20. $\displaystyle\int \frac{du}{u\sqrt{a+bu}} = \begin{cases} \dfrac{1}{\sqrt{a}} \ln\left|\dfrac{\sqrt{a+bu}-\sqrt{a}}{\sqrt{a+bu}+\sqrt{a}}\right| + C & \text{if } a > 0 \\[2ex] \dfrac{2}{\sqrt{-a}} \tan^{-1}\sqrt{\dfrac{a+bu}{-a}} + C & \text{if } a < 0 \end{cases}$

21. $\displaystyle\int \frac{du}{u^n\sqrt{a+bu}} = -\frac{\sqrt{a+bu}}{a(n-1)u^{n-1}} - \frac{b(2n-3)}{2a(n-1)}\int \frac{du}{u^{n-1}\sqrt{a+bu}}$

22. $\displaystyle\int \frac{\sqrt{a+bu}\,du}{u} = 2\sqrt{a+bu} + a\int \frac{du}{u\sqrt{a+bu}}$

23. $\displaystyle\int \frac{\sqrt{a+bu}\,du}{u^n} = -\frac{(a+bu)^{3/2}}{a(n-1)u^{n-1}} - \frac{b(2n-5)}{2a(n-1)}\int \frac{\sqrt{a+bu}\,du}{u^{n-1}}$

Integrals Containing $a^2 \pm u^2$

24. $\displaystyle\int \frac{du}{a^2+u^2} = \frac{1}{a}\tan^{-1}\frac{u}{a} + C$

25. $\displaystyle\int \frac{du}{a^2-u^2} = \frac{1}{2a}\ln\left|\frac{u+a}{u-a}\right| + C = \begin{cases} \dfrac{1}{a}\tanh^{-1}\dfrac{u}{a} + C & \text{if } |u| < a \\[2ex] \dfrac{1}{a}\coth^{-1}\dfrac{u}{a} + C & \text{if } |u| > a \end{cases}$

26. $\displaystyle\int \frac{du}{u^2-a^2} = \frac{1}{2a}\ln\left|\frac{u-a}{u+a}\right| + C = $
$\begin{cases} -\dfrac{1}{a}\tanh^{-1}\dfrac{u}{a} + C & \text{if } |u| < a \\[2ex] -\dfrac{1}{a}\coth^{-1}\dfrac{u}{a} + C & \text{if } |u| > a \end{cases}$

Integrals Containing $\sqrt{u^2 \pm a^2}$

In formulas 27 through 38, we may replace

$\ln(u + \sqrt{u^2 + a^2})$ by $\sinh^{-1}\dfrac{u}{a}$

$\ln|u + \sqrt{u^2 - a^2}|$ by $\cosh^{-1}\dfrac{u}{a}$

$\ln\left|\dfrac{a + \sqrt{u^2 + a^2}}{u}\right|$ by $\sinh^{-1}\dfrac{a}{u}$

27. $\displaystyle\int \frac{du}{\sqrt{u^2 \pm a^2}} = \ln|u + \sqrt{u^2 \pm a^2}| + C$

28. $\displaystyle\int \sqrt{u^2 \pm a^2}\,du = \frac{u}{2}\sqrt{u^2 \pm a^2} \pm \frac{a^2}{2}\ln|u + \sqrt{u^2 \pm a^2}| + C$

29. $\displaystyle\int u^2\sqrt{u^2 \pm a^2}\,du = \frac{u}{8}(2u^2 \pm a^2)\sqrt{u^2 \pm a^2}$
$\qquad\qquad - \dfrac{a^4}{8}\ln|u + \sqrt{u^2 \pm a^2}| + C$

30. $\displaystyle\int \frac{\sqrt{u^2 + a^2}\,du}{u} = \sqrt{u^2 + a^2} - a\ln\left|\frac{a + \sqrt{u^2 + a^2}}{u}\right| + C$

31. $\displaystyle\int \frac{\sqrt{u^2 - a^2}\,du}{u} = \sqrt{u^2 - a^2} - a\sec^{-1}\left|\frac{u}{a}\right| + C$

32. $\displaystyle\int \frac{\sqrt{u^2 \pm a^2}\,du}{u^2} = -\frac{\sqrt{u^2 \pm a^2}}{u} + \ln|u + \sqrt{u^2 \pm a^2}| + C$

33. $\displaystyle\int \frac{u^2\,du}{\sqrt{u^2 \pm a^2}} = \frac{u}{2}\sqrt{u^2 \pm a^2} - \frac{\pm a^2}{2}\ln|u + \sqrt{u^2 \pm a^2}| + C$

34. $\displaystyle\int \frac{du}{u\sqrt{u^2 + a^2}} = -\frac{1}{a}\ln\left|\frac{a + \sqrt{u^2 + a^2}}{u}\right| + C$

35. $\displaystyle\int \frac{du}{u\sqrt{u^2 - a^2}} = \frac{1}{a}\sec^{-1}\frac{u}{a} + C$

36. $\displaystyle\int \frac{du}{u^2\sqrt{u^2 \pm a^2}} = -\frac{\sqrt{u^2 \pm a^2}}{\pm a^2 u} + C$

37. $\displaystyle\int (u^2 \pm a^2)^{3/2}\,du = \frac{u}{8}(2u^2 \pm 5a^2)\sqrt{u^2 \pm a^2}$
$\qquad\qquad + \dfrac{3a^4}{8}\ln|u + \sqrt{u^2 \pm a^2}| + C$

38. $\displaystyle\int \frac{du}{(u^2 \pm a^2)^{3/2}} = \frac{u}{\pm a^2\sqrt{u^2 \pm a^2}} + C$

Integrals Containing $\sqrt{a^2 - u^2}$

39. $\displaystyle\int \frac{du}{\sqrt{a^2 - u^2}} = \sin^{-1}\frac{u}{a} + C$

40. $\displaystyle\int \sqrt{a^2 - u^2}\,du = \frac{u}{2}\sqrt{a^2 - u^2} + \frac{a^2}{2}\sin^{-1}\frac{u}{a} + C$

41. $\displaystyle\int u^2\sqrt{a^2 - u^2}\,du = \frac{u}{8}(2u^2 - a^2)\sqrt{a^2 - u^2} + \frac{a^4}{8}\sin^{-1}\frac{u}{a} + C$

42. $\displaystyle\int \frac{\sqrt{a^2 - u^2}\,du}{u} = \sqrt{a^2 - u^2} - a\ln\left|\frac{a + \sqrt{a^2 - u^2}}{u}\right| + C$
$\qquad\qquad = \sqrt{a^2 - u^2} - a\cosh^{-1}\frac{a}{u} + C$

43. $\displaystyle\int \frac{\sqrt{a^2 - u^2}\,du}{u^2} = -\frac{\sqrt{a^2 - u^2}}{u} - \sin^{-1}\frac{u}{a} + C$

(continued on the back)

Calculus

with analytic geometry

Calculus

with analytic geometry

HOWARD ANTON
Drexel University

THIRD EDITION

JOHN WILEY & SONS
New York   Chichester   Brisbane   Toronto   Singapore

Cover design: Loretta Leiva
Illustration: John Balbalis
Production Supervision: Hudson River Studio

Library of Congress Cataloging in Publication Data:

Anton, Howard.
  Calculus, with analytic geometry.
  Includes index.
  1. Calculus. 2. Geometry, Analytic.   I. Title.
QA303.A53 1988b   515   87-31718
ISBN 0-471-85045-4

Printed in the United States of America

10 9 8 7 6 5 4 3

about the author

Howard Anton was born in Philadelphia, Pennsylvania. He obtained his B.A. from Lehigh University, his M.A. from the University of Illinois, and his Ph.D. from the Polytechnic Institute of Brooklyn, all in mathematics. In the early 1960's he worked for Burroughs Corporation and Avco Corporation at Cape Canaveral, Florida, where he was involved with missile tracking problems for the manned space program. In 1968 he joined the Mathematics Department at Drexel University, where he taught full time until 1983. Since that time he has been an adjunct professor at Drexel and has devoted the majority of his time to textbook writing, development of pedagogical software, and activities for mathematical associations.

He has published numerous research papers in Functional Analysis, Approximation Theory, and Topology, as well as pedagogical papers on applications of mathematics. He is best known for his textbooks in mathematics, which are among the most widely used in the world. There are currently nearly fifty versions of his books including translations into Spanish, Arabic, Portugese, Italian, Indonesian, and Japanese.

Professor Anton was President of the EPADEL Section of the Mathematical Association of America and served on the Board of Governors of that organization. He currently lives in Cherry Hill, New Jersey with his wife, Pat, and his three children, Brian, David, and Lauren. He enjoys traveling and is an avid photographer.

To
My wife Pat
My children Brian, David, and Lauren
My mother Shirley

In memory of
Stephen Girard (1750-1831) benefactor

preface

This text is designed for a standard three-semester or four-quarter calculus course. My intent is to present the material in the clearest possible way. The primary effort has been devoted to *teaching* the subject matter with a level of rigor suitable for the mainstream calculus audience.

Anyone who has taught calculus knows that the course contains so much material that it is impossible for an instructor to spend an adequate amount of classroom time on every topic. For this reason, a calculus textbook bears a major portion of the burden in the teaching process. I am hopeful that this book will be one that can be relied on for sound, clear, and complete explanations, thereby freeing valuable classroom time for the instructor.

FEATURES

Precalculus Material Because of the vast amount of material to be covered, it is desirable to spend as little time as possible on precalculus topics. However, it is a fact that freshmen have a wide variety of educational backgrounds and different levels of preparedness. Therefore, I have included an optional first chapter devoted to precalculus material. It is written in enough detail to enable the instructor to feel confident in moving quickly through these preliminaries.

Trigonometry Deficiencies in trigonometry plague many students throughout the entire calculus sequence. Therefore, I have included a substantial trigonometry review in the appendix. It is more detailed than most such reviews because I feel that students will appreciate having this material readily available. The review is broken into two units: the first to be mastered before reading Section 1.4 and the second before Section 2.8.

Illustrations This calculus text is more heavily illustrated than most. It is my experience that beginners in mathematics frequently have difficulty extracting concepts from mathematical formulas, yet when the right picture is presented the concept becomes clear immediately. For this reason, I have chosen to take full advantage of modern two-color typography and the most up-to-date illustrative techniques. There are over 1200 illustrations designed to clarify and enhance the exposition. (Most calculus books have half this number.)

Pedagogy I have devoted special effort to the explanations of more difficult concepts. In places where students traditionally have trouble, I have tried to give the reader some foothold on the problem. At times I have simply used artwork designed to focus the student's thinking process in the right direction; at other times I have paraphrased ideas informally in a way that cuts through the technical roadblocks; and still at other times I have simply broken the discussion into smaller, more understandable pieces than generally found in most calculus texts.

Rigor Where possible, theory is presented precisely in a style tailored for freshmen. However, in those places where precision conflicts with clarity, I have presented informal intuitive discussions. Whenever this occurs, there is a clear indication that the arguments given are not intended as formal proof. I have tried to make a clear distinction between rigorous mathematics and informal developments. Theory involving $\delta\epsilon$ arguments has been placed in optional sections so it can be avoided if desired.

Flexibility This book is designed to ensure maximum flexibility. There will be no difficulty in permuting chapters in any reasonable way.

Order of Presentation The order in which topics are presented is fairly standard, with two exceptions: derivatives of trigonometric functions are introduced early (Chapter 3) and a discussion of first-order separable and linear differential equations is given in the chapter on logarithmic and exponential functions (Chapter 7). This placement of differential equations material allows us to give some nice applications of logarithms and exponentials immediately, and also helps meet the needs of those engineering and science students who will encounter this material in other courses taken concurrently with calculus.

Applications Standard applications to physics, engineering, biology, population growth, chemistry, and economics appear throughout the text.

Exercises Each exercise set begins with routine drill problems and progresses gradually toward problems of greater difficulty. I have tried to construct well-balanced exercise sets with more variety than is available in most calculus texts. Answers, including art, are given for the odd-numbered exercises, and each chapter contains a set of supplementary exercises to help the student consolidate his or her mastery of the chapter.

Computer Graphics   The use of computers to generate mathematical surfaces is fascinating and growing in importance. This new edition makes even more use of computer-generated surfaces than earlier editions. For many of these surfaces I used the SURFS software package by Mark Bridger.

Reviewing and Class Testing   All of the material in this text has been refined and polished as a result of our extensive experience with the first two editions. In addition to the hundreds of instructors who wrote to me with comments and constructive suggestions, many reviewers worked with me on the third edition to ensure mathematical accuracy and quality exposition. Moreover, we have now had the experience of eight years of classroom use.

Supplements   Numerous supplements are available with this text. Specific descriptions appear later in this preface.

CHANGES FOR THE THIRD EDITION

I am very gratified with the wide acceptance of the first two editions of this text and for the many favorable reviews that have praised the clarity of its mathematical exposition. In this new edition I have continued the process of improving both the exposition and the exercise sets.

Based on a careful and extensive reviewing process, we set four main objectives for the third edition:

- Expand the high end of the exercise sets to include more hard problems, and expand the middle-level problems to include more variety.

- Rewrite the logarithm chapter completely to obtain a better motivation for the definition of the natural logarithm.

- Add more (optional) material on differential equations.

- Revise the chapters on vectors and multivariable calculus to make them clearer and more efficient, thereby allowing the instructor to cover the material at a less hurried pace.

In addition, material has been rewritten and reorganized throughout the text based on reviewer and user suggestions. I think that most users will be quite pleased with these improvements and will find the text even easier to work with than before. Here is a sampling of the changes:

- Continuity is now discussed before differentiability.

- Continuity properties of trigonometric functions now appear in a separate section presented early in the text (Section 2.8).

- Chapter 3 on differentiation was reorganized for improved clarity.

- Section 3.5 on the chain rule was rewritten to improve the exposition.

- Material on velocity and acceleration that was previously scattered among various sections is now presented more cohesively (Sections 4.11, 6.6, and 15.6).

- The logarithm chapter (Chapter 7) was extensively rewritten.

- Vectors in 2-space and 3-space are now presented in parallel rather than in separate sections as before, thereby eliminating redundant exposition.

- The material on vector calculus was rewritten, presenting two- and three-dimensional work in parallel. The definition of curvature in 2-space was modified to make it consistent with the definition in 3-space.

A new chapter on second-order differential equations and their applications was added (Chapter 19).

HOWARD ANTON

acknowledgments

It has been my good fortune to have the advice and guidance of many talented people, whose knowledge and skills have enhanced this book in many ways. Their contributions are diverse: reviewing, creation of exercises, and permissions for access to their work. For their valuable help I would like to thank:

Edith Ainsworth, *University of Alabama*
David Armacost, *Amherst College*
Larry Bates, *University of Calgary*
Marilyn Blockus, *San Jose State University*
David Bolen, *Virginia Military Institute*
George W. Booth, *Brooklyn College*
Mark Bridger, *Northeastern University*
John Brothers, *Indiana University*
Robert C. Bueker, *Western Kentucky University*
Robert Bumcrot, *Hofstra University*
Chris Christensen, *Northern Kentucky University*
David Cohen, *University of California, Los Angeles*
Michael Cohen, *Hofstra University*
Robert Conley, *Precision Visuals*
A.L. Deal, *Virginia Military Institute*
Charles Denlinger, *Millersville State College*
Dennis DeTurck, *University of Pennsylvania*
Jacqueline Dewar, *Loyola Marymount University*
Irving Drooyan, *Los Angeles Pierce College*
Hugh B. Easler, *College of William and Mary*
Joseph M. Egar, *Cleveland State University*
Garret J. Etgen, *University of Houston*
James H. Fife, *University of Richmond*
Barbara Flajnik, *Virginia Military Institute*
Nicholas E. Frangos, *Hofstra University*
Katherine Franklin, *Los Angeles Pierce College*
Michael Frantz, *University of La Verne*
William R. Fuller, *Purdue University*
Raymond Greenwell, *Hofstra University*
Gary Grimes, *Mt. Hood Community College*
Jane Grossman, *University of Lowell*
Michael Grossman, *University of Lowell*
Douglas W. Hall, *Michigan State University*
Nancy A. Harrington, *University of Lowell*
Albert Herr, *Drexel University*
Peter Herron, *Suffolk County Community College*
Robert Higgins, *Quantics Corporation*
Louis F. Hoelzle, *Bucks County Community College*

Harvey B. Keynes, *University of Minnesota*
Paul Kumpel, *SUNY Stony Brook*
Leo Lampone, *Quantics Corporation*
Bruce Landman, *Hofstra University*
Benjamin Levy, *Lexington H.S., Lexington, Mass.*
Phil Locke, *University of Maine, Orono*
Stanley M. Lukawecki, *Clemson University*
Nicholas Macri, *Temple University*
Melvin J. Maron, *University of Louisville*
Thomas McElligott, *University of Lowell*
Judith McKinney, *California State Polytechnic University, Pomona*
Joseph Meier, *Millersville State College*
David Nash, *VP Research, Autofacts, Inc.*
Mark A. Pinsky, *Northeastern University*
William H. Richardson, *Wichita State University*
David Sandell, *U.S. Coast Guard Academy*
Donald R. Sherbert, *University of Illinois*
Wolfe Snow, *Brooklyn College*
Norton Starr, *Amherst College*
Richard B. Thompson, *The University of Arizona*
William F. Trench, *Trinity University*
Walter W. Turner, *Western Michigan University*
Richard C. Vile, *Eastern Michigan University*
James Warner, *Precision Visuals*
Candice A. Weston, *University of Lowell*
Yihren Wu, *Hofstra University*
Richard Yuskaitis, *Precision Visuals*

Proofreading and Problem Solving   I gratefully acknowledge the contributions of some dedicated and hardworking people who helped me with the backbreaking work of proofreading and problem solving. The timely publication of this book is a direct consequence of their efforts: Brian P. Anton, Lilian Brady, Steven Conrad, Michael Dagg, Scott Hecht, Craig Miller, Elizabeth Slikas.

John Wiley and Sons   I would like to take this opportunity to thank Serje Seminoff and Don Ford of John Wiley and Sons for their superb support of this project. They were totally committed to making this a first-rate book, and without hesitation provided me with all the assistance and special help that was required to achieve that goal.

Special thanks are due to my assistant, Mary Parker, who worked on a myriad of important details. She coordinated the complex flow of information and paperwork relating to the new edition and somehow managed to keep me organized and on schedule. I feel most fortunate to have had her valuable help.

Finally, I would like to thank my editor, Robert Pirtle, whose enthusiasm and dedication to this project has been boundless. He has worked tirelessly to ensure the continued high quality of this text, and he deserves much credit for the improvements that have resulted.

supplements

The following supplementary materials for this textbook can be obtained from your bookstore. If they are not in stock ask the bookstore manager to order them for you.

- *The Calculus Companion, Volumes 1 and 2, to Accompany Calculus with Analytic Geometry, Third Edition by Howard Anton*, written by William H. Barker and James E. Ward

   The *Calculus Companion* can help you with course material by serving as a tutorial, a review, and a study aid.

- *Student's Solutions Manual to Accompany Calculus with Analytic Geometry, Third Edition by Howard Anton*, prepared by Albert Herr

   The *Solutions Manual* contains detailed solutions to all odd-numbered exercises.

Information about the following supplementary materials is available to instructors by writing to Valerie Hunter, Mathematics Editor, John Wiley and Sons, 605 Third Avenue, New York, N.Y. 10158. Your request must be sent on your university letterhead.

- *True BASIC Calculus Software* by John G. Kemeny (for the Apple Macintosh, Commodore Amiga, and IBM PC, XT, and AT)

   The *True BASIC Calculus Software* is a useful computational tool for performing various calculus operations and obtaining graphs of functions.

- *True BASIC Study Guide* by Robert G. Phillips

   True BASIC Calculus software written by John Kemeny (for the Apple Macintosh, Commodore Amiga, and IBM PC, XT, and AT, and compatibles) integrates very nicely with this text. A study guide prepared by Robert G. Phillips keys the True BASIC software into the text. It is designed for a calculus course in which a personal computer plays an important role as a computational and pedagogical tool. This supplement uses the computer to reexamine many of the examples and exercises in the text. It also contains additional examples and exercises that are intended to be solved using the True BASIC Calculus software.

Although True BASIC is an item sold in individual copies or in quantity, the publisher has arranged for free copies of the software and the study guide to be made available to those instructors who adopt this text. Contact your Wiley representative for more information.

contents

introduction

Calculus is the mathematical tool used to analyze changes in physical quantities. It was developed in the seventeenth century to study four major classes of scientific and mathematical problems of the time:

1. Find the tangent to a curve at a point.

2. Find the length of a curve, the area of a region, and the volume of a solid.

3. Find the maximum or minimum value of a quantity—for example, the maximum and minimum distances of a planet from the sun, or the maximum range attainable for a projectile by varying its angle of fire.

4. Given a formula for the distance traveled by a body in any specified amount of time, find the velocity and acceleration of the body at any instant. Conversely, given a formula that specifies the acceleration or velocity at any instant, find the distance traveled by the body in a specified period of time.

These problems were attacked by the greatest minds of the seventeenth century, culminating in the crowning achievements of Gottfried Wilhelm Leibniz and Isaac Newton—the creation of calculus.

Gottfried Wilhelm Leibniz (1646–1716)

This gifted genius was one of the last people to have mastered most major fields of knowledge—an impossible accomplishment in our own era of specialization. He was an expert in law, religion, philosophy, literature, politics, geology, metaphysics, alchemy, history, and mathematics.

Leibniz was born in Leipzig, Germany. His father, a professor of moral philosophy at the University of Leipzig, died when Leibniz was six years old. The precocious boy then gained access to his father's library and began reading voraciously on a wide range of subjects, a habit that he maintained throughout his life. At age 15 he entered the University of Leipzig as a law student and by the age of 20 received a doctorate from the University of Altdorf. Subsequently, Leibniz followed a career in law and international politics, serving as counsel to kings and princes.

During his numerous foreign missions, Leibniz came in contact with outstanding mathematicians and scientists who stimulated his interest in mathematics—most notably, the physicist Christian Huygens. In mathematics Leibniz was self-taught, learning the subject by reading papers and journals. As a result of this fragmented mathematical education, Leibniz often duplicated the results of others, and this ultimately led to a raging conflict over the inventor of calculus—Leibniz or Newton? The argument over this question engulfed the scientific circles of England and Europe with most scientists on the continent supporting Leibniz and those in England supporting Newton. The conflict was unfortunate, and both sides suffered in the end. The continent lost the benefit of Newton's discoveries in astronomy and physics for more than 50 years, and for a long period England became a second-rate country mathematically because its mathematicians were hampered by Newton's inferior calculus notation. It is of interest to note that Newton and Leibniz never went to the lengths of vituperation of their advocates—both were sincere admirers of each other's work. The fact is that both men invented calculus independently. Leibniz invented it 10 years after Newton, in 1685, but he published his results 20 years before Newton published his own work on the subject.

Leibniz never married. He was moderate in his habits, quick-tempered, but easily appeased, and charitable in his judgment of other people's work. In spite of his great achievements, Leibniz never received the honors

Gottfried Leibniz
(Culver Pictures)

Isaac Newton
(Culver Pictures)

showered on Newton, and he spent his final years as a lonely embittered man. At his funeral there was one mourner, his secretary. An eyewitness stated, "He was buried more like a robber than what he really was—an ornament of his country."

Isaac Newton (1642–1727)

Newton was born in the village of Woolsthorpe, England. His father died before he was born and his mother raised him on the family farm. As a youth he showed little evidence of his later brilliance, except for an unusual talent with mechanical devices—he apparently built a working water clock and a toy flour mill powered by a mouse. In 1661 he entered Trinity College in Cambridge with a deficiency in geometry. Fortunately, Newton caught the eye of Isaac Barrow, a gifted mathematician and teacher. Under Barrow's guidance Newton immersed himself in mathematics and science, but he graduated without any special distinction. Because the Plague was spreading rapidly through London, Newton returned to his home in Woolsthorpe and stayed there during the years of 1665 and 1666. In those two momentous years the entire framework of modern science was miraculously created in Newton's mind—he discovered calculus, recognized the underlying principles of planetary motion and gravity, and determined that "white" sunlight was composed of all colors, red to violet. For some reason he kept his discoveries to himself. In 1667 he returned to Cambridge to obtain his Master's degree and upon graduation became a teacher at Trinity. Then in 1669 Newton succeeded his teacher, Isaac Barrow, to the Lucasian chair of mathematics at Trinity, one of the most honored chairs of mathematics in the world. Thereafter, brilliant discoveries flowed from Newton steadily. He formulated the law of gravitation and used it to explain the motion of the moon, the planets, and the tides; he formulated basic theories of light, thermodynamics, and hydrodynamics; and he devised and constructed the first modern reflecting telescope.

Throughout his life Newton was hesitant to publish his major discoveries, revealing them only to a select circle of friends, perhaps because of a fear of criticism or controversy. In 1687, only after intense coaxing by the astronomer, Edmond Halley (Halley's comet), did Newton publish his masterpiece, *Philosophae Naturalis Principia Mathematica* (The Mathematical Principles of Natural Philosophy). This work is generally considered to be the most important and influential scientific book ever written. In it Newton explained the workings of the solar system and formulated the basic laws of motion which to this day are fundamental in engineering and physics. However, not even the pleas of his friends could convince Newton to publish his discovery of calculus. Only after Leibniz published his results did Newton relent and publish his own work on calculus.

After 35 years as a professor, Newton suffered depression and a nervous breakdown. He gave up research in 1695 to accept a position as warden and later master of the London mint. During the 25 years that he worked at the mint, he did virtually no scientific or mathematical work. He was knighted in 1705 and on his death was buried in Westminster Abbey with all the honors his country could bestow. It is interesting to note that Newton was a learned theologian who viewed the primary value of his work to be its support of the existence of God. Throughout his life he worked passionately to date biblical events by relating them to astronomical phenomena. He was so consumed with this passion that he spent years searching the Book of Daniel for clues to the end of the world and the geography of hell.

Newton described his brilliant accomplishments as follows, "I seem to have been only like a boy playing on the seashore and diverting myself in now and then finding a smoother pebble or prettier shell than ordinary, whilst the great ocean of truth lay all undiscovered before me."

Calculus
with analytic geometry

1 coordinates, graphs, lines

1.1 REAL NUMBERS, SETS, AND INEQUALITIES (A REVIEW)

Since numbers and their properties play a fundamental role in calculus, we shall use this first section to review some terminology and facts about them.

REAL NUMBERS   The simplest numbers are the *integers:*

$$\ldots, -4, -3, -2, -1, 0, 1, 2, 3, 4, \ldots$$

With the exception that division by zero is ruled out, ratios of integers are called *rational numbers*. Examples are

$$\frac{2}{3}, \frac{0}{9}, \frac{6}{2}, \frac{17}{1000}, -\frac{5}{2}\left(=\frac{-5}{2}=\frac{5}{-2}\right), 3\left(=\frac{3}{1}\right)$$

Observe that every integer is also a rational number because an integer p can be written as the ratio

$$p = \frac{p}{1}$$

Division by zero is ruled out because we would want to express the relationship

$$y = \frac{p}{0}$$

in the alternative form

$$0 \cdot y = p$$

However, if p is different from zero, this equation is contradictory; and if p is equal to zero, this equation is satisfied by any number y, which

1

means that the ratio $p/0$ does not have a unique value—a situation that is mathematically unsatisfactory. For these reasons such symbols as

$$\frac{p}{0} \quad \text{and} \quad \frac{0}{0}$$

are not assigned a value; they are said to be **undefined.**

Figure 1.1.1

The early Greeks believed that the size of every physical quantity could, in theory, be represented by a rational number. They reasoned that the size of a physical quantity must consist of a certain whole number of units plus some fraction m/n of an additional unit. This idea was shattered in the fifth century B.C. by Hippasus of Metapontum,* who demonstrated by geometric methods that the hypotenuse of the triangle in Figure 1.1.1 cannot be expressed as a ratio of integers. This dramatic discovery demonstrated the existence of *irrational numbers;* that is, numbers not expressible as ratios of integers. Other examples of irrational numbers are

$$1 + \sqrt{2}, \quad \sqrt{3}, \quad \sqrt[3]{7}, \quad \pi, \quad \cos 19°$$

The proof that π is irrational is difficult and evaded mathematicians for centuries; it was finally proved in 1761 by J. H. Lambert.†

Rational and irrational numbers can be distinguished by their decimal representations. Rational numbers have *repeating decimals;* that is, from some point the decimal consists entirely of zeros or else a fixed finite string of digits that is repeated over and over. For example,

$$\frac{1}{2} = 0.5000 \ldots \quad \frac{8}{25} = 0.32000 \ldots \quad \frac{12}{4} = 3.000 \ldots$$
$$\frac{4}{3} = 1.3333 \ldots \quad \frac{3}{11} = 0.272727 \ldots \quad \frac{5}{7} = 0.714285714285714285 \ldots$$

Irrational numbers are represented by decimals that are not repeating.

Example 1   Since π is irrational, the decimal

$$\pi = 3.14159265358979323846 \ldots$$

does not begin to repeat from some point. Since the decimal

$$0.101001000100001000001 \ldots$$

---

*HIPPASUS OF METAPONTUM (circa 500 B.C.). A Greek Pythagorean philosopher. According to legend, Hippasus made his discovery at sea and was thrown overboard by fanatic Pythagoreans because his result contradicted their doctrine. The discovery of Hippasus is one of the most fundamental in the entire history of science.

---

†JOHANN HEINRICH LAMBERT (1728–1777). A Swiss-German scientist, Lambert taught at the Berlin Academy of Sciences. In addition to his work on irrational numbers, he wrote landmark books on geometry, the theory of cartography, and perspective in art. His influential book on geometry foreshadowed the discovery of modern non-Euclidean geometry.

is not repeating (the number of zeros between the ones keeps growing), it represents an irrational number.   ◀

COORDINATE LINES

In 1637 René Descartes* published a philosophical work called *Discourse on the Method of Rightly Conducting the Reason.* In the back of that book were three appendices that purported to show how the "method" could be applied to concrete examples. The first two appendices were minor works that endeavored to explain the behavior of lenses and the movement of shooting stars. The third appendix, however, was an inspired stroke of genius; it was described by the nineteenth century British philosopher John Stuart Mill as, "The greatest single step ever made in the progress of the exact sciences." In that appendix René Descartes linked together two branches of mathematics, algebra and geometry. Descartes' work evolved into a new subject called *analytic geometry;* it gave a way of describing algebraic formulas by means of geometric curves and, conversely, geometric curves by algebraic formulas.

In analytic geometry, the key step is to establish a correspondence between real numbers and points on a line. To accomplish this, we arbitrarily choose one direction along the line to be called *positive* and the other *negative.* It is usual to mark the positive direction with an arrowhead, as shown in Figure 1.1.2. Next, we choose an arbitrary reference point on the line to be called the *origin,* and select a unit of length for measuring distances. With each real number we can associate a point on the line as follows:

Figure 1.1.2

- Associate with each positive number r the point that is a distance of r units in the positive direction from the origin.

- Associate with each negative number $-r$ the point that is a distance of r units in the negative direction from the origin.

- Associate the origin with the number 0.

The real number corresponding to a point on the line is called the

---

* RENÉ DESCARTES (1596–1650). Descartes, a French aristocrat, was the son of a government official. He graduated from the University of Poitiers with a law degree at age 20. After a brief probe into the pleasures of Paris he became a military engineer, first for the Dutch Prince of Nassau and then for the German Duke of Bavaria. It was during his service as a soldier that Descartes began to pursue mathematics seriously and develop his analytic geometry. After the wars, he returned to Paris where he stalked the city as an eccentric, wearing a sword in his belt and a plumed hat. He lived in leisure, seldom arose before 11 A.M., and dabbled in the study of human physiology, philosophy, glaciers, meteors, and rainbows. He eventually moved to Holland, where he published his *Discourse on the Method,* and finally to Sweden where he died while serving as tutor to Queen Christina. Descartes is regarded as a genius of the first magnitude. In addition to major contributions in mathematics and philosophy, he is considered, along with William Harvey, to be a founder of modern physiology.

coordinate of the point, and the line is called a *coordinate line* or sometimes the *real line*.

Example 2   In Figure 1.1.3 we have marked the approximate location of the points whose coordinates are -3, -1.75, $-\frac{1}{2}$, $\sqrt{2}$, π, and 4.   ◄

Figure 1.1.3 ▷

It is evident from the way in which real numbers and points on a coordinate line are related that each real number corresponds to a single point and each point corresponds to a single real number. To describe this fact we say that the real numbers and the points on a coordinate line are in *one-to-one correspondence*.

ORDER PROPERTIES   If we traverse a coordinate line in the positive direction, then the numbers increase in size. This is a reflection of the fact that the real numbers are *ordered;* that is, given any two numbers, a and b, exactly one of the following is true:

    a is less than *b*
    b is less than *a*
    a is equal to *b*

To describe the relative size of two real numbers, we use the order symbols $<$ (less than) and $\leq$ (less than or equal to), which are called *inequalities*. These symbols are defined as follows:

1.1.1   DEFINITION. If a and b are real numbers, then

    $a < b$ means $b - a$ is positive
    $a \leq b$ means $a < b$ or $a = b$

The inequality $a < b$, which is read "*a is less than b*," can also be written as $b > a$, which is read "*b is greater than a*"; and the inequality $a \leq b$, which is read "*a is less than or equal to b*," can also be written as $b \geq a$, which is read "*b is greater than or equal to a*."

Geometrically, the inequality $a < b$ states that a is to the left of b on a coordinate line (Figure 1.1.4).

The following definition provides a convenient way of combining two inequalities into a single expression.

Figure 1.1.4

1.1.2   DEFINITION. If a, b, and c are real numbers, we shall write

$$a < b < c$$

when $a < b$ and $b < c$.

Figure 1.1.5

Geometrically, $a < b < c$ states that on a coordinate line, b is to the right of a and c is to the right of b (Figure 1.1.5).

The symbol $a < b \leq c$ means $a < b$ and $b \leq c$. We leave it to the reader to deduce the meanings of such symbols as

$$a \leq b < c, \quad a \leq b \leq c, \quad \text{and} \quad a < b < c < d$$

Example 3   The following inequalities are all correct:

$$3 < 8, \quad -7 < 1.5, \quad -12 \leq -\pi, \quad 5 \leq 5, \quad 0 \leq 2 \leq 4$$
$$8 \geq 3, \quad 1.5 > -7, \quad -\pi > -12, \quad 5 \geq 5, \quad -1 < 0 < 3 \qquad \blacktriangleleft$$

REMARK.   To distinguish verbally between numbers that satisfy $a \geq 0$ and those that satisfy $a > 0$, we shall call a ***nonnegative*** if $a \geq 0$ and ***positive*** if $a > 0$. Thus, a nonnegative number is either positive or zero.

The following properties of inequalities are frequently used in calculus. We omit the proofs.

1.1.3   THEOREM. *Let a, b, c, and d be real numbers.*

(a)   If $a < b$ and $b < c$, then $a < c$.
(b)   If $a < b$, then $a + c < b + c$ and $a - c < b - c$.
(c)   If $a < b$, then $ac < bc$ when c is positive and $ac > bc$ when c is negative.
(d)   If $a < b$ and $c < d$, then $a + c < b + d$.
(e)   If a and b are both positive or both negative and $a < b$, then $1/a > 1/b$.

REMARK.   These five properties remain true if $<$ and $>$ are replaced by $\leq$ and $\geq$, respectively.

If we call the direction in which an inequality points its *sense*, then parts *(b)*–*(e)* of this theorem can be paraphrased informally as follows:

(b)   The sense of an inequality is unchanged if the same number is added to or subtracted from both sides.

(c) *The sense of an inequality is unchanged if both sides are multiplied by the same positive number, but the sense is reversed if both sides are multiplied by the same negative number.*

(d) *Inequalities with the same sense can be added.*

(e) *If both sides of an inequality have the same sign, then the sense of the inequality is reversed by taking the reciprocal of each side.*

Example 4 The statements in Theorem 1.1.3 are illustrated in Table 1.1.1. ◀

Table 1.1.1

STARTING INEQUALITY	OPERATION	RESULTING INEQUALITY
$-2 < 6$	Add 7 to both sides.	$5 < 13$
$-2 < 6$	Subtract 8 from both sides.	$-10 < -2$
$-2 < 6$	Multiply both sides by 3.	$-6 < 18$
$-2 < 6$	Multiply both sides by -3.	$6 > -18$
$3 < 7$	Multiply both sides by 4.	$12 < 28$
$3 < 7$	Multiply both sides by -4.	$-12 > -28$
$3 < 7$	Take reciprocals of both sides.	$\frac{1}{3} > \frac{1}{7}$
$-8 < -6$	Take reciprocals of both sides.	$-\frac{1}{8} > -\frac{1}{6}$
$4 < 5$ and $-7 < 8$	Add corresponding sides.	$-3 < 13$

SETS AND INTERVALS

Since it will be helpful in our later work to use the language of *sets*, let us pause to review some of the relevant terminology and notation.

A set can be viewed as a collection of objects, called **elements** or **members** of the set. In this text we shall be concerned primarily with sets whose members are real numbers. One way to describe a set is to list its elements between braces. Thus, the set of all positive integers less than 5 can be written

$$\{1, 2, 3, 4\}$$

and the set of all positive even integers can be written

$$\{2, 4, 6, \ldots\}$$

where the dots are used to indicate that only a portion of the members have been explicitly listed and the rest of the members are obtained by continuing the pattern.

When it is inconvenient or impossible to list the members, it is sufficient to define a set by stating a property common only to its members. For example,

the set of all rational numbers
the set of all real numbers x such that $2x^2 - 4x + 1 = 0$
the set of all real numbers between 2 and 3

As an alternative to such verbal descriptions of sets, we can use the notation

$$\{x: \underline{\hspace{2cm}}\}$$

which is read, "the set of all x such that _____." Where the line is placed, one would state a property that specifies the set.

Example 5

(a) $\{x:x$ is a rational number$\}$ is read, "the set of all x such that x is a rational number."
(b) $\{x:x$ is a real number satisfying $2x^2 - 4x + 1 = 0\}$ is read, "the set of all x such that x is a real number satisfying $2x^2 - 4x + 1 = 0$."
(c) $\{x:x$ is a real number between 2 and 3$\}$ is read, "the set of all x such that x is a real number between 2 and 3."   ◀

REMARK.   When it is clear that the members of a set are real numbers, we will omit the reference to this fact. Thus, the sets in Example 5 might be written

$$\{x:x \text{ is rational}\}$$
$$\{x:2x^2 - 4x + 1 = 0\}$$
$$\{x:2 < x < 3\}$$

Of special interest are sets of real numbers called *intervals*. Geometrically, an interval is a line segment. For example, if $a < b$, then the *closed interval* from a to b is the set

$$\{x:a \leq x \leq b\}$$

and the *open interval* from a to b is the set

$$\{x:a < x < b\}$$

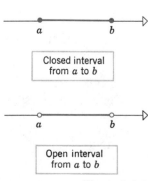

These sets are pictured in Figure 1.1.6.

A closed interval includes both its endpoints (indicated by solid dots in Figure 1.1.6), while an open interval does not include the endpoints (indicated by open dots in Figure 1.1.6). Closed and open intervals are usually denoted by the symbols $[a, b]$ and (a, b), respectively, so

$$[a, b] = \{x:a \leq x \leq b\}$$
$$(a, b) = \{x:a < x < b\}$$

A square bracket, [or], indicates that the endpoint is included, while a rounded bracket, (or), indicates that the endpoint is not included.

Figure 1.1.6

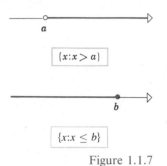

$\{x : x > a\}$

$\{x : x \leq b\}$

Figure 1.1.7

An interval can include one endpoint and not the other. Such intervals are called **half-open** (or sometimes **half-closed**). For example,

$$[a, b) = \{x : a \leq x < b\}$$
$$(a, b] = \{x : a < x \leq b\}$$

An interval can extend indefinitely in either the positive or negative direction (see Figure 1.1.7). The intervals in Figure 1.1.7 are denoted by

$$(a, +\infty) = \{x : x > a\}$$
$$(-\infty, b] = \{x : x \leq b\}$$

where the symbol $+\infty$ ("plus infinity") indicates that the interval extends indefinitely in the positive direction and the symbol $-\infty$ ("minus infinity") indicates that the interval extends indefinitely in the negative direction.
Table 1.1.2 gives a complete list of the kinds of intervals possible.

Table 1.1.2

INTERVAL NOTATION	SET NOTATION	GEOMETRIC PICTURE
$[a, b]$	$\{x : a \leq x \leq b\}$	●——● a b
(a, b)	$\{x : a < x < b\}$	○——○ a b
$[a, b)$	$\{x : a \leq x < b\}$	●——○ a b
$(a, b]$	$\{x : a < x \leq b\}$	○——● a b
$(-\infty, b]$	$\{x : x \leq b\}$	——● b
$(-\infty, b)$	$\{x : x < b\}$	——○ b
$[a, +\infty)$	$\{x : x \geq a\}$	●—— a
$(a, +\infty)$	$\{x : x > a\}$	○—— a
$(-\infty, +\infty)$	$\{x : x \text{ is a real number}\}$	——

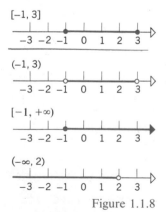

Figure 1.1.8

Example 6   Some intervals and their notation are shown in Figure 1.1.8.   ◄

In many problems we shall be interested in finding points that lie in two or more intervals or points that lie in at least one of several possible intervals. Since such problems can be treated using the notations of union and intersection of sets, we shall review these ideas.
To indicate that an element a is a member of a set A, we write

$$a \in A$$

which is read, "*a* is an element of *A*" or "*a* belongs to *A*." To indicate that the element *a* is *not* a member of the set *A*, we write

$$a \notin A$$

which is read, "*a* is not an element of *A*" or "*a* does not belong to *A*."

Example 7 If $A = \{x : x \text{ is rational}\}$ and $B = \{x : 2 < x < 4\}$, then

$$\tfrac{3}{4} \in A, \quad -2 \in A, \quad \pi \notin A, \quad -\sqrt{2} \notin A$$
$$2.5 \in B, \quad \pi \in B, \quad -1 \notin B, \quad 4 \notin B \quad \blacktriangleleft$$

Sometimes sets arise that have no members. Such a set is denoted by the symbol $\varnothing$ and is called an *empty set* or a *null set*. Thus,

$$\{x : x \text{ is a real number satisfying } x^2 < 0\} = \varnothing$$

1.1.4 DEFINITION. Two sets *A* and *B* are said to be *equal* if they have the same elements, in which case we write $A = B$.

Example 8

$$\{x : x^2 = 1\} = \{-1, 1\}, \quad \{\pi, 0, 3\} = \{3, \pi, 0\}$$
$$\{x : x^2 < 9\} = \{x : -3 < x < 3\} \quad \blacktriangleleft$$

1.1.5 DEFINITION. If every member of a set *A* is also a member of set *B*, then we say *A* is a *subset* of *B* and write $A \subset B$.

REMARK. By convention, the empty set $\varnothing$ is a subset of every set.

Example 9

$$\{-2, 4\} \subset \{-2, 1, 0, 4\}$$
$$\{x : x \text{ is rational}\} \subset \{x : x \text{ is a real number}\}$$
$$\varnothing \subset A \text{ (for every set } A) \quad \blacktriangleleft$$

REMARK. If $A \subset B$ and $B \subset A$, then $A = B$. (Why?)

1.1.6 DEFINITION. If *A* and *B* are two given sets, then
(a) the set of all elements belonging to both *A* and *B* is denoted by $A \cap B$ and is called the *intersection* of *A* and *B*;
(b) the set of all elements belonging to *A* or *B* or both is denoted by $A \cup B$ and is called the *union* of *A* and *B*.

REMARK. There is a convention about the use of the word "or" in mathematics that should be noted. In mathematical parlance, the statement that "*P* is true or *Q* is true" is understood to allow for the possibility that *P* and *Q* are both true. This is called the *inclusive* interpretation of "or." Thus, if *A* and *B* are sets, it is correct to say that $A \cup B$ consists of all points that belong to *A* or *B*. (The possibility that the points may belong to both sets is accounted for in the inclusive interpretation of the "or.")

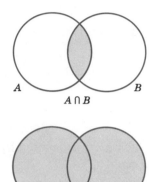

$A \cap B$

$A \cup B$

Figure 1.1.9

Example 10   In Figure 1.1.9, *A* is the set of points inside the left circle and *B* the set of points inside the right circle. The sets $A \cap B$ and $A \cup B$ are shown as shaded regions: ◄

Example 11   Sketch the sets

(a) $[-4, 1) \cup (3, 6)$ (b) $(-3, 2] \cup (1, 7]$.

Solution.   We can write

$$[-4, 1) \cup (3, 6) = \{x: -4 \le x < 1 \quad \text{or} \quad 3 < x < 6\}$$
$$(-3, 2] \cup (1, 7] = \{x: -3 < x \le 2 \quad \text{or} \quad 1 < x \le 7\}$$

These are the sets sketched in Figure 1.1.10.   ◄

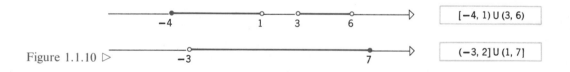

Figure 1.1.10 ▷

Example 12   Sketch the set $(-4, 3] \cap (2, 6)$.

Solution.   The set

$$(-4, 3] \cap (2, 6) = \{x: -4 < x \le 3 \quad \text{and} \quad 2 < x < 6\}$$

consists of all points common to the intervals $(-4, 3]$ and $(2, 6)$. This intersection is the interval $(2, 3]$ shown in Figure 1.1.11.   ◄

Figure 1.1.11 ▷

SOLVING
INEQUALITIES

In the following examples we use Theorem 1.1.3 to solve some inequalities; the solutions are expressed in terms of intervals. The first example points out a subtle problem in logic that we shall encounter throughout the text.

Example 13   Solve $2x - 3 < 7$. (That is, find all real numbers satisfying the inequality.)

Solution.   If x is any solution, then

$$
\begin{array}{ll}
2x - 3 < 7 & \text{[Given.]} \\
2x \quad\ \ < 10 & \text{[We added 3 to both sides.]} \\
x \quad\ \ < 5 & \text{[We multiplied both sides by } \tfrac{1}{2}.]
\end{array}
$$

At this point we are tempted to conclude that the solutions of $2x - 3 < 7$ consist of all x less than 5, that is, all x in the interval $(-\infty, 5)$. While this conclusion is correct, it is premature. To see why, let us examine the logic of our argument. We have proved: *If x is a solution, then $x < 5$.* In other words, if S denotes the set of solutions to $2x - 3 < 7$, we have shown

$$S \subset (-\infty, 5) \tag{1}$$

We have *not* yet shown that

$$S = (-\infty, 5) \tag{2}$$

If we can prove that

$$(-\infty, 5) \subset S \tag{3}$$

then by the remark after Example 9, (2) will follow from (1) and (3).

To prove (3) we must show that every member of $(-\infty, 5)$ is also a member of S. That is, we must assume $x < 5$ and deduce that $2x - 3 < 7$. We do this as follows:

$$
\begin{array}{ll}
x \quad\ \ < 5 & \text{[Given.]} \\
2x \quad\ \ < 10 & \text{[We multiplied both sides by 2.]} \\
2x - 3 < 7 & \text{[We added } -3 \text{ to both sides.]}
\end{array}
$$

Thus, the solutions of $2x - 3 < 7$ form the interval $(-\infty, 5)$.   ◀

REMARK.   In the last example, compare the steps needed to prove (1) with the steps needed to prove (3):

Steps to *Prove* (1)	*Steps to* *Prove* (3)
$2x - 3 < 7$	$x \quad\ \ < 5$
$2x \quad\ \ < 10$	$2x \quad\ \ < 10$
$x \quad\ \ < 5$	$2x - 3 < 7$

In the two parts the steps are the same, but in opposite orders. In any "two-stage" mathematical argument where the steps in the second stage are the same as those in the first stage, but in the opposite order, it is common to give only the steps for the first stage and then conclude the proof by stating that *the steps are reversible*.

Example 14   Solve $3 + 7x \leq 2x - 9$.

Solution.

$$3 + 7x \leq 2x - 9 \quad \text{[Given.]}$$
$$7x \leq 2x - 12 \quad \text{[We added } -3 \text{ to both sides.]}$$
$$5x \leq -12 \quad \text{[We added } -2x \text{ to both sides.]}$$
$$x \leq -\tfrac{12}{5} \quad \text{[We multiplied both sides by } \tfrac{1}{5}.\text{]}$$

Figure 1.1.12

Since the steps are reversible, the set of solutions is $(-\infty, -\tfrac{12}{5}]$; this solution set is shown in Figure 1.1.12.   ◄

Example 15   Solve $7 \leq 2 - 5x < 9$.

Solution.

$$7 \leq 2 - 5x < 9 \quad \text{[Given.]}$$
$$5 \leq \quad - 5x < 7 \quad \text{[We added } -2 \text{ to each member.]}$$
$$-1 \geq \quad\quad x > -\tfrac{7}{5} \quad \text{[We multiplied each member by } -\tfrac{1}{5}.\text{]}$$

Figure 1.1.13

Since the steps are reversible, the set of solutions is $(-\tfrac{7}{5}, -1]$; this solution set is shown in Figure 1.1.13.   ◄

REMARK.   Observe that in Example 15 the inequalities were reversed when we multiplied by the negative quantity $-\tfrac{1}{5}$.

Example 16   Solve $(x + 2)(x - 5) > 0$.

Solution.   The solutions are those values of x for which the factors $(x + 2)$ and $(x - 5)$ have the same sign. From Figure 1.1.14 we see that the

Figure 1.1.14 ▷

Figure 1.1.15

solutions are those x for which $x < -2$ or $x > 5$, since $(x + 2)(x - 5)$ must be positive (> 0). Equivalently, the solution set is $(-\infty, -2) \cup (5, +\infty)$; see Figure 1.1.15. ◀

Example 17 Solve

$$\frac{2x - 5}{x - 2} < 1$$

Solution. Rewrite the inequality as

$$\frac{2x - 5}{x - 2} - 1 < 0 \qquad \text{[We subtracted 1 from both sides.]}$$

$$\frac{(2x - 5) - (x - 2)}{x - 2} < 0 \qquad \text{[We combined terms.]}$$

$$\frac{x - 3}{x - 2} < 0 \qquad \text{[We simplified.]}$$

$- - - - \; 0 + + + + + + + +$

Sign of $x - 2$ (at 2)

$- - - - - - - - - 0 + + + +$

Sign of $x - 3$ (at 3)

$+ + + + \quad - - - 0 + + + +$

Sign of $\frac{x - 3}{x - 2}$ (at 2 and 3)

Figure 1.1.16 ▷

The solutions are those values of x for which the quotient $(x - 3)/(x - 2)$ is negative, and this occurs if and only if the numerator and denominator have opposite signs. From Figure 1.1.16, we see that this is so if and only if x is in the interval $(2, 3)$; see Figure 1.1.17. ◀

Figure 1.1.17

▶ Exercise Set 1.1

1. In each part, sketch on a coordinate line all values of x that satisfy the stated condition.

(a) $x \leq 4$ (b) $x \geq -3$

(c) $-1 \leq x \leq 7$ (d) $x^2 = 9$

(e) $x^2 \leq 9$ (f) $x^2 \geq 9$.

2. In parts (a)–(d), sketch on a coordinate line all values of x, if any, that satisfy the stated conditions.

(a) $x > 4$ and $x \leq 8$

(b) $x \leq 2$ or $x \geq 5$

(c) $x > -2$ and $x \geq 3$

(d) $x \leq 5$ and $x > 7$.

3. Among the terms, *integer*, *rational*, and *irrational*, which ones apply to the given number?

(a) $-\frac{3}{4}$ (b) 0

(c) $\frac{24}{8}$ (d) 0.25

(e) $-\sqrt{16}$ (f) $2^{1/2}$

(g) $0.020202 \ldots$ (h) $7.000 \ldots$

4. Which of the terms, *integer*, *rational*, and *irrational*, apply to the given number?
   (a) 0.31311311131111...   (b) 0.729999...
   (c) 0.376237623762...   (d) $17\frac{4}{5}$.

5. Which of the following are always correct if $a \leq b$?
   (a) $a - 3 \leq b - 3$ (b) $-a \leq -b$
   (c) $3 - a \leq 3 - b$ (d) $6a \leq 6b$
   (e) $a^2 \leq ab$ (f) $a^3 \leq a^2 b$.

6. Which of the following are always correct if $a \leq b$ and $c \leq d$?
   (a) $a + 2c \leq b + 2d$ (b) $a - 2c \leq b - 2d$
   (c) $a - 2c \geq b - 2d$.

7. The repeating decimal $0.137137137\ldots$ can be expressed as a ratio of integers by writing

   $$x = 0.137137137\ldots$$
   $$1000x = 137.137137137\ldots$$

   and subtracting to obtain $999x = 137$ or $x = \frac{137}{999}$. Use this idea, where needed, to express the following decimals as ratios of integers.
   (a) $0.123123123\ldots$ (b) $12.7777\ldots$
   (c) $38.07818181\ldots$ (d) $0.4296000\ldots$

8. Show that the repeating decimal $0.99999\ldots$ represents the number 1. Since $1.000\ldots$ is also a decimal representation of 1, this problem shows that a real number can have two different decimal representations. [*Hint:* Use the technique of Exercise 7.]

9. For what values of a are the following inequalities valid?
   (a) $a \leq a$ (b) $a < a$.

10. If $a \leq b$ and $b \leq a$, what can you say about a and b?

11. (a) If $a < b$ is true, does it follow that $a \leq b$ must also be true?
    (b) If $a \leq b$ is true, does it follow that $a < b$ must also be true?

12. In each part, list the elements in the set.
    (a) $\{x : x^2 - 5x = 0\}$
    (b) $\{x : x$ is an integer satisfying $-2 < x < 3\}$.

13. In each part, express the set in the notation $\{x : \underline{\hspace{1cm}}\}$.
    (a) $\{1, 3, 5, 7, 9, \ldots\}$

(b) the set of even integers
(c) the set of irrational numbers
(d) $\{7, 8, 9, 10\}$.

14. Let $A = \{1, 2, 3\}$. Which of the following are equal to A?
    (a) $\{0, 1, 2, 3\}$ (b) $\{3, 2, 1\}$
    (c) $\{x : (x - 3)(x^2 - 3x + 2) = 0\}$.

15. Which of the following sets are empty?
    (a) $\{x : x^2 + 1 = 0\}$ (b) $\{x : x^2 - 1 = 0\}$
    (c) $\{x : x > 3$ and $x < 3\}$
    (d) $\{x : x \geq 3$ and $x \leq 3\}$.

16. List all subsets of
    (a) $\{a_1, a_2, a_3\}$ (b) $\varnothing$.

17. In Figure 1.1.18, let
    S = the set of points inside the square
    T = the set of points inside the triangle
    C = the set of points inside the circle
    and let a, b, and c be the points shown. Answer the following as true or false.
    (a) $T \subset C$ (b) $T \subset S$
    (c) $a \notin T$ (d) $a \notin S$
    (e) $b \in T$ and $b \in C$ (f) $a \in C$ or $a \in T$
    (g) $c \in T$ and $c \notin C$.

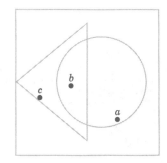

Figure 1.1.18

18. In each part, sketch the set on a coordinate line.
    (a) $[-3, 2] \cup [1, 4]$ (b) $[4, 6] \cup [8, 11]$
    (c) $(-4, 0) \cup (-5, 1)$ (d) $[2, 4) \cup (4, 7)$
    (e) $(-2, 4) \cap (0, 5]$ (f) $[1, 2.3) \cup (1.4, \sqrt{2})$
    (g) $(-\infty, -1) \cup (-3, +\infty)$
    (h) $(-\infty, 5) \cap [0, +\infty)$.

In Exercises 19–40, solve the inequality and sketch the solution on a coordinate line.

19. $3x - 2 < 8$.

20. $\frac{1}{5}x + 6 \geq 14$.

21. $4 + 5x \leq 3x - 7$.

22. $2x - 1 > 11x + 9$.

23. $3 \leq 4 - 2x < 7$.

24. $-2 \geq 3 - 8x \geq -11$.

25. $\dfrac{x}{x - 3} < 4$.

26. $\dfrac{x}{8 - x} \geq -2$.

27. $\dfrac{3x + 1}{x - 2} < 1$.

28. $\dfrac{\frac{1}{2}x - 3}{4 + x} > 1$.

29. $\dfrac{4}{2 - x} \leq 1$.

30. $\dfrac{3}{x - 5} \leq 2$.

31. $x^2 > 9$.

32. $x^2 \leq 5$.

33. $(x - 4)(x + 2) > 0$.

34. $(x - 3)(x + 4) < 0$.

35. $x^2 - 9x + 20 \leq 0$.

36. $2 - 3x + x^2 \geq 0$.

37. $\dfrac{2}{x} < \dfrac{3}{x - 4}$.

38. $\dfrac{1}{x + 1} \geq \dfrac{3}{x - 2}$.

39. $x^3 - x^2 - x - 2 > 0$.

40. $x^3 - 3x + 2 \leq 0$.

In Exercises 41 and 42, find all values of x for which the given expression yields a real number.

41. $\sqrt{x^2 + x - 6}$

42. $\sqrt{\dfrac{x + 2}{x - 1}}$

43. Prove the following results about sums of rational and irrational numbers:
 (a) rational + rational = rational.
 (b) rational + irrational = irrational.

44. Prove the following results about products of rational and irrational numbers:
 (a) rational · rational = rational
 (b) rational · irrational = irrational (provided the rational factor is nonzero).

45. Show that the sum or product of two irrational numbers can be rational or irrational.

46. Classify the following as rational or irrational and justify your conclusion.
 (a) $3 + \pi$ (b) $\sqrt[3]{4}\sqrt{2}$
 (c) $\sqrt{8}\sqrt{2}$ (d) $\sqrt{\pi}$.
 (See Exercises 43 and 44.)

47. Prove: The average of two rational numbers is a rational number, but the average of two irrational numbers can be rational or irrational.

48. Can a rational number satisfy $10^x = 3$?

49. Solve: $8x^3 - 4x^2 - 2x + 1 < 0$.

50. Solve: $12x^3 - 20x^2 \geq -11x + 2$.

51. Prove: If $a, b, c,$ and d are positive numbers such that $a < b$ and $c < d$, then $ac < bd$. (This result gives conditions under which inequalities can be "multiplied together.")

52. Show that rational numbers are represented by repeating decimals. [*Hint:* Examine the long division process that converts a ratio of integers into a decimal.]

1.2 ABSOLUTE VALUE

In this section we shall review the notion of absolute value. This concept plays an important role in algebraic computations involving radicals and in determining the distance between points on a coordinate line.

1.2.1 DEFINITION. The *absolute value* or *magnitude* of a real number a is denoted by $|a|$ and is defined by

$$|a| = a \quad \text{if} \quad a \geq 0$$
$$|a| = -a \quad \text{if} \quad a < 0$$

Example 1.

$$|5| = 5 \qquad \text{[since } 5 \geq 0]$$
$$\left|-\tfrac{4}{7}\right| = -\left(-\tfrac{4}{7}\right) = \tfrac{4}{7} \qquad \text{[since } -\tfrac{4}{7} < 0]$$
$$|0| = 0 \qquad \text{[since } 0 \geq 0] \qquad \blacktriangleleft$$

REMARK.   Note that the effect of taking the absolute value of a number is to strip away the minus sign if the number is negative and to leave the number unchanged if it is nonnegative. Thus, $|a| \geq 0$ for all values of a.

REMARK.   Symbols such as $+a$ and $-a$ are deceptive, since it is tempting to conclude that $+a$ is positive and $-a$ is negative. However, this need not be so, since a itself can represent either a positive or negative number. In fact, if a itself is negative, then $-a$ is positive and $+a$ is negative.

PROPERTIES OF ABSOLUTE VALUE

Recall that a number whose square is a is called a ***square root*** of a. In algebra it is learned that every nonnegative real number has exactly one *nonnegative* square root; we denote this square root by $\sqrt{a}$. For example, the number 9 has two square roots, -3 and 3. Since 3 is the nonnegative square root, we have $\sqrt{9} = 3$.

REMARK.   Readers who are accustomed to writing $\sqrt{9} = \pm 3$ are advised to stop doing so, since it is incorrect.

It is another common error to write $\sqrt{a^2} = a$. Although this equality is correct when a is nonnegative, it is false for negative a. For example, if $a = -4$, then

$$\sqrt{a^2} = \sqrt{(-4)^2} = \sqrt{16} = 4 \neq a$$

A result that is correct for all a is given in the following theorem.

1.2.2   THEOREM. *For any real number a,*

$$\sqrt{a^2} = |a|$$

Proof.   Since $a^2 = (+a)^2 = (-a)^2$, the numbers $+a$ and $-a$ are square roots of a^2. If $a \geq 0$, then $+a$ is the nonnegative square root of a^2, and if $a < 0$, then $-a$ is the nonnegative square root of a^2. Since $\sqrt{a^2}$ denotes the nonnegative square root of a^2, we have

$$\sqrt{a^2} = +a \quad \text{if} \quad a \geq 0$$
$$\sqrt{a^2} = -a \quad \text{if} \quad a < 0$$

That is, $\sqrt{a^2} = |a|$.   ∎

Some basic properties of absolute value are listed in the following theorem.

1.2.3 THEOREM. *If a and b are real numbers, then*

(a) $-|a| \leq a \leq |a|$

(b) $|ab| = |a|\,|b|$

(c) $\left|\dfrac{a}{b}\right| = \dfrac{|a|}{|b|}$

We shall prove parts (a) and (b) only.

Proof (a).   If $a \geq 0$, then we can write $-a \leq a \leq a$ and $|a| = a$ from which it follows that

$$-|a| \leq a \leq |a|$$

If $a < 0$, we can write $a \leq a \leq -a$ and $-a = |a|$ from which it again follows that

$$-|a| \leq a \leq |a|$$

Thus, statement (a) holds in all cases.

Proof (b).   Using Theorem 1.2.2

$$|ab| = \sqrt{(ab)^2} = \sqrt{a^2 b^2} = \sqrt{a^2}\sqrt{b^2} = |a|\,|b|. \quad \blacksquare$$

REMARK.   In words, parts (b) and (c) of this theorem state that *the absolute value of a product is the product of the absolute values and the absolute value of a ratio is the ratio of the absolute values.*

The result in part (b) of Theorem 1.2.3 can be extended to three or more factors. More precisely, for any n real numbers, $a_1, a_2, \ldots, a_n$, it follows that

$$|a_1\, a_2 \cdots a_n| = |a_1|\,|a_2| \cdots |a_n| \tag{1}$$

In the special case where $a_1, a_2, \ldots, a_n$ have the same value, a, it follows from (1) that

$$|a^n| = |a|^n \tag{2}$$

DISTANCE ON A COORDINATE LINE   The notion of absolute values arises naturally in distance problems. On a coordinate line, let A and B be points with coordinates a and b. Because

(a)

(b)

Figure 1.2.1

distance is nonnegative, the distance d between A and B is

$$d = b - a$$

when B is to the right of A (Figure 1.2.1a), and

$$d = a - b$$

when B is to the left of A (Figure 1.2.1b).

In the first case, $b - a$ is positive, so we can write

$$d = b - a = |b - a|$$

and in the second case, $b - a$ is negative, so we can write

$$d = a - b = -(b - a) = |b - a|$$

Thus, regardless of whether B is to the right or left of A, we have the following result:

Distance Formula

> **1.2.4** THEOREM. *If A and B are points on a coordinate line with coordinates a and b, respectively, then the distance d between A and B is*
>
> $$d = |b - a| \tag{3}$$

Formula (3) provides a useful geometric interpretation of some common mathematical expressions:

Table 1.2.1

EXPRESSION	GEOMETRIC INTERPRETATION ON A COORDINATE LINE
$|x - a|$	The distance between x and a.
$|x + a|$	The distance between x and $-a$ (since $|x + a| = |x - (-a)|$).
$|x|$	The distance between x and the origin (since $|x| = |x - 0|$).

The following theorem should be evident from the geometric interpretation of $|x|$ (see Figure 1.2.2). We omit the proof.

> **1.2.5** THEOREM. *For any real numbers x and a and any positive number k*
>
> *(a) $|x - a| < k$ if and only if $a - k < x < a + k$*
>
> *(b) $|x - a| > k$ if and only if $x < a - k$ or $x > a + k$*

REMARK.   Theorem 1.2.5 is also true if we replace the symbol $<$ by $\leq$ and $>$ by $\geq$.

Figure 1.2.2 ▷

SOLVING
INEQUALITIES
INVOLVING
ABSOLUTE VALUE

We shall now illustrate some techniques for solving equations and inequalities involving absolute values.

Example 2 Solve $|x - 3| = 4$.

Solution. Depending on whether $x - 3$ is positive or negative, the equation $|x - 3| = 4$ can be written as

$$x - 3 = 4 \quad \text{or} \quad -(x - 3) = 4$$

Solving these two equations gives $x = 7$ and $x = -1$. ◀

Example 3 Solve $|x - 3| < 4$.

Solution. From part (*a*) of Theorem 1.2.5, the inequality can be rewritten as

$$-4 < x - 3 < 4$$

from which we obtain

$$-4 + 3 < x < 4 + 3$$
$$-1 < x < 7$$

Geometrically, the solution consists of all x in the interval $(-1, 7)$ shown in

Figure 1.2.3 ▷

Example 4 Solve $|x + 4| \geq 2$.

Solution. From part (*b*) of Theorem 1.2.5 with $\geq$ in place of $>$, the solution consists of all x that satisfy

$$x + 4 \leq -2 \quad \text{or} \quad x + 4 \geq 2$$

Solving these inequalities we obtain

$$x \leq -6 \quad \text{or} \quad x \geq -2$$

Geometrically, the solution consists of all x in the set $(-\infty, -6] \cup [-2, +\infty)$ shown in Figure 1.2.4.   ◀

Figure 1.2.4 ▷

REMARK.   The problems in the last two examples were solved algebraically. However, we could also have proceeded geometrically. For example, the solution of $|x - 3| < 4$ consists of all x whose distance from 3 is less than 4 units (Figure 1.2.3), and the solution of $|x + 4| \geq 2$ consists of all x whose distance from -4 is 2 units or more (Figure 1.2.4).

Example 5   Solve

$$\frac{1}{|2x - 3|} > 5$$

Solution.   Observe first that $x = \frac{3}{2}$ is not a solution because it results in a division by zero. Keeping this in mind, we shall begin by applying Theorem 1.1.3e. Taking reciprocals and reversing the inequality yields

$$|2x - 3| < \tfrac{1}{5}$$
$$|2(x - \tfrac{3}{2})| < \tfrac{1}{5} \qquad \text{[Factor out the coefficient of } x.\text{]}$$
$$|2|\,|x - \tfrac{3}{2}| < \tfrac{1}{5} \qquad \text{[Theorem 1.2.3}b.\text{]}$$
$$|x - \tfrac{3}{2}| < \tfrac{1}{10} \qquad \text{[We multiplied both sides by } 1/|2| = 1/2.\text{]}$$
$$-\tfrac{1}{10} < x - \tfrac{3}{2} < \tfrac{1}{10} \qquad \text{[Theorem 1.2.5}a.\text{]}$$
$$\tfrac{7}{5} < x < \tfrac{8}{5} \qquad \text{[We added 3/2 throughout.]}$$

If we now eliminate the value $x = \frac{3}{2}$, we see that the solution consists of all x that satisfy

$$\tfrac{7}{5} < x < \tfrac{3}{2} \quad \text{or} \quad \tfrac{3}{2} < x < \tfrac{8}{5}$$

Figure 1.2.5

Geometrically, the solution consists of all x in the set $(\frac{7}{5}, \frac{3}{2}) \cup (\frac{3}{2}, \frac{8}{5})$ shown in Figure 1.2.5.   ◀

Example 6   The following inequalities, in which a is any real number and δ (Greek "delta") is a positive real number, arise frequently in calculus theory:

$$|x - a| < \delta \tag{4}$$

$$0 < |x - a| < \delta \tag{5}$$

The solution of (4) consists of all x whose distance from a is less than δ units. This is the interval $(a - \delta, a + \delta)$ shown in Figure 1.2.6a. The solution of (5) consists of all x that simultaneously satisfy $0 < |x - a|$ and (4). The former inequality is satisfied by all x except $x = a$ (why?), so the solution of (5) consists of the solutions of (4) with a removed. This is the set $(a - \delta, a) \cup (a, a + \delta)$ shown in Figure 1.2.6b. ◄

Figure 1.2.6 ▷

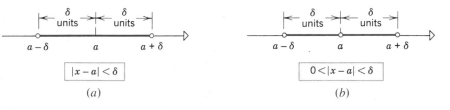

(a)   (b)

REMARK.   It is *not* generally true that $|a + b| = |a| + |b|$. For example, if $a = 2$ and $b = -3$, then $a + b = -1$, so that

$$|a + b| = |-1| = 1$$

whereas

$$|a| + |b| = |2| + |-3| = 2 + 3 = 5$$

The following theorem shows that *the absolute value of a sum is always less than or equal to the sum of the absolute values.*

The Triangle Inequality

1.2.6 THEOREM. *For any real numbers, a and b,*

$$|a + b| \le |a| + |b|$$

Proof.   From Theorem 1.2.3a

$$-|a| \le a \le |a| \quad \text{and} \quad -|b| \le b \le |b|$$

Adding these inequalities yields

$$-(|a| + |b|) \le a + b \le (|a| + |b|) \tag{6}$$

Using Theorem 1.2.5a with $\le$ in place of $<$, with $x = a + b$, and with $k = |a| + |b|$, it follows from (6) that

$$|a + b| \le |a| + |b| \quad \blacksquare$$

▶ Exercise Set 1.2

1. Compute $|x|$ if
   (a) $x = 7$ (b) $x = -\sqrt{2}$
   (c) $x = k^2$ (d) $x = -k^2$.

2. Rewrite $\sqrt{(x-6)^2}$ without using a square root or absolute value sign.

In Exercises 3–10, find all values of x for which the given statement is true.

3. $|x - 3| = 3 - x$.   4. $|x + 2| = x + 2$.

5. $|x^2 + 9| = x^2 + 9$.   6. $|x^2 + 5x| = x^2 + 5x$.

7. $|3x^2 + 2x| = x|3x + 2|$.

8. $|6 - 2x| = 2|x - 3|$.

9. $\sqrt{(x + 5)^2} = x + 5$.

10. $\sqrt{(3x - 2)^2} = 2 - 3x$.

11. Verify $\sqrt{a^2} = |a|$ for $a = 7$ and $a = -7$.

12. Verify the inequalities $-|a| \le a \le |a|$ for $a = 2$ and $a = -5$.

13. Let A and B be points with coordinates a and b. In each part find the distance between A and B.
    (a) $a = 9, b = 7$ (b) $a = 2, b = 3$
    (c) $a = -8, b = 6$ (d) $a = \sqrt{2}, b = -3$
    (e) $a = -11, b = -4$ (f) $a = 0, b = -5$.

14. Is the equality $\sqrt{a^4} = a^2$ valid for all values of a? Explain.

15. Let A and B be points with coordinates a and b. In each part, use the given information to find b.
    (a) $a = -3$, B is to the left of A, and $|b - a| = 6$
    (b) $a = -2$, B is to the right of A, and $|b - a| = 9$
    (c) $a = 5$, $|b - a| = 7$, and $b > 0$.

16. Let E and F be points with coordinates e and f. In each part, determine whether E is to the left or to the right of F on a coordinate line.
    (a) $f - e = 4$ (b) $e - f = 4$
    (c) $f - e = -6$ (d) $e - f = -7$.

In Exercises 17–24 solve for x.

17. $|6x - 2| = 7$.   18. $|3 + 2x| = 11$.

19. $|6x - 7| = |3 + 2x|$.   20. $|4x + 5| = |8x - 3|$.

21. $|9x| - 11 = x$.   22. $2x - 7 = |x + 1|$.

23. $\left|\dfrac{x + 5}{2 - x}\right| = 6$.   24. $\left|\dfrac{x - 3}{x + 4}\right| = 5$.

In Exercises 25–44, solve for x and express the solution in terms of intervals. For Exercises 37–44, use the fact that $|a| < |b|$ (or $\le$) if and only if $a^2 < b^2$ (or $\le$).

25. $|x + 6| < 3$.   26. $|7 - x| \le 5$.

27. $|2x - 3| \le 6$.   28. $|3x + 1| < 4$.

29. $|x + 2| > 1$.   30. $|\frac{1}{2}x - 1| \ge 2$.

31. $|5 - 2x| \ge 4$.   32. $|7x + 1| > 3$.

33. $\dfrac{1}{|x - 1|} < 2$.   34. $\dfrac{1}{|3x + 1|} \ge 5$.

35. $\dfrac{3}{|2x - 1|} \ge 4$.   36. $\dfrac{2}{|x + 3|} < 1$.

37. $|x + 3| < |x - 8|$.   38. $|3x| \le |2x - 5|$.

39. $|4x| \ge |7 - 6x|$.   40. $|2x + 1| > |x - 5|$.

41. $\left|\dfrac{x - \frac{1}{2}}{x + \frac{1}{2}}\right| < 1$.   42. $\left|\dfrac{3 - 2x}{1 + x}\right| \le 4$.

43. $\dfrac{1}{|x - 4|} < \dfrac{1}{|x + 7|}$.

44. $\dfrac{1}{|x - 3|} - \dfrac{1}{|x + 4|} \ge 0$.

45. For which values of x is $\sqrt{(x^2 - 5x + 6)^2} = x^2 - 5x + 6$?

46. Solve $3 \le |x - 2| \le 7$ for x.

47. Solve $|x - 3|^2 - 4|x - 3| = 12$ for x. [*Hint:* First let $z = |x - 3|$.]

48. Prove: If $|x + 2| < 2$, then $|3x - 2| < 14$.

49. Prove: If $|3x - 4| < 5$, then $|x - \frac{1}{3}| < \frac{8}{3}$.

50. Verify the triangle inequality $|a + b| \le |a| + |b|$ (Theorem 1.2.6) for
    (a) $a = 3, b = 4$ (b) $a = -2, b = 6$
    (c) $a = -7, b = -8$ (d) $a = -4, b = 4$.

51. Prove: $|a - b| \le |a| + |b|$.

52. Prove: $|a| - |b| \le |a - b|$.

53. Prove: $\big||a| - |b|\big| \le |a - b|$. [*Hint:* Use Exercise 52.]

54. Find the smallest value of M such that
    $$\left|\frac{1}{x}\right| \le M \text{ for all } x \text{ in the interval } [2, 7].$$

55. Find the smallest value of M such that

$$\left| \frac{1}{x+7} \right| \leq M$$

for all x in the interval $(-4, 2)$.

56. Use the triangle inequality to find a value of M such that

$$|x^3 - 2x + 1| \leq M$$

for all x in the interval $(-2, 3)$.

57. Find a value of M such that

$$\left| \frac{x+3}{x-3} \right| \leq M$$

for all x in the interval $[-\frac{3}{4}, \frac{1}{4}]$.

1.3 COORDINATE PLANES; DISTANCE; GRAPHS

In this section we shall review the basic ideas about rectangular coordinate systems.

RECTANGULAR COORDINATE SYSTEMS

Just as points on a line can be placed in one-to-one correspondence with the real numbers, so points in a plane can be placed in one-to-one correspondence with pairs of real numbers by using two perpendicular coordinate lines that intersect at their origins. Usually, but not always, one of the lines is horizontal with its positive direction to the right and the other is vertical with its positive direction up (Figure 1.3.1*a*). The two lines are called *coordinate axes;* the horizontal line is called the *x-axis,* the vertical line is called the *y-axis,* and the coordinate axes together form

Figure 1.3.1 ▷   (*a*)   (*b*)

what is called a *Cartesian coordinate system* or sometimes a *rectangular coordinate system.* The point of intersection of the coordinate axes is denoted by O and is called the *origin* of the coordinate system (Figure 1.3.1*b*).

REMARK. Throughout this text we will assume that the same unit of measurement is used on both coordinate axes unless stated otherwise (see Exercise 40).

A plane in which a rectangular coordinate system has been introduced is called a *coordinate plane*. Although it is common to label the axes x and y, other letters are also used, especially in applications. A coordinate plane in which x and y are used to label the horizontal and vertical axes, respectively, is called an *xy-plane*. Figure 1.3.2 shows a uv-plane and a ts-plane. The first letter in the name of a coordinate plane refers to the horizontal axis and the second to the vertical axis.

Figure 1.3.2 ▷

Figure 1.3.3

We shall now show how to establish a one-to-one correspondence between points in a coordinate plane and pairs of real numbers. If P is a point in a coordinate plane, then we draw two lines through P, one perpendicular to the x-axis and one perpendicular to the y-axis. If the first line intersects the x-axis at the point with coordinate a and the second line intersects the y-axis at the point with coordinate b, then we associate the pair (a, b) with the point P (Figure 1.3.3). The number a is called the *x-coordinate* or *abscissa* of P and the number b is called the *y-coordinate* or *ordinate* of P; we say that P is the point with *coordinates* (a, b) and denote the point by $P(a, b)$. The process of locating the position of a point in a coordinate plane is called *plotting*.

Example 1 In Figure 1.3.4 we have plotted the points

$$P(2, 5), \quad Q(-4, 3), \quad R(-5, -2), \quad S(4, -3) \quad ◀$$

From the above construction, each point in a coordinate plane determines a unique pair of numbers. Conversely, starting with a pair of numbers (a, b) we can construct lines perpendicular to the x-axis and y-axis at the points with coordinates a and b, respectively; the intersection of

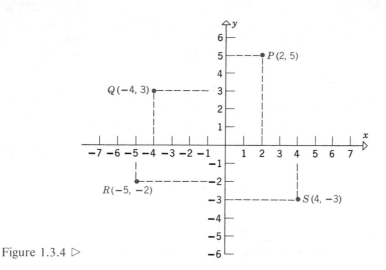

Figure 1.3.4 ▷

these lines determines a unique point P in the plane whose coordinates are (a, b) (Figure 1.3.3). Thus, we have a one-to-one correspondence between pairs of real numbers and points in a coordinate plane.

REMARK.   Since the order in which the members of a set are listed does not matter, the set $\{a, b\}$ and the set $\{b, a\}$ are the same. However, the pair of real numbers (a, b) and the pair of real numbers (b, a) represent different points (unless $a = b$), so that order is important. For this reason, a pair (a, b) of real numbers is called an ***ordered pair***.

The coordinate axes divide the plane into four parts, called ***quadrants***. These quadrants are numbered from one to four as shown in Figure 1.3.5. As shown in Figure 1.3.6 it is easy to determine the quadrant in which a

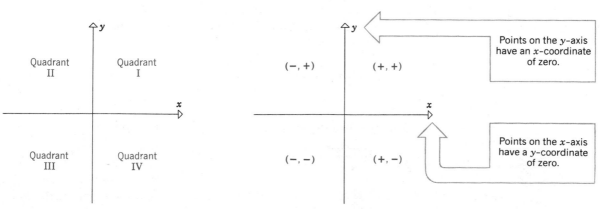

Figure 1.3.5   Figure 1.3.6

point lies from the signs of its coordinates. A point with two positive coordinates $(+, +)$ lies in Quadrant I, a point with a negative x-coordinate and a positive y-coordinate $(-, +)$ lies in Quadrant II, and so on.

DISTANCE BETWEEN POINTS IN A COORDINATE PLANE

In the last section we showed that the distance between points a and b on a coordinate line is $|b - a|$. It follows that the distance d between two points $A(x_1, y)$ and $B(x_2, y)$ on a horizontal line in the xy-plane is $d = |x_2 - x_1|$ (Figure 1.3.7), and the distance d between two points $A(x, y_1)$ and $B(x, y_2)$ on a vertical line in the xy-plane is $d = |y_2 - y_1|$ (Figure 1.3.8).

Figure 1.3.7   Figure 1.3.8

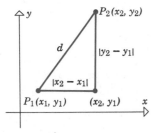

Figure 1.3.9

To find the distance d between any two points $P_1(x_1, y_1)$ and $P_2(x_2, y_2)$ in the xy-plane we apply the Theorem of Pythagoras to the triangle shown in Figure 1.3.9. This yields

$$d = \sqrt{|x_2 - x_1|^2 + |y_2 - y_1|^2}$$

Since $|x_2 - x_1|^2 = (x_2 - x_1)^2$ and $|y_2 - y_1|^2 = (y_2 - y_1)^2$, we are led to the following result.

> **1.3.1**   THEOREM. *The distance d between two points (x_1, y_1) and (x_2, y_2) in a coordinate plane is given by*
>
> $$d = \sqrt{(x_2 - x_1)^2 + (y_2 - y_1)^2} \tag{1}$$

Example 2   The distance between $(-1, 2)$ and $(3, 4)$ is

$$d = \sqrt{(3 - (-1))^2 + (4 - 2)^2} = \sqrt{4^2 + 2^2} = \sqrt{20} \quad \blacktriangleleft$$

THE MIDPOINT FORMULA

It is useful to be able to find the coordinates of the midpoint of the line segment joining two points in the plane. To see how this can be done algebraically, consider two points on a coordinate line with coordinates a and b. If we assume that $a \leq b$, then, as illustrated in Figure 1.3.10a, the distance between a and b is $b - a$, and the point midway between a and b has coordinate

$$a + \tfrac{1}{2}(b - a) = \tfrac{1}{2}a + \tfrac{1}{2}b = \tfrac{1}{2}(a + b)$$

which is the arithmetic average of a and b. The same formula results if $b \leq a$. Now let $P_1(x_1, y_1)$ and $P_2(x_2, y_2)$ be two points in the plane and $M(x, y)$ the midpoint of the line segment joining them (Figure 1.3.10b). Using plane geometry it can be shown that x is midway between x_1 and x_2 on the x-axis and y is midway between y_1 and y_2 on the y-axis, so we have the following result:

$$\left[\begin{array}{l} \textit{The midpoint of the line segment} \\ \textit{joining } P_1(x_1, y_1) \textit{ and } P_2(x_2, y_2) \end{array}\right] = \left(\frac{x_1 + x_2}{2}, \frac{y_1 + y_2}{2}\right) \qquad (2)$$

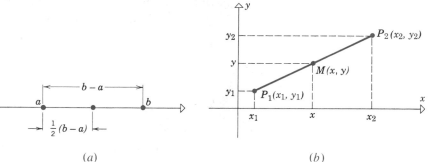

Figure 1.3.10 ▷   (a)   (b)

Example 3   From (2), the midpoint of the line segment joining the points $(3, -4)$ and $(7, 2)$ is

$$\left(\frac{3 + 7}{2}, \frac{-4 + 2}{2}\right) = (5, -1) \qquad ◀$$

GRAPHS   To see how rectangular coordinate systems enable us to describe equations geometrically, assume we have constructed a rectangular coordinate system and are given an equation involving only two variables, x and y; for example,

$$5xy = 2, \quad x^2 + 2y^2 = 7, \quad \text{or} \quad y = \frac{1}{1 - x}$$

We define a *solution* of such an equation to be an ordered pair of real numbers (a, b) such that the equation is satisfied when we substitute $x = a$ and $y = b$.

Example 4   The pair $(3, 2)$ is a solution of

$$6x - 4y = 10$$

since the equation is satisfied when we substitute $x = 3$ and $y = 2$. However, the pair $(2, 0)$ is not a solution, since the equation is not satisfied when we substitute $x = 2$ and $y = 0$.   ◀

The set of all solutions of an equation is called its **solution set,** and the corresponding set of points in a coordinate plane is called the **graph** of the equation.

Example 5   Sketch the graph of $y = x^2$.

Solution.   The solution set of $y = x^2$ has infinitely many members, so that it is impossible to plot them all. However, some sample members of the solution set can be obtained by substituting some arbitrary x values into the right side of $y = x^2$ and solving for the associated values of y. Some typical computations are given in Table 1.3.1.

Table 1.3.1

x	0	1	2	3	-1	-2	-3
$y = x^2$	0	1	4	9	1	4	9
(x, y)	$(0, 0)$	$(1, 1)$	$(2, 4)$	$(3, 9)$	$(-1, 1)$	$(-2, 4)$	$(-3, 9)$

In Figure 1.3.11 the points listed in Table 1.3.1 are connected by a smooth curve. This curve approximates the graph of $y = x^2$.   ◀

REMARK.   It is important to keep in mind that the sketch in Figure 1.3.11 is only an *approximation* to the graph of $y = x^2$ based on plotting a few

Figure 1.3.11   Figure 1.3.12

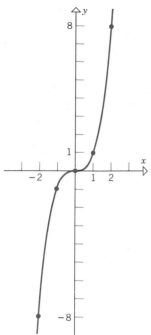

Figure 1.3.13

points. It is conceivable that the smooth curve connecting the plotted points does not accurately describe the true graph. For example, isn't it logically possible that the graph of $y = x^2$ oscillates between the points we have plotted and looks like the dark blue curve in Figure 1.3.12? Although we might look for such oscillations by plotting additional points between those already obtained, we can never resolve the problem with certainty by point plotting, since we will never be sure how the true graph behaves *between* our plotted points. In later sections we will show how calculus can be used to determine the true shape of a graph.

Example 6   Sketch the graph of $y = x^3$.

Solution.   From Table 1.3.2 we obtain the curve in Figure 1.3.13.

Table 1.3.2

x	0	1	2	-1	-2
$y = x^3$	0	1	8	-1	-8
(x, y)	$(0, 0)$	$(1, 1)$	$(2, 8)$	$(-1, -1)$	$(-2, -8)$

Example 7   Sketch the graph of $y = \sqrt{x}$.

Solution.   Since $\sqrt{x}$ is imaginary if $x < 0$, we can only plot points for which $x \geq 0$. From Table 1.3.3 we obtain the curve in Figure 1.3.14. ◀

Table 1.3.3

x	0	1	2	3	4
$y = \sqrt{x}$	0	1	$\sqrt{2} \approx 1.4$	$\sqrt{3} \approx 1.7$	2
(x, y)	$(0, 0)$	$(1, 1)$	$(2, \sqrt{2})$	$(3, \sqrt{3})$	$(4, 2)$

Figure 1.3.14

SYMMETRY   The work required to graph an equation can be reduced if the graph enjoys certain symmetry properties. A curve is called *symmetric about the x-axis* if for each point (x, y) on the curve, the point $(x, -y)$ is also on the curve (Figure 1.3.15a); geometrically, this makes the portion of curve below the x-axis the mirror image of the portion above. Similarly, a curve is *symmetric about the y-axis* if for each point (x, y) on the curve, the point $(-x, y)$ is also on the curve (Figure 1.3.15b). Finally, a curve is *symmetric about the origin* if for each point (x, y) on the curve, the point $(-x, -y)$ is also on the curve (Figure 1.3.15c); geometrically, this means that for each point P on the curve there is a companion point Q on the curve with the origin bisecting the line segment joining P and Q.

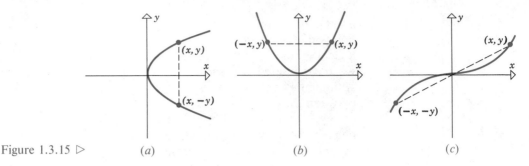

Figure 1.3.15 ▷   (a)   (b)   (c)

Symmetries can often be detected from the equation of a curve. For example, the graph of

$$y = x^2 \tag{3}$$

must be symmetric about the y-axis because for any point (x, y) whose coordinates satisfy (3), the coordinates of the point $(-x, y)$ also satisfy (3), since substituting these values in (3) yields

$$y = (-x)^2 \quad \text{or} \quad y = x^2$$

(Figure 1.3.11). This discussion suggests the following symmetry tests.

Symmetry Tests

> **1.3.2** THEOREM.
>
> (a) *A plane curve is symmetric about the y-axis if replacing x by −x in its equation produces an equivalent equation.*
> (b) *A plane curve is symmetric about the x-axis if replacing y by −y in its equation produces an equivalent equation.*
> (c) *A plane curve is symmetric about the origin if replacing x by −x and y by −y in its equation produces an equivalent equation.*

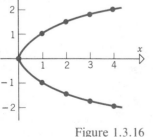

Figure 1.3.16

Example 8   Sketch the graph of $x = y^2$.

Solution. If we solve $x = y^2$ for y in terms of x, we obtain two solutions, $y = \sqrt{x}$ and $y = -\sqrt{x}$. The graph of $y = \sqrt{x}$ is the portion of the curve $x = y^2$ that is on or above the x-axis (since $y = \sqrt{x} \geq 0$), and the graph of $y = -\sqrt{x}$ is the portion that is on or below (since $y = -\sqrt{x} \leq 0$). However, the curve $x = y^2$ is symmetric about the x-axis because substituting $-y$ for y yields $x = (-y)^2$, which is equivalent to the original equation. Thus, we need only graph $y = \sqrt{x}$ (see Figure 1.3.14) and reflect it about the x-axis to complete the graph (Figure 1.3.16).   ◀

Example 9   Sketch the graph of $y = 1/x$.

Solution. Because $1/x$ is undefined when x is zero, we can only plot points for which $x \neq 0$. From Table 1.3.4 we obtain the curve in Figure

Table 1.3.4

x	$y = 1/x$	(x, y)
$\frac{1}{3}$	3	$(\frac{1}{3}, 3)$
$\frac{1}{2}$	2	$(\frac{1}{2}, 2)$
1	1	$(1, 1)$
2	$\frac{1}{2}$	$(2, \frac{1}{2})$
3	$\frac{1}{3}$	$(3, \frac{1}{3})$
$-\frac{1}{3}$	-3	$(-\frac{1}{3}, -3)$
$-\frac{1}{2}$	-2	$(-\frac{1}{2}, -2)$
-1	-1	$(-1, -1)$
-2	$-\frac{1}{2}$	$(-2, -\frac{1}{2})$
-3	$-\frac{1}{3}$	$(-3, -\frac{1}{3})$

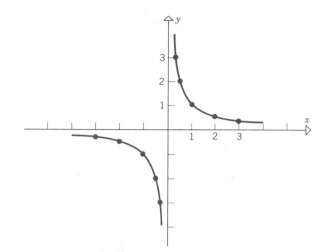

Figure 1.3.17

1.3.17. Note that the curve is symmetric about the origin because replacing x by $-x$ and y by $-y$ yields

$$-y = \frac{1}{-x}$$

which is equivalent to $y = 1/x$. Thus, we could have plotted the points in the first quadrant and obtained those in the third quadrant by symmetry.   ◄

INTERCEPTS   The points where a graph intersects the coordinate axes are of special interest in many problems. Since all points on the x-axis have a y-coordinate of zero, any points where a graph intersects the x-axis have the form $(a, 0)$ (Figure 1.3.18). The number a is called an **x-intercept** of the graph. Similarly, any points where a graph intersects the y-axis have the form $(0, b)$, and the number b is called a **y-intercept** of the graph.

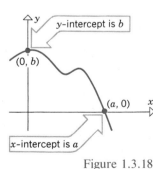

Figure 1.3.18

Example 10   Find all intercepts of

(a) $3x + 2y = 6$ (b) $y = 1/x$.

Solution (a).   To find the x-intercepts we set $y = 0$ and solve for x:

$$3x = 6 \quad \text{or} \quad x = 2$$

To find the y-intercepts we set $x = 0$ and solve for y:

$$2y = 6 \quad \text{or} \quad y = 3$$

The graph is shown in Figure 1.3.19.

Solution (b).   For the x-intercepts we set $y = 0$ and solve for x. This yields

$$\frac{1}{x} = 0$$

which has no solution. (Why?) Thus, there are no x-intercepts. For the y-intercepts we set $x = 0$ and solve for y. But, substituting $x = 0$ leads to a division by zero, which is not allowed, so there are no y-intercepts either. The graph of the equation is shown in Figure 1.3.17.   ◀

We conclude this section with a catalog of basic curves (Figure 1.3.20) that we shall encounter frequently. Some of these curves have been graphed in the examples in this section, while others have not. Try to convince yourself that the graphs are correct by plotting points in each case.

Figure 1.3.19

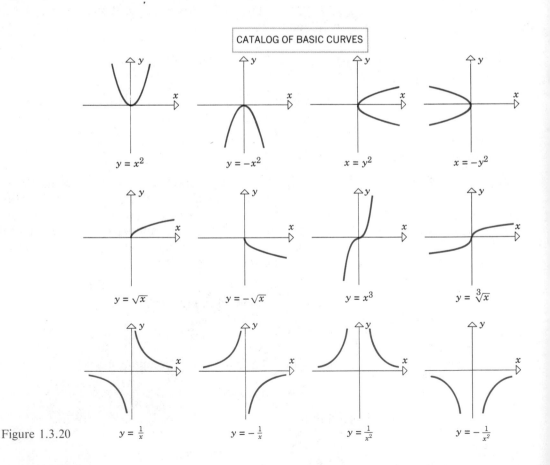

CATALOG OF BASIC CURVES

$y = x^2$ $y = -x^2$ $x = y^2$ $x = -y^2$

$y = \sqrt{x}$ $y = -\sqrt{x}$ $y = x^3$ $y = \sqrt[3]{x}$

$y = \frac{1}{x}$ $y = -\frac{1}{x}$ $y = \frac{1}{x^2}$ $y = -\frac{1}{x^2}$

Figure 1.3.20

▶ Exercise Set 1.3

1. Draw a rectangular coordinate system and locate the points
   (a) $(3, 4)$ (b) $(-2, 5)$
   (c) $(-2.5, -3)$ (d) $(1.7, -2)$
   (e) $(0, -6)$ (f) $(4, 0)$.

In Exercises 2 and 3, draw a rectangular coordinate system and sketch the set of points whose coordinates (x, y) satisfy the given conditions.

2. (a) $x = 0$ (b) $y = 0$
   (c) $y < 0$ (d) $x \geq 1$ and $y \leq 2$
   (e) $x = 3$ (f) $|x| = 5$.

3. (a) $x = 2$ (b) $y = -3$
   (c) $x \geq 0$ (d) $y = x$
   (e) $y \geq x$ (f) $|x| \geq 1$.

4. Find the center of the square whose vertices are $(1, 1)$, $(5, 1)$, $(5, -3)$, and $(1, -3)$.

5. Find the fourth vertex of the rectangle, three of whose vertices are $(-1, 4)$, $(6, 4)$, and $(-1, 9)$.

In Exercises 6 and 7, the points lie on a horizontal or vertical line. Determine whether the line is horizontal or vertical, and find the distance between the points.

6. (a) $A(9, 2)$, $B(7, 2)$
   (b) $A(2, -6)$, $B(3, -6)$
   (c) $A(-8, 1)$, $B(6, 1)$.

7. (a) $A(-4, \sqrt{2})$, $B(-4, -3)$
   (b) $A(3, -11)$, $B(3, -4)$
   (c) $A(0, 0)$, $B(0, -5)$.

In Exercises 8–11, find
(a) the distance between A and B
(b) the midpoint of the line segment joining A and B.

8. $A(2, 5)$, $B(-1, 1)$.

9. $A(7, 1)$, $B(1, 9)$.

10. $A(2, 0)$, $B(-3, 6)$.

11. $A(-2, -6)$, $B(-7, -4)$.

12. Show that the points $(-1, 7)$, $(5, 4)$, $(2, -2)$, and $(-4, 1)$ are vertices of a square.

13. Show that the triangle with vertices $(5, -2)$, $(6, 5)$, $(2, 2)$ is isosceles.

14. Show that $(1, 3)$, $(4, 2)$, and $(-2, -6)$ are vertices

of a right triangle and specify the vertex at which the right angle occurs.

15. Show that $(0, -2)$, $(-4, 8)$, and $(3, 1)$ lie on a circle with center $(-2, 3)$.

16. Show that for all values of t the point $(t, 2t - 6)$ is equidistant from $(0, 4)$ and $(8, 0)$.

17. In each part determine if the given ordered pair (x, y) is a solution of $x^2 - 2x + y = 4$.
    (a) $(0, 4)$ (b) $(-3, 7)$
    (c) $(\frac{1}{2}, \frac{19}{4})$ (d) $(1 + \sqrt{5 - t}, t)$.

In Exercises 18 and 19, determine whether the graph is symmetric about the x-axis, the y-axis, or the origin.

18. (a) $x = 5y^2 + 9$ 19. (a) $x^4 = 2y^3 + y$
    (b) $x^2 - 2y^2 = 3$ (b) $y = \dfrac{x}{3 + x^2}$
    (c) $xy = 5$.   (c) $y^2 = |x| - 5$.

In Exercises 20–29, sketch the graph of the equation. (A calculator will be helpful in some of these problems.)

20. $y = 2x - 3$.   21. $y = 6 - x$.

22. $y = 1 + x^2$.   23. $y = 4 - x^2$.

24. $y = -\sqrt{x + 1}$.   25. $y = \sqrt{x - 4}$.

26. $y = |x|$.   27. $y = |x - 3|$.

28. $xy = -1$.   29. $x^2y = 2$.

In Exercises 30 and 31, sketch the portion of the graph in the first quadrant, and use symmetry to complete the rest of the graph. (A hand calculator will be helpful.)

30. $9x^2 + 4y^2 = 36$.   31. $4x^2 + 16y^2 = 16$.

32. Sketch the graph of $y^2 = 3x$ and explain how this graph is related to the graphs of $y = \sqrt{3x}$ and $y = -\sqrt{3x}$.

33. Sketch the graph of $(x - y)(x + y) = 0$ and explain how it is related to the graphs of $x - y = 0$ and $x + y = 0$.

34. Graph $F = \frac{9}{5}C + 32$ in a CF-coordinate system.

35. Graph $u = 3v^2$ in a uv-coordinate system.

36. Graph $Y = 4X + 5$ in a YX-coordinate system.

37. Find an equation whose graph is the perpendic-

ular bisector of the line segment connecting $(-2, 1)$ and $(4, -3)$.

38. Use the distance formula in Theorem 1.3.1 to prove that $(1, 1)$, $(-2, -8)$, and $(4, 10)$ lie on a straight line.

39. Find k, given that $(2, k)$ is equidistant from $(3, 7)$ and $(9, 1)$.

40. Where in this section did we use the fact that the same unit of measure was used on both coordinate axes?

41. Prove: The midpoint of the hypotenuse of a right triangle is equidistant from the three vertices. [*Hint:* Let a and b denote the lengths of the sides and introduce coordinate axes so that the vertices are $(0, 0)$, $(a, 0)$, and $(0, b)$.]

42. Use Theorem 1.3.2 to show that a graph which is symmetric about the x-axis and y-axis must be symmetric about the origin. Give an example to show that the converse is not true.

1.4 SLOPE OF A LINE

In this section we shall discuss ways to measure the "steepness" or "slope" of a line in the plane. The ideas we develop here will be important when we discuss equations and graphs of straight lines.

This section requires a knowledge of the trigonometry material in Unit 1 of Appendix 1. Readers who need to review that material are advised to do so before starting this section.

In surveying, the *grade* or *slope* of a hill is defined to be the ratio of its rise to its run (Figure 1.4.1). We shall now show how the surveyor's notion of slope can be adapted to measure the steepness of a line in the xy-plane.

$$\text{Slope} = \frac{\text{rise}}{\text{run}} = \frac{2 \text{ ft}}{20 \text{ ft}} = \frac{1}{10}$$

Expressed as a percentage, this is a 10% grade.

Figure 1.4.1 ▷

Consider a particle moving left to right along a *nonvertical* line segment from a point $P_1(x_1, y_1)$ to a point $P_2(x_2, y_2)$. As shown in Figure 1.4.2, the

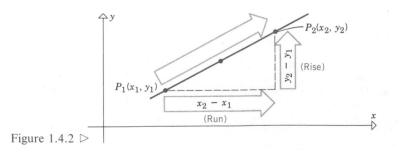

Figure 1.4.2 ▷

particle moves $y_2 - y_1$ units in the y-direction as it travels $x_2 - x_1$ units in the x-direction. The vertical change $y_2 - y_1$ is called the *rise,* and the horizontal change $x_2 - x_1$ the *run.*

By analogy with the surveyor's notion of slope we make the following definition.

1.4.1 DEFINITION. If $P_1(x_1, y_1)$ and $P_2(x_2, y_2)$ are the endpoints of a nonvertical line segment, then the *slope m* of the line segment is defined by

$$m = \frac{\text{rise}}{\text{run}} = \frac{y_2 - y_1}{x_2 - x_1} \qquad (1)$$

For two line segments with the same run, the more steeply inclined segment will have the greater rise and therefore the greater slope (Figure 1.4.3). Thus, the slope m is a numerical measure of steepness.

REMARK. Note that Definition 1.4.1 does not apply to vertical line segments. For such segments we would have $x_2 = x_1$, so (1) would involve a division by zero. The slope of a vertical line segment is *undefined.*

Figure 1.4.3

Example 1 In each part find the slope of the line segment that connects

(a) the points (6, 2) and (9, 8)
(b) the points (2, 9) and (4, 3)
(c) the points (−2, 7) and (5, 7)

Solution (a). If we label (6, 2) as P_1 and (9, 8) as P_2, then from (1) the slope is

$$m = \frac{8 - 2}{9 - 6} = \frac{6}{3} = 2$$

Solution (b). The slope of the line segment connecting (2, 9) and (4, 3) is

$$m = \frac{3 - 9}{4 - 2} = \frac{-6}{2} = -3$$

Solution (c). The slope of the line segment connecting (−2, 7) and (5, 7) is

$$m = \frac{7 - 7}{5 - (-2)} = 0 \qquad \blacktriangleleft$$

REMARK. When applying formula (1), it does not matter which point is used as P_1 and which is used as P_2. For example, had we reversed the roles of P_1 and P_2 in part (a) of Example 1 we would have obtained

$$m = \frac{2 - 8}{6 - 9} = \frac{-6}{-3} = 2$$

which is the same value obtained for the slope using the points in the original order.

It is clear geometrically that two line segments on the same straight line will be equally "steep," and consequently have the same slope. This is the content of the following theorem.

1.4.2   THEOREM. *On a nonvertical line, all line segments have the same slope.*

Proof.   Let $P_1(x_1, y_1)$ and $P_2(x_2, y_2)$ be distinct points on a nonvertical line L and let $P_1'(x_1', y_1')$ and $P_2'(x_2', y_2')$ be another pair of distinct points on L. We shall show that the slope

$$m = \frac{y_2 - y_1}{x_2 - x_1}$$

of the line segment joining P_1 and P_2 is equal to the slope

$$m' = \frac{y_2' - y_1'}{x_2' - x_1'}$$

of the line segment joining P_1' and P_2'. (We shall assume the points are ordered as in Figure 1.4.4. The proofs of the remaining cases are similar and will be omitted for brevity.) The triangles $P_1 Q P_2$ and $P_1' Q' P_2'$ in Figure 1.4.4 are similar, so that the lengths of corresponding sides are propor-

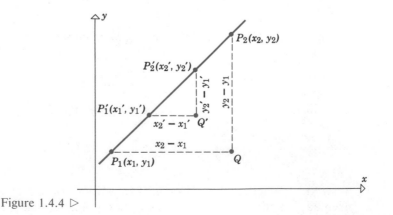

Figure 1.4.4 ▷

tional. Therefore,

$$\frac{y_2 - y_1}{x_2 - x_1} = \frac{y_2' - y_1'}{x_2' - x_1'}$$

or, equivalently, $m = m'$.   ▮

   We define the *slope* of a nonvertical line L to be the common slope of all line segments on L. The slope of a vertical line is **undefined.** Speaking informally, some people say that a vertical line has **infinite slope,** or that it has **no slope.**

Example 2   In applied problems it is essential to include the units in the calculation of slope because changing the units can alter the slope of a line. For example, in Figure 1.4.5 the volume V is measured in cubic centimeters (cm^3) and the temperature T in degrees Kelvin (°K).

V (cm^3)	T (°K)
82.51	223
101.01	273
119.51	323
138.01	373
156.51	423
175.01	473

Figure 1.4.5 ▷

Thus, using the first and third lines in the table that accompanies the graph, we obtain

$$m = \frac{119.51(cm^3) - 82.51(cm^3)}{323(°K) - 223(°K)} = .37 \; cm^3/°K   ◀$$

   The significance of the numerical value of the slope can be seen by rewriting (1) as

$$y_2 - y_1 = m(x_2 - x_1) \tag{2}$$

Because $y_2 - y_1$ represents the rise and $x_2 - x_1$ the run, it follows from (2) that as we travel left to right along a nonvertical line, the rise is proportional to the run and the slope m is the constant of proportionality that relates the two. For this reason the slope m is called the *rate of change of y with respect to x* along the line. As we move from point to point along the graph, through increasing values of the x-coordinate, the slope measures the change in the y-coordinate for each unit change in x.

Figure 1.4.6 ▷

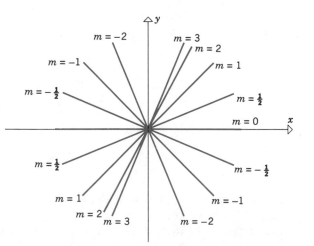

Example 3   Figure 1.4.6 shows the three lines determined by the points in Example 1 and interprets the significance of their slopes.   ◀

As illustrated in the last example, the slope of a line can be positive, negative, or zero. A positive slope means that the line is inclined upward to the right, a negative slope means that it is inclined downward to the right, and a zero slope means that the line is horizontal. An undefined slope means that the line is vertical. In Figure 1.4.7 we have sketched some lines with different slopes.

Figure 1.4.

ANGLE OF   The slope of a line is related to the angle the line makes with the positive
INCLINATION x-axis. To establish this relationship we need the following definition.

1.4.3 DEFINITION. For a line L not parallel to the x-axis, the *angle of inclination* is the smallest angle ϕ measured counterclockwise from the direction of the positive x-axis to L (Figure 1.4.8). For a line parallel to the x-axis, we take $\phi = 0$.

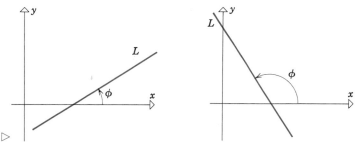

Figure 1.4.8 ▷

REMARK.   In degree measure the angle of inclination satisfies $0° \leq \phi < 180°$ and in radian measure it satisfies $0 \leq \phi < \pi$.

The following theorem relates the slope of a line to its angle of inclination.

1.4.4 THEOREM. *For a nonvertical line, the slope and angle of inclination are related by*

$$m = \tan \phi \tag{3}$$

Proof.   If the line is horizontal, then $m = 0$ and $\phi = 0°$. Thus,

$$\tan \phi = \tan 0 = 0 = m$$

so (3) holds in this case. If the line is not horizontal, let $(x_0, 0)$ be its point of intersection with the x-axis, and construct a circle of radius 1 centered at this point. Because the angle of inclination of the line is ϕ, the line will intersect this circle at the point $(x_0 + \cos \phi, \sin \phi)$ (Figure 1.4.9). (This follows from Theorem 3 in Unit 1 of the trigonometry review in Appendix 1.)

Since the points $(x_0, 0)$ and $(x_0 + \cos \phi, \sin \phi)$ lie on L, we can use them to compute the slope m. This gives

$$m = \frac{\sin \phi - 0}{(x_0 + \cos \phi) - x_0} = \frac{\sin \phi}{\cos \phi} = \tan \phi$$

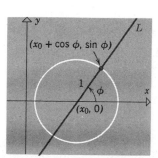

Figure 1.4.9   which proves (3).   ▋

Figure 1.4.10

REMARK.   If the line L is parallel to the y-axis, then $\phi = \frac{1}{2}\pi$ so $\tan\phi$ in (3) is undefined. This agrees with the fact that the slope m is also undefined in this case.

Example 4   Find the angle of inclination for a line of slope $m = 1$ and also for a line of slope $m = -1$.

Solution.   If $m = 1$, then from (3), $\tan\phi = 1$ so that $\phi = \pi/4$ (or in degree measure $\phi = 45°$). If $m = -1$, then from (3), $\tan\phi = -1$. From this equality and the fact that $0 \le \phi < \pi$ we obtain $\phi = 3\pi/4$ or, in degree measure, $\phi = 135°$ (Figure 1.4.10).   ◄

PARALLEL AND PERPENDICULAR LINES

As a consequence of Theorem 1.4.4, we obtain the following basic result.

1.4.5   THEOREM. *Two nonvertical lines are parallel if and only if they have the same slope.*

Proof.   If L_1 and L_2 are nonvertical parallel lines, then their angles of inclination ϕ_1 and ϕ_2 are equal, since two parallel lines cut by a transversal have equal corresponding angles (Figure 1.4.11). Thus,

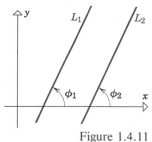

Figure 1.4.11

$$\text{slope } L_1 = \tan\phi_1 = \tan\phi_2 = \text{slope } L_2$$

Conversely, if L_1 and L_2 have the same slope m, then

$$m = \tan\phi_1 = \tan\phi_2 \tag{4}$$

Because $0 \le \phi_1 < \pi$ and $0 \le \phi_2 < \pi$, it follows from (4) that ϕ_1 and ϕ_2 are equal. Thus, L_1 and L_2 are parallel (Figure 1.4.11).   ∎

The next theorem shows how slopes can be used to determine whether two lines are perpendicular.

1.4.6   THEOREM. *Two nonvertical lines are perpendicular if and only if the product of their slopes is* -1; *that is, lines with slopes m_1 and m_2 are perpendicular if and only if $m_1 m_2 = -1$ or equivalently*

$$m_2 = -\frac{1}{m_1} \tag{5}$$

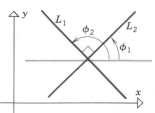

Figure 1.4.12

Proof.   We shall prove first that if L_1 and L_2 are nonvertical perpendicular lines, then (5) holds. Assume that L_1 and L_2 have angles of inclination ϕ_1 and ϕ_2 and that $\phi_1 < \phi_2$ (Figure 1.4.12), so that

$$\phi_2 = \phi_1 + \frac{\pi}{2}$$

Thus,

$$m_2 = \tan \phi_2 = \tan\left(\phi_1 + \frac{\pi}{2}\right) = \frac{\sin\left(\phi_1 + \frac{\pi}{2}\right)}{\cos\left(\phi_1 + \frac{\pi}{2}\right)}$$

$$= \frac{\sin \phi_1 \cos \frac{\pi}{2} + \cos \phi_1 \sin \frac{\pi}{2}}{\cos \phi_1 \cos \frac{\pi}{2} - \sin \phi_1 \sin \frac{\pi}{2}}$$

$$= -\frac{\cos \phi_1}{\sin \phi_1} = -\frac{1}{\sin \phi_1 / \cos \phi_1} = -\frac{1}{\tan \phi_1} = -\frac{1}{m_1}$$

which establishes (5). The proof of the converse is left as an exercise (Exercise 29). ∎

Example 5 Use slopes to show that the points $A(1,3)$, $B(3,7)$, and $C(7,5)$ are vertices of a right triangle.

Solution. The line through A and B has slope

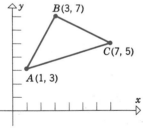

$$m_1 = \frac{7-3}{3-1} = 2$$

and the line through B and C has slope

$$m_2 = \frac{5-7}{7-3} = -\frac{1}{2}$$

Since $m_1 m_2 = -1$, the line through A and B is perpendicular to the line through B and C; thus, ABC is a right triangle (Figure 1.4.13). ◀

Figure 1.4.13

▶ Exercise Set 1.4

1. Find the slope of the line through
   (a) $(-1, 2)$ and $(3, 4)$
   (b) $(5, 3)$ and $(7, 1)$
   (c) $(4, \sqrt{2})$ and $(-3, \sqrt{2})$
   (d) $(-2, -6)$ and $(-2, 12)$.

In Exercises 2 and 3, find the slope of the line whose angle of inclination is given. (You should be able to solve these problems without tables or a calculator.)

2. (a) $45°$ (b) $\frac{2\pi}{3}$ (c) $30°$.

3. (a) $\dfrac{\pi}{6}$ (b) 135°   (c) 60°.

In Exercises 4 and 5, use a calculator or the trigonometric tables in Appendix 3, where necessary, to find (to the nearest degree) the angle of inclination of a line with the given slope.

4. (a) $m = \frac{1}{2}$ (b) $m = -1$
   (c) $m = 2$ (d) $m = -57$.

5. (a) $m = -\frac{1}{2}$ (b) $m = 1$
   (c) $m = -2$ (d) $m = 57$.

6. Draw the line through $(4, 2)$ with slope
   (a) $m = 3$ (b) $m = -2$
   (c) $m = -\frac{3}{4}$.

7. Draw the line through $(-1, -2)$ with slope
   (a) $m = \frac{3}{5}$ (b) $m = -1$
   (c) $m = \sqrt{2}$.

8. Let L be a line with slope $m = 2$. Determine whether the given line L' is parallel to L, perpendicular to L, or neither.
   (a) L' is the line through $(2, 4)$ and $(4, 8)$
   (b) L' is the line through $(2, 4)$ and $(6, 2)$
   (c) L' is the line through $(1, 5)$ and $(2, -3)$.

9. Let L be a line with slope $m = -3$. Determine whether the given line L' is parallel to L, perpendicular to L, or neither.
   (a) L' is the line through $(1, 8)$ and $(2, 5)$
   (b) L' is the line through $(6, 5)$ and $(3, 4)$
   (c) L' is the line through $(1, 0)$ and $(-2, 1)$.

10. A particle, initially at $(7, 5)$, moves along a line of slope $m = -2$ to a new position (x, y).
    (a) Find y if $x = 9$.   (b) Find x if $y = 12$.

11. A particle, initially at $(1, 2)$, moves along a line of slope $m = 3$ to a new position (x, y).
    (a) Find y if $x = 5$.   (b) Find x if $y = -2$.

12. Given that the point $(k, 4)$ is on the line through $(1, 5)$ and $(2, -3)$, find k.

13. Let the point $(3, k)$ lie on the line of slope $m = 5$ through $(-2, 4)$; find k.

14. Find the slopes of the sides of the triangle with vertices $(-1, 2)$, $(6, 5)$, and $(2, 7)$.

15. Use slopes to determine whether the given points lie on the same line.
    (a) $(1, 1)$, $(-2, -5)$, and $(0, -1)$
    (b) $(-2, 4)$, $(0, 2)$, and $(1, 5)$.

16. An equilateral triangle has one vertex at the origin, another on the x-axis, and the third in the first quadrant. Find the slopes of its sides.

17. Find the coordinates of all points P on the x-axis so that the line through $A(1, 2)$ and P is perpendicular to the line through $B(8, 3)$ and P.

18. Use slopes to show that $(3, 1)$, $(6, 3)$, and $(2, 9)$ are vertices of a right triangle.

19. Use slopes to show that $(3, -1)$, $(6, 4)$, $(-3, 2)$, and $(-6, -3)$ are vertices of a parallelogram.

20. Find x and y if the line through $(0, 0)$ and (x, y) has slope $\frac{1}{2}$, and the line through (x, y) and $(7, 5)$ has slope 2.

21. Given two intersecting lines, let L_2 be the line with the larger angle of inclination ϕ_2, and let L_1 be the line with the smaller angle of inclination ϕ_1. We define the **angle θ between L_1 and L_2** by

    $$\theta = \phi_2 - \phi_1$$

    (a) Prove geometrically that θ is the angle pictured in Figure 1.4.14. [*Remark:* The angle θ is the smallest positive angle through which L_1 can be rotated until it coincides with L_2.]

Figure 1.4.14

    (b) Prove: If L_1 and L_2 are not perpendicular, then

    $$\tan \theta = \frac{m_2 - m_1}{1 + m_1 m_2}$$

22. Use the result of Exercise 21 to find the tangent of the angle between the lines whose slopes are
    (a) 1 and 3   (b) -2 and 4
    (c) $-\frac{1}{2}$ and $-\frac{2}{3}$.

23. Use the result of Exercise 21 to find the tangent of the angle between the lines whose slopes are
(a) $\frac{4}{5}$ and $\frac{1}{3}$ (b) -0.7 and 5
(c) -6 and -2.

24. Use the result of Exercise 21 and a calculator or the trigonometric tables in Appendix 3 to find, to the nearest degree, the angle between the lines whose slopes are given in Exercise 22.

25. Repeat the directions of Exercise 24 for the lines whose slopes are given in Exercise 23.

26. Use the result of Exercise 21 to find, to the nearest degree, the interior angles of the triangle whose vertices are $(4, -3)$, $(-3, -1)$, and $(6, 6)$.

27. Use the result of Exercise 21 to find the slope of the line that bisects the angle A of the triangle whose vertices are $A(0, 2)$, $B(-8, 8)$, and $C(8, 6)$.

28. Let L be a line of slope $m = -2$. Use the result of Exercise 21 to find the slope of a line K such that the angle between K and L is $45°$. (Two solutions.)

29. Complete the proof of Theorem 1.4.6 by showing: If L_1 and L_2 are lines whose slopes satisfy $m_1 m_2 = -1$, then L_1 and L_2 are perpendicular.

1.5 EQUATIONS OF STRAIGHT LINES

In this section we shall be concerned with recognizing those equations whose graphs are straight lines and finding equations for lines specified geometrically.

LINES PARALLEL TO THE COORDINATE AXES

A line parallel to the y-axis intersects the x-axis at some point $(a, 0)$. This line consists precisely of those points whose x-coordinate is equal to a (Figure 1.5.1a). Similarly, a line parallel to the x-axis intersects the y-axis at some point $(0, b)$. This line consists precisely of those points whose y-coordinate is equal to b (Figure 1.5.1b). Thus, we have the following theorem.

1.5.1 THEOREM. *The vertical line through $(a, 0)$ and the horizontal line through $(0, b)$ are represented by the equations*

$$x = a \quad and \quad y = b$$

respectively.

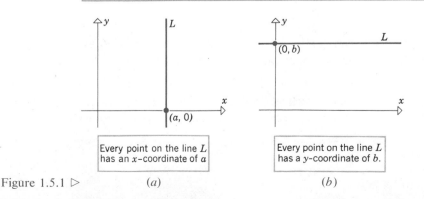

Figure 1.5.1 ▷   (a)   (b)

Example 1   The graph of $x = -5$ is the vertical line through $(-5, 0)$, and the graph of $y = 7$ is the horizontal line through $(0, 7)$ (Figure 1.5.2).   ◀

LINES DETERMINED
BY POINT AND SLOPE

There are infinitely many lines that pass through any given point in the plane. However, if we specify the slope of the line in addition to a point on it, then the point and the slope together determine a unique line (Figure 1.5.3).

Let us now consider how to find an equation of a nonvertical line L that passes through a point $P_1(x_1, y_1)$ and has slope m. If $P(x, y)$ is any point on L, different from P_1, then the slope m can be obtained from the points $P(x, y)$ and $P_1(x_1, y_1)$; this gives

$$m = \frac{y - y_1}{x - x_1}$$

which can be rewritten as

$$y - y_1 = m(x - x_1) \tag{1}$$

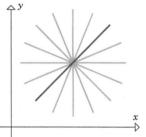

With the possible exception of (x_1, y_1), we have shown that every point on L satisfies (1). But $x = x_1$, $y = y_1$ satisfies (1), so that all points on L satisfy (1). We leave it as an exercise to show that every point satisfying (1) lies on L.

In summary, we have the following theorem.

Figure 1.5.2

1.5.2   THEOREM. *The line passing through $P_1(x_1, y_1)$ and having slope m is given by the equation*

$$y - y_1 = m(x - x_1) \tag{2}$$

*This is called the **point-slope form** of the line.*

A point and slope
determine a unique
line

Figure 1.5.3

Example 2   Find the point-slope form of the line through $(4, -3)$ with slope 5.

Solution.   Substituting the values $x_1 = 4$, $y_1 = -3$, and $m = 5$ in (2) yields the point-slope form $y + 3 = 5(x - 4)$.   ◀

LINES DETERMINED
BY SLOPE AND
y-INTERCEPT

A nonvertical line crosses the y-axis at some point $(0, b)$. If we use this point in the point-slope form of its equation, we obtain

$$y - b = m(x - 0)$$

which we can rewrite as $y = mx + b$. To summarize:

1.5.3 THEOREM. *The line with y-intercept b and slope m is given by the equation*

$$y = mx + b \tag{3}$$

*This is called the **slope-intercept form** of the line.*

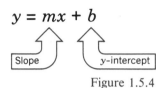

Figure 1.5.4

REMARK.   Note that y is alone on one side of Equation (3). When the equation of a line is written in this way the slope of the line and its y-intercept can be determined by inspection of the equation: the slope is the coefficient of x and the y-intercept is the constant term (Figure 1.5.4).

Example 3

EQUATION	SLOPE	y-INTERCEPT
$y = 3x + 7$	$m = 3$	$b = 7$
$y = -x + \frac{1}{2}$	$m = -1$	$b = \frac{1}{2}$
$y = x$	$m = 1$	$b = 0$
$y = \sqrt{2}x - 8$	$m = \sqrt{2}$	$b = -8$
$y = 2$	$m = 0$	$b = 2$

◄

Example 4   Find the slope-intercept form of the equation of the line that satisfies the stated conditions.

(a)   slope $= -9$; crosses the y-axis at $(0, -4)$
(b)   slope $= 1$; passes through the origin
(c)   passes through $(5, -1)$; perpendicular to $y = 3x + 4$

Solution (a).   From the given conditions $m = -9$ and $b = -4$, so (3) yields $y = -9x - 4$.

Solution (b).   From the given conditions $m = 1$ and the line passes through $(0, 0)$, so $b = 0$. Thus, it follows from (3) that $y = x + 0$ or $y = x$.

Solution (c).   The given line has slope 3, so the line to be determined will have slope $m = -\frac{1}{3}$. Substituting this slope and the given point in (2) then simplifying yields

$$y - (-1) = -\tfrac{1}{3}(x - 5)$$
$$y = -\tfrac{1}{3}x + \tfrac{2}{3} \quad ◄$$

LINES DETERMINED
BY TWO POINTS

If $P_1(x_1, y_1)$ and $P_2(x_2, y_2)$ are distinct points on a nonvertical line, then the slope of the line is

$$m = \frac{y_2 - y_1}{x_2 - x_1}$$

Substituting this expression in (2) we obtain the following result.

1.5.4   THEOREM. *The nonvertical line determined by the points $P_1(x_1, y_1)$ and $P_2(x_2, y_2)$ can be represented by the equation*

$$y - y_1 = \frac{y_2 - y_1}{x_2 - x_1} (x - x_1) \tag{4}$$

*This is called the **two-point form** of the line.*

Example 5   Find the slope-intercept form of the line passing through $(3, 4)$ and $(2, -5)$.

Solution.   Letting $(x_1, y_1) = (3, 4)$ and $(x_2, y_2) = (2, -5)$ and substituting in (4), we obtain the two-point form

$$y - 4 = \frac{-5 - 4}{2 - 3} (x - 3)$$

which can be written as $y - 4 = 9(x - 3)$. Solving for y yields the slope-intercept form

$$y = 9x - 23 \qquad \blacktriangleleft$$

THE GENERAL
EQUATION OF
A LINE

An equation expressible in the form

$$Ax + By + C = 0 \tag{5}$$

where A, B, and C are constants and A and B are not both zero, is called a ***first-degree equation*** in x and y. For example,

$$4x + 6y - 5 = 0$$

is a first-degree equation in x and y since it has form (5) with

$$A = 4, \quad B = 6, \quad C = -5$$

In fact, all the equations of lines studied in this section are first-degree equations in x and y.

The following theorem states that the first-degree equations in x and y are precisely the equations whose graphs in the xy-plane are straight lines. (The proof is left as an exercise.)

1.5.5 THEOREM. *Every first-degree equation in x and y has a straight line as its graph and, conversely, every straight line can be represented by a first-degree equation in x and y.*

Because of this theorem, (5) is sometimes called the **general equation** of a line or a **linear equation** in x and y.

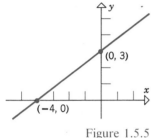

Figure 1.5.5

Example 6   Graph the equation

$$3x - 4y + 12 = 0$$

Solution.   Since this is a linear equation in x and y, its graph is a straight line. Thus, to sketch the graph we need only plot two points on the graph and draw the line through them. It is particularly convenient to plot the points where the line crosses the coordinate axes. These points are $(0, 3)$ and $(-4, 0)$ (verify), so the graph is the line shown in Figure 1.5.5.   ◄

► Exercise Set 1.5

1. Graph the equations
  (a) $2x + 5y = 15$
  (b) $x = 3$
  (c) $y = -2$
  (d) $y = 2x - 7$.

2. Graph the equations
  (a) $\dfrac{x}{3} - \dfrac{y}{4} = 1$
  (b) $x = -8$
  (c) $y = 0$
  (d) $x = 3y + 2$.

3. Graph the equations
  (a) $y = 2x - 1$
  (b) $y = 3$
  (c) $y = -2x$.

4. Graph the equations
  (a) $y = 2 - 3x$
  (b) $y = \tfrac{1}{4}x$
  (c) $y = -\sqrt{3}$.

5. Find the slope and y-intercept of
  (a) $y = 3x + 2$
  (b) $y = 3 - \tfrac{1}{4}x$
  (c) $3x + 5y = 8$
  (d) $y = 1$
  (e) $\dfrac{x}{a} + \dfrac{y}{b} = 1$.

6. Find the slope and y-intercept of
  (a) $y = -4x + 2$
  (b) $x = 3y + 2$
  (c) $\dfrac{x}{2} + \dfrac{y}{3} = 1$
  (d) $y - 3 = 0$
  (e) $a_0 x + a_1 y = 0 \quad (a_1 \neq 0)$.

7. To the nearest degree, find the angle of inclination of
  (a) $y = \sqrt{3}\, x + 2$
  (b) $y + 2x + 5 = 0$.

8. To the nearest degree, find the angle of inclination of
  (a) $3y = 2 - \sqrt{3}\, x$
  (b) $y - 4x + 7 = 0$.

In Exercises 9–24, find the slope-intercept form of the line satisfying the given conditions.

9. Slope $= -2$, y-intercept $= 4$.

10. $m = 5$, $b = -3$.

11. The line is parallel to $y = 4x - 2$ and has y-intercept 7.

12. The line is parallel to $3x + 2y = 5$ and passes through $(-1, 2)$.

13. The line is perpendicular to $y = 5x + 9$ and has y-intercept 6.

14. The line is perpendicular to $x - 4y = 7$ and passes through $(3, -4)$.

15. The line passes through $(2, 4)$ and $(1, -7)$.

16. The line passes through $(-3, 6)$ and $(-2, 1)$.

17. The line is the perpendicular bisector of the line segment joining $(2, 8)$ and $(-4, 6)$.

18. The line is the perpendicular bisector of the line segment joining $(5, -1)$ and $(4, 8)$.

19. The line has an angle of inclination of $\phi = \frac{1}{6}\pi$ and a y-intercept of -3.

20. The line has angle of inclination $\phi = \frac{2}{3}\pi$ and passes through the point $(1, 2)$.

21. The y-intercept is 2 and the x-intercept is -4.

22. The y-intercept is b and the x-intercept is a.

23. The line is perpendicular to the y-axis and passes through $(-4, 1)$.

24. The line is parallel to $y = -5$ and passes through $(-1, -8)$.

25. Find an equation for the line that passes through $(5, -2)$ and has angle of inclination $\phi = \frac{1}{2}\pi$.

26. Find an equation for the line along the y-axis.

27. In each part, classify the lines as parallel, perpendicular, or neither.
    (a) $y = 4x - 7$ and $y = 4x + 9$
    (b) $y = 2x - 3$ and $y = 7 - \frac{1}{2}x$
    (c) $5x - 3y + 6 = 0$ and $10x - 6y + 7 = 0$
    (d) $Ax + By + C = 0$ and $Bx - Ay + D = 0$
    (e) $y - 2 = 4(x - 3)$ and $y - 7 = \frac{1}{4}(x - 3)$.

28. In each part, classify the lines as parallel, perpendicular, or neither.
    (a) $y = -5x + 1$ and $y = 3 - 5x$
    (b) $y - 1 = 2(x - 3)$ and $y - 4 = -\frac{1}{2}(x + 7)$
    (c) $4x + 5y + 7 = 0$ and $5x - 4y + 9 = 0$
    (d) $Ax + By + C = 0$ and $Ax + By + D = 0$
    (e) $y = \frac{1}{2}x$ and $x = \frac{1}{2}y$.

29. Find the area of the triangle formed by the coordinate axes and the line through $(1, 4)$ and $(2, 1)$.

30. Draw the graph of $4x^2 - 9y^2 = 0$. [*Hint:* Factor.]

31. In each part, find the point of intersection of the lines.
    (a) $2x + 3y = 5$ and $y = -1$
    (b) $4x + 3y = -2$ and $5x - 2y = 9$.

32. In each part, find the point of intersection of the lines.
    (a) $6x - 9y = 7$ and $x = -\frac{2}{3}$
    (b) $6x - 2y = -3$ and $-8x + 3y = 5$.

33. Find the distance from the point $(2, 1)$ to the line $4x - 3y + 10 = 0$. [*Hint:* Find the foot of the perpendicular dropped from the point to the line.]

34. Find the distance from the point $(8, 4)$ to the line $5x + 12y - 36 = 0$. [*Hint:* See the hint in Exercise 33.]

35. Use the method described in Exercise 33 to prove that the distance d from (x_0, y_0) to the line $Ax + By + C = 0$ is

$$d = \frac{|Ax_0 + By_0 + C|}{\sqrt{A^2 + B^2}}$$

36. Use the formula in Exercise 35 to solve Exercise 33.

37. Use the formula in Exercise 35 to solve Exercise 34.

38. Prove: For any triangle, the perpendicular bisectors of the sides meet at a point. [*Hint:* Position the triangle with one vertex on the y-axis and the opposite side on the x-axis, so that the vertices are $(0, a)$, $(b, 0)$, and $(c, 0)$.]

39. Find the point on the line $4x - 2y + 3 = 0$ that is equidistant from $(3, 3)$ and $(7, -3)$.

40. In physical problems linear equations involving variables other than x and y often arise. In parts (a)–(g) determine whether the equation is linear.
    (a) $3\alpha - 2\beta = 5$
    (b) $A = 2000(1 + 0.06t)$
    (c) $A = \pi r^2$
    (d) $E = mc^2$ (c constant)
    (e) $V = C(1 - rt)$ (r and C constant)
    (f) $V = \frac{1}{3}\pi r^2 h$ (r constant)
    (g) $V = \frac{1}{3}\pi r^2 h$ (h constant).

41. There are two common systems for measuring temperature, Celsius and Fahrenheit. Water freezes at 0° Celsius and 32° Fahrenheit; it boils at 100° Celsius and 212° Fahrenheit.
(a) Assuming that the Celsius temperature C and the Fahrenheit temperature F are related by a linear equation, find the equation.

(b) What is the slope of the line relating F and C if F is plotted on the horizontal axis?

42. Prove: If (x, y) satisfies Equation (1), then the point $P(x, y)$ lies on the line with slope m passing through $P_1(x_1, y_1)$.

43. Prove Theorem 1.5.5.

1.6 CIRCLES; EQUATIONS OF THE FORM $y = ax^2 + bx + c$

In this section we shall study equations of circles and equations of the form $y = ax^2 + bx + c$. Such equations will be studied in more detail later in the text, but they are introduced here to provide an immediate source of useful examples for subsequent sections.

CIRCLES

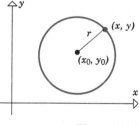

Figure 1.6.1

If (x_0, y_0) is a fixed point in the plane, then the circle of radius r centered at (x_0, y_0) is the set of all points in the plane whose distance from (x_0, y_0) is r (Figure 1.6.1). Thus, a point (x, y) will lie on this circle if and only if

$$\sqrt{(x - x_0)^2 + (y - y_0)^2} = r$$

or equivalently

$$(x - x_0)^2 + (y - y_0)^2 = r^2 \qquad (1)$$

This is called the **standard form of the equation of a circle**.

Example 1

EQUATION OF A CIRCLE	CENTER (x_0, y_0)	RADIUS r
$(x - 2)^2 + (y - 5)^2 = 9$	$(2, 5)$	3
$(x + 7)^2 + (y + 1)^2 = 16$	$(-7, -1)$	4
$x^2 + y^2 = 1$	$(0, 0)$	1
$(x - 4)^2 + y^2 = 5$	$(4, 0)$	$\sqrt{5}$

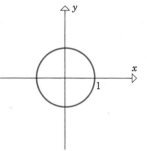

The circle $x^2 + y^2 = 1$, which is centered at the origin and has a radius of 1 (Figure 1.6.2), is often called the **unit circle**. ◀

Figure 1.6.2   **Example 2**   Find an equation for the circle of radius 5 centered at $(-3, 2)$.

Solution.   From (1) with $x_0 = -3$, $y_0 = 2$, and $r = 5$ we obtain

$$(x + 3)^2 + (y - 2)^2 = 25$$

or, if preferred, we can square out the terms to obtain

$$x^2 + y^2 + 6x - 4y - 12 = 0 \quad \blacktriangleleft$$

An alternative version of Equation (1) can be obtained by squaring out the terms and simplifying. This yields an equation of the form

$$x^2 + y^2 + dx + ey + f = 0 \tag{2}$$

Still another version of the equation of a circle can be obtained by multiplying both sides of Equation (2) by a nonzero constant A. This yields an equation of the form

$$Ax^2 + Ay^2 + Dx + Ey + F = 0 \tag{3}$$

where A, D, E, and F are constants and $A \neq 0$.

If the equation of a circle is given in form (2) or (3), then the center and radius can be determined by rewriting the equation in standard form and then reading off the center (x_0, y_0) and radius r from this equation. The following example illustrates how this is done using the algebraic technique of *completing the square.*

Example 3   Find the center and radius of the circle with equation

$$x^2 + y^2 - 8x + 2y + 8 = 0$$

Solution.   First, group the x-terms, group the y-terms, and take the constant to the right side:

$$(x^2 - 8x) + (y^2 + 2y) = -8$$

Next we want to add the appropriate constant within each set of parentheses to complete the square, and add the same constants to the right side to maintain equality. Recall that the appropriate constant is obtained by taking half the coefficient of the first-degree term and squaring it. This yields

$$(x^2 - 8x + 16) + (y^2 + 2y + 1) = -8 + 16 + 1$$

or

$$(x - 4)^2 + (y + 1)^2 = 9$$

Thus from (1), the circle has center $(4, -1)$ and radius 3. $\blacktriangleleft$

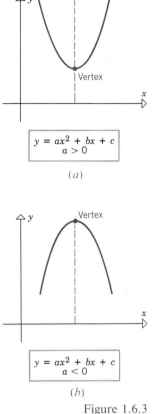

$y = ax^2 + bx + c$
$a > 0$

(a)

$y = ax^2 + bx + c$
$a < 0$

(b)

Figure 1.6.3

REMARK. To find the center and radius of a circle whose equation is of form (3) with $A \neq 1$, we can first divide both sides by A to put the equation in form (2), and then proceed as in Example 3.

An equation of form (3) does not always have a circle as its graph. To see why, suppose we divide both sides of (3) by A and complete the squares to obtain

$$(x - x_0)^2 + (y - y_0)^2 = k$$

• If $k > 0$, then the graph is a circle with center (x_0, y_0) and radius $\sqrt{k}$.

• If $k = 0$, the only solution of the equation is $x = x_0$, $y = y_0$; thus, the graph is the single point (x_0, y_0).

• If $k < 0$, then the equation has no real solutions, so there is no graph.

In summary, we have the following result.

1.6.1   THEOREM. *An equation of the form*

$$Ax^2 + Ay^2 + Dx + Ey + F = 0$$

where $A \neq 0$, represents a circle, or a point, or else has no graph.

The last two cases in Theorem 1.6.1 are called ***degenerate cases***.

THE GRAPH OF
$y = ax^2 + bx + c$

An equation of the form

$$y = ax^2 + bx + c \quad (a \neq 0) \tag{4}$$

is called a ***quadratic equation in x***. Its graph is a curve called a ***parabola***. If a is positive, the parabola will open up as shown in Figure 1.6.3a, and if a is negative, it will open down as in Figure 1.6.3b. In both cases the parabola is symmetric about a vertical line parallel to the y-axis. This line of symmetry cuts the parabola at a point called the ***vertex***. The vertex is the low point on the curve if $a > 0$ and the high point if $a < 0$.

In the exercises we will help the reader show that the x-coordinate of the vertex is given by the formula

$$x = -\frac{b}{2a} \tag{5}$$

With the aid of this formula, a reasonably accurate graph of a quadratic equation in x can be obtained by plotting the vertex and two points on each side of it.

Example 4   Sketch the graph of $y = x^2 - 4x + 5$.

Solution.   The given equation is of form (4), with $a = 1$, $b = -4$, and $c = 5$, so by (5) the x-coordinate of the vertex is

$$x = -\frac{b}{2a} = 2$$

Using this value and two additional values on each side, we obtain Table 1.6.1.

$y = x^2 - 4x + 5$

Table 1.6.1

x	0	1	2	3	4
$y = x^2 - 4x + 5$	5	2	1	2	5

Figure 1.6.4   By plotting the points in this table, we obtain the curve in Figure 1.6.4.
◀

Example 5   Sketch the graph of $y = -x^2 + 2x + 2$.

Solution.   This equation is of form (4), with $a = -1$, $b = 2$, and $c = 2$, so by (5) the x-coordinate of the vertex is

$$x = -\frac{b}{2a} = 1$$

$y = -x^2 + 2x + 2$

Using this value and two additional values on each side, we obtain Table 1.6.2.

Table 1.6.2

x	−1	0	1	2	3
$y = -x^2 + 2x + 2$	−1	2	3	2	−1

Figure 1.6.5   By plotting the points in this table, we obtain the curve in Figure 1.6.5.
◀

Often, the intercepts of a parabola $y = ax^2 + bx + c$ are important to know. The y-intercept, $y = c$, results immediately by setting $x = 0$. However, to obtain the x-intercepts, if any, we must set $y = 0$ and solve the resulting quadratic equation $ax^2 + bx + c = 0$.

Example 6   From Figure 1.6.4, we see that the parabola $y = x^2 - 4x + 5$ has no x-intercepts. This can be verified algebraically by setting $y = 0$ and solving for x. We obtain

$$x^2 - 4x + 5 = 0$$

By the quadratic formula

$$x = \frac{-b \pm \sqrt{b^2 - 4ac}}{2a} = \frac{4 \pm \sqrt{16 - 20}}{2} = 2 \pm \frac{\sqrt{-4}}{2}$$

so the solutions are not real. Thus, there are no x-intercepts.   ◄

Example 7   From Figure 1.6.5 the x-intercepts of the parabola $y = -x^2 + 2x + 2$ appear to lie somewhere in the open intervals $(-1, 0)$ and $(2, 3)$. To find these intercepts we set $y = 0$ to obtain

$$-x^2 + 2x + 2 = 0$$

Solving by the quadratic formula gives

$$x = \frac{-b \pm \sqrt{b^2 - 4ac}}{2a} = \frac{-2 \pm \sqrt{12}}{-2} = 1 \pm \sqrt{3}$$

Thus, the x-intercepts are

$$x = 1 + \sqrt{3} \approx 2.7 \quad \text{and} \quad x = 1 - \sqrt{3} \approx -.7 \quad ◄$$

REMARK.   If x and y are interchanged in (4), the resulting equation,

$$x = ay^2 + by + c$$

is called a *quadratic equation in y*. The graph of such an equation is a parabola with its line of symmetry parallel to the x-axis and its vertex at the point with y-coordinate $y = -b/2a$. Some problems relating to such equations appear in the exercises.

▶ Exercise Set 1.6

In Exercises 1 and 2, find the center and radius of each circle.

1. (a) $x^2 + y^2 = 25$
    (b) $(x - 1)^2 + (y - 4)^2 = 16$
    (c) $(x + 1)^2 + (y + 3)^2 = 5$
    (d) $x^2 + (y + 2)^2 = 1$.

2. (a) $x^2 + y^2 = 9$
    (b) $(x - 3)^2 + (y - 5)^2 = 36$
    (c) $(x + 4)^2 + (y + 1)^2 = 8$
    (d) $(x + 1)^2 + y^2 = 1$.

In Exercises 3–10, find the standard equation of the circle satisfying the given conditions.

3. Center $(3, -2)$; radius $= 4$.

4. Center $(1, 0)$; diameter $= \sqrt{8}$.

5. Center $(-4, 8)$; circle is tangent to the x-axis.

6. Center $(5, 8)$; circle is tangent to the y-axis.

7. Center $(-3, -4)$; circle passes through the origin.

8. Center $(4, -5)$; circle passes through $(1, 3)$.

9. A diameter has endpoints $(2, 0)$ and $(0, 2)$.

10. A diameter has endpoints $(6, 1)$ and $(-2, 3)$.

In Exercises 11–22, determine whether the equation represents a circle, a point, or no graph. If the equation represents a circle, find the center and radius.

11. $x^2 + y^2 - 2x - 4y - 11 = 0$.

12. $x^2 + y^2 + 8x + 8 = 0$.

13. $2x^2 + 2y^2 + 4x - 4y = 0$.

14. $6x^2 + 6y^2 - 6x + 6y = 3$.

15. $x^2 + y^2 + 2x + 2y + 2 = 0$.

16. $x^2 + y^2 - 4x - 6y + 13 = 0$.

17. $9x^2 + 9y^2 = 1$.

18. $\dfrac{x^2}{4} + \dfrac{y^2}{4} = 1$.

19. $x^2 + y^2 + 10y + 26 = 0$.

20. $x^2 + y^2 - 10x - 2y + 29 = 0$.

21. $16x^2 + 16y^2 + 40x + 16y - 7 = 0$.

22. $4x^2 + 4y^2 - 16x - 24y = 9$.

23. Find an equation of
   (a) the bottom half of the circle $x^2 + y^2 = 16$
   (b) the top half of the circle
       $x^2 + y^2 + 2x - 4y + 1 = 0$.

24. Find an equation of
   (a) the right half of the circle $x^2 + y^2 = 9$
   (b) the left half of the circle
       $x^2 + y^2 - 4x + 3 = 0$.

25. Graph
   (a) $y = \sqrt{25 - x^2}$
   (b) $y = \sqrt{5 + 4x - x^2}$.

26. Graph
   (a) $x = -\sqrt{4 - y^2}$
   (b) $x = 3 + \sqrt{4 - y^2}$.

27. Find an equation of the line that is tangent to the circle $x^2 + y^2 = 25$ at the point $(3, 4)$ on the circle.

28. Find an equation of the line that is tangent to the circle at the point P on the circle
   (a) $x^2 + y^2 + 2x = 9$; $P(2, -1)$
   (b) $x^2 + y^2 - 6x + 4y = 13$; $P(4, 3)$.

29. For the circle $x^2 + y^2 = 20$ and the point $P(-1, 2)$:
   (a) Is P inside, outside, or on the circle?
   (b) Find the largest and smallest distances between P and points on the circle.

30. Follow the directions of Exercise 29 for the circle $x^2 + y^2 - 2y - 4 = 0$ and the point $P(3, \frac{3}{2})$.

In Exercises 31–44, graph the parabola and label the coordinates of the vertex and the intersections with the coordinate axes.

31. $y = x^2 + 2$.   32. $y = x^2 - 3$.

33. $y = x^2 + 2x - 3$.   34. $y = x^2 - 3x - 4$.

35. $y = -x^2 + 4x + 5$.   36. $y = -x^2 + x$.

37. $y = (x - 2)^2$.   38. $y = (3 + x)^2$.

39. $x^2 - 2x + y = 0$.   40. $x^2 + 8x + 8y = 0$.

41. $y = 3x^2 - 2x + 1$.   42. $y = x^2 + x + 2$.

43. $x = -y^2 + 2y + 2$.   44. $x = y^2 - 4y + 5$.

45. Find an equation of
   (a) the right half of the parabola $y = 3 - x^2$
   (b) the left half of the parabola $y = x^2 - 2x$.

46. Find an equation of
   (a) the upper half of the parabola $x = y^2 - 5$
   (b) the lower half of the parabola $x = y^2 - y - 2$.

47. Graph
   (a) $y = \sqrt{x + 5}$ (b) $x = -\sqrt{4 - y}$.

48. Graph
   (a) $y = 1 + \sqrt{4 - x}$ (b) $x = 3 + \sqrt{y}$.

49. If a ball is thrown straight up with an initial velocity of 32 ft/sec, then after t seconds, the distance s above its starting height is given by
   $s = 32t - 16t^2$.
   (a) Graph this equation in a ts-coordinate system (t-axis horizontal).
   (b) At what time t will the ball be at its highest point, and how high will it rise?

50. A point (x, y) moves so that its distance to $(2, 0)$ is $\sqrt{2}$ times its distance to $(0, 1)$.
   (a) Show that the point moves along a circle.
   (b) Find the center and radius.

51. A point (x, y) moves so that the sum of the squares of its distances from $(4, 1)$ and $(2, -5)$ is 45.
   (a) Show that the point moves along a circle.
   (b) Find the center and radius.

52. (a) By completing the square, show that

$y = ax^2 + bx + c$ can be rewritten as

$$y = a\left(x + \frac{b}{2a}\right)^2 + \left(c - \frac{b^2}{4a}\right)$$

if $a \neq 0$.
   (b) Use the result in (a) to show that the graph of $y = ax^2 + bx + c$ has its high point at $x = -b/(2a)$ if $a < 0$ and its low point there if $a > 0$.

▶ SUPPLEMENTARY EXERCISES

In Exercises 1–5, use interval notation to describe the set of all values of x (if any) that satisfy the given inequalities.

1. (a) $-3 < x \leq 5$ (b) $-1 < x^2 \leq 9$
   (c) $x^2 \geq \frac{1}{4}$.

2. (a) $|2x + 1| > 5$ (b) $|x^2 - 9| \geq 7$
   (c) $1 \leq |x| \leq 3$.

3. (a) $2x^2 - 5x > 3$ (b) $x^2 - 5x + 4 \leq 0$.

4. (a) $\dfrac{x}{1 - x} \geq 3$ (b) $\dfrac{2x + 3}{x} \geq x$.

5. (a) $\dfrac{|x| - 1}{|x| - 2} \leq 0$ (b) $|x - 1| \leq 2|x + 2|$.

6. Among the terms *integer, rational, irrational,* which ones apply to the given number?
   (a) $\sqrt{4/9}$ (b) 2^{-2}
   (c) $-4^{1/3}$ (d) 0.87
   (e) $-4^{1/2}$ (f) $0.1010010001\ldots$
   (g) 3.222 (h) $3/(-1)$.

7. (a) Find values of a and b such that $a < b$, but $a^2 > b^2$.
   (b) If $a < b$, what additional assumptions on a and b are required to ensure that $a^2 < b^2$?

8. Which of the following are true for all sets A and B?
   (a) $A \subset (A \cap B)$ (b) $(A \cap B) \subset A$
   (c) $\varnothing \subset A$ (d) $A \subset (A \cup B)$
   (e) $(A \cap B) \subset (A \cup B)$
   (f) either $A \subset B$ or $B \subset A$
   (g) $A \in B$.

9. Prove: $|x| \leq \sqrt{x^2 + y^2}$ and $|y| \leq \sqrt{x^2 + y^2}$. Interpret this geometrically.

In Exercises 10–14, draw a rectangular coordinate system and sketch the set of points whose coordinates (x, y) satisfy the given conditions.

10. (a) $y = 0$ and $x > 0$
   (b) $2x - y \leq 3$.

11. (a) $xy = x^2$
   (b) $y(x - 1) = x^2 - 1$.

12. (a) $y = (x^3 - 1)/(x - 1)$
   (b) $y > x^2 - 9$.

13. (a) $y^2 - 6y + x^2 - 2x - 6 \geq 0$
   (b) $x + |y - 2| = 1$.

14. (a) $|x| + |y| = 4$
   (b) $|x| - |y| = 4$.

15. Where does the parabola $y = x^2$ intersect the line $y - 2 = x$?

In Exercises 16–19, sketch the graph of the given equation.

16. $xy + 4 = 0$.   **17.** $y = |x - 2|$.
18. $y = \sqrt{4 - x^2}$.   **19.** $y = x(x - 2)$.

In Exercises 20–24, find the standard equation for the circle satisfying the given conditions.

20. The circle centered at $(3, -2)$ and tangent to the line $y = 1$.

21. The circle centered at $(1, 2)$ and passing through the point $(4, -2)$.

22. The circle centered on the line $x = 2$ and passing through the points $(1, 3)$ and $(3, -11)$.

23. The circle of radius 5 tangent to the lines $y = 7$ and $x = 6$.

24. The circle of radius 13 that passes through the origin and the point $(0, -24)$.

In Exercises 25–28, determine whether the equation represents a circle, a point, or has no graph. If it represents a circle, find the center and radius.

25. $x^2 + y^2 + 4x + 2y + 5 = 0$.

26. $4x^2 + 4y^2 - 4x + 8y + 1 = 0$.

27. $x^2 + y^2 - 3x + 2y + 4 = 0$.

28. $3x^2 + 3y^2 - 5x + 7y + 3 = 0$.

29. In each part, find an equation for the line through A and B, the distance between A and B, and the coordinates of the midpoint of the line segment joining A and B.
    (a) $A(3, 4), B(-3, -4)$ (b) $A(3, 4), B(3, -4)$
    (c) $A(3, 4), B(-3, 4)$ (d) $A(3, 4), B(4, 3)$.

30. Show that the point $(8, 1)$ is *not* on the line through the points $(-3, -2)$ and $(1, -1)$.

31. Where does the circle of radius 5 centered at the origin intersect the line of slope $-3/4$ through the origin?

32. Fahrenheit and Celsius temperatures are related by $F - 32 = 9C/5$. What temperature is the same in both Fahrenheit and Celsius?

33. Find the inclination angle of the line whose equation is
    (a) $x = 3$ (b) $y = -2$
    (c) $2x + 2y = 1$ (d) $\sqrt{3}x - y = 4$.

34. Find the slope of the line whose angle of inclination is

(a) 30° (b) 120° (c) 90°.

In Exercises 35–37, find the slope-intercept form of the line satisfying the stated conditions.

35. The line through $(2, -3)$ and $(4, -3)$.

36. The line with x-intercept -2 and angle of inclination $\phi = 45°$.

37. The line parallel to $x + 2y = 3$ that passes through the origin.

38. Find an equation of the perpendicular bisector of the line segment joining $A(-2, -3)$ and $B(1, 1)$.

In Exercises 39–41, find equations of the lines L and L' and determine their point of intersection.

39. L passes through $(1, 0)$ and $(-1, 4)$.
    L' is perpendicular to L and has y-intercept -3.

40. L passes through $(-2, 0)$ and $(-2, 3)$.
    L' passes through $(-1, 4)$ and is perpendicular to L.

41. L has slope $2/5$ and passes through $(3, 1)$.
    L' has x-intercept $-8/3$ and y-intercept -4.

42. Consider the triangle with vertices $A(5, 2), B(1, -3)$, and $C(-3, 4)$. Find the point-slope form of the line containing:
    (a) the median from C to AB
    (b) the altitude from C to AB.

43. Use slopes to show that the points $(5, 6), (-4, 3)$, $(-3, -2)$, and $(6, 1)$ are vertices of a parallelogram. Is it a rectangle?

44. For what value of k (if any) will the line $2x + ky = 3k$ satisfy the stated condition?
    (a) have slope 3
    (b) have y-intercept 3
    (c) be parallel to the x-axis
    (d) pass through $(1, 2)$.

2 functions and limits

2.1 FUNCTIONS

In this section we shall define one of the most fundamental concepts in mathematics, the notion of a function. We shall discuss the notation used to describe functions and investigate some of their basic properties.

DEFINITION OF A FUNCTION

Historically, the term "function" was first used by Leibniz in 1673 to denote the dependence of one quantity on another. For example:

- The area A of a circle depends on its radius r by the equation $A = \pi r^2$; we say that "A is a function of r."

- The velocity of a ball dropped from a height increases with time until it hits the ground. Thus, the velocity v depends on the time t and we say that "v is a function of t."

- At a fixed point on earth, the wind speed w varies with the time t. Thus, "w is a function of t."

- In a bacteria culture, the number of bacteria present after one hour of growth depends on the number present initially; we say that "the size of the bacteria population after one hour is a function of the initial population size."

In order to describe functions without stating specific formulas, the

Swiss mathematician, Leonhard Euler* (pronounced "oiler") conceived the idea of denoting functions by letters of the alphabet. For example, if we use the letter f to denote a function, then the equation

$$y = f(x) \tag{1}$$

(read "y equals f of x") conveys the idea that y is a function of x. There is nothing special about the letter f; any symbol can be used to denote a function. Thus,

$$y = F(x), \quad y = f_1(x), \quad y = g(x), \quad \text{and} \quad y = \phi(x)$$

$y = f(x)$
(pounds)

x
(in)

Water

Figure 2.1.1

all express the fact that y is a function of x. In fact, by using different symbols for a function we can describe different relationships between y and x. For example, suppose that the bucket in Figure 2.1.1 is filled with water to a height of x inches and placed on a scale. The resulting weight y of the container and liquid in pounds will be some function of x, say $y = f(x)$. However, if we use a liquid other than water, for example mercury or alcohol, this function will be different; mercury will produce some weight function $y = g(x)$ and alcohol still another weight function $y = h(x)$.

---

*LEONHARD EULER (1707–1783). Euler was probably the most prolific mathematician who ever lived. It has been said that, "Euler wrote mathematics as effortlessly as most men breathe." He was born in Basel, Switzerland, and was the son of a Protestant minister who had himself studied mathematics. Euler's genius developed early. He attended the University of Basel, where by age 16 he obtained both a Bachelor of Arts degree and a Master's degree in philosophy. While at Basel, Euler had the good fortune to be tutored one day a week in mathematics by a distinguished mathematician, Johann Bernoulli. At the urging of his father, Euler then began to study theology. The lure of mathematics was too great, however, and by age 18 Euler had begun to do mathematical research. Nevertheless, the influence of his father and his theological studies remained, and throughout his life Euler was a deeply religious, unaffected person. At various times Euler taught at St. Petersburg Academy of Sciences (in Russia), the University of Basel, and the Berlin Academy of Sciences. Euler's energy and capacity for work were virtually boundless. His collected works form more than 100 quarto sized volumes and it is believed that much of his work has been lost. What is particularly astonishing is that Euler was blind for the last 17 years of his life, and this was one of his most productive periods! Euler's flawless memory was phenomenal. Early in his life he memorized the entire *Aeneid* by Virgil and at age 70 could not only recite the entire work, but could also state the first and last sentence on each page of the book from which he memorized the work. His ability to solve problems in his head was beyond belief. He worked out in his head major problems of lunar motion that baffled Isaac Newton and once did a complicated calculation in his head to settle an argument between two students whose computations differed in the fiftieth decimal place.

Following the development of calculus by Leibniz and Newton, results in mathematics developed rapidly in a disorganized way. Euler's genius gave coherence to the mathematical landscape. He was the first mathematician to bring the full power of calculus to bear on problems from physics. He made major contributions to virtually every branch of mathematics as well as to the theory of optics, planetary motion, electricity, magnetism, and general mechanics.

REMARK. It is important to understand that in (1), x and y may represent numerical quantities, but f itself does not represent a numerical quantity; it stands for a "relationship" between y and x.

Since the time of Euler and Leibniz the notion of a function has evolved into the following more precise and general mathematical concept:

2.1.1 DEFINITION. A *function* is a rule that assigns to each element in a set A one and only one element in a set B.

DOMAIN AND RANGE

The element that f associates with x is denoted by $f(x)$.

Figure 2.1.2

The set A in this definition is called the *domain* of the function. If x is an element in the domain of a function f, then the element in B that f associates with x is denoted by the symbol $f(x)$ (read "f of x") and is called the *image of x under f* or the *value of f at x* (Figure 2.1.2). The set of all possible values of $f(x)$ as x varies over the domain of f is called the *range of f*.

In general, the sets A and B need not be sets of real numbers; however, for the time being we will only be concerned with functions for which A and B are both subsets of the real numbers.

To see how Definition 2.1.1 relates to the historical concept of a function, consider the relationship

$$y = f(x)$$

where x and y are the quantities in Figure 2.1.3. Let A be the set of all possible x-values and B the set of all possible y-values. Then with each x in A the function f associates one and only one value y in B. This is precisely the description of a function as given in Definition 2.1.1. To determine the domain and range of this function suppose that

$H =$ the height of the bucket (in)
$y_0 =$ the weight of the empty bucket (lbs)
$y_1 =$ the weight of the bucket when filled with water (lbs)

Then the height x of the water must satisfy

$$0 \leq x \leq H$$

and the corresponding weight y must satisfy

$$y_0 \leq y \leq y_1$$

so the domain of the function f is the interval of real numbers $[0, H] = \{x : 0 \leq x \leq H\}$ and the range is the interval of real numbers $[y_0, y_1] = \{y : y_0 \leq y \leq y_1\}$ (see Figure 2.1.3).

The three most common ways of specifying functions are:

$y_0 =$ weight of the empty bucket

y

$y = f(x)$

$y_1 =$ weight of the full bucket

H

Scale

Water

x

Domain: $0 \leq x \leq H$
Range: $y_0 \leq y \leq y_1$

Figure 2.1.3

- formulas

- equations

- tables

We shall consider each of these methods in this chapter.

FUNCTIONS DEFINED BY FORMULAS

The formula

$$f(x) = x^2$$

describes a function that associates the number x^2 with the number x. Thus,

$$f(3) = 3^2 = 9 \qquad [f \text{ associates 9 with 3}]$$
$$f(-2) = (-2)^2 = 4 \qquad [f \text{ associates 4 with } -2]$$
$$f(0) = 0^2 = 0 \qquad [f \text{ associates 0 with 0}]$$
$$f(\sqrt{2}) = (\sqrt{2})^2 = 2 \qquad [f \text{ associates 2 with } \sqrt{2}]$$

Example 1 If $f(x) = 3x - 4$, then

$$f(0) = 3(0) - 4 = -4, \qquad f(1) = 3(1) - 4 = -1$$
$$f(-3) = 3(-3) - 4 = -13, \quad f(\sqrt{5}) = 3(\sqrt{5}) - 4 = 3\sqrt{5} - 4 \qquad \blacktriangleleft$$

Example 2 If $\phi(x) = 2x^2 - 1$, then

$$\phi(4) = 2(4)^2 - 1 = 31$$
$$\phi(t) = 2t^2 - 1$$
$$\phi(k + 1) = 2(k + 1)^2 - 1 = 2k^2 + 4k + 1$$
$$\phi(k) + 1 = (2k^2 - 1) + 1 = 2k^2 \qquad \blacktriangleleft$$

For reasons that will be discussed later, radian measure of angles is preferred over degree measure in calculus. Therefore, in trigonometric expressions such as $\sin x$, $\cos x$, $\tan x$, $\cot x$, $\sec x$, and $\csc x$ it is understood that x is measured in radians. When degree measure is called for, we shall make this clear by writing $\sin x°$, $\cos x°$, and so forth.

Example 3 If $f(x) = \sin x$, then

$$f\left(\frac{\pi}{2}\right) = \sin \frac{\pi}{2} = 1$$

$$f(1) = \sin(1) \approx .841471 \qquad \left[\begin{array}{l}\text{Use a calculator} \\ \text{set to the radian} \\ \text{mode to obtain} \\ \text{this approximation.}\end{array}\right]$$

$$f(0) = \sin(0) = 0 \qquad \blacktriangleleft$$

REMARK.   If a function is defined by a formula and there is no domain specified, then it is understood that the domain consists of all real numbers for which the formula makes sense and yields a real value for the function.

Example 4   If

$$f(x) = x^2$$

then this formula makes sense and yields a real value for all real values of x. Thus, the domain of f is $\{x: -\infty < x < +\infty\} = (-\infty, +\infty)$. To determine the range it will be helpful to let $y = x^2$. As x varies over the domain of f, the corresponding y values (which must be nonnegative) vary over the set $\{y: 0 \le y < +\infty\} = [0, +\infty)$, which is the range of f.   ◄

Example 5   If

$$g(x) = 2 + \sqrt{x - 1}$$

then $g(x)$ is not a real number if $x < 1$, but is real otherwise. Thus, the domain of g is $\{x: 1 \le x\} = [1, +\infty)$. To determine the range of g, let $y = 2 + \sqrt{x - 1}$. As x varies over the interval $[1, +\infty)$, the value of $\sqrt{x - 1}$ varies over the interval $[0, +\infty)$, so the value of $y = 2 + \sqrt{x - 1}$ varies over the interval $[2, +\infty)$, which is the range of g.   ◄

Example 6   If

$$h(x) = \frac{1}{(x - 1)(x - 3)}$$

Figure 2.1.4

then $h(x)$ is undefined if $x = 1$ or $x = 3$ because division by zero is not allowed; otherwise $h(x)$ has a real value. Thus, the domain of h consists of all real x except $x = 1$ and $x = 3$. In interval notation the domain is $(-\infty, 1) \cup (1, 3) \cup (3, +\infty)$. (See Figure 2.1.4.)   ◄

Example 7   If $f(x) = \tan x$, then $f(x)$ is undefined if

$$x = \pm \frac{\pi}{2}, \pm \frac{3\pi}{2}, \pm \frac{5\pi}{2}, \dots$$

but has a real value otherwise. Thus, the domain consists of all x except

$$x = \pm \frac{\pi}{2}, \pm \frac{3\pi}{2}, \pm \frac{5\pi}{2}, \ldots . \quad \blacktriangleleft$$

The next example shows that care must be exercised when simplifying the formula for a function algebraically to avoid accidentally altering the domain.

Example 8   The function

$$h(x) = \frac{x^2 - 4}{x - 2} \tag{2}$$

has a real value everywhere except at $x = 2$, where we have a division by zero:

$$h(2) = \frac{2^2 - 4}{2 - 2} = \frac{0}{0}$$

Thus, the domain of h consists of all x except $x = 2$. However, if we rewrite (2) as

$$h(x) = \frac{(x - 2)(x + 2)}{(x - 2)} \tag{3}$$

then cancel the common factors, we obtain

$$h(x) = x + 2 \tag{4}$$

which *is* defined at $x = 2$ since

$$h(2) = 2 + 2 = 4$$

Thus, our algebraic simplification has altered the domain of the function. In order to cancel the factors in (3) and not alter the domain of h, we must restrict the domain in (4) and write

$$h(x) = x + 2, \quad x \neq 2 \quad \blacktriangleleft$$

REMARK.   In algebra the reader undoubtedly learned to simplify (3) by writing (4). In elementary problems this often causes no difficulty because the domain of the function is irrelevant to the solution of the problem. However, in more advanced problems the domain is often important, so that the reader should learn to be precise and place the appropriate restrictions on the domain of the function after canceling factors.

Sometimes it is necessary to specify restrictions on the domain, even though the formula for the function makes sense elsewhere.

Example 9   Assume that it costs 12 cents to manufacture a certain computer component, and let $f(x)$ be the cost in dollars for manufacturing x such components. The formula for $f(x)$ is

$$f(x) = 0.12x \tag{5}$$

If we were to specify no restrictions on the domain of this function, then the domain would be $(-\infty, +\infty)$ since (5) makes sense for all real x. Physically, however, x must be a nonnegative integer, so we must restrict the domain to the set $\{0, 1, 2, \ldots\}$. We do this by writing

$$f(x) = 0.12x, \quad x = 0, 1, 2, \ldots \tag{6}$$

From (6) it is to be understood that the formula $f(x) = 0.12x$ only applies when $x = 0, 1, 2, \ldots$; for other values of x, $f(x)$ is undefined. Thus,

$$f(3) = 0.12(3) = 0.36 \quad \text{and} \quad f(\tfrac{3}{2}) \text{ is undefined} \qquad \blacktriangleleft$$

FUNCTIONS DEFINED PIECEWISE

The next example shows how functions can be specified by formulas that have been "pieced together."

Example 10   The cost of a taxi ride in a certain metropolitan area is 75 cents for any ride up to and including one mile. After one mile the rider pays an additional amount at the rate of 50 cents per mile. If $f(x)$ is the total cost in dollars for a ride of x miles, then the value of $f(x)$ is

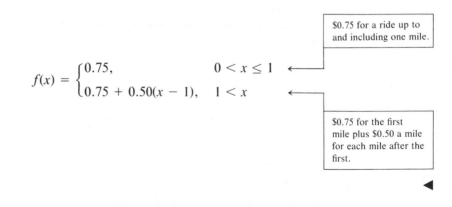

$$f(x) = \begin{cases} 0.75, & 0 < x \le 1 \\ 0.75 + 0.50(x - 1), & 1 < x \end{cases}$$

\$0.75 for a ride up to and including one mile.

\$0.75 for the first mile plus \$0.50 a mile for each mile after the first.

$\blacktriangleleft$

FUNCTIONS DEFINED BY TABLES

Although most functions are specified by formulas, this is not the only possibility. Any description that tells us what values f assigns to the points in its domain will suffice. One possibility is to use a table relating the values of x and $f(x)$.

Example 11   At 10 A.M. on a weekday, a 2-minute call is made between two stations in New Jersey. Table 2.1.1, obtained from a New Jersey telephone directory, shows how the cost of such a call varies with the distance between the stations. If $f(x)$ is the cost in dollars for a call between stations x miles apart, then the function f is completely specified by Table 2.1.1. For example,

$$f(22) = 0.25, \quad f(114) = 0.60, \quad \text{and} \quad f(64) = 0.40 \quad \blacktriangleleft$$

Table 2.1.1

DISTANCE x (MILES)	COST (DOLLARS)
$0 < x \leq 10$	.10
$10 < x \leq 15$	.15
$15 < x \leq 20$	.20
$20 < x \leq 25$	.25
$25 < x \leq 32$	.30
$32 < x \leq 48$	.35
$48 < x \leq 64$	.40
$64 < x \leq 80$	.45
$80 < x \leq 96$	.50
$96 < x \leq 112$	.55
$112 < x$	.60

CLASSIFICATION OF FUNCTIONS

We shall conclude this section by discussing some of the important categories of functions that will occur in this text.

The simplest of all functions are those that assign the same value to every member of the domain. These are called *constant functions*. For example, if f is the constant function defined by $f(x) = 3$, then

$$f(-1) = 3, \quad f(0) = 3, \quad f(\sqrt{2}) = 3, \quad f(9) = 3$$

and so forth.

A function of the form cx^n, where c is a constant and n is a nonnegative integer, is called a *monomial in x*. Examples are:

$$2x^3, \quad \pi x^7, \quad 4x^0 \, (= 4), \quad -6x, \quad \text{and} \quad x^{17}$$

The functions $4x^{1/2}$ and x^{-3} are *not* monomials because the powers of x are not nonnegative integers. A function that is expressible as the sum of finitely many monomials in x is called a *polynomial in x*. Examples are:

$$x^3 + 4x + 7, \quad 3 - 2x^3 + x^{17}, \quad 9, \quad 17 - \tfrac{2}{3}x, \quad x^5$$

The function $(x^2 - 4)^3$ is also a polynomial because it can be expressed as a sum of monomials by performing the cubing operation. Depending on whether one wants the powers written in ascending or descending order, the general formula for a polynomial in x is

$$f(x) = a_0 + a_1 x + a_2 x^2 + \cdots + a_n x^n$$

or

$$f(x) = a_n x^n + a_{n-1} x^{n-1} + \cdots + a_1 x + a_0$$

where n is a nonnegative integer and $a_0, a_1, a_2, \ldots, a_n$ are all constants. The highest power of x that occurs in a nonconstant polynomial is called the *degree* of the polynomial. Thus, $x^3 + 4x + 7$ has degree 3 and $17 - \frac{2}{3}x$ has degree 1. A nonzero constant c has degree zero (it can be written as $c = cx^0$). The constant zero is not assigned a degree. First-, second-, and third-degree polynomials are called *linear, quadratic,* and *cubic,* respectively. These have the following forms:

DESCRIPTION	GENERAL FORMULA
linear polynomial	$a_0 + a_1 x$ (where $a_1 \neq 0$)
quadratic polynomial	$a_0 + a_1 x + a_2 x^2$ (where $a_2 \neq 0$)
cubic polynomial	$a_0 + a_1 x + a_2 x^2 + a_3 x^3$ (where $a_3 \neq 0$)

A function that is expressible as a ratio of two polynomials is called a *rational function.* Examples are:

$$\frac{x^5 - 2x^2 + 1}{x^2 - 4}, \quad \frac{x}{x + 1}, \quad \frac{1}{x^5}$$

In general, f is a rational function if it is expressible in the form

$$f(x) = \frac{a_0 + a_1 x + \cdots + a_n x^n}{b_0 + b_1 x + \cdots + b_m x^m}$$

The domain of f consists of all x where the denominator differs from zero.

The rational functions are part of a broader class of functions called *explicit algebraic functions.* These are functions that can be evaluated using finitely many additions, subtractions, multiplications, divisions, and root extractions. For example,

$$f(x) = x^{2/3} = (\sqrt[3]{x})^2 \quad \text{and} \quad g(x) = \frac{(x - 3)\sqrt[4]{x}}{x^5 + \sqrt{x^2 + 1}}$$

All remaining functions fall into two categories, *implicit algebraic functions* and *transcendental functions.* We shall not define these terms, but in-

stead refer the interested reader to a classic book in calculus, G. H. Hardy,*
A Course of Pure Mathematics, Cambridge Press, 1958 (10th edition).
Among the transcendental functions are those that involve trigonometric
expressions and logarithms.

▶ Exercise Set 2.1

1. Given that $f(x) = 3x^2 + 2$, find
   (a) $f(-2)$ (b) $f(4)$
   (c) $f(0)$ (d) $f(-\sqrt{3})$
   (e) $f(a + 1)$ (f) $f(3t)$.

2. Given that $g(x) = \dfrac{x + 1}{x - 1}$, find
   (a) $g(2)$ (b) $g(-2)$
   (c) $g(\frac{1}{4})$ (d) $g(\pi)$
   (e) $g(a - 1)$ (f) $g(2t + 1)$.

3. Given that
   $$f(x) = \begin{cases} \dfrac{1}{x}, & x > 3 \\[2mm] 2x, & x \le 3 \end{cases}$$
   find
   (a) $f(-4)$ (b) $f(4)$
   (c) $f(0)$ (d) $f(3)$
   (e) $f(2.9)$ (f) $f(t^2 + 5)$.

4. Given that
   $$g(x) = \begin{cases} \sqrt{x + 1}, & x \ge -1 \\ 3, & x < -1 \end{cases}$$
   find
   (a) $g(0)$ (b) $g(-1.1)$
   (c) $g(3)$ (d) $g(-1)$
   (e) $g(-\pi)$ (f) $g(t^2 - 1)$.

In Exercises 5–22, find the domain of the given function.

5. $f(x) = \dfrac{1}{x - 3}$.   6. $f(x) = \dfrac{1}{5x + 7}$.

7. $g(x) = \sqrt{x^2 - 3}$.   8. $g(x) = \sqrt{x^2 + 3}$.

9. $h(x) = \sqrt{\dfrac{x - 1}{x + 2}}$.   10. $h(x) = \sqrt{x - 3x^2}$.

11. $\phi(x) = \dfrac{x}{|x| + 1}$.   12. $\phi(x) = \sqrt{3 - \sqrt{x}}$.

13. $F(x) = \sqrt{x - 5} + \sqrt{8 - x}$.

14. $F(x) = 3\sqrt{x} - \sqrt{x^2 - 4}$.

15. $G(x) = \sqrt{x^2 - 2x + 5}$.

16. $G(x) = \sqrt{\dfrac{x^2 - 4}{x - 4}}$.

17. $f(x) = \dfrac{x}{|x|}$.   18. $f(x) = \dfrac{x^2 - 1}{x + 1}$.

19. $g(x) = \sin \sqrt{x}$.   20. $g(x) = \cos \dfrac{1}{x}$.

21. $h(x) = \dfrac{1}{1 - \sin x}$.   22. $h(x) = \dfrac{3}{2 - \cos x}$.

In Exercises 23–36, find the domain and range of the given function.

23. $f(x) = \sqrt{3 - x}$.   24. $f(x) = \sqrt{3x - 2}$.

G. H. HARDY (1877–1947). Hardy was a world-renowned British mathematician. He taught at Cambridge and Oxford universities and was a prolific researcher who produced over 300 research papers. He received numerous medals and honorary degrees for his accomplishments. Hardy's book, *A Course of Pure Mathematics,* which gave the first rigorous English exposition of functions and limits for the college undergraduate, had a great impact on university teaching. Hardy had a rebellious spirit; he once listed among his most ardent wishes: (1) to prove the Riemann hypothesis (a famous unsolved mathematical problem), (2) to make a brilliant play in a crucial cricket match, (3) to prove the nonexistence of God, and (4) to murder Mussolini.

25. $g(x) = \sqrt{4 - x^2}$.

26. $g(x) = \sqrt{9 - 4x^2}$.

27. $h(x) = 3 + \sqrt{x}$.

28. $h(x) = \dfrac{1}{3 + \sqrt{x}}$.

29. $F(x) = x^2 + 3$.

30. $F(x) = \dfrac{2}{x^2 + 3}$.

31. $G(x) = x^3 + 2$.

32. $G(x) = \dfrac{3}{x}$.

33. $H(x) = 3 \sin x$.

34. $H(x) = \sin^2 \sqrt{x}$.

35. $\phi(x) = 2 + \cos x$.

36. $\phi(x) = \dfrac{5}{3 - \cos 2x}$.

In Exercises 37–40, express the given function in piecewise form without using absolute values.

37. $f(x) = |x| + 3x + 1$.

38. $f(x) = 3 + |2x - 5|$.

39. $g(x) = |x| + |x - 1|$.

40. $g(x) = 3|x - 2| - |x + 1|$.

In Exercises 41–48, find all values of x for which $f(x) = a$.

41. $f(x) = \sqrt{3x - 2}$; $a = 6$.

42. $f(x) = \dfrac{1}{x + 3}$; $a = 5$.

43. $f(x) = x^2 + 5$; $a = 7$.

44. $f(x) = \dfrac{x}{x^2 + 3}$; $a = \frac{1}{4}$.

45. $f(x) = \cos x$; $a = 1$.

46. $f(x) = \sin \dfrac{1}{x}$; $a = 1$.

47. $f(x) = \sin \sqrt{x}$; $a = \frac{1}{2}$.

48. $f(x) = 3 \tan x$; $a = 3$.

49. Express the area A of a circle as a function of its circumference C.

50. Express the area A of an equilateral triangle as a function of
    (a) the length s of each side
    (b) the altitude h.

51. Express the total surface area S of a cube as a function of
    (a) the length x of an edge
    (b) the volume V of the cube.

52. Express the total surface area S of a right circular cylinder with given volume V as a function of its radius r.

53. Criticize the following statement: The function
$$\frac{1 - (1/x)}{1 + (1/x)}$$
can be simplified by multiplying numerator and denominator by x to obtain
$$\frac{1 - (1/x)}{1 + (1/x)} = \frac{x - 1}{x + 1}$$
How would you rewrite the statement to make it accurate?

In Exercises 54–60, simplify the function by canceling factors, but be sure not to alter the domain.

54. $f(x) = \dfrac{x^2 - 4}{x + 2}$

55. $f(x) = \dfrac{(x + 2)(x^2 - 1)}{(x + 2)(x + 1)}$

56. $f(x) = \dfrac{x^2 + x}{x}$

57. $f(x) = \dfrac{x + 1 + \sqrt{x + 1}}{\sqrt{x + 1}}$

58. $f(x) = \dfrac{x^2 - 9}{x - 3}$

59. $f(x) = \dfrac{x^3 + 2x^2 - 3x}{(x - 1)(x + 3)}$

60. $f(x) = \dfrac{x + \sqrt{x}}{\sqrt{x}}$

For the functions in Exercises 61–64, state which of the following term(s) apply: *monomial, polynomial, rational function, explicit algebraic function*.

61. (a) $3x^7$ (b) $4x^{1/7}$ (c) $\dfrac{1}{5x^6}$ (d) $2x^3 - 1$.

62. (a) $2x^{1/3} + 1$ (b) x^{-2} (c) $x^{-1/2}$ (d) $(x - 3)^{12}$.

63. (a) $x^2\sqrt{x^3 - 3}$ (b) $\dfrac{x + 1}{x + 2}$ (c) $\pi^{-2} + 1$ (d) $|x|$.

64. (a) $\sqrt{x^2 + \sqrt{x}}$ (b) $\dfrac{x^3 - 2x + 1}{x^2 - 9}$ (c) $\sqrt{\pi - 3}$ (d) $|x - 2|$.

2.2   GRAPHS OF FUNCTIONS

In this section we shall show how to represent a function geometrically by a graph, and we shall use such graphs to develop additional properties of functions.

We begin with the following definition.

> **2.2.1**   DEFINITION. If f is a function of x, then we define the **graph of f** to be the graph of the equation $y = f(x)$.

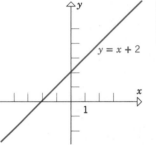

$y = x + 2$

Figure 2.2.1

Example 1   Sketch the graph of $f(x) = x + 2$.

Solution.   By definition, the graph of $f(x) = x + 2$ is the graph of the equation $y = x + 2$, which is a line with slope 1 and y-intercept 2 (Figure 2.2.1).   ◄

Example 2   Sketch the graph of $f(x) = |x|$.

Solution.   By definition, the graph of $f(x) = |x|$ is the graph of $y = |x|$, or equivalently

$$y = \begin{cases} x, & x \geq 0 \\ -x, & x < 0 \end{cases}$$

Graph of $|x|$

Figure 2.2.2

The graph coincides with the line $y = x$ for $x \geq 0$ and with the line $y = -x$ for $x < 0$ (Figure 2.2.2).   ◄

Example 3   Sketch the graph of

$$\phi(x) = \begin{cases} x + 2, & x \neq 2 \\ 6, & x = 2 \end{cases}$$

$y = \begin{cases} x + 2, & x \neq 2 \\ 6, & x = 2 \end{cases}$

Figure 2.2.3

Solution.   The function ϕ is identical to the function f in Example 1, except at $x = 2$, where we have

$$\phi(2) = 6 \quad \text{and} \quad f(2) = 4$$

Thus, the graph of ϕ (Figure 2.2.3) is identical to the graph of f (Figure 2.2.1), except that the graph of ϕ has a point separated from the line at $x = 2$.   ◄

Example 4   Sketch the graph of

$$g(x) = \begin{cases} 1, & x \leq 2 \\ x + 2, & x > 2 \end{cases}$$

Solution.   By definition, the graph of g is the graph of

$$y = \begin{cases} 1, & x \leq 2 \\ x + 2, & x > 2 \end{cases}$$

Thus, for $x \leq 2$, the value of y is always 1 and for $x > 2$, y is given by $y = x + 2$, which is the straight line graphed in Example 1. The graph of g is shown in Figure 2.2.4.   ◀

REMARK.   In Figure 2.2.4, we used the heavy dot and open circle above $x = 2$ to emphasize that the value $g(2) = 1$ lies on the horizontal line and not on the inclined line.

Figure 2.2.4

Example 5   Sketch the graph of

$$h(x) = \frac{x^2 - 4}{x - 2} \tag{1}$$

Solution.   As noted in Example 8 of Section 2.1 this function can be rewritten as

$$h(x) = x + 2, \quad x \neq 2$$

Thus, the function h in (1) is identical to the function f in Example 1, except that h is undefined at $x = 2$. It follows that the graph of h is identical to the graph of f (Figure 2.2.1), except that the graph of h has a hole in it above $x = 2$ (Figure 2.2.5).   ◀

Graph of $\frac{x^2 - 4}{x - 2}$

Figure 2.2.5

TRANSLATIONS OF
GRAPHS

Sometimes the graph of a function can be obtained by translating the graph of a "simpler" function. In Table 2.2.1 we have listed four operations whose effect is to translate the graph of a function f.

Example 6   Sketch the graph of $g(x) = \sqrt{x + 3}$.

Solution.   The function $g(x) = \sqrt{x + 3}$ results when x is replaced by $x + 3$ in the formula $\sqrt{x}$. Thus, from the fourth entry in Table 2.2.1 the graph of g is obtained by translating the graph of $y = \sqrt{x}$ left 3 units (Figure 2.2.6).   ◀

Graph of $\sqrt{x + 3}$

Figure 2.2.6

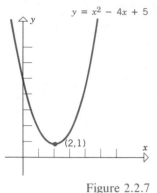

$y = x^2 - 4x + 5$

(2,1)

Figure 2.2.7

Example 7   Sketch the graph of $g(x) = x^2 - 4x + 5$.

Solution.   Complete the square on the first two terms:

$$g(x) = (x^2 - 4x + 4) - 4 + 5 = (x - 2)^2 + 1$$

In this form we see that the graph can be obtained by translating the graph of x^2 right 2 units because of the $x - 2$, and up one unit because of the $+1$ (Figure 2.2.7).

Alternative Solution.   Use the procedure in Example 4 of Section 1.6.   ◄

Table 2.2.1

OPERATION	EFFECT	EXAMPLE
Add a positive constant c to $f(x)$. $(f(x) + c)$	Translates the graph of $f(x)$ up c units.	$y = x^2 + 2$ $y = x^2$ 2
Subtract a positive constant c from $f(x)$. $(f(x) - c)$	Translates the graph of $f(x)$ down c units.	$y = x^2$ $y = x^2 - 2$ -2
Replace x by $x - c$. (c positive)	Translates the graph of $f(x)$ right c units.	$y = (x - 2)^2$ $y = x^2$ 2
Replace x by $x + c$. (c positive)	Translates the graph of $f(x)$ left c units.	$y = (x + 2)^2$ $y = x^2$ -2

In each of the last seven examples we found the graph of a *given* function. We shall now consider the converse problem.

2.2.2 PROBLEM. *Given a curve in the xy-plane, does there exist a function f whose graph is the given curve?*

Sometimes this question can be answered by performing appropriate algebraic computations.

Example 8   Show that the graph of

$$3x^2 - 2y = 1 \tag{2}$$

is also the graph of $f(x)$ for some function f.

Solution.   Equation (2) can be rewritten in the equivalent form

$$y = \tfrac{1}{2}(3x^2 - 1)$$

so that the graph of (2) is also the graph of the function

$$f(x) = \tfrac{1}{2}(3x^2 - 1) \qquad \blacktriangleleft$$

VERTICAL LINE TEST

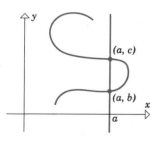

Figure 2.2.8

There is a simple geometric test that can often be used to determine whether a curve in the *xy*-plane is the graph of a function of *x*. To see how it works, consider the curve in Figure 2.2.8 and the vertical line that intersects it at the two points (a, b) and (a, c). This curve cannot be the graph of

$$y = f(x) \tag{3}$$

for any function f. For if it were, then (a, b) and (a, c), being points on the curve, would both have coordinates satisfying (3). Thus, we would have

$$b = f(a) \quad \text{and} \quad c = f(a)$$

But this is impossible since f cannot assign two different values to a. Therefore, there is no function of x whose graph is the curve in Figure 2.2.8.

This discussion illustrates the following general result, which we shall call the ***vertical line test***.

2.2.3   THE VERTICAL LINE TEST. *A curve in the xy-plane is the graph of y = f(x) for some function f if and only if no vertical line intersects the curve more than once.*

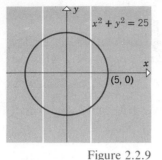

$x^2 + y^2 = 25$

$(5, 0)$

Figure 2.2.9

Example 9   Is the graph of the circle

$$x^2 + y^2 = 25 \tag{4}$$

also the graph of $f(x)$ for some function f?

Solution.   Since some vertical lines intersect the circle more than once (see Figure 2.2.9), the circle is not the graph of any function.

We can also deduce this result algebraically by observing that (4) can be written as

$$y = \pm \sqrt{25 - x^2} \tag{5}$$

But the right side of (5) is not a function of x since it is "multiple-valued." Thus, (4) is not equivalent to an equation of the form $y = f(x)$.   ◄

Sometimes a curve, when considered in its entirety, is not the graph of a function, but some smaller portion of the curve is. For example, we saw above that the circle $x^2 + y^2 = 25$ is not the graph of any single function. However, the upper and lower semicircles are graphs of functions. From (5) the upper semicircle has the equation $y = \sqrt{25 - x^2}$, so it is the graph of $f(x) = \sqrt{25 - x^2}$, and the lower semicircle has the equation $y = -\sqrt{25 - x^2}$, so it is the graph of $g(x) = -\sqrt{25 - x^2}$ (Figure 2.2.10).

$y = \sqrt{25 - x^2}$

$y = -\sqrt{25 - x^2}$

Figure 2.2.10

FUNCTIONS DEFINED BY EQUATIONS

We shall now explain how equations can be used to define functions. For our discussion we shall divide the set of all equations in x and y into three categories:

- equations written in the form $y = f(x)$;

- equations that are not written in the form $y = f(x)$, but which we can rewrite in that form by appropriate algebraic manipulations;

- equations that are not written in the form $y = f(x)$ and cannot be rewritten in that form either because it is mathematically impossible to do so or the algebra required is too complicated.

An equation of the form $y = f(x)$ is said to **define y explicitly as a function of x**. (The function defined is f.)

Example 10   The equation

$$y = x^3 - 2x + 3$$

defines y explicitly as a function of x, the function being $f(x) = x^3 - 2x + 3$.   ◄

An equation that is not of the form $y = f(x)$, but can be written in this form, is said to *define y implicitly as a function of x.* (The function defined is f.)

Example 11   The equation

$$x = y^3 + 3y^2 + 3y + 1$$

defines y implicitly as a function of x because the equation can be rewritten as

$$x = (y + 1)^3 \quad \text{or} \quad y = \sqrt[3]{x} - 1$$

The function of x defined by the equation is $f(x) = \sqrt[3]{x} - 1$.   ◀

For equations that are not of the form $y = f(x)$ and cannot be rewritten in that form, the situation is more complicated. If the graph of the equation is cut at most once by any vertical line, then the graph of the equation is identical to the graph of $y = f(x)$ for some function f. However, because we cannot rewrite the equation in this form, there is no way for us to find an explicit formula for $f(x)$. But even though we cannot produce a formula for the function, we still say that *the equation defines y implicitly as a function of x.* If the graph of the equation is cut more than once by some vertical line, then the graph in its entirety does not coincide with the graph of $y = f(x)$ for any function f. Nevertheless, it may be the case that some portion of the graph is cut at most once by any vertical line (Figure 2.2.11), so that this portion coincides with the graph of $y = f(x)$ for some function f.

To make this idea precise, let C denote the graph of the equation. We will say that the equation *defines y implicitly as a function of x in a neighborhood of* (x_0, y_0) if we can find some rectangle centered at (x_0, y_0), with sides parallel to the coordinate axes, and such that the portion of C contained within the rectangle is cut at most once by any vertical line (Figure 2.2.11).

Example 12   The equation of the unit circle $x^2 + y^2 = 1$ does not define

y as a function of x in a neighborhood of the point $P(1, 0)$, because the portion of the circle within any rectangle centered at P is cut twice by some vertical line (Figure 2.2.12a). The equation does, however, define y as a function of x in a neighborhood of the point $(1/\sqrt{2}, 1/\sqrt{2})$ (Figure 2.2.12b).   ◀

Unfortunately, it is often difficult to determine whether an equation defines y implicitly as a function of x or defines y implicitly as a function of x in a neighborhood of a point. For the most part, we shall leave such questions for a course in advanced calculus.

In an equation of the form $y = f(x)$, we may think of x as a quantity

This portion of C is cut at most once by any vertical line.

Figure 2.2.11

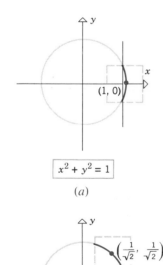

$x^2 + y^2 = 1$

(a)

$x^2 + y^2 = 1$

(b)

Figure 2.2.12

that can be varied arbitrarily over the domain of f and y as a quantity whose value is determined once the value of x is specified. For this reason x is sometimes called the **independent variable** and y the **dependent variable**. Sometimes it is desirable to reverse the roles of x and y, treating y as the independent variable and x as the dependent variable. Thus, if g is a function of y we define the graph of $g(y)$ to be the graph of the equation

$$x = g(y) \tag{6}$$

REMARK.   When graphing equations of the form $y = f(x)$ we have been keeping the x-axis horizontal and the y-axis vertical. Thus, it would seem reasonable to graph (6) with the y-axis horizontal and the x-axis vertical. However, this is rarely done. In this text we shall keep the x-axis horizontal and the y-axis vertical when graphing equations of the form $x = g(y)$ and $y = f(x)$.

Example 13   Sketch the graph of the function $g(y) = y^2$.

Solution.   By definition, the graph of $g(y) = y^2$ is the graph of the equation $x = y^2$ (Figure 2.2.13).   ◀

THE HORIZONTAL
LINE TEST

Just as the graph of a function of x is cut at most once by any vertical line, so the graph of a function of y is cut at most once by any horizontal line. (See Figure 2.2.13, for example.)

An equation of form (6) is said to **define x explicitly as a function of y**. The reader should be able to deduce what it means for an equation to **define x implicitly as a function of y** or **x implicitly as a function of y in a neighborhood of (x_0, y_0).**

Figure 2.2.13

Example 14   The graph of the equation

$$2x + 3y = 6 \tag{7}$$

is cut at most once by any vertical line and is cut at most once by any horizontal line (Figure 2.2.14). Thus, the equation defines y as a function of x and x as a function of y. This can also be seen algebraically by solving (7) for y and x. Solving for y yields

$$y = \tfrac{1}{3}(6 - 2x)$$

and solving for x yields

$$x = \tfrac{1}{2}(6 - 3y)$$

Figure 2.2.14

Thus, the graph of (7) is also the graph of $f(x) = \tfrac{1}{3}(6 - 2x)$ and $g(y) = \tfrac{1}{2}(6 - 3y)$.   ◀

Example 15   The equation

$$3x^2 - 2y = 1 \tag{8}$$

defines y as a function of x because it can be rewritten as

$$y = \tfrac{1}{2}(3x^2 - 1) \tag{9}$$

which has the same graph as the function $f(x) = \tfrac{1}{2}(3x^2 - 1)$. However, the equation does not define x as a function of y because solving (8) for x yields

$$x = \pm \sqrt{\tfrac{1}{3}(1 + 2y)}$$

which is "multiple-valued" and therefore does not define a function of y.
We leave it for the reader to verify these conclusions geometrically by showing that the graph of (9) is cut at most once by any vertical line, but more than once by some horizontal line.   ◀

▶ Exercise Set 2.2

1. Consider the function f graphed in Figure 2.2.15. In each part, find all values of x satisfying the given condition.
  (a) $f(x) = 0$ (b) $f(x) = 3$
  (c) $f(x) \geq 0$ (d) $f(x) \leq 0$.

Figure 2.2.15

In Exercises 2–29, sketch the graph of the function.

2. $f(x) = 2x + 1$.

3. $f(x) = 3x - 2$.

4. $G(x) = x, \quad 1 \leq x \leq 2$.

5. $G(x) = x - 2, \quad -1 \leq x \leq 1$.

6. $h(x) = x^2 - 3$.

7. $h(x) = (x - 2)^2$.

8. $F(x) = \sqrt{x + 1}$.

9. $F(x) = \sqrt{3 - x}$.

10. $f(x) = \sqrt{4 - x^2}$.

11. $f(x) = 2 + \sqrt{9 - x^2}$.

12. $g(x) = \sqrt{4x - x^2}$.

13. $g(x) = \sqrt{7 - 6x - x^2}$.

14. $f(x) = 2 \sin x$.

15. $f(x) = 3 \sin 2x$.

16. $f(x) = \cos 2x$.

17. $f(x) = 1 + \cos x$.

18. $f(x) = \dfrac{x^2 - 4}{x + 2}$.

19. $f(x) = \dfrac{x^2 + 2x}{x}$.

20. $g(x) = \dfrac{x^3 - x^2}{x - 1}$.

21. $g(x) = \dfrac{x - x^3}{x}$.

22. $\phi(x) = \dfrac{x}{|x|}$.

23. $\phi(x) = \dfrac{|x - 2|}{x - 2}$.

24. $g(x) = \begin{cases} x^2, & x \neq 4 \\ 0, & x = 4. \end{cases}$

25. $g(x) = \begin{cases} x - 1, & x \neq 1 \\ 3, & x = 1. \end{cases}$

26. $f(x) = \begin{cases} x + 2, & x \leq 3 \\ x + 4, & x > 3. \end{cases}$

27. $f(x) = \begin{cases} x^2, & x > 1 \\ 2, & x \leq 1. \end{cases}$

28. $h(x) = \begin{cases} 1, & 0 < x \le 1 \\ 3, & 1 < x \le 2 \\ -1, & 2 < x \le 3 \\ 0, & \text{elsewhere.} \end{cases}$

29. $h(x) = \begin{cases} -2, & -2 \le x < -1 \\ 1, & -1 \le x < 0 \\ 2, & 0 \le x < 1 \\ 0, & \text{elsewhere.} \end{cases}$

In Exercises 30–33, express the function in piecewise form without using absolute values and sketch its graph.

30. $f(x) = |x - 3| - x$.

31. $f(x) = 2x + |2 - x|$.

32. $g(x) = |x| + |x - 3|$.

33. $g(x) = |x - 5| - |x - 3|$.

34. A 10 foot long ladder leans against a wall. The base of the ladder is x feet from the wall. Express the distance from the top of the ladder to the ground as a function of x, and sketch the graph of the function.

35. Express the area A enclosed by the graph of

$$f(x) = \begin{cases} 2x, & 0 \le x \le 1 \\ 2, & x > 1 \end{cases}$$

the x-axis, and the vertical line at x ($x \ge 0$) as a function of x.

36. Find a formula for the function f graphed in Figure 2.2.16.

Figure 2.2.16

37. Find a formula for the function g graphed in Figure 2.2.17.

Figure 2.2.17

38. Use Table 2.2.1 and the graph of $y = |x|$ to graph the following:
    (a) $y = |x - 4|$ (b) $y = |x| + 4$
    (c) $y = |x - 4| + 4$ (d) $y = |x + 5| - 2$.

39. Use Table 2.2.1 and the graph of $y = \sqrt{x}$ to graph the following:
    (a) $y = \sqrt{x - 3}$ (b) $y = \sqrt{x} + 3$
    (c) $y = \sqrt{x - 3} + 3$
    (d) $y = \sqrt{x + 1} - 2$.

40. A function f with domain $[-1, 3]$ has the graph shown in Figure 2.2.18. Use this graph to obtain the graphs of
    (a) $y = f(x + 1)$ (b) $y = f(2x)$
    (c) $y = f(-x)$ (d) $y = |f(x)|$.

Figure 2.2.18

41. Graph
    (a) $f(x) = |\cos x|$
    (b) $f(x) = \cos x + |\cos x|$.

In Exercises 42 and 43, determine whether the equation defines y as a function of x, or x as a function of y, or both, or neither.

42. (a) $3x - 4y = 12$ (b) $xy^2 = 1$
    (c) $x^2 + y^2 = 1$ (d) $\dfrac{1 + x}{1 - y} = 2$.

43. (a) $4x + 2y = -8$ (b) $x^2y^3 = 1$
    (c) $3x^2 + 4y^2 = 12$ (d) $\dfrac{xy}{1 - xy} = 1$.

(44.) In each part express x explicitly as a function of y.

(a) $xy - x = 1$ (b) $y = \dfrac{x}{1 + x}$

(c) $x^2 + 2xy + y^2 = 0$.

45. In each part express y explicitly as a function of x.

(a) $x^2y - 1 = 0$ (b) $x = \dfrac{1 - y}{1 + y}$

(c) $y^2 + 2xy + x^2 = 0$.

46. Show that $y^2 + 3xy + x^2 = 0$ does not define y implicitly as a function of x.

47. Show that $y^2 + 4xy + 1 = 0$ does not define y implicitly as a function of x.

48. Find two functions, the union of whose graphs is the graph of the equation in Exercise 47.

49. In Figure 2.2.19, determine whether the curve is the graph of a function of x, a function of y, both, or neither.

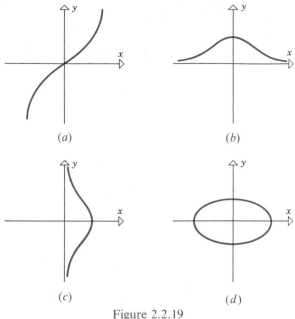

(a) (b)

(c) (d)

Figure 2.2.19

2.3 OPERATIONS ON FUNCTIONS

Just as numbers can be added, subtracted, multiplied, and divided to produce other numbers, so functions can be added, subtracted, multiplied, and divided to produce other functions. In this section we shall discuss these operations and others that have no analogs in ordinary arithmetic.

ARITHMETIC
OPERATIONS ON
FUNCTIONS

2.3.1 DEFINITION. Given functions f and g, their *sum* $f + g$, *difference* $f - g$, *product* $f \cdot g$, and *quotient* f/g, are defined by

$$(f + g)(x) = f(x) + g(x)$$
$$(f - g)(x) = f(x) - g(x)$$
$$(f \cdot g)(x) = f(x) \cdot g(x)$$
$$(f/g)(x) = f(x)/g(x)$$

For the functions $f + g$, $f - g$, and $f \cdot g$ the domain is defined to be the intersection of the domains of f and g, and for f/g the domain is this intersection with the points where $g(x) = 0$ excluded.

Example 1   Let f and g be the functions defined by $f(x) = \sqrt{5 - x}$ and $g(x) = \sqrt{x - 3}$. Then

$$(f + g)(x) = f(x) + g(x) = \sqrt{5 - x} + \sqrt{x - 3}$$
$$(f - g)(x) = f(x) - g(x) = \sqrt{5 - x} - \sqrt{x - 3}$$
$$(f \cdot g)(x) = f(x) \cdot g(x) = \sqrt{5 - x}\sqrt{x - 3}$$
$$(f/g)(x) = f(x)/g(x) = \sqrt{5 - x}/\sqrt{x - 3}$$

Since the domain of f is $(-\infty, 5]$ and the domain of g is $[3, +\infty)$, the domain of $f + g$, $f - g$, and $f \cdot g$ is $[3, 5]$ because this is the intersection of $(-\infty, 5]$ and $[3, +\infty)$. Since $g(x) = 0$ if $x = 3$, this point is not in the domain of f/g; thus, the domain of f/g is $(3, 5]$. ◀

Recall that when a function is given by a formula and there is no mention of the domain, it is understood that the domain consists of all real numbers where the formula makes sense and yields real values. In the foregoing example the formulas for $(f + g)(x)$, $(f - g)(x)$, $(f \cdot g)(x)$, and $(f/g)(x)$ determine the correct domains for the functions $f + g$, $f - g$, $f \cdot g$, and f/g with this convention (verify). However, the following example shows that this is not always the case.

Example 2   Let $f(x) = 3\sqrt{x}$ and $g(x) = \sqrt{x}$. Find $f \cdot g$.

Solution.   The formula for $(f \cdot g)(x)$ is

$$(f \cdot g)(x) = f(x) \cdot g(x) = (3\sqrt{x}) \cdot (\sqrt{x}) = 3x$$

Because $3x$ makes sense and yields a real value for all x, it would determine a domain of $(-\infty, +\infty)$ in absence of any restrictions. However, this is not the correct domain of $f \cdot g$. To see why, observe that both f and g have domain $[0, +\infty)$, so that by definition $f \cdot g$ also has domain $[0, +\infty)$ since this is the intersection of the domains of f and g. Thus, the correct formula for $f \cdot g$ is

$$(f \cdot g)(x) = 3x, \quad x \geq 0 \qquad ◀$$

REMARK.   In light of the foregoing example, we recommend that the reader develop the habit of placing the appropriate restrictions on the domain after performing any operations on functions.

Sometimes we shall write f^2 to denote the product $f \cdot f$. For example, if $f(x) = 8x$, then

$$f^2(x) = (f \cdot f)(x) = f(x) \cdot f(x) = (8x) \cdot (8x) = 64x^2$$

Similarly,

$$f^3 = f \cdot f^2, \quad f^4 = f \cdot f^3, \quad f^5 = f \cdot f^4, \ldots$$

This notation is especially common with trigonometric functions. For example, $(\sin x)^2$ is generally written as $\sin^2 x$.

COMPOSITION OF
FUNCTIONS

Many problems in mathematics are attacked by "decomposing" functions into simpler functions that are evaluated in succession. For example, consider the function h given by

$$h(x) = (x + 1)^2$$

To evaluate $h(x)$ for a given value of x, we would first compute $x + 1$ and then square the result. In other words, if we consider the functions g and f given by

$$g(x) = x + 1 \quad \text{and} \quad f(x) = x^2$$

then

$$h(x) = (x + 1)^2 = [g(x)]^2 = f(g(x))$$

Thus, h has the same effect as g and f evaluated successively. Loosely speaking, h is "composed" of the two functions g and f. This idea is formalized in the following definition.

2.3.2 DEFINITION. Given functions f and g, the *composition of f with g*, denoted by $f \circ g$, is the function defined by

$$(f \circ g)(x) = f(g(x))$$

The domain of $f \circ g$ is defined to consist of all x in the domain of g for which $g(x)$ is in the domain of f.

REMARK. Although the domain of $f \circ g$ may seem complicated at first glance, it is quite natural: To compute $f(g(x))$ one needs x in the domain of g to compute $g(x)$, then one needs $g(x)$ in the domain of f to compute $f(g(x))$.

Example 3 Find $(f \circ g)(x)$ if $f(x) = x^2 + 3$ and $g(x) = \sqrt{x}$.

Solution. The formula for $f(g(x))$ is

$$f(g(x)) = [g(x)]^2 + 3 = (\sqrt{x})^2 + 3 = x + 3$$

Since the domain of g is $[0, +\infty)$ and the domain of f is $(-\infty, +\infty)$, the domain of $f \circ g$ consists of all x in $[0, +\infty)$ such that $g(x) = \sqrt{x}$ lies in $(-\infty, +\infty)$; thus, the domain of $f \circ g$ is $[0, +\infty)$. Therefore,

$$(f \circ g)(x) = x + 3, \quad x \geq 0 \quad \blacktriangleleft$$

Example 4 Let $f(x) = x - 1$ and $g(x) = \sqrt{x}$. Find

(a) $(f \circ g)(x)$ (b) $(g \circ f)(x)$

Solution (a).   The formula for $f(g(x))$ is

$$f(g(x)) = g(x) - 1 = \sqrt{x} - 1$$

Since the domain of g is $[0, +\infty)$ and the domain of f is $(-\infty, +\infty)$, the domain of $f \circ g$ consists of all x in $[0, +\infty)$ such that $g(x) = \sqrt{x}$ lies in $(-\infty, +\infty)$, that is, all x in $[0, +\infty)$. Therefore,

$$(f \circ g)(x) = \sqrt{x} - 1$$

In this formula there is no need to indicate that the domain is $[0, +\infty)$ since this is precisely the set where $\sqrt{x} - 1$ is defined and yields real values.

Solution (b).   The formula for $g(f(x))$ is

$$g(f(x)) = \sqrt{f(x)} = \sqrt{x - 1}$$

The domain of $g \circ f$ consists of all x in the domain of f such that $f(x)$ lies in the domain of g. Since the domain of f is $(-\infty, +\infty)$ and the domain of g is $[0, +\infty)$, the domain of $g \circ f$ consists of all x in $(-\infty, +\infty)$ such that $f(x) = x - 1$ lies in $[0, +\infty)$; thus, the domain is $[1, +\infty)$. Therefore,

$$(g \circ f)(x) = \sqrt{x - 1}$$

[As in part (a), there is no need to indicate specifically that the domain is $[1, +\infty)$, since this is precisely the set where $\sqrt{x - 1}$ is defined and yields real values.]   ◄

REMARK.   Note that $f \circ g$ and $g \circ f$ are different functions in the foregoing example. Thus, the order in which functions are composed can make a difference in the end result.

EXPRESSING A FUNCTION AS A COMPOSITION   Many problems in calculus are solved by expressing a given function as a composition of two simpler functions. Let us begin with an example.

Example 5   Express $h(x) = (x - 4)^5$ as a composition of two functions.

Solution.   To evaluate $h(x)$ for a given value of x one would first compute $x - 4$ and then raise the result to the fifth power. Therefore, if

$$f(x) = x^5 \quad \text{and} \quad g(x) = x - 4$$

then

$$f(g(x)) = [g(x)]^5 = (x - 4)^5$$

so $h(x) = f(g(x))$.   ◄

The thought process in this example suggests a general procedure for decomposing a function h into a composition $h = f \circ g$:

- Think about how you would evaluate $h(x)$ for a specific value of x, trying to break the evaluation into two steps performed successively.
- The first step in the evaluation will determine a function g and the second step a function f.
- The formula for h can then be written as $h(x) = f(g(x))$.

For descriptive purposes, we shall refer to g as the "inside function" and f as the "outside function" in the expression $f(g(x))$.

Example 6   To evaluate $\sin(x^3)$, we would first compute x^3 and then take the sine, so $g(x) = x^3$ is the inside function and $f(x) = \sin x$ the outside function. Therefore,

$$\sin(x^3) = f(g(x)) \quad \text{[where } g(x) = x^3 \text{ and } f(x) = \sin x\text{]}\quad \blacktriangleleft$$

Some more examples are given in Table 2.3.1.

Table 2.3.1

FUNCTION	$g(x)$ INSIDE	$f(x)$ OUTSIDE	COMPOSITION
$(x^2 + 1)^{10}$	$x^2 + 1$	x^{10}	$(x^2 + 1)^{10} = f(g(x))$
$\sin^3 x$	$\sin x$	x^3	$\sin^3 x = f(g(x))$
$\tan(x^5)$	x^5	$\tan x$	$\tan(x^5) = f(g(x))$
$\sqrt{4 - 3x}$	$4 - 3x$	$\sqrt{x}$	$\sqrt{4 - 3x} = f(g(x))$

REMARK.   It should be noted that there is always more than one way to express a function as a composition. For example, here are two ways to express $(x^2 + 1)^{10}$ as a composition that differ from that in Table 2.3.1:

$$(x^2 + 1)^{10} = [(x^2 + 1)^2]^5 = f(g(x)) \quad \text{[}g(x) = (x^2 + 1)^2 \text{ and } f(x) = x^5\text{]}$$

$$(x^2 + 1)^{10} = [(x^2 + 1)^3]^{10/3} = f(g(x)) \quad \text{[}g(x) = (x^2 + 1)^3 \text{ and } f(x) = x^{10/3}\text{]}$$

▶ **Exercise Set 2.3**

1. Let $f(x) = x^2 + 1$. Find
   (a) $f(t)$
   (b) $f(t + 2)$
   (c) $f(x + 2)$
   (d) $f\left(\dfrac{1}{x}\right)$
   (e) $f(x + h)$
   (f) $f(-x)$
   (g) $f(\sqrt{x})$
   (h) $f(3x)$.

2. Let $g(x) = \sqrt{x}$. Find
   (a) $g(5s + 2)$
   (b) $g(\sqrt{x} + 2)$

(c) $3g(5x)$

(d) $\dfrac{1}{g(x)}$

(e) $g(g(x))$

(f) $g^2(x)$

(g) $g\left(\dfrac{1}{\sqrt{x}}\right)$

(h) $g((x-1)^2)$.

In Exercises 3–8, find
  (a) $f+g$
  (b) $f-g$
  (c) $f\cdot g$
  (d) f/g
  (e) $f\circ g$
  (f) $g\circ f$

3. $f(x)=2x,\ g(x)=x^2+1$.

4. $f(x)=3x-2,\ g(x)=|x|$.

5. $f(x)=\sqrt{x+1},\ g(x)=x-2$.

6. $f(x)=\dfrac{x}{1+x^2},\ g(x)=\dfrac{1}{x}$.

7. $f(x)=\sqrt{x-2},\ g(x)=\sqrt{x-3}$.

8. $f(x)=x^3,\ g(x)=\dfrac{1}{\sqrt[3]{x}}$.

9. Given that $f(x)=|x|$ and $g(x)=x$, sketch the graph of
  (a) $(f+g)(x)$
  (b) $(f-g)(x)$
  (c) $(f\cdot g)(x)$
  (d) $(f/g)(x)$.

10. Let $f(x)=\sqrt{2x-10}$ and $g(x)=8x^2+5$. Find $(f\circ g)(x)$ and $(g\circ f)(x)$.

11. Let h be defined by $h(x)=2x-5$. Find
  (a) $h\circ h$
  (b) h^2.

12. Let $g(x)=x^3$ and
$$f(x)=\begin{cases}5x, & x\le 0\\ -x, & 0<x\le 8\\ \sqrt{x}, & x>8\end{cases}$$
Find $(f\circ g)(x)$.

13. Let $f(x)=\dfrac{1}{x}$.
  (a) If $g(x)=x^2+1$, show that $f\circ g$ is defined for all x even though f is not defined when $x=0$.
  (b) Can you find a different function g such that $f\circ g$ is defined for all x?
  (c) What property must a function g have in order for $f\circ g$ to be defined for all x?

14. Is it always true that $f\circ g=g\circ f$? Is it ever true that $f\circ g=g\circ f$?

15. Prove or disprove: For any three functions f, g, and h, $f\circ(g\circ h)=(f\circ g)\circ h$.

In Exercises 16–27, express f as a composition of two functions; that is, find functions g and h such that $f=g\circ h$. (Each exercise has more than one correct solution.)

16. $f(x)=x^2+1$.

17. $f(x)=\sqrt{x+2}$.

18. $f(x)=\dfrac{1}{x-3}$.

19. $f(x)=(x-5)^7$.

20. $f(x)=a+bx$.

21. $f(x)=|x^2-3x+5|$.

22. $f(x)=3\sin(x^2)$.

23. $f(x)=\sin^2 x$.

24. $f(x)=\cos^3 2x$.

25. $f(x)=\dfrac{3}{5+\cos x}$.

26. $f(x)=3\sin^2 x+4\sin x$.

27. $f(x)=\dfrac{\tan x}{3+\tan x}$.

28. Find functions f, g, and h so that
$$f(g(h(x)))=\sin\sqrt{x^2+3x+7}.$$

29. Find functions f, g, and h so that
$$f(g(h(x)))=\sqrt{3-\sin^2 x}.$$

30. Find $(f\circ f)(\pi)$ if $f(x)=\begin{cases}1 & \text{if }x\text{ is rational}\\ 0 & \text{if }x\text{ is irrational}.\end{cases}$

31. Find $f(x)$ if $f(x+1)=x^2+3x+5$. [*Hint:* Let $z=x+1$ and find $f(z)$.]

32. Find $f(x)$ if $f(3x)=\dfrac{x}{x^2+1}$. [*Hint:* Let $z=3x$ and find $f(z)$.]

33. If $f(x)=0$ only for $x=-1$ and $x=2$, and if $g(x)=2x-1$, find x so that $(f\circ g)(x)=0$.

34. Find $g(x)$ if $f(x)=2x-1$ and $(f\circ g)(x)=x^2$.

35. Find $g(x)$ if $f(x)=\sqrt{x+5}$ and $(f\circ g)(x)=3|x|$.

36. Let $f(x)=x^2$, $g(x)=\sin x$, and $h(x)=\cos x$.
  (a) Find the exact numerical value of $f(g(0.3))+f(h(0.3))$
  (b) Show that $f(h(x))-f(g(x))=h(2x)$.

37. If $f(x+y)=f(x)-f(y)$ for all real x and y, show that $f(x)=0$ for all real x.

38. The *greatest integer function*, $[x]$, is defined to be the greatest integer that is less than or equal to x. For example, $[2.7]=2$, $[-2.3]=-3$, $[4]=4$.

Sketch the graph of
(a) $f(x) = [x]$ (b) $f(x) = [x^2]$
(c) $f(x) = [x]^2$ (d) $f(x) = [\sin x]$.

39. A function f is called **even** if $f(-x) = f(x)$ for each x in the domain of f, and **odd** if $f(-x) = -f(x)$ for each such x. In each part, classify the function as even, odd, or neither.
(a) $f(x) = x^2$ (b) $f(x) = x^3$
(c) $f(x) = |x|$ (d) $f(x) = x + 1$
(e) $f(x) = \dfrac{x^5 - x}{1 + x^2}$ (f) $f(x) = 2$.

40. In Figure 2.3.1 we have sketched part of the graph of a function f (the part to the right of the y-axis). Complete the graph assuming
(a) f is an even function
(b) f is an odd function.
(See Exercise 39 for the definitions of even and odd functions.)

Figure 2.3.1

41. Classify the functions graphed in Figure 2.3.2 as even, odd, or neither. (See Exercise 39 for the definitions of even and odd.)

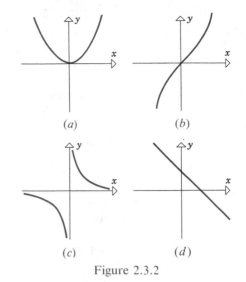

(a)   (b)

(c)   (d)

Figure 2.3.2

42. Can a function be both even and odd? (See Exercise 39 for the definitions of even and odd functions.)

43. Prove that the product of
(a) two even functions is an even function.
(b) two odd functions is an even function.
(c) an even and an odd function is an odd function.
(See Exercise 39 for the definitions of even and odd functions.)

44. Let a be a constant and suppose

$$f(a - x) = f(a + x)$$

for all x. What geometric property must the graph of f have?

2.4 LIMITS (AN INTUITIVE INTRODUCTION)

Figure 2.4.1

Calculus centers around the following two fundamental problems:

THE TANGENT PROBLEM.   Given a function f and a point $P(x_0, y_0)$ on its graph, find an equation of the line tangent to the graph at P (Figure 2.4.1).

THE AREA PROBLEM.   Given a function f, find the area between the graph of f and an interval $[a, b]$ on the x-axis (Figure 2.4.2).

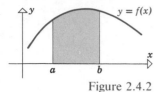

Figure 2.4.2

Traditionally, that portion of calculus arising from the tangent problem is called *differential calculus* and that arising from the area problem is called *integral calculus*. As we shall see, however, the tangent and area problems are closely related, so that the distinction between differential calculus and integral calculus is often hard to discern.

In order to solve the tangent and area problems it is necessary to have a more precise understanding of the concepts of "tangent line" and "area." As we shall explain in this section, both of these concepts rest on a more fundamental concept, known as a "*limit.*"

TANGENT LINES AND LIMITS

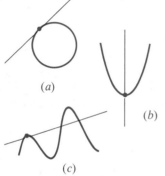

(a)

(b)

(c)

Figure 2.4.3

In plane geometry, a line is called *tangent* to a circle if it meets the circle at precisely one point (Figure 2.4.3a). However, this definition is not satisfactory for other kinds of curves. In Figure 2.4.3b the line meets the curve exactly once, yet is not a tangent, and in Figure 2.4.3c the line is tangent yet meets the curve more than once.

To define the concept of a tangent line so it applies to curves other than circles, we must view tangent lines another way. Consider a point P on a curve in the xy-plane. If Q is any point on the curve different from P, the line through P and Q is called a *secant line* for the curve (Figure 2.4.4). Intuition suggests that if we move the point Q along the curve toward P, the secant line will rotate toward a "limiting" position (Figure 2.4.5). The line T occupying this limiting position we consider to be the *tangent line* at P.

Figure 2.4.4

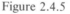

Figure 2.4.5

As suggested by Figure 2.4.6, this new concept of a tangent line coincides with the traditional concept, when applied to circles.

AREA AS A LIMIT

Just as the general notion of a tangent line leads to the concept of a "limit," so does the general notion of area. Areas of some plane regions can be calculated by subdividing them into a *finite* number of rectangles or triangles, then adding the areas of the constituent parts (Figure 2.4.7). However, for many regions a more general approach is needed. Consider the shaded region in Figure 2.4.8a. We can *approximate* the area of this region by inscribing rectangles of equal width under the curve and adding the areas of these rectangles (Figure 2.4.8b). Moreover, intuition suggests

Figure 2.4.6

Figure 2.4.7

that if we repeat the process using more and more rectangles, then the rectangles will tend to fill in the gaps under the curve and our approximations will "approach" the exact area under the curve as a "limiting value" (Figure 2.4.8c).

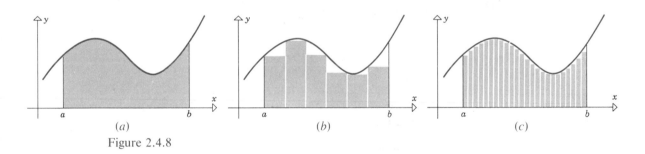

Figure 2.4.8

LIMITS   In the foregoing discussion, we saw that the concepts of a tangent line and the area of a plane region ultimately rest on the notion of a "limit." In this section and the next few we shall investigate the notion of limit in more detail. Our development of limits in this text proceeds in three stages:

- First we discuss limits intuitively.

- Then we discuss methods for computing limits.

- Finally, we give a precise mathematical discussion of limits.

Limits can be used to describe how a function behaves as the independent variable moves toward a certain value. For example, consider the function

$$f(x) = \frac{\sin x}{x}$$

where x is in radians. Although this function is not defined at $x = 0$, it still makes sense to ask what happens to the values of $f(x)$ as x moves along the x-axis toward $x = 0$. To answer this question, we used a calculator set in the radian mode to obtain values of $f(x)$ at a succession of points moving along the *positive* x-axis toward $x = 0$. The results, which appear in Table 2.4.1, suggest that the values of $f(x)$ approach 1. We call the number 1 the **limit** of

$$\frac{\sin x}{x}$$

as x approaches 0 from the right side, and we write

$$\lim_{x \to 0^+} \frac{\sin x}{x} = 1 \tag{1}$$

In this expression, "lim" tells us that we are computing a limit; the symbol $x \to 0$ tells us that we are letting x approach 0; and the label "+" on $x \to 0$ tells us that x is approaching zero from the *right* side. We can also ask what happens to the values of

$$f(x) = \frac{\sin x}{x}$$

as x approaches 0 from the left side. From Table 2.4.2 [or from the fact that $f(-x) = f(x)$] it is evident that the values of $f(x)$ again approach 1. We denote this by writing

$$\lim_{x \to 0^-} \frac{\sin x}{x} = 1 \tag{2}$$

In (2), the label "−" on $x \to 0$ tells us that we are computing the limit as x approaches zero from the *left* side.

REMARK.   It is important to keep in mind that the limits given in (1) and (2) are really just guesses based on numerical evidence. It is conceivable that if we continue the calculator computations in Tables 2.4.1 and 2.4.2 for values of x still closer to 0, the pattern of values for $(\sin x)/x$ might change and approach some value different from 1 or perhaps approach no number at all. Moreover, no matter how much we enlarge our tables, this problem will persist since we can never be certain what happens for values of x not included in the table. To be certain that the limits in (1) and (2) are really correct, we need mathematical proof, and to give a mathematical proof, we need a precise mathematical definition of a limit.

Table 2.4.1

x (RADIANS)	$\dfrac{\sin x}{x}$
1.0	0.84147
0.9	0.87036
0.8	0.89670
0.7	0.92031
0.6	0.94107
0.5	0.95885
0.4	0.97355
0.3	0.98507
0.2	0.99335
0.1	0.99833
0.01	0.99998

Table 2.4.2

x (RADIANS)	$\dfrac{\sin x}{x}$
−1.0	0.84147
−0.9	0.87036
−0.8	0.89670
−0.7	0.92031
−0.6	0.94107
−0.5	0.95885
−0.4	0.97355
−0.3	0.98507
−0.2	0.99335
−0.1	0.99833
−0.01	0.99998

These questions will be taken up later. However, in this section we shall rely completely on our intuition to obtain limits. Our main objective at this time is to introduce the mathematical notation associated with limits and to develop our intuitive understanding of limits geometrically.

2.4.1   NOTATION. If the value of $f(x)$ approaches the number L_1 as x approaches x_0 from the right side, we write

$$\lim_{x \to x_0^+} f(x) = L_1 \qquad\qquad (3)$$

which is read, "the limit of $f(x)$ as x approaches x_0 from the right is equal to L_1."

2.4.2   NOTATION. If the value of $f(x)$ approaches the number L_2 as x approaches x_0 from the left side, we write

$$\lim_{x \to x_0^-} f(x) = L_2 \qquad\qquad (4)$$

which is read, "the limit of $f(x)$ as x approaches x_0 from the left is equal to L_2."

2.4.3   NOTATION. If the limit from the left side is the same as the limit from the right side, that is,

$$\lim_{x \to x_0^-} f(x) = \lim_{x \to x_0^+} f(x) = L$$

then we write

$$\lim_{x \to x_0} f(x) = L \qquad\qquad (5)$$

which is read, "the limit of $f(x)$ as x approaches x_0 is equal to L."

The expressions

$$\lim_{x \to x_0^-} f(x) \quad \text{and} \quad \lim_{x \to x_0^+} f(x)$$

are called the *one-sided limits of $f(x)$ at x_0* and the expression

$$\lim_{x \to x_0} f(x)$$

is called the *two-sided limit of $f(x)$ at x_0*.

Example 1   From (1) and (2)

$$\lim_{x \to 0^+} \frac{\sin x}{x} = \lim_{x \to 0^-} \frac{\sin x}{x} = 1$$

Therefore,

$$\lim_{x \to 0} \frac{\sin x}{x} = 1 \qquad \blacktriangleleft$$

The following examples illustrate limits geometrically.

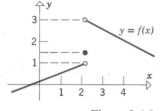

Figure 2.4.9

Example 2   Let f be the function whose graph is shown in Figure 2.4.9. As x approaches 2 from the left, $f(x)$ approaches 1, so

$$\lim_{x \to 2^-} f(x) = 1$$

As x approaches 2 from the right, $f(x)$ approaches 3, so that

$$\lim_{x \to 2^+} f(x) = 3$$

Since $f(2) = 1.5$, this example shows that the value of a function at a point, and the left- and right-hand limits there can all be different. $\blacktriangleleft$

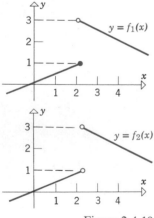

Figure 2.4.10

Example 3   The functions f_1 and f_2 whose graphs are shown in Figure 2.4.10 are identical to the function f in Example 2, except at the point $x = 2$; there we have $f(2) = 1.5$ and $f_1(2) = 1$, while $f_2(2)$ is undefined. However, even though the functions f, f_1, and f_2 behave differently *at* $x = 2$, their limits as x approaches 2 are the same, that is,

$$\lim_{x \to 2^-} f(x) = \lim_{x \to 2^-} f_1(x) = \lim_{x \to 2^-} f_2(x) = 1$$

and

$$\lim_{x \to 2^+} f(x) = \lim_{x \to 2^+} f_1(x) = \lim_{x \to 2^+} f_2(x) = 3 \qquad \blacktriangleleft$$

The foregoing example illustrates a general principle about limits that can be stated loosely as follows:

The limit of a function as the independent variable approaches a point does not depend on the value of the function *at* the point.

Thus, if we alter the value of a function f only at $x = x_0$, then we do not affect

$$\lim_{x \to x_0^-} f(x), \quad \lim_{x \to x_0^+} f(x), \quad \text{or} \quad \lim_{x \to x_0} f(x)$$

Function f_2 in Example 3 illustrates another idea: namely, that the left- and right-hand limits can have values at a point where the function itself is undefined.

Figure 2.4.11

Example 4 Let f be the function whose graph is shown in Figure 2.4.11. From the graph we see that

$$\lim_{x \to 3^-} f(x) = 2 \quad \text{and} \quad \lim_{x \to 3^+} f(x) = 2$$

so that

$$\lim_{x \to 3} f(x) = 2$$

The graph also shows that $f(3) = 2$. Thus, the value of a function at a point and the one-sided and two-sided limits as x approaches that point may be equal.   ◄

There is, in general, no guarantee that a function $f(x)$ actually has a limit as $x \to x_0^+$, $x \to x_0^-$, or $x \to x_0$. If there is no limit, then we say that *the limit does not exist*. Here are some examples.

Example 5 Let f be the function whose graph is shown in Figure 2.4.12. This graph is intended to convey the idea that as x approaches 6 from the right side, $f(x)$ oscillates between -3 and 3 infinitely often and with increasing frequency. Since $f(x)$ approaches no fixed value as x approaches 6 from the right, we say that

$$\lim_{x \to 6^+} f(x) \ does \ not \ exist$$

On the other hand, as x approaches 6 from the left side, $f(x)$ approaches 1, so that

$$\lim_{x \to 6^-} f(x) = 1 \quad ◄$$

Figure 2.4.12 ▷

REMARK.   Recall from 2.4.3 that in order to write $\lim\limits_{x \to x_0} f(x) = L$ we must have $\lim\limits_{x \to x_0^+} f(x) = L$ and $\lim\limits_{x \to x_0^-} f(x) = L$. Thus, if either of the two one-sided limits fails to exist, then the two-sided limit fails to exist. Thus, in the foregoing example $\lim\limits_{x \to 6} f(x)$ does not exist because $\lim\limits_{x \to 6^+} f(x)$ fails to exist.

INFINITE LIMITS

Example 6   Let f be the function whose graph is shown in Figure 2.4.13. As x approaches 0 from the right side, $f(x)$ gets larger and larger without bound and consequently approaches no fixed finite value. Thus,

$$\lim_{x \to 0^+} f(x)$$

does not exist. In this case, we would write

$$\lim_{x \to 0^+} f(x) = +\infty \tag{6}$$

to indicate that the limit fails to exist because $f(x)$ is increasing without bound.

As x approaches 0 from the left side, $f(x)$ becomes more and more negative without bound and consequently approaches no fixed finite value. Thus,

$$\lim_{x \to 0^-} f(x)$$

does not exist. In this case, we would write

$$\lim_{x \to 0^-} f(x) = -\infty \tag{7}$$

to indicate that the limit fails to exist because $f(x)$ is decreasing without bound.   ◀

REMARK.   It should be emphasized that the expressions $\lim\limits_{x \to 0^+} f(x) = +\infty$ and $\lim\limits_{x \to 0^-} f(x) = -\infty$ are simply descriptions of limits that do not exist. The symbols $+\infty$ and $-\infty$ do not represent numbers.

Example 7   Let f and g be the functions with graphs as shown in Figure 2.4.14. From the graph of f we see that

$$\lim_{x \to 2^-} f(x) = -\infty \quad \text{and} \quad \lim_{x \to 2^+} f(x) = -\infty$$

We can indicate these two results more succinctly by writing

$$\lim_{x \to 2} f(x) = -\infty$$

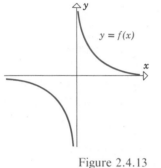

$y = f(x)$

Figure 2.4.13

$y = f(x)$

(a)

$y = g(x)$

-3

(b)

Figure 2.4.14

From the graph of g we see that

$$\lim_{x \to -3^-} g(x) = +\infty \quad \text{and} \quad \lim_{x \to -3^+} g(x) = +\infty$$

More briefly, we shall write

$$\lim_{x \to -3} g(x) = +\infty \quad \blacktriangleleft$$

LIMITS AT INFINITY   So far we have used limits to describe how a function behaves as the independent variable approaches a fixed point on the x-axis. However, limits can also be used to describe how a function behaves as the independent variable moves "indefinitely far" from the origin along the x-axis. If x is allowed to increase without bound, then we write $x \to +\infty$ (read, "x approaches plus infinity"), and if x is allowed to decrease without bound then we write $x \to -\infty$ (read, "x approaches minus infinity").

Example 8 Let f be the function with the graph shown in Figure 2.4.15. As $x \to +\infty$ the graph of f approaches the line $y = 4$ so that the value of $f(x)$ approaches 4. We denote this by writing

$$\lim_{x \to +\infty} f(x) = 4$$

Figure 2.4.15 ▷

As $x \to -\infty$ the graph of f tends toward the line $y = -1$ and so the value of $f(x)$ approaches -1. We denote this by writing

$$\lim_{x \to -\infty} f(x) = -1 \quad \blacktriangleleft$$

Example 9 Let f be the function with the graph shown in Figure 2.4.16.

Figure 2.4.16 ▷

As x approaches $-\infty$, the value of $f(x)$ increases without bound and consequently approaches no fixed value. Thus,

$$\lim_{x \to -\infty} f(x)$$

does not exist. In this case, we would write

$$\lim_{x \to -\infty} f(x) = +\infty$$

to indicate that the limit fails to exist because $f(x)$ is increasing without bound.

As x approaches $+\infty$, the graph of f oscillates but tends toward the line $y = -2$. Thus, the value of $f(x)$ approaches -2 as x approaches $+\infty$ and we write

$$\lim_{x \to +\infty} f(x) = -2 \qquad \blacktriangleleft$$

Example 10   Let f be the function with the graph shown in Figure 2.4.17. As x approaches $-\infty$, the value of $f(x)$ decreases without bound and consequently approaches no fixed finite value. Thus,

$$\lim_{x \to -\infty} f(x)$$

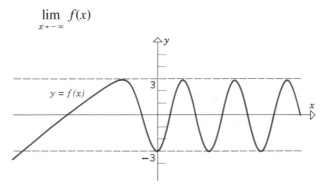

Figure 2.4.17

does not exist. In this case, we would write

$$\lim_{x \to -\infty} f(x) = -\infty$$

to indicate that the limit fails to exist because $f(x)$ is decreasing without bound.

As x approaches $+\infty$, the value of $f(x)$ oscillates between -3 and $+3$ and consequently approaches no fixed value. Thus,

$$\lim_{x \to +\infty} f(x)$$

does not exist. There is no special notation used to describe limits that fail to exist because of oscillation. $\blacktriangleleft$

▶ Exercise Set 2.4

1. For the function f graphed below, find
  (a) $\lim\limits_{x \to 3^-} f(x)$
  (b) $\lim\limits_{x \to 3^+} f(x)$
  (c) $\lim\limits_{x \to 3} f(x)$
  (d) $f(3)$
  (e) $\lim\limits_{x \to -\infty} f(x)$
  (f) $\lim\limits_{x \to +\infty} f(x)$.

4. For the function g graphed below, find
  (a) $\lim\limits_{x \to 0^-} g(x)$
  (b) $\lim\limits_{x \to 0^+} g(x)$
  (c) $\lim\limits_{x \to 0} g(x)$
  (d) $g(0)$
  (e) $\lim\limits_{x \to -\infty} g(x)$
  (f) $\lim\limits_{x \to +\infty} g(x)$.

2. For the function f graphed below, find
  (a) $\lim\limits_{x \to 2^-} f(x)$
  (b) $\lim\limits_{x \to 2^+} f(x)$
  (c) $\lim\limits_{x \to 2} f(x)$
  (d) $f(2)$
  (e) $\lim\limits_{x \to -\infty} f(x)$
  (f) $\lim\limits_{x \to +\infty} f(x)$.

5. For the function F graphed below, find
  (a) $\lim\limits_{x \to -2^-} F(x)$
  (b) $\lim\limits_{x \to -2^+} F(x)$
  (c) $\lim\limits_{x \to -2} F(x)$
  (d) $F(-2)$
  (e) $\lim\limits_{x \to -\infty} F(x)$
  (f) $\lim\limits_{x \to +\infty} F(x)$.

3. For the function g graphed below, find
  (a) $\lim\limits_{x \to 4^-} g(x)$
  (b) $\lim\limits_{x \to 4^+} g(x)$
  (c) $\lim\limits_{x \to 4} g(x)$
  (d) $g(4)$
  (e) $\lim\limits_{x \to -\infty} g(x)$
  (f) $\lim\limits_{x \to +\infty} g(x)$.

6. For the function F graphed below, find
  (a) $\lim\limits_{x \to 3^-} F(x)$
  (b) $\lim\limits_{x \to 3^+} F(x)$
  (c) $\lim\limits_{x \to 3} F(x)$
  (d) $F(3)$
  (e) $\lim\limits_{x \to -\infty} F(x)$
  (f) $\lim\limits_{x \to +\infty} F(x)$.

7. For the function ϕ graphed below, find

    (a) $\lim\limits_{x \to -2^-} \phi(x)$ (b) $\lim\limits_{x \to -2^+} \phi(x)$

    (c) $\lim\limits_{x \to -2} \phi(x)$ (d) $\phi(-2)$

    (e) $\lim\limits_{x \to -\infty} \phi(x)$ (f) $\lim\limits_{x \to +\infty} \phi(x)$.

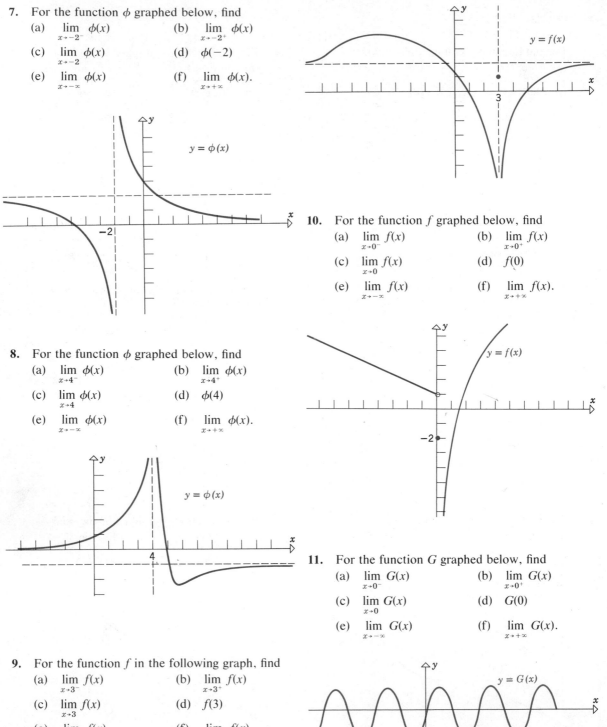

8. For the function ϕ graphed below, find

    (a) $\lim\limits_{x \to 4^-} \phi(x)$ (b) $\lim\limits_{x \to 4^+} \phi(x)$

    (c) $\lim\limits_{x \to 4} \phi(x)$ (d) $\phi(4)$

    (e) $\lim\limits_{x \to -\infty} \phi(x)$ (f) $\lim\limits_{x \to +\infty} \phi(x)$.

9. For the function f in the following graph, find

    (a) $\lim\limits_{x \to 3^-} f(x)$ (b) $\lim\limits_{x \to 3^+} f(x)$

    (c) $\lim\limits_{x \to 3} f(x)$ (d) $f(3)$

    (e) $\lim\limits_{x \to -\infty} f(x)$ (f) $\lim\limits_{x \to +\infty} f(x)$.

10. For the function f graphed below, find

    (a) $\lim\limits_{x \to 0^-} f(x)$ (b) $\lim\limits_{x \to 0^+} f(x)$

    (c) $\lim\limits_{x \to 0} f(x)$ (d) $f(0)$

    (e) $\lim\limits_{x \to -\infty} f(x)$ (f) $\lim\limits_{x \to +\infty} f(x)$.

11. For the function G graphed below, find

    (a) $\lim\limits_{x \to 0^-} G(x)$ (b) $\lim\limits_{x \to 0^+} G(x)$

    (c) $\lim\limits_{x \to 0} G(x)$ (d) $G(0)$

    (e) $\lim\limits_{x \to -\infty} G(x)$ (f) $\lim\limits_{x \to +\infty} G(x)$.

12. For the function G graphed below, find

(a) $\lim\limits_{x \to 0^-} G(x)$ (b) $\lim\limits_{x \to 0^+} G(x)$

(c) $\lim\limits_{x \to 0} G(x)$ (d) $G(0)$

(e) $\lim\limits_{x \to -\infty} G(x)$ (f) $\lim\limits_{x \to +\infty} G(x)$.

13. Consider the function g in the following graph. For what values of x_0 does $\lim\limits_{x \to x_0} g(x)$ exist?

14. Consider the function f graphed below. For what values of x_0 does $\lim\limits_{x \to x_0} f(x)$ exist?

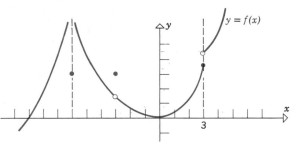

2.5 LIMITS (COMPUTATIONAL TECHNIQUES)

In the last section we concentrated on the graphical interpretation of limits. In this section we shall discuss techniques for finding limits directly from the formula for a function. Again our results will be based on intuition. Mathematical proofs of the results here will have to wait until we have a precise mathematical definition of a limit.

The limits discussed in our first four examples are the basic tools used to solve more complicated limit problems.

Example 1   Since a constant function $f(x) = k$ has the same value k everywhere, it follows that at each point a

$$\lim_{x \to a} k = k$$

and also

$$\lim_{x \to +\infty} k = \lim_{x \to -\infty} k = k$$

(Figure 2.5.1). These results can be summarized by the statement:

The limit of a constant function k at any point is k. ◄

Figure 2.5.1 ▷

Example 2   The limit

$$\lim_{x \to a} x = a$$

is self-evident. For example,

$$\lim_{x \to 5} x = 5, \quad \lim_{x \to 0} x = 0, \quad \lim_{x \to -2} x = -2 \quad ◄$$

Example 3   Find the limits

$$\lim_{x \to 0^+} \frac{1}{x}, \quad \lim_{x \to 0^-} \frac{1}{x}, \quad \lim_{x \to 0} \frac{1}{x}$$

Solution.   It will be evident from some simple calculations that none of these limits exist. As x approaches 0 from the right, the value of $1/x$ gets larger and larger without bound. For example, if x successively assumes values

$$x = 1, \frac{1}{10}, \frac{1}{100}, \frac{1}{1000}, \frac{1}{10,000}, \dots$$

then $1/x$ successively assumes values

$$\frac{1}{x} = 1, 10, 100, 1000, 10,000, \dots$$

Thus,

$$\lim_{x \to 0^+} \frac{1}{x} = +\infty$$

As x approaches 0 from the left, the value of $1/x$ decreases (that is, becomes more and more negative) without bound. For example, if x successively assumes values

$$x = -1, -\frac{1}{10}, -\frac{1}{100}, -\frac{1}{1000}, -\frac{1}{10,000}, \ldots$$

then $1/x$ successively assumes values

$$\frac{1}{x} = -1, -10, -100, -1000, -10,000, \ldots$$

Thus,

$$\lim_{x \to 0^-} \frac{1}{x} = -\infty$$

It follows that

$$\lim_{x \to 0} \frac{1}{x} \text{ does not exist}$$

since the one-sided limits as x approaches zero do not exist. Moreover, we cannot even write

$$\lim_{x \to 0} \frac{1}{x} = +\infty \quad \text{or} \quad \lim_{x \to 0} \frac{1}{x} = -\infty$$

(Why?)

The reader should check that the results in this example are consistent with the graph of $1/x$ (see Figure 1.3.20). ◄

Example 4 Find the limits

$$\lim_{x \to +\infty} \frac{1}{x} \quad \text{and} \quad \lim_{x \to -\infty} \frac{1}{x}$$

Solution. These limits are also evident from some sample calculations. To investigate the case $x \to +\infty$, assume x takes on successive values

$$x = 1, 10, 100, 1000, 10,000, \ldots$$

Then $1/x$ successively assumes values

$$\frac{1}{x} = 1, \frac{1}{10}, \frac{1}{100}, \frac{1}{1000}, \frac{1}{10,000}, \ldots$$

It is clear from these values of $1/x$ that

$$\lim_{x \to +\infty} \frac{1}{x} = 0$$

To investigate the case $x \to -\infty$, assume x takes on successive values

$$x = -1, -10, -100, -1000, -10,000, \ldots$$

Then $1/x$ successively assumes values

$$\frac{1}{x} = -1, -\frac{1}{10}, -\frac{1}{100}, -\frac{1}{1000}, -\frac{1}{10,000}, \ldots$$

It is clear from these values of $1/x$ that

$$\lim_{x \to -\infty} \frac{1}{x} = 0$$

The reader should check that the results in this example are consistent with the graph of $1/x$ (see Figure 1.3.20). ◄

The results in our first four examples, together with the fundamental properties of limits given in the following theorem, can be used to help solve more complicated limit problems. Parts of this theorem are proved in Appendix 2.

2.5.1 THEOREM. *Let* lim *stand for one of the limits* $\lim_{x \to a}$, $\lim_{x \to a^-}$, $\lim_{x \to a^+}$, $\lim_{x \to +\infty}$, *or* $\lim_{x \to -\infty}$. *If* $L_1 = \lim f(x)$ *and* $L_2 = \lim g(x)$ *both exist, then:*

(a) $\lim [f(x) + g(x)] = \lim f(x) + \lim g(x) = L_1 + L_2$.

(b) $\lim [f(x) - g(x)] = \lim f(x) - \lim g(x) = L_1 - L_2$.

(c) $\lim [f(x)g(x)] = \lim f(x) \lim g(x) = L_1 L_2$.

(d) $\lim \dfrac{f(x)}{g(x)} = \dfrac{\lim f(x)}{\lim g(x)} = \dfrac{L_1}{L_2}$ *if* $L_2 \neq 0$.

(e) $\lim \sqrt[n]{f(x)} = \sqrt[n]{\lim f(x)} = \sqrt[n]{L_1}$ *provided* $L_1 > 0$ *if n is even.*

In other words, this theorem states:

(a) *The limit of a sum is the sum of the limits.*
(b) *The limit of a difference is the difference of the limits.*
(c) *The limit of a product is the product of the limits.*
(d) *The limit of a quotient is the quotient of the limits provided the limit of the denominator is nonzero.*
(e) *The limit of an nth root is the nth root of the limit.*

REMARK.   Although results (*a*) and (*c*) are stated for two functions f and g, these results hold as well for any finite number of functions; that is, if

$$\lim f_1(x), \lim f_2(x), \ldots, \lim f_n(x)$$

all exist, then

$$\lim [f_1(x) + f_2(x) + \cdots + f_n(x)] = \lim f_1(x) + \lim f_2(x) + \cdots + \lim f_n(x)$$

and

$$\lim [f_1(x)f_2(x) \cdots f_n(x)] = \lim f_1(x) \lim f_2(x) \cdots \lim f_n(x) \qquad (1)$$

In particular, if $f_1, f_2, \ldots, f_n$ are all the same function f, then (1) reduces to

$$\lim [f(x)]^n = [\lim f(x)]^n \qquad (2)$$

From (2) we obtain the useful result

$$\lim_{x \to a} x^n = [\lim_{x \to a} x]^n = a^n \qquad (3)$$

For example,

$$\lim_{x \to 3} x^4 = 3^4 = 81$$

Another useful result follows from Example 1 and part (c) of Theorem 2.5.1. If k is a constant, then

$$\lim kf(x) = \lim k \lim f(x) = k \lim f(x) \qquad (4)$$

The first and last expressions in (4) tell us:

A constant factor can be moved through a limit sign.

Example 5 Find $\lim\limits_{x \to 5} (x^2 - 4x + 3)$ and justify each step.

Solution.

$$\lim_{x \to 5} (x^2 - 4x + 3) = \lim_{x \to 5} x^2 - \lim_{x \to 5} 4x + \lim_{x \to 5} 3 \qquad \text{[Theorem 2.5.1}(a), (b)]$$

$$= \lim_{x \to 5} x^2 - 4 \lim_{x \to 5} x + \lim_{x \to 5} 3 \qquad \text{[Equation (4)]}$$

$$= 5^2 - 4(5) + 3 \qquad \text{[Equation (3)]}$$

$$= 8 \qquad \blacktriangleleft$$

Example 6   Show that for any polynomial

$$p(x) = c_0 + c_1 x + \cdots + c_n x^n$$

and any real number a,

$$\lim_{x \to a} p(x) = c_0 + c_1 a + \cdots + c_n a^n = p(a)$$

Solution.

$$\lim_{x \to a} p(x) = \lim_{x \to a} (c_0 + c_1 x + \cdots + c_n x^n)$$

$$= \lim_{x \to a} c_0 + \lim_{x \to a} c_1 x + \cdots + \lim_{x \to a} c_n x^n$$

$$= \lim_{x \to a} c_0 + c_1 \lim_{x \to a} x + \cdots + c_n \lim_{x \to a} x^n$$

$$= c_0 + c_1 a + \cdots + c_n a^n = p(a) \quad \blacktriangleleft$$

REMARK.   Example 6 states that for a polynomial $p(x)$, the limit as x approaches a is equal to the value of the polynomial at a. If we apply this result to the limit problem in Example 5, we can bypass the intermediate steps and write immediately

$$\lim_{x \to 5} (x^2 - 4x + 3) = 5^2 - 4(5) + 3 = 8$$

Example 7   Find

$$\lim_{x \to 2} \frac{5x^3 + 4}{x - 3}$$

Solution.   Using the result of Example 6 and part (d) of Theorem 2.5.1, we obtain

$$\lim_{x \to 2} \frac{5x^3 + 4}{x - 3} = \frac{\lim_{x \to 2} (5x^3 + 4)}{\lim_{x \to 2} (x - 3)} = \frac{5 \cdot 2^3 + 4}{2 - 3} = -44 \quad \blacktriangleleft$$

Example 8   Find

$$\lim_{x \to 2} \frac{x^2 - 4}{x - 2}$$

Solution.   The method used in the previous example cannot be used here since part (d) of Theorem 2.5.1 does not apply (the limit of the denominator is 0). However, by factoring the numerator and canceling, we obtain

$$\lim_{x \to 2} \frac{x^2 - 4}{x - 2} = \lim_{x \to 2} \frac{(x - 2)(x + 2)}{x - 2} = \lim_{x \to 2} (x + 2) = 4 \qquad \blacktriangleleft$$

Though correct, the second equality in this calculation needs some justification. We pointed out in Example 8 of Section 2.1 that to cancel the $x - 2$ factors from

$$h(x) = \frac{x^2 - 4}{x - 2} = \frac{(x - 2)(x + 2)}{(x - 2)}$$

we must write

$$h(x) = x + 2, \quad x \neq 2$$

in order not to alter the domain of h. However, the functions

$$f(x) = x + 2 \quad \text{and} \quad h(x) = x + 2, \quad x \neq 2$$

differ only at $x = 2$, so that their limits as x approaches 2 are the same. Thus,

$$\lim_{x \to 2} \frac{(x - 2)(x + 2)}{x - 2} = \lim_{x \to 2} (x + 2)$$

even though

$$\frac{(x - 2)(x + 2)}{x - 2} \quad \text{and} \quad x + 2$$

are different functions.

REMARK. The technique in the foregoing example is extremely important. In any limit problem where the numerator and denominator both approach zero it is often possible to obtain the limit by performing some algebraic manipulations. Sometimes this will not work and more advanced techniques, to be studied later, are required to find the limit.

Example 9 Find

$$\lim_{x \to 4} \frac{2 - x}{(x - 4)(x + 2)}$$

Solution. Again, part (*d*) of Theorem 2.5.1 cannot be applied since the limit of the denominator is zero. There are no common factors in the numerator and denominator as there were in Example 8, so we must use a different approach. This kind of problem is best attacked by looking at the one-sided limits.

First consider

$$\lim_{x \to 4^+} \frac{2 - x}{(x - 4)(x + 2)}$$

As x approaches 4 from the right, the numerator is a negative quantity approaching -2 and the denominator a positive quantity approaching 0. Consequently the ratio is a negative quantity that decreases without bound. That is,

$$\lim_{x \to 4^+} \frac{2 - x}{(x - 4)(x + 2)} = -\infty \tag{5}$$

Now consider

$$\lim_{x \to 4^-} \frac{2 - x}{(x - 4)(x + 2)}$$

As x approaches 4 from the left, the numerator is eventually a negative quantity approaching -2 and the denominator is a negative quantity approaching 0. Consequently, the ratio is a positive quantity that increases without bound; that is,

$$\lim_{x \to 4^-} \frac{2 - x}{(x - 4)(x + 2)} = +\infty \tag{6}$$

From (5) and (6)

$$\lim_{x \to 4} \frac{2 - x}{(x - 4)(x + 2)}$$

does not exist. ◀

Next we shall consider methods for finding limits of rational functions as $x \to +\infty$ or $x \to -\infty$. But first we need some preliminary results.

In Example 4 we obtained the limits

$$\lim_{x \to +\infty} \frac{1}{x} = \lim_{x \to -\infty} \frac{1}{x} = 0$$

Consequently, for any positive integer n,

$$\lim_{x \to +\infty} \frac{1}{x^n} = \left(\lim_{x \to +\infty} \frac{1}{x} \right)^n = 0 \tag{7}$$

and

$$\lim_{x \to -\infty} \frac{1}{x^n} = \left(\lim_{x \to -\infty} \frac{1}{x} \right)^n = 0 \tag{8}$$

Results (7) and (8) are also evident from the graph of $1/x^n$, which has one of the two general shapes shown in Fig. 2.5.2, depending on whether n is even or odd.

If we divide numerator and denominator of a rational function by the highest power of x that occurs in the function, then all the powers of x become constants or powers of $1/x$. The next two examples show how this observation together with (7) and (8) can be used to find $\lim\limits_{x \to +\infty}$ or $\lim\limits_{x \to -\infty}$ for rational functions.

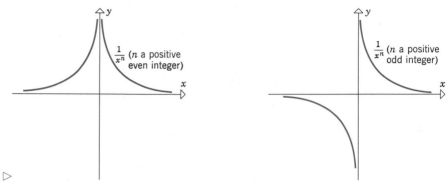

Figure 2.5.2 ▷

Example 10   Find

$$\lim_{x \to +\infty} \frac{3x + 5}{6x - 8}.$$

Solution.   Divide the numerator and denominator by the highest power of x that occurs; this is $x^1 = x$. We obtain

$$\lim_{x \to +\infty} \frac{3x + 5}{6x - 8} = \lim_{x \to +\infty} \frac{3 + 5/x}{6 - 8/x} = \frac{\lim\limits_{x \to +\infty} (3 + 5/x)}{\lim\limits_{x \to +\infty} (6 - 8/x)}$$

$$= \frac{\lim\limits_{x \to +\infty} 3 + \lim\limits_{x \to +\infty} 5/x}{\lim\limits_{x \to +\infty} 6 - \lim\limits_{x \to +\infty} 8/x} = \frac{3 + 5 \lim\limits_{x \to +\infty} 1/x}{6 - 8 \lim\limits_{x \to +\infty} 1/x}$$

$$= \frac{3 + (5 \cdot 0)}{6 - (8 \cdot 0)} = \frac{1}{2} \quad \blacktriangleleft$$

Example 11   Find

$$\lim_{x \to -\infty} \frac{4x^2 - x}{2x^3 - 5}$$

Solution. Divide numerator and denominator by the highest power of x that occurs, namely x^3. We obtain

$$\lim_{x \to -\infty} \frac{4x^2 - x}{2x^3 - 5} = \lim_{x \to -\infty} \frac{4/x - 1/x^2}{2 - 5/x^3} = \frac{\lim\limits_{x \to -\infty} (4/x - 1/x^2)}{\lim\limits_{x \to -\infty} (2 - 5/x^3)}$$

$$= \frac{\lim\limits_{x \to -\infty} 4/x - \lim\limits_{x \to -\infty} 1/x^2}{\lim\limits_{x \to -\infty} 2 - \lim\limits_{x \to -\infty} 5/x^3} = \frac{4 \lim\limits_{x \to -\infty} 1/x - \lim\limits_{x \to -\infty} 1/x^2}{2 - 5 \lim\limits_{x \to -\infty} 1/x^3}$$

$$= \frac{(4 \cdot 0) - 0}{2 - (5 \cdot 0)} = \frac{0}{2} = 0 \quad \blacktriangleleft$$

Example 12 Find

$$\lim_{x \to +\infty} \sqrt[3]{\frac{3x + 5}{6x - 8}}$$

Solution.

$$\lim_{x \to +\infty} \sqrt[3]{\frac{3x + 5}{6x - 8}} \underset{\uparrow}{=} \sqrt[3]{\lim_{x \to +\infty} \frac{3x + 5}{6x - 8}} \underset{\uparrow}{=} \sqrt[3]{\frac{1}{2}} \qquad \blacktriangleleft$$

> Theorem 2.5.1, part (*e*)   Example 10

Example 13 Find

$$\lim_{x \to +\infty} \frac{3 - 2x^4}{x + 1}$$

Solution. Dividing numerator and denominator by x^4, we obtain

$$\lim_{x \to +\infty} \frac{3 - 2x^4}{x + 1} = \lim_{x \to +\infty} \frac{3/x^4 - 2}{1/x^3 + 1/x^4} \tag{9}$$

Since the limit of the denominator in the right side of (9) is zero, we cannot apply Theorem 2.5.1. However, we can argue as follows. In the numerator, $3/x^4$ approaches 0 as $x \to +\infty$ so that

$$\frac{3}{x^4} - 2$$

is eventually negative and approaches -2. As $x \to +\infty$, the denominator is a positive quantity approaching 0. Consequently, as $x \to +\infty$, the ratio is eventually a negative quantity that decreases without bound. That is,

$$\lim_{x \to +\infty} \frac{3 - 2x^4}{x + 1} = \lim_{x \to +\infty} \frac{3/x^4 - 2}{1/x^3 + 1/x^4} = -\infty \qquad \blacktriangleleft$$

Example 14 Find

$$\lim_{x \to +\infty} \frac{\sqrt{x^2 + 2}}{3x - 6}$$

Solution. Since we are letting $x \to +\infty$, it would be helpful to manipulate the function so that the powers of x become powers of $1/x$. To achieve this, we divide numerator and denominator by x. In the numerator, we write x as

$$x = \sqrt{x^2}$$

This is justified because x eventually assumes positive values as $x \to +\infty$, so that

$$\sqrt{x^2} = |x| = x$$

Thus,

$$\lim_{x \to +\infty} \frac{\sqrt{x^2 + 2}}{3x - 6} = \lim_{x \to +\infty} \frac{\sqrt{x^2 + 2}/\sqrt{x^2}}{(3x - 6)/x} = \lim_{x \to +\infty} \frac{\sqrt{1 + 2/x^2}}{3 - 6/x}$$

$$= \frac{\lim\limits_{x \to +\infty} \sqrt{1 + 2/x^2}}{\lim\limits_{x \to +\infty} (3 - 6/x)} = \frac{\sqrt{\lim\limits_{x \to +\infty} (1 + 2/x^2)}}{\lim\limits_{x \to +\infty} (3 - 6/x)}$$

$$= \frac{\sqrt{\lim\limits_{x \to +\infty} 1 + 2 \lim\limits_{x \to +\infty} 1/x^2}}{\lim\limits_{x \to +\infty} 3 - 6 \lim\limits_{x \to +\infty} 1/x} = \frac{\sqrt{1 + (2 \cdot 0)}}{3 - (6 \cdot 0)} = \frac{1}{3} \qquad \blacktriangleleft$$

Example 15 Find

$$\lim_{x \to -\infty} \frac{\sqrt{x^2 + 2}}{3x - 6}$$

Solution. The function here is the same as in Example 14. However, in this problem we want $\lim\limits_{x \to -\infty}$ rather than $\lim\limits_{x \to +\infty}$. As in Example 14, we divide numerator and denominator by x, but in the numerator we write x as

$$x = -\sqrt{x^2}$$

This is justified because x eventually assumes negative values as $x \to -\infty$, so that

$$\sqrt{x^2} = |x| = -x$$

We obtain

$$\lim_{x \to -\infty} \frac{\sqrt{x^2 + 2}}{3x - 6} = \lim_{x \to -\infty} \frac{\sqrt{x^2 + 2}/(-\sqrt{x^2})}{3 - (6/x)}$$

$$= \lim_{x \to -\infty} \frac{-\sqrt{1 + 2/x^2}}{3 - (6/x)} = -\frac{1}{3} \quad \blacktriangleleft$$

Example 16   Find $\lim\limits_{x \to 3} f(x)$ for

$$f(x) = \begin{cases} x^2 - 5, & x \le 3 \\ \sqrt{x + 13}, & x > 3 \end{cases}$$

Solution.   As x approaches 3 from the left, the formula for f is

$$f(x) = x^2 - 5$$

so that

$$\lim_{x \to 3^-} f(x) = \lim_{x \to 3^-} (x^2 - 5) = 3^2 - 5 = 4$$

As x approaches 3 from the right, the formula for f is

$$f(x) = \sqrt{x + 13}$$

so that

$$\lim_{x \to 3^+} f(x) = \lim_{x \to 3^+} \sqrt{x + 13} = \sqrt{\lim_{x \to 3^+} (x + 13)} = \sqrt{16} = 4$$

Therefore,

$$\lim_{x \to 3} f(x) = 4$$

since the one-sided limits are equal. $\blacktriangleleft$

▶ Exercise Set 2.5

Find the limits in Exercises 1–50.

1. $\lim\limits_{x \to 8} 7$.

2. $\lim\limits_{x \to -\infty} (-3)$.

3. $\lim\limits_{x \to 0^+} \pi$.

4. $\lim\limits_{x \to -2} 3x$.

5. $\lim\limits_{y \to 3^+} 12y$.

6. $\lim\limits_{h \to +\infty} (-2h)$.

7. $\lim\limits_{x \to 5} \sqrt{x^3 - 3x - 1}$.

8. $\lim\limits_{x \to 0^-} (x^4 + 12x^3 - 17x + 2)$.

9. $\lim\limits_{y \to -1} (y^6 - 12y + 1)$.

10. $\lim\limits_{x \to 3} \dfrac{x^2 - 2x}{x + 1}$.

11. $\lim\limits_{y \to 2^-} \dfrac{(y - 1)(y - 2)}{y + 1}$.

12. $\lim\limits_{x \to 0} \dfrac{6x - 9}{x^3 - 12x + 3}$.

13. $\lim\limits_{x \to 4} \dfrac{x^2 - 16}{x - 4}$.

14. $\lim\limits_{t \to -2} \dfrac{t^3 + 8}{t + 2}$.

15. $\lim\limits_{x \to 1^+} \dfrac{x^4 - 1}{x - 1}$.

16. $\lim\limits_{x \to 2} \dfrac{x^2 - 4x + 4}{x^2 + x - 6}$.

17. $\lim\limits_{x \to -1} \dfrac{x^2 + 6x + 5}{x^2 - 3x - 4}$.

18. $\lim\limits_{t \to 1} \dfrac{t^3 + t^2 - 5t + 3}{t^3 - 3t + 2}$.

19. $\lim\limits_{x \to +\infty} \dfrac{3x + 1}{2x - 5}$.

20. $\lim\limits_{x \to +\infty} \dfrac{1}{x - 12}$.

21. $\lim\limits_{y \to -\infty} \dfrac{3}{y + 4}$.

22. $\lim\limits_{x \to +\infty} \dfrac{5x^2 + 7}{3x^2 - x}$.

23. $\lim\limits_{x \to -\infty} \dfrac{x - 2}{x^2 + 2x + 1}$.

24. $\lim\limits_{s \to +\infty} \sqrt[3]{\dfrac{3s^7 - 4s^5}{2s^7 + 1}}$.

25. $\lim\limits_{x \to -\infty} \dfrac{\sqrt{5x^2 - 2}}{x + 3}$.

26. $\lim\limits_{x \to +\infty} \dfrac{\sqrt{5x^2 - 2}}{x + 3}$.

27. $\lim\limits_{y \to -\infty} \dfrac{2 - y}{\sqrt{7 + 6y^2}}$.

28. $\lim\limits_{y \to +\infty} \dfrac{2 - y}{\sqrt{7 + 6y^2}}$.

29. $\lim\limits_{x \to -\infty} \dfrac{\sqrt{3x^4 + x}}{x^2 - 8}$.

30. $\lim\limits_{x \to +\infty} \dfrac{\sqrt{3x^4 + x}}{x^2 - 8}$.

31. $\lim\limits_{x \to 3^+} \dfrac{x}{x - 3}$.

32. $\lim\limits_{x \to 3^-} \dfrac{x}{x - 3}$.

33. $\lim\limits_{x \to 3} \dfrac{x}{x - 3}$.

34. $\lim\limits_{x \to 2^+} \dfrac{x}{x^2 - 4}$.

35. $\lim\limits_{x \to 2^-} \dfrac{x}{x^2 - 4}$.

36. $\lim\limits_{x \to 2} \dfrac{x}{x^2 - 4}$.

37. $\lim\limits_{y \to 6^+} \dfrac{y + 6}{y^2 - 36}$.

38. $\lim\limits_{y \to 6^-} \dfrac{y + 6}{y^2 - 36}$.

39. $\lim\limits_{y \to 6} \dfrac{y + 6}{y^2 - 36}$.

40. $\lim\limits_{x \to 4^+} \dfrac{3 - x}{x^2 - 2x - 8}$.

41. $\lim\limits_{x \to 4^-} \dfrac{3 - x}{x^2 - 2x - 8}$.

42. $\lim\limits_{x \to 4} \dfrac{3 - x}{x^2 - 2x - 8}$.

43. $\lim\limits_{x \to +\infty} \dfrac{7 - 6x^5}{x + 3}$.

44. $\lim\limits_{t \to -\infty} \dfrac{5 - 2t^3}{t^2 + 1}$.

45. $\lim\limits_{t \to +\infty} \dfrac{6 - t^3}{7t^3 + 3}$.

46. $\lim\limits_{x \to 0^+} \dfrac{x}{|x|}$.

47. $\lim\limits_{x \to 0^-} \dfrac{x}{|x|}$.

48. $\lim\limits_{x \to 3^-} \dfrac{1}{|x - 3|}$.

49. $\lim\limits_{x \to 9} \dfrac{x - 9}{\sqrt{x} - 3}$.

50. $\lim\limits_{y \to 4} \dfrac{4 - y}{2 - \sqrt{y}}$.

51. Let $f(x) = \begin{cases} x - 1, & x \le 3 \\ 3x - 7, & x > 3 \end{cases}$

Find:

(a) $\lim\limits_{x \to 3^-} f(x)$ (b) $\lim\limits_{x \to 3^+} f(x)$ (c) $\lim\limits_{x \to 3} f(x)$.

52. Let $g(t) = \begin{cases} t^2, & t \ge 0 \\ t - 2, & t < 0 \end{cases}$

Find:

(a) $\lim\limits_{t \to 0^-} g(t)$ (b) $\lim\limits_{t \to 0^+} g(t)$ (c) $\lim\limits_{t \to 0} g(t)$.

53. Find $\lim\limits_{x \to 3} h(x)$ given that

$$h(x) = \begin{cases} x^2 - 2x + 1, & x \ne 3 \\ 7, & x = 3 \end{cases}$$

54. Let $F(x) = \begin{cases} \dfrac{x^2 - 9}{x + 3}, & x \ne -3 \\ k, & x = -3 \end{cases}$

(a) Find k so that $F(-3) = \lim\limits_{x \to -3} F(x)$.

(b) With k assigned the value $\lim\limits_{x \to -3} F(x)$, show that $F(x)$ can be expressed as a polynomial.

55. (a) Explain why the following calculation is incorrect.

$$\lim_{x \to 0^+} \left(\frac{1}{x} - \frac{1}{x^2}\right) = \lim_{x \to 0^+} \frac{1}{x} - \lim_{x \to 0^+} \frac{1}{x^2}$$
$$= +\infty - (+\infty) = 0$$

(b) Show that $\lim_{x \to 0^+} \left(\frac{1}{x} - \frac{1}{x^2}\right) = -\infty$.

56. Let $f(x) = \dfrac{x^3 - 1}{x - 1}$.

(a) Find $\lim_{x \to 1} f(x)$.

(b) Sketch the graph of $y = f(x)$.

Find the limits in Exercises 57–68.

57. $\lim_{x \to 0} \dfrac{\sqrt{x + 3} - \sqrt{3}}{x}$ [*Hint:* Rationalize the numerator.]

58. $\lim_{x \to 0} \dfrac{\sqrt{x^2 + 4} - 2}{x}$.

59. $\lim_{x \to +\infty} (\sqrt{x^2 + ax} - x)$.

60. $\lim_{x \to +\infty} (\sqrt{x^2 + ax} - \sqrt{x^2 + bx})$.

61. $\lim_{x \to +\infty} (\sqrt{x^2 + 3} - x)$.

62. $\lim_{x \to +\infty} (\sqrt{2x^2 + 5} - x)$.

63. $\lim_{x \to 0^+} \sin\left(\dfrac{1}{x}\right)$.

64. $\lim_{x \to 0^+} x \sin\left(\dfrac{1}{x}\right)$.

65. $\lim_{x \to +\infty} \sin x$.

66. $\lim_{x \to +\infty} \dfrac{\sin x}{x}$.

67. $\lim_{x \to +\infty} \cos\left(\dfrac{1}{x}\right)$.

68. $\lim_{x \to +\infty} \dfrac{\cos(1/x)}{x}$.

69. Let $r(x)$ be a rational function. Under what conditions is it true that $\lim_{x \to a} r(x) = r(a)$?

70. Find

$$\lim_{x \to +\infty} \frac{c_0 + c_1 x + \cdots + c_n x^n}{d_0 + d_1 x + \cdots + d_m x^m}$$

where $c_n \neq 0$ and $d_m \neq 0$.
[*Hint:* Your answer will depend on whether $m < n$, $m = n$, or $m > n$.]

2.6   LIMITS: A RIGOROUS APPROACH (OPTIONAL)

This section gives a rigorous treatment of two-sided limits. Readers interested in pursuing the theory of one-sided and infinite limits can turn to Appendix 2, where we give a rigorous treatment of these topics and also prove some of the basic limit theorems.

Let a and L be real numbers. When we write

$$\lim_{x \to a} f(x) = L \tag{1}$$

we mean intuitively that the value of $f(x)$ approaches L as x approaches a from either side. To capture this concept mathematically, we must clarify the meanings of the phrases "x approaches a from either side" and "$f(x)$ approaches L."

The limit in (1) is intended to describe the behavior of f when x is near but *different from* a. The value of f at a is irrelevant to the limit. In fact, we have seen examples such as

$$\lim_{x \to 0} \frac{\sin x}{x} = 1$$

where the function f is not even defined at the point a. Thus, when we say "x approaches a from either side," we mean that x assumes values arbitrarily close to a, on one side or the other, but does not actually take on the value a. On the other hand, when we say that "$f(x)$ approaches L," we want to allow the possibility that $f(x)$ may actually take on the value L.

Let us see if we can phrase these ideas more precisely. To say that $f(x)$ becomes arbitrarily close to L as x approaches a means that if we pick any positive number ϵ (no matter how small), eventually the difference between $f(x)$ and L will be less than ϵ as x approaches a. That is, eventually $f(x)$ will satisfy the condition

$$|f(x) - L| < \epsilon$$

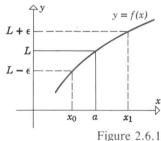

Figure 2.6.1

Figure 2.6.1 illustrates this idea. For the curve in that figure we have

$$\lim_{x \to a} f(x) = L$$

where a and L are the points shown. We picked an arbitrary positive number ϵ and indicated the interval

$$(L - \epsilon, L + \epsilon)$$

on the y-axis. As x approaches a, x will eventually lie between the points x_0 and x_1. When this occurs, $f(x)$ will lie in the interval $(L - \epsilon, L + \epsilon)$, that is, $f(x)$ will satisfy

$$L - \epsilon < f(x) < L + \epsilon$$

or equivalently

$$|f(x) - L| < \epsilon \tag{2}$$

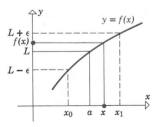

Figure 2.6.2

(Figure 2.6.2). It is evident geometrically that no matter how small we make ϵ, eventually $f(x)$ will satisfy (2) as x approaches a.

For the curve in Figure 2.6.1, we have

Figure 2.6.3

Figure 2.6.4

Figure 2.6.5

$$f(a) = \lim_{x \to a} f(x) = L$$

But the equality of $f(a)$ and L is irrelevant to the limit discussion. As x approaches a, x is not allowed to take on the value a. Thus, if $f(a)$ is undefined, as in Figure 2.6.3, or if $f(a)$ is defined, but unequal to L, as in Figure 2.6.4, it is still true that for an arbitrary positive value of ϵ, $f(x)$ will eventually satisfy (2) as x approaches a. In brief, the condition $|f(x) - L| < \epsilon$ need only be satisfied for x in the set $(x_0, a) \cup (a, x_1)$.

Before we can give a precise definition of limit, we need a preliminary observation. If the condition

$$|f(x) - L| < \epsilon \tag{3}$$

holds for all x in the set

$$(x_0, a) \cup (a, x_1) \tag{4}$$

then this condition holds as well for all x in any *subset* of (4). In particular, if we let δ be any positive number that is smaller than both

$$a - x_0 \quad \text{and} \quad x_1 - a$$

then

$$(a - \delta, a) \cup (a, a + \delta) \tag{5}$$

will be a subset of (4) (Figure 2.6.5) and condition (3) will hold for all x in this subset.

Since the set in (5) consists of all x satisfying

$$0 < |x - a| < \delta$$

we are led to the following definition.

2.6.1 DEFINITION. Let $f(x)$ be defined for all x in some open interval containing the number a, with the possible exception that $f(x)$ may or may not be defined at a. We shall write

$$\lim_{x \to a} f(x) = L$$

if given any number $\epsilon > 0$, we can find a number $\delta > 0$ such that $f(x)$ satisfies

$$|f(x) - L| < \epsilon \quad \text{whenever } x \text{ satisfies} \quad 0 < |x - a| < \delta$$

Example 1   Use Definition 2.6.1 to prove

$$\lim_{x \to 2} (3x - 5) = 1$$

Solution.   We must show that given any positive number ϵ, we can find a positive number δ such that $f(x) = 3x - 5$ satisfies

$$|(3x - 5) - 1| < \epsilon \tag{6}$$

whenever x satisfies

$$0 < |x - 2| < \delta \tag{7}$$

To find δ, we can rewrite (6) as

$$|3x - 6| < \epsilon$$

or equivalently

$$3|x - 2| < \epsilon$$

or

$$|x - 2| < \frac{\epsilon}{3} \tag{8}$$

We must choose δ so that (8) is satisfied whenever (7) is satisfied. We can do this by taking

$$\delta = \frac{\epsilon}{3}$$

To prove that this choice for δ works, assume x satisfies (7). Since we are letting $\delta = \epsilon/3$, it follows from (7) that x satisfies

$$0 < |x - 2| < \frac{\epsilon}{3} \tag{9}$$

Thus, (8) is satisfied since it is just the right-hand inequality in (9). This proves that $\lim_{x \to 2} (3x - 5) = 1$. ◀

As illustrated in Example 1, a limit proof proceeds as follows: We *assume* that an arbitrary positive number ϵ is given to us. Then we try to *find* a positive number δ such that the conditions in the definition are fulfilled. The idea is to find a formula for δ in terms of ϵ so that no matter what ϵ is chosen, we automatically obtain from the formula the required δ. (In Example 1, the formula was $\delta = \epsilon/3$.)

Example 1 is about as easy as a limit proof can get; most limit proofs require a little more algebraic and logical ingenuity. The reader who finds "δ-ϵ" discussions hard going should not become discouraged. The con-

cepts and techniques are intrinsically difficult. In fact, a precise understanding of limits evaded the finest mathematical minds for centuries.

Note that the value of δ in Definition 2.6.1 is not unique. Once one
value of δ is found that fulfills the requirements of the definition, any
smaller positive value for δ will also fulfill these requirements. To see
why, assume we have found a value of δ such that

$$|f(x) - L| < \epsilon \tag{10}$$

whenever x satisfies

$$0 < |x - a| < \delta \tag{11}$$

Let δ_1 be any positive number smaller than δ. If x satisfies

$$0 < |x - a| < \delta_1 \tag{12}$$

then we have

$$0 < |x - a| < \delta_1 < \delta$$

Thus, x satisfies (11) and consequently $f(x)$ satisfies (10). Therefore, δ_1
meets the requirements of Definition 2.6.1 also. For example, we found
in Example 1 that $\delta = \epsilon/3$ fulfills the requirements of Definition 2.6.1.
Consequently, any smaller value for δ, such as $\delta = \epsilon/4$, $\delta = \epsilon/5$, or $\delta = \epsilon/6$,
does also.

Example 2 Prove that $\lim\limits_{x \to 3} x^2 = 9$.

Solution. We must show that given any positive number ϵ we can find
a positive number δ such that

$$|x^2 - 9| < \epsilon \tag{13}$$

whenever x satisfies

$$0 < |x - 3| < \delta \tag{14}$$

Because $|x - 3|$ occurs in (14), it will be helpful to rewrite (13) so that
$|x - 3|$ appears as a factor on the left side. Therefore, we shall rewrite
(13) as

$$|x + 3|\,|x - 3| < \epsilon \tag{15}$$

If we can somehow ensure that when x satisfies (14) the factor $|x + 3|$
remains less than some positive constant, say,

$$|x + 3| < k \tag{16}$$

then on choosing

$$\delta = \frac{\epsilon}{k} \tag{17}$$

it will follow from (14) that

$$0 < |x - 3| < \frac{\epsilon}{k}$$

or

$$0 < k|x - 3| < \epsilon \tag{18}$$

From (16) and the right-hand inequality in (18) we shall then have

$$|x + 3| \, |x - 3| < k|x - 3| < \epsilon$$

so that (15) will be satisfied, and the proof will be complete. As we shall now explain, condition (16) can be obtained by restricting the size of δ. We remarked in the discussion preceding this example that δ is not unique; once a value of δ is found, any *smaller* positive value for δ can also be used. At this point we do not have a formula relating δ and ϵ. However, let us agree in advance that if, for a given ϵ, δ turns out to exceed 1, we shall choose the smaller value $\delta = 1$ instead. Thus, we can assume in our discussion that δ satisfies

$$0 < \delta \leq 1 \tag{19}$$

Assume that x satisfies (14). From (19) and the right side of (14) we obtain

$$|x - 3| < \delta \leq 1$$

so that

$$|x - 3| < 1$$

or equivalently

$$2 < x < 4$$

so

$$5 < x + 3 < 7$$

Therefore,

$$|x + 3| < 7$$

Comparing the last inequality to (16) suggests $k = 7$; and from (17)

$$\delta = \frac{\epsilon}{k} = \frac{\epsilon}{7}.$$

In summary, given $\epsilon > 0$, we choose

$$\delta = \frac{\epsilon}{7}$$

provided $\epsilon/7$ does not exceed 1. If $\epsilon/7$ exceeds 1, we choose

$$\delta = 1$$

In other words, we take δ to be the minimum of the numbers $\epsilon/7$ and 1. This is sometimes written

$$\delta = \min\left(\frac{\epsilon}{7}, 1\right) \qquad \blacktriangleleft$$

In the foregoing example the reader may have wondered how we knew to make the restriction $\delta \leq 1$ as opposed to some other restriction such as $\delta \leq \frac{1}{2}$ or $\delta \leq 5$. Actually, our selection was completely arbitrary; any other restriction of the form $\delta \leq c$ would have worked equally well (Exercise 27).

Example 3   Prove that

$$\lim_{x \to 1/2} \frac{1}{x} = 2$$

Solution.   We must show that given $\epsilon > 0$, there exists a $\delta > 0$ such that

$$\left|\frac{1}{x} - 2\right| < \epsilon \tag{20}$$

whenever x satisfies

$$0 < \left|x - \frac{1}{2}\right| < \delta \tag{21}$$

Because $|x - \frac{1}{2}|$ occurs in (21), it will be helpful to rewrite (20) so that $|x - \frac{1}{2}|$ appears as a factor on the left side. We do this as follows:

$$\left|\frac{1}{x} - 2\right| = \left|\frac{2}{x}\left(\frac{1}{2} - x\right)\right| = \left|\frac{2}{x}\right|\left|\frac{1}{2} - x\right| = \left|\frac{2}{x}\right|\left|x - \frac{1}{2}\right|$$

Thus, (20) is equivalent to

$$\left|\frac{2}{x}\right| \left|x - \frac{1}{2}\right| < \epsilon \tag{22}$$

From here on, the procedure is similar to that used in the last example. By restricting δ we shall try to make the factor $|2/x|$ satisfy

$$\left|\frac{2}{x}\right| < k \tag{23}$$

for some constant k when x satisfies (21). Then on choosing

$$\delta = \frac{\epsilon}{k} \tag{24}$$

we shall have, from (21),

$$0 < \left|x - \frac{1}{2}\right| < \frac{\epsilon}{k}$$

or

$$0 < k \left|x - \frac{1}{2}\right| < \epsilon \tag{25}$$

From (23) and the right-hand inequality in (25) it will then follow that

$$\left|\frac{2}{x}\right| \left|x - \frac{1}{2}\right| < k \left|x - \frac{1}{2}\right| < \epsilon$$

so that (22) will hold and the proof will be complete. Therefore, to finish the proof, we must show that δ can be restricted so that (23) holds for some k when x satisfies (21).

In Figure 2.6.6 we have sketched the graph of $|2/x|$ and marked the set of x values satisfying (21). In part (a) of the figure we took $\delta = \frac{1}{2}$ and in part (b) of the figure we took $\delta < \frac{1}{2}$. Part (a) of this figure makes it clear that if $\delta = \frac{1}{2}$ (or $\delta > \frac{1}{2}$) then the values of $|2/x|$ will have no upper bound,

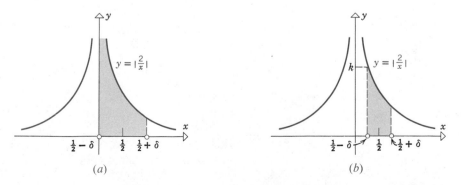

Figure 2.6.6 ▷   (a)   (b)

making it impossible to satisfy (23). However, part (*b*) shows that when $\delta < \frac{1}{2}$, there is an upper bound k to the values of $|2/x|$. Therefore, the arbitrary restriction

$$\delta \leq \frac{1}{4} \tag{26}$$

will be satisfactory to assure that $|2/x|$ has an upper bound.

Assume that x satisfies (21). From (26) and the right side of (21), we obtain

$$\left| x - \frac{1}{2} \right| < \delta \leq \frac{1}{4}$$

so that

$$\left| x - \frac{1}{2} \right| < \frac{1}{4}$$

or equivalently

$$\frac{1}{4} < x < \frac{3}{4} \tag{27}$$

To estimate the size of $|2/x|$ we take reciprocals in (27) and multiply through by 2, which gives

$$8 > \frac{2}{x} > \frac{8}{3}$$

Thus,

$$\left| \frac{2}{x} \right| < 8$$

Comparing the last inequality to (23) suggests $k = 8$; and from (24)

$$\delta = \frac{\epsilon}{k} = \frac{\epsilon}{8}$$

In summary, given $\epsilon > 0$, we can choose

$$\delta = \frac{\epsilon}{8}$$

provided this choice does not violate $\delta \leq \frac{1}{4}$. If it does, we choose the value $\delta = \frac{1}{4}$ instead. In other words,

$$\delta = \min \left(\frac{\epsilon}{8}, \frac{1}{4} \right) \quad \blacktriangleleft$$

Example 4   Let

$$f(x) = \begin{cases} 1, & x > 0 \\ -1, & x < 0 \end{cases}$$

Prove that $\lim_{x \to 0} f(x)$ does not exist.

Solution.   We shall assume that there is a limit and obtain a contradiction. Assume there is a number L such that

$$\lim_{x \to 0} f(x) = L$$

Then given any $\epsilon > 0$, there exists a $\delta > 0$ such that

$$|f(x) - L| < \epsilon \quad \text{whenever} \quad 0 < |x - 0| < \delta$$

In particular, if we take $\epsilon = 1$, there is a $\delta > 0$ such that

$$|f(x) - L| < 1$$

whenever

$$0 < |x - 0| < \delta \tag{28}$$

But $x = \delta/2$ and $x = -\delta/2$ both satisfy (28) so that

$$\left| f\left(\frac{\delta}{2}\right) - L \right| < 1 \quad \text{and} \quad \left| f\left(-\frac{\delta}{2}\right) - L \right| < 1 \tag{29}$$

However, $\delta/2$ is positive and $-\delta/2$ is negative, so

$$f\left(\frac{\delta}{2}\right) = 1 \quad \text{and} \quad f\left(-\frac{\delta}{2}\right) = -1$$

Thus, (29) states that

$$|1 - L| < 1 \quad \text{and} \quad |-1 - L| < 1$$

or equivalently

$$0 < L < 2 \quad \text{and} \quad -2 < L < 0$$

But this is a contradiction, since no number L can satisfy these two conditions. $\blacktriangleleft$

▶ Exercise Set 2.6

In Exercises 1–10, we are told that $\lim_{x \to a} f(x) = L$ and we are given a value of ϵ. In each exercise, find a number δ such that $|f(x) - L| < \epsilon$ whenever $0 < |x - a| < \delta$.

1. $\lim_{x \to 4} 2x = 8$; $\epsilon = 0.1$.

2. $\lim_{x \to -2} \frac{1}{2}x = -1$; $\epsilon = 0.1$.

3. $\lim_{x \to -1} (7x + 5) = -2$; $\epsilon = 0.01$.

4. $\lim_{x \to 3} (5x - 2) = 13$; $\epsilon = 0.01$.

5. $\lim_{x \to 2} \frac{x^2 - 4}{x - 2} = 4$; $\epsilon = 0.05$.

6. $\lim_{x \to -1} \frac{x^2 - 1}{x + 1} = -2$; $\epsilon = 0.05$.

7. $\lim_{x \to 4} x^2 = 16$; $\epsilon = 0.001$.

8. $\lim_{x \to 9} \sqrt{x} = 3$; $\epsilon = 0.001$.

9. $\lim_{x \to 5} \frac{1}{x} = \frac{1}{5}$; $\epsilon = 0.05$.

10. $\lim_{x \to 0} |x| = 0$; $\epsilon = 0.05$.

In Exercises 11–22, use Definition 2.6.1 to prove that the given limit statement is correct.

11. $\lim_{x \to 5} 3x = 15$.

12. $\lim_{x \to 3} (4x - 5) = 7$.

13. $\lim_{x \to 0} \frac{x^2 + x}{x} = 1$.

14. $\lim_{x \to -3} \frac{x^2 - 9}{x + 3} = -6$.

15. $\lim_{x \to 1} 2x^2 = 2$.

16. $\lim_{x \to 3} (x^2 - 5) = 4$.

17. $\lim_{x \to 1/3} \frac{1}{x} = 3$.

18. $\lim_{x \to -2} \frac{1}{x + 1} = -1$.

19. $\lim_{x \to 4} \sqrt{x} = 2$.

20. $\lim_{x \to 6} \sqrt{x + 3} = 3$.

21. $\lim_{x \to 1} f(x) = 3$, where $f(x) = \begin{cases} x + 2, & x \neq 1 \\ 10, & x = 1. \end{cases}$

22. $\lim_{x \to 2} (x^2 + 3x - 1) = 9$.

23. Let $f(x) = \begin{cases} \frac{1}{8}, & x > 0 \\ -\frac{1}{8}, & x < 0. \end{cases}$

Use the method of Example 4 to prove that $\lim_{x \to 0} f(x)$ does not exist.

24. Let $g(x) = \begin{cases} 1 + x, & x > 0 \\ x - 1, & x < 0. \end{cases}$

Prove that $\lim_{x \to 0} g(x)$ does not exist.

25. Prove that $\lim_{x \to 1} \frac{1}{x - 1}$ does not exist.

26. (a) In Definition 2.6.1 there is a condition requiring that $f(x)$ be defined for every x in an open interval containing a, except possibly at a itself. What is the purpose of this requirement?

(b) Why is $\lim_{x \to 0} \sqrt{x} = 0$ an incorrect statement?

(c) Is $\lim_{x \to 0.01} \sqrt{x} = 0.1$ a correct statement?

27. Prove the result in Example 2 under the assumption that $\delta \leq 2$ rather than $\delta \leq 1$.

2.7 CONTINUITY

A moving physical object cannot vanish at some point and reappear someplace else to continue its motion. Thus, we perceive the path of a moving object as a single, unbroken curve without gaps, jumps, or holes. Such curves can be described as "continuous." In this section we shall express this intuitive idea mathematically and develop some properties of continuous curves.

Before we give any formal definitions, let us consider some of the ways in which curves can be "discontinuous." In Figure 2.7.1 we have graphed some curves which, because of their behavior at the point c, are not continuous. The curve in Figure 2.7.1a has a hole at the point c because the function f is undefined there. For the curves in Figures 2.7.1b and 2.7.1c, the function f is defined at c, but

$$\lim_{x \to c} f(x) \tag{1}$$

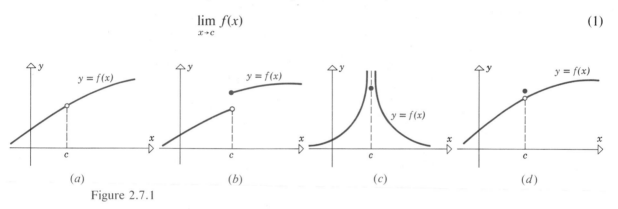

Figure 2.7.1

does not exist, thereby causing a break in the graph. For the curve in Figure 2.7.1d, the function f is defined at c, and the limit in (1) exists, yet the graph still has a break at the point c because

$$\lim_{x \to c} f(x) \neq f(c)$$

Based on this discussion, we see that there is a break or discontinuity in the graph of $y = f(x)$ at a point $x = c$ if any of the following conditions occur:

- The function f is undefined at c.

- The limit $\lim_{x \to c} f(x)$ does not exist.

- The function f is defined at c and the limit $\lim_{x \to c} f(x)$ exists, but the value of the function at c and the value of the limit at c are different.

This suggests the following definition.

2.7.1 DEFINITION. A function f is said to be ***continuous at a point c*** if the following conditions are satisfied:

1. $f(c)$ is defined.

2. $\lim\limits_{x \to c} f(x)$ exists.

3. $\lim\limits_{x \to c} f(x) = f(c)$.

If one or more of the conditions in this definition fails to hold, then f is called ***discontinuous at c*** and c is called a ***point of discontinuity*** of f. If f is continuous at all points of an open interval (a, b), then f is said to be ***continuous on (a, b)***. A function that is continuous on $(-\infty, +\infty)$ is said to be ***continuous everywhere*** or simply ***continuous***.

Example 1 Let

$$f(x) = \frac{x^2 - 4}{x - 2} \quad \text{and} \quad g(x) = \begin{cases} \dfrac{x^2 - 4}{x - 2}, & x \neq 2 \\ 3, & x = 2 \end{cases}$$

Figure 2.7.2

Both f and g are discontinuous at 2 (Figure 2.7.2), the function f because $f(2)$ is undefined, and the function g because $g(2) = 3$, and

$$\lim_{x \to 2} g(x) = \lim_{x \to 2} \frac{x^2 - 4}{x - 2} = \lim_{x \to 2} (x + 2) = 4$$

so that

$$\lim_{x \to 2} g(x) \neq g(2) \qquad \blacktriangleleft$$

REMARK. Some authors define a function to be continuous at c if condition 3 in Definition 2.7.1 holds. This is really equivalent to our definition because if condition 3 holds, then 1 and 2 hold automatically. (Why?) We have stated the three conditions for clarity. However, when we want to show that a function is continuous at a point, we shall just show that condition 3 holds.

Example 2 Show that $f(x) = x^2 - 2x + 1$ is a continuous function.

Solution. We must show that the third condition in Definition 2.7.1 holds for all real c. But

$$\lim_{x \to c} f(x) = \lim_{x \to c} (x^2 - 2x + 1) = c^2 - 2c + 1 = f(c)$$

which shows that the third condition holds.   ◄

The foregoing example is a special case of the following general result.

CONTINUITY OF POLYNOMIALS

> **2.7.2** THEOREM. *Polynomials are continuous functions.*

Proof. If p is a polynomial and c is any real number, then by Example 6 in Section 2.5

$$\lim_{x \to c} p(x) = p(c)$$

which proves the continuity of p at c. Since c is an arbitrary real number, p is continuous everywhere.   ▮

Example 3 Show that $f(x) = |x|$ is a continuous function.

Solution. We can write $f(x)$ as

$$f(x) = |x| = \begin{cases} x & \text{if } x > 0 \\ 0 & \text{if } x = 0 \\ -x & \text{if } x < 0 \end{cases}$$

It follows from Theorem 2.7.2 that $f(x) = |x|$ is continuous if $x > 0$ or $x < 0$ because $|x|$ is identical to the polynomial x in the former case and identical to the polynomial $-x$ in the latter. Thus, $x = 0$ is the only point in question. At this point we have $f(0) = |0| = 0$, so it remains to show that

$$\lim_{x \to 0} f(x) = \lim_{x \to 0} |x| = 0 \tag{2}$$

Because the formula for f changes at 0, it will be helpful to consider the one-sided limits at 0 rather than the two-sided limit. We obtain

$$\lim_{x \to 0^+} |x| = \lim_{x \to 0^+} x = 0$$

$$\lim_{x \to 0^-} |x| = \lim_{x \to 0^-} (-x) = 0$$

Thus, (2) holds and $|x|$ is continuous at $x = 0$.   ◄

The following basic result is an immediate consequence of Theorem 2.5.1.

2.7.3   THEOREM. *If the functions f and g are continuous at c, then*

(a)   *f + g is continuous at c*
(b)   *f − g is continuous at c*
(c)   *f · g is continuous at c*
(d)   *f/g is continuous at c if g(c) ≠ 0 and is discontinuous at c if g(c) = 0.*

We shall prove part (*d*) and leave the remaining proofs as exercises.

Proof of (d).   If $g(c) = 0$, then f/g is discontinuous at c because $f(c)/g(c)$ is undefined.

Assume $g(c) \neq 0$. We must show that

$$\lim_{x \to c} \frac{f(x)}{g(x)} = \frac{f(c)}{g(c)} \tag{3}$$

Since f and g are continuous at c,

$$\lim_{x \to c} f(x) = f(c) \quad \text{and} \quad \lim_{x \to c} g(x) = g(c)$$

Thus, by Theorem 2.5.1(*d*)

$$\lim_{x \to c} \frac{f(x)}{g(x)} = \frac{\lim\limits_{x \to c} f(x)}{\lim\limits_{x \to c} g(x)} = \frac{f(c)}{g(c)}$$

which proves (3).   ∎

CONTINUITY OF RATIONAL FUNCTIONS

Example 4   Where is

$$h(x) = \frac{x^2 - 9}{x^2 - 5x + 6}$$

continuous?

Solution.   The numerator and denominator of h are polynomials and therefore are continuous everywhere by Theorem 2.7.2. Thus, by Theorem 2.7.3(*d*), the ratio is continuous everywhere except at the points where the denominator is zero. Since the solutions of

$$x^2 - 5x + 6 = 0$$

are $x = 2$ and $x = 3$, $h(x)$ is continuous everywhere except at these points.   ◄

The result in the foregoing example is a special case of the following general theorem whose proof is left as an exercise.

2.7.4 THEOREM. *A rational function is continuous everywhere except at the points where the denominator is zero.*

CONTINUITY OF COMPOSITIONS

The next theorem is useful for calculating limits of compositions of functions.

2.7.5 THEOREM. *Let* lim *stand for one of the limits* $\lim_{x \to c}$, $\lim_{x \to c^-}$, $\lim_{x \to c^+}$, $\lim_{x \to +\infty}$, *or* $\lim_{x \to -\infty}$. *If* $\lim g(x) = L$ *and if the function f is continuous at L, then* $\lim f(g(x)) = f(L)$. *That is,* $\lim f(g(x)) = f(\lim g(x))$.

For those who have read Section 2.6, the optional proof of this result appears in Appendix 2.

REMARK. In words, this theorem states that the limit symbol can be moved through a function sign provided the limit of the expression inside the function sign exists and the function is continuous at this limit.

Example 5 We saw in Example 3 that $|x|$ is continuous everywhere, so Theorem 2.7.5 implies that

$$\lim |g(x)| = |\lim g(x)|$$

if $\lim g(x)$ exists. For example,

$$\lim_{x \to 3} |5 - x^2| = |\lim_{x \to 3} (5 - x^2)| = |-4| = 4 \quad \blacktriangleleft$$

The following consequence of Theorem 2.7.5 tells us that compositions of continuous functions are themselves continuous.

2.7.6 THEOREM. *If the function g is continuous at the point c and the function f is continuous at the point g(c), then the composition* $f \circ g$ *is continuous at c.*

Proof. We must show that $f \circ g$ satisfies the third condition of Definition 2.7.1 at c. But this is so since we can write

$$\lim_{x \to c} (f \circ g)(x) = \lim_{x \to c} f(g(x)) = f(\lim_{x \to c} g(x)) = f(g(c)) = (f \circ g)(c)$$

Theorem 2.7.5

g is continuous at c

Example 6   The function $|5 - x^2|$ is continuous because it is the composition of $|x|$ with $5 - x^2$, both of which are continuous.   ◄

Figure 2.7.3 shows the graphs of three functions defined only on a closed interval $[a, b]$. Obviously the function shown in Figure 2.7.3a should be regarded as discontinuous at the left-hand endpoint a, the function in Figure 2.7.3b should be regarded as discontinuous at the right-hand endpoint b, and the function in Figure 2.7.3c should be regarded as continuous at both endpoints. However, the definition of continuity (Definition 2.7.1) does not apply at the endpoints, since the two-sided limits appearing in parts 2 and 3 of the definition make no sense. At the left-hand endpoint the only sensible limit is the one-sided limit

$$\lim_{x \to a^+} f(x)$$

and at the right-hand endpoint the only sensible limit is the one-sided limit

$$\lim_{x \to b^-} f(x)$$

This motivates the following definitions.

(a)

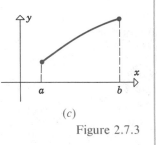

(b)

(c)

Figure 2.7.3

2.7.7   DEFINITION. A function f is called ***continuous from the left at the point c*** if the conditions in the left column below are satisfied, and is called ***continuous from the right at the point c*** if the conditions in the right column are satisfied.

1. $f(c)$ is defined.	1'. $f(c)$ is defined.
2. $\lim\limits_{x \to c^-} f(x)$ exists.	2'. $\lim\limits_{x \to c^+} f(x)$ exists.
3. $\lim\limits_{x \to c^-} f(x) = f(c)$.	3'. $\lim\limits_{x \to c^+} f(x) = f(c)$.

2.7.8   DEFINITION. A function f is said to be ***continuous on a closed interval*** $[a, b]$ if the following conditions are satisfied:

1. f is continuous on (a, b).
2. f is continuous from the right at a.
3. f is continuous from the left at b.

The reader should have no trouble deducing the appropriate definitions of continuity on intervals of the form $[a, +\infty)$, $(-\infty, b]$, $[a, b)$, and $(a, b]$.

Example 7   If f denotes the function graphed in Figure 2.7.3a, then

$$f(a) \neq \lim_{x \to a^+} f(x) \quad \text{and} \quad f(b) = \lim_{x \to b^-} f(x)$$

Thus, f is continuous from the left at b, but is not continuous from the right at a. ◄

Example 8 Show that $f(x) = \sqrt{9 - x^2}$ is continuous on the closed interval $[-3, 3]$.

Solution. Observe that the domain of f is the interval $[-3, 3]$. For c in the interval $(-3, 3)$ we have by Theorem 2.5.1(e)

$$\lim_{x \to c} f(x) = \lim_{x \to c} \sqrt{9 - x^2} = \sqrt{\lim_{x \to c}(9 - x^2)} = \sqrt{9 - c^2} = f(c)$$

so that f is continuous on $(-3, 3)$. Also,

$$\lim_{x \to 3^-} f(x) = \lim_{x \to 3^-} \sqrt{9 - x^2} = 0 = f(3)$$

and

$$\lim_{x \to -3^+} f(x) = \lim_{x \to -3^+} \sqrt{9 - x^2} = 0 = f(-3)$$

so that f is continuous on $[-3, 3]$. ◄

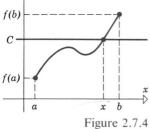

Figure 2.7.4

INTERMEDIATE VALUE THEOREM

If f is continuous on a closed interval $[a, b]$, and we draw a horizontal line crossing the y-axis between the numbers $f(a)$ and $f(b)$ (Figure 2.7.4), then it is geometrically obvious that this line will cross the curve $y = f(x)$ at least once over the interval $[a, b]$. Stated another way, a continuous function must assume every possible value between $f(a)$ and $f(b)$ as x varies from a to b. This idea is formalized in the following theorem.

Intermediate Value Theorem

2.7.9 THEOREM. *If f is continuous on a closed interval $[a, b]$ and C is any number between $f(a)$ and $f(b)$, inclusive, then there is at least one number x in the interval $[a, b]$ such that $f(x) = C$.*

(See Figure 2.7.4.) This theorem, though intuitively obvious, is not easy to prove. The proof can be found in most advanced calculus texts.

The following useful result is an immediate consequence of the Intermediate Value Theorem.

Figure 2.7.5

2.7.10 THEOREM. *If f is continuous on $[a, b]$, and if $f(a)$ and $f(b)$ have opposite signs, then there is at least one solution of the equation $f(x) = 0$ in the interval (a, b).*

The proof is left as an exercise, but the result is illustrated in Figure 2.7.5 in the case where $f(a) > 0$ and $f(b) < 0$.

Figure 2.7.6

Theorem 2.7.10 is the basis for various numerical techniques for approximating solutions of equations of the form $f(x) = 0$.

Example 9   The equation

$$x^3 - x - 1 = 0$$

cannot be solved readily by factoring because the left side has no simple factors. However, from Figure 2.7.6, which was generated on a microcomputer, we see that there is a solution in the interval $(1, 2)$. This can also be seen from Theorem 2.7.10 by letting $f(x) = x^3 - x - 1$, and noting that $f(1) = -1$ and $f(2) = 5$ have opposite signs.

To pinpoint the solution more exactly, we can divide the interval $(1, 2)$ into 10 equal parts and evaluate $f(x)$ at each point of subdivision using a calculator (Table 2.7.1). From the table we see that $f(1.3)$ and $f(1.4)$ have opposite signs, so there is a solution in the subinterval $(1.3, 1.4)$. If we use the midpoint 1.35 as an approximation to the solution, then our error is at most half the subinterval length or .05. If desired, we could reduce the error to at most .005 by subdividing the subinterval $(1.3, 1.4)$ into 10 parts and repeating the sign analysis process. With sufficiently many subdivisions, the solution can be determined to any degree of accuracy. ◀

Table 2.7.1

x	1	1.1	1.2	1.3	1.4	1.5	1.6	1.7	1.8	1.9	2
$f(x)$	-1	$-.77$	$-.47$	$-.10$	.34	.88	1.5	2.2	3.0	3.9	5

We conclude this section by stating an alternative form of the continuity definition that will be useful in later sections.

If we let $h = x - c$ in Definition 2.7.1, then the condition that $x \to c$ is equivalent to $h \to 0$. Thus, Definition 2.7.1 can be restated as follows.

*Alternative Form of the
Continuity Definition*

2.7.11   DEFINITION. A function f is said to be ***continuous at a point c*** if the following conditions are satisfied:

1. $f(c)$ is defined.

2. $\lim_{h \to 0} f(c + h)$ exists.

3. $\lim_{h \to 0} f(c + h) = f(c)$.

REMARK.   As with Definition 2.7.1, conditions 1 and 2 hold automatically if condition 3 holds. Thus, to prove that a function is continuous at a point, it is only necessary to show that condition 3 holds.

► Exercise Set 2.7

In Exercises 1–4, let f be the function whose graph is shown. On which of the following intervals, if any, is f continuous?

(a) $[1, 3]$ (b) $(1, 3)$
(c) $[1, 2]$ (d) $(1, 2)$
(e) $[2, 3]$ (f) $(2, 3)$.

On those intervals where f is discontinuous, state where the discontinuities occur.

1.

2.

3.

4.

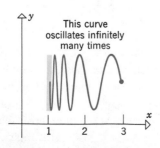

This curve oscillates infinitely many times

In Exercises 5–16, find the points of discontinuity, if any.

5. $f(x) = x^3 - 2x + 3$.

6. $f(x) = (x - 5)^{17}$.

7. $f(x) = \dfrac{x}{x^2 + 1}$.

8. $f(x) = \dfrac{x}{x^2 - 1}$.

9. $f(x) = \dfrac{x - 4}{x^2 - 16}$.

10. $f(x) = \dfrac{3x + 1}{x^2 + 7x - 2}$.

11. $f(x) = \dfrac{x}{|x| - 3}$.

12. $f(x) = \dfrac{5}{x} + \dfrac{2x}{x + 4}$.

13. $f(x) = |x^3 - 2x^2|$.

14. $f(x) = \dfrac{x + 3}{|x^2 + 3x|}$.

15. $f(x) = \begin{cases} 2x + 3, & x \le 4 \\ 7 + \dfrac{16}{x}, & x > 4. \end{cases}$

16. $f(x) = \begin{cases} \dfrac{3}{x - 1}, & x \ne 1 \\ 3, & x = 1. \end{cases}$

17. Find a value for the constant k that will make the function continuous.

(a) $f(x) = \begin{cases} 7x - 2, & x \le 1 \\ kx^2, & x > 1 \end{cases}$

(b) $f(x) = \begin{cases} kx^2, & x \le 2 \\ 2x + k, & x > 2. \end{cases}$

18. On which of the following intervals is

$$f(x) = \dfrac{1}{\sqrt{x - 2}}$$

continuous?

(a) $[2, +\infty)$ (b) $(-\infty, +\infty)$
(c) $(2, +\infty)$ (d) $[1, 2)$.

19. (a) Prove that $f(x) = \sqrt{x}$ is continuous on $[0, +\infty)$.

(b) Prove that if $g(x)$ is continuous and nonnegative, then $\sqrt{g(x)}$ is continuous.

20. Prove that if $g(x)$ is continuous, then $|g(x)|$ is continuous.

21. Find all points of discontinuity of the greatest integer function, $f(x) = [x]$. (See Exercise 38, Section 2.3, for the definition of $[x]$.)

22. A function $f(x)$ has a *removable discontinuity* at $x = c$ if $\lim\limits_{x \to c} f(x)$ exists, but $f(c)$ either is not defined or $f(c) \neq \lim\limits_{x \to c} f(x)$. If $f(x)$ is redefined to have the value $\lim\limits_{x \to c} f(x)$ at $x = c$, then $f(x)$ will be continuous at $x = c$.

Find all points of discontinuity of $f(x)$, and for each such point state if it is removable or not.

(a) $f(x) = \dfrac{|x|}{x}$

(b) $f(x) = \dfrac{x^2 + 3x}{x + 3}$

(c) $f(x) = \dfrac{x - 2}{|x| - 2}$

(d) $f(x) = \dfrac{x^2 - 4}{x^3 - 8}$

(e) $f(x) = \begin{cases} 2x - 3, & x \leq 2 \\ x^2, & x > 2 \end{cases}$

(f) $f(x) = \begin{cases} 3x^2 + 5, & x \neq 1 \\ 6, & x = 1. \end{cases}$

23. Prove:

(a) part (a) of Theorem 2.7.3

(b) part (b) of Theorem 2.7.3

(c) part (c) of Theorem 2.7.3.

24. Prove Theorem 2.7.4.

25. Let f and g be discontinuous at c. Give examples to show:

(a) $f + g$ can be continuous or discontinuous at c

(b) $f \cdot g$ can be continuous or discontinuous at c.

26. **(For students who have read Section 2.6.)** Let f be defined at c. Prove that f is continuous at c if and only if, given $\epsilon > 0$ there exists a $\delta > 0$ such that $|f(x) - f(c)| < \epsilon$ whenever $|x - c| < \delta$.

27. Use Theorem 2.7.9 to prove Theorem 2.7.10.

In Exercises 28 and 29, show that the equation has at least one solution in the given interval.

28. $x^3 - 4x + 1 = 0$; $[1, 2]$

29. $x^3 + x^2 - 2x = 1$; $[-1, 1]$

30. Figure 2.7.7 shows the graph of $y = x^4 + x - 1$ generated on a microcomputer. Use the method of Example 9 to approximate the real solutions of $x^4 + x - 1 = 0$ with an error of at most .05.

$y = x^4 + x - 1$

Figure 2.7.7

31. Figure 2.7.8 shows the graph of $y = 5 - x - x^4$ generated on a microcomputer. Use the method of Example 9 to approximate the real solutions of $5 - x - x^4 = 0$ with an error of at most .05.

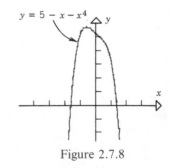

$y = 5 - x - x^4$

Figure 2.7.8

32. For the equation $x^3 - x - 1 = 0$ discussed in Example 9, show that the real solution is approximately 1.325, with an error of at most .005.

33. Use the fact that $\sqrt{5}$ is a solution of $x^2 - 5 = 0$ to approximate $\sqrt{5}$ with an error of at most

(a) .05   (b) .005.

34. Prove: If f and g are continuous on $[a, b]$, and $f(a) > g(a)$, $f(b) < g(b)$, then there is at least one solution of the equation $f(x) = g(x)$ in (a, b). [*Hint:* Consider $f(x) - g(x)$.]

35. Construct an example of a function $f(x)$ that is defined at every point in a closed interval, and whose values at the endpoints have opposite signs, but for which the equation $f(x) = 0$ has no solution in the interval.

36. Prove that the equation

$$\frac{a}{x - 1} + \frac{b}{x - 3} = 0$$

where a and b are both positive real numbers, has at least one solution in the interval $(1, 3)$.

37. Prove: If $p(x)$ is a polynomial of odd degree, then the equation $p(x) = 0$ has at least one real solution.

2.8 LIMITS AND CONTINUITY OF TRIGONOMETRIC FUNCTIONS

In this section we shall derive some basic limits involving trigonometric functions and study the continuity of these functions.

This section requires a knowledge of the trigonometry material in Unit 2 of Appendix 1. Readers who need to review that material are advised to do so before starting this section.

AREA OF A SECTOR

Figure 2.8.1

Before proceeding to the main results in this section, it will be helpful to review the formula for the area of a sector.

Let A denote the area of a sector with radius r and a central angle of θ radians (Figure 2.8.1). If $\theta = 2\pi$, the sector is the entire circle and the area is πr^2. For a general sector the area A is proportional to the central angle θ, so we can write

$$\frac{\text{area of the sector}}{\text{area of the circle}} = \frac{\text{central angle of the sector}}{\text{central angle of the circle}}$$

This yields

$$\frac{A}{\pi r^2} = \frac{\theta}{2\pi}$$

from which we obtain the formula

Area of a Sector with a Central Angle of θ Radians and Radius r

$$A = \frac{1}{2} r^2 \theta \tag{1}$$

REMARK. It is important to keep in mind that the angle θ in the foregoing formula is measured in radians. As we progress through this text you will encounter many calculus formulas that build on (1). All such formulas will also require that angles be measured in radians.

CONTINUITY OF SINE AND COSINE

In trigonometry the graphs of $y = \sin x$ and $y = \cos x$ are drawn as continuous curves (Figures A.24 and A.25 in Appendix 1). Our first objective in this section is to prove that $\sin x$ and $\cos x$ are indeed continuous functions. For this purpose, consider Figure 2.8.2, which shows an angle of h (radians) drawn in standard position with its terminal side intersecting the unit circle at the point $P(\cos h, \sin h)$. It is evident geometrically that as h approaches 0, the point $P(\cos h, \sin h)$ moves toward the point $B(1, 0)$. (Although h was drawn as a positive angle, the same conclusion holds for negative h.) This suggests that

$$\lim_{h \to 0} \cos h = 1 \tag{2a}$$

$$\lim_{h \to 0} \sin h = 0 \tag{2b}$$

Figure 2.8.2

A precise proof of these results will be omitted.

2.8.1   THEOREM. *The functions $\sin x$ and $\cos x$ are continuous.*

Proof.   We shall prove the result for $\sin x$; the proof for $\cos x$ is similar and will be left as an exercise. By the remark following Definition 2.7.11, it suffices to show that the condition

$$\lim_{h \to 0} \sin(c + h) = \sin c \tag{3}$$

is satisfied. Using the addition formula for the sine function we obtain

$$\lim_{h \to 0} \sin(c + h) = \lim_{h \to 0} [\sin c \cos h + \cos c \sin h]$$

$$= \lim_{h \to 0} [\sin c \cos h] + \lim_{h \to 0} [\cos c \sin h] \tag{4}$$

The expressions $\sin c$ and $\cos c$ in (4) do not involve h, so they remain *constant* as $h \to 0$. This allows us to move these expressions through the limit signs and write

$$\lim_{h \to 0} \sin(c + h) = \sin c \lim_{h \to 0} \cos h + \cos c \lim_{h \to 0} \sin h$$

$$= (\sin c)(1) + (\cos c)(0) = \sin c$$

which proves (3).   ∎

CONTINUITY OF OTHER TRIGONOMETRIC FUNCTIONS

The continuity properties of $\tan x$, $\cot x$, $\sec x$, and $\csc x$ can be deduced by expressing these functions in terms of $\sin x$ and $\cos x$. For example, $\tan x = \sin x/\cos x$, so by part (d) of Theorem 2.7.3 the function $\tan x$ is continuous everywhere except at the points where $\cos x = 0$. These points of discontinuity are

$$x = \pm \frac{\pi}{2}, \pm \frac{3\pi}{2}, \pm \frac{5\pi}{2}, \ldots$$

(see Figure A.28 in Appendix 1).

Example 1   Since the functions $\sin x$ and $\cos x$ are continuous everywhere, it follows from Theorem 2.7.5 that we can write

$$\lim \,[\sin\,(g(x))] = \sin\,[\lim\,g(x)]$$

and

$$\lim \,[\cos\,(g(x))] = \cos\,[\lim\,g(x)]$$

if $\lim g(x)$ exists. For example,

$$\lim_{x \to \pi} \left[\sin \left(\frac{x^2}{\pi + x} \right) \right] = \sin \left[\lim_{x \to \pi} \left(\frac{x^2}{\pi + x} \right) \right] = \sin \frac{\pi}{2} = 1$$

$$\lim_{x \to +\infty} \left[\cos \left(\frac{\pi x^2 + 1}{x^2 + 3} \right) \right] = \cos \left[\lim_{x \to +\infty} \left(\frac{\pi x^2 + 1}{x^2 + 3} \right) \right]$$

$$= \cos \left[\lim_{x \to +\infty} \left(\frac{\pi + (1/x^2)}{1 + (3/x^2)} \right) \right] = \cos \pi = -1 \quad \blacktriangleleft$$

Our next objective is to establish two fundamental limits:

$$\lim_{h \to 0} \frac{\sin h}{h} = 1 \quad \text{and} \quad \lim_{h \to 0} \frac{1 - \cos h}{h} = 0 \tag{5}$$

These limits are not at all obvious. For example, in the limit

$$\lim_{h \to 0} \frac{\sin h}{h}$$

the numerator and denominator both approach zero as $h \to 0$. As a result, there are two conflicting influences on the ratio. The numerator approaching 0 drives the magnitude of the ratio toward zero, while the denominator approaching 0 drives the magnitude of the ratio toward $+\infty$. The precise way in which these influences offset one another determine whether the limit exists and what its value is. In a limit problem where the numerator and denominator both approach 0, it is often possible to obtain the limit by some algebraic manipulations. Unfortunately, no such algebraic manipulations will work for the limits in (5); we must develop other techniques. However, some numerical evidence suggesting these limits can be obtained by setting a calculator to the radian mode and computing some sample values as $h \to 0$ (Tables 2.4.1 and 2.4.2 with h replacing x).

OBTAINING LIMITS
BY SQUEEZING

When it is difficult to find the limit of a function directly, it is sometimes possible to obtain the limit indirectly by "squeezing" the function between simpler functions whose limits are known. For example, suppose we are unable to calculate

$$\lim_{x \to a} f(x)$$

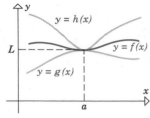

Figure 2.8.3

directly, but we are able to find two functions, g and h, that have the same limit L as $x \to a$ and such that f is "squeezed" between g and h by means of the inequalities

$$g(x) \le f(x) \le h(x)$$

It is evident geometrically that $f(x)$ must also approach L as $x \to a$ because the graph of f lies between the graphs of g and h (Figure 2.8.3).

This idea is formalized in the following theorem, which is called the **Squeezing Theorem** or sometimes the **Pinching Theorem**. We omit the proof.

The Squeezing Theorem

2.8.2 THEOREM. *Let f, g, and h be functions satisfying*

$$g(x) \le f(x) \le h(x)$$

for all x in some open interval containing the point a, with the possible exception that the inequalities need not hold at a. If g and h have the same limit as x approaches a, say

$$\lim_{x \to a} g(x) = \lim_{x \to a} h(x) = L$$

then f also has this limit as x approaches a, that is,

$$\lim_{x \to a} f(x) = L$$

REMARK.   The Squeezing Theorem remains true if $\lim_{x \to a}$ is replaced by $\lim_{x \to a^+}$ or $\lim_{x \to a^-}$. Moreover, for $\lim_{x \to a^+}$ the condition $g(x) \le f(x) \le h(x)$ need only hold on an open interval extending to the right from a, and for $\lim_{x \to a^-}$ the condition need only hold on an open interval extending to the left from a. The Squeezing Theorem can also be extended to limits of the form $\lim_{x \to -\infty}$ and $\lim_{x \to +\infty}$.

Example 2   Use the Squeezing Theorem to evaluate the limit

$$\lim_{x \to 0} x^2 \sin^2 \frac{1}{x}$$

Solution.   If $x \ne 0$, we can write

$$0 \le \sin^2 \frac{1}{x} \le 1$$

Multiplying through by x^2 yields

$$0 \le x^2 \sin^2 \frac{1}{x} \le x^2$$

if $x \ne 0$. But

$$\lim_{x \to 0} 0 = \lim_{x \to 0} x^2 = 0$$

so by the Squeezing Theorem

$$\lim_{x \to 0} x^2 \sin^2 \frac{1}{x} = 0 \qquad \blacktriangleleft$$

SOME IMPORTANT
LIMITS OF
TRIGONOMETRIC
FUNCTIONS

We are now ready to prove the validity of the limits in (5).

2.8.3   THEOREM.

$$\lim_{h \to 0} \frac{\sin h}{h} = 1 \tag{6}$$

Proof.   Assume that h satisfies $0 < h < \pi/2$ and construct the angle h in standard position. The terminal side of h intersects the unit circle at $P(\cos h, \sin h)$ and intersects the vertical line through $B(1, 0)$ at the point $Q(1, \tan h)$ (Figure 2.8.4). From the figure we obtain

$$0 < \text{area of } \triangle\, OBP < \text{area of sector } OBP < \text{area of } \triangle\, OBQ$$

Figure 2.8.4 ▷

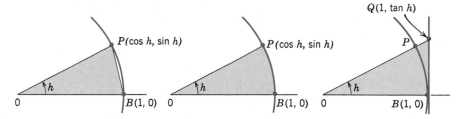

But

$$\text{area } \triangle\, OBP = \tfrac{1}{2}\,\text{base} \cdot \text{altitude} = \tfrac{1}{2} \cdot 1 \cdot \sin h = \tfrac{1}{2} \sin h$$
$$\text{area sector } OBP = \tfrac{1}{2}(1)^2 \cdot h = \tfrac{1}{2} h$$
$$\text{area } \triangle\, OBQ = \tfrac{1}{2}\,\text{base} \cdot \text{altitude} = \tfrac{1}{2} \cdot 1 \cdot \tan h = \tfrac{1}{2} \tan h$$

Therefore,

$$0 < \tfrac{1}{2} \sin h < \tfrac{1}{2} h < \tfrac{1}{2} \tan h$$

Multiplying through by $2/\sin h$ yields

$$1 < \frac{h}{\sin h} < \frac{1}{\cos h}$$

and taking reciprocals yields

$$\cos h < \frac{\sin h}{h} < 1 \tag{7}$$

We have derived (7) under the assumption that $0 < h < \pi/2$. However, the inequalities in (7) are also valid if $-\pi/2 < h < 0$ (Exercise 37) so that (7) holds for all h in the interval $(-\pi/2, \pi/2)$ except $h = 0$.

From (2a) we have

$$\lim_{h \to 0} \cos h = 1$$

and since

$$\lim_{h \to 0} 1 = 1$$

(6) follows by applying the Squeezing Theorem to (7). ■

2.8.4 COROLLARY.

$$\lim_{h \to 0} \frac{1 - \cos h}{h} = 0$$

Proof. With the help of (2a), (2b), and the trigonometric identity $\sin^2 h = 1 - \cos^2 h$ we obtain

$$\lim_{h \to 0} \frac{1 - \cos h}{h} = \lim_{h \to 0} \left[\frac{1 - \cos h}{h} \cdot \frac{1 + \cos h}{1 + \cos h} \right] = \lim_{h \to 0} \frac{\sin^2 h}{h(1 + \cos h)}$$

$$= \left(\lim_{h \to 0} \frac{\sin h}{h} \right) \left(\lim_{h \to 0} \frac{\sin h}{1 + \cos h} \right) = (1) \left(\frac{0}{1 + 1} \right) = 0 \quad ■$$

The following examples illustrate how the limits obtained in this section can be used to calculate other limits involving trigonometric functions.

Example 3   Evaluate $\lim_{x \to 0} \dfrac{\tan x}{x}$.

Solution.

$$\lim_{x \to 0} \frac{\tan x}{x} = \lim_{x \to 0} \left(\frac{\sin x}{x} \cdot \frac{1}{\cos x} \right) = (1)(1) = 1 \quad \blacktriangleleft$$

Example 4   Evaluate $\lim\limits_{\theta \to 0} \dfrac{\sin 2\theta}{\theta}$.

Solution.   Make the substitution $\phi = 2\theta$. Since $\phi \to 0$ as $\theta \to 0$, we can write

$$\lim_{\theta \to 0} \frac{\sin 2\theta}{\theta} = \lim_{\phi \to 0} \frac{\sin \phi}{\frac{1}{2}\phi} = 2 \lim_{\phi \to 0} \frac{\sin \phi}{\phi} = 2(1) = 2 \quad \blacktriangleleft$$

▶ Exercise Set 2.8

In Exercises 1–10, find the points of discontinuity, if any.

1. $f(x) = \sin (x^2 - 2)$.

2. $f(x) = \cos \left(\dfrac{x}{x - \pi} \right)$.

3. $f(x) = \cot x$.

4. $f(x) = \sec x$.

5. $f(x) = \csc x$.

6. $f(x) = \dfrac{1}{1 + \sin^2 x}$.

7. $f(x) = |\cos x|$.

8. $f(x) = \sqrt{2 + \tan^2 x}$.

9. $f(x) = \dfrac{1}{1 - 2 \sin x}$.

10. $f(x) = \dfrac{3}{5 + 2 \cos x}$.

11. Prove that $\sin (g(x))$ is continuous at every point where $g(x)$ is continuous.

12. Use Theorem 2.7.6 to prove that the following functions are continuous.
  (a) $\sin (x^3 + 7x + 1)$ (b) $|\sin x|$
  (c) $\cos^3 (x + 1)$ (d) $\sqrt{3 + \sin 2x}$.

13. Find $\lim\limits_{x \to +\infty} \sin \left(\dfrac{\pi x}{2 - 3x} \right)$.

In Exercises 14–25, find the limit if it exists.

14. $\lim\limits_{h \to 0} \dfrac{\sin h}{2h}$.

15. $\lim\limits_{\theta \to 0} \dfrac{\sin 3\theta}{\theta}$.

16. $\lim\limits_{x \to 0} \dfrac{\sin 6x}{\sin 8x}$.

17. $\lim\limits_{x \to 0} \dfrac{\tan 7x}{\sin 3x}$.

18. $\lim\limits_{\theta \to 0} \dfrac{\sin^2 \theta}{\theta}$.

19. $\lim\limits_{h \to 0} \dfrac{h}{\tan h}$.

20. $\lim\limits_{h \to 0} \dfrac{\sin h}{1 - \cos h}$.

21. $\lim\limits_{\theta \to 0} \dfrac{\theta^2}{1 - \cos \theta}$.

22. $\lim\limits_{x \to 0} \dfrac{x}{\cos (\frac{1}{2}\pi - x)}$.

23. $\lim\limits_{\theta \to 0} \dfrac{\theta}{\cos \theta}$.

24. $\lim\limits_{t \to 0} \dfrac{t^2}{1 - \cos^2 t}$.

25. $\lim\limits_{h \to 0} \dfrac{1 - \cos 5h}{\cos 7h - 1}$.

26. Find a value for the constant k so that

$$f(x) = \begin{cases} \dfrac{\sin 3x}{x}, & x \neq 0 \\[2mm] k, & x = 0 \end{cases}$$

will be continuous at $x = 0$.

27. Find a nonzero value for the constant k so that

$$f(x) = \begin{cases} \dfrac{\tan kx}{x}, & x < 0 \\[2mm] 3x + 2k^2, & x \geq 0 \end{cases}$$

will be continuous at $x = 0$.

28. Is

$$f(x) = \begin{cases} \dfrac{\sin x}{|x|}, & x \neq 0 \\ 1, & x = 0 \end{cases}$$

continuous at $x = 0$?

29. In each part, find the limit by making the indicated substitution.

(a) $\displaystyle\lim_{x \to +\infty} x \sin \dfrac{1}{x}.$ $\left[\text{Let } t = \dfrac{1}{x}.\right]$

(b) $\displaystyle\lim_{x \to -\infty} x \left(1 - \cos \dfrac{1}{x}\right).$ $\left[\text{Let } t = \dfrac{1}{x}.\right]$

(c) $\displaystyle\lim_{x \to \pi} \dfrac{\pi - x}{\sin x}.$ [Let $t = \pi - x$.]

30. Find $\displaystyle\lim_{x \to 2} \dfrac{\cos (\pi/x)}{x - 2}$ by making appropriate substitutions. *Hint: Let $t = \dfrac{\pi}{2} - \dfrac{\pi}{x}$*

31. Find $\displaystyle\lim_{x \to 1} \dfrac{\sin (\pi x)}{x - 1}.$

32. Find $\displaystyle\lim_{x \to \pi/4} \dfrac{\tan x - 1}{x - \pi/4}.$

33. Let f be a function that satisfies

$$1 - x^2 \leq f(x) \leq \cos x$$

for all x in $(-\pi/2, \pi/2)$. Does $\displaystyle\lim_{x \to 0} f(x)$ exist? If so, find the limit. If not, explain why.

34. Let

$$f(x) = \begin{cases} 1 & \text{if } x \text{ is a rational number} \\ 0 & \text{if } x \text{ is an irrational number} \end{cases}$$

Use the Pinching Theorem to prove $\displaystyle\lim_{x \to 0} xf(x) = 0.$

35. Prove: If there are constants L and M such that

$$L \leq f(x) \leq M$$

for all x in some open interval containing 0, with the possible exception that the inequalities may not hold at 0, then

$$\lim_{x \to 0} xf(x) = 0$$

36. State versions of the Pinching Theorem that apply to the limits $\displaystyle\lim_{x \to +\infty} f(x)$ and $\displaystyle\lim_{x \to -\infty} f(x)$. Draw some pictures to illustrate these results.

37. Prove that (7) holds if $-\pi/2 < h < 0$ by substituting $-h$ for h in (7).

38. Prove: If θ is in degrees, then

$$\lim_{\theta \to 0} \dfrac{\sin \theta}{\theta} = \dfrac{\pi}{180}.$$

In Exercises 39 and 40, show that the equation has at least one solution in the given interval.

39. $x = \cos x$; $[0, \pi/2]$.

40. $x + \sin x = 1$; $[0, \pi/6]$.

▶ SUPPLEMENTARY EXERCISES

In Exercises 1–5, find the domain of f and then evaluate f (if defined) at the given values of x.

1. $f(x) = \sqrt{4 - x^2}$; $x = -\sqrt{2}, 0, \sqrt{3}.$

2. $f(x) = 1/\sqrt{(x - 1)^3}$; $x = 0, 1, 2.$

3. $f(x) = (x - 1)/(x^2 + x - 2)$; $x = 0, 1, 2.$

4. $f(x) = \sqrt{|x| - 2}$; $x = -3, 0, 2.$

5. $f(x) = \begin{cases} x^2 - 1, & x \leq 2 \\ \sqrt{x - 1}, & x > 2 \end{cases}$; $x = 0, 2, 4.$

In Exercises 6 and 7, find:

(a) $f(x^2) - (f(x))^2$

(b) $f(x + 3) - [f(x) + f(3)]$

(c) $f(1/x) - 1/f(x)$

(d) $(f \circ f)(x).$

6. $f(x) = \sqrt{3 - x}.$

7. $f(x) = \dfrac{3 - x}{x}.$

In Exercises 8–15, sketch the graph of f and find its domain and range.

8. $f(x) = (x - 2)^2.$

9. $f(x) = -\pi.$

10. $f(x) = |2 - 4x|.$

11. $f(x) = \dfrac{x^2 - 4}{2x + 4}.$

12. $f(x) = \sqrt{-2x}$. **13.** $f(x) = -\sqrt{3x + 1}$.

14. $f(x) = 2 - |x|$. **15.** $f(x) = \dfrac{2x - 4}{x^2 - 4}$.

16. In each part, complete the square, and then find the range of f.
(a) $f(x) = x^2 - 5x + 6$
(b) $f(x) = -3x^2 + 12x - 7$.

17. Express $f(x)$ as a composite function $(g \circ h)(x)$ in two different ways.
(a) $f(x) = x^6 + 3$
(b) $f(x) = \sqrt{x^2 + 1}$
(c) $f(x) = \sin(3x + 2)$.

18. Find $\lim\limits_{x \to k} \dfrac{x^3 - kx^2}{x^2 - k^2}$, where k is a constant.

In Exercises 19 and 20, sketch the graph of f and find the indicated limits of $f(x)$ (if they exist).

19.
$$f(x) = \begin{cases} 1/x, & x < 0 \\ x^2, & 0 \le x < 1 \\ 2, & x = 1 \\ 2 - x, & x > 1 \end{cases}$$
(a) as $x \to -1$ (b) as $x \to 0$
(c) as $x \to 1$ (d) as $x \to 0^+$
(e) as $x \to 0^-$ (f) as $x \to 2^+$
(g) as $x \to -\infty$ (h) as $x \to +\infty$.

20.
$$f(x) = \begin{cases} 2, & x \le -1 \\ -x, & -1 < x < 0 \\ x/(2 - x), & 0 < x < 2 \\ 1, & x \ge 2 \end{cases}$$
(a) as $x \to -1^+$ (b) as $x \to -1^-$
(c) as $x \to -1$ (d) as $x \to 0$
(e) as $x \to 2^+$ (f) as $x \to 2^-$

(g) as $x \to 2$ (h) as $x \to -\infty$.

In Exercises 21–24, find $\lim\limits_{x \to a} f(x)$ (if it exists).

21. $f(x) = \sqrt{2 - x}$;
$a = -2, 1, 2^-, 2^+, -\infty, +\infty$.

22. $f(x) = \begin{cases} (x - 2)/|x - 2|, & x \ne 2 \\ 0, & x = 2 \end{cases}$
$a = 0, 2^-, 2^+, 2, -\infty, +\infty$.

23. $f(x) = (x^2 - 25)/(x - 5)$;
$a = 0, 5^+, -5^-, 5, -5, -\infty, +\infty$.

24. $f(x) = (x + 5)/(x^2 - 25)$;
$a = 0, 5^+, -5^-, -5, 5, -\infty, +\infty$.

In Exercises 25–32, find the indicated limit if it exists.

25. $\lim\limits_{x \to 0} \dfrac{\tan ax}{\sin bx}$ $(a \ne 0, b \ne 0)$.

26. $\lim\limits_{x \to 0} \dfrac{\sin 3x}{\tan 3x}$.

27. $\lim\limits_{\theta \to 0} \dfrac{\sin 2\theta}{\theta^2}$.

28. $\lim\limits_{x \to 0} \dfrac{x \sin x}{1 - \cos x}$.

29. $\lim\limits_{x \to 0^+} \dfrac{\sin x}{\sqrt{x}}$.

30. $\lim\limits_{x \to 0} \dfrac{\sin^2(kx)}{x^2}$, $k \ne 0$.

31. $\lim\limits_{x \to 0} \dfrac{3x - \sin(kx)}{x}$, $k \ne 0$.

32. $\lim\limits_{x \to +\infty} \dfrac{2x + x \sin 3x}{5x^2 - 2x + 1}$.

3 differentiation

3.1 TANGENT LINES AND RATES OF CHANGE

Many physical phenomena involve changing quantities—the speed of a rocket, the inflation of currency, the number of bacteria in a culture, the shock intensity of an earthquake, the voltage of an electrical signal, and so forth. In this section we shall establish a basic relationship between tangent lines and rates of change.

DEFINITION OF A TANGENT LINE

In Section 2.4 we observed informally that if a secant line is drawn between two points P and Q on a curve, and Q is allowed to move along the curve toward P, then we can expect the secant line to rotate toward a "limiting position" which can be regarded as the tangent line to the curve at the point P (Figure 3.1.1). Our first objective in this section is to make this informal idea mathematically precise.

For the moment we shall only consider lines tangent to curves of the form $y = f(x)$. If $P(x_0, y_0)$ and $Q(x_1, y_1)$ are distinct points on such a curve, then the secant line connecting P and Q has slope

$$m_{\text{sec}} = \frac{f(x_1) - f(x_0)}{x_1 - x_0} \tag{1}$$

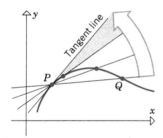

Figure 3.1.1

(See Figure 3.1.2a.) If we let x_1 approach x_0, then Q will approach P along

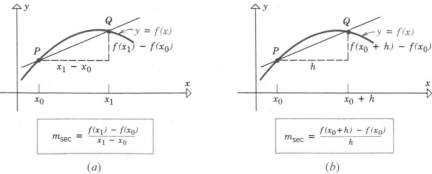

Figure 3.1.2 ▷

$$m_{\text{sec}} = \frac{f(x_1) - f(x_0)}{x_1 - x_0}$$

(a)

$$m_{\text{sec}} = \frac{f(x_0+h) - f(x_0)}{h}$$

(b)

139

the graph of f, and the secant line through P and Q will approach the tangent line at P. Thus, the slope $m_{\sec}$ of the secant line approaches the slope $m_{\tan}$ of the tangent line as x_1 approaches x_0. Therefore, from (1)

$$m_{\tan} = \lim_{x_1 \to x_0} \frac{f(x_1) - f(x_0)}{x_1 - x_0} \tag{2}$$

For many purposes it is desirable to rewrite this expression in an alternative form by letting

$$h = x_1 - x_0$$

Thus (see Figure 3.1.2b), $x_1 = x_0 + h$ and $h \to 0$ as $x_1 \to x_0$, so (2) can be rewritten as

$$m_{\tan} = \lim_{h \to 0} \frac{f(x_0 + h) - f(x_0)}{h}$$

Motivated by the foregoing discussion, we make the following definition.

3.1.1   DEFINITION. If $P(x_0, y_0)$ is a point on the graph of a function f, then the ***tangent line*** to the graph of f at P is defined to be the line through P with slope

$$m_{\tan} = \lim_{h \to 0} \frac{f(x_0 + h) - f(x_0)}{h} \tag{3}$$

For brevity, the tangent line at $P(x_0, y_0)$ is often called the ***tangent line at*** x_0. It follows from the foregoing definition that the point-slope form of the equation of the tangent line at x_0 is

$$y - y_0 = m_{\tan} (x - x_0) \tag{4}$$

REMARK.   It is possible that the limit in (3) may not exist, in which case $m_{\tan}$ will be undefined. We shall consider the geometric significance of this in later sections.

Example 1   Find the slope and the equation of the line tangent to the graph of $f(x) = x^2$ at the point $P(3, 9)$. (See Figure 3.1.3.)

Solution.   We have $x_0 = 3$ and $y_0 = 9$, so from (3)

Figure 3.1.3

$$m_{\tan} = \lim_{h \to 0} \frac{f(3 + h) - f(3)}{h} = \lim_{h \to 0} \frac{(3 + h)^2 - 9}{h}$$

$$= \lim_{h \to 0} \frac{(9 + 6h + h^2) - 9}{h} = \lim_{h \to 0} \frac{6h + h^2}{h}$$

$$= \lim_{h \to 0} (6 + h) = 6$$

Thus, from (4) the point-slope form of the tangent line is

$$y - 9 = 6(x - 3)$$

and the slope-intercept form is $y = 6x - 9$.   ◄

In later problems we shall want to know how the slope of the tangent line varies from point to point along a curve $y = f(x)$. For this purpose we shall need a formula that produces the slope of the tangent line at an arbitrary point $P(x, y)$ on the curve. Such a formula can be obtained by replacing the constant x_0 in (3) by a variable x. This yields

$$m_{\text{tan}} = \lim_{h \to 0} \frac{f(x + h) - f(x)}{h} \tag{5}$$

The limit in (5) occurs so frequently that it has a special notation. We denote it by

$$f'(x) = \lim_{h \to 0} \frac{f(x + h) - f(x)}{h} \tag{6}$$

where $f'(x)$ is read, "f prime of x." This notation emphasizes the fact that the slope of the tangent line at x is a function of x that is "derived" from the function f.

Example 2 Let $f(x) = x^2 + 1$.

(a) Find $f'(x)$.

(b) Use the result in (a) to find the slope of the line tangent to $y = x^2 + 1$ at $x = 2$, $x = 0$, and $x = -2$.

Solution (a). From (6)

$$f'(x) = \lim_{h \to 0} \frac{f(x + h) - f(x)}{h} = \lim_{h \to 0} \frac{[(x + h)^2 + 1] - [x^2 + 1]}{h}$$

$$= \lim_{h \to 0} \frac{x^2 + 2xh + h^2 + 1 - x^2 - 1}{h} = \lim_{h \to 0} \frac{2xh + h^2}{h}$$

$$= \lim_{h \to 0} (2x + h) = 2x$$

Solution (b). From part (a) the slope of the tangent line at any point x

Figure 3.1.4

is

$$f'(x) = 2x$$

Thus, at $x = 2$, $x = 0$, and $x = -2$ the slopes are

$$f'(2) = 2(2) = 4$$

$$f'(0) = 2(0) = 0$$

$$f'(-2) = 2(-2) = -4$$

respectively (Figure 3.1.4).   ◀

At each point the
tangent line has
slope m.

Figure 3.1.5

Example 3   It is obvious geometrically that at each point on a line $y = mx + b$, the tangent line coincides with the line itself and thus has slope m (Figure 3.1.5). Therefore, if $f(x) = mx + b$, we should anticipate that $f'(x) = m$ for all x. The following computation shows that this is so.

$$f'(x) = \lim_{h \to 0} \frac{f(x + h) - f(x)}{h} = \lim_{h \to 0} \frac{[m(x + h) + b] - (mx + b)}{h}$$

$$= \lim_{h \to 0} \frac{mx + mh + b - mx - b}{h} = \lim_{h \to 0} \frac{mh}{h} = \lim_{h \to 0} m = m$$ ◀

AVERAGE AND INSTANTANEOUS VELOCITY

While tangent lines are of interest as a matter of pure geometry, much of the impetus for studying them arose in the seventeenth century when scientists recognized that many problems involving objects moving with varying velocity could be reduced to problems involving tangents. To see why this is so, we need to examine critically the meaning of the word "velocity."

If a car travels 75 miles over a straight road in a 3-hour period, then we say that the average velocity of the car is 25 miles per hour (25 mi/hr). More generally, the *average velocity* of an object moving in *one direction* along a line is

$$\text{average velocity} = \frac{\text{distance traveled}}{\text{time elapsed}}$$

Obviously, if an object travels with an average velocity of 25 mi/hr during a 3-hour trip, it need not travel at a fixed velocity of 25 mi/hr; sometimes it may speed up and sometimes it may slow down.

Although average velocity is useful for some purposes, it is not always significant in physical problems. For example, if a moving car strikes a tree, the damage sustained is not determined by the average velocity up to the time of impact, but rather by the *instantaneous velocity* at the precise moment of impact.

A clear understanding of instantaneous velocity evaded scientists until the advent of calculus in the seventeenth century. The subtlety of this concept was nicely described by Morris Kline* who wrote,

> *In contrasting average velocity with instantaneous velocity we implicitly utilize a distinction between interval and instant. . . . An average velocity is one that concerns what happens over an interval of time—3 hours, 5 seconds, one-half second, and so forth. The interval may be small or large, but it does represent the passage of a definite amount of time. We use the word instant, however, to state the fact that something happens so fast that no time elapses. The event is momentary. When we say, for example, that it is 3 o'clock, we refer to an instant, a precise moment. If the lapse of time is pictured by length along a line, then an interval (of time) is represented by a line segment, whereas an instant corresponds to a point. The notion of an instant, although it is used in everyday life, is strictly a mathematical idealization.*
>
> *Our ways of thinking about real events cause us to speak in terms of instants and velocity at an instant, but closer examination shows that the concept of velocity at an instant presents difficulties. Average velocity, which is simply the distance traveled during some interval of time divided by that amount of time, is easily calculated. Suppose, however, that we try to carry over this process to instantaneous velocity. The distance an automobile travels in one instant is 0 and the time that elapses during one instant is also 0. Hence the distance divided by the time is 0/0, which is meaningless. Thus, although instantaneous velocity is a physical reality, there seems to be a difficulty in calculating it, and unless we can calculate it, we cannot work with it mathematically.*

In order to understand the concept of instantaneous velocity and work with it mathematically we must think in terms of approximating instantaneous velocity by average velocity. For example, suppose we are interested in the instantaneous velocity of a car at a certain instant of time, say, exactly 5 sec after it starts to move along a straight road. Although the velocity of the car may be changing, it is evident that over a short interval of time, say, 0.1 sec, the velocity does not vary much. Thus, we make only a small error if we approximate the instantaneous velocity after 5 sec by the *average velocity* over the *interval* from 5 to 5.1 sec. This average velocity can be calculated by measuring the distance traveled during the time interval between 5 and 5.1 sec and then dividing by the time elapsed, which is 0.1 sec. Thus, even though the instantaneous velocity after 5 sec cannot be calculated directly, it can be approximated

*MORRIS KLINE (1908–   ) American mathematician, scholar, and educator. Kline has made numerous contributions to mathematical thought, written extensively on education, especially mathematics education, and has taught, lectured, and served as a consultant throughout his very active career. He is the author of many popular books including *Mathematical Thought from Ancient to Modern Times* and *Why Johnny Can't Add: The Failure of the New Mathematics*. I wish to thank him for permission to use the above quotation, which is taken from *Calculus: An Intuitive and Physical Approach*. Wiley, New York, 1977, p. 17.

Figure 3.1.6 ▷   Elapsed time

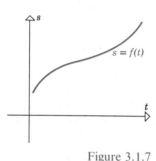

Figure 3.1.7

by an average velocity that can be determined from physical measurements.

Now let us look at these ideas geometrically. Suppose that the car is moving along a straight road and a coordinate line is introduced along the road with its positive direction in the direction of motion of the car. Imagine that a clock is keeping track of the elapsed time, starting with $t = 0$ initially, and that after t seconds have elapsed the car is at a distance of s units from the origin (Figure 3.1.6). Because s changes with t, the position coordinate s is some function of t, say $s = f(t)$. If we graph this function with the t-axis horizontal and the s-axis vertical, we obtain a *position versus time curve* (Figure 3.1.7). Using this curve, we can interpret average velocity and instantaneous velocity geometrically. For example, suppose we are interested in the average velocity of the car over the time interval from t_0 to t_1. At time t_0 the car is at some distance s_0 from the origin and at time t_1 it is at some distance s_1. Over the time interval from t_0 to t_1 the distance traveled is $s_1 - s_0$ and the time elapsed is $t_1 - t_0$ so that the average velocity of the car during the interval is given by

$$\begin{bmatrix} \text{average velocity} \\ \text{from times} \\ t_0 \text{ to } t_1 \end{bmatrix} = \frac{s_1 - s_0}{t_1 - t_0} \tag{7}$$

Figure 3.1.8

However, the points (t_0, s_0) and (t_1, s_1) lie on the position versus time curve, so that expression (7) is also the slope of the secant line connecting these points (Figure 3.1.8). Thus, we conclude:

Geometric Interpretation of Average Velocity
The average velocity of the car between times t_0 and t_1 is represented geometrically by the slope of the secant line connecting (t_0, s_0) and (t_1, s_1) on the position versus time curve.

If we choose t_1 close to t_0, then the average velocity between times t_0 and t_1 closely approximates the instantaneous velocity at time t_0; moreover, as we move t_1 closer and closer to t_0, intuition suggests that these approximations will approach the exact value of the instantaneous velocity at t_0. However, as t_1 moves toward t_0, the point (t_1, s_1) moves toward (t_0, s_0) on the position versus time curve (Figure 3.1.9), so that the slope of the secant line between (t_0, s_0) and (t_1, s_1) approaches the slope of the tangent line at (t_0, s_0). Thus, we conclude:

> **Geometric Interpretation of Instantaneous Velocity**
> The instantaneous velocity of the car at time t_0 is represented geometrically by the slope of the tangent line at (t_0, s_0) on the position versus time curve.

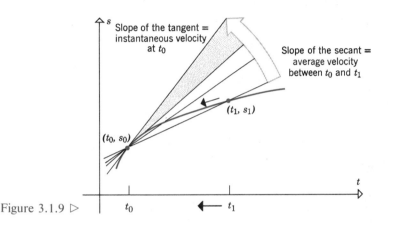

Figure 3.1.9 ▷

AVERAGE AND
INSTANTANEOUS
RATES OF CHANGE

Velocity can be viewed as a *rate of change*—the rate of change of position with time, or in algebraic terms, the rate of change of s with t. Rates of change occur in many applications. For example:

- A geneticist might be interested in the rate at which the number of bacteria in a colony changes with time.

- An engineer might be interested in the rate at which the length of a metal rod changes with temperature.

- An economist might be interested in the rate at which production cost changes with the quantity of a product that is manufactured.

- A medical researcher might be interested in the rate at which the radius of an artery opening changes with the concentration of alcohol in the bloodstream.

In general, if x and y are any quantities related by an equation $y = f(x)$, we can consider the rate at which y varies with x. As with velocity, we distinguish between an average rate of change represented by the slope of a secant line and an instantaneous rate of change represented by the slope of a tangent line. More precisely, we make the following definitions (see Figure 3.1.10).

3.1.2 DEFINITION. If (x_0, y_0) and (x_1, y_1) are points on the graph of $y = f(x)$, then we define

$$m_{\text{sec}} = \frac{y_1 - y_0}{x_1 - x_0} = \frac{f(x_1) - f(x_0)}{x_1 - x_0}$$

to be the **average rate of change of y with respect to x** over the interval $[x_0, x_1]$.

3.1.3 DEFINITION. If (x_0, y_0) is a point on the graph of $y = f(x)$, then we define the **instantaneous rate of change of y with respect to x** at x_0 to be

$$m_{\text{tan}} = f'(x_0)$$

Figure 3.1.10 ▷

Example 4 Consider the curve $y = f(x)$ in Figure 3.1.11. The slope of the secant line joining P and Q is $m_{\text{sec}} = \frac{6}{8} = \frac{3}{4}$. As we travel along that curve from P to Q, y increases slowly at first, then more rapidly and then more slowly again. However, at the end of the trip from P to Q, y has increased by 6 units and x by 8 units. Thus, "on the average" y increases $\frac{6}{8} = \frac{3}{4}$ of a unit for each one unit increase in x between P and Q. ◀

Figure 3.1.11 ▷

Example 5   Suppose that $y = x^2 + 1$. Find

(a)   the average rate of change of y with respect to x over the interval $[3, 5]$;

(b)   the instantaneous rate of change of y with respect to x at the point $x = -4$.

Solution (a).   We will apply Definition 3.1.2 with $x_0 = 3$ and $x_1 = 5$. The y values corresponding to these x values are $y_0 = (3)^2 + 1 = 10$ and $y_1 = (5)^2 + 1 = 26$, so

$$\begin{array}{l}\text{ave. rate of}\\\text{change of } y\\\text{over } [3, 5]\end{array} = \frac{y_1 - y_0}{x_1 - x_0} = \frac{26 - 10}{5 - 3} = \frac{16}{2} = 8$$

Thus, on the average, y increases 8 units per unit increase in x over the interval $[3, 5]$.

Solution (b).   We will apply Definition 3.1.3 with $x_0 = -4$. In Example 2 we found $f'(x) = 2x$ so

$$\begin{array}{l}\text{inst. rate of}\\\text{change of } y\\\text{at } x = -4\end{array} = f'(-4) = 2(-4) = -8$$

Because the instantaneous rate of change is negative, y is *decreasing* at the point $x = -4$; it is decreasing at a rate of 8 units per unit increase in x.   ◀

▶ Exercise Set 3.1

1.   Let $f(x) = \frac{1}{2}x^2$.
   (a)   Find the slope of the secant line between those points on the graph of f for which $x = 3$ and $x = 4$.
   (b)   Use the method of Example 1 to find the slope and equation for the tangent to the graph of f at the point where $x = 3$.
   (c)   Sketch the graph of f together with the secant and tangent lines from (a) and (b).

2.   Let $f(x) = x^3$.
   (a)   Find the slope of the secant line between those points on the graph of f for which $x = 1$ and $x = 2$.
   (b)   Use the method of Example 1 to find the

slope and equation for the tangent to the graph of f at the point where $x = 1$.
   (c)   Sketch the graph of f together with the secant and tangent lines from (a) and (b).

3.   Let $f(x) = 1/x$.
   (a)   Find the slope of the secant line between those points on the graph of f for which $x = 2$ and $x = 3$.
   (b)   Use the method of Example 1 to find the slope and equation for the tangent to the graph of f at the point where $x = 2$.
   (c)   Sketch the graph of f together with the secant and tangent lines from (a) and (b).

4. Let $f(x) = \dfrac{1}{x^2}$.

    (a) Find the slope of the secant line between those points on the graph of f for which $x = 1$ and $x = 2$.

    (b) Use the method of Example 1 to find the slope and equation for the tangent to the graph of f at the point where $x = 1$.

    (c) Sketch the graph of f together with the secant and tangent lines from (a) and (b).

5. Let $f(x) = x^3$.

    (a) Use the method of Example 1 to show that the slope of the tangent to the graph of f at the point where $x = x_0$ is $3x_0^2$.

    (b) Use the result in (a) to find the equation of the tangent to the graph of f at the point where $x = 5$.

    (c) Use the result in (a) to find the equation of the tangent to the graph of f at the point where $x = x_0$.

6. Let $f(x) = \dfrac{1}{x}$.

    (a) Use the method of Example 1 to show that the slope of the tangent to the graph of f at the point where $x = x_0$ is $-1/x_0^2$.

    (b) Use the result in (a) to find the equation of the tangent to the graph of f at the point where $x = -7$.

    (c) Use the result in (a) to find the equation of the tangent to the graph of f at the point where $x = x_0$.

7. Let $f(x) = x^2 + x$.

    (a) Use the method of Example 1 to find the slope of the tangent to the graph of f at the point where $x = x_0$.

    (b) Use the result in (a) to find the equation of the tangent to the graph of f at the point where $x = 2$.

    (c) Use the result in (a) to find the equation of the tangent to the graph of f at the point where $x = x_0$.

8. Follow the directions of Exercise 7 for the function $f(x) = x^2 + 3x + 2$.

9. Figure 3.1.12 shows the position versus time curve for a certain particle moving on a straight line.

    (a) Is the particle moving faster at time t_0 or time t_2? Explain.

    (b) At the origin, the tangent is horizontal. What does this tell us about the initial velocity of the particle?

    (c) Is the particle speeding up or slowing down in the interval $[t_0, t_1]$? Explain.

    (d) Is the particle speeding up or slowing down in the interval $[t_1, t_2]$? Explain.

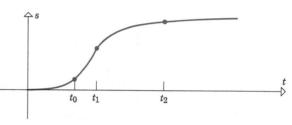

Figure 3.1.12

10. An automobile, initially at rest, begins to move along a straight track. The velocity increases continually until suddenly the driver sees a concrete barrier in the road and applies the brakes sharply at time t_0. The car decelerates rapidly, but it is too late—the car crashes into the barrier at time t_1 and instantaneously comes to rest. Sketch a position versus time curve that might represent the motion of the car.

11. If a particle moves at constant velocity, what can you say about its position versus time curve?

In Exercises 12–15, use the methods of Examples 1 and 5 with $s = f(t)$ to find the average and instantaneous velocity.

12. A rock is dropped from a height of 576 ft and falls toward earth in a straight line. In t sec the rock drops a distance of $s = 16t^2$ ft.

    (a) How many seconds after release does the rock hit the ground?

    (b) What is the average velocity of the rock during the time it is falling?

    (c) What is the average velocity of the rock for the first 3 sec?

(d) What is the instantaneous velocity of the rock when it hits the ground?

13. During the first 40 sec of a rocket flight, the rocket is propelled straight up so that in t sec it reaches a height of $s = 5t^3$ ft.
    (a) How high does the rocket travel in 40 sec?
    (b) What is the average velocity of the rocket during the first 40 sec?
    (c) What is the average velocity of the rocket during the first 135 ft of its flight?
    (d) What is the instantaneous velocity of the rocket at the end of 40 sec?

14. A particle moves on a line away from its initial position so that after t hr it is $s = 3t^2 + t$ mi from its initial position.
    (a) Find the average velocity of the particle over the interval $[1, 3]$.
    (b) Find the instantaneous velocity at $t = 1$.

15. A particle moves in one direction along a straight line so that after t min its distance is $s = 6t^4$ ft from the origin.
    (a) Find the average velocity of the particle over the interval $[2, 4]$.
    (b) Find the instantaneous velocity at $t = 2$.

16. A car is traveling on a straight road that is 120 mi long. For the first 100 mi the car travels at an average velocity of 50 mi/hr. Show that no matter how fast the car travels for the final 20 mi it cannot bring the average velocity up to 60 mi/hr for the entire trip.

17. Let $y = 2x^2 - 1$.
    (a) Find the average rate at which y changes with x over the interval $[1, 4]$.
    (b) Find the instantaneous rate at which y changes with x at the point $x = 1$.

18. Let $y = \dfrac{1}{x^2 + 1}$.
    (a) Find the average rate at which y changes with x over the interval $[-1, 2]$.
    (b) Find the instantaneous rate at which y changes with x at the point $x = -1$.

19. Use the formula $A = \pi r^2$ for the area of a circle to find:
    (a) the average rate at which the area of a circle changes with r as the radius increases from $r = 1$ to $r = 2$;
    (b) the instantaneous rate at which the area changes with r when $r = 2$.

20. Use the formula $V = l^3$ for the volume of a cube of side l to find:
    (a) the average rate at which the volume of a cube changes with l as l increases from $l = 2$ to $l = 4$;
    (b) the instantaneous rate at which the volume of a cube changes with l when $l = 5$.

21. Let $f(x) = x^2$. If we approximate the slope of the tangent at the point $(x_0, f(x_0))$ by the slope of the secant line between $(x_0, f(x_0))$ and $(x_1, f(x_1))$, show that the error is $|x_1 - x_0|$. (By error we mean $|m_{tan} - m_{sec}|$, where m_{tan} is the slope of the tangent line and m_{sec} the slope of the secant line.)

3.2 THE DERIVATIVE

The "slope producing" function f' is of fundamental importance in mathematics. In this section we shall study some of the basic properties of this function.

The function f' defined in the previous section is so important that it has its own name.

> **3.2.1** DEFINITION. The function f' defined by the formula
>
> $$f'(x) = \lim_{h \to 0} \frac{f(x + h) - f(x)}{h} \tag{1}$$
>
> is called the **derivative with respect to x** of the function f. The domain of f' consists of all x for which the limit exists.

Based on our discussion in the previous section, the derivative of a function f can be interpreted two ways:

Geometric Interpretation of the Derivative
f' is the function whose value at x is the slope of the tangent line to $y = f(x)$ at x.

Rate of Change Interpretation of the Derivative
If $y = f(x)$, then f' is the function whose value at x is the instantaneous rate of change of y with respect to x at the point x.

Example 1   Find the derivative of $f(x) = \sqrt{x}$.

Solution.   From Definition 3.2.1,

$$f'(x) = \lim_{h \to 0} \frac{f(x + h) - f(x)}{h} = \lim_{h \to 0} \frac{\sqrt{x + h} - \sqrt{x}}{h}$$

$$= \lim_{h \to 0} \frac{(\sqrt{x + h} - \sqrt{x})(\sqrt{x + h} + \sqrt{x})}{h(\sqrt{x + h} + \sqrt{x})} = \lim_{h \to 0} \frac{(x + h) - x}{h(\sqrt{x + h} + \sqrt{x})}$$

$$= \lim_{h \to 0} \frac{h}{h(\sqrt{x + h} + \sqrt{x})} = \lim_{h \to 0} \frac{1}{\sqrt{x + h} + \sqrt{x}}$$

$$= \frac{1}{\sqrt{x} + \sqrt{x}} = \frac{1}{2\sqrt{x}} \quad \blacktriangleleft$$

Example 2   Find the derivative of

$$f(x) = \frac{x}{x - 9}$$

Solution.

$$f'(x) = \lim_{h \to 0} \frac{f(x + h) - f(x)}{h}$$

$$= \lim_{h \to 0} \frac{1}{h} \left[\frac{x + h}{x + h - 9} - \frac{x}{x - 9} \right]$$

$$= \lim_{h \to 0} \frac{1}{h} \left[\frac{(x + h)(x - 9) - x(x + h - 9)}{(x + h - 9)(x - 9)} \right]$$

$$= \lim_{h \to 0} \frac{(x^2 - 9x + hx - 9h) - (x^2 + hx - 9x)}{h(x + h - 9)(x - 9)}$$

$$= \lim_{h \to 0} \frac{-9h}{h(x + h - 9)(x - 9)}$$

$$= \lim_{h \to 0} \frac{-9}{(x + h - 9)(x - 9)} = - \frac{9}{(x - 9)^2} \quad \blacktriangleleft$$

Example 3 Let $y = \sqrt{x}$, and use the result in Example 1 to find:

(a) the slope of the tangent line to the graph of this equation at $x = 9$;

(b) the instantaneous rate of change of y with respect to x at $x = 5$.

Solution.   If we let $f(x) = \sqrt{x}$, then $f'(x) = 1/(2\sqrt{x})$, so the slope of the tangent line at $x = 9$ is $f'(9) = 1/(2\sqrt{9}) = \frac{1}{6}$, and the instantaneous rate of change of y with respect to x at $x = 5$ is $f'(5) = 1/(2\sqrt{5}) = \sqrt{5}/10$. ◀

DERIVATIVE
NOTATION

The process of finding a derivative is called **differentiation.** It is often useful to think of differentiation as an operation which, when applied to a function f, produces a new function f'. In the case where the independent variable is x, the differentiation operation is often denoted by the symbol

$$\frac{d}{dx} [\quad]$$

which is read, "***the derivative with respect to x of.***" Within the brackets we place the function whose derivative is to be computed. Thus, from Examples 1 and 2 we have

$$\frac{d}{dx} [\sqrt{x}] = \frac{1}{2\sqrt{x}}, \quad \frac{d}{dx} \left[\frac{x}{x - 9} \right] = - \frac{9}{(x - 9)^2} \tag{2}$$

The d/dx notation allows us to write derivatives without using any function symbols like f or f'. This notation also makes it easy to write derivatives of functions involving variables other than x. For example, if we use u as the variable in (2) instead of x, we can write

$$\frac{d}{du} [\sqrt{u}] = \frac{1}{2\sqrt{u}}, \quad \frac{d}{du} \left[\frac{u}{u - 9} \right] = - \frac{9}{(u - 9)^2}$$

(The symbol d/du [ ] is read, "*the derivative with respect to u of.*")

With the d/dx notation it is awkward to denote the value of the derivative at a particular point; the prime notation is better for this purpose. As an illustration, we found in Example 1 that if $f(x) = \sqrt{x}$, then $f'(x) = 1/(2\sqrt{x})$. With this notation the value of the derivative at $x = x_0$ can be written as $f'(x_0) = 1/(2\sqrt{x_0})$. To express this in d/dx notation, we must write

$$\frac{d}{dx}\left[\sqrt{x}\right]\bigg|_{x=x_0} = \frac{1}{2\sqrt{x_0}}$$

which is rather clumsy. In general,

denotes the value of the derivative with respect to x at $x = x_0$.

When we use a dependent variable

$$y = f(x)$$

we will usually denote the derivative as

$$\frac{dy}{dx} \quad \text{or} \quad y'(x)$$

and the value of the derivative at a point $x = x_0$ as

$$\frac{dy}{dx}\bigg|_{x=x_0} \quad \text{or} \quad y'(x_0)$$

For example, if $y = \sqrt{x}$, then

$$y'(x) = \frac{dy}{dx} = \frac{1}{2\sqrt{x}} \quad \text{and} \quad y'(x_0) = \frac{dy}{dx}\bigg|_{x=x_0} = \frac{1}{2\sqrt{x_0}}$$

Similarly, if $y = \sqrt{u}$, then

$$y'(u) = \frac{dy}{du} = \frac{1}{2\sqrt{u}} \quad \text{and} \quad y'(3) = \frac{dy}{du}\bigg|_{u=3} = \frac{1}{2\sqrt{3}}$$

Later, the symbols dy and dx will be defined separately. However, for the time being, dy/dx should not be regarded as a ratio; rather, it should be considered as a single symbol denoting the derivative.

REMARK. Some writers use the notation D_x[ ] to denote the differentiation operation. We shall not use this notation, however.

EXISTENCE OF
DERIVATIVES

The derivative of a function f is defined at those points where the limit in (1) exists. If x_0 is such a point, then we say that *f is differentiable at x_0 or f has a derivative at x_0.* Stated another way, the domain of f' consists of those points where f is differentiable. At points where f is not differentiable we say that *the derivative of f does not exist.* Informally speaking, the most commonly encountered points of nondifferentiability can be classified as:

- corners

- vertical tangents

- points of discontinuity

(see Figure 3.2.1).

Figure 3.2.1

Corner

Vertical tangent

Point of discontinuity

The graph of a function f has a "corner" at a point $P(x_0, f(x_0))$ if f is continuous at P and the limiting position for the secant line joining P and Q depends on whether Q approaches P from the left or the right (Figure 3.2.2). At a corner a tangent line does not exist because the slopes of the secant lines do not have a (two-sided) limit.

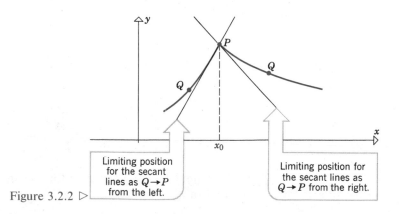

Figure 3.2.2 ▷

Limiting position for the secant lines as $Q \to P$ from the left.

Limiting position for the secant lines as $Q \to P$ from the right.

If the slope of the secant line joining P and Q tends toward $+\infty$ or $-\infty$ as Q approaches P along the graph of f, then f is not differentiable at x_0. Geometrically, such points occur where the secant lines tend toward a vertical limiting position (Figure 3.2.3).

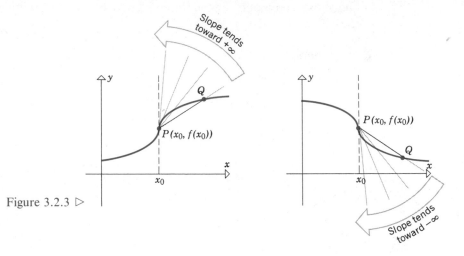

Figure 3.2.3 ▷

RELATIONSHIP BETWEEN DIFFERENTIABILITY AND CONTINUITY

The following major theorem shows that a function must be continuous at each point where it is differentiable.

3.2.2 THEOREM. *If f is differentiable at a point x_0, then f is also continuous at x_0.*

Proof. We shall use Definition 2.7.11 to prove the continuity. We must show that $\lim_{h \to 0} f(x_0 + h) = f(x_0)$ or equivalently

$$\lim_{h \to 0} [f(x_0 + h) - f(x_0)] = 0$$

But

$$\lim_{h \to 0} [f(x_0 + h) - f(x_0)] = \lim_{h \to 0} \left[\frac{f(x_0 + h) - f(x_0)}{h} \cdot h \right]$$

$$= \lim_{h \to 0} \left[\frac{f(x_0 + h) - f(x_0)}{h} \right] \cdot \lim_{h \to 0} h$$

$$= f'(x_0) \cdot 0 = 0 \quad \blacksquare$$

REMARK. It follows from the foregoing theorem that a function cannot be differentiable at a point of discontinuity.

Theorem 3.2.2 shows that differentiability at a point implies continuity at that point. The converse, however, is false—*a function may be continuous at a point but not differentiable there*. Whenever the graph of a function has a corner at a point, but no break or gap there, we have a point where the function is continuous, but not differentiable.

Example 4   The function $f(x) = |x|$ is continuous for all x and consequently continuous at $x = 0$ (Example 3, Section 2.7). Show that $f(x) = |x|$ is not differentiable at $x = 0$.

Figure 3.2.4

Solution.   This result is evident geometrically, since the graph of $|x|$ has a corner at $x = 0$ (Figure 3.2.4). However, an analytic argument can be given as follows. From Definition 3.2.1,

$$f'(0) = \lim_{h \to 0} \frac{f(0 + h) - f(0)}{h} = \lim_{h \to 0} \frac{f(h) - f(0)}{h}$$

$$= \lim_{h \to 0} \frac{|h| - |0|}{h} = \lim_{h \to 0} \frac{|h|}{h}$$

But

$$\frac{|h|}{h} = \begin{cases} 1, & h > 0 \\ -1, & h < 0 \end{cases}$$

so that

$$\lim_{h \to 0^-} \frac{|h|}{h} = -1 \quad \text{and} \quad \lim_{h \to 0^+} \frac{|h|}{h} = 1$$

Thus,

$$f'(0) = \lim_{h \to 0} \frac{|h|}{h}$$

does not exist because the one-sided limits are not equal. Consequently, $f(x) = |x|$ is not differentiable at $x = 0$.   ◄

The relationship between continuity and differentiability was of great historical significance in the development of calculus. In the early nineteenth century mathematicians believed that the graph of a continuous function could not have too many points of nondifferentiability bunched up. They felt that if a continuous function had many points of nondifferentiability, these points, like the tips of a sawblade, would have to be

Figure 3.2.5 ▷

separated from each other and joined by smooth curve segments (Figure 3.2.5). This misconception was shattered by a series of discoveries beginning in 1834. In that year a Bohemian priest, philosopher, and mathematician named Bernhard Bolzano* discovered a procedure for constructing a continuous function that is not differentiable at any point. Later, in 1860, the great German mathematician, Karl Weierstrass**

---

*BERNHARD BOLZANO (1781–1848). Bolzano, the son of an art dealer, was born in Prague, Bohemia (Czechoslovakia). He was educated at the University of Prague, and eventually won enough mathematical fame to be recommended for a mathematics chair there. However, Bolzano became an ordained Roman Catholic priest, and in 1805 he was appointed to a chair of Philosophy at the University of Prague. Bolzano was a man of great human compassion; he spoke out for educational reform, he voiced the right of individual conscience over government demands, and he lectured on the absurdity of war and militarism. His views so disenchanted Emperor Franz I of Austria that the emperor pressed the Archbishop of Prague to have Bolzano recant his statements. Bolzano refused and was then forced to retire in 1824 on a small pension. Bolzano's main contribution to mathematics was philosophical. His work helped convince mathematicians that sound mathematics must ultimately rest on rigorous proof rather than intuition. In addition to his work in mathematics, Bolzano investigated problems concerning space, force, and wave propagation.

**KARL WEIERSTRASS (1815–1897). Weierstrass, the son of a customs officer, was born in Ostenfelde, Germany. As a youth Weierstrass showed outstanding skills in languages and mathematics. However, at the urging of his dominant father, Weierstrass entered the law and commerce program at the University of Bonn. To the chagrin of his family, the rugged and congenial young man concentrated instead on fencing and beer drinking. Four years later he returned home without a degree. In 1839 Weierstrass entered the Academy of Münster to study for a career in secondary education, and he met and studied under an excellent mathematician named Christof Gudermann. Gudermann's ideas greatly influenced the work of Weierstrass. After receiving his teaching certificate, Weierstrass spent the next 15 years in secondary education teaching German, geography, and mathematics. In addition, he taught handwriting to small children. During this period much of Weierstrass's mathematical work was ignored because he was a secondary schoolteacher and not a college professor. Then, in 1854, he published a paper of major importance which created a sensation in the mathematics world and catapulted him to international fame overnight. He was immediately given an honorary Doctorate at the University of Königsberg and began a new career in college teaching at the University of Berlin in 1856. In 1859 the strain of his mathematical research caused a temporary nervous breakdown and led to spells of dizziness that plagued him for the rest of his life. Weierstrass was a brilliant teacher and his classes overflowed with multitudes of auditors. In spite of his fame, he never lost his early beer-drinking congeniality and was always in the company of students, both ordinary and brilliant. Weierstrass was acknowledged as the leading mathematical analyst in the world. He and his students opened the door to the modern school of mathematical analysis.

produced the first formula for such a function. The graphs of such functions are impossible to draw; it is as if they oscillate so wildly that a corner occurs everywhere. Although these pathological functions rarely occur in practical problems,* their discovery was of major importance because it made mathematicians distrustful of their geometric intuition and more reliant on precise mathematical proof.

---

*Important applications of these functions have been discovered recently. See, Bruce Schecter, "A New Geometry of Nature," *Discover Magazine* (June 1982), pp. 66–68.

▶ Exercise Set 3.2

In Exercises 1–12, use Definition 3.2.1 to find $f'(x)$.

1. $f(x) = 3x^2$.   **2.** $f(x) = x^2 - x$.

3. $f(x) = x^3$.   **4.** $f(x) = 2x^3 + 1$.

5. $f(x) = \sqrt{x + 1}$.   **6.** $f(x) = x^4$.

7. $f(x) = \dfrac{1}{x}$.   **8.** $f(x) = \dfrac{1}{x^2}$.

9. $f(x) = ax^2 + b$ (a, b constants).

10. $f(x) = \dfrac{1}{x + 1}$.   **11.** $f(x) = \dfrac{1}{\sqrt{x}}$.

12. $f(x) = x^{1/3}$.

In Exercises 13–18, find $f'(a)$ and the equation of the tangent to the graph of f at the point where $x = a$.

13. f is the function in Exercise 1; $a = 3$.

14. f is the function in Exercise 2; $a = 2$.

15. f is the function in Exercise 3; $a = 0$.

16. f is the function in Exercise 4; $a = -1$.

17. f is the function in Exercise 5; $a = 8$.

18. f is the function in Exercise 6; $a = -2$.

19. Let $y = 4x^2 + 2$. Find

    (a) $\dfrac{dy}{dx}$ (b) $\dfrac{dy}{dx}\bigg|_{x=1}$

20. Let $y = \dfrac{5}{x} + 1$. Find

    (a) $\dfrac{dy}{dx}$ (b) $\dfrac{dy}{dx}\bigg|_{x=-2}$

In Exercises 21–24, use Definition 3.2.1 (with the appropriate change in notation) to obtain the derivative requested.

21. Find $f'(t)$ if $f(t) = 4t^2 + t$.

22. Find $g'(u)$ if $g(u) = 5u + 3$.

23. Find $\dfrac{dA}{d\lambda}$ if $A = 3\lambda^2 - \lambda$.

24. Find $\dfrac{dV}{dr}$ if $V = \frac{4}{3}\pi r^3$.

25. Match the graphs of the functions shown in (a)–(f) with the graphs of their derivatives in (A)–(F).

(a)   (d)

(b)   (e)

(c)   (f)

(A)

(B)

(C)

(D)

(E)

(F)

29.

30.

31.

32.

33.

26. Use the graph of $y = f(x)$ shown in Figure 3.2.6 to estimate the value of $f'(1)$, $f'(3)$, $f'(5)$, and $f'(6)$.

Figure 3.2.6   1

In Exercises 27–34, sketch the graph of the derivative of the function whose graph is shown.

27.   28.

34.

35. Show that $f(x) = \sqrt[3]{x}$ is continuous at $x = 0$ but not differentiable at $x = 0$. Sketch the graph of f.

36. Show that $f(x) = \sqrt[3]{(x - 2)^2}$ is continuous at $x = 2$ but not differentiable at $x = 2$. Sketch the graph of f.

37. Show that

$$f(x) = \begin{cases} x^2 + 1, & x \le 1 \\ 2x, & x > 1 \end{cases}$$

is continuous and differentiable at $x = 1$. Sketch the graph of f.

38. Show that

$$f(x) = \begin{cases} x^2 + 2, & x \le 1 \\ x + 2, & x > 1 \end{cases}$$

is continuous but not differentiable at $x = 1$. Sketch the graph of f.

39. Suppose that the function f is differentiable at $x = 1$ and $\lim\limits_{h \to 0} \dfrac{f(1 + h)}{h} = 5$. Find $f(1)$ and $f'(1)$.

40. Suppose that f is a differentiable function with the property that $f(x + y) = f(x) + f(y) + 5xy$ and $\lim\limits_{h \to 0} \dfrac{f(h)}{h} = 3$. Find $f(0)$ and $f'(x)$.

3.3 TECHNIQUES OF DIFFERENTIATION

Up to now we have obtained derivatives directly from the definition. In this section we shall develop some theorems and formulas that provide more efficient methods.

3.3.1 THEOREM. *If f is a constant function, say $f(x) = c$ for all x, then $f'(x) = 0$.*

Proof.

$$f'(x) = \lim_{h \to 0} \frac{f(x + h) - f(x)}{h} = \lim_{h \to 0} \frac{c - c}{h} = \lim_{h \to 0} 0 = 0 \qquad \blacksquare$$

This result is obvious geometrically. If $f(x) = c$ is a constant function, then the graph of f is a line parallel to the x-axis; consequently, the tangent line at each point is horizontal. Therefore, $f'(x) = m_{tan} = 0$, since a horizontal line has slope 0.

In another notation, Theorem 3.3.1 states

$$\frac{d}{dx}[c] = 0$$

Example 1   If $f(x) = 5$ for all x, then $f'(x) = 0$ for all x; that is,

$$\frac{d}{dx}[5] = 0 \qquad \blacktriangleleft$$

The Power Rule

3.3.2 THEOREM. *If n is a positive integer, then*

$$\frac{d}{dx}[x^n] = nx^{n-1}$$

Proof.   Let $f(x) = x^n$. Then

$$f'(x) = \lim_{h \to 0} \frac{f(x+h) - f(x)}{h} = \lim_{h \to 0} \frac{(x+h)^n - x^n}{h}$$

Expanding $(x + h)^n$ by the binomial theorem, we obtain

$$f'(x) = \lim_{h \to 0} \frac{\left[x^n + nx^{n-1}h + \dfrac{n(n-1)}{2!} x^{n-2}h^2 + \cdots + nxh^{n-1} + h^n \right] - x^n}{h}$$

$$= \lim_{h \to 0} \frac{nx^{n-1}h + \dfrac{n(n-1)}{2!} x^{n-2}h^2 + \cdots + nxh^{n-1} + h^n}{h}$$

Canceling a factor of h we obtain

$$f'(x) = \lim_{h \to 0} \left[nx^{n-1} + \frac{n(n-1)}{2!} x^{n-2}h + \cdots + nxh^{n-2} + h^{n-1} \right]$$

Every term but the first has a factor of h, and so approaches zero as $h \to 0$. Therefore,

$$f'(x) = nx^{n-1} \qquad \blacksquare$$

REMARK.   In words, *to differentiate x to a positive integer power, take the power and multiply it by x to the next lower integer power.*

Example 2

$$\frac{d}{dx}[x^5] = 5x^4, \quad \frac{d}{dx}[x] = 1 \cdot x^0 = 1, \quad \frac{d}{dx}[x^{12}] = 12x^{11} \qquad \blacktriangleleft$$

3.3.3   THEOREM. *Let c be a constant. If f is differentiable at x, then so is cf, and*

$$\frac{d}{dx}[cf(x)] = c\frac{d}{dx}[f(x)]$$

Proof.

$$\frac{d}{dx}[cf(x)] = \lim_{h \to 0} \frac{cf(x+h) - cf(x)}{h}$$

$$= \lim_{h \to 0} c\left[\frac{f(x+h) - f(x)}{h} \right]$$

$$= c \lim_{h \to 0} \frac{f(x + h) - f(x)}{h}$$

A constant factor can be moved through a limit sign.

$$= c \frac{d}{dx}[f(x)] \quad \blacksquare$$

In function notation, Theorem 3.3.3 states

$$(cf)' = cf'$$

REMARK. In words, *a constant factor can be moved through a derivative sign.*

Example 3

$$\frac{d}{dx}[4x^8] = 4 \frac{d}{dx}[x^8] = 4[8x^7] = 32x^7$$

$$\frac{d}{dx}[-x^{12}] = (-1) \frac{d}{dx}[x^{12}] = -12x^{11} \quad \blacktriangleleft$$

3.3.4 THEOREM. *If f and g are differentiable at x, then so is f + g, and*

$$\frac{d}{dx}[f(x) + g(x)] = \frac{d}{dx}[f(x)] + \frac{d}{dx}[g(x)]$$

Proof.

$$\frac{d}{dx}[f(x) + g(x)] = \lim_{h \to 0} \frac{[f(x + h) + g(x + h)] - [f(x) + g(x)]}{h}$$

$$= \lim_{h \to 0} \frac{[f(x + h) - f(x)] + [g(x + h) - g(x)]}{h}$$

$$= \lim_{h \to 0} \frac{f(x + h) - f(x)}{h} + \lim_{h \to 0} \frac{g(x + h) - g(x)}{h}$$

The limit of a sum is the sum of the limits.

$$= \frac{d}{dx}[f(x)] + \frac{d}{dx}[g(x)] \quad \blacksquare$$

Theorem 3.3.4 can be written in function notation as

$$(f + g)' = f' + g'$$

By writing $f - g = f + (-1)g$ and then applying Theorem 3.3.3 and 3.3.4 it follows that

$$\frac{d}{dx}[f(x) - g(x)] = \frac{d}{dx}[f(x)] - \frac{d}{dx}[g(x)]$$

or in function notation

$$(f - g)' = f' - g'$$

REMARK. In words, *the derivative of a sum equals the sum of the derivatives*, and *the derivative of a difference equals the difference of the derivatives*.

Example 4

$$\frac{d}{dx}[x^4 + x^2] = \frac{d}{dx}[x^4] + \frac{d}{dx}[x^2] = 4x^3 + 2x$$

$$\frac{d}{dx}[x^4 - x^2] = \frac{d}{dx}[x^4] - \frac{d}{dx}[x^2] = 4x^3 - 2x$$

$$\frac{d}{dx}[6x^{11} + 9] = \frac{d}{dx}[6x^{11}] + \frac{d}{dx}[9] = 66x^{10} + 0 = 66x^{10} \quad \blacktriangleleft$$

The result in Theorem 3.3.4 can be extended to any finite number of functions. More precisely, if the functions $f_1, f_2, \ldots, f_n$ are all differentiable at x, then their sum is differentiable at x and

$$\frac{d}{dx}[f_1(x) + f_2(x) + \cdots + f_n(x)] = \frac{d}{dx}[f_1(x)] + \frac{d}{dx}[f_2(x)] + \cdots + \frac{d}{dx}[f_n(x)]$$

Example 5

$$\frac{d}{dx}[3x^8 - 2x^5 + 6x + 1] = \frac{d}{dx}[3x^8] + \frac{d}{dx}[-2x^5] + \frac{d}{dx}[6x] + \frac{d}{dx}[1]$$

$$= 24x^7 - 10x^4 + 6 \quad \blacktriangleleft$$

The Product Rule

3.3.5 THEOREM. *If f and g are differentiable at x, then so is the product $f \cdot g$, and*

$$\frac{d}{dx}[f(x)g(x)] = f(x)\frac{d}{dx}[g(x)] + g(x)\frac{d}{dx}[f(x)]$$

Proof.

$$\frac{d}{dx}[f(x)g(x)] = \lim_{h \to 0} \frac{f(x + h) \cdot g(x + h) - f(x) \cdot g(x)}{h}$$

If we add and subtract $f(x + h) \cdot g(x)$ in the numerator, we obtain

$$\frac{d}{dx}[f(x)g(x)] = \lim_{h \to 0} \frac{f(x + h)g(x + h) - f(x + h)g(x) + f(x + h)g(x) - f(x)g(x)}{h}$$

$$= \lim_{h \to 0} \left[f(x + h) \cdot \frac{g(x + h) - g(x)}{h} + g(x) \cdot \frac{f(x + h) - f(x)}{h} \right]$$

$$= \lim_{h \to 0} f(x + h) \cdot \lim_{h \to 0} \frac{g(x + h) - g(x)}{h} + \lim_{h \to 0} g(x) \cdot \lim_{h \to 0} \frac{f(x + h) - f(x)}{h}$$

$$= [\lim_{h \to 0} f(x + h)] \frac{d}{dx}[g(x)] + [\lim_{h \to 0} g(x)] \frac{d}{dx}[f(x)] \tag{1}$$

But

$$\lim_{h \to 0} g(x) = g(x) \tag{2}$$

because $g(x)$ does not involve h and thus remains *constant* as $h \to 0$. Also, it follows from Definition 2.7.11 that

$$\lim_{h \to 0} f(x + h) = f(x) \tag{3}$$

because f is assumed to be differentiable at x and is therefore continuous at x by Theorem 3.2.2. Substituting (2) and (3) into (1) yields

$$\frac{d}{dx}[f(x)g(x)] = f(x)\frac{d}{dx}[g(x)] + g(x)\frac{d}{dx}[f(x)] \quad \blacksquare$$

The product rule can be written in function notation as

$$(f \cdot g)' = f \cdot g' + g \cdot f'$$

REMARK. In words, *the derivative of a product of two functions is the first function times the derivative of the second plus the second function times the derivative of the first.*

WARNING. Note that it is *not* true in general that $(f \cdot g)' = f' \cdot g'$; that is, the derivative of a product is *not* generally the product of the derivatives!

Example 6 Find dy/dx if $y = (4x^2 - 1)(7x^3 + x)$.

Solution. There are two methods. We can either use the product rule or we can multiply out the factors in y and then differentiate. We shall give both methods.

Method I. (*Using the Product Rule.*)

$$\frac{dy}{dx} = \frac{d}{dx}[(4x^2 - 1)(7x^3 + x)]$$

$$= (4x^2 - 1)\frac{d}{dx}[7x^3 + x] + (7x^3 + x)\frac{d}{dx}[4x^2 - 1]$$

$$= (4x^2 - 1)(21x^2 + 1) + (7x^3 + x)(8x)$$

If desired, we can simplify this expression to obtain

$$\frac{dy}{dx} = 140x^4 - 9x^2 - 1$$

Method II. (*Multiplying First.*)

$$y = (4x^2 - 1)(7x^3 + x) = 28x^5 - 3x^3 - x$$

Thus,

$$\frac{dy}{dx} = \frac{d}{dx}[28x^5 - 3x^3 - x] = 140x^4 - 9x^2 - 1$$

which agrees with the result obtained using the product rule. ◄

The Quotient Rule

3.3.6 THEOREM. *If f and g are differentiable at x and $g(x) \neq 0$, then f/g is differentiable at x and*

$$\frac{d}{dx}\left[\frac{f(x)}{g(x)}\right] = \frac{g(x)\dfrac{d}{dx}[f(x)] - f(x)\dfrac{d}{dx}[g(x)]}{[g(x)]^2}$$

Proof.

$$\frac{d}{dx}\left[\frac{f(x)}{g(x)}\right] = \lim_{h \to 0} \frac{\dfrac{f(x+h)}{g(x+h)} - \dfrac{f(x)}{g(x)}}{h}$$

$$= \lim_{h \to 0} \frac{f(x+h) \cdot g(x) - f(x) \cdot g(x+h)}{h \cdot g(x) \cdot g(x+h)}$$

Adding and subtracting $f(x) \cdot g(x)$ in the numerator yields

$$\frac{d}{dx}\left[\frac{f(x)}{g(x)}\right] = \lim_{h \to 0} \frac{f(x+h) \cdot g(x) - f(x) \cdot g(x) - f(x) \cdot g(x+h) + f(x) \cdot g(x)}{h \cdot g(x) \cdot g(x+h)}$$

$$= \lim_{h \to 0} \frac{\left[g(x) \cdot \dfrac{f(x+h) - f(x)}{h}\right] - \left[f(x) \cdot \dfrac{g(x+h) - g(x)}{h}\right]}{g(x) \cdot g(x+h)}$$

$$= \frac{\lim\limits_{h \to 0} g(x) \cdot \lim\limits_{h \to 0} \dfrac{f(x+h) - f(x)}{h} - \lim\limits_{h \to 0} f(x) \cdot \lim\limits_{h \to 0} \dfrac{g(x+h) - g(x)}{h}}{\lim\limits_{h \to 0} g(x) \cdot \lim\limits_{h \to 0} g(x+h)}$$

$$= \frac{[\lim\limits_{h \to 0} g(x)] \cdot \dfrac{d}{dx}[f(x)] - [\lim\limits_{h \to 0} f(x)] \cdot \dfrac{d}{dx}[g(x)]}{\lim\limits_{h \to 0} g(x) \cdot \lim\limits_{h \to 0} g(x+h)} \qquad (4)$$

Since the expressions $f(x)$ and $g(x)$ do not involve h,

$$\lim_{h \to 0} g(x) = g(x) \quad \text{and} \quad \lim_{h \to 0} f(x) = f(x) \qquad (5)$$

Because g is assumed to be differentiable at x, it is continuous at x, and so by Definition 2.7.11

$$\lim_{h \to 0} g(x+h) = g(x) \qquad (6)$$

Substituting (5) and (6) into (4) yields

$$\frac{d}{dx}\left[\frac{f(x)}{g(x)}\right] = \frac{g(x) \dfrac{d}{dx}[f(x)] - f(x) \dfrac{d}{dx}[g(x)]}{[g(x)]^2} \qquad \blacksquare$$

The quotient rule can be written in function notation as

$$\left(\frac{f}{g}\right)' = \frac{g \cdot f' - f \cdot g'}{g^2}$$

REMARK.   In words, *the derivative of a quotient of two functions is the denominator times the derivative of the numerator minus the numerator times the derivative of the denominator, all divided by the denominator squared.*

WARNING.   Note that it is *not* generally true that $(f/g)' = f'/g'$; that is, the derivative of a quotient is *not* generally the quotient of the derivatives.

Example 7   Find dy/dx if

$$y = \frac{x^2 - 1}{x^3 + x}$$

Solution.

$$\frac{dy}{dx} = \frac{d}{dx}\left[\frac{x^2 - 1}{x^3 + x}\right] = \frac{(x^3 + x)\dfrac{d}{dx}[x^2 - 1] - (x^2 - 1)\dfrac{d}{dx}[x^3 + x]}{(x^3 + x)^2}$$

$$= \frac{(x^3 + x)(2x) - (x^2 - 1)(3x^2 + 1)}{(x^3 + x)^2}$$

> The differentiation is complete. The rest is algebraic simplification.

$$= -\frac{x^4 - 4x^2 - 1}{(x^3 + x)^2} \quad \blacktriangleleft$$

The special case of Theorem 3.3.6 in which f is the constant function 1 is of interest in its own right. We leave it as an exercise for the reader to deduce the following result from Theorem 3.3.6.

The Reciprocal Rule

3.3.7   THEOREM. *If g is differentiable at x and g(x) ≠ 0, then 1/g is differentiable at x and*

$$\frac{d}{dx}\left[\frac{1}{g(x)}\right] = -\frac{\dfrac{d}{dx}[g(x)]}{[g(x)]^2}$$

The reciprocal rule can be written in function notation as

$$\left(\frac{1}{g}\right)' = -\frac{g'}{g^2}$$

REMARK.   In words, *the derivative of the reciprocal of a function is the negative of the derivative of the function divided by the function squared.*

Example 8

$$\frac{d}{dx}\left[\frac{1}{x}\right] = -\frac{\dfrac{d}{dx}[x]}{x^2} = -\frac{1}{x^2}$$

$$\frac{d}{dx}\left[\frac{1}{x^3 + 2x - 3}\right] = -\frac{\dfrac{d}{dx}[x^3 + 2x - 3]}{(x^3 + 2x - 3)^2} = -\frac{3x^2 + 2}{(x^3 + 2x - 3)^2} \quad \blacktriangleleft$$

REMARK.   The computations in the foregoing example could have been done using the quotient rule, but this would have been more work. Where it applies, the reciprocal rule is preferable to the quotient rule.

In Theorem 3.3.2 we established the formula

$$\frac{d}{dx}[x^n] = nx^{n-1}$$

for *positive* integer values of n. The following theorem extends this formula so that it applies to *all* integer values of n.

3.3.8   THEOREM. *If n is any integer, then*

$$\frac{d}{dx}[x^n] = nx^{n-1} \tag{7}$$

Proof.   The result has already been established in the case where $n > 0$. If $n < 0$, then let $m = -n$ and let

$$f(x) = x^{-m} = \frac{1}{x^m}$$

From Theorem 3.3.7,

$$f'(x) = \frac{d}{dx}\left[\frac{1}{x^m}\right] = -\frac{\frac{d}{dx}[x^m]}{(x^m)^2}$$

Since $n < 0$, it follows that $m > 0$, so x^m can be differentiated using Theorem 3.3.2. Thus,

$$f'(x) = -\frac{mx^{m-1}}{x^{2m}} = -mx^{m-1-2m} = -mx^{-m-1} = nx^{n-1}$$

which proves (7). In the case $n = 0$ Formula (7) reduces to

$$\frac{d}{dx}[1] = 0 \cdot x^{-1} = 0$$

which is correct by Theorem 3.3.1.   ∎

Example 9

$$\frac{d}{dx}[x^{-9}] = -9x^{-9-1} = -9x^{-10}$$

$$\frac{d}{dx}\left[\frac{1}{x}\right] = \frac{d}{dx}[x^{-1}] = (-1)x^{-1-1} = -x^{-2} = -\frac{1}{x^2}$$

Note that the last result agrees with that obtained in Example 8. ◀

In Example 1 of Section 3.2 we showed that

$$\frac{d}{dx}[\sqrt{x}] = \frac{1}{2\sqrt{x}} \tag{8}$$

If we write this result in exponential notation, we obtain

$$\frac{d}{dx}[x^{1/2}] = \frac{1}{2x^{1/2}} = \frac{1}{2}x^{-1/2}$$

which shows that (7) holds for the rational exponent $n = \frac{1}{2}$. Later, we shall show that (7) actually holds for all rational exponents.

HIGHER DERIVATIVES

If the derivative f' of a function f is itself differentiable, then the derivative of f' is denoted by f'' and is called the **second derivative** of f. As long as we have differentiability, we can continue the process of differentiating derivatives to obtain third, fourth, fifth, and even higher derivatives of f. The successive derivatives of f are denoted by

$$f' \qquad \text{[the first derivative of } f]$$
$$f'' = (f')' \qquad \text{[the second derivative of } f]$$
$$f''' = (f'')' \qquad \text{[the third derivative of } f]$$
$$f^{(4)} = (f''')' \qquad \text{[the fourth derivative of } f]$$
$$f^{(5)} = (f^{(4)})' \qquad \text{[the fifth derivative of } f]$$
$$\vdots \qquad \vdots \qquad \vdots$$

Beyond the third derivative, it is too clumsy to continue using primes, so we switch from primes to integers in parentheses to denote the **order** of the derivative. In this notation it is easy to denote a derivative of arbitrary order by writing

$$f^{(n)} \qquad \text{[the } n\text{th derivative of } f]$$

The significance of the derivatives of order 2, and higher will be discussed later.

Example 10 If $f(x) = 3x^4 - 2x^3 + x^2 - 4x + 2$, then

$$f'(x) = 12x^3 - 6x^2 + 2x - 4$$
$$f''(x) = 36x^2 - 12x + 2$$
$$f'''(x) = 72x - 12$$

$$f^{(4)}(x) = 72$$

$$f^{(5)}(x) = 0$$

$$\vdots$$

$$f^{(n)}(x) = 0 \quad (n \geq 5) \quad \blacktriangleleft$$

Successive applications of the derivative operation are also denoted as follows:

$$f'(x) = \frac{d}{dx} [f(x)]$$

$$f''(x) = \frac{d}{dx} \left[\frac{d}{dx} [f(x)] \right] = \frac{d^2}{dx^2} [f(x)]$$

$$f'''(x) = \frac{d}{dx} \left[\frac{d}{dx} \left[\frac{d}{dx} [f(x)] \right] \right] = \frac{d^3}{dx^3} [f(x)]$$

$$\vdots \qquad\qquad\qquad \vdots$$

In general, we write

$$f^{(n)}(x) = \frac{d^n}{dx^n} [f(x)]$$

which is read, *the nth derivative of f with respect to x.*

When a dependent variable is involved, say,

$$y = f(x)$$

then successive derivatives can be denoted by writing

$$y'(x), \quad y''(x), \quad y'''(x), \quad y^{(4)}(x), \ldots, y^{(n)}(x), \ldots$$

or

$$\frac{dy}{dx}, \quad \frac{d^2y}{dx^2}, \quad \frac{d^3y}{dx^3}, \quad \frac{d^4y}{dx^4}, \ldots, \frac{d^ny}{dx^n}, \ldots$$

The following symbols denote values of derivatives at a particular point x_0; their meanings should be self-evident.

$$y''(x_0), \quad f^{(4)}(x_0), \quad \left. \frac{d^3y}{dx^3} \right|_{x=x_0}, \quad \left. \frac{d^2}{dx^2} [x^7 - x] \right|_{x=x_0}$$

Example 11

$$\frac{d^2}{dx^2} [x^5] = \frac{d}{dx} \left[\frac{d}{dx} (x^5) \right] = \frac{d}{dx} [5x^4] = 20x^3$$

Thus,

$$\frac{d^2}{dx^2}[x^5]\bigg|_{x=2} = 20 \cdot 8 = 160 \qquad \blacktriangleleft$$

▶ Exercise Set 3.3

In Exercises 1–28, use the results of this section to find dy/dx.

1. $y = 4x^7$.

2. $y = -3x^{12}$.

3. $y = 3x^8 + 2x + 1$.

4. $y = \frac{1}{2}(x^4 + 7)$.

5. $y = \pi^3$.

6. $y = \sqrt{2}x + \frac{1}{\sqrt{2}}$.

7. $y = -\frac{1}{3}(x^7 + 2x - 9)$.

8. $y = \frac{x^2 + 1}{5}$.

9. $y = ax^3 + bx^2 + cx + d$
   (a, b, c, d constant).

10. $y = \frac{1}{a}\left(x^2 + \frac{1}{b}x + c\right)$
    (a, b, c constant).

11. $y = -3x^{-8} + 2\sqrt{x}$.

12. $y = 7x^{-6} - 5\sqrt{x}$.

13. $y = x^{-3} + \frac{1}{x^7}$.

14. $y = \sqrt{x} + \frac{1}{x}$.

15. $y = (3x^2 + 6)\left(2x - \frac{1}{4}\right)$.

16. $y = (2 - x - 3x^3)(7 + x^5)$.

17. $y = (x^3 + 7x^2 - 8)(2x^{-3} + x^{-4})$.

18. $y = \left(\frac{1}{x} + \frac{1}{x^2}\right)(3x^3 + 27)$.

19. $y = (3x^2 + 1)^2$.

20. $y = (x^5 + 2x)^2$.

21. $y = \frac{1}{5x - 3}$.

22. $y = \frac{3}{\sqrt{x} + 2}$.

23. $y = \frac{3x}{2x + 1}$.

24. $y = \frac{x^2 + 1}{3x}$.

25. $y = \frac{2x - 1}{x + 3}$.

26. $y = \frac{4x + 1}{x^2 - 5}$.

27. $y = \left(\frac{3x + 2}{x}\right)(x^{-5} + 1)$.

28. $y = (2x^7 - x^2)\left(\frac{x - 1}{x + 1}\right)$.

29. If $f(4) = 3$ and $f'(4) = -5$, find $g'(4)$.
    (a) $g(x) = \sqrt{x} f(x)$
    (b) $g(x) = \frac{f(x)}{x}$.

30. If $f(3) = -2$ and $f'(3) = 4$, find $g'(3)$.
    (a) $g(x) = 3x^2 - 5f(x)$
    (b) $g(x) = \frac{2x + 1}{f(x)}$.

In Exercises 31–36, the functions involve independent variables other than x. Use the results in this section to find the indicated derivative.

31. Find $\frac{d}{dt}[16t^2]$.

32. $c = 2\pi r$; find $\frac{dc}{dr}$.

33. $V(r) = \pi r^3$; find $V'(r)$.

34. Find $\frac{d}{d\alpha}[2\alpha^{-1} + \alpha]$.

35. $s = \frac{t}{t^3 + 7}$; find $\frac{ds}{dt}$.

36. Find $\frac{d}{d\lambda}\left[\frac{\lambda\lambda_0 + \lambda^6}{2 - \lambda_0}\right]$ (λ_0 is constant).

37. Newton's law of gravitation states that the magnitude F of the force exerted by a point with mass M on a point with mass m is

$$F = \frac{GmM}{r^2}$$

where G is a constant and r is the distance between the bodies. Assuming that the points are moving, find a formula for the instantaneous rate of change of F with r.

38. The volume of a sphere is $V = \frac{4}{3}\pi r^3$. Assuming that the radius is changing, find a formula for the instantaneous rate of change of V with r.

39. Find d^2y/dx^2.
  (a) $y = 7x^3 - 5x^2 + x$
  (b) $y = 12x^2 - 2x + 3$
  (c) $y = \dfrac{x+1}{x}$
  (d) $y = (5x^2 - 3)(7x^3 + x)$.

40. Find y''.
  (a) $y = 4x^7 - 5x^3 + 2x$
  (b) $y = 3x + 2$
  (c) $y = \dfrac{3x-2}{5x}$
  (d) $y = (x^3 - 5)(2x + 3)$.

41. Find y'''.
  (a) $y = x^{-5} + x^5$
  (b) $y = \dfrac{1}{x}$
  (c) $y = ax^3 + bx + c$
      (a, b, c constant).

42. Find $\dfrac{d^3y}{dx^3}$.
  (a) $y = 5x^2 - 4x + 7$
  (b) $y = 3x^{-2} + 4x^{-1} + x$
  (c) $y = ax^4 + bx^2 + c$
      (a, b, c constant).

43. Find
  (a) $f'''(2)$, where $f(x) = 3x^2 - 2$
  (b) $\dfrac{d^2y}{dx^2}\bigg|_{x=1}$, where $y = 6x^5 - 4x^2$
  (c) $\dfrac{d^4}{dx^4}[x^{-3}]\bigg|_{x=1}$

44. Find
  (a) $y'''(0)$, where $y = 4x^4 + 2x^3 + 3$
  (b) $\dfrac{d^4y}{dx^4}\bigg|_{x=1}$, where $y = \dfrac{6}{x^4}$.

45. Show that $y = x^3 + 3x + 1$ satisfies the equation $y''' + xy'' - 2y' = 0$.

46. Show that if $x \neq 0$, then $y = 1/x$ satisfies the equation $x^3y'' + x^2y' - xy = 0$.

47. At which point(s) does the graph of the equation $y = \frac{1}{3}x^3 - \frac{3}{2}x^2 + 2x$ have a horizontal tangent line?

48. At which point(s) does the graph of $y = \dfrac{x}{x^2 + 9}$ have a horizontal tangent line?

49. Find an equation of the tangent line to the graph of $y = f(x)$ at the point where $x = -3$ if $f(-3) = 2$ and $f'(-3) = 5$.

50. Find an equation for the line that is tangent to $y = (1 - x)/(1 + x)$ at the point where $x = 2$.

51. Find the values of a and b if the tangent to $y = ax^2 + bx$ at $(1, 5)$ has slope $m_{\tan} = 8$.

52. Find k if the curve $y = x^2 + k$ is tangent to the line $y = 2x$.

53. Find the x-coordinate of the point on the graph of $y = x^2$ where the tangent line is parallel to the secant line that cuts the curve at $x = -1$ and $x = 2$.

54. Find the x-coordinate of the point on the graph of $y = \sqrt{x}$ where the tangent line is parallel to the secant line that cuts the curve at $x = 1$ and $x = 4$.

55. Find the x-coordinate of all points on the graph of $y = 1 - x^2$ at which the tangent line passes through the point $(2, 0)$.

56. Show that any two tangent lines to the parabola $y = ax^2$, $a \neq 0$, intersect at a point that is on the vertical line halfway between the points of tangency.

57. Suppose that L is the tangent line at $x = x_0$ to the graph of the cubic equation $y = ax^3 + bx$. Find the x-coordinate of the point where L intersects the graph a second time.

58. Show that the segment of the tangent line to the graph of $y = 1/x$ that is cut off by the coordinate axes is bisected by the point of tangency.

59. Show that the triangle that is formed by any tangent line to the graph of $y = 1/x$, $x > 0$, and the coordinate axes has an area of 2 square units.

60. Find conditions on $a, b, c,$ and d so that the graph of the polynomial $f(x) = ax^3 + bx^2 + cx + d$ has
  (a) exactly two horizontal tangents

(b)   exactly one horizontal tangent

(c)   no horizontal tangents.

61.   Use Theorems 3.3.3 and 3.3.4 to prove: If the functions f and g are differentiable at x, then $f - g$ is differentiable at x and $(f - g)' = f' - g'$.

62.   (a)   Let the functions f, g, and h be differentiable at x. By applying Theorem 3.3.5 twice, show that the product $f \cdot g \cdot h$ is differentiable at x and

$$(f \cdot g \cdot h)'(x) =$$
$$f(x)g(x)h'(x) + f(x)g'(x)h(x) + f'(x)g(x)h(x)$$

(b)   State a formula for differentiating a product of n functions.

63.   Use the results of Exercise 62 to find:

(a) $\dfrac{d}{dx}\left[(2x + 1)\left(1 + \dfrac{1}{x}\right)(x^{-3} + 7)\right]$

(b) $\dfrac{d}{dx}[x^{-5}(x^2 + 2x)(4 - 3x)(2x^9 + 1)]$

(c) $\dfrac{d}{dx}[(x^7 + 2x - 3)^3]$

(d) $\dfrac{d}{dx}[(x^2 + 1)^{50}]$.

64.   Prove: If the function f is differentiable at x, then $\dfrac{d}{dx}[f^2(x)] = 2f(x)f'(x)$. [*Hint:* Use the product rule.]

65.   Use the result obtained in Exercise 64 to find $\dfrac{d}{dx}(2x^3 - 5x^2 + 7x - 2)^2$.

66.   Let $f(x) = \sqrt{x}$. Assuming that f is differentiable at x, use the result of Exercise 64 to show that $f'(x) = 1/(2\sqrt{x})$.

In Exercises 67–69, you will have to determine whether a function f is differentiable at a point x_0 where the formula for f changes. Use the following result:

Theorem. *Let f be continuous at x_0 and suppose that*

$$\lim_{x \to x_0^+} f'(x) \quad \text{and} \quad \lim_{x \to x_0^-} f'(x)$$

exist. Then f is differentiable at x_0 if and only if these limits are equal. Moreover, in the case of equality

$$f'(x_0) = \lim_{x \to x_0^+} f'(x) = \lim_{x \to x_0^-} f'(x)$$

67.   Let

$$f(x) = \begin{cases} x^2, & x \le 1 \\ \sqrt{x}, & x > 1 \end{cases}$$

Determine whether f is differentiable at $x = 1$. If so, find the value of the derivative there.

68.   Let

$$f(x) = \begin{cases} x^3 + \dfrac{1}{16}, & x < \dfrac{1}{2} \\ \dfrac{3}{4}x^2, & x \ge \dfrac{1}{2} \end{cases}$$

Determine whether f is differentiable at $x = \frac{1}{2}$. If so, find the value of the derivative there.

69.   Find all points where f fails to be differentiable. Justify your answer.

(a) $f(x) = |3x - 2|$

(b) $f(x) = |x^2 - 4|$.

70.   Prove: If f is differentiable at x and $f(x) \ne 0$, then $1/f(x)$ is differentiable at x and

$$\frac{d}{dx}\left[\frac{1}{f(x)}\right] = -\frac{f'(x)}{[f(x)]^2}$$

71.   (a)   Find $f^{(n)}(x)$ if $f(x) = x^n$.

(b)   Find $f^{(n)}(x)$ if $f(x) = x^k$ and $n > k$, where k is a positive integer.

(c)   Find $f^{(n)}(x)$ if $f(x) = a_0 + a_1x + a_2x^2 + \cdots + a_nx^n$.

72.   In each part compute f', f'', f''' and then state the formula for $f^{(n)}$.

(a) $f(x) = \dfrac{1}{x}$

(b) $f(x) = \dfrac{1}{x^2}$.

[*Hint:* The value of $(-1)^n$ is 1 if n is even and -1 if n is odd. Use this expression in your answer.]

73.   (a)   Prove:

$$\frac{d^2}{dx^2}[cf(x)] = c\,\frac{d^2}{dx^2}[f(x)]$$

$$\frac{d^2}{dx^2}[f(x) + g(x)] = \frac{d^2}{dx^2}[f(x)] + \frac{d^2}{dx^2}[g(x)]$$

(b) Do the results in part (a) generalize to nth derivatives? Justify your answer.

74. Prove:

$$(f \cdot g)''(x) = f''(x)g(x) + 2f'(x)g'(x) + f(x)g''(x)$$

75. (a) In our proof of Theorem 3.3.5, we used the fact that

$$\lim_{h \to 0} f(x + h) = f(x)$$

[see Equation (3)]. The argument given used the hypothesis that f is differentiable at x. Find the fallacy in the following proof that makes no assumptions about f. As h approaches 0, $x + h$ approaches x; consequently, $f(x + h)$ approaches $f(x)$; that is,

$$\lim_{h \to 0} f(x + h) = f(x)$$

(b) Let

$$f(x) = \begin{cases} x, & x \neq 1 \\ 3, & x = 1 \end{cases}$$

Show that

$$\lim_{h \to 0} f(x + h) \neq f(x)$$

when $x = 1$.

76. Let $f(x) = x^8 - 2x + 3$ and $x_0 = 2$; find

$$\lim_{h \to 0} \frac{f'(x_0 + h) - f'(x_0)}{h}$$

77. (a) Prove: If $f''(x)$ exists for each x in (a, b), then both f and f' are continuous on (a, b).

(b) What can be said about the continuity of f and its derivatives if $f^{(n)}(x)$ exists for each x in (a, b)?

3.4 DERIVATIVES OF TRIGONOMETRIC FUNCTIONS

The main objective of this section is to obtain formulas for the derivatives of trigonometric functions.

Recall that in the expressions $\sin x$, $\cos x$, $\tan x$, $\cot x$, $\sec x$, and $\csc x$ it is understood that x is measured in radians. Also, we remind the reader of the limits

$$\lim_{h \to 0} \frac{\sin h}{h} = 1 \quad \text{and} \quad \lim_{h \to 0} \frac{1 - \cos h}{h} = 0$$

derived in Section 2.8.

Let us first consider the problem of differentiating $\sin x$. From the definition of a derivative,

$$\frac{d}{dx}[\sin x] = \lim_{h \to 0} \frac{\sin(x + h) - \sin x}{h}$$

$$= \lim_{h \to 0} \frac{\sin x \cos h + \cos x \sin h - \sin x}{h}$$

$$= \lim_{h \to 0} \left[\sin x \left(\frac{\cos h - 1}{h} \right) + \cos x \left(\frac{\sin h}{h} \right) \right]$$

$$= \lim_{h \to 0} \left[\cos x \left(\frac{\sin h}{h} \right) - \sin x \left(\frac{1 - \cos h}{h} \right) \right]$$

Since $\sin x$ and $\cos x$ do not involve h, they remain constant as $h \to 0$; thus,

$$\lim_{h \to 0} (\sin x) = \sin x \quad \text{and} \quad \lim_{h \to 0} (\cos x) = \cos x$$

Consequently,

$$\frac{d}{dx} [\sin x] = \cos x \cdot \lim_{h \to 0} \left(\frac{\sin h}{h} \right) - \sin x \cdot \lim_{h \to 0} \left(\frac{1 - \cos h}{h} \right)$$

$$= \cos x \cdot (1) - \sin x \cdot (0) = \cos x$$

Thus, we have shown that

$$\frac{d}{dx} [\sin x] = \cos x \tag{1}$$

The derivative of $\cos x$ is obtained similarly:

$$\frac{d}{dx} [\cos x] = \lim_{h \to 0} \frac{\cos (x + h) - \cos x}{h}$$

$$= \lim_{h \to 0} \frac{\cos x \cos h - \sin x \sin h - \cos x}{h}$$

$$= \lim_{h \to 0} \left[\cos x \cdot \left(\frac{\cos h - 1}{h} \right) - \sin x \cdot \left(\frac{\sin h}{h} \right) \right]$$

$$= -\cos x \cdot \lim_{h \to 0} \left(\frac{1 - \cos h}{h} \right) - \sin x \cdot \lim_{h \to 0} \left(\frac{\sin h}{h} \right)$$

$$= (-\cos x)(0) - (\sin x)(1) = -\sin x$$

Thus, we have shown that

$$\frac{d}{dx} [\cos x] = -\sin x \tag{2}$$

The derivatives of the remaining trigonometric functions can be obtained using the relationships

$$\tan x = \frac{\sin x}{\cos x} \qquad \cot x = \frac{\cos x}{\sin x} \qquad \sec x = \frac{1}{\cos x} \qquad \csc x = \frac{1}{\sin x}$$

For example,

$$\frac{d}{dx} [\tan x] = \frac{d}{dx} \left[\frac{\sin x}{\cos x} \right]$$

$$= \frac{\cos x \cdot \dfrac{d}{dx}[\sin x] - \sin x \cdot \dfrac{d}{dx}[\cos x]}{\cos^2 x}$$

$$= \frac{\cos x \cdot \cos x - \sin x \cdot (-\sin x)}{\cos^2 x}$$

$$= \frac{\cos^2 x + \sin^2 x}{\cos^2 x} = \frac{1}{\cos^2 x} = \sec^2 x$$

Thus,

$$\frac{d}{dx}[\tan x] = \sec^2 x \tag{3}$$

We leave the remaining formulas for the exercises:

$$\frac{d}{dx}[\cot x] = -\csc^2 x \tag{4}$$

$$\frac{d}{dx}[\sec x] = \sec x \tan x \tag{5}$$

$$\frac{d}{dx}[\csc x] = -\csc x \cot x \tag{6}$$

REMARK.   The derivative formulas for the trigonometric functions should be memorized. An easy way of doing this is discussed in Exercise 34.

Example 1   Find $f'(x)$ if $f(x) = x^2 \tan x$.

Solution.   Using the product rule and formula (3), we obtain

$$f'(x) = x^2 \cdot \frac{d}{dx}[\tan x] + \tan x \cdot \frac{d}{dx}[x^2] = x^2 \sec^2 x + 2x \tan x \qquad \blacktriangleleft$$

Example 2   Find dy/dx if

$$y = \frac{\sin x}{1 + \cos x}$$

Solution.   Using the quotient rule together with formulas (1) and (2) we obtain

$$\frac{dy}{dx} = \frac{(1 + \cos x) \cdot \dfrac{d}{dx}[\sin x] - \sin x \cdot \dfrac{d}{dx}[1 + \cos x]}{(1 + \cos x)^2}$$

$$= \frac{(1 + \cos x)(\cos x) - (\sin x)(-\sin x)}{(1 + \cos x)^2}$$

$$= \frac{\cos x + \cos^2 x + \sin^2 x}{(1 + \cos x)^2} = \frac{\cos x + 1}{(1 + \cos x)^2} = \frac{1}{1 + \cos x} \quad \blacktriangleleft$$

Example 3 Find $y''(\pi/4)$ if $y(x) = \sec x$.

Solution.

$$y'(x) = \sec x \tan x$$

$$y''(x) = \sec x \cdot \frac{d}{dx}[\tan x] + \tan x \cdot \frac{d}{dx}[\sec x]$$

$$= \sec x \cdot \sec^2 x + \tan x \cdot \sec x \tan x$$

$$= \sec^3 x + \sec x \tan^2 x$$

Thus,

$$y''(\pi/4) = \sec^3(\pi/4) + \sec(\pi/4)\tan^2(\pi/4)$$
$$= (\sqrt{2})^3 + (\sqrt{2})(1)^2 = 3\sqrt{2} \quad \blacktriangleleft$$

▶ **Exercise Set 3.4**

In Exercises 1–18, find $f'(x)$.

1. $f(x) = 2 \cos x - 3 \sin x$.

2. $f(x) = \sin x \cos x$.

3. $f(x) = \dfrac{\sin x}{x}$.

4. $f(x) = x^2 \cos x$.

5. $f(x) = x^3 \sin x - 5 \cos x$.

6. $f(x) = \dfrac{\cos x}{x \sin x}$.

7. $f(x) = \sec x - \sqrt{2} \tan x$.

8. $f(x) = (x^2 + 1) \sec x$.

9. $f(x) = \sec x \tan x$.

10. $f(x) = \dfrac{\sec x}{1 + \tan x}$.

11. $f(x) = x - 4 \csc x + 2 \cot x$.

12. $f(x) = \csc x \cot x$.

13. $f(x) = \dfrac{\cot x}{1 + \csc x}$.

14. $f(x) = \dfrac{\csc x}{\tan x}$.

15. $f(x) = \sin^2 x + \cos^2 x$.

16. $f(x) = \dfrac{1}{\cot x}$.

17. $f(x) = \dfrac{\sin x \sec x}{1 + x \tan x}$.

18. $f(x) = \dfrac{(x^2 + 1)\cot x}{3 - \cos x \csc x}$.

In Exercises 19–23, find d^2y/dx^2.

19. $y = x \cos x$.

20. $y = \csc x$.

21. $y = x \sin x - 3 \cos x$.

22. $y = x^2 \cos x + 4 \sin x$.

23. $y = \sin x \cos x$.

In Exercises 24 and 25, find all points where the graph of f has a horizontal tangent line.

24. (a) $f(x) = \sin x$ (b) $f(x) = \tan x$
    (c) $f(x) = \sec x$.

25. (a) $f(x) = \cos x$ (b) $f(x) = \cot x$
    (c) $f(x) = \csc x$.

26. Find the equation of the line tangent to the graph of $\sin x$ at the point where:
(a) $x = 0$ (b) $x = \pi$
(c) $x = \dfrac{\pi}{4}$.

27. Find the equation of the line tangent to the graph of $\tan x$ at the point where:
(a) $x = 0$ (b) $x = \dfrac{\pi}{4}$
(c) $x = -\dfrac{\pi}{4}$.

28. (a) Show that $y = \cos x$ and $y = \sin x$ are solutions of the equation $y'' + y = 0$.
(b) Show that $y = A \sin x + B \cos x$ is a solution for all constants A and B.

29. In each part, determine where f is differentiable.
(a) $f(x) = \sin x$ (b) $f(x) = \cos x$
(c) $f(x) = \tan x$ (d) $f(x) = \cot x$
(e) $f(x) = \sec x$ (f) $f(x) = \csc x$
(g) $f(x) = \dfrac{1}{1 + \cos x}$
(h) $f(x) = \dfrac{1}{\sin x \cos x}$
(i) $f(x) = \dfrac{\cos x}{2 - \sin x}$.

30. Derive the formulas
(a) $\dfrac{d}{dx}[\cot x] = -\csc^2 x$

(b) $\dfrac{d}{dx}[\sec x] = \sec x \tan x$

(c) $\dfrac{d}{dx}[\csc x] = -\csc x \cot x$.

31. Let $f(x) = \cos x$. Find all positive integers n for which $f^{(n)}(x) = \sin x$.

32. (a) Show that $\lim\limits_{h \to 0} \dfrac{\tan h}{h} = 1$.
(b) Use the result in (a) to help derive the formula $(d/dx)[\tan x] = \sec^2 x$ directly from the definition of a derivative.

33. Without using any trigonometric identities, find
$$\lim_{x \to 0} \frac{\tan(x + y) - \tan y}{x}$$

34. Let us agree to call the functions $\cos x$, $\cot x$, and $\csc x$ the **cofunctions** of $\sin x$, $\tan x$, and $\sec x$, respectively. Convince yourself that the derivative of any cofunction can be obtained from the derivative of the corresponding function by introducing a minus sign and replacing each function in the derivative by its cofunction. Memorize the derivatives of $\sin x$, $\tan x$, and $\sec x$ and then use the above observation to deduce the derivatives of the cofunctions.

35. The derivative formulas for $\sin x$, $\cos x$, $\tan x$, $\cot x$, $\sec x$, and $\csc x$ were obtained under the assumption that x is measured in radians. This exercise shows that different (more complicated) formulas result if x is measured in degrees. Prove that if h and x are degree measures, then
(a) $\lim\limits_{h \to 0} \dfrac{\cos h - 1}{h} = 0$
(b) $\lim\limits_{h \to 0} \dfrac{\sin h}{h} = \dfrac{\pi}{180}$
(c) $\dfrac{d}{dx}[\sin x] = \dfrac{\pi}{180} \cos x$.

3.5 THE CHAIN RULE

In this section we shall derive a formula that expresses the derivative of a composition $f \circ g$ in terms of the derivatives of f and g. This formula will enable us to differentiate complicated functions using known derivatives of simpler functions.

3.5.1 PROBLEM. *If we know the derivatives of f and g, how can we use this information to find the derivative of the composition $f \circ g$?*

The key to solving this problem is to introduce dependent variables

$$y = (f \circ g)(x) = f(g(x)) \quad \text{and} \quad u = g(x)$$

so that $y = f(u)$. We are interested in using the known derivatives

$$\frac{dy}{du} = f'(u) \quad \text{and} \quad \frac{du}{dx} = g'(x)$$

to find the unknown derivative

$$\frac{dy}{dx} = \frac{d}{dx}[f(g(x))]$$

Rates of change multiply:
$$\frac{dy}{dx} = \frac{dy}{du} \cdot \frac{du}{dx}$$

Figure 3.5.1

Stated another way, we are interested in using the known rates of change dy/du and du/dx to find the unknown rate of change dy/dx. But intuition suggests that rates of change multiply. For example, if y changes at 4 times the rate of u and u changes at 2 times the rate of x, then y changes at $4 \times 2 = 8$ times the rate of x. Thus, Figure 3.5.1 suggests that

$$\frac{dy}{dx} = \frac{dy}{du} \cdot \frac{du}{dx}$$

These ideas are formalized in the following theorem.

The Chain Rule

3.5.2 THEOREM. *If g is differentiable at the point x and f is differentiable at the point g(x), then the composition $f \circ g$ is differentiable at the point x. Moreover, if*

$$y = f(g(x)) \quad \text{and} \quad u = g(x)$$

then

$$\frac{dy}{dx} = \frac{dy}{du} \cdot \frac{du}{dx} \tag{1}$$

The proof of this result, which is optional, is given in part III of Appendix 2.

Example 1 Find dy/dx if

$$y = 4 \cos (x^3)$$

Solution. Let $u = x^3$ so that

$$y = 4 \cos u$$

By the chain rule,

$$\frac{dy}{dx} = \frac{dy}{du} \cdot \frac{du}{dx} = \frac{d}{du} [4 \cos u] \cdot \frac{d}{dx} [x^3]$$

$$= (-4 \sin u) \cdot (3x^2) = -12x^2 \sin (x^3) \quad \blacktriangleleft$$

REMARK. Formula (1) is easy to remember because the left side is exactly what results if we "cancel" the du's on the right side. This "canceling" device provides a good way to remember the chain rule when variables other than x, y, and u are used.

Example 2 Find dw/dt if

$$w = \tan x \quad \text{and} \quad x = 4t^3 + t$$

Solution. In this case, the chain rule takes the form

$$\frac{dw}{dt} = \frac{dw}{dx} \cdot \frac{dx}{dt} = \frac{d}{dx} [\tan x] \cdot \frac{d}{dt} [4t^3 + t]$$

$$= (\sec^2 x)(12t^2 + 1) = (12t^2 + 1) \sec^2 (4t^3 + t) \quad \blacktriangleleft$$

There are two alternative versions of the chain rule that are useful to know. These can be obtained by writing the derivatives in (1) as

$$\frac{dy}{dx} = \frac{d}{dx} [f(g(x))], \quad \frac{dy}{du} = f'(u) = f'[g(x)], \quad \frac{du}{dx} = g'(x)$$

Substituting these expressions in (1) yields

$$\frac{d}{dx} [f(g(x))] = f'[g(x)]g'(x) \tag{2}$$

and substituting $u = g(x)$ in this formula yields

$$\frac{d}{dx}[f(u)] = f'(u)\frac{du}{dx} \qquad\qquad (3)$$

Formula (2) is important because it involves no dependent variables, and formula (3) is important because it is well-suited for many computations that use the chain rule.

Example 3   Find $\dfrac{d}{dx}[(x^3 + 7x + 1)^{35}]$ using

(a)   formula (2)
(b)   formula (3).

Solution (a).   Let

$$f(x) = x^{35} \quad \text{and} \quad g(x) = x^3 + 7x + 1 \qquad \text{[Thus, } f(g(x)) = (x^3 + 7x + 1)^{35}.]$$

Differentiating these functions yields

$$f'(x) = 35x^{34} \quad \text{and} \quad g'(x) = 3x^2 + 7$$

So from (2)

$$\begin{aligned}
\frac{d}{dx}[(x^3 + 7x + 1)^{35}] &= \frac{d}{dx}[f(g(x))] = f'[g(x)]g'(x) \\
&= 35(g(x))^{34}\, g'(x) \\
&= 35(x^3 + 7x + 1)^{34}\,(3x^2 + 7)
\end{aligned}$$

Solution (b).   If we let

$$u = x^3 + 7x + 1 \quad \text{and} \quad f(u) = u^{35}$$

then

$$f'(u) = 35u^{34} \quad \text{and} \quad \frac{du}{dx} = 3x^2 + 7$$

so from (3)

$$\begin{aligned}
\frac{d}{dx}[(x^3 + 7x + 1)^{35}] &= \frac{d}{dx}[f(u)] = f'(u)\frac{du}{dx} \\
&= 35u^{34}\,(3x^2 + 7) \\
&= 35(x^3 + 7x + 1)^{34}\,(3x^2 + 7) \qquad \blacktriangleleft
\end{aligned}$$

REMARK.   If you take a moment to compare the computations in the two parts of the foregoing example, you will see that there is little difference between the two methods. Some people prefer the first method because no extra variables need to be introduced, and some people the second because they find it easier to work with. The choice is purely a matter of taste.

The procedures illustrated in Example 3 can be summarized as follows:

Applying Formula (2)

Step 1. Express the function to be differentiated in the form $f(g(x))$, where f and g are functions you know how to differentiate.

Step 2. Compute $f'(x)$, and then replace x by $g(x)$ to obtain $f'[g(x)]$.

Step 3. Compute $g'(x)$.

Step 4. Multiply $f'[g(x)]$ and $g'(x)$.

Applying Formula (3)

Step 1. Express the function to be differentiated in the form $f(g(x))$, where f and g are functions you know how to differentiate, and let $u = g(x)$.

Step 2. Compute $f'(u)$.

Step 3. Compute du/dx.

Step 4. Multiply $f'(u)$ and du/dx, and express the result in terms of x.

TECHNIQUES FOR USING THE CHAIN RULE

Eventually, the reader should be able to apply the chain rule without explicitly writing down $f(x)$ and $g(x)$. To reach this goal, it is helpful to express the formula

$$\frac{d}{dx}[f(g(x))] = f'[g(x)]g'(x)$$

in words. For this purpose call f the "outside function" and g the "inside function." Thus, *the derivative of $f(g(x))$ is the derivative of the outside function evaluated at the inside function times the derivative of the inside function.*

Note that in the expression $f(g(x))$, the inside function is evaluated first and the outside second. For example, to evaluate $\sin(x^3)$, we first compute x^3 and then sine, so $u = x^3$ is the inside function and $f(u) = \sin u$ is the outside function. Some more examples are given in the following table.

FUNCTION $f(g(x))$	INSIDE $u = g(x)$	OUTSIDE $f(u)$
$(x^2 + 1)^{10}$	$x^2 + 1$	u^{10}
$\tan^2 x$	$\tan x$	u^2
$\cos (x^2 + 9)$	$x^2 + 9$	$\cos u$
$\sqrt{4 - 3x}$	$4 - 3x$	$\sqrt{u}$

Example 4   Find

$$\frac{d}{dx} [\cos (x^2 + 9)]$$

Solution.   The inside function is $x^2 + 9$ and the outside function is $\cos x$, so

$$\frac{d}{dx} [\cos (x^2 + 9)] = \underbrace{-\sin (x^2 + 9)}_{\substack{\text{derivative of} \\ \text{the outside} \\ \text{evaluated at} \\ \text{the inside}}} \cdot \underbrace{2x}_{\substack{\text{derivative} \\ \text{of the} \\ \text{inside}}} \qquad \blacktriangleleft$$

Example 5   Find

$$\frac{d}{dx} [\tan^2 x]$$

Solution.   The inside function is $\tan x$ and the outside function is x^2, so

$$\frac{d}{dx} [\tan^2 x] = \underbrace{(2 \tan x)}_{\substack{\text{derivative of} \\ \text{the outside} \\ \text{evaluated at} \\ \text{the inside}}} \cdot \underbrace{(\sec^2 x)}_{\substack{\text{derivative} \\ \text{of the} \\ \text{inside}}} = 2 \tan x \sec^2 x \qquad \blacktriangleleft$$

Example 6   Find dy/dx if

$$y = \frac{1}{2x^4 - x^2 + 8}$$

Solution.   First rewrite y as

$$y = (2x^4 - x^2 + 8)^{-1}$$

so the inside function is $2x^4 - x^2 + 8$ and the outside function is x^{-1}.

Thus,

$$\frac{dy}{dx} = \frac{d}{dx}[(2x^4 - x^2 + 8)^{-1}] = -(2x^4 - x^2 + 8)^{-2} \cdot (8x^3 - 2x)$$

$$= \frac{2x - 8x^3}{(2x^4 - x^2 + 8)^2}$$

[Note that this problem can be solved without the chain rule by treating y as a reciprocal and applying Theorem 3.3.7.] ◄

In the following example the chain rule must be applied more than once.

Example 7   Find

$$\frac{d}{dx}[\sin^3 (9x + 1)]$$

Solution.

$$\frac{d}{dx}[\sin^3 (9x + 1)] = 3 \sin^2 (9x + 1) \cdot \frac{d}{dx}[\sin (9x + 1)]$$

$$= 3 \sin^2 (9x + 1) \cdot \cos (9x + 1) \cdot \frac{d}{dx}(9x + 1)$$

$$= 3 \sin^2 (9x + 1) \cdot \cos (9x + 1) \cdot 9$$

$$= 27 \sin^2 (9x + 1) \cdot \cos (9x + 1) \quad ◄$$

Example 8   Find dy/dt if $y = \sqrt{3 + \sec \pi t}$

Solution.   This problem requires several applications of the chain rule, but with t rather than x as the independent variable.

$$\frac{dy}{dt} = \frac{d}{dt}[\sqrt{3 + \sec \pi t}] = \frac{1}{2\sqrt{3 + \sec \pi t}} \cdot \frac{d}{dt}[3 + \sec \pi t]$$

$$= \frac{1}{2\sqrt{3 + \sec \pi t}} \cdot \sec \pi t \tan \pi t \cdot \frac{d}{dt}[\pi t]$$

$$= \frac{\pi \sec \pi t \tan \pi t}{2\sqrt{3 + \sec \pi t}} \quad ◄$$

GENERALIZED DERIVATIVE FORMULAS   We shall conclude this section by discussing another way of using the chain rule. For this purpose consider the three differentiation problems

$$\frac{d}{dx}[\sin 3x], \quad \frac{d}{dx}[\sin \sqrt{x}], \quad \frac{d}{dx}\left[\sin \left(\frac{x}{x + 1}\right)\right] \tag{4}$$

In each case we could use the chain rule to obtain the derivative directly. However, the functions all have the form $\sin u$, where u is a function of x. Thus, an alternative way of obtaining the derivatives is to use the chain rule to find a formula for

$$\frac{d}{dx} [\sin u]$$

then simply substitute the functions

$$u = 3x, \quad u = \sqrt{x}, \quad u = \frac{x}{x + 1}$$

in this formula to obtain the three derivatives. Let us carry out the details. If we let $f(u) = \sin u$, then $f'(u) = \cos u$, so (3) yields

$$\frac{d}{dx} [\sin u] = \cos u \, \frac{du}{dx} \tag{5}$$

Applying this formula to each of the functions in (4), we obtain

$$\frac{d}{dx} [\sin 3x] = \cos 3x \cdot \frac{d}{dx} [3x] = 3 \cos 3x$$

$$\frac{d}{dx} [\sin \sqrt{x}] = \cos \sqrt{x} \cdot \frac{d}{dx} [\sqrt{x}] = \frac{1}{2\sqrt{x}} \cos \sqrt{x}$$

$$\frac{d}{dx} \left[\sin \left(\frac{x}{x + 1} \right) \right] = \cos \left(\frac{x}{x + 1} \right) \cdot \frac{d}{dx} \left[\frac{x}{x + 1} \right] = \frac{1}{(x + 1)^2} \cos \left(\frac{x}{x + 1} \right)$$

The following table contains a list of *generalized derivative formulas* that are consequences of (3).

GENERALIZED DERIVATIVE FORMULAS

$$\frac{d}{dx} [u^n] = nu^{n-1} \frac{du}{dx} \quad (n \text{ an integer}) \qquad \frac{d}{dx} [\sqrt{u}] = \frac{1}{2\sqrt{u}} \frac{du}{dx}$$

$$\frac{d}{dx} [\sin u] = \cos u \, \frac{du}{dx} \qquad\qquad\qquad \frac{d}{dx} [\cos u] = -\sin u \, \frac{du}{dx}$$

$$\frac{d}{dx} [\tan u] = \sec^2 u \, \frac{du}{dx} \qquad\qquad\qquad \frac{d}{dx} [\cot u] = -\csc^2 u \, \frac{du}{dx}$$

$$\frac{d}{dx} [\sec u] = \sec u \, \tan u \, \frac{du}{dx} \qquad\qquad \frac{d}{dx} [\csc u] = -\csc u \, \cot u \, \frac{du}{dx}$$

We conclude with a complicated example to test your mastery of the chain rule.

Example 9

$$\frac{d}{dx}\left[(\cos^2(1-x) + \sqrt{x+3})^5\right]$$

$$= 5(\cos^2(1-x) + \sqrt{x+3})^4 \cdot \frac{d}{dx}\left[\cos^2(1-x) + \sqrt{x+3}\right]$$

$$= 5(\cos^2(1-x) + \sqrt{x+3})^4 \cdot \left[2\cos(1-x)\cdot\frac{d}{dx}\left[\cos(1-x)\right] + \frac{1}{2\sqrt{x+3}}\cdot\frac{d}{dx}(x+3)\right]$$

$$= 5(\cos^2(1-x) + \sqrt{x+3})^4 \cdot \left[-2\cos(1-x)\sin(1-x)\cdot\frac{d}{dx}(1-x) + \frac{1}{2\sqrt{x+3}}\right]$$

$$= 5(\cos^2(1-x) + \sqrt{x+3})^4 \cdot \left[2\cos(1-x)\sin(1-x) + \frac{1}{2\sqrt{x+3}}\right]$$

$$= 5(\cos^2(1-x) + \sqrt{x+3})^4 \left(\sin(2-2x) + \frac{1}{2\sqrt{x+3}}\right)$$

(The last step is just algebraic simplification using the trigonometric double angle formula $\sin 2\alpha = 2\sin\alpha\cos\alpha$.) ◀

▶ Exercise Set 3.5

In Exercises 1–39, find $f'(x)$.

1. $f(x) = (x^3 + 2x)^{37}$.

2. $f(x) = (3x^2 + 2x - 1)^6$.

3. $f(x) = \left(x^3 - \dfrac{7}{x}\right)^{-2}$.

4. $f(x) = \dfrac{1}{(x^5 - x + 1)^9}$.

5. $f(x) = \dfrac{4}{(3x^2 - 2x + 1)^3}$.

6. $f(x) = \sqrt{x^3 - 2x + 5}$.

7. $f(x) = \sqrt{4 + 3\sqrt{x}}$.

8. $f(x) = \sin^3 x$.

9. $f(x) = \sin(x^3)$.

10. $f(x) = \cos^2(3\sqrt{x})$.

11. $f(x) = \tan(4x^2)$.

12. $f(x) = 3\cot^4 x$.

13. $f(x) = 4\cos^5 x$.

14. $f(x) = \csc(x^3)$.

15. $f(x) = \sin\left(\dfrac{1}{x^2}\right)$.

16. $f(x) = \tan^4(x^3)$.

17. $f(x) = 2\sec^2(x^7)$.

18. $f(x) = \cos^3\left(\dfrac{x}{x+1}\right)$.

19. $f(x) = \sqrt{\cos(5x)}$.

20. $f(x) = \sqrt{3x - \sin^2(4x)}$.

21. $f(x) = [x + \csc(x^3 + 3)]^{-3}$.

22. $f(x) = [x^4 - \sec(4x^2 - 2)]^{-4}$.

23. $f(x) = x^2\sqrt{5 - x^2}$.

24. $f(x) = \dfrac{x}{\sqrt{1 - x^2}}$.

25. $f(x) = x^3\sin^2(5x)$.

26. $f(x) = \sqrt{x}\tan^3(\sqrt{x})$.

27. $f(x) = x^5\sec\left(\dfrac{1}{x}\right)$.

28. $f(x) = \dfrac{\sin x}{\sec(3x + 1)}$.

29. $f(x) = \cos(\cos x)$.

30. $f(x) = \sin(\tan 3x)$.

31. $f(x) = \cos^3(\sin 2x)$.

32. $f(x) = \dfrac{1 + \csc(x^2)}{1 - \cot(x^2)}$.

33. $f(x) = (5x + 8)^{13}(x^3 + 7x)^{12}$.

34. $f(x) = (2x - 5)^2(x^2 + 4)^3$.

35. $f(x) = \left(\dfrac{x - 5}{2x + 1}\right)^3$.

36. $f(x) = \left(\dfrac{1 + x^2}{1 - x^2}\right)^{17}$.

37. $f(x) = \dfrac{(4x^2 - 1)^{-8}}{(2x + 1)^{-3}}$.

38. $f(x) = [1 + \sin^3(x^5)]^{12}$.

39. $f(x) = [x \sin 2x + \tan^4(x^7)]^5$.

In Exercises 40–42, find d^2y/dx^2.

40. $y = \sin(3x^2)$.

41. $y = x \cos(5x) - \sin^2 x$.

42. $y = x \tan\left(\dfrac{1}{x}\right)$.

In Exercises 43–46, find an equation for the tangent to the graph at the specified point.

43. $y = x \cos 3x,\ x = \pi$.

44. $y = \sin(1 + x^3),\ x = -3$.

45. $y = \sec^3\left(\dfrac{\pi}{2} - x\right),\ x = -\dfrac{\pi}{2}$.

46. $y = \left(x - \dfrac{1}{x}\right)^3,\ x = 2$.

In Exercises 47–50, find dy/dx in two ways: first by using the chain rule and then by expressing y in terms of x and differentiating directly.

47. $x = 5t + 2,\ y = t^2$.

48. $x = \dfrac{1}{2}\pi - \theta,\ y = \sec \theta$.

49. $x = u^{-1},\ y = 3\sin^3 u$.

50. $x = \dfrac{\lambda}{1 - \lambda},\ y = (1 + \lambda)^{12}$.

In Exercises 51–54, find the indicated derivative.

51. $y = \cot^3(\pi - \theta)$; find $\dfrac{dy}{d\theta}$.

52. $\lambda = \left(\dfrac{au + b}{cu + d}\right)^6$; find $\dfrac{d\lambda}{du}$ $(a, b, c, d$ constants).

53. $\dfrac{d}{d\omega}[a \cos^2 \pi\omega + b \sin^2 \pi\omega]$ $(a, b$ constants).

54. $x = \csc^2\left(\dfrac{\pi}{3} - y\right)$; find $\dfrac{dx}{dy}$.

55. (a)   Show that

$$\frac{d}{dx}(|x|) = \begin{cases} 1, & x > 0 \\ -1, & x < 0 \end{cases}$$

(b)   Use the result in (a) and the chain rule to find

$$\frac{d}{dx}(|\sin x|) \quad \text{and} \quad \frac{d}{dx}(\sin|x|)$$

for nonzero x in the interval $(-\pi, \pi)$.

56. Use the identity

$$\cos x = \sin\left(\frac{\pi}{2} - x\right)$$

and the derivative formula for $\sin x$ to obtain the derivative formula for $\cos x$.

57. Let

$$f(x) = \begin{cases} x \sin\dfrac{1}{x}, & x \neq 0 \\ 0, & x = 0. \end{cases}$$

(a)   Find $f'(x)$ for $x \neq 0$.

(b)   Show that f is continuous at $x = 0$.

(c)   Use Definition 3.2.1 to show that $f'(0)$ does not exist.

58. Let

$$f(x) = \begin{cases} x^2 \sin\dfrac{1}{x}, & x \neq 0 \\ 0, & x = 0. \end{cases}$$

(a)   Find $f'(x)$ for $x \neq 0$.

(b)   Show that f is continuous at $x = 0$.

(c)   Use Definition 3.2.1 to find $f'(0)$.

(d)   Show that f' is not continuous at $x = 0$.

59. Given that $f'(0) = 2$, $g(0) = 0$, and $g'(0) = 3$, find $(f \circ g)'(0)$.

60. Given that $f'(x) = \sqrt{3x + 4}$ and $g(x) = x^2 - 1$, find $F'(x)$ if $F(x) = f(g(x))$.

61. Given that $f'(x) = \dfrac{x}{x^2 + 1}$ and $g(x) = \sqrt{3x - 1}$, find $F'(x)$ if $F(x) = f(g(x))$.

62. Find $f'(x^2)$ if $\dfrac{d}{dx}[f(x^2)] = x^2$.

63. Find $\dfrac{d}{dx}[f(x)]$ if $\dfrac{d}{dx}[f(3x)] = 6x$.

64. A function f is *even* if $f(-x) = f(x)$; *odd* if $f(-x) = -f(x)$, for all x in the domain of f. Assuming that f is differentiable, prove:
(a) f' is odd if f is even

(b) f' is even if f is odd.

65. Find a formula for

$$\frac{d}{dx}[f(g(h(x)))]$$

66. Let $y = f_1(u)$, $u = f_2(v)$, $v = f_3(w)$, and $w = f_4(x)$. Express dy/dx in terms of dy/du, dw/dx, du/dv, and dv/dw.

3.6   IMPLICIT DIFFERENTIATION

In the previous sections of this chapter we showed how to differentiate functions defined explicitly by equations of the form $y = f(x)$. In this section we shall learn to find derivatives of functions that are defined implicitly.

Consider the equation

$$xy = 1 \tag{1}$$

One way to obtain dy/dx is to rewrite this equation as

$$y = \frac{1}{x} \tag{2}$$

from which it follows that

$$\frac{dy}{dx} = \frac{d}{dx}\left[\frac{1}{x}\right] = -\frac{1}{x^2}$$

However, there is another possibility. We can differentiate both sides of (1) *before* solving for y in terms of x, treating y as a (temporarily unspecified) differentiable function of x. With this approach we obtain

$$\frac{d}{dx}[xy] = \frac{d}{dx}[1]$$

$$x\frac{d}{dx}[y] + y\frac{d}{dx}[x] = 0$$

$$x\frac{dy}{dx} + y = 0$$

$$\frac{dy}{dx} = -\frac{y}{x}$$

If we now substitute (2) into the last expression, we obtain

$$\frac{dy}{dx} = -\frac{1}{x^2}$$

which agrees with the previous result. This method of obtaining derivatives is called *implicit differentiation*. It is especially useful when it is inconvenient or impossible to solve explicitly for y in terms of x.

Example 1 By implicit differentiation find dy/dx if $5y^2 + \sin y = x^2$.

Solution. Differentiating both sides with respect to x and treating y as a differentiable function of x, we obtain

$$\frac{d}{dx}[5y^2 + \sin y] = \frac{d}{dx}[x^2]$$

$$5\frac{d}{dx}[y^2] + \frac{d}{dx}[\sin y] = 2x$$

$$5\left(2y\frac{dy}{dx}\right) + (\cos y)\frac{dy}{dx} = 2x \qquad \boxed{\begin{array}{l}\text{The chain rule was}\\\text{used here because } y\\\text{is a function of } x.\end{array}}$$

$$10y\frac{dy}{dx} + (\cos y)\frac{dy}{dx} = 2x$$

Solving for dy/dx, we obtain

$$\frac{dy}{dx} = \frac{2x}{10y + \cos y} \qquad (3)$$

Note that this formula for dy/dx involves the variables x and y. In order to obtain a formula involving x alone we would have to solve the original equation for y in terms of x and substitute in (3). However, it is impossible to do this, so the formula for dy/dx must be left in terms of x and y. ◄

In problems where dy/dx is used to calculate the slope of a tangent line at a point on a curve, the x and y coordinates of the point are often both known, so that there is no problem using a formula for dy/dx that involves both x and y.

Example 2 Find the slope of the tangent line at $(4, 0)$ to the graph of

$$7y^4 + x^3y + x = 4 \qquad (4)$$

[Note that $x = 4$, $y = 0$ satisfies the equation so that $(4, 0)$ is actually a point on the graph.]

Solution. It is difficult to solve (4) for y in terms of x, so we shall differentiate implicitly. We obtain

$$\frac{d}{dx}[7y^4 + x^3y + x] = \frac{d}{dx}[4]$$

$$\frac{d}{dx}[7y^4] + \frac{d}{dx}[x^3y] + \frac{d}{dx}[x] = 0$$

$$\frac{d}{dx}[7y^4] + \left(x^3\frac{dy}{dx} + y\frac{d}{dx}[x^3]\right) + \frac{d}{dx}[x] = 0$$

$$28y^3\frac{dy}{dx} + x^3\frac{dy}{dx} + 3yx^2 + 1 = 0$$

> Note the use of the chain rule in the first term.

Solving for dy/dx yields

$$\frac{dy}{dx} = -\frac{3yx^2 + 1}{28y^3 + x^3} \tag{5}$$

At the point $(4, 0)$ we have $x = 4$ and $y = 0$, so (5) yields

$$m_{\tan} = \frac{dy}{dx}\bigg|_{\substack{x=4\\y=0}} = -\frac{1}{64} \qquad \blacktriangleleft$$

Example 3 Find the slope of the tangent line at $(2, -1)$ to $y^2 - x + 1 = 0$ (Figure 3.6.1).

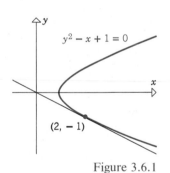

$y^2 - x + 1 = 0$

$(2, -1)$

Figure 3.6.1

Solution. Differentiating implicitly yields

$$\frac{d}{dx}[y^2 - x + 1] = \frac{d}{dx}[0]$$

$$\frac{d}{dx}[y^2] - \frac{d}{dx}[x] + \frac{d}{dx}[1] = \frac{d}{dx}[0]$$

$$2y\frac{dy}{dx} - 1 = 0$$

$$\frac{dy}{dx} = \frac{1}{2y}$$

At $(2, -1)$ we have $y = -1$, so the slope of the tangent line there is

$$m_{\tan} = \frac{dy}{dx}\bigg|_{y=-1} = -\frac{1}{2}$$

Alternative Solution. If we solve $y^2 - x + 1 = 0$ for y in terms of x, we obtain

$$y = \sqrt{x - 1} \quad \text{and} \quad y = -\sqrt{x - 1}$$

$y = -\sqrt{x-1}$

Figure 3.6.2

As shown in Figure 3.6.2, the graph of the first equation is the upper half of the curve $y^2 - x + 1 = 0$ (since $y \geq 0$), and the graph of the second is the lower half (since $y \leq 0$).

Since $(2, -1)$ lies on the lower half of the curve, the slope $m_{\tan}$ of the tangent line at this point can be obtained by evaluating the derivative of $y = -\sqrt{x - 1}$ when $x = 2$; thus,

$$\frac{dy}{dx} = \frac{d}{dx}[-\sqrt{x - 1}] = -\frac{1}{2\sqrt{x - 1}} \cdot \frac{d}{dx}[x - 1] = -\frac{1}{2\sqrt{x - 1}}$$

and

$$m_{\tan} = \frac{dy}{dx}\bigg|_{x=2} = -\frac{1}{2}$$

which agrees with the solution obtained by differentiating implicitly. ◀

Example 4   Use implicit differentiation to find d^2y/dx^2 if $4x^2 - 2y^2 = 9$.

Solution.   Differentiating both sides of $4x^2 - 2y^2 = 9$ implicitly yields

$$8x - 4y\frac{dy}{dx} = 0$$

from which we obtain

$$\frac{dy}{dx} = \frac{2x}{y} \tag{6}$$

Differentiating both sides of (6) implicitly yields

$$\frac{d^2y}{dx^2} = \frac{(y)(2) - (2x)(dy/dx)}{y^2} \tag{7}$$

Substituting (6) into (7) and simplifying, we obtain

$$\frac{d^2y}{dx^2} = \frac{2y - 2x(2x/y)}{y^2} = \frac{2y^2 - 4x^2}{y^3}$$

Finally, using the original equation to simplify further, we obtain

$$\frac{d^2y}{dx^2} = \frac{(-9)}{y^3} = -\frac{9}{y^3} \quad ◀$$

DERIVATIVES OF RATIONAL POWERS OF x

In Section 3.3 we showed that the formula

$$\frac{d}{dx}[x^n] = nx^{n-1} \tag{8}$$

holds for integer values of n and for $n = \frac{1}{2}$. By using implicit differentiation we shall show that this formula holds for any rational power of x. More precisely, we shall show that if r is a rational number, then

$$\frac{d}{dx}[x^r] = rx^{r-1} \tag{9}$$

whenever x^r and x^{r-1} are defined. In our computations we shall assume that x^r is differentiable; the justification for this assumption will be considered later.

Let $y = x^r$. Since r is a rational number, it can be expressed as a ratio of integers $r = m/n$. Thus, $y = x^r = x^{m/n}$ can be written as

$$y^n = x^m$$

so that

$$\frac{d}{dx}[y^n] = \frac{d}{dx}[x^m]$$

On differentiating implicitly with respect to x and using (8), we obtain

$$ny^{n-1}\frac{dy}{dx} = mx^{m-1} \tag{10}$$

But

$$y^{n-1} = [x^{m/n}]^{n-1} = x^{m-(m/n)}$$

Thus, (10) can be written as

$$nx^{m-(m/n)}\frac{dy}{dx} = mx^{m-1}$$

so that

$$\frac{dy}{dx} = \frac{m}{n}x^{(m/n)-1} = rx^{r-1}$$

which establishes (9).

Example 5   From (9)

$$\frac{d}{dx}[x^{4/5}] = \frac{4}{5}x^{(4/5)-1} = \frac{4}{5}x^{-1/5}$$

$$\frac{d}{dx}[x^{-7/8}] = -\frac{7}{8}x^{(-7/8)-1} = -\frac{7}{8}x^{-15/8}$$

$$\frac{d}{dx}[\sqrt[3]{x}] = \frac{d}{dx}[x^{1/3}] = \frac{1}{3}x^{-2/3} = \frac{1}{3\sqrt[3]{x^2}} \quad \blacktriangleleft$$

If u is a differentiable function of x, and r is a rational number, then the chain rule yields the following generalization of (9):

$$\frac{d}{dx}[u^r] = ru^{r-1} \cdot \frac{du}{dx} \tag{11}$$

Example 6

$$\frac{d}{dx}[x^2 - x + 2]^{3/4} = \frac{3}{4}(x^2 - x + 2)^{-1/4} \cdot \frac{d}{dx}[x^2 - x + 2]$$

$$= \frac{3}{4}(x^2 - x + 2)^{-1/4}(2x - 1)$$

$$\frac{d}{dx}[(\sec \pi x)^{-4/5}] = -\frac{4}{5}(\sec \pi x)^{-9/5} \cdot \frac{d}{dx}[\sec \pi x]$$

$$= -\frac{4}{5}(\sec \pi x)^{-9/5} \cdot \sec \pi x \tan \pi x \cdot \pi$$

$$= -\frac{4\pi}{5}(\sec \pi x)^{-4/5} \tan \pi x \quad \blacktriangleleft$$

DIFFERENTIABILITY OF IMPLICIT FUNCTIONS

We conclude this section with some observations about the mathematical assumptions that underlie the method of implicit differentiation.

When differentiating implicitly, it is assumed that y represents a differentiable function of x. If this is not so, the resulting calculations may be nonsense. For example, if we implicitly differentiate the equation

$$x^2 + y^2 + 1 = 0 \tag{12}$$

we obtain

$$2x + 2y\frac{dy}{dx} = 0$$

or

$$\frac{dy}{dx} = -\frac{x}{y} \tag{13}$$

However, the derivative in (13) is meaningless because no function satisfies (12). (The left side of the equation is always greater than zero.) Unfortunately, it can be difficult to determine whether an equation defines y as a function of x and if so, whether the function is differentiable. We leave such questions for a course in advanced calculus. However, the following example should clarify the basic idea.

Example 7   The equation of the unit circle $x^2 + y^2 = 1$ does not define y as a function of x in a neighborhood of the point $P(1, 0)$, because the portion of the circle within any rectangle centered at P is cut twice by some vertical line (Figure 3.6.3a). The equation does, however, define y as a differentiable function of x in a neighborhood of the point $(1/\sqrt{2}, 1/\sqrt{2})$ (Figure 3.6.3b).

The graph of the equation $x^{2/3} + y^{2/3} = 1$, called a *four-cusped hypo-cycloid* (Figure 3.6.3c), defines y as a function of x in a neighborhood of the point $(0, 1)$; however, the function is not differentiable at $x = 0$, because of the corner or "cusp" at $(0, 1)$.   ◀

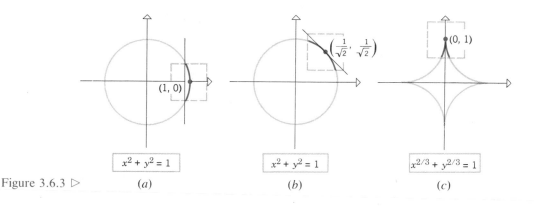

Figure 3.6.3 ▷   (a)   (b)   (c)

▶ Exercise Set 3.6

In Exercises 1–10, find dy/dx.

1. $y = \sqrt[3]{2x - 5}$.

2. $y = \sqrt[3]{2 + \tan(x^2)}$.

3. $y = \left(\dfrac{x - 1}{x + 2}\right)^{3/2}$.

4. $y = \sqrt{\dfrac{x^2 + 1}{x^2 - 5}}$.

5. $y = x^3(5x^2 + 1)^{-2/3}$. **6.** $y = \dfrac{(3 - 2x)^{4/3}}{x^2}$.

7. $y = [\sin(3/x)]^{5/2}$. **8.** $y = [\cos(x^3)]^{-1/2}$.

9. $y = \tan[(2x - 1)^{-1/3}]$.

10. $y = [\tan(2x - 1)]^{-1/3}$.

In Exercises 11–27, find dy/dx by implicit differentiation.

11. $x^2 + y^2 = 100$. **12.** $x^3 - y^3 = 6xy$.

13. $x^2y + 3xy^3 - x = 3$.

14. $x^3y^2 - 5x^2y + x = 1$.

15. $\dfrac{1}{y} + \dfrac{1}{x} = 1$. **16.** $x^2 = \dfrac{x + y}{x - y}$.

17. $\sqrt{x} + \sqrt{y} = 8$. **18.** $\sqrt{xy} + 1 = y$.

19. $(x^2 + 3y^2)^{35} = x$. **20.** $xy^{2/3} + yx^{2/3} = x^2$.

21. $3xy = (x^3 + y^2)^{3/2}$. **22.** $\cos xy = y$.

23. $\sin(x^2y^2) = x$. **24.** $x^2 = \dfrac{\cot y}{1 + \csc y}$.

25. $\tan^3(xy^2 + y) = x$. **26.** $\dfrac{xy^3}{1 + \sec y} = 1 + y^4$.

27. $\sqrt{1 + \sin^3(xy^2)} = y$.

In Exercises 28–32, use implicit differentiation to find

the slope of the tangent to the given curve at the specified point.

28. $x^2y - 5xy^2 + 6 = 0$; (3, 1).

29. $x^3y + y^3x = 10$; (1, 2).

30. $\sin xy = y$; $(\pi/2 , 1)$.

31. $x^{2/3} - y^{2/3} - y = 1$; $(1, -1)$.

32. $\sqrt{3 + \tan xy} - 2 = 0$; $(\pi/12, 3)$.

In Exercises 33–37, find the value of dy/dx at the specified point in two ways: first by solving for y in terms of x and then by implicit differentiation. (See Example 3.)

33. $xy = 8$; (2, 4).

34. $y^2 - x + 1 = 0$; (10, 3).

35. $x^2 + y^2 = 1$; $(1/\sqrt{2} , -1/\sqrt{2})$.

36. $\dfrac{1 - y}{1 + y} = x$; (0, 1).

37. $y^2 - 3xy + 2x^2 = 4$; (3, 2).

In Exercises 38–43, find d^2y/dx^2 by implicit differentiation.

38. $3x^2 - 4y^2 = 7$.

39. $x^3 + y^3 = 1$.

40. $x^3y^3 - 4 = 0$.

41. $2xy - y^2 = 3$.

42. $y + \sin y = x$.

43. $x \cos y = y$.

In Exercises 44–47, use implicit differentiation to find the specified derivative.

44. $\sqrt{u} + \sqrt{v} = 5$; du/dv.

45. $a^4 - t^4 = 6a^2t$; da/dt.

46. $y = \sin x$; dx/dy.

47. $a^2\omega^2 + b^2\lambda^2 = 1$ (a, b constants); $d\omega/d\lambda$.

48. Use implicit differentiation to show that the equation of the tangent to the curve $y^2 = kx$ at (x_0, y_0) is

$$y_0y = \frac{k}{2}(x + x_0)$$

49. Find dy/dx if

$$2y^3t + t^3y = 1 \quad \text{and} \quad \frac{dt}{dx} = \frac{1}{\cos t}$$

50. At what point(s) is the tangent to the curve

$$y^2 = 2x^3$$

perpendicular to the line $4x - 3y + 1 = 0$?

51. Find equations for two lines through the origin that are tangent to the curve

$$x^2 - 4x + y^2 + 3 = 0$$

52. (a) Use implicit differentiation to find the slope of the tangent to the four-cusped hypocycloid $x^{2/3} + y^{2/3} = 1$ at $P(-\frac{1}{4}\sqrt{2}, \frac{1}{4}\sqrt{2})$. (See Figure 3.6.3c.)

(b) Do part (a) by solving explicitly for y as a function of x and then differentiating.

(c) Trace the curve in Figure 3.6.3c and sketch the tangent line at P.

(d) Where does $x^{2/3} + y^{2/3} = 1$ *not* define y as an implicit function of x?

53. The graph of $8(x^2 + y^2)^2 = 100(x^2 - y^2)$, shown in Figure 3.6.4, is called a **lemniscate**.

(a) Where does this equation *not* define y as an implicit function of x?

(b) Use implicit differentiation to find the equation of the tangent to the curve at the point (3, 1).

(c) Trace the curve in Figure 3.6.4 and sketch the tangent at (3, 1).

Figure 3.6.4

54. (a) Show that $f(x) = x^{4/3}$ is differentiable at 0, but not twice differentiable at 0.

(b) Show that $f(x) = x^{7/3}$ is twice differentiable at 0, but not three times differentiable at 0.

(c) Find an exponent k such that $f(x) = x^k$ is $(n - 1)$-times differentiable at 0, but not n-times differentiable at 0.

3.7   Δ-NOTATION; DIFFERENTIALS

Often, the key to solving or understanding a problem is to introduce the right notation. In this section, we shall discuss some powerful notational tools.

INCREMENTS

If the value of a variable changes from one number to another, then the final value minus the initial value is called an ***increment*** in the variable. It is traditional in calculus to denote an increment in a variable x by the symbol Δx ("delta x"). In this notation, "Δx" is not a product of "Δ" and "x." Rather, Δx is a single symbol representing the *change* in the value of x. Similarly, Δy, Δt, and $\Delta \theta$ denote increments in the variables y, t, and θ.

If $y = f(x)$, and x changes from an initial value x_0 to a final value x_1, then there is corresponding change in the value of y from $y_0 = f(x_0)$ to $y_1 = f(x_1)$. Stated another way, the increment

$$\Delta x = x_1 - x_0 \tag{1}$$

produces a corresponding increment

$$\Delta y = y_1 - y_0 = f(x_1) - f(x_0) \tag{2}$$

in y (Figure 3.7.1). Increments can be either positive or negative depending on the relative positions of the initial and final points. In Figure 3.7.1, Δx is positive since the final point x_1 is to the right of the initial point x_0. If x_1 were to the left of x_0, then Δx would be negative. Relation (1) can be rewritten as $x_1 = x_0 + \Delta x$ which states that the final value of x is the initial value of x plus the increment in x. With this notation, (2) can be rewritten as

$$\Delta y = f(x_0 + \Delta x) - f(x_0) \tag{3}$$

Sometimes it is convenient to drop the zero subscript on x_0 and use x to denote the initial value as well as the name of the variable. When this

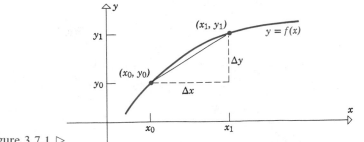

Figure 3.7.1 ▷

is done, $x + \Delta x$ represents the final value of the variable. Similarly, the initial and final values of y may be denoted by y and $y + \Delta y$ rather than y_0 and y_1 (Figure 3.7.2). Moreover, (3) becomes

$$\Delta y = f(x + \Delta x) - f(x) \qquad (4)$$

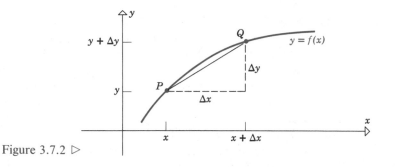

Figure 3.7.2 ▷

The Δ-notation provides a compact way of rewriting the derivative definition,

$$\frac{dy}{dx} = \lim_{h \to 0} \frac{f(x + h) - f(x)}{h} \qquad (5)$$

Referring to Figure 3.7.2, the slope of the secant line joining the points P and Q is $\Delta y / \Delta x$. As $\Delta x \to 0$, Q tends toward P along the curve $y = f(x)$, so that the slope of the secant line approaches the slope of the tangent line at P; that is,

$$\frac{dy}{dx} = \lim_{\Delta x \to 0} \frac{\Delta y}{\Delta x} \qquad (6)$$

This result is not unexpected since $\Delta y = f(x + \Delta x) - f(x)$, so (6) can be written

$$\frac{dy}{dx} = \lim_{\Delta x \to 0} \frac{f(x + \Delta x) - f(x)}{\Delta x} \qquad (7)$$

which is simply a restatement of (5) with Δx used in place of h.
If it is undesirable to introduce a dependent variable y, then we write

$$\Delta f = f(x + \Delta x) - f(x)$$

and (6) as

$$f'(x) = \lim_{\Delta x \to 0} \frac{\Delta f}{\Delta x} \tag{8}$$

Expressions (5), (6), (7), and (8) provide four ways of writing the definition of a derivative. All are commonly used.

DIFFERENTIALS   Up to now we have been viewing the expression dy/dx as a single symbol for the derivative. The symbols "dy" and "dx" in this expression, which are called *differentials,* have had no meaning by themselves. We shall now show how to interpret these symbols so that dy/dx can be regarded as the ratio of dy and dx.

Regard x as fixed and *define dx* to be an independent variable that can be assigned an arbitrary value. If f is differentiable at x, then we *define dy* by the formula

$$dy = f'(x)\,dx \tag{9}$$

If $dx \neq 0$, then we can divide both sides of (9) by dx to obtain

$$\frac{dy}{dx} = f'(x)$$

Thus, we have achieved our goal of defining dy and dx so their ratio is $f'(x)$.

Because

$$\frac{dy}{dx} = f'(x) = m_{\text{tan}}$$

Figure 3.7.3

where m_{tan} is the slope of the tangent to $y = f(x)$ at x, the differentials dy and dx can be interpreted as a corresponding rise and run of this tangent line (Figure 3.7.3).

It is important to understand the distinction between the increment Δy and the differential dy. To see the difference, let us assign the independent variables dx and Δx the same value, so $dx = \Delta x$. Then Δy represents the change in y that occurs when we start at x and travel *along the curve* $y = f(x)$ until we have moved Δx ($= dx$) units in the x-direction, while dy represents the change in y that occurs if we start at x and travel *along the tangent* line until we have moved dx ($= \Delta x$) units in the x-direction (Figure 3.7.4).

Figure 3.7.4

Example 1   If $y = x^2$, then the relation $dy/dx = 2x$ can be written in the *differential form*

$$dy = 2x\,dx$$

When $x = 3$, this becomes

$$dy = 6\,dx$$

This tells us that if we travel along the tangent to the curve $y = x^2$ at $x = 3$, then a change of dx units in x produces a change of $6\,dx$ units in y. For example, if the change in x is $dx = 4$ then the change in y along the tangent is

$$dy = 6(4) = 24 \quad \text{units} \qquad \blacktriangleleft$$

Example 2   Let $y = \sqrt{x}$. Find dy and Δy if $x = 4$ and $dx = \Delta x = 3$. Then make a sketch of $y = \sqrt{x}$, showing dy and Δy in the picture.

Solution.   From (4) with $f(x) = \sqrt{x}$,

$$\Delta y = \sqrt{x + \Delta x} - \sqrt{x} = \sqrt{7} - \sqrt{4} \approx .65$$

If $y = \sqrt{x}$, then

$$\frac{dy}{dx} = \frac{1}{2\sqrt{x}}$$

so

$$dy = \frac{1}{2\sqrt{x}}\,dx = \frac{1}{2\sqrt{4}}\,(3) = .75$$

Figure 3.7.5 shows the curve $y = \sqrt{x}$ together with dy and Δy. $\blacktriangleleft$

Figure 3.7.5 ▷

TANGENT LINE
APPROXIMATIONS

Figure 3.7.6 suggests that if f is differentiable at x_0, then the tangent line to the curve $y = f(x)$ at x_0 is a reasonably good approximation to the curve $y = f(x)$ near x_0. Since the tangent line passes through the point $(x_0, f(x_0))$ and has slope $f'(x_0)$, the point-slope form of its equation is

$$y - f(x_0) = f'(x_0)(x - x_0)$$

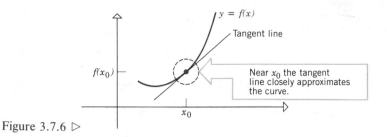

Figure 3.7.6 ▷

or

$$y = f(x_0) + f'(x_0)(x - x_0)$$

For values of x close to x_0, the height y of this tangent line will closely approximate the height $f(x)$ of the curve, which yields the approximation

$$f(x) \approx f(x_0) + f'(x_0)(x - x_0)$$ (10)

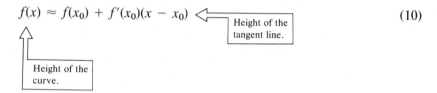

for x near x_0. If we let $\Delta x = x - x_0$, so $x = x_0 + \Delta x$, then (10) can be written in the alternate form

$$f(x_0 + \Delta x) \approx f(x_0) + f'(x_0) \, \Delta x$$ (11)

which is a good approximation when Δx is near zero. This result is called the **linear approximation of f near x_0.**

In the event that $f(x_0 + \Delta x)$ is tedious to calculate, but $f(x_0)$ and $f'(x_0)$ are not, then this formula enables us to use the values of $f(x_0)$ and $f'(x_0)$ to approximate $f(x_0 + \Delta x)$.

Example 3   Use (11) to approximate $\sqrt{1.1}$.

Solution.   If we let $f(x) = \sqrt{x}$, then the problem is to approximate $f(1.1)$. But $f'(x) = 1/(2\sqrt{x})$, so that $f(1)$ and $f'(1)$ are easy to compute. Thus, we shall apply (11) with

$$x_0 + \Delta x = 1.1 \quad \text{and} \quad x_0 = 1$$

It follows that $\Delta x = .1$, so (11) yields

$$f(1.1) \approx f(1) + f'(1)(.1)$$

or

$$\sqrt{1.1} \approx \sqrt{1} + \frac{1}{2\sqrt{1}} (.1) = 1.05. \quad \blacktriangleleft$$

Example 4   Use (11) to approximate $\cos 62°$.

Solution.   We shall take advantage of the fact that $62°$ is close to $60°$, at which point the trigonometric functions are easy to evaluate. However, to apply (11), we must first convert to radian measure because the derivative formulas for the trigonometric functions are based on the assumption that x is in radians. If we let $f(x) = \cos x$ and note that $62° = 31\pi/90$ (radians), then the problem is to approximate $f(31\pi/90)$. Thus, we will apply (11) with

$$x_0 + \Delta x = \frac{31\pi}{90} \quad \text{and} \quad x_0 = \frac{\pi}{3} \ (=60°)$$

It follows that $\Delta x = \pi/90$, so (11) yields

$$f\left(\frac{31\pi}{90}\right) \approx f\left(\frac{\pi}{3}\right) + f'\left(\frac{\pi}{3}\right)\left(\frac{\pi}{90}\right)$$

$$\cos\frac{31\pi}{90} \approx \cos\frac{\pi}{3} - \left(\sin\frac{\pi}{3}\right)\left(\frac{\pi}{90}\right)$$

$$\cos\frac{31\pi}{90} \approx \frac{1}{2} - \frac{\sqrt{3}}{2}\left(\frac{\pi}{90}\right) \approx 0.5 - 0.0302300 = 0.4697700$$

To seven digits, the value of $\cos 62°$ is 0.4694716, so that the error is less than 0.0003. $\blacktriangleleft$

Formula (11) has a useful alternative form, which we shall now derive. Subtract $f(x_0)$ from both sides to obtain

$$f(x_0 + \Delta x) - f(x_0) \approx f'(x_0)\,\Delta x$$

and use (3) to rewrite this as

$$\Delta y \approx f'(x_0)\,\Delta x$$

If we drop the subscript on x_0 and assume that $\Delta x = dx$, then from (9) this can be rewritten as

$$\Delta y \approx dy \tag{12}$$

This formula has applications in the study of ***error propagation***. Suppose a researcher measures a physical quantity. Because of limitations in the instrumentation and other factors, the researcher will not usually obtain the exact value x of the quantity, but rather will obtain $x + \Delta x$, where Δx is a measurement error. This recorded value may then be used to calculate some other quantity y. In this way the measurement error Δx propagates to produce an error Δy in the calculated value of y.

Example 5   The radius of a sphere is measured to be 50 in. with a possible measurement error of ± 0.02 in. Estimate the possible error in the computed volume of the sphere.

Solution.   The volume of the sphere is

$$V = \tfrac{4}{3}\pi r^3 \tag{13}$$

We are given that the error in the radius is $\Delta r = \pm 0.02$, and we want to find the error ΔV in V. If we consider Δr to be small and we let $dr = \Delta r$, then ΔV can be approximated by dV. Thus, from (13),

$$\Delta V \approx dV = 4\pi r^2 \, dr \tag{14}$$

Substituting $r = 50$ and $dr = \pm 0.02$ in (14), we obtain

$$\Delta V \approx 4\pi(2500)(\pm 0.02) \approx \pm 628.32$$

Therefore, the possible error in the calculated volume is approximately ± 628.32 cubic inches (in³).   ◄

REMARK.   In (14), r represents the exact value of the radius. Since the exact value of r was unknown, we substituted the measured value $r = 50$ to obtain ΔV. This is reasonable since the error Δr was assumed to be small.

If the exact value of a quantity is q and a measurement or calculation results in an error Δq, then $\Delta q/q$ is called the **relative error** in the measurement or calculation; when expressed as a percentage, $\Delta q/q$ is called the **percentage error**. As a practical matter, the exact value q is usually unknown, so that the measured or calculated value of q is used instead; and the relative error is approximated by dq/q.

Example 6   For the sphere in Example 5,

$$\text{relative error in } r \approx \frac{dr}{r} = \frac{\pm 0.02}{50} = \pm 0.0004$$

$$\text{relative error in } V \approx \frac{dV}{V} = \frac{4\pi r^2 \, dr}{\tfrac{4}{3}\pi r^3} = 3\,\frac{dr}{r} = \pm 0.0012$$

Thus, the percentage error in the calculated radius is approximately $\pm 0.04\%$, and the percentage error in the calculated volume is approximately $\pm 0.12\%$.   ◀

Example 7   The side of a square is measured with a possible percentage error of $\pm 5\%$. Use differentials to estimate the possible percentage error in the area of the square.

Solution.   The area of a square with side x is given by

$$A = x^2 \tag{15}$$

and the relative errors in A and x are approximately dA/A and dx/x, respectively. From (15),

$$dA = 2x\ dx$$

so

$$\frac{dA}{A} = \frac{2x\ dx}{A} = \frac{2x\ dx}{x^2} = 2\ \frac{dx}{x}$$

We are given that $dx/x \approx \pm 0.05$, so that

$$\frac{dA}{A} \approx \pm 2(0.05) = \pm 0.1$$

Thus, there is a possible percentage error of $\pm 10\%$ in the area of the square.   ◀

DIFFERENTIAL FORMULAS

Just as

$$\frac{d}{dx}[\ \ ]$$

denotes the derivative of the expression inside the brackets, so

$$d[\ \ ]$$

denotes the differential of the expression in the brackets. For example,

$$d[x^3] = 3x^2\ dx$$
$$d[f(x)] = f'(x)\ dx$$

The basic rules of differentiation can be expressed in terms of differentials. In the following table, the differential formulas on the right result when the derivative formulas on the left are multiplied through by dx.

DERIVATIVE FORMULA	DIFFERENTIAL FORMULA
$\dfrac{d}{dx}[c] = 0$	$d[c] = 0$
$\dfrac{d}{dx}[cf] = c\dfrac{df}{dx}$	$d[cf] = c\,df$
$\dfrac{d}{dx}[f+g] = \dfrac{df}{dx} + \dfrac{dg}{dx}$	$d[f+g] = df + dg$
$\dfrac{d}{dx}[fg] = f\dfrac{dg}{dx} + g\dfrac{df}{dx}$	$d[fg] = f\,dg + g\,df$
$\dfrac{d}{dx}\left[\dfrac{f}{g}\right] = \dfrac{g\dfrac{df}{dx} - f\dfrac{dg}{dx}}{g^2}$	$d\left[\dfrac{f}{g}\right] = \dfrac{g\,df - f\,dg}{g^2}$

Example 8 Find dy if $y = x\sin x$.

Solution.

$$dy = d[x\sin x] = x\,d[\sin x] + (\sin x)\,d[x]$$
$$= x(\cos x)\,dx + (\sin x)\,dx \quad \text{[since } d[x] = 1\cdot dx = dx\text{]}$$
$$= (x\cos x + \sin x)\,dx \quad \blacktriangleleft$$

▶ Exercise Set 3.7

1. Let $y = x^2$.
   (a) Find Δy if $\Delta x = 1$ and the initial value of x is $x = 2$.
   (b) Find dy if $dx = 1$ and the initial value of x is $x = 2$.
   (c) Make a sketch of $y = x^2$ and show Δy and dy in the picture.

2. Repeat Exercise 1 with $x = 2$ as the initial value again, but $\Delta x = -1$ and $dx = -1$.

3. Let $y = \dfrac{1}{x}$.
   (a) Find Δy if $\Delta x = 0.5$ and the initial value of x is $x = 1$.
   (b) Find dy if $dx = 0.5$ and the initial value of x is $x = 1$.
   (c) Make a sketch of $y = 1/x$ and show Δy and dy in the picture.

4. Repeat Exercise 3 with $x = 1$ as the initial value again, but $\Delta x = -0.5$ and $dx = -0.5$.

In Exercises 5–8, find general formulas for dy and Δy.

5. $y = x^3$.
6. $y = 8x - 4$.
7. $y = x^2 - 2x + 1$.
8. $y = \sin x$.

In Exercises 9–12, find dy.

9. $y = 4x^3 - 7x^2 + 2x - 1$.

10. $y = \dfrac{1}{x^3 - 1}$.

11. $y = x\cos x$.

12. $y = \dfrac{1 - x^3}{2 - x}$.

In Exercises 13–16, find the limit by expressing it as a derivative.

13. $\lim\limits_{\Delta x \to 0} \dfrac{(x + \Delta x)^2 - x^2}{\Delta x}$.

14. $\lim\limits_{\Delta x \to 0} \dfrac{(3 + \Delta x)^2 - 3^2}{\Delta x}$.

15. $\lim\limits_{\Delta x \to 0} \dfrac{\sin(\pi + \Delta x) - \sin \pi}{\Delta x}$.

16. $\lim\limits_{\Delta x \to 0} \dfrac{5(2 + \Delta x)^4 - 5(2)^4}{\Delta x}$.

In Exercises 17–28, use a linear approximation [formula (11)] to estimate the value of the given quantity.

17. $(3.02)^4$.

18. $(1.97)^3$.

19. $\sqrt{65}$.

20. $\sqrt{24}$.

21. $\sqrt{80.9}$.

22. $\sqrt{36.03}$.

23. $\sqrt[3]{8.06}$.

24. $\sqrt[3]{63.7}$.

25. $\cos 31°$.

26. $\sin 59°$.

27. $\sin 44°$.

28. $\tan 61°$.

In Exercises 29–32, use dy to approximate Δy when x changes as indicated.

29. $y = \sqrt{3x - 2}$; from $x = 2$ to $x = 2.03$.

30. $y = \sqrt{x^2 + 8}$; from $x = 1$ to $x = 0.97$.

31. $y = \dfrac{x}{x^2 + 1}$; from $x = 2$ to $x = 1.96$.

32. $y = x\sqrt{8x + 1}$; from $x = 3$ to $x = 3.05$.

33. The side of a square is measured to be 10 ft, with a possible error of ± 0.1 ft.
   (a) Use differentials to estimate the error in the calculated area.
   (b) Estimate the percentage errors in the side and the area.

34. The side of a cube is measured to be 25 cm, with a possible error of ± 1 cm.
   (a) Use differentials to estimate the error in the calculated volume.
   (b) Estimate the percentage errors in the side and volume.

35. The hypotenuse of a right triangle is known to be 10 in. exactly, and one of the acute angles is measured to be 30°, with a possible error of $\pm 1°$.
   (a) Use differentials to estimate the errors in the sides opposite and adjacent to the measured angle.
   (b) Estimate the percentage errors in the sides.

36. One side of a right triangle is known to be 25 cm exactly. The angle opposite to this side is measured to be 60°, with a possible error of $\pm 0.5°$.
   (a) Use differentials to estimate the errors in the adjacent side and the hypotenuse.
   (b) Estimate the percentage errors in the adjacent side and hypotenuse.

37. The electrical resistance R of a certain wire is given by $R = k/r^2$, where k is a constant and r is the radius of the wire. Assuming that the radius r has a possible error of $\pm 5\%$, use differentials to estimate the percentage error in R. (Assume k is exact.)

38. The side of a square is measured with a possible percentage error of $\pm 1\%$. Use differentials to estimate the percentage error in the area.

39. The side of a cube is measured with a possible percentage error of $\pm 2\%$. Use differentials to estimate the percentage error in the volume.

40. The volume of a sphere is to be computed from a measured value of its radius. Estimate the maximum permissible percentage error in the measurement if the percentage error in the volume must be kept within $\pm 3\%$. [$V = \frac{4}{3}\pi r^3$ is the volume of a sphere of radius r.]

41. The area of a circle is to be computed from a measured value of its diameter. Estimate the maximum permissible percentage error in the measurement if the percentage error in the area must be kept within $\pm 1\%$.

42. A steel cube with 1 in. sides is coated with .01 in. of copper. Use differentials to estimate the volume of copper in the coating. [*Hint:* Let ΔV be the change in the volume of the cube.]

43. A metal rod 15 cm long and 5 cm in diameter is to be covered (except for the ends) with insulation that is .001 cm thick. Use differentials to estimate the volume of insulation. [*Hint:* Let ΔV be the change in volume of the rod.]

44. The time required for one complete oscillation of a pendulum is called its **period**. If the length L of the pendulum is measured in feet and the period P in seconds, then the period is given by $P = 2\pi\sqrt{L/g}$, where g is a constant. Use differentials to show that the percentage error in P is approximately half the percentage error in L.

45. Use differentials to show that $\dfrac{1}{1 + x} \approx 1 - x$ when x is near zero. Calculate, to 4 decimal places, the values of $\dfrac{1}{1 + x}$ and $1 - x$ for $x = 0.1$ and $x = -0.02$.

46. Use differentials to show that $\sqrt{1 + x} \approx 1 + \frac{1}{2}x$ when x is near zero. Calculate, to 4 decimal places, the values of $\sqrt{1 + x}$ and $1 + \frac{1}{2}x$ for $x = 0.1$ and $x = -0.02$.

47. Let $y = x^k$, where k is a positive rational number. Show that the percentage error in y is approximately k times the percentage error in x.

▶ SUPPLEMENTARY EXERCISES

In Exercises 1–4, use Definition 3.2.1 to find $f'(x)$.

1. $f(x) = kx$ (k constant).

2. $f(x) = (x - a)^2$ (a constant).

3. $f(x) = \sqrt{9 - 4x}$.

4. $f(x) = \dfrac{x}{x + 1}$.

5. Use Definition 3.2.1 to find $\dfrac{d}{dx}[|x|^3]\Big|_{x=0}$.

6. Suppose $f(x) = \begin{cases} x^2 - 1, & x \le 1 \\ k(x - 1), & x > 1. \end{cases}$

For what values of k is f
(a) continuous
(b) differentiable?

7. Suppose $f(3) = -1$ and $f'(3) = 5$. Find an equation for the tangent line to the graph of f at $x = 3$.

8. Let $f(x) = x^2$. Show that for any distinct values of a and b, the slope of the tangent line to $y = f(x)$ at $x = \frac{1}{2}(a + b)$ is equal to the slope of the secant line through the points (a, a^2) and (b, b^2).

9. Given the following table of values at $x = 1$ and $x = -2$, find the indicated derivatives in (a)–(l).

x	$f(x)$	$f'(x)$	$g(x)$	$g'(x)$
1	1	3	-2	-1
-2	-2	-5	1	7

(a) $\dfrac{d}{dx}[f^2(x) - 3g(x^2)]\Big|_{x=1}$

(b) $\dfrac{d}{dx}[f(x)g(x)]\Big|_{x=1}$

(c) $\dfrac{d}{dx}\left[\dfrac{f(x)}{g(x)}\right]\Big|_{x=-2}$

(d) $\dfrac{d}{dx}\left[\dfrac{g(x)}{f(x)}\right]\Big|_{x=-2}$

(e) $\dfrac{d}{dx}[f(g(x))]\Big|_{x=1}$

(f) $\dfrac{d}{dx}[f(g(x))]\Big|_{x=-2}$

(g) $\dfrac{d}{dx}[g(f(x))]\Big|_{x=-2}$

(h) $\dfrac{d}{dx}[g(g(x))]\Big|_{x=-2}$

(i) $\dfrac{d}{dx}[f(g(4 - 6x))]\Big|_{x=1}$

(j) $\dfrac{d}{dx}[g^3(x)]\Big|_{x=1}$

(k) $\dfrac{d}{dx}[\sqrt{f(x)}]\Big|_{x=1}$

(l) $\dfrac{d}{dx}[f(-\frac{1}{2}x)]\Big|_{x=-2}$

In Exercises 10–15, find $f'(x)$ and determine those values of x for which $f'(x) = 0$.

10. $f(x) = (2x + 7)^6(x - 2)^5$.

11. $f(x) = \dfrac{(x - 3)^4}{x^2 + 2x}$.

12. $f(x) = \sqrt{3x + 1}\,(x - 1)^2$.

13. $f(x) = \left(\dfrac{3x + 1}{x^2}\right)^3$.

14. $f(x) = \dfrac{3(5x - 1)^{1/3}}{3x - 5}$.

15. $f(x) = \sqrt{x}\,\sqrt[3]{x^2 + x + 1}$.

16. Suppose that $f'(x) = 1/x$ for all $x \ne 0$.
(a) Use the chain rule to show that for any nonzero constant a, $d(f(ax))/dx = d(f(x))/dx$.
(b) If $y = f(\sin x)$ and $v = f(1/x)$, find dy/dx and dv/dx.

In Exercises 17–26, find the indicated derivatives.

17. $\dfrac{d}{dx}\left(\dfrac{\sqrt{2}}{x^2} - \dfrac{2}{5x}\right)$.

18. $\dfrac{dy}{dx}$ if $y = \dfrac{3x^2 + 7}{x^2 - 1}$.

19. $\dfrac{dz}{dr}\Big|_{r=\pi/6}$ if $z = 4 \sin^2 r \cos^2 r$.

20. $g'(2)$ if $g(x) = 1/\sqrt{2x}$.

21. $\dfrac{du}{dx}$ if $u = \left(\dfrac{x}{x-1}\right)^{-2}$.

22. $\dfrac{dw}{dv}$ if $w = \sqrt[5]{v^3} - \sqrt[4]{v}$.

23. $d(\sec^2 x - \tan^2 x)/dx$.

24. $\dfrac{dy}{dx}\Big|_{x=\pi/4}$ if $y = \tan t$ and $t = \cos(2x)$.

25. $F'(x)$ if $F(x) = \dfrac{(1/x) + 2x}{\frac{1}{2}(1/x^2) + 1}$.

26. $\Phi'(x)$ if $\Phi(x) = \dfrac{x^2 - 4x}{5\sqrt{x}}$.

27. Find all values of x for which the tangent to $f(x) = x - (1/x)$ is parallel to the line $2x - y = 5$.

28. Find all values of x for which the tangent to $f(x) = 2x^3 - x^2$ is perpendicular to the line $x + 4y = 10$.

29. Find all values of x for which the tangent to $f(x) = (x + 2)^2$ passes through the origin.

30. Find all values of x for which the tangent to $f(x) = x - \sin 2x$ is horizontal.

31. Find all values of x for which the tangent to $f(x) = 3x - \tan x$ is parallel to the line $y - x = 2$.

In Exercises 32–34, find Δx, Δy, and dy.

32. $y = 1/(x - 1)$; x decreases from 2 to 1.5.

33. $y = \tan x$; x increases from $-\pi/4$ to 0.

34. $y = \sqrt{25 - x^2}$; x increases from 0 to 3.

35. Use a differential to approximate
    (a) $\sqrt[3]{-8.25}$ (b) $\cot 46°$.

36. The area of a right triangle whose hypotenuse is H centimeters is calculated using the formula

$A = \frac{1}{4}H^2 \sin 2\theta$, where θ is one of the acute angles. Use differentials to approximate the error in calculating A if $H = 4$ cm (exactly) and $\theta = 30° \pm 15'$.

37. A 12-ft ladder leaning against a wall makes an angle θ with the floor. If the top of the ladder is h feet up the wall, express h in terms of θ and then use dh to estimate the change in h if θ changes from 60° to 59°.

38. Let V and S denote the volume and surface area of a cube. Find the rate of change of V with respect to S.

39. The amount of water in a tank t minutes after it has started to drain is given by $W = 100(t - 15)^2$ gal.
    (a) At what rate is the water running out at the end of 5 min?
    (b) What is the average rate at which the water flows out during the first 5 min?

In Exercises 40–42, find dy/dx by implicit differentiation and use it to find the equation of the tangent line at the indicated points.

40. $(x + y)^3 + 3xy = -7$; $(-2, 1)$.

41. $xy^2 = \sin(x + 2y)$; $(0, 0)$.

42. $(x + y)^3 - 5x + y = 1$; at the point where the curve intersects the line $x + y = 1$.

43. Show that the curves whose equations are $2x^2 + 3y^2 = 5$ and $y^2 = x^3$ intersect at the point $(1, 1)$ and that their tangent lines are perpendicular there.

44. Show that for any point $P_0(x_0, y_0)$ on the circle $x^2 + y^2 = r^2$, the tangent line at P_0 is perpendicular to the radial line from the origin to P_0.

45. Verify that the function $y = \cos x - 3 \sin x$ satisfies $y''' + y'' + y' + y = 0$.

46. Find d^2y/dx^2 implicitly if
    (a) $y^3 + 3x^2 = 4y$ (b) $\sin y + \cos x = 1$.

4 applications of differentiation

4.1 RELATED RATES

*In this section we shall study **related rates** problems. In such problems one tries to find the rate at which some quantity is changing by relating it to other quantities whose rates of change are known.*

Example 1 Assume that oil spilled from a ruptured tanker spreads in a circular pattern whose radius increases at a constant rate of 2 ft/sec. How fast is the area of the spill increasing when the radius of the spill is 60 ft?

Oil spill

r

Figure 4.1.1

Solution. Let

> t = number of seconds elapsed from the time of the spill
> r = radius of the spill in feet after t seconds
> A = area of the spill in square feet after t seconds

(See Figure 4.1.1.) At each instant the rate at which the radius is increasing with time is dr/dt, and the rate at which the area is increasing with time is dA/dt. We want to find

$$\frac{dA}{dt}\bigg|_{r=60}$$

which is the rate at which the area of the spill is increasing at the instant when $r = 60$.

Since the radius increases at the constant rate of 2 ft/sec, we know

$$\frac{dr}{dt} = 2 \qquad (1)$$

for all t.

From the formula for the area of a circle we obtain

$$A = \pi r^2 \tag{2}$$

Because A and r are functions of t, we can differentiate both sides of (2) with respect to t to obtain

$$\frac{dA}{dt} = 2\pi r \frac{dr}{dt}$$

Substituting (1) yields

$$\frac{dA}{dt} = 2\pi r(2) = 4\pi r$$

Thus, when $r = 60$ the area of the spill is increasing at the rate

$$\frac{dA}{dt}\bigg|_{r=60} = 4\pi(60) = 240\pi \text{ ft}^2/\text{sec}$$

or approximately 754 ft²/sec. ◄

With only minor variations, the method used in Example 1 can be used to solve a variety of related rates problems. The method consists of four steps:

Step 1. Draw a figure and label the quantities that vary.

Step 2. Find an equation relating the quantity with the unknown rate of change to quantities whose rates of change are known.

Step 3. Differentiate both sides of this equation with respect to time and solve for the derivative that will give the unknown rate of change.

Step 4. Evaluate this derivative at the appropriate point.

Figure 4.1.2

In Figure 4.1.2 we have shown a camera mounted at a point 3000 ft from the base of a rocket launching pad. Let us assume that the rocket rises vertically and the camera is to take a series of photographs. Because the rocket will be rising, the elevation angle of the camera will have to vary at just the right rate to keep the rocket in sight. Moreover, because the camera-to-rocket distance will be changing constantly, the camera focusing mechanism will also have to vary at just the right rate to keep the picture sharp. Our next two examples are concerned with these problems.

Example 2   If the rocket shown in Figure 4.1.2 is rising vertically at 880 ft/sec when it is 4000 ft up, how fast is the camera-to-rocket distance changing at that instant?

Solution.   Let

t = number of seconds elapsed from the time of launch
y = camera-to-rocket distance in feet after t seconds
x = height of rocket in feet after t seconds

Figure 4.1.3

At each instant the rate at which the camera-to-rocket distance is changing is dy/dt, and the rate at which the rocket is rising is dx/dt. We want to find

$$\frac{dy}{dt}\bigg|_{x=4000}$$

which is the rate the camera-to-rocket distance is changing at the instant when $x = 4000$.

From the Theorem of Pythagoras and Figure 4.1.3 we have

$$y^2 = x^2 + (3000)^2 \tag{3}$$

Because x and y are functions of t, we can differentiate both sides of (3) with respect to t to obtain

$$2y\frac{dy}{dt} = 2x\frac{dx}{dt} \qquad \boxed{\text{Note the use of the Chain Rule here.}}$$

or

$$\frac{dy}{dt} = \frac{x}{y}\frac{dx}{dt} \tag{4}$$

When $x = 4000$, it follows from (3) that $y = 5000$; moreover, we are given that $dx/dt = 880$ when $x = 4000$ so that from (4)

$$\frac{dy}{dt}\bigg|_{x=4000} = \frac{4000}{5000}(880) = 704 \text{ ft/sec} \qquad \blacktriangleleft$$

Example 3   If the rocket shown in Figure 4.1.4 is rising vertically at 880 ft/sec when it is 4000 ft up, how fast must the camera elevation angle change at that instant to keep the rocket in sight?

Figure 4.1.4

Figure 4.1.5

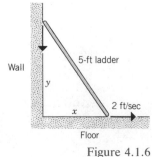

Figure 4.1.6

Solution.   Let
t = number of seconds elapsed from the time of launch
ϕ = camera elevation angle in radians after t seconds
x = height of rocket in feet after t seconds

(See Figure 4.1.4.) At each instant the rate at which the camera elevation angle must change is $d\phi/dt$, and the rate at which the rocket is rising is dx/dt. We want to find

$$\frac{d\phi}{dt}\bigg|_{x=4000}$$

which is the rate the camera elevation angle must change at the instant when $x = 4000$.

From Figure 4.1.4 we see that

$$\tan \phi = \frac{x}{3000} \tag{5}$$

Because ϕ and x are functions of t, we can differentiate both sides of (5) with respect to t to obtain

$$(\sec^2 \phi)\frac{d\phi}{dt} = \frac{1}{3000}\frac{dx}{dt}$$

or

$$\frac{d\phi}{dt} = \frac{1}{3000 \sec^2 \phi}\frac{dx}{dt} \tag{6}$$

When $x = 4000$, it follows that

$$\sec \phi = \frac{5000}{3000} = \frac{5}{3}$$

(See Figure 4.1.5.) Moreover, we are given that $dx/dt = 880$ when $x = 4000$, so that from (6)

$$\frac{d\phi}{dt}\bigg|_{x=4000} = \frac{1}{3000(\frac{5}{3})^2} \cdot 880 = \frac{66}{625} \approx 0.11 \text{ radian/sec} \quad \blacktriangleleft$$

Example 4   A 5-ft ladder, leaning against a wall (Figure 4.1.6), slips so that its base moves away from the wall at a rate of 2 ft/sec. How fast will the top of the ladder be moving down the wall when the base is 4 ft from the wall?

Solution. Let

t = number of seconds after the ladder starts to slip
x = distance in feet from the base of the ladder to the wall
y = distance in feet from the top of the ladder to the floor

(See Figure 4.1.6.) At each instant the rate at which the base moves is dx/dt, and the rate at which the top moves is dy/dt. We want to find

$$\left.\frac{dy}{dt}\right|_{x=4}$$

which is the rate at which the top is moving at the instant the base is 4 ft from the wall.

From the Theorem of Pythagoras we have

$$x^2 + y^2 = 25 \qquad (7)$$

Differentiating both sides of this equation with respect to t yields

$$2x\,\frac{dx}{dt} + 2y\,\frac{dy}{dt} = 0$$

or

$$\frac{dy}{dt} = -\frac{x}{y}\frac{dx}{dt} \qquad (8)$$

When $x = 4$, it follows from (7) that $y = 3$; moreover, we are given that $dx/dt = 2$, so that (8) yields

$$\left.\frac{dy}{dt}\right|_{x=4} = -\frac{4}{3}\,(2) = -\frac{8}{3}\ \text{ft/sec}$$

The negative sign in the answer tells us that y is decreasing, which makes sense physically, since the top of the ladder is moving *down* the wall. ◀

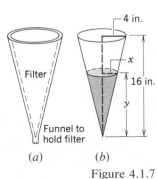

(a)   (b)

Figure 4.1.7

Example 5   Suppose a liquid is to be cleared of sediment by pouring it through a cone-shaped filter as shown in Figure 4.1.7a. Assume the height of the cone is 16 in. and the radius at the base of the cone is 4 in. (Figure 4.1.7b). If the liquid is flowing out of the cone at a constant rate of 2 in³/min, how fast is the depth of the liquid decreasing when the level is 8 in. deep?

Solution.   Let

t = time elapsed from the initial observation (min)
V = volume of liquid in the cone at time t (in³)
y = depth of the liquid in the cone at time t (in)
x = radius of the liquid surface at time t (in)

(See Figure 4.1.7*b*.) At each instant the rate at which the volume of liquid is changing is dV/dt, and the rate at which the depth is changing is dy/dt. We want to find

$$\left. \frac{dy}{dt} \right|_{y=8}$$

which is the rate at which the depth is changing at the instant the depth is 8 in.

From the formula for the volume of a cone, the volume V and the depth y are related by

$$V = \tfrac{1}{3}\pi x^2 y \tag{9}$$

Since we are given that the volume of liquid decreases at a constant rate of 2 in.³/min, we have

$$\frac{dV}{dt} = -2 \tag{10}$$

(We must use a minus sign here because V *decreases* as t increases.) If we differentiate both sides of (9) with respect to t, the right side will involve the quantity dx/dt. Since we have no direct information about dx/dt, it is desirable to eliminate x from (9) before differentiating. This can be done using similar triangles. From Figure 4.1.7*b*, we see that

$$\frac{x}{y} = \frac{4}{16} \quad \text{or} \quad x = \frac{1}{4}y$$

Substituting this expression in (9) gives

$$V = \frac{\pi}{48} y^3 \tag{11}$$

Differentiating both sides of (11) with respect to t we obtain

$$\frac{dV}{dt} = \frac{\pi}{48}\left(3y^2 \frac{dy}{dt}\right)$$

or

$$\frac{dy}{dt} = \frac{16}{\pi y^2} \frac{dV}{dt} \qquad (12)$$

Substituting (10) into (12), and letting $y = 8$ yields

$$\left.\frac{dy}{dt}\right|_{y=8} = \frac{16}{\pi(8)^2}(-2) = -\frac{1}{2\pi} \approx -0.16 \text{ in/min}$$

Thus, when $y = 8$ in., the depth of the liquid is decreasing at an approximate rate of 0.16 in/min. (The minus sign simply means that the depth y is decreasing as t is increasing.) ◀

▶ Exercise Set 4.1

1. Let A be the area of a square whose sides have length x, and assume that x varies with time.
   (a) How are dA/dt and dx/dt related?
   (b) At a certain instant the sides are 3 ft long and growing at a rate of 2 ft/min. How fast is the area growing at that instant?

2. Let A be the area of a circle whose radius is r, and assume that r varies with time.
   (a) How are dA/dt and dr/dt related?
   (b) At a certain instant the radius is 5 in. and is increasing at a rate of 2 in/sec. How fast is the area of the circle increasing at that instant?

3. Let V be the volume of a cylinder having height h and radius r, and assume that h and r vary with time.
   (a) How are dV/dt, dh/dt, and dr/dt related?
   (b) At a certain instant, the height is 6 in. and increasing at 1 in/sec, while the radius is 10 in. and decreasing at 1 in/sec. How fast is the volume changing at that instant? Is the volume increasing or decreasing at that instant?

4. Let l be the length of a diagonal of a rectangle whose sides have lengths x and y, and assume that x and y vary with time.
   (a) How are dl/dt, dx/dt, and dy/dt related?
   (b) If x increases at a constant rate of $\frac{1}{2}$ ft/sec and y decreases at a constant rate of $\frac{1}{4}$ ft/sec,

how fast is the size of the diagonal changing when $x = 3$ ft and $y = 4$ ft? Is the diagonal increasing or decreasing at that instant?

5. Let θ (in radians) be an acute angle in a right triangle, and let x and y, respectively, be the lengths of the sides adjacent and opposite θ. Suppose also that x and y vary with time.
   (a) How are $d\theta/dt$, dx/dt, and dy/dt related?
   (b) At a certain instant, $x = 2$ units and is increasing at 1 unit/sec, while $y = 2$ units and is decreasing at $\frac{1}{4}$ unit/sec. How fast is θ changing at that instant? Is θ increasing or decreasing at that instant?

6. Suppose that $z = x^3y^2$, where both x and y are changing with time. At a certain instant $x = 1$, $y = 2$, x is decreasing at the rate of 2 units/sec, and y is increasing at the rate of 3 units/sec. How fast is z changing at this instant? Is z increasing or decreasing?

7. The minute hand of a certain clock is 4 in. long. Starting from the moment when the hand is pointing straight up, how fast is the area of the sector that is swept out by the hand increasing at any instant during the next revolution of the hand?

8. A stone dropped into a still pond sends out a circular ripple whose radius increases at a constant rate of 3 ft/sec. How rapidly is the area enclosed by the ripple increasing at the end of 10 sec?

9. Oil spilled from a ruptured tanker spreads in a circle whose area increases at a constant rate of 6 mi²/hr. How fast is the radius of the spill increasing when the area is 9 mi²?

10. A spherical balloon is inflated so that its volume is increasing at the rate of 3 ft³/min. How fast is the diameter of the balloon increasing when the radius is 1 ft?

11. A spherical balloon is to be deflated so that its radius decreases at a constant rate of 15 cm/min. At what rate must air be removed when the radius is 9 cm?

12. A 17-ft ladder is leaning against a wall. If the bottom of the ladder is pulled along the ground away from the wall at a constant rate of 5 ft/sec, how fast will the top of the ladder be moving down the wall when it is 8 ft above the ground?

13. A 13-ft ladder is leaning against a wall. If the top of the ladder slips down the wall at a rate of 2 ft/sec, how fast will the foot be moving away from the wall when the top is 5 ft above the ground?

14. A 10-ft plank is leaning against a wall. If at a certain instant the bottom of the plank is 2 ft from the wall and is being pushed toward the wall at the rate of 6 in/sec, how fast is the acute angle that the plank makes with the ground increasing?

15. At a certain instant each edge of a cube is 5 in. long and the volume is increasing at the rate of 2 in³/min. How fast is the surface area of the cube increasing?

16. A rocket, rising vertically, is tracked by a radar station that is on the ground 5 mi from the launch-pad. How fast is the rocket rising when it is 4 mi high and its distance from the radar station is increasing at a rate of 2000 mi/hr?

17. For the camera and rocket shown in Figure 4.1.2, at what rate is the elevation angle changing when the rocket is 3000 ft up and rising vertically at 500 ft/sec?

18. For the camera and rocket shown in Figure 4.1.2, at what rate is the rocket rising when the elevation angle is $\pi/4$ radians and increasing at a rate of .2 radian/sec?

19. A conical water tank with vertex down has a ra-dius of 10 ft at the top and is 24 ft high. If water flows into the tank at a rate of 20 ft³/min, how fast is the depth of the water increasing when the water is 16 ft deep?

20. Grain pouring from a chute at the rate of 8 ft³/min forms a conical pile whose altitude is always twice its radius. How fast is the altitude of the pile increasing at the instant when the pile is 6 ft high?

21. Sand pouring from a chute forms a conical pile whose height is always equal to the diameter. If the height increases at a constant rate of 5 ft/min, at what rate is sand pouring from the chute when the pile is 10 ft high?

22. Wheat is poured through a chute at the rate of 10 ft³/min, and falls in a conical pile whose bottom radius is always half the altitude. How fast will the circumference of the base be increasing when the pile is 8 ft high?

23. An aircraft is climbing at a 30° angle to the horizontal. How fast is the aircraft gaining altitude if its speed is 500 mi/hr?

24. A boat is pulled into a dock by means of a rope attached to a pulley on the dock. The rope is attached to the bow of the boat at a point 10 ft below the pulley. If the rope is pulled through the pulley at a rate of 20 ft/min, at what rate will the boat be approaching the dock when 125 ft of rope is out?

Figure 4.1.8

25. For the boat in Exercise 24, how fast must the rope be pulled if we want the boat to approach the dock at a rate of 12 ft/min at the instant when 125 ft of rope is out?

26. A man 6 ft tall is walking at the rate of 3 ft/sec toward a streetlight 18 ft tall. (See Figure 4.1.9.)
   (a) At what rate is the length of his shadow changing?
   (b) How fast is the tip of his shadow moving?

Figure 4.1.9

27. A beacon which makes one revolution every 10 sec is located on a ship 4 kilometers (km) from a straight shoreline. How fast is the beam moving along the shoreline when it makes an angle of 45° with the shore?

28. An aircraft is flying at a constant altitude with a constant speed of 600 mi/hr. An antiaircraft missile is fired on a straight line perpendicular to the flight path of the aircraft so it will hit the aircraft at a point P. At the instant the aircraft is 2 mi from the impact point P the missile is 4 mi from P and flying at 1200 mi/hr. At that instant, how rapidly is the distance between missile and aircraft decreasing?

Figure 4.1.10

29. Solve Exercise 28 under the assumption that the angle between the flight paths is 120° instead of the assumption that the paths are perpendicular. [*Hint:* Use the law of cosines.]

30. A police helicopter is flying due north at 100 mi/hr, and at a constant altitude of $\frac{1}{2}$ mi. Below, a car is traveling west on a highway at 75 mi/hr. At the moment the helicopter crosses over the highway the car is 2 mi east of the helicopter.
    (a) How fast is the distance between the car and

helicopter changing at the moment the helicopter crosses the highway?
    (b) Is the distance between car and helicopter increasing or decreasing at that moment?

31. A particle is moving along the curve whose equation is

$$\frac{xy^3}{1 + y^2} = \frac{8}{5}$$

Assume that the x-coordinate is increasing at the rate of 6 units/sec when the particle is at the point $(1, 2)$.
    (a) At what rate is the y-coordinate of the point changing at that instant?
    (b) Is the particle rising or falling at that instant?

32. A point P is moving along the curve whose equation is $y = \sqrt{x^3 + 17}$. When P is at $(2, 5)$, y is increasing at the rate of 2 units/sec. How fast is x changing?

33. A point P is moving along the line whose equation is $y = 2x$. How fast is the distance between P and the point $(3, 0)$ changing at the instant when P is at $(3, 6)$ if x is decreasing at the rate of 2 units/sec at that instant?

34. A point P is moving along the curve whose equation is $y = \sqrt{x}$. If x is increasing at the rate of 4 units/sec when $x = 3$,
    (a) how fast is the distance between P and the point $(2, 0)$ changing at this instant?
    (b) how fast is the angle of inclination of the line segment from P to $(2, 0)$ changing at this instant?

35. The *thin lens equation* in physics is

$$\frac{1}{s} + \frac{1}{S} = \frac{1}{f}$$

where s is the object distance from the lens, S is the image distance from the lens, and f is the focal length of the lens. Suppose that a certain lens has a focal length of 6 cm and that an object is moving toward the lens at the rate of 2 cm/sec. How fast is the image distance changing at the instant when the object is 10 cm from the lens? Is the image moving away from the lens or toward the lens?

36. Water is stored in a cone-shaped reservoir (vertex down). Assuming that the water evaporates at a rate proportional to the surface area exposed to the air, show the depth of the water will decrease at a constant rate that does not depend on the dimensions of the reservoir.

37. A meteorite enters the earth's atmosphere and burns up at a rate that, at each instant, is proportional to its surface area. Assuming that the meteorite is always spherical, show that the radius decreases at a constant rate.

38. On a certain clock the minute hand is 4 in. long and the hour hand is 3 in. long. How fast is the distance between the tips of the hands changing at 9 o'clock?

39. Coffee is poured at a uniform rate of 2 cm³/sec into a cup whose inside is shaped like a truncated cone. If the upper and lower radii of the cup are 4 cm and 2 cm and the height of the cup is 6 cm, how fast will the coffee level be rising when the coffee is halfway up? [*Hint:* Extend the cup downward to form a cone.]

Figure 4.1.11

4.2   INTERVALS OF INCREASE AND DECREASE; CONCAVITY

Although point plotting is useful for determining the general shape of a graph, it only provides an approximation because no matter how many points are plotted, we can only guess at the shape of the graph between those points. In this section we shall show how the derivative can be used to resolve such ambiguities.

INCREASING AND DECREASING FUNCTIONS

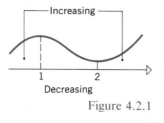

Figure 4.2.1

The terms *increasing* and *decreasing* are used to describe the behavior of a function as we travel *left to right* along its graph. For example, the function graphed in Figure 4.2.1 can be described as increasing on the interval $(-\infty, 1)$, decreasing on the interval $(1, 2)$, and increasing again on the interval $(2, +\infty)$.

The following definition expresses these intuitive ideas precisely.

4.2.1   DEFINITION. Let f be defined on an interval. Then:

(a) f is ***increasing*** on the interval if for any points x_1 and x_2 in the interval such that $x_1 < x_2$, we have $f(x_1) < f(x_2)$. (See Figure 4.2.2*a*.)

(b) f is ***decreasing*** on the interval if for any points x_1 and x_2 in the interval such that $x_1 < x_2$, we have $f(x_2) < f(x_1)$. (See Figure 4.2.2*b*.)

A function is said to be ***strictly monotone*** on a given interval if it is either increasing or decreasing on the interval.

Figure 4.2.3 suggests that if the graph of a function has tangent lines with positive slopes over an interval, then the function is increasing on that interval; and similarly, if the graph has tangent lines with negative

Figure 4.2.2 ▷

$$f(x_1) < f(x_2) \text{ if } x_1 < x_2 \qquad\qquad f(x_2) < f(x_1) \text{ if } x_1 < x_2$$

slopes, the function is decreasing. This intuitive observation suggests the following fundamental theorem whose proof is given in Appendix 2.

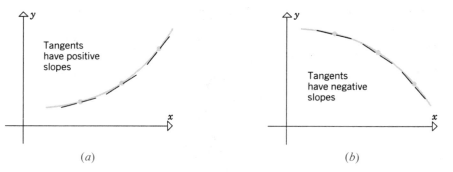

Figure 4.2.3 ▷   (a)   (b)

4.2.2   THEOREM.

(a)   *If $f'(x) > 0$ on an open interval (a, b), then f is increasing on (a, b).*

(b)   *If $f'(x) < 0$ on an open interval (a, b), then f is decreasing on (a, b).*

Sign of $3x$

Sign of $(x - 2)$

Sign of $3x(x - 2)$

Figure 4.2.4

Example 1   On which intervals is the function

$$f(x) = x^3 - 3x^2 + 1$$

increasing? Where is it decreasing?

Solution.   Differentiating f, we obtain

$$f'(x) = 3x^2 - 6x = 3x(x - 2)$$

Thus, f will be increasing when $3x(x - 2) > 0$ and decreasing when $3x(x - 2) < 0$. From the sign analysis in Figure 4.2.4 we see that f is

increasing if $x < 0$
decreasing if $0 < x < 2$
increasing if $x > 2$

The graph of f is shown in Figure 4.2.5. ◀

CONCAVITY

$y = x^3 - 3x^2 + 1$

Figure 4.2.5

Although the derivative of a function f can tell us where the graph of f is increasing or decreasing, it does not reveal where the graph has a downward curvature (Figure 4.2.6a) or an upward curvature (Figure 4.2.6b). To investigate this question, we must study the behavior of the tangent lines shown in the figure.

The curve in part (a) lies below its tangent lines and is called *concave down*. As we travel left to right along this curve, the tangent lines rotate so their slopes *decrease*. In contrast, the curve in part (b) lies above its tangent lines and is called *concave up*. As we travel left to right, its tangent lines rotate so their slopes *increase*. Since f' is the slope of a tangent line to the graph of f, we are led to the following definition.

Concave down

Concave up

Figure 4.2.6 ▷   (a)   (b)

4.2.3 DEFINITION. Let f be differentiable on an open interval.

(a) f is called *concave up* on the interval if f' is increasing on the interval.

(b) f is called *concave down* on the interval if f' is decreasing on the interval.

Concave up "holds water"

Concave down "spills water"

Figure 4.2.7

REMARK.   Informally, a curve that is concave up "holds water" and one that is concave down "spills water" (Figure 4.2.7).

REMARK.   Definition 4.2.3 requires the function f to be differentiable. There are more general definitions of concavity that do not require differentiability. However, we shall have no need for that added generality in this text.

Since f'' is the derivative of f', it follows from Theorem 4.2.2 that f' is increasing on an open interval (a, b) if $f''(x) > 0$ for all x in (a, b), and

f' is decreasing on (a, b) if $f''(x) < 0$ for all x in (a, b). Thus, we have the following result.

4.2.4 THEOREM.

(a) *If $f''(x) > 0$ on an open interval (a, b), then f is concave up on (a, b).*

(b) *If $f''(x) < 0$ on an open interval (a, b), then f is concave down on (a, b).*

Example 2   Where is the function $f(x) = x^3 - 3x^2 + 1$ concave up? Concave down?

Solution.   Calculating the second derivative, we obtain

$$f'(x) = 3x^2 - 6x$$
$$f''(x) = 6x - 6$$

Thus, the curve is concave up where

$$6x - 6 > 0 \tag{1}$$

and concave down where

$$6x - 6 < 0 \tag{2}$$

Since (1) holds if $x > 1$ and (2) holds if $x < 1$, we conclude:

f is concave up if $x > 1$
f is concave down if $x < 1$

The reader can check that these results agree with the graph of f shown in Figure 4.2.8.   ◀

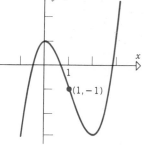

Figure 4.2.8

The point $x = 1$ in Example 2 is of special interest because it is the point where the graph changes the direction of its concavity. There is a name for such points.

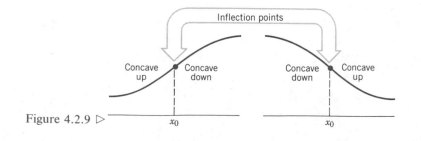

Figure 4.2.9 ▷

> **4.2.5** DEFINITION. If f is continuous on an open interval containing x_0, and if f changes the direction of its concavity at x_0, then the point $(x_0, f(x_0))$ on the graph of f is called an ***inflection point*** of f, and we say that f has an ***inflection point at x_0***. (See Figure 4.2.9.)

Example 3   In Example 2 we showed that $f(x) = x^3 - 3x^2 + 1$ is concave down when $x < 1$ and concave up when $x > 1$. Thus, f has an inflection point at $x = 1$. Since $f(1) = -1$, the inflection point is $(1, -1)$. (See Figure 4.2.8.)   ◄

▶ **Exercise Set 4.2**

Exercises 1 and 2 refer to the function f graphed in Figure 4.2.10.

1. Find the largest open intervals over which f is
   (a) increasing   (b) decreasing
   (c) concave up   (d) concave down.

2. Find all values of x where f has an inflection point.

Figure 4.2.10

In Exercises 3–17, determine open intervals on which f is (a) increasing, (b) decreasing, (c) concave up, (d) concave down, and (e) find the location of all inflection points.

3. $f(x) = x^2 - 5x + 6$.

4. $f(x) = 4 - 3x - x^2$.

5. $f(x) = (x + 2)^3$.

6. $f(x) = 5 + 12x - x^3$.

7. $f(x) = 3x^3 - 4x + 3$.

8. $f(x) = x^4 - 8x^2 + 16$.

9. $f(x) = 3x^4 - 4x^3$.

10. $f(x) = \dfrac{x}{x^2 + 2}$.

11. $f(x) = \cos x,\ 0 < x < 2\pi$.

12. $f(x) = \sin^2 2x,\ 0 < x < \pi$.

13. $f(x) = \tan x,\ -\pi/2 < x < \pi/2$.

14. $f(x) = x^{2/3}$.

15. $f(x) = \sqrt[3]{x + 2}$.

16. $f(x) = x^{4/3} - x^{1/3}$.

17. $f(x) = x^{1/3}(x + 4)$.

18. In each part sketch a continuous curve $y = f(x)$ with the stated properties.
   (a) $f(2) = 4,\ f'(2) = 0,\ f''(x) < 0$ for all x
   (b) $f(2) = 4,\ f'(2) = 0,\ f''(x) > 0$ for $x < 2$, $f''(x) < 0$ for $x > 2$
   (c) $f(2) = 4,\ f''(x) > 0$ for $x \neq 2$ and
   $$\lim_{x \to 2^+} f'(x) = -\infty,\ \lim_{x \to 2^-} f'(x) = +\infty$$

19. In each part sketch a continuous curve $y = f(x)$ with the stated properties.
   (a) $f(2) = 4,\ f'(2) = 0,\ f''(x) > 0$ for all x
   (b) $f(2) = 4,\ f'(2) = 0,\ f''(x) < 0$ for $x < 2$, $f''(x) > 0$ for $x > 2$
   (c) $f(2) = 4,\ f''(x) < 0$ for $x \neq 2$ and
   $$\lim_{x \to 2^+} f'(x) = +\infty,\ \lim_{x \to 2^-} f'(x) = -\infty$$

20. In parts (a)–(c) sketch a continuous curve $y = f(x)$ with the stated properties.
   (a) $f(2) = 4,\ f'(2) = 1,\ f''(x) < 0$ for $x < 2$, $f''(x) > 0$ for $x > 2$
   (b) $f(2) = 4,\ f''(x) > 0$ for $x < 2$, $f''(x) < 0$ for $x > 2$, and

$$\lim_{x \to 2^-} f'(x) = +\infty, \quad \lim_{x \to 2^+} f'(x) = +\infty$$

(c) $f(2) = 4$, $f''(x) < 0$ for $x \neq 2$, and

$$\lim_{x \to 2^-} f'(x) = 1, \quad \lim_{x \to 2^+} f'(x) = -1$$

21. In each part sketch a continuous curve $y = f(x)$ with the stated properties.

   (a) $f(2) = 4$, $f''(x) < 0$ if $x \neq 2$, and

$$\lim_{x \to 2^-} f'(x) = 0, \quad \lim_{x \to 2^+} f'(x) = +\infty$$

   (b) $f(0) = 0$, $f(2) = 4$, $f''(x) = 0$ for all x.

22. Find the inflection points, if any, of the function $f(x) = (x - a)^3$.

23. Find the inflection points, if any, of the function $f(x) = (x - a)^4$.

24. Find the inflection points, if any, of the function $f(x) = ax^2 + bx + c$ ($a \neq 0$).

In Exercises 25–30, use Definition 4.2.1 to prove the given statement.

25. $f(x) = x^2$ is increasing on $[0, +\infty)$.

26. $f(x) = x^2 - 2x$ is decreasing on $(-\infty, 1]$.

27. $f(x) = \sqrt{x}$ is increasing on $[0, +\infty)$.

28. $f(x) = 1/x$ is decreasing on $(0, +\infty)$.

29. If functions f and g are increasing on an interval I, then $f + g$ is increasing on I.

30. If functions f and g are positive and increasing on an interval I, then their product $f \cdot g$ is increasing on I.

31. Give an example of functions f and g that are increasing on an interval I, but $f - g$ is decreasing on I.

32. For the general cubic polynomial

$$f(x) = ax^3 + bx^2 + cx + d \ (a \neq 0)$$

find conditions on a, b, c, and d to assure that f is always increasing or decreasing on $(-\infty, +\infty)$.

33. Prove that a third-degree polynomial

$$f(x) = ax^3 + bx^2 + cx + d \ (a \neq 0)$$

has exactly one inflection point.

34. Prove that an nth-degree polynomial

$$f(x) = a_0x^n + a_1x^{n-1} + \cdots + a_n \ (a_0 \neq 0)$$

has at most $n - 2$ inflection points.

4.3   RELATIVE EXTREMA; FIRST AND SECOND DERIVATIVE TESTS

In this section we shall develop techniques for finding the highest and lowest points on the graph of a function or, equivalently, the largest and smallest values of the function. The methods that we develop here will have important applications in later sections.

RELATIVE MAXIMA
AND MINIMA

The graphs of many functions form hills and valleys. The tops of the hills are called *relative maxima* and the bottoms of the valleys are called *relative minima* (Figure 4.3.1). Just as the top of a hill on the earth's terrain need not be the highest point on earth, so a relative maximum need not be the highest point on the entire graph. However, relative maxima, like

Figure 4.3.1 ▷

tops of hills, are the high points in their *immediate vicinity,* and relative minima, like valley bottoms, are the low points in their *immediate vicinity.* These ideas are captured in the following definitions.

4.3.1 DEFINITION. A function f is said to have a ***relative maximum*** at x_0 if $f(x_0) \geq f(x)$ for all x in some open interval containing x_0.

4.3.2 DEFINITION. A function f is said to have a ***relative minimum*** at x_0 if $f(x_0) \leq f(x)$ for all x in some open interval containing x_0.

4.3.3 DEFINITION. A function f is said to have a ***relative extremum*** at x_0 if it has either a relative maximum or a relative minimum at x_0.

Example 1 The function f graphed in Figure 4.3.2 has a relative maximum at x_0 because there is an open interval containing x_0 on which $f(x_0) \geq f(x)$ holds. (See interval (a, b) in the figure.) ◀

Figure 4.3.2 ▷

CRITICAL POINTS   Relative extrema can be viewed as the transition points that separate the regions where a graph is rising from those where it is falling. As suggested by Figure 4.3.3, the relative extrema of a function f occur either at points where the graph of f has a horizontal tangent or at points where f is not differentiable. This is the content of the following theorem (whose proof is given at the end of this section.)

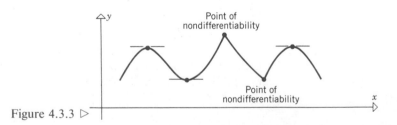

Figure 4.3.3 ▷

4.3.4 THEOREM. *If f has a relative extremum at x_0, then either $f'(x_0) = 0$ or f is not differentiable at x_0.*

Theorem 4.3.4 is of such importance that there is some terminology associated with it.

4.3.5 DEFINITION. A *critical point* for a function f is any value of x in the domain of f at which $f'(x) = 0$ or f is not differentiable; the critical points where $f'(x) = 0$ are called *stationary points* of f.

i.e. $f'(x_0)$ doesn't exist

Example 2   For each of the functions graphed in Figure 4.3.4, x_0 is a critical point. In parts (a), (b), (c), and (d), x_0 is a stationary point because there is a horizontal tangent at x_0, and in the remaining parts x_0 is a critical point, but not a stationary point, because the derivative does not exist there.   ◄

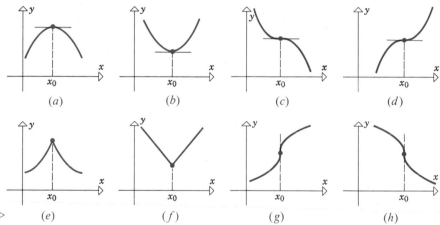

Figure 4.3.4 ▷

(a)   (b)   (c)   (d)

(e)   (f)   (g)   (h)

With the terminology of Definition 4.3.5, Theorem 4.3.4 states that *the relative extrema of a function occur at its critical points*. We emphasize, however, that there need not be a relative extremum at every critical point. Indeed, in parts (c), (d), (g), and (h) of Figure 4.3.4, x_0 is a critical point at which there is no relative extremum. We shall conclude this section with two tests that can be used to determine which critical points correspond to relative extrema and which do not.

FIRST AND SECOND DERIVATIVE TESTS

The nature of a critical point x_0 can often be determined by studying the rise and fall of the curve near x_0. For example, if the curve is increasing to the left of x_0 and decreasing to the right, then there is a relative maximum at x_0 (Figures 4.3.4a, e); if the curve is decreasing to the left of x_0 and increasing to the right, then there is a relative minimum at x_0 (Figures 4.3.4b, f); and if the curve is decreasing on both sides of x_0 or increasing on both sides of x_0, then there is no relative extremum at x_0 (Figures 4.3.4c, d, g, h).

These ideas are stated precisely in the following theorem.

First Derivative Test

4.3.6   THEOREM. *Suppose f is continuous at a critical point x_0.*

(a)   If $f'(x) > 0$ on an open interval extending left from x_0 and $f'(x) < 0$ on an open interval extending right from x_0, then f has a relative maximum at x_0.

(b)   If $f'(x) < 0$ on an open interval extending left from x_0 and $f'(x) > 0$ on an open interval extending right from x_0, then f has a relative minimum at x_0.

(c)   If $f'(x)$ has the same sign [either $f'(x) > 0$ or $f'(x) < 0$] on an open interval extending left from x_0 and on an open interval extending right from x_0, then f does not have a relative extremum at x_0.

REMARK.   To paraphrase this theorem, the relative extrema of a continuous function f occur at those critical points where f' changes sign.

A formal proof of the theorem is given in Appendix 2.

Example 3   Locate the relative extrema of $f(x) = 3x^{5/3} - 15x^{2/3}$.

Solution.

$$f'(x) = 5x^{2/3} - 10x^{-1/3} = 5x^{-1/3}(x - 2)$$

Since $f'(x)$ does not exist when $x = 0$, and $f'(x) = 0$ when $x = 2$, the critical points of f are $x = 0$ and $x = 2$.

As shown in Figure 4.3.5a, the sign of f' changes from positive to negative at 0 and from negative to positive at 2. Thus, there is a relative maximum at 0 and a relative minimum at 2. Although not needed for the solution, the graph of f is shown in Figure 4.3.5b.   ◀

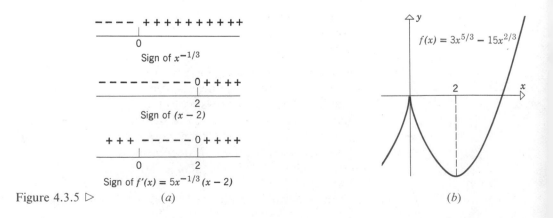

Figure 4.3.5 ▷   (a)   (b)

Example 4 Locate the relative extrema of $f(x) = x^3 - 3x^2 + 3x - 1$.

Solution. Differentiating yields

$$f'(x) = 3x^2 - 6x + 3 = 3(x - 1)^2$$

Solving $f'(x) = 0$ yields $x = 1$ as the only critical point. Since $3(x - 1)^2 \geq 0$ for all x, $f'(x)$ does not change sign at $x = 1$, so f does not have a relative extremum at $x = 1$. Thus, f has no relative extrema. ◄

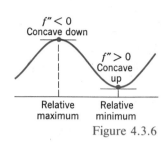

$f'' < 0$
Concave down

$f'' > 0$
Concave up

Relative maximum   Relative minimum

Figure 4.3.6

There is another test for relative extrema that is often easier to apply than the first derivative test. It is based on the geometric observation that a function has a relative maximum where its graph has a horizontal tangent line and is concave down (Figure 4.3.6), and has a relative minimum where its graph has a horizontal tangent line and is concave up (Figure 4.3.6).

Second Derivative Test

> **4.3.7** THEOREM. *Suppose f is twice differentiable at a stationary point x_0.*
>
> *(a) If $f''(x_0) > 0$, then f has a relative minimum at x_0.*
> *(b) If $f''(x_0) < 0$, then f has a relative maximum at x_0.*

if $f''(x_0) = 0$ *can't tell* $\Rightarrow$ *use* 1st *der. test*

A formal proof of this result is given in Appendix 2.

Example 5 Locate and describe the relative extrema of $f(x) = x^4 - 2x^2$.

Solution.

$$f'(x) = 4x^3 - 4x = 4x(x - 1)(x + 1)$$
$$f''(x) = 12x^2 - 4$$

Solving $f'(x) = 0$ yields the stationary points $x = 0$, $x = 1$, and $x = -1$. Since

$$f''(0) = -4 < 0$$
$$f''(1) = 8 > 0$$
$$f''(-1) = 8 > 0$$

there is a relative maximum at $x = 0$ and there are relative minima at $x = 1$ and $x = -1$. ◄

REMARK. If f is not twice differentiable at the critical point x_0, or if $f''(x_0) = 0$, then we must either rely on the first derivative test or devise more imaginative techniques appropriate to the problem.

REMARK. In some problems the first derivative test is easier to apply, and in others the second derivative test is easier. Thus, in each problem some thought should be given to determining the test that involves the least amount of computation.

OPTIONAL

Proof of Theorem 4.3.4. We shall prove this theorem in the case where there is a relative maximum at x_0. The proof in the case of a relative minimum is similar and is left to the reader.

There are two possibilities—either f is differentiable at x_0 or it is not. If it is not, then x_0 is a critical point for f and we are done. If f is differentiable at x_0, then we must show that $f'(x_0) = 0$. We shall do this by showing that $f'(x_0) \geq 0$ and $f'(x_0) \leq 0$, from which it follows that $f'(x_0) = 0$. From the definition of a derivative we have

$$f'(x_0) = \lim_{h \to 0} \frac{f(x_0 + h) - f(x_0)}{h}$$

so that

$$f'(x_0) = \lim_{h \to 0^+} \frac{f(x_0 + h) - f(x_0)}{h} \tag{1}$$

and

$$f'(x_0) = \lim_{h \to 0^-} \frac{f(x_0 + h) - f(x_0)}{h} \tag{2}$$

Because f has a relative maximum at x_0, there is an open interval (a, b) containing x_0 in which $f(x) \leq f(x_0)$ for all x in (a, b).

Assume that h is sufficiently small so that $x_0 + h$ lies in the interval (a, b). Thus,

$$f(x_0 + h) \leq f(x_0)$$

or equivalently

$$f(x_0 + h) - f(x_0) \leq 0$$

Thus, if h is negative

$$\frac{f(x_0 + h) - f(x_0)}{h} \geq 0 \tag{3}$$

and if h is positive

$$\frac{f(x_0 + h) - f(x_0)}{h} \leq 0 \tag{4}$$

But an expression that never assumes negative values cannot approach a negative limit and an expression that never assumes positive values cannot approach a positive limit, so that

$$f'(x_0) = \lim_{h \to 0^-} \frac{f(x_0 + h) - f(x_0)}{h} \geq 0 \quad \text{[from (1) and (3)]}$$

and

$$f'(x_0) = \lim_{h \to 0^+} \frac{f(x_0 + h) - f(x_0)}{h} \leq 0 \quad \text{[from (2) and (4)]}$$

Since $f'(x_0) \geq 0$ and $f'(x_0) \leq 0$, it must be that $f'(x_0) = 0$. ∎

▶ Exercise Set 4.3

In Exercises 1–16, locate the critical points and classify them as stationary points or points of nondifferentiability.

1. $f(x) = x^2 - 5x + 6$.

2. $f(x) = 4x^2 + 2x - 5$.

3. $f(x) = x^3 + 3x^2 - 9x + 1$.

4. $f(x) = 2x^3 - 6x + 7$.

5. $f(x) = x^4 - 6x^2 - 3$.

6. $f(x) = 3x^4 - 4x^3$.   7. $f(x) = \dfrac{x}{x^2 + 2}$.

8. $f(x) = \dfrac{x^2 - 3}{x^2 + 1}$.   9. $f(x) = x^{2/3}$.

10. $f(x) = \sqrt[3]{x + 2}$.   11. $f(x) = \cos 3x$.

12. $f(x) = x \tan x$, $-\pi/2 < x < \pi/2$.

13. $f(x) = \sin^2 2x$, $0 < x < 2\pi$.

14. $f(x) = |\sin x|$.   15. $f(x) = x^{1/3}(x + 4)$.

16. $f(x) = x^{4/3} - 6x^{1/3}$.

In Exercises 17–20, find the relative extrema using (a) the first derivative test; (b) the second derivative test.

17. $f(x) = 1 - 4x - x^2$.

18. $f(x) = 2x^3 - 9x^2 + 12x$.

19. $f(x) = \sin^2 x$, $0 < x < 2\pi$.

20. $f(x) = \frac{1}{2}x - \sin x$, $0 < x < 2\pi$.

In Exercises 21–38, use any method to find the relative extrema.

21. $f(x) = x^3 + 5x - 2$.   22. $f(x) = x^4 - 2x^2 + 7$.

23. $f(x) = x(x - 1)^2$.   24. $f(x) = x^4 + 2x^3$.

25. $f(x) = 2x^2 - x^4$.   26. $f(x) = (2x - 1)^5$.

27. $f(x) = x^{4/5}$.   28. $f(x) = 2x + x^{2/3}$.

29. $f(x) = \dfrac{x^2}{x^2 + 1}$.   30. $f(x) = \dfrac{x}{x + 2}$.

31. $f(x) = |x^2 - 4|$.

32. $f(x) = \begin{cases} 9 - x, & x \leq 3 \\ x^2 - 3, & x > 3. \end{cases}$

33. $f(x) = \cos^2 x$.

34. $f(x) = \sqrt{3}x + 2 \sin x$, $0 < x < 2\pi$.

35. $f(x) = \tan(x^2 + 1)$.

36. $f(x) = \dfrac{\sin x}{2 + \cos x}$, $0 < x < 2\pi$.

37. $f(x) = |\sin 2x|$, $0 < x < 2\pi$.

38. $f(x) = \cos 4x + 2 \sin 2x$, $0 < x < \pi$.

39. Find the value of k so that $x^2 + \dfrac{k}{x}$ will have a relative extremum at $x = 3$.

40. Find the value of k so that $\dfrac{x.}{x^2 + k}$ will have a relative extremum at $x = 2.5$.

41. Recall that the second derivative test (Theorem 4.3.7) does not apply if $f''(x_0) = 0$. Give examples

to show that a function f can have a relative maximum at x_0, a relative minimum at x_0, or neither if $f''(x_0) = 0$. [*Hint:* Try functions of the form $f(x) = x^n$.]

42. Let h and g have relative maxima at x_0. Prove or disprove:
(a) $h + g$ has a relative maximum at x_0
(b) $h - g$ has a relative maximum at x_0.

43. Sketch some curves that show that the three parts of the first derivative test (Theorem 4.3.6) can be false without the assumption that f is continuous at x_0.

4.4 SKETCHING GRAPHS OF POLYNOMIALS AND RATIONAL FUNCTIONS

In this section we shall show how to use the tools developed in the previous section to obtain graphs of polynomials and rational functions.

GRAPHS OF POLYNOMIALS

Polynomials are among the simplest functions to graph; the information obtained in the following steps is usually sufficient to obtain the main elements of the graph.

> ### How to Graph a Polynomial P(x)
>
> **Step 1.** Calculate $P'(x)$ and $P''(x)$.
>
> **Step 2.** From $P'(x)$ determine the stationary points and the intervals where P is increasing and decreasing.
>
> **Step 3.** From $P''(x)$ determine the inflection points and the intervals where P is concave up and concave down.
>
> **Step 4.** Plot a few well-chosen points to provide any additional information needed to complete the shape of the graph. (For example, the points where the graph crosses the axes.)

Example 1 Sketch the graph of $y = x^3 - 3x + 2$.

Solution.

$$\frac{dy}{dx} = 3x^2 - 3 = 3(x - 1)(x + 1)$$

Sign of $(x - 1)$

Sign of $(x + 1)$

Sign of $\dfrac{dy}{dx}$

Figure 4.4.1

$$\frac{d^2y}{dx^2} = 6x$$

From the sign of dy/dx (Figure 4.4.1) the graph of f is increasing if $x < -1$, decreasing if $-1 < x < 1$, and increasing again when $x > 1$; there are stationary points at $x = -1$ and $x = 1$. This information is recorded in Figure 4.4.2a. Since d^2y/dx^2 is positive when $x > 0$ and negative when $x < 0$, the graph of f is concave up when $x > 0$ and concave down when $x < 0$; there is an inflection point at $x = 0$, since f changes concavity there. The concavity information is also recorded in Figure 4.4.2a. From

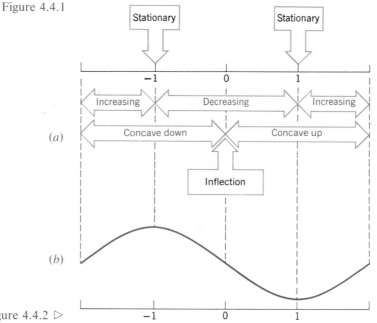

(a)

(b)

Figure 4.4.2 ▷

the information obtained thus far we can deduce the basic shape of the graph (Figure 4.4.2b). All that remains is to refine this sketch by plotting some key points. In this case, the stationary points, the inflection point, and two additional points suffice to give an accurate sketch of the graph. From Table 4.4.1 and Figure 4.4.2 we obtain the final graph shown in Figure 4.4.3.   ◀

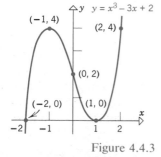

Figure 4.4.3

Table 4.4.1

x	-2	-1	0	1	2
$y = x^3 - 3x + 2$	0	4	2	0	4

BEHAVIOR OF A
POLYNOMIAL AS
$x \to +\infty$ OR $x \to -\infty$

It is often important to know how a polynomial $P(x)$ behaves as $x \to +\infty$ or $x \to -\infty$. For example, it is evident from Figure 4.4.3 that

$$\lim_{x \to +\infty} (x^3 - 3x + 2) = +\infty$$

$$\lim_{x \to -\infty} (x^3 - 3x + 2) = -\infty$$

These results can be obtained algebraically by treating the polynomial as a quotient with 1 as the denominator, then dividing the numerator and denominator by the highest power of x in the polynomial:

$$\lim_{x \to +\infty} (x^3 - 3x + 2) = \lim_{x \to +\infty} \frac{1 - 3/x^2 + 2/x^3}{1/x^3} = +\infty$$

$$\lim_{x \to -\infty} (x^3 - 3x + 2) = \lim_{x \to -\infty} \frac{1 - 3/x^2 + 2/x^3}{1/x^3} = -\infty$$

The first limit results from the fact that the numerator approaches 1 and the denominator approaches 0 through *positive* values, and the second limit results from the fact that the numerator approaches 1 and the denominator approaches 0 through *negative* values.

In the exercises we show how to ascertain the behavior of any polynomial $P(x)$ as $x \to +\infty$ or $x \to -\infty$ by inspection (Exercise 44).

GRAPHS OF
RATIONAL
FUNCTIONS

Recall that if $P(x)$ and $Q(x)$ are polynomials, then their ratio

$$f(x) = \frac{P(x)}{Q(x)}$$

is called a rational function of x. Graphing rational functions is complicated by the fact that discontinuities occur at points where $Q(x) = 0$. In Figure 4.4.4 we have sketched the graph of

$$f(x) = \frac{x}{x - 2}$$

This figure illustrates most of the characteristics that are typical of rational functions. At $x = 2$ the denominator of $f(x)$ is zero, so that a discontinuity occurs in the graph at this point. Approaching the point $x = 2$, we have

$$\lim_{x \to 2^+} \frac{x}{x - 2} = +\infty \quad \text{and} \quad \lim_{x \to 2^-} \frac{x}{x - 2} = -\infty$$

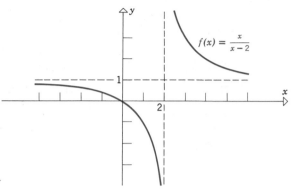

Figure 4.4.4 ▷

The line $x = 2$ is called a *vertical asymptote* for the graph. Also

$$\lim_{x \to +\infty} \frac{x}{x-2} = 1 \quad \text{and} \quad \lim_{x \to -\infty} \frac{x}{x-2} = 1$$

so that the graph tends toward the line $y = 1$ as $x \to +\infty$ and as $x \to -\infty$. The line $y = 1$ is called a *horizontal asymptote* for the graph. More precisely, we make the following definition.

4.4.1 DEFINITION. A line $x = x_0$ is called a ***vertical asymptote*** for the graph of a function f if $f(x) \to +\infty$ or $f(x) \to -\infty$ as x approaches x_0 from the right or from the left. A line $y = L$ is called a ***horizontal asymptote*** for the graph of a function f if $f(x) \to L$ as $x \to +\infty$ or $x \to -\infty$.

In our discussion of graphing rational functions, we shall assume for simplicity that the numerator and denominator have no common factors. For such functions, the information obtained in the following steps is usually sufficient to obtain an accurate sketch of the graph.

How to Graph a Rational Function $f(x) = \dfrac{P(x)}{Q(x)}$

Step 1. Find the x-intercepts of $P(x)$. At these values we have $f(x) = 0$, so that the graph intersects the x-axis at these points.

Step 2. Find the x-intercepts of $Q(x)$. At these values, $f(x)$ approaches $+\infty$ or $-\infty$, and the graph has a vertical asymptote.

Step 3. Compute $\lim_{x \to +\infty} f(x)$ and $\lim_{x \to -\infty} f(x)$. If either limit has a finite value L, then the line $y = L$ is a horizontal asymptote.

Step 4. The only places where $f(x)$ can change sign are at the points where the graph intersects the x-axis or has a vertical asymptote. Calculate a sample value of $f(x)$ in each of the open intervals determined by these points to see whether the graph is above or below the x-axis over the interval.

Step 5. From $f'(x)$ and $f''(x)$ determine the stationary points, inflection points, intervals of increase, decrease, upward concavity, and downward concavity.

Step 6. If needed, plot a few well-chosen points and determine whether the graph crosses any of the horizontal asymptotes.

Example 2   Sketch the graph of

$$f(x) = \frac{x^2 - 1}{x^3}$$

Solution.   Setting $P(x) = x^2 - 1$ equal to zero yields $x = 1$ and $x = -1$ as the x-intercepts.

Setting $Q(x) = x^3$ equal to zero yields the line $x = 0$ as the only vertical asymptote.

From the limits

$$\lim_{x \to +\infty} \frac{x^2 - 1}{x^3} = \lim_{x \to +\infty} \left(\frac{1}{x} - \frac{1}{x^3} \right) = 0 \tag{1}$$

and

$$\lim_{x \to -\infty} \frac{x^2 - 1}{x^3} = \lim_{x \to -\infty} \left(\frac{1}{x} - \frac{1}{x^3} \right) = 0 \tag{2}$$

it follows that the line $y = 0$ is the only horizontal asymptote.

If we mark on the x-axis, the location of the vertical asymptote and the x-intercepts, then these points separate the x-axis into the open intervals

$$(-\infty, -1) \quad (-1, 0) \quad (0, 1) \quad (1, +\infty)$$

Over each interval the graph remains above or below the x-axis. We can determine which is the case by finding the sign of $f(x)$ at an arbitrary sample point in the interval.

At this stage we can make a rough sketch of the graph (Figure 4.4.5). Curve segments A and B in the figure were constructed by using Table 4.4.2 together with limits (1) and (2). Curve segments C and D follow from

Figure 4.4.5 ▷

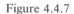

```
+ + + + + + + + + 0 – – –
         |        |
         0       √3
```
Sign of $\sqrt{3} - x$

```
– – – 0 + + + + + + + + + +
  |        |        |
 –√3       0
```
Sign of $\sqrt{3} + x$

```
              ∞
– – – 0 + + + + + + 0 – – –
  |        |        |
 –√3       0       √3
```
Sign of $f'(x)$

Figure 4.4.6

```
– – – – – – – – – 0 + + +
         |        |
         0       √6
```
Sign of $x - \sqrt{6}$

```
– – – 0 + + + + + + + + + +
  |        |        |
 –√6       0
```
Sign of $x + \sqrt{6}$

```
– – – – – – – + + + + + + +
         |        |
         0
```
Sign of x^5

```
– – – 0 + + + – – – 0 + + +
  |        |        |
 –√6       0       √6
```
Sign of $f''(x)$

Figure 4.4.7

Table 4.4.2 and the fact that $x = -1$ and $x = 1$ are x-intercepts; and curve segments E and F follow from Table 4.4.2 and the fact that the y-axis is a vertical asymptote.

Table 4.4.2

INTERVAL	SAMPLE POINT	$f(x)$	CONCLUSION
$(-\infty, -1)$	$x = -2$	$f(-2) = -\dfrac{3}{8} < 0$	The graph is below the x-axis.
$(-1, 0)$	$x = -\dfrac{1}{2}$	$f\left(-\dfrac{1}{2}\right) = 6 > 0$	The graph is above the x-axis.
$(0, 1)$	$x = \dfrac{1}{2}$	$f\left(\dfrac{1}{2}\right) = -6 < 0$	The graph is below the x-axis.
$(1, +\infty)$	$x = 2$	$f(2) = \dfrac{3}{8} > 0$	The graph is above the x-axis.

It is conceivable that our rough sketch may have missed some inflection points or stationary points, but we can investigate this possibility using the first and second derivatives:

$$f'(x) = \frac{x^3(2x) - (x^2 - 1)(3x^2)}{(x^3)^2} = \frac{3 - x^2}{x^4}$$

$$f''(x) = \frac{x^4(-2x) - (3 - x^2)(4x^3)}{(x^4)^2} = \frac{2(x^2 - 6)}{x^5}$$

Since the denominator of $f'(x)$ cannot be negative, the sign of $f'(x)$ is determined by the sign of the numerator. A sign analysis of $f'(x)$ is shown in Figure 4.4.6, and a sign analysis of $f''(x)$ is shown in Figure 4.4.7. From

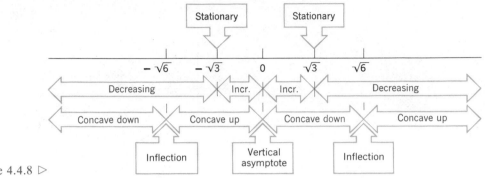

Figure 4.4.8 ▷

Figures 4.4.6 and 4.4.7 we deduce the results recorded in Figure 4.4.8. For our final sketch (Figure 4.4.9) we used our earlier rough sketch, and the data in Figure 4.4.8. We also plotted the stationary points and the inflection points.   ◀

Figure 4.4.9 ▷

REMARK.   Sometimes the work required to graph a polynomial or a rational function can be reduced by taking advantage of its symmetry properties. For example, the curve

$$y = \frac{x^2 - 1}{x^3}$$

in Example 2 is symmetric about the origin, since replacing y by $-y$ and x by $-x$ yields

$$-y = \frac{(-x)^2 - 1}{(-x)^3} \quad \text{or} \quad -y = -\frac{x^2 - 1}{x^3}$$

which reduces to the original equation on multiplying by -1. Thus, we could have proceeded by sketching the portion of the graph to the right of the y-axis, and then constructing the graph in the second and third quadrants by symmetry (Figure 4.4.9).

► Exercise Set 4.4

In Exercises 1–18, use the techniques illustrated in this section to sketch the graph of the given polynomial. Plot the stationary points and the inflection points.

1. $x^2 - 2x - 3$.

2. $1 + x - x^2$.

3. $x^3 - 3x + 1$.

4. $2x^3 - 6x + 4$.

5. $x^3 + 3x^2 + 5$.

6. $x^2 - x^3$.

7. $2x^3 - 3x^2 + 12x + 9$.

8. $x^3 - 3x^2 + 3$.

9. $(x - 1)^4$.

10. $(x - 1)^5$.

11. $x^4 + 2x^3 - 1$.

12. $x^4 - 2x^2 - 12$.

13. $x^4 - 3x^3 + 3x^2 + 1$.

14. $x^5 - 4x^4 + 4x^3$.

15. $3x^5 - 5x^3$.

16. $3x^4 + 4x^3$.

17. $x(x - 1)^3$.

18. $x^5 + 5x^4$.

In Exercises 19–37, use the techniques illustrated in this section to sketch the graph of the given rational function. Show any horizontal and vertical asymptotes, and plot the stationary points and, if reasonable, the inflection points.

19. $\dfrac{2x}{x - 3}$.

20. $\dfrac{x}{x^2 - 1}$.

21. $\dfrac{x^2}{x^2 - 1}$.

22. $\dfrac{1}{(x - 1)^2}$.

23. $\dfrac{x}{1 + x^2}$.

24. $1 - \dfrac{1}{x}$.

25. $\dfrac{x - 1}{x - 2}$.

26. $\dfrac{1}{x^2 + 1}$.

27. $x^2 - \dfrac{1}{x}$.

28. $\dfrac{2x^2 - 1}{x^2}$.

29. $\dfrac{1 - x}{x^2}$.

30. $\dfrac{8}{4 - x^2}$.

31. $\dfrac{x - 1}{x^2 - 4}$.

32. $\dfrac{8(x - 2)}{x^2}$.

33. $\dfrac{(x - 1)^2}{x^2}$.

34. $2 + \dfrac{3}{x} - \dfrac{1}{x^3}$.

35. $3 - \dfrac{4}{x} - \dfrac{4}{x^2}$.

36. $\dfrac{x^2 - 1}{x^2 + 1}$.

37. $\dfrac{x^3 - 1}{x^3 + 1}$.

38. (*Oblique Asymptotes*) If a rational function $P(x)/Q(x)$ is such that the degree of the numerator exceeds the degree of the denominator by *one*, then the graph of $P(x)/Q(x)$ will have an *oblique asymptote*, that is, an asymptote that is neither vertical nor horizontal. To see why, we perform the division of $P(x)$ by $Q(x)$ to obtain

$$\frac{P(x)}{Q(x)} = (ax + b) + \frac{R(x)}{Q(x)}$$

where $ax + b$ is the quotient and $R(x)$ is the remainder. Use the fact that the degree of the remainder $R(x)$ is less than the degree of the divisor $Q(x)$ to help prove:

$$\lim_{x \to +\infty} \left| \frac{P(x)}{Q(x)} - (ax + b) \right| = 0$$

$$\lim_{x \to -\infty} \left| \frac{P(x)}{Q(x)} - (ax + b) \right| = 0$$

These results tell us that the graph of the equation $y = P(x)/Q(x)$ "tends" toward the line (oblique asymptote) $y = ax + b$ as $x \to +\infty$ or $x \to -\infty$ (Figure 4.4.10).

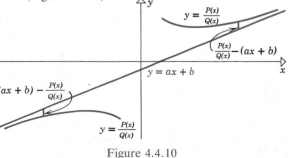

Figure 4.4.10

In Exercises 39–43, sketch the graph of the rational function. Show all vertical, horizontal, and oblique asymptotes (see Exercise 38).

39. $\dfrac{x^2 - 2}{x}$.

40. $\dfrac{x^2 - 2x - 3}{x + 2}$.

41. $\dfrac{(x - 2)^3}{x^2}$.

42. $\dfrac{4 - x^3}{x^2}$.

43. $x + 1 - \dfrac{1}{x} - \dfrac{1}{x^2}$.

44. For an nth degree polynomial

$$P(x) = a_n x^n + a_{n-1} x^{n-1} + \cdots + a_0 \qquad (a_n \neq 0)$$

prove:

(a) If n is even, then

$$\lim_{x \to +\infty} P(x) = \lim_{x \to -\infty} P(x) = +\infty \qquad \text{if } a_n > 0$$

$$\lim_{x \to +\infty} P(x) = \lim_{x \to -\infty} P(x) = -\infty \qquad \text{if } a_n < 0$$

(b) If n is odd, then

$$\lim_{x \to +\infty} P(x) = \lim_{x \to -\infty} P(x) = +\infty \qquad \text{if } a_n < 0$$

$$\lim_{x \to +\infty} P(x) = \lim_{x \to -\infty} P(x) = -\infty \qquad \text{if } a_n > 0$$

4.5   OTHER GRAPHING PROBLEMS

In this section we shall consider functions whose graphs have characteristics not found in the graphs of polynomials and rational functions.

VERTICAL TANGENT LINES

4.5.1   DEFINITION. The graph of a function f is said to have a *vertical tangent line* at x_0 if f is continuous at x_0 and $|f'(x)|$ approaches $+\infty$ as $x \to x_0$.

Example 1   Sketch the graph of $f(x) = \sqrt[3]{x}$.

Solution.   The function $f(x) = \sqrt[3]{x} = x^{1/3}$ is continuous and

$$f'(x) = \frac{1}{3} x^{-2/3} = \frac{1}{3x^{2/3}}$$

$$f''(x) = \frac{1}{3} \left(-\frac{2}{3} \right) x^{-5/3} = -\frac{2}{9x^{5/3}}$$

The first derivative is positive for all x except $x = 0$, so that $f(x)$ is increasing on the intervals $(-\infty, 0)$ and $(0, +\infty)$. There are no stationary points since $f'(x) \neq 0$ for any x. However, $x = 0$ is a critical point since $f'(0)$ is undefined. At this point

$$\lim_{x \to 0} |f'(x)| = \lim_{x \to 0} \left| \frac{1}{3x^{2/3}} \right| = +\infty$$

so the graph of f has a vertical tangent line at $x = 0$.

The second derivative of f is negative if $x > 0$ and positive if $x < 0$,

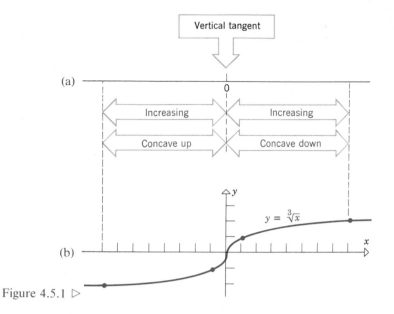

Figure 4.5.1 ▷

so that the graph of f is concave down on the interval $(0, +\infty)$ and concave up on the interval $(-\infty, 0)$ (Figure 4.5.1a).

Since

$$\lim_{x \to +\infty} f(x) = \lim_{x \to +\infty} \sqrt[3]{x} = +\infty$$

$$\lim_{x \to -\infty} f(x) = \lim_{x \to -\infty} \sqrt[3]{x} = -\infty$$

the graph of f has no horizontal asymptotes. Thus, from Figure 4.5.1a and the points plotted from the following table, we obtain the sketch in Figure 4.5.1b.   ◀

x	-8	-1	0	1	8
$\sqrt[3]{x}$	-2	-1	0	1	2

CUSPS

4.5.2 DEFINITION. The graph of a function f is said to have a *cusp* at x_0 if f is continuous at x_0 and $f'(x) \to +\infty$ as x approaches x_0 from one side, while $f'(x) \to -\infty$ as x approaches x_0 from the other side (Figure 4.5.2).

At a cusp there is a vertical tangent since $|f'(x)| \to +\infty$ as $x \to x_0$ from either side.

Figure 4.5.2 ▷

Example 2   Sketch the graph of $f(x) = (x - 4)^{2/3}$.

Solution.   The function f is continuous and

$$f'(x) = \frac{2}{3}(x - 4)^{-1/3} = \frac{2}{3(x - 4)^{1/3}}$$

$$f''(x) = -\frac{2}{9}(x - 4)^{-4/3} = -\frac{2}{9(x - 4)^{4/3}}$$

Since $f'(x) > 0$ when $x > 4$ and $f'(x) < 0$ when $x < 4$, the graph of f is increasing when $x > 4$ and decreasing when $x < 4$. There are no stationary points since $f'(x) \neq 0$ for any value of x. However, $x = 4$ is a critical point since $f'(4)$ does not exist. At this point

$$\lim_{x \to 4^-} f'(x) = \lim_{x \to 4^-} \frac{2}{3(x - 4)^{1/3}} = -\infty$$

$$\lim_{x \to 4^+} f'(x) = \lim_{x \to 4^+} \frac{2}{3(x - 4)^{1/3}} = +\infty$$

so that the graph of f has a cusp at $x = 4$ (Figure 4.5.3).
With the exception of the point $x = 4$, where $f''(x)$ is undefined, $f''(x)$ is negative. Thus, the graph of f is concave down if $x < 4$ and also concave down if $x > 4$ (Figure 4.5.3).

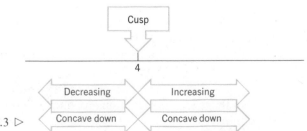

Figure 4.5.3 ▷

Since

$$\lim_{x \to -\infty} f(x) = \lim_{x \to -\infty} (x - 4)^{2/3} = +\infty$$

$$\lim_{x \to +\infty} f(x) = \lim_{x \to +\infty} (x - 4)^{2/3} = +\infty$$

the graph of f has no horizontal asymptotes. Thus, from Figure 4.5.3 and the points plotted from the following table we obtain the graph in Figure 4.5.4. ◄

x	0	3	4	5	8
$y = (x - 4)^{2/3}$	$\sqrt[3]{16} \approx 2.52$	1	0	1	$\sqrt[3]{16} \approx 2.52$

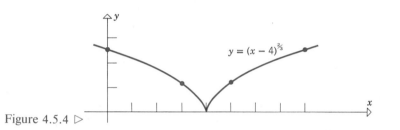

Figure 4.5.4 ▷

$y = (x - 4)^{\frac{2}{3}}$

Example 3 Sketch the graph of $f(x) = 6x^{1/3} + 3x^{4/3}$.

Solution. We have

$$f(x) = 6x^{1/3} + 3x^{4/3} = 3x^{1/3}(2 + x)$$

$$f'(x) = 2x^{-2/3} + 4x^{1/3} = 2x^{-2/3}(1 + 2x)$$

$$f''(x) = -\frac{4}{3}x^{-5/3} + \frac{4}{3}x^{-2/3} = \frac{4}{3}x^{-5/3}(-1 + x)$$

From the formula for f we see that f is continuous and $f(x) = 0$ if $x = 0$ or $x = -2$. From the formulas for f' and f'' we obtain Figure 4.5.5, and from that figure together with the points plotted from the table in Figure 4.5.6, we obtain the graph shown in that figure. ◄

REMARK. Note that in Figure 4.5.6 we had to use different scales on the coordinate axes to keep the size of the figure reasonable. Sometimes this is unavoidable. Also, note how subtle the inflection point is at $x = 1$. It is barely perceptible in our figure, yet we know it is there from our calculations.

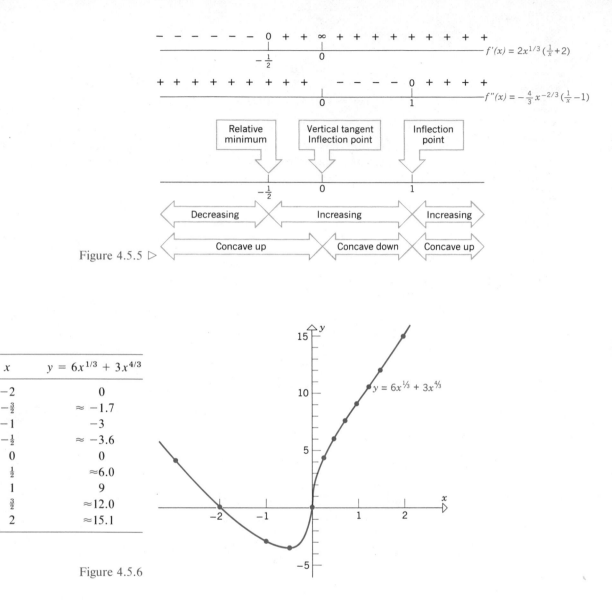

$$f'(x) = 2x^{1/3}\left(\tfrac{1}{x}+2\right)$$

$$f''(x) = -\tfrac{4}{3}x^{-2/3}\left(\tfrac{1}{x}-1\right)$$

Figure 4.5.5 ▷

x	$y = 6x^{1/3} + 3x^{4/3}$
-2	0
$-\frac{3}{2}$	≈ -1.7
-1	-3
$-\frac{1}{2}$	≈ -3.6
0	0
$\frac{1}{2}$	≈ 6.0
1	9
$\frac{3}{2}$	≈ 12.0
2	≈ 15.1

Figure 4.5.6

$y = 6x^{1/3} + 3x^{4/3}$

▶ Exercise Set 4.5

In Exercises 1–24, use the techniques illustrated in this section to sketch the graph of the given function.

1. $(x - 2)^{1/3}$.

2. $x^{1/4}$.

3. $x^{1/5}$.

4. $x^{2/5}$.

5. $x^{4/3}$.

6. $x^{-1/3}$.

7. $1 - x^{2/3}$.

8. $\sqrt{x + 2}$.

9. $\sqrt{x^2 - 1}$.

10. $\sqrt[3]{x^2 - 4}$.

11. $2x + 3x^{2/3}$.

12. $4x - 3x^{4/3}$.

13. $x\sqrt{3 - x}$.

14. $4x^{1/3} - x^{4/3}$.

15. $\dfrac{8(\sqrt{x} - 1)}{x}$.

16. $\dfrac{1 + \sqrt{x}}{1 - \sqrt{x}}$.

17. $\dfrac{\sqrt{x}}{x - 3}$.

18. $x^{2/3}(x - 5)$.

19. $x - \cos x$.

20. $x + \tan x$.

21. $\sin x + \cos x$.

22. $\sqrt{3} \cos x + \sin x$.

23. $\sin^2 x,\ 0 \le x \le 2\pi$.

24. $x \tan x,\ -\dfrac{\pi}{2} < x < \dfrac{\pi}{2}$.

25. (a) How can one obtain the graph of $y = |f(x)|$ from the graph of $y = f(x)$?

    (b) Use the graph of $\sin x$ to obtain the graph of $|\sin x|$.

26. (a) How can one obtain the graph of $y = f(|x|)$ from the graph of $y = f(x)$?

    (b) Use the graph of $\sin x$ to obtain the graph of $\sin |x|$.

4.6 MAXIMUM AND MINIMUM VALUES OF A FUNCTION

Problems concerned with finding the "best" way to perform a task are called **optimization** *problems. A large class of optimization problems can be reduced to finding the largest or smallest value of a function and determining where this value occurs. In this section we shall develop some mathematical tools for solving such problems.*

ABSOLUTE EXTREMA

If we imagine the graph of a function f to be a two-dimensional profile of a mountain range (Figure 4.6.1), then the tops of the mountains correspond to the relative maxima and the bottoms of the valleys to the relative minima. Geologically, these are the high and low points of the terrain in their *immediate vicinity*. However, just as a geologist might be interested in finding the highest mountain and deepest valley in the entire mountain range (Figure 4.6.1), so a mathematician might be interested in finding the largest and smallest values of a function over its *entire domain*. This leads us to the following definitions.

Figure 4.6.1 ▷

Figure 4.6.2

Figure 4.6.3

Figure 4.6.4

Figure 4.6.5

4.6.1   DEFINITION. If $f(x_0) \geq f(x)$ for all x in the domain of f, then $f(x_0)$ is called the *maximum value* or *absolute maximum value* of f.

4.6.2   DEFINITION. If $f(x_0) \leq f(x)$ for all x in the domain of f, then $f(x_0)$ is called the *minimum value* or *absolute minimum value* of f.

4.6.3   DEFINITION. A number that is either the maximum or the minimum value of a function f is called an *extreme value* or *absolute extreme value* of f. Sometimes the terms *extremum* or *absolute extremum* are also used.

Often we shall be concerned with the extreme values of f on some specified interval, rather than on the entire domain of f. The meaning of such terms as *maximum value of f on $[a, b]$* or *minimum value of f on (a, b)* should be clear.

Example 1   The function f graphed in Figure 4.6.2 has no maximum value. Its minimum value is 2, and the minimum occurs when $x = 3$.   ◄

Example 2   The function graphed in Figure 4.6.3 has neither a maximum nor a minimum value. However, on the interval $[-3, 3]$ it has both. The maximum value on $[-3, 3]$ is $f(x) = 3$, which occurs when $x = -2$ and $x = 3$. The minimum value on $[-3, 3]$ is $f(x) = -2$, which occurs when $x = 1$.   ◄

Example 3   The function $f(x) = 2x + 1$ graphed in Figure 4.6.4 has a minimum, but no maximum value on $[0, 3)$. The minimum value is 1, and this occurs when $x = 0$. The reason why $f(x)$ has no maximum value is subtle, but important to understand. If we had been considering the interval $[0, 3]$ rather than $[0, 3)$, then $f(x)$ would have had a maximum value of 7 occurring at $x = 3$. However, the point $x = 3$ does not lie in the interval $[0, 3)$, so that 7 is *not* the maximum value on $[0, 3)$. But no number *less* than 7 can be the maximum either, for if M is any number less than 7, there are values of x in $[0, 3)$ where $f(x) > M$ (see Figure 4.6.5). Thus, $f(x) = 2x + 1$ has no maximum value on $[0, 3)$.   ◄

Given a function f, there are various questions we can ask about its maximum and minimum values; for example:

• Does $f(x)$ have a maximum value?

• If $f(x)$ has a maximum value, what is it?

• If $f(x)$ has a maximum value, where does it occur?

These same questions could, of course, be asked about the minimum value of f. We shall now obtain some results that will help us to answer such questions.

The following theorem gives conditions under which the existence of maximum and minimum values of a function is ensured.

Extreme-Value Theorem

4.6.4 THEOREM. *If a function f is continuous on a closed interval $[a, b]$, then f has both a maximum value and a minimum value on $[a, b]$.*

The proof of this theorem is surprisingly difficult, and will be omitted. However, the result is intuitively obvious if we imagine a particle moving along the graph of a continuous function over a closed interval $[a, b]$; during the trip the particle will have to pass through a highest point and a lowest point (Figure 4.6.6).

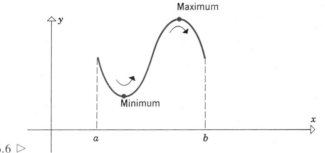

Figure 4.6.6 ▷

In the Extreme-Value Theorem the hypotheses that f is continuous and the interval is closed are essential. If either hypothesis is violated, the existence of maximum or minimum values cannot be guaranteed; this is shown in the next two examples.

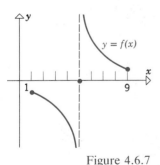

Figure 4.6.7

Example 4 Because $f(x) = 2x + 1$ is a polynomial, it is continuous everywhere. In particular, it is continuous on the interval $[0, 3)$. However, as shown in Example 3, the function f does not have a maximum value on $[0, 3)$. Thus, the closed-interval hypothesis in Theorem 4.6.4 is essential. ◀

Example 5 The function f graphed in Figure 4.6.7 is defined everywhere on the closed interval $[1, 9]$, yet it has neither a maximum nor minimum value on the interval. Since f has a point of discontinuity on the interval $[1, 9]$, this example shows that the continuity hypothesis in Theorem 4.6.4 is essential. ◀

The Extreme-Value Theorem is an example of what mathematicians call an *existence theorem;* it states conditions under which something exists, in this case maximum and minimum values for a function f. However, finding these maximum and minimum values is a separate problem.

Example 6 The polynomial $f(x) = 2x^3 - 15x^2 + 36x$ is continuous everywhere. In particular, it is continuous on the closed interval $[1, 5]$. Thus, the Extreme-Value Theorem tells us that f has both maximum and minimum values on $[1, 5]$. However, this theorem does not tell us what the maximum and minimum values are or where they occur. ◄

The following theorem is the key tool for finding the extreme values of a function.

4.6.5 THEOREM. *If a function f has an extreme value (either a maximum or a minimum) on an open interval (a, b), then the extreme value occurs at a critical point of f.*

Proof. If f has a maximum value on (a, b) at x_0, then $f(x_0)$ is the largest value of f on (a, b) and therefore the largest value of f in the immediate vicinity of x_0. Thus, f has a relative maximum at x_0 and by Theorem 4.3.4, x_0 is a critical point for f. The proof in the case of a minimum value is similar. ∎

FINDING ABSOLUTE EXTREMA

This theorem can be used to locate the extreme values of a continuous function f on a *closed* interval $[a, b]$. For example, consider the possible locations of the maximum. Either the maximum occurs at an endpoint (Figure 4.6.8a) or it occurs in the open interval (a, b), in which case it

(a)   (b)   (c)

Figure 4.6.8

occurs at a critical point (Figures 4.6.8*b* and 4.6.8*c*.) This suggests the following procedure:

> **How to Find the Extreme Values of a Continuous Function f on a Closed Interval [a, b].**
>
> **Step 1.** Find the critical points of f in (a, b).
>
> **Step 2.** Evaluate f at the critical points and the endpoints a and b.
>
> **Step 3.** The largest of the values in Step 2 is the maximum and the smallest is the minimum.

Example 7 Find the maximum and minimum values of the function $f(x) = 2x^3 - 15x^2 + 36x$ on the interval $[1, 5]$, and determine where they occur.

Solution. Since polynomials are differentiable everywhere, f is differentiable everywhere on the interval $(1, 5)$. Thus, if an extreme value occurs on the open interval $(1, 5)$, it must occur at a point where the derivative is zero. Since

$$f'(x) = 6x^2 - 30x + 36$$

the equation $f'(x) = 0$ becomes $6x^2 - 30x + 36 = 0$ or

$$x^2 - 5x + 6 = 0$$
$$(x - 3)(x - 2) = 0$$

Thus, there are two points in the interval $(1, 5)$ where $f'(x) = 0$, namely $x = 2$ and $x = 3$. Evaluating f at these points and the endpoints, we have

$$f(1) = 2(1)^3 - 15(1)^2 + 36(1) = 23$$
$$f(2) = 2(2)^3 - 15(2)^2 + 36(2) = 28$$
$$f(3) = 2(3)^3 - 15(3)^2 + 36(3) = 27$$
$$f(5) = 2(5)^3 - 15(5)^2 + 36(5) = 55$$

Thus, the minimum value is 23 and the maximum value is 55. The minimum occurs at $x = 1$ and the maximum occurs at $x = 5$. ◄

Example 8 Find the extreme values of $f(x) = 6x^{4/3} - 3x^{1/3}$ on the interval $[-1, 1]$ and determine where these values occur.

Solution. Differentiating, we obtain

$$f'(x) = 8x^{1/3} - x^{-2/3} = x^{-2/3}(8x - 1) = \frac{8x - 1}{x^{2/3}}$$

Thus, $f'(x) = 0$ when $x = \frac{1}{8}$ and $f'(x)$ does not exist when $x = 0$. It follows that the critical points of f are $x = 0$, $x = \frac{1}{8}$, both of which lie in the interval $[-1, 1]$. Evaluating f at these critical points and the endpoints we obtain Table 4.6.1. Thus, the minimum value of f on $[-1, 1]$ is $-\frac{9}{8}$, which occurs when $x = \frac{1}{8}$ and the maximum value of f on $[-1, 1]$ is 9, which occurs when $x = -1$. ◄

Table 4.6.1

x	-1	0	$\frac{1}{8}$	1
$f(x)$	9	0	$-\frac{9}{8}$	3

For a continuous function f on a closed interval $[a, b]$, the Extreme-Value Theorem ensures that f has both a maximum and minimum value; and the procedure for finding these values is mechanical—we simply evaluate f at the critical points and the endpoints. For a continuous function on an open interval, half open interval, or infinite interval, the problem of finding maximum and minimum values is more complicated for two reasons:

• The function may not have a maximum or minimum value on such an interval.

• If the function does have maximum or minimum values, it often requires some ingenuity to find them.

It is often possible to determine whether a function has extrema by graphing it. Once it is determined that extrema exist, Theorem 4.6.5 can be applied to help locate them.

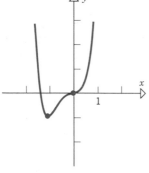

Figure 4.6.9

Example 9 Determine whether the function $f(x) = 3x^4 + 4x^3$ has maximum or minimum values on $(-\infty, +\infty)$ and, if so, find them.

Solution. Using the methods of Section 4.4, the reader should be able to obtain the graph of f shown in Figure 4.6.9. From this graph we see that the function has a minimum value but no maximum. By Theorem 4.6.5 this minimum must occur at a critical point. To locate the critical points, we set $f'(x)$ equal to zero to obtain

$$12x^3 + 12x^2 = 0$$

$$12x^2(x + 1) = 0$$

Thus, the critical points are $x = 0$ and $x = -1$. At $x = 0$ we have an inflection point, and at $x = -1$ we have the desired minimum. Substituting $x = -1$ in $f(x) = 3x^4 + 4x^3$ yields $f(-1) = -1$, which is the minimum value of f on $(-\infty, +\infty)$. ◀

If f is a *continuous* function such that

$$\lim_{x \to +\infty} f(x) = -\infty \quad \text{or} \quad +\infty$$

and

$$\lim_{x \to -\infty} f(x) = -\infty \quad \text{or} \quad +\infty$$

then Table 4.6.2 shows how to ascertain if f has any extrema without actually constructing its graph.

Table 4.6.2

$\lim_{x \to -\infty} f(x)$	$\lim_{x \to +\infty} f(x)$	CONCLUSION (if f is continuous)	GRAPH
$+\infty$	$+\infty$	f has a minimum but no maximum on $(-\infty, +\infty)$.	
$-\infty$	$-\infty$	f has a maximum but no minimum on $(-\infty, +\infty)$.	
$-\infty$	$+\infty$	f has neither a maximum nor a minimum on $(-\infty, +\infty)$.	
$+\infty$	$-\infty$	f has neither a maximum nor a minimum on $(-\infty, +\infty)$.	

Example 10 Find the maximum and minimum values, if any, of $f(x) = x^4 + 2x^3 - 1$ on $(-\infty, +\infty)$.

Solution. Because f is a polynomial it is continuous on $(-\infty, +\infty)$. Moreover,

$$\lim_{x \to +\infty} (x^4 + 2x^3 - 1) = +\infty$$

$$\lim_{x \to -\infty} (x^4 + 2x^3 - 1) = +\infty$$

so f has a minimum but no maximum on $(-\infty, +\infty)$. By Theorem 4.6.5 the minimum occurs at a critical point. Thus, if we evaluate f at each of its critical points, we shall be able to determine where the minimum occurs and find the minimum value. But,

$$f'(x) = 4x^3 + 6x^2 = 2x^2(2x + 3)$$

so that $f'(x) = 0$ yields the critical points $x = 0$ and $x = -\frac{3}{2}$. Evaluating f at these points yields

$$f(0) = -1 \quad \text{and} \quad f(-\tfrac{3}{2}) = -\tfrac{43}{16}$$

so the minimum value is $-\frac{43}{16}$ and this occurs when $x = -\frac{3}{2}$. ◀

There is a variation of the results in Table 4.6.2 that is useful for finding extrema on an open interval. If f is continuous on an open interval (a, b) and

$$\lim_{x \to a^+} f(x) = -\infty \quad \text{or} \quad +\infty$$

$$\lim_{x \to b^-} f(x) = -\infty \quad \text{or} \quad +\infty$$

then Table 4.6.3 shows how to ascertain if f has any extrema on (a, b).

We leave it for the reader to give analogs of the results in Tables 4.6.2 and 4.6.3 for intervals of the form $(-\infty, b)$ and $(a, +\infty)$.

Example 11 Find the maximum and minimum values, if any, of

$$f(x) = \frac{1}{x^2 - x}$$

on the open interval $(0, 1)$.

Solution.   We have

$$\lim_{x \to 0^+} f(x) = \lim_{x \to 0^+} \frac{1}{x^2 - x} = \lim_{x \to 0^+} \frac{1}{x(x-1)} = -\infty$$

$$\lim_{x \to 1^-} f(x) = \lim_{x \to 1^-} \frac{1}{x^2 - x} = \lim_{x \to 1^-} \frac{1}{x(x-1)} = -\infty$$

so f has a maximum, but no minimum on $(0, 1)$. By Theorem 4.6.5 this maximum occurs at a critical point of f. We have

$$f'(x) = -\frac{2x - 1}{(x^2 - x)^2}$$

so that $f'(x) = 0$ if $x = \frac{1}{2}$. Thus, the maximum value of f on $(0, 1)$ occurs at $x = \frac{1}{2}$ and this maximum value is

$$f(\tfrac{1}{2}) = \frac{1}{(\tfrac{1}{2})^2 - \tfrac{1}{2}} = -4 \quad \blacktriangleleft$$

Table 4.6.3

$\lim_{x \to a^+} f(x)$	$\lim_{x \to b^-} f(x)$	CONCLUSION (if f is continuous on (a, b))	GRAPH
$+\infty$	$+\infty$	f has a minimum but no maximum on (a, b).	
$-\infty$	$-\infty$	f has a maximum but no minimum on (a, b).	
$-\infty$	$+\infty$	f has neither a maximum nor a minimum on (a, b).	
$+\infty$	$-\infty$	f has neither a maximum nor a minimum on (a, b).	

Example 12 The function $f(x) = \tan x$ has neither a maximum nor a minimum on $(-\pi/2, \pi/2)$ because $\tan x$ is continuous on this interval and

$$\lim_{x \to -\pi/2^+} \tan x = -\infty, \quad \lim_{x \to \pi/2^-} \tan x = +\infty$$

(See Figure A.28 in Appendix 1.)   ◄

The following theorem can sometimes be helpful in situations where our previous methods do not apply.

4.6.6 THEOREM. *Let f be continuous on an interval I and assume that f has exactly one relative extremum on I, say at x_0.*

(a) If f has a relative minimum at x_0, then $f(x_0)$ is the minimum value of f on the interval I.

(b) If f has a relative maximum at x_0, then $f(x_0)$ is the maximum value of f on the interval I.

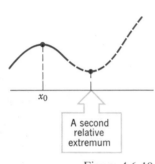

A second relative extremum

Figure 4.6.10

Although we shall omit the proof, the result is easy to visualize. For example, if f has a relative maximum at x_0 and if $f(x_0)$ is *not* the maximum value of f on I, then the graph of f must make an upward turn somewhere on the interval I, thereby introducing a second relative extremum (Figure 4.6.10). Thus, if there is only one relative extremum on I, an absolute maximum value for f must also occur at x_0.

Theorem 4.6.6 reduces the problem of finding extrema to one of finding relative extrema.

Example 13 Find the maximum and minimum values, if any, of $f(x) = x^3 - 3x^2 + 4$ on $(0, +\infty)$.

Solution. We have

$$\lim_{x \to 0^+} (x^3 - 3x^2 + 4) = 4$$

$$\lim_{x \to +\infty} (x^3 - 3x^2 + 4) = +\infty$$

Because the first limit is finite, Tables 4.6.2 and 4.6.3 do not apply. Moreover, the second limit tells us that f definitely has no maximum on $(0, +\infty)$. Thus, there are two possibilities: either f has no extreme values on $(0, +\infty)$ or f has a minimum on $(0, +\infty)$. To determine which is the case, we shall look for critical points. We have

$$f'(x) = 3x^2 - 6x = 3x(x - 2)$$

so that $f'(x) = 0$ yields the critical points $x = 0$ and $x = 2$. However,

$x = 2$ is the only critical point in the interval $(0, +\infty)$, so we shall try to apply Theorem 4.6.6. Since

$$f''(x) = 6x - 6$$

we have $f''(2) = 6 > 0$. Thus, a relative minimum occurs at $x = 2$ by the second derivative test and consequently a minimum value for f occurs at $x = 2$ by Theorem 4.6.6. This minimum value is $f(2) = 0$.   ◀

► Exercise Set 4.6

In Exercises 1–14, find the maximum and minimum values of f on the given closed interval and state where these values occur.

1. $f(x) = 4x^2 - 4x + 1$; $[0, 1]$.

2. $f(x) = 8x - x^2$; $[0, 6]$.

3. $f(x) = (x - 1)^3$; $[0, 4]$.

4. $f(x) = 2x^3 - 3x^2 - 12x$; $[-2, 3]$.

5. $f(x) = \dfrac{3x}{\sqrt{4x^2 + 1}}$; $[-1, 1]$.

6. $f(x) = \dfrac{x}{x^2 + 2}$; $[-1, 4]$.

7. $f(x) = x^{2/3}(20 - x)$; $[-1, 20]$.

8. $f(x) = (x^2 + x)^{2/3}$; $[-2, 3]$.

9. $f(x) = x - \tan x$; $[-\pi/4, \pi/4]$.

10. $f(x) = \sin x - \cos x$; $[0, \pi]$.

11. $f(x) = 2 \sec x - \tan x$; $[0, \pi/4]$.

12. $f(x) = \sin^2 x + \cos x$; $[-\pi, \pi]$.

13. $f(x) = 1 + |9 - x^2|$; $[-5, 1]$.

14. $f(x) = |6 - 4x|$; $[-3, 3]$.

In Exercises 15–26, find the maximum and minimum values of f on the given interval, if they exist.

15. $f(x) = x^2 - 3x - 1$; $(-\infty, +\infty)$.

16. $f(x) = 3 - 4x - 2x^2$; $(-\infty, +\infty)$.

17. $f(x) = 4x^3 - 3x^4$; $(-\infty, +\infty)$.

18. $f(x) = x^4 + 4x$; $(-\infty, +\infty)$.

19. $f(x) = (x^2 - 1)^2$; $(-\infty, +\infty)$.

20. $f(x) = (x - 1)^2(x + 2)^2$; $(-\infty, +\infty)$.

21. $f(x) = x^3 - 3x - 2$; $(-\infty, +\infty)$.

22. $f(x) = x^3 - 9x + 1$; $(-\infty, +\infty)$.

23. $f(x) = 1 + \dfrac{1}{x}$; $(0, +\infty)$.

24. $f(x) = \dfrac{x}{x^2 + 1}$; $[0, +\infty)$.

25. $f(x) = \dfrac{x^2}{x + 1}$; $(-5, -1)$.

26. $f(x) = \dfrac{x + 3}{x - 3}$; $[-5, 5]$.

27. Find the maximum and minimum values of

$$f(x) = 2 \sin 2x + \sin 4x$$

and state where the maximum and minimum values occur. [*Hint*: Since f is periodic, this problem can be solved by first finding the maximum and minimum values on an appropriate closed interval.]

28. Find the maximum and minimum values of

$$f(x) = 3 \cos \frac{x}{3} + 2 \cos \frac{x}{2}$$

[See the hint in the previous problem.]

29. Find the maximum and minimum values of

$$f(x) = \begin{cases} 4x - 2, & x < 1 \\ (x - 2)(x - 3), & x \geq 1 \end{cases}$$

on $[\frac{1}{2}, \frac{7}{2}]$.

30. Let $f(x) = x^2 + px + q$. Find values of p and q such that $f(1) = 3$ is an extreme value of f on $[0, 2]$. Is this value a maximum or minimum?

31. Let $f(x) = (x - a)^p$, where p is an integer greater than 1 and a is any real number. Find the relative extrema of f if

(a) p is even (b) p is odd.

32. What is the smallest possible slope for a tangent to $y = x^3 - 3x^2 + 5x$?

33. (a) Show that

$$f(x) = \frac{64}{\sin x} + \frac{27}{\cos x}$$

has a minimum value, but no maximum value on the interval $(0, \pi/2)$.

(b) Find the minimum value.

34. Prove: For every positive value of t the function $f(x) = x + t/x$ has a minimum value but no maximum value on $(0, +\infty)$.

35. Find a_0, a_1, and a_2 such that the graph of $f(x) = a_0 + a_1 x + a_2 x^2$ passes through the point $(0, 9)$ and f has a minimum value of 1 that occurs when $x = 2$.

36. Prove that $\sin x \leq x$ for all x in the interval $[0, 2\pi]$. [*Hint:* Find the minimum value of $x - \sin x$ on the interval.]

37. Prove that

$$1 - \frac{x^2}{2} \leq \cos x$$

for all x in the interval $[0, 2\pi]$. [See the hint in the previous problem.]

38. (a) Sketch the graph of a function f that is continuous on the open interval $(-1, 1)$ and has both maximum and minimum values on the interval.

(b) Sketch the graph of a function f that is defined everywhere on the open interval $(-1, 1)$ and is not continuous everywhere on $(-1, 1)$, yet has both maximum and minimum values on $(-1, 1)$.

(c) Do the functions in parts (a) and (b) violate the Extreme-Value Theorem? Explain.

39. Prove: If $ax^2 + bx + c = 0$ has two distinct real roots, then the midpoint between these roots is a stationary point for $f(x) = ax^2 + bx + c$.

40. Prove Theorem 4.6.5 in the case where the extreme value is a minimum.

41. Let $f(x) = ax^2 + bx + c$, where $a > 0$. Prove that $f(x) \geq 0$ for all x if and only if $b^2 - 4ac \leq 0$. [*Hint:* Find the minimum of $f(x)$.]

42. Prove that the minimum value of

$$f(x) = x^2 + \frac{16x^2}{(8 - x)^2}, \quad x > 8$$

occurs at $x = 4(2 + \sqrt[3]{2})$.

43. Find the maximum value of the function $\sin^2 \theta \cos \theta$ for $0 \leq \theta \leq \pi/2$.

4.7 APPLIED MAXIMUM AND MINIMUM PROBLEMS

In this section we shall show how the methods developed in the previous section can be used to solve some applied optimization problems.

Example 1 Find the dimensions of a rectangle with perimeter 100 ft whose area is as large as possible.

Solution. Let

x = length of the rectangle (ft)
y = width of the rectangle (ft)
A = area of the rectangle (ft^2)

Then

$$A = xy \tag{1}$$

Since the perimeter of the rectangle is 100 ft, the variables x and y are related by the equation

$$2x + 2y = 100$$

Perimeter
$2x + 2y = 100$

Figure 4.7.1

or

$$y = 50 - x \tag{2}$$

(See Figure 4.7.1.) Substituting (2) in (1) yields

$$A = x(50 - x) = 50x - x^2 \tag{3}$$

Because x represents a length it cannot be negative, and because the two sides of length x cannot have a combined length exceeding the total perimeter of 100 ft, the variable x must satisfy

$$0 \leq x \leq 50 \tag{4}$$

Thus, we have reduced the problem to that of finding the value (or values) of x in $[0, 50]$ for which A is maximum. Since A is a polynomial in x, it is continuous on $[0, 50]$ and so the maximum must occur at an endpoint of this interval or at a critical point.

From (3) we obtain

$$\frac{dA}{dx} = 50 - 2x$$

Setting $dA/dx = 0$ we obtain

$$50 - 2x = 0$$

or $x = 25$. Thus, the maximum occurs at one of the points

$$x = 0, \quad x = 25, \quad x = 50$$

Table 4.7.1

x	0	25	50
A	0	625	0

Substituting these values in (3) yields Table 4.7.1, which tells us that the maximum area of 625 ft^2 occurs when $x = 25$. From (2) the corresponding value of y is $y = 25$, so the rectangle of perimeter 100 ft with greatest area is a square with sides of length 25 ft. ◀

REMARK.   In (4) we included $x = 0$ and $x = 50$ as possible values for x. Because $x = 50$ implies $y = 0$ from (2), these x values correspond to rectangles with two sides of length zero. One can argue that these x values should not be allowed because a "true" rectangle cannot have sides of length zero.

Actually, it is just a matter of the assumptions we choose to make. If we view Example 1 as a purely mathematical problem, then there is nothing wrong with allowing sides of length zero. However, if we view it as a practical problem in which the rectangle is to be constructed from physical material, we would not want to allow $x = 0$ or $x = 50$ and (4) should be replaced by $0 < x < 50$. In this case we no longer have a closed interval to work with and the problem has to be solved by other methods (considered in the next section).

In this section we shall consider only optimization problems over finite closed intervals. In the next section we shall discuss other types of applied optimization problems.

Example 1 illustrates the following five-step procedure for solving many applied maximum and minimum problems.

> **Step 1.** Draw an appropriate figure and label the quantities relevant to the problem.
>
> **Step 2.** Find a formula for the quantity to be maximized or minimized.
>
> **Step 3.** Using the conditions stated in the problem to eliminate variables, express the quantity to be maximized or minimized as a function of one variable.
>
> **Step 4.** Find the interval of possible values for this variable from the physical restrictions in the problem.
>
> **Step 5.** If applicable, use the techniques of the previous section to obtain the maximum or minimum.

Example 2 An open box is to be made from a 16 in. by 30 in. piece of cardboard by cutting out squares of equal size from the four corners and bending up the sides (Figure 4.7.2). What size should the squares be to obtain a box with largest possible volume?

Solution. Let

x = length (in inches) of the sides of the squares to be cut out
V = volume (in cubic inches) of the resulting box

Figure 4.7.2 ▷   (a)   (b)

Because we are removing a square of side x from each corner, the resulting box will have dimensions $16 - 2x$ by $30 - 2x$ by x (Figure 4.7.2b). Since the volume of a box is the product of its dimensions, we have

$$V = (16 - 2x)(30 - 2x)x = 480x - 92x^2 + 4x^3 \qquad (5)$$

The variable x in this expression is subject to certain restrictions. Because x represents a length it cannot be negative and because the width of the cardboard is 16 in., we cannot cut out squares whose sides are more than 8 in. long. Thus, the variable x in (5) must satisfy

$$0 \le x \le 8$$

We have thus reduced our problem to finding the value (or values) of x in the interval $[0, 8]$ for which (5) is maximum. Since the right side of (5) is a polynomial in x, it is continuous on the closed interval $[0, 8]$ and consequently we can use the methods developed in the previous section to find the maximum.

From (5) we obtain

$$\frac{dV}{dx} = 480 - 184x + 12x^2 = 4(120 - 46x + 3x^2)$$

Setting $dV/dx = 0$ yields

$$120 - 46x + 3x^2 = 0$$

Table 4.7.2

x	0	$\frac{10}{3}$	8
V	0	$\frac{19600}{27}$	0

which can be solved by the quadratic formula to obtain the critical points

$$x = \tfrac{10}{3} \quad \text{and} \quad x = 12$$

Since $x = 12$ falls outside the interval $[0, 8]$, the maximum value of V must occur either at the critical point $x = \frac{10}{3}$ or at one of the endpoints $x = 0$, $x = 8$. Substituting these values in (5) yields Table 4.7.2, which tells us that the greatest possible volume $V = \frac{19600}{27}$ in.$^3 \approx 726$ in.^{3} occurs when we cut out squares whose sides have length $\frac{10}{3}$ in. ◀

Figure 4.7.3

Example 3 An offshore oil well is located in the ocean at a point W, which is 5 mi from the closest shorepoint A on a straight shoreline (Figure 4.7.3). The oil is to be piped to a shorepoint B that is 8 mi from A by piping it on a straight line under water from W to some shorepoint P between A and B and then on to B via a pipe along the shoreline. If the cost of laying pipe is $100,000 per mile under water and $75,000 per mile over land, where should the point P be located to minimize the cost of laying the pipe?

REMARK.   Since the shortest distance between two points is a straight line, a pipeline directly from W to B uses the least amount of pipe. However, the pipe, being completely under water, would be expensive to lay. At the other extreme, a pipeline from W to A to B uses the least amount of expensive underwater pipe, but uses the greatest total amount of pipe. Thus, it seems plausible that by piping to some point P between A and B, one might incur less total cost than by piping to either of the extreme locations.

Solution.   Let

x = distance (in miles) between A and P
c = cost (in thousands of dollars) for the entire pipeline

From Figure 4.7.3 the length of pipe under water is the distance between W and P. By the Theorem of Pythagoras, that length is

$$\sqrt{x^2 + 25} \text{ mi} \tag{6}$$

Also from Figure 4.7.3, the length of pipe over land is the distance between P and B, which is

$$8 - x \text{ mi} \tag{7}$$

From (6) and (7) it follows that the total cost c (in thousands of dollars) for the pipeline is

$$c = 100\sqrt{x^2 + 25} + 75(8 - x) \tag{8}$$

Because the distance between A and B is 8 mi, the distance x between A and P must satisfy

$$0 \le x \le 8$$

We have thus reduced our problem to finding the value (or values) of x in the interval $[0, 8]$ for which (8) is a minimum. Since c is a continuous function of x on the closed interval $[0, 8]$, we can use the methods developed in the previous section to find the minimum.
    From (8) we obtain

$$\frac{dc}{dx} = \frac{100x}{\sqrt{x^2 + 25}} - 75 = 25\left(\frac{4x}{\sqrt{x^2 + 25}} - 3\right)$$

Setting $dc/dx = 0$ yields

$$\frac{4x}{\sqrt{x^2 + 25}} - 3 = 0 \tag{9}$$

or

$$4x = 3\sqrt{x^2 + 25}$$

$$16x^2 = 9(x^2 + 25)$$

$$7x^2 = 225$$

$$x = \pm \frac{15}{\sqrt{7}}$$

The number $-15/\sqrt{7}$ is not a solution of (9) and must be discarded, leaving $x = 15/\sqrt{7}$ as the only critical point. Since this point lies in the interval $[0, 8]$, the minimum must occur at one of the points

$$x = 0, \quad x = 15/\sqrt{7}, \quad x = 8$$

Substituting these values in (8) yields Table 4.7.3, which tells us that the least possible cost for the pipeline, approximately $c = 930.719$ (thousand dollars) = \$930,719 occurs when the point P is located at a distance of $15/\sqrt{7} \approx 5.67$ mi from A.   ◄

(a)

(b)

Figure 4.7.4

Table 4.7.3

x	0	$15/\sqrt{7}$	8
c	1100	$600 + 125\sqrt{7} \approx 930.719$	$100\sqrt{89} \approx 943.398$

Example 4 Find the radius and height of the right-circular cylinder of largest volume that can be inscribed in a right-circular cone with radius 6 in. and height 10 in. (Figure 4.7.4*a*).

Solution.   Let

r = radius (in inches) of the cylinder
h = height (in inches) of the cylinder
V = volume (in cubic inches) of the cylinder

The formula for the volume of the inscribed cylinder is

$$V = \pi r^2 h \tag{10}$$

To eliminate one of the variables in (10) we need a relationship between r and h. Using similar triangles (Figure 4.7.4*b*) we obtain

$$\frac{10 - h}{r} = \frac{10}{6}$$

or

$$h = 10 - \tfrac{5}{3}r \qquad (11)$$

Substituting (11) into (10) we obtain

$$V = \pi r^2(10 - \tfrac{5}{3}r) = 10\pi r^2 - \tfrac{5}{3}\pi r^3 \qquad (12)$$

which expresses V in terms of r alone. Because r represents a radius it cannot be negative, and because the radius of the inscribed cylinder cannot exceed the radius of the cone, the variable r must satisfy

$$0 \le r \le 6$$

Thus, we have reduced the problem to that of finding the value (or values) of r in $[0, 6]$ for which (12) is a maximum. Since V is a continuous function of r on $[0, 6]$, the methods developed in the previous section apply.

From (12) we obtain

$$\frac{dV}{dr} = 20\pi r - 5\pi r^2 = 5\pi r(4 - r)$$

Setting $dV/dr = 0$ gives

$$5\pi r(4 - r) = 0$$

so that $r = 0$ and $r = 4$ are critical points. Since these lie in the interval $[0, 6]$, the maximum must occur at one of the points

$$r = 0, \quad r = 4, \quad r = 6$$

Table 4.7.4

r	0	4	6
V	0	$\tfrac{160}{3}\pi$	0

Substituting these values in (12) yields Table 4.7.4, which tells us that the maximum volume $V = \tfrac{160}{3}\pi \approx 168$ in.^{3} occurs when the inscribed cylinder has radius 4 in. When $r = 4$ it follows from (11) that $h = \tfrac{10}{3}$. Therefore, the inscribed cone of largest volume has radius $r = 4$ in. and height $h = \tfrac{10}{3}$ in. ◄

MARGINAL ANALYSIS (AN APPLICATION TO ECONOMICS)

Economists and business people are interested in how changes in such variables as inventory, production, supply, advertising, and price affect other variables such as profit, revenue, demand, inflation, and employment. Such problems are studied using *marginal analysis*. The word, *marginal*, is the economist's term for a rate of change or derivative.

Three functions of importance to an economist or a manufacturer are:

$C(x)$ = total cost of producing x units of a product during some time period.

$R(x)$ = total revenue received from selling x units of the product during the time period.

$P(x)$ = total profit obtained by selling x units of the product during the time period.

These are called, respectively, the *cost function, revenue function,* and *profit function.* If all units produced are sold, then these are related by

$$P(x) = R(x) - C(x) \tag{13}$$

[profit] = [revenue] − [cost]

The derivatives $C'(x)$, $R'(x)$, and $P'(x)$ are called the **marginal cost, marginal revenue,** and **marginal profit.** If all units produced are sold, then a relationship between them is obtained by differentiating (13),

$$P'(x) = R'(x) - C'(x)$$

$$\begin{bmatrix} \text{marginal} \\ \text{profit} \end{bmatrix} = \begin{bmatrix} \text{marginal} \\ \text{revenue} \end{bmatrix} - \begin{bmatrix} \text{marginal} \\ \text{cost} \end{bmatrix}$$

The quantities $P'(x)$, $R'(x)$, and $C'(x)$ represent the instantaneous rates of change of profit, revenue, and cost with respect to x, where x is the amount of the product produced and sold.

In practice, $C'(x)$ is frequently interpreted as the cost of manufacturing the $(x + 1)$-th unit. Although this is not exact, it is usually a good approximation. The justification for this interpretation is based on the fact that x is usually large, so $\Delta x = 1$ can be considered close to zero by comparison. Thus,

$$C'(x) = \lim_{\Delta x \to 0} \frac{C(x + \Delta x) - C(x)}{\Delta x}$$

$$\approx \frac{C(x + 1) - C(x)}{1} = C(x + 1) - C(x)$$

Since $C(x + 1)$ is the cost of producing $x + 1$ units and $C(x)$ is the cost of producing x units, it follows that $C'(x) \approx C(x + 1) - C(x)$ is the approximate cost of producing the $(x + 1)$-th unit. Similarly, $R'(x)$ is the approximate revenue received from selling the $(x + 1)$-th unit and $P'(x)$ is the approximate profit from manufacturing and selling the $(x + 1)$-th unit.

The total cost $C(x)$ of producing x units can be expressed as a sum

$$C(x) = a + M(x) \tag{14}$$

where a is a constant, called **overhead,** and $M(x)$ is a function representing **manufacturing cost.** The overhead, which includes such fixed costs as rent and insurance, does not depend on x; it must be paid even if nothing is produced. On the other hand, the manufacturing cost $M(x)$, which includes such items as cost of materials and labor, depends on the number of items manufactured. It is shown in economics that with suitable simplifying assumptions, $M(x)$ can be expressed in the form

$$M(x) = bx + cx^2$$

Substituting this in (14) yields

$$C(x) = a + bx + cx^2 \tag{15}$$

Example 5 A paint manufacturer determines that the total cost in dollars of producing x gallons of paint per day is

$$C(x) = 5000 + x + 0.001x^2$$

(a) Find the marginal cost when the production level is 500 gal per day.
(b) Use the marginal cost to approximate the cost of producing the 501st gallon.
(c) Find the exact cost of producing the 501st gallon.

Solution.

(a) The marginal cost is $C'(x) = 1 + 0.002x$, so

$$C'(500) = 1 + (0.002)(500) = 2$$

(b) Since $C'(500) = 2$, the cost of producing the 501st gallon is approximately \$2.00.
(c) The total cost of producing 501 gal is

$$C(501) = 5000 + 501 + 0.001(501)^2 = \$5752.001$$

and the total cost of producing 500 gal is

$$C(500) = 5000 + 500 + 0.001(500)^2 = \$5750$$

so the exact cost of producing the 501st gallon is

$$C(501) - C(500) = \$2.001 \quad \blacktriangleleft$$

If a manufacturing firm can sell all the items it produces for p dollars apiece, then its total revenue $R(x)$ (in dollars) will be

$$R(x) = px \tag{16}$$

and its total profit $P(x)$ (in dollars) will be

$$P(x) = [\text{total revenue}] - [\text{total cost}] = R(x) - C(x) = px - C(x)$$

Thus, if the cost function is given by (15),

$$P(x) = px - (a + bx + cx^2) \tag{17}$$

Depending on such factors as number of employees, amount of machinery available, economic conditions, and competition, there will be some upper limit l on the number of items a manufacturer is capable of producing and selling. Thus, during a fixed time period the variable x in (17) will satisfy

$$0 \leq x \leq l$$

By determining the value or values of x in $[0, l]$ that maximize (17), the firm can determine how many units of its product must be manufactured and sold to yield the greatest profit. This is illustrated in the following numerical example.

Example 6   A liquid form of penicillin manufactured by a pharmaceutical firm is sold in bulk at a price of $200 per unit. If the total production cost (in dollars) for x units is

$$C(x) = 500{,}000 + 80x + 0.003x^2$$

and if the production capacity of the firm is at most 30,000 units in a specified time, how many units of penicillin must be manufactured and sold in that time to maximize the profit?

Solution.   Since the total revenue for selling x units is $R(x) = 200x$, the profit $P(x)$ on x units will be

$$P(x) = R(x) - C(x) = 200x - (500{,}000 + 80x + 0.003x^2) \tag{18}$$

Since the production capacity is at most 30,000 units, x must lie in the interval $[0, 30{,}000]$. From (18)

$$\frac{dP}{dx} = 200 - (80 + 0.006x) = 120 - 0.006x$$

Setting $dP/dx = 0$ gives

$$120 - 0.006x = 0$$

$$x = 20,000$$

Since this critical point lies in the interval $[0, 30,000]$, the maximum profit must occur at one of the points

$$x = 0, \quad x = 20,000, \quad \text{or} \quad x = 30,000$$

Substituting these values in (18) yields Table 4.7.5, which tells us that the maximum profit $P = \$700,000$ occurs when $x = 20,000$ units are manufactured and sold in the specified time. ◀

Table 4.7.5

x	0	20,000	30,000
$P(x)$	$-500,000$	700,000	400,000

▶ Exercise Set 4.7

1. Express the number 10 as a sum of two nonnegative terms whose product is as large as possible.

2. How should two nonnegative numbers be chosen so that their sum is 1 and the sum of their squares is:
   (a) as large as possible?
   (b) as small as possible?

3. Find a number in the closed interval $[\frac{1}{2}, \frac{3}{2}]$ such that the sum of the number and its reciprocal is:
   (a) as small as possible;
   (b) as large as possible.

4. A rectangular field is to be bounded by a fence on three sides and by a straight stream on the fourth side. Find the dimensions of the field with maximum area that can be enclosed with 1000 feet of fence.

5. A rectangular plot of land is to be fenced in using two kinds of fencing. Two opposite sides will use heavy-duty fencing selling for $3 a foot, while the remaining two sides will use standard fencing selling for $2 a foot. What are the dimensions of the rectangular plot of greatest area that can be fenced in at a cost of $6000?

6. Show that among all rectangles with perimeter p, the square with side $p/4$ has the maximum area.

7. Find the dimensions of the rectangle with maximum area that can be inscribed in a circle of radius 10.

8. A rectangle is to be inscribed in a right triangle having sides of length 6 in., 8 in., and 10 in. Find the dimensions of the rectangle with greatest area assuming the rectangle is positioned as in Figure 4.7.5a.

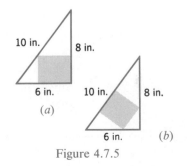

Figure 4.7.5

9. Solve the problem in Exercise 8 assuming the rectangle is positioned as in Figure 4.7.5*b*.

10. A rectangle has its two lower corners on the x-axis and its two upper corners on the curve $y = 16 - x^2$. For all such rectangles, what are the dimensions of the one with largest area?

11. A sheet of cardboard 12 in. square is used to make an open box by cutting squares of equal size from the corners and folding up the sides. What size squares should be cut to obtain a box with largest possible volume?

12. A square sheet of cardboard of side k is used to make an open box by cutting squares of equal size from the corners and folding up the sides. What size squares should be cut from the corners to obtain a box with largest possible volume?

13. An open box is to be made from a 3 ft by 8 ft piece of sheet metal by cutting out squares of equal size from the four corners and bending up the sides. Find the maximum volume that the box can have.

14. The two equal sides of an isosceles triangle are each L units long. Show that the triangle of maximum area is a right triangle.

15. A wire of length 12 in. can be bent into a circle, bent into a square, or cut into two pieces to make both a circle and a square. How much wire should be used for the circle if the total area enclosed by the figure(s) is to be:
(a) a maximum?   (b) a minimum?

16. Assume that the operating cost of a certain truck (excluding driver's wages) is $12 + x/6$ cents per mile when the truck travels at x mi/hr. If the driver earns $6 per hour, what is the most economical speed to operate the truck on a 400-mi turnpike where the minimum speed is 40 mi/hr and the maximum speed is 70 mi/hr?

17. Find the lengths of the sides of the isosceles triangle with perimeter 12 and maximum area.

18. A triangle is inscribed in a semicircle of radius 10 so that one side is along the diameter (Figure 4.7.6). Find the dimensions of the triangle with maximum area.

Figure 4.7.6

19. A box-shaped wire frame consists of two identical wire squares whose vertices are connected by four straight wires of equal length (Figure 4.7.7). If the frame is to be made from a wire of length L, what should the dimensions be to obtain a box of greatest volume?

Figure 4.7.7

20. If air resistance is neglected, then the range R of a cannonball fired from a cannon whose barrel makes an angle θ with the horizontal is

$$R = \left(\frac{v_0^2}{g}\right) \sin 2\theta$$

where the constants v_0 and g are the initial velocity and the acceleration due to gravity. Show that the maximum range is achieved when $\theta = 45°$.

21. (a) A chemical manufacturer sells sulfuric acid in bulk at a price of $100 per unit. If the daily total production cost in dollars for x units is

$$C(x) = 100{,}000 + 50x + 0.0025x^2$$

and if the daily production capacity is at most 7000 units, how many units of sulphuric acid must be manufactured and sold daily to maximize the profit?

(b) Would it benefit the manufacturer to expand the daily production capacity?

22. Find the dimensions of the rectangle of greatest

area that can be inscribed in a semicircle of radius *R*.

23. Find the height and radius of the cone of slant height *L* whose volume is as large as possible.

24. Find the dimensions of the right-circular cylinder of largest volume that can be inscribed in a sphere of radius *R*.

25. Find the dimensions of the right-circular cylinder of greatest surface area that can be inscribed in a sphere of radius *R*.

26. Show that the right-circular cylinder of greatest volume that can be inscribed in a right-circular cone has volume that is $\frac{4}{9}$ the volume of the cone (Figure 4.7.8).

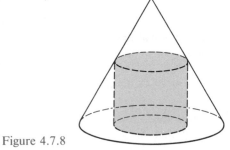

Figure 4.7.8

27. A drainage channel is to be made so that its cross section is a trapezoid with equally sloping sides (Figure 4.7.9). If the sides and bottom all have a length of 5 ft, how should the angle θ be chosen to yield the greatest cross-sectional area? Assume $0 \le \theta \le \pi/2$.

Figure 4.7.9

28. A commercial cattle ranch currently allows 20 steers per acre of grazing land; on the average its steers weigh 2000 lb at market. Estimates by the Agriculture Department indicate that the average market weight per steer will be reduced by 50 lb for each additional steer added per acre of grazing land. How many steers per acre should be allowed in order for the ranch to get the largest possible total market weight for its cattle?

29. A cone is made from a circular sheet of radius *R* by cutting out a sector and gluing the cut edges of the remaining piece together (Figure 4.7.10). What is the maximum volume attainable for the cone?

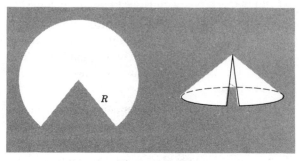

Figure 4.7.10

30. A man is on the bank of a river that is 1 mi wide. He wants to travel to a town on the opposite bank, but 1 mi upstream. He intends to row on a straight line to some point *P* on the opposite bank and then walk the remaining distance along the bank (Figure 4.7.11). To what point should he row in order to reach his destination in the least time if:

(a) he can walk 5 mi/hr and row 3 mi/hr?
(b) he can walk 5 mi/hr and row 4 mi/hr?

Figure 4.7.11

31. The total cost of producing x units of a commodity per week is

$$C(x) = 200 + 4x + 0.1x^2$$

(a) Find the marginal cost when the production level is 100 units.

(b) Use the marginal cost to approximate the cost of producing the 101st unit.

(c) Find the exact cost of producing the 101st unit.

(d) Assuming that the commodity is sold for $10 per unit, find the marginal revenue and marginal profit functions.

32. A firm determines that x units of its product can be sold daily at p dollars per unit, where

$$x = 1000 - p$$

The cost of producing x units per day is

$$C(x) = 3000 + 20x$$

(a) Find the revenue function $R(x)$.

(b) Find the profit function $P(x)$.

(c) Assuming that the production capacity is at most 500 units per day, determine how many units the company must produce and sell each day to maximize the profit.

(d) Find the maximum profit.

(e) What price per unit must be charged to obtain the maximum profit?

33. Prove that $(1, 0)$ is the closest point on the curve $x^2 + y^2 = 1$ to $(2, 0)$.

34. Find all points on the curve $y = \sqrt{x}$ for $0 \le x \le 3$ that are closest to, and at the greatest distance from, the point $(2, 0)$.

35. A church window consisting of a rectangle topped by a semicircle is to have a perimeter p. Find the radius of the semicircle if the area of the window is to be maximum.

36. A triangle is inscribed in a segment of the parabola $y = kx^2$ ($k > 0$), as shown in Figure 4.7.12. Show that the area of the triangle is greatest when $x = (a + b)/2$, and find its area. [*Hint:* The altitude of the triangle will be greatest when the vertical distance between the line and the parabola is greatest. Use the formula in Exercise 35, Section 1.5, to find the altitude of the triangle.]

Figure 4.7.12

37. Suppose that the sum of the surface areas of a sphere and a cube is a constant.

(a) Show that the sum of their volumes is smallest when the diameter of the sphere is equal to the length of an edge of the cube.

(b) When will the sum of their volumes be greatest?

38. We are given a straight fence 100 ft long and we wish to use 200 ft of additional fence to form a rectangular enclosure whose boundary contains all of the original fence. How should we lay out the additional fence to obtain an enclosure of maximum area? (This problem, due to V. L. Klee, Jr., appeared in *Selected Papers on Calculus*, Mathematical Association of America (1969), p. 246.)

39. (a) Find the smallest value of M such that

$$|x^2 - 3x + 2| \le M$$

for all x in the interval $[1, \frac{5}{2}]$.

(b) Find the largest value of m such that

$$|x^2 - 3x + 2| \ge m$$

for all x in the interval $[\frac{3}{2}, \frac{7}{4}]$.

40. At what point(s) in the interval $[0, \pi]$ are the graphs of

$$y = \tfrac{1}{2}x \quad \text{and} \quad y = \sin x$$

farthest apart?

41. Fermat's* (biography on p. 266) principle in optics states that light traveling from one point to another follows that path for which the total travel time is minimum. In a uniform medium, the paths of "minimum time" and "shortest distance" turn out to be the same, so that light, if unobstructed, travels along a straight line. Assume we have a light source, a flat mirror, and an observer in a uniform medium. If a light ray leaves the source, bounces off the mirror, and travels on to the observer, then its path will consist of two line segments, as shown in Figure 4.7.13. According to Fermat's principle, the path will be such that the total travel time t is minimum or, since the medium is uniform, the path will be such that the total distance traveled from A to P to B is as small as possible. Assuming that the minimum occurs when $dt/dx = 0$, show that the light ray will strike the mirror at the point P where the "angle of incidence" θ_1 equals the "angle of reflection" θ_2.

42. Fermat's principle (Exercise 41) also explains why light rays traveling between air and water undergo bending or refraction. Imagine we have two uniform media (such as air and water) and a light ray traveling from a source A in one medium to an observer B in the other medium (Figure 4.7.14). It is known that light travels at a constant speed in a uniform medium, but more slowly in a dense medium (such as water) than in a thin medium (such as air). Consequently, the path of shortest time from A to B is not necessarily a straight line, but rather some broken line path A to P to B allowing the light to take greatest advantage of its higher speed through the thin me-

Figure 4.7.13

*PIERRE DE FERMAT (1601–1665). Fermat, the son of a successful French leather merchant, was a lawyer who practiced mathematics as a hobby. He received a Bachelor of Civil Laws degree from the University of Orleans in 1631 and subsequently held various government positions, including a post as councillor to the Toulouse parliament. Although he was apparently financially successful, confidential documents of that time suggest that his performance in office and as a lawyer was poor, perhaps because he devoted so much time to mathematics. Throughout his life, Fermat fought all efforts to have his mathematical results published. He had the unfortunate habit of scribbling his work in the margins of books and often sent his results to friends without keeping copies for himself. As a result, he never received credit for many major achievements until his name was raised from obscurity in the mid-nineteenth century. It is now know that Fermat, simultaneously and independently of Descartes, developed analytic geometry. Unfortunately, Descartes and Fermat argued bitterly over various problems so that there was never any real cooperation between these two great geniuses.

Fermat solved many fundamental calculus problems. He obtained the first procedure for differentiating polynomials, and solved many important maximization, minimization, area, and tangent problems. His work served to inspire Isaac Newton. Fermat is best known for his work in number theory, the study of properties and relationships between whole numbers. He was the first mathematician to make substantial contributions to this field after the ancient Greek mathematician Diophantus. Unfortunately, none of Fermat's contemporaries appreciated his work in this area, a fact that eventually pushed Fermat into isolation and obscurity in later life.

In addition to his work in calculus and number theory Fermat was one of the founders of probability theory and made major contributions to the theory of optics. Outside mathematics, Fermat was a classical scholar of some note, was fluent in French, Italian, Spanish, Latin, and Greek, and he composed a considerable amount of Latin poetry.

One of the great mysteries of mathematics is shrouded in Fermat's work in number theory. In the margin of a book by Diophantus, Fermat scribbled that for values of n greater than 2, the equation $x^n + y^n = z^n$ has no nonzero integer solutions for x, y, and z. He stated, "I have discovered a truly marvelous proof of this, which however the margin is not large enough to contain." For the past 300 years the greatest mathematical geniuses have been unable to prove this result, even though it seems to be true. This result is now known as "Fermat's last theorem." In 1908 a prize of 100,000 German marks was offered for its solution. Although the prize is worthless because of post-World War I inflation, it has never been won. Whether or not Fermat really proved this theorem is a mystery to this day.

NOTE. A major breakthrough in Fermat's last theorem has just been made by a young West German mathematician named Gerd Faltings (See *Newsweek,* Aug. 1, 1983, p. 66).

dium. Snell's† law of refraction states that the path of the light ray will be such that

$$\frac{\sin \theta_1}{v_1} = \frac{\sin \theta_2}{v_2}$$

where v_1 is the speed of light in the first medium, v_2 is the speed of light in the second medium, and θ_1 and θ_2 are the angles shown in Figure 4.7.14. Show that this follows from the assumption that the path of minimum time occurs when $dt/dx = 0$.

43. A farmer wants to walk at a constant rate from her barn to a straight river, fill her pail, and carry it to her house in the least time.

(a) Explain how this problem relates to Fermat's principle and the light-reflection problem in Exercise 41.

(b) Use the result of Exercise 41 to describe geometrically the best path for the farmer to take.

(c) Use part (b) to determine where the farmer should fill her pail if her house and barn are located as in Figure 4.7.15.

Figure 4.7.14

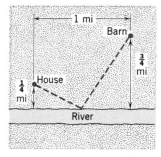

Figure 4.7.15

---

†WILLEBRORD VAN ROIJEN SNELL (1591–1626) Dutch mathematician. Snell, who succeeded his father to the post of Professor of Mathematics at the University of Leiden in 1613, is most famous for the result on light refraction that bears his name. Although this phenomenon was studied as far back as the ancient Greek astronomer Ptolemy, until Snell's work the relationship was incorrectly thought to be $\theta_1/v_1 = \theta_2/v_2$. Snell's law was published by Descartes in 1638 without giving proper credit to Snell. Snell also discovered a method for determining distances by triangulation that founded the modern technique of mapmaking.

4.8   MORE APPLIED MAXIMUM AND MINIMUM PROBLEMS

In the previous section we discussed optimization problems that reduced to maximizing or minimizing a continuous function over a closed interval [a, b]. For such problems, the Extreme-Value Theorem guarantees that a solution exists. In this section we shall consider optimization problems that reduce to maximizing or minimizing continuous functions over open intervals or infinite intervals. For such problems, there is no general guarantee that a solution exists. Thus, part of the problem is to determine whether a solution exists and then to find it if it does.

Example 1   A closed cylindrical can is to hold 1 liter (1000 cm³) of liquid. How should we choose the height and radius to minimize the amount of material needed to manufacture the can?

Solution.   Let

h = height (in cm) of the can
r = radius (in cm) of the can
S = surface area (in cm²) of the can

Assuming there is no waste or overlap, the amount of material needed for manufacture will be the same as the surface area of the can. Since the can can be made from two circular disks of radius r and a rectangular sheet with dimensions h by $2\pi r$ (Figure 4.8.1), the surface area will be

$$S = 2\pi r^2 + 2\pi rh \tag{1}$$

We shall now eliminate one of the variables in (1) so that S will be expressed as a function of one variable. Since the volume of the can is 1000 cm³, it follows from the formula $V = \pi r^2 h$ for the volume of a cylinder

Figure 4.8.1 ▷   Area $2\pi r^2$ Area $2\pi rh$

that

$$1000 = \pi r^2 h \tag{2}$$

or

$$h = \frac{1000}{\pi r^2} \tag{3}$$

Substituting (3) in (1) yields

$$S = 2\pi r^2 + \frac{2000}{r} \tag{4}$$

Because r represents a radius it must be positive;* thus the variable r in (4) must satisfy

$$0 < r < +\infty$$

We have now reduced the problem to that of finding a value of r in $(0, +\infty)$ for which (4) is minimum (provided such a value exists). But S is a continuous function of r on $(0, +\infty)$ and

$$\lim_{r \to 0^+} \left(2\pi r^2 + \frac{2000}{r} \right) = +\infty$$

$$\lim_{r \to +\infty} \left(2\pi r^2 + \frac{2000}{r} \right) = +\infty$$

so (4) has a minimum, but no maximum on $(0, +\infty)$. This minimum must occur at a critical point so we calculate

$$\frac{dS}{dr} = 4\pi r - \frac{2000}{r^2} \tag{5}$$

Setting $dS/dr = 0$ gives

$$4\pi r - \frac{2000}{r^2} = 0$$

$$4\pi r^3 = 2000$$

$$r = \frac{10}{\sqrt[3]{2\pi}} \tag{6}$$

---

*The value $r = 0$ must be excluded; otherwise (2) cannot be satisfied.

Since (6) is the only critical point in the interval $(0, +\infty)$, this value of r yields the minimum value of S. From (3) the value of h corresponding to this r is

$$h = \frac{1000}{\pi(10/\sqrt[3]{2\pi})^2} = \frac{20}{\sqrt[3]{2\pi}}$$

Second Solution.   The conclusion that a minimum occurs at the value of r in (6) can be deduced from Theorem 4.6.6 and the second derivative test by noting that

$$\frac{d^2S}{dr^2} = 4\pi + \frac{4000}{r^3}$$

so that

$$\left.\frac{d^2S}{dr^2}\right|_{r=10/\sqrt[3]{2\pi}} = 4\pi + \frac{4000}{(10/\sqrt[3]{2\pi})^3} = 12\pi$$

Since the second derivative is positive, a relative minimum, and therefore a minimum, occurs at the critical point $r = 10/\sqrt[3]{2\pi}$.

$$S = 2\pi r^2 + \frac{2000}{r}$$

Figure 4.8.2

Third Solution.   Using the methods of Section 4.4, the reader should be able to obtain the graph in Figure 4.8.2. From this graph we see that S has a minimum, and this minimum occurs at the critical point $r = 10/\sqrt[3]{2\pi}$ calculated in our first solution of this problem [see (6)].   ◀

REMARK.   Note that S has no maximum on $(0, +\infty)$. Thus, had we asked for the dimensions of the can requiring the maximum amount of material for its manufacture, there would have been no solution to the problem. Optimization problems with no solution are sometimes called *ill posed*.

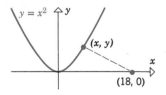

$y = x^2$

(x, y)

$(18, 0)$

Figure 4.8.3

Example 2   Find a point on the curve $y = x^2$ that is closest to the point $(18, 0)$.

Solution.   The distance L between $(18, 0)$ and an arbitrary point (x, y) on the curve $y = x^2$ (Figure 4.8.3) is given by

$$L = \sqrt{(x - 18)^2 + (y - 0)^2}$$

Since (x, y) lies on the curve, x and y satisfy $y = x^2$; thus,

$$L = \sqrt{(x - 18)^2 + x^4} \tag{7}$$

Because there are no restrictions on x, the problem reduces to finding a value of x in $(-\infty, +\infty)$ for which (7) is minimum, provided such a value exists.

In problems of minimizing or maximizing a distance, there is a trick that is helpful for simplifying the computations. It is based on the observation that the distance and the square of the distance have their maximum or minimum at the same point (see Exercise 28). Thus, the minimum value of L in (7) and the minimum value of

$$S = L^2 = (x - 18)^2 + x^4 \tag{8}$$

occur at the same x value.

From (8),

$$\frac{dS}{dx} = 2(x - 18) + 4x^3 = 4x^3 + 2x - 36 \tag{9}$$

so that the critical points satisfy $4x^3 + 2x - 36 = 0$ or equivalently

$$2x^3 + x - 18 = 0 \tag{10}$$

To solve for x we shall begin by searching for integer solutions. This task can be simplified by using the fact that all integer solutions, if there are any, to a polynomial equation with integer coefficients

$$a_n x^n + \cdots + a_1 x + a_0 = 0$$

must be divisors of the constant term a_0. This result is usually proved in algebra courses. Thus, the only possible integer solutions of (10) are the divisors of -18: ± 1, ± 2, ± 3, ± 6, ± 9, ± 18. Successively substituting these values in (10) we find that $x = 2$ is a solution; therefore, $x - 2$ is a factor of the left side of (10). After dividing by the factor $x - 2$ we can rewrite (10) as

$$(x - 2)(2x^2 + 4x + 9) = 0$$

Thus, the remaining solutions of (10) satisfy the quadratic equation

$$2x^2 + 4x + 9 = 0$$

But these solutions are complex numbers (use the quadratic formula) so that $x = 2$ is the only real solution of (10) and consequently the only critical point. To determine the nature of this critical point we shall use the second derivative test. From (9),

$$\frac{d^2S}{dx^2} = 12x^2 + 2$$

so

$$\left. \frac{d^2 S}{dx^2} \right|_{x=2} = 50 > 0$$

which shows that a relative minimum occurs at $x = 2$. Since $x = 2$ is the only relative extremum for L, it follows from Theorem 4.6.6 that an absolute minimum value of L also occurs at $x = 2$. Thus, the point on the curve $y = x^2$ closest to $(18, 0)$ is

$$(x, y) = (x, x^2) = (2, 4) \blacktriangleleft$$

▶ Exercise Set 4.8

1. Find two numbers whose sum is 20 and whose product is
   (a) maximum   (b) minimum.

2. Find the dimensions of the rectangle of area A whose perimeter is
   (a) minimum   (b) maximum.

3. A rectangular area of 3200 ft^2 is to be fenced off. Two opposite sides will use fencing costing $1 per foot and the remaining sides will use fencing costing $2 per foot. Find the dimensions of the rectangle of least cost.

4. A closed rectangular container with a square base is to have a volume of 2250 in.^{3} The material for the top and bottom of the container will cost $2 per in.^{2}, and the material for the sides will cost $3 per in.^{2} Find the dimensions of the container of least cost.

5. A closed rectangular container with a square base is to have a volume of 2000 cm^3. It costs twice as much per square centimeter for the top and bottom as it does for the sides. Find the dimensions of the container of least cost.

6. A container with square base, vertical sides, and open top is to be made from 1000 ft^2 of material. Find the dimensions of the container with greatest volume.

7. A rectangular container with square ends and an open top is to have a volume of V cubic units. Find the dimensions of the container of minimum surface area.

8. A closed cylindrical can is to have a surface area of S square units. Show that the can of maximum volume is achieved when the height is equal to the diameter of the base.

9. A cylindrical can, open at the top, is to hold 500 cm^3 of liquid. Find the height and radius that minimize the amount of material needed to manufacture the can.

10. A rectangular sheet of paper is to contain 72 in.^{2} of printed matter with 2-in. margins at top and bottom and 1-in. margins on each side. What dimensions for the sheet will use the least paper?

11. A cone-shaped paper drinking cup is to hold 10 cm^3 of water. Find the height and radius of the cup that will require the least amount of paper.

12. Find the x-coordinate of the point P on the parabola $y = 1 - x^2$ for $0 < x \leq 1$ where the triangle that is enclosed by the tangent line at P and the coordinate axes has the smallest area.

13. Find all points on the curve $x^2 - y^2 = 1$ closest to $(0, 2)$.

14. Find a point on the curve $x = 2y^2$ closest to $(0, 9)$.

15. Suppose a line L of variable slope passes through $(1, 3)$ and intersects the coordinate axes at the points $(a, 0)$ and $(0, b)$ where a and b are positive. Find the slope of L for which the area of the

triangle with vertices $(a, 0)$, $(0, b)$, and $(0, 0)$ is
(a) maximum   (b) minimum.

16. In a certain chemical manufacturing process, the daily weight y of defective chemical output depends on the total weight x of all output according to the empirical formula

$$y = 0.01x + 0.00003x^2$$

where x and y are in pounds. If the profit is \$100 per pound of nondefective chemical produced and the loss is \$20 per pound of defective chemical produced, how many pounds of chemical should be produced daily to maximize the profit?

17. Two particles, A and B, are in motion in the xy-plane. Their coordinates at each instant of time t $(t \geq 0)$ are given by $x_A = t$, $y_A = 2t$, $x_B = 1 - t$, and $y_B = t$. Find the minimum distance between A and B.

18. Follow the directions of Exercise 17, with $x_A = t$, $y_A = t^2$, $x_B = 2t$, and $y_B = 2$.

19. Find the coordinates of the point P on the curve $y = 1/x^2 (x > 0)$ where the segment of the tangent line at P that is cut off by the coordinate axes will have its shortest length.

20. A lamp is suspended above the center of a round table of radius r. How high above the table should the lamp be placed to achieve maximum illumination at the edge of the table? [Assume that the illumination I is directly proportional to the cosine of the angle of incidence ϕ of the light rays and inversely proportional to the square of the distance l from the light source (Figure 4.8.4).]

Figure 4.8.4

21. Where on the curve

$$y = \frac{1}{1 + x^2}$$

does the tangent line have the greatest slope?

22. Find the dimensions of the isosceles triangle of least area that can be circumscribed about a circle of radius R.

23. Find the height and radius of the right circular cone with least volume that can be circumscribed about a sphere of radius R.

24. If an unknown physical quantity x is measured n times, the measurements $x_1, x_2, \ldots, x_n$ often vary because of uncontrollable factors such as temperature, atmospheric pressure, and so forth. Thus, a scientist is often faced with the problem of using n different observed measurements to obtain an estimate $\bar{x}$ of an unknown quantity x. One method for making such an estimate is based on the *least squares principle*, which states that the estimate $\bar{x}$ should be chosen to minimize

$$s = (x_1 - \bar{x})^2 + (x_2 - \bar{x})^2 + \cdots + (x_n - \bar{x})^2$$

which is the sum of the squares of the deviations between the estimate $\bar{x}$ and the measured values. Show that the estimate resulting from the least squares principle is

$$\bar{x} = \frac{1}{n}(x_1 + x_2 + \cdots + x_n)$$

that is, $\bar{x}$ is the arithmetic average of the observed values.

25. A pipe of negligible diameter is to be carried horizontally around a corner from a hallway 8 ft wide into a hallway 4 ft wide (Figure 4.8.5). What is

Figure 4.8.5

the maximum length that the pipe can have? [An interesting discussion of this problem in the case where the diameter of the pipe is not neglected is given by Norman Miller in the *American Mathematical Monthly*, vol. 56 (1949), pp. 177–179.]

26. A plank is used to reach over a fence 8 ft high to support a wall that is 1 ft behind the fence (Figure 4.8.6). What is the length of the shortest plank that can be used? [*Hint:* Express the length of the plank in terms of the angle θ shown in the figure.]

8 ft

θ

1 ft

Figure 4.8.6

27. The intensity of a point light source is directly proportional to the strength of the source and inversely proportional to the square of the distance from the source. Two point light sources with strengths of S and $8S$ are separated by a distance of 90 cm. Where on the line segment between the two sources is the intensity a minimum?

28. Prove: If $f(x) \geq 0$ on an interval I and if $f(x)$ has a maximum value on I at x_0, then $\sqrt{f(x)}$ also has a maximum value at x_0. Similarly, for minimum values. [*Hint:* Use the fact that $\sqrt{x}$ is an increasing function on the interval $[0, +\infty)$.]

29. Prove: If $P(x_1, y_1)$ is a fixed point, then the minimum distance between P and a point on the line $ax + by + c = 0$ is

$$\frac{|ax_1 + by_1 + c|}{\sqrt{a^2 + b^2}}$$

4.9 NEWTON'S METHOD (OPTIONAL)

In Section 2.7 we showed how to approximate a solution of an equation $f(x) = 0$ to any degree of accuracy using the Intermediate Value Theorem. In this section we shall study a technique, called Newton's Method, that is generally more efficient.

In beginning algebra one learns that the solution of a first-degree equation $ax + b = 0$ is given by the formula $x = -b/a$, and the solutions of a second-degree equation $ax^2 + bx + c = 0$ are given by the quadratic formula. Formulas also exist for the solutions of all third- and fourth-degree equations though they are too complicated to be of practical use. In 1826 it was shown by the Norwegian mathematician Niels Henrik Abel*

---

* NIELS HENRIK ABEL (1802–1829) Norwegian mathematician. Abel was the son of a poor Lutheran minister and a remarkably beautiful mother from whom he inherited strikingly good looks. In his brief life of 26 years Abel lived in virtual poverty and suffered a succession of adversities; yet he managed to prove major results that altered the mathematical landscape forever. At the age of thirteen he was sent away from home to a school whose better days had long passed. By a stroke of luck the school had just hired a teacher named Bernt Michael Holmboe, who quickly discovered that Abel had extraordinary mathematical ability. Together, they studied the calculus texts of Euler and works of Newton and the later French

that it is impossible to construct a formula for the solutions of a *general* fifth degree equation or higher. Thus, for a *specific* fifth-degree polynomial equation such as

$$x^5 - 9x^4 + 2x^3 - 5x^2 + 17x - 8 = 0$$

it may be difficult or impossible to find exact values for all of the solutions. Similar difficulties occur for trigonometric equations such as

$$x - \cos x = 0$$

as well as equations of other types. For such equations the solutions are generally approximated in some way, often by the method we shall now discuss.

NEWTON'S METHOD

We note first that the solutions of $f(x) = 0$ are the values of x where the graph of f crosses the x-axis. Suppose that $x = r$ is the solution we are seeking. Even if we cannot find the value of r exactly, it is usually possible

---

mathematicians. By the time he graduated, Abel was familiar with most of the great mathematical literature. In 1820 his father died, leaving the family in dire financial straits. Abel was able to enter the University of Christiania in Oslo only because he was granted a free room and several professors supported him directly from their salaries. The University had no advanced courses in mathematics, so Abel took a preliminary degree in 1822 and then continued to study mathematics on his own. In 1824 he published at his own expense the proof that it is impossible to solve the general fifth degree polynomial equation algebraically. With the hope that this landmark paper would lead to his recognition and acceptance by the European mathematical community, Abel sent the paper to the great German mathematician Gauss, who casually declared it to be a "monstrosity" and tossed it aside. However, in 1826 Abel's paper on the fifth degree equation and other work was published in the first issue of a new journal, founded by his friend, Leopold Crelle. In the summer of 1826 he completed a landmark work on transcendental functions, which he submitted to the French Academy of Sciences in the hope of establishing himself as a major mathematician, for many young mathematicians had gained quick distinction by having their work accepted by the Academy. However, Abel waited in vain because the paper was either ignored or misplaced by one of the referees, and it did not surface again until two years after his death. That paper was later described by one major mathematician as ". . . the most important mathematical discovery that has been made in our century. . . ." After submitting his paper, Abel returned to Norway, ill with tuberculosis and in heavy debt. While eking out a meager living as a tutor, he continued to produce great work and his fame spread. Soon great efforts were being made to secure a suitable mathematical position for him. Fearing that his great work had been lost by the Academy, he mailed a proof of the main result to Crelle in January of 1829. In April he suffered a violent hemorrhage and died. Two days later Crelle wrote to inform him that an appointment had been secured for him in Berlin and his days of poverty were over! Abel's great paper was finally published by the Academy twelve years after his death.

Figure 4.9.1

to approximate it by graphing f and applying Theorem 2.7.10 to estimate where the graph crosses the x-axis. If we let x_1 denote our initial approximation to r, then we can generally improve on this approximation by moving along the tangent line to $y = f(x)$ at x_1 until we meet the x-axis at a point x_2 (Figure 4.9.1). Usually, x_2 will be closer to r than x_1. To improve the approximation further, we can repeat the process by moving along the tangent line to $y = f(x)$ at x_2 until we meet the x-axis at a point x_3. Continuing in this way we can generate a succession of values $x_1, x_2, x_3, x_4, \ldots$ that will usually get closer and closer to r. This procedure for approximating r is called *Newton's Method*.

To implement Newton's Method analytically, we must derive a formula that will tell us how to calculate each improved approximation from the preceding approximation. For this purpose, we note that the point-slope form of the tangent line to $y = f(x)$ at the initial approximation x_1 is

$$y - f(x_1) = f'(x_1)(x - x_1) \tag{1}$$

If $f'(x_1) \neq 0$, then this line is not parallel to the x-axis and consequently it crosses the x-axis at some point $(x_2, 0)$. Substituting the coordinates of this point in (1) yields

$$-f(x_1) = f'(x_1)(x_2 - x_1)$$

Solving for x_2 we obtain

$$x_2 - x_1 = -\frac{f(x_1)}{f'(x_1)}$$

or

$$x_2 = x_1 - \frac{f(x_1)}{f'(x_1)} \tag{2}$$

The next approximation can be obtained more easily. If we view x_2 as the starting approximation and x_3 the new approximation, we can simply apply (2) with x_2 in place of x_1 and x_3 in place of x_2. This yields

$$x_3 = x_2 - \frac{f(x_2)}{f'(x_2)} \tag{3}$$

provided $f'(x_2) \neq 0$. In general, if x_n is the nth approximation, then it is evident from the pattern in (2) and (3) that the improved approximation x_{n+1} is given by:

Newton's Method

$$x_{n+1} = x_n - \frac{f(x_n)}{f'(x_n)} \quad n = 1, 2, 3, \ldots \tag{4}$$

Example 1   Use Newton's Method to approximate the real solutions of

$$x^3 - x - 1 = 0$$

Solution.   Let $f(x) = x^3 - x - 1$, so $f'(x) = 3x^2 - 1$ and (4) becomes

$$x_{n+1} = x_n - \frac{x_n^3 - x_n - 1}{3x_n^2 - 1}$$

$y = x^3 - x - 1$

or after combining terms and simplifying

$$x_{n+1} = \frac{2x_n^3 + 1}{3x_n^2 - 1} \tag{5}$$

Figure 4.9.2   From the graph of f in Figure 4.9.2, we see that the given equation has only one real solution. This solution lies between 1 and 2 because $f(1) = -1 < 0$ and $f(2) = 5 > 0$. We shall use $x_1 = 1.5$ as our first approximation ($x_1 = 1$ or $x_1 = 2$ would also be reasonable choices.)
   Letting $n = 1$ in (5) and substituting $x_1 = 1.5$ yields

$$x_2 = \frac{2(1.5)^3 + 1}{3(1.5)^2 - 1} = 1.34782609$$

(We used a calculator that displays nine digits.) Next, we let $n = 2$ in (5) and substitute $x_2 = 1.34782609$ to obtain

$$x_3 = \frac{2(1.34782609)^3 + 1}{3(1.34782609)^2 - 1} = 1.32520040$$

If we continue this process until two identical approximations are generated in succession, we shall obtain:

$x_1 = 1.5$
$x_2 = 1.34782609$
$x_3 = 1.32520040$
$x_4 = 1.32471817$
$x_5 = 1.32471796$
$x_6 = 1.32471796$

At this stage there is no need to continue further because we have reached the accuracy limit of our calculator, and all subsequent approximations that it generates will be the same. Thus, the solution is approximately $x \approx 1.32471796$ ◀

Figure 4.9.3

Example 2   It is evident from Figure 4.9.3 that if x is in radians, then the equation

$$\cos x = x$$

has a solution between 0 and 1. Use Newton's Method to approximate it.

Solution.   Rewrite the equation as

$$x - \cos x = 0$$

and apply (4) with $f(x) = x - \cos x$. Since $f'(x) = 1 + \sin x$, (4) becomes

$$x_{n+1} = x_n - \frac{x_n - \cos x_n}{1 + \sin x_n}$$

or after combining terms and simplifying

$$x_{n+1} = \frac{x_n \sin x_n + \cos x_n}{1 + \sin x_n} \tag{6}$$

From Figure 4.9.3, the solution seems closer to $x = 1$ than $x = 0$, so we shall use $x_1 = 1$ (radian) as our initial approximation. Letting $n = 1$ in (6) and substituting $x_1 = 1$ yields

$$x_2 = \frac{1 \cdot \sin 1 + \cos 1}{1 + \sin 1} = .750363868$$

Next, letting $n = 2$ in (6) and substituting this value of x_2 yields

$$x_3 = \frac{(.750363868) \sin (.750363868) + \cos (.750363868)}{1 + \sin (.750363868)} = .739112891$$

If we continue this process until two identical approximations are generated in succession, we obtain:

$$x_1 = 1$$
$$x_2 = .750363868$$
$$x_3 = .739112891$$
$$x_4 = .739085133$$
$$x_5 = .739085133$$

Thus, to the accuracy limit of our calculator, the solution of the equation $\cos x = x$ is $x \approx .739085133$.   ◄

SOME DIFFICULTIES
WITH NEWTON'S
METHOD

x_3 cannot be
generated

Figure 4.9.4

Newton's Method does not always work. For example, if $f'(x_n) = 0$ for some n, then (4) involves a division by zero, making it impossible to generate x_{n+1}. However, this is to be expected because the tangent line to $y = f(x)$ is parallel to the x-axis where $f'(x_n) = 0$, and so this tangent line does not cross the x-axis to generate the next approximation (Figure 4.9.4).

Sometimes the "approximations" produced by Newton's Method do not converge to a solution. For example, consider the equation

$$x^{1/3} = 0$$

which has $x = 0$ as its only solution, and try to approximate this solution by Newton's method with a starting value of $x_0 = 1$. Letting $f(x) = x^{1/3}$, Formula (4) becomes

$$x_{n+1} = x_n - \frac{(x_n)^{1/3}}{\frac{1}{3}(x_n)^{-2/3}} = x_n - 3x_n = -2x_n$$

Beginning with $x_0 = 1$, the successive values generated by this formula are

$$x_0 = 1, \quad x_1 = -2, \quad x_2 = 4, \quad x_3 = -8, \ldots$$

which obviously don't converge to $x = 0$. Figure 4.9.5 illustrates what is happening geometrically.

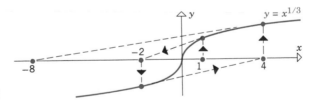

Figure 4.9.5

For a statement of conditions under which Newton's Method works and a discussion of error questions, the reader should consult a book on numerical analysis, for example, Peter Henrici, *Elements of Numerical Analysis,* Wiley, New York, 1964. In situations where Newton's method fails, the method discussed in Section 2.7 can be used; but when Newton's method works, it is the method of choice because it converges more rapidly.

▶ Exercise Set 4.9

In this exercise set, use a hand calculator or micro-computer, and keep as many decimal places as your machine displays.

1. Approximate $\sqrt{2}$ by applying Newton's Method to the equation $x^2 - 2 = 0$.

2. Approximate $\sqrt{7}$ by applying Newton's Method to the equation $x^2 - 7 = 0$.

3. Approximate $\sqrt[3]{6}$ by applying Newton's Method to the equation $x^3 - 6 = 0$.

In Exercises 4–7, the equation has one real solution. Approximate it by Newton's Method.

4. $x^3 + x - 1 = 0$.   5. $x^3 - x + 3 = 0$.

6. $x^5 - x + 1 = 0$.   7. $x^5 + x^4 - 5 = 0$.

In Exercises 8–15, the equation has one solution satisfying the given conditions. Approximate it by Newton's Method.

8. $2x^2 + 4x - 3 = 0$; $x < 0$.

9. $2x^2 + 4x - 3 = 0$; $x > 0$.

10. $x^4 + x - 3 = 0$; $x > 0$.

11. $x^4 + x - 3 = 0$; $x < 0$.

12. $x^5 - 5x^3 = 0$; $x > 0$.

13. $2 \sin x = x$; $x > 0$.

14. $\sin x = x^2$; $x > 0$.

15. $x - \tan x = 0$; $\dfrac{\pi}{2} < x < \dfrac{3\pi}{2}$.

16. Many hand calculators compute reciprocals using the approximation $1/a \approx x_{n+1}$ where

$$x_{n+1} = x_n(2 - ax_n) \qquad n = 1, 2, 3, \ldots$$

and x_1 is an initial approximation to $1/a$. This formula makes it possible to use multiplications and subtractions (which can be done quickly) to perform divisions that would be slow to obtain directly.

(a) Use Newton's Method to derive this approximation.

(b) Use the formula to approximate $\frac{1}{17}$.

17. The *mechanic's rule* for approximating square roots states that $\sqrt{a} \approx x_{n+1}$ where

$$x_{n+1} = \frac{1}{2}\left(x_n + \frac{a}{x_n}\right) \qquad n = 1, 2, 3, \ldots$$

and x_1 is any positive approximation to $\sqrt{a}$.

(a) Use Newton's Method to derive the mechanic's rule.

(b) Use the mechanic's rule to approximate $\sqrt{10}$.

In Exercises 18 and 19, use Newton's Method to approximate the x-coordinate of the point of intersection of the given curves.

18. $y = x^2 + 1$ and $y = x^3$.

19. $y = x^3$ and $y = \frac{1}{2}x - 1$.

4.10 ROLLE'S THEOREM; MEAN-VALUE THEOREM

In this section we shall discuss a result called the Mean-Value Theorem. There are so many major consequences of this theorem that it is regarded as one of the most fundamental results in calculus.

We shall begin with a special case of the Mean-Value Theorem, called ***Rolle's* Theorem***. Geometrically, Rolle's Theorem states that between any two points, *a* and *b*, where a "well-behaved" curve $y = f(x)$ crosses the *x*-axis, there must be at least one place where the tangent line to the curve is horizontal (Figure 4.10.1).

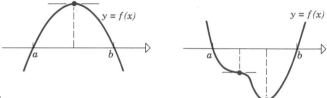

Figure 4.10.1 ▷

The precise statement of this result is as follows.

Rolle's Theorem

4.10.1 THEOREM. *Let f be differentiable on (a, b) and continuous on [a, b]. If $f(a) = f(b) = 0$, then there is at least one point c in (a, b) where $f'(c) = 0$.*

---

* MICHEL ROLLE (1652–1719) French mathematician. Rolle, the son of a shopkeeper, received only an elementary education. He married early and as a young man struggled hard to support his family on the meager wages of a transcriber for notaries and attorneys. In spite of his financial problems and minimal education, Rolle studied algebra and Diophantine analysis (a branch of number theory) on this own. Rolle's fortune changed dramatically in 1682 when he published an elegant solution of a difficult, unsolved problem in Diophantine analysis. The public recognition of his achievement led to a patronage under minister Louvois, a job as an elementary mathematics teacher, and eventually to a short-termed administrative post in the Ministry of War. In 1685 he joined the Académie des Sciences in a low-level position for which he received no regular salary until 1699. He stayed there until he died of apoplexy in 1719.

While Rolle's forté was always Diophantine analysis, his most important work was a book on the algebra of equations, called *Traité d'algèbre,* published in 1690. In that book Rolle firmly established the notation $\sqrt[n]{a}$ [earlier written as $\sqrt{(n)a}$] for the *n*th root of *a*, and proved a polynomial version of the theorem that today bears his name. (Rolle's Theorem was named by Giusto Bellavitis in 1846.) Ironically, Rolle was one of the most vocal early antagonists of calculus. He strove intently to demonstrate that it gave erroneous results and was based on unsound reasoning. He quarreled so vigorously on the subject that the Académie des Sciences was forced to intervene on several occasions. Among his several achievements, Rolle helped advance the currently accepted size order for negative numbers. Descartes, for example, viewed −2 as smaller than −5. Rolle preceded most of his contemporaries by adopting the current convention in 1691.

Proof.   Either $f(x)$ is equal to zero for all x in $[a, b]$ or it is not. If it is, then $f'(x) = 0$ for all x in (a, b), since f is constant on (a, b). Thus, for any c in (a, b)

$$f'(c) = 0$$

If $f(x)$ is not equal to zero for all x in $[a, b]$, then there must be a point x in (a, b) where $f(x) > 0$ or $f(x) < 0$. We shall consider the first case and leave the second as an exercise.

Since f is continuous on $[a, b]$, it follows from the Extreme-Value Theorem that f has a maximum value at some point c in $[a, b]$. Since $f(a) = f(b) = 0$ and $f(x) > 0$ at some point in (a, b), the point c cannot be an endpoint; it must lie in (a, b). By hypothesis, f is differentiable everywhere on (a, b). In particular, it is differentiable at c so that $f'(c) = 0$ by Theorem 4.6.5.   ∎

Example 1   The function

$$f(x) = \sin x$$

is both continuous and differentiable everywhere, hence is continuous on $[0, 2\pi]$ and differentiable on $(0, 2\pi)$. Moreover,

$$f(0) = \sin 0 = 0 \quad \text{and} \quad f(2\pi) = \sin 2\pi = 0$$

so that f satisfies the hypotheses of Rolle's Theorem on the interval $[0, 2\pi]$. Since $f'(c) = \cos c$, Rolle's Theorem guarantees that there is at least one point c in $(0, 2\pi)$ such that

$$\cos c = 0 \tag{1}$$

Solving (1) yields two such values for c, namely $c_1 = \pi/2$ and $c_2 = 3\pi/2$ (Figure 4.10.2).   ◄

REMARK.   In the foregoing example we were able to find c because (1)

Figure 4.10.2 ▷

was easy to solve. However, frequently the equation $f'(c) = 0$ is sufficiently complicated that an exact value for c cannot be obtained.

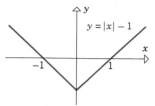

Figure 4.10.3

Example 2   The function

$$f(x) = |x| - 1$$

has the property that $f(-1) = 0$ and $f(1) = 0$, yet there is no point in the
interval $(-1, 1)$ where the graph of f has a horizontal tangent line (Figure
4.10.3). This does *not* contradict Rolle's Theorem because the function
f is not differentiable at every point of the interval $(-1, 1)$.   ◄

Rolle's Theorem is a special case of the **Mean-Value Theorem** which
states that between any two points A and B on a ''well-behaved'' curve
$y = f(x)$, there must be at least one place where the tangent line to the
curve is parallel to the secant line joining A and B (Figure 4.10.4).

Figure 4.10.4 ▷

Noting that the slope of the secant line joining $A(a, f(a))$ and $B(b, f(b))$
is

$$\frac{f(b) - f(a)}{b - a}$$

and the slope of the tangent at c is $f'(c)$, the Mean-Value Theorem can
be stated precisely as follows.

Mean-Value Theorem

4.10.2   THEOREM. *Let f be differentiable on (a, b) and continuous on*
$[a, b]$. Then there is at least one point c in (a, b) where

$$f'(c) = \frac{f(b) - f(a)}{b - a} \qquad (2)$$

Motivation for the Proof of Theorem 4.10.2.   Figure 4.10.4 suggests that
(2) will hold (that is, the tangent line will be parallel to the secant line) at

a point c where the vertical distance between the curve and the secant line is maximum. Thus, to prove the Mean-Value Theorem it is natural to begin by looking for a formula for the vertical distance $v(x)$ between the curve $y = f(x)$ and the secant line joining $(a, f(a))$ and $(b, f(b))$.

Proof of Theorem 4.10.2.   Since the two-point form of the secant line joining $(a, f(a))$ and $(b, f(b))$ is

$$y - f(a) = \frac{f(b) - f(a)}{b - a} (x - a)$$

or equivalently

$$y = \frac{f(b) - f(a)}{b - a} (x - a) + f(a)$$

the difference $v(x)$ between the height of the graph of f and the height of the secant line is

$$v(x) = f(x) - \left[\frac{f(b) - f(a)}{b - a} (x - a) + f(a) \right] \tag{3}$$

Since $f(x)$ is continuous on $[a, b]$ and differentiable on (a, b), so is $v(x)$. Moreover,

$$v(a) = 0 \quad \text{and} \quad v(b) = 0$$

so that $v(x)$ satisfies the hypotheses of Rolle's Theorem on the interval $[a, b]$. Thus, there is a point c in (a, b) such that $v'(c) = 0$. But from Equation (3)

$$v'(x) = f'(x) - \frac{f(b) - f(a)}{b - a}$$

so

$$v'(c) = f'(c) - \frac{f(b) - f(a)}{b - a}$$

Thus, at the point c in (a, b), where $v'(c) = 0$, we have

$$f'(c) = \frac{f(b) - f(a)}{b - a} \quad \blacksquare$$

Example 3   Let $f(x) = x^3 + 1$. Show that f satisfies the hypotheses of the Mean-Value Theorem on the interval $[1, 2]$ and find all values of c in this interval whose existence is guaranteed by the theorem.

Solution.   Because f is a polynomial, f is continuous and differentiable everywhere, hence is continuous on $[1, 2]$ and differentiable on $(1, 2)$. Thus, the hypotheses of the Mean-Value Theorem are satisfied if $a = 1$ and $b = 2$. But

$$f(a) = f(1) = 2, \quad f(b) = f(2) = 9$$
$$f'(x) = 3x^2, \quad\quad f'(c) = 3c^2$$

so that the equation

$$f'(c) = \frac{f(b) - f(a)}{b - a}$$

becomes $3c^2 = 7$, which has two solutions

$$c = \sqrt{7/3} \quad \text{and} \quad c = -\sqrt{7/3}$$

Only the first is in the interval $(1, 2)$ so that $c = \sqrt{7/3}$ is the number whose existence is guaranteed by the Mean-Value Theorem.   ◄

CONSEQUENCES OF
THE MEAN-VALUE
THEOREM

We stated earlier that the Mean-Value Theorem is the starting point for many important results in calculus. We shall conclude this section with two such results.

We know that the derivative of a constant function is zero. Using the Mean-Value Theorem we shall now prove that the converse is also true.

4.10.3   THEOREM. *If $f'(x) = 0$ for all x in an interval, then f is constant on the interval.*

Proof.   Assume $f'(x) = 0$ for all x in an interval I. To prove that f is constant on I it suffices to show that f has the same value at any two points in I. Let a and b be arbitrary points in I with $a < b$. Since f is differentiable on I, it is continuous on I and consequently the hypotheses of the Mean-Value Theorem hold on the interval $[a, b]$. Thus, there is a number c in (a, b) such that

$$f'(c) = \frac{f(b) - f(a)}{b - a}$$

But $f'(c) = 0$ by hypothesis so that $f(b) = f(a)$, which proves that f has the same value at any two points in I.   ∎

4.10.4   COROLLARY. *If $f'(x) = g'(x)$ for all x in an interval I, then f and g differ by a constant on I, that is, there is a constant k such that $f(x) - g(x) = k$ for all x in I.*

Proof.   Let $h(x) = f(x) - g(x)$. Then for every x in I

$$h'(x) = f'(x) - g'(x) = 0$$

Thus, $h(x) = f(x) - g(x)$ is constant on I by Theorem 4.10.3.   ∎

This corollary has a useful geometric interpretation. It states that two functions with the same derivative at each point of an interval have "parallel" graphs over the interval (Figure 4.10.5).

Figure 4.10.5 ▷

▶ Exercise Set 4.10

In Exercises 1–6, verify that the hypotheses of Rolle's Theorem are satisfied on the given interval and find all values of c that satisfy the conclusion of the theorem.

1. $f(x) = x^2 - 6x + 8$; $[2, 4]$.

2. $f(x) = x^3 - 3x^2 + 2x$; $[0, 2]$.

3. $f(x) = \cos x$; $\left[\dfrac{\pi}{2}, \dfrac{3\pi}{2}\right]$.

4. $f(x) = \dfrac{x^2 - 1}{x - 2}$; $[-1, 1]$.

5. $f(x) = \frac{1}{2}x - \sqrt{x}$; $[0, 4]$.

6. $f(x) = \dfrac{1}{x^2} - \dfrac{4}{3x} + \dfrac{1}{3}$; $[1, 3]$.

In Exercises 7–12, verify that the hypotheses of the Mean-Value Theorem are satisfied on the given interval and find all values of c that satisfy the conclusion of the theorem.

7. $f(x) = x^2 + x$; $[-4, 6]$.

8. $f(x) = x^3 + x - 4$; $[-1, 2]$.

9. $f(x) = \sqrt{x + 1}$; $[0, 3]$.

10. $f(x) = x + \dfrac{1}{x}$; $[3, 4]$.

11. $f(x) = \sqrt{25 - x^2}$; $[-5, 3]$.

12. $f(x) = \dfrac{1}{x - 1}$; $[2, 5]$.

13. Let $f(x) = \tan x$.
    (a) Show that there is no point c in $(0, \pi)$ such that $f'(c) = 0$, even though $f(0) = f(\pi) = 0$.
    (b) Explain why the result in part (a) does not violate Rolle's Theorem.

14. Let $f(x) = x^{2/3}$, $a = -1$, and $b = 8$.
    (a) Show that there is no point c in (a, b) such that
    $$f'(c) = \frac{f(b) - f(a)}{b - a}$$
    (b) Explain why the result in part (a) does not violate the Mean-Value Theorem.

In Exercises 15–18, use the Mean-Value Theorem to prove the given statement.

15. $|\sin x - \sin y| \le |x - y|$ for all real values of x and y.

16. $|\tan x + \tan y| \ge |x + y|$ for all real values of x and y in the interval $(-\pi/2, \pi/2)$.

17. If $0 < x < y$, then $\sqrt{xy} < \frac{1}{2}(x + y)$. $\left[\text{Hint: First}\right.$
    show that $\sqrt{y} - \sqrt{x} < \dfrac{y - x}{2\sqrt{x}}$. $\left.\right]$

18. If $f(1) = 0$ and $f'(x) = 1/x$ for all x in $(0, +\infty)$, then $f(x) \le x - 1$ for all x in $(0, +\infty)$. [Hint: Consider the cases, $x > 1$, $0 < x < 1$, and $x = 1$.]

19. Prove: If $f(x) = a_2 x^2 + a_1 x + a_0$ $(a_2 \ne 0)$, then the number c whose existence is guaranteed by the Mean-Value Theorem occurs at the midpoint of the interval $[a, b]$.

In Exercises 20–26, use Rolle's Theorem to prove the given statement.

20. The equation $6x^5 - 4x + 1 = 0$ has at least one solution in the interval $(0, 1)$. (Hint:
    $\dfrac{d}{dx}[x^6 - 2x^2 + x] = 6x^5 - 4x + 1$.)

21. The equation $3ax^2 + 2bx = a + b$ has at least one solution in the interval $(0, 1)$. [Hint: Consider the function $f(x) = ax^3 + bx^2 - (a + b)x$.]

22. If $f'(x)$ exists and $f'(x) \ne 0$ for all x in some open interval I, then $f(x) = 0$ must have fewer than two distinct solutions in I.

23. $x^3 + 4x - 1 = 0$ must have fewer than two distinct real solutions.

24. If $b^2 - 3ac < 0$, then $ax^3 + bx^2 + cx + d = 0$ $(a \ne 0)$, must have fewer than two distinct real solutions.

25. If $f(x) = x^5 + ax^4 + bx^3 + cx^2 + dx + e$, where $2a^2 < 5b$, then the equation $f(x) = 0$ cannot have more than three distinct real solutions. [Hint: Consider $f'''(x)$.]

26. If $f(x) = x^4 - 7x + 2$, then the equation $f(x) = 0$ has exactly two distinct real solutions. [Hint: First find $f(0)$, $f(1)$, and $f(2)$ to show that there are at least two distinct real solutions.]

In Exercises 27 and 28, use Theorem 4.10.3 to prove the given statement.

27. If f and g are functions for which $f'(x) = g(x)$ and $g'(x) = -f(x)$ for all x, then $f^2(x) + g^2(x)$ is a constant.

28. If f and g are functions for which $f'(x) = g(x)$ and $g'(x) = f(x)$ for all x, then $f^2(x) - g^2(x)$ is a constant.

In Exercises 29–32, use Corollary 4.10.4.

29. Let $f(x) = (x - 1)^3$ and $g(x) = (x^2 + 3)(x - 3)$. Show that $f(x) - g(x)$ is a constant. Find the constant.

30. Let $f(x) = \dfrac{x + 2}{3 - x}$ and $g(x) = -\dfrac{5}{x - 3}$ for $x > 3$. Show that $f(x) - g(x)$ is a constant. Find the constant.

31. Let $g(x) = x^3 - 4x + 6$. Find $f(x)$ so that $f'(x) = g'(x)$ and $f(1) = 2$.

32. Let $g(x) = \sqrt{x^2 + 7}$. Find $f(x)$ so that $f'(x) = g'(x)$ and $f(-3) = 1$.

33. Let f and g be continuous on $[a, b]$ and differentiable on (a, b). Prove: If $f(a) = g(a)$ and $f(b) = g(b)$, then there is a point c in (a, b) where $f'(c) = g'(c)$.

34. Prove: If f is continuous on $[a, b]$ and $f'(x) = 0$ for all x in (a, b), then f is constant on $[a, b]$.

35. Prove: If f and g are continuous on $[a, b]$ and $f'(x) = g'(x)$ for all x in (a, b), then f and g differ by a constant on $[a, b]$.

36. (a) Prove: If $f''(x) > 0$ for all x in (a, b), then $f'(x) = 0$ at most once in (a, b).
 (b) Give a geometric interpretation of the result in part (a).

37. Prove: If f is continuous on $[a, b]$ and differentiable on (a, b), and if $f(a) = f(b)$, then there is a point c in (a, b) where $f'(c) = 0$.

38. Let $s = f(t)$ be the position versus time curve for a particle moving in the positive direction along a coordinate line. Prove: If f satisfies the hypotheses of the Mean-Value Theorem on a time interval $[a, b]$, then there will be an instant t_0 in (a, b) where the instantaneous velocity at time t_0 equals the average velocity over the interval $[a, b]$.

39. Complete the proof of Rolle's Theorem by considering the case where $f(x) < 0$ at some point x in (a, b).

40. Use the Mean-Value Theorem to prove:

$$1.71 < \sqrt{3} < 1.75$$

[*Hint:* Let $f(x) = \sqrt{x}$, $a = 3$, and $b = 4$ in the Mean-Value Theorem.]

41. Theorem 4.2.2 gives conditions under which a function is increasing or decreasing on an *open* interval. Prove the following result concerned with *closed* intervals: Let the function f be *continuous* on the closed interval $[a, b]$ and differentiable on the open interval (a, b):
 (a) if $f'(x) > 0$ for all x in (a, b), then f is increasing on $[a, b]$;
 (b) if $f'(x) < 0$ for all x in (a, b), then f is decreasing on $[a, b]$.

4.11   MOTION ALONG A LINE (RECTILINEAR MOTION)

*In this section we shall begin to study the motion of a particle along a line, sometimes called **rectilinear motion**. Examples are: a piston moving up and down in a cylinder, a rock tossed straight up and returning to earth straight down, back and forth vibrations of a spring, and so forth. In later sections we shall study motion along curves in two- and three-dimensional space.*

VELOCITY AND ACCELERATION

To study the motion of a particle along a line, it is usually desirable to coordinatize the line. We do this by selecting an arbitrary origin, positive direction, and unit of length. We also choose a unit for measuring time and let t be the amount of time elapsed from some arbitrary initial observation. Thus, at the initial observation we have $t = 0$.

As the particle moves along the coordinate line its coordinate s will vary as a function of time t. This function, denoted by $s(t)$, is called the **position function** of the particle. The rate at which the particle's coordinate changes with time is called the *velocity* of the particle, and the magnitude or absolute value of the velocity is called the *speed* of the particle. More precisely:

4.11.1 DEFINITION. If $s(t)$ is the position function of a particle moving on a coordinate line, then the *instantaneous velocity* at time t is defined by

$$v(t) = s'(t) = \frac{ds}{dt}$$

and the *instantaneous speed* at time t is defined by

$$\text{speed at time } t = |v(t)| = |s'(t)| = \left|\frac{ds}{dt}\right|$$

If $v(t) > 0$ at a given time t, then the coordinate s of the particle is increasing at that instant, which means that the particle is moving in the positive direction along the line; similarly, if $v(t) < 0$, then s is decreasing at time t and the particle is moving in the negative direction (Figures 4.11.1a and 4.11.1b). The speed of a particle is always nonnegative; it tells how fast the particle is moving, but provides no information about the direction of motion.

Figure 4.11.1 ▷

$s(t)$

$s(t)$ increasing
$v(t) = s'(t) > 0$

(a)

$s(t)$

$s(t)$ decreasing
$v(t) = s'(t) < 0$

(b)

Example 1 Let $s = t^4 - 8t^2$ be the position function of a particle moving on a coordinate line, where t is in seconds and s is in feet. The instantaneous velocity and instantaneous speed of the particle are given by

$$v(t) = \frac{ds}{dt} = 4t^3 - 16t$$

$$|v(t)| = \left|\frac{ds}{dt}\right| = |4t^3 - 16t|$$

Thus, when $t = 1$ sec the particle's velocity is $v(1) = -12$ ft/sec, which means the particle is moving in the negative direction with a speed of $|v(1)| = 12$ ft/sec. When $t = 3$ sec the velocity is $v(3) = 60$ ft/sec, which means that the particle is moving in the positive direction with a speed of $|v(3)| = 60$ ft/sec. When $t = 2$ sec the velocity is $v(2) = 0$ ft/sec, which means that the particle has momentarily stopped.   ◀

If a particle moves along a coordinate line, the rate at which its velocity changes with time is called the *acceleration* of the particle; more precisely:

4.11.2   DEFINITION. If $v(t)$ is the instantaneous velocity at time t of a particle moving on a coordinate line, then the ***instantaneous acceleration*** at time t is defined by

$$a(t) = v'(t) = \frac{dv}{dt}$$

Since $v(t) = s'(t) = ds/dt$, the formula in this definition can also be written as

$$a(t) = s''(t) = \frac{d^2s}{dt^2}$$

Since acceleration is the rate at which velocity changes with time, it is expressed in units of velocity per unit of time. For example, if t is in seconds and s is in feet, then velocity units are feet per second (ft/sec), and acceleration units are feet per second per second, which would be written as

(ft/sec)/sec   or   ft/sec²

Example 2   Let $s = t^3 - 6t^2$, where t is in seconds and s is in feet. Then

$$v(t) = \frac{ds}{dt} = 3t^2 - 12t \quad \text{and} \quad a(t) = \frac{dv}{dt} = 6t - 12$$

At $t = 1$, the acceleration is $a(1) = -6$ ft/sec², which means that the velocity is decreasing at the rate of 6 ft/sec². At $t = 2$ the acceleration is $a(2) = 0$ ft/sec², which means that the velocity is not changing at that instant. At $t = 4$ the acceleration is $a(4) = 12$ ft/sec², which means that the velocity is increasing at the rate of 12 ft/sec².   ◀

REMARK.   We leave it as an exercise to show that the speed of a particle is increasing if its velocity and acceleration have the same sign, and the speed is decreasing if they have opposite signs (Exercise 20). Thus, a particle moving in the positive direction [$v(t) > 0$] is speeding up when its acceleration is positive and slowing down when its acceleration is negative, whereas a particle moving in the negative direction [$v(t) < 0$] is speeding up when its acceleration is negative and slowing down when its acceleration is positive.

Example 3 The position function of a particle moving on a coordinate line is given by $s(t) = 2t^3 - 21t^2 + 60t + 3$, where s is in feet and t is in seconds. Describe the motion of the particle for $t \geq 0$.

Solution. The velocity and acceleration at time t are

$$v(t) = \frac{ds}{dt} = 6t^2 - 42t + 60 = 6(t-2)(t-5)$$

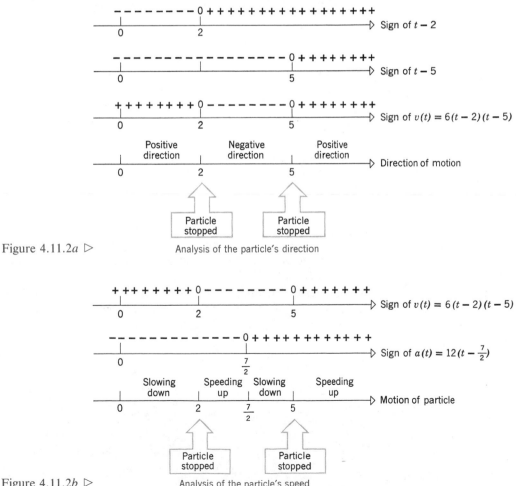

Figure 4.11.2a ▷

Analysis of the particle's direction

Figure 4.11.2b ▷

Analysis of the particle's speed

$$a(t) = \frac{dv}{dt} = 12t - 42 = 12\left(t - \frac{7}{2}\right)$$

At each instant we can determine the direction of motion from the sign of $v(t)$ and whether the particle is speeding up or slowing down from the signs of $v(t)$ and $a(t)$ together (Figures 4.11.2a and 4.11.2b). The motion of the particle is described schematically by the curved line in Figure 4.11.3. At time $t = 0$ the particle is at the point $s(0) = 3$ moving right with velocity $v(0) = 60$ ft/sec, but slowing down with acceleration $a(0) = -42$ ft/sec^2. The particle continues moving right until time $t = 2$, when it stops at the point $s(2) = 55$, reverses direction, and begins to speed up with an acceleration of $a(2) = -18$ ft/sec^2. At time $t = \frac{7}{2}$ the particle begins to slow down, but continues moving left until time $t = 5$, when it stops at the point $s(5) = 28$, reverses direction again, and begins to speed up with acceleration $a(5) = 18$ ft/sec^2. The particle then continues moving right thereafter with increasing speed.   ◄

Figure 4.11.3 ▷

REMARK.   The curved line in Figure 4.11.3 is descriptive only. The actual path of the particle is back and forth on the axis.

GEOMETRIC
INTERPRETATION OF
VELOCITY AND
ACCELERATION

If a particle moving along a coordinate line has position function $s(t)$, instantaneous velocity $v(t)$, and instantaneous acceleration $a(t)$, then $v(t) = s'(t)$ and $a(t) = v'(t)$, so the instantaneous velocity at a time t_0 can be interpreted geometrically as the slope of the tangent line at t_0 to the graph of $s(t)$, and the instantaneous acceleration at time t_0 as the slope of the tangent line at t_0 to the graph of $v(t)$ (Figure 4.11.4).

Figure 4.11.4 ▷

► Exercise Set 4.11

In Exercises 1–16, s is the position in feet, and t is the time in seconds, for a particle moving on a coordinate line.

1. (a) Let $s(t) = t^3 - 6t^2$. Make a table showing the position, velocity, speed, and acceleration at times $t = 1$, $t = 2$, $t = 3$, $t = 4$, and $t = 5$.
   (b) At each of these times specify the direction of motion, if any, and whether the particle is speeding up, slowing down, or neither.

2. Let $s = \dfrac{100}{t^2 + 12}$ for $t \geq 0$. Find the maximum speed of the particle and the direction of motion of the particle when it has this speed.

3. Let $s = 5t^2 - 22t$.
   (a) Find the maximum speed of the particle during the time interval $1 \leq t \leq 3$.
   (b) When, during the time interval $1 \leq t \leq 3$, is the particle farthest from the origin? What is its position at that instant?

In Exercises 4–9, describe the motion of the particle for $t \geq 0$ (as in Example 3) and make a sketch as in Figure 4.11.3.

4. $s = -3t + 2$.

5. $s = 1 + 6t - t^2$.

6. $s = t^3 - 6t^2 + 9t + 1$.

7. $s = t^3 - 9t^2 + 24t$.

8. $s = t + \dfrac{9}{t + 1}$.

9. $s = \begin{cases} \cos t, & 0 \leq t \leq 2\pi \\ 1, & t > 2\pi. \end{cases}$

10. Let $s = t^3 - 6t^2 + 1$.
    (a) Find s and v when $a = 0$.
    (b) Find s and a when $v = 0$.

11. Let $s = 4t^{3/2} - 3t^2$ for $t > 0$.
    (a) Find s and v when $a = 0$.
    (b) Find s and a when $v = 0$.

12. Let $s = \sqrt{2t^2 + 1}$. Find $\lim\limits_{t \to +\infty} v$.

13. (a) Use the chain rule to show that for a particle in rectilinear motion $a = v\dfrac{dv}{ds}$.
    (b) Let $s = \sqrt{3t + 7}$, $t \geq 0$. Find a formula for v in terms of s and use the equation in part (a) to find the acceleration when $s = 5$.

14. If $s = \dfrac{t}{t^2 + 5}$ is the position function of a moving particle for $t \geq 0$, at what instant of time will the particle start to reverse its direction of motion, and where is it at that instant?

15. Suppose that the position functions of two particles, P_1 and P_2, in motion along the same line are given by the formulas $s_1 = \frac{1}{2}t^2 - t + 3$ and $s_2 = -\frac{1}{4}t^2 + t + 1$, respectively, for $t \geq 0$.
    (a) Prove that P_1 and P_2 do not collide.
    (b) How close can P_1 and P_2 get to one another?
    (c) During what intervals of time are they moving in opposite directions?

16. Let $s_A = 15t^2 + 10t + 20$ and $s_B = 5t^2 + 40t$, $t \geq 0$, be the position functions of cars A and B that are moving along parallel straight lanes of a highway.
    (a) How far is car A ahead of car B when $t = 0$?
    (b) At what instant of time are the cars next to one another?
    (c) At what instant of time do they have the same velocity? Which car is ahead at this instant?

17. The position function of a particle moving on a coordinate line is shown in Figure 4.11.5.
    (a) Is the particle moving in the negative direction or positive direction at time t_0?

Figure 4.11.5

(b) Is the acceleration positive or negative at time t_0?

(c) Is the particle speeding up or slowing down at time t_0?

(d) Is the particle speeding up or slowing down at time t_1?

18. For a particle moving on a coordinate line, the *average velocity* v_{ave} and *average acceleration* a_{ave} over a time interval $[t_0, t_1]$ are defined by:

$$v_{ave} = \frac{\text{change in position}}{\text{time elapsed}} = \frac{s(t_1) - s(t_0)}{t_1 - t_0}$$

$$a_{ave} = \frac{\text{change in velocity}}{\text{time elapsed}} = \frac{v(t_1) - v(t_0)}{t_1 - t_0}$$

(a) Interpret v_{ave} geometrically on the graph of $s(t)$.

(b) Interpret a_{ave} geometrically on the graph of $v(t)$.

(c) Find the average velocity and acceleration over the time interval $2 \leq t \leq 4$ for a particle with position function $s(t) = t^3 - 3t^2$.

19. (a) Show that the average velocity v_{ave} over $[t_0, t_1]$ approaches the instantaneous velocity $v(t_0)$ as t_1 approaches t_0. [See Exercise 18 for definitions of v_{ave} and a_{ave}.]

(b) Show that the average acceleration a_{ave} over $[t_0, t_1]$ approaches the instantaneous acceleration $a(t_0)$ as t_1 approaches t_0.

20. Let $r(t) = |v(t)|$ be the speed of a particle moving on a coordinate line. Prove:

(a) $r'(t_0) > 0$ if $a(t_0) > 0$ and $v(t_0) > 0$ or $a(t_0) < 0$ and $v(t_0) < 0$

(b) $r'(t_0) < 0$ if $a(t_0) > 0$ and $v(t_0) < 0$ or $a(t_0) < 0$ and $v(t_0) > 0$.

(To paraphrase: The particle is speeding up if velocity and acceleration have the same sign, and slowing down if they have opposite signs.)

▶ SUPPLEMENTARY EXERCISES

1. For the hollow cylinder shown, assume that R and r are increasing at a rate of 2 meters/sec, and h is decreasing at a rate of 3 meters/sec. At what rate is the volume changing at the instant when $R = 7$ m, $r = 4$ m, and $h = 5$ m?

2. The vessel shown is filled at the rate of 4 ft^3/min. How fast is the fluid level rising when the level is 1 ft?

3. A ball is dropped from a point 10 ft away from a light at the top of a 48-ft pole as shown. When the ball has dropped 16 ft, its velocity (downward) is 32 ft/sec. At what rate is its shadow moving along the ground at that instant?

In Exercises 4–9, find the minimum value m and the maximum value M of f on the indicated interval (if they exist) and state where these extreme values occur.

4. $f(x) = 1/x$; $[-2, -1]$.

5. $f(x) = x^3 - x^4$; $[-1, \frac{3}{2}]$.

6. $f(x) = x^2(x - 2)^{1/3}$; $(0, 3]$.

7. $f(x) = 2x/(x^2 + 3)$; $(0, 2]$.

8. $f(x) = 2x^5 - 5x^4 + 7$; $(-1, 3)$.

9. $f(x) = -|x^2 - 2x|$; $[1, 3]$.

10. Use Newton's Method to approximate the smallest positive solution of $\sin x + \cos x = 0$.

11. Use Newton's Method to approximate all three solutions of $x^3 - 4x + 1 = 0$.

In Exercises 12–19, sketch the graph of f. Use symmetry, where possible, and show all relative extrema, inflection points, and asymptotes.

12. $f(x) = (x^2 - 3)^2$.

13. $f(x) = \dfrac{1}{1 + x^2}$.

14. $f(x) = \dfrac{2x}{1 + x}$.

15. $f(x) = \dfrac{x^3 - 2}{x}$.

16. $f(x) = (1 + x)^{2/3}(3 - x)^{1/3}$.

17. $f(x) = 2\cos^2 x$, $0 \le x \le \pi$.

18. $f(x) = x - \tan x$, $0 \le x \le 2\pi$.

19. $f(x) = \dfrac{3x}{(x + 8)^2}$.

20. Use implicit differentiation to show that a function defined implicitly by $\sin x + \cos y = 2y$ has a critical point whenever $\cos x = 0$. Then use either the first or second derivative test to classify these critical points as relative maxima or minima.

21. Find the equations of the tangent lines at all inflection points of the graph of

$$f(x) = x^4 - 6x^3 + 12x^2 - 8x + 3$$

In Exercises 22–24, find all critical points and use the first derivative test to classify them.

22. $f(x) = x^{1/3}(x - 7)^2$.

23. $f(x) = 2\sin x - \cos 2x$, $0 \le x \le 2\pi$.

24. $f(x) = 3x - (x - 1)^{3/2}$.

In Exercises 25–27, find all critical points and use the second derivative test (if possible) to classify them.

25. $f(x) = x^{-1/2} + \frac{1}{9}x^{1/2}$.

26. $f(x) = x^2 + 8/x$.

27. $f(x) = \sin^2 x - \cos x$, $0 \le x \le 2\pi$.

28. Find two nonnegative numbers whose sum is 20 and such that: (a) the sum of their squares is a maximum, and (b) the product of the square of one and the cube of the other is a maximum.

29. Find the dimensions of the rectangle of maximum area that can be inscribed inside the ellipse $(x/4)^2 + (y/3)^2 = 1$.

30. Find the coordinates of the point on the curve $2y^2 = 5(x + 1)$ that is nearest to the origin. [*Note:* All points $P(x, y)$ on the curve satisfy $x \ge -1$.]

31. A church window consists of a blue semicircular section surmounting a clear rectangular section as shown. The blue glass lets through half as much light per unit area as the clear glass. Find the radius r of the window that admits the most light if the perimeter of the entire window is to be P feet.

32. The cost c (in dollars per hour) to run an ocean liner at a constant speed v (in miles per hour) is given by $c = a + bv^n$ where a, b, and n are positive constants with $n > 1$. Find the speed needed to make the cheapest 3000-mi run.

33. A soup can in the shape of a right-circular cylinder of radius r and height h is to have a prescribed volume V. The top and bottom are cut from squares as shown. If the shaded corners are wasted, but there is no other waste, find the ratio r/h for the can requiring the least material (including waste).

34. If a calculator factory produces x calculators per day, the total daily cost (in dollars) incurred is $0.25x^2 + 35x + 25$. If they are sold for $50 - \frac{1}{2}x$

dollars each, find the value of x that maximizes the daily profit.

In Exercises 35–37, determine if all hypotheses of Rolle's Theorem are satisfied on the stated interval. If not, state which hypotheses fail; if so, find all values of c guaranteed in the conclusion of the theorem.

35. $f(x) = \sqrt{4 - x^2}$ on $[-2, 2]$.

36. $f(x) = x^{2/3} - 1$ on $[-1, 1]$.

37. $f(x) = \sin(x^2)$ on $[0, \sqrt{\pi}]$.

In Exercises 38–41, determine if all hypotheses of the Mean-Value Theorem are satisfied on the stated interval. If not, state which hypotheses fail; if so, find all values of c guaranteed in the conclusion of the theorem.

38. $f(x) = |x - 1|$ on $[-2, 2]$.

39. $f(x) = \sqrt{x}$ on $[0, 4]$.

40. $f(x) = \dfrac{x + 1}{x - 1}$ on $[2, 3]$.

41. $f(x) = \begin{cases} 3 - x^2 & \text{if } x \le 1 \\ 2/x & \text{if } x > 1 \end{cases}$ on $[0, 2]$.

5 integration

5.1 INTRODUCTION

In this chapter we shall study the second major problem of calculus:

> **THE AREA PROBLEM.**   Given a function f that is continuous and non-negative on an interval $[a, b]$, find the area between the graph of f and the interval $[a, b]$ on the x-axis (Figure 5.1.1).

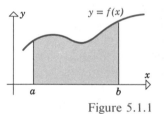

Figure 5.1.1

Area formulas for basic geometric figures such as rectangles, polygons, and circles date back to the earliest written records of mathematics. The first real advance beyond the elementary level of area computation was made by the Greek mathematician, Archimedes,* who devised an inge-

---

ARCHIMEDES (287 B.C.–212 B.C.) Greek mathematician and scientist. Born in Syracuse, Sicily, Archimedes was the son of the astronomer Pheidias and possibly related to Heiron II, king of Syracuse. Most of the facts about his life come from the Roman biographer, Plutarch, who inserted a few tantalizing pages about him in the massive biography of the Roman soldier, Marcellus. In the words of one writer, "the account of Archimedes is slipped like a tissue-thin shaving of ham in a bull-choking sandwich."

Archimedes ranks with Newton and Gauss as one of the three greatest mathematicians who ever lived, and he is certainly the greatest mathematician of antiquity. His mathematical work is so modern in spirit and technique that it is barely distinguishable from that of a seventeenth-century mathematician, yet it was all done without benefit of algebra or a convenient number system. Among his mathematical achievements, Archimedes developed a general method (exhaustion) for finding areas and volumes, and he used the method to find areas bounded by parabolas and spirals and to find volumes of cylinders, paraboloids, and segments of spheres. He gave a procedure for approximating π and bounded its value between $3\frac{1}{7}$ and $3\frac{10}{71}$. In spite of the limitations of the Greek numbering system, he devised methods for finding square roots and invented a method based on the Greek myriad (10,000) for representing numbers as large as 1 followed by 80 million billion zeros.

Of all his mathematical work, Archimedes was most proud of his discovery of the method for finding the volume of a sphere—he showed that the volume of a sphere is two thirds the volume of the smallest cylinder that can contain it. At his request, the figure of a sphere and cylinder was engraved on his tombstone.

In addition to mathematics, Archimedes worked extensively in mechanics and hydro-

(*continued on page 298*)

nious but cumbersome method for obtaining areas, called the *method of exhaustion*. Using this technique Archimedes was able to obtain areas of parabolic regions and spirals. By the early seventeenth century several mathematicians had learned to obtain such areas more simply by calculating appropriate limits. However, both the method of exhaustion and its successor lacked generality. For each different problem one had to devise special procedures or formulas that worked, and more often than not these were difficult or impossible to obtain. The major breakthrough in solving the general area problem was made independently by Newton and Leibniz when they discovered that areas could be obtained by reversing the process of differentiation. This major discovery, which marked the real beginning of calculus, was circulated by Newton in 1669 and then published in 1711 in a paper entitled, *De Analysi per Aequationes Numero Terminorum Infinitas* (*On Analysis by Means of Equations with Infinitely Many Terms*). Independently, Leibniz discovered the same result around 1673 and stated it in an unpublished manuscript dated November 11, 1675.

To obtain a better understanding of this major discovery, consider a continuous curve $y = f(x)$ lying above the x-axis, and let $[a, x]$ be an interval on the x-axis with a fixed left-hand endpoint a and a variable right-hand endpoint. As we move the right-hand endpoint, the area between $y = f(x)$ and the interval $[a, x]$ changes and is therefore a function of x. We denote it by $A(x)$ (Figure 5.1.2). Newton and Leibniz showed

Figure 5.1.2

(continued from page 297)

statics. Nearly every school child knows Archimedes as the absent-minded scientist who, on realizing that a floating object displaces its weight of liquid, leaped from his bath and ran naked through the streets of Syracuse shouting, "Eureka, Eureka!"—(meaning, "I have found it!"). Archimedes actually created the discipline of hydrostatics and used it to find equilibrium positions for various floating bodies. He laid down the fundamental postulates of mechanics, discovered the laws of levers, and calculated centers of gravity for various flat surfaces and solids. In the excitement of discovering the mathematical laws of the lever, he is said to have declared, "Give me a place to stand and I will move the earth."

Although Archimedes was apparently more interested in pure mathematics than its applications, he was an engineering genius. During the second Punic war, when Syracuse was attacked by the Roman fleet under the command of Marcellus, it was reported by Plutarch that Archimedes' military inventions held the fleet at bay for three years. He invented super catapults that showered the Romans with rocks weighing a quarter ton or more, and fearsome mechanical devices with iron "beaks and claws" that reached over the city walls, grasped the ships, and spun them against the rocks. After the first repulse, Marcellus called Archimedes a "geometrical Briareus (a hundred-armed mythological monster) who uses our ships like cups to ladle water from the sea."

Eventually the Roman army was victorious and contrary to Marcellus' specific orders the 75-year-old Archimedes was killed by a Roman soldier. According to one report of the incident, the soldier cast a shadow across the sand in which Archimedes was working on a mathematical problem. When the annoyed Archimedes yelled, "Don't disturb my circles," the soldier flew into a rage and cut the old man down.

With his death the Greek gift of mathematics passed into oblivion, not to be fully resurrected again until the sixteenth century. Unfortunately, there is no known accurate likeness or statue of this great man.

that

$$A'(x) = f(x) \tag{1}$$

Thus, finding the area function reduces to "reversing" the differentiation process and recovering $A(x)$ from its known derivative $f(x)$. Once $A(x)$ is found, the area under $y = f(x)$ over a specific interval $[a, b]$ can be obtained by evaluating $A(x)$ at $x = b$.

The following example illustrates this idea.

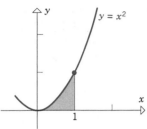

Figure 5.1.3

Example 1   Find the area under the graph of $f(x) = x^2$ over the interval $[0, 1]$. (See Figure 5.1.3.)

Solution.   Let $A(x)$ denote the area under the graph of $f(x) = x^2$ over the interval $[0, x]$. We are interested in finding the value $A(1)$. Since $f(x) = x^2$, it follows from (1) that

$$A'(x) = x^2 \tag{2}$$

Thus, to find $A(x)$ we must look for a function whose derivative is x^2. This is called an ***antidifferentiation*** problem because we are trying to find $A(x)$ by "undoing" a differentiation.

By simply guessing we see that

$$A(x) = \frac{1}{3} x^3$$

is one solution to (2). But this is not the only solution, since

$$A(x) = \frac{1}{3} x^3 + C \tag{3}$$

also satisfies (2) for any constant C. However, there is one additional condition that we can use to single out the appropriate choice for $A(x)$. If $x = 0$, then the interval $[0, x]$ reduces to a single point and consequently the area under the graph of f above this interval is zero. In other words,

$$A(x) = 0 \quad \text{if} \quad x = 0$$

It therefore follows from (3) with $x = 0$ that

$$A(0) = 0 + C = 0 \quad \text{or} \quad C = 0$$

Substituting $C = 0$ in (3) gives

$$A(x) = \frac{1}{3} x^3$$

Thus, the area under the graph of $f(x) = x^2$ over the interval $[0, 1]$ is

$$A(1) = \frac{1}{3} \text{ (square units)} \quad \blacktriangleleft$$

This example shows how certain areas that are impossible to obtain using formulas from plane geometry can be determined using calculus. However, the solution hinged on our ability to *guess* at a function $A(x)$ satisfying $A'(x) = x^2$. In subsequent sections we shall develop a systematic way of finding functions from their derivatives.

5.2 ANTIDERIVATIVES; THE INDEFINITE INTEGRAL

In this section we shall develop some basic results that will ultimately help us to obtain systematic procedures for finding a function from its derivative.

5.2.1 DEFINITION. A function F is called an ***antiderivative*** of a function f if the derivative of F is f.

ANTIDERIVATIVES **Example 1** The functions

$$\frac{1}{3}x^3, \quad \frac{1}{3}x^3 + 2, \quad \frac{1}{3}x^3 - \pi$$

are antiderivatives of $f(x) = x^2$ since the derivative of each is x^2. $\quad \blacktriangleleft$

This example shows that a function can have many antiderivatives. In fact, if $F(x)$ is any antiderivative of $f(x)$ and C is any constant, then

$$F(x) + C$$

is also an antiderivative of $f(x)$ since

$$\frac{d}{dx}[F(x) + C] = \frac{d}{dx}[F(x)] + \frac{d}{dx}[C] = f(x) + 0 = f(x)$$

It is reasonable to ask if there are other antiderivatives of f that cannot be obtained by adding a constant to F. Provided we consider only values of x in an *interval I*, the answer is *no*. To see this, let $G(x)$ be any other antiderivative of $f(x)$; then

$$\frac{d}{dx}[F(x)] = \frac{d}{dx}[G(x)] = f(x)$$

so that by Corollary 4.10.4, F and G differ only by some constant C on I, that is,

$$G(x) = F(x) + C$$

for x in I. The following theorem summarizes these observations.

5.2.2   THEOREM. *If $F(x)$ is any antiderivative of $f(x)$, then for any value of C, the function $F(x) + C$ is also an antiderivative of $f(x)$; moreover on any interval, every antiderivative of $f(x)$ is expressible in the form $F(x)$ plus a constant.*

Example 2   For all x,

$$\frac{d}{dx}[\sin x] = \cos x$$

so that $F(x) = \sin x$ is an antiderivative of $f(x) = \cos x$ on the interval $(-\infty, +\infty)$. Thus, every antiderivative of $\cos x$ on this interval is expressible in the form

$$\sin x + C$$

where C is a constant.   ◀

THE INDEFINITE INTEGRAL

The process of finding antiderivatives is called *antidifferentiation* or *integration*. If

$$\frac{d}{dx}[F(x)] = f(x)$$

then the functions of the form $F(x) + C$ are the antiderivatives of $f(x)$. We denote this by writing

$$\int f(x)\, dx = F(x) + C \tag{1}$$

The symbol $\int$ is called an *integral sign,* and statement (1) is read, "the *indefinite integral* of $f(x)$ equals $F(x)$ plus C." The adjective "indefinite" is used because the right side of (1) is not a definite function, but rather a whole set of possible functions; the constant C is called the *constant of integration.*

REMARK.   The symbol dx in the differentiation operation

$$\frac{d}{dx}[\quad]$$

and in the antidifferentiation operation*

$$\int [\quad] \, dx$$

serves to identify the independent variable. For example, the symbol

$$\int f(t) \, dt$$

denotes a function of t whose derivative with respect to t is $f(t)$.
  The equation

$$\frac{d}{dx} [f(x)] = f'(x)$$

is equivalent to

$$\int f'(x) \, dx = f(x) + C \tag{2}$$

Example 3

DERIVATIVE FORMULA	EQUIVALENT INTEGRATION FORMULA
$\dfrac{d}{dx} [x^3] = 3x^2$	$\displaystyle\int 3x^2 \, dx = x^3 + C$
$\dfrac{d}{dt} [\tan t] = \sec^2 t$	$\displaystyle\int \sec^2 t \, dt = \tan t + C$
$\dfrac{d}{du} [u^{3/2}] = \dfrac{3}{2} u^{1/2}$	$\displaystyle\int \dfrac{3}{2} u^{1/2} \, du = u^{3/2} + C$

◀

Antidifferentiation is guesswork. By looking only at the derivative of a function we try to guess the function itself. It simplifies this guessing process if we keep in mind that every differentiation formula produces a companion integration formula. For example, the differentiation formula

---

*This notation was devised by Leibniz. In his early papers Leibniz used the notation "omn." (an abbreviation for the Latin word "omnes") to denote integration. Then on October 29, 1675 he wrote, "It will be useful to write ∫ for omn., thus ∫ ℓ for omn. ℓ . . . ." Two or three weeks later he refined the notation further and wrote ∫[ ] dx rather than ∫ alone. This notation is so useful and so powerful that its development by Leibniz must be regarded as a major milestone in the history of mathematics and science.

$$\frac{d}{dx}[\sin x] = \cos x$$

gives rise to the integration formula

$$\int \cos x \, dx = \sin x + C$$

Similarly,

$$\frac{d}{dx}[-\cos x] = \sin x \quad \text{yields} \quad \int \sin x \, dx = -\cos x + C$$

and

$$\frac{d}{dx}\left[\frac{x^{r+1}}{r+1}\right] = x^r \quad \text{yields} \quad \int x^r \, dx = \frac{x^{r+1}}{r+1} + C \quad (r \neq -1)$$

Some basic integration formulas are given in Table 5.2.1.

Table 5.2.1

DIFFERENTIATION FORMULA	INTEGRATION FORMULA
1. $\frac{d}{dx}[x] = 1$	$\int 1 \, dx = x + C$
2. $\frac{d}{dx}\left[\frac{x^{r+1}}{r+1}\right] = x^r \ (r \neq -1)$	$\int x^r \, dx = \frac{x^{r+1}}{r+1} + C \ (r \neq -1)$
3. $\frac{d}{dx}[\sin x] = \cos x$	$\int \cos x \, dx = \sin x + C$
4. $\frac{d}{dx}[-\cos x] = \sin x$	$\int \sin x \, dx = -\cos x + C$
5. $\frac{d}{dx}[\tan x] = \sec^2 x$	$\int \sec^2 x \, dx = \tan x + C$
6. $\frac{d}{dx}[-\cot x] = \csc^2 x$	$\int \csc^2 x \, dx = -\cot x + C$
7. $\frac{d}{dx}[\sec x] = \sec x \tan x$	$\int \sec x \tan x \, dx = \sec x + C$
8. $\frac{d}{dx}[-\csc x] = \csc x \cot x$	$\int \csc x \cot x \, dx = -\csc x + C$

Example 4   From the second integration formula in Table 5.2.1 we obtain

$$\int x^2 \, dx = \frac{x^3}{3} + C \quad [r = 2]$$

$$\int x^3 \, dx = \frac{x^4}{4} + C \quad [r = 3]$$

$$\int \frac{1}{x^5} \, dx = \int x^{-5} \, dx = \frac{x^{-5+1}}{-5+1} + C = -\frac{1}{4x^4} + C \quad [r = -5]$$

$$\int \sqrt{x} \, dx = \int x^{\frac{1}{2}} \, dx = \frac{x^{\frac{1}{2}+1}}{\frac{1}{2}+1} + C = \tfrac{2}{3} x^{\frac{3}{2}} + C \quad [r = \tfrac{1}{2}]$$

$$= \tfrac{2}{3}(\sqrt{x})^3 + C \quad \blacktriangleleft$$

PROPERTIES OF THE INDEFINITE INTEGRAL

If we differentiate an antiderivative of $f(x)$, we obtain $f(x)$ back again. Thus,

$$\frac{d}{dx}\left[\int f(x) \, dx\right] = f(x) \tag{3}$$

This result is helpful for proving the following basic properties of anti-derivatives.

5.2.3   THEOREM.

(*a*)   *A constant factor can be moved through an integral sign; that is,*

$$\int cf(x) \, dx = c \int f(x) \, dx$$

(*b*)   *An antiderivative of a sum is the sum of the antiderivatives; that is,*

$$\int [f(x) + g(x)] \, dx = \int f(x) \, dx + \int g(x) \, dx$$

Proof.   To prove (*a*) we must show that $c\int f(x) \, dx$ is an antiderivative of $cf(x)$, and to prove (*b*) we must show that $\int f(x) \, dx + \int g(x) \, dx$ is an anti-derivative of $f(x) + g(x)$. But, these conclusions follow from (3):

$$\frac{d}{dx}\left[c\int f(x) \, dx\right] = c\frac{d}{dx}\left[\int f(x) \, dx\right] = cf(x)$$

$$\frac{d}{dx}\left[\int f(x) \, dx + \int g(x) \, dx\right] = \frac{d}{dx}\left[\int f(x) \, dx\right] + \frac{d}{dx}\left[\int g(x) \, dx\right]$$

$$= f(x) + g(x) \quad \blacksquare$$

Example 5   Evaluate

(a) $\displaystyle\int 4\cos x \, dx$ (b) $\displaystyle\int (x + x^2) \, dx$

Solution (a).

$$\int 4\cos x \, dx = 4 \int \cos x \, dx \quad \text{[Theorem 5.2.3a]}$$

$$= 4\,(\sin x + C) \quad \text{[Table 5.2.1]}$$

$$= 4\sin x + 4C$$

Since C is an arbitrary constant, so is $4C$. However, no purpose is served by keeping the arbitrary constant in this form, so we will replace it by a single letter, say $K = 4C$, and write

$$\int 4\cos x \, dx = 4\sin x + K$$

Solution (b).

$$\int (x + x^2) \, dx = \int x \, dx + \int x^2 \, dx \quad \text{[Theorem 5.2.3b]}$$

$$= \left[\frac{x^2}{2} + C_1\right] + \left[\frac{x^3}{3} + C_2\right] \quad \text{[Table 5.2.1]}$$

$$= \frac{x^2}{2} + \frac{x^3}{3} + C_1 + C_2$$

Since C_1 and C_2 are arbitrary constants, so is $C_1 + C_2$. If we denote this arbitrary constant by the single letter C, we obtain

$$\int (x + x^2) \, dx = \frac{x^2}{2} + \frac{x^3}{3} + C \quad \blacktriangleleft$$

Part (b) of Theorem 5.2.3 can be extended to more than two functions. More precisely,

$$\int [f_1(x) + f_2(x) + \cdots + f_n(x)] \, dx$$

$$= \int f_1(x) \, dx + \int f_2(x) \, dx + \cdots + \int f_n(x) \, dx$$

In addition, we have left it as an exercise to show:

$$\int [f(x) - g(x)] \, dx = \int f(x) \, dx - \int g(x) \, dx$$

Example 6

$$\int (3x^6 - 2x^2 + 7x + 1)\, dx = 3\int x^6\, dx - 2\int x^2\, dx + 7\int x\, dx + \int 1\, dx$$

$$= \frac{3x^7}{7} - \frac{2x^3}{3} + \frac{7x^2}{2} + x + C \quad \blacktriangleleft$$

REMARK. The function $f(x)$ in the expression $\int f(x)\, dx$ is called the *integrand*. Sometimes, for compactness of notation, the dx is incorporated into the integrand. For example:

$$\int 1\, dx \quad \text{can be written as} \quad \int dx$$

$$\int \frac{1}{x^2}\, dx \quad \text{can be written as} \quad \int \frac{dx}{x^2}$$

Sometimes an integrand must be rewritten in a different form before the integration can be performed.

Example 7 Evaluate

$$\int \frac{\cos x}{\sin^2 x}\, dx$$

Solution.

$$\int \frac{\cos x}{\sin^2 x}\, dx = \int \frac{1}{\sin x} \frac{\cos x}{\sin x}\, dx = \int \csc x \cot x\, dx = -\csc x + C$$

Formula 8 in Table 5.2.1 $\quad \blacktriangleleft$

Example 8 Evaluate

$$\int \frac{t^2 - 2t^4}{t^4}\, dt$$

Solution.

$$\int \frac{t^2 - 2t^4}{t^4}\, dt = \int \left(\frac{1}{t^2} - 2 \right) dt = \int (t^{-2} - 2)\, dt$$

$$= \frac{t^{-1}}{-1} - 2t + C = -\frac{1}{t} - 2t + C \quad \blacktriangleleft$$

► Exercise Set 5.2

In Exercises 1–30, evaluate the integrals and check your results by differentiating the answers.

1. $\int x^8 \, dx.$

2. $\int \frac{1}{x^6} \, dx.$

3. $\int x^{5/7} \, dx.$

4. $\int \sqrt[3]{x^2} \, dx.$

5. $\int \frac{4}{\sqrt{t}} \, dt.$

6. $\int \frac{1}{2x^3} \, dx.$

7. $\int x^3 \sqrt{x} \, dx.$

8. $\int (u^3 - 2u + 7) \, du.$

9. $\int (x^{-3} + \sqrt{x} - 3x^{1/4} + x^2) \, dx.$

10. $\int (x^{2/3} - 4x^{-1/5} + 4) \, dx.$

11. $\int \left(\frac{7}{y^{3/4}} - \sqrt[3]{y} + 4\sqrt{y} \right) dy.$

12. $\int (2 + y^2)^2 \, dy.$

13. $\int x(1 + x^3) \, dx.$

14. $\int (1 + x^2)(2 - x) \, dx.$

15. $\int x^{1/3}(2 - x)^2 \, dx.$

16. $\int \frac{1 - 2t^3}{t^3} \, dt.$

17. $\int \frac{x^5 + 2x^2 - 1}{x^4} \, dx.$

18. $\int \left[\frac{1}{t^2} - \cos t \right] dt.$

19. $\int [4 \sin x + 2 \cos x] \, dx.$

20. $\int [4 \sec^2 x + \csc x \cot x] \, dx.$

21. $\int \sec x(\sec x + \tan x) \, dx.$

22. $\int [\sqrt{\theta} - \csc^2 \theta] \, d\theta.$

23. $\int \sec x(\tan x + \cos x) \, dx.$

24. $\int \frac{dy}{\csc y}.$

25. $\int \frac{\sin x}{\cos^2 x} \, dx.$

26. $\int \frac{\sin 2x}{\cos x} \, dx.$

27. $\int [1 + \sin^2 \theta \csc \theta] \, d\theta.$

28. $\int \left[\phi + \frac{2}{\sin^2 \phi} \right] d\phi.$

29. $\int \frac{\cos^3 \theta - 5}{\cos^2 \theta} \, d\theta.$

30. $\int \frac{x^2 \sin x + 2 \sin x}{2 + x^2} \, dx.$

31. Find the antiderivative F of $f(x) = \sqrt[3]{x}$ that satisfies $F(1) = 2.$

32. Find a function f such that $f'(x) + \sin x = 0$ and $f(0) = 2.$

33. Find the general form of a function whose second derivative is $\sqrt{x}$. [*Hint:* Solve the equation $f''(x) = \sqrt{x}$ for $f(x)$ by integrating both sides twice.]

34. Find a function f such that $f''(x) = x + \cos x$ and such that $f(0) = 1$ and $f'(0) = 2.$

[*Hint:* Integrate both sides of the equation twice.]

35. Find $f(x)$ if

$$\int f(x) \, dx = 5x^3 - 3x + C.$$

36. Find $g(t)$ if

$$\int g(t) \, dt = \frac{1}{\sqrt{4 - t^2}} + C.$$

In Exercises 37 and 38, use a trigonometric identity to help evaluate the integral.

37. $\int \tan^2 x \, dx.$

38. $\int \cot^2 x \, dx.$

39. Prove:

$$\int [f(x) - g(x)] \, dx = \int f(x) \, dx - \int g(x) \, dx.$$

40. (a) Show that

$$F(x) = \begin{cases} x, & x > 0 \\ -x, & x < 0 \end{cases}$$

and

$$F_1(x) = \begin{cases} x + 2, & x > 0 \\ -x + 3, & x < 0 \end{cases}$$

are both antiderivatives of

$$f(x) = \begin{cases} 1, & x > 0 \\ -1, & x < 0 \end{cases}$$

but that $F_1(x) \neq F(x)$ plus a constant.

(b)   Does this violate Theorem 5.2.2? Explain.

5.3   INTEGRATION BY SUBSTITUTION

*In this section we shall discuss a technique, called **substitution**, which can often be used to transform complicated integration problems into simpler ones.*

The method of substitution hinges on the following formula in which u stands for a differentiable function of x.

$$\int \left[f(u) \frac{du}{dx} \right] dx = \int f(u) \, du \tag{1}$$

To justify this formula, let F be an antiderivative of f, so that

$$\frac{d}{du} [F(u)] = f(u)$$

or, equivalently,

$$\int f(u) \, du = F(u) + C \tag{2}$$

If u is a differentiable function of x, the chain rule implies that

$$\frac{d}{dx} [F(u)] = \frac{d}{du} [F(u)] \cdot \frac{du}{dx} = f(u) \frac{du}{dx}$$

or, equivalently,

$$\int \left[f(u) \frac{du}{dx} \right] dx = F(u) + C \tag{3}$$

Formula (1) follows from (2) and (3).   ∎

The following example illustrates how Formula (1) is used.

Example 1   Evaluate $\int (x^2 + 1)^{50} \cdot 2x \, dx$.

Solution.   If we let

$$u = x^2 + 1$$

then $du/dx = 2x$, so the given integral can be written as

$$\int (x^2 + 1)^{50} \cdot 2x \, dx = \int \left[u^{50} \frac{du}{dx} \right] dx = \int u^{50} \, du$$

$$\boxed{\int \left[f(u) \frac{du}{dx} \right] dx} \qquad \boxed{\int f(u) \, du}$$

$$= \frac{u^{51}}{51} + C = \frac{(x^2 + 1)^{51}}{51} + C \qquad \blacktriangleleft$$

In general, suppose that we are interested in evaluating

$$\int h(x) \, dx$$

It follows from (1) that if we can express this integral in the form

$$\int h(x) \, dx = \int f(g(x))g'(x) \, dx$$

then the substitution $u = g(x)$ and $du/dx = g'(x)$ will yield

$$\int h(x) \, dx = \int \left[f(u) \frac{du}{dx} \right] dx = \int f(u) \, du$$

With a "good" choice of $u = g(x)$, the integral on the right will be easier to evaluate than the original.

In practice, this substitution process is carried out as follows:

> ### Integration by Substitution
> **Step 1.** Make a choice for u, say $u = g(x)$.
> **Step 2.** Compute $du/dx = g'(x)$.
> **Step 3.** Make the substitution $u = g(x)$, $du = g'(x) \, dx$.
>
> At this stage, the *entire* integral must be in terms of u; no x's should remain. If this is not the case, try a different choice of u.
>
> **Step 4.** Evaluate the resulting integral.
> **Step 5.** Replace u by $g(x)$, so the final answer is in terms of x.

Example 2 Evaluate $\int \sin^2 x \cos x \, dx$.

Solution. If we let

$$u = \sin x$$

then $du/dx = \cos x$, so $du = \cos x \, dx$. Thus,

$$\int \sin^2 x \cos x \, dx = \int u^2 \, du = \frac{u^3}{3} + C = \frac{\sin^3 x}{3} + C \quad \blacktriangleleft$$

Example 3 Evaluate

$$\int \frac{\cos \sqrt{x}}{2\sqrt{x}} \, dx$$

Solution. If we let

$$u = \sqrt{x}$$

then

$$\frac{du}{dx} = \frac{1}{2\sqrt{x}} \quad \text{so} \quad du = \frac{1}{2\sqrt{x}} \, dx$$

Thus,

$$\int \frac{\cos \sqrt{x}}{2\sqrt{x}} \, dx = \int \cos u \, du = \sin u + C = \sin \sqrt{x} + C \quad \blacktriangleleft$$

Example 4 Evaluate $\int 3x^2 \sqrt{x^3 + 1} \, dx$.

Solution. If we let

$$u = x^3 + 1$$

then $du/dx = 3x^2$, so $du = 3x^2 \, dx$. Thus,

$$\int 3x^2 \sqrt{x^3 + 1} \, dx = \int \sqrt{u} \, du = \int u^{1/2} \, du$$

$$= \frac{u^{3/2}}{3/2} + C = \frac{2}{3}(x^3 + 1)^{3/2} + C \quad \blacktriangleleft$$

Example 5 The easiest substitutions occur when the integrand is the derivative of a known function, except for a constant added to or subtracted from the independent variable. For example,

$$\int \sin (x + 9) \, dx = \int \sin u \, du = -\cos u + C = -\cos (x + 9) + C$$

$$u = x + 9$$
$$du = 1 \cdot dx = dx$$

$$\int (x - 8)^{23} \, dx = \int u^{23} \, du = \frac{u^{24}}{24} + C = \frac{(x - 8)^{24}}{24} + C \quad \blacktriangleleft$$

$$u = x - 8$$
$$du = 1 \cdot dx = dx$$

Another easy u-substitution occurs when the integrand is the derivative of a known function, except for a constant that multiplies or divides the independent variable. The following example illustrates two ways to evaluate such integrals.

Example 6 Evaluate $\int \cos 5x \, dx$.

First Solution.

$$\int \cos 5x \, dx = \int (\cos u) \cdot \frac{1}{5} \, du = \frac{1}{5} \int \cos u \, du$$

$$u = 5x$$
$$du = 5 \, dx \text{ or } dx = \tfrac{1}{5} \, du$$

$$= \frac{1}{5} \sin u + C = \frac{1}{5} \sin 5x + C$$

Second Solution. Rather than solve $du = 5 \, dx$ for dx as above, we can replace dx by $5 \, dx$ in the original integrand and compensate by putting a factor of $\frac{1}{5}$ in front of the integral. This yields

$$\int \cos 5x \, dx = \frac{1}{5} \int \cos 5x \cdot 5 \, dx = \frac{1}{5} \int \cos u \, du$$

$$u = 5x$$
$$du = 5 \, dx$$

$$= \frac{1}{5} \sin u + C = \frac{1}{5} \sin 5x + C \quad \blacktriangleleft$$

Example 7

$$\int \frac{dx}{(\frac{1}{3}x - 8)^5} = \int \frac{3\,du}{u^5} = 3 \int u^{-5}\,du$$

$$u = \tfrac{1}{3}x - 8$$
$$du = \tfrac{1}{3}\,dx \text{ or } dx = 3\,du$$

$$= -\frac{3}{4} u^{-4} + C = -\frac{3}{4}\left(\frac{1}{3}x - 8\right)^{-4} + C \quad \blacktriangleleft$$

Example 8   With the help of Theorem 5.2.3, a complicated integral can sometimes be computed by expressing it as a sum of simpler integrals. For example,

$$\int (x + \sec^2 \pi x)\,dx = \int x\,dx + \int \sec^2 \pi x\,dx$$

$$= \frac{x^2}{2} + \int \sec^2 \pi x\,dx$$

$$= \frac{x^2}{2} + \frac{1}{\pi} \int \sec^2 u\,du \qquad \boxed{u = \pi x \\ du = \pi dx \text{ or } dx = \frac{1}{\pi}\,du}$$

$$= \frac{x^2}{2} + \frac{1}{\pi} \tan u + C$$

$$= \frac{x^2}{2} + \frac{1}{\pi} \tan \pi x + C \quad \blacktriangleleft$$

Example 9   Evaluate $\displaystyle\int t^4 \sqrt[3]{3 - 5t^5}\,dt$.

Solution.   After some possible false starts most readers would eventually hit on the following substitution:

$$\int t^4 \sqrt[3]{3 - 5t^5}\,dt = -\frac{1}{25} \int \sqrt[3]{u}\,du = -\frac{1}{25} \int u^{1/3}\,du$$

$$u = 3 - 5t^5$$
$$du = -25t^4\,dt \text{ or } -\tfrac{1}{25}\,du = t^4\,dt$$

$$= -\frac{1}{25}\frac{u^{4/3}}{4/3} + C = -\frac{3}{100}(3 - 5t^5)^{4/3} + C \quad \blacktriangleleft$$

Example 10   Evaluate $\displaystyle\int x^2 \sqrt{x - 1}\,dx$.

Solution. Let

$$u = x - 1 \quad \text{so that} \quad du = dx \tag{4}$$

From the first equality in (4)

$$x^2 = (u + 1)^2 = u^2 + 2u + 1$$

so that

$$\int x^2 \sqrt{x - 1} \, dx = \int (u^2 + 2u + 1)\sqrt{u} \, du$$

$$= \int (u^{5/2} + 2u^{3/2} + u^{1/2}) \, du$$

$$= \frac{2}{7} u^{7/2} + \frac{4}{5} u^{5/2} + \frac{2}{3} u^{3/2} + C$$

$$= \frac{2}{7} (x - 1)^{7/2} + \frac{4}{5} (x - 1)^{5/2} + \frac{2}{3} (x - 1)^{3/2} + C \quad \blacktriangleleft$$

REMARK. If you find integration by substitution hard, don't despair—it is not an easy topic. Work lots of problems.

▶ Exercise Set 5.3

1. Evaluate the integrals by making the indicated substitutions.

(a) $\int 2x(x^2 + 1)^{23} \, dx; \ u = x^2 + 1$

(b) $\int \cos^3 x \sin x \, dx; \ u = \cos x$

(c) $\int \frac{1}{\sqrt{x}} \sin \sqrt{x} \, dx; \ u = \sqrt{x}$

(d) $\int \frac{3x \, dx}{\sqrt{4x^2 + 5}}; \ u = 4x^2 + 5.$

2. Evaluate the integrals by making the indicated substitutions.

(a) $\int \sec^2 (4x + 1) \, dx; \ u = 4x + 1$

(b) $\int y\sqrt{1 + 2y^2} \, dy; \ u = 1 + 2y^2$

(c) $\int \sqrt{\sin \pi\theta} \cos \pi\theta \, d\theta; \ u = \sin \pi\theta$

(d) $\int (2x + 7)(x^2 + 7x + 3)^{4/5} \, dx;$
$\quad u = x^2 + 7x + 3.$

3. Evaluate the integrals by making the indicated substitutions.

(a) $\int \cot x \csc^2 x \, dx; \ u = \cot x$

(b) $\int (1 + \sin t)^9 \cos t \, dt; \ u = 1 + \sin t$

(c) $\int x^2 \sqrt{1 + x} \, dx; \ u = 1 + x$

(d) $\int [\csc (\sin x)]^2 \cos x \, dx; \ u = \sin x.$

In Exercises 4–29, evaluate the integrals.

4. $\int (3x - 1)^5 \, dx.$

5. $\int x(2 - x^2)^3 \, dx.$

6. $\int \sin 3x \, dx.$

7. $\int \cos 8x \, dx.$

8. $\displaystyle\int \sec^2 5x \, dx.$

9. $\displaystyle\int \sec 4x \tan 4x \, dx.$

10. $\displaystyle\int \sqrt{3t + 1} \, dt.$

11. $\displaystyle\int t\sqrt{7t^2 + 12} \, dt.$

12. $\displaystyle\int \frac{x}{\sqrt{4 - 5x^2}} \, dx.$

13. $\displaystyle\int \frac{x^2}{\sqrt{x^3 + 1}} \, dx.$

14. $\displaystyle\int \frac{1}{(1 - 3x)^2} \, dx.$

15. $\displaystyle\int \frac{x}{(4x^2 + 1)^3} \, dx.$

16. $\displaystyle\int x \cos(3x^2) \, dx.$

17. $\displaystyle\int \frac{\sin(5/x)}{x^2} \, dx.$

18. $\displaystyle\int \frac{\sec^2(\sqrt{x})}{\sqrt{x}} \, dx.$

19. $\displaystyle\int x^2 \sec^2(x^3) \, dx.$

20. $\displaystyle\int \cos^3 2t \sin 2t \, dt.$

21. $\displaystyle\int \sin^5 3t \cos 3t \, dt.$

22. $\displaystyle\int \frac{\sin 2\theta}{(5 + \cos 2\theta)^3} \, d\theta.$

23. $\displaystyle\int \cos 4\theta \sqrt{2 - \sin 4\theta} \, d\theta.$

24. $\displaystyle\int \tan^3 5x \sec^2 5x \, dx.$

25. $\displaystyle\int \sec^3 2x \tan 2x \, dx.$

26. $\displaystyle\int [\sin(\sin\theta)] \cos\theta \, d\theta.$

27. $\displaystyle\int [\sec^2(\cos 3\theta)] \sin 3\theta \, d\theta.$

28. $\displaystyle\int \sqrt[n]{a + bx} \, dx \qquad (b \neq 0).$

29. $\displaystyle\int \sin^n (a + bx) \cos(a + bx) \, dx \quad (n > 0, b \neq 0).$

In Exercises 30–36, evaluate the integrals. These are a little trickier than those in the preceding exercises.

30. $\displaystyle\int (4x^2 - 12x + 9)^{2/3} \, dx.$

31. $\displaystyle\int x\sqrt{x - 3} \, dx.$

32. $\displaystyle\int x^2 \sqrt{2 - x} \, dx.$

33. $\displaystyle\int \frac{y \, dy}{\sqrt{y + 1}}.$

34. $\displaystyle\int \sin^3 2\theta \, d\theta.$

[*Hint:* Use the identity $\sin^2 x + \cos^2 x = 1$.]

35. $\displaystyle\int \tan^2 3\theta \, d\theta.$

[*Hint:* Use a trigonometric identity.]

36. $\displaystyle\int \sqrt{1 + x^{-2/3}} \, dx \; (x > 0).$

37. Evaluate the integral in Example 10 by making the substitution $u = \sqrt{x} - 1$. Check your answer against the one obtained in the example.

38. (a) Evaluate the integral $\int \sin x \cos x \, dx$ by two methods: first by letting $u = \sin x$, then by letting $u = \cos x$.
   (b) Explain why the two apparently different answers obtained in part (a) are really equivalent.

39. (a) Evaluate $\int (5x - 1)^2 \, dx$ by two methods: first square and integrate, then let $u = 5x - 1$.
   (b) Explain why the two apparently different answers obtained in part (a) are really equivalent.

40. Find a function f such that $f'(x) = \sqrt{3x + 1}$ and $f(1) = 5$.

41. Find a function f such that $f'(x) = 6 - 5\sin 2x$ and $f(0) = 3$.

In Exercises 42–45 express the given integral in terms of the function f. [*Hint:* Use the fact that $\int f'(u) \, du = f(u) + C$.]

42. $\displaystyle\int f'(5x) \, dx.$

43. $\displaystyle\int f'(3x + 2) \, dx.$

44. $\displaystyle\int xf'(3x^2) \, dx.$

45. $\displaystyle\int \frac{1}{x^2} f'(2/x) \, dx.$

5.4 SIGMA NOTATION

In this section we shall digress from the main theme of this chapter to introduce a notation that can be used to write lengthy sums in a compact form. This material will be helpful when we discuss areas in the next section. The notation uses the uppercase Greek letter Σ (sigma) and is called **sigma notation** *or* **summation notation.**

To illustrate how sigma notation works, consider the sum

$$1^2 + 2^2 + 3^2 + 4^2 + 5^2$$

in which each term is of the form k^2, where k is one of the integers from 1 to 5. In sigma notation this sum can be written

$$\sum_{k=1}^{5} k^2$$

which is read, "the summation of k^2, where k runs from 1 to 5." The notation tells us to form the sum of the terms that result when we substitute successive integers for k in the expression k^2, starting with $k = 1$ and ending with $k = 5$.

More generally, if $f(k)$ is a function of k, and a and b are integers such that $a \le b$, then

$$\sum_{k=a}^{b} f(k) \tag{1}$$

denotes the sum of the terms that result when we substitute successive integers for k, starting with $k = a$ and ending with $k = b$.

Example 1

$$\sum_{k=4}^{8} k^3 = 4^3 + 5^3 + 6^3 + 7^3 + 8^3$$

$$\sum_{k=1}^{5} 2k = 2 \cdot 1 + 2 \cdot 2 + 2 \cdot 3 + 2 \cdot 4 + 2 \cdot 5 = 2 + 4 + 6 + 8 + 10$$

$$\sum_{k=0}^{5} (2k + 1) = 1 + 3 + 5 + 7 + 9 + 11$$

$$\sum_{k=0}^{5} (-1)^k (2k + 1) = 1 - 3 + 5 - 7 + 9 - 11$$

$$\sum_{k=-3}^{1} k^3 = (-3)^3 + (-2)^3 + (-1)^3 + 0^3 + 1^3 = -27 - 8 - 1 + 0 + 1$$

$$\sum_{k=1}^{3} k \sin\left(\frac{k\pi}{5}\right) = \sin\frac{\pi}{5} + 2\sin\frac{2\pi}{5} + 3\sin\frac{3\pi}{5} \quad \blacktriangleleft$$

The numbers a and b in (1) are called, respectively, the **lower** and **upper** **limits of summation;** and the letter k is called the **index of summation.** It is not essential to use k as the index of summation; any letter will do. For example,

$$\sum_{i=1}^{6} \frac{1}{i}, \quad \sum_{j=1}^{6} \frac{1}{j}, \quad \text{and} \quad \sum_{n=1}^{6} \frac{1}{n}$$

all denote the sum

$$1 + \frac{1}{2} + \frac{1}{3} + \frac{1}{4} + \frac{1}{5} + \frac{1}{6}$$

If the upper and lower limits of summation are the same, then the "sum" in (1) reduces to one term. For example,

$$\sum_{k=2}^{2} k^3 = 2^3 \quad \text{and} \quad \sum_{i=1}^{1} \frac{1}{i+2} = \frac{1}{1+2} = \frac{1}{3}$$

In the sums

$$\sum_{i=1}^{5} 2, \quad \sum_{k=3}^{6} 7, \quad \text{and} \quad \sum_{j=0}^{2} x^3$$

the expression to the right of the Σ sign does not involve the index of summation. In such cases, we take all the terms in the sum to be the same, with one term for each allowable value of the summation index. Thus,

$$\sum_{i=1}^{5} 2 = 2 + 2 + 2 + 2 + 2$$

$$\sum_{k=3}^{6} 7 = 7 + 7 + 7 + 7$$

$$\sum_{j=0}^{2} x^3 = x^3 + x^3 + x^3$$

A sum can be written in more than one way with sigma notation by changing the limits of summation. For example, the sum of the first five positive even integers can be written as

$$\sum_{k=1}^{5} 2k = 2 + 4 + 6 + 8 + 10$$

or

$$\sum_{k=0}^{4} (2k + 2) = 2 + 4 + 6 + 8 + 10$$

or

$$\sum_{k=2}^{6} (2k - 2) = 2 + 4 + 6 + 8 + 10$$

CHANGING THE INDEX OF SUMMATION On occasion we shall want to change the sigma notation for a given sum to a sigma notation with different summation limits. The following example illustrates a method for doing this.

Example 2 Express

$$\sum_{k=3}^{7} 5^{k-2}$$

in sigma notation so that the lower limit of summation is 0 rather than 3.

Solution. If we define a new summation index j by means of the formula

$$j = k - 3 \tag{2}$$

then j runs from 0 up to 4 as k runs from 3 up to 7. From (2), $k = j + 3$, so that

$$\sum_{k=3}^{7} 5^{k-2} = \sum_{j=0}^{4} 5^{(j+3)-2} = \sum_{j=0}^{4} 5^{j+1}$$

As a check, the reader can verify that

$$\sum_{j=0}^{4} 5^{j+1} \quad \text{and} \quad \sum_{k=3}^{7} 5^{k-2}$$

both denote the sum $5 + 5^2 + 5^3 + 5^4 + 5^5$. ◄

REMARK. In this solution the summation index was changed from k to j. If it is desirable to keep the same symbol for the summation index, we can change the j back to k *at the very end* and express the final result as

$$\sum_{k=0}^{4} 5^{k+1} \quad \text{instead of} \quad \sum_{j=0}^{4} 5^{j+1}$$

When we want to represent a general sum we shall use letters with subscripts. For example, a general sum with five terms might be written as

$$a_1 + a_2 + a_3 + a_4 + a_5$$

or in sigma notation as

$$\sum_{k=1}^{5} a_k, \quad \sum_{j=1}^{5} a_j, \quad \text{or} \quad \sum_{m=1}^{5} a_m$$

A general sum with n terms might be written as

$$b_1 + b_2 + \cdots + b_n$$

or in sigma notation as

$$\sum_{k=1}^{n} b_k, \quad \sum_{j=1}^{n} b_j, \quad \text{or} \quad \sum_{m=1}^{n} b_m$$

PROPERTIES OF
SIGMA NOTATION

The following properties of sigma notation will help to manipulate sums.

5.4.1 THEOREM.

$(a) \quad \displaystyle\sum_{k=1}^{n} (a_k + b_k) = \sum_{k=1}^{n} a_k + \sum_{k=1}^{n} b_k$

$(b) \quad \displaystyle\sum_{k=1}^{n} (a_k - b_k) = \sum_{k=1}^{n} a_k - \sum_{k=1}^{n} b_k$

$(c) \quad \displaystyle\sum_{k=1}^{n} ca_k = c \sum_{k=1}^{n} a_k$

We shall prove parts (a) and (c) and leave (b) as an exercise.

Proof of (a).

$$\sum_{k=1}^{n} (a_k + b_k) = (a_1 + b_1) + (a_2 + b_2) + \cdots + (a_n + b_n)$$

$$= (a_1 + a_2 + \cdots + a_n) + (b_1 + b_2 + \cdots + b_n)$$

$$= \sum_{k=1}^{n} a_k + \sum_{k=1}^{n} b_k \quad \blacksquare$$

Proof of (c).

$$\sum_{k=1}^{n} ca_k = ca_1 + ca_2 + \cdots + ca_n$$

$$= c(a_1 + a_2 + \cdots + a_n) = c \sum_{k=1}^{n} a_k \quad \blacksquare$$

REMARK. Loosely phrased, this theorem states: *Sigma of a sum equals the sum of the sigmas; sigma of a difference equals the difference of the sigmas; and a constant factor can be moved through a sigma sign.*

SUMMATION
FORMULAS

The following formulas will be used in our later work.

5.4.2 THEOREM.

(a) $$\sum_{k=1}^{n} k = 1 + 2 + 3 + \cdots + n = \frac{n(n + 1)}{2}$$

(b) $$\sum_{k=1}^{n} k^2 = 1^2 + 2^2 + 3^2 + \cdots + n^2 = \frac{n(n + 1)(2n + 1)}{6}$$

(c) $$\sum_{k=1}^{n} k^3 = 1^3 + 2^3 + 3^3 + \cdots + n^3 = \left[\frac{n(n + 1)}{2}\right]^2$$

We shall prove parts (a) and (b) and leave part (c) as an exercise.

Proof of (a). If we write the terms of

$$\sum_{k=1}^{n} k = 1 + 2 + 3 + \cdots + (n - 2) + (n - 1) + n \tag{3}$$

in the opposite order, we obtain

$$\sum_{k=1}^{n} k = n + (n - 1) + (n - 2) + \cdots + 3 + 2 + 1 \tag{4}$$

Adding (3) and (4) term by term yields

$$2\sum_{k=1}^{n} k = \underbrace{(n + 1) + (n + 1) + (n + 1) + \cdots + (n + 1)}_{n \text{ terms}} = n(n + 1)$$

Thus,

$$\sum_{k=1}^{n} k = \frac{n(n + 1)}{2} \qquad \blacksquare$$

Proof of (b).   This proof begins with a trick. Since

$$(k + 1)^3 - k^3 = k^3 + 3k^2 + 3k + 1 - k^3 = 3k^2 + 3k + 1$$

we obtain

$$\sum_{k=1}^{n} [(k + 1)^3 - k^3] = \sum_{k=1}^{n} (3k^2 + 3k + 1) \tag{5}$$

Writing out the left side of (5) yields

$$[2^3 - 1^3] + [3^3 - 2^3] + [4^3 - 3^3] + \cdots + [(n + 1)^3 - n^3] \tag{6}$$

Observe that each term in (6) cancels part of the next term, so that the entire sum collapses like a folding telescope, leaving only $-1^3 + (n + 1)^3$. Thus, (5) can be rewritten as

$$-1 + (n + 1)^3 = \sum_{k=1}^{n} (3k^2 + 3k + 1) \tag{7}$$

or, from Theorem 5.4.1,

$$-1 + (n + 1)^3 = 3 \sum_{k=1}^{n} k^2 + 3 \sum_{k=1}^{n} k + \sum_{k=1}^{n} 1 \tag{8}$$

But

$$\sum_{k=1}^{n} 1 = \underbrace{1 + 1 + \cdots + 1}_{n \text{ terms}} = n$$

and by part (a) of this theorem

$$\sum_{k=1}^{n} k = \frac{n(n + 1)}{2}$$

Thus, (8) can be written

$$-1 + (n + 1)^3 = 3 \sum_{k=1}^{n} k^2 + 3 \frac{n(n + 1)}{2} + n$$

Therefore,

$$\sum_{k=1}^{n} k^2 = \frac{1}{3} \left[(n + 1)^3 - 3 \frac{n(n + 1)}{2} - (n + 1) \right]$$

$$= \frac{n + 1}{6} [2(n + 1)^2 - 3n - 2]$$

$$= \frac{n+1}{6}(2n^2 + n) = \frac{n(n+1)(2n+1)}{6} \qquad \blacksquare$$

Example 3 Evaluate

$$\sum_{k=1}^{30} k(k+1)$$

Solution.

$$\sum_{k=1}^{30} k(k+1) = \sum_{k=1}^{30} (k^2 + k) = \sum_{k=1}^{30} k^2 + \sum_{k=1}^{30} k$$

$$= \frac{30(31)(61)}{6} + \frac{30(31)}{2} = 9920 \quad \text{[Theorem 5.4.2a,b]} \qquad \blacktriangleleft$$

REMARK. In formulas such as

$$\sum_{k=1}^{n} k^2 = \frac{n(n+1)(2n+1)}{6}$$

or

$$1^2 + 2^2 + \cdots + n^2 = \frac{n(n+1)(2n+1)}{6}$$

the left side of the equality is said to express the sum in **open form** and the right side is said to express it in **closed form;** the open form just indicates the terms to be added, while the closed form gives their sum.

Example 4 Express

$$\sum_{k=1}^{n} (3+k)^2$$

in closed form.

Solution.

$$\sum_{k=1}^{n} (3+k)^2 = \sum_{k=1}^{n} (9 + 6k + k^2) = \sum_{k=1}^{n} 9 + 6 \sum_{k=1}^{n} k + \sum_{k=1}^{n} k^2$$

$$= 9n + 6\frac{n(n+1)}{2} + \frac{n(n+1)(2n+1)}{6}$$

$$= \frac{1}{3}n^3 + \frac{7}{2}n^2 + \frac{73}{6}n \qquad \blacktriangleleft$$

▶ Exercise Set 5.4

1. Evaluate

(a) $\displaystyle\sum_{k=1}^{3} k^3$

(b) $\displaystyle\sum_{j=2}^{6} (3j - 1)$

(c) $\displaystyle\sum_{i=-4}^{1} (i^2 - i)$

(d) $\displaystyle\sum_{n=0}^{5} 1$

2. Evaluate

(a) $\displaystyle\sum_{k=1}^{4} k \sin\frac{k\pi}{2}$

(b) $\displaystyle\sum_{j=0}^{5} (-1)^j$

(c) $\displaystyle\sum_{i=7}^{20} \pi$

(d) $\displaystyle\sum_{m=3}^{5} 2^{m+1}$

In Exercises 3–16, express in sigma notation, but do not evaluate.

3. $1 + 2 + 3 + \cdots + 10$.

4. $3 \cdot 1 + 3 \cdot 2 + 3 \cdot 3 + \cdots + 3 \cdot 20$.

5. $1 \cdot 2 + 2 \cdot 3 + 3 \cdot 4 + \cdots + 49 \cdot 50$.

6. $1 + 2 + 2^2 + 2^3 + 2^4$.

7. $2 + 4 + 6 + 8 + \cdots + 20$.

8. $1 + 3 + 5 + 7 + \cdots + 15$.

9. $1 - 3 + 5 - 7 + 9 - 11$.

10. $1 - \dfrac{1}{2} + \dfrac{1}{3} - \dfrac{1}{4} + \dfrac{1}{5}$.

11. $-1 + \dfrac{1}{2} - \dfrac{1}{3} + \dfrac{1}{4} - \dfrac{1}{5}$.

12. $1 + \cos\dfrac{\pi}{7} + \cos\dfrac{2\pi}{7} + \cos\dfrac{3\pi}{7}$.

13. $\sin\dfrac{\pi}{8} + \sin\dfrac{3\pi}{8} + \sin\dfrac{5\pi}{8} + \sin\dfrac{7\pi}{8}$.

14. $2 + 4 + 8 + 16 + 32$.

15. $\dfrac{1}{2} + \dfrac{2}{3} + \dfrac{3}{4} + \dfrac{4}{5} + \dfrac{5}{6}$.

16. $15 + 24 + 35 + \cdots + (n^2 - 1)$.

17. Express in sigma notation.

(a) $a_1 - a_2 + a_3 - a_4 + a_5$

(b) $-b_0 + b_1 - b_2 + b_3 - b_4 + b_5$

(c) $a_0 + a_1 x + a_2 x^2 + \cdots + a_n x^n$

(d) $a^5 + a^4 b + a^3 b^2 + a^2 b^3 + ab^4 + b^5$.

In Exercises 18–25, use Theorem 5.4.2 to evaluate the sums.

18. $\displaystyle\sum_{k=1}^{100} k$.

19. $\displaystyle\sum_{k=3}^{100} k$.

20. $\displaystyle\sum_{k=1}^{100} (7k + 1)$.

21. $\displaystyle\sum_{k=1}^{20} k^2$.

22. $\displaystyle\sum_{k=4}^{20} k^2$.

23. $\displaystyle\sum_{k=1}^{6} (4k^3 - 2k + 1)$.

24. $\displaystyle\sum_{k=1}^{6} (k - k^3)$.

25. $\displaystyle\sum_{k=1}^{30} k(k - 2)(k + 2)$.

When each term of a sum cancels part of the next term, leaving only portions of the first and last terms at the end, the sum is said to *telescope*. In Exercises 26–31, evaluate the telescoping sum.

26. $\displaystyle\sum_{k=1}^{50} \left(\frac{1}{k} - \frac{1}{k + 1}\right)$.

27. $\displaystyle\sum_{k=5}^{17} (3^k - 3^{k-1})$.

28. $\displaystyle\sum_{k=1}^{100} (2^{k+1} - 2^k)$.

29. $\displaystyle\sum_{k=2}^{20} \left(\frac{1}{k^2} - \frac{1}{(k - 1)^2}\right)$.

30. $\displaystyle\sum_{k=1}^{n} (a_k - a_{k+1})$.

31. $\displaystyle\sum_{k=1}^{n} (a_k - a_{k-1})$.

32. Evaluate

(a) $\displaystyle\sum_{j=0}^{m} m$

(b) $\displaystyle\sum_{n=4}^{4} 5$

(c) $\displaystyle\sum_{k=1}^{n} x$

(d) $\displaystyle\sum_{i=1}^{n} i^2 c$.

33. Evaluate

(a) $\displaystyle\sum_{k=1}^{n} n$

(b) $\displaystyle\sum_{i=0}^{0} (-3)$

(c) $\displaystyle\sum_{k=1}^{n} kx$

(d) $\displaystyle\sum_{k=m}^{n} c \; (n \geq m)$.

34. Express $1 + 2 + 2^2 + 2^3 + 2^4 + 2^5$ in sigma notation with:

(a) $j = 0$ as the lower limit of summation
(b) $j = 1$ as the lower limit of summation
(c) $j = 2$ as the lower limit of summation.

35. Express

$$\sum_{k=4}^{18} k(k - 3)$$

in sigma notation with:
(a) $k = 0$ as the lower limit of summation;
(b) $k = 5$ as the lower limit of summation.

36. Express

$$\sum_{k=5}^{9} k2^{k+4}$$

in sigma notation with:
(a) $k = 1$ as the lower limit of summation;
(b) $k = 13$ as the upper limit of summation.

37. Simplify

$$\sum_{k=11}^{28} (k - 10) \sin\left(\frac{\pi}{k - 10}\right)$$

by changing the limits of summation.

38. Which of the following are valid identities?

(a) $\displaystyle\sum_{i=1}^{n} a_i b_i = \sum_{i=1}^{n} a_i \sum_{i=1}^{n} b_i$

(b) $\displaystyle\sum_{i=1}^{n} \frac{a_i}{b_i} = \sum_{i=1}^{n} a_i \Bigg/ \sum_{i=1}^{n} b_i$

(c) $\displaystyle\sum_{i=1}^{n} a_i^2 = \left(\sum_{i=1}^{n} a_i\right)^2$

39. By writing out the sums, determine whether the following are valid identities.

(a) $\displaystyle\int \left[\sum_{i=1}^{n} f_i(x)\right] dx = \sum_{i=1}^{n} \left[\int f_i(x)\, dx\right]$

(b) $\displaystyle\frac{d}{dx}\left[\sum_{i=1}^{n} f_i(x)\right] = \sum_{i=1}^{n} \left[\frac{d}{dx}[f_i(x)]\right]$

40. (a) Evaluate

$$\sum_{k=0}^{n} ar^k - r \sum_{k=0}^{n} ar^k$$

(b) Use the result in part (a) to prove that

$$\sum_{k=0}^{n} ar^k = a + ar + ar^2 + \cdots + ar^n$$

$$= \frac{a - ar^{n+1}}{1 - r} \qquad (r \neq 1)$$

(A sum of this form is called a **geometric sum**.)

41. Use Exercise 40 to evaluate

(a) $\displaystyle\sum_{k=1}^{20} 3^k$ 　　　　 (b) $\displaystyle\sum_{k=5}^{30} 2^k$

(c) $\displaystyle\sum_{k=0}^{100} (-1)^{k+1} \frac{1}{2^k}$.

42. Express the following sums in closed form:

(a) $\displaystyle\sum_{k=1}^{n} (k - \sin^k \theta)$ 　　　 (b) $\displaystyle\sum_{k=1}^{m-3} k^3$.

[*Hint:* Exercise 40 will help in part (a).]

43. Evaluate

$$\sum_{i=1}^{4} \left(\sum_{j=1}^{5} (i + j)\right)$$

44. Let $\bar{x}$ denote the arithmetic average of the n numbers $x_1, x_2, \ldots, x_n$. Use Theorem 5.4.1 to prove

$$\sum_{i=1}^{n} (x_i - \bar{x}) = 0$$

45. Prove part (*b*) of Theorem 5.4.1.

46. Prove part (*c*) of Theorem 5.4.2. [*Hint:* Begin with the difference $(k + 1)^4 - k^4$ and follow the steps used to prove part (*b*) of the theorem.]

5.5 AREAS AS LIMITS

Figure 5.5.1

In previous sections we used the term "area" freely, assuming that its meaning is intuitively clear. However, we must ultimately replace this intuitive concept of area with a precise mathematical definition. It is natural to hope that we might be able to devise a definition that would specify the area of an arbitrary region in the plane (Figure 5.5.1). Surprisingly, there is no way to do this for all regions! It became clear in the late nineteenth century that there are regions in the plane of such complexity that any attempt to assign them areas would ultimately lead to mathematical inconsistencies. While we shall not be concerned with such complicated regions, it is important to know that they exist. In this section we shall show how to obtain areas of certain regions using limits. Although we shall continue to use our intuitive concept of area in this section, the ideas we develop here will form the basis for a precise definition of area in the next section.

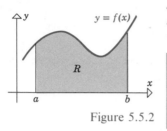

Figure 5.5.2

5.5.1 PROBLEM. Find the area of a region R bounded below by the x-axis, on the sides by the lines $x = a$ and $x = b$, and above by a curve $y = f(x)$, where f is continuous on $[a, b]$ and $f(x) \geq 0$ for all x in $[a, b]$ (Figure 5.5.2).

AREAS AS LIMITS
USING INSCRIBED
RECTANGLES

In Section 5.1 we showed how to obtain the area of this kind of region using antiderivatives. We shall now begin to lay the groundwork for the proof of that result by showing how the area of a region R like that in Figure 5.5.2 can be obtained as a limit. Choose an arbitrary positive integer n and divide the interval $[a, b]$ into n subintervals of width $(b - a)/n$ by introducing points

$$x_1, x_2, \ldots, x_{n-1}$$

equally spaced between a and b (Figure 5.5.3). Next, draw vertical lines through the points $a, x_1, x_2, \ldots, x_{n-1}, b$ to divide the region R into n strips of uniform width. If we approximate each of these strips by a rectangle *inscribed* under the curve $y = f(x)$ (Figure 5.5.3), then the union

Figure 5.5.3 ▷

of these rectangles will form a region R_n, which we can view as an approximation to the entire region R. The area of this approximating region can be calculated by adding the areas of its component rectangles. Moreover, if we allow n to increase, the widths of the rectangles will get smaller, so that the approximation of R by R_n will get better as the smaller rectangles fill in more of the gaps under the curve (Figure 5.5.4). Thus, we can define the *exact* area of R as the limit of the areas of the approximating regions as n goes to plus infinity; that is,

$$A = \text{area } (R) = \lim_{n \to +\infty} [\text{area } (R_n)] \tag{1}$$

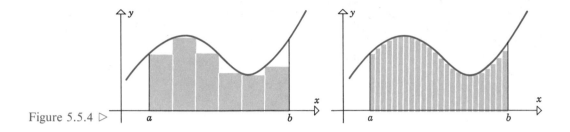

Figure 5.5.4 ▷

REMARK. There is a difference between writing $\lim\limits_{n \to +\infty}$ and writing $\lim\limits_{x \to +\infty}$, where n represents a positive integer, and x has no such restriction. Later we will study limits of the type $\lim\limits_{n \to +\infty}$ in detail, but for now let it suffice to say that the computational techniques we have used for limits of the type $\lim\limits_{x \to +\infty}$ will also work for $\lim\limits_{n \to +\infty}$.

For computational purposes, (1) can be written in a more useful form. If we denote the heights of the inscribed rectangles by $h_1, h_2, \ldots, h_n$ and use the fact that each rectangle has a base of length $(b - a)/n$, then

$$\text{area } (R_n) = h_1 \cdot \frac{b - a}{n} + h_2 \cdot \frac{b - a}{n} + \cdots + h_n \cdot \frac{b - a}{n} \qquad (2)$$

Because f is assumed to be continuous on $[a, b]$, it follows from the Extreme-Value Theorem that f assumes a minimum value on each of the n closed subintervals

$$[a, x_1], [x_1, x_2], \ldots, [x_{n-1}, b]$$

If these minimum values occur at the points $c_1, c_2, \ldots, c_n$, then the heights of the inscribed rectangles are

$$h_1 = f(c_1), h_2 = f(c_2), \ldots, h_n = f(c_n)$$

(Figure 5.5.5), so (2) can be written

Figure 5.5.5 ▷

The height of each inscribed rectangle is the minimum value of $f(x)$ on the subinterval

$$\text{area }(R_n) = f(c_1) \cdot \frac{b-a}{n} + f(c_2) \cdot \frac{b-a}{n} + \cdots + f(c_n) \cdot \frac{b-a}{n} \quad (3)$$

Finally, it will be helpful to write

$$\Delta x = \frac{b-a}{n}$$

for the base dimension of the rectangles, so that (3) becomes

$$\text{area }(R_n) = f(c_1) \cdot \Delta x + f(c_2) \cdot \Delta x + \cdots + f(c_n) \cdot \Delta x$$

or, in sigma notation,

$$\text{area }(R_n) = \sum_{k=1}^{n} f(c_k) \cdot \Delta x$$

With this notation (1) becomes

$$A = \lim_{n \to +\infty} \sum_{k=1}^{n} f(c_k) \cdot \Delta x \quad (4)$$

Example 1   Use inscribed rectangles to find the area under the line $y = x$ over the interval $[1, 2]$.

Solution.   If we subdivide the interval $[1, 2]$ into n equal parts (Figure 5.5.6), then each part will have length

$$\Delta x = \frac{b-a}{n} = \frac{2-1}{n} = \frac{1}{n}$$

and the points of subdivision will be

$$x_1 = 1 + \Delta x = 1 + \frac{1}{n}$$

$$x_2 = 1 + 2\,\Delta x = 1 + \frac{2}{n}$$

$$x_3 = 1 + 3\,\Delta x = 1 + \frac{3}{n}$$

$$\vdots$$

$$x_{n-1} = 1 + (n-1)\,\Delta x = 1 + \frac{n-1}{n}$$

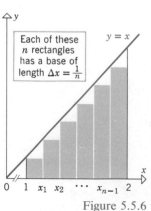

Figure 5.5.6

Because $y = f(x) = x$ is increasing, the minimum value for $f(x)$ on each subinterval occurs at the left endpoint (Figure 5.5.6), so

$$c_1 = 1$$

$$c_2 = x_1 = 1 + \frac{1}{n}$$

$$c_3 = x_2 = 1 + \frac{2}{n}$$

$$c_4 = x_3 = 1 + \frac{3}{n}$$

$$\vdots$$

$$c_n = x_{n-1} = 1 + \frac{n-1}{n}$$

Thus, the inscribed rectangles have areas

$$f(c_1) \cdot \Delta x = c_1 \cdot \Delta x = 1 \cdot \frac{1}{n}$$

$$f(c_2) \cdot \Delta x = c_2 \cdot \Delta x = \left(1 + \frac{1}{n}\right) \cdot \frac{1}{n}$$

$$f(c_3) \cdot \Delta x = c_3 \cdot \Delta x = \left(1 + \frac{2}{n}\right) \cdot \frac{1}{n}$$

$$f(c_4) \cdot \Delta x = c_4 \cdot \Delta x = \left(1 + \frac{3}{n}\right) \cdot \frac{1}{n}$$

$$\vdots$$

$$f(c_n) \cdot \Delta x = c_n \cdot \Delta x = \left(1 + \frac{n-1}{n}\right) \cdot \frac{1}{n}$$

and the sum is

$$\sum_{k=1}^{n} f(c_k) \cdot \Delta x$$

$$= \left[1 + \left(1 + \frac{1}{n}\right) + \left(1 + \frac{2}{n}\right) + \left(1 + \frac{3}{n}\right) + \cdots + \left(1 + \frac{n-1}{n}\right)\right] \cdot \frac{1}{n}$$

$$= \left[n + \left(\frac{1}{n} + \frac{2}{n} + \frac{3}{n} + \cdots + \frac{n-1}{n}\right)\right] \cdot \frac{1}{n}$$

$$= 1 + \frac{1}{n^2}[1 + 2 + 3 + \cdots + (n-1)]$$

$$= 1 + \frac{1}{n^2} \cdot \frac{(n-1)(n)}{2} \qquad \text{[Theorem 5.4.2}a \text{ with } n - 1 \text{ substituted for } n]$$

$$= \frac{3}{2} - \frac{1}{2n}$$

Thus, from (4) the area is

$$A = \lim_{n \to +\infty} \sum_{k=1}^{n} f(c_k) \cdot \Delta x = \lim_{n \to +\infty} \left(\frac{3}{2} - \frac{1}{2n} \right) = \frac{3}{2} - 0 = \frac{3}{2}$$

The region whose area we have computed is a trapezoid with height $h = 1$ and bases $b_1 = 1$ and $b_2 = 2$. From plane geometry, the area of this trapezoid is $A = \frac{1}{2}h(b_1 + b_2) = \frac{1}{2}(1)(1 + 2) = \frac{3}{2}$, which agrees with the result obtained here.   ◄

Example 2   Use inscribed rectangles to find the area under the curve $y = 9 - x^2$ over the interval $[0, 3]$.

Solution.   If we divide the interval $[0, 3]$ into n subintervals of equal length, then each subinterval will have length

$$\Delta x = \frac{b - a}{n} = \frac{3 - 0}{n} = \frac{3}{n}$$

and the points of subdivision will be

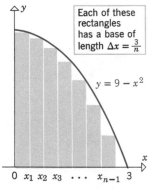

Each of these rectangles has a base of length $\Delta x = \frac{3}{n}$

$y = 9 - x^2$

$0 \quad x_1 \ x_2 \ x_3 \quad \cdots \quad x_{n-1} \quad 3$

Figure 5.5.7

$$x_1 = 0 + \Delta x = \frac{3}{n}$$

$$x_2 = 0 + 2\,\Delta x = 2\left(\frac{3}{n}\right)$$

$$x_3 = 0 + 3\,\Delta x = 3\left(\frac{3}{n}\right)$$

$$\vdots$$

$$x_{n-1} = 0 + (n - 1)\,\Delta x = (n - 1)\left(\frac{3}{n}\right)$$

(Figure 5.5.7). Because $y = f(x) = 9 - x^2$ is decreasing on $[0, 3]$, the minimum value of $f(x)$ on each subinterval occurs at the right endpoint, so

$$c_1 = x_1 = \frac{3}{n}$$

$$c_2 = x_2 = 2 \cdot \frac{3}{n}$$

$$c_3 = x_3 = 3 \cdot \frac{3}{n}$$

$$\vdots$$

$$c_{n-1} = x_{n-1} = (n - 1) \cdot \frac{3}{n}$$

$$c_n = 3$$

In short, $c_k = k \cdot (3/n)$ for $k = 1, 2, \ldots, n$. Thus, the kth inscribed rectangle has area

$$f(c_k) \cdot \Delta x = (9 - c_k{}^2) \cdot \Delta x = \left[9 - k^2 \left(\frac{9}{n^2} \right) \right] \cdot \frac{3}{n} = \frac{27}{n} - \frac{27k^2}{n^3}$$

and the sum of these areas is

$$\sum_{k=1}^{n} f(c_k) \cdot \Delta x = \sum_{k=1}^{n} \left(\frac{27}{n} - \frac{27k^2}{n^3} \right) = \sum_{k=1}^{n} \frac{27}{n} - \sum_{k=1}^{n} \frac{27k^2}{n^3}$$

$$= n \cdot \frac{27}{n} - \frac{27}{n^3} \sum_{k=1}^{n} k^2 = 27 - \frac{27}{n^3} \cdot \frac{n(n+1)(2n+1)}{6}$$

Theorem 5.4.2b

Thus, from (4)

$$A = \lim_{n \to +\infty} \sum_{k=1}^{n} f(c_k) \cdot \Delta x = \lim_{n \to +\infty} \left[27 - \frac{27}{n^3} \cdot \frac{n(n+1)(2n+1)}{6} \right]$$

$$= 27 - \lim_{n \to +\infty} \frac{27}{6} \left(1 + \frac{1}{n} \right) \left(2 + \frac{1}{n} \right) = 27 - \frac{27}{6} (1)(2) = 18 \qquad \blacktriangleleft$$

AREAS AS LIMITS USING CIRCUMSCRIBED RECTANGLES

It may already have occurred to the reader that we could have used *circumscribed* rather than inscribed rectangles in these examples. Indeed, if the *maximum* values of $f(x)$ in the various subintervals occur at the points $d_1, d_2, \ldots, d_n$ then the sum

$$\sum_{k=1}^{n} f(d_k) \cdot \Delta x$$

is an approximation by circumscribed rectangles to the area under the curve (Figure 5.5.8), and the exact area A is

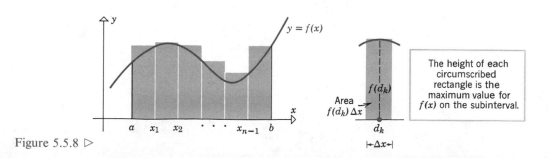

Figure 5.5.8 ▷

$$A = \lim_{n \to +\infty} \sum_{k=1}^{n} f(d_k) \cdot \Delta x \tag{5}$$

since the error (represented by the overlap) diminishes as the rectangles get thinner.

Example 3 Use circumscribed rectangles to find the area under the line $y = x$ over the interval $[1, 2]$.

Solution. As in Example 1, the points

$$x_1 = 1 + \frac{1}{n}$$

$$x_2 = 1 + \frac{2}{n}$$

$$x_3 = 1 + \frac{3}{n}$$

$$\vdots$$

$$x_{n-1} = 1 + \frac{n-1}{n}$$

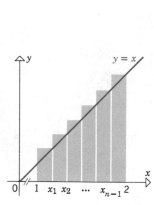

Figure 5.5.9

divide the interval $[1, 2]$ into n subintervals of length $\Delta x = 1/n$. Because $y = f(x) = x$ is increasing, the maximum value for $f(x)$ on each subinterval occurs at the right-hand endpoint (Figure 5.5.9), so that $d_1 = x_1 = 1 + 1/n$, $d_2 = x_2 = 1 + 2/n$, $d_3 = x_3 = 1 + 3/n, \ldots,$ $d_n = 1 + n/n = 2$. Thus, the kth circumscribed rectangle has area

$$f(d_k) \cdot \Delta x = d_k \cdot \Delta x = \left(1 + \frac{k}{n}\right) \cdot \Delta x = \left(1 + \frac{k}{n}\right) \cdot \frac{1}{n}$$

and the sum of the areas of the rectangles is

$$\sum_{k=1}^{n} f(d_k) \cdot \Delta x = \sum_{k=1}^{n} \left[\left(1 + \frac{k}{n}\right) \cdot \frac{1}{n}\right] = \frac{1}{n} \sum_{k=1}^{n} 1 + \frac{1}{n^2} \sum_{k=1}^{n} k$$

$$= \underset{\underset{\boxed{\text{Theorem } 5.4.2a}}{\uparrow}}{\frac{1}{n} \cdot n} + \frac{1}{n^2} \left[\frac{n(n+1)}{2}\right] = \frac{3}{2} + \frac{1}{2n}$$

From (5), the area under the curve is

$$A = \lim_{n \to +\infty} \sum_{k=1}^{n} f(d_k) \cdot \Delta x = \lim_{n \to +\infty} \left(\frac{3}{2} + \frac{1}{2n}\right) = \frac{3}{2}$$

This result agrees with that obtained in Example 1 using inscribed rectangles. ◀

REMARK. Although we shall not do it, it can be proved that, in general, the method of inscribed rectangles [formula (4)] and the method of circumscribed rectangles [formula (5)] both yield the same value for A.

▶ Exercise Set 5.5

In Exercises 1–4, divide the interval $[a, b]$ into $n = 4$ subintervals of equal length and compute: (a) the sum of the areas of the inscribed rectangles, (b) the sum of the areas of the circumscribed rectangles.

1. $y = 3x + 1$; $a = 2$, $b = 6$.
2. $y = 1/x$; $a = 1$, $b = 9$.
3. $y = \cos x$; $a = -\dfrac{\pi}{2}$, $b = \dfrac{\pi}{2}$.
4. $y = 2x - x^2$; $a = 1$, $b = 2$.

In Exercises 5–10, use inscribed rectangles to find the area under the curve $y = f(x)$ over the interval $[a, b]$.

5. $y = \frac{1}{2}x$; $a = 1$, $b = 4$.
6. $y = -x + 5$; $a = 0$, $b = 5$.
7. $y = x^2$; $a = 0$, $b = 1$.
8. $y = 4 - \frac{1}{4}x^2$; $a = 0$, $b = 3$.
9. $y = x^3$; $a = 2$, $b = 6$.
10. $y = 1 - x^3$; $a = -3$, $b = -1$.
11. Use circumscribed rectangles to find the area of the region in Exercise 5.

12. Use circumscribed rectangles to find the area of the region in Exercise 6.
13. Use circumscribed rectangles to find the area of the region in Exercise 7.
14. Use inscribed rectangles to find the area under $y = mx$ over the interval $[a, b]$, where $m > 0$ and $a \geq 0$.
15. (a) Show that the area under $y = x^3$ over the interval $[0, b]$ is $b^4/4$.
    (b) Find a formula for the area under $y = x^3$ over the interval $[a, b]$, where $a \geq 0$.
16. Find the area between the curve $y = \sqrt{x}$ and the interval $0 \leq y \leq 1$ on the y-axis.
17. Assuming that $a > 0$, find the area under $y = x$ over the interval $[a, b]$ three different ways:
    (a) using inscribed rectangles
    (b) using circumscribed rectangles
    (c) using an appropriate area formula from plane geometry.

5.6 THE DEFINITE INTEGRAL

*In mathematics and science there are a variety of concepts, such as length, volume, density, probability, work, and others, whose properties are remarkably similar to properties of area. In this section we shall introduce the concept of a **definite integral**, which is the unifying thread relating these diverse ideas.*

DEFINITION OF AREA   In the last section we gave two equivalent ways of obtaining the area under a continuous curve $y = f(x)$ over an interval $[a, b]$:

$$A = \lim_{n \to +\infty} \sum_{k=1}^{n} f(c_k) \, \Delta x \quad \text{[inscribed rectangles]} \tag{1}$$

and

$$A = \lim_{n \to +\infty} \sum_{k=1}^{n} f(d_k) \, \Delta x \quad \text{[circumscribed rectangles]} \tag{2}$$

However, these are not the only possible formulas for the area A. Instead of choosing the height of the rectangle on the kth subinterval to be the minimum value or the maximum value of f, we can construct a rectangle of some intermediate height. For this purpose, let us select an *arbitrary* point in each subinterval; call these points

$$x_1^*, x_2^*, \ldots, x_n^*$$

Since $f(c_k)$ and $f(d_k)$ are, respectively, the smallest and largest values of f on the kth subinterval, it follows that

$$f(c_k) \le f(x_k^*) \le f(d_k)$$

and

$$f(c_k) \cdot \Delta x \le f(x_k^*) \cdot \Delta x \le f(d_k) \cdot \Delta x$$

so that

$$\sum_{k=1}^{n} f(c_k) \, \Delta x \le \sum_{k=1}^{n} f(x_k^*) \, \Delta x \le \sum_{k=1}^{n} f(d_k) \, \Delta x \tag{3}$$

As $n \to +\infty$, the two outside sums approach A [see (1) and (2)], so the inside sum must also approach A, since it is "squeezed" between the outer sums. Thus, for all possible choices of $x_1^*, x_2^*, \ldots, x_n^*$ we obtain

$$A = \lim_{n \to +\infty} \sum_{k=1}^{n} f(x_k^*) \, \Delta x$$

While rectangles with equal widths are convenient computationally, they are not essential; we can just as well express the area A as the limit of a sum of rectangular areas with different widths (Figure 5.6.1).

Figure 5.6.1 ▷

However, there is a complication here that did not occur before. For rectangles of equal width,

$$\Delta x = \frac{b - a}{n}$$

it follows that $\Delta x \to 0$ as $n \to +\infty$. In other words, we can ensure that the width of *every* rectangle decreases to zero by letting the number of rectangles increase to infinity. For rectangles whose widths vary in size this is not the case. For example, suppose we were to continually divide only the left half of the interval $[a, b]$ into more and more subintervals, but were always to leave the right half of the interval alone (Figure 5.6.2). With this construction the number of rectangles increases to infinity, but the sum of the areas of the rectangles does not approach the area under the curve because there is always an "error" on the right half of the interval. To remedy this problem, we need to ensure that the widths of *all* rectangles decrease to zero as the number of rectangles increases to infinity.

Figure 5.6.2 ▷

Let us suppose that the interval $[a, b]$ is divided into n subintervals whose widths are

$$\Delta x_1, \Delta x_2, \ldots, \Delta x_n$$

and let us denote the largest of these by the symbol

$$\max \Delta x_k$$

(read, "the maximum of the Δx_k's").

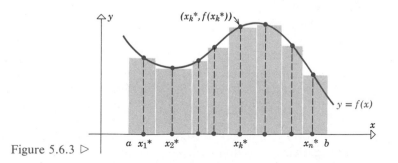

Figure 5.6.3 ▷

If x_k^* is an arbitrary point in the kth subinterval, then

$$f(x_k^*)\,\Delta x_k$$

is the area of a rectangle of height $f(x_k^*)$ and width Δx_k, so

$$\sum_{k=1}^{n} f(x_k^*)\,\Delta x_k \tag{4}$$

is the sum of the shaded rectangular areas in Figure 5.6.3. If we now increase n in such a way that

$$\max \Delta x_k \to 0$$

then the width of *every* rectangle tends to zero because no width exceeds the maximum. Thus, (4) approaches the exact area under the curve as $\max \Delta x_k \to 0$. We denote this by writing

$$A = \lim_{\max \Delta x_k \to 0} \sum_{k=1}^{n} f(x_k^*)\,\Delta x_k$$

We shall give a precise definition of this limit at the end of this section. The preceding informal discussion suggests the following definition.

Area under a Curve

5.6.1   DEFINITION. If the function f is continuous on $[a, b]$ and if $f(x) \geq 0$ for all x in $[a, b]$, then the **area** under the curve $y = f(x)$ over the interval $[a, b]$ is defined by

$$A = \lim_{\max \Delta x_k \to 0} \sum_{k=1}^{n} f(x_k^*)\,\Delta x_k$$

where $x_1^*, x_2^*, \ldots, x_n^*$ are arbitrary points in successive subintervals.

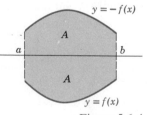

$y = -f(x)$

A

a b

A

$y = f(x)$

Figure 5.6.4

REMARK.   It is proved in advanced calculus that the limit in this definition exists, and that the area A is a nonnegative quantity.

THE DEFINITE INTEGRAL

Definition 5.6.1 applies only to *nonnegative* continuous functions. However, if f is continuous and $f(x) \leq 0$ for all x in $[a, b]$, then $-f$ is continuous and $-f(x) \geq 0$ for all x in $[a, b]$ (Figure 5.6.4). We define the area A between $y = f(x)$ and the interval $[a, b]$ to be the same as the area between $y = -f(x)$ and the interval $[a, b]$.

The limiting process used in Definition 5.6.1 to define area is of fundamental importance in a host of physical and mathematical applications that have no immediate connection with area. In many such applications the function f may have discontinuities and may assume *both* positive and negative values on $[a, b]$. To pave the way for these applications, we shall retrace the three steps leading to Definition 5.6.1, but now we shall drop the assumptions that f is continuous and nonnegative on $[a, b]$; we shall assume only that f is defined on $[a, b]$.

Step 1. Divide the interval $[a, b]$ into n subintervals by choosing arbitrary points $x_1, x_2, \ldots, x_{n-1}$ satisfying

$$a < x_1 < x_2 < \cdots < x_{n-1} < b$$

The points $a, x_1, x_2, \ldots, x_{n-1}, b$ are said to form a **partition** of the interval $[a, b]$. Let

$$\Delta x_1, \Delta x_2, \ldots, \Delta x_n$$

denote the lengths of successive subintervals formed by the partition. The largest of these lengths is called the **norm** of **mesh size** of the partition and is denoted by $\max \Delta x_k$.

Step 2. Choose arbitrary points $x_1^*, x_2^*, \ldots, x_n^*$ in successive subintervals and form the sum

$$f(x_1^*) \, \Delta x_1 + f(x_2^*) \, \Delta x_2 + \cdots + f(x_n^*) \, \Delta x_n$$

$$= \sum_{k=1}^{n} f(x_k^*) \, \Delta x_k$$

The sum in Step 2 is called a **Riemann* sum** (biography on p. 336).

Step 3. Increase n in such a way that the mesh size of the partition approaches zero, and form the limit

$$\lim_{\max \Delta x_k \to 0} \sum_{k=1}^{n} f(x_k^*) \, \Delta x_k$$

if it exists.

The limit in Step 3 is of such importance that there is some notation and terminology associated with it.

5.6.2 DEFINITION. If a function f is defined on a closed interval $[a, b]$, then the **definite integral** of f from a to b, denoted by $\int_a^b f(x) \, dx$, is defined to be

$$\int_a^b f(x) \, dx = \lim_{\max \Delta x_k \to 0} \sum_{k=1}^{n} f(x_k^*) \, \Delta x_k \tag{5}$$

provided the limit exists.

INTEGRABILITY   The numbers a and b in the symbol $\int_a^b f(x)\,dx$ are called, respectively, the *lower* and *upper limits of integration,* and $f(x)$ is called the **integrand**. The reason for the use of an integral sign in the above definition will become clear in the next section where we shall establish a relationship between the definite integral defined by (5) and the indefinite integral studied earlier.

Because the definite integral is defined as a limit, the integral may or may not exist depending on the nature of the integrand f. If the integral does exist, then f is said to be **Riemann integrable** or simply **integrable** on $[a, b]$. The problem of determining precisely which functions are integrable is beyond the scope of this text. However, the following theorem, stated without proof, is useful to know.

5.6.3 THEOREM.

(a) *If f is continuous on $[a, b]$, then f is integrable on $[a, b]$.*

(b) *If f has only finitely many points of discontinuity on $[a, b]$ and if there is a positive number M such that $-M \leq f(x) \leq M$ for all x in $[a, b]$, then f is integrable on $[a, b]$.*

REMARK.   In part (b), the condition that there exists a positive number M such that $-M \leq f(x) \leq M$ for all x in $[a, b]$ is described by stating that

---

*GEORG FRIEDRICH BERNHARD RIEMANN (1826–1866) German mathematician. Bernhard Riemann, as he is commonly known, was the son of a Protestant minister. He received his elementary education from his father and showed brilliance in arithmetic at an early age. In 1846 he enrolled at Göttingen University to study theology and philology, but he soon transferred to mathematics. He studied physics under W. E. Weber and mathematics under Karl Friedrich Gauss, whom some people consider to be the greatest mathematician who ever lived. In 1851 Riemann received his Ph.D. under Gauss, after which he remained at Göttingen to teach. In 1862, one month after his marriage, Riemann suffered an attack of pleuritis, and for the remainder of his life was an extremely sick man. He finally succumbed to tuberculosis in 1866 at age 39.

An interesting story surrounds Riemann's work in geometry. For his introductory lecture prior to becoming an associate professor, Riemann submitted three possible topics to Gauss. Gauss surprised Riemann by choosing the topic Riemann liked the least, the foundations of geometry. The lecture was like a scene from a movie. The old and failing Gauss, a giant in his day, watching intently as his brilliant and youthful protege skillfully pieced together portions of the old man's own work into a complete and beautiful system. Gauss is said to have gasped with delight as the lecture neared its end, and on the way home he marveled at his student's brilliance. Gauss died shortly thereafter. The results presented by Riemann that day eventually evolved into a fundamental tool that Einstein used some 50 years later to develop relativity theory.

In addition to his work in geometry, Riemann made major contributions to the theory of complex functions and mathematical physics. The notion of the definite integral, as it is presented in most basic calculus courses, is due to him. Riemann's early death was a great loss to mathematics, for his mathematical work was brilliant and of fundamental importance.

f is **bounded** on $[a, b]$. Geometrically, this means that over the interval $[a, b]$ the entire graph lies between two horizontal lines, $y = -M$ and $y = M$. A function that approaches $+\infty$ or $-\infty$ somewhere on the interval $[a, b]$ does not satisfy this condition. It can be proved that every Riemann integrable function on $[a, b]$ must be bounded on $[a, b]$.

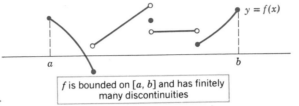

f is bounded on $[a, b]$ and has finitely many discontinuities

Figure 5.6.5 ▷

The function f graphed in Figure 5.6.5 is integrable on $[a, b]$, since it is bounded on the interval $[a, b]$ and has only finitely many points of discontinuity on $[a, b]$.

The following theorem follows from Definitions 5.6.1 and 5.6.2.

5.6.4 THEOREM. *If f is nonnegative and continuous on $[a, b]$, then*

$$A = \begin{bmatrix} \text{area under} \\ y = f(x) \\ \text{over } [a, b] \end{bmatrix} = \int_a^b f(x)\, dx \tag{6}$$

Example 1   In Example 2 of Section 5.5 we showed the area under the curve $y = 9 - x^2$ over the interval $[0, 3]$ to be 18. Thus,

$$\int_0^3 (9 - x^2)\, dx = 18 \qquad ◀$$

GEOMETRIC INTERPRETATION OF THE DEFINITE INTEGRAL

If f is continuous and assumes both positive and negative values on $[a, b]$, then the definite integral can be interpreted as a *difference* of areas. To see why, consider a typical Riemann sum

$$\sum_{k=1}^{n} f(x_k^*)\, \Delta x_k$$

If $f(x_k^*)$ is nonnegative, then the term $f(x_k^*)\, \Delta x_k$ represents the area A_k of a rectangle with height $f(x_k^*)$ and base Δx_k. On the other hand, if $f(x_k^*)$ is negative, then $f(x_k^*)\, \Delta x_k$ is *not* the area of a rectangle, but rather the negative $-A_k$ of such an area. For example, in Figure 5.6.6

$$\sum_{k=1}^{6} f(x_k^*)\, \Delta x_k = f(x_1^*)\, \Delta x_1 + f(x_2^*)\, \Delta x_2 + \cdots + f(x_6^*)\, \Delta x_6$$

$$= A_1 + A_2 - A_3 - A_4 + A_5 + A_6$$

$$= (A_1 + A_2 + A_5 + A_6) - (A_3 + A_4)$$

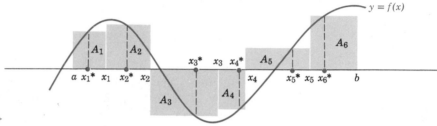

Figure 5.6.6 ▷

It follows that the definite integral

$$\int_a^b f(x)\, dx = \lim_{\max \Delta x_k \to 0} \sum_{k=1}^{n} f(x_k^*)\, \Delta x_k$$

can be interpreted as a difference of areas: the area above the interval $[a, b]$, but below the curve $y = f(x)$, minus the area below the interval $[a, b]$, but above the curve $y = f(x)$. As an illustration, the Riemann sums in Figure 5.6.7 tend toward $A_I - A_{II} + A_{III}$ as a limit, so that

$$\int_a^b f(x)\, dx = (A_I + A_{III}) - A_{II} = \begin{bmatrix} \text{area above} \\ [a, b] \end{bmatrix} - \begin{bmatrix} \text{area below} \\ [a, b] \end{bmatrix}$$

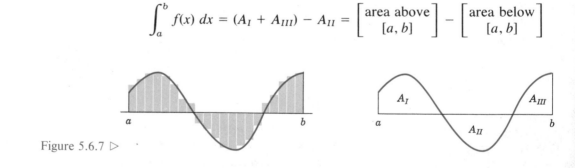

Figure 5.6.7 ▷

In the case where f is continuous and $f(x) \geq 0$ on $[a, b]$, the portion of area below the x-axis is zero so that $\int_a^b f(x)\, dx$ represents the area under $y = f(x)$ over $[a, b]$. On the other hand, if f is continuous and $f(x) \leq 0$ on $[a, b]$, then the portion of area above the x-axis is zero and $\int_a^b f(x)\, dx$ represents the negative of the area above $y = f(x)$ under $[a, b]$. This agrees with our earlier results.

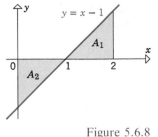

Figure 5.6.8

Example 2   Referring to Figure 5.6.8, the area A_1 is equal to the area A_2 so that

$$\int_0^2 (x-1)\,dx = A_1 - A_2 = 0 \quad \blacktriangleleft$$

The integral in this example was easy to evaluate because the areas involved could be obtained from the area formula for a triangle. In the next section we shall develop techniques for evaluating more complicated integrals.

OPTIONAL

The limits in Definitions 5.6.1 and 5.6.2 are different from those discussed in Chapter 2. Loosely phrased, the expression

$$\lim_{\max \Delta x_k \to 0} \sum_{k=1}^{n} f(x_k^*)\,\Delta x_k = L$$

is intended to convey the idea that we can force the Riemann sums to be as close as we please to L, regardless of how $x_1^*, x_2^*, \ldots, x_n^*$ are chosen, by making the mesh size of the partition sufficiently small. This idea is captured in the following definition.

5.6.5   DEFINITION.   We shall write

$$\lim_{\max \Delta x_k \to 0} \sum_{k=1}^{n} f(x_k^*)\,\Delta x_k = L \qquad (7)$$

if given any number $\epsilon > 0$, there is a number $\delta > 0$ such that

$$\left| \sum_{k=1}^{n} f(x_k^*)\,\Delta x_k - L \right| < \epsilon$$

whenever

$$\max \Delta x_k < \delta$$

and regardless of how $x_1^*, x_2^*, \ldots, x_n^*$ are chosen.

It can be shown that when a number L satisfying (7) exists, it is unique (Exercise 25). We denote this number by $\int_a^b f(x)\,dx$ and write

$$\lim_{\max \Delta x_k \to 0} \sum_{k=1}^{n} f(x_k^*)\,\Delta x_k = \int_a^b f(x)\,dx \qquad (8)$$

REMARK.   Some writers use the symbol $\|\Delta\|$ rather than $\max \Delta x_k$ for the mesh size of the partition, in which case (8) would be written

$$\lim_{\|\Delta\| \to 0} \sum_{k=1}^{n} f(x_k^*) \, \Delta x_k = \int_a^b f(x) \, dx$$

▶ Exercise Set 5.6

In Exercises 1–4, find the value of:

(a) $\displaystyle\sum_{k=1}^{n} f(x_k^*) \, \Delta x_k$ (b) $\max \Delta x_k$.

1. $f(x) = x + 1$; $a = 0$, $b = 4$; $n = 3$; $x_1 = 1$, $x_2 = 2$; $x_1^* = \frac{1}{3}$, $x_2^* = \frac{3}{2}$, $x_3^* = 3$.

2. $f(x) = \cos x$; $a = 0$, $b = 2\pi$; $n = 4$; $x_1 = \pi/2$, $x_2 = 5\pi/4$, $x_3 = 7\pi/4$; $x_1^* = \pi/4$, $x_2^* = \pi$, $x_3^* = 3\pi/2$, $x_4^* = 7\pi/4$.

3. $f(x) = 4 - x^2$; $a = -3$, $b = 4$; $n = 4$; $x_1 = -2$, $x_2 = 0$, $x_3 = 1$; $x_1^* = -\frac{5}{2}$, $x_2^* = -1$, $x_3^* = \frac{1}{4}$, $x_4^* = 3$.

4. $f(x) = x^3$; $a = -3$, $b = 3$; $n = 4$; $x_1 = -1$, $x_2 = 0$, $x_3 = 1$; $x_1^* = -2$, $x_2^* = 0$, $x_3^* = 0$, $x_4^* = 2$.

5. Verify that relationship (3) holds when

$$f(x) = \frac{1}{x}; \ a = 1, \ b = 9; \ n = 4; \ x_1 = 3,$$

$$x_2 = 5, \ x_3 = 7; \ x_1^* = 2, \ x_2^* = 4, \ x_3^* = 6,$$

$$x_4^* = 8.$$

In Exercises 6–8, use the given values of a and b to express the following limits as definite integrals. (Do not evaluate the integrals.)

6. $\displaystyle\lim_{\max \Delta x_k \to 0} \sum_{k=1}^{n} (x_k^*)^3 \, \Delta x_k$; $a = 1$, $b = 2$.

7. $\displaystyle\lim_{\max \Delta x_k \to 0} \sum_{k=1}^{n} 4x_k^*(1 - 3x_k^*) \, \Delta x_k$; $a = -3$, $b = 3$.

8. $\displaystyle\lim_{\max \Delta x_k \to 0} \sum_{k=1}^{n} (\sin^2 x_k^*) \, \Delta x_k$; $a = 0$, $b = \pi/2$.

In Exercises 9–11, express the definite integrals as limits. (Do not evaluate.)

9. $\displaystyle\int_1^2 2x \, dx$.

10. $\displaystyle\int_{-\pi/2}^{\pi/2} (1 + \cos x) \, dx$.

11. $\displaystyle\int_0^1 \frac{x}{x+1} \, dx$.

12. Show on a sketch the area that is represented by

(a) $\displaystyle\int_1^4 \sqrt{x} \, dx$ (b) $\displaystyle\int_1^3 \frac{1}{x} \, dx$

(c) $\displaystyle\int_{-1}^2 \sqrt{9 - x^2} \, dx$ (d) $\displaystyle\int_0^{\pi/2} \sin x \, dx$.

13. Given the areas shown in Figure 5.6.9, find

(a) $\displaystyle\int_a^b f(x) \, dx$ (b) $\displaystyle\int_b^c f(x) \, dx$

(c) $\displaystyle\int_a^c f(x) \, dx$ (d) $\displaystyle\int_a^d f(x) \, dx$.

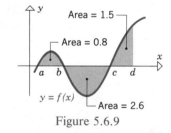

Figure 5.6.9

In Exercises 14–21, evaluate the definite integrals by using appropriate area formulas from plane geometry.

14. (a) $\displaystyle\int_0^3 x \, dx$ (b) $\displaystyle\int_{-2}^{-1} x \, dx$

    (c) $\displaystyle\int_{-1}^4 x \, dx$ (d) $\displaystyle\int_{-5}^5 x \, dx$.

15. Let

$$f(x) = \begin{cases} -1, & x \geq 0 \\ -x, & x < 0 \end{cases}$$

(a) $\displaystyle\int_4^5 f(x) \, dx$ (b) $\displaystyle\int_{-3}^{-2} f(x) \, dx$

(c) $\displaystyle\int_{-5}^2 f(x) \, dx$ (d) $\displaystyle\int_{-k}^k f(x) \, dx \ (k > 0)$.

16. $\displaystyle\int_0^3 |x - 2| \, dx.$ **17.** $\displaystyle\int_{-1}^2 |2x - 3| \, dx.$

18. $\displaystyle\int_0^2 \sqrt{4 - x^2} \, dx.$ **19.** $\displaystyle\int_{-1}^1 \sqrt{1 - x^2} \, dx.$

20. $\displaystyle\int_0^{3/2} \sqrt{9 - 4x^2} \, dx.$ [*Hint:* Factor out the 4 from under the radical.]

21. $\displaystyle\int_0^{10} \sqrt{10x - x^2} \, dx.$ [*Hint:* Complete the square.]

22. Prove that the function

$$f(x) = \begin{cases} 1 & \text{if } x \text{ is rational} \\ 0 & \text{if } x \text{ is irrational} \end{cases}$$

is not integrable on any closed interval $[a, b]$.

23. Prove that the function

$$f(x) = \begin{cases} \dfrac{1}{x^2}, & x > 0 \\ 0, & x = 0 \end{cases}$$

is not integrable on the interval $[0, 1]$.

24. Use Theorem 5.6.3 to show that

$$f(x) = \begin{cases} \sin\dfrac{1}{x}, & x \neq 0 \\ 0, & x = 0 \end{cases}$$

is integrable on the interval $[-1, 1]$.

Exercises 25–27 are for readers who have read the optional material at the end of this section.

25. **(Optional)** Prove: If f is integrable on $[a, b]$, then the value of $\int_a^b f(x) \, dx$ is unique. [*Hint:* Show that if

$$\lim_{\max \Delta x_k \to 0} \sum_{k=1}^n f(x_k^*) \, \Delta x_k = L_1$$

and

$$\lim_{\max \Delta x_k \to 0} \sum_{k=1}^n f(x_k^*) \, \Delta x_k = L_2$$

then $L_1 = L_2$.]

26. **(Optional)** Prove: If f is integrable on $[a, b]$ and c is any constant, then cf is integrable on $[a, b]$ and

$$\int_a^b cf(x) \, dx = c \int_a^b f(x) \, dx$$

27. **(Optional)** Prove: If f and g are integrable on $[a, b]$, then $f + g$ is integrable on $[a, b]$ and

$$\int_a^b [f(x) + g(x)] \, dx = \int_a^b f(x) \, dx + \int_a^b g(x) \, dx$$

28. The function $f(x) = \sqrt{x}$ is continuous on $[0, 4]$ so, by Theorem 5.6.3, the integral $\displaystyle\int_0^4 \sqrt{x} \, dx$ exists. Find its value by using Definition 5.6.2. Use subintervals of unequal length by taking the points

$$x_k = \frac{4k^2}{n^2} \quad (k = 1, 2, \ldots, n - 1) \text{ as the points of}$$

subdivision of $[0, 4]$, and let x_k^* be the right-hand endpoint of the k-th subinterval $(k = 1, 2, \ldots, n)$.

5.7 THE FIRST FUNDAMENTAL THEOREM OF CALCULUS

In Section 5.1 we observed that areas under curves can be obtained by antidifferentiation. In this section we shall prove that result and show that the relationship between areas and antiderivatives is a special case of a more general relationship between definite integrals and antiderivatives.

The following theorem shows how to evaluate the definite integral of a continuous function if we can find an antiderivative for that function.

5.7.1 THEOREM. *If f is continuous on $[a, b]$ and if F is an antiderivative of f on $[a, b]$, then*

$$\int_a^b f(x)\, dx = F(b) - F(a) \tag{1}$$

Proof. Let $x_1, x_2, \ldots, x_{n-1}$ be any points in $[a, b]$ such that

$$a < x_1 < x_2 < \cdots < x_{n-1} < b$$

These points divide $[a, b]$ into n subintervals

$$[a, x_1], [x_1, x_2], \ldots, [x_{n-1}, b] \tag{2}$$

whose lengths, as usual, we denote by

$$\Delta x_1, \Delta x_2, \ldots, \Delta x_n \tag{3}$$

We can write $F(b) - F(a)$ as a telescoping sum:

$$\begin{aligned} F(b) - F(a) = {} & [F(x_1) - F(a)] + [F(x_2) - F(x_1)] \\ & + [F(x_3) - F(x_2)] + \cdots + [F(b) - F(x_{n-1})] \end{aligned} \tag{4}$$

By hypothesis,

$$F'(x) = f(x) \tag{5}$$

for all x in $[a, b]$, so F satisfies the hypotheses of the Mean-Value Theorem (Theorem 4.10.2) on each subinterval in (2). Hence, (4) can be rewritten as

$$\begin{aligned} F(b) - F(a) = {} & F'(x_1^*)(x_1 - a) + F'(x_2^*)(x_2 - x_1) \\ & + F'(x_3^*)(x_3 - x_2) + \cdots + F'(x_n^*)(b - x_{n-1}) \end{aligned} \tag{6}$$

where $x_1^*, x_2^*, \ldots, x_n^*$ are points in successive subintervals.

Using (3) and (5) we can rewrite (6) as

$$F(b) - F(a) = f(x_1^*)\, \Delta x_1 + f(x_2^*)\, \Delta x_2 + f(x_3^*)\, \Delta x_3 + \cdots + f(x_n^*)\, \Delta x_n$$

or in sigma notation,

$$F(b) - F(a) = \sum_{k=1}^{n} f(x_k^*)\, \Delta x_k \tag{7}$$

Let us now increase n in such a way that $\max \Delta x_k \to 0$. Since f is assumed to be continuous, the right side of (7) approaches $\int_a^b f(x)\, dx$, by Theorem 5.6.3a. However, the left side of (7) is a constant that is independent of n; thus,

$$F(b) - F(a) = \lim_{\max \Delta x_k \to 0} \sum_{k=1}^{n} f(x_k^*) \, \Delta x_k = \int_a^b f(x) \, dx \qquad \blacksquare$$

The difference $F(b) - F(a)$ is commonly denoted by $F(x)]_a^b$ so that (1) can be written as

$$\int_a^b f(x) \, dx = F(x) \Big]_a^b \tag{8}$$

Some other common notations are

$$\int_a^b f(x) \, dx = \left[F(x) \right]_a^b \quad \text{and} \quad \int_a^b f(x) \, dx = F(x) \Big]_{x=a}^b$$

The latter notation emphasizes that the limits of integration refer to the variable x. This can be important in problems where more than one variable occurs.

Example 1 Evaluate

$$\int_1^2 x \, dx$$

Solution. The function

$$F(x) = \frac{1}{2} x^2$$

is an antiderivative of $f(x) = x$; thus, from (8)

$$\int_1^2 x \, dx = \frac{1}{2} x^2 \Big]_1^2 = \frac{1}{2} (2)^2 - \frac{1}{2} (1)^2 = 2 - \frac{1}{2} = \frac{3}{2} \qquad \blacktriangleleft$$

When applying the First Fundamental Theorem of Calculus, it does not matter which antiderivative of f is used, for if F is any antiderivative of f on $[a, b]$, then all others have the form

$$F(x) + C \quad \text{[Theorem 5.2.2]}$$

Thus,

$$[F(x) + C]_a^b = [F(b) + C] - [F(a) + C]$$
$$= F(b) - F(a) = F(x)]_a^b$$
$$= \int_a^b f(x) \, dx \tag{9}$$

which shows that all antiderivatives of f on $[a, b]$ yield the same value for $\int_a^b f(x)\,dx$.

Since

$$\int f(x)\,dx = F(x) + C$$

it follows from (9) that

$$\int_a^b f(x)\,dx = \left[\int f(x)\,dx \right]_a^b \tag{10}$$

which relates the definite and indefinite integrals of f.

Example 2 Use the First Fundamental Theorem of Calculus to find the area under the curve $y = \cos x$ over the interval $[0, \pi/2]$. (See Figure 5.7.1.)

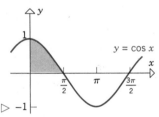

Figure 5.7.1 ▷

Solution. Since $\cos x \geq 0$ for $0 \leq x \leq \pi/2$, the area is

$$A = \int_0^{\pi/2} \cos x\,dx = \left[\int \cos x\,dx \right]_0^{\pi/2}$$

from (10)

$$= \sin x \Big]_0^{\pi/2} = \sin \frac{\pi}{2} - \sin 0 = 1 \quad \blacktriangleleft$$

REMARK. In the third equality of this example we took the constant of integration for the indefinite integral to be $C = 0$. This is justified because we can select any antiderivative of f on $[a, b]$, in particular the one for which $C = 0$.

PROPERTIES OF THE DEFINITE INTEGRAL The following properties of definite integrals follow from (10) and the corresponding properties for indefinite integrals. We shall omit the proofs.

5.7.2 THEOREM. *If f and g are continuous on $[a, b]$ and if c is a constant, then*

(a) $$\int_a^b cf(x)\, dx = c \int_a^b f(x)\, dx$$

(b) $$\int_a^b [f(x) + g(x)]\, dx = \int_a^b f(x)\, dx + \int_a^b g(x)\, dx$$

(c) $$\int_a^b [f(x) - g(x)]\, dx = \int_a^b f(x)\, dx - \int_a^b g(x)\, dx$$

Part (*b*) of Theorem 5.7.2 can be extended to more than two functions. More precisely,

$$\int_a^b [f_1(x) + f_2(x) + \cdots + f_n(x)]\, dx$$

$$= \int_a^b f_1(x)\, dx + \int_a^b f_2(x)\, dx + \cdots + \int_a^b f_n(x)\, dx \quad (11)$$

REMARK. The bracket notation $[F(x)]_a^b$ has some properties in common with the definite integral. We leave it as an exercise to show:

$$[F(x) + G(x)]_a^b = F(x)]_a^b + G(x)]_a^b$$
$$[F(x) - G(x)]_a^b = F(x)]_a^b - G(x)]_a^b$$
$$[cF(x)]_a^b = c[F(x)]_a^b$$

Example 3   Evaluate $\int_0^3 (x^3 - 4x + 1)\, dx$.

Solution.

$$\int_0^3 (x^3 - 4x + 1)\, dx = \int_0^3 x^3\, dx - 4 \int_0^3 x\, dx + \int_0^3 dx$$

$$= \left[\frac{x^4}{4} - 4 \cdot \frac{x^2}{2} + x \right]_0^3$$

$$= \left(\frac{81}{4} - 18 + 3 \right) - (0) = \frac{21}{4} \quad \blacktriangleleft$$

Whenever we speak of the "closed interval $[a, b]$" it is assumed that $a < b$. Thus, our definition of the definite integral

$$\int_a^b f(x)\, dx$$

(Definition 5.6.2) does not allow for the possibilities $a = b$ or $b < a$. The following definition extends the notion of the definite integral to include these cases.

5.7.3 DEFINITION.

(a)   If a is in the domain of f, we define

$$\int_a^a f(x)\, dx = 0$$

(b)   If $b < a$ and if f is integrable on $[b, a]$, then we define

$$\int_b^a f(x)\, dx = -\int_a^b f(x)\, dx$$

The area above
a single point
is zero.

Figure 5.7.2

REMARK.   Geometrically, part (a) states that there is zero area between a curve $y = f(x)$ and a single point a on the x-axis (Figure 5.7.2). In words, part (b) states that interchanging the limits of integration reverses the sign of the integral.

Example 4

(a) $\displaystyle\int_1^1 x^2\, dx = 0$

(b) $\displaystyle\int_4^0 x\, dx = -\int_0^4 x\, dx = -\left.\frac{x^2}{2}\right]_0^4 = -8$ ◄

REMARK.   Part (b) in the foregoing example could have been solved directly without reversing the order of integration by writing

$$\int_4^0 x\, dx = \left.\frac{x^2}{2}\right]_4^0 = \left[\frac{0}{2} - \frac{16}{2}\right] = -8$$

Thus, for the purpose of evaluating a definite integral in which the upper limit is smaller than the lower limit it is not necessary to reverse the order of integration first.

In the next section we shall show how to evaluate definite integrals that require substitutions.

▶ Exercise Set 5.7

In Exercises 1–22, evaluate the definite integrals using the First Fundamental Theorem of Calculus.

1. $\int_2^3 x^3 \, dx.$

2. $\int_{-1}^1 x^4 \, dx.$

3. $\int_{-1}^2 x(1 + x^3) \, dx.$

4. $\int_{-3}^0 (x^2 - 4x + 7) \, dx.$

5. $\int_1^2 (t^2 - 2t + 8) \, dt.$

6. $\int_0^1 (x^5 - x^3 + 2x) \, dx.$

7. $\int_1^3 \frac{1}{x^2} \, dx.$

8. $\int_1^2 \frac{1}{x^6} \, dx.$

9. $\int_1^2 \left(\frac{1}{x^3} - \frac{2}{x^2} + x^{-4} \right) \, dx.$

10. $\int_{-2}^{-1} \left(u^{-4} + 3u^{-2} - \frac{1}{u^5} \right) \, du.$

11. $\int_1^9 \sqrt{x} \, dx.$

12. $\int_1^4 x^{-3/5} \, dx.$

13. $\int_4^9 2y\sqrt{y} \, dy.$

14. $\int_1^8 (5x^{2/3} - 4x^{-2}) \, dx.$

15. $\int_1^4 \left(\frac{3}{\sqrt{x}} - 5\sqrt{x} - x^{-3/2} \right) \, dx.$

16. $\int_4^9 (4y^{-1/2} + 2y^{1/2} + y^{-5/2}) \, dy.$

17. $\int_{-\pi/2}^{\pi/2} \sin \theta \, d\theta.$

18. $\int_0^{\pi/4} \sec^2 \theta \, d\theta.$

19. $\int_{-\pi/4}^{\pi/4} \cos x \, dx.$

20. $\int_0^1 (x - \sec x \tan x) \, dx.$

21. $\int_{\pi/6}^{\pi/2} \left(x + \frac{2}{\sin^2 x} \right) \, dx.$

22. $\int_a^{4a} (a^{1/2} - x^{1/2}) \, dx$ (a a positive constant).

23. Find the area under the curve $y = x^2 + 1$ over the interval $[0, 3]$. Make a sketch of the region.

24. Find the area above the x-axis, but below the curve $y = (1 - x)(x - 2)$. Make a sketch of the region.

25. Find the area under the curve $y = 3 \sin x$ over the interval $[0, 2\pi/3]$. Sketch the region.

26. Find the area below the interval $[-2, -1]$, but above the curve $y = x^3$. Make a sketch of the region.

27. Find the total area that is between the curve $y = x^2 - 3x - 10$ and the interval $[-3, 8]$. Make a sketch of the region. [*Hint:* Find the portion of area above the interval and the portion of area below the interval separately.]

28. Prove:
  (a) $[F(x) + G(x)]_a^b = F(x)]_a^b + G(x)]_a^b$
  (b) $[F(x) - G(x)]_a^b = F(x)]_a^b - G(x)]_a^b$
  (c) $[cF(x)]_a^b = c[F(x)]_a^b.$

29. Assuming the functions involved are all integrable on $[a, b]$, which of the following are always valid?

  (a) $\int_a^b f(x)g(x) \, dx = \int_a^b f(x) \, dx \int_a^b g(x) \, dx$

  (b) $\int_a^b [c_1 f(x) + c_2 g(x)] \, dx$
  $= c_1 \int_a^b f(x) \, dx + c_2 \int_a^b g(x) \, dx$
  (c_1, c_2 constant)

  (c) $\int_a^b \left(\sum_{k=1}^n c_k f_k(x) \right) dx = \sum_{k=1}^n \left[c_k \int_a^b f_k(x) \, dx \right]$
  ($c_1, c_2, \ldots, c_n$ constant)

  (d) $\int_a^b [f(x)]^n \, dx = \left[\int_a^b f(x) \, dx \right]^n$

  (e) $\int_a^b \sqrt{f(x)} \, dx = \sqrt{\int_a^b f(x) \, dx}.$

30. Express Equation (11) in sigma notation.

5.8   EVALUATING DEFINITE INTEGRALS BY SUBSTITUTION

In this section we shall discuss methods of evaluating definite integrals for which a substitution is needed.

In Section 5.3 we showed how to evaluate an indefinite integral

$$\int h(x)\,dx \tag{1}$$

by first expressing it in the form

$$\int h(x)\,dx = \int f(g(x))g'(x)\,dx$$

then making the substitution

$$u = g(x) \quad \text{and} \quad du = g'(x)\,dx$$

to obtain an equivalent, but possibly simpler, indefinite integral

$$\int f(u)\,du$$

We shall now consider two methods for evaluating a definite integral

$$\int_a^b h(x)\,dx$$

by substitution.

Method 1   First evaluate the indefinite integral

$$\int h(x)\,dx$$

by substitution, then use the relationship

$$\int_a^b h(x)\,dx = \left[\int h(x)\,dx\right]_a^b$$

to evaluate the definite integral.

Method 2   Avoid the indefinite integral altogether by first expressing the definite integral in the form

$$\int_a^b h(x)\, dx = \int_a^b f(g(x))g'(x)\, dx \qquad (2)$$

then making the substitution

$$u = g(x) \quad \text{and} \quad du = g'(x)\, dx$$

directly into the definite integral (2). However, to do this we must change the x-limits of integration to corresponding u-limits of integration: Since $u = g(x)$, it follows that

$$u = g(a) \quad \text{if} \quad x = a$$

and

$$u = g(b) \quad \text{if} \quad x = b$$

Thus, if (2) is expressed in terms of u, we obtain

$$\int_a^b h(x)\, dx = \int_{g(a)}^{g(b)} f(u)\, du$$

With a good choice for the substitution, the new definite integral involving u may be easier to evaluate than the original.

Example 1   Use the two methods above to evaluate

$$\int_0^2 2x(x^2 + 1)^3\, dx$$

Method 1   If we let

$$u = x^2 + 1 \quad \text{so that} \quad du = 2x\, dx \qquad (3)$$

then we obtain

$$\int 2x(x^2 + 1)^3\, dx = \int u^3\, du = \frac{u^4}{4} + C = \frac{(x^2 + 1)^4}{4} + C$$

Thus,

$$\int_0^2 2x(x^2 + 1)^3\, dx = \left[\int 2x(x^2 + 1)^3\, dx \right]_{x=0}^{2} = \frac{(x^2 + 1)^4}{4} \Bigg]_{x=0}^{2}$$

$$= \frac{625}{4} - \frac{1}{4} = 156$$

Method 2   For the substitution in (3) we have

$$u = 1 \quad \text{if} \quad x = 0$$
$$u = 5 \quad \text{if} \quad x = 2$$

Thus,

$$\int_0^2 2x(x^2 + 1)^3 \, dx = \int_1^5 u^3 \, du = \frac{u^4}{4}\bigg]_{u=1}^5 = \frac{625}{4} - \frac{1}{4} = 156$$

which agrees with the result obtained by Method 1.   ◀

The choice of methods for evaluating a definite integral by substitution is purely a matter of taste. However, since Method 2 requires some thought to obtain the limits of integration, we shall give a few more examples using that method.

Example 2   Evaluate $\int_0^{\pi/4} \cos (\pi - x) \, dx$

Solution.   Let

$$u = \pi - x \quad \text{so that} \quad du = -dx$$

With this substitution we have

$$u = \pi \qquad \text{if} \quad x = 0$$
$$u = 3\pi/4 \quad \text{if} \quad x = \pi/4$$

so

$$\int_0^{\pi/4} \cos (\pi - x) \, dx = \int_\pi^{3\pi/4} \cos u \, (-du)$$

$$= -\int_\pi^{3\pi/4} \cos u \, du = -\sin u \bigg]_\pi^{3\pi/4}$$

$$= -\left[\sin (3\pi/4) - \sin (\pi) \right]$$

$$= -[1/\sqrt{2} - 0] = -1/\sqrt{2} \quad ◀$$

Example 3   Evaluate $\int_0^{\pi/8} \sin^5 2x \cos 2x \, dx$.

Solution.   Let

$$u = \sin 2x$$

so that

$$du = 2 \cos 2x \, dx \quad \text{or} \quad \tfrac{1}{2}du = \cos 2x \, dx$$

With this substitution we have

$$u = \sin(0) = 0 \quad \text{if} \quad x = 0$$
$$u = \sin(\pi/4) = 1/\sqrt{2} \quad \text{if} \quad x = \pi/8$$

so

$$\int_0^{\pi/8} \sin^5 2x \cos 2x \, dx = \frac{1}{2} \int_0^{1/\sqrt{2}} u^5 \, du = \frac{1}{2} \cdot \frac{u^6}{6} \bigg]_0^{1/\sqrt{2}}$$

$$= \frac{1}{2} \left[\frac{1}{6(\sqrt{2})^6} - 0 \right] = \frac{1}{96} \quad \blacktriangleleft$$

We conclude this section with the theorem that justifies the substitution method used in the foregoing examples.

5.8.1   THEOREM. *If g' is continuous on [a, b] and f is continuous and has an antiderivative on an interval containing the values of g(x) for $a \le x \le b$, then*

$$\int_a^b f(g(x))g'(x) \, dx = \int_{g(a)}^{g(b)} f(u) \, du$$

provided the integrals exist.

Proof.   Let $u = g(x)$ and let F be an antiderivative of f on an interval containing the values of $g(x)$ for $a \le x \le b$. Then by the chain rule

$$\frac{d}{dx} F(g(x)) = \frac{d}{dx} F(u) = \frac{dF}{du}\frac{du}{dx} = f(u)\frac{du}{dx} = f(g(x))g'(x)$$

for each x in $[a, b]$. Thus, $F(g(x))$ is an antiderivative of $f(g(x))g'(x)$ on $[a, b]$. Therefore, by the First Fundamental Theorem of Calculus (Theorem 5.7.1)

$$\int_a^b f(g(x))g'(x) \, dx = F(g(x)) \bigg]_a^b$$

$$= F(g(b)) - F(g(a))$$

$$= \int_{g(a)}^{g(b)} f(u) \, du \quad \blacksquare$$

▶ Exercise Set 5.8

1. In each part express the integral in terms of the variable u, but do not evaluate.

(a) $\displaystyle\int_0^2 (x + 1)^7 \, dx; \; u = x + 1$

(b) $\displaystyle\int_{-1}^2 x\sqrt{8 - x^2} \, dx; \; u = 8 - x^2$

(c) $\displaystyle\int_{-1}^1 \sin(\pi\theta) \, d\theta; \; \theta = u/\pi$

(d) $\displaystyle\int_0^{\pi/4} \tan^2 x \sec^2 x \, dx; \; u = \tan x$

(e) $\displaystyle\int_0^1 x^3\sqrt{x^2 + 3} \, dx; \; u = x^2 + 3$

(f) $\displaystyle\int_0^3 (x + 2)(x - 3)^{20} \, dx; \; u = x - 3.$

In Exercises 2–11, evaluate the integrals two ways: first by a u-substitution in the definite integral, and then by a u-substitution in the corresponding indefinite integral.

2. $\displaystyle\int_1^2 (4x - 2)^3 \, dx.$

3. $\displaystyle\int_0^1 (2x + 1)^4 \, dx.$

4. $\displaystyle\int_1^2 (4 - 3x)^8 \, dx.$

5. $\displaystyle\int_{-1}^0 (1 - 2x)^3 \, dx.$

6. $\displaystyle\int_{-5}^0 x\sqrt{4 - x} \, dx.$

7. $\displaystyle\int_0^8 x\sqrt{1 + x} \, dx.$

8. $\displaystyle\int_0^{\pi/6} 2\cos 3x \, dx.$

9. $\displaystyle\int_0^{\pi/2} 4\sin(x/2) \, dx.$

10. $\displaystyle\int_{1-\pi}^{1+\pi} \sec^2(\tfrac14 x - \tfrac14) \, dx.$

11. $\displaystyle\int_{-2}^{-1} \frac{x}{(x^2 + 2)^3} \, dx.$

In Exercises 12–29, evaluate the integrals by any method.

12. $\displaystyle\int_0^1 \sqrt[3]{a + bx} \, dx \; (b \neq 0).$

13. $\displaystyle\int_0^1 \frac{du}{\sqrt{3u + 1}}.$

14. $\displaystyle\int_1^2 \sqrt{5x - 1} \, dx.$

15. $\displaystyle\int_{-1}^1 \frac{x^2 \, dx}{\sqrt{x^3 + 9}}.$

16. $\displaystyle\int_{-1}^0 6t^2(t^3 + 1)^{19} \, dt.$

17. $\displaystyle\int_1^3 \frac{x + 2}{\sqrt{x^2 + 4x + 7}} \, dx.$

18. $\displaystyle\int_1^2 \frac{dx}{x^2 - 6x + 9}.$

19. $\displaystyle\int_{-3\pi/4}^{-\pi/4} \sin x \cos x \, dx.$

20. $\displaystyle\int_0^{\pi/4} \sqrt{\tan x} \sec^2 x \, dx.$

21. $\displaystyle\int_0^{\sqrt{\pi}} 5x \cos(x^2) \, dx.$

22. $\displaystyle\int_0^{2\pi/t} t^2 \sin tx \, dx \; (t \text{ a positive constant}).$

23. $\displaystyle\int_{\pi^2}^{4\pi^2} \frac{1}{\sqrt{x}} \sin\sqrt{x} \, dx.$

24. $\displaystyle\int_{-\pi/4}^{\pi} \sin\theta \cos\theta \, d\theta.$

25. $\displaystyle\int_0^{\pi/2} \sin^2 3x \cos 3x \, dx.$

26. $\displaystyle\int_0^{\pi/4} \frac{\cos 2x}{\sqrt{7 - 3\sin 2x}} \, dx.$

27. $\displaystyle\int_{\pi/12}^{\pi/9} \sec^2 3\theta \, d\theta.$

28. $\displaystyle\int_{-1}^4 \frac{x \, dx}{\sqrt{5 + x}}.$

29. $\displaystyle\int_0^1 \frac{y^2 \, dy}{\sqrt{4 - 3y}}.$

30. Find the area under the curve $y = 1/(3x + 1)^2$ over the interval $[0, 1]$.

31. Find the area under the curve $y = 3\cos 2x$ over the interval $[0, \pi/8]$. Make a sketch of the region.

In Exercises 32–35, make the indicated substitution and evaluate the resulting definite integral by using an appropriate area formula from plane geometry.

32. $\displaystyle\int_0^{5/3} \sqrt{25 - 9x^2} \, dx; \; \text{let } u = 3x.$

33. $\int_{-3}^{1} \sqrt{3 - 2x - x^2}\, dx$; let $u = x + 1$ after completing the square.

34. $\int_{0}^{2} x\sqrt{16 - x^4}\, dx$; let $u = x^2$.

35. $\int_{\pi/3}^{\pi/2} \sin\theta\sqrt{1 - 4\cos^2\theta}\, d\theta$; let $u = 2\cos\theta$.

36. Find $\int_{0}^{3} f(3x)\, dx$, if $\int_{0}^{9} f(x)\, dx = 5$.

37. Find $\int_{1/2}^{1} \dfrac{1}{x^2} f(1/x)\, dx$, if $\int_{1}^{2} f(x)\, dx = 3$.

38. Prove: If m and n are positive integers, then

$$\int_{0}^{1} x^m(1 - x)^n\, dx = \int_{0}^{1} x^n(1 - x)^m\, dx$$

[*Hint:* Do not evaluate; use a substitution.]

39. Prove: If n is a positive integer, then

$$\int_{0}^{\pi/2} \sin^n x\, dx = \int_{0}^{\pi/2} \cos^n x\, dx.$$

[*Hint:* Do not evaluate; use a trigonometric identity and a substitution.]

40. (a) Prove: If $f(-x) = -f(x)$ for all x in $[-a, a]$

and if f is continuous on $[-a, a]$, then

$$\int_{-a}^{a} f(x)\, dx = 0.$$

(b) Give a geometric explanation of the result in part (a).

41. (a) Let $I = \int_{0}^{a} \dfrac{f(x)}{f(x) + f(a - x)}\, dx$. Show that $I = \dfrac{a}{2}$. [*Hint:* Let $u = a - x$, and then express the integrand as the sum of two fractions.]

(b) Use the result of part (a) to find

$$\int_{0}^{3} \frac{\sqrt{x}}{\sqrt{x} + \sqrt{3 - x}}\, dx.$$

(c) Use the result of part (a) to find

$$\int_{0}^{\pi/2} \frac{\sin x}{\sin x + \cos x}\, dx.$$

42. Let $I = \int_{-1}^{1} \dfrac{1}{1 + x^2}\, dx$. Show that the substitution $x = 1/u$ results in

$$I = -\int_{-1}^{1} \frac{1}{1 + u^2}\, du = -I$$

so $2I = 0$, or $I = 0$, which is impossible (why?). Explain.

5.9 THE MEAN-VALUE THEOREM FOR INTEGRALS; THE SECOND FUNDAMENTAL THEOREM OF CALCULUS

*The First Fundamental Theorem of Calculus tells us how to evaluate the definite integral of a continuous function if we can find an antiderivative for that function. However, the theorem does not address the question of whether every continuous function actually has an antiderivative. That is the purpose of the **Second Fundamental Theorem of Calculus**, which we shall study in this section.*

MORE PROPERTIES OF THE DEFINITE INTEGRAL

Before we can state and prove the Second Fundamental Theorem of Calculus it will be necessary to develop some preliminary results that are important in their own right.

If f is continuous and nonnegative on $[a, b]$ and if c is a point between a and b, then it is evident that the area under $y = f(x)$ over the interval

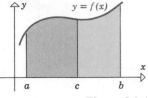

Figure 5.9.1

[a, b] can be split into two parts, the area under the curve from a to c plus the area under the curve from c to b (Figure 5.9.1), that is,

$$\int_a^b f(x)\, dx = \int_a^c f(x)\, dx + \int_c^b f(x)\, dx$$

This is a special case of the following theorem about definite integrals.

5.9.1   THEOREM. *If f is integrable on a closed interval containing the three points a, b, and c, then*

$$\int_a^b f(x)\, dx = \int_a^c f(x)\, dx + \int_c^b f(x)\, dx \qquad (1)$$

no matter how the points are ordered.

(We omit the proof.) This theorem is helpful when the formula for the integrand changes between the limits of integration.

Example 1   Evaluate $\int_0^6 f(x)\, dx$ if

$$f(x) = \begin{cases} x^2, & x \le 2 \\ 3x - 2, & x \ge 2 \end{cases}$$

Solution.

$$\int_0^6 f(x)\, dx = \int_0^2 f(x)\, dx + \int_2^6 f(x)\, dx = \int_0^2 x^2\, dx + \int_2^6 (3x - 2)\, dx$$

$$= \frac{x^3}{3}\Big]_0^2 + \left[\frac{3x^2}{2} - 2x\right]_2^6 = \left(\frac{8}{3} - 0\right) + (42 - 2) = \frac{128}{3} \qquad \blacktriangleleft$$

Example 2   Evaluate $\int_{-1}^2 |x|\, dx$.

Solution.   Since $|x| = x$ when $x \ge 0$ and $|x| = -x$ when $x \le 0$,

$$\int_{-1}^2 |x|\, dx = \int_{-1}^0 |x|\, dx + \int_0^2 |x|\, dx$$

$$= \int_{-1}^0 (-x)\, dx + \int_0^2 x\, dx$$

$$= -\frac{x^2}{2}\Big]_{-1}^{0} + \frac{x^2}{2}\Big]_{0}^{2} = \frac{1}{2} + 2 = \frac{5}{2} \quad \blacktriangleleft$$

The following theorem is useful for comparing the sizes of definite integrals.

5.9.2 THEOREM.

(a) *If f is continuous and $f(x) \geq 0$ for all x in $[a, b]$, then*

$$\int_a^b f(x)\, dx \geq 0$$

(b) *If f and g are continuous and $f(x) \geq g(x)$ for all x in $[a, b]$, then*

$$\int_a^b f(x)\, dx \geq \int_a^b g(x)\, dx$$

Area under $f \geqslant$ area under g

Figure 5.9.2

Proof of (a). Since f is continuous and nonnegative on $[a, b]$, the integral $\int_a^b f(x)\, dx$ represents the area under $y = f(x)$ over $[a, b]$. Thus, the integral is nonnegative by the remark following Definition 5.6.1.

Proof of (b). It follows from the hypotheses that $f - g$ is continuous and $f(x) - g(x) \geq 0$ for all x in $[a, b]$. Thus by part (a),

$$\int_a^b [f(x) - g(x)]\, dx = \int_a^b f(x)\, dx - \int_a^b g(x)\, dx \geq 0$$

from which it follows that

$$\int_a^b f(x)\, dx \geq \int_a^b g(x)\, dx \quad \blacksquare$$

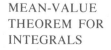

Figure 5.9.3

In the case where $f(x) \geq g(x) \geq 0$ for all x in $[a, b]$, part (b) of this theorem states that the area under the "higher" curve $y = f(x)$ is at least as large as the area under the "lower" curve $y = g(x)$ (Figure 5.9.2).

MEAN-VALUE
THEOREM FOR
INTEGRALS

Our next objective is to obtain an important result called the *Mean-Value Theorem for Integrals*. As we shall see, this theorem is not only the cornerstone for the Second Fundamental Theorem of Calculus, but can also be used to define the notion of "average value" for continuous functions.

Let f be a continuous nonnegative function on $[a, b]$, and let m and M be the minimum and maximum values of $f(x)$ on this interval. Consider the rectangles of heights m and M over the interval $[a, b]$ (Figure 5.9.3). It is clear geometrically that the area

$$A = \int_a^b f(x)\, dx$$

is at least as large as the area of the rectangle of height m, and no larger than the area of the rectangle of height M. It seems reasonable, therefore, that somewhere between m and M there is an appropriate height $f(x^*)$ such that the rectangle of this height over $[a, b]$ has precisely area A; that is,

$$\int_a^b f(x)\, dx = f(x^*)(b - a)$$

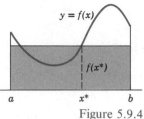

Figure 5.9.4

(Figure 5.9.4). This is a special case of the following result.

The Mean-Value Theorem for Integrals

5.9.3   THEOREM. *If f is continuous on a closed interval $[a, b]$, then there is at least one number x^* in $[a, b]$ such that*

$$\int_a^b f(x)\, dx = f(x^*)(b - a) \qquad (2)$$

Proof.   By the Extreme-Value Theorem (4.6.4), f assumes a maximum value M and a minimum value m on $[a, b]$. Thus, for all x in $[a, b]$,

$$m \le f(x) \le M \qquad (3)$$

and from Theorem 5.9.2*b*

$$\int_a^b m\, dx \le \int_a^b f(x)\, dx \le \int_a^b M\, dx$$

or

$$m(b - a) \le \int_a^b f(x)\, dx \le M(b - a)$$

or

$$m \le \frac{1}{b - a} \int_a^b f(x)\, dx \le M \qquad (4)$$

Since (4) states that

$$\frac{1}{b - a} \int_a^b f(x)\, dx \qquad (5)$$

is a number between m and M, and since $f(x)$ assumes the values m and

M on $[a, b]$, it follows from the Intermediate-Value Theorem (2.7.9) that $f(x)$ must assume the value (5) at some point x^* in $[a, b]$, that is,

$$\frac{1}{b-a} \int_a^b f(x)\, dx = f(x^*) \quad \text{or} \quad \int_a^b f(x)\, dx = f(x^*)(b-a) \qquad \blacksquare$$

Example 3 Since $f(x) = x^2$ is continuous on the interval $[1, 4]$, the Mean-Value Theorem for Integrals guarantees that there is a number x^* in $[1, 4]$ such that

$$\int_1^4 x^2\, dx = f(x^*)(4 - 1) = (x^*)^2(4 - 1) = 3(x^*)^2$$

But

$$\int_1^4 x^2\, dx = \frac{x^3}{3}\bigg]_1^4 = 21$$

so that

$$3(x^*)^2 = 21 \quad \text{or} \quad (x^*)^2 = 7 \quad \text{or} \quad x^* = \pm\sqrt{7}$$

Thus, $x^* = \sqrt{7} \approx 2.65$ is the number in the interval $[1, 4]$ whose existence is guaranteed by the Mean-Value Theorem for Integrals. ◀

AVERAGE VALUE The number $f(x^*)$ in Theorem 5.9.3 is closely related to the familiar notion of an *arithmetic average*. To see this, divide the interval $[a, b]$ into n subintervals of equal length

$$\Delta x = \frac{b-a}{n} \tag{6}$$

and choose arbitrary points $x_1^*, x_2^*, \ldots, x_n^*$ in successive subintervals. Then the arithmetic average of the numbers $f(x_1^*), f(x_2^*), \ldots, f(x_n^*)$ is

$$\text{ave} = \frac{1}{n}[f(x_1^*) + f(x_2^*) + \cdots + f(x_n^*)]$$

or from (6)

$$\text{ave} = \frac{1}{b-a}[f(x_1^*)\,\Delta x + f(x_2^*)\,\Delta x + \cdots + f(x_n^*)\,\Delta x]$$

$$= \frac{1}{b-a}\sum_{k=1}^n f(x_k^*)\,\Delta x$$

Taking the limit as $n \to +\infty$ yields

$$\lim_{n \to +\infty} \frac{1}{b-a} \sum_{k=1}^{n} f(x_k^*)\, \Delta x = \frac{1}{b-a} \int_a^b f(x)\, dx \qquad (7)$$

Since this equation describes what happens when we compute the average of "more and more" values of $f(x)$, we are led to the following definition.

5.9.4 DEFINITION. If f is integrable on $[a, b]$, then the **average value** (or **mean value**) of $f(x)$ on $[a, b]$ is defined to be

$$f(x)_{\text{ave}} = \frac{1}{b-a} \int_a^b f(x)\, dx$$

Example 4 Find the average value of $f(x) = x^2$ on the interval $[1, 4]$.

Solution.

$$f(x)_{\text{ave}} = \frac{1}{b-a} \int_a^b f(x)\, dx = \frac{1}{4-1} \int_1^4 x^2\, dx = \frac{1}{3}(21) = 7 \qquad \blacktriangleleft$$

REMARK. In light of Definition 5.9.4, the quantity $f(x^*)$ in the Mean-Value Theorem for Integrals (5.9.3) is just the average value of $f(x)$ over $[a, b]$. Thus, the Mean-Value Theorem for Integrals can be stated in the form

$$\int_a^b f(x)\, dx = (b-a)f(x)_{\text{ave}}$$

Before turning to the question of "existence" of antiderivatives, it will be helpful to consider some notational matters.

Sometimes it is convenient to use a letter other than x for the variable of integration in a definite integral. For example,

$$\int_a^b f(t)\, dt, \quad \int_a^b f(u)\, du, \quad \int_a^b f(y)\, dy$$

These integrals all have the same value. In general, we have the following result:

The value of a definite integral is unaffected if we change the letter used for the variable of integration, but do not change the limits of integration.

For example, the following integrals have the same limits of integration, but different variables of integration:

$$\int_1^3 x^2 \, dx = \left. \frac{x^3}{3} \right]_{x=1}^{3} = \frac{27}{3} - \frac{1}{3} = \frac{26}{3}$$

$$\int_1^3 t^2 \, dt = \left. \frac{t^3}{3} \right]_{t=1}^{3} = \frac{27}{3} - \frac{1}{3} = \frac{26}{3}$$

$$\int_1^3 u^2 \, du = \left. \frac{u^3}{3} \right]_{u=1}^{3} = \frac{27}{3} - \frac{1}{3} = \frac{26}{3}$$

Because the letter used for the variable of integration has no effect on the final value of the definite integral, it is sometimes called a *dummy variable*.

In the work to follow we shall consider definite integrals of the form

$$\int_a^x \underline{\qquad\qquad}$$

where the upper limit x is allowed to vary. For such integrals we shall use a letter different from x (often t) for the variable of integration; thus, we would write

$$\int_a^x f(t) \, dt \quad \text{rather than} \quad \int_a^x f(x) \, dx$$

This avoids using x in two different ways (as a variable of integration and as a limit of integration), which can cause errors.

Example 5   Evaluate $\int_2^x t^2 \, dt$

Solution.

$$\int_2^x t^2 \, dt = \left. \frac{t^3}{3} \right]_{t=2}^{x} = \frac{x^3}{3} - \frac{8}{3} \qquad \blacktriangleleft$$

Note that the final expression in the foregoing example is a function of x alone. In general, an expression of the form

$$\int_a^x f(t) \, dt$$

represents a function of x—the variable t does not enter into the final result.

We now turn to the major result in this section.

5.9.5 THEOREM. *Let f be continuous on an open interval I, and let a be any point in I. If F is defined by*

$$F(x) = \int_a^x f(t)\, dt \qquad (8)$$

then $F'(x) = f(x)$ *at each point x in the interval I.*

Proof. From the definition of a derivative,

$$F'(x) = \lim_{h \to 0} \frac{F(x + h) - F(x)}{h}$$

$$= \lim_{h \to 0} \frac{1}{h} \left[\int_a^{x+h} f(t)\, dt - \int_a^x f(t)\, dt \right]$$

$$= \lim_{h \to 0} \frac{1}{h} \left[\int_a^{x+h} f(t)\, dt + \int_x^a f(t)\, dt \right]$$

$$= \lim_{h \to 0} \frac{1}{h} \int_x^{x+h} f(t)\, dt \qquad \boxed{\text{Theorem 5.9.1}}$$

Applying the Mean-Value Theorem for Integrals (5.9.3) to the last expression, we obtain

$$F'(x) = \lim_{h \to 0} \frac{1}{h} [f(t^*) \cdot h] = \lim_{h \to 0} f(t^*) \qquad (9)$$

where t^* is some number between x and $x + h$. Because t^* is between x and $x + h$, it follows that $t^* \to x$ as $h \to 0$. Thus, $f(t^*) \to f(x)$ as $h \to 0$, since f is assumed continuous at x. Therefore, from (9) $F'(x) = f(x)$. ∎

The Second Fundamental Theorem of Calculus ensures that if f is continuous on an open interval, then f has an antiderivative on that interval, namely, $\int_a^x f(t)\, dt$. However, because we can add an arbitrary constant to an antiderivative of f and still have an antiderivative of f, it follows that a function that is continuous on an open interval has infinitely many antiderivatives on the interval. Note, however, that (8) singles out that antiderivative of f whose value at $x = a$ is zero because

$$F(a) = \int_a^a f(t)\, dt = 0$$

Example 6   Use (8) to find that antiderivative of $f(x) = x$ on $(-\infty, +\infty)$ whose value is zero at $x = 2$.

Solution.   The function $f(x) = x^2$ is continuous on $(-\infty, +\infty)$. Thus, by the Second Fundamental Theorem of Calculus and the preceding remarks, the function

$$F(x) = \int_2^x f(t)\, dt = \int_2^x t^2\, dt$$

must be that antiderivative of $f(x) = x^2$ on $(-\infty, +\infty)$ with a value of zero at $x = 2$. This is indeed the case since

$$F(x) = \frac{x^3}{3} - \frac{8}{3} \tag{10}$$

(see Example 5) so that

$$F'(x) = x^2 \quad \text{and} \quad F(2) = \frac{8}{3} - \frac{8}{3} = 0 \quad \blacktriangleleft$$

Example 7   Because $f(x) = 1/x$ is continuous on the interval $(0, +\infty)$,

$$F(x) = \int_1^x \frac{1}{t}\, dt$$

is the antiderivative of $1/x$ on $(0, +\infty)$ whose value at $x = 1$ is zero. That is,

$$F'(x) = \frac{1}{x} \quad \text{and} \quad F(1) = 0 \quad \blacktriangleleft$$

Example 8   Find

$$\frac{d}{dx} \int_1^x \frac{\sin t}{t}\, dt$$

Solution.   Let

$$F(x) = \int_1^x \frac{\sin t}{t}\, dt$$

so that $f(t) = \dfrac{\sin t}{t}$ in (8). Thus, from the Second Fundamental Theorem of Calculus

$$\frac{d}{dx} \int_1^x \frac{\sin t}{t}\, dt = \frac{d}{dx}(F(x)) = F'(x) = f(x) = \frac{\sin x}{x} \quad \blacktriangleleft$$

As a result of the Second Fundamental Theorem of Calculus, we now know that continuous functions have antiderivatives. However, there are two possible impediments that can prevent us from applying the formula

$$\int_a^b f(x)\,dx = F(b) - F(a)$$

even if f is continuous:

- We may not be clever enough to find a usable formula for an antiderivative F.

- The antiderivative F may not have a formula in terms of familiar functions.

To illustrate the second situation, consider the function $f(x) = 1/x$. Because this function is continuous on $(0, +\infty)$, we know that it has an antiderivative on this interval. In fact, we showed in Example 7 that the function

$$F(x) = \int_1^x \frac{1}{t}\,dt \tag{11}$$

is such an antiderivative. Although this formula is interesting from a theoretical viewpoint, it would be desirable to have a more elementary formula that uses familiar functions and does not involve an integral. It can be proved, however, that no antiderivative of $f(x) = 1/x$ can be expressed by an elementary formula involving finitely many polynomials, rational functions, or trigonometric functions. In short, our repertoire of basic functions is too limited, at present, to produce an antiderivative of $f(x) = 1/x$ which is simpler than (11). In Chapter 7 we shall use the Second Fundamental Theorem of Calculus to expand our repertoire of basic functions.

▶ Exercise Set 5.9

In Exercises 1–6, use Theorem 5.9.1 to evaluate the integrals.

1. $\int_0^2 |2x - 3|\,dx.$

2. $\int_1^5 |x - 2|\,dx.$

3. $\int_0^{3\pi/4} |\cos x|\,dx.$

4. $\int_{-1}^2 \sqrt{2 + |x|}\,dx.$

5. $\int_{-2}^3 f(x)\,dx,$ where $f(x) = \begin{cases} -x, & x \geq 0 \\ x^2, & x < 0. \end{cases}$

6. $\int_0^9 g(x)\,dx,$ where $g(x) = \begin{cases} 1, & 0 \leq x < 1 \\ x^3, & 1 \leq x < 4 \\ \sqrt{x}, & 4 \leq x \leq 9. \end{cases}$

7. Prove: If f is continuous and $f(x) \leq 0$ for all x in $[a, b]$, then $\int_a^b f(x)\,dx \leq 0.$

8. (a) Find f_{ave} of $f(x) = 2x$ over $[0, 4]$.
   (b) Find a point x^* in $[0, 4]$ such that $f(x^*) = f_{\text{ave}}$.

(c)   Sketch the graph of $f(x) = 2x$ over $[0, 4]$ and construct a rectangle over the interval whose area is the same as the area under the graph of f over the interval.

9. (a)   Find f_{ave} of $f(x) = x^2$ over $[0, 2]$.

(b)   Find a point x^* in $[0, 2]$ such that $f(x^*) = f_{ave}$.

(c)   Sketch the graph of $f(x) = x^2$ over $[0, 2]$ and construct a rectangle over the interval whose area is the same as the area under the graph of f over the interval.

In Exercises 10–13, find the average value of $f(x)$ over the interval and find all values of x^* described in the Mean-Value Theorem for Integrals.

10. $f(x) = 1/x^2$; $[1, 3]$.

11. $f(x) = \sqrt{x}$; $[0, 9]$.

12. $f(x) = \sin x$; $[-\pi, \pi]$.

13. $f(x) = \alpha x + \beta$; $[x_0, x_1]$.

14.   Let $s(t)$ be the position function of a particle moving on a coordinate line. In Section 4.11 (Exercise 18) we defined the average velocity v_{ave} over a time interval $[t_0, t_1]$ by

$$v_{ave} = \frac{\text{change in position}}{\text{time elapsed}} = \frac{s(t_1) - s(t_0)}{t_1 - t_0}$$

(a)   Show that

$$v_{ave} = \frac{1}{t_1 - t_0} \int_{t_0}^{t_1} v(t)\, dt$$

so that v_{ave} can also be interpreted as the average value of the velocity function over the time interval $[t_0, t_1]$.

(b)   Use the result in (a) to compute the average velocity over the time interval $[0, 5]$ for a particle moving on a coordinate line with velocity function $v(t) = 32t$.

15.   For a particle moving on a coordinate line, show that the average acceleration a_{ave} over a time interval $[t_0, t_1]$ (as defined in Exercise 18 of Section 4.11) can be written as

$$a_{ave} = \frac{1}{t_1 - t_0} \int_{t_0}^{t_1} a(t)\, dt$$

which is the average of the acceleration function over the time interval $[t_0, t_1]$.

16.   Consider a particle moving on a coordinate line. Use Exercises 14 and 15 to find:

(a) v_{ave} for $1 \le t \le 4$ if $v(t) = 3t^3 + 2$

(b) a_{ave} for $2 \le t \le 9$ if $a(t) = t^{1/2}$

(c) v_{ave} for $t_0 \le t \le t_1$ if $v(t) = 32t + v_0$
        (v_0 constant).

17.   Water is run at a constant rate of 1 ft^3/min to fill a cylindrical tank of radius 3 ft and height 5 ft. Assuming the tank is initially empty, find the average force on the bottom over the time period required to fill the tank (density of water = 62.4 lb/ft^3).

18.   Prove: If $f(x) = k$ is constant on $[a, b]$, then $f_{ave} = k$ on $[a, b]$.

19.   (a)   Prove: If f is continuous on $[a, b]$, then

$$\int_a^b [f(x) - f_{ave}]\, dx = 0$$

(b)   Does there exist a constant $c \ne f_{ave}$ such that

$$\int_a^b [f(x) - c]\, dx = 0?$$

20.   Define $F(x)$ by

$$F(x) = \int_{\pi/4}^x \cos 2t\, dt$$

(a)   Use the Second Fundamental Theorem of Calculus to find $F'(x)$.

(b)   Check the result in (a) by first integrating and then differentiating.

21.   Define $F(x)$ by

$$F(x) = \int_1^x (t^3 + 1)\, dt$$

(a)   Use the Second Fundamental Theorem of Calculus to find $F'(x)$.

(b)   Check the result in (a) by first integrating and then differentiating.

In Exercises 22–25, use the Second Fundamental Theorem of Calculus to find the derivative.

22. $\dfrac{d}{dx} \displaystyle\int_0^x \dfrac{dt}{1 + \sqrt{t}}$.

23. $\dfrac{d}{dx} \displaystyle\int_1^x \sin(\sqrt{t})\, dt$.

24. $\dfrac{d}{dx} \displaystyle\int_0^x \dfrac{t}{\cos t}\, dt$.

25. $\dfrac{d}{dx} \displaystyle\int_0^x |t|\, dt$.

In Exercises 26–29, express the antiderivatives as integrals.

26. The antiderivative of $1/(1 + x^2)$ on the interval $(-\infty, +\infty)$ whose value at $x = 1$ is 0.

27. The antiderivative of $1/(x - 1)$ on the interval $(1, +\infty)$ whose value at $x = 2$ is 0.

28. The antiderivative of $1/(x - 1)$ on the interval $(-\infty, 1)$ whose value at $x = -3$ is 0.

29. The antiderivative of $1/(x - 1)$ on the interval $(-\infty, 1)$ whose value at $x = 0$ is 0.

30. (a) Over what open interval does the formula
$$F(x) = \int_1^x \frac{1}{t^2 - 9} dt$$
represent an antiderivative of
$$f(x) = \frac{1}{x^2 - 9}?$$
(b) Find a point where the graph of F crosses the x-axis.

31. (a) Over what open interval does the formula
$$F(x) = \int_1^x \frac{dt}{t}$$
represent an antiderivative of $f(x) = 1/x$?
(b) Find a point where the graph of F crosses the x-axis.

32. Let $F(x) = \int_2^x \sqrt{3t^2 + 1}\ dt$. Find
(a) $F(2)$ (b) $F'(2)$ (c) $F''(2)$.

33. Let $F(x) = \int_0^x \frac{\cos t}{t^2 + 3} dt$. Find
(a) $F(0)$ (b) $F'(0)$ (c) $F''(0)$.

34. Let $F(x) = \int_0^x \frac{t - 3}{t^2 + 7} dt$ for $-\infty < x < +\infty$.
(a) Find the value of x where F attains its minimum value.
(b) Find open intervals over which F is only increasing or only decreasing.
(c) Find open intervals over which F is only concave up or only concave down.

In Exercises 35–37, express $F(x)$ in a piecewise form that does not employ integrals.

35. $F(x) = \int_{-1}^x |t|\ dt$.

36. $F(x) = \int_0^x f(t)\ dt$, where $f(x) = \begin{cases} x, & 0 \le x \le 2 \\ 2, & x > 2. \end{cases}$

37. $F(x) = \int_{-1}^x f(t)\ dt$, where $f(x) = \begin{cases} x^2, & x \le 0 \\ 2x, & x > 0. \end{cases}$

38. Use the Second Fundamental Theorem of Calculus and the chain rule to show that
$$\frac{d}{dx} \int_a^{g(x)} f(t)\ dt = f(g(x))g'(x)$$

In Exercises 39 and 40, use the result in Exercise 38 to perform the differentiation.

39. $\dfrac{d}{dx} \displaystyle\int_1^{x^3} \frac{1}{t}\ dt$.
40. $\dfrac{d}{dx} \displaystyle\int_3^{\sin x} \frac{1}{1 + t^2}\ dt$.

41. Prove that the function
$$F(x) = \int_0^x \frac{1}{1 + t^2} dt + \int_0^{1/x} \frac{1}{1 + t^2} dt$$
is constant on the interval $(0, +\infty)$.

42. Use Exercise 38 and Theorem 5.9.1 to show that
$$\frac{d}{dx} \int_{h(x)}^{g(x)} f(t)\ dt = f(g(x))g'(x) - f(h(x))h'(x)$$

43. Use the result in Exercise 42 to perform the following differentiations:
(a) $\dfrac{d}{dx} \displaystyle\int_{x^2}^{x^3} \sin^2 t\ dt$
(b) $\dfrac{d}{dx} \displaystyle\int_{-x}^x \frac{1}{1 + t}\ dt$.

44. Prove that the function
$$F(x) = \int_x^{3x} \frac{1}{t}\ dt$$
is constant on the interval $(0, +\infty)$ by using Exercise 42 to find $F'(x)$.

45. Prove: If f is continuous on an open interval I and b is any point in I, then at each point in I
$$\frac{d}{dx} \int_x^b f(t)\ dt = -f(x)$$

46. Prove: If f is continuous on an open interval I and a is any point in I, then
$$F(x) = \int_a^x f(t)\ dt$$
is continuous on I.

▶ SUPPLEMENTARY EXERCISES

In Exercises 1–10, evaluate the integrals and check your results by differentiation.

1. $\int \left[\dfrac{1}{x^3} + \dfrac{1}{\sqrt{x}} - 5 \sin x \right] dx.$

2. $\int \dfrac{2t^4 - t + 2}{t^3} \, dt.$ 3. $\int \dfrac{(\sqrt{x} + 2)^8}{\sqrt{x}} \, dx.$

4. $\int x^3 \cos (2x^4 - 1) \, dx.$

5. $\int \dfrac{x \sin \sqrt{2x^2 - 5}}{\sqrt{2x^2 - 5}} \, dx.$

6. $\int \sqrt{\cos \theta} \, \sin (2\theta) \, d\theta.$

7. $\int \sqrt{x}(3 + \sqrt[3]{x^4}) \, dx.$

8. $\int \dfrac{x^{1/3} \, dx}{x^{8/3} + 2x^{4/3} + 1}.$

9. $\int \sec^2 (\sin 5t) \cos 5t \, dt.$

10. $\int \dfrac{\cot^2 x}{\sin^2 x} \, dx.$

11. Evaluate $\int y(y^2 + 2)^2 \, dy$ two ways: (a) by multiplying out and integrating term by term; and (b) by using the substitution $u = y^2 + 2$. Show that your answers differ by a constant.

In Exercises 12–17, evaluate the definite integral by making the indicated substitution and changing the x-limits of integration to u-limits.

12. $\int_1^0 \sqrt[5]{1 - 2x} \, dx, \ u = 1 - 2x.$

13. $\int_0^{\pi/2} \sin^4 x \cos x \, dx, \ u = \sin x.$

14. $\int_0^{-3} \dfrac{x \, dx}{\sqrt{x^2 + 16}}, \ u = x^2 + 16.$

15. $\int_2^5 \dfrac{x - 2}{\sqrt{x - 1}} \, dx, \ u = x - 1.$

16. $\int_{\pi/6}^{\pi/4} \dfrac{\sin 2x \, dx}{\sqrt{1 - \frac{3}{2} \cos 2x}}, \ u = 1 - \frac{3}{2} \cos 2x.$

17. $\int_1^4 \dfrac{1}{\sqrt{x}} \cos \left(\dfrac{\pi\sqrt{x}}{2} \right) dx, \ u = \dfrac{\pi\sqrt{x}}{2}.$

In Exercises 18 and 19, evaluate $\int_{-2}^2 f(x) \, dx.$

18. $f(x) = \begin{cases} x^3 \text{ for } x \geq 0 \\ -x \text{ for } x < 0. \end{cases}$

19. $f(x) = |2x - 1|.$

In Exercises 20–22, solve for x.

20. $\int_1^x \dfrac{1}{\sqrt{t}} \, dt = 3.$

21. $\int_0^x \dfrac{1}{(3t + 1)^2} \, dt = \dfrac{1}{6}.$

22. $\int_2^x (4t - 1) \, dt = 9.$

23. Evaluate:

(a) $\displaystyle\sum_{i=3}^6 5$ (b) $\displaystyle\sum_{i=n}^{n+3} 2$

(c) $\displaystyle\sum_{i=n}^{n+3} n$ (d) $\displaystyle\sum_{k=1}^3 \left(\dfrac{k - 1}{k + 3} \right)$

(e) $\displaystyle\sum_{k=2}^4 \dfrac{6}{k^2}$ (f) $\displaystyle\sum_{n=4}^4 (2n + 1)$

(g) $\displaystyle\sum_{k=0}^4 \sin (k\pi/4)$ (h) $\displaystyle\sum_{k=1}^4 \sin^k (\pi/4).$

24. Express in sigma notation and evaluate:
(a) $3 \cdot 1 + 4 \cdot 2 + 5 \cdot 3 + \cdots + 102 \cdot 100$
(b) $200 + 198 + \cdots + 4 + 2.$

25. Express in sigma notation, first starting with $k = 1$, and then with $k = 2$. Do not evaluate.

(a) $\dfrac{1}{4} - \dfrac{4}{9} + \dfrac{9}{16} - \cdots - \dfrac{64}{81} + \dfrac{81}{100}$

(b) $\dfrac{\pi^2}{1} - \dfrac{\pi^3}{2} + \dfrac{\pi^4}{3} - \cdots + \dfrac{\pi^{12}}{11}.$

In Exercises 26–29, use the partition of $[a, b]$ into n subintervals of equal length, and find a closed form for the sum of the areas of: (a) the inscribed rectangles and (b) the circumscribed rectangles. (c) Use your answer in either (a) or (b) to find the area under the curve $y = f(x)$ over the interval $[a, b]$. (Check your answer by integration.)

26. $f(x) = 6 - 2x$; $a = 1$, $b = 3$.

27. $f(x) = 16 - x^2$; $a = 0$, $b = 4$.

28. $f(x) = x^2 + 2$; $a = 1$, $b = 4$.

29. $f(x) = 6$; $a = -1$, $b = 1$.

30. (a) Prove: If $f(-x) = f(x)$ for all x in $[-a, a]$ and f is integrable on the interval $[-a, a]$, then

$$\int_{-a}^{a} f(x)\,dx = 2 \int_{0}^{a} f(x)\,dx.$$

   (b) Give a geometric explanation of the result in part (a).

31. In each part evaluate the definite integral geometrically, that is, by interpreting it as an area and using geometric reasoning to find the area.

   (a) $\int_{0}^{2} (2 - 4x)\,dx$ (b) $\int_{-3}^{0} \sqrt{9 - t^2}\,dt$

   (c) $\int_{-4}^{2} \frac{|x|}{x}\,dx$ (d) $\int_{-\pi/3}^{\pi/3} \sin x\,dx$

   (e) $\int_{0}^{\pi} \cos x\,dx$ (f) $\int_{-10}^{-5} 6\,dx$

   (g) $\int_{-5}^{4} f(x)\,dx$ if $f(x) = \begin{cases} -4, & x \le -3 \\ 2, & -3 < x < 0 \\ 1, & x \ge 0. \end{cases}$

32. Given that

$$\int_{-1}^{0} f(x)\,dx = 3, \quad \int_{-1}^{2} f(x)\,dx = -1,$$

and

$$\int_{-1}^{2} g(x)\,dx = 2,$$

evaluate the following:

   (a) $\int_{0}^{-1} 8f(x)\,dx$ (b) $\int_{-1}^{2} f(y)\,dy$

   (c) $\int_{2}^{-2} g(x)\,dx$ (d) $\int_{0}^{2} f(x)\,dx$

   (e) $\int_{-1}^{2} [f(x) - 2g(x)]\,dx.$

33. Given that

$$\int_{1}^{5} P(x)\,dx = -1, \quad \int_{3}^{5} P(x)\,dx = 3$$

and

$$\int_{3}^{5} Q(x)\,dx = 4,$$

evaluate the following:

   (a) $\int_{3}^{5} [2P(x) + Q(x)]\,dx$ (b) $\int_{5}^{1} P(t)\,dt$

   (c) $\int_{-3}^{-5} Q(-x)\,dx$ (d) $\int_{3}^{1} P(x)\,dx.$

34. Suppose that f is continuous and $x^2 \le f(x) \le 6$ for all x in $[-1, 2]$. Find values of A and B such that

$$A \le \int_{-1}^{2} f(x)\,dx \le B.$$

In Exercises 35–38, find: (a) the average value of $f(x)$ over the indicated interval, and (b) all values of x^* described in the Mean-Value Theorem for Integrals.

35. $f(x) = 3x^2$; $[-2, -1]$.

36. $f(x) = \dfrac{x}{\sqrt{x^2 + 9}}$; $[0, 4]$.

37. $f(x) = 2 + |x|$; $[-3, 1]$.

38. $f(x) = \sin^2 x$; $[0, \pi]$
    [*Hint:* $\sin^2 x = \frac{1}{2}(1 - \cos 2x)$.]

6 applications of the definite integral

6.1 AREA BETWEEN TWO CURVES

In this section we shall discuss methods for calculating the area between curves in the plane.

> FIRST AREA PROBLEM. Suppose that f and g are continuous functions on an interval $[a, b]$ and
>
> $$f(x) \geq g(x) \quad \text{for} \quad a \leq x \leq b$$
>
> (This means that the curve $y = f(x)$ does not cross under the curve $y = g(x)$ over $[a, b]$.) Find the area A of the region bounded above by $y = f(x)$, below by $y = g(x)$, and on the sides by $x = a$ and $x = b$ (Figure 6.1.1a).

If f and g are nonnegative on $[a, b]$, then as illustrated in Figure 6.1.1b, we have

$$A = [\text{area under } f] - [\text{area under } g]$$

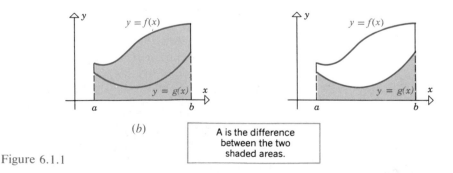

Figure 6.1.1

or equivalently

$$A = \int_a^b f(x)\, dx - \int_a^b g(x)\, dx = \int_a^b [f(x) - g(x)]\, dx$$

This formula can be extended to the case where g has negative values by translating the curves $y = f(x)$ and $y = g(x)$ upward until both are above the x-axis. To do this, let $-m$ be the minimum value of $g(x)$ on $[a, b]$ (Figure 6.1.2a). Since $g(x) \geq -m$, it follows that

$$g(x) + m \geq 0$$

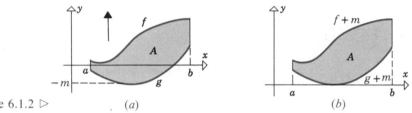

Figure 6.1.2 ▷   (a)   (b)

so that the functions $g + m$ and $f + m$ are nonnegative on $[a, b]$ (Figure 6.1.2b). It is intuitively clear that the area of a region is unchanged by translation, so the area A between f and g is the same as the area between $f + m$ and $g + m$. Thus,

$$A = [\text{area under } f + m] - [\text{area under } g + m]$$

or equivalently

$$A = \int_a^b [f(x) + m]\, dx - \int_a^b [g(x) + m]\, dx = \int_a^b [f(x) - g(x)]\, dx$$

which is the same formula we obtained above.

The foregoing discussion can be summarized as follows:

Area between
$y = f(x)$ *and* $y = g(x)$

6.1.1   AREA FORMULA. If f and g are continuous functions on the interval $[a, b]$, and if $f(x) \geq g(x)$ for all x in $[a, b]$, then the area of the region bounded above by $y = f(x)$, below by $y = g(x)$, on the left by the line $x = a$, and on the right by the line $x = b$ is

$$A = \int_a^b [f(x) - g(x)]\, dx \tag{1}$$

When the region is complicated, it may require some careful thought to determine the integrand and limits of integration in (1). Here is a systematic procedure that you can follow to set up this formula.

> **Step 1.** Sketch the region and then draw a vertical line segment through the region at an arbitrary point x, connecting the top and bottom boundaries (Figure 6.1.3a).
>
> **Step 2.** The top endpoint of the line segment sketched in Step 1 will be $f(x)$, the bottom one $g(x)$, and the length of the line segment will be $f(x) - g(x)$. This is the integrand in (1).
>
> **Step 3.** To determine the limits of integration, imagine moving the line segment left and then right. The leftmost position at which the line segment intersects the region is $x = a$ and the rightmost is $x = b$ (Figures 6.1.3b and 6.1.3c).

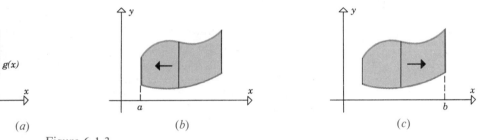

(a) (b) (c)

Figure 6.1.3

Example 1 Find the area of the region bounded above by $y = x + 6$, bounded below by $y = x^2$, and bounded on the sides by the lines $x = 0$ and $x = 2$.

Figure 6.1.4

Solution. The region and a vertical line segment through it are shown in Figure 6.1.4. The line segment extends from $f(x) = x + 6$ on the top to $g(x) = x^2$ on the bottom. If the line segment is moved through the region, its leftmost position will be $x = 0$ and its rightmost will be $x = 2$. Thus, from (1)

$$A = \int_0^2 [(x + 6) - x^2]\, dx = \left[\frac{x^2}{2} + 6x - \frac{x^3}{3} \right]_0^2 = \frac{34}{3} - 0 = \frac{34}{3} \quad \blacktriangleleft$$

Sometimes the upper curve $y = f(x)$ intersects the lower curve $y = g(x)$ at either the left-hand boundary $x = a$, the right-hand boundary $x = b$, or both. When this happens the side of the region where the upper and lower curves intersect reduces to a point, rather than a vertical line segment (Figure 6.1.5).

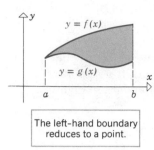

The left-hand boundary reduces to a point.

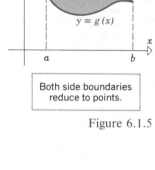

Both side boundaries reduce to points.

Figure 6.1.5

Example 2 Find the area of the region enclosed between the curves $y = x^2$ and $y = x + 6$.

Solution. A sketch of the region (Figure 6.1.6) shows that the lower boundary is $y = x^2$ and the upper boundary is $y = x + 6$. At the endpoints of the region, the upper and lower boundaries have the same y-coordinates; thus, to find the endpoints we equate

$$y = x^2 \quad \text{and} \quad y = x + 6 \tag{2}$$

This yields

$$x^2 = x + 6 \quad \text{or} \quad x^2 - x - 6 = 0 \quad \text{or} \quad (x + 2)(x - 3) = 0$$

from which we obtain

$$x = -2, \, x = 3$$

Although the y-coordinates of the endpoints are not essential to our solution, they may be obtained from (2) by substituting $x = -2$ and $x = 3$ in either equation. This yields $y = 4$ and $y = 9$, so the upper and lower boundaries intersect at $(-2, 4)$ and $(3, 9)$.

From (1) with $f(x) = x + 6$, $g(x) = x^2$, $a = -2$, and $b = 3$, we obtain the area

$$A = \int_{-2}^{3} [(x + 6) - x^2] \, dx = \left[\frac{x^2}{2} + 6x - \frac{x^3}{3} \right]_{-2}^{3}$$

$$= \frac{27}{2} - \left(-\frac{22}{3} \right) = \frac{125}{6} \quad \blacktriangleleft$$

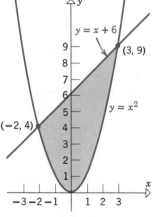

Figure 6.1.6

Example 3 Find the area of the region enclosed by $x = y^2$ and $y = x - 2$.

Solution. To make an accurate sketch of the region we need to know where the curves $x = y^2$ and $y = x - 2$ intersect. In Example 2 we found intersections by equating the expressions for y. Here it is easier to rewrite the latter equation as $x = y + 2$ and equate the expressions for x, namely,

$$x = y^2 \quad \text{and} \quad x = y + 2 \tag{3}$$

This yields

$$y^2 = y + 2 \quad \text{or} \quad y^2 - y - 2 = 0 \quad \text{or} \quad (y + 1)(y - 2) = 0$$

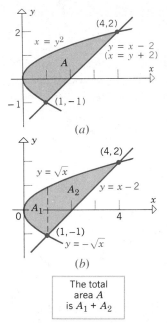

(a)

(b)

The total
area A
is $A_1 + A_2$

Figure 6.1.7

from which we obtain $y = -1$, $y = 2$. Substituting these values in either equation in (3) we see that the corresponding x-values are $x = 1$ and $x = 4$, respectively, so the points of intersection are $(1, -1)$ and $(4, 2)$ (Figure 6.1.7a).

To apply formula (1), the equations of the boundaries must be written so y is expressed explicitly as a function of x. The upper boundary can be written as $y = \sqrt{x}$ (rewrite $x = y^2$ as $y = \pm\sqrt{x}$ and choose the $+$ for the upper portion of the curve). The lower portion of the boundary consists of two parts: $y = -\sqrt{x}$ for $0 \le x \le 1$ and $y = x - 2$ for $1 \le x \le 4$ (Figure 6.1.7b). Because of this change in the formula for the lower boundary, it is necessary to divide the region into two parts and find the area of each part separately.

From (1) with $f(x) = \sqrt{x}$, $g(x) = -\sqrt{x}$, $a = 0$, and $b = 1$, we obtain

$$A_1 = \int_0^1 [\sqrt{x} - (-\sqrt{x})] \, dx = 2 \int_0^1 \sqrt{x} \, dx$$

$$= 2 \left[\frac{2}{3} x^{3/2} \right]_0^1 = \frac{4}{3} - 0 = \frac{4}{3}$$

From (1) with $f(x) = \sqrt{x}$, $g(x) = x - 2$, $a = 1$, and $b = 4$, we obtain

$$A_2 = \int_1^4 [\sqrt{x} - (x - 2)] \, dx = \int_1^4 (\sqrt{x} - x + 2) \, dx$$

$$= \left[\frac{2}{3} x^{3/2} - \frac{1}{2} x^2 + 2x \right]_1^4$$

$$= \left(\frac{16}{3} - 8 + 8 \right) - \left(\frac{2}{3} - \frac{1}{2} + 2 \right) = \frac{19}{6}$$

Thus, the area of the entire region is

$$A = A_1 + A_2 = \tfrac{4}{3} + \tfrac{19}{6} = \tfrac{9}{2} \quad \blacktriangleleft$$

Sometimes it is possible to avoid splitting a region into parts by integrating with respect to y rather than x. We shall now show how this can be done.

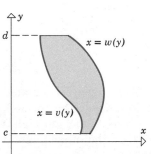

SECOND AREA PROBLEM.   Suppose that w and v are continuous functions on an interval $[c, d]$ and that

$$w(y) \ge v(y) \quad \text{for} \quad c \le y \le d$$

(This means that the curve $x = w(y)$ does not cross to the left of the curve $x = v(y)$ over $[c, d]$.) Find the area A of the region bounded on the left by $x = v(y)$, on the right by $x = w(y)$, and above and below by the lines $y = d$ and $y = c$ (Figure 6.1.8).

Proceeding as in the derivation of (1), but with the roles of x and y reversed, yields the following result.

6.1.2   AREA FORMULA. If w and v are continuous functions and if $w(y) \geq v(y)$ for all y in $[c, d]$, then the area of the region bounded on the left by $x = v(y)$, on the right by $x = w(y)$, below by $y = c$, and above by $y = d$ is

$$A = \int_c^d [w(y) - v(y)] \, dy \qquad (4)$$

Figure 6.1.9

The procedure for finding the integrand and limits of integration in (4) is similar to that used for (1): Draw a horizontal line segment through the region from boundary to boundary at an arbitrary point y (Figure 6.1.9). The right-hand endpoint of the line segment will be $w(y)$, the left-hand endpoint $v(y)$, and the length of the line segment will be $w(y) - v(y)$, which is the integrand in (4). To determine the limits of integration, imagine moving the line segment down and then up (Figure 6.1.9). The lowest position where the line segment intersects the region is $y = c$ and the highest is $y = d$.

In Example 3, where we integrated with respect to x to find the area of the region enclosed by $x = y^2$ and $y = x - 2$, we had to split the region into parts and evaluate two integrals. In the next example we shall see that by integrating with respect to y no splitting of the region is necessary.

Example 4   Find the area of the region enclosed by $x = y^2$ and $y = x - 2$, integrating with respect to y.

Solution.   From Figure 6.1.7 the left boundary is $x = y^2$, the right boundary is $y = x - 2$, and the region extends over the interval $-1 \leq y \leq 2$. However, to apply (4) the equations for the boundaries must be written so x is expressed explicitly as a function of y. Thus, we rewrite $y = x - 2$ as $x = y + 2$. It now follows from (4) that

$$A = \int_{-1}^2 [(y + 2) - y^2] \, dy = \left[\frac{y^2}{2} + 2y - \frac{y^3}{3} \right]_{-1}^2 = \frac{9}{2}$$

which agrees with the result obtained in Example 3.   ◀

REMARK.   The choice between formulas (1) and (4) is generally dictated by the shape of the region, and one would usually choose the formula that requires the least amount of splitting. However, if the integral(s) resulting by one method are difficult to evaluate, then the other method might be preferable, even if it requires more splitting.

► Exercise Set 6.1

1. Find the area of the region enclosed by the curves $y = x^2$ and $y = 4x$ by integrating
    (a) with respect to x
    (b) with respect to y.

2. Find the area of the region enclosed by the curves $y^2 = 4x$ and $y = 2x - 4$ by integrating
    (a) with respect to x
    (b) with respect to y.

3. Find the area of the region enclosed by the curves $y^2 = 2x$ and $y = 2x - 2$ by integrating
    (a) with respect to x
    (b) with respect to y.

In Exercises 4–23, sketch the region enclosed by the curves and find its area by any method.

4. $y = x^3$, $y = x$, $x = 0$, $x = 1/2$.

5. $y = x^2$, $y = \sqrt{x}$, $x = 1/4$, $x = 1$.

6. $y = x^3 - 4x$, $y = 0$, $x = 0$, $x = 2$.

7. $y = \cos 2x$, $y = 0$, $x = \pi/4$, $x = \pi/2$.

8. $y = x^3 - 4x^2 + 3x$, $y = 0$, $x = 0$, $x = 3$.

9. $x = y^2 - 4y$, $x = 0$, $y = 0$, $y = 4$.

10. $x = \sin y$, $x = 0$, $y = \pi/4$, $y = 3\pi/4$.

11. $y = \sec^2 x$, $y = 2$, $x = -\pi/4$, $x = \pi/4$.

12. $x^2 = y$, $x = y - 2$.

13. $y = x^2 + 4$, $x + y = 6$.

14. $y = x^3$, $y = -x$, $y = 8$.

15. $y^2 = -x$, $y = x - 6$, $y = -1$, $y = 4$.

16. $y = x$, $y = 4x$, $y = -x + 2$.

17. $y = 2 + |x - 1|$, $y = -\frac{1}{5}x + 7$.

18. $y = x^3 - 4x$, $y = 0$, $x = -2$, $x = 2$.

19. $x = y^3 - y$, $x = 0$.

20. $y = x^3 - 2x^2$, $y = 2x^2 - 3x$, $x = 0$, $x = 3$.

21. $y = \sin x$, $y = \cos x$, $x = 0$, $x = 2\pi$.

22. $y = \sqrt{x + 2}$, $y = x$, $y = 0$.

23. $y = 1/x^2$, $y = x$, $y = 4$.

24. Find the area between the curve $y = \sin x$ and the line segment that joins the points $(0, 0)$ and $(5\pi/6, 1/2)$ on the curve.

25. Find the area enclosed by the curve $y = \sqrt{x}$, the tangent to the curve at $x = 4$, and the y-axis.

26. Let A be the area enclosed by $y = 1/x^2$, $y = 0$, $x = 1$, and $x = b$ $(b > 1)$.
    (a) Find A.   (b) Find $\lim\limits_{b \to +\infty} A$.

27. Let A be the area enclosed by $y = 1/\sqrt{x}$, $y = 0$, $x = 1$, and $x = b$ $(b > 1)$.
    (a) Find A.   (b) Find $\lim\limits_{b \to +\infty} A$.

28. Find a vertical line $x = k$ that divides the area enclosed by $x = \sqrt{y}$, $x = 2$, and $y = 0$ into two equal parts.

29. Find a horizontal line $y = k$ that divides the area between $y = x^2$ and $y = 9$ into two equal parts.

30. Find the area enclosed between the curve $x^{1/2} + y^{1/2} = a^{1/2}$ and the coordinate axes.

31. Suppose that f and g are integrable on $[a, b]$, but neither $f(x) \geq g(x)$ nor $g(x) \geq f(x)$ holds for all x in $[a, b]$. (That is, the curves $y = f(x)$ and $y = g(x)$ are intertwined.)
    (a) What is the geometric significance of the integral

$$\int_a^b [f(x) - g(x)] \, dx?$$

    (b) What is the geometric significance of the integral

$$\int_a^b |f(x) - g(x)| \, dx?$$

6.2   VOLUMES BY SLICING; DISKS AND WASHERS

In this section we shall use definite integrals to find volumes of three-dimensional solids.

A right-circular cylinder (Figure 6.2.1, top) can be generated by translating a plane circular disk along a line perpendicular to the disk. In general, we define a *right cylinder* to be any solid that can be generated by translating a plane region along a line or *axis* perpendicular to the region. All of the solids in Figure 6.2.1 are right cylinders, the top one being a *right circular cylinder.* Observe that all cross sections of a right cylinder taken perpendicular to the axis are identical in size and shape.

> Cross sections perpendicular to the axis are identical in size and shape.

Figure 6.2.1 ▷

THE METHOD OF SLICING

If a right cylinder is generated by moving a plane region of area A through a distance h (Figure 6.2.2), then the volume V of the cylinder is *defined* to be

$$V = A \cdot h$$

that is, *the volume is the cross-sectional area times the height.*

Volumes of solids that are neither right cylinders nor composed of finitely many right cylinders can be obtained by a technique called "slicing." To illustrate the idea, suppose that a solid S extends along the x-axis and is bounded on the left and right by planes perpendicular to the x-axis at

Volume = $A \cdot h$

Figure 6.2.2

$x = a$ and $x = b$ (Figure 6.2.3). Because S is not assumed to be a right cylinder, its cross sections perpendicular to the x-axis can vary from point to point; we will denote by $A(x)$ the area of the cross section at x (Figure 6.2.3).

Let us divide the interval $[a, b]$ into n subintervals with widths

$$\Delta x_1, \Delta x_2, \ldots, \Delta x_n$$

by inserting points

$$x_1, x_2, \ldots, x_{n-1}$$

Figure 6.2.3 ▷

between a and b, and let us pass a plane perpendicular to the x-axis through each of these points. As illustrated in Figure 6.2.4, these planes cut the solid S into n slices

Figure 6.2.4 ▷

$$S_1, S_2, \ldots, S_n$$

The cross section here has area $A(x_k{}^*)$

Figure 6.2.5

Consider a typical slice S_k. In general, this slice may not be a right cylinder because its cross section can vary. However, if the slice is very thin, the cross section will not vary much. Therefore, if we choose an arbitrary point x_k^* in the kth subinterval, each cross section of slice S_k will be approximately the same as the cross section at x_k^*, and we can approximate slice S_k by a right cylinder of thickness Δx_k and cross-sectional area $A(x_k^*)$ (Figure 6.2.5).

Thus, the volume V_k of slice S_k is approximately the volume of this cylinder, namely,

$$V_k \approx A(x_k^*) \, \Delta x_k$$

and the volume V of the entire solid is approximately

$$V = V_1 + V_2 + \cdots + V_n \approx \sum_{k=1}^{n} A(x_k^*)\, \Delta x_k \tag{1}$$

If we now increase the number of slices in such a way that max $\Delta x_k \to 0$, then the slices will become thinner and thinner and our approximations will get better and better. Thus, intuition suggests that approximation (1) will approach the exact value of the volume V as max $\Delta x_k \to 0$, that is,

$$V = \lim_{\max \Delta x_k \to 0} \sum_{k=1}^{n} A(x_k^*)\, \Delta x_k \tag{2}$$

Since the right side of (2) is just the definite integral

$$\int_a^b A(x)\, dx$$

we are led to the following result.

Volumes by Cross Sections
Perpendicular to the x-Axis

> **6.2.1**   VOLUME FORMULA. Let S be a solid bounded by two parallel planes perpendicular to the x-axis at $x = a$ and $x = b$. If, for each x in $[a, b]$, the cross-sectional area of S perpendicular to the x-axis is $A(x)$, then the volume of the solid is
>
> $$V = \int_a^b A(x)\, dx \tag{3}$$
>
> provided $A(x)$ is integrable.

There is a similar result for cross sections perpendicular to the y-axis.

Volumes by Cross Sections
Perpendicular to the y-Axis

> **6.2.2**   VOLUME FORMULA. Let S be a solid bounded by two parallel planes perpendicular to the y-axis at $y = c$ and $y = d$. If, for each y in $[c, d]$, the cross-sectional area of S perpendicular to the y-axis is $A(y)$, then the volume of the solid is
>
> $$V = \int_c^d A(y)\, dy \tag{4}$$
>
> provided $A(y)$ is integrable.

Example 1   Derive the formula for the volume of a right pyramid whose altitude is h and whose base is a square with sides of length a.

Solution.   As illustrated in Figure 6.2.6*a*, we introduce a rectangular coordinate system so that the *y*-axis passes through the apex, and the *x*-axis passes through the base and is parallel to a side of the base.

At any point *y* in the interval $[0, h]$ on the *y*-axis the cross section perpendicular to the *y*-axis is a square. If *s* denotes the length of a side of this square, then by similar triangles (Figure 6.2.6*b*)

Figure 6.2.6 ▷

(*a*)

(*b*)

(*a*)

(*b*)

Figure 6.2.7

$$\frac{\frac{1}{2}s}{\frac{1}{2}a} = \frac{h - y}{h} \quad \text{or} \quad s = \frac{a}{h}(h - y)$$

Thus, the area $A(y)$ of the cross section at *y* is

$$A(y) = s^2 = \frac{a^2}{h^2}(h - y)^2$$

and by (4) the volume is

$$V = \int_0^h A(y)\,dy = \int_0^h \frac{a^2}{h^2}(h - y)^2\,dy = \frac{a^2}{h^2}\int_0^h (h - y)^2\,dy$$

$$= \frac{a^2}{h^2}\left[-\frac{1}{3}(h - y)^3 \right]_0^h = \frac{a^2}{h^2}\left[0 + \frac{1}{3}h^3 \right] = \frac{1}{3}a^2 h \quad \blacktriangleleft$$

VOLUMES OF SOLIDS OF REVOLUTION; THE METHODS OF DISKS AND WASHERS

Let *f* be nonnegative and continuous on $[a, b]$, and let *R* be the region bounded above by the graph of *f*, below by the *x*-axis, and on the sides by the lines $x = a$ and $x = b$ (Figure 6.2.7*a*). When this region is revolved about the *x*-axis, it generates a solid having circular cross sections (Figure 6.2.7*b*). Since the cross section at *x* has radius $f(x)$, the cross-sectional area is

$$A(x) = \pi[f(x)]^2$$

Therefore, from (3), the volume of the solid is

$$V = \int_a^b \pi[f(x)]^2 \, dx \tag{5}$$

Because the cross sections are circular or disk shaped, the application of this formula is called the **method of disks.**

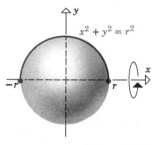

$y = \sqrt{x}$

Figure 6.2.8

Example 2　Find the volume of the solid obtained when the region under the curve $y = \sqrt{x}$ over the interval $[1, 4]$ is revolved about the x-axis (Figure 6.2.8).

Solution.　From (5), the volume is

$$V = \int_a^b \pi[f(x)]^2 \, dx = \int_1^4 \pi x \, dx = \left. \frac{\pi x^2}{2} \right]_1^4 = 8\pi - \frac{\pi}{2} = \frac{15\pi}{2} \quad \blacktriangleleft$$

Example 3　Derive the formula for the volume of a sphere of radius r.

Solution.　As indicated in Figure 6.2.9, a sphere of radius r can be generated by revolving the upper half of the circle

$$x^2 + y^2 = r^2$$

about the x-axis. Since the upper half of this circle is the graph of $y = f(x) = \sqrt{r^2 - x^2}$, it follows from (5) that the volume of the sphere is

$x^2 + y^2 = r^2$

Figure 6.2.9

$$V = \int_a^b \pi[f(x)]^2 \, dx = \int_{-r}^r \pi(r^2 - x^2) \, dx$$

$$= \pi\left[r^2 x - \frac{x^3}{3} \right]_{-r}^r = \frac{4}{3}\pi r^3 \quad \blacktriangleleft$$

We shall now consider more general solids of revolution. Suppose that f and g are nonnegative continuous functions such that

$$g(x) \le f(x) \quad \text{for} \quad a \le x \le b$$

and let R be the region enclosed between the graphs of these functions and the lines $x = a$ and $x = b$ (Figure 6.2.10a). When this region is revolved about the x-axis, it generates a solid having annular or washer-shaped cross sections (Figure 6.2.10b). Since the cross section at x has inner radius $g(x)$ and outer radius $f(x)$, its area is

$$A(x) = \pi[f(x)]^2 - \pi[g(x)]^2 = \pi([f(x)]^2 - [g(x)]^2)$$

Figure 6.2.10 ▷   (a)   (b)

Therefore, from (3), the volume of the solid is

Volumes by Washers
Perpendicular to the x-Axis

$$V = \int_a^b \pi([f(x)]^2 - [g(x)]^2)\, dx \qquad (6)$$

The application of this formula is called the ***method of washers.***

Example 4   Find the volume of the solid generated when the region between the graphs of $f(x) = \frac{1}{2} + x^2$ and $g(x) = x$ over the interval $[0, 2]$ is revolved about the x-axis (Figure 6.2.11).

Solution.   From (6) the volume is

$$V = \int_a^b \pi([f(x)]^2 - [g(x)]^2)\, dx = \int_0^2 \pi([\tfrac{1}{2} + x^2]^2 - x^2)\, dx$$

$$= \int_0^2 \pi\left(\frac{1}{4} + x^4\right) dx = \pi\left[\frac{x}{4} + \frac{x^5}{5}\right]_0^2 = \frac{69\pi}{10} \qquad \blacktriangleleft$$

The methods of disks and washers have analogs for regions revolved about the y-axis. If the region of Figure 6.2.12 is revolved about the

Figure 6.2.11

Figure 6.2.12 ▷   (a)   (b)

y-axis, the cross sections of the resulting solid taken perpendicular to the y-axis are disks, and it follows from Formula 6.2.2 that the volume is

Volumes by Disks
Perpendicular to the y-Axis

$$V = \int_c^d \pi[u(y)]^2 \, dy \tag{7}$$

(Verify.)

Also, if the region of Figure 6.2.13a is revolved about the y-axis, the cross sections taken perpendicular to the y-axis are washers, and it follows from Formula 6.2.2 that the volume of the solid in Figure 6.2.13b is

Volumes by Washers
Perpendicular to the y-Axis

$$V = \int_c^d \pi([u(y)]^2 - [v(y)]^2) \, dy \tag{8}$$

(Verify.)

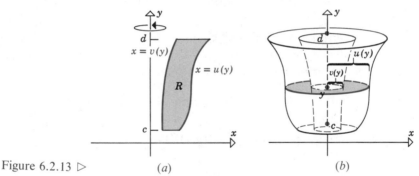

Figure 6.2.13 ▷   (a)   (b)

Example 5   Find the volume of the solid generated when the region enclosed by $y = \sqrt{x}$, $y = 2$, and $x = 0$ is revolved about the y-axis (Figure 6.2.14).

Cross section at y is
a circular disk

Figure 6.2.14 ▷

Solution. The cross sections taken perpendicular to the y-axis are disks, so we shall apply (7). But first we must rewrite $y = \sqrt{x}$ as $x = y^2$. Thus, from (7) with $u(y) = y^2$, the volume is

$$V = \int_c^d \pi[u(y)]^2 \, dy = \int_0^2 \pi y^4 \, dy = \frac{\pi y^5}{5}\Big]_0^2 = \frac{32\pi}{5} \qquad \blacktriangleleft$$

▶ Exercise Set 6.2

In Exercises 1–12, find the volume of the solid that results when the region enclosed by the given curves is revolved about the x-axis.

1. $y = x^2$, $x = 0$, $x = 2$, $y = 0$.
2. $y = \sec x$, $x = \pi/4$, $x = \pi/3$, $y = 0$.
3. $y = 1 + x^3$, $x = 1$, $x = 2$, $y = 0$.
4. $y = 1/x$, $x = 1$, $x = 4$, $y = 0$.
5. $y = 9 - x^2$, $y = 0$.
6. $y = \sqrt{\cos x}$, $x = \pi/4$, $x = \pi/2$, $y = 0$.
7. $y = x^2$, $y = 4x$.
8. $y = x^2$, $y = 9$.
9. $y = \sin x$, $y = \cos x$, $x = 0$, $x = \pi/4$. [*Hint:* Use the identity $\cos 2x = \cos^2 x - \sin^2 x$.]
10. $y = x^2 + 1$, $y = x + 3$.
11. $y = \sqrt{x}$, $y = x$.
12. $y = x^2$, $y = x^3$.

In Exercises 13–24, find the volume of the solid that results when the region enclosed by the given curves is revolved about the y-axis.

13. $y = x^3$, $x = 0$, $y = 1$.
14. $x = 1 - y^2$, $x = 0$.
15. $x = \sqrt{1 + y}$, $x = 0$, $y = 3$.
16. $x = \sqrt{\cos y}$, $y = 0$, $y = \pi/2$, $x = 0$.
17. $x = \csc y$, $y = \pi/4$, $y = 3\pi/4$, $x = 0$.
18. $y = 2/x$, $y = 1$, $y = 3$, $x = 0$.
19. $x = \sqrt{9 - y^2}$, $y = 1$, $y = 3$, $x = 0$.
20. $y = x^2 - 1$, $x = 2$, $y = 0$.
21. $y = 1 + x^3$, $x = 1$, $y = 9$.
22. $y = x^2$, $x = y^2$.
23. $x = y^2$, $x = y + 2$.

24. $x = 1 - y^2$, $x = 2 + y^2$, $y = -1$, $y = 1$.
25. Find the volume of the solid that results when the region enclosed by the semicircle $y = \sqrt{25 - x^2}$ and the line $y = 3$ is revolved about the x-axis.
26. Find the volume of the torus that results when the region enclosed by the circle of radius r with center at $(h, 0)$, $h > r$, is revolved about the y-axis. [*Hint:* Use an appropriate formula from plane geometry to help evaluate the definite integral.]
27. Find the volume of the solid that results when the region enclosed by $y = \sqrt{x}$, $y = 0$, and $x = 9$ is revolved about the line $x = 9$.
28. Find the volume of the solid that results when the region in Exercise 27 is revolved about the line $y = 3$.
29. Find the volume of the solid that results when the region enclosed by $x = y^2$ and $x = y$ is revolved about the line $y = -1$.
30. Find the volume of the solid that results when the region in Exercise 29 is revolved about the line $x = -1$.
31. Find the volume of the solid that results when the region above the x-axis and below the curve

$$\frac{x^2}{a^2} + \frac{y^2}{b^2} = 1 \qquad (a > 0, b > 0)$$

is revolved about the x-axis.
32. Find the volume of the solid generated when the region enclosed by $y = \sqrt{x}$, $y = 6 - x$, and $y = 0$ is revolved about the x-axis. [*Hint:* Split the solid into two parts.]
33. Find the volume of the solid generated when the

region enclosed by $y = \sqrt{x + 1}$, $y = \sqrt{2x}$, and $y = 0$ is revolved about the x-axis. [*Hint:* Split the solid into two parts.]

34. Let R_1, R_2, R_3, and R_4 be the regions indicated in Figure 6.2.15. Express the following volumes as definite integrals:
    (a) The volume of the solid generated when R_1 is revolved about the x-axis.
    (b) The volume of the solid generated when R_2 is revolved about the x-axis.
    (c) The volume of the solid generated when R_3 is revolved about the y-axis.
    (d) The volume of solid generated when R_4 is revolved about the y-axis.

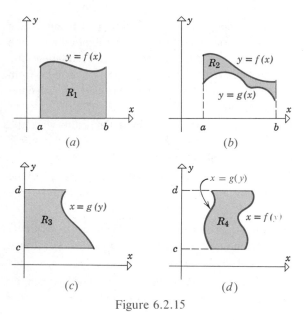

(a)   (b)

(c)   (d)

Figure 6.2.15

35. Derive the formula for the volume of a right-circular cone with radius r and height h.

36. A hole of radius $r/2$ is drilled through the center of a sphere of radius r. Find the volume of the remaining solid.

37. A cylindrical hole is drilled all the way through the center of a sphere (Figure 6.2.16). Show that the volume of the remaining solid depends only on the length L of the hole, and not on the size of the sphere.

Figure 6.2.16

38. (a) A vat, shaped like a hemisphere of radius r ft, is filled with a fluid to a depth of h ft. Find the volume of the fluid.
    (b) If fluid enters a hemispherical vat of radius 10 ft at a rate of $\frac{1}{2}$ ft³/min, how fast will the fluid be rising when the depth is 5 ft?

39. A cocktail glass with a bowl shaped like a hemisphere of diameter 8 cm contains a cherry with a diameter of 2 cm (Figure 6.2.17). If the glass is filled to a depth of h cm, what is the volume of liquid it contains? [*Hint:* First consider the case where the cherry is partially submerged, then the case where it is totally submerged.]

Figure 6.2.17

40. A *general cylinder* is any solid that can be generated by translating a plane region along a line passing through but not contained in the region. Derive a formula for the volume of a general cylinder with base area A and height h (Figure 6.2.18).

Figure 6.2.18

41. A nose cone for a space reentry vehicle is designed so that a cross section, taken x feet from the tip and perpendicular to the axis of symmetry, is a circle of radius $\frac{1}{4}x^2$ ft. Find the volume of the nose cone given that its length is 20 ft.

42. A certain solid is 1 ft high, and a horizontal cross section taken x ft above the bottom of the solid is an annulus of inner radius x^2 and outer radius $\sqrt{x}$. Find the volume of the solid.

43. The base of a certain solid is the circle $x^2 + y^2 = 9$ and each cross section perpendicular to the x-axis is an equilateral triangle with one side across the base. Find the volume of the solid.

44. The base of a certain solid is the region enclosed by $y = \sqrt{x}$, $y = 0$, and $x = 4$. Every cross section perpendicular to the x-axis is a semicircle with its diameter across the base. Find the volume of the solid.

45. The base of a certain solid is the region enclosed by $y = \sin x$, $y = 0$, $x = \pi/4$, and $x = 3\pi/4$; and every cross section perpendicular to the x-axis is a square with one side across the base. Find the volume of the solid. [*Hint:* To help with the integration, use the identity $\sin^2 x = \frac{1}{2}(1 - \cos 2x)$.]

46. The base of a certain solid is the region enclosed by $y = 1/x$, $y = 0$, $x = 1$, and $x = 3$. Every cross section perpendicular to the x-axis is an isosceles right triangle with its hypotenuse across the base. Find the volume of the solid.

47. A wedge is cut from a right-circular cylinder of radius r by two planes, one perpendicular to the axis of the cylinder and the other making an angle θ with the first. Find the volume of the wedge by slicing perpendicular to the y-axis that is shown in Figure 6.2.19.

48. Find the volume of wedge described in Exercise 47 by slicing perpendicular to the x-axis.

49. Two right-circular cylinders of radius r have axes that intersect at right angles. Find the volume of the solid common to the two cylinders. [*Hint:* One eighth of the solid is sketched in Figure 6.2.20.]

Figure 6.2.20

50. In 1635 Bonaventura Cavalieri, a student of Galileo, stated the following result, called *Cavalieri's Principle: If two solids have the same height, and if the areas of their cross sections taken parallel to and at equal distances from their bases are always equal, then the solids have the same volume.* Prove this result.

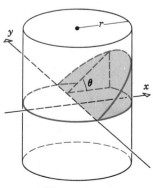

Figure 6.2.19

VOLUMES BY CYLINDRICAL SHELLS

The methods discussed so far for computing volumes of solids depend on our ability to compute cross-sectional areas of the solid. In this section we shall develop an alternative technique for computing volumes that can sometimes be applied when it is inconvenient or difficult to determine the cross-sectional areas or if the integration is too difficult.

Figure 6.3.1

A *cylindrical shell* is a solid enclosed by two concentric right-circular cylinders (Figure 6.3.1). The volume V of a cylindrical shell having inner radius r_1, outer radius r_2, and height h can be written as

$$V = [\text{area of cross section}] \cdot [\text{height}] = (\pi r_2^2 - \pi r_1^2)h$$

$$= \pi(r_2 + r_1)(r_2 - r_1)h = 2\pi\left(\frac{r_2 + r_1}{2}\right)h(r_2 - r_1)$$

Thickness of the shell

Average Radius

Thus,

Volume of a Cylindrical Shell

$$V = 2\pi \cdot [\text{average radius}] \cdot [\text{height}] \cdot [\text{thickness}] \qquad (1)$$

We shall now show how this formula can be used to find the volume of a solid of revolution.

Let R be a plane region bounded above by a continuous curve $y = f(x)$, bounded below by the x-axis, and bounded on the left and right, respectively, by the lines $x = a$ and $x = b$. Let S be the solid generated by revolving the region R about the y-axis (Figure 6.3.2.)

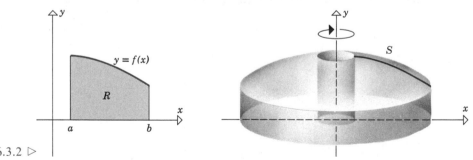

Figure 6.3.2 ▷

To find the volume of S, let us divide the interval $[a, b]$ into n subintervals with widths

$$\Delta x_1, \Delta x_2, \ldots, \Delta x_n$$

by inserting points

$$x_1, x_2, \ldots, x_{n-1}$$

between a and b, and let us draw a vertical line through each of these points to divide the region R into n strips $R_1, R_2, \ldots, R_n$ (Figure 6.3.3a).

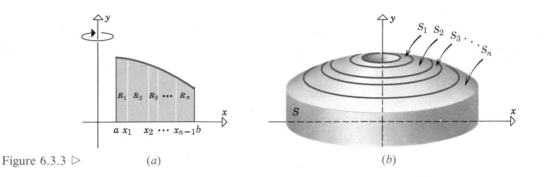

Figure 6.3.3 ▷   (a)   (b)

These strips, when revolved about the y-axis, generate solids $S_1, S_2, \ldots, S_n$. As illustrated in Figure 6.3.3b, these solids are nested one inside the other and together form the entire solid S. Thus, the volume of the solid S can be obtained by adding together the volumes of solids $S_1, S_2, \ldots, S_n$:

$$V(S) = V(S_1) + V(S_2) + \cdots + V(S_n) \tag{2}$$

Consider a typical strip R_k and the solid S_k that it generates (Figure 6.3.4). Although solid S_k resembles a cylindrical shell, it will not, in general, be a cylindrical shell because it can have a curved upper surface.

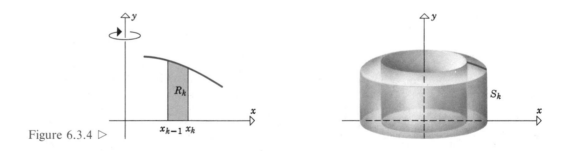

Figure 6.3.4 ▷

However, if the interval width

$$\Delta x_k = x_k - x_{k-1}$$

is small, we can obtain a good approximation to the region R_k by a rectangle of width Δx_k and height $f(x_k^*)$, where

$$x_k^* = \frac{x_k + x_{k-1}}{2}$$

is the midpoint of the interval $[x_{k-1}, x_k]$ (Figure 6.3.5a). This rectangle, when revolved about the y-axis, generates a cylindrical shell, which is a good approximation to the solid S_k (Figure 6.3.5b).

Figure 6.3.5 ▷　　　　　　　　(a)　　　　　　　　　　　　　　　　　　　　　　　　(b)

This approximating cylindrical shell has thickness Δx_k, height $f(x_k^*)$, and average radius x_k^*, so that by (1)

$$V(S_k) \approx \text{volume of approximating cylindrical shell} = 2\pi x_k^* f(x_k^*)\,\Delta x_k$$

Thus, from (2), the volume, $V(S)$, of the entire solid S is approximately

$$\sum_{k=1}^{n} 2\pi x_k^* f(x_k^*)\,\Delta x_k \tag{3}$$

If we now divide $[a, b]$ into more and more subintervals in such a way that $\max \Delta x_k \to 0$, then intuition suggests that our approximations will tend to get better and (3) will approach the exact value of the volume, that is,

$$V(S) = \lim_{\max \Delta x_k \to 0} \sum_{k=1}^{n} 2\pi x_k^* f(x_k^*)\,\Delta x_k \tag{4}$$

Because the right side of (4) is just the definite integral

$$\int_a^b 2\pi x f(x)\,dx$$

we are led to the following result.

Volumes by Cylindrical Shells Centered on the y-Axis

6.3.1 VOLUME FORMULA. Let R be a plane region bounded above by a continuous curve $y = f(x)$, below by the x-axis, and on the left and right, respectively, by the lines $x = a$ and $x = b$. Then the volume of the solid generated by revolving R about the y-axis is given by

$$V = \int_a^b 2\pi x f(x)\,dx \tag{5}$$

Example 1 Use cylindrical shells to find the volume of the solid generated when the region enclosed between $y = \sqrt{x}$, $x = 1$, $x = 4$, and the x-axis is revolved about the y-axis (Figure 6.3.6).

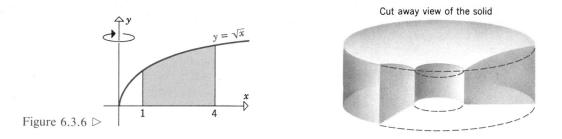

Cut away view of the solid

Figure 6.3.6 ▷

Solution. Since $f(x) = \sqrt{x}$, $a = 1$, and $b = 4$, Formula (5) yields

$$V = \int_1^4 2\pi x \sqrt{x}\, dx = 2\pi \int_1^4 x^{3/2}\, dx$$

$$= \left[2\pi \cdot \frac{2}{5} x^{5/2} \right]_1^4 = \frac{4\pi}{5} [32 - 1] = \frac{124\pi}{5} \qquad ◀$$

There is a way of thinking about (5) that is sometimes useful. At each point x in $[a, b]$ the vertical line through x cuts the region R in a line segment that we can view as the vertical "cross section" of R at x (Figure 6.3.7a). When the region R is revolved about the y-axis, the vertical cross

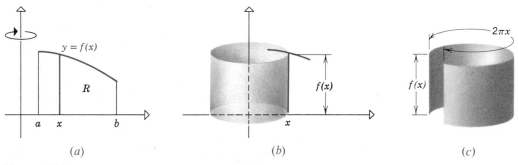

(a)   (b)   (c)

Figure 6.3.7

section at x generates the *surface* of a right-circular cylinder having height $f(x)$ and radius x (Figure 6.3.7b). The area of this surface is

$$2\pi x f(x)$$

(see Figure 6.3.7c), which is precisely the integrand in (5). Thus, 6.3.1 can be paraphrased as follows:

> *The volume V by cylindrical shells is the integral of the surface area generated by an arbitrary cross section of R taken parallel to the axis about which R is revolved.*

The foregoing result is helpful for adapting the method of cylindrical shells to problems that are not exactly of the form required in 6.3.1.

Example 2   Use cylindrical shells to find the volume of the solid generated when the region R in the first quadrant enclosed between $y = x$ and $y = x^2$ is revolved about the y-axis (Figure 6.3.8).

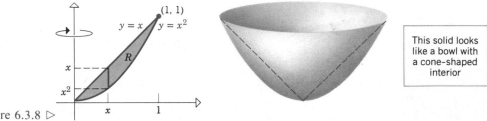

Figure 6.3.8 ▷

This solid looks like a bowl with a cone-shaped interior

Solution.   At each x in $[0, 1]$ the cross section of R parallel to the y-axis generates a cylindrical surface of height $x - x^2$ and radius x. Since the area of this surface is

$$2\pi x(x - x^2)$$

the volume of the solid is

$$V = \int_0^1 2\pi x(x - x^2)\, dx = 2\pi \int_0^1 (x^2 - x^3)\, dx$$

$$= 2\pi \left[\frac{x^3}{3} - \frac{x^4}{4} \right]_0^1 = 2\pi \left[\frac{1}{3} - \frac{1}{4} \right] = \frac{\pi}{6} \quad ◀$$

Example 3   Use cylindrical shells to find the volume of the solid generated when the region R under $y = x^2$ over the interval $[0, 2]$ is revolved about the x-axis (Figure 6.3.9).

Solution.   At each y in the interval $0 \leq y \leq 4$, the cross section of R parallel to the x-axis generates a cylindrical surface of height $2 - \sqrt{y}$ and radius y. Since the area of this surface is

$$2\pi y(2 - \sqrt{y})$$

the volume of the solid is

$$V = \int_0^4 2\pi y(2 - \sqrt{y})\, dy = 2\pi \int_0^4 (2y - y^{3/2})\, dy$$

$$= 2\pi \left[y^2 - \frac{2}{5} y^{5/2} \right]_0^4 = \frac{32\pi}{5} \quad \blacktriangleleft$$

Figure 6.3.9 ▷

The volume in the foregoing example can also be obtained by the method of disks [Formula (5) of Section 6.2]. The computations are

$$V = \int_0^2 \pi(x^2)^2\, dx = \int_0^2 \pi x^4\, dx = \frac{\pi x^5}{5} \Bigg]_0^2 = \frac{32}{5}\pi$$

▶ Exercise Set 6.3

In Exercises 1–8, use cylindrical shells to find the volume of the solid generated when the region enclosed by the given curves is revolved about the y-axis.

1. $y = x^3$, $x = 1$, $y = 0$.

2. $y = \sqrt{x}$, $x = 4$, $x = 9$, $y = 0$.

3. $x^2 + y^3 = 4$, $x = 0$, $x = 4$, $y = 0$.

4. $y = \cos(x^2)$, $x = 0$, $x = \frac{1}{2}\sqrt{\pi}$, $y = 0$.

5. $y = 2x - 1$, $y = -2x + 3$, $x = 2$.

6. $x = y^2$, $y = x^2$.

7. $y = 1/x$, $y = 0$, $x = 1$, $x = 3$.

8. $y = 2x - x^2$, $y = 0$.

In Exercises 9–12, use cylindrical shells to find the volume of the solid generated when the region enclosed by the given curves is revolved about the x-axis.

9. $y^2 = x$, $y = 1$, $x = 0$.

10. $x = 2y$, $y = 2$, $y = 3$, $x = 0$.

11. $y = x^2$, $x = 1$, $y = 0$.

12. $xy = 4$, $x + y = 5$.

13. (a) Use cylindrical shells to find the volume of the solid generated when the region under the curve $y = x^3 - 3x^2 + 2x$ over $[0, 1]$ is revolved about the y-axis.

(b) For this problem, is the method of cylindrical shells easier or harder than the method of slicing discussed in the last section? Explain.

14. Use cylindrical shells to find the volume of the solid generated when the region that is enclosed by $y = 1/x^3$, $x = 1$, $x = 2$, $y = 0$ is revolved about the line $x = -1$.

15. Use cylindrical shells to find the volume of the solid generated when the region that is enclosed by $y = x^3$, $y = 1$, $x = 0$ is revolved about the line $y = 1$.

16. Let R_1 and R_2 be regions of the form shown in Figure 6.3.10. Use cylindrical shells to find a formula for the volume of the solid that results when:
    (a) region R_1 is revolved about the y-axis
    (b) region R_2 is revolved about the x-axis.

17. Use cylindrical shells to find the volume of the cone generated when the triangle with vertices $(0,0)$, $(0,r)$, $(h,0)$, where $r > 0$ and $h > 0$ is revolved about the x-axis.

18. The region enclosed between the curve $y^2 = kx$ and the line $x = \frac{1}{4}k$ is revolved about the line $x = \frac{1}{2}k$. Use cylindrical shells to find the volume of the resulting solid. Assume $k > 0$.

19. A round hole of radius a is drilled through the center of a solid sphere of radius r. Use cylindrical shells to find the volume of the portion removed. (Assume $r > a$.)

20. Use cylindrical shells to find the volume of the torus obtained by revolving the circle

$$x^2 + y^2 = a^2$$

about the line $x = b$, where $b > a$. [*Hint:* It may help in the integration to think of an integral as an area.]

21. Let V_x and V_y be the volumes of the solids that result when the region enclosed by $y = 1/x$, $y = 0$, $x = \frac{1}{2}$, and $x = b$ ($b > \frac{1}{2}$) is revolved about the x-axis and y-axis, respectively. Is there a value of b for which $V_x = V_y$?

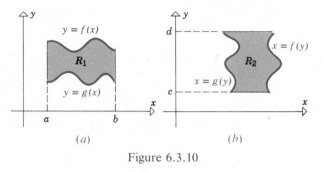

Figure 6.3.10

6.4 LENGTH OF A PLANE CURVE

In this section we shall use definite integrals to find arc lengths of plane curves. To start, we shall consider only curves that are graphs of functions. In a later section, we shall extend our results to more general curves.

If f' is continuous on an interval, we shall say that $y = f(x)$ is a *smooth curve* (or f is a *smooth function*) on that interval. We shall restrict our discussion of arc length to smooth curves in order to eliminate some complications that would otherwise occur.

Figure 6.4.1

6.4.1   ARC LENGTH PROBLEM. Suppose f is smooth on the interval $[a, b]$. Find the arc length L of the curve $y = f(x)$ over the interval $[a, b]$ (Figure 6.4.1).

In order to solve Problem 6.4.1 we must first define the term "arc length" precisely. For motivation, consider the graph of a smooth curve $y = f(x)$ over an interval $[a, b]$, and as shown in Figure 6.4.2, divide the interval

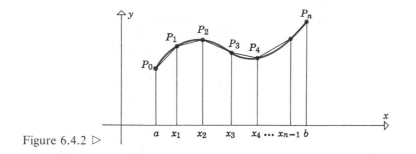

Figure 6.4.2 ▷

[a, b] into n subintervals with widths

$$\Delta x_1, \Delta x_2, \ldots, \Delta x_n$$

by inserting points

$$x_1, x_2, \ldots, x_{n-1}$$

between a and b. Let $P_0, P_1, \ldots, P_n$ be the points on the curve whose x-coordinates are $a, x_1, x_2, \ldots, x_{n-1}, b$ and join these points with straight-line segments. These segments form a ***polygonal path*** that we can regard as an approximation to the curve $y = f(x)$. Intuition suggests that the length of the approximating polygonal path will approach the length of the curve if we increase the number of points in such a way that the lengths of the line segments in the polygonal path approach zero.

Figure 6.4.3

To examine this idea more closely, let us isolate a typical subinterval, say the kth (Figure 6.4.3). As suggested by this figure, the length L_k of the kth line segment in the polygonal path is given by

$$L_k = \sqrt{(\Delta x_k)^2 + [f(x_k) - f(x_{k-1})]^2} \qquad (1)$$

By the Mean-Value Theorem (4.10.2) there is a point x_k^* between x_{k-1} and x_k such that

$$\frac{f(x_k) - f(x_{k-1})}{x_k - x_{k-1}} = f'(x_k^*)$$

or

$$f(x_k) - f(x_{k-1}) = f'(x_k^*)\,\Delta x_k$$

Thus, (1) can be rewritten as

$$L_k = \sqrt{1 + [f'(x_k^*)]^2}\,\Delta x_k$$

which means that the length of the *entire* polygonal path is

$$\sum_{k=1}^{n} L_k = \sum_{k=1}^{n} \sqrt{1 + [f'(x_k^*)]^2}\,\Delta x_k$$

If we now increase the number of subintervals in such a way that $\max \Delta x_k \to 0$, then the length of the polygonal path will approach the arc length L of the curve $y = f(x)$ over $[a, b]$. Thus,

$$L = \lim_{\max \Delta x_k \to 0} \sum_{k=1}^{n} \sqrt{1 + [f'(x_k^*)]^2} \, \Delta x_k \qquad (2)$$

Since the right side of (2) is just the definite integral

$$\int_a^b \sqrt{1 + [f'(x)]^2} \, dx$$

we are led to the following result.

6.4.2 ARC LENGTH FORMULAS. If f is a smooth function on $[a, b]$, then the **arc length** L of the curve $y = f(x)$ from $x = a$ to $x = b$ is given by

$$L = \int_a^b \sqrt{1 + [f'(x)]^2} \, dx \qquad (3)$$

or equivalently,

$$L = \int_a^b \sqrt{1 + \left(\frac{dy}{dx}\right)^2} \, dx \qquad (3a)$$

Similarly, for a curve expressed in the form $x = g(y)$, where g' is continuous on $[c, d]$, the arc length L from $y = c$ to $y = d$ is given by

$$L = \int_c^d \sqrt{1 + [g'(y)]^2} \, dy \qquad (4)$$

or equivalently,

$$L = \int_c^d \sqrt{1 + \left(\frac{dx}{dy}\right)^2} \, dy \qquad (4a)$$

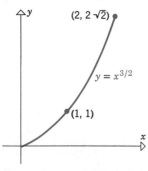

Figure 6.4.4

Example 1 Find the arc length of the curve $y = x^{3/2}$ from $(1, 1)$ to $(2, 2\sqrt{2})$ (Figure 6.4.4) using

(a) Formula (3)   (b) Formula (4).

Solution (a). Since $f(x) = x^{3/2}$,

$$f'(x) = \frac{3}{2}x^{1/2}$$

Thus, from (3), the arc length from $x = 1$ to $x = 2$ is

$$L = \int_1^2 \sqrt{1 + \frac{9}{4}x}\; dx$$

To evaluate this integral we make the *u*-substitution

$$u = 1 + \frac{9}{4}x, \quad du = \frac{9}{4}dx$$

and change the *x*-limits ($x = 1, x = 2$) to the corresponding *u*-limits ($u = \frac{13}{4}, u = \frac{22}{4}$):

$$L = \frac{4}{9} \int_{13/4}^{22/4} u^{1/2}\; du = \frac{8}{27} u^{3/2} \Big]_{13/4}^{22/4} = \frac{8}{27}\left[\left(\frac{22}{4}\right)^{3/2} - \left(\frac{13}{4}\right)^{3/2} \right]$$

$$= \frac{22\sqrt{22} - 13\sqrt{13}}{27}$$

Solution (b). Solving $y = x^{3/2}$ for x in terms of y, we obtain $x = y^{2/3}$. Hence $g(y) = y^{2/3}$ and

$$g'(y) = \frac{2}{3}y^{-1/3}$$

Thus, from (4), the arc length from $y = 1$ to $y = 2\sqrt{2}$ is

$$L = \int_1^{2\sqrt{2}} \sqrt{1 + \frac{4}{9}y^{-2/3}}\; dy = \frac{1}{3}\int_1^{2\sqrt{2}} y^{-1/3}\sqrt{9y^{2/3} + 4}\; dy$$

To evaluate this integral we make the *u*-substitution

$$u = 9y^{2/3} + 4, \quad du = 6y^{-1/3}\; dy$$

and change the *y*-limits ($y = 1, y = 2\sqrt{2}$) to the corresponding *u*-limits ($u = 13, u = 22$). This gives

$$L = \frac{1}{18}\int_{13}^{22} u^{1/2}\; du = \frac{1}{27}u^{3/2}\Big]_{13}^{22}$$

$$= \frac{1}{27}[(22)^{3/2} - (13)^{3/2}]$$

$$= \frac{22\sqrt{22} - 13\sqrt{13}}{27}$$

This result agrees with that in part (a); however, the integration here is more tedious. In problems where there is a choice between using (3) or (4), it is sometimes worthwhile to determine which one leads to the simpler integral. ◀

▶ Exercise Set 6.4

1. Find the arc length of the curve $y = 2x$ from $(1, 2)$ to $(2, 4)$ using
   (a) formula (3) (b) formula (4)
   (c) the Theorem of Pythagoras.

2. Find the arc length of the curve $y = mx + b$ from $x = k_1$ to $x = k_2$ $(m \neq 0, k_2 > k_1)$ using
   (a) formula (3) (b) formula (4)
   (c) the Theorem of Pythagoras.

3. Find the arc length of the curve $y = 3x^{3/2} - 1$ from $x = 0$ to $x = 1$.

4. Find the arc length of the curve $x = \frac{1}{3}(y^2 + 2)^{3/2}$ from $y = 0$ to $y = 1$.

5. Find the arc length of the curve $y = x^{2/3}$ from $x = 1$ to $x = 8$.

6. Find the arc length of the curve $y = \frac{x^4}{16} + \frac{1}{2x^2}$ from $x = 2$ to $x = 3$.

7. Find the arc length of the curve $24xy = y^4 + 48$ from $y = 2$ to $y = 4$.

8. Find the arc length of the curve $x = \frac{1}{8}y^4 + \frac{1}{4}y^{-2}$ from $y = 1$ to $y = 4$.

9. Consider the curve $y = x^{2/3}$.

(a) Sketch the portion of the curve between $x = -1$ and $x = 8$.
(b) Explain why (3) cannot be used to find the arc length of the curve sketched in (a).
(c) Find the arc length of the curve sketched in (a).

10. Find the arc length in the second quadrant of the curve $x^{2/3} + y^{2/3} = a^{2/3}$ from the point $x = -a$ to $x = -\frac{1}{8}a$ $(a > 0)$.

11. Let $y = f(x)$ be a smooth curve and suppose $f'(x) \geq 0$ on the closed interval $[a, b]$.
    (a) Prove: There are numbers m and M such that $m \leq f'(x) \leq M$ for all x in $[a, b]$.
    (b) Prove: The arc length L of $y = f(x)$ over the closed interval $[a, b]$ satisfies the inequalities
    $$(b - a)\sqrt{1 + m^2} \leq L \leq (b - a)\sqrt{1 + M^2}$$

12. Use the result of Exercise 11 to show that the arc length L of $y = \sin x$ over the interval $0 \leq x \leq \frac{\pi}{4}$ satisfies
    $$\frac{\pi}{4}\sqrt{\frac{3}{2}} \leq L \leq \frac{\pi}{4}\sqrt{2}$$

6.5   AREA OF A SURFACE OF REVOLUTION

In this section we shall apply the definite integral to the problem of finding the area of a surface of revolution.

6.5.1 SURFACE AREA PROBLEM. Let f be a smooth, nonnegative function on $[a, b]$. Find the area of the surface generated by revolving the portion of the curve $y = f(x)$ between $x = a$ and $x = b$ about the x-axis (Figure 6.5.1).

Figure 6.5.1 ▷

In order to solve this problem we must first define the term "surface area" precisely. For motivation, let us approximate the curve $y = f(x)$ by a polygonal path of straight line segments connecting the points on the curve that have x-coordinates

$$a, x_1, x_2, \ldots, x_{n-1}, b$$

(Figure 6.5.2a). As usual, let

$$\Delta x_1, \Delta x_2, \ldots, \Delta x_n$$

be the widths of the subintervals determined by these x-coordinates. If these widths are small, then the surface generated by revolving the po-

Figure 6.5.2 ▷   (a)   (b)

Figure 6.5.3

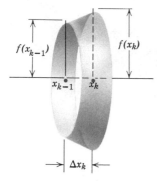

Figure 6.5.4

lygonal path about the x-axis will have approximately the same area as the surface generated by revolving the curve $y = f(x)$ about the x-axis (Figure 6.5.2b). Observe that the surface generated by the polygonal path is made of parts, each of which is a frustum of a cone. Thus, the lateral area of each of these parts can be obtained from the formula

$$S = \pi(r_1 + r_2)l \tag{1}$$

for the lateral area S of a frustum of slant height l and base radii r_1 and r_2 (Figure 6.5.3). A derivation of this formula is discussed in Exercise 17. Intuition suggests that the area of the approximating surface will approach the desired area of the surface if we increase the number of subdivisions in such a way that the lengths of the line segments in the polygonal path approach zero.

To examine this idea more closely, let us isolate a typical section of the approximating surface, say the kth (Figure 6.5.4). Applying (1), we obtain as the lateral area S_k of this kth section

$$S_k = \pi[f(x_{k-1}) + f(x_k)]\sqrt{(\Delta x_k)^2 + [f(x_k) - f(x_{k-1})]^2} \tag{2}$$

By the Mean-Value Theorem (4.10.2) there is a point x_k^* between x_{k-1} and x_k such that

$$\frac{f(x_k) - f(x_{k-1})}{x_k - x_{k-1}} = f'(x_k^*) \quad \text{or} \quad f(x_k) - f(x_{k-1}) = f'(x_k^*)\,\Delta x_k$$

Thus, (2) can be rewritten as

$$S_k = \pi[f(x_{k-1}) + f(x_k)]\sqrt{1 + [f'(x_k^*)]^2}\,\Delta x_k \tag{3}$$

Since the arithmetic average of two numbers lies between those numbers, $\frac{1}{2}[f(x_{k-1}) + f(x_k)]$ is between $f(x_{k-1})$ and $f(x_k)$. Thus, because f is continuous on the interval $[x_{k-1}, x_k]$, the Intermediate-Value Theorem (2.7.9) implies that there exists a point x_k^{**} in this interval such that

$$\frac{1}{2}[f(x_{k-1}) + f(x_k)] = f(x_k^{**})$$

Thus, (3) can be rewritten

$$S_k = 2\pi f(x_k^{**})\sqrt{1 + [f'(x_k^*)]^2}\,\Delta x_k$$

so that the area of the *entire* polygonal surface is

$$\sum_{k=1}^{n} S_k = \sum_{k=1}^{n} 2\pi f(x_k^{**})\sqrt{1 + [f'(x_k^*)]^2}\,\Delta x_k$$

If we now increase the number of subintervals in such a way that $\max \Delta x_k \to 0$, then the area of the approximating polygonal surface will approach the exact surface area S. Thus,

$$S = \lim_{\max \Delta x_k \to 0} \sum_{k=1}^{n} 2\pi f(x_k^{**})\sqrt{1 + [f'(x_k^*)]^2}\,\Delta x_k \tag{4}$$

If it were the case that $x_k^{**} = x_k^*$, then the right side of (4) would be the definite integral

$$\int_a^b 2\pi f(x)\sqrt{1 + [f'(x)]^2}\,dx \tag{5}$$

However, it is proved in advanced calculus that this is true even if $x_k^* \neq x_k^{**}$ because of the continuity of f and f'. In light of (4) and (5) we are led to the following result.

Surface Area of a
Surface of Revolution

6.5.2 SURFACE AREA FORMULAS. Let f be a smooth, nonnegative function on $[a, b]$. Then the *surface area S* generated by revolving the portion of the curve $y = f(x)$ between $x = a$ and $x = b$ about the x-axis is

$$S = \int_a^b 2\pi f(x)\sqrt{1 + [f'(x)]^2}\,dx \tag{6}$$

For a curve expressed in the form $x = g(y)$, where g' is continuous on $[c, d]$, and $g(y) \geq 0$ for $c \leq y \leq d$, the surface area S generated by revolving the portion of the curve from $y = c$ to $y = d$ about the y-axis is given by

$$S = \int_c^d 2\pi g(y)\sqrt{1 + [g'(y)]^2}\,dy \tag{7}$$

Figure 6.5.5

Example 1 Find the surface area generated by revolving the curve

$$y = \sqrt{1 - x^2}, \quad 0 \leq x \leq \tfrac{1}{2}$$

about the x-axis (Figure 6.5.5).

Solution. Since $f(x) = \sqrt{1 - x^2}$,

$$f'(x) = -\frac{x}{\sqrt{1 - x^2}}$$

Thus, (6) yields

$$S = \int_0^{1/2} 2\pi\sqrt{1-x^2}\ \sqrt{1 + \frac{x^2}{1-x^2}}\ dx = \int_0^{1/2} 2\pi\ dx = 2\pi x\bigg]_0^{1/2} = \pi$$

◀

Figure 6.5.6

Example 2   Find the surface area generated by revolving the curve

$$y = \sqrt[3]{3x}, \quad 0 \le y \le 2$$

about the y-axis (Figure 6.5.6).

Solution.   We shall apply (7) after rewriting $y = \sqrt[3]{3x}$ as $x = g(y) = \frac{1}{3}y^3$. Thus,

$$g'(y) = y^2$$

so that from (7) we obtain

$$S = \int_0^2 2\pi\left(\frac{1}{3}y^3\right)\sqrt{1 + y^4}\ dy = \frac{2\pi}{3}\int_0^2 y^3\sqrt{1 + y^4}\ dy \tag{8}$$

The u-substitution

$$u = 1 + y^4, \quad du = 4y^3\ dy$$

yields

$$\int y^3\sqrt{1 + y^4}\ dy = \frac{1}{4}\int \sqrt{u}\ du = \frac{1}{4}\cdot\frac{2}{3}u^{3/2} + C = \frac{1}{6}(1 + y^4)^{3/2} + C$$

Thus, from (8),

$$S = \frac{2\pi}{3}\left[\frac{1}{6}(1 + y^4)^{3/2}\right]_0^2 = \frac{\pi}{9}(17^{3/2} - 1)$$

◀

▶ **Exercise Set 6.5**

In Exercises 1–6, find the area of the surface generated by revolving the given curve about the x-axis.

1. $y = 7x, 0 \le x \le 1$.

2. $y = \sqrt{x}, 1 \le x \le 4$.

3. $y = \sqrt{4 - x^2}, -1 \le x \le 1$.

4. $x = \sqrt[3]{y}, 1 \le y \le 8$.

5. $y = \sqrt{x} - \frac{1}{3}x^{3/2}, 1 \le x \le 3$.

6. $y = \frac{1}{3}x^3 + \frac{1}{4}x^{-1}, 1 \le x \le 2$.

In Exercises 7–12, find the area of the surface generated by revolving the given curve about the y-axis.

7. $x = 9y + 1, 0 \le y \le 2$.

8. $x = y^3, 0 \le y \le 1$.

9. $x = \sqrt{9 - y^2}, -2 \le y \le 2$.

10. $x = 2\sqrt{1 - y}, -1 \le y \le 0$.

11. $8xy^2 = 2y^6 + 1,\ 1 \le y \le 2$.

12. $x = |y - 11|,\ 0 \le y \le 2$.

13. The lateral area S of a right-circular cone with height h and base radius r is $S = \pi r \sqrt{r^2 + h^2}$. Obtain this result using (6).

14. Show that the area of the surface of a sphere of radius r is $4\pi r^2$. [*Hint:* Revolve the semicircle $y = \sqrt{r^2 - x^2}$ about the x-axis.]

15. The portion of the surface of a sphere between two parallel planes that cut the sphere and are h units apart is called a *zone* of altitude h. Show that the area of the surface of a zone depends only on the radius r of the sphere and the altitude h of the zone, and not on the location of the zone. [*Hint:* Let $-r \le a \le r - h$. Revolve the portion of $y = \sqrt{r^2 - x^2}$ for $a \le x \le a + h$ about the x-axis.]

16. Assume $f(x) \ge 0$ for $a \le x \le b$. Derive a formula for the surface area generated when the curve $y = f(x),\ a \le x \le b$ is revolved about the line $y = -k\ (k > 0)$.

17. (*Formula for Surface Area of a Frustum*)
    (a) If a cone of slant height l and base radius r is cut along a lateral edge and laid flat, it becomes a sector of a circle of radius l (Figure 6.5.7). Use the formula $\frac{1}{2}l^2\theta$ for the area of a sector with radius l and central angle θ (in radians) to show that the lateral surface area of the cone is $\pi r l$.

(b) Use the result in (a) to obtain Formula (1) for the lateral surface area of a frustum.

Figure 6.5.7

18. Let $y = f(x)$ be a smooth curve on the interval $[a, b]$ and assume $f(x) \ge 0$ for $a \le x \le b$. By the Extreme-Value Theorem (4.6.4), the function f has a maximum value K and a minimum value k on $[a, b]$. Prove: If L is the arc length of the curve $y = f(x)$ between $x = a$ and $x = b$ and if S is the area of the surface that is generated by revolving this curve about the x-axis, then

$$2\pi k L \le S \le 2\pi K L$$

19. Let $y = f(x)$ be a smooth curve on $[a, b]$ and assume $f(x) \ge 0$ for $a \le x \le b$. Let A be the area under the curve $y = f(x)$ between $x = a$ and $x = b$ and let S be the area of the surface obtained when this section of curve is revolved about the x-axis.
    (a) Prove that $2\pi A \le S$.
    (b) For what functions f is $2\pi A = S$?

6.6 APPLICATION OF INTEGRATION TO RECTILINEAR MOTION

In Section 4.11 we used the derivative to define the notions of instantaneous velocity and acceleration for a particle moving along a line. In this section we shall resume the study of such motion, using the integration tools developed in the previous chapter.

Recall from Definitions 4.11.1 and 4.11.2 that if a particle moving along a coordinate line has position function $s(t)$, then its instantaneous velocity and acceleration are given by

$$v(t) = s'(t) = \frac{ds}{dt}$$

$$a(t) = v'(t) = \frac{dv}{dt}$$

It follows from these formulas that $s(t)$ is an antiderivative of $v(t)$ and $v(t)$ is an antiderivative of $a(t)$, that is,

$$s(t) = \int v(t)\, dt \tag{1}$$

$$v(t) = \int a(t)\, dt \tag{2}$$

Thus, if we have sufficient information to determine the constants of integration, the velocity can be determined from the acceleration, and the position can be determined from the velocity.

Example 1   Find the position function of a particle moving with velocity $v(t) = \cos \pi t$ along a straight line, assuming the particle is at $s = 4$ when $t = 0$.

Solution.   The position function is

$$s(t) = \int v(t)\, dt = \int \cos \pi t\, dt = \frac{1}{\pi} \sin \pi t + C$$

Since $s = 4$ when $t = 0$, it follows that

$$4 = s(0) = \frac{1}{\pi} \sin 0 + C = C$$

Thus,

$$s(t) = \frac{1}{\pi} \sin \pi t + 4 \qquad \blacktriangleleft$$

MOTION NEAR THE EARTH'S SURFACE   It is a fact of physics that an object moving on a vertical line near the earth's surface and subject only to the force of gravity moves with constant acceleration.* This constant, denoted by the letter g, is approximately 32 ft/sec^2 or 9.8 m/sec^2, depending on whether distance is measured in feet or meters.

---

*Strictly speaking, the acceleration changes with the distance from the earth's center. However, near the surface of the earth the acceleration is approximately constant.

From this simple principle it is possible to describe completely the vertical motion of a particle moving freely near the earth's surface, provided the initial position and velocity of the particle are known. To see how, assume that a coordinate line has been superimposed on the line of motion with the origin at the surface of the earth and the positive direction upward (Figure 6.6.1). Moreover, let us assume that time is measured in seconds, distance is measured in feet or meters, and that

the position at time $t = 0$ is s_0

the velocity at time $t = 0$ is v_0

Figure 6.6.1

where the initial displacement s_0 and initial velocity v_0 are known. Recall that a particle speeds up when its velocity and acceleration have the same sign and slows down when they have opposite signs. Because we have chosen the positive direction to be up, and because a particle moving up (positive velocity) slows down under the force of gravity and a particle moving down (negative velocity) speeds up under the force of gravity, it follows that the acceleration a of the particle must always be negative, that is,

$$a(t) = -g$$

Since the acceleration of the particle is constant, we obtain

$$v(t) = \int a(t)\, dt = \int -g\, dt = -gt + C_1 \tag{3}$$

To determine the constant of integration we use the fact that the velocity is v_0 when $t = 0$. Thus,

$$v_0 = v(0) = -g \cdot 0 + C_1 = C_1$$

Substituting this in (3) yields

$$v(t) = -gt + v_0 \tag{4}$$

Since v_0 is a constant, it follows that

$$s(t) = \int v(t)\, dt = \int (-gt + v_0)\, dt = -\frac{1}{2} gt^2 + v_0 t + C_2 \tag{5}$$

To determine the constant C_2 we use the fact that the position is s_0 when $t = 0$. Thus,

$$s_0 = s(0) = -\frac{1}{2} g \cdot 0 + v_0 \cdot 0 + C_2 = C_2$$

Substituting this in (5) yields

$$s(t) = -\frac{1}{2} gt^2 + v_0 t + s_0 \tag{6}$$

Example 2   A rock, initially at rest, is dropped from a height of 400 ft. Assuming gravity is the only force acting, how long does it take for the rock to hit the ground and what is its speed at the time of impact?

Solution.   Since distance is in feet, we take $g = 32$ ft/sec^2. Initially, we have $s_0 = 400$ and $v_0 = 0$, so that from (6)

$$s(t) = -16t^2 + 400$$

Impact occurs when $s(t) = 0$. Solving this equation for t, we obtain

$$-16t^2 + 400 = 0$$
$$t^2 = 25$$
$$t = \pm 5$$

where t is in seconds. Since the rock is released at $t = 0$, the time of impact is positive, so that we can discard the negative solution and conclude that it takes 5 sec for the rock to hit the ground. Substituting $t = 5$ and $v_0 = 0$ in (4), we obtain the velocity at time of impact,

$$v(5) = -32(5) + 0 = -160 \text{ ft/sec}$$

Thus, the speed at impact is $|v(5)| = 160$ ft/sec   ◀

Example 3   A ball is thrown directly upward from a point 8 meters above the ground with an initial velocity of 49 meters/sec. Assuming gravity is the only force acting on the ball after its release, how high will the ball travel?

Solution.   Since distance is in meters, we take $g = 9.8$ meters/sec^2. Initially, $s_0 = 8$ and $v_0 = 49$, so that from (4) and (6)

$$v(t) = -9.8t + 49$$
$$s(t) = -4.9t^2 + 49t + 8$$

The ball will rise until $v(t) = 0$, that is, until $-9.8t + 49 = 0$ or $t = 5$. At this instant the height will be

$$s(5) = -4.9(5)^2 + 49(5) + 8 = 130.5 \text{ meters}$$ ◀

DISTANCE TRAVELED IN RECTILINEAR MOTION   We conclude this section with an application of the First Fundamental Theorem of Calculus (Theorem 5.7.1) to rectilinear motion. It follows from the First Fundamental Theorem of Calculus and (1) that

$$\int_{t_1}^{t_2} v(t)\, dt = s(t) \Big]_{t_1}^{t_2} = s(t_2) - s(t_1) \tag{7}$$

Since $s(t_1)$ is the position of the particle at time t_1 and $s(t_2)$ is the position at time t_2, the difference, $s(t_2) - s(t_1)$, is the change in position or **displacement** of the particle during the time interval $[t_1, t_2]$. A positive displacement means that the particle is farther to the right at time t_2 than at time t_1; a negative displacement means that the particle is farther to the left. In the case where $v(t) \geq 0$ throughout the time interval $[t_1, t_2]$, the particle moves in the positive direction only; thus, the displacement $s(t_2) - s(t_1)$ is the same as the distance traveled by the particle (Figure 6.6.2).

Figure 6.6.2 ▷

In the case where $v(t) \leq 0$ throughout the time interval $[t_1, t_2]$, the particle moves in the negative direction only; thus, the displacement $s(t_2) - s(t_1)$ is the negative of the distance traveled by the particle (Figure 6.6.3). In the case where $v(t)$ assumes both positive and negative values during the time interval $[t_1, t_2]$, the particle moves back and forth and the displacement is the distance traveled in the positive direction minus the distance traveled in the negative direction.

Figure 6.6.3 ▷

If we want to find the total distance traveled in this case (distance traveled in the positive direction *plus* the distance traveled in the negative direction), we must integrate the absolute value of the velocity function, that is,

$$\begin{bmatrix} \text{total distance} \\ \text{traveled during} \\ \text{time interval} \\ [t_1, t_2] \end{bmatrix} = \int_{t_1}^{t_2} |v(t)|\, dt \tag{8}$$

Example 4 A particle moves on a coordinate line so that its velocity at time t is $v(t) = t^2 - 2t$ meters/sec. Find

(a)   the displacement of the particle during the time interval $0 \leq t \leq 3$;

(b)   the distance traveled by the particle during the time interval $0 \leq t \leq 3$.

Solution (a).   From (7) the displacement is

$$\int_0^3 v(t) \, dt = \int_0^3 (t^2 - 2t) \, dt = \left[\frac{t^3}{3} - t^2 \right]_0^3 = 0$$

Thus, the particle is at the same position at time $t = 3$ as at $t = 0$.

Solution (b).   The velocity can be written as $v(t) = t^2 - 2t = t(t - 2)$ from which we see that $v(t) \leq 0$ for $0 \leq t \leq 2$ and $v(t) \geq 0$ for $2 \leq t \leq 3$. Thus, it follows from (8) that the distance traveled is

$$\int_0^3 |v(t)| \, dt = \int_0^2 -v(t) \, dt + \int_2^3 v(t) \, dt$$

$$= \int_0^2 -(t^2 - 2t) \, dt + \int_2^3 (t^2 - 2t) \, dt$$

$$= -\left[\frac{t^3}{3} - t^2 \right]_0^2 + \left[\frac{t^3}{3} - t^2 \right]_2^3$$

$$= \frac{4}{3} + \frac{4}{3} = \frac{8}{3} \text{ meters} \qquad \blacktriangleleft$$

▶ Exercise Set 6.6

In Exercises 1–8, use the given information to find the position function of the particle.

1. $v(t) = 2t - 3$; $s(1) = 5$.

2. $v(t) = 3t^2$; $s(0) = 0$.

3. $v(t) = t^3 - 2t^2 + 1$; $s(0) = 1$.

4. $v(t) = 1 + \sin t$; $s(0) = -3$.

5. $a(t) = 4$; $v(0) = 1$, $s(0) = 0$.

6. $a(t) = t^2 - 3t + 1$; $v(0) = 0$; $s(0) = 0$.

7. $a(t) = 4 \cos 2t$; $v(0) = -1$; $s(0) = -3$.

8. $a(t) = 1/\sqrt{2t + 3}$, $t \geq 0$; $v(3) = 1$, $s(3) = 0$.

9.   In each part use the given information to find the position, velocity, speed, and acceleration at time $t = 1$.
(a) $v = \sin \frac{1}{2} \pi t$; $s = 0$ when $t = 0$.
(b) $a = -3t$; $s = 1$ and $v = 0$ when $t = 0$.

10.   A car traveling 60 mi/hr along a straight road decelerates at a constant rate of 10 ft/sec².

(a)   How long will it take until the speed is 45 mi/hr? [*Note:* 60 mi/hr = 88 ft/sec.]

(b)   How far will the car travel before coming to a stop?

11.   A car traveling 60 mi/hr skids 180 ft after its brakes are applied. Find the acceleration of the car, assuming that it is constant.

12.   A particle moving along a straight line is accelerating at a constant rate of 3 meters/sec². Find the initial velocity if the particle moves 40 meters in the first 4 sec.

In Exercises 13–23, assume that the only force acting is the earth's gravity, and apply formulas (4) and (6) to solve the problems.

13.   A projectile is launched vertically upward from ground level with an initial velocity of 112 ft/sec.
(a)   Find the velocity at $t = 3$ sec and $t = 5$ sec.
(b)   How high will the projectile rise?

(c) Find the speed of the projectile when it hits the ground.

14. A projectile fired downward from a height of 112 ft reaches the ground in 2 sec. What is its initial velocity?

15. A projectile is fired vertically upward from ground level with an initial velocity of 16 ft/sec.
(a) How long will it take for the projectile to hit the ground?
(b) How long will the projectile be moving upward?

16. A rock is dropped from the top of the Washington Monument, which is 555 ft high.
(a) How long will it take for the rock to hit the ground?
(b) What is the speed of the rock at impact?

17. A helicopter pilot drops a package when the helicopter is 200 ft above the ground and rising at a speed of 20 ft/sec.
(a) How long will it take for the package to hit the ground?
(b) What will be its speed at impact?

18. A stone is thrown downward with an initial speed of 96 ft/sec from a height of 112 ft.
(a) How long will it take for the stone to hit the ground?
(b) What will be its speed at impact?

19. A projectile is fired vertically upward with an initial velocity of 49 meters/sec from a tower 150 meters high.
(a) How long will it take for the projectile to reach its maximum height?
(b) What is the maximum height?
(c) How long will it take for the projectile to pass its starting point on the way down?
(d) What is the velocity when it passes the starting point on the way down?
(e) How long will it take for the projectile to hit the ground?
(f) What will be its speed at impact?

20. A man drops a stone from a bridge. How high is the bridge if:
(a) the stone hits the water 4 sec later?
(b) the sound of the splash reaches the man 4 sec later? (Take 1080 ft/sec as the speed of sound.)

21. A projectile fired upward from ground level is to reach a height of 1000 ft. What must its initial velocity be?

22. A stone is released from rest from a point 40 ft above the ground. Find a formula for v in terms of s.

23. A projectile is fired upward from ground level. Find the initial velocity if it hits the ground 8 sec later. How high does it go?

In Exercises 24–29, $v(t)$ is the velocity (meters/sec) of a particle moving on a coordinate line. Find the displacement and the distance traveled by the particle during the given time interval.

24. $v(t) = 2t - 4;\ 0 \le t \le 4$.

25. $v(t) = t^2 + t - 2;\ 0 \le t \le 2$.

26. $v(t) = |t - 3|;\ 0 \le t \le 5$.

27. $v(t) = \cos t;\ 0 \le t \le \pi$.

28. $v(t) = 3 \sin t;\ \pi/4 \le t \le \pi$.

29. $v(t) = t^3 - 3t^2 + 2t;\ 0 \le t \le 3$.

In Exercises 30–33, $a(t)$ is the acceleration (meters/sec²) of a particle moving on a coordinate line, and v_0 is its velocity at time $t = 0$. Find the displacement and the distance traveled by the particle during the given time interval.

30. $a(t) = -2;\ v_0 = 3;\ 1 \le t \le 4$.

31. $a(t) = t - 2;\ v_0 = 0;\ 1 \le t \le 5$.

32. $a(t) = \sin t;\ v_0 = 1;\ \dfrac{\pi}{4} \le t \le \dfrac{\pi}{2}$.

33. $a(t) = \dfrac{1}{\sqrt{5t + 1}};\ v_0 = 2;\ 0 \le t \le 3$.

6.7   WORK

In this section we shall use the integration tools developed in the previous chapter to study the concept of work that arises in physics and engineering.

If an object moves a distance d along a line while subjected to a *constant* force F applied in the direction of motion, then physicists and engineers define the **work** W done on the object by the force F to be

$$W = F \cdot d$$
$$[\text{work}] = [\text{force}] \cdot [\text{distance}]$$

(1)

If force is measured in pounds and distance in feet, then the unit of work is foot-pounds. In the metric system force is measured in **dynes** (D) (the force required to give a mass of 1 gm an acceleration of 1 cm/sec²) or **newtons** (N) (the force required to give a mass of 1 kg an acceleration of 1 m/sec²). In this system the most common units of work are newton-meters and dyne-centimeters. One newton-meter is called a **joule** (J). One foot-pound is approximately 1.36 J.

Example 1   An object moves 5 ft, while subjected to a constant force of 100 lb along its direction of motion. The work done is

$$W = F \cdot d = 100 \cdot 5 = 500 \text{ ft-lb} \quad \blacktriangleleft$$

Calculus is needed when it is required to calculate the work done by a *variable* force. We shall consider the following problem.

> **6.7.1**   PROBLEM. Suppose an object moves in the positive direction along a coordinate line while subject to a force $F(x)$, in the direction of motion, whose magnitude depends on the coordinate x. Find the work done by the force when the object moves over an interval $[a, b]$.

For example, Figure 6.7.1 shows a block subjected to the force of a compressed spring. As the block moves from a to b the spring expands and the force it applies diminishes. Thus, the force $F(x)$ applied by the spring varies with x.

Before we can solve Problem 6.7.1 we must define precisely what is

Figure 6.7.1 ▷

meant by the work done by a variable force. To motivate this definition, let us divide the interval $[a, b]$ into n subintervals with widths

$$\Delta x_1, \Delta x_2, \ldots, \Delta x_n$$

by inserting points

$$x_1, x_2, \ldots, x_{n-1}$$

between a and b. If we denote by W_k the work done when the object moves across the kth subinterval, then total work W done when the object moves across the entire interval $[a, b]$ will be

$$W = W_1 + W_2 + \cdots + W_n$$

Let us estimate W_k. If the kth subinterval is short and if $F(x)$ is continuous, then the force will not vary much over this subinterval; it will be almost constant. We can approximate this nearly constant force by $F(x_k^*)$ where x_k^* is any point in the kth subinterval. Thus, from (1), the work done over the kth subinterval is approximately

$$W_k \approx F(x_k^*) \, \Delta x_k$$

and the work W done over the entire interval $[a, b]$ is approximately

$$\sum_{k=1}^{n} F(x_k^*) \, \Delta x_k$$

If we now increase the number of subintervals in such a way that $\max \Delta x_k \to 0$, then intuition suggests that our approximations will tend to get better; hence,

$$W = \lim_{\max \Delta x_k \to 0} \sum_{k=1}^{n} F(x_k^*) \, \Delta x_k$$

Since the limit on the right is just the definite integral

$$\int_a^b F(x) \, dx$$

we are led to the following definition.

Work by a Variable Force

6.7.2 DEFINITION. If an object moves in the positive direction over the interval $[a, b]$ while subjected to a variable force $F(x)$ in the direction of motion, then the **work** done by the force is

$$W = \int_a^b F(x) \, dx \tag{2}$$

Hooke's law [Robert Hooke (1635–1703)-English physicist] states that under appropriate conditions a spring stretched x units beyond its equilibrium position pulls back with a force

$$F(x) = kx$$

where k is a constant (called the *spring constant*). The value of k depends on such factors as the thickness of the spring, the material used in its composition, and the units of force and distance.

Example 2   A spring whose natural length is 24 in. exerts a force of 5 lb when stretched 10 in. beyond its natural length.

(a)   Find the spring constant k.

(b)   How much work is required to stretch the spring from its natural length to a length of 42 in.?

Solution (a).   From Hooke's law,

$$F(x) = kx$$

From the data, $F(x) = 5$ lb when $x = 10$ in. so that

$$5 = k \cdot 10$$

Thus, the spring constant is

$$k = \frac{5}{10} = \frac{1}{2}$$

This means that the force $F(x)$ required to stretch the spring x in. is

$$F(x) = \frac{1}{2}x \tag{3}$$

Solution (b).   Place the spring along a coordinate line as shown in Figure 6.7.2. We want to find the work W required to stretch the spring over the interval from $x = 0$ to $x = 18$. From (2) and (3) the work W required is

$$W = \int_a^b F(x)\,dx = \int_0^{18} \frac{1}{2}x\,dx = \frac{x^2}{4}\Bigg]_0^{18} = 81 \text{ in-lb} \quad \blacktriangleleft$$

Figure 6.7.2 ▷

Example 3   A cylindrical water tank of radius 10 ft and height 30 ft is half filled with water. How much work is required to pump all the water over the upper rim of the tank?

Solution.   Introduce a coordinate line as shown in Figure 6.7.3. Imagine

Figure 6.7.3 ▷

the water to be divided into n thin layers with thicknesses

$$\Delta x_1, \Delta x_2, \ldots, \Delta x_n$$

The force required to move the kth layer equals the weight of the layer, which can be found by multiplying its volume by the density of water (62.4 lb/ft^3). Since the kth layer is a cylinder of radius $r = 10$ ft and height Δx_k, the force required to move it is

$$\begin{bmatrix} \text{Force to move} \\ \text{the } k\text{th layer} \end{bmatrix} = (\pi r^2 \, \Delta x_k) \cdot [\text{density of water}]$$

$$= (\pi (10)^2 \, \Delta x_k)(62.4) = 6240\pi \, \Delta x_k$$

Because the kth layer has a finite thickness, the upper and lower surfaces are at different distances from the origin. However, if the layer is thin, the difference in these distances is small, and we can reasonably assume that the entire layer is concentrated at a single distance x_k^* from the origin (Figure 6.7.3). With this assumption, the work W_k required to pump the kth layer over the rim will be approximately

$$W_k \approx \underbrace{(30 - x_k^*)}_{\text{distance}} \cdot \underbrace{6240\pi \, \Delta x_k}_{\text{force}}$$

and the work W required to pump all n layers will be approximately

$$W = \sum_{k=1}^{n} W_k \approx \sum_{k=1}^{n} (30 - x_k^*)(6240\pi) \, \Delta x_k$$

To find the *exact* value of the work we take the limit as max $\Delta x_k \to 0$. This yields

$$W = \lim_{\max \Delta x_k \to 0} \sum_{k=1}^{n} (30 - x_k^*)(6240\pi)\, \Delta x_k = \int_0^{15} (30 - x)(6240\pi)\, dx$$

$$= 6240\pi \left(30x - \frac{x^2}{2}\right)\Bigg]_0^{15} = 2{,}106{,}000\pi \text{ ft-lb} \qquad \blacktriangleleft$$

▶ Exercise Set 6.7

1. Find the work done when
   (a) a constant force of 30 lb along the x-axis moves an object from $x = -2$ to $x = 5$ in.;
   (b) a variable force of $F(x) = 1/x^2$ lb along the x-axis moves an object from $x = 1$ to $x = 6$ in.

2. A spring whose natural length is 15 in. exerts a force of 45 lb when stretched to a length of 20 in.
   (a) Find the spring constant.
   (b) Find the work done in stretching the spring 3 in. beyond its natural length.
   (c) Find the work done in stretching the spring from a length of 20 in. to a length of 25 in.

3. A spring exerts a force of $\frac{1}{2}$ ton when stretched 5 ft beyond its natural length. How much work is required to stretch the spring 6 ft beyond its natural length?

4. Assume a force of 6 N (newtons) is required to compress a spring from a natural length of 4 m (meters) to a length of $3\frac{1}{2}$ m. Find the work required to compress the spring from its natural length to a length of 2 m. (Hooke's law applies to compression as well as extension.)

5. Assume 10 ft-lb of work is required to stretch a spring 1 ft beyond its natural length. What is the spring constant?

6. A cylindrical tank of radius 5 ft and height 9 ft is two-thirds filled with water. Find the work required to pump all the water over the upper rim.

7. Solve Problem 6 assuming that the tank is two-thirds filled with a liquid of density ρ lb/ft³.

8. A cone-shaped water reservoir is 20 ft in diameter across the top and 15 ft deep. If the reservoir is filled to a depth of 10 ft, how much work is required to pump all the water to the top of the reservoir?

9. A swimming pool is built in the shape of a rectangular parallelepiped 10 ft deep, 15 ft wide, and 20 ft long.
   (a) If the pool is filled 1 ft below the top, how much work is required to pump all the water into a drain at the top edge of the pool?
   (b) If a one-horsepower motor can do 550 ft-lb of work per second, what size motor is required to empty the pool in one hour?

10. A water tower in the shape of a hemisphere of radius 10 ft is to be filled by pumping water over the top edge from a lake 200 ft below the top. How much work is required to fill the tank? [*Hint:* Solve without calculus.]

11. A 100-ft length of steel chain weighing 15 lb/ft is dangling from a pulley. How much work is required to wind the chain onto the pulley?

12. A rocket weighing 3 tons is filled with 40 tons of liquid fuel. In the initial part of the flight, fuel is burned off at a constant rate of 2 tons per 1000 ft of vertical height. How much work is done in lifting the rocket to 3000 ft?

13. (**Satellite Problem**) The weight of an object is the force exerted on it by the earth's gravity. Thus, if a person weighs 100 lb on the surface of the earth, the earth's gravity is pulling on that person with a force of 100 lb. It is a fundamental law of physics that the force the earth exerts on an object varies inversely as the square of its distance from the earth's center. Thus an object's weight $F(x)$ is related to its distance x from the earth's center

by a formula of the form

$$F(x) = k/x^2$$

where k is a constant of proportionality depending on the mass of the object and the units of force and distance.

(a) Assuming that the earth is a sphere of radius 4000 mi, find the constant k in the formula above for a satellite that weighs 6000 lb on the earth's surface.

(b) How much work must be performed to lift this satellite to an orbital position 1000 miles above the earth's surface?

14. **(Coulomb's law)** It follows from Coulomb's law in physics that two like electrostatic charges repel each other with a force inversely proportional to the square of the distance between them. Suppose that two charges A and B repel with a force of k pounds when they are positioned at points $A(-a, 0)$ and $B(a, 0)$. Find the work W required to move charge A along the x-axis to the origin if charge B remains stationary.

6.8 LIQUID PRESSURE AND FORCE

In this section we shall use the definite integral to calculate the force exerted by a liquid on a submerged surface.

If a flat surface of area A is submerged horizontally at a depth h in a container of fluid, then the weight of the fluid above exerts a force F on the surface given by

$$F = \rho h A \qquad (1)$$

where ρ is the density of the fluid (weight per unit volume). It is a physical fact that the force F in (1) does not depend on the shape or size of the container. Thus, if the three containers in Figure 6.8.1 have bases of the same area and are filled with a fluid to the same depth, h, then each container will have the same fluid force on its base.

Figure 6.8.1 ▷

The units in (1) must be compatible. For example, if height is in feet and force in pounds, then the density ρ must be expressed in pounds per cubic foot. The density of water is approximately 62.4 lb/ft^3.

Pressure is defined to be force per unit area. Thus, from (1), the pressure p exerted at each point of a flat surface of area A submerged horizontally at a depth h is given by

$$p = \frac{F}{A} = \rho h \qquad (2)$$

Example 1 If a flat circular plate of radius $r = 2$ ft is submerged horizontally in water so the top surface is at a depth of 3 ft, then the force on the top surface of the plate is

$$F = \rho h A = \rho h (\pi r^2) = (62.4)(3)(4\pi) = 748.8\pi \text{ lb}$$

and the pressure at each point on the plate surface is

$$p = \rho h = (62.4)(3) = 187.2 \text{ lb/ft}^2 \quad \blacktriangleleft$$

Pascal's* principle in physics states that *fluid pressure is the same in all directions*. Thus, if a flat surface is submerged vertically (or at any angle at all), the pressure at a point of depth h is exactly the same as the pressure at a point of depth h on a horizontal surface (Figure 6.8.2). However, (1) cannot be applied to obtain the total force acting on a flat surface that is not submerged horizontally, because the depth h can vary from point to point on the surface. It is in such problems that calculus comes into play.

Suppose we want to compute the total fluid force against a submerged vertical surface. As shown in Figure 6.8.3, let us introduce a vertical

By Pascal's Principle the fluid pressure at points A, B and C is the same.

Figure 6.8.2 ▷

---

*BLAISE PASCAL (1623–1662) French mathematician and scientist. Pascal's mother died when he was three years old and his father, a highly educated magistrate, personally provided the boy's early education. Although Pascal showed an inclination for science and mathematics, his father refused to tutor him in those subjects until he mastered Latin and Greek. Pascal's sister and primary biographer claimed that he independently discovered the first thirty-two propositions of Euclid without ever reading a book on geometry. (However, it is generally agreed that the story is apocryphal.) Nevertheless, the precocious Pascal published a highly respected essay on conic sections by the time he was sixteen years old. Descartes, who read the essay, thought it so brilliant that he could not believe that it was written by such a young man. By age 18 his health began to fail and until his death he was in frequent pain. However, his creativity was unimpaired.

Pascal's contributions to physics include the discovery that air pressure decreases with altitude and the principle of fluid pressure that bears his name. However, the originality of his work is questioned by some historians. Pascal made major contributions to a branch of mathematics called "projective geometry," and he helped to develop probability theory through a series of letters with Fermat.

In 1646, Pascal's health problems resulted in a deep emotional crisis that led him to become increasingly concerned with religious matters. Although born a Catholic, he converted to a religious doctrine called Jansenism and spent most of his final years writing on religion and philosophy.

x-axis whose positive direction is downward and whose origin is at any convenient point.

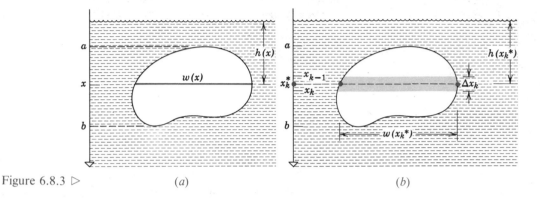

Figure 6.8.3 ▷   (a)   (b)

As indicated in Figure 6.8.3*a*, let us assume that the submerged portion of the plate extends from $x = a$ to $x = b$ on the *x*-axis. Also, suppose that a point x on the axis lies $h(x)$ units below the surface and that the surface and that the cross section of the plate at x has width $w(x)$.

Next, divide the interval $[a, b]$ into n subintervals with lengths

$$\Delta x_1, \Delta x_2, \ldots, \Delta x_n$$

and in each subinterval choose an arbitrary point x_k^*. Following a familiar pattern, we approximate the section of plate along the *k*th subinterval by a rectangle of length $w(x_k^*)$ and width Δx_k^* (Figure 6.8.3*b*). Because the upper and lower edges of this rectangle are at different depths, (1) cannot be used to calculate the force on this rectangle. However, if Δx_k is small, the difference in depth between the upper and lower edges is small, and we can reasonably assume that the entire rectangle is concentrated at a single depth $h(x_k^*)$ below the surface. With this assumption, (1) can be used to *approximate* the force F_k on the *k*th rectangle. We obtain

$$F_k \approx \rho \underbrace{h(x_k^*)}_{\text{depth}} \cdot \underbrace{w(x_k^*) \Delta x_k}_{\text{area of rectangle}}$$

Thus, the total force F on the plate is approximately

$$F = \sum_{k=1}^{n} F_k \approx \sum_{k=1}^{n} \rho h(x_k^*) w(x_k^*) \, \Delta x_k$$

To find the *exact* value of the force we take the limit as $\max \Delta x_k \to 0$. This yields

$$F = \lim_{\max \Delta x_k \to 0} \sum_{k=1}^{n} \rho h(x_k^*) w(x_k^*) \, \Delta x_k = \int_a^b \rho h(x) w(x) \, dx$$

which suggests the following result.

6.8.1 FORMULA FOR FLUID FORCE. Assume that a plate is immersed vertically in a liquid of density ρ and that the submerged portion extends from $x = a$ to $x = b$ on a vertical x-axis. For $a \leq x \leq b$, let $w(x)$ be the width of the plate at x and let $h(x)$ be the depth of the point x. Then the total **fluid force** on the plate is

$$F = \int_a^b \rho h(x) w(x) \, dx \qquad (3)$$

Figure 6.8.4

Example 2 The face of a dam is a vertical rectangle of height 100 ft and width 200 ft (Figure 6.8.4). Find the total fluid force exerted on the face when the water surface is level with the top of the dam.

Solution. Introduce an x-axis with its origin at the water surface as shown in Figure 6.8.4. At a point x on this axis, the width of the dam in feet is $w(x) = 200$ and the depth in feet is $h(x) = x$. Thus, from (3) with $\rho = 62.4$ lb/ft^3 (the density of water) we obtain as the total force on the face

$$F = \int_0^{100} (62.4)(x)(200) \, dx = 12{,}480 \int_0^{100} x \, dx$$

$$= 12{,}480 \frac{x^2}{2} \Big]_0^{100} = 62{,}400{,}000 \text{ lb} \qquad \blacktriangleleft$$

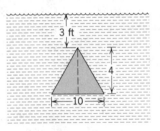

(a)

Example 3 A plate in the form of an isosceles triangle with base 10 ft and altitude 4 ft is submerged vertically in oil as shown in Figure 6.8.5a. Find the fluid force F against a surface of the plate if the oil has density $\rho = 30$ lb/ft^3.

Solution. Introduce an x-axis as shown in Figure 6.8.5b. By similar triangles, the width of the plate, in feet, at a depth of $h(x) = (3 + x)$ ft satisfies

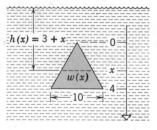

(b)

Figure 6.8.5

$$\frac{w(x)}{10} = \frac{x}{4} \quad \text{so} \quad w(x) = \frac{5}{2} x$$

Thus, it follows from (3) that the force on the plate is

$$F = \int_a^b \rho h(x) w(x) \, dx = \int_0^4 (30)(3 + x) \left(\frac{5}{2} x \right) dx$$

$$= 75 \int_0^4 (3x + x^2) \, dx = 75 \left[\frac{3x^2}{2} + \frac{x^3}{3} \right]_0^4 = 3400 \text{ lb} \qquad \blacktriangleleft$$

▶ Exercise Set 6.8

1. A flat square plate with 3-ft sides and negligible thickness is submerged horizontally in a liquid. Find the force and pressure on a surface of the plate if
   (a) the liquid is water and the plate is at a depth of 5 ft;
   (b) the liquid has density 40 lb/ft^3 and the plate is at a depth of 10 ft.

In Exercises 2–7, the flat surfaces shown are submerged vertically in water. Find the fluid force against the surface.

2.

3.

4.

5.

6.

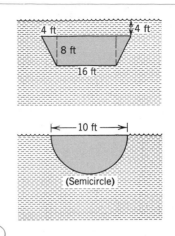

7.

8. An oil tank is shaped like a right-circular cylinder of diameter 4 ft. Find the total fluid force against one end when the axis is horizontal and the tank is half filled with oil of density 50 lb/ft^3.

9. A square plate of side a ft is dipped in a liquid of density ρ lb/ft^3. Find the fluid force on the plate if a vertex is at the surface and a diagonal is perpendicular to the surface.

10. Figure 6.8.6 shows a dam whose face is an inclined rectangle. Find the fluid force on the face when the water is level with the top of this dam.

Figure 6.8.6

11. Figure 6.8.7 shows a rectangular swimming pool whose bottom is an inclined plane. Find the fluid force on the bottom when the pool is filled to the top.

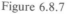

Figure 6.8.7

12. An observation window on a submarine is a square with 2-ft sides. Using ρ_0 for the density of seawater, find the fluid force on the window when the submarine has descended vertically so that the top of the window is at a depth of h feet.

13. (a) Show: If the submarine in Exercise 12 descends vertically at a constant rate, then the fluid force on the window increases at a constant rate.

(b) At what rate is the force on the window increasing if the submarine is descending vertically at 20 ft/min?

▶ SUPPLEMENTARY EXERCISES

In Exercises 1–3, set up, but do not evaluate, an integral or sum of integrals that gives the area of the region R. (Set up the integral with respect to x or y as directed.)

1. R is the region in the first quadrant enclosed by $y = x^2$, $y = 2 + x$, and $x = 0$.
   (a) Integrate with respect to x.
   (b) Integrate with respect to y.

2. R is enclosed by $x = 4y - y^2$ and $y = \frac{1}{2}x$.
   (a) Integrate with respect to x.
   (b) Integrate with respect to y.

3. R is enclosed by $x = 9$ and $x = y^2$.
   (a) Integrate with respect to x.
   (b) Integrate with respect to y.

In Exercises 4–9, set up, but do not evaluate, an integral or sum of integrals that gives the stated volume. (Set up the integral with respect to x or y as directed.)

4. The volume generated by revolving the region in Exercise 1 about the x-axis.
   (a) Integrate with respect to x.
   (b) Integrate with respect to y.

5. The volume generated by revolving the region in Exercise 1 about the y-axis.
   (a) Integrate with respect to x.
   (b) Integrate with respect to y.

6. The volume generated by revolving the region in Exercise 2 about the x-axis.
   (a) Integrate with respect to x.
   (b) Integrate with respect to y.

7. The volume generated by revolving the region in Exercise 2 about the y-axis.
   (a) Integrate with respect to x.
   (b) Integrate with respect to y.

8. The volume generated by revolving the region in Exercise 3 about the x-axis.
   (a) Integrate with respect to x.
   (b) Integrate with respect to y.

9. The volume generated by revolving the region in Exercise 3 about the y-axis.
   (a) Integrate with respect to x.
   (b) Integrate with respect to y.

In Exercises 10 and 11, find:
   (a) the area of the region described;
   (b) the volume generated by revolving the region about the indicated line.

10. The region in the first quadrant enclosed by $y = \sin x$, $y = \cos x$, and $x = 0$; revolved about the x-axis. [*Hint:* $\cos^2 x - \sin^2 x = \cos 2x$.]

11. The region enclosed by the x-axis, the y-axis, and $x = \sqrt{4 - y}$; revolved about the y-axis.

12. Set up a sum of definite integrals that represents the total shaded area between the curves $y = f(x)$ and $y = g(x)$ below.

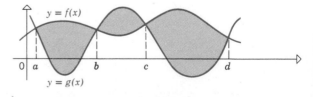

13. Find the *total* area bounded between $y = x^3$ and $y = x$ over the interval $[-1, 2]$. (See previous exercise.)

14. Find the volume of the solid whose base is the region bounded between the curves $y = x$ and $y = x^2$, and whose cross sections perpendicular to the x-axis are squares.

15. Find the volume of the solid whose base is the triangular region with vertices $(0, 0)$, $(a, 0)$, and $(0, b)$, where $a > 0$ and $b > 0$, and whose cross sections perpendicular to the y-axis are semicircles.

In Exercises 16 and 17, find the volume generated by revolving the region described about the axis indicated.

16. The region bounded above by the curve $y = \cos x^2$, on the left by the y-axis, and below by the x-axis; revolved about the y-axis.

17. The region bounded by $y = \sqrt{x}$, $x = 4$, and $y = 0$; revolved about:
(a) the line $x = 4$ (b) the line $y = 2$.

18. A football has the shape of the solid generated by revolving the region bounded between the x-axis and the parabola $y = 4R(x^2 - \frac{1}{4}L^2)/L^2$ about the x-axis. Find its volume.

In Exercises 19–22, find the arc length of the indicated curve.

19. $8y^2 = x^3$ between $(0, 0)$ and $(2, 1)$.

20. $y = \dfrac{1}{3}(x^2 + 2)^{3/2}$, $0 \le x \le 3$.

21. $y = \dfrac{1}{10}x^5 + \dfrac{1}{6}x^{-3}$, $1 \le x \le 2$.

22. $y = \dfrac{1}{3}x^3 + \dfrac{1}{4x}$, $1 \le x \le 2$.

In Exercises 23–28, find the area of the surface generated by revolving the given curve about the indicated axis.

23. $y = x^3$ between $(1, 1)$ and $(2, 8)$; x-axis.

24. $y^2 = 12x$ between $(0, 0)$ and $(3, 6)$; x-axis.

25. $y = \frac{2}{3}x^{3/2} - \frac{1}{2}x^{1/2}$ between $(0, 0)$ and $(9, 33/2)$; y-axis.

26. The curve in Exercise 25 revolved about the line $x = 9$.

27. $3y = \sqrt{x}\,(3 - x)$ between $(0, 0)$ and $(3, 0)$; y-axis.

28. $y = \sqrt{2x - x^2}$ between $(\frac{1}{2}, \sqrt{3}/2)$ and $(1, 1)$; x-axis.

29. The natural length of a spring is 6 in. If a force of 2 lb is needed to hold it at a length of 10 in., find the work done in stretching it from 8 in. to 10 in.

30. Find the spring constant if 180 in-lbs of work are required to stretch a spring 3 in. from its natural length.

31. A 250-lb weight is suspended from a ledge by a uniform 40-ft cable weighing 30 lb. How much work is required to bring the weight up to the ledge?

32. A tank in the shape of a right-circular cone has a 6-ft diameter at the top and a height of 5 ft. It is filled with a liquid of density 64 lb/ft³. How much work can be done by the liquid if it runs out of the bottom of the tank?

33. A vessel has the shape obtained by revolving about the y-axis the part of the parabola $y = 2(x^2 - 4)$ lying below the x-axis. If x and y are in feet, how much work is required to pump all the water in the full vessel to a point 4 ft above its top?

34. Two like magnetic poles repel each other with a force $F = k/x^2$ newtons. Express the work needed to move them along a line from D meters apart to $D/3$ meters apart.

In Exercises 35 and 36, the flat surface shown is submerged vertically in a liquid of density ρ lb/ft³. Find the fluid force against the surface.

35.

36.

(Semicircle)

7 logarithm and exponential functions

7.1 INVERSE FUNCTIONS

In spite of its seeming simplicity, the integral

$$\int \frac{1}{x}\, dx$$

cannot be evaluated in terms of finitely many polynomials, rational functions, or trigonometric functions. The same difficulty occurs with such basic integrals as

$$\int \tan x\, dx \quad and \quad \int \sec x\, dx$$

It is the primary purpose of this chapter to define new functions that will enable us to evaluate these and other important integrals. In this initial section we shall discuss some preliminary results that will be essential to our work in this chapter and the next.

The functions $f(x) = 2x$ and $g(x) = \frac{1}{2}x$ have the property that each cancels out the effect of the other in the sense that

$$f(g(x)) = f(\tfrac{1}{2}x) = 2(\tfrac{1}{2}x) = x$$
$$g(f(x)) = g(2x) = \tfrac{1}{2}(2x) = x$$

Similarly, the functions $f(x) = x^{1/3}$ and $g(x) = x^3$ cancel the effect of one another since

$$f(g(x)) = f(x^3) = (x^3)^{1/3} = x$$
$$g(f(x)) = g(x^{1/3}) = (x^{1/3})^3 = x$$

Pairs of functions that cancel the effect of one another are of such importance that there is some terminology associated with them.

419

7.1.1 DEFINITION. If the functions f and g satisfy the two conditions

$f(g(x)) = x$ for every x in the domain of g

$g(f(x)) = x$ for every x in the domain of f

then we say that *f is an inverse of g* and *g is an inverse of f*. We also say that *f and g are inverses of one another.*

Example 1 In the terminology of the foregoing definition, the functions $f(x) = 2x$ and $g(x) = \frac{1}{2}x$ are inverses of one another, as are $f(x) = x^{1/3}$ and $g(x) = x^3$. ◀

It can be shown that a function cannot have two different inverses. Thus, if f has an inverse, we are entitled to talk about *the* inverse of f. The inverse of f is commonly denoted by f^{-1} (read, "f inverse"); thus,

$$f(f^{-1}(x)) = x \quad \text{and} \quad f^{-1}(f(x)) = x \tag{1}$$

WARNING. The symbol f^{-1} does not mean $1/f$.

GRAPHS OF INVERSE FUNCTIONS

There is an important relationship between the graphs of f and f^{-1} that can be seen by graphing the inverse functions in Example 1 (see Figure 7.1.1). That figure shows that the graphs of $y = 2x$ and $y = \frac{1}{2}x$ are reflections ("mirror images") of one another about the line $y = x$. The same is true of the graphs of $y = x^3$ and $y = x^{1/3}$. This is not accidental. In general, *the graphs of any two inverse functions are reflections of one another about the line $y = x$.*

To see that this is so we need some preliminary results. It is easy to prove using plane geometry that the points (a, b) and (b, a) are symmetrically positioned about the line $y = x$; that is, the line $y = x$ is the per-

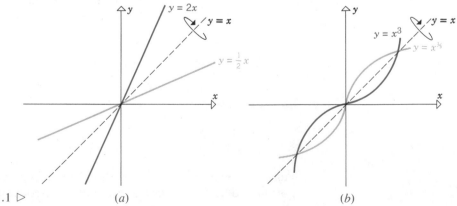

Figure 7.1.1 ▷   (a)   (b)

pendicular bisector of the line segment joining these points (Figure 7.1.2a). Thus, interchanging the x- and y-coordinates of a point causes the point to be reflected about the line $y = x$. Consequently, interchanging the x and y variables in an equation causes the graph of that equation to be reflected about this line. Thus, to prove that the graphs of inverse functions are reflections of one another about the line $y = x$ we must show that if a function f has an inverse, and (a, b) is any point on the graph of f, then the reflection of this point about the line $y = x$, namely (b, a), lies on the graph of f^{-1}, and conversely (Figure 7.1.2b). We argue as follows: Because (a, b) lies on the graph of f, the coordinates of this point satisfy the equation $y = f(x)$, so

$$b = f(a)$$
$$f^{-1}(b) = f^{-1}(f(a)) \quad \text{[Take } f^{-1} \text{ of both sides.]}$$
$$f^{-1}(b) = a \quad \text{[From (1).]}$$

The last equation shows that the point (b, a) lies on the graph of $y = f^{-1}(x)$, which is what we wanted to show. The proof of the converse is similar.

DOMAIN AND RANGE OF INVERSE FUNCTIONS

Figure 7.1.2 provides some geometric insight into the relationship between the domains and ranges of f and f^{-1}, for if (a, b) is a point on the graph of f, then *a lies in the domain of f and b lies in the range of f.* Similarly, because (b, a) is a point on the graph of f^{-1}, *a lies in the range of f^{-1} and b lies in the domain of f^{-1}.* This suggests the following result:

$$\text{range of } f^{-1} = \text{domain of } f \tag{2}$$
$$\text{domain of } f^{-1} = \text{range of } f$$

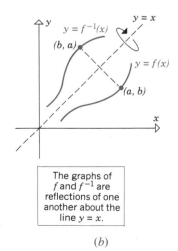

The graphs of f and f^{-1} are reflections of one another about the line $y = x$.

Figure 7.1.2 ▷   (*a*)   (*b*)

EXISTENCE OF
INVERSE FUNCTIONS

Given a function f, we shall be interested in two important questions:

1. Does f have an inverse?
2. If so, how do we find it?

Let us consider the first of these questions. In order for a function f to have an inverse, the reflection of the graph of f about the line $y = x$ would have to pass the vertical line test (2.2.3), because this reflection would be the graph of the function f^{-1}. However, reflection about $y = x$ causes horizontal lines to become vertical and vertical lines to become horizontal (Figure 7.1.3). Thus, for the reflection of the graph of f to pass the vertical line test, the graph of f itself would have to be cut at most once by any *horizontal* line. This yields the following result.

The vertical line V reflects into the horizontal line H and conversely.

Figure 7.1.3

7.1.2   THE HORIZONTAL LINE TEST. A function f has an inverse if and only if no horizontal line intersects its graph more than once.

Example 2   The function $f(x) = x^2$ has no inverse because it does not pass the horizontal line test (Figure 7.1.4).   ◄

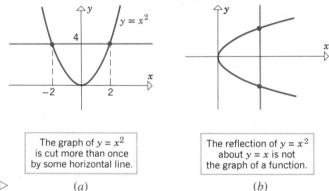

The graph of $y = x^2$ is cut more than once by some horizontal line.

The reflection of $y = x^2$ about $y = x$ is not the graph of a function.

Figure 7.1.4 ▷ (a) (b)

Referring to Figure 7.1.5, it is evident that if the graph of a function f has multiple intersections with a horizontal line, then those intersections lie above points on the x-axis at which f has the same value. Thus, to state that a function f is cut at most once by any horizontal line is equivalent to stating that f does not have the same value at two distinct points in its domain. A function f with this property is said to be *one-to-one*.

7.1.3   DEFINITION. A function f is ***one-to-one***, written 1–1, if its graph is cut at most once by any horizontal line, or equivalently, if f does not have the same value at two distinct points in its domain.

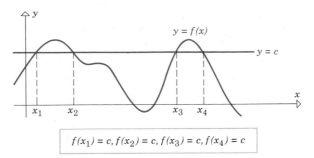

Figure 7.1.5

$$f(x_1) = c, f(x_2) = c, f(x_3) = c, f(x_4) = c$$

Using the terminology of this definition, we can rephrase 7.1.2 as follows.

7.1.4 THEOREM. *A function f has an inverse if and only if it is one-to-one.*

Figure 7.1.6

Example 3   The function $f(x) = x^3$ is one-to-one since two different numbers cannot have the same cube, that is, if $x_1 \neq x_2$, then $x_1^3 \neq x_2^3$. Geometrically, no horizontal line cuts the graph of $f(x) = x^3$ more than once (Figure 7.1.6).   ◄

Example 4   We saw in Example 2 that the function $f(x) = x^2$ is not one-to-one because there are horizontal lines that cut its graph more than once (Figure 7.1.4a). From an algebraic viewpoint, $f(x) = x^2$ is not one-to-one because f can have the same value at two distinct points; for example, $f(-2) = 4$ and $f(2) = 4$.   ◄

Example 5   Determine whether $f(x) = \sin x$ has an inverse.

Solution.   There exist horizontal lines that cut the graph of $\sin x$ more than once (Figure 7.1.7), so $f(x) = \sin x$ is not one-to-one and therefore has no inverse.   ◄

Figure 7.1.7 ▷

The following theorem describes a major class of functions with inverses.

7.1.5 THEOREM. *If the domain of f is an interval, and if f is either an increasing function or a decreasing function on that interval, then f has an inverse.*

Proof. If f is increasing and $x_1 < x_2$ are distinct points in its domain, then $f(x_1) < f(x_2)$ (Definition 4.2.1), so that $f(x_1) \neq f(x_2)$. Thus, f is one-to-one and therefore has an inverse by Theorem 7.1.4. The proof for decreasing functions is similar. ∎

Recall from Theorem 4.2.2 that f is increasing on an open interval if $f'(x) > 0$ on the interval and is decreasing if $f'(x) < 0$. This result makes it easy to apply Theorem 7.1.5 to differentiable functions.

Example 6   Show that $f(x) = x^5 + 7x^3 + 4x + 1$ has an inverse.

Solution.   The function f is increasing on $(-\infty, +\infty)$ because

$$f'(x) = 5x^4 + 21x^2 + 4 > 0$$

for all x. Thus, f has an inverse by Theorem 7.1.5.   ◀

FINDING A FORMULA
FOR THE INVERSE

We now turn to the problem of actually finding a formula for the inverse of a one-to-one function f. If we let

$$y = f^{-1}(x) \tag{3}$$

then it follows that

$$f(y) = f(f^{-1}(x))$$

or

$$f(y) = x \tag{4}$$

Observe that (4) expresses x explicitly as a function of y and (3) expresses y explicitly as a function of x. Thus, if we start with Equation (4) and solve this equation for y as a function of x, we will obtain (3) from which we can read off the formula for f^{-1}.

Example 7   Find the inverse of $f(x) = 4x - 5$.

Solution.   Replacing x by y yields

$$f(y) = 4y - 5$$

from which we obtain our starting equation

$$x = 4y - 5 \quad [x = f(y)]$$

Solving for y, we obtain

$$y = \tfrac{1}{4}(x + 5) \quad [y = f^{-1}(x)]$$

so

$$f^{-1}(x) = \tfrac{1}{4}(x + 5)$$

The graphs of f and f^{-1} are shown in Figure 7.1.8. As expected, they are reflections of one another about $y = x$.   ◀

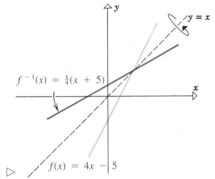

Figure 7.1.8 ▷

Example 8   Find the inverse of

$$f(x) = \frac{x}{x - 2}$$

Solution.   From Figure 4.4.4, the graph of f is cut at most once by any horizontal line, so we are guaranteed that f has an inverse. To find f^{-1}, replace x by y to obtain

$$f(y) = \frac{y}{y - 2}$$

from which we obtain the starting equation

$$x = \frac{y}{y - 2}$$

Solving for y, we obtain

$$y = \frac{2x}{x - 1}$$

so

$$f^{-1}(x) = \frac{2x}{x-1} \qquad \blacktriangleleft$$

Example 9 In Example 6 we proved that $f(x) = x^5 + 7x^3 + 4x + 1$ has an inverse. However, to find a formula for f^{-1} we would have to solve the equation

$$x = y^5 + 7y^3 + 4y + 1$$

for y in terms of x, which is too complicated to do. Thus, we cannot produce a formula for f^{-1} even though we know the inverse exists. $\blacktriangleleft$

DERIVATIVES OF INVERSE FUNCTIONS

We conclude this section by investigating the differentiability of inverse functions and the relationship between the derivative of a function and the derivative of its inverse. If, as in Examples 7 and 8, we can obtain an explicit formula for f^{-1}, then we can simply work with that formula to investigate the differentiability of f^{-1} and to find its derivative. However, if as in Example 9, no explicit formula for f^{-1} can be found, then the following theorem is helpful.

7.1.6 THEOREM. *Suppose that f has an inverse and is differentiable on an open interval I. If $f^{-1}(x)$ is a point in I at which f' is nonzero, then f^{-1} is differentiable at the point x and*

$$(f^{-1})'(x) = \frac{1}{f'(f^{-1}(x))} \qquad (5)$$

REMARK. If we let

$$y = f^{-1}(x) \quad \text{so that} \quad x = f(y)$$

then

$$\frac{dy}{dx} = (f^{-1})'(x) \quad \text{and} \quad \frac{dx}{dy} = f'(y) = f'(f^{-1}(x))$$

so that (5) can be written in more compact form as

$$\frac{dy}{dx} = \frac{1}{dx/dy} \qquad (6)$$

Although we shall not give a formal proof of this theorem, we can

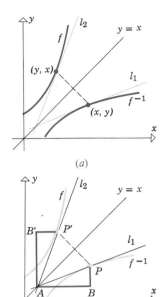

(a)

(b)

Figure 7.1.9

motivate it by considering Figure 7.1.9a, which shows the graphs of a function f and its inverse. For f^{-1} to be differentiable at x its graph must have a nonvertical tangent line l_1 at (x, y). This will occur if the graph of f has a *nonhorizontal* tangent line l_2 at the reflected point (y, x), or equivalently if $f'(y) \neq 0$. But $y = f^{-1}(x)$, so if $f'(y) = f'(f^{-1}(x)) \neq 0$, then the graph of f^{-1} will have a nonvertical tangent line at x and f^{-1} will be differentiable at x.

To find the relationship between $(f^{-1})'$ and f', consider Figure 7.1.9b in which the slopes m_1 and m_2 of tangent lines l_1 and l_2 are

$$m_1 = \frac{\overline{BP}}{\overline{AB}} \quad \text{and} \quad m_2 = \frac{\overline{AB'}}{\overline{B'P'}} \tag{7}$$

But triangles ABP and $AB'P'$ are congruent (verify), so that $\overline{BP} = \overline{B'P'}$ and $\overline{AB} = \overline{AB'}$. From these equalities and (7) it follows that

$$m_1 = \frac{1}{m_2} \tag{8}$$

Since $m_1 = (f^{-1})'(x)$ and $m_2 = f'(y)$, (8) yields

$$(f^{-1})'(x) = \frac{1}{f'(y)} \tag{9}$$

But (x, y) is on the graph of f^{-1}, so $y = f^{-1}(x)$. Consequently, (9) can also be written as

$$(f^{-1})'(x) = \frac{1}{f'(f^{-1}(x))}$$

which is precisely Formula (5) in Theorem 7.1.6.

Example 10 In Example 6 we saw that $f(x) = x^5 + 7x^3 + 4x + 1$ has an inverse.

(a) Show that this inverse is differentiable for all x in the interval $(-\infty, +\infty)$.

(b) Find the derivative of f^{-1} by using Formula (6) and also by implicit differentiation.

Solution (a). From the formula for f we have

$$f'(x) = 5x^4 + 21x^2 + 4 > 0$$

so f' is nonzero for all x. In particular, if x is a point in the domain of f^{-1}, then f' is nonzero at $f^{-1}(x)$. Thus, by Theorem 7.1.6 f^{-1} is differentiable at all x in the domain of f^{-1}. But the domain of f^{-1} is the range of f, which is $(-\infty, +\infty)$ (verify).

Solution (b).   Let $y = f^{-1}(x)$, so that $f(y) = f(f^{-1}(x)) = x$, or equivalently

$$x = y^5 + 7y^3 + 4y + 1 \tag{10}$$

Thus,

$$\frac{dx}{dy} = 5y^4 + 21y^2 + 4$$

so

$$\frac{dy}{dx} = \frac{1}{dx/dy} = \frac{1}{5y^4 + 21y^2 + 4} \tag{11}$$

Because (10) is too complicated to solve for y in terms of x, we must leave (11) in terms of y.

Alternative Solution.   Instead of using (6), we can differentiate (10) implicitly with respect to x. We obtain

$$\frac{d}{dx}[x] = \frac{d}{dx}[y^5 + 7y^3 + 4y + 1]$$

$$1 = 5y^4\frac{dy}{dx} + 21y^2\frac{dy}{dx} + 4\frac{dy}{dx}$$

$$1 = (5y^4 + 21y^2 + 4)\frac{dy}{dx}$$

which yields (11).   ◀

▶ Exercise Set 7.1

1 In (a)–(d), determine whether f and g are inverse functions.

(a) $f(x) = 4x$, $g(x) = \dfrac{1}{4}x$

(b) $f(x) = 3x + 1$, $g(x) = 3x - 1$

(c) $f(x) = \sqrt[3]{x - 2}$, $g(x) = x^3 + 2$

(d) $f(x) = x^4$, $g(x) = \sqrt[4]{x}$.

In Exercises 2–14, determine whether the function has an inverse.

2. $f(x) = 1 - x$.   **3.** $f(x) = 3x + 2$.

4. $f(x) = x^2 - 2x + 1$.   **5.** $f(x) = 2 - x - x^2$.

6. $f(x) = x^3 - 3x + 2$.   **7.** $f(x) = x^3 - x - 1$.

8. $f(x) = x^3 - 3x^2 + 3x - 1$.

9. $f(x) = x^3 + 3x^2 + 3x + 1$.

10. $f(x) = x^5 + 8x^3 + 2x - 1$.

11. $f(x) = 2x^5 + x^3 + 3x + 2$.

12. $f(x) = x + \dfrac{1}{x}$, $x > 0$.

13. $f(x) = \sin x$, $-\dfrac{\pi}{2} < x < \dfrac{\pi}{2}$.

14. $f(x) = \tan x$, $-\dfrac{\pi}{2} < x < \dfrac{\pi}{2}$.

In Exercises 15–25, find $f^{-1}(x)$.

15. $f(x) = x^5$.

16. $f(x) = 6x$.

17. $f(x) = 7x - 6.$

18. $f(x) = \dfrac{x + 1}{x - 1}.$

19. $f(x) = 3x^3 - 5.$

20. $f(x) = \sqrt[5]{4x + 2}.$

21. $f(x) = \sqrt[3]{2x - 1}.$

22. $f(x) = 5/(x^2 + 1), \; x \geq 0.$

23. $f(x) = 3/x^2, \; x < 0.$

24. $f(x) = \begin{cases} 2x, & x \leq 0 \\ x^2, & x > 0. \end{cases}$

25. $f(x) = \begin{cases} 5/2 - x, & x < 2 \\ 1/x, & x \geq 2. \end{cases}$

In Exercises 26–32, use Formula (6) to find the derivative of f^{-1}, and check your work by differentiating implicitly.

26. $f(x) = 2x^3 + 5x + 3.$

27. $f(x) = 5x^3 + x - 7.$

28. $f(x) = 1/x^2, \; x > 0.$

29. $f(x) = \tan 2x, \; -\pi/4 < x < \pi/4.$

30. $f(x) = 5x - \sin 2x.$

31. $f(x) = 2x^5 + x^3 + 1.$

32. $f(x) = x^7 + 2x^5 + x^3.$

33. Let $f(x) = x^2, \; x > 1$ and $g(x) = \sqrt{x}.$
   (a) Show that $f(g(x)) = x, x > 1$, and $g(f(x)) = x, x > 1.$
   (b) Show that f and g are *not* inverses of one another by showing that the graphs of $y = f(x)$ and $y = g(x)$ are not reflections of one another about $y = x.$
   (c) Do parts (a) and (b) contradict Definition 7.1.1? Explain.

34. Let $f(x) = ax^2 + bx + c, \; a > 0.$ Find f^{-1} if
   (a) $x \geq -b/(2a)$ (b) $x \leq -b/(2a).$

In Exercises 35–39, find $f^{-1}(x)$ and its domain.

35. $f(x) = (x + 2)^4, \; x \geq 0.$

36. $f(x) = \sqrt{x + 3}.$

37. $f(x) = -\sqrt{3 - 2x}.$

38. $f(x) = 3x^2 + 5x - 2, \; x \geq 0.$

39. $f(x) = x - 5x^2, \; x \geq 1.$

40. Prove that if $d = -a$, then the graph of
$$f(x) = \frac{ax + b}{cx + d}$$
is symmetric about the line $y = x.$

41. (a) Show that $f(x) = (3 - x)/(1 - x)$ is its own inverse.
   (b) What does the result in (a) tell you about the graph of f?

42. Suppose that a line of nonzero slope m intersects the x-axis at $(x_0, 0)$. Find an equation for the reflection of this line about $y = x.$

43. (a) Show that $f(x) = x^3 - 3x^2 + 2x$ is not one-to-one on $(-\infty, +\infty).$
   (b) Find the largest value of k such that f is one-to-one on the interval $(-k, k).$

44. If f is continuous and one-to-one, do you think that f^{-1} must also be continuous? Justify your conclusion.

45. Let $f(x) = \int_1^x \sqrt[3]{1 + t^2}\, dt.$
   (a) Without integrating, show that f is one-to-one on the interval $(-\infty, +\infty).$
   (b) Find $\dfrac{d}{dx}[f^{-1}(x)]\Big|_{x=0}.$

46. (a) Prove: If f and g are one-to-one, then so is the composition $f \circ g.$
   (b) Prove: If f and g are one-to-one, then
$$(f \circ g)^{-1} = g^{-1} \circ f^{-1}$$

47. Sketch the graph of a function that is one-to-one on $(-\infty, +\infty)$, yet not strictly monotone on $(-\infty, +\infty).$

48. Prove: A one-to-one function f cannot have two different inverses.

49. Let $F(x) = f(2g(x))$ where $f(x) = x^4 + x^3 + 1$ for $0 \leq x \leq 2$, and $g(x) = f^{-1}(x).$ Find $F'(3).$

7.2   LOGARITHMS AND IRRATIONAL EXPONENTS (AN OVERVIEW)

In this section we shall first review some basic facts about exponents and logarithms, then we shall discuss some ideas about irrational exponents, and explain why certain logarithms, called natural logarithms, play a special role in calculus. Our work in this section is informal and intuitive. In subsequent sections we shall give formal definitions and proofs.

IRRATIONAL EXPONENTS   In algebra, integer exponents and general rational exponents are defined by

$$a^n = a \cdot a \cdot \cdots \cdot a \quad (n \text{ factors})$$

$$a^{-n} = \frac{1}{a^n}$$

$$a^0 = 1$$

$$a^{p/q} = \sqrt[q]{a^p} = (\sqrt[q]{a})^p$$

$$a^{-p/q} = \frac{1}{a^{p/q}}$$

(a)

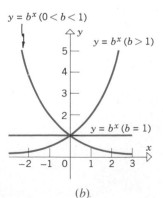

(b)

Figure 7.2.1

One of the objectives of this chapter is to define irrational exponents, thereby giving meaning to such expressions as 2^{π}, $3^{\sqrt{2}}$, and $\sqrt{2}^{\sin(\pi/8)}$. To motivate some reasonable guidelines for such a definition, let b be a *positive* constant, and let us consider the graph of the function

$$f(x) = b^x \qquad (1)$$

which for now is defined only for rational values of x. Depending on whether $0 < b < 1$, $b = 1$, or $b > 1$, the graph of (1) will have one of the shapes shown in Figure 7.2.1a. Because $f(x) = b^x$ is not yet defined for irrational values of x, each of the graphs consists of densely packed dots separated by holes at the irrational values of x. One of our objectives in defining b^x for irrational x will be to fill in the holes in these graphs in such a way that the graphs become continuous curves as in Figure 7.2.1b. As a second objective, we shall want to make sure that standard properties of exponents such as $a^k \cdot a^l = a^{k+l}$ and $(a^k)^l = a^{kl}$ continue to hold.

If we accept, for the moment, that it will be possible to define irrational exponents so the above objectives are met, then we can give a simple method for approximating b^x for irrational x. For example, if we want to approximate the value of

$$3^{\sqrt{2}}$$

then we can start by approximating the exponent $\sqrt{2}$ with a rational number. For example, a calculator yields

$$\sqrt{2} \approx 1.414214$$

from which we obtain

$$1.414210 < \sqrt{2} < 1.414220$$

Since 3^x is an increasing function (Figure 7.2.1b), we have

$$3^{1.414210} < 3^{\sqrt{2}} < 3^{1.414220}$$

or with the help of a calculator again

$$4.728786 < 3^{\sqrt{2}} < 4.728838$$

If we round the decimals on the left and right to four decimal places, we obtain 4.7288 for both. This suggests that

$$3^{\sqrt{2}} \approx 4.7288$$

to four decimal places.

REVIEW OF
LOGARITHMS

Recall from algebra that a logarithm is an exponent. More precisely, if b is a positive number other than 1, then

$$\log_b x$$

(read, "the logarithm to the base b of x") represents that power to which b must be raised to produce x. Thus,

$$\log_{10} 100 = 2$$

because 10 must be raised to the second power to produce 100. Similarly,

$\log_2 8 = 3$ since $2^3 = 8$

$\log_{10} \dfrac{1}{1000} = -3$ since $10^{-3} = \dfrac{1}{1000}$

$\log_{10} 1 = 0$ since $10^0 = 1$

$\log_3 81 = 4$ since $3^4 = 81$

In general,

$$y = \log_b x \quad \text{and} \quad x = b^y$$

are equivalent statements. It follows from these equations that

$$f(x) = b^x \quad \text{and} \quad g(x) = \log_b x$$

are inverse functions (verify). Thus,

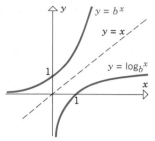

Figure 7.2.2

$$\log_b b^x = x \qquad (2)$$

$$b^{\log_b x} = x \qquad (3)$$

and the graph of $\log_b x$ is the reflection about $y = x$ of the graph of b^x. (See Figure 7.2.2 for the graph of $\log_b x$ in the case where $b > 1$.)

REMARK. It is important to note that for every base b, $\log_b x$ is only defined for $x > 0$. This is because $y = \log_b x$ is equivalent to $x = b^y$, and $b^y > 0$ for all real values of y (since b is positive).

The reader should already be familiar with the properties of logarithms listed in the following theorem.

7.2.1 THEOREM.

(a) $\log_b 1 = 0$
(b) $\log_b b = 1$
(c) $\log_b ac = \log_b a + \log_b c$
(d) $\log_b \dfrac{a}{c} = \log_b a - \log_b c$
(e) $\log_b a^r = r \log_b a$
(f) $\log_b \dfrac{1}{c} = -\log_b c$

We shall prove (a) and (c) and leave the remaining proofs as exercises.

Proof of (a).   Since $b^0 = 1$, it follows that $\log_b 1 = 0$.

Proof of (c).   Let

$$x = \log_b a \quad \text{and} \quad y = \log_b c \qquad (4)$$

so

$$b^x = a \quad \text{and} \quad b^y = c$$

Therefore,

$$ac = b^x b^y = b^{x+y}$$

or equivalently,

$$\log_b ac = x + y$$

Thus, from (4)

$$\log_b ac = \log_b a + \log_b c \qquad \blacksquare$$

THE NUMBER e
NATURAL
LOGARITHMS

The most important logarithms are those with base 10, called *common logarithms,* and those with the base $e \approx 2.71828 \ldots$, called *natural logarithms*. The base e is an irrational number which was discovered and named by Leonhard Euler (see p. 58) who suggested its use as a base for logarithms in an unpublished paper he wrote in 1728. In later sections we shall formally define the number e and show that it can be expressed as

$$e = \lim_{x \to 0} (1 + x)^{1/x} \tag{5}$$

or as

$$e = \lim_{x \to +\infty} \left(1 + \frac{1}{x}\right)^x \tag{6}$$

On some hand calculators an approximate value of e can be obtained by depressing a single key. However, e can also be approximated from either (5) or (6). For example, to use (6) we would compute

$$\left(1 + \frac{1}{x}\right)^x$$

for large values of x—the greater the value of x, the more accurate the approximation. Some sample computations, generated with a calculator, are shown in Table 7.2.1. The last value in the table is correct to five decimal places.

Table 7.2.1

x	$1 + \dfrac{1}{x}$	$\left(1 + \dfrac{1}{x}\right)^x$
1	2	2.00000
10	1.1	2.59374
100	1.01	2.70481
1000	1.001	2.71692
10,000	1.0001	2.71815
100,000	1.00001	2.71827
1,000,000	1.000001	2.71828

It has become fairly standard to denote a natural logarithm by the symbol ln (read "ell-en"). Thus,

$$\log_e x \quad \text{and} \quad \ln x$$

both denote the natural logarithm of x. In words, $\ln x$ can be interpreted as that power to which e must be raised to produce x.

Example 1

$$\ln 1 = 0 \quad (\text{since } e^0 = 1)$$
$$\ln e = 1 \quad (\text{since } e^1 = e)$$
$$\ln \frac{1}{e} = -1 \quad \left(\text{since } e^{-1} = \frac{1}{e}\right)$$
$$\ln (e^2) = 2 \quad \blacktriangleleft$$

For reference, values of the natural logarithm are given in Table 3 of Appendix 3.

To help explain why the natural logarithm plays a special role in calculus, let us consider the problem of differentiating the function

$$f(x) = \log_b x$$

where b is some arbitrary base. Since our goal in this section is motivation and not formal proof, we shall assume without proof that $\log_b x$ is differentiable and consequently continuous for $x > 0$. (The proof of this fact will be given later.) Using the derivative definition we obtain

$$\frac{d}{dx}[\log_b x] = \lim_{h \to 0} \frac{\log_b (x + h) - \log_b x}{h}$$

$$= \lim_{h \to 0} \frac{1}{h} \log_b \left(\frac{x + h}{x}\right) \qquad \text{[Theorem 7.2.1}d\text{]}$$

$$= \lim_{h \to 0} \frac{1}{h} \log_b \left(1 + \frac{h}{x}\right)$$

If we now make the substitution $t = h/x$ in the first and last expressions and use the fact that $t \to 0$ as $h \to 0$, then we obtain

$$\frac{d}{dx}[\log_b x] = \lim_{t \to 0} \frac{1}{tx} \log_b (1 + t)$$

$$= \frac{1}{x} \lim_{t \to 0} \frac{1}{t} \log_b (1 + t) \qquad \boxed{\begin{array}{l} 1/x \text{ does not} \\ \text{vary with } t \text{ and} \\ \text{so can be moved} \\ \text{through the} \\ \text{limit sign.} \end{array}}$$

$$= \frac{1}{x} \lim_{t \to 0} \log_b (1 + t)^{1/t} \qquad \boxed{\text{Theorem 7.2.1}e.}$$

$$= \frac{1}{x} \log_b \left[\lim_{t \to 0} (1 + t)^{1/t}\right] \qquad \boxed{\begin{array}{l} \text{See the remark} \\ \text{following} \\ \text{Theorem 2.7.5.} \end{array}}$$

$$= \frac{1}{x} \log_b e$$

To summarize,

$$\frac{d}{dx}[\log_b x] = \frac{1}{x} \log_b e \qquad (7)$$

In the special case where $b = e$, we have

$$\log_b e = \log_e e = 1$$

so (7) simplifies to

$$\frac{d}{dx}[\log_e x] = \frac{1}{x}$$

or in our alternative notation

$$\frac{d}{dx}[\ln x] = \frac{1}{x} \qquad (8)$$

Thus, among all possible choices for the base b, the one that yields the simplest derivative formula for $\log_b x$ is $b = e$. This is one of the main reasons why the natural logarithm plays such a special role in calculus.

► Exercise Set 7.2

1. Without using a calculator or tables, find the exact value of the following logarithms.
   (a) $\log_2 16$
   (b) $\log_2 (\frac{1}{32})$
   (c) $\log_4 4$
   (d) $\log_9 3$
   (e) $\log_{10} (.001)$
   (f) $\log_{10} (10^4)$
   (g) $\ln (e^3)$
   (h) $\ln (\sqrt{e})$.

In Exercises 2–11, solve for x without using a calculator or tables.

2. $\log_{10} (1 + x) = 3$.

3. $\log_{10} (\sqrt{x}) = -1$.

4. $\ln (x^2) = 4$.

5. $\ln (1/x) = -2$.

6. $\log_3 (3^x) = 7$.

7. $\log_5 (5^{2x}) = 8$.

8. $\log_{10} x^2 + \log_{10} x = 30$.

9. $\log_{10} x^{3/2} - \log_{10} \sqrt{x} = 5$.

10. $\ln 4x - 3 \ln (x^2) = \ln 2$.

11. $\ln (1/x) + \ln (2x^3) = \ln 3$.

12. In each part, sketch the graph of $f(x)$.
    (a) $f(x) = 2^x$
    (b) $f(x) = (1/2)^x$
    (c) $f(x) = 3^x$
    (d) $f(x) = (1/3)^x$.

13. Prove:
    (a) $b^{\log_b x} = x$ for $x > 0$
    (b) $\log_b (b^x) = x$ for $-\infty < x < +\infty$.

14. (a) Prove part (b) of Theorem 7.2.1.
    (b) Prove part (d) of Theorem 7.2.1.
    (c) Prove part (e) of Theorem 7.2.1.
    (d) Prove part (f) of Theorem 7.2.1.

15. Prove the following change of base formula for logarithms:

$$\log_b a = \frac{1}{\log_a b} \qquad \text{if } a > 0 \text{ and } b > 0$$

16. The intensity of sound is measured using the decibel scale. On this scale the magnitude L of a sound is given by $L = 10 \log_{10}(I/I_0)$, where I is the intensity (in watts per square meter) of the sound being measured, and I_0 is the threshold intensity (the point where the sound can just barely be heard).

(a) If one sound is three times as intense as another, how much greater is its decibel level?

(b) According to one source, the noise inside a moving automobile is about 70 decibels, while an electric blender generates 93 decibels. How much more intense is the noise of the blender than that of the automobile?

17. Magnitudes of earthquakes are measured using the ***Richter scale***. On this scale the magnitude R of an earthquake is given by $R = \log_{10}(I/I_0)$ where I_0 is a fixed standard intensity used for comparison, and I is the intensity of the earthquake being measured.

(a) Show that if an earthquake measures $R = 3$ on the Richter scale, then its intensity I is 1000 times the standard, that is, $I = 1000 I_0$.

(b) The San Francisco earthquake of 1906 registered $R = 8.2$ on the Richter scale. Express its intensity in terms of the standard intensity.

(c) How many times more intense is an earthquake measuring $R = 8$ than one measuring $R = 4$?

7.3   THE NATURAL LOGARITHM

In the previous section we gave an informal discussion of irrational exponents and natural logarithms. In this section we shall put those concepts on a sounder mathematical footing.

DEFINITION OF THE
NATURAL
LOGARITHM

Our immediate objective is to give precise definitions of irrational exponents, the number e, and the natural logarithm. This can be done in various ways. One approach is to define irrational exponents as limits of rational exponents, to define e as either limit (5) or limit (6) from the previous section, and to define $\ln x$ as the inverse function of e^x. Although this approach seems "natural" to many people, proofs about differentiability and continuity become cumbersome when it is used. For this reason mathematicians have developed a "less natural" approach to these definitions, but one which is ultimately easier to work with. That approach, which we shall use here, proceeds as follows:

1. Define $\ln x$ in such a way that irrational exponents are not used.

2. Define e using properties of $\ln x$.

3. Define irrational exponents using properties of e and $\ln x$.

Let us now investigate how we might define $\ln x$ without using irrational exponents. Recall from Example 1 and Formula (8) of the previous section that $\ln x$ must satisfy the conditions

$$\ln 1 = 0 \quad \text{and} \quad \frac{d}{dx} [\ln x] = \frac{1}{x}$$

Thus, for positive values of x, $\ln x$ can be viewed as that antiderivative of $1/x$ that has a value of 0 when $x = 1$. However, by the Second Fundamental Theorem of Calculus (Theorem 5.9.5),

$$F(x) = \int_1^x \frac{1}{t} dt$$

is such a function. (Also, see Example 7 in Section 5.9.) Thus, we are led to the following definition of $\ln x$ which makes no use of irrational exponents.

The Natural Logarithm Function

7.3.1 DEFINITION. The **natural logarithm** function is defined by the formula

$$\ln x = \int_1^x \frac{1}{t} dt, \quad x > 0 \tag{1}$$

In the next section we shall use this definition as our starting point for rigorously defining irrational exponents and the number e. In the remainder of this section we shall use this definition to derive some basic properties of $\ln x$.

It follows from (1) and the Second Fundamental Theorem of Calculus (Theorem 5.9.5) that $\ln x$ is differentiable for $x > 0$ and

$$\frac{d}{dx} [\ln x] = \frac{1}{x}, \quad x > 0 \tag{2}$$

If $u(x) > 0$, and if the function u is differentiable at x, then it follows from the chain rule (Formula (3) in Section 3.5) that

$$\frac{d}{dx} [\ln u] = \frac{1}{u} \cdot \frac{du}{dx} \tag{3}$$

Example 1   Find $\frac{d}{dx} [\ln (x^2 + 1)]$.

Solution.   From (3) with $u = x^2 + 1$,

$$\frac{d}{dx} [\ln (x^2 + 1)] = \frac{1}{x^2 + 1} \cdot \frac{d}{dx} [x^2 + 1] = \frac{1}{x^2 + 1} \cdot 2x = \frac{2x}{x^2 + 1} \quad \blacktriangleleft$$

Example 2   Find $\dfrac{d}{dx}[\ln|x|]$.

Solution.   We shall consider the cases where $x > 0$ and $x < 0$ separately. [The case where $x = 0$ is excluded because $\ln 0$ is undefined; see (1).] If $x > 0$, then $|x| = x$, so that

$$\frac{d}{dx}[\ln|x|] = \frac{d}{dx}[\ln x] = \frac{1}{x}$$

If $x < 0$, then $|x| = -x$, so that from (3) with $u = -x$,

$$\frac{d}{dx}[\ln|x|] = \frac{d}{dx}[\ln(-x)] = \frac{1}{(-x)} \cdot \frac{d}{dx}[-x] = \frac{1}{x}$$

Thus,

$$\frac{d}{dx}[\ln|x|] = \frac{1}{x} \qquad \text{if } x \neq 0 \qquad \blacktriangleleft \tag{4}$$

Formula (4) states that the function $\ln|x|$ is an antiderivative of $1/x$ everywhere except at $x = 0$. In contrast, the function $\ln x$ is an antiderivative of $1/x$ only for $x > 0$. The companion integration formula for (4) is

$$\int \frac{1}{x}\,dx = \ln|x| + C \tag{5}$$

Example 3   From (4) and the chain rule,

$$\frac{d}{dx}[\ln|\sin x|] = \frac{1}{\sin x} \cdot \frac{d}{dx}[\sin x] = \frac{\cos x}{\sin x} = \cot x \qquad \blacktriangleleft$$

Example 4   Evaluate

$$\int \frac{3x^2}{x^3 + 5}\,dx$$

Solution.   Make the substitution

$$u = x^3 + 5, \quad du = 3x^2\,dx$$

so that

$$\int \frac{3x^2}{x^3 + 5}\, dx = \int \frac{1}{u}\, du = \ln|u| + C = \ln|x^3 + 5| + C \quad \blacktriangleleft$$

Formula (5)

REMARK. The foregoing example illustrates an important point: any integral of the form

$$\int \frac{g'(x)}{g(x)}\, dx$$

(where the numerator of the integrand is the derivative of the denominator) can be evaluated by the u-substitution $u = g(x)$, $du = g'(x)\, dx$, since this substitution yields

$$\int \frac{g'(x)}{g(x)}\, dx = \int \frac{du}{u} = \ln|u| + C = \ln|g(x)| + C$$

Example 5 Evaluate

$$\int \tan x\, dx$$

Solution.

$$\int \tan x\, dx = \int \frac{\sin x}{\cos x}\, dx = -\int \frac{1}{u}\, du = -\ln|u| + C = -\ln|\cos x| + C \quad \blacktriangleleft$$

$u = \cos x$
$du = -\sin x\, dx$

PROPERTIES OF THE NATURAL LOGARITHM

The following theorem shows that $\ln x$, as defined in Definition 7.3.1, has the basic properties of a logarithm, thereby justifying the use of the term "logarithm" in its name.

study

7.3.2 THEOREM. *For any positive numbers a and c and any rational number r,*

(a) $\ln 1 = 0$
(b) $\ln ac = \ln a + \ln c$
(c) $\ln \dfrac{a}{c} = \ln a - \ln c$
(d) $\ln a^r = r \ln a$
(e) $\ln \dfrac{1}{c} = -\ln c$

Proof of (a).   From (1) with $x = 1$,

$$\ln 1 = \int_1^1 \frac{1}{t} \, dt = 0$$

Proof of (b).   Consider the function $f(x) = \ln ax$. Treating a as constant and differentiating with respect to x, we obtain

$$f'(x) = \frac{d}{dx}[\ln ax] = \frac{1}{ax} \cdot \frac{d}{dx}[ax] = \frac{1}{ax} \cdot a = \frac{1}{x}$$

which shows that $\ln ax$ and $\ln x$ have the same derivative on $(0, +\infty)$. Thus, by Corollary 4.10.4, there is a constant k such that

$$\ln ax - \ln x = k \tag{6}$$

If we let $x = 1$ in this equation and use the fact that $\ln 1 = 0$, we obtain

$$\ln a = k$$

so that (6) can be written as

$$\ln ax - \ln x = \ln a$$

In particular, if $x = c$ we obtain

$$\ln ac - \ln c = \ln a$$

or

$$\ln ac = \ln a + \ln c$$

Proofs of (c) and (e).   Using parts (a) and (b), we can write

$$0 = \ln 1 = \ln\left(c \cdot \frac{1}{c}\right) = \ln c + \ln \frac{1}{c}$$

so that

$$\ln \frac{1}{c} = -\ln c$$

which proves (e). To prove (c) we write

$$\ln \frac{a}{c} = \ln\left(a \cdot \frac{1}{c}\right) = \ln a + \ln \frac{1}{c} = \ln a - \ln c$$

Proof of (d).   The functions $\ln x^r$ and $r \ln x$ have the same derivative on $(0, +\infty)$, since

$$\frac{d}{dx}[\ln x^r] = \frac{1}{x^r} \cdot \frac{d}{dx}[x^r] = \frac{1}{x^r} \cdot rx^{r-1} = \frac{r}{x}$$

and

$$\frac{d}{dx}[r \ln x] = r\frac{d}{dx}[\ln x] = \frac{r}{x}$$

Thus, there is a constant k such that

$$\ln x^r - r \ln x = k \tag{7}$$

If we let $x = 1$ in this equation and use the fact that $\ln 1 = 0$, we obtain $k = 0$ (verify) so that (7) can be written as

$$\ln x^r - r \ln x = 0 \quad \text{or} \quad \ln x^r = r \ln x$$

Letting $x = a$ completes the proof. ∎

Example 6   Property (d) in the above theorem yields the following useful result:

$$\ln \sqrt[n]{a} = \ln a^{1/n} = \frac{1}{n} \ln a \quad ◀$$

THE GRAPH OF $\ln x$ We now turn to the problem of graphing the function $\ln x$. For this purpose it will be helpful to interpret the values of $\ln x$ as areas. If $x > 1$, then

$$\ln x = \int_1^x \frac{1}{t}\,dt$$

can be viewed as the area under the curve $y = 1/t$ between the points 1 and x on the t-axis (Figure 7.3.1a). On the other hand, if $0 < x < 1$, then by writing

$$\ln x = -\int_x^1 \frac{1}{t}\,dt$$

we can view $\ln x$ as the *negative* of the area under the curve $y = 1/t$ between the points x and 1 on the t-axis (Figure 7.3.1b). Therefore,

$$\begin{array}{lll} \ln x > 0 & \text{if} & x > 1 \\ \ln x < 0 & \text{if} & 0 < x < 1 \\ \ln x = 0 & \text{if} & x = 1 \end{array}$$

Since $\ln x$ is differentiable for $x > 0$, it is continuous on the interval $(0, +\infty)$. The first derivative

Figure 7.3.1

$$\frac{d}{dx}[\ln x] = \frac{1}{x}$$

is positive for all x in $(0, +\infty)$, so that $\ln x$ is *increasing* on this interval, and the second derivative

$$\frac{d^2}{dx^2}[\ln x] = \frac{d}{dx}\left[\frac{1}{x}\right] = -\frac{1}{x^2}$$

is negative for all x in $(0, +\infty)$, so that the graph of $\ln x$ is *concave down* on this interval.

To complete the picture of the graph, we must investigate the behavior of $\ln x$ as $x \to 0^+$ and $x \to +\infty$. For this purpose we shall need an estimate of $\ln 2$. From the Mean-Value Theorem for integrals (5.9.3) we obtain

$$\ln 2 = \int_1^2 \frac{1}{t}\, dt = (2-1) \cdot \frac{1}{t^*} = \frac{1}{t^*} \tag{8}$$

for some t^* satisfying

$$1 \le t^* \le 2 \tag{9}$$

Taking reciprocals in (9) yields

$$\frac{1}{2} \le \frac{1}{t^*} \le 1$$

so that from (8),

$$\frac{1}{2} \le \ln 2 \le 1 \tag{10}$$

We shall now use this estimate to deduce the following results.

7.3.3   THEOREM.

(a) $\displaystyle\lim_{x \to +\infty} \ln x = +\infty$

(b) $\displaystyle\lim_{x \to 0^+} \ln x = -\infty$

Proof of (a).   We shall show that for any positive number N (no matter how large) the values of $\ln x$ eventually exceed N as $x \to +\infty$. It will then follow that $\displaystyle\lim_{x \to +\infty} \ln x = +\infty$.

Since $\ln x$ is an increasing function, it follows from (10) that for $x > 2^{2N}$ we must have

$$\ln x > \ln 2^{2N} = 2N \ln 2 \geq 2N\left(\frac{1}{2}\right) = N$$

Therefore, $\lim\limits_{x \to +\infty} \ln x = +\infty$.

Proof of (b).   If we let $v = 1/x$ then $v \to +\infty$ as $x \to 0^+$, so that from part (*a*) of this theorem and Theorem 7.3.2*e* we obtain

$$\lim_{x \to 0^+} \ln x = \lim_{x \to 0^+} \left(-\ln \frac{1}{x}\right) = \lim_{v \to +\infty} (-\ln v) = -\lim_{v \to +\infty} \ln v = -\infty \quad \blacksquare$$

Figure 7.3.2

From the foregoing theorem and the fact that $y = \ln x$ is increasing and concave down on $(0, +\infty)$, the graph of $\ln x$ is as shown in Figure 7.3.2. The graph crosses the *x*-axis at $x = 1$ since $\ln 1 = 0$. A more accurate picture can be obtained by plotting some points and sketching some tangent lines. For this purpose we shall use the approximation,

$$\ln 2 \approx 0.6931 \qquad \boxed{\text{Appendix 3, Table 3}}$$

which is accurate to four decimal places. (We shall show how to obtain this approximation later.) Using this value of $\ln 2$ and applying Theorem 7.3.2*d* we can estimate the natural logarithm of any power of 2. For example,

$$\ln 8 = \ln 2^3 = 3 \ln 2 \approx 2.0793$$
$$\ln 4 = \ln 2^2 = 2 \ln 2 \approx 1.3862$$
$$\ln \tfrac{1}{4} = \ln 2^{-2} = -2 \ln 2 \approx -1.3862$$
$$\ln \tfrac{1}{8} = \ln 2^{-3} = -3 \ln 2 \approx -2.0793$$
$$\ln \tfrac{1}{2} = -\ln 2 \approx -0.6931$$

In Figure 7.3.3 we have plotted these points and drawn some tangent lines to obtain a more accurate sketch of the curve $y = \ln x$. The slopes of the tangent lines were obtained from the derivative of $\ln x$.

When possible, the properties of logarithms in Theorem 7.3.2 should be used to convert products, quotients, and exponents into sums, differ-

Figure 7.3.3

ences, and constant multiples *before* differentiating a function involving natural logarithms.

Example 7

$$\frac{d}{dx}\left[\ln\left(\frac{x^2\sin x}{\sqrt{1+x}}\right)\right] = \frac{d}{dx}\left[2\ln x + \ln(\sin x) - \frac{1}{2}\ln(1+x)\right]$$

$$= \frac{2}{x} + \frac{\cos x}{\sin x} - \frac{1}{2(1+x)}$$

$$= \frac{2}{x} + \cot x - \frac{1}{2+2x} \qquad \blacktriangleleft$$

LOGARITHMIC DIFFERENTIATION

We conclude this section by discussing a technique, called *logarithmic differentiation,* that is useful for differentiating functions that are composed of products and quotients.

Example 8   The derivative of

$$y = \frac{x^2\sqrt[3]{7x-14}}{(1+x^2)^4} \tag{11}$$

is messy to calculate directly. However, if we first take ln of both sides and use properties of the natural logarithm, we can write

$$\ln y = 2\ln x + \frac{1}{3}\ln(7x-14) - 4\ln(1+x^2)$$

Differentiating both sides with respect to x yields

$$\frac{1}{y}\frac{dy}{dx} = \frac{2}{x} + \frac{7/3}{7x-14} - \frac{8x}{1+x^2}$$

Thus, on solving for dy/dx and using (11) we obtain

$$\frac{dy}{dx} = \frac{x^2\sqrt[3]{7x-14}}{(1+x^2)^4}\left[\frac{2}{x} + \frac{1}{3x-6} - \frac{8x}{1+x^2}\right] \qquad \blacktriangleleft$$

Since $\ln y$ is defined only for $y > 0$, results obtained by logarithmic differentiation are valid only at points where this condition is satisfied. The derivative formula obtained in the last example is valid for $x > 2$ since $y > 0$ for such x [see (11)]. As the next example shows, logarithmic differentiation can sometimes be used with functions having negative values by first taking absolute values.

Example 9   The derivative of

$$y = \frac{x \sqrt[3]{x} - 5}{1 + \sin^3 x}$$

can be obtained by writing

$$\ln|y| = \ln\left|\frac{x \sqrt[3]{x} - 5}{1 + \sin^3 x}\right|$$

$$\ln|y| = \ln|x| + \frac{1}{3}\ln|x - 5| - \ln|1 + \sin^3 x|$$

$$\frac{1}{y}\frac{dy}{dx} = \frac{1}{x} + \frac{1}{3(x - 5)} - \frac{3\sin^2 x \cos x}{1 + \sin^3 x}$$

$$\frac{dy}{dx} = \frac{x \sqrt[3]{x} - 5}{1 + \sin^3 x}\left[\frac{1}{x} + \frac{1}{3(x - 5)} - \frac{3\sin^2 x \cos x}{1 + \sin^3 x}\right] \quad \blacktriangleleft$$

▶ Exercise Set 7.3

1. Let $r = \ln 2$ and $s = \ln 3$. Express the following in terms of r and s:
   (a) $\ln 6$
   (b) $\ln 1.5$
   (c) $\ln \frac{1}{2}$
   (d) $\ln 9$
   (e) $\ln \sqrt[5]{3}$
   (f) $\ln \frac{1}{36}$.

2. Assume that $\ln x_0 = 1$. Solve the following equations for x in terms of x_0:
   (a) $\ln x = -1$
   (b) $\ln x = 2$
   (c) $\ln \sqrt[3]{x} = -1/2$.

3. (a) On graph paper, make an accurate sketch of the curve $y = \ln x$ by first plotting the points where $x = \frac{1}{9}, \frac{1}{3}, 1, 3, 9$, then sketching tangent lines at these points, and finally drawing a smooth curve through the points. [Use $\ln 3 \approx 1.1$.]
   (b) Use your sketch to estimate $\ln 2$.
   (c) Use your sketch to estimate x_0 such that $\ln x_0 = 1$.

In Exercises 4–7, draw the graph. Avoid point plotting.

4. $y = \ln|x|$.

5. $y = \ln \frac{1}{x}$.

6. $y = \ln \sqrt{x}$.

7. $y = \ln (x - 1)$.

In Exercises 8–28, find dy/dx.

8. $y = 4 \ln x$.

9. $y = \ln 2x$.

10. $y = \ln (x^3)$.

11. $y = (\ln x)^2$.

12. $y = \ln (\sin x)$.

13. $y = \ln |\tan x|$.

14. $y = \ln (2 + \sqrt{x})$.

15. $y = \ln \left(\frac{x}{1 + x^2}\right)$.

16. $y = \ln (\ln x)$.

17. $y = \ln |x^3 - 7x^2 - 3|$.

18. $y = x^3 \ln x$.

19. $y = \sqrt{\ln x}$.

20. $y = \cos (\ln x)$.

21. $y = \sin \left(\frac{5}{\ln x}\right)$.

22. $y = \sqrt{1 + \ln^2 x}$.

23. $y = x^3 \ln (3 - 2x)$.

24. $y = x[\ln (x^2 - 2x)]^3$.

25. $y = (x^2 + 1)[\ln (x^2 + 1)]^2$.

26. $y = \frac{\ln x}{1 + \ln x}$.

27. $y = \frac{x^2}{1 + \ln x}$.

28. $y = \ln \left|\frac{1 - \cos \pi x}{1 + \cos \pi x}\right|$.

29. Find dy/dx by implicit differentiation if $y + \ln xy = 1$.

30. Find dy/dx by implicit differentiation if $y = \ln (x \tan y)$.

In Exercises 31–42, evaluate the indefinite integrals.

31. $\int \frac{dx}{2x}$.

32. $\int \frac{5x^4}{x^5 + 1} dx$.

33. $\displaystyle\int \frac{x^2}{x^3 - 4}\, dx.$ **34.** $\displaystyle\int \frac{t + 1}{t}\, dt.$

35. $\displaystyle\int \frac{\sec^2 x}{\tan x}\, dx.$ **36.** $\displaystyle\int \cot x\, dx.$

37. $\displaystyle\int \frac{\sin 3\theta}{1 + \cos 3\theta}\, d\theta.$ **38.** $\displaystyle\int \frac{dx}{x \ln x}.$

39. $\displaystyle\int \frac{x^3}{x^2 + 1}\, dx.$ **40.** $\displaystyle\int \frac{1}{x}\cos(\ln x)\, dx.$

41. $\displaystyle\int \frac{1}{y}(\ln y)^3\, dy.$ **42.** $\displaystyle\int \frac{dx}{\sqrt{x}(1 - 2\sqrt{x})}.$

In Exercises 43–46, evaluate the definite integral.

43. $\displaystyle\int_0^1 \frac{1}{3x + 2}\, dx.$ **44.** $\displaystyle\int_1^4 \frac{3}{1 - 2x}\, dx.$

45. $\displaystyle\int_{-1}^0 \frac{x}{x^2 + 5}\, dx.$ **46.** $\displaystyle\int_1^4 \frac{1}{\sqrt{x}(1 + \sqrt{x})}\, dx.$

47. Let $f(x) = \ln(x^2)$.
  (a) Find the domain of f.
  (b) For what values of x does $f(x) = 2 \ln x$?

In Exercises 48–53, use Theorem 7.3.2 to help perform the indicated differentiations and integrations.

48. $\displaystyle\frac{d}{dx}\left[\ln \sqrt{\frac{x - 1}{x + 1}}\, \right].$

49. $\displaystyle\frac{d}{dx}[\ln(\sqrt{x}\sqrt[3]{x} + 3\sqrt[5]{(3x - 2)})].$

50. $\displaystyle\frac{d}{dx}\left[\ln\left(\frac{\sqrt{x}\sqrt[3]{x} + 1}{\sin x \sec x} \right) \right].$

51. $\displaystyle\int \frac{1}{x}\ln(x^3)\, dx.$ **52.** $\displaystyle\int \frac{dx}{x \ln \sqrt{x}}.$

53. $\displaystyle\int \frac{\ln(1/x)}{x}\, dx.$

54. Use an area interpretation of $\ln x$ to obtain (10).

In Exercises 55–58, obtain dy/dx by logarithmic differentiation.

55. $y = x\sqrt[3]{1 + x^2}.$

56. $y = \sqrt[5]{\dfrac{x - 1}{x + 1}}.$

57. $y = \dfrac{(x^2 - 8)^{1/3}\sqrt{x^3 + 1}}{x^6 - 7x + 5}.$

58. $y = \dfrac{\sin x \cos x \tan^3 x}{\sqrt{x}}.$

59. (a) Make an appropriate u-substitution to show that

$$\int_a^{ab} \frac{1}{t}\, dt = \int_1^b \frac{1}{t}\, dt$$

  (b) Use this equality to obtain another proof of $\ln(ab) = \ln a + \ln b$.

60. If the interval $[1, x]$ on the t-axis is divided into n equal subintervals and A_1 and A_2 denote, respectively, the total areas of the inscribed and circumscribed rectangles shown in Figure 7.3.4, then

$$A_1 \leq \int_1^x \frac{1}{t}\, dt \leq A_2$$

Use this to solve the following problems.
  (a) Find upper and lower estimates for $\ln 2$ using $n = 4$ subintervals.
  (b) Find upper and lower estimates for $\ln 2$ using $n = 10$ subintervals.
  (c) Compare your answers to the value listed in Appendix 3, Table 3.

Figure 7.3.4

61. Prove: For all $x > 0$

$$1 - \frac{1}{x} \leq \ln x \leq x - 1$$

[*Hint:* Review the derivation of (10).]

62. Use the result of Exercise 61 to prove that

$$\lim_{x \to 0} \frac{\ln(x + 1)}{x} = 1$$

63. (a) Use the result of Exercise 61 to prove that $0 < \ln x < x$ when $x > 1$.
  (b) Use the result in (a) to prove that

$$0 < \frac{\ln x}{x} < \frac{2}{\sqrt{x}}$$

when $x > 1$. [*Hint:* $\ln \sqrt{x} < \sqrt{x}$ for $x > 1$.]

(c) Use the result in (b) to prove that

$$\lim_{x \to +\infty} \frac{\ln x}{x} = 0$$

(d) Use the result in (c) to prove that

$$\lim_{x \to 0^+} x \ln x = 0$$

[*Remark:* This problem was suggested by an article published by D. S. Greenstein in the *American Mathematical Monthly*, vol. 72 (1965), p. 767.]

64. Sketch the graph of $y = x \ln x$. [*Remark:* The limits in Exercise 63 may be helpful.]

65. **Boyle's law** in physics states that under appropriate conditions the pressure p exerted by a gas is related to its volume v by $pv = c$, where c is a constant depending on the units and various physical factors. Prove: When such a gas increases in volume from v_0 to v_1 the average pressure it exerts is

$$p_{ave} = \frac{c}{v_1 - v_0} \ln \frac{v_1}{v_0}$$

66. Prove: $\ln x < \sqrt{x}$. [*Hint:* Show that $\sqrt{x} - \ln x$ is always positive.]

67. Find the minimum value of $x^2 - \ln x$.

68. Find $\displaystyle\lim_{n \to +\infty} \sum_{k=1}^{n} \frac{1}{n + k}$. [*Hint:* Factor $\frac{1}{n}$ out of $\frac{1}{n + k}$ and identify the result as a Riemann sum.]

69. Find the volume of the solid that is generated when the region enclosed by $y = 1/\sqrt{x}$, $y = 0$, $x = 1$, and $x = 4$ is revolved about the x-axis.

70. Find the volume of the solid that is generated when the region enclosed by $y = 1/x^2$, $y = 4$, and $x = 3$ is revolved about the y-axis.

71. Find the area of the region enclosed by $y = \tan x$, $y = 0$, and $x = \pi/3$.

7.4 IRRATIONAL EXPONENTS; THE NUMBER e; THE FUNCTIONS a^x AND e^x

Recall from Section 7.2 that we had two objectives in defining irrational exponents: we wanted the standard properties of exponents to remain valid and the function $f(x) = b^x$ ($b > 0$) to be continuous. In this section we shall give such a definition.

DEFINITION OF e

We saw in the previous section that $\ln x$ is an increasing, continuous function on the interval $(0, +\infty)$ and that its values vary from $-\infty$ to $+\infty$ as x varies over this interval. This being the case, it follows that the curve $y = \ln x$ crosses every horizontal exactly once and that the intersection occurs over the interval $(0, +\infty)$. (See Figure 7.4.1.) Thus, we have the following result.

7.4.1 THEOREM. *For every real number c there is a unique real number x such that $\ln x = c$. Moreover, the number x is positive.*

Figure 7.4.1

It follows as a special case of this theorem that there is a unique positive real number x such that $\ln x = 1$. We shall denote this number by e. Thus,

$$\ln e = 1 \tag{1}$$

(Figure 7.4.2). Note that (1) agrees with the result obtained informally in Example 1 of Section 7.2.

IRRATIONAL EXPONENTS

Figure 7.4.2

We now turn to the problem of defining irrational exponents. Observe first that if a is a positive real number and r is a *rational* number, then Theorem 7.3.2d states that

$$\ln a^r = r \ln a \tag{2}$$

This seemingly innocuous equation, together with Theorem 7.4.1, gives us an alternative definition of rational exponents. For positive a we can, if desired, redefine a^r to be the unique real number whose natural logarithm is $r \ln a$. Motivated by this alternative definition of rational exponents, we make the following definition, which applies to all real exponents, both rational and irrational.

7.4.2 DEFINITION. If a is any positive real number and k is any real number, then a^k is defined to be that real number whose natural logarithm has value $k \ln a$; that is,

$$\ln a^k = k \ln a \tag{3}$$

Example 1   By definition, 2^π is that real number whose natural logarithm is

$$\ln 2^\pi = \pi \ln 2 \approx (3.142)(.6931) \approx 2.1778 \quad \text{[Appendix 3, Table 3]}$$

From Appendix 3, Table 3, we see that the value of 2^π is between 8.8 and 8.9.   ◄

The following theorem shows that the familiar laws for rational exponents are valid for all real exponents.

7.4.3   THEOREM. *If a and b are positive real numbers, then for all real numbers k and l the following laws of exponents hold.*

(a) $a^0 = 1$ (e) $\dfrac{a^k}{a^l} = a^{k-l}$

(b) $a^1 = a$ (f) $a^k \cdot b^k = (ab)^k$

(c) $a^k \cdot a^l = a^{k+l}$ (g) $\dfrac{a^k}{b^k} = \left(\dfrac{a}{b}\right)^k$

(d) $a^{-l} = \dfrac{1}{a^l}$ (h) $(a^k)^l = a^{kl}$

Proof.   We shall prove (*c*) and leave the rest as exercises. We can write

$$\ln(a^k \cdot a^l) = \ln a^k + \ln a^l = k \ln a + l \ln a$$
$$= (k + l) \ln a = \ln(a^{k+l})$$

Thus, we have $a^k \cdot a^l = a^{k+l}$ since the two sides have the same natural logarithm.   ∎

If b is a positive real constant, then the formula

$$f(x) = b^x \tag{4}$$

defines a function of x on the interval $-\infty < x < +\infty$, called the *exponential function with base b.* Some examples are

$$2^x, \quad \pi^x, \quad \text{and} \quad e^x$$

EXPONENTIAL
FUNCTIONS

The function $f(x) = e^x$, called the *natural exponential function,* or more simply the *exponential function,* has special importance because of the following theorem, which states that the natural logarithm and exponential functions are inverses.

7.4.4   THEOREM.

(*a*) $\ln e^x = x$ *for* $-\infty < x < +\infty$
(*b*) $e^{\ln x} = x$ *for* $x > 0$

Proof of (a).

$$\ln e^x = x \ln e = x$$

Proof of (b).   Let $y = e^{\ln x}$. Then

$$\ln y = \ln(e^{\ln x}) = (\ln x)(\ln e) = \ln x$$

Since y and x have the same natural logarithm, it follows that $y = x$ or, equivalently, $e^{\ln x} = x$.   ∎

In words, part (*b*) of this theorem states that $\ln x$ is that power to which e must be raised to produce x. Thus, we have shown that

$$\log_e x = \ln x$$

that is, the exponential and integral definitions of the natural logarithm are equivalent.

The derivative formula for e^x is provided by the following theorem.

7.4.5   THEOREM. *The function $f(x) = e^x$ is differentiable on $(-\infty, +\infty)$ and*

$$\frac{d}{dx}[e^x] = e^x \tag{5}$$

Proof.   Let

$$f(x) = \ln x$$

so that

$$f^{-1}(x) = e^x$$

If x is a point in $(-\infty, +\infty)$, then $f^{-1}(x) = e^x$ is a point in $(0, +\infty)$, since e^x assumes only positive values. Moreover, f' is nonzero on $(0, +\infty)$ since

$$f'(x) = \frac{1}{x} > 0$$

In particular, f' is nonzero at the point $f^{-1}(x) = e^x$. Thus, f^{-1} is differentiable at x by Theorem 7.1.6. Since x is any point in $(-\infty, +\infty)$, we have shown that e^x is differentiable everywhere.

To obtain (5) we shall use Formula (5) in Theorem 7.1.6 with $f(x) = \ln x$ and $f^{-1}(x) = e^x$. This yields

$$\frac{d}{dx}[e^x] = (f^{-1})'(x) = \frac{1}{f'(f^{-1}(x))} = \frac{1}{f'(e^x)} = \frac{1}{1/e^x} = e^x \quad \blacksquare$$

If u is a differentiable function of x, then it follows from (5) and the chain rule that

$$\frac{d}{dx}[e^u] = e^u \cdot \frac{du}{dx} \tag{6}$$

Example 2

$$\frac{d}{dx}[e^{-2x}] = e^{-2x} \cdot \frac{d}{dx}[-2x] = -2e^{-2x}$$

$$\frac{d}{dx}[e^{x^3}] = e^{x^3} \cdot \frac{d}{dx}[x^3] = 3x^2 e^{x^3}$$

$$\frac{d}{dx}[e^{\cos x}] = e^{\cos x} \cdot \frac{d}{dx}[\cos x] = -(\sin x)e^{\cos x} \quad \blacktriangleleft$$

Because $\ln x$ and e^x are inverse functions, it follows that the graph of $y = e^x$ is the reflection of the graph $y = \ln x$ about the line $y = x$ (Figure 7.4.3).

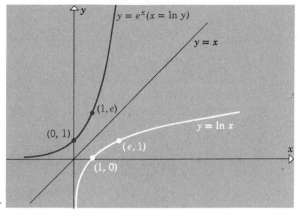

Figure 7.4.3 ▷

Earlier, we proved that $f(x) = e^x$ is everywhere differentiable and consequently everywhere continuous. However, the continuity should also be intuitively clear since the graph of $y = e^x$ is the reflection of the continuous curve $y = \ln x$.

The following properties are evident from the graph of e^x (proofs are discussed in the exercises):

7.4.6 THEOREM.

(a) $e^x > 0$ *for all* x
(b) $\displaystyle\lim_{x \to +\infty} e^x = +\infty$
(c) $\displaystyle\lim_{x \to -\infty} e^x = 0$

The following very useful theorem enables us to express a^x in terms of the natural logarithm and exponential functions.

7.4.7 THEOREM. *If a is a positive real number and x is any real number, then*

$$a^x = e^{x \ln a} \tag{7}$$

Proof. From (3),

$$\ln a^x = x \ln a$$

so that

$$e^{\ln a^x} = e^{x \ln a}$$

or by Theorem 7.4.4*b*

$$a^x = e^{x \ln a} \qquad \blacksquare$$

Because e^x and $x \ln a$ are differentiable on $(-\infty, +\infty)$, it follows from the chain rule (Theorem 3.5.2) that the composition

$$a^x = e^{x \ln a}$$

is differentiable on $(-\infty, +\infty)$ and

$$\frac{d}{dx}[a^x] = \frac{d}{dx}[e^{x \ln a}] = e^{x \ln a} \cdot \frac{d}{dx}[x \ln a] = e^{x \ln a} \cdot \ln a = a^x \ln a$$

In summary,

$$\frac{d}{dx}[a^x] = a^x \ln a \qquad (8)$$

Example 3

$$\frac{d}{dx}[\pi^x] = \pi^x \ln \pi \qquad \blacktriangleleft$$

If u is a differentiable function of x, then it follows from (8) and the chain rule that

$$\frac{d}{dx}[a^u] = a^u \ln a \cdot \frac{du}{dx} \qquad (9)$$

Example 4   From (9) with $a = 2$ and $u = \sin x$

$$\frac{d}{dx}[2^{\sin x}] = (2^{\sin x})(\ln 2) \cdot \frac{d}{dx}[\sin x] = (2^{\sin x})(\ln 2)(\cos x) \qquad \blacktriangleleft$$

The following theorem extends a familiar derivative formula so it applies to arbitrary real exponents.

> **7.4.8**   THEOREM. *If x is a positive real number and r is any real exponent, then*
>
> $$\frac{d}{dx}[x^r] = rx^{r-1}$$

Proof.   From Theorem 7.4.7

$$x^r = e^{r \ln x}$$

so that

$$\frac{d}{dx}[x^r] = \frac{d}{dx}[e^{r \ln x}] = e^{r \ln x} \cdot \frac{d}{dx}[r \ln x] = x^r \cdot \frac{r}{x} = rx^{r-1} \qquad \blacksquare$$

Example 5

$$\frac{d}{dx}[x^\pi] = \pi x^{\pi - 1} \qquad \blacktriangleleft$$

REMARK.   It is important to distinguish between the derivative formulas for a^x (variable exponent) and x^a (constant exponent). Compare, for example, the derivative of x^π in Example 5 and the derivative of π^x in Example 3.

Associated with derivatives (5) and (8) are the companion integration formulas

$$\int e^x \, dx = e^x + C \tag{10}$$

$$\int a^x \, dx = \frac{a^x}{\ln a} + C \tag{11}$$

Example 6

$$\int 2^x \, dx = \frac{2^x}{\ln 2} + C \qquad \blacktriangleleft$$

Example 7   To evaluate

$$\int e^{5x} \, dx$$

let $u = 5x$ so that $du = 5\,dx$ or $dx = \frac{1}{5}\,du$, which yields

$$\int e^{5x}\,dx = \int e^u\left(\frac{1}{5}\,du\right) = \frac{1}{5}\int e^u\,du = \frac{1}{5}e^u + C$$

$$= \frac{1}{5}e^{5x} + C \qquad \blacktriangleleft$$

Example 8

$$\int e^{-x}\,dx = -\int e^u\,du = -e^u + C = -e^{-x} + C \qquad \blacktriangleleft$$

$$u = -x$$
$$du = -dx$$

Example 9

$$\int x^2 e^{x^3}\,dx = \int \frac{1}{3}e^u\,du = \frac{1}{3}e^u + C = \frac{1}{3}e^{x^3} + C \qquad \blacktriangleleft$$

$$u = x^3$$
$$du = 3x^2\,dx$$

Example 10   Evaluate

$$\int_0^{\ln 3} e^x(1 + e^x)^{1/2}\,dx$$

Solution.   Make the u-substitution

$$u = 1 + e^x, \quad du = e^x\,dx$$

and change the x-limits ($x = 0, x = \ln 3$) to the u-limits ($u = 1 + e^0 = 2$, $u = 1 + e^{\ln 3} = 1 + 3 = 4$).

$$\int_0^{\ln 3} e^x(1 + e^x)^{1/2}\,dx = \int_2^4 u^{1/2}\,du = \frac{2}{3}u^{3/2}\Big]_2^4 = \frac{2}{3}[4^{3/2} - 2^{3/2}]$$

$$= \frac{16}{3} - \frac{4\sqrt{2}}{3} \qquad \blacktriangleleft$$

REMARK.   For ease of printing, the exponential function is often denoted by $\exp x$ rather than e^x. For example, in this notation the statements

$$e^{x_1 + x_2} = e^{x_1}e^{x_2}$$

$$e^{\ln x} = x$$

$$\ln(e^x) = x$$

would be written as

$$\exp(x_1 + x_2) = \exp x_1 \cdot \exp x_2$$
$$\exp(\ln x) = x$$
$$\ln(\exp x) = x$$

▶ Exercise Set 7.4

1. Simplify the expression and state the values of x for which your simplification is valid.
   - (a) $e^{-\ln x}$
   - (b) $e^{\ln x^2}$
   - (c) $\ln(e^{-x^2})$
   - (d) $\ln(1/e^x)$
   - (e) $\exp(3\ln x)$
   - (f) $\ln(xe^x)$
   - (g) $\ln(e^{x-\sqrt[3]{x}})$
   - (h) $e^{x-\ln x}$.

2. Solve for x:
   - (a) $\ln(x^2) = 5$
   - (b) $e^{-4x} = 3$
   - (c) $\ln(\ln x) = 0$
   - (d) $e^x + e^{-x} = 2$.

3. Solve for x:
   - (a) $\ln(\sqrt{x}) + \ln(x^{3/2}) = 1$
   - (b) $e^{2\pi x} = \sqrt{2}$
   - (c) $\ln(\cos x) = 0, \quad 0 \le x \le \pi, \quad x \ne \pi/2$
   - (d) $e^{2x} + 2e^x + 1 = 9$.

4. Where does the graph of $y = e^{2x} - 3e^x - 4$ intersect the x-axis?

5. Use Theorem 7.4.7 to approximate the following. (Round off each computation to two significant digits.)
   - (a) $2^{1.7}$
   - (b) $5^{\sqrt{3}}$
   - (c) 3^π.

In Exercises 6–21, find dy/dx.

6. $y = e^{7x}$.

7. $y = e^{-5x^2}$.

8. $y = e^{1/x}$.

9. $y = x^3 e^x$.

10. $y = \sin(e^x)$.

11. $y = \dfrac{e^x - e^{-x}}{e^x + e^{-x}}$.

12. $y = \dfrac{e^x}{\ln x}$.

13. $y = e^{x\tan x}$.

14. $y = \exp(\sqrt{1 + 5x^3})$.

15. $y = e^{(x - e^{3x})}$.

16. $y = \ln(\cos e^x)$.

17. $y = \ln(1 - xe^{-x})$.

18. $y = \sqrt{1 + e^x}$.

19. $y = e^{ax}\cos bx$ (a, b constant).

20. $y = \dfrac{a}{1 + be^{-x}}$ (a, b constant).

21. $y = e^{\ln(x^3 + 1)}$.

In Exercises 22–26, find $f'(x)$ by Formula (9) and then by logarithmic differentiation.

22. $f(x) = 2^x$.

23. $f(x) = 3^{-x}$.

24. $f(x) = \pi^{\sin x}$.

25. $f(x) = \pi^{x\tan x}$.

26. $f(x) = (\sqrt{2})^{x\ln x}$.

27. (a) Explain why Formula (8) cannot be used to find $(d/dx)[x^x]$.
    (b) Find this derivative by logarithmic differentiation.

In Exercises 28–33, find dy/dx by logarithmic differentiation.

28. $y = x^{\sin x}$.

29. $y = (x^3 - 2x)^{\ln x}$.

30. $y = (x^2 + 3)^{\ln x}$.

31. $y = (\ln x)^{\tan x}$.

32. $y = (1 + x)^{1/x}$.

33. $y = x^{(e^x)}$.

34. Show that $y = e^{3x}$ and $y = e^{-3x}$ both satisfy the equation $y'' - 9y = 0$.

35. Show that for any constants A and B, the function $y = Ae^{2x} + Be^{-4x}$ satisfies the equation $y'' + 2y' - 8y = 0$.

36. Show that for any constants A and k, the function $y = Ae^{kt}$ satisfies the equation $dy/dt = ky$.

37. Let $f(x) = e^{kx}$ and $g(x) = e^{-kx}$. Find
    - (a) $f^{(n)}(x)$
    - (b) $g^{(n)}(x)$.

38. Find dy/dt if $y = e^{-\lambda t}(A\sin\omega t + B\cos\omega t)$, where A, B, λ, and ω are constants.

39. Find $f'(x)$ if

$$f(x) = \frac{1}{\sqrt{2\pi}\sigma}\exp\left[-\frac{1}{2}\left(\frac{x - \mu}{\sigma}\right)^2\right]$$

where μ and σ are constants.

In Exercises 40–67, evaluate the integrals.

40. $\displaystyle\int \frac{dx}{e^x}.$ 41. $\displaystyle\int e^{-5x}\,dx.$

42. $\displaystyle\int e^{\tan x}\sec^2 x\,dx.$ 43. $\displaystyle\int e^{\sin x}\cos x\,dx.$

44. $\displaystyle\int x^3 e^{x^4}\,dx.$ 45. $\displaystyle\int x^2 e^{-2x^3}\,dx.$

46. $\displaystyle\int \frac{e^x + e^{-x}}{e^x - e^{-x}}\,dx.$ 47. $\displaystyle\int \frac{e^x}{1 + e^x}\,dx.$

48. $\displaystyle\int \sqrt{e^x}\,dx.$ 49. $\displaystyle\int e^{2t}\sqrt{1 + e^{2t}}\,dt.$

50. $\displaystyle\int (x + 3)\exp(x^2 + 6x)\,dx.$

51. $\displaystyle\int \cos x \exp(\sin x)\,dx.$

52. $\displaystyle\int e^x \sin(1 + e^x)\,dx.$

53. $\displaystyle\int e^{-x}\sec^2(2 - e^{-x})\,dx.$

54. $\displaystyle\int 2^{5x}\,dx.$ 55. $\displaystyle\int \pi^{\sin x}\cos x\,dx.$

56. $\displaystyle\int \left[ex^2 + \left(\frac{1}{2}\ln 2\right)\sin x\right]dx.$

57. $\displaystyle\int \left(x \ln 3 - 4\pi e^2 \cos x\right)dx.$

58. $\displaystyle\int e^{2\ln x}\,dx.$

59. $\displaystyle\int \left[\ln(e^x) + \ln(e^{-x})\right]dx.$

60. $\displaystyle\int \frac{dy}{\sqrt{y}e^{\sqrt{y}}}.$ 61. $\displaystyle\int \frac{e^{\sqrt{y}}}{\sqrt{y}}\,dy.$

62. $\displaystyle\int_0^{\ln 2} e^{-3x}\,dx.$ 63. $\displaystyle\int_0^{\ln 5} e^x(3 - 4e^x)\,dx.$

64. $\displaystyle\int_1^{\sqrt{2}} x 4^{-x^2}\,dx.$ 65. $\displaystyle\int_1^2 (3 - e^x)\,dx.$

66. $\displaystyle\int_0^e \frac{dx}{x + e}.$ 67. $\displaystyle\int_{-\ln 3}^{\ln 3} \frac{e^x}{e^x + 4}\,dx.$

68. Show that the equation $e^{1/x} - e^{-1/x} = 0$ has no solution.

69. Prove: If $f'(x) = f(x)$ for all x in $(-\infty, +\infty)$, then $f(x)$ has the form $f(x) = ke^x$ for some constant k. [*Hint*: Let $g(x) = e^{-x}f(x)$ and consider $g'(x)$.]

70. A particle moves along the x-axis so that its x-coordinate at time t is given by $x = ae^{kt} + be^{-kt}$. Show that its acceleration is proportional to x.

71. Find the intersections of the curves $y = 2^x$ and $y = 3^{x+1}$.

72. Find a point on the graph of $y = e^{3x}$ at which the tangent line passes through the origin.

73. Find $f'(x)$ if $f(x) = x^e$.

74. Show that
   (a) $y = xe^{-x}$ satisfies the equation
   $$xy' = (1 - x)y$$
   (b) $y = xe^{-x^2/2}$ satisfies the equation
   $$xy' = (1 - x^2)y.$$

75. The equilibrium constant k of a balanced chemical reaction changes with the absolute temperature T according to the law
$$k = k_0 \exp\left(-\frac{q}{2}\frac{T - T_0}{T_0 T}\right)$$
where k_0, q, and T_0 are constants. Find the rate of change of k with respect to T.

76. Find a function $y = f(x)$ such that $e^y - e^{-y} = x$.

77. Evaluate
$$\int \frac{e^{2x}}{e^x + 3}\,dx$$

In Exercises 78–81, find $f^{-1}(x)$.

78. $f(x) = e^{2x+1}.$ 79. $f(x) = e^{1/x}.$

80. $f(x) = 4\ln(x + 1).$ 81. $f(x) = 1 - \ln(3x).$

82. Find $\dfrac{dy}{dx}$:
   (a) $y = \log_x e$ (b) $y = \log_x 2.$

83. Use the Mean-Value Theorem (4.10.2) to prove that
   (a) $e^x \geq 1 + x$ if $x \geq 0$
   (b) $e^{-x} \geq 1 - x$ if $x \geq 0.$

84. Prove that $\lim\limits_{x \to +\infty} e^x = +\infty$ by showing that for any $N > 0$, there is a point x_0 such that $e^x > N$ whenever $x > x_0$.

85. Prove that $\lim\limits_{x \to -\infty} e^x = 0$ by showing that for any $\epsilon > 0$ there is a point x_0 such that $0 < e^x < \epsilon$ whenever $x < x_0$.

86. Sketch the graph of $y = x^{1/\ln x}$. [*Hint:* First find a simpler form for $x^{1/\ln x}$.]

87. Find the minimum value of x^x, $x > 0$.

88. Find the maximum value of $x^{1/x}$, $x > 0$.

89. (a) Give a geometric argument to show that

$$\frac{1}{x+1} < \int_x^{x+1} \frac{1}{t}\, dt < \frac{1}{x}, \quad x > 0$$

(b) Use the result in part (a) to prove that

$$\frac{1}{x+1} < \ln\left(1 + \frac{1}{x}\right) < \frac{1}{x}, \quad x > 0$$

(c) Use the result in part (b) to prove that

$$e^{\frac{x}{x+1}} < \left(1 + \frac{1}{x}\right)^x < e, \quad x > 0$$

and hence that

$$\lim_{x \to +\infty} \left(1 + \frac{1}{x}\right)^x = e$$

(d) Use the inequality in part (c) to prove that

$$\left(1 + \frac{1}{x}\right)^x < e < \left(1 + \frac{1}{x}\right)^{x+1}, \quad x > 0$$

7.5 LIMITS AND GRAPHS INVOLVING EXPONENTIALS AND LOGARITHMS

In this section we shall describe some basic limits involving the natural exponential and logarithm functions, and we shall use these results in some graphing problems.

Let us begin by recalling the following limits that were previously stated in Theorems 7.3.3 and 7.4.6.

$$\lim_{x \to +\infty} e^x = +\infty \tag{1}$$

$$\lim_{x \to -\infty} e^x = 0 \tag{2}$$

$$\lim_{x \to +\infty} \ln x = +\infty \tag{3}$$

$$\lim_{x \to 0^+} \ln x = -\infty \tag{4}$$

The following limits are consequences of (1) and (2):

$$\lim_{x \to +\infty} e^{-x} = \lim_{x \to +\infty} \frac{1}{e^x} = 0 \tag{5}$$

$$\lim_{x \to -\infty} e^{-x} = \lim_{x \to -\infty} \frac{1}{e^x} = +\infty \tag{6}$$

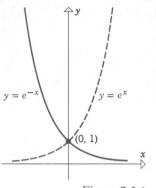

Figure 7.5.1

Example 1   Sketch the graph of $y = e^{-x}$.

Solution.   The effect of replacing x by $-x$ in the equation $y = e^x$ is to reflect the graph of this equation about the y-axis. Thus, the graph of $y = e^{-x}$ is as shown in Figure 7.5.1. Note that limits (5) and (6) are evident geometrically from this graph.   ◀

Example 2   Sketch the graph of $f(x) = e^{-x^2/2}$.

Solution.

$$f'(x) = e^{-x^2/2} \frac{d}{dx}\left[-\frac{x^2}{2}\right] = -xe^{-x^2/2}$$

$$f''(x) = -x\frac{d}{dx}[e^{-x^2/2}] + e^{-x^2/2}\frac{d}{dx}[-x]$$

$$= x^2 e^{-x^2/2} - e^{-x^2/2}$$

$$= (x^2 - 1)e^{-x^2/2}$$

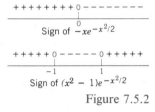

Figure 7.5.2

Since $e^{-x^2/2} > 0$ for all x, the signs of $f'(x)$ and $f''(x)$ are the same as those of $-x$ and $x^2 - 1$, respectively (Figure 7.5.2). The information conveyed by these signs is summarized in Figure 7.5.3a. Since the curve is concave down at $x = 0$, there is a relative maximum at this stationary point. The values of f at this point and at the inflection points, $x = 1$ and $x = -1$, are

$$f(0) = e^0 = 1$$
$$f(1) = e^{-1/2} \approx .61 \qquad \text{[Appendix 3, Table 2]}$$
$$f(-1) = e^{-1/2} \approx .61$$

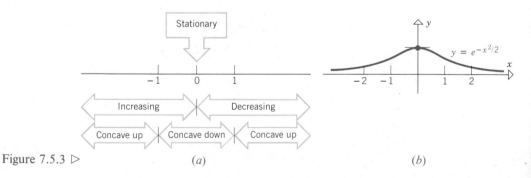

Figure 7.5.3 ▷   (a)   (b)

Since $e^{-x^2/2} > 0$ for all x, it follows that the graph of $y = e^{-x^2/2}$ is entirely above the x-axis. Finally, since $x^2/2 \to +\infty$ as $x \to +\infty$ or $x \to -\infty$, it follows from (5) and (6) that

$$\lim_{x \to +\infty} e^{-x^2/2} = \lim_{x \to -\infty} e^{-x^2/2} = 0$$

Piecing together all of our information yields the graph in Figure 7.5.3b. Note that the graph is symmetric about the y-axis. This could have been anticipated from the fact that the equation $y = e^{-x^2/2}$ is unaltered when x is replaced by $-x$.   ◄

Formulas (1) and (3) tell us that as x increases e^x and $\ln x$ both tend toward $+\infty$. However, the two functions increase in very different ways— the increase in e^x is extremely rapid and that of $\ln x$ is extremely slow (Figure 7.4.3). For example, as x increases from 0 to 10, e^x increases from $e^0 = 1$ to $e^{10} \approx 22{,}026$ (Appendix 3, Table 2). Thus, an increase of only 10 units in x results in an increase of more than 22,000 units in e^x. The situation is reversed for $\ln x$. If x increases from 1 to $e^{10} \approx 22{,}026$, then $\ln x$ increases from $\ln 1 = 0$ to $\ln e^{10} = 10 \ln e = 10$. Thus, an increase in x of more than 22,000 units produces a mere 10-unit increase in $\ln x$.

Mathematicians often use integer powers of x as a "measuring stick" for describing how rapidly a function grows. For example, it is shown in the exercises (Exercises 41 and 42 of Section 7.5) that if n is any positive integer power of x, then

$$\lim_{x \to +\infty} \frac{e^x}{x^n} = +\infty \tag{7}$$

$$\lim_{x \to +\infty} \frac{\ln x}{x^n} = 0 \tag{8}$$

Comparing (1) and (7) tells us that e^x increases so rapidly that no matter how large we choose the integer n, division by x^n does not prevent the growth toward $+\infty$. Thus, we say that e^x *increases more rapidly than any positive integer power of x*. On the other hand, comparing (3) and (8) tells us that $\ln x$ increases so slowly that division by x^n for any positive integer n prevents the growth toward $+\infty$. Thus, we say that $\ln x$ *increases more slowly than any positive integer power of x*.

The following alternative forms of (7) and (8) can be obtained by taking reciprocals

$$\lim_{x \to +\infty} \frac{x^n}{e^x} = 0 \tag{9}$$

$$\lim_{x \to +\infty} \frac{x^n}{\ln x} = +\infty \tag{10}$$

Example 3   Sketch the graph of

$$f(x) = \frac{\ln x}{x}$$

Solution.

$$f'(x) = \frac{x\left(\dfrac{1}{x}\right) - (\ln x)(1)}{x^2} = \frac{1 - \ln x}{x^2}$$

$$f''(x) = \frac{x^2\left(-\dfrac{1}{x}\right) - (1 - \ln x)(2x)}{x^4} = \frac{2x \ln x - 3x}{x^4} = \frac{2 \ln x - 3}{x^3}$$

Observe first that $\ln x$ is defined only for $x > 0$, so the same is true of $f(x)$. From the formula for $f'(x)$, there is a stationary point if $1 - \ln x = 0$, which occurs if $x = e$. Moreover, since $x^2 > 0$, the sign of f' is determined by its numerator. Thus, $f'(x) > 0$ if $\ln x < 1$, or equivalently if $x < e$, and $f'(x) < 0$ if $\ln x > 1$, or equivalently if $x > e$. Thus, the graph of f is increasing if $x < e$ and decreasing if $x > e$ (Figure 7.5.4a).

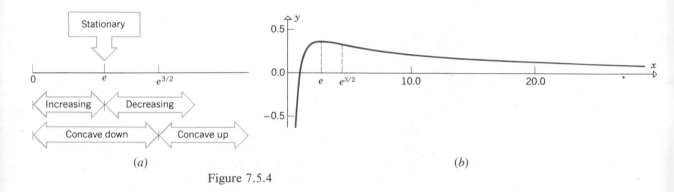

Figure 7.5.4

Since we are only concerned with positive values of x, the sign of f'' is determined by its numerator. Thus, $f''(x) > 0$ if

$$2 \ln x - 3 > 0$$

$$\ln x > \tfrac{3}{2}$$

$$e^{\ln x} > e^{3/2} \qquad \boxed{\text{Because } e^x \text{ is an increasing function}}$$

$$x > e^{3/2}$$

Similarly, $f''(x) < 0$ if $x < e^{3/2}$. Thus, the graph is concave down if $x < e^{3/2}$, concave up if $x > e^{3/2}$, and has an inflection point at $x = e^{3/2} \approx 4.48$ (Figure 7.5.4b). The values of f at the stationary and inflection points are

$$f(e) = \frac{\ln e}{e} = \frac{1}{e} \approx .37$$

$$f(e^{3/2}) = \frac{\ln(e^{3/2})}{e^{3/2}} = \frac{\frac{3}{2}\ln e}{e^{3/2}} = \frac{3}{2e^{3/2}} \approx .33$$

Finally,

$$\lim_{x \to +\infty} \frac{\ln x}{x} = 0 \qquad \text{[Formula (8)]}$$

$$\lim_{x \to 0^+} \frac{\ln x}{x} = -\infty \qquad \boxed{\text{A direct consequence of (4)}}$$

Piecing together all our information yields the graph in Figure 7.5.4b. (For visual clarity we have used different scales on the coordinate axes.)   ◄

► Exercise Set 7.5

In Exercises 1–6, find the limits.

1. (a) $\lim\limits_{x \to +\infty} 4e^{3x}$ (b) $\lim\limits_{x \to -\infty} 4e^{3x}$.

2. (a) $\lim\limits_{x \to +\infty} 3e^{-2x}$ (b) $\lim\limits_{x \to -\infty} 3e^{-2x}$.

3. (a) $\lim\limits_{x \to +\infty} (e^x + e^{-x})$ (b) $\lim\limits_{x \to -\infty} (e^x + e^{-x})$.

4. (a) $\lim\limits_{x \to +\infty} (e^x - e^{-x})$ (b) $\lim\limits_{x \to -\infty} (e^x - e^{-x})$.

5. (a) $\lim\limits_{x \to +\infty} (1 - e^{-x^2})$ (b) $\lim\limits_{x \to -\infty} (1 - e^{-x^2})$.

6. (a) $\lim\limits_{x \to +\infty} e^{1/x}$ (b) $\lim\limits_{x \to -\infty} e^{1/x}$.

In Exercises 7–12, sketch the graph of f and label the relative extreme points and points of inflection.

7. $f(x) = 4e^{3x}$.   8. $f(x) = 3e^{-2x}$.

9. $f(x) = e^x + e^{-x}$.   10. $f(x) = e^x - e^{-x}$.

11. $f(x) = 1 - e^{-x^2}$.   12. $f(x) = e^{1/x}$.

13. Let $f(x) = e^{|x|}$.
    (a) Is f continuous at $x = 0$?
    (b) Is f differentiable at $x = 0$?
    (c) Sketch the graph of f.

14. In each part determine whether the limit exists. If so, find it.
    (a) $\lim\limits_{x \to +\infty} e^x \cos x$ (b) $\lim\limits_{x \to -\infty} e^x \cos x$.

15. Sketch the graphs of e^x, $-e^x$, and $e^x \cos x$, all in the same figure, and label any points of intersection.

In Exercises 16–19, find the limits.

16. $\lim\limits_{x \to +\infty} \dfrac{e^x + e^{-x}}{e^x - e^{-x}}$.   17. $\lim\limits_{x \to -\infty} \dfrac{e^x + e^{-x}}{e^x - e^{-x}}$.

18. $\lim\limits_{x \to +\infty} \dfrac{2 + e^x}{1 + 3e^x}$.   19. $\lim\limits_{x \to +\infty} e^{(-e^x)}$.

20. One of the fundamental functions of mathematical statistics is

$$f(x) = \frac{1}{\sqrt{2\pi}\sigma} \exp\left[-\frac{1}{2}\left(\frac{x - \mu}{\sigma}\right)^2\right]$$

where μ and σ are constants such that $\sigma > 0$ and $-\infty < \mu < +\infty$.
    (a) Locate the inflection points and relative extreme points.
    (b) Find $\lim\limits_{x \to +\infty} f(x)$ and $\lim\limits_{x \to -\infty} f(x)$.
    (c) Sketch the graph of f.

21. By writing the derivative

$$\frac{d}{dx}[e^x]\bigg|_{x=0} = 1$$

as a limit, prove that

$$\lim_{h \to 0} \frac{e^h - 1}{h} = 1$$

In Exercises 22–25, use the method of Exercise 21 to find the limits.

22. $\lim\limits_{x\to 0} \dfrac{e^{2x} - e^{x}}{x}$.

23. $\lim\limits_{x\to 0} \dfrac{1 - e^{-x}}{x}$.

24. $\lim\limits_{x\to a} \dfrac{e^{x} - e^{a}}{x - a}$.

25. $\lim\limits_{x\to +\infty} x(e^{1/x} - 1)$.

In Exercise 40 of this set, it will be shown that

$$\lim_{x\to +\infty} xe^{-x} = 0$$

Use this result, where needed, in Exercises 26–30.

26. (a) Find $\lim\limits_{x\to +\infty} xe^{-2x}$ and $\lim\limits_{x\to -\infty} xe^{-2x}$.
    (b) Sketch the graph of $y = xe^{-2x}$ and label all relative extrema and inflection points.

27. (a) Find $\lim\limits_{x\to +\infty} xe^{x}$ and $\lim\limits_{x\to -\infty} xe^{x}$.
    (b) Sketch the graph of $y = xe^{x}$ and label all relative extrema and inflection points.

28. (a) Find $\lim\limits_{x\to +\infty} x^{2}e^{2x}$ and $\lim\limits_{x\to -\infty} x^{2}e^{2x}$.
    (b) Sketch the graph of $y = x^{2}e^{2x}$ and label all relative extrema and inflection points.

29. (a) Find $\lim\limits_{x\to +\infty} x^{2}/e^{2x}$ and $\lim\limits_{x\to -\infty} x^{2}/e^{2x}$.
    (b) Sketch the graph of $y = x^{2}/e^{2x}$ and label all relative extrema and inflection points.

30. (a) Find $\lim\limits_{x\to +\infty} e^{x}/x$, $\lim\limits_{x\to -\infty} e^{x}/x$, $\lim\limits_{x\to 0^{+}} e^{x}/x$, and $\lim\limits_{x\to 0^{-}} e^{x}/x$.
    (b) Sketch the graph of $y = e^{x}/x$ and label all relative extrema and inflection points.

31. Let c_1, c_2, k_1, and k_2 be positive constants. In each part make a sketch that shows the relative positions of the curves $y = c_1 e^{k_1 x}$ and $y = c_2 e^{k_2 x}$ when:
    (a) $c_1 = c_2$ and $k_1 < k_2$
    (b) $c_1 < c_2$ and $k_1 = k_2$
    (c) $c_1 < c_2$ and $k_2 < k_1$.

32. Repeat Exercise 31 for the curves $y = c_1 e^{-k_1 x}$ and $y = c_2 e^{-k_2 x}$.

33. (a) Sketch the curve $y = 2^{x}$.
    (b) Sketch the curve $y = a^{x}$ assuming $a > 1$.
    (c) Make a sketch that shows the relative positions of the curve $y = a_1^{x}$ and $y = a_2^{x}$ if $a_2 > a_1 > 1$.

34. Find the volume of the solid generated when the region bounded by $y = e^{x}$, $x = 0$, $x = \ln 3$, and $y = 0$ is revolved about the x-axis.

35. Find the area of the region enclosed by the curve $y = e^{-x}$ and the line through the points $(0, 1)$ and $(1, 1/e)$.

36. Evaluate $\int_1^5 \ln x \, dx$. [*Hint:* Interpret the integral as an area and integrate with respect to y.]

37. Evaluate $\int_1^3 x \ln x \, dx$. [*Hint:* Relate the integral to the volume of a solid by the method of shells and use the method of washers to find the volume.]

In Exercises 38 and 39, sketch the graph of f. (The limits in parts (c) and (d) of Exercise 63, Section 7.3, will be helpful.)

38. $f(x) = x^{2} \ln x$.

39. $f(x) = \dfrac{\ln x}{x^{2}}$.

40. It was proved in Exercise 63 of Section 7.3 that $0 < \ln x < x$ for $x > 1$.
    (a) Use this result to prove that $\ln x < 2\sqrt{x}$ for $x > 1$.
    (b) Use the result in (a) to obtain

    $$\frac{1}{e^{x}} < \frac{x}{e^{x}} < e^{2\sqrt{x}-x}$$

    for $x > 1$.
    (c) Use the result in (b) to show that

    $$\lim_{x\to +\infty} xe^{-x} = 0$$

    (d) By inspection, explain why

    $$\lim_{x\to -\infty} xe^{-x} = -\infty$$

41. Use the result in part (a) of Exercise 40 to obtain

    $$\frac{e^{x}}{x^{n}} > e^{x-2n\sqrt{x}}$$

    where n is any positive integer and $x > 1$. Use this result to show that

    $$\lim_{x\to +\infty} \frac{e^{x}}{x^{n}} = +\infty$$

42. Use the result in part (a) of Exercise 40 to show that

    $$\lim_{x\to +\infty} \frac{\ln x}{x^{n}} = 0$$

    for any positive integer n.

7.6   THE HYPERBOLIC FUNCTIONS

*In this section we shall study certain combinations of e^x and e^{-x}, called **hyperbolic functions**. These functions have numerous engineering applications and arise naturally in many mathematical problems. It will become evident as we progress that the hyperbolic functions have many properties in common with the trigonometric functions. This similarity is reflected in the names of the hyperbolic functions.*

7.6.1   DEFINITION. The **hyperbolic sine** and **hyperbolic cosine** functions, denoted by **sinh** and **cosh**, respectively, are defined by

$$\sinh x = \frac{e^x - e^{-x}}{2}$$

$$\cosh x = \frac{e^x + e^{-x}}{2}$$

REMARK.   We note that sinh rhymes with "cinch" and cosh rhymes with "gosh."

The graph of $\cosh x = \frac{1}{2}e^x + \frac{1}{2}e^{-x}$ can be obtained by separately graphing $\frac{1}{2}e^x$ and $\frac{1}{2}e^{-x}$ and then adding the y-coordinates together at each point (Figure 7.6.1a). This graphing technique, called **addition of ordinates,** can also be used to graph $\sinh x = \frac{1}{2}e^x - \frac{1}{2}e^{-x}$ (Figure 7.6.1b).

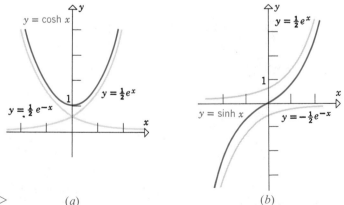

Figure 7.6.1 ▷   (a)   (b)

As an illustration of how hyperbolic functions occur in physical problems, consider a homogeneous flexible cable hanging suspended between two points (for example, an electrical transmission line suspended between two poles). The cable forms a curve called a *catenary* (from the Latin "catena" meaning chain). If a coordinate system is introduced in such a way that the low point of the cable occurs on the y-axis at $(0, a)$ where $a > 0$, then it can be shown using principles of physics that the equation of the curve formed by the cable is

$$y = a \cosh\left(\frac{x}{a}\right)$$

Figure 7.6.2

where a depends on the tension and physical properties of the cable (Figure 7.6.2).

The remaining hyperbolic functions, *hyperbolic tangent, hyperbolic cotangent, hyperbolic secant,* and *hyperbolic cosecant* are defined in terms of sinh and cosh as follows.

$$\tanh x = \frac{\sinh x}{\cosh x} = \frac{e^x - e^{-x}}{e^x + e^{-x}}$$

$$\coth x = \frac{\cosh x}{\sinh x} = \frac{e^x + e^{-x}}{e^x - e^{-x}}$$

$$\operatorname{sech} x = \frac{1}{\cosh x} = \frac{2}{e^x + e^{-x}}$$

$$\operatorname{csch} x = \frac{1}{\sinh x} = \frac{2}{e^x - e^{-x}}$$

The graphs of these functions are shown in Figure 7.6.3 (see Exercises 39–42).

HYPERBOLIC IDENTITIES   The hyperbolic functions satisfy various identities similar to the identities for the trigonometric functions. The most fundamental of these is

$$\cosh^2 x - \sinh^2 x = 1 \tag{1}$$

which can be proved by writing

$$\cosh^2 x - \sinh^2 x = \left(\frac{e^x + e^{-x}}{2}\right)^2 - \left(\frac{e^x - e^{-x}}{2}\right)^2$$

$$= \frac{1}{4}(e^{2x} + 2e^0 + e^{-2x}) - \frac{1}{4}(e^{2x} - 2e^0 + e^{-2x})$$

$$= 1$$

If we divide (1) by $\cosh^2 x$ we obtain

$$1 - \tanh^2 x = \operatorname{sech}^2 x \tag{2}$$

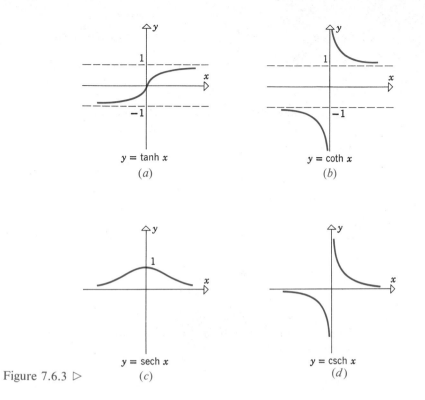

$$y = \tanh x$$
$$(a)$$

$$y = \coth x$$
$$(b)$$

$$y = \operatorname{sech} x$$
$$(c)$$

$$y = \operatorname{csch} x$$
$$(d)$$

Figure 7.6.3 ▷

and if we divide (1) by $\sinh^2 x$ we obtain

$$\coth^2 x - 1 = \operatorname{csch}^2 x \tag{3}$$

The addition formulas for sinh and cosh are

$$\sinh (x + y) = \sinh x \cosh y + \cosh x \sinh y \tag{4a}$$
$$\cosh (x + y) = \cosh x \cosh y + \sinh x \sinh y \tag{4b}$$

These are easier to prove than the corresponding identities for the trigonometric functions. For example, to prove (4a) we use the relations

$$\cosh x + \sinh x = e^x \tag{5a}$$
$$\cosh x - \sinh x = e^{-x} \tag{5b}$$

which follow immediately from Definition 7.6.1. From (5a) and (5b) it follows that

$$\sinh (x + y) = \frac{e^{(x+y)} - e^{-(x+y)}}{2} = \frac{e^x e^y - e^{-x} e^{-y}}{2}$$

$$= \frac{1}{2}[(\cosh x + \sinh x)(\cosh y + \sinh y)$$

$$- (\cosh x - \sinh x)(\cosh y - \sinh y)]$$

$$= \sinh x \cosh y + \cosh x \sinh y$$

which proves (4a). The proof of (4b) is similar.

If we let $x = y$ in (4a) and (4b) we obtain the following analogs of the trigonometric double angle formulas:

$$\sinh 2x = 2 \sinh x \cosh x \tag{6a}$$
$$\cosh 2x = \cosh^2 x + \sinh^2 x \tag{6b}$$

By using (1) and (6b) the reader should be able to show that

$$\cosh 2x = 2 \sinh^2 x + 1 \tag{7a}$$
$$\cosh 2x = 2 \cosh^2 x - 1 \tag{7b}$$

To obtain subtraction formulas for sinh and cosh we shall use the addition formulas and the relations

$$\cosh (-x) = \cosh x \tag{8a}$$
$$\sinh (-x) = -\sinh x \tag{8b}$$

which follow immediately from Definition 7.6.1. If we replace y by $-y$ in (4a) and (4b) and then apply (8a) and (8b) we obtain

$$\sinh (x - y) = \sinh x \cosh y - \cosh x \sinh y \tag{9a}$$
$$\cosh (x - y) = \cosh x \cosh y - \sinh x \sinh y \tag{9b}$$

DERIVATIVE AND INTEGRAL FORMULAS

Derivative formulas for $\sinh x$ and $\cosh x$ follow easily from Definition 7.6.1. For example,

$$\frac{d}{dx}[\cosh x] = \frac{d}{dx}\left[\frac{e^x + e^{-x}}{2}\right] = \frac{1}{2}\left(\frac{d}{dx}[e^x] + \frac{d}{dx}[e^{-x}]\right)$$

$$= \frac{e^x - e^{-x}}{2} = \sinh x$$

Similarly,

$$\frac{d}{dx}[\sinh x] = \cosh x$$

The derivatives of the remaining hyperbolic functions can be obtained by first expressing them in terms of sinh and cosh. For example,

$$\frac{d}{dx}[\tanh x] = \frac{d}{dx}\left[\frac{\sinh x}{\cosh x}\right] = \frac{\cosh x \dfrac{d}{dx}[\sinh x] - \sinh x \dfrac{d}{dx}[\cosh x]}{\cosh^2 x}$$

$$= \frac{\cosh^2 x - \sinh^2 x}{\cosh^2 x} = \frac{1}{\cosh^2 x} = \operatorname{sech}^2 x$$

Derivations of differentiation formulas for the remaining hyperbolic functions are exercises. For reference we summarize the results.

$$\frac{d}{dx}[\sinh x] = \cosh x$$

$$\frac{d}{dx}[\cosh x] = \sinh x$$

$$\frac{d}{dx}[\tanh x] = \operatorname{sech}^2 x$$

$$\frac{d}{dx}[\coth x] = -\operatorname{csch}^2 x$$

$$\frac{d}{dx}[\operatorname{sech} x] = -\operatorname{sech} x \tanh x$$

$$\frac{d}{dx}[\operatorname{csch} x] = -\operatorname{csch} x \coth x$$

REMARK.   Except for a difference in the pattern of signs, these formulas are similar to the formulas for the derivatives of the trigonometric functions.

These differentiation formulas produce the following companion integration formulas.

$$\int \sinh x \, dx = \cosh x + C$$

$$\int \cosh x \, dx = \sinh x + C$$

$$\int \operatorname{sech}^2 x \, dx = \tanh x + C$$

$$\int \operatorname{csch}^2 x \, dx = -\coth x + C$$

$$\int \operatorname{sech} x \tanh x \, dx = -\operatorname{sech} x + C$$

$$\int \operatorname{csch} x \coth x \, dx = -\operatorname{csch} x + C$$

Example 1

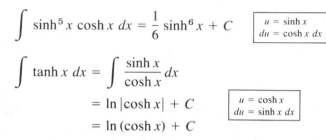

$$\frac{d}{dx}[\cosh(x^3)] = \sinh(x^3) \cdot \frac{d}{dx}[x^3] = 3x^2 \sinh(x^3)$$

$$\frac{d}{dx}[\ln(\tanh x)] = \frac{1}{\tanh x} \cdot \frac{d}{dx}[\tanh x] = \frac{\text{sech}^2 x}{\tanh x} \quad \blacktriangleleft$$

Example 2

Figure 7.6.4

$$\int \sinh^5 x \cosh x \, dx = \frac{1}{6} \sinh^6 x + C \qquad \boxed{\begin{array}{l} u = \sinh x \\ du = \cosh x \, dx \end{array}}$$

$$\int \tanh x \, dx = \int \frac{\sinh x}{\cosh x} \, dx$$

$$= \ln|\cosh x| + C \qquad \boxed{\begin{array}{l} u = \cosh x \\ du = \sinh x \, dx \end{array}}$$

$$= \ln(\cosh x) + C$$

We are justified in writing $|\cosh x| = \cosh x$ because $\cosh x$ is positive for all x. In fact, $\cosh x \geq 1$ for all x. This is geometrically clear from Figure 7.6.1a and can be proved using (1) (Exercise 41).   ◀

We conclude this section with a brief explanation of why the adjective "hyperbolic" is used for hyperbolic functions.

If t is any real number, then the point $(\cos t, \sin t)$ lies on the circle $x^2 + y^2 = 1$ because

$$\cos^2 t + \sin^2 t = 1$$

(Figure 7.6.4). For this reason sine and cosine are called *circular functions*. Analogously, for any real number t the point $(\cosh t, \sinh t)$ lies on the curve $x^2 - y^2 = 1$ because

$$\cosh^2 t - \sinh^2 t = 1$$

Figure 7.6.5

(Figure 7.6.5). As we shall see later, this curve is called a hyperbola and accordingly sinh and cosh are called *hyperbolic functions*.

▶ Exercise Set 7.6

1. In each part, a value for one of the hyperbolic functions is given at x_0. Find the values of the remaining five hyperbolic functions at x_0.

(a) $\sinh x_0 = -2$

(b) $\cosh x_0 = \frac{5}{4}$, $x_0 < 0$

(c) $\tanh x_0 = -\frac{4}{5}$

(d) $\coth x_0 = 2$

(e) $\text{sech} \, x_0 = \frac{15}{17}$, $x_0 > 0$

(f) $\text{csch} \, x_0 = -1$.

In Exercises 2–14, establish the identities.

2. $\cosh(x + y) = \cosh x \cosh y + \sinh x \sinh y.$

3. $\cosh 2x = 2 \sinh^2 x + 1.$

4. $\cosh 2x = 2 \cosh^2 x - 1.$

5. $\cosh(-x) = \cosh x.$

6. $\sinh(-x) = -\sinh x.$

7. $\tanh(x + y) = \dfrac{\tanh x + \tanh y}{1 + \tanh x \tanh y}.$

8. $\tanh(x - y) = \dfrac{\tanh x - \tanh y}{1 - \tanh x \tanh y}.$

9. $\tanh 2x = \dfrac{2 \tanh x}{1 + \tanh^2 x}.$

10. $\cosh \dfrac{1}{2} x = \sqrt{\dfrac{1}{2}(\cosh x + 1)}.$

11. $\sinh \dfrac{1}{2} x = \pm\sqrt{\dfrac{1}{2}(\cosh x - 1)}.$

12. $\sinh x + \sinh y = 2 \sinh\left(\dfrac{x + y}{2}\right) \cosh\left(\dfrac{x - y}{2}\right).$

13. $\cosh x + \cosh y = 2 \cosh\left(\dfrac{x + y}{2}\right) \cosh\left(\dfrac{x - y}{2}\right).$

14. $\cosh 3x = 4 \cosh^3 x - 3 \cosh x.$

15. Derive the differentiation formula for
    (a) $\sinh x$ (b) $\coth x$
    (c) $\operatorname{sech} x$ (d) $\operatorname{csch} x.$

In Exercises 16–25, find dy/dx.

16. $y = \cosh(x^4).$ 17. $y = \sinh(4x - 8).$

18. $y = \ln(\tanh 2x).$ 19. $y = \coth(\ln x).$

20. $y = \operatorname{sech}(e^{2x}).$ 21. $y = \operatorname{csch}(1/x).$

22. $y = \sinh^3(2x).$

23. $y = \sqrt{4x + \cosh^2(5x)}.$

24. $y = \sinh(\cos 3x).$ 25. $y = x^3 \tanh^2(\sqrt{x}).$

In Exercises 26–35, evaluate the integral.

26. $\displaystyle\int \cosh(2x - 3)\, dx.$ 27. $\displaystyle\int \sinh^6 x \cosh x\, dx.$

28. $\displaystyle\int \operatorname{csch}^2(3x)\, dx.$ 29. $\displaystyle\int \sqrt{\tanh x}\, \operatorname{sech}^2 x\, dx.$

30. $\displaystyle\int \coth^2 x\, \operatorname{csch}^2 x\, dx.$

31. $\displaystyle\int \tanh x\, dx.$

32. $\displaystyle\int \dfrac{e^x - e^{-x}}{e^x + e^{-x}}\, dx.$ 33. $\displaystyle\int \tanh x\, \operatorname{sech}^3 x\, dx.$

34. $\displaystyle\int \dfrac{\sinh 2x}{3 + 5 \cosh 2x}\, dx.$

35. $\displaystyle\int \dfrac{\cosh(\sqrt{x})}{\sqrt{x}}\, dx.$

36. In each part, find the exact numerical value:
    (a) $\sinh(\ln 3)$ (b) $\cosh(-\ln 2)$
    (c) $\tanh(2 \ln 5)$ (d) $\sinh(-3 \ln 2).$

37. In each part, express as a rational function:
    (a) $\cosh(\ln x)$ (b) $\sinh(\ln x)$
    (c) $\tanh(2 \ln x)$ (d) $\cosh(-\ln x).$

38. Prove the following facts about $\tanh x$:
    (a) $-1 < \tanh x < 1$ for all x
    (b) $\displaystyle\lim_{x \to +\infty} \tanh x = 1$
    (c) $\displaystyle\lim_{x \to -\infty} \tanh x = -1.$

39. Determine the intervals on which $y = \tanh x$ is positive, negative, increasing, decreasing, concave up, and concave down. Then use these results and Exercise 38 to obtain the graph of $y = \tanh x.$

40. Use the results found in Exercises 38 and 39 to help obtain the graph of $y = \coth x.$

41. (a) Prove: $\cosh x \geq 1$ for all $x.$
    (b) Prove: $0 < \operatorname{sech} x \leq 1$ for all $x.$
    (c) Obtain the graph of $y = \operatorname{sech} x.$

42. (a) Show that $\displaystyle\lim_{x \to +\infty} \operatorname{csch} x = \lim_{x \to -\infty} \operatorname{csch} x = 0.$
    (b) Show that
    $$\lim_{x \to 0^+} \operatorname{csch} x = +\infty \quad \text{and} \quad \lim_{x \to 0^-} \operatorname{csch} x = -\infty$$
    (c) Obtain the graph of $y = \operatorname{csch} x.$

43. Prove: $(\sinh x + \cosh x)^n = \sinh nx + \cosh nx.$

44. (a) It is sometimes said that $\sinh x$ and $\cosh x$ "behave like $\frac{1}{2}e^x$" for x large and positive. Give an informal justification of this statement.
    (b) Make a sketch showing the graphs of $y = \sinh x$, $y = \cosh x$, and $y = \frac{1}{2}e^x$, all in the same figure.

45. Find the arc length of $y = \cosh x$ between $x = 0$ and $x = \ln 2.$

46. Find the arc length of the catenary $y = a \cosh(x/a)$ between $x = 0$ and $x = x_1$ $(x_1 > 0)$.

47. Find the volume of the solid that is generated when the region enclosed by $y = \cosh 2x$, $y = \sinh 2x$, $x = 0$, and $x = 5$ is revolved about the x-axis.

7.7   FIRST-ORDER DIFFERENTIAL EQUATIONS AND APPLICATIONS

*In this section we shall begin to study equations that involve an unknown function and its derivatives. These are called **differential equations**. Such equations govern many of the fundamental principles in science and engineering, and their study constitutes a major field of mathematics. Our work in this section is limited to the most basic types of differential equations. In the final chapter of this text we shall go a little further into this topic.*

Some examples of differential equations are

$$\frac{dy}{dx} = 3y \tag{1}$$

$$\frac{d^2y}{dx^2} - 6\frac{dy}{dx} + 8y = 0 \tag{2}$$

$$y' - y = e^{2x} \tag{3}$$

$$\frac{d^3y}{dt^3} - t\frac{dy}{dt} + (t^2 - 1)y = e^t \tag{4}$$

In the first three equations, $y = y(x)$ is an unknown function of x and in the last equation $y = y(t)$ is an unknown function of t. The **order** of a differential equation is the order of the highest derivative that appears in the equation. Thus, (1) and (3) are first-order equations, (2) is second order, and (4) is third order.

A function $y = y(x)$ is a **solution** of a differential equation if the equation is satisfied when $y(x)$ and its derivatives are substituted. For example,

$$y = e^{2x} \tag{5}$$

is a solution of the equation

$$\frac{dy}{dx} - y = e^{2x} \tag{6}$$

since

$$\frac{dy}{dx} - y = 2e^{2x} - e^{2x} = e^{2x}$$

Similarly,

$$y = e^x + e^{2x} \tag{7}$$

is a solution. More generally,

$$y = Ce^x + e^{2x} \tag{8}$$

is a solution for any constant C (verify). This solution is of special importance since it can be proved that *all* solutions of (6) can be obtained by substituting values for the arbitrary constant C. For example, $C = 0$ yields solution (5) and $C = 1$ yields solution (7). A solution of a differential equation from which all other solutions can be derived by substituting values for arbitrary constants is called the **general solution** of the equation. Usually, the general solution of an nth-order equation contains n arbitrary constants. Thus, (8), which is the general solution of the first-order equation (6), contains one arbitrary constant.

Often, solutions of differential equations are expressed as implicitly defined functions. For example,

$$\ln y = xy + C \tag{9}$$

defines a solution of

$$\frac{dy}{dx} = \frac{y^2}{1 - xy} \tag{10}$$

for any value of the constant C, since implicit differentiation of (9) yields

$$\frac{1}{y}\frac{dy}{dx} = x\frac{dy}{dx} + y$$

or

$$\frac{dy}{dx} - xy\frac{dy}{dx} = y^2$$

from which (10) follows.

FIRST-ORDER SEPARABLE EQUATIONS   A first-order differential equation is called *separable* if it is expressible in the form

$$\frac{dy}{dx} = \frac{g(x)}{h(y)}$$

To solve such an equation we rewrite it in the differential form

$$h(y)\, dy = g(x)\, dx \tag{11}$$

and integrate both sides, thereby obtaining the general solution

$$\int h(y)\, dy = \int g(x)\, dx + C$$

where C is an arbitrary constant.

In (11), the x and y variables are "separated" from each other, hence the term *separable* differential equation.

Example 1   Solve the equation

$$\frac{dy}{dx} = \frac{x}{y^2}$$

Solution.   Changing to differential form and integrating yields

$$y^2\, dy = x\, dx$$

$$\int y^2\, dy = \int x\, dx$$

$$\frac{y^3}{3} = \frac{x^2}{2} + C$$

This expresses the solution implicitly. If desired, we can solve explicitly for y to obtain

$$y = \left(\frac{3}{2} x^2 + 3C \right)^{1/3}$$

or

$$y = \left(\frac{3}{2} x^2 + K \right)^{1/3}$$

where $K(= 3C)$ is an arbitrary constant.   ◀

Example 2   Solve the equation

$$x(y - 1)\frac{dy}{dx} = y$$

Solution.   Separating the variables and integrating yields

$$x(y - 1)\, dy = y\, dx$$

$$\frac{y - 1}{y}\, dy = \frac{dx}{x}$$

$$\int \frac{y - 1}{y}\, dy = \int \frac{dx}{x}$$

$$\int \left(1 - \frac{1}{y}\right) dy = \int \frac{dx}{x}$$

$$y - \ln|y| = \ln|x| + C$$

or, using properties of logarithms,

$$y = \ln|xy| + C \quad \blacktriangleleft$$

INITIAL-VALUE
PROBLEMS

When a physical problem leads to a differential equation, there are usually conditions in the problem that determine specific values for the arbitrary constants in the general solution of the equation. For a first-order equation, a condition that specifies the value of the unknown function $y(x)$ at some point $x = x_0$ is called an *initial condition*. A first-order differential equation together with one initial condition constitutes a *first-order initial-value problem*.

Example 3 Solve the initial-value problem

$$\frac{dy}{dx} = -4xy^2, \quad y(0) = 1$$

Solution. We first solve the differential equation:

$$\frac{dy}{y^2} = -4x \, dx$$

$$\int \frac{dy}{y^2} = \int -4x \, dx$$

$$-\frac{1}{y} = -2x^2 + C_1$$

or on multiplying by -1, taking reciprocals, and writing C in place of $-C_1$,

$$y = \frac{1}{2x^2 + C} \tag{12}$$

The initial condition, $y(0) = 1$, requires that $y = 1$ when $x = 0$. Substituting these values in (12) yields $C = 1$. Thus, the solution of the initial-value problem is

$$y = \frac{1}{2x^2 + 1} \quad \blacktriangleleft$$

FIRST-ORDER LINEAR
EQUATIONS

Not every first-order differential equation is separable. For example, it is impossible to separate the variables in the equation

$$\frac{dy}{dx} + x^2y = e^x$$

However, this equation can be solved by a different method that we shall now consider.

A first-order differential equation is called *linear* if it is expressible in the form

$$\frac{dy}{dx} + p(x)y = q(x) \tag{13}$$

where the functions $p(x)$ and $q(x)$ may or may not be constant. Some examples are

$$\frac{dy}{dx} + x^2y = e^x \qquad\qquad [p(x) = x^2, q(x) = e^x]$$

$$y' - 3e^xy = 0 \qquad\qquad [p(x) = -3e^x, q(x) = 0]$$

$$\frac{dy}{dx} + 5y = 2 \qquad\qquad [p(x) = 5, q(x) = 2]$$

$$\frac{dy}{dx} + (\sin x)y + x^3 = 0 \qquad [p(x) = \sin x, q(x) = -x^3]$$

One procedure for solving (13) is based on the observation that if we define $\rho = \rho(x)$ by

$$\rho = e^{\int p(x)\,dx}$$

then

$$\frac{d\rho}{dx} = e^{\int p(x)\,dx} \cdot \frac{d}{dx}\int p(x)\,dx = \rho p(x)$$

Thus,

$$\frac{d}{dx}(\rho y) = \rho\frac{dy}{dx} + \frac{d\rho}{dx}y = \rho\frac{dy}{dx} + \rho p(x)y \tag{14}$$

If (13) is multiplied through by ρ, it becomes

$$\rho\frac{dy}{dx} + \rho p(x)y = \rho q(x)$$

or from (14),

$$\frac{d}{dx}(\rho y) = \rho q(x)$$

This equation can be solved by integrating both sides to obtain

$$\rho y = \int \rho q(x) \, dx + C$$

or

$$y = \frac{1}{\rho} \left[\int \rho q(x) \, dx + C \right]$$

To summarize, (13) can be solved in three steps:

Step 1. Calculate

$$\rho = e^{\int p(x) \, dx}$$

This is called the *integrating factor.* Since any ρ will suffice, we can take the constant of integration to be zero in this step.

Step 2. Multiply both sides of (13) by ρ and express the result as

$$\frac{d}{dx}(\rho y) = \rho q(x)$$

Step 3. Integrate both sides of the equation obtained in Step 2 and then solve for y. Be sure to include a constant of integration in this step.

Example 4 Solve the equation

$$\frac{dy}{dx} - 4xy = x \tag{15}$$

Solution. Since $p(x) = -4x$, the integrating factor is

$$\rho = e^{\int(-4x) \, dx} = e^{-2x^2}$$

If we multiply (15) by ρ we obtain

$$\frac{d}{dx}(e^{-2x^2}y) = xe^{-2x^2}$$

Integrating both sides of this equation yields

$$e^{-2x^2}y = \int xe^{-2x^2} \, dx = -\frac{1}{4}e^{-2x^2} + C$$

and then multiplying both sides by e^{2x^2} yields

$$y = -\frac{1}{4} + Ce^{2x^2} \qquad \blacktriangleleft$$

Example 5 Solve the equation

$$x\frac{dy}{dx} - y = x \quad (x > 0)$$

Solution. To put the equation in form (13), we divide through by x to obtain

$$\frac{dy}{dx} - \frac{1}{x}y = 1 \tag{16}$$

Since $p(x) = -1/x$, the integrating factor is

$$\rho = e^{\int -(1/x)\,dx} = e^{-\ln|x|} = \frac{1}{|x|} = \frac{1}{x}$$

(The absolute value was dropped because of the assumption that $x > 0$.) If we multiply (16) by ρ, we obtain

$$\frac{d}{dx}\left(\frac{1}{x}y\right) = \frac{1}{x}$$

Integrating both sides of this equation yields

$$\frac{1}{x}y = \int \frac{1}{x}\,dx = \ln x + C$$

or

$$y = x\ln x + Cx \qquad \blacktriangleleft$$

APPLICATIONS   We conclude this section with some applications of first-order differential equations.

Example 6 (*Geometry*)   Find a curve in the xy-plane that passes through $(0, 3)$ and whose tangent line at a point (x, y) has slope $2x/y^2$.

Solution. Since the slope of the tangent line is dy/dx, we have

$$\frac{dy}{dx} = \frac{2x}{y^2} \tag{17}$$

and, since the curve passes through $(0, 3)$, we have the initial condition

$$y(0) = 3 \tag{18}$$

Equation (17) is separable and can be written as

$$y^2\,dy = 2x\,dx$$

so

$$\int y^2 \, dy = \int 2x \, dx$$

or

$$\frac{1}{3} y^3 = x^2 + C$$

From the initial condition, (18), it follows that $C = 9$, and so the curve has the equation

$$\frac{1}{3} y^3 = x^2 + 9 \qquad \text{or} \qquad y = (3x^2 + 27)^{1/3} \qquad \blacktriangleleft$$

Example 7 (*Mixing Problems*)   At time $t = 0$, a tank contains 4 lb of salt dissolved in 100 gal of water. Suppose that brine containing 2 lb of salt per gallon of water is allowed to enter the tank at a rate of 5 gal/min and that the mixed solution is drained from the tank at the same rate. Find a formula for the amount of salt in the tank after 10 min.

Solution.   Let $y(t)$ be the amount of salt (in pounds) at time t. We are interested in finding $y(10)$, the amount of salt at time $t = 10$. We will begin by finding an expression for dy/dt, the rate of change of the amount of salt in the tank at time t. Clearly,

$$\frac{dy}{dt} = \text{rate in} - \text{rate out} \qquad\qquad (19)$$

where *rate in* is the rate at which salt enters the tank and *rate out* is the rate at which salt leaves the tank. But

$$\text{rate in} = (2 \text{ lb/gal}) \cdot (5 \text{ gal/min}) = 10 \text{ lb/min}$$

At time t, the mixture contains $y(t)$ pounds of salt in 100 gallons of water; thus, the concentration of salt at time t is $y(t)/100$ lb per gal and

$$\text{rate out} = \left(\frac{y(t)}{100} \text{ lb/gal} \right) \cdot (5 \text{ gal/min}) = \frac{y(t)}{20} \text{ lb/min}$$

Therefore, (19) can be written as

$$\frac{dy}{dt} = 10 - \frac{y}{20}$$

or

$$\frac{dy}{dt} + \frac{y}{20} = 10 \qquad\qquad (20)$$

which is a first-order linear differential equation. We also have the initial condition

$$y(0) = 4 \tag{21}$$

since the tank contains 4 lb of salt at time $t = 0$.

Multiplying both sides of (20) by the integrating factor

$$\rho = e^{\int (1/20)\, dt} = e^{t/20}$$

yields

$$\frac{d}{dt}\left(e^{t/20}y\right) = 10e^{t/20}$$

so

$$e^{t/20}y = \int 10e^{t/20}\, dt = 200e^{t/20} + C$$

or

$$y(t) = 200 + Ce^{-t/20}$$

From the initial condition, (21), it follows that

$$4 = 200 + C$$

or $C = -196$, so

$$y(t) = 200 - 196e^{-t/20}$$

Thus, after 10 min ($t = 10$), the amount of salt in the tank is

$$y(10) = 200 - 196e^{-0.5} \approx 81.1 \text{ lb} \qquad \blacktriangleleft$$

EXPONENTIAL
GROWTH

Many quantities increase or decrease with time in proportion to the amount of the quantity present. Some examples are human population, bacteria in a culture, drug concentration in the bloodstream, radioactivity, and the values of certain kinds of investments. We shall show how differential equations can be used to study the growth and decay of such quantities.

7.7.1 DEFINITION. A quantity is said to have an *exponential growth* *(decay) model* if at each instant of time its rate of increase (decrease) is proportional to the amount of the quantity present.

Consider a quantity with an exponential growth or decay model and denote by $y(t)$ the amount of the quantity present at time t. We assume that $y(t) > 0$ for all t. Since the rate of change of $y(t)$ is proportional to the amount present, it follows that $y(t)$ satisfies

$$\frac{dy}{dt} = ky \tag{22}$$

where k is a constant of proportionality.

The constant k in (22) is called the **growth constant** if $k > 0$ and the **decay constant** if $k < 0$. If $k > 0$, then $dy/dt > 0$, so that y is increasing with time (a growth model). If $k < 0$, then $dy/dt < 0$, so that y is decreasing with time (a decay model).

REMARK.   In problems of exponential growth and decay, the constant k is often expressed as a percentage. Thus, a growth constant of 3% means $k = 0.03$ and a decay constant of -500% means $k = -5$. We also note that the constant k is often called the **growth rate**. Strictly speaking, this is not correct; as indicated by (22), the growth rate dy/dt is not k, but rather ky. However, the description of k as a growth rate is so common that we will use this standard terminology in this section.

If we are given a quantity with an exponential growth or decay model, and if we know the amount y_0 present at some initial time $t = 0$, then we can find the amount present at any time t by solving the initial-value problem

$$\frac{dy}{dt} = ky, \quad y(0) = y_0 \tag{23}$$

This differential equation is both separable and linear and thus can be solved by either of the methods we have studied in this section. We shall treat it as a linear equation. Rewriting the differential equation as

$$\frac{dy}{dt} - ky = 0$$

and multiplying through by the integrating factor

$$\rho = e^{\int -k\,dt} = e^{-kt}$$

yields

$$\frac{d}{dt}(e^{-kt}y) = 0$$

After integrating,

$$e^{-kt}y = C$$

or

$$y = Ce^{kt}$$

From the initial condition, $y(0) = y_0$, it follows that $C = y_0$; thus the solution of (23) is

$$y(t) = y_0 e^{kt} \tag{24}$$

Exponential models have proved useful in studies of population growth. Although populations (e.g., people, bacteria, and flowers) grow in discrete steps, we can apply the results of this section if we are willing to approximate the population graph by a continuous curve $y = y(t)$ (Figure 7.7.1).

Example 8 (*Population Growth*)   According to United Nations data, the world population at the beginning of 1975 was approximately 4 billion and growing at a rate of about 2% per year. Assuming an exponential growth model, estimate the world population at the beginning of the year 2000.

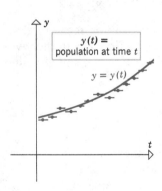

Figure 7.7.1

Solution. Let

t = time elapsed from the beginning of 1975 (in years)

y = world population (in billions)

Since the beginning of 1975 corresponds to $t = 0$, it follows from the given data that

$$y_0 = y(0) = 4 \text{ (billion)}$$

Since the growth rate is 2% ($k = 0.02$), it follows from (24) that the world population at time t will be

$$y(t) = y_0 e^{kt} = 4e^{0.02t} \tag{25}$$

Since the beginning of the year 2000 corresponds to an elapsed time of $t = 25$ years ($2000 - 1975 = 25$), it follows from (25) that the world population by the year 2000 will be

$$y(25) = 4e^{0.02(25)} = 4e^{0.5}$$

or from Appendix 3, Table 2,

$$y(25) = 4(1.6487) = 6.5948 \text{ (billion)}$$

which is a population of approximately 6.6 billion.   ◄

If a quantity has an exponential growth model, then the time required for it to double in size is called the **doubling time**. Similarly, if a quantity has an exponential decay model, then the time required for it to reduce in value by half is called the **halving time**. As it turns out, doubling and halving times depend only on the growth rate and not on the amount present initially. To see why, suppose y has an exponential growth model so that

$$y = y_0 e^{kt} \quad (k > 0)$$

At any fixed time t_1 let

$$y_1 = y_0 e^{kt_1} \tag{26}$$

be the value of y, and let T denote the amount of time required for y to double in size. Thus, at time $t_1 + T$ the value of y will be $2y_1$ so that

$$2y_1 = y_0 e^{k(t_1+T)} = y_0 e^{kt_1} e^{kT}$$

or, from (26),

$$2y_1 = y_1 e^{kT}$$

Thus,

$$2 = e^{kT} \quad \text{and} \quad \ln 2 = kT$$

Therefore, the doubling time T is

Doubling Time
$$T = \frac{1}{k} \ln 2 \tag{27}$$

which does not depend on y_0 or t_1. We leave it as an exercise to show that the halving time for a quantity with an exponential decay model $(k < 0)$ is

Halving Time
$$T = -\frac{1}{k} \ln 2 \tag{28}$$

Example 9   It follows from (27) that at the current 2% annual growth rate, the doubling time for the world population is

$$T = \frac{1}{0.02} \ln 2 \approx \frac{1}{0.02}(0.6931) = 34.655 \quad \boxed{\text{Appendix 3, Table 3}}$$

or approximately 35 years. Thus, with a continued 2% annual growth rate the population of 4 billion in 1975 will double to 8 billion by the year 2010 and will double again to 16 billion by 2045.   ◄

Radioactive elements continually undergo a process of disintegration called *radioactive decay*. It is a physical fact that at each instant of time the rate of decay is proportional to the amount of the element present. Consequently, the amount of any radioactive element has an exponential decay model. For radioactive elements, halving time is called *half-life*.

Example 10 (*Radioactive Decay*)   The radioactive element carbon-14 has a half-life of 5750 years. If 100 grams of this element are present initially, how much will be left after 1000 years?

Solution.   From (28) the decay constant is

$$k = -\frac{1}{T} \ln 2 \approx -\frac{1}{5750}(0.6931) \approx -0.00012 \quad \boxed{\text{Appendix 3, Table 3}}$$

Thus, if we take $t = 0$ to be the present time, then $y_0 = y(0) = 100$, so that (24) implies that the amount of carbon-14 after 1000 years will be

$$y(1000) = 100e^{-0.00012(1000)} = 100e^{-0.12} \approx 100(0.88692) = 88.692$$

Thus, about 88.69 grams of carbon-14 will remain.   ◄

► Exercise Set 7.7

In Exercises 1–6, solve the given separable differential equation. Where convenient, express the solution explicitly as a function of x.

1. $\dfrac{dy}{dx} = \dfrac{y}{x}$.

2. $\dfrac{dy}{dx} = \dfrac{x^3}{(1 + x^4)y}$.

3. $\sqrt{1 + x^2}\, y' + x(1 + y) = 0$.

4. $3 \tan y - \dfrac{dy}{dx} \sec x = 0$.

5. $e^{-y} \sin x - y' \cos^2 x = 0$.

6. $\dfrac{dy}{dx} = 1 - y + x^2 - yx^2$.

In Exercises 7–12, solve the given first-order linear differential equation. Where convenient, express the solution explicitly as a function of x.

7. $\dfrac{dy}{dx} + 3y = e^{-2x}$.

8. $\dfrac{dy}{dx} - \dfrac{5}{x}y = x \quad (x > 0)$.

9. $y' + y = \cos(e^x)$.

10. $2\dfrac{dy}{dx} + 4y = 1$.

11. $x^2 y' + 3xy + 2x^5 = 0 \quad (x > 0)$.

12. $\dfrac{dy}{dx} + y - \dfrac{1}{1 + e^x} = 0$.

In Exercises 13–18, solve the initial-value problems.

13. $\dfrac{dy}{dx} - xy = x, \quad y(0) = 3$.

14. $2y\dfrac{dy}{dx} = 3x^2(x^3 + 1)^{-1/2}, \quad y(2) = 1$.

15. $\dfrac{dy}{dt} + y = 2, \quad y(0) = 1$.

16. $y' - xe^y = 2e^y, \quad y(0) = 0$.

17. $y^2 t\dfrac{dy}{dt} - t + 1 = 0, \quad y(1) = 3$.

18. $y' \cosh x + y \sinh x = \cosh^2 x, \quad y(0) = \dfrac{1}{4}$.

19. Find a curve in the xy-plane that passes through the point $(1, 2)$ and whose tangent at a point (x, y) has slope $2 - (y/x)$.

20. Find a curve in the xy-plane that passes through the point $(1, 1)$ and whose normal at a point (x, y) has slope $-2y/(3x^2)$.

21. The line normal to a curve at any point (x, y) passes through $(3, 0)$. Given that the curve passes through $(3, 2)$, find its equation.

22. A function f with positive values has the property that its value at every point is twice the slope

of the tangent at the point. Given that $f(0) = 2$, find f.

In Exercises 23–40, use a calculator or the tables in the appendix where needed.

23. At time $t = 0$, a tank contains 25 lb of salt dissolved in 50 gal of water. Then brine containing 4 lb of salt per gallon of water is allowed to enter the tank at a rate of 2 gal/min and the mixed solution is drained from the tank at the same rate.
    (a) How much salt is in the tank at an arbitrary time t?
    (b) How much salt is in the tank after 25 min?

24. A tank initially contains 200 gal of pure water. Then at time $t = 0$ brine containing 5 lb of salt per gallon of water is allowed to enter the tank at a rate of 10 gal/min and the mixed solution is drained from the tank at the same rate.
    (a) How much salt is in the tank at an arbitrary time t?
    (b) How much salt is in the tank after 30 min?

25. A tank with a 1000-gal capacity initially contains 500 gal of brine containing 50 lb of salt. At time $t = 0$, pure water is added at a rate of 20 gal/min and the mixed solution is drained off at a rate of 10 gal/min. How much salt is in the tank when it reaches the point of overflowing?

26. The number of bacteria in a certain culture grows exponentially at a rate of 1% per hour. Assuming that 10,000 bacteria are present initially, find
    (a) the number of bacteria present at any time t;
    (b) the number of bacteria present after 5 hr;
    (c) the time required for the number of bacteria to reach 45,000.

27. Polonium-210 is a radioactive element with a half-life of 140 days. Assume a sample weighs 10 mg initially.
    (a) Find a formula for the amount that will remain after t days.
    (b) How much will remain after 10 weeks?

28. In a certain chemical reaction a substance decomposes at a rate proportional to the amount present. Tests show that under appropriate conditions 15,000 grams will reduce to 5000 grams in 10 hours.

    (a) Find a formula for the amount that will remain from a 15,000 g sample after t hours.
    (b) How long will it take for 50% of an initial sample of y_0 grams to decompose?

29. One hundred fruit flies are placed in a breeding container that can support a population of at most 5000 flies. If the population grows exponentially at a rate of 2% per day, how long will it take for the container to reach capacity?

30. In 1960 the American scientist W. F. Libby won the Nobel prize for his discovery of carbon dating, a method for determining the age of certain fossils. Carbon dating is based on the fact that nitrogen is converted to radioactive carbon-14 by cosmic radiation in the upper atmosphere. This radioactive carbon is absorbed by plant and animal tissue through the life processes while the plant or animal lives. However, when the plant or animal dies the absorption process stops and the amount of carbon-14 decreases through radioactive decay. Suppose that tests on a fossil show that 70% of its carbon-14 has decayed. Estimate the age of the fossil, assuming a half-life of 5750 years for carbon-14.

31. Forty percent of a radioactive substance decays in 5 years. Find the half-life of the substance.

32. Assume that if the temperature is constant, then the atmospheric pressure p varies with the altitude h (above sea level) in such a way that

$$\frac{dp}{dh} = kp$$

where k is a constant.
    (a) Find a formula for p in terms of k, h, and the atmospheric pressure p_0 at sea level.
    (b) Given that p measures 15 lb/in.^{2} at sea level and 12 lb/in.^{2} at 5000 ft above sea level, find the pressure at 10,000 ft (assuming temperature is constant).

33. The town of Grayrock had a population of 10,000 in 1960 and 12,000 in 1970.
    (a) Assuming an exponential growth model, estimate the population in 1980.
    (b) What is the doubling time for the town's population?

34. Prove: If a quantity A has an exponential growth or decay model and A has values A_1 and A_2 at times t_1 and t_2, respectively, then the growth rate k is

$$k = \frac{1}{t_1 - t_2} \ln\left(\frac{A_1}{A_2}\right)$$

35. Newton's law of cooling states that the rate at which an object cools is proportional to the difference in temperature between the object and the surrounding medium. Show that if C is the constant temperature of a surrounding medium, then the temperature $T(t)$ at time t of a cooling object is given by

$$T(t) = (T_0 - C)e^{kt} + C$$

where T_0 is the temperature of the object at $t = 0$ and k is a negative constant.

36. A liquid with initial temperature $200°$ is surrounded by a body of air at a constant temperature of $80°$. If the liquid cools to $120°$ in 30 min, what will the temperature be after 1 hr? (Use the result of Exercise 35.)

37. Suppose P dollars is invested at an annual interest rate of $r \times 100\%$. If the accumulated interest is credited to the account at the end of the year, then the interest is said to be *compounded annually;* if it is credited at the end of each six-month period then it is said to be *compounded semiannually;* and if it is credited at the end of each three-month period, then it is said to be *compounded quarterly.* The more frequently the interest is compounded, the better it is for the investor since more of the interest is itself earning interest.

   (a)   Show that if interest is compounded n times a year at equally spaced intervals, then the value A of the investment after t years is

$$A = P\left(1 + \frac{r}{n}\right)^{nt}$$

   (b)   One can imagine interest to be compounded each day, each hour, each minute, and so forth. Carried to the limit one can conceive of interest compounded at each instant of time; this is called *continuous compounding.* Thus, from (a), the value A of P dollars after t years when invested at an annual rate of $r \times 100\%$, compounded continuously, is

$$A = \lim_{n \to +\infty} P\left(1 + \frac{r}{n}\right)^{nt}$$

   Use the fact that $\lim_{x \to 0}(1 + x)^{1/x} = e$ to prove that

$$A = Pe^{rt}$$

   (c)   Use the result in (b) to show that money invested at continuous compound interest increases at a rate proportional to the amount present.

38. (a)   If \$1000 is invested at 8% per year compounded continuously (Exercise 37) what will the investment be worth after 5 years?

   (b)   If it is desired that an investment at 8% per year compounded continuously should have a value of \$10,000 after 10 years, how much should be invested now?

   (c)   How long does it take for an investment at 8% per year compounded continuously to double in value?

39. Derive Formula (28) for halving time.

40. Let a quantity have an exponential growth model with growth rate k. How long does it take for the quantity to triple in size?

▶ SUPPLEMENTARY EXERCISES

1. In each part determine whether f and g are inverse functions.

   (a) $f(x) = mx$ $g(x) = 1/(mx)$
   (b) $f(x) = 3/(x + 1)$ $g(x) = (3 - x)/x$
   (c) $f(x) = x^3 - 8$ $g(x) = \sqrt[3]{x} + 2$
   (d) $f(x) = x^3 - 1$ $g(x) = \sqrt[3]{x} + 1$
   (e) $f(x) = \sqrt{e^x}$ $g(x) = 2 \ln x.$

In Exercises 2–6, find $f^{-1}(x)$ if it exists.

2. $f(x) = 8x^3 - 1$.

3. $f(x) = x^2 - 2x + 1$.

4. $f(x) = x^2 - 2x + 1$, $x \geq 1$.

5. $f(x) = (e^x)^2 + 1$.

6. $f(x) = \exp(x^2) + 1$.

7. Let $f(x) = (ax + b)/(cx + d)$.
    What conditions on a, b, c, d guarantee that f^{-1} exists? Find $f^{-1}(x)$.

8. Show that $f(x) = (x + 2)/(x - 1)$ is its own inverse.

9. Find the largest open interval containing the origin on which f is one-to-one.
    (a) $f(x) = |2x - 5|$ (b) $f(x) = x^2 + 4x$
    (c) $f(x) = \cos(x - 2\pi/3)$.

In Exercises 10–13, find $f^{-1}(x)$, and then use Formula (5) of Section 7.1 to obtain $(f^{-1})'(x)$. Check your work by differentiating $f^{-1}(x)$ directly.

10. $f(x) = x^3 - 8$.

11. $f(x) = 3/(x + 1)$.

12. $f(x) = mx + b$ $(m \neq 0)$.

13. $f(x) = \sqrt{e^x}$.

14. Prove that the line $y = x$ is the perpendicular bisector of the line segment joining (a, b) and (b, a).

15. If $r = \ln 2$ and $s = \ln 3$, express the following in terms of r and s:
    (a) $\ln(1/12)$ (b) $\ln(9/\sqrt{8})$
    (c) $\ln(\sqrt[4]{8/3})$.

16. Simplify.
    (a) $e^{2 - \ln x}$ (b) $\exp(\ln x^2 - 2\ln y)$
    (c) $\ln[x^3 \exp(-x^2)]$.

17. Solve for x in terms of $\ln 3$ and $\ln 5$.
    (a) $25^x = 3^{1-x}$ (b) $\sinh x = \frac{1}{4}\cosh x$.

18. Express the following as a rational function of x:
    $3\ln(e^{2x}(e^x)^3) + 2\exp(\ln 1)$.

19. If $\sinh x = -3/5$, find:
    (a) $\cosh x$ (b) $\tanh x$
    (c) $\sinh(2x)$.

20. Use appendix tables to approximate the following quantities to two decimal places.
    (a) 2^e (b) $(\sqrt{2})^\pi$

In Exercises 21–42, find dy/dx. When appropriate, use implicit or logarithmic differentiation.

21. $y = 1/\sqrt{e^x}$.

22. $y = 1/e^{\sqrt{x}}$.

23. $y = x/\ln x$.

24. $y = e^x \ln(1/x)$.

25. $y = x/e^{\ln x}$.

26. $y = \ln\sqrt{x^2 + 2x}$.

27. $y = \ln(10^x/\sin x)$.

28. $y = \cos(e^{-2x})$.

29. $y = e^{\tan x}e^{4\ln x}$.

30. $y = \ln\left|\dfrac{a + x}{a - x}\right|$.

31. $y = \ln|x + \sqrt{x^2 + a^2}|$.

32. $y = \ln|\tan 3x + \sec 3x|$.

33. $y = [\exp(x^2)]^3$.

34. $y = \ln(x^3/\sqrt{5 + \sin x})$.

35. $y = \sqrt{\ln(\sqrt{x})}$.

36. $y = e^{5x} + (5x)^e$.

37. $y = \pi^x x^\pi$.

38. $y = 4(e^x)^3/\sqrt{\exp(5x)}$.

39. $y = \sinh[\tanh(5x)]$.

40. $x^4 + e^{xy} - y^2 = 20$.

41. $y = e^{3x}(1 + e^{-x})^2$.

42. $y = (\cosh x)^{x^3}$.

43. Show that the function $y = e^{ax}\sin bx$ satisfies $y'' - 2ay' + (a^2 + b^2)y = 0$ for any real constants a and b.

44. Assuming that u and v are differentiable functions of x, use logarithmic differentiation to derive a formula for $d(u^v)/dx$ in terms of u, v, du/dx, and dv/dx. What formula results when the base u is constant? When the exponent v is constant?

45. Find dy if
    (a) $y = e^{-x}$ (b) $y = \ln(1 + x)$ (c) $y = 2^{x^2}$.

46. If $y = 3^{2x}5^{7x}$, show that dy/dx is proportional to y.

47. Use the chain rule and the Second Fundamental Theorem of Calculus to find the derivative.
    (a) $\dfrac{d}{dx}\left(\displaystyle\int_0^{\ln x} \dfrac{dt}{\sqrt{4 + e^t}}\right)$
    (b) $\dfrac{d}{dx}\left(\displaystyle\int_1^{e^{5x}} \sqrt{\ln u + u}\; du\right)$.
    [*Hint:* See Exercise 38, Section 5.9.]

48. Suppose $F(-\pi) = 0$ and that $y = F(x)$ satisfies $dy/dx = (3\sin^2 2x + 2\cos^2 3x)^{1/2}$. Express $F(x)$ as an integral.

49. If $y = Ce^{kt}$ (C, k constant) and $Y = \ln y$, show that the graph of Y versus t is a straight line.

50. Evaluate $\int e^{2x}(4 + e^{2x})\,dx$ two ways: (a) using the substitution $u = 4 + e^{2x}$, and (b) expanding the integrand. Verify that the antiderivatives in (a) and (b) differ by a constant.

In Exercises 51–66, evaluate the indicated integral.

51. $\int \dfrac{e^x}{1 + e^x} \, dx.$

52. $\int \dfrac{1 + e^x}{e^x} \, dx.$

53. $\int x^e \, dx.$

54. $\int \dfrac{x^2}{5 - 2x^3} \, dx.$

55. $\int \dfrac{4x^2 - 3x}{x^3} \, dx.$

56. $\int \dfrac{(\ln x^2)^2}{x} \, dx.$

57. $\int \dfrac{\sec x \tan x}{2 \sec x - 1} \, dx.$

58. $\int \dfrac{e^{5x}}{3 + e^{5x}} \, dx.$

59. $\int (\cos 2x) \exp(\sin 2x) \, dx.$

60. $\int \tanh(3x + 1) \, dx.$

61. $\int \operatorname{sech}^2 x \tanh x \, dx.$

62. $\int e^{2x}(4 + e^{-3x}) \, dx.$

63. $\int_e^{e^2} \dfrac{dx}{x \ln x}.$

64. $\int_0^1 \dfrac{dx}{\sqrt{e^x}}.$

65. $\int_0^{\pi/4} \dfrac{2^{\tan x}}{\cos^2 x} \, dx.$

66. $\int_1^4 \dfrac{dx}{\sqrt{x} e^{\sqrt{x}}}.$

67. Show that $\int e^{kx} \, dx = (e^{kx}/k) + C$ for any nonzero constant k.

68. In each part find the indicated limit by interpreting the expression as a derivative and evaluating the derivative.

(a) $\displaystyle\lim_{h \to 0} \dfrac{10^h - 1}{h}$

(b) $\displaystyle\lim_{h \to 0} \dfrac{e^{(3+h)^2} - e^9}{h}$

(c) $\displaystyle\lim_{h \to 0} \dfrac{\ln(e^2 + h) - 2}{h}$

(d) $\displaystyle\lim_{x \to 1} \dfrac{2^x - 2}{x - 1}.$

69. Sketch $y = x^3 e^{-x}$, showing all relative extrema and inflection points. You may assume that $\lim\limits_{x \to +\infty} x^3 e^{-x} = 0$.

70. Use the second derivative test to determine the nature of the critical points of $f(x) = x^2 e^{-x}$.

71. Show that if $y = f(x)$ satisfies $y' = e^y + 2y + x$, then every critical point of y is a relative minimum.

72. Sketch the curve $y = e^{-x/2} \sin 2x$ on $[-\pi/2, 3\pi/2]$. Show all x-intercepts and all points where $y = \pm e^{-x/2}$.

73. Let A denote the area under the curve $y = e^{-2x}$ over the interval $[0, b]$. Express A as a function of b, and then find the value of b for which $A = \frac{1}{4}$. What is the limiting value of A as $b \to +\infty$.

74. Let R be the region bounded by the curve $y = 4(2x - 1)^{-1/2}$, the x-axis, and the lines $x = 1$ and $x = 3$. Find the volume of the solid generated by revolving R about the x-axis.

75. Find the surface area generated when the curve $y = \cosh x$, $0 \le x \le 1$, is revolved about the x-axis.

[*Hint:* Equation (7b) of Section 7.6 will help to evaluate the integral.]

76. Suppose that a crystal dissolves at a rate proportional to the amount *un*dissolved. If 9 grams are undissolved initially and 6 grams remain undissolved after 1 min, how many grams remain undissolved after 3 min?

77. The population of the United States was 205 million in 1970. Assuming an annual growth rate of 1.8%, find: (a) the population in the year 2000, (b) the year in which the population will reach one billion.

8 inverse trigonometric and hyperbolic functions

8.1   INVERSE TRIGONOMETRIC FUNCTIONS

Because the six basic trigonometric functions are periodic, each of their values is repeated infinitely many times. Thus, the trigonometric functions are not one-to-one, and consequently have no inverses. However, we shall show in this section that it is possible to restrict the domains of the trigonometric functions in such a way that the restricted functions do have inverses.

INVERSE SINE   In Figure 8.1.1 we have sketched the graph of

$$f(x) = \sin x, \quad -\frac{\pi}{2} \le x \le \frac{\pi}{2} \tag{1}$$

Figure 8.1.1 ▷

Although $\sin x$ is defined on the entire interval $(-\infty, +\infty)$, the function $f(x)$ in (1) is defined only on the interval $[-\pi/2, \pi/2]$. Since no horizontal line cuts the graph of $f(x)$ more than once, this function has an inverse. This inverse is called the **inverse sine function,** and is denoted by

$$\sin^{-1} x$$

The symbol $\sin^{-1} x$ is *never* used to denote $1/\sin x$. If desired, $1/\sin x$

can be rewritten as $(\sin x)^{-1}$ or $\csc x$. In the older literature, $\sin^{-1} x$ is called the *arcsine* function and is denoted by $\arcsin x$. We shall not use this terminology or notation.

REMARK. To define $\sin^{-1} x$ we restricted the domain of $\sin x$ to the interval $[-\pi/2, \pi/2]$ to obtain a one-to-one function. There are other ways to restrict the domain of $\sin x$ to obtain one-to-one functions; for example, we might have required that $\pi/2 \le x \le 3\pi/2$ or $-3\pi/2 \le x \le -\pi/2$. However, the choice $-\pi/2 \le x \le \pi/2$ is customary.

Figure 8.1.2

The graph of $\sin^{-1} x$ is obtained by reflecting the graph of (1) about the line $y = x$ (Figure 8.1.2). If we let $y = \sin^{-1} x$, then it is evident from this graph that $-1 \le x \le 1$ and $-\pi/2 \le y \le \pi/2$, so the domain of $\sin^{-1} x$ is the interval $[-1, 1]$ and the range is $[-\pi/2, \pi/2]$.

Because $\sin x$ (restricted) and $\sin^{-1} x$ are inverses of one another, it follows that

$$\sin^{-1}(\sin y) = y \quad \text{if} \quad -\frac{\pi}{2} \le y \le \frac{\pi}{2} \qquad (2a)$$

$$\sin(\sin^{-1} x) = x \quad \text{if} \quad -1 \le x \le 1 \qquad (2b)$$

From these equations we obtain the following important result.

8.1.1 THEOREM. *If* $-1 \le x \le 1$ *and* $-\pi/2 \le y \le \pi/2$, *then*

$$y = \sin^{-1} x \quad \text{and} \quad \sin y = x \qquad (3)$$

are equivalent statements.

Proof. If $y = \sin^{-1} x$ and $-1 \le x \le 1$, then it follows from (2b) that $\sin y = \sin(\sin^{-1} x) = x$. Conversely, if $\sin y = x$ and $-\pi/2 \le y \le \pi/2$, then $\sin^{-1}(\sin y) = \sin^{-1} x$, or from (2a) $y = \sin^{-1} x$. ∎

REMARK. If we think of y as an angle in radian measure, then it follows from (3) that $y = \sin^{-1} x$ is that angle between $-\pi/2$ and $\pi/2$ whose sine is x.

Example 1   Find

(a) $\sin^{-1}(\tfrac{1}{2})$ (b) $\sin^{-1}(-1/\sqrt{2})$

Solution (a). Let $y = \sin^{-1}(\tfrac{1}{2})$. From Theorem 8.1.1 this equation is equivalent to

$$\sin y = \tfrac{1}{2}, \quad -\pi/2 \le y \le \pi/2$$

The only value of y satisfying these conditions is $y = \pi/6$, so $\sin^{-1}\left(\frac{1}{2}\right) = \pi/6$.

Solution (b).   Let $y = \sin^{-1}(-1/\sqrt{2})$. This is equivalent to

$$\sin y = -1/\sqrt{2}, \quad -\pi/2 \leq y \leq \pi/2$$

which is satisfied only by $y = -\pi/4$. Thus, $\sin^{-1}(-1/\sqrt{2}) = -\pi/4$.   ◄

   In the next section we shall encounter functions that are compositions of trigonometric and inverse trigonometric functions. The following example illustrates a technique for simplifying such functions.

Example 2   Simplify the function $\cos(\sin^{-1}x)$.

Solution.   The idea is to express cosine in terms of sine in order to take advantage of the simplification $\sin(\sin^{-1}x) = x$. Thus, we start with the identity

$$\cos^2\theta = 1 - \sin^2\theta$$

and substitute $\theta = \sin^{-1}x$ to obtain

$$\cos^2(\sin^{-1}x) = 1 - \sin^2(\sin^{-1}x)$$

or on taking square roots

$$\left|\cos(\sin^{-1}x)\right| = \sqrt{1 - \sin^2(\sin^{-1}x)}$$

or

$$\left|\cos(\sin^{-1}x)\right| = \sqrt{1 - x^2}$$

Since $-\pi/2 \leq \sin^{-1}x \leq \pi/2$, it follows that $\cos(\sin^{-1}x)$ is nonnegative; thus, we can drop the absolute value sign and write

$$\cos(\sin^{-1}x) = \sqrt{1 - x^2}   \quad ◄ \tag{4}$$

Figure 8.1.3

   Formula (4) can also be motivated geometrically by first constructing a right triangle with an acute angle of $\theta = \sin^{-1}x$ (Figure 8.1.3). For this triangle, $\sin\theta = x$, so we can assume that the side opposite θ has length x and the hypotenuse has length 1. Thus,

$$\sin\theta = \frac{\text{opposite side}}{\text{hypotenuse}} = \frac{x}{1} = x$$

By the Theorem of Pythagoras, the side that is adjacent to θ has length $\sqrt{1 - x^2}$, so

$$\cos\theta = \frac{\text{adjacent side}}{\text{hypotenuse}} = \frac{\sqrt{1 - x^2}}{1} = \sqrt{1 - x^2}$$

768°

or

$$\cos(\sin^{-1} x) = \sqrt{1 - x^2}$$

INVERSE TANGENT The *inverse tangent* function, $\tan^{-1} x$, is defined to be the inverse of the restricted tangent function,

$$f(x) = \tan x, \quad -\frac{\pi}{2} < x < \frac{\pi}{2}$$

(Figure 8.1.4*a*). The graph of $\tan^{-1} x$ is obtained by reflecting the graph of f about the line $y = x$ (Figure 8.1.4*b*).

If we let $y = \tan^{-1} x$, then from Figure 8.1.4*b* we see that $-\infty < x < +\infty$ and $-\pi/2 < y < \pi/2$, so the domain of $\tan^{-1} x$ is $(-\infty, +\infty)$ and the range is $(-\pi/2, \pi/2)$.

$$f(x) = \tan x$$
$$-\frac{\pi}{2} < x < \frac{\pi}{2}$$

$$y = \tan^{-1} x$$

Figure 8.1.4 ▷   (*a*)   (*b*)

Because $\tan x$ (restricted) and $\tan^{-1} x$ are inverses of one another, it follows that

$$\tan^{-1}(\tan y) = y \quad \text{if} \quad -\pi/2 < y < \pi/2$$
$$\tan(\tan^{-1} x) = x \quad \text{if} \quad -\infty < x < +\infty$$

By imitating the proof of Theorem 8.1.1, the reader should be able to prove the following result.

8.1.2 THEOREM. *If* $-\infty < x < +\infty$ *and* $-\pi/2 < y < \pi/2$, *then*

$$y = \tan^{-1} x \quad \text{and} \quad \tan y = x$$

are equivalent statements.

Example 3   Simplify the function $\sec^2(\tan^{-1}x)$.

Solution.   The object is to express secant in terms of tangent to take advantage of the simplification $\tan(\tan^{-1}x) = x$. Thus, if we let $\theta = \tan^{-1}x$ in the identity

$$\sec^2\theta = 1 + \tan^2\theta$$

we obtain

$$\sec^2(\tan^{-1}x) = 1 + \tan^2(\tan^{-1}x)$$

or

$$\sec^2(\tan^{-1}x) = 1 + x^2 \qquad \blacktriangleleft \tag{5}$$

Formula (5) can also be motivated from the triangle in Figure 8.1.5. The tangent of the acute angle $\tan^{-1}x$ is x, so we can assume that the side opposite to this angle has length x and the adjacent side length 1. The hypotenuse is determined by the Theorem of Pythagoras. From the triangle we obtain

Figure 8.1.5

$$\sec(\tan^{-1}x) = \frac{\text{hypotenuse}}{\text{adjacent side}} = \frac{\sqrt{1 + x^2}}{1} = \sqrt{1 + x^2}$$

or

$$\sec^2(\tan^{-1}x) = 1 + x^2$$

INVERSE SECANT   The *inverse secant* function, $\sec^{-1}x$, is defined to be the inverse of the restricted secant function

$$f(x) = \sec x, \quad 0 \le x < \pi/2 \quad \text{or} \quad \pi \le x < 3\pi/2$$

(Figure 8.1.6a). The graph of $\sec^{-1}x$ is obtained by reflecting the graph of f about the line $y = x$ (Figure 8.1.6b).

REMARK.   There is no universal agreement among mathematicians about the definition of $\sec^{-1}x$. For example, some writers define $\sec^{-1}x$ by restricting x so $0 \le x < \pi/2$ or $\pi/2 < x \le \pi$. Although there are some advantages to this definition, our definition will produce simpler derivative and integration formulas in the next section.

If we let $y = \sec^{-1}x$, then from Figure 8.1.6b we see that $x \le -1$ or $x \ge 1$ and $0 \le y < \pi/2$ or $\pi \le y < 3\pi/2$. Thus, the domain of $\sec^{-1}x$ is $(-\infty, -1] \cup [1, +\infty)$ and the range is $[0, \pi/2) \cup [\pi, 3\pi/2)$.

$$y = \sec x,$$
$$0 \le x < \frac{\pi}{2}$$
or
$$\pi \le x < \frac{3\pi}{2}$$

$$y = \sec^{-1} x$$

Figure 8.1.6 ▷   (a)   (b)

Because $\sec x$ (restricted) and $\sec^{-1} x$ are inverses of one another, it follows that

$$\sec^{-1}(\sec y) = y \quad \text{if} \quad 0 \le y < \pi/2 \quad \text{or} \quad \pi \le y < 3\pi/2$$
$$\sec(\sec^{-1} x) = x \quad \text{if} \quad x \le -1 \quad \text{or} \quad x \ge 1$$

Moreover, the following result parallels Theorems 8.1.1 and 8.1.2.

8.1.3 THEOREM. *If* $|x| \ge 1$ *and if* $0 \le y < \pi/2$ *or* $\pi \le y < 3\pi/2$, *then*

$$y = \sec^{-1} x \quad \text{and} \quad \sec y = x$$

are equivalent statements.

INVERSE COSINE,
COTANGENT, AND
COSECANT

The inverse cosine, cotangent, and cosecant functions are of lesser importance, so we shall summarize their properties briefly.

Table 8.1.1

$y = \cos^{-1} x$	is equivalent to	$x = \cos y$	if $\begin{array}{l} 0 \le y \le \pi \\ -1 \le x \le 1 \end{array}$
$y = \cot^{-1} x$	is equivalent to	$x = \cot y$	if $\begin{array}{l} 0 < y < \pi \\ -\infty < x < +\infty \end{array}$
$y = \csc^{-1} x$	is equivalent to	$x = \csc y$	if $\begin{cases} 0 < y \le \pi/2 \\ \quad\text{or} \\ -\pi < y \le -\pi/2 \end{cases}$ $|x| \ge 1$

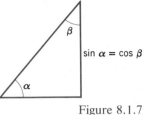

Figure 8.1.7

We leave it as an exercise to find the graph of these three inverse trigonometric functions (Exercise 20).

In the definitions of the six inverse trigonometric functions, the restrictions on the domains are designed not only to produce one-to-one functions, but also to ensure that certain "natural" identities hold for the inverse trigonometric functions. For example, if α and β are acute complementary angles, then from basic trigonometry, $\sin \alpha$ and $\cos \beta$ are equal (Figure 8.1.7). Let us write $x = \sin \alpha = \cos \beta$ so that

$$\alpha = \sin^{-1} x \quad \text{and} \quad \beta = \cos^{-1} x$$

since $\alpha + \beta = \pi/2$ we obtain the identity

$$\sin^{-1} x + \cos^{-1} x = \frac{\pi}{2} \tag{6}$$

(Because α and β were assumed to be nonnegative acute angles, this derivation is only valid for $0 \le x \le 1$; for a derivation valid for all x in $[-1, 1]$ see Exercise 25.) Similarly, we can obtain the identities

$$\tan^{-1} x + \cot^{-1} x = \frac{\pi}{2} \tag{7}$$

$$\sec^{-1} x + \csc^{-1} x = \frac{\pi}{2} \tag{8}$$

The proofs of the following identities, which are evident from Figures 8.1.2, 8.1.4b, and 8.1.6b, are left as exercises.

$$\sin^{-1} (-x) = -\sin^{-1} x \tag{9}$$

$$\tan^{-1} (-x) = -\tan^{-1} x \tag{10}$$

$$\sec^{-1} (-x) = \pi + \sec^{-1} x \quad \text{if } x \ge 1 \tag{11}$$

▶ Exercise Set 8.1

1. Find the exact value of
   (a) $\sin^{-1}(-1)$
   (b) $\cos^{-1}(-1)$
   (c) $\tan^{-1}(-1)$
   (d) $\cot^{-1}(1)$
   (e) $\sec^{-1}(1)$
   (f) $\csc^{-1}(1)$.

2. Find the exact value of
   (a) $\sin^{-1}(\tfrac{1}{2}\sqrt{3})$
   (b) $\cos^{-1}(\tfrac{1}{2})$

   (c) $\tan^{-1}(1)$
   (d) $\cot^{-1}(-1)$
   (e) $\sec^{-1}(-2)$
   (f) $\csc^{-1}(-2)$.

3. Given that $\theta = \sin^{-1}(-\tfrac{1}{2}\sqrt{3})$, find the exact values of $\cos \theta$, $\tan \theta$, $\cot \theta$, $\sec \theta$, and $\csc \theta$.

4. Given that $\theta = \cos^{-1}(\tfrac{1}{2})$, find the exact values of $\sin \theta$, $\tan \theta$, $\cot \theta$, $\sec \theta$, and $\csc \theta$.

5. Given that $\theta = \tan^{-1}(\frac{4}{3})$, find the exact values of $\sin\theta$, $\cos\theta$, $\cot\theta$, $\sec\theta$, and $\csc\theta$.

6. Make a table that lists the six inverse trigonometric functions together with their domains and ranges.

7. Find the exact value of

(a) $\sin^{-1}\left(\sin\dfrac{\pi}{7}\right)$ (b) $\sin^{-1}(\sin\pi)$

(c) $\sin^{-1}\left(\sin\dfrac{5\pi}{7}\right)$ (d) $\sin^{-1}(\sin 630)$.

8. Find the exact value of

(a) $\cos^{-1}\left(\cos\dfrac{\pi}{7}\right)$ (b) $\cos^{-1}(\cos\pi)$

(c) $\cos^{-1}\left(\cos\dfrac{12\pi}{7}\right)$ (d) $\cos^{-1}(\cos 200)$.

9. For which values of x is it true that
(a) $\cos^{-1}(\cos x) = x$ (b) $\cos(\cos^{-1}x) = x$
(c) $\tan^{-1}(\tan x) = x$ (d) $\tan(\tan^{-1}x) = x$
(e) $\csc^{-1}(\csc x) = x$ (f) $\csc(\csc^{-1}x) = x$.

In Exercises 10–15, find the exact value of the given quantity.

10. $\sec\left[\sin^{-1}\left(-\dfrac{3}{4}\right)\right]$.   **11.** $\sin\left[2\cos^{-1}\left(\dfrac{3}{5}\right)\right]$.

12. $\tan^{-1}\left[\sin\left(-\dfrac{\pi}{2}\right)\right]$.   **13.** $\sin^{-1}\left[\cot\left(\dfrac{\pi}{4}\right)\right]$.

14. $\sin\left[\sin^{-1}\left(\dfrac{2}{3}\right) + \cos^{-1}\left(\dfrac{1}{3}\right)\right]$.

15. $\tan\left[2\sec^{-1}\left(\dfrac{3}{2}\right)\right]$.

16. Prove:

$$\tan^{-1}x + \tan^{-1}y = \tan^{-1}\left(\frac{x+y}{1-xy}\right)$$

provided $-\pi/2 < \tan^{-1}x + \tan^{-1}y < \pi/2$. [*Hint:* Use an identity for $\tan(\alpha+\beta)$.]

17. Use the result in Exercise 16 to show:

(a) $\tan^{-1}\dfrac{1}{2} + \tan^{-1}\dfrac{1}{3} = \pi/4$

(b) $2\tan^{-1}\dfrac{1}{3} + \tan^{-1}\dfrac{1}{7} = \pi/4$.

Following Example 2 in this section, we derived the identity $\cos(\sin^{-1}x) = \sqrt{1 - x^2}$ by considering a right triangle with an acute angle of $\sin^{-1}x$ (see Figure 8.1.3). In Exercises 18 and 19, complete the identity by constructing an appropriate triangle.

18. (a) $\sin(\cos^{-1}x) = ?$ (b) $\tan(\cos^{-1}x) = ?$
(c) $\csc(\tan^{-1}x) = ?$ (d) $\sin(\tan^{-1}x) = ?$

19. (a) $\cos(\tan^{-1}x) = ?$ (b) $\tan(\cot^{-1}x) = ?$
(c) $\sin(\sec^{-1}x) = ?$ (d) $\cot(\csc^{-1}x) = ?$

20. Sketch the graphs of $\cos^{-1}x$, $\cot^{-1}x$, and $\csc^{-1}x$.

21. Sketch the graphs of

(a) $y = \sin^{-1}2x$ (b) $y = \tan^{-1}\dfrac{1}{2}x$.

22. Sketch the graphs of

(a) $y = \cos^{-1}\dfrac{1}{3}x$ (b) $y = 2\cot^{-1}2x$.

23. Prove:
(a) $\sin^{-1}(-x) = -\sin^{-1}x$
(b) $\tan^{-1}(-x) = -\tan^{-1}x$
(c) $\sec^{-1}(-x) = \pi + \sec^{-1}x$, if $x \geq 1$.

24. Prove: $\cos^{-1}(-x) = \pi - \cos^{-1}x$.

25. In the text we proved that

$$\sin^{-1}x + \cos^{-1}x = \frac{\pi}{2}$$

for $0 \leq x \leq 1$. Use this result together with Exercises 23(a) and 24 to prove this identity for $-1 \leq x \leq 1$.

26. Prove Theorem 8.1.2.

27. Prove Theorem 8.1.3.

28. A camera is positioned x feet from the base of a missile launching pad. If a missile of length a feet is launched vertically, show that when the base of the missile is b feet above the camera lens, the angle θ subtended at the lens by the missile is

$$\theta = \cot^{-1}\frac{x}{a+b} - \cot^{-1}\frac{x}{b}$$

29. In the computer programming language, BASIC, the instruction ATN(X) causes the computer to calculate $\tan^{-1}(X)$. According to the programming manual for a popular microcomputer, this instruction can be used to calculate $\sin^{-1}(X)$ and $\cos^{-1}(X)$ by means of the formulas

(a) $\sin^{-1}(X) = \text{ATN}\left(\dfrac{X}{\sqrt{1-X^2}}\right)$

(b) $\cos^{-1}(X) = -\text{ATN}\left(\dfrac{X}{\sqrt{1-X^2}}\right) + 1.5708.$

Derive these results.

8.2 DERIVATIVES AND INTEGRALS INVOLVING INVERSE TRIGONOMETRIC FUNCTIONS

In this section we shall obtain the derivatives of the inverse trigonometric functions and the related integration formulas.

8.2.1 THEOREM.

$$\frac{d}{dx}[\sin^{-1}x] = \frac{1}{\sqrt{1-x^2}} \tag{1}$$

$$\frac{d}{dx}[\cos^{-1}x] = -\frac{1}{\sqrt{1-x^2}} \tag{2}$$

$$\frac{d}{dx}[\tan^{-1}x] = \frac{1}{1+x^2} \tag{3}$$

$$\frac{d}{dx}[\cot^{-1}x] = -\frac{1}{1+x^2} \tag{4}$$

$$\frac{d}{dx}[\sec^{-1}x] = \frac{1}{x\sqrt{x^2-1}} \tag{5}$$

$$\frac{d}{dx}[\csc^{-1}x] = -\frac{1}{x\sqrt{x^2-1}} \tag{6}$$

Proof. To obtain (1), let

$$y = \sin^{-1}x$$

so

$$x = \sin y, \quad -\frac{\pi}{2} \le y \le \frac{\pi}{2}$$

Thus, from (6) of Section 7.1 and (4) of Section 8.1,

$$\frac{d}{dx}[\sin^{-1}x] = \frac{dy}{dx} = \frac{1}{dx/dy} = \frac{1}{\cos y} = \frac{1}{\cos(\sin^{-1}x)} = \frac{1}{\sqrt{1-x^2}}$$

Formula (3) is obtained similarly; let

$$y = \tan^{-1}x$$

so

$$x = \tan y, \quad -\frac{\pi}{2} < y < \frac{\pi}{2}$$

Thus, from (6) of Section 7.1 and (5) of Section 8.1,

$$\frac{d}{dx}[\tan^{-1}x] = \frac{dy}{dx} = \frac{1}{dx/dy} = \frac{1}{\sec^2 y} = \frac{1}{\sec^2(\tan^{-1}x)} = \frac{1}{1+x^2}$$

To obtain (5), let

$$y = \sec^{-1}x$$

so

$$x = \sec y, \quad 0 \le y < \frac{\pi}{2} \quad \text{or} \quad \pi \le y < \frac{3\pi}{2}$$

Thus, from (6) of Section 7.1 and Figure 8.2.1,

$$\frac{d}{dx}[\sec^{-1}x] = \frac{dy}{dx} = \frac{1}{dx/dy} = \frac{1}{\sec y \tan y} = \frac{1}{\sec(\sec^{-1}x)\tan(\sec^{-1}x)}$$

$$= \frac{1}{x\sqrt{x^2-1}}$$

Figure 8.2.1 ▷ $\boxed{\tan(\sec^{-1}x) = \sqrt{x^2-1}}$

The remaining formulas can all be obtained from the derivatives of $\sin^{-1}$, $\tan^{-1}$, and $\sec^{-1}$ by using appropriate identities. For example, using identity (6) of Section 8.1 yields

$$\frac{d}{dx}[\cos^{-1}x] = \frac{d}{dx}\left[\frac{\pi}{2} - \sin^{-1}x\right] = -\frac{d}{dx}[\sin^{-1}x] = -\frac{1}{\sqrt{1-x^2}}$$

Similarly, the derivatives of $\cot^{-1} x$ and $\csc^{-1} x$ follow from identities (7) and (8) of Section 8.1.   ∎

If u is a differentiable function of x, then applying the chain rule to (1)–(6) yields the following generalized derivative formulas:

$$\frac{d}{dx}[\sin^{-1} u] = \frac{1}{\sqrt{1 - u^2}}\frac{du}{dx} \tag{1a}$$

$$\frac{d}{dx}[\cos^{-1} u] = -\frac{1}{\sqrt{1 - u^2}}\frac{du}{dx} \tag{2a}$$

$$\frac{d}{dx}[\tan^{-1} u] = \frac{1}{1 + u^2}\frac{du}{dx} \tag{3a}$$

$$\frac{d}{dx}[\cot^{-1} u] = -\frac{1}{1 + u^2}\frac{du}{dx} \tag{4a}$$

$$\frac{d}{dx}[\sec^{-1} u] = \frac{1}{u\sqrt{u^2 - 1}}\frac{du}{dx} \tag{5a}$$

$$\frac{d}{dx}[\csc^{-1} u] = -\frac{1}{u\sqrt{u^2 - 1}}\frac{du}{dx} \tag{6a}$$

Example 1   Find dy/dx if $y = \sin^{-1}(x^3)$.

Solution.   From (1a),

$$\frac{dy}{dx} = \frac{1}{\sqrt{1 - (x^3)^2}}(3x^2) = \frac{3x^2}{\sqrt{1 - x^6}} \quad \blacktriangleleft$$

Example 2   Find dy/dx if $y = \sec^{-1}(e^x)$.

Solution.   From (5a),

$$\frac{dy}{dx} = \frac{1}{e^x\sqrt{(e^x)^2 - 1}}(e^x) = \frac{1}{\sqrt{e^{2x} - 1}} \quad \blacktriangleleft$$

Differentiation formulas (1)–(6) yield useful integration formulas. Those most commonly needed are:

$$\int \frac{dx}{\sqrt{1 - x^2}} = \sin^{-1} x + C \tag{7}$$

$$\int \frac{dx}{1 + x^2} = \tan^{-1} x + C \tag{8}$$

$$\int \frac{dx}{x\sqrt{x^2 - 1}} = \sec^{-1}x + C \tag{9}$$

Example 3   Evaluate

$$\int \frac{dx}{1 + 3x^2}$$

Solution.   Substituting

$$u = \sqrt{3}\, x, \quad du = \sqrt{3}\, dx$$

yields

$$\int \frac{dx}{1 + 3x^2} = \frac{1}{\sqrt{3}} \int \frac{du}{1 + u^2} = \frac{1}{\sqrt{3}} \tan^{-1}u + C$$

$$= \frac{1}{\sqrt{3}} \tan^{-1}(\sqrt{3}\, x) + C \quad \blacktriangleleft$$

Example 4   Evaluate

$$\int \frac{e^x}{\sqrt{1 - e^{2x}}}\, dx$$

Solution.   Substituting

$$u = e^x, \quad du = e^x\, dx$$

yields

$$\int \frac{e^x}{\sqrt{1 - e^{2x}}}\, dx = \int \frac{du}{\sqrt{1 - u^2}} = \sin^{-1}u + C = \sin^{-1}(e^x) + C$$

$$\blacktriangleleft$$

Example 5   Evaluate

$$\int \frac{dx}{a^2 + x^2}$$

where $a \neq 0$ is a constant.

Solution.   If we had a 1 in place of a^2, we could use (8). Thus, we look for a u-substitution that will enable us to replace the a^2 with a 1. If we let

$$x = au, \quad dx = a\, du$$

then

$$\int \frac{dx}{a^2 + x^2} = \int \frac{a\,du}{a^2 + a^2 u^2} = \frac{1}{a} \int \frac{du}{1 + u^2}$$

$$= \frac{1}{a} \tan^{-1} u + C = \frac{1}{a} \tan^{-1} \frac{x}{a} + C \quad \blacktriangleleft$$

The method of Example 5 leads to the following generalizations of (7), (8), and (9).

$$\int \frac{dx}{\sqrt{a^2 - x^2}} = \sin^{-1} \frac{x}{a} + C \qquad (a > 0) \qquad (10)$$

$$\int \frac{dx}{a^2 + x^2} = \frac{1}{a} \tan^{-1} \frac{x}{a} + C \qquad (a \neq 0) \qquad (11)$$

$$\int \frac{dx}{x\sqrt{x^2 - a^2}} = \frac{1}{a} \sec^{-1} \frac{x}{a} + C \quad (a > 0) \qquad (12)$$

Example 6   Evaluate

$$\int \frac{dx}{\sqrt{2 - x^2}}$$

Solution.   Applying (10) with $a = \sqrt{2}$ yields

$$\int \frac{dx}{\sqrt{2 - x^2}} = \sin^{-1} \frac{x}{\sqrt{2}} + C \quad \blacktriangleleft$$

▶ Exercise Set 8.2

In Exercises 1–11, find dy/dx.

1. (a) $y = \sin^{-1}(\tfrac{1}{3}x)$
    (b) $y = \cos^{-1}(2x + 1)$.

2. (a) $y = \tan^{-1}(x^2)$
    (b) $y = \cot^{-1}(\sqrt{x})$.

3. (a) $y = \sec^{-1}(x^7)$
    (b) $y = \csc^{-1}(e^x)$.

4. (a) $y = (\tan x)^{-1}$
    (b) $y = \dfrac{1}{\tan^{-1} x}$.

5. (a) $y = \sin^{-1}\left(\dfrac{1}{x}\right)$
    (b) $y = \cos^{-1}(\cos x)$.

6. (a) $y = \ln(\cos^{-1} x)$
    (b) $y = \sqrt{\cot^{-1} x}$.

7. (a) $y = e^x \sec^{-1} x$
    (b) $y = x^2(\sin^{-1} x)^3$.

8. (a) $y = \sin^{-1} x + \cos^{-1} x$
    (b) $y = \sec^{-1} x + \csc^{-1} x$.

9. (a) $y = \tan^{-1}\left(\dfrac{1 - x}{1 + x}\right)$
    (b) $y = (1 + x \csc^{-1} x)^{10}$.

10. (a) $y = \sin^{-1}(e^{-3x})$
    (b) $y = \tan^{-1}(xe^{2x})$.

11. (a) $y = \tan^{-1} \sqrt{\dfrac{1 - x}{1 + x}}$
    (b) $y = \sin^{-1}(x^2 \ln x)$.

In Exercises 12 and 13, find dy/dx by implicit differentiation.

12. $x^3 + x \tan^{-1} y = e^y$.

13. $\sin^{-1}(xy) = \cos^{-1}(x - y)$.

In Exercises 14–27, evaluate the integral.

14. $\displaystyle\int_0^{1/\sqrt{2}} \frac{dx}{\sqrt{1 - x^2}}$.

15. $\displaystyle\int_{-1}^{1} \frac{dx}{1 + x^2}$.

16. $\displaystyle\int_{\sqrt{2}}^{2} \frac{dx}{x\sqrt{x^2 - 1}}$.

17. $\displaystyle\int_{-\sqrt{2}}^{-2/\sqrt{3}} \frac{dx}{x\sqrt{x^2 - 1}}$.

18. $\displaystyle\int \frac{dx}{\sqrt{1 - 4x^2}}$.

19. $\displaystyle\int \frac{dx}{1 + 16x^2}$.

20. $\displaystyle\int \frac{dx}{x\sqrt{9x^2 - 1}}$.

21. $\displaystyle\int \frac{e^x}{1 + e^{2x}} dx$.

22. $\displaystyle\int_{\ln 2}^{\ln (2/\sqrt{3})} \frac{e^{-x} dx}{\sqrt{1 - e^{-2x}}}$.

23. $\displaystyle\int_1^3 \frac{dx}{\sqrt{x}(x + 1)}$.

24. $\displaystyle\int \frac{t}{t^4 + 1} dt$.

25. $\displaystyle\int \frac{\sec^2 x\, dx}{\sqrt{1 - \tan^2 x}}$.

26. $\displaystyle\int \frac{\sin\theta}{\cos^2\theta + 1} d\theta$.

27. $\displaystyle\int \frac{dx}{x\sqrt{1 - (\ln x)^2}}$.

28. Derive integration formulas (10) and (12).

In Exercises 29 and 30, use (10), (11), and (12) to evaluate the integrals.

29. (a) $\displaystyle\int \frac{dx}{\sqrt{9 - x^2}}$ (b) $\displaystyle\int \frac{dx}{5 + x^2}$

(c) $\displaystyle\int \frac{dx}{x\sqrt{x^2 - \pi}}$.

30. (a) $\displaystyle\int \frac{e^x}{4 + e^{2x}} dx$ (b) $\displaystyle\int \frac{dx}{\sqrt{9 - 4x^2}}$

(c) $\displaystyle\int \frac{dy}{y\sqrt{5y^2 - 3}}$.

31. Find the area of the region enclosed by the graphs of $y = 1/\sqrt{1 - 9x^2}$, $y = 0$, $x = 0$, and $x = 1/6$.

32. Find the area of the region enclosed by the graphs of $y = \sin^{-1} x$, $x = 0$, and $y = \pi/2$.

33. Find the volume of the solid generated when the region bounded by $x = 2$, $x = -2$, $y = 0$, and $y = 1/\sqrt{4 + x^2}$ is revolved about the x-axis.

34. (a) Find the volume V of the solid generated when the region bounded by $y = 1/(1 + x^4)$, $y = 0$, $x = 1$, and $x = b$ ($b > 1$) is revolved about the y-axis.

(b) Find $\displaystyle\lim_{b \to +\infty} V$.

35. Evaluate $\int_0^1 \sin^{-1} x\, dx$. [*Hint:* Interpret the integral as the area of a region in the xy-plane, and integrate with respect to y.]

36. In Exercise 28 of Section 8.1, how far from the launching pad should the camera be positioned to maximize the angle θ subtended at the lens by the missile?

37. Given points $A(2, 1)$ and $B(5, 4)$, find the point P in the interval $[2, 5]$ on the x-axis that maximizes angle APB.

38. An aircraft is flying at an altitude of 4 mi at a speed of 800 mi/hr in a direction away from a tracking station on the ground. How fast is the angle of elevation changing when the aircraft is over a point 10 mi from the station?

39. A lighthouse is located 3 mi off a straight shore. If the light revolves at 2 revolutions per minute, how fast is the beam moving along the coastline at a point 2 mi down the coast?

40. A 25-ft ladder leans against a vertical wall. If the bottom of the ladder slides away from the base of the wall at the rate of 4 ft/sec, how fast is the angle between the ladder and the wall changing when the top of the ladder is 20 ft above the ground?

41. The lower edge of a painting, 10 ft in height, is 2 ft above an observer's eye level. Assuming the best view is obtained when the angle subtended at the observer's eye by the painting is maximum, how far from the wall should the observer stand?

42. Use Corollary 4.10.4 to prove that
(a) $2 \sin^{-1} \sqrt{x} = \sin^{-1}(2x - 1) + \pi/2$ for $0 < x < 1$.
(b) $\sin^{-1}(\tanh x) = \tan^{-1}(\sinh x)$.

43. Use Theorem 4.10.2 (the Mean-Value Theorem) to prove that

$$\frac{x}{1 + x^2} < \tan^{-1} x < x \text{ for } x > 0$$

44. Find $\displaystyle\lim_{n \to +\infty} \sum_{k=1}^{n} \frac{n}{n^2 + k^2}$. [*Hint:* Factor $\dfrac{1}{n}$ out of $\dfrac{n}{n^2 + k^2}$ and identify the result as a Riemann sum.]

8.3 INVERSE HYPERBOLIC FUNCTIONS

In this section we shall discuss inverses of hyperbolic functions. We shall be interested primarily in the integration formulas that these functions produce.

Because the hyperbolic sine function is increasing on the interval $(-\infty, +\infty)$ (Figure 7.6.1*b*), it follows that $\sinh x$ is one-to-one on this interval and consequently has an inverse function, $\sinh^{-1} x$. Thus, for all real values of x and y

$$\sinh^{-1}(\sinh y) = y$$
$$\sinh(\sinh^{-1} x) = x$$

Moreover,

$$y = \sinh^{-1} x \quad \text{and} \quad \sinh y = x$$

are equivalent statements. The graph of $\sinh^{-1} x$ can be obtained by reflecting the graph of $\sinh x$ about the line $y = x$ (Figure 8.3.1).

Because $\sinh x$ is expressible in terms of e^x, one might suspect that $\sinh^{-1} x$ is expressible in terms of natural logarithms. To see that this is so, we write $y = \sinh^{-1} x$ in the alternative form

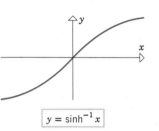

$y = \sinh^{-1} x$

Figure 8.3.1

$$x = \sinh y = \frac{e^y - e^{-y}}{2}$$

so that

$$e^y - 2x - e^{-y} = 0$$

or, on multiplying through by e^y

$$e^{2y} - 2xe^y - 1 = 0$$

Applying the quadratic formula yields

$$e^y = \frac{2x \pm \sqrt{4x^2 + 4}}{2} = x \pm \sqrt{x^2 + 1}$$

Since $e^y > 0$, the solution involving the minus sign is extraneous and must be discarded. Thus,

$$e^y = x + \sqrt{x^2 + 1}$$

Taking natural logarithms yields

$$y = \ln(x + \sqrt{x^2 + 1})$$

or

$$\sinh^{-1} x = \ln(x + \sqrt{x^2 + 1}) \tag{1}$$

This formula and a table of natural logarithms can be used to evaluate $\sinh^{-1} x$. Moreover, we can use it to obtain the derivative formula for $\sinh^{-1} x$ by writing

$$\frac{d}{dx}[\sinh^{-1} x] = \frac{d}{dx}[\ln(x + \sqrt{x^2 + 1})]$$

$$= \frac{1}{x + \sqrt{x^2 + 1}}\left(1 + \frac{x}{\sqrt{x^2 + 1}}\right)$$

$$= \frac{\sqrt{x^2 + 1} + x}{(x + \sqrt{x^2 + 1})(\sqrt{x^2 + 1})}$$

or, on canceling

$$\frac{d}{dx}[\sinh^{-1} x] = \frac{1}{\sqrt{x^2 + 1}} \tag{2}$$

Derivatives of the remaining inverse hyperbolic functions are obtained in a similar manner with the exception of $\cosh^{-1}$ and sech^{-1}. As evidenced by Figures 7.6.1a and 7.6.3, $\cosh x$ and $\mathrm{sech}\, x$ are not one-to-one, so that we must restrict the domains of these functions appropriately before we can obtain inverses. For example, if we let

$$f(x) = \cosh x, \ x \geq 0$$

then f is one-to-one (Figure 8.3.2a) and consequently has an inverse. We call this inverse $\cosh^{-1} x$. The graph of $y = \cosh^{-1} x$ is the reflection of the graph of f about the line $y = x$ (Figure 8.3.2b). The domain of $\cosh^{-1}$ is the interval $[1, +\infty)$, the range is $[0, +\infty)$, and

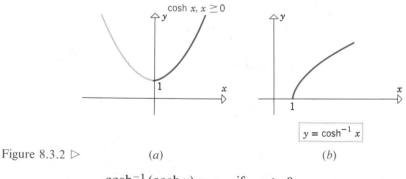

Figure 8.3.2 ▷   (a)   (b)

$$\cosh^{-1}(\cosh y) = y \quad \text{if} \quad y \geq 0$$
$$\cosh(\cosh^{-1} x) = x \quad \text{if} \quad x \geq 1$$

Moreover, if $x \geq 1$ and $y \geq 0$, then the statements

$$y = \cosh^{-1} x \quad \text{and} \quad x = \cosh y$$

are equivalent.

By following the derivations (1) and (2), the reader should be able to show that

$$\cosh^{-1} x = \ln (x + \sqrt{x^2 - 1})$$

and

$$\frac{d}{dx}[\cosh^{-1} x] = \frac{1}{\sqrt{x^2 - 1}}$$

if $x > 1$.

For reference, we shall summarize the definitions, logarithmic expressions, and derivatives of the inverse hyperbolic functions.

8.3.1 DEFINITION.

$y = \sinh^{-1} x$ is equivalent to $x = \sinh y$ for all x, y

$y = \cosh^{-1} x$ is equivalent to $x = \cosh y$ if $\begin{cases} x \geq 1 \\ y \geq 0 \end{cases}$

$y = \tanh^{-1} x$ is equivalent to $x = \tanh y$ if $\begin{cases} -1 < x < 1 \\ -\infty < y < +\infty \end{cases}$

$y = \coth^{-1} x$ is equivalent to $x = \coth y$ if $\begin{cases} |x| > 1 \\ y \neq 0 \end{cases}$

$y = \operatorname{sech}^{-1} x$ is equivalent to $x = \operatorname{sech} y$ if $\begin{cases} 0 < x \leq 1 \\ y \geq 0 \end{cases}$

$y = \operatorname{csch}^{-1} x$ is equivalent to $x = \operatorname{csch} y$ if $\begin{cases} x \neq 0 \\ y \neq 0 \end{cases}$

8.3.2 THEOREM.

$$\sinh^{-1} x = \ln (x + \sqrt{x^2 + 1}) \qquad -\infty < x < +\infty$$
$$\cosh^{-1} x = \ln (x + \sqrt{x^2 - 1}) \qquad x \geq 1$$
$$\tanh^{-1} x = \frac{1}{2} \ln \frac{1 + x}{1 - x} \qquad -1 < x < 1$$
$$\coth^{-1} x = \frac{1}{2} \ln \frac{x + 1}{x - 1} \qquad |x| > 1$$
$$\operatorname{sech}^{-1} x = \ln \left(\frac{1 + \sqrt{1 - x^2}}{x} \right) \qquad 0 < x \leq 1$$
$$\operatorname{csch}^{-1} x = \ln \left(\frac{1}{x} + \frac{\sqrt{1 + x^2}}{|x|} \right) \qquad x \neq 0$$

8.3.3 THEOREM.

$$\frac{d}{dx}[\sinh^{-1}x] = \frac{1}{\sqrt{1+x^2}}$$

$$\frac{d}{dx}[\cosh^{-1}x] = \frac{1}{\sqrt{x^2-1}}$$

$$\frac{d}{dx}[\tanh^{-1}x] = \frac{1}{1-x^2} \quad (|x|<1)$$

$$\frac{d}{dx}[\coth^{-1}x] = \frac{1}{1-x^2} \quad (|x|>1)$$

$$\frac{d}{dx}[\text{sech}^{-1}x] = -\frac{1}{x\sqrt{1-x^2}}$$

$$\frac{d}{dx}[\text{csch}^{-1}x] = -\frac{1}{|x|\sqrt{1+x^2}}$$

REMARK.   Care must be exercised in differentiating $\tanh^{-1}x$ and $\coth^{-1}x$. In both cases the expression for the derivative is

$$\frac{1}{1-x^2} \tag{3}$$

However, the domain of $\tanh^{-1}x$ is the interval $-1<x<1$ (verify this), so that the derivative formula for $\tanh^{-1}x$ applies only when $|x|<1$. Similarly, the derivative formula for $\coth^{-1}x$ applies only on the domain of $\coth^{-1}x$, that is, when $|x|>1$. This distinction becomes important when we integrate (3); for then we must write

$$\int \frac{1}{1-x^2}\,dx = \begin{cases} \tanh^{-1}x + C & \text{if } |x|<1 \\ \coth^{-1}x + C & \text{if } |x|>1 \end{cases}$$

The following integration formulas follow from Theorem 8.3.3.

8.3.4 THEOREM.

$$\int \frac{dx}{\sqrt{1+x^2}} = \sinh^{-1}x + C$$

$$\int \frac{dx}{\sqrt{x^2-1}} = \cosh^{-1}x + C \quad x>1$$

$$\int \frac{dx}{1-x^2} = \begin{cases} \tanh^{-1}x + C & \text{if } |x|<1 \\ \coth^{-1}x + C & \text{if } |x|>1 \end{cases}$$

$$\int \frac{dx}{x\sqrt{1 - x^2}} = -\operatorname{sech}^{-1}|x| + C$$

$$\int \frac{dx}{x\sqrt{1 + x^2}} = -\operatorname{csch}^{-1}|x| + C$$

The first three integration formulas follow immediately from the corresponding differentiation formulas. The last two require additional work (see Exercises 27 and 28). By using Theorem 8.3.2 these formulas can also be expressed in terms of the natural logarithm. In particular, we leave it as an exercise to show that the third formula can be written:

$$\int \frac{dx}{1 - x^2} = \frac{1}{2} \ln \left| \frac{1 + x}{1 - x} \right| + C \tag{4}$$

If u is a differentiable function of x, then applying the chain rule to the formulas in Theorem 8.3.3 yields the following generalized derivative formulas:

$$\frac{d}{dx}[\sinh^{-1} u] = \frac{1}{\sqrt{1 + u^2}} \frac{du}{dx}$$

$$\frac{d}{dx}[\cosh^{-1} u] = \frac{1}{\sqrt{u^2 - 1}} \frac{du}{dx} \quad (u > 1)$$

$$\frac{d}{dx}[\tanh^{-1} u] = \frac{1}{1 - u^2} \frac{du}{dx} \quad (|u| < 1)$$

$$\frac{d}{dx}[\coth^{-1} u] = \frac{1}{1 - u^2} \frac{du}{dx} \quad (|u| > 1)$$

$$\frac{d}{dx}[\operatorname{sech}^{-1} u] = -\frac{1}{u\sqrt{1 - u^2}} \frac{du}{dx}$$

$$\frac{d}{dx}[\operatorname{csch}^{-1} u] = -\frac{1}{|u|\sqrt{1 + u^2}} \frac{du}{dx}$$

▶ Exercise Set 8.3
_____

1. (a) Prove: $\cosh^{-1} x = \ln(x + \sqrt{x^2 - 1}), x \geq 1$.
    (b) Use (a) to obtain the derivative of $\cosh^{-1} x$.

2. (a) Prove: $\tanh^{-1} x = \dfrac{1}{2} \ln \dfrac{1 + x}{1 - x}, -1 < x < 1$.
    (b) Use (a) to obtain the derivative of $\tanh^{-1} x$.

3. (a) Prove:

$$\operatorname{sech}^{-1} x = \cosh^{-1}(1/x), \quad 0 < x \leq 1$$

$$\coth^{-1} x = \tanh^{-1}(1/x), \quad |x| > 1$$

$$\operatorname{csch}^{-1} x = \sinh^{-1}(1/x), \quad x \neq 0$$

(b)   Use (a) and the derivatives of $\sinh^{-1}x$, $\cosh^{-1}x$, and $\tanh^{-1}x$ to find the derivatives of $\text{sech}^{-1}x$, $\coth^{-1}x$, and $\text{csch}^{-1}x$.

(c)   Use (a) and the logarithmic expressions for $\sinh^{-1}x$, $\cosh^{-1}x$, and $\tanh^{-1}x$ to derive the logarithmic expressions for $\text{sech}^{-1}x$, $\coth^{-1}x$, and $\text{csch}^{-1}x$.

4.   Without referring to the text, state the domains of the six inverse hyperbolic functions.

In Exercises 5 and 6, find the logarithmic equivalent of each of the given expressions.

5.   (a) $\cosh^{-1}(3)$ (b) $\sinh^{-1}(-2)$.

6.   (a) $\tanh^{-1}(3/4)$ (b) $\coth^{-1}(-5/4)$.

In Exercises 7–15, find dy/dx.

7.   (a) $y = \sinh^{-1}(\tfrac{1}{3}x)$
     (b) $y = \cosh^{-1}(2x + 1)$.

8.   (a) $y = \tanh^{-1}(x^2)$ (b) $y = \coth^{-1}(\sqrt{x})$.

9.   (a) $y = \text{sech}^{-1}(x^7)$ (b) $y = \text{csch}^{-1}(e^x)$.

10.   (a) $y = (\tanh^{-1}x)^2$ (b) $y = \dfrac{1}{\tanh^{-1}x}$.

11.   (a) $y = \sinh^{-1}\left(\dfrac{1}{x}\right)$
      (b) $y = \cosh^{-1}(\cosh x)$.

12.   (a) $y = \ln(\cosh^{-1}x)$ (b) $y = \sqrt{\coth^{-1}x}$.

13.   (a) $y = e^x \text{sech}^{-1}x$ (b) $y = x^2 (\sinh^{-1}x)^3$.

14.   (a) $y = \sinh^{-1}(\tanh x)$
      (b) $y = \cosh^{-1}(\sinh^{-1}x)$.

15.   (a) $y = \tanh^{-1}\left(\dfrac{1 - x}{1 + x}\right)$
      (b) $y = (1 + x\, \text{csch}^{-1}x)^{10}$.

In Exercises 16–21, evaluate the integral.

16. $\displaystyle\int \frac{dx}{\sqrt{1 + 9x^2}}$.   17. $\displaystyle\int \frac{dx}{\sqrt{x^2 - 2}}$.

18. $\displaystyle\int \frac{dx}{\sqrt{9x^2 - 25}}$.   19. $\displaystyle\int \frac{dx}{\sqrt{1 - e^{2x}}}$.

20. $\displaystyle\int \frac{\sin\theta\, d\theta}{\sqrt{1 + \cos^2\theta}}$.   21. $\displaystyle\int \frac{dx}{x\sqrt{1 + x^6}}$.

In Exercises 22–24, use the natural logarithm table in Appendix 3 to obtain a numerical value for the integral.

22. $\displaystyle\int_0^{1/2} \frac{dx}{1 - x^2}$.   23. $\displaystyle\int_2^3 \frac{dx}{1 - x^2}$.

24. $\displaystyle\int_0^{\sqrt{3}} \frac{dt}{\sqrt{t^2 + 1}}$.

25.   Sketch the graphs of
      (a) $\tanh^{-1}x$ (b) $\text{csch}^{-1}x$.

26.   Sketch the graphs of
      (a) $\coth^{-1}x$ (b) $\text{sech}^{-1}x$.

27.   Show that
$$\frac{d}{dx}[\text{sech}^{-1}|x|] = -\frac{1}{x\sqrt{1 - x^2}}$$

28.   Show that
$$\frac{d}{dx}[\text{csch}^{-1}|x|] = -\frac{1}{x\sqrt{1 + x^2}}$$

29.   Derive integration formula (4).

30.   Let a be a positive constant. Derive an integration formula for

(a) $\displaystyle\int \frac{du}{\sqrt{a^2 + u^2}}$ (b) $\displaystyle\int \frac{du}{\sqrt{u^2 - a^2}}$

(c) $\displaystyle\int \frac{du}{a^2 - u^2}$.

[*Hint:* See Example 5 of Section 8.2.]

31.   Our derivation of the differentiation formula for $\sinh^{-1}x$ used the logarithmic expression for this function. Show that the derivative can also be obtained by the method we used in Section 8.2 to obtain the derivative formula for $\sin^{-1}x$.

32.   Find
(a) $\displaystyle\lim_{x \to +\infty} \sinh^{-1}x$ (b) $\displaystyle\lim_{x \to +\infty} \coth^{-1}x$
(c) $\displaystyle\lim_{x \to 0^+} \text{csch}^{-1}x$ (d) $\displaystyle\lim_{x \to +\infty} (\cosh^{-1}x - \ln x)$.

33.   Show that
$$\int \frac{1}{\sqrt{x^2 - 1}}\, dx = -\cosh^{-1}(-x) + C$$
$$\text{if } x < -1$$

34.   Use the result in Exercise 33 to show that
$$\int \frac{1}{\sqrt{x^2 - 1}}\, dx = \ln|x + \sqrt{x^2 - 1}| + C$$
$$\text{if } x < -1$$

▶ SUPPLEMENTARY EXERCISES

In Exercises 1–3, find the exact value.

1. (a) $\cos^{-1}(-1/2)$ (b) $\cot^{-1}[\cot(3/4)]$
   (c) $\cos[\sin^{-1}(4/5)]$
   (d) $\cos[\sin^{-1}(-4/5)]$.

2. (a) $\tan^{-1}(-1)$ (b) $\csc^{-1}(-2/\sqrt{3})$
   (c) $\cos^{-1}[\cos(-\pi/3)]$
   (d) $\sin[-\sec^{-1}(2/\sqrt{3})]$.

3. (a) $\sin^{-1}(1/\sqrt{2})$ (b) $\sin^{-1}[\sin(5\pi/4)]$
   (c) $\tan(\sec^{-1}5)$ (d) $\tan^{-1}[\cot(\pi/6)]$.

4. Use a double angle formula to convert the given expression to an algebraic function of x.
   (a) $\sin(2\csc^{-1}x)$, $|x| \geq 1$
   (b) $\cos(2\sin^{-1}x)$, $|x| \leq 1$
   (c) $\sin(2\tan^{-1}x)$.

5. Simplify:
   (a) $\cos[\cos^{-1}(4/5) + \sin^{-1}(5/13)]$
   (b) $\sin[\sin^{-1}(4/5) + \cos^{-1}(5/13)]$
   (c) $\tan[\tan^{-1}(1/3) + \tan^{-1}(2)]$.

6. If $u = \operatorname{csch}^{-1}(-5/12)$, find $\coth u$, $\sinh u$, $\cosh u$, and $\sinh(2u)$.

7. If $u = \tanh^{-1}(-3/5)$, find $\cosh u$, $\sinh u$, and $\cosh(2u)$.

In Exercises 8 and 9, sketch the graph of f.

8. (a) $f(x) = 3\sin^{-1}(x/2)$
   (b) $f(x) = \cos^{-1}x - \pi/2$.

9. (a) $f(x) = 2\tan^{-1}(-3x)$
   (b) $f(x) = \cos^{-1}x + \sin^{-1}x$.

In Exercises 10–23, find dy/dx, using implicit or logarithmic differentiation where convenient.

10. $y = \sin^{-1}(e^x) + 2\tan^{-1}(3x)$.

11. $y = \dfrac{1}{\sec^{-1}x^2}$.

12. $y = x\sin^{-1}x + \sqrt{1 - x^2}$.

13. $y = \cosh^{-1}(\sec x)$.   14. $\tan^{-1}y = \sin^{-1}x$.

15. $y = x\tanh^{-1}(\ln x)$.   16. $y = \tan^{-1}\left(\dfrac{2x}{1 - x^2}\right)$.

17. $y = \sqrt{\sin^{-1}3x}$.   18. $y = (\sin^{-1}2x)^{-1}$.

19. $y = \exp(\sec^{-1}x)$.   20. $y = (\tan^{-1}x)/\ln x$.

21. $y = \pi^{\sin^{-1}x}$.   22. $y = (\sinh^{-1}x)^{\pi}$.

23. $y = \tanh^{-1}\left(\dfrac{1}{\coth x}\right)$.

24. Let $f(x) = \tan^{-1}x + \tan^{-1}(1/x)$.
   (a) By considering $f'(x)$, show that $f(x) = C_1$ on $(-\infty, 0)$ and $f(x) = C_2$ on $(0, +\infty)$, where C_1 and C_2 are constants.
   (b) Find C_1 and C_2 in part (a).

25. Show that $y = \tan^{-1}x$ satisfies
   $$y'' = -2\sin y\cos^3 y.$$

In Exercises 26–35, evaluate the integral.

26. $\displaystyle\int \frac{dx}{\sqrt{9 - 4x^2}}$.

27. $\displaystyle\int \frac{e^x\,dx}{1 - e^{2x}}$.

28. $\displaystyle\int \frac{\cot x\,dx}{\sqrt{1 - \sin^2 x}}$.

29. $\displaystyle\int \frac{dx}{x\sqrt{(\ln x)^2 - 1}}$.

30. $\displaystyle\int \frac{dx}{e^x\sqrt{1 - e^{-2x}}}$.

31. $\displaystyle\int_{2\sqrt{3}/9}^{2/3} \frac{dx}{3x\sqrt{9x^2 - 1}}$.

32. $\displaystyle\int_0^{\sqrt{2}} \frac{x\,dx}{4 + x^4}$.

33. $\displaystyle\int \frac{dx}{x^{1/2} + x^{3/2}}$.

34. $\displaystyle\int \frac{x^2\,dx}{\sqrt{1 + x^6}}$.

35. $\displaystyle\int_{1/4}^{1/2} \frac{dx}{\sqrt{x}\sqrt{1 - x}}$.

36. The hypotenuse of a right triangle is growing at a rate of a cm/sec and one leg is decreasing at a rate of b cm/sec. How fast is the acute angle between the hypotenuse and the other leg changing at the instant when both legs are 1 cm?

37. For $f(x) = \tan^{-1}(2x) - \tan^{-1}x$ find the value(s) of x at which $f(x)$ assumes its maximum value on the interval $[0, +\infty)$.

38. Find the area between the curve $y = 1/(9 + x^2)$, the x-axis, and the lines $x = \pm\sqrt{3}$.

9 techniques of integration

9.1 A BRIEF REVIEW

Although we can now integrate a wide variety of functions, there remain many important kinds of integrals that we cannot yet evaluate. The purpose of this chapter is to develop some additional techniques of integration and also to systematize the procedure of integration.

In this section we shall review the integration formulas studied in previous sections. We shall restate those formulas with u rather than x as the variable of integration because this is the form that commonly arises when substitution is used. However, the reader should be prepared to encounter these formulas with other letters denoting the variable of integration as well.

CONSTANTS, POWERS, EXPONENTIALS

See Sections 5.2, 7.3, and 7.4

1. $\displaystyle\int du = u + C$

4. $\displaystyle\int \frac{du}{u} = \ln|u| + C$

2. $\displaystyle\int a\, du = a \int du = au + C$

5. $\displaystyle\int e^u\, du = e^u + C$

3. $\displaystyle\int u^r\, du = \frac{u^{r+1}}{r+1} + C, r \neq -1$

6. $\displaystyle\int a^u\, du = \frac{a^u}{\ln a} + C$

TRIGONOMETRIC FUNCTIONS

See Sections 5.2 and 7.3

7. $\displaystyle\int \sin u\, du = -\cos u + C$

9. $\displaystyle\int \sec^2 u\, du = \tan u + C$

8. $\displaystyle\int \cos u\, du = \sin u + C$

10. $\displaystyle\int \csc^2 u\, du = -\cot u + C$

11. $\displaystyle\int \sec u \tan u \, du = \sec u + C$ **13.** $\displaystyle\int \tan u \, du = -\ln|\cos u| + C$

12. $\displaystyle\int \csc u \cot u \, du = -\csc u + C$ **14.** $\displaystyle\int \cot u \, du = \ln|\sin u| + C$

HYPERBOLIC
FUNCTIONS

See Section 7.6

15. $\displaystyle\int \sinh u \, du = \cosh u + C$ **18.** $\displaystyle\int \operatorname{csch}^2 u \, du = -\coth u + C$

16. $\displaystyle\int \cosh u \, du = \sinh u + C$ **19.** $\displaystyle\int \operatorname{sech} u \tanh u \, du = -\operatorname{sech} u + C$

17. $\displaystyle\int \operatorname{sech}^2 u \, du = \tanh u + C$ **20.** $\displaystyle\int \operatorname{csch} u \coth u \, du = -\operatorname{csch} u + C$

ALGEBRAIC
FUNCTIONS

See Sections 8.1 and 8.2

21. $\displaystyle\int \frac{du}{\sqrt{1 - u^2}} = \sin^{-1} u + C$

22. $\displaystyle\int \frac{du}{1 + u^2} = \tan^{-1} u + C$

23. $\displaystyle\int \frac{du}{u\sqrt{u^2 - 1}} = \sec^{-1} u + C$

24. $\displaystyle\int \frac{du}{\sqrt{1 + u^2}} = \sinh^{-1} u + C = \ln(u + \sqrt{u^2 + 1}) + C$

25. $\displaystyle\int \frac{du}{\sqrt{u^2 - 1}} = \cosh^{-1} u + C = \ln(u + \sqrt{u^2 - 1}) + C \qquad (u > 1)$

26. $\displaystyle\int \frac{du}{1 - u^2} = \begin{cases} \tanh^{-1} u + C & \text{if } |u| < 1 \\ \coth^{-1} u + C & \text{if } |u| > 1 \end{cases} = \frac{1}{2} \ln\left|\frac{1 + u}{1 - u}\right| + C$

27. $\displaystyle\int \frac{du}{u\sqrt{1 - u^2}} = -\operatorname{sech}^{-1}|u| + C = -\ln\left(\frac{1 + \sqrt{1 - u^2}}{u}\right) + C$

28. $\displaystyle\int \frac{du}{u\sqrt{1 + u^2}} = -\operatorname{csch}^{-1}|u| + C = -\ln\left(\frac{1}{u} + \frac{\sqrt{1 + u^2}}{|u|}\right) + C$

REMARK.   Readers who did not cover Section 8.3 should ignore Formulas 24–28. We shall study alternative methods for evaluating these integrals in this chapter.

▶ Exercise Set 9.1

Without looking at the text, complete the following integration formulas.

Constants, Powers, Exponentials

1. $\int du =$

2. $\int a\ du =$

3. $\int u^r\ du =$

4. $\int \dfrac{du}{u} =$

5. $\int e^u\ du =$

6. $\int a^u\ du =$

Trigonometric Functions

7. $\int \sin u\ du =$

8. $\int \cos u\ du =$

9. $\int \sec^2 u\ du =$

10. $\int \csc^2 u\ du =$

11. $\int \sec u \tan u\ du =$

12. $\int \csc u \cot u\ du =$

13. $\int \tan u\ du =$

14. $\int \cot u\ du =$

Hyperbolic Functions

15. $\int \sinh u\ du =$

16. $\int \cosh u\ du =$

17. $\int \text{sech}^2 u\ du =$

18. $\int \text{csch}^2 u\ du =$

19. $\int \text{sech}\ u \tanh u\ du =$

20. $\int \text{csch}\ u \coth u\ du =$

Algebraic Functions

21. $\int \dfrac{du}{\sqrt{1 - u^2}} =$

22. $\int \dfrac{du}{1 + u^2} =$

23. $\int \dfrac{du}{u\sqrt{u^2 - 1}} =$

24. $\int \dfrac{du}{\sqrt{1 + u^2}} =$

25. $\int \dfrac{du}{\sqrt{u^2 - 1}} =$

26. $\int \dfrac{du}{1 - u^2} =$

27. $\int \dfrac{du}{u\sqrt{1 - u^2}} =$

28. $\int \dfrac{du}{u\sqrt{1 + u^2}} =$

9.2 INTEGRATION BY PARTS

In this section, we shall develop a technique that will help us to evaluate a wide variety of integrals that do not fit any of the basic integration formulas.

If f and g are differentiable functions, then by the rule for differentiating products

$$\frac{d}{dx}[f(x)g(x)] = f(x)g'(x) + g(x)f'(x)$$

Integrating both sides we obtain

$$\int \frac{d}{dx}[f(x)g(x)]\ dx = \int f(x)g'(x)\ dx + \int g(x)f'(x)\ dx$$

or

$$f(x)g(x) + C = \int f(x)g'(x)\ dx + \int g(x)f'(x)\ dx$$

or

$$\int f(x)g'(x) \, dx = f(x)g(x) - \int g(x)f'(x) \, dx + C$$

Since the integral on the right will produce another constant of integration, there is no need to keep the C in this last equation; thus, we obtain

$$\int f(x)g'(x) \, dx = f(x)g(x) - \int f'(x)g(x) \, dx \qquad (1)$$

which is called the formula for **integration by parts**. By using this formula we can sometimes reduce a hard integration problem to an easier one.

In practice, it is usual to rewrite (1) by letting

$$u = f(x) \quad du = f'(x) \, dx$$
$$v = g(x) \quad dv = g'(x) \, dx$$

This yields the following formula.

Integration by Parts for Indefinite Integrals

$$\int u \, dv = uv - \int v \, du \qquad (2a)$$

For definite integrals the corresponding formula is

Integration by Parts for Definite Integrals

$$\int_a^b u \, dv = uv \Big]_a^b - \int_a^b v \, du \qquad (2b)$$

where a and b are the limits of integration for the variable x.

Example 1   Evaluate

$$\int x e^x \, dx$$

Solution.   To apply (2a) we must write the integral in the form

$$\int u \, dv$$

One way to do this is to let

$$u = x \quad \text{and} \quad dv = e^x \, dx$$

so that

$$du = dx \quad \text{and} \quad v = \int e^x \, dx = e^x$$

Thus, from (2a)

$$\int \underset{u}{\underbrace{x}} \, \underset{dv}{\underbrace{e^x \, dx}} = \underset{u}{\underbrace{x}} \, \underset{v}{\underbrace{e^x}} - \int \underset{v}{\underbrace{e^x}} \, \underset{du}{\underbrace{dx}}$$

or

$$\int xe^x \, dx = xe^x - e^x + C \quad \blacktriangleleft$$

REMARK.   In the calculation of v from dv above, we omitted the constant of integration and wrote $v = \int e^x \, dx = e^x$. Had we included a constant of integration and written $v = \int e^x \, dx = e^x + C_1$, the constant C_1 would have eventually canceled out (Exercise 56(a)). This is always the case in integration by parts (Exercise 56(b)), so we shall usually omit the constant when calculating v from dv.

To use integration by parts successfully, the choice of u and dv must be made so that the new integral is easier than the original. For example, had we decided above to let

$$u = e^x \qquad dv = x \, dx$$

$$du = e^x \, dx \qquad v = \int x \, dx = \frac{x^2}{2}$$

then we would have obtained

$$\int xe^x \, dx = \int u \, dv = uv - \int v \, du = \frac{x^2}{2} e^x - \frac{1}{2} \int x^2 e^x \, dx$$

For this choice of u and dv the new integral is actually more complicated than the original. It is difficult to give hard and fast rules for choosing u and dv. It is a matter of experience that comes with lots of practice.

The next example shows that it is sometimes necessary to use integration by parts more than once in the same problem.

Example 2   Evaluate

$$\int x^2 e^{-x} \, dx$$

Solution.   Let

$$u = x^2 \qquad dv = e^{-x}\, dx$$

$$du = 2x\, dx \qquad v = \int e^{-x}\, dx = -e^{-x}$$

so that

$$\int x^2 e^{-x}\, dx = \int u\, dv = uv - \int v\, du = -x^2 e^{-x} + 2 \int x e^{-x}\, dx \qquad (3)$$

The last integral is similar to the original except that we have replaced x^2 by x. Another integration by parts applied to $\int x e^{-x}\, dx$ will complete the problem. We let

$$u = x \qquad dv = e^{-x}\, dx$$

$$du = dx \qquad v = \int e^{-x}\, dx = -e^{-x}$$

so that

$$\int x e^{-x}\, dx = \int u\, dv = uv - \int v\, du$$

$$= -x e^{-x} + \int e^{-x}\, dx$$

$$= -x e^{-x} - e^{-x} + C_1$$

Substituting in (3) we obtain

$$\int x^2 e^{-x}\, dx = -x^2 e^{-x} + 2(-x e^{-x} - e^{-x} + C_1)$$

$$= -x^2 e^{-x} - 2x e^{-x} - 2e^{-x} + 2C_1$$

$$= -(x^2 + 2x + 2)e^{-x} + C$$

where $C = 2C_1$.   ◄

Example 3   Evaluate

$$\int \ln x\, dx$$

Solution.   Let

$$u = \ln x \qquad dv = dx$$

$$du = \frac{1}{x} \, dx \qquad v = \int dx = x$$

so that

$$\int \ln x \, dx = \int u \, dv = uv - \int v \, du = x \ln x - \int x \left(\frac{1}{x} \right) dx$$

$$= x \ln x - \int dx = x \ln x - x + C \qquad \blacktriangleleft$$

The next example illustrates how integration by parts can be used to integrate the inverse trigonometric functions.

Example 4 Evaluate

$$\int_0^1 \tan^{-1} x \, dx$$

Solution. Let

$$u = \tan^{-1} x \qquad dv = dx$$

$$du = \frac{1}{1 + x^2} \, dx \qquad v = \int dx = x$$

Thus,

$$\int_0^1 \tan^{-1} x \, dx = \int_0^1 u \, dv = uv \Big]_0^1 - \int_0^1 v \, du$$

$$= x \tan^{-1} x \Big]_0^1 - \int_0^1 \frac{x}{1 + x^2} \, dx$$

> The limits of integration refer to x; that is, $x = 0$ and $x = 1$.

But

$$\int_0^1 \frac{x}{1 + x^2} \, dx = \frac{1}{2} \int_0^1 \frac{2x}{1 + x^2} \, dx = \frac{1}{2} \ln (1 + x^2) \Big]_0^1 = \frac{1}{2} \ln 2$$

so

$$\int_0^1 \tan^{-1} x \, dx = x \tan^{-1} x \Big]_0^1 - \frac{1}{2} \ln 2 = \left(\frac{\pi}{4} - 0 \right) - \frac{1}{2} \ln 2$$

$$= \frac{\pi}{4} - \ln \sqrt{2} \qquad \blacktriangleleft$$

Example 5   Evaluate

$$\int e^x \cos x \, dx$$

Solution.   Let

$$u = e^x \qquad dv = \cos x \, dx$$

$$du = e^x \, dx \qquad v = \int \cos x \, dx = \sin x$$

Thus,

$$\int e^x \cos x \, dx = \int u \, dv = uv - \int v \, du$$

$$= e^x \sin x - \int e^x \sin x \, dx \qquad (4)$$

Since the integral $\int e^x \sin x \, dx$ is similar in form to the original integral $\int e^x \cos x \, dx$, it seems that nothing has been accomplished. However, let us integrate this new integral by parts; we let

$$u = e^x \qquad dv = \sin x \, dx$$

$$du = e^x \, dx \qquad v = \int \sin x \, dx = -\cos x$$

Thus,

$$\int e^x \sin x \, dx = \int u \, dv = uv - \int v \, du$$

$$= -e^x \cos x + \int e^x \cos x \, dx$$

Substituting in (4) yields

$$\int e^x \cos x \, dx = e^x \sin x - \left[-e^x \cos x + \int e^x \cos x \, dx \right]$$

or

$$\int e^x \cos x \, dx = e^x \sin x + e^x \cos x - \int e^x \cos x \, dx$$

which is an equation we can solve for the unknown integral. We obtain

$$2 \int e^x \cos x \, dx = e^x \sin x + e^x \cos x$$

or

$$\int e^x \cos x \, dx = \frac{1}{2} e^x \sin x + \frac{1}{2} e^x \cos x + C \qquad \blacktriangleleft$$

REDUCTION FORMULAS

If n is a positive integer, then

$$\int \sin^n x \, dx \quad \text{and} \quad \int \cos^n x \, dx$$

can be evaluated by using *reduction formulas*. These are formulas that express the given integral in terms of a similar integral involving a *lower* power. For example, we can obtain a reduction formula for $\int \cos^n x \, dx$ by writing $\cos^n x$ as $\cos^{n-1} x \cdot \cos x$ and letting

$$u = \cos^{n-1} x \qquad\qquad dv = \cos x \, dx$$

$$v = \int \cos x \, dx = \sin x \qquad du = (n-1) \cos^{n-2} x(-\sin x) \, dx$$
$$= -(n-1) \cos^{n-2} x \sin x \, dx$$

so that

$$\int \cos^n x \, dx = \int \cos^{n-1} x \cos x \, dx = \int u \, dv = uv - \int v \, du$$

$$= \cos^{n-1} x \sin x + (n-1) \int \sin^2 x \cos^{n-2} x \, dx$$

$$= \cos^{n-1} x \sin x + (n-1) \int (1 - \cos^2 x) \cos^{n-2} x \, dx$$

$$= \cos^{n-1} x \sin x + (n-1) \int \cos^{n-2} x \, dx - (n-1) \int \cos^n x \, dx$$

Transposing the last term on the right to the left side yields

$$n \int \cos^n x \, dx = \cos^{n-1} x \sin x + (n-1) \int \cos^{n-2} x \, dx$$

or

$$\int \cos^n x \, dx = \frac{1}{n} \cos^{n-1} x \sin x + \frac{n-1}{n} \int \cos^{n-2} x \, dx \qquad (5)$$

This reduction formula reduces the exponent by 2. Thus, if we apply it repeatedly, we can eventually express $\int \cos^n x \, dx$ in terms of

$$\int \cos x \; dx = \sin x + C$$

if n is odd, or

$$\int \cos^0 x \; dx = \int dx = x + C$$

if n is even.

Example 6 Evaluate

(a) $\displaystyle\int \cos^3 x \; dx$ (b) $\displaystyle\int \cos^4 x \; dx$

Solution (a). From (5) with $n = 3$

$$\int \cos^3 x \; dx = \frac{1}{3} \cos^2 x \sin x + \frac{2}{3} \int \cos x \; dx$$

$$= \frac{1}{3} \cos^2 x \sin x + \frac{2}{3} \sin x + C$$

Solution (b). From (5) with $n = 4$

$$\int \cos^4 x \; dx = \frac{1}{4} \cos^3 x \sin x + \frac{3}{4} \int \cos^2 x \; dx$$

and from (5) with $n = 2$

$$\int \cos^2 x \; dx = \frac{1}{2} \cos x \sin x + \frac{1}{2} \int dx = \frac{1}{2} \cos x \sin x + \frac{1}{2} x + C_1$$

so that

$$\int \cos^4 x \; dx = \frac{1}{4} \cos^3 x \sin x + \frac{3}{4} \left(\frac{1}{2} \cos x \sin x + \frac{1}{2} x + C_1 \right)$$

$$= \frac{1}{4} \cos^3 x \sin x + \frac{3}{8} \cos x \sin x + \frac{3}{8} x + C$$

where $C = \frac{3}{4} C_1$ ◀

We leave it as an exercise to derive the following companion formula to (5).

$$\int \sin^n x \; dx = -\frac{1}{n} \sin^{n-1} x \cos x + \frac{n-1}{n} \int \sin^{n-2} x \; dx \qquad (6)$$

► Exercise Set 9.2

In Exercises 1–28, evaluate the integral.

1. $\int xe^{-x}\,dx.$

2. $\int xe^{3x}\,dx.$

3. $\int \ln(2x+3)\,dx.$

4. $\int x\ln x\,dx.$

5. $\int x\ln\sqrt{x}\,dx.$

6. $\int \sin^{-1}x\,dx.$

7. $\int \cos^{-1}(2x)\,dx.$

8. $\int x^2 e^{-2x}\,dx.$

9. $\int x^2 e^x\,dx.$

10. $\int x^3 e^{-x}\,dx.$

11. $\int e^x \sin x\,dx.$

12. $\int e^{-3\theta}\sin 3\theta\,d\theta.$

13. $\int \dfrac{\cos 2\pi x}{e^{2\pi x}}\,dx.$

14. $\int e^{2x}\cos 3x\,dx.$

15. $\int e^{ax}\sin bx\,dx.$

16. $\int x^2 \ln x\,dx.$

17. $\int x^2 \cos x\,dx.$

18. $\int x\tan^{-1}x\,dx.$

19. $\int x\sin(3x+1)\,dx.$

20. $\int x\sinh x\,dx.$

21. $\int x\sec^2 x\,dx.$

22. $\int \dfrac{x^3\,dx}{\sqrt{1-x^2}}.$

23. $\int \cos(\ln x)\,dx.$

24. $\int \sin(3\ln x)\,dx.$

25. $\int \sin(\ln x)\,dx.$

26. $\int (\ln x)^2\,dx.$

27. $\int x\tan^2 x\,dx.$

28. $\int \dfrac{xe^x}{(x+1)^2}\,dx.$

In Exercises 29–41, evaluate the definite integral.

29. $\displaystyle\int_0^1 xe^{-5x}\,dx.$

30. $\displaystyle\int_0^2 xe^{2x}\,dx.$

31. $\displaystyle\int_1^e x^2 \ln x\,dx.$

32. $\displaystyle\int_{\sqrt{e}}^{e} \dfrac{\ln x}{x^2}\,dx.$

33. $\displaystyle\int_{-2}^{2} \ln(x+3)\,dx.$

34. $\displaystyle\int_0^{1/2} \sin^{-1}x\,dx.$

35. $\displaystyle\int_2^4 \sec^{-1}\sqrt{\theta}\,d\theta.$

36. $\displaystyle\int_1^2 x\sec^{-1}x\,dx.$

37. $\displaystyle\int_0^{\pi/2} x\sin 4x\,dx.$

38. $\displaystyle\int_0^{\pi} (x+x\cos x)\,dx.$

39. $\displaystyle\int_1^3 \sqrt{x}\,\tan^{-1}\sqrt{x}\,dx.$

40. $\displaystyle\int_0^2 \ln(x^2+1)\,dx.$

41. $\displaystyle\int_0^1 \dfrac{x^3}{\sqrt{x^2+1}}\,dx.$

42. Solve Exercise 41 without using integration by parts.

43. (a) Find the area of the region enclosed by $y=\ln x$, the line $x=e$, and the x-axis.
    (b) Find the volume of the solid generated when the region in part (a) is revolved about the x-axis.

44. Find the area of the region enclosed by $y=x\sin x$, $y=x$, $x=0$, and $x=\pi/2$.

45. Find the volume of the solid generated when the region enclosed by $y=\sin x$, $y=0$, $x=0$, and $x=\pi$ is revolved about the y-axis.

46. Find the volume of the solid generated when the region enclosed by $y=\sin x\cos x$, $y=0$, $x=0$, and $x=\pi/2$ is revolved about the y-axis.

47. Use reduction formula (6) to evaluate

    (a) $\displaystyle\int \sin^3 x\,dx$ (b) $\displaystyle\int_0^{\pi/4} \sin^4 x\,dx.$

48. Use reduction formula (5) to evaluate

    (a) $\displaystyle\int \cos^5 x\,dx$ (b) $\displaystyle\int_0^{\pi/2} \cos^6 x\,dx.$

49. Use reduction formula (5) to help evaluate

    (a) $\displaystyle\int \cos^3 5x\,dx$ (b) $\displaystyle\int x\cos^4(x^2)\,dx.$

    [Hint: First make a substitution.]

50. Use reduction formula (6) to help evaluate

    (a) $\displaystyle\int \sin^4 2x\,dx$ (b) $\displaystyle\int \dfrac{\sin^3\sqrt{x}}{\sqrt{x}}\,dx.$

    [Hint: First make a substitution.]

51. Derive reduction formula (6).

In Exercises 52 and 53, derive the reduction formula in part (a) and use it to evaluate the integral in part (b).

52. (a) $\displaystyle\int \sec^n x \, dx =$

$$\frac{\sec^{n-2} x \tan x}{n - 1} + \frac{n - 2}{n - 1} \int \sec^{n-2} x \, dx$$

(b) $\displaystyle\int \sec^4 x \, dx.$

53. (a) $\displaystyle\int x^n e^x \, dx = x^n e^x - n \int x^{n-1} e^x \, dx$

(b) $\displaystyle\int x^3 e^x \, dx.$

54. Use the reduction formula in part (a) of Exercise 53 to help evaluate

(a) $\displaystyle\int x^2 e^{3x} \, dx$ (b) $\displaystyle\int_0^1 x e^{-\sqrt{x}} \, dx.$

[*Hint:* First make a substitution.]

55. Let f be a function whose second derivative is continuous on $[-1, 1]$. Show that

$$\int_{-1}^1 x f''(x) \, dx = f'(1) + f'(-1) + f(-1) - f(1)$$

56. (a) In Example 1, let

$$u = x \qquad dv = e^x \, dx$$

$$du = dx \qquad v = \int e^x \, dx = e^x + C_1$$

and show that the constant C_1 cancels out, thus leading to the same solution we obtained by omitting C_1.

(b) Show that

$$uv - \int v \, du = u(v + C_1) - \int (v + C_1) \, du$$

thereby justifying the omission of the constant of integration when calculating v in integration by parts.

57. Use integration by parts on $\displaystyle\int \frac{1}{x} \, dx$ with $u = 1/x$ and $dv = dx$. Explain.

9.3   INTEGRATING POWERS OF SINE AND COSINE

In this section we shall study methods for evaluating integrals of the form

$$\int \sin^m x \cos^n x \, dx$$

where m and n are nonnegative integers.

Integrals of the form

$$\int \sin^m x \, dx \quad (n = 0) \qquad \text{and} \qquad \int \cos^n x \, dx \quad (m = 0)$$

can be treated using reduction formulas (5) and (6) of the previous section.

Example 1   Show that

$$\int \sin^2 x \, dx = \frac{1}{2}x - \frac{1}{4}\sin 2x + C \tag{1}$$

$$\int \cos^2 x \, dx = \frac{1}{2}x + \frac{1}{4}\sin 2x + C \tag{2}$$

Solution. Letting $n = 2$ in reduction formulas (5) and (6) of Section 9.2 yields

$$\int \sin^2 x \, dx = -\frac{1}{2}\sin x \cos x + \frac{1}{2}x + C = \frac{1}{2}x - \frac{1}{4}\sin 2x + C$$

and

$$\int \cos^2 x \, dx = \frac{1}{2}\cos x \sin x + \frac{1}{2}x + C = \frac{1}{2}x + \frac{1}{4}\sin 2x + C$$

Alternative Solution. Apply the identities

$$\sin^2 x = \frac{1}{2}(1 - \cos 2x) \tag{3a}$$

$$\cos^2 x = \frac{1}{2}(1 + \cos 2x) \tag{3b}$$

which follow from the double-angle formulas $\cos 2x = 1 - 2\sin^2 x$ and $\cos 2x = 2\cos^2 x - 1$. We obtain

$$\int \sin^2 x \, dx = \frac{1}{2}\int (1 - \cos 2x) \, dx = \frac{1}{2}x - \frac{1}{4}\sin 2x + C$$

$$\int \cos^2 x \, dx = \frac{1}{2}\int (1 + \cos 2x) \, dx = \frac{1}{2}x + \frac{1}{4}\sin 2x + C \quad \blacktriangleleft$$

Example 2   The formulas

$$\int \sin^4 x \, dx = \frac{3}{8}x - \frac{1}{4}\sin 2x + \frac{1}{32}\sin 4x + C \tag{4}$$

$$\int \cos^4 x \, dx = \frac{3}{8}x + \frac{1}{4}\sin 2x + \frac{1}{32}\sin 4x + C \tag{5}$$

can be obtained from the reduction formulas and basic trigonometric identities. However, a more direct approach is as follows. From (3b)

$$\int \cos^4 x \, dx = \int (\cos^2 x)^2 \, dx = \int \left[\frac{1}{2}(1 + \cos 2x) \right]^2 dx$$

$$= \frac{1}{4} \int (1 + 2 \cos 2x + \cos^2 2x) \, dx$$

To finish, we apply (3b) again and write

$$\cos^2 2x = \frac{1}{2}(1 + \cos 4x) = \frac{1}{2} + \frac{1}{2} \cos 4x$$

which gives

$$\int \cos^4 x \, dx = \frac{1}{4} \int \left(\frac{3}{2} + 2 \cos 2x + \frac{1}{2} \cos 4x \right) dx$$

$$= \frac{3}{8}x + \frac{1}{4} \sin 2x + \frac{1}{32} \sin 4x + C$$

A similar procedure, using (3a), will yield (4). ◀

Example 3 Show that

$$\int \sin^3 x \, dx = -\cos x + \frac{1}{3} \cos^3 x + C \qquad (6)$$

$$\int \cos^3 x \, dx = \sin x - \frac{1}{3} \sin^3 x + C \qquad (7)$$

Solution. In the exercises we ask the reader to derive these results from reduction formulas. An alternative approach is as follows.

$$\int \sin^3 x \, dx = \int \sin^2 x \sin x \, dx$$

$$= \int (1 - \cos^2 x) \sin x \, dx$$

$$= \int \sin x \, dx - \int \cos^2 x \sin x \, dx$$

To integrate, let $u = \cos x$, $du = -\sin x \, dx$

$$= -\cos x + \frac{1}{3} \cos^3 x + C$$

$$\int \cos^3 x \, dx = \int \cos^2 x \cos x \, dx$$

$$= \int (1 - \sin^2 x) \cos x \, dx$$

$$= \int \cos x \, dx - \int \sin^2 x \cos x \, dx$$

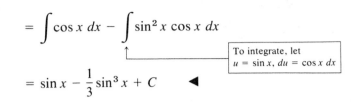

To integrate, let
$u = \sin x$, $du = \cos x \, dx$

$$= \sin x - \frac{1}{3} \sin^3 x + C \quad \blacktriangleleft$$

If m and n are both positive integers, then the integral

$$\int \sin^m x \cos^n x \, dx$$

can be evaluated by one of three procedures, depending on whether m and n are odd or even. The procedures are outlined in Table 9.3.1.

Table 9.3.1

CASE	PROCEDURE	RELEVANT IDENTITIES
n odd	Substitute $u = \sin x$	$\cos^2 x = 1 - \sin^2 x$
m odd	Substitute $u = \cos x$	$\sin^2 x = 1 - \cos^2 x$
$\begin{cases} m \text{ even} \\ n \text{ even} \end{cases}$	Use identities to reduce the powers on sin and cos	$\begin{cases} \sin^2 x = \frac{1}{2}(1 - \cos 2x) \\ \cos^2 x = \frac{1}{2}(1 + \cos 2x) \end{cases}$

Example 4   Evaluate

$$\int \sin^4 x \cos^5 x \, dx$$

Solution.   Since $n = 5$ is odd, we shall follow the first procedure in Table 9.3.1 and make the substitution

$$u = \sin x \quad du = \cos x \, dx$$

To form the du, we shall first split off a factor of $\cos x$.

$$\int \sin^4 x \cos^5 x \, dx = \int \sin^4 x \cos^4 x \cos x \, dx$$

$$= \int \sin^4 x (1 - \sin^2 x)^2 \cos x \, dx$$

$$= \int u^4 (1 - u^2)^2 \, du$$

$$= \int (u^4 - 2u^6 + u^8) \, du$$

$$= \frac{1}{5} u^5 - \frac{2}{7} u^7 + \frac{1}{9} u^9 + C$$

$$= \frac{1}{5}\sin^5 x - \frac{2}{7}\sin^7 x + \frac{1}{9}\sin^9 x + C \quad \blacktriangleleft$$

Example 5   Evaluate

$$\int \sin^3 x \cos^2 x \, dx$$

Solution.   Since $m = 3$ is odd, we shall follow the second procedure in Table 9.3.1 and make the substitution

$$u = \cos x \quad du = -\sin x \, dx$$

To form the du, we shall split off a factor of $\sin x$.

$$\int \sin^3 x \cos^2 x \, dx = \int \sin^2 x \cos^2 x \sin x \, dx$$

$$= \int (1 - \cos^2 x) \cos^2 x \sin x \, dx$$

$$= -\int (1 - u^2) u^2 \, du$$

$$= \int (u^4 - u^2) \, du$$

$$= \frac{1}{5}u^5 - \frac{1}{3}u^3 + C$$

$$= \frac{1}{5}\cos^5 x - \frac{1}{3}\cos^3 x + C \quad \blacktriangleleft$$

Example 6   Evaluate

$$\int \sin^4 x \cos^4 x \, dx$$

Solution.   Since $m = n = 4$, we shall follow the third procedure in Table 9.3.1.

$$\int \sin^4 x \cos^4 x \, dx = \int (\sin^2 x)^2 (\cos^2 x)^2 \, dx$$

$$= \int (\tfrac{1}{2}[1 - \cos 2x])^2 (\tfrac{1}{2}[1 + \cos 2x])^2 \, dx$$

$$= \frac{1}{16} \int (1 - \cos^2 2x)^2 \, dx$$

$$= \frac{1}{16} \int \sin^4 2x \, dx$$

To finish, we shall let $u = 2x$, $du = 2\,dx$ and then use (4).

$$\int \sin^4 x \cos^4 x\, dx = \frac{1}{32} \int \sin^4 u\, du$$

$$= \frac{1}{32}\left(\frac{3}{8}u - \frac{1}{4}\sin 2u + \frac{1}{32}\sin 4u\right) + C$$

$$= \frac{3}{128}x - \frac{1}{128}\sin 4x + \frac{1}{1024}\sin 8x + C \quad \blacktriangleleft$$

Integrals of the form

$$\int \sin mx \cos nx\, dx, \quad \int \sin mx \sin nx\, dx, \quad \int \cos mx \cos nx\, dx$$

can be found using the product formulas from trigonometry (see 19a, 19b, and 19c of the Unit I trigonometry review in Appendix 1).

Example 7 Evaluate

$$\int \sin 7x \cos 3x\, dx$$

Solution. Since

$$\sin 7x \cos 3x = \frac{1}{2}(\sin 4x + \sin 10x)$$

we can write

$$\int \sin 7x \cos 3x\, dx = \frac{1}{2}\int (\sin 4x + \sin 10x)\, dx$$

$$= -\frac{1}{8}\cos 4x - \frac{1}{20}\cos 10x + C \quad \blacktriangleleft$$

▶ **Exercise Set 9.3**

In Exercises 1–30, perform the indicated integration.

1. $\int \cos^5 x \sin x\, dx.$

2. $\int \sin^4 3x \cos 3x\, dx.$

3. $\int \sin ax \cos ax\, dx \quad (a \neq 0).$

4. $\int \cos^2 3x\, dx.$

5. $\int \sin^2 5\theta\, d\theta.$

6. $\int \cos^3 at\, dt \quad (a \neq 0).$

7. $\int \cos^4 \left(\frac{x}{4}\right) dx.$

8. $\int \sin^5 x\, dx.$

9. $\int \cos^5 \theta\, d\theta.$

10. $\int \sin^3 x \cos^3 x\, dx.$

11. $\int \sin^2 2t \cos^3 2t\, dt.$

12. $\displaystyle\int \sin^4 x \cos^5 x \, dx.$ **13.** $\displaystyle\int \cos^4 x \sin^3 x \, dx.$

14. $\displaystyle\int \sin^3 2x \cos^2 2x \, dx.$

15. $\displaystyle\int \sin^5 \theta \cos^4 \theta \, d\theta.$

16. $\displaystyle\int \cos^{1/5} x \sin x \, dx.$ **17.** $\displaystyle\int \sin^2 x \cos^2 x \, dx.$

18. $\displaystyle\int \sin^2 x \cos^4 x \, dx.$ **19.** $\displaystyle\int \sin x \cos 2x \, dx.$

20. $\displaystyle\int \sin 3\theta \cos 2\theta \, d\theta.$ **21.** $\displaystyle\int \sin x \cos\left(\frac{x}{2}\right) dx.$

22. $\displaystyle\int \sin ax \cos bx \, dx$ $(a > 0, b > 0, a \neq b).$

23. $\displaystyle\int \frac{\sin x}{\cos^8 x} \, dx.$ **24.** $\displaystyle\int \sqrt{\cos \theta} \sin \theta \, d\theta.$

25. $\displaystyle\int_0^{\pi/4} \cos^3 x \, dx.$ **26.** $\displaystyle\int_{-\pi}^{\pi} \cos^2 5\theta \, d\theta.$

27. $\displaystyle\int_0^{\pi/3} \sin^4 3x \cos^3 3x \, dx.$

28. $\displaystyle\int_0^{\pi/2} \sin^2 \frac{x}{2} \cos^2 \frac{x}{2} \, dx.$

29. $\displaystyle\int_0^{\pi/6} \sin 2x \cos 4x \, dx.$

30. $\displaystyle\int_0^{2\pi} \sin^2 kx \, dx$ $(k \neq 0).$

31. Let m, n be distinct nonnegative integers. Prove:

(a) $\displaystyle\int_0^{2\pi} \sin mx \cos nx \, dx = 0$

(b) $\displaystyle\int_0^{2\pi} \cos mx \cos nx \, dx = 0$

(c) $\displaystyle\int_0^{2\pi} \sin mx \sin nx \, dx = 0.$

32. The region bounded below by the x-axis and above by the portion of $y = \sin x$ from $x = 0$ to $x = \pi$ is revolved about the x-axis. Find the volume of the resulting solid.

33. Find the volume of the solid that results when the region enclosed by $y = \cos x$, $y = \sin x$, $x = 0$, and $x = \pi/4$ is revolved about the x-axis.

34. (a) Use Formula (6) in Section 9.2 to show that

$$\int_0^{\pi/2} \sin^n x \, dx = \frac{n-1}{n} \int_0^{\pi/2} \sin^{n-2} x \, dx$$

(b) Use this result to derive the **Wallis sine formulas:**

$$\int_0^{\pi/2} \sin^n x \, dx = \frac{\pi}{2} \cdot \frac{1 \cdot 3 \cdot 5 \cdots (n-1)}{2 \cdot 4 \cdot 6 \cdots n} \quad (n \text{ even})$$

$$\int_0^{\pi/2} \sin^n x \, dx = \frac{2 \cdot 4 \cdot 6 \cdots (n-1)}{1 \cdot 3 \cdot 5 \cdots n} \quad \left(\begin{matrix} n \text{ odd} \\ \text{and} \geq 3 \end{matrix}\right)$$

35. Use the Wallis formulas in Exercise 34 to evaluate:

(a) $\displaystyle\int_0^{\pi/2} \sin^3 x \, dx$ (b) $\displaystyle\int_0^{\pi/2} \sin^4 x \, dx$

(c) $\displaystyle\int_0^{\pi/2} \sin^5 x \, dx$ (d) $\displaystyle\int_0^{\pi/2} \sin^6 x \, dx.$

36. Use Formula (5) in Section 9.2 and the method of Exercise 34 to derive the **Wallis cosine formulas:**

$$\int_0^{\pi/2} \cos^n x \, dx = \frac{2 \cdot 4 \cdot 6 \cdots (n-1)}{3 \cdot 5 \cdot 7 \cdots n} \quad \left(\begin{matrix} n \text{ odd} \\ \text{and} \geq 3 \end{matrix}\right)$$

$$\int_0^{\pi/2} \cos^n x \, dx = \frac{\pi}{2} \cdot \frac{1 \cdot 3 \cdot 5 \cdots (n-1)}{2 \cdot 4 \cdot 6 \cdots n} \quad (n \text{ even})$$

37. Derive (6) and (7) using reduction formulas and appropriate trigonometric identities.

9.4   INTEGRATING POWERS OF SECANT AND TANGENT

In this section we shall discuss methods for evaluating integrals of the form

$$\int \tan^m x \, \sec^n x \, dx$$

where m and n are nonnegative integers.

We shall begin with the integrals

$$\int \tan x \, dx \quad (m = 1, n = 0)$$

and

$$\int \sec x \, dx \quad (m = 0, n = 1)$$

The first integral is evaluated by writing

$$\int \tan x \, dx = \int \frac{\sin x}{\cos x} \, dx$$

from which it follows ($u = \cos x$, $du = -\sin x \, dx$) that

$$\int \tan x = -\ln |\cos x| + C \tag{1a}$$

or since $-\ln|\cos x| = \ln(1/|\cos x|) = \ln|\sec x|$,

$$\int \tan x \, dx = \ln |\sec x| + C \tag{1b}$$

The second integral requires a trick. We write

$$\int \sec x \, dx = \int \sec x \left(\frac{\sec x + \tan x}{\sec x + \tan x} \right) dx$$

$$= \int \frac{\sec^2 x + \sec x \tan x}{\sec x + \tan x} \, dx$$

$$= \int \frac{du}{u} \qquad \boxed{\begin{aligned} u &= \sec x + \tan x \\ du &= (\sec^2 x + \sec x \tan x) \, dx \end{aligned}}$$

$$= \ln |u| + C$$

from which it follows that

$$\int \sec x \, dx = \ln |\sec x + \tan x| + C \tag{2}$$

Higher powers of secant and tangent can be evaluated using the reduction formulas

$$\int \sec^n x \, dx = \frac{\sec^{n-2} x \tan x}{n-1} + \frac{n-2}{n-1} \int \sec^{n-2} x \, dx \tag{3}$$

$$\int \tan^m x \, dx = \frac{\tan^{m-1} x}{m-1} - \int \tan^{m-2} x \, dx \tag{4}$$

Formula (3) was discussed in Exercise 52 of Section 9.2, and Formula (4) can be obtained from the identity

$$1 + \tan^2 x = \sec^2 x$$

by writing

$$\int \tan^m x \, dx = \int \tan^{m-2} x \tan^2 x \, dx = \int \tan^{m-2} x \, (\sec^2 x - 1) \, dx$$

$$= \int \tan^{m-2} x \sec^2 x \, dx - \int \tan^{m-2} x \, dx$$

$$= \frac{\tan^{m-1} x}{m-1} - \int \tan^{m-2} x \, dx \qquad \boxed{\begin{array}{l} \text{To integrate,} \\ \text{let } u = \tan x \\ du = \sec^2 x \, dx \end{array}}$$

Example 1   Evaluate

$$\int \sec^3 x \, dx$$

Solution.   From (3) with $n = 3$,

$$\int \sec^3 x \, dx = \frac{\sec x \tan x}{2} + \frac{1}{2} \int \sec x \, dx$$

$$= \frac{1}{2} \sec x \tan x + \frac{1}{2} \ln |\sec x + \tan x| + C \qquad \blacktriangleleft$$

Example 2   Evaluate

$$\int \tan^5 x \, dx$$

Solution. We shall use Formula (4) twice.

$$\int \tan^5 x \, dx = \frac{\tan^4 x}{4} - \int \tan^3 x \, dx$$

$$= \frac{\tan^4 x}{4} - \left[\frac{\tan^2 x}{2} - \int \tan x \, dx \right]$$

$$= \frac{1}{4} \tan^4 x - \frac{1}{2} \tan^2 x - \ln |\cos x| + C \quad \blacktriangleleft$$

If *m* and *n* are positive integers, then the integral

$$\int \tan^m x \, \sec^n x \, dx$$

can be evaluated by one of the three procedures in Table 9.4.1.

Table 9.4.1

CASE	PROCEDURE	RELEVANT IDENTITIES
n even	Substitute $u = \tan x$	$\sec^2 x = \tan^2 x + 1$
m odd	Substitute $u = \sec x$	$\tan^2 x = \sec^2 x - 1$
$\begin{cases} m \text{ even} \\ n \text{ odd} \end{cases}$	Reduce to powers of sec *x* alone	$\tan^2 x = \sec^2 x - 1$

Example 3 Evaluate

$$\int \tan^2 x \, \sec^4 x \, dx$$

Solution. Since *n* = 4 is even, we shall follow the first procedure in Table 9.4.1 and make the substitution

$$u = \tan x, \quad du = \sec^2 x \, dx$$

To form the *du*, we shall split off a factor of $\sec^2 x$.

$$\int \tan^2 x \, \sec^4 x \, dx = \int \tan^2 x \, \sec^2 x \, \sec^2 x \, dx$$

$$= \int \tan^2 x \, (\tan^2 x + 1) \, \sec^2 x \, dx$$

$$= \int u^2 (u^2 + 1) \, du$$

$$= \frac{1}{5} u^5 + \frac{1}{3} u^3 + C = \frac{1}{5} \tan^5 x + \frac{1}{3} \tan^3 x + C \quad \blacktriangleleft$$

Example 4 Evaluate

$$\int \tan^3 x \sec^3 x \, dx$$

Solution. Since $m = 3$ is odd, we shall follow the second procedure in Table 9.4.1 and make the substitution

$$u = \sec x, \quad du = \sec x \tan x \, dx$$

To form the du, we shall split off the product $\sec x \tan x$.

$$
\begin{aligned}
\int \tan^3 x \sec^3 x \, dx &= \int \tan^2 x \sec^2 x (\sec x \tan x) \, dx \\
&= \int (\sec^2 x - 1) \sec^2 x (\sec x \tan x) \, dx \\
&= \int (u^2 - 1) u^2 \, du \\
&= \frac{1}{5} u^5 - \frac{1}{3} u^3 + C = \frac{1}{5} \sec^5 x - \frac{1}{3} \sec^3 x + C \quad \blacktriangleleft
\end{aligned}
$$

Example 5 Evaluate

$$\int \tan^2 x \sec x \, dx$$

Solution. Since $m = 2$ and $n = 1$, we shall follow the third procedure in Table 9.4.1.

$$
\begin{aligned}
\int \tan^2 x \sec x \, dx & \\
&= \int (\sec^2 x - 1) \sec x \, dx = \int \sec^3 x \, dx - \int \sec x \, dx
\end{aligned}
$$

See Example 1

$$= \frac{1}{2} \sec x \tan x + \frac{1}{2} \ln |\sec x + \tan x| - \ln |\sec x + \tan x| + C$$

$$= \frac{1}{2} \sec x \tan x - \frac{1}{2} \ln |\sec x + \tan x| + C \quad \blacktriangleleft$$

Example 6 Instead of using reduction formula (3), the first procedure of Table 9.4.1 can be used to integrate an even power of $\sec x$. For example,

$$\int \sec^6 x \; dx = \int \sec^4 x \, \sec^2 x \; dx = \int (\sec^2 x)^2 \sec^2 x \; dx$$

$$= \int (\tan^2 x + 1)^2 \sec^2 x \; dx = \int (u^2 + 1)^2 \; du \qquad \boxed{\begin{array}{l} \text{To integrate,} \\ \text{let } u = \tan x \\ du = \sec^2 x \; dx \end{array}}$$

$$= \int (u^4 + 2u^2 + 1) \; du = \frac{1}{5} u^5 + \frac{2}{3} u^3 + u + C$$

$$= \frac{1}{5} \tan^5 x + \frac{2}{3} \tan^3 x + \tan x + C \qquad \blacktriangleleft$$

REMARK.   With the aid of the identity

$$1 + \cot^2 x = \csc^2 x$$

some of the techniques in this section can be adapted to treat integrals of the form

$$\int \cot^m x \, \csc^n x \; dx$$

▶ Exercise Set 9.4

In Exercises 1–34, perform the indicated integration.

1. $\displaystyle\int \sec^2 (3x + 1) \; dx.$ 2. $\displaystyle\int \tan 5x \; dx.$

3. $\displaystyle\int e^{-2x} \tan (e^{-2x}) \; dx.$ 4. $\displaystyle\int \cot 3x \; dx.$

5. $\displaystyle\int \sec 2x \; dx.$ 6. $\displaystyle\int \frac{\sec (\sqrt{x})}{\sqrt{x}} \; dx.$

7. $\displaystyle\int \tan^2 x \, \sec^2 x \; dx.$ 8. $\displaystyle\int \tan^5 x \, \sec^4 x \; dx.$

9. $\displaystyle\int \tan^3 4x \, \sec^4 4x \; dx.$

10. $\displaystyle\int \tan^4 \theta \, \sec^4 \theta \; d\theta.$

11. $\displaystyle\int \sec^5 x \, \tan^3 x \; dx.$ 12. $\displaystyle\int \tan^5 \theta \, \sec \theta \; d\theta.$

13. $\displaystyle\int \tan^4 x \, \sec x \; dx.$ 14. $\displaystyle\int \tan^2 \frac{x}{2} \, \sec^3 \frac{x}{2} \; dx.$

15. $\displaystyle\int \tan 2t \, \sec^3 2t \; dt.$ 16. $\displaystyle\int \tan x \, \sec^5 x \; dx.$

17. $\displaystyle\int \sec^4 x \; dx.$ 18. $\displaystyle\int \sec^5 x \; dx.$

19. $\displaystyle\int \sec^6 (\pi x) \; dx.$ 20. $\displaystyle\int \tan^3 4x \; dx.$

21. $\displaystyle\int \tan^4 x \; dx.$ 22. $\displaystyle\int \tan^7 \theta \; d\theta.$

23. $\displaystyle\int x \tan^2 (x^2) \sec^2 (x^2) \; dx.$

24. $\displaystyle\int \tan^2 (1 - 2x) \sec (1 - 2x) \; dx.$

25. $\displaystyle\int \cot^3 x \, \csc^3 x \; dx.$ 26. $\displaystyle\int \cot^2 3t \, \sec 3t \; dt.$

27. $\displaystyle\int \cot^3 x \; dx.$ 28. $\displaystyle\int \csc^4 x \; dx.$

29. $\displaystyle\int \sqrt{\tan x} \, \sec^4 x \; dx.$ 30. $\displaystyle\int \tan x \, \sec^{3/2} x \; dx.$

31. $\displaystyle\int_0^{\pi/6} \tan^2 2x \; dx.$ 32. $\displaystyle\int_0^{\pi/6} \sec^3 \theta \, \tan \theta \; d\theta.$

33. $\displaystyle\int_{0}^{\pi/2} \tan^5 \frac{x}{2}\, dx.$ **34.** $\displaystyle\int_{\pi/4}^{\pi/2} \csc^3 x \cot x \, dx.$ $\displaystyle\int \csc x \, dx = \ln|\tan \tfrac{1}{2}x| + C.$

35. Find the arc length of the curve $y = \ln(\cos x)$ over the interval $[0, \pi/4]$.

36. Find the volume of the solid generated when the region enclosed by $y = \tan x$, $y = 1$, and $x = 0$ is revolved about the x-axis.

37. (a) Show that

$$\int \csc x \, dx = -\ln|\csc x + \cot x| + C$$

(b) Show that the result in (a) can also be written

$$\int \csc x \, dx = \ln|\csc x - \cot x| + C$$

and

38. Rewrite $\sin x + \cos x$ in the form

$$A \sin(x + \phi)$$

and use your result together with Exercise 37 to evaluate

$$\int \frac{dx}{\sin x + \cos x}$$

39. Use the method of Exercise 38 to evaluate

$$\int \frac{dx}{a \sin x + b \cos x} \qquad (a, b \text{ not both zero})$$

40. Use integration by parts and Formula (2) to evaluate $\int \sec^3 x \, dx$.

9.5   TRIGONOMETRIC SUBSTITUTIONS

In this section we shall show how to evaluate integrals that contain expressions of the form

$$\sqrt{a^2 - x^2}, \quad \sqrt{x^2 + a^2}, \quad and \quad \sqrt{x^2 - a^2}$$

$(a > 0)$ *by making substitutions involving trigonometric functions.*

The basic idea for evaluating an integral that involves one of the radicals described above is to make a substitution that will eliminate the radical. For example, substituting

$$x = a \sin \theta, \quad -\pi/2 \leq \theta \leq \pi/2 \tag{1}$$

in $\sqrt{a^2 - x^2}$ yields

$$\sqrt{a^2 - x^2} = \sqrt{a^2 - a^2 \sin^2 \theta} = \sqrt{a^2(1 - \sin^2 \theta)}$$
$$= a\sqrt{\cos^2 \theta} = a\,|\cos \theta| = a \cos \theta$$

> $\cos \theta \geq 0$ since $-\pi/2 \leq \theta \leq \pi/2$

Thus, we have eliminated the radical by this substitution. The purpose of the restriction $-\pi/2 \le \theta \le \pi/2$ in (1) is twofold. First it enables us to rewrite (1) as

$$\theta = \sin^{-1}\left(\frac{x}{a}\right)$$

if desired (see Section 8.1), and second this restriction enables us to replace $|\cos \theta|$ by the simpler expression $\cos \theta$ in the resulting calculations.

Example 1   Evaluate

$$\int \frac{dx}{x^2\sqrt{4 - x^2}}$$

Solution.   To eliminate the radical we make the substitution

$$x = 2 \sin \theta, \quad -\pi/2 \le \theta \le \pi/2$$

so that

$$\frac{dx}{d\theta} = 2 \cos \theta \quad \text{or} \quad dx = 2 \cos \theta \, d\theta$$

This yields

$$\int \frac{dx}{x^2\sqrt{4 - x^2}} = \int \frac{2 \cos \theta \, d\theta}{(2 \sin \theta)^2 \sqrt{4 - 4 \sin^2 \theta}}$$

$$= \int \frac{2 \cos \theta \, d\theta}{(2 \sin \theta)^2 (2 \cos \theta)} = \frac{1}{4}\int \frac{d\theta}{\sin^2 \theta}$$

$$= \frac{1}{4}\int \csc^2 \theta \, d\theta = -\frac{1}{4}\cot \theta + C$$

$x = 2 \sin \theta$

Figure 9.5.1

To complete the solution we must express $\cot \theta$ in terms of x. This can be done using trigonometric identities or more simply by displaying the substitution $x = 2 \sin \theta$ ($\sin \theta = x/2$) as in Figure 9.5.1. From the figure we obtain

$$\cot \theta = \frac{\sqrt{4 - x^2}}{x}$$

so that

$$\int \frac{dx}{x^2\sqrt{4 - x^2}} = -\frac{1}{4}\cot \theta + C = -\frac{1}{4}\frac{\sqrt{4 - x^2}}{x} + C \quad \blacktriangleleft$$

To eliminate the radical in $\sqrt{x^2 + a^2}$ ($a > 0$), we can make the substitution

$$x = a \tan \theta, \quad -\pi/2 < \theta < \pi/2 \tag{2}$$

to obtain

$$\sqrt{x^2 + a^2} = \sqrt{a^2 \tan^2 \theta + a^2} = \sqrt{a^2 (1 + \tan^2 \theta)}$$
$$= \sqrt{a^2 \sec^2 \theta} = a \, |\sec \theta| = a \sec \theta$$

$$\boxed{\text{sec } \theta > 0 \text{ since } -\pi/2 < \theta < \pi/2}$$

As before, the restriction on θ in (2) enables us to write

$$\theta = \tan^{-1}\left(\frac{x}{a}\right)$$

if needed, and also eliminates the absolute value sign in the resulting computation.

Example 2   Evaluate

$$\int \frac{dx}{\sqrt{x^2 + a^2}}$$

Solution.   To eliminate the radical we make the substitution

$$x = a \tan \theta, \quad -\frac{\pi}{2} < \theta < \frac{\pi}{2}$$

so that

$$\frac{dx}{d\theta} = a \sec^2 \theta \quad \text{or} \quad dx = a \sec^2 \theta \, d\theta$$

This yields

$$\int \frac{dx}{\sqrt{x^2 + a^2}} = \int \frac{a \sec^2 d\theta}{\sqrt{a^2 \tan^2 \theta + a^2}} = \int \frac{a \sec^2 \theta \, d\theta}{a \sec \theta}$$
$$= \int \sec \theta \, d\theta = \ln |\sec \theta + \tan \theta| + C$$

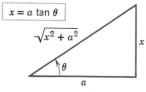

Figure 9.5.2

To express the solution in terms of x, we use Figure 9.5.2, which yields

$$\int \frac{dx}{\sqrt{x^2 + a^2}} = \ln \left| \frac{\sqrt{x^2 + a^2}}{a} + \frac{x}{a} \right| + C$$

or if we prefer we can rewrite the expression on the right as

$$\ln\left|\sqrt{x^2 + a^2} + x\right| - \ln a + C$$

and combine the constant $\ln a$ with the constant of integration to obtain

$$\int \frac{dx}{\sqrt{x^2 + a^2}} = \ln\left|\sqrt{x^2 + a^2} + x\right| + C'$$

Moreover, $\sqrt{x^2 + a^2} + x > 0$ for all x, so that we can drop the absolute value sign and write

$$\int \frac{dx}{\sqrt{x^2 + a^2}} = \ln\left(\sqrt{x^2 + a^2} + x\right) + C' \tag{3}$$

◄

The integral in the last example can also be evaluated by making the substitution

$$x = au \quad \text{or} \quad u = \frac{x}{a}$$

so that

$$dx = a\, du$$

which yields

$$\int \frac{dx}{\sqrt{x^2 + a^2}} = \int \frac{a\, du}{\sqrt{a^2 u^2 + a^2}} = \int \frac{du}{\sqrt{u^2 + 1}} = \sinh^{-1} u + C$$

or

$$\int \frac{dx}{\sqrt{x^2 + a^2}} = \sinh^{-1}\left(\frac{x}{a}\right) + C \tag{4}$$

Using the logarithmic expression for $\sinh^{-1}$ in Theorem 8.3.2, the reader can show that (3) and (4) are equivalent.

To eliminate the radical $\sqrt{x^2 - a^2}$, we can make the substitution

$$x = a \sec\theta, \quad 0 \le \theta < \pi/2 \quad \text{or} \quad \pi \le \theta < 3\pi/2 \tag{5}$$

to obtain

$$\sqrt{x^2 - a^2} = \sqrt{a^2 \sec^2\theta - a^2}$$
$$= \sqrt{a^2(\sec^2\theta - 1)} = \sqrt{a^2 \tan^2\theta}$$
$$= a\left|\tan\theta\right| = a\tan\theta$$

The removal of the absolute value sign is justified because the restriction on θ in (5) implies that $\tan \theta \geq 0$. This restriction also enables us to write

$$\theta = \sec^{-1}\left(\frac{x}{a}\right)$$

when needed (Theorem 8.1.3).

Example 3   Evaluate

$$\int \frac{\sqrt{x^2 - 25}}{x}\, dx$$

Solution.   To eliminate the radical, we make the substitution

$$x = 5 \sec \theta, \quad 0 \leq \theta < \pi/2 \quad \text{or} \quad \pi \leq \theta < 3\pi/2$$

so that

$$\frac{dx}{d\theta} = 5 \sec \theta \tan \theta \quad \text{or} \quad dx = 5 \sec \theta \tan \theta\, d\theta$$

Thus,

$$\int \frac{\sqrt{x^2 - 25}}{x}\, dx = \int \frac{\sqrt{25 \sec^2 \theta - 25}}{5 \sec \theta} (5 \sec \theta \tan \theta)\, d\theta$$

$$= \int \frac{5 \tan \theta}{5 \sec \theta} (5 \sec \theta \tan \theta)\, d\theta$$

$$= 5 \int \tan^2 \theta\, d\theta$$

$$= 5 \int (\sec^2 \theta - 1)\, d\theta$$

$$= 5 \tan \theta - 5\theta + C$$

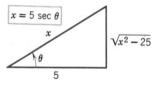

Figure 9.5.3

But $\theta = \sec^{-1}(x/5)$ and from Figure 9.5.3,

$$\tan \theta = \frac{\sqrt{x^2 - 25}}{5}$$

so that

$$\int \frac{\sqrt{x^2 - 25}}{x}\, dx = \sqrt{x^2 - 25} - 5 \sec^{-1}\left(\frac{x}{5}\right) + C \quad \blacktriangleleft$$

The integral in the next example will arise frequently in later sections.

Example 4 Evaluate

$$\int_{-a}^{a} \sqrt{a^2 - x^2}\, dx \quad (a > 0)$$

Figure 9.5.4

Solution. This integral can be evaluated by the substitution $x = a \sin \theta$, but the computations are tedious. A better approach is to observe that the integral represents the area of a semicircle of radius a (Figure 9.5.4). Thus,

$$\int_{-a}^{a} \sqrt{a^2 - x^2}\, dx = \frac{1}{2} \pi a^2 \quad \blacktriangleleft$$

▶ **Exercise Set 9.5**

In Exercises 1–30, perform the indicated integration.

1. $\displaystyle\int \sqrt{4 - x^2}\, dx.$

2. $\displaystyle\int \sqrt{1 - 4x^2}\, dx.$

3. $\displaystyle\int \frac{x^2}{\sqrt{9 - x^2}}\, dx.$

4. $\displaystyle\int \frac{dx}{x^2\sqrt{16 - x^2}}.$

5. $\displaystyle\int \frac{dx}{(4 + x^2)^2}.$

6. $\displaystyle\int \frac{dx}{(x^2 + 1)^{3/2}}.$

7. $\displaystyle\int \frac{\sqrt{x^2 - 9}}{x}\, dx.$

8. $\displaystyle\int \frac{dx}{x^2\sqrt{x^2 - 16}}.$

9. $\displaystyle\int \frac{x^3}{\sqrt{2 - x^2}}\, dx.$

10. $\displaystyle\int x^3\sqrt{5 - x^2}\, dx.$

11. $\displaystyle\int \frac{dx}{(3 + x^2)^{3/2}}.$

12. $\displaystyle\int \frac{x^2}{\sqrt{5 + x^2}}\, dx.$

13. $\displaystyle\int \frac{dx}{x^2\sqrt{4x^2 - 9}}.$

14. $\displaystyle\int \frac{\sqrt{1 + t^2}}{t}\, dt.$

15. $\displaystyle\int \frac{dx}{(1 - x^2)^{3/2}}.$

16. $\displaystyle\int \frac{dx}{x^2\sqrt{x^2 + 25}}.$

17. $\displaystyle\int \frac{x^2}{1 + x^2}\, dx.$

18. $\displaystyle\int \frac{dx}{1 + 2x^2 + x^4}.$

19. $\displaystyle\int \frac{dx}{x^2\sqrt{9 - 4x^2}}.$

20. $\displaystyle\int \frac{x^2}{\sqrt{x^2 - 25}}\, dx.$

21. $\displaystyle\int \frac{dx}{(9x^2 - 1)^{3/2}}.$

22. $\displaystyle\int \frac{\cos \theta}{\sqrt{2 - \sin^2 \theta}}\, d\theta.$

23. $\displaystyle\int e^x\sqrt{1 - e^{2x}}\, dx.$

24. $\displaystyle\int_{0}^{1/3} \frac{dx}{(4 - 9x^2)^2}.$

25. $\displaystyle\int_{0}^{4} x^3\sqrt{16 - x^2}\, dx.$

26. $\displaystyle\int_{\sqrt{2}}^{2} \frac{\sqrt{2x^2 - 4}}{x}\, dx.$

27. $\displaystyle\int_{\sqrt{2}}^{2} \frac{dx}{x^2\sqrt{x^2 - 1}}.$

28. $\displaystyle\int_{-1/\sqrt{2}}^{1/\sqrt{2}} (1 - 2x^2)^{3/2}\, dx.$

29. $\displaystyle\int_{1}^{3} \frac{dx}{x^4\sqrt{x^2 + 3}}.$

30. $\displaystyle\int_{0}^{3} \frac{x^3}{(3 + x^2)^{5/2}}\, dx.$

31. The integral

$$\int \frac{x}{x^2 + 4}\, dx$$

can be evaluated either by a trigonometric substitution or by the substitution $u = x^2 + 4$. Do it both ways and show that the results are equivalent.

32. By integrating, prove that the area of a circle of radius r is πr^2. [*Hint:* $x^2 + y^2 = r^2$ is the equation of such a circle.]

33. Find the arc length of the curve $y = \ln x$ from $x = 1$ to $x = 2$.

34. Find the arc length of the curve $y = x^2$ from $x = 0$ to $x = 1$.

35. Find the area of the surface generated when the curve in Exercise 34 is revolved about the x-axis.

36. Find the volume of the solid generated when the region enclosed by $x = y(1 - y^2)^{1/4}$, $y = 0$, $y = 1$, and $x = 0$ is revolved about the y-axis.

In cases where the trigonometric substitutions $x = a \sec \theta$ and $x = a \tan \theta$ lead to difficult integrals, it is sometimes possible to use the **hyperbolic substitutions:**

$x = a \sinh u$ for integrals involving $\sqrt{x^2 + a^2}$

$x = a \cosh u$ for integrals involving $\sqrt{x^2 - a^2}$

These substitutions are useful because in each case the hyperbolic identity

$$a^2 \cosh^2 u - a^2 \sinh^2 u = a^2$$

removes the radical.

37. (a) Evaluate

$$\int \frac{dx}{\sqrt{x^2 + 9}}$$

using the hyperbolic substitution suggested above.

(b) Evaluate the integral in (a) by a trigonometric substitution and show that the results in (a) and (b) agree.

38. Follow the directions of Exercise 37 for the integral

$$\int \sqrt{x^2 - 1} \, dx, \quad x \geq 1$$

39. For the substitution $x = a \sin \theta$, $-\pi/2 \leq \theta \leq \pi/2$, the triangle below suggests the relationships

(a) $\cos \theta = \dfrac{1}{a} \sqrt{a^2 - x^2}$

(b) $\tan \theta = \dfrac{x}{\sqrt{a^2 - x^2}}$.

Use trigonometric identities to prove that these results are correct.

9.6 INTEGRALS INVOLVING $ax^2 + bx + c$

In this section we shall discuss integrals whose integrands contain a quadratic expression $ax^2 + bx + c$.

Integrals that involve a quadratic expression $ax^2 + bx + c$, where $b \neq 0$ can often be evaluated by first completing the square as follows:

$$ax^2 + bx + c = a\left(x^2 + \frac{b}{a}x\right) + c$$

$$= a\left(x^2 + \frac{b}{a}x + \frac{b^2}{4a^2}\right) + c - \frac{b^2}{4a}$$

$$= a\left(x + \frac{b}{2a}\right)^2 + c - \frac{b^2}{4a}$$

At this point, the substitution

$$u = x + \frac{b}{2a}$$

will reduce the original expression $ax^2 + bx + c$ to the simpler form $au^2 + d$, where $d = c - (b^2/4a)$. Once this simplification is made, methods discussed in previous sections can be applied.

Example 1   Evaluate

$$\int \frac{dx}{x^2 - 2x + 5}$$

Solution.   Completing the square yields

$$x^2 - 2x + 5 = (x^2 - 2x + 1) + 5 - 1 = (x - 1)^2 + 4$$

Thus,

$$\int \frac{dx}{x^2 - 2x + 5} = \int \frac{dx}{(x - 1)^2 + 4} = \int \frac{du}{u^2 + 4}$$

$$\begin{array}{c} u = x - 1 \\ du = dx \end{array}$$

$$= \frac{1}{2} \tan^{-1} \frac{u}{2} + C = \frac{1}{2} \tan^{-1} \left(\frac{x - 1}{2} \right) + C$$

Formula (11) of Section 8.2

◀

Example 2   Evaluate

$$\int \frac{dx}{\sqrt{5 - 4x - 2x^2}}$$

Solution.   Completing the square yields

$$\begin{aligned} 5 - 4x - 2x^2 &= 5 - 2(x^2 + 2x) \\ &= 5 - 2(x^2 + 2x + 1) + 2 \\ &= 5 - 2(x + 1)^2 + 2 = 7 - 2(x + 1)^2 \end{aligned}$$

Thus,

$$\int \frac{dx}{\sqrt{5 - 4x - 2x^2}} = \int \frac{dx}{\sqrt{7 - 2(x + 1)^2}}$$

$$= \int \frac{du}{\sqrt{7 - 2u^2}} \qquad \begin{array}{c} u = x + 1 \\ du = dx \end{array}$$

$$= \frac{1}{\sqrt{2}} \int \frac{du}{\sqrt{(7/2) - u^2}}$$

$$= \frac{1}{\sqrt{2}} \sin^{-1}\left(\frac{u}{\sqrt{7/2}}\right) + C \qquad \boxed{\begin{array}{l} \text{Formula (10),} \\ \text{Section 8.2} \\ \text{with } a = \sqrt{7/2} \end{array}}$$

$$= \frac{1}{\sqrt{2}} \sin^{-1}(\sqrt{2/7}\,u) + C$$

$$= \frac{1}{\sqrt{2}} \sin^{-1}(\sqrt{2/7}(x + 1)) + C \qquad \blacktriangleleft$$

Example 3   Evaluate

$$\int \frac{x}{x^2 - 4x + 8}\,dx$$

Solution.   Completing the square yields

$$x^2 - 4x + 8 = (x^2 - 4x + 4) + 8 - 4 = (x - 2)^2 + 4$$

Thus, the substitution

$$u = x - 2, \quad du = dx$$

yields

$$\int \frac{x}{x^2 - 4x + 8}\,dx = \int \frac{x}{(x - 2)^2 + 4}\,dx = \int \frac{u + 2}{u^2 + 4}\,du$$

$$= \int \frac{u}{u^2 + 4}\,du + 2\int \frac{du}{u^2 + 4}$$

$$= \frac{1}{2} \int \frac{2u}{u^2 + 4}\,du + 2\int \frac{du}{u^2 + 4}$$

$$= \frac{1}{2} \ln (u^2 + 4) + 2\left(\frac{1}{2}\right) \tan^{-1}\frac{u}{2} + C$$

$$= \frac{1}{2} \ln [(x - 2)^2 + 4] + \tan^{-1}\left(\frac{x - 2}{2}\right) + C \qquad \blacktriangleleft$$

▶ **Exercise Set 9.6**

Evaluate the integrals in Exercises 1–15.

1. $\displaystyle\int \frac{dx}{x^2 - 4x + 13}.$ **2.** $\displaystyle\int \frac{dx}{\sqrt{2x - x^2}}.$ **7.** $\displaystyle\int \sqrt{3 - 2x - x^2}\,dx.$

3. $\displaystyle\int \frac{dx}{\sqrt{8 + 2x - x^2}}.$ **4.** $\displaystyle\int \frac{dx}{16x^2 + 16x + 5}.$ **8.** $\displaystyle\int \frac{e^x}{\sqrt{1 + e^x + e^{2x}}}\,dx.$

5. $\displaystyle\int \frac{dx}{\sqrt{x^2 - 6x + 10}}.$ **6.** $\displaystyle\int \frac{x}{x^2 + 6x + 10}\,dx.$ **9.** $\displaystyle\int \frac{dx}{2x^2 + 4x + 7}.$

10. $\displaystyle\int \frac{\cos\theta}{\sin^2\theta - 6\sin\theta + 12}\,d\theta.$

13. $\displaystyle\int \frac{x + 3}{\sqrt{x^2 + 2x + 2}}\,dx.$

11. $\displaystyle\int \frac{2x + 5}{x^2 + 2x + 5}\,dx.$ **12.** $\displaystyle\int \frac{2x + 3}{4x^2 + 4x + 5}\,dx.$

14. $\displaystyle\int_1^2 \frac{dx}{\sqrt{4x - x^2}}.$

15. $\displaystyle\int_0^1 \sqrt{x(4 - x)}\,dx.$

9.7 INTEGRATING RATIONAL FUNCTIONS; PARTIAL FRACTIONS

Recall that a rational function is the quotient of two polynomials. Some examples are:

$$\frac{3x}{x^2 + 1}, \quad \frac{6x^2 - x + 2}{(x - 1)(x^2 + 4)}, \quad \frac{\frac{1}{2}x^3 + 2}{x^3 - \frac{1}{3}}, \quad \frac{7x^4 - 8x^2 + 2x + 1}{3x^3 + 2x - 5} \tag{1}$$

In this section we shall discuss a method for integrating rational functions.

In algebra we learn to combine two or more fractions into a single fraction; for example,

$$\frac{1}{x} + \frac{3}{x - 1} + \frac{2}{x + 2} = \frac{(x - 1)(x + 2) + 3x(x + 2) + 2x(x - 1)}{x(x - 1)(x + 2)}$$

$$= \frac{6x^2 + 5x - 2}{x^3 + x^2 - 2x} \tag{2}$$

However, the left side of (2) is easier to integrate than the right. Thus, it would be helpful if we knew how to obtain the left side of the equation starting with the right. The procedure for doing this is called **partial fraction decomposition**.

A rational function

$$\frac{P(x)}{Q(x)} \tag{3}$$

is called **proper** if the degree of $P(x)$ is less than the degree of $Q(x)$; otherwise it is **improper**. Thus, in (1), the first two rational functions are proper and the last two are improper. It can be proved that any proper rational function is expressible as a sum of terms (called **partial fractions**) having the form:

$$\frac{A}{(ax + b)^k} \quad \text{or} \quad \frac{Bx + C}{(ax^2 + bx + c)^k}$$

The exact number of terms of each type depends on how the denominator $Q(x)$ in (3) factors. In theory, a polynomial $Q(x)$ with real coefficients can always be factored into a product of linear and quadratic factors with real coefficients. For example, the polynomial

$$Q(x) = x^3 - 3x^2 + x - 3$$

factors into

$$Q(x) = (x - 3)(x^2 + 1)$$

The quadratic factor $x^2 + 1$ cannot be further decomposed into linear factors without using imaginary numbers $[x^2 + 1 = (x - i)(x + i)]$. Such quadratic factors are said to be *irreducible.*

The first step in the partial fraction decomposition of $P(x)/Q(x)$ is to completely factor the denominator $Q(x)$ into linear and irreducible quadratic factors and then collect all repeated factors so that $Q(x)$ is expressed as a product of *distinct* factors of the form

$$(ax + b)^m \quad \text{and} \quad (ax^2 + bx + c)^m$$

where $ax^2 + bx + c$ is irreducible. Once this is done, the structure of the partial fraction decomposition of $P(x)/Q(x)$ is determined as follows:

Linear Factors
For each factor of the form $(ax + b)^m$, introduce the m terms

$$\frac{A_1}{ax + b} + \frac{A_2}{(ax + b)^2} + \cdots + \frac{A_m}{(ax + b)^m}$$

where $A_1, A_2, \ldots, A_m$ are constants to be determined.

Irreducible Quadratic Factors
For each factor of the form $(ax^2 + bx + c)^m$, introduce m terms

$$\frac{A_1 x + B_1}{ax^2 + bx + c} + \frac{A_2 x + B_2}{(ax^2 + bx + c)^2} + \cdots + \frac{A_m x + B_m}{(ax^2 + bx + c)^m}$$

where $A_1, A_2, \ldots, A_m, B_1, B_2, \ldots, B_m$ are constants to be determined.

Example 1   Evaluate

$$\int \frac{dx}{x^2 + x - 2}$$

Solution.   The integrand can be written as

$$\frac{1}{x^2 + x - 2} = \frac{1}{(x - 1)(x + 2)}$$

According to the rule above for linear factors, the factor $x - 1$ introduces one term (since $m = 1$) of the form

$$\frac{A}{x - 1}$$

and the factor $x + 2$ introduces one term of the form

$$\frac{B}{x + 2}$$

so that the partial fraction decomposition is

$$\frac{1}{(x - 1)(x + 2)} = \frac{A}{x - 1} + \frac{B}{x + 2} \tag{4}$$

where A and B are constants to be determined so that (4) becomes an *identity*. To find these constants we can multiply both sides of (4) by $(x - 1)(x + 2)$ to obtain

$$1 = A(x + 2) + B(x - 1) \tag{5}$$

Next, we substitute values of x to make the various terms zero. Setting $x = -2$ in (5) yields

$$1 = -3B \quad \text{or} \quad B = -\frac{1}{3}$$

and setting $x = 1$ in (5) yields

$$1 = 3A \quad \text{or} \quad A = \frac{1}{3}$$

Thus, (4) becomes

$$\frac{1}{(x - 1)(x + 2)} = \frac{1/3}{x - 1} + \frac{-1/3}{x + 2}$$

and

$$\int \frac{dx}{(x - 1)(x + 2)} = \frac{1}{3} \int \frac{dx}{x - 1} - \frac{1}{3} \int \frac{dx}{x + 2}$$

$$= \frac{1}{3} \ln|x - 1| - \frac{1}{3} \ln|x + 2| + C$$

$$= \frac{1}{3} \ln \left| \frac{x - 1}{x + 2} \right| + C$$

Alternative Solution. The constants A and B in (5) can also be determined by collecting like terms:

$$1 = (A + B)x + (2A - B) \tag{6}$$

and then equating corresponding coefficients on both sides:

$$A + B = 0$$
$$2A - B = 1$$

and then solving these equations simultaneously to obtain $A = \frac{1}{3}$, $B = -\frac{1}{3}$. This method is justified because (6) is an *identity* holding for all x, and two polynomials are equal for all x if and only if their corresponding coefficients are equal (Exercise 52). ◄

Example 2 Evaluate

$$\int \frac{2x + 4}{x^3 - 2x^2} \, dx$$

Solution. The integrand can be rewritten as

$$\frac{2x + 4}{x^3 - 2x^2} = \frac{2x + 4}{x^2(x - 2)}$$

Although x^2 is a quadratic factor, it is *not* irreducible since $x^2 = xx$. Thus, by the rule for linear factors, x^2 introduces two terms (since $m = 2$) of the form

$$\frac{A}{x} + \frac{B}{x^2}$$

and the factor $x - 2$ introduces one term (since $m = 1$) of the form

$$\frac{C}{x - 2}$$

so the partial fraction decomposition is

$$\frac{2x + 4}{x^2(x - 2)} = \frac{A}{x} + \frac{B}{x^2} + \frac{C}{x - 2} \tag{7}$$

Multiplying by $x^2(x - 2)$ yields

$$2x + 4 = Ax(x - 2) + B(x - 2) + Cx^2 \tag{8}$$

To determine A, B, and C we shall follow the alternative method used in Example 1. Multiplying and collecting like terms in (8) yields

$$2x + 4 = (A + C)x^2 + (-2A + B)x - 2B$$

Equating corresponding coefficients gives

$$A \qquad + C = 0$$
$$-2A + B \qquad = 2$$
$$-2B \qquad = 4$$

and solving this system yields

$$A = -2, \quad B = -2, \quad C = 2$$

so (7) becomes

$$\frac{2x + 4}{x^2(x - 2)} = \frac{-2}{x} + \frac{-2}{x^2} + \frac{2}{x - 2}$$

Thus,

$$\int \frac{2x + 4}{x^2(x - 2)} \, dx = -2 \int \frac{dx}{x} - 2 \int \frac{dx}{x^2} + 2 \int \frac{dx}{x - 2}$$

$$= -2 \ln |x| + \frac{2}{x} + 2 \ln |x - 2| + C$$

$$= 2 \ln \left| \frac{x - 2}{x} \right| + \frac{2}{x} + C \qquad \blacktriangleleft$$

Example 3 Evaluate

$$\int \frac{x^2 + x - 2}{3x^3 - x^2 + 3x - 1} \, dx$$

Solution. The denominator in the integrand can be factored by grouping:

$$\frac{x^2 + x - 2}{3x^3 - x^2 + 3x - 1} = \frac{x^2 + x - 2}{x^2(3x - 1) + (3x - 1)} = \frac{x^2 + x - 2}{(3x - 1)(x^2 + 1)}$$

By the rule for linear factors, the factor $3x - 1$ introduces one term:

$$\frac{A}{3x - 1}$$

and by the rule for irreducible quadratic factors, the factor $x^2 + 1$ introduces one term:

$$\frac{Bx + C}{x^2 + 1}$$

Thus, the partial fraction decomposition is

$$\frac{x^2 + x - 2}{(3x - 1)(x^2 + 1)} = \frac{A}{3x - 1} + \frac{Bx + C}{x^2 + 1} \tag{9}$$

Multiplying by $(3x - 1)(x^2 + 1)$ yields

$$x^2 + x - 2 = A(x^2 + 1) + (Bx + C)(3x - 1)$$

To determine A, B, and C, we multiply out and collect like terms:

$$x^2 + x - 2 = (A + 3B)x^2 + (-B + 3C)x + (A - C)$$

Equating corresponding coefficients gives

$$
\begin{aligned}
A + 3B \quad\;\; &= \quad 1 \\
- \; B + 3C &= \quad 1 \\
A \qquad\;\; - \; C &= -2
\end{aligned}
$$

To solve this system, subtract the third equation from the first to eliminate A. Then use the resulting equation together with the second equation to solve for B and C. Finally, determine A from the first or third equation. This yields (verify):

$$A = -\frac{7}{5}, \quad B = \frac{4}{5}, \quad C = \frac{3}{5}$$

Thus, (9) becomes

$$\frac{x^2 + x - 2}{(3x - 1)(x^2 + 1)} = \frac{-\frac{7}{5}}{3x - 1} + \frac{\frac{4}{5}x + \frac{3}{5}}{x^2 + 1}$$

and

$$
\begin{aligned}
\int \frac{x^2 + x - 2}{(3x - 1)(x^2 + 1)}\, dx \\
= -\frac{7}{5} \int \frac{dx}{3x - 1} + \frac{4}{5} \int \frac{x}{x^2 + 1}\, dx + \frac{3}{5} \int \frac{dx}{x^2 + 1} \\
= -\frac{7}{15} \ln|3x - 1| + \frac{2}{5} \ln(x^2 + 1) + \frac{3}{5} \tan^{-1} x + C \quad \blacktriangleleft
\end{aligned}
$$

Example 4 Evaluate

$$\int \frac{3x^4 + 4x^3 + 16x^2 + 20x + 9}{(x + 2)(x^2 + 3)^2}\, dx$$

Solution. By the rule for linear factors, the factor $x + 2$ introduces one term:

$$\frac{A}{x + 2}$$

and by the rule for irreducible quadratic factors, the factor $(x^2 + 3)^2$ introduces two terms (since $m = 2$):

$$\frac{Bx + C}{x^2 + 3} + \frac{Dx + E}{(x^2 + 3)^2}$$

Thus, the partial fraction decomposition of the integrand is

$$\frac{3x^4 + 4x^3 + 16x^2 + 20x + 9}{(x + 2)(x^2 + 3)^2} = \frac{A}{x + 2} + \frac{Bx + C}{x^2 + 3} + \frac{Dx + E}{(x^2 + 3)^2} \quad (10)$$

Multiplying by $(x + 2)(x^2 + 3)^2$ yields

$$3x^4 + 4x^3 + 16x^2 + 20x + 9$$
$$= A(x^2 + 3)^2 + (Bx + C)(x^2 + 3)(x + 2) + (Dx + E)(x + 2) \quad (11)$$

Multiplying out and collecting terms on the right side of (11) yields

$$(A + B)x^4 + (2B + C)x^3 + (6A + 3B + 2C + D)x^2$$
$$+ (6B + 3C + 2D + E)x + (9A + 6C + 2E)$$

and equating corresponding coefficients with the left side of (11) yields

$$\begin{aligned} A + B &= 3 \\ 2B + C &= 4 \\ 6A + 3B + 2C + D &= 16 \\ 6B + 3C + 2D + E &= 20 \\ 9A + 6C + 2E &= 9 \end{aligned} \quad (12)$$

This system of five equations in five unknowns is tedious to solve. However, we can reduce the work considerably by substituting $x = -2$ in (11) to obtain

$$49 = 49A \quad \text{or} \quad A = 1$$

(The choice of $x = -2$ is suggested by the fact that all terms on the right side of (11) except one have $x + 2$ as a factor, so all but one drop out with this substitution.) Substituting $A = 1$ in the first equation of (12) gives $B = 2$ and substituting this value of B in the second equation of (12) gives $C = 0$; continuing in this way we are led to the values (verify):

$$A = 1, \quad B = 2, \quad C = 0, \quad D = 4, \quad E = 0$$

Thus, (10) becomes

$$\frac{3x^4 + 4x^3 + 16x^2 + 20x + 9}{(x + 2)(x^2 + 3)^2} = \frac{1}{x + 2} + \frac{2x}{x^2 + 3} + \frac{4x}{(x^2 + 3)^2}$$

and

$$\int \frac{3x^4 + 4x^3 + 16x^2 + 20x + 9}{(x + 2)(x^2 + 3)^2} \, dx$$

$$= \int \frac{dx}{x + 2} + \int \frac{2x}{x^2 + 3} \, dx + 4 \int \frac{x}{(x^2 + 3)^2} \, dx$$

$$= \ln|x + 2| + \ln(x^2 + 3) - \frac{2}{x^2 + 3} + C$$

(The third integral on the right was evaluated by the substitution $u = x^2 + 3$.)   ◄

As noted in the beginning of this section, partial fraction decomposition only applies to *proper* rational functions. However, the next example shows that improper rational functions can be integrated by first performing a long division, and then working with the remainder term.

Example 5   Evaluate

$$\int \frac{3x^4 + 3x^3 - 5x^2 + x - 1}{x^2 + x - 2} \, dx$$

Solution.   Since the integrand is an improper rational function, we cannot use a partial fraction decomposition directly. However, if we perform the long division

$$
\begin{array}{r}
3x^2 + 1 \\
x^2 + x - 2 \enclose{longdiv}{3x^4 + 3x^3 - 5x^2 + x - 1} \\
\underline{3x^4 + 3x^3 - 6x^2} \\
x^2 + x - 1 \\
\underline{x^2 + x - 2} \\
1
\end{array}
$$

we can write the integrand as the quotient plus the remainder over the divisor, that is,

$$\frac{3x^4 + 3x^3 - 5x^2 + x - 1}{x^2 + x - 2} = (3x^2 + 1) + \frac{1}{x^2 + x - 2}$$

Thus,

$$\int \frac{3x^4 + 3x^3 - 5x^2 + x - 1}{x^2 + x - 2} \, dx = \int (3x^2 + 1) \, dx + \int \frac{dx}{x^2 + x - 2}$$

The second integral on the right now involves a proper rational function and can thus be evaluated by a partial fraction decomposition. Using the result of Example 1 we obtain

$$\int \frac{3x^4 + 3x^3 - 5x^2 + x - 1}{x^2 + x - 2} \, dx = x^3 + x + \frac{1}{3} \ln \left| \frac{x - 1}{x + 2} \right| + C \qquad \blacktriangleleft$$

FACTORING POLYNOMIALS—OPTIONAL DISCUSSION

The method of partial fractions depends on our ability to carry out the necessary factorization. This is not always easy to do. The following results, usually proved in algebra courses, are helpful to know.

Factor Theorem

9.7.1 THEOREM. *If $p(x)$ is a polynomial and r is a solution of the equation $p(x) = 0$, then $x - r$ is a factor of $p(x)$.*

9.7.2 THEOREM. *Let*

$$p(x) = a_0 x^n + a_1 x^{n-1} + \cdots + a_{n-1} x + a_n$$

be a polynomial with integer coefficients.

(a) *If r is an integer solution of $p(x) = 0$, then the constant term a_n is an integer multiple of r.*

(b) *If c/d is a rational solution of $p(x) = 0$, and if c/d is expressed in lowest terms, then the constant term a_n is an integer multiple of c; and the leading coefficient a_0 is an integer multiple of d.*

Example 6 For the equation

$$x^3 + x^2 - 10x + 8 = 0$$

it follows from part (a) of the above theorem that the only possible integer solutions are ± 1, ± 2, ± 4, ± 8. By substitution, or by using synthetic division, the reader can show that 1, 2, -4 are solutions and the rest are not. It follows that

$$x^3 + x^2 - 10x + 8 = (x - 1)(x - 2)(x + 4) \qquad \blacktriangleleft$$

Example 7 For the equation

$$2x^3 + x^2 - 6x - 3 = 0$$

the only possible numerators for rational solutions are ± 1, ± 3, and the only possible denominators are ± 1, ± 2; thus, the only possible rational solutions are

$$\pm 1, \ \pm 3, \ \pm \tfrac{1}{2}, \ \pm \tfrac{3}{2}$$

By substitution, or by using synthetic division, the reader can show that $-\tfrac{1}{2}$ is a solution, but the rest are not. It follows by division that

$$2x^3 + x^2 - 6x - 3 = (x + \tfrac{1}{2})(2x^2 - 6) = 2(x + \tfrac{1}{2})(x + \sqrt{3})(x - \sqrt{3})$$

Therefore, the solutions of the given equation are $x = -\tfrac{1}{2}$, $x = -\sqrt{3}$, and $x = \sqrt{3}$. ◀

▶ Exercise Set 9.7

In Exercises 1–38, perform the integrations.

1. $\displaystyle\int \frac{dx}{x^2 + 3x - 4}.$

2. $\displaystyle\int \frac{dx}{x^2 + 8x + 7}.$

3. $\displaystyle\int \frac{x}{x^2 - 5x + 6}\,dx.$

4. $\displaystyle\int \frac{5x - 4}{x^2 - 4x}\,dx.$

5. $\displaystyle\int \frac{11x + 17}{2x^2 + 7x - 4}\,dx.$

6. $\displaystyle\int \frac{5x - 5}{3x^2 - 8x - 3}\,dx.$

7. $\displaystyle\int \frac{dx}{(x - 1)(x + 2)(x - 3)}.$

8. $\displaystyle\int \frac{dx}{x(x^2 - 1)}.$

9. $\displaystyle\int \frac{2x^2 - 9x - 9}{x^3 - 9x}\,dx.$

10. $\displaystyle\int \frac{2x^2 + 4x - 8}{x^3 - 4x}\,dx.$

11. $\displaystyle\int \frac{x^2 + 2}{x + 2}\,dx.$

12. $\displaystyle\int \frac{x^2 - 4}{x - 1}\,dx.$

13. $\displaystyle\int \frac{3x^2 - 10}{x^2 - 4x + 4}\,dx.$

14. $\displaystyle\int \frac{x^2}{x^2 - 3x + 2}\,dx.$

15. $\displaystyle\int \frac{x^3}{x^2 - 3x + 2}\,dx.$

16. $\displaystyle\int \frac{x^3}{x^2 - x - 6}\,dx.$

17. $\displaystyle\int \frac{x^5 + 2x^2 + 1}{x^3 - x}\,dx.$

18. $\displaystyle\int \frac{2x^5 - x^3 - 1}{x^3 - 4x}\,dx.$

19. $\displaystyle\int \frac{2x^2 + 3}{x(x - 1)^2}\,dx.$

20. $\displaystyle\int \frac{3x^2 - x + 1}{x^3 - x^2}\,dx.$

21. $\displaystyle\int \frac{x^2 + x - 16}{(x + 1)(x - 3)^2}\,dx.$

22. $\displaystyle\int \frac{2x^2 - 2x - 1}{x^3 - x^2}\,dx.$

23. $\displaystyle\int \frac{x^2}{(x + 2)^3}\,dx.$

24. $\displaystyle\int \frac{2x^2 + 3x + 3}{(x + 1)^3}\,dx.$

25. $\displaystyle\int \frac{2x^2 - 1}{(4x - 1)(x^2 + 1)}\,dx.$

26. $\displaystyle\int \frac{dx}{x(x^2 + x + 1)}.$

27. $\displaystyle\int \frac{dx}{x^4 - 16}.$

28. $\displaystyle\int \frac{dx}{x^3 + x}.$

29. $\displaystyle\int \frac{x^3 + 3x^2 + x + 9}{(x^2 + 1)(x^2 + 3)}\,dx.$

30. $\displaystyle\int \frac{x^3 + x^2 + x + 2}{(x^2 + 1)(x^2 + 2)}\,dx.$

31. $\displaystyle\int \frac{x^3 - 3x^2 + 2x - 3}{x^2 + 1}\,dx.$

32. $\displaystyle\int \frac{x^4 + 6x^3 + 10x^2 + x}{x^2 + 6x + 10}\,dx.$

33. $\displaystyle\int \frac{x^2 + 1}{(x^2 + 2x + 3)^2}\,dx.$

34. $\displaystyle\int \frac{x^5 + x^4 + 4x^3 + 4x^2 + 4x + 4}{(x^2 + 2)^3}\,dx.$

35. $\displaystyle\int \frac{\cos \theta}{\sin^2 \theta + 4 \sin \theta - 5}\,d\theta.$

36. $\displaystyle\int \frac{e^t}{e^{2t} - 4}\,dt.$

37. $\displaystyle\int \frac{dx}{1 + e^x}.$

38. $\displaystyle\int \frac{\sec^2 \theta}{\tan^3 \theta - \tan^2 \theta}\,d\theta.$

39. (a) Find constants a and b such that

$$x^4 + 1 = (x^2 + ax + 1)(x^2 + bx + 1)$$

(b) Use the result in (a) to show that

$$\int_0^1 \frac{x}{x^4 + 1}\, dx = \frac{\pi}{8}$$

40. Find the area of the region enclosed by $y = (x - 3)/(x^3 + x^2)$, $y = 0$, $x = 1$, and $x = 2$.

41. Find the volume of the solid generated when the region enclosed by $y = x^2/(9 - x^2)$, $y = 0$, $x = 0$, and $x = 2$ is revolved about the x-axis.

In Exercises 42–45, solve the differential equations.

42. $\dfrac{dy}{dx} = y^2 + y$.

43. $\dfrac{dy}{dx} = y^2 - 5y + 6$.

44. $\dfrac{dy}{dt} = t^2y^2 - 4t^2y$.

45. $t(t - 1)\dfrac{dy}{dt} - (y^2 + y) = 0$.

46. The differential equation

$$\frac{dy}{dt} = ay - by^2 \quad (a > 0, b > 0)$$

which is called the **logistic equation,** first arose in the study of human population growth. By solving the equation, show that its general solution is

$$y = \frac{a}{b + Ce^{-at}}$$

where C is an arbitrary constant.

In Exercises 47–50, use Theorems 9.7.1 and 9.7.2.

47. Find all rational solutions, if any, and use your results to factor the polynomial into a product of linear and irreducible quadratic factors.
(a) $x^3 - 6x^2 + 11x - 6 = 0$
(b) $x^3 - 3x^2 + x - 20 = 0$
(c) $x^4 - 5x^3 + 7x^2 - 5x + 6 = 0$.

48. Find all rational solutions, if any, and use your results to factor the polynomial into a product of linear and irreducible quadratic factors.
(a) $8x^3 + 4x^2 - 2x - 1 = 0$
(b) $6x^4 - 7x^3 + 6x^2 - 1 = 0$
(c) $9x^4 - 56x^3 + 57x^2 + 98x - 24 = 0$.

49. Evaluate

$$\int \frac{dx}{x^4 - 3x^3 - 7x^2 + 27x - 18}$$

50. Evaluate

$$\int \frac{dx}{16x^3 - 4x^2 + 4x - 1}$$

51. Use Theorem 9.7.2 to prove:
(a) $\sqrt{2}$ is irrational
(b) If a is a positive integer, then $\sqrt{a}$ is either an integer or is irrational.

52. (a) Prove: $a_0x^n + a_1x^{n-1} + \cdots + a_n \equiv 0$ if and only if $a_0 = a_1 = \cdots = a_n = 0$.
(b) Use the result in (a) to prove:
$a_0x^n + a_1x^{n-1} + \cdots + a_n \equiv$
$\qquad\qquad b_0x^n + b_1x^{n-1} + \cdots + b_n$
if and only if $a_0 = b_0, a_1 = b_1, \ldots, a_n = b_n$.
(*Note.* The symbol $\equiv$ means that the two sides are equal for all values of x.)

9.8 MISCELLANEOUS SUBSTITUTIONS (OPTIONAL)

In this section we shall consider some integrals that do not fit into any of the categories previously studied.

INTEGRALS INVOLVING RATIONAL EXPONENTS

Integrals involving rational powers of x can often be simplified by substituting

$$u = x^{1/n}$$

where n is the least common multiple of the denominators of the exponents. The effect of this substitution is to replace fractional exponents with integer exponents, which are easier to work with.

Example 1   Evaluate

$$\int \frac{\sqrt{x}}{1 + \sqrt[3]{x}}\, dx$$

Solution.   The integrand involves $x^{1/2}$ and $x^{1/3}$, so we make the substitution

$$u = x^{1/6}$$

or

$$x = u^6 \quad \text{and} \quad dx = 6u^5\, du$$

Thus,

$$\int \frac{\sqrt{x}}{1 + \sqrt[3]{x}}\, dx = \int \frac{(u^6)^{1/2}}{1 + (u^6)^{1/3}}\, (6u^5)\, du = 6 \int \frac{u^8}{1 + u^2}\, du$$

By long division

$$\frac{u^8}{1 + u^2} = u^6 - u^4 + u^2 - 1 + \frac{1}{1 + u^2}$$

Thus,

$$\int \frac{\sqrt{x}}{1 + \sqrt[3]{x}}\, dx = 6 \int \left(u^6 - u^4 + u^2 - 1 + \frac{1}{1 + u^2} \right) du$$

$$= \frac{6}{7}u^7 - \frac{6}{5}u^5 + 2u^3 - 6u + 6 \tan^{-1} u + C$$

$$= \frac{6}{7}x^{7/6} - \frac{6}{5}x^{5/6} + 2x^{1/2} - 6x^{1/6} + 6 \tan^{-1} (x^{1/6}) + C \quad \blacktriangleleft$$

Example 2   Evaluate

$$\int \frac{dx}{2 + 2\sqrt{x}}$$

Solution.   The integrand contains $\sqrt{x} = x^{1/2}$, so we make the substitution

$$u = x^{1/2}$$

or

$$x = u^2 \quad \text{and} \quad dx = 2u\, du$$

This yields

$$\int \frac{dx}{2 + 2\sqrt{x}} = \int \frac{2u}{2 + 2u}\, du = \int \left(1 - \frac{1}{1 + u}\right) du$$

$$= u - \ln|1 + u| + C = \sqrt{x} - \ln|1 + \sqrt{x}| + C \quad \blacktriangleleft$$

The following example illustrates a variation on the above idea.

Example 3 Evaluate

$$\int \sqrt{1 + e^x}\, dx$$

Solution. The substitution

$$u^2 = 1 + e^x$$

will eliminate the square root. To express dx in terms of du, it is helpful to solve this equation for x and then differentiate. We obtain

$$e^x = u^2 - 1$$
$$x = \ln(u^2 - 1)$$
$$\frac{dx}{du} = \frac{2u}{u^2 - 1}, \quad dx = \frac{2u}{u^2 - 1}\, du$$

Thus,

$$\int \sqrt{1 + e^x}\, dx = \int u\left(\frac{2u}{u^2 - 1}\right) du$$

$$= \int \frac{2u^2}{u^2 - 1}\, du$$

$$= \int \left(2 + \frac{2}{u^2 - 1}\right) du \quad \text{[long division]}$$

$$= 2u + \int \left(\frac{1}{u - 1} - \frac{1}{u + 1}\right) du \quad \text{[partial fractions]}$$

$$= 2u + \ln|u - 1| - \ln|u + 1| + C$$

$$= 2u + \ln\left|\frac{u - 1}{u + 1}\right| + C$$

$$= 2\sqrt{1 + e^x} + \ln\left|\frac{\sqrt{1 + e^x} - 1}{\sqrt{1 + e^x} + 1}\right| + C \quad \blacktriangleleft$$

INTEGRALS
CONTAINING
RATIONAL
EXPRESSIONS IN
SIN x AND COS x

Functions such as

$$\frac{\sin x + 3 \cos^2 x}{\cos x + 4 \sin x}, \quad \frac{\sin x}{1 + \cos x - \cos^2 x}, \quad \frac{3 \sin^5 x}{1 + 4 \sin x}$$

are called *rational expressions in sin x and cos x.* Such expressions consist of finitely many sums, differences, products, and quotients of $\sin x$ and $\cos x$. If an integrand is a rational expression in $\sin x$ and $\cos x$, then the substitution

$$u = \tan (x/2), \quad -\pi < x < \pi \tag{1}$$

will transform the integrand into a rational function of u (which can then be integrated by methods already discussed). To see this, observe that

$$\cos (x/2) = \frac{1}{\sec (x/2)} = \frac{1}{\sqrt{1 + \tan^2 (x/2)}} = \frac{1}{\sqrt{1 + u^2}}$$

$$\sin (x/2) = \tan (x/2) \cos (x/2) = u \left(\frac{1}{\sqrt{1 + u^2}} \right)$$

Therefore,

$$\sin x = 2 \sin (x/2) \cos (x/2) = 2u \left(\frac{1}{\sqrt{1 + u^2}} \right) \left(\frac{1}{\sqrt{1 + u^2}} \right)$$

or

$$\sin x = \frac{2u}{1 + u^2} \tag{2}$$

Also,

$$\cos x = 1 - 2 \sin^2 \left(\frac{x}{2} \right) = 1 - \frac{2u^2}{1 + u^2}$$

so

$$\cos x = \frac{1 - u^2}{1 + u^2} \tag{3}$$

Moreover, from (1)

$$\frac{x}{2} = \tan^{-1} u \quad \text{so that} \quad \frac{dx}{du} = \frac{2}{1 + u^2}$$

or

$$dx = \frac{2}{1 + u^2} du \qquad (4)$$

It follows from (2), (3), and (4) that substitution (1) will yield an integrand that is a rational function of u.

Example 4 Evaluate

$$\int \frac{dx}{1 + \sin x}$$

Solution. Making substitution (1), it follows from (2) and (4) above that

$$\int \frac{dx}{1 + \sin x} = \int \frac{1}{1 + \left(\dfrac{2u}{1 + u^2}\right)} \left(\frac{2}{1 + u^2}\right) du = \int \frac{2}{1 + 2u + u^2} du$$

$$= \int \frac{2}{(1 + u)^2} du = -\frac{2}{1 + u} + C = -\frac{2}{1 + \tan\left(\dfrac{x}{2}\right)} + C \qquad \blacktriangleleft$$

REMARK. While substitution (1) is a useful tool, it can lead to cumbersome partial fraction decompositions. Consequently, this method should be used only after looking for simpler methods.

► Exercise Set 9.8

In Exercises 1–26, perform the integrations.

1. $\displaystyle\int x\sqrt{x - 2}\, dx.$

2. $\displaystyle\int_0^8 \frac{x}{\sqrt{x + 1}}\, dx.$

3. $\displaystyle\int_4^8 \frac{\sqrt{x - 4}}{x}\, dx.$

4. $\displaystyle\int_0^9 \frac{\sqrt{x}}{x + 9}\, dx.$

5. $\displaystyle\int_0^4 \frac{1}{3 + \sqrt{x}}\, dx.$

6. $\displaystyle\int \frac{x^5}{\sqrt{x^3 + 1}}\, dx.$

7. $\displaystyle\int x^5 \sqrt{x^3 + 1}\, dx.$

8. $\displaystyle\int \frac{1}{x\sqrt{x^3 - 1}}\, dx.$

9. $\displaystyle\int \frac{dx}{\sqrt{x} + \sqrt[3]{x}}.$

10. $\displaystyle\int \frac{dx}{x - x^{3/5}}.$

11. $\displaystyle\int \frac{dv}{v(1 - v^{1/4})}.$

12. $\displaystyle\int \frac{x^{2/3}}{x + 1}\, dx.$

13. $\displaystyle\int \frac{dt}{t^{1/2} - t^{1/3}}.$

14. $\displaystyle\int \frac{1 + \sqrt{x}}{1 - \sqrt{x}}\, dx.$

15. $\displaystyle\int \frac{x^3}{\sqrt{1 + x^2}}\, dx.$

16. $\displaystyle\int \frac{x}{(x + 3)^{1/5}}\, dx.$

17. $\displaystyle\int \sin \sqrt{x}\, dx.$

18. $\displaystyle\int e^{\sqrt{x}}\, dx.$

19. $\displaystyle\int \frac{1}{\sqrt{e^x} + 1}\, dx.$

20. $\displaystyle\int_0^{\ln 2} \sqrt{e^x - 1}\, dx.$

21. $\displaystyle\int \frac{dx}{1 + \sin x + \cos x}.$

22. $\displaystyle\int \frac{dx}{2 + \sin x}.$

23. $\displaystyle\int_{\pi/2}^{\pi} \frac{d\theta}{1 - \cos \theta}.$

24. $\displaystyle\int \frac{dx}{4 \sin x - 3 \cos x}.$

25. $\displaystyle\int \frac{\cos x}{2 - \cos x}\, dx.$

26. $\displaystyle\int \frac{dx}{\sin x + \tan x}.$

27. (a) Use the substitution $u = \tan(x/2)$ to show that

$$\int \sec x \, dx = \ln \left| \frac{1 + \tan\frac{1}{2}x}{1 - \tan\frac{1}{2}x} \right| + C$$

(b) Use the result in (a) to show that

$$\int \sec x \, dx = \ln \left| \tan\left(\frac{\pi}{4} + \frac{x}{2}\right) \right| + C$$

(c) Show that this agrees with (2) of Section 9.4.

28. (a) Use the substitution $u = \tan(x/2)$ to show that

$$\int \csc x \, dx = \frac{1}{2} \ln \left[\frac{1 - \cos x}{1 + \cos x} \right] + C$$

(b) Show that this agrees with Exercise 37(a) of Section 9.4.

29. Develop a substitution that can be used to integrate rational functions of $\sinh x$ and $\cosh x$ and use your substitution to evaluate

$$\int \frac{dx}{2 \cosh x + \sinh x}$$

without expressing the integrand in terms of e^x and e^{-x}.

In Exercises 30–33, use the substitution $x = 1/u$ to help evaluate the integral. (Assume that $x > 0$.)

30. $\displaystyle\int \frac{\sqrt{4 - x^2}}{x^4} \, dx.$ 31. $\displaystyle\int \frac{1}{x^2\sqrt{3 - x^2}} \, dx.$

32. $\displaystyle\int \frac{1}{x^2\sqrt{x^2 + 1}} \, dx.$ 33. $\displaystyle\int \frac{\sqrt{x^2 - 5}}{x^4} \, dx.$

9.9 NUMERICAL INTEGRATION; SIMPSON'S RULE

If an antiderivative of f cannot be found, then the integral

$$\int_a^b f(x) \, dx$$

cannot be evaluated by the First Fundamental Theorem of Calculus. In such cases, the value of the integral can be approximated using methods we shall study in this section.

ENDPOINT APPROXIMATIONS

Suppose we want to approximate

$$\int_a^b f(x) \, dx$$

where f is continuous on $[a, b]$. Using the definition of a definite integral (Definition 5.6.2), we can write this integral as

$$\int_a^b f(x) \, dx = \lim_{\max \Delta x_k \to 0} \sum_{k=1}^n f(x_k^*) \, \Delta x_k \tag{1}$$

If we divide the interval $[a, b]$ into n subintervals of equal length, then

$$\Delta x_1 = \Delta x_2 = \cdots = \Delta x_n = \frac{b-a}{n}$$

so that we can rewrite (1) as

$$\int_a^b f(x)\,dx = \lim_{n \to +\infty} \sum_{k=1}^n f(x_k^*)\left(\frac{b-a}{n}\right)$$

Thus, we expect the following approximation to be good when n is large:

$$\int_a^b f(x)\,dx \approx \sum_{k=1}^n f(x_k^*)\left(\frac{b-a}{n}\right) = \frac{b-a}{n}\sum_{k=1}^n f(x_k^*) \tag{2}$$

where x_k^* denotes an arbitrary point in the kth subinterval. Expression (2) yields various approximation formulas, depending on how the points $x_1^*, x_2^*, \ldots, x_n^*$ are chosen. For example, let

$$y_0, y_1, \ldots, y_n$$

be the values of f at the endpoints of the subintervals (Figure 9.9.1).

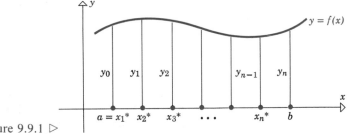

Figure 9.9.1 ▷

If we choose $x_1^*, x_2^*, \ldots, x_n^*$ to be the left-hand endpoints of the subintervals, then

$$y_0 = f(x_1^*), \; y_1 = f(x_2^*), \ldots, \; y_{n-1} = f(x_n^*)$$

so that (2) yields:

Left-hand Endpoint Approximation

$$\int_a^b f(x)\,dx \approx \left(\frac{b-a}{n}\right)[y_0 + y_1 + \cdots + y_{n-1}] \tag{3}$$

(Figure 9.9.2a). Similarly, if $x_1^*, x_2^*, \ldots, x_n^*$ are the right-hand endpoints of the subintervals, then (2) yields:

Right-hand Endpoint Approximation

$$\int_a^b f(x)\, dx \approx \left(\frac{b-a}{n}\right)[y_1 + y_2 + \cdots + y_n] \tag{4}$$

TRAPEZOIDAL
APPROXIMATION

(Figure 9.9.2*b*). If we take the average of the left-hand and right-hand endpoint approximations, we obtain:

Trapezoidal Approximation

$$\int_a^b f(x)\, dx \approx \left(\frac{b-a}{2n}\right)[y_0 + 2y_1 + \cdots + 2y_{n-1} + y_n] \tag{5}$$

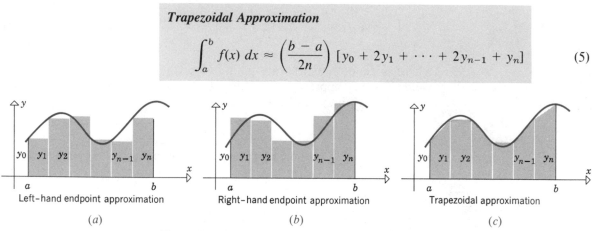

Left–hand endpoint approximation

(*a*)

Right–hand endpoint approximation

(*b*)

Trapezoidal approximation

(*c*)

Figure 9.9.2

The name, *trapezoidal approximation,* can be explained by considering the case where $f(x) \geq 0$ on $[a, b]$, so that $\int_a^b f(x)\, dx$ represents the area under $f(x)$ over $[a, b]$. Geometrically, the trapezoidal approximation formula is what results if we approximate this area by the sum of the trapezoidal areas shown in Figure 9.9.2*c* (Exercise 8). By comparison, the left-hand and right-hand approximation formulas result when we approximate the area using the rectangles shown in Figures 9.9.2*a* and 9.9.2*b*.

Example 1   Below, we have approximated

$$\int_1^2 \frac{1}{x}\, dx$$

by three methods, left-hand endpoint approximation, right-hand endpoint approximation, and trapezoidal approximation. In each case, we used $n = 10$ subdivisions of the interval $[1, 2]$ so that the subintervals have width

$$\frac{b-a}{n} = \frac{2-1}{10} = \frac{1}{10} = 0.1$$

Left-hand Endpoint Approximation

i	Endpoint x_i	$y_i = f(x_i) = 1/x_i$
0	1.0	1.0000
1	1.1	0.9091
2	1.2	0.8333
3	1.3	0.7692
4	1.4	0.7143
5	1.5	0.6667
6	1.6	0.6250
7	1.7	0.5882
8	1.8	0.5556
9	1.9	0.5263
		Total
		7.1877

Right-hand Endpoint Approximation

i	Endpoint x_i	$y_i = f(x_i) = 1/x_i$
1	1.1	0.9091
2	1.2	0.8333
3	1.3	0.7692
4	1.4	0.7143
5	1.5	0.6667
6	1.6	0.6250
7	1.7	0.5882
8	1.8	0.5556
9	1.9	0.5263
10	2.0	0.5000
		Total
		6.6877

From (3)

$$\int_1^2 \frac{1}{x}\,dx \approx (0.1)(7.1877) = 0.71877$$

From (4)

$$\int_1^2 \frac{1}{x}\,dx \approx (0.1)(6.6877) = 0.66877$$

Trapezoidal Approximation

i	Endpoint x_i	$y_i = f(x_i) = 1/x_i$	Multiplier w_i	$w_i y_i$
0	1.0	1.0000	1	1.0000
1	1.1	0.9091	2	1.8182
2	1.2	0.8333	2	1.6666
3	1.3	0.7692	2	1.5384
4	1.4	0.7143	2	1.4286
5	1.5	0.6667	2	1.3334
6	1.6	0.6250	2	1.2500
7	1.7	0.5882	2	1.1764
8	1.8	0.5556	2	1.1112
9	1.9	0.5263	2	1.0526
10	2.0	0.5000	1	0.5000
				Total
				13.8754

Since

$$\frac{b-a}{2n} = \frac{2-1}{2(10)} = \frac{1}{20} = 0.05$$

we obtain from (5)

$$\int_1^2 \frac{1}{x}\, dx \approx (0.05)(13.8754) = 0.69377 \quad \blacktriangleleft$$

REMARK. Because $1/x$ is decreasing on the interval $[1, 2]$, we would expect the left-hand endpoint approximation to be too large and the right-hand endpoint approximation to be too small. This is in fact the case since

$$\int_1^2 \frac{1}{x}\, dx = \ln 2 \approx 0.6931$$

to four-decimal-place accuracy.

Figure 9.9.3

Intuition suggests that we might improve on the endpoint approximations and the trapezoidal approximation by replacing the rectangular and trapezoidal strips of Figure 9.9.2 with strips having curved upper boundaries chosen to match the shape of the curve $y = f(x)$. This is the idea behind **Simpson's* rule**, which uses curves of the form

$$y = ax^2 + bx + c$$

to approximate sections of the curve $y = f(x)$. As noted in Section 1.6, these curves are parabolas with axes of symmetry parallel to the y-axis (Figure 9.9.3).

To simplify the description of Simpson's rule we shall assume $f(x) \geq 0$ on $[a, b]$ so we can interpret $\int_a^b f(x)\, dx$ as an area. However, the method is valid without this assumption.

The heart of Simpson's rule is the formula

---

* THOMAS SIMPSON (1710–1761) English mathematician. Simpson was the son of a weaver. He was trained to follow in his father's footsteps and had little formal education in his early life. His interest in science and mathematics was aroused in 1724, when he witnessed an eclipse of the sun and received two books from a peddler, one on astrology and the other on arithmetic. Simpson quickly absorbed their contents and soon became a successful local fortune teller. His improved financial situation enabled him to give up weaving and marry his landlady, an older woman. Then in 1733 some mysterious "unfortunate incident" forced him to move. He settled in Derby, where he taught in an evening school and worked at weaving during the day. In 1736 he moved to London and published his first mathematical work in a periodical called the *Ladies' Diary* (of which he later became the editor). In 1737 he published a successful calculus textbook that enabled him to give up weaving completely and concentrate on textbook writing and teaching. His fortunes improved further in 1740 when one Robert Heath accused him of plagiarism. The publicity was marvelous, and Simpson proceeded to dash off a succession of best-selling textbooks, *Algebra* (ten editions plus translations), *Geometry* (twelve editions plus translations), *Trigonometry* (five editions plus translations), and numerous others.

It is interesting to note that Simpson did not discover the rule that bears his name. It was a well-known result by Simpson's time.

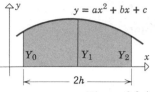

Figure 9.9.4

$$A = \frac{h}{3}[Y_0 + 4Y_1 + Y_2] \tag{6}$$

which gives the area under the curve

$$y = ax^2 + bx + c$$

over an arbitrary interval of width $2h$. In this formula Y_0, Y_1, and Y_2 represent the y values at the left-hand endpoint, the midpoint, and the right-hand endpoint of the interval (Figure 9.9.4).

To derive Formula (6), denote the midpoint of the interval by m. Since the interval has length $2h$, the left-hand endpoint will be $m - h$ and the right-hand endpoint $m + h$. Thus, the area A under $y = ax^2 + bx + c$ over this interval will be

$$A = \int_{m-h}^{m+h} (ax^2 + bx + c)\, dx$$

$$= \frac{a}{3}x^3 + \frac{b}{2}x^2 + cx \Big]_{m-h}^{m+h}$$

$$= \frac{a}{3}[(m + h)^3 - (m - h)^3] + \frac{b}{2}[(m + h)^2 - (m - h)^2]$$

$$+ c[(m + h) - (m - h)]$$

or on simplifying

$$A = \frac{h}{3}[a(6m^2 + 2h^2) + b(6m) + 6c] \tag{7}$$

But the values of $y = ax^2 + bx + c$ at the left-hand endpoint, the midpoint, and the right-hand endpoint are, respectively,

$$Y_0 = a(m - h)^2 + b(m - h) + c$$
$$Y_1 = am^2 + bm + c$$
$$Y_2 = a(m + h)^2 + b(m + h) + c$$

from which it follows that

$$Y_0 + 4Y_1 + Y_2 = a(6m^2 + 2h^2) + b(6m) + 6c \tag{8}$$

Thus, (6) follows from (7) and (8).

Simpson's rule is obtained by dividing the interval $[a, b]$ into an *even* number of equal subintervals of width h and applying formula (6) to approximate the area under $y = f(x)$ over successive pairs of subintervals. The sum of these approximations then serves as an estimate of $\int_a^b f(x)\, dx$. More precisely, let $[a, b]$ be divided into n subintervals of width h (n even) and let

$$y_0, y_1, \ldots, y_n$$

be the values of $y = f(x)$ at the subinterval endpoints

$$a = x_0, x_1, \ldots, x_n = b$$

By (6) the area under $y = f(x)$ over the first two subintervals is approximately

$$\frac{h}{3}[y_0 + 4y_1 + y_2]$$

and the area over the second pair of subintervals is approximately

$$\frac{h}{3}[y_2 + 4y_3 + y_4]$$

and the area over the last pair of subintervals is approximately

$$\frac{h}{3}[y_{n-2} + 4y_{n-1} + y_n]$$

Adding all the approximations, collecting terms, and replacing h by $(b - a)/n$ yields:

Simpson's Rule

$$\int_a^b f(x)\, dx \approx \left(\frac{b - a}{3n}\right)[y_0 + 4y_1 + 2y_2 + 4y_3 + 2y_4 + \cdots$$
$$+ 2y_{n-2} + 4y_{n-1} + y_n]$$

Example 2   Below, we have approximated

$$\int_1^2 \frac{1}{x}\, dx$$

by Simpson's rule using $n = 10$ subdivisions so that the subintervals have width

$$\frac{b - a}{n} = \frac{2 - 1}{10} = \frac{1}{10} = 0.1$$

Since

$$\frac{b - a}{3n} = \frac{2 - 1}{3(10)} = \frac{1}{30}$$

we obtain from Simpson's rule

Simpson's Rule

	Endpoint		Multiplier	
i	x_i	$y_i = f(x_i) = 1/x_i$	w_i	$w_i y_i$
0	1.0	1.0000	1	1.0000
1	1.1	0.9091	4	3.6364
2	1.2	0.8333	2	1.6666
3	1.3	0.7692	4	3.0768
4	1.4	0.7143	2	1.4286
5	1.5	0.6667	4	2.6668
6	1.6	0.6250	2	1.2500
7	1.7	0.5882	4	2.3528
8	1.8	0.5556	2	1.1112
9	1.9	0.5263	4	2.1052
10	2.0	0.5000	1	0.5000
				Total
				20.7944

$$\int_1^2 \frac{1}{x}\, dx \approx \frac{1}{30}(20.7944) \approx 0.69315$$

To six decimal places, the value of $\ln 2 = \int_1^2 (1/x)\, dx$ is 0.693147 so that with $n = 10$ Simpson's rule is more accurate than the two methods of Example 1.   ◄

With all the methods studied in this section, there are two sources of error, the *intrinsic* or *truncation error* due to the approximation formula, and the *roundoff* error introduced in the calculations. In general, increasing n reduces the truncation error but increases the roundoff error, since more computations are required for larger n. In practical applications, it is important to know how large n must be taken to ensure a specified degree of accuracy (see Exercise 10). Such problems are studied in a branch of mathematics called **numerical analysis**.

▶ Exercise Set 9.9

In Exercises 1–6, use the given value of n to approximate the integral by:

   (a) the left-hand endpoint rule
   (b) the right-hand endpoint rule
   (c) the trapezoidal rule
   (d) Simpson's rule.

Round off all computations to four decimal places and use a calculator or the appendix tables of trigonometric functions and exponentials where needed.

1. $\int_0^1 \frac{dx}{x + 1}$, $n = 6$.   **2.** $\int_0^4 x^2\, dx$, $n = 8$.

3. $\displaystyle\int_0^\pi \sin x \, dx, \quad n = 4.$

4. $\displaystyle\int_0^1 \frac{dx}{\sqrt{1 + x^2}}, \quad n = 4.$

5. $\displaystyle\int_0^2 e^{-x^2} \, dx, \quad n = 4.$

6. $\displaystyle\int_1^4 \sqrt{1 + x^3} \, dx, \quad n = 6.$

7. Use the relationship

$$\frac{\pi}{4} = \tan^{-1}(1) = \int_0^1 \frac{dx}{1 + x^2}$$

and Simpson's rule with $n = 10$ to estimate π. Round off all computations to four decimal places.

8. Derive the trapezoidal rule by summing the areas of the trapezoids in Figure 9.9.2c.

9. If, in (2), x_k^* is taken as the midpoint of the kth subinterval, we obtain the **midpoint rule** for approximating $\int_a^b f(x) \, dx$. Use the midpoint rule with $n = 10$ to approximate

$$\int_1^2 \frac{1}{x} \, dx$$

(Compare your results to those obtained in Examples 1 and 2.)

10. Let K be any number such that $|f''(x)| \leq K$, and let M be any number such that $|f^{(4)}(x)| \leq M$ for

all x in $[a, b]$. It is shown in advanced calculus that the absolute value of the error resulting from Simpson's rule with n subintervals is at most

$$\frac{M(b - a)^5}{180n^4}$$

and the absolute value of the error from the trapezoidal rule is at most

$$\frac{K(b - a)^3}{12n^2}$$

(a) In Examples 1 and 2, $\int_1^2 (1/x) \, dx$ was estimated by the trapezoidal rule and Simpson's rule using $n = 10$. Show that the absolute value of the error for the trapezoidal rule is at most

$$\frac{2}{12 \times 10^2} \approx 0.00167$$

and the absolute value of the error for Simpson's rule is at most

$$\frac{24}{180 \times 10^4} \approx 0.0000133$$

(b) How large should n be taken to ensure that the absolute value of the error in the approximation of $\int_1^2 (1/x) \, dx$ by Simpson's rule is at most 0.0002?

(c) Answer (b) for the trapezoidal rule.

▶ **SUPPLEMENTARY EXERCISES**

In Exercises 1–64, evaluate the integrals.

1. $\displaystyle\int x \cos 2x \, dx.$

2. $\displaystyle\int x \cos x^2 \, dx.$

3. $\displaystyle\int \tan^3 x \sec x \, dx.$

4. $\displaystyle\int \sin^3 x \cos^2 x \, dx.$

5. $\displaystyle\int \tan^2 3t \sec^2 3t \, dt.$

6. $\displaystyle\int \cot 2x \csc^3 2x \, dx.$

7. $\displaystyle\int \frac{\sin^2 x \, dx}{1 + \cos x}.$

8. $\displaystyle\int \frac{\sin 2x \, dx}{\cos x \, (1 + \cos x)}.$

9. $\displaystyle\int x^2 \cos^2 x \, dx.$

10. $\displaystyle\int \sin^2 2x \cos^2 2x \, dx.$

11. $\displaystyle\int \sec^5 x \sin x \, dx.$

12. $\displaystyle\int \tan^5 2x \, dx.$

13. $\displaystyle\int \sin^4 x \cos^2 x \, dx.$

14. $\displaystyle\int \frac{dx}{\sec^4 x}.$

15. $\displaystyle\int_0^{\pi/4} \sin 5x \sin 3x \, dx.$

16. $\displaystyle\int_{-\pi/10}^0 \sin 2x \cos 3x \, dx.$

17. $\displaystyle\int_0^1 \sin^2 \pi x \, dx.$ **18.** $\displaystyle\int_0^{\pi/3} \sin^3 3x \, dx.$

19. $\displaystyle\int_0^{\sqrt{\pi/2}} x \sec^2 (x^2) \, dx.$

20. $\displaystyle\int_0^{\pi/4} \frac{\sec^2 x \, dx}{\sqrt{1 + 3 \tan x}}.$

21. $\displaystyle\int \frac{\sin (\cot^{-1} x) \, dx}{1 + x^2}.$ **22.** $\displaystyle\int \frac{e^{\tan 3x} \, dx}{\cos^2 3x}.$

23. $\displaystyle\int e^x \sec (e^x) \, dx.$ **24.** $\displaystyle\int x \sec^2 3x \, dx.$

25. $\displaystyle\int \frac{e^{2x}}{\sqrt{e^{2x} + 1}} \, dx.$ **26.** $\displaystyle\int \frac{e^x}{\sqrt{e^{2x} + 1}} \, dx.$

27. $\displaystyle\int e^{3x} \sin 2x \, dx.$ **28.** $\displaystyle\int \ln (a^2 + x^2) \, dx.$

29. $\displaystyle\int_1^2 \sin^{-1} (x/2) \, dx.$ **30.** $\displaystyle\int (\ln x)^2 \, dx.$

31. $\displaystyle\int (\ln x)/x^2 \, dx.$ **32.** $\displaystyle\int x^3 e^{-x^2} \, dx.$

33. $\displaystyle\int \frac{x \, dx}{\sqrt{x^2 - 9}}.$ **34.** $\displaystyle\int_1^2 \frac{\sqrt{4x^2 - 1}}{x} \, dx.$

35. $\displaystyle\int_1^3 \frac{\sqrt{9 - x^2}}{x} \, dx.$ **36.** $\displaystyle\int_{-\sqrt{8}}^{-\sqrt{8/3}} \frac{ds}{s\sqrt{s^2 + 8}}.$

37. $\displaystyle\int \frac{x^2 \, dx}{\sqrt{2x + 3}}.$ **38.** $\displaystyle\int \frac{dx}{\sqrt{x^2 + 16}}.$

39. $\displaystyle\int \frac{dt}{\sqrt{3 - 4t - 4t^2}}.$ **40.** $\displaystyle\int_4^7 \frac{dx}{\sqrt{x^2 - 2x - 8}}.$

41. $\displaystyle\int \frac{dx}{x^2\sqrt{a^2 - x^2}}.$ **42.** $\displaystyle\int \frac{x^3}{(x^2 + 4)^{1/3}} \, dx.$

43. $\displaystyle\int \sqrt{a^2 - x^2} \, dx.$ **44.** $\displaystyle\int x\sqrt{a^2 - x^2} \, dx.$

45. $\displaystyle\int \frac{x - 2}{\sqrt{4x - x^2}} \, dx.$ **46.** $\displaystyle\int_1^3 \frac{dx}{x^2 - 2x + 5}.$

47. $\displaystyle\int \frac{dx}{2x^2 + 3x + 1}.$ **48.** $\displaystyle\int \frac{dx}{(x^2 + 4)^2}.$

49. $\displaystyle\int \frac{x + 1}{x^3 + x^2 - 6x} \, dx.$

50. $\displaystyle\int \frac{x^3 + 1}{x - 2} \, dx.$

51. $\displaystyle\int \frac{x^2 - 1}{x^3 - 3x} \, dx.$ **52.** $\displaystyle\int \frac{x - 3}{x^3 - 1} \, dx.$

53. $\displaystyle\int \frac{2x^2 + 5}{x^4 - 1} \, dx.$

54. $\displaystyle\int \frac{x^4 - x^3 - x - 1}{x^3 - x^2} \, dx.$

55. $\displaystyle\int \frac{dx}{(x^2 + 4)(x - 3)}.$ **56.** $\displaystyle\int \frac{x \, dx}{(x + 1)^3}.$

57. $\displaystyle\int \frac{3x^2 + 12x + 2}{(x^2 + 4)^2} \, dx.$

58. $\displaystyle\int \frac{(4x + 2) \, dx}{x^4 + 2x^3 + x^2}.$

59. $\displaystyle\int \frac{x \, dx}{x^2 + 2x + 5}.$ **60.** $\displaystyle\int \frac{6x \, dx}{(x^2 + 9)^3}.$

61. $\displaystyle\int \frac{dx}{\sqrt{3 - 2x^2}}.$ **62.** $\displaystyle\int \frac{1 + t}{\sqrt{t}} \, dt.$

63. $\displaystyle\int \frac{\sqrt{t} \, dt}{1 + t}.$ **64.** $\displaystyle\int \frac{\sqrt{1 - x^2}}{x^2} \, dx.$

Exercises 65–70 relate to the optional Section 9.8. Evaluate the integrals.

65. $\displaystyle\int_0^1 \frac{x^{2/3}}{1 + x^{1/3}} \, dx.$ **66.** $\displaystyle\int \frac{dx}{x^{1/2} + x^{1/4}}.$

67. $\displaystyle\int \frac{dx}{1 - \tan x}.$ **68.** $\displaystyle\int \frac{dx}{3 \cos x + 5}.$

69. $\displaystyle\int \frac{dx}{\sin x - \tan x}.$ **70.** $\displaystyle\int_0^{\pi/3} \frac{dx}{5 \sec x - 3}.$

71. Use partial fractions to show that

$$\int \frac{dx}{x^2 - a^2} = \frac{1}{2a} \ln \left| \frac{x - a}{x + a} \right| + C \quad (a \neq 0)$$

72. Find the arc length of: (a) the parabola $y = x^2/2$ from $(0, 0)$ to $(2, 2)$, (b) the curve $y = \ln (\sec x)$ from $(0, 0)$ to $(\pi/4, \frac{1}{2} \ln 2)$.

73. Let R be the region bounded by the curve $y = 1/(4 + x^2)$ and the lines $x = 0$, $y = 0$, and $x = 2$. Find: (a) the area of R, (b) the volume of the solid obtained by revolving R about the x-axis, (c) the volume of the solid obtained by revolving R about the y-axis.

74. Derive the following reduction formulas for $a \neq 0$:

(a) $\displaystyle \int x^n e^{ax} \, dx = \frac{x^n e^{ax}}{a} - \frac{n}{a} \int x^{n-1} e^{ax} \, dx$

(b) $\displaystyle \int x^n \sin ax \, dx =$

$$\frac{-x^n \cos ax}{a} + \frac{n}{a} \int x^{n-1} \cos ax \, dx$$

$$\int x^n \cos ax \, dx =$$

$$\frac{x^n \sin ax}{a} - \frac{n}{a} \int x^{n-1} \sin ax \, dx$$

(c) $\displaystyle \int \sin^n ax \cos^m ax \, dx = -\frac{\sin^{n-1} ax \cos^{m+1} ax}{a(m+n)}$

$$+ \frac{n-1}{m+n} \int \sin^{n-2} ax \cos^m ax \, dx$$

$$= \frac{\sin^{n+1} ax \cos^{m-1} ax}{a(m+n)}$$

$$+ \frac{m-1}{m+n} \int \sin^n ax \cos^{m-2} ax \, dx.$$

75. Use Exercise 74 to evaluate the following integrals.

(a) $\displaystyle \int x^3 e^{2x} \, dx$ (b) $\displaystyle \int_0^{\pi/10} x^2 \sin 5x \, dx$

(c) $\displaystyle \int \sin^2 x \cos^4 x \, dx.$

76. Evaluate the following integrals assuming that $a \neq 0$.

(a) $\displaystyle \int x^n \ln ax \, dx \quad (n \neq -1)$

(b) $\displaystyle \int \sec^n ax \tan ax \, dx \quad (n \geq 1).$

77. Find $\int (\sin^3 \theta / \cos^5 \theta) \, d\theta$ two ways: (a) letting $u = \cos \theta$ and (b) expressing the integrand in terms of $\sec \theta$ and $\tan \theta$. Show that your answers differ by a constant.

In Exercises 78–81, approximate the integral using the given value of n and (a) the trapezoidal rule, (b) Simpson's rule. Use a calculator and express the answer to four decimal places.

78. $\displaystyle \int_0^1 \sqrt{x} \, dx, \, n = 4.$ **79.** $\displaystyle \int_{-4}^2 e^{-x} \, dx, \, n = 6.$

80. $\displaystyle \int_0^4 \sinh x \, dx, \, n = 4.$ **81.** $\displaystyle \int_4^{5.2} \ln x \, dx, \, n = 6.$

In Exercises 82 and 83, use Simpson's rule with $n = 10$ to approximate the given integral. Use a calculator and express the answer to five decimal places.

82. $\displaystyle \int_0^2 \cos(\sinh x) \, dx.$ **83.** $\displaystyle \int_1^2 \sin(\ln x) \, dx.$

84. (a) Show that if $f(x)$ is continuous for $0 \leq x \leq 1$, then

$$\int_0^\pi x f(\sin x) \, dx = \frac{\pi}{2} \int_0^\pi f(\sin x) \, dx$$

[*Hint:* Let $x = \pi - u.$]

(b) Use the result in part (a) to find

$$\int_0^\pi \frac{x \sin x}{2 - \sin^2 x} \, dx$$

In Exercises 85–94, evaluate the integrals.

85. $\displaystyle \int \frac{1}{e^{ax} + 1} \, dx, \, a \neq 0.$

86. $\displaystyle \int \frac{(x-2)^3}{\sqrt{4x - x^2}} \, dx.$

87. $\displaystyle \int \frac{\sqrt{1+x} + \sqrt{1-x}}{\sqrt{1+x} - \sqrt{1-x}} \, dx.$

88. $\displaystyle \int (\cos^{32} x \sin^{30} x - \cos^{30} x \sin^{32} x) \, dx.$

89. $\displaystyle \int \frac{\sqrt{x+1}}{(x-1)^{5/2}} \, dx.$ [*Hint:* Let $x - 1 = 1/u.$]

90. $\displaystyle \int \frac{1}{x^{10} + x} \, dx.$ [*Hint:* Rewrite the denominator as $x^{10}(1 + x^{-9})$ and let $u = 1 + x^{-9}.$]

91. $\displaystyle \int \frac{1}{x(3x^5 + 2)} \, dx.$ [*Hint:* See Exercise 90.]

92. $\displaystyle \int \frac{3x^6 - 2}{x(2x^6 + 5)} \, dx.$ [*Hint:* Rewrite as the sum of two integrals and see Exercise 90.]

93. $\displaystyle \int \sqrt{x - \sqrt{x^2 - 4}} \, dx.$

[*Hint:* $\frac{1}{2}(\sqrt{x+2} - \sqrt{x-2})^2 = ?$]

94. $\displaystyle \int_0^1 \sqrt{1 + \sqrt{1 - x^2}} \, dx.$

[*Hint:* $\frac{1}{2}(\sqrt{1+x} + \sqrt{1-x})^2 = ?$]

10

improper integrals; L'Hôpital's rule

10.1 IMPROPER INTEGRALS

In the definition of $\int_a^b f(x)\,dx$, it is assumed that the interval $[a, b]$ is finite. Moreover, in order for the limit

$$\lim_{\max \Delta x_k \to 0} \sum_{k=1}^{n} f(x_k^*)\,\Delta x_k = \int_a^b f(x)\,dx$$

to exist, the function f must be bounded on the interval $[a, b]$. (See the remark following Theorem 5.6.3.) For example, the integral

$$\int_0^3 \frac{1}{(x-2)^{2/3}}\,dx$$

does not exist, since the integrand approaches $+\infty$ at the point $x = 2$ in the interval of integration.

  In this section we shall extend the concept of the definite integral to include:

- *integrals with infinite intervals of integration;*

- *integrals in which the integrand becomes infinite within the interval of integration.*

*These are called **improper integrals**.*

INTEGRALS OVER
INFINITE INTERVALS

If f is continuous on the interval $[a, +\infty)$, then we define the improper integral $\int_a^{+\infty} f(x)\,dx$ as a limit in the following way:

$$\int_a^{+\infty} f(x)\,dx = \lim_{l \to +\infty} \int_a^l f(x)\,dx \tag{1}$$

567

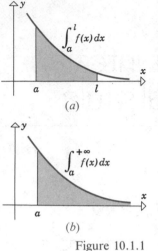

$\int_a^l f(x)\,dx$

(a)

$\int_a^{+\infty} f(x)\,dx$

(b)

Figure 10.1.1

If this limit exists, the improper integral is said to *converge,* and the value of the limit is the value assigned to the integral. If the limit does not exist, then the improper integral is said to *diverge,* in which case it is not assigned a value.

If f is nonnegative and continuous on $[a, +\infty)$, then (1) has an important geometric interpretation. For each value of $l > a$, the definite integral $\int_a^l f(x)\,dx$ represents the area under the curve $y = f(x)$ over the finite interval $[a, l]$ (Figure 10.1.1a). As we let $l \to +\infty$, this area tends toward the area under $y = f(x)$ over the entire interval $[a, +\infty)$ (Figure 10.1.1b). Thus, $\int_a^{+\infty} f(x)\,dx$ can be regarded as the area under $y = f(x)$ over the interval $[a, +\infty)$.

Example 1   Find

$$\int_1^{+\infty} \frac{dx}{x^2}$$

Solution.   We begin by replacing the infinite upper limit with a finite upper limit l:

$$\int_1^l \frac{dx}{x^2} = -\frac{1}{x}\Big]_1^l = -\frac{1}{l} - (-1) = 1 - \frac{1}{l}$$

Thus,

$$\int_1^{+\infty} \frac{dx}{x^2} = \lim_{l \to +\infty} \int_1^l \frac{dx}{x^2} = \lim_{l \to +\infty} \left(1 - \frac{1}{l}\right) = 1$$

so the given integral converges to 1.   ◀

Example 2   Evaluate

$$\int_1^{+\infty} \frac{dx}{x}$$

Solution.

$$\int_1^{+\infty} \frac{dx}{x} = \lim_{l \to +\infty} \int_1^l \frac{dx}{x} = \lim_{l \to +\infty} \Big[\ln|x|\Big]_1^l = \lim_{l \to +\infty} \ln|l| = +\infty$$

Thus, the integral diverges.   ◀

$y = 1/x^2$

$y = 1/x$

Figure 10.1.2

It is worthwhile to reflect on the results of Examples 1 and 2. Why is it that $\int_1^{+\infty} dx/x^2$ converges while $\int_1^{+\infty} dx/x$ diverges when, in fact, the graphs of $1/x$ and $1/x^2$ look similar over the interval $[1, +\infty)$ (Figure 10.1.2)?

The explanation is that $1/x^2$ approaches zero more rapidly than $1/x$ as $x \to +\infty$, so that area accumulates under the curve $y = 1/x^2$ less rapidly than under the curve $y = 1/x$. When we calculate

$$\lim_{l \to +\infty} \int_1^l \frac{dx}{x} \quad \text{and} \quad \lim_{l \to +\infty} \int_1^l \frac{dx}{x^2}$$

the difference is enough that the first limit is infinite while the second is finite.

If f is continuous on the interval $(-\infty, b]$, then we define the improper integral $\int_{-\infty}^{b} f(x)\, dx$ as a limit in the following way:

$$\int_{-\infty}^{b} f(x)\, dx = \lim_{l \to -\infty} \int_{l}^{b} f(x)\, dx$$

As before, the improper integral is said to **converge** if the limit exists and **diverge** if it does not. If f is nonnegative, then the integral represents the area under $y = f(x)$ over the interval $(-\infty, b]$.

Example 3 Evaluate

$$\int_{-\infty}^{0} e^x\, dx$$

Solution. We begin by replacing the infinite lower limit with a finite lower limit l:

$$\int_{l}^{0} e^x\, dx = e^x \Big]_{l}^{0} = e^0 - e^l = 1 - e^l$$

Thus,

$$\int_{-\infty}^{0} e^x\, dx = \lim_{l \to -\infty} \int_{l}^{0} e^x\, dx = \lim_{l \to -\infty} (1 - e^l) = 1 \quad \blacktriangleleft$$

If the two improper integrals $\int_{-\infty}^{0} f(x)\, dx$ and $\int_{0}^{+\infty} f(x)\, dx$ both converge, then we say that $\int_{-\infty}^{+\infty} f(x)\, dx$ **converges** and we define

$$\int_{-\infty}^{+\infty} f(x)\, dx = \int_{-\infty}^{0} f(x)\, dx + \int_{0}^{+\infty} f(x)\, dx \tag{2}$$

If either integral on the right side of (2) diverges, then we say that $\int_{-\infty}^{+\infty} f(x)\, dx$ **diverges.** If f is nonnegative, then the integral $\int_{-\infty}^{+\infty} f(x)\, dx$ represents the area under $y = f(x)$ over the interval $(-\infty, +\infty)$.

Example 4   Evaluate

$$\int_{-\infty}^{+\infty} \frac{dx}{1 + x^2}$$

Solution.

$$\int_0^{+\infty} \frac{dx}{1 + x^2} = \lim_{l \to +\infty} \int_0^l \frac{dx}{1 + x^2} = \lim_{l \to +\infty} \left[\tan^{-1} x \right]_0^l = \lim_{l \to +\infty} (\tan^{-1} l) = \frac{\pi}{2}$$

$$\int_{-\infty}^0 \frac{dx}{1 + x^2} = \lim_{l \to -\infty} \int_l^0 \frac{dx}{1 + x^2} = \lim_{l \to -\infty} \left[\tan^{-1} x \right]_l^0 = \lim_{l \to -\infty} (-\tan^{-1} l) =$$

$$- \left(-\frac{\pi}{2} \right) = \frac{\pi}{2}$$

Thus,

$$\int_{-\infty}^{+\infty} \frac{dx}{1 + x^2} = \int_{-\infty}^0 \frac{dx}{1 + x^2} + \int_0^{+\infty} \frac{dx}{1 + x^2} = \frac{\pi}{2} + \frac{\pi}{2} = \pi \qquad \blacktriangleleft$$

REMARK.   In (2), the decision to split the integral $\int_{-\infty}^{+\infty} f(x) \, dx$ at $x = 0$ is arbitrary. We can just as well make the split at any other point $x = c$ without affecting the convergence, the divergence, or the value of the integral; that is,

$$\int_{-\infty}^{+\infty} f(x) \, dx = \int_{-\infty}^c f(x) \, dx + \int_c^{+\infty} f(x) \, dx$$

(We omit the proof.)

INTEGRALS WHOSE INTEGRANDS BECOME INFINITE

Next, we consider improper integrals in which the integrand approaches $+\infty$ or $-\infty$ somewhere in the interval of integration. We start with the cases where the integrand becomes infinite at one of the endpoints of integration.

If f is continuous on the interval $[a, b)$, but $f(x) \to +\infty$ or $f(x) \to -\infty$ as x approaches b from the left, then we define the improper integral $\int_a^b f(x) \, dx$ as a limit in the following way:

$$\int_a^b f(x) \, dx = \lim_{l \to b^-} \int_a^l f(x) \, dx \tag{3}$$

To visualize this definition geometrically, consider the case where f is nonnegative on the interval $[a, b)$. For each number l satisfying $a \le l < b$, the integral $\int_a^l f(x) \, dx$ represents the area under $y = f(x)$ over the interval $[a, l]$ (Figure 10.1.3a). As we let l approach b from the left, these areas

(a)

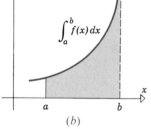

(b)

Figure 10.1.3

tend to fill out the entire area under $y = f(x)$ over the interval $[a, b)$ (Figure 10.1.3b). As before, the improper integral $\int_a^b f(x)\, dx$ is said to *converge* or *diverge* depending on whether the limit in (3) does or does not exist.

Example 5 Evaluate

$$\int_0^1 \frac{dx}{\sqrt{1-x}}$$

Solution. The integral is improper because the integrand approaches $+\infty$ as x approaches the upper limit 1 from the left. From (3),

$$\int_0^1 \frac{dx}{\sqrt{1-x}} = \lim_{l \to 1^-} \int_0^l \frac{dx}{\sqrt{1-x}} = \lim_{l \to 1^-} \left[-2\sqrt{1-x} \right]_0^l$$

$$= \lim_{l \to 1^-} [-2\sqrt{1-l} + 2] = 2 \quad ◄$$

If f is continuous on the interval $(a, b]$, but $f(x) \to +\infty$ or $f(x) \to -\infty$ as x approaches a from the right, then we define the improper integral $\int_a^b f(x)\, dx$ as:

$$\int_a^b f(x)\, dx = \lim_{l \to a^+} \int_l^b f(x)\, dx$$

The improper integral *converges* or *diverges* depending on whether the limit does or does not exist.

Example 6 The integral

$$\int_1^2 \frac{dx}{1-x}$$

is improper because the integrand approaches $-\infty$ as x approaches the lower limit 1 from the right. The integral diverges because

$$\int_1^2 \frac{dx}{1-x} = \lim_{l \to 1^+} \int_l^2 \frac{dx}{1-x} = \lim_{l \to 1^+} \left[-\ln|1-x| \right]_l^2$$

$$= \lim_{l \to 1^+} [-\ln|-1| + \ln|1-l|]$$

$$= \lim_{l \to 1^+} \ln|1-l| = -\infty \quad ◄$$

Let f be continuous on the interval $[a, b]$ with the exception that at some point c satisfying $a < c < b$, $f(x)$ becomes infinite (tends to $+\infty$ or $-\infty$) as x approaches c from the left or the right. If the two improper integrals $\int_a^c f(x)\, dx$ and $\int_c^b f(x)\, dx$ both converge, then we say that the improper integral $\int_a^b f(x)\, dx$ **converges,** and we define:

$$\int_a^b f(x)\, dx = \int_a^c f(x)\, dx + \int_c^b f(x)\, dx \tag{4}$$

If either integral on the right side of (4) diverges, then we say that $\int_a^b f(x)\, dx$ **diverges.**

Example 7 Evaluate

$$\int_1^4 \frac{dx}{(x-2)^{2/3}}$$

Solution. The integrand approaches $+\infty$ as $x \to 2$, so we use (4) to write

$$\int_1^4 \frac{dx}{(x-2)^{2/3}} = \int_1^2 \frac{dx}{(x-2)^{2/3}} + \int_2^4 \frac{dx}{(x-2)^{2/3}} \tag{5}$$

But

$$\int_1^2 \frac{dx}{(x-2)^{2/3}} = \lim_{l \to 2^-} \int_1^l \frac{dx}{(x-2)^{2/3}} = \lim_{l \to 2^-} [3(l-2)^{1/3} - 3(1-2)^{1/3}] = 3$$

$$\int_2^4 \frac{dx}{(x-2)^{2/3}} = \lim_{l \to 2^+} \int_l^4 \frac{dx}{(x-2)^{2/3}} = \lim_{l \to 2^+} [3(4-2)^{1/3} - 3(l-2)^{1/3}] = 3\sqrt[3]{2}$$

Thus, from (5)

$$\int_1^4 \frac{dx}{(x-2)^{2/3}} = 3 + 3\sqrt[3]{2} \quad \blacktriangleleft$$

It is sometimes tempting to apply the First Fundamental Theorem of Calculus directly to an improper integral without taking the appropriate limits. To illustrate what can go wrong with this procedure, suppose we ignore the fact that the integral

$$\int_0^2 \frac{dx}{(x-1)^2} \tag{6}$$

is improper and write

$$\int_0^2 \frac{dx}{(x-1)^2} = -\frac{1}{x-1} \Big]_0^2 = -1 - (1) = -2$$

This result is clearly nonsense because the integrand is never negative and consequently the integral cannot be negative! To evaluate (6) correctly we should write

$$\int_0^2 \frac{dx}{(x-1)^2} = \int_0^1 \frac{dx}{(x-1)^2} + \int_1^2 \frac{dx}{(x-1)^2} \tag{7}$$

But

$$\int_0^1 \frac{dx}{(x-1)^2} = \lim_{l \to 1^-} \int_0^l \frac{dx}{(x-1)^2} = \lim_{l \to 1^-} \left[-\frac{1}{l-1} - 1 \right] = +\infty$$

so that (6) diverges.

▶ Exercise Set 10.1

In Exercises 1–32, evaluate the integrals that converge.

1. $\displaystyle\int_0^{+\infty} e^{-x} \, dx.$

2. $\displaystyle\int_1^{+\infty} \frac{dx}{x^3}.$

3. $\displaystyle\int_1^{+\infty} \frac{dx}{\sqrt{x}}.$

4. $\displaystyle\int_{-1}^{+\infty} \frac{x}{1+x^2} \, dx.$

5. $\displaystyle\int_4^{+\infty} \frac{2}{x^2-1} \, dx.$

6. $\displaystyle\int_0^{+\infty} xe^{-x^2} \, dx.$

7. $\displaystyle\int_e^{+\infty} \frac{1}{x \ln^3 x} \, dx.$

8. $\displaystyle\int_2^{+\infty} \frac{1}{x\sqrt{\ln x}} \, dx.$

9. $\displaystyle\int_a^{+\infty} \frac{x \, dx}{(x^2+1)^2}.$

10. $\displaystyle\int_0^{+\infty} \frac{dx}{a^2+b^2x^2}; \quad a > 0, \, b > 0.$

11. $\displaystyle\int_{-\infty}^0 \frac{dx}{(2x-1)^3}.$

12. $\displaystyle\int_{-\infty}^2 \frac{dx}{x^2+4}.$

13. $\displaystyle\int_{-\infty}^0 e^{3x} \, dx.$

14. $\displaystyle\int_{-\infty}^0 \frac{e^x \, dx}{3-2e^x}.$

15. $\displaystyle\int_{-\infty}^{+\infty} x^3 \, dx.$

16. $\displaystyle\int_{-\infty}^{+\infty} \frac{x}{\sqrt{x^2+2}} \, dx.$

17. $\displaystyle\int_{-\infty}^{+\infty} \frac{x}{(x^2+3)^2} \, dx.$

18. $\displaystyle\int_{-\infty}^{+\infty} \frac{e^{-t}}{1+e^{-2t}} \, dt.$

19. $\displaystyle\int_3^4 \frac{dx}{(x-3)^2}.$

20. $\displaystyle\int_0^8 \frac{dx}{\sqrt[3]{x}}.$

21. $\displaystyle\int_0^{\pi/2} \tan x \, dx.$

22. $\displaystyle\int_0^9 \frac{dx}{\sqrt{9-x}}.$

23. $\displaystyle\int_0^1 \frac{dx}{\sqrt{1-x^2}}.$

24. $\displaystyle\int_{-3}^1 \frac{x \, dx}{\sqrt{9-x^2}}.$

25. $\displaystyle\int_0^{\pi/6} \frac{\cos x}{\sqrt{1-2\sin x}} \, dx.$

26. $\displaystyle\int_0^{\pi/4} \frac{\sec^2 x}{1-\tan x} \, dx.$

27. $\displaystyle\int_0^3 \frac{dx}{x-2}.$

28. $\displaystyle\int_{-2}^2 \frac{dx}{x^2}.$

29. $\displaystyle\int_{-1}^8 x^{-1/3} \, dx.$

30. $\displaystyle\int_0^4 \frac{dx}{(x-2)^{2/3}}.$

31. $\displaystyle\int_0^{+\infty} \frac{dx}{\sqrt{x}(x+4)}.$

32. $\displaystyle\int_1^{+\infty} \frac{dx}{x\sqrt{x^2-1}}.$

33. Find a positive value of a such that

(a) $\displaystyle\int_0^{+\infty} e^{-ax} \, dx = 5$

(b) $\displaystyle\int_0^{+\infty} \frac{1}{x^2+a^2} \, dx = 1.$

34. Given that

$$\int_0^{+\infty} e^{-x^2} \, dx = \frac{\sqrt{\pi}}{2}$$

find

$$\int_0^{+\infty} e^{-a^2x^2} \, dx \quad \text{(where } a \neq 0)$$

35. (a) It is possible for an improper integral to diverge without becoming infinite. Show that $\int_0^{+\infty} \cos x \, dx$ does this.

(b) Evaluate $\int_0^{+\infty} e^{-x} \cos x \, dx.$

36. Show that $\displaystyle\int_1^{+\infty} \frac{dx}{x^p}$ converges if $p > 1$ and diverges if $p \le 1$.

37. Show that $\displaystyle\int_0^1 \frac{dx}{x^p}$ converges if $p < 1$ and diverges if $p \ge 1$.

38. Find the area of the region between the x-axis and the curve $y = 8/(x^2 - 4)$ for $x \ge 3$.

39. Find the area of the region bounded by $y = 0$, $x = 0$, $x = 4$, and $y = (1 - x)^{-2}$.

40. (a)   Show that

$$\int_{-1}^0 \frac{dx}{x} = -\infty \quad \text{and} \quad \int_0^1 \frac{dx}{x} = +\infty$$

(b)   Evaluate

$$\int_{-1}^1 \frac{dx}{x}.$$

41. It can be shown that if f and g are continuous and $0 \le f(x) \le g(x)$ for all $x \ge a$, then $\int_a^{+\infty} f(x)\, dx$ converges if $\int_a^{+\infty} g(x)\, dx$ converges. Moreover

$$\int_a^{+\infty} f(x)\, dx \le \int_a^{+\infty} g(x)\, dx$$

and $\int_a^{+\infty} g(x)\, dx$ is an upper bound for $\int_a^{+\infty} f(x)\, dx$. Use this result to obtain an upper bound for each of the following improper integrals.

(a) $\displaystyle\int_2^{+\infty} \frac{x}{x^5 + 1}\, dx$ (b) $\displaystyle\int_1^{+\infty} e^{-x^2}\, dx$

42. It can be shown that if f and g are continuous and $0 \le f(x) \le g(x)$ for all $x \ge a$, then $\int_a^{+\infty} g(x)\, dx$ diverges if $\int_a^{+\infty} f(x)\, dx$ diverges. Use this result to show that each of the following improper integrals diverges.

(a) $\displaystyle\int_2^{+\infty} \frac{\sqrt{x^3 + 1}}{x}\, dx$ (b) $\displaystyle\int_0^{+\infty} \frac{e^x}{2x + 1}\, dx$

43. Let R be the region to the right of $x = 1$ that is bounded by the x-axis and the curve $y = 1/x$.
(a)   Show that the solid obtained by revolving R about the x-axis has a finite volume.
(b)   Show that the solid of part (a) has an infinite surface area. [*Hint:* See Exercise 42.]
[It has been suggested that by filling this solid with paint and letting it seep through to the surface one could paint an infinite surface area with a finite amount of paint!]

44. For what values of p does $\int_0^{+\infty} e^{px}\, dx$ converge?

45. In electromagnetic theory, the magnetic potential at a point on the axis of a circular coil is given by

$$u = \frac{2\pi N I r}{k} \int_a^{+\infty} \frac{dx}{(r^2 + x^2)^{3/2}}$$

where N, I, r, k, and a are constants. Find u.

46. Sketch the region whose area is $\displaystyle\int_0^{+\infty} \frac{dx}{1 + x^2}$, and use your sketch to show that

$$\int_0^{+\infty} \frac{dx}{1 + x^2} = \int_0^1 \sqrt{\frac{1 - y}{y}}\, dy$$

47. (**Satellite Problem**). In Exercise 13 of Section 6.7, we determined the work required to lift a 6000-lb satellite to a specified orbital position. The result in part (a) of that problem will be needed here.
(a)   Find a definite integral that represents the work required to lift a 6000-lb satellite to a position l miles above the earth's surface.
(b)   Find a definite integral that represents the work required to lift a 6000-lb satellite an "infinite distance" above the earth's surface. Evaluate the integral. [The result obtained here is sometimes called the work required to "escape" the earth's gravity.]

10.2   L'HÔPITAL'S RULE (INDETERMINATE FORMS OF TYPE 0/0)

In this section we shall develop an important new technique for finding limits of functions.

In each of the limits

$$\lim_{x \to 2} \frac{x^2 - 4}{x - 2} \quad \text{and} \quad \lim_{x \to 0} \frac{\sin x}{x} \tag{1}$$

the numerator and denominator both approach zero. It is customary to describe such limits as **indeterminate forms of type 0/0.** As we shall see, a limit of this type can have any real number whatsoever as its value or can diverge to $+\infty$ or $-\infty$. The value of such a limit, if it converges, is not generally evident by inspection, so the term "indeterminate" is used to convey the idea that the limit cannot be determined without some additional work.

In Example 8 of Section 2.5 we evaluated the first limit in (1) by canceling the common factor $x - 2$ from the numerator and denominator, and in Theorem 2.8.3 we resorted to a rather intricate geometric argument to obtain the second limit in (1). Because geometric arguments and the technique of canceling factors apply only to a limited range of problems, it is desirable to have a general method for handling indeterminate forms. This is provided by **L'Hôpital's* rule**, which we now discuss.

L'Hôpital's Rule
for Form 0/0

10.2.1   THEOREM. *Let* lim *stand for one of the limits* $\lim_{x \to a}$, $\lim_{x \to a^+}$, $\lim_{x \to a^-}$, $\lim_{x \to +\infty}$, *or* $\lim_{x \to -\infty}$, *and suppose that* $\lim f(x) = 0$ *and* $\lim g(x) = 0$. *If* $\lim [f'(x)/g'(x)]$ *has a finite value* L, *or if this limit is* $+\infty$ *or* $-\infty$, *then*

$$\lim \frac{f(x)}{g(x)} = \lim \frac{f'(x)}{g'(x)}$$

GUILLAUME FRANCOIS ANTOINE DE L'HÔPITAL (1661–1704) French mathematician. L'Hôpital, born to parents of the French high nobility, held the title of Marquis de Sainte-Mesme Comte d'Autrement. He showed mathematical talent quite early and at age 15 solved a difficult problem about cycloids posed by Pascal. As a young man he served briefly as a cavalry officer, but resigned because of nearsightedness. In his own time he gained fame as the author of the first textbook ever published on differential calculus, L'Analyse des Infiniment Petits pour l'Intelligence des Lignes Courbes (1696). L'Hôpital's rule appeared for the first time in that book. Actually, L'Hôpital's rule and most of the material in the calculus text were due to John Bernoulli, who was L'Hôpital's teacher. L'Hôpital dropped his plans for a book on integral calculus when Leibniz informed him that he intended to write such a text.

L'Hôpital was apparently generous and personable and his many contacts with major mathematicians provided the vehicle for disseminating major discoveries in calculus throughout Europe.

REMARK. There are some hypotheses implicit in this theorem. For example, in the case where $x \to a$, the statement $\lim_{x \to a} [f'(x)/g'(x)] = L$ requires that f'/g' be defined in some open interval I containing a (except possibly at a). This implies that f and g are differentiable and $g'(x) \neq 0$ in I (except possibly at a). Similar hypotheses are implicit in the other cases.

In essence, L'Hôpital's rule enables us to replace one limit problem with another that may be simpler. In each of the following examples we shall employ the following three-step process:

Step 1. Check that $\lim f(x)/g(x)$ is an indeterminate form. If it is not, then L'Hôpital's rule cannot be used.

Step 2. Differentiate f and g separately.

Step 3. Find $\lim f'(x)/g'(x)$. If this limit is finite, $+\infty$, or $-\infty$, then it is equal to $\lim f(x)/g(x)$.

Example 1   Use L'Hôpital's rule to evaluate

(a) $\displaystyle\lim_{x \to 2} \frac{x^2 - 4}{x - 2}$ (b) $\displaystyle\lim_{x \to 0} \frac{\sin 2x}{x}$

Solution (a).   Since

$$\lim_{x \to 2} (x^2 - 4) = 0 \quad \text{and} \quad \lim_{x \to 2} (x - 2) = 0$$

the given limit is an indeterminate form of type 0/0. Thus, L'Hôpital's rule applies and we can write

$$\lim_{x \to 2} \frac{x^2 - 4}{x - 2} = \lim_{x \to 2} \frac{\dfrac{d}{dx}[x^2 - 4]}{\dfrac{d}{dx}[x - 2]} = \lim_{x \to 2} \frac{2x}{1} = 4$$

Observe that this agrees with the result obtained in Example 8 of Section 2.5 by factoring.

Solution (b).   Since

$$\lim_{x \to 0} \sin 2x = 0 \quad \text{and} \quad \lim_{x \to 0} x = 0$$

the given limit is an indeterminate form of type 0/0. Thus, L'Hôpital's rule applies and we can write

$$\lim_{x \to 0} \frac{\sin 2x}{x} = \lim_{x \to 0} \frac{\dfrac{d}{dx}[\sin 2x]}{\dfrac{d}{dx}[x]} = \lim_{x \to 0} \frac{2 \cos 2x}{1} = 2$$

Observe that this agrees with the result obtained in Example 4 of Section 2.8 by substitution.   ◄

REMARK.   To be rigorous, in each of the foregoing examples the first equality is not justified until the limit on the right is shown to exist. However, for simplicity we shall usually arrange the computations as shown when applying L'Hôpital's rule.

Example 2   Evaluate

$$\lim_{x \to \pi/2} \frac{1 - \sin x}{\cos x}$$

Solution.   Since

$$\lim_{x \to \pi/2} (1 - \sin x) = \lim_{x \to \pi/2} \cos x = 0$$

the given limit is an indeterminate form of type 0/0. Thus, by L'Hôpital's rule

$$\lim_{x \to \pi/2} \frac{1 - \sin x}{\cos x} = \lim_{x \to \pi/2} \frac{\dfrac{d}{dx}[1 - \sin x]}{\dfrac{d}{dx}[\cos x]} = \lim_{x \to \pi/2} \frac{-\cos x}{-\sin x} = \frac{0}{-1} = 0 \qquad ◄$$

Example 3   Evaluate

$$\lim_{x \to 0} \frac{e^x - 1}{x^3}$$

Solution.   Since

$$\lim_{x \to 0} (e^x - 1) = \lim_{x \to 0} x^3 = 0$$

the given limit is an indeterminate form of type 0/0. Thus, by L'Hôpital's rule

$$\lim_{x \to 0} \frac{e^x - 1}{x^3} = \lim_{x \to 0} \frac{\frac{d}{dx}[e^x - 1]}{\frac{d}{dx}[x^3]} = \lim_{x \to 0} \frac{e^x}{3x^2} = +\infty \quad \blacktriangleleft$$

As the following example shows, it is sometimes necessary to apply L'Hôpital's rule more than once in the same problem.

Example 4 Evaluate

$$\lim_{x \to 0} \frac{1 - \cos x}{x^2}$$

Solution. Since

$$\lim_{x \to 0} (1 - \cos x) = \lim_{x \to 0} x^2 = 0$$

the given limit is an indeterminate form of type 0/0. Thus, by L'Hôpital's rule

$$\lim_{x \to 0} \frac{1 - \cos x}{x^2} = \lim_{x \to 0} \frac{\sin x}{2x}$$

However, the new limit is also an indeterminate form of type 0/0, so we apply L'Hôpital's rule again. This yields

$$\lim_{x \to 0} \frac{1 - \cos x}{x^2} = \lim_{x \to 0} \frac{\sin x}{2x} = \lim_{x \to 0} \frac{\cos x}{2} = \frac{1}{2} \quad \blacktriangleleft$$

WARNING. We warn the reader about two possible sources of errors. First, when applying L'Hôpital's rule to $\lim f(x)/g(x)$, the derivatives of $f(x)$ and $g(x)$ are taken separately to yield the new limit, $\lim [f'(x)/g'(x)]$. Do not make the mistake of differentiating $f(x)/g(x)$ according to the quotient rule. Second, make sure that $\lim [f(x)/g(x)]$ is an indeterminate form; otherwise, L'Hôpital's rule does not apply.

Example 5 Evaluate

$$\lim_{x \to 0} \frac{e^x}{x^2}$$

Solution.

$$\lim_{x \to 0} e^x = 1 \quad \text{and} \quad \lim_{x \to 0} x^2 = 0$$

so the given problem is not an indeterminate form of type 0/0. By inspection

$$\lim_{x \to 0} \frac{e^x}{x^2} = +\infty \quad \blacktriangleleft$$

Example 6   Evaluate

$$\lim_{x \to +\infty} \frac{x^{-4/3}}{\sin (1/x)}$$

Solution.   Since

$$\lim_{x \to +\infty} x^{-4/3} = \lim_{x \to +\infty} \sin (1/x) = 0$$

the given limit is an indeterminate form of type 0/0. Thus, by L'Hôpital's rule

$$\lim_{x \to +\infty} \frac{x^{-4/3}}{\sin (1/x)} = \lim_{x \to +\infty} \frac{-\frac{4}{3}x^{-7/3}}{(-1/x^2) \cos (1/x)} = \lim_{x \to +\infty} \frac{\frac{4}{3}x^{-1/3}}{\cos (1/x)} = \frac{0}{1} = 0$$

$$\blacktriangleleft$$

Example 7   Evaluate

$$\lim_{x \to 0^-} \frac{\tan x}{x^2}$$

Solution.   Since

$$\lim_{x \to 0^-} \tan x = \lim_{x \to 0^-} x^2 = 0$$

the given limit is an indeterminate form of type 0/0. Thus, by L'Hôpital's rule

$$\lim_{x \to 0^-} \frac{\tan x}{x^2} = \lim_{x \to 0^-} \frac{\sec^2 x}{2x} = -\infty \quad \blacktriangleleft$$

OPTIONAL

THEORY BEHIND
L'HÔPITAL'S RULE

The proof of L'Hôpital's rule depends on the following result, called the ***Extended Mean-Value Theorem*** or sometimes the ***Cauchy* Mean-Value Theorem*** (biography on p. 580).

10.2.2 THEOREM. *Let the functions f and g be differentiable on (a, b) and continuous on $[a, b]$. If $g'(x) \neq 0$ for any x in (a, b), then there is at least one point c in (a, b) such that*

$$\frac{f'(c)}{g'(c)} = \frac{f(b) - f(a)}{g(b) - g(a)} \tag{2}$$

Proof. Observe first that $g(b) - g(a) \neq 0$, since otherwise it would follow from the Mean-Value Theorem (4.10.2) that $g'(x) = 0$ at some point x in (a, b), contradicting our hypothesis. For convenience, introduce a new function F defined by

$$F(x) = [f(b) - f(a)]g(x) - [g(b) - g(a)]f(x) \tag{3}$$

It follows from our assumptions about f and g that F is continuous on $[a, b]$ and differentiable on (a, b). Moreover, $F(b) = F(a)$ (verify), so that the Mean-Value Theorem (4.10.2) implies that there is at least one point c in (a, b) where $F'(c) = 0$. Thus, from (3)

$$[f(b) - f(a)]g'(c) - [g(b) - g(a)]f'(c) = 0$$

or

---

*AUGUSTIN LOUIS CAUCHY (1789–1857) French mathematician. Cauchy's early education was acquired from his father, a barrister and master of the classics. Cauchy entered L'Ecole Polytechnique in 1805 to study engineering, but because of poor health, was advised to concentrate on mathematics. His major mathematical work began in 1811 with a series of brilliant solutions to some difficult outstanding problems.

In 1814 he wrote a treatise on integrals that was to become the basis for modern complex variable theory; in 1816 there followed a classic paper on wave propagation in liquids that won a prize from the French Academy; and in 1822 he wrote a paper that formed the basis of modern elasticity theory.

Cauchy's mathematical contributions for the next 35 years were brilliant and staggering in quantity, over 700 papers filling 26 modern volumes. Cauchy's work initiated the era of modern analysis. He brought to mathematics standards of precision and rigor undreamed of by Leibniz and Newton.

Cauchy's life was inextricably tied to the political upheavals of the time. A strong partisan of the Bourbons, he left his wife and children in 1830 to follow the Bourbon king Charles X into exile. For his loyalty he was made a baron by the ex-king. Cauchy eventually returned to France, but refused to accept a university position until the government waived its requirement that he take a loyalty oath.

It is difficult to get a clear picture of the man. Devoutly Catholic, he sponsored charitable work for unwed mothers, criminals, and relief for Ireland. Yet other aspects of his life cast him in an unfavorable light. The Norwegian mathematician Abel described him as, "mad, infinitely Catholic, and bigoted." Some writers praise his teaching, yet others say he rambled incoherently and, according to a report of the day, he once devoted an entire lecture to extracting the square root of seventeen to ten decimal places by a method well known to his students. In any event, Cauchy is undeniably one of the greatest minds in the history of science.

$$\frac{f'(c)}{g'(c)} = \frac{f(b) - f(a)}{g(b) - g(a)} \qquad \blacksquare$$

Note that in the special case where $g(x) = x$, Formula (2) reduces to

$$f'(c) = \frac{f(b) - f(a)}{b - a}$$

which is precisely the conclusion of the Mean-Value Theorem (4.10.2). Thus, the Extended Mean-Value Theorem is, in fact, an extension of the Mean-Value Theorem.

We shall now prove one of the cases of L'Hôpital's rule.

Proof of Theorem 10.2.1 (L'Hôpital's Rule). We shall consider only the case where L is finite and lim is the two-sided limit $\lim\limits_{x \to a}$, with a finite.

By hypothesis,

$$\lim_{x \to a} \frac{f'(x)}{g'(x)} = L \tag{4}$$

As noted in the remark following the statement of Theorem 10.2.1, (4) implies that there is an interval $(a, r]$ extending to the right of a and an interval $[l, a)$ extending to the left of a on which $f'(x)$ and $g'(x)$ are defined and $g'(x) \neq 0$. For convenience, define two new functions F and G by

$$F(x) = \begin{cases} f(x), & x \neq a \\ 0, & x = a \end{cases} \qquad G(x) = \begin{cases} g(x), & x \neq a \\ 0, & x = a \end{cases}$$

Both F and G are continuous on $[l, r]$; they are continuous on the intervals $[l, a)$ and $(a, r]$ because on these intervals $F(x) = f(x)$, $G(x) = g(x)$, and f and g are continuous (since they are differentiable). The continuity of F and G at a follows from the fact that $\lim\limits_{x \to a} f(x)/g(x)$ is an indeterminate form of type 0/0, so

$$\lim_{x \to a} F(x) = \lim_{x \to a} f(x) = 0$$

$$\lim_{x \to a} G(x) = \lim_{x \to a} g(x) = 0$$

Thus, F and G satisfy the hypotheses of the Extended Mean-Value Theorem (10.2.2) on $[l, a]$ and $[a, r]$. Moreover, the definitions of F and G imply that

$$F'(c) = f'(c) \quad \text{and} \quad G'(c) = g'(c) \tag{5}$$

at any point c (different from a) in one of these intervals. If we choose a point $x \neq a$ in one of the intervals $[l, a]$ or $[a, r]$ and apply the Extended Mean-Value Theorem to $[x, a]$ (or $[a, x]$), we conclude that there is a number c between a and x such that

$$\frac{F(x) - F(a)}{G(x) - G(a)} = \frac{F'(c)}{G'(c)} \tag{6}$$

From (5) together with the fact that $F(a) = G(a) = 0$ and $F(x) = f(x)$, $G(x) = g(x)$, we can rewrite (6) as

$$\frac{f(x)}{g(x)} = \frac{f'(c)}{g'(c)}$$

Thus,

$$\lim_{x \to a} \frac{f(x)}{g(x)} = \lim_{x \to a} \frac{f'(c)}{g'(c)} \tag{7}$$

Since c is between a and x, it follows that $c \to a$ as $x \to a$. This fact together with (4) yields

$$\lim_{x \to a} \frac{f'(c)}{g'(c)} = \lim_{c \to a} \frac{f'(c)}{g'(c)} = L$$

Thus, from (7)

$$\lim_{x \to a} \frac{f(x)}{g(x)} = L \quad \blacksquare$$

▶ Exercise Set 10.2

In Exercises 1–32, find the limits.

1. $\displaystyle \lim_{x \to 1} \frac{\ln x}{x - 1}$.

2. $\displaystyle \lim_{x \to 0} \frac{\sin 2x}{\sin 5x}$.

3. $\displaystyle \lim_{x \to 0} \frac{e^x - 1}{\sin x}$.

4. $\displaystyle \lim_{x \to 3} \frac{x - 3}{3x^2 - 13x + 12}$.

5. $\displaystyle \lim_{\theta \to 0} \frac{\tan \theta}{\theta}$.

6. $\displaystyle \lim_{t \to 0} \frac{te^t}{1 - e^t}$.

7. $\displaystyle \lim_{x \to 1} \frac{\ln x}{\tan \pi x}$.

8. $\displaystyle \lim_{x \to c} \frac{x^{1/3} - c^{1/3}}{x - c}$.

9. $\displaystyle \lim_{x \to \pi^+} \frac{\sin x}{x - \pi}$.

10. $\displaystyle \lim_{x \to 0^+} \frac{\sin x}{x^2}$.

11. $\displaystyle \lim_{x \to \frac{1}{2}\pi^-} \frac{\cos x}{\sqrt{\frac{1}{2}\pi - x}}$.

12. $\displaystyle \lim_{x \to 0^+} \frac{1 - \cos x}{x^3}$.

13. $\displaystyle \lim_{x \to 0} \frac{e^x + e^{-x} - 2}{1 - \cos 2x}$.

14. $\displaystyle \lim_{x \to 0} \frac{2 \cosh x - 2}{1 - \cos 2x}$.

15. $\displaystyle \lim_{x \to 0} \frac{\sin 3x}{\sinh 2x}$.

16. $\displaystyle \lim_{x \to 0} \frac{1 - e^{-2x}}{x^2 + 3x}$.

17. $\displaystyle \lim_{x \to \pi/2} \frac{\sin 2x}{4x^2 - \pi^2}$.

18. $\displaystyle \lim_{x \to 2} \frac{\ln (5x - 9)}{x^3 - 8}$.

19. $\displaystyle \lim_{x \to 0} \frac{x - \ln (x + 1)}{1 - \cos 2x}$.

20. $\displaystyle \lim_{x \to 0} \frac{x - \tan^{-1} x}{x^3}$.

21. $\displaystyle \lim_{x \to 0} \frac{2 - x^2 - 2 \cos x}{x^4}$.

22. $\displaystyle \lim_{x \to +\infty} \frac{\ln (1 + 3/x)}{\sin (2/x)}$.

23. $\displaystyle \lim_{x \to 0} \frac{x - \sin x}{x^3}$.

24. $\displaystyle \lim_{x \to -1} \frac{x^2 - 1}{\ln (3x + 4)}$.

25. $\displaystyle \lim_{x \to 0} \frac{e^{ax} - e^{bx}}{x}$.

26. $\displaystyle \lim_{x \to 0} \frac{a^x - 1}{x}, \quad a > 0$.

27. $\displaystyle \lim_{x \to 0} \frac{x - \tan x}{\sin x - x}$.

28. $\displaystyle \lim_{x \to \pi} \frac{\sin^2 x}{1 + \cos 3x}$.

29. $\displaystyle \lim_{x \to +\infty} \frac{\frac{1}{2}\pi - \tan^{-1} x}{\ln \left(1 + \frac{1}{x^2}\right)}$.

30. $\displaystyle \lim_{x \to 0^+} \frac{\tan x}{\tan 2x}$.

31. $\displaystyle\lim_{x \to 0^+} \frac{\ln (\cos x)}{\ln (\cos 3x)}$.

32. $\displaystyle\lim_{\theta \to 0} \frac{\sin^2 \theta - \sin (\theta^2)}{\theta^4}$.

33. (a) Find the error in the following calculation:

$$\lim_{x \to 1} \frac{x^3 - x^2 + x - 1}{x^3 - x^2} = \lim_{x \to 1} \frac{3x^2 - 2x + 1}{3x^2 - 2x}$$

$$= \lim_{x \to 1} \frac{6x - 2}{6x - 2} = 1$$

(b) Find the correct answer.

34. Find $\displaystyle\lim_{x \to 1} \frac{x^4 - 4x^3 + 6x^2 - 4x + 1}{x^4 - 3x^3 + 3x^2 - x}$.

35. Find all values of k and l such that

$$\lim_{x \to 0} \frac{k + \cos lx}{x^2} = -4$$

36. Evaluate

$$\lim_{x \to 0} \frac{(2 + x) \ln (1 - x)}{(1 - e^x) \cos x}$$

37. Evaluate

$$\lim_{x \to +\infty} x \ln \left(\frac{x + 1}{x - 1}\right)$$

[*Hint:* Rewrite the problem as an indeterminate form of type 0/0.]

38. (a) Explain why L'Hôpital's rule does not apply to the problem

$$\lim_{x \to 0} \frac{x^2 \sin (1/x)}{\sin x}$$

(b) Find the limit.

39. Find

$$\lim_{x \to 0^+} \frac{x \sin (1/x)}{\sin x}$$

if it exists.

40. (a) Given that $k \neq 0$ and $x > 0$, show that

$$\int_1^x \frac{1}{t^{1-k}} \, dt = \frac{x^k - 1}{k}$$

(b) Use the result in (a) to make a guess at the value of the limit

$$\lim_{k \to 0} \frac{x^k - 1}{k}$$

(c) Use L'Hôpital's rule to substantiate your guess.

[*Remark:* This exercise was motivated by an article by Henry C. Finlayson, which appeared in the *American Math. Monthly,* Vol 94, No. 5, May 1987, p. 450.]

41. In Figure 10.2.1, let $T(\theta)$ be the area of the right triangle ABC, and $S(\theta)$ the area of the segment of the circle formed by chord AB and the arc of the unit circle subtended by the central angle θ. Find $\displaystyle\lim_{\theta \to 0^+} T(\theta)/S(\theta)$.

Figure 10.2.1

10.3 OTHER INDETERMINATE FORMS $\left(\frac{\infty}{\infty}, 0 \cdot \infty, 0^0, \infty^0, 1^\infty, \infty - \infty\right)$

In this section we shall continue our study of indeterminate forms and introduce another version of L'Hôpital's rule.

In Figure 10.3.1 we have illustrated four ways in which a function can become "infinite at a."

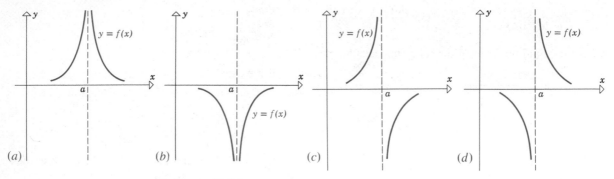

Figure 10.3.1

- Figure 10.3.1a: $\lim\limits_{x \to a} f(x) = +\infty$

- Figure 10.3.1b: $\lim\limits_{x \to a} f(x) = -\infty$

- Figure 10.3.1c: $\begin{cases} \lim\limits_{x \to a^+} f(x) = -\infty \\ \lim\limits_{x \to a^-} f(x) = +\infty \end{cases}$

- Figure 10.3.1d: $\begin{cases} \lim\limits_{x \to a^+} f(x) = +\infty \\ \lim\limits_{x \to a^-} f(x) = -\infty \end{cases}$

Sometimes we shall want to indicate that a function becomes infinite at a point a without actually specifying which of these four possibilities occurs. To do so, we shall write

$$\lim_{x \to a} f(x) = \infty$$

Similarly, we shall write:

$$\lim_{x \to +\infty} f(x) = \infty \quad \text{if} \quad \lim_{x \to +\infty} f(x) = +\infty \text{ or } \lim_{x \to +\infty} f(x) = -\infty$$

$$\lim_{x \to -\infty} f(x) = \infty \quad \text{if} \quad \lim_{x \to -\infty} f(x) = +\infty \text{ or } \lim_{x \to -\infty} f(x) = -\infty$$

INDETERMINATE FORMS OF TYPE ∞/∞

An *indeterminate form of type* ∞/∞ is a limit $\lim f(x)/g(x)$ in which $\lim f(x) = \infty$ and $\lim g(x) = \infty$. Some examples are

$$\lim_{x \to 0^+} \frac{\ln x}{\csc x} \quad (\text{numerator} \to -\infty, \text{ denominator} \to +\infty)$$

$$\lim_{x \to +\infty} \frac{x}{e^x} \quad (\text{numerator} \to +\infty, \text{ denominator} \to +\infty)$$

$$\lim_{x \to 0} \frac{1 + \dfrac{1}{x}}{\cot x} \quad \text{(numerator} \to \infty, \text{ denominator} \to \infty)$$

The following version of L'Hôpital's rule, usually proved in advanced courses, is applicable to problems like these.

L'Hôpital's Rule for Form $\frac{\infty}{\infty}$

10.3.1   THEOREM.   *Let* lim *stand for one of the limits* $\lim\limits_{x \to a}$, $\lim\limits_{x \to a^+}$, $\lim\limits_{x \to a^-}$, $\lim\limits_{x \to +\infty}$, *or* $\lim\limits_{x \to -\infty}$, *and suppose that* $\lim f(x) = \infty$ *and* $\lim g(x) = \infty$. *If* $\lim [f'(x)/g'(x)]$ *has a finite value L, or if this limit is* $+\infty$ *or* $-\infty$, *then*

$$\lim \frac{f(x)}{g(x)} = \lim \frac{f'(x)}{g'(x)}$$

REMARK.   The remark about hypotheses, which follows Theorem 10.2.1, is also applicable here.

Example 1   Evaluate $\lim\limits_{x \to +\infty} \dfrac{x}{e^x}$

Solution.

$$\lim_{x \to +\infty} x = \lim_{x \to +\infty} e^x = +\infty$$

so that the given limit is an indeterminate form of type ∞/∞. Thus, by L'Hôpital's rule

$$\lim_{x \to +\infty} \frac{x}{e^x} = \lim_{x \to +\infty} \frac{\dfrac{d}{dx}[x]}{\dfrac{d}{dx}[e^x]} = \lim_{x \to +\infty} \frac{1}{e^x} = 0 \quad \blacktriangleleft$$

Example 2   Evaluate $\lim\limits_{x \to 0^+} \dfrac{\ln x}{\csc x}$

Solution.

$$\lim_{x \to 0^+} \ln x = -\infty \quad \text{and} \quad \lim_{x \to 0^+} \csc x = +\infty$$

so that the given limit is an indeterminate form of type ∞/∞. Thus, by L'Hôpital's rule

$$\lim_{x \to 0^+} \frac{\ln x}{\csc x} = \lim_{x \to 0^+} \frac{1/x}{-\csc x \cot x} \qquad (1)$$

This last limit is again an indeterminate form of type ∞/∞. Moreover, any additional applications of L'Hôpital's rule will yield powers of $1/x$ in the numerator and expressions involving $\csc x$ and $\cot x$ in the denominator; thus, repeated application of L'Hôpital's rule simply produces new indeterminate forms. We must try something else. The last limit in (1) can be rewritten as

$$\lim_{x \to 0^+} \left(-\frac{\sin x}{x} \tan x \right) = -\lim_{x \to 0^+} \frac{\sin x}{x} \cdot \lim_{x \to 0^+} \tan x = -(1)(0) = 0$$

Thus,

$$\lim_{x \to 0^+} \frac{\ln x}{\csc x} = 0 \qquad \blacktriangleleft$$

INDETERMINATE
FORMS OF TYPE $0 \cdot \infty$

A limit of a product, $\lim f(x)g(x)$, is called an **indeterminate form of type** $\mathbf{0 \cdot \infty}$ if $\lim f(x) = 0$ and $\lim g(x) = \infty$. Frequently, limit problems of this type can be converted to the form $0/0$ by writing

$$f(x)g(x) = \frac{f(x)}{1/g(x)}$$

or to the form ∞/∞ by writing

$$f(x)g(x) = \frac{g(x)}{1/f(x)}$$

The limit can then be treated by L'Hôpital's rule.

Example 3   Evaluate $\lim\limits_{x \to 0^+} x \ln x$.

Solution.   Since

$$\lim_{x \to 0^+} x = 0 \quad \text{and} \quad \lim_{x \to 0^+} \ln x = -\infty$$

the given problem is an indeterminate form of type $0 \cdot \infty$. We shall convert the problem to the form ∞/∞ and apply L'Hôpital's rule as follows:

$$\lim_{x \to 0^+} x \ln x = \lim_{x \to 0^+} \frac{\ln x}{1/x} = \lim_{x \to 0^+} \frac{1/x}{-1/x^2} = \lim_{x \to 0^+} (-x) = 0 \qquad \blacktriangleleft$$

In this example we could have converted the problem to form $0/0$ by writing

$$x \ln x = \frac{x}{1/\ln x}$$

However, this is a less desirable first choice because of the relatively complicated derivative of $1/\ln x$.

WARNING. It is tempting to argue that an indeterminate form of type $0 \cdot \infty$ has value 0 since "zero times anything is zero." However, this is fallacious since $0 \cdot \infty$ is not a product of numbers, but rather a statement about limits. The following example gives an indeterminate form of type $0 \cdot \infty$ whose value is not zero.

Example 4 Evaluate $\lim_{x \to \pi/4} (1 - \tan x) \sec 2x$.

Solution. The given problem is an indeterminate form of type $0 \cdot \infty$. We convert it to type 0/0 and apply L'Hôpital's rule as follows:

$$\lim_{x \to \pi/4} (1 - \tan x) \sec 2x = \lim_{x \to \pi/4} \frac{1 - \tan x}{1/\sec 2x} = \lim_{x \to \pi/4} \frac{1 - \tan x}{\cos 2x}$$

$$= \lim_{x \to \pi/4} \frac{-\sec^2 x}{-2 \sin 2x} = \frac{-2}{-2} = 1 \quad \blacktriangleleft$$

INDETERMINATE FORMS OF TYPE $0^0, \infty^0, 1^\infty$

Limits of the form

$$\lim_{x \to a} f(x)^{g(x)}$$

give rise to *indeterminate forms of the types 0^0, ∞^0, and 1^∞.* (At this stage the meaning of these symbols should be clear.) All three types are treated by first introducing a dependent variable.

$$y = f(x)^{g(x)}$$

and then calculating

$$\lim_{x \to a} \ln y = \lim_{x \to a} [\ln (f(x)^{g(x)})] = \lim_{x \to a} [g(x) \ln f(x)]$$

Once the value of $\lim_{x \to a} \ln y$ is known, it is a simple matter to determine $\lim_{x \to a} y = \lim_{x \to a} f(x)^{g(x)}$ as our examples will show.

In the next example we shall derive the two limit formulas for e stated in Section 7.2. (See Formulas (5) and (6) in that section.)

Example 5   Show that

$$\text{(a)} \quad \lim_{x \to 0} (1 + x)^{1/x} = e \qquad \text{(b)} \quad \lim_{x \to +\infty} \left(1 + \frac{1}{x}\right)^x = e$$

Solution (a).   Since

$$\lim_{x \to 0} (1 + x) = 1 \quad \text{and} \quad \lim_{x \to 0} \frac{1}{x} = \infty$$

the given limit is an indeterminate form of type 1^∞. As discussed above, we introduce a dependent variable

$$y = (1 + x)^{1/x}$$

and take the natural logarithm of both sides

$$\ln y = \ln (1 + x)^{1/x} = \frac{1}{x} \ln (1 + x) = \frac{\ln (1 + x)}{x}$$

The limit

$$\lim_{x \to 0} \ln y = \lim_{x \to 0} \frac{\ln (1 + x)}{x}$$

is an indeterminate form of type 0/0, so by L'Hôpital's rule,

$$\lim_{x \to 0} \ln y = \lim_{x \to 0} \frac{\ln (1 + x)}{x} = \lim_{x \to 0} \frac{1/(1 + x)}{1} = 1 \qquad (2)$$

To complete the calculation, we shall use the known limit, $\lim_{x \to 0} \ln y = 1$, to find the unknown limit, $\lim_{x \to 0} y = \lim_{x \to 0} (1 + x)^{1/x}$. We argue as follows: As $x \to 0$, we have

$$\ln y \to 1 \quad \text{[from (2)]}$$

so

$$e^{\ln y} \to e^1 \quad \text{[since } e^x \text{ is a continuous function]}$$

and thus,

$$y \to e$$

Therefore,

$$\lim_{x \to 0} (1 + x)^{1/x} = e \qquad (3)$$

Solution (b).   The formula in this part can be derived by the same procedure used in (a). Alternatively, we can simply let $x = 1/h$ in (3) to obtain

$$\lim_{h \to +\infty} \left(1 + \frac{1}{h}\right)^h = e$$

which is the same as the stated formula except for the notation used for the variable. ◀

INDETERMINATE FORMS OF TYPE $\infty - \infty$

If a limit problem such as

$$\lim [f(x) - g(x)] \quad \text{or} \quad \lim [f(x) + g(x)]$$

leads to one of the expressions

$$(+\infty) - (+\infty), \quad (-\infty) - (-\infty)$$

$$(+\infty) + (-\infty), \quad (-\infty) + (+\infty)$$

then one term tends to make the expression large and the other tends to make it small, resulting in an indeterminate form. These are called *inde-terminate forms of type $\infty - \infty$*. Such forms can sometimes be treated by combining the two terms into one and manipulating the result into one of the previous forms.

Example 6 Evaluate

$$\lim_{x \to 0^+} \left(\frac{1}{x} - \frac{1}{\sin x}\right)$$

Solution. Since

$$\lim_{x \to 0^+} \frac{1}{x} = +\infty \quad \text{and} \quad \lim_{x \to 0^+} \frac{1}{\sin x} = +\infty$$

the given limit is an indeterminate form of type $\infty - \infty$. Combining terms yields

$$\lim_{x \to 0^+} \left(\frac{1}{x} - \frac{1}{\sin x}\right) = \lim_{x \to 0^+} \left(\frac{\sin x - x}{x \sin x}\right)$$

which is an indeterminate form of type 0/0. Applying L'Hôpital's rule twice yields

$$\lim_{x \to 0^+} \left(\frac{\sin x - x}{x \sin x}\right) = \lim_{x \to 0^+} \frac{\cos x - 1}{\sin x + x \cos x}$$

$$= \lim_{x \to 0^+} \frac{-\sin x}{\cos x + \cos x - x \sin x}$$

$$= \frac{0}{2} = 0 \quad ◀$$

Example 7   Evaluate

$$\lim_{x \to 0^+} (\cot x - \ln x)$$

Solution.   Since

$$\lim_{x \to 0^+} \cot x = +\infty \quad \text{and} \quad \lim_{x \to 0^+} \ln x = -\infty$$

we are dealing with a problem of the type $(+\infty) - (-\infty)$. This is not an indeterminate form. The first term tends to make the limit large and because of the subtraction, the second term also tends to make the limit large. Thus,

$$\lim_{x \to 0^+} (\cot x - \ln x) = +\infty \qquad \blacktriangleleft$$

▶ Exercise Set 10.3

In Exercises 1–42, find the limits.

1. $\displaystyle \lim_{x \to +\infty} \frac{\ln x}{x}$.

2. $\displaystyle \lim_{x \to +\infty} \frac{e^{3x}}{x^2}$.

3. $\displaystyle \lim_{x \to 0^+} \frac{\cot x}{\ln x}$.

4. $\displaystyle \lim_{x \to 0^+} \frac{1 - \ln x}{e^{1/x}}$.

5. $\displaystyle \lim_{x \to +\infty} \frac{x \ln x}{x + \ln x}$.

6. $\displaystyle \lim_{x \to +\infty} \frac{x^3 - 2x + 1}{4x^3 + 2}$.

7. $\displaystyle \lim_{x \to +\infty} \frac{x^{100}}{e^x}$.

8. $\displaystyle \lim_{x \to 0^+} \frac{\ln (\sin x)}{\ln (\tan x)}$.

9. $\displaystyle \lim_{x \to +\infty} x e^{-x}$.

10. $\displaystyle \lim_{x \to \pi^-} (x - \pi) \tan \tfrac{1}{2}x$.

11. $\displaystyle \lim_{x \to +\infty} x \sin \frac{\pi}{x}$.

12. $\displaystyle \lim_{x \to 0^+} \tan x \ln x$.

13. $\displaystyle \lim_{x \to +\infty} x(e^{\sin (2/x)} - 1)$.

14. $\displaystyle \lim_{x \to 1} x^{1/(1-x)}$.

15. $\displaystyle \lim_{x \to +\infty} (1 - 3/x)^x$.

16. $\displaystyle \lim_{x \to 0} (1 + 2x)^{-3/x}$.

17. $\displaystyle \lim_{x \to 0} (e^x + x)^{1/x}$.

18. $\displaystyle \lim_{x \to +\infty} (1 + a/x)^{bx}$.

19. $\displaystyle \lim_{x \to +\infty} (1 + 1/x^2)^x$.

20. $\displaystyle \lim_{x \to +\infty} \left(\frac{x + 1}{x + 2}\right)^x$.

21. $\displaystyle \lim_{x \to +\infty} (1 + 1/x)^{x^2}$.

22. $\displaystyle \lim_{x \to 0} (1 + \sin 2x)^{1/x}$.

23. $\displaystyle \lim_{x \to 1} (2 - x)^{\tan (\pi/2) x}$.

24. $\displaystyle \lim_{x \to +\infty} [\cos (2/x)]^{x^2}$.

25. $\displaystyle \lim_{x \to 0^+} x^{\sin x}$.

26. $\displaystyle \lim_{x \to 0^+} x^x$.

27. $\displaystyle \lim_{x \to 0^+} (\sin x)^{3/\ln x}$.

28. $\displaystyle \lim_{x \to 0^+} (e^{2x} - 1)^{1/\ln x}$.

29. $\displaystyle \lim_{x \to (1/2)\pi^-} (\tan x)^{\cos x}$.

30. $\displaystyle \lim_{x \to +\infty} (\ln x)^{1/x}$.

31. $\displaystyle \lim_{x \to +\infty} (1 + x^2)^{1/\ln x}$.

32. $\displaystyle \lim_{x \to +\infty} (3^x + 5^x)^{1/x}$.

33. $\displaystyle \lim_{\theta \to 0} \left(\frac{1}{1 - \cos \theta} - \frac{2}{\sin^2 \theta}\right)$.

34. $\displaystyle \lim_{x \to 0} \left(\frac{1}{x^2} - \frac{\cos 3x}{x^2}\right)$.

35. $\displaystyle \lim_{x \to 0} (\csc x - 1/x)$.

36. $\displaystyle \lim_{x \to 0} \left(\frac{1}{x} - \frac{1}{e^x - 1}\right)$.

37. $\displaystyle \lim_{x \to 0} (\cot x - \csc x)$.

38. $\displaystyle \lim_{x \to +\infty} [\ln x - \ln (1 + x)]$.

39. $\displaystyle \lim_{x \to +\infty} [x - \ln (x^2 + 1)]$.

40. $\displaystyle \lim_{x \to +\infty} [x - \ln (1 + 2e^x)]$.

41. $\displaystyle \lim_{x \to 0^+} \frac{\cot x}{\cot 2x}$.

42. $\displaystyle \lim_{x \to (1/2)\pi^-} \frac{4 \tan x}{1 + \sec x}$.

43. Show that for any positive integer n:

    (a) $\displaystyle \lim_{x \to +\infty} \frac{x^n}{e^x} = 0$ (b) $\displaystyle \lim_{x \to +\infty} \frac{e^x}{x^n} = +\infty$.

44. Show that for any positive integer n:

    (a) $\displaystyle \lim_{x \to +\infty} \frac{\ln x}{x^n} = 0$ (b) $\displaystyle \lim_{x \to +\infty} \frac{x^n}{\ln x} = +\infty$.

45. Limits of the type

$$\frac{0}{\infty}, \frac{\infty}{0}, 0^\infty, \infty \cdot \infty, +\infty + (+\infty),$$

$$+\infty - (-\infty), \quad -\infty + (-\infty), \quad -\infty - (+\infty)$$

are *not* indeterminate forms. Find the following limits by inspection.

(a) $\lim\limits_{x \to 0^+} \dfrac{x}{\ln x}$ (b) $\lim\limits_{x \to +\infty} \dfrac{x^3}{e^{-x}}$

(c) $\lim\limits_{x \to (1/2)\pi^-} (\cos x)^{\tan x}$ (d) $\lim\limits_{x \to 0^+} (\ln x) \cot x$

(e) $\lim\limits_{x \to (1/2)\pi^-} \left(\dfrac{1}{\frac{1}{2}\pi - x} + \tan x \right)$

(f) $\lim\limits_{x \to 0^+} \left(\dfrac{1}{x} - \ln x \right)$

(g) $\lim\limits_{x \to -\infty} (x + x^3)$ (h) $\lim\limits_{x \to +\infty} \left(\ln \left(\dfrac{1}{x} \right) - e^x \right)$.

46. Find $\lim\limits_{x \to +\infty} (e^x - x^2)$.

47. Sketch the graph of x^x, $x > 0$.

48. Sketch the graph of $(1/x) \tan x$, $-\pi/2 < x < \pi/2$.

In Exercises 49–51, evaluate the improper integral.

49. $\displaystyle\int_0^1 \ln x \, dx$.   **50.** $\displaystyle\int_1^{+\infty} \dfrac{\ln x}{x^2} \, dx$.

51. $\displaystyle\int_0^{+\infty} xe^{-3x} \, dx$.

52. (a) Show that $\int_1^{+\infty} e^{t^2} \, dt = +\infty$

       [*Hint:* Don't try to carry out the integration. Instead, compare the sizes of $\int_1^l e^t \, dt$ and $\int_1^l e^{t^2} \, dt$.]

    (b) Use L'Hôpital's rule to evaluate

$$\lim_{x \to +\infty} \frac{1}{x} \int_1^x e^{t^2} \, dt$$

53. (a) Show that

$$\int_0^{+\infty} \sqrt{1 + t^3} \, dt = +\infty.$$

       [*Hint:* $\sqrt{1 + t^3} \geq t^{3/2}$ for $t \geq 0$.]

    (b) Use L'Hôpital's rule to evaluate

$$\lim_{x \to +\infty} \frac{\displaystyle\int_0^{2x} \sqrt{1 + t^3} \, dt}{x^{5/2}}.$$

In Exercises 54–56, find the limit.

54. $\lim\limits_{x \to +\infty} \dfrac{2x - \sin x}{3x + \sin x}$.   **55.** $\lim\limits_{x \to +\infty} \dfrac{x(2 + \sin x)}{x + 1}$.

56. $\lim\limits_{x \to +\infty} \dfrac{x(2 + \sin x)}{x^2 + 1}$.

57. There is a myth that circulates among beginning calculus students which states that all indeterminate forms of types 0^0, ∞^0, and 1^∞ have value 1 because "anything to the zero power is 1" and "1 to any power is 1." The fallacy is that 0^0, ∞^0, and 1^∞ are not powers of numbers, but rather descriptions of limits. The following examples, which were transmitted to me by the late Professor Jack Staib of Drexel University, show that such indeterminate forms can have any positive real value:

    (a) $\lim\limits_{x \to 0^+} [x^{(\ln a)/(1 + \ln x)}] = 0^0 = a$.

    (b) $\lim\limits_{x \to +\infty} [x^{(\ln a)/(1 + \ln x)}] = \infty^0 = a$.

    (c) $\lim\limits_{x \to 0} [(x + 1)^{(\ln a)/x}] = 1^\infty = a$.

Prove these results.

▶ SUPPLEMENTARY EXERCISES

In Exercises 1–14, evaluate the integrals that converge.

1. $\displaystyle\int_{-\infty}^{+\infty} \dfrac{dx}{x^2 + 4}$.   **2.** $\displaystyle\int_4^6 \dfrac{dx}{4 - x}$.   **7.** $\displaystyle\int_{-\infty}^{+\infty} xe^{-x^2} \, dx$.   **8.** $\displaystyle\int_{-\infty}^0 xe^x \, dx$.

3. $\displaystyle\int_0^1 \dfrac{x \, dx}{\sqrt{1 - x^2}}$.   **4.** $\displaystyle\int_0^5 \dfrac{dx}{\sqrt{25 - x^2}}$.   **9.** $\displaystyle\int_0^{\pi/2} \cot x \, dx$.   **10.** $\displaystyle\int_0^{+\infty} \dfrac{dx}{x^5}$.

5. $\displaystyle\int_{-1}^1 \dfrac{dx}{\sqrt{x^2}}$.   **6.** $\displaystyle\int_0^{\pi/2} \sec^2 x \, dx$.   **11.** $\displaystyle\int_e^{+\infty} \dfrac{dx}{x(\ln x)^2}$.   **12.** $\displaystyle\int_0^1 \sqrt{x} \ln x \, dx$.

13. $\displaystyle\int_0^{+\infty} \frac{dx}{x^2 + 2x + 2}$.   14. $\displaystyle\int_0^4 \frac{e^{-\sqrt{x}}}{\sqrt{x}}\, dx$.

15. For what values of n does $\displaystyle\int_0^1 x^n \ln x\, dx$ converge?
    What is its value?

In Exercises 16–31, find the limit.

16. $\displaystyle\lim_{x \to 1} \frac{\ln x}{x - 1}$.

17. $\displaystyle\lim_{x \to 0} \frac{xe^{3x} - x}{1 - \cos 2x}$.

18. $\displaystyle\lim_{x \to +\infty} \frac{\ln (\ln x)}{\sqrt{x}}$.

19. $\displaystyle\lim_{x \to 0^+} x^2 e^{1/x}$.

20. $\displaystyle\lim_{x \to +\infty} (\sqrt{x^2 + x} - x)$.

21. $\displaystyle\lim_{x \to 0^-} x^2 e^{1/x}$.

22. $\displaystyle\lim_{x \to 0^-} (1 - x)^{2/x}$.

23. $\displaystyle\lim_{\theta \to 0} \left(\frac{\csc \theta}{\theta} - \frac{1}{\theta^2} \right)$.

24. $\displaystyle\lim_{x \to 0} \frac{x - \tan^{-1} x}{x^4}$.

25. $\displaystyle\lim_{x \to 2} \frac{x - 1 - e^{x-2}}{1 - \cos 2\pi x}$.

26. $\displaystyle\lim_{x \to 0} \frac{9^x - 3^x}{x}$.

27. $\displaystyle\lim_{x \to 0} \frac{\displaystyle\int_0^x \sin t^2\, dt}{\sin x^2}$.

28. $\displaystyle\lim_{x \to +\infty} x^{1/x}$.

29. $\displaystyle\lim_{x \to +\infty} \frac{(\ln x)^3}{x}$.

30. $\displaystyle\lim_{x \to +\infty} \left(\frac{x}{x - 3} \right)^x$.

31. $\displaystyle\lim_{x \to 0^+} (1 + x)^{\ln x}$.

32. Find the area of the region in the first quadrant between the curves $y = e^{-x}$ and $y = (e^{-x})^2$.

In Exercises 33 and 34, find:
(a) the area of the region R;
(b) the volume obtained by revolving R about the x-axis.

33. The region R bounded between the x-axis and the curve $y = x^{-2/3}$, $x \geq 8$.

34. The region R bounded between the x-axis and the curve $y = x^{-1/3}$, $0 \leq x \leq 1$.

35. Find the total arc length of the graph of $f(x) = \sqrt{x - x^2} - \sin^{-1} \sqrt{x}$. [*Hint:* First find the domain of f.]

36. The *Gamma function*, $\Gamma(x)$, is defined as

$$\Gamma(x) = \int_0^{+\infty} t^{x-1} e^{-t}\, dt$$

It can be shown that this improper integral converges if and only if $x > 0$.
(a) Find $\Gamma(1)$.
(b) Prove: $\Gamma(x + 1) = x\Gamma(x)$ for all $x > 0$. [*Hint:* Use integration by parts.]
(c) Use the results in parts (a) and (b) to find $\Gamma(2)$, $\Gamma(3)$, and $\Gamma(4)$.
(d) Find $\Gamma(\frac{1}{2})$, given that $\int_0^{+\infty} e^{-x^2}\, dx = \sqrt{\pi}/2$.
(e) Use the results in parts (b) and (d) to find $\Gamma(\frac{3}{2})$ and $\Gamma(\frac{5}{2})$.

11 infinite series

11.1 SEQUENCES

This chapter is concerned with the study of "infinite series," which, loosely speaking, are sums with infinitely many terms. The material in this chapter has far-reaching applications in engineering and science and is the cornerstone for many branches of mathematics. In this initial section we shall develop some preliminary results that are important in their own right.

In everyday language, we use the term "sequence" to suggest a succession of objects or events given in a specified order. *Informally* speaking, the term "sequence" in mathematics is used to describe an unending succession of numbers. Some possibilities are

$$1, 2, 3, 4, \ldots$$
$$2, 4, 6, 8, \ldots$$
$$1, \tfrac{1}{2}, \tfrac{1}{3}, \tfrac{1}{4}, \ldots$$
$$1, -1, 1, -1, \ldots$$

In each case, the three dots are used to suggest that the sequence continues indefinitely, following the obvious pattern. The numbers in a sequence are called the *terms* of the sequence. The terms may be described according to the positions they occupy. Thus, a sequence has a *first term,* a *second term,* a *third term,* and so forth. Because a sequence continues indefinitely, there is no last term.

The most common way to specify a sequence is to give a formula for the terms. To illustrate the idea, we have listed the terms in the sequence $2, 4, 6, 8, \ldots$ together with their term numbers:

term number	1	2	3	4	$\ldots$
term		2	4	6	8 $\ldots$

There is a clear relationship here; each term is twice its term number.

Thus, for each positive integer n the nth term in the sequence 2, 4, 6, 8, ... is given by the formula $2n$. This is denoted by writing

$$2, 4, 6, 8, \ldots, 2n, \ldots$$

or more compactly in **bracket notation** as

$$\{2n\}_{n=1}^{+\infty}$$

From the bracket notation, the terms in the sequence can be generated by successively substituting the integer values $n = 1, 2, 3, \ldots$ into the formula $2n$.

Example 1   List the first five terms of the sequence $\{2^n\}_{n=1}^{+\infty}$

Solution.   Substituting $n = 1, 2, 3, 4, 5$ into the formula 2^n yields

$$2^1, 2^2, 2^3, 2^4, 2^5, \ldots$$

or, equivalently,

$$2, 4, 8, 16, 32, \ldots \quad \blacktriangleleft$$

Example 2   Express the following sequences in bracket notation.

(a) $\dfrac{1}{2}, \dfrac{2}{3}, \dfrac{3}{4}, \dfrac{4}{5}, \ldots$

(b) $\dfrac{1}{2}, \dfrac{1}{4}, \dfrac{1}{8}, \dfrac{1}{16}, \ldots$

(c) $1, -1, 1, -1, \ldots$

(d) $\dfrac{1}{2}, -\dfrac{2}{3}, \dfrac{3}{4}, -\dfrac{4}{5}, \ldots$

(e) $1, 3, 5, 7, \ldots$

Solution.
(a)   Begin by comparing terms and term numbers:

term number	1	2	3	4	...
term	$\frac{1}{2}$	$\frac{2}{3}$	$\frac{3}{4}$	$\frac{4}{5}$	...

In each term, the numerator is the same as the term number, and the denominator is one greater than the term number. Thus, the nth term is $n/(n + 1)$ and the sequence can be written as

$$\left\{ \frac{n}{n + 1} \right\}_{n=1}^{+\infty}$$

(b) Observe that the sequence can be rewritten as

$$\frac{1}{2}, \frac{1}{2^2}, \frac{1}{2^3}, \frac{1}{2^4}, \dots$$

and construct a table comparing terms and term numbers:

term number	1	2	3	4	$\cdots$
term	$\frac{1}{2}$	$\frac{1}{2^2}$	$\frac{1}{2^3}$	$\frac{1}{2^4}$	$\cdots$

From the table we see that the nth term is $1/2^n$, so the sequence can be written as

$$\left\{\frac{1}{2^n}\right\}_{n=1}^{+\infty}$$

(c) Observe first that $(-1)^r$ is either 1 or -1 according to whether r is an even integer or an odd integer. In the sequence $1, -1, 1, -1, \dots$ the odd-numbered terms are 1's and the even-numbered terms are -1's. Thus, a formula for the nth term can be obtained by raising -1 to a power that will be even when n is odd and odd when n is even. This is accomplished by the formula $(-1)^{n+1}$, so that the sequence can be written as

$$\{(-1)^{n+1}\}_{n=1}^{+\infty}$$

(d) Combining the results in parts (a) and (c), we can write this sequence as

$$\left\{(-1)^{n+1}\frac{n}{n+1}\right\}_{n=1}^{+\infty}$$

(e) Begin by comparing terms and term numbers:

term number	1	2	3	4	$\cdots$
term	1	3	5	7	$\cdots$

From the table we see that each term is one less than twice the term number. Thus, the nth term is $2n - 1$, and the sequence can be written as

$$\{2n - 1\}_{n=1}^{+\infty} \quad \blacktriangleleft$$

Frequently we shall want to write down a sequence without specifying the numerical values of the terms. We do this by writing

$$a_1, a_2, \dots, a_n, \dots$$

or in bracket notation

$$\{a_n\}_{n=1}^{+\infty}$$

or sometimes simply

$$\{a_n\}$$

(There is nothing special about the letter a; any other letter may be used.)

In the beginning of this section we stated informally that a sequence is an unending succession of numbers. However, this is not a satisfactory mathematical definition since the word "succession" is itself an undefined term. The time has come to formally define the term "sequence." When we write a sequence such as

$$2, 4, 6, 8, \ldots, 2n, \ldots$$

in bracket notation

$$\{2n\}_{n=1}^{+\infty} \tag{1}$$

we are specifying a rule that tells us how to associate a numerical value, namely $2n$, with each positive integer n. Stated another way, $\{2n\}_{n=1}^{+\infty}$ may be regarded as a formula for a function whose independent variable n ranges over the positive integers. Indeed, we could rewrite (1) in functional notation as

$$f(n) = 2n, \quad n = 1, 2, 3, \ldots$$

From this point of view, the notation

$$2, 4, 6, 8, \ldots, 2n, \ldots$$

represents a listing of the function values

$$f(1), f(2), f(3), \ldots, f(n), \ldots$$

This suggests the following definition.

11.1.1   DEFINITION. A *sequence* (or *infinite sequence*) is a function whose domain is the set of positive integers.

Because sequences are functions, we may inquire about the graph of a sequence. For example, the graph of the sequence

$$\left\{\frac{1}{n}\right\}_{n=1}^{+\infty}$$

is the graph of the equation

$$y = \frac{1}{n}, \quad n = 1, 2, 3, \ldots$$

Because the right side of this equation is defined only for positive integer values of n, the graph consists of a succession of isolated points (Figure 11.1.1a). This is in marked distinction to the graph of

$$y = \frac{1}{x}, \quad x \geq 1$$

which is a continuous curve (Figure 11.1.1b).

Figure 11.1.1 ▷   (a)   (b)

In Figure 11.1.2 we have sketched the graphs of four sequences:

$$\{n + 1\}_{n=1}^{+\infty}, \quad \{(-1)^{n+1}\}_{n=1}^{+\infty}, \quad \left\{\frac{n}{n+1}\right\}_{n=1}^{+\infty}, \quad \left\{1 + \left(-\frac{1}{2}\right)^n\right\}_{n=1}^{+\infty}$$

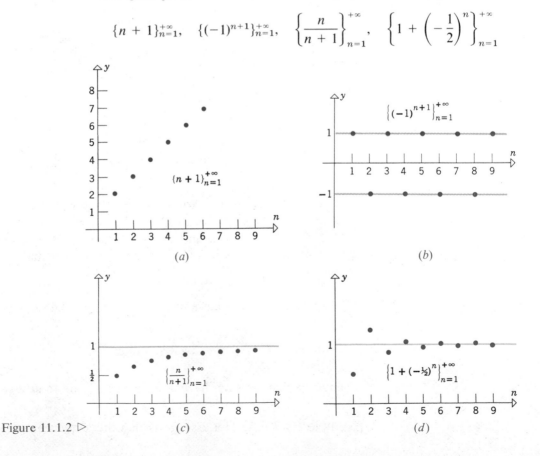

Figure 11.1.2 ▷   (c)   (d)

Each of these sequences behaves differently as n gets larger and larger. In the sequence $\{n + 1\}$, the terms grow without bound; in the sequence $\{(-1)^{n+1}\}$ the terms oscillate between 1 and -1; in the sequence $\left\{\dfrac{n}{n + 1}\right\}$ the terms increase toward a "limit" of 1; and finally, in the sequence $\{1 + (-\frac{1}{2})^n\}$ the terms also tend toward a "limit" of 1, but do so in an oscillatory fashion.

LIMIT OF A
SEQUENCE

Let us try to make the term "limit" more precise. To say that a sequence $\{a_n\}_{n=1}^{+\infty}$ approaches a limit L as n gets large is intended to mean that eventually the terms in the sequence become arbitrarily close to the number L. Thus, if we choose *any* positive number ϵ, the terms in the sequence will eventually be within ϵ units of L. Geometrically, this means that if we sketch the lines $y = L + \epsilon$ and $y = L - \epsilon$, the terms in the sequence will eventually be trapped within the band between these lines, and thus be within ϵ units of L (Figure 11.1.3).

Figure 11.1.3 ▷

The following definition expresses this idea precisely.

11.1.2   DEFINITION. A sequence $\{a_n\}_{n=1}^{+\infty}$ is said to have the *limit L* if given any $\epsilon > 0$ there is a positive integer N such that $|a_n - L| < \epsilon$ when $n \geq N$.

If a sequence $\{a_n\}_{n=1}^{+\infty}$ has a limit L, we say that the sequence *converges* to L and write

$$\lim_{n \to +\infty} a_n = L$$

A sequence that does not have a finite limit is said to *diverge*.

Example 3   Figure 11.1.2 suggests that the sequences $\{n + 1\}_{n=1}^{+\infty}$ and

$\{(-1)^{n+1}\}_{n=1}^{+\infty}$ diverge, while the sequences

$$\left\{\frac{n}{n+1}\right\}_{n=1}^{+\infty} \quad \text{and} \quad \left\{1 + \left(-\frac{1}{2}\right)^n\right\}_{n=1}^{+\infty}$$

converge to 1; that is,

$$\lim_{n \to +\infty} \frac{n}{n+1} = 1 \quad \text{and} \quad \lim_{n \to +\infty} \left(1 + \left(-\frac{1}{2}\right)^n\right) = 1 \quad \blacktriangleleft$$

Many familiar properties of limits apply to limits of sequences.

11.1.3   THEOREM. *Suppose that the sequences $\{a_n\}$ and $\{b_n\}$ converge to limits L_1 and L_2, respectively, and c is a constant. Then*

(a) $\displaystyle\lim_{n \to +\infty} c = c$

(b) $\displaystyle\lim_{n \to +\infty} ca_n = c \lim_{n \to +\infty} a_n = cL_1$

(c) $\displaystyle\lim_{n \to +\infty} (a_n + b_n) = \lim_{n \to +\infty} a_n + \lim_{n \to +\infty} b_n = L_1 + L_2$

(d) $\displaystyle\lim_{n \to +\infty} (a_n - b_n) = \lim_{n \to +\infty} a_n - \lim_{n \to +\infty} b_n = L_1 - L_2$

(e) $\displaystyle\lim_{n \to +\infty} (a_n b_n) = \lim_{n \to +\infty} a_n \cdot \lim_{n \to +\infty} b_n = L_1 L_2$

(f) $\displaystyle\lim_{n \to +\infty} \left(\frac{a_n}{b_n}\right) = \frac{\displaystyle\lim_{n \to +\infty} a_n}{\displaystyle\lim_{n \to +\infty} b_n} = \frac{L_1}{L_2}$ (if $L_2 \neq 0$)

(We omit the proof.)

Example 4   In each part, determine whether the given sequence converges or diverges. If it converges, find the limit.

(a) $\left\{\dfrac{n}{2n+1}\right\}_{n=1}^{+\infty}$ (b) $\left\{(-1)^{n+1} \dfrac{n}{2n+1}\right\}_{n=1}^{+\infty}$

(c) $\left\{(-1)^{n+1} \dfrac{1}{n}\right\}_{n=1}^{+\infty}$ (d) $\{8 - 2n\}_{n=1}^{+\infty}$ (e) $\left\{\dfrac{n}{e^n}\right\}_{n=1}^{+\infty}$

Solution.
(a)   Dividing numerator and denominator by n yields

$$\lim_{n \to +\infty} \frac{n}{2n + 1} = \lim_{n \to +\infty} \frac{1}{\left(2 + \dfrac{1}{n}\right)} = \frac{\lim\limits_{n \to +\infty} 1}{\lim\limits_{n \to +\infty} \left(2 + \dfrac{1}{n}\right)} = \frac{\lim\limits_{n \to +\infty} 1}{\lim\limits_{n \to +\infty} 2 + \lim\limits_{n \to +\infty} \dfrac{1}{n}}$$

$$= \frac{1}{2 + 0} = \frac{1}{2}$$

Thus, $\left\{ \dfrac{n}{2n + 1} \right\}_{n=1}^{+\infty}$ converges to $\dfrac{1}{2}$.

(b) From part (a),

$$\lim_{n \to +\infty} \frac{n}{2n + 1} = \frac{1}{2}$$

Thus, since $(-1)^{n+1}$ oscillates between $+1$ and -1, the product $(-1)^{n+1} \dfrac{n}{2n + 1}$ oscillates between positive and negative values, with the odd-numbered terms approaching $\frac{1}{2}$ and the even-numbered terms approaching $-\frac{1}{2}$. Therefore, the sequence $\left\{ (-1)^{n+1} \dfrac{n}{2n + 1} \right\}_{n=1}^{+\infty}$ approaches no limit—it diverges.

(c) Since $\lim\limits_{n \to +\infty} 1/n = 0$, the product $(-1)^{n+1} (1/n)$ oscillates between positive and negative values, with the odd-numbered terms approaching 0 through positive values and the even-numbered terms approaching 0 through negative values. Thus,

$$\lim_{n \to +\infty} (-1)^{n+1} \frac{1}{n} = 0$$

so that the sequence converges to 0.

(d) $\lim\limits_{n \to +\infty} (8 - 2n) = -\infty$, so the sequence $\{8 - 2n\}_{n=1}^{+\infty}$ diverges.

(e) We want to find $\lim\limits_{n \to +\infty} n/e^n$, which is an indeterminate form of type ∞/∞. Unfortunately, we cannot apply L'Hôpital's rule directly since e^n and n are not differentiable functions (n assumes only integer values). However, we can apply L'Hôpital's rule to the related problem $\lim\limits_{x \to +\infty} x/e^x$ to obtain

$$\lim_{x \to +\infty} \frac{x}{e^x} = \lim_{x \to +\infty} \frac{1}{e^x} = 0$$

We conclude from this that $\lim\limits_{n \to +\infty} n/e^n = 0$ since the values of n/e^n and x/e^x are the same when x is a positive integer.   ◄

Example 5 Show

$$\lim_{n \to +\infty} \sqrt[n]{n} = 1$$

Solution. With the aid of L'Hôpital's rule, $\lim_{n \to +\infty} \dfrac{1}{n} \ln n = 0$. Thus,

$$\lim_{n \to +\infty} \sqrt[n]{n} = \lim_{n \to +\infty} n^{1/n} = \lim_{n \to +\infty} e^{(1/n)\ln n} = e^0 = 1 \quad \blacktriangleleft$$

▶ Exercise Set 11.1

In Exercises 1–18, show the first five terms of the sequence, determine whether the sequence converges, and if so find the limit. (When writing out the terms of the sequence, you need not find numerical values; leave the terms in the first form you obtain.)

1. $\left\{ \dfrac{n}{n+2} \right\}_{n=1}^{+\infty}$.

2. $\left\{ \dfrac{n^2}{2n+1} \right\}_{n=1}^{+\infty}$.

3. $\{2\}_{n=1}^{+\infty}$.

4. $\left\{ \ln\left(\dfrac{1}{n}\right) \right\}_{n=1}^{+\infty}$.

5. $\left\{ \dfrac{\ln n}{n} \right\}_{n=1}^{+\infty}$.

6. $\left\{ n \sin \dfrac{\pi}{n} \right\}_{n=1}^{+\infty}$.

7. $\{1 + (-1)^n\}_{n=1}^{+\infty}$.

8. $\left\{ \dfrac{(-1)^{n+1}}{n^2} \right\}_{n=1}^{+\infty}$.

9. $\left\{ (-1)^n \dfrac{2n^3}{n^3+1} \right\}_{n=1}^{+\infty}$.

10. $\left\{ \dfrac{n}{2^n} \right\}_{n=1}^{+\infty}$.

11. $\left\{ \dfrac{(n+1)(n+2)}{2n^2} \right\}_{n=1}^{+\infty}$.

12. $\left\{ \dfrac{\pi^n}{4^n} \right\}_{n=1}^{+\infty}$.

13. $\{\cos(3/n)\}_{n=1}^{+\infty}$.

14. $\left\{ \cos \dfrac{\pi n}{2} \right\}_{n=1}^{+\infty}$.

15. $\{n^2 e^{-n}\}_{n=1}^{+\infty}$.

16. $\{\sqrt{n^2 + 3n} - n\}_{n=1}^{+\infty}$.

17. $\left\{ \left(\dfrac{n+3}{n+1}\right)^n \right\}_{n=1}^{+\infty}$.

18. $\left\{ \left(1 - \dfrac{2}{n}\right)^n \right\}_{n=1}^{+\infty}$.

In Exercises 19–26, express the sequence in the notation $\{a_n\}_{n=1}^{+\infty}$, determine whether the sequence converges, and if so find its limit.

19. $\dfrac{1}{2}, \dfrac{3}{4}, \dfrac{5}{6}, \dfrac{7}{8}, \ldots$

20. $0, \dfrac{1}{2^2}, \dfrac{2}{3^2}, \dfrac{3}{4^2}, \ldots$

21. $\dfrac{1}{3}, \dfrac{1}{9}, \dfrac{1}{27}, \dfrac{1}{81}, \ldots$

22. $-1, 2, -3, 4, -5, \ldots$

23. $\left(1 - \dfrac{1}{2}\right), \left(\dfrac{1}{2} - \dfrac{1}{3}\right), \left(\dfrac{1}{3} - \dfrac{1}{4}\right), \left(\dfrac{1}{4} - \dfrac{1}{5}\right), \ldots$

24. $3, \dfrac{3}{2}, \dfrac{3}{2^2}, \dfrac{3}{2^3}, \ldots$

25. $(\sqrt{2} - \sqrt{3}), (\sqrt{3} - \sqrt{4}), (\sqrt{4} - \sqrt{5}), \ldots$

26. $\dfrac{1}{3^5}, -\dfrac{1}{3^6}, \dfrac{1}{3^7}, -\dfrac{1}{3^8}, \ldots$

27. (a) Let $\{a_n\}$ be a sequence for which $a_1 = 3$ and $a_n = 2a_{n-1}$ when $n \geq 2$. Find the first eight terms.

   (b) Let $\{a_n\}$ be a sequence for which $a_1 = 1$, $a_2 = 1$, and $a_n = a_{n-1} + a_{n-2}$ when $n \geq 3$. Find the first eight terms.

28. The nth term a_n of the sequence $1, 2, 1, 4, 1, 6, \ldots$ is best written in the form

$$a_n = \begin{cases} 1, & \text{if } n \text{ is odd} \\ n, & \text{if } n \text{ is even} \end{cases}$$

since it would be tedious to find one formula applicable to all terms. By considering even and odd terms separately, find a formula for the nth term of the sequence.

   (a) $1, \dfrac{1}{2^2}, 3, \dfrac{1}{2^4}, 5, \dfrac{1}{2^6}, \ldots$

   (b) $1, \dfrac{1}{3}, \dfrac{1}{3}, \dfrac{1}{5}, \dfrac{1}{5}, \dfrac{1}{7}, \dfrac{1}{7}, \dfrac{1}{9}, \dfrac{1}{9}, \ldots$

29. Consider the sequence $\{a_n\}_{n=1}^{+\infty}$ where

$$a_n = \frac{1}{n^2} + \frac{2}{n^2} + \cdots + \frac{n}{n^2}$$

(a) Write out the first four terms of the sequence.

(b) Find the limit of the sequence.

[*Hint:* Sum up the terms in the formula for a_n.]

30. Follow the directions in Exercise 29 with

$$a_n = \frac{1^2}{n^3} + \frac{2^2}{n^3} + \cdots + \frac{n^2}{n^3}$$

31. If we accept the fact that the sequence $\{1/n\}_{n=1}^{+\infty}$ converges to the limit $L = 0$, then according to Definition 11.1.2, for every $\epsilon > 0$, there exists an integer N such that $|a_n - L| = |(1/n) - 0| < \epsilon$ when $n \geq N$. In each part, find the smallest possible value of N for the given value of ϵ.

(a) $\epsilon = 0.5$ (b) $\epsilon = 0.1$ (c) $\epsilon = 0.001$.

32. If we accept the fact that the sequence $\left\{\dfrac{n}{n+1}\right\}_{n=1}^{+\infty}$ converges to the limit $L = 1$, then according to Definition 11.1.2, for every $\epsilon > 0$, there exists an integer N such that $|a_n - L| = \left|\dfrac{n}{n+1} - 1\right| < \epsilon$ when $n \geq N$. In each part, find the smallest value of N for the given value of ϵ.

(a) $\epsilon = 0.25$ (b) $\epsilon = 0.1$ (c) $\epsilon = 0.001$.

33. Prove:

(a) The sequence $\left\{\dfrac{1}{n}\right\}_{n=1}^{+\infty}$ converges to 0.

(b) The sequence $\left\{\dfrac{n}{n+1}\right\}_{n=1}^{+\infty}$ converges to 1.

34. Consider the sequence $\{a_n\}_{n=1}^{+\infty}$ whose nth term is

$$a_n = \sum_{k=0}^{n-1} \frac{1}{1 + \dfrac{k}{n}} \cdot \frac{1}{n}$$

Show that $\lim\limits_{n \to +\infty} a_n = \ln 2$. [*Hint:* Interpret $\lim\limits_{n \to +\infty} a_n$ as a definite integral.]

35. (a) Show that a polygon with n equal sides inscribed in a circle of radius r has perimeter $p_n = 2rn \sin(\pi/n)$.

(b) By finding the limit of the sequence $\{p_n\}_{n=1}^{+\infty}$, show that the perimeters approach the circumference of the circle as n increases.

36. Find $\lim\limits_{n \to +\infty} r^n$, where r is a real number. [*Hint:* Consider the cases $|r| < 1$, $|r| > 1$, $r = 1$, and $r = -1$ separately.]

37. Find the limit of the sequence

$$\{(2^n + 3^n)^{1/n}\}_{n=1}^{+\infty}$$

11.2 MONOTONE SEQUENCES

Sometimes the critical information about a sequence is whether it converges or not, with the limit being of little or no importance. In this section we shall discuss results that are used to study convergence of sequences.

We begin with some terminology.

11.2.1   DEFINITION. A sequence $\{a_n\}$ is called

increasing if	$a_1 < a_2 < a_3 < \cdots < a_n < \cdots$
nondecreasing if	$a_1 \leq a_2 \leq a_3 \leq \cdots \leq a_n \leq \cdots$
decreasing if	$a_1 > a_2 > a_3 > \cdots > a_n > \cdots$
nonincreasing if	$a_1 \geq a_2 \geq a_3 \geq \cdots \geq a_n \geq \cdots$

A sequence that is either nondecreasing or nonincreasing is called *mono-tone,* and a sequence that is increasing or decreasing is called *strictly monotone.* Observe that a strictly monotone sequence is monotone, but not conversely. (Why?)

Example 1

$$\frac{1}{2}, \frac{2}{3}, \frac{3}{4}, \ldots, \frac{n}{n+1}, \ldots \qquad \text{is increasing}$$

$$1, \frac{1}{2}, \frac{1}{3}, \ldots, \frac{1}{n}, \ldots \qquad \text{is decreasing}$$

$$1, 1, 2, 2, 3, 3, \ldots \qquad \text{is nondecreasing}$$

$$1, 1, \frac{1}{2}, \frac{1}{2}, \frac{1}{3}, \frac{1}{3}, \ldots \qquad \text{is nonincreasing}$$

All four of these sequences are monotone, but the sequence

$$1, -\frac{1}{2}, \frac{1}{3}, -\frac{1}{4}, \ldots, (-1)^{n+1}\frac{1}{n}, \ldots$$

is not. The first and second sequences are strictly monotone. ◀

In order for a sequence to be increasing, *all* pairs of successive terms, a_n and a_{n+1}, must satisfy $a_n < a_{n+1}$, or equivalently, $a_n - a_{n+1} < 0$. More generally, monotone sequences can be classified as follows:

Difference Between Successive Terms	Classification
$a_n - a_{n+1} < 0$	Increasing
$a_n - a_{n+1} > 0$	Decreasing
$a_n - a_{n+1} \leq 0$	Nondecreasing
$a_n - a_{n+1} \geq 0$	Nonincreasing

Frequently, one can *guess* whether a sequence is increasing, decreasing, nondecreasing, or nonincreasing after writing out some of the initial terms. However, to be certain that the guess is correct, a precise mathematical proof is needed. The following example illustrates a method for doing this.

Example 2   Show that

$$\frac{1}{2}, \frac{2}{3}, \frac{3}{4}, \ldots, \frac{n}{n+1}, \ldots$$

is an increasing sequence.

Solution. It is intuitively clear that the sequence is increasing. To prove that this is so, let

$$a_n = \frac{n}{n + 1}$$

We can obtain a_{n+1} by replacing n by $n + 1$ in this formula. This yields

$$a_{n+1} = \frac{n + 1}{(n + 1) + 1} = \frac{n + 1}{n + 2}$$

Thus, for $n \geq 1$

$$a_n - a_{n+1} = \frac{n}{n + 1} - \frac{n + 1}{n + 2} = \frac{n^2 + 2n - n^2 - 2n - 1}{(n + 1)(n + 2)}$$

$$= -\frac{1}{(n + 1)(n + 2)} < 0$$

This proves that the sequence is increasing.　◄

If a_n and a_{n+1} are any successive terms in an increasing sequence, then $a_n < a_{n+1}$. If the terms in the sequence are all positive, then we can divide both sides of this inequality by a_n to obtain $1 < a_{n+1}/a_n$ or equivalently $a_{n+1}/a_n > 1$. More generally, monotone sequences with *positive* terms can be classified as follows:

Ratio of Successive Terms	Classification
$a_{n+1}/a_n > 1$	Increasing
$a_{n+1}/a_n < 1$	Decreasing
$a_{n+1}/a_n \geq 1$	Nondecreasing
$a_{n+1}/a_n \leq 1$	Nonincreasing

Example 3 Show that the sequence in Example 2 is increasing by examining the ratio of successive terms.

Solution. As shown in the solution of Example 2,

$$a_n = \frac{n}{n + 1} \quad \text{and} \quad a_{n+1} = \frac{n + 1}{n + 2}$$

Thus,

$$\frac{a_{n+1}}{a_n} = \frac{(n + 1)/(n + 2)}{n/(n + 1)} = \frac{n + 1}{n + 2} \cdot \frac{n + 1}{n} = \frac{n^2 + 2n + 1}{n^2 + 2n} \tag{1}$$

Since the numerator in (1) exceeds the denominator, the ratio exceeds 1, that is, $a_{n+1}/a_n > 1$ for $n \geq 1$. This proves that the sequence is increasing.

◀

In our subsequent work we will encounter sequences involving *facto-rials*. The reader will recall that if n is a positive integer, then $n!$ (n factorial) is the product of the first n positive integers, that is, $n! = 1 \cdot 2 \cdot 3 \cdots n$. Furthermore, it is agreed that $0! = 1$.

Example 4 Show that the sequence

$$\frac{e}{2!}, \frac{e^2}{3!}, \frac{e^3}{4!}, \ldots, \frac{e^n}{(n+1)!}, \ldots$$

is decreasing.

Solution. We shall examine the ratio of successive terms. Since

$$a_n = \frac{e^n}{(n+1)!}$$

it follows on replacing n by $n + 1$ that

$$a_{n+1} = \frac{e^{n+1}}{[(n+1)+1]!} = \frac{e^{n+1}}{(n+2)!}$$

Thus,

$$\frac{a_{n+1}}{a_n} = \frac{e^{n+1}/(n+2)!}{e^n/(n+1)!} = \frac{e^{n+1}}{e^n} \cdot \frac{(n+1)!}{(n+2)!} = \frac{e}{n+2}$$

For $n \geq 1$, it follows that $n + 2 \geq 3 > e$ ($\approx 2.718\ldots$), so

$$\frac{a_{n+1}}{a_n} = \frac{e}{n+2} < 1$$

for $n \geq 1$. This proves that the sequence is decreasing. ◀

The following example illustrates still a third technique for determining whether a sequence is increasing or decreasing.

Example 5 In Examples 2 and 3 we proved that the sequence

$$\frac{1}{2}, \frac{2}{3}, \frac{3}{4}, \ldots, \frac{n}{n+1}, \ldots$$

is increasing by considering the difference and ratio of successive terms.

Alternatively, we can proceed as follows. Let

$$f(x) = \frac{x}{x + 1}$$

so the nth term in the given sequence is $a_n = f(n)$. The function f is increasing for $x \geq 1$ since

$$f'(x) = \frac{(x + 1)(1) - x(1)}{(x + 1)^2} = \frac{1}{(x + 1)^2} > 0$$

Thus,

$$a_n = f(n) < f(n + 1) = a_{n+1}$$

which proves that the given sequence is increasing.   ◄

In general, if $f(n) = a_n$ is the nth term of a sequence, and if f is differentiable for $x \geq 1$, then we have the following results:

Derivative of f for $x \geq 1$	Classification of the Sequence with nth Term $a_n = f(n)$
$f'(x) > 0$	Increasing
$f'(x) < 0$	Decreasing
$f'(x) \geq 0$	Nondecreasing
$f'(x) \leq 0$	Nonincreasing

We omit the proof.

The following two theorems, whose optional proofs are discussed at the end of this section, show that a monotone sequence either converges or it becomes infinite—divergence by oscillation cannot occur.

11.2.2   THEOREM. *If $a_1 \leq a_2 \leq a_3 \leq \cdots \leq a_n \leq \cdots$ is a nondecreasing sequence, then there are two possibilities:*

(a)   *There is a constant M such that $a_n \leq M$ for all n, in which case the sequence converges to a limit L satisfying $L \leq M$.*

(b)   *No such constant exists, in which case $\lim\limits_{n \to +\infty} a_n = +\infty$.*

11.2.3   THEOREM. *If $a_1 \geq a_2 \geq a_3 \geq \cdots \geq a_n \geq \cdots$ is a nonincreasing sequence, then there are two possibilities:*

(a)   *There is a constant M such that $a_n \geq M$ for all n, in which case the sequence converges to a limit L satisfying $L \geq M$.*

(b)   *No such constant exists, in which case $\lim\limits_{n \to +\infty} a_n = -\infty$.*

It should be noted that these results do not give a method for obtaining limits; they tell us only whether a limit exists. To prove these theorems we need a preliminary result that takes us to the very foundations of the real number system. In this text we have not been concerned with a logical development of the real numbers; our approach has been to accept the familiar properties of real numbers and to work with them. Indeed, we have not even attempted to define the term "real number." However, by the late nineteenth century, the study of limits and functions in calculus necessitated a precise axiomatic formulation of the real numbers in much the same way that Euclidean geometry is developed from axioms. While we will not attempt to pursue this development, we shall have need for the following axiom about real numbers.

The Completeness Axiom

11.2.4 AXIOM. *If S is a nonempty set of real numbers, and if there is some real number that is greater than or equal to every number in S, then there is a smallest real number that is greater than or equal to every number in S.*

For example, let S be the set of numbers in the interval $(1, 3)$. It is true that there exists a number u greater than or equal to every number in S; some examples are $u = 10$, $u = 100$, and $u = 3.2$. The smallest number u that is greater than or equal to every number in S is $u = 3$.

There is an alternative phrasing of the completeness axiom that is useful to know. Let us call a number u an ***upper bound*** for a set S if u is greater than or equal to every number in S; and if S has a smallest upper bound, call it the ***least upper bound*** of S. Using this terminology the Completeness Axiom states:

The Completeness Axiom
(Alternative Form)

11.2.5 AXIOM. *If a nonempty set S of real numbers has an upper bound, then S has a least upper bound.*

Example 6   As shown in Examples 2 and 3, the sequence $\left\{\dfrac{n}{n + 1}\right\}_{n=1}^{+\infty}$ is increasing (hence nondecreasing). Since

$$a_n = \frac{n}{n + 1} < 1, \quad n = 1, 2, \ldots$$

the terms in the sequence have $M = 1$ as an upper bound. By Theorem 11.2.2 the sequence must converge to a limit $L \le M$. This is indeed the case since

$$\lim_{n \to +\infty} \frac{n}{n + 1} = \lim_{n \to +\infty} \frac{1}{1 + \dfrac{1}{n}} = 1 \quad \blacktriangleleft$$

Because the limit of a sequence $\{a_n\}$ describes the behavior of the terms as n gets *large*, one can alter or even delete a *finite* number of terms in a sequence without affecting either the convergence or the value of the limit. That is, the original and the modified sequence will both converge or both diverge, and in the case of convergence both will have the same limit. We omit the proof.

Example 7   Show that the sequence

$$\left\{\frac{5^n}{n!}\right\}_{n=1}^{+\infty}$$

converges.

Solution.   It is tedious to determine convergence directly from the limit

$$\lim_{n \to +\infty} \frac{5^n}{n!}$$

Thus, we shall proceed indirectly. If we let

$$a_n = \frac{5^n}{n!}$$

then

$$a_{n+1} = \frac{5^{n+1}}{(n + 1)!}$$

so that

$$\frac{a_{n+1}}{a_n} = \frac{5^{n+1}/(n + 1)!}{5^n/n!} = \frac{5^{n+1}}{5^n} \cdot \frac{n!}{(n + 1)!} = \frac{5}{n + 1}$$

For $n = 1$, 2, and 3, the value of a_{n+1}/a_n is greater than 1 so $a_{n+1} > a_n$. Thus,

$$a_1 < a_2 < a_3 < a_4$$

For $n = 4$ the value of a_{n+1}/a_n is 1, so

$$a_4 = a_5$$

For $n \geq 5$ the value of a_{n+1}/a_n is less than 1, so

$$a_5 > a_6 > a_7 > a_8 > \cdots$$

Thus, if we discard the first four terms of the given sequence (which will not affect convergence) the resulting sequence will be decreasing. Moreover, each term in the sequence is positive, so that by Theorem 11.2.3, the sequence converges to some limit that is ≥ 0. ◄

OPTIONAL

Proof of Theorem 11.2.2.

(a)   Assume there exists a number M such that $a_n \leq M$ for $n = 1, 2, \ldots$ Then M is an upper bound for the set of terms in the sequence. By the Completeness Axiom there is a least upper bound for the terms, call it L. Now let ϵ be any positive number. Since L is the least upper bound for the terms, $L - \epsilon$ is not an upper bound for the terms, which means that there is at least one term a_N such that

$$a_N > L - \epsilon$$

Moreover, since $\{a_n\}$ is a nondecreasing sequence, we must have

$$a_n \geq a_N > L - \epsilon \tag{2}$$

when $n \geq N$. But a_n cannot exceed L since L is an upper bound for the terms. This observation together with (2) tells us that $L \geq a_n > L - \epsilon$ for $n \geq N$, so that all terms from the Nth on are within ϵ units of L. This is exactly the requirement to have

$$\lim_{n \to +\infty} a_n = L$$

Finally, $L \leq M$ since M is an upper bound for the terms and L is the least upper bound. This proves (a).

(b)   If there is no number M such that $a_n \leq M$ for $n = 1, 2, \ldots$, then no matter how large we choose M, there is a term a_N such that

$$a_N > M$$

and, since the sequence is nondecreasing,

$$a_n \geq a_N > M$$

when $n \geq N$. Thus, the terms in the sequence become arbitrarily large as n increases. That is,

$$\lim_{n \to +\infty} a_n = +\infty \qquad \blacksquare$$

The proof of Theorem 11.2.3 will be omitted since it is similar to 11.2.2.

► Exercise Set 11.2

In Exercises 1–6, determine whether the given sequence $\{a_n\}$ is monotone by examining $a_n - a_{n+1}$. If so, classify it as increasing, decreasing, nonincreasing, or nondecreasing.

1. $\left\{\dfrac{1}{n}\right\}_{n=1}^{+\infty}$.

2. $\left\{1 - \dfrac{1}{n}\right\}_{n=1}^{+\infty}$.

3. $\left\{\dfrac{n}{2n + 1}\right\}_{n=1}^{+\infty}$.

4. $\left\{\dfrac{n}{4n - 1}\right\}_{n=1}^{+\infty}$.

5. $\{n - 2^n\}_{n=1}^{+\infty}$.

6. $\{n - n^2\}_{n=1}^{+\infty}$.

In Exercises 7–18, determine whether the given sequence $\{a_n\}$ is monotone by examining a_{n+1}/a_n. If so, classify it as increasing, decreasing, nonincreasing, or nondecreasing.

7. $\left\{\dfrac{n}{2n + 1}\right\}_{n=1}^{+\infty}$.

8. $\left\{\dfrac{n}{2^n}\right\}_{n=1}^{+\infty}$.

9. $\{ne^{-n}\}_{n=1}^{+\infty}$.

10. $\left\{\dfrac{n^2}{3^n}\right\}_{n=1}^{+\infty}$.

11. $\left\{\dfrac{2^n}{n!}\right\}_{n=1}^{+\infty}$.

12. $\left\{\dfrac{e^n}{n!}\right\}_{n=1}^{+\infty}$.

13. $\left\{\dfrac{n!}{3^n}\right\}_{n=1}^{+\infty}$.

14. $\left\{\dfrac{n^2}{n!}\right\}_{n=1}^{+\infty}$.

15. $\left\{\dfrac{10^n}{(2n)!}\right\}_{n=1}^{+\infty}$.

16. $\left\{\dfrac{2^n}{1 + 2^n}\right\}_{n=1}^{+\infty}$.

17. $\left\{\dfrac{n^n}{n!}\right\}_{n=1}^{+\infty}$.

18. $\left\{\dfrac{10^n}{2^{(n^2)}}\right\}_{n=1}^{+\infty}$.

In Exercises 19–24, use differentiation to show that the sequence is strictly monotone and classify it as increasing or decreasing.

19. $\left\{\dfrac{n}{2n + 1}\right\}_{n=1}^{+\infty}$.

20. $\left\{3 - \dfrac{1}{n}\right\}_{n=1}^{+\infty}$.

21. $\left\{\dfrac{1}{n + \ln n}\right\}_{n=1}^{+\infty}$.

22. $\{ne^{-2n}\}_{n=1}^{+\infty}$.

23. $\left\{\dfrac{\ln (n + 2)}{n + 2}\right\}_{n=1}^{+\infty}$.

24. $\{\tan^{-1} n\}_{n=1}^{+\infty}$.

In Exercises 25–30, show that the sequence is monotone and apply Theorem 11.2.2 or 11.2.3 to determine whether it converges.

25. $\left\{\dfrac{n}{5^n}\right\}_{n=1}^{+\infty}$.

26. $\left\{\dfrac{2^n}{(n + 1)!}\right\}_{n=1}^{+\infty}$.

27. $\left\{n - \dfrac{1}{n}\right\}_{n=1}^{+\infty}$.

28. $\left\{\cos \dfrac{\pi}{2n}\right\}_{n=1}^{+\infty}$.

29. $\left\{2 + \dfrac{1}{n}\right\}_{n=1}^{+\infty}$.

30. $\left\{\dfrac{4n - 1}{5n + 2}\right\}_{n=1}^{+\infty}$.

31. Find the limit (if it exists) of the sequence.

(a) $1, -1, 1, \dfrac{1}{2}, \dfrac{1}{3}, \dfrac{1}{4}, \dfrac{1}{5}, \ldots$

(b) $-\dfrac{1}{2}, 0, 0, 0, 1, 2, 3, 4, \ldots$

32. (a) Is $\left\{\dfrac{100^n}{n!}\right\}_{n=1}^{+\infty}$ a monotone sequence?

(b) Is it a convergent sequence? Justify your answer.

33. Show that $\left\{\dfrac{3^n}{1 + 3^{2n}}\right\}_{n=1}^{+\infty}$ is a decreasing sequence.

34. Show that $\left\{\dfrac{1 \cdot 3 \cdot 5 \cdots (2n - 1)}{n!}\right\}_{n=1}^{+\infty}$ is an increasing sequence.

35. (a) Show that if $\{a_n\}_{n=1}^{+\infty}$ is a nonincreasing sequence, then $\{-a_n\}_{n=1}^{+\infty}$ is a nondecreasing sequence.

(b) Use part (a) and Theorem 11.2.2 to help prove Theorem 11.2.3.

36. Show that $n! > \dfrac{n^n}{e^{n-1}}$, $n > 1$.

[*Hint:* $\ln n! = \ln 1 + \ln 2 + \cdots + \ln n > \int_1^n \ln x \, dx$.]

37. (a) Show that $\left\{\dfrac{n^n}{n!e^n}\right\}_{n=1}^{+\infty}$ is a decreasing sequence. [*Hint:* From part (d) of Exercise 89, Section 7.4, $(1 + 1/x)^x < e$ for $x > 0$.]

(b) Does the sequence converge? Justify your answer.

11.3 INFINITE SERIES

The purpose of this section is to discuss sums

$$u_1 + u_2 + u_3 + \cdots + u_k + \cdots$$

that contain infinitely many terms. The most familiar examples of such sums occur in the decimal representation of real numbers. For example, when we write $\frac{1}{3}$ in the decimal form $\frac{1}{3} = 0.3333 \ldots$ we mean

$$\frac{1}{3} = 0.3 + 0.03 + 0.003 + 0.0003 + \cdots$$

$$= \frac{3}{10} + \frac{3}{10^2} + \frac{3}{10^3} + \frac{3}{10^4} + \cdots$$

Our first objective in this section is to define what is meant by a sum with infinitely many terms. Since it is impossible to "add up" infinitely many numbers, we shall deal with infinite sums by means of a limiting process involving sequences.

Definition of an Infinite Series

11.3.1 DEFINITION. An *infinite series* is an expression of the form

$$u_1 + u_2 + u_3 + \cdots + u_k + \cdots$$

or in sigma notation

$$\sum_{k=1}^{\infty} u_k$$

The numbers $u_1, u_2, u_3, \ldots$ are called the *terms* of the series.

SUM OF AN INFINITE SERIES

Informally speaking, the expression $\sum_{k=1}^{\infty} u_k$ directs us to obtain the "sum" of the terms $u_1, u_2, u_3, \ldots$ To carry out this summation process we proceed as follows: Let s_n denote the sum of the first n terms of the series. Thus,

$$s_1 = u_1$$
$$s_2 = u_1 + u_2$$
$$s_3 = u_1 + u_2 + u_3$$
$$\vdots$$
$$s_n = u_1 + u_2 + u_3 + \cdots + u_n = \sum_{k=1}^{n} u_k$$

The number s_n is called the ***nth partial sum*** of the series and the sequence $\{s_n\}_{n=1}^{+\infty}$ is called the ***sequence of partial sums.***

WARNING.   In everyday English the words "sequence" and "series" are often used interchangeably. However, this is not so in mathematics—mathematically, a sequence is a *succession* and a series is a *sum*. It is essential that you keep this distinction in mind.

Example 1   For the infinite series

$$\frac{3}{10} + \frac{3}{10^2} + \frac{3}{10^3} + \frac{3}{10^4} + \cdots$$

the partial sums are

$$s_1 = \frac{3}{10}$$

$$s_2 = \frac{3}{10} + \frac{3}{10^2} = \frac{33}{100}$$

$$s_3 = \frac{3}{10} + \frac{3}{10^2} + \frac{3}{10^3} = \frac{333}{1000}$$

$$s_4 = \frac{3}{10} + \frac{3}{10^2} + \frac{3}{10^3} + \frac{3}{10^4} = \frac{3333}{10000}$$

$$\vdots \qquad\qquad\qquad\qquad\qquad \blacktriangleleft$$

As n increases, the partial sum $s_n = u_1 + u_2 + \cdots + u_n$ includes more and more terms of the series. Thus, if s_n tends toward a limit as $n \to +\infty$, it is reasonable to view this limit as the sum of *all* the terms in the series. This suggests the following definition.

11.3.2   DEFINITION. Let $\{s_n\}$ be the sequence of partial sums of the series $\sum_{k=1}^{\infty} u_k$. If the sequence $\{s_n\}$ converges to a limit S, then the series is said to ***converge*** and S is called the ***sum*** of the series. We denote this by writing

$$S = \sum_{k=1}^{\infty} u_k$$

If the sequence of partial sums diverges, then the series is said to ***diverge.*** A divergent series has no sum.

Example 2   If Definition 11.3.2 is to be reasonable, it should be the case that

$$\frac{1}{3} = \frac{3}{10} + \frac{3}{10^2} + \frac{3}{10^3} + \cdots + \frac{3}{10^k} + \cdots$$

Let us verify that this is indeed the case. The nth partial sum is

$$s_n = \frac{3}{10} + \frac{3}{10^2} + \cdots + \frac{3}{10^n} \tag{1}$$

The problem of calculating $\lim_{n \to +\infty} s_n$ is complicated by the fact that the number of terms in (1) changes with n. For purposes of calculation, it is desirable to rewrite (1) in closed form (see the remark following Example 3 in Section 5.4). To do this, we multiply both sides of (1) by $\frac{1}{10}$ to obtain

$$\frac{1}{10} s_n = \frac{3}{10^2} + \frac{3}{10^3} + \cdots + \frac{3}{10^n} + \frac{3}{10^{n+1}} \tag{2}$$

and then subtract (2) from (1) to obtain:

$$s_n - \frac{1}{10} s_n = \frac{3}{10} - \frac{3}{10^{n+1}}$$

$$\frac{9}{10} s_n = \frac{3}{10}\left(1 - \frac{1}{10^n}\right)$$

$$s_n = \frac{1}{3}\left(1 - \frac{1}{10^n}\right) \tag{3}$$

Since $1/10^n \to 0$ as $n \to +\infty$, it follows from (3) that $S = \lim_{n \to +\infty} s_n = \frac{1}{3}$. Thus,

$$\frac{1}{3} = \frac{3}{10} + \frac{3}{10^2} + \frac{3}{10^3} + \cdots + \frac{3}{10^n} + \cdots \qquad \blacktriangleleft$$

Example 3   Determine whether the series

$$1 - 1 + 1 - 1 + 1 - 1 + \cdots$$

converges or diverges. If it converges, find the sum.

Solution.   The partial sums are

$$s_1 = 1$$

$$s_2 = 1 - 1 = 0$$

$$s_3 = 1 - 1 + 1 = 1$$
$$s_4 = 1 - 1 + 1 - 1 = 0$$

and so forth. Thus, the sequence of partial sums is

$$1, 0, 1, 0, 1, 0, \ldots$$

Since this is a divergent sequence, the given series diverges and consequently has no sum. ◄

GEOMETRIC SERIES   The series in Examples 2 and 3 are examples of **geometric series.** A geometric series is one of the form

$$a + ar + ar^2 + ar^3 + \cdots + ar^{k-1} + \cdots \quad (a \neq 0)$$

where each term is obtained by multiplying the previous one by a constant r. The multiplier r is called the **ratio** for the series. Some examples of geometric series are:

$$1 + 2 + 4 + 8 + \cdots + 2^{k-1} + \cdots \qquad [a = 1, r = 2]$$

$$3 + \frac{3}{10} + \frac{3}{10^2} + \frac{3}{10^3} + \cdots + \frac{3}{10^{k-1}} + \cdots \qquad [a = 3, r = \tfrac{1}{10}]$$

$$\frac{1}{2} - \frac{1}{4} + \frac{1}{8} - \frac{1}{16} + \cdots + (-1)^{k+1}\frac{1}{2^k} + \cdots \qquad [a = \tfrac{1}{2}, r = -\tfrac{1}{2}]$$

$$1 + 1 + 1 + \cdots + 1 + \cdots \qquad [a = 1, r = 1]$$

$$1 - 1 + 1 - 1 + \cdots + (-1)^{k+1} + \cdots \qquad [a = 1, r = -1]$$

The following theorem is the fundamental result on convergence of geometric series.

11.3.3   THEOREM. *A geometric series*

$$a + ar + ar^2 + \cdots + ar^{k-1} + \cdots \quad (a \neq 0)$$

converges if $|r| < 1$ and diverges if $|r| \geq 1$. If the series converges the sum is

$$\frac{a}{1 - r} = a + ar + ar^2 + \cdots + ar^{k-1} + \cdots$$

Proof.   Let us treat the case $|r| = 1$ first. If $r = 1$, then the series is

$$a + a + a + \cdots + a + \cdots$$

so that the nth partial sum is $s_n = na$ and $\lim\limits_{n \to +\infty} s_n = \lim\limits_{n \to +\infty} na = \pm\infty$ (the sign depending on whether a is positive or negative). This proves divergence. If $r = -1$, the series is

$$a - a + a - a + \cdots$$

so the sequence of partial sums is

$$a, 0, a, 0, a, 0, \ldots$$

which diverges.

Now let us consider the case where $|r| \neq 1$. The nth partial sum of the series is

$$s_n = a + ar + ar^2 + \cdots + ar^{n-1} \tag{4}$$

Multiplying both sides of (4) by r yields

$$rs_n = ar + ar^2 + \cdots + ar^{n-1} + ar^n \tag{5}$$

and subtracting (5) from (4) gives

$$s_n - rs_n = a - ar^n$$

or

$$(1 - r)s_n = a - ar^n \tag{6}$$

Since $r \neq 1$ in the case we are considering, this can be rewritten as

$$s_n = \frac{a - ar^n}{1 - r} = \frac{a}{1 - r} - \frac{ar^n}{1 - r} \tag{7}$$

If $|r| < 1$, then $\lim\limits_{n \to +\infty} r^n = 0$, so that $\{s_n\}$ converges. From (7)

$$\lim\limits_{n \to +\infty} s_n = \frac{a}{1 - r}$$

If $|r| > 1$, then either $r > 1$ or $r < -1$. In the case $r > 1$, $\lim\limits_{n \to +\infty} r^n = +\infty$, and in the case $r < -1$, r^n oscillates between positive and negative values that grow in magnitude, so $\{s_n\}$ diverges in both cases. ∎

Example 4   The series

$$5 + \frac{5}{4} + \frac{5}{4^2} + \cdots + \frac{5}{4^{k-1}} + \cdots$$

is a geometric series with $a = 5$ and $r = \frac{1}{4}$. Since $|r| = \frac{1}{4} < 1$, the series converges and the sum is

$$\frac{a}{1-r} = \frac{5}{1-\frac{1}{4}} = \frac{20}{3} \qquad \blacktriangleleft$$

Example 5 Find the rational number represented by the repeating decimal

$$0.784784784 \ldots$$

Solution. We can write

$$0.784784784 \ldots = 0.784 + 0.000784 + 0.000000784 + \cdots$$

so the given decimal is the sum of a geometric series with $a = 0.784$ and $r = 0.001$. Thus,

$$0.784784784 \ldots = \frac{a}{1-r} = \frac{0.784}{1 - 0.001} = \frac{0.784}{0.999} = \frac{784}{999} \qquad \blacktriangleleft$$

Example 6 Determine whether the series

$$\sum_{k=1}^{\infty} \frac{1}{k(k+1)} = \frac{1}{1 \cdot 2} + \frac{1}{2 \cdot 3} + \frac{1}{3 \cdot 4} + \frac{1}{4 \cdot 5} + \cdots$$

converges or diverges. If it converges, find the sum.

Solution. The nth partial sum of the series is

$$s_n = \sum_{k=1}^{n} \frac{1}{k(k+1)} = \frac{1}{1 \cdot 2} + \frac{1}{2 \cdot 3} + \frac{1}{3 \cdot 4} + \cdots + \frac{1}{n(n+1)}$$

To calculate $\lim_{n \to +\infty} s_n$ we shall rewrite s_n in closed form. This may be accomplished by using the method of partial fractions to obtain (verify):

$$\frac{1}{k(k+1)} = \frac{1}{k} - \frac{1}{k+1}$$

from which it follows that

$$s_n = \sum_{k=1}^{n} \left(\frac{1}{k} - \frac{1}{k+1} \right)$$

$$= \left(1 - \frac{1}{2} \right) + \left(\frac{1}{2} - \frac{1}{3} \right) + \left(\frac{1}{3} - \frac{1}{4} \right) + \cdots + \left(\frac{1}{n} - \frac{1}{n+1} \right)$$

$$= 1 + \left(-\frac{1}{2} + \frac{1}{2} \right) + \left(-\frac{1}{3} + \frac{1}{3} \right) + \cdots + \left(-\frac{1}{n} + \frac{1}{n} \right) - \frac{1}{n+1} \qquad (8)$$

The sum in (8) is an example of a **telescoping sum,** which means that each term cancels part of the next term, thereby collapsing the sum (like a folding telescope) into only two terms. After the cancellation, (8) can be written as

$$s_n = 1 - \frac{1}{n+1}$$

so

$$\lim_{n \to +\infty} s_n = \lim_{n \to +\infty} \left(1 - \frac{1}{n+1}\right) = 1$$

and therefore

$$1 = \sum_{k=1}^{\infty} \frac{1}{k(k+1)} \qquad \blacktriangleleft$$

HARMONIC SERIES   One of the most famous and important of all diverging series is the **harmonic series,**

$$\sum_{k=1}^{\infty} \frac{1}{k} = 1 + \frac{1}{2} + \frac{1}{3} + \frac{1}{4} + \frac{1}{5} + \cdots$$

which arises in connection with the overtones produced by a vibrating musical string. It is not immediately evident that this series diverges. However, the divergence will become apparent when we examine the partial sums in detail. Because the terms in the series are all positive, the partial sums

$$s_1 = 1, \quad s_2 = 1 + \frac{1}{2}, \quad s_3 = 1 + \frac{1}{2} + \frac{1}{3}, \quad s_4 = 1 + \frac{1}{2} + \frac{1}{3} + \frac{1}{4}, \ldots$$

form an increasing sequence

$$s_1 < s_2 < s_3 < \cdots < s_n < \cdots$$

Thus, by Theorem 11.2.2 we can prove divergence by demonstrating that there is no constant M that is greater than or equal to *every* partial sum. To this end, we shall consider some selected partial sums, namely s_2, s_4, s_8, s_{16}, $s_{32}, \ldots$. Note that the subscripts are successive powers of 2, so that these are the partial sums of the form s_{2^n}. These partial sums satisfy the inequalities

$$s_2 = 1 + \frac{1}{2} > \frac{1}{2} + \frac{1}{2} = \frac{2}{2}$$

$$s_4 = s_2 + \frac{1}{3} + \frac{1}{4} > s_2 + \left(\frac{1}{4} + \frac{1}{4}\right) = s_2 + \frac{1}{2} > \frac{3}{2}$$

$$s_8 = s_4 + \frac{1}{5} + \frac{1}{6} + \frac{1}{7} + \frac{1}{8} > s_4 + \left(\frac{1}{8} + \frac{1}{8} + \frac{1}{8} + \frac{1}{8}\right) = s_4 + \frac{1}{2} > \frac{4}{2}$$

$$s_{16} = s_8 + \frac{1}{9} + \frac{1}{10} + \frac{1}{11} + \frac{1}{12} + \frac{1}{13} + \frac{1}{14} + \frac{1}{15} + \frac{1}{16}$$

$$\vdots \quad > s_8 + \left(\frac{1}{16} + \frac{1}{16} + \frac{1}{16} + \frac{1}{16} + \frac{1}{16} + \frac{1}{16} + \frac{1}{16} + \frac{1}{16}\right) = s_8 + \frac{1}{2} > \frac{5}{2}$$

$$s_{2^n} > \frac{n+1}{2}$$

Now if M is any constant, we can certainly find a positive integer n such that $(n + 1)/2 > M$. But for this n

$$s_{2^n} > \frac{n+1}{2} > M$$

so that no constant M is greater than or equal to *every* partial sum of the harmonic series. This proves divergence.

▶ Exercise Set 11.3

1. In each part, find the first four partial sums; find a closed form for the nth partial sum; determine whether the series converges, and if so give the sum.

    (a) $\displaystyle\sum_{k=1}^{\infty} \frac{2}{5^{k-1}}$ (b) $\displaystyle\sum_{k=1}^{\infty} \frac{1}{(k+1)(k+2)}$

    (c) $\displaystyle\sum_{k=1}^{\infty} \frac{2^{k-1}}{4}$.

In Exercises 2–16 determine whether the series converges or diverges. If it converges, find the sum.

2. $\displaystyle\sum_{k=1}^{\infty} \frac{1}{5^k}$.

3. $\displaystyle\sum_{k=1}^{\infty} \left(-\frac{3}{4}\right)^{k-1}$.

4. $\displaystyle\sum_{k=1}^{\infty} \left(\frac{2}{3}\right)^{k+2}$.

5. $\displaystyle\sum_{k=1}^{\infty} (-1)^{k-1} \frac{7}{6^{k-1}}$.

6. $\displaystyle\sum_{k=1}^{\infty} 4^{k-1}$.

7. $\displaystyle\sum_{k=1}^{\infty} \left(-\frac{3}{2}\right)^{k+1}$.

8. $\displaystyle\sum_{k=1}^{\infty} \left(\frac{1}{k+3} - \frac{1}{k+4}\right)$.

9. $\displaystyle\sum_{k=1}^{\infty} \frac{1}{(k+2)(k+3)}$.

10. $\displaystyle\sum_{k=1}^{\infty} \left(\frac{1}{2^k} - \frac{1}{2^{k+1}}\right)$.

11. $\displaystyle\sum_{k=1}^{\infty} \frac{1}{9k^2 + 3k - 2}$.

12. $\displaystyle\sum_{k=2}^{\infty} \frac{1}{k^2 - 1}$.

13. $\displaystyle\sum_{k=1}^{\infty} \frac{4^{k+2}}{7^{k-1}}$.

14. $\displaystyle\sum_{k=1}^{\infty} \left(\frac{e}{\pi}\right)^{k-1}$.

15. $\displaystyle\sum_{k=1}^{\infty} \left(-\frac{1}{2}\right)^k$.

16. $\displaystyle\sum_{k=3}^{\infty} \frac{5}{k-2}$.

In Exercises 17–22, express the repeating decimal as a fraction.

17. 0.4444 . . .   18. 0.9999 . . .

19. 5.373737 . . .   20. 0.159159159 . . .

21. 0.782178217821 . . .   22. 0.451141414 . . .

23. Find a closed form for the nth partial sum of the series

$$\ln\frac{1}{2} + \ln\frac{2}{3} + \ln\frac{3}{4} + \cdots + \ln\frac{n}{n+1} + \cdots$$

and determine whether the series converges.

24. A ball is dropped from a height of 10 meters. Each time it strikes the ground it bounces vertically to a height that is three-fourths of the previous height. Find the total distance the ball will travel if it is allowed to bounce indefinitely.

25. Show: $\displaystyle\sum_{k=2}^{\infty} \ln(1 - 1/k^2) = -\ln 2$.

26. Show: $\displaystyle\sum_{k=1}^{\infty} \frac{\sqrt{k+1} - \sqrt{k}}{\sqrt{k^2 + k}} = 1$.

27. Show: $\displaystyle\sum_{k=1}^{\infty} \left(\frac{1}{k} - \frac{1}{k+2}\right) = \frac{3}{2}$.

28. Show: $\displaystyle\sum_{k=1}^{\infty} \frac{6^k}{(3^{k+1} - 2^{k+1})(3^k - 2^k)} = 2$.

$\left[\begin{array}{l} \textit{Hint: Find } A \text{ and } B \text{ so that} \\[4pt] \dfrac{6^k}{(3^{k+1} - 2^{k+1})(3^k - 2^k)} = \dfrac{2^k A}{3^k - 2^k} + \dfrac{2^k B}{3^{k+1} - 2^{k+1}}. \end{array}\right]$

29. Use geometric series to show:

(a) $\displaystyle\sum_{k=0}^{\infty} (-1)^k x^k = \frac{1}{1+x}$ if $-1 < x < 1$

(b) $\displaystyle\sum_{k=0}^{\infty} (x-3)^k = \frac{1}{4-x}$ if $2 < x < 4$

(c) $\displaystyle\sum_{k=0}^{\infty} (-1)^k x^{2k} = \frac{1}{1+x^2}$ if $-1 < x < 1$.

In Exercises 30–33, find all values of x for which the series converges, and for these values find its sum.

30. $x - x^3 + x^5 - x^7 + x^9 - \cdots$

31. $\dfrac{1}{x^2} + \dfrac{2}{x^3} + \dfrac{4}{x^4} + \dfrac{8}{x^5} + \dfrac{16}{x^6} + \cdots$

32. $e^{-x} + e^{-2x} + e^{-3x} + e^{-4x} + e^{-5x} + \cdots$

33. $\sin x - \frac{1}{2}\sin^2 x + \frac{1}{4}\sin^3 x - \frac{1}{8}\sin^4 x + \cdots$

34. Prove the following decimal equality assuming $a_n \neq 9$:
$$0.a_1 a_2 \ldots a_n 9999 \ldots = 0.a_1 a_2 \ldots (a_n + 1)0000 \ldots$$

35. Let a_1 be any real number and define $a_{n+1} = \frac{1}{2}(a_n + 1)$ for $n = 1, 2, 3, \ldots$. Show that the sequence $\{a_n\}_{n=1}^{+\infty}$ converges and find its limit. [*Hint:* Express a_n in terms of a_1.]

36. Lines L_1 and L_2 form an angle θ, $0 < \theta < \pi/2$, at their point of intersection P (Figure 11.3.1). A point P_0 is chosen that is on L_1 and a units from P. Starting from P_0 a zig-zag path is constructed by successively going back and forth between L_1 and L_2 along a perpendicular from one line to the other. Find the following sums in terms of θ:

(a) $P_0 P_1 + P_1 P_2 + P_2 P_3 + \cdots$

(b) $P_0 P_1 + P_2 P_3 + P_4 P_5 + \cdots$

(c) $P_1 P_2 + P_3 P_4 + P_5 P_6 + \cdots$.

Figure 11.3.1

11.4   CONVERGENCE; THE INTEGRAL TEST

In the previous section we found sums of series and investigated convergence by first writing the nth partial sum s_n in closed form and then examining the limit $\lim\limits_{n \to +\infty} s_n$. Unfortunately, it is relatively rare that the nth partial sum of a series can be written in closed form; for most series, convergence or divergence is determined by using convergence tests, some of which we shall introduce in this section. Once it is established that a series converges, the sum of the series can always be approximated to any degree of accuracy by a partial sum with sufficiently many terms.

Our first theorem states that the terms of an infinite series must tend toward zero if the series is to converge.

11.4.1   THEOREM. *If the series Σu_k converges, then $\lim\limits_{k \to +\infty} u_k = 0$.*

Proof.   The term u_k can be written

$$u_k = s_k - s_{k-1} \tag{1}$$

where s_k is the sum of the first k terms and s_{k-1} is the sum of the first $k - 1$ terms. If S denotes the sum of the series, then $\lim\limits_{k \to +\infty} s_k = S$, and since $(k - 1) \to +\infty$ as $k \to +\infty$, we also have $\lim\limits_{k \to +\infty} s_{k-1} = S$. Thus, from (1)

$$\lim_{k \to +\infty} u_k = \lim_{k \to +\infty} (s_k - s_{k-1}) = S - S = 0 \quad \blacksquare$$

The following result is just an alternative phrasing of the above theorem and needs no additional proof.

The Divergence Test

11.4.2   THEOREM. *If $\lim\limits_{k \to +\infty} u_k \neq 0$, then the series Σu_k diverges.*

Example 1   The series

$$\sum_{k=1}^{\infty} \frac{k}{k+1} = \frac{1}{2} + \frac{2}{3} + \frac{3}{4} + \cdots + \frac{k}{k+1} + \cdots$$

diverges since

$$\lim_{k \to +\infty} \frac{k}{k+1} = \lim_{k \to +\infty} \frac{1}{1 + 1/k} = 1 \neq 0 \quad \blacktriangleleft$$

WARNING.   The converse of Theorem 11.4.1 is false. To prove that a series converges it does not suffice to show that $\lim\limits_{k \to +\infty} u_k = 0$, since this property may hold for divergent as well as convergent series. For example, the kth term of the divergent harmonic series $1 + 1/2 + 1/3 + \cdots + 1/k + \cdots$ tends to zero as $k \to +\infty$, and the kth term of the convergent geometric series $1/2 + 1/2^2 + \cdots + 1/2^k + \cdots$ tends to zero as $k \to +\infty$.

ALGEBRAIC
PROPERTIES OF
INFINITE SERIES

For brevity, the proof of the following result is left for the exercises.

11.4.3   THEOREM.

(a)   *If Σu_k and Σv_k are convergent series, then $\Sigma(u_k + v_k)$ and $\Sigma(u_k - v_k)$ are convergent series and the sums of these series are related by*

$$\sum_{k=1}^{\infty} (u_k + v_k) = \sum_{k=1}^{\infty} u_k + \sum_{k=1}^{\infty} v_k$$

$$\sum_{k=1}^{\infty} (u_k - v_k) = \sum_{k=1}^{\infty} u_k - \sum_{k=1}^{\infty} v_k$$

(b)   *If c is a nonzero constant, then the series Σu_k and Σcu_k both converge or both diverge. In the case of convergence, the sums are related by*

$$\sum_{k=1}^{\infty} cu_k = c \sum_{k=1}^{\infty} u_k$$

(c)   *Convergence or divergence is unaffected by deleting a finite number of terms from the beginning of a series; that is, for any positive integer K, the series*

$$\sum_{k=1}^{\infty} u_k = u_1 + u_2 + u_3 + \cdots$$

and

$$\sum_{k=K}^{\infty} u_k = u_K + u_{K+1} + u_{K+2} + \cdots$$

both converge or both diverge.

REMARK.   Do not read too much into part (c) of this theorem. Although the convergence is not affected when a finite number of terms is deleted from the beginning of a convergent series, the *sum* of the series is changed by the removal of these terms.

Example 2   Find the sum of the series

$$\sum_{k=1}^{\infty} \left(\frac{3}{4^k} - \frac{2}{5^{k-1}} \right)$$

Solution.   The series

$$\sum_{k=1}^{\infty} \frac{3}{4^k} = \frac{3}{4} + \frac{3}{4^2} + \frac{3}{4^3} + \cdots$$

is a convergent geometric series ($a = \frac{3}{4}, r = \frac{1}{4}$), and the series

$$\sum_{k=1}^{\infty} \frac{2}{5^{k-1}} = 2 + \frac{2}{5} + \frac{2}{5^2} + \frac{2}{5^3} + \cdots$$

is also a convergent geometric series ($a = 2, r = \frac{1}{5}$). Thus, from Theorems 11.4.3(a) and 11.3.3 the given series converges and

$$\sum_{k=1}^{\infty} \left(\frac{3}{4^k} - \frac{2}{5^{k-1}} \right) = \sum_{k=1}^{\infty} \frac{3}{4^k} - \sum_{k=1}^{\infty} \frac{2}{5^{k-1}} = \frac{\frac{3}{4}}{1 - \frac{1}{4}} - \frac{2}{1 - \frac{1}{5}}$$

$$= 1 - \frac{5}{2} = -\frac{3}{2} \quad \blacktriangleleft$$

Example 3   The series

$$\sum_{k=1}^{\infty} \frac{5}{k} = 5 + \frac{5}{2} + \frac{5}{3} + \cdots + \frac{5}{k} + \cdots$$

diverges by part (b) of Theorem 11.4.3, since

$$\sum_{k=1}^{\infty} \frac{5}{k} = \sum_{k=1}^{\infty} 5 \left(\frac{1}{k} \right)$$

so each term is a constant times the corresponding term of the divergent harmonic series.   ◀

Example 4   The series

$$\sum_{k=10}^{\infty} \frac{1}{k} = \frac{1}{10} + \frac{1}{11} + \frac{1}{12} + \cdots$$

diverges by part (c) of Theorem 11.4.3, since this series results by deleting the first nine terms from the divergent harmonic series.   ◀

CONVERGENCE
TESTS

If an infinite series $u_1 + u_2 + u_3 + \cdots + u_k + \cdots$ has *positive terms*, then the partial sums $s_1 = u_1$, $s_2 = u_1 + u_2$, $s_3 = u_1 + u_2 + u_3, \ldots$ form an increasing sequence, that is

$$s_1 < s_2 < s_3 < \cdots < s_n < \cdots$$

If there is a finite constant M such that $s_n \leq M$ for all n, then according to Theorem 11.2.2, the sequence of partial sums will converge to a limit S, satisfying $S \leq M$. If no such constant exists, then $\lim_{n \to +\infty} s_n = +\infty$. This yields the following theorem.

11.4.4   THEOREM. *If Σu_k is a series with positive terms, and if there is a constant M such that*

$$s_n = u_1 + u_2 + \cdots + u_n \leq M$$

for every n, then the series converges and the sum S satisfies $S \leq M$. If no such M exists then the series diverges.

If we have a series with positive terms, say

$$\sum_{k=1}^{\infty} \frac{1}{k^2}$$

and if we form the improper integral

$$\int_1^{+\infty} \frac{1}{x^2}\, dx$$

whose integrand is obtained by replacing the summation index k by x, then there is a relationship between convergence of the series and convergence of the improper integral.

The Integral Test

11.4.5   THEOREM. *Let Σu_k be a series with positive terms, and let $f(x)$ be the function that results when k is replaced by x in the formula for u_k. If f is decreasing and continuous for $x \geq 1$, then*

$$\sum_{k=1}^{\infty} u_k \quad and \quad \int_1^{+\infty} f(x)\, dx$$

both converge or both diverge.

We shall defer the proof to the end of the section and proceed with some examples.

Example 5   Determine whether

$$\sum_{k=1}^{\infty} \frac{1}{k^2}$$

converges or diverges.

Solution.   If we replace k by x in the formula for u_k, we obtain the function

$$f(x) = \frac{1}{x^2}$$

which satisfies the hypotheses of the integral test. (Verify.) Since

$$\int_1^{+\infty} \frac{1}{x^2} \, dx = \lim_{l \to +\infty} \int_1^l \frac{dx}{x^2} = \lim_{l \to +\infty} \left[-\frac{1}{x} \right]_1^l = \lim_{l \to +\infty} \left[1 - \frac{1}{l} \right] = 1$$

the integral converges and consequently the series converges.   ◄

REMARK.   In the above example, do *not* erroneously conclude that $\sum_{k=1}^{\infty} \frac{1}{k^2} = 1$ from the fact that $\int_1^{+\infty} \frac{1}{x^2} \, dx = 1$. (If you write out the terms of the series, it will be evident that the sum exceeds 1.)

Example 6   The integral test provides another way to demonstrate divergence of the harmonic series $\sum_{k=1}^{\infty} \frac{1}{k}$. If we replace k by x in the formula for u_k we obtain the function $f(x) = 1/x$, which satisfies the hypotheses of the integral test. (Verify.) Since

$$\int_1^{+\infty} \frac{1}{x} \, dx = \lim_{l \to +\infty} \int_1^l \frac{1}{x} \, dx = \lim_{l \to +\infty} [\ln l - \ln 1] = +\infty$$

the integral diverges and consequently so does the series.   ◄

Example 7   Determine whether the series

$$\frac{1}{e} + \frac{2}{e^4} + \frac{3}{e^9} + \cdots + \frac{k}{e^{k^2}} + \cdots$$

converges or diverges.

Solution.   If we replace k by x in the formula for u_k, we obtain the function

$$f(x) = \frac{x}{e^{x^2}} = xe^{-x^2}$$

For $x \geq 1$, this function has positive values and is continuous. Moreover, for $x \geq 1$ the derivative

$$f'(x) = e^{-x^2} - 2x^2 e^{-x^2} = e^{-x^2}(1 - 2x^2)$$

is negative, so that f is decreasing for $x \geq 1$. Thus, the hypotheses of the integral test are met. But

$$\int_{1}^{+\infty} x e^{-x^2}\, dx = \lim_{l \to +\infty} \int_{1}^{l} x e^{-x^2}\, dx = \lim_{l \to +\infty} \left[-\frac{1}{2} e^{-x^2} \right]_{1}^{l}$$

$$= \left(-\frac{1}{2} \right) \lim_{l \to +\infty} \left[e^{-l^2} - e^{-1} \right] = \frac{1}{2e}$$

Thus, the improper integral and the series converge. ◀

p-SERIES   The harmonic series and the series in Example 5 are special cases of a class of series called **p-series** or **hyperharmonic series**. A *p*-series is an infinite series of the form

$$\sum_{k=1}^{\infty} \frac{1}{k^p} = 1 + \frac{1}{2^p} + \frac{1}{3^p} + \cdots + \frac{1}{k^p} + \cdots$$

where $p > 0$. Examples of *p*-series are

$$\sum_{k=1}^{\infty} \frac{1}{k} = 1 + \frac{1}{2} + \frac{1}{3} + \cdots + \frac{1}{k} + \cdots \qquad \boxed{p = 1}$$

$$\sum_{k=1}^{\infty} \frac{1}{k^2} = 1 + \frac{1}{2^2} + \frac{1}{3^2} + \cdots + \frac{1}{k^2} + \cdots \qquad \boxed{p = 2}$$

$$\sum_{k=1}^{\infty} \frac{1}{\sqrt{k}} = 1 + \frac{1}{\sqrt{2}} + \frac{1}{\sqrt{3}} + \cdots + \frac{1}{\sqrt{k}} + \cdots \qquad \boxed{p = \tfrac{1}{2}}$$

The following theorem tells when a *p*-series converges.

Convergence of p-Series

11.4.6   THEOREM.

$$\sum_{k=1}^{\infty} \frac{1}{k^p} = 1 + \frac{1}{2^p} + \frac{1}{3^p} + \cdots + \frac{1}{k^p} + \cdots$$

converges if $p > 1$ and diverges if $0 < p \leq 1$.

Proof.   To establish this result when $p \neq 1$, we shall use the integral test.

$$\int_{1}^{+\infty} \frac{1}{x^p}\, dx = \lim_{l \to +\infty} \int_{1}^{l} x^{-p}\, dx$$

$$= \lim_{l \to +\infty} \frac{x^{1-p}}{1-p} \Bigg]_1^l$$

$$= \lim_{l \to +\infty} \left[\frac{l^{1-p}}{1-p} - \frac{1}{1-p} \right]$$

For $p > 1$, $1 - p < 0$ and $l^{1-p} \to 0$ as $l \to +\infty$, so the integral and the series converge. For $0 < p < 1$, $1 - p > 0$ and $l^{1-p} \to +\infty$ as $l \to +\infty$, so the integral and the series diverge. The case $p = 1$ is the harmonic series, which was previously shown to diverge. ∎

Example 8

$$1 + \frac{1}{\sqrt[3]{2}} + \frac{1}{\sqrt[3]{3}} + \cdots + \frac{1}{\sqrt[3]{k}} + \cdots$$

diverges since it is a p-series with $p = \frac{1}{3} < 1$.   ◀

PROOF OF THE INTEGRAL TEST   We conclude this section by proving Theorem 11.4.5 (the integral test).

Proof.   Let $f(x)$ satisfy the hypotheses of the theorem. Since

$$f(1) = u_1, \ f(2) = u_2, \ldots, \ f(n) = u_n, \ldots$$

the rectangles in Figures 11.4.1a and 11.4.1b have areas $u_1, u_2, \ldots, u_n$ as indicated. In Figure 11.4.1a, the total area of the rectangles is less than the area under the curve from $x = 1$ to $x = n$, so

$$u_2 + u_3 + \cdots + u_n < \int_1^n f(x)\,dx$$

Therefore,

$$u_1 + u_2 + u_3 + \cdots + u_n < u_1 + \int_1^n f(x)\,dx \tag{2}$$

In Figure 11.4.1b, the total area of the rectangles is greater than the area under the curve from $x = 1$ to $x = n + 1$, so that

$$\int_1^{n+1} f(x)\,dx < u_1 + u_2 + \cdots + u_n \tag{3}$$

If we let $s_n = u_1 + u_2 + \cdots + u_n$ be the nth partial sum of the series, then (2) and (3) yield

$$\int_1^{n+1} f(x)\,dx < s_n < u_1 + \int_1^n f(x)\,dx \tag{4}$$

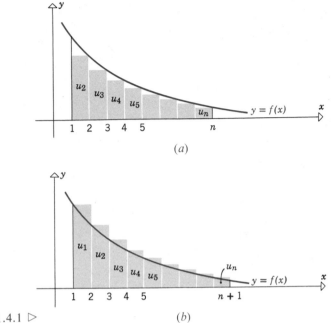

Figure 11.4.1 ▷   (b)

If the integral $\int_1^{+\infty} f(x)\,dx$ converges to a finite value L, then from the right-hand inequality in (4)

$$s_n < u_1 + \int_1^n f(x)\,dx < u_1 + \int_1^{+\infty} f(x)\,dx = u_1 + L$$

Thus, each partial sum is less than the finite constant $u_1 + L$, and the series converges by Theorem 11.4.4. On the other hand, if the integral $\int_1^{+\infty} f(x)\,dx$ diverges, then

$$\lim_{n \to +\infty} \int_1^{n+1} f(x)\,dx = +\infty$$

so that from the left-hand inequality in (4), $\lim_{n \to +\infty} s_n = +\infty$. This means that the series also diverges. ∎

REMARK. If the summation index in a series Σu_k does not begin with $k = 1$, a variation of the integral test may still apply. It can be shown that

$$\sum_{k=K}^{\infty} u_k \quad \text{and} \quad \int_K^{+\infty} f(x)\,dx$$

both converge or both diverge provided the hypotheses of Theorem 11.4.5 hold for $x \geq K$.

► Exercise Set 11.4

In Exercises 1–4, use Theorem 11.4.3 to find the sum of the series.

1. $\sum\limits_{k=1}^{\infty} \left[\dfrac{1}{2^k} + \dfrac{1}{4^k} \right].$ **2.** $\sum\limits_{k=1}^{\infty} \left[\dfrac{1}{5^k} - \dfrac{1}{k(k+1)} \right].$

3. $\sum\limits_{k=2}^{\infty} \left[\dfrac{1}{k^2-1} - \dfrac{7}{10^{k-1}} \right].$

4. $\sum\limits_{k=1}^{\infty} \left[\dfrac{7}{3^k} + \dfrac{6}{(k+3)(k+4)} \right].$

5. In each part, determine whether the given p-series converges or diverges.

(a) $\sum\limits_{k=1}^{\infty} \dfrac{1}{k^3}$ (b) $\sum\limits_{k=1}^{\infty} \dfrac{1}{\sqrt{k}}$

(c) $\sum\limits_{k=1}^{\infty} k^{-1}$ (d) $\sum\limits_{k=1}^{\infty} k^{-2/3}$

(e) $\sum\limits_{k=1}^{\infty} k^{-4/3}$ (f) $\sum\limits_{k=1}^{\infty} \dfrac{1}{\sqrt[4]{k}}$

(g) $\sum\limits_{k=1}^{\infty} \dfrac{1}{\sqrt[3]{k^5}}$ (h) $\sum\limits_{k=1}^{\infty} \dfrac{1}{k^\pi}.$

In Exercises 6–8, use the divergence test to show that the series diverges.

6. (a) $\sum\limits_{k=1}^{\infty} \dfrac{k+1}{k+2}$ (b) $\sum\limits_{k=1}^{\infty} \ln k.$

7. (a) $\sum\limits_{k=1}^{\infty} \dfrac{k^2+k+3}{2k^2+1}$ (b) $\sum\limits_{k=1}^{\infty} \left(1 + \dfrac{1}{k} \right)^k.$

8. (a) $\sum\limits_{k=1}^{\infty} \cos k\pi$ (b) $\sum\limits_{k=1}^{\infty} \dfrac{e^k}{k}.$

In Exercises 9–30, determine whether the series converges or diverges.

9. $\sum\limits_{k=1}^{\infty} \dfrac{1}{k+6}.$ **10.** $\sum\limits_{k=1}^{\infty} \dfrac{3}{5k}.$

11. $\sum\limits_{k=1}^{\infty} \dfrac{1}{5k+2}.$ **12.** $\sum\limits_{k=1}^{\infty} \dfrac{k}{1+k^2}.$

13. $\sum\limits_{k=1}^{\infty} \dfrac{1}{1+9k^2}.$ **14.** $\sum\limits_{k=1}^{\infty} \dfrac{1}{(4+2k)^{3/2}}.$

15. $\sum\limits_{k=1}^{\infty} \dfrac{1}{\sqrt{k+5}}.$ **16.** $\sum\limits_{k=1}^{\infty} \dfrac{1}{\sqrt[k]{e}}.$

17. $\sum\limits_{k=1}^{\infty} \dfrac{1}{\sqrt[3]{2k-1}}.$ **18.** $\sum\limits_{k=3}^{\infty} \dfrac{\ln k}{k}.$

19. $\sum\limits_{k=1}^{\infty} \dfrac{k}{\ln(k+1)}.$ **20.** $\sum\limits_{k=1}^{\infty} ke^{-k^2}.$

21. $\sum\limits_{k=1}^{\infty} \dfrac{1}{(k+1)[\ln(k+1)]^2}.$

22. $\sum\limits_{k=1}^{\infty} \dfrac{k^2+1}{k^2+3}.$ **23.** $\sum\limits_{k=1}^{\infty} \left(1 + \dfrac{1}{k} \right)^k.$

24. $\sum\limits_{k=1}^{\infty} \dfrac{1}{\sqrt{k^2+1}}.$ **25.** $\sum\limits_{k=1}^{\infty} \dfrac{\tan^{-1} k}{1+k^2}.$

26. $\sum\limits_{k=1}^{\infty} \operatorname{sech}^2 k.$ **27.** $\sum\limits_{k=5}^{\infty} 7k^{-p} \ (p>1).$

28. $\sum\limits_{k=1}^{\infty} 7(k+5)^{-p} \ (p \le 1).$

29. $\sum\limits_{k=1}^{\infty} k^2 \sin^2 \left(\dfrac{1}{k} \right).$ **30.** $\sum\limits_{k=1}^{\infty} k^2 e^{-k^3}.$

31. Prove: $\sum\limits_{k=2}^{\infty} \dfrac{1}{k(\ln k)^p}$ converges if $p>1$ and diverges if $p \le 1.$

32. Prove: $\sum\limits_{k=3}^{\infty} \dfrac{1}{k(\ln k)[\ln(\ln k)]^p}$ converges if $p>1$ and diverges if $p \le 1.$

33. Prove: If Σu_k converges and Σv_k diverges, then $\Sigma(u_k + v_k)$ diverges and $\Sigma(u_k - v_k)$ diverges. [*Hint:* Assume $\Sigma(u_k + v_k)$ converges and use Theorem 11.4.3 to obtain a contradiction. Similarly, for $\Sigma(u_k - v_k)$.]

34. Find examples to show that $\Sigma(u_k + v_k)$ and $\Sigma(u_k - v_k)$ may converge or may diverge if Σu_k and Σv_k both diverge.

35. With the help of Exercise 33, determine whether the given series in (a)–(d) converge or diverge.

(a) $\sum\limits_{k=1}^{\infty} \left[\left(\dfrac{2}{3} \right)^{k-1} + \dfrac{1}{k} \right]$

(b) $\sum\limits_{k=1}^{\infty} \left[\dfrac{k^2}{1+k^2} + \dfrac{1}{k(k+1)} \right]$

(c) $\sum\limits_{k=1}^{\infty} \left[\dfrac{1}{3k+2} + \dfrac{1}{k^{3/2}} \right]$

(d) $\displaystyle\sum_{k=2}^{\infty} \left[\frac{1}{k(\ln k)^2} - \frac{1}{k^2} \right].$

36. The harmonic series $\displaystyle\sum_{k=1}^{\infty} \frac{1}{k}$ diverges because the sequence of partial sums $\{s_n\}_{n=1}^{+\infty}$ increases without bound. However, s_n increases very slowly. Use inequality (4) to obtain an upper bound for $s_{1,000,000}$.

37. If the sum S of a convergent series $\displaystyle\sum_{k=1}^{\infty} u_k$ of positive terms is approximated by the nth partial sum s_n, then the error in the approximation is $S - s_n$. Let $f(x)$ be the function that results when k is replaced by x in the formula for u_k.

(a) Show that $S - s_n < \int_n^{+\infty} f(x)\,dx$ if f is decreasing for $x \geq n$.

(b) Use (a) to obtain an upper bound on the error if $\displaystyle\sum_{k=1}^{\infty} \frac{1}{k^3}$ is approximated by s_{10}.

(c) Use (a) to estimate n to assure that s_n will approximate $\displaystyle\sum_{k=1}^{\infty} \frac{1}{k^4}$ with an error less than 10^{-5}.

11.5 ADDITIONAL CONVERGENCE TESTS

In this section we shall develop some additional convergence tests for series with positive terms.

Our first result is not only a practical test for convergence, but is also a theoretical tool that will be used to develop other convergence tests.

The Comparison Test

11.5.1 THEOREM. *Let Σa_k and Σb_k be series with positive terms and suppose*

$$a_1 \leq b_1,\ a_2 \leq b_2,\ a_3 \leq b_3, \ldots,\ a_k \leq b_k, \ldots$$

(a) *If the "bigger series" Σb_k converges, then the "smaller series" Σa_k also converges.*

(b) *On the other hand, if the "smaller series" Σa_k diverges, then the "bigger series" Σb_k also diverges.*

Proof of (a). Suppose that the series Σb_k converges and its sum is B. Then for all n

$$b_1 + b_2 + \cdots + b_n < \sum_{k=1}^{\infty} b_k = B$$

From our hypothesis it follows that

$$a_1 + a_2 + \cdots + a_n \leq b_1 + b_2 + \cdots + b_n$$

so that

$$a_1 + a_2 + \cdots + a_n < B$$

Thus, each partial sum of the series Σa_k is less than B, so that Σa_k converges by Theorem 11.4.4.

Proof of (b). This part is really just an alternative phrasing of part (a). If Σa_k diverges, then Σb_k must diverge since convergence of Σb_k would imply convergence of Σa_k, contrary to the hypothesis. ∎

Since the comparison test requires a little ingenuity to use, we shall wait until the next section before applying it. For now, we shall use the comparison test to develop some other tests that are easier to apply.

The Ratio Test

11.5.2 THEOREM. *Let Σu_k be a series with positive terms and suppose*

$$\lim_{k \to +\infty} \frac{u_{k+1}}{u_k} = \rho$$

(a) *If $\rho < 1$, the series converges.*
(b) *If $\rho > 1$ or $\rho = +\infty$, the series diverges.*
(c) *If $\rho = 1$, the series may converge or diverge, so that another test must be tried.*

Proof of (a). Assume $\rho < 1$, and let $r = \frac{1}{2}(1 + \rho)$. Thus, $\rho < r < 1$, since r is the midpoint between 1 and ρ. It follows that the number

$$\epsilon = r - \rho \tag{1}$$

is positive. Since

$$\rho = \lim_{k \to +\infty} \frac{u_{k+1}}{u_k}$$

it follows that for k sufficiently large, say $k \geq K$, the ratios u_{k+1}/u_k are within ϵ units of ρ. Thus, we will have

$$\frac{u_{k+1}}{u_k} < \rho + \epsilon \quad \text{when} \quad k \geq K$$

or on substituting (1)

$$\frac{u_{k+1}}{u_k} < r \quad \text{when} \quad k \geq K$$

that is,

$$u_{k+1} < ru_k \quad \text{when} \quad k \geq K$$

This yields the inequalities

$$u_{K+1} < ru_K$$
$$u_{K+2} < ru_{K+1} < r^2 u_K$$
$$u_{K+3} < ru_{K+2} < r^3 u_K$$
$$u_{K+4} < ru_{K+3} < r^4 u_K \tag{2}$$
$$\vdots$$

But $|r| < 1$ (why?) so that

$$ru_K + r^2 u_K + r^3 u_K + \cdots$$

is a convergent geometric series. From the inequalities in (2) and the comparison test it follows that

$$u_{K+1} + u_{K+2} + u_{K+3} + \cdots$$

must also be a convergent series. Thus, $u_1 + u_2 + u_3 + \cdots + u_k + \cdots$ converges by Theorem 11.4.3c.

Proof of (b).   Assume $\rho > 1$. Thus,

$$\epsilon = \rho - 1 \tag{3}$$

is a positive number. Since

$$\rho = \lim_{k \to +\infty} \frac{u_{k+1}}{u_k}$$

it follows that for k sufficiently large, say $k \geq K$, the ratio u_{k+1}/u_k is within ϵ units of ρ. Thus,

$$\frac{u_{k+1}}{u_k} > \rho - \epsilon \quad \text{when} \quad k \geq K$$

or on substituting (3)

$$\frac{u_{k+1}}{u_k} > 1 \quad \text{when} \quad k \geq K$$

that is,

$$u_{k+1} > u_k \quad \text{when} \quad k \geq K$$

This yields the inequalities

$$u_{K+1} > u_K$$

$$u_{K+2} > u_{K+1} > u_K$$

$$u_{K+3} > u_{K+2} > u_K$$

$$u_{K+4} > u_{K+3} > u_K \qquad (4)$$

$$\vdots$$

Since $u_K > 0$, it follows from the inequalities in (4) that $\lim\limits_{k \to +\infty} u_k \neq 0$, so $u_1 + u_2 + \cdots + u_k + \cdots$ diverges by Theorem 11.4.2. The proof in the case where $\rho = +\infty$ is omitted.

Proof of (c).   The series

$$\sum_{k=1}^{\infty} \frac{1}{k} \quad \text{and} \quad \sum_{k=1}^{\infty} \frac{1}{k^2}$$

both have $\rho = 1$ (verify). Since the first is the divergent harmonic series and the second is a convergent p-series, the ratio test does not distinguish between convergence and divergence when $\rho = 1$.   ∎

Example 1   The series

$$\sum_{k=1}^{\infty} \frac{1}{k!}$$

converges by the ratio test since

$$\rho = \lim_{k \to +\infty} \frac{u_{k+1}}{u_k} = \lim_{k \to +\infty} \frac{1/(k+1)!}{1/k!} = \lim_{k \to +\infty} \frac{k!}{(k+1)!} = \lim_{k \to +\infty} \frac{1}{k+1} = 0$$

so that $\rho < 1$.   ◀

Example 2   The series

$$\sum_{k=1}^{\infty} \frac{k}{2^k}$$

converges by the ratio test since

$$\rho = \lim_{k \to +\infty} \frac{u_{k+1}}{u_k} = \lim_{k \to +\infty} \frac{k+1}{2^{k+1}} \cdot \frac{2^k}{k} = \frac{1}{2} \lim_{k \to +\infty} \frac{k+1}{k} = \frac{1}{2}$$

so that $\rho < 1$.   ◀

Example 3   The series

$$\sum_{k=1}^{\infty} \frac{k^k}{k!}$$

diverges by the ratio test since

$$\rho = \lim_{k \to +\infty} \frac{u_{k+1}}{u_k} = \lim_{k \to +\infty} \frac{(k+1)^{k+1}}{(k+1)!} \cdot \frac{k!}{k^k}$$

$$= \lim_{k \to +\infty} \frac{(k+1)^k}{k^k}$$

$$= \lim_{k \to +\infty} \left(1 + \frac{1}{k}\right)^k = e \qquad \boxed{\text{See Section 7.2} \\ \text{Formula (6).}}$$

Since $\rho = e > 1$, the series diverges. ◀

Example 4   Determine whether the series

$$1 + \frac{1}{3} + \frac{1}{5} + \frac{1}{7} + \cdots + \frac{1}{2k-1} + \cdots$$

converges or diverges.

Solution.   The ratio test is of no help since

$$\rho = \lim_{k \to +\infty} \frac{u_{k+1}}{u_k} = \lim_{k \to +\infty} \frac{1}{2(k+1)-1} \cdot \frac{2k-1}{1} = \lim_{k \to +\infty} \frac{2k-1}{2k+1} = 1$$

However, the integral test proves that the series diverges since

$$\int_{1}^{+\infty} \frac{dx}{2x-1} = \lim_{l \to +\infty} \int_{1}^{l} \frac{dx}{2x-1} = \lim_{l \to +\infty} \frac{1}{2} \ln(2x-1) \Big]_{1}^{l} = +\infty \qquad ◀$$

Example 5   The series

$$\frac{2!}{4} + \frac{4!}{4^2} + \frac{6!}{4^3} + \cdots + \frac{(2k)!}{4^k} + \cdots$$

diverges since

$$\rho = \lim_{k \to +\infty} \frac{u_{k+1}}{u_k} = \lim_{k \to +\infty} \frac{[2(k+1)]!}{4^{k+1}} \cdot \frac{4^k}{(2k)!} = \lim_{k \to +\infty} \left(\frac{(2k+2)!}{(2k)!} \cdot \frac{1}{4}\right)$$

$$= \frac{1}{4} \lim_{k \to +\infty} (2k+2)(2k+1) = +\infty \qquad ◀$$

Sometimes the following result is easier to apply than the ratio test.

The Root Test

11.5.3 THEOREM. *Let Σu_k be a series with positive terms and suppose*

$$\rho = \lim_{k \to +\infty} \sqrt[k]{u_k} = \lim_{k \to +\infty} (u_k)^{1/k}$$

(a) *If $\rho < 1$, the series converges.*
(b) *If $\rho > 1$, or $\rho = +\infty$, the series diverges.*
(c) *If $\rho = 1$, the series may converge or diverge, so that another test must be tried.*

Since the proof of the root test is similar to the proof of the ratio test, we shall omit it.

Example 6   The series

$$\sum_{k=1}^{\infty} \left(\frac{4k - 5}{2k + 1} \right)^k$$

diverges by the root test since

$$\rho = \lim_{k \to +\infty} (u_k)^{1/k} = \lim_{k \to +\infty} \frac{4k - 5}{2k + 1} = 2 > 1 \quad \blacktriangleleft$$

Example 7   The series

$$\sum_{k=1}^{\infty} \frac{1}{(\ln (k + 1))^k}$$

converges by the root test, since

$$\lim_{k \to +\infty} (u_k)^{1/k} = \lim_{k \to +\infty} \frac{1}{\ln (k + 1)} = 0 < 1 \quad \blacktriangleleft$$

COMMENTS ON NOTATION

We conclude this section with a remark about notation. Until now we have written most of our infinite series in the form

$$\sum_{k=1}^{\infty} u_k \tag{5}$$

with the summation index beginning at 1. If the summation index begins at some other integer, it is always possible to rewrite the series in form (5). Thus, for example, the series

$$\sum_{k=0}^{\infty} \frac{2^k}{k!} = 1 + 2 + \frac{2^2}{2!} + \frac{2^3}{3!} + \cdots \qquad (6)$$

can be written as

$$\sum_{k=1}^{\infty} \frac{2^{k-1}}{(k-1)!} = 1 + 2 + \frac{2^2}{2!} + \frac{2^3}{3!} + \cdots \qquad (7)$$

However, for purposes of applying the convergence tests, it is not necessary that the series have form (5). For example, we can apply the ratio test to (6) without converting to the more complicated form (7). Doing so yields

$$\rho = \lim_{k \to +\infty} \frac{u_{k+1}}{u_k} = \lim_{k \to +\infty} \frac{2^{k+1}}{(k+1)!} \cdot \frac{k!}{2^k} = \lim_{k \to +\infty} \frac{2}{k+1} = 0$$

which shows that the series converges since $\rho < 1$.

▶ Exercise Set 11.5

In Exercises 1–6, apply the ratio test. According to the test, does the series converge, does the series diverge, or are the results inconclusive?

1. $\displaystyle\sum_{k=1}^{\infty} \frac{3^k}{k!}$.

2. $\displaystyle\sum_{k=1}^{\infty} \frac{4^k}{k^2}$.

3. $\displaystyle\sum_{k=2}^{\infty} \frac{1}{5k}$.

4. $\displaystyle\sum_{k=1}^{\infty} k\left(\frac{1}{2}\right)^k$.

5. $\displaystyle\sum_{k=1}^{\infty} \frac{k!}{k^3}$.

6. $\displaystyle\sum_{k=1}^{\infty} \frac{k}{k^2+1}$.

In Exercises 7–10, apply the root test. According to the test, does the series converge, does the series diverge, or are the results inconclusive?

7. $\displaystyle\sum_{k=1}^{\infty} \left(\frac{3k+2}{2k-1}\right)^k$.

8. $\displaystyle\sum_{k=1}^{\infty} \left(\frac{k}{100}\right)^k$.

9. $\displaystyle\sum_{k=1}^{\infty} \frac{k}{5^k}$.

10. $\displaystyle\sum_{k=1}^{\infty} (1 + e^{-k})^k$.

In Exercises 11–32, use any appropriate test to determine whether the series converges.

11. $\displaystyle\sum_{k=1}^{\infty} \frac{2^k}{k^3}$.

12. $\displaystyle\sum_{k=1}^{\infty} \frac{1}{k^2}$.

13. $\displaystyle\sum_{k=0}^{\infty} \frac{7^k}{k!}$.

14. $\displaystyle\sum_{k=1}^{\infty} \frac{1}{2k+1}$.

15. $\displaystyle\sum_{k=1}^{\infty} \frac{k^2}{5^k}$.

16. $\displaystyle\sum_{k=1}^{\infty} \frac{k!\,10^k}{3^k}$.

17. $\displaystyle\sum_{k=1}^{\infty} k^{50} e^{-k}$.

18. $\displaystyle\sum_{k=1}^{\infty} \frac{k^2}{k^3+1}$.

19. $\displaystyle\sum_{k=1}^{\infty} k\left(\frac{2}{3}\right)^k$.

20. $\displaystyle\sum_{k=1}^{\infty} k^k$.

21. $\displaystyle\sum_{k=2}^{\infty} \frac{1}{k \ln k}$.

22. $\displaystyle\sum_{k=1}^{\infty} \frac{2^k}{k^3+1}$.

23. $\displaystyle\sum_{k=1}^{\infty} \left(\frac{4}{7k-1}\right)^k$.

24. $\displaystyle\sum_{k=1}^{\infty} \frac{(k!)^2 2^k}{(2k+2)!}$.

25. $\displaystyle\sum_{k=0}^{\infty} \frac{(k!)^2}{(2k)!}$.

26. $\displaystyle\sum_{k=1}^{\infty} \frac{1}{k^2+25}$.

27. $\displaystyle\sum_{k=1}^{\infty} \frac{1}{1+\sqrt{k}}$.

28. $\displaystyle\sum_{k=1}^{\infty} \frac{k^k}{k!}$.

29. $\displaystyle\sum_{k=1}^{\infty} \frac{\ln k}{e^k}$.

30. $\displaystyle\sum_{k=1}^{\infty} \frac{k!}{e^{k^2}}$.

31. $\displaystyle\sum_{k=0}^{\infty} \frac{(k+4)!}{4!k!4^k}$.

32. $\displaystyle\sum_{k=1}^{\infty} \left(\frac{k}{k+1}\right)^{k^2}$.

33. Determine whether the following series converges:

$$1 + \frac{1 \cdot 2}{1 \cdot 3} + \frac{1 \cdot 2 \cdot 3}{1 \cdot 3 \cdot 5} + \frac{1 \cdot 2 \cdot 3 \cdot 4}{1 \cdot 3 \cdot 5 \cdot 7} + \cdots$$

34. For which positive values of α does $\sum_{k=1}^{\infty} \frac{\alpha^k}{k^\alpha}$ converge?

35. (a) Show $\lim_{k \to +\infty} (\ln k)^{1/k} = 1$ [*Hint:* Let $y = (\ln x)^{1/x}$ and find $\lim_{x \to +\infty} \ln y$.]

(b) Use the result in (a) and the root test to show that $\sum_{k=1}^{\infty} \frac{\ln k}{3^k}$ converges.

(c) Show that the series converges using the ratio test.

36. Prove: $\lim_{k \to +\infty} k!/k^k = 0$ [*Hint:* Exploit Theorem 11.4.1.]

37. Prove: $\lim_{k \to +\infty} \frac{a^k}{k!} = 0$ for every real number a. [See Hint to Exercise 36.]

11.6 APPLYING THE COMPARISON TEST

In this section we shall discuss procedures for applying the comparison test and we shall state an alternative version of this test which is easier to work with. Before starting, we remind the reader that the comparison test applies only to series with positive terms.

There are two basic steps required to apply the comparison to a series Σu_k of positive terms:

- Guess at whether the series Σu_k converges or diverges.

- Find a series that proves the guess to be correct. Thus, if the guess is divergence we must find a divergent series whose terms are "smaller" than the corresponding terms of Σu_k, and if the guess is convergence we must find a convergent series whose terms are "bigger" than the corresponding terms of Σu_k.

To help with the guessing process in the first step we have formulated some principles that sometimes *suggest* whether a series is likely to converge or diverge. We have called these "informal principles" because they are not intended as formal theorems. In fact, we shall not guarantee that they *always* work. However, they work often enough to be useful as a starting point for the comparison test.

> **11.6.1** INFORMAL PRINCIPLE. Constant terms in the denominator of u_k can usually be deleted without affecting the convergence or divergence of the series.

Example 1 Use the above principle to help guess whether the following series converge or diverge.

(a) $\displaystyle\sum_{k=1}^{\infty} \frac{1}{2^k + 1}$ (b) $\displaystyle\sum_{k=5}^{\infty} \frac{1}{\sqrt{k} - 2}$ (c) $\displaystyle\sum_{k=1}^{\infty} \frac{1}{(k + \frac{1}{2})^3}$

Solution.
(a)   Deleting the constant 1 suggests that

$$\sum_{k=1}^{\infty} \frac{1}{2^k + 1} \quad \text{behaves like} \quad \sum_{k=1}^{\infty} \frac{1}{2^k}$$

The modified series is a convergent geometric series, so the given series is likely to converge.

(b)   Deleting the -2 suggests that

$$\sum_{k=5}^{\infty} \frac{1}{\sqrt{k} - 2} \quad \text{behaves like} \quad \sum_{k=5}^{\infty} \frac{1}{\sqrt{k}}$$

The modified series is a portion of a divergent p-series ($p = \frac{1}{2}$), so the given series is likely to diverge.

(c)   Deleting the $\frac{1}{2}$ suggests that

$$\sum_{k=1}^{\infty} \frac{1}{(k + \frac{1}{2})^3} \quad \text{behaves like} \quad \sum_{k=1}^{\infty} \frac{1}{k^3}$$

The modified series is a convergent p-series ($p = 3$), so the given series is likely to converge.   ◀

11.6.2   INFORMAL PRINCIPLE. *If a polynomial in k appears as a factor in the numerator or denominator of u_k, all but the highest power of k in the polynomial may usually be deleted without affecting the convergence or divergence of the series.*

Example 2   Use the above principle to help guess whether the following series converge or diverge.

(a) $\displaystyle\sum_{k=1}^{\infty} \frac{1}{\sqrt{k^3 + 2k}}$ (b) $\displaystyle\sum_{k=1}^{\infty} \frac{6k^4 - 2k^3 + 1}{k^5 + k^2 - 2k}$

Solution.
(a)   Deleting the term $2k$ suggests that

$$\sum_{k=1}^{\infty} \frac{1}{\sqrt{k^3 + 2k}} \quad \text{behaves like} \quad \sum_{k=1}^{\infty} \frac{1}{\sqrt{k^3}} = \sum_{k=1}^{\infty} \frac{1}{k^{3/2}}$$

Since the modified series is a convergent p-series ($p = \frac{3}{2}$), the given series is likely to converge.

(b) Deleting all but the highest powers of k in the numerator and also in the denominator suggests that

$$\sum_{k=1}^{\infty} \frac{6k^4 - 2k^3 + 1}{k^5 + k^2 - 2k} \quad \text{behaves like} \quad \sum_{k=1}^{\infty} \frac{6k^4}{k^5} = 6 \sum_{k=1}^{\infty} \frac{1}{k}$$

Since the modified series is a constant times the divergent harmonic series, the given series is likely to diverge. ◀

Once it is decided whether a series is likely to converge or diverge, the second step in applying the comparison test is to produce a series with which the given series can be compared to substantiate the guess. Let us consider the case of convergence first. To prove Σa_k converges by the comparison test, we must find a *convergent* series Σb_k such that

$$a_k \le b_k$$

for all k. Frequently, b_k is derived from the formula for a_k by either increasing the numerator of a_k, or decreasing the denominator of a_k, or both.

Example 3 Use the comparison test to determine whether

$$\sum_{k=1}^{\infty} \frac{1}{2k^2 + k}$$

converges or diverges.

Solution. Using Principle 11.6.2, the given series behaves like the series

$$\sum_{k=1}^{\infty} \frac{1}{2k^2} = \frac{1}{2} \sum_{k=1}^{\infty} \frac{1}{k^2}$$

which is a constant times a convergent p-series. Thus, the given series is likely to converge. To prove the convergence, observe that when we discard the k from the denominator of $1/(2k^2 + k)$, the denominator decreases and the ratio increases, so that

$$\frac{1}{2k^2 + k} < \frac{1}{2k^2}$$

for $k = 1, 2, \ldots$. Since

$$\sum_{k=1}^{\infty} \frac{1}{2k^2} = \frac{1}{2} \sum_{k=1}^{\infty} \frac{1}{k^2}$$

converges, so does $\displaystyle\sum_{k=1}^{\infty} \frac{1}{2k^2 + k}$ by the comparison test.   ◀

Example 4   Use the comparison test to determine whether

$$\sum_{k=1}^{\infty} \frac{1}{2k^2 - k}$$

converges or diverges.

Solution.   Using Principle 11.6.2, the series behaves like the convergent series

$$\sum_{k=1}^{\infty} \frac{1}{2k^2} = \frac{1}{2} \sum_{k=1}^{\infty} \frac{1}{k^2}$$

Thus, the given series is likely to converge. However, if we discard the k from the denominator of $1/(2k^2 - k)$, the denominator increases and the ratio decreases, so that

$$\frac{1}{2k^2 - k} > \frac{1}{2k^2}$$

Unfortunately, this inequality is in the wrong direction to prove convergence of the given series. A different approach is needed; we must do something to decrease the denominator, not increase it. We accomplish this by replacing k by k^2 to obtain

$$\frac{1}{2k^2 - k} \le \frac{1}{2k^2 - k^2} = \frac{1}{k^2}$$

Since $\displaystyle\sum_{k=1}^{\infty} \frac{1}{k^2}$ is a convergent p-series, the given series converges by the comparison test.   ◀

To prove that a series Σa_k diverges by the comparison test, we must produce a divergent series Σb_k of positive terms such that $a_k \ge b_k$ for all k.

Example 5   Use the comparison test to determine whether

$$\sum_{k=1}^{\infty} \frac{1}{k - \frac{1}{4}}$$

converges or diverges.

Solution.   Using Principle 11.6.1, the series behaves like the divergent harmonic series

$$\sum_{k=1}^{\infty} \frac{1}{k}$$

Thus, the given series is likely to diverge. Since

$$\frac{1}{k - \frac{1}{4}} > \frac{1}{k} \quad \text{for } k = 1, 2, \ldots$$

and since $\sum_{k=1}^{\infty} \frac{1}{k}$ diverges, the given series diverges by the comparison test.   ◄

Example 6   Use the comparison test to determine whether

$$\sum_{k=1}^{\infty} \frac{1}{\sqrt{k} + 5}$$

converges or diverges.

Solution.   Using Principle 11.6.1, the series behaves like the divergent *p*-series

$$\sum_{k=1}^{\infty} \frac{1}{\sqrt{k}}$$

Thus, the given series is likely to diverge. For $k \geq 25$ we have

$$\frac{1}{\sqrt{k} + 5} \geq \frac{1}{\sqrt{k} + \sqrt{k}} = \frac{1}{2\sqrt{k}}$$

and since

$$\sum_{k=25}^{\infty} \frac{1}{2\sqrt{k}}$$

diverges (why?), the series

$$\sum_{k=25}^{\infty} \frac{1}{\sqrt{k} + 5}$$

diverges by the comparison test; consequently, the given series diverges by Theorem 11.4.3(c). ◀

As the last four examples show, some tricky work with inequalities may be required to apply the comparison test. Fortunately, there is an alternative version of this test, which avoids this complication. We shall state the result, but defer its proof to the end of the section, so we can proceed immediately to some examples.

The Limit Comparison Test

11.6.3 THEOREM. *Let* Σa_k *and* Σb_k *be series with positive terms and suppose*

$$\rho = \lim_{k \to +\infty} \frac{a_k}{b_k}$$

If ρ *is finite and* $\rho \neq 0$, *then the series both converge or both diverge.*

To illustrate how this test works we shall reconsider the problems in Examples 4 and 5.

Example 7   Use the limit comparison test to determine whether the following series converge or diverge:

(a) $\displaystyle\sum_{k=1}^{\infty} \frac{1}{2k^2 - k}$ (b) $\displaystyle\sum_{k=1}^{\infty} \frac{1}{k - \frac{1}{4}}$

Solution (*a*).   As in Example 4, we guess that the given series behaves like the convergent series

$$\sum_{k=1}^{\infty} \frac{1}{2k^2} \tag{1}$$

Thus, from Theorem 11.6.3 with

$$a_k = \frac{1}{2k^2 - k} \quad \text{and} \quad b_k = \frac{1}{2k^2}$$

we obtain

$$\rho = \lim_{k \to +\infty} \frac{a_k}{b_k} = \lim_{k \to +\infty} \frac{2k^2}{2k^2 - k} = \lim_{k \to +\infty} \frac{2}{2 - 1/k} = 1$$

Since (1) converges, so does the given series since ρ is finite and positive.

Solution (*b*).   As in Example 5, we guess that the given series behaves like the divergent series

$$\sum_{k=1}^{\infty} \frac{1}{k} \tag{2}$$

Thus, from Theorem 11.6.3 with

$$a_k = \frac{1}{k - \frac{1}{4}} \quad \text{and} \quad b_k = \frac{1}{k}$$

we obtain

$$\rho = \lim_{k \to +\infty} \frac{a_k}{b_k} = \lim_{k \to +\infty} \frac{k}{k - \frac{1}{4}} = \lim_{k \to +\infty} \frac{1}{1 - \frac{1}{4k}} = 1$$

Since (2) diverges, so does the given series since ρ is finite and non-zero. ◄

Example 8   Use the limit comparison test to determine whether

$$\sum_{k=1}^{\infty} \frac{3k^3 - 2k^2 + 4}{k^5 - k^3 + 2}$$

converges or diverges.

Solution.   From Principle 11.6.2, the series behaves like

$$\sum_{k=1}^{\infty} \frac{3k^3}{k^5} = \sum_{k=1}^{\infty} \frac{3}{k^2} \tag{3}$$

which converges since it is a constant times a convergent p-series. Thus, the given series is likely to converge. To substantiate this, we apply the limit comparison test to series (3) and the given series. We obtain

$$\rho = \lim_{k \to +\infty} \frac{\dfrac{3k^3 - 2k^2 + 4}{k^5 - k^3 + 2}}{\dfrac{3}{k^2}} = \lim_{k \to +\infty} \frac{3k^5 - 2k^4 + 4k^2}{3k^5 - 3k^3 + 6} = 1$$

Since $\rho \neq 0$, the given series converges because series (3) converges. ◄

Unlike the comparison test, the limit comparison test does not require any tricky manipulation of inequalities. However, the test only applies when $0 < \rho < +\infty$. We note, though, that if $\rho = 0$ or $\rho = +\infty$, conclusions about convergence or divergence may be drawn in certain cases (see Exercise 44).

REMARK.   As a practical matter you should generally try the limit comparison test before the comparison test because it is easier to apply.

OPTIONAL

We conclude this section with a proof of the limit comparison test.

Proof of Theorem 11.6.3.   We must show that Σb_k converges when Σa_k converges and conversely. To this end, let

$$\epsilon = \frac{\rho}{2} \tag{4}$$

so that $\epsilon > 0$ because $\rho > 0$ by hypothesis. From the assumption that

$$\rho = \lim_{k \to +\infty} \frac{a_k}{b_k}$$

it follows that for k sufficiently large, say $k \geq K$, the ratio a_k/b_k will be within ϵ units of ρ. Thus,

$$\rho - \epsilon < \frac{a_k}{b_k} < \rho + \epsilon \quad \text{when} \quad k \geq K$$

or on substituting (4)

$$\frac{1}{2}\rho < \frac{a_k}{b_k} < \frac{3}{2}\rho \quad \text{when} \quad k \geq K$$

or

$$\frac{1}{2}\rho b_k < a_k < \frac{3}{2}\rho b_k \quad \text{when} \quad k \geq K \tag{5}$$

If $\sum_{k=1}^{\infty} a_k$ converges, then $\sum_{k=K}^{\infty} a_k$ converges by Theorem 11.4.3(c). It follows from the comparison test and the left-hand inequality in (5), that $\sum_{k=K}^{\infty} \frac{1}{2}\rho b_k$ converges. Thus, $\sum_{k=K}^{\infty} b_k$ converges (Theorem 11.4.3(b)) and $\sum_{k=1}^{\infty} b_k$ converges (Theorem 11.4.3(c)).

Conversely, if $\sum_{k=1}^{\infty} b_k$ converges, then $\sum_{k=K}^{\infty} \frac{3}{2}\rho b_k$ converges (Theorem 11.4.3(b),(c)), so that $\sum_{k=K}^{\infty} a_k$ converges by the comparison test and the right-hand inequality in (5). Thus, $\sum_{k=1}^{\infty} a_k$ converges.   ∎

▶ Exercise Set 11.6

In Exercises 1–6, prove that the series converges by the comparison test.

1. $\displaystyle\sum_{k=1}^{\infty} \frac{1}{3^k + 5}$.

2. $\displaystyle\sum_{k=1}^{\infty} \frac{2}{k^4 + k}$.

3. $\displaystyle\sum_{k=1}^{\infty} \frac{1}{5k^2 - k}$.

4. $\displaystyle\sum_{k=1}^{\infty} \frac{k}{8k^3 + 2k^2 - 1}$.

5. $\displaystyle\sum_{k=1}^{\infty} \frac{2^k - 1}{3^k + 2k}$.

6. $\displaystyle\sum_{k=1}^{\infty} \frac{5 \sin^2 k}{k!}$.

In Exercises 7–12, prove that the series diverges by the comparison test.

7. $\displaystyle\sum_{k=1}^{\infty} \frac{3}{k - \frac{1}{4}}$.

8. $\displaystyle\sum_{k=1}^{\infty} \frac{1}{\sqrt{k + 8}}$.

9. $\displaystyle\sum_{k=1}^{\infty} \frac{9}{\sqrt{k} + 1}$.

10. $\displaystyle\sum_{k=2}^{\infty} \frac{k + 1}{k^2 - k}$.

11. $\displaystyle\sum_{k=1}^{\infty} \frac{k^{4/3}}{8k^2 + 5k + 1}$.

12. $\displaystyle\sum_{k=1}^{\infty} \frac{k^{-1/2}}{2 + \sin^2 k}$.

In Exercises 13–18, use the limit comparison test to determine whether the series converges or diverges.

13. $\displaystyle\sum_{k=1}^{\infty} \frac{4k^2 - 2k + 6}{8k^7 + k - 8}$.

14. $\displaystyle\sum_{k=1}^{\infty} \frac{1}{9k + 6}$.

15. $\displaystyle\sum_{k=1}^{\infty} \frac{5}{3^k + 1}$.

16. $\displaystyle\sum_{k=1}^{\infty} \frac{k(k + 3)}{(k + 1)(k + 2)(k + 5)}$.

17. $\displaystyle\sum_{k=1}^{\infty} \frac{1}{\sqrt[3]{8k^2 - 3k}}$.

18. $\displaystyle\sum_{k=1}^{\infty} \frac{1}{(2k + 3)^{17}}$.

In Exercises 19–34, use any method to determine whether the series converges or diverges. In some cases, you may have to use tests from earlier sections.

19. $\displaystyle\sum_{k=1}^{\infty} \frac{1}{k^3 + 2k + 1}$.

20. $\displaystyle\sum_{k=1}^{\infty} \frac{1}{(3 + k)^{2/5}}$.

21. $\displaystyle\sum_{k=1}^{\infty} \frac{1}{9k - 2}$.

22. $\displaystyle\sum_{k=1}^{\infty} \frac{\ln k}{k}$.

23. $\displaystyle\sum_{k=1}^{\infty} \frac{\sqrt{k}}{k^3 + 1}$.

24. $\displaystyle\sum_{k=1}^{\infty} \frac{4}{2 + 3^k k}$.

25. $\displaystyle\sum_{k=1}^{\infty} \frac{1}{\sqrt{k(k + 1)}}$.

26. $\displaystyle\sum_{k=1}^{\infty} \frac{2 + (-1)^k}{5^k}$.

27. $\displaystyle\sum_{k=1}^{\infty} \frac{2 + \sqrt{k}}{(k + 1)^3 - 1}$.

28. $\displaystyle\sum_{k=1}^{\infty} \frac{4 + |\cos k|}{k^3}$.

29. $\displaystyle\sum_{k=1}^{\infty} \frac{1}{4 + 2^{-k}}$.

30. $\displaystyle\sum_{k=1}^{\infty} \frac{\sqrt{k} \ln k}{k^3 + 1}$.

31. $\displaystyle\sum_{k=1}^{\infty} \frac{\tan^{-1} k}{k^2}$.

32. $\displaystyle\sum_{k=1}^{\infty} \frac{5^k + k}{k! + 3}$.

33. $\displaystyle\sum_{k=1}^{\infty} \frac{\ln k}{k\sqrt{k}}$.

34. $\displaystyle\sum_{k=1}^{\infty} \frac{\cos(1/k)}{k^2}$.

35. Use the limit comparison test to show that $\displaystyle\sum_{k=1}^{\infty} (1 - \cos(1/k))$ converges. [*Hint:* Compare with the series $\displaystyle\sum_{k=1}^{\infty} \frac{1}{k^2}$.]

36. Use the limit comparison test to show that $\displaystyle\sum_{k=1}^{\infty} \sin\left(\frac{\pi}{k}\right)$ diverges. [*Hint:* Compare with the series $\displaystyle\sum_{k=1}^{\infty} \frac{\pi}{k}$.]

37. Use the comparison test to determine whether $\displaystyle\sum_{k=1}^{\infty} \frac{\ln k}{k^2}$ converges or diverges.
[*Hint:* $\ln x < 2\sqrt{x}$ for $x > 1$ by Exercise 63 of Section 7.3.]

38. Determine whether $\displaystyle\sum_{k=2}^{\infty} \frac{1}{(\ln k)^2}$ converges or diverges. [*Hint:* See Hint to Exercise 37.]

39. Let a, b, and p be positive constants. For which values of p does the series $\displaystyle\sum_{k=1}^{\infty} \frac{1}{(a + bk)^p}$ converge?

40. (a) Show that $k^k \geq k!$ and use this result to prove that the series $\displaystyle\sum_{k=1}^{\infty} k^{-k}$ converges by the comparison test.
(b) Prove convergence using the root test.

41. Use the limit comparison test to investigate convergence of $\displaystyle\sum_{k=1}^{\infty} \frac{(k+1)^2}{(k+2)!}$.

42. Use the limit comparison test to investigate convergence of the series $1 + \frac{1}{3} + \frac{1}{5} + \frac{1}{7} + \cdots$

43. Prove that $\displaystyle\sum_{k=1}^{\infty} \frac{1}{k!}$ converges by comparison with a suitable geometric series.

44. Let Σa_k and Σb_k be series with positive terms. Prove:

(a) If $\displaystyle\lim_{k \to +\infty} (a_k/b_k) = 0$ and Σb_k converges, then Σa_k converges.

(b) If $\displaystyle\lim_{k \to +\infty} (a_k/b_k) = +\infty$ and Σb_k diverges, then Σa_k diverges.

11.7 ALTERNATING SERIES; CONDITIONAL CONVERGENCE

So far our emphasis has been on series with positive terms. In this section we shall discuss series containing negative terms.

ALTERNATING SERIES

Of special importance are series whose terms are alternately positive and negative. These are called **alternating series**. Such series have one of two possible forms:

$$a_1 - a_2 + a_3 - a_4 + \cdots + (-1)^{k+1} a_k + \cdots$$

or

$$-a_1 + a_2 - a_3 + a_4 - \cdots + (-1)^{k} a_k + \cdots$$

where the a_k's are all positive.

The following theorem is the key result on convergence of alternating series.

Alternating Series Test

11.7.1 THEOREM. *An alternating series*

$$\sum_{k=1}^{\infty} (-1)^{k+1} a_k \quad or \quad \sum_{k=1}^{\infty} (-1)^{k} a_k$$

converges if the following two conditions are satisfied:

(a) $a_1 \geq a_2 \geq a_3 \geq \cdots \geq a_k \geq \cdots$

(b) $\displaystyle\lim_{k \to +\infty} a_k = 0$

To prove this result, we shall need the following fact about sequences:

If the even-numbered terms of a sequence tend toward a limit L, and if the odd-numbered terms of the sequence tend toward the same limit L, then the entire sequence tends toward the limit L.

This result should be intuitively obvious; we omit the proof.

Proof of 11.7.1.   We shall consider the case

$$a_1 - a_2 + a_3 - a_4 + \cdots + (-1)^{k+1} a_k + \cdots$$

The other case is left as an exercise. Consider the partial sums

$$s_2 = a_1 - a_2$$
$$s_4 = (a_1 - a_2) + (a_3 - a_4)$$
$$s_6 = (a_1 - a_2) + (a_3 - a_4) + (a_5 - a_6)$$
$$s_8 = (a_1 - a_2) + (a_3 - a_4) + (a_5 - a_6) + (a_7 - a_8)$$
$$\vdots$$

By hypothesis (*a*), each of the differences appearing in the parentheses is nonnegative so that

$$s_2 \leq s_4 \leq s_6 \leq s_8 \leq \cdots$$

Moreover, the terms in this sequence are all less than or equal to a_1 since we can write

$$s_2 = a_1 - a_2$$
$$s_4 = a_1 - (a_2 - a_3) - a_4$$
$$s_6 = a_1 - (a_2 - a_3) - (a_4 - a_5) - a_6$$
$$s_8 = a_1 - (a_2 - a_3) - (a_4 - a_5) - (a_6 - a_7) - a_8$$
$$\vdots$$

Thus, the sequence

$$s_2, s_4, s_6, s_8, \ldots, s_{2n}, \ldots$$

converges to some limit S, by Theorem 11.2.2. That is,

$$\lim_{n \to +\infty} s_{2n} = S \tag{1}$$

We shall show next that the sequence

$$s_1, s_3, s_5, \ldots, s_{2n-1}, \ldots$$

also has limit S. Since the $(2n)$-th term in the given alternating series is $-a_{2n}$, it follows that $s_{2n} - s_{2n-1} = -a_{2n}$, which can be written as

$$s_{2n-1} = s_{2n} + a_{2n} \tag{2}$$

But $2n \to +\infty$ as $n \to +\infty$ so, $\lim\limits_{n \to +\infty} a_{2n} = 0$ by hypothesis (b). Thus, from (1) and (2)

$$\lim_{n \to +\infty} s_{2n-1} = \lim_{n \to +\infty} s_{2n} + \lim_{n \to +\infty} a_{2n} = S + 0 = S \tag{3}$$

From (1) and (3), the sequence $s_1, s_2, s_3, \ldots, s_n, \ldots$ converges to S, so the series converges. ∎

Example 1 The series

$$1 - \frac{1}{2} + \frac{1}{3} - \frac{1}{4} + \cdots + (-1)^{k+1} \frac{1}{k} + \cdots$$

is called the **alternating harmonic series**. Since

$$a_k = \frac{1}{k} > \frac{1}{k+1} = a_{k+1}$$

and

$$\lim_{k \to +\infty} a_k = \lim_{k \to +\infty} \frac{1}{k} = 0$$

this series converges by the alternating series test. ◄

Example 2 Determine whether the alternating series

$$\sum_{k=1}^{\infty} (-1)^{k+1} \frac{k+3}{k(k+1)}$$

converges or diverges.

Solution. Requirement (b) of the alternating series test is satisfied since

$$\lim_{k \to +\infty} a_k = \lim_{k \to +\infty} \frac{k+3}{k(k+1)} = \lim_{k \to +\infty} \frac{\dfrac{1}{k} + \dfrac{3}{k^2}}{1 + \dfrac{1}{k}} = 0$$

To see if requirement (a) is met, we must determine whether the sequence

$$\{a_k\}_{k=1}^{+\infty} = \left\{ \frac{k+3}{k(k+1)} \right\}_{k=1}^{+\infty}$$

is nonincreasing. Since

$$\frac{a_{k+1}}{a_k} = \frac{k+4}{(k+1)(k+2)} \cdot \frac{k(k+1)}{k+3} = \frac{k^2 + 4k}{k^2 + 5k + 6}$$

$$= \frac{k^2 + 4k}{(k^2 + 4k) + (k+6)} < 1$$

we have $a_k > a_{k+1}$, so the series converges by the alternating series test. ◄

REMARK.   If an alternating series violates condition (b) of the alternating series test, then the series must diverge by the divergence test (11.4.2). However, if condition (b) is satisfied, but (a) is not, the series may either converge or diverge.*

Figure 11.7.1 provides some insight into the way in which an alternating series

$$a_1 - a_2 + a_3 - a_4 + \cdots + (-1)^{k+1} a_k + \cdots$$

converges to its sum S when the hypotheses of the alternating test are satisfied.

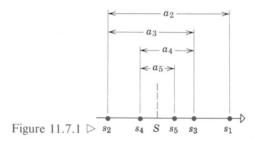

Figure 11.7.1 ▷

In the figure we have plotted the successive partial sums on the x-axis. Because

$$a_1 \geq a_2 \geq a_3 \geq a_4 \geq \cdots$$

and

$$\lim_{k \to +\infty} a_k = 0$$

the successive partial sums oscillate in smaller and smaller steps, closing in on the sum S. It is of interest to note that the even-numbered partial sums are less than or equal to S and the odd-numbered partial sums are

*The interested reader will find some nice examples in an article by R. Lariviere, "On a Convergence Test for Alternating Series," *Mathematics Magazine*, Vol. 29 (1956), p. 88.

greater than or equal to S. Thus, the sum S falls between any two successive partial sums; that is, for any positive integer n

$$s_n \leq S \leq s_{n+1} \quad \text{or} \quad s_{n+1} \leq S \leq s_n$$

depending on whether n is even or odd. In either case,

$$|S - s_n| \leq |s_{n+1} - s_n| \tag{4}$$

But $s_{n+1} - s_n = \pm a_{n+1}$ (the sign depending on whether n is even or odd) so (4) yields

$$|S - s_n| \leq a_{n+1} \tag{5}$$

Since $|S - s_n|$ represents the magnitude of error that results when we approximate the sum of the entire series by the sum of the first n terms, (5) tells us that this error is less than or equal to the magnitude of the $(n + 1)$-st term in the series. The reader can check that (5) also holds for alternating series of the form

$$-a_1 + a_2 - a_3 + a_4 - \cdots + (-1)^k a_k + \cdots$$

In summary, we have the following result.

11.7.2 THEOREM. *If an alternating series satisfies the hypotheses of the alternating series test, and if the sum S of the series is approximated by the nth partial sum s_n, then the absolute value of the error is less than or equal to a_{n+1}.*

Example 3   As shown in Example 1, the alternating harmonic series

$$1 - \frac{1}{2} + \frac{1}{3} - \frac{1}{4} + \cdots + (-1)^{k+1} \frac{1}{k} + \cdots$$

satisfies the hypotheses of the alternating series test. If we approximate the sum of the series by

$$1 - \frac{1}{2} + \frac{1}{3} - \frac{1}{4} = \frac{7}{12}$$

then the absolute value of the error is at most $\frac{1}{5} = 0.2$, and if we approximate the sum by

$$1 - \frac{1}{2} + \frac{1}{3} - \frac{1}{4} + \frac{1}{5} - \frac{1}{6} + \frac{1}{7} = \frac{319}{420}$$

then the absolute value of the error is at most $\frac{1}{8} = 0.125$.   ◄

ABSOLUTE AND
CONDITIONAL
CONVERGENCE

The series

$$1 - \frac{1}{2} - \frac{1}{2^2} + \frac{1}{2^3} + \frac{1}{2^4} - \frac{1}{2^5} - \frac{1}{2^6} + \cdots$$

does not fit in any of the categories studied so far—it has mixed signs, but is not alternating. We shall now develop some convergence tests that can be applied to such series.

11.7.3 DEFINITION. A series

$$\sum_{k=1}^{\infty} u_k = u_1 + u_2 + \cdots + u_k + \cdots$$

is said to *converge absolutely* if the series of absolute values

$$\sum_{k=1}^{\infty} |u_k| = |u_1| + |u_2| + \cdots + |u_k| + \cdots$$

converges.

Example 4   The series

$$1 - \frac{1}{2} - \frac{1}{2^2} + \frac{1}{2^3} + \frac{1}{2^4} - \frac{1}{2^5} - \frac{1}{2^6} + \cdots$$

converges absolutely since the series of absolute values

$$1 + \frac{1}{2} + \frac{1}{2^2} + \frac{1}{2^3} + \frac{1}{2^4} + \frac{1}{2^5} + \frac{1}{2^6} + \cdots$$

is a convergent geometric series. On the other hand, the alternating harmonic series

$$1 - \frac{1}{2} + \frac{1}{3} - \frac{1}{4} + \frac{1}{5} - \cdots$$

does not converge absolutely since the series of absolute values

$$1 + \frac{1}{2} + \frac{1}{3} + \frac{1}{4} + \frac{1}{5} + \cdots$$

diverges.   ◄

Absolute convergence is of importance because of the following theorem.

11.7.4 THEOREM. *If the series*

$$\sum_{k=1}^{\infty} |u_k| = |u_1| + |u_2| + \cdots + |u_k| + \cdots$$

converges, then so does the series

$$\sum_{k=1}^{\infty} u_k = u_1 + u_2 + \cdots + u_k + \cdots$$

In other words, if a series converges absolutely, then it converges.

Proof. Our proof is based on a trick. We shall show that the series

$$\sum_{k=1}^{\infty} (u_k + |u_k|) \tag{6}$$

converges. Since $\Sigma|u_k|$ is assumed to converge, it will then follow from Theorem 11.4.3(*a*) that Σu_k converges, since

$$\sum_{k=1}^{\infty} u_k = \sum_{k=1}^{\infty} [(u_k + |u_k|) - |u_k|]$$

For all k, the value of $u_k + |u_k|$ is either 0 or $2|u_k|$, depending on whether u_k is negative or not. Thus, for all values of k

$$0 \le u_k + |u_k| \le 2|u_k| \tag{7}$$

But $\Sigma 2|u_k|$ is a convergent series since it is a constant times the convergent series $\Sigma|u_k|$. Thus, from (7), series (6) converges by the comparison test. ∎

Example 5   In Example 4 we showed that

$$1 - \frac{1}{2} - \frac{1}{2^2} + \frac{1}{2^3} + \frac{1}{2^4} - \frac{1}{2^5} - \frac{1}{2^6} + \cdots$$

converges absolutely. It follows from Theorem 11.7.4 that the series converges.   ◄

Example 6   Show that the series

$$\sum_{k=1}^{\infty} \frac{\cos k}{k^2}$$

converges.

Solution.   Since $|\cos k| \le 1$ for all k

$$\left| \frac{\cos k}{k^2} \right| \le \frac{1}{k^2}$$

Thus,

$$\sum_{k=1}^{\infty} \left| \frac{\cos k}{k^2} \right|$$

converges by the comparison test, and consequently

$$\sum_{k=1}^{\infty} \frac{\cos k}{k^2}$$

converges.

If $\Sigma|u_k|$ *diverges,* no conclusion can be drawn about the convergence or divergence of Σu_k. For example, consider the two series

$$1 - \frac{1}{2} + \frac{1}{3} - \frac{1}{4} + \cdots + (-1)^{k+1} \frac{1}{k} + \cdots \qquad (8)$$

$$-1 - \frac{1}{2} - \frac{1}{3} - \frac{1}{4} - \cdots - \frac{1}{k} - \cdots \qquad (9)$$

Series (8), the alternating harmonic series, converges; while series (9), being a constant times the harmonic series, diverges. Yet in each case the series of absolute values is

$$1 + \frac{1}{2} + \frac{1}{3} + \cdots + \frac{1}{k} + \cdots$$

which diverges. A series such as (8), which is convergent, but not absolutely convergent, is called ***conditionally convergent.***

The following version of the ratio test is useful for investigating absolute convergence.

The Ratio Test for Absolute Convergence

11.7.5   THEOREM. *Let Σu_k be a series with nonzero terms and suppose*

$$\lim_{k \to +\infty} \frac{|u_{k+1}|}{|u_k|} = \rho$$

(a)   *If $\rho < 1$, the series Σu_k converges absolutely.*
(b)   *If $\rho > 1$ or if $\rho = +\infty$, then the series Σu_k diverges.*
(c)   *If $\rho = 1$, no conclusion about convergence can be drawn from this test.*

The proof is discussed in the exercises.

Example 7 The series

$$\sum_{k=1}^{\infty} (-1)^k \frac{2^k}{k!}$$

converges absolutely since

$$\rho = \lim_{k \to +\infty} \frac{|u_{k+1}|}{|u_k|} = \lim_{k \to +\infty} \frac{2^{k+1}}{(k+1)!} \cdot \frac{k!}{2^k} = \lim_{k \to +\infty} \frac{2}{k+1} = 0 < 1 \qquad \blacktriangleleft$$

Example 8 We proved earlier (Theorem 11.3.3) that a geometric series

$$a + ar + ar^2 + \cdots + ar^{k-1} + \cdots$$

converges if $|r| < 1$ and diverges if $|r| \geq 1$. However, a stronger statement can be made—the series converges *absolutely* if $|r| < 1$. This follows from Theorem 11.7.5 since

$$\rho = \lim_{k \to +\infty} \frac{|u_{k+1}|}{|u_k|} = \lim_{k \to +\infty} \frac{|ar^k|}{|ar^{k-1}|} = \lim_{k \to +\infty} |r| = |r|$$

so that $\rho < 1$ if $|r| < 1$. $\qquad \blacktriangleleft$

The following review is included as a ready reference to convergence tests.

Review of Convergence Tests

NAME	STATEMENT	COMMENTS
Divergence Test (11.4.2)	If $\lim\limits_{k \to +\infty} u_k \neq 0$, then Σu_k diverges.	If $\lim\limits_{k \to +\infty} u_k = 0$, Σu_k may or may not converge.
Integral Test (11.4.5)	Let Σu_k be a series with positive terms and let $f(x)$ be the function that results when k is replaced by x in the formula for u_k. If f is decreasing and continuous for $x \geq 1$, then $$\sum_{k=1}^{\infty} u_k \quad \text{and} \quad \int_1^{+\infty} f(x)\, dx$$ both converge or both diverge.	Use this test when $f(x)$ is easy to integrate. This test only applies to series that have positive terms.

Review of Convergence Tests (Continued)

NAME	STATEMENT	COMMENTS
Comparison Test (11.5.1)	Let Σa_k and Σb_k be series with positive terms such that $$a_1 \leq b_1,\ a_2 \leq b_2, \ldots,\ a_k \leq b_k, \ldots$$ If Σb_k converges, then Σa_k converges, and if Σa_k diverges, then Σb_k diverges.	Use this test as a last resort. Other tests are often easier to apply. This test only applies to series with positive terms.
Ratio Test (11.5.2)	Let Σu_k be a series with positive terms and suppose $$\lim_{k \to +\infty} \frac{u_{k+1}}{u_k} = \rho$$ (a) Series converges if $\rho < 1$. (b) Series diverges if $\rho > 1$ or $\rho = +\infty$. (c) No conclusion if $\rho = 1$.	Try this test when u_k involves factorials or kth powers.
Root Test (11.5.3)	Let Σu_k be a series with positive terms such that $$\rho = \lim_{k \to +\infty} \sqrt[k]{u_k}$$ (a) Series converges if $\rho < 1$. (b) Series diverges if $\rho > 1$ or $\rho = +\infty$. (c) No conclusion if $\rho = 1$.	Try this test when u_k involves kth powers.
Limit Comparison Test (11.6.3)	Let Σa_k and Σb_k be series with positive terms such that $$\rho = \lim_{k \to +\infty} \frac{a_k}{b_k}$$ If $0 < \rho < +\infty$, then both series converge or both diverge.	This is easier to apply than the comparison test, but still requires some skill in choosing the series Σb_k for comparison.
Alternating Series Test (11.7.1)	The series $$a_1 - a_2 + a_3 - a_4 + \cdots$$ and $$-a_1 + a_2 - a_3 + a_4 - \cdots$$ converge if (a) $a_1 \geq a_2 \geq a_3 \geq \cdots$ (b) $\lim_{k \to +\infty} a_k = 0$	This test applies only to alternating series.

Review of Convergence Tests (Continued)

NAME	STATEMENT	COMMENTS
Ratio Test for Absolute Convergence (11.7.5)	Let Σu_k be a series with nonzero terms such that $$\rho = \lim_{k \to +\infty} \frac{|u_{k+1}|}{|u_k|}$$ (a) Series converges absolutely if $\rho < 1$. (b) Series diverges if $\rho > 1$ or $\rho = +\infty$. (c) No conclusion if $\rho = 1$.	The series need not have positive terms and need not be alternating to use this test.

▶ Exercise Set 11.7

In Exercises 1–6, use the alternating series test to determine whether the series converges or diverges.

1. $\displaystyle\sum_{k=1}^{\infty} \frac{(-1)^{k+1}}{2k+1}$.

2. $\displaystyle\sum_{k=1}^{\infty} (-1)^{k+1} \frac{k}{3^k}$.

3. $\displaystyle\sum_{k=1}^{\infty} (-1)^{k+1} \frac{k+1}{3k+1}$.

4. $\displaystyle\sum_{k=1}^{\infty} (-1)^{k+1} \frac{k+4}{k^2+k}$.

5. $\displaystyle\sum_{k=1}^{\infty} (-1)^{k+1} e^{-k}$.

6. $\displaystyle\sum_{k=3}^{\infty} (-1)^k \frac{\ln k}{k}$.

In Exercises 7–12, use the ratio test for absolute convergence to determine whether the series converges absolutely or diverges.

7. $\displaystyle\sum_{k=1}^{\infty} \left(-\frac{3}{5}\right)^k$.

8. $\displaystyle\sum_{k=1}^{\infty} (-1)^{k+1} \frac{2^k}{k!}$.

9. $\displaystyle\sum_{k=1}^{\infty} (-1)^{k+1} \frac{3^k}{k^2}$.

10. $\displaystyle\sum_{k=1}^{\infty} (-1)^k \left(\frac{k}{5^k}\right)$.

11. $\displaystyle\sum_{k=1}^{\infty} (-1)^k \left(\frac{k^3}{e^k}\right)$.

12. $\displaystyle\sum_{k=1}^{\infty} (-1)^{k+1} \frac{k^k}{k!}$.

In Exercises 13–30, classify the series as: absolutely convergent, conditionally convergent, or divergent.

13. $\displaystyle\sum_{k=1}^{\infty} \frac{(-1)^{k+1}}{3k}$.

14. $\displaystyle\sum_{k=1}^{\infty} \frac{(-1)^{k+1}}{k^{4/3}}$.

15. $\displaystyle\sum_{k=1}^{\infty} \frac{(-4)^k}{k^2}$.

16. $\displaystyle\sum_{k=1}^{\infty} \frac{(-1)^{k+1}}{k!}$.

17. $\displaystyle\sum_{k=1}^{\infty} \frac{\cos k\pi}{k}$.

18. $\displaystyle\sum_{k=3}^{\infty} \frac{(-1)^k \ln k}{k}$.

19. $\displaystyle\sum_{k=1}^{\infty} (-1)^{k+1} \left(\frac{k+2}{3k-1}\right)^k$.

20. $\displaystyle\sum_{k=1}^{\infty} \frac{(-1)^{k+1}}{k^2+1}$.

21. $\displaystyle\sum_{k=1}^{\infty} (-1)^{k+1} \frac{k+2}{k(k+3)}$.

22. $\displaystyle\sum_{k=1}^{\infty} \frac{(-1)^{k+1}k^2}{k^3+1}$.

23. $\displaystyle\sum_{k=1}^{\infty} \sin \frac{k\pi}{2}$.

24. $\displaystyle\sum_{k=1}^{\infty} \frac{\sin k}{k^3}$.

25. $\displaystyle\sum_{k=2}^{\infty} \frac{(-1)^k}{k \ln k}$.

26. $\displaystyle\sum_{k=1}^{\infty} \frac{(-1)^k}{\sqrt{k(k+1)}}$.

27. $\displaystyle\sum_{k=2}^{\infty} \left(-\frac{1}{\ln k}\right)^k$.

28. $\displaystyle\sum_{k=1}^{\infty} \frac{(-1)^{k+1}}{\sqrt{k+1}+\sqrt{k}}$.

29. $\displaystyle\sum_{k=2}^{\infty} \frac{(-1)^k(k^2+1)}{k^3+2}$.

30. $\displaystyle\sum_{k=1}^{\infty} \frac{k \cos k\pi}{k^2+1}$.

In Exercises 31–34, the given series satisfies the hypotheses of the alternating series test. For the stated value of n, estimate the error that results if the sum of the series is approximated by the nth partial sum.

31. $\displaystyle\sum_{k=1}^{\infty} \frac{(-1)^{k+1}}{k}$; $n = 7$.

32. $\displaystyle\sum_{k=1}^{\infty} \frac{(-1)^{k+1}}{k!}$; $n = 5$.

33. $\displaystyle\sum_{k=1}^{\infty}\frac{(-1)^{k+1}}{\sqrt{k}}$; $n = 99$.

34. $\displaystyle\sum_{k=1}^{\infty}\frac{(-1)^{k+1}}{(k+1)\ln(k+1)}$; $n = 3$.

In Exercises 35–38, the given series satisfies the hypotheses of the alternating series test. Find the smallest value of n for which the nth partial sum approximates the sum of the series to the stated accuracy.

35. $\displaystyle\sum_{k=1}^{\infty}\frac{(-1)^{k+1}}{k}$; $|\text{error}| < 0.0001$.

36. $\displaystyle\sum_{k=1}^{\infty}\frac{(-1)^{k+1}}{k!}$; $|\text{error}| < 0.00001$.

37. $\displaystyle\sum_{k=1}^{\infty}\frac{(-1)^{k+1}}{\sqrt{k}}$; $|\text{error}| < 0.005$.

38. $\displaystyle\sum_{k=1}^{\infty}\frac{(-1)^{k+1}}{(k+1)\ln(k+1)}$; $|\text{error}| < 0.1$.

39. Prove: If Σa_k converges absolutely, then Σa_k^2 converges.

40. Show that the converse of the result in Exercise 39 is false by finding a series for which Σa_k^2 converges, but $\Sigma |a_k|$ diverges.

41. Prove Theorem 11.7.1 for series of the form

$$-a_1 + a_2 - a_3 + a_4 - \cdots + (-1)^k a_k + \cdots$$

42. Prove Theorem 11.7.5. [*Hint:* Theorem 11.7.4 will help in part (*a*). For part (*b*), it may help to review the proof of Theorem 11.5.2.]

43. The sum of an absolutely convergent series is independent of the order in which the terms are added, but the terms of a conditionally convergent series can be rearranged to converge to any given value, or even diverge. For example, let S be the sum of the conditionally convergent alternating harmonic series,

$$S = 1 - \frac{1}{2} + \frac{1}{3} - \frac{1}{4} + \frac{1}{5} - \frac{1}{6} + \cdots$$

Rearrange the terms in this series to get

$$\left(1 - \frac{1}{2} - \frac{1}{4}\right) + \left(\frac{1}{3} - \frac{1}{6} - \frac{1}{8}\right)$$
$$+ \left(\frac{1}{5} - \frac{1}{10} - \frac{1}{12}\right) + \cdots$$

Show that this rearrangement results in a series that converges to $S/2$. [*Hint:* Add the first two terms within each pair of parentheses.]

11.8   POWER SERIES

In previous sections we studied series with constant terms. In this section we shall consider series whose terms involve variables. Such series are of fundamental importance in many branches of mathematics and the physical sciences.

POWER SERIES IN x If $c_0, c_1, c_2, \ldots$ are constants and x is a variable, then a series of the form

$$\sum_{k=0}^{\infty} c_k x^k = c_0 + c_1 x + c_2 x^2 + \cdots + c_k x^k + \cdots$$

is called a ***power series*** in x. Some examples are

$$\sum_{k=0}^{\infty} x^k = 1 + x + x^2 + x^3 + \cdots$$

$$\sum_{k=0}^{\infty} \frac{x^k}{k!} = 1 + x + \frac{x^2}{2!} + \frac{x^3}{3!} + \cdots$$

$$\sum_{k=0}^{\infty} (-1)^k \frac{x^{k+1}}{k+1} = x - \frac{x^2}{2} + \frac{x^3}{3} - \frac{x^4}{4} + \cdots$$

$$\sum_{k=0}^{\infty} (-1)^k \frac{x^{2k}}{(2k)!} = 1 - \frac{x^2}{2!} + \frac{x^4}{4!} - \frac{x^6}{6!} + \cdots$$

$$\sum_{k=0}^{\infty} (-1)^k \frac{x^{2k+1}}{(2k+1)!} = x - \frac{x^3}{3!} + \frac{x^5}{5!} - \frac{x^7}{7!} + \cdots$$

If a numerical value is substituted for x in a power series $\Sigma c_k x^k$, then we obtain a series of constants that may either converge or diverge. This leads to the following basic problem.

> **A Fundamental Problem.** For what values of x does a given power series, $\Sigma c_k x^k$, converge?

The following theorem is the fundamental result on convergence of power series. The proof can be found in most advanced calculus texts.

> **11.8.1** THEOREM. *For any power series in x, exactly one of the following is true:*
>
> (a) *The series converges only for $x = 0$.*
>
> (b) *The series converges absolutely for all x.*
>
> (c) *The series converges absolutely for all x in some finite open interval $(-R, R)$, and diverges if $x < -R$ or $x > R$ (Figure 11.8.1). At the points $x = R$ and $x = -R$ the series may converge absolutely, converge conditionally, or diverge, depending on the particular series.*

| Series diverges | Series converges absolutely | Series diverges |

Figure 11.8.1 ▷ $-R$ 0 R

RADIUS AND INTERVAL OF CONVERGENCE

In case (c), where the power series converges absolutely for $|x| < R$ and diverges for $|x| > R$, we call R the *radius of convergence*. In case (a), where the series converges only for $x = 0$, we define the radius of convergence to be $R = 0$; and in case (b), where the series converges absolutely for all x, we define the radius of convergence to be $R = +\infty$. The set of all values of x for which a power series converges is called the *interval of convergence*.

Example 1 Find the interval of convergence and radius of convergence of the power series

$$\sum_{k=0}^{\infty} x^k = 1 + x + x^2 + \cdots + x^k + \cdots$$

Solution. For every x, the given series is a geometric series with ratio $r = x$. Thus, by Example 8 of Section 11.7, the series converges absolutely if $-1 < x < 1$ and diverges if $|x| \geq 1$. Therefore, the interval of convergence is $(-1, 1)$ and the radius of convergence is $R = 1$. ◀

Example 2 Find the interval of convergence and radius of convergence of

$$\sum_{k=0}^{\infty} \frac{x^k}{k!}$$

Solution. We shall apply the ratio test for absolute convergence (11.7.5). For every real number x,

$$\rho = \lim_{k \to +\infty} \left| \frac{u_{k+1}}{u_k} \right| = \lim_{k \to +\infty} \left| \frac{x^{k+1}}{(k+1)!} \cdot \frac{k!}{x^k} \right| = \lim_{k \to +\infty} \left| \frac{x}{k+1} \right| = 0$$

Since $\rho < 1$ for all x, the series converges absolutely for all x. Thus, the interval of convergence is $(-\infty, +\infty)$ and the radius of convergence is $R = +\infty$. ◀

REMARK. There is a useful consequence of Example 2. Since

$$\sum_{k=0}^{\infty} \frac{x^k}{k!}$$

converges for all x, Theorem 11.4.1 implies that for all values of x

$$\lim_{k \to +\infty} \frac{x^k}{k!} = 0 \tag{1}$$

We shall need this result later.

Example 3 Find the interval of convergence and radius of convergence of

$$\sum_{k=0}^{\infty} k!x^k$$

Solution.   If $x = 0$, the series has only one nonzero term and therefore converges. If $x \neq 0$, the ratio test yields

$$\rho = \lim_{k \to +\infty} \left| \frac{u_{k+1}}{u_k} \right| = \lim_{k \to +\infty} \left| \frac{(k + 1)! x^{k+1}}{k! x^k} \right| = \lim_{k \to +\infty} |(k + 1)x| = +\infty$$

Therefore, the series converges if $x = 0$, but diverges for all other x. Consequently, the interval of convergence is the single point $x = 0$ and the radius of convergence is $R = 0$.   ◄

Example 4   Find the interval of convergence and radius of convergence of

$$\sum_{k=0}^{\infty} \frac{(-1)^k x^k}{3^k (k + 1)}$$

Solution.   Since $|(-1)^k| = |(-1)^{k+1}| = 1$, we obtain

$$\rho = \lim_{k \to +\infty} \left| \frac{u_{k+1}}{u_k} \right| = \lim_{k \to +\infty} \left| \frac{x^{k+1}}{3^{k+1}(k + 2)} \cdot \frac{3^k (k + 1)}{x^k} \right|$$

$$= \lim_{k \to +\infty} \left[\frac{|x|}{3} \cdot \left(\frac{k + 1}{k + 2} \right) \right]$$

$$= \frac{|x|}{3} \lim_{k \to +\infty} \left(\frac{1 + 1/k}{1 + 2/k} \right) = \frac{|x|}{3}$$

The ratio test for absolute convergence implies that the series converges absolutely if

$$|x| < 3$$

and diverges if

$$|x| > 3$$

The ratio test fails when

$$|x| = 3$$

so the cases $x = -3$ and $x = 3$ need separate analyses. Substituting $x = -3$ in the given series yields

$$\sum_{k=0}^{\infty} \frac{(-1)^k (-3)^k}{3^k (k + 1)} = \sum_{k=0}^{\infty} \frac{(-1)^k (-1)^k 3^k}{3^k (k + 1)} = \sum_{k=0}^{\infty} \frac{1}{k + 1}$$

which is the divergent harmonic series $1 + \frac{1}{2} + \frac{1}{3} + \frac{1}{4} + \cdots$. Substituting $x = 3$ in the given series yields

$$\sum_{k=0}^{\infty} \frac{(-1)^k 3^k}{3^k(k+1)} = \sum_{k=0}^{\infty} \frac{(-1)^k}{k+1}$$

which is the conditionally convergent alternating harmonic series $1 - \frac{1}{2} + \frac{1}{3} - \frac{1}{4} + \cdots$. Thus, the interval of convergence for the given series is $(-3, 3]$ and the radius of convergence is $R = 3$.   ◄

Example 5   Find the interval and radius of convergence of the series

$$\sum_{k=0}^{\infty} (-1)^k \frac{x^{2k}}{(2k)!}$$

Solution.   Since $|(-1)^k| = |(-1)^{k+1}| = 1$, we have

$$\rho = \lim_{k \to +\infty} \left| \frac{u_{k+1}}{u_k} \right| = \lim_{k \to +\infty} \left| \frac{x^{2(k+1)}}{[2(k+1)]!} \cdot \frac{(2k)!}{x^{2k}} \right| = \lim_{k \to +\infty} \left| \frac{x^{2k+2}}{(2k+2)!} \frac{(2k)!}{x^{2k}} \right|$$

$$= \lim_{k \to +\infty} \left| \frac{x^2}{(2k+2)(2k+1)} \right| = x^2 \lim_{k \to +\infty} \frac{1}{(2k+2)(2k+1)} = x^2 \cdot 0 = 0$$

Thus, $\rho < 1$ for all x, which means that the interval of convergence is $(-\infty, +\infty)$ and the radius of convergence is $R = +\infty$.   ◄

POWER SERIES IN
$x - a$

In addition to power series in x, we shall be interested in series of the form

$$\sum_{k=0}^{\infty} c_k(x-a)^k = c_0 + c_1(x-a) + c_2(x-a)^2 + \cdots + c_k(x-a)^k + \cdots$$

where $c_0, c_1, c_2, \ldots$ and a are constants. Such a series is called a *power series in $x - a$*. Some examples are

$$\sum_{k=0}^{\infty} \frac{(x-1)^k}{k+1} = 1 + \frac{(x-1)}{2} + \frac{(x-1)^2}{3} + \frac{(x-1)^3}{4} + \cdots$$

$$\sum_{k=0}^{\infty} \frac{(-1)^k(x+3)^k}{k!} = 1 - (x+3) + \frac{(x+3)^2}{2!} - \frac{(x+3)^3}{3!} + \cdots$$

The first is a power series in $x - 1$ ($a = 1$) and the second a power series in $x + 3$ ($a = -3$).

The convergence properties of a power series $\Sigma c_k(x-a)^k$ may be obtained from Theorem 11.8.1 by substituting $X = x - a$, to obtain

$$\sum_{k=0}^{\infty} a_k X^k$$

which is a power series in X. There are three possibilities for this series; it converges only when $X = 0$, or equivalently only when $x = a$; it converges absolutely for all values of X, or equivalently for all values of x; or finally, it converges absolutely for all X satisfying

$$-R < X < R \qquad (2)$$

and diverges when $X < -R$ or $X > R$. But (2) can be written as

$$-R < x - a < R$$

or

$$a - R < x < a + R$$

Thus, we are led to the following result.

11.8.2 THEOREM. *For a power series $\Sigma c_k(x - a)^k$, exactly one of the following is true:*

(a) *The series converges only for $x = a$.*

(b) *The series converges absolutely for all x.*

(c) *The series converges absolutely for all x in some finite open interval $(a - R, a + R)$ and diverges if $x < a - R$ or $x > a + R$ (Figure 11.8.2). At the points $x = a - R$ and $x = a + R$, the series may converge absolutely, converge conditionally, or diverge, depending on the particular series.*

Series diverges	Series converges absolutely	Series diverges
$a - R$	a	$a + R$

Figure 11.8.2 ▷

In cases (a), (b), and (c) of Theorem 11.8.2 the series is said to have *radius of convergence*, 0, $+\infty$, and R, respectively. The set of all values of x for which the series converges is called the *interval of convergence*.

Example 6 Find the interval of convergence and radius of convergence of the series

$$\sum_{k=1}^{\infty} \frac{(x - 5)^k}{k^2}$$

Solution. We apply the ratio test for absolute convergence.

$$\rho = \lim_{k \to +\infty} \left| \frac{u_{k+1}}{u_k} \right| = \lim_{k \to +\infty} \left| \frac{(x - 5)^{k+1}}{(k + 1)^2} \cdot \frac{k^2}{(x - 5)^k} \right|$$

$$= \lim_{k \to +\infty} \left[|x - 5| \left(\frac{k}{k + 1} \right)^2 \right]$$

$$= |x - 5| \lim_{k \to +\infty} \left(\frac{1}{1 + 1/k} \right)^2 = |x - 5|$$

Thus, the series converges absolutely if $|x - 5| < 1$, or $-1 < x - 5 < 1$, or $4 < x < 6$. The series diverges if $x < 4$ or $x > 6$.

To determine the convergence behavior at the endpoints $x = 4$ and $x = 6$, we substitute these values in the given series. If $x = 6$ the series becomes

$$\sum_{k=1}^{\infty} \frac{1^k}{k^2} = \sum_{k=1}^{\infty} \frac{1}{k^2} = 1 + \frac{1}{2^2} + \frac{1}{3^2} + \frac{1}{4^2} + \cdots$$

which is a convergent p-series ($p = 2$). If $x = 4$ the series becomes

$$\sum_{k=1}^{\infty} \frac{(-1)^k}{k^2} = -1 + \frac{1}{2^2} - \frac{1}{3^2} + \frac{1}{4^2} - \cdots$$

Since this series converges absolutely, the interval of convergence for the given series is $[4, 6]$. The radius of convergence is $R = 1$ (Figure 11.8.3).   ◀

Figure 11.8.3 ▷

Series diverges   Series converges absolutely   Series diverges

4                          $a = 5$ 6

|← $R = 1$ →|← $R = 1$ →|

▶ Exercise Set 11.8

In Exercises 1–24, find the radius of convergence and the interval of convergence.

1. $\displaystyle\sum_{k=0}^{\infty} \frac{x^k}{k + 1}$.

2. $\displaystyle\sum_{k=0}^{\infty} 3^k x^k$.

3. $\displaystyle\sum_{k=0}^{\infty} \frac{(-1)^k x^k}{k!}$.

4. $\displaystyle\sum_{k=0}^{\infty} \frac{k!}{2^k} x^k$.

5. $\displaystyle\sum_{k=1}^{\infty} \frac{5^k}{k^2} x^k$.

6. $\displaystyle\sum_{k=2}^{\infty} \frac{x^k}{\ln k}$.

7. $\displaystyle\sum_{k=1}^{\infty} \frac{x^k}{k(k + 1)}$.

8. $\displaystyle\sum_{k=0}^{\infty} \frac{(-2)^k x^{k+1}}{k + 1}$.

9. $\displaystyle\sum_{k=1}^{\infty} (-1)^{k-1} \frac{x^k}{\sqrt{k}}$.

10. $\displaystyle\sum_{k=0}^{\infty} \frac{(-1)^k x^{2k}}{(2k)!}$.

11. $\displaystyle\sum_{k=0}^{\infty} (-1)^k \frac{x^{2k+1}}{(2k + 1)!}$.

12. $\displaystyle\sum_{k=1}^{\infty} (-1)^k \frac{x^{3k}}{k^{3/2}}$.

13. $\displaystyle\sum_{k=0}^{\infty} \frac{3^k}{k!} x^k$.

14. $\displaystyle\sum_{k=2}^{\infty} (-1)^{k+1} \frac{x^k}{k(\ln k)^2}$.

15. $\displaystyle\sum_{k=0}^{\infty} \frac{x^k}{1 + k^2}$.

16. $\displaystyle\sum_{k=0}^{\infty} \frac{(x - 3)^k}{2^k}$.

17. $\displaystyle\sum_{k=1}^{\infty} (-1)^{k+1} \frac{(x + 1)^k}{k}$.

18. $\displaystyle\sum_{k=0}^{\infty} (-1)^k \frac{(x - 4)^k}{(k + 1)^2}$.

19. $\displaystyle\sum_{k=0}^{\infty} \left(\frac{3}{4} \right)^k (x + 5)^k$.

20. $\displaystyle\sum_{k=1}^{\infty} \frac{(2k + 1)!}{k^3} (x - 2)^k$.

21. $\displaystyle\sum_{k=1}^{\infty} (-1)^k \frac{(x + 1)^{2k+1}}{k^2 + 4}$.

22. $\displaystyle\sum_{k=1}^{\infty} \frac{(\ln k)(x - 3)^k}{k}$.

23. $\displaystyle\sum_{k=0}^{\infty} \frac{\pi^k(x - 1)^{2k}}{(2k + 1)!}$. **24.** $\displaystyle\sum_{k=0}^{\infty} \frac{(2x - 3)^k}{4^{2k}}$.

25. Use the root test to find the interval of convergence of $\displaystyle\sum_{k=2}^{\infty} \frac{x^k}{(\ln k)^k}$.

26. Find the radius of convergence of

$$\sum_{k=1}^{\infty} (-1)^k \frac{1 \cdot 2 \cdot 3 \cdots k}{1 \cdot 3 \cdot 5 \cdots (2k - 1)} x^{2k+1}$$

27. Find the interval of convergence of

$$\sum_{k=0}^{\infty} \frac{(x - a)^k}{b^k}$$

where $b > 0$.

28. Find the radius of convergence of the power series $\displaystyle\sum_{k=0}^{\infty} \frac{(pk)!}{(k!)^p} x^k$, where p is a positive integer.

29. Find the radius of convergence of the power series $\displaystyle\sum_{k=0}^{\infty} \frac{(k + p)!}{k!(k + q)!} x^k$, where p and q are positive integers.

30. Prove: If $\displaystyle\lim_{k \to +\infty} |c_k|^{1/k} = L$, where $L \neq 0$, then $1/L$ is the radius of convergence of the power series $\displaystyle\sum_{k=0}^{\infty} c_k x^k$.

31. Prove: If the power series $\displaystyle\sum_{k=0}^{\infty} c_k x^k$ has radius of convergence R, then the series $\displaystyle\sum_{k=0}^{\infty} c_k x^{2k}$ has radius of convergence $\sqrt{R}$.

32. Prove: If the interval of convergence of the series $\displaystyle\sum_{k=0}^{\infty} c_k(x - a)^k$ is $(a - R, a + R]$, then the series converges conditionally at $a + R$.

11.9 TAYLOR AND MACLAURIN SERIES

One of the early applications of calculus was the calculation of values for functions such as $\sin x$, $\ln x$, and e^x. The basic idea is to approximate the given function by a polynomial in such a way that the resulting error is within some specified tolerance. In this section we shall study the approximation of functions by polynomials and introduce an important class of power series.

APPROXIMATING FUNCTIONS BY POLYNOMIALS

Suppose we are interested in approximating a function f by a polynomial

$$p(x) = c_0 + c_1 x + \cdots + c_n x^n \tag{1}$$

over an interval centered at $x = 0$. Because $p(x)$ has $n + 1$ coefficients, it seems reasonable that we will be able to impose $n + 1$ conditions on this polynomial. We shall assume that the first n derivatives of f exist at $x = 0$, and we shall choose these $n + 1$ conditions to be:

$$f(0) = p(0), \ f'(0) = p'(0), \ f''(0) = p''(0), \ldots, \ f^{(n)}(0) = p^{(n)}(0) \quad (2)$$

These conditions require that the value of $p(x)$ and its first n derivatives match the value of $f(x)$ and its first n derivatives at $x = 0$. By forcing this high degree of "match" at $x = 0$, it is reasonable to hope that $f(x)$ and $p(x)$ will remain close over some interval (possibly small) centered at $x = 0$. Since

$$
\begin{aligned}
p(x) &= c_0 + c_1 x + c_2 x^2 + c_3 x^3 + \cdots + c_n x^n \\
p'(x) &= c_1 + 2c_2 x + 3c_3 x^2 + \cdots + nc_n x^{n-1} \\
p''(x) &= 2c_2 + 3 \cdot 2c_3 x + \cdots + n(n-1)c_n x^{n-2} \\
p'''(x) &= 3 \cdot 2c_3 + \cdots + n(n-1)(n-2)c_n x^{x-3} \\
&\vdots \\
p^{(n)}(x) &= n(n-1)(n-2) \cdots (1)c_n
\end{aligned}
$$

we obtain on substituting $x = 0$

$$
\begin{aligned}
p(0) &= c_0 \\
p'(0) &= c_1 \\
p''(0) &= 2c_2 = 2!c_2 \\
p'''(0) &= 3 \cdot 2c_3 = 3!c_3 \\
&\vdots \\
p^{(n)}(0) &= n(n-1)(n-2) \cdots (1)c_n = n!c_n
\end{aligned}
$$

Thus, from (2),

$$
\begin{aligned}
f(0) &= c_0 \\
f'(0) &= c_1 \\
f''(0) &= 2!c_2 \\
f'''(0) &= 3!c_3 \\
&\vdots \\
f^{(n)}(0) &= n!c_n
\end{aligned}
$$

so

$$c_0 = f(0), \ c_1 = f'(0), \ c_2 = \frac{f''(0)}{2!}, \ c_3 = \frac{f'''(0)}{3!}, \ldots, \ c_n = \frac{f^{(n)}(0)}{n!}$$

MACLAURIN
POLYNOMIALS

Substituting these values in (1) yields a polynomial, called the nth *Maclaurin* polynomial for f.*

11.9.1 DEFINITION. If f can be differentiated n times at 0, then we define the nth **Maclaurin polynomial** for f to be

$$p_n(x) = f(0) + f'(0)x + \frac{f''(0)}{2!}x^2 + \frac{f'''(0)}{3!}x^3 + \cdots + \frac{f^{(n)}(0)}{n!}x^n \quad (3)$$

This polynomial has the property that its value and the values of its first n derivatives match the value of $f(x)$ and its first n derivatives when $x = 0$.

Example 1 Find the Maclaurin polynomials p_0, p_1, p_2, p_3, and p_n for e^x.

Solution. Let $f(x) = e^x$. Thus,

$$f'(x) = f''(x) = f'''(x) = \cdots = f^{(n)}(x) = e^x$$

and

$$f(0) = f'(0) = f''(0) = f'''(0) = \cdots = f^{(n)}(0) = e^0 = 1$$

Therefore,

$$p_0(x) = f(0) = 1$$
$$p_1(x) = f(0) + f'(0)x = 1 + x$$
$$p_2(x) = f(0) + f'(0)x + \frac{f''(0)}{2!}x^2 = 1 + x + \frac{x^2}{2!} = 1 + x + \frac{1}{2}x^2$$

---

* COLIN MACLAURIN (1698–1746) Scottish mathematician. Maclaurin's father, a minister, died when the boy was only six months old, and his mother when he was nine years old. He was then raised by an uncle who was also a minister.

Maclaurin entered Glasgow University as a divinity student, but transferred to mathematics after one year. He received his Master's degree at age 17 and, in spite of his youth, began teaching at Marischal College in Aberdeen, Scotland.

Maclaurin met Isaac Newton during a visit to London in 1719 and from that time on he became Newton's disciple. During that era, some of Newton's analytic methods were bitterly attacked by major mathematicians and much of Maclaurin's important mathematical work resulted from his efforts to defend Newton's ideas geometrically. Maclaurin's work, *A Treatise of Fluxions* (1742), was the first systematic formulation of Newton's methods. The treatise was so carefully done that it was a standard of mathematical rigor in calculus until the work of Cauchy in 1821.

Maclaurin was an outstanding experimentalist. He devised numerous ingenious mechanical devices, made important astronomical observations, performed actuarial computations for insurance societies, and helped to improve maps of the islands around Scotland.

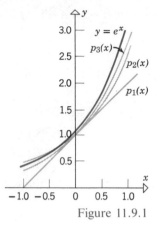

Figure 11.9.1

$$p_3(x) = f(0) + f'(0)x + \frac{f''(0)}{2!}x^2 + \frac{f'''(0)}{3!}x^3$$

$$= 1 + x + \frac{x^2}{2!} + \frac{x^3}{3!} = 1 + x + \frac{1}{2}x^2 + \frac{1}{6}x^3$$

$$p_n(x) = f(0) + f'(0)x + \frac{f''(0)}{2!}x^2 + \cdots + \frac{f^{(n)}(0)}{n!}x^n$$

$$= 1 + x + \frac{x^2}{2!} + \cdots + \frac{x^n}{n!} \quad \blacktriangleleft$$

In Figure 11.9.1 we have sketched the graphs of e^x and its Maclaurin polynomials of degree 1, 2, and 3. Note that the graphs of e^x and $p_3(x)$ are virtually indistinguishable over the interval from -0.5 to $+0.5$. (In the next section we shall investigate in detail the accuracy of Maclaurin polynomial approximations.)

Example 2 Find the nth Maclaurin polynomial for $\ln(x + 1)$.

Solution. Let $f(x) = \ln(x + 1)$ and arrange the computations as follows:

$$f(x) = \ln(x + 1) \qquad\qquad f(0) = \ln 1 = 0$$

$$f'(x) = \frac{1}{x + 1} \qquad\qquad f'(0) = 1$$

$$f''(x) = -\frac{1}{(x + 1)^2} \qquad\qquad f''(0) = -1$$

$$f'''(x) = \frac{2}{(x + 1)^3} \qquad\qquad f'''(0) = 2$$

$$f^{(4)}(x) = -\frac{3 \cdot 2}{(x + 1)^4} \qquad\qquad f^{(4)}(0) = -3!$$

$$f^{(5)}(x) = \frac{4 \cdot 3 \cdot 2}{(x + 1)^5} \qquad\qquad f^{(5)}(0) = 4!$$

$$\vdots \qquad\qquad\qquad\qquad \vdots$$

$$f^{(n)}(x) = (-1)^{n+1}\frac{(n - 1)!}{(x + 1)^n} \qquad f^{(n)}(0) = (-1)^{n+1}(n - 1)!$$

Substituting these values in (3) yields

$$p_n(x) = x - \frac{x^2}{2} + \frac{x^3}{3} - \cdots + (-1)^{n+1}\frac{x^n}{n} \quad \blacktriangleleft$$

Example 3 In the Maclaurin polynomials for $\sin x$, only the odd powers of x appear explicitly. To see this, let $f(x) = \sin x$, thus,

$$f(x) \quad = \sin x \qquad f(0) \quad = 0$$
$$f'(x) \quad = \cos x \qquad f'(0) \quad = 1$$
$$f''(x) \quad = -\sin x \qquad f''(0) \quad = 0$$
$$f'''(x) = -\cos x \qquad f'''(0) = -1$$

Since $f^{(4)}(x) = \sin x = f(x)$, the pattern $0, 1, 0, -1$ will repeat over and over as we evaluate successive derivatives at 0. Therefore, the successive Maclaurin polynomials for $\sin x$ are

$$p_1(x) = 0 + x = x$$

$$p_2(x) = 0 + x + 0 = x$$

$$p_3(x) = 0 + x + 0 - \frac{x^3}{3!} = x - \frac{x^3}{3!}$$

$$p_4(x) = 0 + x + 0 - \frac{x^3}{3!} + 0 = x - \frac{x^3}{3!}$$

$$p_5(x) = 0 + x + 0 - \frac{x^3}{3!} + 0 + \frac{x^5}{5!} = x - \frac{x^3}{3!} + \frac{x^5}{5!}$$

$$p_6(x) = 0 + x + 0 - \frac{x^3}{3!} + 0 + \frac{x^5}{5!} + 0 = x - \frac{x^3}{3!} + \frac{x^5}{5!}$$

$$p_7(x) = 0 + x + 0 - \frac{x^3}{3!} + 0 + \frac{x^5}{5!} + 0 - \frac{x^7}{7!} = x - \frac{x^3}{3!} + \frac{x^5}{5!} - \frac{x^7}{7!}$$

$$\vdots$$

In general, the Maclaurin polynomials for $\sin x$ are

$$p_{2n+1}(x) = p_{2n+2}(x) = x - \frac{x^3}{3!} + \frac{x^5}{5!} - \frac{x^7}{7!} + \cdots + (-1)^n \frac{x^{2n+1}}{(2n+1)!}$$

$(n = 0, 1, 2, \ldots)$.   ◀

Example 4 In the Maclaurin polynomials for $\cos x$, only the even powers of x appear explicitly. Using computations similar to those in Example 3, the reader should be able to show that the successive Maclaurin polynomials for $\cos x$ are

$$p_0(x) = p_1(x) = 1$$

$$p_2(x) = p_3(x) = 1 - \frac{x^2}{2!}$$

$$p_4(x) = p_5(x) = 1 - \frac{x^2}{2!} + \frac{x^4}{4!}$$

$$p_6(x) = p_7(x) = 1 - \frac{x^2}{2!} + \frac{x^4}{4!} - \frac{x^6}{6!}$$

$$p_8(x) = p_9(x) = 1 - \frac{x^2}{2!} + \frac{x^4}{4!} - \frac{x^6}{6!} + \frac{x^8}{8!}$$

In general, the Maclaurin polynomials for $\cos x$ are

$$p_{2n}(x) = p_{2n+1}(x) = 1 - \frac{x^2}{2!} + \frac{x^4}{4!} - \cdots + (-1)^n \frac{x^{2n}}{(2n)!}$$

$(n = 0, 1, 2, \ldots)$.   ◀

TAYLOR POLYNOMIALS

If we are interested in a polynomial approximation to $f(x)$ on an interval centered at $x = a$, then the idea is to choose the polynomial $p(x)$ so that the values of $p(x)$ and its first n derivatives match the values of $f(x)$ and its first n derivatives at $x = a$. The computations are simplest if the approximating polynomial is expressed in the form

$$p(x) = c_0 + c_1(x - a) + c_2(x - a)^2 + \cdots + c_n(x - a)^n \tag{4}$$

We leave it as an exercise for the reader to calculate the first n derivatives of $p(x)$ and show

$$p(a) = c_0, \ p'(a) = c_1, \ p''(a) = 2!c_2, \ p'''(a) = 3!c_3, \ldots, p^{(n)}(a) = n!c_n$$

Thus, if we want the values of $p(x)$ and its first n derivatives to match the values of $f(x)$ and its first n derivatives at $x = a$, we must have

$$c_0 = f(a), \ c_1 = f'(a), \ c_2 = \frac{f''(a)}{2!}, \ c_3 = \frac{f'''(a)}{3!}, \ldots, c_n = \frac{f^{(n)}(a)}{n!}$$

Substituting these values in (4) we obtain a polynomial called the *nth Taylor* polynomial about $x = a$ for f.*

---

*BROOK TAYLOR (1685–1731) English mathematician. Taylor was born of well-to-do parents. Musicians and artists were entertained frequently in the Taylor home, which undoubtedly had a lasting influence on young Brook. In later years, Taylor published a definitive work on the mathematical theory of perspective and obtained major mathematical results about the vibrations of strings. There also exists an unpublished work, *On Musick,* that was intended to be part of a joint paper with Isaac Newton.

Taylor's life was scarred with unhappiness, illness, and tragedy. Because his first wife was not rich enough to suit his father, the two men argued bitterly and parted ways. Subsequently, his wife died in childbirth. Then, after he remarried, his second wife also died in childbirth, though his daughter survived.

Taylor's most productive period was from 1714 to 1719, during which time he wrote on a wide range of subjects—magnetism, capillary action, thermometers, perspective, and calculus. In his final years, Taylor devoted his writing efforts to religion and philosophy. According to Taylor, the results that bear his name were motivated by coffeehouse conversations about works of Newton on planetary motion and works of Halley ("Halley's comet") on roots of polynomials.

Taylor's writing style was so terse and hard to understand that he never received credit for many of his innovations.

11.9.2 DEFINITION. *If f can be differentiated n times at a, then we define the **nth Taylor polynomial for f about x = a** to be*

$$p_n(x) = f(a) + f'(a)(x - a) + \frac{f''(a)}{2!}(x - a)^2$$

$$+ \frac{f'''(a)}{3!}(x - a)^3 + \cdots + \frac{f^{(n)}(a)}{n!}(x - a)^n \quad (5)$$

Observe that the Taylor polynomials include the Maclaurin polynomials as a special case (let $a = 0$).

Example 5 Find the Taylor polynomials $p_1(x)$, $p_2(x)$, and $p_3(x)$ for $\sin x$ about $x = \pi/3$.

Solution.   Let $f(x) = \sin x$. Thus,

$$f(x) \;\; = \sin x \qquad f(\pi/3) \;\; = \sin\frac{\pi}{3} = \frac{\sqrt{3}}{2}$$

$$f'(x) = \cos x \qquad f'(\pi/3) = \cos\frac{\pi}{3} = \frac{1}{2}$$

$$f''(x) = -\sin x \qquad f''(\pi/3) = -\sin\frac{\pi}{3} = -\frac{\sqrt{3}}{2}$$

$$f'''(x) = -\cos x \qquad f'''(\pi/3) = -\cos\frac{\pi}{3} = -\frac{1}{2}$$

Substituting in (5) with $a = \pi/3$ yields

$$p_1(x) = f(\pi/3) + f'(\pi/3)\left(x - \frac{\pi}{3}\right) = \frac{\sqrt{3}}{2} + \frac{1}{2}\left(x - \frac{\pi}{3}\right)$$

$$p_2(x) = f(\pi/3) + f'(\pi/3)\left(x - \frac{\pi}{3}\right) + \frac{f''(\pi/3)}{2!}\left(x - \frac{\pi}{3}\right)^2$$

$$= \frac{\sqrt{3}}{2} + \frac{1}{2}\left(x - \frac{\pi}{3}\right) - \frac{\sqrt{3}}{2\cdot2!}\left(x - \frac{\pi}{3}\right)^2$$

$$p_3(x) = f(\pi/3) + f'(\pi/3)\left(x - \frac{\pi}{3}\right) + \frac{f''(\pi/3)}{2!}\left(x - \frac{\pi}{3}\right)^2 + \frac{f'''(\pi/3)}{3!}\left(x - \frac{\pi}{3}\right)^3$$

$$= \frac{\sqrt{3}}{2} + \frac{1}{2}\left(x - \frac{\pi}{3}\right) - \frac{\sqrt{3}}{2\cdot2!}\left(x - \frac{\pi}{3}\right)^2 - \frac{1}{2\cdot3!}\left(x - \frac{\pi}{3}\right)^3 \quad \blacktriangleleft$$

Frequently, it is convenient to express the defining formula for the Taylor polynomial in sigma notation. To do this, we use the notation $f^{(k)}(a)$ to denote the kth derivative of f at $x = a$, and we make the added convention that $f^{(0)}(a)$ denotes $f(a)$. This enables us to write

$$\sum_{k=0}^{n} \frac{f^{(k)}(a)}{k!}(x-a)^k = f(a) + f'(a)(x-a)$$
$$+ \frac{f''(a)}{2!}(x-a)^2 + \cdots + \frac{f^{(n)}(a)}{n!}(x-a)^n$$

In particular, the nth Maclaurin polynomial for $f(x)$ may be written

$$\sum_{k=0}^{n} \frac{f^{(k)}(0)}{k!}x^k = f(0) + f'(0)x + \frac{f''(0)}{2!}x^2 + \cdots + \frac{f^{(n)}(0)}{n!}x^n$$

TAYLOR SERIES Because the value of f and its first n derivatives match the value of the Taylor polynomial and its first n derivatives at $x = a$, we might hope that as n increases the Taylor polynomials for f about $x = a$ will become better and better approximations to $f(x)$—at least in some interval centered at $x = a$. This raises the problem of finding those values of x for which the Taylor polynomials converge to $f(x)$ as $n \to +\infty$. In other words, for which values of x is it true that

$$f(x) = \lim_{n \to +\infty} \sum_{k=0}^{n} \frac{f^{(k)}(a)}{k!}(x-a)^k = \sum_{k=0}^{\infty} \frac{f^{(k)}(a)}{k!}(x-a)^k \qquad (6)$$

The power series in (6) is called the *Taylor series* about $x = a$ for the function f.

11.9.3 DEFINITION. If f has derivatives of all orders at a, then we define the **Taylor series for f about x = a** to be

$$\sum_{k=0}^{\infty} \frac{f^{(k)}(a)}{k!}(x-a)^k = f(a) + f'(a)(x-a)$$
$$+ \frac{f''(a)}{2!}(x-a)^2 + \cdots + \frac{f^{(k)}(a)}{k!}(x-a)^k + \cdots \qquad (7)$$

11.9.4 DEFINITION. If f has derivatives of all orders at 0, then we define the **Maclaurin series for f** to be

$$\sum_{k=0}^{\infty} \frac{f^{(k)}(0)}{k!}x^k = f(0) + f'(0)x + \frac{f''(0)}{2!}x^2 + \cdots + \frac{f^{(k)}(0)}{k!}x^k + \cdots \qquad (8)$$

Observe that the Maclaurin series for f is just the Taylor series for f about $a = 0$.

Example 6 In Example 1, we found the nth Maclaurin polynomial for e^x to be

$$\sum_{k=0}^{n} \frac{x^k}{k!} = 1 + x + \frac{x^2}{2!} + \cdots + \frac{x^n}{n!}$$

Thus, the Maclaurin series for e^x is

$$\sum_{k=0}^{\infty} \frac{x^k}{k!} = 1 + x + \frac{x^2}{2!} + \frac{x^3}{3!} + \cdots + \frac{x^n}{n!} + \cdots$$

Similarly, from Example 3 the Maclaurin series for $\sin x$ is

$$\sum_{k=0}^{\infty} (-1)^k \frac{x^{2k+1}}{(2k+1)!} = x - \frac{x^3}{3!} + \frac{x^5}{5!} - \frac{x^7}{7!} + \cdots$$

and from Example 4 the Maclaurin series for $\cos x$ is

$$\sum_{k=0}^{\infty} (-1)^k \frac{x^{2k}}{(2k)!} = 1 - \frac{x^2}{2!} + \frac{x^4}{4!} - \frac{x^6}{6!} + \cdots \qquad \blacktriangleleft$$

Example 7 Find the Taylor series about $x = 1$ for $1/x$.

Solution. Let $f(x) = 1/x$ so that

$$f(x) \quad = \frac{1}{x} \qquad\qquad f(1) \quad = 1$$

$$f'(x) \quad = -\frac{1}{x^2} \qquad\qquad f'(1) \quad = -1$$

$$f''(x) \quad = \frac{2}{x^3} \qquad\qquad f''(1) \quad = 2!$$

$$f'''(x) \quad = -\frac{3 \cdot 2}{x^4} \qquad\qquad f'''(1) \quad = -3!$$

$$f^{(4)}(x) = \frac{4 \cdot 3 \cdot 2}{x^5} \qquad\qquad f^{(4)}(1) = 4!$$

$$\vdots \qquad\qquad\qquad \vdots$$

$$f^{(k)}(x) = (-1)^k \frac{k!}{x^{k+1}} \qquad f^{(k)}(1) = (-1)^k k!$$

$$\vdots \qquad\qquad\qquad \vdots$$

Thus, substituting in (7) with $a = 1$ yields

$$\sum_{k=0}^{\infty} \frac{(-1)^k k!}{k!} (x - 1)^k = \sum_{k=0}^{\infty} (-1)^k (x - 1)^k$$

$$= 1 - (x - 1) + (x - 1)^2 - (x - 1)^3 + \cdots \qquad \blacktriangleleft$$

Convergence of Taylor and Maclaurin series will be considered in the next section.

▶ Exercise Set 11.9

In Exercises 1–12, find the fourth Maclaurin polynomial ($n = 4$) for the given function.

1. e^{-2x}.

2. $\dfrac{1}{1 + x}$.

3. $\sin 2x$.

4. $e^x \cos x$.

5. $\tan x$.

6. $x^3 - x^2 + 2x + 1$.

7. xe^x.

8. $\tan^{-1} x$.

9. $\sec x$.

10. $\sqrt{1 + x}$.

11. $\ln(3 + 2x)$.

12. $\sinh x$.

In Exercises 13–22, find the third Taylor polynomial ($n = 3$) about $x = a$ for the given function.

13. e^x; $a = 1$.

14. $\ln x$; $a = 1$.

15. $\sqrt{x}$; $a = 4$.

16. $x^4 + x - 3$; $a = -2$.

17. $\cos x$; $a = \dfrac{\pi}{4}$.

18. $\tan x$; $a = \dfrac{\pi}{3}$.

19. $\sin \pi x$; $a = -\dfrac{1}{3}$.

20. $\csc x$; $a = \dfrac{\pi}{2}$.

21. $\tan^{-1} x$; $a = 1$.

22. $\cosh x$; $a = \ln 2$.

In Exercises 23–31, find the Maclaurin series for the given function. Express your answer in sigma notation.

23. e^{-x}.

24. e^{ax}.

25. $\dfrac{1}{1 + x}$.

26. xe^x.

27. $\ln(1 + x)$.

28. $\sin \pi x$.

29. $\cos\left(\dfrac{x}{2}\right)$.

30. $\sinh x$.

31. $\cosh x$.

In Exercises 32–39, find the Taylor series about $x = a$ for the given function. Express your answer in sigma notation.

32. $\dfrac{1}{x}$; $a = 3$.

33. $\dfrac{1}{x}$; $a = -1$.

34. e^x; $a = 2$.

35. $\ln x$; $a = 1$.

36. $\cos x$; $a = \dfrac{\pi}{2}$.

37. $\sin \pi x$; $a = \dfrac{1}{2}$.

38. $\dfrac{1}{x + 2}$; $a = 3$.

39. $\sinh x$; $a = \ln 4$.

40. Prove: The value of $p_n(x) = \displaystyle\sum_{k=0}^{n} \dfrac{f^{(k)}(a)}{k!}(x - a)^k$ and its first n derivatives match the value of $f(x)$ and its first n derivatives at $x = a$.

11.10 TAYLOR FORMULA WITH REMAINDER; CONVERGENCE OF TAYLOR SERIES

In this section we shall analyze the error that results when a function f is approximated by a Taylor or Maclaurin polynomial. This work will enable us to investigate convergence properties of Taylor and Maclaurin series.

If we approximate a function f by its nth Taylor polynomial p_n, then the error at a point x is represented by the difference $f(x) - p_n(x)$. This difference is commonly called the **nth remainder** and is denoted by

$$R_n(x) = f(x) - p_n(x)$$

The following theorem gives an important explicit formula for this remainder.

Taylor's Theorem

11.10.1 THEOREM. *Suppose that a function f can be differentiated $n + 1$ times at each point in an interval containing the point a, and let*

$$p_n(x) = f(a) + f'(a)(x - a) + \frac{f''(a)}{2!}(x - a)^2 + \cdots + \frac{f^{(n)}(a)}{n!}(x - a)^n$$

be the nth Taylor polynomial about $x = a$ for f. Then for each x in the interval, there is at least one point c between a and x such that

$$R_n(x) = f(x) - p_n(x) = \frac{f^{(n+1)}(c)}{(n + 1)!}(x - a)^{n+1} \tag{1}$$

In the foregoing theorem, the statement that c is "between" a and x means that c is in the interval (a, x) if $a < x$, or in (x, a) if $x < a$, or $c = a = x$ if $a = x$.

Rewriting (1) as

$$f(x) = p_n(x) + \frac{f^{(n+1)}(c)}{(n + 1)!}(x - a)^{n+1}$$

then substituting the expression for $p_n(x)$ stated in the theorem yields the following result, called **Taylor's formula with remainder.**

$$f(x) = f(a) + f'(a)(x - a) + \frac{f''(a)}{2!}(x - a)^2 + \cdots$$
$$+ \frac{f^{(n)}(a)}{n!}(x - a)^n + \frac{f^{(n+1)}(c)}{(n + 1)!}(x - a)^{n+1} \tag{2}$$

This formula expresses $f(x)$ as the sum of its nth Taylor polynomial about $x = a$ plus a remainder or error term,

$$R_n(x) = \frac{f^{(n+1)}(c)}{(n+1)!}(x-a)^{n+1} \qquad (3)$$

in which c is some unspecified number between a and x. The value of c depends on a, x, and n. Historically, (3) was not due to Taylor, but rather to Joseph Louis Lagrange.* For this reason, (3) is frequently called *Lagrange's form of the remainder.* Other formulas for the remainder can be found in any advanced calculus text. The proof of Taylor's Theorem will be deferred until the end of the section.

For Maclaurin polynomials ($a = 0$), (2) and (3) become:

$$f(x) = f(0) + f'(0)x + \frac{f''(0)}{2!}x^2 + \cdots + \frac{f^{(n)}(0)}{n!}x^n + \frac{f^{(n+1)}(c)}{(n+1)!}x^{n+1} \qquad (2a)$$

$$R_n(x) = \frac{f^{(n+1)}(c)}{(n+1)!}x^{n+1} \quad \text{where } c \text{ is between 0 and } x \qquad (3a)$$

---

*JOSEPH LOUIS LAGRANGE (1736–1813) French–Italian mathematician and astronomer. Lagrange, the son of a public official, was born in Turin, Italy. (Baptismal records list his name as Giuseppe Lodovico Lagrangia.) Although his father wanted him to be a lawyer, Lagrange was attracted to mathematics and astronomy after reading a memoir by the astronomer Halley. At age 16 he began to study mathematics on his own and by age 19 was appointed to a professorship at the Royal Artillery School in Turin. The following year Lagrange sent Euler solutions to some famous problems using new methods that eventually blossomed into a branch of mathematics called calculus of variations. These methods and Lagrange's applications of them to problems in celestial mechanics were so monumental that by age 25 he was regarded by many of his contemporaries as the greatest living mathematician.

In 1776, on the recommendation of Euler, he was chosen to succeed Euler as the director of the Berlin Academy. During his stay in Berlin, Lagrange distinguished himself not only in celestial mechanics, but also in algebraic equations and the theory of numbers. After twenty years in Berlin, he moved to Paris at the invitation of Louis XVI. He was given apartments in the Louvre and treated with great honor, even during the revolution.

Napoleon was a great admirer of Lagrange and showered him with honors—count, senator, and Legion of Honor. The years Lagrange spent in Paris were devoted primarily to didactic treatises summarizing his mathematical conceptions. One of Lagrange's most famous works is a memoir, *Mécanique Analytique,* in which he reduced the theory of mechanics to a few general formulas from which all other necessary equations could be derived.

It is an interesting historical fact that Lagrange's father speculated unsuccessfully in several financial ventures, so his family was forced to live quite modestly. Lagrange himself stated that if his family had money, he would not have made mathematics his vocation.

In spite of his fame, Lagrange was always a shy and modest man. On his death, he was buried with honor in the Pantheon.

Example 1 From Example 1 of the previous section, the nth Maclaurin polynomial for $f(x) = e^x$ is

$$1 + x + \frac{x^2}{2!} + \cdots + \frac{x^n}{n!}$$

Since $f^{(n+1)}(x) = e^x$, it follows that

$$f^{(n+1)}(c) = e^c$$

Thus, from Taylor's formula with remainder (2a)

$$e^x = 1 + x + \frac{x^2}{2!} + \cdots + \frac{x^n}{n!} + \frac{e^c}{(n+1)!} x^{n+1} \qquad (4)$$

where c is between 0 and x. Note that (4) is valid for all real values of x since the hypotheses of Taylor's Theorem are satisfied on the interval $(-\infty, +\infty)$ (verify). ◀

Example 2 In Example 5 of the previous section, we found the Taylor polynomial of degree 3 about $x = \pi/3$ for $f(x) = \sin x$ to be

$$\frac{\sqrt{3}}{2} + \frac{1}{2}\left(x - \frac{\pi}{3}\right) - \frac{\sqrt{3}}{4}\left(x - \frac{\pi}{3}\right)^2 - \frac{1}{12}\left(x - \frac{\pi}{3}\right)^3$$

Since $f^{(4)}(x) = \sin x$ (verify) it follows that $f^{(4)}(c) = \sin c$. Thus, from Taylor's formula with remainder (2) in the case $a = \pi/3$ and $n = 3$ we obtain

$$\sin x = \frac{\sqrt{3}}{2} + \frac{1}{2}\left(x - \frac{\pi}{3}\right) - \frac{\sqrt{3}}{4}\left(x - \frac{\pi}{3}\right)^2 - \frac{1}{12}\left(x - \frac{\pi}{3}\right)^3$$

$$+ \frac{\sin c}{4!}\left(x - \frac{\pi}{3}\right)^4$$

where c is some number between $\pi/3$ and x. ◀

CONVERGENCE OF TAYLOR SERIES In the previous section we raised the problem of finding those values of x for which the Taylor series for f about $x = a$ converges to $f(x)$, that is

$$f(x) = f(a) + f'(a)(x - a) + \frac{f''(a)}{2!}(x - a)^2$$

$$+ \cdots + \frac{f^{(k)}(a)}{k!}(x - a)^k + \cdots \qquad (5)$$

To solve this problem, let us write (5) in sigma notation:

$$f(x) = \sum_{k=0}^{\infty} \frac{f^{(k)}(a)}{k!} (x - a)^k$$

This equality is equivalent to

$$f(x) = \lim_{n \to +\infty} \sum_{k=0}^{n} \frac{f^{(k)}(a)}{k!} (x - a)^k$$

or

$$\lim_{n \to +\infty} \left[f(x) - \sum_{k=0}^{n} \frac{f^{(k)}(a)}{k!} (x - a)^k \right] = 0 \tag{6}$$

But the bracketed expression in (6) is $R_n(x)$, since it is the difference between $f(x)$ and its nth Taylor polynomial about $x = a$. Thus, (6) can be written as

$$\lim_{n \to +\infty} R_n(x) = 0$$

which leads us to the following result.

11.10.2   THEOREM. *The equality*

$$f(x) = \sum_{k=0}^{\infty} \frac{f^{(k)}(a)}{k!} (x - a)^k$$

holds if and only if $\lim_{n \to +\infty} R_n(x) = 0.$

To paraphrase this theorem, the *Taylor series for f converges to f(x) at precisely those points where the remainder approaches zero.*

Example 3   Show that the Maclaurin series for e^x converges to e^x for all x.

Solution.   We want to show

$$e^x = 1 + x + \frac{x^2}{2!} + \frac{x^3}{3!} + \cdots + \frac{x^n}{n!} + \cdots \tag{7}$$

holds for all x. From (4),

$$e^x = 1 + x + \frac{x^2}{2!} + \cdots + \frac{x^n}{n!} + R_n(x)$$

where

$$R_n(x) = \frac{e^c}{(n+1)!} x^{n+1}$$

and c is between 0 and x. Thus, we must show that for all x

$$\lim_{n \to +\infty} R_n(x) = \lim_{n \to +\infty} \frac{e^c}{(n+1)!} x^{n+1} = 0 \qquad (8)$$

Our proof of this result will hinge on the limit

$$\lim_{n \to +\infty} \frac{x^{n+1}}{(n+1)!} = 0 \qquad (9)$$

which follows from Formula (1) of Section 11.8. We shall consider three cases. If $x > 0$, then

$$0 < c < x$$

from which it follows that

$$0 < e^c < e^x$$

and consequently

$$0 < \frac{e^c}{(n+1)!} x^{n+1} < \frac{e^x}{(n+1)!} x^{n+1} \qquad (10)$$

From (9) we obtain

$$\lim_{n \to +\infty} \frac{e^x}{(n+1)!} x^{n+1} = e^x \lim_{n \to +\infty} \frac{x^{n+1}}{(n+1)!} = e^x \cdot 0 = 0$$

Thus, from (10) and the Squeezing Theorem (2.8.2), result (8) follows.

In the case $x < 0$, we have $c < 0$ since c is between 0 and x. Thus, $0 < e^c < 1$ and consequently

$$0 < e^c \left| \frac{x^{n+1}}{(n+1)!} \right| < \left| \frac{x^{n+1}}{(n+1)!} \right|$$

or

$$0 < \left| \frac{e^c}{(n+1)!} x^{n+1} \right| < \left| \frac{x^{n+1}}{(n+1)!} \right|$$

or

$$0 < |R_n(x)| < \left| \frac{x^{n+1}}{(n+1)!} \right|$$

From (9) and the Squeezing Theorem it follows that $\lim\limits_{n\to+\infty} |R_n(x)| = 0$, and so $\lim\limits_{n\to+\infty} R_n(x) = 0$.

Convergence in the case $x = 0$ is obvious since (7) reduces to

$$e^0 = 1 + 0 + 0 + 0 + \cdots \qquad \blacktriangleleft$$

Example 4 Show that the Maclaurin series for $\sin x$ converges to $\sin x$ for all x.

Solution. We must show that

$$\sin x = x - \frac{x^3}{3!} + \frac{x^5}{5!} - \frac{x^7}{7!} + \cdots$$

holds for all x. Let $f(x) = \sin x$, so that for all x either

$$f^{(n+1)}(x) = \pm\cos x \quad \text{or} \quad f^{(n+1)}(x) = \pm\sin x$$

In all of these cases $|f^{(n+1)}(x)| \le 1$, so that for all possible values of c

$$|f^{(n+1)}(c)| \le 1$$

Hence,

$$0 \le |R_n(x)| = \left| \frac{f^{(n+1)}(c)}{(n+1)!} x^{n+1} \right| \le \frac{|x|^{n+1}}{(n+1)!} \tag{11}$$

From (9) with $|x|$ replacing x,

$$\lim_{n\to+\infty} \frac{|x|^{n+1}}{(n+1)!} = 0$$

so that from (11) and the Squeezing Theorem

$$\lim_{n\to+\infty} |R_n(x)| = 0$$

and consequently

$$\lim_{n\to+\infty} R_n(x) = 0$$

for all x. $\blacktriangleleft$

Example 5 Find the Taylor series for $\sin x$ about $x = \pi/2$ and show that the series converges to $\sin x$ for all x.

Solution.   Let $f(x) = \sin x$. Thus,

$$f(x) = \sin x \qquad f\left(\frac{\pi}{2}\right) = \sin\frac{\pi}{2} = 1$$

$$f'(x) = \cos x \qquad f'\left(\frac{\pi}{2}\right) = \cos\frac{\pi}{2} = 0$$

$$f''(x) = -\sin x \qquad f''\left(\frac{\pi}{2}\right) = -\sin\frac{\pi}{2} = -1$$

$$f'''(x) = -\cos x \qquad f'''\left(\frac{\pi}{2}\right) = -\cos\frac{\pi}{2} = 0$$

Since $f^{(4)}(x) = \sin x$, the pattern $1, 0, -1, 0$ will repeat over and over as we evaluate successive derivatives at $\pi/2$. Thus, the Taylor series representation about $x = \pi/2$ for $\sin x$ is

$$\sin x = 1 - \frac{1}{2!}\left(x - \frac{\pi}{2}\right)^2 + \frac{1}{4!}\left(x - \frac{\pi}{2}\right)^4 - \frac{1}{6!}\left(x - \frac{\pi}{2}\right)^6 + \cdots \quad (12)$$

which we are trying to show is valid for all x. As in Example 4, $|f^{(n+1)}(c)| \leq 1$ so that

$$0 \leq |R_n(x)| = \left|\frac{f^{(n+1)}(c)}{(n+1)!}\left(x - \frac{\pi}{2}\right)^{n+1}\right| \leq \frac{\left|x - \frac{\pi}{2}\right|^{n+1}}{(n+1)!}$$

From (9) with $|x - \pi/2|$ replacing x it follows that

$$\lim_{n \to +\infty} \frac{\left|x - \frac{\pi}{2}\right|^{n+1}}{(n+1)!} = 0$$

so that by the same argument given in Example 4, $\lim_{n \to +\infty} R_n(x) = 0$ for all x. This shows that (12) is valid for all x.   ◀

REMARK.   It can be shown that the Taylor series for e^x, $\sin x$, and $\cos x$ about any point $x = a$ converges to these functions for all x.

For reference, we have listed in Table 11.10.1 the Maclaurin series for a number of important functions, and we have indicated the interval over which the series converges to the function. It should be noted that the intervals of convergence stated for $\ln(1 + x)$ and $\tan^{-1} x$ are somewhat difficult to obtain directly. However, these intervals can be obtained by indirect methods that we shall study in the last section of this chapter.

Table 11.10.1

MACLAURIN SERIES	INTERVAL OF VALIDITY
$e^x = \displaystyle\sum_{k=0}^{\infty} \frac{x^k}{k!} = 1 + x + \frac{x^2}{2!} + \frac{x^3}{3!} + \frac{x^4}{4!} + \cdots$	$-\infty < x < +\infty$
$\sin x = \displaystyle\sum_{k=0}^{\infty} (-1)^k \frac{x^{2k+1}}{(2k+1)!} = x - \frac{x^3}{3!} + \frac{x^5}{5!} - \frac{x^7}{7!} + \cdots$	$-\infty < x < +\infty$
$\cos x = \displaystyle\sum_{k=0}^{\infty} (-1)^k \frac{x^{2k}}{(2k)!} = 1 - \frac{x^2}{2!} + \frac{x^4}{4!} - \frac{x^6}{6!} + \cdots$	$-\infty < x < +\infty$
$\ln(1+x) = \displaystyle\sum_{k=0}^{\infty} (-1)^k \frac{x^{k+1}}{k+1} = x - \frac{x^2}{2} + \frac{x^3}{3} - \frac{x^4}{4} + \cdots$	$-1 < x \le 1$
$\tan^{-1} x = \displaystyle\sum_{k=0}^{\infty} (-1)^k \frac{x^{2k+1}}{2k+1} = x - \frac{x^3}{3} + \frac{x^5}{5} - \frac{x^7}{7} + \cdots$	$-1 \le x \le 1$
$\dfrac{1}{1-x} = \displaystyle\sum_{k=0}^{\infty} x^k = 1 + x + x^2 + x^3 + \cdots$	$-1 < x < 1$
$\sinh x = \displaystyle\sum_{k=0}^{\infty} \frac{x^{2k+1}}{(2k+1)!} = x + \frac{x^3}{3!} + \frac{x^5}{5!} + \frac{x^7}{7!} + \cdots$	$-\infty < x < +\infty$
$\cosh x = \displaystyle\sum_{k=0}^{\infty} \frac{x^{2k}}{(2k)!} = 1 + \frac{x^2}{2!} + \frac{x^4}{4!} + \frac{x^6}{6!} + \cdots$	$-\infty < x < +\infty$

CONSTRUCTING MACLAURIN SERIES BY SUBSTITUTION

Sometimes Maclaurin series can be obtained by substituting in other Maclaurin series.

Example 6   Using the Maclaurin series

$$e^x = 1 + x + \frac{x^2}{2!} + \frac{x^3}{3!} + \frac{x^4}{4!} + \cdots \quad -\infty < x < +\infty$$

we can derive the Maclaurin series for e^{-x} by substituting $-x$ for x to obtain

$$e^{-x} = 1 + (-x) + \frac{(-x)^2}{2!} + \frac{(-x)^3}{3!} + \frac{(-x)^4}{4!} + \cdots \quad -\infty < -x < +\infty$$

or

$$e^{-x} = 1 - x + \frac{x^2}{2!} - \frac{x^3}{3!} + \frac{x^4}{4!} - \cdots \quad -\infty < x < +\infty$$

From the Maclaurin series for e^x and e^{-x} we can obtain the Maclaurin series for $\cosh x$ by writing

$$\cosh x = \frac{1}{2}(e^x + e^{-x}) = \frac{1}{2}\left(\left[1 + x + \frac{x^2}{2!} + \frac{x^3}{3!} + \frac{x^4}{4!} + \cdots\right]\right.$$
$$\left. + \left[1 - x + \frac{x^2}{2!} - \frac{x^3}{3!} + \frac{x^4}{4!} + \cdots\right]\right)$$

or

$$\cosh x = 1 + \frac{x^2}{2!} + \frac{x^4}{4!} + \cdots \qquad -\infty < x < +\infty \qquad \blacktriangleleft$$

We could have also derived this Maclaurin series directly. However, sometimes indirect methods, such as those in this example, are useful when it is messy to calculate the higher derivatives required for a Maclaurin series. There is, however, a loose thread in the logic of this example. We have produced a power series in x that converges to $\cosh x$ for all x. But isn't it conceivable that we have produced a power series in x *different* from the Maclaurin series? In Section 11.12 we shall show that if a power series in $x - a$ converges to $f(x)$ on some interval containing a, then the series must be the Taylor series for f about $x = a$. Thus, we are assured the series obtained for $\cosh x$ is, in fact, the Maclaurin series.

Example 7   Using the Maclaurin series

$$\frac{1}{1-x} = 1 + x + x^2 + x^3 + \cdots \qquad -1 < x < 1$$

we can derive the Maclaurin series for $1/(1 - 2x^2)$ by substituting $2x^2$ for x to obtain

$$\frac{1}{1 - 2x^2} = 1 + (2x^2) + (2x^2)^2 + (2x^2)^3 + \cdots \qquad -1 < 2x^2 < 1$$

or

$$\frac{1}{1 - 2x^2} = 1 + 2x^2 + 4x^4 + 8x^6 + \cdots = \sum_{k=0}^{\infty} 2^k x^{2k} \qquad -1 < 2x^2 < 1$$

Since $2x^2 \geq 0$ for all x, the convergence condition $-1 < 2x^2 < 1$ can be written in the equivalent form $0 \leq 2x^2 < 1$ or $0 \leq x^2 < 1/2$ or $-1/\sqrt{2} < x < 1/\sqrt{2}$. $\blacktriangleleft$

OPTIONAL

We conclude this section with a proof of Taylor's Theorem.

Proof of Theorem 11.10.1.   By hypothesis, f can be differentiated $n + 1$ times at each point in an interval containing the point a. Choose any point b in this interval. To be specific, we shall assume $b > a$. (The cases $b < a$ and $b = a$ are left to the reader.) Let $p_n(x)$ be the nth Taylor polynomial for $f(x)$ about $x = a$ and define

$$h(x) = f(x) - p_n(x) \tag{13}$$

$$g(x) = (x - a)^{n+1} \tag{14}$$

Because $f(x)$ and $p_n(x)$ have the same value and the same first n derivatives at $x = a$, it follows that

$$h(a) = h'(a) = h''(a) = \cdots = h^{(n)}(a) = 0 \tag{15}$$

Also, we leave it for the reader to show that

$$g(a) = g'(a) = g''(a) = \cdots = g^{(n)}(a) = 0 \tag{16}$$

and that $g(x)$ and its first n derivatives are nonzero when $x \neq a$.

It is straightforward to check that h and g satisfy the hypotheses of the Extended Mean-Value Theorem (10.2.2) on the interval $[a, b]$, so that there is a point c_1 with $a < c_1 < b$ such that

$$\frac{h(b) - h(a)}{g(b) - g(a)} = \frac{h'(c_1)}{g'(c_1)} \tag{17}$$

or, from (15) and (16),

$$\frac{h(b)}{g(b)} = \frac{h'(c_1)}{g'(c_1)} \tag{18}$$

If we now apply the Extended Mean-Value Theorem to h' and g' over the interval $[a, c_1]$, we may deduce that there is a point c_2 with $a < c_2 < c_1 < b$ such that

$$\frac{h'(c_1) - h'(a)}{g'(c_1) - g'(a)} = \frac{h''(c_2)}{g''(c_2)}$$

or, from (15) and (16),

$$\frac{h'(c_1)}{g'(c_1)} = \frac{h''(c_2)}{g''(c_2)}$$

which, when combined with (18), yields

$$\frac{h(b)}{g(b)} = \frac{h''(c_2)}{g''(c_2)}$$

It should now be clear that if we continue in this way, applying the

Extended Mean-Value Theorem to the successive derivatives of h and g, we shall eventually obtain a relationship of the form

$$\frac{h(b)}{g(b)} = \frac{h^{(n+1)}(c_{n+1})}{g^{(n+1)}(c_{n+1})} \tag{19}$$

where $a < c_{n+1} < b$. However, $p_n(x)$ is a polynomial of degree n, so that its $(n + 1)$-st derivative is zero. Thus, from (13)

$$h^{(n+1)}(c_{n+1}) = f^{(n+1)}(c_{n+1}) \tag{20}$$

Also, from (14), the $(n + 1)$-st derivative of $g(x)$ is the constant $(n + 1)!$, so that

$$g^{(n+1)}(c_{n+1}) = (n + 1)! \tag{21}$$

Substituting (20) and (21) in (19) yields

$$\frac{h(b)}{g(b)} = \frac{f^{(n+1)}(c_{n+1})}{(n + 1)!}$$

Letting $c = c_{n+1}$, and using (13) and (14), it follows that

$$f(b) - p_n(b) = \frac{f^{(n+1)}(c)}{(n + 1)!}(b - a)^{n+1}$$

But this is precisely Formula (1) in Taylor's Theorem, with the exception that the variable here is b rather than x. Thus, to finish we need only replace b by x. ∎

▶ Exercise Set 11.10

For the functions in Exercises 1–14, find Lagrange's form of the remainder for the given values of a and n.

1. e^{2x}; $a = 0$; $n = 5$.

2. $\cos x$; $a = 0$; $n = 8$.

3. $\dfrac{1}{x + 1}$; $a = 0$; $n = 4$.

4. $\tan x$; $a = 0$; $n = 2$.

5. xe^x; $a = 0$; $n = 3$.

6. $\ln(1 + x)$; $a = 0$; $n = 5$.

7. $\tan^{-1} x$; $a = 0$; $n = 2$.

8. $\sinh x$; $a = 0$; $n = 6$.

9. $\sqrt{x}$; $a = 4$; $n = 3$.

10. $\dfrac{1}{x}$; $a = 1$; $n = 5$.

11. $\sin x$; $a = \dfrac{\pi}{6}$; $n = 4$.

12. $\cos \pi x$; $a = \dfrac{1}{2}$; $n = 2$.

13. $\dfrac{1}{(1 + x)^2}$; $a = -2$; $n = 5$.

14. $\csc x$; $a = \dfrac{\pi}{2}$; $n = 1$.

In Exercises 15–18, find Lagrange's form of the remainder $R_n(x)$ when $a = 0$.

15. $f(x) = \dfrac{1}{1 - x}$.

16. $f(x) = e^{-x}$.

17. $f(x) = e^{2x}$.

18. $f(x) = \ln(1 + x)$.

19. Prove: The Maclaurin series for $\cos x$ converges to $\cos x$ for all x.

20. Prove: The Taylor series for $\sin x$ about $x = \pi/4$ converges to $\sin x$ for all x.

21. Prove: The Taylor series for e^x about $x = 1$ converges to e^x for all x.

22. (a) Prove: The Maclaurin series for $\ln(1 + x)$ converges to $\ln(1 + x)$ if $0 \leq x \leq 1$. [*Remark:* The Maclaurin series actually converges to $\ln(1 + x)$ on the interval $(-1, 1]$, but the Lagrange form of the remainder is not strong enough to establish this fact.]

(b) Use $x = 1$ in the Maclaurin series for $\ln(1 + x)$, Table 11.10.1, to show that

$$\ln 2 = 1 - \frac{1}{2} + \frac{1}{3} - \frac{1}{4} + \cdots$$

23. Prove: The Taylor series for e^x about any point $x = a$ converges to e^x for all x.

24. Prove: The Taylor series for $\sin x$ about any point $x = a$ converges to $\sin x$ for all x.

25. Prove: The Taylor series for $\cos x$ about any point $x = a$ converges to $\cos x$ for all x.

In Exercises 26–38, use the known Maclaurin series for e^x, $\sin x$, $\cos x$, and $1/(1 - x)$ to find the Maclaurin series for the given function. In each case, specify the interval on which your computations are valid.

26. xe^x.

27. e^{-2x}.

28. e^{x^2}.

29. $\dfrac{1}{1 + x}$.

30. $\dfrac{1}{1 - 4x^2}$.

31. $\dfrac{x^2}{1 + 3x}$.

32. $\sin^2 x$. $\left[Hint: \sin^2 x = \dfrac{1}{2}(1 - \cos 2x). \right]$

33. $\cos^2 x$. $\left[Hint: \cos^2 x = \dfrac{1}{2}(1 + \cos 2x). \right]$

34. $\sinh x$.

35. $\sin(2x)$.

36. $\cos(2x)$.

37. $\cos(x^2)$.

38. $\sin(x^2)$.

39. Use the Maclaurin series for $1/(1 - x)$ to express $1/x$ in powers of $x - 1$. Find the interval of con-

vergence. $\left[Hint: \dfrac{1}{x} = \dfrac{1}{1 + (x - 1)}. \right]$

In Exercises 40–43, use the known Maclaurin series for e^x, $\sin x$, and $\cos x$ to help find the sum of the given series.

40. $2 + \dfrac{4}{2!} + \dfrac{8}{3!} + \dfrac{16}{4!} + \cdots$

41. $\pi - \dfrac{\pi^3}{3!} + \dfrac{\pi^5}{5!} - \dfrac{\pi^7}{7!} + \cdots$

42. $1 - \dfrac{e^2}{2!} + \dfrac{e^4}{4!} - \dfrac{e^6}{6!} + \cdots$

43. $1 - \ln 3 + \dfrac{(\ln 3)^2}{2!} - \dfrac{(\ln 3)^3}{3!} + \cdots$

44. (a) Use the Maclaurin series for $\dfrac{1}{1 - x}$ to help find the Maclaurin series for the function

$$f(x) = \dfrac{x}{1 - x^2}.$$

(b) Use the result in part (a) to help find $f^{(5)}(0)$ and $f^{(6)}(0)$. [*Hint:* From Theorem 11.10.2, what is the coefficient of x^k?]

45. (a) Use the Maclaurin series for $\cos x$ to find the Maclaurin series for $f(x) = x^2 \cos 2x$.

(b) Use the result in part (a) to help find $f^{(5)}(0)$. [See the hint in part (b) of Exercise 44.]

46. The purpose of this exercise is to show that the Taylor series of a function f may possibly converge to a value different from $f(x)$ for certain x. Let

$$f(x) = \begin{cases} e^{-1/x^2}, & x \neq 0 \\ 0, & x = 0 \end{cases}$$

(a) Use the definition of a derivative to show that $f'(0) = 0$.

(b) With some difficulty it can be shown that $f^{(n)}(0) = 0$ for $n \geq 2$. Accepting this fact, show that the Maclaurin series of f converges for all x, but converges to $f(x)$ only at the point $x = 0$.

11.11 COMPUTATIONS USING TAYLOR SERIES

In this section we shall show how Taylor and Maclaurin series can be used to obtain approximate values for trigonometric functions and logarithms.

In the previous section we showed that for all x

$$e^x = 1 + x + \frac{x^2}{2!} + \frac{x^3}{3!} + \frac{x^4}{4!} + \cdots$$

In particular, if we let $x = 1$, we obtain the following expression for e as the sum of an infinite series

$$e = 1 + 1 + \frac{1}{2!} + \frac{1}{3!} + \frac{1}{4!} + \cdots$$

Thus, we can approximate e to any degree of accuracy using an appropriate partial sum

$$e \approx 1 + 1 + \frac{1}{2!} + \frac{1}{3!} + \cdots + \frac{1}{n!} \tag{1}$$

The value of n in this formula is at our disposal, the larger we choose n the more accurate the approximation. For practical applications one would be interested in approximating e to a specified degree of accuracy, in which case it would be important to determine how large n should be chosen to attain the desired accuracy. For example, how large should n be taken in (1) to ensure an error of at most 0.00005? We shall now show how Lagrange's remainder formula can be used to answer such questions.

According to Taylor's theorem, the absolute value of the error that results when $f(x)$ is approximated by its nth Taylor polynomial $p_n(x)$ about $x = a$ is

$$|R_n(x)| = |f(x) - p_n(x)| = \left| \frac{f^{(n+1)}(c)}{(n+1)!}(x-a)^{n+1} \right| \tag{2}$$

In this formula, c is an unknown number between a and x, so that the value of $f^{(n+1)}(c)$ usually cannot be determined. However, it is frequently possible to determine an upper bound on the size of $|f^{(n+1)}(c)|$; that is, we can find a constant M such that $|f^{(n+1)}(c)| \le M$. For such an M, it follows from (2) that

$$|R_n(x)| = |f(x) - p_n(x)| \le \frac{M}{(n+1)!}|x-a|^{n+1} \tag{3}$$

which gives an upper bound on the magnitude of the error $R_n(x)$. The examples to follow illustrate the usefulness of this result, but first let us introduce some terminology.

An approximation is said to be **accurate to n decimal places** if the magnitude of the error is less than 0.5×10^{-n}. For example,

DESCRIPTION	MAGNITUDE OF THE ERROR IS LESS THAN
1 decimal place accuracy	$0.05 \quad = 0.5 \times 10^{-1}$
2 decimal place accuracy	$0.005 \; = 0.5 \times 10^{-2}$
3 decimal place accuracy	$0.0005 = 0.5 \times 10^{-3}$
$\vdots$	$\vdots$

APPROXIMATING e **Example 1**   Use (1) to approximate e to four decimal place accuracy.

Solution.   From Taylor's formula with remainder (see Example 1, Section 11.10)

$$e^x = 1 + x + \frac{x^2}{2!} + \cdots + \frac{x^n}{n!} + \frac{e^c}{(n+1)!} x^{n+1}$$

where c is between 0 and x. Thus, in the case $x = 1$ we obtain

$$e = 1 + 1 + \frac{1}{2!} + \cdots + \frac{1}{n!} + \frac{e^c}{(n+1)!}$$

where c is between 0 and 1. This tells us that the magnitude of the error in approximation (1) is

$$|R_n| = \left| \frac{e^c}{(n+1)!} \right| = \frac{e^c}{(n+1)!} \tag{4}$$

where $0 < c < 1$. Since $c < 1$, it follows that

$$e^c < e^1 = e$$

so that from (4)

$$|R_n| < \frac{e}{(n+1)!} \tag{5}$$

This inequality provides an upper bound on the magnitude of the error R_n. Unfortunately, this inequality is not very useful since the right side involves the quantity e, which we are trying to estimate. However, if we use the fact that $e < 3$ (Table 7.2.1), then we can replace (5) with the more useful result

$$|R_n| < \frac{3}{(n+1)!}$$

It follows that if we choose n so that

$$|R_n| < \frac{3}{(n+1)!} < 0.5 \times 10^{-4} = 0.00005 \tag{6}$$

then approximation (1) will be accurate to four decimal places. An appropriate value for n may be found by trial and error. For example, using a hand-held calculator or computer one can evaluate $3/(n+1)!$ for $n = 0, 1, 2, \ldots$ until a value of n satisfying (6) is obtained. We leave it for the reader to show that $n = 8$ is the first positive integer satisfying (6). Thus, to four decimal place accuracy

$$e \approx 1 + 1 + \frac{1}{2!} + \frac{1}{3!} + \frac{1}{4!} + \frac{1}{5!} + \frac{1}{6!} + \frac{1}{7!} + \frac{1}{8!} \approx 2.7183 \quad \blacktriangleleft$$

APPROXIMATING TRIGONOMETRIC FUNCTIONS

Example 2   Use the Maclaurin series for $\sin x$ to approximate $\sin 3°$ to five decimal place accuracy.

Solution.   In the Maclaurin series

$$\sin x = x - \frac{x^3}{3!} + \frac{x^5}{5!} - \frac{x^7}{7!} + \cdots \tag{7}$$

the angle x is assumed to be in radians (because the differentiation formulas for the trigonometric functions were derived with this assumption). Since $3° = \pi/60$ radians, it follows from (7) that

$$\sin 3° = \sin \frac{\pi}{60} = \left(\frac{\pi}{60}\right) - \frac{\left(\frac{\pi}{60}\right)^3}{3!} + \frac{\left(\frac{\pi}{60}\right)^5}{5!} - \frac{\left(\frac{\pi}{60}\right)^7}{7!} + \cdots \tag{8}$$

We must now decide how many terms in this series must be kept in order to obtain five decimal place accuracy. We shall consider two possible approaches, one using Lagrange's remainder formula, and the other exploiting the fact that (8) satisfies the hypotheses of the alternating series test.

If we let $f(x) = \sin x$, then the magnitude of the error that results when $\sin x$ is approximated by its nth Maclaurin polynomial is

$$|R_n| = \left| \frac{f^{(n+1)}(c)}{(n+1)!} x^{n+1} \right|$$

where c is between 0 and x. Since $f^{(n+1)}(c)$ is either $\pm \sin c$ or $\pm \cos c$, it follows that $|f^{(n+1)}(c)| \le 1$, so

$$|R_n| \leq \frac{|x|^{n+1}}{(n+1)!}$$

In particular, if $x = \pi/60$, then

$$|R_n| \leq \frac{\left(\dfrac{\pi}{60}\right)^{n+1}}{(n+1)!}$$

Thus, for five decimal place accuracy, we must choose n so that

$$\frac{\left(\dfrac{\pi}{60}\right)^{n+1}}{(n+1)!} < 0.5 \times 10^{-5} = 0.000005$$

By trial and error with the help of a calculator or computer, the reader can check that $n = 3$ is the smallest n that works. Thus, in (8) we need only keep terms up to the third power for five decimal place accuracy, that is,

$$\sin 3° \approx \left(\frac{\pi}{60}\right) - \frac{\left(\dfrac{\pi}{60}\right)^3}{3!} \approx 0.05234 \qquad (9)$$

An alternative approach to determining n uses the fact that (8) satisfies the hypotheses of the alternating series test, Theorem 11.7.1. (Verify.) Thus, by Theorem 11.7.2, if we use only those terms up to and including

$$\pm \frac{\left(\dfrac{\pi}{60}\right)^m}{m!} \qquad [m \text{ is an odd positive integer.}]$$

then the magnitude of the error will be at most

$$\frac{\left(\dfrac{\pi}{60}\right)^{m+2}}{(m+2)!}$$

Thus, for five decimal place accuracy, we look for the first positive odd integer m such that

$$\frac{\left(\dfrac{\pi}{60}\right)^{m+2}}{(m+2)!} < 0.5 \times 10^{-5} = 0.000005$$

By trial and error, $m = 3$ is the first such integer. Thus, to five decimal place accuracy

$$\sin 3° \approx \left(\frac{\pi}{60}\right) - \frac{\left(\frac{\pi}{60}\right)^3}{3!} \approx 0.05234 \tag{10}$$

which is the same as our earlier result. ◄

REMARK. It should be noted that there are two types of errors that result when computing with series. The first, called *truncation error,* is the error that results when the entire series is approximated by a partial sum. The second kind of error, called *round-off error,* results when decimal approximations are used. For example, (9) involves a truncation error of at most 0.5×10^{-5} (five decimal place accuracy). However, to obtain the numerical value (10) it was necessary to approximate π, thereby introducing a round-off error. Round-off error is not limited to approximating irrational numbers. Even a fraction like $\frac{1}{3}$ must be approximated to represent it by a decimal in a calculator or a computer. If one seeks n decimal place accuracy in the final result, then it is common procedure to use $n + 1$ decimal places in each decimal approximation and then round off to n decimal places at the end of all computations. Although this procedure may occasionally not produce n decimal place accuracy, it works often enough that we shall use it in this text. Better procedures are considered in a branch of mathematics called *numerical analysis.*

Example 3 Approximate $\sin 92°$ to five decimal place accuracy.

Solution. We could proceed, as in the previous example, using the Maclaurin series

$$\sin x = x - \frac{x^3}{3!} + \frac{x^5}{5!} - \frac{x^7}{7!} + \cdots \tag{11}$$

with $x = 92° = \frac{23}{45}\pi$ (radians). However, a better procedure is to work with the Taylor series about $x = \pi/2$ for $\sin x$ (Example 5, Section 11.10):

$$\sin x = 1 - \frac{1}{2!}\left(x - \frac{\pi}{2}\right)^2 + \frac{1}{4!}\left(x - \frac{\pi}{2}\right)^4 - \frac{1}{6!}\left(x - \frac{\pi}{2}\right)^6 + \cdots \tag{12}$$

The advantage of this series over the Maclaurin series is its faster convergence at $x = 92° = \frac{23}{45}\pi$. In general, the rate of convergence of a Taylor series about $x = a$ decreases as one moves farther away from the point a. That is, the farther x becomes from a, the more terms that are required to achieve a given degree of accuracy. Therefore, for the purpose of approximating $\sin 92° = \sin \frac{23}{45}\pi$, series (12) is preferable to series (11) since $a = \pi/2$ is closer to $\frac{23}{45}\pi$ than $a = 0$.

Substituting $x = \frac{23}{45}\pi$ in (12) yields

$$\sin 92° = \sin \frac{23}{45}\pi = 1 - \frac{1}{2!}\left(\frac{\pi}{90}\right)^2 + \frac{1}{4!}\left(\frac{\pi}{90}\right)^4 - \frac{1}{6!}\left(\frac{\pi}{90}\right)^6 + \cdots \quad (13)$$

If we let $f(x) = \sin x$, then the magnitude of the error in approximating $\sin x$ by its nth Taylor polynomial about $a = \pi/2$ is

$$|R_n| = \left|\frac{f^{(n+1)}(c)}{(n+1)!}\left(x - \frac{\pi}{2}\right)^{n+1}\right|$$

where c is between $\pi/2$ and x. As was the case in the previous example, $|f^{(n+1)}(c)| \le 1$, so that

$$|R_n| \le \frac{\left|x - \dfrac{\pi}{2}\right|^{n+1}}{(n+1)!}$$

In particular, if $x = \frac{23}{45}\pi$, then

$$|R_n| \le \frac{\left(\dfrac{\pi}{90}\right)^{n+1}}{(n+1)!}$$

For five decimal place accuracy, we must choose n so that

$$\frac{\left(\dfrac{\pi}{90}\right)^{n+1}}{(n+1)!} < 0.5 \times 10^{-5} = 0.000005$$

By trial and error, the smallest such integer is $n = 3$. Thus, in (13) we need only keep terms up to the third power for five decimal place accuracy, that is,

$$\sin 92° \approx 1 + 0 \cdot \left(\frac{\pi}{90}\right) - \frac{1}{2!}\left(\frac{\pi}{90}\right)^2 + 0 \cdot \left(\frac{\pi}{90}\right)^3$$

$$= 1 - \frac{1}{2!}\left(\frac{\pi}{90}\right)^2 \approx 0.99939$$

As in Example 2, we could also have obtained this result by exploiting the fact that (13) is an alternating series.   ◄

APPROXIMATING   The Maclaurin series
LOGARITHMS

$$\ln(1 + x) = x - \frac{x^2}{2} + \frac{x^3}{3} - \frac{x^4}{4} + \cdots \quad -1 < x \le 1 \quad (14)$$

is the starting point for the approximation of natural logarithms. Unfortunately, the usefulness of this series is limited because of its slow convergence and the restriction $-1 < x \le 1$. However, if we replace x by $-x$ in this series, we obtain

$$\ln(1 - x) = -x - \frac{x^2}{2} - \frac{x^3}{3} - \frac{x^4}{4} - \cdots \qquad -1 \le x < 1 \qquad (15)$$

and on subtracting (15) from (14) we obtain

$$\ln \frac{1 + x}{1 - x} = 2 \left(x + \frac{x^3}{3} + \frac{x^5}{5} + \frac{x^7}{7} + \cdots \right) \qquad -1 < x < 1 \qquad (16)$$

Series (16), first obtained by James Gregory* in 1668, can be used to compute the natural logarithm of any positive number y by letting

$$y = \frac{1 + x}{1 - x}$$

or equivalently

$$x = \frac{y - 1}{y + 1} \qquad (17)$$

For example, to compute $\ln 2$ we let $y = 2$ in (17), which yields $x = 1/3$. Substituting this value in (16) gives

$$\ln 2 = 2 \left[(1/3) + \frac{(1/3)^3}{3} + \frac{(1/3)^5}{5} + \frac{(1/3)^7}{7} + \frac{(1/3)^9}{9} + \cdots \right]$$

$$= 2 \left[\frac{1}{3} + \frac{1}{81} + \frac{1}{1,215} + \frac{1}{15,309} + \frac{1}{177,147} + \cdots \right]$$

$$\approx 2[0.33333 + 0.01235 + 0.00082 + 0.00007 + 0.00001 + \cdots]$$

Adding the five terms shown and then rounding to four decimal places at the end yields

$$\ln 2 \approx 2[0.34657] = 0.69314 \approx 0.6931$$

It is of interest to note that in the case $x = 1$, series (14) yields

$$\ln 2 = 1 - \frac{1}{2} + \frac{1}{3} - \frac{1}{4} + \frac{1}{5} - \cdots$$

---

*JAMES GREGORY (1638–1675) Scottish mathematician and astronomer. Gregory, the son of a minister, was famous in his time as the inventor of the Gregorian reflecting telescope, so named in his honor. Although he is not generally ranked with the great mathematicians, much of his work relating to calculus was studied by Leibniz and Newton and undoubtedly influenced some of their discoveries. There is a manuscript, discovered posthumously, which shows that Gregory had anticipated Taylor series well before Taylor.

This result is noteworthy because it is the first time we have been able to obtain the sum of the alternating harmonic series. However, this series converges too slowly to be of any computational value.

APPROXIMATING π If we let $x = 1$ in the Maclaurin series

$$\tan^{-1} x = x - \frac{x^3}{3} + \frac{x^5}{5} - \frac{x^7}{7} + \cdots \qquad -1 \leq x \leq 1 \tag{18}$$

we obtain

$$\frac{\pi}{4} = \tan^{-1} 1 = 1 - \frac{1}{3} + \frac{1}{5} - \frac{1}{7} + \cdots$$

or

$$\pi = 4 \left[1 - \frac{1}{3} + \frac{1}{5} - \frac{1}{7} + \cdots \right]$$

This famous series, obtained by Leibniz in 1674, converges too slowly to be of computational importance. A more practical procedure for approximating π uses the identity

$$\frac{\pi}{4} = \tan^{-1} \frac{1}{2} + \tan^{-1} \frac{1}{3} \tag{19}$$

By using this identity and series (18) to approximate $\tan^{-1} \frac{1}{2}$ and $\tan^{-1} \frac{1}{3}$, the value of π can be effectively approximated to any degree of accuracy.

▶ Exercise Set 11.11

In this exercise set, a hand-held calculator will be useful.

1.   How large should n be taken in (1) to approximate e to:
   (a)   five decimal place accuracy?
   (b)   ten decimal place accuracy?

In Exercises 2–8, apply Lagrange's form of the remainder.

2.   Use $x = -1$ in the Maclaurin series for e^x to approximate $1/e$ to three decimal place accuracy.

3.   Use $x = \frac{1}{2}$ in the Maclaurin series for e^x to approximate $\sqrt{e}$ to four decimal place accuracy.

4.   Use the Maclaurin series for $\sin x$ to approximate $\sin 4°$ to five decimal place accuracy.

5.   Use the Maclaurin series for $\cos x$ to approximate $\cos (\pi/20)$ to four decimal place accuracy.

6.   Use an appropriate Taylor series for $\sin x$ to approximate $\sin 85°$ to four decimal place accuracy.

7.   Use an appropriate Taylor series for $\cos x$ to approximate $\cos 58°$ to four decimal place accuracy.

8.   Use an appropriate Taylor series for $\sin x$ to approximate $\sin 35°$ to four decimal place accuracy.

9.   Use series (16) to approximate $\ln 1.25$ to three decimal place accuracy.

10.   Use series (16) to approximate $\ln 3$ to four decimal place accuracy.

11.   Use the Maclaurin series for $\tan^{-1} x$ to approxi-

mate $\tan^{-1} 0.1$ to three decimal place accuracy. [*Hint:* Use the fact that (18) is an alternating series.]

12. Use the Maclaurin series for $\sinh x$ to approximate $\sinh 0.5$ to three decimal place accuracy.

13. Use the Maclaurin series for $\cosh x$ to approximate $\cosh 0.1$ to four decimal place accuracy.

14. Use an appropriate Taylor series for $\sqrt[3]{x}$ to approximate $\sqrt[3]{28}$ to three decimal place accuracy.

15. For what range of x values can $\sin x$ be approximated by $x - x^3/3!$ with three decimal place accuracy?

16. Estimate the maximum error in the approximation $\cos x \approx 1 - x^2/2! + x^4/4!$ if $-0.2 \leq x \leq 0.2$.

17. For what range of x values can e^x be approximated by $1 + x + x^2/2!$ if the allowable error is at most 0.0005?

18. Estimate the maximum error in the approximation $\ln(1 + x) \approx x$ if $|x| < 0.01$.

19. (a) How many terms in the alternating harmonic series are needed to approximate $\ln 2$ to six decimal place accuracy?

    (b) How many terms in series (16) with $x = \frac{1}{3}$ are needed to approximate $\ln 2$ to six decimal place accuracy?

20. Prove identity (19).

21. Approximate $\tan^{-1}\frac{1}{2}$ and $\tan^{-1}\frac{1}{3}$ to three decimal place accuracy and then use identity (19) to approximate π.

22. The curve $y = \cos x$ is to be approximated by the parabola $x^2 = 2(1 - y)$ over the interval $[-0.2, 0.2]$. Estimate the maximum difference in the y-coordinates of the curves over this interval.

11.12 DIFFERENTIATION AND INTEGRATION OF POWER SERIES

If a function f is expressed as a power series

$$f(x) = \sum_{k=0}^{\infty} c_k(x - a)^k$$

*for all x in some interval, then we say that the series **represents** $f(x)$ on that interval. Thus, for example, the geometric series $1 + x + x^2 + x^3 + \cdots$ represents the function $1/(1 - x)$ on the interval $(-1, 1)$ since*

$$\frac{1}{1 - x} = 1 + x + x^2 + x^3 + \cdots \qquad -1 < x < 1$$

In this section we shall show how a power series representation of a function f can be used to obtain power series representations of $f'(x)$ and $\int f(x)\, dx$, and we shall discuss some applications of these results.

The following theorem tells us that a power series representation of $f'(x)$ or $\int f(x)\, dx$ can be obtained by differentiating or integrating a power series representation of $f(x)$ term by term. We omit the proofs.

11.12.1   THEOREM. *If the series* $\sum_{k=0}^{\infty} c_k(x - a)^k$ *has a nonzero radius of convergence R, and if for each x in the interval* $(a - R, a + R)$ *we have* $f(x) = \sum_{k=0}^{\infty} c_k(x - a)^k$, *then:*

(*a*)   *The series* $\sum_{k=0}^{\infty} \dfrac{d}{dx}[c_k(x - a)^k] = \sum_{k=1}^{\infty} kc_k(x - a)^{k-1}$ *has radius of convergence R, and for all x in the interval* $(a - R, a + R)$

$$f'(x) = \sum_{k=0}^{\infty} \frac{d}{dx}[c_k(x - a)^k]$$

(*b*)   *The series* $\sum_{k=0}^{\infty} \left[\int c_k(x - a)^k \, dx \right] = \sum_{k=0}^{\infty} \dfrac{c_k}{k + 1}(x - a)^{k+1}$ *has radius of convergence R, and for all x in the interval* $(a - R, a + R)$

$$\int f(x) \, dx = \sum_{k=0}^{\infty} \left[\int c_k(x - a)^k \, dx \right] + C$$

(*c*)   *For all* α *and* β *in the interval* $(a - R, a + R)$, *the series*

$$\sum_{k=0}^{\infty} \left[\int_{\alpha}^{\beta} c_k(x - a)^k \, dx \right]$$

converges absolutely and

$$\int_{\alpha}^{\beta} f(x) \, dx = \sum_{k=0}^{\infty} \left[\int_{\alpha}^{\beta} c_k(x - a)^k \, dx \right]$$

REMARK.   Note that in part (*b*) a separate constant of integration is not introduced for each term in the series $\sum_{k=0}^{\infty} \left[\int c_k(x - a)^k \, dx \right]$; rather, a single constant C is added to the entire series.

Example 1   To illustrate Theorem 11.12.1, we shall obtain the familiar results

$$\frac{d}{dx}[\sin x] = \cos x \quad \text{and} \quad \int \cos x \, dx = \sin x + C$$

using power series. Since

$$\sin x = x - \frac{x^3}{3!} + \frac{x^5}{5!} - \frac{x^7}{7!} + \cdots \qquad -\infty < x < +\infty$$

$$\cos x = 1 - \frac{x^2}{2!} + \frac{x^4}{4!} - \frac{x^6}{6!} + \cdots \qquad -\infty < x < +\infty$$

it follows from part (*a*) of Theorem 11.12.1 that

$$\frac{d}{dx}[\sin x] = \frac{d}{dx}\left[x - \frac{x^3}{3!} + \frac{x^5}{5!} - \frac{x^7}{7!} + \cdots\right]$$

$$= 1 - 3\frac{x^2}{3!} + 5\frac{x^4}{5!} - 7\frac{x^6}{7!} + \cdots$$

$$= 1 - \frac{x^2}{2!} + \frac{x^4}{4!} - \frac{x^6}{6!} + \cdots$$

$$= \cos x$$

Also, from part (*b*) of Theorem 11.12.1

$$\int \cos x \, dx = \int \left[1 - \frac{x^2}{2!} + \frac{x^4}{4!} - \frac{x^6}{6!} + \cdots\right] dx$$

$$= \left[x - \frac{x^3}{3(2!)} + \frac{x^5}{5(4!)} - \frac{x^7}{7(6!)} + \cdots\right] + C$$

$$= \left[x - \frac{x^3}{3!} + \frac{x^5}{5!} - \frac{x^7}{7!} + \cdots\right] + C$$

$$= \sin x + C \qquad \blacktriangleleft$$

Example 2   The integral

$$\int_0^1 e^{-x^2} \, dx$$

cannot be evaluated directly because there is no elementary antiderivative of e^{-x^2}. However, it is possible to approximate the integral by some numerical technique such as Simpson's rule. Still, another possibility is to represent e^{-x^2} by its Maclaurin series and then integrate term by term in accordance with part (*c*) of Theorem 11.12.1. This produces a series that converges to the integral.

The simplest way to obtain the Maclaurin series for e^{-x^2} is to replace x by $-x^2$ in the Maclaurin series

$$e^x = 1 + x + \frac{x^2}{2!} + \frac{x^3}{3!} + \frac{x^4}{4!} + \cdots$$

to obtain

$$e^{-x^2} = 1 - x^2 + \frac{x^4}{2!} - \frac{x^6}{3!} + \frac{x^8}{4!} - \cdots$$

Therefore,

$$\int_0^1 e^{-x^2}\, dx = \int_0^1 \left[1 - x^2 + \frac{x^4}{2!} - \frac{x^6}{3!} + \frac{x^8}{4!} - \cdots \right] dx$$

$$= \left[x - \frac{x^3}{3} + \frac{x^5}{5(2!)} - \frac{x^7}{7(3!)} + \frac{x^9}{9(4!)} - \cdots \right]_0^1$$

$$= 1 - \frac{1}{3} + \frac{1}{5 \cdot 2!} - \frac{1}{7 \cdot 3!} + \frac{1}{9 \cdot 4!} - \cdots \tag{1}$$

Thus, we have found a series that converges to the value of the integral $\int_0^1 e^{-x^2}\, dx$. Using the first three terms in this series we obtain the approximation

$$\int_0^1 e^{-x^2}\, dx \approx 1 - \frac{1}{3} + \frac{1}{10} = \frac{23}{30} = 0.766 \ldots$$

Since series (1) satisfies the hypotheses of the alternating series test, it follows from Theorem 11.7.2 that the magnitude of the error in this approximation is at most $1/(7 \cdot 3!) = 1/42 \approx 0.024$. Greater accuracy can be obtained by using more terms in the series.   ◀

The following example illustrates how Theorem 11.12.1 can sometimes be used to obtain new Taylor series from known Taylor series.

Example 3   Derive the Maclaurin series for $\tan^{-1} x$.

Solution.   We could calculate this Maclaurin series directly. However, we can also exploit Theorem 11.12.1 by first writing

$$\int \frac{1}{1 + x^2}\, dx = \tan^{-1} x + C$$

and then integrating the Maclaurin series for $1/(1 + x^2)$ term by term. Since

$$\frac{1}{1 - x} = 1 + x + x^2 + x^3 + x^4 + \cdots$$

it follows on replacing x by $-x^2$ that

$$\frac{1}{1 + x^2} = 1 - x^2 + x^4 - x^6 + x^8 - \cdots \tag{2}$$

Thus,

$$\tan^{-1} x + C = \int \frac{1}{1 + x^2}\, dx = \int [1 - x^2 + x^4 - x^6 + x^8 - \cdots]\, dx$$

or

$$\tan^{-1} x = \left[x - \frac{x^3}{3} + \frac{x^5}{5} - \frac{x^7}{7} + \frac{x^9}{9} - \cdots \right] - C$$

The constant of integration may be evaluated by substituting $x = 0$ and using the condition $\tan^{-1} 0 = 0$. This gives $C = 0$, so that

$$\tan^{-1} x = x - \frac{x^3}{3} + \frac{x^5}{5} - \frac{x^7}{7} + \frac{x^9}{9} - \cdots \tag{3}$$

◀

Although it is conceivable at this point that series (3) is not the Maclaurin series for $\tan^{-1} x$, but rather some other power series converging to $\tan^{-1} x$, the following theorem tells us that this is not the case.

11.12.2 THEOREM. *If*

$$f(x) = c_0 + c_1(x - a) + c_2(x - a)^2 + \cdots + c_n(x - a)^n + \cdots$$

for all x in some open interval containing a, then the series is the Taylor series for f about a.

Proof. By repeated application of Theorem 11.12.1(a) we obtain

$$f(x) = c_0 + c_1(x - a) + c_2(x - a)^2 + c_3(x - a)^3 + c_4(x - a)^4 + \cdots$$
$$f'(x) = c_1 + 2c_2(x - a) + 3c_3(x - a)^2 + 4c_4(x - a)^3 + \cdots$$
$$f''(x) = 2!c_2 + (3 \cdot 2)c_3(x - a) + (4 \cdot 3)c_4(x - a)^2 + \cdots$$
$$f'''(x) = 3!c_3 + (4 \cdot 3 \cdot 2)c_4(x - a) + \cdots$$
$$\vdots$$

On substituting $x = a$, all the powers of $x - a$ drop out leaving

$$
\begin{array}{lll}
f(a) = c_0 & & c_0 = f(a) \\[4pt]
f'(a) = c_1 & & c_1 = f'(a) \\[4pt]
f''(a) = 2!c_2 & \text{or} & c_2 = \dfrac{f''(a)}{2!} \\[8pt]
f'''(a) = 3!c_3 & & c_3 = \dfrac{f'''(a)}{3!} \\[8pt]
\vdots & &
\end{array}
$$

which shows that the coefficients $c_0, c_1, c_2, c_3, \ldots$ are precisely the coefficients in the Taylor series about a for $f(x)$.   ∎

REMARK.   The foregoing theorem tells us that no matter how we arrive at a power series in $x - a$ converging to $f(x)$, be it by substitution, by integration, by differentiation, or by algebraic manipulation, the resulting series will be the Taylor series about a for $f(x)$.

REMARK.   Parts (a) and (b) of Theorem 11.12.1 say nothing about the behavior of the differentiated and integrated series at the endpoints $a - R$ and $a + R$. Indeed, by differentiating termwise, convergence may be lost at one or both endpoints; and by integrating termwise, convergence may be gained at one or both endpoints. As an illustration, in Example 3 we derived the series (3) for $\tan^{-1} x$ by integrating series (2) for $1/(1 + x^2)$. We leave it as an exercise to show that the series for $\tan^{-1} x$ converges to $\tan^{-1} x$ on the interval $[-1, 1]$, while the series for $1/(1 + x^2)$ converges to $1/(1 + x^2)$ only on $(-1, 1)$. Thus, convergence was gained at both endpoints by integrating.

MISCELLANEOUS TECHNIQUES FOR OBTAINING TAYLOR SERIES

We conclude this section with some tricks for obtaining Taylor series that would be messy to obtain directly. Our objective is to make the reader aware of these techniques; we shall not attempt to justify the results.

Example 4   Find the first three terms that occur in the Maclaurin series for $e^{-x^2} \tan^{-1} x$.

Solution.   Using the series for e^{-x^2} and $\tan^{-1} x$ obtained in Examples 2 and 3 gives

$$e^{-x^2} \tan^{-1} x = \left(1 - x^2 + \frac{x^4}{2} - \cdots\right)\left(x - \frac{x^3}{3} + \frac{x^5}{5} + \cdots\right)$$

We now multiply out, following a familiar format from elementary algebra:

$$
\begin{array}{r}
1 - x^2 + \dfrac{x^4}{2} - \cdots \\[2mm]
\times \quad x - \dfrac{x^3}{3} + \dfrac{x^5}{5} + \cdots \\[1mm]
\hline
x - x^3 + \dfrac{x^5}{2} - \cdots \\[2mm]
- \dfrac{x^3}{3} + \dfrac{x^5}{3} - \dfrac{x^7}{6} + \cdots \\[2mm]
\dfrac{x^5}{5} - \dfrac{x^7}{5} + \cdots \\[1mm]
\hline
x - \tfrac{4}{3}x^3 + \tfrac{31}{30}x^5 - \cdots
\end{array}
$$

Thus,

$$e^{-x^2} \tan^{-1} x = x - \frac{4}{3}x^3 + \frac{31}{30}x^5 - \cdots$$

If desired, more terms in the series can be obtained by including more terms in the factors. ◄

Example 5 Find the first three terms in the Maclaurin series for $\tan x$.

Solution. Instead of computing the series directly, we write

$$\tan x = \frac{\sin x}{\cos x} = \frac{x - \dfrac{x^3}{3!} + \dfrac{x^5}{5!} - \cdots}{1 - \dfrac{x^2}{2!} + \dfrac{x^4}{4!} - \cdots}$$

and follow the familiar "long division" process to obtain

$$
\begin{array}{r}
x + \dfrac{x^3}{3} + \dfrac{2x^5}{15} + \cdots \\[2mm]
1 - \dfrac{x^2}{2} + \dfrac{x^4}{24} - \cdots \,\Big)\, x - \dfrac{x^3}{6} + \dfrac{x^5}{120} - \cdots \\[2mm]
x - \dfrac{x^3}{2} + \dfrac{x^5}{24} - \cdots \\[2mm]
\dfrac{x^3}{3} - \dfrac{x^5}{30} + \cdots \\[2mm]
\dfrac{x^3}{3} - \dfrac{x^5}{6} + \cdots \\[2mm]
\dfrac{2x^5}{15} + \cdots
\end{array}
$$

Thus,

$$\tan x = x + \frac{x^3}{3} + \frac{2x^5}{15} + \cdots \qquad ◄$$

BINOMIAL SERIES   If m is a real number, then the Maclaurin series for $(1 + x)^m$ is called the *binomial series;* it is given by (verify)

$$1 + mx + \frac{m(m-1)}{2!}x^2 + \frac{m(m-1)(m-2)}{3!}x^3 + \cdots$$

It is proved in advanced calculus that the binomial series converges to $(1 + x)^m$ if $|x| < 1$. Thus, for such values of x

$$(1 + x)^m = 1 + mx + \frac{m(m-1)}{2!}x^2 + \cdots$$

$$+ \frac{m(m-1)(m-2)\cdots(m-k+1)}{k!}x^k + \cdots \quad (4)$$

or in sigma notation

$$(1 + x)^m = 1 + \sum_{k=1}^{\infty} \frac{m(m-1)\cdots(m-k+1)}{k!}x^k \quad \text{if } |x| < 1 \quad (4a)$$

REMARK.   If m is a positive integer, then $f(x) = (1 + x)^m$ is a polynomial of degree m so that

$$f^{(m+1)}(0) = f^{(m+2)}(0) = f^{(m+3)}(0) = \cdots = 0$$

and the binomial series reduces to the familiar binomial expansion

$$(1 + x)^m = 1 + mx + \frac{m(m-1)}{2!}x^2 + \frac{m(m-1)(m-2)}{3!}x^3 + \cdots + x^m$$

Example 6   Express $1/\sqrt{1 + x}$ as a binomial series.

Solution.   Substituting $m = -\frac{1}{2}$ in (4) yields

$$\frac{1}{\sqrt{1 + x}} = 1 - \frac{1}{2}x + \frac{(-\frac{1}{2})(-\frac{1}{2} - 1)}{2!}x^2 + \frac{(-\frac{1}{2})(-\frac{1}{2} - 1)(-\frac{1}{2} - 2)}{3!}x^3$$

$$+ \cdots + \frac{(-\frac{1}{2})(-\frac{3}{2})(-\frac{5}{2})\cdots(-\frac{1}{2} - k + 1)}{k!}x^k + \cdots$$

$$= 1 - \frac{1}{2}x + \frac{1 \cdot 3}{2^2 \cdot 2!}x^2 - \frac{1 \cdot 3 \cdot 5}{2^3 \cdot 3!}x^3 + \cdots$$

$$+ (-1)^k \frac{1 \cdot 3 \cdot 5 \cdots (2k-1)}{2^k k!}x^k + \cdots \quad \blacktriangleleft$$

▶ Exercise Set 11.12

In Exercises 1–4, obtain the stated results by differentiating or integrating Maclaurin series term by term.

1. (a) $\dfrac{d}{dx}[e^x] = e^x$ (b) $\displaystyle\int e^x \, dx = e^x + C.$

2. (a) $\dfrac{d}{dx}[\cos x] = -\sin x$

(b) $\displaystyle\int \sin x \, dx = -\cos x + C.$

3. (a) $\dfrac{d}{dx}[\sinh x] = \cosh x$

(b) $\displaystyle\int \sinh x \, dx = \cosh x + C.$

4. (a) $\dfrac{d}{dx}[\ln(1+x)] = \dfrac{1}{1+x}$

(b) $\displaystyle\int \dfrac{1}{1+x}\,dx = \ln(1+x) + C.$

5. Derive the Maclaurin series for $1/(1+x)^2$ by differentiating an appropriate Maclaurin series term by term.

6. By differentiating an appropriate series, show that

$$\sum_{k=1}^{\infty} kx^k = \dfrac{x}{(1-x)^2}$$

7. By integrating an appropriate series, show that

$$\sum_{k=1}^{\infty} \dfrac{x^k}{k} = \ln\left(\dfrac{1}{1-x}\right)$$

8. Use the result of Exercise 6 to find the sum of the series

$$\dfrac{1}{3} + \dfrac{2}{3^2} + \dfrac{3}{3^3} + \dfrac{4}{3^4} + \cdots$$

9. Use the result of Exercise 7 to find the sum of the series

$$\dfrac{1}{4} + \dfrac{1}{2(4^2)} + \dfrac{1}{3(4^3)} + \dfrac{1}{4(4^4)} + \cdots$$

10. Find the sum

$$\sum_{k=0}^{\infty} \dfrac{k+1}{k!} = 1 + 2 + \dfrac{3}{2!} + \dfrac{4}{3!} + \dfrac{5}{4!} + \cdots$$

[*Hint:* Differentiate the Maclaurin series for xe^x.]

11. Find the function to which the series

$$2 + 6x + 12x^2 + 20x^3 + \cdots$$

converges. [*Hint:* Find the second derivative of the Maclaurin series for $1/(1-x)$.]

12. Find the sum

$$\sum_{k=1}^{\infty} \dfrac{k^2}{4^k}$$

[*Hint:* Differentiate the Maclaurin series for $1/(1-x)$, multiply by x, differentiate, and multiply by x again.]

13. Let $\displaystyle f(x) = \sum_{k=0}^{\infty} (-1)^k \dfrac{x^{k+1}}{k+1}$

$$= x - \dfrac{x^2}{2} + \dfrac{x^3}{3} - \dfrac{x^4}{4} + \cdots$$

(a) Use the ratio test to show that the series converges for all x in the interval $(-1, 1)$.

(b) Use part (a) of Theorem 11.12.1 to find a power series for $f'(x)$. What is its interval of convergence?

(c) From the series obtained in part (b) deduce that $f'(x) = 1/(1+x)$ and hence that

$$f(x) = \ln(1+x) \text{ for } -1 < x < 1.$$

[*Remark:* The Lagrange form of the remainder can be used to show that the Maclaurin series for $\ln(1+x)$ converges to $\ln(1+x)$ for $x = 1$ as well.]

In Exercises 14–19, use a binomial series to find the first four nonzero terms in the Maclaurin series for the function and give the radius of convergence.

14. $\sqrt{1+x}.$

15. $(1-x)^{1/3}.$

16. $(1+x^2)^{-1/4}.$

17. $\dfrac{x}{\sqrt[3]{1-2x}}.$

18. $\dfrac{1}{(1+2x)^3}.$

19. $\sqrt{1-4x^2}.$

In Exercises 20–27, use series to approximate the value of the integral to three decimal place accuracy.

20. $\displaystyle\int_0^1 \sin x^2\,dx.$

21. $\displaystyle\int_0^1 \cos\sqrt{x}\,dx.$

22. $\displaystyle\int_0^{0.1} \dfrac{\sin x}{x}\,dx.$

23. $\displaystyle\int_0^{1/2} \dfrac{dx}{1+x^4}.$

24. $\displaystyle\int_0^{1/2} \tan^{-1} 2x^2\,dx.$

25. $\displaystyle\int_0^{0.1} e^{-x^3}\,dx.$

26. $\displaystyle\int_0^{0.2} \sqrt[3]{1+x^4}\,dx.$

27. $\displaystyle\int_0^{1/2} \dfrac{dx}{\sqrt[4]{x^2+1}}.$

In Exercises 28–35, use any method to find the first four nonzero terms in the Maclaurin series of the given function.

28. $x^4 e^x.$

29. $e^{-x^2}\cos x.$

30. $\dfrac{x^2}{1+x^4}.$

31. $\dfrac{\sin x}{e^x}.$

32. $\tanh x$.

33. $x \ln(1 - x^2)$.

34. $\dfrac{\ln(1 + x)}{1 - x}$.

35. $x^2 e^{4x}\sqrt{1 + x}$.

36. Obtain the familiar result, $\lim\limits_{x \to 0}(\sin x)/x = 1$, by finding a power series for $(\sin x)/x$, and taking the limit term by term.

37. Use the method of Exercise 36 to find the limits:

  (a) $\lim\limits_{x \to 0} \dfrac{1 - \cos x}{\sin x}$

  (b) $\lim\limits_{x \to 0} \dfrac{\ln\sqrt{1 + x} - \sin 2x}{x}$.

38. (a) Use the relationship

$$\int \frac{1}{\sqrt{1 - x^2}}\, dx = \sin^{-1} x + C$$

     to find the first four nonzero terms in the Maclaurin series for $\sin^{-1} x$.

  (b) Express the series in sigma notation.

  (c) What is the radius of convergence?

39. (a) Use the relationship

$$\int \frac{1}{\sqrt{1 + x^2}}\, dx = \sinh^{-1} x + C$$

     to find the first four nonzero terms in the Maclaurin series for $\sinh^{-1} x$.

  (b) Express the series in sigma notation.

  (c) What is the radius of convergence?

40. Estimate the maximum error in the approximation $\sqrt{1 + x} \approx 1 + (x/2)$ if $|x| < 0.001$.

41. Derive (4). (Do not try to prove convergence.)

42. Prove: If the power series $\sum\limits_{k=0}^{\infty} a_k x^k$ and $\sum\limits_{k=0}^{\infty} b_k x^k$ have the same sum on an interval $(-r, r)$, then $a_k = b_k$ for all values of k.

43. If m is any real number, and k is a nonnegative integer, then we define the **binomial coefficients** $\dbinom{m}{k}$ by the formulas $\dbinom{m}{0} = 1$ and

$$\binom{m}{k} = \frac{m(m - 1)(m - 2) \cdots (m - k + 1)}{k!}$$

for $k \geq 1$.

Express Formula (4a) in terms of binomial coefficients.

In Exercises 44–49, use the formula obtained in Exercise 43 to express the Maclaurin series for the given function in sigma notation. State the radius of convergence of the series.

44. $(1 + x)^{1/3}$.

45. $(1 - x)^{1/3}$.

46. $(1 - x^2)^{2/3}$.

47. $(1 + x^2)^{2/3}$.

48. $\dfrac{1}{\sqrt{1 + x}}$.

49. $\dfrac{x}{\sqrt{1 + x}}$.

▶ SUPPLEMENTARY EXERCISES

In Exercises 1–6, find $L = \lim\limits_{n \to +\infty} a_n$ if it exists.

1. $a_n = (-1)^n/e^n$.

2. $a_n = e^{1/n}$.

3. $a_n = \dfrac{1}{\sqrt{n}} - \dfrac{1}{\sqrt{n + 1}}$.

4. $a_n = \sin(\pi n)$.

5. $a_n = \sin\left(\dfrac{(2n - 1)\pi}{2}\right)$.

6. $a_n = \dfrac{n + 1}{n(n + 2)}$.

7. Which of the sequences $\{a_n\}_{n=1}^{+\infty}$ in Exercises 1–6 are (a) decreasing? (b) nondecreasing? (c) alternating?

8. Suppose $f(x)$ satisfies

$$f'(x) > 0 \quad \text{and} \quad f(x) \leq 1 - e^{-x}$$

for all $x \geq 1$. What can you conclude about the convergence of $\{a_n\}$ if $a_n = f(n)$, $n = 1, 2, \dots$?

9. Use your knowledge of geometric series and p-series to determine all values of q for which the following series converge.

  (a) $\sum\limits_{k=0}^{\infty} \pi^k/q^{2k}$

  (b) $\sum\limits_{k=1}^{\infty} (1/k^q)^3$

  (c) $\sum\limits_{k=2}^{\infty} 1/(\ln q^k)$

  (d) $\sum\limits_{k=2}^{\infty} 1/(\ln q)^k$.

10. (a) Use a suitable test to find all values of q for which $\sum_{k=2}^{\infty} 1/[k \, (\ln k)^q]$ converges.

(b) Why can't you use the integral test for the series $\sum_{k=1}^{\infty} (2 + \cos k)/k^2$? Test for convergence using a test that does apply.

11. Express $1.3636\ldots$ as (a) an infinite series in sigma notation, (b) a ratio of integers.

12. In parts (a)–(d), use the comparison test to determine whether the series converges.

(a) $\sum_{k=1}^{\infty} \dfrac{2k-1}{3k^2-k}$ (b) $\sum_{k=1}^{\infty} \dfrac{2k+1}{3k^2+k}$

(c) $\sum_{k=1}^{\infty} \dfrac{2k-1}{3k^3-k^2}$ (d) $\sum_{k=1}^{\infty} \dfrac{2k+1}{3k^3+k^2}$.

13. Find the sum of the series (if it converges).

(a) $\sum_{k=1}^{\infty} \dfrac{2^k + 3^k}{6^{k+1}}$ (b) $\sum_{k=2}^{\infty} \ln\left(1 + \dfrac{1}{k}\right)$

(c) $\sum_{k=1}^{\infty} [k^{-1/2} - (k+1)^{-1/2}]$.

In Exercises 14–21, determine whether the series converges or diverges. You may use the following limits without proof:

$$\lim_{k \to +\infty} (1 + 1/k)^k = e, \quad \lim_{k \to +\infty} \sqrt[k]{k} = 1, \quad \lim_{k \to +\infty} \sqrt[k]{a} = 1$$

14. $\sum_{k=0}^{\infty} e^{-k}$.

15. $\sum_{k=1}^{\infty} k e^{-k^2}$.

16. $\sum_{k=1}^{\infty} \dfrac{k}{k^2 + 2k + 7}$.

17. $\sum_{k=1}^{\infty} \dfrac{\sqrt{k}}{k^2 + 7}$.

18. $\sum_{k=1}^{\infty} \left(\dfrac{k}{k+1}\right)^k$.

19. $\sum_{k=0}^{\infty} \dfrac{3^k k!}{(2k)!}$.

20. $\sum_{k=0}^{\infty} \dfrac{k^6 3^k}{(k+1)!}$.

21. $\sum_{k=1}^{\infty} \left(\dfrac{5k}{2k+1}\right)^{3k}$.

In Exercises 22–25, determine whether the given series is absolutely convergent, conditionally convergent, or divergent.

22. $\sum_{k=1}^{\infty} (-1)^k/e^{1/k}$.

23. $\sum_{k=0}^{\infty} (-2)^k/(3^k + 1)$.

24. $\sum_{k=0}^{\infty} (-1)^k/(2k+1)$.

25. $\sum_{k=0}^{\infty} (-1)^k 3^k/2^{k+1}$.

26. Find the smallest value of n for which the nth partial sum approximates the sum of the series to the stated accuracy:

(a) $\sum_{k=1}^{\infty} \dfrac{(-1)^k}{k^2 + 1}$; $|\text{error}| < 0.0001$.

(b) $\sum_{k=1}^{\infty} \dfrac{(-1)^k}{5^k + 1}$; $|\text{error}| < 0.00005$.

In Exercises 27–32, determine the radius of convergence and the interval of convergence of the given power series.

27. $\sum_{k=1}^{\infty} \dfrac{(x-1)^k}{k\sqrt{k}}$.

28. $\sum_{k=1}^{\infty} \dfrac{(2x)^k}{3k}$.

29. $\sum_{k=1}^{\infty} \dfrac{(1-x)^{2k}}{4^k k}$.

30. $\sum_{k=1}^{\infty} \dfrac{k^2(x+2)^k}{(k+1)!}$.

31. $\sum_{k=1}^{\infty} \dfrac{k!(x-1)^k}{5^k}$.

32. $\sum_{k=1}^{\infty} \dfrac{(2k)! x^k}{(2k+1)!}$.

In Exercises 33–35, find:
(a) the nth Taylor polynomial for f about $x = a$ (for the stated values of n and a);
(b) Lagrange's form of $R_n(x)$ (for the stated values of n and a);
(c) an upper bound on the absolute value of the error if $f(x)$ is approximated over the given interval by the Taylor polynomial obtained in part (a).

33. $f(x) = \ln(x-1)$; $a = 2$; $n = 3$; $[\frac{3}{2}, 2]$.

34. $f(x) = e^{x/2}$; $a = 0$; $n = 4$; $[-1, 0]$.

35. $f(x) = \sqrt{x}$; $a = 1$; $n = 2$; $[\frac{4}{9}, 1]$.

36. (a) Use the identity $a - x = a(1 - x/a)$ to find the Maclaurin series for $1/(a - x)$ from the geometric series. What is its radius of convergence?

(b) Find the Maclaurin series and radius of convergence of $1/(3 + x)$.

(c) Find the Maclaurin series and radius of convergence of $2x/(4 + x^2)$.

(d) Use partial fractions to find the Maclaurin series and radius of convergence of
$$\dfrac{1}{(1 - x)(2 - x)}.$$

37. Use the known Maclaurin series for $\ln(1 + x)$ to find the Maclaurin series and radius of convergence of $\ln(a + x)$ for $a > 0$.

38. Use the identity $x = a + (x - a)$ and the known Maclaurin series for e^x, $\sin x$, $\cos x$, and $1/(1 - x)$ to find the Taylor series about $x = a$ for: (a) e^x, (b) $\sin x$, and (c) $1/x$.

39. Use the series of Example 6 in Section 11.12 to find the Maclaurin series and radius of convergence of $1/\sqrt{9 + x}$.

In Exercises 40–45, use any method to find the first three nonzero terms of the Maclaurin series.

40. $e^{\tan x}$.

41. $\sec x$.

42. $(\sin x)/(e^x - x)$.

43. $\sqrt{\cos x}$.

44. $e^x \ln (1 - x)$.

45. $\ln (1 + \sin x)$.

46. Find a power series for $\dfrac{1 - \cos 3x}{x^2}$ and use it to evaluate $\lim\limits_{x \to 0} \dfrac{1 - \cos 3x}{x^2}$.

47. Find a power series for $\dfrac{\ln (1 - 2x)}{x}$ and use it to evaluate $\lim\limits_{x \to 0} \dfrac{\ln (1 - 2x)}{x}$.

48. How many decimal places of accuracy can be guaranteed if we approximate $\cos x$ by $1 - x^2/2$ for $-0.1 < x < 0.1$?

49. For what values of x can $\sin x$ be replaced by $x - x^3/6 + x^5/120$ with an assured accuracy of 6×10^{-4}?

In Exercises 50–52, approximate the indicated quantity to three decimal place accuracy.

50. $\cos (10°)$.

51. $\displaystyle\int_0^1 \frac{(1 - e^{-t/2})}{t} \, dt$.

52. $\displaystyle\int_0^1 \frac{\sin x}{\sqrt{x}} \, dx$.

53. Show that $y = \displaystyle\sum_{n=0}^{\infty} k^n x^n/n!$ satisfies $y' - ky = 0$ for any fixed k.

12 topics in analytic geometry

12.1 INTRODUCTION TO THE CONIC SECTIONS

Axis

Vertex

Right circular cone

Figure 12.1.1

The surface shown in Figure 12.1.1 is called a *double right circular cone* or sometimes simply a *cone*. It is the three-dimensional surface generated by a line that revolves about a fixed *axis* in such a way that the line always passes through a fixed point on the axis, called the *vertex,* and always makes the same angle with the axis. The cone consists of two parts, or *nappes,* that intersect at the vertex.

The curves that can be obtained by intersecting a cone with a plane are called *conics* or *conic sections,* the most important of which are circles, ellipses, parabolas, and hyperbolas. A *circle* is obtained by intersecting a cone with a plane that is perpendicular to the axis and does not contain the vertex (Figure 12.1.2a). If the cutting plane is tilted slightly and intersects only one nappe, the resulting intersection is an *ellipse* (Figure 12.1.2b). If the cutting plane is tilted still further so that it is parallel to a line on the surface of the cone, but intersects only one nappe, the resulting intersection is a *parabola* (Figure 12.1.2c). Finally, if the plane intersects both nappes, but does not contain the vertex, the resulting intersection is a *hyperbola* (Figure 12.1.2d). By choosing the cutting plane to pass through the vertex, it is possible to obtain a point or a pair of lines for the intersection. These are called *degenerate conic sections.*

According to the Alexandrian geographer and astronomer Eratosthenes, the conic sections were first discovered by Menaechmus, a geometer and astronomer in Plato's academy. While it is not known for certain what motivated the discovery of the conic sections, it is commonly believed that they resulted from the study of construction problems. Menaechmus, for example, used them to solve the problem of ''doubling the cube,'' that is, constructing a cube whose volume is twice that of a given cube. Another theory suggests that the conic sections may have originated as a result of work on sundials. With the advent of analytic geometry and calculus, conic sections gained importance in the physical sciences. In 1609 Johannes Kepler published a book known as *Astronomia Nova* in which he presented his landmark discovery that the path of each planet

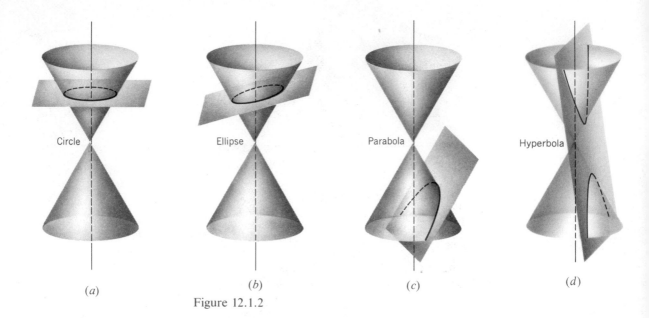

Figure 12.1.2

about the sun is an ellipse. Galileo and Newton showed that objects subject to gravitational forces can also move along paths that are parabolas or hyperbolas.

Today, properties of conic sections are used in the construction of telescopes, radar antennas, and navigational systems, and in the determination of satellite orbits. In this chapter we shall develop the basic geometric properties and equations of conic sections. For simplicity, our working definitions of the conic sections will be based on their geometric properties rather than their interpretation as intersections of a plane with a cone. This will enable us to keep our work in a two-dimensional setting.

12.2   THE PARABOLA; TRANSLATION OF COORDINATE AXES

In this section we shall discuss properties of parabolas.

12.2.1   DEFINITION. A *parabola* is the set of all points in the plane that are equidistant from a given line and a given point not on the line.

The given line is called the *directrix* of the parabola, and the given point the *focus* (Figure 12.2.1). A parabola is symmetric about the line that passes through the focus at right angles to the directrix. This line of symmetry, called the *axis* of the parabola, meets the parabola at a point called the *vertex*.

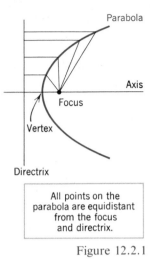

Parabola

Axis

Focus

Vertex

Directrix

All points on the parabola are equidistant from the focus and directrix.

Figure 12.2.1

The equation of a parabola is simplest if the coordinate axes are positioned so that the vertex is at the origin and the axis of symmetry is along the x-axis or y-axis. The four possible such orientations are shown in Table 12.2.1. In the first two orientations, the focus is assumed to be at the point $(p, 0)$, where p is positive in one case and negative in the other. Since the vertex is equidistant from focus and directrix, it follows

Table 12.2.1

ORIENTATION	DESCRIPTION	EQUATION
$x = -p$, $(p, 0)$	Vertex at the origin. Parabola opens in the positive x-direction. Symmetric about the x-axis.	$y^2 = 4px$ (p positive)
$(p, 0)$, $x = -p$	Vertex at the origin. Parabola opens in the negative x-direction. Symmetric about the x-axis.	$y^2 = 4px$ (p negative)
$(0, p)$, $y = -p$	Vertex at the origin. Parabola opens in the positive y-direction. Symmetric about the y-axis.	$x^2 = 4py$ (p positive)
$y = -p$, $(0, p)$	Vertex at the origin. Parabola opens in the negative y-direction. Symmetric about the y-axis.	$x^2 = 4py$ (p negative)

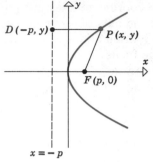

Figure 12.2.2

that the directrix has equation $x = -p$ in these cases. In the last two orientations, the focus is assumed to be at $(0, p)$; and the directrix has equation $y = -p$, where again p may be positive or negative.

To illustrate how the equations in Table 12.2.1 are obtained, let us begin with the first entry in the table—the parabola with focus at $(p, 0)$, directrix $x = -p$, and p positive.

Let $P(x, y)$ be any point on the parabola. Since P is equidistant from focus and directrix, the distances PF and PD in Figure 12.2.2 are equal; that is,

$$PF = PD \tag{1}$$

where $D(-p, y)$ is the foot of the perpendicular from P to the directrix. From the distance formula, the distances PF and PD are

$$PF = \sqrt{(x - p)^2 + y^2} \quad \text{and} \quad PD = \sqrt{(x + p)^2} \tag{2}$$

Substituting in (1) and squaring yields

$$(x - p)^2 + y^2 = (x + p)^2 \tag{3}$$

and after simplifying

$$y^2 = 4px \tag{4}$$

Conversely, any point $P(x, y)$ satisfying (4) also satisfies (3) (reverse the steps in the simplification); thus, from (2), $PF = PD$, which shows that P is equidistant from focus and directrix. Therefore, each point satisfying (4) lies on the parabola.

The remaining equations in Table 12.2.1 have similar derivations; they are left as exercises.

Example 1   Find the focus and directrix of the parabola with equation $y^2 = -8x$.

Solution.   The equation is of the form $y^2 = 4px$ with $4p = -8$ or $p = -2$. Since y occurs only to an even power, the parabola is symmetric about the x-axis. Thus, the focus is on the x-axis at the point $(-2, 0)$ and the directrix has equation $x = 2$ (Figure 12.2.3).   ◀

Example 2   Find an equation for the parabola that is symmetric about the y-axis, has its vertex at the origin, and passes through the point $(5, 2)$.

Solution.   Since the parabola is symmetric about the y-axis and has its vertex at the origin, the equation is of the form

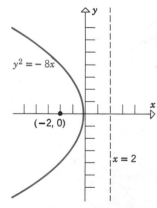

Figure 12.2.3

$$x^2 = 4py \tag{5}$$

Since the parabola passes through $(5, 2)$ we must have $5^2 = 4p(2)$ or $4p = \frac{25}{2}$. Therefore, (5) becomes

$$x^2 = \frac{25}{2} y \quad \blacktriangleleft$$

If a parabola has its axis of symmetry parallel to one of the coordinate axes, but its vertex is not at the origin, then the location of the focus and directrix may be found by introducing an auxiliary coordinate system with its origin at the vertex and its axes parallel to the original axes. Before we can discuss the details, we need some preliminary results.

TRANSLATION OF AXES

In Figure 12.2.4a we have translated the axes of an xy-coordinate system to obtain a new $x'y'$-coordinate system whose origin O' is at the point $(x, y) = (h, k)$. As a result, a point P in the plane will have both (x, y)-coordinates and (x', y')-coordinates. As suggested by Figure 12.2.4b, these coordinates are related by

$$x' = x - h, \quad y' = y - k \tag{6}$$

or, equivalently,

$$x = x' + h, \quad y = y' + k \tag{7}$$

These are called the *translation equations*.

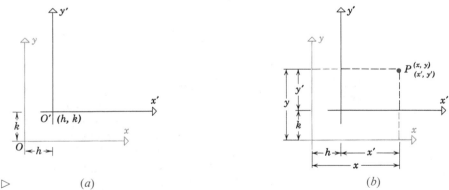

Figure 12.2.4 ▷   (a)   (b)

As an illustration, if the new origin is at $(h, k) = (4, -1)$ and the xy-coordinates of a point P are $(2, 5)$, then the $x'y'$-coordinates of P are

$$x' = x - h = 2 - 4 = -2 \quad \text{and} \quad y' = y - k = 5 - (-1) = 6$$

TRANSLATED PARABOLAS

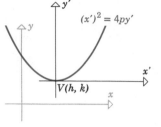

$(x')^2 = 4py'$

$V(h, k)$

Figure 12.2.5

Let us now consider a parabola with vertex V at (h, k) and axis of symmetry parallel to the y-axis. If, as in Figure 12.2.5, we translate the axes so that the vertex V is at the origin of an $x'y'$-coordinate system, then in $x'y'$-coordinates the equation of the parabola will be

$$(x')^2 = 4py'$$

and from the translation equations (6), the corresponding equation in xy-coordinates will be:

> **Parabola with Vertex (h, k) and Axis Parallel to y-axis**
>
> $$(x - h)^2 = 4p(y - k) \tag{8}$$

Thus, in xy-coordinates, an equation of form (8) represents a parabola with vertex at (h, k) and axis of symmetry parallel to the y-axis. The parabola opens in the positive or negative y-direction according to whether p is positive or negative. Similarly,

> **Parabola with Vertex (h, k) and Axis Parallel to x-axis**
>
> $$(y - k)^2 = 4p(x - h) \tag{9}$$

represents a parabola with vertex at (h, k) and axis of symmetry parallel to the x-axis. The parabola opens in the positive or negative x-direction according to whether p is positive or negative.

Example 3   Sketch the parabola

$$(y - 3)^2 = 8(x + 4)$$

and locate the focus and directrix.

Solution.   From (9), the equation represents a parabola with vertex at $(h, k) = (-4, 3)$ and axis of symmetry parallel to the x-axis. Moreover, $4p = 8$ or $p = 2$. Since p is positive, the parabola opens in the positive x-direction. This places the focus 2 units to the right of the vertex or at the point $(-2, 3)$. The directrix is 2 units to the left of the vertex (and parallel to the y-axis), so its equation is $x = -6$. The parabola is sketched in Figure 12.2.6.   ◄

Sometimes (8) and (9) appear in expanded form, in which case some algebraic manipulations are required to identify the parabola.

Vertex (−4, 3)   Focus (−2, 3)

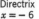
Directrix $x = -6$

Figure 12.2.6   **Example 4**   Show that the curve

$$y = 6x^2 - 12x + 8$$

is a parabola.

Solution.   Because the equation involves x to the second power and y to the first power, and because the expanded form of (8) has the same property, we shall try to rewrite the equation in form (8). To do this, we first divide by the coefficient of x^2 and collect all the x-terms on one side:

$$\frac{1}{6}y - \frac{8}{6} = x^2 - 2x$$

Next, we complete the square on the x-terms by adding 1 to both sides:

$$\frac{1}{6}y - \frac{2}{6} = (x - 1)^2$$

Finally, we factor out the coefficient of the y-term to obtain

$$(x - 1)^2 = \frac{1}{6}(y - 2)$$

which has form (8) with

$$h = 1, \quad k = 2, \quad 4p = \frac{1}{6}, \quad p = \frac{1}{24}$$

Thus, the curve is a parabola with vertex at $(1, 2)$ and axis of symmetry parallel to the y-axis. Since $p = \frac{1}{24}$ is positive, the parabola opens in the positive y-direction, the focus is $\frac{1}{24}$ unit above the vertex, and the directrix $\frac{1}{24}$ unit below the vertex. Thus, the focus is at $(1, \frac{49}{24})$ and the directrix has equation $y = \frac{47}{24}$ (Figure 12.2.7).   ◀

Figure 12.2.7 ▷

The procedure used in Example 4 can be used to prove the following general result.

12.2.2 THEOREM. *The graph of*

$$y = Ax^2 + Bx + C \quad (A \neq 0)$$

is a parabola with axis of symmetry parallel to the y-axis; the parabola opens in the positive y-direction if $A > 0$ and in the negative y-direction if $A < 0$. The graph of

$$x = Ay^2 + By + C \quad (A \neq 0)$$

is a parabola with axis of symmetry parallel to the x-axis; the parabola opens in the positive x-direction if $A > 0$ and in the negative x-direction if $A < 0$.

Example 5 Find an equation for the parabola with vertex $(1, 2)$ and focus $(4, 2)$.

Solution. Since the focus and vertex are on a horizontal line, and since the focus is to the right of the vertex, the parabola opens to the right and its equation has the form

$$(y - k)^2 = 4p(x - h) \quad (p \text{ positive})$$

Since the vertex and focus are three units apart we have $p = 3$, and since the vertex is at $(h, k) = (1, 2)$ we obtain

$$(y - 2)^2 = 12(x - 1) \quad \blacktriangleleft$$

REFLECTION PROPERTIES OF PARABOLAS

Parabolas have important applications in the design of telescopes, radar antennas, and lighting systems. This is due to the following property of parabolas (Exercise 44).

A Geometric Property of Parabolas

12.2.3 THEOREM. *The tangent line at a point P on a parabola makes equal angles with the line through P parallel to the axis of symmetry and the line through P and the focus (Figure 12.2.8).*

It is a principle of physics that when light is reflected from a point P on a surface, the angle of incidence equals the angle of reflection; that is, the angle between the incoming ray and the tangent line at P equals the angle between the outgoing ray and the tangent line at P. Therefore, if the reflecting surface has parabolic cross sections with a common focus, then it follows from Theorem 12.2.3 that all light rays entering parallel to the axis will be reflected through the focus (Figure 12.2.9a). In reflecting

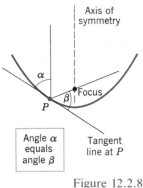

Axis of symmetry

α

Focus

β

P

| Angle α equals angle β | Tangent line at P |

Figure 12.2.8

telescopes, this principle is used to reflect the (approximately) parallel rays of light from the stars or planets off a parabolic mirror to an eyepiece at the focus of the parabola. Conversely, if a light source is located at the focus of a parabolic reflector, it follows from Theorem 12.2.3 that the reflected rays will form a beam parallel to the axis (Figure 12.2.9b). The parabolic reflectors in flashlights and automobile headlights utilize this principle. The optical principles just discussed also apply to radar signals, which explains the parabolic shape of many radar antennas.

Figure 12.2.9   (a)   (b)

▶ Exercise Set 12.2

In Exercises 1–16, sketch the parabola. Show the focus, vertex, and directrix.

1. $y^2 = 6x$.

2. $y^2 = -10x$.

3. $x^2 = -9y$.

4. $x^2 = 4y$.

5. $5y^2 = 12x$.

6. $x^2 - 40y = 0$.

7. $(y - 3)^2 = 6(x - 2)$.

8. $(y + 1)^2 = -7(x - 4)$.

9. $(x + 2)^2 = -(y + 2)$.

10. $(x - \frac{1}{2})^2 = 2(y - 1)$.

11. $x^2 - 4x + 2y = 1$.

12. $y^2 - 6y - 2x + 1 = 0$.

13. $x = y^2 - 4y + 2$.

14. $y = 4x^2 + 8x + 5$.

15. $-y^2 + 2y + x = 0$.

16. $y = 1 - 4x - x^2$.

In Exercises 17–31, find an equation for the parabola satisfying the given conditions.

17. Vertex $(0, 0)$; focus $(3, 0)$.

18. Vertex $(0, 0)$; focus $(0, -4)$.

19. Vertex $(0, 0)$; directrix $x = 7$.

20. Vertex $(0, 0)$; directrix $y = \frac{1}{2}$.

21. Vertex $(0, 0)$; symmetric about the x-axis; passes through $(2, 2)$.

22. Vertex $(0, 0)$; symmetric about the y-axis; passes through $(-1, 3)$.

23. Focus $(0, -3)$; directrix $y = 3$.

24. Focus $(6, 0)$; directrix $x = -6$.

25. Axis $y = 0$; passes through $(3, 2)$ and $(2, -3)$.

26. Axis $x = 0$; passes through $(2, -1)$ and $(-4, 5)$.

27. Focus $(3, 0)$; directrix $x = 0$.

28. Vertex $(4, -5)$; focus $(1, -5)$.

29. Vertex $(1, 1)$; directrix $y = -2$.

30. Focus $(-1, 4)$; directrix $x = 5$.

31. Vertex $(5, -3)$; axis parallel to the y-axis; passes through $(9, 5)$.

32. Use the definition of a parabola (12.2.1) to find

the equation of the parabola with focus (2, 1) and directrix $x + y + 1 = 0$. [*Hint:* Use the result of Exercise 35, Section 1.5.]

33. (a) Find an equation for the parabola with axis parallel to the y-axis and passing through (0, 3), (2, 0), and (3, 2).

(b) Find an equation for the parabola with axis parallel to the x-axis and passing through the points in (a).

34. Prove: The line tangent to the parabola $x^2 = 4py$ at (x_0, y_0) is

$$y = \frac{x_0}{2p}x - y_0$$

35. Find the vertex, focus, and directrix of the parabola $y = Ax^2 + Bx + C$ $(A \neq 0)$.

36. (a) Find an equation for the following parabolic arch with base b and height h.

(b) Find the area under the arch.

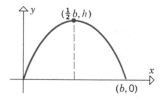

37. A parabolic arch spans a road 40 feet wide. How high is the arch if a center section of the road 20 feet wide has a minimum clearance of 12 feet?

38. Let C be a circle of radius r and L a line that does not intersect C and is in the same plane as C. Show that the centers of all circles that do not enclose C and are tangent to both C and L lie on a parabola. Specify the location of the focus and directrix of the parabola.

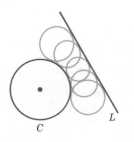

39. Prove: The vertex is the closest point on a parabola to the focus.

40. A comet moves in a parabolic orbit with the sun at its focus. When the comet is 40 million miles from the sun, the line from the sun to the comet makes an angle of 60° with the axis of the parabola. How close will the comet come to the sun? [*Hint:* See Exercise 39.]

41. How far from the vertex should a light source be placed on the axis of the following parabolic reflector to produce a beam of parallel rays?

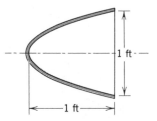

42. Derive the equation $x^2 = 4py$ (p positive) in Table 12.2.1.

43. Derive the equation $y^2 = 4px$ (p negative) in Table 12.2.1.

44. Prove Theorem 12.2.3. [*Hint:* Choose coordinate axes so the parabola has the equation $x^2 = 4py$. Show that the tangent line at $P(x_0, y_0)$ intersects the y-axis at $Q(0, -y_0)$ and that the triangle whose three vertices are at P, Q, and the focus is isosceles.]

12.3 THE ELLIPSE

In this section we shall discuss properties of ellipses.

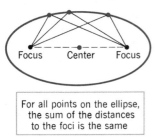

For all points on the ellipse, the sum of the distances to the foci is the same

Figure 12.3.1

> **12.3.1** DEFINITION. An *ellipse* is the set of all points in the plane, the sum of whose distances from two fixed points is a given positive constant.

The two fixed points are called the *foci* (plural of "focus"), and the midpoint of the line segment joining the foci is called the *center* of the ellipse (Figure 12.3.1).

To help visualize Definition 12.3.1, imagine that two ends of a string are tacked to the foci and a pencil traces a curve as it is held tight against the string (Figure 12.3.2). The resulting curve will be an ellipse since the sum of the distances to the foci is a constant, namely the total length of the string. Note that if the foci coincide, the ellipse reduces to a circle.

The line segment through the foci and across the ellipse is called the *major axis* (Figure 12.3.3), while the line segment across the ellipse, through the center, and perpendicular to the major axis is called the *minor axis*. It is traditional in the study of ellipses to denote the length of the major axis by $2a$, the length of the minor axis by $2b$, and the distance between the foci by $2c$ (Figure 12.3.4). The numbers a and b are called the *semiaxes* of the ellipse.

There is a basic relationship between the numbers a, b, and c that can be obtained by considering a point P at the end of the major axis and a point Q at the end of the minor axis. Because P and Q both lie on the ellipse, the sum of the distances from each of them to the foci will be the same. If we express this fact as an equation we obtain (see Figure 12.3.5):

$$2\sqrt{b^2 + c^2} = (a - c) + (a + c)$$

Figure 12.3.2

or

$$a = \sqrt{b^2 + c^2} \tag{1}$$

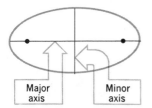

Major axis Minor axis

Figure 12.3.3

Figure 12.3.4

Figure 12.3.5

Figure 12.3.6

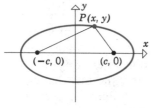

Figure 12.3.7

Relationship (1) shows that the distance from a focus to an end of the minor axis is a (Figure 12.3.6), which implies that for *all* points on the ellipse the sum of the distances to the foci is $2a$. (Why?)

It also follows from (1) that $a \geq b$ with the equality holding only when $c = 0$. Geometrically, this means that the major axis of an ellipse is at least as large as the minor axis, and that the two axes have equal length only when the foci coincide, in which case the ellipse is a circle.

The equation of an ellipse is simplest if the coordinate axes are positioned so the center of the ellipse is at the origin and the foci are on the x-axis or y-axis. The two possible such orientations are shown in Table 12.3.1.

To illustrate how the equations in Table 12.3.1 are derived, let us consider the first entry in the table. Let $P(x, y)$ be any point on the ellipse. Since the sum of the distances from P to the foci is $2a$ it follows (Figure 12.3.7) that

$$\sqrt{(x + c)^2 + y^2} + \sqrt{(x - c)^2 + y^2} = 2a$$

Transposing the second radical to the right side of the equation and squaring yields

$$(x + c)^2 + y^2 = 4a^2 - 4a\sqrt{(x - c)^2 + y^2} + (x - c)^2 + y^2$$

and, on simplifying,

$$\sqrt{(x - c)^2 + y^2} = a - \frac{c}{a}x$$

Squaring again and simplifying yields

$$\frac{x^2}{a^2} + \frac{y^2}{a^2 - c^2} = 1$$

which, by virtue of (1), can be written as

$$\frac{x^2}{a^2} + \frac{y^2}{b^2} = 1 \tag{2}$$

Conversely, it can be shown that any point whose coordinates satisfy (2) has $2a$ as the sum of its distances from the foci, so that such a point is on the ellipse.

The second equation in Table 12.3.1 has a similar derivation.

Example 1   Does the ellipse

$$\frac{x^2}{9} + \frac{y^2}{4} = 1$$

have its major axis on the x-axis or the y-axis?

Table 12.3.1

ORIENTATION	DESCRIPTION	EQUATION
(ellipse with major axis on x-axis; points $(0,b)$, $(0,-b)$, $(-a,0)$, $(a,0)$, $(-c,0)$, $(c,0)$)	Foci and major axis on the x-axis. Minor axis on the y-axis. Center at the origin.	$\dfrac{x^2}{a^2} + \dfrac{y^2}{b^2} = 1$
(ellipse with major axis on y-axis; points $(0,a)$, $(0,-a)$, $(0,c)$, $(0,-c)$, $(-b,0)$, $(b,0)$)	Foci and major axis on the y-axis. Minor axis on the x-axis. Center at the origin.	$\dfrac{x^2}{b^2} + \dfrac{y^2}{a^2} = 1$

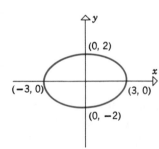

Figure 12.3.8

Solution.   Let us locate the intersections of the ellipse with the coordinate axes. Setting $y = 0$ yields $x^2 = 9$, or $x = \pm 3$, so the ellipse intersects the x-axis at $(-3, 0)$ and $(3, 0)$. Setting $x = 0$ yields $y^2 = 4$, or $y = \pm 2$, so the ellipse intersects the y-axis at $(0, -2)$ and $(0, 2)$. Plotting the points of intersection and making a rough sketch shows that the major axis is along the x-axis (Figure 12.3.8).

Alternative Solution.   Since it is always true that $a^2 \geq b^2$, it follows that $a^2 = 9$ and $b^2 = 4$, so that the given equation is of the form

$$\frac{x^2}{a^2} + \frac{y^2}{b^2} = 1$$

which is an ellipse with major axis along the x-axis (Table 12.3.1).   ◀

Example 2   Show that the graph of the equation

$$2x^2 + y^2 = 4$$

is an ellipse, and locate the foci.

Solution.   Dividing through by 4 yields

$$\frac{x^2}{2} + \frac{y^2}{4} = 1 \tag{3}$$

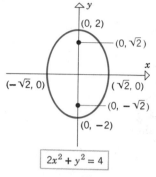

$$2x^2 + y^2 = 4$$

Figure 12.3.9

which is the equation of an ellipse having $a^2 = 4$ ($a = 2$) and $b^2 = 2$ ($b = \sqrt{2}$). Since the a^2 is under the y^2 in (3), the major axis is along the y-axis. From relationship (1)

$$c = \sqrt{a^2 - b^2} = \sqrt{4 - 2} = \sqrt{2}$$

so the foci are $(0, \sqrt{2})$ and $(0, -\sqrt{2})$. The ellipse is shown in Figure 12.3.9.   ◀

Example 3   Find an equation for the ellipse with foci $(0, \pm 2)$ and major axis with endpoints $(0, \pm 4)$.

Solution.   From Table 12.3.1, the equation has the form

$$\frac{x^2}{b^2} + \frac{y^2}{a^2} = 1$$

and from the given information, $a = 4$ and $c = 2$. It follows from (1) that

$$b^2 = a^2 - c^2 = 16 - 4 = 12$$

so the equation of the ellipse is

$$\frac{x^2}{12} + \frac{y^2}{16} = 1   ◀$$

TRANSLATED
ELLIPSES

If the axes of an ellipse are parallel to the coordinate axes but the center is not at the origin, then the equation of the ellipse may be determined by translation of axes. Specifically, if the center of the ellipse is at the point (h, k), and if we translate the xy-axes so the origin of the $x'y'$-system is at the center (h, k), then the equation of the ellipse in the $x'y'$-system will be

$$\frac{(x')^2}{a^2} + \frac{(y')^2}{b^2} = 1 \quad \text{or} \quad \frac{(x')^2}{b^2} + \frac{(y')^2}{a^2} = 1$$

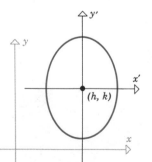

Figure 12.3.10

depending on the orientation of the major and minor axes (Figure 12.3.10). Thus, from the translation equations $x' = x - h$, $y' = y - k$, the equation of the ellipse in the xy-system will be one of the following:

Ellipse with Center (h, k) and Major Axis Parallel to x-axis

$$\frac{(x - h)^2}{a^2} + \frac{(y - k)^2}{b^2} = 1 \tag{4}$$

> **Ellipse with Center (h, k) and Major Axis Parallel to y-axis**
>
> $$\frac{(x - h)^2}{b^2} + \frac{(y - k)^2}{a^2} = 1 \tag{5}$$

Sometimes (4) and (5) appear in expanded form, in which case it is necessary to complete the squares before the ellipse can be analyzed.

Example 4   Show that the curve

$$16x^2 + 9y^2 - 64x - 54y + 1 = 0$$

is an ellipse. Sketch the ellipse and show the location of the foci.

Solution.   Our objective is to rewrite the equation in one of the forms, (4) or (5). To do this, group the x-terms and y-terms and take the constant to the right side:

$$(16x^2 - 64x) + (9y^2 - 54y) = -1$$

Next, factor out the coefficients of x^2 and y^2 and complete the squares:

$$16(x^2 - 4x + 4) + 9(y^2 - 6y + 9) = -1 + 64 + 81$$

or

$$16(x - 2)^2 + 9(y - 3)^2 = 144$$

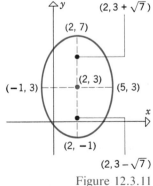

(2, 7)

(2, 3 + $\sqrt{7}$)

(2, 3)

(−1, 3)

(5, 3)

(2, −1)

(2, 3 − $\sqrt{7}$)

Figure 12.3.11

Finally, divide through by 144 to introduce a 1 on the right side:

$$\frac{(x - 2)^2}{9} + \frac{(y - 3)^2}{16} = 1$$

This is an equation of form (5), with $h = 2$, $k = 3$, $a^2 = 16$, and $b^2 = 9$. Thus, the given equation is an ellipse with center $(2, 3)$ and major axis parallel to the y-axis. Since $a = 4$, the major axis extends 4 units above and 4 units below the center, so its endpoints are $(2, 7)$ and $(2, -1)$ (Figure 12.3.11). Since $b = 3$, the minor axis extends 3 units to the left and 3 units to the right of the center, so its endpoints are $(-1, 3)$ and $(5, 3)$. Since

$$c = \sqrt{a^2 - b^2} = \sqrt{16 - 9} = \sqrt{7}$$

the foci lie $\sqrt{7}$ units above and below the center, placing them at the points $(2, 3 + \sqrt{7})$ and $(2, 3 - \sqrt{7})$.   ◄

REFLECTION PROPERTIES OF ELLIPSES

In the exercises the reader is asked to prove the following result.

A Geometric Property of Ellipses

12.3.2 THEOREM. *A line tangent to an ellipse at a point P makes equal angles with the lines through P and the foci (Figure 12.3.12).*

Tangent line at P

Angle α equals angle β

Figure 12.3.12

It follows from Theorem 12.3.2 that a ray of light emanating from one focus of an ellipse will be reflected through the other focus. This property of ellipses is used in "whispering galleries." Such rooms have ceilings whose cross sections are elliptical in shape with common foci. As a result, if a person standing at one focus whispers, the sound waves are reflected by the ceiling to the other focus, making it possible for a person at that focus to hear the whispered sound.

▶ Exercise Set 12.3

In Exercises 1–14, sketch the ellipse. Label the foci and the ends of the major and minor axes.

1. $\dfrac{x^2}{16} + \dfrac{y^2}{9} = 1$.

2. $\dfrac{x^2}{4} + \dfrac{y^2}{25} = 1$.

3. $9x^2 + y^2 = 9$.

4. $4x^2 + 9y^2 = 36$.

5. $x^2 + 3y^2 = 2$.

6. $16x^2 + 4y^2 = 1$.

7. $9(x - 1)^2 + 16(y - 3)^2 = 144$.

8. $(x + 3)^2 + 4(y - 5)^2 = 16$.

9. $3(x + 2)^2 + 4(y + 1)^2 = 12$.

10. $\frac{1}{4}x^2 + \frac{1}{9}(y + 2)^2 - 1 = 0$.

11. $x^2 + 9y^2 + 2x - 18y + 1 = 0$.

12. $9x^2 + 4y^2 + 18x - 24y + 9 = 0$.

13. $4x^2 + y^2 + 8x - 10y = -13$.

14. $5x^2 + 9y^2 - 20x + 54y = -56$.

In Exercises 15–26, find an equation for the ellipse satisfying the given conditions.

15. Ends of major axis $(\pm 3, 0)$; ends of minor axis $(0, \pm 2)$.

16. Ends of major axis $(0, \pm\sqrt{5})$; ends of minor axis $(\pm 1, 0)$.

17. Length of major axis 26; foci $(\pm 5, 0)$.

18. Length of minor axis 16; foci $(0, \pm 6)$.

19. Foci $(\pm 1, 0)$; $b = \sqrt{2}$.

20. Foci $(\pm 3, 0)$; $a = 4$.

21. $c = 2\sqrt{3}$; $a = 4$; center at the origin; foci on a coordinate axis (two answers).

22. $b = 3$; $c = 4$; center at the origin; foci on a coordinate axis (two answers).

23. Ends of major axis $(\pm 6, 0)$; passes through $(2, 3)$.

24. Center at $(0, 0)$; major and minor axes along the coordinate axes; passes through $(3, 2)$ and $(1, 6)$.

25. Foci $(1, 2)$ and $(1, 4)$; minor axis of length 2.

26. Foci $(2, 1)$ and $(2, -3)$; major axis of length 6.

27. Find an equation of the ellipse traced by a point that moves so the sum of its distances to $(4, 1)$ and $(4, 5)$ is 12.

28. Find an equation of the ellipse traced by a point that moves so the sum of its distances to $(0, 0)$ and $(1, 1)$ is 4.

In Exercises 29–32, find all intersections of the given curves and make a sketch of the curves that shows the points of intersection.

29. $x^2 + 4y^2 = 40$ and $x + 2y = 8$.

30. $y^2 = 2x$ and $x^2 + 2y^2 = 12$.

31. $x^2 + 9y^2 = 36$ and $x^2 + y^2 = 20$.

32. $16x^2 + 9y^2 = 36$ and $5x^2 + 18y^2 = 45$.

33. Find two values of k such that the line $x + 2y = k$ is tangent to the ellipse $x^2 + 4y^2 = 8$. Find the points of tangency.

34. Prove: The line that is tangent to the ellipse $x^2/a^2 + y^2/b^2 = 1$ at (x_0, y_0) has equation $xx_0/a^2 + yy_0/b^2 = 1$.

35. Find the area that is enclosed by the ellipse $b^2x^2 + a^2y^2 = a^2b^2$.

36. Find the volume of the solid generated when the region enclosed by the ellipse $b^2x^2 + a^2y^2 = a^2b^2$ is
  (a) revolved about the x-axis;
  (b) revolved about the y-axis.

37. Find the area of the square that can be inscribed in the ellipse $x^2/a^2 + y^2/b^2 = 1$.

38. A semielliptic arch spans a highway 50 feet wide. How high is the arch if a center section of the highway 30 feet wide has a minimum clearance of 14 feet?

39. The distance between a point $P(x, y)$ and the point $(4, 0)$ is 4/5 of the distance between P and the line $x = 25/4$. Show that the set of all such points is an ellipse. Find the center and the lengths of the major and minor axes.

40. Suppose that you want to draw an ellipse having given values for the lengths of the major and minor axes by using the method shown in Figure 12.3.2. Assuming that the axes are drawn, explain how a compass can be used to locate the positions for the tacks.

41. Let C_1 and C_2 be circles, in the same plane, with radii r_1 and r_2, respectively. Assume that $r_1 < r_2$ and that C_1 lies entirely inside C_2, but that C_1 and C_2 do not have the same center. Show that the centers of all circles that are outside C_1 and inside C_2, and are tangent to both C_1 and C_2, lie on an ellipse whose foci are the centers of C_1 and C_2. Find the length of the major axis and the location of the center of the ellipse.

Figure 12.3.13

42. Show that the curve of intersection of a plane and a right circular cylinder is an ellipse, assuming that the plane is neither perpendicular to nor parallel to the axis of the cylinder. [*Hint:* Let θ be the angle that the plane makes with a circular cross section of the cylinder. Introduce xy- and $x'y'$-axes as shown in Figure 12.3.14, and find x' and y' in terms of x and y.]

Figure 12.3.14

43. The shape of an ellipse depends on the relative sizes of a and c. The ratio c/a is called the *eccentricity* and is denoted by e, so that $e = c/a$.
  (a) Show that $0 < e < 1$ for an ellipse.
  (b) If the major axis of an ellipse is kept constant, what happens to the ellipse as e approaches 0? As e approaches 1?

44. The earth's orbit around the sun is approximately an ellipse with the sun at one of its foci. The smallest and largest distances between the earth and the sun occur at the vertices.
  (a) If the ratio of the smallest to the largest distance is 59/61, what is the eccentricity of the orbit. [Refer to Exercise 43 for the definition of eccentricity.]
  (b) Given that the semimajor axis of the earth's orbit has a length of 93 million miles, find the shortest distance between the earth and the sun.

45. Derive the equation $x^2/b^2 + y^2/a^2 = 1$ in Table 12.3.1.

46. Prove Theorem 12.3.2. [*Hint:* Introduce coordinate axes so the ellipse has the equation $x^2/a^2 + y^2/b^2 = 1$, and use the result of Exercise 21 of Section 1.4.]

12.4 THE HYPERBOLA

In this section we shall discuss properties of hyperbolas.

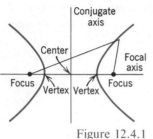

Figure 12.4.1

> **12.4.1** DEFINITION. A *hyperbola* is the set of all points in the plane, the difference of whose distances from two fixed points is a given positive constant.

In this definition the "difference" of the distances is understood to mean the distance to the farther point minus the distance to the closer point.

The two fixed points are called the *foci,* and the midpoint of the line segment joining the foci is called the **center** of the hyperbola (Figure 12.4.1). The line through the foci is called the *focal axis* (or *transverse axis*) and the line through the center and perpendicular to the focal axis is called the *conjugate axis.* The hyperbola intersects the focal axis at two points, called *vertices.* The two separate parts of a hyperbola are called the *branches.*

It is traditional in the study of hyperbolas to denote the distance between the vertices by $2a$, the distance between the foci by $2c$ (Figure 12.4.2), and to define the quantity b as

Figure 12.4.2

$$b = \sqrt{c^2 - a^2} \tag{1}$$

(See Figure 12.4.3.)

If V is one vertex of a hyperbola, then, as illustrated in Figure 12.4.4, the distance from V to the farther focus minus the distance from V to the closer focus is

$$[(c - a) + 2a] - (c - a) = 2a$$

Thus, for *all* points on a hyperbola, the distance to the farther focus minus the distance to the closer focus is $2a$. (Why?)

The equation of a hyperbola is simplest if the coordinate axes are positioned so the center of the hyperbola is at the origin and the foci are on the x-axis or y-axis. The two possible such orientations are shown in Table 12.4.1.

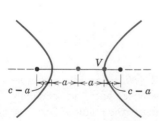

Figure 12.4.3

To illustrate how the equations in Table 12.4.1 are derived, let us consider the first entry in the table. Let $P(x, y)$ be any point on the hyperbola, so that the distance from P to the farther focus minus the distance from P to the closer focus is $2a$. Depending on which focus is farther from P, this condition leads either to the equation

$$\sqrt{(x + c)^2 + y^2} - \sqrt{(x - c)^2 + y^2} = 2a \tag{2a}$$

Figure 12.4.4 or to

Table 12.4.1

ORIENTATION	DESCRIPTION	EQUATION
	Foci on the x-axis. Conjugate axis on the y-axis. Center at the origin.	$\dfrac{x^2}{a^2} - \dfrac{y^2}{b^2} = 1$
	Foci on the y-axis. Conjugate axis on the x-axis. Center at the origin.	$\dfrac{y^2}{a^2} - \dfrac{x^2}{b^2} = 1$

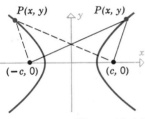

$$\sqrt{(x - c)^2 + y^2} - \sqrt{(x + c)^2 + y^2} = 2a \qquad (2b)$$

(See Figure 12.4.5.) Rewriting (2a) yields

$$\sqrt{(x + c)^2 + y^2} = 2a + \sqrt{(x - c)^2 + y^2}$$

then squaring both sides yields

Figure 12.4.5 ▷ $(x + c)^2 + y^2 = 4a^2 + 4a\sqrt{(x - c)^2 + y^2} + (x - c)^2 + y^2$

Simplifying, then isolating the radical, then squaring again to remove the radical, ultimately yields (verify)

$$\frac{x^2}{a^2} - \frac{y^2}{c^2 - a^2} = 1$$

which, by virtue of (1), can be written as

$$\frac{x^2}{a^2} - \frac{y^2}{b^2} = 1 \qquad (3)$$

We leave it as an exercise to show that (2b) also simplifies to (3). Thus, each point $P(x, y)$ on the hyperbola satisfies (3), regardless of the branch on which it lies. Conversely, it can be shown that for any point P whose coordinates satisfy (3), the distance from P to the farther focus minus the distance from P to the closer focus is $2a$, so that such a point must lie on the hyperbola.

The derivation of the second equation in Table 12.4.1 is similar.

Example 1   Does the hyperbola

$$\frac{x^2}{4} - \frac{y^2}{9} = 1$$

have its focal axis on the x-axis or the y-axis?

Solution.   Unlike the case of the ellipse, the orientation of a hyperbola is not determined by examining the relative sizes of a^2 and b^2, but rather by noting where the minus sign occurs in the equation. From Table 12.4.1 we see that the focal axis is on the x-axis when the minus precedes the y^2-term, and it is on the y-axis when the minus precedes the x^2-term. In the given equation the minus precedes the y^2-term so the focal axis is on the x-axis.

Alternatively, the orientation of a hyperbola can be obtained by locating its vertices. If we let $x = 0$ in the given equation, we obtain

$$-\frac{y^2}{9} = 1 \quad \text{or} \quad y^2 = -9$$

This equation has no real solutions, so the hyperbola does not intersect the y-axis. This tells us that the focal axis is along the x-axis. The same conclusion can be reached by setting $y = 0$ in the given equation to obtain

$$\frac{x^2}{4} = 1 \quad \text{or} \quad x = \pm 2$$

This tells us that the vertices are at the points $(-2, 0)$ and $(2, 0)$, so again we conclude that the focal axis is on the x-axis (Figure 12.4.6).   ◄

$(-2, 0)$ $(2, 0)$

$$\frac{x^2}{4} - \frac{y^2}{9} = 1$$

Figure 12.4.6

ASYMPTOTES OF A
HYPERBOLA

Associated with every hyperbola is a pair of lines called the **asymptotes** of the hyperbola. These lines intersect at the center of the hyperbola and have the property that as a point $P(x, y)$ moves along the hyperbola so $x \to +\infty$ or $x \to -\infty$, the vertical distance between P and one of the asymp-

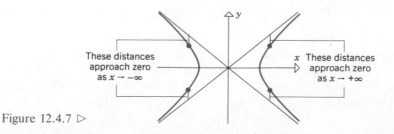

These distances approach zero as $x \to -\infty$

These distances approach zero as $x \to +\infty$

Figure 12.4.7 ▷

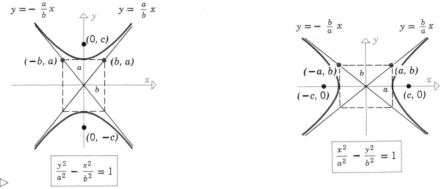

Figure 12.4.8 ▷

The following theorem proves that hyperbolas have asymptotes and gives their equations (Figure 12.4.8).

Asymptotes of Hyperbolas

12.4.2 THEOREM.

(*a*) *The hyperbola*

$$\frac{x^2}{a^2} - \frac{y^2}{b^2} = 1$$

has asymptotes

$$y = \frac{b}{a}x \quad and \quad y = -\frac{b}{a}x$$

(*b*) *The hyperbola*

$$\frac{y^2}{a^2} - \frac{x^2}{b^2} = 1$$

has asymptotes

$$y = \frac{a}{b}x \quad and \quad y = -\frac{a}{b}x$$

Proof. We shall prove part (*a*); the proof of (*b*) is similar. If we write the equation $x^2/a^2 - y^2/b^2 = 1$ in the form

$$y^2 = \frac{b^2}{a^2}(x^2 - a^2)$$

$y = \frac{b}{a}\sqrt{x^2 - a^2}$

y ↑ $y = \frac{b}{a}x$

Figure 12.4.9

then, in the first quadrant, the vertical distance between the line $y = (b/a)x$ and the hyperbola can be written (Figure 12.4.9) as

$$\frac{b}{a}x - \frac{b}{a}\sqrt{x^2 - a^2}$$

But this distance tends to zero as $x \to +\infty$ since

$$\lim_{x \to +\infty} \left(\frac{b}{a}x - \frac{b}{a}\sqrt{x^2 - a^2} \right) = \lim_{x \to +\infty} \frac{b}{a}(x - \sqrt{x^2 - a^2})$$

$$= \lim_{x \to +\infty} \frac{b}{a}\frac{(x - \sqrt{x^2 - a^2})(x + \sqrt{x^2 - a^2})}{x + \sqrt{x^2 - a^2}}$$

$$= \lim_{x \to +\infty} \frac{ab}{x + \sqrt{x^2 - a^2}}$$

$$= 0$$

The analysis in the remaining quadrants is similar.   ∎

REMARKS.   There is a trick that can be used to avoid memorizing the equations of the asymptotes of a hyperbola. They can be obtained, when needed, by substituting 0 for the 1 on the right side of the hyperbola equation, and then solving for y in terms of x. For example, for the hyperbola

$$\frac{x^2}{a^2} - \frac{y^2}{b^2} = 1$$

we would write

$$\frac{x^2}{a^2} - \frac{y^2}{b^2} = 0 \quad \text{or} \quad y^2 = \frac{b^2}{a^2}x^2 \quad \text{or} \quad y = \pm\frac{b}{a}x$$

which are the equations for the asymptotes. As indicated in Figure 12.4.8, the asymptotes of a hyperbola are along the diagonals of a box extending a units on each side of the origin along the focal axis, and b units on each side of the origin along the conjugate axis.

Example 2   Find the foci, vertices, and asymptotes of the hyperbola

$$\frac{y^2}{16} - \frac{x^2}{9} = 1$$

and sketch the graph.

Solution.   Since the minus precedes the x^2-term, this hyperbola has its foci on the y-axis. Also, $a^2 = 16$ and $b^2 = 9$. Since $a = 4$, the vertices

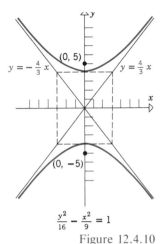

$$\frac{y^2}{16} - \frac{x^2}{9} = 1$$

Figure 12.4.10

are at $(0, -4)$ and $(0, 4)$. Moreover, from (1), $c = \sqrt{a^2 + b^2} = \sqrt{25} = 5$, so the foci are at $(0, -5)$ and $(0, 5)$.

To obtain the asymptotes, we substitute 0 for 1 in the given equation, which yields

$$\frac{y^2}{16} - \frac{x^2}{9} = 0 \quad \text{or} \quad y = \pm\frac{4}{3}x$$

With the aid of a box extending $a = 4$ units above and below the origin in the y-direction and $b = 3$ units to the left and right of the origin in the x-direction, we obtain the sketch in Figure 12.4.10. ◀

Example 3 Find the equation of the hyperbola with vertices $(0, \pm 8)$ and asymptotes $y = \pm\frac{4}{3}x$.

Solution. Since the vertices are on the y-axis, the equation of the hyperbola has the form $y^2/a^2 - x^2/b^2 = 1$ and the asymptotes are

$$y = \pm\frac{a}{b}x$$

From the location of the vertices we have $a = 8$, so that the given equations of the asymptotes yield

$$y = \pm\frac{a}{b}x = \pm\frac{8}{b}x = \pm\frac{4}{3}x$$

from which it follows that $b = 6$. Thus, the hyperbola has the equation

$$\frac{y^2}{64} - \frac{x^2}{36} = 1 \quad ◀$$

TRANSLATED HYPERBOLAS   If the center of a hyperbola is at the point (h, k) and its focal and conjugate axes are parallel to the coordinate axes, then its equation has one of the forms:

Hyperbola with Center (h, k) and Focal Axis Parallel to x-axis.

$$\frac{(x - h)^2}{a^2} - \frac{(y - k)^2}{b^2} = 1 \qquad (4)$$

Hyperbola with Center (h, k) and Focal Axis Parallel to y-axis.

$$\frac{(y - k)^2}{a^2} - \frac{(x - h)^2}{b^2} = 1 \qquad (5)$$

The derivations are similar to those for the parabola and ellipse and will be omitted. As before, the equations of the asymptotes can be obtained by replacing the 1 by a 0 on the right side of (4) or (5), and then solving for y in terms of x.

Example 4   Show that the curve

$$x^2 - y^2 - 4x + 8y - 21 = 0$$

is a hyperbola. Sketch the hyperbola and show the foci, vertices, and asymptotes.

Solution.   Our objective is to rewrite the equation in form (4) or (5) by completing the squares. To do this, we first group the x-terms and y-terms and take the constant to the right side:

$$(x^2 - 4x) - (y^2 - 8y) = 21$$

Next, complete the squares:

$$(x^2 - 4x + 4) - (y^2 - 8y + 16) = 21 + 4 - 16$$

or

$$(x - 2)^2 - (y - 4)^2 = 9$$

Finally, divide by 9 to introduce a 1 on the right side:

$$\frac{(x - 2)^2}{9} - \frac{(y - 4)^2}{9} = 1 \tag{6}$$

This is an equation of form (4) with $h = 2$, $k = 4$, $a^2 = 9$, and $b^2 = 9$. Thus, the equation represents a hyperbola with center $(2, 4)$ and focal axis parallel to the x-axis. Since $a = 3$, the vertices are located 3 units to the left and 3 units to the right of the center, or at the points $(-1, 4)$ and $(5, 4)$. From (1), $c = \sqrt{a^2 + b^2} = \sqrt{9 + 9} = 3\sqrt{2}$, so that the foci are located $3\sqrt{2}$ units to the left and right of the center or at the points $(2 - 3\sqrt{2}, 4)$ and $(2 + 3\sqrt{2}, 4)$.

The equations of the asymptotes may be found by substituting 0 for 1 in (6) to obtain

$$\frac{(x - 2)^2}{9} - \frac{(y - 4)^2}{9} = 0 \quad \text{or} \quad y - 4 = \pm(x - 2)$$

which yields the asymptotes

$$y = x + 2 \quad \text{and} \quad y = -x + 6$$

With the aid of a box extending $a = 3$ units left and right of the center and $b = 3$ units above and below the center we obtain the sketch in Figure 12.4.11.   ◄

Figure 12.4.11

A hyperbola (like that in the last example) is called *equilateral* if $a = b$. For such hyperbolas, the asymptotes are perpendicular. (Verify.)

HYPERBOLIC
NAVIGATION

By measuring the difference in reception times of synchronized radio signals from two widely spaced transmitters, a ship's electronic equipment can determine the difference $2a$ in its distances from the two transmitters. This information places the ship somewhere on the hyperbola whose foci are at the transmitters and whose points have $2a$ as the difference in their distances from the foci. By using two pairs of transmitters, the position of a ship can be determined as the intersection of two hyperbolas (Figure 12.4.12).

Figure 12.4.12

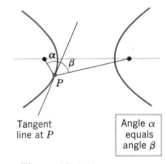

Tangent
line at P

Angle α equals angle β

Figure 12.4.13

REFLECTION
PROPERTIES OF
HYPERBOLAS

In the exercises, the reader is asked to prove the following result.

A Geometric Property of Hyperbolas

12.4.3   THEOREM. *A line tangent to a hyperbola at a point P makes equal angles with the lines through P and the foci (Figure 12.4.13).*

It follows from Theorem 12.4.3 that a ray of light emanating from one focus of a hyperbola will be reflected back along the line from the opposite focus (Figure 12.4.14). Light reflection properties of hyperbolas are used to advantage in the design of high-quality telescopes.

Figure 12.4.14 ▷

▶ Exercise Set 12.4

In Exercises 1–16, sketch the hyperbola. Find the co-ordinates of the vertices and foci, and find equations for the asymptotes.

1. $\dfrac{x^2}{16} - \dfrac{y^2}{4} = 1.$ **2.** $\dfrac{y^2}{9} - \dfrac{x^2}{25} = 1.$

3. $9y^2 - 4x^2 = 36.$ **4.** $16x^2 - 25y^2 = 400.$

5. $8x^2 - y^2 = 8.$ **6.** $3x^2 - y^2 = -9.$

7. $x^2 - y^2 = 1.$ **8.** $4y^2 - 4x^2 = 1.$

9. $\dfrac{(x - 2)^2}{9} - \dfrac{(y - 4)^2}{4} = 1.$

10. $\dfrac{(y + 4)^2}{3} - \dfrac{(x - 2)^2}{5} = 1.$

11. $(y + 3)^2 - 9(x + 2)^2 = 36.$

12. $16(x + 1)^2 - 8(y - 3)^2 = 16.$

13. $x^2 - 4y^2 + 2x + 8y - 7 = 0.$

14. $4x^2 - 9y^2 + 16x + 54y - 29 = 0.$

15. $16x^2 - y^2 - 32x - 6y = 57.$

16. $4y^2 - x^2 + 40y - 4x = -60.$

In Exercises 17–30, find an equation for the hyperbola satisfying the given conditions.

17. Vertices $(\pm 2, 0)$; foci $(\pm 3, 0)$.

18. Vertices $(0, \pm 3)$; foci $(0, \pm 5)$.

19. Vertices $(\pm 1, 0)$; asymptotes $y = \pm 2x$.

20. Vertices $(0, \pm 3)$; asymptotes $y = \pm x$.

21. Asymptotes $y = \pm \frac{3}{2}x$; $b = 4$.

22. Foci $(0, \pm 5)$; asymptotes $y = \pm 2x$.

23. Passes through $(5, 9)$; asymptotes $y = \pm x$.

24. Asymptotes $y = \pm \frac{3}{4}x$; $c = 5$.

25. Vertices $(\pm 2, 0)$; passes through $(4, 2)$.

26. Foci $(\pm 3, 0)$; asymptotes $y = \pm 2x$.

27. Vertices $(4, -3)$ and $(0, -3)$; foci 6 units apart.

28. Foci $(1, 8)$ and $(1, -12)$; vertices 4 units apart.

29. Vertices $(2, 4)$ and $(10, 4)$; foci 10 units apart.

30. Asymptotes $y = 2x + 1$ and $y = -2x + 3$; passes through the origin.

31. Find the equation of the hyperbola traced by a point that moves so the difference between its distances to $(0, 0)$ and $(1, 1)$ is 1.

32. Find the equation of the hyperbola traced by a point that moves so the difference between its distances to $(4, -3)$ and $(-2, 5)$ is 6.

In Exercises 33–36, find all intersections of the given curves, and make a sketch of the curves that shows the points of intersection.

33. $x^2 - 4y^2 = 36$ and $x - 2y - 20 = 0.$

34. $y^2 - 8x^2 = 5$ and $y - 2x^2 = 0.$

35. $3x^2 - 7y^2 = 5$ and $9y^2 - 2x^2 = 1.$

36. $x^2 - y^2 = 1$ and $x^2 + y^2 = 7.$

37. A point moves so that the product of its distances from the lines $y = mx$ and $y = -mx$ is a constant k^2. Show that the point moves along a hyperbola having these lines as asymptotes.

38. Prove: The line tangent to the hyperbola $x^2/a^2 - y^2/b^2 = 1$ at (x_0, y_0) has the equation $xx_0/a^2 - yy_0/b^2 = 1.$

39. A line tangent to the hyperbola $4x^2 - y^2 = 36$ intersects the y-axis at the point $(0, 4)$. Find the point(s) of tangency.

40. Let R be the region enclosed between the hyperbola $b^2x^2 - a^2y^2 = a^2b^2$ and the line $x = c$ through the focus $(c, 0)$. Find the volume of the solid generated by revolving R about
(a)   the x-axis   (b)   the y-axis.

41. Find the coordinates of all points on the hyperbola $4x^2 - y^2 = 4$ where the two lines that pass through the point and the foci are perpendicular.

42. The distance between a point $P(x, y)$ and the point $(5, 0)$ is 5/3 of the distance between P and the line $x = 9/5$. Show that the set of all such points is a hyperbola.

43. The shape of a hyperbola depends on the relative sizes of a and c. The ratio c/a is called the *eccentricity* and is denoted by e, so $e = c/a$.
(a)   Show that $e > 1$ for a hyperbola.
(b)   If the distance between vertices is kept constant, what happens to the hyperbola as e approaches 1? As e gets larger without bound?

44. Let C_1 and C_2 be circles in the same plane with unequal radii r_1 and r_2 respectively. Assume that C_1 and C_2 do not intersect and that neither circle is inside the other. Show that the centers of all circles that are outside C_1 and C_2, and are tangent to both C_1 and C_2, lie on a branch of a hyperbola whose foci are the centers of C_1 and C_2. Where is the center of the hyperbola?

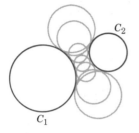

45. Suppose that the report of a gun is heard first by one observer and then t seconds later by a second observer in another location. Assuming that the speed of sound is constant, show that the gun was fired somewhere on a branch of a hyperbola whose foci coincide with the observers.

46. Show that Equation (2b) simplifies to (3).

47. Derive the equation $y^2/a^2 - x^2/b^2 = 1$ in Table 12.4.1.

48. Prove Theorem 12.4.3. [*Hint:* Introduce coordinate axes so the hyperbola has the equation $x^2/a^2 - y^2/b^2 = 1$, and use the result of Exercise 21 of Section 1.4.]

12.5 ROTATION OF AXES; SECOND DEGREE EQUATIONS

In previous sections we obtained equations of conic sections with axes parallel to the coordinate axes. In this section we shall study the equations of conics that are "tilted" relative to the coordinate axes. This will lead us to investigate rotations of coordinate axes.

Each of the equations

$$(y - k)^2 = 4p(x - h) \qquad (x - h)^2 = 4p(y - k)$$

$$\frac{(x - h)^2}{a^2} + \frac{(y - k)^2}{b^2} = 1 \qquad \frac{(x - h)^2}{b^2} + \frac{(y - k)^2}{a^2} = 1$$

$$\frac{(x - h)^2}{a^2} - \frac{(y - k)^2}{b^2} = 1 \qquad \frac{(y - k)^2}{a^2} - \frac{(x - h)^2}{b^2} = 1$$

represents a conic section with axis or axes parallel to the coordinate axes. By squaring out the quadratic terms and simplifying, each of these equations can be rewritten in the form

$$Ax^2 + Cy^2 + Dx + Ey + F = 0 \qquad\qquad (1)$$

For example, $(y - k)^2 = 4p(x - h)$ can be written as

$$y^2 - 4px - 2ky + (k^2 + 4ph) = 0$$

which is of form (1) with

$$A = 0, \quad C = 1, \quad D = -4p, \quad E = -2k, \quad F = k^2 + 4ph$$

Equation (1) is a special case of the more general equation

$$Ax^2 + Bxy + Cy^2 + Dx + Ey + F = 0 \tag{2}$$

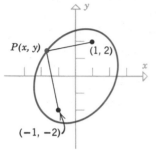

Figure 12.5.1

which, if A, B, and C are not all zero, is called a ***second degree equation*** or ***quadratic equation*** in x and y. We shall show later that the graph of any second degree equation is a conic section (or possibly a special case of a conic section). If $B = 0$, then (2) reduces to (1) and the conic section has its axis or axes parallel to the coordinate axes. However, if $B \neq 0$ then (2) contains a "cross product" term Bxy, and the graph of the conic section represented by the equation has its axis or axes "tilted" relative to the coordinate axes. As an illustration, consider the ellipse with foci $F_1(1, 2)$ and $F_2(-1, -2)$ and such that the sum of the distances from each point $P(x, y)$ on the ellipse to the foci is 6 units. Expressing this condition as an equation, we obtain (Figure 12.5.1):

$$\sqrt{(x - 1)^2 + (y - 2)^2} + \sqrt{(x + 1)^2 + (y + 2)^2} = 6$$

Squaring both sides, then isolating the remaining radical, then squaring again, ultimately yields

$$8x^2 - 4xy + 5y^2 = 36$$

as the equation of the ellipse. This is an equation of form (2) with $A = 8$, $B = -4$, $C = 5$, $D = 0$, $E = 0$, $F = -36$.

ROTATION OF AXES

To study conics that are tilted relative to the coordinate axes, it is frequently helpful to rotate the coordinate axes, so that the rotated coordinate axes are parallel to the axes of the conic. Before we can discuss the details, we need to develop some ideas about rotation of coordinate axes.

In Figure 12.5.2a the axes of an xy-coordinate system have been rotated about the origin through an angle θ to produce a new $x'y'$-coordinate system. As shown in the figure, each point P in the plane has coordinates (x', y') as well as coordinates (x, y). To see how the two are related, let r be the distance from the common origin to the point P, and let α be the angle shown in Figure 12.5.2b. It follows that

Figure 12.5.2 ▷   (a)   (b)

$$x = r \cos(\theta + \alpha), \quad y = r \sin(\theta + \alpha) \tag{3}$$

and

$$x' = r \cos \alpha, \quad y' = r \sin \alpha \tag{4}$$

Using familiar trigonometric identities, the relationships in (3) can be written as

$$x = r \cos \theta \cos \alpha - r \sin \theta \sin \alpha$$
$$y = r \sin \theta \cos \alpha + r \cos \theta \sin \alpha$$

and on substituting (4) in these equations we obtain the following relationships:

> **The Rotation Equations**
>
> $$x = x' \cos \theta - y' \sin \theta$$
> $$y = x' \sin \theta + y' \cos \theta$$
$$\tag{5}$$

Example 1 Suppose the axes of an xy-coordinate system are rotated through an angle of $\theta = 45°$ to obtain an $x'y'$-coordinate system. Find the equation of the curve

$$x^2 - xy + y^2 - 6 = 0$$

in $x'y'$-coordinates.

Solution. Substituting the values $\sin \theta = \sin 45° = 1/\sqrt{2}$ and $\cos \theta = \cos 45° = 1/\sqrt{2}$ in (5) yields the rotation equations

$$x = \frac{x'}{\sqrt{2}} - \frac{y'}{\sqrt{2}}$$

$$y = \frac{x'}{\sqrt{2}} + \frac{y'}{\sqrt{2}}$$

Substituting these into the given equation yields

$$\left(\frac{x'}{\sqrt{2}} - \frac{y'}{\sqrt{2}}\right)^2 - \left(\frac{x'}{\sqrt{2}} - \frac{y'}{\sqrt{2}}\right)\left(\frac{x'}{\sqrt{2}} + \frac{y'}{\sqrt{2}}\right) + \left(\frac{x'}{\sqrt{2}} + \frac{y'}{\sqrt{2}}\right)^2 - 6 = 0$$

or

$$\frac{x'^2 - 2x'y' + y'^2 - x'^2 + y'^2 + x'^2 + 2x'y' + y'^2}{2} = 6$$

or

Figure 12.5.3

$$\frac{x'^2}{12} + \frac{y'^2}{4} = 1$$

which is the equation of an ellipse (Figure 12.5.3). ◀

If the rotation equations (5) are solved for x' and y' in terms of x and y, one obtains (Exercise 14):

$$x' = x \cos \theta + y \sin \theta \qquad (6)$$
$$y' = -x \sin \theta + y \cos \theta$$

Example 2 Find the new coordinates of the point $(2, 4)$ if the coordinate axes are rotated through an angle of $\theta = 30°$.

Solution. From (6) with $x = 2$, $y = 4$, $\cos \theta = \cos 30° = \sqrt{3}/2$, and $\sin \theta = \sin 30° = 1/2$, we obtain

$$x' = 2(\sqrt{3}/2) + 4(1/2) = \sqrt{3} + 2$$
$$y' = -2(1/2) + 4(\sqrt{3}/2) = -1 + 2\sqrt{3}$$

Thus, the new coordinates are $(\sqrt{3} + 2, -1 + 2\sqrt{3})$. ◀

In Example 1, we were able to identify the curve $x^2 - xy + y^2 - 6 = 0$ as an ellipse because the rotation of axes eliminated the xy-term, thereby reducing the equation to a familiar form. This occurred because the new $x'y'$-axes were aligned with the axes of the ellipse. The following theorem tells how to determine an appropriate rotation of axes to eliminate the cross-product term of a second-degree equation in x and y.

12.5.1 THEOREM. *If the equation*

$$Ax^2 + Bxy + Cy^2 + Dx + Ey + F = 0 \qquad (7)$$

is such that $B \neq 0$, and if an $x'y'$-coordinate system is obtained by rotating the xy-axes through an angle θ satisfying

$$\cot 2\theta = \frac{A - C}{B} \qquad (8)$$

then, in $x'y'$-coordinates, equation (7) will have the form

$$A'x'^2 + C'y'^2 + D'x' + E'y' + F' = 0$$

Proof. Substituting (5) into (7) and simplifying yields,

$$A'x'^2 + B'x'y' + C'y'^2 + D'x' + E'y' + F' = 0$$

where

$$A' = A \cos^2 \theta + B \cos \theta \sin \theta + C \sin^2 \theta$$
$$B' = B (\cos^2 \theta - \sin^2 \theta) + 2(C - A) \sin \theta \cos \theta$$
$$C' = A \sin^2 \theta - B \sin \theta \cos \theta + C \cos^2 \theta \qquad (9)$$
$$D' = D \cos \theta + E \sin \theta$$
$$E' = -D \sin \theta + E \cos \theta$$
$$F' = F$$

(Verify.) To complete the proof we must show that $B' = 0$ if

$$\cot 2\theta = \frac{A - C}{B}$$

or equivalently

$$\frac{\cos 2\theta}{\sin 2\theta} = \frac{A - C}{B} \qquad (10)$$

However, by using the trigonometric double angle formulas, we can re-write B' in the form

$$B' = B \cos 2\theta - (A - C) \sin 2\theta$$

Thus, $B' = 0$ if θ satisfies (10). ∎

REMARK.   It is always possible to satisfy (8) with an angle θ in the range $0 < \theta < \pi/2$. We shall always use such a value of θ.

Example 3   Identify and sketch the curve $xy = 1$.

Solution.   As a first step, we shall rotate the coordinate axes to eliminate the cross-product term. Comparing the given equation to (7), we have

$$A = 0, \quad B = 1, \quad C = 0$$

Thus, the desired angle of rotation must satisfy

$$\cot 2\theta = \frac{A - C}{B} = \frac{0 - 0}{1} = 0$$

This condition can be met by taking $2\theta = \pi/2$ or $\theta = \pi/4 = 45°$. Substituting $\cos \theta = \cos 45° = 1/\sqrt{2}$ and $\sin \theta = \sin 45° = 1/\sqrt{2}$ in (5) yields

$$x = \frac{x'}{\sqrt{2}} - \frac{y'}{\sqrt{2}}$$
$$\qquad (11)$$
$$y = \frac{x'}{\sqrt{2}} + \frac{y'}{\sqrt{2}}$$

and substituting these in the equation $xy = 1$ yields

$$\left(\frac{x'}{\sqrt{2}} - \frac{y'}{\sqrt{2}}\right)\left(\frac{x'}{\sqrt{2}} + \frac{y'}{\sqrt{2}}\right) = 1$$

or on simplifying

$$\frac{x'^2}{2} - \frac{y'^2}{2} = 1 \tag{12}$$

This is an equation of the form

$$\frac{x'^2}{a^2} - \frac{y'^2}{b^2} = 1$$

with $a = b = \sqrt{2}$. With the exception that the variables are x' and y' rather than x and y, this is the familiar equation of a hyperbola (see Table 12.4.1). This hyperbola has its focal axis on the x'-axis, its conjugate axis on the y'-axis. Moreover, $c = \sqrt{a^2 + b^2} = 2$, so that the vertices, foci, and asymptotes are:

vertices: $(x', y') = (\pm a, 0) = (\pm\sqrt{2}, 0)$

foci: $(x', y') = (\pm c, 0) = (\pm 2, 0)$

asymptotes: $y' = \pm\dfrac{b}{a}x'$ or $y' = \pm x'$

The vertices in xy-coordinates can be found by substituting $x' = \sqrt{2}$, $y' = 0$ and $x' = -\sqrt{2}$, $y' = 0$ in (11), and the foci may be obtained by substituting $x' = 2$, $y' = 0$ and $x' = -2$, $y' = 0$. The results are

vertices: $(x, y) = (1, 1)$ and $(x, y) = (-1, -1)$

foci: $(x, y) = (\sqrt{2}, \sqrt{2})$ and $(x, y) = (-\sqrt{2}, -\sqrt{2})$

To obtain the xy-equations of the asymptotes, it is desirable to use form (6) of the rotation equations. Substituting $\cos \theta = \cos 45° = 1/\sqrt{2}$ and $\sin \theta = \sin 45° = 1/\sqrt{2}$ in (6) yields

$$x' = \frac{1}{\sqrt{2}}x + \frac{1}{\sqrt{2}}y$$

$$y' = -\frac{1}{\sqrt{2}}x + \frac{1}{\sqrt{2}}y \tag{13}$$

Substituting (13) in the asymptote equations $y' = x'$ and $y' = -x'$ and

Figure 12.5.4

simplifying shows that the asymptotes are

$$x = 0 \quad \text{and} \quad y = 0$$

The curve $xy = 1$ is sketched in Figure 12.5.4.   ◀

In problems where it is inconvenient to solve

$$\cot 2\theta = \frac{A - C}{B}$$

for θ, the values of $\sin \theta$ and $\cos \theta$ needed for the rotation equations may be obtained by first calculating $\cos 2\theta$ and then computing $\sin \theta$ and $\cos \theta$ from the identities

$$\sin \theta = \sqrt{\frac{1 - \cos 2\theta}{2}} \quad \text{and} \quad \cos \theta = \sqrt{\frac{1 + \cos 2\theta}{2}}$$

Figure 12.5.5

Example 4   Identify and sketch the curve

$$153x^2 - 192xy + 97y^2 - 30x - 40y - 200 = 0$$

Solution.   We have $A = 153$, $B = -192$, and $C = 97$ so that

$$\cot 2\theta = \frac{A - C}{B} = -\frac{56}{192} = -\frac{7}{24}$$

Since θ is to be chosen in the range $0 < \theta < \pi/2$, this relationship is represented by the triangle in Figure 12.5.5. From that triangle we obtain

$$\cos 2\theta = -\frac{7}{25}$$

which implies that

$$\cos \theta = \sqrt{\frac{1 + \cos 2\theta}{2}} = \sqrt{\frac{1 - 7/25}{2}} = \frac{3}{5}$$

$$\sin \theta = \sqrt{\frac{1 - \cos 2\theta}{2}} = \sqrt{\frac{1 + 7/25}{2}} = \frac{4}{5}$$

Substituting these values in (5) yields the rotation equations

$$x = \frac{3}{5}x' - \frac{4}{5}y'$$

$$y = \frac{4}{5}x' + \frac{3}{5}y'$$

Substituting these expressions in the given equation yields

$$\frac{153}{25}(3x' - 4y')^2 - \frac{192}{25}(3x' - 4y')(4x' + 3y') + \frac{97}{25}(4x' + 3y')^2$$
$$- \frac{30}{5}(3x' - 4y') - \frac{40}{5}(4x' + 3y') - 200 = 0$$

which simplifies to

$$25x'^2 + 225y'^2 - 50x' - 200 = 0$$

or

$$x'^2 + 9y'^2 - 2x' - 8 = 0$$

Completing the square yields

$$\frac{(x' - 1)^2}{9} + y'^2 = 1$$

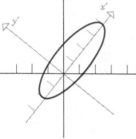

Figure 12.5.6

This is an ellipse whose center is $(1, 0)$ in $x'y'$-coordinates (Figure 12.5.6). ◄

OPTIONAL (THE DISCRIMINANT)

It is possible to describe the graph of a second-degree equation without rotating coordinate axes.

12.5.2 THEOREM. *Consider a second-degree equation*

$$Ax^2 + Bxy + Cy^2 + Dx + Ey + F = 0 \tag{14}$$

(a) *If $B^2 - 4AC < 0$, the equation represents an ellipse, a circle, a point, or else has no graph.*

(b) *If $B^2 - 4AC > 0$, the equation represents a hyperbola or a pair of intersecting lines.*

(c) *If $B^2 - 4AC = 0$, the equation represents a parabola, a line, a pair of parallel lines, or else has no graph.*

The quantity $B^2 - 4AC$ in this theorem is called the ***discriminant*** of the quadratic equation. To see why Theorem 12.5.2 is true, we need a fact about the discriminant. It can be shown (Exercise 19) that if the coordinate axes are rotated through any angle θ, and if

$$A'x'^2 + B'x'y' + C'y'^2 + D'x' + E'y' + F' = 0 \tag{15}$$

is the equation resulting from (14) after rotation, then

$$B^2 - 4AC = B'^2 - 4A'C' \tag{16}$$

In other words, the discriminant of a quadratic equation is not altered by rotating the coordinate axes. For this reason the discriminant is said to be *invariant* under a rotation of coordinate axes. In particular, if we choose the angle of rotation to eliminate the cross-product term, then (15) becomes

$$A'x'^2 + C'y'^2 + D'x' + E'y' + F' = 0 \tag{17}$$

and since $B' = 0$, (16) tells us that

$$B^2 - 4AC = -4A'C' \tag{18}$$

Part (a) of Theorem 12.5.2 can now be proved as follows. If $B^2 - 4AC < 0$, then from (18), $A'C' > 0$, so that (17) may be divided through by $A'C'$ and written in the form

$$\frac{1}{C'}\left(x'^2 + \frac{D'}{A'}x'\right) + \frac{1}{A'}\left(y'^2 + \frac{D'}{C'}y'\right) = -\frac{F'}{A'C'}$$

By completing the squares, this equation may be rewritten in the form

$$\frac{1}{C'}(x' - h)^2 + \frac{1}{A'}(y' - k)^2 = K \tag{19}$$

But A' and C' have the same sign (since $A'C' > 0$), so that by multiplying (19) through by -1, if necessary, we can assume A' and C' to be positive. Thus, (19) can be written as

$$\frac{(x' - h)^2}{(\sqrt{C'})^2} + \frac{(y' - k)^2}{(\sqrt{A'})^2} = K \tag{20}$$

If $K < 0$, this equation has no graph (the left side is nonnegative for all x' and y'); if $K = 0$ the equation is satisfied only by $x' = h$ and $y' = k$, so the graph is the single point (h, k); finally, if $K > 0$, we can rewrite (20) in the form

$$\frac{(x' - h)^2}{(\sqrt{C'}\sqrt{K})^2} + \frac{(y' - k)^2}{(\sqrt{A'}\sqrt{K})^2} = 1$$

which is an ellipse or a circle. The proofs of parts (b) and (c) require a similar kind of analysis. ∎

Example 5 Use the discriminant to identify the graph of

$$8x^2 - 3xy + 5y^2 - 7x + 6 = 0$$

Solution. We have

$$B^2 - 4AC = (-3)^2 - 4(8)(5) = -151$$

Since the discriminant is negative, the equation represents an ellipse, a point, or else has no graph. (Why can't the graph be a circle?) ◄

In cases where a quadratic equation represents a point, a pair of parallel lines, a pair of intersecting lines or has no graph, we say that equation represents a ***degenerate conic section***. Thus, if we allow for possible degeneracy, it follows from Theorem 12.5.2 that *every quadratic equation has a conic section as its graph.*

► Exercise Set 12.5

1. Let an $x'y'$-coordinate system be obtained by rotating an xy-coordinate system through an angle of $\theta = 60°$.
    (a) Find the $x'y'$-coordinates of the point whose xy-coordinates are $(-2, 6)$.
    (b) Find an equation of the curve
        $\sqrt{3}xy + y^2 = 6$ in $x'y'$-coordinates.
    (c) Sketch the curve in (b), showing both xy-axes and $x'y'$-axes.

2. Let an $x'y'$-coordinate system be obtained by rotating an xy-coordinate system through an angle of $\theta = 30°$.
    (a) Find the $x'y'$-coordinates of the point whose xy-coordinates are $(1, -\sqrt{3})$.
    (b) Find an equation of the curve
        $2x^2 + 2\sqrt{3}xy = 3$ in $x'y'$-coordinates.
    (c) Sketch the curve in (b), showing both xy-axes and $x'y'$-axes.

In Exercises 3–12, rotate the coordinate axes to remove the xy-term. Then name the conic and sketch its graph.

3. $xy = -9$.

4. $x^2 - xy + y^2 - 2 = 0$.

5. $x^2 + 4xy - 2y^2 - 6 = 0$.

6. $31x^2 + 10\sqrt{3}xy + 21y^2 - 144 = 0$.

7. $x^2 + 2\sqrt{3}xy + 3y^2 + 2\sqrt{3}x - 2y = 0$.

8. $34x^2 - 24xy + 41y^2 - 25 = 0$.

9. $9x^2 - 24xy + 16y^2 - 80x - 60y + 100 = 0$.

10. $5x^2 - 6xy + 5y^2 - 8\sqrt{2}x + 8\sqrt{2}y = 8$.

11. $52x^2 - 72xy + 73y^2 + 40x + 30y - 75 = 0$.

12. $6x^2 + 24xy - y^2 - 12x + 26y + 11 = 0$.

13. Let an $x'y'$-coordinate system be obtained by rotating an xy-coordinate system through an angle θ. Prove: For every choice of θ, the equation $x^2 + y^2 = r^2$ becomes $x'^2 + y'^2 = r^2$. Give a geometric explanation of this result.

14. Solve the rotation equations (5) for x' and y' in terms of x and y.

15. Let an $x'y'$-coordinate system be obtained by rotating an xy-coordinate system through an angle of 45°. Use (6) to find an equation in xy-coordinates of the curve $3x'^2 + y'^2 = 6$.

16. Let an $x'y'$-coordinate system be obtained by rotating an xy-coordinate system through an angle of 30°. Use (5) to find an equation in $x'y'$-coordinates of the curve $y = x^2$.

17. Show that the graph of the equation

    $$\sqrt{x} + \sqrt{y} = 1$$

    is a portion of a parabola. [*Hint:* First rationalize the equation and then perform a rotation of axes.]

18. Derive the expression for B' in (9).

19. Use (9) to prove that $B^2 - 4AC = B'^2 - 4A'C'$ for all values of θ.

20. Use (9) to prove that $A + C = A' + C'$ for all values of θ.

21. Prove: If $A = C$ in (7), then the cross-product term can be eliminated by rotating through $45°$.

22. Prove: If $B \neq 0$, then the graph of $x^2 + Bxy + F = 0$ is a hyperbola if $F \neq 0$ and two intersecting lines if $F = 0$.

In Exercises 23–27, use the discriminant to identify the graph of the given equation.

23. $x^2 - xy + y^2 - 2 = 0$.

24. $x^2 + 4xy - 2y^2 - 6 = 0$.

25. $x^2 + 2\sqrt{3}xy + 3y^2 + 2\sqrt{3}x - 2y = 0$.

26. $6x^2 + 24xy - y^2 - 12x + 26y + 11 = 0$.

27. $34x^2 - 24xy + 41y^2 - 25 = 0$.

28. Each of the following represents a degenerate conic section. Where possible, sketch the graph.
  (a) $x^2 - y^2 = 0$
  (b) $x^2 + 3y^2 + 7 = 0$
  (c) $8x^2 + 7y^2 = 0$
  (d) $x^2 - 2xy + y^2 = 0$
  (e) $9x^2 + 12xy + 4y^2 - 36 = 0$
  (f) $x^2 + y^2 - 2x - 4y = -5$.

29. Prove parts (b) and (c) of Theorem 12.5.2.

▶ SUPPLEMENTARY EXERCISES

In Exercises 1–8, identify the curve as a parabola, ellipse, or hyperbola, and give the following information:

Parabola: the coordinates of the vertex and focus, and the equation of the directrix.
Ellipse: the coordinates of the center and foci, and the lengths of the major and minor axes.
Hyperbola: the coordinates of the center, foci, and vertices, and equations of the asymptotes.

1. $y^2 + 12x - 6y + 33 = 0$.

2. $x^2 - 4y^2 = -1$.

3. $9x^2 + 4y^2 + 36x - 8y + 4 = 0$.

4. $6x + 8y - x^2 - 4y^2 = 12$.

5. $x^2 - 9y^2 - 4x + 18y - 14 = 0$.

6. $4y = x^2 + 2x - 7$.

7. $3x + 2y^2 - 4y - 7 = 0$.

8. $4x^2 = y^2 - 4y$.

In Exercises 9–16, find an equation for the curve described.

9. The parabola with vertex at $(1, 3)$ and directrix $x = -3$.

10. The ellipse with major axis of length 12 and foci at $(2, 7)$ and $(2, -1)$.

11. The hyperbola with foci $(0, \pm 5)$ and vertices 6 units apart.

12. The parabola with axis $y = 1$, vertex $(2, 1)$, and passing through $(3, -1)$.

13. The ellipse with foci $(\pm 3, 0)$ and such that the distances from the foci to $P(x, y)$ on the ellipse add up to 10 units.

14. The hyperbola with vertices $(0, \pm 2)$ and asymptotes $y = \pm 3x$.

15. The curve C with the property that the distance between the point $(3, 4)$ and any point $P(x, y)$ on C is equal to the distance between P and the line $y = 2$.

16. The hyperbola with vertices $(-3, 2)$ and $(1, 2)$ and perpendicular asymptotes.

In Exercises 17–20, sketch the curve whose equation is given in the stated exercise.

17. Exercise 1.   **18.** Exercise 2.

19. Exercise 3.   **20.** Exercise 6.

In Exercises 21–26, find the rotation angle θ needed to remove the xy-term; then name the conic and give its equation in $x'y'$-coordinates after the xy-term is removed.

21. $3x^2 - 2xy + 3y^2 = 4$.

22. $7x^2 - 8xy + y^2 = 9$.

23. $11x^2 + 10\sqrt{3}xy + y^2 = 4$.

24. $x^2 + 4xy + 4y^2 - 2\sqrt{5}x + \sqrt{5}y = 0.$

25. $16x^2 - 24xy + 9y^2 - 60x - 80y + 100 = 0.$

26. $73x^2 - 72xy + 52y^2 - 100 = 0.$

In Exercises 27–29, sketch the graph of the equation in the stated exercise, showing the xy- and $x'y'$-axes.

27. Exercise 21.   **28.** Exercise 23.

29. Exercise 24.

30. For the curve $xy + x - y = 1$,
   (a) find the rotation angle necessary to remove the xy-term;
   (b) find the equation in $x'y'$-coordinates obtained by rotating the xy-coordinate axes through an angle of $\theta = \tan^{-1}(3/4)$ radians and verify the identities $A' + C' = A + C$ and $B'^2 - 4A'C' = B^2 - 4AC$.

13
polar coordinates and parametric equations

13.1 POLAR COORDINATES

*Up to now, we have specified the location of points in the plane by means of coordinates relative to a rectangular coordinate system consisting of two perpendicular coordinate axes. In this section, we shall introduce a new kind of coordinate system, called a **polar coordinate system,** that will be easier to work with in many problems.*

Figure 13.1.1

To form a polar coordinate system in a plane, we pick a fixed point O, called the *origin* or *pole;* and using the origin as an endpoint, we construct a ray, called the *polar axis.* After selecting a unit of measurement, we may associate with any point P in the plane a pair of *polar coordinates*

$$(r, \theta)$$

where r is the distance from P to the origin, and θ measures the angle from the polar axis to the line segment OP (Figure 13.1.1). The number r is called the *radial distance* of P and θ is called a *polar angle* of P. In Figure 13.1.2, the points $(6, 45°)$, $(3, 225°)$, $(5, 120°)$, and $(4, 330°)$ are plotted in polar coordinate systems.

Figure 13.1.2 ▷   (a)   (b)

The polar coordinates of a point are not unique. For example, the polar coordinates

$$(1, 315°), \quad (1, -45°), \quad \text{and} \quad (1, 675°)$$

all represent the same point (Figure 13.1.3). In general, if a point P has polar coordinates (r, θ), then for any integer $n = 0, 1, 2, \ldots$

$$(r, \theta + n \cdot 360°) \quad \text{and} \quad (r, \theta - n \cdot 360°)$$

are also polar coordinates of P.

(1, 315°)   (1, -45°)   (1, 675°)

All three pairs of polar coordinates
represent the same point.

Figure 13.1.3 ▷

In the case where P is the origin, the line segment OP in Figure 13.1.1 reduces to a point, since $r = 0$. Because there is no clearly defined polar angle in this case, we will agree that an arbitrary polar angle θ may be used. Thus, for arbitrary θ, the point $(0, \theta)$ is the origin.

RELATIONSHIP
BETWEEN POLAR
AND RECTANGULAR
COORDINATES

Frequently, it is helpful to use both polar and rectangular coordinates in the same problem. To do this, we let the positive x-axis of the rectangular coordinate system serve as the polar axis for the polar coordinate system. When this is done, each point P has both polar coordinates (r, θ) and rectangular coordinates (x, y). As suggested by Figure 13.1.4, these coordinates are related by the equations

Figure 13.1.4

$$x = r \cos \theta$$
$$y = r \sin \theta$$

(1)

and

$$r^2 = x^2 + y^2$$

(2a)

$$\tan \theta = \frac{y}{x}$$

(2b)

Example 1   Find the rectangular coordinates of the point P whose polar coordinates are $(6, 135°)$.

Solution.   Substituting the polar coordinates $r = 6$ and $\theta = 135°$ in (1) yields

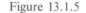

The point P has polar coordinates $(6, 135°)$ and rectangular coordinates $(-3\sqrt{2}, 3\sqrt{2})$

Figure 13.1.5

(a)

(b)

Figure 13.1.6

$$x = 6 \cos 135° = 6\left(-\frac{\sqrt{2}}{2}\right) = -3\sqrt{2}$$

$$y = 6 \sin 135° = 6\left(\frac{\sqrt{2}}{2}\right) = 3\sqrt{2}$$

Thus, the rectangular coordinates of P are $(-3\sqrt{2}, 3\sqrt{2})$ (Figure 13.1.5). ◄

When we start graphing curves in polar coordinates, it will be desirable to allow negative values for r. This will require a special definition. For motivation, consider the point P with polar coordinates $(3, 225°)$. We can reach this point by rotating the polar axis 225° and then moving *forward* from the origin 3 units along the terminal side of the angle (Figure 13.1.6a). On the other hand, we can also reach the point P by rotating the polar axis 45° and then moving *backward* 3 units from the origin along the *extension* of the terminal side of the angle (Figure 13.1.6b). This suggests that the point $(3, 225°)$ might also be denoted by $(-3, 45°)$, with the minus sign serving to indicate that the point is on the extension of the angle's terminal side rather than on the terminal side itself.

Since the terminal side of the angle $\theta + 180°$ is the extension of the terminal side of the angle θ, we shall define

$$(-r, \theta) \quad \text{and} \quad (r, \theta + 180°) \tag{3}$$

to be polar coordinates for the same point. With $r = 3$ and $\theta = 45°$ in (3) it follows that $(-3, 45°)$ and $(3, 225°)$ represent the same point.

In the exercises we ask the reader to show that relationships (1) and (2) remain valid when r is negative.

REMARK. For many purposes it does not matter whether polar angles are represented in degrees or radians. However, later on, when we are concerned with differentiation problems, it will be essential to measure polar angles in radians, since the derivative formulas for the trigonometric functions were derived under this assumption. From here on, we shall use radian measure.

Example 2   Find polar coordinates of the point P whose rectangular coordinates are $(-2, 2\sqrt{3})$.

Solution.   We will find polar coordinates (r, θ) of P such that $r > 0$ and $0 \leq \theta < 2\pi$. From (2a),

$$r^2 = x^2 + y^2 = (-2)^2 + (2\sqrt{3})^2 = 4 + 12 = 16$$

so $r = 4$. From (2b),

$$\tan \theta = \frac{y}{x} = \frac{2\sqrt{3}}{-2} = -\sqrt{3}$$

From this and the fact that $(-2, 2\sqrt{3})$ lies in the second quadrant, it follows that $\theta = 2\pi/3$. Thus, $(4, 2\pi/3)$ are polar coordinates of P. All other polar coordinates of P have the form

$$\left(4, \frac{2\pi}{3} + 2n\pi\right) \quad \text{or} \quad \left(-4, \frac{5\pi}{3} + 2n\pi\right) \text{ where } n \text{ is an integer.} \quad \blacktriangleleft$$

Polar coordinates provide a new way of graphing certain equations. As an example, consider the equation

$$r = \sin \theta \tag{4}$$

By substituting values for θ at increments of $\pi/6$ $(=30°)$, and calculating r we can construct the following table:

Table 13.1.1

θ (radians)	0	$\dfrac{\pi}{6}$	$\dfrac{\pi}{3}$	$\dfrac{\pi}{2}$	$\dfrac{2\pi}{3}$	$\dfrac{5\pi}{6}$	π	$\dfrac{7\pi}{6}$	$\dfrac{4\pi}{3}$	$\dfrac{3\pi}{2}$	$\dfrac{5\pi}{3}$	$\dfrac{11\pi}{6}$	2π
$r = \sin \theta$	0	$\dfrac{1}{2}$	$\dfrac{\sqrt{3}}{2}$	1	$\dfrac{\sqrt{3}}{2}$	$\dfrac{1}{2}$	0	$-\dfrac{1}{2}$	$-\dfrac{\sqrt{3}}{2}$	-1	$-\dfrac{\sqrt{3}}{2}$	$-\dfrac{1}{2}$	0

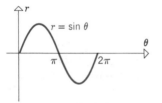

Figure 13.1.7

The pairs of values listed in this table may be plotted in one of two ways. One possibility is to view each pair (r, θ) as *rectangular* coordinates of a point and plot each point in the $r\theta$-plane. This yields points on the familiar sine curve (Figure 13.1.7). Alternatively, we can view each pair (r, θ) as *polar* coordinates of a point and plot each point in a polar coordinate system (Figure 13.1.8). Note that there are 13 pairs listed in Table 13.1.1, but only 6 points plotted in Figure 13.1.8. This is because the pairs from $\theta = \pi$ on yield duplicates of the preceding points. For example, $(-1/2, 7\pi/6)$ and $(1/2, \pi/6)$ represent the same point.

The points in Figure 13.1.8 appear to lie on a circle. That this is indeed the case may be seen by expressing (4) in terms of x and y. To do this we first multiply (4) through by r to obtain

$$r^2 = r \sin \theta$$

which may be rewritten using (1) and (2) as

$$x^2 + y^2 = y \quad \text{or} \quad x^2 + y^2 - y = 0$$

Figure 13.1.8   or on completing the square

$$x^2 + \left(y - \frac{1}{2}\right)^2 = \frac{1}{4}$$

This is a circle of radius $\frac{1}{2}$ centered at the point $(0, \frac{1}{2})$ in the xy-plane.

Example 3 Sketch the curve $r = \sin 2\theta$ in polar coordinates.

Solution. A rough sketch of the graph may be obtained without plotting points by observing how the value of r changes with θ.

If θ increases from 0 to $\pi/4$, then 2θ increases from 0 to $\pi/2$ and $r = \sin 2\theta$ increases from 0 to 1 (Figure 13.1.9a).

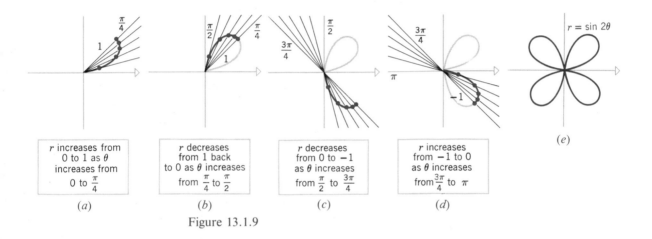

r increases from 0 to 1 as θ increases from 0 to $\frac{\pi}{4}$	r decreases from 1 back to 0 as θ increases from $\frac{\pi}{4}$ to $\frac{\pi}{2}$	r decreases from 0 to -1 as θ increases from $\frac{\pi}{2}$ to $\frac{3\pi}{4}$	r increases from -1 to 0 as θ increases from $\frac{3\pi}{4}$ to π
(a)	(b)	(c)	(d)

Figure 13.1.9

If θ increases from $\pi/4$ to $\pi/2$, then 2θ increases from $\pi/2$ to π and $r = \sin 2\theta$ decreases from 1 back to 0. Thus, the curve completes a loop (Figure 13.1.9b).

If θ increases from $\pi/2$ to $3\pi/4$, then 2θ increases from π to $3\pi/2$ and r decreases from 0 to -1. Because r is negative, the resulting portion of the graph is in the fourth quadrant although the values of θ are in the second quadrant (Figure 13.1.9c).

If θ increases from $3\pi/4$ to π, then 2θ increases from $3\pi/2$ to 2π and $r = \sin 2\theta$ increases from -1 to 0, and the curve completes a loop in the fourth quadrant (Figure 13.1.9d).

We leave it for the reader to continue the analysis for θ in the range π to 2π and show that the remaining portion of the curve consists of two loops in the second and third quadrants (Figure 13.1.9e). No new points are generated by allowing θ to vary past 2π or through negative values (verify) so that Figure 13.1.9e is the complete graph. ◄

▶ Exercise Set 13.1

1. Plot the following points in polar coordinates.
   (a) $(3, \pi/4)$ (b) $(5, 2\pi/3)$
   (c) $(1, \pi/2)$ (d) $(4, 7\pi/6)$
   (e) $(2, 4\pi/3)$ (f) $(0, \pi)$.

2. Plot the following points in polar coordinates.
   (a) $(2, -\pi/3)$ (b) $(3/2, -7\pi/4)$
   (c) $(-3, 3\pi/2)$ (d) $(-5, -\pi/6)$
   (e) $(-6, -\pi)$ (f) $(-1, 9\pi/4)$.

3. Find rectangular coordinates of the points whose polar coordinates are given.
   (a) $(6, \pi/6)$ (b) $(7, 2\pi/3)$
   (c) $(8, 9\pi/4)$ (d) $(5, 0)$
   (e) $(7, 17\pi/6)$ (f) $(0, \pi)$.

4. Find rectangular coordinates of the points whose polar coordinates are given.
   (a) $(-8, \pi/4)$ (b) $(7, -\pi/4)$
   (c) $(-6, -5\pi/6)$ (d) $(0, -\pi)$
   (e) $(-2, -3\pi/2)$ (f) $(-5, 0)$.

5. The following points are given in rectangular coordinates. Express the points in polar coordinates with $r \geq 0$ and $0 \leq \theta < 2\pi$.
   (a) $(-5, 0)$ (b) $(2\sqrt{3}, -2)$
   (c) $(0, -2)$ (d) $(-8, -8)$
   (e) $(-3, 3\sqrt{3})$ (f) $(1, 1)$.

6. Express the points in Exercise 5 in polar coordinates with $r \geq 0$ and $-\pi < \theta \leq \pi$.

7. Express the points in Exercise 5 in polar coordinates with $r \leq 0$ and $0 \leq \theta < 2\pi$.

In Exercises 8–15, identify the curve by transforming to rectangular coordinates.

8. $r = 3$.

9. $r \sin \theta = 4$.

10. $r = 2 \sin \theta$.

11. $r = \dfrac{6}{2 - \cos \theta}$.

12. $r = 5 \sec \theta$.

13. $r + 4 \cos \theta = 0$.

14. $r = \dfrac{2}{1 - 3 \sin \theta}$.

15. $r = \dfrac{6}{3 \cos \theta + 2 \sin \theta}$.

In Exercises 16–23, express the equations in polar coordinates.

16. $x^2 + y^2 = 9$.

17. $x = 7$.

18. $4xy = 9$.

19. $(x^2 + y^2)^2 = 16(x^2 - y^2)$.

20. $x^2 - y^2 = 4$.

21. $x^2 = 9y$.

22. $2x - 5y = 3$.

23. $x^2 + y^2 - 6y = 0$.

In Exercises 24–27, use the method of Example 3 to sketch the curve in polar coordinates.

24. $r = \cos 2\theta$.

25. $r = 2(1 + \sin \theta)$.

26. $r = 4 \cos 3\theta$.

27. $r = 1 - \cos \theta$.

28. Prove that the distance between the points with polar coordinates (r_1, θ_1) and (r_2, θ_2) is
$$d = \sqrt{r_1^2 + r_2^2 - 2r_1 r_2 \cos(\theta_1 - \theta_2)}$$

29. Prove: In polar coordinates, the equation $r = a \sin \theta + b \cos \theta$ represents a circle.

30. Prove: The area of the triangle whose vertices have polar coordinates $(0, 0)$, (r_1, θ_1), and (r_2, θ_2) is $A = \frac{1}{2} r_1 r_2 \sin(\theta_2 - \theta_1)$. [Assume $0 \leq \theta_1 < \theta_2 \leq \pi$ and r_1 and r_2 are positive.]

31. Prove: Equations (1) and (2) hold if r is negative.

32. Prove that
$$r r_1 \sin(\theta - \theta_1) + r r_2 \sin(\theta_2 - \theta) + r_1 r_2 \sin(\theta_1 - \theta_2) = 0$$
is the polar equation of the line through the points (r_1, θ_1) and (r_2, θ_2).

13.2   GRAPHS IN POLAR COORDINATES

In this section we shall graph some curves in polar coordinates. Since we shall be working with both rectangular xy-coordinates and polar co-ordinates, we shall assume throughout that the positive x-axis coincides with the polar axis.

LINES IN POLAR
COORDINATES

A line perpendicular to the x-axis and passing through the point with xy-coordinates $(a, 0)$ has the equation

$$x = a$$

To express this equation in polar coordinates we substitute $x = r \cos \theta$ [see (1) in the previous section]; this yields

$$r \cos \theta = a \tag{1}$$

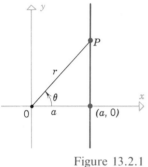

Figure 13.2.1

This result makes sense geometrically since each point $P(r, \theta)$ on this line will yield the value a for $r \cos \theta$ (Figure 13.2.1).

A line parallel to the x-axis that meets the y-axis at the point with xy-coordinates $(0, b)$ has the equation

$$y = b$$

Substituting $y = r \sin \theta$ yields

$$r \sin \theta = b \tag{2}$$

as the polar equation of this line. This makes sense geometrically since each point $P(r, \theta)$ on this line will yield the value b for $r \sin \theta$ (Figure 13.2.2).

For any constant θ_0, the equation

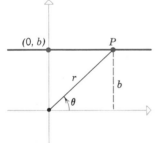

Figure 13.2.2

$$\theta = \theta_0 \tag{3}$$

is satisfied by the coordinates of all points of the form $P(r, \theta_0)$, regardless of the value of r. Thus, the equation represents the line through the origin making an angle of θ_0 (radians) with the polar axis (Figure 13.2.3).

Example 1   Sketch the graphs of the following equations in polar co-ordinates.

Figure 13.2.3

(a) $r \cos \theta = 3$ (b) $r \sin \theta = -2$ (c) $\theta = \dfrac{3\pi}{4}$

Solution. From (1), (2), and (3), each graph is a line; they are shown in Figure 13.2.4.   ◄

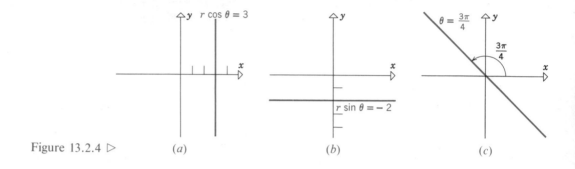

Figure 13.2.4 ▷   (a)   (b)   (c)

By substituting $x = r \cos \theta$ and $y = r \sin \theta$ in the equation $Ax + By + C = 0$, we obtain the **general polar form of a line,**

$$r(A \cos \theta + B \sin \theta) + C = 0$$

CIRCLES IN POLAR COORDINATES

Let us try to find the polar equation of a circle whose radius is a and whose center has polar coordinates (r_0, θ_0) (Figure 13.2.5). If we let $P(r, \theta)$ be an arbitrary point on the circle, and if we apply the law of cosines to the triangle OCP in Figure 13.2.5 we obtain

$$r^2 - 2rr_0 \cos(\theta - \theta_0) + r_0{}^2 = a^2 \qquad (4)$$

This general equation is rarely used since the corresponding equation in rectangular coordinates is easier to work with. However, we will now consider some special cases of (4) that arise frequently.

A circle of radius a, centered at the origin, has an especially simple polar equation. If we let $r_0 = 0$ in (4), we obtain

Figure 13.2.5

$$r^2 = a^2$$

or since $a \geq 0$

$$r = a \qquad (5)$$

This equation makes sense geometrically since the circle of radius a, centered at the origin, consists of all points $P(r, \theta)$ for which $r = a$, regardless of the value of θ (Figure 13.2.6a).

If a circle of radius a has its center on the x-axis and passes through the origin, then the polar coordinates of the center are either

$$(a, 0) \quad \text{or} \quad (a, \pi)$$

(a)

(b)

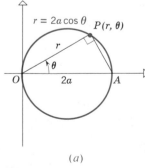

(c)

Figure 13.2.6

depending on whether the center is to the right or left of the origin (Figure 13.2.6b). For the circle to the right of the origin we have $r_0 = a$ and $\theta_0 = 0$, so (4) reduces to

$$r^2 - 2ra \cos \theta = 0$$

or

$$r = 2a \cos \theta \qquad (6a)$$

For the circle to the left of the origin, $r_0 = a$ and $\theta_0 = \pi$, so (4) reduces to

$$r = -2a \cos \theta \qquad (6b)$$

because $\cos (\theta - \pi) = -\cos \theta$.

If a circle of radius a passes through the origin and has its center on the y-axis, then the polar coordinates of the center are $(a, \pi/2)$ or $(a, -\pi/2)$, and (4) reduces to

$$r = 2a \sin \theta \qquad (7a)$$

or

$$r = -2a \sin \theta \qquad (7b)$$

depending on whether the center is above or below the origin (Figure 13.2.6c).

Equations (6a) and (7a) are interpreted geometrically in Figures 13.2.7a and 13.2.7b. In each figure part, OPA is a right triangle (why?) so that for each point $P(r, \theta)$ on the circle, we have $r = 2a \cos \theta$ or $r = 2a \sin \theta$, depending on the orientation of the circle.

(a)

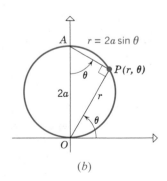

(b)

Figure 13.2.7

Example 2   Sketch the graphs of the following equations in polar coordinates

(a) $r = 4 \cos \theta$ (b) $r = -5 \sin \theta$ (c) $r = 3$

Solution.   These equations are of forms (6a), (7b), and (5), respectively, and therefore represent circles. The graphs are shown in Figure 13.2.8.   ◄

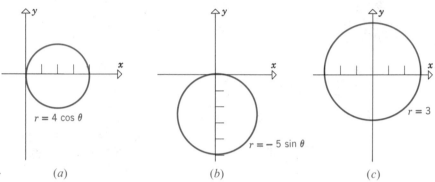

Figure 13.2.8 ▷   (a)   (b)   (c)

The points with polar coordinates (r, θ) and $(r, -\theta)$ are symmetric about the x-axis (Figure 13.2.9a). Thus, if a polar equation is unchanged on replacing θ by $-\theta$, then its graph is symmetric about the x-axis. The following tests for symmetry will be useful.

Symmetry Tests

13.2.1   THEOREM.

(a)   *A curve in polar coordinates is symmetric about the x-axis if replacing θ by $-\theta$ in its equation produces an equivalent equation (Figure 13.2.9a).*

(b)   *A curve in polar coordinates is symmetric about the y-axis if replacing θ by $\pi - \theta$ in its equation produces an equivalent equation (Figure 13.2.9b).*

(c)   *A curve in polar coordinates is symmetric about the origin if replacing r by $-r$ in its equation produces an equivalent equation (Figure 13.2.9c).*

CARDIOIDS AND   Equations of the form
LIMAÇONS

$$r = a + b \sin \theta \qquad r = a - b \sin \theta \qquad \text{(8a–8b)}$$
$$r = a + b \cos \theta \qquad r = a - b \cos \theta \qquad \text{(8c–8d)}$$

produce polar curves called *limaçons* (from the Latin word ''limax,'' for

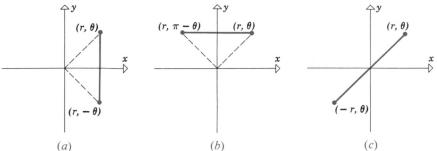

Figure 13.2.9 ▷   (a)   (b)   (c)

a slug, a snail-like creature). There are four possible shapes for a limaçon*
that can be determined from the ratio a/b when $a > 0$ and $b > 0$ (Figure
13.2.10).

The orientation of the limaçon relative to the polar axis depends on
whether $\sin \theta$ or $\cos \theta$ appears in the equation and whether the $+$ or $-$
occurs. Because of the heart-shaped appearance of the curve in the case
$a = b$, limaçons of this type are called *cardioids* (from the Greek word
"kardia," for heart).

Example 3 Assuming a to be a positive constant, sketch the curve
$r = a(1 - \cos \theta)$ in polar coordinates.

Solution. This is an equation of form (8d) with $a = b$ and, therefore,
represents a cardioid. Since $\cos(-\theta) = \cos \theta$ the equation is unaltered
when θ is replaced by $-\theta$; thus, the cardioid is symmetric about the
x-axis. This being the case, we can obtain the entire curve by first sketching the portion of the cardioid above the x-axis and then reflecting this
portion about the x-axis.

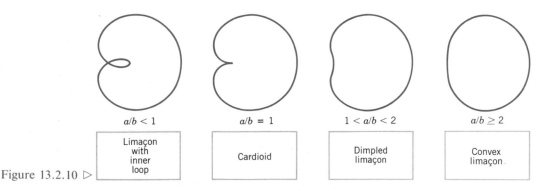

Limaçon with inner loop	Cardioid	Dimpled limaçon	Convex limaçon
$a/b < 1$	$a/b = 1$	$1 < a/b < 2$	$a/b \geq 2$

Figure 13.2.10 ▷

*For a detailed discussion of limaçons and their shapes, the reader is referred to the article
by Jane T. Grossman and Michael P. Grossman, "Dimple or No Dimple," *Two Year College
Mathematics Journal*, Vol. 13, No. 1 (1982), pp. 52–55.

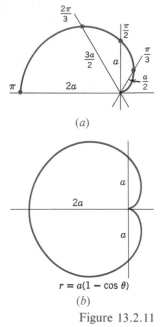

(a)

$r = a(1 - \cos\theta)$

(b)

Figure 13.2.11

As θ varies from 0 to π, $\cos\theta$ decreases steadily from 1 to -1, and $1 - \cos\theta$ increases steadily from 0 to 2. Thus, as θ varies from 0 to π, the value of $r = a(1 - \cos\theta)$ will increase steadily from an initial value of $r = 0$ to a final value of $r = 2a$. Using this information, and by plotting the points in Table 13.2.1, we obtain the graph in Figure 13.2.11a.

Table 13.2.1

θ	0	$\dfrac{\pi}{3}$	$\dfrac{\pi}{2}$	$\dfrac{2\pi}{3}$	π
$r = a(1 - \cos\theta)$	0	$\dfrac{a}{2}$	a	$\dfrac{3a}{2}$	$2a$

On reflecting the curve in Figure 13.2.11a about the x-axis, we obtain the entire cardioid (Figure 13.2.11b). ◀

REMARK. We have sketched the cardioid in Figure 13.2.11a with the line $\theta = 0$ tangent to the curve at the origin. While it is not evident from our discussion that this is the case, we will prove in Section 13.5 that if a polar curve passes through the origin when $\theta = \theta_0$, then the line $\theta = \theta_0$ is tangent to the curve at the origin.

LEMNISCATES   If $a > 0$ then equations of the form

$$r^2 = a^2 \cos 2\theta \qquad r^2 = -a^2 \cos 2\theta \qquad \text{(9a–9b)}$$
$$r^2 = a^2 \sin 2\theta \qquad r^2 = -a^2 \sin 2\theta \qquad \text{(9c–9d)}$$

A lemniscate

Figure 13.2.12

represent propeller-shaped curves, called *lemniscates* (Figure 13.2.12). The lemniscates are centered at the origin, but the orientation relative to the polar axis depends on the sign preceding the a^2 and whether $\sin 2\theta$ or $\cos 2\theta$ appears in the equation.

Example 4   Sketch the curve

$$r^2 = 4 \cos 2\theta \qquad \text{(10)}$$

in polar coordinates.

Solution. The equation is of type (9a) with $a = 2$, and thus represents a lemniscate. Solving (10) for r in terms of θ yields two functions of θ:

$$r = 2\sqrt{\cos 2\theta} \qquad \text{(11a)}$$

and

$$r = -2\sqrt{\cos 2\theta} \qquad \text{(11b)}$$

We leave it for the reader to apply the symmetry tests in Theorem 13.2.1 and show that the graph of each of these equations is symmetric about the x-axis and the y-axis. Therefore, we can obtain each graph by first sketching the portion of the graph in the range $0 \leq \theta \leq \pi/2$ and then reflecting that portion about the x- and y-axes.

We shall consider the graph of (11a) first. As θ varies from 0 to $\pi/4$, the value of $\cos 2\theta$ decreases steadily from 1 to 0, so that $r = 2\sqrt{\cos 2\theta}$ decreases steadily from 2 to 0. With this information and the points plotted from Table 13.2.2, we obtain the sketch in Figure 13.2.13a. (Note that the curve passes through the origin when $\theta = \pi/4$, so that the line $\theta = \pi/4$ is tangent to the curve at the origin by the remark following Example 3.)

For θ in the range $\pi/4 < \theta < \pi/2$, the quantity $\cos 2\theta$ is negative, so there are no real values of r satisfying (11a). Thus, there are no points on the graph for such θ. The entire graph is obtained by reflecting the curve in Figure 13.2.13a about the x-axis and then reflecting the resulting curve about the y-axis (Figure 13.2.13b).

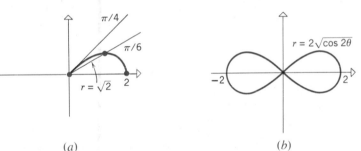

Figure 13.2.13 ▷   (a)   (b)

Table 13.2.2

θ	0	$\dfrac{\pi}{6}$	$\dfrac{\pi}{4}$
$r = 2\sqrt{\cos 2\theta}$	2	$\sqrt{2}$	0

Figure 13.2.14

To graph $r = -2\sqrt{\cos 2\theta}$, we simply note that (11a) and (11b) differ only in the sign of r, so the graph of (11b) is identical to that in Figure 13.2.13b, but traced in a "diagonally opposite" manner. For example, as θ varies from 0 to $\pi/4$, the portion of the graph generated is in the third quadrant because r is negative (Figure 13.2.14). To summarize, the graph of (10) actually consists of two superimposed lemniscates, one being the graph of (11a) and the other the graph of (11b). ◀

SPIRALS   A curve that "winds around the origin" infinitely many times in such a way that r increases (or decreases) steadily as θ increases is called a *spiral*.

The most common example is the *spiral of Archimedes,* which has an equation of the form

$$r = a\theta \quad (\theta \geq 0) \tag{12a}$$

or

$$r = a\theta \quad (\theta \leq 0) \tag{12b}$$

In these equations, θ is in radians and a is positive.

Example 5   Sketch the curve

$$r = \theta \quad (\theta \geq 0)$$

in polar coordinates.

Solution.   This is an equation of form (12a) with $a = 1$ and, thus, represents an Archimedean spiral. Since $r = 0$ when $\theta = 0$, the origin is on the curve and the polar axis is tangent to the spiral.

A reasonably accurate sketch may be obtained by plotting the intersections of the spiral with the x- and y-axes and noting that r increases steadily as θ increases. The intersections with the x-axis occur when

$$\theta = 0, \pi, 2\pi, 3\pi, \ldots$$

at which points r has the values

$$r = 0, \pi, 2\pi, 3\pi, \ldots$$

and the intersections with the y-axis occur when

$$\theta = \frac{\pi}{2}, \frac{3\pi}{2}, \frac{5\pi}{2}, \frac{7\pi}{2}, \ldots$$

at which points r has the values

$$r = \frac{\pi}{2}, \frac{3\pi}{2}, \frac{5\pi}{2}, \frac{7\pi}{2}, \ldots$$

The graph is shown in Figure 13.2.15.   ◀

Starting from the origin, the spiral in Figure 13.2.15 loops counterclockwise around the origin. This is true of all Archimedean spirals of form (12a); Archimedean spirals of form (12b) loop clockwise around the origin.

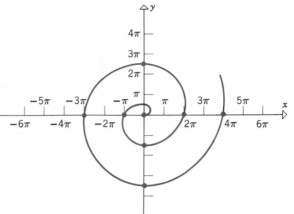

Figure 13.2.15 ▷

ROSE CURVES   Equations of the form

$$r = a \sin n\theta \tag{13a}$$
$$r = a \cos n\theta \tag{13b}$$

represent flower-shaped curves called *roses*. The rose has n equally spaced petals or loops if n is odd and $2n$ equally spaced petals if n is even (Figure 13.2.16). The orientation of the rose relative to the polar axis depends on the sign of the constant a and whether $\sin \theta$ or $\cos \theta$ appears in the equation. (A rose curve occurred in Example 3 of the previous section.)

If $n = 1$ in (13a) or (13b) then we obtain the equation of a circle, which can be regarded as a one-petal rose.

REMARK. It can be shown that a rose with an even number petals is traced out exactly once as θ varies over the interval $0 \le \theta < 2\pi$ (Example 3 in the previous section), and a rose with an odd number of petals is traced out exactly once as θ varies over the interval $0 \le \theta < \pi$ (Exercise 52).

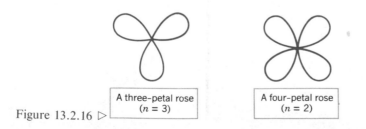

A three-petal rose
($n = 3$)

A four-petal rose
($n = 2$)

Figure 13.2.16 ▷

▶ Exercise Set 13.2

In Exercises 1–36, sketch the curve in polar coordinates and give its name.

1. $\theta = \pi/6$.

2. $\theta = -3\pi/4$.

3. $r = 5$.

4. $r = 4 \sin \theta$.

5. $r = -6 \cos \theta$.

6. $r = -3 \sin \theta$.

7. $2r = \cos \theta$.

8. $r = 1 + \sin \theta$.

9. $r = 3(1 - \sin \theta)$.

10. $r - 2 = 2 \cos \theta$.

11. $r = 4 - 4 \cos \theta$.

12. $r = -5 + 5 \sin \theta$.

13. $r = -1 - \cos \theta$.

14. $r = 1 + 2 \sin \theta$.

15. $r = 1 - 2 \cos \theta$.

16. $r = 4 + 3 \cos \theta$.

17. $r = 3 + 2 \sin \theta$.

18. $r = 3 - \cos \theta$.

19. $r = 2 + \sin \theta$.

20. $r - 5 = 3 \sin \theta$.

21. $r = 3 + 4 \cos \theta$.

22. $r = -3 - 4 \sin \theta$.

23. $r = 5 - 2 \cos \theta$.

24. $r^2 = 9 \cos 2\theta$.

25. $r^2 = -9 \cos 2\theta$.

26. $r^2 = \sin 2\theta$.

27. $r^2 = -16 \sin 2\theta$.

28. $r = 4\theta \ (\theta \geq 0)$.

29. $r = 4\theta \ (\theta \leq 0)$.

30. $r = 4\theta$.

31. $r = \cos 2\theta$.

32. $r = 3 \sin 2\theta$.

33. $r = \sin 3\theta$.

34. $r = 2 \cos 3\theta$.

35. $r = 9 \sin 4\theta$.

36. $r = \cos 5\theta$.

In Exercises 37–46, sketch the curve in polar coordinates.

37. $r = 4 \cos \theta + 4 \sin \theta$.

38. $r = e^{\theta}$ (logarithmic spiral).

39. $r = 4 \tan \theta$ (kappa curve).

40. $r = 2 + 2 \sec \theta$ (conchoid of Nichomedes).

41. $r = \sin (\theta/2)$.

42. $r^2 = \theta$ (parabolic spiral).

43. $r = 2 \sin \theta \tan \theta$ (cissoid).

44. $r = 1/\theta \ (\theta > 0)$ (hyperbolic spiral). [*Hint:* Find $\lim\limits_{\theta \to 0^+} y$ to show that there is a horizontal asymptote.]

45. $r = 1/\sqrt{\theta} \ (\theta > 0)$. [*Hint:* Find $\lim\limits_{\theta \to 0^+} y$ to show that there is a horizontal asymptote.]

46. $r = 1/\theta^2 \ (\theta > 0)$. Show that $\lim\limits_{\theta \to 0^+} y = +\infty$.

47. Find the maximum value of the y-coordinate of points on the cardioid $r = 1 + \cos \theta$ for θ in the interval $[0, \pi]$.

48. (a) Show that the maximum value of the y-coordinate of points on the curve $r = 1/\sqrt{\theta}$ for θ in the interval $(0, \pi]$ occurs when θ satisfies the equation $\tan \theta = 2\theta$.

   (b) Use Newton's Method to solve the equation in part (a) for θ to at least four decimal place accuracy.

   (c) Use the result of part (b) to approximate the maximum value of y for $0 < \theta \leq \pi$.

49. Find the minimum value of the x-coordinate of points on the cardioid $r = 1 + \cos \theta$ for θ in the interval $[0, \pi]$.

50. Show that the minimum value of the x-coordinate of points on the curve $r = a + b \cos \theta$, $a > 0$, $b > 0$, for θ in the interval $[0, \pi]$ is $-a^2/(4b)$ if $a < 2b$ and is $b - a$ if $a \geq 2b$.

51. Use Definition 12.2.1 to derive an equation in polar coordinates of the parabola with its focus at the pole and its directrix perpendicular to the polar axis and k units ($k > 0$) to the left of the pole.

52. Prove that a rose with an even number of petals is traced out exactly once as θ varies over the interval $0 \leq \theta < 2\pi$ and a rose with an odd number of petals is traced out exactly once as θ varies over the interval $0 \leq \theta < \pi$.

13.3 AREA IN POLAR COORDINATES

In this section we shall show how to find areas of regions that are bounded by curves whose equations are given in polar coordinates.

The Area Problem in Polar Coordinates

13.3.1 PROBLEM. Find the area of the region R between a polar curve $r = f(\theta)$ and two lines, $\theta = \alpha$ and $\theta = \beta$ (Figure 13.3.1).

$r = f(\theta)$

$\theta = \beta$

R

$\theta = \alpha$

Figure 13.3.1 ▷

We shall assume that $\alpha < \beta$ and $f(\theta)$ is continuous and nonnegative for $\alpha \le \theta \le \beta$. As a first step toward finding the area of the region R, let θ_1, $\theta_2, \ldots, \theta_{n-1}$ be any numbers such that

$$\alpha < \theta_1 < \theta_2 < \cdots < \theta_{n-1} < \beta$$

and construct the lines

$$\theta = \theta_1, \theta = \theta_2, \ldots, \theta = \theta_{n-1}$$

These lines divide the angle from α to β into n subangles, and they divide the region R into n subregions (Figure 13.3.2). As indicated in this figure, we shall denote the subangles by

$\theta = \theta_2$

$\theta = \theta_{n-1}$

$\theta = \theta_1$

$\theta = \beta$

A_n

A_2

$\Delta\theta_2$

A_1

$\theta = \alpha$

$\Delta\theta_n$

$\Delta\theta_1$

Figure 13.3.2 ▷

$$\Delta\theta_1, \Delta\theta_2, \ldots, \Delta\theta_n$$

and the areas of the subregions by

$$A_1, A_2, \ldots, A_n$$

Then the area A of the entire region may be written as

$$A = A_1 + A_2 + \cdots + A_n = \sum_{k=1}^{n} A_k$$

Let us now concentrate on the areas $A_1, A_2, \ldots, A_n$ of the subregions. If $\Delta\theta_k$ is not too large, we can approximate the area A_k by the area of a *sector* having central angle $\Delta\theta_k$ and radius $f(\theta_k^*)$, where θ_k^* is an arbitrary angle terminating in the kth subregion (Figure 13.3.3). Thus, from the area formula for a sector [Formula (1), Section 2.8],

$(f(\theta_k^*), \theta_k^*)$

$r = f(\theta)$

$\Delta\theta_k$

θ_k^*

The shaded sector has radius $f(\theta_k^*)$ and central angle $\Delta\theta_k$

Figure 13.3.3 ▷

$$A_k \approx \frac{1}{2}[f(\theta_k^*)]^2 \, \Delta\theta_k$$

and

$$A = \sum_{k=1}^{n} A_k \approx \sum_{k=1}^{n} \frac{1}{2}[f(\theta_k^*)]^2 \, \Delta\theta_k \tag{1}$$

If we now increase n in such a way that $\max \Delta\theta_k \to 0$, then the sectors will become better and better approximations to the subregions and (1) will approach the exact value of the area A; that is,

$$A = \lim_{\max \Delta\theta_k \to 0} \sum_{k=1}^{n} \frac{1}{2}[f(\theta_k^*)]^2 \, \Delta\theta_k \tag{2}$$

Since the limit on the right side of (2) is the definite integral

$$\int_{\alpha}^{\beta} \frac{1}{2}[f(\theta)]^2 \, d\theta$$

we are led to the following result.

Area in Polar Coordinates

13.3.2 AREA FORMULA   If $f(\theta)$ is continuous and nonnegative for $\alpha \leq \theta \leq \beta$, then the area A enclosed by the polar curve $r = f(\theta)$ and the lines $\theta = \alpha$ and $\theta = \beta$ is

$$A = \int_{\alpha}^{\beta} \frac{1}{2} [f(\theta)]^2 \, d\theta \qquad (3a)$$

or equivalently

$$A = \int_{\alpha}^{\beta} \frac{1}{2} r^2 \, d\theta \qquad (3b)$$

The hardest part of applying (3) is determining the limits of integration. This can be done as follows:

Step 1. Sketch the region R whose area is to be determined.

Step 2. Draw an arbitrary "radial line" from the origin to the boundary curve $r = f(\theta)$.

Step 3. Ask, "Over what interval of values must θ vary in order for the radial line to sweep out the region R?"

Step 4. Your answer in Step 3 will determine the lower and upper limits of integration.

Example 1   Find the area of the region in the first quadrant within the cardioid $r = 1 - \cos \theta$.

Solution.   The region and a typical radial line are shown in Figure 13.3.4. For the radial line to sweep out the region, θ must vary from 0 to $\pi/2$. Thus, from (3b) with $\alpha = 0$ and $\beta = \pi/2$,

$$A = \int_{0}^{\pi/2} \frac{1}{2} r^2 \, d\theta = \frac{1}{2} \int_{0}^{\pi/2} (1 - \cos \theta)^2 \, d\theta$$

$$= \frac{1}{2} \int_{0}^{\pi/2} (1 - 2 \cos \theta + \cos^2 \theta) \, d\theta$$

With the help of the identity $\cos^2 \theta = \frac{1}{2}(1 + \cos 2\theta)$, this may be rewritten as

$$A = \frac{1}{2} \int_{0}^{\pi/2} \left(\frac{3}{2} - 2 \cos \theta + \frac{1}{2} \cos 2\theta \right) d\theta$$

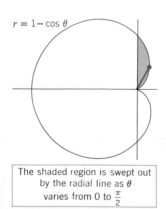

$r = 1 - \cos \theta$

The shaded region is swept out by the radial line as θ varies from 0 to $\frac{\pi}{2}$

Figure 13.3.4

$$= \frac{1}{2} \left[\frac{3}{2}\theta - 2 \sin \theta + \frac{1}{4} \sin 2\theta \right]_0^{\pi/2}$$

$$= \tfrac{3}{8}\pi - 1 \qquad \blacktriangleleft$$

Example 2 Find the entire area within the cardioid of Example 1.

Solution. For the radial line to sweep out the entire cardioid, θ must vary from 0 to 2π. Thus, from (3b) with $\alpha = 0$ and $\beta = 2\pi$

$$A = \int_0^{2\pi} \frac{1}{2} r^2 \, d\theta = \frac{1}{2} \int_0^{2\pi} (1 - \cos \theta)^2 \, d\theta$$

If we proceed as in Example 1, this reduces to

$$A = \frac{1}{2} \int_0^{2\pi} \left(\frac{3}{2} - 2 \cos \theta + \frac{1}{2} \cos 2\theta \right) d\theta = \frac{3\pi}{2}$$

Alternative Solution. Since the cardioid is symmetric about the x-axis, we can calculate the portion of area above the x-axis and double the result. In the portion of the cardioid above the x-axis, θ ranges from 0 to π, so that

$$A = 2 \int_0^{\pi} \frac{1}{2} r^2 \, d\theta = \int_0^{\pi} (1 - \cos \theta)^2 \, d\theta = \frac{3\pi}{2} \qquad \blacktriangleleft$$

It is important to keep in mind that Area Formula 13.3.2 requires $r = f(\theta)$ to be nonnegative. For example, suppose we were interested in finding the area of the region enclosed by the rose $r = \sin 2\theta$ shown in Figure 13.1.9. Although this region is swept out by a radial line as θ varies from 0 to 2π, the value of

$$r = \sin 2\theta$$

is negative over portions of this interval (if $\pi/2 < \theta < \pi$ or $3\pi/2 < \theta < 2\pi$). Thus, for this area problem Formulas (3a) and (3b) are not applicable over the entire interval $0 \le \theta \le 2\pi$. The following example shows how to avoid this difficulty by calculating a portion of the area and using symmetry.

Example 3 Find the area of the region enclosed by the rose curve $r = \sin 2\theta$.

Solution. The petal in the first quadrant is swept out as θ varies over the interval $0 \le \theta \le \pi/2$ (Figures 13.1.9a and b), and for these values of θ, the value of $r = \sin 2\theta$ is nonnegative, so Formula (3a) or (3b) can be used to find the area of this petal. By symmetry, the area enclosed by one petal is one-fourth of the entire area A, so

$$A = 4 \int_0^{\pi/2} \frac{1}{2} r^2 \, d\theta = 2 \int_0^{\pi/2} \sin^2 2\theta \, d\theta$$

$$= 2 \int_0^{\pi/2} \frac{1}{2} (1 - \cos 4\theta) \, d\theta = \int_0^{\pi/2} (1 - \cos 4\theta) \, d\theta$$

$$= \left[\theta - \frac{1}{4} \sin 4\theta \right]_0^{\pi/2} = \pi/2 \quad \blacktriangleleft$$

As the following example shows, it is sometimes convenient to allow θ to vary through negative angles.

Example 4 Find the area of the region that is inside the cardioid $r = 4 + 4 \cos \theta$ and outside the circle $r = 6$.

Solution. To sketch the region, we need to know where the circle and cardioid intersect. To find these points, we equate the given expressions for r. This yields

$$4 + 4 \cos \theta = 6$$

from which we obtain $\cos \theta = \frac{1}{2}$ or

$$\theta = \frac{\pi}{3} \quad \text{and} \quad \theta = -\frac{\pi}{3}$$

The region is sketched in Figure 13.3.5. The desired area can be obtained by subtracting the shaded areas in parts (*b*) and (*c*) of Figure 13.3.5. Thus,

$$A = \int_{-\pi/3}^{\pi/3} \frac{1}{2} (4 + 4 \cos \theta)^2 \, d\theta - \int_{-\pi/3}^{\pi/3} \frac{1}{2} (6)^2 \, d\theta$$

$$= \int_{-\pi/3}^{\pi/3} \frac{1}{2} [(4 + 4 \cos \theta)^2 - 36] \, d\theta$$

$$= \int_{-\pi/3}^{\pi/3} (16 \cos \theta + 8 \cos^2 \theta - 10) \, d\theta \tag{4}$$

$$= \left[16 \sin \theta + (4\theta + 2 \sin 2\theta) - 10\theta \right]_{-\pi/3}^{\pi/3}$$

$$= 18\sqrt{3} - 4\pi$$

Alternative Solution. Using symmetry, we may calculate the portion of area above the x-axis and double it to obtain the entire area A. This yields

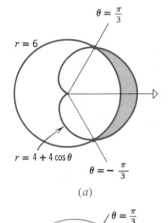

$\theta = \frac{\pi}{3}$

$r = 6$

$r = 4 + 4 \cos \theta$

$\theta = -\frac{\pi}{3}$

(*a*)

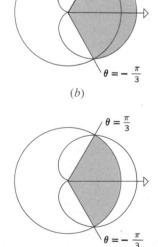

$\theta = \frac{\pi}{3}$

$\theta = -\frac{\pi}{3}$

(*b*)

$\theta = \frac{\pi}{3}$

$\theta = -\frac{\pi}{3}$

(*c*)

Figure 13.3.5

$$A = 2 \int_0^{\pi/3} \frac{1}{2} [(4 + 4 \cos \theta)^2 - 36] \, d\theta$$

rather than (4). However, the end result will be the same.   ◄

WARNING.   In the last example we determined the intersections of the curves $r = 4 + 4 \cos \theta$ and $r = 6$ by equating the right-hand sides and solving for θ. However, it may not be possible to find *all* the intersections of two polar curves, $r = f_1(\theta)$ and $r = f_2(\theta)$, by solving the equation $f_1(\theta) = f_2(\theta)$ for θ. This is because every point has infinitely many pairs of polar coordinates. It is possible, therefore, that a point of intersection may have no single pair of polar coordinates that satisfies both equations. For example, the cardioids

$$r = 1 - \cos \theta \quad \text{and} \quad r = 1 + \cos \theta \tag{5}$$

intersect at three points, the origin, the point $(1, \pi/2)$, and the point $(1, 3\pi/2)$ (Figure 13.3.6). Equating the right-hand sides of the equations in (5) yields $1 - \cos \theta = 1 + \cos \theta$ or $\cos \theta = 0$, so

$$\theta = \frac{\pi}{2} + k\pi, \quad k = 0, \pm 1, \pm 2, \ldots$$

Substituting any of these values in (5) yields $r = 1$, so that we have found only two distinct points of intersection, $(1, \pi/2)$ and $(1, 3\pi/2)$; the origin has been missed.

When calculating intersections in polar coordinates, it is a good idea to make a sketch of the curves to determine how many intersections there should be.

Figure 13.3.6 ▷

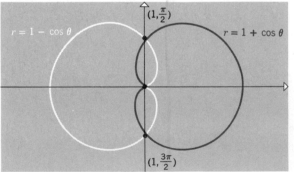

▶ Exercise Set 13.3

In Exercises 1–21, find the area of the region described.

1. The region in the first quadrant enclosed by the first loop of the spiral $r = \theta$ ($\theta \geq 0$), and the lines $\theta = \pi/6$ and $\theta = \pi/3$.

2. The region in the first quadrant within the cardioid $r = 1 + \sin\theta$.

3. The region that is enclosed by the cardioid $r = 2 + 2\cos\theta$.

4. The region outside the cardioid $r = 2 - 2\cos\theta$ and inside the circle $r = 4$.

5. The region inside the circle $r = 5\sin\theta$ and outside the limaçon $r = 2 + \sin\theta$.

6. The region enclosed by the inner loop of the limaçon $r = 1 + 2\cos\theta$.

7. The region inside the cardioid $r = 2 + 2\cos\theta$ and outside the circle $r = 3$.

8. The region inside the circle $r = 2a\sin\theta$.

9. The region enclosed by the curve $r^2 = \sin 2\theta$.

10. The region common to the circles $r = 4\cos\theta$ and $r = 4\sin\theta$.

11. The region enclosed by the rose $r = 4\cos 3\theta$.

12. The region inside the rose $r = 2a\cos 2\theta$ and outside the circle $r = a\sqrt{2}$.

13. The region common to the circle $r = 3\cos\theta$ and the cardioid $r = 1 + \cos\theta$.

14. The region between the loops of the limaçon $r = \tfrac{1}{2} + \cos\theta$.

15. The region inside the circle $r = 10$ and to the right of the line $r\cos\theta = 6$.

16. The region inside the cardioid $r = a(1 + \sin\theta)$ and outside the circle $r = a\sin\theta$.

17. The region enclosed by $r = a\sec^2\tfrac{1}{2}\theta$ and the rays whose polar angles are $\theta = 0$ and $\theta = \tfrac{1}{2}\pi$.

18. The region inside the cardioid $r = 2 + 2\cos\theta$ and to the right of the line $r\cos\theta = 3/2$.

19. The region swept out by a radial line from the origin to the curve $r = 3e^{-2\theta}$ as θ varies over the interval $0 \leq \theta \leq \pi$.

20. The region swept out by a radial line from the origin to the curve $r = 2/\theta$ as θ varies over the interval $1 \leq \theta \leq 3$.

21. The region swept out by a radial line from the origin to the curve $r = 1/\sqrt{\theta}$ as θ varies over the interval $1/9 \leq \theta \leq 4$.

22. (a) Find the error: The area inside the lemniscate $r^2 = a^2 \cos 2\theta$ is

$$A = \int_0^{2\pi} \frac{1}{2} r^2 \, d\theta = \int_0^{2\pi} \frac{1}{2} a^2 \cos 2\theta \, d\theta$$

$$= \frac{1}{4} a^2 \sin 2\theta \Big]_0^{2\pi} = 0$$

(b) Find the correct area.

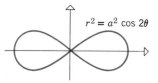

$r^2 = a^2 \cos 2\theta$

23. Find the area inside the lemniscate $r^2 = 4\cos 2\theta$ and outside the circle $r = \sqrt{2}$.

24. A radial line is drawn from the origin to the spiral $r = a\theta$ ($a > 0$ and $\theta \geq 0$). Find the area swept out during the second revolution of the radial line that was not swept out during the first revolution.

13.4   PARAMETRIC EQUATIONS

To study a curve using calculus, we need some way to represent the curve mathematically. Often, the representation is an equation $y = f(x)$ whose graph is the given curve. However, graphs of functions are specialized curves in the sense that no vertical line can cut such a curve more than once. In order to represent more general curves different methods are required. In this section we shall discuss one of these methods.

Curves that are not graphs of functions can often be specified by using a pair of equations

$$x = x(t)$$
$$y = y(t)$$

to express the coordinates of a point (x, y) on the curve as functions of an auxiliary variable t. These are called **parametric equations** for the curve, and the variable t is called a **parameter.**

Parametric equations arise naturally if one imagines a plane curve C to be traced by a moving point. If we use the parameter t to denote time, then the parametric equations $x = x(t)$, $y = y(t)$ specify how the x- and y-coordinates of the moving point vary with time.

Example 1   A particle moves in the plane so its x- and y-coordinates vary with time according to the equations

$$x = \tfrac{1}{2}t^3 - 6t$$
$$y = \tfrac{1}{2}t^2$$

Sketch the path of the particle over the time interval $0 \le t \le 4$.

Solution.   The path of the particle, shown in Figure 13.4.1, was found by calculating the x- and y-coordinates of the particle for $t = 0, 1, 2, 3, 4$ (Table 13.4.1), plotting the points (x, y), and connecting successive points with a smooth curve. The arrows on the curve serve to indicate the direction of motion.   ◀

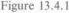
Figure 13.4.1

Table 13.4.1

t	0	1	2	3	4
$x = \frac{1}{2}t^3 - 6t$	0	$-\frac{11}{2}$	-8	$-\frac{9}{2}$	8
$y = \frac{1}{2}t^2$	0	$\frac{1}{2}$	2	$\frac{9}{2}$	8
(x, y)	$(0, 0)$	$(-\frac{11}{2}, \frac{1}{2})$	$(-8, 2)$	$(-\frac{9}{2}, \frac{9}{2})$	$(8, 8)$

Although time is a common parameter, it is not the only possibility. Other parameters are illustrated in the following examples.

Example 2   Let C be a circle of radius a centered at the origin. If $P(x, y)$ is any point on the circle and θ is the angle measured counterclockwise from the positive x-axis to the line segment OP (Figure 13.4.2a), then

$$x = a \cos \theta \quad \text{and} \quad y = a \sin \theta$$

As θ varies from 0 to 2π, the point $P(x, y)$ makes one complete revolution. Thus,

$$\begin{array}{l} x = a \cos \theta \\ y = a \sin \theta \end{array}, \quad 0 \le \theta \le 2\pi$$

are parametric equations for the circle.

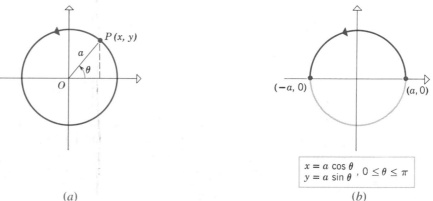

Figure 13.4.2 ▷ (a) (b)

By changing the interval over which θ is allowed to vary, we can obtain different portions of the circle. For example,

$$\begin{array}{l} x = a \cos \theta \\ y = a \sin \theta \end{array}, \quad 0 \le \theta \le \pi$$

represents just the upper half of the circle (Figure 13.4.2b). ◀

A curve that is represented parametrically is traced in a certain direction as the parameter increases. This is called the **direction of increasing pa-**

rameter or sometimes the *orientation* of the curve. For example, the circle

$$x = a \cos \theta \atop y = a \sin \theta, \quad 0 \le \theta \le 2\pi$$

is traced counterclockwise as θ increases from 0 to 2π, so this is the direction of increasing parameter. Phrased another way, the circle is oriented counterclockwise. We have indicated this in Figure 13.4.2a with an arrow on the circle.

Sometimes a curve given parametrically can be recognized by eliminating the parameter.

Example 3   Sketch the curve

$$x = 2t - 3$$
$$y = 6t - 7$$

Solution.   Solving each equation for t, we obtain

$$t = \frac{x + 3}{2}$$

$$t = \frac{y + 7}{6}$$

Thus,

$$\frac{x + 3}{2} = \frac{y + 7}{6}$$

or

$$y = 3x + 2$$

which is the straight line shown in Figure 13.4.3. The direction of increasing parameter, shown in the figure, was obtained from the original parametric equations by observing that x and y both increase as t increases.   ◄

REMARK.   In the last example, there were no conditions imposed on the parameter t. When no conditions are stated explicitly, it is understood that the parameter varies over the interval $(-\infty, +\infty)$.

Example 4   Sketch the curve

$$x = a \cos t \atop y = b \sin t, \quad 0 \le t \le 2\pi$$

where $a > 0$ and $b > 0$.

Figure 13.4.3

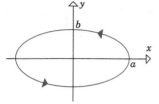

Figure 13.4.4

Solution. We can eliminate t by writing

$$\frac{x}{a} = \cos t \quad \text{and} \quad \frac{y}{b} = \sin t$$

from which it follows that

$$\frac{x^2}{a^2} + \frac{y^2}{b^2} = 1$$

This is the equation of the ellipse shown in Figure 13.4.4. We leave it for the reader to show that the ellipse has counterclockwise orientation. ◀

Example 5   Let C be the parabola

$$y = x^2$$

and let $m = dy/dx$ be the slope of the tangent to C at the point (x, y). Find parametric equations for C using m as a parameter.

Solution. Our objective is to express the coordinates (x, y) of an arbitrary point on the parabola in terms of the parameter m. But

$$\frac{dy}{dx} = 2x$$

so

$$x = \frac{1}{2}\frac{dy}{dx} = \frac{1}{2}m$$

and

$$y = x^2 = \left(\frac{1}{2}m\right)^2 = \frac{1}{4}m^2$$

which leads to the parametric equations

$$x = \frac{1}{2}m$$

$$y = \frac{1}{4}m^2$$

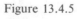

Figure 13.4.5

The direction of increasing parameter, indicated in Figure 13.4.5, is obtained by noting that x increases as the slope m increases, since $x = \frac{1}{2}m$. ◀

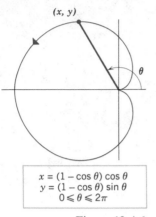

$$x = (1 - \cos \theta) \cos \theta$$
$$y = (1 - \cos \theta) \sin \theta$$
$$0 \leqslant \theta \leqslant 2\pi$$

Figure 13.4.6

Example 6 If $r = f(\theta)$ is a polar curve, we can obtain parametric equations for the curve in terms of θ by substituting $r = f(\theta)$ in the relationships

$$x = r \cos \theta$$
$$y = r \sin \theta$$

This yields

$$x = f(\theta) \cos \theta$$
$$y = f(\theta) \sin \theta$$

For example, the cardioid $r = 1 - \cos \theta$ can be represented by the parametric equations

$$x = (1 - \cos \theta) \cos \theta$$
$$y = (1 - \cos \theta) \sin \theta$$

If we impose the added restriction, $0 \leq \theta \leq 2\pi$, then the point (x, y) will traverse the cardioid exactly once as θ varies over $[0, 2\pi]$ (Figure 13.4.6). ◄

If the parameter is eliminated from a pair of parametric equations, the graph of the resulting equation in x and y may possibly extend beyond the graph of the original parametric equations. For example, consider the curve with parametric equations

$$x = \cosh t$$
$$y = \sinh t \tag{1}$$

We may eliminate the parameter by using the identity

$$\cosh^2 t - \sinh^2 t = 1$$

(Formula (1) of Section 7.6) to obtain

$$x^2 - y^2 = 1 \tag{2}$$

which is the equation of the hyperbola shown in Figure 13.4.7. However, the original parametric equations represent only the right branch of this hyperbola since $x = \cosh t$ is never less than 1. To exclude the unwanted left branch we would have to replace (2) by

$$x^2 - y^2 = 1, \quad x \geq 1$$

or more simply

$$x = \sqrt{1 + y^2}$$

To summarize, whenever the parameter is eliminated from parametric equations, it is important to check that the graph of the resulting equation

$x = -\sqrt{1 + y^2}$ $x = \sqrt{1 + y^2}$

$$x^2 - y^2 = 1$$

Figure 13.4.7

does not include too much. If it does, added conditions must be imposed.

While parametric equations are of special importance for curves that are not graphs of functions, curves of the form $y = f(x)$ or $x = g(y)$ may also be represented parametrically. This can always be done by introducing a parameter t that is equal to the independent variable.

Example 7   The curves $y = 4x^2 - 1$ and $x = 5y^3 - y$ can be represented parametrically as

$$x = t \qquad \qquad x = 5t^3 - 1$$
$$\text{and}$$
$$y = 4t^2 - 1 \qquad \qquad y = t$$

respectively.

REMARK.   A given curve can always be represented parametrically in infinitely many different ways. For example, a parametrization of $y = 4x^2 - 1$ different from that in Example 7 can be obtained by making the substitution $t = s + 1$. This yields new parametric equations

$$x = s + 1$$
$$y = 4(s + 1)^2 - 1 = 4s^2 + 8s + 3$$

Different substitutions would yield different parametric equations for the curve.

Sometimes it is difficult to eliminate the parameter from a pair of parametric equations and express y directly in terms of x. Nevertheless, by using the chain rule, it is possible to find the derivative of y with respect to x without eliminating the parameter. We start with the relationship

$$\frac{dy}{dt} = \frac{dy}{dx}\frac{dx}{dt}$$

If $dx/dt \neq 0$, this can be rewritten as

$$\frac{dy}{dx} = \frac{dy/dt}{dx/dt} \tag{3}$$

Example 8   Given the curve

$$x = t^2$$
$$y = t^3$$

find dy/dx and d^2y/dx^2 at the point $(1, 1)$ without eliminating the parameter t.

Solution.

$$\frac{dy}{dx} = \frac{dy/dt}{dx/dt} = \frac{3t^2}{2t} = \frac{3}{2}t \tag{4a}$$

To find the second derivative, denote dy/dx by y' and use (3) to write

$$\frac{d^2y}{dx^2} = \frac{dy'}{dx} = \frac{dy'/dt}{dx/dt} = \frac{3/2}{2t} = \frac{3}{4t} \tag{4b}$$

Note that these calculations express dy/dx and d^2y/dx^2 in terms of the parameter t. Since the point $(1, 1)$ on the curve corresponds to $t = 1$, it follows from (4a) and (4b) that

$$\left.\frac{dy}{dx}\right|_{t=1} = \frac{3}{2}(1) = \frac{3}{2}$$

and

$$\left.\frac{d^2y}{dx^2}\right|_{t=1} = \frac{3}{4}$$

The curve and the tangent at $(1, 1)$ are shown in Figure 13.4.8. Note that the tangent line has slope $\frac{3}{2}$ at $(1, 1)$ and the curve is concave up at $(1, 1)$, in keeping with our calculations.   ◄

Figure 13.4.8 ▷

REMARK.   Formula (3) does not apply if $dx/dt = 0$. However, at points where $dx/dt = 0$ and $dy/dt \neq 0$ there is generally a **vertical tangent** line. Points (x, y) where both $dx/dt = 0$ and $dy/dt = 0$ are called **singular points**. The behavior of a curve at a singular point must be determined on a case-by-case basis.

In Section 6.4 we obtained the following formula for the arc length L of a curve $y = f(x)$ from $x = a$ to $x = b$:

$$L = \int_a^b \sqrt{1 + [f'(x)]^2} \, dx = \int_a^b \sqrt{1 + \left(\frac{dy}{dx}\right)^2} \, dx \qquad (5)$$

This formula is a special case of the following general result about arc length of parametric curves:

Arc Length of Parametric Curves

13.4.1 THEOREM. *If $x'(t)$ and $y'(t)$ are continuous functions for $a \le t \le b$, then the curve*

$$x = x(t) \atop y = y(t)}, \quad a \le t \le b$$

has arc length L given by

$$L = \int_a^b \sqrt{[x'(t)]^2 + [y'(t)]^2} \, dt \qquad (6)$$

or equivalently

$$L = \int_a^b \sqrt{\left(\frac{dx}{dt}\right)^2 + \left(\frac{dy}{dt}\right)^2} \, dt \qquad (7)$$

The derivation of these formulas is similar to the derivation of (5) and will be omitted. Observe that (5) follows from (7) by expressing the curve $y = f(x)$ in the parametric form

$$x = t$$
$$y = f(t)$$

in which case

$$\frac{dx}{dt} = 1 \quad \text{and} \quad \frac{dy}{dt} = f'(t) = f'(x) = \frac{dy}{dx}$$

Example 9 Use (7) to find the arc length of the circle

$$x = a \cos t \atop y = a \sin t}, \quad 0 \le t \le 2\pi \quad (a > 0)$$

Solution.

$$L = \int_0^{2\pi} \sqrt{\left(\frac{dx}{dt}\right)^2 + \left(\frac{dy}{dt}\right)^2} \, dt$$

$$= \int_0^{2\pi} \sqrt{(-a \sin t)^2 + (a \cos t)^2} \, dt$$

$$= \int_0^{2\pi} a \, dt = at \bigg]_0^{2\pi} = 2\pi a \quad \blacktriangleleft$$

OPTIONAL

___

If a wheel rolls along a straight line without slipping, then a point on the rim of the wheel traces a curve called a *cycloid* (Figure 13.4.9). The cycloid

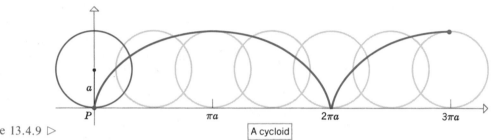

Figure 13.4.9 ▷

A cycloid

is of special interest because it provides the solution of two famous mathematical problems: the **brachistochrone problem** (from Greek words meaning ''shortest time'') and the **tautochrone problem** (from Greek words meaning ''equal time''). In June of 1696, Johann Bernoulli* posed the

___

*BERNOULLI. An amazing Swiss family that included several generations of outstanding mathematicians and scientists. Nikolaus Bernoulli (1623–1708), a druggist, fled from Antwerp to escape religious persecution and ultimately settled in Basel, Switzerland. There he had three sons, Jakob I (also called Jacques or James), Nikolaus, and Johann I (also called Jean or John). The Roman numerals are used to distinguish family members with identical names (continued on p. 775).

Figure 13.4.10

brachistochrone problem in the form of a challenge to other mathematicians. The problem was to determine the shape of a wire down which a bead might slide from a point P to another point Q, not directly below, in the *shortest time*. At first, one might conjecture that the wire should form a straight line, since that shape yields the shortest distance from P to Q. However, the correct answer is half of one arch of an inverted cycloid (Figure 13.4.10). In essence, this shape allows the bead to fall rapidly at first, building up sufficient initial speed to reach Q in the shortest time, even though the path does not provide the shortest distance from P to Q. This problem was solved by Newton, Leibniz, L'Hôpital, Johann Bernoulli, and his older brother Jakob Bernoulli. It was formulated and incorrectly solved years earlier by Galileo, who gave the arc of a circle as the answer.

In the tautochrone problem the object is to find the shape of a wire

Following Newton and Leibniz, the Bernoulli brothers, Jakob I and Johann I, are considered by some to be the two most important founders of calculus. Jakob I was self-taught in mathematics. His father wanted him to study for the ministry, but he turned to mathematics and in 1686 became a professor at the University of Basel. When he started working in mathematics, he knew nothing of Newton's and Leibniz' work. He eventually became familiar with Newton's results, but because so little of Leibniz' work was published, Jakob duplicated many of Leibniz' results.

Jakob's younger brother Johann I was urged to enter into business by his father. Instead, he turned to medicine and studied mathematics under the guidance of his older brother. He eventually became a mathematics professor at Gröningen in Holland and then, when Jakob died in 1705, Johann succeeded him as mathematics professor at Basel. Throughout their lives, Jakob I and Johann I had a mutual passion for criticizing each other's work, which frequently erupted into ugly confrontations. Leibniz tried to mediate the disputes, but Jakob, who resented Leibniz' superior intellect, accused him of siding with Johann and thus Leibniz became entangled in the arguments. The brothers often worked on common problems that they posed as challenges to one another. Johann, interested in gaining fame, often used unscrupulous means to make himself appear the originator of his brother's results; Jakob occasionally retaliated. Thus, it is often difficult to determine who deserves credit for many results. However, both men made major contributions to the development of calculus. In addition to his work on calculus, Jakob helped establish fundamental principles in probability, including the Law of Large Numbers, which is a cornerstone of modern probability theory. Johann was Euler's teacher at Basel and L'Hôpital's tutor in France.

Among the other members of the Bernoulli family, Daniel, son of Johann I, is the most famous. He was a professor of mathematics at St. Petersburg Academy in Russia and subsequently a professor of anatomy and then physics at Basel. He did work in calculus and probability, but is best known for his work in physics. A basic law of fluid flow, called Bernoulli's principle, is named in his honor. He won the annual prize of the French Academy 10 times for work on vibrating strings, tides of the sea, and kinetic theory of gases.

Johann II succeeded his father as professor of mathematics at Basel. His research was on the theory of heat and sound. Nikolaus I was a mathematician and law scholar who worked on probability and series. On the recommendation of Leibniz, he was appointed professor of mathematics at Padua and then went to Basel as a professor of logic and then law. Nikolaus II was professor of jurisprudence in Switzerland and then professor of mathematics at St. Petersburg Academy. Johann III was a professor of mathematics and astronomy in Berlin and Jakob II succeeded his uncle Daniel as professor of mathematics at St. Petersburg Academy in Russia. Truly an incredible family!

from P to Q such that two beads started at *any* points on the wire between P and Q reach Q in the same amount of time. Again, the answer is half of one arch of a cycloid.

We conclude this section by deriving parametric equations for a cycloid.

Example 10   Let P be a fixed point on the rim of a wheel of radius a. Suppose that the wheel rolls along the positive x-axis of a rectangular coordinate system, and that P is initially at the origin. When the wheel rolls a given distance, the radial line from the center C to the point P rotates through an angle ϕ (Figure 13.4.11), which we shall use as our parameter.

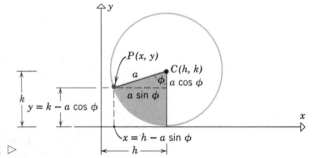

Figure 13.4.11 ▷

(It is customary to regard ϕ as a positive angle, even though it is generated by a clockwise rotation.) Our objective is to express the coordinates of $P(x, y)$ in terms of the angle ϕ. Figure 13.4.11 suggests that the coordinates of $P(x, y)$ and the coordinates of the wheel's center $C(h, k)$ are related by

$$x = h - a \sin \phi \qquad (8)$$
$$y = k - a \cos \phi$$

Since the height of the wheel's center is the radius of the wheel, we have $k = a$; and since the distance h moved by the center is the same as the circular arc length subtended by ϕ (why?), we have $h = a\phi$. Therefore, (8) yields

$$\begin{array}{l} x = a\phi - a \sin \phi \\ y = a - a \cos \phi \end{array}, \ 0 \le \phi < +\infty$$

These are the parametric equations of the cycloid. ◀

▶ Exercise Set 13.4

1. (a) Sketch the curve $x = t$, $y = t^2$ by plotting the points corresponding to $t = -3, -2, -1, 0, 1, 2, 3$.
   (b) Show that the curve is a parabola by eliminating the parameter t.

2. Sketch the curve $x = t^3 - 4t$, $y = 2t$, $-3 \le t \le 3$ by plotting some appropriate points.

In Exercises 3–15, sketch the curve by eliminating the parameter t, and label the direction of increasing t.

3. $x = \cos t$, $y = \sin t$, $0 \le t \le 2\pi$.

4. $x = 1 + \cos t$, $y = 3 - \sin t$, $0 \le t \le 2\pi$.

5. $x = 3t - 4$, $y = 6t + 2$.

6. $x = t - 3$, $y = 3t - 7$, $0 \le t \le 3$.

7. $x = 2 \cos t$, $y = 5 \sin t$, $0 \le t \le 2\pi$.

8. $x = \sqrt{t}$, $y = 2t + 4$.

9. $x = 3 + 2 \cos t$, $y = 2 + 4 \sin t$, $0 \le t \le 2\pi$.

10. $x = 2 \cosh t$, $y = 4 \sinh t$.

11. $x = 4 \sin 2\pi t$, $y = 4 \cos 2\pi t$, $0 \le t \le 1$.

12. $x = \sec t$, $y = \tan t$, $\pi \le t < \dfrac{3\pi}{2}$.

13. $x = \cos 2t$, $y = \sin t$, $-\dfrac{\pi}{2} \le t \le \dfrac{\pi}{2}$.

14. $x = 4t + 3$, $y = 16t^2 - 9$.

15. $x = t^2$, $y = 2 \ln t$, $t \ge 1$.

In Exercises 16–23, sketch the curve.

16. $x = 3e^{-t} - 2$, $y = 4e^{-t} - 1$, $t \ge 0$.

17. $x = 3e^{-2t} - 1$, $y = e^{-t}$, $t \ge 0$.

18. $x = \sec^2 t$, $y = \tan^2 t$.

19. $x = 2 \sin^2 t$, $y = 3 \cos^2 t$.

20. $x = \cos 2t$, $y = 2 \cos t$.

21. $x = \sin^2 t$, $y = 1 + \cos t$.

22. $x = \cos 3t$, $y = 2 \sin^2 3t - 1$.

23. $x = \cos(e^{-t})$, $y = \sin(e^{-t})$, $t \ge 0$.

24. Find parametric equations of the semicircle $y = \sqrt{a^2 - x^2}$ ($a > 0$) in terms of the parameter t, where t is the distance between $(-a, 0)$ and any point on the semicircle.

25. Find parametric equations of the curve $y = \sqrt{x}$ in terms of the parameter t, using for t the x-coordinate of the point where the tangent line to the curve crosses the x-axis.

In Exercises 26–32, find dy/dx at the point corresponding to the given value of the parameter without eliminating the parameter.

26. $x = t^2 + 4$, $y = 8t$; $t = 2$.

27. $x = \cos t$, $y = \sin t$; $t = 3\pi/4$.

28. $x = t + 5$, $y = 5t - 7$; $t = 1$.

29. $x = \sqrt{t}$, $y = 2t + 4$; $t = 9$.

30. $x = \sec \theta$, $y = \tan \theta$; $\theta = \pi/3$.

31. $x = 4 \cos 2\pi s$, $y = 3 \sin 2\pi s$; $s = -1/4$.

32. $x = \sinh t$, $y = \cosh t$; $t = 0$.

In Exercises 33–36, find d^2y/dx^2 at the point corresponding to the given value of the parameter without eliminating the parameter.

33. $x = \tfrac{1}{2}t^2$, $y = \tfrac{1}{3}t^3$; $t = 2$.

34. $x = \cos \phi$, $y = \sin \phi$; $\phi = \pi/4$.

35. $x = \sqrt{t}$, $y = 2t + 4$; $t = 1$.

36. $x = \sec t$, $y = \tan t$; $t = \pi/3$.

In Exercises 37–45, find the arc length of the curve.

37. $x = 4t + 3$, $y = 3t - 2$, $0 \le t \le 2$.

38. $x = \cos^3 t$, $y = \sin^3 t$, $0 \le t \le \pi/2$.

39. $x = \tfrac{1}{3}t^3$, $y = \tfrac{1}{2}t^2$, $0 \le t \le 1$.

40. $x = \tfrac{1}{3}t^3$, $y = \tfrac{1}{2}t^2$, $-1 \le t \le 0$.

41. $x = \cos 2t$, $y = \sin 2t$, $0 \le t \le \pi/2$.

42. $x = e^t(\sin t + \cos t)$, $y = e^t(\cos t - \sin t)$, $1 \le t \le 4$.

43. $x = (1 + t)^2$, $y = (1 + t)^3$, $0 \le t \le 1$.

44. $x = e^t \cos t$, $y = e^t \sin t$, $0 \le t \le \pi/2$.

45. One arch of the cycloid
$x = a(t - \sin t)$, $y = a(1 - \cos t)$.

46. Find parametric equations for the rose $r = 2 \cos 2\theta$ using θ as the parameter.

47. Find parametric equations for the limaçon $r = 2 + 3 \sin \theta$ using θ as the parameter.

48. Find the equation of the tangent line to the curve $x = 2t + 4$, $y = 8t^2 - 2t + 4$ at the point where $t = 1$.

49. Find the equation of the tangent line to the curve $x = e^t$, $y = e^{-t}$ at the point where $t = 2$.

50. Find all values of t at which the curve $x = 2 \cos t$, $y = 4 \sin t$ has a tangent that is
  (a) horizontal   (b) vertical.

51. Find all values of the parameter t at which the curve $x = 2t^3 - 15t^2 + 24t + 7$, $y = t^2 + t + 1$ has a tangent line that is
  (a) horizontal   (b) vertical.

52. Show that the curve $x = t^3 - 4t$, $y = t^2$ intersects itself at the point $(0, 4)$, and find equations for two tangent lines to the curve at the point of intersection.

53. Show that the curve with parametric equations $x = t^2 - 3t + 5$, $y = t^3 + t^2 - 10t + 9$ intersects itself at the point $(3, 1)$, and find equations for two tangent lines to the curve at the point of intersection.

54. A point traces the circle $x^2 + y^2 = 25$ so that $dx/dt = 8$ when the point reaches $(4, 3)$. Find dy/dt there.

55. Describe the curve whose parametric equations are $x = a \cos t + h$, $y = b \sin t + k$, $0 \le t \le 2\pi$.

56. A *hypocycloid* is a curve traced by a point P on the circumference of a circle that rolls inside a larger fixed circle. Suppose that the fixed circle has radius a, the rolling circle has radius b, and the fixed circle is centered at the origin. Let ϕ be the angle shown in the following figure, and assume that the point P is at $(a, 0)$ when $\phi = 0$. Show that the hypocycloid generated is given by

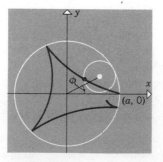

the parametric equations

$$x = (a - b) \cos \phi + b \cos \left(\frac{a - b}{b} \phi \right)$$

$$y = (a - b) \sin \phi - b \sin \left(\frac{a - b}{b} \phi \right)$$

57. If $b = \frac{1}{4}a$ in Exercise 56, then the resulting curve is called a four-cusped hypocycloid.
  (a) Sketch this curve.
  (b) Show that the curve is given by the parametric equations $x = a \cos^3 \phi$, $y = a \sin^3 \phi$.
  (c) Show that the curve is given by the equation $x^{2/3} + y^{2/3} = a^{2/3}$ in rectangular coordinates.

Exercises 58–63 require the formulas developed in the following discussion: If $x'(t)$ and $y'(t)$ are continuous functions for $a \le t \le b$, then it can be shown that the area of the surface generated by revolving the curve

$$x = x(t), \qquad a \le t \le b$$
$$y = y(t)$$

about the *x*-axis is

$$S = \int_a^b 2\pi y(t) \sqrt{[x'(t)]^2 + [y'(t)]^2} \, dt$$

and the area of the surface generated by revolving the curve about the *y*-axis is

$$S = \int_a^b 2\pi x(t) \sqrt{[x'(t)]^2 + [y'(t)]^2} \, dt$$

(The derivations are similar to those used to obtain Formulas (6) and (7) in Section 6.5.)

58. Find the area of the surface generated by revolving $x = t^2$, $y = 2t$, $0 \le t \le 4$ about the *x*-axis.

59. Find the area of the surface generated by revolving $x = e^t \cos t$, $y = e^t \sin t$, $0 \le t \le \pi/2$ about the *x*-axis.

60. Find the area of the surface generated by revolving $x = \cos^2 t$, $y = \sin^2 t$, $0 \le t \le \pi/2$ about the *y*-axis.

61. Find the area of the surface generated by revolving $x = t$, $y = 2t^2$, $0 \le t \le 1$ about the *y*-axis.

62. By revolving the semicircle $x = r \cos t$, $y = r \sin t$, $0 \le t \le \pi$ about the x-axis, show that the surface area of a sphere of radius r is $4\pi r^2$.

63. The equations

$$x = a\phi - a \sin \phi, \quad y = a - a \cos \phi, \quad 0 \le \phi \le 2\pi$$

represent one arch of a cycloid. Show that the surface area generated by revolving this curve about the x-axis is $S = 64\pi a^2/3$.

13.5 TANGENT LINES AND ARC LENGTH IN POLAR COORDINATES (OPTIONAL)

In this section we shall use our results on parametric equations to derive formulas for arc length and slopes of tangent lines to polar curves.

TANGENT LINES   Let

$$r = f(\theta)$$

be a curve in polar coordinates, for which f is a differentiable function of θ. We can obtain parametric equations for this curve in terms of θ by substituting $r = f(\theta)$ in the relationships

$$x = r \cos \theta$$
$$y = r \sin \theta$$

This yields the parametric equations

$$x = f(\theta) \cos \theta$$
$$y = f(\theta) \sin \theta \tag{1}$$

If we assume the existence of a tangent line to the curve $r = f(\theta)$ at the point $P(r, \theta)$, then the slope of this tangent line is

$$m = \tan \phi = \frac{dy}{dx}$$

where ϕ is the angle of inclination. From (1) we have

$$\frac{dx}{d\theta} = -f(\theta) \sin \theta + f'(\theta) \cos \theta$$
$$\frac{dy}{d\theta} = f(\theta) \cos \theta + f'(\theta) \sin \theta \tag{2}$$

Thus, if $dx/d\theta \ne 0$, it follows from Equation (3) of Section 13.4 that

$$m = \tan \phi = \frac{dy}{dx} = \frac{dy/d\theta}{dx/d\theta} \tag{3}$$

or

$$m = \tan \phi = \frac{f(\theta)\cos\theta + f'(\theta)\sin\theta}{-f(\theta)\sin\theta + f'(\theta)\cos\theta} = \frac{r\cos\theta + \sin\theta\dfrac{dr}{d\theta}}{-r\sin\theta + \cos\theta\dfrac{dr}{d\theta}} \qquad (4)$$

which is a formula for the slope of the tangent line to $r = f(\theta)$ at $P(r, \theta)$.

Formula (4) was derived by assuming that $dx/d\theta \neq 0$. If $dx/d\theta = 0$ and $dy/d\theta \neq 0$, we will agree that the curve has a **vertical tangent**. This is reasonable since $\tan\phi$ becomes infinite in this case [see (3)], indicating that $\phi = \pi/2$. Points where both $dx/d\theta = 0$ and $dy/d\theta = 0$ are called **singular points**. No general statement can be made about the behavior of a polar curve at a singular point. Each case requires its own analysis.

If $\cos\theta \neq 0$, we may divide the numerator and denominator of (4) by $\cos\theta$ to obtain the following alternative formula for the slope of a tangent line.

$$m = \tan\phi = \frac{r + \tan\theta\dfrac{dr}{d\theta}}{-r\tan\theta + \dfrac{dr}{d\theta}} \qquad (5)$$

Example 1   Find the slope of the tangent line to the circle

$$r = 4\cos\theta \qquad (6)$$

at the point where $\theta = \pi/4$.

Solution.   Differentiating (6) yields

$$\frac{dr}{d\theta} = -4\sin\theta \qquad (7)$$

If $\theta = \pi/4$ it follows from (6) and (7) that

$$r = 4\cos\frac{\pi}{4} = 2\sqrt{2} \quad \text{and} \quad \frac{dr}{d\theta} = -4\sin\frac{\pi}{4} = -2\sqrt{2}$$

Substituting these values and $\tan\theta = \tan\dfrac{\pi}{4} = 1$ in (5) yields

$$\tan\phi = \frac{2\sqrt{2} + (1)(-2\sqrt{2})}{-(2\sqrt{2})(1) + (-2\sqrt{2})} = 0$$

Thus, the circle has a horizontal tangent line when $\theta = \pi/4$ (Figure 13.5.1).   ◄

Figure 13.5.1

Example 2 At what points does the cardioid $r = 1 - \cos\theta$ have a vertical tangent line?

Solution. A vertical tangent line will occur at points where $dx/d\theta = 0$ and $dy/d\theta \neq 0$. The cardioid is given parametrically by the equations

$$x = r\cos\theta = (1 - \cos\theta)\cos\theta$$
$$y = r\sin\theta = (1 - \cos\theta)\sin\theta \quad, \quad 0 \leq \theta \leq 2\pi$$

Differentiating and then simplifying, we obtain (verify)

$r = 1 - \cos\theta$

$$\frac{dx}{d\theta} = \sin\theta(2\cos\theta - 1) \tag{8}$$

$$\frac{dy}{d\theta} = (1 - \cos\theta)(1 + 2\cos\theta) \tag{9}$$

From (8), $dx/d\theta = 0$ if $\sin\theta = 0$ or $\cos\theta = \frac{1}{2}$; this occurs where $\theta = 0$, π, $\pi/3$, $5\pi/3$, or 2π. But from (9), $dy/d\theta \neq 0$ if $\theta = \pi$, $\pi/3$, and $5\pi/3$, so vertical tangents occur at these points. Since $dy/d\theta = 0$ if $\theta = 0$ or $\theta = 2\pi$, these are singular points. Although we shall not prove it, the cardioid has a horizontal tangent at the origin (Figure 13.5.2). ◀

Figure 13.5.2

Another way to find the tangent line to a polar curve at a point $P(r, \theta)$ is to find the angle ψ between the **radial line** OP and the tangent line (Figure 13.5.3).

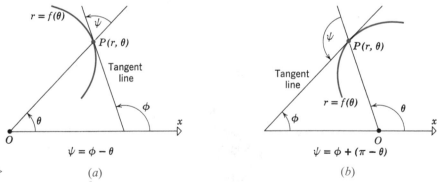

Figure 13.5.3 ▷   (a)   (b)

By definition, ψ is measured counterclockwise from the radial line OP to the tangent line, and is selected so

$$0 \leq \psi < \pi$$

If we choose the polar angle θ in the range $0 \leq \theta < 2\pi$, there is a simple algebraic relationship between θ, ϕ, and ψ that depends on the relative sizes of ϕ and θ. If $\phi \geq \theta$, then

$$\psi = \phi - \theta \tag{10}$$

(Figure 13.5.3a), and if $\phi < \theta$, then the relationship is

$$\psi = \phi + (\pi - \theta) = (\phi - \theta) + \pi \tag{11}$$

(Figure 13.5.3b). In either case, we may write

$$\tan \psi = \tan (\phi - \theta) \tag{12}$$

from which it follows that

$$\tan \psi = \frac{\tan \phi - \tan \theta}{1 + \tan \phi \tan \theta} \tag{13}$$

Substituting (5) into (13) yields

$$\tan \psi = \frac{\left[\left(r + \tan \theta \dfrac{dr}{d\theta}\right) \Big/ \left(-r \tan \theta + \dfrac{dr}{d\theta}\right)\right] - \tan \theta}{1 + \left[\left(r + \tan \theta \dfrac{dr}{d\theta}\right) \Big/ \left(-r \tan \theta + \dfrac{dr}{d\theta}\right)\right] \tan \theta}$$

$$= \frac{r + \tan \theta \dfrac{dr}{d\theta} + r \tan^2 \theta - \tan \theta \dfrac{dr}{d\theta}}{-r \tan \theta + \dfrac{dr}{d\theta} + r \tan \theta + \tan^2 \theta \dfrac{dr}{d\theta}}$$

$$= \frac{r(1 + \tan^2 \theta)}{(1 + \tan^2 \theta) \dfrac{dr}{d\theta}}$$

or

$$\tan \psi = \frac{r}{dr/d\theta} \tag{14}$$

Example 3 For the cardioid $r = 1 - \cos \theta$, find the angle ψ between the tangent line and radial line at the points where the cardioid crosses the y-axis.

Solution. For the given cardioid we have

$$\tan \psi = \frac{r}{dr/d\theta} = \frac{1 - \cos \theta}{\sin \theta}$$

so that from the trigonometric identity

$$\tan \frac{\theta}{2} = \frac{1 - \cos \theta}{\sin \theta}$$

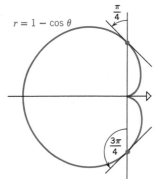

$r = 1 - \cos\theta$

Figure 13.5.4

(derive) it follows that

$$\tan\psi = \tan\frac{\theta}{2}$$

If $0 \le \theta < 2\pi$, this implies that

$$\psi = \frac{\theta}{2}$$

since $0 \le \psi < \pi$. In particular, where the cardioid crosses the y-axis we have $\theta = \pi/2$ and $\theta = 3\pi/2$, so the angles from the radial lines to the tangent lines at these points are $\psi = \pi/4$ and $\psi = 3\pi/4$, respectively (Figure 13.5.4).   ◄

In Section 13.2 we remarked informally that if a polar curve $r = f(\theta)$ passes through the origin where $\theta = \theta_0$, then the line $\theta = \theta_0$ is tangent to the curve at the origin. To see that this is so, we need only observe that in order for the curve $r = f(\theta)$ to pass through the origin when $\theta = \theta_0$, we must have $r = 0$ for this value of θ. Substituting $r = 0$ in (14) yields

$$\tan\psi = 0$$

(provided $dr/d\theta \ne 0$), which tells us that the angle ψ between the radial line $\theta = \theta_0$ and the tangent line is zero. Thus, the tangent line coincides with the line $\theta = \theta_0$.

ARC LENGTH

The arc length formula for a parametric curve can be used to obtain a formula for the arc length of a polar curve.

Arc Length of a Polar Curve

> **13.5.1   THEOREM.** *If a curve has the polar equation $r = f(\theta)$, where $f'(\theta)$ is continuous for $\alpha \le \theta \le \beta$, then its arc length L from $\theta = \alpha$ to $\theta = \beta$ is*
>
> $$L = \int_{\alpha}^{\beta} \sqrt{[f(\theta)]^2 + [f'(\theta)]^2}\, d\theta \qquad (15)$$
>
> *or equivalently*
>
> $$L = \int_{\alpha}^{\beta} \sqrt{r^2 + \left(\frac{dr}{d\theta}\right)^2}\, d\theta \qquad (16)$$

Proof.   It follows from (2) that (verify):

$$\left(\frac{dx}{d\theta}\right)^2 + \left(\frac{dy}{d\theta}\right)^2 = [f(\theta)]^2 + [f'(\theta)]^2$$

Thus, from (7) of Theorem 13.4.1 with θ in place of t, we obtain (15). ∎

Example 4   Find the arc length of the spiral $r = e^{\theta}$ between $\theta = 0$ and $\theta = 1$.

Solution.

$$L = \int_{\alpha}^{\beta} \sqrt{r^2 + \left(\frac{dr}{d\theta}\right)^2}\, d\theta = \int_{0}^{1} \sqrt{(e^{\theta})^2 + (e^{\theta})^2}\, d\theta$$

$$= \int_{0}^{1} \sqrt{2}\, e^{\theta}\, d\theta = \sqrt{2} e^{\theta}\Big]_{0}^{1} = \sqrt{2}(e - 1) \qquad \blacktriangleleft$$

Example 5   Find the total arc length of the cardioid $r = 1 + \cos\theta$.

Solution.   The cardioid is traced out once as θ varies from $\theta = 0$ to $\theta = 2\pi$. Thus,

$$L = \int_{\alpha}^{\beta} \sqrt{r^2 + \left(\frac{dr}{d\theta}\right)^2}\, d\theta = \int_{0}^{2\pi} \sqrt{(1 + \cos\theta)^2 + (-\sin\theta)^2}\, d\theta$$

$$= \sqrt{2} \int_{0}^{2\pi} \sqrt{1 + \cos\theta}\, d\theta$$

From the identity $1 + \cos\theta = 2\cos^2\tfrac{1}{2}\theta$, we obtain

$$L = 2 \int_{0}^{2\pi} \sqrt{\cos^2\tfrac{1}{2}\theta}\, d\theta = 2 \int_{0}^{2\pi} |\cos\tfrac{1}{2}\theta|\, d\theta \qquad (17)$$

Since

$$\cos\tfrac{1}{2}\theta \geq 0 \quad \text{when} \quad 0 \leq \theta \leq \pi$$

and

$$\cos\tfrac{1}{2}\theta \leq 0 \quad \text{when} \quad \pi \leq \theta \leq 2\pi$$

we may rewrite (17) as

$$L = 2\left[\int_{0}^{\pi} \cos\tfrac{1}{2}\theta\, d\theta - \int_{\pi}^{2\pi} \cos\tfrac{1}{2}\theta\, d\theta\right]$$

$$= 4\sin\tfrac{1}{2}\theta\Big]_{0}^{\pi} - 4\sin\tfrac{1}{2}\theta\Big]_{\pi}^{2\pi} = 8$$

Alternative Solution. Since the cardioid is symmetric about the *x*-axis (Figure 13.5.5), the entire arc length is twice the arc length from $\theta = 0$ to $\theta = \pi$. Thus, proceeding as above, we obtain

$$L = 2\sqrt{2} \int_0^\pi \sqrt{1 + \cos\theta}\; d\theta = 4 \int_0^\pi \cos\tfrac{1}{2}\theta\; d\theta = 8 \quad \blacktriangleleft$$

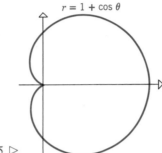

$r = 1 + \cos\theta$

Figure 13.5.5 ▷

► Exercise Set 13.5

In Exercises 1–6, find the slope of the tangent to the curve at the point with the given value of θ.

1. $r = 2 \cos\theta$; $\theta = \pi/3$.

2. $r = 1 + \sin\theta$; $\theta = \pi/4$.

3. $r = 1/\theta$; $\theta = 2$.

4. $r = a \sec 2\theta$; $\theta = \pi/6$.

5. $r = \cos 3\theta$; $\theta = 3\pi/4$.

6. $r = 4 - 3 \sin\theta$; $\theta = \pi$.

In Exercises 7–12, find $\tan\psi$ at the point on the curve with the given value of θ.

7. $r = 3(1 - \cos\theta)$; $\theta = \pi/2$.

8. $r = 2\theta$; $\theta = 1$.

9. $r = 5 \sin\theta$; $\theta = \pi$.

10. $r = \tan\theta$; $\theta = 3\pi/4$.

11. $r = 4 \sin^2\theta$; $\theta = 5\pi/6$.

12. $r = \sin\theta/(1 + \cos\theta)$; $\theta = \pi/3$.

In Exercises 13–19, find the arc length of the curve.

13. $r = e^{3\theta}$ from $\theta = 0$ to $\theta = 2$.

14. The entire circle $r = a$.

15. The entire circle $r = 2a \cos\theta$.

16. $r = \sin^2(\theta/2)$ from $\theta = 0$ to $\theta = \pi$.

17. $r = a\theta^2$ from $\theta = 0$ to $\theta = \pi$.

18. $r = \sin^3(\theta/3)$ from $\theta = 0$ to $\theta = \pi/2$.

19. The entire cardioid $r = a(1 - \cos\theta)$.
[*Hint:* See Example 5.]

20. Find all points on the cardioid $r = a(1 + \cos\theta)$ where the tangent is
(a) horizontal   (b) vertical.

21. Find all points on the limaçon $r = 1 - 2 \sin\theta$ where the tangent is horizontal.

22. Prove: If the polar curves $r = f_1(\theta)$ and $r = f_2(\theta)$ intersect at a point P, and if β is the smallest nonnegative angle between the tangent lines at P, then

$$\tan\beta = \left| \frac{\tan\psi_2 - \tan\psi_1}{1 + \tan\psi_1 \tan\psi_2} \right|$$

where ψ_1 and ψ_2 are evaluated at P.

23. Show that the curves $r = \sin 2\theta$ and $r = \cos\theta$ intersect at $(\sqrt{3}/2, \pi/6)$ and use the result of Exercise 22 to find the smallest nonnegative angle between their tangent lines at the point of intersection.

24. Find all points of intersection of the curves $r = \cos\theta$ and $r = 1 - \cos\theta$, and use the result of Exercise 22 to find the smallest nonnegative angle between the tangent lines at each point of intersection.

25. Prove that ψ is the same at each point of the curve $r = e^{a\theta}$.

26. Prove: At all points of intersection of the cardioids $r = a(1 + \cos\theta)$ and $r = b(1 - \cos\theta)$ (excluding the origin), the tangent lines are perpendicular.

▶ SUPPLEMENTARY EXERCISES

Find the rectangular coordinates of the points with the given polar coordinates.

1. (a) $(-2, 4\pi/3)$ (b) $(2, -\pi/2)$
   (c) $(0, -\pi)$ (d) $(-\sqrt{2}, -\pi/4)$
   (e) $(3, \pi)$ (f) $(1, \tan^{-1}(-\tfrac{4}{3}))$.

2. In parts (a)–(c), points are given in rectangular coordinates. Express them in polar coordinates in three ways:

   (i) with $r \geq 0$ and $0 \leq \theta < 2\pi$;
   (ii) with $r \geq 0$ and $-\pi < \theta \leq \pi$;
   (iii) with $r \leq 0$ and $0 \leq \theta < 2\pi$.

   (a) $(-\sqrt{3}, -1)$ (b) $(-3, 0)$ (c) $(1, -1)$.

3. Sketch the region in polar coordinates determined by the given inequalities.
   (a) $1 \leq r \leq 2$, $\cos\theta \leq 0$
   (b) $-1 \leq r \leq 1$, $\pi/4 \leq \theta \leq \pi/2$.

In Exercises 4–11, identify the curve by transforming to rectangular coordinates.

4. $r = 2/(1 - \cos\theta)$.   **5.** $r^2 \sin(2\theta) = 1$.
6. $r = \pi/2$.   **7.** $r = -4\csc\theta$.
8. $r = 6/(3 - \sin\theta)$.   **9.** $\theta = \pi/3$.
10. $r = 2\sin\theta + 3\cos\theta$.   **11.** $r = 0$.

In Exercises 12–15, express the given equation in polar coordinates.

12. $x^2 + y^2 = kx$.   **13.** $x = -3$.
14. $y^2 = 4x$.   **15.** $y = 3x$.

In Exercises 16–23, sketch the curve in polar coordinates.

16. $r = -4\sin 3\theta$.   **17.** $r = -1 - 2\cos\theta$.

18. $r = 5\cos\theta$.   **19.** $r = 4 - \sin\theta$.
20. $r = 3(\cos\theta - 1)$.   **21.** $r = \theta/\pi$ $(\theta \geq 0)$.
22. $r = \sqrt{2}\cos(\theta/2)$.   **23.** $r = e^{-\theta/\pi}$ $(\theta \geq 0)$.

In Exercises 24–26, sketch the curves in the same polar coordinate system, and find all points of intersection.

24. $r = 3\cos\theta$, $r = 1 + \cos\theta$.

25. $r = a\cos(2\theta)$, $r = a/2$ $(a > 0)$.

26. $r = 2\sin\theta$, $r = 2 + 2\cos\theta$.

In Exercises 27–29, set up, but *do not evaluate*, definite integrals for the stated area and arc length.

27. (a) The area inside both the circle and cardioid in Exercise 24.
   (b) The arc length of that part of the cardioid outside the circle in Exercise 24.

28. (a) The area inside the rose and outside the circle in Exercise 25.
   (b) The arc length of that part of the rose lying inside the circle in Exercise 25.

29. (a) The area inside the circle and outside the cardioid in Exercise 26.
   (b) The arc length of that portion of the circle lying inside the cardioid in Exercise 26.

In Exercises 30–33, find the area of the region described.

30. One petal of the rose $r = a\sin 3\theta$.

31. The region outside the circle $r = a$ and inside the lemniscate $r^2 = 2a^2\cos 2\theta$.

32. The region in part (a) of Exercise 28.

33. The region in part (a) of Exercise 29.

In Exercises 34–37:

(a) Sketch the curve and label the direction of increasing parameter.

(b) Use the parametric equations to find dy/dx, d^2y/dx^2, and the equation of the tangent line at the point on the curve corresponding to the parameter value t_0 (or θ_0).

34. $x = 3 - t^2$, $y = 2 + t$, $0 \le t \le 3$; $t_0 = 1$.

35. $x = 1 + 3 \cos \theta$, $y = -1 + 2 \sin \theta$, $0 \le \theta \le \pi$; $\theta_0 = \pi/2$.

36. $x = 2 \tan \theta$, $y = \sec \theta$, $-\pi/2 < \theta < \pi/2$; $\theta_0 = \pi/3$.

37. $x = 1/t$, $y = \ln t$, $1 \le t \le e$; $t_0 = 2$.

In Exercises 38–43, find the arc length of the curve described.

38. $x = 2t^3$, $y = 3t^2$, $-4 \le t \le 4$.

39. $x = \ln \cos 2t$, $y = 2t$, $0 \le t \le \pi/6$.

40. $x = 3 \cos t - 1$, $y = 3 \sin t + 4$, $0 \le t \le \pi$.

41. $x = 3t^2$, $y = t^3 - 3t$, $0 \le t \le 1$.

42. $x = 1 - \cos t$, $y = t - \sin t$, $-\pi \le t \le \pi$.

43. $r = e^\theta$, $0 \le \theta \le 2\pi$.

In Exercises 44–46, $x(t)$ and $y(t)$ describe the motion of a particle. Find the coordinates of the particle when the instantaneous direction of motion is: (a) horizontal, (b) vertical.

44. $x = -2t^2$, $y = t^3 - 3t + 5$.

45. $x = 1 - 2 \sin t$, $y = t + 2 \cos t$, $0 \le t \le \pi$.

46. $x = \ln t$, $y = t^2 - 4t$, $t > 0$.

47. At what instant does the trajectory described in Exercise 46 have a point of inflection?

48. Find a set of parametric equations for
(a) the line $y = 2x + 3$
(b) the ellipse $4(x - 2)^2 + y^2 = 4$.

Exercises 49–50 refer to the optional Section 13.5.

49. For the cardioid $r = 2(1 + \cos \theta)$, find the slope of the tangent line and the angle ψ when $\theta = \pi/2$.

50. For the circle $r = 4 \sin \theta$, show that $\psi = \theta$ if $0 \le \theta < \pi$ and find a formula for the inclination angle ϕ as a function of θ for $0 \le \theta < \pi/2$.

14 three-dimensional space; vectors

14.1 RECTANGULAR COORDINATES IN 3-SPACE; SPHERES; CYLINDRICAL SURFACES

In this section we shall discuss coordinate systems in three-dimensional space and some basic facts about surfaces in three dimensions.

RECTANGULAR
COORDINATE
SYSTEMS

Just as points in a plane can be placed in one-to-one correspondence with pairs of real numbers by using two perpendicular coordinate lines, so points in three-dimensional space can be placed in one-to-one correspondence with triples of real numbers by using three mutually perpendicular coordinate lines. To obtain this correspondence, we choose the coordinate lines so they intersect at their origins, and we call these lines the *x-axis*, the *y-axis*, and the *z-axis* (Figure 14.1.1*a*).

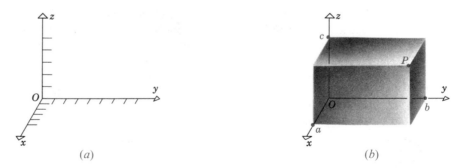

Figure 14.1.1   (*a*)   (*b*)

The three coordinate axes form a three-dimensional *rectangular* or *Cartesian coordinate system* and the point of intersection of the coordinate axes is called the *origin* of the coordinate system.

Each pair of coordinate axes determines a plane called a *coordinate plane*. These are referred to as the *xy-plane*, the *xz-plane*, and the *yz-plane*.

To each point P in 3-space we assign a triple of numbers (a, b, c), called the ***coordinates of P,*** by passing three planes through P parallel to the coordinate planes, and letting a, b, and c be the coordinates of the intersections of these planes, with the x, y, and z axes, respectively (Figure 14.1.1b). The notation $P(a, b, c)$ will sometimes be used to denote a point P with coordinates (a, b, c).

In Figure 14.1.2 we have constructed the points whose coordinates are $(4, 5, 6)$ and $(-3, 2, -4)$.

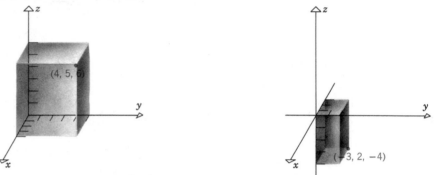

Figure 14.1.2 ▷

In this text we shall call three-dimensional space ***3-space,*** two-dimensional space (a plane) ***2-space,*** and one-dimensional space (a line) ***1-space.***

Rectangular coordinate systems in 3-space fall into two categories: ***left-handed*** and ***right-handed.*** A right-handed system has the property that when the fingers of the right hand are cupped so they curve from the positive x-axis toward the positive y-axis, the thumb points (roughly) in the direction of the positive z-axis (Figure 14.1.3a). A system that is not right-handed is called left-handed (Figure 14.1.3b). We shall use only right-handed coordinate systems.

Just as the coordinate axes in a two-dimensional coordinate system divide 2-space into four quadrants, so the coordinate planes of a three-dimensional coordinate system divide 3-space into eight parts, called ***octants*** (count them). Those points having three positive coordinates form the ***first octant;*** the remaining octants have no standard numbering.

The reader should be able to visualize the following results about three-dimensional rectangular coordinate systems:

Right-handed
(a)

Left-handed
(b)

Figure 14.1.3

REGION	DESCRIPTION
xy-plane	consists of all points of the form $(x, y, 0)$
xz-plane	consists of all points of the form $(x, 0, z)$
yz-plane	consists of all points of the form $(0, y, z)$
x-axis	consists of all points of the form $(x, 0, 0)$
y-axis	consists of all points of the form $(0, y, 0)$
z-axis	consists of all points of the form $(0, 0, z)$

DISTANCE FORMULA IN 3-SPACE

To obtain a formula for the distance between two points in 3-space, we first consider a box (rectangular parallelepiped) with dimensions a, b, and c (Figure 14.1.4a). By the Theorem of Pythagoras, the length of the diagonal of the base is $\sqrt{a^2 + b^2}$. Moreover, the diagonal of the box is the hypotenuse of a right triangle having a diagonal of the base for one side and a vertical edge for the other side (Figure 14.1.4b). Thus, by the Theorem of Pythagoras, the length of the diagonal of the box is the square root of $[\sqrt{a^2 + b^2}]^2 + c^2$ or more simply

$$\begin{bmatrix} \text{the length of a} \\ \text{diagonal of a} \\ \text{box} \end{bmatrix} = \sqrt{a^2 + b^2 + c^2} \tag{1}$$

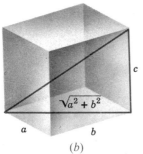

(a)

(b)

Figure 14.1.4

Suppose now that $P_1(x_1, y_1, z_1)$ and $P_2(x_2, y_2, z_2)$ are points in 3-space. As shown in Figure 14.1.5, these points are on diagonally opposite corners of a box whose dimensions are

$$|x_2 - x_1|, \quad |y_2 - y_1|, \quad \text{and} \quad |z_2 - z_1|$$

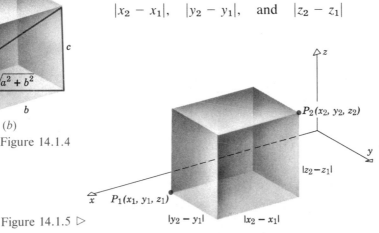

Figure 14.1.5 ▷

Thus, from (1) the distance d between P_1 and P_2 is

$$d = \sqrt{|x_2 - x_1|^2 + |y_2 - y_1|^2 + |z_2 - z_1|^2}$$

or equivalently

$$d = \sqrt{(x_2 - x_1)^2 + (y_2 - y_1)^2 + (z_2 - z_1)^2} \tag{2}$$

Example 1 The distance between the points $(4, -1, 3)$ and $(2, 3, -1)$ is

$$d = \sqrt{(4 - 2)^2 + (-1 - 3)^2 + (3 + 1)^2} = \sqrt{36} = 6 \quad \blacktriangleleft$$

In 2-space the midpoint of the line segment joining $P_1(x_1, y_1)$ and $P_2(x_2, y_2)$ is

$$\left(\frac{x_1 + x_2}{2}, \frac{y_1 + y_2}{2}\right)$$

In 3-space the formula is similar:

$$\left[\begin{array}{c}\text{The midpoint of the} \\ \text{line segment joining} \\ P_1(x_1, y_1, z_1) \text{ and } P_2(x_2, y_2, z_2)\end{array}\right] = \left(\frac{x_1 + x_2}{2}, \frac{y_1 + y_2}{2}, \frac{z_1 + z_2}{2}\right) \quad (3)$$

(We shall prove this result later.)

Example 2   The midpoint of the line segment joining the points $(-1, 3, -8)$ and $(3, 1, 0)$ is

$$\left(\frac{-1 + 3}{2}, \frac{3 + 1}{2}, \frac{-8 + 0}{2}\right) = (1, 2, -4) \quad \blacktriangleleft$$

SPHERES   In 2-space, the graph of an equation relating the variables x and y is the set of all points (x, y) whose coordinates satisfy the equation. Usually, such graphs are curves. Similarly, in 3-space the graph of an equation relating x, y, and z is the set of all points (x, y, z) whose coordinates satisfy the equation. Usually, such graphs are surfaces in 3-space. For example, in 3-space, a sphere with center (x_0, y_0, z_0) and radius r consists of all points (x, y, z) at a distance of r units from (x_0, y_0, z_0) (Figure 14.1.6). Thus, from (2)

$$\sqrt{(x - x_0)^2 + (y - y_0)^2 + (z - z_0)^2} = r$$

or equivalently

$$(x - x_0)^2 + (y - y_0)^2 + (z - z_0)^2 = r^2 \quad (4)$$

Figure 14.1.6 ▷

This equation represents the sphere with center (x_0, y_0, z_0) and radius r; it is called the **standard form of the equation** of a sphere. Some examples are:

EQUATION	GRAPH
$(x - 3)^2 + (y - 2)^2 + (z - 1)^2 = 9$	Sphere with center $(3, 2, 1)$ and radius 3
$(x + 1)^2 + y^2 + (z + 4)^2 = 5$	Sphere with center $(-1, 0, -4)$ and radius $\sqrt{5}$
$x^2 + y^2 + z^2 = 1$	Sphere with center $(0, 0, 0)$ and radius 1

Example 3   Describe the graph of

$$x^2 + y^2 + z^2 - 2x - 4y + 8z + 17 = 0$$

Solution.   We can put the equation in the form of (4) by completing the squares:

$$(x^2 - 2x) + (y^2 - 4y) + (z^2 + 8z) = -17$$
$$(x^2 - 2x + 1) + (y^2 - 4y + 4) + (z^2 + 8z + 16) = -17 + 21$$
$$(x - 1)^2 + (y - 2)^2 + (z + 4)^2 = 4$$

Thus, the graph is a sphere of radius 2 centered at $(1, 2, -4)$.   ◀

By squaring out and collecting terms in (4), we can rewrite the equation of a sphere in the form:

$$x^2 + y^2 + z^2 + Gx + Hy + Iz + J = 0 \tag{5}$$

However, it is not true that every equation of form (5) has a sphere as its graph; for if we begin with an equation of form (5) and complete the squares as in Example 2, we shall obtain an equation of the form

$$(x - x_0)^2 + (y - y_0)^2 + (z - z_0)^2 = k$$

where k is a constant. If $k > 0$, then the equation represents a sphere of radius $\sqrt{k}$. However, if $k = 0$ the graph is the single point (x_0, y_0, z_0), and if $k < 0$ the equation is not satisfied by any real values of x, y, and z; thus, there is no graph. To summarize:

14.1.1   THEOREM. *An equation of the form*

$$x^2 + y^2 + z^2 + Gx + Hy + Iz + J = 0$$

represents either a sphere or a point, or else has no graph.

CYLINDRICAL
SURFACES

Now that we have introduced graphs in 3-space, an equation in x and y such as

$$x + y = 1 \tag{6}$$

becomes ambiguous. Should we graph this equation in 2-space (Figure 14.1.7*a*) or should we take the position that (6) is really the equation

$$x + y + 0z = 1 \tag{7}$$

in which case it should be graphed in 3-space? If we take the latter position, then the graph is the plane shown in Figure 14.1.7*b* since (x, y, z) will satisfy (7) for *arbitrary z* as long as x and y satisfy (6).

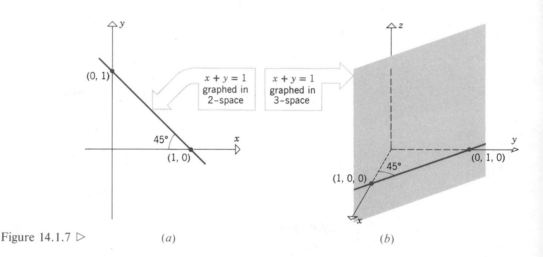

Figure 14.1.7 ▷   (*a*)   (*b*)

Actually, it is correct to graph (6) either in 2-space or 3-space—the appropriate choice will usually be clear from the context in which the equation arises.

A three-dimensional surface generated by a line traversing a plane curve and moving parallel to a fixed line is sometimes called a ***cylindrical surface;*** the moving line is called the ***generator*** of the surface. In general, if an equation in x and y has a curve C for its graph in the xy-plane, then the same equation graphed in 3-space produces the cylindrical surface that is traced out as a generator parallel to the z-axis traverses the curve C in the xy-plane. Similarly, an equation in x and z only represents a cylindrical surface in 3-space with generator parallel to the y-axis; and an equation in y and z alone represents a cylindrical surface in 3-space with generator parallel to the x-axis. In brief:

> *An equation containing only two of the three variables x, y, and z represents a cylindrical surface in 3-space. The generator of the surface is parallel to the axis corresponding to the missing variable.*

Example 4   Sketch the graph of $y = x^2$ in 3-space.

Solution.   (Figure 14.1.8)   ◀

Example 5   Sketch the graph of $x^2 + z^2 = 1$ in 3-space.

Solution.   Since y does not appear in this equation, the graph is a cylindrical surface with generator parallel to the y-axis. It is helpful to begin with a sketch of the equation in 2-space. In the xz-plane the curve $x^2 + z^2 = 1$ is a circle (Figure 14.1.9a). Thus, in 3-space this equation represents a right-circular cylinder parallel to the y-axis (14.1.9b).   ◀

Figure 14.1.8

Figure 14.1.9 ▷   (a)   (b)

Example 6   Sketch the graph of $z = \sin y$.

Solution.   (Figure 14.1.10)   ◀

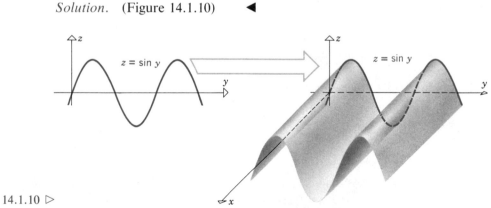

Figure 14.1.10 ▷

▶ Exercise Set 14.1

1. Plot the points P and Q in a right-handed coordinate system. Then, find the distance between them and the midpoint of the line segment joining them.
   (a) $P(0, 0, 0)$; $Q(2, 1, 3)$
   (b) $P(5, 2, 3)$; $Q(4, 1, 6)$
   (c) $P(-2, -1, 3)$; $Q(3, 0, 5)$
   (d) $P(-1, -1, -3)$; $Q(4, 3, -2)$.

2. A cube of side 4 has its geometric center at the origin and its faces parallel to the coordinate planes. Sketch the cube and give the coordinates of the vertices.

3. A rectangular parallelepiped has its faces parallel to the coordinate planes and has $(4, 2, -2)$ and $(-6, 1, 1)$ as endpoints of a diagonal. Sketch the parallelepiped and give the coordinates of the vertices.

4. Show that $(4, 5, 2)$, $(1, 7, 3)$, and $(2, 4, 5)$ are vertices of an equilateral triangle.

5. (a) Show that $(2, 1, 6)$, $(4, 7, 9)$, and $(8, 5, -6)$ are the vertices of a right triangle.
   (b) Which vertex is at the 90° angle?
   (c) Find the area of the triangle.

6. Find the distance from the point $(-5, 2, -3)$ to the:
   (a) xy-plane   (b) xz-plane
   (c) yz-plane   (d) x-axis
   (e) y-axis   (f) z-axis.

7. Show that the distance from a point (x_0, y_0, z_0) to the z-axis is $\sqrt{x_0^2 + y_0^2}$, and find the distances from the point to the x- and y-axes.

In Exercises 8–11, find an equation for the sphere with center C and radius r.

8. $C(0, 0, 0)$; $r = 8$.   9. $C(-2, 4, -1)$; $r = 6$.
10. $C(5, -2, 4)$; $r = \sqrt{7}$. 11. $C(0, 1, 0)$; $r = 3$.

12. In each part find an equation for the sphere with center $(-3, 5, -4)$ and satisfying the given condition:
   (a) tangent to the xy-plane
   (b) tangent to the xz-plane
   (c) tangent to the yz-plane.

13. In each part, find an equation for the sphere with center $(2, -1, -3)$ and satisfying the given condition.
   (a) tangent to the xy-plane
   (b) tangent to the xz-plane
   (c) tangent to the yz-plane.

In Exercises 14–19, find the standard equation of the sphere satisfying the given conditions.

14. Center $(1, 0, -1)$; diameter $= 8$.

15. A diameter has endpoints $(-1, 2, 1)$ and $(0, 2, 3)$.

16. Center $(-1, 3, 2)$ and passing through the origin.

17. Center $(3, -2, 4)$ and passing through $(7, 2, 1)$.

18. Center $(-3, 5, -4)$; tangent to the sphere of radius 1 centered at the origin (two answers).

19. Center $(0, 0, 0)$; tangent to the sphere of radius 1 centered at $(3, -2, 4)$ (two answers).

In Exercises 20–25, describe the surface whose equation is given.

20. $x^2 + y^2 + z^2 - 2x - 6y - 8z + 1 = 0$.

21. $x^2 + y^2 + z^2 + 10x + 4y + 2z - 19 = 0$.

22. $x^2 + y^2 + z^2 - y = 0$.

23. $2x^2 + 2y^2 + 2z^2 - 2x - 3y + 5z - 2 = 0$.

24. $x^2 + y^2 + z^2 + 2x - 2y + 2z + 3 = 0$.

25. $x^2 + y^2 + z^2 - 3x + 4y - 8z + 25 = 0$.

26. Find the largest and smallest distances between the point $P(1, 1, 1)$ and the sphere $x^2 + y^2 + z^2 - 2y + 6z - 6 = 0$.

27. Find the largest and smallest distances between the origin and the sphere $x^2 + y^2 + z^2 + 2x - 2y - 4z - 3 = 0$.

28. Describe the set of all points in 3-space whose coordinates satisfy the inequality $x^2 + y^2 + z^2 - 2x + 8z \leq 8$.

29. Describe the set of all points in 3-space whose coordinates satisfy the inequality $y^2 + z^2 + 6y - 4z > 3$.

30. The distance between a point $P(x, y, z)$ and the point $A(1, -2, 0)$ is twice the distance between P and the point $B(0, 1, 1)$. Show that the set of all

such points is a sphere, and find the center and radius of the sphere.

31. A bowling ball of radius R is placed inside a box just large enough to hold it, and it is secured for shipping by packing a styrofoam sphere into each corner of the box. Find the radius of the largest styrofoam sphere that can be used. [*Hint:* Take the origin of a Cartesian coordinate system at a corner of the box with the coordinate axes along the edges.]

(a)

(b)

(c)

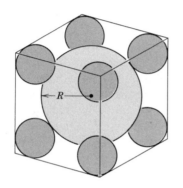

In Exercises 32–39, sketch the surface whose equation is given.

32. (a) $y = x$ (b) $y = z$
    (c) $x = z$.

33. (a) $x^2 + y^2 = 25$ (b) $y^2 + z^2 = 25$
    (c) $x^2 + z^2 = 25$.

34. (a) $y = x^2$ (b) $z = x^2$
    (c) $y = z^2$.

35. (a) $y = e^x$ (b) $x = \ln z$
    (c) $yz = 1$.

36. (a) $2x + 3y = 6$ (b) $2x + z = 3$.

37. (a) $y = \sin x$ (b) $z = \cos x$.

38. (a) $z = 1 - y^2$ (b) $z = \sqrt{3 - x}$.

39. (a) $4x^2 + 9z^2 = 36$ (b) $y^2 - 4z^2 = 4$.

40. Find an equation for the right-circular cylinder of radius a shown.

41. Show that for all values of θ and ϕ, the point $(a \sin \phi \cos \theta, a \sin \phi \sin \theta, a \cos \phi)$ lies on the sphere $x^2 + y^2 + z^2 = a^2$.

42. Consider the equation

$$x^2 + y^2 + z^2 + Gx + Hy + Iz + J = 0$$

and let $K = G^2 + H^2 + I^2 - 4J$.

(a) Prove the equation represents a sphere, a point, or no graph, according to whether $K > 0$, $K = 0$, or $K < 0$.

(b) In the case where $K > 0$, find the center and radius of the sphere.

14.2   VECTORS

*Many physical quantities such as area, length, mass, and temperature are completely described once the magnitude of the quantity is given. Such quantities are called scalars. Other physical quantities, called **vectors**, are not completely determined until both a magnitude and a direction are specified. For example, wind movement is usually described by giving the speed and the direction, say 20 mph northeast. The wind speed and wind direction together form a vector quantity called the wind velocity. Other examples of vectors are force and displacement. In this section we shall develop the basic mathematical properties of vectors.*

VECTORS VIEWED
GEOMETRICALLY

(a)   (b)

Figure 14.2.1

Vectors can be represented geometrically as directed line segments or arrows in two- or three-dimensional space; the direction of the arrow specifies the direction of the vector and the length of the arrow describes its magnitude. The tail of the arrow is called the ***initial point*** of the vector, and the tip of the arrow the ***terminal point***. We shall denote vectors by lowercase boldface type like **a**, **k**, **v**, **w**, and **x**. When discussing vectors, we shall refer to numbers as ***scalars***. Scalars will be denoted by lowercase italic type like a, k, v, w, and x.

If, as in Figure 14.2.1a, the initial point of a vector **v** is A and the terminal point is B, we write

$$\mathbf{v} = \overrightarrow{AB}$$

Vectors having the same length and same direction, such as those in Figure 14.2.1b, are called ***equivalent***. Since we want a vector to be determined solely by its length and direction, equivalent vectors are regarded as ***equal*** even though they may be located in different positions. If **v** and **w** are equivalent we write

$$\mathbf{v} = \mathbf{w}$$

14.2.1   DEFINITION. If **v** and **w** are any two vectors, then the sum **v** + **w** is the vector determined as follows. Position the vector **w** so that its initial point coincides with the terminal point of **v**. The vector **v** + **w** is represented by the arrow from the initial point of **v** to the terminal point of **w** (Figure 14.2.2a).

In Figure 14.2.2b we have constructed two sums, **v** + **w** (blue arrows) and **w** + **v** (white arrows). It is evident that

$$\mathbf{v} + \mathbf{w} = \mathbf{w} + \mathbf{v}$$

and that the sum coincides with the diagonal of the parallelogram deter-

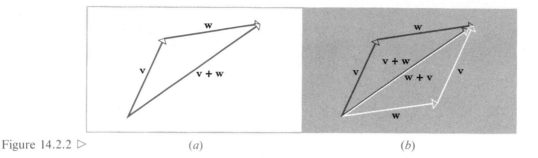

Figure 14.2.2 ▷   (*a*)   (*b*)

mined by **v** and **w** when these vectors are located so they have the same initial point.

The vector of length zero is called the *zero vector* and is denoted by **0**. We define

$$\mathbf{0} + \mathbf{v} = \mathbf{v} + \mathbf{0} = \mathbf{v}$$

Figure 14.2.3

Since there is no natural direction for the zero vector, we shall agree that it can be assigned any direction that is convenient for the problem at hand.

If **v** is any nonzero vector, then −**v**, the *negative* of **v**, is defined to be the vector having the same magnitude as **v**, but oppositely directed (Figure 14.2.3). This vector has the property

$$\mathbf{v} + (-\mathbf{v}) = \mathbf{0}$$

(Why?) In addition, we define −**0** = **0**.

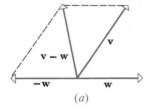

(*a*)

14.2.2 DEFINITION. If **v** and **w** are any two vectors, then *subtraction* of **w** from **v** is defined by

$$\mathbf{v} - \mathbf{w} = \mathbf{v} + (-\mathbf{w})$$

(See Figure 14.2.4*a*.)

To obtain **v** − **w** without constructing −**w**, position **v** and **w** so their initial points coincide; the vector from the terminal point of **w** to the terminal point of **v** is then the vector **v** − **w** (Figure 14.2.4*b*).

(*b*)

Figure 14.2.4

14.2.3 DEFINITION. If **v** is a nonzero vector and k is a nonzero real number (scalar), then the *product* $k\mathbf{v}$ is defined to be the vector whose length is $|k|$ times the length of **v** and whose direction is the same as that of **v** if $k > 0$ and opposite to that of **v** if $k < 0$. We define $k\mathbf{v} = \mathbf{0}$ if $k = 0$ or $\mathbf{v} = \mathbf{0}$.

Figure 14.2.5

Figure 14.2.5 illustrates the relationship between a vector **v** and the vectors $\frac{1}{2}$**v**, (-1)**v**, 2**v**, and (-3)**v**. Note that the vector (-1)**v** has the same length as **v** but is oppositely directed. Thus, (-1)**v** is just the negative of **v**; that is,

$$(-1)\mathbf{v} = -\mathbf{v}$$

VECTORS IN COORDINATE SYSTEMS

Up to this point everything we have said about vectors applies equally well to vectors in 2-space or 3-space. However, problems involving vectors can often be simplified by introducing a rectangular coordinate system. For this purpose it will be necessary for us to distinguish between vectors in 2-space (the plane) and vectors in 3-space.

If **v** is a vector in 2-space or 3-space with its initial point at the origin of a rectangular coordinate system (Figure 14.2.6), then the coordinates (v_1, v_2) or (v_1, v_2, v_3) of the terminal point are called the ***components*** of **v** and we write

$$\mathbf{v} = \langle v_1, v_2 \rangle \quad \text{or} \quad \mathbf{v} = \langle v_1, v_2, v_3 \rangle$$

depending on whether the vector is in 2-space or 3-space.

Because the zero vector **0** has length zero, its terminal point and initial point coincide. Thus,

$$\mathbf{0} = \langle 0, 0 \rangle \quad \text{(in 2-space)}$$
$$\mathbf{0} = \langle 0, 0, 0 \rangle \quad \text{(in 3-space)}$$

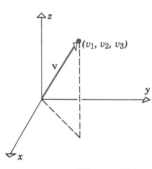

Figure 14.2.6

If equivalent vectors, **v** and **w**, are located so their initial points fall at the origin, then it is obvious that their terminal points must coincide (since the vectors have the same length and direction). The vectors thus have the same components. Conversely, vectors with the same components are equivalent, since they have the same length and same direction. In summary, the vectors

$$\mathbf{v} = \langle v_1, v_2 \rangle \quad \text{and} \quad \mathbf{w} = \langle w_1, w_2 \rangle$$

in 2-space are equivalent (that is, **v** = **w**) if and only if

$$v_1 = w_1 \quad \text{and} \quad v_2 = w_2$$

and the vectors

$$\mathbf{v} = \langle v_1, v_2, v_3 \rangle \quad \text{and} \quad \mathbf{w} = \langle w_1, w_2, w_3 \rangle$$

in 3-space are equivalent if and only if

$$v_1 = w_1, \quad v_2 = w_2, \quad \text{and} \quad v_3 = w_3$$

ARITHMETIC
OPERATIONS ON
VECTORS

The following theorem shows how to perform arithmetic operations on vectors using components.

14.2.4 THEOREM. *If* $\mathbf{v} = \langle v_1, v_2 \rangle$ *and* $\mathbf{w} = \langle w_1, w_2 \rangle$ *are vectors in 2-space and k is any scalar, then*

$$\mathbf{v} + \mathbf{w} = \langle v_1 + w_1, v_2 + w_2 \rangle \qquad (1a)$$

$$\mathbf{v} - \mathbf{w} = \langle v_1 - w_1, v_2 - w_2 \rangle \qquad (1b)$$

$$k\mathbf{v} = \langle kv_1, kv_2 \rangle \qquad (1c)$$

Similarly, if $\mathbf{v} = \langle v_1, v_2, v_3 \rangle$ *and* $\mathbf{w} = \langle w_1, w_2, w_3 \rangle$ *are vectors in 3-space and k is any scalar, then*

$$\mathbf{v} + \mathbf{w} = \langle v_1 + w_1, v_2 + w_2, v_3 + w_3 \rangle \qquad (2a)$$

$$\mathbf{v} - \mathbf{w} = \langle v_1 - w_1, v_2 - w_2, v_3 - w_3 \rangle \qquad (2b)$$

$$k\mathbf{v} = \langle kv_1, kv_2, kv_3 \rangle \qquad (2c)$$

We shall not prove this theorem. However, results (1a) and (1c) should be evident from Figure 14.2.7. Similar figures in 3-space can be used to motivate (2a) and (2c). Formulas (1b) and (2b) can be obtained from parts (a) and (c) by writing

$$\mathbf{v} - \mathbf{w} = \mathbf{v} + (-1)\mathbf{w}$$

Figure 14.2.7 ▷

Example 1   If $\mathbf{v} = \langle 1, -2 \rangle$ and $\mathbf{w} = \langle 7, 6 \rangle$ then

$$\mathbf{v} + \mathbf{w} = \langle 1, -2 \rangle + \langle 7, 6 \rangle = \langle 1 + 7, -2 + 6 \rangle = \langle 8, 4 \rangle$$

$$4\mathbf{v} = 4\langle 1, -2 \rangle = \langle 4(1), 4(-2) \rangle = \langle 4, -8 \rangle$$

$$-\mathbf{v} = (-1)\mathbf{v} = (-1)\langle 1, -2 \rangle = \langle -1, 2 \rangle$$
$$\mathbf{v} - \mathbf{w} = \langle 1, -2 \rangle - \langle 7, 6 \rangle = \langle 1 - 7, -2 - 6 \rangle = \langle -6, -8 \rangle \quad \blacktriangleleft$$

Example 2 If $\mathbf{v} = \langle -2, 0, 1 \rangle$ and $\mathbf{w} = \langle 3, 5, -4 \rangle$, then

$$\mathbf{v} + \mathbf{w} = \langle -2, 0, 1 \rangle + \langle 3, 5, -4 \rangle = \langle 1, 5, -3 \rangle$$
$$-3\mathbf{v} = \langle 6, 0, -3 \rangle$$
$$-\mathbf{w} = \langle -3, -5, 4 \rangle$$
$$\mathbf{w} - 2\mathbf{v} = \langle 3, 5, -4 \rangle - \langle -4, 0, 2 \rangle = \langle 7, 5, -6 \rangle \quad \blacktriangleleft$$

REMARK. It should be evident from Theorem 14.2.4 and the foregoing examples that, except for the number of components, there is no difference between arithmetic computations on vectors in 2-space and 3-space.

VECTORS WITH
INITIAL POINT NOT
AT THE ORIGIN

Sometimes a vector is positioned so that its initial point is not at the origin (Figure 14.2.8).

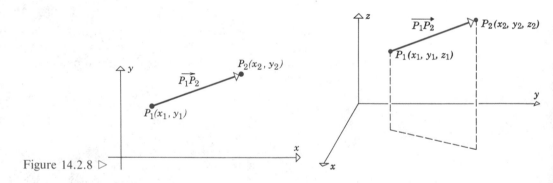

Figure 14.2.8 ▷

If the coordinates of the initial and terminal points of a vector are known, then the components of the vector can be obtained using the following theorem.

14.2.5 THEOREM. *If $\overrightarrow{P_1P_2}$ is a vector in 2-space with initial point $P_1(x_1, y_1)$ and terminal point $P_2(x_2, y_2)$, then*

$$\overrightarrow{P_1P_2} = \langle x_2 - x_1, y_2 - y_1 \rangle \tag{3a}$$

Similarly, if $\overrightarrow{P_1P_2}$ is a vector in 3-space with initial point $P_1(x_1, y_1, z_1)$ and terminal point $P_2(x_2, y_2, z_2)$, then

$$\overrightarrow{P_1P_2} = \langle x_2 - x_1, y_2 - y_1, z_2 - z_1 \rangle \tag{3b}$$

Proof. We shall give the proof in 2-space. The proof in 3-space is similar. The vector P_1P_2 is the difference of vectors $\overrightarrow{OP_2}$ and $\overrightarrow{OP_1}$ (Figure 14.2.9). Thus,

$$\overrightarrow{P_1P_2} = \overrightarrow{OP_2} - \overrightarrow{OP_1} = \langle x_2, y_2 \rangle - \langle x_1, y_1 \rangle = \langle x_2 - x_1, y_2 - y_1 \rangle$$ ▪

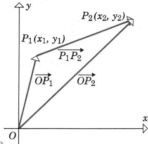

Figure 14.2.9 ▷

To paraphrase the foregoing theorem loosely, *the components of any vector are the coordinates of its terminal point minus the coordinates of its initial point.*

Example 3   In 2-space the vector with initial point $P_1(1, 3)$ and terminal point $P_2(4, -2)$ is

$$\overrightarrow{P_1P_2} = \langle 4 - 1, -2 - 3 \rangle = \langle 3, -5 \rangle$$

and in 3-space the vector with initial point $A(0, -2, 5)$ and terminal point $B(3, 4, -1)$ is

$$\overrightarrow{AB} = \langle 3 - 0, 4 - (-2), -1 - 5 \rangle = \langle 3, 6, -6 \rangle$$ ◀

In the previous section we stated without proof that the midpoint of the line segment joining the points $P_1(x_1, y_1, z_1)$ and $P_2(x_2, y_2, z_2)$ has coordinates $M(\bar{x}, \bar{y}, \bar{z})$ given by

$$\bar{x} = \frac{x_1 + x_2}{2}, \quad \bar{y} = \frac{y_1 + y_2}{2}, \quad \bar{z} = \frac{z_1 + z_2}{2} \tag{4}$$

To see that this is so we need only observe that

$$\overrightarrow{P_1M} = \tfrac{1}{2}\overrightarrow{P_1P_2}$$

(Figure 14.2.10), so that

$$\langle \bar{x} - x_1, \bar{y} - y_1, \bar{z} - z_1 \rangle = \tfrac{1}{2}\langle x_2 - x_1, y_2 - y_1, z_2 - z_1 \rangle$$

$P_1(x_1, y_1, z_1)$

M

$P_2(x_2, y_2, z_2)$

$\overrightarrow{P_1M} = \tfrac{1}{2}\overrightarrow{P_1P_2}$

Figure 14.2.10   or, on equating components,

$$\bar{x} - x_1 = \tfrac{1}{2}(x_2 - x_1), \quad \bar{y} - y_1 = \tfrac{1}{2}(y_2 - y_1), \quad \bar{z} - z_1 = \tfrac{1}{2}(z_2 - z_1)$$

from which (4) follows.

RULES OF VECTOR
ARITHMETIC

The following theorem shows that many of the familiar rules of ordinary arithmetic also hold for vector arithmetic.

14.2.6   THEOREM. *For any vectors* **u**, **v**, *and* **w** *and any scalars k and l, the following relationships hold:*

(a) $\mathbf{u} + \mathbf{v} = \mathbf{v} + \mathbf{u}$
(b) $(\mathbf{u} + \mathbf{v}) + \mathbf{w} = \mathbf{u} + (\mathbf{v} + \mathbf{w})$
(c) $\mathbf{u} + \mathbf{0} = \mathbf{0} + \mathbf{u} = \mathbf{u}$
(d) $\mathbf{u} + (-\mathbf{u}) = \mathbf{0}$
(e) $k(l\mathbf{u}) = (kl)\mathbf{u}$
(f) $k(\mathbf{u} + \mathbf{v}) = k\mathbf{u} + k\mathbf{v}$
(g) $(k + l)\mathbf{u} = k\mathbf{u} + l\mathbf{u}$
(h) $1\mathbf{u} = \mathbf{u}$

Before discussing the proof, we note that we have developed two approaches to vectors: *geometric,* in which vectors are represented by arrows or directed line segments, and *analytic,* in which vectors are represented by pairs of numbers called components. As a consequence, the results in this theorem can be established either geometrically or analytically. As an illustration, we shall prove part (b) both ways. The remaining proofs are left as exercises.

Proof of (b) (analytic in 2-space).   Let $\mathbf{u} = \langle u_1, u_2 \rangle$, $\mathbf{v} = \langle v_1, v_2 \rangle$, and $\mathbf{w} = \langle w_1, w_2 \rangle$. Then

$$
\begin{aligned}
(\mathbf{u} + \mathbf{v}) + \mathbf{w} &= (\langle u_1, u_2 \rangle + \langle v_1, v_2 \rangle) + \langle w_1, w_2 \rangle \\
&= \langle u_1 + v_1, u_2 + v_2 \rangle + \langle w_1, w_2 \rangle \\
&= \langle (u_1 + v_1) + w_1, (u_2 + v_2) + w_2 \rangle \\
&= \langle u_1 + (v_1 + w_1), u_2 + (v_2 + w_2) \rangle \\
&= \langle u_1, u_2 \rangle + \langle v_1 + w_1, v_2 + w_2 \rangle \\
&= \mathbf{u} + (\mathbf{v} + \mathbf{w})
\end{aligned}
$$

Proof of (b) (geometric).   Let **u**, **v**, and **w** be represented by $\overrightarrow{PQ}$, $\overrightarrow{QR}$, and $\overrightarrow{RS}$ as shown in Figure 14.2.11. Then

$$\mathbf{v} + \mathbf{w} = \overrightarrow{QS} \quad \text{and} \quad \mathbf{u} + (\mathbf{v} + \mathbf{w}) = \overrightarrow{PS}$$
$$\mathbf{u} + \mathbf{v} = \overrightarrow{PR} \quad \text{and} \quad (\mathbf{u} + \mathbf{v}) + \mathbf{w} = \overrightarrow{PS}$$

Therefore,

$$(\mathbf{u} + \mathbf{v}) + \mathbf{w} = \mathbf{u} + (\mathbf{v} + \mathbf{w}) \quad \blacksquare$$

REMARK.   In light of part (*b*) of this theorem, the symbol **u** + **v** + **w** is unambiguous since the same result is obtained no matter where parentheses are inserted. Moreover, if the vectors **u**, **v**, and **w** are placed "tip to tail," then the sum **u** + **v** + **w** is the vector from the initial point of **u** to the terminal point of **w** (Figure 14.2.11).

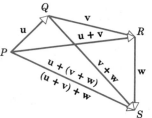

Figure 14.2.11 ▷

LENGTH OF A VECTOR

Geometrically, the *length* of a vector **v**, also called the *norm* of **v**, is the distance between its initial and terminal points. The length (or norm) of **v** is denoted by ‖**v**‖. It follows from the distance formulas in 2-space and 3-space that the norm of a vector $\mathbf{v} = \langle v_1, v_2 \rangle$ in 2-space is given by

$$\|\mathbf{v}\| = \sqrt{v_1^2 + v_2^2} \tag{5a}$$

and the norm of a vector $\mathbf{v} = \langle v_1, v_2, v_3 \rangle$ in 3-space is given by

$$\|\mathbf{v}\| = \sqrt{v_1^2 + v_2^2 + v_3^2} \tag{5b}$$

(Figure 14.2.12).

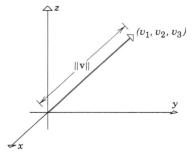

Figure 14.2.12 ▷

Example 4   Find the norm of

(a) $\mathbf{v} = \langle -2, 3 \rangle$ (b) $\mathbf{w} = \langle 2, 3, 6 \rangle$

Solution.   From (5a) and (5b)

$$\|\mathbf{v}\| = \sqrt{(-2)^2 + 3^2} = \sqrt{13}$$
$$\|\mathbf{w}\| = \sqrt{2^2 + 3^2 + 6^2} = \sqrt{49} = 7 \qquad \blacktriangleleft$$

Recall from Definition 14.2.3 that the length of $k\mathbf{v}$ is $|k|$ times the length of $\mathbf{v}$. Expressed as an equation, this statement says that

$$\|k\mathbf{v}\| = |k|\,\|\mathbf{v}\| \tag{6}$$

This formula applies to both vectors in 2-space and 3-space.

UNIT VECTORS   It follows from (6) that if a nonzero vector $\mathbf{v}$ is multiplied by $1/\|\mathbf{v}\|$ (the reciprocal of its length), then the result is a vector of length 1 in the same direction as $\mathbf{v}$. (Why?) This process of multiplying $\mathbf{v}$ by $1/\|\mathbf{v}\|$ to obtain a vector of length 1 is called *normalizing* $\mathbf{v}$.

Example 5   A vector of length 1 in the same direction as $\mathbf{v} = \langle 3, 4 \rangle$ is

$$\frac{1}{\|\mathbf{v}\|}\mathbf{v} = \frac{1}{\sqrt{3^2 + 4^2}}\langle 3, 4 \rangle = \frac{1}{5}\langle 3, 4 \rangle = \left\langle \frac{3}{5}, \frac{4}{5} \right\rangle \quad \blacktriangleleft$$

A vector of length 1 is called a *unit vector*. Of special importance are the unit vectors that run along the positive coordinate axes of a rectangular coordinate system. In both 2-space and 3-space the unit vectors along the x and y axes are denoted by $\mathbf{i}$ and $\mathbf{j}$, respectively, and in 3-space the unit vector along the z-axis is denoted by $\mathbf{k}$. Thus,

$$\mathbf{i} = \langle 1, 0 \rangle, \qquad \mathbf{j} = \langle 0, 1 \rangle \qquad \text{[in 2-space]}$$

$$\mathbf{i} = \langle 1, 0, 0 \rangle, \quad \mathbf{j} = \langle 0, 1, 0 \rangle, \quad \mathbf{k} = \langle 0, 0, 1 \rangle \qquad \text{[in 3-space]}$$

(See Figure 14.2.13.)

Figure 14.2.13 $\triangleright$

Every vector $\mathbf{v} = \langle v_1, v_2 \rangle$ in 2-space is expressible in terms of $\mathbf{i}$ and $\mathbf{j}$, since we can write

$$\mathbf{v} = \langle v_1, v_2 \rangle = \langle v_1, 0 \rangle + \langle 0, v_2 \rangle = v_1\langle 1, 0 \rangle + v_2\langle 0, 1 \rangle = v_1\mathbf{i} + v_2\mathbf{j}$$

and similarly every vector $\mathbf{v} = \langle v_1, v_2, v_3 \rangle$ in 3-space is expressible in terms of $\mathbf{i}$, $\mathbf{j}$, and $\mathbf{k}$ since we can write

$$\mathbf{v} = \langle v_1, v_2, v_3 \rangle = v_1\langle 1, 0, 0 \rangle + v_2\langle 0, 1, 0 \rangle + v_3\langle 0, 0, 1 \rangle = v_1\mathbf{i} + v_2\mathbf{j} + v_3\mathbf{k}$$

Example 6

$$\langle 2, 3 \rangle = 2\mathbf{i} + 3\mathbf{j}$$
$$\langle 5, -4 \rangle = 5\mathbf{i} + (-4)\mathbf{j} = 5\mathbf{i} - 4\mathbf{j}$$
$$\langle 0, 2 \rangle = 0\mathbf{i} + 2\mathbf{j} = 2\mathbf{j}$$
$$\langle -4, 0 \rangle = -4\mathbf{i} + 0\mathbf{j} = -4\mathbf{i}$$
$$\langle 0, 0 \rangle = 0\mathbf{i} + 0\mathbf{j} = \mathbf{0}$$

$$(3\mathbf{i} + 2\mathbf{j}) + (4\mathbf{i} + \mathbf{j}) = 7\mathbf{i} + 3\mathbf{j}$$
$$5(6\mathbf{i} - 2\mathbf{j}) = 30\mathbf{i} - 10\mathbf{j}$$
$$\|2\mathbf{i} - 3\mathbf{j}\| = \sqrt{2^2 + (-3)^2} = \sqrt{13}$$

Example 7

$$\langle 2, -3, 4 \rangle = 2\mathbf{i} - 3\mathbf{j} + 4\mathbf{k}$$
$$\langle 0, 3, 0 \rangle = 3\mathbf{j}$$
$$(3\mathbf{i} + 2\mathbf{j} - \mathbf{k}) - (4\mathbf{i} - \mathbf{j} + 2\mathbf{k}) = -\mathbf{i} + 3\mathbf{j} - 3\mathbf{k}$$
$$2(\mathbf{i} + \mathbf{j} - \mathbf{k}) + 4(\mathbf{i} - \mathbf{j}) = 6\mathbf{i} - 2\mathbf{j} - 2\mathbf{k}$$
$$\|\mathbf{i} + 2\mathbf{j} - 3\mathbf{k}\| = \sqrt{1^2 + 2^2 + (-3)^2} = \sqrt{14} \quad \blacktriangleleft$$

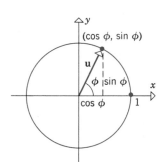

Figure 14.2.14

Example 8 If $\mathbf{u}$ is a unit vector in 2-space and if ϕ is the angle from the positive x-axis to $\mathbf{u}$, then

$$\mathbf{u} = \langle \cos \phi, \sin \phi \rangle = (\cos \phi)\mathbf{i} + (\sin \phi)\mathbf{j}$$

(Figure 14.2.14); and if $\mathbf{v}$ is an arbitrary vector making an angle ϕ with the positive x-axis, then $\mathbf{v}$ has the same direction as $\mathbf{u}$, but is $\|\mathbf{v}\|$ times as long. Thus,

$$\mathbf{v} = \|\mathbf{v}\| \langle \cos \phi, \sin \phi \rangle$$

or

$$\mathbf{v} = \|\mathbf{v}\| \cos \phi \, \mathbf{i} + \|\mathbf{v}\| \sin \phi \, \mathbf{j}$$

For example, the vector of length 2 making an angle of $\pi/4$ with the positive x-axis is

$$\mathbf{v} = 2 \cos \frac{\pi}{4} \mathbf{i} + 2 \sin \frac{\pi}{4} \mathbf{j} = \sqrt{2}\mathbf{i} + \sqrt{2}\mathbf{j} \quad \blacktriangleleft$$

▶ Exercise Set 14.2

In Exercises 1–4, sketch the vectors with the initial point at the origin.

1. (a) $\langle 2, 5 \rangle$
    (b) $\langle -5, -4 \rangle$
    (c) $\langle 2, 0 \rangle$.

2. (a) $\langle -3, 7 \rangle$
    (b) $\langle 6, -2 \rangle$
    (c) $\langle 0, -8 \rangle$.

3. (a) $\langle 1, -2, 2 \rangle$
    (b) $\langle 2, 2, -1 \rangle$.

4. (a) $\langle -1, 3, 2 \rangle$
    (b) $\langle 3, 4, 2 \rangle$.

In Exercises 5–8, sketch the vectors with the initial point at the origin.

5. (a) $-5\mathbf{i} + 3\mathbf{j}$
    (b) $3\mathbf{i} - 2\mathbf{j}$
    (c) $-6\mathbf{j}$.

6. (a) $4\mathbf{i} + 2\mathbf{j}$
    (b) $-2\mathbf{i} - \mathbf{j}$
    (c) $4\mathbf{i}$.

7. (a) $-\mathbf{i} + 2\mathbf{j} + 3\mathbf{k}$
    (b) $2\mathbf{i} + 3\mathbf{j} - \mathbf{k}$.

8. (a) $2\mathbf{j} - \mathbf{k}$
    (b) $\mathbf{i} - \mathbf{j} + 2\mathbf{k}$.

In Exercises 9–12, find the components of the vector $\overrightarrow{P_1P_2}$.

9. (a) $P_1(3, 5)$, $P_2(2, 8)$
    (b) $P_1(7, -2)$, $P_2(0, 0)$
    (c) $P_1(-6, -2)$, $P_2(-4, -1)$
    (d) $P_1(0, 0)$, $P_2(-8, 7)$.

10. (a) $P_1(1, 3)$, $P_2(4, 1)$
    (b) $P_1(6, -4)$, $P_2(0, 0)$
    (c) $P_1(-8, -1)$, $P_2(-3, -2)$
    (d) $P_1(0, 0)$, $P_2(-3, -5)$.

11. (a) $P_1(5, -2, 1)$, $P_2(2, 4, 2)$
    (b) $P_1(-1, 3, 5)$, $P_2(0, 0, 0)$.

12. (a) $P_1(0, 0, 0)$, $P_2(-1, 6, 1)$
    (b) $P_1(4, 1, -3)$, $P_2(9, 1, -3)$.

13. Find the terminal point of $\mathbf{v} = 3\mathbf{i} - 2\mathbf{j}$ if the initial point is $(1, -2)$.

14. Find the terminal point of $\mathbf{v} = \langle 7, 6 \rangle$ if the initial point is $(2, -1)$.

15. Find the initial point of $\mathbf{v} = \langle -2, 4 \rangle$ if the terminal point is $(2, 0)$.

16. Find the terminal point of $\mathbf{v} = \mathbf{i} + 2\mathbf{j} - 3\mathbf{k}$ if the initial point is $(-2, 1, 4)$.

17. Find the initial point of $\mathbf{v} = \langle -3, 1, 2 \rangle$ if the terminal point is $(5, 0, -1)$.

18. Let $\mathbf{u} = \langle 1, 3 \rangle$, $\mathbf{v} = \langle 2, 1 \rangle$, and $\mathbf{w} = \langle 4, -1 \rangle$. Find:
    (a) $\mathbf{u} - \mathbf{w}$
    (b) $7\mathbf{v} + 3\mathbf{w}$

    (c) $-\mathbf{w} + \mathbf{v}$
    (d) $3(\mathbf{u} - 7\mathbf{v})$
    (e) $-3\mathbf{v} - 8\mathbf{w}$
    (f) $2\mathbf{v} - (\mathbf{u} + \mathbf{w})$.

19. Let $\mathbf{u} = 2\mathbf{i} + 3\mathbf{j}$, $\mathbf{v} = 4\mathbf{i}$, $\mathbf{w} = -\mathbf{i} - 2\mathbf{j}$. Find:
    (a) $\mathbf{w} - \mathbf{v}$
    (b) $6\mathbf{u} + 4\mathbf{w}$
    (c) $-\mathbf{v} - 2\mathbf{w}$
    (d) $4(3\mathbf{u} + \mathbf{v})$
    (e) $-8(\mathbf{v} + \mathbf{w}) + 2\mathbf{u}$
    (f) $3\mathbf{w} - (\mathbf{v} - \mathbf{w})$.

20. Let $\mathbf{u} = \langle 2, -1, 3 \rangle$, $\mathbf{v} = \langle 4, 0, -2 \rangle$, $\mathbf{w} = \langle 1, 1, 3 \rangle$. Find:
    (a) $\mathbf{u} - \mathbf{w}$
    (b) $7\mathbf{v} + 3\mathbf{w}$
    (c) $-\mathbf{w} + \mathbf{v}$
    (d) $3(\mathbf{u} - 7\mathbf{v})$
    (e) $-3\mathbf{v} - 8\mathbf{w}$
    (f) $2\mathbf{v} - (\mathbf{u} + \mathbf{w})$.

21. Let $\mathbf{u} = 3\mathbf{i} - \mathbf{k}$, $\mathbf{v} = \mathbf{i} - \mathbf{j} + 2\mathbf{k}$, $\mathbf{w} = 3\mathbf{j}$. Find:
    (a) $\mathbf{w} - \mathbf{v}$
    (b) $6\mathbf{u} + 4\mathbf{w}$
    (c) $-\mathbf{v} - 2\mathbf{w}$
    (d) $4(3\mathbf{u} + \mathbf{v})$
    (e) $-8(\mathbf{v} + \mathbf{w}) + 2\mathbf{u}$
    (f) $3\mathbf{w} - (\mathbf{v} - \mathbf{w})$.

In Exercises 22–25, compute the norm of $\mathbf{v}$.

22. (a) $\mathbf{v} = \langle 3, 4 \rangle$
    (b) $\mathbf{v} = -\mathbf{i} + 7\mathbf{j}$
    (c) $\mathbf{v} = -3\mathbf{j}$.

23. (a) $\mathbf{v} = \langle 1, -1 \rangle$
    (b) $\mathbf{v} = \langle 2, 0 \rangle$
    (c) $\mathbf{v} = \sqrt{2}\mathbf{i} - \sqrt{7}\mathbf{j}$.

24. (a) $\mathbf{v} = \mathbf{i} + \mathbf{j} + \mathbf{k}$
    (b) $\mathbf{v} = \langle -1, 2, 4 \rangle$.

25. (a) $\mathbf{v} = -3\mathbf{i} + 2\mathbf{j} + \mathbf{k}$
    (b) $\mathbf{v} = \langle 0, -3, 0 \rangle$.

26. Let $\mathbf{u} = \langle 1, -3 \rangle$, $\mathbf{v} = \langle 1, 1 \rangle$, and $\mathbf{w} = \langle 2, -4 \rangle$. Find:
    (a) $\|\mathbf{u} + \mathbf{v}\|$
    (b) $\|\mathbf{u}\| + \|\mathbf{v}\|$
    (c) $\|-2\mathbf{u}\| + 2\|\mathbf{v}\|$
    (d) $\|3\mathbf{u} - 5\mathbf{v} + \mathbf{w}\|$.

27. Let $\mathbf{u} = 2\mathbf{i} - 5\mathbf{j}$, $\mathbf{v} = 2\mathbf{i}$, and $\mathbf{w} = 3\mathbf{i} + 4\mathbf{j}$. Find:
    (a) $\|\mathbf{v} + \mathbf{w}\|$
    (b) $\|\mathbf{v}\| + \|\mathbf{w}\|$
    (c) $\|-3\mathbf{u}\| + 4\|\mathbf{v}\|$
    (d) $\|\mathbf{u} - \mathbf{v} - \mathbf{w}\|$
    (e) $\dfrac{1}{\|\mathbf{w}\|}\mathbf{w}$
    (f) $\left\|\dfrac{1}{\|\mathbf{w}\|}\mathbf{w}\right\|$.

28. Let $\mathbf{u} = \langle 2, -1, 0 \rangle$ and $\mathbf{v} = \langle 0, 1, -1 \rangle$. Find:
    (a) $\|\mathbf{u} + \mathbf{v}\|$
    (b) $\|\mathbf{u}\| + \|\mathbf{v}\|$
    (c) $\|3\mathbf{u}\|$
    (d) $\|2\mathbf{u} - 3\mathbf{v}\|$.

29. Let $\mathbf{u} = \mathbf{i} - 3\mathbf{j} + 2\mathbf{k}$, $\mathbf{v} = \mathbf{i} + \mathbf{j}$, and $\mathbf{w} = 2\mathbf{i} + 2\mathbf{j} - 4\mathbf{k}$. Find:
    (a) $\|\mathbf{u} + \mathbf{v}\|$
    (b) $\|\mathbf{u}\| + \|\mathbf{v}\|$
    (c) $\|-2\mathbf{u}\| + 2\|\mathbf{v}\|$
    (d) $\|3\mathbf{u} - 5\mathbf{v} + \mathbf{w}\|$
    (e) $\dfrac{1}{\|\mathbf{w}\|}\mathbf{w}$
    (f) $\left\|\dfrac{1}{\|\mathbf{w}\|}\mathbf{w}\right\|$

30. Let $\mathbf{u} = \langle -1, 1 \rangle$, $\mathbf{v} = \langle 0, 1 \rangle$, and $\mathbf{w} = \langle 3, 4 \rangle$. Find the vector $\mathbf{x}$ that satisfies $\mathbf{u} - 2\mathbf{x} = \mathbf{x} - \mathbf{w} + 3\mathbf{v}$.

31. Let $\mathbf{u} = \langle 1, 3 \rangle$, $\mathbf{v} = \langle 2, 1 \rangle$, $\mathbf{w} = \langle 4, -1 \rangle$. Find the vector $\mathbf{x}$ that satisfies $2\mathbf{u} - \mathbf{v} + \mathbf{x} = 7\mathbf{x} + \mathbf{w}$.

32. Find $\mathbf{u}$ and $\mathbf{v}$ if $\mathbf{u} + \mathbf{v} = \langle 2, -3 \rangle$ and $3\mathbf{u} + 2\mathbf{v} = \langle -1, 2 \rangle$.

33. Find $\mathbf{u}$ and $\mathbf{v}$ if $\mathbf{u} + 2\mathbf{v} = 3\mathbf{i} - \mathbf{k}$ and $3\mathbf{u} - \mathbf{v} = \mathbf{i} + \mathbf{j} + \mathbf{k}$.

34. Give a geometric argument to show that if $\mathbf{u}$ and $\mathbf{v}$ are nonzero and $\mathbf{u}$ is not parallel to $\mathbf{v}$, then any vector $\mathbf{w}$ in the plane of $\mathbf{u}$ and $\mathbf{v}$ can be written as $\mathbf{w} = c_1\mathbf{u} + c_2\mathbf{v}$ for a suitable choice of scalars c_1 and c_2.

35. Give a geometric argument to show that if $\mathbf{u}$, $\mathbf{v}$, and $\mathbf{w}$ are not coplanar, then any vector $\mathbf{z}$ can be written as $\mathbf{z} = c_1\mathbf{u} + c_2\mathbf{v} + c_3\mathbf{w}$ for a suitable choice of scalars c_1, c_2, and c_3.

36. Let $\mathbf{u} = 2\mathbf{i} - \mathbf{j}$ and $\mathbf{v} = 4\mathbf{i} + 2\mathbf{j}$. Find scalars c_1 and c_2 such that $c_1\mathbf{u} + c_2\mathbf{v} = -4\mathbf{j}$.

37. Let $\mathbf{u} = \langle 1, -3 \rangle$ and $\mathbf{v} = \langle -2, 6 \rangle$. Show that there do not exist scalars c_1 and c_2 such that $c_1\mathbf{u} + c_2\mathbf{v} = \langle 3, 5 \rangle$.

38. Let $\mathbf{u} = \langle 1, 0, 1 \rangle$, $\mathbf{v} = \langle 3, 2, 0 \rangle$, and $\mathbf{w} = \langle 0, 1, 1 \rangle$. Find scalars c_1, c_2, and c_3 such that $c_1\mathbf{u} + c_2\mathbf{v} + c_3\mathbf{w} = \langle -1, 1, 5 \rangle$.

39. Let $\mathbf{u} = \mathbf{i} - \mathbf{j}$, $\mathbf{v} = 3\mathbf{i} + \mathbf{k}$, and $\mathbf{w} = 4\mathbf{i} - \mathbf{j} + \mathbf{k}$. Show that there do not exist scalars c_1, c_2, and c_3 such that $c_1\mathbf{u} + c_2\mathbf{v} + c_3\mathbf{w} = 2\mathbf{i} + \mathbf{j} - \mathbf{k}$.

40. Let $\mathbf{v} = 4\mathbf{i} - 3\mathbf{j}$. Find all scalars k such that $\|k\mathbf{v}\| = 3$.

41. Verify parts (b), (e), (f), and (g) of Theorem 14.2.6 for $\mathbf{u} = \langle 1, -3 \rangle$, $\mathbf{v} = \langle 6, 6 \rangle$, $\mathbf{w} = \langle -8, 1 \rangle$, $k = 3$, and $l = 6$.

42. Find a unit vector having the same direction as $-\mathbf{i} + 4\mathbf{j}$.

43. Find a unit vector oppositely directed to $3\mathbf{i} - 4\mathbf{j}$.

44. Find a unit vector having the same direction as $2\mathbf{i} - \mathbf{j} - 2\mathbf{k}$.

45. Find a unit vector oppositely directed to $6\mathbf{i} - 4\mathbf{j} + 2\mathbf{k}$.

46. Find a unit vector having the same direction as the vector from the point $A(-3, 2)$ to the point $B(1, -1)$.

47. Find a unit vector having the same direction as the vector from the point $A(-1, 0, 2)$ to the point $B(3, 1, 1)$.

48. Find a vector with the same direction as $\mathbf{v} = -2\mathbf{i} + 3\mathbf{j}$ but with three times the length of $\mathbf{v}$.

49. Find a vector oppositely directed to $\mathbf{v} = \langle 3, -4 \rangle$ but with half the length of $\mathbf{v}$.

50. Find a vector with the same direction as $\mathbf{v} = \langle 7, 0, -6 \rangle$ but with twice the length of $\mathbf{v}$.

51. Find a vector oppositely directed to $\mathbf{v} = -3\mathbf{i} + 4\mathbf{j} + \mathbf{k}$, but with twice the length of $\mathbf{v}$.

52. Let $\mathbf{r} = \langle x, y \rangle$. Describe the set of points (x, y) for which $\|\mathbf{r}\| = 1$.

53. Let $\mathbf{r}_0 = \langle x_0, y_0 \rangle$ and $\mathbf{r} = \langle x, y \rangle$. Describe the set of all points (x, y) for which $\|\mathbf{r} - \mathbf{r}_0\| = 1$.

54. Let $\mathbf{r}_1 = \langle x_1, y_1 \rangle$, $\mathbf{r}_2 = \langle x_2, y_2 \rangle$, and $\mathbf{r} = \langle x, y \rangle$. Describe the set of all points (x, y) for which $\|\mathbf{r} - \mathbf{r}_1\| + \|\mathbf{r} - \mathbf{r}_2\| = k$, where $k > \|\mathbf{r}_2 - \mathbf{r}_1\|$.

55. Let $\mathbf{r}_0 = \langle x_0, y_0, z_0 \rangle$ and $\mathbf{r} = \langle x, y, z \rangle$. Describe the set of all points (x, y, z) for which
(a) $\|\mathbf{r}\| = 2$ (b) $\|\mathbf{r} - \mathbf{r}_0\| = 3$
(c) $\|\mathbf{r} - \mathbf{r}_0\| \le 1$.

56. Find two unit vectors in 2-space parallel to the line $y = 3x + 2$.

57. (a) Find two unit vectors in 2-space parallel to the line $x + y = 4$.
(b) Find two unit vectors in 2-space perpendicular to the line in (a).

58. Let P be the point $(2, 3)$ and Q the point $(7, -4)$. Use vectors to find the point on the line segment joining P and Q that is $\frac{3}{4}$ of the way from P to Q.

59. For the points P and Q in Exercise 58, use vectors to find the point on the line segment joining P and Q that is $\frac{3}{4}$ of the way from Q to P.

60. Find a unit vector in 2-space making an angle of $135°$ with the x-axis.

61. Find:
(a) a unit vector in 2-space making an angle of $\pi/3$ with the positive x-axis;
(b) a vector of length 4 in 2-space making an angle of $3\pi/4$ with the positive x-axis.

62. Use vectors to find the length of the diagonal of the parallelogram determined by $\mathbf{i} + \mathbf{j}$ and $\mathbf{i} - 2\mathbf{j}$.

63. Use vectors to find the fourth vertex of a parallelogram, three of whose vertices are $(0,0)$, $(1,3)$, and $(2,4)$. [*Note:* There is more than one answer.]

64. Prove: $\|\mathbf{u} + \mathbf{v}\| \leq \|\mathbf{u}\| + \|\mathbf{v}\|$ geometrically.

65. Prove parts (a), (c), and (e) of Theorem 14.2.6 analytically in 2-space.

66. Prove parts (d), (g), and (h) of Theorem 14.2.6 analytically in 2-space.

67. Prove part (f) of Theorem 14.2.6 geometrically.

68. Use vectors to prove that the midpoints of the sides of a quadrilateral are the vertices of a parallelogram.

69. Use vectors to prove that the line segment joining the midpoints of two sides of a triangle is parallel to the third side and half as long.

70. Find $\overrightarrow{AB} + \overrightarrow{BC} + \overrightarrow{CA}$, where A, B, and C are the vertices of a triangle.

In Exercises 71–73, let A, B, C, and D be any four points in 3-space.

71. If M is the midpoint of BC, show that $\overrightarrow{AB} + \overrightarrow{AC} = 2\overrightarrow{AM}$.

72. If M and N are the midpoints of AC and BD, show that $\overrightarrow{AB} + \overrightarrow{CD} = 2\overrightarrow{MN}$.

73. If M and N are the midpoints of AC and BD, show that $\overrightarrow{AB} + \overrightarrow{AD} + \overrightarrow{CB} + \overrightarrow{CD} = 4\overrightarrow{MN}$.

74. Let A and B be distinct points on a straight line L. If a point P different from B is on L, then $\overrightarrow{AP} = t\overrightarrow{PB}$ for some value of the scalar t. Let $\mathbf{a}$, $\mathbf{b}$, and $\mathbf{r}$ be vectors from the origin to the points A, B, and P, respectively. Show that $\mathbf{r} = (\mathbf{a} + t\mathbf{b})/(1 + t)$.

75. Let $\mathbf{r}_1, \mathbf{r}_2, \ldots, \mathbf{r}_n$ be vectors from the origin to points $P_1, P_2, \ldots, P_n$, respectively. The *centroid* of points $P_1, P_2, \ldots, P_n$ is defined as the point P for which $\sum_{k=1}^{n} \overrightarrow{PP_k} = \mathbf{0}$. Let $\mathbf{r}$ be the vector from the origin to P. Show that $\mathbf{r} = \dfrac{1}{n} \sum_{k=1}^{n} \mathbf{r}_k$. [*Hint:* Write $\overrightarrow{PP_k}$ as a difference of vectors $\mathbf{r}$ and $\mathbf{r}_k$.]

76. Let $P_1, P_2, \ldots, P_n$ be consecutive vertices of a polygon in 2-space, all of whose interior angles are less than π. Let $\mathbf{r}_k$ be the vector from the origin to the point P_k for $k = 1, 2, \ldots, n$. From Exercise 75, the endpoint of the vector $\mathbf{r} = \dfrac{1}{n} \sum_{k=1}^{n} \mathbf{r}_k$ drawn from the origin is the centroid of the points $P_1, P_2, \ldots, P_n$. It can be shown that the centroid is the balance point of the polygon.
   (a) Find the centroid of the triangle with vertices $P_1(1, 1)$, $P_2(3, 3)$, and $P_3(5, 0)$.
   (b) Cut the polygon described in part (a) out of cardboard and show that it balances when the tip of a pencil is placed at the centroid.

77. Follow the directions of Exercise 76 for the quadrilateral with vertices $P_1(-1, 2)$, $P_2(2, 3)$, $P_3(5, -2)$, and $P_4(0, -1)$.

14.3 DOT PRODUCT; PROJECTIONS

In this section we shall introduce a type of multiplication of vectors in 2-space and 3-space. We shall also discuss the arithmetic properties of this multiplication and give some of its applications.

ANGLE BETWEEN VECTORS

Let $\mathbf{u}$ and $\mathbf{v}$ be two nonzero vectors in 2-space or 3-space, and assume these vectors have been positioned so their initial points coincide. By the *angle between* $\mathbf{u}$ *and* $\mathbf{v}$, we shall mean the angle θ determined by $\mathbf{u}$ and $\mathbf{v}$ that satisfies $0 \leq \theta \leq \pi$ (Figure 14.3.1).

Figure 14.3.1 ▷

14.3.1 DEFINITION. If **u** and **v** are vectors in 2-space or 3-space and θ is the angle between **u** and **v**, then the *dot product* or *Euclidean inner product* **u** · **v** is defined by

$$\mathbf{u} \cdot \mathbf{v} = \begin{cases} \|\mathbf{u}\| \, \|\mathbf{v}\| \cos \theta, & \text{if } \mathbf{u} \neq \mathbf{0} \text{ and } \mathbf{v} \neq \mathbf{0} \\ 0, & \text{if } \mathbf{u} = \mathbf{0} \text{ or } \mathbf{v} = \mathbf{0} \end{cases}$$

Figure 14.3.2

Example 1 As shown in Figure 14.3.2, the angle between vectors $\mathbf{u} = \langle 0, 2 \rangle$ and $\mathbf{v} = \langle 1, 1 \rangle$ is 45°. Thus,

$$\mathbf{u} \cdot \mathbf{v} = \|\mathbf{u}\| \, \|\mathbf{v}\| \cos \theta = \sqrt{0^2 + 2^2} \, \sqrt{1^2 + 1^2} \cos 45° = (2)(\sqrt{2}) \frac{1}{\sqrt{2}} = 2$$

◀

FORMULA FOR THE
DOT PRODUCT

For purposes of computation, it is desirable to have a formula that expresses the dot product of two vectors in terms of the components of the vectors. We shall derive such a formula for vectors in 3-space and just state the corresponding formula for vectors in 2-space.

Let $\mathbf{u} = \langle u_1, u_2, u_3 \rangle$ and $\mathbf{v} = \langle v_1, v_2, v_3 \rangle$ be two nonzero vectors. If, as in Figure 14.3.3, θ is the angle between **u** and **v**, then the law of cosines yields

$$\|\overrightarrow{PQ}\|^2 = \|\mathbf{u}\|^2 + \|\mathbf{v}\|^2 - 2\|\mathbf{u}\| \, \|\mathbf{v}\| \cos \theta \tag{1}$$

Figure 14.3.3

Since $\overrightarrow{PQ} = \mathbf{v} - \mathbf{u}$, we can rewrite (1) as

$$\|\mathbf{u}\| \, \|\mathbf{v}\| \cos \theta = \tfrac{1}{2}(\|\mathbf{u}\|^2 + \|\mathbf{v}\|^2 - \|\mathbf{v} - \mathbf{u}\|^2)$$

or

$$\mathbf{u} \cdot \mathbf{v} = \tfrac{1}{2}(\|\mathbf{u}\|^2 + \|\mathbf{v}\|^2 - \|\mathbf{v} - \mathbf{u}\|^2)$$

Substituting

$$\|\mathbf{u}\|^2 = u_1{}^2 + u_2{}^2 + u_3{}^2, \quad \|\mathbf{v}\|^2 = v_1{}^2 + v_2{}^2 + v_3{}^2$$

and

$$\|\mathbf{v} - \mathbf{u}\|^2 = (v_1 - u_1)^2 + (v_2 - u_2)^2 + (v_3 - u_3)^2$$

we obtain, after simplifying,

$$\mathbf{u} \cdot \mathbf{v} = u_1 v_1 + u_2 v_2 + u_3 v_3 \tag{2a}$$

If $\mathbf{u} = \langle u_1, u_2 \rangle$ and $\mathbf{v} = \langle v_1, v_2 \rangle$ are two vectors in 2-space, then the formula corresponding to (2a) is

$$\mathbf{u} \cdot \mathbf{v} = u_1 v_1 + u_2 v_2 \tag{2b}$$

If $\mathbf{u}$ and $\mathbf{v}$ are nonzero vectors, the formula in Definition 14.3.1 can be written as

$$\cos \theta = \frac{\mathbf{u} \cdot \mathbf{v}}{\|\mathbf{u}\| \, \|\mathbf{v}\|} \tag{3}$$

Example 2   Consider the vectors

$$\mathbf{u} = 2\mathbf{i} - \mathbf{j} + \mathbf{k} \quad \text{and} \quad \mathbf{v} = \mathbf{i} + \mathbf{j} + 2\mathbf{k}$$

Find $\mathbf{u} \cdot \mathbf{v}$ and determine the angle θ between $\mathbf{u}$ and $\mathbf{v}$.

Solution.

$$\mathbf{u} \cdot \mathbf{v} = u_1 v_1 + u_2 v_2 + u_3 v_3 = (2)(1) + (-1)(1) + (1)(2) = 3$$

For the given vectors, $\|\mathbf{u}\| = \|\mathbf{v}\| = \sqrt{6}$, so that

$$\cos \theta = \frac{3}{\sqrt{6}\sqrt{6}} = \frac{1}{2}$$

Thus, $\theta = 60°$ ◀

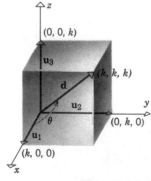

Figure 14.3.4

Example 3   Find the angle between a diagonal of a cube and one of its edges.

Solution.   Let k be the length of an edge and let us introduce a coordinate system as shown in Figure 14.3.4.

If we let $\mathbf{u}_1 = \langle k, 0, 0 \rangle$, $\mathbf{u}_2 = \langle 0, k, 0 \rangle$, and $\mathbf{u}_3 = \langle 0, 0, k \rangle$, then the vector

$$\mathbf{d} = \langle k, k, k \rangle = \mathbf{u}_1 + \mathbf{u}_2 + \mathbf{u}_3$$

is a diagonal of the cube. The angle θ between $\mathbf{d}$ and the edge $\mathbf{u}_1$ satisfies

$$\cos\theta = \frac{\mathbf{u}_1 \cdot \mathbf{d}}{\|\mathbf{u}_1\| \, \|\mathbf{d}\|} = \frac{k^2}{(k)(\sqrt{3k^2})} = \frac{1}{\sqrt{3}}$$

Thus,

$$\theta = \cos^{-1}\frac{1}{\sqrt{3}} \approx 54°44' \quad \blacktriangleleft$$

The sign of the dot product provides useful information about the angle between two vectors.

14.3.2   THEOREM. *If* **u** *and* **v** *are nonzero vectors in 2-space or 3-space, and if* θ *is the angle between them, then*

θ *is acute*   *if and only if* $\mathbf{u} \cdot \mathbf{v} > 0$

θ *is obtuse*   *if and only if* $\mathbf{u} \cdot \mathbf{v} < 0$

$\theta = \pi/2$ *if and only if* $\mathbf{u} \cdot \mathbf{v} = 0$

Proof.   Since **u** and **v** are nonzero vectors, $\|\mathbf{u}\| > 0$ and $\|\mathbf{v}\| > 0$. Thus,

$$\mathbf{u} \cdot \mathbf{v} = \|\mathbf{u}\| \, \|\mathbf{v}\| \cos\theta$$

is positive, negative, or zero according to whether $\cos\theta$ is positive, negative, or zero. Since $0 \le \theta \le \pi$, it follows that θ is acute if and only if $\cos\theta > 0$; θ is obtuse if and only if $\cos\theta < 0$, and $\theta = \pi/2$ if and only if $\cos\theta = 0$.   ∎

Example 4   If $\mathbf{u} = \mathbf{i} - 2\mathbf{j} + 3\mathbf{k}$, $\mathbf{v} = -3\mathbf{i} + 4\mathbf{j} + 2\mathbf{k}$, and $\mathbf{w} = 3\mathbf{i} + 6\mathbf{j} + 3\mathbf{k}$, then

$$\mathbf{u} \cdot \mathbf{v} = (1)(-3) + (-2)(4) + (3)(2) = -5$$
$$\mathbf{v} \cdot \mathbf{w} = (-3)(3) + (4)(6) + (2)(3) = 21$$
$$\mathbf{u} \cdot \mathbf{w} = (1)(3) + (-2)(6) + (3)(3) = 0$$

Therefore **u** and **v** make an obtuse angle, **v** and **w** make an acute angle, and **u** and **w** are perpendicular.   ◄

ORTHOGONAL VECTORS   Perpendicular vectors are also called ***orthogonal*** vectors. In light of Theorem 14.3.2, two nonzero vectors are orthogonal if and only if their dot product is zero. If we agree to consider **u** and **v** to be perpendicular when either or both of these vectors is **0**, then we can state without exception that two vectors **u** and **v** are orthogonal (perpendicular) if and only if $\mathbf{u} \cdot \mathbf{v} = 0$.

Example 5   Show that in 2-space the vector $a\mathbf{i} + b\mathbf{j}$ is perpendicular to the line $ax + by + c = 0$.

Solution.   Let $P_1(x_1, y_1)$ and $P_2(x_2, y_2)$ be distinct points on the line, so that

$$ax_1 + by_1 + c = 0 \tag{4}$$
$$ax_2 + by_2 + c = 0$$

Since the vector $\overrightarrow{P_1P_2} = (x_2 - x_1)\mathbf{i} + (y_2 - y_1)\mathbf{j}$ runs along the line, we need only show that $a\mathbf{i} + b\mathbf{j}$ and $\overrightarrow{P_1P_2}$ are perpendicular. But on subtracting the equations in (4), we obtain

$$a(x_2 - x_1) + b(y_2 - y_1) = 0$$

which can be expressed in the form

$$(a\mathbf{i} + b\mathbf{j}) \cdot [(x_2 - x_1)\mathbf{i} + (y_2 - y_1)\mathbf{j}] = 0$$

or

$$(a\mathbf{i} + b\mathbf{j}) \cdot \overrightarrow{P_1P_2} = 0$$

so that $a\mathbf{i} + b\mathbf{j}$ and $\overrightarrow{P_1P_2}$ are perpendicular.   ◀

Figure 14.3.5

Of special interest are the angles α, β, and γ that a vector $\mathbf{u}$ in 3-space makes with the vectors $\mathbf{i}$, $\mathbf{j}$, and $\mathbf{k}$ (Figure 14.3.5). These are called the **direction angles** of $\mathbf{u}$. The numbers $\cos\alpha$, $\cos\beta$, and $\cos\gamma$ are called the **direction cosines** of $\mathbf{u}$. Formulas for the direction cosines follow easily from (3).

14.3.3   THEOREM. *The three direction cosines of a nonzero vector* $\mathbf{u} = u_1\mathbf{i} + u_2\mathbf{j} + u_3\mathbf{k}$ *in 3-space are*

$$\cos\alpha = \frac{u_1}{\|\mathbf{u}\|}, \quad \cos\beta = \frac{u_2}{\|\mathbf{u}\|}, \quad \cos\gamma = \frac{u_3}{\|\mathbf{u}\|}$$

Proof.   Since $\mathbf{u} \cdot \mathbf{i} = (u_1)(1) + (u_2)(0) + (u_3)(0) = u_1$, it follows that

$$\cos\alpha = \frac{\mathbf{u} \cdot \mathbf{i}}{\|\mathbf{u}\| \|\mathbf{i}\|} = \frac{u_1}{\|\mathbf{u}\|}$$

Similarly for $\cos\beta$ and $\cos\gamma$.   ∎

The direction cosines of a vector $\mathbf{u} = u_1\mathbf{i} + u_2\mathbf{j} + u_3\mathbf{k}$ can be obtained by simply reading off the components of the unit vector $\mathbf{u}/\|\mathbf{u}\|$ since

$$\frac{\mathbf{u}}{\|\mathbf{u}\|} = \frac{u_1}{\|\mathbf{u}\|}\mathbf{i} + \frac{u_2}{\|\mathbf{u}\|}\mathbf{j} + \frac{u_3}{\|\mathbf{u}\|}\mathbf{k} = (\cos\alpha)\mathbf{i} + (\cos\beta)\mathbf{j} + (\cos\gamma)\mathbf{k}$$

Example 6 Find the direction cosines of the vector $\mathbf{u} = 2\mathbf{i} - 4\mathbf{j} + 4\mathbf{k}$, and estimate the direction angles to the nearest degree.

Solution. $\|\mathbf{u}\| = \sqrt{4 + 16 + 16} = 6$, so that $\mathbf{u}/\|\mathbf{u}\| = \frac{1}{3}\mathbf{i} - \frac{2}{3}\mathbf{j} + \frac{2}{3}\mathbf{k}$. Thus,

$$\cos\alpha = \frac{1}{3}, \quad \cos\beta = -\frac{2}{3}, \quad \cos\gamma = \frac{2}{3}$$

With the help of trigonometric tables or a calculator that can compute inverse trigonometric functions, one obtains

$$\alpha = \cos^{-1}\left(\frac{1}{3}\right) \approx 71°, \quad \beta = \cos^{-1}\left(-\frac{2}{3}\right) \approx 132°, \quad \gamma = \cos^{-1}\left(\frac{2}{3}\right) \approx 48°$$

◀

PROPERTIES OF THE DOT PRODUCT   The following arithmetic properties of the dot product are useful in calculations involving vectors.

14.3.4   THEOREM. *If* $\mathbf{u}$, $\mathbf{v}$, *and* $\mathbf{w}$ *are vectors in 2- or 3-space and k is a scalar, then*

(*a*) $\mathbf{u} \cdot \mathbf{v} = \mathbf{v} \cdot \mathbf{u}$
(*b*) $\mathbf{u} \cdot (\mathbf{v} + \mathbf{w}) = \mathbf{u} \cdot \mathbf{v} + \mathbf{u} \cdot \mathbf{w}$
(*c*) $k(\mathbf{u} \cdot \mathbf{v}) = (k\mathbf{u}) \cdot \mathbf{v} = \mathbf{u} \cdot (k\mathbf{v})$
(*d*) $\mathbf{v} \cdot \mathbf{v} = \|\mathbf{v}\|^2$

Proof.   We shall prove parts (*c*) and (*d*) for vectors in 3-space and omit the remaining proofs.

(*c*)   Let $\mathbf{u} = \langle u_1, u_2, u_3 \rangle$ and $\mathbf{v} = \langle v_1, v_2, v_3 \rangle$; then

$$k(\mathbf{u} \cdot \mathbf{v}) = k(u_1v_1 + u_2v_2 + u_3v_3)$$
$$= (ku_1)v_1 + (ku_2)v_2 + (ku_3)v_3$$
$$= (k\mathbf{u}) \cdot \mathbf{v}$$

Similarly, $k(\mathbf{u} \cdot \mathbf{v}) = \mathbf{u} \cdot (k\mathbf{v})$.

(*d*) $\mathbf{v} \cdot \mathbf{v} = v_1v_1 + v_2v_2 + v_3v_3 = v_1^2 + v_2^2 + v_3^2 = \|\mathbf{v}\|^2$.   ∎

Part (*d*) of the last theorem is sometimes expressed in the following alternative form.

$$\|\mathbf{v}\| = \sqrt{\mathbf{v} \cdot \mathbf{v}}$$

(5)

ORTHOGONAL
PROJECTIONS OF
VECTORS

In many applications it is of interest to "decompose" a vector $\mathbf{u}$ into a sum of two terms, one parallel to a specified nonzero vector $\mathbf{a}$ and the other perpendicular to $\mathbf{a}$. If $\mathbf{u}$ and $\mathbf{a}$ are positioned so their initial points coincide at a point Q, we may decompose the vector $\mathbf{u}$ as follows (Figure 14.3.6): Drop a perpendicular from the tip of $\mathbf{u}$ to the line through $\mathbf{a}$, and construct the vector $\mathbf{w}_1$ from Q to the foot of this perpendicular. Next form the difference

$$\mathbf{w}_2 = \mathbf{u} - \mathbf{w}_1$$

Figure 14.3.6 ▷

As indicated in Figure 14.3.6, the vector $\mathbf{w}_1$ is parallel to $\mathbf{a}$, the vector $\mathbf{w}_2$ is perpendicular to $\mathbf{a}$, and

$$\mathbf{w}_1 + \mathbf{w}_2 = \mathbf{w}_1 + (\mathbf{u} - \mathbf{w}_1) = \mathbf{u}$$

The vector $\mathbf{w}_1$ is called the ***orthogonal projection of*** $\mathbf{u}$ ***on*** $\mathbf{a}$ or sometimes the ***vector component of*** $\mathbf{u}$ ***along*** $\mathbf{a}$. It is denoted by

$$\text{proj}_{\mathbf{a}}\, \mathbf{u} \tag{6}$$

The vector $\mathbf{w}_2$ is called the ***vector component of*** $\mathbf{u}$ ***orthogonal to*** $\mathbf{a}$. Since $\mathbf{w}_2 = \mathbf{u} - \mathbf{w}_1$, this vector may be written in notation (6) as

$$\mathbf{w}_2 = \mathbf{u} - \text{proj}_{\mathbf{a}}\, \mathbf{u}$$

The following theorem gives formulas for calculating the vectors $\text{proj}_{\mathbf{a}}\, \mathbf{u}$ and $\mathbf{u} - \text{proj}_{\mathbf{a}}\, \mathbf{u}$.

14.3.5   THEOREM. *If* $\mathbf{u}$ *and* $\mathbf{a}$ *are vectors in 2-space or 3-space and if* $\mathbf{a} \neq \mathbf{0}$, *then*

$$\text{proj}_{\mathbf{a}}\, \mathbf{u} = \frac{\mathbf{u} \cdot \mathbf{a}}{\|\mathbf{a}\|^2}\, \mathbf{a} \quad \textit{(vector component of } \mathbf{u} \textit{ along } \mathbf{a}\textit{)}$$

$$\mathbf{u} - \text{proj}_{\mathbf{a}}\, \mathbf{u} = \mathbf{u} - \frac{\mathbf{u} \cdot \mathbf{a}}{\|\mathbf{a}\|^2}\, \mathbf{a} \quad \begin{array}{l}\textit{(vector component of } \mathbf{u} \\ \textit{orthogonal to } \mathbf{a}\textit{)}\end{array}$$

Proof.   Let $\mathbf{w}_1 = \text{proj}_{\mathbf{a}}\, \mathbf{u}$ and $\mathbf{w}_2 = \mathbf{u} - \text{proj}_{\mathbf{a}}\, \mathbf{u}$. Since $\mathbf{w}_1$ is parallel to $\mathbf{a}$, it must be a scalar multiple of $\mathbf{a}$, so it can be written in the form $\mathbf{w}_1 = k\mathbf{a}$. Thus,

$$\mathbf{u} = \mathbf{w}_1 + \mathbf{w}_2 = k\mathbf{a} + \mathbf{w}_2 \tag{7}$$

Taking the dot product of both sides of (7) with $\mathbf{a}$ and using Theorem 14.3.4 yields

$$\mathbf{u} \cdot \mathbf{a} = (k\mathbf{a} + \mathbf{w}_2) \cdot \mathbf{a} = k\|\mathbf{a}\|^2 + \mathbf{w}_2 \cdot \mathbf{a} \tag{8}$$

But $\mathbf{w}_2 \cdot \mathbf{a} = 0$ since $\mathbf{w}_2$ is perpendicular to $\mathbf{a}$; thus, (8) yields

$$k = \frac{\mathbf{u} \cdot \mathbf{a}}{\|\mathbf{a}\|^2}$$

Since $\text{proj}_a \mathbf{u} = \mathbf{w}_1 = k\mathbf{a}$, we obtain

$$\text{proj}_a \mathbf{u} = \frac{\mathbf{u} \cdot \mathbf{a}}{\|\mathbf{a}\|^2} \mathbf{a} \qquad ∎$$

Example 7 Let $\mathbf{u} = 2\mathbf{i} - \mathbf{j} + 3\mathbf{k}$ and $\mathbf{a} = 4\mathbf{i} - \mathbf{j} + 2\mathbf{k}$. Find the vector component of $\mathbf{u}$ along $\mathbf{a}$ and the vector component of $\mathbf{u}$ orthogonal to $\mathbf{a}$.

Solution.

$$\mathbf{u} \cdot \mathbf{a} = (2)(4) + (-1)(-1) + (3)(2) = 15$$
$$\|\mathbf{a}\|^2 = 4^2 + (-1)^2 + 2^2 = 21$$

Thus, the vector component of $\mathbf{u}$ along $\mathbf{a}$ is

$$\text{proj}_a \mathbf{u} = \frac{\mathbf{u} \cdot \mathbf{a}}{\|\mathbf{a}\|^2} \mathbf{a} = \frac{15}{21}(4\mathbf{i} - \mathbf{j} + 2\mathbf{k}) = \frac{20}{7}\mathbf{i} - \frac{5}{7}\mathbf{j} + \frac{10}{7}\mathbf{k}$$

and the vector component of $\mathbf{u}$ orthogonal to $\mathbf{a}$ is

$$\mathbf{u} - \text{proj}_a \mathbf{u} = (2\mathbf{i} - \mathbf{j} + 3\mathbf{k}) - \left(\frac{20}{7}\mathbf{i} - \frac{5}{7}\mathbf{j} + \frac{10}{7}\mathbf{k}\right)$$
$$= -\frac{6}{7}\mathbf{i} - \frac{2}{7}\mathbf{j} + \frac{11}{7}\mathbf{k}$$

As a check, the reader may wish to verify that the vectors $\mathbf{u} - \text{proj}_a \mathbf{u}$ and $\mathbf{a}$ are perpendicular by showing that their dot product is zero. ◀

A formula for the length of the vector component of $\mathbf{u}$ along $\mathbf{a}$ may be obtained by writing

$$\|\text{proj}_a \mathbf{u}\| = \left\| \frac{\mathbf{u} \cdot \mathbf{a}}{\|\mathbf{a}\|^2}\mathbf{a} \right\| = \left| \frac{\mathbf{u} \cdot \mathbf{a}}{\|\mathbf{a}\|^2} \right| \|\mathbf{a}\| \quad \left[\text{Since } \frac{\mathbf{u} \cdot \mathbf{a}}{\|\mathbf{a}\|^2} \text{ is a scalar.} \right]$$
$$= \frac{|\mathbf{u} \cdot \mathbf{a}|}{\|\mathbf{a}\|^2} \|\mathbf{a}\| \qquad [\text{Since } \|\mathbf{a}\|^2 > 0.]$$

which yields

$$\|\text{proj}_a \mathbf{u}\| = \frac{|\mathbf{u} \cdot \mathbf{a}|}{\|\mathbf{a}\|} \tag{9}$$

If θ denotes the angle between $\mathbf{u}$ and $\mathbf{a}$, then $\mathbf{u} \cdot \mathbf{a} = \|\mathbf{u}\| \|\mathbf{a}\| \cos \theta$, so that (9) can also be written as

$$\|\text{proj}_a \mathbf{u}\| = \|\mathbf{u}\| |\cos \theta| \tag{10}$$

(Verify.) A geometric interpretation of this result is given in Figure 14.3.7.

Figure 14.3.7 ▷

As an application of these formulas we shall use vector methods to derive a formula for the distance from a point in the plane to a line.

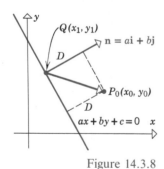

Figure 14.3.8

Example 8   Find a formula for the distance D between the point $P_0(x_0, y_0)$ and the line $ax + by + c = 0$.

Solution.   Let $Q(x_1, y_1)$ be any point on the line and position the vector

$$\mathbf{n} = a\mathbf{i} + b\mathbf{j}$$

so that its initial point is at Q.

By virtue of Example 5, the vector $\mathbf{n}$ is perpendicular to the line (Figure 14.3.8). As indicated in the figure, the distance D is equal to the length of the orthogonal projection of the vector $\overrightarrow{QP_0}$ on $\mathbf{n}$; thus, from (9),

$$D = \|\text{proj}_{\mathbf{n}} \overrightarrow{QP_0}\| = \frac{|\overrightarrow{QP_0} \cdot \mathbf{n}|}{\|\mathbf{n}\|}$$

But

$$\overrightarrow{QP_0} = \langle x_0 - x_1, y_0 - y_1 \rangle$$
$$\overrightarrow{QP_0} \cdot \mathbf{n} = a(x_0 - x_1) + b(y_0 - y_1)$$
$$\|\mathbf{n}\| = \sqrt{a^2 + b^2}$$

so that

$$D = \frac{|a(x_0 - x_1) + b(y_0 - y_1)|}{\sqrt{a^2 + b^2}} \qquad (11)$$

Since the point $Q(x_1, y_1)$ lies on the line, its coordinates satisfy the equation of the line, so

$$ax_1 + by_1 + c = 0$$

or

$$c = -ax_1 - by_1$$

Substituting this expression in (11) yields the formula

$$D = \frac{|ax_0 + by_0 + c|}{\sqrt{a^2 + b^2}} \qquad \blacktriangleleft \qquad (12)$$

As an illustration of Formula (12), the distance D from the point $(1, -2)$ to the line $3x + 4y - 6 = 0$ is

$$D = \frac{|(3)(1) + 4(-2) - 6|}{\sqrt{3^2 + 4^2}} = \frac{|-11|}{\sqrt{25}} = \frac{11}{5}$$

WORK   Recall from Section 6.7 that the work W done by a constant force $\mathbf{F}$ of magnitude $\|\mathbf{F}\|$ acting in the direction of motion on a particle moving from P to Q on a line is

$$W = (\text{force}) \cdot (\text{distance}) = \|\mathbf{F}\| \, \|\overrightarrow{PQ}\|$$

If the force $\mathbf{F}$ is constant, but makes an angle θ with the direction of motion (Figure 14.3.9), then we define the work done by $\mathbf{F}$ to be

$$W = \mathbf{F} \cdot \overrightarrow{PQ} = (\|\mathbf{F}\| \cos \theta) \, \|\overrightarrow{PQ}\| \qquad (13)$$

The quantity $\|\mathbf{F}\| \cos \theta$ is the "component" of force in the direction of motion and $\|\overrightarrow{PQ}\|$ is the distance traveled by the particle.

Figure 14.3.9

Example 9   A wagon is pulled horizontally by exerting a continual force

Figure 14.3.10 ▷

of 10 lb on the handle at an angle of 60° with the horizontal. How much work is done in moving the wagon 50 ft?

Solution.   Introduce an xy-coordinate system so the wagon moves from $P(0, 0)$ to $Q(50, 0)$ along the x-axis (Figure 14.3.10). In this coordinate system

$$\overrightarrow{PQ} = 50\mathbf{i}$$

and

$$\mathbf{F} = (10 \cos 60°)\mathbf{i} + (10 \sin 60°)\mathbf{j} = 5\mathbf{i} + 5\sqrt{3}\mathbf{j}$$

so that the work done is

$$W = \mathbf{F} \cdot \overrightarrow{PQ} = (5\mathbf{i} + 5\sqrt{3}\mathbf{j}) \cdot (50\mathbf{i}) = 250 \text{ (foot-pounds)} \quad \blacktriangleleft$$

▶ Exercise Set 14.3

1. Find $\mathbf{u} \cdot \mathbf{v}$.
   (a) $\mathbf{u} = \mathbf{i} + 2\mathbf{j}$, $\mathbf{v} = 6\mathbf{i} - 8\mathbf{j}$
   (b) $\mathbf{u} = \langle -7, -3 \rangle$, $\mathbf{v} = \langle 0, 1 \rangle$
   (c) $\mathbf{u} = \mathbf{i} - 3\mathbf{j} + 7\mathbf{k}$, $\mathbf{v} = 8\mathbf{i} - 2\mathbf{j} - 2\mathbf{k}$
   (d) $\mathbf{u} = \langle -3, 1, 2 \rangle$, $\mathbf{v} = \langle 4, 2, -5 \rangle$.

2. In each part of Exercise 1, find the cosine of the angle θ between $\mathbf{u}$ and $\mathbf{v}$.

3. Determine whether $\mathbf{u}$ and $\mathbf{v}$ make an acute angle, an obtuse angle, or are orthogonal.
   (a) $\mathbf{u} = 7\mathbf{i} + 3\mathbf{j} + 5\mathbf{k}$, $\mathbf{v} = -8\mathbf{i} + 4\mathbf{j} + 2\mathbf{k}$
   (b) $\mathbf{u} = 6\mathbf{i} + \mathbf{j} + 3\mathbf{k}$, $\mathbf{v} = 4\mathbf{i} - 6\mathbf{k}$
   (c) $\mathbf{u} = \langle 1, 1, 1 \rangle$, $\mathbf{v} = \langle -1, 0, 0 \rangle$
   (d) $\mathbf{u} = \langle 4, 1, 6 \rangle$, $\mathbf{v} = \langle -3, 0, 2 \rangle$.

4. Find the orthogonal projection of $\mathbf{u}$ on $\mathbf{a}$.
   (a) $\mathbf{u} = 2\mathbf{i} + \mathbf{j}$, $\mathbf{a} = -3\mathbf{i} + 2\mathbf{j}$
   (b) $\mathbf{u} = \langle 2, 6 \rangle$, $\mathbf{a} = \langle -9, 3 \rangle$
   (c) $\mathbf{u} = -7\mathbf{i} + \mathbf{j} + 3\mathbf{k}$, $\mathbf{a} = 5\mathbf{i} + \mathbf{k}$
   (d) $\mathbf{u} = \langle 0, 0, 1 \rangle$, $\mathbf{a} = \langle 8, 3, 4 \rangle$.

5. In each part of Exercise 4, find the vector component of $\mathbf{u}$ orthogonal to $\mathbf{a}$.

6. Find $\|\text{proj}_\mathbf{a} \mathbf{u}\|$
   (a) $\mathbf{u} = 2\mathbf{i} - \mathbf{j}$, $\mathbf{a} = 3\mathbf{i} + 4\mathbf{j}$
   (b) $\mathbf{u} = \langle 4, 5 \rangle$, $\mathbf{a} = \langle 1, -2 \rangle$
   (c) $\mathbf{u} = 2\mathbf{i} - \mathbf{j} + 3\mathbf{k}$, $\mathbf{a} = \mathbf{i} + 2\mathbf{j} + 2\mathbf{k}$
   (d) $\mathbf{u} = \langle 4, -1, 7 \rangle$, $\mathbf{a} = \langle 2, 3, -6 \rangle$.

7. Verify Theorem 14.3.4(c) for $\mathbf{u} = 6\mathbf{i} - \mathbf{j} + 2\mathbf{k}$, $\mathbf{v} = 2\mathbf{i} + 7\mathbf{j} + 4\mathbf{k}$, and $k = -5$.

8. Find two vectors in 2-space of norm 1 that are orthogonal to $3\mathbf{i} - 2\mathbf{j}$.

9. Let $\mathbf{u} = \langle 1, 2 \rangle$, $\mathbf{v} = \langle 4, -2 \rangle$, and $\mathbf{w} = \langle 6, 0 \rangle$. Find:
   (a) $\mathbf{u} \cdot (7\mathbf{v} + \mathbf{w})$ (b) $\|(\mathbf{u} \cdot \mathbf{w})\mathbf{w}\|$
   (c) $\|\mathbf{u}\|(\mathbf{v} \cdot \mathbf{w})$ (d) $(\|\mathbf{u}\|\mathbf{v}) \cdot \mathbf{w}$.

10. Explain why each of the following expressions makes no sense.
    (a) $\mathbf{u} \cdot (\mathbf{v} \cdot \mathbf{w})$ (b) $(\mathbf{u} \cdot \mathbf{v}) + \mathbf{w}$
    (c) $\|\mathbf{u} \cdot \mathbf{v}\|$ (d) $k \cdot (\mathbf{u} + \mathbf{v})$.

11. Use vectors to find the cosines of the interior angles of the triangle with vertices $(-1, 0)$, $(2, -1)$, and $(1, 4)$.

12. Find two unit vectors in 2-space that make an angle of 45° with $4\mathbf{i} + 3\mathbf{j}$.

13. Show that $A(2, -1, 1)$, $B(3, 2, -1)$, and $C(7, 0, -2)$ are vertices of a right triangle. At which vertex is the right angle?

14. Find k so that the vector from the point $A(1, -1, 3)$ to the point $B(3, 0, 5)$ is perpendicular to the vector from A to the point $P(k, k, k)$.

15. Let $\mathbf{r}_0 = \langle x_0, y_0 \rangle$ and $\mathbf{r} = \langle x, y \rangle$. Describe the set of all points (x, y) for which

(a) $\mathbf{r} \cdot \mathbf{r}_0 = 0$ (b) $(\mathbf{r} - \mathbf{r}_0) \cdot \mathbf{r}_0 = 0$.
(c) $\mathbf{r} \cdot (\mathbf{r} - \mathbf{r}_0) = 0$.

16. Suppose $\mathbf{a} \cdot \mathbf{b} = \mathbf{a} \cdot \mathbf{c}$ and $\mathbf{a} \neq \mathbf{0}$. Does it follow that $\mathbf{b} = \mathbf{c}$? Explain.

17. Let $\mathbf{a} = k\mathbf{i} + \mathbf{j}$ and $\mathbf{b} = 4\mathbf{i} + 3\mathbf{j}$. Find k so that
(a) $\mathbf{a}$ and $\mathbf{b}$ are orthogonal
(b) the angle between $\mathbf{a}$ and $\mathbf{b}$ is $\pi/4$
(c) the angle between $\mathbf{a}$ and $\mathbf{b}$ is $\pi/6$
(d) $\mathbf{a}$ and $\mathbf{b}$ are parallel.

18. Find the direction cosines of $\mathbf{u}$ and estimate the direction angles to the nearest degree.
(a) $\mathbf{u} = \mathbf{i} + \mathbf{j} - \mathbf{k}$ (b) $\mathbf{u} = 2\mathbf{i} - 2\mathbf{j} + \mathbf{k}$
(c) $\mathbf{u} = 3\mathbf{i} - 2\mathbf{j} - 6\mathbf{k}$ (d) $\mathbf{u} = 3\mathbf{i} - 4\mathbf{k}$.

19. Prove: The direction cosines of a vector satisfy the equation $\cos^2 \alpha + \cos^2 \beta + \cos^2 \gamma = 1$.

20. Prove: Two nonzero vectors $\mathbf{u}_1$ and $\mathbf{u}_2$ are perpendicular if and only if their direction cosines satisfy

$$\cos \alpha_1 \cos \alpha_2 + \cos \beta_1 \cos \beta_2 + \cos \gamma_1 \cos \gamma_2 = 0$$

21. Use Formula (12) to calculate the distance between the point and the line.
(a) $3x + 4y + 7 = 0;\ (1, -2)$
(b) $y = -2x + 1;\ (-3, 5)$
(c) $2x + y = 8;\ (2, 6)$.

22. Given the points $A(1, 1, 0)$, $B(-2, 3, -4)$, and $P(-3, 1, 2)$,
(a) find $\|\mathrm{proj}_{\overrightarrow{AB}}\, \overrightarrow{AP}\|$
(b) use the Pythagorean Theorem and the result of part (a) to find the distance from P to the line through A and B.

23. Follow the directions of Exercise 22 for the points $A(2, 1, -3)$, $B(0, 2, -1)$, and $P(4, 3, 0)$.

24. A boat travels 100 meters due north while the wind exerts a force of 50 newtons toward the northeast. How much work does the wind do?

25. Find the work done by a force $\mathbf{F} = -3\mathbf{j}$ (pounds) applied to a point that moves on a line from $(1, 3)$ to $(4, 7)$. Assume distance is measured in feet.

26. Let $\mathbf{u}$ and $\mathbf{v}$ determine a parallelogram. Use vectors to prove that the diagonals of the parallelogram are perpendicular if and only if the sides are equal in length.

27. Let $\mathbf{u}$ and $\mathbf{v}$ determine a parallelogram. Use vectors to prove that the parallelogram is a rectangle if and only if the diagonals are equal in length.

28. Prove: $\|\mathbf{u} + \mathbf{v}\|^2 + \|\mathbf{u} - \mathbf{v}\|^2 = 2\|\mathbf{u}\|^2 + 2\|\mathbf{v}\|^2$.

29. Prove: $\mathbf{u} \cdot \mathbf{v} = \frac{1}{4}\|\mathbf{u} + \mathbf{v}\|^2 - \frac{1}{4}\|\mathbf{u} - \mathbf{v}\|^2$.

30. Find, to the nearest degree, the angle between the diagonal of a cube and a diagonal of one of its faces.

31. Find, to the nearest degree, the acute angle formed by two diagonals of a cube.

32. Find, to the nearest degree, the angles that a diagonal of a box with dimensions 10 in. by 15 in. by 25 in. makes with the edges of the box.

33. Prove: If vectors $\mathbf{v}_1$, $\mathbf{v}_2$, and $\mathbf{v}_3$ are nonzero and mutually perpendicular, then any vector $\mathbf{v}$ can be written as

$$\mathbf{v} = c_1\mathbf{v}_1 + c_2\mathbf{v}_2 + c_3\mathbf{v}_3$$

where $c_i = (\mathbf{v} \cdot \mathbf{v}_i)/\|\mathbf{v}_i\|^2,\ i = 1, 2, 3$.

34. Show that the vectors $\mathbf{v}_1 = 3\mathbf{i} - \mathbf{j} + 2\mathbf{k}$, $\mathbf{v}_2 = \mathbf{i} + \mathbf{j} - \mathbf{k}$, and $\mathbf{v}_3 = \mathbf{i} - 5\mathbf{j} - 4\mathbf{k}$ are mutually perpendicular. Use the result of Exercise 33 to find scalars c_1, c_2, and c_3 so that $c_1\mathbf{v}_1 + c_2\mathbf{v}_2 + c_3\mathbf{v}_3 = \mathbf{i} - \mathbf{j} + \mathbf{k}$.

35. Prove: If $\mathbf{v}$ is orthogonal to $\mathbf{w}_1$ and $\mathbf{w}_2$, then $\mathbf{v}$ is orthogonal to $k_1\mathbf{w}_1 + k_2\mathbf{w}_2$ for all scalars k_1 and k_2.

36. Let $\mathbf{u}$ and $\mathbf{v}$ be nonzero vectors, and let $k = \|\mathbf{u}\|$ and $l = \|\mathbf{v}\|$. Prove that

$$\mathbf{w} = l\mathbf{u} + k\mathbf{v}$$

bisects the angle between $\mathbf{u}$ and $\mathbf{v}$.

14.4   CROSS PRODUCT

In many applications of vectors to problems in geometry, physics, and engineering, it is of interest to construct a vector in 3-space that is perpendicular to two given vectors. In this section we shall introduce a type of vector multiplication that facilitates this construction.

DETERMINANTS   We begin with some notation. If a, b, c, and d are real numbers, then the symbol

$$\begin{vmatrix} a & b \\ c & d \end{vmatrix}$$

called a **2 × 2 (*two-by-two*) determinant,** denotes the number

$$\begin{vmatrix} a & b \\ c & d \end{vmatrix} = ad - bc$$

For example,

$$\begin{vmatrix} 3 & -2 \\ 4 & 5 \end{vmatrix} = (3)(5) - (-2)(4) = 15 + 8 = 23$$

A **3 × 3 *determinant***

$$\begin{vmatrix} a_1 & a_2 & a_3 \\ b_1 & b_2 & b_3 \\ c_1 & c_2 & c_3 \end{vmatrix}$$

is defined in terms of 2 × 2 determinants by the formula

$$\begin{vmatrix} a_1 & a_2 & a_3 \\ b_1 & b_2 & b_3 \\ c_1 & c_2 & c_3 \end{vmatrix} = a_1 \begin{vmatrix} b_2 & b_3 \\ c_2 & c_3 \end{vmatrix} - a_2 \begin{vmatrix} b_1 & b_3 \\ c_1 & c_3 \end{vmatrix} + a_3 \begin{vmatrix} b_1 & b_2 \\ c_1 & c_2 \end{vmatrix}$$

The right side of this formula is easily remembered by noting that a_1, a_2, and a_3 are the entries in the first "row" of the left side, and the 2 × 2 determinants on the right side arise by deleting the first row and an appropriate column from the left side. The pattern is as follows:

$$\begin{vmatrix} a_1 & a_2 & a_3 \\ b_1 & b_2 & b_3 \\ c_1 & c_2 & c_3 \end{vmatrix} = a_1 \begin{vmatrix} \cancel{a_1} & a_2 & a_3 \\ b_1 & b_2 & b_3 \\ c_1 & c_2 & c_3 \end{vmatrix} - a_2 \begin{vmatrix} a_1 & \cancel{a_2} & a_3 \\ b_1 & b_2 & b_3 \\ c_1 & c_2 & c_3 \end{vmatrix} + a_3 \begin{vmatrix} a_1 & a_2 & \cancel{a_3} \\ b_1 & b_2 & b_3 \\ c_1 & c_2 & c_3 \end{vmatrix}$$

For example,

$$\begin{vmatrix} 3 & -2 & -5 \\ 1 & 4 & -4 \\ 0 & 3 & 2 \end{vmatrix} = 3 \begin{vmatrix} 4 & -4 \\ 3 & 2 \end{vmatrix} - (-2) \begin{vmatrix} 1 & -4 \\ 0 & 2 \end{vmatrix} + (-5) \begin{vmatrix} 1 & 4 \\ 0 & 3 \end{vmatrix}$$

$$= 3(20) + 2(2) - 5(3) = 49$$

CROSS PRODUCT

14.4.1 DEFINITION. If $\mathbf{u} = \langle u_1, u_2, u_3 \rangle$ and $\mathbf{v} = \langle v_1, v_2, v_3 \rangle$ are vectors in 3-space, then the *cross product* $\mathbf{u} \times \mathbf{v}$ is the vector defined by

$$\mathbf{u} \times \mathbf{v} = \begin{vmatrix} u_2 & u_3 \\ v_2 & v_3 \end{vmatrix} \mathbf{i} - \begin{vmatrix} u_1 & u_3 \\ v_1 & v_3 \end{vmatrix} \mathbf{j} + \begin{vmatrix} u_1 & u_2 \\ v_1 & v_2 \end{vmatrix} \mathbf{k} \tag{1}$$

Formula (1) can be remembered by writing it in the form,

$$\mathbf{u} \times \mathbf{v} = \begin{vmatrix} \mathbf{i} & \mathbf{j} & \mathbf{k} \\ u_1 & u_2 & u_3 \\ v_1 & v_2 & v_3 \end{vmatrix} \tag{2}$$

However, this is just a mnemonic device since the entries in a determinant must be numbers, not vectors.

Example 1 Find $\mathbf{u} \times \mathbf{v}$, where $\mathbf{u} = \langle 1, 2, -2 \rangle$ and $\mathbf{v} = \langle 3, 0, 1 \rangle$.

Solution.

$$\mathbf{u} \times \mathbf{v} = \begin{vmatrix} \mathbf{i} & \mathbf{j} & \mathbf{k} \\ 1 & 2 & -2 \\ 3 & 0 & 1 \end{vmatrix}$$

$$= \begin{vmatrix} 2 & -2 \\ 0 & 1 \end{vmatrix} \mathbf{i} - \begin{vmatrix} 1 & -2 \\ 3 & 1 \end{vmatrix} \mathbf{j} + \begin{vmatrix} 1 & 2 \\ 3 & 0 \end{vmatrix} \mathbf{k} = 2\mathbf{i} - 7\mathbf{j} - 6\mathbf{k} \quad \blacktriangleleft$$

Observe that the cross product of two vectors is another vector, whereas the dot product of two vectors is a scalar.

The following theorem gives an important relationship between dot product and cross product and also shows that $\mathbf{u} \times \mathbf{v}$ is orthogonal to both $\mathbf{u}$ and $\mathbf{v}$.

14.4.2 THEOREM. *If $\mathbf{u}$ and $\mathbf{v}$ are vectors in 3-space, then:*

(a) $\mathbf{u} \cdot (\mathbf{u} \times \mathbf{v}) = 0$ ($\mathbf{u} \times \mathbf{v}$ *is orthogonal to* $\mathbf{u}$)

(b) $\mathbf{v} \cdot (\mathbf{u} \times \mathbf{v}) = 0$ ($\mathbf{u} \times \mathbf{v}$ *is orthogonal to* $\mathbf{v}$)

(c) $\|\mathbf{u} \times \mathbf{v}\|^2 = \|\mathbf{u}\|^2 \|\mathbf{v}\|^2 - (\mathbf{u} \cdot \mathbf{v})^2$ (*Lagrange's identity*)

Proof.   Let $\mathbf{u} = \langle u_1, u_2, u_3 \rangle$ and $\mathbf{v} = \langle v_1, v_2, v_3 \rangle$.

(*a*)   By definition

$$\mathbf{u} \times \mathbf{v} = \begin{vmatrix} u_2 & u_3 \\ v_2 & v_3 \end{vmatrix} \mathbf{i} - \begin{vmatrix} u_1 & u_3 \\ v_1 & v_3 \end{vmatrix} \mathbf{j} + \begin{vmatrix} u_1 & u_2 \\ v_1 & v_2 \end{vmatrix} \mathbf{k}$$

which may be rewritten as

$$\mathbf{u} \times \mathbf{v} = \langle u_2 v_3 - u_3 v_2, \; u_3 v_1 - u_1 v_3, \; u_1 v_2 - u_2 v_1 \rangle \tag{3}$$

so that

$$\mathbf{u} \cdot (\mathbf{u} \times \mathbf{v}) = u_1(u_2 v_3 - u_3 v_2) + u_2(u_3 v_1 - u_1 v_3) \\ + u_3(u_1 v_2 - u_2 v_1) = 0$$

(*b*)   Similar to (*a*).

(*c*)   From (3) we obtain

$$\|\mathbf{u} \times \mathbf{v}\|^2 = (u_2 v_3 - u_3 v_2)^2 + (u_3 v_1 - u_1 v_3)^2 + (u_1 v_2 - u_2 v_1)^2 \tag{4}$$

Moreover,

$$\|\mathbf{u}\|^2 \|\mathbf{v}\|^2 - (\mathbf{u} \cdot \mathbf{v})^2 = (u_1{}^2 + u_2{}^2 + u_3{}^2)(v_1{}^2 + v_2{}^2 + v_3{}^2) \\ - (u_1 v_1 + u_2 v_2 + u_3 v_3)^2 \tag{5}$$

Lagrange's identity can be established by "multiplying out" the right sides of (4) and (5) and verifying their equality.   ∎

Example 2   Let $\mathbf{u} = \langle 1, 2, -2 \rangle$ and $\mathbf{v} = \langle 3, 0, 1 \rangle$. In Example 1 we showed that

$$\mathbf{u} \times \mathbf{v} = \langle 2, -7, -6 \rangle$$

Thus,

$$\mathbf{u} \cdot (\mathbf{u} \times \mathbf{v}) = (1)(2) + (2)(-7) + (-2)(-6) = 0$$

and

$$\mathbf{v} \cdot (\mathbf{u} \times \mathbf{v}) = (3)(2) + (0)(-7) + (1)(-6) = 0$$

so that $\mathbf{u} \times \mathbf{v}$ is orthogonal to both $\mathbf{u}$ and $\mathbf{v}$ as guaranteed by Theorem 14.4.2.   ◄

If $\mathbf{u}$ and $\mathbf{v}$ are nonzero vectors in 3-space, then the length of $\mathbf{u} \times \mathbf{v}$ has a useful geometric interpretation. Lagrange's identity, given in Theorem 14.4.2, states that

$$\|\mathbf{u} \times \mathbf{v}\|^2 = \|\mathbf{u}\|^2 \|\mathbf{v}\|^2 - (\mathbf{u} \cdot \mathbf{v})^2 \tag{6}$$

If θ denotes the angle between $\mathbf{u}$ and $\mathbf{v}$, then $\mathbf{u} \cdot \mathbf{v} = \|\mathbf{u}\| \, \|\mathbf{v}\| \cos \theta$, so (6) can be rewritten as

$$\begin{aligned} \|\mathbf{u} \times \mathbf{v}\|^2 &= \|\mathbf{u}\|^2 \, \|\mathbf{v}\|^2 - \|\mathbf{u}\|^2 \, \|\mathbf{v}\|^2 \cos^2 \theta \\ &= \|\mathbf{u}\|^2 \, \|\mathbf{v}\|^2 \, (1 - \cos^2 \theta) \\ &= \|\mathbf{u}\|^2 \, \|\mathbf{v}\|^2 \sin^2 \theta \end{aligned}$$

Since $0 \leq \theta \leq \pi$, it follows that $\sin \theta \geq 0$, so

$$\|\mathbf{u} \times \mathbf{v}\| = \|\mathbf{u}\| \, \|\mathbf{v}\| \sin \theta \tag{7}$$

Figure 14.4.1

But $\|\mathbf{v}\| \sin \theta$ is the altitude of the parallelogram determined by $\mathbf{u}$ and $\mathbf{v}$ (Figure 14.4.1). Thus, from (7), the area A of this parallelogram is given by

$$A = (\text{base})(\text{altitude}) = \|\mathbf{u}\| \, \|\mathbf{v}\| \sin \theta = \|\mathbf{u} \times \mathbf{v}\|$$

In other words, the length of $\mathbf{u} \times \mathbf{v}$ is numerically equal to the area of the parallelogram determined by $\mathbf{u}$ and $\mathbf{v}$.

Example 3 Find the area of the triangle that is determined by the points $P_1(2, 2, 0)$, $P_2(-1, 0, 2)$, and $P_3(0, 4, 3)$.

Figure 14.4.2

Solution. The area A of the triangle is half the area of the parallelogram determined by the vectors $\overrightarrow{P_1P_2}$ and $\overrightarrow{P_1P_3}$ (Figure 14.4.2). But $\overrightarrow{P_1P_2} = \langle -3, -2, 2 \rangle$ and $\overrightarrow{P_1P_3} = \langle -2, 2, 3 \rangle$ so that

$$\overrightarrow{P_1P_2} \times \overrightarrow{P_1P_3} = \langle -10, 5, -10 \rangle$$

(verify) and consequently

$$A = \frac{1}{2} \|\overrightarrow{P_1P_2} \times \overrightarrow{P_1P_3}\| = \frac{15}{2} \quad \blacktriangleleft$$

It follows from (7) that $\mathbf{u} \times \mathbf{v} = \mathbf{0}$ if and only if

$$\mathbf{u} = \mathbf{0}, \quad \text{or} \quad \mathbf{v} = \mathbf{0}, \quad \text{or} \quad \sin \theta = 0$$

In all three cases the vectors $\mathbf{u}$ and $\mathbf{v}$ are parallel. For the first two cases this is true because $\mathbf{0}$ is parallel to every vector, and in the third case, $\sin \theta = 0$ implies that the angle θ between $\mathbf{u}$ and $\mathbf{v}$ is $\theta = 0$ or $\theta = \pi$. In summary, we have the following result.

> **14.4.3** THEOREM. *If* $\mathbf{u}$ *and* $\mathbf{v}$ *are vectors in 3-space, then* $\mathbf{u} \times \mathbf{v} = \mathbf{0}$ *if and only if* $\mathbf{u}$ *and* $\mathbf{v}$ *are parallel vectors.*

The main arithmetic properties of the cross product are listed in the next theorem.

14.4.4   THEOREM. *If* **u**, **v**, *and* **w** *are any vectors in 3-space and k is any scalar, then:*

(*a*)   **u** × **v** = −(**v** × **u**)
(*b*)   **u** × (**v** + **w**) = (**u** × **v**) + (**u** × **w**)
(*c*)   (**u** + **v**) × **w** = (**u** × **w**) + (**v** × **w**)
(*d*)   *k*(**u** × **v**) = (*k***u**) × **v** = **u** × (*k***v**)
(*e*)   **u** × **0** = **0** × **u** = **0**
(*f*)   **u** × **u** = **0**

We shall prove (*a*) and leave the remaining proofs as exercises.

Proof (a).   First note that interchanging the rows of a 2 × 2 determinant changes the sign of the determinant, since

$$\begin{vmatrix} c & d \\ a & b \end{vmatrix} = bc - ad = -(ad - bc) = -\begin{vmatrix} a & b \\ c & d \end{vmatrix}$$

It follows that interchanging **u** and **v** in (1) interchanges the rows of the three determinants on the right side of (1) and thereby changes the sign of each component in the cross product. Thus, **u** × **v** = −(**v** × **u**).   ∎

Cross products of the unit vectors **i**, **j**, and **k** are of special interest. We obtain, for example,

$$\mathbf{i} \times \mathbf{j} = \begin{vmatrix} \mathbf{i} & \mathbf{j} & \mathbf{k} \\ 1 & 0 & 0 \\ 0 & 1 & 0 \end{vmatrix} = \begin{vmatrix} 0 & 0 \\ 1 & 0 \end{vmatrix}\mathbf{i} - \begin{vmatrix} 1 & 0 \\ 0 & 0 \end{vmatrix}\mathbf{j} + \begin{vmatrix} 1 & 0 \\ 0 & 1 \end{vmatrix}\mathbf{k} = \mathbf{k}$$

The reader should have no trouble obtaining the following list of cross products:

$$\begin{array}{lll} \mathbf{i} \times \mathbf{j} = \mathbf{k} & \mathbf{j} \times \mathbf{k} = \mathbf{i} & \mathbf{k} \times \mathbf{i} = \mathbf{j} \\ \mathbf{j} \times \mathbf{i} = -\mathbf{k} & \mathbf{k} \times \mathbf{j} = -\mathbf{i} & \mathbf{i} \times \mathbf{k} = -\mathbf{j} \\ \mathbf{i} \times \mathbf{i} = \mathbf{0} & \mathbf{j} \times \mathbf{j} = \mathbf{0} & \mathbf{k} \times \mathbf{k} = \mathbf{0} \end{array}$$

The following diagram is helpful for remembering these results.

In this diagram, the cross product of two consecutive vectors going clockwise is the next vector around, and the cross product of two consecutive vectors going counterclockwise is the negative of the next vector around.

WARNING.   It is *not* true in general that $\mathbf{u} \times (\mathbf{v} \times \mathbf{w}) = (\mathbf{u} \times \mathbf{v}) \times \mathbf{w}$. For example,

$$\mathbf{i} \times (\mathbf{j} \times \mathbf{j}) = \mathbf{i} \times \mathbf{0} = \mathbf{0}$$

and

$$(\mathbf{i} \times \mathbf{j}) \times \mathbf{j} = \mathbf{k} \times \mathbf{j} = -\mathbf{i}$$

so that

$$\mathbf{i} \times (\mathbf{j} \times \mathbf{j}) \neq (\mathbf{i} \times \mathbf{j}) \times \mathbf{j}$$

Figure 14.4.3

We know from Theorem 14.4.2 that $\mathbf{u} \times \mathbf{v}$ is orthogonal to both $\mathbf{u}$ and $\mathbf{v}$. It can be shown that if $\mathbf{u}$ and $\mathbf{v}$ are nonzero vectors, then the direction of $\mathbf{u} \times \mathbf{v}$ can be determined using the following "right-hand rule"* (Figure 14.4.3). Let θ be the angle between $\mathbf{u}$ and $\mathbf{v}$, and suppose $\mathbf{u}$ is rotated through the angle θ until it coincides with $\mathbf{v}$. If the fingers of the right hand are cupped so they point in the direction of rotation, then the thumb indicates (roughly) the direction of $\mathbf{u} \times \mathbf{v}$. The reader may find it instructive to practice this rule with the products

$$\mathbf{i} \times \mathbf{j} = \mathbf{k} \qquad \mathbf{j} \times \mathbf{k} = \mathbf{i} \qquad \mathbf{k} \times \mathbf{i} = \mathbf{j}$$

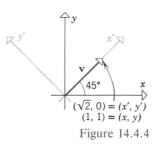

$(\sqrt{2}, 0) = (x', y')$
$(1, 1) = (x, y)$

Figure 14.4.4

Initially, we defined a vector to be a directed line segment or arrow in 2-space or 3-space; coordinate systems and components were introduced later in order to simplify computations with vectors. Thus, a vector has a "mathematical existence" regardless of whether a coordinate system has been introduced. Furthermore, the components of a vector are not determined by the vector alone; they depend as well on the coordinate system chosen. For example, in Figure 14.4.4 we have indicated a fixed-plane vector $\mathbf{v}$ and two different coordinate systems. In the xy-coordinate system the components of $\mathbf{v}$ are $\langle 1, 1 \rangle$, and in the $x'y'$-system they are $\langle \sqrt{2}, 0 \rangle$.

This raises an important question about our definition of cross product. Since we defined the cross product $\mathbf{u} \times \mathbf{v}$ in terms of the components of $\mathbf{u}$ and $\mathbf{v}$, and since these components depend on the coordinate system chosen, it seems possible that two *fixed* vectors $\mathbf{u}$ and $\mathbf{v}$ might yield different vectors for $\mathbf{u} \times \mathbf{v}$ in different coordinate systems. Fortunately, this is not the case. To see this, we need only recall that:

(i) $\mathbf{u} \times \mathbf{v}$ is perpendicular to both $\mathbf{u}$ and $\mathbf{v}$.

(ii)   The orientation of $\mathbf{u} \times \mathbf{v}$ is determined by the right-hand rule.

(iii) $\|\mathbf{u} \times \mathbf{v}\| = \|\mathbf{u}\| \, \|\mathbf{v}\| \sin \theta$.

---

*Recall that we agreed to consider only right-handed coordinate systems in this text. Had we used left-handed systems instead, a "left-hand rule" would apply here.

These three properties completely determine the vector $\mathbf{u} \times \mathbf{v}$. Properties (i) and (ii) determine the direction, and property (iii) determines the length. Since these properties depend only on the lengths and relative positions of $\mathbf{u}$ and $\mathbf{v}$ and not on the particular right-handed coordinate system being used, the vector $\mathbf{u} \times \mathbf{v}$ will remain unchanged if a different right-handed coordinate system is introduced. Thus, we say that the definition of $\mathbf{u} \times \mathbf{v}$ is *coordinate free*. This result is of importance to physicists and engineers who often work with many coordinate systems in the same problem.

TRIPLE SCALAR
PRODUCTS

If $\mathbf{a} = \langle a_1, a_2, a_3 \rangle$, $\mathbf{b} = \langle b_1, b_2, b_3 \rangle$, and $\mathbf{c} = \langle c_1, c_2, c_3 \rangle$ are vectors in 3-space, then the number

$$\mathbf{a} \cdot (\mathbf{b} \times \mathbf{c})$$

is called the *triple scalar product* of $\mathbf{a}$, $\mathbf{b}$, and $\mathbf{c}$. The triple scalar product may be conveniently calculated from the formula

$$\mathbf{a} \cdot (\mathbf{b} \times \mathbf{c}) = \begin{vmatrix} a_1 & a_2 & a_3 \\ b_1 & b_2 & b_3 \\ c_1 & c_2 & c_3 \end{vmatrix} \tag{8}$$

The validity of this formula may be seen by writing

$$\mathbf{a} \cdot (\mathbf{b} \times \mathbf{c}) = \mathbf{a} \cdot \left(\begin{vmatrix} b_2 & b_3 \\ c_2 & c_3 \end{vmatrix} \mathbf{i} - \begin{vmatrix} b_1 & b_3 \\ c_1 & c_3 \end{vmatrix} \mathbf{j} + \begin{vmatrix} b_1 & b_2 \\ c_1 & c_2 \end{vmatrix} \mathbf{k} \right)$$

$$= \begin{vmatrix} b_2 & b_3 \\ c_2 & c_3 \end{vmatrix} a_1 - \begin{vmatrix} b_1 & b_3 \\ c_1 & c_3 \end{vmatrix} a_2 + \begin{vmatrix} b_1 & b_2 \\ c_1 & c_2 \end{vmatrix} a_3$$

$$= \begin{vmatrix} a_1 & a_2 & a_3 \\ b_1 & b_2 & b_3 \\ c_1 & c_2 & c_3 \end{vmatrix}$$

Example 4   Calculate the triple scalar product $\mathbf{a} \cdot (\mathbf{b} \times \mathbf{c})$ of the vectors

$$\mathbf{a} = 3\mathbf{i} - 2\mathbf{j} - 5\mathbf{k}, \quad \mathbf{b} = \mathbf{i} + 4\mathbf{j} - 4\mathbf{k}, \quad \mathbf{c} = 3\mathbf{j} + 2\mathbf{k}$$

Solution.

$$\mathbf{a} \cdot (\mathbf{b} \times \mathbf{c}) = \begin{vmatrix} 3 & -2 & -5 \\ 1 & 4 & -4 \\ 0 & 3 & 2 \end{vmatrix}$$

$$= 3 \begin{vmatrix} 4 & -4 \\ 3 & 2 \end{vmatrix} - (-2) \begin{vmatrix} 1 & -4 \\ 0 & 2 \end{vmatrix} + (-5) \begin{vmatrix} 1 & 4 \\ 0 & 3 \end{vmatrix}$$

$$= 60 + 4 - 15 = 49 \quad \blacktriangleleft$$

REMARK.   The symbol $(\mathbf{a} \cdot \mathbf{b}) \times \mathbf{c}$ makes no sense since we cannot form the cross product of a scalar and a vector. Thus, no ambiguity arises if we write $\mathbf{a} \cdot \mathbf{b} \times \mathbf{c}$ rather than $\mathbf{a} \cdot (\mathbf{b} \times \mathbf{c})$. However, for clarity we shall usually keep the parentheses.

From the following calculations, we see that interchanging the rows of a 2×2 determinant changes the sign of its value, and that converting the rows into columns has no effect on the value.

$$\begin{vmatrix} c & d \\ a & b \end{vmatrix} = bc - ad = -(ad - bc) = - \begin{vmatrix} a & b \\ c & d \end{vmatrix}$$

$$\begin{vmatrix} a & c \\ b & d \end{vmatrix} = ad - bc = \begin{vmatrix} a & b \\ c & d \end{vmatrix}$$

The following analogous result holds for 3×3 determinants. We omit the proof.

14.4.5   THEOREM.

(a)   Interchanging two rows of a 3×3 determinant changes the sign of its value.

$$(b) \quad \begin{vmatrix} a_1 & a_2 & a_3 \\ b_1 & b_2 & b_3 \\ c_1 & c_2 & c_3 \end{vmatrix} = \begin{vmatrix} a_1 & b_1 & c_1 \\ a_2 & b_2 & c_2 \\ a_3 & b_3 & c_3 \end{vmatrix}$$

Figure 14.4.5

It follows from part (a) of this theorem that

$$\mathbf{a} \cdot (\mathbf{b} \times \mathbf{c}) = \mathbf{c} \cdot (\mathbf{a} \times \mathbf{b}) = \mathbf{b} \cdot (\mathbf{c} \times \mathbf{a}) \tag{9}$$

since the 3×3 determinants that represent these products can be obtained from one another by *two* row interchanges. (Verify.) These relationships may be remembered by moving the vectors $\mathbf{a}$, $\mathbf{b}$, and $\mathbf{c}$ clockwise around the vertices of the triangle in Figure 14.4.5.

The triple scalar product $\mathbf{a} \cdot (\mathbf{b} \times \mathbf{c})$ has a useful geometric interpretation. If we assume, for the moment, that the vectors $\mathbf{a}$, $\mathbf{b}$, $\mathbf{c}$ do not all lie in the same plane, then the three vectors form adjacent sides of a parallelepiped (Figure 14.4.6) when they are positioned with a common initial point. If the parallelogram determined by $\mathbf{b}$ and $\mathbf{c}$ is regarded as the base of the parallelepiped, then the area of the base is $\|\mathbf{b} \times \mathbf{c}\|$, and the height h is the length of the orthogonal projection of $\mathbf{a}$ on $\mathbf{b} \times \mathbf{c}$ (Figure 14.4.6). Therefore, by (9) of Section 14.3

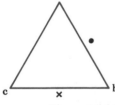

Figure 14.4.6

$$h = \|\text{proj}_{\mathbf{b} \times \mathbf{c}} \, \mathbf{a}\| = \frac{|\mathbf{a} \cdot (\mathbf{b} \times \mathbf{c})|}{\|\mathbf{b} \times \mathbf{c}\|}$$

It follows that the volume V of the parallelepiped is

$$V = (\text{area of base}) \cdot \text{height} = \|\mathbf{b} \times \mathbf{c}\| \frac{|\mathbf{a} \cdot (\mathbf{b} \times \mathbf{c})|}{\|\mathbf{b} \times \mathbf{c}\|}$$

or, on simplifying,

$$V = \begin{bmatrix} \text{volume of parallelepiped with} \\ \text{adjacent sides } \mathbf{a}, \mathbf{b}, \text{ and } \mathbf{c} \end{bmatrix} = |\mathbf{a} \cdot (\mathbf{b} \times \mathbf{c})| \qquad (10)$$

REMARK.   It follows from this formula that

$$\mathbf{a} \cdot (\mathbf{b} \times \mathbf{c}) = \pm V$$

where the $+$ or $-$ results depending on whether $\mathbf{a}$ makes an acute or obtuse angle with $\mathbf{b} \times \mathbf{c}$ (Theorem 14.3.2).

If the vectors $\mathbf{a}$, $\mathbf{b}$, and $\mathbf{c}$ do not lie in the same plane, then they determine a parallelepiped of positive volume. Thus, from (10), $\mathbf{a} \cdot (\mathbf{b} \times \mathbf{c}) \neq 0$ if $\mathbf{a}$, $\mathbf{b}$, and $\mathbf{c}$ do not lie in the same plane. It follows, therefore, that if $\mathbf{a} \cdot (\mathbf{b} \times \mathbf{c}) = 0$, then $\mathbf{a}$, $\mathbf{b}$, and $\mathbf{c}$ lie in the same plane. Conversely, it can be shown that if $\mathbf{a}$, $\mathbf{b}$, and $\mathbf{c}$ lie in the same plane, then $\mathbf{a} \cdot (\mathbf{b} \times \mathbf{c}) = 0$. In summary, we have the following result.

14.4.6   THEOREM. *If the vectors* $\mathbf{a} = \langle a_1, a_2, a_3 \rangle$, $\mathbf{b} = \langle b_1, b_2, b_3 \rangle$, *and* $\mathbf{c} = \langle c_1, c_2, c_3 \rangle$ *have the same initial point, then they lie in the same plane if and only if*

$$\mathbf{a} \cdot (\mathbf{b} \times \mathbf{c}) = \begin{vmatrix} a_1 & a_2 & a_3 \\ b_1 & b_2 & b_3 \\ c_1 & c_2 & c_3 \end{vmatrix} = 0$$

▶ Exercise Set 14.4

In Exercises 1–4, find $\mathbf{a} \times \mathbf{b}$.

1. $\mathbf{a} = \langle 1, 2, -3 \rangle$, $\mathbf{b} = \langle -4, 1, 2 \rangle$.

2. $\mathbf{a} = 3\mathbf{i} + 2\mathbf{j} - \mathbf{k}$, $\mathbf{b} = -\mathbf{i} - 3\mathbf{j} + \mathbf{k}$.

3. $\mathbf{a} = \langle 0, 1, -2 \rangle$, $\mathbf{b} = \langle 3, 0, -4 \rangle$.

4. $\mathbf{a} = 4\mathbf{i} + \mathbf{k}$, $\mathbf{b} = 2\mathbf{i} - \mathbf{j}$.

5. Let $\mathbf{u} = \langle 2, -1, 3 \rangle$, $\mathbf{v} = \langle 0, 1, 7 \rangle$, and $\mathbf{w} = \langle 1, 4, 5 \rangle$. Find:
(a) $\mathbf{u} \times (\mathbf{v} \times \mathbf{w})$ (b) $(\mathbf{u} \times \mathbf{v}) \times \mathbf{w}$
(c) $\mathbf{u} \times (\mathbf{v} - 2\mathbf{w})$ (d) $(\mathbf{u} \times \mathbf{v}) - 2\mathbf{w}$
(e) $(\mathbf{u} \times \mathbf{v}) \times (\mathbf{v} \times \mathbf{w})$ (f) $(\mathbf{v} \times \mathbf{w}) \times (\mathbf{u} \times \mathbf{v})$.

6. Find a vector orthogonal to both $\mathbf{u}$ and $\mathbf{v}$.
(a) $\mathbf{u} = -7\mathbf{i} + 3\mathbf{j} + \mathbf{k}$, $\mathbf{v} = 2\mathbf{i} + 4\mathbf{k}$
(b) $\mathbf{u} = \langle -1, -1, -1 \rangle$, $\mathbf{v} = \langle 2, 0, 2 \rangle$.

7. Verify Theorem 14.4.2 for the vectors $\mathbf{u} = \mathbf{i} - 5\mathbf{j} + 6\mathbf{k}$ and $\mathbf{v} = 2\mathbf{i} + \mathbf{j} + 2\mathbf{k}$.

8. Verify Theorem 14.4.4 for $\mathbf{u} = \langle 2, 0, -1 \rangle$, $\mathbf{v} = \langle 6, 7, 4 \rangle$, $\mathbf{w} = \langle 1, 1, 1 \rangle$, and $k = -3$.

9. Find all unit vectors parallel to the yz-plane that are perpendicular to the vector $3\mathbf{i} - \mathbf{j} + 2\mathbf{k}$.

10. Prove: If θ is the angle between $\mathbf{u}$ and $\mathbf{v}$ and $\mathbf{u} \cdot \mathbf{v} \neq 0$, then $\tan \theta = \|\mathbf{u} \times \mathbf{v}\|/(\mathbf{u} \cdot \mathbf{v})$.

11. Simplify $(\mathbf{u} + \mathbf{v}) \times (\mathbf{u} - \mathbf{v})$.

12. Find the area of the parallelogram determined by the vectors $\mathbf{u}$ and $\mathbf{v}$:
(a) $\mathbf{u} = \mathbf{i} - \mathbf{j} + 2\mathbf{k}, \mathbf{v} = 3\mathbf{j} + \mathbf{k}$
(b) $\mathbf{u} = 2\mathbf{i} + 3\mathbf{j}, \mathbf{v} = -\mathbf{i} + 2\mathbf{j} - 2\mathbf{k}$.

13. Find the area of the triangle having vertices P, Q, and R.
(a) $P(1, 5, -2), Q(0, 0, 0), R(3, 5, 1)$
(b) $P(2, 0, -3), Q(1, 4, 5), R(7, 2, 9)$.

14. Use the cross product to find the sine of the angle between the vectors $\mathbf{a} = 2\mathbf{i} + 3\mathbf{j} - 6\mathbf{k}$ and $\mathbf{b} = 2\mathbf{i} + 3\mathbf{j} + 6\mathbf{k}$.

15. (a) Find the area of the triangle having vertices $A(1, 0, 1), B(0, 2, 3)$, and $C(2, 1, 0)$.
(b) Use the result of part (a) to find the length of the altitude from vertex C to side AB.

16. Show that if $\mathbf{u}$ is a vector from any point on a line to a point P not on the line, and $\mathbf{v}$ is a vector parallel to the line, then the distance between P and the line is given by $\|\mathbf{u} \times \mathbf{v}\|/\|\mathbf{v}\|$.

17. Use the result of Exercise 16 to find the distance between the point P and the line through the points A and B:
(a) $P(-3, 1, 2), A(1, 1, 0), B(-2, 3, -4)$
(b) $P(4, 3, 0), A(2, 1, -3), B(0, 2, -1)$.

18. Find all unit vectors in the plane of $\mathbf{u} = 3\mathbf{i} + \mathbf{k}$ and $\mathbf{v} = \mathbf{i} - \mathbf{j} - \mathbf{k}$ that are perpendicular to the vector $\mathbf{w} = \mathbf{i} + 2\mathbf{j}$.

19. What is wrong with the expression $\mathbf{u} \times \mathbf{v} \times \mathbf{w}$?

In Exercises 20–23, find $\mathbf{a} \cdot (\mathbf{b} \times \mathbf{c})$.

20. $\mathbf{a} = \langle 1, -2, 2 \rangle, \mathbf{b} = \langle 0, 3, 2 \rangle, \mathbf{c} = \langle -4, 1, -3 \rangle$.

21. $\mathbf{a} = 2\mathbf{i} - 3\mathbf{j} + \mathbf{k}, \mathbf{b} = 4\mathbf{i} + \mathbf{j} - 3\mathbf{k}, \mathbf{c} = \mathbf{j} + 5\mathbf{k}$.

22. $\mathbf{a} = \langle 2, 1, 0 \rangle, \mathbf{b} = \langle 1, -3, 1 \rangle, \mathbf{c} = \langle 4, 0, 1 \rangle$.

23. $\mathbf{a} = \mathbf{i}, \mathbf{b} = \mathbf{i} + \mathbf{j}, \mathbf{c} = \mathbf{i} + \mathbf{j} + \mathbf{k}$.

24. Suppose $\mathbf{u} \cdot (\mathbf{v} \times \mathbf{w}) = 3$. Find
(a) $\mathbf{u} \cdot (\mathbf{w} \times \mathbf{v})$
(b) $(\mathbf{v} \times \mathbf{w}) \cdot \mathbf{u}$
(c) $\mathbf{w} \cdot (\mathbf{u} \times \mathbf{v})$
(d) $\mathbf{v} \cdot (\mathbf{u} \times \mathbf{w})$
(e) $(\mathbf{u} \times \mathbf{w}) \cdot \mathbf{v}$
(f) $\mathbf{v} \cdot (\mathbf{w} \times \mathbf{w})$.

25. Find the volume of the parallelepiped with sides $\mathbf{a}$, $\mathbf{b}$, and $\mathbf{c}$.
(a) $\mathbf{a} = \langle 2, -6, 2 \rangle, \mathbf{b} = \langle 0, 4, -2 \rangle,$ $\mathbf{c} = \langle 2, 2, -4 \rangle$
(b) $\mathbf{a} = 3\mathbf{i} + \mathbf{j} + 2\mathbf{k}, \mathbf{b} = 4\mathbf{i} + 5\mathbf{j} + \mathbf{k},$ $\mathbf{c} = \mathbf{i} + 2\mathbf{j} + 4\mathbf{k}$.

26. Determine whether $\mathbf{u}$, $\mathbf{v}$, and $\mathbf{w}$ lie in the same plane.
(a) $\mathbf{u} = \langle 1, -2, 1 \rangle, \mathbf{v} = \langle 3, 0, -2 \rangle,$ $\mathbf{w} = \langle 5, -4, 0 \rangle$
(b) $\mathbf{u} = 5\mathbf{i} - 2\mathbf{j} + \mathbf{k}, \mathbf{v} = 4\mathbf{i} - \mathbf{j} + \mathbf{k},$ $\mathbf{w} = \mathbf{i} - \mathbf{j}$
(c) $\mathbf{u} = \langle 4, -8, 1 \rangle, \mathbf{v} = \langle 2, 1, -2 \rangle,$ $\mathbf{w} = \langle 3, -4, 12 \rangle$.

27. Consider the parallelepiped with sides
$$\mathbf{a} = 3\mathbf{i} + 2\mathbf{j} + \mathbf{k}$$
$$\mathbf{b} = \mathbf{i} + \mathbf{j} + 2\mathbf{k}$$
$$\mathbf{c} = \mathbf{i} + 3\mathbf{j} + 3\mathbf{k}$$
(a) Find the volume.
(b) Find the area of the face determined by $\mathbf{a}$ and $\mathbf{c}$.
(c) Find the angle between $\mathbf{a}$ and the plane containing the face determined by $\mathbf{b}$ and $\mathbf{c}$.

28. Find a vector $\mathbf{n}$ perpendicular to the plane determined by the points $A(0, -2, 1), B(1, -1, -2)$, and $C(-1, 1, 0)$.

29. Prove:
(a) $(\mathbf{u} + k\mathbf{v}) \times \mathbf{v} = \mathbf{u} \times \mathbf{v}$
(b) $\mathbf{u} \cdot \mathbf{v} \times \mathbf{z} = \mathbf{u} \times \mathbf{v} \cdot \mathbf{z}$.

30. Let $\mathbf{u}$, $\mathbf{v}$, and $\mathbf{w}$ be vectors in 3-space with the same initial point and such that no two are collinear. Prove:
(a) $\mathbf{u} \times (\mathbf{v} \times \mathbf{w})$ lies in the plane determined by $\mathbf{v}$ and $\mathbf{w}$;
(b) $(\mathbf{u} \times \mathbf{v}) \times \mathbf{w}$ lies in the plane determined by $\mathbf{u}$ and $\mathbf{v}$.

31. Prove parts (b) and (c) of Theorem 14.4.4.

32. Prove parts (d), (e), and (f) of Theorem 14.4.4.

33. Prove that $\mathbf{x} \times (\mathbf{y} \times \mathbf{z}) = (\mathbf{x} \cdot \mathbf{z})\mathbf{y} - (\mathbf{x} \cdot \mathbf{y})\mathbf{z}$.
[*Hint:* First prove the result for the case $\mathbf{z} = \mathbf{i}$, then for $\mathbf{z} = \mathbf{j}$, and then for $\mathbf{z} = \mathbf{k}$. Finally, prove it for an arbitrary vector $\mathbf{z} = z_1\mathbf{i} + z_2\mathbf{j} + z_3\mathbf{k}$.]

34. Let $\mathbf{a} = \mathbf{i} + 3\mathbf{j} - \mathbf{k}, \quad \mathbf{b} = \mathbf{i} + \mathbf{j} + 2\mathbf{k}, \quad$ and

$\mathbf{c} = 3\mathbf{i} - \mathbf{j} + 2\mathbf{k}$. Calculate $\mathbf{a} \times (\mathbf{b} \times \mathbf{c})$ using Exercise 33 and then check your result by calculating directly.

35. Prove: If $\mathbf{a}$, $\mathbf{b}$, $\mathbf{c}$, and $\mathbf{d}$ lie in the same plane, then

$$(\mathbf{a} \times \mathbf{b}) \times (\mathbf{c} \times \mathbf{d}) = \mathbf{0}.$$

36. It is a theorem of solid geometry that the volume of a tetrahedron is $\frac{1}{3}$(area of base) · (height). Use this result to prove that the volume of a tetra-

hedron whose sides are the vectors $\mathbf{a}$, $\mathbf{b}$, and $\mathbf{c}$ is $\frac{1}{6}|\mathbf{a} \cdot (\mathbf{b} \times \mathbf{c})|$.

37. Use the result of Exercise 36 to find the volume of the tetrahedron with vertices P, Q, R, and S.
   (a) $P(-1, 2, 0)$, $Q(2, 1, -3)$, $R(1, 0, 1)$, $S(3, -2, 3)$.
   (b) $P(0, 0, 0)$, $Q(1, 2, -1)$, $R(3, 4, 0)$, $S(-1, -3, 4)$.

14.5   PARAMETRIC EQUATIONS OF LINES

In this section we shall discuss parametric equations of lines in 2-space and 3-space. In 3-space, parametric equations of lines are especially important because they are generally the most convenient form for representing such lines algebraically.

LINES DETERMINED BY A POINT AND A VECTOR

A line in 2-space or 3-space can be determined uniquely by specifying a point on the line and a nonzero vector parallel to the line (Figure 14.5.1).

The following theorem gives parametric equations of the line through a point P_0 and parallel to a nonzero vector $\mathbf{v}$.

Parametric Equations for Lines in 2-space and 3-space

14.5.1   THEOREM.

(a)   *The line in 2-space that passes through the point $P_0(x_0, y_0)$ and is parallel to the nonzero vector $\mathbf{v} = \langle a, b \rangle = a\mathbf{i} + b\mathbf{j}$ has parametric equations*

$$\begin{aligned} x &= x_0 + at \\ y &= y_0 + bt \end{aligned} \tag{1a}$$

(b)   *The line in 3-space that passes through the point $P_0(x_0, y_0, z_0)$ and is parallel to the nonzero vector $\mathbf{v} = \langle a, b, c \rangle = a\mathbf{i} + b\mathbf{j} + c\mathbf{k}$ has parametric equations*

$$\begin{aligned} x &= x_0 + at \\ y &= y_0 + bt \\ z &= z_0 + ct \end{aligned} \tag{1b}$$

Proof.   We shall prove part (b). The proof of (a) is similar.

If L is the line in 3-space passing through the point $P_0(x_0, y_0, z_0)$ and

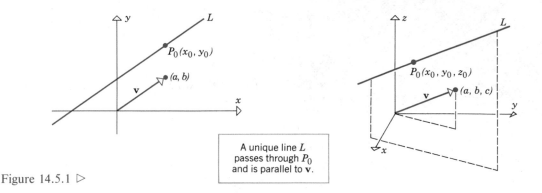

A unique line L
passes through P_0
and is parallel to $\mathbf{v}$.

Figure 14.5.1 ▷

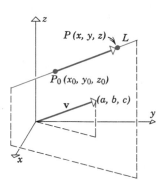

Figure 14.5.2

parallel to the nonzero vector $\mathbf{v} = \langle a, b, c \rangle$, it is clear (Figure 14.5.2) that L consists precisely of those points $P(x, y, z)$ for which the vector $\overrightarrow{P_0 P}$ is parallel to $\mathbf{v}$. In other words, the point $P(x, y, z)$ is on L if and only if $\overrightarrow{P_0 P}$ is a scalar multiple of $\mathbf{v}$, say

$$\overrightarrow{P_0 P} = t\mathbf{v}$$

This equation can be written as

$$\langle x - x_0, y - y_0, z - z_0 \rangle = \langle ta, tb, tc \rangle$$

Equating components yields

$$x - x_0 = ta, \quad y - y_0 = tb, \quad z - z_0 = tc$$

from which (1b) follows.   ∎

REMARK.   Although it is not stated explicitly, it is understood in (1a) and (1b) that $-\infty < t < +\infty$, which reflects the fact that lines extend indefinitely.

Example 1   Find parametric equations of the line

(a)   passing through $(4, 2)$ and parallel to $\mathbf{v} = \langle -1, 5 \rangle$;
(b)   passing through $(1, 2, -3)$ and parallel to $\mathbf{v} = 4\mathbf{i} + 5\mathbf{j} - 7\mathbf{k}$;
(c)   passing through the origin in 3-space and parallel to $\mathbf{v} = \langle 1, 1, 1 \rangle$.

Solution (a).   From (1a) with $x_0 = 4$, $y_0 = 2$, $a = -1$, and $b = 5$ we obtain

$$x = 4 - t$$
$$y = 2 + 5t$$

Solution (b).   From (1b) we obtain

$$x = 1 + 4t$$
$$y = 2 + 5t$$
$$z = -3 - 7t$$

Solution (c).   From (1b) with $x_0 = 0$, $y_0 = 0$, $z_0 = 0$, $a = 1$, $b = 1$, and $c = 1$ we obtain

$$x = t$$
$$y = t$$
$$z = t \quad \blacktriangleleft$$

Example 2

(a)   Find parametric equations of the line L passing through the points $P_1(2, 4, -1)$ and $P_2(5, 0, 7)$.

(b)   Where does the line intersect the xy-plane?

Solution (a).   Since the vector $\overrightarrow{P_1P_2} = \langle 3, -4, 8 \rangle$ is parallel to L and $P_1(2, 4, -1)$ lies on L, the line L is given by

$$x = 2 + 3t$$
$$y = 4 - 4t$$
$$z = -1 + 8t$$

Solution (b).   The line intersects the xy-plane at the point where $z = -1 + 8t = 0$; that is, when $t = \frac{1}{8}$. Substituting this value of t in the parametric equations for L yields the point of intersection

$$(x, y, z) = \left(\frac{19}{8}, \frac{7}{2}, 0 \right) \quad \blacktriangleleft$$

Example 3   Let L_1 and L_2 be the lines

$$
\begin{array}{ll}
 & x = 1 + 4t \qquad\qquad x = 2 + 8t \\
L_1: & y = 5 - 4t \qquad L_2: \; y = 4 - 3t \\
 & z = -1 + 5t \qquad\qquad z = 5 + t
\end{array}
$$

(a)   Are the lines parallel?

(b)   Do the lines intersect?

Solution (a).   The line L_1 is parallel to the vector $4\mathbf{i} - 4\mathbf{j} + 5\mathbf{k}$, and the line L_2 is parallel to the vector $8\mathbf{i} - 3\mathbf{j} + \mathbf{k}$. These vectors are not parallel since neither is a scalar multiple of the other. Thus, the lines are not parallel.

Solution (b).   In order for the lines to intersect at some point (x_0, y_0, z_0)

these coordinates would have to satisfy the equations of both L_1 and L_2. In other words, there would have to exist values t_1 and t_2 for the parameters such that

$$
\begin{aligned}
x_0 &= 1 + 4t_1 & x_0 &= 2 + 8t_2 \\
y_0 &= 5 - 4t_1 \quad \text{and} \quad & y_0 &= 4 - 3t_2 \\
z_0 &= -1 + 5t_1 & z_0 &= 5 + t_2
\end{aligned}
$$

This leads to three conditions on t_1 and t_2,

$$
\begin{aligned}
1 + 4t_1 &= 2 + 8t_2 \\
5 - 4t_1 &= 4 - 3t_2 \\
-1 + 5t_1 &= 5 + t_2
\end{aligned}
\tag{2}
$$

We shall try to solve these equations for t_1 and t_2. If we obtain a solution, then the lines intersect; if we find that there is no solution, then the lines do not intersect, since the three conditions cannot be satisfied.

The first two equations in (2) may be solved by adding them together to obtain

$$ 6 = 6 + 5t_2 $$

or $t_2 = 0$. Substituting $t_2 = 0$ in the first equation yields

$$ 1 + 4t_1 = 2 $$

or $t_1 = \frac{1}{4}$. However, the values $t_1 = \frac{1}{4}$, $t_2 = 0$ do not satisfy the third equation in (2), so that there is no simultaneous solution to the three equations. Thus, the lines do not intersect. ◀

Two lines in 3-space that are not parallel and do not intersect (such as those in Example 3) are called *skew* lines. As illustrated in Figure 14.5.3, any two skew lines lie in parallel planes.

Parallel planes containing skew lines L_1 and L_2 may be determined by translating each line until it intersects the other.

Figure 14.5.3

LINE SEGMENTS

Sometimes one is not interested in an entire line, but rather some *segment* of a line. Parametric equations of a line segment can be obtained by finding parametric equations for the entire line, then restricting the parameter appropriately so that only the desired segment is generated as the parameter varies.

Example 4 Find parametric equations for the line segment joining the points $P_1(2, 4, -1)$ and $P_2(5, 0, 7)$.

Solution. From Example 2, the line through P_1 and P_2 has parametric equations $x = 2 + 3t$, $y = 4 - 4t$, $z = -1 + 8t$. With these equations, the point P_1 corresponds to $t = 0$ and P_2 to $t = 1$. Thus, the line segment from P_1 to P_2 is given by

$$x = 2 + 3t$$
$$y = 4 - 4t, \qquad 0 \le t \le 1$$
$$z = -1 + 8t \qquad \blacktriangleleft$$

VECTOR EQUATIONS OF LINES

We shall now show how vector notation can be used to express the parametric equations of a line in a more compact form. Because two vectors are equal if and only if their components are equal, (1a) and (1b) can be written as

$$\langle x, y \rangle = \langle x_0 + at, y_0 + bt \rangle$$

and

$$\langle x, y, z \rangle = \langle x_0 + at, y_0 + bt, z_0 + ct \rangle$$

or, equivalently, as

$$\langle x, y \rangle = \langle x_0, y_0 \rangle + t \langle a, b \rangle \qquad (3a)$$

and

$$\langle x, y, z \rangle = \langle x_0, y_0, z_0 \rangle + t \langle a, b, c \rangle \qquad (3b)$$

In 2-space let us introduce the vectors $\mathbf{r}$, $\mathbf{r}_0$, and $\mathbf{v}$ given by

$$\mathbf{r} = \langle x, y \rangle, \quad \mathbf{r}_0 = \langle x_0, y_0 \rangle, \quad \mathbf{v} = \langle a, b \rangle \qquad (4a)$$

and in 3-space

$$\mathbf{r} = \langle x, y, z \rangle, \quad \mathbf{r}_0 = \langle x_0, y_0, z_0 \rangle, \quad \mathbf{v} = \langle a, b, c \rangle \qquad (4b)$$

Substituting (4a) and (4b) in (3a) and (3b), respectively, yields the equation

$$\mathbf{r} = \mathbf{r}_0 + t\mathbf{v} \qquad (5)$$

in both cases. We call this the *vector equation of a line* in 2-space or 3-space. In this equation $\mathbf{v}$ is a nonzero vector parallel to the line and $\mathbf{r}_0$ is a vector whose components are the coordinates of a point on the line.

Example 5   The equation

$$\langle x, y, z \rangle = \langle -1, 0, 2 \rangle + t \langle 1, 5, -4 \rangle$$

is of form (5) with

$$\mathbf{r}_0 = \langle -1, 0, 2 \rangle \quad \text{and} \quad \mathbf{v} = \langle 1, 5, -4 \rangle$$

$$r = r_0 + tv$$

Figure 14.5.4

Thus, the equation represents the line in 3-space that passes through the point $(-1, 0, 2)$ and is parallel to the vector $\langle 1, 5, -4 \rangle$.   ◀

Figure 14.5.4 illustrates how Equation (5) can be interpreted geometrically: r_0 can be viewed as a vector from the origin to a point P_0 on L, tv is a scalar multiple of a nonzero vector v that is parallel to L, and $r = r_0 + tv$ can be interpreted as a vector from the origin to a point on L. As the parameter t varies from $-\infty$ to $+\infty$, the terminal point of r traces out the line L.

Example 6   Find a vector equation of the line in 3-space that passes through the points $P_1(2, 4, -1)$ and $P_2(5, 0, 7)$.

Solution.   The vector

$$\overrightarrow{P_1 P_2} = \langle 3, -4, 8 \rangle$$

is parallel to the line and so can be used as v in (5). For r_0 we can use either the vector from the origin to P_1 or the vector from the origin to P_2. Using the former yields

$$r_0 = \langle 2, 4, -1 \rangle$$

Thus, a vector equation of the line through P_1 and P_2 is

$$\langle x, y, z \rangle = \langle 2, 4, -1 \rangle + t \langle 3, -4, 8 \rangle$$

(Note that equating corresponding components on the two sides of this vector equation yields the parametric equations for the line obtained in Example 2.)   ◀

Line segments in 2-space or 3-space can be represented by vector equations with restrictions on the parameter.

Example 7   The vector equation

$$\langle x, y \rangle = \langle -3, 2 \rangle + t \langle 1, 1 \rangle, \quad 0 \le t \le 3$$

represents a segment of the line in 2-space that passes through the point $(-3, 2)$ and is parallel to the vector $\langle 1, 1 \rangle$. Since

$$\langle x, y \rangle = \langle -3, 2 \rangle \quad \text{if } t = 0$$

and

$$\langle x, y \rangle = \langle -3, 2 \rangle + \langle 3, 3 \rangle = \langle 0, 5 \rangle \quad \text{if } t = 3$$

the equation represents the portion of the line between the points $(-3, 2)$ and $(0, 5)$ (Figure 14.5.5).   ◀

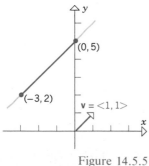

Figure 14.5.5

▶ Exercise Set 14.5

In Exercises 1–8, find parametric equations for the line through P_1 and P_2.

1. $P_1(3, -2)$, $P_2(5, 1)$.   2. $P_1(0, 1)$, $P_2(-3, -4)$.

3. $P_1(4, 1)$, $P_2(4, 3)$.   4. $P_1(5, 2)$, $P_2(3, 7)$.

5. $P_1(5, -2, 1)$, $P_2(2, 4, 2)$.

6. $P_1(-1, 3, 5)$, $P_2(-1, 3, 2)$.

7. $P_1(0, 0, 0)$, $P_2(-1, 6, 1)$.

8. $P_1(4, 0, 7)$, $P_2(-1, -1, 2)$.

In Exercises 9–16, find parametric equations for the line segment joining P_1 and P_2.

9. $P_1(3, -2)$, $P_2(5, 1)$.   10. $P_1(0, 1)$, $P_2(-3, -4)$.

11. $P_1(4, 1)$, $P_2(4, 3)$.   12. $P_1(5, 2)$, $P_2(3, 7)$.

13. $P_1(5, -2, 1)$, $P_2(2, 4, 2)$.

14. $P_1(-1, 3, 5)$, $P_2(-1, 3, 2)$.

15. $P_1(0, 0, 0)$, $P_2(-1, 6, 1)$.

16. $P_1(4, 0, 7)$, $P_2(-1, -1, 2)$.

In Exercises 17–25, find parametric equations for the line.

17. The line through $(-5, 2)$ and parallel to $2\mathbf{i} - 3\mathbf{j}$.

18. The line through $(0, 3)$ and parallel to the line $x = -5 + t$, $y = 1 - 2t$.

19. The line tangent to the circle $x^2 + y^2 = 25$ at the point $(3, -4)$.

20. The line tangent to the parabola $y = x^2$ at the point $(-2, 4)$.

21. The line through $(-1, 2, 4)$ and parallel to $3\mathbf{i} - 4\mathbf{j} + \mathbf{k}$.

22. The line through $(2, -1, 5)$ and parallel to $\langle -1, 2, 7 \rangle$.

23. The line through $(-2, 0, 5)$ and parallel to the line $x = 1 + 2t$, $y = 4 - t$, $z = 6 + 2t$.

24. The line through the origin and parallel to the line $x = t$, $y = -1 + t$, $z = 2$.

25. The line through $(3, 7, 0)$ and parallel to the x-axis.

26. Where does the line $x = 1 + 3t$, $y = 2 - t$ intersect
    (a) the x-axis   (b) the y-axis?

27. Where does the line $x = 2t$, $y = 3 + 4t$ intersect the parabola $y = x^2$?

28. Where does the line $x = -1 + 2t$, $y = 3 + t$, $z = 4 - t$ intersect
    (a) the xy-plane   (b) the xz-plane
    (c) the yz-plane?

29. Where does the line $x = -2$, $y = 4 + 2t$, $z = -3 + t$ intersect
    (a) the xy-plane   (b) the xz-plane
    (c) the yz-plane?

30. Where does the line $x = 2 - t$, $y = 3t$, $z = -1 + 2t$ intersect the plane $2y + 3z = 6$?

31. Where does the line $x = 1 + t$, $y = 3 - t$, $z = 2t$ intersect the cylinder $x^2 + y^2 = 16$?

32. Find parametric equations for the line through (x_0, y_0, z_0) and (x_1, y_1, z_1).

33. Find parametric equations for the line through (x_1, y_1, z_1) and parallel to the line $x = x_0 + at$, $y = y_0 + bt$, $z = z_0 + ct$.

34. Prove: If a, b, and c are nonzero, then each point on the line $x = x_0 + at$, $y = y_0 + bt$, $z = z_0 + ct$ satisfies

$$\frac{x - x_0}{a} = \frac{y - y_0}{b} = \frac{z - z_0}{c}$$

and conversely, each point (x, y, z) satisfying these equations lies on the line. (These are called the *symmetric equations* of the line.)

35. Show that the lines

$$\begin{array}{ll} x = 2 + t & x = 2 + t \\ y = 2 + 3t \quad \text{and} & y = 3 + 4t \\ z = 3 + t & z = 4 + 2t \end{array}$$

intersect and find the point of intersection.

36. Show that the lines

$$\begin{array}{ll} x + 1 = 4t & x + 13 = 12t \\ y - 3 = t \quad \text{and} & y - 1 = 6t \\ z - 1 = 0 & z - 2 = 3t \end{array}$$

intersect and find the point of intersection.

37. Show that the lines

$$x = 1 + 7t$$
$$y = 3 + t \quad \text{and}$$
$$z = 5 - 3t$$

$$x = 4 - t$$
$$y = 6$$
$$z = 7 + 2t$$

are skew.

38. Show that the lines

$$x = 2 + 8t$$
$$y = 6 - 8t \quad \text{and}$$
$$z = 10t$$

$$x = 3 + 8t$$
$$y = 5 - 3t$$
$$z = 6 + t$$

are skew.

39. Determine whether P_1, P_2, and P_3 lie on the same line.
(a) $P_1(6, 9, 7)$, $P_2(9, 2, 0)$, $P_3(0, -5, -3)$
(b) $P_1(1, 0, 1)$, $P_2(3, -4, -3)$, $P_3(4, -6, -5)$.

40. Find k_1 and k_2 so that the point $(k_1, 1, k_2)$ lies on the line passing through $(0, 2, 3)$ and $(2, 7, 5)$.

41. Find the point on the line segment joining $P_1(1, 4, -3)$ and $P_2(1, 5, -1)$ that is $\frac{2}{3}$ of the way from P_1 to P_2.

42. In each part, determine whether the lines are parallel.

(a) $x = 3 - 2t$
$y = 4 + t \quad$ and
$z = 6 - t$

$x = 5 - 4t$
$y = -2 + 2t$
$z = 7 - 2t$

(b) $x = 5 + 3t$
$y = 4 - 2t \quad$ and
$z = -2 + 3t$

$x = -1 + 9t$
$y = 5 - 6t$
$z = 3 + 8t$.

43. Show that the equations

$$x = 3 - t$$
$$y = 1 + 2t \quad \text{and}$$

$$x = -1 + 3t$$
$$y = 9 - 6t$$

represent the same line.

44. Show that the equations

$$x = 1 + 3t$$
$$y = -2 + t \quad \text{and}$$
$$z = 2t$$

$$x = 4 - 6t$$
$$y = -1 - 2t$$
$$z = 2 - 4t$$

represent the same line.

45. Find a vector equation of the line in 2-space that passes through the points P_1 and P_2.
(a) $P_1(2, -1)$, $P_2(-5, 3)$
(b) $P_1(0, 3)$, $P_2(4, 3)$.

46. Find a vector equation of the line in 3-space that passes through the points P_1 and P_2.
(a) $P_1(3, -1, 2)$, $P_2(0, 1, 1)$
(b) $P_1(2, 4, 1)$, $P_2(2, -1, 1)$.

47. Describe the line segment represented by the vector equation $\langle x, y \rangle = \langle 1, 0 \rangle + t \langle -2, 3 \rangle$, $0 \le t \le 2$.

48. Describe the line segment represented by the vector equation $\langle x, y, z \rangle = \langle -2, 1, 4 \rangle + t \langle 3, 0, -1 \rangle$, $0 \le t \le 3$.

49. Find the distance between the point $(-2, 1, 1)$ and the line $x = 3 - t$, $y = t$, $z = 1 + 2t$. [*Hint:* See Exercise 16, Section 14.4.]

50. Find the distance between the point $(1, 4, -3)$ and the line $x = 2 + t$, $y = -1 - t$, $z = 3t$.

51. Find an equation of the line that passes through the point $(0, 2, 1)$ and intersects the line $x = 2t$, $y = 1 - t$, $z = 2 + t$ at a right angle.

PLANES IN 3-SPACE

In this section we shall use vectors to derive equations of planes in 3-space, and we shall use these equations to solve some basic geometric problems.

PLANES DETERMINED BY A POINT AND A NORMAL VECTOR

A plane in 3-space is uniquely determined by specifying a point in the plane and a vector perpendicular to the plane (Figure 14.6.1). A vector perpendicular to a plane is called a **normal** to the plane.

The shaded plane is uniquely determined by the point P and the vector $\mathbf{n}$ perpendicular to the plane.

Figure 14.6.1

Figure 14.6.2

Suppose we want the equation of the plane passing through the point $P_0(x_0, y_0, z_0)$ and perpendicular to the nonzero vector $\mathbf{n} = \langle a, b, c \rangle$. It is evident from Figure 14.6.2 that the plane consists precisely of those points $P(x, y, z)$ for which the vector $\overrightarrow{P_0P}$ is perpendicular to $\mathbf{n}$; or, phrased as an equation,

$$\mathbf{n} \cdot \overrightarrow{P_0P} = 0 \tag{1}$$

Since $P_0P = \langle x - x_0, y - y_0, z - z_0 \rangle$, (1) can be rewritten as

$$a(x - x_0) + b(y - y_0) + c(z - z_0) = 0 \tag{2}$$

We shall call this the **point-normal form** of the equation of a plane.

Example 1   Find an equation of the plane passing through the point $(3, -1, 7)$ and perpendicular to the vector $\mathbf{n} = \langle 4, 2, -5 \rangle$.

Solution.   From (2), a point-normal form of the equation is

$$4(x - 3) + 2(y + 1) - 5(z - 7) = 0 \quad \blacktriangleleft$$

By multiplying out and collecting terms, (2) can be rewritten in the form

$$ax + by + cz + d = 0 \tag{3}$$

where a, b, c, and d are constants, and a, b, and c are not all zero. To illustrate, the equation in Example 1 can be rewritten as

$$4x + 2y - 5z + 25 = 0$$

As our next theorem shows, every equation of form (3) represents a plane in 3-space.

> **14.6.1**   THEOREM. *If a, b, c, and d are constants and a, b, and c are not all zero, then the graph of the equation*
>
> $$ax + by + cz + d = 0$$
>
> *is a plane having the vector $\mathbf{n} = \langle a, b, c \rangle$ as a normal.*

Proof.   By hypothesis, a, b, and c are not all zero. Assume, for the moment, that $a \neq 0$. The equation $ax + by + cz + d = 0$ can be rewritten as $a[x + (d/a)] + by + cz = 0$. But this is a point-normal form of the plane passing through the point $(-d/a, 0, 0)$ and having $\mathbf{n} = \langle a, b, c \rangle$ as a normal.

If $a = 0$, then either $b \neq 0$ or $c \neq 0$. A straightforward modification of the above argument will handle these other cases.   ▮

The equation in Theorem 14.6.1 is called the *general form* of the equation of a plane.

Example 2 Determine whether the planes

$$3x - 4y + 5z = 0$$

and

$$-6x + 8y - 10z - 4 = 0$$

are parallel.

Solution. It is clear geometrically that two planes are parallel if and only if their normals are parallel vectors. A normal to the first plane is

$$\mathbf{n}_1 = \langle 3, -4, 5 \rangle$$

and a normal to the second plane is

$$\mathbf{n}_2 = \langle -6, 8, -10 \rangle$$

Since $\mathbf{n}_2$ is a scalar multiple of $\mathbf{n}_1$, the normals are parallel. Thus, the planes are also parallel. ◄

A plane cannot be specified by giving a point on it and *one* vector parallel to it, since there are infinitely many such planes (Figure 14.6.3*a*). However, as shown in Figure 14.6.3*b*, a plane is uniquely determined by giving a point in the plane and *two* nonparallel vectors that are parallel to the plane. A plane is also uniquely determined by specifying three noncollinear points in the plane (Figure 14.6.3*c*).

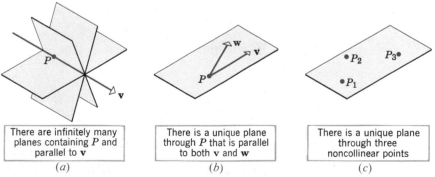

There are infinitely many planes containing P and parallel to $\mathbf{v}$	There is a unique plane through P that is parallel to both $\mathbf{v}$ and $\mathbf{w}$	There is a unique plane through three noncollinear points
(*a*)	(*b*)	(*c*)

Figure 14.6.3 ▷

Example 3 Find an equation of the plane through the points $P_1(1, 2, -1)$, $P_2(2, 3, 1)$, and $P_3(3, -1, 2)$.

Solution.   Since the points P_1, P_2, and P_3 lie in the plane, the vectors $\overrightarrow{P_1P_2} = \langle 1, 1, 2 \rangle$ and $\overrightarrow{P_1P_3} = \langle 2, -3, 3 \rangle$ are parallel to the plane. Therefore,

$$\overrightarrow{P_1P_2} \times \overrightarrow{P_1P_3} = \begin{vmatrix} \mathbf{i} & \mathbf{j} & \mathbf{k} \\ 1 & 1 & 2 \\ 2 & -3 & 3 \end{vmatrix} = 9\mathbf{i} + \mathbf{j} - 5\mathbf{k}$$

is normal to the plane, since it is perpendicular to both $\overrightarrow{P_1P_2}$ and $\overrightarrow{P_1P_3}$. By using this normal and the point $P_1(1, 2, -1)$ in the plane, we obtain the point-normal form

$$9(x - 1) + (y - 2) - 5(z + 1) = 0$$

which may be rewritten as

$$9x + y - 5z - 16 = 0 \quad \blacktriangleleft$$

Example 4   Determine whether the line

$$x = 3 + 8t$$
$$y = 4 + 5t$$
$$z = -3 - t$$

is parallel to the plane $x - 3y + 5z = 12$.

Solution.   The vector $\mathbf{v} = \langle 8, 5, -1 \rangle$ is parallel to the line and the vector $\mathbf{n} = \langle 1, -3, 5 \rangle$ is normal to the plane. In order for the line and plane to be parallel, the vectors $\mathbf{v}$ and $\mathbf{n}$ must be perpendicular. But this is not so, since the dot product

$$\mathbf{v} \cdot \mathbf{n} = (8)(1) + (5)(-3) + (-1)(5) = -12$$

is nonzero. Thus, the line and plane are not parallel. $\blacktriangleleft$

Example 5   Find the intersection of the line and plane in Example 4.

Solution.   If we let (x_0, y_0, z_0) be the point of intersection, then the coordinates of this point satisfy the equation of the plane and the parametric equations of the line. Thus,

$$x_0 - 3y_0 + 5z_0 = 12 \tag{4}$$

and for some value of t, say $t = t_0$,

$$x_0 = 3 + 8t_0$$
$$y_0 = 4 + 5t_0 \tag{5}$$
$$z_0 = -3 - t_0$$

(a)

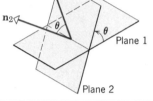

In this figure the angle between n_1 and n_2 is θ. If the direction of one of the normals is reversed, then the angle between the normals would be $180° - \theta$.

(b)

Figure 14.6.4

Substituting (5) in (4) yields

$$(3 + 8t_0) - 3(4 + 5t_0) + 5(-3 - t_0) = 12$$

Solving for t_0 yields

$$t_0 = -3$$

and on substituting this value in (5), we obtain

$$(x_0, y_0, z_0) = (-21, -11, 0) \qquad \blacktriangleleft$$

Two intersecting planes determine two angles of intersection, an acute angle θ ($0 \leq \theta \leq 90°$) and its supplement $180° - \theta$ (Figure 14.6.4a). If n_1 and n_2 are normals to the planes, then the angle between n_1 and n_2 is θ or $180° - \theta$, depending on the directions of the normals (Figure 14.6.4b). Thus, the angles of intersection of two planes may be determined from the normals.

Example 6   Find the acute angle of intersection of the two planes $2x - 4y + 4z = 7$ and $6x + 2y - 3z = 2$.

Solution.   From the given equations, we obtain the normals $n_1 = \langle 2, -4, 4 \rangle$ and $n_2 = \langle 6, 2, -3 \rangle$. Since the dot product $n_1 \cdot n_2 = -8$ is negative, the normals make an obtuse angle. To obtain normals making an acute angle, we can reverse the direction of either n_1 or n_2. To be specific, let us reverse the direction of n_1. Thus, the acute angle θ between the planes will be the angle between $-n_1 = \langle -2, 4, -4 \rangle$ and $n_2 = \langle 6, 2, -3 \rangle$. From (3) of Section 14.3 we obtain

$$\cos \theta = \frac{(-n_1) \cdot n_2}{\|-n_1\| \, \|n_2\|} = \frac{8}{\sqrt{36}\sqrt{49}} = \frac{4}{21}$$

With the aid of trigonometric tables or a calculator that can compute inverse trigonometric functions one obtains

$$\theta = \cos^{-1}\left(\frac{4}{21}\right) \approx 79° \qquad \blacktriangleleft$$

DISTANCE PROBLEMS INVOLVING PLANES

We conclude this section by discussing three basic "distance problems" in 3-space:

- Find the distance between a point and a plane.

- Find the distance between two parallel planes.

- Find the distance between two skew lines.

(a)

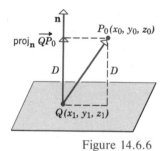

(b)

Figure 14.6.5

The three problems are related. If we can find the distance between a point and a plane, then we can find the distance between parallel planes by computing the distance between one of the planes and an arbitrary point P_0 in the other (Figure 14.6.5a). Moreover, we can find the distance between two skew lines by computing the distance between parallel planes containing them (Figure 14.6.5b).

14.6.2 THEOREM. *The distance D between a point $P_0(x_0, y_0, z_0)$ and the plane $ax + by + cz + d = 0$ is*

$$D = \frac{|ax_0 + by_0 + cz_0 + d|}{\sqrt{a^2 + b^2 + c^2}} \qquad (6)$$

Proof. Let $Q(x_1, y_1, z_1)$ be any point in the plane, and position the normal $\mathbf{n} = \langle a, b, c \rangle$ so that its initial point is at Q. As illustrated in Figure 14.6.6, the distance D is equal to the length of the orthogonal projection of QP_0 on $\mathbf{n}$. Thus, from (9) of Section 14.3,

$$D = \|\text{proj}_{\mathbf{n}} \overrightarrow{QP_0}\| = \frac{|\overrightarrow{QP_0} \cdot \mathbf{n}|}{\|\mathbf{n}\|}$$

But

$$\overrightarrow{QP_0} = \langle x_0 - x_1, y_0 - y_1, z_0 - z_1 \rangle$$
$$\overrightarrow{QP_0} \cdot \mathbf{n} = a(x_0 - x_1) + b(y_0 - y_1) + c(z_0 - z_1)$$
$$\|\mathbf{n}\| = \sqrt{a^2 + b^2 + c^2}$$

Figure 14.6.6

Thus,

$$D = \frac{|a(x_0 - x_1) + b(y_0 - y_1) + c(z_0 - z_1)|}{\sqrt{a^2 + b^2 + c^2}} \qquad (7)$$

Since the point $Q(x_1, y_1, z_1)$ lies in the plane, its coordinates satisfy the equation of the plane, so that

$$ax_1 + by_1 + cz_1 + d = 0$$

or

$$d = -ax_1 - by_1 - cz_1$$

Substituting this expression in (7) yields (6). ∎

REMARK. Note the similarity between (6) and the formula for the distance between a point and a line in 2-space [(12) of Section 14.3].

Example 7   Find the distance D between the point $(1, -4, -3)$ and the plane $2x - 3y + 6z = -1$.

Solution.   To apply (6), we first rewrite the equation of the plane in the form

$$2x - 3y + 6z + 1 = 0$$

Then

$$D = \frac{|(2)(1) + (-3)(-4) + 6(-3) + 1|}{\sqrt{2^2 + (-3)^2 + 6^2}} = \frac{|-3|}{7} = \frac{3}{7} \qquad \blacktriangleleft$$

Example 8   The planes

$$x + 2y - 2z = 3 \quad \text{and} \quad 2x + 4y - 4z = 7$$

are parallel since their normals, $\langle 1, 2, -2 \rangle$ and $\langle 2, 4, -4 \rangle$, are parallel vectors. Find the distance between these planes.

Solution.   To find the distance D between the planes, we may select an arbitrary point in one of the planes and compute its distance to the other plane. By setting $y = z = 0$ in the equation $x + 2y - 2z = 3$, we obtain the point $P_0(3, 0, 0)$ in this plane. From (6), the distance from P_0 to the plane $2x + 4y - 4z = 7$ is

$$D = \frac{|(2)(3) + 4(0) + (-4)(0) - 7|}{\sqrt{2^2 + 4^2 + (-4)^2}} = \frac{1}{6} \qquad \blacktriangleleft$$

Example 9   It was shown in Example 3 of Section 14.5 that the lines

$$L_1: \begin{array}{l} x = 1 + 4t \\ y = 5 - 4t \\ z = -1 + 5t \end{array} \qquad L_2: \begin{array}{l} x = 2 + 8t \\ y = 4 - 3t \\ z = 5 + t \end{array}$$

are skew. Find the distance between them.

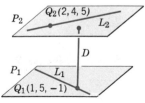

Figure 14.6.7

Solution.   Let P_1 and P_2 denote parallel planes containing L_1 and L_2, respectively (Figure 14.6.7). To find the distance D between L_1 and L_2, we shall calculate the distance from a point in P_1 to the plane P_2. Setting $t = 0$ in the equations of L_1 yields the point $Q_1(1, 5, -1)$ in plane P_1. (The value $t = 0$ was chosen for simplicity. Any value of t will suffice.) Next we shall obtain an equation for plane P_2.

The vector $\mathbf{u}_1 = \langle 4, -4, 5 \rangle$ is parallel to line L_1, and therefore also parallel to plane P_1. Similarly, $\mathbf{u}_2 = \langle 8, -3, 1 \rangle$ is parallel to L_2 and P_2. Therefore, the cross product

$$n = u_1 \times u_2 = \begin{vmatrix} i & j & k \\ 4 & -4 & 5 \\ 8 & -3 & 1 \end{vmatrix} = 11i + 36j + 20k$$

is normal to both P_1 and P_2. Using this normal and the point $Q_2(2, 4, 5)$ found by setting $t = 0$ in the equations of L_2, we obtain an equation for P_2:

$$11(x - 2) + 36(y - 4) + 20(z - 5) = 0$$

or

$$11x + 36y + 20z - 266 = 0$$

The distance between $Q_1(1, 5, -1)$ and this plane is

$$D = \frac{|(11)(1) + (36)(5) + (20)(-1) - 266|}{\sqrt{11^2 + 36^2 + 20^2}} = \frac{95}{\sqrt{1817}}$$

This is also the distance between L_1 and L_2.   ◀

▶ Exercise Set 14.6

In Exercises 1–4, find an equation of the plane passing through P and having n as a normal.

1. $P(2, 6, 1)$; $n = \langle 1, 4, 2 \rangle$.

2. $P(-1, -1, 2)$; $n = \langle -1, 7, 6 \rangle$.

3. $P(1, 0, 0)$; $n = \langle 0, 0, 1 \rangle$.

4. $P(0, 0, 0)$; $n = \langle 2, -3, -4 \rangle$.

5. Find an equation of the plane passing through the given points.
   (a) $(-2, 1, 1)$, $(0, 2, 3)$, and $(1, 0, -1)$
   (b) $(3, 2, 1)$, $(2, 1, -1)$, and $(-1, 3, 2)$.

6. Determine whether the planes are parallel.
   (a) $3x - 2y + z = 4$ and $6x - 4y + 3z = 7$
   (b) $2x - 8y - 6z - 2 = 0$ and
       $-x + 4y + 3z - 5 = 0$
   (c) $y = 4x - 2z + 3$ and $x = \frac{1}{4}y + \frac{1}{2}z$.

7. Determine whether the line and plane are parallel.
   (a) $x = 4 + 2t$, $y = -t$, $z = -1 - 4t$;
       $3x + 2y + z - 7 = 0$
   (b) $x = t$, $y = 2t$, $z = 3t$; $x - y + 2z = 5$.

8. Determine whether the planes are perpendicular.
   (a) $x - y + 3z - 2 = 0$, $2x + z = 1$
   (b) $3x - 2y + z = 1$, $4x + 5y - 2z = 4$.

9. Determine whether the line and plane are perpendicular.
   (a) $x = -1 + 2t$, $y = 4 + t$, $z = 1 - t$;
       $4x + 2y - 2z = 7$
   (b) $x = 3 - t$, $y = 2 + t$, $z = 1 - 3t$;
       $2x + 2y - 5 = 0$.

10. Find the point of intersection of the line and plane.
    (a) $x = t$, $y = t$, $z = t$; $3x - 2y + z - 5 = 0$
    (b) $x = 1 + t$, $y = -1 + 3t$, $z = 2 + 4t$;
        $x - y + 4z = 7$
    (c) $x = 2 - t$, $y = 3 + t$, $z = t$;
        $2x + y + z = 1$.

11. Find the acute angle of intersection of the planes (to the nearest degree).
    (a) $x = 0$ and $2x - y + z - 4 = 0$
    (b) $x + 2y - 2z = 5$ and $6x - 3y + 2z = 8$.

12. Find an equation of the plane through $(-1, 4, -3)$ and perpendicular to the line $x - 2 = t$, $y + 3 = 2t$, $z = -t$.

13. Find an equation of
    (a) the xy-plane   (b) the xz-plane
    (c) the yz-plane.

14. Find an equation of the plane that contains the point (x_0, y_0, z_0) and is
    (a) parallel to the xy-plane
    (b) parallel to the yz-plane
    (c) parallel to the xz-plane.

15. Find an equation of the plane through the origin that is parallel to the plane $4x - 2y + 7z + 12 = 0$.

16. Find an equation of the plane containing the line $x = -2 + 3t$, $y = 4 + 2t$, $z = 3 - t$ and perpendicular to the plane $x - 2y + z = 5$.

17. Find an equation of the plane through $(-1, 4, 2)$ and containing the line of intersection of the planes $4x - y + z - 2 = 0$ and $2x + y - 2z - 3 = 0$.

18. Show that the points $(1, 0, -1)$, $(0, 2, 3)$, $(-2, 1, 1)$, and $(4, 2, 3)$ lie in the same plane.

19. Find parametric equations of the line through $(5, 0, -2)$ that is parallel to the planes $x - 4y + 2z = 0$ and $2x + 3y - z + 1 = 0$.

20. Find an equation of the plane through $(-1, 2, -5)$ and perpendicular to the planes $2x - y + z = 1$ and $x + y - 2z = 3$.

21. Find an equation of the plane through $(1, 2, -1)$ and perpendicular to the line of intersection of the planes $2x + y + z = 2$ and $x + 2y + z = 3$.

22. Find a plane through the points $P_1(-2, 1, 4)$, $P_2(1, 0, 3)$ and perpendicular to the plane $4x - y + 3z = 2$.

23. Show that the lines

    $$x = -2 + t \qquad x = 3 - t$$
    $$y = 3 + 2t \quad \text{and} \quad y = 4 - 2t$$
    $$z = 4 - t \qquad z = t$$

    are parallel and find an equation of the plane they determine.

24. Find an equation of the plane that contains the point $(2, 0, 3)$ and the line $x = -1 + t$, $y = t$, $z = -4 + 2t$.

25. Find an equation of the plane, each of whose points is equidistant from $(2, -1, 1)$ and $(3, 1, 5)$.

26. Find an equation of the plane containing the line $x = 3t$, $y = 1 + t$, $z = 2t$ and parallel to the intersection of the planes $2x - y + z = 0$ and $y + z + 1 = 0$.

27. Show that the line $x = 0$, $y = t$, $z = t$
    (a) lies in the plane $6x + 4y - 4z = 0$
    (b) is parallel to and below the plane $5x - 3y + 3z = 1$
    (c) is parallel to and above the plane $6x + 2y - 2z = 3$.

28. Show that the lines

    $$x + 1 = 4t \qquad x + 13 = 12t$$
    $$y - 3 = t \quad \text{and} \quad y - 1 = 6t$$
    $$z - 1 = 0 \qquad z - 2 = 3t$$

    intersect and find an equation of the plane they determine.

29. Find parametric equations of the line of intersection of the planes
    (a) $-2x + 3y + 7z + 2 = 0$ and
        $x + 2y - 3z + 5 = 0$
    (b) $3x - 5y + 2z = 0$ and $z = 0$.

30. Show that the plane whose intercepts with the coordinate axes are $x = a$, $y = b$, and $z = c$ has equation

    $$\frac{x}{a} + \frac{y}{b} + \frac{z}{c} = 1$$

    provided a, b, and c are nonzero.

In Exercises 31–33, find the distance between the point and the plane.

31. $(1, -2, 3)$; $2x - 2y + z = 4$.

32. $(0, 1, 5)$; $3x + 6y - 2z - 5 = 0$.

33. $(7, 2, -1)$; $20x - 4y - 5z = 0$.

In Exercises 34–36, find the distance between the given parallel planes.

34. $2x - 3y + 4z = 7$, $4x - 6y + 8z = 3$.

35. $-2x + y + z = 0$, $6x - 3y - 3z - 5 = 0$.

36. $x + y + z = 1$, $x + y + z = -1$.

In Exercises 37–39, find the distance between the given skew lines.

37. $x = 1 + 7t$, $y = 3 + t$, $z = 5 - 3t$; $x = 4 - t$, $y = 6$, $z = 7 + 2t$.

38. $x = 3 - t, y = 4 + 4t, z = 1 + 2t; x = t, y = 3,$ $z = 2t.$

39. $x = 2 + 4t, \quad y = 6 - 4t, \quad z = 5t; \quad x = 3 + 8t,$ $y = 5 - 3t, z = 6 + t.$

40. Show that the line $x = -1 + t, y = 3 + 2t, z = -t$ and the plane $2x - 2y - 2z + 3 = 0$ are parallel, and find the distance between them.

41. Prove: The planes $a_1x + b_1y + c_1z = d_1$ and $a_2x + b_2y + c_2z = d_2$ are perpendicular if and only if $a_1a_2 + b_1b_2 + c_1c_2 = 0$.

42. Let $\mathbf{r}_0 = \langle x_0, y_0, z_0 \rangle$ and $\mathbf{r} = \langle x, y, z \rangle$. Describe the set of all points (x, y, z) for which
   (a) $\mathbf{r} \cdot \mathbf{r}_0 = 0$ (b) $(\mathbf{r} - \mathbf{r}_0) \cdot \mathbf{r}_0 = 0$

43. Find an equation of the sphere with its center at $(2, 1, -3)$ and tangent to the plane $x - 3y + 2z = 4$.

44. Find where the line that passes through the point $(3, 1, 0)$ and is normal to the plane $2x + y - z = 0$ intersects the plane.

14.7 QUADRIC SURFACES

In this section we shall study an important class of surfaces that are the three-dimensional analogs of the conic sections.

Figure 14.7.1

In 2-space the general shape of a curve can be obtained by plotting points. However, for surfaces in 3-space, point plotting is not generally helpful since too many points are needed to obtain even a crude picture of the surface. It is better to build up the shape of the surface by using curves of intersection with some well-chosen planes. The curve of intersection of a plane and a surface is called the *trace* of the surface in the plane (Figure 14.7.1). In Figure 14.7.2, we get an excellent picture of the surface $z = x^3 - 3xy^2$ from the traces shown. (This surface is called a "monkey saddle" because a monkey sitting astride the x-axis has a place for its two legs and tail.)

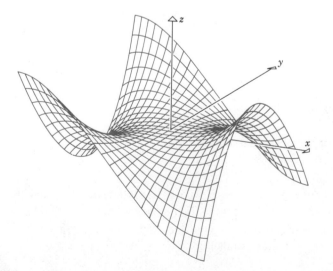

Figure 14.7.2

Earlier in the text, we saw that the graph in two dimensions of a second-degree equation in x and y,

$$Ax^2 + Bxy + Cy^2 + Dx + Ey + F = 0$$

is a conic section. In three dimensions, the graph of a second-degree equation in x, y, and z,

$$Ax^2 + By^2 + Cz^2 + Dxy + Exz + Fyz + Gx + Hy + Iz + J = 0$$

is called a *quadric surface* or a *quadric*.

Figure 14.7.3 shows the equations and graphs of six important quadric surfaces. In those equations, a, b, and c are positive constants. The stated equations apply only when the quadric surfaces are in the positions shown. If the quadric surfaces are rotated or translated from these positions, then the equations change. We shall consider this point in more detail later.

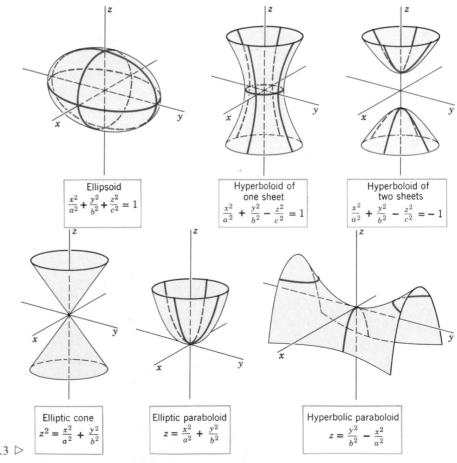

Ellipsoid
$$\frac{x^2}{a^2} + \frac{y^2}{b^2} + \frac{z^2}{c^2} = 1$$

Hyperboloid of one sheet
$$\frac{x^2}{a^2} + \frac{y^2}{b^2} - \frac{z^2}{c^2} = 1$$

Hyperboloid of two sheets
$$\frac{x^2}{a^2} + \frac{y^2}{b^2} - \frac{z^2}{c^2} = -1$$

Elliptic cone
$$z^2 = \frac{x^2}{a^2} + \frac{y^2}{b^2}$$

Elliptic paraboloid
$$z = \frac{x^2}{a^2} + \frac{y^2}{b^2}$$

Hyperbolic paraboloid
$$z = \frac{y^2}{b^2} - \frac{x^2}{a^2}$$

Figure 14.7.3 ▷

The traces of the quadric surfaces in planes parallel to the coordinate planes are described in Table 14.7.1.

To study the traces of a quadric surface, we will need equations of planes parallel to the coordinate planes. The plane parallel to the xy-plane and passing through $(0, 0, k)$ (Figure 14.7.4) consists of all points (x, y, z) with $z = k$. Thus, its equation is

$$z = k$$

Similarly, $x = k$ represents a plane parallel to the yz-plane and passing through $(k, 0, 0)$, while $y = k$ represents a plane parallel to the xz-plane and passing through $(0, k, 0)$. The equation of the trace of a surface in the plane $z = k$ is obtained by substituting $z = k$ in the equation of the surface. Similarly for traces in the planes $x = k$ and $y = k$.

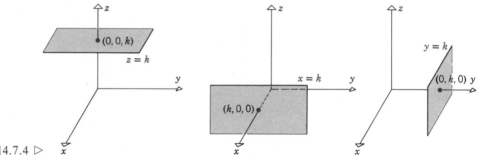

Figure 14.7.4 ▷

We shall show how to obtain the results in Table 14.7.1 for elliptic cones. (See pages 852 and 853.) The derivations of the remaining results in the table are left as exercises.

For the elliptic cone

$$z^2 = \frac{x^2}{a^2} + \frac{y^2}{b^2} \qquad (a > 0, b > 0)$$

the trace in the xy-plane is

$$\frac{x^2}{a^2} + \frac{y^2}{b^2} = 0 \qquad (z = 0)$$

which implies that $x = 0$, $y = 0$, $z = 0$. Thus, the trace is the single point $(0, 0, 0)$. In the plane $z = k$ parallel to the xy-plane, the trace is

$$\frac{x^2}{a^2} + \frac{y^2}{b^2} = k^2$$

or

$$\frac{x^2}{(ak)^2} + \frac{y^2}{(bk)^2} = 1$$

which is an ellipse. As $|k|$ increases, so do $(ak)^2$ and $(bk)^2$. Thus, the dimensions of these ellipses increase as the planes containing them recede from the xy-plane.

The trace in the xz-plane is

$$z^2 = \frac{x^2}{a^2} \qquad (y = 0)$$

or

$$\left(z - \frac{x}{a}\right)\left(z + \frac{x}{a}\right) = 0$$

Thus, the trace is the pair of lines

$$z = \frac{x}{a} \quad \text{and} \quad z = -\frac{x}{a} \qquad (y = 0)$$

which intersect at the origin. Similarly, the trace in the yz-plane is the pair of intersecting lines

$$z = \frac{y}{b} \quad \text{and} \quad z = -\frac{y}{b} \qquad (x = 0)$$

The trace in the plane $y = k$ parallel to the xz-plane is

$$z^2 = \frac{x^2}{a^2} + \frac{k^2}{b^2}$$

or

$$z^2 - \frac{x^2}{a^2} = \frac{k^2}{b^2}$$

or

$$\frac{z^2}{k^2/b^2} - \frac{x^2}{a^2 k^2/b^2} = 1$$

which is a hyperbola in the plane $y = k$ opening along a line parallel to the z-axis. Similarly, the trace in the plane $x = k$ is

$$z^2 = \frac{k^2}{a^2} + \frac{y^2}{b^2}$$

or

$$z^2 - \frac{y^2}{b^2} = \frac{k^2}{a^2}$$

Table 14.7.1

SURFACE	NAME	EQUATION
The traces in the coordinate planes are ellipses, as are the traces in planes parallel to the coordinate planes.	ELLIPSOID	$\dfrac{x^2}{a^2} + \dfrac{y^2}{b^2} + \dfrac{z^2}{c^2} = 1$
The trace in the xy-plane is an ellipse, as are the traces in planes parallel to the xy-plane. The traces in the yz-plane and xz-plane are hyperbolas, as are the traces in planes parallel to these.	HYPERBOLOID OF ONE SHEET	$\dfrac{x^2}{a^2} + \dfrac{y^2}{b^2} - \dfrac{z^2}{c^2} = 1$
There is no trace in the xy-plane. In planes parallel to the xy-plane, which intersect the surface, the traces are ellipses. In the yz- and xz-planes, the traces are hyperbolas, as are the traces in planes parallel to these.	HYPERBOLOID OF TWO SHEETS	$\dfrac{x^2}{a^2} + \dfrac{y^2}{b^2} - \dfrac{z^2}{c^2} = -1$

Table 14.7.1 *(continued)*

SURFACE	NAME	EQUATION
	ELLIPTIC CONE	$z^2 = \dfrac{x^2}{a^2} + \dfrac{y^2}{b^2}$

The trace in the xy-plane is a point (the origin), and the traces in planes parallel to the xy-plane are ellipses. The traces in the yz- and xz-planes are pairs of lines intersecting at the origin. The traces in planes parallel to these are hyperbolas.

SURFACE	NAME	EQUATION
	ELLIPTIC PARABOLOID	$z = \dfrac{x^2}{a^2} + \dfrac{y^2}{b^2}$

The trace in the xy-plane is a point (the origin), and the traces in planes parallel to and above the xy-plane are ellipses. The traces in the yz- and xz-planes are parabolas, as are the traces in planes parallel to these.

SURFACE	NAME	EQUATION
	HYPERBOLIC PARABOLOID	$z = \dfrac{y^2}{b^2} - \dfrac{x^2}{a^2}$

The trace in the xy-plane is a pair of lines intersecting at the origin. The traces in planes parallel to the xy-plane are hyperbolas. The hyperbolas above the xy-plane open in the y-direction, and those below in the x-direction. The traces in the yz- and xz-planes are parabolas, as are the traces in planes parallel to these.

or

$$\frac{z^2}{k^2/a^2} - \frac{y^2}{b^2k^2/a^2} = 1$$

which is a hyperbola in the plane $x = k$ opening along a line parallel to the z-axis.

We shall now discuss the problem of sketching quadric surfaces.

ELLIPSOIDS   A fairly accurate graph of the ellipsoid

$$\frac{x^2}{a^2} + \frac{y^2}{b^2} + \frac{z^2}{c^2} = 1 \quad (a > 0, b > 0, c > 0) \tag{1}$$

can be obtained by first plotting the intercepts with the coordinate axes. The x-intercepts occur where $y = z = 0$. Substituting these values in (1) yields $x^2/a^2 = 1$ or $x = \pm a$. Thus, the x-intercepts are

$$(a, 0, 0) \quad \text{and} \quad (-a, 0, 0)$$

The y-intercepts (obtained by setting $x = z = 0$) are $(0, \pm b, 0)$, and the z-intercepts (obtained by setting $x = y = 0$) are $(0, 0, \pm c)$. The ellipsoid is sketched in Figure 14.7.5.

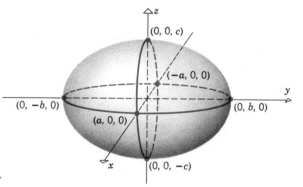

Figure 14.7.5 ▷

Observe that if $a = b = c$, then the ellipsoid represented by (1) reduces to a sphere of radius a centered at the origin.

HYPERBOLOIDS OF   A fairly accurate graph of the hyperboloid of one sheet
ONE SHEET

$$\frac{x^2}{a^2} + \frac{y^2}{b^2} - \frac{z^2}{c^2} = 1 \quad (a > 0, b > 0, c > 0) \tag{2}$$

can be obtained by first sketching its traces in the coordinate planes. The trace in the xy-plane is the ellipse,

$$\frac{x^2}{a^2} + \frac{y^2}{b^2} = 1 \quad (z = 0)$$

which intersects the x-axis at $(\pm a, 0, 0)$ and the y-axis at $(0, \pm b, 0)$ (Figure 14.7.6). The trace in the yz-plane is the hyperbola

$$\frac{y^2}{b^2} - \frac{z^2}{c^2} = 1 \quad (x = 0)$$

which intersects the y-axis at $(0, \pm b, 0)$ and has asymptotes

$$y = \pm \frac{b}{c} z$$

The trace in the xz-plane is the hyperbola

$$\frac{x^2}{a^2} - \frac{z^2}{c^2} = 1 \quad (y = 0)$$

which intersects the x-axis at $(\pm a, 0, 0)$ and has asymptotes

$$x = \pm \frac{a}{c} z$$

Figure 14.7.6

HYPERBOLOIDS OF TWO SHEETS

A fairly accurate graph of the hyperboloid of two sheets,

$$\frac{x^2}{a^2} + \frac{y^2}{b^2} - \frac{z^2}{c^2} = -1 \quad (a > 0, b > 0, c > 0) \tag{3}$$

can be obtained by plotting the intercepts with the z-axis and sketching the traces in the coordinate planes.

The z-intercepts occur where $x = y = 0$. Substituting these values in (3) yields

$$-\frac{z^2}{c^2} = -1$$

or $z = \pm c$. Thus, the z-intercepts are $(0, 0, \pm c)$ (Figure 14.7.7). The trace in the yz-plane is the hyperbola

$$\frac{z^2}{c^2} - \frac{y^2}{b^2} = 1 \quad (x = 0)$$

which intersects the z-axis at $(0, 0, \pm c)$ and has asymptotes

$$z = \pm \frac{c}{b} y$$

Figure 14.7.7

The trace in the xz-plane is the hyperbola,

$$\frac{z^2}{c^2} - \frac{x^2}{a^2} = 1 \quad (y = 0)$$

which intersects the z-axis at the points $(0, 0, \pm c)$ and has asymptotes

$$z = \pm \frac{c}{a} x$$

ELLIPTIC CONES   A fairly accurate graph of the elliptic cone

$$z^2 = \frac{x^2}{a^2} + \frac{y^2}{b^2} \quad (a > 0, b > 0) \tag{4}$$

can be obtained by first sketching the traces in the planes $z = 1$ and $z = -1$ (Figure 14.7.8). In both planes, the traces are ellipses with the equation

$$\frac{x^2}{a^2} + \frac{y^2}{b^2} = 1$$

If $a = b$, then all traces of the cone in planes parallel to the xy-plane are circles, so the surface is called a *circular cone*.

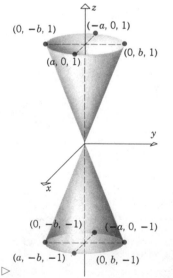

Figure 14.7.8 ▷

ELLIPTIC
PARABOLOIDS   A fairly accurate graph of the elliptic paraboloid,

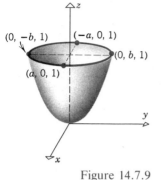

Figure 14.7.9

$$z = \frac{x^2}{a^2} + \frac{y^2}{b^2} \quad (a > 0, b > 0) \tag{5a}$$

can be obtained by first sketching its trace in the plane $z = 1$ (Figure 14.7.9). This trace is the ellipse with the equation

$$\frac{x^2}{a^2} + \frac{y^2}{b^2} = 1$$

If $a = b$, then all traces of the paraboloid in planes parallel to the xy-plane are circles, so the surface is called a *circular paraboloid.*

Later, we shall encounter equations of the form

$$z = -\left(\frac{x^2}{a^2} + \frac{y^2}{b^2}\right) \tag{5b}$$

An equation of this type represents an elliptic paraboloid opening in the negative z-direction.

HYPERBOLIC
PARABOLOIDS

The hyperbolic paraboloid

$$z = \frac{y^2}{b^2} - \frac{x^2}{a^2} \quad (a > 0, b > 0) \tag{6}$$

is a difficult surface to draw accurately. Fortunately, a rough sketch, simply showing the position of the surface, suffices for many purposes.

TRANSLATION OF
AXES IN 3-SPACE

Let an $x'y'z'$-coordinate system be obtained by translating an xyz-coordinate system so the $x'y'z'$-origin is at the point whose xyz-coordinates are $(x, y, z) = (h, k, l)$ (Figure 14.7.10). It can be shown that the $x'y'z'$-coordinates and xyz-coordinates of a point P are related by

$$x' = x - h, \quad y' = y - k, \quad z' = z - l \tag{7}$$

[Compare this to (6) in Section 12.2.]

Figure 14.7.10

Example 1 Sketch the surface $z = 1 - x^2 - y^2$.

Solution. Rewrite the equation in the form

$$z - 1 = -(x^2 + y^2) \tag{8}$$

If we translate the coordinate axes so the new origin is at the point $(h, k, l) = (0, 0, 1)$, then the translation equations (7) are

$$x' = x, \quad y' = y, \quad z' = z - 1$$

so that in $x'y'z'$-coordinates (8) becomes

$$z' = -(x'^2 + y'^2)$$

which is of form (5b) with $a = b = 1$. Thus, the surface is a circular paraboloid, opening down, with vertex $(0, 0, 1)$ in xyz-coordinates (Figure 14.7.11).   ◄

Figure 14.7.11 ▷

Example 2   Sketch the surface

$$4x^2 + 4y^2 + z^2 + 8y - 4z = -4$$

Solution.   Completing the squares yields

$$4x^2 + 4(y + 1)^2 + (z - 2)^2 = -4 + 4 + 4$$

or

$$x^2 + (y + 1)^2 + \frac{(z - 2)^2}{4} = 1 \tag{9}$$

If we translate the coordinate axes so the new origin is at the point $(h, k, l) = (0, 0, 1)$, then the translation equations (7) are

$$x' = x, \quad y' = y + 1, \quad z' = z - 2$$

so that in $x'y'z'$-coordinates (9) becomes

$$x'^2 + y'^2 + \frac{z'^2}{4} = 1$$

which is of form (1) with $a = 1$, $b = 1$, and $c = 2$. Thus, the surface is the ellipsoid shown in Figure 14.7.12.   ◄

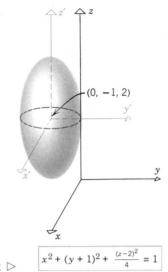

$$x^2 + (y + 1)^2 + \frac{(z-2)^2}{4} = 1$$

Figure 14.7.12 ▷

► Exercise Set 14.7

In Exercises 1–12, name and sketch the quadric surfaces.

1. $x^2 + \dfrac{y^2}{4} + \dfrac{z^2}{9} = 1.$ **2.** $\dfrac{z^2}{4} - \dfrac{y^2}{9} - \dfrac{x^2}{4} = 1.$

3. $\dfrac{x^2}{4} + \dfrac{y^2}{9} - \dfrac{z^2}{16} = 1.$ **4.** $z = x^2 + y^2.$

5. $z^2 = \dfrac{x^2}{4} + \dfrac{y^2}{9}.$ **6.** $z - y^2 + x^2 = 0.$

7. $16z + x^2 + 4y^2 = 0.$ **8.** $36 - x^2 - 4y^2 = 9z^2.$

9. $4x^2 + y^2 - z^2 = 16.$ **10.** $9z^2 - 4y^2 - x^2 = 36.$

11. $9x^2 - 4y^2 = 36z.$ **12.** $2x^2 - 8z + y^2 = 0.$

13. The following equations represent quadric surfaces with orientations different from those in Figure 14.7.3. Name and sketch the surface.

(a) $\dfrac{z^2}{c^2} - \dfrac{y^2}{b^2} + \dfrac{x^2}{a^2} = 1$ (b) $\dfrac{x^2}{a^2} - \dfrac{y^2}{b^2} - \dfrac{z^2}{c^2} = 1$

(c) $x = \dfrac{y^2}{b^2} + \dfrac{z^2}{c^2}$ (d) $x^2 = \dfrac{y^2}{b^2} + \dfrac{z^2}{c^2}$

(e) $y = \dfrac{z^2}{c^2} - \dfrac{x^2}{a^2}$

(f) $y = -\left(\dfrac{x^2}{a^2} + \dfrac{z^2}{c^2}\right).$

In Exercises 14–17, sketch the graph of the equation.

14. $z = \sqrt{1 - x^2 - y^2}.$ **15.** $z = \sqrt{x^2 + y^2}.$

16. $z = \sqrt{1 + x^2 + y^2}.$ **17.** $z = \sqrt{x^2 + y^2 - 1}.$

In Exercises 18–23, name and sketch the surface.

18. $\dfrac{(x-1)^2}{4} + \dfrac{(y-2)^2}{9} + \dfrac{(z-4)^2}{16} = 1.$

19. $z = (x + 2)^2 + (y - 3)^2 - 9.$

20. $4x^2 - y^2 + 16(z - 2)^2 = 100.$

21. $9x^2 + y^2 + 4z^2 - 18x + 2y + 16z = 10.$

22. $z^2 = 4x^2 + y^2 + 8x - 2y + 4z.$

23. $z = 4 - x^2 - y^2 - 2y.$

24. Obtain the results in Table 14.7.1 for the ellipsoid, $x^2/a^2 + y^2/b^2 + z^2/c^2 = 1.$

25. Obtain the results in Table 14.7.1 for the hyperboloid of one sheet, $x^2/a^2 + y^2/b^2 - z^2/c^2 = 1$.

26. Obtain the results in Table 14.7.1 for the hyperboloid of two sheets, $x^2/a^2 + y^2/b^2 - z^2/c^2 = -1$.

27. Obtain the results in Table 14.7.1 for the elliptic paraboloid $z = x^2/a^2 + y^2/b^2$.

28. Obtain the results in Table 14.7.1 for the hyperbolic paraboloid $z = y^2/b^2 - x^2/a^2$.

29. For the elliptic paraboloid

$$z = \frac{x^2}{9} + \frac{y^2}{4}$$

find

(a) the focus and vertex of the (parabolic) trace in the plane $x = k$;

(b) the foci and the endpoints of the major and minor axes of the (elliptic) trace in the plane $z = k$.

30. Use the method of slicing to find the volume of the ellipsoid

$$\frac{x^2}{a^2} + \frac{y^2}{b^2} + \frac{z^2}{c^2} = 1$$

[*Hint:* The area of the ellipse $x^2/a^2 + y^2/b^2 = 1$ is πab.]

31. Find an equation of the surface consisting of points

$P(x, y, z)$ for which the distance between P and the plane $z = -1$ is equal to the distance between P and the point $(0, 0, 1)$. Identify the surface.

32. Find an equation of the surface consisting of points $P(x, y, z)$ for which the distance between P and the plane $z = -1$ is twice the distance between P and the point $(0, 0, 1)$. Identify the surface.

33. (a) Show that the lines $x = 3 + t$, $y = 2 + t$, $z = 5 + 2t$ and $x = 3 + t$, $y = 2 - t$, $z = 5 + 10t$ both lie completely on the hyperbolic paraboloid $z = x^2 - y^2$.

(b) Let $P_0(x_0, y_0, z_0)$ be any point on the surface $z = x^2 - y^2$. Show that it is always possible to find two lines with equations of the form $x = x_0 + t$, $y = y_0 + at$, $z = z_0 + bt$ that pass through P_0 and lie completely on the surface.

34. (a) Show that the lines $x = 2 + \frac{3}{5}t$, $y = 1 + \frac{4}{5}t$, $z = 2 + t$ and $x = 2 + t$, $y = 1$, $z = 2 + t$ both lie completely on the hyperboloid $x^2 + y^2 - z^2 = 1$.

(b) Let $P_0(x_0, y_0, z_0)$ be any point on the surface $x^2 + y^2 - z^2 = 1$. Show that it is always possible to find two lines with equations of the form $x = x_0 + at$, $y = y_0 + bt$, $z = z_0 + t$ that pass through P_0 and lie completely on the surface.

14.8 SPHERICAL AND CYLINDRICAL COORDINATES

In this section we shall discuss two new types of coordinate systems in three dimensions that are extremely important in applied problems.

CYLINDRICAL
COORDINATES

The **cylindrical coordinates** (r, θ, z) of a point $P(x, y, z)$ are obtained by replacing x and y by polar coordinates and keeping z the same (Figure 14.8.1). For simplicity, we shall require that

$$r \geq 0 \quad \text{and} \quad 0 \leq \theta < 2\pi$$

From the familiar relationship between rectangular and polar coordinates in the xy-plane (Formula (1) of Section 13.1), it follows that the rectangular and cylindrical coordinates of a point in 3-space are related by

$$x = r \cos \theta, \quad y = r \sin \theta, \quad z = z \tag{1}$$

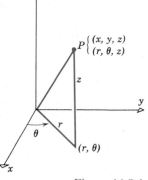

Figure 14.8.1

or expressed another way

$$r = \sqrt{x^2 + y^2}, \quad \tan\theta = \frac{y}{x}, \quad z = z \tag{2}$$

Example 1   The point whose cylindrical coordinates are $(r, \theta, z) = (4, \pi/3, -3)$ has rectangular coordinates

$$x = 4\cos\frac{\pi}{3} = 2, \quad y = 4\sin\frac{\pi}{3} = 2\sqrt{3}, \quad z = -3 \quad \blacktriangleleft$$

Example 2   The equation in cylindrical coordinates of the surface $z = x^2 + y^2 - 2x + y$ is

$$z = r^2 - 2r\cos\theta + r\sin\theta \quad \blacktriangleleft$$

Example 3   The equation in rectangular coordinates of the surface

$$r = 4\cos\theta$$

can be obtained by multiplying both sides by r to get $r^2 = 4r\cos\theta$, from which it follows that

$$x^2 + y^2 = 4x$$

or equivalently

$$(x - 2)^2 + y^2 = 4$$

This is a right-circular cylinder parallel to the z-axis. $\blacktriangleleft$

In cylindrical coordinates, surfaces of the forms

$$r = r_0, \quad \theta = \theta_0, \quad \text{and} \quad z = z_0$$

(r_0, θ_0, and z_0 constants) are of special interest. These surfaces are shown in Figure 14.8.2.

Figure 14.8.2 $\triangleright$

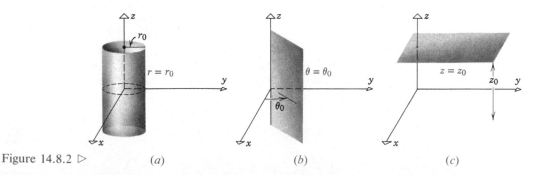

(a)   (b)   (c)

The surface $r = r_0$ is a right-circular cylinder of radius r_0 centered on the z-axis. At each point (r, θ, z) on this cylinder, r has the value r_0, but θ and z are unrestricted except for our blanket assumption that $0 \le \theta < 2\pi$.

The surface $\theta = \theta_0$ is a half plane attached along the z-axis and making an angle θ_0 with the positive x-axis. At each point (r, θ, z) on this surface, θ has the value θ_0, but r and z are unrestricted except for our blanket assumption that $r \ge 0$.

The surface $z = z_0$ is a horizontal plane. At each point (r, θ, z) on this plane, z has the value z_0, but r and θ are unrestricted except for our blanket assumptions.

SPHERICAL
COORDINATES

The **spherical coordinates** (ρ, θ, ϕ) of a point P in 3-space are illustrated in Figure 14.8.3.

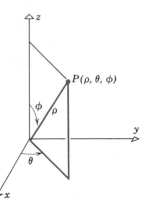

Figure 14.8.3 ▷

The coordinate ρ is the distance from P to the origin, the coordinate θ is the same as in polar coordinates, and the coordinate ϕ is the angle from the positive z-axis to the line segment joining P and the origin. We shall require that

$$\rho \ge 0, \quad 0 \le \theta < 2\pi, \quad 0 \le \phi \le \pi$$

The following relationships between spherical and cylindrical coordinates of a point P are immediate from Figure 14.8.4:

$$r = \rho \sin \phi, \quad \theta = \theta, \quad z = \rho \cos \phi \qquad (3)$$

Combining these with the equations $x = r \cos \theta$ and $y = r \sin \theta$, we obtain the following relationships between the spherical and rectangular coordinates of P:

$$x = \rho \sin \phi \cos \theta, \quad y = \rho \sin \phi \sin \theta, \quad z = \rho \cos \phi \qquad (4)$$

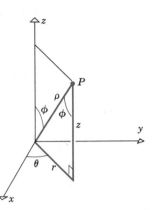

Figure 14.8.4 ▷

Moreover, since ρ is the distance between the origin and P, we have

$$\rho = \sqrt{x^2 + y^2 + z^2} \tag{5}$$

Example 4 The point with spherical coordinates $(\rho, \theta, \phi) = (4, \pi/3, \pi/4)$ has rectangular coordinates

$$x = \rho \sin \phi \cos \theta = 4 \sin \frac{\pi}{4} \cos \frac{\pi}{3} = \sqrt{2}$$

$$y = \rho \sin \phi \sin \theta = 4 \sin \frac{\pi}{4} \sin \frac{\pi}{3} = \sqrt{6}$$

$$z = \rho \cos \phi = 4 \cos \frac{\pi}{4} = 2\sqrt{2} \quad \blacktriangleleft$$

Example 5 Find the equation of the paraboloid $z = x^2 + y^2$ in spherical coordinates

Solution. Substituting (4) in this equation yields

$$\rho \cos \phi = \rho^2 \sin^2 \phi \cos^2 \theta + \rho^2 \sin^2 \phi \sin^2 \theta$$
$$\rho \cos \phi = \rho^2 \sin^2 \phi \, (\cos^2 \theta + \sin^2 \theta)$$
$$\rho \cos \phi = \rho^2 \sin^2 \phi$$

which simplifies to $\rho \sin^2 \phi = \cos \phi$ $\quad \blacktriangleleft$

In spherical coordinates, surfaces of the forms

$$\rho = \rho_0, \quad \theta = \theta_0, \quad \phi = \phi_0$$

(ρ_0, θ_0, ϕ_0 constants) are of special interest. These surfaces are shown in Figure 14.8.5.

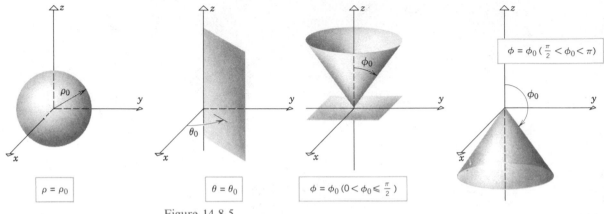

Figure 14.8.5

The surface $\rho = \rho_0$ consists of all points whose distance ρ from the origin is ρ_0. Assuming ρ_0 to be nonnegative, this is a sphere of radius ρ_0 centered at the origin.

As in cylindrical coordinates, the surface $\theta = \theta_0$ is a half plane attached along the z-axis, making an angle of θ_0 with the positive x-axis.

The surface $\phi = \phi_0$ consists of all points from which a line segment to the origin makes an angle of ϕ_0 with the positive z-axis. Depending on whether $0 < \phi_0 < \pi/2$, or $\pi/2 < \phi_0 < \pi$, or $\phi_0 = \pi/2$ this will be a cone opening up, a cone opening down, or the xy-plane.

SPHERICAL COORDINATES IN NAVIGATION

Spherical coordinates are related to longitude and latitude coordinates used in navigation. Let us construct a right-hand rectangular coordinate system with origin at the earth's center, positive z-axis passing through the north pole, and positive x-axis passing through the prime meridian (Figure 14.8.6). If we assume the earth to be a perfect sphere of radius

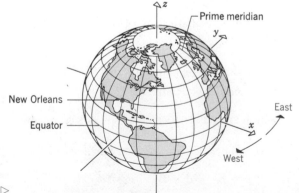

Figure 14.8.6 ▷

$\rho = 4000$ miles, then each point on the earth has spherical coordinates of the form $(4000, \theta, \phi)$ where ϕ and θ determine the latitude and longitude of the point. It is usual to specify longitudes in degrees east or west of the prime meridian and latitudes in degrees north or south of the equator. However, it is a simple matter to determine ϕ and θ from such data.

Example 6 The city of New Orleans is located at 90° West longitude and 30° North latitude. Find its spherical and rectangular coordinates relative to the coordinate axes of Figure 14.8.6. (Take miles as the unit of distance.)

Solution. A longitude of 90° West corresponds to $\theta = 360° - 90° = 270°$ or $\theta = 3\pi/2$ radians; and a latitude of 30° North corresponds to $\phi = 90° - 30° = 60°$ or $\phi = \pi/3$ radians. Thus, the spherical coordinates of New Orleans are $(4000, 3\pi/2, \pi/3)$.
From (4), the rectangular coordinates of New Orleans are

$$x = 4000 \sin\frac{\pi}{3}\cos\frac{3\pi}{2} = 4000\frac{\sqrt{3}}{2}(0) = 0 \text{ miles}$$

$$y = 4000 \sin\frac{\pi}{3}\sin\frac{3\pi}{2} = 4000\frac{\sqrt{3}}{2}(-1) = -2000\sqrt{3} \text{ miles}$$

$$z = 4000 \cos\frac{\pi}{3} = 4000\left(\frac{1}{2}\right) = 2000 \text{ miles} \qquad \blacktriangleleft$$

▶ Exercise Set 14.8

1. Convert from rectangular to cylindrical coordinates.
   (a) $(4\sqrt{3}, 4, -4)$ (b) $(-5, 5, 6)$
   (c) $(0, 2, 0)$ (d) $(4, -4\sqrt{3}, 6)$
   (e) $(\sqrt{2}, -\sqrt{2}, 1)$ (f) $(0, 0, 1)$.

2. Convert from cylindrical to rectangular coordinates.
   (a) $(4, \pi/6, 3)$ (b) $(8, 3\pi/4, -2)$
   (c) $(5, 0, 4)$ (d) $(7, \pi, -9)$
   (e) $(6, 5\pi/3, 7)$ (f) $(1, \pi/2, 0)$.

3. Convert from rectangular to spherical coordinates.
   (a) $(1, \sqrt{3}, -2)$ (b) $(1, -1, \sqrt{2})$
   (c) $(0, 3\sqrt{3}, 3)$ (d) $(-5\sqrt{3}, 5, 0)$
   (e) $(4, 4, 4\sqrt{6})$ (f) $(1, -\sqrt{3}, -2)$.

4. Convert from spherical to rectangular coordinates.

   (a) $(5, \pi/6, \pi/4)$ (b) $(7, 0, \pi/2)$
   (c) $(1, \pi, 0)$ (d) $(2, 3\pi/2, \pi/2)$
   (e) $(1, 2\pi/3, 3\pi/4)$ (f) $(3, 7\pi/4, 5\pi/6)$.

5. Convert from cylindrical to spherical coordinates.
   (a) $(\sqrt{3}, \pi/6, 3)$ (b) $(1, \pi/4, -1)$
   (c) $(2, 3\pi/4, 0)$ (d) $(6, 1, -2\sqrt{3})$
   (e) $(4, 5\pi/6, 4)$ (f) $(2, 0, -2)$.

6. Convert from spherical to cylindrical coordinates.
   (a) $(5, \pi/4, 2\pi/3)$ (b) $(1, 7\pi/6, \pi)$
   (c) $(3, 0, 0)$ (d) $(4, \pi/6, \pi/2)$
   (e) $(5, \pi/2, 0)$ (f) $(6, 0, 3\pi/4)$.

In Exercises 7–14, an equation is given in cylindrical coordinates. Express the equation in rectangular coordinates and sketch the graph.

7. $r = 3$.

8. $\theta = \pi/4$.

9. $z = r^2$.

10. $z = r \cos \theta$.

11. $r = 4 \sin \theta$.

12. $r = 2 \sec \theta$.

13. $r^2 + z^2 = 1$.

14. $r^2 \cos 2\theta = z$.

In Exercises 15–22, an equation is given in spherical coordinates. Express the equation in rectangular coordinates and sketch the graph.

15. $\rho = 3$.

16. $\theta = \pi/3$.

17. $\phi = \pi/4$.

18. $\rho = 2 \sec \phi$.

19. $\rho = 4 \cos \phi$.

20. $\rho \sin \phi = 1$.

21. $\rho \sin \phi = 2 \cos \theta$.

22. $\rho - 2 \sin \phi \cos \theta = 0$.

In Exercises 23–28, express the equation in: (a) cylindrical coordinates, (b) spherical coordinates.

23. $x^2 + y^2 + z^2 = 9$.

24. $z^2 = x^2 - y^2$.

25. $2x + 3y + 4z = 1$.

26. $x^2 + y^2 - z^2 = 1$.

27. $x^2 = 16 - z^2$.

28. $x^2 + y^2 + z^2 = 2z$.

29. Leningrad, Russia, is located at 30° East longitude and 60° North latitude. Find its spherical and rectangular coordinates relative to the coordinate axes of Figure 14.8.6. Take miles as the unit of distance and assume the earth to be a sphere of radius 4000 miles.

30. (a)   Show that the curve of intersection of the surfaces $z = \sin \theta$ and $r = a$ (cylindrical coordinates) is an ellipse.
  (b)   Sketch a portion of the surface $z = \sin \theta$ for $0 \le \theta \le \pi/2$.

31. Sketch the surface whose equation in spherical coordinates is $\rho = a(1 - \cos \phi)$. [*Hint:* The surface is shaped like a familiar fruit.]

▶ SUPPLEMENTARY EXERCISES

1. Find $\overrightarrow{P_1P_2}$ and $\|\overrightarrow{P_1P_2}\|$.
  (a) $P_1(2, 3)$, $P_2(5, -1)$ (b) $P_1(2, -1)$, $P_2(1, 3)$.

In Exercises 2–7, find all vectors satisfying the given conditions.

2. A vector of length 1 in 2-space that is perpendicular to the line $x + y = -1$.

3. The vector oppositely directed to $3\mathbf{i} - 4\mathbf{j}$, and having the same length.

4. The vector obtained by rotating $\mathbf{i}$ counterclockwise through an angle θ in 2-space.

5. The vector with initial point $(1, 2)$ and a terminal point that is 3/5 of the way from $(1, 2)$ to $(5, 5)$.

6. A vector of length 2 that is parallel to the tangent to the curve $y = x^2$ at $(-1, 1)$.

7. The vector of length 12 in 2-space that makes an angle of 120° with the x-axis.

8. Solve for c_1 and c_2 given that $c_1\langle -2, 5\rangle + 3c_2\langle 1, 3\rangle = \langle -6, -51\rangle$.

9. Solve for $\mathbf{u}$ if $3\mathbf{u} - (\mathbf{i} + \mathbf{j}) = \mathbf{i} + \mathbf{u}$.

10. Solve for $\mathbf{u}$ and $\mathbf{v}$ if $3\mathbf{u} - 4\mathbf{v} = 3\mathbf{v} - 2\mathbf{u} = \langle 1, 2\rangle$.

11. Two forces $\mathbf{F}_1 = 2\mathbf{i} - \mathbf{j}$ and $\mathbf{F}_2 = -3\mathbf{i} - 4\mathbf{j}$ are applied at a point. What force $\mathbf{F}_3$ must be applied at the point to cancel the effect of $\mathbf{F}_1$ and $\mathbf{F}_2$?

12. Given the points $P(3, 4)$, $Q(1, 1)$, and $R(5, 2)$, use vector methods to find the coordinates of the fourth vertex of the parallelogram whose adjacent sides are $\overrightarrow{PQ}$ and $\overrightarrow{QR}$.

In Exercises 13 and 14, find
  (a) $\|\mathbf{a}\|$ (b) $\mathbf{a} \cdot \mathbf{b}$
  (c) $\mathbf{a} \times \mathbf{b}$ (d) $\mathbf{b} \times \mathbf{a}$
  (e)   the area of the triangle with sides $\mathbf{a}$ and $\mathbf{b}$
  (f) $3\mathbf{a} - 2\mathbf{b}$.

13. $\mathbf{a} = \langle 1, 2, -1\rangle$, $\mathbf{b} = \langle 2, -1, 3\rangle$.

14. $\mathbf{a} = \langle 1, -2, 2\rangle$, $\mathbf{b} = \langle 3, 4, -5\rangle$.

In Exercises 15 and 16, find:
  (a) $\|\text{proj}_{\mathbf{b}}\, \mathbf{a}\|$ (b) $\|\text{proj}_{\mathbf{a}}\, \mathbf{b}\|$
  (c)   the angle between $\mathbf{a}$ and $\mathbf{b}$
  (d)   the direction cosines of $\mathbf{a}$.

15. $\mathbf{a} = 3\mathbf{i} - 4\mathbf{j}$, $\mathbf{b} = 2\mathbf{i} + 2\mathbf{j} - \mathbf{k}$.

16. $\mathbf{a} = -\mathbf{j}$, $\mathbf{b} = \mathbf{i} + \mathbf{j}$.

17. Verify the identity $\mathbf{a} \times (\mathbf{b} \times \mathbf{c}) = (\mathbf{a} \cdot \mathbf{c})\mathbf{b} - (\mathbf{a} \cdot \mathbf{b})\mathbf{c}$ for $\mathbf{a} = \mathbf{i} + \mathbf{j}$, $\mathbf{b} = 2\mathbf{i} - \mathbf{k}$, $\mathbf{c} = \mathbf{j} - \mathbf{k}$.

18. Find the vector with length 5 and direction angles $\alpha = 60°$, $\beta = 120°$, $\gamma = 135°$.

19. Find the vector with length 3 and direction cosines $-1/\sqrt{2}$, 0, and $1/\sqrt{2}$.

20. For the points $P(6, 5, 7)$ and $Q(7, 3, 9)$ find:
 (a) the midpoint of the line segment PQ
 (b) the length and direction cosines of $\overrightarrow{PQ}$.

21. If $\mathbf{u} = \mathbf{i} + 2\mathbf{j} - 3\mathbf{k}$ and $\mathbf{v} = \mathbf{i} + \mathbf{j} + 2\mathbf{k}$, find:
 (a) the vector component of $\mathbf{u}$ along $\mathbf{v}$,
 (b) the vector component of $\mathbf{u}$ orthogonal to $\mathbf{v}$.

22. Find the vector component of $\mathbf{i}$ along $3\mathbf{i} - 2\mathbf{j} + \mathbf{k}$.

23. A diagonal of a box makes angles of 50° and 70° with two of its edges. Find to the nearest degree the angle that it makes with the third edge.

24. Consider the points $O(0, 0, 0)$, $A(0, a, a)$, and $B(-3, 4, 2)$. Find all nonzero values of a that make $\overrightarrow{OA}$ orthogonal to $\overrightarrow{AB}$.

25. Under what conditions are $\mathbf{u} + \mathbf{v}$ and $\mathbf{u} - \mathbf{v}$ orthogonal?

26. If $M(3, -1, 5)$ is the midpoint of the line segment PQ and if the coordinates of P are $(1, 2, 3)$, find the coordinates of Q.

27. Find all possible vectors of length 1 orthogonal to both $\mathbf{a} = \langle 3, -2, 1 \rangle$ and $\mathbf{b} = \langle -2, 1, -3 \rangle$.

28. Find the distance from the point $P(2, 3, 4)$ to the plane containing the points $A(0, 0, 1)$, $B(1, 0, 0)$, and $C(0, 2, 0)$.

In Exercises 29–32, find an equation for the plane that satisfies the given conditions.

29. The plane through $A(1, 2, 3)$ and $B(2, 4, 2)$ that is parallel to $\mathbf{v} = \langle -3, -1, -2 \rangle$.

30. The plane through $P_0(-1, 2, 3)$ that is perpendicular to the planes $2x - 3y + 5 = 0$ and $3x - y - 4z + 6 = 0$.

31. The plane that passes through $P(1, 1, 1)$, $Q(2, 3, 0)$, and $R(2, 1, 2)$.

32. The plane with intercepts $x = 2$, $y = -3$, $z = 10$.

33. Let L be the line through $P(1, 2, 8)$ that is parallel to $\mathbf{v} = \langle 3, -1, -4 \rangle$.
 (a) For what values of k and l will the point $Q(k, 3, l)$ be on L?
 (b) If L' has parametric equations $x = -8 - 3t$,

$y = 5 + t$, $z = 0$, show that L' intersects L and find the point of intersection.
 (c) Find the point at which L intersects the plane through $R(-4, 0, 3)$ having a normal vector $\langle 3, -2, 6 \rangle$.

34. Consider the lines L_1 and L_2 with symmetric equations,

$$L_1: \frac{x - 1}{2} = \frac{y + \frac{3}{2}}{1} = \frac{z + 1}{2}$$

$$L_2: \frac{4 - x}{1} = \frac{3 - y}{2} = \frac{4 + z}{2}$$

(see Exercise 34, Section 14.5).
 (a) Are L_1 and L_2 parallel? Perpendicular?
 (b) Find parametric equations for L_1 and L_2.
 (c) Do L_1 and L_2 intersect? If so, where?

35. Find parametric equations for the line through P_1 and P_2.
 (a) $P_1(1, -1, 2)$, $P_2(3, 2, -1)$
 (b) $P_1(1, -3, 4)$, $P_2(1, 2, -3)$.

36. For points $A(1, -1, 2)$, $B(2, -3, 0)$, $C(-1, -2, 0)$, and $D(2, 1, -1)$ find:
 (a) $\overrightarrow{AB} \times \overrightarrow{AC}$ (b) the area of triangle ABC
 (c) the volume of the parallelepiped determined by the vectors $\overrightarrow{AB}$, $\overrightarrow{AC}$, $\overrightarrow{AD}$
 (d) the distance from D to the plane containing A, B, and C.

37. (a) Find parametric equations for the intersection of the planes $2x + y - z = 3$ and $x + 2y + z = 3$.
 (b) Find the acute angle between the two planes.

In Exercises 38–40, describe the region satisfying the given conditions:

38. (a) $x^2 + 9y^2 + 4z^2 > 36$
 (b) $x^2 + y^2 + z^2 - 6x + 2y - 6 < 0$.

39. (a) $z > 4x^2 + 9y^2$
 (b) $x^2 + 4y^2 + z^2 = 0$.

40. (a) $y^2 + 4z^2 = 4$, $0 \le x \le 2$
 (b) $9x^2 + 4y^2 + 36x - 8y = -60$.

In Exercises 41–45, identify the quadric surface whose equation is given.

41. $100x^2 + 225y^2 - 36z^2 = 0$.

42. $x^2 - z^2 + y = 0$.

43. $400x^2 + 25y^2 + 16z^2 = 400$.

44. $4x^2 - y^2 + 4z^2 = 4$.

45. $-16x^2 - 100y^2 + 25z^2 = 400$.

46. Identify the surface by completing the squares:
    (a) $x^2 + 4y^2 - z^2 - 6x + 8y + 4z = 0$
    (b) $x^2 + y^2 + z^2 + 6x - 4y + 12z = 0$.

47. Find the work done by a constant force $\mathbf{F} = 3\mathbf{i} - 4\mathbf{j} + \mathbf{k}$ (pounds) acting on a particle that moves along the line segment from $P(5, 7, 0)$ to $Q(6, 6, 6)$ (units in feet).

48. Two forces $\mathbf{F}_1 = \mathbf{i} - 3\mathbf{j} + \mathbf{k}$ and $\mathbf{F}_2 = \mathbf{i} + 2\mathbf{j} + 2\mathbf{k}$ (pounds) act on a particle as it moves in a straight line from $P(-1, -2, 3)$ to $Q(0, 2, 0)$ (units in feet). How much work is done?

49. Express the spherical coordinates ρ, θ, ϕ in terms of cylindrical coordinates r, θ, z.

50. Convert from rectangular coordinates to: (i) cylindrical coordinates, (ii) spherical coordinates.
    (a) $(2, 2, 2\sqrt{6})$ (b) $(1, \sqrt{3}, 0)$.

51. Convert $(\sqrt{2}, \pi/4, 1)$ from cylindrical coordinates to
    (a) rectangular coordinates
    (b) spherical coordinates.

52. Express the equation in terms of: (i) cylindrical coordinates, (ii) spherical coordinates:
    (a) $x^2 + y^2 = z$ (b) $x^2 - y^2 - z^2 = 0$.

53. Express the equation in terms of rectangular coordinates:
    (a) $z = r^2 \cos 2\theta$
    (b) $\rho^2 \sin \phi \cos \phi \cos \theta = 1$.

54. Sketch the set of points defined by the given conditions.
    (a) $0 \le \theta \le \pi/2$, $0 \le r \le \cos \theta$, $0 \le z \le 2$ (cylindrical coordinates)
    (b) $0 \le \theta \le \pi/2$, $0 \le \phi \le \pi/4$, $0 \le \rho \le 2 \sec \phi$ (spherical coordinates)
    (c) $r = 2 \sin \theta$, $0 \le z \le 2$ (cylindrical coordinates)
    (d) $\rho = 2 \cos \phi$ (spherical coordinates).

15 vector-valued functions

15.1 INTRODUCTION TO VECTOR-VALUED FUNCTIONS

In this section we shall show how vectors can be used to express parametric equations in a more compact form. As part of our work we shall discuss functions that associate vectors with real numbers. This new category of functions has important applications in science and engineering.

VECTOR-VALUED
FUNCTIONS

In Section 2.1 we defined a function to be a rule that assigns to each element in its domain one and only one element in its range. Up to now, both the domain and range have consisted of real numbers. Such functions are called ***real-valued functions of a real variable*** or sometimes simply ***real-valued functions.*** In this section we shall need to consider functions for which the domain consists of real numbers, and the range consists of vectors in 2-space or 3-space. Such functions are called ***vector-valued functions of a real variable*** or more simply ***vector-valued functions.*** In 2-space such functions can be expressed in the form

$$\mathbf{f}(t) = \langle x(t), y(t) \rangle = x(t)\mathbf{i} + y(t)\mathbf{j}$$

and in 3-space in the form

$$\mathbf{f}(t) = \langle x(t), y(t), z(t) \rangle = x(t)\mathbf{i} + y(t)\mathbf{j} + z(t)\mathbf{k}$$

where $x(t)$, $y(t)$, and $z(t)$ are real-valued functions of the real variable t. These real-valued functions are called the ***component functions*** or ***components*** of **f**. As a matter of notation, we shall denote vector-valued functions with boldface type (for example, $\mathbf{f}(t)$, $\mathbf{g}(t)$, and $\mathbf{r}(t)$) and real-valued functions, as usual, with lightface type ($f(t)$, $g(t)$, and $r(t)$).

Example 1   If

$$\mathbf{f}(t) = (\ln t)\mathbf{i} + \sqrt{t^2 + 2}\,\mathbf{j} + (\cos t\pi)\mathbf{k}$$

then the component functions are

$$x(t) = \ln t, \quad y(t) = \sqrt{t^2 + 2}, \quad \text{and} \quad z(t) = \cos t\pi$$

The vector that $\mathbf{f}(t)$ associates with $t = 1$ is

$$\mathbf{f}(1) = (\ln 1)\mathbf{i} + \sqrt{3}\,\mathbf{j} + (\cos \pi)\mathbf{k} = \sqrt{3}\,\mathbf{j} - \mathbf{k}$$

The function $\mathbf{f}$ is undefined if $t \le 0$ because $\ln t$ is undefined for such t.   ◄

In general, if the domain of a vector-valued function is not stated explicitly, then it is understood to consist of all real numbers for which every component is defined and yields a real value. Stated another way, the domain of a vector-valued function is the intersection of the domains of its components.

Vector-valued functions can be used to express parametric equations in a more compact form. To see how this can be done, let C be a curve in 2-space represented by the parametric equations

$$x = x(t)$$
$$y = y(t)$$

Because two vectors are equivalent if and only if their corresponding components are equal, this pair of equations can be replaced by the single vector equation

$$x\mathbf{i} + y\mathbf{j} = x(t)\mathbf{i} + y(t)\mathbf{j} \tag{1}$$

We call this a *vector equation* of C. Similarly, a curve C in 3-space represented by the parametric equations

$$x = x(t)$$
$$y = y(t)$$
$$z = z(t)$$

can be replaced by the single vector equation

$$x\mathbf{i} + y\mathbf{j} + z\mathbf{k} = x(t)\mathbf{i} + y(t)\mathbf{j} + z(t)\mathbf{k} \tag{2}$$

Equations (1) and (2) can be written more compactly by introducing vectors $\mathbf{r}$ and $\mathbf{f}(t)$ defined by

$$\mathbf{r} = x\mathbf{i} + y\mathbf{j}, \quad \mathbf{f}(t) = x(t)\mathbf{i} + y(t)\mathbf{j}$$

and

$$\mathbf{r} = x\mathbf{i} + y\mathbf{j} + z\mathbf{k}, \quad \mathbf{f}(t) = x(t)\mathbf{i} + y(t)\mathbf{j} + z(t)\mathbf{k}$$

in 2-space and 3-space, respectively. With this notation both (1) and (2) can be rewritten as

$$\mathbf{r} = \mathbf{f}(t) \tag{3}$$

Thus, in both 2-space and 3-space, parametric equations can be expressed as a single vector equation of form (3), and conversely, every vector equation of this form can be rewritten as parametric equations by equating the corresponding components on the two sides of the equation. In Equation (3) the vector $\mathbf{r}$ can be viewed as a dependent variable whose value is determined by the independent variable t.

Example 2 Express the given parametric equations as a single vector equation.

(a) $x = t^2$ (b) $x = \cos t$

     $y = 3t$ $y = \sin t$

                     $z = t$

Solution (a). Using the two sides of the equations as components of a vector yields

$$x\mathbf{i} + y\mathbf{j} = t^2\mathbf{i} + 3t\mathbf{j}$$

which is of form (3) with

$$\mathbf{r} = x\mathbf{i} + y\mathbf{j} \quad \text{and} \quad \mathbf{f}(t) = t^2\mathbf{i} + 3t\mathbf{j}$$

Solution (b). Proceeding as in part (a) yields

$$x\mathbf{i} + y\mathbf{j} + z\mathbf{k} = (\cos t)\mathbf{i} + (\sin t)\mathbf{j} + t\mathbf{k}$$

which is of form (3) with

$$\mathbf{r} = x\mathbf{i} + y\mathbf{j} + z\mathbf{k}, \quad \mathbf{f}(t) = (\cos t)\mathbf{i} + (\sin t)\mathbf{j} + t\mathbf{k} \quad \blacktriangleleft$$

Example 3 Find the parametric equations that correspond to the vector equation

$$x\mathbf{i} + y\mathbf{j} + z\mathbf{k} = (t^3 + 1)\mathbf{i} + 3\mathbf{j} + e^t\mathbf{k}$$

Solution. Equating corresponding components yields

$$x = t^3 + 1$$
$$y = 3$$
$$z = e^t \quad \blacktriangleleft$$

REMARK.   When it is desirable to limit the number of different letters occurring in a problem, we shall write $\mathbf{r} = \mathbf{r}(t)$ rather than $\mathbf{r} = \mathbf{f}(t)$. We did the same thing with the parametric equations $x = x(t)$, $y = y(t)$, and $z = z(t)$, where we used the same letter for the function name and the dependent variable.

GRAPHS OF
VECTOR-VALUED
FUNCTIONS

We shall now consider some geometric aspects of vector equations and vector-valued functions. We start with some basic definitions.

15.1.1   DEFINITION. The *graph* of a vector equation $\mathbf{r} = \mathbf{f}(t)$ in 2-space or 3-space is defined to be the graph of the corresponding parametric equations.

Example 4   Describe the graph of the vector equation

$$x\mathbf{i} + y\mathbf{j} + z\mathbf{k} = (-2 + t)\mathbf{i} + 3t\mathbf{j} + (5 - 4t)\mathbf{k}$$

Solution.   The corresponding parametric equations are

$$x = -2 + t$$
$$y = 3t$$
$$z = 5 - 4t$$

From Theorem 14.5.1, the graph of these equations is the line in 3-space that passes through the point $(-2, 0, 5)$ and is parallel to the vector $\mathbf{i} + 3\mathbf{j} - 4\mathbf{k}$.   ◄

Example 5   Describe the graph of the vector equation

$$\mathbf{r} = 2\mathbf{i} + 3\mathbf{j} - \mathbf{k}$$

Solution.   Equating the components on the right side of the equation with the corresponding components of $\mathbf{r} = x\mathbf{i} + y\mathbf{j} + z\mathbf{k}$ yields

$$x = 2$$
$$y = 3$$
$$z = -1$$

Since the right sides of these equations do not vary with the parameter t the graph is the single point $(2, 3, -1)$.   ◄

REMARK. In general, if $\mathbf{c} = x_0\mathbf{i} + y_0\mathbf{j} + z_0\mathbf{k}$ is a constant vector (that is, does not depend on a parameter), then the graph of $\mathbf{r} = \mathbf{c}$ is the single point (x_0, y_0, z_0). Similarly, in 2-space, if $\mathbf{c} = x_0\mathbf{i} + y_0\mathbf{j}$, then the graph of $\mathbf{r} = \mathbf{c}$ is the single point (x_0, y_0).

Recall that when we first began to study functions in Chapter 2 we defined the graph of a function $f(x)$ to be the graph of the equation $y = f(x)$. Similarly, for vector-valued functions we define the **graph of $\mathbf{f}(t)$** to be the graph of the vector equation $\mathbf{r} = \mathbf{f}(t)$.

Example 6 Sketch the graph of the vector-valued function

$$\mathbf{f}(t) = (\cos t)\mathbf{i} + (\sin t)\mathbf{j}, \quad 0 \le t \le 2\pi$$

Solution. The graph of $\mathbf{f}(t)$ is the graph of the vector equation

$$x\mathbf{i} + y\mathbf{j} = (\cos t)\mathbf{i} + (\sin t)\mathbf{j}, \quad 0 \le t \le 2\pi$$

or equivalently, it is the graph of the parametric equations

$$x = \cos t \atop y = \sin t, \quad 0 \le t \le 2\pi$$

This is a circle of radius 1 centered at the origin. The direction of increasing t is counterclockwise (Figure 15.1.1). ◀

$\mathbf{f}(t) = (\cos t)\,\mathbf{i} + (\sin t)\,\mathbf{j}$

Figure 15.1.1

Example 7 Sketch the graph of the vector-valued function

$$\mathbf{f}(t) = (\cos t)\mathbf{i} + (\sin t)\mathbf{j} + 2\mathbf{k}, \quad 0 \le t \le 2\pi$$

Solution. The graph of $\mathbf{f}(t)$ is the graph of the vector equation

$$x\mathbf{i} + y\mathbf{j} + z\mathbf{k} = (\cos t)\mathbf{i} + (\sin t)\mathbf{j} + 2\mathbf{k}, \quad 0 \le t \le 2\pi$$

or, equivalently, it is the graph of the parametric equations

$$x = \cos t$$
$$y = \sin t, \quad 0 \le t \le 2\pi$$
$$z = 2$$

The first two equations tell us that as the parameter t varies, the point (x, y, z) moves in a path directly over the circle in Figure 15.1.1. The last equation tells us that this circle lies in the plane $z = 2$ (Figure 15.1.2). ◀

$\mathbf{r} = (\cos t)\,\mathbf{i} + (\sin t)\,\mathbf{j} + 2\mathbf{k}$

Figure 15.1.2

Example 8   Sketch the graph of the vector-valued function

$$\mathbf{f}(t) = (a \cos t)\mathbf{i} + (a \sin t)\mathbf{j} + ct\mathbf{k}$$

where a and c are positive constants.

Solution.   The graph of $\mathbf{f}(t)$ is the same as the graph of the parametric equations

$$x = a \cos t$$
$$y = a \sin t$$
$$z = ct$$

As the parameter t increases, the value of $z = ct$ also increases, so the point (x, y, z) moves upward. However, as t increases, the point (x, y, z) also moves in a path directly over the circle

$$x = a \cos t$$
$$y = a \sin t$$

in the xy-plane. The combination of these upward and circular motions produces a corkscrew-shaped curve that wraps around a right circular cylinder of radius a centered on the z-axis (Figure 15.1.3). This curve is called a ***circular helix.***   ◄

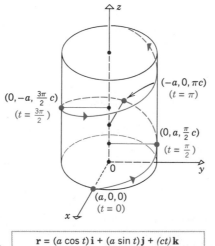

Figure 15.1.3 ▷

$$\mathbf{r} = (a \cos t)\mathbf{i} + (a \sin t)\mathbf{j} + (ct)\mathbf{k}$$

We shall now describe a way of thinking about the graph of a vector equation that is purely "vectorial" in the sense that it does not require the vector equation to be split into parametric equations. Denote the graph of the vector equation $\mathbf{r} = \mathbf{f}(t)$ by C and position the vector $\mathbf{r}$ (or equivalently $\mathbf{f}(t)$) so its initial point is at the origin. As illustrated in Figure 15.1.4, the terminal point of $\mathbf{r}$ will lie on the curve C. When positioned in this way the vector **r** is called a ***radius vector*** or ***position vector*** for the curve C. As the parameter t varies, the terminal point of **r** will trace out the curve C.

As t varies, the tip
of the radius vector **r**
traces out the curve C.

Figure 15.1.4 ▷

NORM OF A
VECTOR-VALUED
FUNCTION

If $\mathbf{f}(t)$ is a vector-valued function, then for each value of the parameter t, the expression $\|\mathbf{f}(t)\|$ is a real-valued function of t because the norm (or length) of $\mathbf{f}(t)$ is a real number. For example, if

$$\mathbf{f}(t) = t\mathbf{i} + (t - 1)\mathbf{j}$$

then

$$\|\mathbf{f}(t)\| = \sqrt{t^2 + (t - 1)^2} = \sqrt{2t^2 - 2t + 1}$$

which is a real-valued function of t.

Example 9   In Example 7 we saw that the graph of

$$\mathbf{f}(t) = (\cos t)\mathbf{i} + (\sin t)\mathbf{j} + 2\mathbf{k}, \quad 0 \le t \le 2\pi$$

is a circle of radius 1 centered on the z-axis and lying in the plane $z = 2$. This circle lies on the surface of a sphere of radius $\sqrt{5}$ because for each value of t

$$\|\mathbf{f}(t)\| = \sqrt{\cos^2 t + \sin^2 t + 4} = \sqrt{1 + 4} = \sqrt{5}$$

which shows that each point on the circle is at a distance of $\sqrt{5}$ units from the origin (Figure 15.1.5).   ◀

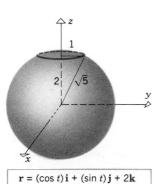

$\mathbf{r} = (\cos t)\mathbf{i} + (\sin t)\mathbf{j} + 2\mathbf{k}$

Figure 15.1.5

▶ Exercise Set 15.1

In Exercises 1–4, find the domain of **f** and the value of $\mathbf{f}(t_0)$.

1. $\mathbf{f}(t) = (\cos t)\mathbf{i} - 3t\mathbf{j}$; $t_0 = \pi$.

2. $\mathbf{f}(t) = \langle \sqrt{3t + 1}, t^2 \rangle$; $t_0 = 1$.

3. $\mathbf{f}(t) = (\cos \pi t)\mathbf{i} - (\ln t)\mathbf{j} + \sqrt{t - 2}\,\mathbf{k}$; $t_0 = 3$.

4. $\mathbf{f}(t) = \langle 2e^{-t}, \sin^{-1} t, \ln(1 - t) \rangle$; $t_0 = 0$.

In Exercises 5–8, express the parametric equations as a single vector equation $\mathbf{r} = x(t)\mathbf{i} + y(t)\mathbf{j} + z(t)\mathbf{k}$.

5. $x = 3 \cos t$, $y = t + \sin t$.

6. $x = t^2 + 1$, $y = e^{-2t}$.

7. $x = 2t$, $y = 2 \sin 3t$, $z = 5 \cos 3t$.

8. $x = t \sin t$, $y = \ln t$, $z = \cos^2 t$.

In Exercises 9–12, find the parametric equations that correspond to the given vector equation.

9. $\mathbf{r} = 3t^2\mathbf{i} - 2\mathbf{j}$.

10. $\mathbf{r} = (\sin^2 t)\mathbf{i} + (1 - \cos 2t)\mathbf{j}$.

11. $\mathbf{r} = (2t - 1)\mathbf{i} - 3\sqrt{t}\,\mathbf{j} + (\sin 3t)\mathbf{k}$.

12. $\mathbf{r} = te^{-t}\mathbf{i} - 5t^2\mathbf{k}$.

In Exercises 13–16, describe the graph of the vector equation.

13. $\mathbf{r} = (2 - 3t)\mathbf{i} - 4t\mathbf{j}$.

14. $\mathbf{r} = (3 \sin 2t)\mathbf{i} + (3 \cos 2t)\mathbf{j}$.

15. $\mathbf{r} = 2t\mathbf{i} - 3\mathbf{j} + (1 + 3t)\mathbf{k}$.

16. $\mathbf{r} = 3\mathbf{i} + (2 \cos t)\mathbf{j} + (2 \sin t)\mathbf{k}$.

17. Find the slope of the line in 2-space that is represented by the vector equation

$$\mathbf{r} = (1 - 2t)\mathbf{i} - (2 - 3t)\mathbf{j}.$$

18. Find the y-intercept of the line in 2-space that is represented by the vector equation

$$\mathbf{r} = (3 + 2t)\mathbf{i} + 5t\mathbf{j}.$$

19. Find the coordinates of the point where the line $\mathbf{r} = (2 + t)\mathbf{i} + (1 - 2t)\mathbf{j} + 3t\mathbf{k}$ intersects the xz-plane.

20. Find the coordinates of the point where the line $\mathbf{r} = t\mathbf{i} + (1 + 2t)\mathbf{j} - 3t\mathbf{k}$ intersects the plane $3x - y - z = 2$.

In Exercises 21–32, sketch the graph of $\mathbf{r}(t)$ and show the direction of increasing t.

21. $\mathbf{r}(t) = 2\mathbf{i} + t\mathbf{j}$.

22. $\mathbf{r}(t) = \langle 3t - 4, 6t + 2 \rangle$.

23. $\mathbf{r}(t) = (1 + \cos t)\mathbf{i} + (3 - \sin t)\mathbf{j}$, $0 \le t \le 2\pi$.

24. $\mathbf{r}(t) = \langle 2 \cos t, 5 \sin t \rangle$, $0 \le t \le 2\pi$.

25. $\mathbf{r}(t) = (\cosh t)\mathbf{i} + (\sinh t)\mathbf{j}$.

26. $\mathbf{r}(t) = \sqrt{t}\,\mathbf{i} + (2t + 4)\mathbf{j}$.

27. $\mathbf{r}(t) = t\mathbf{i} + t\mathbf{j} + t\mathbf{k}$.

28. $\mathbf{r}(t) = (1 + 3t)\mathbf{i} + (-1 + t)\mathbf{j} + 2t\mathbf{k}$.

29. $\mathbf{r}(t) = (2 \cos t)\mathbf{i} + (2 \sin t)\mathbf{j} + t\mathbf{k}$.

30. $\mathbf{r}(t) = (9 \cos t)\mathbf{i} + (4 \sin t)\mathbf{j} + t\mathbf{k}$.

31. $\mathbf{r}(t) = t\mathbf{i} + t^2\mathbf{j} + 2\mathbf{k}$.

32. $\mathbf{r}(t) = t\mathbf{i} + t\mathbf{j} + (\sin t)\mathbf{k}$, $0 \le t \le 2\pi$.

33. For the helix $\mathbf{r} = (a \cos t)\mathbf{i} + (a \sin t)\mathbf{j} + ct\mathbf{k}$, find c $(c > 0)$ so that the helix will make one complete turn in a distance of 3 units measured along a line parallel to the z-axis.

34. How many revolutions will the circular helix $\mathbf{r} = (a \cos t)\mathbf{i} + (a \sin t)\mathbf{j} + 0.2t\mathbf{k}$ make in a distance of 10 units measured along a line parallel to the z-axis?

35. Show that the curve $\mathbf{r} = (t \cos t)\mathbf{i} + (t \sin t)\mathbf{j} + t\mathbf{k}$, $t \ge 0$, lies completely on the cone $z = \sqrt{x^2 + y^2}$. Describe the curve.

36. Describe the curve $\mathbf{r} = (a \cos t)\mathbf{i} + (b \sin t)\mathbf{j} + ct$ where a, b, and c are positive constants such that $a \ne b$.

15.2 LIMITS AND DERIVATIVES OF VECTOR-VALUED FUNCTIONS

In this section we shall define limits and derivatives of vector-valued functions and give some basic interpretations and applications of these concepts.

LIMITS OF
VECTOR-VALUED
FUNCTIONS

The limit of a vector-valued function is defined to be the vector that results by taking the limit of each component. More precisely, we make the following definition.

15.2.1 DEFINITION. Let lim stand for any one of the limit symbols $\lim\limits_{t \to a}$, $\lim\limits_{t \to a^+}$, $\lim\limits_{t \to a^-}$, $\lim\limits_{t \to +\infty}$, or $\lim\limits_{t \to -\infty}$.

(a) For a vector-valued function

$$\mathbf{f}(t) = x(t)\mathbf{i} + y(t)\mathbf{j}$$

in 2-space, we define $\lim \mathbf{f}(t)$ by

$$\lim \mathbf{f}(t) = (\lim x(t))\mathbf{i} + (\lim y(t))\mathbf{j}$$

provided $\lim x(t)$ and $\lim y(t)$ exist.

(b) For a vector-valued function

$$\mathbf{f}(t) = x(t)\mathbf{i} + y(t)\mathbf{j} + z(t)\mathbf{k}$$

in 3-space, we define $\lim \mathbf{f}(t)$ by

$$\begin{aligned} \lim \mathbf{f}(t) &= \lim \left[x(t)\mathbf{i} + y(t)\mathbf{j} + z(t)\mathbf{k} \right] \\ &= (\lim x(t))\mathbf{i} + (\lim y(t))\mathbf{j} + (\lim z(t))\mathbf{k} \end{aligned}$$

provided $\lim x(t)$, $\lim y(t)$, and $\lim z(t)$ exist.

Limits of vector-valued functions have many of the same basic properties as limits of real-valued functions (Exercise 52).

Example 1   Let $\mathbf{f}(t) = t^2\mathbf{i} + 3t\mathbf{j}$. Find $\lim\limits_{t \to 2} \mathbf{f}(t)$.

Solution.

$$\lim_{t \to 2} \mathbf{f}(t) = \lim_{t \to 2}(t^2\mathbf{i} + 3t\mathbf{j}) = (\lim_{t \to 2} t^2)\mathbf{i} + (\lim_{t \to 2} 3t)\mathbf{j} = 4\mathbf{i} + 6\mathbf{j} \quad \blacktriangleleft$$

Example 2   Let $\mathbf{f}(t) = e^t\mathbf{i} + (\cos \pi t)\mathbf{j} + (\ln t)\mathbf{k}$. Find $\lim\limits_{t \to 0^+} \mathbf{f}(t)$.

Solution.   For a vector-valued function to have a limit, each component must have a limit. But

$$\lim_{t \to 0^+} \ln t = -\infty$$

so $\lim\limits_{t \to 0^+} \mathbf{f}(t)$ does not exist.   ◄

It follows from Definition 15.2.1 that

$$\lim_{t \to a} \mathbf{f}(t) = \mathbf{L}$$

if and only if the components of $\mathbf{f}(t)$ approach the components of $\mathbf{L}$ as $t \to a$. Geometrically, this is equivalent to stating that the length and direction of $\mathbf{f}(t)$ approach the length and direction of $\mathbf{L}$ as $t \to a$ (Figure 15.2.1).

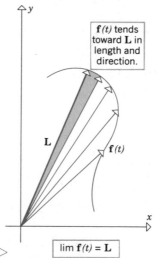

$$\boxed{\mathbf{f}(t) \text{ tends toward } \mathbf{L} \text{ in length and direction.}}$$

$\mathbf{L}$

$\mathbf{f}(t)$

$$\boxed{\lim \mathbf{f}(t) = \mathbf{L}}$$

Figure 15.2.1 ▷

CONTINUITY OF
VECTOR-VALUED
FUNCTIONS

The definition of continuity for vector-valued functions is similar to that for real-valued functions. We shall say that $\mathbf{f}$ is *continuous at* t_0 if:

1. $\mathbf{f}(t_0)$ is defined.
2. $\lim\limits_{t \to t_0} \mathbf{f}(t)$ exists.
3. $\lim\limits_{t \to t_0} \mathbf{f}(t) = \mathbf{f}(t_0)$.

It can be shown that $\mathbf{f}$ is continuous at t_0 if and only if each component of $\mathbf{f}$ is continuous at t_0 (Exercise 53). As with real-valued functions, we shall call **f** *continuous everywhere* or simply *continuous* if **f** is continuous at all real values of t. Geometrically, the graph of a continuous vector-valued function is an unbroken curve.

DERIVATIVES OF VECTOR-VALUED FUNCTIONS

Derivatives played a fundamental role in defining and determining tangent lines to graphs of real-valued functions. We shall now extend the notion of a derivative to vector-valued functions. This definition will enable us to define the notion of a tangent line to the graph of a vector-valued function. However, as was the case for derivatives of real-valued functions, derivatives of vector-valued functions have applications far beyond the tangent-line problem.

The definition of a derivative for vector-valued functions is analogous to the definition for real-valued functions.

15.2.2 DEFINITION. The derivative $\mathbf{f}'(t)$ of a vector-valued function $\mathbf{f}(t)$ is defined by

$$\mathbf{f}'(t) = \lim_{h \to 0} \frac{\mathbf{f}(t + h) - \mathbf{f}(t)}{h}$$

provided this limit exists.

For computational purposes the following theorem is extremely useful. It states that the derivative of a vector-valued function can be computed by differentiating each component.

15.2.3 THEOREM.

(a) *If $\mathbf{f}(t) = x(t)\mathbf{i} + y(t)\mathbf{j}$ is a vector-valued function in 2-space, and if $x(t)$ and $y(t)$ are differentiable, then*

$$\mathbf{f}'(t) = x'(t)\mathbf{i} + y'(t)\mathbf{j}$$

(b) *If $\mathbf{f}(t) = x(t)\mathbf{i} + y(t)\mathbf{j} + z(t)\mathbf{k}$ is a vector-valued function in 3-space, and if $x(t)$, $y(t)$, and $z(t)$ are differentiable, then*

$$\mathbf{f}'(t) = x'(t)\mathbf{i} + y'(t)\mathbf{j} + z'(t)\mathbf{k}$$

Proof. We shall prove part (a). The proof of (b) is similar. From Definition 15.2.2,

$$\mathbf{f}'(t) = \lim_{h \to 0} \frac{\mathbf{f}(t + h) - \mathbf{f}(t)}{h}$$

$$= \lim_{h \to 0} \frac{[x(t + h)\mathbf{i} + y(t + h)\mathbf{j}] - [x(t)\mathbf{i} + y(t)\mathbf{j}]}{h}$$

$$= \lim_{h \to 0} \frac{[x(t + h) - x(t)]}{h} \mathbf{i} + \lim_{h \to 0} \frac{[y(t + h) - y(t)]}{h} \mathbf{j}$$

$$= x'(t)\mathbf{i} + y'(t)\mathbf{j} \quad \blacksquare$$

As with real-valued functions, there are various notations for the derivative of a vector-valued function. If $\mathbf{r} = \mathbf{f}(t)$, then some possibilities are

$$\frac{d}{dt}[\mathbf{f}(t)], \quad \frac{d\mathbf{f}}{dt}, \quad \frac{d\mathbf{r}}{dt}, \quad \text{and} \quad \mathbf{r}'$$

Example 3   Let $\mathbf{f}(t) = t^2\mathbf{i} + t^3\mathbf{j}$. Find $\mathbf{f}'(t)$ and $\mathbf{f}'(1)$.

Solution.

$$\mathbf{f}'(t) = \frac{d}{dt}[t^2]\mathbf{i} + \frac{d}{dt}[t^3]\mathbf{j} = 2t\mathbf{i} + 3t^2\mathbf{j}$$

Substituting $t = 1$ yields

$$\mathbf{f}'(1) = 2\mathbf{i} + 3\mathbf{j} \quad \blacktriangleleft$$

TANGENT LINES AND TANGENT VECTORS

The derivative of a vector-valued function has an important geometric interpretation which can be stated *informally* as follows:

If C is the graph of the vector-valued function $\mathbf{f}(t)$, *then for each value of t where* $\mathbf{f}'(t)$ *exists and is nonzero, the vector* $\mathbf{f}'(t)$ *is tangent to the curve C when this vector is positioned so its initial point is at the terminal point of the radius vector* $\mathbf{f}(t)$. *Moreover,* $\mathbf{f}'(t)$ *points in the direction of increasing parameter (Figure 15.2.2).*

To make this result plausible, let C be the graph of $\mathbf{f}(t)$, and for a fixed value of the parameter t construct the position vectors $\mathbf{f}(t)$ and $\mathbf{f}(t + h)$. If $h > 0$, then the tip of the vector $\mathbf{f}(t + h)$ is in the direction of increasing parameter from the tip of $\mathbf{f}(t)$, and if $h < 0$ it is the other way around (Figure 15.2.3). In either case, the difference $\mathbf{f}(t + h) - \mathbf{f}(t)$ coincides with the secant line through the tips of the vectors $\mathbf{f}(t)$ and $\mathbf{f}(t + h)$. Moreover, since h is a scalar, the vector

$$\frac{1}{h}[\mathbf{f}(t + h) - \mathbf{f}(t)] = \frac{\mathbf{f}(t + h) - \mathbf{f}(t)}{h} \tag{1}$$

also coincides with this secant line. If $h < 0$, then vector (1) is oppositely directed to $\mathbf{f}(t + h) - \mathbf{f}(t)$, and if $h > 0$ these vectors have the same direction. In both cases, vector (1) points in the direction of increasing parameter. As $h \to 0$, the secant lines through the tips of $\mathbf{f}(t + h)$ and $\mathbf{f}(t)$

Figure 15.2.2

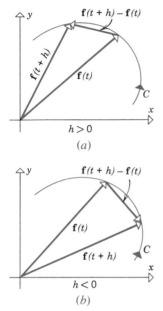

$$\mathbf{f}(t + h) - \mathbf{f}(t)$$

$h > 0$

(a)

$$\mathbf{f}(t + h) - \mathbf{f}(t)$$

$h < 0$

(b)

Figure 15.2.3

tend toward the tangent line to C at the tip of $\mathbf{f}(t)$. Thus, the vector

$$\mathbf{f}'(t) = \lim_{h \to 0} \frac{\mathbf{f}(t + h) - \mathbf{f}(t)}{h}$$

is tangent to the curve C at the tip of $\mathbf{f}(t)$ and points in the direction of increasing parameter (Figure 15.2.3).

Motivated by the foregoing discussion, we shall now extend the definition of a tangent line to graphs of vector-valued functions.

15.2.4 DEFINITION. Let P be a point on the graph of a vector-valued function $\mathbf{f}(t)$, and let $\mathbf{f}(t_0)$ be the radius vector from the origin to P (Figure 15.2.4). If $\mathbf{f}'(t_0)$ exists and $\mathbf{f}'(t_0) \neq \mathbf{0}$, then we call $\mathbf{f}'(t_0)$ the *tangent vector* to the graph of $\mathbf{f}$ at $\mathbf{f}(t_0)$. If $\mathbf{f}'(t_0)$ is positioned with its initial point at P, then the line through $\mathbf{f}'(t_0)$ is called the *tangent line to the graph of $\mathbf{f}(t)$ at $f(t_0)$.*

REMARK.   Recall that for a real-valued function f, a tangent line does not exist at points of discontinuity or corners on the graph of f. The same is true for vector-valued functions.

It follows from Definition 15.2.4 and Formula (5) of Section 14.5 that a vector equation of the tangent line to the graph of $\mathbf{f}$ at $\mathbf{f}(t_0)$ is

$$\mathbf{r} = \mathbf{f}(t_0) + t\mathbf{f}'(t_0) \tag{2}$$

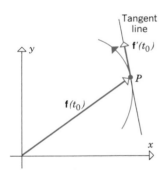

Figure 15.2.4

Example 4   Find parametric equations of the tangent line to the circular helix

$$x = \cos t$$
$$y = \sin t$$
$$z = t$$

at the point where $t = \pi/6$.

Solution.   We shall first use Formula (2) to find a vector equation of the tangent line, then we shall equate components to obtain the parametric equations.

A vector equation $\mathbf{r} = \mathbf{f}(t)$ of the helix is

$$x\mathbf{i} + y\mathbf{j} + z\mathbf{k} = (\cos t)\mathbf{i} + (\sin t)\mathbf{j} + t\mathbf{k}$$

Thus,

$$\mathbf{f}(t) = (\cos t)\mathbf{i} + (\sin t)\mathbf{j} + t\mathbf{k}$$
$$\mathbf{f}'(t) = (-\sin t)\mathbf{i} + (\cos t)\mathbf{j} + \mathbf{k}$$

$$\mathbf{f}\left(\frac{\pi}{6}\right) = \frac{\sqrt{3}}{2}\mathbf{i} + \frac{1}{2}\mathbf{j} + \frac{\pi}{6}\mathbf{k}$$

$$\mathbf{f}'\left(\frac{\pi}{6}\right) = -\frac{1}{2}\mathbf{i} + \frac{\sqrt{3}}{2}\mathbf{j} + \mathbf{k}$$

so from (2) a vector equation of the tangent line is

$$\mathbf{r} = \mathbf{f}\left(\frac{\pi}{6}\right) + t\mathbf{f}'\left(\frac{\pi}{6}\right)$$

or

$$\mathbf{r} = \left(\frac{\sqrt{3}}{2}\mathbf{i} + \frac{1}{2}\mathbf{j} + \frac{\pi}{6}\mathbf{k}\right) + t\left(-\frac{1}{2}\mathbf{i} + \frac{\sqrt{3}}{2}\mathbf{j} + \mathbf{k}\right)$$

Equating corresponding components yields the desired parametric equations

$$x = \frac{\sqrt{3}}{2} - \frac{1}{2}t$$

$$y = \frac{1}{2} + \frac{\sqrt{3}}{2}t$$

$$z = \frac{\pi}{6} + t \quad \blacktriangleleft$$

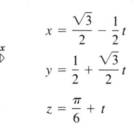

$\mathbf{f}(t) = t^2\mathbf{i} + t^3\mathbf{j}$

Figure 15.2.5

Example 5 The graph of $\mathbf{f}(t) = t^2\mathbf{i} + t^3\mathbf{j}$ is called a *semicubical parabola* (Figure 15.2.5). Find a vector equation of the tangent line to the graph of $\mathbf{f}(t)$ at:

(a)   the point $(1, 1)$ (b)   the point $(0, 0)$

Solution (a).   The point $(1, 1)$ on the graph of $\mathbf{f}$ corresponds to $t = 1$, so from (2) a vector equation of the tangent line at this point is

$$\mathbf{r} = \mathbf{f}(1) + t\mathbf{f}'(1) \tag{3}$$

But

$$\mathbf{f}(t) = t^2\mathbf{i} + t^3\mathbf{j}$$
$$\mathbf{f}'(t) = 2t\mathbf{i} + 3t^2\mathbf{j}$$
$$\mathbf{f}(1) = \mathbf{i} + \mathbf{j} \tag{4}$$
$$\mathbf{f}'(1) = 2\mathbf{i} + 3\mathbf{j}$$

so from (3)

$$\mathbf{r} = (\mathbf{i} + \mathbf{j}) + t(2\mathbf{i} + 3\mathbf{j})$$

is a vector-equation of the tangent line.

Solution (b).   The point $(0,0)$ on the graph of $\mathbf{f}$ corresponds to $t = 0$. But from (4) we have $\mathbf{f}'(0) = \mathbf{0}$, so there is no tangent line to the graph of $\mathbf{f}(t)$ at this point.   ◄

If $\mathbf{f}$ is a vector-valued function in 2-space or 3-space, then we call $\mathbf{f}$ a *smooth function* and its graph a *smooth curve* if the components of $\mathbf{f}$ have continuous derivatives and $\mathbf{f}'(t) \neq \mathbf{0}$ for any value of t. Thus, in 2-space

$$\mathbf{f}(t) = x(t)\mathbf{i} + y(t)\mathbf{j}$$

is smooth if $x'(t)$ and $y'(t)$ are continuous and

$$\mathbf{f}'(t) = x'(t)\mathbf{i} + y'(t)\mathbf{j} \neq \mathbf{0}$$

for any value of t. Similarly, in 3-space. The helix in Example 4 is smooth, but the semicubical parabola in Example 5 is not.

PROPERTIES OF
DERIVATIVES

Because derivatives of vector-valued functions are computed by components, most basic theorems on differentiating real-valued functions carry over to vector-valued functions.

Rules of Differentiation

15.2.5 THEOREM. *In either 2-space or 3-space let* $\mathbf{r}(t)$, $\mathbf{r}_1(t)$, *and* $\mathbf{r}_2(t)$ *be vector-valued functions,* $f(t)$ *a real-valued function,* k *a scalar, and* $\mathbf{c}$ *a fixed (constant) vector. Then the following rules of differentiation hold:*

$$\frac{d}{dt}[\mathbf{c}] = \mathbf{0}$$

$$\frac{d}{dt}[k\mathbf{r}(t)] = k\frac{d}{dt}[\mathbf{r}(t)]$$

$$\frac{d}{dt}[\mathbf{r}_1(t) + \mathbf{r}_2(t)] = \frac{d}{dt}[\mathbf{r}_1(t)] + \frac{d}{dt}[\mathbf{r}_2(t)]$$

$$\frac{d}{dt}[\mathbf{r}_1(t) - \mathbf{r}_2(t)] = \frac{d}{dt}[\mathbf{r}_1(t)] - \frac{d}{dt}[\mathbf{r}_2(t)]$$

$$\frac{d}{dt}[f(t)\mathbf{r}(t)] = f(t)\frac{d}{dt}[\mathbf{r}(t)] + \mathbf{r}(t)\frac{d}{dt}[f(t)]$$

The proofs are left as exercises.
In addition to the rules listed in the foregoing theorem, we have the following rules for differentiating dot products in 2-space or 3-space and cross products in 3-space:

$$\frac{d}{dt}[\mathbf{r}_1(t) \cdot \mathbf{r}_2(t)] = \mathbf{r}_1(t) \cdot \frac{d\mathbf{r}_2}{dt} + \frac{d\mathbf{r}_1}{dt} \cdot \mathbf{r}_2(t) \tag{5}$$

$$\frac{d}{dt}[\mathbf{r}_1(t) \times \mathbf{r}_2(t)] = \mathbf{r}_1(t) \times \frac{d\mathbf{r}_2}{dt} + \frac{d\mathbf{r}_1}{dt} \times \mathbf{r}_2(t) \tag{6}$$

The proofs of (5) and (6) are left as exercises.

REMARK.   In (5), the order of the factors in each term on the right does not matter, but in (6) it does.

The following theorem will be very useful in our later work.

15.2.6   THEOREM. *If* $\mathbf{r}(t)$ *is a vector-valued function in 2-space or 3-space and* $\|\mathbf{r}(t)\|$ *is constant for all t, then*

$$\mathbf{r}(t) \cdot \mathbf{r}'(t) = 0 \tag{7}$$

that is, $\mathbf{r}(t)$ *and* $\mathbf{r}'(t)$ *are orthogonal vectors for all t.*

Proof.   It follows from (5) with $\mathbf{r}_1(t) = \mathbf{r}_2(t) = \mathbf{r}(t)$ that

$$\frac{d}{dt}[\mathbf{r}(t) \cdot \mathbf{r}(t)] = \mathbf{r}(t) \cdot \frac{d\mathbf{r}}{dt} + \frac{d\mathbf{r}}{dt} \cdot \mathbf{r}(t)$$

or, equivalently,

$$\frac{d}{dt}[\|\mathbf{r}(t)\|^2] = 2\mathbf{r}(t) \cdot \frac{d\mathbf{r}}{dt} \tag{8}$$

But $\|\mathbf{r}(t)\|^2$ is constant, so its derivative is zero. Thus

$$2\mathbf{r}(t) \cdot \frac{d\mathbf{r}}{dt} = 0$$

from which (7) follows.   ∎

Figure 15.2.6

Example 6   In plane geometry one learns that a tangent line to a circle is perpendicular to the radius at the point of tangency. Similarly, in 3-space a tangent line to a curve on the surface of a sphere is perpendicular to the radius at the point of tangency (Figure 15.2.6). To see that this is so suppose that the graph of $\mathbf{r}(t)$ lies on the surface of the sphere of radius $k > 0$ centered at the origin. For each value of t we have $\|\mathbf{r}(t)\| = k$, so by Theorem 15.2.6

$$\mathbf{r}(t) \cdot \mathbf{r}'(t) = 0$$

which implies that the radius vector $\mathbf{r}(t)$ and the tangent vector $\mathbf{r}'(t)$ are perpendicular. This completes the argument because the tangent line, where it exists, is parallel to the tangent vector. ◄

If $\mathbf{r}(t)$ is a vector-valued function, and if the parameter t is a function of another parameter w, then the following version of the chain rule holds (Exercise 51).

The Chain Rule

15.2.7 THEOREM. *If* $\mathbf{u} = \mathbf{r}(t)$ *is a differentiable function of* t, *and* t *is a differentiable function of* w, *then*

$$\frac{d\mathbf{u}}{dw} = \frac{d\mathbf{u}}{dt}\frac{dt}{dw}$$

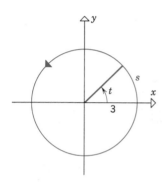

Figure 15.2.7

Example 7 The graph of

$$\mathbf{u} = (3\cos t)\mathbf{i} + (3\sin t)\mathbf{j}, \quad 0 \le t \le 2\pi$$

is a circle of radius 3 centered at the origin and oriented counterclockwise. For each value of t, let s be the arc length on the circle subtended by the angle t (Figure 15.2.7), so that

$$s = 3t \quad \text{or} \quad t = \frac{s}{3}$$

Thus,

$$\mathbf{u} = 3\cos\left(\frac{s}{3}\right)\mathbf{i} + 3\sin\left(\frac{s}{3}\right)\mathbf{j}, \quad 0 \le s \le 6\pi \tag{9}$$

is a representation of the circle with arc length s as the parameter. From (9) the derivative of $\mathbf{u}$ with respect to s is

$$\frac{d\mathbf{u}}{ds} = \left[-\sin\left(\frac{s}{3}\right)\right]\mathbf{i} + \left[\cos\left(\frac{s}{3}\right)\right]\mathbf{j}$$

This same result can be obtained from the chain rule and the given formula for $\mathbf{u}$ in terms of t by writing

$$\frac{d\mathbf{u}}{ds} = \frac{d\mathbf{u}}{dt}\frac{dt}{ds} = [(-3\sin t)\mathbf{i} + (3\cos t)\mathbf{j}]\left(\frac{1}{3}\right)$$

$$= (-\sin t)\mathbf{i} + (\cos t)\mathbf{j}$$

$$= \left[-\sin\left(\frac{s}{3}\right)\right]\mathbf{i} + \left[\cos\left(\frac{s}{3}\right)\right]\mathbf{j} \quad ◄$$

▶ Exercise Set 15.2

In Exercises 1–8, find the limit.

1. $\lim\limits_{t \to 3} (t^2 \mathbf{i} + 2t\mathbf{j})$.

2. $\lim\limits_{t \to \pi/4} \langle \cos t, \sin t \rangle$.

3. $\lim\limits_{t \to 0^+} \left(\sqrt{t}\mathbf{i} + \dfrac{\sin t}{t}\mathbf{j} \right)$.

4. $\lim\limits_{t \to +\infty} \left\langle \dfrac{t^2 + 1}{3t^2 + 2}, \dfrac{1}{t} \right\rangle$.

5. $\lim\limits_{t \to 2} (t\mathbf{i} - 3\mathbf{j} + t^2\mathbf{k})$.

6. $\lim\limits_{t \to \pi} \langle \cos 3t, e^{-t}, \sqrt{t} \rangle$.

7. $\lim\limits_{t \to +\infty} \left(\tan^{-1} t\,\mathbf{i} + \dfrac{t}{t^2 + 3}\mathbf{j} + \cos\dfrac{2}{t}\mathbf{k} \right)$.

8. $\lim\limits_{t \to 1} \left\langle \dfrac{3}{t^2}, \dfrac{\ln t}{t^2 - 1}, \sin 2t \right\rangle$.

In Exercises 9 and 10, prove that $\mathbf{f}$ is continuous at t_0.

9. $\mathbf{f}(t) = (3 \sin t)\mathbf{i} - 2t\mathbf{j}$; $t_0 = \pi/2$.

10. $\mathbf{f}(t) = 5\mathbf{i} - \sqrt{3t + 1}\mathbf{j} + e^{2t}\mathbf{k}$; $t_0 = 1$.

In Exercises 11–14, find $\mathbf{f}'(t)$.

11. $\mathbf{f}(t) = (4 + 5t)\mathbf{i} + (t - t^2)\mathbf{j}$.

12. $\mathbf{f}(t) = 4\mathbf{i} - (\cos t)\mathbf{j}$.

13. $\mathbf{f}(t) = \dfrac{1}{t}\mathbf{i} + (\tan t)\mathbf{j} + e^{2t}\mathbf{k}$.

14. $\mathbf{f}(t) = (\tan^{-1} t)\mathbf{i} + (t \cos t)\mathbf{j} - \sqrt{t}\mathbf{k}$.

In Exercises 15–22, find $\mathbf{r}'(t_0)$; then sketch the graph of $\mathbf{r}(t)$ and the tangent vector $\mathbf{r}'(t_0)$.

15. $\mathbf{r}(t) = \langle t, t^2 \rangle$; $t_0 = 2$.

16. $\mathbf{r}(t) = (\cos t)\mathbf{i} + (\sin t)\mathbf{j}$; $t_0 = 3\pi/4$.

17. $\mathbf{r}(t) = \langle e^{-t}, e^{2t} \rangle$; $t_0 = \ln 2$.

18. $\mathbf{r}(t) = (\cos 2t)\mathbf{i} - (4 \sin t)\mathbf{j}$; $t_0 = \pi$.

19. $\mathbf{r}(t) = (2 \sin t)\mathbf{i} + \mathbf{j} + (2 \cos t)\mathbf{k}$; $t_0 = \pi/2$.

20. $\mathbf{r}(t) = (\cos t)\mathbf{i} + (\sin t)\mathbf{j} + t\mathbf{k}$; $t_0 = \pi/4$.

21. $\mathbf{r}(t) = 3\mathbf{i} + t\mathbf{j} + (2 - t^2)\mathbf{k}$; $t_0 = 1$.

22. $\mathbf{r}(t) = t\mathbf{i} + 2t\mathbf{j} + t^2\mathbf{k}$; $t_0 = 2$.

In Exercises 23–26, find parametric equations of the line tangent to the graph of $\mathbf{r}(t)$ at the point where $t = t_0$.

23. $\mathbf{r}(t) = t^2\mathbf{i} + (2 - \ln t)\mathbf{j}$; $t_0 = 1$.

24. $\mathbf{r}(t) = e^{2t}\mathbf{i} - (2 \cos 3t)\mathbf{j}$; $t_0 = 0$.

25. $\mathbf{r}(t) = (2 \cos \pi t)\mathbf{i} + (2 \sin \pi t)\mathbf{j} + 3t\mathbf{k}$; $t_0 = \frac{1}{3}$.

26. $\mathbf{r}(t) = (\ln t)\mathbf{i} + e^{-t}\mathbf{j} + t^3\mathbf{k}$; $t_0 = 2$.

In Exercises 27–30, find a vector equation of the line tangent to the graph of $\mathbf{r}(t)$ at the point P_0 on the curve.

27. $\mathbf{r}(t) = (2t - 1)\mathbf{i} + \sqrt{3t + 4}\mathbf{j}$; $P_0(-1, 2)$.

28. $\mathbf{r}(t) = (4 \cos t)\mathbf{i} - 3t\mathbf{j}$; $P_0(2, -\pi)$.

29. $\mathbf{r}(t) = t^2\mathbf{i} - \dfrac{1}{t + 1}\mathbf{j} + (4 - t^2)\mathbf{k}$; $P_0(4, 1, 0)$.

30. $\mathbf{r}(t) = (\sin t)\mathbf{i} + (\cosh t)\mathbf{j} + (\tan^{-1} t)\mathbf{k}$; $P_0(0, 1, 0)$.

31. Find an equation of the plane that is perpendicular to the curve $\mathbf{r} = (3 \sin t)\mathbf{i} - (2 \cos t)\mathbf{j} + t\mathbf{k}$ at the point where $t = \pi/2$. [*Note:* A plane is considered to be perpendicular to a curve at a point if it is perpendicular to the tangent line at that point.]

32. Find an equation of the plane that is perpendicular to the curve $\mathbf{r} = 3t^2\mathbf{i} + \sqrt{t + 5}\mathbf{j} - 2t\mathbf{k}$ at the point $P(3, 2, 2)$ on the curve. [See the note in Exercise 31.]

33. (a) Find the points where the curve $\mathbf{r} = t\mathbf{i} + t^2\mathbf{j} - 3t\mathbf{k}$ intersects the plane $2x - y + z = -2$.

   (b) For the curve and plane in part (a), find, to the nearest degree, the acute angle that the tangent line to the curve makes with the plane at each point of intersection.

34. Find where the tangent line to the curve $\mathbf{r} = e^{-2t}\mathbf{i} + (\cos t)\mathbf{j} + (3 \sin t)\mathbf{k}$ at the point $(1, 1, 0)$ intersects the yz-plane.

35. Show that the graphs of $\mathbf{r}_1(t) = t\mathbf{i} + t^2\mathbf{j}$ and $\mathbf{r}_2(t) = t^3\mathbf{i} + t^6\mathbf{j}$ are the same, but that $\mathbf{r}_1'(t)$ is never zero, whereas $\mathbf{r}_2'(t) = \mathbf{0}$ for some value of t.

36. Show that the graphs of $\mathbf{r}_1(t) = (\cos t)\mathbf{i} + (\sin t)\mathbf{j} + t\mathbf{k}$ and $\mathbf{r}_2(t) = \cos (t^3)\mathbf{i} + \sin (t^3)\mathbf{j} + t^3\mathbf{k}$ are the same, but that $\mathbf{r}_1'(t)$ is never zero, whereas $\mathbf{r}_2'(t) = \mathbf{0}$ for some value of t.

37. Calculate $(d/dt)\,[\mathbf{r}_1(t) \cdot \mathbf{r}_2(t)]$ two ways: first using Formula (5), and then differentiating $\mathbf{r}_1(t) \cdot \mathbf{r}_2(t)$ directly.

   (a) $\mathbf{r}_1(t) = 2t\mathbf{i} + 3t^2\mathbf{j} + t^3\mathbf{k}$, $\mathbf{r}_2(t) = t^4\mathbf{k}$

   (b) $\mathbf{r}_1(t) = 3 \sec t\,\mathbf{i} - t\mathbf{j} + \ln t\mathbf{k}$, $\mathbf{r}_2(t) = 4t\mathbf{i} - \sin t\mathbf{k}$.

38. Calculate $\dfrac{d}{dt}[r_1(t) \times r_2(t)]$ two ways: first using

Formula (6), and then by differentiating $r_1(t) \times r_2(t)$ directly.

(a) $r_1(t) = 2t\mathbf{i} + 3t^2\mathbf{j} + t^3\mathbf{k}$, $r_2(t) = t^4\mathbf{k}$

(b) $r_1(t) = 3 \sec t\mathbf{i} - t\mathbf{j} + \ln t\mathbf{k}$,
$r_2(t) = 4t\mathbf{i} - \sin t\mathbf{k}$.

In Exercises 39–41, calculate du/dw by the chain rule, and then check your result by expressing $\mathbf{u}$ in terms of w and differentiating.

39. $\mathbf{u} = t\mathbf{i} + t^2\mathbf{j}$; $t = 4w + 1$.

40. $\mathbf{u} = \langle 3 \cos t, 3 \sin t \rangle$; $t = \pi w$.

41. $\mathbf{u} = e^t\mathbf{i} + 4e^{-t}\mathbf{j}$; $t = w^2$.

42. Let $r(t) = (a \cos t)\mathbf{i} + (a \sin t)\mathbf{j}$ $(a > 0)$.
(a) Show that $r(t) \cdot r'(t) = 0$.
(b) Give a geometric explanation of the result in part (a).

43. Prove: $(d/dt)[r(t) \times r'(t)] = r(t) \times r''(t)$.   [Use Formula (6).]

44. Let $\mathbf{u} = \mathbf{u}(t)$, $\mathbf{v} = \mathbf{v}(t)$, and $\mathbf{w} = \mathbf{w}(t)$ be differentiable vector-valued functions. Use Formulas (5) and (6) to show that

$$\frac{d}{dt}[\mathbf{u} \cdot (\mathbf{v} \times \mathbf{w})] =$$

$$\frac{d\mathbf{u}}{dt} \cdot [\mathbf{v} \times \mathbf{w}] + \mathbf{u} \cdot \left[\frac{d\mathbf{v}}{dt} \times \mathbf{w}\right] + \mathbf{u} \cdot \left[\mathbf{v} \times \frac{d\mathbf{w}}{dt}\right]$$

45. Let $u_1, u_2, u_3, v_1, v_2, v_3, w_1, w_2$, and w_3 be differentiable functions of t. Use Exercise 44 to show that:

$$\frac{d}{dt}\begin{vmatrix} u_1 & u_2 & u_3 \\ v_1 & v_2 & v_3 \\ w_1 & w_2 & w_3 \end{vmatrix}$$

$$= \begin{vmatrix} u_1' & u_2' & u_3' \\ v_1 & v_2 & v_3 \\ w_1 & w_2 & w_3 \end{vmatrix} + \begin{vmatrix} u_1 & u_2 & u_3 \\ v_1' & v_2' & v_3' \\ w_1 & w_2 & w_3 \end{vmatrix} + \begin{vmatrix} u_1 & u_2 & u_3 \\ v_1 & v_2 & v_3 \\ w_1' & w_2' & w_3' \end{vmatrix}$$

46. Let $\mathbf{r} = r(t)$. Show that

$$\frac{d}{dt}[\|\mathbf{r}\|] = \frac{1}{\|\mathbf{r}\|} \mathbf{r} \cdot \mathbf{r}'$$

[*Hint:* Consider $\mathbf{r} \cdot \mathbf{r}$.]

47. Use Theorem 15.2.5 and Exercise 46 to derive the formula

$$\frac{d}{dt}\left[\frac{\mathbf{r}}{\|\mathbf{r}\|}\right] = \frac{1}{\|\mathbf{r}\|} \mathbf{r}' - \frac{\mathbf{r} \cdot \mathbf{r}'}{\|\mathbf{r}\|^3} \mathbf{r}$$

48. Prove Theorem 15.2.5 for 2-space.

49. Derive Formulas (5) and (6) for 3-space.

50. Let $\mathbf{f}(t) = x(t)\mathbf{i} + y(t)\mathbf{j}$. Prove: If $\mathbf{f}'(t) = \mathbf{0}$ for all t in an interval, then $\mathbf{f}$ is constant on the interval. [*Hint:* See Theorem 4.10.3.]

51. Prove Theorem 15.2.7 for 2-space.

52. Let lim stand for any one of the limit symbols $\lim_{t \to a}$, $\lim_{t \to a^+}$, $\lim_{t \to a^-}$, $\lim_{t \to +\infty}$, or $\lim_{t \to -\infty}$. Prove for 2-space:
(a) If k is a scalar and $\lim \mathbf{f}(t)$ exists, then $\lim k\mathbf{f}(t) = k \lim \mathbf{f}(t)$.
(b) If $\lim \mathbf{f}(t)$ and $\lim \mathbf{g}(t)$ exist, then:
$\lim [\mathbf{f}(t) + \mathbf{g}(t)] = \lim \mathbf{f}(t) + \lim \mathbf{g}(t)$
$\lim [\mathbf{f}(t) - \mathbf{g}(t)] = \lim \mathbf{f}(t) - \lim \mathbf{g}(t)$.

53. Prove for 2-space: $\mathbf{f}$ is continuous at t_0 if and only if each component of $\mathbf{f}$ is continuous at t_0.

15.3   INTEGRATION OF VECTOR-VALUED FUNCTIONS; ARC LENGTH

In this section we shall define integrals of vector-valued functions and give some of their basic properties. We shall also derive some important formulas for the arc length of the graph of a vector-valued function.

Because derivatives of vector-valued functions are calculated by differentiating components, it is natural to define integrals of vector-valued functions in terms of components.

15.3.1   DEFINITION.

(a)   If $\mathbf{f}(t) = x(t)\mathbf{i} + y(t)\mathbf{j}$ is a vector-valued function in 2-space, then we define

$$\int \mathbf{f}(t)\, dt = \left(\int x(t)\, dt \right)\mathbf{i} + \left(\int y(t)\, dt \right)\mathbf{j}$$

$$\int_a^b \mathbf{f}(t)\, dt = \left(\int_a^b x(t)\, dt \right)\mathbf{i} + \left(\int_a^b y(t)\, dt \right)\mathbf{j}$$

(b)   If $\mathbf{f}(t) = x(t)\mathbf{i} + y(t)\mathbf{j} + z(t)\mathbf{k}$ is a vector-valued function in 3-space, then we define

$$\int \mathbf{f}(t)\, dt = \int [x(t)\mathbf{i} + y(t)\mathbf{j} + z(t)\mathbf{k}]\, dt$$

$$= \left(\int x(t)\, dt \right)\mathbf{i} + \left(\int y(t)\, dt \right)\mathbf{j} + \left(\int z(t)\, dt \right)\mathbf{k}$$

$$\int_a^b \mathbf{f}(t)\, dt = \int_a^b [x(t)\mathbf{i} + y(t)\mathbf{j} + z(t)\mathbf{k}]\, dt$$

$$= \left(\int_a^b x(t)\, dt \right)\mathbf{i} + \left(\int_a^b y(t)\, dt \right)\mathbf{j} + \left(\int_a^b z(t)\, dt \right)\mathbf{k}$$

Example 1   Let $\mathbf{f}(t) = 2t\mathbf{i} + 3t^2\mathbf{j}$. Find:

(a) $\displaystyle\int \mathbf{f}(t)\, dt$ (b) $\displaystyle\int_0^2 \mathbf{f}(t)\, dt$

Solution (a).

$$\int \mathbf{f}(t)\, dt = \int (2t\mathbf{i} + 3t^2\mathbf{j})\, dt$$

$$= \left(\int 2t\, dt \right)\mathbf{i} + \left(\int 3t^2\, dt \right)\mathbf{j} = (t^2 + C_1)\mathbf{i} + (t^3 + C_2)\mathbf{j}$$

$$= (t^2\mathbf{i} + t^3\mathbf{j}) + C_1\mathbf{i} + C_2\mathbf{j} = t^2\mathbf{i} + t^3\mathbf{j} + \mathbf{C}$$

where $\mathbf{C} = C_1\mathbf{i} + C_2\mathbf{j}$ is an arbitrary vector constant of integration.

Solution (b). $\displaystyle\int_0^2 \mathbf{f}(t)\, dt = \int_0^2 (2t\mathbf{i} + 3t^2\mathbf{j})\, dt$

$$= \left(\int_0^2 2t\, dt \right)\mathbf{i} + \left(\int_0^2 3t^2\, dt \right)\mathbf{j} = \left[t^2 \right]_0^2 \mathbf{i} + \left[t^3 \right]_0^2 \mathbf{j} = 4\mathbf{i} + 8\mathbf{j} \quad \blacktriangleleft$$

PROPERTIES OF INTEGRALS

In the exercises we ask the reader to show that the standard properties of integrals carry over to vector-valued functions in 2-space and 3-space:

$$\int c\mathbf{f}(t)\, dt = c \int \mathbf{f}(t)\, dt \tag{1}$$

$$\int [\mathbf{f}(t) + \mathbf{g}(t)]\, dt = \int \mathbf{f}(t)\, dt + \int \mathbf{g}(t)\, dt \tag{2}$$

$$\int [\mathbf{f}(t) - \mathbf{g}(t)]\, dt = \int \mathbf{f}(t)\, dt - \int \mathbf{g}(t)\, dt \tag{3}$$

These properties also hold for definite integrals of vector-valued functions. In addition, we leave it for the reader to show that if $\mathbf{f}$ is a vector-valued function in 2-space or 3-space, then

$$\frac{d}{dt}\left[\int \mathbf{f}(t)\, dt \right] = \mathbf{f}(t) \tag{4}$$

This shows that an indefinite integral of $\mathbf{f}(t)$ is, in fact, an antiderivative of $\mathbf{f}(t)$, just as for real-valued functions. We also leave it as an exercise to show that if $\mathbf{F}(t)$ is any antiderivative of $\mathbf{f}(t)$ in the sense that $\mathbf{F}'(t) = \mathbf{f}(t)$, then

$$\int \mathbf{f}(t)\, dt = \mathbf{F}(t) + \mathbf{C} \tag{5}$$

where $\mathbf{C}$ is an arbitrary vector constant of integration. Moreover,

$$\int_a^b \mathbf{f}(t)\, dt = \mathbf{F}(t) \Bigg]_{t=a}^b = \mathbf{F}(b) - \mathbf{F}(a) \tag{6}$$

which is the extension of the First Fundamental Theorem of Calculus (Theorem 5.7.1) to vector-valued functions.

Example 2 In part (a) of Example 1 we showed that $\mathbf{F}(t) = t^2\mathbf{i} + t^3\mathbf{j}$ is an antiderivative of $\mathbf{f}(t) = 2t\mathbf{i} + 3t^2\mathbf{j}$. (It is the antiderivative that results when $\mathbf{C} = \mathbf{0}$.) Thus, from (6)

$$\int_0^2 \mathbf{f}(t)\, dt = \int_0^2 (2t\mathbf{i} + 3t^2\mathbf{j})\, dt = \left[\mathbf{F}(t) \right]_0^2$$

$$= \left[t^2\mathbf{i} + t^3\mathbf{j} \right]_0^2 = (4\mathbf{i} + 8\mathbf{j}) - (0\mathbf{i} + 0\mathbf{j})$$

$$= 4\mathbf{i} + 8\mathbf{j}$$

which agrees with the result obtained in Example 1 by integrating term by term. ◄

In Theorem 13.4.1 we showed that if $x'(t)$ and $y'(t)$ are continuous for $a \leq t \leq b$, then the curve given by the parametric equations

$$\begin{array}{c} x = x(t) \\ y = y(t) \end{array}, \quad a \leq t \leq b \tag{7}$$

has arc length

$$L = \int_a^b \sqrt{\left(\frac{dx}{dt}\right)^2 + \left(\frac{dy}{dt}\right)^2}\, dt \tag{8}$$

This result generalizes to curves in 3-space exactly as one would expect: If $x'(t)$, $y'(t)$, and $z'(t)$ are continuous for $a \leq t \leq b$, then the curve given by the parametric equations

$$\begin{array}{c} x = x(t) \\ y = y(t) \,, \quad a \leq t \leq b \\ z = z(t) \end{array} \tag{9}$$

has arc length

$$L = \int_a^b \sqrt{\left(\frac{dx}{dt}\right)^2 + \left(\frac{dy}{dt}\right)^2 + \left(\frac{dz}{dt}\right)^2}\, dt \tag{10}$$

Example 3   Find the arc length of that portion of the circular helix

$$\begin{array}{l} x = \cos t \\ y = \sin t \\ z = t \end{array}$$

from $t = 0$ to $t = \pi$.

Solution.   From (10) the arc length is

$$\begin{aligned} L &= \int_0^\pi \sqrt{\left(\frac{dx}{dt}\right)^2 + \left(\frac{dy}{dt}\right)^2 + \left(\frac{dz}{dt}\right)^2}\, dt \\ &= \int_0^\pi \sqrt{(-\sin t)^2 + (\cos t)^2 + 1}\, dt \\ &= \int_0^\pi \sqrt{2}\, dt = \sqrt{2}\, \pi \quad ◄ \end{aligned}$$

ARC LENGTH AS A
PARAMETER

Figure 15.3.1

For many purposes the best parameter to use for representing a curve in 2-space or 3-space parametrically is the length of arc measured along the curve from some fixed reference point. This can be done as follows:

> **Step 1.** Select an arbitrary point on the curve C to serve as a *reference point*.
>
> **Step 2.** Starting from the reference point, choose one direction along the curve to be the *positive direction* and the other to be the *negative direction*.
>
> **Step 3.** If P is a point on the curve, let s be the "signed" arc length along C from the reference point to P, where s is positive if P is in the positive direction from the reference point, and s is negative if P is in the negative direction. Figure 15.3.1 illustrates this idea.

By this procedure, a unique point P on the curve is determined when a value for s is given. For example, $s = 2$ determines the point that is 2 units along the curve in the positive direction from the reference point, and $s = -\frac{3}{2}$ determines the point that is $\frac{3}{2}$ units along the curve in the negative direction from the reference point.

Let us now treat s as a variable. As the value of s changes, the corresponding point P moves along C and the coordinates of P become functions of s. Thus, in 2-space the coordinates of P are $(x(s), y(s))$, and in 3-space they are $(x(s), y(s), z(s))$. Therefore, in 2-space the curve C is given by the parametric equations

$$x = x(s)$$
$$y = y(s)$$

and in 3-space by

$$x = x(s)$$
$$y = y(s)$$
$$z = z(s)$$

REMARK.   When defining the parameter s, the choice of positive and negative directions is arbitrary. However, it may be that the curve C is already specified in terms of some other parameter t, in which case we shall agree always to take the direction of increasing t as the positive direction for the parameter s. By so doing, s will increase as t increases and vice versa.

The following theorem gives a formula for computing an arc-length parameter s when the curve C is expressed in terms of some other parameter t. This result will be used when we want to change the parametrization for C from t to s.

15.3.2   THEOREM.

(a)   *Let a curve C in 2-space be given parametrically by*

$$x = x(t)$$
$$y = y(t)$$

where $x'(t)$ and $y'(t)$ are continuous functions. If an arc-length parameter s is introduced with its reference point at $(x(t_0), y(t_0))$, then the parameters s and t are related by

$$s = \int_{t_0}^{t} \sqrt{\left(\frac{dx}{du}\right)^2 + \left(\frac{dy}{du}\right)^2}\, du \qquad (11a)$$

(b)   *Let a curve C in 3-space be given parametrically by*

$$x = x(t)$$
$$y = y(t)$$
$$z = z(t)$$

where $x'(t)$, $y'(t)$, and $z'(t)$ are continuous functions. If an arc-length parameter s is introduced with its reference point at $(x(t_0), y(t_0), z(t_0))$, then the parameters s and t are related by

$$s = \int_{t_0}^{t} \sqrt{\left(\frac{dx}{du}\right)^2 + \left(\frac{dy}{du}\right)^2 + \left(\frac{dz}{du}\right)^2}\, du \qquad (11b)$$

Proof.   We shall prove (a) and leave the proof of (b) as an exercise.

If $t > t_0$, then from (8) (with u as the variable of integration rather than t) it follows that

$$\int_{t_0}^{t} \sqrt{\left(\frac{dx}{du}\right)^2 + \left(\frac{dy}{du}\right)^2}\, du \qquad (12)$$

represents the arc length of that portion of the curve C that lies between $(x(t_0), y(t_0))$ and $(x(t), y(t))$. If $t < t_0$, then (12) is the negative of this arc length. In either case, integral (12) represents the "signed" arc length s between these points, which proves (11a).   ∎

It follows from Formulas (11a) and (11b) and the Second Fundamental Theorem of Calculus (Theorem 5.9.5) that in 2-space

$$\frac{ds}{dt} = \frac{d}{dt}\left[\int_{t_0}^{t}\sqrt{\left(\frac{dx}{du}\right)^2 + \left(\frac{dy}{du}\right)^2}\,du\right] = \sqrt{\left(\frac{dx}{dt}\right)^2 + \left(\frac{dy}{dt}\right)^2}$$

and in 3-space

$$\frac{ds}{dt} = \frac{d}{dt}\left[\int_{t_0}^{t}\sqrt{\left(\frac{dx}{du}\right)^2 + \left(\frac{dy}{du}\right)^2 + \left(\frac{dz}{du}\right)^2}\,dt\right]$$

$$= \sqrt{\left(\frac{dx}{dt}\right)^2 + \left(\frac{dy}{dt}\right)^2 + \left(\frac{dz}{dt}\right)^2}$$

Thus, in 2-space and 3-space, respectively,

$$\frac{ds}{dt} = \sqrt{\left(\frac{dx}{dt}\right)^2 + \left(\frac{dy}{dt}\right)^2} \tag{13a}$$

$$\frac{ds}{dt} = \sqrt{\left(\frac{dx}{dt}\right)^2 + \left(\frac{dy}{dt}\right)^2 + \left(\frac{dz}{dt}\right)^2} \tag{13b}$$

REMARK.   Formulas (13a) and (13b) reveal two facts worth noting. First, ds/dt does not depend on t_0; that is, the value of ds/dt is independent of where the reference point for the parameter s is located. This is to be expected since changing the position of the reference point shifts each value of s by a constant (the arc length between the reference points), and this constant drops out when we differentiate. The second fact to be noted from (13a) and (13b) is that $ds/dt \geq 0$ for all t. This is also to be expected since s increases with t by the remark preceding Theorem 15.3.2.

Example 4   Find parametric equations for the line

$$x = 2t + 1 \tag{14}$$
$$y = 3t - 2$$

using arc length s as a parameter, where the reference point for s is the point $(1, -2)$.

Solution.   In Formula (11a) we used u as the variable of integration because t was needed as a limit of integration. To apply (11a), we first rewrite the given parametric equations with u in place of t; this gives

$$x = 2u + 1$$
$$y = 3u - 2$$

from which we obtain

$$\frac{dx}{du} = 2, \quad \frac{dy}{du} = 3$$

From (14) we see that the reference point $(1, -2)$ corresponds to $t = t_0 = 0$, so (11a) yields

$$s = \int_{t_0}^{t} \sqrt{\left(\frac{dx}{du}\right)^2 + \left(\frac{dy}{du}\right)^2}\, du = \int_{0}^{t} \sqrt{13}\, du = \sqrt{13}\,u \Bigg]_{u=0}^{t} = \sqrt{13}\,t$$

Therefore,

$$t = \frac{1}{\sqrt{13}} s$$

Substituting this expression in the given parametric equations yields

$$x = 2\left(\frac{1}{\sqrt{13}} s\right) + 1 = \frac{2}{\sqrt{13}} s + 1$$

$$y = 3\left(\frac{1}{\sqrt{13}} s\right) - 2 = \frac{3}{\sqrt{13}} s - 2 \quad \blacktriangleleft$$

Example 5   Find parametric equations for the circle

$$\begin{array}{l} x = a \cos t \\ y = a \sin t \end{array}, \quad 0 \le t \le 2\pi$$

using arc length s as a parameter, with the reference point for s being $(a, 0)$, where $a > 0$.

Solution.   Proceeding as in Example 4, we first replace t by u in the given equations so that

$$x = a \cos u, \quad y = a \sin u$$

and

$$\frac{dx}{du} = -a \sin u, \quad \frac{dy}{du} = a \cos u$$

Since the reference point $(a, 0)$ corresponds to $t = 0$, we obtain from (11a),

$$s = \int_{t_0}^{t} \sqrt{\left(\frac{dx}{du}\right)^2 + \left(\frac{dy}{du}\right)^2}\, du = \int_{0}^{t} \sqrt{(-a \sin u)^2 + (a \cos u)^2}\, du$$

$$= \int_{0}^{t} a\, du = au \Bigg]_{u=0}^{u=t} = at$$

Solving for t in terms of s yields

$$t = \frac{s}{a}$$

Substituting this in the given parametric equations and using the fact that $s = at$ ranges from 0 to $2\pi a$ as t ranges from 0 to 2π, we obtain

$$x = a \cos\left(\frac{s}{a}\right)$$
$$y = a \sin\left(\frac{s}{a}\right) \quad , \quad 0 \le s \le 2\pi a$$

◀

VECTOR FORMULAS FOR ARC LENGTH

Some of the formulas in this section have alternative formulations in vector notation that are important to know. To derive these we shall, as usual, let

$$\mathbf{r} = x(t)\mathbf{i} + y(t)\mathbf{j}$$

and

$$\mathbf{r} = x(t)\mathbf{i} + y(t)\mathbf{j} + z(t)\mathbf{k}$$

be vector equations in 2-space and 3-space, respectively. Differentiating these equations with respect to t yields

$$\frac{d\mathbf{r}}{dt} = \left(\frac{dx}{dt}\right)\mathbf{i} + \left(\frac{dy}{dt}\right)\mathbf{j}$$
$$\frac{d\mathbf{r}}{dt} = \left(\frac{dx}{dt}\right)\mathbf{i} + \left(\frac{dy}{dt}\right)\mathbf{j} + \left(\frac{dz}{dt}\right)\mathbf{k}$$

from which we obtain

$$\left\|\frac{d\mathbf{r}}{dt}\right\| = \sqrt{\left(\frac{dx}{dt}\right)^2 + \left(\frac{dy}{dt}\right)^2} \tag{15a}$$

$$\left\|\frac{d\mathbf{r}}{dt}\right\| = \sqrt{\left(\frac{dx}{dt}\right)^2 + \left(\frac{dy}{dt}\right)^2 + \left(\frac{dz}{dt}\right)^2} \tag{15b}$$

in 2-space and 3-space, respectively. Comparing these to (13a) and (13b), we see that in both 2-space and 3-space

$$\left\|\frac{d\mathbf{r}}{dt}\right\| = \frac{ds}{dt} \tag{16}$$

Substituting (15a) in arc-length formula (8) and (15b) in arc-length formula (10), we see that in both 2-space and 3-space the formula for arc length can be written as

$$L = \int_a^b \left\| \frac{d\mathbf{r}}{dt} \right\| dt \tag{17}$$

Still another formula for arc length can be obtained by substituting (16) in (17) to obtain

$$L = \int_a^b \left(\frac{ds}{dt} \right) dt \tag{18}$$

All of these formulas will have applications in subsequent sections.

▶ Exercise Set 15.3

In Exercises 1–12, evaluate the integral.

1. $\displaystyle\int (3\mathbf{i} + 4t\mathbf{j})\, dt$.

2. $\displaystyle\int [(\cos t)\mathbf{i} + (\sin t)\mathbf{j}]\, dt$.

3. $\displaystyle\int_0^{\pi/3} \langle \cos 3t, -\sin 3t \rangle\, dt$.

4. $\displaystyle\int_0^1 (t^2\mathbf{i} + t^3\mathbf{j})\, dt$.

5. $\displaystyle\int_1^9 (t^{1/2}\mathbf{i} + t^{-1/2}\mathbf{j})\, dt$.

6. $\displaystyle\int [(t \sin t)\mathbf{i} + \mathbf{j}]\, dt$.   7. $\displaystyle\int \langle te^t, \ln t \rangle\, dt$.

8. $\displaystyle\int_0^2 \|t\mathbf{i} + t^2\mathbf{j}\|\, dt$.

9. $\displaystyle\int \left[t^2\mathbf{i} - 2t\mathbf{j} + \frac{1}{t}\mathbf{k} \right] dt$.

10. $\displaystyle\int \langle e^{-t}, e^t, 3t^2 \rangle\, dt$.

11. $\displaystyle\int_0^1 (e^{2t}\mathbf{i} + e^{-t}\mathbf{j} + t\mathbf{k})\, dt$.

12. $\displaystyle\int_{-3}^3 \langle (3 - t)^{3/2}, (3 + t)^{3/2}, 1 \rangle\, dt$.

13. Suppose that a particle moves through 3-space along the curve $\mathbf{r} = t\mathbf{i} - 3t^2\mathbf{j} + \mathbf{k}$ and that it is subjected to a force of $\mathbf{F} = 3x\mathbf{i} - 2\mathbf{j} + yz\mathbf{k}$ when it is at the point (x, y, z).
    (a) Find $\mathbf{F}$ in terms of t for points on the path.
    (b) Find $\displaystyle\int_0^2 \mathbf{F} \cdot \frac{d\mathbf{r}}{dt}\, dt$. [Note: Later, we shall see that this is the work done by the force as the particle moves along the curve from the point where $t = 0$ to the point where $t = 2$.]

14. Find $\mathbf{r}(t)$ given that $\mathbf{r}'(t) = t^2\mathbf{i} + 2t\mathbf{j}$ and $\mathbf{r}(0) = \mathbf{i} + \mathbf{j}$.

15. Find $\mathbf{r}(t)$ given that $\mathbf{r}'(t) = (\cos t)\mathbf{i} + (\sin t)\mathbf{j}$ and $\mathbf{r}(0) = \mathbf{i} - \mathbf{j}$.

16. Find $\mathbf{r}(t)$ given that $\mathbf{r}''(t) = \mathbf{i} + e^t\mathbf{j}$, $\mathbf{r}(0) = 2\mathbf{i}$, and $\mathbf{r}'(0) = \mathbf{j}$.

17. Find $\mathbf{r}(t)$ given that $\mathbf{r}''(t) = 12t^2\mathbf{i} - 2\mathbf{j}$, $\mathbf{r}'(0) = \mathbf{0}$ and $\mathbf{r}(0) = 2\mathbf{i} - 4\mathbf{j}$.

18. Find $\mathbf{r}(t)$ given that $\mathbf{r}'(t) = e^{-2t}\mathbf{i} + (\cos t)\mathbf{j} - \mathbf{k}$ and $\mathbf{r}(0) = 3\mathbf{j} + 2\mathbf{k}$.

19. Find $\mathbf{r}(t)$ given that $\mathbf{r}'(t) = 2\mathbf{i} + \dfrac{t}{t^2 + 1}\mathbf{j} + t\mathbf{k}$ and $\mathbf{r}(1) = \mathbf{0}$.

20. Find $\mathbf{r}(t)$ given that $\mathbf{r}''(t) = (4 \sin 2t)\mathbf{i} + 6t\mathbf{j} + e^{-t}\mathbf{k}$, $\mathbf{r}(0) = 2\mathbf{i}$, and $\mathbf{r}'(0) = \mathbf{k}$.

In Exercises 21–26, find the arc length of the curve.

21. $r(t) = (4 + 3t)i + (2 - 2t)j + (5 + t)k$;
$3 \le t \le 4$.

22. $r(t) = 3 \cos t\, i + 3 \sin t\, j + t\, k$; $0 \le t \le 2\pi$.

23. $r(t) = t^3 i + t j + \frac{1}{2}\sqrt{6}t^2 k$; $1 \le t \le 3$.

24. $x = \cos^3 t$, $y = \sin^3 t$, $z = 2$; $0 \le t \le \pi/2$.

25. $r(t) = \langle e^t, e^{-t}, \sqrt{2}t \rangle$; $0 \le t \le 1$.

26. $x = \frac{1}{2}t$, $y = \frac{1}{3}(1 - t)^{3/2}$, $z = \frac{1}{3}(1 + t)^{3/2}$;
$-1 \le t \le 1$.

27. Find the arc length of the circular helix $x = a \cos t$, $y = a \sin t$, $z = ct$ for $0 \le t \le t_0$.

28. Copper tubing with an outside diameter of $\frac{1}{2}$ in. is to be wrapped in a circular helix around a cylindrical core that has a 12 in. diameter. What length of tubing will make one complete turn around the cylinder in a distance of 20 in. measured along the axis of the cylinder?

Enlarged cross-section

12 in.

20 in.

$\frac{1}{2}$ in.

In Exercises 29–39, find parametric equations for the curve using arc length s as a parameter. Use the point on the curve where $t = 0$ as the reference point.

29. $r(t) = (3t - 2)i + (4t + 3)j$.

30. $r(t) = (3 \cos 2t)i + (3 \sin 2t)j$; $0 \le t \le \pi$.

31. $r(t) = (3 + \cos t)i + (2 + \sin t)j$; $0 \le t \le 2\pi$.

32. $r(t) = (\cos^3 t)i + (\sin^3 t)j$; $0 \le t \le \pi/2$.

33. $r(t) = \frac{1}{3}t^3 i + \frac{1}{2}t^2 j$; $t \ge 0$.

34. $r(t) = (1 + t)^2 i + (1 + t)^3 j$; $0 \le t \le 1$.

35. $r(t) = (e^t \cos t)i + (e^t \sin t)j$; $0 \le t \le \pi/2$.

36. $r(t) = (a \cos t)i + (a \sin t)j + ct k$; $t \ge 0$.

37. $r(t) = (1 - 3t)i + 5t j + (2 + t)k$.

38. $r(t) = \sin(e^t)i + \cos(e^t)j + \sqrt{3}e^t k$; $t \ge 0$.

39. $r(t) = (t \cos t)i + (t \sin t)j + \frac{2}{3}\sqrt{2}t^{3/2}k$; $t \ge 0$.

40. Find parametric equations for the cycloid

$$x = at - a \sin t$$
$$y = a - a \cos t$$
, $0 \le t \le 2\pi$

using arc length as the parameter. Take $(0, 0)$ as the reference point.

41. A thread with negligible thickness is unwound from a spool of radius a, with the unwound piece always held straight.

(a) Show that, in terms of the parameter θ ($\theta \ge 0$) as shown in the figure, the curve traced out by the free end of the thread has parametric equations

$$x = a(\cos \theta + \theta \sin \theta)$$
$$y = a(\sin \theta - \theta \cos \theta)$$

(b) Find parametric equations for the curve of part (a) using arc length as the parameter, where $\theta = 0$ is the reference point.

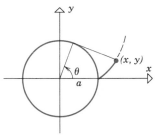

42. Show that in cylindrical coordinates a curve given by the parametric equations $r = r(t)$, $\theta = \theta(t)$, $z = z(t)$, for $a \le t \le b$, has arc length

$$L = \int_a^b \sqrt{\left(\frac{dr}{dt}\right)^2 + r^2\left(\frac{d\theta}{dt}\right)^2 + \left(\frac{dz}{dt}\right)^2}\, dt$$

[*Hint:* Use the relationships $x = r \cos \theta$, $y = r \sin \theta$.]

43. Use the formula in Exercise 42 to find the arc length of the following curves.
(a) $r = e^{2t}$, $\theta = t$, $z = e^{2t}$; $0 \le t \le \ln 2$.
(b) $r = t^2$, $\theta = \ln t$, $z = \frac{1}{3}t^3$; $1 \le t \le 2$.

44. Show that in spherical coordinates a curve given by the parametric equations $\rho = \rho(t)$, $\theta = \theta(t)$, $\phi = \phi(t)$, for $a \le t \le b$, has arc length

$$L = \int_a^b \sqrt{\left(\frac{d\rho}{dt}\right)^2 + \rho^2 \sin^2 \phi \left(\frac{d\theta}{dt}\right)^2 + \rho^2 \left(\frac{d\phi}{dt}\right)^2}\; dt$$

[*Hint:* Use the relationships $x = \rho \sin \phi \cos \theta$, $y = \rho \sin \phi \sin \theta$, $z = \rho \cos \phi$.]

45. Use the formula in Exercise 44 to find the arc length of the following curves.

(a) $\rho = e^{-t}$, $\theta = 2t$, $\phi = \dfrac{\pi}{4}$; $0 \le t \le 2$.

(b) $\rho = 2t$, $\theta = \ln t$, $\phi = \dfrac{\pi}{6}$; $1 \le t \le 5$.

46. Let $\mathbf{r}(t) = t\mathbf{i} + (\cosh t)\mathbf{j}$. Find:

(a) $\|\mathbf{r}'(t)\|$ (b) $\dfrac{ds}{dt}$

(c) $\displaystyle\int_0^{\ln 3} \|\mathbf{r}'(t)\|\; dt$.

47. Let $\mathbf{r}(t) = (\ln t)\mathbf{i} + 2t\mathbf{j} + t^2\mathbf{k}$. Find:

(a) $\|\mathbf{r}'(t)\|$ (b) $\dfrac{ds}{dt}$

(c) $\displaystyle\int_1^3 \|\mathbf{r}'(t)\|\; dt$.

48. Let $x = \cos t$, $y = \sin t$, $z = t^{3/2}$. Find:

$$\int_0^2 \left(\frac{ds}{dt}\right) dt$$

49. Prove for 2-space: If $\mathbf{F}'(t) = \mathbf{f}(t)$, then

(a) $\displaystyle\int \mathbf{f}(t)\; dt = \mathbf{F}(t) + \mathbf{C}$, where $\mathbf{C}$ is an arbitrary vector constant;

(b) $\displaystyle\int_a^b \mathbf{f}(t)\; dt = \mathbf{F}(b) - \mathbf{F}(a)$.

50. Prove for 2-space:

(a) $\displaystyle\int c\mathbf{f}(t)\; dt = c \int \mathbf{f}(t)\; dt$, where c is a scalar constant;

(b) $\displaystyle\int [\mathbf{f}(t) + \mathbf{g}(t)]\; dt = \int \mathbf{f}(t)\; dt + \int \mathbf{g}(t)\; dt$.

51. Prove part (b) of Theorem 15.3.2.

15.4 UNIT TANGENT AND NORMAL VECTORS

In this section we shall discuss some geometric properties of vector-valued functions. Our work here will have important applications to the study of motion along a curved path in 2-space or 3-space.

UNIT TANGENT VECTORS

Recall from Section 15.2 that if $\mathbf{r}(t)$ is a vector-valued function in 2-space or 3-space with graph C, then the vector $\mathbf{r}'(t)$ is tangent to the graph of C if this vector is positioned so its initial point is at the terminal point of $\mathbf{r}(t)$. Moreover, $\mathbf{r}'(t)$ points in the direction of increasing t. Thus, if $\mathbf{r}'(t) \ne \mathbf{0}$, then the vector $\mathbf{T}(t)$ defined by

$$\mathbf{T}(t) = \frac{\mathbf{r}'(t)}{\|\mathbf{r}'(t)\|}$$

will be tangent to C, point in the direction of increasing t, and have length 1 (Figure 15.4.1). This suggests the following definition.

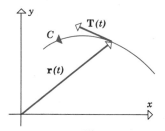

Figure 15.4.1

15.4.1 DEFINITION. Let $\mathbf{r}(t)$ be a vector-valued function in 2-space or 3-space with graph C. The *unit tangent vector* $\mathbf{T} = \mathbf{T}(t)$ to C is defined by

$$\mathbf{T}(t) = \frac{\mathbf{r}'(t)}{\|\mathbf{r}'(t)\|}$$

if $\mathbf{r}'(t) \neq \mathbf{0}$.

Observe that $\mathbf{T}(t)$ is defined only at those points where $\mathbf{r}'(t) \neq \mathbf{0}$. At each such point $\mathbf{T}(t)$ can be regarded as an indicator of the "direction" of the curve C.

Example 1 Find the unit tangent vector to the graph of $\mathbf{r}(t) = t^2\mathbf{i} + t^3\mathbf{j}$ at the point where $t = 2$.

Solution. Since

$$\mathbf{r}'(t) = 2t\mathbf{i} + 3t^2\mathbf{j}$$

we obtain

$$\mathbf{T}(t) = \frac{\mathbf{r}'(t)}{\|\mathbf{r}'(t)\|} = \frac{2t\mathbf{i} + 3t^2\mathbf{j}}{\|2t\mathbf{i} + 3t^2\mathbf{j}\|} = \frac{2t\mathbf{i} + 3t^2\mathbf{j}}{\sqrt{4t^2 + 9t^4}}$$

Thus,

$$\mathbf{T}(2) = \frac{4\mathbf{i} + 12\mathbf{j}}{\sqrt{160}} = \frac{4\mathbf{i} + 12\mathbf{j}}{4\sqrt{10}} = \frac{1}{\sqrt{10}}\mathbf{i} + \frac{3}{\sqrt{10}}\mathbf{j}$$

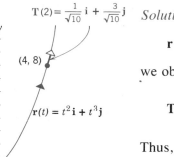

Figure 15.4.2

The graph of $\mathbf{r}(t)$ and the vector $\mathbf{T}(2)$ are shown in Figure 15.4.2. ◄

REMARK. Unless stated otherwise, we shall always assume that the unit tangent vector $\mathbf{T}(t)$ is positioned with its initial point at the terminal point of $\mathbf{r}(t)$ as in Figure 15.4.1. This will ensure that $\mathbf{T}(t)$ is actually tangent to the graph of $\mathbf{r}(t)$ and not simply parallel to a tangent vector.

T EXPRESSED IN TERMS OF AN ARC LENGTH PARAMETER

If $\mathbf{r}(t)$ is a smooth vector-valued function in 2-space or 3-space and s is an arc-length parameter for the graph of $\mathbf{r}(t)$, then the formula for a unit tangent vector has the following simple form in terms of s,

$$\mathbf{T} = \frac{d\mathbf{r}}{ds} \tag{1}$$

This result can be obtained from Formula (16) of Section 15.3 and the chain rule as follows:

$$\mathbf{T} = \frac{1}{\|\mathbf{r}'(t)\|} \mathbf{r}'(t) = \frac{1}{ds/dt} \frac{d\mathbf{r}}{dt} = \frac{1}{ds/dt} \frac{d\mathbf{r}}{ds} \frac{ds}{dt} = \frac{d\mathbf{r}}{ds}$$

Example 2   In Example 5 of Section 15.3 we obtained the parametric equations

$$\begin{matrix} x = a \cos(s/a) \\ y = a \sin(s/a) \end{matrix}, \quad 0 \leq s \leq 2\pi a$$

for a circle of radius a centered at the origin. In vector notation, this circle is the graph of

$$\mathbf{r}(s) = a \cos\left(\frac{s}{a}\right)\mathbf{i} + a \sin\left(\frac{s}{a}\right)\mathbf{j}, \quad 0 \leq s \leq 2\pi a$$

so that for each value of s, the unit tangent vector to the circle is

$$\mathbf{T} = \mathbf{T}(s) = \frac{d\mathbf{r}}{ds} = -\sin\left(\frac{s}{a}\right)\mathbf{i} + \cos\left(\frac{s}{a}\right)\mathbf{j} \quad \blacktriangleleft$$

Recall that the graph C of a function $\mathbf{r}(t)$ in 2-space or 3-space is called smooth if the components of $\mathbf{r}'(t)$ are continuous and $\mathbf{r}'(t) \neq \mathbf{0}$ for any value of t. Smooth curves are sometimes described as having "continuously turning tangents." This is because the tangent vector $\mathbf{T}(t)$ for a smooth curve varies "continuously" without abrupt changes in direction as t increases. The function $\mathbf{r}(t) = t^2\mathbf{i} + t^3\mathbf{j}$ in Example 1 is not smooth since $\mathbf{r}'(t) = 2t\mathbf{i} + 3t^2\mathbf{j}$ is zero when $t = 0$. (Note the abrupt change in the direction of $\mathbf{T}(t)$ that occurs at the origin.)

UNIT NORMAL VECTORS

Let C be the graph of a vector-valued function $\mathbf{r}(t)$ in 2-space or 3-space. In 2-space there are two unit vectors that are perpendicular to the unit tangent vector $\mathbf{T}(t)$, and in 3-space there are infinitely many such vectors (Figure 15.4.3). In both 2-space and 3-space the following definition singles out a unique unit vector that is perpendicular to $\mathbf{T}(t)$.

15.4.2   DEFINITION. Let $\mathbf{r}(t)$ be a vector-valued function in 2-space or 3-space with graph C. The ***principal unit normal vector*** $\mathbf{N} = \mathbf{N}(t)$ to C is defined by

$$\mathbf{N}(t) = \frac{\mathbf{T}'(t)}{\|\mathbf{T}'(t)\|}$$

if $\mathbf{T}'(t) \neq \mathbf{0}$.

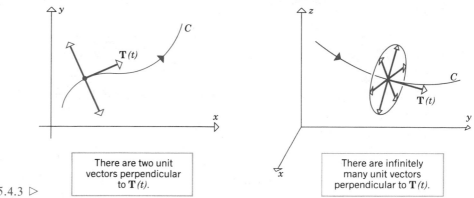

Figure 15.4.3 ▷

There are two unit vectors perpendicular to **T**(t).

There are infinitely many unit vectors perpendicular to **T**(t).

For simplicity, we shall often refer to **N**(t) as simply the **unit normal vector**. It is evident that this vector has length 1 because it is obtained by normalizing **T**'(t). The following theorem shows that **N** and **T** are actually perpendicular.

15.4.3 THEOREM. *If* **r**(t) *is a vector-valued function in 2-space or 3-space, then for each value of t at which* **T**(t) *and* **N**(t) *exist, these vectors are perpendicular.*

Proof. The proof will be complete if we can show that $\mathbf{T}(t) \cdot \mathbf{N}(t) = 0$. But

$$\mathbf{T}(t) \cdot \mathbf{N}(t) = \mathbf{T}(t) \cdot \frac{\mathbf{T}'(t)}{\|\mathbf{T}'(t)\|} = \frac{1}{\|\mathbf{T}'(t)\|} \left(\mathbf{T}(t) \cdot \mathbf{T}'(t)\right)$$

so it suffices to show that $\mathbf{T}(t) \cdot \mathbf{T}'(t) = 0$. But this follows from Theorem 15.2.6 (with **T**(t) replacing **r**(t)) and the fact that $\|\mathbf{T}(t)\| = 1$ for all t. ∎

REMARK. Unless stated otherwise, we shall always assume that **N**(t) is positioned with its initial point at the terminal point of **r**(t) as shown in Figure 15.4.4.

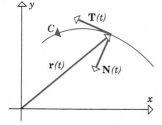

Figure 15.4.4

Example 3 Find **T**(t) and **N**(t) for the circular helix

$$x = a \cos t$$
$$y = a \sin t \quad (a > 0)$$
$$z = ct$$

Solution. The position vector for the helix is

$$\mathbf{r}(t) = (a \cos t)\mathbf{i} + (a \sin t)\mathbf{j} + (ct)\mathbf{k}$$

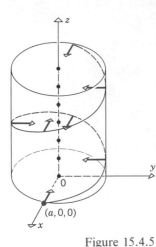

Figure 15.4.5

Thus,

$$\mathbf{r}'(t) = (-a \sin t)\mathbf{i} + (a \cos t)\mathbf{j} + c\mathbf{k}$$

$$\|\mathbf{r}'(t)\| = \sqrt{(-a \sin t)^2 + (a \cos t)^2 + c^2} = \sqrt{a^2 + c^2}$$

$$\mathbf{T}(t) = \frac{\mathbf{r}'(t)}{\|\mathbf{r}'(t)\|} = -\frac{a \sin t}{\sqrt{a^2 + c^2}}\mathbf{i} + \frac{a \cos t}{\sqrt{a^2 + c^2}}\mathbf{j} + \frac{c}{\sqrt{a^2 + c^2}}\mathbf{k}$$

$$\mathbf{T}'(t) = -\frac{a \cos t}{\sqrt{a^2 + c^2}}\mathbf{i} - \frac{a \sin t}{\sqrt{a^2 + c^2}}\mathbf{j}$$

$$\|\mathbf{T}'(t)\| = \sqrt{\left(-\frac{a \cos t}{\sqrt{a^2 + c^2}}\right)^2 + \left(-\frac{a \sin t}{\sqrt{a^2 + c^2}}\right)^2} = \sqrt{\frac{a^2}{a^2 + c^2}}$$

$$= \frac{a}{\sqrt{a^2 + c^2}}$$

$$\mathbf{N}(t) = \frac{\mathbf{T}'(t)}{\|\mathbf{T}'(t)\|} = (-\cos t)\mathbf{i} - (\sin t)\mathbf{j}$$

Because the $\mathbf{k}$ component of $\mathbf{N}(t)$ is zero, this vector lies in a horizontal plane for every value of t. Moreover, in the exercises we ask the reader to show that $\mathbf{N}(t)$ points directly toward the z-axis for all t (Figure 15.4.5). ◀

Figure 15.4.6

If $\mathbf{T}$ is a unit tangent vector to a curve C in 2-space, then, as noted earlier, there are two unit vectors perpendicular to $\mathbf{T}$ (Figure 15.4.3). To determine which of these is the principal unit normal $\mathbf{N}$, we shall let $\phi = \phi(t)$ be the counterclockwise angle from the direction of the positive x-axis to $\mathbf{T} = \mathbf{T}(t)$ (Figure 15.4.6). Since $\mathbf{T}$ has length 1, we may write (see Example 8 of Section 14.2)

$$\mathbf{T} = (\cos \phi)\mathbf{i} + (\sin \phi)\mathbf{j} \tag{2}$$

It follows from this formula and the chain rule that

$$\mathbf{T}'(t) = \frac{d\mathbf{T}}{dt} = \frac{d\mathbf{T}}{d\phi}\frac{d\phi}{dt} = [(-\sin \phi)\mathbf{i} + (\cos \phi)\mathbf{j}]\frac{d\phi}{dt} = \mathbf{n}(t)\phi'(t) \tag{3}$$

where

$$\mathbf{n} = \mathbf{n}(t) = (-\sin \phi)\mathbf{i} + (\cos \phi)\mathbf{j} \tag{4}$$

Using the identities

$$\cos(\phi + \pi/2) = -\sin \phi \quad \text{and} \quad \sin(\phi + \pi/2) = \cos \phi$$

we can rewrite (4) as

$$\mathbf{n} = \mathbf{n}(t) = \cos\left(\phi + \frac{\pi}{2}\right)\mathbf{i} + \sin\left(\phi + \frac{\pi}{2}\right)\mathbf{j} \qquad (5)$$

Comparing (2) and (5) we see that $\mathbf{n}$ is 90° counterclockwise from $\mathbf{T}$. But from (3), $\mathbf{T}'(t)$ has the same direction as $\mathbf{n}(t)$ if $\phi'(t) > 0$ and the opposite direction if $\phi'(t) < 0$. The same is true of $\mathbf{N} = \mathbf{T}'(t)/\|\mathbf{T}'(t)\|$ (why?). (See Figure 15.4.7.)

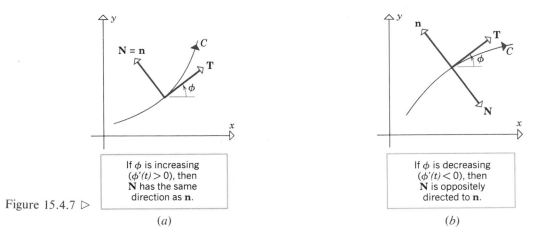

Figure 15.4.7 ▷

If ϕ is increasing ($\phi'(t) > 0$), then $\mathbf{N}$ has the same direction as $\mathbf{n}$.

(a)

If ϕ is decreasing ($\phi'(t) < 0$), then $\mathbf{N}$ is oppositely directed to $\mathbf{n}$.

(b)

It follows from the foregoing discussion that regardless of whether $\phi(t)$ is increasing or decreasing, the principal unit normal $\mathbf{N}$ points "inward" toward the concave side of the curve. For this reason $\mathbf{N}$ is called the *inward unit normal* in 2-space.

▶ Exercise Set 15.4

In Exercises 1–8, find the unit tangent vector $\mathbf{T}$ and the unit normal vector $\mathbf{N}$ for the given value of t. Then sketch the vectors and a portion of the curve containing the point of tangency.

1. $\mathbf{r}(t) = (5 \cos t)\mathbf{i} + (5 \sin t)\mathbf{j}; \ t = \pi/3.$

2. $\mathbf{r}(t) = 2t\mathbf{i} + 4t^2\mathbf{j}; \ t = 1.$

3. $\mathbf{r}(t) = (t^2 - 1)\mathbf{i} + t\mathbf{j}; \ t = 1.$

4. $\mathbf{r}(t) = e^t\mathbf{i} + e^{-t}\mathbf{j}; \ t = 0.$

5. $\mathbf{r}(t) = \frac{1}{2}t^2\mathbf{i} + \frac{1}{3}t^3\mathbf{j}; \ t = 1.$

6. $\mathbf{r}(t) = (\ln t)\mathbf{i} + t\mathbf{j}; \ t = e.$

7. $\mathbf{r}(t) = (4 \cos t)\mathbf{i} + (9 \sin t)\mathbf{j}; \ t = \pi/4.$

8. $\mathbf{r}(t) = (\ln \sin t)\mathbf{i} + (\ln \cos t)\mathbf{j}; \ t = \pi/6.$

In Exercises 9–16, find the unit tangent vector $\mathbf{T}$ and the unit normal vector $\mathbf{N}$ for the given value of t.

9. $\mathbf{r}(t) = 4 \cos t\mathbf{i} + 4 \sin t\mathbf{j} + t\mathbf{k}; \ t = \pi/2.$

10. $x = e^t, \ y = e^{-t}, \ z = t; \ t = 0.$

11. $\mathbf{r}(t) = t\mathbf{i} + \frac{1}{2}t^2\mathbf{j} + \frac{1}{3}t^3\mathbf{k}; \ t = 0.$

12. $x = \sin t, \ y = \cos t, \ z = \frac{1}{2}t^2; \ t = 0.$

13. $\mathbf{r}(t) = 3 \cos t\mathbf{i} + 4 \sin t\mathbf{j} + t\mathbf{k}; \ t = \pi/2.$

14. $x = e^t \cos t, \ y = e^t \sin t, \ z = e^t; \ t = 0.$

15. $r(t) = i + tj + t^2k; t = 1.$

16. $x = \cosh t, y = \sinh t, z = t; t = \ln 2.$

17. Let C be the curve given parametrically in terms of arc length by

$$\begin{array}{l} x = \frac{3}{5}s + 1 \\ y = \frac{4}{5}s - 2 \end{array}, \quad 0 \le s \le 10$$

(a) Find $T = T(s)$.
(b) Sketch the curve and the vector $T(5)$.

18. Let C be the curve given parametrically in terms of arc length by

$$\begin{array}{l} x = 2 \cos\left(\dfrac{s}{2}\right) \\ y = 2 \sin\left(\dfrac{s}{2}\right) \end{array}, \quad 0 \le s \le 4\pi$$

(a) Find $T = T(s)$ and $N = N(s)$.
(b) Sketch the curve and the vectors $T(\pi/2)$ and $N(\pi/2)$.

19. Show that the vector $N(t) = (-\cos t)i - (\sin t)j$ found in Example 3 points directly toward the z-axis.

20. Replace r by r' in the formula of Exercise 47, Section 15.2, to show that N points in the direction of the vector $v = \|r'\|^2 r'' - (r' \cdot r'')r'$ and hence that N can also be obtained from the formula $N = v/\|v\|$.

In Exercises 21–28, use the formula in Exercise 20 to find N for the indicated problem.

21. Exercise 1.   **22.** Exercise 2.

23. Exercise 3.   **24.** Exercise 4.

25. Exercise 9.   **26.** Exercise 10.

27. Exercise 11.   **28.** Exercise 12.

29. Show that if arc length is used as the parameter, then $N(s) = r''(s)/\|r''(s)\|$.

30. The *binormal vector* B at a point on the graph of $r(t)$ is defined by

$$B = T \times N$$

(a) Show that B has length 1 and that $T, N,$ and B are mutually perpendicular.
(b) Find the binormal at a general point on the helix $x = a \cos t, y = a \sin t, z = bt.$

15.5   CURVATURE

In this section we shall consider the problem of obtaining a numerical measure of how sharply a curve in 2-space or 3-space bends. Our results will have applications in geometry and in the study of motion along a curved path.

Let s be an arc-length parameter and $r(s)$ a vector-valued function in 2-space or 3-space whose graph is a smooth curve C. The "sharpness" with which the curve bends can be ascertained by observing the behavior of the unit tangent vector T as the parameter s increases. If C is a straight line (no bend), then the direction of T remains constant (Figure 15.5.1a). If C bends slightly, then T undergoes a gradual change of direction (Figure 15.5.1b); and if C bends sharply, then T undergoes a rapid change of direction (Figure 15.5.1c). In summary, the "sharpness" of the bend is closely related to the rate of change of T with respect to s. This suggests the following numerical measure of curvature.

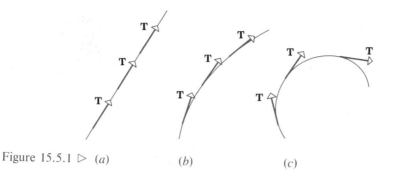

Figure 15.5.1 ▷ (a)   (b)   (c)

15.5.1   DEFINITION. If s is an arc-length parameter and $\mathbf{r}(s)$ is a vector-valued function in 2-space or 3-space whose graph is a smooth curve C, then the ***curvature*** $\kappa = \kappa(s)$ (κ = Greek "kappa") of C is defined by

$$\kappa = \left\| \frac{d\mathbf{T}}{ds} \right\| \tag{1}$$

INTERPRETATION
OF CURVATURE IN
2-SPACE

A simple geometric interpretation of curvature in 2-space can be obtained by considering the angle $\phi = \phi(s)$ measured counterclockwise from the direction of the positive x-axis to $\mathbf{T}$ (Figure 15.4.6). As noted in Formula (2) of Section 15.4, we can write

$$\mathbf{T} = (\cos \phi)\mathbf{i} + (\sin \phi)\mathbf{j}$$

from which we obtain

$$\frac{d\mathbf{T}}{d\phi} = (-\sin \phi)\mathbf{i} + (\cos \phi)\mathbf{j}$$

By the chain rule

$$\frac{d\mathbf{T}}{ds} = \frac{d\mathbf{T}}{d\phi} \frac{d\phi}{ds}$$

so

$$\left\| \frac{d\mathbf{T}}{ds} \right\| = \left| \frac{d\phi}{ds} \right| \left\| \frac{d\mathbf{T}}{d\phi} \right\| = \left| \frac{d\phi}{ds} \right| \sqrt{(-\sin \phi)^2 + \cos^2 \phi} = \left| \frac{d\phi}{ds} \right|$$

It now follows from (1) that

$$\kappa = \left| \frac{d\phi}{ds} \right| \tag{2}$$

Thus, in 2-space curvature can be interpreted as the magnitude of the rate of the rate of change of ϕ with respect to s (Figure 15.5.2).

In 2-space, κ is the magnitude of the rate of change of ϕ with respect to s.

Figure 15.5.2 ▷

FORMULAS FOR CURVATURE

For computational purposes, Formula (1) is not useful unless the curve happens to be expressed in terms of an arc-length parameter. We shall now derive a formula that works for any parameter t.

By the chain rule

$$\frac{d\mathbf{T}}{dt} = \frac{d\mathbf{T}}{ds}\frac{ds}{dt}$$

Thus,

$$\kappa = \left\|\frac{d\mathbf{T}}{ds}\right\| = \left\|\frac{d\mathbf{T}/dt}{ds/dt}\right\| \tag{3}$$

But, by Formula (16) of Section 15.3 $ds/dt = \|d\mathbf{r}/dt\|$, so (3) yields

$$\kappa = \frac{\|d\mathbf{T}/dt\|}{\|d\mathbf{r}/dt\|}$$

or in an alternative notation

$$\kappa = \kappa(t) = \frac{\|\mathbf{T}'(t)\|}{\|\mathbf{r}'(t)\|} \tag{4}$$

Figure 15.5.3

Example 1   It is evident geometrically that a line in 2-space has zero curvature at every point because the angle $\phi = \phi(s)$ is constant (Figure 15.5.3). Thus,

$$\kappa = \left|\frac{d\phi}{ds}\right| = 0$$

An analytic proof that lines in both 2-space and 3-space have zero curvature at every point can be obtained by writing the equation of a line in the vector form $\mathbf{r} = \mathbf{r}_0 + t\mathbf{v}$ (see Formula (5) of Section 14.5). Thus,

$$\mathbf{r}'(t) = \frac{d\mathbf{r}}{dt} = \mathbf{v} \quad [\mathbf{r}_0 \text{ is constant.}]$$

$$\|\mathbf{r}'(t)\| = \|\mathbf{v}\|$$

$$\mathbf{T}(t) = \frac{\mathbf{r}'(t)}{\|\mathbf{r}'(t)\|} = \frac{\mathbf{v}}{\|\mathbf{v}\|}$$

$$\mathbf{T}'(t) = \mathbf{0} \quad [\text{Because T}(t) \text{ is constant.}]$$

Thus, (4) yields

$$\kappa(t) = \frac{\|\mathbf{T}'(t)\|}{\|\mathbf{r}'(t)\|} = \frac{0}{\|\mathbf{v}\|} = 0 \quad \blacktriangleleft$$

Example 2 Show that a circle of radius a has curvature $1/a$ at every point.

Solution. Let the circle be centered at the origin, so it is represented by the parametric equations

$$\begin{aligned} x &= a \cos t \\ y &= a \sin t \end{aligned}, \quad 0 \leq t \leq 2\pi$$

or equivalently by the vector equation

$$\mathbf{r} = (a \cos t)\mathbf{i} + (a \sin t)\mathbf{j}, \quad 0 \leq t \leq 2\pi$$

Thus,

$$\mathbf{r}'(t) = (-a \sin t)\mathbf{i} + (a \cos t)\mathbf{j}$$

$$\|\mathbf{r}'(t)\| = \sqrt{(-a \sin t)^2 + (a \cos t)^2} = \sqrt{a^2(\sin^2 t + \cos^2 t)} = a$$

$$\mathbf{T}(t) = \frac{\mathbf{r}'(t)}{\|\mathbf{r}'(t)\|} = (-\sin t)\mathbf{i} + (\cos t)\mathbf{j}$$

$$\mathbf{T}'(t) = (-\cos t)\mathbf{i} + (-\sin t)\mathbf{j}$$

$$\|\mathbf{T}'(t)\| = \sqrt{(-\cos t)^2 + (-\sin t)^2} = 1$$

$$\kappa = \frac{\|\mathbf{T}'(t)\|}{\|\mathbf{r}'(t)\|} = \frac{1}{a} \quad \blacktriangleleft$$

There are numerous formulas that can be used to compute curvature, and choosing the right one can make a big difference in the complexity of the computations. The formula stated in the following theorem is generally a good first choice.

15.5.2 THEOREM. *If $\mathbf{r}(t)$ is a vector-valued function in 3-space whose graph is a smooth curve, then*

$$\kappa(t) = \frac{\|\mathbf{r}'(t) \times \mathbf{r}''(t)\|}{\|\mathbf{r}'(t)\|^3} \qquad (5)$$

Proof. From Definition 15.4.1 and Formula (16) of Section 15.3 we have

$$\mathbf{T}(t) = \frac{\mathbf{r}'(t)}{\|\mathbf{r}'(t)\|} = \frac{\mathbf{r}'(t)}{ds/dt}$$

so

$$\mathbf{r}'(t) = \frac{ds}{dt}\mathbf{T}(t)$$

Differentiating both sides of this equation yields

$$\mathbf{r}''(t) = \left(\frac{d^2s}{dt^2}\right)\mathbf{T}(t) + \frac{ds}{dt}\mathbf{T}'(t)$$

Therefore,

$$\mathbf{r}'(t) \times \mathbf{r}''(t) = \left(\frac{ds}{dt}\right)\left(\frac{d^2s}{dt^2}\right)[\mathbf{T}(t) \times \mathbf{T}(t)] + \left(\frac{ds}{dt}\right)^2[\mathbf{T}(t) \times \mathbf{T}'(t)]$$

From Formula (16) of Section 15.3 and the fact that $\mathbf{T}(t) \times \mathbf{T}(t) = \mathbf{0}$, it follows that

$$\mathbf{r}'(t) \times \mathbf{r}''(t) = \left(\frac{ds}{dt}\right)^2[\mathbf{T}(t) \times \mathbf{T}'(t)] = \|\mathbf{r}'(t)\|^2[\mathbf{T}(t) \times \mathbf{T}'(t)]$$

and consequently

$$\|\mathbf{r}'(t) \times \mathbf{r}''(t)\| = \|\mathbf{r}'(t)\|^2\|\mathbf{T}(t) \times \mathbf{T}'(t)\| \qquad (6)$$

From Lagrange's identity (Theorem 14.4.2), and the fact that $\mathbf{T}(t) \cdot \mathbf{T}'(t) = 0$ (Theorem 15.2.6 with $\mathbf{T}(t)$ replacing $\mathbf{r}(t)$), we obtain

$$\|\mathbf{T}(t) \times \mathbf{T}'(t)\| = \|\mathbf{T}(t)\|\,\|\mathbf{T}'(t)\| = \|\mathbf{T}'(t)\|$$

so from (4)

$$\|\mathbf{T}(t) \times \mathbf{T}'(t)\| = \kappa\|\mathbf{r}'(t)\|$$

Thus (6) can be rewritten as

$$\|\mathbf{r}'(t) \times \mathbf{r}''(t)\| = \kappa \|\mathbf{r}'(t)\|^3$$

from which (5) follows.   ∎

Example 3   Find $\kappa(t)$ for the circular helix

$$x = a \cos t$$
$$y = a \sin t \quad (a > 0)$$
$$z = ct$$

Solution.   The position vector for the helix is

$$\mathbf{r} = (a \cos t)\mathbf{i} + (a \sin t)\mathbf{j} + (ct)\mathbf{k}$$

Thus,

$$\mathbf{r}'(t) = (-a \sin t)\mathbf{i} + (a \cos t)\mathbf{j} + c\mathbf{k}$$
$$\mathbf{r}''(t) = (-a \cos t)\mathbf{i} + (-a \sin t)\mathbf{j}$$

and

$$\mathbf{r}'(t) \times \mathbf{r}''(t) = \begin{vmatrix} \mathbf{i} & \mathbf{j} & \mathbf{k} \\ -a \sin t & a \cos t & c \\ -a \cos t & -a \sin t & 0 \end{vmatrix}$$
$$= (ac \sin t)\mathbf{i} - (ac \cos t)\mathbf{j} + a^2\mathbf{k}$$

Therefore,

$$\|\mathbf{r}'(t)\| = \sqrt{(-a \sin t)^2 + (a \cos t)^2 + c^2} = \sqrt{a^2 + c^2}$$

and

$$\|\mathbf{r}'(t) \times \mathbf{r}''(t)\| = \sqrt{(ac \sin t)^2 + (-ac \cos t)^2 + a^4}$$
$$= \sqrt{a^2 c^2 + a^4} = a\sqrt{a^2 + c^2}$$

so

$$\kappa(t) = \frac{\|\mathbf{r}'(t) \times \mathbf{r}''(t)\|}{\|\mathbf{r}'(t)\|^3} = \frac{a\sqrt{a^2 + c^2}}{(\sqrt{a^2 + c^2})^3} = \frac{a}{a^2 + c^2}$$

Note that κ does not depend on t, which means that the helix has constant curvature.   ◀

Although Formula (5) was derived for curves in 3-space, it can be used to calculate curvature in 2-space by thinking of a 2-space function $\mathbf{r}(t) = x(t)\mathbf{i} + y(t)\mathbf{j}$ as the 3-space function $\mathbf{r}(t) = x(t)\mathbf{i} + y(t)\mathbf{j} + 0\mathbf{k}$ with a zero $\mathbf{k}$ component.

Example 4   The parametric equations

$$x = 2 \cos t \atop y = 3 \sin t, \quad 0 \le t \le 2\pi$$

represent the ellipse

$$\frac{x^2}{4} + \frac{y^2}{9} = 1$$

with counterclockwise orientation (Figure 15.5.4). Find the curvature at the point $(2, 0)$ and at the point $(0, 3)$.

Solution.   A vector equation of the ellipse is

$$\mathbf{r} = (2 \cos t)\mathbf{i} + (3 \sin t)\mathbf{j}, \quad 0 \le t \le 2\pi$$

Thus,

$$\mathbf{r}'(t) = (-2 \sin t)\mathbf{i} + (3 \cos t)\mathbf{j}$$
$$\mathbf{r}''(t) = (-2 \cos t)\mathbf{i} + (-3 \sin t)\mathbf{j}$$

and

$$\mathbf{r}'(t) \times \mathbf{r}''(t) = \begin{vmatrix} \mathbf{i} & \mathbf{j} & \mathbf{k} \\ -2 \sin t & 3 \cos t & 0 \\ -2 \cos t & -3 \sin t & 0 \end{vmatrix}$$

$$= [(6 \sin^2 t) + (6 \cos^2 t)]\mathbf{k} = 6\mathbf{k}$$

Therefore,

$$\|\mathbf{r}'(t)\| = \sqrt{(-2 \sin t)^2 + (3 \cos t)^2} = \sqrt{4 \sin^2 t + 9 \cos^2 t}$$

and

$$\|\mathbf{r}'(t) \times \mathbf{r}''(t)\| = 6$$

so

$$\kappa = \kappa(t) = \frac{\|\mathbf{r}'(t) \times \mathbf{r}''(t)\|}{\|\mathbf{r}'(t)\|^3} = \frac{6}{[4 \sin^2 t + 9 \cos^2 t]^{3/2}} \tag{7}$$

The point $(2, 0)$ on the ellipse corresponds to $t = 0$. Substituting $t = 0$ in (7) yields

$$\kappa = \kappa(0) = \frac{6}{9^{3/2}} = \frac{6}{27} = \frac{2}{9}$$

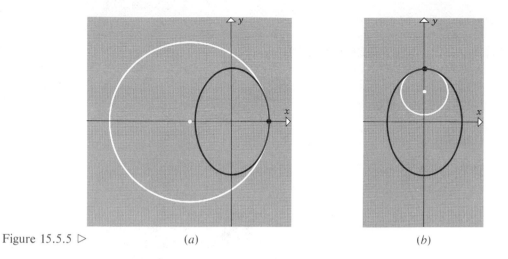

Figure 15.5.5 ▷   (a)   (b)

Similarly, the point $(0, 3)$ corresponds to $t = \pi/2$, so the curvature there is

$$\kappa = \kappa\left(\frac{\pi}{2}\right) = \frac{6}{4^{3/2}} = \frac{3}{4}$$

Note that the curvature at $(2, 0)$ (Figure 15.5.5a) is smaller than the curvature at $(0, 3)$ (Figure 15.5.5b), as it should be. ◀

RADIUS OF CURVATURE

If a curve C in 2-space has nonzero curvature κ at a point P, then the circle of radius

$$\rho = \frac{1}{\kappa}$$

sharing a common tangent with C at P, and centered on the concave side of the curve at P is called the *circle of curvature* or *osculating circle* at P. The radius ρ is called the *radius of curvature* at P, and the center of the circle is called the *center of curvature* at P (Figure 15.5.6).

From Example 2, the osculating circle at P has curvature $1/\rho = \kappa$. Thus, the osculating circle and the curve C not only touch at P but they have equal curvatures at that point. In this sense, the osculating circle is the circle that best approximates the curve C near P.

Figure 15.5.6

Example 5   In Example 4 we found the curvature of the ellipse

$$x = 2 \cos t \atop y = 3 \sin t , \quad 0 \leq t \leq 2\pi$$

to be 2/9 at $(2, 0)$ and 3/4 at $(0, 3)$. Thus, the radius of curvature is 9/2 at $(2, 0)$ and 4/3 at $(0, 3)$. The osculating circles at these points are shown in Figures 15.5.5a and 15.5.5b. ◀

▶ Exercise Set 15.5

In Exercises 1–14, use formula (5) to find the curvature at the indicated point.

1. $\mathbf{r}(t) = t^2\mathbf{i} + t^3\mathbf{j}$; $t = \frac{1}{2}$.

2. $\mathbf{r}(t) = (4 \cos t)\mathbf{i} + (\sin t)\mathbf{j}$; $t = \pi/2$.

3. $\mathbf{r}(t) = e^{3t}\mathbf{i} + e^{-t}\mathbf{j}$; $t = 0$.

4. $x = 1 - t^3$, $y = t - t^2$; $t = 1$.

5. $x = t \cos t$, $y = t \sin t$; $t = 1$.

6. $x = 2abt$, $y = b^2t^2$; $t = 1$ ($a > 0, b > 0$).

7. $\mathbf{r}(t) = 4 \cos t\mathbf{i} + 4 \sin t\mathbf{j} + t\mathbf{k}$; $t = \pi/2$.

8. $x = e^t$, $y = e^{-t}$, $z = t$; $t = 0$.

9. $\mathbf{r}(t) = t\mathbf{i} + \frac{1}{2}t^2\mathbf{j} + \frac{1}{3}t^3\mathbf{k}$; $t = 0$.

10. $x = \sin t$, $y = \cos t$, $z = \frac{1}{2}t^2$; $t = 0$.

11. $\mathbf{r}(t) = 3 \cos t\mathbf{i} + 4 \sin t\mathbf{j} + t\mathbf{k}$; $t = \pi/2$.

12. $x = e^t \cos t$, $y = e^t \sin t$, $z = e^t$; $t = 0$.

13. $\mathbf{r}(t) = \mathbf{i} + t\mathbf{j} + t^2\mathbf{k}$; $t = 1$.

14. $x = \cosh t$, $y = \sinh t$, $z = t$; $t = \ln 2$.

15. Use Formula (5) to show that for a plane curve described by $y = f(x)$ the curvature $\kappa(x)$ is

$$\kappa(x) = \frac{\left|\dfrac{d^2y}{dx^2}\right|}{\left[1 + \left(\dfrac{dy}{dx}\right)^2\right]^{3/2}}$$

[*Hint:* Let x be the parameter, so $\mathbf{r}(x) = x\mathbf{i} + y\mathbf{j} = x\mathbf{i} + f(x)\mathbf{j}$.]

16. Show that the formula for κ in Exercise 15 can also be written as $\kappa = |y'' \cos^3 \phi|$, where ϕ is the angle of inclination of the tangent line to the graph of $y = f(x)$.

In Exercises 17–22, use the formula of Exercise 15 to find the curvature at the indicated point.

17. $y = \sin x$; $x = \pi/2$.   18. $y = x^3/3$; $x = 0$.

19. $y = 1/x$; $x = 1$.   20. $y = e^{-x}$; $x = 1$.

21. $y = \tan x$; $x = \pi/4$.   22. $y^2 - 4x^2 = 9$; $(2, 5)$.

23. Use the formula of Exercise 15 to find the curvature of $y = \ln \cos x$ for $-\pi/2 < x < \pi/2$. For what value of x is the curvature maximum?

24. Use Formula (5) to show that for a plane curve described by $\mathbf{r} = x(t)\mathbf{i} + y(t)\mathbf{j}$ the curvature is

$$\kappa = \frac{|x'y'' - y'x''|}{(x'^2 + y'^2)^{3/2}}$$

where a prime denotes differentiation with respect to t.

In Exercises 25–30, use the formula of Exercise 24 to find the curvature in the indicated problem.

25. Exercise 1.   26. Exercise 2.

27. Exercise 3.   28. Exercise 4.

29. Exercise 5.   30. Exercise 6.

31. Use the formula of Exercise 24 to find the curvature at $t = 0$ and $t = \pi/2$ of the curve $x = a \cos t$, $y = b \sin t$ ($a > 0, b > 0$).

32. Use Formula (5) to show that for a curve in polar coordinates described by $r = f(\theta)$ the curvature is

$$\kappa = \frac{\left|r^2 + 2\left(\dfrac{dr}{d\theta}\right)^2 - r\dfrac{d^2r}{d\theta^2}\right|}{\left[r^2 + \left(\dfrac{dr}{d\theta}\right)^2\right]^{3/2}}$$

[*Hint:* Let θ be the parameter and use the relationships $x = r \cos \theta$, $y = r \sin \theta$.]

In Exercises 33–36, use the formula of Exercise 32 to find the curvature at the indicated point.

33. $r = 2 \sin \theta$; $\theta = \pi/6$.

34. $r = \theta$; $\theta = 1$.

35. $r = a(1 + \cos \theta)$; $\theta = \pi/2$.

36. $r = e^{2\theta}$; $\theta = 1$.

37. Find $\kappa(t)$ for the cycloid $x = a(t - \sin t)$, $y = a(1 - \cos t)$ for $0 < t < \pi$ and sketch the graph of $\kappa(t)$.

38. Find $\kappa(t)$ for the curve $x = e^{-t} \cos t$, $y = e^{-t} \sin t$ and sketch the graph of $\kappa(t)$.

In Exercises 39–45, sketch the curve, calculate the radius of curvature at the indicated point, and sketch the osculating circle.

39. $y = \sin x$; $x = \pi/2$.   40. $x = t$, $y = t^2$; $t = 1$.

41. $y = \frac{1}{2}x^2$; $x = -1$.   42. $x = t$, $y = \sqrt{t}$; $t = 2$.

43. $y = \ln x$; $x = 1$.

44. $x = 2t$, $y = 4/t$; $t = 1$.

45. $x = t - \sin t$, $y = 1 - \cos t$; $t = \pi$.

46. Find the radius of curvature of $y = \cos x$ at $x = 0$ and $x = \pi$. Sketch the osculating circles at those points.

47. Find the radius of curvature of the ellipse $x = 2\cos t$, $y = \sin t$, $0 \le t \le 2\pi$ at $t = 0$ and $t = \pi/2$. Sketch the osculating circles at those points.

48. Consider the curve $y = x^4 - 2x^2$.
    (a) Find the radius of curvature at each relative extremum.
    (b) Sketch the curve and show the osculating circles at the relative extrema.

49. Find the radius of curvature of the parabola $y^2 = 4px$ at $(0, 0)$.

50. At what point(s) does $y = e^x$ have maximum curvature?

51. At what point(s) does $4x^2 + 9y^2 = 36$ have minimum radius of curvature?

52. Find the value of x, $x > 0$, where $y = x^3$ has maximum curvature.

53. Find the maximum and minimum values of the radius of curvature for the curve $x = \cos t$, $y = \sin t$, $z = \cos t$.

54. Find the minimum value of the radius of curvature for the curve $x = e^t$, $y = e^{-t}$, $z = \sqrt{2}t$.

55. Show that the curvature of the polar curve $r = e^{a\theta}$ is inversely proportional to r.

56. Show that the curvature of the polar curve $r^2 = a^2 \cos 2\theta$ is directly proportional to r $(r > 0)$.

57. Given that $\dfrac{d\mathbf{T}}{ds} = 0.03\mathbf{i} - 0.04\mathbf{j}$ at a certain point P on a curve, find the value of $\left|\dfrac{d\phi}{ds}\right|$, where ϕ is the angle of inclination of the tangent line to the curve at P. Assuming that s is measured in cm, express your answer in units of °/cm.

58. Assuming that arc length is measured in inches along the curve $y = x^3$, find the magnitude of the rate of change with respect to arc length of the angle of inclination of the tangent line to the curve in units of °/in. at the point where $x = 1$.

In Exercises 59–62, we shall say that a *smooth transition* occurs at $x = x_0$ on a curve $y = f(x)$ if the curvature κ is continuous at x_0 and f does not have an inflection point at x_0 if $\kappa(x_0) \ne 0$.

59. Show that the transition at $x = 0$ from the horizontal line $y = 0$ for $x \le 0$ to the parabola $y = x^2$ for $x > 0$ is not smooth, whereas the transition to $y = x^3$ for $x > 0$ is smooth.

60. (a) Sketch the graph of the curve defined piecewise by $y = x^2$ for $x < 0$, $y = x^4$ for $x \ge 0$.
    (b) Show that for the curve in part (a) the transition at $x = 0$ is not smooth.

61. The figure below shows the arc of a circle of radius r with center at $(0, r)$. Find the value of a so that there is a smooth transition from the circle to the parabola $y = ax^2$ at the point where $x = 0$.

62. Find a, b, and c so that there is a smooth transition at $x = 0$ from the curve $y = e^x$ for $x \le 0$ to the parabola $y = ax^2 + bx + c$ for $x > 0$.

63. Show that if arc length is used as the parameter in 3-space, then

$$\kappa = \sqrt{\left(\frac{d^2x}{ds^2}\right)^2 + \left(\frac{d^2y}{ds^2}\right)^2 + \left(\frac{d^2z}{ds^2}\right)^2}$$

15.6   MOTION ALONG A CURVE

Earlier in this text we studied the motion of a particle moving along a line (Sections 4.11 and 6.6). In this section we shall study the motion of a particle moving along a curve in 2-space or 3-space.

VELOCITY, ACCELERATION, AND SPEED

Recall from Section 4.11 that if a particle moves along a coordinate line so its position function is $s(t)$, then its instantaneous velocity, acceleration, and speed are defined by

$$\text{velocity} = v(t) = \frac{ds}{dt}$$

$$\text{acceleration} = a(t) = \frac{dv}{dt} = \frac{d^2s}{dt^2}$$

$$\text{speed} = |v(t)| = \left|\frac{ds}{dt}\right|$$

Because there are only two possible directions for the motion, the sign of $v(t)$ serves to distinguish the direction of motion—a positive velocity means that the particle is moving in the positive direction of the coordinate line, and a negative velocity means that it is moving in the negative direction. For a particle moving along a curve C in 2-space or 3-space, the situation is much more complicated. At each instant, the direction of motion is considered to be tangent to C, so there are infinitely many possible directions for the motion (Figure 15.6.1). Vectors provide the mathematical tool for dealing with this problem. At each instant the speed of the particle is taken to be the instantaneous rate at which the arc length traveled by the particle is changing with time. The speed and direction of motion together form a vector quantity called the "velocity" of the particle. This velocity can be represented geometrically by an arrow that

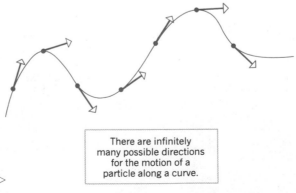

There are infinitely many possible directions for the motion of a particle along a curve.

Figure 15.6.1 ▷

The velocity vector **v** points
in the direction of motion
and ‖**v**‖ is equal to
the speed of the particle

Figure 15.6.2 ▷

points in the direction of motion and has magnitude equal to the speed of the particle (Figure 15.6.2).

The following definition makes these ideas precise.

15.6.1 DEFINITION. If a particle moves along a curve C in 2-space or 3-space so its position vector at time t is $\mathbf{r}(t)$, then the *instantaneous velocity, instantaneous acceleration,* and *instantaneous speed* of the particle are defined by

$$\text{velocity} = \mathbf{v}(t) = \frac{d\mathbf{r}}{dt} \tag{1}$$

$$\text{acceleration} = \mathbf{a}(t) = \frac{d\mathbf{v}}{dt} = \frac{d^2\mathbf{r}}{dt^2} \tag{2}$$

$$\text{speed} = \|\mathbf{v}(t)\| \tag{3}$$

The definition of velocity given here agrees with our initial informal discussion, since $\mathbf{v} = d\mathbf{r}/dt$ is tangent to C and points in the direction of motion. Moreover, the speed, $\|\mathbf{v}(t)\| = \|d\mathbf{r}/dt\| = ds/dt$, is the rate at which arc length traveled changes with time. For reference, we state the formula

$$\text{speed} = \|\mathbf{v}(t)\| = \frac{ds}{dt} \tag{4}$$

It follows from formulas (1), (2), and (3) that if

$$\mathbf{r}(t) = x(t)\mathbf{i} + y(t)\mathbf{j}$$

is the position vector of a particle moving along a curve in the plane (2-space), then

$$\mathbf{v}(t) = \frac{d\mathbf{r}}{dt} = x'(t)\mathbf{i} + y'(t)\mathbf{j} \tag{4a}$$

$$\mathbf{a}(t) = \frac{d^2\mathbf{r}}{dt^2} = x''(t)\mathbf{i} + y''(t)\mathbf{j} \tag{4b}$$

$$\|\mathbf{v}(t)\| = \sqrt{[x'(t)]^2 + [y'(t)]^2} \tag{4c}$$

Similarly, if

$$\mathbf{r}(t) = x(t)\mathbf{i} + y(t)\mathbf{j} + z(t)\mathbf{k}$$

is the position vector of a particle moving along a curve in 3-space, then

$$\mathbf{v}(t) = \frac{d\mathbf{r}}{dt} = x'(t)\mathbf{i} + y'(t)\mathbf{j} + z'(t)\mathbf{k} \tag{5a}$$

$$\mathbf{a}(t) = \frac{d^2\mathbf{r}}{dt^2} = x''(t)\mathbf{i} + y''(t)\mathbf{j} + z''(t)\mathbf{k} \tag{5b}$$

$$\|\mathbf{v}(t)\| = \sqrt{[x'(t)]^2 + [y'(t)]^2 + [z'(t)]^2} \tag{5c}$$

Example 1   A particle moves around a circle so that its x- and y-coordinates at time t are

$$x = 2\cos t, \quad y = 2\sin t$$

Find the velocity, acceleration, and speed when $t = \pi/4$. Sketch the path of the particle and the velocity and acceleration vectors at $t = \pi/4$.

Solution.   At time t, the position, velocity, acceleration, and speed are

$$\mathbf{r}(t) = 2\cos t\,\mathbf{i} + 2\sin t\,\mathbf{j}$$

$$\mathbf{v}(t) = \frac{d\mathbf{r}}{dt} = -2\sin t\,\mathbf{i} + 2\cos t\,\mathbf{j}$$

$$\mathbf{a}(t) = \frac{d\mathbf{v}}{dt} = -2\cos t\,\mathbf{i} - 2\sin t\,\mathbf{j}$$

$$\|\mathbf{v}(t)\| = \sqrt{(-2\sin t)^2 + (2\cos t)^2} = 2$$

Thus, at $t = \pi/4$, the velocity, acceleration, and speed are

$$v(\pi/4) = -2 \sin \frac{\pi}{4} i + 2 \cos \frac{\pi}{4} j = -\sqrt{2}i + \sqrt{2}j$$

$$a(\pi/4) = -2 \cos \frac{\pi}{4} i - 2 \sin \frac{\pi}{4} j = -\sqrt{2}i - \sqrt{2}j$$

$$\|v(\pi/4)\| = 2$$

Note that the particle moves around the circle with constant speed since $\|v(t)\| = 2$ does not depend on t. The path of the particle and the velocity and acceleration vectors at $t = \pi/4$ are shown in Figure 15.6.3.   ◄

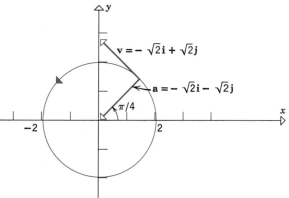

Figure 15.6.3 ▷

REMARK.   Except for some applications that appear later in this section, we shall omit the units on length, velocity, acceleration, and speed to keep the presentation uncluttered.

Example 2   A particle moves along a curve in 3-space so that its coordinates at time t are

$$x = t, \quad y = \frac{1}{2}t^2, \quad z = \frac{1}{3}t^3$$

Find the velocity, speed, and acceleration at time $t = 1$.

Solution.   At time t, the position, velocity, acceleration, and speed are

$$r(t) = ti + \frac{1}{2}t^2 j + \frac{1}{3}t^3 k$$

$$v(t) = \frac{dr}{dt} = i + tj + t^2 k$$

$$a(t) = \frac{dv}{dt} = j + 2tk$$

$$\|v(t)\| = \sqrt{1 + t^2 + t^4}$$

Thus, at time $t = 1$, the velocity, acceleration, and speed are

$$\mathbf{v}(1) = \mathbf{i} + \mathbf{j} + \mathbf{k}$$
$$\mathbf{a}(1) = \mathbf{j} + 2\mathbf{k}$$
$$\|\mathbf{v}(1)\| = \sqrt{3} \quad \blacktriangleleft$$

It follows from (1) and (2) that

$$\mathbf{r}(t) = \int \mathbf{v}(t) \, dt \tag{6}$$

$$\mathbf{v}(t) = \int \mathbf{a}(t) \, dt \tag{7}$$

Thus, if the acceleration $\mathbf{a}(t)$ is known, it is possible to determine the velocity $\mathbf{v}(t)$ and the position $\mathbf{r}(t)$ by integration, provided sufficient information is available to determine the constants of integration.

Example 3   For a particle moving in a plane, find the position $\mathbf{r}(t)$ and the velocity $\mathbf{v}(t)$ given that the acceleration is $\mathbf{a}(t) = 6t\mathbf{i} + 12t^2\mathbf{j}$ and the position and velocity at time $t = 0$ are $\mathbf{r}(0) = \mathbf{i} + \mathbf{j}$ and $\mathbf{v}(0) = \mathbf{0}$.

Solution.   From (7)

$$\mathbf{v}(t) = \int (6t\mathbf{i} + 12t^2\mathbf{j}) \, dt = 3t^2\mathbf{i} + 4t^3\mathbf{j} + \mathbf{c}_1 \tag{8}$$

where $\mathbf{c}_1$ is a vector constant of integration. Since $\mathbf{v}(0) = \mathbf{0}$, it follows on substituting $t = 0$ in (8) that $\mathbf{c}_1 = \mathbf{0}$. Thus,

$$\mathbf{v}(t) = 3t^2\mathbf{i} + 4t^3\mathbf{j}$$

From (6)

$$\mathbf{r}(t) = \int (3t^2\mathbf{i} + 4t^3\mathbf{j}) \, dt = t^3\mathbf{i} + t^4\mathbf{j} + \mathbf{c}_2 \tag{9}$$

Since $\mathbf{r}(0) = \mathbf{i} + \mathbf{j}$, it follows on substituting $t = 0$ in (9) that $\mathbf{c}_2 = \mathbf{i} + \mathbf{j}$. Thus,

$$\mathbf{r}(t) = (t^3 + 1)\mathbf{i} + (t^4 + 1)\mathbf{j} \quad \blacktriangleleft$$

DISTANCE TRAVELED   It follows from Formula (17) of Section 15.3 that if $\mathbf{r}(t)$ is the position vector of a particle moving along a curve C in 2-space or 3-space, then the length of arc L traversed by the particle (that is, the distance it travels) during the time interval $t_1 \leq t \leq t_2$ is given by

$$L = \int_{t_1}^{t_2} \left\| \frac{d\mathbf{r}}{dt} \right\| \, dt$$

or from (1),

$$L = \int_{t_1}^{t_2} \|\mathbf{v}(t)\| \, dt \qquad (10)$$

In words, *the length of arc traversed by a particle during a time interval is obtained by integrating the speed of the particle over that time interval.* (See Formula (8) of Section 6.6 for the comparable result in rectilinear motion.)

Example 4 Suppose that a particle travels along a circular helix in 3-space so that its position vector at time t is

$$\mathbf{r}(t) = (4 \cos t)\mathbf{i} + (4 \sin t)\mathbf{j} + t\mathbf{k}$$

Find the distance traveled by the particle during the time interval $0 \leq t \leq 5$.

Solution. We have

$$\mathbf{v}(t) = \frac{d\mathbf{r}}{dt} = (-4 \sin t)\mathbf{i} + (4 \cos t)\mathbf{j} + \mathbf{k}$$

$$\|\mathbf{v}(t)\| = \sqrt{(-4 \sin t)^2 + (4 \cos t)^2 + 1} = \sqrt{17}$$

Thus, from (10), the distance traveled is

$$L = \int_0^5 \sqrt{17} \, dt = 5\sqrt{17} \qquad \blacktriangleleft$$

DISPLACEMENT Recall that in rectilinear motion we distinguished between the distance traveled by a particle and its change in position or *displacement*. A similar

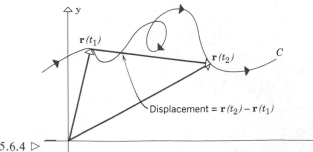

Figure 15.6.4 ▷

distinction exists for motion along a curve in 2-space or 3-space. If the position vector of a particle is $\mathbf{r}(t)$, then its ***displacement*** over the time interval $t_1 \leq t \leq t_2$ is defined to be the vector $\mathbf{r}(t_2) - \mathbf{r}(t_1)$ (Figure 15.6.4).

NORMAL AND
TANGENTIAL
COMPONENTS OF
ACCELERATION

If a particle moves along a curve in 2-space or 3-space, then Formulas (4a), (4b), (5a), and (5b) express $\mathbf{v}(t)$ and $\mathbf{a}(t)$ in terms of components parallel to the coordinate axes. For many purposes, however, it is desirable to resolve the velocity and acceleration vectors into components parallel to the unit tangent and unit normal vectors to the curve C. The following theorem explains how this can be done.

15.6.2   THEOREM. *For a particle moving along a smooth curve C in 2-space or 3-space, the velocity and acceleration vectors can be written as*

$$\mathbf{v} = \frac{ds}{dt}\mathbf{T} \tag{11}$$

$$\mathbf{a} = \frac{d^2 s}{dt^2}\mathbf{T} + \kappa\left(\frac{ds}{dt}\right)^2 \mathbf{N} \tag{12}$$

where s is an arc-length parameter for the curve, and $\mathbf{T}$, $\mathbf{N}$, and κ denote the unit tangent vector, unit normal vector, and curvature (Figure 15.6.5).

Proof.   To prove (11) we write

$$\mathbf{v} = \frac{d\mathbf{r}}{dt} = \left\|\frac{d\mathbf{r}}{dt}\right\|\mathbf{T} = \frac{ds}{dt}\mathbf{T}$$

| Definition 15.4.1 | Formula (16), Section 15.3 |

Figure 15.6.5 ▷

To prove (12) we shall differentiate this expression for **v** using the product rule for derivatives (Theorem 15.2.5):

$$\mathbf{a} = \frac{d\mathbf{v}}{dt} = \frac{ds}{dt}\frac{d\mathbf{T}}{dt} + \mathbf{T}\frac{d^2s}{dt^2} = \frac{ds}{dt}\left\|\frac{d\mathbf{T}}{dt}\right\|\mathbf{N} + \mathbf{T}\frac{d^2s}{dt^2}$$

$$\boxed{\text{Definition 15.4.2}}$$

$$= \frac{ds}{dt}\left(\kappa\left\|\frac{d\mathbf{r}}{dt}\right\|\right)\mathbf{N} + \mathbf{T}\frac{d^2s}{dt^2} \quad \boxed{\begin{array}{l}\text{Formula (4)}\\\text{of Section 15.5}\end{array}}$$

$$= \frac{ds}{dt}\left(\kappa\frac{ds}{dt}\right)\mathbf{N} + \mathbf{T}\frac{d^2s}{dt^2} \quad \boxed{\begin{array}{l}\text{Formula (16)}\\\text{of Section 15.3}\end{array}}$$

$$= \frac{d^2s}{dt^2}\mathbf{T} + \kappa\left(\frac{ds}{dt}\right)^2\mathbf{N} \quad \blacksquare$$

Formula (11) is in keeping with the fact that the velocity vector is tangent to C and has magnitude equal to the speed ds/dt of the particle. Formula (12) should make sense from your experience as an automobile passenger. If a car increases its speed rapidly, the effect is to press the passenger back against the seat. This occurs because the rapid increase in speed results in a large value for

$$\frac{d}{dt}\left(\frac{ds}{dt}\right) = \frac{d^2s}{dt^2}$$

which produces a large tangential component of acceleration in Formula (12). On the other hand, if the car rounds a turn in the road, the passenger is thrown to the side. The greater the curvature in the road or the greater the speed of the car, the greater the force with which the passenger is thrown. This occurs because ds/dt (speed) and κ (curvature) both enter as factors in the normal component of acceleration in (12).

The coefficients of **T** and **N** in (12) are commonly denoted by

$$a_T = \frac{d^2s}{dt^2} \quad \text{and} \quad a_N = \kappa\left(\frac{ds}{dt}\right)^2 \tag{13}$$

so that

$$\mathbf{a} = a_T\mathbf{T} + a_N\mathbf{N} \tag{14}$$

The scalars a_T and a_N are called the *scalar tangential component of acceleration* and the *scalar normal component of acceleration*, respectively. The vectors $a_T\mathbf{T}$ and $a_N\mathbf{N}$ are called the *vector tangential component of acceleration* and *vector normal component of acceleration*, respectively.

It follows from the Theorem of Pythagoras (Figure 15.6.6) that

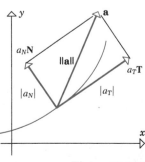

Figure 15.6.6

$$\|\mathbf{a}\|^2 = a_T{}^2 + a_N{}^2 \tag{15}$$

Also, observe from Formula (13) that a_N is nonnegative because κ is nonnegative. Thus, for motion in 2-space the vector $a_N\mathbf{N}$ points "inward" toward the concave side of the curve because $\mathbf{N}$ does. (See the last paragraph of Section 15.4.) For this reason, the vector $a_N\mathbf{N}$ is also called the ***inward component of acceleration*** in 2-space. Finally, observe that it follows from (14) that in 3-space the acceleration $\mathbf{a}$ always lies in the plane determined by $\mathbf{T}$ and $\mathbf{N}$. (Why?)

Example 5   A particle moves so that its position vector at time t is $\mathbf{r}(t) = t\mathbf{i} + \frac{1}{2}t^2\mathbf{j}$. Find the scalar tangential and normal components of acceleration at time $t = 1$.

Solution.   We have

$$\mathbf{v} = \frac{d\mathbf{r}}{dt} = \mathbf{i} + t\mathbf{j}$$

$$\mathbf{a} = \frac{d\mathbf{v}}{dt} = \mathbf{j} \tag{16}$$

$$\frac{ds}{dt} = \|\mathbf{v}\| = \sqrt{1 + t^2}$$

$$\frac{d^2s}{dt^2} = \frac{d}{dt}[\sqrt{1 + t^2}] = \frac{t}{\sqrt{1 + t^2}}$$

Thus, it follows from (13) that

$$a_T = \frac{d^2s}{dt^2} = \frac{t}{\sqrt{1 + t^2}} \tag{17}$$

We could calculate a_N from (13), but this would involve calculating the curvature κ. A better procedure is to use Formula (15). From (16)

$$\|\mathbf{a}\| = \|\mathbf{j}\| = 1$$

so from (15)

$$a_N = \sqrt{\|\mathbf{a}\|^2 - a_T{}^2} = \sqrt{1 - \frac{t^2}{1 + t^2}} = \frac{1}{\sqrt{1 + t^2}} \tag{18}$$

From (17) and (18) the scalar components of acceleration at time $t = 1$ are

$$a_T = \frac{1}{\sqrt{2}} \quad \text{and} \quad a_N = \frac{1}{\sqrt{2}} \quad \blacktriangleleft$$

MOTION OF A
PROJECTILE

We shall conclude this section by showing how our work can be used to study the motion of an object moving near the surface of the earth in a vertical plane. The key result we need from physics is **Newton's Second Law of Motion,** which states that when an object with constant mass m is subjected to a force **F**, it undergoes an acceleration **a** satisfying

$$\mathbf{F} = m\mathbf{a} \tag{19}$$

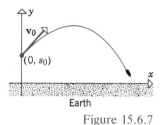

Figure 15.6.7

Consider an xy-coordinate system whose origin is on the surface of the earth and whose positive y-axis points vertically up. From an initial point s_0 units up on the y-axis, let a projectile of mass m be fired, thrown, or released at time $t = 0$ with an initial velocity vector of $\mathbf{v_0}$ (Figure 15.6.7). We shall obtain formulas for the position and velocity of the projectile as functions of t, where t is the amount of time elapsed (in seconds) from the initial firing or release.

We shall assume that once the projectile is fired or released, the only force acting on it is the downward force of the earth's gravity. For a projectile of mass m, it is shown in physics that the downward force of gravity is

$$\mathbf{F} = -mg\mathbf{j} \tag{20}$$

where g is a constant approximately equal to 32 ft/sec² if distance is measured in feet and 9.8 m/sec² if distance is measured in meters. By Newton's Second Law of Motion (19), this force produces an acceleration $\mathbf{a} = \mathbf{a}(t)$ satisfying

$$\mathbf{F} = m\mathbf{a}$$

or from (20)

$$-mg\mathbf{j} = m\mathbf{a} \tag{21}$$

or on canceling m

$$\mathbf{a} = -g\mathbf{j} \tag{22}$$

From (22) it follows that

$$\mathbf{v}(t) = \int -g\mathbf{j}\, dt$$

or since g is constant

$$\mathbf{v}(t) = -gt\mathbf{j} + \mathbf{c_1} \tag{23}$$

where $\mathbf{c_1}$ is a vector constant of integration. Since the initial velocity at $t = 0$ is $\mathbf{v_0}$, it follows on substituting $t = 0$ in (23) that

$$\mathbf{v}_0 = \mathbf{c}_1$$

so that (23) may be written as

$$\mathbf{v}(t) = -gt\mathbf{j} + \mathbf{v}_0 \tag{24}$$

This formula specifies the velocity of the object at any time t.

The position function $\mathbf{r}(t)$ can be obtained by integrating $\mathbf{v}(t)$. This yields

$$\mathbf{r}(t) = \int (-gt\mathbf{j} + \mathbf{v}_0)\, dt$$

or since $\mathbf{v}_0$ (the initial velocity) is constant

$$\mathbf{r}(t) = -\frac{1}{2} gt^2 \mathbf{j} + \mathbf{v}_0 t + \mathbf{c}_2 \tag{25}$$

Since the object is initially s_0 units up on the y-axis, the position vector at $t = 0$ is $\mathbf{r}(0) = s_0 \mathbf{j}$. It follows on substituting $t = 0$ in (25) that

$$s_0 \mathbf{j} = \mathbf{c}_2$$

so that (25) can be written as

$$\mathbf{r}(t) = \left(-\frac{1}{2} gt^2 + s_0\right)\mathbf{j} + \mathbf{v}_0 t \tag{26}$$

This formula specifies the position of the object at any time t.

REMARK. Because the mass m cancels in (21), the mass of the projectile does not enter into the final formulas for velocity and position. Physically, this means that the mass has no influence on the path or the velocity of the projectile—these are completely determined by the initial position and velocity. This explains the famous observation of Galileo, that two objects of different mass, released from the same height, will reach the ground at the same time.

Example 6 A shell, fired from a cannon, has a muzzle speed (the speed as it leaves the barrel) of 800 ft/sec. The barrel makes an angle of 45° with the horizontal and, for simplicity, the barrel opening is assumed to be at ground level.

Figure 15.6.8

(a) Find parametric equations for the shell's trajectory relative to the coordinate system in Figure 15.6.8.

(b) How high does the shell rise?

(c) How far does the shell travel horizontally?

(d) What is the speed of the shell at its point of impact with the ground?

Figure 15.6.9

Solution (*a*). From the given data, the initial velocity vector $\mathbf{v}_0$ has a magnitude of 800 and makes an angle of 45° with the positive x-axis. Thus,

$$\mathbf{v}_0 = (800 \cos 45°)\mathbf{i} + (800 \sin 45°)\mathbf{j} = 400\sqrt{2}\,\mathbf{i} + 400\sqrt{2}\,\mathbf{j}$$

(Figure 15.6.9). From the given data, the initial y-coordinate of the shell is $s_0 = 0$. Substituting s_0 and $\mathbf{v}_0$ in (26) and taking $g = 32$ (since distance is measured in feet), we obtain the position vector of the shell after t seconds:

$$\mathbf{r}(t) = -16t^2\mathbf{j} + (400\sqrt{2}\,\mathbf{i} + 400\sqrt{2}\,\mathbf{j})t$$

or

$$\mathbf{r}(t) = (400\sqrt{2})t\mathbf{i} + (400\sqrt{2}\,t - 16t^2)\mathbf{j} \qquad (27)$$

which yields the parametric equations

$$x = 400\sqrt{2}\,t \qquad\qquad (28a)$$
$$y = 400\sqrt{2}\,t - 16t^2 , \quad t \ge 0 \qquad (28b)$$

(The reader should be able to verify that the trajectory is a parabola.)

Solution (*b*). The shell is rising when $dy/dt > 0$ (y is increasing with t) and is falling when $dy/dt < 0$ (y is decreasing with t). Consequently, the peak occurs when $dy/dt = 0$, since this is the transition between rise and fall. Thus, from (28b) the peak occurs when

$$400\sqrt{2} - 32t = 0$$

or

$$t = \frac{25\sqrt{2}}{2}$$

From (28b), the value of y at this time is

$$y = 5000 \text{ (feet)}$$

Solution (*c*). The shell will hit the ground when $y = 0$. From (28b), this occurs when

$$400\sqrt{2}\,t - 16t^2 = 0$$

or

$$t(400\sqrt{2} - 16t) = 0$$

The solution $t = 0$ corresponds to the initial position of the shell and the solution $t = 25\sqrt{2}$ to the time of impact. Substituting the latter value in (28a) yields

$$x = 20,000 \text{ ft}$$

as the horizontal distance traveled by the shell.

Solution (d).   The velocity at time t is obtained by differentiating (27):

$$\mathbf{v}(t) = 400\sqrt{2}\,\mathbf{i} + (400\sqrt{2} - 32t)\mathbf{j}$$

From (c), impact occurs when $t = 25\sqrt{2}$, so that the velocity vector at this point is

$$\mathbf{v}(25\sqrt{2}) = 400\sqrt{2}\,\mathbf{i} + [400\sqrt{2} - 32(25\sqrt{2})]\mathbf{j} = 400\sqrt{2}\,\mathbf{i} - 400\sqrt{2}\,\mathbf{j}$$

Thus, the speed at impact is

$$\|\mathbf{v}(25\sqrt{2})\| = \sqrt{(400\sqrt{2})^2 + (-400\sqrt{2})^2} = 800 \text{ ft/sec} \quad \blacktriangleleft$$

▶ Exercise Set 15.6

In Exercises 1–6, $\mathbf{r}(t)$ is the position vector of a particle moving in the plane. Find the velocity, acceleration, and speed at an arbitrary time t; then sketch the path of the particle together with the velocity and acceleration vectors at the indicated time t.

1. $\mathbf{r}(t) = 3\cos t\,\mathbf{i} + 3\sin t\,\mathbf{j}$; $t = \pi/3$.

2. $\mathbf{r}(t) = t\,\mathbf{i} + t^2\,\mathbf{j}$; $t = 2$.

3. $\mathbf{r}(t) = e^t\,\mathbf{i} + e^{-t}\,\mathbf{j}$; $t = 0$.

4. $\mathbf{r}(t) = (2 + 4t)\,\mathbf{i} + (1 - t)\,\mathbf{j}$; $t = 1$.

5. $\mathbf{r}(t) = \cosh t\,\mathbf{i} + \sinh t\,\mathbf{j}$; $t = \ln 2$.

6. $\mathbf{r}(t) = 4\cos t\,\mathbf{i} + 9\sin t\,\mathbf{j}$; $t = \pi/2$.

In Exercises 7–12, find the velocity, speed, and acceleration at the given time t of a particle moving along the given curve.

7. $\mathbf{r}(t) = t\,\mathbf{i} + \frac{1}{2}t^2\,\mathbf{j} + \frac{1}{3}t^3\,\mathbf{k}$; $t = 1$.

8. $x = 1 + 3t$, $y = 2 - 4t$, $z = 7 + t$; $t = 2$.

9. $\mathbf{r}(t) = 2\cos t\,\mathbf{i} + 2\sin t\,\mathbf{j} + t\,\mathbf{k}$; $t = \pi/4$.

10. $\mathbf{r}(t) = 3t\,\mathbf{i} + 2t^2\,\mathbf{j} + \ln t\,\mathbf{k}$; $t = 1$.

11. $\mathbf{r}(t) = e^t\sin t\,\mathbf{i} + e^t\cos t\,\mathbf{j} + t\,\mathbf{k}$; $t = \pi/2$.

12. $\mathbf{r}(t) = 2t\,\mathbf{i} + t^2\,\mathbf{j} + \ln t\,\mathbf{k}$; $t = 2$.

In Exercises 13–20, use the given information to find the position and velocity vectors of the particle.

13. $\mathbf{a}(t) = -32\mathbf{j}$; $\mathbf{v}(0) = \mathbf{0}$; $\mathbf{r}(0) = \mathbf{0}$.

14. $\mathbf{a}(t) = t\mathbf{j}$; $\mathbf{v}(0) = \mathbf{i} + \mathbf{j}$; $\mathbf{r}(0) = \mathbf{0}$.

15. $\mathbf{a}(t) = -\cos t\,\mathbf{i} - \sin t\,\mathbf{j}$; $\mathbf{v}(0) = \mathbf{i}$; $\mathbf{r}(0) = \mathbf{j}$.

16. $\mathbf{a}(t) = \mathbf{i} + e^{-t}\mathbf{j}$; $\mathbf{v}(0) = 2\mathbf{i} + \mathbf{j}$; $\mathbf{r}(0) = \mathbf{i} - \mathbf{j}$.

17. $\mathbf{a}(t) = \mathbf{i} + t\mathbf{k}$; $\mathbf{v}(0) = \mathbf{0}$; $\mathbf{r}(0) = \mathbf{j}$.

18. $\mathbf{a}(t) = (\sin 2t)\mathbf{j}$; $\mathbf{v}(0) = \mathbf{i} - \mathbf{k}$; $\mathbf{r}(0) = \mathbf{0}$.

19. $\mathbf{a}(t) = \sin t\,\mathbf{i} + \cos t\,\mathbf{j} + e^t\mathbf{k}$; $\mathbf{v}(0) = \mathbf{k}$; $\mathbf{r}(0) = -\mathbf{i} + \mathbf{k}$.

20. $\mathbf{a}(t) = (t + 1)^{-2}\mathbf{j} - e^{-2t}\mathbf{k}$, $t > -1$; $\mathbf{v}(0) = 3\mathbf{i} - \mathbf{j}$; $\mathbf{r}(0) = 2\mathbf{k}$.

In Exercises 21–30, find the scalar tangential and normal components of acceleration at the indicated time t.

21. $\mathbf{r}(t) = 2\cos t\,\mathbf{i} + 2\sin t\,\mathbf{j}$; $t = \pi/3$.

22. $\mathbf{r}(t) = t\,\mathbf{i} + t^2\,\mathbf{j}$; $t = 1$.

23. $\mathbf{r}(t) = e^{-t}\mathbf{i} + e^t\mathbf{j}$; $t = 0$.

24. $\mathbf{r}(t) = \cos t^2\,\mathbf{i} + \sin t^2\,\mathbf{j}$; $t = \sqrt{\pi}/2$.

25. $\mathbf{r}(t) = (t^3 - 2t)\mathbf{i} + (t^2 - 4)\mathbf{j}$; $t = 1$.

26. $\mathbf{r}(t) = e^t\cos t\,\mathbf{i} + e^t\sin t\,\mathbf{j}$; $t = \pi/4$.

27. $\mathbf{r}(t) = t\,\mathbf{i} + t^2\,\mathbf{j} + t^3\,\mathbf{k}$; $t = 1$.

28. $\mathbf{r}(t) = e^t\mathbf{i} + e^{-2t}\mathbf{j} + t\,\mathbf{k}$; $t = 0$.

29. $\mathbf{r}(t) = 3\sin t\,\mathbf{i} + 2\cos t\,\mathbf{j} - \sin 2t\,\mathbf{k}$; $t = \pi/2$.

30. $\mathbf{r}(t) = 2\mathbf{i} + t^3\,\mathbf{j} - 16\ln t\,\mathbf{k}$; $t = 1$.

In Exercises 31–34, $\mathbf{v}$ and $\mathbf{a}$ are given at a certain instant of time. Find a_T, a_N, $\mathbf{T}$, and $\mathbf{N}$ at this instant.

31. $\mathbf{v} = -4\mathbf{j}$, $\mathbf{a} = 2\mathbf{i} + 3\mathbf{j}$.

32. $\mathbf{v} = \mathbf{i} + 2\mathbf{j}$, $\mathbf{a} = 3\mathbf{i}$.

33. $\mathbf{v} = 2\mathbf{i} + 2\mathbf{j} + \mathbf{k}$, $\mathbf{a} = \mathbf{i} + 2\mathbf{k}$.

34. $\mathbf{v} = 3\mathbf{i} - 4\mathbf{k}$, $\mathbf{a} = \mathbf{i} - \mathbf{j} + 2\mathbf{k}$.

35. Prove: If a particle moves along the curve $y = f(x)$ so x increases with time, then the acceleration vector is tangent to the curve at a point of inflection.

36. Prove:

(a) $a_T = \dfrac{d}{dt}\|\mathbf{v}\|$ (b) $a_N = \dfrac{\|\mathbf{v} \times \mathbf{a}\|}{\|\mathbf{v}\|}$.

37. The nuclear accelerator at the Enrico Fermi Laboratory is circular with a radius of 1 kilometer. Find the scalar normal component of acceleration of a proton moving around the accelerator with a constant speed of 3×10^5 kilometers per second.

38. Prove: If the speed of a particle is constant, then the acceleration and velocity vectors are orthogonal.

39. Prove: If the acceleration of a particle is zero for all t, then the particle moves on a straight line.

40. (a) Show that the speed of a particle is constant if $\mathbf{r} = 3 \sin 2t\mathbf{i} - 3 \cos 2t\mathbf{j} - 8t\mathbf{k}$.
    (b) Show that $\mathbf{v}$ and $\mathbf{a}$ are orthogonal at each point on the path of part (a), and thus verify the result in Exercise 38.

41. (a) If $\mathbf{r} = (2 - t^3)\mathbf{i} - t^3\mathbf{j} + (t^3 - 5)\mathbf{k}$, show that the acceleration is zero at only one instant of time.
    (b) Show that the path of part (a) is a line. Does the fact that the acceleration is zero at only one instant of time contradict the result in Exercise 39?

42. The motion of a particle is described by $\mathbf{r} = (t - t^2)\mathbf{i} - t^2\mathbf{j}$. Find the minimum speed of the particle and its location when it has this speed.

43. Find the maximum and minimum speeds of a particle whose motion is described by the formula $\mathbf{r} = \sin 3t\mathbf{i} - 2 \cos 3t\mathbf{j}$.

44. Find the maximum and minimum speeds of a par-

ticle whose motion is described by the formula $\mathbf{r} = 3 \cos 2t\mathbf{i} + \sin 2t\mathbf{j} + 4t\mathbf{k}$.

45. Find, to the nearest degree, the angle between $\mathbf{v}$ and $\mathbf{a}$ for $\mathbf{r} = t^3\mathbf{i} + t^2\mathbf{j}$ when $t = 1$.

46. Show that the angle between $\mathbf{v}$ and $\mathbf{a}$ is constant for $\mathbf{r} = e^t \cos t\mathbf{i} + e^t \sin t\mathbf{j}$. Find the angle.

47. Where on the path

$$\mathbf{r} = (t^2 - 5t)\mathbf{i} + (2t + 1)\mathbf{j} + 3t^2\mathbf{k}$$

are the velocity and acceleration vectors orthogonal?

48. Show that for a particle in motion over the time interval $[t_1, t_2]$:
    (a) the displacement $\mathbf{r}(t_2) - \mathbf{r}(t_1)$ is given by

$$\mathbf{r}(t_2) - \mathbf{r}(t_1) = \int_{t_1}^{t_2} \mathbf{v}(t)\, dt$$

    (b) the change in velocity $\mathbf{v}(t_2) - \mathbf{v}(t_1)$ is given by

$$\mathbf{v}(t_2) - \mathbf{v}(t_1) = \int_{t_1}^{t_2} \mathbf{a}(t)\, dt$$

In Exercises 49–54, find the displacement and the distance traveled over the indicated time interval.

49. $\mathbf{r}(t) = t^2\mathbf{i} + \frac{1}{3}t^3\mathbf{j}$; $1 \le t \le 3$.

50. $\mathbf{r}(t) = (1 - 3 \sin t)\mathbf{i} + 3 \cos t\mathbf{j}$; $0 \le t \le 3\pi/2$.

51. $\mathbf{r}(t) = 2 \sin 3t\mathbf{i} + 2 \cos 3t\mathbf{j}$; $0 \le t \le 2\pi$.

52. $\mathbf{r}(t) = 3 \sin 2t\mathbf{i} - 3 \cos 2t\mathbf{j} + 8t\mathbf{k}$; $0 \le t \le 3\pi/4$.

53. $\mathbf{r}(t) = e^t\mathbf{i} + e^{-t}\mathbf{j} + \sqrt{2}t\mathbf{k}$; $0 \le t \le \ln 3$.

54. $\mathbf{r}(t) = \cos 2t\mathbf{i} + (1 - \cos 2t)\mathbf{j} + (3 + \frac{1}{2} \cos 2t)\mathbf{k}$; $0 \le t \le \pi$.

55. A shell is fired from ground level with a muzzle speed of 320 ft/sec and elevation angle of 60°. Find:
    (a) parametric equations for the shell's trajectory;
    (b) the maximum height reached by the shell;
    (c) the horizontal distance traveled by the shell;
    (d) the speed of the shell at impact.

56. Solve Exercise 55 assuming that the muzzle speed is 980 m/sec and the elevation angle is 45°.

57. A rock is thrown downward from the top of a building, 168 ft high, at an angle of 60° with the

horizontal. How far from the base of the building will the rock land if its initial speed is 80 ft/sec?

58. Solve Exercise 57 assuming that the rock is thrown horizontally at a speed of 80 ft/sec.

59. A shell is to be fired from ground level at an elevation angle of 30°. What should the muzzle speed be in order for the maximum height of the shell to be 2500 ft?

60. A shell, fired from ground level at an elevation angle of 45°, hits the ground 24,500 m away. Calculate the muzzle speed of the shell.

61. Find two elevation angles that will enable a shell, fired from ground level, with a muzzle speed of 800 ft/sec, to hit a ground-level target 10,000 ft away.

62. A ball rolls off a table 4 ft high while moving at a constant speed of 5 ft/sec.
  (a) How long does it take for the ball to hit the floor after it leaves the table?
  (b) At what speed does the ball hit the floor?
  (c) If a ball were dropped from table height (initial speed 0) at the same time the rolling ball leaves the table, which ball would hit the ground first?

63. A shell is fired from ground level at an elevation angle of α and with a muzzle speed of v_0.
  (a) Find parametric equations for the shell's trajectory.
  (b) Show that the maximum height reached by the shell is

$$\text{maximum height} = \frac{(v_0 \sin \alpha)^2}{2g}$$

  (c) With v_0 kept fixed, what elevation angle will enable the shell to travel the largest possible horizontal distance?

64. Suppose that a particle in 3-space undergoes a constant acceleration of $-g\mathbf{k}$, and assume that at the initial time, $t = 0$, the position vector is $\mathbf{r}_0$ and the velocity vector is $\mathbf{v}_0$. By integration, show that the position of the particle at any time t is $\mathbf{r}(t) = -\frac{1}{2}gt^2\mathbf{k} + t\mathbf{v}_0 + \mathbf{r}_0$.

65. Use the result of Exercise 64 to find the position, velocity, and speed of a particle having the given acceleration, initial position, and initial velocity.
  (a) $\mathbf{a}(t) = -32\mathbf{k}$; $\mathbf{v}_0 = \mathbf{0}$; $\mathbf{r}_0 = \mathbf{0}$
  (b) $\mathbf{a}(t) = -32\mathbf{k}$; $\mathbf{v}_0 = \mathbf{0}$; $\mathbf{r}_0 = \mathbf{i} + 2\mathbf{j} + \mathbf{k}$
  (c) $\mathbf{a}(t) = -32\mathbf{k}$; $\mathbf{v}_0 = \mathbf{i} + \mathbf{k}$; $\mathbf{r}_0 = 3\mathbf{i} - \mathbf{j} + 4\mathbf{k}$.

▶ SUPPLEMENTARY EXERCISES

In Exercises 1–3,
(a) find $\mathbf{v} = d\mathbf{r}/dt$ and $\mathbf{a} = d^2\mathbf{r}/dt^2$;
(b) sketch the graph of $\mathbf{r}(t)$, showing the direction of increasing t, and find the vectors $\mathbf{r}'(t)$ and $\mathbf{r}''(t)$ at the points corresponding to $t = t_0$ and $t = t_1$.

1. $\mathbf{r}(t) = \sqrt{t + 4}\,\mathbf{i} + 2t\mathbf{j}$; $t_0 = -3$, $t_1 = 0$.

2. $\mathbf{r}(t) = \langle 2 + \cosh t, 1 - 2\sinh t \rangle$; $t_0 = 0$, $t_1 = \ln 2$.

3. $\mathbf{r}(t) = \langle 2t^3 - 1, t^3 + 1 \rangle$; $t_0 = 0$, $t_1 = -\frac{1}{2}$.

4. Find the limits.
  (a) $\lim\limits_{t \to e} \langle t + \ln t^2, \ln t + t^2 \rangle$
  (b) $\lim\limits_{t \to \pi/6} (\cos 2t\mathbf{i} - 3t\mathbf{j})$.

5. Evaluate the integrals.
  (a) $\displaystyle\int (k\mathbf{i} + m\mathbf{j})\, dt$

  (b) $\displaystyle\int_0^{\ln 3} \langle e^{2t}, 2e^t \rangle\, dt$

  (c) $\displaystyle\int_0^2 \|\cos t\mathbf{i} + \sin t\mathbf{j}\|\, dt$

  (d) $\displaystyle\int \frac{d}{dt}[\sqrt{t^2 + 3}\,\mathbf{i} + \ln(\sin t)\mathbf{j}]\, dt$.

In Exercises 6–8, find: (a) ds/dt, (b) parametric equations for the curve with arc length s as a parameter assuming that the point corresponding to t_0 is the reference point.

6. $\mathbf{r}(t) = (3e^t + 2)\mathbf{i} + (e^t - 1)\mathbf{j}$; $t_0 = 0$.

7. $\mathbf{r}(t) = \left\langle \dfrac{t^2 + 1}{t}, \ln t^2 \right\rangle$, where $t > 0$; $t_0 = 1$.

8. $\mathbf{r}(t) = \langle t^3, t^2 \rangle$, where $t \geq 0$; $t_0 = 0$.

In Exercises 9 and 10, sketch the graph of the curve, showing the direction of increasing t.

9. $\mathbf{r}(t) = \langle t, t^2 + 1, 1 \rangle$.

10. $x = t$, $y = t$, $z = 2\cos(\pi t/2)$, $0 \le t \le 2$.

11. Find the velocity, speed, acceleration, unit tangent vector, unit normal vector, and curvature when $t = 0$ for the motion given by $x = a\sin t$, $y = a\cos t$, $z = a\ln(\cos t)$ $(a > 0)$.

12. The position vector of a particle is
$\mathbf{r}(t) = \sin(2t)\mathbf{i} + \cos(2t)\mathbf{j} + 2e^t\mathbf{k}$.
    (a) Find the velocity, acceleration, and speed as functions of t.
    (b) Find the scalar tangential and normal components of acceleration, and the curvature when $t = 0$.

In Exercises 13 and 14, find the arc length of the curve.

13. $x = 2t$, $y = 4\sin 3t$, $z = 4\cos 3t$, $0 \le t \le 2\pi$.

14. $\mathbf{r}(t) = \langle e^{-t}, \sqrt{2}t, e^t \rangle$, $0 \le t \le \ln 2$.

15. Consider the curve whose position vector is

$$\mathbf{r}(t) = \langle e^{-t}, e^{2t}, t^3 + 1 \rangle$$

Find parametric equations for the tangent line to the curve at the point where $t = 0$.

16. Prove: If the tangential component of acceleration of a moving particle is zero at every point, then $d\mathbf{r}/dt$ has constant length.

17. If $\mathbf{u} = \langle 2t, 3, -t^2 \rangle$ and $\mathbf{v} = \langle 0, t^2, t \rangle$, find:

    (a) $\displaystyle\int_0^3 \mathbf{u}\,dt$ (b) $d(\mathbf{u} \times \mathbf{v})/dt$.

18. If $\mathbf{r}(t) = \langle \cos(\pi e^t), \sin(\pi e^t), \pi t \rangle$, find the angle between the acceleration $\mathbf{a}$ and the velocity $\mathbf{v}$ when $t = 0$.

For the curves given in Exercises 19–21, find:
(a) the unit tangent vector $\mathbf{T}$ at P_0;
(b) the unit normal vector $\mathbf{N}$ at P_0;
(c) the curvature κ at P_0.

19. $\mathbf{r}(t) = (t^2 + 1)\mathbf{i} + (1/t)\mathbf{j}$; $P_0(2, 1)$.

20. $y = \ln x$; $P_0(1, 0)$.

21. $x = (y - 1)^2$; $P_0(0, 1)$.

In Exercises 22–25, find the curvature κ of the given curve at P_0.

22. $xy^2 = 1$; $P_0(1, 1)$.

23. $\mathbf{r}(t) = (t + t^3)\mathbf{i} + (t + t^2)\mathbf{j}$; $P_0(2, 2)$.

24. $y = a\cosh(x/a)$; $P_0(a, a\cosh 1)$.

25. $e^x = \sec y$; $P_0(0, 0)$.

26. Find the smallest radius of curvature and the point at which it occurs for $\mathbf{r}(t) = \langle e^{2t}, e^{-2t} \rangle$.

27. Find the equation of the osculating circle for the parabola $y = (x - 1)^2$ at the point $(1, 0)$. Verify that y' and y'' for the parabola are the same as y' and y'' for the osculating circle at $(1, 0)$.

In Exercises 28 and 29, calculate $d\mathbf{u}/dw$ by the chain rule, and check the result by first expressing $\mathbf{u}$ in terms of w.

28. $\mathbf{u} = \langle \sin t, 2\cos 2t \rangle$; $t = e^{w/2}$.

29. $\mathbf{u} = \langle e^t - 1, 2e^{2t} \rangle$; $t = \ln w$.

In Exercises 30 and 31, find the scalar tangential and normal components of acceleration.

30. $\mathbf{r}(t) = \langle \cosh 2t, \sinh 2t \rangle$, $t \ge 0$.

31. $\mathbf{r}(t) = \langle \sin t - t\cos t, \cos t + t\sin t \rangle$, $t \ge 0$.

For the motion described in Exercises 32 and 33:
(a) find $\mathbf{v}$, $\mathbf{a}$, and ds/dt at P_0;
(b) find $\mathbf{T}$, $\mathbf{N}$, and κ at P_0;
(c) find a_T and a_N at P_0;
(d) describe the trajectory;
(e) find the center of the osculating circle at P_0.

32. $\mathbf{r}(t) = (1 - t^2)\mathbf{i} + 2t\mathbf{j}$; $P_0(0, 2)$.

33. $\mathbf{r}(t) = \langle e^{-t}, e^t \rangle$; $P_0(1, 1)$.

34. At $t = 0$, a particle of mass m is located at the point $(-2/m, 0)$ and has a velocity of $(2\mathbf{i} - 3\mathbf{j})/m$. Find the position function $\mathbf{r}(t)$ if the particle is acted upon by a force $\mathbf{F} = \langle 2\cos t, 3\sin t \rangle$ for $t \ge 0$.

35. The force acting on a particle of unit mass $(m = 1)$ is $\mathbf{F} = (\sin t)\mathbf{i} + (4e^{2t})\mathbf{j}$. If the particle starts at the origin with an initial velocity $\mathbf{v}_0 = \mathbf{i} + 2\mathbf{j}$, find the position function $\mathbf{r}(t)$ at any $t \ge 0$.

36. A curve in a railroad track has the shape of the parabola $x = y^2/100$. If a train is loaded so that its scalar normal component of acceleration cannot exceed 25 units/sec^2, what is its maximum possible speed as it rounds the curve at $(0, 0)$?

37. A particle moves along the parabola $y = 2x - x^2$ with a constant x-component of velocity of 4 ft/sec in. Find the scalar tangential and normal components of acceleration at the points (a) $(1, 1)$ and (b) $(0, 0)$.

38. If a particle moves along the curve $y = 2x^2$ with constant speed $ds/dt = 10$, what are a_T and a_N at $P(x, 2x^2)$?

39. A child twirls a weight at the end of a 2-meter string at a rate of 1 revolution/second. Find a_T and a_N for the motion of the weight.

For the motion described in Exercises 40 and 41, find (a) ds/dt and (b) the distance traveled over the interval described.

40. $\mathbf{r}(t) = \langle 2 \sinh t, \sinh^2 t \rangle$, $0 \le t \le 1$.

41. $\mathbf{r}(t) = e^t \langle \sin 2t, \cos 2t \rangle$, $0 \le t \le \ln 3$.

16 partial derivatives

16.1   FUNCTIONS OF TWO OR MORE VARIABLES

In previous sections we studied real-valued functions of a real variable and vector-valued functions of a real variable. In this section we shall consider real-valued functions of two or more real variables.

NOTATION AND TERMINOLOGY

There are many familiar formulas in which a given variable depends on two or more other variables. For example, the area A of a triangle depends on the base length b and height h by the formula

$$A = \tfrac{1}{2}bh,$$

the volume V of a rectangular box depends on the length l, the width w, and the height h by the formula

$$V = lwh,$$

and the arithmetic average $\bar{x}$ of n real numbers, $x_1, x_2, \ldots, x_n$ depends on those numbers by the formula

$$\bar{x} = \frac{1}{n}(x_1 + x_2 + \cdots + x_n).$$

Thus, we say that:

A is a function of the two variables b and h.

V is a function of the three variables l, w, and h.

$\bar{x}$ is a function of the n variables $x_1, x_2, \ldots, x_n$.

The terminology and notation for functions of two or more variables is similar to that used for functions of one variable. For example, the expression

$$z = f(x, y)$$

means that z is a function of x and y in the sense that a unique value of the *dependent variable* z is determined by specifying values for the *independent variables* x and y. Similarly,

$$w = f(x, y, z)$$

expresses w as a function of x, y, and z and

$$u = f(x_1, x_2, \ldots, x_n)$$

expresses u as a function of $x_1, x_2, \ldots, x_n$.

The functional relationship

$$z = f(x, y) \tag{1}$$

has a useful geometric interpretation. When values of the independent variables x and y are specified, a point (x, y) in the xy-plane is determined. Thus, the dependent variable z in (1) may be viewed as a numerical value associated with the point (x, y). Similarly, the functional relationship

$$w = f(x, y, z)$$

associates the numerical value w with the point (x, y, z) in 3-space.

The following definitions summarize this discussion.

16.1.1   DEFINITION. A *function f of two real variables,* x and y, is a rule that assigns a unique real number $f(x, y)$ to each point (x, y) in some set D of the xy-plane.

16.1.2   DEFINITION. A *function f of three variables,* x, y, and z, is a rule that assigns a unique real number $f(x, y, z)$ to each point (x, y, z) in some set D of three-dimensional space.

The set D in these definitions is the *domain* of the function; it is the set of points at which the function is defined. If a function f is specified by a formula and the domain of f is not stated explicitly, then it is understood that the domain consists of all points at which the formula produces only real numbers and has no divisions by zero.

REMARK.   In more advanced courses the notion of "n-dimensional space" for $n > 3$ is defined, and a *function f of n real variables,* $x_1, x_2, \ldots, x_n$, is regarded as a rule that assigns a unique real number $f(x_1, x_2, \ldots, x_n)$ to each "point" $(x_1, x_2, \ldots, x_n)$ in some set of n-dimensional space. However, we shall not pursue this idea in this text.

Example 1   Let

$$f(x, y) = 3x^2\sqrt{y} - 1$$

Find $f(1, 4)$, $f(0, 9)$, $f(t^2, t)$, $f(ab, 9b)$, and the domain of f.

Solution.   By substitution

$$f(1, 4) = 3(1)^2\sqrt{4} - 1 = 5$$
$$f(0, 9) = 3(0)^2\sqrt{9} - 1 = -1$$
$$f(t^2, t) = 3(t^2)^2\sqrt{t} - 1 = 3t^4\sqrt{t} - 1$$
$$f(ab, 9b) = 3(ab)^2\sqrt{9b} - 1 = 9a^2b^2\sqrt{b} - 1$$

$y \geq 0$

The heavy boundary line is included in the domain.

Figure 16.1.1

Because of the $\sqrt{y}$, we must have $y \geq 0$ to avoid imaginary values for $f(x, y)$. Thus, the domain of f consists of all points in the xy-plane that are on or above the x-axis. (See Figure 16.1.1.)   ◄

Example 2   Sketch the domain of the function $f(x, y) = \ln(x^2 - y)$.

Solution. $\ln(x^2 - y)$ is defined only when $0 < x^2 - y$ or $y < x^2$. To sketch this region, we use the fact that the curve $y = x^2$ separates the region where $y < x^2$ from the region where $y > x^2$. To determine the region where $y < x^2$ holds, we can select an arbitrary "test point" off the boundary $y = x^2$ and determine whether $y < x^2$ or $y > x^2$ at the test point. For example, if we choose the test point $(x, y) = (0, 1)$, then $x^2 = 0$, $y = 1$, so that this point lies in the region where $y > x^2$. Thus, the region where $y < x^2$ is the one that does *not* contain the test point (Figure 16.1.2).   ◄

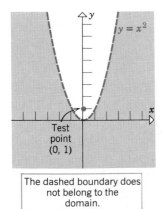

$y = x^2$

Test point (0, 1)

The dashed boundary does not belong to the domain.

Figure 16.1.2

Example 3   Let

$$f(x, y, z) = \sqrt{1 - x^2 - y^2 - z^2}$$

Find $f(0, \frac{1}{2}, -\frac{1}{2})$ and the domain of f.

Solution.   By substitution,

$$f(0, \tfrac{1}{2}, -\tfrac{1}{2}) = \sqrt{1 - (0)^2 - (\tfrac{1}{2})^2 - (-\tfrac{1}{2})^2} = \sqrt{\tfrac{1}{2}}$$

Because of the square root sign, we must have $0 \leq 1 - x^2 - y^2 - z^2$ in order to have a real value for $f(x, y, z)$. Rewriting this inequality in the form

$$x^2 + y^2 + z^2 \leq 1$$

we see that the domain of f consists of all points on or within the sphere $x^2 + y^2 + z^2 = 1$.   ◄

GRAPHS OF
FUNCTIONS OF TWO
VARIABLES

Recall that for a function f of one variable, the graph of $f(x)$ was defined to be the graph of the equation $y = f(x)$. Similarly, if f is a function of two variables, we define the **graph** of $f(x, y)$ to be the graph of the equation $z = f(x, y)$. In general, such a graph will be a surface in 3-space.

Example 4 Describe the graph of the function $f(x, y) = 1 - x - \frac{1}{2}y$.

Solution. By definition, the graph of the given function is the graph of the equation

$$z = 1 - x - \tfrac{1}{2}y$$

or equivalently

$$x + \tfrac{1}{2}y + z = 1$$

which is a plane. A triangular portion of the plane can be sketched by plotting the intersections with the coordinate axes and joining them with line segments (Figure 16.1.3). ◀

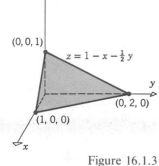

Figure 16.1.3

Example 5 Sketch the graph of the function $f(x, y) = \sqrt{1 - x^2 - y^2}$.

Solution. By definition, the graph of the given function is the graph of the equation

$$z = \sqrt{1 - x^2 - y^2} \qquad (2)$$

After squaring both sides, this can be rewritten as

$$x^2 + y^2 + z^2 = 1$$

which represents a sphere of radius 1, centered at the origin. Since (2) imposes the added condition that $z \geq 0$, the graph is just the upper hemisphere (Figure 16.1.4). ◀

Figure 16.1.4

Example 6 Sketch the graph of the function $f(x, y) = -\sqrt{x^2 + y^2}$.

Solution. The graph of the given function is the graph of the equation

$$z = -\sqrt{x^2 + y^2} \qquad (3)$$

After squaring, we obtain

$$z^2 = x^2 + y^2$$

which is the equation of a right circular cone [(4) of Section 14.7]. Since (3) imposes the condition that $z \leq 0$, the graph is just the lower nappe of the cone (Figure 16.1.5). ◀

Figure 16.1.5

Figure 16.1.6

$z = x^2 + \frac{1}{4}y^2$

Example 7   Sketch the graph of the function $f(x, y) = x^2 + \frac{1}{4}y^2$.

Solution.   The graph of f is the graph of the equation

$$z = x^2 + \frac{1}{4}y^2$$

As discussed in Section 14.7 [Equation (5a)], this is an elliptic paraboloid (Figure 16.1.6).   ◄

When it is necessary to introduce a dependent variable for a function $f(x, y, z)$ of three variables, we shall usually use the letter w and write $w = f(x, y, z)$. Because this equation involves four variables, it cannot be graphed in three dimensions—"four dimensions" are needed. Thus, there is no *direct* way to represent a function of three or more variables geometrically. However, we shall now discuss some alternative methods for representing functions geometrically that can be applied to functions of three variables.

LEVEL CURVES AND SURFACES

We are all familiar with topographic (or contour) maps in which a three-dimensional landscape, such as a mountain range, is represented by two-dimensional contour lines or curves of constant elevation. Consider, for example, the model hill and its contour map shown in Figure 16.1.7. The contour map is constructed by passing planes of constant elevation through the hill, projecting the resulting contours onto a flat surface, and labeling the contours with their elevations. In Figure 16.1.7, note how the two gullies appear as indentations in the contour lines and how the curves are close together on the contour map where the hill has a steep slope and become more widely spaced where the slope is gradual.

Contour maps are useful for studying functions of two variables. If the surface

$$z = f(x, y) \tag{4}$$

is cut by the horizontal plane

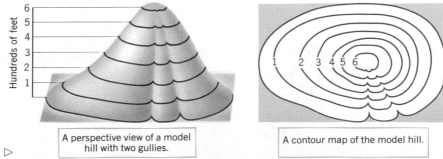

Figure 16.1.7 ▷

Hundreds of feet

A perspective view of a model hill with two gullies.

A contour map of the model hill.

$$z = k \tag{5}$$

and if the resulting curve is projected onto the xy-plane, then we obtain what is called the *level curve of height k* (or the *level curve with constant k*) for the function. It follows from (4) and (5) that this level curve has the equation $f(x, y) = k$ (see Figure 16.1.8).

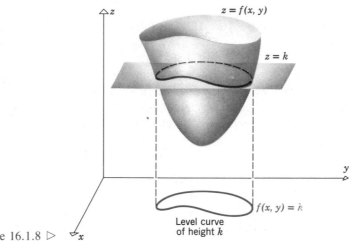

Figure 16.1.8 ▷

Level curve
of height k

Example 8 The graph of the function $f(x, y) = x^2 + \frac{1}{4}y^2$ is the paraboloid shown in Figure 16.1.9. The level curves have equations of the form

$$x^2 + \tfrac{1}{4}y^2 = k \tag{6}$$

For $k > 0$ these are ellipses; for $k = 0$, it is the single point $(0, 0)$; and for $k < 0$ there are no level curves, since (6) is not satisfied by any real values of x and y. Some sample level curves are shown in Figure 16.1.9.   ◀

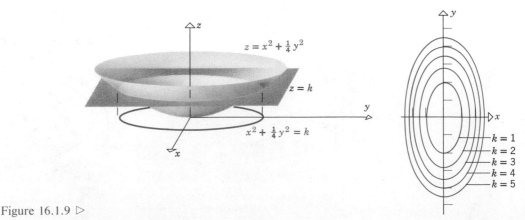

Figure 16.1.9 ▷

Example 9   The graph of the function $f(x, y) = 2 - x - y$ is the plane shown in Figure 16.1.10a. The level curves have equations of the form $2 - x - y = k$, or $y = -x + (2 - k)$. These form a family of parallel lines of slope -1 (Figure 16.1.10b).   ◀

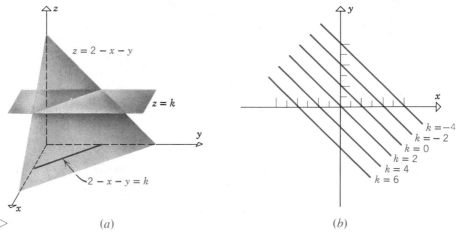

Figure 16.1.10 ▷   (a)   (b)

Example 10   The graph of the function $f(x, y) = y^2 - x^2$ is the hyperbolic paraboloid (saddle curve) shown in Figure 16.1.11a. The level curves have equations of the form $y^2 - x^2 = k$. For $k > 0$ these curves are hyperbolas opening along lines parallel to the y-axis; for $k < 0$ they are hyperbolas opening along lines parallel to the x-axis; and for $k = 0$ the level curve consists of the intersecting lines $y + x = 0$ and $y - x = 0$ (Figure 16.1.11b).   ◀

The concept of a level curve for a function of two variables can be extended to functions of three variables. If k is a constant, then an equation of the form $f(x, y, z) = k$ will, in general, represent a surface in three-dimensional space. (For example, $x^2 + y^2 + z^2 = 1$ represents a sphere.) The graph of this surface is called the *level surface with constant k* for the function f.

REMARK.   The term "level surface" can be confusing. A level surface need *not* be level in the sense of being horizontal. It is simply a surface on which all values of f are the same.

Example 11   Describe the level surfaces of $f(x, y, z) = x^2 + y^2 + z^2$.

Solution.   The level surfaces have equations of the form

$$x^2 + y^2 + z^2 = k$$

For $k > 0$, the graph of this equation is a sphere of radius $\sqrt{k}$, centered

Figure 16.1.11   (a)   (b)

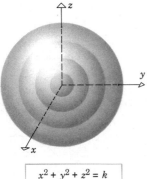

$x^2 + y^2 + z^2 = k$
$(k \geqslant 0)$

Figure 16.1.12

at the origin; for $k = 0$, the graph is the single point $(0, 0, 0)$; and for $k < 0$ there is no level surface (Figure 16.1.12).   ◄

Example 12   Describe the level surfaces of $f(x, y, z) = z^2 - x^2 - y^2$.

Solution.   The level surfaces have equations of the form

$$z^2 - x^2 - y^2 = k$$

As discussed in Section 14.7, this equation represents a cone if $k = 0$, a hyperboloid of two sheets if $k > 0$, and a hyperboloid of one sheet if $k < 0$ (Figure 16.1.13).   ◄

OPEN AND CLOSED
SETS

In our study of functions of one variable, the domains of the functions we encountered were generally intervals. For functions of two or three variables the situation is more complicated, so we shall need to discuss some terminology about sets in 2-space and 3-space that will be helpful when we want to accurately describe the domain of a function of two or three variables.

If D is a set of points in 2-space, then a point (x_0, y_0) is called an *interior*

$k>0 \quad k=0 \quad k<0$

Level surfaces of
$f(x, y, z) = z^2 - x^2 - y^2$

Figure 16.1.13

point of D if there is *some* circular disk with positive radius, centered at (x_0, y_0), and containing only points in D (Figure 16.1.14). A point (x_0, y_0) is called a **boundary point** of D if *every* circular disk with positive radius and centered at (x_0, y_0) contains both points in D and points not in D (Figure 16.1.14). Similarly, if D is a set of points in 3-space, then a point (x_0, y_0, z_0) is called an **interior point** of D if there is some spherical ball with positive radius, centered at (x_0, y_0, z_0), and containing only points in D (Figure 16.1.15). A point (x_0, y_0, z_0) is called a **boundary point** of D if *every* spherical ball with positive radius and centered at (x_0, y_0, z_0) contains both points in D and points not in D (Figure 16.1.15).

An interior point

A boundary point

A boundary point

An interior point

Figure 16.1.14   Figure 16.1.15

For a set D in either 2-space or 3-space, the set of all boundary points of D is called the **boundary** of D and the set of all interior points of D is called the **interior** of D.

Recall that an open interval (a, b) on a coordinate line contains *neither* of its endpoints and a closed interval $[a, b]$ contains *both* of its endpoints. Analogously, a set D in 2-space or 3-space is called **open** if it contains *none* of its boundary points and **closed** if it contains *all* of its boundary points.

Example 13   Let D be the set of points in the xy-plane that are inside or on the circle of radius 1 centered at the origin (Figure 16.1.16a). In set notation

$$D = \{(x, y) : x^2 + y^2 \leq 1\}$$

The interior of D is the set

$$I = \{(x, y) : x^2 + y^2 < 1\}$$

and the boundary of D is the set

$$B = \{(x, y) : x^2 + y^2 = 1\}$$

(Figures 16.1.16b and 16.1.16c). The set D is closed and the set I is open.   ◄

Figure 16.1.16 ▷   (a)   (b)   (c)

BOUNDED SETS   Just as we distinguished between finite intervals and infinite intervals on the real line, so we shall want to distinguish between regions of "finite extent" and regions of "infinite extent" in 2-space and 3-space. A set of points in 2-space is called **bounded** if the entire set can be contained within some rectangle, and is called **unbounded** if there is no rectangle that contains all the points of the set. Similarly, a set of points in 3-space is **bounded** if the entire set can be contained within some box, and is unbounded otherwise (Figure 16.1.17).

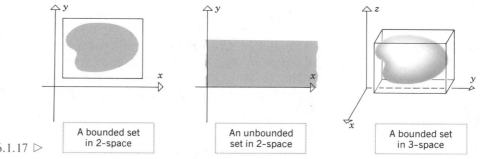

Figure 16.1.17 ▷

A bounded set in 2-space

An unbounded set in 2-space

A bounded set in 3-space

COMPUTER GRAPHICS   In recent years computers have been used to generate graphic representations of solids and surfaces in three dimensions that would be difficult or impossible to draw accurately without computer assistance.

Representations of Solids   Computers can be used to generate various kinds of solids (Figure 16.1.18).

Representations of Surfaces   A portion of a surface $z = f(x, y)$ can be viewed in perspective and delineated by a rectangular mesh over a rectangular portion of the xy-plane. By altering program parameters, it is possible to change the section of

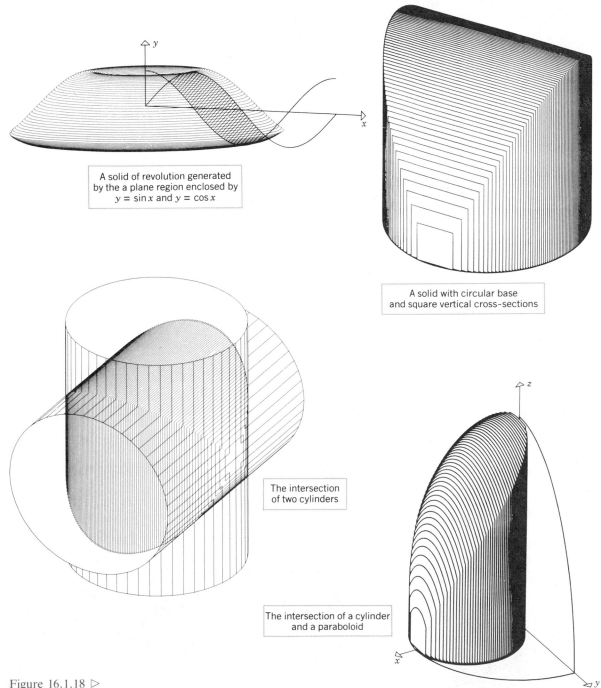

A solid of revolution generated
by the a plane region enclosed by
$y = \sin x$ and $y = \cos x$

A solid with circular base
and square vertical cross-sections

The intersection
of two cylinders

The intersection of a cylinder
and a paraboloid

Figure 16.1.18 ▷

surface shown, the point of view, and the apparent distance from the observer to the origin (Figure 16.1.19).

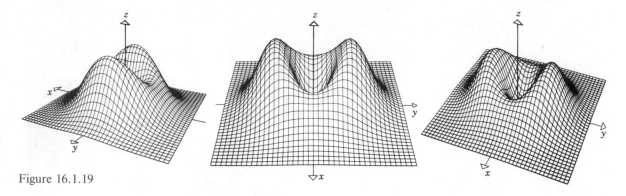

Figure 16.1.19

It is also possible to "zoom in" on interesting features and to generate a two-dimensional map of level curves pictured either head-on or in a perspective view that corresponds to the orientation of the surface (Figures 16.1.20 and 16.1.21).

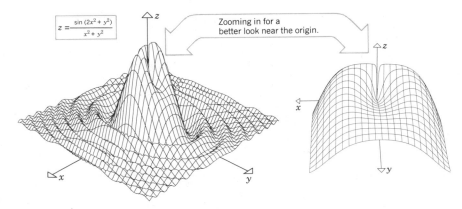

$$z = \frac{\sin(2x^2 + y^2)}{x^2 + y^2}$$

Zooming in for a better look near the origin.

Figure 16.1.20 ▷

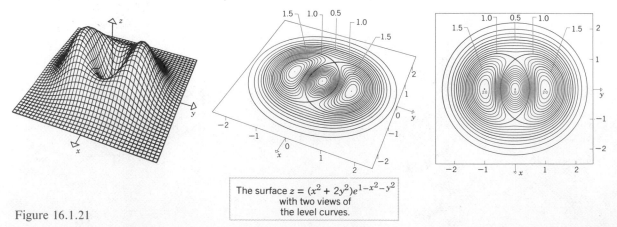

The surface $z = (x^2 + 2y^2)e^{1-x^2-y^2}$ with two views of the level curves.

Figure 16.1.21

We conclude this section with a sampling of surfaces in 3-space that illustrates the variety and complexity that such surfaces can have.

A GALLERY OF SURFACES IN 3-SPACE

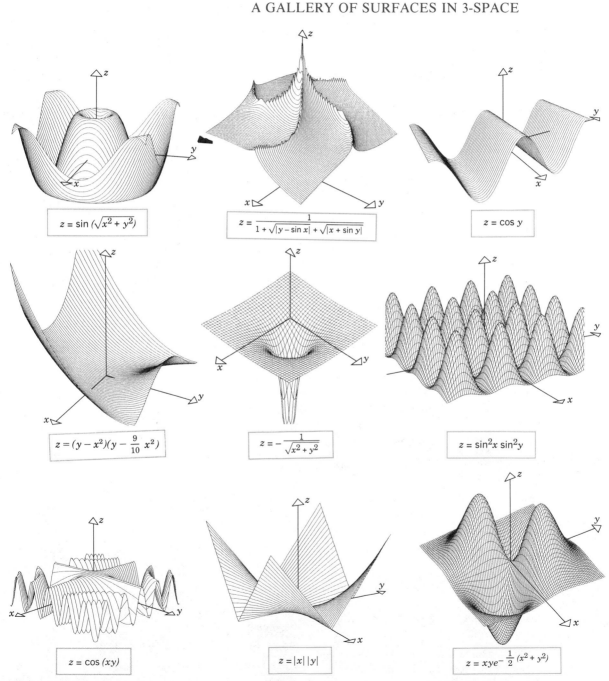

$z = \sin\left(\sqrt{x^2 + y^2}\right)$

$z = \dfrac{1}{1 + \sqrt{|y - \sin x|} + \sqrt{|x + \sin y|}}$

$z = \cos y$

$z = (y - x^2)\left(y - \frac{9}{10}x^2\right)$

$z = -\dfrac{1}{\sqrt{x^2 + y^2}}$

$z = \sin^2 x \, \sin^2 y$

$z = \cos(xy)$

$z = |x|\,|y|$

$z = xy e^{-\frac{1}{2}(x^2 + y^2)}$

Figure 16.1.22

▶ Exercise Set 16.1

1. Let $f(x, y) = x^2y + 1$. Find:
   (a) $f(2, 1)$ (b) $f(1, 2)$
   (c) $f(0, 0)$ (d) $f(1, -3)$
   (e) $f(3a, a)$ (f) $f(ab, a - b)$.

2. Let $f(x, y) = x + \sqrt[3]{xy}$. Find:
   (a) $f(t, t^2)$ (b) $f(x, x^2)$
   (c) $f(2y^2, 4y)$.

3. Let $f(x, y) = xy + 3$. Find:
   (a) $f(x + y, x - y)$ (b) $f(xy, 3x^2y^3)$.

4. Let $g(x) = x \sin x$. Find:
   (a) $g\left(\dfrac{x}{y}\right)$ (b) $g(xy)$
   (c) $g(x - y)$.

5. Find $F(g(x), h(y))$ if $F(x, y) = xe^{xy}$, $g(x) = x^3$, and $h(y) = 3y + 1$.

6. Find $g(u(x, y), v(x, y))$ if $g(x, y) = y \sin(x^2y)$, $u(x, y) = x^2y^3$, and $v(x, y) = \pi xy$.

7. Let $f(x, y) = x + 3x^2y^2$, $x(t) = t^2$, and $y(t) = t^3$. Find:
   (a) $f(x(t), y(t))$ (b) $f(x(0), y(0))$
   (c) $f(x(2), y(2))$.

8. Let $g(x, y) = ye^{-3x}$, $x(t) = \ln(t^2 + 1)$, and $y(t) = \sqrt{t}$. Find $g(x(t), y(t))$.

9. Let $f(x, y, z) = xy^2z^3 + 3$. Find:
   (a) $f(2, 1, 2)$ (b) $f(-3, 2, 1)$
   (c) $f(0, 0, 0)$ (d) $f(a, a, a)$
   (e) $f(t, t^2, -t)$ (f) $f(a + b, a - b, b)$.

10. Let $f(x, y, z) = zxy + x$. Find:
    (a) $f(x + y, x - y, x^2)$ (b) $f(xy, y/x, xz)$.

11. Find $F(f(x), g(y), h(z))$ if $F(x, y, z) = ye^{xyz}$, $f(x) = x^2$, $g(y) = y + 1$, and $h(z) = z^2$.

12. Find $g(u(x, y, z), v(x, y, z), w(x, y, z))$ if $g(x, y, z) = z \sin xy$, $u(x, y, z) = x^2z^3$, $v(x, y, z) = \pi xyz$, and $w(x, y, z) = xy/z$.

13. Let $f(x, y, z) = x^2y^2z^4$, $x(t) = t^3$, $y(t) = t^2$, and $z(t) = t$. Find:
    (a) $f(x(t), y(t), z(t))$ (b) $f(x(0), y(0), z(0))$
    (c) $f(x(2), y(2), z(2))$.

In Exercises 14–19, sketch the domain of f. Use solid lines for portions of the boundary included in the domain and dashed lines for portions not included.

14. $f(x, y) = xy\sqrt{y - 1}$.

15. $f(x, y) = \ln(1 - x^2 - y^2)$.

16. $f(x, y) = \sqrt{x^2 + y^2 - 4}$.

17. $f(x, y) = \dfrac{1}{x - y^2}$.

18. $f(x, y) = \ln xy$.

19. $f(x, y) = \sqrt{\dfrac{x^2 + y^2}{x^2 - y^2}}$.

In Exercises 20–27, describe the domain of f.

20. $f(x, y) = \sin^{-1}(x + y)$.

21. $f(x, y) = xe^{-\sqrt{y+2}}$.

22. $f(x, y) = \dfrac{\sqrt{4 - x^2}}{y^2 + 3}$.

23. $f(x, y) = \ln(y - 2x)$.

24. $f(x, y, z) = \sqrt{25 - x^2 - y^2 - z^2}$.

25. $f(x, y, z) = \dfrac{xyz}{x + y + z}$.   26. $f(x, y, z) = e^{xyz}$.

27. $f(x, y, z) = z + \ln(1 - x^2 - y^2)$.

In Exercises 28–39, sketch the graph of f.

28. $f(x, y) = 3$.

29. $f(x, y) = 4 - 2x - 4y$.

30. $f(x, y) = \sqrt{9 - x^2 - y^2}$.

31. $f(x, y) = \sqrt{x^2 + y^2}$.

32. $f(x, y) = x^2 + y^2$.

33. $f(x, y) = x^2 - y^2$.

34. $f(x, y) = 4 - x^2 - y^2$.

35. $f(x, y) = -\sqrt{1 - x^2/4 - y^2/9}$.

36. $f(x, y) = \sqrt{x^2 + y^2 - 1}$.

37. $f(x, y) = \sqrt{x^2 + y^2 + 1}$.

38. $f(x, y) = x^2$.

39. $f(x, y) = y + 1$.

In Exercises 40–47, sketch the level curve $z = k$ for the specified values of k.

40. $z = 3x + y$; $k = -2, -1, 0, 1, 2$.

41. $z = x^2 + y^2$; $k = 0, 1, 2, 3, 4$.

42. $z = y/x$; $k = -2, -1, 0, 1, 2$.

43. $z = x^2 + y$; $k = -2, -1, 0, 1, 2$.

44. $z = x^2 + 9y^2$; $k = 0, 1, 2, 3, 4$.

45. $z = x^2 - y^2$; $k = -2, -1, 0, 1, 2$.

46. $z = y \csc x$; $k = -2, -1, 0, 1, 2$.

47. $z = \sqrt{\dfrac{x+y}{x-y}}$; $k = 0, 1, 2, 3, 4$.

In Exercises 48–51, sketch the level surface $f(x, y, z) = k$.

48. $f(x, y, z) = 4x - 2y + z$; $k = 1$.

49. $f(x, y, z) = 4x^2 + y^2 + 4z^2$; $k = 16$.

50. $f(x, y, z) = x^2 + y^2 - z^2$; $k = 0$.

51. $f(x, y, z) = z - x^2 - y^2 + 4$; $k = 7$.

In Exercises 52–55, describe the level surfaces.

52. $f(x, y, z) = 3x - y + 2z$.

53. $f(x, y, z) = (x - 2)^2 + y^2 + z^2$.

54. $f(x, y, z) = z - x^2 - y^2$.

55. $f(x, y, z) = x^2 + z^2$.

56. Let $f(x, y) = yx^2 + 1$. Find an equation of the level curve that passes through the point
(a) $(1, 2)$ (b) $(-2, 4)$ (c) $(0, 0)$.

57. Let $f(x, y) = x^2 - 2x^3 + 3xy$. Find an equation of the level curve that passes through the point
(a) $(-1, 1)$ (b) $(0, 0)$ (c) $(2, -1)$.

58. Let $f(x, y) = ye^x$. Find an equation of the level curve that passes through the point
(a) $(\ln 2, 1)$ (b) $(0, 3)$ (c) $(1, -2)$.

59. Let $f(x, y, z) = x^2 + y^2 - z$. Find an equation of the level surface that passes through the point
(a) $(1, -2, 0)$ (b) $(1, 0, 3)$ (c) $(0, 0, 0)$.

60. Let $f(x, y, z) = xyz + 3$. Find an equation of the level surface that passes through the point
(a) $(1, 0, 2)$ (b) $(-2, 4, 1)$ (c) $(0, 0, 0)$.

61. If $V(x, y)$ is the voltage or potential at a point (x, y) in the xy-plane, then the level curves of V are called *equipotential curves*. Along such a curve, the voltage remains constant. Given that

$$V(x, y) = \frac{8}{\sqrt{16 + x^2 + y^2}}$$

sketch the equipotential curves at which $V = 2.0$, $V = 1.0$, and $V = 0.5$.

62. If $T(x, y)$ is the temperature at a point (x, y) on a thin metal plate in the xy-plane, then the level curves of T are called *isothermal curves*. All points on such a curve are at the same temperature. Suppose that a plate occupies the first quadrant and $T(x, y) = xy$.
(a) Sketch the isothermal curves on which

$$T = 1, \quad T = 2, \quad T = 3$$

(b) An ant, initially at $(1, 4)$, wants to walk on the plate so that the temperature along its path remains constant. What path should the ant take?

In Exercises 63–65, use the contour map shown in the figure (all elevations in hundreds of feet).

63. For points A and B:
(a) which one is higher?
(b) which one is on the steeper slope?

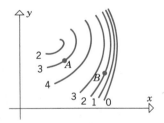

64. Starting at point A, will the elevation begin to increase or decrease if we travel so that
(a) y remains constant and x increases
(b) y remains constant and x decreases
(c) x remains constant and y increases
(d) x remains constant and y decreases?

65. Starting at point B, will the elevation begin to increase or decrease if we travel so that
(a) y remains constant and x increases
(b) y remains constant and x decreases
(c) x remains constant and y increases
(d) x remains constant and y decreases?

In Exercises 66 and 67, classify the given set of points as open, closed, or neither.

66. The set of points in the xy-plane that satisfies the inequality
   (a) $1 \le x^2 + y^2 \le 3$
   (b) $0 \le y < 2x + 1$ for $0 \le x \le 1$
   (c) $\sqrt{y} < x < 1$
   (d) $0 < y < e^x$ for $0 < x < 2$.

67. The set of points in 3-space that satisfies the inequality
   (a) $x^2 + y^2 + z^2 < 3$
   (b) $0 \le z \le x^2 + y^2$ for $x^2 + y^2 < 1$
   (c) $0 \le z \le 5 - 3x - 2y$ for $x \ge 0$ and $y \ge 0$
   (d) $y^2 + z^2 \le 5$ for $1 \le x \le 2$.

In Exercises 68 and 69, classify the given set as bounded or unbounded.

68. The set of points in the xy-plane that satisfies the inequality
   (a) $x^2 + y^2 < 100$ (b) $x^2 + y^2 \ge 100$
   (c) $-1 < y < 2$
   (d) $y \le 3 - x$ for $-1 < x < 1$.

69. The set of points in 3-space that satisfies the inequality
   (a) $x^2 + y^2 + z^2 \le 3$
   (b) $0 < z < 2 - y$ for $0 < x < 1$
   (c) $x^2 + y^2 < 4$ (d) $z \ge 2 - x - 3y$.

16.2   LIMITS AND CONTINUITY

In this section we shall introduce the notions of limit and continuity for functions of two or more variables. We shall not go into great detail; our objective is to develop the basic ideas accurately, and to obtain results needed in later sections. A more extensive study of these topics is usually given in advanced calculus.

LIMITS OF FUNCTIONS OF TWO VARIABLES

Recall that for a function of one variable, the statement

$$\lim_{x \to x_0} f(x) = L$$

This circular region with the center removed consists of all points (x, y) that satisfy
$$0 < \sqrt{(x - x_0)^2 + (y - y_0^2)} < \delta$$

For each point (x, y) within this region we have
$$|f(x, y) - L| < \varepsilon$$

In this particular figure, this condition is also satisfied at (x_0, y_0), although this is not relevant to the limit.

$$\lim_{(x, y) \to (x_0, y_0)} f(x, y) = L$$

$z = f(x, y)$

Figure 16.2.1

means that the value of $f(x)$ can be made arbitrarily close to L by making x sufficiently close to (but different from) x_0. Similarly, the statement

$$\lim_{(x,y)\to(x_0,y_0)} f(x, y) = L$$

is intended to mean that the value of $f(x, y)$ can be made arbitrarily close to L by making the point (x, y) sufficiently close to (but different from) (x_0, y_0). To phrase this idea another way, for every positive number ϵ we want to be able to guarantee that $f(x, y)$ is within ϵ units of L by restricting (x, y) to some sufficiently small circle of radius δ centered at (x_0, y_0), with the possible exception that $f(x, y)$ need not be within ϵ units of L if (x, y) is at the center of the circle. The following definition makes this informal idea precise (Figure 16.2.1).

16.2.1 DEFINITION. Let f be a function of two variables, and suppose that f is defined at each point in some circular region centered at (x_0, y_0), except possibly at (x_0, y_0) itself. We shall write

$$\lim_{(x,y)\to(x_0,y_0)} f(x, y) = L \qquad (1)$$

if given any number $\epsilon > 0$, we can find a number $\delta > 0$ such that $f(x, y)$ satisfies

$$|f(x, y) - L| < \epsilon$$

whenever the distance between (x, y) and (x_0, y_0) satisfies

$$0 < \sqrt{(x - x_0)^2 + (y - y_0)^2} < \delta$$

LIMITS OF FUNCTIONS OF THREE VARIABLES

The following definition extends the notion of a limit to functions of three variables. Note how closely it parallels Definition 16.2.1.

16.2.2 DEFINITION. Let f be a function of three variables, and suppose that f is defined at each point in some spherical region centered at (x_0, y_0, z_0), except possibly at (x_0, y_0, z_0) itself. We shall write

$$\lim_{(x,y,z)\to(x_0,y_0,z_0)} f(x, y, z) = L \qquad (2)$$

if given any number $\epsilon > 0$, we can find a number $\delta > 0$ such that $f(x, y, z)$ satisfies

$$|f(x, y, z) - L| < \epsilon$$

whenever the distance between (x, y, z) and (x_0, y_0, z_0) satisfies

$$0 < \sqrt{(x - x_0)^2 + (y - y_0)^2 + (z - z_0)^2} < \delta$$

When convenient, (1) and (2) can also be written in the alternative notations

$$f(x, y) \to L \quad \text{as} \quad (x, y) \to (x_0, y_0)$$

and

$$f(x, y, z) \to L \quad \text{as} \quad (x, y, z) \to (x_0, y_0, z_0)$$

REMARK.   Limits can also be defined for functions of more than three variables, but we shall leave this for more advanced courses.

PROPERTIES OF
LIMITS

We note without proof that the basic properties of limits hold for functions of two or three variables.

16.2.3   THEOREM. *If*

$$\lim_{(x,y) \to (x_0,y_0)} f(x, y) = L_1 \quad and \quad \lim_{(x,y) \to (x_0,y_0)} g(x, y) = L_2$$

then

(a) $\lim\limits_{(x,y) \to (x_0,y_0)} [cf(x, y)] = cL_1 \quad$ (*if c is constant*)

(b) $\lim\limits_{(x,y) \to (x_0,y_0)} [f(x, y) + g(x, y)] = L_1 + L_2$

(c) $\lim\limits_{(x,y) \to (x_0,y_0)} [f(x, y) - g(x, y)] = L_1 - L_2$

(d) $\lim\limits_{(x,y) \to (x_0,y_0)} [f(x, y)g(x, y)] = L_1 L_2$

(e) $\lim\limits_{(x,y) \to (x_0,y_0)} \dfrac{f(x, y)}{g(x, y)} = \dfrac{L_1}{L_2} \quad$ (*if* $L_2 \neq 0$)

Similarly, for functions of three variables.

As with functions of one variable, results (*b*) and (*d*) can be extended to any finite number of functions; that is,

• the limit of a sum is the sum of the limits;

• the limit of a product is the product of the limits.

Example 1   It can be proved using Definitions 16.2.1 and 16.2.2 that

$$\lim_{(x,y) \to (x_0,y_0)} c = c \quad \text{[c a constant]}$$

$$\lim_{(x,y) \to (x_0,y_0)} x = x_0$$

$$\lim_{(x,y) \to (x_0,y_0)} y = y_0$$

with similar results for limits of functions with three variables. Thus,

$$\lim_{(x,y)\to(2,3)} x = 2, \qquad \lim_{(x,y)\to(2,3)} y = 3$$

$$\lim_{(x,y,z)\to(0,1,-2)} z = -2, \qquad \lim_{(x,y)\to(0,0)} 7 = 7 \qquad \blacktriangleleft$$

Example 2

$$\lim_{(x,y)\to(1,4)} [5x^3y^2 - 9] = \lim_{(x,y)\to(1,4)} [5x^3y^2] - \lim_{(x,y)\to(1,4)} 9$$

$$= 5\left[\lim_{(x,y)\to(1,4)} x\right]^3 \left[\lim_{(x,y)\to(1,4)} y\right]^2 - 9$$

$$= 5(1)^3(4)^2 - 9 = 71 \qquad \blacktriangleleft$$

CONTINUITY   Later in this section we shall describe some more techniques for evaluating limits, but most of the limits we shall need in this text will follow from the notion of continuity, which we shall discuss next.

The definition of continuity for functions of two or three variables is similar to the definition for functions of one variable (see Definition 2.7.1).

16.2.4   DEFINITION. A function f of two variables is called *continuous* at the point (x_0, y_0) if

1. $f(x_0, y_0)$ is defined.
2. $\displaystyle\lim_{(x,y)\to(x_0,y_0)} f(x,y)$ exists.
3. $\displaystyle\lim_{(x,y)\to(x_0,y_0)} f(x,y) = f(x_0,y_0)$.

16.2.5   DEFINITION. A function f of three variables is called *continuous at a point* (x_0, y_0, z_0) if

1. $f(x_0, y_0, z_0)$ is defined.
2. $\displaystyle\lim_{(x,y,z)\to(x_0,y_0,z_0)} f(x, y, z)$ exists.
3. $\displaystyle\lim_{(x,y,z)\to(x_0,y_0,z_0)} f(x, y, z) = f(x_0, y_0, z_0)$.

A function f that is continuous at each point of a region R in 2-space or 3-space is said to be *continuous on R*. A function that is continuous at every point in 2-space or 3-space is called *continuous everywhere* or simply *continuous*. As with functions of one variable, the first and second conditions in the above definitions are consequences of the third condition, so only the third condition needs to be verified to prove continuity.

Intuitively, we can imagine the graph of a continuous function of two variables to be constructed from a thin sheet of clay that has been hollowed and pinched into peaks and valleys without creating tears or pinholes: the requirement that $f(x_0, y_0)$ be defined eliminates the possibility of a hole in the surface $z = f(x, y)$ above the point (x_0, y_0); the requirement that $\lim\limits_{(x,y)\to(x_0,y_0)} f(x, y)$ exists ensures that $z = f(x, y)$ does not become "infinite" at (x_0, y_0) or oscillate wildly; and the requirement that $\lim\limits_{(x,y)\to(x_0,y_0)} f(x, y) = f(x_0, y_0)$ ensures that the surface does not have a vertical jump or step above the point (x_0, y_0). (See Figure 16.2.2.)

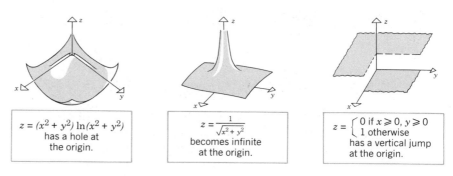

Figure 16.2.2

$z = (x^2 + y^2) \ln(x^2 + y^2)$
has a hole at
the origin.

$z = \dfrac{1}{\sqrt{x^2 + y^2}}$
becomes infinite
at the origin.

$z = \begin{cases} 0 \text{ if } x \geqslant 0, y \geqslant 0 \\ 1 \text{ otherwise} \end{cases}$
has a vertical jump
at the origin.

COMPOSITIONS OF CONTINUOUS FUNCTIONS

The following theorem, which we state without proof, will help us to identify continuous functions of two variables.

16.2.6   THEOREM.

(a) *If g and h are continuous functions of one variable, then $f(x, y) = g(x)h(y)$ is a continuous function of x and y.*

(b) *If g is a continuous function of one variable and h is a continuous function of two variables, then their composition $f(x, y) = g\big(h(x, y)\big)$ is a continuous function of x and y.*

Example 3   The function $f(x, y) = 3x^2 y^5$ is continuous because it is the product of the continuous functions $g(x) = 3x^2$ and $h(y) = y^5$.

In general, any function of the form $f(x, y) = Ax^m y^n$ (*m* and *n* non-negative integers) is continuous because it is the product of the continuous functions Ax^m and y^n.   ◄

Example 4   The function $g(x) = \sin x$ is a continuous function of one variable and the function $h(x, y) = xy^2$ is a continuous function of two variables, so $g\big(h(x, y)\big) = g(xy^2) = \sin(xy^2)$ is a continuous function of *x* and *y*. By a similar argument, each of the following is continuous:

$$(x^4 y^5)^{1/3}, \quad e^{xy}, \quad \cosh(x^3 y) \qquad ◄$$

Example 5 By Example 4, e^{xy} is continuous. Thus, $\cos(e^{xy})$ is continuous by part (*b*) of Theorem 16.2.6. ◄

Theorem 16.2.6 is one of a whole class of theorems about continuity of functions in any number of variables. The content of these theorems can be summarized informally with three basic principles:

- A composition of continuous functions is continuous.

- A sum, difference, or product of continuous functions is continuous.

- A quotient of continuous functions is continuous, except where the denominator is zero.

Example 6 The following functions are continuous since they are sums, differences, products, and compositions of continuous functions:

$$3 - 2x^2yz + 9x^4y^8z^3, \quad e^{xy}\cos(xy^2 + 1), \quad (3z + ye^x)^{17} \qquad ◄$$

Example 7 Since the function

$$f(x, y) = \frac{x^3y^2}{1 - xy}$$

is a quotient of continuous functions, it is continuous except where $1 - xy = 0$. Thus, $f(x, y)$ is continuous everywhere except on the hyperbola $xy = 1$. ◄

Example 8 Evaluate

$$\lim_{(x,y)\to(-1,2)} \frac{xy}{x^2 + y^2}$$

Solution. Since $f(x, y) = xy/(x^2 + y^2)$ is continuous at $(-1, 2)$ (why?), it follows from part 3 of Definition 16.2.4 that

$$\lim_{(x,y)\to(-1,2)} \frac{xy}{x^2 + y^2} = \frac{(-1)(2)}{(-1)^2 + (2)^2} = -\frac{2}{5} \qquad ◄$$

LIMITS AT POINTS OF DISCONTINUITY At a point of discontinuity the method of Example 8 cannot be used. However, sometimes such limits can be obtained by converting the given function to polar coordinates.

Example 9   Find

$$\lim_{(x,y)\to(0,0)} (x^2 + y^2)\ln(x^2 + y^2)$$

Solution.   Let (r, θ) be the coordinates of the point (x, y) with $r \geq 0$. Then we have

$$x = r\cos\theta, \quad y = r\sin\theta, \quad r^2 = x^2 + y^2$$

Moreover, since $r \geq 0$ we have $r = \sqrt{x^2 + y^2}$, so that $r \to 0^+$ if and only if $(x, y) \to (0, 0)$. Thus, we can rewrite the given limit as

$$\lim_{(x,y)\to(0,0)} (x^2 + y^2)\ln(x^2 + y^2) = \lim_{r\to 0^+} r^2 \ln r^2$$

$$= \lim_{r\to 0^+} \frac{2\ln r}{1/r^2} \qquad \boxed{\text{This converts the limit to an indeterminate form of type } \infty/\infty.}$$

$$= \lim_{r\to 0^+} \frac{2/r}{-2/r^3} \qquad \boxed{\text{L'Hôpital's Rule}}$$

$$= \lim_{r\to 0^+} (-r^2)$$

$$= 0 \qquad \blacktriangleleft$$

REMARK.   The function $f(x, y) = (x^2 + y^2)\ln(x^2 + y^2)$ from this example is graphed in Figure 16.2.2.

LIMITS ALONG CURVE

For a function of one variable we considered the one-sided limits,

$$\lim_{x\to x_0^+} f(x) \quad \text{and} \quad \lim_{x\to x_0^-} f(x)$$

which reflected the fact that there are two directions from which x can approach x_0—the left or the right. For functions of two variables the situation is much more complicated because there are infinitely many different curves along which a point (x, y) can approach a point (x_0, y_0) (Figure 16.2.3).

There are infinitely many different curves along which *(x, y)* can approach *(x₀, y₀)*.

Figure 16.2.3 ▷

Example 10   Let

$$f(x, y) = \frac{xy}{x^2 + y^2}$$

(See Figure 16.2.4.) Find the limit of $f(x, y)$ as $(x, y) \to (0, 0)$ along

(a) the x-axis;   (b) the y-axis;

(c) the parabola $y = x^2$;   (d) the line $y = x$.

Solution (a). If $(x, y) \to (0, 0)$ along the x-axis, then $y = 0$, so

$$\lim_{\substack{(x,y)\to(0,0) \\ (\text{along } y=0)}} f(x, y) = \lim_{(x,y)\to(0,0)} f(x, 0) = \lim_{(x,y)\to(0,0)} \frac{0}{x^2} = \lim_{(x,y)\to(0,0)} 0 = 0$$

Solution (b). If $(x, y) \to (0, 0)$ along the y-axis, then $x = 0$, so

$$\lim_{\substack{(x,y)\to(0,0) \\ (\text{along } x=0)}} f(x, y) = \lim_{(x,y)\to(0,0)} f(0, y) = \lim_{(x,y)\to(0,0)} \frac{0}{y^2} = \lim_{(x,y)\to(0,0)} 0 = 0$$

Solution (c). If $(x, y) \to (0, 0)$ along $y = x^2$, then

$$\lim_{\substack{(x,y)\to(0,0) \\ (\text{along } y=x^2)}} f(x, y) = \lim_{(x,y)\to(0,0)} f(x, x^2) = \lim_{(x,y)\to(0,0)} \frac{x^3}{x^2 + x^4} = \lim_{(x,y)\to(0,0)} \frac{x}{1 + x^2} = 0$$

Solution (d). If $(x, y) \to (0, 0)$ along $y = x$, then

$$\lim_{\substack{(x,y)\to(0,0) \\ (\text{along } y=x)}} f(x, y) = \lim_{(x,y)\to(0,0)} f(x, x) = \lim_{(x,y)\to(0,0)} \frac{x^2}{2x^2} = \frac{1}{2} \quad \blacktriangleleft$$

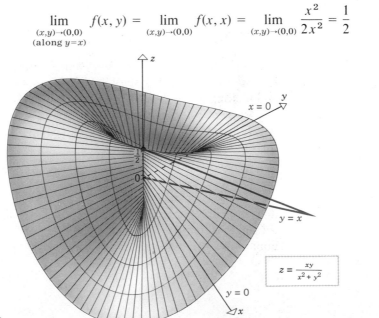

$$z = \frac{xy}{x^2 + y^2}$$

Figure 16.2.4 ▷

Recall that for a function of one variable we have

$$\lim_{x \to x_0} f(x) = L$$

if and only if

$$\lim_{x \to x_0^+} f(x) = L \quad \text{and} \quad \lim_{x \to x_0^-} f(x) = L$$

If either one-sided limit fails to exist, or if the two exist but are unequal, then the two-sided limit does not exist. An analogous result holds for functions of two variables. In order to have

$$\lim_{(x,y) \to (x_0,y_0)} f(x, y) = L$$

the function $f(x, y)$ must approach L as $(x, y) \to (x_0, y_0)$ along every continuous curve. Thus, if we find two different continuous curves of approach toward (x_0, y_0) along which $f(x, y)$ has different limits, or if we find some continuous curve of approach along which $f(x, y)$ has no limit, then

$$\lim_{(x,y) \to (x_0,y_0)} f(x, y)$$

does not exist. Similarly, for functions of three variables.

Example 11   The limit

$$\lim_{(x,y) \to (0,0)} \frac{xy}{x^2 + y^2}$$

does not exist because we showed in Example 10, that the limit does not have the same value along all paths to $(0,0)$. (The limit is $\frac{1}{2}$ along $y = x$ and 0 along the x-axis, for example.)   ◄

REMARK.   Calculating limits along specific curves is a useful technique for showing that a limit

$$\lim_{(x,y) \to (x_0,y_0)} f(x, y)$$

does not exist because it is only necessary to find two curves along which the limits differ. However, this method is of no use for showing that the limit exists because it is impossible to examine all possible curves.

▶ Exercise Set 16.2

In Exercises 1–8, sketch the region where f is continuous.

1. $f(x, y) = y \ln (1 + x)$.

2. $f(x, y) = \sqrt{x - y}$.

3. $f(x, y) = \dfrac{x^2 y}{\sqrt{25 - x^2 - y^2}}$.

4. $f(x, y) = \ln (2x - y + 1)$.

5. $f(x, y) = \cos \left(\dfrac{xy}{1 + x^2 + y^2} \right)$.

6. $f(x, y) = e^{(1-xy)}$.

7. $f(x, y) = \sin^{-1} (xy)$.

8. $f(x, y) = \tan^{-1} (y - x)$.

In Exercises 9–12, describe the region where f is continuous.

9. $f(x, y, z) = 3x^2 e^{yz} \cos (xyz)$.

10. $f(x, y, z) = \ln (4 - x^2 - y^2 - z^2)$.

11. $f(x, y, z) = \dfrac{y + 1}{x^2 + z^2 - 1}$.

12. $f(x, y, z) = \sin \sqrt{x^2 + y^2 + 3z^2}$.

In Exercises 13–35, find the limit, if it exists.

13. $\lim\limits_{(x,y) \to (1,3)} (4xy^2 - x)$.

14. $\lim\limits_{(x,y) \to (1/2, \pi)} (xy^2 \sin xy)$.

15. $\lim\limits_{(x,y) \to (-1,2)} \dfrac{xy^3}{x + y}$.

16. $\lim\limits_{(x,y) \to (1,-3)} e^{2x-y^2}$.

17. $\lim\limits_{(x,y) \to (0,0)} \ln (1 + x^2 y^3)$.

18. $\lim\limits_{(x,y) \to (4,-2)} x\sqrt[3]{y^3 + 2x}$.

19. $\lim\limits_{(x,y) \to (0,0)} \dfrac{x - y}{x^2 + y^2}$. [*Hint:* Let $(x, y) \to (0, 0)$ along the line $y = 0$.]

20. $\lim\limits_{(x,y) \to (0,0)} \dfrac{3}{x^2 + 2y^2}$.

21. $\lim\limits_{(x,y) \to (0,0)} \dfrac{\sin(x^2 + y^2)}{x^2 + y^2}$.

22. $\lim\limits_{(x,y) \to (0,0)} \dfrac{1 - \cos(x^2 + y^2)}{x^2 + y^2}$.

23. $\lim\limits_{(x,y) \to (0,0)} \dfrac{x^4 - y^4}{x^2 + y^2}$.

24. $\lim\limits_{(x,y) \to (0,0)} \dfrac{x^4 - 16y^4}{x^2 + 4y^2}$.

25. $\lim\limits_{(x,y) \to (0,0)} \dfrac{xy}{3x^2 + 2y^2}$.

26. $\lim\limits_{(x,y) \to (0,0)} \dfrac{1 - x^2 - y^2}{x^2 + y^2}$.

27. $\lim\limits_{(x,y) \to (0,0)} e^{-1/(x^2+y^2)}$.

28. $\lim\limits_{(x,y) \to (0,0)} \dfrac{e^{-1/\sqrt{x^2+y^2}}}{\sqrt{x^2 + y^2}}$.

29. $\lim\limits_{(x,y) \to (0,0)} y \ln (x^2 + y^2)$.

30. $\lim\limits_{(x,y) \to (0,0)} x \ln (|x| + |y|)$.

31. $\lim\limits_{(x,y,z) \to (2,-1,2)} \dfrac{xz^2}{\sqrt{x^2 + y^2 + z^2}}$.

32. $\lim\limits_{(x,y,z) \to (2,0,-1)} \ln (2x + y - z)$.

33. $\lim\limits_{(x,y,z) \to (0,0,0)} \dfrac{\sin (x^2 + y^2 + z^2)}{\sqrt{x^2 + y^2 + z^2}}$.

34. $\lim\limits_{(x,y,z) \to (0,0,0)} \dfrac{\sin \sqrt{x^2 + y^2 + z^2}}{x^2 + y^2 + z^2}$.

35. $\lim\limits_{(x,y,z) \to (0,0,0)} \dfrac{yz}{x^2 + y^2 + z^2}$.
    [*Hint:* First let $(x, y, z) \to (0, 0, 0)$ along the z-axis and then along the line $x = t, y = t, z = t$.]

36. Show that $\dfrac{xy}{x^2 + y^2}$ can be made to approach any value in the interval $[-\frac{1}{2}, \frac{1}{2}]$ by letting $(x, y) \to (0, 0)$ along some line $y = mx$.

37. (a) Show that $\dfrac{x^2 y}{x^4 + y^2}$ approaches zero as $(x, y) \to (0, 0)$ along any straight line $y = mx$.

    (b) Show that $\lim\limits_{(x,y) \to (0,0)} \dfrac{x^2 y}{x^4 + y^2}$ does not exist by letting $(x, y) \to (0, 0)$ along the parabola $y = x^2$.

38. (a) Show that $\dfrac{x^3 y}{2x^6 + y^2}$ approaches 0 as $(x, y) \to (0, 0)$ along any straight line $y = mx$, or along any parabola $y = kx^2$.

    (b) Show that $\lim\limits_{(x,y) \to (0,0)} \dfrac{x^3 y}{2x^6 + y^2}$ does not exist by letting $(x, y) \to (0, 0)$ along the curve $y = x^3$.

39. (a) Show that $\dfrac{xyz}{x^2 + y^4 + z^4}$ approaches 0 as $(x, y, z) \to (0, 0, 0)$ along any line $x = at$, $y = bt$, $z = ct$.

(b) Show that $\displaystyle\lim_{(x,y,z)\to(0,0,0)} \dfrac{xyz}{x^2 + y^4 + z^4}$ does not exist by letting $(x, y, z) \to (0, 0, 0)$ along the curve $x = t^2$, $y = t$, $z = t$.

40. Find

$$\lim_{(x,y)\to(0,1)} \tan^{-1}\left[\frac{x^2 + 1}{x^2 + (y - 1)^2}\right].$$

41. Find

$$\lim_{(x,y)\to(0,1)} \tan^{-1}\left[\frac{x^2 - 1}{x^2 + (y - 1)^2}\right].$$

42. Let $f(x, y) = \begin{cases} \dfrac{\sin(x^2 + y^2)}{x^2 + y^2}, & (x, y) \neq (0, 0) \\ 1, & (x, y) = (0, 0) \end{cases}$

Show that f is continuous at $(0, 0)$.

43. Let $f(x, y) = \dfrac{x^2}{x^2 + y^2}$. Is it possible to define $f(0, 0)$ so that f will be continuous at $(0, 0)$?

44. Let $f(x, y) = xy \ln(x^2 + y^2)$. Is it possible to define $f(0, 0)$ so that f will be continuous at $(0, 0)$?

In Exercises 45 and 46, use Definition 16.2.1 to prove the given statement. [*Hint:* Let $r = \sqrt{x^2 + y^2}$.]

45. $\displaystyle\lim_{(x,y)\to(0,0)} (x^2 + y^2) = 0$.

46. $\displaystyle\lim_{(x,y)\to(0,0)} \dfrac{x^2 y^2}{\sqrt{x^2 + y^2}} = 0$.

In Exercises 47 and 48, use Definition 16.2.2 to prove the given statement. [*Hint:* Let $\rho = \sqrt{x^2 + y^2 + z^2}$.]

47. $\displaystyle\lim_{(x,y,z)\to(0,0,0)} (x^2 + y^2 + z^2) = 0$.

48. $\displaystyle\lim_{(x,y,z)\to(0,0,0)} e^{\sqrt{x^2+y^2+z^2}} = 1$.

16.3 PARTIAL DERIVATIVES

*If f is a function of two or more independent variables and all but one of those variables are held fixed, then the derivative of f with respect to that one remaining independent variable is called a **partial derivative** of f. In this section we shall show how to compute partial derivatives and discuss their geometric significance.*

PARTIAL
DERIVATIVES OF
FUNCTIONS OF TWO
VARIABLES

Let f be a function of x and y. If we hold y constant, say $y = y_0$, and view x as a variable, then $f(x, y_0)$ is a function of x alone. If this function is differentiable at $x = x_0$, then the value of this derivative is denoted by

$$f_x(x_0, y_0) \tag{1}$$

and is called the ***partial derivative of f with respect to x*** at the point (x_0, y_0). Similarly, if we hold x constant, say $x = x_0$, then $f(x_0, y)$ is a function of y alone. If this function is differentiable at $y = y_0$, then the value of this derivative is denoted by

$$f_y(x_0, y_0) \tag{2}$$

and is called the ***partial derivative of f with respect to y*** at (x_0, y_0).

The partial derivatives of $f(x, y)$ have a simple geometric interpretation. Let P be a point on the surface

$$z = f(x, y)$$

If y is held constant, say $y = y_0$, and x is allowed to vary, then the point P moves along the curve C_1 that is the intersection of the surface with the vertical plane $y = y_0$ (Figure 16.3.1a). Thus, the partial derivative $f_x(x_0, y_0)$ can be interpreted as the slope (change in z per unit increase in x) of the tangent line to the curve C_1 at the point (x_0, y_0). Similarly, if x is held constant, say $x = x_0$, and y is allowed to vary, then the point P moves along the curve C_2 that is the intersection of the surface with the vertical plane $x = x_0$. Thus, the partial derivative $f_y(x_0, y_0)$ can be interpreted as the slope of the tangent line (change in z per unit increase in y) to the curve C_2 at the point (x_0, y_0) (Figure 16.3.1b).

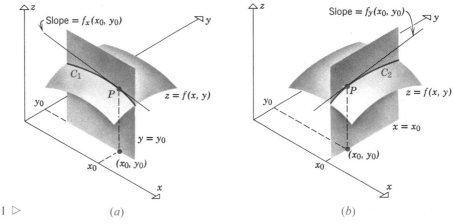

Figure 16.3.1 ▷ (a) (b)

REMARK.   In this text we shall only consider partial derivatives at *interior points* of the domain of f. Partial derivatives at boundary points lead to complications that are best left for more advanced courses.

The values of $f_x(x_0, y_0)$ and $f_y(x_0, y_0)$ are usually obtained by finding expressions for $f_x(x, y)$ and $f_y(x, y)$ at a general point (x, y) and then substituting $x = x_0$ and $y = y_0$ in these expressions. To obtain $f_x(x, y)$ we differentiate $f(x, y)$ with respect to x, *treating y as a constant;* and to obtain $f_y(x, y)$ we differentiate $f(x, y)$ with respect to y, *treating x as a constant.*

Example 1   Find $f_x(1, 2)$ and $f_y(1, 2)$ if

$$f(x, y) = 2x^3y^2 + 2y + 4x$$

Solution.   Treating y as a constant and differentiating with respect to x, we obtain

$$f_x(x, y) = 6x^2y^2 + 4$$

Treating x as a constant and differentiating with respect to y, we obtain

$$f_y(x, y) = 4x^3y + 2$$

Substituting $x = 1$ and $y = 2$ in these partial-derivative formulas yields

$$f_x(1, 2) = 6(1)^2(2)^2 + 4 = 28$$
$$f_y(1, 2) = 4(1)^3(2) + 2 = 10 \quad \blacktriangleleft$$

The partial derivatives f_x and f_y are also denoted by the symbols*

$$\frac{\partial f}{\partial x} \quad \text{and} \quad \frac{\partial f}{\partial y}$$

and if a dependent variable $z = f(x, y)$ is introduced, then the symbols

$$\frac{\partial z}{\partial x} \quad \text{and} \quad \frac{\partial z}{\partial y}$$

may be used. Some typical notations for the partial derivatives at a point (x_0, y_0) are

$$\left.\frac{\partial f}{\partial x}\right|_{x=x_0,\, y=y_0} \qquad \left.\frac{\partial z}{\partial y}\right|_{(x_0,\, y_0)} \qquad \left.\frac{\partial f}{\partial x}\right|_{(x_0,\, y_0)} \qquad \frac{\partial f}{\partial x}(x_0, y_0)$$

Example 2   Find $\partial z/\partial x$ and $\partial z/\partial y$ if

$$z = x^4 \sin(xy^3)$$

Solution.

$$\frac{\partial z}{\partial x} = \frac{\partial}{\partial x}[x^4 \sin(xy^3)] = x^4 \frac{\partial}{\partial x}[\sin(xy^3)] + \sin(xy^3) \cdot \frac{\partial}{\partial x}(x^4)$$
$$= x^4 \cos(xy^3) \cdot y^3 + \sin(xy^3) \cdot 4x^3 = x^4y^3 \cos(xy^3) + 4x^3 \sin(xy^3)$$

$$\frac{\partial z}{\partial y} = \frac{\partial}{\partial y}[x^4 \sin(xy^3)] = x^4 \frac{\partial}{\partial y}[\sin(xy^3)] + \sin(xy^3) \cdot \frac{\partial}{\partial y}(x^4)$$
$$= x^4 \cos(xy^3) \cdot 3xy^2 + \sin(xy^3) \cdot 0 = 3x^5y^2 \cos(xy^3) \quad \blacktriangleleft$$

Example 3   A point Q moves along the intersection of the sphere $x^2 + y^2 + z^2 = 1$ with the plane $x = \frac{2}{3}$. At what rate is z changing with y when the point is at $P(\frac{2}{3}, \frac{1}{3}, \frac{2}{3})$?

_____

*The symbol ∂ is called a partial derivative sign. It is a letter from the Cyrillic alphabet.

Figure 16.3.2

Solution. Since the z-coordinate of the point $P(\frac{2}{3}, \frac{1}{3}, \frac{2}{3})$ is positive, this point lies on the upper hemisphere

$$z = \sqrt{1 - x^2 - y^2} \tag{3}$$

(Figure 16.3.2). Since $x = \frac{2}{3}$, $y = \frac{1}{3}$ at P, the rate at which z changes with y at this point (as Q moves along the circle of intersection) is

$$\frac{\partial z}{\partial y}\bigg|_{x=\frac{2}{3}, y=\frac{1}{3}}$$

From (3) we obtain

$$\frac{\partial z}{\partial y} = \frac{\partial}{\partial y}[(1 - x^2 - y^2)^{1/2}]$$

$$= \frac{1}{2}(1 - x^2 - y^2)^{-1/2}(-2y)$$

$$= -\frac{y}{\sqrt{1 - x^2 - y^2}}$$

so that

$$\frac{\partial z}{\partial y}\bigg|_{x=\frac{2}{3}, y=\frac{1}{3}} = -\frac{\frac{1}{3}}{\sqrt{1 - (\frac{2}{3})^2 - (\frac{1}{3})^2}} = -\frac{1}{2}$$

Alternative Solution. Instead of solving $x^2 + y^2 + z^2 = 1$ explicitly for z as a function of x and y, we can obtain $\partial z/\partial y$ by implicit differentiation. Differentiating both sides of $x^2 + y^2 + z^2 = 1$ with respect to y and treating z as a function of x and y yields

$$\frac{\partial}{\partial y}[x^2 + y^2 + z^2] = \frac{\partial}{\partial y}[1]$$

$$2y + 2z\frac{\partial z}{\partial y} = 0$$

$$\frac{\partial z}{\partial y} = -\frac{y}{z}$$

Substituting the y- and z-coordinates of the point $(\frac{2}{3}, \frac{1}{3}, \frac{2}{3})$ yields

$$\frac{\partial z}{\partial y} = -\frac{\frac{1}{3}}{\frac{2}{3}} = -\frac{1}{2}$$

which agrees with our previous result. ◀

Example 4 According to the ideal gas law of physics, the pressure P exerted by a confined gas is related to its volume V and its temperature T by

$$P = k\frac{T}{V}$$

where k is a constant depending on the amount of gas present and the units of measurement. Show that if the temperature of such a gas remains constant, then the pressure and volume satisfy the condition

$$\frac{\partial P}{\partial V} = -\frac{P}{V} \tag{4}$$

Solution. Treating T as a constant we obtain

$$\frac{\partial P}{\partial V} = \frac{\partial}{\partial V}\left(k\frac{T}{V}\right) = (kT)\left(-\frac{1}{V^2}\right) = -k\frac{T}{V}\cdot\frac{1}{V} = -\frac{P}{V} \quad \blacktriangleleft$$

Example 5 The diagonal D of a rectangle is given by $D = \sqrt{x^2 + y^2}$, where x and y are the lengths of the sides.

(a) Find a formula for the instantaneous rate of change of D with respect to x if x varies while y remains constant.

(b) Suppose $y = 4$ in. Find the rate of change of D with respect to x at the instant when $x = 3$ in.

Solution (a).

$$\frac{\partial D}{\partial x} = \frac{1}{2}(x^2 + y^2)^{-1/2}(2x) = \frac{x}{\sqrt{x^2 + y^2}}$$

Solution (b).

$$\frac{\partial D}{\partial x}\bigg|_{x=3,\,y=4} = \frac{3}{\sqrt{3^2 + 4^2}} = \frac{3}{5}$$

Thus, the diagonal D is increasing at a rate of $\frac{3}{5}$ inches per inch increase in x. $\quad \blacktriangleleft$

HIGHER-ORDER PARTIAL DERIVATIVES

Since the partial derivatives $\partial f/\partial x$ and $\partial f/\partial y$ are functions of x and y, each can have partial derivatives. This gives rise to four possible *second-order* partial derivatives of f, which are defined by

$$\frac{\partial^2 f}{\partial x^2} = \frac{\partial}{\partial x}\left(\frac{\partial f}{\partial x}\right), \quad \frac{\partial^2 f}{\partial y^2} = \frac{\partial}{\partial y}\left(\frac{\partial f}{\partial y}\right)$$

$$\frac{\partial^2 f}{\partial x\,\partial y} = \frac{\partial}{\partial x}\left(\frac{\partial f}{\partial y}\right), \quad \frac{\partial^2 f}{\partial y\,\partial x} = \frac{\partial}{\partial y}\left(\frac{\partial f}{\partial x}\right)$$

Henceforth, we shall call $\partial f / \partial x$ and $\partial f / \partial y$ the *first-order* partial derivatives of f.

Example 6 Find the second-order partial derivatives of the function $f(x, y) = x^2 y^3 + x^4 y$.

Solution. We have

$$\frac{\partial f}{\partial x} = 2xy^3 + 4x^3 y$$

$$\frac{\partial f}{\partial y} = 3x^2 y^2 + x^4$$

so that

$$\frac{\partial^2 f}{\partial x^2} = \frac{\partial}{\partial x}\left(\frac{\partial f}{\partial x}\right) = \frac{\partial}{\partial x}(2xy^3 + 4x^3 y) = 2y^3 + 12x^2 y$$

$$\frac{\partial^2 f}{\partial y^2} = \frac{\partial}{\partial y}\left(\frac{\partial f}{\partial y}\right) = \frac{\partial}{\partial y}(3x^2 y^2 + x^4) = 6x^2 y$$

$$\frac{\partial^2 f}{\partial x\, \partial y} = \frac{\partial}{\partial x}\left(\frac{\partial f}{\partial y}\right) = \frac{\partial}{\partial x}(3x^2 y^2 + x^4) = 6xy^2 + 4x^3$$

$$\frac{\partial^2 f}{\partial y\, \partial x} = \frac{\partial}{\partial y}\left(\frac{\partial f}{\partial x}\right) = \frac{\partial}{\partial y}(2xy^3 + 4x^3 y) = 6xy^2 + 4x^3 \quad \blacktriangleleft$$

REMARK. The derivatives

$$\frac{\partial^2 f}{\partial y\, \partial x} \quad \text{and} \quad \frac{\partial^2 f}{\partial x\, \partial y}$$

are called the *mixed second-order partial derivatives* or *mixed second-partials*. For most functions that arise in applications, these mixed partial derivatives are equal (as in the last example). In the next section we shall state precise conditions under which equality holds.

By successively differentiating, we can obtain third-order partial derivatives and higher. Some possibilities are

$$\frac{\partial^3 f}{\partial x^3} = \frac{\partial}{\partial x}\left(\frac{\partial^2 f}{\partial x^2}\right), \quad \frac{\partial^3 f}{\partial y^2\, \partial x} = \frac{\partial}{\partial y}\left(\frac{\partial^2 f}{\partial y\, \partial x}\right)$$

$$\frac{\partial^3 f}{\partial y\, \partial x^2} = \frac{\partial}{\partial y}\left(\frac{\partial^2 f}{\partial x^2}\right), \quad \frac{\partial^4 f}{\partial y^2\, \partial x^2} = \frac{\partial}{\partial y}\left(\frac{\partial^3 f}{\partial y\, \partial x^2}\right)$$

Higher-order partial derivatives can be denoted more compactly with subscript notation. For example,

$$\frac{\partial^2 f}{\partial y\, \partial x} = \frac{\partial}{\partial y}\left(\frac{\partial f}{\partial x}\right) = \frac{\partial}{\partial y}(f_x) = (f_x)_y$$

It is usual to drop the parentheses and write simply

$$\frac{\partial^2 f}{\partial y\, \partial x} = f_{xy}$$

Note that in "∂" notation the sequence of differentiations is obtained by reading from right to left, but in the subscript notation it is left to right. Some other examples are

$$f_{xx} = \frac{\partial^2 f}{\partial x^2}, \quad f_{yyx} = \frac{\partial^3 f}{\partial x\, \partial y^2}, \quad f_{xxyy} = \frac{\partial^4 f}{\partial y^2\, \partial x^2}$$

Example 7   Let $f(x, y) = y^2 e^x + y$. Find f_{xyy}.

Solution.

$$f_{xyy} = \frac{\partial^3 f}{\partial y^2\, \partial x} = \frac{\partial^2}{\partial y^2}\left(\frac{\partial f}{\partial x}\right) = \frac{\partial^2}{\partial y^2}(y^2 e^x) = \frac{\partial}{\partial y}(2ye^x) = 2e^x \quad \blacktriangleleft$$

PARTIAL DERIVATIVES OF FUNCTIONS WITH MORE THAN TWO VARIABLES

For a function $f(x, y, z)$ of three variables, there are three *partial derivatives:*

$$f_x(x, y, z), \quad f_y(x, y, z), \quad f_z(x, y, z)$$

The partial derivative f_x is calculated by holding y and z constant, and differentiating with respect to x. For f_y, the variables x and z are held constant, and for f_z the variables x and y are held constant. If a dependent variable

$$w = f(x, y, z)$$

is used, then the three partial derivatives of f may be denoted by

$$\frac{\partial w}{\partial x}, \quad \frac{\partial w}{\partial y}, \quad \text{and} \quad \frac{\partial w}{\partial z}$$

Example 8   If $f(x, y, z) = x^3 y^2 z^4 + 2xy + z$, then

$$f_x(x, y, z) = 3x^2 y^2 z^4 + 2y$$
$$f_y(x, y, z) = 2x^3 yz^4 + 2x$$
$$f_z(x, y, z) = 4x^3 y^2 z^3 + 1$$
$$f_z(-1, 1, 2) = 4(-1)^3(1)^2(2)^3 + 1 = -31 \quad \blacktriangleleft$$

Example 9   If $f(\rho, \theta, \phi) = \rho^2 \cos\phi \sin\theta$ then

$$f_\rho(\rho, \theta, \phi) = 2\rho \cos\phi \sin\theta$$
$$f_{\rho\phi}(\rho, \theta, \phi) = -2\rho \sin\phi \sin\theta$$
$$f_{\rho\phi\theta}(\rho, \theta, \phi) = -2\rho \sin\phi \cos\theta \quad \blacktriangleleft$$

In general, if $f(v_1, v_2, \ldots, v_n)$ is a function of n variables, there are n partial derivatives of f, each of which is obtained by holding $n - 1$ of the variables fixed and differentiating with respect to the remaining variable. If $w = f(v_1, v_2, \ldots, v_n)$, then these partial derivatives are denoted by

$$\frac{\partial w}{\partial v_1}, \frac{\partial w}{\partial v_2}, \ldots, \frac{\partial w}{\partial v_n}$$

where $\partial w / \partial v_i$ is obtained by holding all variables except v_i fixed and differentiating with respect to v_i.

Example 10   Find

$$\frac{\partial}{\partial x_i} [\sqrt{x_1^2 + x_2^2 + \cdots + x_n^2}]$$

for $i = 1, 2, \ldots, n$.

Solution.   For each $i = 1, 2, \ldots, n$ we obtain

$$\frac{\partial}{\partial x_i}[\sqrt{x_1^2 + x_2^2 + \cdots + x_n^2}] = \frac{1}{2\sqrt{x_1^2 + x_2^2 + \cdots + x_n^2}} \cdot \frac{\partial}{\partial x_i}[x_1^2 + x_2^2 + \cdots + x_n^2]$$

$$= \frac{1}{2\sqrt{x_1^2 + x_2^2 + \cdots + x_n^2}}[2x_i] \qquad \boxed{\text{All terms except } x_i^2 \text{ are constant.}}$$

$$= \frac{x_i}{\sqrt{x_1^2 + x_2^2 + \cdots + x_n^2}} \quad \blacktriangleleft$$

► Exercise Set 16.3

In Exercises 1–6, find $\partial z/\partial x$ and $\partial z/\partial y$.

1. $z = 3x^3 y^2$.

2. $z = 4x^2 - 2y + 7x^4 y^5$.

3. $z = 4e^{x^2 y^3}$.

4. $z = \cos(x^5 y^4)$.

5. $z = x^3 \ln(1 + xy^{-3/5})$.

6. $z = e^{xy} \sin 4y^2$.

In Exercises 7–12, find $f_x(x, y)$ and $f_y(x, y)$.

7. $f(x, y) = \sqrt{3x^5 y - 7x^3 y}$.

8. $f(x, y) = \dfrac{x + y}{x - y}$.

9. $f(x, y) = y^{-3/2} \tan^{-1}(x/y)$.

10. $f(x, y) = x^3 e^{-y} + y^3 \sec \sqrt{x}$.

11. $f(x, y) = (y^2 \tan x)^{-4/3}$.

12. $f(x, y) = \cosh(\sqrt{x}) \sinh^2(xy^2)$.

13. Given $f(x, y) = 9 - x^2 - 7y^3$, find
  (a) $f_x(3, 1)$ (b) $f_y(3, 1)$.

14. Given $f(x, y) = x^2 y e^{xy}$, find
  (a) $\left.\dfrac{\partial f}{\partial x}\right|_{(1, 1)}$ (b) $\left.\dfrac{\partial f}{\partial y}\right|_{(1, 1)}$.

15. Given $z = \sqrt{x^2 + 4y^2}$, find
  (a) $\left.\dfrac{\partial z}{\partial x}\right|_{(1, 2)}$ (b) $\left.\dfrac{\partial z}{\partial y}\right|_{(1, 2)}$.

16. Given $w = x^2 \cos xy$, find
  (a) $\dfrac{\partial w}{\partial x}(\tfrac{1}{2}, \pi)$ (b) $\dfrac{\partial w}{\partial y}(\tfrac{1}{2}, \pi)$.

In Exercises 17–20, calculate $\partial z/\partial x$ and $\partial z/\partial y$ using implicit differentiation. Leave your answers in terms of x, y, and z.

17. $(x^2 + y^2 + z^2)^{3/2} = 1$.

18. $\ln(2x^2 + y - z^3) = x$.

19. $x^2 + z \sin xyz = 0$.

20. $e^{xy} \sinh z - z^2 x + 1 = 0$.

In Exercises 21–26, find f_{xx}, f_{yy}, f_{xy}, and f_{yx}.

21. $f(x, y) = 4x^2 - 8xy^4 + 7y^5 - 3$.

22. $f(x, y) = \sqrt{x^2 + y^2}$. **23.** $f(x, y) = e^x \cos y$.

24. $f(x, y) = e^{x-y^2}$.   **25.** $f(x, y) = \ln(4x - 5y)$.

26. $f(x, y) = (x^2 - y^2)/(x^2 + y^2)$.

27. Given $f(x, y) = x^3 y^5 - 2x^2 y + x$, find
  (a) f_{xxy} (b) f_{yxy} (c) f_{yyy}.

28. Given $z = (2x - y)^5$, find
  (a) $\dfrac{\partial^3 z}{\partial y \, \partial x \, \partial y}$ (b) $\dfrac{\partial^3 z}{\partial x^2 \, \partial y}$ (c) $\dfrac{\partial^4 z}{\partial x^2 \, \partial y^2}$.

29. Given $f(x, y) = y^3 e^{-5x}$, find
  (a) $f_{xyy}(0, 1)$ (b) $f_{xxx}(0, 1)$ (c) $f_{yyxx}(0, 1)$.

30. Given $w = e^y \cos x$, find
  (a) $\left.\dfrac{\partial^3 w}{\partial y^2 \, \partial x}\right|_{(\pi/4, 0)}$ (b) $\left.\dfrac{\partial^3 w}{\partial x^2 \, \partial y}\right|_{(\pi/4, 0)}$.

31. Express the following derivatives in "∂" notation.
  (a) f_{xxx} (b) f_{xyy}
  (c) f_{yyxx} (d) f_{xyyy}.

32. Express the following derivatives in "subscript" notation.
  (a) $\dfrac{\partial^3 f}{\partial y^2 \, \partial x}$ (b) $\dfrac{\partial^4 f}{\partial x^4}$
  (c) $\dfrac{\partial^4 f}{\partial y^2 \, \partial x^2}$ (d) $\dfrac{\partial^5 f}{\partial x^2 \, \partial y^3}$.

In Exercises 33–37, find $\partial w/\partial x$, $\partial w/\partial y$, and $\partial w/\partial z$.

33. $w = x^2 y^4 z^3 + xy + z^2 + 1$.

34. $w = ye^z \sin x$.   **35.** $w = \dfrac{x^2 - y^2}{y^2 + z^2}$.

36. $w = y^3 e^{2x+3z}$.   **37.** $w = \sqrt{x^2 + y^2 + z^2}$.

In Exercises 38–42, find f_x, f_y, and f_z.

38. $f(x, y, z) = z \ln(x^2 y \cos z)$.

39. $f(x, y, z) = \tan^{-1}\left(\dfrac{1}{xy^2 z^3}\right)$.

40. $f(x, y, z) = y^{-3/2} \sec\left(\dfrac{xz}{y}\right)$.

41. $f(x, y, z) = \cosh(\sqrt{z}) \sinh^2(x^2 yz)$.

42. $f(x, y, z) = \left(\dfrac{xz}{1 - z^2 - y^2}\right)^{-3/4}$.

43. Let $f(x, y, z) = 5x^2 yz^3$. Find
  (a) $f_x(1, -1, 2)$ (b) $f_y(1, -1, 2)$
  (c) $f_z(1, -1, 2)$.

44. Let $f(x, y, z) = y^2 e^{xz}$. Find
  (a) $\left.\dfrac{\partial f}{\partial x}\right|_{(1, 1, 1)}$ (b) $\left.\dfrac{\partial f}{\partial y}\right|_{(1, 1, 1)}$ (c) $\left.\dfrac{\partial f}{\partial z}\right|_{(1, 1, 1)}$.

45. Let $w = \sqrt{x^2 + 4y^2 - z^2}$. Find
  (a) $\left.\dfrac{\partial w}{\partial x}\right|_{(2, 1, -1)}$ (b) $\left.\dfrac{\partial w}{\partial y}\right|_{(2, 1, -1)}$
  (c) $\left.\dfrac{\partial w}{\partial z}\right|_{(2, 1, -1)}$.

46. Let $w = x \sin xyz$. Find
  (a) $\dfrac{\partial w}{\partial x}(1, \tfrac{1}{2}, \pi)$ (b) $\dfrac{\partial w}{\partial y}(1, \tfrac{1}{2}, \pi)$
  (c) $\dfrac{\partial w}{\partial z}(1, \tfrac{1}{2}, \pi)$.

In Exercises 47–50, find $\partial w/\partial x$, $\partial w/\partial y$, and $\partial w/\partial z$ using implicit differentiation. Leave your answers in terms of x, y, z, and w.

47. $(x^2 + y^2 + z^2 + w^2)^{3/2} = 4$.

48. $\ln(2x^2 + y - z^3 + 3w) = z$.

49. $w^2 + w \sin xyz = 0$.

50. $e^{xy} \sinh w - z^2 w + 1 = 0$.

51. Let $f(x, y, z) = x^3 y^5 z^7 + xy^2 + y^3 z$. Find
(a) f_{xy} (b) f_{yz} (c) f_{xz} (d) f_{zz}
(e) f_{zyy} (f) f_{xxy} (g) f_{zyx} (h) f_{xxyz}.

52. Let $w = (4x - 3y + 2z)^5$. Find
(a) $\dfrac{\partial^2 w}{\partial x \, \partial z}$ (b) $\dfrac{\partial^3 w}{\partial x \, \partial y \, \partial z}$ (c) $\dfrac{\partial^4 w}{\partial z^2 \, \partial y \, \partial x}$.

53. Show that the following functions satisfy

$$\frac{\partial^2 f}{\partial x^2} + \frac{\partial^2 f}{\partial y^2} = 0$$

(This is called *Laplace's equation*.)
(a) $f(x, y) = e^x \sin y + e^y \cos x$
(b) $f(x, y) = \ln(x^2 + y^2)$
(c) $f(x, y) = \tan^{-1} \dfrac{2xy}{x^2 - y^2}$.

54. Show that $u(x, y)$ and $v(x, y)$ satisfy

$$\frac{\partial u}{\partial x} = \frac{\partial v}{\partial y} \quad \text{and} \quad \frac{\partial u}{\partial y} = -\frac{\partial v}{\partial x}$$

(These are called the *Cauchy-Riemann equations*.)
(a) $u = x^2 - y^2$, $\quad v = 2xy$
(b) $u = e^x \cos y$, $\quad v = e^x \sin y$
(c) $u = \ln(x^2 + y^2)$, $\quad v = 2 \tan^{-1}(y/x)$.

55. A point moves along the intersection of the paraboloid $z = x^2 + 3y^2$ and the plane $x = 2$. At what rate is z changing with y when the point is at $(2, 1, 7)$?

56. A point moves along the intersection of the surface $z = \sqrt{29 - x^2 - y^2}$ and the plane $y = 3$. At what rate is z changing with x when the point is at $(4, 3, 2)$?

57. Find the slope of the tangent line at $(-1, 1, 5)$ to the curve of intersection of the surface $z = x^2 + 4y^2$ and
(a) the plane $x = -1$ (b) the plane $y = 1$.

58. Find the slope of the tangent line at $(2, 1, 2)$ to the curve of intersection of the surface $x^2 + y^2 + z^2 = 9$ and
(a) the plane $x = 2$ (b) the plane $y = 1$.

59. The volume V of a right-circular cylinder is given by $V = \pi r^2 h$, where r is the radius and h is the height.

(a) Find a formula for the instantaneous rate of change of V with respect to r if h remains constant.
(b) Find a formula for the instantaneous rate of change of V with respect to h if r remains constant.
(c) Suppose h has a constant value of 4 in., but r varies. Find the rate of change of V with respect to r at the instant when $r = 6$ in.
(d) Suppose r has a constant value of 8 in., but h varies. Find the instantaneous rate of change of V with respect to h at the instant when $h = 10$ in.

60. The volume V of a right-circular cone is given by

$$V = \frac{\pi}{24} d^2 \sqrt{4s^2 - d^2}$$

where s is the slant height and d is the diameter of the base.
(a) Find a formula for the instantaneous rate of change of V with respect to s if d remains constant.
(b) Find a formula for the instantaneous rate of change of V with respect to d if s remains constant.
(c) Suppose d has a constant value of 16 cm, but s varies. Find the rate of change of V with respect to s at the instant when $s = 10$ cm.
(d) Suppose s has a constant value of 10 cm, but d varies. Find the rate of change of V with respect to d at the instant when $d = 16$ cm.

61. According to the ideal gas law (Example 4), the pressure, temperature, and volume of a gas are related by $P = kT/V$. Suppose that for a certain gas, $k = 10$.
(a) Find the instantaneous rate of change of pressure (lb/in²) with respect to temperature if the temperature is $80°K$ and the volume remains fixed at 50 in³.
(b) Find the instantaneous rate of change of volume with respect to pressure if the volume is 50 in³ and the temperature remains fixed at $80°K$.

62. Find parametric equations for the tangent line at $(1, 3, 3)$ to the curve of intersection of the surface $z = x^2 y$ and
   (a) the plane $x = 1$
   (b) the plane $y = 3$.

63. Suppose

$$\sin(x + z) + \sin(x - y) = 1$$

Use implicit differentiation to find $\partial z/\partial x$, $\partial z/\partial y$, and $\partial^2 z/\partial x \partial y$ in terms of x, y, and z.

64. The volume of a right-circular cone of radius r and height h is $V = \frac{1}{3}\pi r^2 h$. Show that if the height remains constant while the radius changes, then the volume satisfies

$$\frac{\partial V}{\partial r} = \frac{2V}{r}$$

65. The temperature at a point (x, y) on a metal plate in the xy-plane is $T(x, y) = x^3 + 2y^2 + x$ degrees. Assume that distance is measured in centimeters and find the rate at which temperature changes with distance if we start at the point $(1, 2)$ and move
   (a) to the right and parallel to the x-axis
   (b) upward and parallel to the y-axis.

66. When two resistors having resistances R_1 ohms and R_2 ohms are connected in parallel, their combined resistance R in ohms is

$$R = R_1 R_2/(R_1 + R_2)$$

Show that

$$\frac{\partial^2 R}{\partial R_1^2}\frac{\partial^2 R}{\partial R_2^2} = \frac{4R^2}{(R_1 + R_2)^4}$$

67. Prove: If $u(x, y)$ and $v(x, y)$ each have equal mixed second-partials, and if u and v satisfy the Cauchy-Riemann equations (Exercise 54), then u and v both satisfy Laplace's equation (Exercise 53).

68. Recall that for a function of one variable the derivative $f'(x)$ can be expressed as the limit

$$f'(x) = \lim_{h \to 0} \frac{f(x + h) - f(x)}{h}$$

Express $f_x(x, y)$ and $f_y(x, y)$ as limits.

In Exercises 69 and 70, the figures show some of the isotherms (curves of constant temperature) of a temperature function $T(x, y)$ for a thin metal plate in the xy-plane. Based on the figures, determine at the point P

   (a) the signs of $\dfrac{\partial T}{\partial x}$ and $\dfrac{\partial T}{\partial y}$, and which is largest in absolute value.

   (b) the signs of $\dfrac{\partial^2 T}{\partial x^2}$, $\dfrac{\partial^2 T}{\partial y^2}$, $\dfrac{\partial^2 T}{\partial y \partial x}$, and $\dfrac{\partial^2 T}{\partial x \partial y}$.

69.

70.

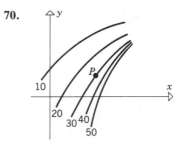

16.4 DIFFERENTIABILITY AND CHAIN RULES FOR FUNCTIONS OF TWO VARIABLES

In this section we shall extend the notion of differentiability to functions of two variables and derive versions of the chain rule for these functions. We have restricted the discussion in this section to functions of two variables because some of the results we shall discuss have geometric interpretations that only apply to such functions. In a later section we shall extend the concepts developed here to functions of three or more variables.

DIFFERENTIABILITY OF FUNCTIONS OF TWO VARIABLES

Recall that a function f of one variable is called differentiable at x_0 if it has a derivative at x_0, or, in other words, if the limit

$$f'(x_0) = \lim_{\Delta x \to 0} \frac{f(x_0 + \Delta x) - f(x_0)}{\Delta x} \tag{1}$$

exists. A function f that is differentiable at a point x_0 enjoys two important properties:

- $f(x)$ is continuous at x_0.

- The curve $y = f(x)$ has a nonvertical tangent line at x_0.

Our primary objective in this section is to extend the notion of differentiability to functions of two variables in such a way that the natural analogs of these two properties hold. More precisely, when $f(x, y)$ is differentiable at (x_0, y_0), we shall want it to be the case that

- $f(x, y)$ is continuous at (x_0, y_0).

- The surface $z = f(x, y)$ has a nonvertical tangent plane at (x_0, y_0) (Figure 16.4.1). (A precise definition of a tangent plane will be given later.)

$z = f(x, y)$

(x_0, y_0)

Figure 16.4.1

It would not be unreasonable to guess that a function f of two variables should be called differentiable at (x_0, y_0) if the two partial derivatives $f_x(x_0, y_0)$ and $f_y(x_0, y_0)$ exist at (x_0, y_0). Unfortunately, this condition is not strong enough to meet our objectives, since there are functions that have partial derivatives at a point, but are not continuous at that point. For example, the function

$$f(x, y) = \begin{cases} -1 & \text{if } x \text{ and } y \text{ are positive} \\ 0 & \text{otherwise} \end{cases}$$

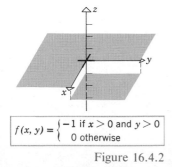

$f(x, y) = \begin{cases} -1 \text{ if } x > 0 \text{ and } y > 0 \\ 0 \text{ otherwise} \end{cases}$

Figure 16.4.2

is discontinuous at $(0, 0)$, but has partial derivatives at $(0, 0)$. To be precise, $f_x(0, 0) = 0$ and $f_y(0, 0) = 0$. These facts should be evident from Figure 16.4.2.

To motivate the definition of differentiability for functions of two variables, we shall reformulate (1) in another form. Using the notation

$$\Delta f = f(x_0 + \Delta x) - f(x_0)$$

(1) can be rewritten as

$$f'(x_0) = \lim_{\Delta x \to 0} \frac{\Delta f}{\Delta x} \tag{2}$$

Let us define ϵ as

$$\epsilon = \frac{\Delta f}{\Delta x} - f'(x_0) \tag{3}$$

so that

$$\Delta f = f'(x_0)\,\Delta x + \epsilon\,\Delta x \tag{4}$$

Figure 16.4.3

The quantities in this equation are pictured in Figure 16.4.3. The term Δf represents the change in height that results when a point moves along the graph of f so that the x-coordinate changes from x_0 to $x_0 + \Delta x$; the term $f'(x_0)\,\Delta x$ represents the change in height that results when a point moves along the tangent line at $(x_0, f(x_0))$ so that the x-coordinate changes from x_0 to $x_0 + \Delta x$; finally, the term $\epsilon\,\Delta x$ represents the difference between Δf and $f'(x_0)\,\Delta x$. It is evident from Figure 16.4.3 that $\epsilon\,\Delta x \to 0$ as $\Delta x \to 0$. However, it is also true that $\epsilon \to 0$ as $\Delta x \to 0$. This is not at all evident from Figure 16.4.3, but follows immediately from (2) and (3).

If f is a function of x and y, then the symbol Δf, called the **increment** of f, denotes the change in the value of $f(x, y)$ that results when (x, y) varies from some initial position (x_0, y_0) to some new position $(x_0 + \Delta x, y_0 + \Delta y)$; thus,

$$\Delta f = f(x_0 + \Delta x, y_0 + \Delta y) - f(x_0, y_0)$$

(See Figure 16.4.4.) If a dependent variable $z = f(x, y)$ is used, then we shall sometimes write Δz rather than Δf.

Figure 16.4.4 ▷

Motivated by the form of (4), we now make the following definition of differentiability for functions of two variables.

16.4.1   DEFINITION. A function f of two variables is said to be *differentiable* at (x_0, y_0) if $f_x(x_0, y_0)$ and $f_y(x_0, y_0)$ exist and Δf can be written in the form

$$\Delta f = f_x(x_0, y_0) \cdot \Delta x + f_y(x_0, y_0) \, \Delta y + \epsilon_1 \, \Delta x + \epsilon_2 \, \Delta y$$

where ϵ_1 and ϵ_2 are functions of Δx and Δy such that $\epsilon_1 \to 0$ and $\epsilon_2 \to 0$ as $(\Delta x, \Delta y) \to (0, 0)$.

A function is called *differentiable on a region R* of the xy-plane if it is differentiable at each point of R. A function that is differentiable on the entire xy-plane is called *everywhere differentiable* or simply *differentiable*.

REMARK.   Before proceeding further, we want to reemphasize an important point. For functions of one variable, the terms "differentiable" and "has a derivative" are synonymous. However, for functions of two variables differentiability is a more stringent requirement than the mere existence of partial derivatives.

RELATIONSHIP
BETWEEN
DIFFERENTIABILITY
AND CONTINUITY

Earlier, we set two goals for our definition of differentiability—first, we wanted a function f that is differentiable at (x_0, y_0) to be continuous at (x_0, y_0), and second we wanted the graph of f to have a nonvertical tangent plane at (x_0, y_0). The next theorem shows that the continuity criterion is met; the existence of a nonvertical tangent plane will be demonstrated in the next section.

16.4.2   THEOREM. *If f is differentiable at (x_0, y_0), then f is continuous at (x_0, y_0).*

Proof.   In order to prove that

$$\lim_{(x,y)\to(x_0,y_0)} f(x, y) = f(x_0, y_0)$$

we shall show that

$$\lim_{(x,y)\to(x_0,y_0)} [f(x, y) - f(x_0, y_0)] = 0 \tag{5}$$

Let $\Delta x = x - x_0$ and $\Delta y = y - y_0$, so that $x = x_0 + \Delta x$, $y = y_0 + \Delta y$. Since $\Delta x \to 0$ as $x \to x_0$ and $\Delta y \to 0$ as $y \to y_0$, (5) can be rewritten in the form

$$\lim_{(\Delta x, \Delta y) \to (0,0)} [f(x_0 + \Delta x, y_0 + \Delta y) - f(x_0, y_0)] = 0$$

or more briefly

$$\lim_{(\Delta x, \Delta y) \to (0,0)} \Delta f = 0 \tag{6}$$

Let us now prove (6). By hypothesis,

$$\Delta f = f_x(x_0, y_0) \Delta x + f_y(x_0, y_0) \Delta y + \epsilon_1 \Delta x + \epsilon_2 \Delta y$$

where $\epsilon_1 \to 0$, $\epsilon_2 \to 0$ as $(\Delta x, \Delta y) \to (0,0)$. Thus,

$$\lim_{(\Delta x, \Delta y) \to (0,0)} \Delta f = \lim_{(\Delta x, \Delta y) \to (0,0)} [f_x(x_0, y_0) \Delta x + f_y(x_0, y_0) \Delta y + \epsilon_1 \Delta x + \epsilon_2 \Delta y] = 0$$

which completes the proof. ∎

The next theorem gives simple conditions for a function of two variables to be differentiable at a point.

16.4.3 THEOREM. *If f has first-order partial derivatives at each point in some circular region centered at (x_0, y_0), and if these partial derivatives are continuous at (x_0, y_0), then f is differentiable at (x_0, y_0).*

The proof is optional and can be found at the end of this section.

Example 1   Show that $f(x, y) = x^3 y^4$ is a differentiable function.

Solution.   The partial derivatives $f_x = 3x^2 y^4$ and $f_y = 4x^3 y^3$ are defined and continuous everywhere in the xy-plane. Thus, the hypotheses of Theorem 16.4.3 are satisfied at each point (x_0, y_0) in the xy-plane, so $f(x, y) = x^3 y^4$ is everywhere differentiable.   ◀

The following important corollary follows from Theorems 16.4.2 and 16.4.3.

16.4.4 COROLLARY. *If f has first-order partial derivatives at each point of some circular region centered at (x_0, y_0), and if these partial derivatives are continuous at (x_0, y_0), then f is continuous at (x_0, y_0).*

In addition to guaranteeing differentiability and continuity of $f(x, y)$, continuity of the first-order partial derivatives also ensures that the mixed second-order partial derivatives of f are equal. This is the content of the following theorem which we state without proof.

Equality of Mixed Partials

16.4.5   THEOREM. *Let f be a function of two variables. If f_x, f_y, f_{xy}, and f_{yx} are continuous on an open set, then $f_{xy} = f_{yx}$ at each point of the set.*

Example 2   Let $f(x, y) = 2e^{xy} \sin y$. It should be evident from the form of this function and from Theorem 16.2.6 that f and all its partial derivatives are continuous everywhere. Thus, Theorem 16.4.5 guarantees that $f_{xy} = f_{yx}$ everywhere. This is, in fact, the case since

$$f_x(x, y) = 2ye^{xy} \sin y = (2y \sin y)e^{xy}$$

$$f_{xy}(x, y) = (2y \sin y)(xe^{xy}) + e^{xy}(2y \cos y + 2 \sin y)$$

$$= 2e^{xy}(xy \sin y + y \cos y + \sin y)$$

$$f_y(x, y) = 2e^{xy} \cos y + 2xe^{xy} \sin y$$

$$f_{yx}(x, y) = 2ye^{xy} \cos y + 2xye^{xy} \sin y + 2e^{xy} \sin y$$

$$= 2e^{xy}(xy \sin y + y \cos y + \sin y)$$

so $f_{xy}(x, y) = f_{yx}(x, y)$ for all (x, y).   ◄

In general, the order of differentiation in an n-th order partial derivative can be changed without affecting the final result whenever the function and all its partial derivatives of order n or less are continuous. For example, if f and its partial derivatives of the first, second, and third orders are continuous on an open set, then at each point of that set,

$$f_{xyy} = f_{yxy} = f_{yyx}$$

or in another notation,

$$\frac{\partial^3 f}{\partial y^2 \, \partial x} = \frac{\partial^3 f}{\partial y \, \partial x \, \partial y} = \frac{\partial^3 f}{\partial x \, \partial y^2}$$

CHAIN RULES   If y is a differentiable function of x and x is a differentiable function of t, then the chain rule for functions of one variable states that

$$\frac{dy}{dt} = \frac{dy}{dx}\frac{dx}{dt}$$

We shall now derive versions of the chain rule for functions of two variables.

Assume that z is a function of x and y, say

$$z = f(x, y) \tag{7}$$

and suppose that x and y, in turn, are functions of a single variable t, say

$$x = x(t), \quad y = y(t)$$

On substituting these functions of t in (7), we obtain the relationship

$$z = f(x(t), y(t))$$

which expresses z as a function of the single variable t. Thus, we may ask for the derivative dz/dt, and we may inquire about its relationship to the derivatives $\partial z/\partial x$, $\partial z/\partial y$, dx/dt, and dy/dt.

Chain Rule

16.4.6   THEOREM. *If $x = x(t)$ and $y = y(t)$ are differentiable at t, and if $z = f(x, y)$ is differentiable at the point $(x(t), y(t))$, then $z = f(x(t), y(t))$ is differentiable at t, and*

$$\frac{dz}{dt} = \frac{\partial z}{\partial x}\frac{dx}{dt} + \frac{\partial z}{\partial y}\frac{dy}{dt} \tag{8}$$

Proof.   From the derivative definition for functions of one variable

$$\frac{dz}{dt} = \lim_{\Delta t \to 0} \frac{\Delta z}{\Delta t} \tag{9}$$

Since $z = f(x, y)$ is differentiable at the point $(x, y) = (x(t), y(t))$, we can express Δz in the form

$$\Delta z = \frac{\partial z}{\partial x}\Delta x + \frac{\partial z}{\partial y}\Delta y + \epsilon_1 \Delta x + \epsilon_2 \Delta y \tag{10}$$

where the partial derivatives are evaluated at $(x(t), y(t))$ and $\epsilon_1 \to 0$, $\epsilon_2 \to 0$ as $(\Delta x, \Delta y) \to (0, 0)$. Thus, from (9) and (10),

$$\frac{dz}{dt} = \lim_{\Delta t \to 0} \frac{\Delta z}{\Delta t} = \lim_{\Delta t \to 0} \left[\frac{\partial z}{\partial x}\frac{\Delta x}{\Delta t} + \frac{\partial z}{\partial y}\frac{\Delta y}{\Delta t} + \epsilon_1 \frac{\Delta x}{\Delta t} + \epsilon_2 \frac{\Delta y}{\Delta t} \right] \tag{11}$$

But

$$\lim_{\Delta t \to 0} \frac{\Delta x}{\Delta t} = \frac{dx}{dt} \quad \text{and} \quad \lim_{\Delta t \to 0} \frac{\Delta y}{\Delta t} = \frac{dy}{dt}$$

Therefore, if we can show that $\epsilon_1 \to 0$, $\epsilon_2 \to 0$ as $\Delta t \to 0$, then the proof will be complete, since (11) will reduce to (8). But $\Delta x \to 0$ and $\Delta y \to 0$ as $\Delta t \to 0$, since

$$\lim_{\Delta t \to 0} \Delta x = \lim_{\Delta t \to 0} \frac{\Delta x}{\Delta t}\Delta t = \frac{dx}{dt} \cdot 0 = 0$$

and similarly for Δy. Thus, as Δt tends to zero, $(\Delta x, \Delta y) \to (0, 0)$, which implies that $\epsilon_1 \to 0$, $\epsilon_2 \to 0$. ∎

Example 3 Suppose

$$z = x^2y, \quad x = t^2, \quad y = t^3$$

Use the chain rule to find dz/dt, and check the result by expressing z as a function of t and differentiating directly.

Solution. By the chain rule

$$\frac{dz}{dt} = \frac{\partial z}{\partial x}\frac{dx}{dt} + \frac{\partial z}{\partial y}\frac{dy}{dt} = (2xy)(2t) + (x^2)(3t^2)$$
$$= (2t^5)(2t) + (t^4)(3t^2) = 7t^6$$

Alternatively, we may express z directly as a function of t,

$$z = x^2y = (t^2)^2(t^3) = t^7$$

and then differentiate to obtain $dz/dt = 7t^6$. However, this procedure is not always convenient. ◄

Example 4 Suppose

$$z = \sqrt{xy + y}, \quad x = \cos \theta, \quad y = \sin \theta$$

Use the chain rule to find $dz/d\theta$ when $\theta = \pi/2$.

Solution. From the chain rule with θ in place of t,

$$\frac{dz}{d\theta} = \frac{\partial z}{\partial x}\frac{dx}{d\theta} + \frac{\partial z}{\partial y}\frac{dy}{d\theta}$$

we obtain

$$\frac{dz}{d\theta} = \frac{1}{2}(xy + y)^{-1/2}(y)(-\sin \theta) + \frac{1}{2}(xy + y)^{-1/2}(x + 1)(\cos \theta)$$

When $\theta = \pi/2$, we have

$$x = \cos\frac{\pi}{2} = 0, \quad y = \sin\frac{\pi}{2} = 1$$

Substituting $x = 0$, $y = 1$, $\theta = \pi/2$ in the formula for $dz/d\theta$ yields

$$\left.\frac{dz}{d\theta}\right|_{\theta=\pi/2} = \frac{1}{2}(1)(1)(-1) + \frac{1}{2}(1)(1)(0) = -\frac{1}{2} \quad ◄$$

REMARK. There are many variations in derivative notations, each of which gives the chain rule a different look. If $z = f(x, y)$, where x and y are functions of t, then some possibilities are

$$\frac{dz}{dt} = f_x \frac{dx}{dt} + f_y \frac{dy}{dt}$$

$$\frac{df}{dt} = \frac{\partial f}{\partial x} \frac{dx}{dt} + \frac{\partial f}{\partial y} \frac{dy}{dt}$$

$$\frac{df}{dt} = f_x x'(t) + f_y y'(t)$$

The reader may be able to construct other variations as well.

In the special case where $z = F(x, y)$ and y is a differentiable function of x, chain-rule formula (8) yields

$$\frac{dz}{dx} = \frac{\partial F}{\partial x} \frac{dx}{dx} + \frac{\partial F}{\partial y} \frac{dy}{dx} = \frac{\partial F}{\partial x} + \frac{\partial F}{\partial y} \frac{dy}{dx} \tag{12}$$

This result can be used to find derivatives of functions that are defined implicitly. Suppose that the equation

$$F(x, y) = 0 \tag{13}$$

defines y implicitly as a differentiable function of x, and we are interested in finding dy/dx. Differentiating both sides of (13) with respect to x and applying (12) yields

$$\frac{\partial F}{\partial x} + \frac{\partial F}{\partial y} \frac{dy}{dx} = 0$$

or

$$\frac{dy}{dx} = -\frac{\partial F/\partial x}{\partial F/\partial y} \tag{14}$$

provided $\partial F/\partial y \neq 0$.

Example 5   Given that

$$x^3 + y^2 x - 3 = 0$$

find dy/dx using (14) and check the result using implicit differentiation.

Solution.   By (14)

$$\frac{dy}{dx} = -\frac{\partial F/\partial x}{\partial F/\partial y} = -\frac{3x^2 + y^2}{2yx}$$

On the other hand, differentiating the given equation implicitly yields

$$3x^2 + y^2 + x\left(2y\,\frac{dy}{dx}\right) - 0 = 0$$

or

$$\frac{dy}{dx} = -\frac{3x^2 + y^2}{2yx}$$

which agrees with the result obtained by (14).   ◄

In Theorem 16.4.6 the variables x and y are each functions of a single variable t. We now consider the case where x and y are each functions of two variables. Let

$$z = f(x, y) \tag{15}$$

and suppose x and y are functions of u and v, say

$$x = x(u, v), \quad y = y(u, v)$$

On substituting these functions of u and v into (15) we obtain the relationship

$$z = f(x(u, v), y(u, v))$$

which expresses z as a function of the two variables u and v. Thus, we may ask for the partial derivatives $\partial z/\partial u$ and $\partial z/\partial v$; and we may inquire about the relationship between these derivatives and the derivatives $\partial z/\partial x$, $\partial z/\partial y$, $\partial x/\partial u$, $\partial x/\partial v$, $\partial y/\partial u$, and $\partial y/\partial v$.

Chain Rule

16.4.7   THEOREM. *If $x = x(u, v)$ and $y = y(u, v)$ have first-order partial derivatives at the point (u, v) and if $z = f(x, y)$ is differentiable at the point $(x(u, v), y(u, v))$, then $z = f(x(u, v), y(u, v))$ has first-order partial derivatives at (u, v), given by*

$$\frac{\partial z}{\partial u} = \frac{\partial z}{\partial x}\frac{\partial x}{\partial u} + \frac{\partial z}{\partial y}\frac{\partial y}{\partial u}$$

$$\frac{\partial z}{\partial v} = \frac{\partial z}{\partial x}\frac{\partial x}{\partial v} + \frac{\partial z}{\partial y}\frac{\partial y}{\partial v}$$

Proof.   If v is held fixed, then $x = x(u, v)$ and $y = y(u, v)$ become functions of u alone. Thus, we are back to the case of Theorem 16.4.6. If we apply that theorem with u in place of t and if we use ∂ rather than d to indicate that the variable v is fixed, we obtain

$$\frac{\partial z}{\partial u} = \frac{\partial z}{\partial x}\frac{\partial x}{\partial u} + \frac{\partial z}{\partial y}\frac{\partial y}{\partial u}$$

The formula for $\partial z/\partial v$ is derived similarly.   ∎

Example 6 Given that

$$z = e^{xy}, \quad x = 2u + v, \quad y = u/v$$

find $\partial z/\partial u$ and $\partial z/\partial v$ using the chain rule.

Solution.

$$\frac{\partial z}{\partial u} = \frac{\partial z}{\partial x}\frac{\partial x}{\partial u} + \frac{\partial z}{\partial y}\frac{\partial y}{\partial u} = (ye^{xy})(2) + (xe^{xy})(1/v) = \left[2y + \frac{x}{v}\right]e^{xy}$$

$$= \left[\frac{2u}{v} + \frac{2u + v}{v}\right]e^{(2u+v)(u/v)} = \left[\frac{4u}{v} + 1\right]e^{(2u+v)(u/v)}$$

$$\frac{\partial z}{\partial v} = \frac{\partial z}{\partial x}\frac{\partial x}{\partial v} + \frac{\partial z}{\partial y}\frac{\partial y}{\partial v} = (ye^{xy})(1) + (xe^{xy})\left(-\frac{u}{v^2}\right)$$

$$= \left[y - x\left(\frac{u}{v^2}\right)\right]e^{xy} = \left[\frac{u}{v} - (2u + v)\left(\frac{u}{v^2}\right)\right]e^{(2u+v)(u/v)}$$

$$= -\frac{2u^2}{v^2}e^{(2u+v)(u/v)}   ◀$$

The chain rules are useful in related rates problems.

Example 7 At what rate is the area of a rectangle changing if its length is 15 ft and increasing at 3 ft/sec while its width is 6 ft and increasing at 2 ft/sec?

Solution. Let

$x =$ length of the rectangle in feet
$y =$ width of the rectangle in feet
$A =$ area of the rectangle in square feet
$t =$ time in seconds

We are given that

$$\frac{dx}{dt} = 3 \quad \text{and} \quad \frac{dy}{dt} = 2 \tag{16}$$

at the instant when

$$x = 15, \quad y = 6 \tag{17}$$

We want to find dA/dt at that instant.

From the area formula $A = xy$, we obtain

$$\frac{dA}{dt} = \frac{\partial A}{\partial x}\frac{dx}{dt} + \frac{\partial A}{\partial y}\frac{dy}{dt} = y\frac{dx}{dt} + x\frac{dy}{dt}$$

Substituting (16) and (17) in this equation yields

$$\frac{dA}{dt} = 6(3) + 15(2) = 48$$

Thus, the area is increasing at a rate of 48 ft²/sec at the given instant.   ◄

OPTIONAL

Proof of Theorem 16.4.3.   We must show that

$$\Delta f = f(x_0 + \Delta x, y_0 + \Delta y) - f(x_0, y_0)$$

can be expressed in the form

$$\Delta f = f_x(x_0, y_0)\, \Delta x + f_y(x_0, y_0)\, \Delta y + \epsilon_1\, \Delta x + \epsilon_2\, \Delta y$$

where $\epsilon_1 \to 0$, $\epsilon_2 \to 0$ as $(\Delta x, \Delta y) \to (0, 0)$. In our discussion, we shall restrict Δx and Δy to be sufficiently small so that the points

$$B(x_0 + \Delta x, y_0) \quad \text{and} \quad C(x_0 + \Delta x, y_0 + \Delta y)$$

Figure 16.4.5

lie inside a circle where f_x and f_y exist (Figure 16.4.5). The increment Δf represents the change in the value of f as (x, y) varies from the initial point $A(x_0, y_0)$ to the final point $C(x_0 + \Delta x, y_0 + \Delta y)$. Let us resolve Δf into two terms

$$\Delta f = \Delta f_1 + \Delta f_2 \tag{18}$$

where Δf_1 is the change in the value of f as (x, y) varies from A to B and Δf_2 is the change in the value of f as (x, y) varies from B to C. Thus,

$$\Delta f_1 = f(x_0 + \Delta x, y_0) - f(x_0, y_0) \tag{19}$$

$$\Delta f_2 = f(x_0 + \Delta x, y_0 + \Delta y) - f(x_0 + \Delta x, y_0) \tag{20}$$

Along the line segment from A to B, y has a constant value of y_0, so that $f(x, y) = f(x, y_0)$ is a function of x alone. Let us denote this function by

$$f_1(x) = f(x, y_0)$$

Similarly, along the line segment from B to C, x has a constant value of $x = x_0 + \Delta x$, so that $f(x, y) = f(x_0 + \Delta x, y)$ is a function of y alone. Let us denote this function by

$$f_2(y) = f(x_0 + \Delta x, y)$$

With these definitions of f_1 and f_2, (19) and (20) can be written as

$$\Delta f_1 = f_1(x_0 + \Delta x) - f_1(x_0) \tag{21}$$

$$\Delta f_2 = f_2(y_0 + \Delta y) - f_2(y_0) \tag{22}$$

The reader should be able to show that f_1 and f_2 satisfy the hypotheses of the Mean-Value Theorem (4.10.2). Accepting this, we can rewrite (21) and (22) in the form

$$\Delta f_1 = f_1'(x^*) \, \Delta x \tag{23}$$

$$\Delta f_2 = f_2'(y^*) \, \Delta y \tag{24}$$

where x^* is between x_0 and $x_0 + \Delta x$ and y^* is between y_0 and $y_0 + \Delta y$. From the definitions of $f_1(x)$ and $f_2(y)$

$$f_1'(x) = \frac{d}{dx} [f_1(x)] = f_x(x, y_0)$$

$$f_2'(y) = \frac{d}{dy} [f_2(y)] = f_y(x_0 + \Delta x, y)$$

so that (23) and (24) can be rewritten as

$$\Delta f_1 = f_x(x^*, y_0) \, \Delta x \tag{25}$$

$$\Delta f_2 = f_y(x_0 + \Delta x, y^*) \, \Delta y \tag{26}$$

If we define

$$\epsilon_1 = f_x(x^*, y_0) - f_x(x_0, y_0) \tag{27}$$

$$\epsilon_2 = f_y(x_0 + \Delta x, y^*) - f_y(x_0, y_0) \tag{28}$$

then (25) and (26) can be rewritten as

$$\Delta f_1 = f_x(x_0, y_0) \, \Delta x + \epsilon_1 \, \Delta x$$

$$\Delta f_2 = f_y(x_0, y_0) \, \Delta y + \epsilon_2 \, \Delta y$$

and substituting these in (18) yields

$$\Delta f = f_x(x_0, y_0) \, \Delta x + f_y(x_0, y_0) \, \Delta y + \epsilon_1 \, \Delta x + \epsilon_2 \, \Delta y$$

To finish, we must show that $\epsilon_1 \to 0$, $\epsilon_2 \to 0$ as $(\Delta x, \Delta y) \to (0, 0)$.

Since x^* is between x_0 and $x_0 + \Delta x$ and y^* is between y_0 and $y_0 + \Delta y$, it follows that

$$x^* \to x_0 \quad \text{as} \quad \Delta x \to 0$$
$$y^* \to y_0 \quad \text{as} \quad \Delta y \to 0$$

But this, together with the hypothesis that f_x and f_y are continuous at (x_0, y_0), implies that

$$f_x(x^*, y_0) \to f_x(x_0, y_0)$$
$$f_y(x_0 + \Delta x, y^*) \to f_y(x_0, y_0)$$

as $(\Delta x, \Delta y) \to (0, 0)$. Thus, from (27) and (28) it follows that $\epsilon_1 \to 0$ and $\epsilon_2 \to 0$ as $(\Delta x, \Delta y) \to (0, 0)$. ∎

▶ Exercise Set 16.4

1. Find Δf given that $f(x, y) = x^2 y$, $(x_0, y_0) = (1, 3)$, $\Delta x = 0.1$, and $\Delta y = 0.2$.

2. Find Δz given that $z = 3x^2 - 2y$, $(x_0, y_0) = (-2, 4)$, $\Delta x = 0.02$, and $\Delta y = -0.03$.

3. Find the increment of $f(x, y) = x/y$ as (x, y) varies from $(-1, 2)$ to $(3, 1)$.

4. Find the increment of $g(u, v) = 2uv - v^3$ as (u, v) varies from $(0, 1)$ to $(4, -2)$.

In Exercises 5–8, use Definition 16.4.1 to establish the differentiability of the given function. [*Remark:* ϵ_1 and ϵ_2 are not unique.]

5. $f(x, y) = xy$.   **6.** $f(x, y) = x^2 + y^2$.

7. $f(x, y) = x^2 y$.   **8.** $f(x, y) = 3x + y^2$.

9. Let $f(x, y) = \sqrt{x^2 + y^2}$.
  (a) Show that f is continuous at $(0, 0)$.
  (b) Show that $f_x(0, 0)$ does not exist, and hence that f is not differentiable at $(0, 0)$. [*Hint:* Express $f_x(0, 0)$ as a limit (see Exercise 68, Section 16.3).]

10. Let

$$f(x, y) = \begin{cases} 5 - 3x - 2y, & x \ge 0 \text{ or } y \ge 0 \\ 0, & x < 0 \text{ and } y < 0 \end{cases}$$

Show that $f_x(0, 0)$ and $f_y(0, 0)$ exist, but f is not continuous at $(0, 0)$.

11. Let

$$f(x, y) = \begin{cases} \dfrac{xy}{x^2 + y^2}, & (x, y) \ne (0, 0) \\ 0, & (x, y) = (0, 0) \end{cases}$$

Prove: $f_x(0, 0)$ and $f_y(0, 0)$ exist, but f is not continuous at $(0, 0)$. [*Hint:* Show that $f_x(0, 0) = 0$ and $f_y(0, 0) = 0$ by expressing these derivatives as limits (see Exercise 68, Section 16.3). To prove that f is not continuous at $(0, 0)$, show that

$$\lim_{(x,y) \to (0,0)} f(x, y)$$

does not exist by letting $(x, y) \to (0, 0)$ along $y = 0$ and along $y = x$.]

In Exercises 12–16, find f_{xy} and f_{yx} and verify their equality.

12. $f(x, y) = 2xy - 3y^2$.

13. $f(x, y) = 4x^3 y + 3x^2 y$.

14. $f(x, y) = \dfrac{x^3}{y}$.

15. $f(x, y) = \sin(x^2 + y^3)$.

16. $f(x, y) = \sqrt{x^2 + y^2 - 1}$.

17. Let f be a function of two variables with continuous third- and fourth-order partial derivatives.
  (a) How many of the third-order partial derivatives can be distinct?
  (b) How many of the fourth order?

18. Let $f(x, y) = e^{xy^2}$. Find f_{xyx}, f_{xxy}, and f_{yxx} and verify their equality.

In Exercises 19–24, find dz/dt using the chain rule.

19. $z = 3x^2 y^3$; $x = t^4$, $y = t^2$.

20. $z = \ln(2x^2 + y)$; $x = \sqrt{t}$, $y = t^{2/3}$.

21. $z = 3\cos x - \sin xy$; $x = 1/t$, $y = 3t$.

22. $z = \sqrt{1 + x - 2xy^4}$; $x = \ln t$, $y = t$.

23. $z = e^{1-xy}$; $x = t^{1/3}$, $y = t^3$.

24. $z = \cosh^2 xy$; $x = t/2$, $y = e^t$.

In Exercises 25–31, find $\partial z/\partial u$ and $\partial z/\partial v$ by the chain rule.

25. $z = 8x^2y - 2x + 3y$; $x = uv$, $y = u - v$.

26. $z = x^2 - y \tan x$; $x = u/v$, $y = u^2v^2$.

27. $z = x/y$; $x = 2 \cos u$, $y = 3 \sin v$.

28. $z = 3x - 2y$; $x = u + v \ln u$, $y = u^2 - v \ln v$.

29. $z = e^{x^2y}$; $x = \sqrt{uv}$, $y = 1/v$.

30. $z = \cos x \sin y$; $x = u - v$, $y = u^2 + v^2$.

31. $z = \tan^{-1}(x^2 + y^2)$; $x = e^u \sin v$; $y = e^u \cos v$.

32. Let $w = rs/(r^2 + s^2)$; $r = uv$, $s = u - 2v$. Use the chain rule to find $\partial w/\partial u$ and $\partial w/\partial v$.

33. Let $T = x^2y - xy^3 + 2$; $x = r \cos \theta$, $y = r \sin \theta$. Use the chain rule to find $\partial T/\partial r$ and $\partial T/\partial \theta$.

34. Let $R = e^{2s-t^2}$; $s = 3\phi$, $t = \phi^{1/2}$. Use the chain rule to find $dR/d\phi$.

35. Let $t = u/v$; $u = x^2 - y^2$, $v = 4xy^3$. Use the chain rule to find $\partial t/\partial x$ and $\partial t/\partial y$.

36. Use the chain rule to find $\dfrac{dz}{dt}\Big|_{t=3}$ if $z = x^2y$; $x = t^2$, $y = t + 7$.

37. Use the chain rule to find the value of $\dfrac{dw}{ds}\Big|_{s=1/4}$ if $w = r^2 - r \tan \theta$; $r = \sqrt{s}$, $\theta = \pi s$.

38. Use the chain rule to find the value of

$$\frac{\partial f}{\partial u}\Big|_{u=1, v=-2} \quad \text{and} \quad \frac{\partial f}{\partial v}\Big|_{u=1, v=-2}$$

if $f(x, y) = x^2y^2 - x + 2y$; $x = \sqrt{u}$, $y = uv^3$.

39. Use the chain rule to find the value of

$$\frac{\partial z}{\partial r}\Big|_{r=2, \theta=\pi/6} \quad \text{and} \quad \frac{\partial z}{\partial \theta}\Big|_{r=2, \theta=\pi/6}$$

if $z = xye^{x/y}$; $x = r \cos \theta$, $y = r \sin \theta$.

In Exercises 40–43, use (14) to find dy/dx and check your result using implicit differentiation.

40. $x^2y^3 + \cos y = 0$. **41.** $x^3 - 3xy^2 + y^3 = 5$.

42. $e^{xy} + ye^y = 1$. **43.** $x - \sqrt{xy} + 3y = 4$.

44. The portion of a tree usable for lumber may be viewed as a right-circular cylinder. If the height of a tree increases 24 in. per year and the diameter increases 3 in. per year, how fast is the volume of usable lumber increasing when the tree is 240 in. high and the diameter is 30 in.?

45. Two straight roads intersect at right angles. Car A, moving on one of the roads, approaches the intersection at 25 mi/hr and car B, moving on the other road, approaches the intersection at 30 mi/hr. At what rate is the distance between the cars changing when A is 0.3 miles from the intersection and B is 0.4 miles from the intersection?

46. Use the ideal gas law (Example 4, Section 16.3) with $k = 10$ to find the rate at which the temperature of a gas is changing when the volume is 200 in.^{3} and increasing at the rate of 4 in.^{3}/sec, while the pressure is 5 lb/in.^{2} and decreasing at the rate of 1 lb/in.^{2} per sec.

47. Two sides of a triangle have lengths $a = 4$ cm and $b = 3$ cm, but are increasing at the rate of 1 cm/sec. If the area of the triangle remains constant, at what rate is the angle θ between a and b changing when $\theta = \pi/6$?

48. Two sides of a triangle have lengths $a = 5$ cm and $b = 10$ cm, and the included angle is $\theta = \pi/3$. If a is increasing at a rate of 2 cm/sec, b is increasing at a rate of 1 cm/sec, and θ remains constant, at what rate is the third side changing? Is it increasing or decreasing? [*Hint:* Use the law of cosines.]

49. Suppose that a particle moving along a metal plate in the xy-plane has velocity $\mathbf{v} = \mathbf{i} - 4\mathbf{j}$ (cm/sec) at the point $(3, 2)$. If the temperature of the plate at points in the xy-plane is $T(x, y) = y^2 \ln x$, $x \geq 1$, in degrees Celsius, find dT/dt at $(3, 2)$.

In Exercises 50–67, you may assume that the derivatives satisfy all continuity requirements needed to solve the problem.

50. Let $z = f(u)$ where $u = g(x, y)$. Express $\partial z/\partial x$ and $\partial z/\partial y$ in terms of dz/du, $\partial u/\partial x$, and $\partial u/\partial y$.

51. Let $z = f(x^2 - y^2)$. Show that $y\partial z/\partial x + x\partial z/\partial y = 0$.

52. Let $z = f(xy)$. Show that $x \partial z/\partial x - y \partial z/\partial y = 0$.

53. Let $z = f(u)$ where $u = g(x, y)$. Show that

(a) $\dfrac{\partial^2 z}{\partial x^2} = \dfrac{dz}{du}\dfrac{\partial^2 u}{\partial x^2} + \dfrac{d^2 z}{du^2}\left(\dfrac{\partial u}{\partial x}\right)^2$

and $\dfrac{\partial^2 z}{\partial y^2} = \dfrac{dz}{du}\dfrac{\partial^2 u}{\partial y^2} + \dfrac{d^2 z}{du^2}\left(\dfrac{\partial u}{\partial y}\right)^2$

(b) $\dfrac{\partial^2 z}{\partial y\,\partial x} = \dfrac{dz}{du}\dfrac{\partial^2 u}{\partial y\,\partial x} + \dfrac{d^2 z}{du^2}\dfrac{\partial u}{\partial x}\dfrac{\partial u}{\partial y}$.

54. Let $r = \sqrt{x^2 + y^2}$. Show that

(a) $\dfrac{\partial r}{\partial x} = \dfrac{x}{r}$ (b) $\dfrac{\partial r}{\partial y} = \dfrac{y}{r}$

(c) $\dfrac{\partial^2 r}{\partial x^2} = \dfrac{y^2}{r^3}$ (d) $\dfrac{\partial^2 r}{\partial y^2} = \dfrac{x^2}{r^3}$.

55. Let $z = f(r)$ where $r = \sqrt{x^2 + y^2}$, $r \neq 0$.

(a) Use the formulas in Exercises 53 and 54 to show that if z satisfies Laplace's equation,
$$\dfrac{\partial^2 z}{\partial x^2} + \dfrac{\partial^2 z}{\partial y^2} = 0, \text{ then } r\dfrac{d^2 z}{dr^2} + \dfrac{dz}{dr} = 0.$$

(b) Use the result of part (a) to find the general solution of Laplace's equation for $z = f(r)$.

[*Hint:* Write $r\dfrac{d^2 z}{dr^2} + \dfrac{dz}{dr}$ as the derivative of a product.]

56. Let $z = f(x - y, y - x)$. Show that
$$\partial z/\partial x + \partial z/\partial y = 0$$

57. Let $z = f(y + cx) + g(y - cx)$, where $c \neq 0$. Show that
$$\dfrac{\partial^2 z}{\partial x^2} = c^2 \dfrac{\partial^2 z}{\partial y^2}$$

58. Let $z = f(x, y)$ where $x = g(t)$ and $y = h(t)$.

(a) Show that
$$\dfrac{d}{dt}\left(\dfrac{\partial z}{\partial x}\right) = \dfrac{\partial^2 z}{\partial x^2}\dfrac{dx}{dt} + \dfrac{\partial^2 z}{\partial y\,\partial x}\dfrac{dy}{dt}$$

and
$$\dfrac{d}{dt}\left(\dfrac{\partial z}{\partial y}\right) = \dfrac{\partial^2 z}{\partial x\,\partial y}\dfrac{dx}{dt} + \dfrac{\partial^2 z}{\partial y^2}\dfrac{dy}{dt}$$

(b) Use the formulas in part (a) to help find a formula for $d^2 z/dt^2$.

59. Let $z = f(x, y)$ where $x = g(u, v)$ and $y = h(u, v)$. Show that
$$\dfrac{\partial}{\partial u}\left(\dfrac{\partial z}{\partial x}\right) = \dfrac{\partial^2 z}{\partial x^2}\dfrac{\partial x}{\partial u} + \dfrac{\partial^2 z}{\partial y\,\partial x}\dfrac{\partial y}{\partial u}$$

60. Let $z = f(x, y)$ where $x = u + v$ and $y = u - v$. Show that
$$\dfrac{\partial^2 z}{\partial v\,\partial u} = \dfrac{\partial^2 z}{\partial x^2} - \dfrac{\partial^2 z}{\partial y^2}$$

61. The equations $x = r\cos\theta$ and $y = r\sin\theta$, which relate Cartesian and polar coordinates, define r and θ implicitly as functions of x and y.

(a) Use implicit differentiation with respect to the variable x on both equations to show that
$$\dfrac{\partial r}{\partial x} = \cos\theta \text{ and } \dfrac{\partial\theta}{\partial x} = -\dfrac{\sin\theta}{r}.$$

(b) Use implicit differentiation with respect to y on both equations to show that
$$\dfrac{\partial r}{\partial y} = \sin\theta \text{ and } \dfrac{\partial\theta}{\partial y} = \dfrac{\cos\theta}{r}.$$

62. Let $z = f(r, \theta)$, where r and θ are defined implicitly as functions of x and y by the equations $x = r\cos\theta$ and $y = r\sin\theta$. Use the results in Exercise 61 to show that

(a) $\dfrac{\partial z}{\partial x} = \dfrac{\partial z}{\partial r}\cos\theta - \dfrac{1}{r}\dfrac{\partial z}{\partial\theta}\sin\theta$

(b) $\dfrac{\partial z}{\partial y} = \dfrac{\partial z}{\partial r}\sin\theta + \dfrac{1}{r}\dfrac{\partial z}{\partial\theta}\cos\theta$.

63. Use the formulas in Exercise 62 to show that
$$\left(\dfrac{\partial z}{\partial x}\right)^2 + \left(\dfrac{\partial z}{\partial y}\right)^2 = \left(\dfrac{\partial z}{\partial r}\right)^2 + \dfrac{1}{r^2}\left(\dfrac{\partial z}{\partial\theta}\right)^2$$

64. Use the formulas in Exercises 61 and 62 to show that Laplace's equation
$$\dfrac{\partial^2 z}{\partial x^2} + \dfrac{\partial^2 z}{\partial y^2} = 0$$

becomes
$$\dfrac{\partial^2 z}{\partial r^2} + \dfrac{1}{r^2}\dfrac{\partial^2 z}{\partial\theta^2} + \dfrac{1}{r}\dfrac{\partial z}{\partial r} = 0$$

in polar coordinates.

65. A function $f(x, y)$ is said to be **homogeneous of degree n** if $f(tx, ty) = t^n f(x, y)$ for $t > 0$. Show that the following functions are homogeneous and find the degree.

(a) $f(x, y) = 3x^2 + y^2$

(b) $f(x, y) = \sqrt{x^2 + y^2}$

(c) $f(x, y) = x^2 y - 2y^3$

(d) $f(x, y) = \dfrac{5}{(x^2 + 2y^2)^2}$, $(x, y) \neq (0, 0)$.

66. Show that for a homogeneous function $f(x, y)$ of degree n (see Exercise 65)

$$x \frac{\partial f}{\partial x} + y \frac{\partial f}{\partial y} = nf$$

[*Hint:* Let $u = tx$ and $v = ty$ in $f(tx, ty)$, and differentiate both sides of $f(u, v) = t^n f(x, y)$ with respect to t.]

67. Let $w = f(x, y)$ where $y = g(x, z)$. Taking x and z as the independent variables, express each of the following in terms of $\partial f/\partial x$, $\partial f/\partial y$, $\partial y/\partial x$, and $\partial y/\partial z$.

(a) $\dfrac{\partial w}{\partial x}$

(b) $\dfrac{\partial w}{\partial z}$.

68. Show that if $u(x, y)$ and $v(x, y)$ satisfy the Cauchy-Riemann equations (Exercise 54, Section 16.3), and if $x = r \cos \theta$ and $y = r \sin \theta$, then

$$\frac{\partial u}{\partial r} = \frac{1}{r} \frac{\partial v}{\partial \theta} \quad \text{and} \quad \frac{\partial v}{\partial r} = -\frac{1}{r} \frac{\partial u}{\partial \theta}$$

69. Prove: If f, f_x, and f_y are continuous on a circular region containing $A(x_0, y_0)$ and $B(x_1, y_1)$, then there is a point (x^*, y^*) on the line segment joining A and B such that

$$f(x_1, y_1) - f(x_0, y_0)$$
$$= f_x(x^*, y^*)(x_1 - x_0) + f_y(x^*, y^*)(y_1 - y_0)$$

This result is the two-dimensional version of the Mean-Value Theorem. [*Hint:* Express the line segment joining A and B in parametric form and use the Mean-Value Theorem for functions of one variable.]

70. Prove: If $f_x(x, y) = 0$ and $f_y(x, y) = 0$ throughout a circular region, then $f(x, y)$ is constant on that region. [*Hint:* Use the result of Exercise 69.]

16.5   TANGENT PLANES; TOTAL DIFFERENTIALS FOR FUNCTIONS OF TWO VARIABLES

In this section we shall discuss tangent planes to surfaces in three-dimensional space. We are concerned with three main questions: What is a tangent plane? When do tangent planes exist? How do we find equations of tangent planes? We shall use our results on tangent planes to extend the concept of a differential to functions of two variables.

TANGENT PLANES

Recall that if C is a smooth curve in 3-space, then the tangent line to C at a point P_0 is the line through P_0 along the unit tangent vector to C at P_0 (Figure 16.5.1). The concept of a *tangent plane* builds on this definition. If $P_0(x_0, y_0, z_0)$ is a point on a surface S, and if the tangent lines at P_0 to all smooth curves that pass through P_0 and lie on the surface S all lie in a common plane, then we shall regard that plane to be the **tangent plane** to the surface at P_0 (Figure 16.5.2).

The following theorem states conditions that ensure the existence of a tangent plane and tell us how to find its equation.

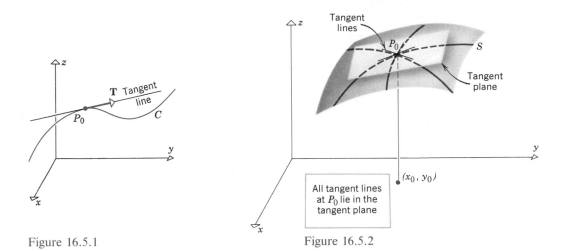

Figure 16.5.1

Figure 16.5.2

16.5.1   THEOREM. *Let $P_0(x_0, y_0, z_0)$ be any point on the surface $z = f(x, y)$. If $f(x, y)$ is differentiable at (x_0, y_0), then the surface has a tangent plane at P_0, and this plane has the equation*

$$f_x(x_0, y_0)(x - x_0) + f_y(x_0, y_0)(y - y_0) - (z - z_0) = 0 \qquad (1)$$

Proof.   To prove the existence of a tangent plane at P_0, we must show that all smooth curves on the surface $z = f(x, y)$ that pass through P_0 have tangent lines that lie in a common plane. We shall do this by showing that these curves all have unit tangent vectors at P_0 that are perpendicular to the vector

$$\mathbf{n} = \langle f_x(x_0, y_0), f_y(x_0, y_0), -1 \rangle \qquad (2)$$

This will force all the tangent lines at P_0 to lie in the plane through P_0 with $\mathbf{n}$ as a normal. Moreover, it will follow from (2) that the point-normal equation of this plane is (1), thereby completing the proof.

Assume that C is any smooth curve that lies on the surface $z = f(x, y)$ and passes through $P_0(x_0, y_0, z_0)$. Moreover, assume that C has parametric equations

$$x = x(s)$$

$$y = y(s)$$

$$z = z(s)$$

where s is an arc-length parameter, and assume that $P_0(x_0, y_0, z_0)$ is the point on C that corresponds to the parameter value $s = s_0$. Thus, $x_0 = x(s_0)$,

$y_0 = y(s_0)$, and $z_0 = z(s_0)$. Because C lies on the surface $z = f(x, y)$, every point $(x(s), y(s), z(s))$ on C must satisfy this equation for all s, so

$$z(s) = f(x(s), y(s))$$

for all s. If we differentiate both sides of this equation and apply the version of the chain rule in Theorem 16.4.6 with s replacing t, we obtain

$$\frac{dz}{ds} = \frac{\partial f}{\partial x}\frac{dx}{ds} + \frac{\partial f}{\partial y}\frac{dy}{ds}$$

or equivalently

$$\frac{\partial f}{\partial x}\frac{dx}{ds} + \frac{\partial f}{\partial y}\frac{dy}{ds} - \frac{dz}{ds} = 0$$

The left side of this equation can be rewritten as a dot product of vectors:

$$\left\langle \frac{\partial f}{\partial x}, \frac{\partial f}{\partial y}, -1 \right\rangle \cdot \left\langle \frac{dx}{ds}, \frac{dy}{ds}, \frac{dz}{ds} \right\rangle = 0$$

or in an alternative notation

$$\langle f_x(x, y), f_y(x, y), -1 \rangle \cdot \langle x'(s), y'(s), z'(s) \rangle = 0$$

In particular, if $s = s_0$, we have

$$\langle f_x(x_0, y_0), f_y(x_0, y_0), -1 \rangle \cdot \langle x'(s_0), y'(s_0), z'(s_0) \rangle = 0 \tag{3}$$

But the second vector in the dot product is the unit tangent vector to C at the point $P_0(x_0, y_0, z_0)$ (see Formula 1 of Section 15.4), so that from (3) the unit tangent vector to C at P_0 is perpendicular to the vector

$$\mathbf{n} = \langle f_x(x_0, y_0), f_y(x_0, y_0), -1 \rangle$$

which completes the proof. ∎

If $f(x, y)$ is differentiable at (x_0, y_0), then the vector

$$\mathbf{n} = \langle f_x(x_0, y_0), f_y(x_0, y_0), -1 \rangle \tag{4}$$

is called a ***normal vector*** to the surface $z = f(x, y)$ at $P_0(x_0, y_0, z_0)$ and the line through P_0 parallel to $\mathbf{n}$ is called the ***normal line*** to the surface at P_0 (Figure 16.5.3). Parametric equations of the normal line are

$$\begin{aligned} x &= x_0 + f_x(x_0, y_0)t \\ y &= y_0 + f_y(x_0, y_0)t \\ z &= z_0 - t \end{aligned} \tag{5}$$

Figure 16.5.3 ▷

Example 1   Find equations of the tangent plane and normal line to the surface $z = x^2 y$ at the point $(2, 1, 4)$.

Solution.   Since $f(x, y) = x^2 y$, it follows that

$$f_x(x, y) = 2xy \quad \text{and} \quad f_y(x, y) = x^2$$

so that with $x = 2$, $y = 1$,

$$f_x(2, 1) = 4 \quad \text{and} \quad f_y(2, 1) = 4$$

Thus, a vector normal to the surface at $(2, 1, 4)$ is

$$\mathbf{n} = f_x(2, 1)\mathbf{i} + f_y(2, 1)\mathbf{j} - \mathbf{k} = 4\mathbf{i} + 4\mathbf{j} - \mathbf{k}$$

Therefore, the tangent plane has the equation

$$4(x - 2) + 4(y - 1) - (z - 4) = 0$$

or

$$4x + 4y - z = 8$$

and the normal line has equations

$$x = 2 + 4t$$
$$y = 1 + 4t$$
$$z = 4 - t \quad \blacktriangleleft$$

REMARK.   In the previous section we set two goals for the definition of differentiability of a function $f(x, y)$ of two variables at a point (x_0, y_0)—we wanted f to be continuous at (x_0, y_0) and we wanted the surface

$z = f(x, y)$ to have a nonvertical tangent plane at (x_0, y_0). Both of these goals have now been achieved, for we showed in Theorem 16.4.2 that differentiability implies continuity and now Theorem 16.5.1 shows that differentiability implies the existence of a nonvertical tangent plane. (The tangent plane given by (1) is nonvertical because the third component of the normal vector $\mathbf{n}$ in (2) is not zero.)

DIFFERENTIALS   Recall that if $y = f(x)$ is a function of one variable, then the differential

$$dy = f'(x_0)\, dx$$

represents the change in y along the *tangent line* at (x_0, y_0) produced by a change dx in x and

$$\Delta y = f(x_0 + \Delta x) - f(x_0)$$

represents the change in y along the *curve* $y = f(x)$ produced by a change Δx in x. Analogously, if $z = f(x, y)$ is a function of two variables, we shall define dz to be the change in z along the *tangent plane* at (x_0, y_0, z_0) to the surface $z = f(x, y)$ produced by changes dx and dy in x and y, respectively. This is in contrast to

$$\Delta z = f(x_0 + \Delta x, y_0 + \Delta y) - f(x_0, y_0)$$

which represents the change in z *along the surface* produced by changes Δx and Δy in x and y. A comparison of dz and Δz is shown in Figure 16.5.4 in the case where $dx = \Delta x$ and $dy = \Delta y$.

Figure 16.5.4 ▷

To derive a formula for dz, let $P_0(x_0, y_0, z_0)$ be a fixed point on the surface $z = f(x, y)$. If we assume f to be differentiable at (x_0, y_0), then the surface has a tangent plane at P_0, given by the equation

$$f_x(x_0, y_0)(x - x_0) + f_y(x_0, y_0)(y - y_0) - (z - z_0) = 0$$

or

$$z = z_0 + f_x(x_0, y_0)(x - x_0) + f_y(x_0, y_0)(y - y_0) \tag{6}$$

From (6) it follows that the tangent plane has height z_0 when $x = x_0$, $y = y_0$, and it has height

$$z_0 + f_x(x_0, y_0) \, dx + f_y(x_0, y_0) \, dy \tag{7}$$

when $x = x_0 + dx$, $y = y_0 + dy$. Thus, the change dz in the height of the tangent plane as (x, y) varies from (x_0, y_0) to $(x_0 + dx, y_0 + dy)$ is obtained by subtracting z_0 from expression (7). This yields

$$dz = f_x(x_0, y_0) \, dx + f_y(x_0, y_0) \, dy$$

This quantity is called the **total differential** of z at (x_0, y_0). Often, we omit the subscripts on x_0 and y_0 and write

$$dz = f_x(x, y) \, dx + f_y(x, y) \, dy \tag{8}$$

In this formula, dx and dy are usually viewed as variables and x and y fixed numbers. Formula (8) may also be written using df in place of dz.

If $z = f(x, y)$ is differentiable at the point (x, y), then the increment Δz can be written as

$$\Delta z = f_x(x, y) \, \Delta x + f_y(x, y) \, \Delta y + \epsilon_1 \, \Delta x + \epsilon_2 \, \Delta y \tag{9}$$

where $\epsilon_1 \to 0$, $\epsilon_2 \to 0$ as $(\Delta x, \Delta y) \to (0, 0)$. In the case where $dx = \Delta x$ and $dy = \Delta y$, it follows from (8) and (9) that

$$\Delta z = dz + \epsilon_1 \, \Delta x + \epsilon_2 \, \Delta y$$

Thus, when $\Delta x = dx$ and $\Delta y = dy$ are small, we can approximate Δz by

$$\Delta z \approx dz \tag{10}$$

Geometrically, this approximation tells us that the change in z along the surface and the change in z along the tangent plane are approximately equal when $\Delta x = dx$ and $\Delta y = dy$ are small (see Figure 16.5.4).

Example 2 Let $z = 4x^3y^2$. Find dz.

Solution. Since $f(x, y) = 4x^3y^2$,

$$f_x(x, y) = 12x^2y^2 \quad \text{and} \quad f_y(x, y) = 8x^3y$$

so

$$dz = 12x^2y^2 \, dx + 8x^3y \, dy \quad \blacktriangleleft$$

Example 3   Let $f(x, y) = \sqrt{x^2 + y^2}$. Use a total differential to approximate the change in $f(x, y)$ as (x, y) varies from the point $(3, 4)$ to the point $(3.04, 3.98)$.

Solution.   We shall approximate Δf (the change in f) by

$$df = f_x(x, y)\, dx + f_y(x, y)\, dy = \frac{x}{\sqrt{x^2 + y^2}}\, dx + \frac{y}{\sqrt{x^2 + y^2}}\, dy$$

Since $x = 3$, $y = 4$, $dx = 0.04$, and $dy = -0.02$, we obtain

$$\Delta f \approx df = \frac{3}{\sqrt{3^2 + 4^2}}(0.04) + \frac{4}{\sqrt{3^2 + 4^2}}(-0.02)$$

$$= \frac{3}{5}(0.04) - \frac{4}{5}(0.02) = 0.008$$

The reader may want to use a hand calculator to show that the true value of Δf to five decimal places is

$$\Delta f = \sqrt{(3.04)^2 + (3.98)^2} - \sqrt{3^2 + 4^2} \approx 0.00819 \qquad \blacktriangleleft$$

Example 4   The volume V of a right-circular cone of radius r and height h is given by $V = \frac{1}{3}\pi r^2 h$. Suppose the height increases from 10 cm to 10.01 cm, while the radius decreases from 12 cm to 11.95 cm. Use a total differential to approximate the change ΔV in volume.

Solution.   We shall approximate ΔV by

$$dV = \frac{\partial V}{\partial r}\, dr + \frac{\partial V}{\partial h}\, dh = \frac{2}{3}\pi r h\, dr + \frac{1}{3}\pi r^2\, dh$$

With $h = 10$, $dh = 0.01$, $r = 12$, and $dr = -0.05$, we obtain

$$\Delta V \approx dV = \frac{2}{3}\pi(12)(10)(-0.05) + \frac{1}{3}\pi(12)^2(0.01)$$

or

$$\Delta V \approx -3.52\pi \approx -11.06$$

Thus, the volume decreases by approximately 11.06 cm^3. $\blacktriangleleft$

Example 5   The radius of a right-circular cylinder is measured with an error of at most 2%, and the height is measured with an error of at most 4%. Estimate the maximum possible percentage error in the calculated volume V.

Solution.   Let r, h, and V be the true radius, height, and volume of the cylinder, and let Δr, Δh, and ΔV be the errors in these quantities. We are given that

$$\left|\frac{\Delta r}{r}\right| \le 0.02 \quad \text{and} \quad \left|\frac{\Delta h}{h}\right| \le 0.04$$

We want to find the maximum possible value of $|\Delta V/V|$. Since the volume of the cylinder is $V = \pi r^2 h$, it follows from (8) that

$$dV = \frac{\partial V}{\partial r}\, dr + \frac{\partial V}{\partial h}\, dh = 2\pi r h\, dr + \pi r^2 dh$$

If we choose $dr = \Delta r$ and $dh = \Delta h$, then we can use the approximations

$$\Delta V \approx dV \quad \text{and} \quad \frac{\Delta V}{V} \approx \frac{dV}{V}$$

But

$$\frac{dV}{V} = \frac{2\pi r h\, dr + \pi r^2 dh}{\pi r^2 h} = 2\frac{dr}{r} + \frac{dh}{h}$$

so by the triangle inequality (Theorem 1.2.6)

$$\left|\frac{dV}{V}\right| = \left|2\frac{dr}{r} + \frac{dh}{h}\right| \le 2\left|\frac{dr}{r}\right| + \left|\frac{dh}{h}\right| \le 2(0.02) + (0.04) = 0.08$$

Thus, the maximum percentage error in V is approximately 8%.   ◀

▶ Exercise Set 16.5

In Exercises 1–8, find equations for the tangent plane and normal line to the given surface at the point P.

1. $z = 4x^3y^2 + 2y$; $P(1, -2, 12)$.

2. $z = \frac{1}{2}x^7y^{-2}$; $P(2, 4, 4)$.

3. $z = xe^{-y}$; $P(1, 0, 1)$.

4. $z = \ln\sqrt{x^2 + y^2}$; $P(-1, 0, 0)$.

5. $z = e^{3y}\sin 3x$; $P(\pi/6, 0, 1)$.

6. $z = x^{1/2} + y^{1/2}$; $P(4, 9, 5)$.

7. $x^2 + y^2 + z^2 = 25$; $P(-3, 0, 4)$.

8. $x^2y - 4z^2 = -7$; $P(-3, 1, -2)$.

In Exercises 9–12, find dz.

9. $z = 7x - 2y$.

10. $z = 5x^2y^5 - 2x + 4y + 7$.

11. $z = \tan^{-1}xy$.   12. $z = \sec^2(x - 3y)$.

In Exercises 13–16, use a total differential to approximate the change in $f(x, y)$ as (x, y) varies from P to Q.

13. $f(x, y) = x^2 + 2xy - 4x$; $P(1, 2)$, $Q(1.01, 2.04)$.

14. $f(x, y) = x^{1/3}y^{1/2}$; $P(8, 9)$, $Q(7.78, 9.03)$.

15. $f(x, y) = \dfrac{x + y}{xy}$; $P(-1, -2)$, $Q(-1.02, -2.04)$.

16. $f(x, y) = \ln\sqrt{1 + xy}$; $P(0, 2)$, $Q(-0.09, 1.98)$.

17. Find all points on the surface at which the tangent plane is horizontal.
     (a) $z = x^3y^2$
     (b) $z = x^2 - xy + y^2 - 2x + 4y$.

18. Find a point on the surface $z = 3x^2 - y^2$ at which the tangent plane is parallel to the plane $6x + 4y - z = 5$.

19. Find a point on the surface $z = 8 - 3x^2 - 2y^2$ at which the tangent plane is perpendicular to the line, $x = 2 - 3t$, $y = 7 + 8t$, $z = 5 - t$.

20. Show that the surfaces $z = \sqrt{x^2 + y^2}$ and $z = \frac{1}{10}(x^2 + y^2) + \frac{5}{2}$ intersect at $(3, 4, 5)$ and have a common tangent plane at that point.

21. Show that the surfaces $z = \sqrt{16 - x^2 - y^2}$ and $z = \sqrt{x^2 + y^2}$ intersect at $(2, 2, 2\sqrt{2})$ and have tangent planes that are perpendicular at that point.

22. (a) Show that every line normal to the cone $z = \sqrt{x^2 + y^2}$ passes through the z-axis.
 (b) Show that every tangent plane intersects the cone in a line passing through the origin.

23. One leg of a right triangle increases from 3 cm to 3.2 cm, while the other leg decreases from 4 cm to 3.96 cm. Use a total differential to approximate the change in the length of the hypotenuse.

24. The volume V of a right-circular cone of radius r and height h is given by $V = \frac{1}{3}\pi r^2 h$. Suppose the height decreases from 20 in. to 19.95 in., while the radius increases from 4 in. to 4.05 in. Use a total differential to approximate the change in volume.

25. The length and width of a rectangle are measured with errors of at most 3% and 5%, respectively. Use differentials to estimate the maximum percentage error in the calculated area.

26. The radius and height of a right-circular cone are measured with errors of at most 1% and 4%, respectively. Use differentials to estimate the maximum percentage error in the calculated volume.

27. The length and width of a rectangle are measured with errors of at most $r\%$. Use differentials to estimate the maximum percentage error in the calculated length of the diagonal.

28. The legs of a right triangle are measured to be 3 cm and 4 cm, with a maximum error of 0.05 cm in each measurement. Use differentials to estimate the maximum possible error in the calculated value of (a) the hypotenuse and (b) the area of the triangle.

29. The total resistance R of two resistances R_1 and R_2, connected in parallel, is

$$R = \frac{R_1 R_2}{R_1 + R_2}$$

Suppose R_1 and R_2 are measured to be 200 ohms and 400 ohms, respectively, with a maximum error of 2% in each. Use differentials to estimate the maximum percentage error in the calculated value of R.

30. According to the ideal gas law, the pressure, temperature, and volume of a confined gas are related by $P = kT/V$ where k is a constant. Use differentials to estimate the percentage change in pressure if the temperature of a gas is increased 3% and the volume is increased 5%.

31. An angle θ of a right triangle is calculated by the formula

$$\theta = \sin^{-1}\frac{a}{c}$$

where a is the length of the side opposite to θ, and c is the length of the hypotenuse. Suppose that the measurements $a = 3$ in. and $c = 5$ in. each have a maximum possible error of 0.01 in. Use differentials to estimate the maximum possible error in the calculated value of θ.

32. An open cylindrical can has an inside radius of 2 cm and an inside height of 5 cm. Use differentials to estimate the volume of metal in the can if it is 0.01 cm thick. [*Hint:* The volume of metal is the difference, ΔV, in the volumes of two cylinders.]

33. The period T of a simple pendulum with small oscillations is calculated from the formula $T = 2\pi\sqrt{L/g}$. The values of L and g have errors of at most 0.5% and 0.1%, respectively. Use differentials to estimate the maximum percentage error in the calculated value of T.

34. The angle of elevation from a point on the ground to the top of a building is measured as 60° with a maximum possible error of 0.2°. The distance from the point to the building is measured as 100 ft with a maximum possible error of 2 in. Use differentials to estimate the maximum possible error in the calculated height of the building.

35. Suppose that x and y have errors of at most $r\%$ and $s\%$, respectively. For each of the following formulas in x and y, use differentials to estimate the maximum possible error in the calculated result.

(a) xy

(b) x/y

(c) x^2y^3

(d) $x^3\sqrt{y}$.

36. Show that the volume of the solid bounded by the coordinate planes and the plane tangent to the portion of the surface $xyz = k$, $k > 0$ in the first octant does not depend on the point of tangency.

37. (a) Find all points of intersection of the line $x = -1 + t$, $y = 2 + t$, $z = 2t + 7$, and the surface $z = x^2 + y^2$.

(b) At each point of intersection, find the cosine of the acute angle between the given line and the line normal to the surface.

38. Show that if f is differentiable and $z = xf(x/y)$, then all tangent planes to the graph of this equation pass through a common point.

39. Show that the equation of the plane tangent to the ellipsoid

$$\frac{x^2}{a^2} + \frac{y^2}{b^2} + \frac{z^2}{c^2} = 1$$

at (x_0, y_0, z_0) can be written in the form

$$\frac{x_0 x}{a^2} + \frac{y_0 y}{b^2} + \frac{z_0 z}{c^2} = 1$$

40. Show that the equation of the plane tangent to the paraboloid

$$z = \frac{x^2}{a^2} + \frac{y^2}{b^2}$$

at (x_0, y_0, z_0) can be written in the form

$$z + z_0 = \frac{2x_0 x}{a^2} + \frac{2y_0 y}{b^2}$$

41. Prove: If $z = f(x, y)$ and $z = g(x, y)$ intersect at $P(x_0, y_0, z_0)$ and if f and g are differentiable at (x_0, y_0), then the normal lines at P are perpendicular if and only if

$$f_x(x_0, y_0)g_x(x_0, y_0) + f_y(x_0, y_0)g_y(x_0, y_0) = -1$$

16.6 DIRECTIONAL DERIVATIVES AND GRADIENTS FOR FUNCTIONS OF TWO VARIABLES

The partial derivatives $f_x(x, y)$ and $f_y(x, y)$ represent the rates of change of $f(x, y)$ in directions parallel to the x- and y-axes. In this section we shall investigate rates of change of $f(x, y)$ in other directions.

DIRECTIONAL DERIVATIVES

As shown in Figure 16.6.1, let $P_0(x_0, y_0)$ be a fixed point in the xy-plane, let l be a line in the xy-plane that passes through P_0, and let f be a function

Figure 16.6.1 ▷

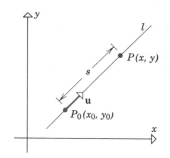

Figure 16.6.2

of two variables that is differentiable at (x_0, y_0). If a point $P(x, y)$ moves along the line l, then, as illustrated in the figure, there is a companion point Q directly above it, that moves along the surface $z = f(x, y)$, tracing out some curve C. We shall be interested in finding the rate at which the z-coordinate of Q changes with the distance s between P_0 and P. To make this idea more precise, consider the *unit vector*

$$\mathbf{u} = \langle u_1, u_2 \rangle$$

with initial point $P_0(x_0, y_0)$ and pointing in the direction of motion of $P(x, y)$. (See Figure 16.6.2.) Let s be the (signed) distance along l from $P_0(x_0, y_0)$ to $P(x, y)$, where s is nonnegative if P is in the direction of $\mathbf{u}$ from P_0 and is negative otherwise. The vector $\overrightarrow{P_0 P}$ is related to the vector $\mathbf{u}$ by

$$\overrightarrow{P_0 P} = s\mathbf{u}$$

or

$$\langle x - x_0, y - y_0 \rangle = \langle su_1, su_2 \rangle$$

(Figure 16.6.2). By equating corresponding components, we obtain the following parametric equations of l:

$$\begin{aligned} x &= x_0 + su_1 \\ y &= y_0 + su_2 \end{aligned} \tag{1}$$

Since the point Q is directly above $P(x, y)$ and lies on the surface, the z-coordinate of Q is

$$z = f(x, y)$$

or, in terms of s,

$$z = f(x_0 + su_1, y_0 + su_2)$$

The rate at which z changes with s can now be calculated using the chain rule. We obtain

$$\frac{dz}{ds} = f_x(x, y) \frac{dx}{ds} + f_y(x, y) \frac{dy}{ds}$$

or from (1)

$$\frac{dz}{ds} = f_x(x, y)u_1 + f_y(x, y)u_2 \tag{2}$$

where x and y in this formula are expressed in terms of s by equations (1). But $P_0(x_0, y_0)$ is the point on l that corresponds to $s = 0$ (see (1)), so setting $s = 0$ in (2) yields the instantaneous rate of change of z with respect

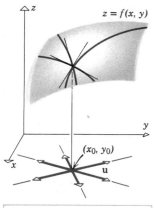

The rate at which z changes with s at (x_0, y_0) depends on the direction of **u**.

Figure 16.6.3

to s at the point $P_0(x_0, y_0)$:

$$\frac{dz}{ds}\bigg|_{s=0} = f_x(x_0, y_0)u_1 + f_y(x_0, y_0)u_2 \tag{3}$$

This quantity is commonly denoted by $D_{\mathbf{u}}z(x_0, y_0)$ or $D_{\mathbf{u}}f(x_0, y_0)$. We make the following definition.

16.6.1 DEFINITION. If $f(x, y)$ is differentiable at (x_0, y_0) and if $\mathbf{u} = \langle u_1, u_2 \rangle$ is a unit vector, then the *directional derivative* of f at (x_0, y_0) in the direction of **u** is defined by

$$D_{\mathbf{u}}f(x_0, y_0) = f_x(x_0, y_0)u_1 + f_y(x_0, y_0)u_2 \tag{4}$$

It should be kept in mind that there are infinitely many directional derivatives of $z = f(x, y)$ at a point (x_0, y_0), one for each possible choice of the direction vector **u** (Figure 16.6.3).

REMARK. The directional derivative $D_{\mathbf{u}}f(x_0, y_0)$ can be interpreted algebraically as the instantaneous rate of change in the direction of **u** at (x_0, y_0) of $z = f(x, y)$ with respect to the distance parameter s described above, or geometrically as the rise over the run of the tangent line to the curve C at the point Q_0 shown in Figure 16.6.4.

Figure 16.6.4

Example 1 Find the directional derivative of $f(x, y) = 3x^2y$ at the point $(1, 2)$ in the direction of the vector $\mathbf{a} = 3\mathbf{i} + 4\mathbf{j}$.

Solution. The partial derivatives of f are

$$f_x(x, y) = 6xy, \quad f_y(x, y) = 3x^2$$

so that

$$f_x(1, 2) = 12, \quad f_y(1, 2) = 3$$

Thus, the directional derivative at $(1, 2)$ is

$$D_{\mathbf{u}}f(1, 2) = f_x(1, 2)u_1 + f_y(1, 2)u_2 = 12u_1 + 3u_2 \qquad (5)$$

where u_1, u_2 are components of the *unit* vector in the direction of differentiation. We obtain these components by normalizing $\mathbf{a} = 3\mathbf{i} + 4\mathbf{j}$:

$$\mathbf{u} = \frac{\mathbf{a}}{\|\mathbf{a}\|} = \frac{1}{\sqrt{25}}(3\mathbf{i} + 4\mathbf{j}) = \frac{3}{5}\mathbf{i} + \frac{4}{5}\mathbf{j}$$

Thus, from (5)

$$D_{\mathbf{u}}f(1, 2) = 12\left(\frac{3}{5}\right) + 3\left(\frac{4}{5}\right) = \frac{48}{5} \qquad \blacktriangleleft$$

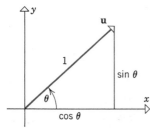

Figure 16.6.5

If $\mathbf{u} = u_1\mathbf{i} + u_2\mathbf{j}$ is a unit vector making an angle θ with the positive x-axis, then

$$u_1 = \cos\theta \quad \text{and} \quad u_2 = \sin\theta$$

(see Figure 16.6.5) so that (4) can be written in the form

$$D_{\mathbf{u}}f(x_0, y_0) = f_x(x_0, y_0)\cos\theta + f_y(x_0, y_0)\sin\theta \qquad (6)$$

Example 2 Find the directional derivative of e^{xy} at $(-2, 0)$ in the direction of the unit vector $\mathbf{u}$ that makes an angle of $\pi/3$ with the positive x-axis.

Solution. Let $f(x, y) = e^{xy}$, so that

$$f_x(x, y) = ye^{xy}, \quad f_y(x, y) = xe^{xy}$$
$$f_x(-2, 0) = 0, \quad f_y(-2, 0) = -2$$

From (6)

$$D_{\mathbf{u}}f(-2, 0) = f_x(-2, 0)\cos\frac{\pi}{3} + f_y(-2, 0)\sin\frac{\pi}{3}$$

$$= 0\left(\frac{1}{2}\right) + (-2)\left(\frac{\sqrt{3}}{2}\right) = -\sqrt{3} \qquad \blacktriangleleft$$

GRADIENT   As might be expected, directional derivatives in the directions of the positive coordinate axes reduce to partial derivatives:

$$D_{\mathbf{i}}f(x_0, y_0) = f_x(x_0, y_0)(1) + f_y(x_0, y_0)(0) = f_x(x_0, y_0)$$

$$D_{\mathbf{j}}f(x_0, y_0) = f_x(x_0, y_0)(0) + f_y(x_0, y_0)(1) = f_y(x_0, y_0)$$

Moreover, directional derivatives in the directions of the negative coordinate axes reduce to negatives of partial derivatives:

$$D_{-\mathbf{i}}f(x_0, y_0) = f_x(x_0, y_0)(-1) + f_y(x_0, y_0)(0) = -f_x(x_0, y_0)$$

$$D_{-\mathbf{j}}f(x_0, y_0) = f_x(x_0, y_0)(0) + f_y(x_0, y_0)(-1) = -f_y(x_0, y_0)$$

The directional derivative formula

$$D_{\mathbf{u}}f(x, y) = f_x(x, y)u_1 + f_y(x, y)u_2$$

can be expressed in the form of a dot product by writing

$$D_{\mathbf{u}}f(x, y) = (f_x(x, y)\mathbf{i} + f_y(x, y)\mathbf{j}) \cdot (u_1\mathbf{i} + u_2\mathbf{j}) \tag{7}$$

The second vector in the dot product is **u**. However, the first vector is something new; it is called the *gradient of f* and is denoted by the symbol ∇f or $\nabla f(x, y)$.*

16.6.2 DEFINITION. If f is a function of x and y, then the ***gradient of f*** is defined by

$$\nabla f(x, y) = f_x(x, y)\mathbf{i} + f_y(x, y)\mathbf{j} \tag{8}$$

With the gradient notation, Formula (7) for the directional derivative can be written in the following compact form:

$$D_{\mathbf{u}}f(x, y) = \nabla f(x, y) \cdot \mathbf{u} \tag{9}$$

In words, the dot product of the gradient of f with a unit vector **u** produces the directional derivative of f in the direction of **u**.

Example 3 Find the gradient of $f(x, y) = 3x^2y$ at the point $(1, 2)$ and use it to calculate the directional derivative of f at $(1, 2)$ in the direction of the vector $\mathbf{a} = 3\mathbf{i} + 4\mathbf{j}$.

Solution. From (8)

$$\nabla f(x, y) = f_x(x, y)\mathbf{i} + f_y(x, y)\mathbf{j} = 6xy\mathbf{i} + 3x^2\mathbf{j}$$

so that the gradient of f at $(1, 2)$ is

---

*The symbol ∇ (read, "del") is an inverted delta. In older books this symbol is sometimes called a "nabla" because of its similarity in form to an ancient Hebrew ten-stringed harp of that name.

$$\nabla f(1, 2) = 12\mathbf{i} + 3\mathbf{j}$$

The unit vector in the direction of **a** is

$$\mathbf{u} = \frac{\mathbf{a}}{\|\mathbf{a}\|} = \frac{1}{5}(3\mathbf{i} + 4\mathbf{j}) = \tfrac{3}{5}\mathbf{i} + \tfrac{4}{5}\mathbf{j}$$

Thus, from (9)

$$D_{\mathbf{u}}f(1, 2) = \nabla f(1, 2) \cdot \mathbf{u} = (12\mathbf{i} + 3\mathbf{j}) \cdot (\tfrac{3}{5}\mathbf{i} + \tfrac{4}{5}\mathbf{j}) = \tfrac{48}{5}$$

which agrees with the result obtained in Example 1.   ◀

PROPERTIES OF THE GRADIENT

The gradient is not merely a notational device to simplify the formula for the directional derivative; the length and direction of the gradient ∇f provide important information about the function f.

16.6.3   THEOREM. *Let f be a function of two variables that is differentiable at (x_0, y_0).*

(a)   *If $\nabla f(x_0, y_0) = \mathbf{0}$, then all directional derivatives of f at (x_0, y_0) are zero.*

(b)   *If $\nabla f(x_0, y_0) \neq \mathbf{0}$, then among all possible directional derivatives of f at (x_0, y_0), the derivative in the direction of $\nabla f(x_0, y_0)$ has the largest value. The value of that directional derivative is $\|\nabla f(x_0, y_0)\|$.*

(c)   *If $\nabla f(x_0, y_0) \neq \mathbf{0}$, then among all possible directional derivatives of f at (x_0, y_0), the derivative in the direction opposite to that of $\nabla f(x_0, y_0)$ has the smallest value. The value of that directional derivative is $-\|\nabla f(x_0, y_0)\|$.*

Proof of (a).   If $\nabla f(x_0, y_0) = \mathbf{0}$, then for all choices of **u** we have

$$D_{\mathbf{u}}f(x_0, y_0) = \nabla f(x_0, y_0) \cdot \mathbf{u} = \mathbf{0} \cdot \mathbf{u} = 0$$

Proofs of (b) and (c).   Assume $\nabla f(x_0, y_0) \neq \mathbf{0}$ and let θ be the angle between $\nabla f(x_0, y_0)$ and an arbitrary unit vector **u**. By the definition of dot product,

$$D_{\mathbf{u}}f(x_0, y_0) = \nabla f(x_0, y_0) \cdot \mathbf{u} = \|\nabla f(x_0, y_0)\| \, \|\mathbf{u}\| \cos \theta$$

or, since $\|\mathbf{u}\| = 1$,

$$D_{\mathbf{u}}f(x_0, y_0) = \|\nabla f(x_0, y_0)\| \cos \theta$$

Thus, the maximum value of $D_{\mathbf{u}}f(x_0, y_0)$ is $\|\nabla f(x_0, y_0)\|$, and this occurs when $\cos \theta = 1$, or when **u** has the same direction as $\nabla f(x_0, y_0)$ (since $\theta = 0$). The minimum value of $D_{\mathbf{u}}f(x_0, y_0)$ is $-\|\nabla f(x_0, y_0)\|$, and this occurs when $\cos \theta = -1$ or when $\mathbf{u}$ and $\nabla f(x_0, y_0)$ are oppositely directed (since $\theta = \pi$). ∎

Example 4 For the function $f(x, y) = x^2 e^y$, find the maximum value of a directional derivative at $(-2, 0)$, and give a unit vector in the direction of the maximum.

Solution. Since

$$\nabla f(x, y) = f_x(x, y)\mathbf{i} + f_y(x, y)\mathbf{j} = 2xe^y\mathbf{i} + x^2 e^y\mathbf{j}$$

the gradient of f at $(-2, 0)$ is

$$\nabla f(-2, 0) = -4\mathbf{i} + 4\mathbf{j}$$

By Theorem 16.6.3, the maximum value of the directional derivative is

$$\|\nabla f(-2, 0)\| = \sqrt{(-4)^2 + 4^2} = \sqrt{32} = 4\sqrt{2}$$

This maximum occurs in the direction of $\nabla f(-2, 0)$. A unit vector in this direction is

$$\frac{\nabla f(-2, 0)}{\|\nabla f(-2, 0)\|} = \frac{1}{4\sqrt{2}}(-4\mathbf{i} + 4\mathbf{j}) = -\frac{1}{\sqrt{2}}\mathbf{i} + \frac{1}{\sqrt{2}}\mathbf{j} \quad ◀$$

There is an important geometric relationship between the gradient $\nabla f(x, y)$ and the level curves of $f(x, y)$. If (x_0, y_0) is any point in the domain of f, and if $f(x_0, y_0) = c$, then the equation

$$f(x, y) = c$$

defines a level curve of f that passes through the point (x_0, y_0) (Figure 16.6.6a). Let us recall that the function f has its maximum rate of increase at (x_0, y_0) in the direction of $\nabla f(x_0, y_0)$ and has its maximum rate of decrease in the direction of $-\nabla f(x_0, y_0)$. The level curve through (x_0, y_0) is the "dividing line" between these extremes in the sense that $f(x, y)$ neither

Figure 16.6.6 ▷ (a) (b)

increases nor decreases along this curve (Figure 16.6.6b). The following theorem shows that the gradient of f at (x_0, y_0) is perpendicular to the level curve of f through (x_0, y_0). (A vector is *perpendicular* or *normal* to a curve at a point if it is perpendicular to a nonzero tangent vector at that point.)

> **16.6.4** THEOREM. *If f is differentiable at (x_0, y_0) and $\nabla f(x_0, y_0) \neq \mathbf{0}$, then $\nabla f(x_0, y_0)$ is perpendicular to the level curve of f through (x_0, y_0).*

To see why this result is true, let

$$f(x, y) = c \tag{10}$$

be the level curve through (x_0, y_0). It is proved in advanced calculus that with the given hypotheses, this level curve can be represented parametrically by the equations

$$x = x(t)$$

$$y = y(t)$$

so that the level curve has a nonzero tangent vector at (x_0, y_0). More precisely,

$$x'(t_0)\mathbf{i} + y'(t_0)\mathbf{j} \neq \mathbf{0}$$

where t_0 is the value of the parameter corresponding to (x_0, y_0).

Differentiating both sides of (10) with respect to t, and applying the chain rule yields

$$f_x\big(x(t), y(t)\big)x'(t) + f_y\big(x(t), y(t)\big)y'(t) = 0$$

Substituting $t = t_0$, and using the fact that $x(t_0) = x_0$ and $y(t_0) = y_0$, we obtain

$$f_x(x_0, y_0)x'(t_0) + f_y(x_0, y_0)y'(t_0) = 0$$

which can be written as

$$\nabla f(x_0, y_0) \cdot (x'(t_0)\mathbf{i} + y'(t_0)\mathbf{j}) = 0$$

This equation tells us that $\nabla f(x_0, y_0)$ is perpendicular to the tangent vector $x'(t_0)\mathbf{i} + y'(t_0)\mathbf{j}$ and, therefore, is perpendicular to the level curve through (x_0, y_0).

Example 5 For the function $f(x, y) = x^2 + y^2$, sketch the level curve through the point $(3, 4)$, and draw the gradient vector at this point.

Solution. Since $f(3, 4) = 25$, the level curve through the point $(3, 4)$ has

Figure 16.6.7

the equation $f(x, y) = 25$, which is the circle

$$x^2 + y^2 = 25$$

Since

$$\nabla f(x, y) = f_x(x, y)\mathbf{i} + f_y(x, y)\mathbf{j} = 2x\mathbf{i} + 2y\mathbf{j}$$

the gradient vector at $(3, 4)$ is

$$\nabla f(3, 4) = 6\mathbf{i} + 8\mathbf{j}$$

(Figure 16.6.7). Note that the gradient vector is perpendicular to the circle at $(3, 4)$, as guaranteed by Theorem 16.6.4.   ◄

▶ Exercise Set 16.6

In Exercises 1–4, find ∇z.

1. $z = 4x - 8y$.

2. $z = e^{-3y} \cos 4x$.

3. $z = \ln \sqrt{x^2 + y^2}$.

4. $z = e^{-5x} \sec x^2 y$.

In Exercises 5–8, find the gradient of f at the indicated point.

5. $f(x, y) = (x^2 + xy)^3$; $(-1, -1)$.

6. $f(x, y) = (x^2 + y^2)^{-1/2}$; $(3, 4)$.

7. $f(x, y) = y \ln (x + y)$; $(-3, 4)$.

8. $f(x, y) = y^2 \tan^3 x$; $(\pi/4, -3)$.

In Exercises 9–12, find $D_{\mathbf{u}} f$ at P.

9. $f(x, y) = (1 + xy)^{3/2}$; $P(3, 1)$; $\mathbf{u} = \dfrac{1}{\sqrt{2}}\mathbf{i} + \dfrac{1}{\sqrt{2}}\mathbf{j}$.

10. $f(x, y) = e^{2xy}$; $P(4, 0)$; $\mathbf{u} = -\tfrac{3}{5}\mathbf{i} + \tfrac{4}{5}\mathbf{j}$.

11. $f(x, y) = \ln (1 + x^2 + y)$; $P(0, 0)$;

$$\mathbf{u} = -\frac{1}{\sqrt{10}}\mathbf{i} - \frac{3}{\sqrt{10}}\mathbf{j}.$$

12. $f(x, y) = \dfrac{cx + dy}{x - y}$; $P(3, 4)$; $\mathbf{u} = \dfrac{4}{5}\mathbf{i} + \dfrac{3}{5}\mathbf{j}$.

In Exercises 13–18, find the directional derivative of f at P in the direction of $\mathbf{a}$.

13. $f(x, y) = 4x^3 y^2$; $P(2, 1)$; $\mathbf{a} = 4\mathbf{i} - 3\mathbf{j}$.

14. $f(x, y) = x^2 - 3xy + 4y^3$; $P(-2, 0)$; $\mathbf{a} = \mathbf{i} + 2\mathbf{j}$.

15. $f(x, y) = y^2 \ln x$; $P(1, 4)$; $\mathbf{a} = -3\mathbf{i} + 3\mathbf{j}$.

16. $f(x, y) = e^x \cos y$; $P(0, \pi/4)$; $\mathbf{a} = 5\mathbf{i} - 2\mathbf{j}$.

17. $f(x, y) = \tan^{-1}\left(\dfrac{y}{x}\right)$; $P(-2, 2)$; $\mathbf{a} = -\mathbf{i} - \mathbf{j}$.

18. $f(x, y) = xe^y - ye^x$; $P(0, 0)$; $\mathbf{a} = 5\mathbf{i} - 2\mathbf{j}$.

In Exercises 19–22, find the directional derivative of f at P in the direction of a vector making the angle θ with the positive x-axis.

19. $f(x, y) = \sqrt{xy}$; $P(1, 4)$; $\theta = \pi/3$.

20. $f(x, y) = \dfrac{x - y}{x + y}$; $P(-1, -2)$; $\theta = \pi/2$.

21. $f(x, y) = \tan (2x + y)$; $P(\pi/6, \pi/3)$; $\theta = 7\pi/4$.

22. $f(x, y) = \sinh x \cosh y$; $P(0, 0)$; $\theta = \pi$.

In Exercises 23–26, sketch the level curve of $f(x, y)$ that passes through P and draw the gradient vector at P.

23. $f(x, y) = 4x - 2y + 3$; $P(1, 2)$.

24. $f(x, y) = y/x^2$; $P(-2, 2)$.

25. $f(x, y) = x^2 + 4y^2$; $P(-2, 0)$.

26. $f(x, y) = x^2 - y^2$; $P(2, -1)$.

In Exercises 27–30, find a unit vector in the direction in which f increases most rapidly at P; and find the rate of change of f at P in that direction.

27. $f(x, y) = 4x^3 y^2$; $P(-1, 1)$.

28. $f(x, y) = 3x - \ln y$; $P(2, 4)$.

29. $f(x, y) = \sqrt{x^2 + y^2}$; $P(4, -3)$.

30. $f(x, y) = \dfrac{x}{x + y}$; $P(0, 2)$.

In Exercises 31–34, find a unit vector in the direction in which f decreases most rapidly at P; and find the rate of change of f at P in that direction.

31. $f(x, y) = 20 - x^2 - y^2$; $P(-1, -3)$.

32. $f(x, y) = e^{xy}$; $P(2, 3)$.

33. $f(x, y) = \cos(3x - y)$; $P\left(\dfrac{\pi}{6}, \dfrac{\pi}{4}\right)$.

34. $f(x, y) = \sqrt{\dfrac{x - y}{x + y}}$; $P(3, 1)$.

35. Find the directional derivative of $f(x, y) = \dfrac{x}{x + y}$ at $P(1, 0)$ in the direction to $Q(-1, -1)$.

36. Find the directional derivative of $f(x, y) = e^{-x} \sec y$ at $P(0, \pi/4)$ in the direction of the origin.

37. Find the directional derivative of the function $f(x, y) = \sqrt{xy}\, e^y$ at $P(1, 1)$ in the direction of the negative y-axis.

38. Let $f(x, y) = \dfrac{y}{x + y}$. Find a unit vector $\mathbf{u}$ for which $D_{\mathbf{u}} f(2, 3) = 0$.

39. Find a unit vector $\mathbf{u}$ that is perpendicular at $P(1, -2)$ to the level curve of $f(x, y) = 4x^2y$ through P.

40. Find a unit vector $\mathbf{u}$ that is perpendicular at $P(2, -3)$ to the level curve of $f(x, y) = 3x^2y - xy$ through P.

41. Given that $D_{\mathbf{u}} f(1, 2) = -5$ if $\mathbf{u} = \frac{3}{5}\mathbf{i} - \frac{4}{5}\mathbf{j}$ and $D_{\mathbf{v}} f(1, 2) = 10$ if $\mathbf{v} = \frac{4}{5}\mathbf{i} + \frac{3}{5}\mathbf{j}$, find
(a) $f_x(1, 2)$ (b) $f_y(1, 2)$
(c) the directional derivative of f at $(1, 2)$ in the direction of the origin.

42. Given that $f_x(-5, 1) = -3$ and $f_y(-5, 1) = 2$, find the directional derivative of f at $P(-5, 1)$ in the direction to $Q(-4, 3)$.

43. Given that $\nabla f(4, -5) = 2\mathbf{i} - \mathbf{j}$, find the directional derivative of f at $(4, -5)$ in the direction of $\mathbf{a} = 5\mathbf{i} + 2\mathbf{j}$.

44. Given that $\nabla f(x_0, y_0) = \mathbf{i} - 2\mathbf{j}$ and $D_{\mathbf{u}} f(x_0, y_0) = -2$, find $\mathbf{u}$.

45. Let $z = 3x^2 - y^2$. Find all points where $\|\nabla z\| = 6$.

46. Given that $z = 3x + y^2$, find the maximum value

of $d\|\nabla z\|/ds$ at the point $(5, 2)$ and a unit vector in the direction in which the maximum is attained.

47. A particle moves along a path C given by $x = t$ and $y = -t^2$. If $z = x^2 + y^2$, find dz/ds along C at the instant when the particle is at the point $(2, -4)$.

48. The temperature at a point (x, y) on a metal plate in the xy-plane is $T(x, y) = \dfrac{xy}{1 + x^2 + y^2}$ degrees Celsius.
(a) Find the rate of change of temperature at $(1, 1)$ in the direction of $\mathbf{a} = 2\mathbf{i} - \mathbf{j}$.
(b) An ant at $(1, 1)$ wants to walk in the direction in which the temperature drops most rapidly. Find a unit vector in that direction.

49. If the electric potential at a point (x, y) in the xy-plane is $V(x, y)$, then the **electric intensity vector** at (x, y) is $\mathbf{E} = -\nabla V(x, y)$. Suppose $V(x, y) = e^{-2x} \cos 2y$.
(a) Find the electric intensity vector at $(\pi/4, 0)$.
(b) Show that at each point in the plane, the electric potential decreases most rapidly in the direction of $\mathbf{E}$.

50. On a certain mountain, the elevation z above a point (x, y) in a horizontal xy-plane at sea level is $z = 2000 - 2x^2 - 4y^2$ ft. The positive x-axis points east, and the positive y-axis north. A climber is at the point $(-20, 5, 1100)$.
(a) If the climber uses a compass reading to walk due west, will he begin to ascend or descend?
(b) If the climber uses a compass reading to walk northeast, will he ascend or descend? At what rate?
(c) In what compass direction should the climber walk to travel a level path?

51. Let $r = \sqrt{x^2 + y^2}$.
(a) Show that $\nabla r = \dfrac{\mathbf{r}}{r}$, where $\mathbf{r} = x\mathbf{i} + y\mathbf{j}$.
(b) Show that $\nabla f(r) = f'(r)\nabla r = \dfrac{f'(r)}{r}\mathbf{r}$.

52. Use the formula of part (b) in Exercise 51 to find
(a) $\nabla f(r)$ if $f(r) = re^{-3r}$
(b) $f(r)$ if $\nabla f(r) = 3r^2\mathbf{r}$ and $f(2) = 1$.

53. Let $\mathbf{u}_r$ be a unit vector making an angle θ with the positive x-axis, and let $\mathbf{u}_\theta$ be a unit vector $90°$ counterclockwise from $\mathbf{u}_r$. Show that if $z = f(x, y)$, $x = r \cos \theta$, and $y = r \sin \theta$, then

$$\nabla z = \frac{\partial z}{\partial r} \mathbf{u}_r + \frac{1}{r} \frac{\partial z}{\partial \theta} \mathbf{u}_\theta$$

[*Hint:* Use parts (a) and (b) of Exercise 62, Section 16.4.]

54. Prove: If f and g are differentiable, then
(a) $\nabla(f + g) = \nabla f + \nabla g$
(b) $\nabla(cf) = c\nabla f$ (c constant)
(c) $\nabla(fg) = f\nabla g + g\nabla f$
(d) $\nabla\left(\dfrac{f}{g}\right) = \dfrac{g\nabla f - f\nabla g}{g^2}$
(e) $\nabla(f^p) = pf^{p-1}\nabla f$.

55. Prove: If $x = x(t)$ and $y = y(t)$ are differentiable at t, and if $z = f(x, y)$ is differentiable at the point $(x(t), y(t))$, then

$$\frac{dz}{dt} = \nabla z \cdot \mathbf{r}'(t)$$

where $\mathbf{r}(t) = x(t)\mathbf{i} + y(t)\mathbf{j}$.

56. Prove: If f, f_x, and f_y are continuous on a circular region, and if $\nabla f(x, y) = \mathbf{0}$ throughout the region, then $f(x, y)$ is constant on the region. [*Hint:* Exercise 70, Section 16.4.]

57. Prove: If $D_{\mathbf{u}} f(x, y) = 0$ in two nonparallel directions, then $D_{\mathbf{u}} f(x, y) = 0$ in all directions.

16.7 DIFFERENTIABILITY, DIRECTIONAL DERIVATIVES, AND GRADIENTS FOR FUNCTIONS OF THREE VARIABLES

In this section we shall extend most of the results obtained in the last two sections to functions of three variables. The main difference between functions of two and three variables is geometric: The graph of $z = f(x, y)$ represents a surface in 3-space, whereas $w = f(x, y, z)$ has no analogous interpretation.

DIFFERENTIABILITY

The definition of differentiability for functions of three variables is a direct generalization of Definition 16.4.1 for functions of two variables.

> **16.7.1** DEFINITION. A function f of three variables is defined to be **differentiable** at the point (x_0, y_0, z_0) if $f_x(x_0, y_0, z_0)$, $f_y(x_0, y_0, z_0)$, and $f_z(x_0, y_0, z_0)$ exist and
>
> $$\Delta f = f(x_0 + \Delta x, y_0 + \Delta y, z_0 + \Delta z) - f(x_0, y_0, z_0)$$
>
> can be written in the form
>
> $$\Delta f = f_x(x_0, y_0, z_0) \, \Delta x + f_y(x_0, y_0, z_0) \, \Delta y \\ + f_z(x_0, y_0, z_0) \, \Delta z + \epsilon_1 \, \Delta x + \epsilon_2 \, \Delta y + \epsilon_3 \, \Delta z$$
>
> where ϵ_1, ϵ_2, and ϵ_3 are functions of Δx, Δy, and Δz such that $\epsilon_1 \to 0$, $\epsilon_2 \to 0$, and $\epsilon_3 \to 0$ as $(\Delta x, \Delta y, \Delta z) \to (0, 0, 0)$.

The meaning of the terms *differentiable on a region R* and *differentiable* (i.e., *everywhere differentiable*) should be clear.

As in the cases of one and two variables, a function $f(x, y, z)$ that is differentiable at a point must also be continuous at that point. Moreover, the following generalization of Theorem 16.4.3, which we state without proof, gives simple conditions for $f(x, y, z)$ to be differentiable at a point.

16.7.2 THEOREM. *If f has first-order partial derivatives at each point of some spherical region centered at (x_0, y_0, z_0), and if these partial derivatives are continuous at (x_0, y_0, z_0), then f is differentiable at (x_0, y_0, z_0).*

The reader should have no trouble modifying the proof of Theorem 16.4.6 to obtain the following version of the chain rule.

Chain Rule

16.7.3 THEOREM. *If $x = x(t)$, $y = y(t)$, and $z = z(t)$ are differentiable at t and $w = f(x, y, z)$ is differentiable at the point $(x(t), y(t), z(t))$, then $w = f(x(t), y(t), z(t))$ is differentiable at t, and*

$$\frac{dw}{dt} = \frac{\partial w}{\partial x}\frac{dx}{dt} + \frac{\partial w}{\partial y}\frac{dy}{dt} + \frac{\partial w}{\partial z}\frac{dz}{dt}$$

Example 1 Suppose

$$w = x^3 y^2 z, \quad x = t^2, \quad y = t^3, \quad z = t^4$$

Use the chain rule to find dw/dt.

Solution. By the chain rule,

$$\frac{dw}{dt} = \frac{\partial w}{\partial x}\frac{dx}{dt} + \frac{\partial w}{\partial y}\frac{dy}{dt} + \frac{\partial w}{\partial z}\frac{dz}{dt}$$

$$= (3x^2 y^2 z)(2t) + (2x^3 yz)(3t^2) + (x^3 y^2)(4t^3)$$

$$= (3t^{14})(2t) + (2t^{13})(3t^2) + (t^{12})(4t^3) = 16t^{15} \quad \blacktriangleleft$$

Other variations of the chain rule for functions of three variables will be considered in the next section.

DIRECTIONAL DERIVATIVES

We shall now show how to define directional derivatives for functions of three variables. Suppose $w = f(x, y, z)$, where f is differentiable at the point $P_0(x_0, y_0, z_0)$. Let l be a line passing through $P_0(x_0, y_0, z_0)$ and

$$\mathbf{u} = \langle u_1, u_2, u_3 \rangle$$

a unit vector along l. Let s be the (signed) distance from $P_0(x_0, y_0, z_0)$ to

Figure 16.7.1

a point $P(x, y, z)$ on l, where s is nonnegative if P is in the direction of $\mathbf{u}$ from P_0 and negative otherwise (Figure 16.7.1).

Since $\overrightarrow{P_0P} = s\mathbf{u}$, it follows that l is given parametrically by

$$
\begin{aligned}
x &= x_0 + su_1 \\
y &= y_0 + su_2 \\
z &= z_0 + su_3
\end{aligned}
\tag{1}
$$

As the parameter s varies, the point $P(x, y, z)$ moves along l and the value of $w = f(x, y, z)$ changes. The reader should have no trouble modifying the derivation of Formula (3) of Section 16.6 to show that the instantaneous rate of change of w with s in the direction of $\mathbf{u}$ at (x_0, y_0, z_0) (i.e., when $s = 0$) is

$$
\left. \frac{dw}{ds} \right|_{s=0} = f_x(x_0, y_0, z_0)u_1 + f_y(x_0, y_0, z_0)u_2 + f_z(x_0, y_0, z_0)u_3
$$

This quantity is denoted by $D_\mathbf{u}w(x_0, y_0, z_0)$ or $D_\mathbf{u}f(x_0, y_0, z_0)$.

16.7.4 DEFINITION. If f is differentiable at (x_0, y_0, z_0), and if $\mathbf{u} = \langle u_1, u_2, u_3 \rangle$ is a unit vector, then the ***directional derivative*** of f at (x_0, y_0, z_0) in the direction of $\mathbf{u}$ is defined by

$$
D_\mathbf{u}f(x_0, y_0, z_0) = f_x(x_0, y_0, z_0)u_1 + f_y(x_0, y_0, z_0)u_2 + f_z(x_0, y_0, z_0)u_3 \tag{2}
$$

GRADIENT

The directional derivative $D_\mathbf{u}f(x_0, y_0, z_0)$ represents the instantaneous rate at which $w = f(x, y, z)$ changes with distance as (x, y, z) moves from (x_0, y_0, z_0) in the direction of the unit vector $\mathbf{u}$.

Note the similarity between the formula for $D_\mathbf{u}f(x_0, y_0, z_0)$ and the formula for $D_\mathbf{u}f(x_0, y_0)$ in Definition 16.6.1. If we now define the ***gradient*** of f by

$$
\nabla f(x, y, z) = f_x(x, y, z)\mathbf{i} + f_y(x, y, z)\mathbf{j} + f_z(x, y, z)\mathbf{k} \tag{3}
$$

then the directional derivative at (x, y, z),

$$
D_\mathbf{u}f(x, y, z) = f_x(x, y, z)u_1 + f_y(x, y, z)u_2 + f_z(x, y, z)u_3
$$

can be expressed in the form

$$
D_\mathbf{u}f(x, y, z) = \nabla f(x, y, z) \cdot \mathbf{u} \tag{4}
$$

which parallels (9) of Section 16.6. Using this result, the reader should have no trouble proving the following extension of Theorem 16.6.3.

16.7.5   THEOREM. *Let f be a function of three variables that is differentiable at (x_0, y_0, z_0).*

(a) *If $\nabla f(x_0, y_0, z_0) = \mathbf{0}$, then all directional derivatives of f at (x_0, y_0, z_0) are zero.*

(b) *If $\nabla f(x_0, y_0, z_0) \neq \mathbf{0}$, then among all possible directional derivatives of f at (x_0, y_0, z_0), the derivative in the direction of $\nabla f(x_0, y_0, z_0)$ has the largest value. The value of that directional derivative is $\|\nabla f(x_0, y_0, z_0)\|$.*

(c) *If $\nabla f(x_0, y_0, z_0) \neq \mathbf{0}$, then among all possible directional derivatives of f at (x_0, y_0, z_0), the derivative in the direction opposite to that of $\nabla f(x_0, y_0, z_0)$ has the smallest value. The value of that directional derivative is $-\|\nabla f(x_0, y_0, z_0)\|$.*

Example 2   Find the directional derivative of $f(x, y, z) = x^2 y - yz^3 + z$ at the point $P(1, -2, 0)$ in the direction of the vector $\mathbf{a} = 2\mathbf{i} + \mathbf{j} - 2\mathbf{k}$, and find the maximum rate of increase of f at P.

Solution.   Since

$$f_x(x, y, z) = 2xy, \quad f_y(x, y, z) = x^2 - z^3, \quad f_z(x, y, z) = -3yz^2 + 1$$

it follows that

$$\nabla f(x, y, z) = 2xy\mathbf{i} + (x^2 - z^3)\mathbf{j} + (-3yz^2 + 1)\mathbf{k}$$
$$\nabla f(1, -2, 0) = -4\mathbf{i} + \mathbf{j} + \mathbf{k}$$

A unit vector in the direction of $\mathbf{a}$ is

$$\mathbf{u} = \frac{\mathbf{a}}{\|\mathbf{a}\|} = \frac{1}{\sqrt{9}}(2\mathbf{i} + \mathbf{j} - 2\mathbf{k}) = \tfrac{2}{3}\mathbf{i} + \tfrac{1}{3}\mathbf{j} - \tfrac{2}{3}\mathbf{k}$$

Therefore

$$D_\mathbf{u} f(1, -2, 0) = \nabla f(1, -2, 0) \cdot \mathbf{u} = (-4)(\tfrac{2}{3}) + (1)(\tfrac{1}{3}) + (1)(-\tfrac{2}{3}) = -3$$

The maximum rate of increase of f at P is

$$\|\nabla f(1, -2, 0)\| = \sqrt{(-4)^2 + (1)^2 + (1)^2} = 3\sqrt{2} \qquad \blacktriangleleft$$

If $P_0(x_0, y_0, z_0)$ is a point in the domain of the function f and $f(x_0, y_0, z_0) = c$, then the equation

$$f(x, y, z) = c$$

defines a level surface of f that passes through the point $P_0(x_0, y_0, z_0)$. The following theorem states that the gradient of f at P_0 is perpendicular

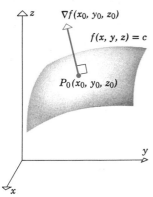

Figure 16.7.2

to this surface at P_0 (Figure 16.7.2). (Recall that for functions of two variables, the gradient at a point is perpendicular to the level *curve* through the point.)

16.7.6   THEOREM. *If f is differentiable at (x_0, y_0, z_0) and $\nabla f(x_0, y_0, z_0) \neq \mathbf{0}$, then $\nabla f(x_0, y_0, z_0)$ is perpendicular to the level surface of $f(x, y, z)$ through (x_0, y_0, z_0).*

We omit the proof.

Example 3   Find an equation of the tangent plane to the ellipsoid $x^2 + 4y^2 + z^2 = 18$ at the point $(1, 2, -1)$.

Solution.   If we let $F(x, y, z) = x^2 + 4y^2 + z^2$, then the given equation has the form $F(x, y, z) = 18$, which may be viewed as the equation of a level surface for F. Thus, the vector $\nabla F(1, 2, -1)$ is normal to the ellipsoid at the point $(1, 2, -1)$. To find this vector we write

$$\nabla F(x, y, z) = \frac{\partial F}{\partial x}\mathbf{i} + \frac{\partial F}{\partial y}\mathbf{j} + \frac{\partial F}{\partial z}\mathbf{k} = 2x\mathbf{i} + 8y\mathbf{j} + 2z\mathbf{k}$$

$$\nabla F(1, 2, -1) = 2\mathbf{i} + 16\mathbf{j} - 2\mathbf{k}$$

Using this normal and the point $(1, 2, -1)$, we obtain as the equation of the tangent plane

$$2(x - 1) + 16(y - 2) - 2(z + 1) = 0$$

or

$$x + 8y - z = 18$$

Alternative Solution.   Solving the given equation for z, we obtain $z = \pm\sqrt{18 - x^2 - 4y^2}$. Since the point $(1, 2, -1)$ has a negative z-coordinate, this point lies on the lower half of the ellipsoid, which is given by

$$z = -\sqrt{18 - x^2 - 4y^2}$$

Using this expression and Theorem 16.5.1, we can obtain the equation of the tangent plane as follows:

$$\frac{\partial z}{\partial x} = \frac{x}{\sqrt{18 - x^2 - 4y^2}}, \quad \frac{\partial z}{\partial y} = \frac{4y}{\sqrt{18 - x^2 - 4y^2}}$$

so that

$$\left.\frac{\partial z}{\partial x}\right|_{x=1,\, y=2} = 1, \quad \left.\frac{\partial z}{\partial y}\right|_{x=1,\, y=2} = 8$$

Thus, from Formula (1) of Section 16.5 the tangent plane at $(1, 2, -1)$ is given by

$$(1)(x - 1) + 8(y - 2) - (z + 1) = 0$$

or

$$x + 8y - z = 18$$

which agrees with our previous solution.   ◀

REMARK.   The first method in the last example is useful when it is inconvenient or impossible to express z explicitly as a function of x and y.

TOTAL
DIFFERENTIALS

If $w = f(x, y, z)$, then we define the *increment* Δw (also written Δf) to be

$$\Delta w = f(x + \Delta x, y + \Delta y, z + \Delta z) - f(x, y, z)$$

and we define the ***total differential*** dw (also written df) to be

$$dw = f_x(x, y, z)\, dx + f_y(x, y, z)\, dy + f_z(x, y, z)\, dz$$

where Δx, Δy, Δz, dx, dy, and dz are all variables representing changes in the values of x, y, and z.

The increment Δw represents the change in the value of $w = f(x, y, z)$ when x, y, and z are changed by amounts Δx, Δy, and Δz, respectively. However, for functions of three variables, the total differential dw has no natural geometric interpretation.

If we let

$$dx = \Delta x, \quad dy = \Delta y, \quad dz = \Delta z$$

and if we assume that $w = f(x, y, z)$ is differentiable at (x, y, z), then it follows from Definition 16.7.1 that

$$\Delta w = dw + \epsilon_1 \Delta x + \epsilon_2 \Delta y + \epsilon_3 \Delta z \tag{5}$$

where $\epsilon_1 \to 0$, $\epsilon_2 \to 0$, $\epsilon_3 \to 0$ as $(\Delta x, \Delta y, \Delta z) \to (0, 0, 0)$. Thus, when Δx, Δy, and Δz are small, it follows from (5) that

$$\Delta w \approx dw$$

Example 4   The length, width, and height of a rectangular box are each measured with an error of at most 5%. Estimate the maximum possible percentage error that results if these quantities are used to calculate the diagonal of the box.

Solution.   Let x, y, z, and D be the true length, width, height, and diagonal of the box, respectively; and let Δx, Δy, Δz, and ΔD be the errors in these quantities. We are given that

$$\left| \frac{\Delta x}{x} \right| \leq 0.05, \quad \left| \frac{\Delta y}{y} \right| \leq 0.05, \quad \left| \frac{\Delta z}{z} \right| \leq 0.05$$

We want to estimate $|\Delta D/D|$. Since the diagonal D is related to the length, width, and height by

$$D = \sqrt{x^2 + y^2 + z^2}$$

it follows that

$$dD = \frac{\partial D}{\partial x} dx + \frac{\partial D}{\partial y} dy + \frac{\partial D}{\partial z} dz$$

$$= \frac{x}{\sqrt{x^2 + y^2 + z^2}} dx + \frac{y}{\sqrt{x^2 + y^2 + z^2}} dy + \frac{z}{\sqrt{x^2 + y^2 + z^2}} dz$$

If we choose $\Delta x = dx$, $\Delta y = dy$, $\Delta z = dz$, then we can use the approximation $\Delta D/D \approx dD/D$. But,

$$\frac{dD}{D} = \frac{x}{x^2 + y^2 + z^2} dx + \frac{y}{x^2 + y^2 + z^2} dy + \frac{z}{x^2 + y^2 + z^2} dz$$

or

$$\frac{dD}{D} = \frac{x^2}{x^2 + y^2 + z^2} \frac{dx}{x} + \frac{y^2}{x^2 + y^2 + z^2} \frac{dy}{y} + \frac{z^2}{x^2 + y^2 + z^2} \frac{dz}{z}$$

Thus,

$$\left| \frac{dD}{D} \right| = \left| \frac{x^2}{x^2 + y^2 + z^2} \frac{dx}{x} + \frac{y^2}{x^2 + y^2 + z^2} \frac{dy}{y} + \frac{z^2}{x^2 + y^2 + z^2} \frac{dz}{z} \right|$$

$$\leq \left| \frac{x^2}{x^2 + y^2 + z^2} \frac{dx}{x} \right| + \left| \frac{y^2}{x^2 + y^2 + z^2} \frac{dy}{y} \right| + \left| \frac{z^2}{x^2 + y^2 + z^2} \frac{dz}{z} \right|$$

$$\leq \frac{x^2}{x^2 + y^2 + z^2} (0.05) + \frac{y^2}{x^2 + y^2 + z^2} (0.05) + \frac{z^2}{x^2 + y^2 + z^2} (0.05)$$

$$= 0.05$$

Therefore, the maximum percentage error in D is approximately 5%.   ◀

▶ Exercise Set 16.7

In Exercises 1–4, find dw/dt using the chain rule.

1. $w = 5x^2y^3z^4$; $x = t^2$, $y = t^3$, $z = t^5$.

2. $w = \ln(3x^2 - 2y + 4z^3)$; $x = t^{1/2}$, $y = t^{2/3}$, $z = t^{-2}$.

3. $w = 5\cos xy - \sin xz$; $x = 1/t$, $y = t$, $z = t^3$.

4. $w = \sqrt{1 + x - 2yz^4x}$; $x = \ln t$, $y = t$, $z = 4t$.

5. Use the chain rule to find $\left.\dfrac{dw}{dt}\right|_{t=1}$ if $w = x^3y^2z^4$; $x = t^2$, $y = t + 2$, $z = 2t^4$.

6. Use the chain rule to find $\left.\dfrac{dw}{dt}\right|_{t=0}$ if $w = x\sin yz^2$; $x = \cos t$, $y = t^2$, $z = e^t$.

In Exercises 7–10, find the gradient of f at P, and then use the gradient to calculate $D_{\mathbf{u}}f$ at P.

7. $f(x, y, z) = 4x^5y^2z^3$; $P(2, -1, 1)$;
$\mathbf{u} = \dfrac{1}{3}\mathbf{i} + \dfrac{2}{3}\mathbf{j} - \dfrac{2}{3}\mathbf{k}$.

8. $f(x, y, z) = ye^{xz} + z^2$; $P(0, 2, 3)$;
$\mathbf{u} = \dfrac{2}{7}\mathbf{i} - \dfrac{3}{7}\mathbf{j} + \dfrac{6}{7}\mathbf{k}$.

9. $f(x, y, z) = \ln(x^2 + 2y^2 + 3z^2)$; $P(-1, 2, 4)$;
$\mathbf{u} = -\dfrac{3}{13}\mathbf{i} - \dfrac{4}{13}\mathbf{j} - \dfrac{12}{13}\mathbf{k}$.

10. $f(x, y, z) = \sin xyz$; $P(\frac{1}{2}, \frac{1}{3}, \pi)$;
$\mathbf{u} = \dfrac{1}{\sqrt{3}}\mathbf{i} - \dfrac{1}{\sqrt{3}}\mathbf{j} + \dfrac{1}{\sqrt{3}}\mathbf{k}$.

In Exercises 11–14, find the directional derivative of f at P in the direction of $\mathbf{a}$.

11. $f(x, y, z) = x^3z - yx^2 + z^2$; $P(2, -1, 1)$;
$\mathbf{a} = 3\mathbf{i} - \mathbf{j} + 2\mathbf{k}$.

12. $f(x, y, z) = y - \sqrt{x^2 + z^2}$; $P(-3, 1, 4)$;
$\mathbf{a} = 2\mathbf{i} - 2\mathbf{j} - \mathbf{k}$.

13. $f(x, y, z) = \dfrac{z - x}{z + y}$; $P(1, 0, -3)$;
$\mathbf{a} = -6\mathbf{i} + 3\mathbf{j} - 2\mathbf{k}$.

14. $f(x, y, z) = e^{x+y+3z}$; $P(-2, 2, -1)$;
$\mathbf{a} = 20\mathbf{i} - 4\mathbf{j} + 5\mathbf{k}$.

In Exercises 15–18, find a unit vector in the direction in which f increases most rapidly at P, and find the rate of increase of f in that direction.

15. $f(x, y, z) = x^3z^2 + y^3z + z - 1$; $P(1, 1, -1)$.

16. $f(x, y, z) = \sqrt{x - 3y + 4z}$; $P(0, -3, 0)$.

17. $f(x, y, z) = \dfrac{x}{z} + \dfrac{z}{y^2}$; $P(1, 2, -2)$.

18. $f(x, y, z) = \tan^{-1}\left(\dfrac{x}{y + z}\right)$; $P(4, 2, 2)$.

In Exercises 19 and 20, find a unit vector in the direction in which f decreases most rapidly at P, and find the rate of change of f in that direction.

19. $f(x, y, z) = \dfrac{x + z}{z - y}$; $P(5, 7, 6)$.

20. $f(x, y, z) = 4e^{xy}\cos z$; $P(0, 1, \pi/4)$.

21. Find the directional derivative of the function
$f(x, y, z) = \dfrac{y}{x + z}$ at $P(2, 1, -1)$ in the direction
from P to $Q(-1, 2, 0)$.

22. Find the directional derivative of the function
$f(x, y, z) = x^3y^2z^5 - 2xz + yz + 3x$ at $P(-1, -2, 1)$
in the direction of the negative z-axis.

23. Given that the directional derivative of $f(x, y, z)$
at the point $(3, -2, 1)$ in the direction of
$\mathbf{a} = 2\mathbf{i} - \mathbf{j} - 2\mathbf{k}$ is -5 and that $\|\nabla f(3, -2, 1)\| = 5$,
find $\nabla f(3, -2, 1)$.

24. The temperature (in degrees Celsius) at a point
(x, y, z) in a metal solid is

$$T(x, y, z) = \dfrac{xyz}{1 + x^2 + y^2 + z^2}$$

(a) Find the rate of change of temperature at
$(1, 1, 1)$ in the direction of the origin.

(b) Find the direction in which the temperature
rises most rapidly at $(1, 1, 1)$. (Express your
answer as a unit vector.)

(c) Find the rate at which the temperature rises
moving from $(1, 1, 1)$ in the direction obtained in (b).

In Exercises 25–28, find equations for the tangent plane
and the line that is normal to the given surface at the
point P.

25. $x^2 + y^2 + z^2 = 49$; $P(-3, 2, -6)$.

26. $xz - yz^3 + yz^2 = 2$; $P(2, -1, 1)$.

27. $\sqrt{\dfrac{z + x}{y - 1}} = z^2$; $P(3, 5, 1)$.

28. $\sin xz - 4 \cos yz = 4$; $P(\pi, \pi, 1)$.

29. Show that every line that is normal to the sphere $x^2 + y^2 + z^2 = 1$ passes through the origin.

30. Find all points on the ellipsoid $2x^2 + 3y^2 + 4z^2 = 9$ at which the tangent plane is parallel to the plane $x - 2y + 3z = 5$.

31. Find all points on the surface $x^2 + y^2 - z^2 = 1$ at which the normal line is parallel to the line through $P(1, -2, 1)$ and $Q(4, 0, -1)$.

32. Show that the ellipsoid $2x^2 + 3y^2 + z^2 = 9$ and the sphere

$$x^2 + y^2 + z^2 - 6x - 8y - 8z + 24 = 0$$

have a common tangent plane at the point $(1, 1, 2)$.

In Exercises 33–36, find dw.

33. $w = 8x - 3y + 4z$.

34. $w = 4x^2y^3z^7 - 3xy + z + 5$.

35. $w = \tan^{-1}(xyz)$.

36. $w = \sqrt{x} + \sqrt{y} + \sqrt{z}$.

37. Use a total differential to approximate the change in $f(x, y, z) = 2xy^2z^3$ as (x, y, z) varies from $P(1, -1, 2)$ to $Q(0.99, -1.02, 2.02)$.

38. Use a total differential to approximate the change in $f(x, y, z) = xyz/(x + y + z)$ as (x, y, z) varies from $P(-1, -2, 4)$ to $Q(-1.04, -1.98, 3.97)$.

39. The length, width, and height of a rectangular box are measured to be 3 cm, 4 cm, and 5 cm, respectively, with a maximum error of 0.05 cm in each measurement. Use differentials to approximate the maximum error in the calculated volume.

40. The total resistance R of three resistances R_1, R_2, and R_3, connected in parallel, is given by

$$\frac{1}{R} = \frac{1}{R_1} + \frac{1}{R_2} + \frac{1}{R_3}$$

Suppose that R_1, R_2, and R_3 are measured to be 100 ohms, 200 ohms, and 500 ohms, respectively, with a maximum error of 10% in each. Use differentials to approximate the maximum percentage error in the calculated value of R.

41. The area of a triangle is to be computed from the formula $A = \frac{1}{2}ab \sin \theta$, where a and b are the lengths of two sides and θ is the included angle. Suppose that a, b, and θ are measured to be 40 ft, 50 ft, and 30°, respectively. Use differentials to approximate the maximum error in the calculated value of A if the maximum errors in a, b, and θ are $\frac{1}{2}$ ft, $\frac{1}{4}$ ft, and 2°, respectively.

42. The length, width, and height of a rectangular box are measured with errors of at most $r\%$. Use differentials to approximate the maximum percentage error in the computed value of the volume.

43. Use differentials to approximate the maximum percentage error in $w = xy^2z^3$ if x, y, and z have errors of at most 1%, 2%, and 3%, respectively.

44. Two surfaces are said to be **orthogonal** at a point of intersection if their normal lines are perpendicular at that point. Prove that the surfaces $f(x, y, z) = 0$ and $g(x, y, z) = 0$ are orthogonal at a point of intersection, (x_0, y_0, z_0), if and only if

$$f_x g_x + f_y g_y + f_z g_z = 0$$

at (x_0, y_0, z_0). (Assume that $\nabla f(x_0, y_0, z_0) \neq \mathbf{0}$ and $\nabla g(x_0, y_0, z_0) \neq \mathbf{0}$.)

45. Use the result of Exercise 44 to show that the sphere $x^2 + y^2 + z^2 = a^2$ and the cone $z^2 = x^2 + y^2$ are orthogonal at every point of intersection.

16.8   FUNCTIONS OF n VARIABLES; MORE ON THE CHAIN RULE

In this section we shall discuss functions involving more than three variables. Our main objective is to develop forms of the chain rule for such functions.

All the definitions and theorems we have stated for functions of two and three variables can be extended to functions of four or more variables. Recall that if

$$w = f(v_1, v_2, \ldots, v_n)$$

is a function of n variables, then there are n partial derivatives

$$\frac{\partial w}{\partial v_1}, \frac{\partial w}{\partial v_2}, \ldots, \frac{\partial w}{\partial v_n}$$

each of which is calculated by holding $n - 1$ of the variables fixed and differentiating with respect to the remaining variable. In order to define directional derivatives for a function of n variables, it is first necessary to define the notion of a vector in "n-dimensional space." This topic is studied in a branch of mathematics called *linear algebra,* and will not be considered in this text.

TOTAL DIFFERENTIALS   If $w = f(v_1, v_2, \ldots, v_n)$, we define the *increment* Δw and the *total differential* dw by

$$\Delta w = f(v_1 + \Delta v_1, v_2 + \Delta v_2, \ldots, v_n + \Delta v_n) - f(v_1, v_2, \ldots, v_n) \quad (1)$$

and

$$dw = \frac{\partial w}{\partial v_1} dv_1 + \frac{\partial w}{\partial v_2} dv_2 + \cdots + \frac{\partial w}{\partial v_n} dv_n \quad (2)$$

where $\Delta v_1, \Delta v_2, \ldots, \Delta v_n$ and $dv_1, dv_2, \ldots, dv_n$ are variables representing changes in the values of $v_1, v_2, \ldots, v_n$.

If $v_1, v_2, \ldots, v_n$ are functions of a single variable t, then $w = f(v_1, v_2, \ldots, v_n)$ is a function of t, and a chain rule formula for dw/dt is

$$\frac{dw}{dt} = \frac{\partial w}{\partial v_1} \frac{dv_1}{dt} + \frac{\partial w}{\partial v_2} \frac{dv_2}{dt} + \cdots + \frac{\partial w}{\partial v_n} \frac{dv_n}{dt} \quad (3)$$

which is a natural extension of the chain rule formulas in Theorems 16.4.6 and 16.7.3. Observe that (3) results if we formally divide both sides of (2) by dt.

CHAIN RULES   Other forms of the chain rule arise, depending on the number of variables involved. For example, in Theorem 16.4.7 we obtained the chain rule formulas

$$\frac{\partial z}{\partial u} = \frac{\partial z}{\partial x} \frac{\partial x}{\partial u} + \frac{\partial z}{\partial y} \frac{\partial y}{\partial u} \quad (4)$$

$$\frac{\partial z}{\partial v} = \frac{\partial z}{\partial x} \frac{\partial x}{\partial v} + \frac{\partial z}{\partial y} \frac{\partial y}{\partial v} \quad (5)$$

for the case where z is a function of two variables $z = f(x, y)$; and x and y in turn are functions of two other variables, $x = x(u, v)$, $y = y(u, v)$.

Formulas (4) and (5) can be represented schematically by a "tree diagram" constructed as follows (Figure 16.8.1). Starting with z at the top of the diagram and moving downward, join each variable by lines (or branches) to those variables that depend *directly* on it. Thus, z is joined to x and y and these in turn are each joined to u and v. Next, label each branch with a derivative whose "numerator" contains the variable at the top end of that branch, and whose "denominator" contains the variable at the bottom end of that branch. This completes the "tree." To find the formula for $\partial z/\partial u$ trace all paths through the tree that start at z and end at u. Each such path produces one of the terms in the formula for $\partial z/\partial u$ (Figure 16.8.1a). Similarly, each term in the formula for $\partial z/\partial v$ corresponds to a path starting at z and ending at v (Figure 16.8.1b).

The following examples illustrate how tree diagrams can be used to construct other forms of the chain rule.

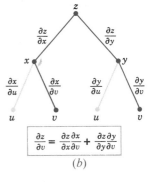

Figure 16.8.1 ▷ (a) (b)

Example 1   Suppose

$$w = e^{xyz}, \quad x = 3r + s, \quad y = 3r - s, \quad z = r^2 s$$

Use appropriate forms of the chain rule to find $\partial w/\partial r$ and $\partial w/\partial s$.

Solution.   From the tree diagram

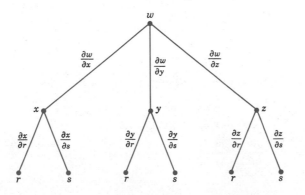

we obtain the formulas

$$\frac{\partial w}{\partial r} = \frac{\partial w}{\partial x}\frac{\partial x}{\partial r} + \frac{\partial w}{\partial y}\frac{\partial y}{\partial r} + \frac{\partial w}{\partial z}\frac{\partial z}{\partial r}$$

$$\frac{\partial w}{\partial s} = \frac{\partial w}{\partial x}\frac{\partial x}{\partial s} + \frac{\partial w}{\partial y}\frac{\partial y}{\partial s} + \frac{\partial w}{\partial z}\frac{\partial z}{\partial s}$$

from which we obtain

$$\frac{\partial w}{\partial r} = yze^{xyz}(3) + xze^{xyz}(3) + xye^{xyz}(2rs) = e^{xyz}(3yz + 3xz + 2xyrs)$$

and

$$\frac{\partial w}{\partial s} = yze^{xyz}(1) + xze^{xyz}(-1) + xye^{xyz}(r^2) = e^{xyz}(yz - xz + xyr^2)$$

If desired, we can express $\partial w/\partial r$ and $\partial w/\partial s$ in terms of r and s alone by replacing x, y, and z by their expressions in terms of r and s. ◄

Example 2 Suppose $w = x^2 + y^2 - z^2$ and

$$x = \rho \sin\phi \cos\theta, \quad y = \rho \sin\phi \sin\theta, \quad z = \rho \cos\phi$$

Use appropriate forms of the chain rule to find $\partial w/\partial\rho$ and $\partial w/\partial\theta$.

Solution. From the tree diagram

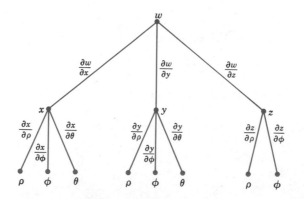

we obtain the formulas

$$\frac{\partial w}{\partial \rho} = \frac{\partial w}{\partial x}\frac{\partial x}{\partial \rho} + \frac{\partial w}{\partial y}\frac{\partial y}{\partial \rho} + \frac{\partial w}{\partial z}\frac{\partial z}{\partial \rho}$$

$$\frac{\partial w}{\partial \theta} = \frac{\partial w}{\partial x}\frac{\partial x}{\partial \theta} + \frac{\partial w}{\partial y}\frac{\partial y}{\partial \theta}$$

from which we obtain

$$\frac{\partial w}{\partial \rho} = (2x) \sin \phi \cos \theta + 2y \sin \phi \sin \theta - 2z \cos \phi$$

$$= 2\rho \sin^2 \phi \cos^2 \theta + 2\rho \sin^2 \phi \sin^2 \theta - 2\rho \cos^2 \phi$$
$$= 2\rho \sin^2 \phi (\cos^2 \theta + \sin^2 \theta) - 2\rho \cos^2 \phi$$
$$= 2\rho (\sin^2 \phi - \cos^2 \phi)$$
$$= -2\rho \cos 2\phi$$

$$\frac{\partial w}{\partial \theta} = (2x)(-\rho \sin \phi \sin \theta) + (2y) \rho \sin \phi \cos \theta$$

$$= -2\rho^2 \sin^2 \phi \sin \theta \cos \theta + 2\rho^2 \sin^2 \phi \sin \theta \cos \theta$$
$$= 0$$

This result is explained by the fact that w does not vary with θ. We may see this directly by expressing w in terms of ρ, ϕ, and θ. If this is done, the expressions involving θ will cancel, leaving w as a function of ρ and ϕ alone. (Verify.) ◀

In many applications of the chain rule, some of the variables in the function $w = f(v_1, v_2, \ldots, v_n)$ are functions of the remaining variables. Tree diagrams are especially helpful in such situations.

Example 3 Suppose

$$w = xy + yz, \quad y = \sin x, \quad z = e^x$$

Use an appropriate form of the chain rule to find dw/dx.

Solution. From the tree diagram

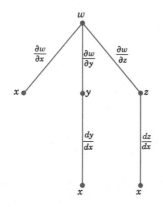

we obtain the formula

$$\frac{dw}{dx} = \frac{\partial w}{\partial x} + \frac{\partial w}{\partial y}\frac{dy}{dx} + \frac{\partial w}{\partial z}\frac{dz}{dx}$$

from which it follows that

$$\frac{dw}{dx} = y + (x + z)\cos x + ye^x$$

$$= \sin x + (x + e^x)\cos x + e^x \sin x$$

This same result can be obtained by first expressing w explicitly in terms of x,

$$w = x \sin x + e^x \sin x$$

and then differentiating with respect to x; however, such direct substitution is not always convenient.   ◄

REMARK.   Unlike the differential dz, a partial symbol ∂z has no meaning of its own. For example, if we were to "cancel" partial symbols in the chain rule formula

$$\frac{\partial z}{\partial u} = \frac{\partial z}{\partial x}\frac{\partial x}{\partial u} + \frac{\partial z}{\partial y}\frac{\partial y}{\partial u}$$

we would obtain

$$\frac{\partial z}{\partial u} = \frac{\partial z}{\partial u} + \frac{\partial z}{\partial u}$$

which is nonsense in cases where $\partial z/\partial u \neq 0$.

In each of the expressions

$$z = \sin xy, \quad z = \frac{xy}{1 + xy}, \quad z = e^{xy}$$

the independent variables occur only in the combination xy, so the substitution $t = xy$ reduces the expression to a function of one variable

$$z = \sin t, \quad z = \frac{t}{1 + t}, \quad z = e^t$$

Conversely, if we begin with a function of one variable $z = f(t)$ and substitute $t = xy$, we obtain a function $z = f(xy)$ in which the variables appear only in the combination xy. Functions whose variables occur in fixed combinations arise frequently in applications.

Example 4   Show that a function of the form $z = f(xy)$ satisfies the equation

$$x\frac{\partial z}{\partial x} - y\frac{\partial z}{\partial y} = 0$$

(assuming the derivatives exist).

Solution.   Let $t = xy$, so that $z = f(t)$. From the tree diagram

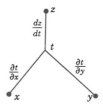

we obtain

$$\frac{\partial z}{\partial x} = \frac{dz}{dt}\frac{\partial t}{\partial x} = y\frac{dz}{dt}$$

$$\frac{\partial z}{\partial y} = \frac{dz}{dt}\frac{\partial t}{\partial y} = x\frac{dz}{dt}$$

Thus,

$$x\frac{\partial z}{\partial x} - y\frac{\partial z}{\partial y} = xy\frac{dz}{dt} - yx\frac{dz}{dt} = 0 \quad \blacktriangleleft$$

▶ Exercise Set 16.8

1.   Let $f(v, w, x, y) = 4v^2w^3x^4y^5$. Find $\partial f/\partial v$, $\partial f/\partial w$, $\partial f/\partial x$, and $\partial f/\partial y$.

2.   Let $w = r\cos st + e^u \sin ur$. Find $\partial w/\partial r$, $\partial w/\partial s$, $\partial w/\partial t$, $\partial w/\partial u$.

3.   Let $f(v_1, v_2, v_3, v_4) = \dfrac{v_1^2 - v_2^2}{v_3^2 + v_4^2}$. Find $\partial f/\partial v_1$, $\partial f/\partial v_2$, $\partial f/\partial v_3$, $\partial f/\partial v_4$.

4.   Let $V = xe^{2x-y} + we^{zw} + yw$. Find V_x, V_y, V_z, and V_w.

5.   Let $f(v, w, x, y) = 2v^{1/2}w^4x^{1/2}y^{2/3}$. Find $f_v(1, -2, 4, 8)$, $f_w(1, -2, 4, 8)$, $f_x(1, -2, 4, 8)$, and $f_y(1, -2, 4, 8)$.

6.   Let $u(w, x, y, z) = xe^{yw} \sin^2 z$. Find

(a) $\dfrac{\partial u}{\partial x}(0, 0, 1, \pi)$

(b) $\dfrac{\partial u}{\partial y}(0, 0, 1, \pi)$

(c) $\dfrac{\partial u}{\partial w}(0, 0, 1, \pi)$

(d) $\dfrac{\partial u}{\partial z}(0, 0, 1, \pi)$

(e) $\dfrac{\partial^4 u}{\partial x\, \partial y\, \partial w\, \partial z}$

(f) $\dfrac{\partial^4 u}{\partial w\, \partial z\, \partial y^2}$.

In Exercises 7–13, use appropriate forms of the chain rule to find the derivatives.

7.   Let $v = 7w^2x^3y^4z^5$, where $w = t^4$, $x = t^3$, $y = t^2$, $z = t$. Find dv/dt.

8. Let $z = u^7$, where $u = 5x^2 - 2y^3$. Find $\partial z/\partial x$ and $\partial z/\partial y$.

9. Let $z = \ln(x^2 + 1)$, where $x = r\cos\theta$. Find $\partial z/\partial r$ and $\partial z/\partial\theta$.

10. Let $u = rs^2 \ln t$, $r = x^2$, $s = 4y + 1$, $t = xy^3$. Find $\partial u/\partial x$ and $\partial u/\partial y$.

11. Let $w = 4x^2 + 4y^2 + z^2$, $x = \rho\sin\phi\cos\theta$, $y = \rho\sin\phi\sin\theta$, $z = \rho\cos\phi$. Find $\partial w/\partial\rho$, $\partial w/\partial\phi$, and $\partial w/\partial\theta$.

12. Let $w = 3xy^2z^3$, $y = 3x^2 + 2$, $z = \sqrt{x - 1}$. Find dw/dx.

13. Let $w = \sqrt{x^2 + y^2 + z^2}$, $x = \cos 2y$, $z = \sqrt{y}$. Find dw/dy.

14. The length, width, and height of a rectangular box are increasing at rates of 1 in./sec, 2 in./sec and 3 in./sec, respectively.
    (a) At what rate is the volume increasing when the length is 2 in., the width is 3 in., and the height is 6 in.?
    (b) At what rate is the diagonal increasing at that instant?

15. Angle A of triangle ABC is increasing at a rate of $\pi/60$ radians/sec, side AB is increasing at a rate of 2 cm/sec, and side AC is increasing at a rate of 4 cm/sec. At what rate is the length of BC changing when angle A is $\pi/3$ radians, $AB = 20$ cm and $AC = 10$ cm? Is the length of BC increasing or decreasing? [*Hint:* Use the law of cosines.]

16. The area A of a triangle is given by $A = \frac{1}{2}ab\sin\theta$, where a and b are the lengths of two sides and θ is the angle between these sides. Suppose $a = 5$, $b = 10$, and $\theta = \pi/3$. Find
    (a) the rate at which A changes with a if b and θ are held constant
    (b) the rate at which A changes with θ if a and b are held constant
    (c) the rate at which b changes with a if A and θ are held constant.

17. Suppose $x^2 + 4xz + z^2 - 3yz + 5 = 0$. Find $\partial z/\partial x$ and $\partial z/\partial y$ by implicit differentiation.

18. Suppose $e^{xy}\cos yz - e^{yz}\sin xz + 2 = 0$. Find $\partial z/\partial x$ and $\partial z/\partial y$ by implicit differentiation.

19. Let $f(w, x, y, z) = wz\tan^{-1}\dfrac{x}{y} + 5w$. Show that

$$f_{ww} + f_{xx} + f_{yy} + f_{zz} = 0$$

In the remaining exercises, you may assume that all derivatives mentioned exist.

20. Let f be a function of one variable, and let $z = f(x + 2y)$. Show that

$$2\frac{\partial z}{\partial x} - \frac{\partial z}{\partial y} = 0$$

21. Let f be a function of one variable and let $z = f(x^2 + y^2)$. Show that

$$y\frac{\partial z}{\partial x} - x\frac{\partial z}{\partial y} = 0$$

22. Let f be a function of one variable, and let $w = f(\rho)$, where $\rho = (x^2 + y^2 + z^2)^{1/2}$. Show that

$$\left(\frac{\partial w}{\partial x}\right)^2 + \left(\frac{\partial w}{\partial y}\right)^2 + \left(\frac{\partial w}{\partial z}\right)^2 = \left(\frac{dw}{d\rho}\right)^2$$

23. Let f be a function of three variables and let $w = f(x - y, y - z, z - x)$. Show that

$$\frac{\partial w}{\partial x} + \frac{\partial w}{\partial y} + \frac{\partial w}{\partial z} = 0$$

24. Let $w = f(x, y, z)$, $x = \rho\sin\phi\cos\theta$, $y = \rho\sin\phi\sin\theta$, and $z = \rho\cos\phi$. Express $\partial w/\partial\rho$, $\partial w/\partial\phi$, and $\partial w/\partial\theta$ in terms of $\partial w/\partial x$, $\partial w/\partial y$, and $\partial w/\partial z$.

25. Assume that $F(x, y, z) = 0$ defines z implicitly as a function of x and y. Show that

$$\frac{\partial z}{\partial x} = -\frac{\partial F/\partial x}{\partial F/\partial z} \quad\text{and}\quad \frac{\partial z}{\partial y} = -\frac{\partial F/\partial y}{\partial F/\partial z}$$

In Exercises 26–28, use the formulas in Exercise 25 to find $\partial z/\partial x$ and $\partial z/\partial y$.

26. $x^2 - 3yz^2 + xyz - 2 = 0$.

27. $ye^x - 5\sin 3z = 3z$.

28. $\ln(1 + z) + xy^2 + z = 1$.

29. Given that $u = u(x, y, z)$, $v = v(x, y, z)$, and $w = w(x, y, z)$, show that

$$\nabla f(u, v, w) = \frac{\partial f}{\partial u}\nabla u + \frac{\partial f}{\partial v}\nabla v + \frac{\partial f}{\partial w}\nabla w$$

30. Let $w = f(x, y, z)$ where $z = g(x, y)$. Taking x and y as the independent variables, express each of

the following in terms of $\partial f/\partial x$, $\partial f/\partial y$, $\partial f/\partial z$, $\partial z/\partial x$, and $\partial z/\partial y$.

(a) $\dfrac{\partial w}{\partial x}$ (b) $\dfrac{\partial w}{\partial y}$

31. Let $w = \ln{(e^r + e^s + e^t + e^u)}$. Show that

$$w_{rstu} = -6e^{r+s+t+u-4w}$$

[*Hint:* Take advantage of the relationship $e^w = e^r + e^s + e^t + e^u$.]

32. Suppose w is a function of x_1, x_2, and x_3, and

$$x_1 = a_1 y_1 + b_1 y_2$$
$$x_2 = a_2 y_1 + b_2 y_2$$
$$\dot{x}_3 = a_3 y_1 + b_3 y_2$$

where the a's and b's are constants. Express $\partial w/\partial y_1$ and $\partial w/\partial y_2$ in terms of $\partial w/\partial x_1$, $\partial w/\partial x_2$, and $\partial w/\partial x_3$.

33. (a) Let w be a function of x_1, x_2, x_3, and x_4, and let each x_i be a function of t. Find a chain rule formula for dw/dt.

(b) Let w be a function of x_1, x_2, x_3, and x_4, and let each x_i be a function of v_1, v_2, and v_3. Find chain rule formulas for $\partial w/\partial v_1$, $\partial w/\partial v_2$, and $\partial w/\partial v_3$.

34. Let $w = (x_1{}^2 + x_2{}^2 + \cdots + x_n{}^2)^k$, where $n > 2$. For what values of k does

$$\frac{\partial^2 w}{\partial x_1{}^2} + \frac{\partial^2 w}{\partial x_2{}^2} + \cdots + \frac{\partial^2 w}{\partial x_n{}^2} = 0$$

hold?

35. Show that

$$\frac{d}{dx}\left[\int_{a(x)}^{b(x)} f(t)\, dt\right] = f(b(x))b'(x) - f(a(x))a'(x)$$

This result is called *Leibniz' rule*. [*Hint:* Let $u = a(x)$, $v = b(x)$, and

$$F(u, v) = \int_u^v f(t)\, dt$$

Then use a chain rule and Theorem 5.9.5.]

36. Use the result of Exercise 35 to compute the following derivatives without performing the integrations.

(a) $\dfrac{d}{dx}\displaystyle\int_x^{x^2} e^{t^2}\, dt$

(b) $\dfrac{d}{dx}\displaystyle\int_{\sin x}^{\cos x} (t^3 + 2)^{2/3}\, dt$

(c) $\dfrac{d}{dx}\displaystyle\int_{3x}^{x^3} \sin^5 t\, dt$

(d) $\dfrac{d}{dx}\displaystyle\int_{e^x}^{e^{2x}} (\ln t)^4\, dt$.

16.9 MAXIMA AND MINIMA OF FUNCTIONS OF TWO VARIABLES

Earlier in this text we learned how to find maximum and minimum values of a function of one variable. In this section we shall develop similar techniques for functions of two variables.

If we imagine the graph of a function f of two variables to be a portion of the earth's terrain (Figure 16.9.1), then the mountaintops, which are the high points in their immediate vicinity, are called *relative maxima* of f, and the valley bottoms, which are the low points in their immediate vicinity, are called *relative minima* of f.

Just as a geologist might be interested in finding the highest mountain and deepest valley in an entire mountain range, so a mathematician might be interested in finding the largest and smallest values of $f(x, y)$ over the *entire* domain of f. These are called the *absolute maximum* and *absolute minimum values* of f. The following definitions make these informal ideas precise.

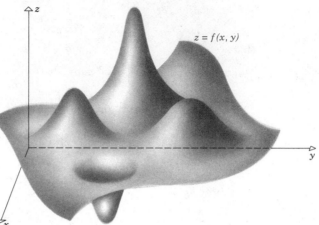

$z = f(x, y)$

Figure 16.9.1 ▷

16.9.1   DEFINITION. A function f of two variables is said to have a *relative maximum* at a point (x_0, y_0) if there is a circle centered at (x_0, y_0) such that $f(x_0, y_0) \geq f(x, y)$ for all points (x, y) inside the circle, and f is said to have an *absolute maximum* at (x_0, y_0) if $f(x_0, y_0) \geq f(x, y)$ for all points (x, y) in the domain of f.

16.9.2   DEFINITION. A function f of two variables is said to have a *relative minimum* at a point (x_0, y_0) if there is a circle centered at (x_0, y_0) such that $f(x_0, y_0) \leq f(x, y)$ for all points (x, y) inside the circle, and f is said to have an *absolute minimum* at (x_0, y_0) if $f(x_0, y_0) \leq f(x, y)$ for all points (x, y) in the domain of f.

If f has a relative maximum or a relative minimum at (x_0, y_0), then we say that f has a *relative extremum* at (x_0, y_0), and if f has an absolute maximum or absolute minimum at (x_0, y_0), then we say that f has an *absolute extremum* at (x_0, y_0).

In Figure 16.9.2 we have sketched the graph of a function f whose domain is the closed square region in the xy-plane whose points satisfy the inequalities $0 \leq x \leq 1, 0 \leq y \leq 1$. The function f has relative minima at the points A and C and a relative maximum at B. There is an absolute minimum at A and an absolute maximum at D.

For functions of two variables we shall be concerned with two important questions:

• Are there any relative or absolute extrema?

• If so, where are they located?

For functions of one variable the Extreme-Value Theorem (4.6.4) an-

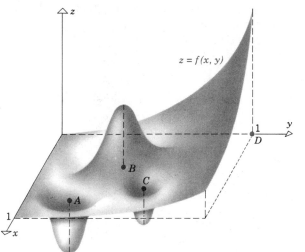

Figure 16.9.2 ▷

swered the existence question for absolute extrema. The following theorem, which we state without proof, is the corresponding result for functions of two variables.

Extreme-Value Theorem

16.9.3   THEOREM. *If $f(x, y)$ is continuous on a closed and bounded set R, then f has both an absolute maximum and an absolute minimum at points of R.*

Example 1   The square region R whose points satisfy the inequalities

$$0 \leq x \leq 1 \quad \text{and} \quad 0 \leq y \leq 1$$

is a closed and bounded set in the xy-plane. The function f whose graph is sketched in Figure 16.9.2 is continuous on R, and so is guaranteed to have an absolute maximum and minimum on R by the foregoing theorem. These occur at points D and A in the figure.   ◀

REMARK.   If any of the conditions in the Extreme-Value Theorem fail to hold, then there is no guarantee that an absolute maximum or absolute minimum exists on the region R. Thus, a discontinuous function on a closed and bounded set need not have any absolute extrema and a continuous function on a set that is not closed and bounded also need not have any absolute extrema. (See Exercise 49.)

Recall that if a function g of one variable has a relative extremum at a point x_0 where g is differentiable, then $g'(x_0) = 0$ (Theorem 4.3.4). To obtain the analog of this result for functions of two variables, suppose that $f(x, y)$ has a relative maximum at a point (x_0, y_0), and that the partial derivatives of f exist at (x_0, y_0). It seems plausible geometrically that the traces of the surface $z = f(x, y)$ on the planes $x = x_0$ and $y = y_0$ have horizontal tangent lines at (x_0, y_0) (Figure 16.9.3), so

$$f_x(x_0, y_0) = 0 \quad \text{and} \quad f_y(x_0, y_0) = 0$$

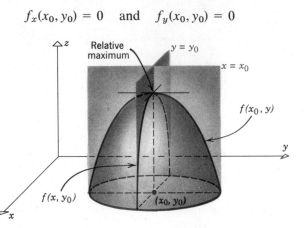

Figure 16.9.3

Similarly, if f has a relative minimum at (x_0, y_0).

This suggests the following result.

16.9.4 THEOREM. *If f has a relative extremum at a point (x_0, y_0), and if the first-order partial derivatives of f exist at this point, then*

$$f_x(x_0, y_0) = 0 \quad \text{and} \quad f_y(x_0, y_0) = 0$$

Proof. Assume that f has a relative maximum at (x_0, y_0) and that both partial derivatives of f exist at (x_0, y_0). We leave it as an exercise to show that

$$G(x) = f(x, y_0)$$

has a relative maximum at $x = x_0$ and

$$H(y) = f(x_0, y)$$

has a relative maximum at $y = y_0$ (Figure 16.9.3), so

$$G'(x_0) = f_x(x_0, y_0) = 0$$

and

$$H'(y_0) = f_y(x_0, y_0) = 0$$

(Exercise 48). The proof for a relative minimum is similar. ∎

Recall that for a function f of one variable we defined a critical point to be a point x_0 at which $f'(x_0) = 0$ or at which $f'(x_0)$ does not exist. Analogously, for a function f of two variables we define a *critical point* of f to be a point (x_0, y_0) in the interior of the domain of f at which

$$f_x(x_0, y_0) = 0 \quad \text{and} \quad f_y(x_0, y_0) = 0$$

or at which one or both of the partial derivatives does not exist.

For a function of one variable, the condition $f'(x_0) = 0$ is *not* sufficient to guarantee that f has a relative extremum at x_0. (The graph of f may have an inflection point with a horizontal tangent line at x_0.) Similarly, the conditions $f_x(x_0, y_0) = 0$ and $f_y(x_0, y_0) = 0$ are not sufficient to guarantee that a function f of two variables has a relative extremum at (x_0, y_0).

Example 2   The graphs of the functions

$$z = f(x, y) = x^2 + y^2 \qquad \text{(paraboloid)}$$
$$z = g(x, y) = 1 - x^2 - y^2 \quad \text{(paraboloid)}$$
$$z = h(x, y) = y^2 - x^2 \qquad \text{(hyperbolic paraboloid)}$$

are the quadric surfaces graphed in Figure 16.9.4. We have

$$f_x(x, y) = 2x, \qquad f_y(x, y) = 2y$$
$$g_x(x, y) = -2x, \quad g_y(x, y) = -2y$$
$$h_x(x, y) = -2x, \quad h_y(x, y) = 2y$$

$z = x^2 + y^2$

$z = 1 - x^2 - y^2$

$z = y^2 - x^2$

Figure 16.9.4

so in all three cases the partial derivatives are zero at $(0, 0)$, which means that $(0, 0)$ is a critical point for all three functions. The function f has a relative minimum at $(0, 0)$ and the function g a relative maximum. However, the function h has neither. To see this, observe that inside any circle in the xy-plane centered at $(0, 0)$, there exist points where $h(x, y)$ is positive (points on the y-axis) and there exist points where $h(x, y)$ is negative (points on the x-axis). Thus, $h(0, 0) = 0$ is neither the largest nor the smallest value of $h(x, y)$ in the circle.   ◀

A critical point at which a function does not have a relative extremum is called a *saddle point* of the function. Thus, the point $(0, 0)$ is a saddle

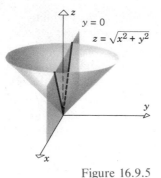

$y = 0$

$z = \sqrt{x^2 + y^2}$

Figure 16.9.5

point of the function $h(x, y) = y^2 - x^2$. (More advanced books distinguish between various types of saddle points. We shall not do this, however.)

Example 3   The graph of the function

$$f(x, y) = \sqrt{x^2 + y^2}$$

is the circular cone shown in Figure 16.9.5. The point $(0, 0)$ is a critical point of f because the partial derivatives do not both exist there. It is evident geometrically that $f_x(0, 0)$ does not exist because the trace of the cone in the plane $y = 0$ has a corner at the origin. The fact that $f_x(0, 0)$ does not exist can also be seen algebraically by noting that $f_x(0, 0)$ can be interpreted as the derivative with respect to x of the function

$$f(x, 0) = \sqrt{x^2 + 0} = |x|$$

at $x = 0$. But $|x|$ is not differentiable at $x = 0$, so $f_x(0, 0)$ does not exist. Similarly, $f_y(0, 0)$ does not exist.   ◄

For functions of one variable the Second-Derivative Test (Theorem 4.3.7) was used to determine the behavior of a function at a critical point. The following theorem, which is usually proved in advanced calculus, is the analog of that theorem for functions of two variables.

The Second-Partials Test

16.9.5   THEOREM. *Let f be a function of two variables with continuous second-order partial derivatives in some circle centered at a critical point (x_0, y_0); and let*

$$D = f_{xx}(x_0, y_0)f_{yy}(x_0, y_0) - f_{xy}{}^2(x_0, y_0)$$

(a)   If $D > 0$ and $f_{xx}(x_0, y_0) > 0$, then f has a relative minimum at (x_0, y_0).
(b)   If $D > 0$ and $f_{xx}(x_0, y_0) < 0$, then f has a relative maximum at (x_0, y_0).
(c)   If $D < 0$, then f has a saddle point at (x_0, y_0).
(d)   If $D = 0$, then no conclusion can be drawn.

Example 4   Locate all relative extrema and saddle points of

$$f(x, y) = 3x^2 - 2xy + y^2 - 8y$$

Solution.   Since $f_x(x, y) = 6x - 2y$ and $f_y(x, y) = -2x + 2y - 8$, the critical points of f satisfy the equations

$$6x - 2y = 0$$
$$-2x + 2y - 8 = 0$$

Solving these for x and y yields $x = 2$, $y = 6$ (verify), so $(2, 6)$ is the only critical point. To apply Theorem 16.9.5 we need the second-order partial derivatives

$$f_{xx}(x, y) = 6 \quad f_{yy}(x, y) = 2, \quad f_{xy}(x, y) = -2$$

At the point $(2, 6)$ we have

$$D = f_{xx}(2, 6)f_{yy}(2, 6) - f_{xy}{}^2(2, 6) = (6)(2) - (-2)^2 = 8 > 0$$

and

$$f_{xx}(2, 6) = 6 > 0$$

so that f has a relative minimum at $(2, 6)$ by part (a) of the second-partials test.   ◀

Example 5   Locate all relative extrema and saddle points of

$$f(x, y) = 4xy - x^4 - y^4$$

Solution.   Since

$$f_x(x, y) = 4y - 4x^3 \tag{1}$$
$$f_y(x, y) = 4x - 4y^3$$

the critical points of f have coordinates satisfying the equations

$$\begin{matrix} 4y - 4x^3 = 0 \\ 4x - 4y^3 = 0 \end{matrix} \quad \text{or} \quad \begin{matrix} y = x^3 \\ x = y^3 \end{matrix} \tag{2}$$

Substituting the top equation in the bottom yields $x = (x^3)^3$ or $x^9 - x = 0$ or $x(x^8 - 1) = 0$, which has solutions $x = 0$, $x = 1$, $x = -1$. Substituting these values in the top equation of (2) we obtain the corresponding y values $y = 0$, $y = 1$, $y = -1$. Thus, the critical points of f are $(0, 0)$, $(1, 1)$, and $(-1, -1)$.
   From (1),

$$f_{xx}(x, y) = -12x^2, \quad f_{yy}(x, y) = -12y^2, \quad f_{xy}(x, y) = 4$$

which yields the following table:

CRITICAL POINT (x_0, y_0)	$f_{xx}(x_0, y_0)$	$f_{yy}(x_0, y_0)$	$f_{xy}(x_0, y_0)$	$D = f_{xx}f_{yy} - f_{xy}{}^2$
$(0, 0)$	0	0	4	-16
$(1, 1)$	-12	-12	4	128
$(-1, -1)$	-12	-12	4	128

At the points $(1, 1)$ and $(-1, -1)$, we have $D > 0$ and $f_{xx} < 0$ so that relative maxima occur at these critical points. At $(0, 0)$ there is a saddle point since $D < 0$. The surface and its contour map are shown in Figure 16.9.6.   ◀

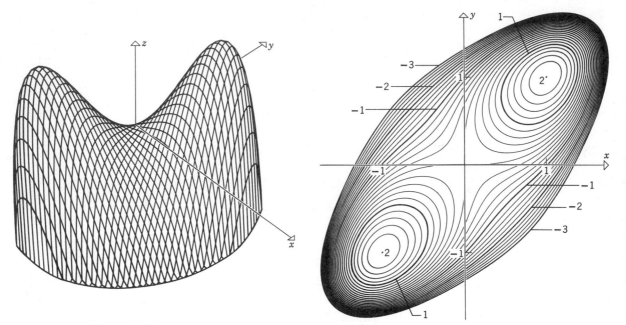

Figure 16.9.6

The following analog of Theorem 4.6.5 is a key result for locating the absolute extreme values of a function of two variables.

16.9.6   THEOREM.   *If a function f of two variables has an absolute extremum (either an absolute maximum or an absolute minimum) at an interior point of its domain, then this extremum occurs at a critical point.*

Proof.   We shall prove the result for an absolute maximum. The proof for an absolute minimum is similar.

If f has an absolute maximum at the point (x_0, y_0) in the interior of the domain of f, then f has a relative maximum at (x_0, y_0) (why?). If both partial derivatives exist at (x_0, y_0), then

$$f_x(x_0, y_0) = 0 \quad \text{and} \quad f_y(x_0, y_0) = 0$$

by Theorem 16.9.4, so (x_0, y_0) is a critical point of f. If either partial derivative does not exist, then again (x_0, y_0) is a critical point, so (x_0, y_0) is a critical point in all cases.   ▮

FINDING ABSOLUTE
EXTREMA ON
CLOSED AND
BOUNDED SETS

If $f(x, y)$ is continuous on a closed and bounded set R, then the Extreme-Value Theorem (16.9.3) guarantees the existence of an absolute maximum and an absolute minimum of f on R. These absolute extrema can occur either on the boundary of R or in the interior of R, but if an absolute extremum occurs in the interior, then it occurs at a critical point by Theorem 16.9.6. Thus, we are led to the following procedure for finding absolute extrema:

> **How to Find the Absolute Extrema of a Continuous Function f of Two Variables on a Closed and Bounded Set R.**
>
> **Step 1.** Find the critical points of f that lie in the interior of R.
>
> **Step 2.** Find all boundary points at which the absolute extrema can occur.
>
> **Step 3.** Evaluate $f(x, y)$ at the points obtained in the previous steps. The largest of these values is the absolute maximum and the smallest the absolute minimum.

Example 6   Find the absolute maximum and minimum values of

$$f(x, y) = 3xy - 6x - 3y + 7 \tag{3}$$

on the closed triangular region R with vertices $(0, 0)$, $(3, 0)$, and $(0, 5)$.

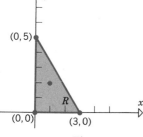

(0, 5)

R

(0, 0)   (3, 0)

Figure 16.9.7

Solution.   The region R is shown in Figure 16.9.7. We have

$$\frac{\partial f}{\partial x} = 3y - 6 \quad \text{and} \quad \frac{\partial f}{\partial y} = 3x - 3$$

so all critical points occur where

$$3y - 6 = 0 \quad \text{and} \quad 3x - 3 = 0$$

Solving these equations yields $x = 1$ and $y = 2$, so $(1, 2)$ is the only critical point. As shown in Figure 16.9.7, this critical point is in the interior of R.

Next, we want to determine the location of the points on the boundary of R at which the absolute extrema might occur. The boundary of R consists of three line segments, each of which we shall treat separately:

The line segment between $(0, 0)$ and $(3, 0)$.   On this line segment we have $y = 0$, so (3) simplifies to a function of the single variable x,

$$u(x) = f(x, 0) = -6x + 7, \quad 0 \le x \le 3$$

This function has no critical points because $u'(x) = -6$ is nonzero for all *x*. Thus the extreme values of $u(x)$ occur at the endpoints $x = 0$ and $x = 3$, which correspond to the points $(0, 0)$ and $(3, 0)$ of R.

The line segment between $(0, 0)$ *and* $(0, 5)$.   On this line segment we have $x = 0$, so (3) simplifies to a function of the single variable y,

$$v(y) = f(0, y) = -3y + 7, \quad 0 \le y \le 5$$

This function has no critical points because $v'(y) = -3$ is nonzero for all y. Thus, the extreme values of $v(y)$ occur at the endpoints $y = 0$ and $y = 5$, which correspond to the points $(0, 0)$ and $(0, 5)$ of R.

The line segment between $(3, 0)$ *and* $(0, 5)$.   In the *xy*-plane, an equation for this line segment is

$$y = -\tfrac{5}{3}x + 5, \quad 0 \le x \le 3 \tag{4}$$

so (3) simplifies to a function of the single variable x,

$$\begin{aligned} w(x) = f(x, -\tfrac{5}{3}x + 5) &= 3x(-\tfrac{5}{3}x + 5) - 6x - 3(-\tfrac{5}{3}x + 5) + 7 \\ &= -5x^2 + 14x - 8, \quad 0 \le x \le 3 \end{aligned}$$

Since $w'(x) = -10x + 14$, the equation $w'(x) = 0$ yields $x = \tfrac{7}{5}$ as the only critical point of w. Thus, the extreme values of w occur either at the critical point $x = \tfrac{7}{5}$ or at the endpoints $x = 0$ and $x = 3$. The endpoints correspond to the points $(0, 5)$ and $(3, 0)$ of R, and from (4) the critical point corresponds to $(\tfrac{7}{5}, \tfrac{8}{3})$.

Finally, Table 16.9.1 lists the values of $f(x, y)$ at the interior critical point and at the points on the boundary where an absolute extremum can occur. From the table we conclude that the absolute maximum value of f is $f(0, 0) = 7$ and the absolute minimum value is $f(3, 0) = -11$. ◄

Table 16.9.1

(x, y)	$(0, 0)$	$(3, 0)$	$(0, 5)$	$(\tfrac{7}{5}, \tfrac{8}{3})$	$(1, 2)$
$f(x, y)$	7	-11	-8	$\tfrac{9}{5}$	1

Example 7   Determine the dimensions of a rectangular box, open at the top, having a volume of 32 ft^3, and requiring the least amount of material for its construction.

Solution.   Let

$x =$ length of the box (in feet)
$y =$ width of the box (in feet)
$z =$ height of the box (in feet)
$S =$ surface area of the box (in square feet)

Two sides have area xz.
Two sides have area yz.
One side has area xy.

Figure 16.9.8

We may reasonably assume that the box with least surface area requires the least amount of material, so that our objective is to minimize the surface area

$$S = xy + 2xz + 2yz \qquad (5)$$

(see Figure 16.9.8) subject to the volume requirement

$$xyz = 32 \qquad (6)$$

From (6) we obtain $z = 32/xy$, so (5) can be rewritten as

$$S = xy + \frac{64}{y} + \frac{64}{x} \qquad (7)$$

which expresses S as a function of two variables. The dimensions x and y in this formula must be positive, but otherwise have no limitation, so our problem reduces to finding the absolute minimum value of S over the region for which $x > 0$ and $y > 0$ (Figure 16.9.9). Because this region is not bounded, we have no mathematical guarantee at this stage that an absolute minimum exists. However, if it does, then it occurs at a critical point of S, so we shall begin by finding the critical points. Differentiating (7) we obtain,

Figure 16.9.9

$$\frac{\partial S}{\partial x} = y - \frac{64}{x^2}, \quad \frac{\partial S}{\partial y} = x - \frac{64}{y^2} \qquad (8)$$

so that the coordinates of the critical points of S satisfy

$$y - \frac{64}{x^2} = 0$$

$$x - \frac{64}{y^2} = 0$$

Solving the first equation for y yields

$$y = \frac{64}{x^2} \qquad (9)$$

and substituting this expression in the second equation yields

$$x - \frac{64}{(64/x^2)^2} = 0$$

which can be rewritten as

$$x\left(1 - \frac{x^3}{64}\right) = 0$$

The solutions of this equation are $x = 0$ and $x = 4$. Since we require $x > 0$, the only solution of significance is $x = 4$. Substituting this value in (9) yields $y = 4$. To see that we have located a relative minimum, we use the second-partials test. From (8),

$$\frac{\partial^2 S}{\partial x^2} = \frac{128}{x^3}, \quad \frac{\partial^2 S}{\partial y^2} = \frac{128}{y^3}, \quad \frac{\partial^2 S}{\partial y \, \partial x} = 1$$

Thus, when $x = 4$ and $y = 4$, we have

$$\frac{\partial^2 S}{\partial x^2} = 2, \quad \frac{\partial^2 S}{\partial y^2} = 2, \quad \frac{\partial^2 S}{\partial y \, \partial x} = 1$$

and

$$D = \frac{\partial^2 S}{\partial x^2}\frac{\partial^2 S}{\partial y^2} - \left(\frac{\partial^2 S}{\partial y \, \partial x}\right)^2 = (2)(2) - (1)^2 = 3$$

Since $\partial^2 S/\partial x^2 > 0$ and $D > 0$, it follows from the second-partials test that a relative minimum occurs when $x = y = 4$. Substituting these values in (6) yields $z = 2$, so that the box using least material has a height of 2 ft and a square base whose sides are 4 ft long.   ◄

REMARK.   Strictly speaking, the solution in the last example is incomplete since we have not shown that an *absolute minimum* for S occurs when $x = y = 4$ and $z = 2$, only a relative minimum. The problem of showing that a relative extremum is also an absolute extremum can be difficult for functions of two or more variables and will not be considered in this text. However, in applied problems such as this it is often obvious from physical or geometric considerations that an absolute extremum has been found.

Definitions of relative extrema can be given for functions of three or more variables. For example, if f is a function of three variables, then f has a **relative maximum** at (x_0, y_0, z_0) if $f(x_0, y_0, z_0) \geq f(x, y, z)$ for all points (x, y, z) within some sphere centered at (x_0, y_0, z_0), and for a **relative minimum** the inequality is reversed. If $f(x, y, z)$ has first partial derivatives, then the relative extrema occur at **critical points,** that is, points where

$$f_x(x_0, y_0, z_0) = f_y(x_0, y_0, z_0) = f_z(x_0, y_0, z_0) = 0$$

The extension of the second-partials test (Theorem 16.9.5) to functions of three or more variables is given in advanced calculus texts.

▶ Exercise Set 16.9

In Exercises 1–20, locate all relative maxima, relative minima, and saddle points.

1. $f(x, y) = 3x^2 + 2xy + y^2$.

2. $f(x, y) = x^3 - 3xy - y^3$.

3. $f(x, y) = y^2 + xy + 3y + 2x + 3$.

4. $f(x, y) = x^2 + xy - 2y - 2x + 1$.

5. $f(x, y) = x^2 + xy + y^2 - 3x$.

6. $f(x, y) = xy - x^3 - y^2$.

7. $f(x, y) = x^2 + 2y^2 - x^2y$.

8. $f(x, y) = 2x^2 - 4xy + y^4 + 2$.

9. $f(x, y) = x^2 + y^2 + 2/(xy)$.

10. $f(x, y) = x^3 + y^3 - 3x - 3y$.

11. $f(x, y) = x^2 + y - e^y$.

12. $f(x, y) = xe^y$.

13. $f(x, y) = e^x \sin y$.

14. $f(x, y) = xy + 2/x + 4/y$.

15. $f(x, y) = 2y^2x - yx^2 + 4xy$.

16. $f(x, y) = y \sin x$.

17. $f(x, y) = e^{-(x^2+y^2+2x)}$.

18. $f(x, y) = xy + \dfrac{a^3}{x} + \dfrac{b^3}{y}$ $(a \neq 0, b \neq 0)$.

19. $f(x, y) = \sin x + \sin y$, $0 < x < \pi$, $0 < y < \pi$.

20. $f(x, y) = \sin x + \sin y + \sin(x + y)$, $0 < x < \pi/2$, $0 < y < \pi/2$.

21. (a) Show that the second-partials test provides no information about the critical points of $f(x, y) = x^4 + y^4$.
    (b) Classify all critical points as relative maxima, relative minima, or saddle points.

22. (a) Show that the second-partials test provides no information about the critical points of $f(x, y) = x^4 - y^4$.
    (b) Classify all critical points as relative maxima, relative minima, or saddle points.

23. If f is a function of one variable, and f is continuous on an interval I and has exactly one relative extremum on I, say at x_0, then f has an absolute extremum at x_0 (Theorem 4.6.6). This exercise

shows that a similar result does not hold for functions of two variables.

(a) Show that $f(x, y) = 3xe^y - x^3 - e^{3y}$ has only one critical point and that a relative maximum occurs there (see Figure 16.9.10).

(b) Show that f does not have an absolute maximum.

[*Remark:* This exercise is based on the article "The Only Critical Point in Town" Test by Ira Rosenholtz and Lowell Smylie, *Mathematics Magazine*, Vol. 58, No. 3, May 1985, pp. 149–150.]

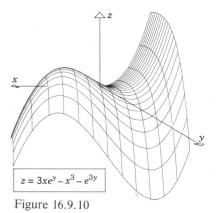

$$z = 3xe^y - x^3 - e^{3y}$$

Figure 16.9.10

24. Let f be a function of two variables that is continuous everywhere. It has been said that if f has relative maxima at two points, then f must have another critical point because it is impossible to have two mountains without some sort of valley in between. This exercise shows that this is not true.
Let $f(x, y) = 4x^2e^y - 2x^4 - e^{4y}$.
(a) Show that f has exactly two critical points and that a relative maximum occurs at each one (see Figure 16.9.11, page 1030).
(b) Show that f does not have an absolute maximum.

[*Remark:* This exercise is based on the problem Two Mountains Without a Valley, proposed and solved by Ira Rosenholtz, *Mathematics Magazine*, Vol. 60, No. 1, February 1987, p. 48.]

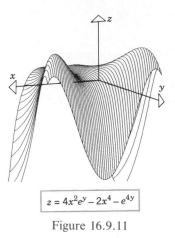

$$z = 4x^2 e^y - 2x^4 - e^{4y}$$

Figure 16.9.11

In Exercises 25–30, find the absolute extrema of the given function on the indicated closed and bounded set R.

25. $f(x, y) = xy - x - 3y$; R is the triangular region with vertices $(0, 0)$, $(0, 4)$, and $(5, 0)$.

26. $f(x, y) = xy - 2x$; R is the triangular region with vertices $(0, 0)$, $(0, 4)$, and $(4, 0)$.

27. $f(x, y) = x^2 - 3y^2 - 2x + 6y$; R is the square region with vertices $(0, 0)$, $(0, 2)$, $(2, 2)$, and $(2, 0)$.

28. $f(x, y) = xe^y - x^2 - e^y$; R is the rectangular region with vertices $(0, 0)$, $(0, 1)$, $(2, 1)$, and $(2, 0)$.

29. $f(x, y) = x^2 + 2y^2 - x$; R is the circular region $x^2 + y^2 \leq 4$.

30. $f(x, y) = xy^2$; R is the region that satisfies the inequalities $x \geq 0$, $y \geq 0$, and $x^2 + y^2 \leq 1$.

31. Show that $f(x, y) = y^2 - 2xy + x^2$ has an absolute minimum at each point on the line $y = x$.

32. Find three positive numbers whose sum is 48 and such that their product is as large as possible.

33. Find three positive numbers whose sum is 27 and such that the sum of their squares is as small as possible.

34. Find all points on the plane $x + y + z = 5$ in the first octant at which $f(x, y, z) = xy^2z^2$ has a maximum value.

35. Find the points on the surface $x^2 - yz = 5$ that are closest to the origin.

36. Find the dimensions of the rectangular box of maximum volume that can be inscribed in a sphere of radius a.

37. Find the maximum volume of a rectangular box with three faces in the coordinate planes and a vertex in the first octant on the plane $x + y + z = 1$.

38. A manufacturer makes two models of an item, standard and deluxe. It costs \$40 to manufacture the standard model and \$60 for the deluxe. A market research firm estimates that if the standard model is priced at x dollars and the deluxe at y dollars, then the manufacturer will sell $500(y - x)$ of the standard items and $45{,}000 + 500(x - 2y)$ of the deluxe each year. How should the items be priced to maximize the profit?

39. A closed rectangular box with a volume of 16 ft^3 is made from two kinds of materials. The top and bottom are made of material costing 10¢ per square foot and the sides from material costing 5¢ per square foot. Find the dimensions of the box so that the cost of materials is minimized.

40. Use the methods of this section to find the distance between the lines

$$
\begin{array}{lll}
x = 3t & & x = 2t \\
y = 2t & \text{and} & y = 2t + 3 \\
z = t & & z = 2t
\end{array}
$$

41. Use the methods of this section to find the distance from the point $(-1, 3, 2)$ to the plane $x - 2y + z = 4$.

42. Show that among all parallelograms with perimeter l, a square with sides of length $l/4$ has maximum area. [*Hint:* The area of a parallelogram is $A = ab \sin \alpha$, where a and b are the lengths of two adjacent sides and α is the angle between them.]

43. Determine the dimensions of a rectangular box, open at the top, having volume V, and requiring the least amount of material for its construction.

44. A length of sheet metal 27 in. wide is to be made into a water trough by bending up two sides as shown in Figure 16.9.12. Find x and ϕ so that the

Figure 16.9.12

trapezoid-shaped cross section has a maximum area.

45. A common problem in experimental work is to obtain a mathematical relationship $y = f(x)$ between two variables x and y by "fitting" a curve to points in the plane corresponding to various experimentally determined values of x and y, say

$$(x_1, y_1), (x_2, y_2), \ldots, (x_n, y_n)$$

Based on theoretical considerations, or simply on the pattern of the points, one decides on the general form of the curve $y = f(x)$ to be fitted. Often, the "curve" to be fitted is a straight line, $y = ax + b$. However, because of experimental errors in the data it is often impossible to find a line that passes through all of the points (Figure 16.9.13), so one looks for a line that "best fits" the data. One criterion for selecting a line of "best fit" is to choose a and b to minimize the function

$$f(a, b) = \sum_{k=1}^{n} (ax_k + b - y_k)^2$$

Figure 16.9.13

Geometrically, $|ax_k + b - y_k|$ is the vertical distance between the data point (x_k, y_k) and the line $y = ax + b$, so in effect, minimizing $f(a, b)$ minimizes the sum of the squares of these vertical distances (Figure 16.9.13). This procedure is called the method of **least squares**.

(a) Show that the conditions $\dfrac{\partial f}{\partial a} = 0$ and $\dfrac{\partial f}{\partial b} = 0$ result in the equations

$$(\Sigma x_k^2)a + (\Sigma x_k)b = \Sigma x_k y_k$$
$$(\Sigma x_k)a + nb = \Sigma y_k$$

where $\Sigma = \sum\limits_{k=1}^{n}$.

(b) Solve the equations in part (a) for a and b to show that

$$a = \frac{n(\Sigma x_k y_k) - (\Sigma x_k)(\Sigma y_k)}{n(\Sigma x_k^2) - (\Sigma x_k)^2}$$

and

$$b = \frac{(\Sigma x_k^2)(\Sigma y_k) - (\Sigma x_k)(\Sigma x_k y_k)}{n(\Sigma x_k^2) - (\Sigma x_k)^2}$$

46. This exercise shows that the values of a and b obtained in Exercise 45 result in the absolute minimum value of $f(a, b)$.

(a) Given that the arithmetic average $\bar{x} = \dfrac{1}{n}\Sigma x_k$ minimizes $\Sigma(x_k - \bar{x})^2$ (see Exercise 24, Section 4.8), show that $n(\Sigma x_k^2) - (\Sigma x_k)^2 > 0$. [*Note:* $\Sigma(x_k - \bar{x})^2 > 0$ if the x_k's are not all the same.]

(b) Find $f_{aa}(a, b)$, $f_{bb}(a, b)$, and $f_{ab}(a, b)$.

(c) Use Theorem 16.9.5 and the results of parts (a) and (b) to show that f has a relative minimum at the critical point found in Exercise 45.

(d) Based on the result of part (c) and the nature of the function $f(a, b)$, show that f takes on its absolute minimum value at the critical point.

47. Use the formulas in part (b) of Exercise 45 to find the equation of the least squares line $y = ax + b$ for the following data:

x	1	2	3	4
y	1.5	1.6	2.1	3.0

48. Prove: If f has a relative maximum at (x_0, y_0), and $f_x(x_0, y_0)$ exists, then $f(x, y_0)$ has a relative maximum at $x = x_0$, and $f_x(x_0, y_0) = 0$.

49. Find an example to show that a function that is discontinuous on a closed and bounded set need not have any absolute extrema and an example to show that a continuous function on a set that is not closed and bounded also need not have any absolute extrema.

16.10 LAGRANGE MULTIPLIERS

In this section we shall study a powerful method for solving certain types of optimization problems that is sometimes simpler to apply than the methods studied in the last section.

In the previous section, we considered the problem of minimizing

$$S = xy + 2xz + 2yz$$

subject to the constraint

$$xyz - 32 = 0$$

This is a special case of the following general problem, which we shall study in this section:

> **Three-Variable Extremum Problem with One Constraint**
> Maximize or minimize the function $f(x, y, z)$ subject to the constraint $g(x, y, z) = 0$.

We shall also be interested in the two-variable version of this problem.

> **Two-Variable Extremum Problem with One Constraint**
> Maximize or minimize the function $f(x, y)$ subject to the constraint $g(x, y) = 0$.

One way to attack these problems is to solve the constraint equation for one of the variables in terms of the rest and substitute the result into f. The resulting function of one or two variables can then be maximized or minimized by finding its critical points. (See the solution of Example 7 in the previous section.) However, if the constraint equation is too complicated to solve for one of the variables in terms of the rest, then other techniques must be used. We shall discuss one such technique, called *the method of Lagrange multipliers*. Since a rigorous discussion of this topic requires results from advanced calculus, we shall not be too concerned about all the technical details; we shall emphasize computational techniques.

Let us begin with the two-variable problem of maximizing or minimizing $f(x, y)$ subject to the constraint $g(x, y) = 0$. The graph of $g(x, y) = 0$ is some curve C in the xy-plane. Geometrically, we are concerned with finding the maximum or minimum value of $f(x, y)$ as (x, y) varies over the constraint curve C. If (x_0, y_0) is a point on the constraint curve C, then we shall say that $f(x, y)$ has a *constrained relative maximum* at (x_0, y_0) if there is a circle centered at (x_0, y_0) such that

$$f(x_0, y_0) \geq f(x, y) \tag{1}$$

for all points (x, y) on C within the circle (Figure 16.10.1). For a *constrained relative minimum* at (x_0, y_0), the inequality in (1) is reversed. We shall say that f has a *constrained relative extremum* at (x_0, y_0) if f has either a constrained relative maximum or a constrained relative minimum at (x_0, y_0). The following result is the key to finding constrained relative extrema.

The Constrained-Extremum Principle for Two Variables

16.10.1   THEOREM. *Let f and g be functions of two variables with continuous first partial derivatives on some open set containing the constraint curve $g(x, y) = 0$, and assume that $\nabla g \neq 0$ at any point on this curve. If f has a constrained relative extremum, then this extremum occurs at a point (x_0, y_0) on the constraint curve where the gradient vectors $\nabla f(x_0, y_0)$ and $\nabla g(x_0, y_0)$ are parallel, that is,*

$$\nabla f(x_0, y_0) = \lambda \nabla g(x_0, y_0) \tag{2}$$

for some number λ.

At a constrained relative maximum, the inequality $f(x_0, y_0) \geq f(x, y)$ holds at all points of C within some circle centered at (x_0, y_0).

Figure 16.10.1

The number λ in (2) is called a *Lagrange multiplier*.

To see why the constrained-extremum principle is true, let C denote the constraint curve $g(x, y) = 0$, and suppose that a constrained relative extremum occurs at the point (x_0, y_0) on C. It is proved in advanced calculus that under the hypotheses of this theorem, the curve C can be represented parametrically by the equations

$$x = x(t)$$
$$y = y(t)$$

such that C has a nonzero tangent vector at (x_0, y_0). More precisely,

$$x'(t_0)\mathbf{i} + y'(t_0)\mathbf{j} \neq 0$$

where t_0 is the value of the parameter corresponding to (x_0, y_0). In terms of the parameter t, the value of f at a point (x, y) on the curve C is $f(x, y) = f(x(t), y(t))$. Since f has a constrained relative extremum at the point (x_0, y_0), and since this point corresponds to $t = t_0$, it follows that the function

$$F(t) = f(x(t), y(t))$$

has a relative extremum at $t = t_0$, that is,

$$F'(t_0) = 0 \tag{3}$$

By the chain rule

$$F'(t) = \frac{d}{dt}[f(x(t), y(t))] = f_x(x, y)\frac{dx}{dt} + f_y(x, y)\frac{dy}{dt}$$

so that condition (3) yields the equation

$$f_x(x_0, y_0)x'(t_0) + f_y(x_0, y_0)y'(t_0) = 0$$

which may be expressed in the form

$$\nabla f(x_0, y_0) \cdot [x'(t_0)\mathbf{i} + y'(t_0)\mathbf{j}] = 0 \tag{4}$$

Since the vector $x'(t_0)\mathbf{i} + y'(t_0)\mathbf{j}$ is tangent to the curve C at (x_0, y_0), it follows from (4) that $\nabla f(x_0, y_0)$ is perpendicular to C at (x_0, y_0). But $\nabla g(x_0, y_0)$ is also perpendicular to C at (x_0, y_0) since C is the level curve $g(x, y) = 0$ for the function g (see Theorem 16.6.4). Since $\nabla f(x_0, y_0)$ and $\nabla g(x_0, y_0)$ are both perpendicular to C at (x_0, y_0), these vectors are parallel.

Example 1 At what point on the circle $x^2 + y^2 = 1$ is the product xy a maximum?

Solution. We want to maximize $f(x, y) = xy$ subject to the constraint

$$g(x, y) = x^2 + y^2 - 1 = 0 \tag{5}$$

At a constrained relative extremum we must have

$$\nabla f = \lambda \nabla g$$

or

$$y\mathbf{i} + x\mathbf{j} = \lambda(2x\mathbf{i} + 2y\mathbf{j})$$

which is equivalent to the pair of equations

$$y = 2x\lambda \quad \text{and} \quad x = 2y\lambda \tag{6}$$

Since the maximum value of xy on the circle $x^2 + y^2 = 1$ is obviously greater than zero, we must have $x \neq 0$ and $y \neq 0$ at a constrained maximum. Thus, the equations in (6) can be rewritten as

$$\lambda = \frac{y}{2x} \quad \text{and} \quad \lambda = \frac{x}{2y}$$

from which we obtain

$$\frac{y}{2x} = \frac{x}{2y}$$

or

$$y^2 = x^2 \tag{7}$$

Substituting this in (5) yields

$$2x^2 - 1 = 0$$

or

$$x = \frac{1}{\sqrt{2}} \quad \text{and} \quad x = -\frac{1}{\sqrt{2}}$$

Substituting $x = 1/\sqrt{2}$ in (7) yields $y = \pm 1/\sqrt{2}$ and substituting $x = -1/\sqrt{2}$ in (7) yields $y = \pm 1/\sqrt{2}$, so that there are four candidates for the location of a maximum:

$$\left(\frac{1}{\sqrt{2}}, \frac{1}{\sqrt{2}}\right), \ \left(\frac{1}{\sqrt{2}}, -\frac{1}{\sqrt{2}}\right), \ \left(-\frac{1}{\sqrt{2}}, \frac{1}{\sqrt{2}}\right), \ \left(-\frac{1}{\sqrt{2}}, -\frac{1}{\sqrt{2}}\right)$$

At the first and fourth points the function $f(x, y) = xy$ has value $\frac{1}{2}$, while at the second and third points the value is $-\frac{1}{2}$. Thus, the constrained maximum value of $\frac{1}{2}$ occurs at $(1/\sqrt{2}, 1/\sqrt{2})$ and $(-1/\sqrt{2}, -1/\sqrt{2})$. ◀

REMARK.   If c is a constant, then the functions $g(x, y)$ and $g(x, y) - c$ have the same gradient since the constant c drops out when we differentiate. Consequently, it is *not* essential to rewrite a constraint of the form $g(x, y) = c$ as $g(x, y) - c = 0$ in order to apply the constrained-extremum principle. Thus, in the last example, we could have kept the constraint in the form $x^2 + y^2 = 1$ and taken $g(x, y) = x^2 + y^2$ rather than $g(x, y) = x^2 + y^2 - 1$.

In Exercise 6 of Section 4.7, it was stated that among all rectangles of perimeter p, a square of side $p/4$ has maximum area. This result can be obtained using Lagrange multipliers.

Example 2   Find the dimensions of a rectangle having perimeter p and maximum area.

Solution.   Let

$x = $ length of the rectangle
$y = $ width of the rectangle
$A = $ area of the rectangle

We want to maximize $A = xy$ subject to the perimeter constraint

$$2x + 2y = p \tag{8}$$

From the constrained extremum principle with $f(x, y) = xy$ and $g(x, y) = 2x + 2y$, we must have $\nabla f = \lambda \nabla g$, or

$$y\mathbf{i} + x\mathbf{j} = \lambda(2\mathbf{i} + 2\mathbf{j})$$

at a constrained relative maximum. This is equivalent to the two equations

$$y = 2\lambda \quad \text{and} \quad x = 2\lambda$$

Eliminating λ from these equations we obtain

$$x = y$$

Using this condition and constraint (8), we obtain $x = p/4$, $y = p/4$. ◄

Lagrange multipliers can also be used in the three-variable problem of maximizing or minimizing $f(x, y, z)$ subject to the constraint $g(x, y, z) = 0$. The graph of $g(x, y, z) = 0$ is some surface S in 3-space. Geometrically we are concerned with finding the maximum or minimum of the function $f(x, y, z)$ as (x, y, z) varies over the surface S. We shall say that $f(x, y, z)$ has a *constrained relative maximum* at (x_0, y_0, z_0) if there is a sphere centered at (x_0, y_0, z_0) such that

$$f(x_0, y_0, z_0) \geq f(x, y, z)$$

for all points (x, y, z) on S within the sphere (Figure 16.10.2). The meaning of the terms *constrained relative minimum* and *constrained relative extremum* should be clear. It can be shown that a constrained relative extremum can only occur at a point (x_0, y_0, z_0) where $\nabla f(x_0, y_0, z_0)$ and $\nabla g(x_0, y_0, z_0)$ are parallel, that is,

$$\nabla f(x_0, y_0, z_0) = \lambda \nabla g(x_0, y_0, z_0)$$

for some number λ.

(x_0, y_0, z_0)

S

$g(x, y, z) = 0$

At a constrained relative maximum, the inequality $f(x_0, y_0, z_0) \geqslant f(x, y, z)$ holds at all points of S inside some sphere centered at (x_0, y_0, z_0).

Figure 16.10.2

Example 3   Find the points on the sphere $x^2 + y^2 + z^2 = 36$ that are closest to and farthest from $(1, 2, 2)$.

Solution.   To avoid radicals, we shall find points on the sphere that minimize and maximize the *square* of the distance to $(1, 2, 2)$. Thus, we want to extremize

$$f(x, y, z) = (x - 1)^2 + (y - 2)^2 + (z - 2)^2$$

subject to the constraint

$$x^2 + y^2 + z^2 = 36 \tag{9}$$

Therefore, with $g(x, y, z) = x^2 + y^2 + z^2$, we must have $\nabla f(x, y, z) = \lambda \nabla g(x, y, z)$ at a constrained relative extremum; that is,

$$2(x - 1)\mathbf{i} + 2(y - 2)\mathbf{j} + 2(z - 2)\mathbf{k} = \lambda(2x\mathbf{i} + 2y\mathbf{j} + 2z\mathbf{k})$$

which leads to the equations

$$\begin{aligned}
2(x - 1) &= 2x\lambda \\
2(y - 2) &= 2y\lambda \\
2(z - 2) &= 2z\lambda
\end{aligned} \tag{10}$$

It is clear geometrically that the points on the sphere $x^2 + y^2 + z^2 = 36$ that are closest to and farthest from $(1, 2, 2)$ do not lie in the coordinate planes (make a sketch). Thus, we can assume that x, y, and z are nonzero, and we can rewrite (10) as

$$\frac{x - 1}{x} = \lambda$$

$$\frac{y - 2}{y} = \lambda$$

$$\frac{z - 2}{z} = \lambda$$

The first two equations imply that

$$\frac{x - 1}{x} = \frac{y - 2}{y}$$
$$xy - y = xy - 2x$$
$$y = 2x \tag{11}$$

Similarly, the first and third equations imply that

$$z = 2x \tag{12}$$

Substituting (11) and (12) in the constraint equation (9), we obtain

$$9x^2 = 36$$

or

$$x = \pm 2$$

Substituting these values in (11) and (12) yields two points

$$(2, 4, 4) \quad \text{and} \quad (-2, -4, -4)$$

Since $f(2, 4, 4) = 9$ and $f(-2, -4, -4) = 81$, it follows that $(2, 4, 4)$ is the point on the sphere closest to $(1, 2, 2)$, and $(-2, -4, -4)$ is the point that is farthest (Figure 16.10.3). ◀

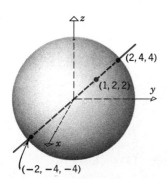

Figure 16.10.3

Next we shall use Lagrange multipliers to solve the problem of Example 7 in the previous section.

Example 4   Use Lagrange multipliers to determine the dimensions of a rectangular box, open at the top, having a volume of 32 ft^3, and requiring the least amount of material for its construction.

Solution.   With the notation of Example 7 in Section 16.9, the problem is to minimize the surface area

$$S = xy + 2xz + 2yz$$

subject to the volume constraint

$$xyz = 32 \tag{13}$$

If we let $f(x, y, z) = xy + 2xz + 2yz$ and $g(x, y, z) = xyz$, then we must have $\nabla f = \lambda \nabla g$, that is,

$$(y + 2z)\mathbf{i} + (x + 2z)\mathbf{j} + (2x + 2y)\mathbf{k} = \lambda(yz\mathbf{i} + xz\mathbf{j} + xy\mathbf{k})$$

This condition yields the three equations

$$y + 2z = \lambda yz$$
$$x + 2z = \lambda xz$$
$$2x + 2y = \lambda xy$$

Because of the volume constraint $xyz = 32$, we know that x, y, and z are nonzero, so that these equations can be rewritten as

$$\frac{1}{z} + \frac{2}{y} = \lambda$$

$$\frac{1}{z} + \frac{2}{x} = \lambda$$

$$\frac{2}{y} + \frac{2}{x} = \lambda$$

From the first two equations,

$$y = x \tag{14}$$

and from the second and third equations, $z = \frac{1}{2}y$. This and (14) imply that

$$z = \tfrac{1}{2}x \tag{15}$$

Substituting (14) and (15) in the volume constraint (13) yields

$$\tfrac{1}{2}x^3 = 32$$

This equation, together with (14) and (15), yields

$$x = 4, \quad y = 4, \quad z = 2$$

which agrees with the result obtained in Example 7 of the previous section.   ◀

Lagrange multipliers can also be used in problems involving two or more constraints. However, we shall not pursue this topic here.

▶ Exercise Set 16.10

In Exercises 1–7, use Lagrange multipliers to find the maximum and minimum values of f subject to the given constraint. Also, find the points at which these extreme values occur.

1. $f(x, y) = xy; 4x^2 + 8y^2 = 16.$

2. $f(x, y) = x^2 - y; x^2 + y^2 = 25.$

3. $f(x, y) = 4x^3 + y^2; 2x^2 + y^2 = 1.$

4. $f(x, y) = x - 3y - 1; x^2 + 3y^2 = 16.$

5. $f(x, y, z) = 2x + y - 2z; x^2 + y^2 + z^2 = 4.$

6. $f(x, y, z) = 3x + 6y + 2z; 2x^2 + 4y^2 + z^2 = 70.$

7. $f(x, y, z) = xyz; x^2 + y^2 + z^2 = 1.$

In Exercises 8–16, solve using Lagrange multipliers.

8. Find the point on the line $2x - 4y = 3$ that is closest to the origin.

9. Find the point on the line $y = 2x + 3$ that is closest to $(4, 2)$.

10. Find the point on the plane $x + 2y + z = 1$ that is closest to the origin.

11. Find the point on the plane $4x + 3y + z = 2$ that is closest to $(1, -1, 1)$.

12. Find the points on the surface $xy - z^2 = 1$ that are closest to the origin.

13. Find the maximum value of $\sin x \sin y$, where x and y denote the acute angles of a right triangle.

14. Find a vector in 3-space whose length is 5 and whose components have the largest possible sum.

15. Find the points on the circle $x^2 + y^2 = 45$ that are closest to and farthest from $(1, 2)$.

16. The temperature at a point (x, y) on a metal plate is $T(x, y) = 4x^2 - 4xy + y^2$. An ant, walking on the plate, traverses a circle of radius 5 centered at the origin. What are the highest and lowest temperatures encountered by the ant?

In Exercises 17–24, use Lagrange multipliers to solve the indicated problems from Section 16.9.

17. Exercise 33.

18. Exercise 34.

19. Exercise 35.

20. Exercise 36.

21. Exercise 39.

22. Exercise 41.

23. Exercise 42.

24. Exercise 43.

▶ SUPPLEMENTARY EXERCISES

1. Let $f(x, y) = e^x \ln y$. Find:
   (a) $f(0, e)$
   (b) $f(\ln y, e^x)$
   (c) $f(r + s, rs)$.

2. Sketch the domain of f using solid lines for portions of the boundary included in the domain and dashed lines for portions not included.

(a) $f(x, y) = \sqrt{x - y}/(2x - y)$
(b) $f(x, y) = \ln(xy - 1)$
(c) $f(x, y) = (\sin^{-1} x)/e^y$.

3. Describe the graph of f.
   (a) $f(x, y) = \sqrt{x^2 + 4y^2}$
   (b) $f(x, y) = 1 - x/a - y/b$.

4. Find f_x, f_y, and f_z if $f(x, y, z) = x^2/(y^2 + z^2)$.

5. Find $\partial w/\partial r$ if $w = \ln(xy)/\sin yz$, $x = r + s$, $y = s$, $z = 3r - s$.

6. Find $\partial f/\partial x$, $\partial f/\partial y$, and $\partial f/\partial u$ if

$$f(x, y, z, u) = (e^{yz}/x) + \ln(u - x).$$

7. Find f_x, f_y, f_{yx}, and f_{yzx} at $(0, \pi/2, 1)$ if

$$f(x, y, z) = e^{xy} \sin yz.$$

8. Find $g_{xy}(0, 3)$ and $g_{yy}(2, 0)$ if

$$g(x, y) = \sin(xy) + xe^y.$$

9. Find $\partial w/\partial\theta|_{r=1, \theta=0}$ if $w = \ln(x^2 + y^2)$, $x = re^\theta$, $y = \tan(r\theta)$.

10. Find $\partial w/\partial r$ if $w = x \cos y + y \sin x$, $x = rs^2$, $y = r + s$.

In Exercises 12–15, verify the assertion.

12. If $w = \tan(x^2 + y^2) + x\sqrt{y}$, then $w_{xy} = w_{yx}$.

13. If $w = \ln(3x - 3y) + \cos(x + y)$, then $\partial^2 w/\partial x^2 = \partial^2 w/\partial y^2$.

14. If $F(x, y, z) = 2z^3 - 3(x^2 + y^2)z$, then $F_{xx} + F_{yy} + F_{zz} = 0$.

15. If $f(x, y, z) = xyz + x^2 + \ln(y/z)$, then $f_{xyzx} = f_{zxxy}$.

16. Find the slope of the tangent line at $(1, -2, -3)$ to the curve of intersection of the surface $z = 5 - 4x^2 - y^2$ with
(a) the plane $x = 1$ (b) the plane $y = -2$.

17. The pressure in newtons/m^2 of a gas in a cylinder is given by $P = 10T/V$ with T in °K and V in m^3.
(a) If T is increasing at a rate of 3°K/min with V held fixed at 2.5 m^3, find the rate at which the pressure is changing when $T = 50$°C.
(b) If T is held fixed at 50°K while V is decreasing at the rate of 3 m^3/min, find the rate at which the pressure is changing when $V = 2.5$ m^3.

In Exercises 18 and 19:
(a) Find the limit of $f(x, y)$ as $(x, y) \to (0, 0)$ if it exists.
(b) Determine if f is continuous at $(0, 0)$.

18. $f(x, y) = \dfrac{x^4 - x + y - x^3y}{x - y}$.

19. $f(x, y) = \begin{cases} \dfrac{x^4 - y^4}{x^2 + y^2} & \text{if } (x, y) \neq (0, 0) \\ 0 & \text{if } (x, y) = (0, 0). \end{cases}$

20. Find dw/dt using the chain rule.
(a) $w = \sin xy + y\ln xz + z$, $x = e^t$, $y = t^2$, $z = 1$
(b) $w = \sqrt{xy - e^z}$, $x = \sin t$, $y = 3t$, $z = \cos t$.

21. Use Formula (14) of Section 16.4 to find dy/dx.
(a) $3x^2 - 5xy + \tan xy = 0$
(b) $x \ln y + \sin(x - y) = \pi$.

22. If $F(x, y) = 0$, find a formula for d^2y/dx^2 in terms of partial derivatives of F. [*Hint:* Use Formula (14) of Section 16.4.]

23. The voltage V across a fixed resistance R in series with a variable resistance r is $V = RE/(r + R)$, where E is the source voltage. Express dV/dt in terms of dE/dt and dr/dt.

24. Let $f(x, y, z) = 1/(z - x^2 - 4y^2)$.
(a) Describe the domain of f.
(b) Describe the level surface $f(x, y, z) = 2$.
(c) Find $f(3t, uv, e^{3t})$.

In Exercises 25–29, find:
(a) the gradient of f at P_0
(b) the directional derivative at P_0 in the indicated direction.

25. $f(x, y) = x^2y^5$, $P_0(3, 1)$; from P_0 toward $P_1(4, -3)$.

26. $f(x, y, z) = ye^x \sin z$, $P_0(\ln 2, 2, \pi/4)$; in the direction of $\mathbf{a} = \langle 1, -2, 2 \rangle$.

27. $f(x, y, z) = \ln(xyz)$, $P_0(3, 2, 6)$; $\mathbf{u} = \langle -1, 1, 1 \rangle/\sqrt{3}$.

28. $f(x, y) = x^2y + 2xy^2$, $P_0(1, 2)$; $\mathbf{u}$ makes an angle of 60° with the positive x-axis.

29. $f(x, y, z) = xy + yz + zx$, $P_0(1, -1, 2)$; from P_0 toward $P_1(11, 10, 0)$.

30. Let $f(x, y, z) = (x + y)^2 + (y + z)^2 + (z + x)^2$. Find the maximum rate of decrease of f at $P_0(2, -1, 2)$ and the direction in which this rate of decrease occurs.

31. Find all unit vectors $\mathbf{u}$ such that $D_{\mathbf{u}}f = 0$ at P_0.
(a) $f(x, y) = x^3y^3 - xy$, $P_0(1, -1)$
(b) $f(x, y) = xe^y$, $P_0(-2, 0)$.

32. The directional derivative $D_{\mathbf{u}}f$ at (x_0, y_0) is known to be 2 when $\mathbf{u}$ makes an angle of $30°$ with the positive x-axis, and 8 when this angle is $150°$. Find $D_{\mathbf{u}}f(x_0, y_0)$ in the direction of the vector $\sqrt{3}\mathbf{i} + 2\mathbf{j}$.

33. At the point $(1, 2)$, the directional derivative $D_{\mathbf{u}}f$ is $2\sqrt{2}$ toward $P_1(2, 3)$ and -3 toward $P_2(1, 0)$. Find $D_{\mathbf{u}}f(1, 2)$ toward the origin.

In Exercises 34 and 35:

(a) Find a normal vector $\mathbf{N}$ at $P_0(x_0, y_0, f(x_0, y_0))$.

(b) Find an equation for the tangent plane at P_0.

34. $f(x, y) = 4x^2 + y^2 + 1$; $P_0(1, 2, 9)$.

35. $f(x, y) = 2\sqrt{x^2 + y^2}$; $P_0(4, -3, 10)$.

36. Find equations for the tangent plane and normal line to the given surface at P_0.

(a) $z = x^2 e^{2y}$; $P_0(1, \ln 2, 4)$

(b) $x^2 y^3 z^4 + xyz = 2$; $P_0(2, 1, -1)$.

37. Find all points P_0 on the surface $z = 2 - xy$ at which the normal line passes through the origin.

38. Show that for all tangent planes to the surface $x^{2/3} + y^{2/3} + z^{2/3} = 1$, the sum of the squares of the x-, y-, and z-intercepts is 1.

39. Find all points on the paraboloid $z = 9x^2 + 4y^2$ at which the normal line is parallel to the line through the points $P(4, -2, 5)$ and $Q(-2, -6, 4)$.

40. If $w = x^2 y - 2xy + y^2 x$, find the increment Δw and the differential dw if (x, y) varies from $(1, 0)$ to $(1.1, -0.1)$.

41. Use differentials to estimate the change in the volume $V = \frac{1}{3}x^2 h$ of a pyramid with a square base when its height h is increased from 2 to 2.2 m, while its base dimension x is decreased from 1 to 0.9 m. Compare this to ΔV.

42. If $f(x, y, z) = x^2 y^4/(1 + z^2)$, use differentials to estimate $f(4.996, 1.003, 1.995)$.

In Exercises 43–45, locate all relative minima, relative maxima, and saddle points.

43. $f(x, y) = x^2 + 3xy + 3y^2 - 6x + 3y$.

44. $f(x, y) = x^2 y - 6y^2 - 3x^2$.

45. $f(x, y) = x^3 - 3xy + \frac{1}{2}y^2$.

Solve Exercises 46 and 47 two ways:

(a) Use the constraint to eliminate a variable.

(b) Use Lagrange multipliers.

46. Find all relative extrema of $x^2 y^2$ subject to the constraint $4x^2 + y^2 = 8$.

47. Find the dimensions of the rectangular box of maximum volume that can be inscribed in the ellipsoid $(x/a)^2 + (y/b)^2 + (z/c)^2 = 1$.

In Exercises 48 and 49, use Lagrange multipliers.

48. Find the points on the curve $5x^2 - 6xy + 5y^2 = 8$ whose distance from the origin is (i) minimum (ii) maximum.

49. A current I branches into currents I_1, I_2, and I_3 through resistors R_1, R_2, and R_3 (see figure) in such a way that the total energy to the three resistors is a minimum. If the energy delivered to R_i is $I_i^2 R_i$ ($i = 1, 2, 3$), find the ratios $I_1 : I_2 : I_3$.

17

multiple integrals

17.1 DOUBLE INTEGRALS

The notion of a definite integral can be extended to functions of two or more variables. In this section we shall discuss the double integral, which is the extension to functions of two variables.

DEFINITION OF A
DOUBLE INTEGRAL

To set the groundwork, let us review the steps needed to define the definite integral $\int_a^b f(x)\, dx$ of a function of one variable.

Step 1. Divide the interval $[a, b]$ into n subintervals having lengths

$$\Delta x_1, \Delta x_2, \ldots, \Delta x_n$$

Step 2. In each subinterval, choose an arbitrary point, and let these points be denoted by

$$x_1^*, x_2^*, \ldots, x_n^*$$

Step 3. Form the Riemann sum

$$\sum_{k=1}^{n} f(x_k^*)\, \Delta x_k$$

Step 4. Repeat this process with more and more subdivisions, so that the length of each subinterval approaches zero, and n, the number of subintervals, approaches $+\infty$. Define

$$\int_a^b f(x)\, dx = \lim_{n \to +\infty} \sum_{k=1}^{n} f(x_k^*)\, \Delta x_k$$

Recall that when $f(x)$ is nonnegative on the interval $[a, b]$, the Riemann sum

$$\sum_{k=1}^{n} f(x_k^*) \, \Delta x_k$$

can be interpreted as an approximation by rectangles to the area under the curve $y = f(x)$ over the interval $[a, b]$ (Figure 17.1.1) and the integral $\int_a^b f(x) \, dx$ can be interpreted as the exact area under the curve. In the case where $f(x)$ assumes both positive and negative values on $[a, b]$, the integral $\int_a^b f(x) \, dx$ can be interpreted as a difference of areas, the area above the x-axis between $[a, b]$ and $y = f(x)$ minus the area below the x-axis between $[a, b]$ and $y = f(x)$.

Figure 17.1.1

Figure 17.1.2

The definition of an integral for a function of two variables is a natural extension of these ideas. Whereas the integration of $f(x)$ takes place over a closed interval on the x-axis, the integration of $f(x, y)$ takes place over a closed region R in the xy-plane. Later, we shall place more restrictions on the region R, but for now let us just assume that the entire region R can be enclosed within some suitably large rectangle with sides parallel to the coordinate axes. This ensures that R does not extend indefinitely in any direction (Figure 17.1.2).

To define a double integral we proceed as follows:

Step 1. Using lines parallel to the coordinate axes, divide the rectangle enclosing the region R into subrectangles, and exclude from consideration all those subrectangles that contain any points outside of R. This leaves only rectangles that are subsets of R (Figure 17.1.3). Denote the areas of these rectangles by

$$\Delta A_1, \Delta A_2, \ldots, \Delta A_n$$

Step 2. Choose an arbitrary point in each of these subrectangles, and denote them by

$$(x_1^*, y_1^*), (x_2^*, y_2^*), \ldots, (x_n^*, y_n^*)$$

(see Figure 17.1.3).

Step 3. Form the sum

$$\sum_{k=1}^{n} f(x_k^*, y_k^*) \, \Delta A_k$$

This is called a ***Riemann sum.***

Step 4. Repeat this process with more and more subdivisions, so that the length and width of each rectangle approach zero; and n, the number of rectangles, approaches $+\infty$. Define

$$\iint\limits_{R} f(x, y) \, dA = \lim_{n \to +\infty} \sum_{k=1}^{n} f(x_k^*, y_k^*) \, \Delta A_k \qquad (1)$$

The symbol $\iint\limits_{R} f(x, y) \, dA$ is called the ***double integral*** of $f(x, y)$ over R.

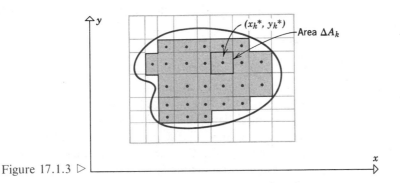

Figure 17.1.3 ▷

GEOMETRIC INTERPRETATION OF DOUBLE INTEGRALS

A precise definition of expression (1) (in terms of ϵ's and δ's) and conditions under which the double integral exists are studied in advanced calculus. However, for our purposes it suffices to say that existence is ensured when f is continuous on R and the region R is not too "complicated."

In the special case where $f(x, y)$ is *nonnegative* on the region R, the double integral may be interpreted as the volume of the solid S bounded

$z = f(x, y)$

Height $f(x_k{}^*, y_k{}^*)$

S

x

Area ΔA_k

$(x_k{}^*, y_k{}^*)$

y

Figure 17.1.4 ▷

above by the surface $z = f(x, y)$ and below by the region R (Figure 17.1.4). To see this, observe that the product

$$f(x_k^*, y_k^*)\, \Delta A_k$$

represents the volume of a rectangular parallelepiped whose height is $f(x_k^*, y_k^*)$ and whose base has area ΔA_k (Figure 17.1.4). Consequently, the Riemann sum

$$\sum_{k=1}^{n} f(x_k^*, y_k^*)\, \Delta A_k$$

may be interpreted as an approximation by rectangular parallelepipeds to the volume of the solid S. There are two sources of error in this approximation. First, the bases of the parallelepipeds will not, in general, cover the region R exactly and, second, the approximating parallelepipeds have flat tops, whereas the upper surface of the solid may be curved. However, it is intuitively clear that as we use more and more subrectangles of decreasing size, both errors diminish, so (1) gives the exact volume.

In the case where $f(x, y)$ has both positive and negative values on the region R, the double integral of f over R may be interpreted as a difference of volumes: the volume above the xy-plane between $z = f(x, y)$ and R minus the volume below the xy-plane between $z = f(x, y)$ and R. (Explain why.)

PROPERTIES OF DOUBLE INTEGRALS   Double integrals enjoy many properties analogous to those of single integrals:

$$\iint_R cf(x, y)\, dA = c \iint_R f(x, y)\, dA \quad (c \text{ a constant})$$

$$\iint_R [f(x, y) + g(x, y)]\, dA = \iint_R f(x, y)\, dA + \iint_R g(x, y)\, dA$$

$$\iint_R [f(x, y) - g(x, y)]\, dA = \iint_R f(x, y)\, dA - \iint_R g(x, y)\, dA$$

Moreover, if the region R is subdivided into two regions R_1 and R_2 (Figure 17.1.5), then

$$\iint_R f(x, y)\, dA = \iint_{R_1} f(x, y)\, dA + \iint_{R_2} f(x, y)\, dA$$

We omit the proofs.

Figure 17.1.5 ▷

EVALUATING
DOUBLE INTEGRALS

Except in the simplest cases, it is impractical to obtain the value of a double integral from the limit in (1). However, we shall now show how to evaluate double integrals by calculating two successive single integrals. For the remainder of this section, we shall limit our discussion to the case where R is a rectangle. In the next section we shall consider double integrals over more complicated regions.

The partial derivatives of a function $f(x, y)$ are calculated by holding one of the variables fixed and differentiating with respect to the other variable. Let us consider the reverse of this process, *partial integration.* The symbol

$$\int_a^b f(x, y)\, dx$$

is a *partial definite integral with respect to x.* It is evaluated by holding y fixed and integrating with respect to x. Similarly, the *partial definite integral with respect to y,*

$$\int_{c}^{d} f(x, y) \, dy$$

is evaluated by holding x fixed and integrating with respect to y.

Example 1

$$\int_{0}^{1} xy^2 \, dx = y^2 \int_{0}^{1} x \, dx = \frac{y^2 x^2}{2} \bigg]_{x=0}^{1} = \frac{y^2}{2}$$

$$\int_{0}^{1} xy^2 \, dy = x \int_{0}^{1} y^2 \, dy = \frac{xy^3}{3} \bigg]_{y=0}^{1} = \frac{x}{3} \quad \blacktriangleleft$$

As this example shows, an integral of the form $\int_a^b f(x, y) \, dx$ produces a function of y as the result, while an integral of the form $\int_c^d f(x, y) \, dy$ produces a function of x. This being the case, we can consider the following types of calculations:

$$\int_{c}^{d} \left[\int_{a}^{b} f(x, y) \, dx \right] dy \tag{2a}$$

$$\int_{a}^{b} \left[\int_{c}^{d} f(x, y) \, dy \right] dx \tag{2b}$$

In (2a), the inside integration, $\int_a^b f(x, y) \, dx$, yields a function of y, which is then integrated over the interval $c \le y \le d$. In (2b), the integration, $\int_c^d f(x, y) \, dy$, yields a function of x, which is then integrated over the interval $a \le x \le b$. Expressions (2a) and (2b) are called *iterated* (or *repeated*) *integrals*. Often the brackets are omitted and these expressions are written as

$$\int_{c}^{d} \int_{a}^{b} f(x, y) \, dx \, dy = \int_{c}^{d} \left[\int_{a}^{b} f(x, y) \, dx \right] dy \tag{3a}$$

$$\int_{a}^{b} \int_{c}^{d} f(x, y) \, dy \, dx = \int_{a}^{b} \left[\int_{c}^{d} f(x, y) \, dy \right] dx \tag{3b}$$

Example 2   Evaluate

(a) $\displaystyle\int_{0}^{3} \int_{1}^{2} (1 + 8xy) \, dy \, dx$ (b) $\displaystyle\int_{1}^{2} \int_{0}^{3} (1 + 8xy) \, dx \, dy$

Solution (a).

$$\int_0^3 \int_1^2 (1 + 8xy)\, dy\, dx = \int_0^3 \left[\int_1^2 (1 + 8xy)\, dy \right] dx$$

$$= \int_0^3 \left[y + 4xy^2 \right]_{y=1}^{2} dx$$

$$= \int_0^3 [(2 + 16x) - (1 + 4x)]\, dx$$

$$= \int_0^3 (1 + 12x)\, dx$$

$$= x + 6x^2 \Big|_0^3 = 57$$

Solution (b).

$$\int_1^2 \int_0^3 (1 + 8xy)\, dx\, dy = \int_1^2 \left[\int_0^3 (1 + 8xy)\, dx \right] dy$$

$$= \int_1^2 \left[x + 4x^2y \right]_{x=0}^{3} dy$$

$$= \int_1^2 (3 + 36y)\, dy$$

$$= 3y + 18y^2 \Big|_1^2 = 57 \quad \blacktriangleleft$$

It is no accident that the two iterated integrals in the last example have the same value; it is a consequence of the following theorem.

17.1.1   THEOREM. *Let R be the rectangle defined by the inequalities*

$$a \le x \le b, \quad c \le y \le d$$

If $f(x, y)$ is continuous on this rectangle, then

$$\iint\limits_R f(x, y)\, dA = \int_c^d \int_a^b f(x, y)\, dx\, dy = \int_a^b \int_c^d f(x, y)\, dy\, dx$$

This major theorem enables us to evaluate a double integral over a rectangle by calculating an iterated integral. Moreover, the theorem tells us that the order of integration in the iterated integral does not matter. We shall not formally prove this result. However, we offer the following geometric argument for the case where $f(x, y)$ is nonnegative on R.

If $f(x, y)$ is nonnegative on R, the double integral

$$\iint\limits_{R} f(x, y) \, dA$$

represents the volume of the solid S bounded above by the surface $z = f(x, y)$ and below by the rectangle R. However, as discussed in Section 6.2 (Formula 6.2.2), the volume of the solid S is also given by

$$\text{Vol}\,(S) = \int_{c}^{d} A(y) \, dy \tag{4}$$

where $A(y)$ represents the area of the cross section perpendicular to the y-axis taken at the point y (Figure 17.1.6). But consider how we might compute the cross-sectional area $A(y)$. For each *fixed* y in the interval $c \le y \le d$, the function $f(x, y)$ is a function of x alone, and $A(y)$ may be viewed as the area under the graph of this function along the interval $a \le x \le b$ (see the blue shaded area in Figure 17.1.6). Thus,

$$A(y) = \int_{a}^{b} f(x, y) \, dx$$

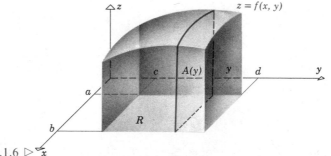

Figure 17.1.6 ▷

Substituting this expression in (4) yields

$$\text{Vol}\,(S) = \int_{c}^{d} \left[\int_{a}^{b} f(x, y) \, dx \right] dy = \int_{c}^{d} \int_{a}^{b} f(x, y) \, dx \, dy \tag{5}$$

The volume of the solid S can also be obtained using cross sections perpendicular to the x-axis. By Formula 6.2.1,

$$\text{Vol}(S) = \int_a^b A(x)\, dx \tag{6}$$

where $A(x)$ is the area of the cross section perpendicular to the x-axis taken at the point x (Figure 17.1.7). For each *fixed* x in the interval $a \le x \le b$, the function $f(x, y)$ is a function of y alone, so that the area $A(x)$ is given by

$$A(x) = \int_c^d f(x, y)\, dy$$

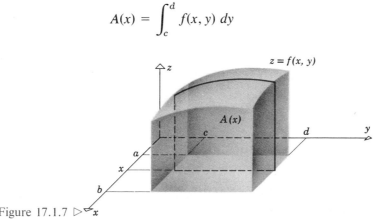

Figure 17.1.7 ▷

Substituting this expression in (6) yields

$$\text{Vol}(S) = \int_a^b \left[\int_c^d f(x, y)\, dy \right] dx = \int_a^b \int_c^d f(x, y)\, dy\, dx \tag{7}$$

Since the volume of S is also given by the double integral $\displaystyle\iint_R f(x, y)\, dA$, it follows from (5) and (7) that

$$\iint_R f(x, y)\, dA = \int_c^d \int_a^b f(x, y)\, dx\, dy = \int_a^b \int_c^d f(x, y)\, dy\, dx$$

which is what we intended to show.

Example 3 Evaluate the double integral

$$\iint_R y^2 x\, dA$$

over the rectangle $R = \{(x, y)\colon -3 \le x \le 2,\, 0 \le y \le 1\}$.

Solution.   In view of Theorem 17.1.1, the value of the double integral may be obtained from either of the iterated integrals

$$\int_{-3}^{2} \int_{0}^{1} y^2 x \, dy \, dx \quad \text{or} \quad \int_{0}^{1} \int_{-3}^{2} y^2 x \, dx \, dy \tag{8}$$

Using the first of these, we obtain

$$\iint_{R} y^2 x \, dA = \int_{-3}^{2} \int_{0}^{1} y^2 x \, dy \, dx = \int_{-3}^{2} \left[\frac{1}{3} y^3 x \right]_{y=0}^{1} dx$$

$$= \int_{-3}^{2} \frac{1}{3} x \, dx = \left. \frac{x^2}{6} \right]_{-3}^{2} = -\frac{5}{6}$$

The reader can check this result by evaluating the second integral in (8). ◀

$z = 4 - x - y$

$(1, 2)$

Figure 17.1.8

Example 4   Use a double integral to find the volume of the solid that is bounded above by the plane $z = 4 - x - y$ and below by the rectangle $R = \{(x, y): 0 \le x \le 1, 0 \le y \le 2\}$ (Figure 17.1.8).

Solution.

$$V = \iint_{R} (4 - x - y) \, dA = \int_{0}^{2} \int_{0}^{1} (4 - x - y) \, dx \, dy$$

$$= \int_{0}^{2} \left[4x - \frac{x^2}{2} - xy \right]_{x=0}^{1} dy = \int_{0}^{2} \left(\frac{7}{2} - y \right) dy$$

$$= \left[\frac{7}{2} y - \frac{y^2}{2} \right]_{0}^{2} = 5$$

The volume can also be obtained by first integrating with respect to y and then with respect to x.   ◀

▶ Exercise Set 17.1

In Exercises 1–14, evaluate the iterated integrals.

1. $\displaystyle\int_{0}^{1} \int_{0}^{2} (x + 3) \, dy \, dx.$

2. $\displaystyle\int_{1}^{3} \int_{-1}^{1} (2x - 4y) \, dy \, dx.$

3. $\displaystyle\int_{2}^{4} \int_{0}^{1} x^2 y \, dx \, dy.$

4. $\displaystyle\int_{-2}^{0} \int_{-1}^{2} (x^2 + y^2) \, dx \, dy.$

5. $\displaystyle\int_{0}^{\ln 3} \int_{0}^{\ln 2} e^{x+y} \, dy \, dx.$

6. $\displaystyle\int_{0}^{2} \int_{0}^{1} y \sin x \, dy \, dx.$

7. $\displaystyle\int_0^3 \int_0^1 x(x^2 + y)^{1/2} \, dx \, dy.$

8. $\displaystyle\int_{-1}^2 \int_2^4 (2x^2y + 3xy^2) \, dx \, dy.$

9. $\displaystyle\int_{-1}^0 \int_2^5 dx \, dy.$ **10.** $\displaystyle\int_4^6 \int_{-3}^7 dy \, dx.$

11. $\displaystyle\int_0^1 \int_0^1 \frac{x}{(xy + 1)^2} \, dy \, dx.$

12. $\displaystyle\int_{\pi/2}^\pi \int_1^2 x \cos xy \, dy \, dx.$

13. $\displaystyle\int_0^{\ln 2} \int_0^1 xy \, e^{y^2x} \, dy \, dx.$

14. $\displaystyle\int_3^4 \int_1^2 \frac{1}{(x + y)^2} \, dy \, dx.$

In Exercises 15–19, evaluate the double integral over the rectangular region R.

15. $\displaystyle\iint_R 4xy^3 \, dA;$

$R = \{(x, y): -1 \leq x \leq 1, -2 \leq y \leq 2\}.$

16. $\displaystyle\iint_R \frac{xy}{\sqrt{x^2 + y^2 + 1}} \, dA;$

$R = \{(x, y): 0 \leq x \leq 1, 0 \leq y \leq 1\}.$

17. $\displaystyle\iint_R x\sqrt{1 - x^2} \, dA;$

$R = \{(x, y): 0 \leq x \leq 1, 2 \leq y \leq 3\}.$

18. $\displaystyle\iint_R (x \sin y - y \sin x) \, dA;$

$R = \{(x, y): 0 \leq x \leq \pi/2, 0 \leq y \leq \pi/3\}.$

19. $\displaystyle\iint_R \cos (x + y) \, dA;$

$R = \{(x, y): -\pi/4 \leq x \leq \pi/4, 0 \leq y \leq \pi/4\}.$

In Exercises 20–25, the iterated integral represents the volume of a solid. Make an accurate sketch of the solid. (You do *not* have to find the volume.)

20. $\displaystyle\int_0^5 \int_1^2 4 \, dx \, dy.$

21. $\displaystyle\int_0^1 \int_0^1 (2 - x - y) \, dy \, dx.$

22. $\displaystyle\int_2^3 \int_3^4 y \, dx \, dy.$

23. $\displaystyle\int_0^3 \int_0^4 \sqrt{25 - x^2 - y^2} \, dy \, dx.$

24. $\displaystyle\int_{-2}^2 \int_{-2}^2 (x^2 + y^2) \, dx \, dy.$

25. $\displaystyle\int_0^1 \int_{-1}^1 \sqrt{4 - x^2} \, dy \, dx.$

In Exercises 26–30, use a double integral to find the volume.

26. The volume under the plane $z = 2x + y$ and over the rectangle $R = \{(x, y): 3 \leq x \leq 5, 1 \leq y \leq 2\}$.

27. The volume under the surface $z = 3x^3 + 3x^2y$ and over the rectangle $R = \{(x, y): 1 \leq x \leq 3, 0 \leq y \leq 2\}$.

28. The volume in the first octant bounded by the coordinate planes, the plane $y = 4$, and the plane $x/3 + z/5 = 1$.

29. The volume of the solid in the first octant enclosed by the surface $z = x^2$ and the planes $x = 2, y = 3$, $y = 0$, and $z = 0$.

30. The volume of the solid in the first octant that is enclosed by the planes $x = 0$, $z = 0$, $x = 5$, $z - y = 0$, and $z = -2y + 6$. [*Hint:* Break the solid into two parts.]

31. Suppose that $f(x, y) = g(x)h(y)$ and $R = \{(x, y): a \leq x \leq b, c \leq y \leq d\}$. Show that

$$\iint_R f(x, y) \, dA = \left[\int_a^b g(x) \, dx\right]\left[\int_c^d h(y) \, dy\right]$$

32. Evaluate

$$\iint_R x \cos (xy) \cos^2 \pi x \, dA$$

where $R = \{(x, y): 0 \leq x \leq \frac{1}{2}, 0 \leq y \leq \pi\}$. [*Hint:* One order of integration leads to a simpler solution than the other.]

17.2   DOUBLE INTEGRALS OVER NONRECTANGULAR REGIONS

In this section we shall show how to evaluate double integrals over regions other than rectangles.

Until now we have considered only iterated integrals with constant limits of integration. However, our work in this section will lead us to iterated integrals of the following types:

$$\int_a^b \int_{g_1(x)}^{g_2(x)} f(x, y)\, dy\, dx = \int_a^b \left[\int_{g_1(x)}^{g_2(x)} f(x, y)\, dy \right] dx \tag{1a}$$

$$\int_c^d \int_{h_1(y)}^{h_2(y)} f(x, y)\, dx\, dy = \int_c^d \left[\int_{h_1(y)}^{h_2(y)} f(x, y)\, dx \right] dy \tag{1b}$$

Example 1   Evaluate

$$\int_0^2 \int_{x^2}^x y^2 x\, dy\, dx$$

Solution.

$$\int_0^2 \int_{x^2}^x y^2 x\, dy\, dx = \int_0^2 \left[\int_{x^2}^x y^2 x\, dy \right] dx = \int_0^2 \left[\frac{y^3 x}{3} \right]_{y=x^2}^x dx$$

$$= \int_0^2 \left(\frac{x^4}{3} - \frac{x^7}{3} \right) dx = \left[\frac{x^5}{15} - \frac{x^8}{24} \right]_0^2$$

$$= \frac{32}{15} - \frac{256}{24} = -\frac{128}{15} \quad \blacktriangleleft$$

Example 2   Evaluate

$$\int_0^\pi \int_0^{\cos y} x \sin y\, dx\, dy$$

Solution.

$$\int_0^\pi \int_0^{\cos y} x \sin y\, dx\, dy = \int_0^\pi \left[\int_0^{\cos y} x \sin y\, dx \right] dy$$

$$= \int_0^\pi \left[\frac{x^2}{2} \sin y \right]_{x=0}^{\cos y} dy$$

$$= \int_0^\pi \frac{1}{2} \cos^2 y \sin y\, dy$$

$$= \left[-\frac{1}{6} \cos^3 y \right]_0^\pi = \frac{1}{3} \quad \blacktriangleleft$$

We shall be concerned with evaluating double integrals over two types of closed regions, which we shall call *type I* and *type II* (Figure 17.2.1).

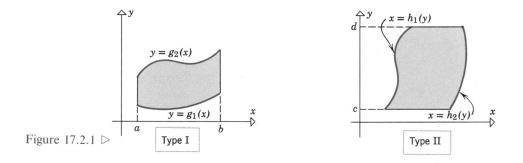

Figure 17.2.1 ▷

Type I

Type II

A type I region is bounded on the left and right by vertical lines $x = a$ and $x = b$ and is bounded below and above by smooth curves $y = g_1(x)$ and $y = g_2(x)$, where $g_1(x) \le g_2(x)$ for $a \le x \le b$.

A type II region is bounded below and above by horizontal lines $y = c$ and $y = d$ and is bounded on the left and right by smooth curves $x = h_1(y)$ and $x = h_2(y)$ satisfying $h_1(y) \le h_2(y)$ for $c \le y \le d$.

The following theorem will enable us to evaluate double integrals over type I and type II regions using iterated integrals.

17.2.1   THEOREM.

(a)   *If R is a type I region on which $f(x, y)$ is continuous, then*

$$\iint\limits_{R} f(x, y)\, dA = \int_a^b \int_{g_1(x)}^{g_2(x)} f(x, y)\, dy\, dx \qquad (2a)$$

(b)   *If R is a type II region on which $f(x, y)$ is continuous, then*

$$\iint\limits_{R} f(x, y)\, dA = \int_c^d \int_{h_1(y)}^{h_2(y)} f(x, y)\, dx\, dy \qquad (2b)$$

Although we shall not prove this theorem, the results are easy to visualize geometrically. Suppose $f(x, y)$ is nonnegative on a type I region R, so that the integral

$$\iint\limits_{R} f(x, y)\, dA \tag{3}$$

represents the volume of the solid S bounded above by the surface $z = f(x, y)$ and below by the region R. By the method of cross sections, the volume of S is also given by

$$\text{Vol}(S) = \int_{a}^{b} A(x)\, dx \tag{4}$$

where $A(x)$ is the area of the cross section at the fixed point x. As shown in Figure 17.2.2, this cross-sectional area extends from $g_1(x)$ to $g_2(x)$ in the y-direction, so

$$A(x) = \int_{g_1(x)}^{g_2(x)} f(x, y)\, dy$$

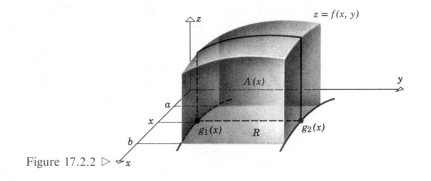

Figure 17.2.2 ▷

Substituting this in (4) we obtain

$$\text{Vol}(S) = \int_{a}^{b} \int_{g_1(x)}^{g_2(x)} f(x, y)\, dy\, dx$$

Since the volume of S is also given by (3), we obtain

$$\iint\limits_{R} f(x, y)\, dA = \int_{a}^{b} \int_{g_1(x)}^{g_2(x)} f(x, y)\, dy\, dx$$

Part (b) of Theorem 17.2.1 can be motivated in a similar way by using cross sections parallel to the x-axis.

To apply Theorem 17.2.1, it is usual to start with a two-dimensional sketch of the region R. [It is not necessary to graph $f(x, y)$.] For a type I region, the limits of integration in the formula

$$\iint\limits_{R} f(x, y)\, dA = \int_{a}^{b} \int_{g_1(x)}^{g_2(x)} f(x, y)\, dy\, dx \tag{5}$$

may be obtained as follows:

Step 1. Since x is held fixed for the first integration, we draw a vertical line through the region R at an arbitrary fixed point x (Figure 17.2.3). This line crosses the boundary of R twice. The lower point of intersection is on the curve $y = g_1(x)$ and the higher point is on the curve $y = g_2(x)$. These intersections determine the y-limits of integration in Formula (5).

Step 2. Imagine moving the line drawn in Step 1 first to the left and then to the right (Figure 17.2.3). The leftmost position where the line intersects the region R is $x = a$ and the rightmost position where the line intersects the region R is $x = b$. This yields the limits for the x-integration in Formula (5).

   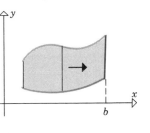

Figure 17.2.3

Example 3 Evaluate

$$\iint\limits_{R} xy\, dA$$

over the region R enclosed between $y = \tfrac{1}{2}x$, $y = \sqrt{x}$, $x = 2$, and $x = 4$.

Solution. We view R as a type I region. The region R and a vertical line corresponding to a fixed x are shown in Figure 17.2.4. This line meets the region R at the lower boundary $y = \tfrac{1}{2}x$ and the upper boundary $y = \sqrt{x}$. These are the y-limits of integration. Moving this line first left and then right yields the x-limits of integration, $x = 2$ and $x = 4$. Thus,

$$\iint\limits_{R} xy\, dA = \int_{2}^{4} \int_{x/2}^{\sqrt{x}} xy\, dy\, dx = \int_{2}^{4} \left[\frac{xy^2}{2} \right]_{y=x/2}^{\sqrt{x}} dx = \int_{2}^{4} \left(\frac{x^2}{2} - \frac{x^3}{8} \right) dx$$

$$= \left[\frac{x^3}{6} - \frac{x^4}{32} \right]_{2}^{4} = \left(\frac{64}{6} - \frac{256}{32} \right) - \left(\frac{8}{6} - \frac{16}{32} \right) = \frac{11}{6} \quad \blacktriangleleft$$

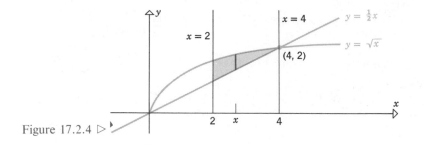

Figure 17.2.4 ▷

If R is a type II region, then the limits of integration in the formula

$$\iint_R f(x, y) \, dA = \int_c^d \int_{h_1(y)}^{h_2(y)} f(x, y) \, dx \, dy \tag{6}$$

may be obtained as follows:

Step 1. Since y is held fixed for the first integration, we draw a horizontal line through the region R at a fixed point y (Figure 17.2.5). This line crosses the boundary of R twice. The leftmost point of intersection is on the curve $x = h_1(y)$ and the rightmost point is on the curve $x = h_2(y)$. These intersections determine x-limits of integration in (6).

Step 2. Imagine moving the line drawn in Step 1 first down and then up (Figure 17.2.5). The lowest position where the line intersects the region R is $y = c$, and the highest position where the line intersects the region R is $y = d$. This yields the y-limits of integration in (6).

Figure 17.2.5

Figure 17.2.6

Example 4   Evaluate

$$\iint_R (2x - y^2)\, dA$$

over the triangular region R enclosed between the lines $y = -x + 1$, $y = x + 1$, and $y = 3$.

Solution.   We view R as a type II region. The region R and a horizontal line corresponding to a fixed y are shown in Figure 17.2.6. This line meets the region R at its left-hand boundary $x = 1 - y$ and its right-hand boundary $x = y - 1$. These are the x-limits of integration. Moving this line first down and then up yields the y-limits, $y = 1$ and $y = 3$. Thus,

$$\iint_R (2x - y^2)\, dA = \int_1^3 \int_{1-y}^{y-1} (2x - y^2)\, dx\, dy = \int_1^3 \left[x^2 - y^2 x \right]_{x=1-y}^{y-1} dy$$

$$= \int_1^3 [(1 - 2y + 2y^2 - y^3) - (1 - 2y + y^3)]\, dy$$

$$= \int_1^3 (2y^2 - 2y^3)\, dy = \left[\frac{2y^3}{3} - \frac{y^4}{2} \right]_1^3 = -\frac{68}{3} \quad \blacktriangleleft$$

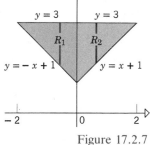

Figure 17.2.7

REMARK.   To integrate over a type II region, the left- and right-hand boundaries must be expressed in the form $x = h_1(y)$ and $x = h_2(y)$. This is why we rewrote the boundary equations $y = -x + 1$ and $y = x + 1$ as $x = 1 - y$ and $x = y - 1$ in the last example.

In Example 4 we could have treated R as a type I region, but with an added complication. Viewed as a type I region, the upper boundary of R is the line $y = 3$ (Figure 17.2.7) and the lower boundary consists of two parts, the line $y = -x + 1$ to the left of the origin and the line $y = x + 1$ to the right of the origin. To carry out the integration it is necessary to decompose the region R into two parts, R_1 and R_2, as shown in Figure 17.2.7, and write:

$$\iint_R (2x - y^2)\, dA = \iint_{R_1} (2x - y^2)\, dA + \iint_{R_2} (2x - y^2)\, dA$$

$$= \int_{-2}^0 \int_{-x+1}^3 (2x - y^2)\, dy\, dx + \int_0^2 \int_{x+1}^3 (2x - y^2)\, dy\, dx$$

This will yield the same result that was obtained in Example 4.

REVERSING THE ORDER OF INTEGRATION

Sometimes the evaluation of an iterated integral can be simplified by reversing the order of integration. The next example illustrates how this is done.

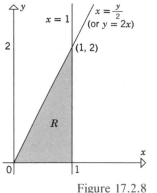

Figure 17.2.8

Example 5   Since there is no elementary antiderivative of e^{x^2}, the integral

$$\int_0^2 \int_{y/2}^1 e^{x^2} \, dx \, dy$$

cannot be evaluated by performing the x-integration first. Evaluate this integral by expressing it as an equivalent iterated integral with the order of integration reversed.

Solution.   For the inside integration, y is fixed and x varies from the line $x = y/2$ to the line $x = 1$ (Figure 17.2.8). For the outside integration, y varies from 0 to 2, so the given iterated integral is equal to a double integral over the triangular region R in Figure 17.2.8.

To reverse the order of integration, we treat R as a type I region, which enables us to write the given integral as,

$$\int_0^2 \int_{y/2}^1 e^{x^2} \, dx \, dy = \iint_R e^{x^2} \, dA = \int_0^1 \int_0^{2x} e^{x^2} \, dy \, dx = \int_0^1 \left[e^{x^2} y \right]_{y=0}^{2x} dx$$

$$= \int_0^1 2x e^{x^2} \, dx = e^{x^2} \bigg]_0^1 = e - 1 \quad \blacktriangleleft$$

The double integral

$$\iint_R 1 \, dA = \iint_R dA$$

represents the volume of a solid of constant height 1 above the region R. Numerically, this is the same as the area of the region R. Thus,

Area Calculated as a Double Integral.

$$\text{area of } R = \iint_R dA \tag{7}$$

Example 6   Use a double integral to find the area of the region R enclosed between the parabola $y = \frac{1}{2}x^2$ and the line $y = 2x$.

Solution.   The region R may be treated equally well as type I (Figure 17.2.9a) or type II (Figure 17.2.9b). Treating R as type I yields,

$$\text{area of } R = \iint_R dA = \int_0^4 \int_{x^2/2}^{2x} dy \, dx = \int_0^4 \left[y \right]_{y=x^2/2}^{2x} dx$$

$$= \int_0^4 \left(2x - \frac{1}{2}x^2 \right) dx = \left[x^2 - \frac{x^3}{6} \right]_0^4 = \frac{16}{3}$$

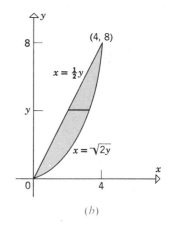

Figure 17.2.9 ▷   (a)   (b)

Treating R as type II yields

$$\text{area of } R = \iint_R dA = \int_0^8 \int_{y/2}^{\sqrt{2y}} dx \, dy = \int_0^8 \left[x \right]_{x=y/2}^{\sqrt{2y}} dy$$

$$= \int_0^8 \left(\sqrt{2y} - \frac{1}{2}y \right) dy = \left[\frac{2\sqrt{2}}{3} y^{3/2} - \frac{y^2}{4} \right]_0^8 = \frac{16}{3} \quad \blacktriangleleft$$

Figure 17.2.10

Example 7 Use a double integral to find the volume of the tetrahedron bounded by the coordinate planes and the plane $z = 4 - 4x - 2y$.

Solution. The tetrahedron in question is bounded above by the plane

$$z = 4 - 4x - 2y \tag{8}$$

and below by the triangular region R shown in Figure 17.2.10. Thus, the volume is given by

$$V = \iint_R (4 - 4x - 2y) \, dA$$

The region R is bounded by the x-axis, the y-axis, and the line $y = 2 - 2x$ [set $z = 0$ in (8)], so that treating R as a type I region yields

$$V = \iint_R (4 - 4x - 2y) \, dA = \int_0^1 \int_0^{2-2x} (4 - 4x - 2y) \, dy \, dx$$

$$= \int_0^1 \left[4y - 4xy - y^2 \right]_{y=0}^{2-2x} dx = \int_0^1 (4 - 8x + 4x^2) \, dx = \frac{4}{3} \quad \blacktriangleleft$$

Figure 17.2.11

Example 8   Find the volume of the solid bounded by the cylinder $x^2 + y^2 = 4$ and the planes $y + z = 4$ and $z = 0$.

Solution.   The solid shown in Figure 17.2.11 is bounded above by the plane $z = 4 - y$ and below by the region R within the circle $x^2 + y^2 = 4$. The volume is given by

$$V = \iint_R (4 - y) \, dA$$

Treating R as a type I region we obtain

$$V = \int_{-2}^{2} \int_{-\sqrt{4-x^2}}^{\sqrt{4-x^2}} (4 - y) \, dy \, dx = \int_{-2}^{2} \left[4y - \frac{1}{2} y^2 \right]_{y=-\sqrt{4-x^2}}^{\sqrt{4-x^2}} dx$$

$$= \int_{-2}^{2} 8\sqrt{4 - x^2} \, dx = 8(2\pi) = 16\pi \quad \boxed{\text{See Example 4 of Section 9.5.}} \quad ◄$$

► Exercise Set 17.2

In Exercises 1–12, evaluate the iterated integral.

1. $\displaystyle\int_0^1 \int_{x^2}^x xy^2 \, dy \, dx.$ **2.** $\displaystyle\int_1^2 \int_y^{3-y} y \, dx \, dy.$

3. $\displaystyle\int_0^3 \int_0^{\sqrt{9-y^2}} y \, dx \, dy.$ **4.** $\displaystyle\int_{1/4}^1 \int_{x^2}^x \sqrt{\frac{x}{y}} \, dy \, dx.$

5. $\displaystyle\int_{\sqrt{\pi}}^{\sqrt{2\pi}} \int_0^{x^3} \sin\frac{y}{x} \, dy \, dx.$

6. $\displaystyle\int_{-1}^1 \int_{-x^2}^{x^2} (x^2 - y) \, dy \, dx.$

7. $\displaystyle\int_{\pi/2}^{\pi} \int_0^{x^2} \frac{1}{x}\cos\frac{y}{x} \, dy \, dx.$

8. $\displaystyle\int_0^{\pi/2} \int_0^{\sin y} e^x \cos y \, dx \, dy.$

9. $\displaystyle\int_0^a \int_0^{\sqrt{a^2-x^2}} (x + y) \, dy \, dx.$

10. $\displaystyle\int_1^2 \int_0^{y^2} e^{x/y^2} \, dx \, dy.$

11. $\displaystyle\int_0^1 \int_0^x y\sqrt{x^2 - y^2} \, dy \, dx.$

12. $\displaystyle\int_0^1 \int_0^x e^{x^2} \, dy \, dx.$

In Exercises 13–28, evaluate the double integral.

13. $\displaystyle\iint_R 6xy \, dA$; R is the region bounded by $y = 0$, $x = 2$, and $y = x^2$.

14. $\displaystyle\iint_R xy \, dA$; R is the region bounded by the trapezoid with vertices $(1, 3)$, $(5, 3)$, $(2, 1)$, and $(4, 1)$.

15. $\displaystyle\iint_R x \cos xy \, dA$; R is the region enclosed by $x = 1$, $x = 2$, $y = \pi/2$, and $y = 2\pi/x$.

16. $\displaystyle\iint_R (x + y) \, dA$; R is the region enclosed between the curves $y = x^2$ and $y = \sqrt{x}$.

17. $\displaystyle\iint_R x^2 \, dA$; R is the region bounded by $y = 16/x$, $y = x$, and $x = 8$.

18. $\displaystyle\iint_R xy^2 \, dA$; R is the region enclosed by $y = 1$, $y = 2$, $x = 0$, and $y = x$.

19. $\displaystyle\iint_R x(1 + y^2)^{-1/2} \, dA$; R is the region in the first quadrant enclosed by $y = x^2$, $y = 4$, and $x = 0$. [*Hint:* Choose your order of integration carefully.]

20. $\displaystyle\iint_R x \cos y \, dA$; R is the triangular region bounded by $y = x$, $y = 0$, and $x = \pi$.

21. $\displaystyle\iint_R (3x - 2y) \, dA$; R is the region enclosed by the circle $x^2 + y^2 = 1$.

22. $\displaystyle\iint_R y \, dA$; R is the region in the first quadrant enclosed between the circle $x^2 + y^2 = 25$ and the line $x + y = 5$.

23. $\displaystyle\iint_R \frac{1}{1 + x^2} \, dA$; R is the triangular region with vertices $(0, 0)$, $(1, 1)$, and $(0, 1)$.

24. $\displaystyle\iint_R (x^2 - xy) \, dA$; R is the region enclosed by $y = x$ and $y = 3x - x^2$.

25. $\displaystyle\iint_R xy \, dA$; R is the region enclosed by $y = \sqrt{x}$, $y = 6 - x$, and $y = 0$.

26. $\displaystyle\iint_R x \, dA$; R is the region enclosed by $y = \sin^{-1} x$, $x = 1/\sqrt{2}$, and $y = 0$.

27. $\displaystyle\iint_R (x - 1) \, dA$; R is the region enclosed between $y = x$ and $y = x^3$.

28. $\displaystyle\iint_R x^2 \, dA$; R is the region in the first quadrant enclosed by $xy = 1$, $y = x$, and $y = 2x$.

In Exercises 29–34, use double integration to find the area of the plane region enclosed by the given curves.

29. $x + y = 5$, $x = 0$, and $y = 0$.

30. $y = x^2$ and $y = 4x$.

31. $y = \sin x$ and $y = \cos x$, for $0 \le x \le \pi/4$.

32. $y^2 = -x$ and $3y - x = 4$.

33. $y^2 = 9 - x$ and $y^2 = 9 - 9x$.

34. $y = \cosh x$, $y = \sinh x$, $x = 0$, and $x = 1$.

In Exercises 35–48, use double integration to find the volume of each solid.

35. The tetrahedron in the first octant that is bounded by the three coordinate planes and the plane $z = 5 - 2x - y$.

36. The solid bounded by the cylinder $x^2 + y^2 = 9$ and the planes $z = 0$ and $z = 3 - x$.

37. The solid bounded above by the plane $z = x + 2y + 2$, below by the xy-plane, and laterally by $y = 0$ and $y = 1 - x^2$.

38. The solid in the first octant bounded above by the paraboloid $z = x^2 + 3y^2$, below by the plane $z = 0$, and laterally by $y = x^2$ and $y = x$.

39. The solid bounded above by the paraboloid $z = 9x^2 + y^2$, below by the plane $z = 0$, and laterally by the planes $x = 0$, $y = 0$, $x = 3$, and $y = 2$.

40. The solid enclosed by $y^2 = x$, $z = 0$, and $x + z = 1$.

41. The wedge cut from the cylinder $4x^2 + y^2 = 9$ by the planes $z = 0$ and $z = y + 3$.

42. The solid in the first octant bounded above by $z = 9 - x^2$, below by $z = 0$, and laterally by $y^2 = 3x$.

43. The solid in the first octant bounded by the coordinate planes and the planes $x + 2y = 4$ and $x + 8y - 4z = 0$.

44. The solid in the first octant bounded by the surface $z = e^{y-x}$, the plane $x + y = 1$, and the coordinate planes.

45. The solid bounded above by the paraboloid $z = 1 - x^2 - y^2$ and below by the xy-plane. [*Hint:*

Use a trigonometric substitution to evaluate the integral.]

46. The solid in the first octant bounded by the paraboloid $z = x^2 + y^2$ and the cylinder $x^2 + y^2 = 4$. [*Hint:* Use a trigonometric substitution to evaluate the integral.]

47. The solid common to the cylinders $x^2 + y^2 = 25$ and $x^2 + z^2 = 25$.

48. The solid bounded above by the paraboloid $z = x^2 + y^2$, laterally by the cylinder $x^2 + (y - 1)^2 = 1$, and below by the xy-plane.

In Exercises 49–56, express the integral as an equivalent integral with the order of integration reversed.

49. $\displaystyle\int_0^2 \int_0^{\sqrt{x}} f(x, y)\, dy\, dx.$

50. $\displaystyle\int_0^4 \int_{2y}^8 f(x, y)\, dx\, dy.$

51. $\displaystyle\int_0^2 \int_1^{e^y} f(x, y)\, dx\, dy.$

52. $\displaystyle\int_1^e \int_0^{\ln x} f(x, y)\, dy\, dx.$

53. $\displaystyle\int_{-2}^2 \int_{-\sqrt{1-(x^2/4)}}^{\sqrt{1-(x^2/4)}} f(x, y)\, dy\, dx.$

54. $\displaystyle\int_0^1 \int_{y^2}^{\sqrt{y}} f(x, y)\, dx\, dy.$

55. $\displaystyle\int_0^1 \int_{\sin^{-1}y}^{\pi/2} f(x, y)\, dx\, dy.$

56. $\displaystyle\int_{-3}^1 \int_{x^2+6x}^{4x+3} f(x, y)\, dy\, dx.$

In Exercises 57–62, evaluate the integral by first reversing the order of integration.

57. $\displaystyle\int_0^1 \int_{4x}^4 e^{-y^2}\, dy\, dx.$

58. $\displaystyle\int_0^2 \int_{y/2}^1 \cos(x^2)\, dx\, dy.$

59. $\displaystyle\int_0^4 \int_{\sqrt{y}}^2 e^{x^3}\, dx\, dy.$

60. $\displaystyle\int_1^3 \int_0^{\ln x} x\, dy\, dx.$

61. $\displaystyle\int_0^1 \int_0^{\cos^{-1}x} x\, dy\, dx.$

62. $\displaystyle\int_0^1 \int_{\sin^{-1}y}^{\pi/2} \sec^2(\cos x)\, dx\, dy.$

63. Evaluate $\displaystyle\iint_R \sin(y^3)\, dA$, where R is the region bounded by $y = \sqrt{x}$, $y = 2$, and $x = 0$. [*Hint:* Choose the order of integration carefully.]

64. Evaluate $\displaystyle\iint_R x\, dA$, where R is the region bounded by $x = \ln y$, $x = 0$, and $y = e$ [*Hint:* Choose the order of integration carefully.]

65. In each part evaluate $\displaystyle\iint_R xy^2\, dA$.

(a)

(b)

66. Let R be a region consisting of subregions R_1 and R_2 that are symmetric with respect to the y-axis, and suppose that $\displaystyle\iint_{R_1} f(x, y)\, dA = 3$.

(a) Find $\displaystyle\iint_R f(x, y)\, dA$ if $f(-x, y) = f(x, y)$.

(b) Find $\iint\limits_{R} f(x, y)\, dA$ if $f(-x, y) = -f(x, y)$.

68. $\displaystyle\int_{0}^{2} \int_{x-2}^{2-x} \sin(xy^3)\, dy\, dx.$

In Exercises 67–69, evaluate the integral without performing an integration. [*Hint:* See Exercise 66.]

69. $\displaystyle\iint\limits_{R} x^3 y\, dA$, where R is the semicircular region bounded by $y = \sqrt{4 - x^2}$ and $x = 0$.

67. $\displaystyle\int_{-1}^{1} \int_{0}^{\sqrt{1-x^2}} (2 + x\sqrt{9 - y^2})\, dy\, dx.$

17.3   DOUBLE INTEGRALS IN POLAR COORDINATES

In this section we shall study double integrals in which the integrand and the region of integration are expressed in polar coordinates. Such integrals are important for two reasons: First, they arise naturally in many applications, and second, many double integrals in rectangular coordinates are more easily evaluated if they are converted to polar coordinates.

DEFINITION OF DOUBLE INTEGRALS IN POLAR COORDINATES

Our first objective is to define integrals of the type

$$\iint\limits_{R} f(r, \theta)\, dA$$

where R is a closed region in the xy-plane expressed in polar coordinates, and (r, θ) are polar coordinates of a point in the xy-plane. If we introduce a dependent variable

$$z = f(r, \theta)$$

then this equation represents a three-dimensional surface expressed in cylindrical coordinates (Figure 17.3.1). For simplicity, we shall assume that the region R has the form shown in Figure 17.3.2a. The region is

Figure 17.3.1

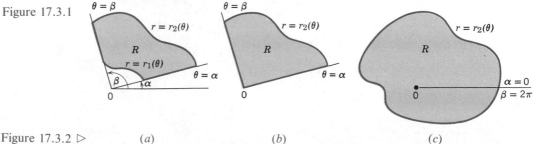

Figure 17.3.2 ▷   (a)   (b)   (c)

enclosed between two rays $\theta = \alpha$ and $\theta = \beta$ ($\beta > \alpha$) and between two smooth polar curves $r = r_1(\theta)$ and $r = r_2(\theta)$, which satisfy $r_1(\theta) \leq r_2(\theta)$ for $\alpha \leq \theta \leq \beta$.

In the special case where $r_1(\theta)$ is identically zero, the graph of $r_1(\theta)$ reduces to a single point, the origin, and the region R has a form like that in Figure 17.3.2b. If, in addition, $\alpha = 0$ and $\beta = 2\pi$, then R has a form like that in Figure 17.3.2c.

If $f(r, \theta)$ is defined on a region R having one of the above forms, then we define the *double integral* of $f(r, \theta)$ over R as follows:

Step 1. Cover the region R with a grid of circular arcs centered at the origin and rays emanating from the origin (Figure 17.3.3). The blocks formed by this grid are called ***polar rectangles***. Exclude from consideration all polar rectangles that contain any points outside of R. This leaves only polar rectangles that are subsets of R. Denote the areas of these polar rectangles by

$$\Delta A_1, \Delta A_2, \ldots, \Delta A_n$$

Step 2. Choose an arbitrary point in each of these polar rectangles, and denote them by

$$(r_1^*, \theta_1^*), (r_2^*, \theta_2^*), \ldots, (r_n^*, \theta_n^*)$$

(Figure 17.3.3).

Figure 17.3.3 ▷

Step 3. Form the ***Riemann sum***

$$\sum_{k=1}^{n} f(r_k^*, \theta_k^*) \, \Delta A_k$$

Step 4. Repeat this process using more and more subdivisions in such a way that the dimensions of all the polar rectangles tend to zero. Thus n, the number of interior polar rectangles, will tend to $+\infty$. Define

$$\iint_R f(r, \theta)\, dA = \lim_{n \to +\infty} \sum_{k=1}^{n} f(r_k^*, \theta_k^*)\, \Delta A_k$$

The symbol $\displaystyle\iint_R f(r, \theta)\, dA$ is called the **polar double integral** of $f(r, \theta)$ over R.

GEOMETRIC
INTERPRETATION OF
POLAR DOUBLE
INTEGRALS

In the special case where $f(r, \theta)$ is nonnegative on the region R, the polar double integral may be interpreted as the volume of the solid S bounded above by the surface $z = f(r, \theta)$ and below by the region R (Figure 17.3.4). To see this, observe that the product

$$f(r_k^*, \theta_k^*)\, \Delta A_k \tag{1}$$

represents the volume of a cylinder whose base is the kth polar rectangle and whose height is $f(r_k^*, \theta_k^*)$. Thus, (1) is an approximation to the volume

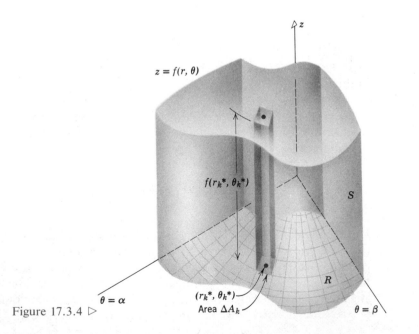

Figure 17.3.4 ▷

of the portion of S above the kth polar rectangle; and the sum

$$\sum_{k=1}^{n} f(r_k^*, \theta_k^*) \, \Delta A_k$$

is an approximation to the entire volume of S. There are two sources of error in this approximation. First, the polar rectangles may not fill up the region R exactly and, second, the approximating solids have flat tops, whereas the upper surface of the solid may be curved. However, as we use more and more polar rectangles with decreasing dimensions, it is intuitively clear that the errors in the approximations diminish and the limit in Step 4 above gives the exact volume.

In the case where $f(r, \theta)$ has both positive and negative values on the region R, the double polar integral of $f(r, \theta)$ over R may be interpreted as a difference of volumes: the volume above the xy-plane between $z = f(r, \theta)$ and R minus the volume below the xy-plane between $z = f(r, \theta)$ and R.

EVALUATION OF
POLAR DOUBLE
INTEGRALS

The following theorem will enable us to evaluate double polar integrals by using iterated integrals.

17.3.1   THEOREM. *If R is a region of the type shown in Figure 17.3.2 and if $f(r, \theta)$ is continuous on R, then*

$$\iint\limits_{R} f(r, \theta) \, dA = \int_{\alpha}^{\beta} \int_{r_1(\theta)}^{r_2(\theta)} f(r, \theta) r \, dr \, d\theta \qquad (2)$$

REMARK.   In (2) note that the dA in the double integral becomes $r \, dr \, d\theta$ in the iterated integral.

We shall not formally prove this theorem. However, the appearance of the factor r in the iterated integral may be explained by returning to the definition

$$\iint\limits_{R} f(r, \theta) \, dA = \lim_{n \to +\infty} \sum_{k=1}^{n} f(r_k^*, \theta_k^*) \, \Delta A_k \qquad (3)$$

In the Riemann sum $\sum_{k=1}^{n} f(r_k^*, \theta_k^*) \, \Delta A_k$, suppose the arbitrary point (r_k^*, θ_k^*) is chosen at the "center" of the kth polar rectangle, that is, at the point halfway between the bounding circular arcs on the ray that bisects them (Figure 17.3.5). Suppose also that this polar rectangle has a central angle $\Delta \theta_k$ and a "radial thickness" Δr_k. Thus, the inner radius of this polar rectangle is $r_k^* - \frac{1}{2}\Delta r_k$ and the outer radius is $r_k^* + \frac{1}{2}\Delta r_k$. Treating

Figure 17.3.5

the area ΔA_k of this polar rectangle as the difference in area of two sectors, we obtain

$$\Delta A_k = \frac{1}{2}\left(r_k^* + \frac{1}{2}\Delta r_k\right)^2 \Delta \theta_k - \frac{1}{2}\left(r_k^* - \frac{1}{2}\Delta r_k\right)^2 \Delta \theta_k$$

which simplifies to

$$\Delta A_k = r_k^* \, \Delta r_k \, \Delta \theta_k \tag{4}$$

(Verify.) Substituting this expression in (3) yields

$$\iint\limits_R f(r, \theta) \, dA = \lim_{n \to +\infty} \sum_{k=1}^{n} f(r_k^*, \theta_k^*) r_k^* \, \Delta r_k \, \Delta \theta_k$$

which explains the form of the integrand in (2).

To apply Theorem 17.3.1, we start with a sketch of the region R. From this sketch the limits of integration in the formula

$$\iint\limits_R f(r, \theta) \, dA = \int_\alpha^\beta \int_{r_1(\theta)}^{r_2(\theta)} f(r, \theta) r \, dr \, d\theta \tag{5}$$

can be obtained as follows:

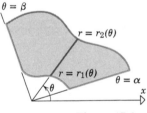

Figure 17.3.

Step 1. Since θ is held fixed for the first integration, draw a radial line from the origin through the region R at a fixed angle θ (Figure 17.3.6). This line crosses the boundary of R at most twice. The innermost point of intersection is on the curve $r = r_1(\theta)$ and the outermost point is on the curve $r = r_2(\theta)$. These intersections determine the r-limits of integration in (5).

Step 2. Imagine rotating a ray along the positive x-axis one revolution counterclockwise about the origin. The smallest angle at which this ray intersects the region R is $\theta = \alpha$ and the largest angle is $\theta = \beta$. This yields the θ-limits of integration.

Example 1 Evaluate

$$\iint\limits_R \sin \theta \, dA$$

where R is the region in the first quadrant that is outside the circle $r = 2$ and inside the cardioid $r = 2(1 + \cos \theta)$.

Figure 17.3.7

Solution. The region R is sketched in Figure 17.3.7. Following the two steps outlined above we obtain

$$\iint\limits_R \sin\theta \, dA = \int_0^{\pi/2} \int_2^{2(1+\cos\theta)} (\sin\theta)r \, dr \, d\theta$$

$$= \int_0^{\pi/2} \frac{1}{2}r^2 \sin\theta \Bigg]_{r=2}^{2(1+\cos\theta)} d\theta$$

$$= 2\int_0^{\pi/2} [(1+\cos\theta)^2 \sin\theta - \sin\theta] \, d\theta$$

$$= 2\left[-\frac{1}{3}(1+\cos\theta)^3 + \cos\theta \right]_0^{\pi/2}$$

$$= 2\left[-\frac{1}{3} - \left(-\frac{5}{3}\right) \right] = \frac{8}{3} \quad \blacktriangleleft$$

Example 2 In cylindrical coordinates, the equation $r^2 + z^2 = a^2$ represents a sphere of radius a centered at the origin. (It is the sphere whose equation in rectangular coordinates is $x^2 + y^2 + z^2 = a^2$.) Use a double polar integral to calculate the volume of this sphere.

Solution. The upper hemisphere is given by the (cylindrical) equation

$$z = \sqrt{a^2 - r^2}$$

so that the volume enclosed by the entire sphere is

$$V = 2\iint\limits_R \sqrt{a^2 - r^2} \, dA$$

Figure 17.3.8   where R is the circular region shown in Figure 17.3.8. Thus,

$$V = 2\iint\limits_R \sqrt{a^2 - r^2} \, dA = \int_0^{2\pi} \int_0^a \sqrt{a^2 - r^2} \, (2r) \, dr \, d\theta$$

$$= \int_0^{2\pi} \left[-\frac{2}{3}(a^2 - r^2)^{3/2} \right]_{r=0}^a d\theta = \int_0^{2\pi} \frac{2}{3}a^3 \, d\theta$$

$$= \left[\frac{2}{3}a^3\theta \right]_0^{2\pi} = \frac{4}{3}\pi a^3 \quad \blacktriangleleft$$

The double polar integral

$$\iint_R 1 \, dA = \iint_R dA$$

represents the volume of a solid of constant height 1 above the region R. Numerically, this volume is the same as the area of the region R. Thus,

$$\text{area of } R = \iint_R dA \tag{6}$$

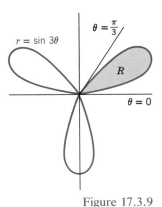

$\theta = \dfrac{\pi}{3}$

$r = \sin 3\theta$

R

$\theta = 0$

Figure 17.3.9

Example 3 Use a double polar integral to find the area enclosed by the three-petaled rose $r = \sin 3\theta$.

Solution. The rose is sketched in Figure 17.3.9. We shall calculate the area of the petal R in the first quadrant and multiply by three.

$$A = 3 \iint_R dA = 3 \int_0^{\pi/3} \int_0^{\sin 3\theta} r \, dr \, d\theta$$

$$= \frac{3}{2} \int_0^{\pi/3} \sin^2 3\theta \, d\theta = \frac{3}{4} \int_0^{\pi/3} (1 - \cos 6\theta) \, d\theta$$

$$= \left[\frac{3}{4} \theta - \frac{3}{24} \sin 6\theta \right]_0^{\pi/3} = \frac{1}{4} \pi \quad \blacktriangleleft$$

CONVERTING
DOUBLE INTEGRALS
FROM RECTANGULAR
TO POLAR
COORDINATES

Sometimes a double integral

$$\iint_R f(x, y) \, dA$$

that is hard to evaluate in rectangular coordinates is easy to evaluate when the integrand and the region R are expressed in polar coordinates. This is especially true when the integrand or the boundary of R involve expressions of the form

$$x^2 + y^2 \quad \text{or} \quad \sqrt{x^2 + y^2}$$

since these simplify to

$$r^2 \quad \text{and} \quad r$$

in polar coordinates.

Example 4   Use polar coordinates to evaluate

$$\int_{-1}^{1} \int_{0}^{\sqrt{1-x^2}} (x^2 + y^2)^{3/2} \, dy \, dx$$

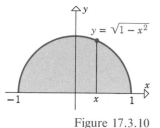

Figure 17.3.10

Solution. The first step is to express this iterated integral as a double integral in rectangular coordinates. For fixed x, the y-integration runs from the lower boundary $y = 0$ up to the semicircle $y = \sqrt{1 - x^2}$. The fixed x can extend from -1 on the left to $+1$ on the right, so

$$\int_{-1}^{1} \int_{0}^{\sqrt{1-x^2}} (x^2 + y^2)^{3/2} \, dy \, dx = \iint_{R} (x^2 + y^2)^{3/2} \, dA$$

where R is the region shown in Figure 17.3.10. Expressing this double integral as a double integral in polar coordinates yields

$$\int_{-1}^{1} \int_{0}^{\sqrt{1-x^2}} (x^2 + y^2)^{3/2} \, dy \, dx = \iint_{R} (r^2)^{3/2} \, dA$$

$$= \int_{0}^{\pi} \int_{0}^{1} (r^3) r \, dr \, d\theta$$

$$= \int_{0}^{\pi} \frac{1}{5} \, d\theta = \frac{\pi}{5} \quad \blacktriangleleft$$

▶ Exercise Set 17.3

In Exercises 1–6, evaluate the iterated integral.

1. $\displaystyle\int_{0}^{\pi/2} \int_{0}^{\sin\theta} r \cos\theta \, dr \, d\theta.$

2. $\displaystyle\int_{0}^{\pi} \int_{0}^{1+\cos\theta} r \, dr \, d\theta.$

3. $\displaystyle\int_{-\pi/2}^{\pi/2} \int_{0}^{a\sin\theta} r^2 \, dr \, d\theta.$

4. $\displaystyle\int_{0}^{\pi/3} \int_{0}^{\cos 3\theta} r \, dr \, d\theta.$

5. $\displaystyle\int_{0}^{\pi} \int_{0}^{1-\sin\theta} r^2 \cos\theta \, dr \, d\theta.$

6. $\displaystyle\int_{0}^{\pi} \int_{0}^{\cos\theta} r^3 \, dr \, d\theta.$

In Exercises 7–12, use a double integral in polar coordinates to find the area of the region described.

7. The region enclosed by the cardioid $r = 1 - \cos\theta$.

8. The region enclosed by the rose $r = \sin 2\theta$.

9. The region in the first quadrant bounded by $r = 1$ and $r = \sin 2\theta$, with $\pi/4 \le \theta \le \pi/2$.

10. The region inside the circle $x^2 + y^2 = 4$ and to the right of the line $x = 1$.

11. The region inside the circle $r = 4\sin\theta$ and outside the circle $r = 2$.

12. The region inside the circle $r = 1$ and outside the cardioid $r = 1 + \cos\theta$.

In Exercises 13–18, use a double integral in polar coordinates to find the volume of the solid.

13. The solid common to the sphere $x^2 + y^2 + z^2 = 9$ and the cylinder $x^2 + y^2 = 1$.

14. The solid common to the sphere $r^2 + z^2 = 4$ and the cylinder $r = 2\cos\theta$.

15. The solid bounded above by the cone $z = \sqrt{x^2 + y^2}$, below by the xy-plane, and laterally by the cylinder $x^2 + y^2 = 2y$.

16. The solid bounded above by the surface $z = (x^2 + y^2)^{-1/2}$, below by the xy-plane, and enclosed between the cylinders $x^2 + y^2 = 1$ and $x^2 + y^2 = 9$.

17. The solid bounded above by the paraboloid $z = 1 - x^2 - y^2$, below by the xy-plane, and laterally by the cylinder $x^2 + y^2 - x = 0$.

18. The solid in the first octant bounded above by the plane $z = r \sin \theta$, below by the xy-plane, and laterally by the plane $x = 0$ and the cylinder $r = 3 \sin \theta$.

In Exercises 19–22, use polar coordinates to evaluate the double integral.

19. $\displaystyle\iint_R e^{-(x^2+y^2)} \, dA$, where R is the region enclosed by the circle $x^2 + y^2 = 1$.

20. $\displaystyle\iint_R \sqrt{9 - x^2 - y^2} \, dA$, where R is the region in the first quadrant within the circle $x^2 + y^2 = 9$.

21. $\displaystyle\iint_R \frac{1}{1 + x^2 + y^2} \, dA$, where R is the sector in the first quadrant bounded by $y = 0$, $y = x$, and $x^2 + y^2 = 4$.

22. $\displaystyle\iint_R 2y \, dA$, where R is the region in the first quadrant, bounded above by the circle $(x - 1)^2 + y^2 = 1$ and below by the line $y = x$.

In Exercises 23–30, evaluate the iterated integral by converting to polar coordinates.

23. $\displaystyle\int_0^1 \int_0^{\sqrt{1-x^2}} (x^2 + y^2) \, dy \, dx.$

24. $\displaystyle\int_{-2}^2 \int_{-\sqrt{4-y^2}}^{\sqrt{4-y^2}} e^{-(x^2+y^2)} \, dx \, dy.$

25. $\displaystyle\int_0^2 \int_0^{\sqrt{2x-x^2}} \sqrt{x^2 + y^2} \, dy \, dx.$

26. $\displaystyle\int_0^1 \int_0^{\sqrt{1-y^2}} \cos(x^2 + y^2) \, dx \, dy.$

27. $\displaystyle\int_0^a \int_0^{\sqrt{a^2-x^2}} \frac{dy \, dx}{(1 + x^2 + y^2)^{3/2}}.$

28. $\displaystyle\int_0^1 \int_y^{\sqrt{y}} \sqrt{x^2 + y^2} \, dx \, dy.$

29. $\displaystyle\int_0^{\sqrt{2}} \int_y^{\sqrt{4-y^2}} \frac{1}{\sqrt{1 + x^2 + y^2}} \, dx \, dy.$

30. $\displaystyle\int_0^4 \int_3^{\sqrt{25-x^2}} dy \, dx.$

31. Use polar coordinates to find the volume of the solid bounded above by the ellipsoid $x^2/a^2 + y^2/a^2 + z^2/c^2 = 1$, below by the xy-plane, and laterally by the cylinder $x^2 + y^2 - ay = 0$.

32. Find the area of the region enclosed by the lemniscate $r^2 = 2a^2 \cos 2\theta$.

33. Find the area in the first quadrant inside the circle $r = 4 \sin \theta$ and outside the lemniscate $r^2 = 8 \cos 2\theta$.

34. Show that the shaded area in the following figure is $a^2\phi - \frac{1}{2}a^2 \sin 2\phi$.

35. The integral $\displaystyle\int_0^{+\infty} e^{-x^2} \, dx$

which arises in probability theory, can be evaluated using a trick. Let the value of the integral be I. Thus,

$$I = \int_0^{+\infty} e^{-x^2} \, dx = \int_0^{+\infty} e^{-y^2} \, dy$$

since the letter used for the variable of integration in a definite integral does not matter.

(a) Show that

$$I^2 = \int_0^{+\infty} \int_0^{+\infty} e^{-(x^2+y^2)} \, dx \, dy$$

(b) Evaluate the iterated integral in (a) by converting to polar coordinates.

(c) Use the result in (b) to find I.

36. In each part evaluate $\iint x^2\, dA$.

(a)

(b)

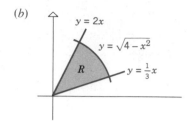

17.4   SURFACE AREA

In Section 6.5 we showed how to find the surface area of a surface of revolution. In this section we shall consider more general surface-area problems.

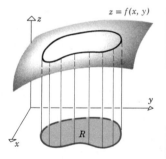

17.4.1   PROBLEM. Let f be a function defined on a closed region R of the xy-plane. Find the area of that portion of the surface $z = f(x, y)$ whose projection on the xy-plane is the region R (Figure 17.4.1).

In the special case where the surface is a plane and the region R is rectangular, we have the following result.

17.4.2   THEOREM. *Let R be a closed rectangular region in the xy-plane. If R has sides of length l and w, then the surface area S of that portion of the plane $z = ax + by + c$ that projects onto the region R is given by*

$$S = \sqrt{a^2 + b^2 + 1}\; lw \tag{1}$$

Proof. The portion of the plane that projects onto the region R is a parallelogram (Figure 17.4.2). Thus, if we can find vectors **a** and **b** forming adjacent sides of this parallelogram, then we can obtain its surface area S from the formula

$$S = \|\mathbf{a} \times \mathbf{b}\| \tag{2}$$

(see Section 14.4).

To find the vectors **a** and **b**, suppose that the four corners of the rectangle R are

$$A(x_0, y_0), \quad B(x_0 + l, y_0), \quad C(x_0, y_0 + w), \quad D(x_0 + l, y_0 + w)$$

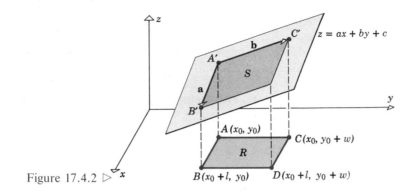

Figure 17.4.2 ▷

(Figure 17.4.2). Then those points in the plane $z = ax + by + c$ that lie above the corners A, B, and C are

$A'(x_0, y_0, ax_0 + by_0 + c)$

$B'(x_0 + l, y_0, a[x_0 + l] + by_0 + c)$

$C'(x_0, y_0 + w, ax_0 + b[y_0 + w] + c)$

Thus, the vectors

$\mathbf{a} = \overrightarrow{A'B'} = l\,\mathbf{i} + 0\,\mathbf{j} + al\,\mathbf{k}$

$\mathbf{b} = \overrightarrow{A'C'} = 0\,\mathbf{i} + w\,\mathbf{j} + bw\,\mathbf{k}$

form adjacent sides of the parallelogram in question. Since

$$\mathbf{a} \times \mathbf{b} = \begin{vmatrix} \mathbf{i} & \mathbf{j} & \mathbf{k} \\ l & 0 & al \\ 0 & w & bw \end{vmatrix} = -alw\,\mathbf{i} - lbw\,\mathbf{j} + lw\,\mathbf{k}$$

it follows from (2) that

$$S = \|\mathbf{a} \times \mathbf{b}\| = \sqrt{(-alw)^2 + (-lbw)^2 + (lw)^2} = \sqrt{a^2 + b^2 + 1}\, lw \qquad \blacksquare$$

Theorem 17.4.2 leads to the following more general result.

Surface Area

17.4.3 AREA FORMULA. If f has continuous first partial derivatives on a closed region R of the xy-plane, then the area S of that portion of the surface $z = f(x, y)$ that projects onto R is

$$S = \iint\limits_{R} \sqrt{\left(\frac{\partial z}{\partial x}\right)^2 + \left(\frac{\partial z}{\partial y}\right)^2 + 1}\; dA \qquad (3)$$

Formula (3) can be motivated as follows:

Step 1.   Enclose R within a rectangle whose sides are parallel to the x- and y-axes. Using lines parallel to the x- and y-axes, divide this rectangle into subrectangles, and exclude from consideration all those subrectangles that contain any points outside of R. This leaves only rectangles that are subsets of R (Figure 17.4.3a). Let these rectangles be denoted by

$$R_1, R_2, \ldots, R_n$$

and suppose that the sides of rectangle R_k have lengths

$$\Delta x_k \quad \text{and} \quad \Delta y_k$$

Step 2.   When projected up to the surface $z = f(x, y)$, each subrectangle determines a patch of area on the surface (Figure 17.4.3a). If we denote the areas of these patches by

$$S_1, S_2, \ldots, S_n$$

then the entire surface area S can be approximated by adding the areas of these patches:

$$S \approx S_1 + S_2 + \cdots + S_n \tag{4}$$

(This is only an approximation since the subrectangles may not completely fill out the region R.)

Figure 17.4.3 ▷   (a)   (b)

Step 3. We shall now approximate each of the areas S_1, $S_2, \ldots, S_n$. Let

$$(x_k^*, y_k^*)$$

be an arbitrary point in the kth subrectangle and above this point construct the tangent plane to the surface $z = f(x, y)$ (Figure 17.4.3b). It follows from Theorem 16.5.1 that the equation of this tangent plane can be written in the form

$$z = f_x(x_k^*, y_k^*)x + f_y(x_k^*, y_k^*)y + c$$

where c is an appropriate constant. If the rectangle R_k is small, then we can reasonably approximate the area S_k of the kth patch on the surface by the portion of area on the tangent plane over R_k (Figure 17.4.3b). Thus, by Theorem 17.4.2

$$S_k \approx \sqrt{f_x(x_k^*, y_k^*)^2 + f_y(x_k^*, y_k^*)^2 + 1} \ \Delta x_k \ \Delta y_k$$

Substituting this expression in (4) and writing $\Delta A_k = \Delta x_k \ \Delta y_k$ for the area of the kth subrectangle, we obtain the Riemann sum

$$S \approx \sum_{k=1}^{n} \sqrt{f_x(x_k^*, y_k^*)^2 + f_y(x_k^*, y_k^*)^2 + 1} \ \Delta A_k \quad (5)$$

Step 4. There are two sources of error in approximation (5). First, the rectangles $R_1, R_2, \ldots, R_n$ may not fill up the region R completely, and second, we have approximated patches of area on the surface by areas on tangent planes. However, let us repeat this approximation process using more and more rectangles of decreasing dimensions. It is intuitively clear that both kinds of errors diminish and the exact surface area S is given by

$$S = \lim_{n \to +\infty} \sum_{k=1}^{n} \sqrt{f_x(x_k^*, y_k^*)^2 + f_y(x_k^*, y_k^*)^2 + 1} \ \Delta A_k$$

or equivalently

$$S = \iint\limits_{R} \sqrt{f_x(x, y)^2 + f_y(x, y)^2 + 1} \ dA$$

This is the motivation for Definition 17.4.3.

Example 1   Find the surface area of the portion of the cylinder $x^2 + z^2 = 4$ above the rectangle $R = \{(x, y): 0 \le x \le 1, 0 \le y \le 4\}$ in the xy-plane.

Solution.   The surface is shown in Figure 17.4.4. The portion of the cylinder, $x^2 + z^2 = 4$, that lies above the xy-plane is given by the equation $z = \sqrt{4 - x^2}$. Thus, from (3)

$$S = \iint_R \sqrt{\left(\frac{\partial z}{\partial x}\right)^2 + \left(\frac{\partial z}{\partial y}\right)^2 + 1} \, dA$$

$$= \iint_R \sqrt{\left(-\frac{x}{\sqrt{4 - x^2}}\right)^2 + 0 + 1} \, dA = \int_0^4 \int_0^1 \frac{2}{\sqrt{4 - x^2}} \, dx \, dy$$

$$= 2 \int_0^4 \left[\sin^{-1}\left(\frac{1}{2}x\right) \right]_{x=0}^1 \, dy = 2 \int_0^4 \frac{\pi}{6} \, dy = \frac{4}{3}\pi$$

Formula (10)
Section 8.3

◀

Figure 17.4.4 ▷

Example 2   Find the surface area of the portion of the paraboloid $z = x^2 + y^2$ below the plane $z = 1$.

Figure 17.4.5

Solution.   The surface is shown in Figure 17.4.5. From the equations $z = 1$ and $z = x^2 + y^2$, we see that the plane and the paraboloid intersect in a circle whose projection on the xy-plane has the equation $x^2 + y^2 = 1$. Consequently, the surface whose area we seek projects onto the region R enclosed by this circle. Since the surface has the equation

$$z = x^2 + y^2$$

it follows from (3) that

$$S = \iint_R \sqrt{4x^2 + 4y^2 + 1} \, dA$$

This integral is best evaluated in polar coordinates. Replacing $x^2 + y^2$ by r^2 and substituting $r\,dr\,d\theta$ for dA we obtain

$$S = \int_0^{2\pi} \int_0^1 \sqrt{4r^2 + 1}\ r\ dr\ d\theta = \int_0^{2\pi} \left[\frac{1}{12}(4r^2 + 1)^{3/2} \right]_{r=0}^1 d\theta$$

$$= \int_0^{2\pi} \frac{1}{12}(5\sqrt{5} - 1)\ d\theta = \frac{1}{6}\pi(5\sqrt{5} - 1) \quad \blacktriangleleft$$

▶ Exercise Set 17.4

1. Find the surface area of the portion of the cylinder $y^2 + z^2 = 9$ above the rectangle $R = \{(x, y): 0 \le x \le 2, -3 \le y \le 3\}$.

2. By integration, find the surface area of the portion of the plane $2x + 2y + z = 8$ in the first octant that is cut off by the three coordinate planes.

3. Find the surface area of the portion of the cone $z^2 = 4x^2 + 4y^2$ that is above the region in the first quadrant bounded by the line $y = x$ and the parabola $y = x^2$.

4. Find the surface area of the portion of the cone $z = \sqrt{x^2 + y^2}$ that lies inside the cylinder $x^2 + y^2 = 2x$.

5. Find the surface area of the portion of the paraboloid $z = 1 - x^2 - y^2$ that is above the xy-plane.

6. Find the area of the portion of the surface $z = 2x + y^2$ that is above the triangular region with vertices $(0, 0)$, $(0, 1)$, and $(1, 1)$.

7. Find the area of the portion of the surface $z = xy$ that is above the sector in the first quadrant bounded by the lines $y = x/\sqrt{3}$, $y = 0$, and the circle $x^2 + y^2 = 9$.

8. Find the surface area of the portion of the paraboloid $2z = x^2 + y^2$ that is inside the cylinder $x^2 + y^2 = 8$.

9. Find the surface area of the portion of the sphere $x^2 + y^2 + z^2 = 16$ between the planes $z = 1$ and $z = 2$.

10. Find the surface area of the portion of the sphere $x^2 + y^2 + z^2 = 8$ that is cut out by the cone $z = \sqrt{x^2 + y^2}$.

11. Find the total surface area of the portion of the sphere $x^2 + y^2 + z^2 = a^2$ inside the cylinder $x^2 + y^2 = ay$.

12. Use a double integral to derive the formula for the surface area of a sphere of radius a.

13. Find the surface area of the portion of the cylinder $x^2 + z^2 = 16$ that lies inside the cylinder $x^2 + y^2 = 16$. [*Hint:* Find the area in the first octant and use symmetry.]

14. Find the surface area of the portion of the cylinder $x^2 + z^2 = 5x$ that lies inside the sphere $x^2 + y^2 + z^2 = 25$.

15. The portion of the surface

$$z = \frac{h}{a}\sqrt{x^2 + y^2} \qquad (a, h > 0)$$

between the xy-plane and the plane $z = h$ is a right-circular cone of height h and radius a. Use a double integral to show that the lateral surface area of this cone is $S = \pi a \sqrt{a^2 + h^2}$.

17.5   TRIPLE INTEGRALS

Earlier we defined the notion of a double integral for functions of two variables. In this section we shall discuss triple integrals for functions of three variables.

DEFINITION OF A
TRIPLE INTEGRAL

Whereas a double integral

$$\iint\limits_{R} f(x, y)\ dA$$

is evaluated over a closed region R in the xy-plane, a triple integral of a function $f(x, y, z)$ is evaluated over a closed three-dimensional region G. We shall assume that G can be enclosed within some suitably large box (rectangular parallelepiped) with sides parallel to the coordinate planes (Figure 17.5.1). This ensures that the region does not extend indefinitely in any direction.

To define a triple integral, we proceed as follows:

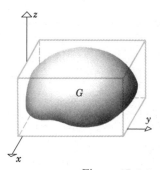

Figure 17.5.1

Step 1.   Enclose G in a box with sides parallel to the co-ordinate planes, and use planes parallel to the co-ordinate planes to divide this box into subboxes. Exclude from consideration all subboxes that contain any points outside of G. This leaves only boxes that are subsets of G (Figure 17.5.2). Denote the volumes of these boxes by

$$\Delta V_1, \Delta V_2, \dots, \Delta V_n$$

Step 2.   Choose an arbitrary point in each box, and let these be denoted by

$$(x_1^*, y_1^*, z_1^*), (x_2^*, y_2^*, z_2^*), \dots, (x_n^*, y_n^*, z_n^*)$$

(See Figure 17.5.2.)

Step 3.   Form the sum

$$\sum_{k=1}^{n} f(x_k^*, y_k^*, z_k^*)\ \Delta V_k$$

This is called a ***Riemann sum.***

Step 4.   Repeat this process with more and more subdivi-sions, so that the length, width, and height of each box approach zero, and n, the number of boxes, approaches $+\infty$. Define

$$\iiint\limits_{G} f(x, y, z) \, dV = \lim_{n \to +\infty} \sum_{k=1}^{n} f(x_k^*, y_k^*, z_k^*) \, \Delta V_k \qquad (1)$$

The symbol

$$\iiint\limits_{G} f(x, y, z) \, dV$$

is called the *triple integral* of $f(x, y, z)$ over the region G. Conditions under which the triple integral exists are studied in advanced calculus. However, for our purposes it suffices to say that existence is ensured when f is continuous on G and the region G is not too "complicated."

Triple integrals have a number of physical interpretations, some of which we shall consider in the next section. In the special case where $f(x, y, z) = 1$ the triple integral over G represents the volume of the solid G, that is,

Volume Calculated as a Triple Integral

$$\text{volume of } G = \iiint\limits_{G} dV \qquad (2)$$

[Compare this to (7) of Section 17.2.] To obtain (2), let $f(x, y, z) = 1$ in (1). This yields

$$\iiint\limits_{G} dV = \lim_{n \to +\infty} \sum_{k=1}^{n} \Delta V_k$$

As suggested by Figure 17.5.2, the sum $\Sigma \, \Delta V_k$ represents the total volume of the boxes interior to the solid G. As n increases and the dimensions of these boxes approach zero, the boxes tend to fill up the solid G, so that their total volume approaches the volume of G, that is,

$$\iiint\limits_{G} dV = \lim_{n \to +\infty} \sum_{k=1}^{n} \Delta V_k = \text{volume of } G$$

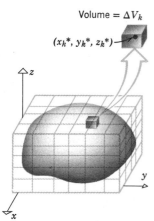

Volume $= \Delta V_k$

(x_k^*, y_k^*, z_k^*)

Figure 17.5.2

PROPERTIES OF TRIPLE INTEGRALS

Triple integrals enjoy many properties of single and double integrals:

$$\iiint\limits_{G} cf(x, y, z) \, dV = c \iiint\limits_{G} f(x, y, z) \, dV \quad (c \text{ a constant})$$

$$\iiint\limits_{G} [f(x, y, z) + g(x, y, z)]\, dV = \iiint\limits_{G} f(x, y, z)\, dV + \iiint\limits_{G} g(x, y, z)\, dV$$

$$\iiint\limits_{G} [f(x, y, z) - g(x, y, z)]\, dV = \iiint\limits_{G} f(x, y, z)\, dV - \iiint\limits_{G} g(x, y, z)\, dV$$

G_1 G_2

G

Figure 17.5.3

Moreover, if the region G is subdivided into two subregions G_1 and G_2 (Figure 17.5.3), then

$$\iiint\limits_{G} f(x, y, z)\, dV = \iiint\limits_{G_1} f(x, y, z)\, dV + \iiint\limits_{G_2} f(x, y, z)\, dV$$

We omit the proofs.

EVALUATING TRIPLE INTEGRALS

Just as a double integral can be evaluated by two single integrations, a triple integral can be evaluated by three single integrations.

17.5.1   THEOREM. *Let G be the rectangular box defined by the inequalities*

$$a \leq x \leq b, \quad c \leq y \leq d, \quad k \leq z \leq l$$

If f is continuous on the region G, then

$$\iiint\limits_{G} f(x, y, z)\, dV = \int_a^b \int_c^d \int_k^l f(x, y, z)\, dz\, dy\, dx \qquad (3)$$

Moreover, the iterated integral on the right can be replaced with any of the five other iterated integrals that result by altering the order of integration.

We omit the proof.

Example 1   Evaluate the triple integral

$$\iiint\limits_{G} 12xy^2z^3\, dV$$

over the rectangular box G defined by the inequalities: $-1 \leq x \leq 2$, $0 \leq y \leq 3$, $0 \leq z \leq 2$.

Solution.   Of the six possible iterated integrals we might use, we shall choose the one in (3). Thus, we shall first integrate with respect to z, holding x and y fixed, then with respect to y holding x fixed, and finally with respect to x.

$$\iiint\limits_{G} 12xy^2z^3\, dV = \int_{-1}^{2}\int_{0}^{3}\int_{0}^{2} 12xy^2z^3\, dz\, dy\, dx$$

$$= \int_{-1}^{2}\int_{0}^{3}\left[3xy^2z^4\right]_{z=0}^{2} dy\, dx = \int_{-1}^{2}\int_{0}^{3} 48xy^2\, dy\, dx$$

$$= \int_{-1}^{2}\left[16xy^3\right]_{y=0}^{3} dx = \int_{-1}^{2} 432x\, dx$$

$$= 216x^2\Big]_{-1}^{2} = 648 \quad \blacktriangleleft$$

We shall also be concerned with evaluating triple integrals over solid regions other than rectangular boxes. For simplicity, we shall restrict our discussion to solid regions constructed as follows. Let R be a closed region in the xy-plane and let $g_1(x, y)$ and $g_2(x, y)$ be continuous functions satisfying

$$g_1(x, y) \le g_2(x, y)$$

for all (x, y) in R. Geometrically, this condition states that the surface $z = g_2(x, y)$ does not dip below the surface $z = g_1(x, y)$ over R (Figure 17.5.4a). We shall call $z = g_1(x, y)$ the **lower surface** and $z = g_2(x, y)$ the **upper surface**. Let G be the solid consisting of all points above or below the region R that lie between the upper surface and the lower surface (Figure 17.5.4b). A solid G constructed in this way will be called a **simple solid,** and the region R will be called the **projection** of G on the xy-plane.

The following theorem, which we state without proof, will enable us to evaluate triple integrals over simple solids.

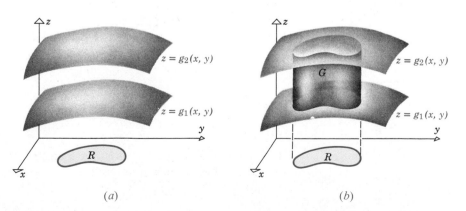

Figure 17.5.4   (a)   (b)

17.5.2   THEOREM. *Let G be a simple solid with upper surface* $z = g_2(x, y)$ *and lower surface* $z = g_1(x, y)$; *and let R be the projection of G on the xy-plane. If* $f(x, y, z)$ *is continuous on G, then*

$$\iiint_G f(x, y, z)\, dV = \iint_R \left[\int_{g_1(x, y)}^{g_2(x,y)} f(x, y, z)\, dz \right] dA \qquad (4)$$

In (4), the first integration is with respect to z, after which a function of x and y remains. This function of x and y is then integrated over the region R in the xy-plane. To apply (4), it is usual to begin with a three-dimensional sketch of the solid G, from which the limits of integration can be obtained as follows:

Step 1.   Find an equation $z = g_2(x, y)$ for the upper surface and an equation $z = g_1(x, y)$ for the lower surface of G. The functions $g_1(x, y)$ and $g_2(x, y)$ determine the z-limits of integration.

Step 2.   Make a two-dimensional sketch of the projection R of the solid on the xy-plane. From this sketch determine the limits of integration for the double integral over R in (4).

Example 2   Let G be the wedge in the first octant cut from the cylinder $y^2 + z^2 = 1$ by the planes $y = x$ and $x = 0$. Evaluate

$$\iiint_G z\, dV$$

Solution.   The solid G and its projection R on the xy-plane are shown in Figure 17.5.5. The upper surface of the solid is formed by the cylinder and the lower surface by the xy-plane. Since the portion of the cylinder $y^2 + z^2 = 1$ that lies above the xy-plane has the equation $z = \sqrt{1 - y^2}$, and the xy-plane has the equation $z = 0$, it follows from (4) that

$$\iiint_G z\, dV = \iint_R \left[\int_0^{\sqrt{1-y^2}} z\, dz \right] dA \qquad (5)$$

For the double integral over R, the x and y integrations can be performed in either order, since R is both a type I and type II region. We shall integrate with respect to x first. With this choice, (5) yields

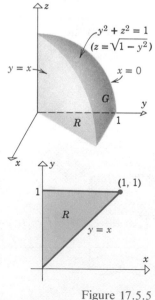

Figure 17.5.5

$$\iiint_G z \, dV = \int_0^1 \int_0^y \int_0^{\sqrt{1-y^2}} z \, dz \, dx \, dy = \int_0^1 \int_0^y \frac{1}{2}z^2 \Big]_{z=0}^{\sqrt{1-y^2}} dx \, dy$$

$$= \int_0^1 \int_0^y \frac{1}{2}(1 - y^2) \, dx \, dy = \frac{1}{2} \int_0^1 (1 - y^2)x \Big]_{x=0}^y dy$$

$$= \frac{1}{2} \int_0^1 (y - y^3) \, dy = \frac{1}{2}\left[\frac{1}{2}y^2 - \frac{1}{4}y^4\right]_0^1 = \frac{1}{8} \quad \blacktriangleleft$$

Example 3 Use a triple integral to find the volume of the solid enclosed between the cylinder $x^2 + y^2 = 9$ and the planes $z = 1$ and $x + z = 5$.

Solution. The solid G and its projection R on the xy-plane are shown in Figure 17.5.6. The lower surface of the solid is the plane $z = 1$; and the upper surface is the plane $x + z = 5$, or equivalently, $z = 5 - x$. Thus, from (2) and (4)

$$\text{volume of } G = \iiint_G dV = \iint_R \left[\int_1^{5-x} dz\right] dA \tag{6}$$

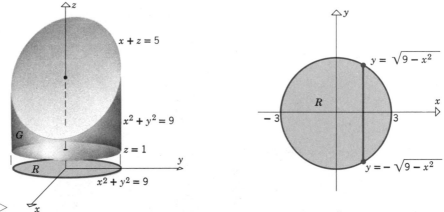

Figure 17.5.6 ▷

For the double integral over R, we shall integrate with respect to y first. Thus, (6) yields

$$\text{volume of } G = \int_{-3}^3 \int_{-\sqrt{9-x^2}}^{\sqrt{9-x^2}} \int_1^{5-x} dz \, dy \, dx = \int_{-3}^2 \int_{-\sqrt{9-x^2}}^{\sqrt{9-x^2}} z \Big]_{z=1}^{5-x} dy \, dx$$

$$= \int_{-3}^3 \int_{-\sqrt{9-x^2}}^{\sqrt{9-x^2}} (4 - x) \, dy \, dx = \int_{-3}^3 (8 - 2x)\sqrt{9 - x^2} \, dx$$

$$= 8 \int_{-3}^{3} \sqrt{9 - x^2} \, dx - \int_{-3}^{3} 2x\sqrt{9 - x^2} \, dx \qquad \left[\begin{array}{l} \text{For the first integral,} \\ \text{see Example 4, Section} \\ \text{9.5.} \end{array} \right]$$

$$= 8\left(\frac{9}{2}\pi\right) - \int_{-3}^{3} 2x\sqrt{9 - x^2} \, dx \qquad \left[\begin{array}{l} \text{Let } u = 9 - x^2, \text{ or better, apply} \\ \text{the result in Exercise 40} \\ \text{of Section 5.8.} \end{array} \right]$$

$$= 8\left(\frac{9}{2}\pi\right) - 0 = 36\pi \qquad \blacktriangleleft$$

Example 4   Find the volume of the solid enclosed by the paraboloids

$$z = 5x^2 + 5y^2 \quad \text{and} \quad z = 6 - 7x^2 - y^2$$

Solution.   The solid G and its projection R on the xy-plane are shown in Figure 17.5.7. The projection R is obtained by solving the given equations simultaneously to determine where the paraboloids intersect. We obtain

$$5x^2 + 5y^2 = 6 - 7x^2 - y^2$$

or

$$2x^2 + y^2 = 1 \tag{7}$$

which tells us that the paraboloids intersect in a curve on the elliptic cylinder given by (7).

The projection of this intersection on the xy-plane is an ellipse with this same equation. Therefore,

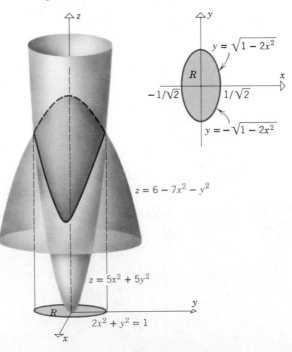

Figure 17.5.7 ▷

$$\text{volume of } G = \iiint_G dV = \iint_R \left[\int_{5x^2+5y^2}^{6-7x^2-y^2} dz \right] dA$$

$$\text{volume of } G = \int_{-1/\sqrt{2}}^{1/\sqrt{2}} \int_{-\sqrt{1-2x^2}}^{\sqrt{1-2x^2}} \int_{5x^2+5y^2}^{6-7x^2-y^2} dz\, dy\, dx$$

$$= \int_{-1/\sqrt{2}}^{1/\sqrt{2}} \int_{-\sqrt{1-2x^2}}^{\sqrt{1-2x^2}} (6 - 12x^2 - 6y^2)\, dy\, dx$$

$$= \int_{-1/\sqrt{2}}^{1/\sqrt{2}} \left[6(1 - 2x^2)y - 2y^3 \right]_{y=-\sqrt{1-2x^2}}^{\sqrt{1-2x^2}} dx$$

$$= 8 \int_{-1/\sqrt{2}}^{1/\sqrt{2}} (1 - 2x^2)^{3/2}\, dx = \frac{8}{\sqrt{2}} \int_{-\pi/2}^{\pi/2} \cos^4 \theta\, d\theta = \frac{3\pi}{\sqrt{2}}$$

Let $x = \dfrac{1}{\sqrt{2}} \sin \theta$.

Formula (5), Section 9.3

◀

For certain regions, triple integrals are best evaluated by integrating first with respect to x or y rather than z. For example, if the solid G is bounded on the left and right by the surfaces $y = g_1(x, z)$ and $y = g_2(x, z)$ and bounded laterally by a cylinder extending in the y-direction (Figure 17.5.8a), then

$$\iiint_G f(x, y, z)\, dV = \iint_R \left[\int_{g_1(x,z)}^{g_2(x,z)} f(x, y, z)\, dy \right] dA$$

where R is the projection of the solid G on the xz-plane.

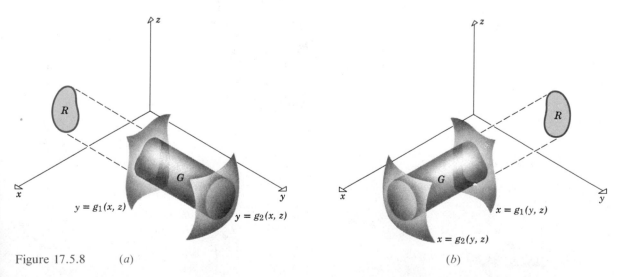

Figure 17.5.8　　(a)　　　　　　　　　　　　　　　　　(b)

Similarly, if the solid G is bounded in the back and front by the surfaces $x = g_1(y, z)$ and $x = g_2(y, z)$ and bounded laterally by a cylinder extending in the x-direction, then

$$\iiint_G f(x, y, z)\, dV = \iint_R \left[\int_{g_1(y,z)}^{g_2(y,z)} f(x, y, z)\, dx \right] dA$$

where R is the projection of the solid on the yz-plane (Figure 17.5.8b).

Example 5   In Example 2, we evaluated

$$\iiint_G z\, dV$$

over the wedge in Figure 17.5.5 by integrating first with respect to z. Evaluate this integral by integrating first with respect to x.

Figure 17.5.9

Solution.   The solid is bounded in the back by the plane $x = 0$ and in the front by the plane $x = y$, so

$$\iiint_G z\, dV = \iint_R \left[\int_0^y z\, dx \right] dA$$

where R is the projection of G on the yz-plane (Figure 17.5.9). The integration over R can be performed first with respect to z and then y or vice versa. Performing the z-integration first yields

$$\iiint_G z\, dV = \int_0^1 \int_0^{\sqrt{1-y^2}} \int_0^y z\, dx\, dz\, dy = \int_0^1 \int_0^{\sqrt{1-y^2}} zx \Big]_{x=0}^y dz\, dy$$

$$= \int_0^1 \int_0^{\sqrt{1-y^2}} zy\, dz\, dy = \int_0^1 \frac{1}{2} z^2 y \Big]_{z=0}^{\sqrt{1-y^2}} = \int_0^1 \frac{1}{2}(1 - y^2)y\, dy = \frac{1}{8}$$

which agrees with the result in Example 2.   ◀

▶ Exercise Set 17.5

In Exercises 1–8, evaluate the iterated integral.

1. $\displaystyle \int_{-1}^1 \int_0^2 \int_0^1 (x^2 + y^2 + z^2)\, dx\, dy\, dz.$

2. $\displaystyle \int_{1/3}^{1/2} \int_0^\pi \int_0^1 zx \sin xy\, dz\, dy\, dx.$

3. $\displaystyle \int_0^2 \int_{-1}^{y^2} \int_1^z yz\, dx\, dz\, dy.$

4. $\displaystyle \int_0^{\pi/4} \int_0^1 \int_0^{x^2} x \cos y\, dz\, dx\, dy.$

5. $\displaystyle\int_0^3 \int_0^{\sqrt{9-z^2}} \int_0^x xy \, dy \, dx \, dz.$

6. $\displaystyle\int_1^3 \int_x^{x^2} \int_0^{\ln z} xe^y \, dy \, dz \, dx.$

7. $\displaystyle\int_0^2 \int_0^{\sqrt{4-x^2}} \int_{-5+x^2+y^2}^{3-x^2-y^2} x \, dz \, dy \, dx.$

8. $\displaystyle\int_1^2 \int_z^2 \int_0^{\sqrt{3}y} \frac{y}{x^2+y^2} \, dx \, dy \, dz.$

In Exercises 9–12, evaluate the triple integral.

9. $\displaystyle\iiint_G xy \sin yz \, dV,$ where G is the rectangular box defined by the inequalities $0 \le x \le \pi,$ $0 \le y \le 1, 0 \le z \le \pi/6.$

10. $\displaystyle\iiint_G y \, dV,$ where G is the solid enclosed by the plane $z = y$, the xy-plane, and the parabolic cylinder $y = 1 - x^2.$

11. $\displaystyle\iiint_G xyz \, dV,$ where G is the solid in the first octant bounded by the parabolic cylinder $z = 2 - x^2$ and the planes $z = 0, y = x,$ and $y = 0.$

12. $\displaystyle\iiint_G \cos(z/y) \, dV,$ where G is the solid defined by the inequalities, $\pi/6 \le y \le \pi/2, y \le x \le \pi/2,$ $0 \le z \le xy.$

In Exercises 13–21, use a triple integral to find the volume of the solid.

13. The solid in the first octant bounded by the coordinate planes and the plane $3x + 6y + 4z = 12.$

14. The solid bounded by the surface $z = \sqrt{y}$ and the planes $x + y = 1$ and $z = 0.$

15. The solid bounded by the surface $y = x^2$ and the planes $y + z = 4, x = 0,$ and $z = 0.$

16. The wedge in the first octant cut from the cylinder $y^2 + z^2 = 1$ by the planes $y = x$ and $x = 0.$

17. The solid enclosed between the elliptic cylinder $x^2 + 9y^2 = 9$ and the planes $z = 0$ and $z = x + 3.$

18. The solid common to the cylinders $x^2 + y^2 = 1$ and $x^2 + z^2 = 1.$

19. The solid bounded by the paraboloid $z = 4x^2 + y^2$ and the parabolic cylinder $z = 4 - 3y^2.$

20. The solid enclosed between the paraboloids $z = 8 - x^2 - y^2$ and $z = 3x^2 + y^2.$

21. The solid that is enclosed between the sphere $x^2 + y^2 + z^2 = 2a^2$ and the paraboloid $az = x^2 + y^2 \ (a > 0).$

22. In each part sketch the solid whose volume is given by the integral.

(a) $\displaystyle\int_0^3 \int_{x^2}^9 \int_0^2 dz \, dy \, dx$

(b) $\displaystyle\int_0^2 \int_0^{2-y} \int_0^{2-x-y} dz \, dx \, dy.$

23. In each part sketch the solid whose volume is given by the integral.

(a) $\displaystyle\int_{-1}^1 \int_{-\sqrt{1-x^2}}^{\sqrt{1-x^2}} \int_0^{y+1} dz \, dy \, dx$

(b) $\displaystyle\int_0^9 \int_0^{y/3} \int_0^{\sqrt{y^2-9x^2}} dz \, dx \, dy.$

24. In each part sketch the solid whose volume is given by the integral.

(a) $\displaystyle\int_0^1 \int_0^{\sqrt{1-x^2}} \int_0^2 dy \, dz \, dx$

(b) $\displaystyle\int_{-2}^2 \int_0^{4-y^2} \int_0^2 dx \, dz \, dy.$

25. Let G be the tetrahedron in the first octant bounded by the coordinate planes and the plane

$$\frac{x}{a} + \frac{y}{b} + \frac{z}{c} = 1 \qquad (a > 0, b > 0, c > 0)$$

(a) List six different iterated integrals that represent the volume of G.

(b) Evaluate any one of the six to show that the volume of G is $\frac{1}{6}abc.$

26. In parts (a)–(c), express the integral as an equivalent integral in which the z-integration is performed first, the y-integration second, and the x-integration last.

(a) $\displaystyle\int_0^3 \int_0^{\sqrt{9-z^2}} \int_0^{\sqrt{9-y^2-z^2}} f(x, y, z) \, dx \, dy \, dz$

(b) $\displaystyle\int_0^4 \int_0^2 \int_0^{x/2} f(x, y, z)\, dy\, dz\, dx$

(c) $\displaystyle\int_0^4 \int_0^{4-y} \int_0^{\sqrt{z}} f(x, y, z)\, dx\, dz\, dy.$

27. Use a triple integral to find the volume of the tetrahedron with vertices $(0, 0, 0)$, $(a, a, 0)$, $(a, 0, 0)$ and $(a, 0, a)$.

28. Let G be the rectangular box defined by the inequalities $a \le x \le b, c \le y \le d, k \le z \le l$. Show that

$$\iiint_G f(x)g(y)h(z)\, dV$$

$$= \left[\int_a^b f(x)\, dx\right]\left[\int_c^d g(y)\, dy\right]\left[\int_k^l h(z)\, dz\right]$$

29. Use the result of Exercise 28 to evaluate

(a) $\displaystyle\iiint_G xy^2 \sin z\, dV$, where G is the set of points satisfying

$$-1 \le x \le 1, 0 \le y \le 1, 0 \le z \le \frac{\pi}{2}$$

(b) $\displaystyle\iiint_G e^{2x+y-z}\, dV$, where G is the set of points satisfying

$$0 \le x \le 1, 0 \le y \le \ln 3, 0 \le z \le \ln 2$$

30. Use a triple integral to find the volume of the ellipsoid

$$\frac{x^2}{a^2} + \frac{y^2}{b^2} + \frac{z^2}{c^2} = 1$$

31. Let G be a region consisting of subregions G_1 and G_2 that are symmetric with respect to the xy-plane, and suppose that $\displaystyle\iiint_{G_1} f(x, y, z)\, dV = 5$.

(a) Find $\displaystyle\iiint_G f(x, y, z)\, dV$ given that

$$f(x, y, -z) = f(x, y, z).$$

(b) Find $\displaystyle\iiint_G f(x, y, z)\, dV$ given that

$$f(x, y, -z) = -f(x, y, z).$$

In Exercises 32–34, evaluate the integral without performing an integration. [*Hint:* See Exercise 31.]

32. $\displaystyle\int_{-2}^2 \int_0^{\sqrt{4-x^2}} \int_{-\sqrt{4-x^2-y^2}}^{\sqrt{4-x^2-y^2}} [z^3 \cos(xyz) - 3]\, dz\, dy\, dx.$

33. $\displaystyle\int_{-3}^3 \int_{-\sqrt{9-y^2}}^{\sqrt{9-y^2}} \int_{x^2+y^2}^9 x^2 y^3 e^z\, dz\, dx\, dy.$

34. $\displaystyle\iiint_G e^{xy} \sin z\, dV$, where G is the sphere $x^2 + y^2 + z^2 = R^2$.

17.6   CENTROIDS, CENTERS OF GRAVITY, THEOREM OF PAPPUS

Suppose that a physical body is acted on by a gravitational field. Because the body is composed of many particles, each of which is affected by gravity, the action of the gravitational field on the body consists of a large number of forces distributed over the entire body. However, it is a physical fact that a body subjected to gravity behaves as if the entire force of gravity acts at a single point. This point is called the **center of gravity** *or* **center of mass** *of the body. In this section we shall show how double and triple integrals can be used to locate centers of gravity.*

DENSITY OF A LAMINA
To begin, let us consider an idealized flat object that is sufficiently thin to be viewed as a two-dimensional solid (Figure 17.6.1). Such a solid is

The thickness of a lamina is negligible

Figure 17.6.1

called a *lamina*. A lamina is called *homogeneous* if its composition and structure are uniform throughout, and is called *inhomogeneous* otherwise.

For a *homogeneous* lamina of mass M and area A, we define the *density* δ of the lamina by

$$\delta = \frac{M}{A}$$

Thus, δ measures the mass per unit area.

For an inhomogeneous lamina the composition may vary from point to point. If the lamina is placed in the xy-plane, then the "density" of the lamina at a general point (x, y) can be specified by a function $\delta(x, y)$, called the *density function* for the lamina. Informally, the density function can be visualized as follows. Construct a small rectangle centered at (x, y) and let ΔM and ΔA be the mass and area of the portion of lamina enclosed by this rectangle (Figure 17.6.2). If the ratio $\Delta M/\Delta A$ tends toward a limiting value as the dimensions of the rectangle tend to zero, then this limit is considered to be the density of the lamina at (x, y). Symbolically,

$$\delta(x, y) = \lim \frac{\Delta M}{\Delta A} \tag{1}$$

Area $= \Delta A$
Mass $= \Delta M$

Figure 17.6.2

From this relationship we obtain the approximation

$$\Delta M \approx \delta(x, y)\, \Delta A \tag{2}$$

which relates the mass and area of a small rectangular portion of lamina centered at (x, y). It is assumed that as the dimensions of the rectangle tend to zero, the error in this approximation also tends to zero.

MASS OF A LAMINA

We can now obtain a formula for the mass of a lamina.

17.6.1 DEFINITION. If a lamina with density function $\delta(x, y)$ occupies a region R in the xy-plane, then its total mass M is defined to be

$$M = \iint\limits_{R} \delta(x, y)\, dA \tag{3}$$

To motivate this result, let a rectangle enclosing the region R be divided into subrectangles by lines parallel to the coordinate axes, and exclude from consideration all those rectangles that contain any points outside of R. Let

$$\Delta A_1, \Delta A_2, \ldots, \Delta A_n$$

be the areas of the remaining rectangles and let

$$\Delta M_1, \Delta M_2, \ldots, \Delta M_n$$

be the masses of the portions of lamina covered by these rectangles. We can approximate the total mass of the lamina by adding the masses of these pieces

$$M = \text{total mass} \approx \sum_{k=1}^{n} \Delta M_k \qquad (4)$$

If we now let (x_k^*, y_k^*) be the center of the kth rectangle, then it follows from (2) that

$$\Delta M_k \approx \delta(x_k^*, y_k^*) \, \Delta A_k$$

Substituting this in (4) yields

$$M \approx \sum_{k=1}^{n} \delta(x_k^*, y_k^*) \, \Delta A_k \qquad (5)$$

If we now increase n in such a way that the dimensions of the rectangles tend to zero, then the errors in our approximations will diminish so that

$$M = \lim_{n \to +\infty} \sum_{k=1}^{n} \delta(x_k^*, y_k^*) \, \Delta A_k = \iint_R \delta(x, y) \, dA$$

Example 1 A triangular lamina with vertices $(0,0)$, $(0,1)$, and $(1,0)$ has density function $\delta(x, y) = xy$. Find its total mass.

Solution. Referring to (3) and Figure 17.6.3, the mass M of the lamina is

$$M = \iint_R \delta(x, y) \, dA = \iint_R xy \, dA = \int_0^1 \int_0^{-x+1} xy \, dy \, dx$$

$$= \int_0^1 \left[\frac{1}{2} xy^2 \right]_{y=0}^{-x+1} dx = \int_0^1 \left[\frac{1}{2} x^3 - x^2 + \frac{1}{2} x \right] dx$$

$$= \frac{1}{24} \text{ (units of mass)} \qquad \blacktriangleleft$$

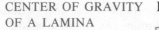
Figure 17.6.3

CENTER OF GRAVITY OF A LAMINA Let us now consider the following problem.

17.6.2 PROBLEM. Suppose that a lamina with density function $\delta(x, y)$ occupies a region R in a horizontal xy-plane. Find the coordinates $(\bar{x}, \bar{y})$ of the center of gravity.

To motivate the solution, consider what happens if we try to balance the lamina on a knife-edge parallel to the x-axis. Suppose the lamina in Figure 17.6.4 is placed on a knife-edge along a line $y = c$ that does not pass through the center of gravity. Because the lamina behaves as if its entire mass is concentrated at the center of gravity $(\bar{x}, \bar{y})$, the lamina will be rotationally unstable and the force of gravity will cause a rotation about $y = c$. Similarly, the lamina will undergo a rotation if placed on a knife-edge along $y = d$. However, if the knife-edge runs along the line $y = \bar{y}$ through the center of gravity, the lamina will be in perfect balance. Similarly, the lamina will be in perfect balance on a knife-edge along the line $x = \bar{x}$ through the center of gravity.

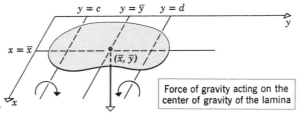

Figure 17.6.4 ▷

Force of gravity acting on the center of gravity of the lamina

This discussion suggests that the center of gravity of a lamina can be determined as the intersection of two lines of balance, one parallel to the x-axis and the other parallel to the y-axis. In order to find these lines of balance, we shall need some preliminary results about rotations.

Children on a seesaw learn by experience that a lighter child can balance a heavier one by sitting farther from the fulcrum or pivot point. This is because the tendency for a mass to produce rotation is proportional not only to the magnitude of the mass but also to the distance between the mass and the fulcrum. To be precise, if a point-mass m is located on a coordinate axis at a point x, then the tendency for that mass to produce a rotation about a point a on the axis is measured by the following quantity, called the **moment of m about a,**

$$\begin{bmatrix} \text{moment of } m \\ \text{about } a \end{bmatrix} = m(x - a)$$

The number $x - a$ is called the **lever arm.** Depending on whether the mass is to the right or left of a, the lever arm is either the distance between x and a or the negative of this distance (Figure 17.6.5). Positive lever arms

positive moment
about a
(clockwise rotation)

negative moment
about a
(counterclockwise rotation)

Figure 17.6.5 ▷

result in positive moments and clockwise rotations, while negative lever arms result in negative moments and counterclockwise rotations.

Suppose masses $m_1, m_2, \ldots, m_n$ are located at points $x_1, x_2, \ldots, x_n$ on a coordinate axis and a fulcrum is positioned at the point a (Figure 17.6.6). Depending on whether the sum of the moments about a,

$$\sum_{k=1}^{n} m_k(x_k - a) = m_1(x_1 - a) + m_2(x_2 - a) + \cdots + m_n(x_n - a)$$

is positive, negative, or zero, the axis will rotate clockwise about a, rotate counterclockwise about a, or balance perfectly. In the last case, the system of masses is said to be in *equilibrium*.

Figure 17.6.6 ▷

Example 2   Where should a fulcrum be placed so that the following system of masses is in equilibrium?

Solution.   Introduce an x-axis with origin at the mass m_1, and let a be the unknown coordinate of the fulcrum (Figure 17.6.7). (Our choice of location for the origin is arbitrary. Any convenient location will suffice.) We shall determine a so that the total moment about a is zero. The sum of the moments about a is

$$\text{total moment} = 20(0 - a) + 10(3 - a) + 40(15 - a) = 630 - 70a$$

Figure 17.6.7 ▷

For equilibrium, a should be chosen so that

$$630 - 70a = 0$$

or $a = 9$. Thus, the fulcrum should be located 9 feet to the right of mass m_1.   ◄

The ideas we have discussed can be extended to masses distributed in two-dimensional space. If a mass m is located at a point (x, y) in a horizontal xy-plane, then we define the **moment of m about the line $x = a$** and the **moment of m about the line $y = c$** by

$$\begin{bmatrix} \text{moment of } m \\ \text{about the} \\ \text{line } x = a \end{bmatrix} = m(x - a) \qquad \begin{pmatrix} \text{mass times} \\ \text{lever arm} \end{pmatrix} \tag{6}$$

$$\begin{bmatrix} \text{moment of } m \\ \text{about the} \\ \text{line } y = c \end{bmatrix} = m(y - c) \qquad \begin{pmatrix} \text{mass times} \\ \text{lever arm} \end{pmatrix} \tag{7}$$

(See Figure 17.6.8.)

Figure 17.6.8 ▷

If a number of masses are distributed in the xy-plane, and if we imagine the xy-plane to be a weightless sheet, then this sheet will balance on the line $x = a$ (or $y = c$) if the sum of the moments about that line is zero.

Suppose now that we are dealing not with isolated point masses, but with a lamina in the xy-plane. How should we define the moments about the lines $x = a$ and $y = c$? To motivate the appropriate definitions, assume the lamina occupies a region R in the xy-plane and has density function $\delta(x, y)$. Following a now familiar procedure, let R be partitioned into subregions and let R_k be a typical rectangle interior to R. Let ΔA_k be the area of R_k, let ΔM_k be the mass of that portion of the lamina covered by R_k, and let (x_k^*, y_k^*) be the center of the rectangle R_k. From (2), the mass ΔM_k may be approximated by

$$\Delta M_k \approx \delta(x_k^*, y_k^*) \, \Delta A_k$$

Moreover, if the rectangle R_k is small, there should be little error in assuming that the entire mass ΔM_k is concentrated at the center (x_k^*, y_k^*) of the rectangle. Thus, from (6) and (7), the moment of the mass ΔM_k about the line $x = a$ is approximately

$$(x_k^* - a) \, \Delta M_k \approx (x_k^* - a)\delta(x_k^*, y_k^*) \, \Delta A_k$$

and the moment about the line $y = c$ is approximately

$$(y_k^* - c) \, \Delta M_k \approx (y_k^* - c)\delta(x_k^*, y_k^*) \, \Delta A_k$$

Therefore, the total moment about $x = a$ produced by all the rectangular regions interior to R is approximately

$$\sum_{k=1}^{n} (x_k^* - a)\delta(x_k^*, y_k^*)\, \Delta A_k \qquad (8)$$

and the total moment about $y = c$ is approximately

$$\sum_{k=1}^{n} (y_k^* - c)\delta(x_k^*, y_k^*)\, \Delta A_k \qquad (9)$$

If we assume that the errors in our approximations diminish as the dimensions of the rectangles tend to zero, then (8) and (9) suggest the following definitions:

$$\begin{bmatrix} \text{moment of} \\ \text{lamina } R \\ \text{about the} \\ \text{line } x = a \end{bmatrix} = \lim_{n \to +\infty} \sum_{k=1}^{n} (x_k^* - a)\delta(x_k^*, y_k^*)\, \Delta A_k = \iint_{R} (x - a)\delta(x, y)\, dA \qquad (10)$$

$$\begin{bmatrix} \text{moment of} \\ \text{lamina } R \\ \text{about the} \\ \text{line } y = c \end{bmatrix} = \lim_{n \to +\infty} \sum_{k=1}^{n} (y_k^* - c)\delta(x_k^*, y_k^*)\, \Delta A_k = \iint_{R} (y - c)\delta(x, y)\, dA \qquad (11)$$

Of special importance are the moments about the y-axis ($a = 0$) and about the x-axis ($c = 0$). These moments are denoted by

$$M_y = \begin{bmatrix} \text{moment of lamina } R \\ \text{about the } y\text{-axis} \end{bmatrix} = \iint_{R} x\delta(x, y)\, dA \qquad (12)$$

$$M_x = \begin{bmatrix} \text{moment of lamina } R \\ \text{about the } x\text{-axis} \end{bmatrix} = \iint_{R} y\delta(x, y)\, dA \qquad (13)$$

We are now in a position to obtain formulas for the center of gravity $(\bar{x}, \bar{y})$ of a lamina. Since a lamina balances on the lines $x = \bar{x}$ and $y = \bar{y}$ through the center of gravity, the moments of the lamina about these lines must be zero. Thus, from (10) and (11)

$$\iint_{R} (x - \bar{x})\delta(x, y)\, dA = 0$$

$$\iint_{R} (y - \bar{y})\delta(x, y)\, dA = 0$$

Since $\bar{x}$ and $\bar{y}$ are constant, these equations can be rewritten as

$$\iint\limits_R x\delta(x,y)\,dA = \bar{x}\iint\limits_R \delta(x,y)\,dA$$

$$\iint\limits_R y\delta(x,y)\,dA = \bar{y}\iint\limits_R \delta(x,y)\,dA$$

from which we obtain the following formulas:

Center of gravity of a lamina

$$\bar{x} = \frac{\displaystyle\iint\limits_R x\delta(x,y)\,dA}{\displaystyle\iint\limits_R \delta(x,y)\,dA} \qquad \bar{y} = \frac{\displaystyle\iint\limits_R y\delta(x,y)\,dA}{\displaystyle\iint\limits_R \delta(x,y)\,dA} \qquad (14a)$$

By virtue of (3), (12), and (13), these formulas can be written as

$$\bar{x} = \frac{\begin{bmatrix}\text{moment about}\\\text{the }y\text{-axis}\end{bmatrix}}{\begin{bmatrix}\text{mass of the}\\\text{lamina}\end{bmatrix}} = \frac{M_y}{M} \qquad \bar{y} = \frac{\begin{bmatrix}\text{moment about}\\\text{the }x\text{-axis}\end{bmatrix}}{\begin{bmatrix}\text{mass of the}\\\text{lamina}\end{bmatrix}} = \frac{M_x}{M} \qquad (14b)$$

Example 3 Find the center of gravity of the triangular lamina with vertices $(0,0)$, $(0,1)$, and $(1,0)$ and density function $\delta(x,y) = xy$.

Solution. The lamina is shown in Figure 17.6.3. In Example 1 we found the mass of the lamina to be

$$M = \iint\limits_R \delta(x,y)\,dA = \iint\limits_R xy\,dA = \frac{1}{24}$$

The moment of the lamina about the y-axis is

$$M_y = \iint\limits_R x\delta(x,y)\,dA = \iint\limits_R x^2 y\,dA = \int_0^1 \int_0^{-x+1} x^2 y\,dy\,dx$$

$$= \int_0^1 \left[\frac{1}{2}x^2 y^2\right]_{y=0}^{-x+1} dx = \int_0^1 \left(\frac{1}{2}x^4 - x^3 + \frac{1}{2}x^2\right) dx = \frac{1}{60}$$

and the moment about the x-axis is

$$M_x = \iint\limits_R y\delta(x, y)\, dA = \iint\limits_R xy^2\, dA = \int_0^1 \int_0^{-x+1} xy^2\, dy\, dx$$

$$= \int_0^1 \left[\frac{1}{3}xy^3\right]_{y=0}^{-x+1} dx = \int_0^1 \left(-\frac{1}{3}x^4 + x^3 - x^2 + \frac{1}{3}x\right) dx = \frac{1}{60}$$

From (14b),

$$\bar{x} = \frac{M_y}{M} = \frac{1/60}{1/24} = \frac{2}{5}, \quad \bar{y} = \frac{M_x}{M} = \frac{1/60}{1/24} = \frac{2}{5}$$

so the center of gravity is $(2/5, 2/5)$.   ◄

CENTROIDS   In the special case of a *homogeneous* lamina, the center of gravity is called the **centroid of the lamina** or sometimes the **centroid of the region R**. Because the density function δ is constant for a homogeneous lamina, the factor δ may be moved through the integral signs in (14a) and canceled. Thus, the centroid $(\bar{x}, \bar{y})$ of a region R is given by the formulas

---

Centroid of a region R

$$\bar{x} = \frac{\displaystyle\iint\limits_R x\, dA}{\displaystyle\iint\limits_R dA} = \frac{1}{\text{area of } R} \iint\limits_R x\, dA \tag{15a}$$

$$\bar{y} = \frac{\displaystyle\iint\limits_R y\, dA}{\displaystyle\iint\limits_R dA} = \frac{1}{\text{area of } R} \iint\limits_R y\, dA \tag{15b}$$

Example 4   Find the centroid of the following semicircular region:

Solution. By symmetry, $\bar{x} = 0$ since the y-axis is obviously a line of balance. From (15b),

$$\bar{y} = \frac{\displaystyle\iint_R y \, dA}{\text{area of } R} = \frac{\displaystyle\iint_R y \, dA}{\frac{1}{2}\pi a^2} \tag{16}$$

The integral in (16) is best evaluated in polar coordinates:

$$\iint_R y \, dA = \int_0^\pi \int_0^a (r \sin \theta) r \, dr \, d\theta = \int_0^\pi \left[\frac{1}{3} r^3 \sin \theta \right]_{r=0}^a d\theta$$

$$= \frac{1}{3} a^3 \int_0^\pi \sin \theta \, d\theta = \frac{2}{3} a^3$$

Thus, from (16)

$$\bar{y} = \frac{\frac{2}{3} a^3}{\frac{1}{2}\pi a^2} = \frac{4a}{3\pi}$$

so the centroid is $\left(0, \dfrac{4a}{3\pi} \right)$. ◄

CENTER OF GRAVITY
OF A SOLID

For three-dimensional solids the formulas for moments, centers of gravity, and centroids are similar to the formulas for laminas. If a solid G of mass M and volume V is homogeneous in composition, we define the *density* δ of the solid to be

$$\delta = \frac{M}{V}$$

which measures the mass per unit volume. If the solid G is not homogeneous, then the density at a general point (x, y, z) is specified by a *density function* $\delta(x, y, z)$ that may be viewed as a limit

$$\delta(x, y, z) = \lim \frac{\Delta M}{\Delta V}$$

where ΔM and ΔV represent the mass and volume of a rectangular parallelepiped, centered at (x, y, z), whose dimensions tend to zero (Figure 17.6.9).

The following formulas generalize the corresponding results for laminas:

$$\boxed{\text{mass of } G = \iiint_G \delta(x, y, z) \, dV} \tag{17}$$

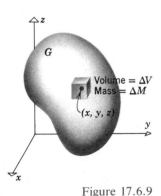

Figure 17.6.9

Center of gravity $(\bar{x}, \bar{y}, \bar{z})$ *of a solid G*

$$\bar{x} = \frac{\iiint\limits_{G} x\delta(x, y, z)\, dV}{\iiint\limits_{G} \delta(x, y, z)\, dV} = \frac{1}{\text{mass of } G} \iiint\limits_{G} x\delta(x, y, z)\, dV \qquad (18a)$$

$$\bar{y} = \frac{\iiint\limits_{G} y\delta(x, y, z)\, dV}{\iiint\limits_{G} \delta(x, y, z)\, dV} = \frac{1}{\text{mass of } G} \iiint\limits_{G} y\delta(x, y, z)\, dV \qquad (18b)$$

$$\bar{z} = \frac{\iiint\limits_{G} z\delta(x, y, z)\, dV}{\iiint\limits_{G} \delta(x, y, z)\, dV} = \frac{1}{\text{mass of } G} \iiint\limits_{G} z\delta(x, y, z)\, dV \qquad (18c)$$

Centroid $(\bar{x}, \bar{y}, \bar{z})$ *of a solid G*

$$\bar{x} = \frac{\iiint\limits_{G} x\, dV}{\iiint\limits_{G} dV} = \frac{\iiint\limits_{G} x\, dV}{\text{volume of } G} \qquad (19a)$$

$$\bar{y} = \frac{\iiint\limits_{G} y\, dV}{\iiint\limits_{G} dV} = \frac{\iiint\limits_{G} y\, dV}{\text{volume of } G} \qquad (19b)$$

$$\bar{z} = \frac{\iiint\limits_{G} z\, dV}{\iiint\limits_{G} dV} = \frac{\iiint\limits_{G} z\, dV}{\text{volume of } G} \qquad (19c)$$

Figure 17.6.10

Example 5   Find the mass and the center of gravity of a cylindrical solid (Figure 17.6.10), assuming that the density at each point is proportional to the distance between the point and the base of the solid.

Solution.   Since the density is proportional to the distance z from the base, the density function has the form $\delta(x, y, z) = kz$ where k is some (unknown) constant of proportionality. From (17) the mass of the solid is

$$M = \iiint_G \delta(x, y\,z)\,dV = \int_{-a}^{a} \int_{-\sqrt{a^2-x^2}}^{\sqrt{a^2-x^2}} \int_{0}^{h} kz\,dz\,dy\,dx$$

$$= k \int_{-a}^{a} \int_{-\sqrt{a^2-x^2}}^{\sqrt{a^2-x^2}} \tfrac{1}{2}h^2 dy\,dx$$

$$= kh^2 \int_{-a}^{a} \sqrt{a^2 - x^2}\,dx$$

$$= \tfrac{1}{2}kh^2\pi a^2 \qquad \left[\begin{array}{l}\text{See Example 4}\\\text{of Section 9.5.}\end{array}\right]$$

Without additional information, the constant k cannot be determined. However, as we shall now see, the value of k does not affect the center of gravity.
    From (18),

$$\bar{z} = \dfrac{\displaystyle\iiint_G z\delta(x, y, z)\,dV}{\text{mass of } G} = \dfrac{\displaystyle\iiint_G z\delta(x, y, z)\,dV}{\tfrac{1}{2}kh^2\pi a^2}$$

But

$$\iiint_G z\delta(x, y, z)\,dV = \int_{-a}^{a} \int_{-\sqrt{a^2-x^2}}^{\sqrt{a^2-x^2}} \int_{0}^{h} z(kz)\,dz\,dy\,dx$$

$$= k \int_{-a}^{a} \int_{-\sqrt{a^2-x^2}}^{\sqrt{a^2-x^2}} \tfrac{1}{3}h^3\,dy\,dx$$

$$= \tfrac{2}{3}kh^3 \int_{-a}^{a} \sqrt{a^2 - x^2}\,dx$$

$$= \tfrac{1}{3}kh^3\pi a^2$$

so that

$$\bar{z} = \frac{\tfrac{1}{3}kh^3\pi a^2}{\tfrac{1}{2}kh^2\pi a^2} = \tfrac{2}{3}h$$

Similar calculations using (18) will yield $\bar{x} = \bar{y} = 0$. However, this is evident by inspection, since it follows from the symmetry of the solid and the form of its density function that the center of gravity is on the z-axis. Thus, the center of gravity is $(0, 0, \frac{2}{3}h)$.   ◄

The next theorem, due to the Greek mathematician Pappus,* gives an interesting and useful relationship between the centroid of a plane region R and the volume of the solid generated when the region is revolved about a line.

Theorem of Pappus

17.6.3   THEOREM. *If R is a plane region and L is a line that lies in the plane of R, but does not intersect R, then the volume of the solid formed by revolving R about L is given by*

$$volume = (area\ of\ R) \cdot \left(\begin{array}{c} distance\ traveled \\ by\ the\ centroid \end{array} \right)$$

Proof.   Introduce an xy-coordinate system so that L is along the y-axis and the region R is in the first quadrant (Figure 17.6.11). Let R be partitioned into subregions in the usual way and let R_k be a typical rectangle interior to R. If (x_k^*, y_k^*) is the center of R_k and if the area of R_k is $\Delta A_k = \Delta x_k \, \Delta y_k$, then the volume generated by R_k as it revolves about L is

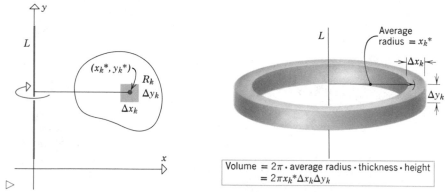

Figure 17.6.11 ▷

*PAPPUS OF ALEXANDRIA (4th century A.D.) Greek mathematician. Pappus lived during the early Christian era when mathematical activity was in a period of decline. His main contributions to mathematics appeared in a series of eight books called *The Collection* (written about 340 A.D.). This work, which survives only partially, contained some original results, but was devoted mostly to statements, refinements, and proofs of results by earlier mathematicians. Pappus' Theorem, stated without proof in Book VII of *The Collection*, was probably known and proved in earlier times. This result is sometimes called Guldin's Theorem in recognition of the Swiss mathematician, Paul Guldin (1577–1643), who rediscovered it independently.

$$2\pi x_k^* \, \Delta x_k \, \Delta y_k = 2\pi x_k^* \, \Delta A_k$$

[See Figure 17.6.11 and Formula (1) of Section 6.3.] Therefore, the total volume of the solid is approximately

$$V \approx \sum_{k=1}^{n} 2\pi x_k^* \, \Delta A_k$$

from which it follows that the exact volume is

$$V = \iint_R 2\pi x \, dA = 2\pi \iint_R x \, dA$$

Thus, from (15a)

$$V = 2\pi \cdot \bar{x} \cdot [\text{area of } R]$$

This completes the proof since $2\pi\bar{x}$ is the distance traveled by the centroid when R is revolved about the y-axis. ∎

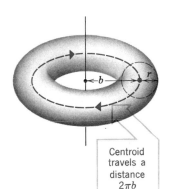

Centroid
travels a
distance
$2\pi b$

Figure 17.6.12

Example 6 Use Pappus' Theorem to find the volume V of the torus generated by revolving a circular region of radius r about a line at a distance b (greater than r) from the center of the circle (Figure 17.6.12).

Solution. By symmetry, the centroid of a circular region is its center. Thus, the distance traveled by the centroid is $2\pi b$. Since the area of a circle is πr^2, it follows from Pappus' Theorem that the volume of the torus is

$$V = (2\pi b)(\pi r^2) = 2\pi^2 b r^2 \quad \blacktriangleleft$$

▶ Exercise Set 17.6

1. Where should the fulcrum be placed so the see-saw is in equilibrium?

m_1 m_2 m_3

5 10 20

|← 5 ft →|← 10 ft →|

2. A rectangular lamina with density $\delta(x, y) = xy^2$ has vertices $(0,0)$, $(0,2)$, $(3,0)$, $(3,2)$. Find its mass and center of gravity.

3. A lamina with density $\delta(x, y) = x + y$ is bounded by the x-axis, the line $x = 1$, and the curve $y = \sqrt{x}$. Find its mass and center of gravity.

4. A lamina with density $\delta(x, y) = y$ is bounded by $y = \sin x$, $y = 0$, $x = 0$, and $x = \pi$. Find its mass and center of gravity.

5. A lamina with density $\delta(x, y) = xy$ is in the first quadrant and is bounded by the circle $x^2 + y^2 = a^2$ and the coordinate axes. Find its mass and center of gravity.

6. A lamina with density $\delta(x, y) = x^2 + y^2$ is bounded by the x-axis and the upper half of the circle $x^2 + y^2 = 1$. Find its mass and center of gravity.

In Exercises 7–13, find the centroid of the region.

7. The triangular region bounded by $y = x$, $x = 1$, and the x-axis.

8. The region bounded by $y = x^2$, $x = 1$, and the x-axis.

9. The region enclosed between $y = x$ and $y = 2 - x^2$.

10. The region enclosed by the parabolas $x = 3y^2 - 6y$ and $x = 2y - y^2$.

11. The region above the x-axis and between the circles $x^2 + y^2 = a^2$ and $x^2 + y^2 = b^2$ $(a < b)$.

12. The region enclosed between the y-axis and the right half of the circle $x^2 + y^2 = a^2$.

13. The region enclosed between $y = |x|$ and the line $y = 4$.

14. A rectangular box defined by the inequalities $0 \le x \le 1$, $0 \le y \le 1$, $0 \le z \le 1$ has density $\delta(x, y, z) = 3xyz$. Find its mass and center of gravity.

15. A cube defined by the inequalities $0 \le x \le a$, $0 \le y \le a$, $0 \le z \le a$ has density $\delta(x, y, z) = a - x$. Find its mass and center of gravity.

16. The cylindrical solid enclosed by $x^2 + y^2 = a^2$, $z = 0$, and $z = h$ has density $\delta(x, y, z) = h - z$. Find its mass and center of gravity.

17. The solid enclosed by the surface $z = 1 - y^2$ (where $y \ge 0$) and the planes $z = 0$, $x = -1$, $x = 1$ has density $\delta(x, y, z) = yz$. Find its mass and center of gravity.

18. The solid enclosed by the surface $y = 9 - x^2$ (where $x \ge 0$) and the planes $y = 0$, $z = 0$, and $z = 1$ has density $\delta(x, y, z) = xz$. Find its mass and center of gravity.

In Exercises 19–23, find the centroid of the solid.

19. The tetrahedron in the first octant enclosed by the coordinate planes and the plane $x + y + z = 1$.

20. The solid enclosed by the xy-plane and the hemisphere $z = \sqrt{a^2 - x^2 - y^2}$.

21. The solid bounded by the surface $z = y^2$, and the planes $x = 0$, $x = 1$, and $z = 1$.

22. The solid in the first octant bounded by the surface $z = xy$ and the planes $z = 0$, $x = 2$, and $y = 2$.

23. The solid in the first octant bounded by the sphere $x^2 + y^2 + z^2 = a^2$ and the coordinate planes.

24. Find the center of gravity of the square lamina with vertices $(0, 0)$, $(1, 0)$, $(0, 1)$, and $(1, 1)$ if
    (a) the density is proportional to the square of the distance from the origin;
    (b) the density is proportional to the distance from the y-axis.

25. Find the mass of a circular lamina of radius a whose density at a point is proportional to the distance from the center. (Let k denote the constant of proportionality.)

26. Find the center of gravity of the cube determined by the inequalities $0 \le x \le 1$, $0 \le y \le 1$, $0 \le z \le 1$ if
    (a) the density is proportional to the square of the distance to the origin;
    (b) the density is proportional to the sum of the distances to the faces that lie in the coordinate planes.

27. Show that in polar coordinates the formulas for the centroid $(\bar{x}, \bar{y})$ of a region R are

$$\bar{x} = \frac{1}{\text{area of } R} \iint\limits_R r^2 \cos \theta \, dr \, d\theta$$

$$\bar{y} = \frac{1}{\text{area of } R} \iint\limits_R r^2 \sin \theta \, dr \, d\theta$$

28. Use the result of Exercise 27 to find the centroid $(\bar{x}, \bar{y})$ of the region enclosed by the cardioid $r = a(1 + \sin \theta)$.

29. Use the result of Exercise 27 to find the centroid $(\bar{x}, \bar{y})$ of the petal of the rose $r = \sin 2\theta$ that lies in the first quadrant.

30. Let R be the rectangle bounded by the lines $x = 0$, $x = 3$, $y = 0$, and $y = 2$. By inspection, find the centroid of R and use it to evaluate

$$\iint\limits_R x \, dA \quad \text{and} \quad \iint\limits_R y \, dA$$

31. Use the Theorem of Pappus and the fact that the volume of a sphere of radius a is $V = \frac{4}{3}\pi a^3$ to show that the centroid of the lamina bounded by the x-axis and the semicircle $y = \sqrt{a^2 - x^2}$ is $(0, 4a/(3\pi))$. (This problem was solved directly in Example 4.)

32. Use the Theorem of Pappus and the result of Exercise 31 to find the volume of the solid generated when the region bounded by the x-axis and the semicircle $y = \sqrt{a^2 - x^2}$ is revolved about
(a) the line $y = -a$
(b) the line $y = x - a$.

33. Use the Theorem of Pappus and the fact that the area of an ellipse with semiaxes a and b is πab to find the volume of the elliptical torus generated by revolving the ellipse

$$\frac{(x-k)^2}{a^2} + \frac{y^2}{b^2} = 1$$

about the y-axis. Assume $k > a$.

34. Use the Theorem of Pappus to find the volume of the solid generated when the region enclosed by $y = x^2$ and $y = 8 - x^2$ is revolved about the x-axis.

35. Use the Theorem of Pappus to find the centroid of the triangular region with vertices $(0, 0)$, $(a, 0)$, and $(0, b)$, where $a > 0$ and $b > 0$. [*Hint:* Revolve the region about the x-axis to obtain $\bar{y}$ and about the y-axis to obtain $\bar{x}$.]

36. If a lamina with density function $\delta(x, y)$ occupies a region R in the xy-plane, then the *second moments* (or *moments of inertia*) of the lamina about the x-axis, y-axis, and z-axis, respectively, are defined by

$$I_x = \iint_R y^2 \delta \, dA, \qquad I_y = \iint_R x^2 \delta \, dA,$$

$$I_z = \iint_R (x^2 + y^2)\delta \, dA.$$

The corresponding definitions for a solid with density function $\delta(x, y, z)$ that occupies a region G are:

$$I_x = \iiint_G (y^2 + z^2)\delta(x, y, z) \, dV$$

$$I_y = \iiint_G (x^2 + z^2)\delta(x, y, z) \, dV$$

$$I_z = \iiint_G (x^2 + y^2)\delta(x, y, z) \, dV$$

Find I_x, I_y, and I_z for the homogeneous (constant δ) rectangular lamina described by the inequalities $0 \le x \le a$ and $0 \le y \le b$. Express your answers in terms of the mass M of the lamina. [*Note:* Moments of inertia are important in problems involving rotational motion.]

In Exercises 37–44, a homogeneous lamina or solid is given. Use the formulas in Exercise 36 to find the indicated moment of inertia. Express your answers in terms of the mass M of the lamina or solid.

37. The moment I_z for the rectangular lamina $-a/2 \le x \le a/2$, $-b/2 \le y \le b/2$.

38. The moment I_x for the rectangular lamina $-a/2 \le x \le a/2$, $-b/2 \le y \le b/2$.

39. The circular lamina $x^2 + y^2 \le R^2$; I_y.

40. The circular lamina $x^2 + y^2 \le R^2$; I_z.

41. The circular lamina $x^2 + (y - R)^2 \le R^2$; I_z.

42. The triangular lamina $0 \le y \le 2 - 2x$, $x \ge 0$; I_y.

43. The solid $0 \le x \le a$, $0 \le y \le a$, $0 \le z \le a$; I_z.

44. The rectangular solid $0 \le x \le a$, $-b/2 \le y \le b/2$, $-c/2 \le z \le c/2$; I_x.

17.7   TRIPLE INTEGRALS IN CYLINDRICAL AND SPHERICAL COORDINATES

In Section 17.3 we saw that some double integrals are easier to evaluate in polar coordinates. Similarly, some triple integrals are easier to evaluate in cylindrical or spherical coordinates. In this section we shall study triple integrals in these coordinate systems.

TRIPLE INTEGRALS IN CYLINDRICAL COORDINATES

In rectangular coordinates the triple integral of a function f over a solid region G is defined as

$$\iiint\limits_{G} f(x, y, z)\, dV = \lim_{n \to +\infty} \sum_{k=1}^{n} f(x_k^*, y_k^*, z_k^*)\, \Delta V_k$$

Figure 17.7.1

where ΔV_k denotes the volume of a rectangular parallelepiped interior to G and (x_k^*, y_k^*, z_k^*) is a point in this parallelepiped (Figure 17.7.1). Triple integrals in cylindrical and spherical coordinates are defined similarly, except that the region G is not divided into rectangular parallelepipeds, but into regions more appropriate to these coordinate systems.

In cylindrical coordinates, the simplest equations are of the form

$$r = \text{constant}, \quad \theta = \text{constant}, \quad z = \text{constant}$$

As indicated in Figure 14.8.2, the first equation represents a right-circular cylinder centered on the z-axis, the second a vertical half plane hinged on the z-axis, and the third a horizontal plane. These surfaces can be paired up to determine solids called **cylindrical wedges** or **cylindrical elements of volume.** To be precise, a cylindrical wedge is a solid enclosed between six surfaces of the following type:

two cylinders $r = r_1, r = r_2$ $(r_1 < r_2)$

two half planes $\theta = \theta_1, \theta = \theta_2$ $(\theta_1 < \theta_2)$

two planes $z = z_1, z = z_2$ $(z_1 < z_2)$

(Figure 17.7.2). The dimensions $\theta_2 - \theta_1$, $r_2 - r_1$, and $z_2 - z_1$ are called the **central angle, thickness,** and **height** of the wedge.

To define the triple integral over G of a function $f(r, \theta, z)$ in cylindrical coordinates, we proceed as follows:

Step 1. Construct a three-dimensional grid consisting of right-circular cylinders centered on the z-axis, half planes hinged on the z-axis, and horizontal planes. The blocks determined by this grid will be cylindrical wedges. Using this grid, divide the region G

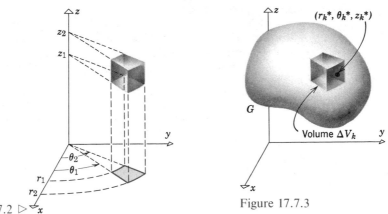

Figure 17.7.2 ▷

Figure 17.7.3

into subregions and exclude from consideration subregions that contain any portion of the surface of the solid G. This leaves only cylindrical wedges interior to G (Figure 17.7.3). Denote the volumes of these interior cylindrical wedges by

$$\Delta V_1, \Delta V_2, \ldots, \Delta V_n$$

Step 2. Choose an arbitrary point in each interior cylindrical wedge, and let these be denoted by

$$(r_1^*, \theta_1^*, z_1^*), (r_2^*, \theta_2^*, z_2^*), \ldots, (r_n^*, \theta_n^*, z_n^*)$$

(See Figure 17.7.3.)

Step 3. Form the sum

$$\sum_{k=1}^{n} f(r_k^*, \theta_k^*, z_k^*) \, \Delta V_k$$

Step 4. Repeat this process with more and more subdivisions, so that the height, thickness, and central angle of each interior cylindrical wedge approaches zero, and n, the number of such wedges, approaches $+\infty$. Define

$$\iiint\limits_{G} f(r, \theta, z) \, dV = \lim_{n \to +\infty} \sum_{k=1}^{n} f(r_k^*, \theta_k^*, z_k^*) \, \Delta V_k$$

In Section 17.5, we defined simple solids. The following theorem will enable us to evaluate triple integrals in cylindrical coordinates over such solids by repeated integration.

17.7.1   THEOREM. *Let G be a simple solid whose upper surface has the equation $z = g_2(r, \theta)$ and whose lower surface has the equation $z = g_1(r, \theta)$ in cylindrical coordinates. If R is the projection of the solid on the xy-plane and if $f(r, \theta, z)$ is continuous on G, then*

$$\iiint_G f(r, \theta, z)\, dV = \iint_R \left[\int_{g_1(r,\theta)}^{g_2(r,\theta)} f(r, \theta, z)\, dz \right] dA \qquad (1)$$

where the double integral over R is evaluated in polar coordinates. In particular, if the projection R is as shown in Figure 17.7.4, then (1) can be written as

$$\iiint_G f(r, \theta, z)\, dV = \int_{\theta_1}^{\theta_2} \int_{r_1(\theta)}^{r_2(\theta)} \int_{g_1(r,\theta)}^{g_2(r,\theta)} f(r, \theta, z) r\, dz\, dr\, d\theta \qquad (2)$$

The type of solid to which formula (2) applies is illustrated in Figure 17.7.4.

Figure 17.7.4

Figure 17.7.5

We shall not formally prove Theorem 17.7.1. However, the appearance of the factor r in the integrand of (2) can be explained by returning to the definition

$$\iiint_G f(r, \theta, z)\, dV = \lim_{n \to +\infty} \sum_{k=1}^{n} f(r_k^*, \theta_k^*, z_k^*)\, \Delta V_k \qquad (3)$$

In this formula, the volume ΔV_k of the kth cylindrical wedge can be written

$$\Delta V_k = [\text{area of base}] \cdot [\text{height}] \qquad (4)$$

If we denote the thickness, central angle, and height of this wedge by Δr_k, $\Delta \theta_k$, and Δz_k, and if we choose the arbitrary point $(r_k^*, \theta_k^*, z_k^*)$ to lie above the "center" of the base (Figures 17.3.5 and 17.7.5), then it follows from (4) of Section 17.3 that the base has area $r_k^* \, \Delta r_k \, \Delta \theta_k$. Thus, (4) can be written as

$$\Delta V_k = r_k^* \, \Delta r_k \, \Delta \theta_k \, \Delta z_k$$

Substituting this expression in (3) yields

$$\iiint_G f(r, \theta, z)\, dV = \lim_{n \to +\infty} \sum_{k=1}^{n} f(r_k^*, \theta_k^*, z_k^*) r_k^* \, \Delta z_k \, \Delta r_k \, \Delta \theta_k$$

which explains the form of the integrand in (2).

To apply (1) and (2) it is best to begin with a three-dimensional sketch of the solid G, from which the limits of integration can be obtained as follows:

> **Step 1.** Identify the upper surface $z = g_2(r, \theta)$ and the lower surface $z = g_1(r, \theta)$ of the solid. The functions $g_1(r, \theta)$ and $g_2(r, \theta)$ determine the z-limits of integration. (If the upper and lower surfaces are given in rectangular coordinates, convert them to cylindrical coordinates.)
>
> **Step 2.** Make a two-dimensional sketch of the projection R of the solid on the xy-plane. From this sketch the r- and θ-limits of integration may be obtained exactly as with double integrals in polar coordinates.

Example 1 Use triple integration in cylindrical coordinates to find the volume and the centroid of the solid G that is bounded above by the hemisphere $z = \sqrt{25 - x^2 - y^2}$, below by the xy-plane, and laterally by the cylinder $x^2 + y^2 = 9$.

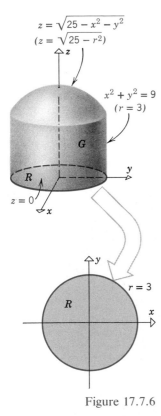

$z = \sqrt{25 - x^2 - y^2}$
$(z = \sqrt{25 - r^2})$

$x^2 + y^2 = 9$
$(r = 3)$

G

R

$z = 0$

$r = 3$

R

Figure 17.7.6

Solution. The solid G and its projection R on the xy-plane are shown in Figure 17.7.6. In cylindrical coordinates, the upper surface of G is the hemisphere $z = \sqrt{25 - r^2}$ and the lower surface is the plane $z = 0$. Thus, from (1), the volume of G is

$$V = \iiint_G dV = \iint_R \left[\int_0^{\sqrt{25 - r^2}} dz \right] dA$$

For the double integral over R, we use polar coordinates:

$$V = \int_0^{2\pi} \int_0^3 \int_0^{\sqrt{25 - r^2}} r \, dz \, dr \, d\theta = \int_0^{2\pi} \int_0^3 \left[rz \right]_{z=0}^{\sqrt{25 - r^2}} dr \, d\theta$$

$$= \int_0^{2\pi} \int_0^3 r\sqrt{25 - r^2} \, dr \, d\theta = \int_0^{2\pi} \left[-\frac{1}{3}(25 - r^2)^{3/2} \right]_{r=0}^{3} d\theta$$

$$= \int_0^{2\pi} \frac{61}{3} \, d\theta = \frac{122}{3} \pi$$

$$u = 25 - r^2$$
$$du = -2r \, dr.$$

From this result and Formula (19c) of Section 17.6,

$$\bar{z} = \frac{1}{V} \iiint_G z\, dV = \frac{3}{122\pi} \iiint_G z\, dV = \frac{3}{122\pi} \iint_R \left[\int_0^{\sqrt{25-r^2}} z\, dz \right] dA$$

$$= \frac{3}{122\pi} \int_0^{2\pi} \int_0^3 \int_0^{\sqrt{25-r^2}} zr\, dz\, dr\, d\theta = \frac{3}{122\pi} \int_0^{2\pi} \int_0^3 \left[\frac{1}{2} rz^2 \right]_{z=0}^{\sqrt{25-r^2}} dr\, d\theta$$

$$= \frac{3}{244\pi} \int_0^{2\pi} \int_0^3 (25r - r^3)\, dr\, d\theta = \frac{3}{244\pi} \int_0^{2\pi} \frac{369}{4}\, d\theta = \frac{1107}{488}$$

By symmetry, the centroid $(\bar{x}, \bar{y}, \bar{z})$ of G lies on the z-axis, so $\bar{x} = \bar{y} = 0$. Thus, the centroid is at the point $(0, 0, 1107/488)$.   ◄

Frequently, a triple integral that is hard to evaluate in rectangular coordinates is easy to evaluate when the integrand and the limits of integration are expressed in cylindrical coordinates. This is especially true when the integrand or the limits of integration involve expressions of the form

$$x^2 + y^2 \quad \text{or} \quad \sqrt{x^2 + y^2}$$

since these simplify to

$$r^2 \quad \text{and} \quad r$$

in cylindrical coordinates.

Example 2   Use cylindrical coordinates to evaluate

$$\int_{-3}^3 \int_{-\sqrt{9-x^2}}^{\sqrt{9-x^2}} \int_0^{9-x^2-y^2} x^2\, dz\, dy\, dx$$

Solution.   In problems of this type, it is helpful to sketch the region of integration G and its projection R on the xy-plane. From the z-limits of integration, the upper surface of G is the paraboloid $z = 9 - x^2 - y^2$ and the lower surface is the xy-plane, $z = 0$. From the x- and y-limits of integration, the projection R is the region in the xy-plane enclosed by the circle $x^2 + y^2 = 9$ (Figure 17.7.7). Thus,

$$\int_{-3}^3 \int_{-\sqrt{9-x^2}}^{\sqrt{9-x^2}} \int_0^{9-x^2-y^2} x^2\, dz\, dy\, dx = \iiint_G x^2\, dV$$

$$= \iiint_G (r\cos\theta)^2\, dV = \iint_R \left[\int_0^{9-r^2} r^2 \cos^2\theta\, dz \right] dA$$

Figure 17.7.7

$$= \int_0^{2\pi} \int_0^3 \int_0^{9-r^2} r^3 \cos^2 \theta \, dz \, dr \, d\theta = \int_0^{2\pi} \int_0^3 \left[zr^3 \cos^2 \theta \right]_{z=0}^{9-r^2} dr \, d\theta$$

$$= \int_0^{2\pi} \int_0^3 (9r^3 - r^5) \cos^2 \theta \, dr \, d\theta = \int_0^{2\pi} \left[\left(\frac{9r^4}{4} - \frac{r^6}{6} \right) \cos^2 \theta \right]_{r=0}^3 d\theta$$

$$= \frac{243}{4} \int_0^{2\pi} \cos^2 \theta \, d\theta = \frac{243}{4} \int_0^{2\pi} \frac{1}{2}(1 + \cos 2\theta) \, d\theta = \frac{243\pi}{4} \quad \blacktriangleleft$$

TRIPLE INTEGRALS IN SPHERICAL COORDINATES

In spherical coordinates, the simplest equations are of the form

$$\rho = \text{constant}, \quad \theta = \text{constant}, \quad \phi = \text{constant}$$

As indicated in Figure 14.8.5, the first equation represents a sphere centered at the origin, the second a half plane hinged on the z-axis, and the third a right-circular cone with vertex at the origin and centered on the z-axis. By a *spherical wedge* or *spherical element of volume* we mean a solid enclosed between six surfaces of the following form:

two spheres	$\rho = \rho_1, \rho = \rho_2$	$(\rho_1 < \rho_2)$
two half planes	$\theta = \theta_1, \theta = \theta_2$	$(\theta_1 < \theta_2)$
two right-circular cones	$\phi = \phi_1, \phi = \phi_2$	$(\phi_1 < \phi_2)$

(Figure 17.7.8). We shall refer to the numbers $\rho_2 - \rho_1$, $\theta_2 - \theta_1$, and $\phi_2 - \phi_1$ as the *dimensions* of a spherical wedge.

If G is a solid region in three-dimensional space, then the triple integral over G of a function $f(\rho, \theta, \phi)$ in spherical coordinates is similar in definition to the triple integral in cylindrical coordinates, except that the solid G is partitioned into *spherical* wedges by a three-dimensional grid consisting of spheres centered at the origin, half planes hinged on the z-axis,

Figure 17.7.8 ▷

and right-circular cones with vertices at the origin and opening along the z-axis (Figure 17.7.9).

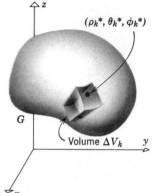

Figure 17.7.9 ▷

The defining equation of a triple integral in spherical coordinates is

$$\iiint\limits_{G} f(\rho, \theta, \phi) \, dV = \lim_{n \to +\infty} \sum_{k=1}^{n} f(\rho_k^*, \theta_k^*, \phi_k^*) \, \Delta V_k \tag{5}$$

where ΔV_k is the volume of the kth spherical wedge interior to G, $(\rho_k^*, \theta_k^*, \phi_k^*)$ is an arbitrary point in this wedge, and n increases in such a way that the dimensions of each interior spherical wedge tend to zero.

In the exercises we shall help the reader to show that if the point $(\rho_k^*, \theta_k^*, \phi_k^*)$ is suitably chosen, then the volume ΔV_k in (5) can be written

$$\Delta V_k = \rho_k^{*2} \sin \phi_k^* \, \Delta \rho_k \, \Delta \phi_k \, \Delta \theta_k$$

where $\Delta \rho_k$, $\Delta \phi_k$, and $\Delta \theta_k$ are the dimensions of the wedge. Substituting this in (5) we obtain

$$\iiint\limits_{G} f(\rho, \theta, \phi) \, dV = \lim_{n \to +\infty} \sum_{k=1}^{n} f(\rho_k^*, \theta_k^*, \phi_k^*) \rho_k^{*2} \sin \phi_k^* \, \Delta \rho_k \, \Delta \phi_k \, \Delta \theta_k$$

which suggests the following formula for evaluating a triple integral in spherical coordinates by repeated integration:

$$\iiint\limits_{G} f(\rho, \theta, \phi) \, dV = \iiint f(\rho, \theta, \phi) \rho^2 \sin \phi \, d\rho \, d\phi \, d\theta \tag{6}$$

appropriate
limits of
integration

REMARK.   Note the extra factor of $\rho^2 \sin \phi$ that appears in the integrand of the iterated integral. This is analogous to the extra factor of r that appeared when we integrated in cylindrical coordinates.

In (6) we omitted the limits of integration. Instead, we shall give some examples that explain how to obtain such limits. In all our examples we shall use the same order of integration: first with respect to ρ, then ϕ, and then θ.

Suppose we want to integrate $f(\rho, \theta, \phi)$ over the spherical solid G enclosed by the sphere $\rho = \rho_0$. The basic idea is to choose the limits of integration so that every point of the solid is accounted for in the integration process. Figure 17.7.10 illustrates one way of doing this. Holding θ and ϕ fixed for the first integration, we let ρ vary from 0 to ρ_0. This covers a radial line from the origin to the surface of the sphere. Next, keeping θ fixed, we let ϕ vary from 0 to π so that the radial line sweeps out a fan-shaped region. Finally, we let θ vary from 0 to 2π so that the fan-shaped region makes a complete revolution, thereby sweeping out the entire sphere. Thus, the triple integral of $f(\rho, \theta, \phi)$ over the spherical solid G may be evaluated by writing

$$\iiint\limits_{G} f(\rho, \theta, \phi) \, dV = \int_0^{2\pi} \int_0^{\pi} \int_0^{\rho_0} f(\rho, \theta, \phi)\rho^2 \sin \phi \, d\rho \, d\phi \, d\theta$$

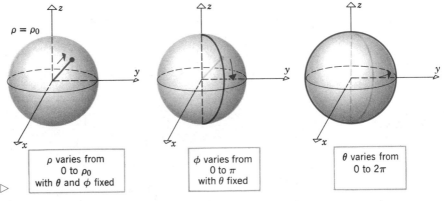

$\rho = \rho_0$

| ρ varies from 0 to ρ_0 with θ and ϕ fixed | ϕ varies from 0 to π with θ fixed | θ varies from 0 to 2π |

Figure 17.7.10 ▷

Table 17.7.1 (pages 1115-1116) suggests how the limits of integration in spherical coordinates can be obtained for some other common solids.

Example 3   Use spherical coordinates to find the volume and the centroid of the solid G bounded above by the sphere $x^2 + y^2 + z^2 = 16$ and below by the cone $z = \sqrt{x^2 + y^2}$.

Solution.   The solid G is sketched in Figure 17.7.11.

$x^2 + y^2 + z^2 = 16$
$(\rho = 4)$

$z = \sqrt{x^2 + y^2}$

$(\phi = \frac{\pi}{4})$

Figure 17.7.11 $\triangleright$

In spherical coordinates, the equation of the sphere $x^2 + y^2 + z^2 = 16$ is $\rho = 4$ and the equation of the cone $z = \sqrt{x^2 + y^2}$ is

$$\rho \cos \phi = \sqrt{\rho^2 \sin^2 \phi \cos^2 \theta + \rho^2 \sin^2 \phi \sin^2 \theta}$$

which simplifies to

$$\rho \cos \phi = \rho \sin \phi$$

or on dividing both sides by $\rho \cos \phi$

$$\tan \phi = 1$$

Thus, $\phi = \pi/4$, so the volume of G is

$$V = \iiint\limits_{G} dV = \int_0^{2\pi} \int_0^{\pi/4} \int_0^4 \rho^2 \sin \phi \, d\rho \, d\phi \, d\theta$$

$$= \int_0^{2\pi} \int_0^{\pi/4} \left[\frac{\rho^3}{3} \sin \phi \right]_{\rho=0}^{4} d\phi \, d\theta$$

$$= \int_0^{2\pi} \int_0^{\pi/4} \frac{64}{3} \sin \phi \, d\phi \, d\theta$$

$$= \frac{64}{3} \int_0^{2\pi} \left[-\cos \phi \right]_{\phi=0}^{\pi/4} d\theta = \frac{64}{3} \int_0^{2\pi} \left(1 - \frac{\sqrt{2}}{2} \right) d\theta$$

$$= \frac{64\pi}{3} (2 - \sqrt{2})$$

By symmetry, the centroid $(\bar{x}, \bar{y}, \bar{z})$ is on the z-axis, so $\bar{x} = \bar{y} = 0$. From Formula (19c) of Section 17.6 and the volume calculated above

$$\bar{z} = \frac{1}{V} \iiint\limits_{G} z \, dV = \frac{1}{V} \int_0^{2\pi} \int_0^{\pi/4} \int_0^4 (\rho \cos \phi)\rho^2 \sin \phi \, d\rho \, d\phi \, d\theta$$

$$= \frac{1}{V} \int_0^{2\pi} \int_0^{\pi/4} \left[\frac{\rho^4}{4} \cos \phi \sin \phi \right]_{\rho=0}^{4} d\phi \, d\theta$$

$$= \frac{64}{V} \int_0^{2\pi} \int_0^{\pi/4} \sin \phi \cos \phi \, d\phi \, d\theta = \frac{64}{V} \int_0^{2\pi} \left[\frac{1}{2} \sin^2 \phi \right]_{\phi=0}^{\pi/4} d\theta$$

$$= \frac{16}{V} \int_0^{2\pi} d\theta = \frac{32\pi}{V} = \frac{3}{2(2 - \sqrt{2})}$$

With the help of a calculator, $\bar{z} \approx 2.56$ (to two decimal places), so the approximate location of the centroid is $(0, 0, 2.56)$.   ◀

Table 17.7.1

DETERMINATION OF LIMITS	INTEGRAL

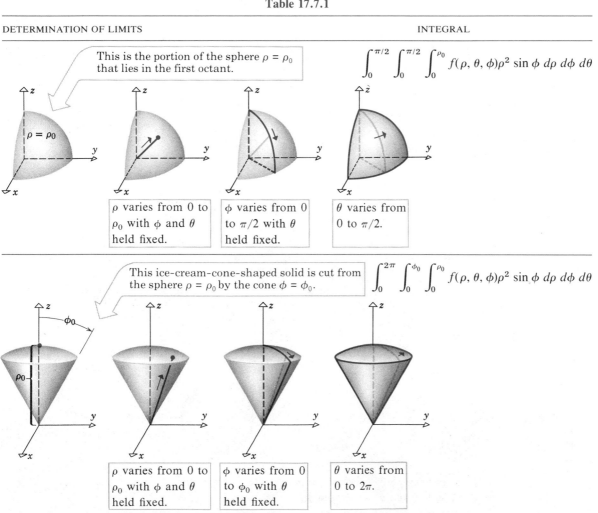

This is the portion of the sphere $\rho = \rho_0$ that lies in the first octant.

$$\int_0^{\pi/2} \int_0^{\pi/2} \int_0^{\rho_0} f(\rho, \theta, \phi) \rho^2 \sin \phi \, d\rho \, d\phi \, d\theta$$

ρ varies from 0 to ρ_0 with ϕ and θ held fixed.

ϕ varies from 0 to $\pi/2$ with θ held fixed.

θ varies from 0 to $\pi/2$.

This ice-cream-cone-shaped solid is cut from the sphere $\rho = \rho_0$ by the cone $\phi = \phi_0$.

$$\int_0^{2\pi} \int_0^{\phi_0} \int_0^{\rho_0} f(\rho, \theta, \phi) \rho^2 \sin \phi \, d\rho \, d\phi \, d\theta$$

ρ varies from 0 to ρ_0 with ϕ and θ held fixed.

ϕ varies from 0 to ϕ_0 with θ held fixed.

θ varies from 0 to 2π.

This solid is cut from the sphere $\rho = \rho_0$ by two cones $\phi = \phi_1$ and $\phi = \phi_2$.

$$\int_0^{2\pi} \int_{\phi_1}^{\phi_2} \int_0^{\rho_0} f(\rho, \theta, \phi)\rho^2 \sin\phi \, d\rho \, d\phi \, d\theta$$

ρ varies from 0 to ρ_0 with ϕ and θ held fixed.

ϕ varies from ϕ_1 to ϕ_2 with θ held fixed.

θ varies from 0 to 2π.

This solid is enclosed laterally by the cone $\phi = \phi_0$ and on top by the horizontal plane $z = a$.

$$\int_0^{2\pi} \int_0^{\phi_0} \int_0^{a\,\sec\phi} f(\rho, \theta, \phi)\rho^2 \sin\phi \, d\rho \, d\phi \, d\theta$$

ρ varies from 0 to $a \sec\phi$ with ϕ and θ held fixed.

ϕ varies from 0 to ϕ_0 with θ held fixed.

θ varies from 0 to 2π.

This solid is enclosed between two concentric spheres, $\rho = \rho_1$, $\rho = \rho_2$.

$$\int_0^{2\pi} \int_0^{\pi} \int_{\rho_1}^{\rho_2} f(\rho, \theta, \phi)\rho^2 \sin\phi \, d\rho \, d\phi \, d\theta$$

ρ varies from ρ_1 to ρ_2 with ϕ and θ held fixed.

ϕ varies from 0 to π with θ held fixed.

θ varies from 0 to 2π.

Example 4   Use spherical coordinates to evaluate

$$\int_{-2}^{2} \int_{-\sqrt{4-x^2}}^{\sqrt{4-x^2}} \int_{0}^{\sqrt{4-x^2-y^2}} z^2 \sqrt{x^2 + y^2 + z^2} \; dz \; dy \; dx$$

Solution. In problems like this, it is helpful to begin (when possible) with a sketch of the region G of integration. From the z-limits of integration, the upper surface of G is the hemisphere $z = \sqrt{4 - x^2 - y^2}$, and the lower surface is the *xy*-plane, $z = 0$. From the x- and y-limits of integration, the projection of the solid G on the *xy*-plane is the region enclosed by the circle $x^2 + y^2 = 4$. From this information we obtain the sketch of G in Figure 17.7.12.

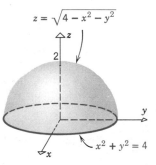

Figure 17.7.12

Thus,

$$\int_{-2}^{2} \int_{-\sqrt{4-x^2}}^{\sqrt{4-x^2}} \int_{0}^{\sqrt{4-x^2-y^2}} z^2 \sqrt{x^2 + y^2 + z^2} \; dz \; dy \; dx$$

$$= \iiint_{G} z^2 \sqrt{x^2 + y^2 + z^2} \; dV$$

$$= \iiint_{G} (\rho^2 \cos^2 \phi)\rho \; dV$$

$$= \int_{0}^{2\pi} \int_{0}^{\pi/2} \int_{0}^{2} \rho^5 \cos^2 \phi \sin \phi \; d\rho \; d\phi \; d\theta$$

$$= \int_{0}^{2\pi} \int_{0}^{\pi/2} \frac{32}{3} \cos^2 \phi \sin \phi \; d\phi \; d\theta$$

$$= \frac{32}{3} \int_{0}^{2\pi} \left[-\frac{1}{3} \cos^3 \phi \right]_{\phi=0}^{\pi/2} d\theta$$

$$= \frac{32}{9} \int_{0}^{2\pi} d\theta = \frac{64}{9} \pi \qquad \blacktriangleleft$$

▶ **Exercise Set 17.7**

In Exercises 1–4, evaluate the iterated integral.

1. $\displaystyle\int_0^{2\pi}\int_0^1\int_0^{\sqrt{1-r^2}} zr\,dz\,dr\,d\theta.$

2. $\displaystyle\int_0^{\pi/2}\int_0^{\cos\theta}\int_0^{r^2} r\sin\theta\,dz\,dr\,d\theta.$

3. $\displaystyle\int_0^{\pi/2}\int_0^{\pi/2}\int_0^1 \rho^3\sin\phi\cos\phi\,d\rho\,d\phi\,d\theta.$

4. $\displaystyle\int_0^{2\pi}\int_0^{\pi/4}\int_0^{a\sec\phi} \rho^2\sin\phi\,d\rho\,d\phi\,d\theta.$

In Exercises 5–9, use cylindrical coordinates to find the volume of the solid.

5. The solid bounded by the paraboloid $z = x^2 + y^2$ and the plane $z = 9$.

6. The solid bounded above and below by the sphere $x^2 + y^2 + z^2 = 9$ and laterally by the cylinder $x^2 + y^2 = 4$.

7. The solid enclosed between the surface $r^2 + z^2 = 20$ and the surface $z = r^2$.

8. The solid bounded by the cone $z = \dfrac{h}{a}r$ and $z = h$.

9. The solid in the first octant below the sphere $x^2 + y^2 + z^2 = 16$ and inside the cylinder $x^2 + y^2 = 4x$.

In Exercises 10–14, use spherical coordinates to find the volume of the solid.

10. The solid bounded above by the sphere $\rho = 4$ and below by the cone $\phi = \pi/3$.

11. The solid in the first octant bounded by the sphere $\rho = 2$, the coordinate planes, and the cones $\phi = \pi/6$ and $\phi = \pi/3$.

12. The solid within the cone $\phi = \pi/4$ and between the spheres $\rho = 1$ and $\rho = 2$.

13. The solid within the sphere $x^2 + y^2 + z^2 = 9$, outside the cone $z = \sqrt{x^2 + y^2}$, and above the xy-plane.

14. The solid bounded by the sphere $x^2 + y^2 + z^2 = 4a^2$ and the planes $z = 0$ and $z = a$.

15. Find the volume of $x^2 + y^2 + z^2 = a^2$ using
(a) cylindrical coordinates
(b) spherical coordinates.

In Exercises 16–18, use cylindrical coordinates.

16. Find the mass of the solid bounded by the cone $z = \sqrt{x^2 + y^2}$ and $z = 3$ if the density is $\delta(x, y, z) = 3 - z$.

17. Find the mass of the solid in the first octant bounded above by the paraboloid $z = 4 - x^2 - y^2$, below by the plane $z = 0$, and laterally by the cylinder $x^2 + y^2 = 2x$ and the planes $x = 0$ and $y = 0$, assuming the density to be $\delta(x, y, z) = z$. [*Hint:* Formula (5) of Section 9.2 and Formula (5) of Section 9.3 will help with the integration.]

18. Find the mass of a right-circular cylinder of radius a and height h if the density is proportional to the distance from the base. (Let k be the constant of proportionality.)

In Exercises 19–21, use spherical coordinates.

19. Find the mass of the solid common to the sphere $x^2 + y^2 + z^2 = 1$ and the cone $z = \sqrt{x^2 + y^2}$ if the density is $\delta(x, y, z) = \sqrt{x^2 + y^2 + z^2}$.

20. Find the mass of the solid enclosed between the spheres $x^2 + y^2 + z^2 = 1$ and $x^2 + y^2 + z^2 = 4$ if the density is $\delta(x, y, z) = (x^2 + y^2 + z^2)^{-1/2}$.

21. Find the mass of a spherical solid of radius a if the density is proportional to the distance from the center. (Let k be the constant of proportionality.)

In Exercises 22–24, use cylindrical coordinates to find the centroid of the solid.

22. The solid bounded by the cone $z = \sqrt{x^2 + y^2}$ and the plane $z = 2$.

23. The solid bounded above by the sphere $x^2 + y^2 + z^2 = 2$ and below by the paraboloid $z = x^2 + y^2$.

24. The solid bounded above by the paraboloid $z = x^2 + y^2$, below by the plane $z = 0$, and laterally by the cylinder $(x - 1)^2 + y^2 = 1$. [*Hint:* Formula (5) of Section 9.2 and Formula (5) of Section 9.3 will help with the integration.]

In Exercises 25–27, use spherical coordinates to find the centroid.

25. The solid in the first octant bounded by the coordinate planes and the sphere $x^2 + y^2 + z^2 = a^2$.

26. The solid bounded above by the sphere $\rho = 4$ and below by the cone $\phi = \pi/3$.

27. The solid that is enclosed by the hemispheres $y = \sqrt{9 - x^2 - z^2}$, $y = \sqrt{4 - x^2 - z^2}$, and the plane $y = 0$.

In Exercises 28–31, use cylindrical or spherical coordinates to evaluate the integral.

28. $\int_0^a \int_0^{\sqrt{a^2-x^2}} \int_0^{a^2-x^2-y^2} x^2 \, dz \, dy \, dx$.

29. $\int_{-1}^1 \int_0^{\sqrt{1-x^2}} \int_0^{\sqrt{1-x^2-y^2}} e^{-(x^2+y^2+z^2)^{3/2}} \, dz \, dy \, dx$.

30. $\int_0^2 \int_0^{\sqrt{4-y^2}} \int_{\sqrt{x^2+y^2}}^{\sqrt{8-x^2-y^2}} z^2 \, dz \, dx \, dy$.

31. $\int_{-3}^3 \int_{-\sqrt{9-y^2}}^{\sqrt{9-y^2}} \int_{-\sqrt{9-x^2-y^2}}^{\sqrt{9-x^2-y^2}} \sqrt{x^2 + y^2 + z^2} \, dz \, dx \, dy$.

32. Let G be the solid in the first octant bounded by the sphere $x^2 + y^2 + z^2 = 4$ and the coordinate planes. Evaluate

$$\iiint_G xyz \, dV$$

using
(a) rectangular coordinates
(b) cylindrical coordinates
(c) spherical coordinates.

Solve Exercises 33–38 using either cylindrical or spherical coordinates, whichever seems appropriate.

33. Find the center of gravity of the solid hemisphere bounded by $z = \sqrt{a^2 - x^2 - y^2}$ and $z = 0$ if the density is proportional to the distance from the origin.

34. Find the center of gravity of the solid in the first octant bounded by the cylinder $x^2 + y^2 = a^2$, the coordinate planes, and the plane $z = a$ if the density is $\delta(x, y, z) = xyz$.

35. Find the center of gravity of the solid bounded by

the paraboloid $z = 1 - x^2 - y^2$ and the xy-plane if the density is $\delta(x, y, z) = x^2 + y^2 + z^2$.

36. Find the center of gravity of the solid bounded by the cylinder $x^2 + y^2 = 1$, the cone $z = \sqrt{x^2 + y^2}$, and the xy-plane if the density is $\delta(x, y, z) = z$.

37. Find the volume common to the spheres $x^2 + y^2 + z^2 = 9$ and $x^2 + y^2 + (z - 2)^2 = 4$.

38. Suppose the density at a point on a spherical planet is assumed to be

$$\delta = \delta_0 e^{[(\rho/R^3)-1]}$$

where δ_0 is a constant, R is the radius of the planet, and ρ is the distance from the point to the planet's center. Calculate the mass of the planet.

39. In this exercise we shall obtain a formula for the volume of the spherical wedge in Figure 17.7.8.
(a) Use a triple integral in cylindrical coordinates to show that the volume of the solid bounded above by a sphere $\rho = \rho_0$, below by a cone $\phi = \phi_0$, and on the sides by $\theta = \theta_1$ and $\theta = \theta_2$ $(\theta_1 < \theta_2)$ is

$$V = \frac{1}{3}\rho_0^3(1 - \cos \phi_0)(\theta_2 - \theta_1)$$

[*Hint:* In cylindrical coordinates, the sphere has the equation $r^2 + z^2 = \rho_0^2$ and the cone has the equation $z = r \cot \phi_0$. For simplicity, consider only the case $0 < \phi_0 < \pi/2$.]
(b) Subtract appropriate volumes and use the result in (a) to deduce that the volume ΔV of the spherical wedge is

$$\Delta V = \frac{\rho_2^3 - \rho_1^3}{3} (\cos \phi_1 - \cos \phi_2)(\theta_2 - \theta_1)$$

(c) Apply the Mean-Value Theorem to the functions $\cos \phi$ and ρ^3 to deduce that the formula in (b) can be written as

$$\Delta V = \rho^{*2} \sin \phi^* \, \Delta \rho \, \Delta \phi \, \Delta \theta$$

where ρ^* is between ρ_1 and ρ_2, ϕ^* is between ϕ_1 and ϕ_2, and $\Delta \rho = \rho_2 - \rho_1$, $\Delta \phi = \phi_2 - \phi_1$, $\Delta \theta = \theta_2 - \theta_1$.

In Exercises 40–44, use the formulas in Exercise 36, Section 17.6, to find the indicated moment of inertia for the given homogeneous solid. Express your answer in terms of the mass M of the solid.

40. I_z for the solid cylinder $x^2 + y^2 \leq R^2, 0 \leq z \leq h$.

41. I_y for the solid cylinder $x^2 + y^2 \leq R^2, 0 \leq z \leq h$.

42. I_z for the hollow cylinder $R_1^2 \leq x^2 + y^2 \leq R_2^2$, $0 \leq z \leq h$.

43. I_z for the solid sphere $x^2 + y^2 + z^2 \leq R^2$.

44. I_z for the solid cone $\dfrac{h}{R}\sqrt{x^2 + y^2} \leq z \leq h$.

45. If d is the distance between two point masses m and m', then according to Newton's *inverse-square law of gravitational attraction* each point mass attracts the other with a force of magnitude kmm'/d^2, where k is a constant. Because force is a vector, the attraction between a lamina or solid and a point mass can be found by considering the components of the force. For example, assume that a solid G attracts a unit point mass at the origin

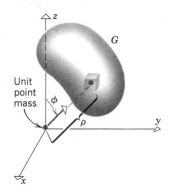

(see figure). The component in the z-direction of the force that G exerts on the unit mass can be found by approximating the corresponding components exerted by small subregions, summing, and taking the limit. This leads to the formula, using the spherical coordinates ρ and ϕ,

$$\iiint\limits_{G} \frac{k\delta \cos \phi}{\rho^2}\, dV$$

for the component of force in the z-direction.

(a) Use cylindrical coordinates to find the component in the z-direction of the force of attraction by the homogeneous solid cylinder $x^2 + y^2 \leq R^2, 0 \leq a \leq z \leq a + h$ on a unit mass at the origin.

(b) Explain why the force in part (a) is the total force that acts on the unit mass.

In Exercises 46–48, use the formula in Exercise 45 to find the magnitude of the force of attraction on a unit mass at the origin by the given homogeneous lamina or solid. (The regions are all such that the force in the z-direction is the only force of attraction.)

46. The circular lamina $x^2 + y^2 \leq R^2, z = a > 0$. (Use polar coordinates with dV replaced by dA.)

47. The solid cone $\dfrac{h}{R}r \leq z \leq h$. (Use spherical coordinates.)

48. The solid sphere of radius R with its center on the z-axis at $z = a > R$. (Use spherical coordinates.) Show that the force is the same as if the total mass of the sphere were concentrated at its center.

▶ SUPPLEMENTARY EXERCISES

In Exercises 1–4, evaluate the iterated integrals.

1. $\displaystyle\int_{1/2}^{1} \int_{0}^{2x} \cos(\pi x^2)\, dy\, dx$.

2. $\displaystyle\int_{0}^{2} \int_{-y}^{2y} xe^{y^3}\, dx\, dy$.

3. $\displaystyle\int_{-1}^{0} \int_{0}^{y^2} \int_{xy}^{1} 2y\, dz\, dx\, dy$.

4. $\displaystyle\int_{0}^{1} \int_{0}^{z} \int_{0}^{\sqrt{yz}} x\, dx\, dy\, dz$.

In Exercises 5 and 6, express the iterated integral as an equivalent integral with the order of integration reversed.

5. $\displaystyle\int_{0}^{2} \int_{0}^{x/2} e^x e^y\, dy\, dx$.

6. $\displaystyle\int_{0}^{\pi} \int_{y}^{\pi} \frac{\sin x}{x}\, dx\, dy$.

7. Use a double integral to find the area of the region bounded by $y = 2x^3, 2x + y = 4$, and the x-axis.

8. Use a double integral to find the area of the region bounded by $x = y^2$ and $x = 4y - y^2$.

9. Sketch the region R whose area is given by the iterated integral.

(a) $\displaystyle\int_0^{\pi/2} \int_{\tan(x/2)}^{\sin x} dy \, dx.$

(b) $\displaystyle\int_{\pi/6}^{\pi/2} \int_a^{a(1+\cos\theta)} r \, dr \, d\theta \qquad (a > 0).$

In Exercises 10–12, evaluate the double integral over R using either rectangular or polar coordinates.

10. $\displaystyle\iint_R xy \, dA$; R is the region bounded by $y = \sqrt{x}$, $y = 2 - \sqrt{x}$, and the y-axis.

11. $\displaystyle\iint_R x^2 \sin y^2 \, dA$; R is the region bounded by $y = x^3$, $y = -x^3$, and $y = 8$.

12. $\displaystyle\iint_R (4 - x^2 - y^2) \, dA$; R is the sector in the first quadrant bounded by the circle $x^2 + y^2 = 4$ and the coordinate axes.

In Exercises 13–15, use a double integral in rectangular or polar coordinates to find the volume of the solid.

13. The solid in the first octant bounded by the coordinate planes and the plane $3x + 2y + z = 6$.

14. The solid enclosed by the cylinders $y = 3x + 4$ and $y = x^2$, and such that $0 \le z \le \sqrt{y}$.

15. The solid $G = \{(x, y, z) : 2 \le x^2 + y^2 \le 4 \text{ and } 0 \le z \le 1/(x^2 + y^2)^2\}$.

16. Convert to polar coordinates and evaluate:

$$\int_0^{\sqrt{2}} \int_x^{\sqrt{4-x^2}} 4xy \, dy \, dx.$$

17. Convert to rectangular coordinates and evaluate:

$$\int_0^{\pi/2} \int_0^{2a \sin\theta} r \sin 2\theta \, dr \, d\theta.$$

In Exercises 18–20, find the area of the region using a double integral in polar coordinates.

18. The region outside the circle $r = \sqrt{2}a$ and inside the lemniscate $r^2 = 4a^2 \cos 2\theta$.

19. The region enclosed by the rose $r = \cos 3\theta$.

20. The region inside the circle $r = 2\sqrt{3} \sin\theta$ and outside the circle $r = 3$.

In Exercises 21–23, find the area of the surface described.

21. The part of the paraboloid $z = 3x^2 + 3y^2 - 3$ below the xy-plane.

22. The part of the plane $2x + 2y + z = 7$ in the first octant.

23. The part of the cone $z^2 = x^2 + y^2$ between the planes $z = 1$ and $z = 4$.

24. Evaluate $\displaystyle\iiint_G x^2yz \, dV$, where G is the set of points satisfying $0 \le x \le 2$, $-x \le y \le x^2$, $0 \le z \le x + y$.

25. Evaluate $\displaystyle\iiint_G \sqrt{x^2 + y^2} \, dV$, where G is the set of points satisfying $x^2 + y^2 \le 16, 0 \le z \le 4 - y$.

26. If $G = \{(x, y, z) : x^2 + y^2 \le z \le 4x\}$, express the volume of G as a triple integral in (a) rectangular coordinates, (b) cylindrical coordinates.

27. In each part find an equivalent integral of the form

$$\iiint_G (\ ) \, dx \, dz \, dy.$$

(a) $\displaystyle\int_0^1 \int_0^{(1-x)/2} \int_0^{1-x-2y} z \, dz \, dy \, dx.$

(b) $\displaystyle\int_0^2 \int_{x^2}^4 \int_0^{4-y} 3 \, dz \, dy \, dx.$

28. (a) Change to cylindrical coordinates and then evaluate:

$$\int_{-2}^2 \int_{-\sqrt{4-x^2}}^{\sqrt{4-x^2}} \int_{(x^2+y^2)^2}^{16} x^2 \, dz \, dy \, dx$$

(b) Change to spherical coordinates and evaluate:

$$\int_0^1 \int_0^{\sqrt{1-x^2}} \int_0^{\sqrt{1-x^2-y^2}} \frac{1}{1 + x^2 + y^2 + z^2} \, dz \, dy \, dx.$$

29. If G is the region bounded above by the sphere $\rho = a$ and below by the cone $\phi = \pi/3$, express

$$\iiint\limits_{G} (x^2 + y^2) \, dV$$ as an iterated integral in (a)

spherical coordinates, (b) cylindrical coordinates, (c) Cartesian coordinates.

In Exercises 30–33, find the volume of G.

30. $G = \{(r, \theta, z): 0 \le r \le 2 \sin \theta, 0 \le z \le r \sin \theta\}$.

31. G is the solid enclosed by the "inverted apple" $\rho = a(1 + \cos \phi)$.

32. G is the solid enclosed between the surfaces $x = y^2 + z^2$ and $x = 1 - y^2$.

33. G is the solid bounded below by the upper nappe of the cone $\phi = \pi/6$ and above by the plane $z = a$.

In Exercises 34–36, find the centroid $(\bar{x}, \bar{y})$ of the plane region R.

34. The region R is the upper half of the ellipse $(x/a)^2 + (y/b)^2 = 1$.

35. The region R is enclosed by the cardioid $r = a(1 + \sin \theta)$.

36. The region R is bounded by $y^2 = 4x$ and $y^2 = 8(x - 2)$.

In Exercises 37 and 38, find the center of gravity of the lamina with density δ.

37. The triangular lamina with vertices $(a, 0)$, $(-a, 0)$, and $(0, b)$, where $a > 0$ and $b > 0$; and $\delta(x, y)$ is proportional to the distance from (x, y) to the y-axis.

38. The lamina enclosed by the circle $r = 3 \cos \theta$, but outside the cardioid $r = 1 + \cos \theta$, and with $\delta(r, \theta)$ proportional to the distance from (r, θ) to the x-axis.

In Exercises 39 and 40, find the mass of the solid G if its density is δ.

39. The solid G is the part of the first octant under the plane $x/a + y/b + z/c = 1$, where a, b, c are positive, and $\delta(x, y, z) = kz$.

40. The spherical solid G is bounded by $\rho = a$ and $\delta(x, y, z)$ is twice the distance from (x, y, z) to the origin.

In Exercises 41–43, find the centroid of G.

41. The solid G bounded by $y = x^2$, $y = 4$, $z = 0$, and $y + z = 4$.

42. The solid G is the part of the sphere $\rho \le a$ lying within the cone $\phi \le \phi_0$, where $\phi_0 \le \pi/2$.

43. The solid G is bounded by the cone with vertex $(0, 0, h)$, and base $x^2 + y^2 \le R^2$ in the xy-plane.

18

topics in vector calculus

18.1 LINE INTEGRALS

In previous chapters we considered three kinds of integrals in rectangular coordinates: single integrals over intervals, double integrals over two-dimensional regions, and triple integrals over three-dimensional regions. In this section we shall discuss line integrals, which are integrals over curves in two- or three-dimensional space.

18.1.1 DEFINITION. Let C be a smooth plane curve given parametrically by the equations

$$x = x(t) \atop y = y(t) \quad , \quad a \le t \le b$$

and let $f(x, y)$ and $g(x, y)$ be continuous on some open region containing the curve C. Then we define

$$\int_C f(x, y)\, dx = \int_a^b f(x(t), y(t))x'(t)\, dt \tag{1}$$

$$\int_C g(x, y)\, dy = \int_a^b g(x(t), y(t))y'(t)\, dt \tag{2}$$

The symbol $\int_C f(x, y)\, dx$ is called the ***line integral of f along C with respect to x*** and $\int_C g(x, y)\, dy$ is called the ***line integral of g along C with respect to y.***

REMARK. In words, the line integrals $\int_C f(x, y)\, dx$ and $\int_C g(x, y)\, dy$ are evaluated by replacing x and y by their expressions in terms of t, writing

dx and dy as $dx = x'(t)\, dt$ and $dy = y'(t)\, dt$, and then integrating in the usual way over the interval $a \le t \le b$.

Frequently, the line integrals in (1) and (2) occur in combination, in which case we dispense with one of the integral signs and write:

$$\int_C f(x, y)\, dx + g(x, y)\, dy = \int_C f(x, y)\, dx + \int_C g(x, y)\, dy \qquad (3)$$

The expression $f(x, y)\, dx + g(x, y)\, dy$ is called a **differential form**.

Example 1   Evaluate

$$\int_C 2xy\, dx + (x^2 + y^2)\, dy$$

over the circular arc C given by $x = \cos t$, $y = \sin t$, $0 \le t \le \pi/2$ (Figure 18.1.1).

$x = \cos t$, $0 \le t \le \pi/2$
$y = \sin t$

Figure 18.1.1

Solution.   From (1) and (2)

$$\int_C 2xy\, dx = \int_0^{\pi/2} (2 \cos t \sin t) \left[\frac{d}{dt} (\cos t) \right] dt$$

$$= -2 \int_0^{\pi/2} \sin^2 t \cos t\, dt = -\frac{2}{3} \sin^3 t \Big]_0^{\pi/2} = -\frac{2}{3}$$

$$\int_C (x^2 + y^2)\, dy = \int_0^{\pi/2} (\cos^2 t + \sin^2 t) \left[\frac{d}{dt} (\sin t) \right] dt$$

$$= \int_0^{\pi/2} \cos t\, dt = \sin t \Big]_0^{\pi/2} = 1$$

Thus, from (3)

$$\int_C 2xy\, dx + (x^2 + y^2)\, dy = \int_C 2xy\, dx + \int_C (x^2 + y^2)\, dy$$

$$= -\frac{2}{3} + 1 = \frac{1}{3} \qquad \blacktriangleleft$$

Since the parametric equations of a curve are used to evaluate line integrals over that curve, it seems possible that two different parametrizations of a curve C might produce different values for the same line integral over C. Fortunately, this is not the case.

Independence of
Parametrization

18.1.2   THEOREM. *If two different sets of parametric equations rep-resent the same curve C, and if the curve is traced in the same direction in both cases, then the two sets of equations produce the same value for a given line integral over C.*

Reversal of Orientation

18.1.3   THEOREM. *If two different sets of parametric equations rep-resent the same curve C, but the curve is traced in opposite directions in the two cases, then the two sets of equations produce values for a line integral over C that are negatives of one another.*

We omit the proofs. Instead, we shall illustrate these facts with some examples.

Example 2   Evaluate

$$\int_C 2xy \, dx + (x^2 + y^2) \, dy$$

over the curve C given parametrically by

(a) $x = \cos 2t$, $y = \sin 2t$, where $0 \leq t \leq \dfrac{\pi}{4}$

(b) $x = \cos(t^2)$, $y = \sin(t^2)$, where $0 \leq t \leq \sqrt{\pi/2}$

(c) $x = \cos(-t)$, $y = \sin(-t)$, where $\dfrac{3\pi}{2} \leq t \leq 2\pi$

(d) $x = t$, $y = \sqrt{1 - t^2}$, where $0 \leq t \leq 1$

Solution.   In parts (a) and (b), the curve C is the quarter circle shown in Figure 18.1.1, traced in the counterclockwise direction. Thus, from Example 1 and Theorem 18.1.2,

$$\int_C 2xy \, dx + (x^2 + y^2) \, dy = \frac{1}{3}$$

in both cases. In parts (c) and (d), the curve is the same quarter circle, but traced in the clockwise direction. Thus,

$$\int_C 2xy \, dx + (x^2 + y^2) \, dy = -\frac{1}{3}$$

These results should be verified using (1), (2), and (3) to evaluate the integrals directly.   ◄

If a curve C is traced in a certain direction, then the same curve traced in the opposite direction is often denoted by the symbol $-C$. Thus, it follows from Theorem 18.1.3 that

$$\int_{-C} f(x, y)\, dx + g(x, y)\, dy = -\int_{C} f(x, y)\, dx + g(x, y)\, dy \qquad (4)$$

In the definition of a line integral, we required the curve C to be smooth. However, the definition can be extended to curves formed by finitely many smooth curves $C_1, C_2, \ldots, C_n$ joined end to end. Such a curve is called *piecewise smooth* (Figure 18.1.2). We define a line integral over a piecewise smooth curve C to be the sum of the integrals over the pieces:

$$\int_C = \int_{C_1} + \int_{C_2} + \cdots + \int_{C_n}$$

Figure 18.1.2 ▷

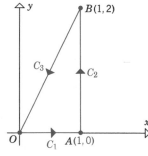

Figure 18.1.3

Example 3   Evaluate

$$\int_C x^2 y\, dx + x\, dy$$

in a counterclockwise direction around the triangular path shown in Figure 18.1.3.

Solution.   We shall integrate over C_1, C_2, and C_3 separately and add the results. For each of the three integrals we must find parametric equations that trace the path of integration in the correct direction.

The horizontal line segment C_1 can be represented as follows using $x = t$ as the parameter:

$$\begin{aligned} x &= t \\ y &= 0 \end{aligned}, \quad 0 \le t \le 1$$

The vertical line segment C_2 can be represented as follows using $y = t$ as the parameter:

$$\begin{aligned} x &= 1 \\ y &= t \end{aligned}, \quad 0 \le t \le 2$$

To find parametric equations for C_3 we shall use Formula (5) of Section 14.5 with $\mathbf{v} = \overrightarrow{BO} = \langle -1, -2 \rangle$ and $\mathbf{r}_0 = \langle 1, 2 \rangle$ to obtain the vector equation

$$\langle x, y \rangle = \langle 1, 2 \rangle + t \langle -1, -2 \rangle$$

for the line through O and B traced in the direction of C_3. Equating corresponding components and using the fact that the line segment C_3 is traced as t varies from 0 to 1 yields the parametric equations

$$\begin{array}{l} x = 1 - t \\ y = 2 - 2t \end{array}, \quad 0 \le t \le 1$$

We can now carry out the calculations.

$$\int_{C_1} x^2 y \, dx + x \, dy = \int_0^1 (t^2)(0) \frac{d}{dt} [t] \, dt + \int_0^1 (t) \frac{d}{dt} [0] \, dt = 0$$

$$\int_{C_2} x^2 y \, dx + x \, dy = \int_0^2 (1^2)(t) \frac{d}{dt} [1] \, dt + \int_0^2 (1) \frac{d}{dt} [t] \, dt = 0 + 2 = 2$$

$$\int_{C_3} x^2 y \, dx + x \, dy = \int_0^1 (1 - t)^2 (2 - 2t) \frac{d}{dt} [1 - t] \, dt$$

$$+ \int_0^1 (1 - t) \frac{d}{dt} [2 - 2t] \, dt$$

$$= 2 \int_0^1 (t - 1)^3 \, dt + 2 \int_0^1 (t - 1) \, dt$$

$$= -\frac{1}{2} - 1 = -\frac{3}{2}$$

Thus,

$$\int_C x^2 y \, dx + x \, dy = 0 + 2 + \left(-\frac{3}{2} \right) = \frac{1}{2} \quad \blacktriangleleft$$

For many applications it is helpful to express the line integral

$$\int_C f(x, y) \, dx + g(x, y) \, dy \tag{5}$$

in vector notation. To do this, we introduce the vector-valued function

$$\mathbf{F}(x, y) = f(x, y)\mathbf{i} + g(x, y)\mathbf{j} \tag{6}$$

and express the curve C: $x = x(t)$, $y = y(t)$ in terms of its position vector

$$\mathbf{r}(t) = x(t)\mathbf{i} + y(t)\mathbf{j}$$

If we formally multiply both sides of the equation

$$\frac{d\mathbf{r}}{dt} = \frac{dx}{dt}\mathbf{i} + \frac{dy}{dt}\mathbf{j}$$

by dt, we are led to the following convenient notation

$$d\mathbf{r} = \frac{d\mathbf{r}}{dt}dt = dx\mathbf{i} + dy\mathbf{j} \tag{7}$$

If we now calculate the dot product of (6) and (7), we obtain

$$\mathbf{F}(x, y) \cdot d\mathbf{r} = f(x, y)\, dx + g(x, y)\, dy$$

which is precisely the integrand in (5). This suggests the notation

$$\int_C f(x, y)\, dx + g(x, y)\, dy = \int_C \mathbf{F}(x, y) \cdot d\mathbf{r} \tag{8}$$

Moreover, the equation

$$\int_C f(x, y)\, dx + g(x, y)\, dy = \int_a^b \left[f(x(t), y(t))\frac{dx}{dt} + g(x(t), y(t))\frac{dy}{dt} \right] dt$$

can be written as

$$\int_C \mathbf{F}(x, y) \cdot d\mathbf{r} = \int_a^b \left[\mathbf{F}(x(t), y(t)) \cdot \frac{d\mathbf{r}}{dt} \right] dt \tag{9}$$

WORK   One of the most important applications of line integrals occurs in the study of work. Recall that if a constant force $\mathbf{F}$ acts on a particle moving along a straight line from a point P to a point Q, then the work done by $\mathbf{F}$ is

$$W = \mathbf{F} \cdot \overrightarrow{PQ} = (\|\mathbf{F}\| \cos \theta)\|\overrightarrow{PQ}\| \tag{10}$$

where $\|\mathbf{F}\| \cos \theta$ is the component of force in the direction of motion and $\|\overrightarrow{PQ}\|$ is the distance traveled by the particle [see (13), Section 14.3].

This elementary notion of work is inadequate for many applications. For example, if a rocket moves through the earth's gravitational field, the path is usually a curve rather than a straight line; and the force of gravity on the rocket does not remain constant, but changes with the position of the rocket relative to the earth's center. Thus, we must consider the more general notion of work done on a particle moving along a curve C while subject to a force $\mathbf{F}(x, y)$ that varies from point to point along the curve (Figure 18.1.4).

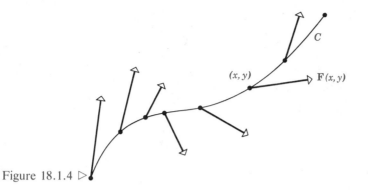

Figure 18.1.4 ▷

Let us assume that C is a smooth curve given parametrically by the equations

$$x = x(t) \atop y = y(t), \quad a \le t \le b$$

and that the force vector at a point (x, y) on the curve is

$$\mathbf{F}(x, y) = f(x, y)\mathbf{i} + g(x, y)\mathbf{j}$$

where $f(x, y)$ and $g(x, y)$ are continuous functions. Let $\mathbf{r}(t) = x(t)\mathbf{i} + y(t)\mathbf{j}$ $(a \le t \le b)$ denote the position vector for C, and let us define the function $W(t)$ by

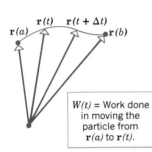

$W(t)$ = Work done in moving the particle from $\mathbf{r}(a)$ to $\mathbf{r}(t)$.

Figure 18.1.5

$$W(t) = \begin{bmatrix} \text{the work done as the particle moves from} \\ \text{the initial point } \mathbf{r}(a) \text{ to the point } \mathbf{r}(t) \end{bmatrix}$$

(See Figure 18.1.5.) Thus, the work done as the particle moves from $\mathbf{r}(a)$ to $\mathbf{r}(t + \Delta t)$ is $W(t + \Delta t)$, and the work done as the particle moves from $\mathbf{r}(t)$ to $\mathbf{r}(t + \Delta t)$ is $W(t + \Delta t) - W(t)$.

If Δt is small, then we may reasonably approximate the portion of curve between $\mathbf{r}(t)$ and $\mathbf{r}(t + \Delta t)$ by a straight line. If the force vector $\mathbf{F}$ does not vary much over this small section of curve, we may assume the force to be constant between $\mathbf{r}(t)$ and $\mathbf{r}(t + \Delta t)$ and equal to its value at $\mathbf{r}(t)$. Thus, (10) implies that the work done as the particle moves from $\mathbf{r}(t)$ to $\mathbf{r}(t + \Delta t)$ can be approximated in the following way if Δt is small:

$$W(t + \Delta t) - W(t) \approx \mathbf{F}\big(x(t), y(t)\big) \cdot [\mathbf{r}(t + \Delta t) - \mathbf{r}(t)]$$

If we assume that the error in this approximation diminishes to zero as $\Delta t \to 0$, then it follows on dividing by Δt and taking the limit that

$$\lim_{\Delta t \to 0} \frac{W(t + \Delta t) - W(t)}{\Delta t} = \lim_{\Delta t \to 0} \mathbf{F}\big(x(t), y(t)\big) \cdot \frac{\mathbf{r}(t + \Delta t) - \mathbf{r}(t)}{\Delta t}$$

or

$$W'(t) = \mathbf{F}(x(t), y(t)) \cdot \mathbf{r}'(t) \tag{11}$$

Since $W(t)$ represents the total work done as the particle moves from $\mathbf{r}(a)$ to $\mathbf{r}(t)$, it follows that

$$W(a) = 0$$

Thus, the total work W done as the particle moves from $\mathbf{r}(a)$ to $\mathbf{r}(b)$ can be written as

$$W = W(b) = W(b) - W(a) = \int_a^b W'(t) \, dt$$

or from (11)

$$W = \int_a^b \mathbf{F}(x(t), y(t)) \cdot \mathbf{r}'(t) \, dt$$

Since this is precisely the line integral over C of the force $\mathbf{F}$ [see (9)], we are led to the following definition of the work W done by a force $\mathbf{F}(x, y)$ acting on a particle as it moves along a curve C:

$$W = \int_C \mathbf{F}(x, y) \cdot d\mathbf{r} \tag{12}$$

Example 4   Find the work done if a particle moves from $(-2, 4)$ to $(1, 1)$ along the parabola $y = x^2$, while subject to the force

$$\mathbf{F}(x, y) = x^3 y \mathbf{i} + (x - y) \mathbf{j}$$

Solution.   If we use $x = t$ as the parameter, the path of the particle is represented by

$$\begin{array}{l} x = t \\ y = t^2 \end{array}, \quad -2 \le t \le 1$$

or

$$\mathbf{r}(t) = t\mathbf{i} + t^2 \mathbf{j} \quad (-2 \le t \le 1)$$

Since $\mathbf{F}(x, y) = x^3 y \mathbf{i} + (x - y)\mathbf{j}$, it follows that

$$\mathbf{F}(x(t), y(t)) = (t^3)(t^2)\mathbf{i} + (t - t^2)\mathbf{j} = t^5 \mathbf{i} + (t - t^2)\mathbf{j}$$

Thus, from (12) and (9)

$$W = \int_C \mathbf{F}(x, y) \cdot d\mathbf{r} = \int_a^b \left[\mathbf{F}(x(t), y(t)) \cdot \frac{d\mathbf{r}}{dt} \right] dt$$

$$= \int_{-2}^{1} [t^5\mathbf{i} + (t - t^2)\mathbf{j}] \cdot [\mathbf{i} + 2t\mathbf{j}] \, dt$$

$$= \int_{-2}^{1} (t^5 + 2t^2 - 2t^3) \, dt$$

$$= \frac{1}{6}t^6 + \frac{2}{3}t^3 - \frac{1}{2}t^4 \Big]_{-2}^{1}$$

$$= 3 \quad \text{(units of work)} \quad \blacktriangleleft$$

LINE INTEGRALS IN THREE DIMENSIONS The notion of a line integral can be extended to three dimensions. If C is a smooth curve in three dimensions given parametrically by

$$x = x(t)$$
$$y = y(t), \quad a \le t \le b$$
$$z = z(t)$$

and if $f(x, y, z)$, $g(x, y, z)$ and $h(x, y, z)$ are continuous on some region containing C, then we define

$$\int_C f(x, y, z) \, dx = \int_a^b f(x(t), y(t), z(t))x'(t) \, dt$$

$$\int_C g(x, y, z) \, dy = \int_a^b g(x(t), y(t), z(t))y'(t) \, dt \qquad (13)$$

$$\int_C h(x, y, z) \, dz = \int_a^b h(x(t), y(t), z(t))z'(t) \, dt$$

$$\int_C f(x, y, z) \, dx + g(x, y, z) \, dy + h(x, y, z) \, dz$$

$$= \int_C f(x, y, z) \, dx + \int_C g(x, y, z) \, dy + \int_C h(x, y, z) \, dz \quad (14)$$

If we let

$$\mathbf{F}(x, y, z) = f(x, y, z)\mathbf{i} + g(x, y, z)\mathbf{j} + h(x, y, z)\mathbf{k}$$
$$\mathbf{r}(t) = x(t)\mathbf{i} + y(t)\mathbf{j} + z(t)\mathbf{k}$$

$$d\mathbf{r} = \frac{d\mathbf{r}}{dt} dt = dx\mathbf{i} + dy\mathbf{j} + dz\mathbf{k}$$

then (14) can be written as

$$\int_C f(x, y, z)\, dx + g(x, y, z)\, dy + h(x, y, z)\, dz = \int_C \mathbf{F}(x, y, z) \cdot d\mathbf{r} \quad (15)$$

Moreover, it follows from (13) that

$$\int_C \mathbf{F}(x, y, z) \cdot d\mathbf{r} = \int_a^b \left[\mathbf{F}(x(t), y(t), z(t)) \cdot \frac{d\mathbf{r}}{dt} \right] dt \quad (16)$$

Finally, if a particle moves along a curve C in three dimensions while subject to a force $\mathbf{F}(x, y, z) = f(x, y, z)\mathbf{i} + g(x, y, z)\mathbf{j} + h(x, y, z)\mathbf{k}$, then we define the work W done by the force on the particle to be

$$W = \int_C \mathbf{F}(x, y, z) \cdot d\mathbf{r} \quad (17)$$

Example 5   Evaluate $\int_C \mathbf{F} \cdot d\mathbf{r}$ if $\mathbf{F}(x, y, z) = yz\mathbf{i} + xz\mathbf{j} + xy\mathbf{k}$ and C is the curve $\mathbf{r}(t) = t\mathbf{i} + t^2\mathbf{j} + t^3\mathbf{k}$ $(0 \le t \le 1)$.

Solution.   From (16)

$$\int_C \mathbf{F} \cdot d\mathbf{r} = \int_0^1 \left[\mathbf{F}(x(t), y(t), z(t)) \cdot \frac{d\mathbf{r}}{dt} \right] dt$$

$$= \int_0^1 (t^2 t^3 \mathbf{i} + t t^3 \mathbf{j} + t t^2 \mathbf{k}) \cdot (\mathbf{i} + 2t\mathbf{j} + 3t^2\mathbf{k})\, dt$$

$$= \int_0^1 (t^5 + 2t^5 + 3t^5)\, dt = \int_0^1 6t^5\, dt = 1 \quad \blacktriangleleft$$

▶ Exercise Set 18.1

1. Let C be the curve $x = t$, $y = t^2$, $0 \le t \le 1$. Find

    (a) $\displaystyle\int_C (2x + y)\, dx$ (b) $\displaystyle\int_C (x^2 - y)\, dy$

    (c) $\displaystyle\int_C (2x + y)\, dx + (x^2 - y)\, dy.$

2. Let $\mathbf{F} = (3x + 2y)\mathbf{i} + (2x - y)\mathbf{j}$. Evaluate $\int_C \mathbf{F} \cdot d\mathbf{r}$, where C is
    (a) the line segment from $(0, 0)$ to $(1, 1)$
    (b) the parabolic arc $y = x^2$ from $(0, 0)$ to $(1, 1)$
    (c) the curve $y = \sin(\pi x/2)$ from $(0, 0)$ to $(1, 1)$
    (d) the curve $x = y^3$ from $(0, 0)$ to $(1, 1)$.

In Exercises 3–10, evaluate the line integral.

3. $\displaystyle\int_C y\, dx - x^2\, dy$, where C is the curve $x = t$, $y = \frac{1}{2}t^2$, $0 \le t \le 2$.

4. $\displaystyle\int_C e^{y-x}\, dx + xy\, dy$, where C is the curve $x = 2 + t$, $y = 3 - t$, $0 \le t \le 1$.

5. $\displaystyle\int_C (x + 2y)\, dx + (x - y)\, dy$, where C is the curve $x = 2\cos t$, $y = 4\sin t$, $0 \le t \le \pi/4$.

6. $\int_C (x^2 - y^2)\, dx + x\, dy$, where C is the curve

$x = t^{2/3}, y = t, -1 \le t \le 1$.

7. $\int_C -y\, dx + x\, dy$ along $y^2 = 3x$ from the point

$(3, 3)$ to the point $(0, 0)$.

8. $\int_C (y - x)\, dx + x^2 y\, dy$ along $y^2 = x^3$ from the

point $(1, -1)$ to the point $(1, 1)$.

9. $\int_C (x^2 + y^2)\, dx - x\, dy$ along the quarter circle

$x^2 + y^2 = 1$ from $(1, 0)$ to $(0, 1)$.

10. $\int_C (y - x)\, dx + xy\, dy$ along the line segment from

$(3, 4)$ to $(2, 1)$.

11. Find $\int_C \mathbf{F} \cdot d\mathbf{r}$, where $\mathbf{F}(x, y) = x^2 \mathbf{i} + xy\mathbf{j}$ and

C is the semicircle given by $\mathbf{r}(t) = 2 \cos t\mathbf{i} + 2 \sin t\mathbf{j}$,

$0 \le t \le \pi$.

12. Find $\int_C \mathbf{F} \cdot d\mathbf{r}$, where $\mathbf{F}(x, y) = x^2 y\mathbf{i} + 4\mathbf{j}$ and C

is the curve $\mathbf{r}(t) = e^t \mathbf{i} + e^{-t}\mathbf{j}, 0 \le t \le 1$.

13. Find $\int_C \mathbf{F} \cdot d\mathbf{r}$, where

$\mathbf{F}(x, y) = (x^2 + y^2)^{-3/2}(x\mathbf{i} + y\mathbf{j})$ and C is the curve

$\mathbf{r}(t) = e^t \sin t\mathbf{i} + e^t \cos t\mathbf{j}, 0 \le t \le 1$.

14. Evaluate $\int_C y\, dx - x\, dy$ over each of the follow-

ing curves:

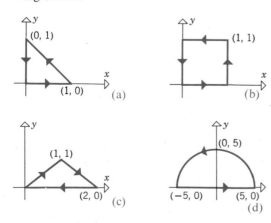

In Exercises 15–18, evaluate the three-dimensional line integral.

15. $\int_C yz\, dx - xz\, dy + xy\, dz$, where C is the curve

$x = e^t, y = e^{3t}, z = e^{-t}, 0 \le t \le 1$.

16. $\int_C y\, dx + z\, dy - x\, dz$, where C is the line segment

from $(1, 1, 1)$ to $(-3, 2, 0)$.

17. $\int_C \mathbf{F} \cdot d\mathbf{r}$, where $\mathbf{F}(x, y, z) = z\mathbf{i} + x\mathbf{j} + y\mathbf{k}$ and C

is the curve $\mathbf{r}(t) = \sin t\mathbf{i} + 3 \sin t\mathbf{j} + \sin^2 t\mathbf{k}$,

$0 \le t \le \pi/2$.

18. $\int_C x^2 z\, dx - yx^2\, dy + 3xz\, dz$, where C is the tri-

angular path from $(0, 0, 0)$ to $(1, 1, 0)$ to $(1, 1, 1)$

to $(0, 0, 0)$.

19. Find the work done if a particle moves along the

parabolic arc $x = y^2$ from $(0, 0)$ to $(1, 1)$ while

subject to the force $\mathbf{F}(x, y) = xy\mathbf{i} + x^2 \mathbf{j}$.

20. Find the work done if a particle moves along the

curve $x = t, y = 1/t, 1 \le t \le 3$ while subject to

the force $\mathbf{F}(x, y) = (x^2 + xy)\mathbf{i} + (y - x^2 y)\mathbf{j}$.

21. Find the work done by the force

$$\mathbf{F}(x, y) = \frac{1}{x^2 + y^2}\mathbf{i} + \frac{4}{x^2 + y^2}\mathbf{j}$$

acting on a particle that moves along the curve

C shown.

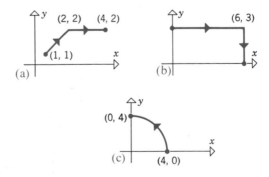

22. Find the work done by a force $\mathbf{F}(x, y, z) =$

$xy\mathbf{i} + yz\mathbf{j} + xz\mathbf{k}$ acting on a particle that moves

along the curve $\mathbf{r}(t) = t\mathbf{i} + t^2 \mathbf{j} + t^3 \mathbf{k}, 0 \le t \le 1$.

23. Find the work that is done by a force $F(x, y, z) = (x + y)\mathbf{i} + xy\mathbf{j} - z^2\mathbf{k}$ acting on a particle that moves along the line segment from $(0, 0, 0)$ to $(1, 3, 1)$ and then along the line segment from $(1, 3, 1)$ to $(2, -1, 4)$.

24. Evaluate $\displaystyle\int_{-C} \frac{x\,dy - y\,dx}{x^2 + y^2}$, where C is the circle $x^2 + y^2 = a^2$ traversed counterclockwise.

25. A particle moves from the point $(0, 0)$ to the point $(1, 0)$ along the curve $x = t$, $y = \lambda t(1 - t)$ while subject to the force $F(x, y) = xy\mathbf{i} + (x - y)\mathbf{j}$. For what value of λ is the work done equal to 1?

26. (a) Show that $\displaystyle\int_C f(x, y)\,dx = 0$ if C is a vertical line segment in the xy-plane.

(b) Show that $\displaystyle\int_C f(x, y)\,dy = 0$ if C is a horizontal line segment in the xy-plane.

27. A particle moves counterclockwise along the upper half of the circle $x^2 + y^2 = a^2$ from $(a, 0)$ to $(-a, 0)$ while subject to the force

$$F(x, y) = \frac{k\mathbf{r}}{\|\mathbf{r}\|^3}$$

where k is a constant and $\mathbf{r} = x\mathbf{i} + y\mathbf{j}$. Find the work done.

28. A farmer carries a sack of grain weighing 20 lb up a circular helical staircase encircling a silo of radius 25 ft. As the farmer climbs, grain leaks from the sack at a rate of 1 lb per 10 ft of ascent. How much work is performed by the farmer on the sack of grain if he climbs 60 ft in exactly four revolutions?

18.2   LINE INTEGRALS INDEPENDENT OF PATH

In general, the value of a line integral $\int_C \mathbf{F} \cdot d\mathbf{r}$ depends on the curve C. However, we shall show in this section that when the integrand satisfies appropriate conditions, the value of the integral depends only on the location of the endpoints of the curve C and not on the shape of the curve joining the endpoints. In such cases the evaluation of the line integral is greatly simplified.

INDEPENDENCE OF PATH

Let us begin with an example.

Figure 18.2.1

Example 1   Let $F(x, y) = y\mathbf{i} + x\mathbf{j}$. Evaluate the line integral $\int_C \mathbf{F} \cdot d\mathbf{r} = \int_C y\,dx + x\,dy$ over the following curves (Figure 18.2.1):

(a) the line segment $y = x$ from $(0, 0)$ to $(1, 1)$

(b) the parabola $y = x^2$ from $(0, 0)$ to $(1, 1)$

(c) the cubic $y = x^3$ from $(0, 0)$ to $(1, 1)$

Solution (a).   With $x = t$ as the parameter, the path of integration is given by $\mathbf{r}(t) = t\mathbf{i} + t\mathbf{j}$ $(0 \leq t \leq 1)$. Since $F(x, y) = y\mathbf{i} + x\mathbf{j}$, it follows that $F(x(t), y(t)) = t\mathbf{i} + t\mathbf{j}$. Thus,

$$\int_C \mathbf{F} \cdot d\mathbf{r} = \int_0^1 \left[F(x(t), y(t)) \cdot \frac{d\mathbf{r}}{dt} \right] dt = \int_0^1 (t\mathbf{i} + t\mathbf{j}) \cdot (\mathbf{i} + \mathbf{j})\,dt$$

$$= \int_0^1 2t \, dt = 1$$

Solution (b). With $x = t$ as the parameter, $\mathbf{r}(t) = t\mathbf{i} + t^2\mathbf{j}$ ($0 \le t \le 1$), and $\mathbf{F}(x(t), y(t)) = t^2\mathbf{i} + t\mathbf{j}$. Thus,

$$\int_C \mathbf{F} \cdot d\mathbf{r} = \int_0^1 \left[\mathbf{F}(x(t), y(t)) \cdot \frac{d\mathbf{r}}{dt} \right] dt$$

$$= \int_0^1 (t^2\mathbf{i} + t\mathbf{j}) \cdot (\mathbf{i} + 2t\mathbf{j}) \, dt$$

$$= \int_0^1 3t^2 \, dt = 1$$

Solution (c). With $x = t$ as the parameter, $\mathbf{r}(t) = t\mathbf{i} + t^3\mathbf{j}$ ($0 \le t \le 1$), and $\mathbf{F}(x(t), y(t)) = t^3\mathbf{i} + t\mathbf{j}$. Thus,

$$\int_C \mathbf{F} \cdot d\mathbf{r} = \int_0^1 \left[\mathbf{F}(x(t), y(t)) \cdot \frac{d\mathbf{r}}{dt} \right] dt$$

$$= \int_0^1 (t^3\mathbf{i} + t\mathbf{j}) \cdot (\mathbf{i} + 3t^2\mathbf{j}) \, dt$$

$$= \int_0^1 4t^3 \, dt = 1 \qquad \blacktriangleleft$$

In this example we obtained the same value for the line integral even though we integrated over three different paths joining $(0, 0)$ to $(1, 1)$. This is not accidental. The following theorem shows that this is the case because $\mathbf{F}(x, y) = y\mathbf{i} + x\mathbf{j}$ is the gradient of some function ϕ [specifically $\mathbf{F}(x, y) = \nabla\phi$, where $\phi(x, y) = xy$].

The Fundamental Theorem of
Line Integrals

18.2.1 THEOREM. *Suppose that* $\mathbf{F}(x, y) = f(x, y)\mathbf{i} + g(x, y)\mathbf{j}$, *where* f *and* g *are continuous in some open region containing the points* (x_0, y_0) *and* (x_1, y_1). *If*

$$\mathbf{F}(x, y) = \nabla\phi(x, y)$$

at each point of this region, then for any piecewise smooth curve C *that starts at* (x_0, y_0), *ends at* (x_1, y_1), *and lies entirely in the region,*

$$\int_C \mathbf{F}(x, y) \cdot d\mathbf{r} = \phi(x_1, y_1) - \phi(x_0, y_0) \tag{1}$$

Proof. We shall give the proof for a smooth curve C. The proof for piecewise smooth curves is obtained by considering each piece separately. The details are left for the exercises.

If C is given parametrically by $x = x(t)$, $y = y(t)$ ($a \le t \le b$), then the initial and final points of the curve C are

$$(x_0, y_0) = (x(a), y(a))$$
$$(x_1, y_1) = (x(b), y(b))$$

Since $\mathbf{F}(x, y) = \nabla\phi$, it follows that

$$\mathbf{F}(x, y) = \frac{\partial\phi}{\partial x}\mathbf{i} + \frac{\partial\phi}{\partial y}\mathbf{j}$$

so

$$\int_C \mathbf{F}(x, y) \cdot d\mathbf{r} = \int_C \frac{\partial\phi}{\partial x}\,dx + \frac{\partial\phi}{\partial y}\,dy = \int_a^b \left[\frac{\partial\phi}{\partial x}\frac{dx}{dt} + \frac{\partial\phi}{\partial y}\frac{dy}{dt}\right] dt$$

$$= \int_a^b \frac{d}{dt}\left[\phi(x(t), y(t))\right] dt = \phi[x(t), y(t)]\Big]_{t=a}^{b}$$

$$= \phi(x(b), y(b)) - \phi(x(a), y(a))$$

$$= \phi(x_1, y_1) - \phi(x_0, y_0) \qquad \blacksquare$$

CONSERVATIVE
FUNCTIONS

Since the right side of (1) involves only the values of ϕ at the endpoints (x_0, y_0) and (x_1, y_1), it follows that the integral on the left has the same value for *every* piecewise smooth curve C joining these endpoints. We say that the integral is ***independent of the path C***. If $\mathbf{F}(x, y)$ is the gradient of a function $\phi(x, y)$ in some open region, then $\mathbf{F}$ is said to be ***conservative*** in that region, and ϕ is called a ***potential function*** for $\mathbf{F}$ in that region. In brief, Theorem 18.2.1 tells us that if the function $\mathbf{F}$ is conservative in a region, and C is a path (curve) in that region, then the line integral of $\mathbf{F} \cdot d\mathbf{r}$ is independent of the path and the value of the integral can be determined from the values of a potential function at the endpoints of the path.

Example 2   The function $\mathbf{F}(x, y) = y\mathbf{i} + x\mathbf{j}$ is conservative in the entire xy-plane since it is the gradient of $\phi(x, y) = xy$. Thus, for any piecewise smooth curve C running from $(0, 0)$ to $(1, 1)$, it follows from (1) that

$$\int_C y\,dx + x\,dy = \int_C \mathbf{F} \cdot d\mathbf{r} = \phi(1, 1) - \phi(0, 0) = 1 - 0 = 1$$

which explains the results obtained in Example 1.   ◀

Figure 18.2.2

A curve C given by $\mathbf{r}(t) = x(t)\mathbf{i} + y(t)\mathbf{j}$ $(a \leq t \leq b)$ is called ***closed*** if the initial point $\mathbf{r}(a) = \langle x_0, y_0 \rangle$ and the terminal point $\mathbf{r}(b) = \langle x_1, y_1 \rangle$ coincide (Figure 18.2.2). Line integrals over smooth closed curves require no new computational techniques. However, if $\mathbf{F}(x, y) = f(x, y)\mathbf{i} + g(x, y)\mathbf{j}$ is the gradient of $\phi(x, y)$, then it follows from the Fundamental Theorem of Line Integrals (18.2.1) that

$$\int_C \mathbf{F} \cdot d\mathbf{r} = \int_C f(x, y)\, dx + g(x, y)\, dy = \phi(x_1, y_1) - \phi(x_0, y_0) = 0$$

Thus, we have the following result.

18.2.2   THEOREM. *Suppose that* $\mathbf{F}(x, y) = f(x, y)\mathbf{i} + g(x, y)\mathbf{j}$, *where* f *and* g *are continuous in some open region. If* $\mathbf{F}$ *is conservative in that region, then*

$$\int_C \mathbf{F} \cdot d\mathbf{r} = 0$$

for every piecewise smooth closed curve C *in the region.*

Theorem 18.2.1 raises two basic questions that must be answered if Formula (1) is to be useful for calculations.

- How can we tell whether $\mathbf{F}(x, y) = f(x, y)\mathbf{i} + g(x, y)\mathbf{j}$ is the gradient of some function $\phi(x, y)$?

- If $\mathbf{F}(x, y)$ is the gradient of some function $\phi(x, y)$, how do we find ϕ?

To answer these questions, we shall need some terminology. A plane curve $\mathbf{r} = \mathbf{r}(t)$ ($a \le t \le b$) is said to be *simple* if it does not intersect itself anywhere between its endpoints. A simple curve may or may not be closed (Figure 18.2.3). A plane region whose boundary consists of *one* simple closed curve is called *simply connected*. As illustrated in Figure 18.2.4, a

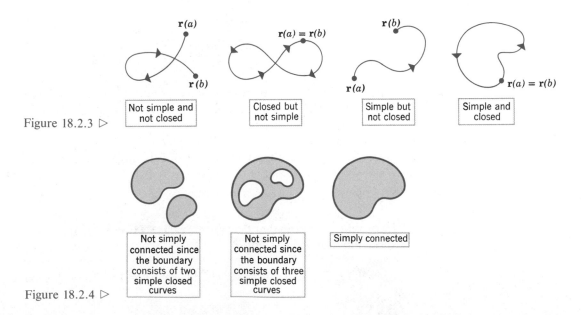

Figure 18.2.3 ▷

Figure 18.2.4 ▷

simply connected region cannot consist of two or more separated pieces and cannot contain any holes.

The following theorem will help us to determine whether a function $\mathbf{F}(x, y)$ is conservative.

18.2.3 THEOREM. *Let* $\mathbf{F}(x, y) = f(x, y)\mathbf{i} + g(x, y)\mathbf{j}$, *where f and g have continuous first partial derivatives in an open simply connected region. Then* $\mathbf{F}(x, y)$ *is the gradient of some function* $\phi(x, y)$ *on that region if and only if*

$$\frac{\partial f}{\partial y} = \frac{\partial g}{\partial x} \tag{2}$$

at each point of the region.

Proof. We shall prove that if $\mathbf{F}$ is the gradient of some function, then condition (2) holds. The proof of the converse is omitted since it requires results from advanced calculus.

Suppose that $\mathbf{F}$ is the gradient of some function ϕ, that is,

$$\mathbf{F}(x, y) = f(x, y)\mathbf{i} + g(x, y)\mathbf{j} = \frac{\partial \phi}{\partial x}\mathbf{i} + \frac{\partial \phi}{\partial y}\mathbf{j} = \nabla \phi$$

It follows that

$$f(x, y) = \frac{\partial \phi}{\partial x} \quad \text{and} \quad g(x, y) = \frac{\partial \phi}{\partial y}$$

so

$$\frac{\partial f}{\partial y} = \frac{\partial^2 \phi}{\partial y \, \partial x} \quad \text{and} \quad \frac{\partial g}{\partial x} = \frac{\partial^2 \phi}{\partial x \, \partial y}$$

Since $\partial f/\partial y$ and $\partial g/\partial x$ were assumed continuous, it follows from Theorem 16.4.5 that the two mixed partial derivatives of ϕ are equal. Thus, $\partial f/\partial y = \partial g/\partial x$. ∎

WARNING. In (2), the **i**-component of **F** is differentiated with respect to y and the **j**-component with respect to x. It is easy to get this backwards by mistake.

Example 3 Show that $\mathbf{F}(x, y) = 2xy^3\mathbf{i} + (1 + 3x^2y^2)\mathbf{j}$ is the gradient of some function ϕ, and find ϕ.

Solution. Since $f(x, y) = 2xy^3$ and $g(x, y) = 1 + 3x^2y^2$ we have

$$\frac{\partial f}{\partial y} = 6xy^2 = \frac{\partial g}{\partial x}$$

so that (2) holds. Thus, **F** is the gradient of some function ϕ, that is,

$$\mathbf{F}(x, y) = 2xy^3\mathbf{i} + (1 + 3x^2y^2)\mathbf{j} = \frac{\partial \phi}{\partial x}\mathbf{i} + \frac{\partial \phi}{\partial y}\mathbf{j} = \nabla\phi \qquad (3)$$

To find ϕ, we begin by equating components in (3):

$$\frac{\partial \phi}{\partial x} = 2xy^3 \qquad (4)$$

$$\frac{\partial \phi}{\partial y} = 1 + 3x^2y^2 \qquad (5)$$

Integrating (4) with respect to x (and treating y as constant), we obtain

$$\phi = \int 2xy^3 \, dx = x^2y^3 + k(y) \qquad (6)$$

where $k(y)$ represents the "constant" of integration. We are justified in treating the constant of integration as a function of y, since y is held constant in the integration process, so

$$\frac{\partial}{\partial x}[k(y)] = 0$$

To find $k(y)$, we differentiate (6) with respect to y:

$$\frac{\partial \phi}{\partial y} = 3x^2y^2 + k'(y)$$

Equating this with (5) yields

$$1 + 3x^2y^2 = 3x^2y^2 + k'(y)$$

Thus,

$$k'(y) = 1 \quad \text{or} \quad k(y) = \int 1 \, dy = y + K$$

where K is a (numerical) constant of integration. Substituting in (6) we obtain

$$\phi = x^2y^3 + y + K$$

The appearance of the arbitrary constant K tells us that ϕ is not unique. The reader may want to verify that $\nabla\phi = \mathbf{F}$.

Alternative Solution.   Instead of integrating (4) with respect to x, as above, we can begin by integrating (5) with respect to y (treating x as a constant in the integration process). We obtain

$$\phi = \int (1 + 3x^2y^2)\, dy = y + x^2y^3 + k(x) \tag{7}$$

where $k(x)$ is the "constant" of integration. Differentiating this expression with respect to x yields

$$\frac{\partial \phi}{\partial x} = 2xy^3 + k'(x)$$

and equating this with (4) gives

$$2xy^3 = 2xy^3 + k'(x)$$

so $k'(x) = 0$ and $k(x) = K$, where K is a (numerical) constant of integration. Substituting this in (7) gives

$$\phi = y + x^2y^3 + K$$

which agrees with our previous solution.   ◄

If a curve C runs from the initial point (x_0, y_0) to the terminal point (x_1, y_1), and if the line integral $\int_C \mathbf{F} \cdot d\mathbf{r}$ is independent of the path joining these points, then we write

$$\int_C \mathbf{F} \cdot d\mathbf{r} = \int_{(x_0,y_0)}^{(x_1,y_1)} \mathbf{F} \cdot d\mathbf{r}$$

or, equivalently,

$$\int_C f(x, y)\, dx + g(x, y)\, dy = \int_{(x_0,y_0)}^{(x_1,y_1)} f(x, y)\, dx + g(x, y)\, dy$$

Example 4   Evaluate

$$\int_{(1,4)}^{(3,1)} 2xy^3\, dx + (1 + 3x^2y^2)\, dy$$

Solution.   From Example 3, $\mathbf{F}(x, y) = 2xy^3\mathbf{i} + (1 + 3x^2y^2)\mathbf{j}$ is the gradient of $\phi(x, y) = y + x^2y^3 + K$. Thus, from Theorem 18.2.1,

$$\int_{(1,4)}^{(3,1)} 2xy^3\, dx + (1 + 3x^2y^2)\, dy = \int_{(1,4)}^{(3,1)} \mathbf{F} \cdot d\mathbf{r}$$

$$= \phi(3, 1) - \phi(1, 4)$$

$$= (10 + K) - (68 + K) = -58 \quad ◄$$

REMARK.   Note that the constant K drops out. In future integration problems we shall omit K from the computations.

Example 5   A particle moves over the semicircle C: $\mathbf{r}(t) = \cos t\mathbf{i} + \sin t\mathbf{j}$ ($0 \le t \le \pi$) while subject to the force $\mathbf{F}(x, y) = e^y\mathbf{i} + xe^y\mathbf{j}$. Find the work done.

Solution.   The work done is

$$W = \int_C \mathbf{F} \cdot d\mathbf{r} = \int_C e^y\, dx + xe^y\, dy \tag{8}$$

If we try to evaluate this integral directly, we must deal with the complicated integral

$$W = \int_C \mathbf{F} \cdot d\mathbf{r} = \int_0^\pi \left[\mathbf{F}(x(t), y(t)) \cdot \frac{d\mathbf{r}}{dt} \right] dt$$

$$= \int_0^\pi [e^{\sin t}\mathbf{i} + (\cos t)e^{\sin t}\mathbf{j}] \cdot [-\sin t\mathbf{i} + \cos t\mathbf{j}]\, dt$$

$$= \int_0^\pi [(-\sin t)e^{\sin t} + (\cos^2 t)e^{\sin t}]\, dt$$

To avoid this difficulty, we shall try an alternative approach.
The force $\mathbf{F}$ is the gradient of some function ϕ, since

$$\frac{\partial}{\partial y}(e^y) = e^y = \frac{\partial}{\partial x}(xe^y)$$

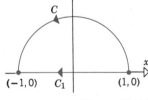

Figure 18.2.5

Thus, the integral in (8) is independent of the path and we are free to replace C by a straight line segment C_1 joining the endpoint $(1, 0)$ to the endpoint $(-1, 0)$ (Figure 18.2.5). This will simplify the computations.
The straight line segment C_1 can be represented parametrically by

$$\begin{matrix} x = 1 - 2t \\ y = 0 \end{matrix}, \quad 0 \le t \le 1$$

(Verify.) Thus,

$$\mathbf{r}(t) = (1 - 2t)\mathbf{i} + 0\mathbf{j} = (1 - 2t)\mathbf{i} \quad (0 \le t \le 1)$$

and

$$\frac{d\mathbf{r}}{dt} = -2\mathbf{i}$$

Therefore, the work done is

$$W = \int_{C_1} \mathbf{F} \cdot d\mathbf{r} = \int_{C_1} \left[(e^y\mathbf{i} + xe^y\mathbf{j}) \cdot \frac{d\mathbf{r}}{dt} \right] dt$$

$$= \int_0^1 [e^0\mathbf{i} + (1 - 2t)e^0\mathbf{j}] \cdot (-2\mathbf{i}) \, dt$$

$$= \int_0^1 [\mathbf{i} + (1 - 2t)\mathbf{j}] \cdot (-2\mathbf{i}) \, dt$$

$$= \int_0^1 -2dt = -2 \quad \text{(units of work)}$$

Second Solution.   Since the line segment C_1 has initial point $(1, 0)$ and terminal point $(-1, 0)$, the line segment $-C_1$ has initial point $(-1, 0)$ and terminal point $(1, 0)$. Thus, $-C_1$ can be represented parametrically by

$$\begin{matrix} x = t \\ y = 0 \end{matrix}, \quad -1 \le t \le 1$$

It follows that

$$\mathbf{r}(t) = t\mathbf{i} + 0\mathbf{j} = t\mathbf{i} \quad (-1 \le t \le 1)$$

and

$$\frac{d\mathbf{r}}{dt} = \mathbf{i}$$

From (4) of Section 18.1,

$$W = \int_{C_1} \mathbf{F} \cdot d\mathbf{r} = -\int_{-C_1} \mathbf{F} \cdot d\mathbf{r} = -\int_{-C_1} \left[(e^y\mathbf{i} + xe^y\mathbf{j}) \cdot \frac{d\mathbf{r}}{dt} \right] dt$$

$$= -\int_{-1}^1 [(e^0\mathbf{i} + te^0\mathbf{j}) \cdot \mathbf{i}] \, dt = -\int_{-1}^1 [(\mathbf{i} + t\mathbf{j}) \cdot \mathbf{i}] \, dt = -\int_{-1}^1 dt = -2$$

Third Solution.   Since $\mathbf{F}(x, y)$ is the gradient of some function ϕ, we can proceed, as in Example 4, by finding ϕ and applying (1). Since $\nabla\phi = \mathbf{F}(x, y) = e^y\mathbf{i} + xe^y\mathbf{j}$,

$$\frac{\partial\phi}{\partial x} = e^y \quad \text{and} \quad \frac{\partial\phi}{\partial y} = xe^y \tag{9}$$

Thus,

$$\phi = \int e^y \, dx = xe^y + k(y) \tag{10}$$

and consequently

$$\frac{\partial \phi}{\partial y} = xe^y + k'(y)$$

Equating this with the expression for $\partial \phi/\partial y$ in (9), we obtain $k'(y) = 0$ or $k(y) = K$. Thus, from (10),

$$\phi = xe^y + K$$

Since the curve C begins at $(1, 0)$ and ends at $(-1, 0)$, it follows that

$$W = \int_C \mathbf{F} \cdot d\mathbf{r} = \int_{(1,0)}^{(-1,0)} e^y \, dx + xe^y \, dy = \phi(-1, 0) - \phi(1, 0)$$

$$= (-1)e^0 - (1)e^0 = -2 \quad \blacktriangleleft$$

Example 6 A particle moves around the circle C given by $r(t) = \cos t\mathbf{i} + \sin t\mathbf{j}$ $(0 \leq t \leq 2\pi)$ while subject to the force $\mathbf{F}(x, y) = e^y\mathbf{i} + xe^y\mathbf{j}$. Find the work done.

Solution. In Example 5 we showed that $\mathbf{F} \cdot d\mathbf{r} = e^y \, dx + xe^y \, dy$ is conservative. Since C is a closed curve, it follows that

$$W = \int_C \mathbf{F} \cdot d\mathbf{r} = \int_C e^y \, dx + xe^y \, dy = 0 \quad \blacktriangleleft$$

▶ Exercise Set 18.2

In Exercises 1–6, determine whether $\mathbf{F}$ is conservative. If it is, find a potential function for it.

1. $\mathbf{F}(x, y) = x\mathbf{i} + y\mathbf{j}$.

2. $\mathbf{F}(x, y) = 3y^2\mathbf{i} + 6xy\mathbf{j}$.

3. $\mathbf{F}(x, y) = x^2y\mathbf{i} + 5xy^2\mathbf{j}$.

4. $\mathbf{F}(x, y) = e^x \cos y\mathbf{i} - e^x \sin y\mathbf{j}$.

5. $\mathbf{F}(x, y) = (\cos y + y \cos x)\mathbf{i} + (\sin x - x \sin y)\mathbf{j}$.

6. $\mathbf{F}(x, y) = x \ln y\mathbf{i} + y \ln x\mathbf{j}$.

7. Show that $\displaystyle\int_{(-1, 2)}^{(1, 3)} y^2 \, dx + 2xy \, dy$ is independent of the path and evaluate the integral by:
    (a) using Theorem 18.2.1

    (b) integrating along the line segment from $(-1, 2)$ to $(1, 3)$.

8. Show that $\displaystyle\int_{(0, 1)}^{(\pi, -1)} y \sin x \, dx - \cos x \, dy$ is independent of the path, and evaluate the integral by:
    (a) using Theorem 18.2.1
    (b) integrating along the line segment from $(0, 1)$ to $(\pi, -1)$.

In Exercises 9–14, show that the integral is independent of the path, and find its value by any method.

9. $\displaystyle\int_{(1, 2)}^{(4, 0)} 3y \, dx + 3x \, dy$.

10. $\displaystyle\int_{(0,0)}^{(1,\pi/2)} e^x \sin y \, dx + e^x \cos y \, dy.$

11. $\displaystyle\int_{(0,0)}^{(3,2)} 2xe^y \, dx + x^2 e^y \, dy.$

12. $\displaystyle\int_{(-1,2)}^{(0,1)} (3x - y + 1) \, dx - (x + 4y + 2) \, dy.$

13. $\displaystyle\int_{(2,-2)}^{(-1,0)} 2xy^3 \, dx + 3y^2 x^2 \, dy.$

14. $\displaystyle\int_{(1,1)}^{(3,3)} \left(e^x \ln y - \frac{e^y}{x}\right) dx + \left(\frac{e^x}{y} - e^y \ln x\right) dy,$

where x and y are positive.

In Exercises 15–18, find the work done by the conservative force F as it acts on a particle moving from P to Q.

15. $F(x, y) = xy^2 i + x^2 y j$; $P(1, 1)$, $Q(0, 0)$.

16. $F(x, y) = ye^{xy} i + xe^{xy} j$; $P(-1, 1)$, $Q(2, 0)$.

17. $F(x, y) = \dfrac{y}{x^2 + y^2} i - \dfrac{x}{x^2 + y^2} j$; $P(0, 2)$, $Q(3, 3)$.

[Assume $y > 0$.]

18. $F(x, y) = e^{-y} \cos x i - e^{-y} \sin x j$; $P(\pi/2, 1)$, $Q(-\pi/2, 0)$.

19. Let

$$F(x, y) = (e^y + ye^x)i + (xe^y + e^x)j$$

Find the work done by the force F if it acts on a particle that moves

(a) from $(a, 0)$ to $(-a, 0)$ along the x-axis
(b) from $(a, 0)$ to $(-a, 0)$ along the upper half of the circle $x^2 + y^2 = a^2$
(c) from $(a, 0)$ to $(-a, 0)$ along the upper half of the ellipse

$$\frac{x^2}{a^2} + \frac{y^2}{b^2} = 1$$

(d) once around the circle $x^2 + y^2 = a^2$.

20. Evaluate

$$\int_C (\sin y \sinh x + \cos y \cosh x) \, dx$$
$$+ (\cos y \cosh x - \sin y \sinh x) \, dy$$

where C is the line segment

$$x = t \atop y = \tfrac{1}{2}\pi(t - 1), \qquad 1 \le t \le 2.$$

21. Let

$$F(x, y) = \frac{k}{\|r\|^3} r$$

where $r = xi + yj$. Show that F is conservative on any region that does not contain the origin.

22. Find a function h for which

$$F(x, y) = h(x)[x \sin y + y \cos y]i$$
$$+ h(x)[x \cos y - y \sin y]j$$

is conservative.

23. (a) Prove: $F(x, y) = f(x, y)i + g(x, y)j$ is conservative in a region if and only if there is a function $\phi(x, y)$ such that

$$d\phi = f(x, y) \, dx + g(x, y) \, dy$$

on the region.

(b) Prove: If $F(x, y)$ is conservative on a simply connected open region containing (x_0, y_0) and (x_1, y_1), then

$$\int_{(x_0,y_0)}^{(x_1,y_1)} d\phi = \phi(x_1, y_1) - \phi(x_0, y_0)$$

where ϕ is the function in part (a).

24. Prove Theorem 18.2.1 in the case where C is a piecewise smooth curve composed of smooth curves $C_1, C_2, \ldots, C_n$.

18.3 GREEN'S THEOREM

In this section we shall discuss a remarkable and beautiful theorem that expresses the double integral over a plane region in terms of a line integral over the boundary.

Green's* Theorem

18.3.1 THEOREM. *Let R be a simply connected plane region whose boundary is a simple, closed piecewise smooth curve C traversed counterclockwise. If $f(x, y)$ and $g(x, y)$ have continuous first partial derivatives on some open set containing R, then*

$$\int_C f(x, y)\, dx + g(x, y)\, dy = \iint_R \left(\frac{\partial g}{\partial x} - \frac{\partial f}{\partial y} \right) dA \tag{1}$$

Proof.   For simplicity, we shall prove the theorem only for regions that are simultaneously type I and type II (see Section 17.2). Such a region is shown in Figure 18.3.1. The crux of the proof is to show that

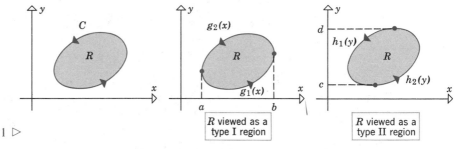

Figure 18.3.1 ▷

R viewed as a type I region

R viewed as a type II region

---

*GEORGE GREEN (1793–1841), English mathematician and physicist. Green left school at an early age to work in his father's bakery and consequently had little early formal education. When his father opened a mill, the boy used the top room as a study in which he taught himself physics and mathematics from library books. In 1828 Green published his most important work, *An Essay on the Application of Mathematical Analysis to the Theories of Electricity and Magnetism.* Although Green's Theorem appeared in that paper, the result went virtually unnoticed because of the small pressrun and local distribution. Following the death of his father in 1829, Green was urged by friends to seek a college education. After four years of self-study to close the gaps in his elementary education, Green was admitted to Caius College, Cambridge in 1833. He graduated four years later, but with a disappointing performance on his final examinations—possibly because he was more interested in his own research. After a succession of works on light and sound, he was named to be Perse Fellow at Caius College. Two years later he died. In 1845, four years after his death, his paper of 1828 was published and the theories developed therein by this obscure, self-taught baker's son helped pave the way to the modern theories of electricity and magnetism.

$$\int_C f(x, y)\, dx = -\iint_R \frac{\partial f}{\partial y}\, dA \qquad (2a)$$

and

$$\int_C g(x, y)\, dy = \iint_R \frac{\partial g}{\partial x}\, dA \qquad (2b)$$

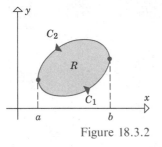

Figure 18.3.2

Formula (1) will then follow by adding (2a) and (2b).

To prove (2a), view R as a type I region and let C_1 and C_2 be the lower and upper boundary curves, oriented as in Figure 18.3.2. Then

$$\int_C f(x, y)\, dx = \int_{C_1} f(x, y)\, dx + \int_{C_2} f(x, y)\, dx$$

or, equivalently,

$$\int_C f(x, y)\, dx = \int_{C_1} f(x, y)\, dx - \int_{-C_2} f(x, y)\, dx \qquad (3)$$

(This step will help simplify our calculations since C_1 and $-C_2$ are both oriented left to right.) With $x = t$ as the parameter, C_1 and $-C_2$ are represented by

$$C_1: x = t,\ y = g_1(t) \quad (a \le t \le b)$$

$$-C_2: x = t,\ y = g_2(t) \quad (a \le t \le b)$$

Thus, (3) yields

$$\int_C f(x, y)\, dx = \int_a^b f(t, g_1(t)) \left(\frac{dx}{dt}\right) dt - \int_a^b f(t, g_2(t)) \left(\frac{dx}{dt}\right) dt$$

$$= \int_a^b f(t, g_1(t))\, dt - \int_a^b f(t, g_2(t))\, dt$$

$$= -\int_a^b [f(t, g_2(t)) - f(t, g_1(t))]\, dt$$

$$= -\int_a^b \left[f(t, y) \right]_{y=g_1(t)}^{y=g_2(t)} dt$$

$$= -\int_a^b \left[\int_{g_1(t)}^{g_2(t)} \frac{\partial f}{\partial y}\, dy \right] dt$$

$$= -\int_a^b \int_{g_1(x)}^{g_2(x)} \frac{\partial f}{\partial y}\, dy\, dx \quad \text{[Since } x = t\text{]}$$

$$= -\iint_R \frac{\partial f}{\partial y}\, dA$$

The proof of (2b) is obtained similarly by treating R as a type II region. The details are omitted. ∎

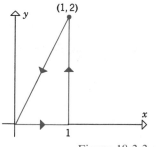

(1, 2)

Figure 18.3.3

Example 1 Use Green's Theorem to evaluate

$$\int_C x^2 y \, dx + x \, dy$$

over the triangular path shown in Figure 18.3.3.

Solution. Since $f(x, y) = x^2 y$ and $g(x, y) = x$, it follows from (1) that

$$\int_C x^2 y \, dx + x \, dy = \iint\limits_R \left[\frac{\partial}{\partial x}(x) - \frac{\partial}{\partial y}(x^2 y) \right] dA$$

$$= \int_0^1 \int_0^{2x} (1 - x^2) \, dy \, dx$$

$$= \int_0^1 (2x - 2x^3) \, dx$$

$$= \left[x^2 - \frac{x^4}{2} \right]_0^1 = \frac{1}{2}$$

This agrees with the result obtained in Example 3 of Section 18.1, where we evaluated the line integral directly. Note how much simpler this solution is. ◀

Example 2 Suppose that a particle travels once around the unit circle in the counterclockwise direction, while subject to the force $\mathbf{F}(x, y) = (e^x - y^3)\mathbf{i} + (\cos y + x^3)\mathbf{j}$. Use Green's Theorem to find the work done by the force.

Solution.

$$W = \int_C (e^x - y^3) \, dx + (\cos y + x^3) \, dy$$

$$= \iint\limits_R \left[\frac{\partial}{\partial x}(\cos y + x^3) - \frac{\partial}{\partial y}(e^x - y^3) \right] dA$$

$$= \iint\limits_R (3x^2 + 3y^2) \, dA = 3 \iint\limits_R (x^2 + y^2) \, dA$$

$$= 3 \int_0^{2\pi} \int_0^1 (r^2) r \, dr \, d\theta = \frac{3}{4} \int_0^{2\pi} d\theta = \frac{3\pi}{2} \quad \blacktriangleleft$$

We converted to polar coordinates.

COMPUTING AREAS USING GREEN'S THEOREM

Green's Theorem leads to some new formulas for area. Letting $f(x, y) = 0$ and $g(x, y) = x$ in (1) yields

$$\int_C x \, dy = \iint_R dA = \text{area of } R \tag{4a}$$

Letting $f(x, y) = -y$ and $g(x, y) = 0$ in (1) yields

$$\int_C -y \, dx = \iint_R dA = \text{area of } R \tag{4b}$$

Adding (4a) and (4b) and dividing the result by 2 yields

$$\frac{1}{2} \int_C -y \, dx + x \, dy = \text{area of } R \tag{4c}$$

Formulas (4a), (4b), and (4c) each express the area of R in terms of a line integral taken counterclockwise around the boundary of R. These formulas apply to those regions that satisfy the hypotheses of Green's Theorem. Although (4c) is more complicated than (4a) or (4b), it frequently results in an easier integral to evaluate.

Example 3 Use (4c) to find the area enclosed by the ellipse $x^2/a^2 + y^2/b^2 = 1$.

Solution. The ellipse, with counterclockwise orientation, can be represented parametrically by

$$\begin{aligned} x &= a \cos t \\ y &= b \sin t \end{aligned}, \quad 0 \le t \le 2\pi$$

If we denote this curve by C, then the area enclosed by the ellipse is

$$A = \frac{1}{2} \int_C -y \, dx + x \, dy$$

$$= \frac{1}{2} \int_0^{2\pi} [(-b \sin t)(-a \sin t) + (a \cos t)(b \cos t)] \, dt$$

$$= \frac{1}{2} ab \int_0^{2\pi} (\sin^2 t + \cos^2 t) \, dt$$

$$= \frac{1}{2} ab \int_0^{2\pi} dt = \pi ab$$

This result can also be obtained from Formula (4a) or (4b).   ◀

REMARK.   Green's Theorem provides another viewpoint about some earlier results. If $\partial f/\partial y = \partial g/\partial x$ throughout a simply connected open region R with boundary C, then it follows from Green's Theorem that

$$\int_C f \, dx + g \, dy = \iint_R \left(\frac{\partial g}{\partial x} - \frac{\partial f}{\partial y} \right) dA = 0$$

However, this is to be expected since the condition $\partial f/\partial y = \partial g/\partial x$ implies that $\mathbf{F}(x, y) = f(x, y)\mathbf{i} + g(x, y)\mathbf{j}$ is conservative in R (Theorem 18.2.3) and we already know that a line integral around a closed curve has value zero in this case (Theorem 18.2.2).

▶ Exercise Set 18.3

In Exercises 1 and 2, evaluate the line integral using Green's Theorem and check the answer by evaluating directly.

1. $\int_C y^2 \, dx + x^2 \, dy$, where C is the square with vertices $(0, 0)$, $(1, 0)$, $(1, 1)$, and $(0, 1)$ traversed counterclockwise.

2. $\int_C y \, dx + x \, dy$, where C is the unit circle traversed counterclockwise.

In Exercises 3–13, use Green's Theorem to evaluate the integral. In each exercise, assume that the curve C is traversed counterclockwise.

3. $\int_C 3xy \, dx + 2xy \, dy$, where C is the rectangle bounded by $x = -2$, $x = 4$, $y = 1$, and $y = 2$.

4. $\int_C (x^2 - y^2) \, dx + x \, dy$, where C is the circle $x^2 + y^2 = 9$.

5. $\int_C x \cos y \, dx - y \sin x \, dy$, where C is the square with vertices $(0, 0)$, $(0, \pi/2)$, $(\pi/2, \pi/2)$, and $(\pi/2, 0)$.

6. $\int_C y \tan^2 x \, dx + \tan x \, dy$, where C is the circle $x^2 + (y + 1)^2 = 1$.

7. $\int_C (x^2 - y) \, dx + x \, dy$, where C is the circle $x^2 + y^2 = 4$.

8. $\int_C (e^x + y^2) \, dx + (e^y + x^2) \, dy$, where C is the boundary of the region between $y = x^2$ and $y = x$.

9. $\int_C \ln(1 + y) \, dx - \frac{xy}{1 + y} \, dy$, where C is the triangle with vertices $(0, 0)$, $(2, 0)$, and $(0, 4)$.

10. $\int_C x^2 y \, dx - y^2 x \, dy$, where C is the boundary of the region in the first quadrant, enclosed between the coordinate axes and the circle $x^2 + y^2 = 16$.

11. $\int_C \tan^{-1} y \, dx - \dfrac{y^2 x}{1 + y^2} \, dy$, where C is the square with vertices $(0, 0)$, $(0, 1)$, $(1, 1)$, and $(1, 0)$.

12. $\int_C \cos x \sin y \, dx + \sin x \cos y \, dy$, where C is the triangle with vertices $(0, 0)$, $(3, 3)$, and $(0, 3)$.

13. $\int_C x^2 y \, dx + (y + xy^2) \, dy$, where C is the boundary of the region enclosed by $y = x^2$ and $x = y^2$.

14. Let C be the boundary of the region enclosed between $y = x^2$ and $y = 2x$. Assuming that C is traversed counterclockwise, evaluate the following integrals by Green's Theorem:

(a) $\int_C (6xy - y^2) \, dx$ (b) $\int_C (6xy - y^2) \, dy$.

15. Find the area of the ellipse in Example 3 using
(a)   Formula (4a)   (b)   Formula (4b).

16. Use a line integral to find the area of the region enclosed by

$$x = a \cos^3 t$$
$$y = a \sin^3 t , \quad 0 \le t \le 2\pi$$

17. Use a line integral to find the area of the triangle with vertices $(0, 0)$, $(a, 0)$, and $(0, b)$, where a and b are positive.

18. Use a line integral to find the area of the region in the first quadrant enclosed by $y = x$, $y = 1/x$, and $y = x/9$.

19. A particle, starting at $(5, 0)$, traverses the upper semicircle $x^2 + y^2 = 25$ and returns to its starting point along the x-axis. Use Green's Theorem to find the work done on the particle by a force $F(x, y) = xy\mathbf{i} + (\frac{1}{2}x^2 + yx)\mathbf{j}$.

20. A particle which is subject to a force of $F(x, y) = \sqrt{y}\mathbf{i} + \sqrt{x}\mathbf{j}$ moves counterclockwise around the closed curve formed by $y = 0$, $x = 2$, and $y = x^3/4$. Use Green's Theorem to find the work done.

21. (a)   Let R be a plane region with area A whose boundary is a piecewise smooth simple closed curve C. Use Green's Theorem to prove that the centroid $(\bar{x}, \bar{y})$ of R is given by

$$\bar{x} = \frac{1}{2A} \int_C x^2 \, dy, \qquad \bar{y} = -\frac{1}{2A} \int_C y^2 \, dx$$

(b)   Use the result in (a) to find the centroid of the region enclosed between the x-axis and the upper half of the circle $x^2 + y^2 = a^2$.

22. Evaluate $\int_C y \, dx - x \, dy$, where C is the cardioid $r = a(1 + \cos \theta)$, $0 \le \theta \le 2\pi$.

23. (a)   Let C be the line segment from a point (a, b) to a point (c, d). Show that

$$\int_C x \, dy - y \, dx = ad - bc$$

(b)   Use the result in (a) to show that the area A of a triangle with successive vertices (x_1, y_1), (x_2, y_2), and (x_3, y_3) going counterclockwise is

$$A = \tfrac{1}{2}[(x_1 y_2 - x_2 y_1) + (x_2 y_3 - x_3 y_2) + (x_3 y_1 - x_1 y_3)]$$

(c)   Find a formula for the area of a polygon with successive vertices (x_1, y_1), (x_2, y_2), ..., (x_n, y_n) going counterclockwise.

(d)   Use the result in (c) to find the area of a quadrilateral with vertices $(0, 0)$, $(3, 4)$, $(-2, 2)$, $(-1, 0)$.

24. Find a simple closed curve C that maximizes the value of

$$\int_C \frac{1}{3} y^3 \, dx + \left(x - \frac{1}{3} x^3 \right) dy$$

25. (a)   Let R be the region enclosed by the circle, $C: x = \cos t, y = \sin t, 0 \le t \le 2\pi$. By evaluating directly, show that

$$\int_C -\frac{y}{x^2 + y^2} \, dx + \frac{x}{x^2 + y^2} \, dy = 2\pi$$

(b)   Let f and g be the functions

$$f(x, y) = -\frac{y}{x^2 + y^2}, \quad g(x, y) = \frac{x}{x^2 + y^2}$$

Show that $\partial g/\partial x = \partial f/\partial y$.

(c)   Find the error in the following argument: By

Green's Theorem

$$\int_C -\frac{y}{x^2 + y^2}\,dx + \frac{x}{x^2 + y^2}\,dy$$

$$= \int_C f(x, y)\,dx + g(x, y)\,dy$$

$$= \iint_R \left(\frac{\partial g}{\partial x} - \frac{\partial f}{\partial y}\right) dA = 0$$

[which contradicts the result in (a)].

18.4 INTRODUCTION TO SURFACE INTEGRALS

In previous sections we considered four kinds of integrals—integrals over intervals, double integrals over two-dimensional regions, triple integrals over three-dimensional solids, and line integrals over curves in two- or three-dimensional space. In this section we shall discuss integrals over surfaces in three-dimensional space. Such integrals occur in problems of fluid and heat flow, electricity, magnetism, mass, and center of gravity.

MASS OF A
CURVED LAMINA

We shall motivate the definition of a surface integral by considering a problem about mass. In Section 17.6 we defined a lamina to be an idealized flat object that is sufficiently thin to be viewed as a two-dimensional solid. If we imagine a flat lamina to be bent into some curved shape, the result is a *curved lamina* (Figure 18.4.1). The density of a curved lamina at a point (x, y, z) can be specified by a function $\delta(x, y, z)$, called the *density function* for the lamina. Informally, the density function can be visualized as follows: Consider a small section of the lamina containing the point (x, y, z), and let ΔM and ΔS be the mass and surface area of this section (Figure 18.4.2). If the ratio $\Delta M/\Delta S$ tends toward a limiting value when the small section of lamina is allowed to shrink down to the point (x, y, z), then this limit is the density of the lamina at (x, y, z). Symbolically,

$$\delta(x, y, z) = \lim \frac{\Delta M}{\Delta S}$$

The thickness of a curved
lamina is negligible

Figure 18.4.1

(x,y,z)

Figure 18.4.2

From this equation we obtain the approximation

$$\Delta M \approx \delta(x, y, z)\,\Delta S \tag{1}$$

which relates the mass and surface area of a small section of lamina containing the point (x, y, z).

We shall now use (1) to obtain a formula for the mass of a curved lamina.

18.4.1 DEFINITION. If a curved lamina with density $\delta(x, y, z)$ has the equation $z = f(x, y)$, and if the projection of this lamina on the xy-plane is the region R (Figure 18.4.3), then the mass M of the lamina is defined to be

$$M = \iint\limits_{R} \delta[x, y, f(x, y)] \sqrt{f_x(x, y)^2 + f_y(x, y)^2 + 1}\, dA \qquad (2a)$$

or in an alternative notation

$$M = \iint\limits_{R} \delta[x, y, f(x, y)] \sqrt{\left(\frac{\partial z}{\partial x}\right)^2 + \left(\frac{\partial z}{\partial y}\right)^2 + 1}\, dA \qquad (2b)$$

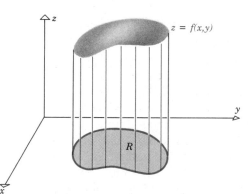

Figure 18.4.3 ▷

We shall motivate this formula in four steps that closely parallel the four steps in the motivation of Definition 17.4.3 for surface area. (The reader should review the discussion following that definition before continuing.)

Step 1. As shown in Figure 18.4.4, subdivide a rectangle containing R into subrectangles and discard all those subrectangles that contain any points outside of R. This leaves only rectangles $R_1, R_2, \ldots, R_n$ that are subsets of R. We shall denote the area of R_k by ΔA_k.

Step 2. As illustrated in Figure 18.4.4, project each subrectangle onto the lamina, let the surface areas of the resulting sections of lamina be

$$\Delta S_1, \Delta S_2, \ldots, \Delta S_n$$

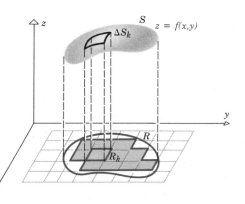

Figure 18.4.4 ▷

and let $(x_k^*, y_k^*, z_k^*) = [x_k^*, y_k^*, f(x_k^*, y_k^*)]$ be an arbitrary point in the kth section.

Step 3. As in Step 3 of the surface area discussion, the surface area ΔS_k can be approximated by

$$\Delta S_k \approx \sqrt{f_x(x_k^*, y_k^*)^2 + f_y(x_k^*, y_k^*)^2 + 1}\, \Delta A_k \quad (3)$$

It follows from (1) that the mass ΔM_k of the kth section of lamina is given by the approximation

$$\Delta M_k \approx \delta(x_k^*, y_k^*, z_k^*)\, \Delta S_k = \delta[x_k^*, y_k^*, f(x_k^*, y_k^*)]\, \Delta S_k \quad (4)$$

Substituting (3) into (4) and adding the masses of the individual sections yields the following approximation to the total mass M of the lamina

$$M \approx \sum_{k=1}^{n} \delta[x_k^*, y_k^*, f(x_k^*, y_k^*)]\sqrt{f_x(x_k^*, y_k^*)^2 + f_y(x_k^*, y_k^*)^2 + 1}\, \Delta A_k$$

Step 4. If we repeat the subdivision process using more and more rectangles with dimensions that decrease to zero as $n \to +\infty$, then it is reasonable to assume that the errors in the approximations diminish to zero and the exact mass of the lamina is

$$M = \lim_{n \to +\infty} \sum_{k=1}^{n} \delta[x_k^*, y_k^*, f(x_k^*, y_k^*)]\sqrt{f_x(x_k^*, y_k^*)^2 + f_y(x_k^*, y_k^*)^2 + 1}\ \Delta A_k$$

or, equivalently,

$$M = \iint_R \delta[x, y, f(x, y)]\sqrt{f_x(x, y)^2 + f_y(x, y)^2 + 1}\ dA$$

which agrees with (2a).

Example 1   A curved lamina is the portion of the paraboloid $z = x^2 + y^2$ below the plane $z = 1$ (Figure 18.4.5) and has constant density $\delta(x, y, z) = \delta_0$. Find the mass of the lamina.

Figure 18.4.5

Solution.   Since $z = f(x, y) = x^2 + y^2$, it follows that

$$\frac{\partial z}{\partial x} = 2x \quad \text{and} \quad \frac{\partial z}{\partial y} = 2y$$

Substituting these expressions and $\delta(x, y, z) = \delta(x, y, f(x, y)) = \delta_0$ into (2b) yields

$$M = \iint_R \delta_0 \sqrt{(2x)^2 + (2y)^2 + 1}\ dA = \delta_0 \iint_R \sqrt{4x^2 + 4y^2 + 1}\ dA \quad (5)$$

where R is the circular region enclosed by $x^2 + y^2 = 1$. To evaluate (5) we use polar coordinates:

$$M = \delta_0 \int_0^{2\pi} \int_0^1 \sqrt{4r^2 + 1}\ r\ dr\ d\theta = \frac{\delta_0}{12} \int_0^{2\pi} (4r^2 + 1)^{3/2} \Big]_{r=0}^{1} d\theta$$

$$= \frac{\delta_0}{12} \int_0^{2\pi} (5^{3/2} - 1)\ d\theta = \frac{\pi\delta_0}{6}(5\sqrt{5} - 1) \quad \blacktriangleleft$$

SURFACE INTEGRALS   We shall now generalize the procedure used to derive Formulas (2a) and (2b) to develop the concept of a *surface integral*.

Let σ be a surface with finite surface area and $g(x, y, z)$ a continuous function defined on σ. We shall *not* assume in this discussion that σ can be represented by an equation of the form $z = f(x, y)$. Thus, the surface might be a sphere $x^2 + y^2 + z^2 = a^2$ or perhaps a surface represented by an equation of the form $y = f(x, z)$ or $x = f(y, z)$.

Subdivide σ into n parts $\sigma_1, \sigma_2, \ldots, \sigma_n$ with surface areas $\Delta S_1, \Delta S_2, \ldots, \Delta S_n$ and form the sum

$$\sum_{k=1}^{n} g(x_k^*, y_k^*, z_k^*)\ \Delta S_k \qquad (6)$$

(x_k^*, y_k^*, z_k^*)

σ_k

σ

Figure 18.4.6

where (x_k^*, y_k^*, z_k^*) is an arbitrary point in the kth part σ_k (Figure 18.4.6).

Now, repeat the subdivision process dividing σ into more and more parts, so that each part shrinks to a point as $n \to +\infty$. If (6) tends toward a limit as $n \to +\infty$, and if this limit does not depend on the way the subdivisions are made or how the points (x_k^*, y_k^*, z_k^*) are chosen, then this limit is called the *surface integral* of $g(x, y, z)$ over σ, and is denoted by

$$\iint_\sigma g(x, y, z)\, dS = \lim_{n \to +\infty} \sum_{k=1}^{n} g(x_k^*, y_k^*, z_k^*)\, \Delta S_k \tag{7}$$

Example 2 In the special case where $g(x, y, z) = 1$, the integral in (7) is the surface area of σ, that is

$$S = \text{surface area of } \sigma = \iint_\sigma dS \tag{8}$$

To see this, observe that for every subdivision of σ into parts with surface areas $\Delta S_1, \Delta S_2, \ldots, \Delta S_n$, these surface areas add up to the total surface area S, that is,

$$S = \sum_{k=1}^{n} \Delta S_k$$

Since the right side has the value S for all n, it follows that

$$S = \lim_{n \to +\infty} \sum_{k=1}^{n} \Delta S_k$$

Formula (8) now follows from this and (7). ◀

Example 3 If σ is a curved lamina with density $\delta(x, y, z)$ and mass M, then

$$M = \iint_\sigma \delta(x, y, z)\, dS \tag{9}$$

To see this, let σ be subdivided into n parts with surface areas $\Delta S_1, \Delta S_2, \ldots, \Delta S_n$ and for $k = 1, 2, \ldots, n$ let (x_k^*, y_k^*, z_k^*) be an arbitrary point in the kth part. It follows from (1) that

$$\delta(x_k^*, y_k^*, z_k^*)\, \Delta S_k$$

is approximately the mass of the kth part, and so the entire mass M can be approximated by

$$M \approx \sum_{k=1}^{n} \delta(x_k^*, y_k^*, z_k^*) \, \Delta S_k$$

If we assume that the error in this approximation tends to zero as the number of subdivisions tends to infinity and the parts shrink to points, then the exact mass is

$$M = \lim_{n \to +\infty} \sum_{k=1}^{n} \delta(x_k^*, y_k^*, z_k^*) \, \Delta S_k = \iint_{\sigma} \delta(x, y, z) \, dS \quad \blacktriangleleft$$

If σ is a curved lamina with density $\delta(x, y, z)$ and equation $z = f(x, y)$, then (2b) expresses the mass of the lamina as a double integral and (9) as a surface integral. Equating (2b) and (9) yields the following relationship between the surface integral and the double integral:

$$\iint_{\sigma} \delta(x, y, z) \, dS = \iint_{R} \delta[x, y, f(x, y)] \sqrt{\left(\frac{\partial z}{\partial x}\right)^2 + \left(\frac{\partial z}{\partial y}\right)^2 + 1} \, dA$$

This is a special case of the following general result that can be used to evaluate many surface integrals.

18.4.2   THEOREM.

(a)   *Let σ be a surface with equation $z = f(x, y)$ and R its projection on the xy-plane. If f has continuous first partial derivatives on R and $g(x, y, z)$ is continuous on σ, then*

$$\iint_{\sigma} g(x, y, z) \, dS = \iint_{R} g[x, y, f(x, y)] \sqrt{\left(\frac{\partial z}{\partial x}\right)^2 + \left(\frac{\partial z}{\partial y}\right)^2 + 1} \, dA \quad (10)$$

(b)   *Let σ be a surface with equation $y = f(x, z)$ and let R be its projection on the xz-plane. If f has continuous first partial derivatives on R and $g(x, y, z)$ is continuous on σ, then*

$$\iint_{\sigma} g(x, y, z) \, dS = \iint_{R} g[x, f(x, z), z] \sqrt{\left(\frac{\partial y}{\partial x}\right)^2 + \left(\frac{\partial y}{\partial z}\right)^2 + 1} \, dA \quad (11)$$

(c)   *Let σ be a surface with equation $x = f(y, z)$ and let R be its projection on the yz-plane. If f has continuous first partial derivatives on R and $g(x, y, z)$ is continuous on σ, then*

$$\iint_{\sigma} g(x, y, z) \, dS = \iint_{R} g[f(y, z), y, z] \sqrt{\left(\frac{\partial x}{\partial y}\right)^2 + \left(\frac{\partial x}{\partial z}\right)^2 + 1} \, dA \quad (12)$$

We omit the proof.

Example 4   Evaluate the surface integral

$$\iint_\sigma xz \, dS$$

where σ is the part of the plane $x + y + z = 1$ that lies in the first octant.

Solution.   Since the equation of the plane can be written as

$$z = 1 - x - y$$

which is of the form $z = f(x, y)$, we can apply Formula (10) with $z = f(x, y) = 1 - x - y$ and $g(x, y, z) = xz$. Thus,

$$\frac{\partial z}{\partial x} = -1 \quad \text{and} \quad \frac{\partial z}{\partial y} = -1$$

so (10) becomes

$$\iint_\sigma xz \, dS = \iint_R x(1 - x - y)\sqrt{(-1)^2 + (-1)^2 + 1} \, dA \tag{13}$$

where R is the projection of σ on the xy-plane (Figure 18.4.7a). Rewriting the double integral in (13) as an iterated integral yields

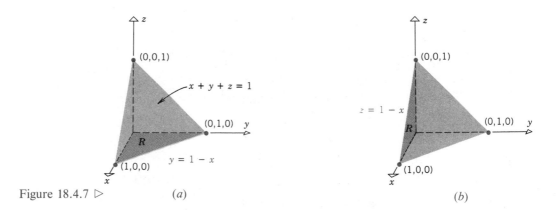

Figure 18.4.7 ▷   (a)   (b)

$$\iint_\sigma xz \, dS = \sqrt{3} \int_0^1 \int_0^{1-x} (x - x^2 - xy) \, dy \, dx$$

$$= \sqrt{3} \int_0^1 xy - x^2y - \frac{xy^2}{2} \Big]_{y=0}^{1-x} dx$$

$$= \sqrt{3} \int_0^1 \left(\frac{x}{2} - x^2 + \frac{x^3}{2} \right) dx$$

$$= \sqrt{3} \left[\frac{x^2}{4} - \frac{x^3}{3} + \frac{x^4}{8} \right]_0^1 = \frac{\sqrt{3}}{24}$$

Alternative Solution.   Since the equation of the plane can be written as

$$y = 1 - x - z$$

which is of the form $y = f(x, z)$, we can apply Formula (11) with $y = f(x, z) = 1 - x - z$. Thus,

$$\frac{\partial y}{\partial x} = -1 \quad \text{and} \quad \frac{\partial y}{\partial z} = -1$$

so (11) yields

$$\iint_\sigma xz \, dS = \iint_R xz \sqrt{(-1)^2 + (-1)^2 + 1} \, dA \tag{14}$$

where R is the projection of σ on the xz-plane (Figure 18.4.7*b*). Rewriting the double integral in (14) as an iterated integral yields

$$\iint_\sigma xz \, dS = \sqrt{3} \int_0^1 \int_0^{1-x} xz \, dz \, dx = \sqrt{3} \int_0^1 \frac{xz^2}{2} \Big]_{z=0}^{1-x} dx$$

$$= \frac{\sqrt{3}}{2} \int_0^1 (x - 2x^2 + x^3) \, dx$$

$$= \frac{\sqrt{3}}{2} \left[\frac{x^2}{2} - \frac{2x^3}{3} + \frac{x^4}{4} \right]_0^1 = \frac{\sqrt{3}}{24} \quad \blacktriangleleft$$

Example 5   Evaluate the surface integral

$$\iint_\sigma y^2 z^2 \, dS$$

where σ is the part of the cone $z = \sqrt{x^2 + y^2}$ between the planes $z = 1$ and $z = 2$ (Figure 18.4.8).

Solution.   We shall apply Formula (10) with $z = f(x, y) = \sqrt{x^2 + y^2}$ and $g(x, y, z) = y^2 z^2$. Thus,

$$\frac{\partial z}{\partial x} = \frac{x}{\sqrt{x^2 + y^2}} \quad \text{and} \quad \frac{\partial z}{\partial y} = \frac{y}{\sqrt{x^2 + y^2}}$$

$z = 2$

$z = 1$

$z = \sqrt{x^2 + y^2}$

R

Figure 18.4.8 ▷

so

$$\sqrt{\left(\frac{\partial z}{\partial x}\right)^2 + \left(\frac{\partial z}{\partial y}\right)^2 + 1} = \sqrt{2}$$

(verify), and (10) yields

$$\iint_{\sigma} y^2 z^2 \, dS = \iint_{R} y^2 (\sqrt{x^2 + y^2})^2 \sqrt{2} \, dA = \sqrt{2} \iint_{R} y^2 (x^2 + y^2) \, dA$$

where R is the annulus enclosed between $x^2 + y^2 = 1$ and $x^2 + y^2 = 4$ (Figure 18.4.8). Using polar coordinates to evaluate this double integral over the annulus R yields

$$\iint_{\sigma} y^2 z^2 \, dS = \sqrt{2} \int_{0}^{2\pi} \int_{1}^{2} (r \sin \theta)^2 (r^2) r \, dr \, d\theta$$

$$= \sqrt{2} \int_{0}^{2\pi} \int_{1}^{2} r^5 \sin^2 \theta \, dr \, d\theta$$

$$= \sqrt{2} \int_{0}^{2\pi} \frac{r^6}{6} \sin^2 \theta \Big]_{r=1}^{2} \, d\theta$$

$$= \frac{21}{\sqrt{2}} \int_{0}^{2\pi} \sin^2 \theta \, d\theta$$

$$= \frac{21}{\sqrt{2}} \left[\frac{1}{2}\theta - \frac{1}{4}\sin 2\theta\right]_{0}^{2\pi} = \frac{21\pi}{\sqrt{2}} \qquad \boxed{\begin{array}{l}\text{See Example 1,}\\ \text{Section 9.3.}\end{array}} \qquad ◀$$

Sometimes Formulas (10), (11), and (12) cannot be applied because the partial derivatives in these formulas do not exist at every point of the region R. However, it may still be possible to compute surface integrals in such cases by means of a limiting procedure. The next example illustrates this.

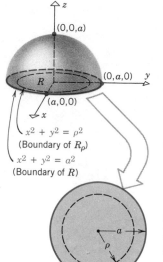

z

$(0,0,a)$

R

$(0,a,0)$ y

$(a,0,0)$

x

$x^2 + y^2 = \rho^2$
(Boundary of R_ρ)

$x^2 + y^2 = a^2$
(Boundary of R)

a

ρ

Figure 18.4.9

Example 6 Evaluate the surface integral

$$\iint_\sigma dS \qquad (15)$$

where σ is the upper hemisphere of radius a given by $z = \sqrt{a^2 - x^2 - y^2}$ (Figure 18.4.9).

Solution. The partial derivatives

$$\frac{\partial z}{\partial x} = \frac{-x}{\sqrt{a^2 - x^2 - y^2}} \quad \text{and} \quad \frac{\partial z}{\partial y} = \frac{-y}{\sqrt{a^2 - x^2 - y^2}}$$

do not exist on the boundary of the region R since $x^2 + y^2 = a^2$ there. In order to overcome this problem we evaluate the double integral over a slightly smaller circular region R_ρ of radius ρ, and then let ρ approach a. The computations are as follows:

$$\iint_\sigma dS = \lim_{\rho \to a} \iint_{R_\rho} \sqrt{\left(\frac{\partial z}{\partial x}\right)^2 + \left(\frac{\partial z}{\partial y}\right)^2 + 1} \; dA$$

$$= \lim_{\rho \to a} \iint_{R_\rho} \sqrt{\frac{x^2}{a^2 - x^2 - y^2} + \frac{y^2}{a^2 - x^2 - y^2} + 1} \; dA$$

$$= \lim_{\rho \to a} \iint_{R_\rho} \frac{a}{\sqrt{a^2 - x^2 - y^2}} \, dA$$

$$= \lim_{\rho \to a} \int_0^{2\pi} \int_0^{\rho} \frac{a}{\sqrt{a^2 - r^2}} r \, dr \, d\theta$$

$$= \lim_{\rho \to a} \int_0^{2\pi} -a\sqrt{a^2 - r^2} \, \Big]_{r=0}^{\rho} \, d\theta$$

$$= \lim_{\rho \to a} \int_0^{2\pi} (a^2 - a\sqrt{a^2 - \rho^2}) \, d\theta$$

$$= \lim_{\rho \to a} 2\pi(a^2 - a\sqrt{a^2 - \rho^2}) = 2\pi a^2 \qquad \blacktriangleleft$$

Note that (15) is the surface area of the hemisphere σ [see (8)], so we have shown that the surface area of a *sphere* of radius a is $4\pi a^2$.

We conclude this section by noting without proof that the standard properties of integrals also hold for surface integrals. More precisely:

$$\iint_\sigma (f + g) \, dS = \iint_\sigma f \, dS + \iint_\sigma g \, dS$$

$$\iint_{\sigma} (f - g)\, dS = \iint_{\sigma} f\, dS - \iint_{\sigma} g\, dS$$

$$\iint_{\sigma} kf\, dS = k \iint_{\sigma} f\, dS \qquad (k \text{ constant})$$

Figure 18.4.10

and, finally, if a surface σ is subdivided into finitely many parts, then a surface integral over σ can be computed by computing it over each part and adding the results. Thus, if σ is subdivided into two parts, σ_1 and σ_2 (Figure 18.4.10), then

$$\iint_{\sigma} f\, dS = \iint_{\sigma_1} f\, dS + \iint_{\sigma_2} f\, dS$$

▶ Exercise Set 18.4

In Exercises 1–4, find the mass of the given lamina assuming the density to be a constant δ_0.

1. The lamina that is the portion of the paraboloid $z = 1 - x^2 - y^2$ above the xy-plane.

2. The lamina that is the portion of the plane $2x + 2y + z = 8$ in the first octant.

3. The lamina that is the portion of the cylinder $x^2 + z^2 = 4$ that lies above the rectangle $R = \{(x, y) : 0 \le x \le 1, \ 0 \le y \le 4\}$ in the xy-plane.

4. The lamina that is the portion of the paraboloid $2z = x^2 + y^2$ inside the cylinder $x^2 + y^2 = 8$.

In Exercises 5–10, evaluate the surface integrals.

5. $\displaystyle\iint_{\sigma} z^2\, dS$, where σ is the portion of the cone $z = \sqrt{x^2 + y^2}$ between the planes $z = 1$ and $z = 2$.

6. $\displaystyle\iint_{\sigma} xyz\, dS$, where σ is the portion of the plane $x + y + z = 1$ lying in the first octant.

7. $\displaystyle\iint_{\sigma} x^2 y\, dS$, where σ is the portion of the cylinder $x^2 + z^2 = 1$ between the planes $y = 0$, $y = 1$, and above the xy-plane.

8. $\displaystyle\iint_{\sigma} (x^2 + y^2)z\, dS$, where σ is the portion of the sphere $x^2 + y^2 + z^2 = 4$ above the plane $z = 1$.

9. $\displaystyle\iint_{\sigma} (x + y + z)\, dS$, where σ is the portion of the plane $x + y = 1$ in the first octant between $z = 0$ and $z = 1$.

10. $\displaystyle\iint_{\sigma} (x + y)\, dS$, where σ is the portion of the plane $z = 6 - 2x - 3y$ in the first octant.

11. Evaluate

$$\iint_{\sigma} (x + y + z)\, dS$$

over the surface of the cube defined by the inequalities, $0 \leq x \leq 1, 0 \leq y \leq 1, 0 \leq z \leq 1$. [*Hint:* Integrate over each surface separately.]

12. Evaluate

$$\iint_\sigma zx^2 \, dS$$

over the portion of the cylinder $x^2 + y^2 = 1$ between $z = 0$ and $z = 1$. [*Hint:* Divide the surface into two parts and evaluate by projecting on the xz-plane.]

13. Evaluate

$$\iint_\sigma \sqrt{x^2 + y^2 + z^2} \, dS$$

over the portion of the cone $z = \sqrt{x^2 + y^2}$ below the plane $z = 1$.

14. Evaluate

$$\iint_\sigma (z + 1) \, dS$$

where σ is the upper hemisphere

$$z = \sqrt{1 - x^2 - y^2}.$$

15. Evaluate

$$\iint_\sigma (x^2 + y^2) \, dS$$

over the surface of the sphere $x^2 + y^2 + z^2 = a^2$. [*Hint:* Divide the sphere into two parts and integrate over each part separately.]

16. Find the mass of the lamina that is the portion of the cone

$$z = \sqrt{x^2 + y^2}$$

between $z = 1$ and $z = 4$ if the density is

$$\delta(x, y, z) = x^2 z$$

17. Find the mass of the lamina that is the portion of the surface $y^2 = 4 - z$ between the planes $x = 0$, $x = 3$, $y = 0$, and $y = 3$, if the density is $\delta(x, y, z) = y$.

18. Find the mass of the lamina that is the portion of the paraboloid $z = x^2 + y^2$ below the plane $z = 1$ if the density is $\delta(x, y, z) = \sqrt{x^2 + y^2}$.

In Exercises 19 and 20, set up, but do not evaluate, an iterated integral equal to the given surface integral by projecting σ on (a) the xy-plane, (b) the yz-plane, and (c) the xz-plane.

19. $\iint_\sigma xyz \, dS$, where σ is the portion of the plane

$$2x + 3y + 4z = 12 \text{ in the first octant.}$$

20. $\iint_\sigma xz \, dS$, where σ is the portion of the sphere

$$x^2 + y^2 + z^2 = a^2 \text{ in the first octant.}$$

In Exercises 21 and 22, set up, but do not evaluate, two different iterated integrals equal to the given surface integral.

21. $\iint_\sigma xyz \, dS$, where σ is the portion of the surface

$$y^2 = x \text{ between the planes } z = 0, z = 4, y = 1,$$
and $y = 2$.

22. $\iint_\sigma x^2 y \, dS$, where σ is the portion of the cylinder

$$y^2 + z^2 = a^2 \text{ in the first octant between the planes}$$
$x = 0$, $x = 9$, $z = y$, and $z = 2y$.

23. Prove: If a curved lamina has constant density δ_0, then its mass is the density times the surface area.

24. Show that the mass of the spherical lamina $x^2 + y^2 + z^2 = a^2$ is $2\pi a^3$ if the density at any point is equal to the distance from the point to the xy-plane.

18.5   SURFACE INTEGRALS OF VECTOR FUNCTIONS

In this section we shall study surface integrals with integrands that involve vector functions. Applications of the results we obtain here will be given in later sections.

ORIENTED SURFACES

Figure 18.5.1

If a surface σ has a nonzero normal vector at a point (x, y, z), then there are two oppositely directed *unit normal* vectors at that point (Figure 18.5.1). These vectors are described by various names, depending on the signs of their components. For example, if the unit normal vector **n** has a positive **k** component, then **n** points roughly in the upward direction and is called an ***upward unit normal*** (Figure 18.5.2). If **n** has a negative **k** component, then **n** points roughly downward and is called a ***downward unit normal***.

Table 18.5.1 explains the terminology used to describe unit normal vectors.

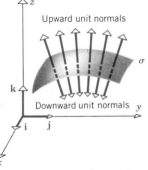

Figure 18.5.2

Example 1   If the vector

$$\mathbf{n} = \frac{1}{\sqrt{3}}\mathbf{i} - \frac{1}{\sqrt{3}}\mathbf{j} + \frac{1}{\sqrt{3}}\mathbf{k}$$

is normal to a surface σ, then **n** is an upward unit normal (positive **k** component), a left unit normal (negative **j** component), and a forward unit normal (positive **i** component).   ◄

To compute the unit normals to a surface σ given by an equation $z = z(x, y)$, we first rewrite this equation as

$$z - z(x, y) = 0$$

which is a level surface for the function

$$G(x, y, z) = z - z(x, y)$$

By Theorem 16.7.9, the gradient

$$\nabla G = -\frac{\partial z}{\partial x}\mathbf{i} - \frac{\partial z}{\partial y}\mathbf{j} + \mathbf{k} \tag{1}$$

is normal to this surface. To obtain a *unit* normal vector, we must normalize (1). Since

$$\|\nabla G\| = \sqrt{\left(\frac{\partial z}{\partial x}\right)^2 + \left(\frac{\partial z}{\partial y}\right)^2 + 1}$$

it follows that

Table 18.5.1

TERMINOLOGY	MATHEMATICAL MEANING	PICTURE
Upward unit normal	Positive $\mathbf{k}$ component	
Downward unit normal	Negative $\mathbf{k}$ component	
Right unit normal	Positive $\mathbf{j}$ component	
Left unit normal	Negative $\mathbf{j}$ component	
Forward unit normal	Positive $\mathbf{i}$ component	
Backward unit normal	Negative $\mathbf{i}$ component	

$$\frac{\nabla G}{\|\nabla G\|} = \frac{-\dfrac{\partial z}{\partial x}\mathbf{i} - \dfrac{\partial z}{\partial y}\mathbf{j} + \mathbf{k}}{\sqrt{\left(\dfrac{\partial z}{\partial x}\right)^2 + \left(\dfrac{\partial z}{\partial y}\right)^2 + 1}} \qquad (2)$$

is a unit normal vector to σ at the point (x, y, z). Moreover, because the $\mathbf{k}$ component is positive, (2) is the upward unit normal. To find the downward unit normal, we must multiply this vector by -1. This yields

$$-\frac{\nabla G}{\|\nabla G\|} = \frac{\frac{\partial z}{\partial x}\mathbf{i} + \frac{\partial z}{\partial y}\mathbf{j} - \mathbf{k}}{\sqrt{\left(\frac{\partial z}{\partial x}\right)^2 + \left(\frac{\partial z}{\partial y}\right)^2 + 1}} \tag{3}$$

For surfaces expressed in the form $y = y(x, z)$ or $x = x(y, z)$, the unit normal vectors are obtained by normalizing the gradients of the functions $y - y(x, z)$ and $x - x(y, z)$, respectively. The resulting formulas are listed in Table 18.5.2.

Table 18.5.2

EQUATION OF σ	NORMALS TO σ	
$z = z(x, y)$	Upward unit normal $$\frac{-\frac{\partial z}{\partial x}\mathbf{i} - \frac{\partial z}{\partial y}\mathbf{j} + \mathbf{k}}{\sqrt{\left(\frac{\partial z}{\partial x}\right)^2 + \left(\frac{\partial z}{\partial y}\right)^2 + 1}}$$	Downward unit normal $$\frac{\frac{\partial z}{\partial x}\mathbf{i} + \frac{\partial z}{\partial y}\mathbf{j} - \mathbf{k}}{\sqrt{\left(\frac{\partial z}{\partial x}\right)^2 + \left(\frac{\partial z}{\partial y}\right)^2 + 1}}$$
$y = y(x, z)$	Right unit normal $$\frac{-\frac{\partial y}{\partial x}\mathbf{i} + \mathbf{j} - \frac{\partial y}{\partial z}\mathbf{k}}{\sqrt{\left(\frac{\partial y}{\partial x}\right)^2 + \left(\frac{\partial y}{\partial z}\right)^2 + 1}}$$	Left unit normal $$\frac{\frac{\partial y}{\partial x}\mathbf{i} - \mathbf{j} + \frac{\partial y}{\partial z}\mathbf{k}}{\sqrt{\left(\frac{\partial y}{\partial x}\right)^2 + \left(\frac{\partial y}{\partial z}\right)^2 + 1}}$$
$x = x(y, z)$	Forward unit normal $$\frac{\mathbf{i} - \frac{\partial x}{\partial y}\mathbf{j} - \frac{\partial x}{\partial z}\mathbf{k}}{\sqrt{\left(\frac{\partial x}{\partial y}\right)^2 + \left(\frac{\partial x}{\partial z}\right)^2 + 1}}$$	Backward unit normal $$\frac{-\mathbf{i} + \frac{\partial x}{\partial y}\mathbf{j} + \frac{\partial x}{\partial z}\mathbf{k}}{\sqrt{\left(\frac{\partial x}{\partial y}\right)^2 + \left(\frac{\partial x}{\partial z}\right)^2 + 1}}$$

Example 2 Find the unit normal vector at the point $(1, 1, 1)$ to $x + 2y + 3z = 6$ shown in Figure 18.5.3.

Solution. Since the vector points above the xy-plane, to the front of the yz-plane, and to the right of the xz-plane, it can be calculated from the formula for an upward normal, a right normal, or a forward normal. For illustrative purposes we shall calculate the vector two ways—as an upward normal and as a forward normal.

To use the formula for an upward normal we rewrite the equation of the plane as

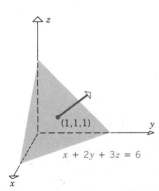

Figure 18.5.3

$$z = 2 - \frac{1}{3}x - \frac{2}{3}y$$

from which we obtain

$$\frac{\partial z}{\partial x} = -\frac{1}{3} \quad \text{and} \quad \frac{\partial z}{\partial y} = -\frac{2}{3}.$$

Thus, from Table 18.5.2, the upward unit normal is

$$\mathbf{n} = \frac{\frac{1}{3}\mathbf{i} + \frac{2}{3}\mathbf{j} + \mathbf{k}}{\sqrt{\left(-\frac{1}{3}\right)^2 + \left(-\frac{2}{3}\right)^2 + 1}} = \frac{3}{\sqrt{14}}\left(\frac{1}{3}\mathbf{i} + \frac{2}{3}\mathbf{j} + \mathbf{k}\right)$$

$$= \frac{1}{\sqrt{14}}\mathbf{i} + \frac{2}{\sqrt{14}}\mathbf{j} + \frac{3}{\sqrt{14}}\mathbf{k}$$

To use the formula for a forward unit normal we rewrite the equation of the plane as

$$x = 6 - 2y - 3z$$

from which we obtain

$$\frac{\partial x}{\partial y} = -2 \quad \text{and} \quad \frac{\partial x}{\partial z} = -3$$

Thus, from Table 18.5.2, the forward unit normal is

$$\mathbf{n} = \frac{\mathbf{i} + 2\mathbf{j} + 3\mathbf{k}}{\sqrt{(-2)^2 + (-3)^2 + 1}} = \frac{1}{\sqrt{14}}\mathbf{i} + \frac{2}{\sqrt{14}}\mathbf{j} + \frac{3}{\sqrt{14}}\mathbf{k}$$

which agrees with the result obtained above.   ◄

A surface σ is said to be **oriented** if a unit normal vector is constructed at each point of the surface in such a way that the vectors vary continuously (have no abrupt changes in direction) as we traverse any curve on the surface. The unit normal vectors are then said to form an **orientation** of the surface. For example, the surface σ in Figure 18.5.4a can be oriented by constructing an upward unit normal at each point (Figure 18.5.4b) or

Figure 18.5.4   (a)   (b)   (c)   (d)

a downward unit normal at each point (Figure 18.5.4c). However, the unit normals in Figure 18.5.4d do not form an orientation because the directions of these vectors change abruptly as we cross the curve drawn on the surface.

Most common surfaces can be oriented by constructing the unit normal vectors appropriately; such surfaces are called *orientable*. (A famous non-orientable surface, called a Möbius strip, is discussed in Exercise 22.) It is proved in advanced courses that an orientable surface has only two possible orientations. For example, the surface σ in Figure 18.5.4 can be oriented by upward normals or by downward normals, but any mixture of the two is not an orientation since abrupt direction changes in the normals are introduced.

Example 3 The two possible orientations of a sphere are by *inward normals* or *outward normals* (Figure 18.5.5).   ◄

Figure 18.5.5 ▷   Inward normals   Outward normals

SURFACE INTEGRALS   We now turn to the main topic of this section, surface integrals of vector-valued functions.

18.5.1   DEFINITION. If $\mathbf{F}(x, y, z) = f(x, y, z)\mathbf{i} + g(x, y, z)\mathbf{j} + h(x, y, z)\mathbf{k}$ has continuous components on the oriented surface σ, and if $\mathbf{n} = \mathbf{n}(x, y, z)$ is the unit normal vector of the orientation at (x, y, z), then

$$\iint_{\sigma} \mathbf{F} \cdot \mathbf{n} \, dS$$

is called the *flux integral of F over σ*, or the *surface integral of F over σ*, or the *surface integral of the normal component of F over σ*.

In Section 18.8, we shall see that surface integrals of vector-valued functions have important applications and physical interpretations. However, for now we shall concentrate on computing them.

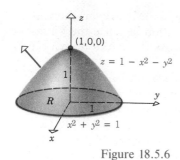

Figure 18.5.6

Example 4   Let σ be the portion of the surface $z = 1 - x^2 - y^2$ above the xy-plane oriented by upward normals (Figure 18.5.6), and let $F(x, y, z) = x\mathbf{i} + y\mathbf{j} + z\mathbf{k}$. Evaluate

$$\iint_\sigma \mathbf{F} \cdot \mathbf{n}\, dS$$

Solution.   From the first entry in Table 18.5.2, the upward unit normal to the surface

$$z = 1 - x^2 - y^2 \tag{4}$$

is given by

$$\mathbf{n} = \frac{-\dfrac{\partial z}{\partial x}\mathbf{i} - \dfrac{\partial z}{\partial y}\mathbf{j} + \mathbf{k}}{\sqrt{\left(\dfrac{\partial z}{\partial x}\right)^2 + \left(\dfrac{\partial z}{\partial y}\right)^2 + 1}}$$

Thus, from Formula (10) in Theorem 18.4.2,

$$\iint_\sigma \mathbf{F} \cdot \mathbf{n}\, dS = \iint_R \mathbf{F} \cdot \left[\frac{-\dfrac{\partial z}{\partial x}\mathbf{i} - \dfrac{\partial z}{\partial y}\mathbf{j} + \mathbf{k}}{\sqrt{\left(\dfrac{\partial z}{\partial x}\right)^2 + \left(\dfrac{\partial z}{\partial y}\right)^2 + 1}}\right]\sqrt{\left(\dfrac{\partial z}{\partial x}\right)^2 + \left(\dfrac{\partial z}{\partial y}\right)^2 + 1}\, dA$$

or, on canceling square roots,

$$\iint_\sigma \mathbf{F} \cdot \mathbf{n}\, dS = \iint_R \mathbf{F} \cdot \left(-\frac{\partial z}{\partial x}\mathbf{i} - \frac{\partial z}{\partial y}\mathbf{j} + \mathbf{k}\right) dA \tag{5}$$

where R is the region shown in Figure 18.5.6. From (4) and the given formula for $\mathbf{F}$, it follows that

$$\mathbf{F} \cdot \left(-\frac{\partial z}{\partial x}\mathbf{i} - \frac{\partial z}{\partial y}\mathbf{j} + \mathbf{k}\right) = (x\mathbf{i} + y\mathbf{j} + z\mathbf{k}) \cdot (2x\mathbf{i} + 2y\mathbf{j} + \mathbf{k})$$

$$= 2x^2 + 2y^2 + z$$
$$= 2x^2 + 2y^2 + (1 - x^2 - y^2)$$
$$= x^2 + y^2 + 1$$

Substituting this expression in (5) yields

$$\iint_\sigma \mathbf{F} \cdot \mathbf{n}\, dS = \iint_R (x^2 + y^2 + 1)\, dA$$

$$= \int_0^{2\pi} \int_0^1 (r^2 + 1)r\, dr\, d\theta \quad \left[\begin{array}{l}\text{Using polar coordinates} \\ \text{to evaluate the integral.}\end{array}\right]$$

$$= \int_0^{2\pi} \left(\frac{3}{4}\right) d\theta = \frac{3\pi}{2} \quad \blacktriangleleft$$

EVALUATING
SURFACE INTEGRALS

The procedure used to derive Formula (5) in the last example can be used to obtain the following general results.

> **18.5.2** THEOREM. *If the surface σ is given by an equation of the form $z = z(x, y)$, and if R is the projection of the surface on the xy-plane, then*
>
> (a) $$\iint_\sigma \mathbf{F} \cdot \mathbf{n}\, dS = \iint_R \mathbf{F} \cdot \left(-\frac{\partial z}{\partial x}\mathbf{i} - \frac{\partial z}{\partial y}\mathbf{j} + \mathbf{k}\right) dA \qquad (6)$$
>
> *if σ is oriented by upward normals.*
>
> (b) $$\iint_\sigma \mathbf{F} \cdot \mathbf{n}\, dS = \iint_R \mathbf{F} \cdot \left(\frac{\partial z}{\partial x}\mathbf{i} + \frac{\partial z}{\partial y}\mathbf{j} - \mathbf{k}\right) dA \qquad (7)$$
>
> *if σ is oriented by downward normals.*

Similar formulas that apply to surfaces that are expressed in the form $x = x(y, z)$ and $y = y(x, z)$ are discussed in the exercises.

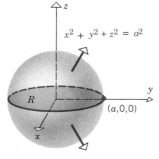

$x^2 + y^2 + z^2 = a^2$

R

$(a, 0, 0)$

Figure 18.5.7

Example 5 Let σ be the sphere $x^2 + y^2 + z^2 = a^2$ oriented by outward normals (Figure 18.5.7), and let $\mathbf{F}(x, y, z) = z\mathbf{k}$. Evaluate

$$\iint_\sigma \mathbf{F} \cdot \mathbf{n}\, dS$$

Solution. On the upper hemisphere the outward unit normal is upward, while on the lower hemisphere it is downward. Since different formulas for these normals apply on the two hemispheres, it is desirable to write

$$\iint_\sigma \mathbf{F} \cdot \mathbf{n}\, dS = \iint_{\sigma_1} \mathbf{F} \cdot \mathbf{n}\, dS + \iint_{\sigma_2} \mathbf{F} \cdot \mathbf{n}\, dS \qquad (8)$$

where σ_1 is the upper hemisphere and σ_2 is the lower hemisphere.
The upper hemisphere σ_1 has the equation

$$z = \sqrt{a^2 - x^2 - y^2} \qquad (9)$$

so that (6) yields

$$\iint_{\sigma_1} \mathbf{F} \cdot \mathbf{n} \, dS = \iint_R (z\mathbf{k}) \cdot \left(\frac{x}{\sqrt{a^2 - x^2 - y^2}} \mathbf{i} + \frac{y}{\sqrt{a^2 - x^2 - y^2}} \mathbf{j} + \mathbf{k} \right) dA$$

$$= \iint_R z \, dA = \iint_R \sqrt{a^2 - x^2 - y^2} \, dA \qquad \begin{bmatrix} R \text{ is the region shown} \\ \text{in Figure 18.5.7.} \end{bmatrix}$$

$$= \int_0^{2\pi} \int_0^a \sqrt{a^2 - r^2} \, r \, dr \, d\theta$$

$$= \int_0^{2\pi} -\frac{1}{3}(a^2 - r^2)^{3/2} \Big]_0^a \, d\theta$$

$$= \int_0^{2\pi} \frac{1}{3} a^3 \, d\theta = \frac{2\pi a^3}{3}$$

The lower hemisphere σ_2 has the equation

$$z = -\sqrt{a^2 - x^2 - y^2}$$

so that (7) yields

$$\iint_{\sigma_2} \mathbf{F} \cdot \mathbf{n} \, dS = \iint_R (z\mathbf{k}) \cdot \left(\frac{x}{\sqrt{a^2 - x^2 - y^2}} \mathbf{i} + \frac{y}{\sqrt{a^2 - x^2 - y^2}} \mathbf{j} - \mathbf{k} \right) dA$$

$$= \iint_R -z \, dA = \iint_R \sqrt{a^2 - x^2 - y^2} \, dA$$

$$= \frac{2\pi a^3}{3} \qquad \begin{bmatrix} \text{The computations} \\ \text{are identical to} \\ \text{those above.} \end{bmatrix}$$

Thus, from (8)

$$\iint_{\sigma} \mathbf{F} \cdot \mathbf{n} \, dS = \frac{2\pi a^3}{3} + \frac{2\pi a^3}{3} = \frac{4\pi a^3}{3} \qquad \blacktriangleleft$$

▶ Exercise Set 18.5

1. Use three different formulas in Table 18.5.2 to calculate the unit normal to $2x + 3y + 4z = 9$ at $(1, 1, 1)$ that has positive components.

2. Use three different formulas in Table 18.5.2 to calculate the unit normal to $x^2 + y^2 + z^2 = 9$ at $(2, 1, -2)$ that points below the xy-plane.

3. In parts (a)–(c), use any appropriate formula to calculate the indicated unit normal.

   (a) The unit normal to $z = x^2 + y^2$ at $(1, 2, 5)$ that points toward the z-axis;

   (b) the unit normal to $z = \sqrt{x^2 + y^2}$ at $(-3, 4, 5)$ that points toward the xz-plane;

(c) the unit normal to the cylinder $x^2 + z^2 = 25$ at $(3, 2, -4)$ that points away from the xy-plane.

4. In each part, use any appropriate formula to calculate the indicated unit normal.
   (a) The unit normal to the surface $y^2 = x$ at $(1, 1, 2)$ that points toward the xz-plane;
   (b) the unit normal to the hyperbolic paraboloid $y = z^2 - x^2$ at $(1, 3, 2)$ that points toward the yz-plane;
   (c) the unit normal to the cone $x^2 = y^2 + z^2$ at $(\sqrt{2}, -1, -1)$ that points away from the xy-plane.

In Exercises 5–15, evaluate $\iint\limits_{\sigma} \mathbf{F} \cdot \mathbf{n}\, dS$.

5. $\mathbf{F}(x, y, z) = x\mathbf{i} + y\mathbf{j} + 2z\mathbf{k}$; σ is the portion of the surface $z = 1 - x^2 - y^2$ above the xy-plane oriented by upward normals.

6. $\mathbf{F}(x, y, z) = (x + y)\mathbf{i} + (y + z)\mathbf{j} + (z + x)\mathbf{k}$; σ is the portion of the plane $x + y + z = 1$ in the first octant oriented by unit normals with positive components.

7. $\mathbf{F}(x, y, z) = z^2\mathbf{k}$; σ is the upper hemisphere $z = \sqrt{1 - x^2 - y^2}$ oriented by upward unit normals.

8. $\mathbf{F}(x, y, z) = x^2\mathbf{i} + yx\mathbf{j} + zx\mathbf{k}$; σ is the portion of the plane $6x + 3y + 2z = 6$ in the first octant oriented by unit normals with positive components.

9. $\mathbf{F}(x, y, z) = x\mathbf{i} + y\mathbf{j} + z\mathbf{k}$; σ is the upper hemisphere $z = \sqrt{9 - x^2 - y^2}$ oriented by upward unit normals.

10. $\mathbf{F}(x, y, z) = \mathbf{i} + \mathbf{j} + \mathbf{k}$; σ is the portion of the cone $z = \sqrt{x^2 + y^2}$ below the plane $z = 1$, oriented by downward unit normals.

11. $\mathbf{F}(x, y, z) = x\mathbf{i} + y\mathbf{j} + 2z\mathbf{k}$; σ is the portion of the cone $z^2 = x^2 + y^2$ between the planes $z = 1$ and $z = 2$, oriented by upward unit normals.

12. $\mathbf{F}(x, y, z) = y\mathbf{j} + \mathbf{k}$; σ is the portion of the paraboloid $z = x^2 + y^2$ below the plane $z = 4$, oriented by downward unit normals.

13. $\mathbf{F}(x, y, z) = x\mathbf{k}$; σ is the portion of the paraboloid $z = x^2 + y^2$ below the plane $z = y$, oriented by downward unit normals.

14. $\mathbf{F}(x, y, z) = x\mathbf{i} + y\mathbf{j} + z\mathbf{k}$; σ is the portion of the cylinder $z^2 = 1 - x^2$ between the planes $y = 1$ and $y = -2$, oriented by outward unit normals.

15. $\mathbf{F}(x, y, z) = x\mathbf{i} + y\mathbf{j} + z\mathbf{k}$; σ is the sphere $x^2 + y^2 + z^2 = a^2$ oriented by outward unit normals.

16. Let σ be the surface of the cube bounded by the planes $x = \pm 1, y = \pm 1, z = \pm 1$, oriented by outward unit normals. In each part, evaluate the flux integral of $\mathbf{F}$ over σ.
   (a) $\mathbf{F}(x, y, z) = x\mathbf{i}$
   (b) $\mathbf{F}(x, y, z) = x\mathbf{i} + y\mathbf{j} + z\mathbf{k}$
   (c) $\mathbf{F}(x, y, z) = x^2\mathbf{i} + y^2\mathbf{j} + z^2\mathbf{k}$

17. Show that reversing the orientation of σ reverses the sign of
$$\iint\limits_{\sigma} \mathbf{F} \cdot \mathbf{n}\, dS$$

18. Derive Formulas (6) and (7) in Theorem 18.5.2. [*Hint:* Review the derivation of (5) in Example 4.]

19. Obtain the analogs of Formulas (6) and (7) for:
   (a) surfaces of the form $x = x(y, z)$;
   (b) surfaces of the form $y = y(x, z)$.

20. Evaluate $\iint\limits_{\sigma} \mathbf{F} \cdot \mathbf{n}\, dS$ if σ is the portion of the paraboloid $x = y^2 + z^2$ with $x \leq 1$ and $z \geq 0$ oriented by backward unit normals and $\mathbf{F}(x, y, z) = y\mathbf{i} - z\mathbf{j} + 8\mathbf{k}$. [*Hint:* See Exercise 19(a).]

21. Evaluate $\iint\limits_{\sigma} \mathbf{F} \cdot \mathbf{n}\, dS$ if σ is the hemisphere $y = \sqrt{1 - x^2 - z^2}$ oriented by right unit normals and $\mathbf{F}(x, y, z) = x\mathbf{i} + y\mathbf{j} + z\mathbf{k}$. [*Hint:* See Exercise 19(b).]

22. The best-known example of a nonorientable surface is the *Möbius strip*, which can be visualized by taking a band of paper, twisting it once, and gluing the ends together (Figures 18.5.8a and 18.5.8b). In Figure 18.5.8c, we have tried to construct unit normal vectors whose directions vary continuously moving counterclockwise around the dashed curve starting and finishing on the vertical line.
   (a) Explain why these vectors are not part of an orientation of the surface.
   (b) Explain why the surface is not orientable.

(a)

(b)

(c)

Figure 18.5.8

18.6   THE DIVERGENCE THEOREM

In this section we shall discuss an important three-dimensional analog of Green's Theorem.

We begin with some terminology.

18.6.1   DEFINITION.   If $\mathbf{F}(x, y, z) = f(x, y, z)\mathbf{i} + g(x, y, z)\mathbf{j} + h(x, y, z)\mathbf{k}$, then we define the **divergence of F,** written div **F**, by

$$\operatorname{div} \mathbf{F} = \frac{\partial f}{\partial x} + \frac{\partial g}{\partial y} + \frac{\partial h}{\partial z}$$

Example 1   The divergence of

$$\mathbf{F}(x, y, z) = xy^2z^2\mathbf{i} + (3xz^2 + y^4)\mathbf{j} + x^3y^2z\mathbf{k}$$

is

$$\operatorname{div} \mathbf{F} = \frac{\partial}{\partial x}(xy^2z^2) + \frac{\partial}{\partial y}(3xz^2 + y^4) + \frac{\partial}{\partial z}(x^3y^2z)$$

$$= y^2z^2 + 4y^3 + x^3y^2 \quad \blacktriangleleft$$

The following result, called the **Divergence Theorem** or **Gauss'* Theorem,**

---

*KARL FRIEDRICH GAUSS (1777–1855)—German mathematician and scientist. Sometimes called the "prince of mathematicians," Gauss ranks with Newton and Archimedes as one of the three greatest mathematicians who ever lived. His father, a laborer, was an uncouth but honest man who would have liked Gauss to take up a trade such as gardening or bricklaying; but the boy's genius for mathematics was not be be denied. In the entire history of mathematics there may never have been a child so precocious as Gauss—by his own account he worked out the rudiments of arithmetic before he could talk. One day, before he was even three years old, his genius became apparent to his parents in a very dramatic way.

His father was preparing the weekly payroll for the laborers under his charge while the boy watched quietly from a corner. At the end of the long and tedious calculation, Gauss informed his father that there was an error in the result and stated the answer, which he had worked out in his head. To the astonishment of his parents, a check of the computations showed Gauss to be correct!

For his elementary education Gauss was enrolled in a squalid school run by a man named Büttner whose main teaching technique was thrashing. Büttner was in the habit of assigning long addition problems which, unknown to his students, were arithmetic progressions that he could sum up using formulas. On the first day that Gauss entered the arithmetic class, the students were asked to sum the numbers from 1 to 100. But no sooner had Büttner stated the problem than Gauss turned over his slate and exclaimed in his peasant dialect, "Ligget se'." (Here it lies.) For nearly an hour Büttner glared at Gauss, who sat with folded hands while his classmates toiled away. When Büttner examined the slates at the end of the period, Gauss' slate contained a single number, 5050—the only correct solution in the class.

To his credit, Büttner recognized the genius of Gauss and with the help of his assistant, John Bartels, had him brought to the attention of Karl Wilhelm Ferdinand, Duke of Brunswick. The shy and awkward boy, who was then fourteen, so captivated the Duke that he subsidized him through preparatory school, college, and the early part of his career.

From 1795 to 1798 Gauss studied mathematics at the University of Göttingen, receiving his degree in absentia from the University of Helmstadt. For his dissertation, he gave the first complete proof of the fundamental theorem of algebra, which states that every polynomial equation has as many solutions as its degree. At age 19 he solved a problem that baffled Euclid, inscribing a regular polygon of seventeen sides in a circle using straightedge and compass; and in 1801, at age 24, he published his first masterpiece, *Disquisitiones Arithmeticae,* considered by many to be one of the most brilliant achievements in mathematics. In that paper Gauss systematized the study of number theory (properties of the integers) and formulated the basic concepts that form the foundation of that subject.

In the same year that the *Disquisitiones* was published, Gauss again applied his phenomenal computational skills in a dramatic way. The astronomer Giuseppi Piazzi had observed the asteroid Ceres for $\frac{1}{40}$ of its orbit, but lost it in the sun. Using only three observations and the "method of least squares" that he had developed in 1795, Gauss computed the orbit with such accuracy that astronomers had no trouble relocating it the following year. This achievement brought him instant recognition as the premier mathematician in Europe, and in 1807 he was made Professor of Astronomy and head of the astronomical observatory at Göttingen.

In the years that followed, Gauss revolutionized mathematics by bringing to it standards of precision and rigor undreamed of by his predecessors. He had a passion for perfection that drove him to polish and rework his papers rather than publish less finished work in greater numbers—his favorite saying was "Pauca, sed matura" (Few, but ripe). As a result, many of his important discoveries were squirreled away in diaries that remained unpublished until years after his death.

Among his myriad achievements, Gauss discovered the Gaussian or "bell-shaped" error curve fundamental in probability, gave the first geometric interpretation of complex numbers and established their fundamental role in mathematics, developed methods of characterizing surfaces intrinsically by means of the curves that they contain, developed the theory of conformal (angle-preserving) maps, and discovered non-Euclidean geometry 30 years before the ideas were published by others. In physics he made major contributions to the theory of lenses and capillary action, and with Wilhelm Weber he did fundamental work in electromagnetism. Gauss invented the heliotrope, bifilar magnetometer, and an electrotelegraph.

Gauss was deeply religious and aristocratic in demeanor. He mastered foreign languages with ease, read extensively, and enjoyed minerology and botany as hobbies. He disliked teaching and was usually cool and discouraging to other mathematicians, possibly because he had already anticipated their work. It has been said that if Gauss had published all of his discoveries, the current state of mathematics would be advanced by 50 years. He was without a doubt the greatest mathematician of the modern era.

Divergence Theorem

18.6.2   THEOREM. *Let G be a solid with surface σ oriented by outward unit normals. If*

$$\mathbf{F}(x, y, z) = f(x, y, z)\mathbf{i} + g(x, y, z)\mathbf{j} + h(x, y, z)\mathbf{k}$$

where f, g, and h have continuous first partial derivatives on some open set containing G, then

$$\iint_{\sigma} \mathbf{F} \cdot \mathbf{n} \, dS = \iiint_{G} \operatorname{div} \mathbf{F} \, dV \tag{1}$$

The proof of this theorem for a general solid G is too difficult to present here. However, we can give a proof for the special case where G is a simple solid (see Figure 17.5.4 and the discussion preceding it).

Proof (for simple solids).   Let G be a simple solid with upper surface $z = g_2(x, y)$, lower surface $z = g_1(x, y)$, and projection R on the xy-plane. Let σ_1 denote the lower surface, σ_2 the upper surface, and σ_3 the lateral surface (Figure 18.6.1*a*). If the upper surface and lower surface meet as in Figure 18.6.1*b*, then there is no lateral surface σ_3. Our proof will allow for both cases shown in those figures.

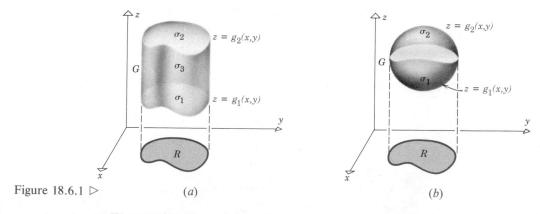

Figure 18.6.1 ▷   (a)   (b)

We want to show that

$$\iint_{\sigma} \mathbf{F} \cdot \mathbf{n} \, dS = \iiint_{G} \operatorname{div} \mathbf{F} \, dV$$

or equivalently

$$\iint_{\sigma} [f(x, y, z)\mathbf{i} + g(x, y, z)\mathbf{j} + h(x, y, z)\mathbf{k}] \cdot \mathbf{n} \, dS = \iiint_{G} \left(\frac{\partial f}{\partial x} + \frac{\partial g}{\partial y} + \frac{\partial h}{\partial z} \right) dV$$

To prove this, it suffices to prove the following three equalities:

$$\iint_{\sigma} [f(x, y, z)\mathbf{i} \cdot \mathbf{n}] \, dS = \iiint_{G} \frac{\partial f}{\partial x} \, dV \tag{2a}$$

$$\iint_{\sigma} [g(x, y, z)\mathbf{j} \cdot \mathbf{n}] \, dS = \iiint_{G} \frac{\partial g}{\partial y} \, dV \tag{2b}$$

$$\iint_{\sigma} [h(x, y, z)\mathbf{k} \cdot \mathbf{n}] \, dS = \iiint_{G} \frac{\partial h}{\partial z} \, dV \tag{2c}$$

Since the proofs of all three formulas are similar, we shall prove only the third.

It follows from Theorem 17.5.2 that

$$\iiint_{G} \frac{\partial h}{\partial z} \, dV = \iint_{R} \left[\int_{g_1(x,y)}^{g_2(x,y)} \frac{\partial h}{\partial z} \, dz \right] dA = \iint_{R} \left[h(x, y, z) \right]_{z = g_1(x,y)}^{g_2(x,y)} dA$$

so

$$\iiint_{G} \frac{\partial h}{\partial z} \, dV = \iint_{R} [h(x, y, g_2(x, y)) - h(x, y, g_1(x, y))] \, dA \tag{3}$$

We shall evaluate the surface integral in (2c) by integrating over each surface of G separately. If there is a lateral surface σ_3, then at each point of this surface $\mathbf{n} \cdot \mathbf{k} = 0$ since $\mathbf{n}$ is horizontal and $\mathbf{k}$ is vertical. Thus,

$$\iint_{\sigma_3} [h(x, y, z)\mathbf{k} \cdot \mathbf{n}] \, dS = 0$$

Therefore, regardless of whether G has a lateral surface or not, we can write

$$\iint_{\sigma} [h(x, y, z)\mathbf{k} \cdot \mathbf{n}] \, dS = \iint_{\sigma_1} [h(x, y, z)\mathbf{k} \cdot \mathbf{n}] \, dS + \iint_{\sigma_2} [h(x, y, z)\mathbf{k} \cdot \mathbf{n}] \, dS \tag{4}$$

On the upper surface σ_2, the outer normal is an upward normal, and on the lower surface σ_1 the outer normal is a downward normal. Thus, Theorem 18.5.2 implies that

$$\iint\limits_{\sigma_2} [h(x, y, z)\mathbf{k} \cdot \mathbf{n}] \, dS = \iint\limits_{R} \left[h(x, y, g_2(x, y))\mathbf{k} \cdot \left(-\frac{\partial z}{\partial x}\mathbf{i} - \frac{\partial z}{\partial y}\mathbf{j} + \mathbf{k} \right) \right] dA$$

$$= \iint\limits_{R} [h(x, y, g_2(x, y))] \, dA \tag{5}$$

and

$$\iint\limits_{\sigma_1} [h(x, y, z)\mathbf{k} \cdot \mathbf{n}] \, dS = \iint\limits_{R} \left[h(x, y, g_1(x, y))\mathbf{k} \cdot \left(\frac{\partial z}{\partial x}\mathbf{i} + \frac{\partial z}{\partial y}\mathbf{j} - \mathbf{k} \right) \right] dA$$

$$= -\iint\limits_{R} [h(x, y, g_1(x, y))] \, dA \tag{6}$$

Substituting (5) and (6) into (4) and combining the terms into a single integral yields

$$\iint\limits_{\sigma} [h(x, y, z)\mathbf{k} \cdot \mathbf{n}] \, dS = \iint\limits_{R} [h(x, y, g_2(x, y)) - h(x, y, g_1(x, y))] \, dA \tag{7}$$

Equation (2c) now follows from (3) and (7).   ■

Example 2   Let σ be the sphere $x^2 + y^2 + z^2 = a^2$ oriented by outward normals and let $\mathbf{F}(x, y, z) = z\mathbf{k}$. Use the Divergence Theorem to evaluate

$$\iint\limits_{\sigma} \mathbf{F} \cdot \mathbf{n} \, dS$$

Solution.   Let G be the spherical solid enclosed by σ. Since

$$\operatorname{div} \mathbf{F} = \frac{\partial z}{\partial z} = 1$$

it follows from (1) that

$$\iint\limits_{\sigma} \mathbf{F} \cdot \mathbf{n} \, dS = \iiint\limits_{G} dV = \text{volume of } G = \frac{4\pi a^3}{3} \quad \blacktriangleleft$$

REMARK.   Notice how much simpler this solution is than the direct cal-culation of $\iint\limits_{\sigma} \mathbf{F} \cdot \mathbf{n} \, dS$ in Example 5 of Section 18.5.

Figure 18.6.2

Example 3   Let G be the cube in the first octant shown in Figure 18.6.2, and let σ be its surface. Use the Divergence Theorem to compute

$$\iint_{\sigma} \mathbf{F} \cdot \mathbf{n}\, dS$$

where $\mathbf{F}(x, y, z) = 2x\mathbf{i} + 3y\mathbf{j} + z^2\mathbf{k}$, and $\mathbf{n}$ is the outward unit normal.

Solution.   Since

$$\operatorname{div} \mathbf{F} = \frac{\partial}{\partial x}(2x) + \frac{\partial}{\partial y}(3y) + \frac{\partial}{\partial z}(z^2) = 5 + 2z$$

it follows from (1) that

$$\iint_{\sigma} \mathbf{F} \cdot \mathbf{n}\, dS = \iiint_{G} (5 + 2z)\, dV = \int_0^1 \int_0^1 \int_0^1 (5 + 2z)\, dz\, dy\, dx$$

$$= \int_0^1 \int_0^1 \left[\frac{1}{4}(5 + 2z)^2 \right]_{z=0}^{1} dy\, dx = \int_0^1 \int_0^1 6\, dy\, dx = 6 \quad \blacktriangleleft$$

Example 4   Let G be the cylindrical solid bounded by $x^2 + y^2 = 9$, $z = 0$, and $z = 2$ (Figure 18.6.3), and let σ be its surface. Use the Divergence Theorem to evaluate

$$\iint_{\sigma} \mathbf{F} \cdot \mathbf{n}\, dS$$

Figure 18.6.3 ▷

where $\mathbf{F}(x, y, z) = x^3\mathbf{i} + y^3\mathbf{j} + z^2\mathbf{k}$, and $\mathbf{n}$ is the outward unit normal to σ.

Solution.   Since

$$\operatorname{div} \mathbf{F} = \frac{\partial}{\partial x}(x^3) + \frac{\partial}{\partial y}(y^3) + \frac{\partial}{\partial z}(z^2) = 3x^2 + 3y^2 + 2z$$

it follows from (1) that

$$\iint_\sigma \mathbf{F} \cdot \mathbf{n} \, dS = \iiint_G (3x^2 + 3y^2 + 2z) \, dV$$

We shall use cylindrical coordinates to evaluate the triple integral. We obtain

$$\iint_\sigma \mathbf{F} \cdot \mathbf{n} \, dS = \int_0^{2\pi} \int_0^3 \int_0^2 (3r^2 + 2z)r \, dz \, dr \, d\theta$$

$$= \int_0^{2\pi} \int_0^3 \left[3r^3 z + z^2 r \right]_{z=0}^2 dr \, d\theta$$

$$= \int_0^{2\pi} \int_0^3 (6r^3 + 4r) \, dr \, d\theta$$

$$= \int_0^{2\pi} \left[\frac{3r^4}{2} + 2r^2 \right]_0^3 d\theta$$

$$= \int_0^{2\pi} \frac{279}{2} \, d\theta = 279\pi \quad \blacktriangleleft$$

$z = \sqrt{a^2 - x^2 - y^2}$

Figure 18.6.4

Example 5 Let G be the solid bounded above by the hemisphere $z = \sqrt{a^2 - x^2 - y^2}$ and below by the plane $z = 0$ (Figure 18.6.4). Find

$$\iint_\sigma \mathbf{F} \cdot \mathbf{n} \, dS$$

where $\mathbf{F}(x, y, z) = x^3\mathbf{i} + y^3\mathbf{j} + z^3\mathbf{k}$, σ is the surface of G, and $\mathbf{n}$ is the outer unit normal.

Solution. Since

$$\text{div } \mathbf{F} = \frac{\partial}{\partial x}(x^3) + \frac{\partial}{\partial y}(y^3) + \frac{\partial}{\partial z}(z^3) = 3x^2 + 3y^2 + 3z^2$$

it follows from (1) that

$$\iint_\sigma \mathbf{F} \cdot \mathbf{n} \, dS = \iiint_G (3x^2 + 3y^2 + 3z^2) \, dV$$

We shall use spherical coordinates to evaluate the triple integral. We obtain

$$\iint_\sigma \mathbf{F} \cdot \mathbf{n} \, dS = \int_0^{2\pi} \int_0^{\pi/2} \int_0^a (3\rho^2)\rho^2 \sin\phi \, d\rho \, d\phi \, d\theta$$

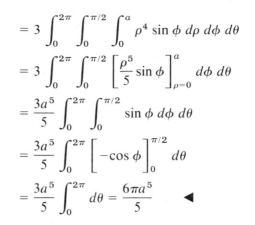

$$= 3 \int_0^{2\pi} \int_0^{\pi/2} \int_0^a \rho^4 \sin \phi \, d\rho \, d\phi \, d\theta$$

$$= 3 \int_0^{2\pi} \int_0^{\pi/2} \left[\frac{\rho^5}{5} \sin \phi \right]_{\rho=0}^a d\phi \, d\theta$$

$$= \frac{3a^5}{5} \int_0^{2\pi} \int_0^{\pi/2} \sin \phi \, d\phi \, d\theta$$

$$= \frac{3a^5}{5} \int_0^{2\pi} \left[-\cos \phi \right]_0^{\pi/2} d\theta$$

$$= \frac{3a^5}{5} \int_0^{2\pi} d\theta = \frac{6\pi a^5}{5} \quad \blacktriangleleft$$

▶ Exercise Set 18.6

In Exercises 1–5, find the divergence of **F**.

1. $\mathbf{F}(x, y, z) = xz^3\mathbf{i} + 2y^4x^2\mathbf{j} + 5z^2y\mathbf{k}$.

2. $\mathbf{F}(x, y, z) = 7y^3z^2\mathbf{i} - 8x^2z^5\mathbf{j} - 3xy^4\mathbf{k}$.

3. $\mathbf{F}(x, y, z) = e^{xy}\mathbf{i} - \cos y\mathbf{j} + \sin^2 z\mathbf{k}$.

4. $\mathbf{F}(x, y, z) = \dfrac{1}{\sqrt{x^2 + y^2 + z^2}}(x\mathbf{i} + y\mathbf{j} + z\mathbf{k})$.

5. $\mathbf{F}(x, y, z) = \ln x\mathbf{i} + e^{xyz}\mathbf{j} + \tan^{-1}\left(\dfrac{z}{x}\right)\mathbf{k}$.

In Exercises 6–17, use the Divergence Theorem to evaluate $\displaystyle\iint_\sigma \mathbf{F} \cdot \mathbf{n} \, dS$, where **n** is the outer unit normal to σ.

6. $\mathbf{F}(x, y, z) = 2x\mathbf{i} + 2y\mathbf{j} + 2z\mathbf{k}$; σ is the sphere $x^2 + y^2 + z^2 = 9$.

7. $\mathbf{F}(x, y, z) = 4x\mathbf{i} - 3y\mathbf{j} + 7z\mathbf{k}$; σ is the surface of the cube bounded by the coordinate planes and the planes $x = 1$, $y = 1$, and $z = 1$.

8. $\mathbf{F}(x, y, z) = z^3\mathbf{i} - x^3\mathbf{j} + y^3\mathbf{k}$; σ is the sphere $x^2 + y^2 + z^2 = a^2$.

9. $\mathbf{F}(x, y, z) = (x - z)\mathbf{i} + (y - x)\mathbf{j} + (z - y)\mathbf{k}$; σ is the surface of the cylindrical solid bounded by $x^2 + y^2 = a^2$, $z = 0$, and $z = 1$.

10. $\mathbf{F}(x, y, z) = x\mathbf{i} + y\mathbf{j} + z\mathbf{k}$; σ is the surface of the solid bounded by the paraboloid $z = 1 - x^2 - y^2$ and the xy-plane.

11. $\mathbf{F}(x, y, z) = x^3\mathbf{i} + y^3\mathbf{j} + z^3\mathbf{k}$; σ is the surface of the cylindrical solid bounded by $x^2 + y^2 = 4$, $z = 0$, and $z = 3$.

12. $\mathbf{F}(x, y, z) = (x^3 - e^y)\mathbf{i} + (y^3 + \sin z)\mathbf{j} + (z^3 - xy)\mathbf{k}$; σ is the surface of the solid bounded by $z = \sqrt{4 - x^2 - y^2}$ and the xy-plane. [*Hint:* Use spherical coordinates.]

13. $\mathbf{F}(x, y, z) = (x^2 + y)\mathbf{i} + xy\mathbf{j} - (2xz + y)\mathbf{k}$; σ is the surface of the tetrahedron in the first octant bounded by $x + y + z = 1$ and the coordinate planes.

14. $\mathbf{F}(x, y, z) = 2xz\mathbf{i} + yz\mathbf{j} + z^2\mathbf{k}$; σ is the surface of the hemispherical solid bounded by $z = \sqrt{a^2 - x^2 - y^2}$ and the xy-plane.

15. $\mathbf{F}(x, y, z) = x^2\mathbf{i} + y^2\mathbf{j} + z^2\mathbf{k}$; σ is the surface of the conical solid bounded by $z = \sqrt{x^2 + y^2}$ and $z = 1$.

16. $\mathbf{F}(x, y, z) = x^2y\mathbf{i} - xy^2\mathbf{j} + (z + 2)\mathbf{k}$; σ is the surface of the solid bounded above by the plane $z = 2x$ and below by the paraboloid $z = x^2 + y^2$.

17. $\mathbf{F}(x, y, z) = x^3\mathbf{i} + x^2y\mathbf{j} + xy\mathbf{k}$; σ is the surface of the solid bounded by $z = 4 - x^2$, $y + z = 5$, $z = 0$, and $y = 0$.

In Exercises 18–20, verify the Divergence Theorem.

18. $\mathbf{F}(x, y, z) = x^2\mathbf{i} + y^2\mathbf{j} + z^2\mathbf{k}$; σ is the surface of the cube bounded by $x = 1$, $y = 1$, $z = 1$, and the coordinate planes.

19. $\mathbf{F}(x, y, z) = x\mathbf{i} + y\mathbf{j} + z\mathbf{k}$; σ is the sphere $x^2 + y^2 + z^2 = a^2$.

20. $\mathbf{F}(x, y, z) = (x + y)\mathbf{i} + (y + z)\mathbf{j} + (z + x)\mathbf{k}$; σ is the surface of the cylindrical solid bounded by $x^2 + y^2 = 9$, $z = 0$, and $z = 5$.

21. Prove the following properties of divergence:
    (a) $\operatorname{div}(\mathbf{F} + \mathbf{G}) = \operatorname{div}\mathbf{F} + \operatorname{div}\mathbf{G}$;
    (b) $\operatorname{div}(f\mathbf{F}) = f \operatorname{div}\mathbf{F} + \nabla f \cdot \mathbf{F}$, where $f(x, y, z)$ is a scalar-valued function and $\mathbf{F}(x, y, z)$ a vector-valued function.

22. Prove that if σ is the surface of a solid G oriented by outward unit normals, then the volume of G is given by

$$\operatorname{vol}(G) = \frac{1}{3} \iint_{\sigma} \mathbf{F} \cdot \mathbf{n} \, dS$$

where $\mathbf{F}(x, y, z) = x\mathbf{i} + y\mathbf{j} + z\mathbf{k}$ and $\mathbf{n}$ is an outer unit normal to σ.

23. Use the formula in Exercise 22 to find the volume of a right-circular cylinder of radius a and height h.

24. Prove: If $\mathbf{F}(x, y, z) = a\mathbf{i} + b\mathbf{j} + c\mathbf{k}$, where a, b, and c are constant, and σ is the surface of a solid G, then

$$\iint_{\sigma} \mathbf{F} \cdot \mathbf{n} \, dS = 0$$

where $\mathbf{n}$ is an outer unit normal to σ.

18.7 STOKES' THEOREM

*In this section we shall discuss a generalization of Green's Theorem to three dimensions, called **Stokes'* Theorem**. In two dimensions, Green's Theorem expresses a double integral over a plane region in terms of a line integral over the boundary of the region. In three dimensions, Stokes' Theorem expresses a surface integral in terms of a line integral over the boundary of the surface.*

---

* GEORGE GABRIEL STOKES (1819–1903), Irish mathematician and physicist. Born in Skreen, Ireland, Stokes came from a family deeply rooted in the Church of Ireland. His father was a rector, his mother the daughter of a rector, and three of his brothers took holy orders. He received his early education from his father and a local parish clerk. In 1837, he entered Pembroke College and after graduating with top honors accepted a fellowship at the college. In 1847 he was appointed Lucasian professor of mathematics at Cambridge, a position once held by Isaac Newton, but one that had lost its esteem through the years. By virtue of his accomplishments, Stokes ultimately restored the position to the eminence it once held. Unfortunately, the position paid very little and Stokes was forced to teach at the Government School of Mines during the 1850s to supplement his income.

Stokes was one of several outstanding nineteenth century scientists who helped turn the physical sciences in a more empirical direction. He systematically studied hydrodynamics, elasticity of solids, behavior of waves in elastic solids, and diffraction of light. For Stokes, mathematics was a tool for his physical studies. He wrote classic papers on the motion of viscous fluids that laid the foundation for modern hydrodynamics; he elaborated on the wave theory of light; and he wrote papers on gravitational variation that established him as a founder of the modern science of geodesy.

Stokes was honored in his later years with degrees, medals, and memberships in foreign societies. He was knighted in 1889. Throughout his life, Stokes gave generously of his time to learned societies and readily assisted those who sought his help in solving problems. He was deeply religious and vitally concerned with the relationship between science and religion.

We begin with some terminology.

18.7.1 DEFINITION. If $\mathbf{F}(x, y, z) = f(x, y, z)\mathbf{i} + g(x, y, z)\mathbf{j} + h(x, y, z)\mathbf{k}$, then we define the ***curl of* F** by

$$\operatorname{curl} \mathbf{F} = \left(\frac{\partial h}{\partial y} - \frac{\partial g}{\partial z}\right)\mathbf{i} + \left(\frac{\partial f}{\partial z} - \frac{\partial h}{\partial x}\right)\mathbf{j} + \left(\frac{\partial g}{\partial x} - \frac{\partial f}{\partial y}\right)\mathbf{k} \tag{1}$$

REMARK. This formula can be remembered by writing it in the determinant form

$$\operatorname{curl} \mathbf{F} = \begin{vmatrix} \mathbf{i} & \mathbf{j} & \mathbf{k} \\ \dfrac{\partial}{\partial x} & \dfrac{\partial}{\partial y} & \dfrac{\partial}{\partial z} \\ f & g & h \end{vmatrix} \tag{2}$$

The reader should verify that Formula (1) results if we compute this determinant by interpreting a "product" such as $(\partial/\partial x)(f)$ to mean $\partial f/\partial x$. We note, however, that (2) is just a mnemonic device and not a true determinant, since the entries in a determinant must be numbers, not vectors and partial derivative symbols.

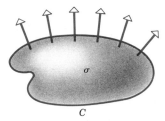

Example 1 If $\mathbf{F}(x, y, z) = x^2 y\mathbf{i} + 2y^3 z\mathbf{j} + 3z\mathbf{k}$, then

$$\operatorname{curl} \mathbf{F} = \begin{vmatrix} \mathbf{i} & \mathbf{j} & \mathbf{k} \\ \dfrac{\partial}{\partial x} & \dfrac{\partial}{\partial y} & \dfrac{\partial}{\partial z} \\ x^2 y & 2y^3 z & 3z \end{vmatrix}$$

$$= \left[\frac{\partial}{\partial y}(3z) - \frac{\partial}{\partial z}(2y^3 z)\right]\mathbf{i} - \left[\frac{\partial}{\partial x}(3z) - \frac{\partial}{\partial z}(x^2 y)\right]\mathbf{j}$$

$$+ \left[\frac{\partial}{\partial x}(2y^3 z) - \frac{\partial}{\partial y}(x^2 y)\right]\mathbf{k}$$

$$= -2y^3\mathbf{i} - x^2\mathbf{k} \quad \blacktriangleleft$$

Positive direction of C

Negative direction of C

Figure 18.7.1

To state Stokes' Theorem we need some terminology about oriented surfaces. If σ is an oriented surface bounded by a curve C (Figure 18.7.1a), then the orientation of σ induces an orientation or direction along C as follows. If we imagine a woman walking along the curve C with her head in the (general) direction of the normals that orient σ, then the woman is walking in the ***positive direction*** of C if the surface σ is on the woman's left and in the ***negative direction*** if σ is on her right (Figures 18.7.1b and 18.7.1c).

Stokes' Theorem

18.7.2 THEOREM. *Let σ be an oriented surface, bounded by a curve C. If the components of* $\mathbf{F}(x, y, z) = f(x, y, z)\mathbf{i} + g(x, y, z)\mathbf{j} + h(x, y, z)\mathbf{k}$ *have continuous first partial derivatives on some open set containing σ, then*

$$\int_C \mathbf{F} \cdot d\mathbf{r} = \iint_\sigma (\text{curl } \mathbf{F}) \cdot \mathbf{n} \, dS \tag{3}$$

where the line integral is taken in the positive direction of C.

The proof of Stokes' Theorem is too difficult for a first course and will be omitted.

Recall that if s is an arc-length parameter for the curve C, and if the positive direction of C is in the direction of increasing s, then

$$\frac{d\mathbf{r}}{ds} = \mathbf{T}$$

where $\mathbf{T}$ is the unit tangent vector to C that points in the positive direction of C. This suggests the notation

$$d\mathbf{r} = \mathbf{T} \, ds$$

which enables us to write (3) as

$$\int_C \mathbf{F} \cdot \mathbf{T} \, ds = \iint_\sigma (\text{curl } \mathbf{F}) \cdot \mathbf{n} \, dS \tag{4}$$

Example 2 Verify Stokes' Theorem if σ is the portion of the paraboloid $z = 4 - x^2 - y^2$ for which $z \geq 0$ and $\mathbf{F}(x, y, z) = 2z\mathbf{i} + 3x\mathbf{j} + 5y\mathbf{k}$.

Solution. If σ is oriented by outward normals, then the positive direction for the boundary curve C is as shown in Figure 18.7.2. Since the curve C is a circle in the xy-plane with radius 2 and traversed counterclockwise looking down the z-axis, it is represented (with the proper direction) by the parametric equations

$$x = 2 \cos t$$
$$y = 2 \sin t, \quad 0 \leq t \leq 2\pi \tag{5}$$
$$z = 0$$

or equivalently by the position function

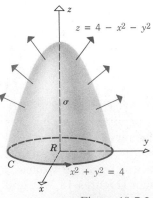

Figure 18.7.2

$$\mathbf{r}(t) = 2 \cos t\mathbf{i} + 2 \sin t\mathbf{j} + 0 = 2 \cos t\mathbf{i} + 2 \sin t\mathbf{j}$$

where $0 \leq t \leq 2\pi$. Since $\mathbf{F}(x, y, z) = 2z\mathbf{i} + 3x\mathbf{j} + 5y\mathbf{k}$, it follows from (5) that

$$\mathbf{F}(x(t), y(t), z(t)) = 2(0)\mathbf{i} + 3(2 \cos t)\mathbf{j} + 5(2 \sin t)\mathbf{k} = 6 \cos t\mathbf{j} + 10 \sin t\mathbf{k}$$

Thus,

$$
\begin{aligned}
\int_C \mathbf{F} \cdot d\mathbf{r} &= \int_0^{2\pi} \left[\mathbf{F}(x(t), y(t), z(t)) \cdot \frac{d\mathbf{r}}{dt} \right] dt \\
&= \int_0^{2\pi} (6 \cos t\mathbf{j} + 10 \sin t\mathbf{k}) \cdot (-2 \sin t\mathbf{i} + 2 \cos t\mathbf{j}) \, dt \\
&= \int_0^{2\pi} 12 \cos^2 t \, dt = 12 \left[\frac{1}{2}t + \frac{1}{4} \sin 2t \right]_0^{2\pi} = 12\pi
\end{aligned}
$$

On the other hand,

$$
\operatorname{curl} \mathbf{F} =
\begin{vmatrix}
\mathbf{i} & \mathbf{j} & \mathbf{k} \\
\dfrac{\partial}{\partial x} & \dfrac{\partial}{\partial y} & \dfrac{\partial}{\partial z} \\
2z & 3x & 5y
\end{vmatrix}
= 5\mathbf{i} + 2\mathbf{j} + 3\mathbf{k}
$$

Since the outward normals are upward normals, and the paraboloid is given by $z = z(x, y) = 4 - x^2 - y^2$, it follows from Formula (6) of Section 18.5 (with curl $\mathbf{F}$ replacing $\mathbf{F}$) that

$$
\begin{aligned}
\iint_\sigma (\operatorname{curl} \mathbf{F}) \cdot \mathbf{n} \, dS &= \iint_R (\operatorname{curl} \mathbf{F}) \cdot \left(-\frac{\partial z}{\partial x}\mathbf{i} - \frac{\partial z}{\partial y}\mathbf{j} + \mathbf{k} \right) dA \\
&= \iint_R (5\mathbf{i} + 2\mathbf{j} + 3\mathbf{k}) \cdot (2x\mathbf{i} + 2y\mathbf{j} + \mathbf{k}) \, dA \\
&= \iint_R (10x + 4y + 3) \, dA \\
&= \int_0^{2\pi} \int_0^2 (10r \cos \theta + 4r \sin \theta + 3) r \, dr \, d\theta \\
&= \int_0^{2\pi} \left[\frac{10r^3}{3} \cos \theta + \frac{4r^3}{3} \sin \theta + \frac{3r^2}{2} \right]_{r=0}^2 d\theta \\
&= \int_0^{2\pi} \left(\frac{80}{3} \cos \theta + \frac{32}{3} \sin \theta + 6 \right) d\theta \\
&= \left[\frac{80}{3} \sin \theta - \frac{32}{3} \cos \theta + 6\theta \right]_0^{2\pi} = 12\pi
\end{aligned}
$$

which is the same as the value of the line integral.   ◀

REMARK.   Had we oriented σ by inward normals in this example, then the surface integral and the line integral would have had value -12π. (Why?)

Figure 18.7.3

Example 3   Let C be the rectangle in the plane $z = y$ oriented as in Figure 18.7.3, and let $\mathbf{F}(x, y, z) = x^2\mathbf{i} + 4xy^3\mathbf{j} + y^2x\mathbf{k}$. Find

$$\int_C \mathbf{F} \cdot d\mathbf{r}$$

Solution.   To evaluate the integral directly would require four separate integrations, one over each side of the rectangle. Instead, we shall apply Stokes' Theorem, so we need only evaluate a single surface integral over the rectangular surface σ bounded by C. For the positive direction of C to be as shown in Figure 18.7.3, the surface σ must be oriented by downward normals.

Since the surface σ has equation $z = y$ and

$$\text{curl } \mathbf{F} = \begin{vmatrix} \mathbf{i} & \mathbf{j} & \mathbf{k} \\ \dfrac{\partial}{\partial x} & \dfrac{\partial}{\partial y} & \dfrac{\partial}{\partial z} \\ x^2 & 4xy^3 & y^2x \end{vmatrix} = 2yx\mathbf{i} - y^2\mathbf{j} + 4y^3\mathbf{k}$$

it follows from Formula (7) of Theorem 18.5.2 that

$$\iint_\sigma (\text{curl } \mathbf{F}) \cdot \mathbf{n} \, dS = \iint_R (\text{curl } \mathbf{F}) \cdot \left(\frac{\partial z}{\partial x}\mathbf{i} + \frac{\partial z}{\partial y}\mathbf{j} - \mathbf{k} \right) dA$$

$$= \iint_R (2yx\mathbf{i} - y^2\mathbf{j} + 4y^3\mathbf{k}) \cdot (0\mathbf{i} + \mathbf{j} - \mathbf{k}) \, dA$$

$$= \int_0^1 \int_0^3 (-y^2 - 4y^3) \, dy \, dx$$

$$= -\int_0^1 \left[\frac{y^3}{3} + y^4 \right]_{y=0}^3 dx$$

$$= -\int_0^1 90 \, dx = -90 \quad ◀$$

▶ Exercise Set 18.7

In Exercises 1–5, find the curl of **F**.

1. $\mathbf{F}(x, y, z) = x\mathbf{i} + y\mathbf{j} + z\mathbf{k}$.

2. $\mathbf{F}(x, y, z) = xz^3\mathbf{i} + 2y^4x^2\mathbf{j} + 5z^2y\mathbf{k}$.

3. $\mathbf{F}(x, y, z) = e^{xy}\mathbf{i} - \cos y\mathbf{j} + \sin^2 z\mathbf{k}$.

4. $\mathbf{F}(x, y, z) = 7y^3z^2\mathbf{i} - 8x^2z^5\mathbf{j} - 3xy^4\mathbf{k}$.

5. $\mathbf{F}(x, y, z) = \ln x\mathbf{i} + e^{xyz}\mathbf{j} + \tan^{-1}\left(\dfrac{z}{x}\right)\mathbf{k}$.

In Exercises 6–12, use Stokes' Theorem to evaluate the integral $\displaystyle\int_C \mathbf{F} \cdot d\mathbf{r}$.

6. $\mathbf{F}(x, y, z) = xz\mathbf{i} + 3x^2y^2\mathbf{j} + yx\mathbf{k}$; C is the rectangle in the plane $z = y$ shown in Figure 18.7.3.

7. $\mathbf{F}(x, y, z) = 3z\mathbf{i} + 4x\mathbf{j} + 2y\mathbf{k}$; C is the boundary of the paraboloid shown in Figure 18.7.2.

8. $\mathbf{F}(x, y, z) = -3y^2\mathbf{i} + 4z\mathbf{j} + 6x\mathbf{k}$; C is the triangle in the plane $z = \frac{1}{2}y$ with vertices $(2, 0, 0)$, $(0, 2, 1)$, and $(0, 0, 0)$ with a counterclockwise orientation looking down the positive z-axis.

9. $\mathbf{F}(x, y, z) = xy\mathbf{i} + x^2\mathbf{j} + z^2\mathbf{k}$; C is the intersection of the paraboloid $z = x^2 + y^2$ and the plane $z = y$ with a counterclockwise orientation looking down the positive z-axis.

10. $\mathbf{F}(x, y, z) = xy\mathbf{i} + yz\mathbf{j} + zx\mathbf{k}$; C is the triangle in the plane $x + y + z = 1$ with vertices $(1, 0, 0)$, $(0, 1, 0)$, and $(0, 0, 1)$ with a counterclockwise orientation looking from the first octant toward the origin.

11. $\mathbf{F}(x, y, z) = (x - y)\mathbf{i} + (y - z)\mathbf{j} + (z - x)\mathbf{k}$; C is the circle $x^2 + y^2 = a^2$ in the xy-plane with counterclockwise orientation looking down the positive z-axis.

12. $\mathbf{F}(x, y, z) = (z + \sin x)\mathbf{i} + (x + y^2)\mathbf{j} + (y + e^z)\mathbf{k}$; C is the intersection of the sphere $x^2 + y^2 + z^2 = 1$ and the cone $z = \sqrt{x^2 + y^2}$ with counterclockwise orientation looking down the positive z-axis.

In Exercises 13–16, verify Stokes' Theorem by computing the line and surface integrals in (3) and showing that they are equal.

13. $\mathbf{F}(x, y, z) = (x - y)\mathbf{i} + (y - z)\mathbf{j} + (z - x)\mathbf{k}$; σ is the portion of the plane $x + y + z = 1$ in the first octant.

14. $\mathbf{F}(x, y, z) = x^2\mathbf{i} + y^2\mathbf{j} + z^2\mathbf{k}$; σ is the portion of the cone $z = \sqrt{x^2 + y^2}$ below the plane $z = 1$.

15. $\mathbf{F}(x, y, z) = x\mathbf{i} + y\mathbf{j} + z\mathbf{k}$; σ is the upper hemisphere $z = \sqrt{a^2 - x^2 - y^2}$.

16. $\mathbf{F}(x, y, z) = (z - y)\mathbf{i} + (z + x)\mathbf{j} - (x + y)\mathbf{k}$; σ is the portion of the paraboloid $z = 9 - x^2 - y^2$ above the xy-plane.

17. Use Stokes' Theorem to evaluate

$$\int_C z^2\, dx + 2x\, dy - y^3\, dz$$

over the circle $x^2 + y^2 = 1$ in the xy-plane traversed counterclockwise looking down the positive z-axis. [*Hint:* Find a surface σ whose boundary is C.]

18. Use Stokes' Theorem to evaluate

$$\int_C y^2\, dx + z^2\, dy + x^2\, dz$$

over the triangle with vertices $P(a, 0, 0)$, $Q(0, a, 0)$, $R(0, 0, a)$ traversed counterclockwise looking toward the origin from the first octant.

19. Prove: If the components of **F** and their first- and second-order partial derivatives are continuous, then

$$\text{div}\,(\text{curl}\,\mathbf{F}) = 0$$

20. Use Exercise 19 and the Divergence Theorem to prove that

$$\iint_\sigma (\text{curl}\,\mathbf{F}) \cdot \mathbf{n}\, dS = 0$$

if σ is a sphere, **n** is the unit outer normal to σ, and the components of **F** satisfy the conditions stated in Exercise 19.

21. Prove: If f and the components of **F** are continuous and have continuous second-order partial derivatives, then
    (a) $\text{curl}\,(\nabla f) = \mathbf{0}$;
    (b) $\text{curl}\,(\nabla f + \text{curl}\,\mathbf{F}) = \text{curl}\,(\text{curl}\,\mathbf{F})$. [*Hint:* You may use the property $\text{curl}\,(\mathbf{F}_1 + \mathbf{F}_2) = \text{curl}\,\mathbf{F}_1 + \text{curl}\,\mathbf{F}_2$ without proof.]

22. (a) Use Stokes' Theorem to prove: If σ_1 and σ_2 are oriented surfaces with the same boundary C, and if C has the same positive orientation relative to each surface, then

$$\iint_{\sigma_1} (\text{curl } \mathbf{F}) \cdot \mathbf{n} \, dS = \iint_{\sigma_2} (\text{curl } \mathbf{F}) \cdot \mathbf{n} \, dS$$

   (b) Use the result in (a) to evaluate

$$\iint_{\sigma} (\text{curl } \mathbf{F}) \cdot \mathbf{n} \, dS$$

where $\mathbf{F}(x, y, z) = x^3 \mathbf{i} + y \mathbf{j} + z^2 e^{xy} \mathbf{k}$ and the surface σ is the upper hemisphere $z = \sqrt{a^2 - x^2 - y^2}$ oriented by upward unit normals. [*Hint:* Find a circular disk in the xy-plane that shares a common boundary with the hemisphere.]

23. Use Stokes' Theorem to prove the result in Exercise 20. [*Hint:* See Exercise 22(a) and note that the upper and lower hemispheres share a common boundary.]

18.8   APPLICATIONS

In this section we shall discuss applications of Stokes' Theorem and the Divergence Theorem to problems of fluid flow. These theorems also have applications to problems in electricity, magnetism, and mechanics which we shall not discuss here.

FLUX   Consider a body of fluid flowing in a region of three-dimensional space, and suppose that at each point in the region the velocity of the fluid is given by

$$\mathbf{F}(x, y, z) = f(x, y, z)\mathbf{i} + g(x, y, z)\mathbf{j} + h(x, y, z)\mathbf{k}$$

The function $\mathbf{F}(x, y, z)$ is called the *flow field* for the fluid motion (Figure 18.8.1).

For simplicity, we shall make two assumptions:

Figure 18.8.1

1. The velocity of the fluid at each point of the region does not vary with time. Such a flow is said to be *steady* over the region.

2. The density (mass per unit volume) of the fluid is constant throughout the region. Such a fluid is called *incompressible* over the region. To keep our formulas simple, we shall assume this density to be 1.

Figure 18.8.2

If σ is an oriented surface within such a three-dimensional body of fluid, then at each point (x, y, z) on the surface the velocity vector $\mathbf{F}(x, y, z)$ can be resolved into two components, a component $\mathbf{F} \cdot \mathbf{T}$ tangent to the surface, and a component $\mathbf{F} \cdot \mathbf{n}$ along the unit normal vector $\mathbf{n}$ in the orientation of σ (Figure 18.8.2). The component $\mathbf{F} \cdot \mathbf{T}$ is the component of flow *along* the surface, and $\mathbf{F} \cdot \mathbf{n}$ is the component *across* the surface. Fluid crosses σ in the direction of orientation where $\mathbf{F} \cdot \mathbf{n}$ is positive and opposite to that direction where $\mathbf{F} \cdot \mathbf{n}$ is negative. (Why?) In a specified time interval, the volume of fluid crossing σ in the direction of orientation minus the volume crossing in the opposite direction is called the ***net volume*** of fluid crossing σ during the time interval, and the surface integral

$$\iint_{\sigma} \mathbf{F} \cdot \mathbf{n} \, dS$$

is called the ***flux of F across σ;*** it represents the net volume of fluid that crosses the oriented surface σ in one unit of time. To see this, subdivide σ into n parts $\sigma_1, \sigma_2, \ldots, \sigma_n$ with areas

$$\Delta S_1, \Delta S_2, \ldots, \Delta S_n$$

If the parts are small and the flow is not too erratic, it is reasonable to assume that the velocity does not vary much on each part. Thus, if (x_k^*, y_k^*, z_k^*) is any point in the kth part, we can assume that $\mathbf{F}(x, y, z)$ is constant and equal to $\mathbf{F}(x_k^*, y_k^*, z_k^*)$ throughout the part, and that the component of flow across the surface σ_k is

$$\mathbf{F}(x_k^*, y_k^*, z_k^*) \cdot \mathbf{n}(x_k^*, y_k^*, z_k^*) \tag{1}$$

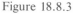

Figure 18.8.3

where $\mathbf{n}(x_k^*, y_k^*, z_k^*)$ is the unit normal at (x_k^*, y_k^*, z_k^*) in the orientation of σ (Figure 18.8.3).

If we observe the flow across σ_k for one unit of time, then the section of fluid initially on the surface will move to some new position and the volume of fluid that crosses σ_k will form a solid (Figure 18.8.4) with a volume that can be approximated by

Figure 18.8.4

$$\text{volume} \approx \begin{bmatrix} \text{distance traveled by} \\ \text{the fluid in one unit} \\ \text{of time} \end{bmatrix} \Delta S_k \tag{2}$$

But (1) is the component of fluid velocity across the surface and therefore represents either the distance traveled by the fluid in one unit of time or the negative of that distance, depending on whether $\mathbf{F} \cdot \mathbf{n}$ is positive or negative. Thus, from (2)

$$\mathbf{F}(x_k^*, y_k^*, z_k^*) \cdot \mathbf{n}(x_k^*, y_k^*, z_k^*) \, \Delta S_k$$

is either the approximate volume of fluid crossing σ_k in one unit of time or the negative of that approximate volume. It is positive if the fluid is crossing in the same direction as the orienting vector $\mathbf{n}(x_k^*, y_k^*, z_k^*)$ and negative if it is crossing in the opposite direction. Thus,

$$\sum_{k=1}^{n} \mathbf{F}(x_k^*, y_k^*, z_k^*) \cdot \mathbf{n}(x_k^*, y_k^*, z_k^*) \, \Delta S_k \tag{3}$$

is approximately the volume of fluid crossing σ in the direction of orientation minus the volume crossing in the opposite direction.

If we now repeat the subdivision process, dividing σ into more and more parts, in such a way that each part shrinks to a point, then we can assume that the errors in our approximations diminish to zero, and

$$\lim_{n \to +\infty} \sum_{k=1}^{n} \mathbf{F}(x_k^*, y_k^*, z_k^*) \cdot \mathbf{n}(x_k^*, y_k^*, z_k^*) \, \Delta S_k = \iint_{\sigma} \mathbf{F}(x, y, z) \cdot \mathbf{n}(x, y, z) \, dS$$

represents the *exact* net volume of fluid crossing σ in one unit of time.

In summary, we have the following result:

Formula for Flux

$$\begin{bmatrix} \text{flux of } \mathbf{F} \\ \text{across } \sigma \end{bmatrix} = \begin{bmatrix} \text{net volume of fluid crossing} \\ \sigma \text{ in the direction of orientation in one unit of time.} \end{bmatrix} = \iint_{\sigma} \mathbf{F} \cdot \mathbf{n} \, dS \quad (4)$$

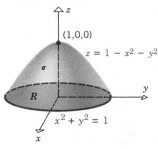

Figure 18.8.5

Example 1   Let σ be the portion of the surface $z = 1 - x^2 - y^2$ above the xy-plane oriented by upward normals (Figure 18.8.5), and suppose the flow field is

$$\mathbf{F}(x, y, z) = x\mathbf{i} + y\mathbf{j} + z\mathbf{k}$$

Find the net volume of fluid that crosses σ in the direction of orientation in one unit of time.

Solution.   From (4), the net volume of fluid crossing σ in one unit of time is

$$\iint_{\sigma} \mathbf{F} \cdot \mathbf{n} \, dS$$

We showed in Example 4 of Section 18.5 that

$$\iint_{\sigma} \mathbf{F} \cdot \mathbf{n} \, dS = \iint_{\sigma} (x^2 + y^2 + 1) \, dA = \frac{3\pi}{2}$$

so there is a net volume of $3\pi/2$ cubic units of fluid crossing σ in one unit of time.   ◀

PHYSICAL INTERPRETATION OF DIVERGENCE

We now have the background to give a physical interpretation of the divergence of a flow field

$$\mathbf{F}(x, y, z) = f(x, y, z)\mathbf{i} + g(x, y, z)\mathbf{j} + h(x, y, z)\mathbf{k}$$

If G is a *small* spherical region centered at (x, y, z) and div $\mathbf{F}$ is continuous on G, then div $\mathbf{F}$ will not vary much from its value at the center (x, y, z); and we can assume that div $\mathbf{F}$ is constant over G. If σ is the surface of G oriented by outward normals, then the Divergence Theorem (18.6.2) yields

$$\iint_{\sigma} \mathbf{F} \cdot \mathbf{n} \, dS = \iiint_{G} \text{div } \mathbf{F} \, dV \approx \text{div } \mathbf{F}(x, y, z) \iiint_{G} dV$$

where div $\mathbf{F}(x, y, z)$ is the divergence at the center of the sphere. Therefore,

$$\iint_{\sigma} \mathbf{F} \cdot \mathbf{n} \, dS \approx \text{div } \mathbf{F}(x, y, z) \, [\text{volume } (G)] \tag{5}$$

or

$$\text{div } \mathbf{F}(x, y, z) \approx \frac{1}{\text{vol } (G)} \iint_{\sigma} \mathbf{F} \cdot \mathbf{n} \, dS = \frac{\text{flux of } \mathbf{F} \text{ across } \sigma}{\text{vol } (G)} \tag{6}$$

If we assume that the errors in our approximations diminish to zero over smaller and smaller spheres, then it follows from (6) that

$$\text{div } \mathbf{F}(x, y, z) = \lim_{\text{vol } (G) \to 0} \frac{\text{flux of } \mathbf{F} \text{ across } \sigma}{\text{vol } (G)}$$

The quantity

$$\frac{\text{flux of } \mathbf{F} \text{ across } \sigma}{\text{vol } (G)}$$

is called the *flux density of F over the sphere G,* and its limit as vol $(G) \to 0$ is called the *flux density of F* at the point (x, y, z). Thus, we have the following physical description of the divergence:

$$\text{div } \mathbf{F}(x, y, z) = \text{flux density of } \mathbf{F} \text{ at } (x, y, z) \tag{7}$$

The flux density at (x, y, z) is a measure of the rate at which fluid is approaching or departing from the point (x, y, z). To see why, consider a sphere G centered at (x, y, z) with its surface σ oriented by outward unit normals. If the sphere is small, then the volume of fluid approaching or departing from (x, y, z) in one unit of time is approximately the volume of fluid crossing the surface σ in that time. But (4) and (5) imply that the net volume of fluid crossing σ in one unit of time is directly proportional to the divergence of $\mathbf{F}$ at (x, y, z).

SOURCES AND SINKS   We leave it for the reader to show that the net fluid flow is away from (x, y, z) if div $\mathbf{F}(x, y, z) > 0$, and toward (x, y, z) if div $\mathbf{F}(x, y, z) < 0$. Points where div $\mathbf{F}(x, y, z) > 0$ are called *sources*, and points where div $\mathbf{F}(x, y, z) < 0$ are called *sinks*. At a source, fluid is entering the flow, and at a sink it is leaving the flow. If div $\mathbf{F}(x, y, z) = 0$ throughout the flow, then there are no sources or sinks.

Example 2   The fluid with flow field $\mathbf{F}(x, y, z) = yz\mathbf{i} + xz\mathbf{j} + xy\mathbf{k}$ has no sources or sinks since

$$\text{div } \mathbf{F}(x, y, z) = \frac{\partial}{\partial x}(yz) + \frac{\partial}{\partial y}(xz) + \frac{\partial}{\partial z}(xy) = 0 \quad \blacktriangleleft$$

We are now in a position to give an important physical interpretation of the Divergence Theorem. Let σ be an oriented surface that encloses a finite three-dimensional region G inside a moving body of fluid (Figure 18.8.6). If $\mathbf{F}(x, y, z)$ is the flow field, then it follows from the Divergence Theorem (Formula (1), Section 18.6) that

$$\iint_{\sigma} \mathbf{F} \cdot \mathbf{n} \, dS = \iiint_{G} \text{div } \mathbf{F} \, dV$$

Figure 18.8.6

But from (4) and (7) this equation states that

$$\begin{bmatrix} \text{flux of } \mathbf{F} \\ \text{across } \sigma \end{bmatrix} = \iiint_{G} [\text{flux density}] \, dV$$

In other words, the Divergence Theorem asserts that *the net volume of fluid that crosses σ in one unit of time is obtained by integrating the flux density over the region enclosed by σ.*

PHYSICAL
INTERPRETATION OF
CURL

We conclude this section by obtaining a physical interpretation of the curl of a flow field

$$\mathbf{F}(x, y, z) = f(x, y, z)\mathbf{i} + g(x, y, z)\mathbf{j} + h(x, y, z)\mathbf{k}$$

Let σ_a be a *small* oriented disk-shaped region of radius a, centered at (x, y, z), and having boundary C_a (Figure 18.8.7a). If the components of curl $\mathbf{F}$ are continuous on σ_a, then curl $\mathbf{F}$ will not vary much from its "value" at the center (x, y, z), and we can assume curl $\mathbf{F}$ to be approximately constant on σ_a. Since σ_a is a flat surface, the unit normal vectors that orient σ_a are all the same as the unit normal $\mathbf{n} = \mathbf{n}(x, y, z)$ at the center (x, y, z) (Figure 18.8.7b). Thus, curl $\mathbf{F} \cdot \mathbf{n}$ is also constant on σ_a, so Equation (4) of Section 18.7 and Stokes' Theorem (18.7.2) yield

$$\int_{C_a} \mathbf{F} \cdot \mathbf{T} \, ds = \iint_{\sigma_a} (\text{curl } \mathbf{F}) \cdot \mathbf{n} \, dS \approx \text{curl } \mathbf{F}(x, y, z) \cdot \mathbf{n} \iint_{\sigma_a} dS$$

(a)

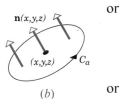

(b)

Figure 18.8.7

where the line integral is taken in the positive direction of C_a, and $\mathbf{T}$ is the unit tangent vector to C_a pointing in the positive direction. Therefore,

$$\int_{C_a} \mathbf{F} \cdot \mathbf{T}\, ds \approx [\text{curl } \mathbf{F}(x, y, z) \cdot \mathbf{n}]\ [\text{surface area of } \sigma_a]$$

or

$$\int_{C_a} \mathbf{F} \cdot \mathbf{T}\, ds \approx \pi a^2[\text{curl } \mathbf{F}(x, y, z) \cdot \mathbf{n}]$$

or

$$\text{curl } \mathbf{F}(x, y, z) \cdot \mathbf{n} \approx \frac{1}{\pi a^2} \int_{C_a} \mathbf{F} \cdot \mathbf{T}\, ds \tag{8}$$

If we assume that the errors in our approximations diminish to zero over smaller and smaller disks, then it follows from (8) that

$$\text{curl } \mathbf{F}(x, y, z) \cdot \mathbf{n} = \lim_{a \to 0} \frac{1}{\pi a^2} \int_{C_a} \mathbf{F} \cdot \mathbf{T}\, ds \tag{9}$$

Figure 18.8.8

The integral in (9) has a physical interpretation that we shall now explain. At each point of C_a, the flow field $\mathbf{F}$ has a component $\mathbf{F} \cdot \mathbf{u}$ along the outward normal to C_a and a component $\mathbf{F} \cdot \mathbf{T}$ tangent to C_a (Figure 18.8.8). When $\mathbf{F}$ is closely aligned with $\mathbf{T}$, the magnitude of $\mathbf{F} \cdot \mathbf{T}$ is large compared to the magnitude of $\mathbf{F} \cdot \mathbf{u}$, so the fluid tends to move along the circle C_a rather than pass through it. Thus, physicists call the integral

$$\int_{C_a} \mathbf{F} \cdot \mathbf{T}\, ds$$

(a)

(b)

Figure 18.8.9

the *circulation of $\mathbf{F}$ around C_a* and use it as a measure of the tendency for fluid to circulate in the positive direction of C_a.

If the flow is normal to the circle at each point of C_a, then $\mathbf{F} \cdot \mathbf{T} = 0$ so

$$\int_{C_a} \mathbf{F} \cdot \mathbf{T}\, ds = 0$$

and there is no tendency for the fluid to circulate around C_a (Figure 18.8.9a). However, the more closely $\mathbf{F}$ is aligned with $\mathbf{T}$, the larger the value of $\mathbf{F} \cdot \mathbf{T}$ and the larger the value of the circulation. The maximum circulation around C_a occurs when $\mathbf{F}$ is aligned with $\mathbf{T}$ at each point of C_a (Figure 18.8.9b).

The quantity

$$\frac{1}{\pi a^2} \int_{C_a} \mathbf{F} \cdot \mathbf{T} \, ds = \frac{\text{circulation of } \mathbf{F} \text{ around } C_a}{\text{area of the disk } \sigma_a}$$

is called the *specific circulation* of $\mathbf{F}$ over the disk σ_a and its limit as $a \to 0$ is called the *rotation of* $\mathbf{F}$ *about* $\mathbf{n}$ at (x, y, z). Thus, from (9),

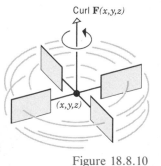

Curl $\mathbf{F}(x,y,z)$

(x,y,z)

Figure 18.8.10

$$\text{curl } \mathbf{F}(x, y, z) \cdot \mathbf{n} = \text{rotation of } \mathbf{F} \text{ about } \mathbf{n} \text{ at } (x, y, z)$$

Since curl $\mathbf{F} \cdot \mathbf{n}$ has its maximum value when $\mathbf{n}$ is in the same direction as curl $\mathbf{F}(x, y, z)$, it follows that the rotation of $\mathbf{F}$ at (x, y, z) has its maximum value about this $\mathbf{n}$. Thus, in the vicinity of the point (x, y, z) the maximum circulation occurs around small circles in the plane normal to curl $\mathbf{F}$. Physically, if a paddle wheel is immersed in the fluid so the pivot point is at (x, y, z), then the paddles will turn most quickly when the spindle is aligned with curl $\mathbf{F}(x, y, z)$ (Figure 18.8.10).

▶ Exercise Set 18.8

In Exercises 1–4, a flow field $\mathbf{F}(x, y, z)$ is given. Find the net volume of fluid that crosses the surface σ in the direction of orientation in one unit of time.

1. $\mathbf{F}(x, y, z) = x\mathbf{i} + y\mathbf{j} + 2z\mathbf{k}$; σ is the portion of the surface $z = 1 - x^2 - y^2$ oriented by upward unit normals.

2. $\mathbf{F}(x, y, z) = x^2\mathbf{i} + yx\mathbf{j} + zx\mathbf{k}$; σ is the portion of the plane $6x + 3y + 2z = 6$ in the first octant oriented by upward unit normals.

3. $\mathbf{F}(x, y, z) = (x + y)\mathbf{i} + (y + z)\mathbf{j} + (z + x)\mathbf{k}$; σ is the upper hemisphere $z = \sqrt{9 - x^2 - y^2}$ oriented by upward unit normals.

4. $\mathbf{F}(x, y, z) = x\mathbf{i} + y\mathbf{j} + 2z\mathbf{k}$; σ is the portion of the cone $z^2 = x^2 + y^2$ between the planes $z = 1$ and $z = 2$, oriented by upward unit normals.

In Exercises 5–8, a flow field $\mathbf{F}(x, y, z)$ is given. Use the Divergence Theorem to find the net volume of fluid that crosses the surface σ in the direction of orientation in one unit of time.

5. $\mathbf{F}(x, y, z) = x\mathbf{i} + y\mathbf{j} + z\mathbf{k}$; σ is the sphere $x^2 + y^2 + z^2 = a^2$ oriented by outward unit normals.

6. $\mathbf{F}(x, y, z) = xy\mathbf{i} + yz\mathbf{j} + xz\mathbf{k}$; σ is the cube bounded by the coordinate planes and the planes $x = 2$,

$y = 2$, $z = 2$, and oriented by outward unit normals.

7. $\mathbf{F}(x, y, z) = (x^3 - y)\mathbf{i} + (y^3 + z)\mathbf{j} + (z^3 - 2xy)\mathbf{k}$; σ is the surface of the cylinder (including top and bottom) bounded by $x^2 + y^2 = 4$, $z = 0$, and $z = 3$, oriented by outward unit normals.

8. $\mathbf{F}(x, y, z) = x^2\mathbf{i} + xy\mathbf{j} - 2xz\mathbf{k}$; σ is the tetrahedron bounded by the coordinate planes and the plane $x + y + z = 1$, oriented by outward unit normals.

In Exercises 9–12, determine whether the flow field $\mathbf{F}(x, y, z)$ is free of sources and sinks. If it is not, find the location of all sources and sinks.

9. $\mathbf{F}(x, y, z) = (y + z)\mathbf{i} - xz^3\mathbf{j} + (x^2 \sin y)\mathbf{k}$.

10. $\mathbf{F}(x, y, z) = xy\mathbf{i} - xy\mathbf{j} + y^2\mathbf{k}$.

11. $\mathbf{F}(x, y, z) = x^3\mathbf{i} + y^3\mathbf{j} + z^3\mathbf{k}$.

12. $\mathbf{F}(x, y, z) = (x^3 - x)\mathbf{i} + (y^3 - y)\mathbf{j} + (z^3 - z)\mathbf{k}$.

13. For the flow field that is given by the formula $\mathbf{F}(x, y, z) = (x - z)\mathbf{i} + (y - x)\mathbf{j} + (z - xy)\mathbf{k}$:
    (a) use Stokes' Theorem to find the circulation around the triangle with vertices $A(1, 0, 0)$, $B(0, 2, 0)$, and $C(0, 0, 1)$ oriented counterclockwise looking from the origin toward the

first octant;

(b) find the rotation of F about the vector k at the origin;

(c) find the unit vector n such that the rotation of F about n at the origin is maximum.

▶ SUPPLEMENTARY EXERCISES

In Exercises 1–6, evaluate the line integral by any method.

1. $\displaystyle\int_C 2y\,dx + 3\,dy$ along $y = \sin x$ from the point $(0,0)$ to the point $(\pi, 0)$.

2. $\displaystyle\int_C x^5\,dy$ along the curve C given by $x = 1/t, y = 4t^2$, $1 \le t \le 2$.

3. $\displaystyle\int_C \langle 2e^y, -x\rangle \cdot d\mathbf{r}$ along $y = \ln x$ from the point $(1,0)$ to the point $(3, \ln 3)$.

4. $\displaystyle\int_C (y\mathbf{i} + z\mathbf{j} + x\mathbf{k}) \cdot d\mathbf{r}$ along the curve C given by $\mathbf{r} = \langle t^2 - 2t, -2t, t - 2\rangle,\ 0 \le t \le 2$.

5. $\displaystyle\int_C z\,dx + y\,dy - x\,dz$ along the line segment from $(0,0,0)$ to $(1,2,3)$.

6. $\displaystyle\int_C x \sin xy\,dx - y \sin xy\,dy$ along the line segment from $(0,0)$ to $(1, \pi)$.

In Exercises 7–9, determine whether F is conservative. If it is, find a potential function for it.

7. $F(x, y) = y \sin xy\mathbf{i} - x \cos xy\mathbf{j}$.

8. $F(x, y) = 2x(\ln y - 1)\mathbf{i} + \left(\dfrac{x^2}{y} - 3y^2\right)\mathbf{j}$.

9. $F(x, y) = \left(3x^2 - \dfrac{y^2}{x^2}\right)\mathbf{i} + \left(\dfrac{2y}{x} + 4y\right)\mathbf{j}$.

10. Let $F(x, y) = y\mathbf{i} - 2x\mathbf{j}$. Find functions $h(x)$ and $g(y)$ such that $h(x)F(x, y)$ and $g(y)F(x, y)$ are each conservative.

In Exercises 11–14, determine whether the integral is independent of the path. If it is, find its value by any method.

11. $\displaystyle\int_{(1,\pi/4)}^{(2,\pi/4)} (\cos 2y - 3x^2y^2)\,dx + (\cos 2y - 2x \sin 2y - 2x^3y)\,dy.$

12. $\displaystyle\int_{(0,0)}^{(2,-2)} x^2y^4\,dx + y^2x^4\,dy.$

13. $\displaystyle\int_{(1,2)}^{(2,1)} \dfrac{y\,dx - x\,dy}{y^2}.$

14. $\displaystyle\int_{(0,1)}^{(1,0)} (ye^{xy} - 1)\,dx + (xe^{xy})\,dy.$

In Exercises 15–18, evaluate the line integral counterclockwise over C using Green's Theorem.

15. $\displaystyle\int_C 2xy\,dx + xy^2\,dy$ along the triangle with vertices $(0,0)$, $(2,0)$, $(2,4)$.

16. $\displaystyle\int_C 3y\,dx - 4x\,dy$ along the boundary of the semicircular region $\{(x, y): x^2 + y^2 \le 2 \text{ and } 0 \le x\}$.

17. $\displaystyle\int_C -2y\,dx + (x - 4)\,dy$; C runs from $(0,2)$ to $(0, -2)$ on the y-axis, then to $(0, 2)$ along the curve $x = 4 - y^2$.

18. $\displaystyle\int_C 5xy\,dx + x^3\,dy$ along the curve $y = x^2$ from $(0,0)$ to $(2,4)$, then back to $(0,0)$ along $y = 2x$.

19. Use Green's Theorem to evaluate $\displaystyle\int_C y^3\,dx - x^3\,dy$ counterclockwise around the boundary of the region $R = \{(x, y): 1 \le x^2 + y^2 \le 4 \text{ and } 0 \le y\}$.

20. Find the area of the circle $r = 2 \cos \theta$ using Formula (4c) of Section 18.3.

In Exercises 21–23, use any method to find the work done by the force F as it acts on a particle moving along C.

21. $\mathbf{F} = (2x + 3y)\mathbf{i} + (xy)\mathbf{j}$ along $y = x^2$ from $(0, 0)$ to $(1, 1)$.

22. $\mathbf{F} = 3x^2y^3\mathbf{i} + 3x^3y^2\mathbf{j}$ clockwise once around the cardioid $r = 3 + 3 \cos\theta$.

23. $\mathbf{F} = z\mathbf{i} + y\mathbf{j} - x\mathbf{k}$ on the line segment from $(0, 0, 0)$ to $(1, 2, 3)$.

24. Assuming that $\phi(x, y)$ has continuous first- and second-order partial derivatives, show that
    (a) $\text{curl}(\nabla\phi) = \mathbf{0}$ (b) $\text{div}(\nabla\phi) = \phi_{xx} + \phi_{yy}$.

The centroid of a curved lamina σ is defined by

$$\bar{x} = \frac{\displaystyle\iint_\sigma x\, dS}{\text{area of }\sigma}, \quad \bar{y} = \frac{\displaystyle\iint_\sigma y\, dS}{\text{area of }\sigma}, \quad \bar{z} = \frac{\displaystyle\iint_\sigma z\, dS}{\text{area of }\sigma}$$

In Exercises 25 and 26, find the centroid of the lamina.

25. The lamina that is the portion of the paraboloid $z = \frac{1}{2}(x^2 + y^2)$ below the plane $z = 4$.

26. The lamina that is the portion of the sphere $x^2 + y^2 + z^2 = 4$ above the plane $z = 1$.

27. Let σ be the portion of the sphere $x^2 + y^2 + z^2 = 1$ that lies in the first octant, $\mathbf{n}$ an outward unit normal (with respect to the sphere), and $D_\mathbf{n} f$ the directional derivative of

$$f(x, y, z) = \ln\sqrt{x^2 + y^2 + z^2}$$

Evaluate the surface integral

$$\iint_\sigma D_\mathbf{n} f\, dS$$

28. Let σ be the sphere $x^2 + y^2 + z^2 = 1$, $\mathbf{n}$ an inward unit normal, and $D_\mathbf{n} f$ the directional derivative of $f(x, y, z) = x^2 + y^2 + z^2$. Evaluate the surface integral

$$\iint_\sigma D_\mathbf{n} f\, dS$$

29. Let G be a solid with surface σ oriented by outward unit normals, suppose that ϕ has continuous first and second partial derivatives in some open set containing G, and let $D_\mathbf{n}\phi$ be the directional derivative of ϕ. Show that

$$\iint_\sigma D_\mathbf{n}\phi\, dS = \iiint_G \left[\frac{\partial^2\phi}{\partial x^2} + \frac{\partial^2\phi}{\partial y^2} + \frac{\partial^2\phi}{\partial z^2}\right] dV$$

30. Show that Green's Theorem is a special case of Stokes' Theorem by applying Stokes' Theorem to $\mathbf{F}(x, y) = f(x, y)\mathbf{i} + g(x, y)\mathbf{j}$.

19 second-order differential equations

19.1 SECOND-ORDER LINEAR HOMOGENEOUS DIFFERENTIAL EQUATIONS WITH CONSTANT COEFFICIENTS

In Section 7.7 we showed how to solve first-order linear differential equations. In this section we shall show how to solve certain second-order linear differential equations.

SECOND-ORDER LINEAR DIFFERENTIAL EQUATIONS

Recall from Section 7.7 that a first-order linear differential equation has the form

$$\frac{dy}{dx} + p(x)y = q(x)$$

In this section we shall be concerned with the general *second-order linear differential equation*

$$\frac{d^2y}{dx^2} + p(x)\frac{dy}{dx} + q(x)y = r(x) \tag{1}$$

or in an alternative notation

$$y'' + p(x)y' + q(x)y = r(x)$$

If $r(x) = 0$ for all x, then (1) reduces to

$$\frac{d^2y}{dx^2} + p(x)\frac{dy}{dx} + q(x)y = 0$$

which is called the general second-order linear *homogeneous* differential equation. If $r(x)$ is not identically zero, then (1) is said to be *nonhomogeneous*. Some examples of second-order linear differential equations are

$$\frac{d^2y}{dx^2} + x^2\frac{dy}{dx} - xy = e^x \qquad [p(x) = x^2, \quad q(x) = -x, \quad r(x) = e^x]$$

$$y'' + y' - 3y = \sin x \qquad [p(x) = 1, \quad q(x) = -3, \quad r(x) = \sin x]$$

$$y'' + e^x y = 3 \qquad [p(x) = 0, \quad q(x) = e^x, \quad r(x) = 3]$$

$$\frac{d^2y}{dx^2} - \frac{dy}{dx} + 2y = 0 \qquad [p(x) = -1, \quad q(x) = 2, \quad r(x) = 0]$$

The last equation is homogeneous and the first three are nonhomogeneous.

In Section 7.7 we gave a general procedure for solving first-order linear differential equations by integration. For second-order linear differential equations the situation is more complicated and simple general procedures for solving such equations can only be given in special cases. Before we can pursue this matter further, we shall need some preliminary results.

First, some terminology. Two functions f and g are said to be *linearly dependent* if one is a constant multiple of the other. If neither is a constant multiple of the other, then they are called *linearly independent*. Thus,

$$f(x) = \sin x \quad \text{and} \quad g(x) = 3 \sin x$$

are linearly dependent, but

$$f(x) = x \quad \text{and} \quad g(x) = x^2$$

are linearly independent.

The following theorem is central to the study of second-order linear differential equations.

19.1.1 THEOREM. *If $y_1 = y_1(x)$ and $y_2 = y_2(x)$ are linearly independent solutions of the homogeneous equation*

$$\frac{d^2y}{dx^2} + p(x)\frac{dy}{dx} + q(x)y = 0 \qquad (2)$$

then

$$y(x) = c_1 y_1(x) + c_2 y_2(x) \qquad (3)$$

is the general solution of (2) in the sense that every solution of (2) can be obtained from (3) by choosing appropriate values for the arbitrary constants c_1 and c_2, and conversely, (3) is a solution of (2) for all choices of c_1 and c_2.

(Readers interested in a proof of this theorem are referred to *Elementary Differential Equations and Boundary Value Problems,* John Wiley and Sons, 1969, by William E. Boyce and Richard C. DiPrima.)

REMARK. The expression on the right side of (3) is called a *linear combination* of $y_1(x)$ and $y_2(x)$. Thus, Theorem 19.1.1 tells us that once we find two linearly independent solutions of (2), we essentially know all the solutions because every other solution can be expressed as a linear combination of those two.

CONSTANT
COEFFICIENTS

For the remainder of this section we shall restrict our attention to second-order linear homogeneous equations in which $p(x)$ and $q(x)$ are constants, p and q. Such equations have the form

$$\frac{d^2y}{dx^2} + p\frac{dy}{dx} + qy = 0 \tag{4}$$

Our objective is to find two linearly independent solutions of this equation. To start, we note that the function e^{mx} has the property that its derivatives are constant multiples of itself. This suggests the possibility that

$$y = e^{mx} \tag{5}$$

might be a solution of (4) if the constant m is suitably chosen. Since

$$\frac{dy}{dx} = me^{mx}, \quad \frac{d^2y}{dx^2} = m^2e^{mx} \tag{6}$$

substituting (5) and (6) into (4) yields

$$(m^2 + pm + q)e^{mx} = 0 \tag{7}$$

which is satisfied if and only if

$$m^2 + pm + q = 0 \tag{8}$$

since $e^{mx} \neq 0$ for any x.

Equation (8), which is called the *auxiliary equation* for (4), can be obtained from (4) by replacing d^2y/dx^2 by m^2, dy/dx by m, and y by 1. The solutions, m_1 and m_2, of the auxiliary equation can be obtained by factoring or by the quadratic formula. These solutions are

$$m_1 = \frac{-p + \sqrt{p^2 - 4q}}{2}, \quad m_2 = \frac{-p - \sqrt{p^2 - 4q}}{2} \tag{9}$$

Depending on whether $p^2 - 4q$ is positive, zero, or negative, these roots will be distinct and real, equal and real, or complex conjugates.* We shall consider each of these cases separately.

---

*Recall that the complex solutions of a polynomial equation, and in particular of a quadratic equation, occur as conjugate pairs $a + bi$ and $a - bi$.

DISTINCT REAL ROOTS   If m_1 and m_2 are distinct real roots, then (4) has the two solutions

$$y_1 = e^{m_1 x}, \quad y_2 = e^{m_2 x}$$

Neither of the functions $e^{m_1 x}$ and $e^{m_2 x}$ is a constant multiple of the other (Exercise 29), so the general solution of (4) in this case is

$$y(x) = c_1 e^{m_1 x} + c_2 e^{m_2 x} \tag{10}$$

Example 1   Find the general solution of

$$y'' - y' - 6y = 0$$

Solution.   The auxiliary equation

$$m^2 - m - 6 = 0$$

can be rewritten as

$$(m + 2)(m - 3) = 0$$

so its roots are $m = -2$, $m = 3$. Thus, from (10) the general solution of the differential equation is

$$y = c_1 e^{-2x} + c_2 e^{3x}$$

where c_1 and c_2 are arbitrary constants.   ◀

EQUAL REAL ROOTS   If m_1 and m_2 are equal real roots, say $m_1 = m_2 (= m)$, then the auxiliary equation yields only one solution of (4):

$$y_1(x) = e^{mx}$$

We shall now show that

$$y_2(x) = x e^{mx} \tag{11}$$

is a second linearly independent solution. To see that this is so, note that $p^2 - 4q = 0$ in (9) since the roots are equal. Thus,

$$m = m_1 = m_2 = -p/2$$

and (11) becomes

$$y_2(x) = x e^{(-p/2)x}$$

Differentiating yields

$$y_2'(x) = \left(1 - \frac{p}{2} x\right) e^{(-p/2)x}$$

$$y_2''(x) = \left(\frac{p^2}{4}x - p\right) e^{-(p/2)x}$$

so

$$y_2''(x) + py_2'(x) + qy_2(x) = \left[\left(\frac{p^2}{4}x - p\right) + p\left(1 - \frac{p}{2}x\right) + qx\right]e^{(-p/2)x}$$

$$= \left[-\frac{p^2}{4} + q\right] xe^{(-p/2)x}$$

(12)

But $p^2 - 4q = 0$ implies that $(-p^2/4) + q = 0$, so (12) becomes

$$y_2''(x) + py_2'(x) + qy_2(x) = 0$$

which tells us that $y_2(x)$ is a solution of (4). It can be shown that

$$y_1(x) = e^{mx} \quad \text{and} \quad y_2(x) = xe^{mx}$$

are linearly independent (Exercise 29), so the general solution of (4) in this case is

$$y = c_1 e^{mx} + c_2 x e^{mx}$$

(13)

Example 2   Find the general solution of

$$y'' - 8y' + 16y = 0$$

Solution.   The auxiliary equation

$$m^2 - 8m + 16 = 0$$

can be rewritten as

$$(m - 4)^2 = 0$$

so $m = 4$ is the only root. Thus, from (13) the general solution of the differential equation is

$$y = c_1 e^{4x} + c_2 x e^{4x} \quad \blacktriangleleft$$

COMPLEX ROOTS   If the auxiliary equation has complex roots $m_1 = a + bi$ and $m_2 = a - bi$, then $y_1(x) = e^{ax}\cos bx$ and $y_2(x) = e^{ax}\sin bx$ are linearly independent solutions of (4) and

$$y = e^{ax}(c_1 \cos bx + c_2 \sin bx)$$

(14)

is the general solution. The proof is discussed in the exercises (Exercise 30).

Example 3   Find the general solution of

$$y'' + y' + y = 0$$

Solution.   The auxiliary equation

$$m^2 + m + 1 = 0$$

has roots

$$m_1 = \frac{-1 + \sqrt{1 - 4}}{2} = -\frac{1}{2} + \frac{\sqrt{3}}{2}i$$

$$m_2 = \frac{-1 - \sqrt{1 - 4}}{2} = -\frac{1}{2} - \frac{\sqrt{3}}{2}i$$

Thus, from (14) with $a = -1/2$ and $b = \sqrt{3}/2$, the general solution of the differential equation is

$$y = e^{-x/2}\left(c_1 \cos \frac{\sqrt{3}}{2}x + c_2 \sin \frac{\sqrt{3}}{2}x\right) \quad \blacktriangleleft$$

INITIAL VALUE PROBLEMS

When a physical problem leads to a second-order differential equation, there are usually two conditions in the problem that determine specific values for the two arbitrary constants in the general solution of the equation. Conditions that specify the value of the solution $y(x)$ and its derivative $y'(x)$ at some point $x = x_0$ are called *initial conditions*. A second-order differential equation with initial conditions is called a *second-order initial-value problem.*

Example 4   Solve the initial-value problem

$$y'' - y = 0, \quad y(0) = 1, \quad y'(0) = 0$$

Solution.   We must first solve the differential equation. The auxiliary equation

$$m^2 - 1 = 0$$

has distinct real roots $m_1 = 1$, $m_2 = -1$, so from (10) the general solution is

$$y(x) = c_1 e^x + c_2 e^{-x} \tag{15}$$

and the derivative of this solution is

$$y'(x) = c_1 e^x - c_2 e^{-x} \tag{16}$$

Substituting $x = 0$ in (15) and (16) and using the initial conditions $y(0) = 1$ and $y'(0) = 0$ yields the system of equations

$$c_1 + c_2 = 1$$
$$c_1 - c_2 = 0$$

Solving this system yields $c_1 = \frac{1}{2}$, $c_2 = \frac{1}{2}$, so from (15) the solution of the initial-value problem is

$$y(x) = \tfrac{1}{2}e^x + \tfrac{1}{2}e^{-x} \quad \blacktriangleleft$$

The following summary is included as a ready reference for the solution of second-order homogeneous linear differential equations with constant coefficients.

Summary

EQUATION: $y'' + py' + qy = 0$
AUXILIARY EQUATION: $m^2 + pm + q = 0$

CASE	GENERAL SOLUTION
Distinct real roots m_1, m_2 to the auxiliary equation.	$y = c_1 e^{m_1 x} + c_2 e^{m_2 x}$
Equal real roots $m_1 = m_2 (= m)$ to the auxiliary equation.	$y = c_1 e^{mx} + c_2 x e^{mx}$
Complex roots $m_1 = a + bi$, $m_2 = a - bi$ to the auxiliary equation.	$y = e^{ax}(c_1 \cos bx + c_2 \sin bx)$

▶ Exercise Set 19.1

1. Verify that the following are solutions of the differential equation $y'' - y' - 2y = 0$ by substituting these functions into the equation:
   (a) e^{2x} and e^{-x}
   (b) $c_1 e^{2x} + c_2 e^{-x}$ (c_1, c_2 constants).

2. Verify that the following are solutions of the differential equation $y'' + 4y' + 4y = 0$ by substituting these functions into the equation:
   (a) e^{-2x} and $x e^{-2x}$
   (b) $c_1 e^{-2x} + c_2 x e^{-2x}$ (c_1, c_2 constants).

In Exercises 3–16, find the general solution of the differential equation.

3. $y'' + 3y' - 4y = 0$.   4. $y'' + 6y' + 5y = 0$.

5. $y'' - 2y' + y = 0$.   6. $y'' + 6y' + 9y = 0$.

7. $y'' + 5y = 0$.   8. $y'' + y = 0$.

9. $\dfrac{d^2 y}{dx^2} - \dfrac{dy}{dx} = 0$.   10. $\dfrac{d^2 y}{dx^2} + 3\dfrac{dy}{dx} = 0$.

11. $\dfrac{d^2 y}{dt^2} + 4\dfrac{dy}{dt} + 4y = 0$.

12. $\dfrac{d^2y}{dt^2} - 10\dfrac{dy}{dt} + 25y = 0.$

13. $\dfrac{d^2y}{dx^2} - 4\dfrac{dy}{dx} + 13y = 0.$

14. $\dfrac{d^2y}{dx^2} - 6\dfrac{dy}{dx} + 25y = 0.$

15. $8y'' - 2y' - 1 = 0.$ **16.** $9y'' - 6y' + 1 = 0.$

In Exercises 17–22, solve the initial-value problem.

17. $y'' + 2y' - 3y = 0$; $y(0) = 1,\ y'(0) = 5.$

18. $y'' - 6y' - 7y = 0$; $y(0) = 5,\ y'(0) = 3.$

19. $y'' - 6y' + 9y = 0$; $y(0) = 2,\ y'(0) = 1.$

20. $y'' + 4y' + y = 0$; $y(0) = 5,\ y'(0) = 4.$

21. $y'' + 4y' + 5y = 0$; $y(0) = -3,\ y'(0) = 0.$

22. $y'' - 6y' + 13y = 0$; $y(0) = -1,\ y'(0) = 1.$

23. In each part find a second-order linear homogeneous differential equation with constant coefficients that has the given functions as solutions.
   (a) $y_1 = e^{5x},\ y_2 = e^{-2x}$
   (b) $y_1 = e^{4x},\ y_2 = xe^{4x}$
   (c) $y_1 = e^{-x}\cos 4x,\ y_2 = e^{-x}\sin 4x.$

24. Show that if e^x and e^{-x} are solutions of a second-order linear homogeneous differential equation, then so are $\cosh x$ and $\sinh x.$

25. Find all values of k for which the equation $y'' + ky' + ky = 0$ has a general solution of the given form.
   (a) $y = c_1 e^{ax} + c_2 e^{bx}$ (b) $y = c_1 e^{ax} + c_2 x e^{ax}$
   (c) $y = c_1 e^{ax}\cos bx + c_2 e^{ax}\sin bx.$

26. The equation

$$x^2\frac{d^2y}{dx^2} + px\frac{dy}{dx} + qy = 0$$

where p and q are constants, is called *Euler's equidimensional equation*. Show that the substitution $|x| = e^z$ transforms this equation into the equation

$$\frac{d^2y}{dz^2} + (p - 1)\frac{dy}{dz} + qy = 0$$

27. Use the result in Exercise 26 to find the general solution of:
   (a) $x^2\dfrac{d^2y}{dx^2} + 3x\dfrac{dy}{dx} + 2y = 0$

   (b) $x^2\dfrac{d^2y}{dx^2} - x\dfrac{dy}{dx} - 2y = 0.$

28. Let $y(x)$ be a solution of $y'' + py' + qy = 0$. Prove: If p and q are positive constants, then $\lim\limits_{x \to +\infty} y(x) = 0.$

29. The *Wronskian* of two functions y_1 and y_2 is denoted by $W(y_1, y_2)$ and is defined to be the function

$$W(y_1, y_2) = y_1 y_2' - y_1' y_2 = \begin{vmatrix} y_1 & y_2 \\ y_1' & y_2' \end{vmatrix}$$

The value of $W(y_1, y_2)$ at a point x is denoted by $W(y_1, y_2)(x)$ or often more simply by $W(x)$. It can be proved that two solutions, $y_1 = y_1(x)$ and $y_2 = y_2(x)$, of Equation (2) are linearly dependent if and only if $W(x) = 0$ for all x. Equivalently, the functions are linearly independent if and only if $W(x) \neq 0$ for at least one value of x. Use this result to prove that the following solutions of Equation (4) are linearly independent:
   (a) $y_1 = e^{m_1 x},\ y_2 = e^{m_2 x}$ $(m_1 \neq m_2)$
   (b) $y_1 = e^{mx},\ y_2 = xe^{mx}.$

30. Prove: If the auxiliary equation of $y'' + py' + qy = 0$ has complex roots, $a + bi$ and $a - bi$, then the general solution of this differential equation is $y(x) = e^{ax}(c_1\cos bx + c_2\sin bx)$. [*Hint:* By substitution, verify that $y_1 = e^{ax}\cos bx$ and $y_2 = e^{ax}\sin bx$ are solutions of the differential equation. Then use Exercise 29 to prove that y_1 and y_2 are linearly independent.]

31. Suppose that the auxiliary equation of the differential equation $y'' + py' + qy = 0$ has distinct real roots μ and m.
   (a) Show that the function

$$g_\mu(x) = \frac{e^{\mu x} - e^{mx}}{\mu - m}$$

   is a solution of the differential equation.
   (b) Use L'Hôpital's Rule to show that

$$\lim_{\mu \to m} g_\mu(x) = xe^{mx}$$

   [*Note:* Can you see how the result in part (b) makes it plausible that the function $y(x) = xe^{mx}$ is a solution of $y'' + py' + qy = 0$

when m is a repeated root of the auxiliary equation?]

32. Consider the problem of solving the differential equation

$$y'' + \lambda y = 0$$

subject to the conditions $y(0) = 0$, $y(\pi) = 0$.

(a) Show that if $\lambda \leq 0$, then $y = 0$ is the only solution.

(b) Show that if $\lambda > 0$, then the solution is $y = c \sin \sqrt{\lambda} x$, where c is an arbitrary constant, if $\lambda = 1, 2^2, 3^2, 4^2, \ldots$ and the solution is $y = 0$ otherwise.

19.2 SECOND-ORDER LINEAR NONHOMOGENEOUS DIFFERENTIAL EQUATIONS WITH CONSTANT COEFFICIENTS; UNDETERMINED COEFFICIENTS

In this section we shall study techniques for solving second-order linear differential equations that are not homogeneous.

The following theorem is the key result for solving a second-order *nonhomogeneous* linear differential equation with constant coefficients

$$y'' + py' + qy = r(x) \tag{1}$$

where p and q are constants and $r(x)$ is a continuous function of x.

19.2.1 THEOREM. *The general solution of* (1) *is*

$$y(x) = c_1 y_1(x) + c_2 y_2(x) + y_p(x)$$

where $c_1 y_1(x) + c_2 y_2(x)$ *is the general solution of the homogeneous equation*

$$y'' + py' + qy = 0 \tag{2}$$

and $y_p(x)$ *is any solution of* (1).

The proof is deferred to the end of the section.

REMARK.   Equation (2) is called the ***complementary equation*** to (1) and $y_p(x)$ is called a ***particular solution*** of (1). Thus, Theorem 19.2.1 states that *the general solution of* (1) *is obtained by adding a particular solution of* (1) *to the general solution of the complementary equation.*

THE METHOD OF UNDETERMINED COEFFICIENTS

Since we already know how to obtain the general solution of (2), we shall focus our attention on the problem of obtaining a particular solution of (1). In this section we shall discuss a procedure for doing this called the method of ***undetermined coefficients***, and in the next section we shall discuss a second procedure called *variation of parameters.*

The first step in the method of undetermined coefficients is to make a reasonable guess about the form of the particular solution. This guess will involve one or more unknown coefficients which we shall then determine from conditions obtained by substituting the proposed solution into the differential equation.

We shall begin with some examples that illustrate the basic idea.

Example 1   Find a particular solution of the differential equation

$$y'' + 2y' - 8y = e^{3x} \tag{3}$$

Solution.   It should be clear that such functions as $\sin x$, $\cos x$, $\ln x$, or x^3 are not reasonable possibilities for $y_p(x)$ because the derivatives of such functions do not yield expressions involving e^{3x}. Since the obvious choice for $y_p(x)$ is an expression involving e^{3x}, we shall *guess* that $y_p(x)$ has the form

$$y_p(x) = Ae^{3x} \tag{4}$$

where A is an unknown constant to be determined. It follows that

$$y_p'(x) = 3Ae^{3x} \tag{5}$$

$$y_p''(x) = 9Ae^{3x} \tag{6}$$

Substituting expressions (4), (5), and (6) for y, y', and y'' in (3) yields

$$9Ae^{3x} + 2(3Ae^{3x}) - 8(Ae^{3x}) = e^{3x}$$

or

$$7Ae^{3x} = e^{3x}$$

Thus, $A = \frac{1}{7}$. From this result and (4), a particular solution of (3) is

$$y_p(x) = \tfrac{1}{7}e^{3x} \qquad \blacktriangleleft$$

Example 2   Find a particular solution of the differential equation

$$y'' - y' - 6y = e^{3x} \tag{7}$$

Solution.   As in Example 1, a reasonable guess is $y_p(x) = Ae^{3x}$. However, if we substitute (4), (5), and (6) in (7) we obtain

$$9Ae^{3x} - 3Ae^{3x} - 6Ae^{3x} = e^{3x}$$

or

$$0 = e^{3x}$$

which is contradictory. The problem here is that $y_p(x) = Ae^{3x}$ is a solution of the complementary equation

$$y'' - y' - 6y = 0$$

which makes it impossible for y_p to satisfy (7), no matter how A is selected. How should we proceed in this case? Experience has shown that multiplying (4) by x produces the correct form for the solution. Thus, we shall try

$$y_p(x) = Axe^{3x} \tag{8}$$

It follows that

$$y_p'(x) = 3Axe^{3x} + Ae^{3x} \tag{9}$$
$$y_p''(x) = 9Axe^{3x} + 6Ae^{3x} \tag{10}$$

Substituting expressions (8), (9), and (10) for y, y', and y'' in (7) yields

$$(9Axe^{3x} + 6Ae^{3x}) - (3Axe^{3x} + Ae^{3x}) - 6(Axe^{3x}) = e^{3x}$$

or

$$5Ae^{3x} = e^{3x}$$

Thus, $A = \frac{1}{5}$, so (8) implies that a particular solution of (7) is

$$y_p(x) = \tfrac{1}{5}xe^{3x} \qquad \blacktriangleleft$$

REMARK.   Had it turned out that Axe^{3x} was also a solution of the complementary equation, then we would have tried $y_p(x) = Ax^2e^{3x}$ as our candidate for a particular solution.

In summary, a particular solution of an equation of the form

$$y'' + py' + qy = ke^{ax}$$

can be obtained as follows:

Step 1.   Start with $y_p = Ae^{ax}$ as an initial guess.

Step 2.   Determine if the initial guess is a solution of the complementary equation $y'' + py' + qy = 0$.

Step 3.   If the initial guess is not a solution of the complementary equation, then $y_p = Ae^{ax}$ is the correct form of a particular solution.

Step 4.   If the initial guess is a solution of the complementary equation, then multiply it by the smallest positive integer power of x required to produce a function that is not a solution of the complementary equation. This will yield either $y_p = Axe^{ax}$ or $y_p = Ax^2 e^{ax}$.

Table 19.2.1, which we provide without proof, restates the preceeding procedure more compactly and also explains how to find a particular solution of (1) when $r(x)$ is a polynomial or a combination of sine and cosine functions.

Table 19.2.1

EQUATION	INITIAL GUESS FOR y_p
$y'' + py' + qy = ke^{ax}$	$y_p = Ae^{ax}$
$y'' + py' + qy = a_0 + a_1 x + \cdots + a_n x^n$	$y_p = A_0 + A_1 x + \cdots + A_n x^n$
$y'' + py' + qy = a_1 \cos bx + a_2 \sin bx$	$y_p = A_1 \cos bx + A_2 \sin bx$

MODIFICATION RULE

If any term in the initial guess is a solution of the complementary equation, then the correct form for y_p is obtained by multiplying the initial guess by the smallest positive integer power of x required so that no term is a solution of the complementary equation.

Example 3   Find the general solution of

$$y'' + y' = 4x^2 \tag{11}$$

Solution.   We begin by finding the general solution of the complementary equation

$$y'' + y' = 0$$

The auxiliary equation for this homogeneous equation is

$$m^2 + m = 0$$

which has roots $m_1 = 0$, $m_2 = -1$. Thus, the general solution, $y_c(x)$, of the complementary equation is

$$y_c(x) = c_1 e^{0x} + c_2 e^{-x} = c_1 + c_2 e^{-x}$$

Since the right-hand side of (11) is a second-degree polynomial, our initial guess for y_p is

$$y_p = A_0 + A_1 x + A_2 x^2$$

But A_0 is a solution of the complementary equation (take $c_1 = A_0, c_2 = 0$), so our second guess for y_p is

$$y_p = x(A_0 + A_1x + A_2x^2) = A_0x + A_1x^2 + A_2x^3 \tag{12}$$

Since no term of this function is a solution of the complementary equation, this is the correct form for y_p. Differentiating (12) we obtain

$$y_p'(x) = A_0 + 2A_1x + 3A_2x^2 \tag{13}$$

$$y_p''(x) = 2A_1 + 6A_2x \tag{14}$$

Substituting (13) and (14) in (11) yields

$$(2A_1 + 6A_2x) + (A_0 + 2A_1x + 3A_2x^2) = 4x^2$$

or

$$(A_0 + 2A_1) + (2A_1 + 6A_2)x + 3A_2x^2 = 4x^2$$

Equating corresponding coefficients on the two sides of this equation yields the system of equations

$$A_0 + 2A_1 = 0$$
$$2A_1 + 6A_2 = 0$$
$$3A_2 = 4$$

from which we obtain

$$A_2 = \tfrac{4}{3}, \quad A_1 = -4, \quad A_0 = 8$$

Substituting these values in (12) yields the particular solution

$$y_p(x) = 8x - 4x^2 + \tfrac{4}{3}x^3$$

Thus, the general solution of (11) is

$$y(x) = y_c(x) + y_p(x) = c_1 + c_2e^{-x} + 8x - 4x^2 + \tfrac{4}{3}x^3 \qquad \blacktriangleleft$$

Example 4   Find the general solution of

$$y'' - 2y' + y = \sin 2x \tag{15}$$

Solution.   We begin by finding the general solution of the complementary equation

$$y'' - 2y' + y = 0$$

The auxiliary equation for this homogeneous equation is

$$m^2 - 2m + 1 = 0$$

which has the repeated root $m_1 = 1$, $m_2 = 1$. Thus, the general solution of the complementary equation is

$$y_c(x) = c_1 e^x + c_2 x e^x \tag{16}$$

Since $\sin 2x$ has the form $a_1 \cos bx + a_2 \sin bx$ $(a_1 = 0, \ a_2 = 1, \ b = 2)$, our initial guess for y_p is

$$y_p(x) = A_1 \cos 2x + A_2 \sin 2x \tag{17}$$

Since no term of y_p is a solution of the complementary equation [see (16)], this is the correct form for a particular solution. Differentiating (17) we obtain

$$y_p'(x) = -2A_1 \sin 2x + 2A_2 \cos 2x \tag{18}$$

$$y_p''(x) = -4A_1 \cos 2x - 4A_2 \sin 2x \tag{19}$$

Substituting (17), (18), and (19) in (15) yields

$$(-4A_1 \cos 2x - 4A_2 \sin 2x) - 2(-2A_1 \sin 2x + 2A_2 \cos 2x)$$
$$+ (A_1 \cos 2x + A_2 \sin 2x) = \sin 2x$$

or

$$(-3A_1 - 4A_2) \cos 2x + (4A_1 - 3A_2) \sin 2x = \sin 2x$$

Equating the coefficients of $\sin 2x$ and $\cos 2x$ on the two sides of this equation yields the system of equations

$$-3A_1 - 4A_2 = 0$$
$$4A_1 - 3A_2 = 1$$

from which we obtain

$$A_1 = \tfrac{4}{25}, \quad A_2 = -\tfrac{3}{25}$$

(Verify.) From this result and (17), a particular solution of (15) is

$$y_p(x) = \tfrac{4}{25} \cos 2x - \tfrac{3}{25} \sin 2x$$

Thus, the general solution of (15) is

$$y(x) = y_c(x) + y_p(x) = c_1 e^x + c_2 x e^x + \tfrac{4}{25} \cos 2x - \tfrac{3}{25} \sin 2x \qquad \blacktriangleleft$$

We conclude this section with a proof of Theorem 19.2.1.

OPTIONAL

Proof of Theorem 19.2.1.   To prove that

$$y(x) = c_1 y_1(x) + c_2 y_2(x) + y_p(x) \tag{20}$$

is the general solution of (1) we must prove two results: first, that for all choices of c_1 and c_2 the function $y(x)$ defined by (20) satisfies (1); and second, that every solution of (1) can be obtained from (20) by choosing appropriate values for the constants c_1 and c_2.

To prove the first statement, let c_1 and c_2 have any real values. Then differentiating (20) yields

$$y'(x) = c_1 y_1'(x) + c_2 y_2'(x) + y_p'(x)$$
$$y''(x) = c_1 y_1''(x) + c_2 y_2''(x) + y_p''(x)$$

so

$$
\begin{aligned}
y''(x) + py'(x) + qy(x) = {} & c_1(y_1''(x) + py_1'(x) + qy_1(x)) \\
& + c_2(y_2''(x) + py_2'(x) + qy_2(x)) \\
& + y_p''(x) + py_p'(x) + qy_p(x)
\end{aligned}
\tag{21}
$$

But $y_1(x)$ and $y_2(x)$ satisfy (2) and $y_p(x)$ satisfies (1), so (21) reduces to

$$y''(x) + py'(x) + qy(x) = 0 + 0 + r(x) = r(x)$$

which proves that $y(x)$ satisfies (1).

To prove the second statement, let $y(x)$ be any solution of (1). We must find values of c_1 and c_2 such that

$$y(x) = c_1 y_1(x) + c_2 y_2(x) + y_p(x) \tag{22}$$

But $y(x) - y_p(x)$ satisfies (2) since

$$
\begin{aligned}
(y(x) - y_p(x))'' &+ p(y(x) - y_p(x))' + q(y(x) - y_p(x)) \\
&= [y''(x) + py'(x) + qy(x)] - [y_p''(x) + py_p'(x) + qy_p(x)] \\
&= r(x) - r(x) = 0
\end{aligned}
$$

Thus, since $c_1 y_1(x) + c_2 y_2(x)$ is the general solution of (2), there exist values of c_1 and c_2 such that the solution $y(x) - y_p(x)$ can be written as

$$y(x) - y_p(x) = c_1 y_1(x) + c_2 y_2(x)$$

from which (22) follows. ∎

▶ Exercise Set 19.2

In Exercises 1–24, use the method of undetermined coefficients to find the general solution of the differential equation.

1. $y'' + 6y' + 5y = 2e^{3x}$. 2. $y'' + 3y' - 4y = 5e^{7x}$.

3. $y'' - 9y' + 20y = -3e^{5x}$.

4. $y'' + 7y' - 8y = 7e^x$.

5. $y'' + 2y' + y = e^{-x}$.

6. $y'' + 4y' + 4y = 4e^{-2x}$.

7. $y'' + y' - 12y = 4x^2$.

8. $y'' - 4y' - 5y = -6x^2$.

9. $y'' - 6y' = x - 1$.   10. $y'' + 3y' = 2x + 2$.

11. $y'' - x^3 + 1 = 0$.   12. $y'' + 3x^3 + x = 0$.

13. $y'' - y' - 2y = 10 \cos x$.

14. $y'' - 3y' - 4y = 2 \sin x.$

15. $y'' - 4y = 2 \sin 2x + 3 \cos 2x.$

16. $y'' - 9y = \cos 3x - \sin 3x.$

17. $y'' + y = \sin x.$ **18.** $y'' + 4y = \cos 2x.$

19. $y'' - 3y' + 2y = x.$

20. $y'' + 4y' + 4y = 3x + 3.$

21. $y'' + 4y' + 9y = x^2 + 3x.$

22. $y'' - y = 1 + x + x^2.$

23. $y'' + 4y = \sin x \cos x.$

24. $y'' + 4y = \cos^2 x - \sin^2 x.$

25. (a) Prove: If $y_1(x)$ is a solution of

$$y'' + p(x)y' + q(x)y = r_1(x)$$

and $y_2(x)$ is a solution of

$$y'' + p(x)y' + q(x)y = r_2(x)$$

then $y_1(x) + y_2(x)$ is a solution of

$$y'' + p(x)y' + q(x)y = r_1(x) + r_2(x)$$

(b) Use the result in part (a) to find a particular solution of the equation $y'' + 3y' - 4y = x + e^x.$

(c) State a generalization of the result in (a) that is applicable to the equation

$$y'' + p(x)y' + q(x)y = r_1(x)$$
$$+ r_2(x) + \cdots + r_n(x)$$

In Exercises 26–34, use the results in Exercise 25 to find the general solution of the differential equation.

26. $y'' - y' - 2y = x + e^{-x}.$

27. $y'' - y = 1 + e^x.$

28. $y'' - 4y' + 3y = 2 \cos x + 4 \sin x.$

29. $y'' + 4y = 1 + x + \sin x.$

30. $y'' + 2y' + y = 2 + 3x + 3e^x + 2 \cos 2x.$

31. $y'' - 2y' + y = \sinh x.$ [*Hint:* $\sinh x = \frac{1}{2}(e^x - e^{-x}).$]

32. $y'' + 4y' - 5y = \cosh x.$
[*Hint:* $\cosh x = \frac{1}{2}(e^x + e^{-x}).$]

33. $y'' + y = 12 \cos^2 x.$ [*Hint:* $\cos^2 x = \frac{1}{2}(1 + \cos 2x).$]

34. $y'' + 2y' + y = \sin^2 x.$
[*Hint:* $\sin^2 x = \frac{1}{2}(1 - \cos 2x).$]

35. (a) Find the general solution of

$$y'' + \mu^2 y = a \sin bx$$

where a is an arbitrary constant and μ and b are positive constants such that $\mu \neq b.$

(b) Use part (a) and Exercise 25 to find the general solution of

$$y'' + \mu^2 y = \sum_{k=1}^{n} a_k \sin k\pi x$$

where $\mu > 0$ and $\mu \neq k\pi,\ k = 1, 2, \ldots, n.$

36. Find the general solution of

$$y'' + \lambda^2 y = \sum_{k=1}^{n} a_k \cos k\pi x$$

where $\lambda > 0$ and $\lambda \neq k\pi,\ k = 1, 2, \ldots, n.$ [*Hint:* See Exercise 35.]

37. Find all solutions of the equation

$$y'' - y' = 4 - 4x$$

(if any) with the property that $y'(x_0) = y''(x_0) = 0$ at some point $x_0.$

19.3 VARIATION OF PARAMETERS

In this section* we shall consider an alternative method for finding a particular solution of a second-order linear nonhomogeneous equation that often works when the method of undetermined coefficients cannot be applied.

---

*It is assumed in this section that the reader is familiar with Cramer's Rule, which is discussed in Unit VI of Appendix 2.

As in the previous section, we shall be concerned with equations of the form

$$y'' + py' + qy = r(x) \tag{1}$$

where p and q are constants and $r(x)$ is a continuous function of x. The method we shall discuss, called **variation of parameters**, is based on the (not very obvious) fact that if

$$y_c(x) = c_1 y_1(x) + c_2 y_2(x) \tag{2}$$

is the general solution of the complementary equation

$$y'' + py' + qy = 0$$

then it is possible to find functions $u(x)$ and $v(x)$ such that

$$y_p(x) = u(x)y_1(x) + v(x)y_2(x) \tag{3}$$

is a particular solution of (1).

To see how the functions $u(x)$ and $v(x)$ can be found, consider the derivative of (3):

$$y_p' = (uy_1' + vy_2') + (u'y_1 + v'y_2) \tag{4}$$

If we now differentiate y_p', we shall introduce second derivatives of the unknown functions $u(x)$ and $v(x)$. To avoid this complication we shall require that $u(x)$ and $v(x)$ satisfy the condition

$$u'y_1 + v'y_2 = 0 \tag{5}$$

so (4) simplifies to

$$y_p' = uy_1' + vy_2'$$

Thus,

$$y_p'' = uy_1'' + u'y_1' + vy_2'' + v'y_2' \tag{6}$$

On substituting (3), (4), and (6) in (1) and rearranging terms we obtain

$$u(y_1'' + py_1' + qy_1) + v(y_2'' + py_2' + qy_2) + u'y_1' + v'y_2' = r(x) \tag{7}$$

Since y_1 and y_2 are solutions of the complementary equation, we have

$$y_1'' + py_1' + qy_1 = 0 \quad \text{and} \quad y_2'' + py_2' + qy_2 = 0$$

so (7) simplifies to

$$u'y_1' + v'y_2' = r(x) \tag{8}$$

Equations (5) and (8) together yield two equations in the two unknowns

$u'(x)$ and $v'(x)$:

$$u'y_1 + v'y_2 = 0$$
$$u'y_1' + v'y_2' = r(x) \tag{9}$$

It can be shown (see Boyce and DiPrima, *Elementary Differential Equations and Boundary Value Problems,* John Wiley and Sons, N.Y., 1969) that this system has a unique solution for u' and v'. Once this solution is found, u and v can be obtained by integration.

Example 1   Find the general solution of

$$y'' + y = \sec x \tag{10}$$

Solution.   We begin by finding the general solution of the complementary equation

$$y'' + y = 0$$

The auxiliary equation for this homogeneous equation is

$$m^2 + 1 = 0$$

which has roots $m_1 = i$ and $m_2 = -i$. Thus, the general solution of the complementary equation is

$$y_c(x) = c_1 \cos x + c_2 \sin x$$

Comparing this to (2) yields $y_1(x) = \cos x$ and $y_2(x) = \sin x$, so we shall look for a particular solution of the form

$$y_p(x) = u(x) \cos x + v(x) \sin x \tag{11}$$

Substituting $y_1 = \cos x$, $y_2 = \sin x$, and $r(x) = \sec x$ in (9) yields

$$u' \cos x + v' \sin x = 0$$
$$-u' \sin x + v' \cos x = \sec x \tag{12}$$

Solving this system for u' and v' by Cramer's Rule we obtain

$$u' = \frac{\begin{vmatrix} 0 & \sin x \\ \sec x & -\cos x \end{vmatrix}}{\begin{vmatrix} \cos x & \sin x \\ -\sin x & \cos x \end{vmatrix}} = \frac{-\sec x \sin x}{\cos^2 x + \sin^2 x} = \frac{-\tan x}{1} = -\tan x$$

$$v' = \frac{\begin{vmatrix} \cos x & 0 \\ -\sin x & \sec x \end{vmatrix}}{\begin{vmatrix} \cos x & \sin x \\ -\sin x & \cos x \end{vmatrix}} = \frac{\cos x \sec x}{\cos^2 x + \sin^2 x} = \frac{1}{1} = 1$$

Integrating u' and v' yields

$$u(x) = \int -\tan x \, dx = \ln|\cos x|$$

$$v(x) = \int 1 \, dx = x$$

(We set the constants of integration equal to zero because any functions $u(x)$ and $v(x)$ satisfying (12) will suffice.) Substituting these functions in (11) yields

$$y_p(x) = (\ln|\cos x|) \cos x + x \sin x$$

Thus, the general solution of (10) is

$$y(x) = y_c(x) + y_p(x) = c_1 \cos x + c_2 \sin x + (\ln|\cos x|) \cos x + x \sin x$$

◀

▶ Exercise Set 19.3

In Exercises 1–6, use variation of parameters to find the general solution of the differential equation and check your result using undetermined coefficients.

1. $y'' + y = x^2$.

2. $y'' + 9y = 3x$.

3. $y'' + y' - 2y = 2e^x$.

4. $y'' + 5y' + 6y = e^{-x}$.

5. $y'' + 4y = \sin 2x$.

6. $y'' + 9y = \cos 3x$.

In Exercises 7–28, use variation of parameters to find the general solution of the differential equation.

7. $y'' + y = \tan x$.

8. $y'' + y = \cot x$.

9. $y'' - 2y' + y = \dfrac{1}{x} e^x$.

10. $y'' - 4y' + 4y = \dfrac{1}{x} e^{2x}$.

11. $y'' + y = 3 \sin^2 x$.

12. $y'' + y = 6 \cos^2 x$.

13. $y'' + y = \csc x$.

14. $y'' + 9y = 6 \sec 3x$.

15. $y'' + y = \sec x \tan x$.

16. $y'' + y = \csc x \cot x$.

17. $y'' + 2y' + y = e^{-x}/x^2$.

18. $y'' - y = x^2 e^x$.

19. $y'' + 4y' + 4y = xe^{-x}$.

20. $y'' + 4y' + 4y = xe^{2x}$.

21. $y'' + y = \sec^2 x$.

22. $y'' + y = \sec^3 x$.

23. $y'' - 2y' + y = e^x/x^2$.

24. $y'' - 2y' + y = x^3 e^x$.

25. $y'' - y = e^x \cos x$.

26. $y'' - 2y' + 2y = e^{2x} \sin x$.

27. $y'' + 2y' + y = e^{-x} \ln|x|$.

28. $y'' - 3y' + 2y = \dfrac{e^x}{1 + e^x}$.

29. Use the method of variation of parameters to show that the general solution of $y'' + y = r(x)$ is

$$y = c_1 \cos x + c_2 \sin x$$

$$- \left(\int r(x) \sin x \, dx \right) \cos x$$

$$+ \left(\int r(x) \cos x \, dx \right) \sin x$$

30. Use the method of variation of parameters to show that if y_1 and y_2 are linearly independent solutions of $y'' + py' + qy = 0$, then a particular solution of $y'' + py' + qy = r(x)$ is given by

$$y_p(x) = -y_1(x) \int \frac{y_2(x)r(x)}{W(x)} \, dx + y_2(x) \int \frac{y_1(x)r(x)}{W(x)} \, dx$$

where $W(x) = y_1(x)y_2'(x) - y_1'(x)y_2(x)$.

19.4 VIBRATION OF A SPRING

In this section we shall use second-order linear differential equations with constant coefficients to study the vibration of a spring.

> **19.4.1**   THE VIBRATING SPRING PROBLEM. As shown in Figure 19.4.1, let a mass attached to a vertical spring be allowed to settle into an equilibrium position. Then, let the spring be stretched (or compressed) by pulling (or pushing) the mass, and finally, let the mass be released, thereby causing it to undergo a vibratory motion. We shall be interested in finding a formula for the position of the mass at any time t.

Figure 19.4.1 ▷

To solve this problem we shall need three results from physics:

> HOOKE'S LAW   *If a spring is stretched (or compressed) l units beyond its natural position, then it pulls back (or pushes) with a force of magnitude*
>
> $F = kl$
>
> *where k is a positive constant, called the **spring constant**. This constant depends on such factors as the thickness of the spring, the material from which it is made, and the units of force and distance.*

NEWTON'S SECOND LAW OF MOTION   *If an object with mass **M** is subjected to a force **F**, then it undergoes an acceleration **a** satisfying*

$$\mathbf{F} = M\mathbf{a}$$

WEIGHT   *The gravitational force exerted by the earth on an object is called the **weight** (more precisely, **earth weight**) of the object. It follows from Newton's Second Law of Motion that an object with mass M has a weight whose magnitude w given by*

$$w = Mg \tag{1}$$

*where g is a constant, called the **acceleration due to gravity**. Near the surface of the earth an approximate value of g is*

$$g \approx 32 \text{ ft/sec}^2$$

if length is measured in feet and time in seconds, or

$$g \approx 980 \text{ cm/sec}^2$$

if length is measured in centimeters and time in seconds.

In any problem involving length, mass, time, and force it is important that the units of measurement be consistent. The most important systems of measurement are summarized in Table 19.4.1.

Table 19.4.1

SYSTEM OF MEASUREMENT	LENGTH	TIME	MASS	FORCE
Engineering System	foot (ft)	second (sec)	slug	pound (lb)
Mks System	meter (m)	second (sec)	kilogram (Kg)	newton (N)
Cgs System	centimeter (cm)	second (sec)	gram (g)	dyne

Figure 19.4.2

To solve the spring problem posed above, we introduce a coordinate axis (a y-axis) with the positive direction up and the origin at the bottom end of the spring when the mass is in its equilibrium position (Figure 19.4.2). If we take $t = 0$ to be the time at which the mass is released, then at each subsequent time t, the end of the spring has a position $y(t)$, a velocity $y'(t)$, and an acceleration $y''(t)$.* Because the positive direction of the y-axis is up, force, velocity, and acceleration are positive when directed up and negative when directed down.

---

*For convenience of terminology we shall refer to $y(t)$, $y'(t)$, and $y''(t)$ as the position, velocity, and acceleration of the mass, respectively.

Mass in
equilibrium

Figure 19.4.3

Before we can solve the spring problem, it will be necessary to derive a preliminary result from the equilibrium conditions for the mass. Let us assume that the spring constant is k, the mass of the object is M, and the spring is stretched l units beyond its natural length when the mass is in equilibrium (Figure 19.4.3). When the mass is in its equilibrium position its downward weight, $-Mg$, is balanced exactly by the upward force, kl, of the spring, so the sum of these forces must be zero*; that is,

$$kl - Mg = 0$$

or

$$kl = Mg \qquad (2)$$

The basic strategy for solving the spring problem is to find the total force $F(t)$ that acts on the mass at time t. Then, since the acceleration of the mass at time t is $y''(t)$, it will follow from Newton's Second Law that

$$My''(t) = F(t)$$

or

$$y''(t) = F(t)/M \qquad (3)$$

which is a second-order linear differential equation that can be solved for the position function $y(t)$.

At each instant, the force $F(t)$ in (3) consists of four possible components:

$F_g(t)$ = the force of gravity (weight of the mass)

$F_s(t)$ = the force of the spring

$F_d(t)$ = the *damping force* or frictional force exerted on the mass by the surrounding medium (air, water, oil, etc.)

$F_e(t)$ = external forces due to such factors as movement in the spring support, magnetic forces acting on the mass, and so on.

UNDAMPED FREE
VIBRATIONS

The simplest case occurs when there is no damping ($F_d = 0$) and the mass is *free* of external forces ($F_e = 0$). In this case the only forces acting on the mass are the force of gravity, F_g, and the spring force, F_s. Let us try to calculate these forces when the mass is at an arbitrary point $y(t)$.

When the mass is at the point $y(t)$, the spring is stretched (or compressed) $l - y(t)$ units from its natural length (Figure 19.4.4), so by Hooke's

---

*We shall assume that the weight of the spring is small relative to the weight of the mass and can be neglected.

Figure 19.4.4

law the spring exerts a force of

$$F_s(t) = k(l - y(t))$$

on the mass. Adding this to the force of gravity,

$$F_g(t) = -Mg$$

acting on the mass yields the total force acting on the mass:

$$F_s(t) + F_g(t) = k(l - y(t)) - Mg$$

But $kl = Mg$ from (2), so

$$F_s(t) + F_g(t) = -ky(t) \qquad (4)$$

Substituting this expression for $F(t)$ in (3) yields

$$y''(t) = -\frac{k}{M} y(t)$$

or

$$y''(t) + \frac{k}{M} y(t) = 0 \qquad (5)$$

which is a second-order linear differential equation with constant coefficients.

Because the mass is *released* (i.e., has zero initial velocity) at time $t = 0$, we have $y'(0) = 0$. If we assume, in addition, that the position of the mass at time $t = 0$ is $y(0) = y_0$, then we have two initial conditions that can be combined with (5) to yield an initial-value problem for $y(t)$:

$$y'' + \frac{k}{M} y = 0$$
$$y(0) = y_0, \quad y'(0) = 0 \qquad (6)$$

The auxiliary equation for the differential equation in (6) is

$$m^2 + \frac{k}{M} = 0$$

which has roots $m_1 = \sqrt{\dfrac{k}{M}} i$, $m_2 = -\sqrt{\dfrac{k}{M}} i$ (since k and M are positive), so the general solution of the differential equation is

$$y(t) = c_1 \cos\left(\sqrt{\frac{k}{M}} t\right) + c_2 \sin\left(\sqrt{\frac{k}{M}} t\right)$$

From the initial conditions $y(0) = y_0$ and $y'(0) = 0$ it follows that $c_1 = 0$, $c_2 = y_0$ (verify), so the solution of (6) is

$$y(t) = y_0 \cos\left(\sqrt{\frac{k}{M}}\,t\right) \tag{7}$$

This formula describes a periodic vibration with an *amplitude* of $|y_0|$ and *period T* given by

$$T = \frac{2\pi}{\sqrt{k/M}} = 2\pi\sqrt{\frac{M}{k}} \tag{8}$$

(Figure 19.4.5). The *frequency f* of the vibration is the number of cycles per unit time, so

$$f = \frac{1}{T} = \frac{1}{2\pi}\sqrt{\frac{k}{M}} \tag{9}$$

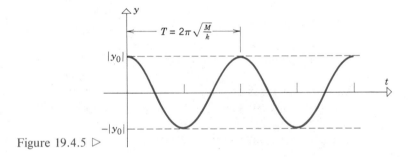

Figure 19.4.5 ▷

REMARK.   In the preceding problem, the spring oscillates indefinitely with constant amplitude because there are no frictional forces to dissipate the energy in the mass-spring system.

Example 1   Suppose that the top of a spring is fixed to a ceiling and a mass attached to the bottom end stretches the spring $\frac{1}{2}$ foot. If the mass is pulled 2 feet below its equilibrium position and released, find:

(a)   a formula for the position of the mass at any time t;

(b)   the amplitude, period, and frequency of the motion.

Solution (a).   The appropriate formula is (7). Although we are not given the mass M or the spring constant k, it does not matter because we are given that the mass stretches the spring $l = \frac{1}{2}$ ft and we know that

$g = 32\,\text{ft/sec}^2$, so (2) implies that

$$\frac{k}{M} = \frac{g}{l} = \frac{32}{\frac{1}{2}} = 64$$

Also, the mass is initially 2 units *below* its equilibrium position, so $y_0 = -2$. Therefore, from (7) the formula for the position of the mass at time t is

$$y(t) = -2\cos 8t$$

Solution (b). From (8) and (9)

$$\text{amplitude} = |y_0| = |-2| = 2\ \text{ft}$$

$$\text{period} = T = 2\pi\,\sqrt{\frac{M}{k}} = 2\pi\,\sqrt{\frac{1}{64}} = \frac{\pi}{4}\ \text{sec/cycle}$$

$$\text{frequency} = f = \frac{1}{T} = \frac{4}{\pi}\ \text{cycles/sec} \quad \blacktriangleleft$$

DAMPED FREE
VIBRATIONS

We shall now consider the solution of the spring problem in the case where damping cannot be neglected; that is, $F_d \neq 0$.

Physicists have shown that under appropriate conditions the damping force F_d is opposite to the direction of motion of the mass (i.e., tends to slow the mass down) and has a magnitude that is proportional to the speed of the mass (the greater the speed, the greater the effect of friction). This type of damping force is described by an equation of the form

$$F_d(t) = -cy'(t) \tag{10}$$

where c is a positive constant, called the **damping constant.** The damping constant depends on the viscosity of the surrounding medium.

It follows from (4) and (10) that

$$F_s(t) + F_g(t) + F_d(t) = -ky(t) - cy'(t)$$

If we substitute this expression in (3) for $F(t)$, we obtain

$$y''(t) = -\frac{k}{M}y(t) - \frac{c}{M}y'(t)$$

or

$$y''(t) + \frac{c}{M}y'(t) + \frac{k}{M}y(t) = 0$$

Combining this equation with the initial conditions $y(0) = y_0$, $y'(0) = 0$ yields the following initial-value problem whose solution describes the motion of the mass subject to damping:

$$y'' + \frac{c}{M} y' + \frac{k}{M} y = 0$$
$$y(0) = y_0, \quad y'(0) = 0$$

(11)

The form of the solution to (11) will depend on whether the auxiliary equation

$$m^2 + \frac{c}{M} m + \frac{k}{M} = 0$$

(12)

has distinct real roots, equal real roots, or complex roots. We leave it for the reader to show that the roots are distinct and real if $c^2 > 4kM$, equal and real if $c^2 = 4kM$, and complex if $c^2 < 4kM$ (Exercise 22).

The cases $c^2 > 4kM$ and $c^2 = 4kM$ are called *overdamped* and *critically damped*, respectively. In these cases vibration in the usual sense does not occur. In the overdamped case, the mass passes through the equilibrium position once, reaches a peak displacement, and then gradually drifts back toward the equilibrium position (Figure 19.4.6*a*); and in the critically damped case, the mass simply drifts slowly toward its equilibrium position without ever passing through it (Figure 19.4.6*b*). We shall leave the analysis of these cases for the exercises and concentrate on the case $c^2 < 4kM$ in which true vibratory motion occurs. This is called the *underdamped case*.

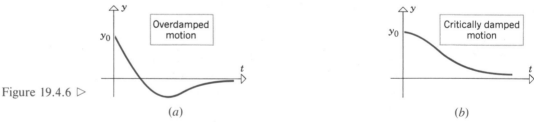

Figure 19.4.6 ▷

(a)

(b)

Underdamped Vibrations
($c^2 < 4kM$) In this case the roots of the auxiliary equation (12) are

$$m_1 = -\frac{c}{2M} + \frac{\sqrt{4Mk - c^2}}{2M} i, \quad m_2 = -\frac{c}{2M} - \frac{\sqrt{4Mk - c^2}}{2M} i$$

(Verify.) For convenience, let

$$\alpha = \frac{c}{2M}, \quad \beta = \frac{\sqrt{4Mk - c^2}}{2M}$$

(13)

so the solution of the differential equation in (11) is

$$y(t) = e^{-\alpha t}(c_1 \cos \beta t + c_2 \sin \beta t) \tag{14}$$

It follows that

$$y'(t) = -\alpha e^{-\alpha t}(c_1 \cos \beta t + c_2 \sin \beta t) + \beta e^{-\alpha t}(-c_1 \sin \beta t + c_2 \cos \beta t)$$

so the initial conditions $y(0) = y_0$, $y'(0) = 0$ yield the equations

$$c_1 = y_0$$
$$-\alpha c_1 + \beta c_2 = 0$$

which can be solved to obtain

$$c_1 = y_0, \quad c_2 = \alpha y_0/\beta$$

Substituting these values in (14) yields the solution of (11):

$$y(t) = \frac{y_0}{\beta} e^{-\alpha t}(\beta \cos \beta t + \alpha \sin \beta t) \tag{15}$$

In the exercises (Exercise 23) we ask the reader to use the cosine addition formula to show that this solution can be rewritten in the alternative form

$$y(t) = \frac{y_0 \sqrt{\alpha^2 + \beta^2}}{\beta} e^{-\alpha t} \cos(\beta t - \omega)$$

where

$$\omega = \tan^{-1}\left(\frac{\alpha}{\beta}\right), \quad 0 \le \omega < \pi/2 \tag{16}$$

Since $\cos(\beta t - \omega)$ has values between $+1$ and -1, it follows from (16) that the graph of $y(t)$ oscillates between the curves

$$y = \frac{y_0 \sqrt{\alpha^2 + \beta^2}}{\beta} e^{-\alpha t} \quad \text{and} \quad y = -\frac{y_0 \sqrt{\alpha^2 + \beta^2}}{\beta} e^{-\alpha t}$$

Thus, the graph of $y(t)$ resembles a cosine curve, but with decreasing amplitude (Figure 19.4.7). Strictly speaking, the function $y(t)$ is not periodic. However, $\cos(\beta t - \omega)$ has a period of $2\pi/\beta$, so the displacement $y(t)$ reaches a relative maximum at times spaced $2\pi/\beta$ units apart. Thus, we shall define the *period T* of this motion as

$$T = \frac{2\pi}{\beta} = \frac{4M\pi}{\sqrt{4Mk - c^2}} \tag{17}$$

(Figure 19.4.7) and the **frequency** to be

$$f = \frac{1}{T} = \frac{\sqrt{4Mk - c^2}}{4M\pi} \tag{18}$$

If we think of a relative maximum as marking the end of one cycle of motion and the start of the next, then the period T is the time required for the completion of one cycle, and the frequency is the number of cycles that occur per unit time.

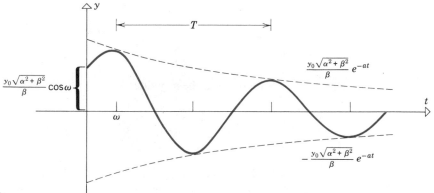

Figure 19.4.7 ▷

Example 2 A spring with a spring constant of $k = 4$ is attached to a ceiling and a 64-pound weight is attached to the bottom end. The weight is pushed 2 feet above its equilibrium position and released with an initial velocity of 0. Assuming that the damping constant due to air resistance is $c = 4$, find:

(a) a formula for the position of the weight at any time t;

(b) the period and frequency of the vibration.

Solution (a). We shall first use (13) to find α and β, after which we can use either (15) or (16). The attached weight is 64 pounds, so from (1) its mass is

$$M = \frac{w}{g} = \frac{64}{32} = 2 \text{ slugs}$$

Thus,

$$\alpha = \frac{c}{2M} = \frac{4}{4} = 1, \quad \beta = \frac{\sqrt{4Mk - c^2}}{2M} = \frac{4}{4} = 1$$

Since the weight is initially 2 feet above its equilibrium position, we have

$y_0 = 2$, so from (15) the position function of the weight is

$$y(t) = 2e^{-t}(\cos t + \sin t)$$

Alternatively, we can apply (16), but we must first calculate ω. Since

$$\omega = \tan^{-1}\left(\frac{\alpha}{\beta}\right) = \tan^{-1}(1) = \frac{\pi}{4}$$

(16) yields the alternative formula

$$y(t) = 2\sqrt{2}e^{-t} \cos\left(t - \frac{\pi}{4}\right)$$

Solution (b). From (17) and (18)

$$T = \frac{2\pi}{\beta} = 2\pi \quad \text{sec/cycle}$$

$$f = \frac{1}{T} = \frac{1}{2\pi} \quad \text{cycles/sec} \quad \blacktriangleleft$$

FORCED VIBRATIONS If there are external forces such as movement of the spring support or magnetic forces acting on the mass, then $F_e \neq 0$ and the vibrations are said to be *forced.* The study of forced vibrations leads to important phenomena such as *resonance* and *beats,* but this is beyond the scope of this section. Readers interested in this topic are referred to *Elementary Differential Equations and Boundary Value Problems,* John Wiley and Sons, N.Y., 1969 by William E. Boyce and Richard C. DiPrima.

▶ Exercise Set 19.4

In this exercise set assume that the y-axis is oriented as in Figure 19.4.2.

1. A weight of 64 lb is attached to a vertical spring with spring constant $k = 8$. The weight is pushed 1 ft above its equilibrium position and released.
    (a) Find an initial-value problem whose solution $y(t)$ is the position function of the weight, assuming that there is no damping.
    (b) Solve the initial-value problem.
    (c) Check the solution in (b) using Formula (7).

2. A mass of 1000 grams is attached to a vertical spring with spring constant $k = 25$. The mass is pushed 50 cm above its equilibrium position and released.
    (a) Find an initial-value problem whose solution $y(t)$ is the position function of the mass, assuming that there is no damping.
    (b) Solve the initial-value problem.
    (c) Check the solution in (b) using Formula (7).

3. A mass attached to a vertical spring stretches the spring 5 cm. The mass is pulled 10 cm below its equilibrium position and released.
    (a) Find an initial-value problem whose solution $y(t)$ is the position function of the mass, assuming that there is no damping.

(b)   Solve the initial-value problem.
(c)   Check the solution in (b) using Formula (7).

4.   A mass attached to a vertical spring stretches the spring 2 ft. The mass is pulled 4 ft below its equilibrium position and released.
(a)   Find an initial-value problem whose solution $y(t)$ is the position function of the mass, assuming that there is no damping.
(b)   Solve the initial-value problem.
(c)   Check the solution in (b) using Formula (7).

5.   A weight of $\frac{1}{2}$ lb is attached to a vertical spring with spring constant $k = 1$. The weight is pushed 2 ft above its equilibrium position and released. Use Formulas (7), (8), and (9) to find:
(a)   the position function of the weight;
(b)   the amplitude of the vibration;
(c)   the period of the vibration;
(d)   the frequency of the vibration.

6.   A mass of 2 slugs is attached to a vertical spring with spring constant $k = 4$. The mass is pushed 1 ft above its equilibrium position and released. Use Formulas (7), (8), and (9) to find:
(a)   the position function of the mass;
(b)   the amplitude of the vibration;
(c)   the period of the vibration;
(d)   the frequency of the vibration.

7.   A mass attached to a vertical spring stretches the spring 1 in. The mass is pulled 3 in. below its equilibrium position and released. Use Formulas (7), (8), and (9) to find:
(a)   the position function of the mass;
(b)   the amplitude of the vibration;
(c)   the period of the vibration;
(d)   the frequency of the vibration.

8.   A mass attached to a vertical spring stretches the spring 8 meters. The mass is pulled 2 meters below its equilibrium position and released. Use Formulas (7), (8), and (9) to find:
(a)   the position function of the mass;
(b)   the amplitude of the vibration;
(c)   the period of the vibration;
(d)   the frequency of the vibration.

9.   A weight of 32 lb is attached to a vertical spring with spring constant $k = 8$. The surrounding medium has a damping constant of $c = 4$. The weight is pulled 3 ft below its equilibrium position and released.
(a)   Find an initial-value problem whose solution $y(t)$ is the position function of the weight.
(b)   Solve the initial-value problem.
(c)   Check the solution in (b) using Formula (15).
(d)   Express the solution in the form of (16).
(e)   Find the period of the vibration.
(f)   Find the frequency of the vibration.

10.   A mass of 3 slugs is attached to a vertical spring with spring constant $k = 9$. The surrounding medium has a damping constant of $c = 6$. The mass is pulled 1 ft below its equilibrium position and released.
(a)   Find an initial-value problem whose solution $y(t)$ is the position function of the mass.
(b)   Solve the initial-value problem.
(c)   Check the solution in (b) using Formula (15).
(d)   Express the solution in the form of (16).
(e)   Find the period of the vibration.
(f)   Find the frequency of the vibration.

11.   A mass of 25 grams is attached to a vertical spring with spring constant $k = 3$. The surrounding medium has a damping constant of $c = 10$. The mass is pushed 5 cm above its equilibrium position and released.
(a)   Use Formula (16) to find the position function of the mass.
(b)   Find the period of the vibration.
(c)   Find the frequency of the vibration.

12.   A mass of 490 grams is attached to a vertical spring with spring constant $k = 40$. The surrounding medium has a damping constant of $c = 140$. The mass is pushed 20 cm above its equilibrium position and released.
(a)   Use Formula (16) to find the position function of the mass.
(b)   Find the period of the vibration.
(c)   Find the frequency of the vibration.

If the object in Figure 19.4.1 is given an initial velocity v_0 rather than being released with initial velocity 0, then in (6) and (11) the initial conditions become $y(0) = y_0$, $y'(0) = v_0$. Use this fact in Exercises 13–16.

13. A weight of 3 lb attached to a vertical spring stretches the spring $\frac{1}{2}$ ft. While in its equilibrium position, the weight is struck to give it a downward initial velocity of 2 ft/sec. Assuming that there is no damping, find:
    (a) the position function of the weight;
    (b) the amplitude of the vibration;
    (c) the period of the vibration;
    (d) the frequency of the vibration.

14. A mass of 64 slugs attached to a vertical spring stretches the spring $1\frac{1}{2}$ in. The mass is pulled 4 in. below its equilibrium position and struck to give it a downward initial velocity of 8 ft/sec. Assuming that there is no damping, find:
    (a) the position function of the weight;
    (b) the amplitude of the vibration;
    (c) the period of the vibration;
    (d) the frequency of the vibration.

15. A weight of 4 lb is attached to a vertical spring with spring constant $k = 6\frac{1}{4}$. The weight is pushed $\frac{1}{3}$ ft above its equilibrium position and struck to give it a downward initial velocity of 5 ft/sec. Assuming that the damping constant is $c = \frac{1}{4}$, find the position function of the weight.

16. A weight of 49 dynes is attached to a vertical spring with spring constant $k = \frac{1}{4}$. The weight is pushed 1 cm above its equilibrium position and struck to give it an upward initial velocity of 2 cm/sec. Assuming that the damping constant is $c = \frac{1}{5}$, find the position function of the weight.

17. A weight of w pounds is attached to a spring, then pulled below its equilibrium position and released, thereby causing it to vibrate with a period of 3 sec. When 4 additional pounds are added, the period becomes 5 sec. Assuming there is no damping, find:
    (a) the spring constant k;
    (b) the weight w.

18. As illustrated in the following figure, let a toy cart of mass M be attached to a wall by a spring with spring constant k, and let an x-axis be introduced as shown with its origin at the point where the cart is in equilibrium. Suppose that the cart is pulled or pushed horizontally, then released. Find

a differential equation for the position function of the cart if:
(a) there is no damping;
(b) there is a damping constant of c.

19. A cylindrical buoy of height h and radius r floats in the water with its axis vertical. By **Archimedes principle** the water exerts an upward force on the buoy (the buoyancy force) with magnitude equal to the weight of the water displaced. Neglecting all forces except the buoyancy force and the force of gravity, determine the period with which the buoy will vibrate vertically if it is depressed slightly from its equilibrium position and released. Take ρ lb/in.^{3} to be the density of water and δ lb/in.^{3} to be the density of the buoy material (density = weight per unit of volume).

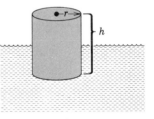

20. Suppose that the mass of the object in Figure 19.4.1 is M and the spring constant is k. Show that if the mass is displaced y_0 units from its equilibrium position and given an initial velocity of v_0 rather than being released with velocity 0, then its position function is given by

$$y(t) = y_0 \cos\left(\sqrt{\frac{k}{M}}\,t\right) + v_0\sqrt{\frac{M}{k}}\sin\left(\sqrt{\frac{k}{M}}\,t\right)$$

21. A mass attached to a vertical spring is displaced from its equilibrium position and released, thereby causing it to vibrate with amplitude $|y_0|$ and period T (no damping).

(a) Show that the velocity of the mass has maximum magnitude $2\pi|y_0|/T$ and the maximum occurs when the mass is at its equilibrium position.

(b) Show that the acceleration of the mass has maximum magnitude $4\pi^2|y_0|/T^2$ and the maximum occurs when the mass is at a top or bottom point of its motion.

22. Prove that the roots of Equation (12) are distinct and real, equal and real, or complex according to whether $c^2 > 4kM$, $c^2 = 4kM$, or $c^2 < 4kM$.

23. Use the addition formula for cosine to show that Formula (15) can be rewritten as (16).

24. (Overdamped Motion)
(a) Prove that in the overdamped case $(c^2 > 4kM)$ the solution of (11) is

$$y(t) = \frac{y_0}{m_2 - m_1}(m_2 e^{m_1 t} - m_1 e^{m_2 t})$$

(b) Prove that $\lim_{t \to +\infty} y(t) = 0$. [*Hint:* Show that m_1 and m_2 are negative.]

(c) Prove that $y(t) = 0$ for exactly one positive value of t.

25. (Critically Damped Motion)
(a) Prove that in the critically damped case $(c^2 = 4kM)$ the solution of (11) is

$$y(t) = y_0 e^{-\alpha t}(1 + \alpha t)$$

(b) Prove that $\lim_{t \to +\infty} y(t) = 0$.

(c) Prove that $y(t) \neq 0$ for any positive value of t (assuming that the initial displacement y_0 is not zero).

Appendix 1

trigonometry review

UNIT I TRIGONOMETRIC FUNCTIONS AND IDENTITIES

In this section we shall review the definitions and basic properties of trigonometric functions.

ANGLES We shall be concerned with angles in the plane that are generated by rotating a ray (or half-line) about its endpoint. The starting position of the ray is called the ***initial side*** of the angle and the final position of the ray is called the ***terminal side***. The initial and terminal sides meet at a point called the ***vertex*** of the angle (Figure A.1).

An angle is considered ***positive*** if it is generated by a counterclockwise rotation and ***negative*** if it is generated by a clockwise rotation. In a rectangular coordinate system, an angle is said to be in ***standard position*** if its vertex is at the origin and its initial side is along the positive x-axis

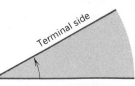

Figure A.1 Vertex   Initial side

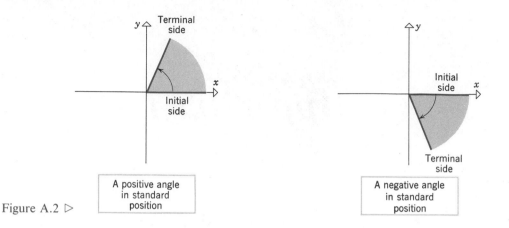

Figure A.2 ▷

(Figure A.2). As shown in Figure A.3, an angle may be generated by making more than one complete revolution.

Figure A.3 ▷

TRIGONOMETRIC FUNCTIONS

The *sine, cosine, tangent, cosecant, secant,* and *cotangent* of a positive acute angle θ can be defined as ratios of the sides of a right triangle. Using the notation from Figure A.4, these definitions take the form:

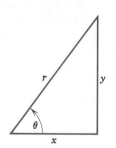

Figure A.4

$$\sin \theta = \frac{\text{side opposite } \theta}{\text{hypotenuse}} = \frac{y}{r} \tag{1}$$

$$\cos \theta = \frac{\text{side adjacent to } \theta}{\text{hypotenuse}} = \frac{x}{r} \tag{2}$$

$$\tan \theta = \frac{\text{side opposite } \theta}{\text{side adjacent to } \theta} = \frac{y}{x} \tag{3}$$

$$\csc \theta = \frac{\text{hypotenuse}}{\text{side opposite } \theta} = \frac{r}{y} \tag{4}$$

$$\sec \theta = \frac{\text{hypotenuse}}{\text{side adjacent to } \theta} = \frac{r}{x} \tag{5}$$

$$\cot \theta = \frac{\text{side adjacent to } \theta}{\text{side opposite to } \theta} = \frac{x}{y} \tag{6}$$

We shall call sin, cos, tan, csc, sec, and cot the ***trigonometric functions***. Because similar triangles have proportional sides, the values of the trigonometric functions depend only on the size of θ and not on the particular right triangle used to compute the ratios.

Example 1   Recall from high school geometry that the two legs of a 45°–45°–90° triangle are of equal size. This fact and the Theorem of Pythagoras yield Figure A.5*a*. Recall also that the hypotenuse of a 30°–60°–90° triangle is twice the size of the shorter leg and the shorter leg is opposite the 30° angle. These facts and the Theorem of Pythagoras yield Figure A.5*b*.

Figure A.5   (*a*)   (*b*)

From Figure A.5 we obtain

$$\sin 45° = \frac{1}{\sqrt{2}} \quad \cos 45° = \frac{1}{\sqrt{2}} \quad \tan 45° = 1$$

$$\csc 45° = \sqrt{2} \quad \sec 45° = \sqrt{2} \quad \cot 45° = 1$$

$$\sin 30° = \frac{1}{2} \quad \cos 30° = \frac{\sqrt{3}}{2} \quad \tan 30° = \frac{1}{\sqrt{3}}$$

$$\csc 30° = 2 \quad \sec 30° = \frac{2}{\sqrt{3}} \quad \cot 30° = \sqrt{3}$$

$$\sin 60° = \frac{\sqrt{3}}{2} \quad \cos 60° = \frac{1}{2} \quad \tan 60° = \sqrt{3}$$

$$\csc 60° = \frac{2}{\sqrt{3}} \quad \sec 60° = 2 \quad \cot 60° = \frac{1}{\sqrt{3}} \quad ◀$$

Since a right triangle cannot have an angle larger than 90°, Formulas (1)–(6) are not applicable if θ is obtuse. To obtain definitions of the trigonometric functions that apply to all angles, we take the following approach: Given an angle θ, introduce a coordinate system so that the angle is in standard position. Then construct a circle of arbitrary radius r centered at the origin and let $P(x, y)$ be the point where the terminal side of

Figure A.6

the angle intersects the circle (Figure A.6). We make the following definition:

DEFINITION 1.

$$\sin \theta = \frac{y}{r}, \quad \cos \theta = \frac{x}{r}, \quad \tan \theta = \frac{y}{x}$$

$$\csc \theta = \frac{r}{y}, \quad \sec \theta = \frac{r}{x}, \quad \cot \theta = \frac{x}{y}$$

The formulas in Definition 1 agree with Formulas (1)–(6). However, Definition 1 applies to all angles—positive, negative, acute, or obtuse, whereas (1)–(6) apply only to positive acute angles. Note that $\tan \theta$ and $\sec \theta$ are undefined if the terminal side of θ is on the y-axis (since $x = 0$), while $\csc \theta$ and $\cot \theta$ are undefined if the terminal side of θ is on the x-axis (since $y = 0$). Moreover,

$$\csc \theta = \frac{1}{\sin \theta}, \quad \sec \theta = \frac{1}{\cos \theta}, \quad \cot \theta = \frac{1}{\tan \theta} \qquad (7a)$$

$$\tan \theta = \frac{\sin \theta}{\cos \theta}, \quad \cot \theta = \frac{\cos \theta}{\sin \theta} \qquad (7b)$$

(Verify.)

Example 2   In Figure A.7,

$$x = -3, \quad y = 4, \quad \text{and} \quad r = 5$$

so that for the angle θ shown,

$$\sin \theta = \frac{4}{5}, \quad \cos \theta = -\frac{3}{5}, \quad \tan \theta = -\frac{4}{3}$$

$$\csc \theta = \frac{5}{4}, \quad \sec \theta = -\frac{5}{3}, \quad \cot \theta = -\frac{3}{4} \qquad ◄$$

Figure A.7

Since the value of a trigonometric function is determined by the point where the terminal side of the angle intersects a circle of radius r, it follows that the value of a trigonometric function will be the same for any two angles with the same terminal side (like those in Figure A.3).

Using properties of similar triangles, it can be shown that the values of the trigonometric functions in Definition 1 are the same for every value r. In particular, for the *unit circle* (the circle of radius $r = 1$ centered at the origin) the formulas for $\sin \theta$ and $\cos \theta$ in Definition 1 become

$$x = \cos \theta, \quad y = \sin \theta$$

Thus, we have the following geometric interpretation of $\sin \theta$ and $\cos \theta$ (Figure A.8):

Geometric Interpretation of $\sin \theta$ and $\cos \theta$

> THEOREM **2.**   *If an angle θ is placed in standard position, then its terminal side intersects the unit circle at the point $(\cos \theta, \sin \theta)$.*

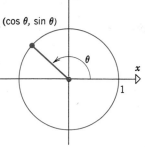

Figure A.8

Example 3   Evaluate the trigonometric functions of $\theta = 150°$.

Solution.   Construct a unit circle and place the angle $\theta = 150°$ in standard position (Figure A.9). Since angle AOP is 30° and $\triangle OAP$ is a 30°–60°–90° triangle, the leg AP has length $\frac{1}{2}$ (half the hypotenuse) and the leg OA has length $\sqrt{3}/2$ by the Theorem of Pythagoras. Thus, the coordinates of P are $(-\sqrt{3}/2, 1/2)$ and Theorem 2 yields

$$\sin 150° = \frac{1}{2}, \quad \cos 150° = -\frac{\sqrt{3}}{2}$$

Moreover, from Formulas (7a) and (7b)

$$\tan 150° = \frac{\sin 150°}{\cos 150°} = \frac{1/2}{-\sqrt{3}/2} = -\frac{1}{\sqrt{3}}$$

$$\cot 150° = \frac{1}{\tan 150°} = -\sqrt{3}$$

$$\sec 150° = \frac{1}{\cos 150°} = -\frac{2}{\sqrt{3}}$$

$$\csc 150° = \frac{1}{\sin 150°} = 2 \quad \blacktriangleleft$$

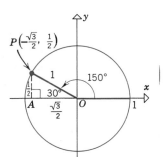

Figure A.9

The following useful theorem is suggested by Figure A.10. We omit the proof.

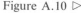

Figure A.10 ▷

> **THEOREM 3.**   *If an angle θ is generated by a ray with endpoint (x_0, y_0) and initial side extending in the positive x-direction, then the intersection of the terminal side with the circle of radius r centered at (x_0, y_0) is*
>
> $$(x_0 + r \cos \theta, y_0 + r \sin \theta)$$

RADIAN MEASURE

Figure A.11

In calculus, angles are measured in *radians* rather than degrees, minutes, and seconds because it simplifies many important formulas. Before we can define radian measure, we need some preliminary ideas.

Let Q be an arbitrary point on a circle of radius r and imagine that a particle, initially at Q, moves around the circle and stops at a point P (Figure A.11). During its motion the point will travel a distance d along the circle. To distinguish between clockwise and counterclockwise motions we introduce the *signed distance* or *signed arc length* s traveled by the point; it is defined by:

$$s = d \qquad \text{if the motion is counterclockwise}$$
$$s = -d \quad \text{if the motion is clockwise}$$
$$s = 0 \qquad \text{if there is no motion}$$

For example, if $s = 3$, the point has traveled 3 units in the counterclockwise direction from Q; and, if $s = -\pi$, the point has traveled π units in the clockwise direction from Q.

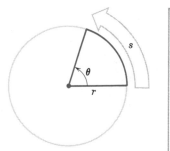

Figure A.12

> DEFINITION 4.   To define the radian measure of an angle, we place the vertex of the angle at the center of a circle of arbitrary radius r. As a ray sweeps out the angle, the intersection of the ray with the circle moves some signed distance s along the circle (Figure A.12). We define
>
> $$\theta = \frac{s}{r} \qquad\qquad (8)$$
>
> to be the *radian measure* of the angle.

REMARK.   We leave it as an exercise to show that the radian measure of an angle depends only on the size of the angle, and not on the radius r selected for the circle.

Example 4   Express the angle 90° in radian measure.

Solution.   An angle of 90° placed with its vertex at the center of a circle of radius r intercepts one-fourth of the circumference (Figure A.13). Since

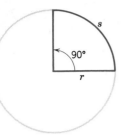

Figure A.13

the full circumference is $2\pi r$, we obtain

$$s = \tfrac{1}{4}(2\pi r) = \tfrac{1}{2}\pi r$$

Thus, the radian measure of the angle is

$$\theta = \frac{s}{r} = \frac{\tfrac{1}{2}\pi r}{r} = \frac{\pi}{2} \quad \text{(radians)} \quad \blacktriangleleft$$

Example 5   From the fact that 90° corresponds to $\pi/2$ radians, the reader should be able to obtain the relationships in the following table:

Degrees	30°	45°	60°	90°	120°	135°	150°	180°	270°	360°
Radians	$\dfrac{\pi}{6}$	$\dfrac{\pi}{4}$	$\dfrac{\pi}{3}$	$\dfrac{\pi}{2}$	$\dfrac{2\pi}{3}$	$\dfrac{3\pi}{4}$	$\dfrac{5\pi}{6}$	π	$\dfrac{3\pi}{2}$	2π

Some angles and their radian measure are shown in the following figure. $\blacktriangleleft$

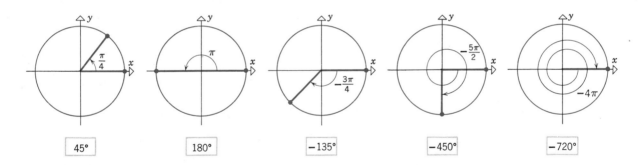

The following table will be useful in the next example.

$$360° = 2\pi \text{ radians}$$
$$180° = \pi \text{ radians} \approx 3.14159 \ldots \text{ radians}$$
$$\left(\frac{180}{\pi}\right)^{\circ} = 1 \text{ radian} \approx 57° \ 17' \ 44.8''$$
$$1° = \frac{\pi}{180} \text{ radian} \approx 0.01745 \text{ radian}$$

Example 6

(a)   Express 146° in radians.

(b)   Express 3 radians in degrees.

Solution (a).   Since

$$1° = \frac{\pi}{180} \text{ radians}$$

it follows that

$$146° = \frac{\pi}{180} \cdot 146 \text{ radians} = \frac{73\pi}{90} \text{ radians}$$

Solution (b).   Since

$$1 \text{ radian} = \left(\frac{180}{\pi}\right)°$$

it follows that

$$3 \text{ radians} = \left(3 \cdot \frac{180}{\pi}\right)° = \left(\frac{540}{\pi}\right)° \approx 171.9° \quad \blacktriangleleft$$

REMARK.   By convention, an angle written without any units is under-stood to be measured in radians. Thus, the statement $\theta = 3.2$ means $\theta = 3.2$ radians, *not* $\theta = 3.2°$.

EVALUATING
TRIGONOMETRIC
FUNCTIONS

Example 7   Evaluate the trigonometric functions at $\theta = 5\pi/6$.

Solution.   Since $5\pi/6 = 150°$, this problem is equivalent to that of Ex-ample 3. From that example we obtain

$$\sin\frac{5\pi}{6} = \frac{1}{2}, \quad \cos\frac{5\pi}{6} = -\frac{\sqrt{3}}{2}, \quad \tan\frac{5\pi}{6} = -\frac{1}{\sqrt{3}}$$

$$\csc\frac{5\pi}{6} = 2, \quad \sec\frac{5\pi}{6} = -\frac{2}{\sqrt{3}}, \quad \cot\frac{5\pi}{6} = -\sqrt{3} \quad \blacktriangleleft$$

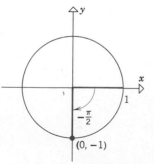

Example 8   Evaluate the trigonometric functions at $\theta = -\pi/2$.

Solution.   Construct a unit circle and place the angle $\theta = -\pi/2 = -90°$ in standard position (Figure A.14). The terminal side of θ intersects the circle at $P(0, -1)$ so that from Theorem 2 we obtain

$$\sin\left(-\frac{\pi}{2}\right) = -1, \quad \cos\left(-\frac{\pi}{2}\right) = 0$$

Figure A.14   and from Formulas (7a) and (7b)

$$\tan\left(-\frac{\pi}{2}\right) = \frac{\sin\left(-\dfrac{\pi}{2}\right)}{\cos\left(-\dfrac{\pi}{2}\right)} = \frac{-1}{0} = \text{undefined}$$

$$\cot\left(-\frac{\pi}{2}\right) = \frac{\cos\left(-\dfrac{\pi}{2}\right)}{\sin\left(-\dfrac{\pi}{2}\right)} = \frac{0}{-1} = 0$$

$$\sec\left(-\frac{\pi}{2}\right) = \frac{1}{\cos\left(-\dfrac{\pi}{2}\right)} = \frac{1}{0} = \text{undefined}$$

$$\csc\left(-\frac{\pi}{2}\right) = \frac{1}{\sin\left(-\dfrac{\pi}{2}\right)} = \frac{1}{-1} = -1 \qquad \blacktriangleleft$$

It follows from Theorem 2 that $\cos\theta$ is positive if θ terminates in the first or fourth quadrant, and is negative if θ terminates in the second or third quadrant. (To visualize this, look at Figure A.8.) Similarly, $\sin\theta$ is positive if θ terminates in the first or second quadrant, and is negative if θ terminates in the third or fourth quadrant. Moreover, since $\tan\theta = \sin\theta/\cos\theta$, it follows that $\tan\theta$ is positive where $\sin\theta$ and $\cos\theta$ have the same sign (first and third quadrants) and negative where they have opposite signs (second and fourth quadrants). These results are summarized in the adjacent figure.

The signs of $\sin\theta$, $\cos\theta$, and $\tan\theta$ can be remembered from the adjacent diagram in which each quadrant is labeled with the trigonometric functions that are positive there.

It follows from Formulas (7a) and (7b) that $\csc\theta$ has the same sign as $\sin\theta$, $\cot\theta$ the same sign as $\tan\theta$, and $\sec\theta$ the same sign as $\cos\theta$.

Example 9   Find θ if $\sin\theta = \frac{1}{2}$.

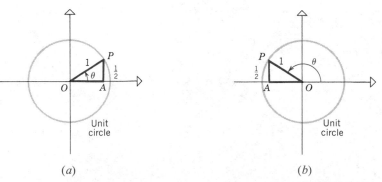

Figure A.15 ▷   (a)   (b)

Solution.   We shall begin by looking for positive angles that satisfy the equation.

Because $\sin \theta$ is positive, the angle θ must terminate in the first or second quadrant. If it terminates in the first quadrant, then the hypotenuse of $\triangle OAP$ in Figure A.15a is double the leg AP, so

$$\theta = 30° = \frac{\pi}{6} \text{ radians}$$

If θ terminates in the second quadrant, then the hypotenuse of $\triangle OAP$ is double the leg AP, so $\angle AOP = 30°$, which implies that

$$\theta = 180° - 30° = 150° = \frac{5\pi}{6} \text{ radians}$$

Now that we have found these two solutions, all other solutions are obtained by adding or subtracting multiples of $360°$ (2π radians) to them. Thus, the entire set of solutions is given by the formulas

$$\theta = 30° \pm n \cdot 360°, \quad n = 0, 1, 2, \ldots$$

and

$$\theta = 150° \pm n \cdot 360°, \quad n = 0, 1, 2, \ldots$$

or in radian measure

$$\theta = \frac{\pi}{6} \pm n \cdot 2\pi, \quad n = 0, 1, 2, \ldots$$

and

$$\theta = \frac{5\pi}{6} \pm n \cdot 2\pi, \quad n = 0, 1, 2, \ldots \qquad \blacktriangleleft$$

Table 1 lists the values of the trigonometric functions for some frequently used angles. A more extensive table is given in Appendix 3. Dashes denote undefined quantities.

TRIGONOMETRIC IDENTITIES   Earlier, we obtained the relationships

$$\csc \theta = \frac{1}{\sin \theta}, \quad \sec \theta = \frac{1}{\cos \theta}, \quad \cot \theta = \frac{1}{\tan \theta}$$

and

$$\tan \theta = \frac{\sin \theta}{\cos \theta}, \quad \cot \theta = \frac{\cos \theta}{\sin \theta}$$

Table 1

	$\theta = 0$	$\dfrac{\pi}{6}$	$\dfrac{\pi}{4}$	$\dfrac{\pi}{3}$	$\dfrac{\pi}{2}$	$\dfrac{2\pi}{3}$	$\dfrac{3\pi}{4}$	$\dfrac{5\pi}{6}$	π	$\dfrac{3\pi}{2}$	2π
$\sin\theta$	0	1/2	$1/\sqrt{2}$	$\sqrt{3}/2$	1	$\sqrt{3}/2$	$1/\sqrt{2}$	1/2	0	-1	0
$\cos\theta$	1	$\sqrt{3}/2$	$1/\sqrt{2}$	1/2	0	$-1/2$	$-1/\sqrt{2}$	$-\sqrt{3}/2$	-1	0	1
$\tan\theta$	0	$1/\sqrt{3}$	1	$\sqrt{3}$	—	$-\sqrt{3}$	-1	$-1/\sqrt{3}$	0	—	0
$\csc\theta$	—	2	$\sqrt{2}$	$2/\sqrt{3}$	1	$2/\sqrt{3}$	$\sqrt{2}$	2	—	-1	—
$\sec\theta$	1	$2/\sqrt{3}$	$\sqrt{2}$	2	—	-2	$-\sqrt{2}$	$-2/\sqrt{3}$	-1	—	1
$\cot\theta$	—	$\sqrt{3}$	1	$1/\sqrt{3}$	0	$-1/\sqrt{3}$	-1	$-\sqrt{3}$	—	0	—

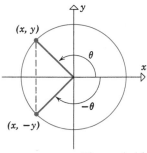

Figure A.16

These results are examples of *identities,* which means that the equalities hold for all values of θ where both sides are defined. One of the most important identities in trigonometry can be derived by applying the Theorem of Pythagoras to the triangle in Figure A.6 to obtain

$$x^2 + y^2 = r^2$$

Dividing both sides by r^2 and using the definitions of $\sin\theta$ and $\cos\theta$ (Definition 1), we obtain the following fundamental result:

$$\sin^2\theta + \cos^2\theta = 1 \qquad (9)$$

Figure A.16 shows two angles, θ and $-\theta$, having equal magnitude but opposite sign. By symmetry, the terminal sides of these angles intersect a circle centered at the origin at points whose x-coordinates are equal and whose y-coordinates differ only in sign. Thus,

$$\sin(-\theta) = \frac{-y}{r} = -\frac{y}{r} = -\sin\theta$$

$$\cos(-\theta) = \frac{x}{r} = \cos\theta$$

In summary, we have the identities

$$\sin(-\theta) = -\sin\theta \qquad (10a)$$
$$\cos(-\theta) = \cos\theta \qquad (10b)$$

As previously noted, a trigonometric function has the same value for two angles with the same terminal side. In radian measure, the angles θ, $\theta + 2\pi$, and $\theta - 2\pi$ all have the same terminal side, so

$$\sin \theta = \sin (\theta + 2\pi) = \sin (\theta - 2\pi)$$
$$\cos \theta = \cos (\theta + 2\pi) = \cos (\theta - 2\pi)$$

(11)

More generally, any multiple of 2π can be added to or subtracted from an angle θ in radians without changing the terminal side. Thus,

$$\sin \theta = \sin (\theta \pm 2n\pi), \quad n = 0, 1, 2, \ldots$$
$$\cos \theta = \cos (\theta \pm 2n\pi), \quad n = 0, 1, 2, \ldots$$

(12)

The following result, called the *law of cosines*, generalizes the Theorem of Pythagoras:

Law of Cosines

> **THEOREM 5.**   *If the sides of a triangle have lengths a, b, and c, and if θ is the angle between the sides with lengths a and b, then*
>
> $$c^2 = a^2 + b^2 - 2ab \cos \theta$$

Proof.   Introduce a coordinate system so that θ is in standard position and the side of length a falls along the positive x-axis (Figure A.17). As shown in Figure A.17, the side of length a extends from the origin to $(a, 0)$ and the side of length b extends from the origin to some point (x, y). From the definition of $\sin \theta$ and $\cos \theta$ we have

$$\sin \theta = \frac{y}{b}, \quad \cos \theta = \frac{x}{b}$$

Figure A.17 ▷

so

$$y = b \sin \theta, \quad x = b \cos \theta$$

(13)

From the distance formula in Theorem 1.3.1 we obtain

$$c^2 = (x - a)^2 + (y - 0)^2$$

so that, from (13),

$$c^2 = (b \cos \theta - a)^2 + b^2 \sin^2 \theta$$

$$= a^2 + b^2 (\cos^2 \theta + \sin^2 \theta) - 2ab \cos \theta$$
$$= a^2 + b^2 - 2ab \cos \theta$$

which completes the proof.   ∎

Using the law of cosines we shall now be able to establish the following identities, called the **addition formulas** for sine and cosine.

$$\sin(\alpha - \beta) = \sin \alpha \cos \beta - \cos \alpha \sin \beta \qquad (14a)$$
$$\cos(\alpha - \beta) = \cos \alpha \cos \beta + \sin \alpha \sin \beta \qquad (14b)$$

$$\sin(\alpha + \beta) = \sin \alpha \cos \beta + \cos \alpha \sin \beta \qquad (15a)$$
$$\cos(\alpha + \beta) = \cos \alpha \cos \beta - \sin \alpha \sin \beta \qquad (15b)$$

Identities (15a) and (15b) can be obtained by substituting $-\beta$ for β in (14a) and (14b) and using the identities

$$\sin(-\beta) = -\sin \beta, \quad \cos(-\beta) = \cos \beta$$

We shall derive (14b) and then deduce (14a) from (14b). In our derivation of (14b), we shall assume that $0 \le \beta < \alpha < 2\pi$ (Figure A.18). The proofs for the remaining cases will be omitted.

Figure A.18 ▷

As shown in Figure A.18, the terminal sides of α and β intersect the unit circle at the points $P_1(\cos \alpha, \sin \alpha)$ and $P_2(\cos \beta, \sin \beta)$. If we denote the lengths of the sides of triangle OP_1P_2 by OP_1, P_1P_2, and OP_2, then $OP_1 = OP_2 = 1$ and, from the distance formula in Theorem 1.3.1,

$$(P_1P_2)^2 = (\cos \beta - \cos \alpha)^2 + (\sin \beta - \sin \alpha)^2$$
$$= (\sin^2 \alpha + \cos^2 \alpha) + (\sin^2 \beta + \cos^2 \beta)$$
$$- 2(\cos \alpha \cos \beta + \sin \alpha \sin \beta)$$
$$= 2 - 2(\cos \alpha \cos \beta + \sin \alpha \sin \beta)$$

But angle $P_2OP_1 = \alpha - \beta$, so that the law of cosines yields

$$(P_1P_2)^2 = (OP_1)^2 + (OP_2)^2 - 2(OP_1)(OP_2) \cos(\alpha - \beta)$$
$$= 2 - 2 \cos(\alpha - \beta)$$

Equating the two expressions for $(P_1P_2)^2$ and simplifying, we obtain

$$\cos(\alpha - \beta) = \cos \alpha \cos \beta + \sin \alpha \sin \beta$$

which completes the derivation of (14b).

To derive identity (14a), we shall need the identities

$$\cos\left(\frac{\pi}{2} - \alpha\right) = \sin \alpha \qquad (16a)$$

$$\sin\left(\frac{\pi}{2} - \alpha\right) = \cos \alpha \qquad (16b)$$

which state that *the sine of an angle is the cosine of its complement* and *the cosine of an angle is the sine of its complement*. The derivations of (16a) and (16b) are left as exercises.

We can now derive (14a) as follows:

$$\sin(\alpha - \beta) = \cos\left[\frac{\pi}{2} - (\alpha - \beta)\right]$$

$$= \cos\left[\left(\frac{\pi}{2} - \alpha\right) + \beta\right]$$

$$= \cos\left[\left(\frac{\pi}{2} - \alpha\right) - (-\beta)\right]$$

$$= \cos\left(\frac{\pi}{2} - \alpha\right) \cos(-\beta) + \sin\left(\frac{\pi}{2} - \alpha\right) \sin(-\beta)$$

$$= \cos\left(\frac{\pi}{2} - \alpha\right) \cos \beta - \sin\left(\frac{\pi}{2} - \alpha\right) \sin \beta$$

$$= \sin \alpha \cos \beta - \cos \alpha \sin \beta$$

In the special case where $\alpha = \beta$, (15a) and (15b) reduce to the identities

$$\sin 2\alpha = 2 \sin \alpha \cos \alpha \qquad (17a)$$
$$\cos 2\alpha = \cos^2 \alpha - \sin^2 \alpha \qquad (17b)$$

These are called the **double-angle formulas**. By using the identity $\sin^2 \alpha + \cos^2 \alpha = 1$, (17b) can be rewritten in the alternative forms

$$\cos 2\alpha = 2 \cos^2 \alpha - 1 \qquad (17c)$$
$$\cos 2\alpha = 1 - 2 \sin^2 \alpha \qquad (17d)$$

If we replace α by $\alpha/2$ in (17c) and (17d) and use some algebra, we obtain the *half-angle formulas*

$$\cos^2 \frac{\alpha}{2} = \frac{1 + \cos \alpha}{2} \tag{18a}$$

$$\sin^2 \frac{\alpha}{2} = \frac{1 - \cos \alpha}{2} \tag{18b}$$

Identities (14a) and (15a) and some algebra yield the first of the following *product formulas*:

$$\sin \alpha \cos \beta = \frac{1}{2}[\sin(\alpha - \beta) + \sin(\alpha + \beta)] \tag{19a}$$

$$\sin \alpha \sin \beta = \frac{1}{2}[\cos(\alpha - \beta) - \cos(\alpha + \beta)] \tag{19b}$$

$$\cos \alpha \cos \beta = \frac{1}{2}[\cos(\alpha - \beta) + \cos(\alpha + \beta)] \tag{19c}$$

The derivations of (19b) and (19c) are left as exercises.

We can deduce some important identities involving tangent, cotangent, secant, and cosecant from (9). The first of the identities

$$\tan^2 \theta + 1 = \sec^2 \theta \tag{20a}$$
$$1 + \cot^2 \theta = \csc^2 \theta \tag{20b}$$

can be obtained by dividing

$$\sin^2 \theta + \cos^2 \theta = 1$$

by $\cos^2 \theta$ and the second by dividing by $\sin^2 \theta$.

In the exercises we have asked the reader to deduce the first of the identities

$$\tan(\alpha + \beta) = \frac{\tan \alpha + \tan \beta}{1 - \tan \alpha \tan \beta} \tag{21a}$$

$$\tan(\alpha - \beta) = \frac{\tan \alpha - \tan \beta}{1 + \tan \alpha \tan \beta} \tag{21b}$$

from the relationship $\tan(\alpha + \beta) = \sin(\alpha + \beta)/\cos(\alpha + \beta)$. Identity (21b) is obtained from (21a) by substituting $-\beta$ for β and using the relationship

$$\tan(-\beta) = \frac{\sin(-\beta)}{\cos(-\beta)} = \frac{-\sin \beta}{\cos \beta} = -\tan \beta \tag{22}$$

If we let $\alpha = \beta$ in (21a), we obtain the double-angle formula

$$\tan 2\alpha = \frac{2 \tan \alpha}{1 - \tan^2 \alpha} \tag{23}$$

If we let $\beta = \pi$ in (21a) and (21b) and use the fact that $\tan \pi = 0$, we obtain the identities

$$\tan(\alpha + \pi) = \tan \alpha \tag{24a}$$
$$\tan(\alpha - \pi) = \tan \alpha \tag{24b}$$

The trigonometric identities obtained in this section are the ones most commonly used. In later sections other identities will be derived as they are needed.

▶ Exercises

In Exercises 1 and 2, express the angles in radians.

1. (a) $270°$ (b) $390°$ (c) $20°$
    (d) $138°$ (e) $\left(\dfrac{\pi}{10}\right)°$.

2. (a) $150°$ (b) $420°$ (c) $15°$
    (d) $117°$ (e) $165°$.

In Exercises 3 and 4, express the angles in degrees.

3. (a) $\dfrac{\pi}{15}$ (b) $\dfrac{3\pi}{2}$ (c) 4.5 (d) $\dfrac{8\pi}{5}$
    (e) 5π.

4. (a) $\dfrac{\pi}{10}$ (b) 3.2 (c) $\dfrac{\pi}{6}$ (d) $\dfrac{2\pi}{5}$
    (e) 2.

5. Sketch the following angles in standard position.
    (a) $\dfrac{\pi}{3}$ (b) $-\dfrac{5\pi}{6}$ (c) $\dfrac{9\pi}{4}$ (d) $-\pi$
    (e) 4π.

In Exercises 6 and 7, evaluate $\sin \theta$, $\cos \theta$, $\tan \theta$, $\csc \theta$, $\sec \theta$, $\cot \theta$ for the given values of θ.

6. (a) $\theta = \dfrac{\pi}{3}$ (b) $\theta = -\dfrac{5\pi}{6}$ (c) $\theta = \dfrac{9\pi}{4}$
    (d) $\theta = -\pi$ (e) $\theta = 4\pi$.

7. (a) $\theta = \dfrac{\pi}{6}$ (b) $\theta = -\dfrac{7\pi}{3}$ (c) $\theta = -\dfrac{5\pi}{4}$

(d) $\theta = -3\pi$ (e) $\theta = \pi$.

8. Find all values of θ (in radians) such that
    (a) $\sin \theta = 1$ (b) $\cos \theta = 1$ (c) $\tan \theta = 1$
    (d) $\csc \theta = 1$ (e) $\sec \theta = 1$ (f) $\cot \theta = 1$.

9. Find all values of θ (in radians) such that
    (a) $\sin \theta = 0$ (b) $\cos \theta = 0$ (c) $\tan \theta = 0$
    (d) $\csc \theta$ is undefined
    (e) $\sec \theta$ is undefined
    (f) $\cot \theta$ is undefined.

In Exercises 10–25, find all values of θ (in radians) that satisfy the given equation. Do not use any tables.

10. $\cos \theta = -\dfrac{1}{\sqrt{2}}$.   11. $\sin \theta = -\dfrac{1}{\sqrt{2}}$.

12. $\tan \theta = -1$.   13. $\cos \theta = \dfrac{1}{2}$.

14. $\sin \theta = -\dfrac{1}{2}$.   15. $\tan \theta = \sqrt{3}$.

16. $\tan \theta = \dfrac{1}{\sqrt{3}}$.   17. $\sin \theta = -\dfrac{\sqrt{3}}{2}$.

18. $\sin \theta = -1$.   19. $\cos \theta = -1$.

20. $\cot \theta = 1$.   21. $\cot \theta = \sqrt{3}$.

22. $\sec \theta = -2$.   23. $\csc \theta = -2$.

24. $\csc \theta = \dfrac{2}{\sqrt{3}}$.   **25.** $\sec \theta = \dfrac{2}{\sqrt{3}}$.

26. Find $\sin \theta$, $\cos \theta$, $\tan \theta$, $\csc \theta$, $\sec \theta$, $\cot \theta$ for the angles sketched in Figures A.19a and A.19b.

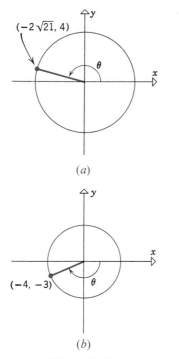

$(-2\sqrt{21}, 4)$

(a)

$(-4, -3)$

(b)

Figure A.19

27. Use identities and Table 1 to find the values of
(a) $\cos 15°$ (b) $\sin 22.5°$ (c) $\sin 37.5°$.

28. Use identities and Table 1 to find the values of
(a) $\sin 75°$ (b) $\tan 75°$.

29. (a) Let θ be an acute angle such that $\sin \theta = a/3$. Express the remaining trigonometric functions in terms of a by using the triangle in Figure A.20.

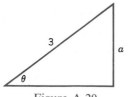

Figure A.20

(b) Let θ be an acute angle such that $\tan \theta = a/5$. Construct an appropriate triangle, and use it to express the remaining trigonometric functions in terms of a.

(c) Let θ be an acute angle such that $\sec \theta = a$. Construct an appropriate triangle, and use it to express the remaining trigonometric functions in terms of a.

30. How could you use a ruler and protractor to estimate $\sin 17°$ and $\cos 17°$?

31. Let ABC be a triangle whose angles at A and B are 30° and 45°. If the side opposite the angle B has length 9, find the lengths of the remaining sides and the size of the angle C.

Exercises 32–34 refer to an arbitrary triangle ABC in which side a is opposite angle A, side b is opposite angle B, and side c is opposite angle C.

32. Prove: The area of a triangle ABC can be written as

$$\text{area} = \frac{1}{2}bc \sin A$$

Find two other similar formulas for the area.

33. Prove the *law of sines:* In any triangle, the ratios of the sides to the sines of the opposite angles are equal. That is,

$$\frac{a}{\sin A} = \frac{b}{\sin B} = \frac{c}{\sin C}$$

34. (a) Use Exercises 32 and 33 to prove that the area of a triangle ABC can be written as

$$\text{area} = \frac{b^2 \sin A \sin C}{2 \sin B}$$

(b) Find two other similar formulas for the area.

35. Prove: If θ is positive and measured in radians, then the area of a sector with angle θ and radius r is

$$\text{area} = \frac{1}{2}r^2\theta$$

36. Derive the identities

$$\cos\left(\frac{\pi}{2} - \alpha\right) = \sin \alpha \qquad \sin\left(\frac{\pi}{2} - \alpha\right) = \cos \alpha$$

37. From the identities in Exercise 36, obtain the identities

$$\cos\left(\frac{\pi}{2} + \alpha\right) = -\sin\alpha$$

$$\sin\left(\frac{\pi}{2} + \alpha\right) = \cos\alpha$$

38. Derive identities (18a) and (18b).

39. Derive identities (19b) and (19c).

40. Derive identity (21a).

41. Express

$$\sin\left(\frac{3\pi}{2} - \theta\right) \quad \text{and} \quad \cos\left(\frac{3\pi}{2} + \theta\right)$$

in terms of $\sin\theta$ and $\cos\theta$.

In Exercises 42–53, verify the given identities.

42. $\tan\dfrac{\theta}{2} = \dfrac{1 - \cos\theta}{\sin\theta}$.

43. $\tan\dfrac{\theta}{2} = \dfrac{\sin\theta}{1 + \cos\theta}$.

44. $2\csc 2\theta = \sec\theta\csc\theta$.

45. $\tan\theta + \cot\theta = 2\csc 2\theta$.

46. $\sin 3\theta + \sin\theta = 2\sin 2\theta\cos\theta$.

47. $\sin 3\theta - \sin\theta = 2\cos 2\theta\sin\theta$.

48. $\dfrac{\cos\theta\sec\theta}{1 + \tan^2\theta} = \cos^2\theta$.

49. $\dfrac{\cos\theta\tan\theta + \sin\theta}{\tan\theta} = 2\cos\theta$.

50. $\dfrac{\sin 2\theta}{\sin\theta} - \dfrac{\cos 2\theta}{\cos\theta} = \sec\theta$.

51. $\dfrac{\sin\theta + \cos 2\theta - 1}{\cos\theta - \sin 2\theta} = \tan\theta$.

52. $\cos\left(\dfrac{\pi}{3} + \theta\right) + \cos\left(\dfrac{\pi}{3} - \theta\right) = \cos\theta$.

53. $\sin\left(\dfrac{3\pi}{2} - \theta\right) + \sin\left(\dfrac{3\pi}{2} + \theta\right) = -2\cos\theta$.

54. Beginning with identity $\sin^2\theta + \cos^2\theta = 1$, obtain the identity

$$1 + \cot^2\theta = \csc^2\theta$$

55. Express $\sin 3\theta$ and $\cos 3\theta$ in terms of $\sin\theta$ and $\cos\theta$.

56. (a) Express $3\sin\alpha + 5\cos\alpha$ in the form $C\sin(\alpha + \phi)$.
(b) Show that a sum of the form

$$A\sin\alpha + B\cos\alpha$$

can be rewritten in the form $C\sin(\alpha + \phi)$.

57. Show that the length of the diagonal of the parallelogram in Figure A.21 is

$$d = \sqrt{a^2 + b^2 + 2ab\cos\theta}$$

Figure A.21

58. Prove that the values of the trigonometric functions, as defined in Definition 1, do not depend on the value of r.

59. Prove that the ratio s/r in Formula (8) depends only on the angle and not on the radius r selected for the circle. [*Hint:* Use the result from plane geometry that states that the arc subtended on a circle by a central angle is proportional to the angle.]

UNIT II   GRAPHS OF TRIGONOMETRIC FUNCTIONS

In this section we shall review the basic properties of graphs of trigonometric functions.

Recall that an angle written without any units is understood to be measured in radians. Thus, the expression

$$\sin 3$$

means, the sine of 3 radians. More generally, for any real number x, the expression

$$\sin x \qquad\qquad (1)$$

means, the sine of x radians (i.e., the sine of an angle whose radian measure is x).

REMARK.   The use of the letter x in expression (1) rather than θ, α, β, or some other symbol is dangerous but virtually unavoidable in many problems. The danger is that x had a different meaning in the definition of the trigonometric functions (Definition 1 of Unit I). While it is obviously not correct to give a symbol two different meanings in the *same* problem, there is nothing wrong with using a symbol differently in *different* problems as long as we clearly understand how it is being used.

The expression $\sin x$ is defined for every real number x so that

$$f(x) = \sin x$$

defines a function f with domain $(-\infty, +\infty)$. By definition, the graph of $\sin x$ is the graph of the equation

$$y = \sin x$$

The general shape of this graph can be obtained geometrically by constructing an angle of x radians in standard position and recalling from the previous section that the terminal side of this angle intersects the unit circle at the point

$$(\cos x, \sin x)$$

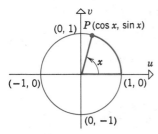

Figure A.22

This is illustrated in Figure A.22, where we have used u and v to label the coordinate axes so we can use x for the angle. Thus, by observing the ordinate of the point P in Figure A.22, we can see how the value of $\sin x$ varies with x. When $x = 0$, P is the point $(1, 0)$, so that $\sin x = 0$. As x increases from 0 to $\pi/2$, P moves counterclockwise along the circle from $(1, 0)$ to the point $(0, 1)$; thus, $\sin x$ increases from 0 to 1 (Figure

Figure A.23 ▷

(a)   (b)   (c)   (d)

A.23*a*). As *x* increases from $\pi/2$ to π, *P* moves counterclockwise along the circle from $(0, 1)$ to $(-1, 0)$; thus, $\sin x$ decreases from 1 to 0 (Figure A.23*b*). As *x* increases from π to $3\pi/2$, *P* moves counterclockwise along the circle from $(-1, 0)$ to $(0, -1)$; thus, $\sin x$ decreases from 0 to -1 (Figure A.23*c*). As *x* increases from $3\pi/2$ to 2π, *P* completes one full revolution by moving counterclockwise from $(0, -1)$ to $(1, 0)$; thus, $\sin x$ increases from -1 to 0 (Figure A.23*d*). As *x* increases past 2π, the point *P* begins a second revolution around the circle so that the values of $\sin x$ begin to repeat. In fact, the graph of

$$y = \sin x$$

will repeat itself every 2π units, since *P* makes a complete revolution whenever *x* changes by 2π. For negative values of *x*, the graph of $\sin x$ can be obtained by observing the ordinate of *P* during a *clockwise* rotation around the unit circle. Alternatively, we can use the identity

$$\sin(-x) = -\sin x$$

to observe that the values of the sine function at *x* and $-x$ differ only in sign. Either way we are led to the graph in Figure A.24.

   The fact that the graph of $y = \sin x$ repeats itself every 2π units could have been anticipated from the identities

$$\sin(x + 2\pi) = \sin x, \quad \sin(x - 2\pi) = \sin x$$

It is this *periodic* or repetitive nature of trigonometric functions that makes

Figure A.24 ▷

Graph of $y = \sin x$

them useful for studying repetitive physical phenomena such as vibrations and wave motion.

By observing the abscissa of the point P in Figure A.22, we can see how the value of $\cos x$ varies with x. An analysis similar to the one we used to graph $\sin x$ yields the graph of $y = \cos x$ shown in Figure A.25.

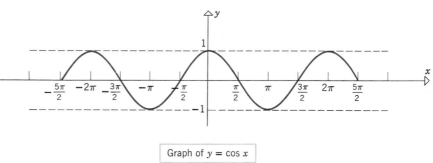

Figure A.25 ▷

Graph of $y = \cos x$

Observe that the graph of $\cos x$ is symmetric about the y-axis. This could have been anticipated from the identity

$$\cos(-x) = \cos x$$

which shows that for all x the cosine function assigns the same value to $-x$ and x.

Figure A.26

The graph of

$$y = \tan x$$

can also be obtained geometrically. To begin, let us assume that

$$0 \le x < \frac{\pi}{2}$$

If we construct the angle x in standard position and let Q be the intersection of its terminal side with the vertical line L through $(1, 0)$ (Figure A.26), then the coordinates of Q are

$(1, \tan x)$

To see this, observe that triangles OAP and OBQ are similar, so that

$$\frac{BQ}{OB} = \frac{AP}{OA} \qquad\qquad (2)$$

But $OB = 1$, $OA = \cos x$, and $AP = \sin x$, so that from (2)

$$BQ = \frac{\sin x}{\cos x} = \tan x$$

Thus, as shown in Figure A.26, the ordinate of Q is $\tan x$.

As x increases from 0 toward $\pi/2$, the point Q travels from $(1, 0)$ up the vertical line L getting higher and higher without any bound. Thus, as x increases from 0 toward $\pi/2$, the value of $\tan x$ increases without bound from an initial value of zero (Figure A.27a).

Figure A.27 ▷

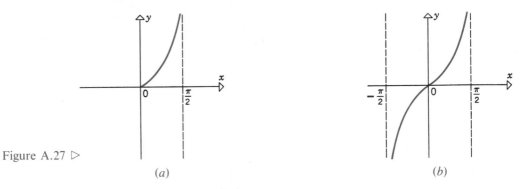

(a)   (b)

From the identity

$$\tan(-x) = -\tan x$$

and Figure A.27a, we can obtain the portion of the graph shown in Figure A.27b. Finally, from the identities

$$\tan(x + \pi) = \tan x \quad \text{and} \quad \tan(x - \pi) = \tan x$$

we see that the graph of $y = \tan x$ repeats itself every π units. This fact together with Figure A.27b leads us to the complete graph of $y = \tan x$ shown in Figure A.28.

REMARK. Unlike the sine and cosine functions that have domain $(-\infty, +\infty)$, the tangent function is not defined for all x. Since $\tan x = \sin x/\cos x$, the tangent function is undefined whenever $\cos x = 0$; this occurs when x is an odd integer multiple of $\pi/2$, that is, when

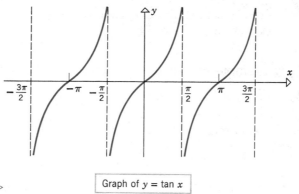

Figure A.28 ▷   Graph of $y = \tan x$

$$x = \pm \frac{\pi}{2}, \pm \frac{3\pi}{2}, \pm \frac{5\pi}{2}, \ldots$$

DEFINITION 1.   A function f is called **periodic** if there is a positive number p such that

$$f(x + p) = f(x) \tag{3}$$

whenever x and $x + p$ lie in the domain of f. We call p a **period** of the function. The smallest positive period is called the **fundamental period** of f or sometimes the **period** of f.

In words, (3) states that the values of f repeat themselves every p units.

Example 1   For the sine and cosine functions, 2π is a period since

$$\sin(x + 2\pi) = \sin x, \quad \cos(x + 2\pi) = \cos x$$

Also, 4π, 6π, 8π, and so on, are periods for sine and cosine since

$$\sin(x + 4\pi) = \sin x, \sin(x + 6\pi) = \sin x, \sin(x + 8\pi) = \sin x$$
$$\cos(x + 4\pi) = \cos x, \cos(x + 6\pi) = \cos x, \cos(x + 8\pi) = \cos x$$

and so forth. The fundamental period of sine and cosine is 2π.
    For the tangent function, π is a period since

$$\tan(x + \pi) = \tan x$$

Also, 2π, 3π, 4π, and so on, are periods; but π is the fundamental period.   ◀

Functions of the form

$$a \sin bx \quad \text{and} \quad a \cos bx \tag{4}$$

occur frequently in applications. The graphs of such functions are considered in the following examples.

Example 2   Sketch the graphs of

(a) $y = 3 \sin x$ (b) $y = \sin 4x$ (c) $y = 3 \sin 4x$

Solution (a).   The factor of 3 in

$$y = 3 \sin x$$

has the effect of multiplying each y-coordinate in the graph of $y = \sin x$ by 3, thereby stretching the graph of $y = \sin x$ vertically by a factor of 3 (Figure A.29a).

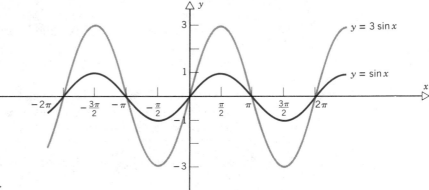

Figure A.29(a) ▷

Solution (b).   The graph of $y = \sin x$ repeats itself when x changes by 2π. Thus, the graph of $y = \sin 4x$ repeats itself whenever $4x$ changes by 2π, or equivalently when x changes by $\frac{1}{4}(2\pi) = \pi/2$. Thus, the factor of 4 in

$$y = \sin 4x$$

has the effect of reducing the fundamental period by a factor of $\frac{1}{4}$. This leads to the graph in Figure A.29b.

Figure A.29(b) ▷

Solution (c).   The graph of

$$y = 3 \sin 4x$$

combines both the stretching effect discussed in (a) and the period effect discussed in (b). The graph is shown in Figure A.29c.   ◀

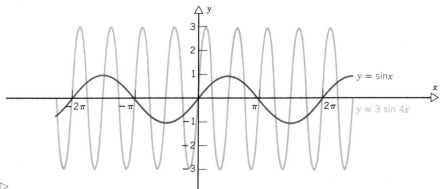

Figure A.29(c) ▷

The last example illustrates the following general result.

> **THEOREM 2.** *If $a \neq 0$ and $b \neq 0$, then the functions $a \sin bx$ and $a \cos bx$ have fundamental period $2\pi/|b|$ and their graphs oscillate between $-a$ and a.*

We leave the proof as an exercise.

Example 3   The graph of the function $6 \cos 7x$ oscillates between -6 and 6 and has fundamental period $2\pi/7$; the function $\frac{1}{3} \sin \frac{1}{4}x$ oscillates between $-\frac{1}{3}$ and $\frac{1}{3}$ and has fundamental period 8π.   ◀

We leave it as an exercise to prove the following result.

> **THEOREM 3.** *If $b \neq 0$, then the function $\tan bx$ has fundamental period $\pi/|b|$.*

▶ Exercises

1. Find the fundamental period of

   (a) $\sin 5x$
   (b) $\cos \frac{1}{3}x$
   (c) $\tan \frac{1}{8}x$

   (d) $\tan 7x$
   (e) $\sin \pi x$
   (f) $\sec \frac{\pi}{5}x$

   (g) $\cot \pi x$
   (h) $\csc \frac{x}{k}$.

   As stated in Theorem 2 of this section, the graphs of $a \sin bx$ and $a \cos bx$ oscillate between $-a$ and a. The number $|a|$ is called the *amplitude* of the function. In Exercises 2 and 3, find the amplitudes of the functions.

2. (a) $4 \sin 6x$ (b) $-8 \cos \pi x$
    (c) $\frac{1}{2} \sin x$ (d) $-\frac{1}{8} \cos 2x.$

3. (a) $5 \cos 2x$ (b) $\frac{1}{3} \sin 4x$
    (c) $-\frac{1}{2} \cos x$ (d) $-\sin \pi x.$

In Exercises 4–11, sketch the graph of the equation.

4. $y = \sin 3x.$ 5. $y = 2 \sin 3x.$

6. $y = 2 \sin x.$ 7. $y = \dfrac{1}{2} \cos \pi x.$

8. $y = -4 \cos x.$ 9. $y = \cos(-4x).$

10. $y = \sin(2\pi x).$ 11. $y = -2 \sin(-2x).$

12. (a) How are the graphs of $y = \cos x$ and $y = \cos(-x)$ related?
    (b) How are the graphs of $y = \sin x$ and $y = \sin(-x)$ related?

13. Use the graphs of $\sin x$, $\cos x$, $\tan x$, and the relationships

$$\cot x = \frac{1}{\tan x}, \quad \sec x = \frac{1}{\cos x}, \quad \csc x = \frac{1}{\sin x}$$

to help sketch the graphs of
    (a) $\cot x$ (b) $\sec x$ (c) $\csc x.$

In Exercises 14–17, sketch the graph of the equation.

14. $y = 3 \cot 2x.$ 15. $y = \dfrac{1}{4} \sec \pi x.$

16. $y = 2 \csc \dfrac{x}{3}.$ 17. $y = 2 \cot \dfrac{x}{4}.$

In Exercises 18–32, sketch the graph of the equation.

18. $y = \sin(x + \pi).$ 19. $y = \cos(x - \pi).$

20. $y = \cos\left(x - \dfrac{\pi}{4}\right).$ 21. $y = \sin\left(x + \dfrac{\pi}{2}\right).$

22. $y = 2 - \cos x.$ 23. $y = 1 + \sin x.$

24. $y = 2 \cos\left(4x + \dfrac{\pi}{2}\right).$ 25. $y = 3 \sin\left(2x + \dfrac{\pi}{4}\right).$

26. $y = \tan\left(\pi x - \dfrac{1}{2}\right).$ 27. $y = 2 \tan(2\pi x + 1).$

28. $y = |\tan x|.$ 29. $y = |\cos x|.$

30. $y = \sin x - \cos x.$ 31. $y = \sin x + \cos x.$

32. $y = 2 \sin x + 3 \cos 2x.$

33. Classify the following functions as even, odd, or neither. (See Exercise 39 in Section 2.3.)
    (a) $\sin x$ (b) $\cos x$ (c) $\tan x$
    (d) $|\sec x|$ (e) $\cot(x^2)$ (f) $3 \sec(-2x)$
    (g) $-4 \csc x$ (h) $\dfrac{\sin x}{x}.$

34. Prove: If $0 < x < \pi/2$, then

$$\sin x < x < \tan x$$

    [*Hint:* In Figure A.26, compare the areas of triangle OAP, sector OBP, and triangle OBQ. The area of a sector is discussed in Exercise 35 of Unit I of the trigonometry review.]

35. Prove: For all real numbers x,

$$|\sin x| \leq |x|$$

    [*Hint:* First consider the cases $|x| \geq \pi/2$ and $x = 0$; these are easy. Then use the left-hand inequality in Exercise 34 to complete the proof.]

36. Prove Theorem 2 of this section.

37. Prove Theorem 3 of this section.

Appendix 2

supplementary material

I. ONE-SIDED AND INFINITE LIMITS: A RIGOROUS APPROACH

In this section we shall give precise definitions of one-sided and infinite limits and illustrate how these definitions are used to prove results about limits.

In Section 2.4 we discussed the one-sided limits

$$\lim_{x \to a^-} f(x) = L \quad \text{and} \quad \lim_{x \to a^+} f(x) = L$$

from an intuitive viewpoint. We interpreted the first statement to mean that the value of $f(x)$ approaches L as x approaches a from the left side, and we interpreted the second statement to mean that the value of $f(x)$ approaches L as x approaches a from the right side. We can formalize these ideas mathematically as follows:

DEFINITION 1.   Let f be defined on some open interval extending to the left from a. We shall write

$$\lim_{x \to a^-} f(x) = L$$

if given any number $\epsilon > 0$, we can find a number $\delta > 0$ such that $f(x)$ satisfies $|f(x) - L| < \epsilon$ whenever x satisfies $a - \delta < x < a$.

DEFINITION 2.   Let f be defined on some open interval extending to the right from a. We shall write

$$\lim_{x \to a^+} f(x) = L$$

if given any number $\epsilon > 0$, we can find a number $\delta > 0$ such that $f(x)$ satisfies $|f(x) - L| < \epsilon$ whenever x satisfies $a < x < a + \delta$.

Before discussing some examples, let us compare Definitions 1 and 2 to the definition of $\lim_{x \to a} f(x) = L$ as stated in Definition 2.6.1. Given $\epsilon > 0$, the definition of $\lim_{x \to a} f(x) = L$ requires that we be able to find a $\delta > 0$ such that

$$|f(x) - L| < \epsilon \tag{1}$$

whenever x satisfies

$$0 < |x - a| < \delta$$

or equivalently, whenever x lies in the set

$$(a - \delta, a) \cup (a, a + \delta)$$

(See Figure A.30a.)

$$|f(x) - L| < \epsilon$$
$$\text{if}$$
$$0 < |x - a| < \delta$$

(a)

Figure A.30

$$|f(x) - L| < \epsilon$$
$$\text{if}$$
$$a - \delta < x < a$$

(b)

$$|f(x) - L| < \epsilon$$
$$\text{if}$$
$$a < x < a + \delta$$

(c)

On the other hand, the definition of $\lim_{x \to a^-} f(x) = L$ requires only that (1) hold for x in the set

$$(a - \delta, a)$$

(Figure A.30b), while the definition of $\lim_{x \to a^+} f(x) = L$ requires only that (1) be satisfied on the set

$$(a, a + \delta)$$

(See Figure A.30c.)

Example 1   Because the function $f(x) = \sqrt{x}$ is undefined for $x < 0$, we cannot consider the limits

$$\lim_{x \to 0} \sqrt{x} \quad \text{or} \quad \lim_{x \to 0^-} \sqrt{x}$$

The expression

$$\lim_{x \to 0^+} \sqrt{x}$$

is the only one that makes sense. Prove that

$$\lim_{x \to 0^+} \sqrt{x} = 0$$

Solution.   Let $\epsilon > 0$ be given. We must find a $\delta > 0$ such that $f(x) = \sqrt{x}$ satisfies

$$|\sqrt{x} - 0| < \epsilon \tag{2}$$

whenever x satisfies

$$0 < x < 0 + \delta$$

or equivalently,

$$0 < x < \delta \tag{3}$$

To find δ, we rewrite (2) as

$$\sqrt{x} < \epsilon \tag{4}$$

(since $\sqrt{x} = |\sqrt{x}|$). We must choose δ so that (4) holds whenever (3) does. We can do this by taking

$$\delta = \epsilon^2$$

To see that this choice of δ works, assume that x satisfies (3). Since we are letting $\delta = \epsilon^2$, it follows from (3) that

$$0 < x < \epsilon^2$$

or equivalently,

$$0 < \sqrt{x} < \epsilon \tag{5}$$

Thus, (4) holds since it is just the right-hand inequality in (5). This proves that

$$\lim_{x \to 0^+} \sqrt{x} = 0 \quad \blacktriangleleft$$

Example 2   Let

$$f(x) = \begin{cases} 2x, & x < 1 \\ x^3, & x \geq 1 \end{cases}$$

Prove that

$$\lim_{x \to 1^-} f(x) = 2$$

Solution.   Let $\epsilon > 0$ be given. We must find a $\delta > 0$ such that $f(x)$ satisfies

$$|f(x) - 2| < \epsilon \tag{6}$$

whenever x satisfies

$$1 - \delta < x < 1 \tag{7}$$

Because we are considering only x satisfying (7), it follows from the formula for $f(x)$ that

$$f(x) = 2x$$

Thus, we can rewrite (6) as

$$|2x - 2| < \epsilon$$

or

$$|x - 1| < \frac{\epsilon}{2}$$

or

$$1 - \frac{\epsilon}{2} < x < 1 + \frac{\epsilon}{2} \tag{8}$$

We must choose δ so that (8) is satisfied whenever (7) is satisfied. We can do this by taking

$$\delta = \frac{\epsilon}{2}$$

To see that this choice of δ works, assume that x satisfies (7). With $\delta = \epsilon/2$, it follows from (7) that

$$1 - \frac{\epsilon}{2} < x < 1 \tag{9}$$

and this implies that

$$1 - \frac{\epsilon}{2} < x < 1 + \frac{\epsilon}{2}$$

since ϵ is positive.

Thus, we have proved

$$\lim_{x \to 1^-} f(x) = 2 \quad \blacktriangleleft$$

In Section 2.4 we discussed the limits

$$\lim_{x \to -\infty} f(x) = L \quad \text{and} \quad \lim_{x \to +\infty} f(x) = L$$

from an intuitive viewpoint. We interpreted the first statement to mean that the value of $f(x)$ approaches L as x moves indefinitely far from the origin along the negative x-axis, and we interpreted the second statement to mean that the value of $f(x)$ approaches L as x moves indefinitely far from the origin along the positive x-axis.

INFINITE LIMITS   The phrase, "x moves indefinitely far from the origin along the positive x-axis" is intended to convey the idea that given *any* positive number N (no matter how large) eventually x is larger than N (Figure A.31a). Similarly, the phrase "x moves indefinitely far from the origin along the negative x-axis" is intended to convey the idea that given any negative number N (no matter how small) eventually x is smaller than N (Figure A.31b). Thus, when we say that $\lim_{x \to +\infty} f(x) = L$, we mean that for any $\epsilon > 0$, we can find a positive number N (sufficiently large) so that $f(x)$ satisfies

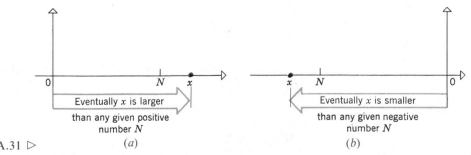

Figure A.31 ▷

(a)   Eventually x is larger than any given positive number N

(b)   Eventually x is smaller than any given negative number N

Figure A.32 ▷

$$f(x) \text{ satisfies } |f(x) - L| < \epsilon$$
$$\text{when } x > N.$$

$$|f(x) - L| < \epsilon$$

when x is greater than N (Figure A.32). Similarly, when we say that $\lim\limits_{x \to -\infty} f(x) = L$, we mean that for any $\epsilon > 0$, we can find a negative number N so that $f(x)$ satisfies

$$|f(x) - L| < \epsilon$$

when x is smaller than N (Figure A.33).

Figure A.33 ▷

$$f(x) \text{ satisfies } |f(x) - L| < \epsilon$$
$$\text{when } x < N.$$

These ideas suggest the following formal definitions.

DEFINITION 3.   Let f be a function that is defined on some infinite open interval $(x_0, +\infty)$. We shall write

$$\lim_{x \to +\infty} f(x) = L$$

if given any number $\epsilon > 0$, there corresponds a positive number N such that

$$|f(x) - L| < \epsilon$$

whenever $x > N$.

DEFINITION 4.   Let f be a function that is defined on some infinite open interval $(-\infty, x_0)$. We shall write

$$\lim_{x \to -\infty} f(x) = L$$

if given any number $\epsilon > 0$, there corresponds a negative number N such that

$$|f(x) - L| < \epsilon$$

whenever x satisfies $x < N$.

Example 3   Prove that

$$\lim_{x \to +\infty} \frac{1}{x} = 0$$

Solution.   We must show that for any $\epsilon > 0$ we can find a positive number N such that $f(x) = 1/x$ satisfies

$$\left| \frac{1}{x} - 0 \right| < \epsilon \tag{10}$$

whenever x satisfies

$$x > N \tag{11}$$

To find N, rewrite (10) as

$$\frac{1}{|x|} < \epsilon$$

or

$$|x| > \frac{1}{\epsilon} \tag{12}$$

Thus, we must find a positive number N such that x satisfies (12) whenever x satisfies (11). We can do this by taking

$$N = \frac{1}{\epsilon}$$

To prove that this choice of N works, assume that x satisfies (11). Since we are letting $N = 1/\epsilon$, (11) becomes

$$x > \frac{1}{\epsilon} \tag{13}$$

But $1/\epsilon$ is positive (since $\epsilon > 0$) so that the x in (13) is positive. Thus, (13) can be written as

$$|x| > \frac{1}{\epsilon}$$

which is precisely inequality (12). Thus, we have proved

$$\lim_{x \to +\infty} \frac{1}{x} = 0 \qquad \blacktriangleleft$$

In Section 2.4 we discussed the limits

$$\lim_{x \to a} f(x) = +\infty \qquad \lim_{x \to a} f(x) = -\infty \tag{14}$$

$$\lim_{x \to a^+} f(x) = +\infty \qquad \lim_{x \to a^+} f(x) = -\infty \tag{15}$$

$$\lim_{x \to a^-} f(x) = +\infty \qquad \lim_{x \to a^-} f(x) = -\infty \tag{16}$$

$$\lim_{x \to +\infty} f(x) = +\infty \qquad \lim_{x \to +\infty} f(x) = -\infty \tag{17}$$

$$\lim_{x \to -\infty} f(x) = +\infty \qquad \lim_{x \to -\infty} f(x) = -\infty \tag{18}$$

from an intuitive viewpoint. Recall that each of these expressions describes a particular way in which the limit fails to exist. The $+\infty$ to the right of the $=$ sign indicates that the limit fails to exist because $f(x)$ is increasing without bound, and the $-\infty$ indicates that the limit fails to exist because $f(x)$ is decreasing without bound.

To make these ideas mathematically precise, we must clarify the meanings of the phrases "$f(x)$ is increasing without bound" and "$f(x)$ is decreasing without bound." We shall limit our discussion to the limits in (14); formal definitions of the remaining limits are discussed in the exercises.

When we say that $f(x)$ increases without bound as x approaches a from either side, we mean that given any positive number N (no matter how large) the value of $f(x)$ eventually exceeds N as x approaches a from either side. Figure A.34 illustrates this idea. For the curve in that figure we have

Figure A.34

$$\lim_{x \to a} f(x) = +\infty$$

We picked an arbitrary positive number N and marked it on the y-axis. As x approaches a from either side, x will eventually lie between the points

$$a - \delta \quad \text{and} \quad a + \delta$$

shown in the figure. When this occurs the value of $f(x)$ will satisfy

$$f(x) > N \tag{19}$$

It is evident from the figure that no matter how large we make N, we shall always be able to find a sufficiently small positive number δ such that $f(x)$ satisfies (19) when x is between $a - \delta$ and $a + \delta$ (but different from a). This suggests the following definition.

DEFINITION 5.   Let f be defined in some open interval containing the number a, except that f need not be defined at a. We shall write

$$\lim_{x \to a} f(x) = +\infty$$

if given any positive number N, we can find a number $\delta > 0$ such that $f(x)$ satisfies

$$f(x) > N$$

whenever x satisfies $0 < |x - a| < \delta$.

When we say that $f(x)$ decreases without bound as x approaches a from either side, we mean that given any negative number N (no matter how small) the value of $f(x)$ will eventually be smaller than N as x approaches a from either side. Figure A.35 illustrates this idea. For the curve in that figure we have

$$\lim_{x \to a} f(x) = -\infty$$

Figure A.35 ▷

We picked an arbitrary negative number N and marked it on the y-axis. As x approaches a from either side, x will eventually lie between the points

$$a - \delta \quad \text{and} \quad a + \delta$$

shown in the figure. When this occurs the value of $f(x)$ will satisfy

$$f(x) < N$$

This suggests the following definition.

DEFINITION 6.   Let f be defined in some open interval containing the number a, except that f need not be defined at a. We shall write

$$\lim_{x \to a} f(x) = -\infty$$

if given any negative number N we can find a number $\delta > 0$ such that $f(x)$ satisfies

$$f(x) < N$$

whenever x satisfies $0 < |x - a| < \delta$.

Example 4   Prove that

$$\lim_{x \to 0} \frac{1}{x^2} = +\infty$$

Solution.   We must show that given any positive number N, we can find a positive number δ such that $f(x) = 1/x^2$ satisfies

$$\frac{1}{x^2} > N \tag{20}$$

whenever x satisfies

$$0 < |x - 0| < \delta$$

or equivalently,

$$0 < |x| < \delta \tag{21}$$

To find δ we shall rewrite (20) as

$$x^2 < \frac{1}{N} \tag{22}$$

Thus, we must find a positive number δ such that x satisfies (22) whenever x satisfies (21). We can do this by taking

$$\delta = \frac{1}{\sqrt{N}}$$

To prove that this choice of δ works, assume that x satisfies (21). Since we are letting $\delta = 1/\sqrt{N}$, (21) becomes

$$0 < |x| < \frac{1}{\sqrt{N}}$$

or equivalently

$$0 < |x|^2 < \frac{1}{N}$$

or since $|x|^2 = x^2$,

$$0 < x^2 < \frac{1}{N} \tag{23}$$

Thus, (22) is satisfied, since it is just the right-hand inequality in (23). This proves that

$$\lim_{x \to +\infty} \frac{1}{x^2} = 0 \quad \blacktriangleleft$$

▶ Exercises

In Exercises 1–6, use Definitions 1 and 2 to prove that the stated limit is correct.

1. $\lim_{x \to 2^+} (x + 1) = 3$.

2. $\lim_{x \to 1^-} (3x + 2) = 5$.

3. $\lim_{x \to 4^+} \sqrt{x - 4} = 0$.

4. $\lim_{x \to 0^-} \sqrt{-x} = 0$.

5. $\lim_{x \to 2^+} f(x) = 2$, where $f(x) = \begin{cases} x, & x > 2 \\ 3x, & x \le 2. \end{cases}$

6. $\lim_{x \to 2^-} f(x) = 6$, where $f(x) = \begin{cases} x, & x > 2 \\ 3x, & x \le 2. \end{cases}$

In Exercises 7–14, use Definitions 3 and 4 to prove that the stated limit is correct.

7. $\lim_{x \to +\infty} \frac{1}{x^2} = 0$.

8. $\lim_{x \to -\infty} \frac{1}{x} = 0$.

9. $\lim_{x \to -\infty} \frac{1}{x + 2} = 0$.

10. $\lim_{x \to +\infty} \frac{1}{x + 2} = 0$.

11. $\lim_{x \to +\infty} \frac{x}{x + 1} = 1$.

12. $\lim_{x \to -\infty} \frac{x}{x + 1} = 1$.

13. $\lim_{x \to -\infty} \frac{4x - 1}{2x + 5} = 2$.

14. $\lim_{x \to +\infty} \frac{4x - 1}{2x + 5} = 2$.

In Exercises 15–20, use Definitions 5 and 6 to prove that the stated limit is correct.

15. $\lim_{x \to 3} \frac{1}{(x - 3)^2} = +\infty$.

16. $\lim_{x \to 3} \frac{-1}{(x - 3)^2} = -\infty$.

17. $\lim_{x \to 0} \frac{1}{|x|} = +\infty$.

18. $\lim_{x \to 1} \frac{1}{|x - 1|} = +\infty$.

19. $\lim_{x \to 0} \left(-\frac{1}{x^4} \right) = -\infty$.

20. $\lim_{x \to 0} \frac{1}{x^4} = +\infty$.

21. Define limit statements (15) and (16).

22. Use the definitions in Exercise 21 to prove:

   (a) $\lim_{x \to 0^+} \frac{1}{x} = +\infty$

   (b) $\lim_{x \to 0^-} \frac{1}{x} = -\infty$.

23. Use the definitions in Exercise 21 to prove:

   (a) $\lim_{x \to 1^+} \frac{1}{1 - x} = -\infty$

   (b) $\lim_{x \to 1^-} \frac{1}{1 - x} = +\infty$.

24. Define limit statements (17) and (18).

25. Use the definitions in Exercise 24 to prove:

(a) $\lim_{x \to +\infty} (x + 1) = +\infty$

(b) $\lim_{x \to -\infty} (x + 1) = -\infty$.

26. Use the definitions in Exercise 24 to prove:

(a) $\lim_{x \to +\infty} (x^2 - 3) = +\infty$

(b) $\lim_{x \to -\infty} (x^3 + 5) = -\infty$.

II. SOME PROOFS OF LIMIT THEOREMS

In this section we shall prove a number of results stated in Section 2.5, as well as some other basic properties of limits.

Our first result states that the limit of a function, if it exists, is unique; that is, there cannot be two distinct limits.

> **THEOREM 1.** *If* $\lim_{x \to a} f(x) = L_1$ *and* $\lim_{x \to a} f(x) = L_2$, *then* $L_1 = L_2$.

REMARK.   In this section we shall only consider the cases where a and the limits are finite (i.e., not equal to $+\infty$ or $-\infty$). Other cases are considered in the exercises.

Proof.   We shall assume that $L_1 \neq L_2$ and show that this assumption leads to a contradiction. It will then follow that $L_1 = L_2$. Let

$$\epsilon = \frac{|L_1 - L_2|}{2} \tag{1}$$

Since we are assuming $L_1 \neq L_2$, it follows that $\epsilon > 0$.
   From the assumption

$$\lim_{x \to a} f(x) = L_1$$

we can find $\delta_1 > 0$ such that

$$|f(x) - L_1| < \epsilon \tag{2}$$

whenever

$$0 < |x - a| < \delta_1 \tag{3}$$

and since

$$\lim_{x \to a} f(x) = L_2 \tag{4}$$

we can find $\delta_2 > 0$ such that

$$|f(x) - L_2| < \epsilon \tag{5}$$

whenever

$$0 < |x - a| < \delta_2 \tag{6}$$

Let δ be the smaller of the numbers δ_1 and δ_2, and choose any x satisfying

$$0 < |x - a| < \delta$$

Since $\delta \leq \delta_1$ and $\delta \leq \delta_2$, this x satisfies both (3) and (6) and, therefore, $f(x)$ satisfies both (2) and (5). But

$$
\begin{aligned}
2\epsilon &= |L_1 - L_2| && \text{[From (1)]} \\
&= |L_1 - f(x) + f(x) - L_2| \\
&= |(L_1 - f(x)) + (f(x) - L_2)| \\
&\leq |L_1 - f(x)| + |f(x) - L_2| && \text{[The triangle inequality]} \\
&< \epsilon + \epsilon && \text{[(2) and (5)]} \\
&= 2\epsilon
\end{aligned}
$$

Thus, we have shown that

$$2\epsilon < 2\epsilon$$

which is a contradiction.   ∎

Next, we prove the result discussed in Example 1 of Section 2.5.

THEOREM 2.   *For any constant k,*

$$\lim_{x \to a} k = k$$

Proof.   Let $\epsilon > 0$ be given. We must find $\delta > 0$ such that

$$|k - k| < \epsilon \tag{7}$$

whenever

$$0 < |x - a| < \delta \tag{8}$$

But (7) can be written as

$$0 < \epsilon$$

which holds for all values of x. Thus, for *any* positive δ whatever, (7) will be satisfied when (8) is satisfied.   ∎

Next we shall prove part (*a*) of Theorem 2.5.1.

Theorem 2.5.1a

> THEOREM 3.   *If* $\lim\limits_{x \to a} f(x) = L_1$ *and* $\lim\limits_{x \to a} g(x) = L_2$, *then*
>
> $$\lim_{x \to a} [\,f(x) + g(x)] = \lim_{x \to a} f(x) + \lim_{x \to a} g(x) = L_1 + L_2$$

Proof.   Let $\epsilon > 0$ be given. To prove that

$$\lim_{x \to a} [\,f(x) + g(x)] = L_1 + L_2$$

we must show that given $\epsilon > 0$ we can find $\delta > 0$ such that

$$|[\,f(x) + g(x)] - [L_1 + L_2]| < \epsilon \tag{9}$$

whenever

$$0 < |x - a| < \delta \tag{10}$$

Since

$$\lim_{x \to a} f(x) = L_1$$

we can find $\delta_1 > 0$ such that

$$|f(x) - L_1| < \frac{\epsilon}{2} \tag{11}$$

whenever

$$0 < |x - a| < \delta_1 \tag{12}$$

and since

$$\lim_{x \to a} g(x) = L_2$$

we can find $\delta_2 > 0$ such that

$$|g(x) - L_2| < \frac{\epsilon}{2} \tag{13}$$

whenever

$$0 < |x - a| < \delta_2 \tag{14}$$

Let δ be the smaller of the numbers δ_1 and δ_2. If x satisfies (10), then x will also satisfy both (12) and (14) since $\delta \leq \delta_1$ and $\delta \leq \delta_2$. Consequently,

$f(x)$ and $g(x)$ will satisfy both (11) and (13). Therefore,

$$|[f(x) + g(x)] - [L_1 + L_2]| = |[f(x) - L_1] + [g(x) - L_2]|$$

$$\leq |f(x) - L_1| + |g(x) - L_2|$$

$$< \frac{\epsilon}{2} + \frac{\epsilon}{2} = \epsilon$$

Thus, for the δ we have selected, (9) is satisfied whenever (10) is satisfied. This completes the proof.   ∎

Next we shall prove part (c) of Theorem 2.5.1. This proof is a little more complicated than the previous proofs.

Theorem 2.5.1c

THEOREM 4.   *If* $\lim\limits_{x \to a} f(x) = L_1$ *and* $\lim\limits_{x \to a} g(x) = L_2$, *then*

$$\lim_{x \to a} [f(x)g(x)] = \lim_{x \to a} f(x) \lim_{x \to a} g(x) = L_1 L_2$$

Proof.   Let $\epsilon > 0$ be given. We must find $\delta > 0$ such that

$$|f(x)g(x) - L_1 L_2| < \epsilon \tag{15}$$

whenever

$$0 < |x - a| < \delta \tag{16}$$

To find δ we shall express (15) in a different form. We can write

$$f(x) = L_1 + [f(x) - L_1] \quad \text{and} \quad g(x) = L_2 + [g(x) - L_2]$$

When we multiply these expressions and subtract $L_1 L_2$, we obtain

$$f(x)g(x) - L_1 L_2 = L_1[g(x) - L_2] + L_2[f(x) - L_1]$$
$$+ [f(x) - L_1][g(x) - L_2]$$

so that (15) can be rewritten

$$|L_1[g(x) - L_2] + L_2[f(x) - L_1] + [f(x) - L_1][g(x) - L_2]| < \epsilon \tag{17}$$

Thus, we must find $\delta > 0$ such that (17) holds whenever (16) does.
Since ϵ is positive, the numbers

$$\sqrt{\epsilon/3}, \quad \frac{\epsilon}{3(1 + |L_1|)}, \quad \frac{\epsilon}{3(1 + |L_2|)}$$

are also positive. Therefore, since

$$\lim_{x \to a} f(x) = L_1 \quad \text{and} \quad \lim_{x \to a} g(x) = L_2$$

we can find positive numbers δ_1, δ_2, δ_3, and δ_4 such that

$$
\left.
\begin{array}{lll}
|f(x) - L_1| < \sqrt{\epsilon/3} & \text{when} & 0 < |x - a| < \delta_1 \\[2mm]
|f(x) - L_1| < \dfrac{\epsilon}{3(1 + |L_2|)} & \text{when} & 0 < |x - a| < \delta_2 \\[3mm]
|g(x) - L_2| < \sqrt{\epsilon/3} & \text{when} & 0 < |x - a| < \delta_3 \\[2mm]
|g(x) - L_2| < \dfrac{\epsilon}{3(1 + |L_1|)} & \text{when} & 0 < |x - a| < \delta_4
\end{array}
\right\} \tag{18}
$$

Let δ be the smallest of the numbers δ_1, δ_2, δ_3, and δ_4. Then $\delta \leq \delta_1$, $\delta \leq \delta_2$, $\delta \leq \delta_3$, and $\delta \leq \delta_4$. Thus, if x satisfies (16), then x will also satisfy the four conditions on the right side of (18). Consequently, $f(x)$ and $g(x)$ will satisfy the four conditions on the left side of (18). Therefore,

$$
\begin{aligned}
& |L_1[g(x) - L_2] + L_2[f(x) - L_1] + [f(x) - L_1][g(x) - L_2]| \\[2mm]
& \quad \leq |L_1[g(x) - L_2]| + |L_2[f(x) - L_1]| + |[f(x) - L_1][g(x) - L_2]| \\[2mm]
& \quad = |L_1|\,|g(x) - L_2| + |L_2|\,|f(x) - L_1| + |f(x) - L_1|\,|g(x) - L_2| \\[2mm]
& \quad < |L_1|\frac{\epsilon}{3(1 + |L_1|)} + |L_2|\frac{\epsilon}{3(1 + |L_2|)} + \sqrt{\epsilon/3}\sqrt{\epsilon/3} \quad \text{[From (18)]} \\[2mm]
& \quad = \frac{\epsilon}{3}\frac{|L_1|}{1 + |L_1|} + \frac{\epsilon}{3}\frac{|L_2|}{1 + |L_2|} + \frac{\epsilon}{3} \\[2mm]
& \quad < \frac{\epsilon}{3} + \frac{\epsilon}{3} + \frac{\epsilon}{3} = \epsilon
\end{aligned}
$$

$$
\boxed{\text{Since } \frac{|L_1|}{1 + |L_1|} < 1 \text{ and } \frac{|L_2|}{1 + |L_2|} < 1}
$$

Thus, for the δ we have selected, (17) holds whenever (16) does. This completes the proof. ∎

RELATIONSHIP BETWEEN ONE-SIDED AND TWO-SIDED LIMITS

In Section 2.4 we interpreted the statement

$$
\lim_{x \to a} f(x) = L
$$

to mean

$$
\lim_{x \to a^+} f(x) = L \quad \text{and} \quad \lim_{x \to a^-} f(x) = L
$$

The following theorem shows that this interpretation is consistent with our formal limit definitions.

THEOREM 5. *If* $\lim\limits_{x \to a^+} f(x) = \lim\limits_{x \to a^-} f(x) = L$, *then* $\lim\limits_{x \to a} f(x) = L$ *and conversely.*

Proof.   Assume

$$\lim_{x \to a^+} f(x) = \lim_{x \to a^-} f(x) = L \tag{19}$$

To prove

$$\lim_{x \to a} f(x) = L$$

we must show that given $\epsilon > 0$ there exists a $\delta > 0$ such that

$$|f(x) - L| < \epsilon \tag{20}$$

whenever

$$0 < |x - a| < \delta \tag{21}$$

Because of (19), there exists a number $\delta_1 > 0$ such that (20) is satisfied whenever

$$a < x < a + \delta_1 \tag{22}$$

and there exists a number $\delta_2 > 0$ such that (20) is satisfied whenever

$$a - \delta_2 < x < a \tag{23}$$

Figure A.36

Let δ be the smaller of δ_1 and δ_2, so that $\delta \leq \delta_1$ and $\delta \leq \delta_2$. Thus, if x satisfies (21), x will satisfy both (22) and (23) (Figure A.36) and consequently $f(x)$ will satisfy (20). Thus, for the δ we have selected, (20) holds whenever (21) does. This completes this part of the proof. The proof of the converse is left as an exercise.   ∎

► Exercises

1. Prove: For any constant k, $\lim_{x \to +\infty} k = k$.

2. Prove: For any constant k, $\lim_{x \to -\infty} k = k$.

3. Prove: If $\lim_{x \to -\infty} f(x) = L_1$ and $\lim_{x \to -\infty} g(x) = L_2$, then
$$\lim_{x \to -\infty} [f(x) + g(x)] = L_1 + L_2.$$

4. Prove: If $\lim_{x \to +\infty} f(x) = L_1$ and $\lim_{x \to +\infty} g(x) = L_2$, then
$$\lim_{x \to +\infty} [f(x) + g(x)] = L_1 + L_2.$$

5. Use Theorems 2, 3, and 4 of this section to prove:
If $\lim_{x \to a} f(x) = L_1$ and $\lim_{x \to a} g(x) = L_2$, then

$$\lim_{x \to a} [f(x) - g(x)] = \lim_{x \to a} f(x) - \lim_{x \to a} g(x)$$
$$= L_1 - L_2.$$

6. Suppose $\lim_{x \to a} f(x) = +\infty$ and $\lim_{x \to a} g(x) = +\infty$.

(a) Prove: $\lim_{x \to a} [f(x) + g(x)] = +\infty$.

(b) Is it true that $\lim_{x \to a} [f(x) - g(x)] = 0$?

7. Suppose $\lim_{x \to a} f(x) = -\infty$ and $\lim_{x \to a} g(x) = +\infty$.

(a) Prove: $\lim_{x \to a} [f(x) - g(x)] = -\infty$.

(b) Is it true that $\lim_{x \to a} [f(x) + g(x)] = 0$?

8. Use Theorems 2 and 4 of this section to prove: If $\lim_{x \to a} f(x) = L$ and k is a constant, then
$$\lim_{x \to a} [kf(x)] = kL.$$

9. Prove: $\lim_{x \to a} f(x) = L$ if and only if

$$\lim_{x \to a} [f(x) - L] = 0.$$

10. Prove: If $\lim_{x \to a} f(x) = L$, then $\lim_{x \to a} |f(x)| = |L|$.

11. Finish the proof of Theorem 5 of this section.

III. PROOF OF THE LIMIT PROPERTY OF CONTINUITY

In this section we shall prove Theorem 2.7.5 for the case of the two-sided limit $\lim_{x \to c}$.

Theorem 2.7.5

THEOREM 1. *If* $\lim_{x \to c} g(x) = L$ *and if the function f is continuous at L, then*

$$\lim_{x \to c} f(g(x)) = f(L)$$

that is,

$$\lim_{x \to c} f(g(x)) = f\left(\lim_{x \to c} g(x)\right)$$

Proof. To prove that

$$\lim_{x \to c} f(g(x)) = f(L)$$

we must show that for every $\epsilon > 0$ there is a $\delta > 0$ such that $f(g(x))$ satisfies

$$|f(g(x)) - f(L)| < \epsilon \tag{1}$$

whenever x satisfies

$$0 < |x - c| < \delta \tag{2}$$

Since f is continuous at L,

$$\lim_{x \to L} f(x) = f(L)$$

If we use u as the variable rather than x, then this limit can be written as

$$\lim_{u \to L} f(u) = f(L)$$

Thus, given $\epsilon > 0$ we can find a $\delta_1 > 0$ such that $f(u)$ satisfies

$$|f(u) - f(L)| < \epsilon \tag{3}$$

whenever u satisfies

$$|u - L| < \delta_1 \tag{4}$$

Since

$$\lim_{x \to c} g(x) = L$$

there exists $\delta > 0$ such that

$$|g(x) - L| < \delta_1 \tag{5}$$

whenever x satisfies (2). To complete the proof let

$$u = g(x)$$

and assume that x satisfies (2). This implies that $g(x)$ satisfies (5), which implies that u satisfies (4). Therefore, $f(u)$ satisfies (3), which implies that $f(g(x))$ satisfies (1). Thus, (1) is satisfied whenever (2) is satisfied.   ∎

IV.   PROOF OF THE CHAIN RULE

In this section we shall prove the chain rule (Theorem 3.5.2).

We begin with a preliminary result.

THEOREM **1.**   *If f is differentiable at x and if $y = f(x)$, then*

$$\Delta y = f'(x)\,\Delta x + \epsilon\,\Delta x$$

where $\epsilon \to 0$ as $\Delta x \to 0$.

Proof.   Define

$$\epsilon = \begin{cases} \dfrac{f(x + \Delta x) - f(x)}{\Delta x} - f'(x) & \text{if } \Delta x \neq 0 \\[2mm] 0 & \text{if } \Delta x = 0 \end{cases} \tag{1}$$

If $\Delta x \neq 0$, it follows from (1) that

$$\epsilon\,\Delta x = [f(x + \Delta x) - f(x)] - f'(x)\,\Delta x \tag{2}$$

But,

$$\Delta y = f(x + \Delta x) - f(x) \tag{3}$$

so (2) can be written as

$$\epsilon \, \Delta x = \Delta y - f'(x) \, \Delta x$$

or

$$\Delta y = f'(x) \, \Delta x + \epsilon \, \Delta x \tag{4}$$

If $\Delta x = 0$, then (4) still holds (why?), so (4) is valid for all values of Δx. It remains to show that $\epsilon \to 0$ as $\Delta x \to 0$. But, this follows from the assumption that f is differentiable at x, since

$$\lim_{\Delta x \to 0} \epsilon = \lim_{\Delta x \to 0} \left[\frac{f(x + \Delta x) - f(x)}{\Delta x} - f'(x) \right] = f'(x) - f'(x) = 0 \quad \blacksquare$$

We are now ready to prove the chain rule.

Theorem 3.5.2
The Chain Rule

THEOREM 2. *If g is differentiable at the point x and f is differentiable at the point g(x), then the composition f ∘ g is differentiable at the point x. Moreover, if y = f(g(x)) and u = g(x), then*

$$\frac{dy}{dx} = \frac{dy}{du} \cdot \frac{du}{dx}$$

Proof. Since g is differentiable at x and $u = g(x)$, it follows from Theorem 1 that

$$\Delta u = g'(x) \, \Delta x + \epsilon_1 \, \Delta x \tag{5}$$

where $\epsilon_1 \to 0$ as $\Delta x \to 0$. And since $y = f(g(x)) = f(u)$ is differentiable at $u = g(x)$, it follows from Lemma 1 that

$$\Delta y = f'(u) \, \Delta u + \epsilon_2 \, \Delta u \tag{6}$$

where $\epsilon_2 \to 0$ as $\Delta u \to 0$.

Factoring out the Δu in (6) and then substituting (5) yields

$$\Delta y = [f'(u) + \epsilon_2][g'(x) \, \Delta x + \epsilon_1 \, \Delta x]$$

or

$$\Delta y = [f'(u) + \epsilon_2][g'(x) + \epsilon_1] \, \Delta x$$

or if $\Delta x \neq 0$,

$$\frac{\Delta y}{\Delta x} = [f'(u) + \epsilon_2][g'(x) + \epsilon_1] \tag{7}$$

Since $\epsilon_1 \to 0$ and $\epsilon_2 \to 0$ as $\Delta x \to 0$, it follows from (7) that

$$\lim_{\Delta x \to 0} \frac{\Delta y}{\Delta x} = f'(u)g'(x)$$

or

$$\frac{dy}{dx} = f'(u)g'(x) = \frac{dy}{du} \cdot \frac{du}{dx} \quad \blacksquare$$

V.   PROOFS OF KEY RESULTS USING THE MEAN-VALUE THEOREM

In this section we shall prove three fundamental results that rest on the Mean-Value Theorem: Theorem 4.2.2, Theorem 4.3.6 (the first derivative test), and Theorem 4.3.7 (the second derivative test).

Theorem 4.2.2

THEOREM **1.**

(a)   *If $f'(x) > 0$ on an open interval (a, b), then f is increasing on (a, b).*

(b)   *If $f'(x) < 0$ on an open interval (a, b), then f is decreasing on (a, b).*

Proof of (a).   Let x_1 and x_2 be points in (a, b) such that $x_1 < x_2$. We must show that $f(x_1) < f(x_2)$. Since f is differentiable on (a, b), it is continuous on (a, b). Therefore, f is continuous on $[x_1, x_2]$ and differentiable on (x_1, x_2), since the interval $[x_1, x_2]$ is contained inside the interval (a, b). Thus, we can apply the Mean-Value Theorem over the interval $[x_1, x_2]$ and conclude that there is a number c in the interval (x_1, x_2) such that

$$\frac{f(x_2) - f(x_1)}{x_2 - x_1} = f'(c)$$

or equivalently

$$f(x_2) - f(x_1) = f'(c)(x_2 - x_1) \tag{1}$$

But $x_2 - x_1 > 0$ since $x_1 < x_2$, and $f'(c) > 0$ follows from the hypothesis; thus, from (1)

$$f(x_2) - f(x_1) > 0$$

or equivalently

$$f(x_1) < f(x_2) \quad \blacksquare$$

The proof of part (b) is similar and is left as an exercise.

Theorem 4.3.6

> **THEOREM 2.**   *Suppose f is continuous at a critical point x_0.*
>
> (a)   *If $f'(x) > 0$ on an open interval extending left from x_0 and $f'(x) < 0$ on an open interval extending right from x_0, then f has a relative maximum at x_0.*
>
> (b)   *If $f'(x) < 0$ on an open interval extending left from x_0 and $f'(x) > 0$ on an open interval extending right from x_0, then f has a relative minimum at x_0.*
>
> (c)   *If $f'(x)$ has the same sign [either $f'(x) > 0$ or $f'(x) < 0$] on an open interval extending left from x_0 and on an open interval extending right from x_0, then f does not have a relative extremum at x_0.*

Proof of (a).   In accordance with the hypothesis, assume that $f'(x) > 0$ on the interval (a, x_0) and $f'(x) < 0$ on the interval (x_0, b). To prove that f has a relative maximum at x_0, we shall show that

$$f(x_0) \geq f(x) \tag{2}$$

for all x in (a, b). First, let x be any point in the interval (a, x_0). Because f is differentiable on the interval (a, x_0) and because of the hypothesis that f is continuous at x_0, it follows that the hypotheses of the Mean-Value Theorem are satisfied on the interval $[x, x_0]$. Thus, there is a point c in the interval (x, x_0) such that

$$\frac{f(x_0) - f(x)}{x_0 - x} = f'(c)$$

or equivalently

$$f(x_0) - f(x) = (x_0 - x)f'(c) \tag{3}$$

But $x_0 - x > 0$ since $x_0 > x$, and $f'(c) > 0$ holds since f' is positive everywhere on the interval (a, x_0). Thus, from (3)

$$f(x_0) - f(x) > 0$$

which shows that (2) holds for all x in the interval (a, x_0).

Next, let x be any point in the interval (x_0, b). As above, the hypotheses of the Mean-Value Theorem hold on the interval $[x_0, x]$, so there is a point c in the interval (x_0, x) such that

$$\frac{f(x) - f(x_0)}{x - x_0} = f'(c)$$

or equivalently

$$f(x) - f(x_0) = (x - x_0)f'(c) \tag{4}$$

But $x - x_0 > 0$ since $x > x_0$, and $f'(c) < 0$ holds since f' is negative everywhere on the interval (x_0, b). Thus, from (4)

$$f(x) - f(x_0) < 0$$

which shows that (2) holds for all x in the interval (x_0, b).

Since (2) obviously holds when $x = x_0$, we have shown that it holds for all x in (a, b). ∎

The proofs of parts (b) and (c) are left as exercises.

Theorem 4.3.7

> **THEOREM 3.**   *Suppose f is twice differentiable at a stationary point x_0.*
>
> (a)   *If $f''(x_0) > 0$, then f has a relative minimum at x_0.*
> (b)   *If $f''(x_0) < 0$, then f has a relative maximum at x_0.*

Proof of (a).   Using the definition of a derivative we can write

$$f''(x_0) = \lim_{h \to 0} \frac{f'(x_0 + h) - f'(x_0)}{h} \tag{5}$$

By hypothesis, $f''(x_0) > 0$ so that we can use $\epsilon = \frac{1}{2}f''(x_0)$ in the definition of a limit and deduce from (5) that there exists a $\delta > 0$ such that

$$\left| \frac{f'(x_0 + h) - f'(x_0)}{h} - f''(x_0) \right| < \frac{1}{2}f''(x_0) \tag{6}$$

whenever h satisfies

$$0 < |h| < \delta \tag{7}$$

To prove that f has a relative minimum at x_0, we shall show that

$$f'(x) > 0 \quad \text{for all } x \text{ in } (x_0, x_0 + \delta) \tag{8}$$

and

$$f'(x) < 0 \quad \text{for all } x \text{ in } (x_0 - \delta, x_0) \tag{9}$$

It will then follow from the first derivative test (Theorem 2) that f has a relative minimum at x_0.

From (6) and (7) it follows that

$$\frac{1}{2}f''(x_0) < \frac{f'(x_0 + h) - f'(x_0)}{h} < \frac{3}{2}f''(x_0) \tag{10}$$

whenever h satisfies

$$0 < |h| < \delta$$

By hypothesis, x_0 is a stationary point for f, so that $f'(x_0) = 0$. From this, the fact that $f''(x_0) > 0$, and the left-hand inequality in (10), it follows that

$$0 < \frac{f'(x_0 + h)}{h} \qquad (11)$$

whenever h satisfies (7).

To prove (8), let x be any point in $(x_0, x_0 + \delta)$. If we let

$$h = x - x_0 \qquad (12)$$

then $h > 0$ and h satisfies (7), so that (11) holds.

Multiplying (11) by h yields

$$0 < f'(x_0 + h) \qquad (13)$$

then substituting (12) in (13) yields $0 < f'(x)$, which establishes (8). To obtain (9) let x be any point in $(x_0 - \delta, x_0)$. If we let

$$h = x - x_0 \qquad (14)$$

then $h < 0$ and h satisfies (7) so that (11) holds.

Multiplying (11) by h yields

$$f'(x_0 + h) < 0 \qquad (15)$$

then substituting (14) in (15) yields $f'(x) < 0$, which establishes (9).   ∎

The proof of part (b) is similar and is left as an exercise.

▶ Exercises

1. Prove part (b) of Theorem 4.2.2.

2. Prove part (b) of Theorem 4.3.6.

3. Prove part (c) of Theorem 4.3.6.

4. Prove part (b) of Theorem 4.3.7.

5. A function f is called *nondecreasing* on a given interval if

$$f(x_2) \geq f(x_1) \text{ whenever } x_2 > x_1$$

where x_1 and x_2 are points in the interval. Similarly, f is called **nonincreasing** on the interval if

$$f(x_2) \leq f(x_1) \text{ whenever } x_2 > x_1$$

(a) Sketch the graph of a function that is non-decreasing, yet is not increasing.

(b) Sketch the graph of a function that is non-increasing, yet is not decreasing.

(c) Prove: If $f'(x) \geq 0$ on an open interval (a, b), then f is nondecreasing on (a, b).

(d) Prove: If $f'(x) \leq 0$ on an open interval (a, b), then f is nonincreasing on (a, b).

[REMARK. Unfortunately, terminology is not always consistent in the mathematical literature. Some writers use the terms "strictly increasing" and "strictly decreasing" where we have used the terms increasing and decreasing. These writers then use the terms "increasing" and "decreasing" where we have used the terms non-decreasing and nonincreasing.]

6. Prove: if $f'(x) < g'(x)$ for all x in (a, b), then for all points x_1, x_2 in (a, b), where $x_2 > x_1$,

$$f(x_2) - f(x_1) < g(x_2) - g(x_1)$$

7. (a) Prove: If f is continuous on (a, b) and $f'(x) > 0$, except at a single point x_0 in (a, b), then f is increasing on (a, b).

(b) Does this result remain true if the condition $f'(x) > 0$ fails to hold at any finite number of points in (a, b)? Justify your answer.

(c) Use part (a) to show that $f(x) = x^3$ is increasing on $(-\infty, +\infty)$.

8. (a) Prove: If $f'(x_0) > 0$ and f' is continuous at x_0, then there is an open interval containing x_0 on which f is increasing.

(b) Prove: If $f'(x_0) < 0$ and f' is continuous at x_0, then there is an open interval containing x_0 on which f is decreasing.

9. (a) Use Exercise 8(a) to prove: If $f''(x_0) > 0$ and f'' is continuous at x_0, then there is an open interval containing x_0 on which f is concave up.

(b) Use Exercise 8(b) to prove: If $f''(x_0) < 0$ and f'' is continuous at x_0, then there is an open interval containing x_0 on which f is concave down.

10. Prove: If f is differentiable at each point of the interval $[a, b]$ and f' is continuous at a and b, and $f'(a)f'(b) < 0$, then there is a number c in (a, b) such that $f'(c) = 0$. [*Hint:* Use the results in Exercise 8.]

VI. CRAMER'S RULE

In this section we shall show how determinants can be used to solve systems of two linear equations in two unknowns and three linear equations in three unknowns. We assume that the reader has read the text through Section 14.6.

Recall that a **solution** of a system of two linear equations in two unknowns

$$a_1 x + b_1 y = k_1$$
$$a_2 x + b_2 y = k_2$$
(1)

is a value for x and a value for y that satisfy *both* equations. The graphs of these equations are lines, which we shall denote by L_1 and L_2. Since a point (x, y) lies on a line if and only if the numbers x and y satisfy the

(a)

(b)

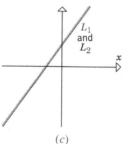

(c)

Figure A.37

equation of the line, the solutions of the system will correspond to points of intersection of L_1 and L_2. There are three possibilities (Figure A.37).

(a)   The lines L_1 and L_2 are parallel and distinct, in which case there is no intersection and, consequently, no solution to the system.

(b)   The lines L_1 and L_2 intersect at only one point, in which case the system has exactly one solution.

(c)   The lines L_1 and L_2 coincide, in which case there are infinitely many points of intersection, and consequently infinitely many solutions to the system.

Example 1   If we multiply the second equation of the system

$$x + y = 4$$
$$2x + 2y = 6$$

by $\frac{1}{2}$, it becomes evident that there is no solution since the two equations in the resulting system

$$x + y = 4$$
$$x + y = 3$$

contradict each other. Geometrically, the lines $x + y = 4$ and $2x + 2y = 6$ are distinct and parallel.   ◄

Example 2   Since the second equation of the system

$$x + y = 4$$
$$2x + 2y = 8$$

is a multiple of the first equation, it is evident that any solution of the first equation will satisfy the second equation automatically. But the first equation, $x + y = 4$, has infinitely many solutions since we may assign x an arbitrary value and determine y from the relationship $y = 4 - x$. (Some possibilities are $x = 0, y = 4; x = 1, y = 3; x = -1, y = 5$.) Thus, the system has infinitely many solutions. Geometrically, the lines $x + y = 4$ and $2x + 2y = 8$ coincide.   ◄

Example 3   If we add the equations of the system

$$x + y = 3$$
$$2x - y = 6$$

we obtain $3x = 9$ or $x = 3$, and if we substitute this value in the first equation we obtain $y = 0$, so that this system has the unique solution

$x = 3, y = 0$. Geometrically, the lines $x + y = 3$ and $2x - y = 6$ intersect only at the point $(3, 0)$.   ◄

A *solution* of a system of three linear equations in three unknowns

$$a_1 x + b_1 y + c_1 z = k_1$$
$$a_2 x + b_2 y + c_2 z = k_2 \qquad\qquad (2)$$
$$a_3 x + b_3 y + c_3 z = k_3$$

consists of values of x, y, and z that satisfy all three equations. The graphs of these equations are planes in 3-space, and the solutions of this system correspond to points where all three planes intersect. It follows that system (2), like system (1), has zero, one, or infinitely many solutions, depending on the relative orientation of the planes (Figure A.38).

If a system of two linear equations in two unknowns or three linear equations in three unknowns has a unique solution, the solution can be expressed as a ratio of determinants by using the following theorems, which are cases of a general result called *Cramer's* Rule*.

---

*GABRIEL CRAMER (1704–1752) Swiss mathematician. Although Cramer does not rank with the great mathematicians of his time, his contributions as a disseminator of mathematical ideas have earned him a well-deserved place in the history of mathematics. The son of a physician, Cramer was born and educated in Geneva, Switzerland. At age 20 he competed for, but failed to secure, the chair of philosophy at the Académie de Calvin at Geneva. However, the awarding magistrates were sufficiently impressed with Cramer and a fellow competitor to create a new chair of mathematics for both men to share. Alternately, each assumed the full responsibility and salary associated with the chair for two or three years while the other traveled. During his travels Cramer met many of the great mathematicians and scientists of his day the Bernoullis, Euler, Halley, D'Alembert, and others. Many of these contacts and friendships led to extensive correspondence in which information about new mathematical discoveries was transmitted. Eventually, Cramer became sole possessor of the mathematics chair and the chair of philosophy as well.

Cramer's mathematical work was primarily in geometry and probability; he had relatively little knowledge of calculus and did not use it to any great extent in his work. In 1730 he finished second to Johann I Bernoulli in a competition for a prize offered by the Paris Academy to explain properties of planetary orbits.

Cramer's most widely known work, *Introduction à l'analyse des lignes courbes algébriques* (1750) was a study and classification of algebraic curves; Cramer's Rule appeared in the appendix. Although the rule bears his name, variations of the basic idea were formulated earlier by Leibniz (and even earlier by Chinese mathematicians). However, Cramer's superior notation helped clarify and popularize the technique.

Perhaps Cramer's most important contributions stemmed from his work as an editor of the mathematical creations of others. He edited and published the works of Jacob I Bernoulli and Leibniz.

Overwork combined with a fall from a carriage eventually led to his death in 1752. Cramer was apparently a good-natured and pleasant person, though he never married. His interests were broad. He wrote on philosophy of law and government and the history of mathematics. He served in public office, participated in artillery and fortifications activities for the government, instructed workers on techniques of cathedral repair, and undertook excavations of cathedral archives. Cramer received numerous honors for his activities.

| 3 parallel planes No solutions | No solutions | No solutions | 3 coincident planes Infinitely many solutions | 3 planes intersecting in a line Infinitely many solutions | 3 planes intersecting at a point One solution |

Figure A.38

Cramer's Rule for Two Unknowns

THEOREM **1.**   *If the system of equations*

$$a_1x + b_1y = k_1$$
$$a_2x + b_2y = k_2$$

is such that $\begin{vmatrix} a_1 & b_1 \\ a_2 & b_2 \end{vmatrix} \neq 0$

then the system has a unique solution. This solution is

$$x = \frac{\begin{vmatrix} k_1 & b_1 \\ k_2 & b_2 \end{vmatrix}}{\begin{vmatrix} a_1 & b_1 \\ a_2 & b_2 \end{vmatrix}}, \quad y = \frac{\begin{vmatrix} a_1 & k_1 \\ a_2 & k_2 \end{vmatrix}}{\begin{vmatrix} a_1 & b_1 \\ a_2 & b_2 \end{vmatrix}}$$

Cramer's Rule for Three Unknowns

THEOREM **2.**   *If the system of equations*

$$a_1x + b_1y + c_1z = k_1$$
$$a_2x + b_2y + c_2z = k_2$$
$$a_3x + b_3y + c_3z = k_3$$

is such that $\begin{vmatrix} a_1 & b_1 & c_1 \\ a_2 & b_2 & c_2 \\ a_3 & b_3 & c_3 \end{vmatrix} \neq 0$

then the system has a unique solution. This solution is

$$x = \frac{\begin{vmatrix} k_1 & b_1 & c_1 \\ k_2 & b_2 & c_2 \\ k_3 & b_3 & c_3 \end{vmatrix}}{\begin{vmatrix} a_1 & b_1 & c_1 \\ a_2 & b_2 & c_2 \\ a_3 & b_3 & c_3 \end{vmatrix}}, \quad y = \frac{\begin{vmatrix} a_1 & k_1 & c_1 \\ a_2 & k_2 & c_2 \\ a_3 & k_3 & c_3 \end{vmatrix}}{\begin{vmatrix} a_1 & b_1 & c_1 \\ a_2 & b_2 & c_2 \\ a_3 & b_3 & c_3 \end{vmatrix}}, \quad z = \frac{\begin{vmatrix} a_1 & b_1 & k_1 \\ a_2 & b_2 & k_2 \\ a_3 & b_3 & k_3 \end{vmatrix}}{\begin{vmatrix} a_1 & b_1 & c_1 \\ a_2 & b_2 & c_2 \\ a_3 & b_3 & c_3 \end{vmatrix}}$$

REMARK. There is a pattern to the formulas in these theorems. In each formula the determinant in the denominator is formed from the coefficients of the unknowns; and the determinant in the numerator differs from the determinant in the denominator in that the coefficients of the unknown being calculated are replaced by the k's. It is assumed in these theorems that the system is written so that like unknowns are aligned vertically and the constants appear by themselves on the right side of each equation.

We shall prove Cramer's Rule for the case of three unknowns. However, let us first look at some examples.

Example 4 Use Cramer's Rule to solve

$$5x - 2y = -1$$
$$2x + 3y = 3$$

Solution.

$$x = \frac{\begin{vmatrix} -1 & -2 \\ 3 & 3 \end{vmatrix}}{\begin{vmatrix} 5 & -2 \\ 2 & 3 \end{vmatrix}} = \frac{3}{19}; \quad y = \frac{\begin{vmatrix} 5 & -1 \\ 2 & 3 \end{vmatrix}}{\begin{vmatrix} 5 & -2 \\ 2 & 3 \end{vmatrix}} = \frac{17}{19} \quad \blacktriangleleft$$

Example 5 Use Cramer's Rule to solve

$$x \quad\quad + 2z = 6$$
$$-3x + 4y + 6z = 30$$
$$-x - 2y + 3z = 8$$

Solution.

$$x = \frac{\begin{vmatrix} 6 & 0 & 2 \\ 30 & 4 & 6 \\ 8 & -2 & 3 \end{vmatrix}}{\begin{vmatrix} 1 & 0 & 2 \\ -3 & 4 & 6 \\ -1 & -2 & 3 \end{vmatrix}} = \frac{-10}{11}; \quad y = \frac{\begin{vmatrix} 1 & 6 & 2 \\ -3 & 30 & 6 \\ -1 & 8 & 3 \end{vmatrix}}{\begin{vmatrix} 1 & 0 & 2 \\ -3 & 4 & 6 \\ -1 & -2 & 3 \end{vmatrix}} = \frac{18}{11};$$

$$z = \frac{\begin{vmatrix} 1 & 0 & 6 \\ -3 & 4 & 30 \\ -1 & -2 & 8 \end{vmatrix}}{\begin{vmatrix} 1 & 0 & 2 \\ -3 & 4 & 6 \\ -1 & -2 & 3 \end{vmatrix}} = \frac{38}{11} \quad \blacktriangleleft$$

Proof of Cramer's Rule for Three Unknowns. Since two vectors are equal if and only if their corresponding components are equal, the system

$$a_1x + b_1y + c_1z = k_1$$
$$a_2x + b_2y + c_2z = k_2 \tag{3}$$
$$a_3x + b_3y + c_3z = k_3$$

is equivalent to the single vector equation

$$\langle a_1x + b_1y + c_1z, a_2x + b_2y + c_2z, a_3x + b_3y + c_3z \rangle = \langle k_1, k_2, k_3 \rangle$$

or, equivalently,

$$x\langle a_1, a_2, a_3 \rangle + y\langle b_1, b_2, b_3 \rangle + z\langle c_1, c_2, c_3 \rangle = \langle k_1, k_2, k_3 \rangle \tag{4}$$

If we define

$$\mathbf{a} = \langle a_1, a_2, a_3 \rangle, \ \mathbf{b} = \langle b_1, b_2, b_3 \rangle, \ \mathbf{c} = \langle c_1, c_2, c_3 \rangle, \ \mathbf{k} = \langle k_1, k_2, k_3 \rangle$$

then (4) can be written as

$$x\mathbf{a} + y\mathbf{b} + z\mathbf{c} = \mathbf{k} \tag{5}$$

Therefore, solving system (3) is equivalent to solving the vector equation (5) for the unknown scalars x, y, and z. If we take the dot product of both sides of (5) with $\mathbf{b} \times \mathbf{c}$, we obtain

$$(x\mathbf{a} + y\mathbf{b} + z\mathbf{c}) \cdot (\mathbf{b} \times \mathbf{c}) = \mathbf{k} \cdot (\mathbf{b} \times \mathbf{c})$$

or

$$x(\mathbf{a} \cdot \mathbf{b} \times \mathbf{c}) + y(\mathbf{b} \cdot \mathbf{b} \times \mathbf{c}) + z(\mathbf{c} \cdot \mathbf{b} \times \mathbf{c}) = \mathbf{k} \cdot \mathbf{b} \times \mathbf{c} \tag{6}$$

By Theorem 15.4.2,

$$\mathbf{b} \cdot \mathbf{b} \times \mathbf{c} = 0 \quad \text{and} \quad \mathbf{c} \cdot \mathbf{b} \times \mathbf{c} = 0$$

so that (6) reduces to

$$x(\mathbf{a} \cdot \mathbf{b} \times \mathbf{c}) = \mathbf{k} \cdot \mathbf{b} \times \mathbf{c}$$

Expressing these triple scalar products as determinants yields

$$x \begin{vmatrix} a_1 & a_2 & a_3 \\ b_1 & b_2 & b_3 \\ c_1 & c_2 & c_3 \end{vmatrix} = \begin{vmatrix} k_1 & k_2 & k_3 \\ b_1 & b_2 & b_3 \\ c_1 & c_2 & c_3 \end{vmatrix}$$

Solving for x and applying Theorem 14.4.5(*b*) yields

$$x = \frac{\begin{vmatrix} k_1 & b_1 & c_1 \\ k_2 & b_2 & c_2 \\ k_3 & b_3 & c_3 \end{vmatrix}}{\begin{vmatrix} a_1 & b_1 & c_1 \\ a_2 & b_2 & c_2 \\ a_3 & b_3 & c_3 \end{vmatrix}}$$

The formulas for y and z have similar derivations.

As a matter of logic, we have shown that *if there exists a solution* of system (3), then the solution is given by the formulas in Cramer's Rule. However, it is logically possible that the system has no solution, even though Cramer's Rule generates values of x, y, and z. This can be ruled out by substituting these values into (3) and verifying that the equations are satisfied. The details are tedious and will be omitted.   ∎

▶ Exercises

In Exercises 1–8, solve the system using Cramer's Rule.

1. $3x - 4y = -5$
$2x + y = 4.$

2. $-x + 3y = 8$
$2x + 5y = 7.$

3. $2x_1 - 5x_2 = -2$
$4x_1 + 6x_2 = 1.$

4. $3a + 2b = 4$
$-a + b = 7.$

5. $x + 2y + z = 3$
$2x + y - z = 0$
$x - y + z = 6.$

6. $x - 3y + z = 4$
$2x - y = -2$
$4x - 3z = 0.$

7. $x_1 + x_2 - 2x_3 = 1$
$2x_1 - x_2 + x_3 = 2$
$x_1 - 2x_2 - 4x_3 = -4.$

8. $r + s + t = 2$
$r - s - 2t = 0$
$-r + 2s + t = 4.$

9. Use Cramer's Rule to solve the rotation equations

$x = x' \cos\theta - y' \sin\theta$
$y = x' \sin\theta + y' \cos\theta$

for x' and y' in terms of x and y.

10. Solve the following system of equations for the unknown angles α, β, and γ, where $0 \le \alpha \le 2\pi$, $0 \le \beta \le 2\pi$, and $0 \le \gamma < \pi$:

$2 \sin\alpha - \cos\beta + 3 \tan\gamma = 3$
$4 \sin\alpha + 2 \cos\beta - 2 \tan\gamma = 2$
$6 \sin\alpha - 3 \cos\beta + \tan\gamma = 9.$

[*Hint:* First solve for $\sin\alpha$, $\cos\beta$, and $\tan\gamma$.]

Appendix 3

tables

Table 1
Table of Trigonometric Functions

DEGREES	RADIANS	SIN	COS	TAN	DEGREES	RADIANS	SIN	COS	TAN
0°	0.000	0.000	1.000	0.000					
1°	0.017	0.017	1.000	0.017	21°	0.367	0.358	0.934	0.384
2°	0.035	0.035	0.999	0.035	22°	0.394	0.375	0.927	0.404
3°	0.052	0.052	0.999	0.052	23°	0.401	0.391	0.921	0.424
4°	0.070	0.070	0.998	0.070	24°	0.419	0.407	0.914	0.445
5°	0.087	0.087	0.996	0.087	25°	0.436	0.423	0.906	0.466
6°	0.105	0.105	0.995	0.105	26°	0.454	0.438	0.899	0.488
7°	0.122	0.122	0.993	0.123	27°	0.471	0.454	0.891	0.510
8°	0.140	0.139	0.990	0.141	28°	0.489	0.469	0.883	0.532
9°	0.157	0.156	0.988	0.158	29°	0.506	0.485	0.875	0.554
10°	0.175	0.174	0.985	0.176	30°	0.524	0.500	0.866	0.577
11°	0.192	0.191	0.982	0.194	31°	0.541	0.515	0.857	0.601
12°	0.209	0.208	0.978	0.213	32°	0.559	0.530	0.848	0.625
13°	0.227	0.225	0.974	0.231	33°	0.576	0.545	0.839	0.649
14°	0.244	0.242	0.970	0.249	34°	0.593	0.559	0.829	0.675
15°	0.262	0.259	0.966	0.268	35°	0.611	0.574	0.819	0.700
16°	0.279	0.276	0.961	0.287	36°	0.628	0.588	0.809	0.727
17°	0.297	0.292	0.956	0.306	37°	0.646	0.602	0.799	0.754
18°	0.314	0.309	0.951	0.325	38°	0.663	0.616	0.788	0.781
19°	0.332	0.326	0.946	0.344	39°	0.681	0.629	0.777	0.810
20°	0.349	0.342	0.940	0.364	40°	0.698	0.643	0.766	0.839

Table 1
Table of Trigonometric Functions (Continued)

DEGREES	RADIANS	SIN	COS	TAN	DEGREES	RADIANS	SIN	COS	TAN
41°	0.716	0.656	0.755	0.869	66°	1.152	0.914	0.407	2.246
42°	0.733	0.669	0.743	0.900	67°	1.169	0.921	0.391	2.356
43°	0.750	0.682	0.731	0.933	68°	1.187	0.927	0.375	2.475
44°	0.768	0.695	0.719	0.966	69°	1.204	0.934	0.358	2.605
45°	0.785	0.707	0.707	1.000	70°	1.222	0.940	0.342	2.748
46°	0.803	0.719	0.695	1.036	71°	1.239	0.946	0.326	2.904
47°	0.820	0.731	0.682	1.072	72°	1.257	0.951	0.309	3.078
48°	0.838	0.743	0.669	1.111	73°	1.274	0.956	0.292	3.271
49°	0.855	0.755	0.656	1.150	74°	1.292	0.961	0.276	3.487
50°	0.873	0.766	0.643	1.192	75°	1.309	0.966	0.259	3.732
51°	0.890	0.777	0.629	1.235	76°	1.326	0.970	0.242	4.011
52°	0.908	0.788	0.616	1.280	77°	1.344	0.974	0.225	4.332
53°	0.925	0.799	0.602	1.327	78°	1.361	0.978	0.208	4.705
54°	0.942	0.809	0.588	1.376	79°	1.379	0.982	0.191	5.145
55°	0.960	0.819	0.574	1.428	80°	1.396	0.985	0.174	5.671
56°	0.977	0.829	0.559	1.483	81°	1.414	0.988	0.156	6.314
57°	0.995	0.839	0.545	1.540	82°	1.431	0.990	0.139	7.115
58°	1.012	0.848	0.530	1.600	83°	1.449	0.993	0.122	8.144
59°	1.030	0.857	0.515	1.664	84°	1.466	0.995	0.105	9.514
60°	1.047	0.866	0.500	1.732	85°	1.484	0.996	0.087	11.43
61°	1.065	0.875	0.485	1.804	86°	1.501	0.998	0.070	14.30
62°	1.082	0.883	0.469	1.881	87°	1.518	0.999	0.052	19.08
63°	1.100	0.891	0.454	1.963	88°	1.536	0.999	0.035	28.64
64°	1.117	0.899	0.438	2.050	89°	1.553	1.000	0.017	57.29
65°	1.134	0.906	0.423	2.145	90°	1.571	1.000	0.000	——

Table 2
Table of Exponential and Hyperbolic Functions

x	e^x	e^{-x}	SINH x	COSH x	TANH x	x	e^x	e^{-x}	SINH x	COSH x	TANH x
.00	1.0000	1.00000	.0000	1.0000	.00000	.40	1.4918	.67032	.4108	1.0811	.37995
.01	1.0101	.99005	.0100	1.0001	.01000	.41	1.5068	.66365	.4216	1.0852	.38847
.02	1.0202	.98020	.0200	1.0002	.02000	.42	1.5220	.65705	.4325	1.0895	.39693
.03	1.0305	.97045	.0300	1.0005	.02999	.43	1.5373	.65051	.4434	1.0939	.40532
.04	1.0408	.96079	.0400	1.0008	.03998	.44	1.5527	.64404	.4543	1.0984	.41364
.05	1.0513	.95123	.0500	1.0013	.04996	.45	1.5683	.63763	.4653	1.1030	.42190
.06	1.0618	.94176	.0600	1.0018	.05993	.46	1.5841	.63128	.4764	1.1077	.43008
.07	1.0725	.93239	.0701	1.0025	.06989	.47	1.6000	.62500	.4875	1.1125	.43820
.08	1.0833	.92312	.0801	1.0032	.07983	.48	1.6161	.61878	.4986	1.1174	.44624
.09	1.0942	.91393	.0901	1.0041	.08976	.49	1.6323	.61263	.5098	1.1225	.45422
.10	1.1052	.90484	.1002	1.0050	.09967	.50	1.6487	.60653	.5211	1.1276	.46212
.11	1.1163	.89583	.1102	1.0061	.10956	.51	1.6653	.60050	.5324	1.1329	.46995
.12	1.1275	.88692	.1203	1.0072	.11943	.52	1.6820	.59452	.5438	1.1383	.47770
.13	1.1388	.87809	.1304	1.0085	.12927	.53	1.6989	.58860	.5552	1.1438	.48538
.14	1.1503	.86936	.1405	1.0098	.13909	.54	1.7160	.58275	.5666	1.1494	.49299
.15	1.1618	.86071	.1506	1.0113	.14889	.55	1.7333	.57695	.5782	1.1551	.50052
.16	1.1735	.85214	.1607	1.0128	.15865	.56	1.7507	.57121	.5897	1.1609	.50798
.17	1.1853	.84366	.1708	1.0145	.16838	.57	1.7683	.56553	.6014	1.1669	.51536
.18	1.1972	.83527	.1810	1.0162	.17808	.58	1.7860	.55990	.6131	1.1730	.52267
.19	1.2092	.82696	.1911	1.0181	.18775	.59	1.8040	.55433	.6248	1.1792	.52990
.20	1.2214	.81873	.2013	1.0201	.19738	.60	1.8221	.54881	.6367	1.1855	.53705
.21	1.2337	.81058	.2115	1.0221	.20697	.61	1.8404	.54335	.6485	1.1919	.54413
.22	1.2461	.80252	.2218	1.0243	.21652	.62	1.8589	.53794	.6605	1.1984	.55113
.23	1.2586	.79453	.2320	1.0266	.22603	.63	1.8776	.53259	.6725	1.2051	.55805
.24	1.2712	.78663	.2423	1.0289	.23550	.64	1.8965	.52729	.6846	1.2119	.56490
.25	1.2840	.77880	.2526	1.0314	.24492	.65	1.9155	.52205	.6967	1.2188	.57167
.26	1.2969	.77105	.2629	1.0340	.25430	.66	1.9348	.51685	.7090	1.2258	.57836
.27	1.3100	.76338	.2733	1.0367	.26362	.67	1.9542	.51171	.7213	1.2330	.58498
.28	1.3231	.75578	.2837	1.0395	.27291	.68	1.9739	.50662	.7336	1.2402	.59152
.29	1.3364	.74826	.2941	1.0423	.28213	.69	1.9937	.50158	.7461	1.2476	.59798
.30	1.3499	.74082	.3045	1.0453	.29131	.70	2.0138	.49659	.7586	1.2552	.60437
.31	1.3634	.73345	.3150	1.0484	.30044	.71	2.0340	.49164	.7712	1.2628	.61068
.32	1.3771	.72615	.3255	1.0516	.30951	.72	2.0544	.48675	.7838	1.2706	.61691
.33	1.3910	.71892	.3360	1.0549	.31852	.73	2.0751	.48191	.7966	1.2785	.62307
.34	1.4049	.71177	.3466	1.0584	.32748	.74	2.0959	.47711	.8094	1.2865	.62915
.35	1.4191	.70469	.3572	1.0619	.33638	.75	2.1170	.47237	.8223	1.2947	.63515
.36	1.4333	.69768	.3678	1.0655	.34521	.76	2.1383	.46767	.8353	1.3030	.64108
.37	1.4477	.69073	.3785	1.0692	.35399	.77	2.1598	.46301	.8484	1.3114	.64693
.38	1.4623	.68386	.3892	1.0731	.36271	.78	2.1815	.45841	.8615	1.3199	.65271
.39	1.4770	.67706	.4000	1.0770	.37136	.79	2.2034	.45384	.8748	1.3286	.65841

Table 2
Table of Exponential and Hyperbolic Functions (Continued)

x	e^x	e^{-x}	SINH x	COSH x	TANH x	x	e^x	e^{-x}	SINH x	COSH x	TANH x
.80	2.2255	.44933	.8881	1.3374	.66404	3.00	20.086	.04979	10.018	10.068	.99505
.81	2.2479	.44486	.9015	1.3464	.66959	3.10	22.198	.04505	11.076	11.122	.99595
.82	2.2705	.44043	.9150	1.3555	.67507	3.20	24.533	.04076	12.246	12.287	.99668
.83	2.2933	.43605	.9286	1.3647	.68048	3.30	27.113	.03688	13.538	13.575	.99728
.84	2.3164	.43171	.9423	1.3740	.68581	3.40	29.964	.03337	14.965	14.999	.99777
.85	2.3396	.42741	.9561	1.3835	.69107	3.50	33.115	.03020	16.543	16.573	.99818
.86	2.3632	.42316	.9700	1.3932	.69626	3.60	36.598	.02732	18.286	18.313	.99851
.87	2.3869	.41895	.9840	1.4029	.70137	3.70	40.447	.02472	20.211	20.236	.99878
.88	2.4109	.41478	.9981	1.4128	.70642	3.80	44.701	.02237	22.339	22.362	.99900
.89	2.4351	.41066	1.0122	1.4229	.71139	3.90	49.402	.02024	24.691	24.711	.99918
.90	2.4596	.40657	1.0265	1.4331	.71630	4.00	54.598	.01832	27.290	27.308	.99933
.91	2.4843	.40252	1.0409	1.4434	.72113	4.10	60.340	.01657	30.162	30.178	.99945
.92	2.5093	.39852	1.0554	1.4539	.72590	4.20	66.686	.01500	33.336	33.351	.99955
.93	2.5345	.39455	1.0700	1.4645	.73059	4.30	73.700	.01357	36.843	36.857	.99963
.94	2.5600	.39063	1.0847	1.4753	.73522	4.40	81.451	.01227	40.719	40.732	.99970
.95	2.5857	.38674	1.0995	1.4862	.73978	4.50	90.017	.01111	45.003	45.014	.99975
.96	2.6117	.38289	1.1144	1.4973	.74428	4.60	99.484	.01005	49.737	49.747	.99980
.97	2.6379	.37908	1.1294	1.5085	.74870	4.70	109.95	.00910	54.969	54.978	.99983
.98	2.6645	.37531	1.1446	1.5199	.75307	4.80	121.51	.00823	60.751	60.759	.99986
.99	2.6912	.37158	1.1598	1.5314	.75736	4.90	134.29	.00745	67.141	67.149	.99989
1.00	2.7183	.36788	1.1752	1.5431	.76159	5.00	148.41	.00674	74.203	74.210	.99991
1.10	3.0042	.33287	1.3356	1.6685	.80050	5.10	164.02	.00610	82.008	82.014	.99993
1.20	3.3201	.30119	1.5095	1.8107	.83365	5.20	181.27	.00552	90.633	90.639	.99994
1.30	3.6693	.27253	1.6984	1.9709	.86172	5.30	200.34	.00499	100.17	100.17	.99995
1.40	4.0552	.24660	1.9043	2.1509	.88535	5.40	221.41	.00452	110.70	110.71	.99996
1.50	4.4817	.22313	2.1293	2.3524	.90515	5.50	244.69	.00409	122.34	122.35	.99997
1.60	4.9530	.20190	2.3756	2.5775	.92167	5.60	270.43	.00370	135.21	135.22	.99997
1.70	5.4739	.18268	2.6456	2.8283	.93541	5.70	298.87	.00335	149.43	149.44	.99998
1.80	6.0496	.16530	2.9422	3.1075	.94681	5.80	330.30	.00303	165.15	165.15	.99998
1.90	6.6859	.14957	3.2682	3.4177	.95624	5.90	365.04	.00274	182.52	182.52	.99998
2.00	7.3891	.13534	3.6269	3.7622	.96403	6.00	403.43	.00248	201.71	201.72	.99999
2.10	8.1662	.12246	4.0219	4.1443	.97045	6.25	518.01	.00193	259.01	259.01	.99999
2.20	9.0250	.11080	4.4571	4.5679	.97574	6.50	665.14	.00150	332.57	332.57	1.0000
2.30	9.9742	.10026	4.9370	5.0372	.98010	6.75	854.06	.00117	427.03	427.03	1.0000
2.40	11.023	.09072	5.4662	5.5569	.98367	7.00	1096.6	.00091	548.32	548.32	1.0000
2.50	12.182	.08208	6.0502	6.1323	.98661	7.50	1808.0	.00055	904.02	904.02	1.0000
2.60	13.464	.07427	6.6947	6.7690	.98903	8.00	2981.0	.00034	1490.5	1490.5	1.0000
2.70	14.880	.06721	7.4063	7.4735	.99101	8.50	4914.8	.00020	2457.4	2457.4	1.0000
2.80	16.445	.06081	8.1919	8.2527	.99263	9.00	8103.1	.00012	4051.5	4051.5	1.0000
2.90	18.174	.05502	9.0596	9.1146	.99396	9.50	13360	.00007	6679.9	6679.9	1.0000
						10.00	22026	.00005	11013	11013	1.0000

Table 3
Natural Logarithms

n	$\ln n$	n	$\ln n$	n	$\ln n$
0.0	—	4.5	1.5041	9.0	2.1972
0.1	−2.3026	4.6	1.5261	9.1	2.2083
0.2	−1.6094	4.7	1.5476	9.2	2.2192
0.3	−1.2040	4.8	1.5686	9.3	2.2300
0.4	−0.9163	4.9	1.5892	9.4	2.2407
0.5	−0.6931	5.0	1.6094	9.5	2.2513
0.6	−0.5108	5.1	1.6292	9.6	2.2618
0.7	−0.3567	5.2	1.6487	9.7	2.2721
0.8	−0.2231	5.3	1.6677	9.8	2.2824
0.9	−0.1054	5.4	1.6864	9.9	2.2925
1.0	0.0000	5.5	1.7047	10	2.3026
1.1	0.0953	5.6	1.7228	11	2.3979
1.2	0.1823	5.7	1.7405	12	2.4849
1.3	0.2624	5.8	1.7579	13	2.5649
1.4	0.3365	5.9	1.7750	14	2.6391
1.5	0.4055	6.0	1.7918	15	2.7081
1.6	0.4700	6.1	1.8083	16	2.7726
1.7	0.5306	6.2	1.8245	17	2.8332
1.8	0.5878	6.3	1.8405	18	2.8904
1.9	0.6419	6.4	1.8563	19	2.9444
2.0	0.6931	6.5	1.8718	20	2.9957
2.1	0.7419	6.6	1.8871	25	3.2189
2.2	0.7885	6.7	1.9021	30	3.4012
2.3	0.8329	6.8	1.9169	35	3.5553
2.4	0.8755	6.9	1.9315	40	3.6889
2.5	0.9163	7.0	1.9459	45	3.8067
2.6	0.9555	7.1	1.9601	50	3.9120
2.7	0.9933	7.2	1.9741	55	4.0073
2.8	1.0296	7.3	1.9879	60	4.0943
2.9	1.0647	7.4	2.0015	65	4.1744
3.0	1.0986	7.5	2.0149	70	4.2485
3.1	1.1314	7.6	2.0281	75	4.3175
3.2	1.1632	7.7	2.0412	80	4.3820
3.3	1.1939	7.8	2.0541	85	4.4427
3.4	1.2238	7.9	2.0669	90	4.4998
3.5	1.2528	8.0	2.0794	95	4.5539
3.6	1.2809	8.1	2.0919	100	4.6052
3.7	1.3083	8.2	2.1041	200	5.2983
3.8	1.3350	8.3	2.1163	300	5.7038
3.9	1.3610	8.4	2.1282	400	5.9915
4.0	1.3863	8.5	2.1401	500	6.2146
4.1	1.4110	8.6	2.1518	600	6.3069
4.2	1.4351	8.7	2.1633	700	6.5511
4.3	1.4586	8.8	2.1748	800	6.6846
4.4	1.4816	8.9	2.1861	900	6.8024

answers to odd-numbered exercises

▶ Exercise Set 1.1 (Page 13)

1. (a)

(b)

(c)

(d)

(e)

(f)

3. (a) rational (b) integer, rational
(c) integer, rational (d) rational
(e) integer, rational (f) irrational
(g) rational (h) integer, rational

5. a, d, f

7. (a) $\frac{41}{333}$ (b) $\frac{115}{9}$ (c) $\frac{20943}{550}$ (d) $\frac{537}{1250}$

9. (a) all values (b) none

11. (a) yes (b) no

13. (a) $\{x: x \text{ is a positive odd integer}\}$
(b) $\{x: x \text{ is an even integer}\}$
(c) $\{x: x \text{ is irrational}\}$
(d) $\{x: x \text{ is an integer and } 7 \le x \le 10\}$

15. a, c

17. (a) false (b) true (c) true (d) false
(e) true (f) true (g) true

19. $(-\infty, \frac{10}{3})$

21. $(-\infty, -\frac{11}{2}]$

23. $(-\frac{3}{2}, \frac{1}{2}]$

25. $(-\infty, 3) \cup (4, +\infty)$

27. $(-\frac{3}{2}, 2)$

29. $(-\infty, -2] \cup (2, +\infty)$

31. $(-\infty, -3) \cup (3, +\infty)$

33. $(-\infty, -2) \cup (4, +\infty)$

35. $[4, 5]$

37. $(-8, 0) \cup (4, +\infty)$

39. $(2, +\infty)$

41. $(-\infty, -3] \cup [2, +\infty)$ **49.** $(-\infty, -\frac{1}{2})$

▶ Exercise Set 1.2 (Page 22)

1. (a) 7   (b) $\sqrt{2}$ (c) k^2 (d) k^2

3. $x \leq 3$ **5.** all real x

7. $x \geq 0$ or $x = -\frac{2}{3}$ **9.** $x \geq -5$

13. (a) 2   (b) 1   (c) 14
(d) $3 + \sqrt{2}$ (e) 7   (f) 5

15. (a) -9 (b) 7   (c) 12

17. $-\frac{5}{6}, \frac{3}{2}$ **19.** $\frac{1}{2}, \frac{5}{2}$ **21.** $-\frac{11}{10}, \frac{11}{8}$

23. $1, \frac{17}{5}$ **25.** $(-9, -3)$ **27.** $[-\frac{3}{2}, \frac{9}{2}]$

29. $(-\infty, -3) \cup (-1, +\infty)$ **31.** $(-\infty, \frac{1}{2}] \cup [\frac{9}{2}, +\infty)$

33. $(-\infty, \frac{1}{2}) \cup (\frac{3}{2}, +\infty)$ **35.** $[\frac{1}{8}, \frac{1}{2}) \cup (\frac{1}{2}, \frac{7}{8}]$

37. $(-\infty, \frac{5}{2})$ **39.** $[\frac{7}{10}, \frac{7}{2}]$ **41.** $(0, +\infty)$

43. $(-\infty, -7) \cup (-7, -\frac{3}{2})$

45. $x \in (-\infty, 2] \cup [3, +\infty)$

47. $-3, 9$ **55.** $\frac{1}{3}$

57. $\frac{13}{11}$ (Other answers are possible.)

▶ Exercise Set 1.3 (Page 33)

1.

3. (a)   (b)   (c)

(d)   (e)   (f)

5. $(6, 9)$

7. (a) vertical; $3 + \sqrt{2}$
(b) vertical; 7   (c) vertical; 5

9. (a) 10   (b) $(4, 5)$

11. (a) $\sqrt{29}$ (b) $(-\frac{9}{2}, -5)$

17. (a) yes   (b) no   (c) yes   (d) yes

19. (a) y-axis   (b) origin
(c) x-axis, y-axis, origin

21.   **23.**

25.   **27.**   **29.**   **31.**

33. The union of the graphs of $x - y = 0$ and $x + y = 0$.

35.

37. $3x - 2y - 5 = 0$

39. 0

► Exercise Set 1.4 (Page 41)

1. (a) $\frac{1}{2}$ (b) -1 (c) 0 (d) not defined

3. (a) $\dfrac{1}{\sqrt{3}}$ (b) -1 (c) $\sqrt{3}$

5. (a) $153°$ (b) $45°$ (c) $117°$ (d) $89°$

7.

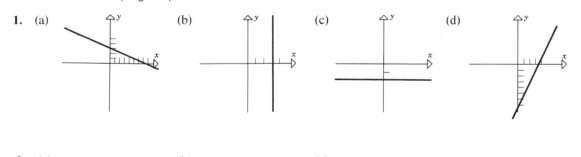

9. (a) parallel   (b) perpendicular   (c) neither

11. (a) 14   (b) $-\frac{1}{3}$ **13.** 29

15. (a) yes   (b) no   **17.** $(2, 0), (7, 0)$

23. (a) $\frac{7}{19}$ (b) $\frac{57}{25}$ (c) $\frac{4}{13}$

25. (a) $20°$ (b) $66°$ (c) $17°$ **27.** 11.09

► Exercise Set 1.5 (Page 47)

1. (a)   (b)   (c)   (d)

3. (a)   (b)   (c)

5.

	(a)	(b)	(c)	(d)	(e)
Slope	3	$-1/4$	$-3/5$	0	$-b/a$
y-intercept	2	3	8/5	1	b

7. (a) 60°   (b) 117°   **9.** $y = -2x + 4$

11. $y = 4x + 7$ **13.** $y = -\frac{1}{5}x + 6$

15. $y = 11x - 18$ **17.** $y = -3x + 4$

19. $y = \dfrac{1}{\sqrt{3}}x - 3$ **21.** $y = \frac{1}{2}x + 2$

23. $y = 1$ **25.** $x = 5$

27. (a) parallel   (b) perpendicular   **29.** $\frac{49}{6}$
(c) parallel   (d) perpendicular
(e) neither

31. (a) $(4, -1)$ (b) $(1, -2)$

33. 3   **37.** 4   **39.** $(-\frac{29}{8}, -\frac{23}{4})$

41. (a) $F = \frac{9}{5}C + 32$ (b) $\frac{5}{9}$

► Exercise Set 1.6 (Page 53)

1. (a) $(0,0)$; 5   (b) $(1,4)$; 4
(c) $(-1, -3)$; $\sqrt{5}$ (d) $(0, -2)$; 1

3. $(x - 3)^2 + (y + 2)^2 = 16$

5. $(x + 4)^2 + (y - 8)^2 = 64$

7. $(x + 3)^2 + (y + 4)^2 = 25$

9. $(x - 1)^2 + (y - 1)^2 = 2$

11. circle; center $(1, 2)$, radius 4

13. circle; center $(-1, 1)$, radius $\sqrt{2}$

15. the point $(-1, -1)$

17. circle; center $(0, 0)$, radius $\frac{1}{3}$

19. no graph

21. circle; center $(-\frac{5}{4}, -\frac{1}{2})$, radius $\frac{3}{2}$

23. (a) $y = -\sqrt{16 - x^2}$
(b) $y = 2 + \sqrt{3 - 2x - x^2}$

25. (a)   (b)

27. $y = -\frac{3}{4}x + \frac{25}{4}$

29. (a) inside
(b) largest $3\sqrt{5}$, smallest $\sqrt{5}$

31.   **33.**   **35.**   **37.**

39.   **41.**   **43.**

45. (a) $x = \sqrt{3 - y}$ (b) $x = 1 - \sqrt{y + 1}$

47. (a)   (b)

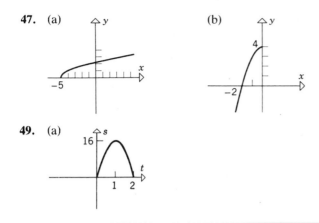

51. (a) equation of path: $2x^2 + 2y^2 - 12x + 8y + 1 = 0$

(b) center $(3, -2)$, radius $\dfrac{5}{\sqrt{2}}$

49. (a)

▶ Chapter 1 Supplementary Exercises (Page 55)

1. (a) $(-3, 5]$ (b) $[-3, 3]$
    (c) $(-\infty, -\frac{1}{2}] \cup [\frac{1}{2}, +\infty)$

3. (a) $(-\infty, -\frac{1}{2}) \cup (3, +\infty)$ (b) $[1, 4]$

5. (a) $(-2, -1] \cup [1, 2)$
    (b) $(-\infty, -5] \cup [-1, +\infty)$

7. (a) Take $a = -2$, $b = 1$.
    (b) $a + b > 0$

9. Both legs of a right triangle are no longer than the hypotenuse.

11. (a) all points on the y-axis or the line $y = x$
    (b) all points on the line $x = 1$ or the line $y = x + 1$

13. (a) all points on or outside the circle of radius 4 about $(1, 3)$
    (b)

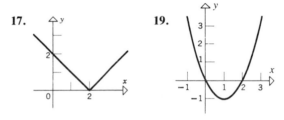

15. at $(2, 4)$ and $(-1, 1)$

17.

19.

21. $(x - 1)^2 + (y - 2)^2 = 25$

23. four circles: $(x - h)^2 + (y - k)^2 = 25$, where $h = 1$ or 11 and $k = 2$ or 12

25. point $(-2, -1)$ **27.** no graph

29. (a) $y = 4x/3$; 10; $(0, 0)$
    (b) $x = 3$; 8; $(3, 0)$
    (c) $y = 4$; 6; $(0, 4)$
    (d) $y = 7 - x$; $\sqrt{2}$; $(\frac{7}{2}, \frac{7}{2})$

31. $(4, -3)$ and $(-4, 3)$

33. (a) $90°$ (b) $0°$ (c) $135°$ (d) $60°$

35. $y = -3$ **37.** $y = -\frac{1}{2}x$

39. $L: y - 0 = (-2)(x - 1)$
    $L': y = \frac{1}{2}x - 3$; $(2, -2)$

41. $L: y - 1 = \frac{2}{5}(x - 3)$ **43.** no
    $L': 3x + 2y = -8$; $(-2, -1)$

▶ Exercise Set 2.1 (Page 66)

1. (a) 14 (b) 50 (c) 2
    (d) 11 (e) $3a^2 + 6a + 5$ (f) $27t^2 + 2$

3. (a) -8 (b) $\frac{1}{4}$ (c) 0
    (d) 6 (e) 5.8 (f) $\dfrac{1}{t^2 + 5}$

5. $(-\infty, 3) \cup (3, +\infty)$

7. $(-\infty, -\sqrt{3}] \cup [\sqrt{3}, +\infty)$

9. $(-\infty, -2) \cup [1, +\infty)$ **11.** $(-\infty, +\infty)$

13. $[5, 8]$ **15.** $(-\infty, +\infty)$

17. $(-\infty, 0) \cup (0, +\infty)$ **19.** $[0, +\infty)$

21. $x \neq \dfrac{\pi}{2} + 2k\pi,\ k = 0, \pm 1, \pm 2, \ldots$

23. domain $(-\infty, 3]$,   **25.** domain $[-2, 2]$,
range $[0, +\infty)$ range $[0, 2]$

27. domain $[0, +\infty)$,   **29.** domain $(-\infty, +\infty)$,
range $[3, +\infty)$ range $[3, +\infty)$

31. domain $(-\infty, +\infty)$,   **33.** domain $(-\infty, +\infty)$,
range $(-\infty, +\infty)$ range $[-3, 3]$

35. domain $(-\infty, +\infty)$,
range $[1, 3]$

37. $f(x) = \begin{cases} 2x + 1, & x < 0 \\ 4x + 1, & x \geq 0 \end{cases}$

39. $g(x) = \begin{cases} 1 - 2x, & x < 0 \\ 1, & 0 \leq x < 1 \\ 2x - 1, & x \geq 1 \end{cases}$

41. $\frac{38}{3}$ **43.** $\pm\sqrt{2}$

45. $2k\pi,\ k = 0, \pm 1, \pm 2, \ldots$

47. $(\frac{1}{6} + 2k)^2\pi^2$ or $(\frac{5}{6} + 2k)^2\pi^2$ for $k = 0, 1, 2, \ldots$

49. $A = \dfrac{C^2}{4\pi}$

51. (a) $S = 6x^2$ (b) $S = 6V^{2/3}$

53. $\dfrac{1 - (1/x)}{1 + (1/x)} = \dfrac{x - 1}{x + 1},\ x \neq 0$

55. $x - 1,\ x \neq -1$ or -2

57. $\sqrt{x + 1} + 1,\ x \neq -1$

59. $x,\ x \neq -3$ or 1

61. (a) monomial, polynomial, rational, explicit algebraic
(b) explicit algebraic
(c) rational, explicit algebraic
(d) polynomial, rational, explicit algebraic

63. (a) explicit algebraic
(b) rational, explicit algebraic
(c) monomial, polynomial, rational, explicit algebraic
(d) explicit algebraic

▶ Exercise Set 2.2 (Page 75)

1. (a) $-4, -3, -2, 2, 3$ (b) $0, 4$
(c) $-4 \leq x \leq -3, -2 \leq x \leq 2, x \geq 3$
(d) $x \leq -4, -3 \leq x \leq -2, 2 \leq x \leq 3$

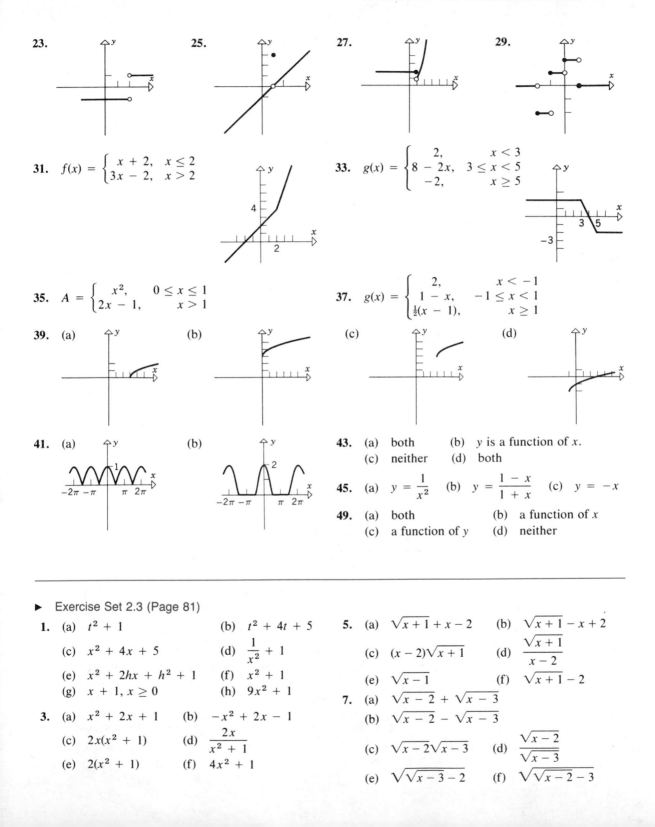

23.

25.

27.

29.

31. $f(x) = \begin{cases} x + 2, & x \le 2 \\ 3x - 2, & x > 2 \end{cases}$

33. $g(x) = \begin{cases} 2, & x < 3 \\ 8 - 2x, & 3 \le x < 5 \\ -2, & x \ge 5 \end{cases}$

35. $A = \begin{cases} x^2, & 0 \le x \le 1 \\ 2x - 1, & x > 1 \end{cases}$

37. $g(x) = \begin{cases} 2, & x < -1 \\ 1 - x, & -1 \le x < 1 \\ \frac{1}{2}(x - 1), & x \ge 1 \end{cases}$

39. (a)   (b)   (c)   (d)

41. (a)   (b)

43. (a) both   (b) y is a function of x.
       (c) neither   (d) both

45. (a) $y = \dfrac{1}{x^2}$ (b) $y = \dfrac{1 - x}{1 + x}$ (c) $y = -x$

49. (a) both   (b) a function of x
       (c) a function of y (d) neither

▶ **Exercise Set 2.3 (Page 81)**

1. (a) $t^2 + 1$
   (b) $t^2 + 4t + 5$
   (c) $x^2 + 4x + 5$
   (d) $\dfrac{1}{x^2} + 1$
   (e) $x^2 + 2hx + h^2 + 1$
   (f) $x^2 + 1$
   (g) $x + 1, x \ge 0$
   (h) $9x^2 + 1$

3. (a) $x^2 + 2x + 1$ (b) $-x^2 + 2x - 1$
   (c) $2x(x^2 + 1)$ (d) $\dfrac{2x}{x^2 + 1}$
   (e) $2(x^2 + 1)$ (f) $4x^2 + 1$

5. (a) $\sqrt{x + 1} + x - 2$ (b) $\sqrt{x + 1} - x + 2$
   (c) $(x - 2)\sqrt{x + 1}$ (d) $\dfrac{\sqrt{x + 1}}{x - 2}$
   (e) $\sqrt{x - 1}$ (f) $\sqrt{x + 1} - 2$

7. (a) $\sqrt{x - 2} + \sqrt{x - 3}$
   (b) $\sqrt{x - 2} - \sqrt{x - 3}$
   (c) $\sqrt{x - 2}\sqrt{x - 3}$ (d) $\dfrac{\sqrt{x - 2}}{\sqrt{x - 3}}$
   (e) $\sqrt{\sqrt{x - 3} - 2}$ (f) $\sqrt{\sqrt{x - 2} - 3}$

9. (a)   (b)   (c)   (d)

11. (a) $4x - 15$ (b) $4x^2 - 20x + 25$

17. $g(x) = \sqrt{x}, h(x) = x + 2$

19. $g(x) = x^7, h(x) = x - 5$

21. $g(x) = |x|, h(x) = x^2 - 3x + 5$

23. $g(x) = x^2, h(x) = \sin x$

25. $g(x) = \dfrac{3}{5 + x}, h(x) = \cos x$

27. $g(x) = \dfrac{x}{3 + x}, h(x) = \tan x$

29. $f(x) = \sqrt{x}, g(x) = 3 - x^2, h(x) = \sin x$

31. $f(x) = x^2 + x + 3$ 33. 0, $\frac{3}{2}$

35. $g(x) = 9x^2 - 5$

39. (a) even   (b) odd   (c) even
    (d) neither (e) odd   (f) even

41. (a) even   (b) odd   (c) odd   (d) neither

► Exercise Set 2.4 (Page 93)

1. (a) -1 (b) 3
   (c) does not exist   (d) 1
   (e) -1 (f) 3

3. (a) 1   (b) 1
   (c) 1   (d) 1
   (e) $-\infty$ (f) $+\infty$

5. (a) 0   (b) 0
   (c) 0   (d) 3
   (e) $+\infty$ (f) $+\infty$

7. (a) $-\infty$ (b) $+\infty$
   (c) does not exist   (d) not defined
   (e) 2   (f) 0

9. (a) $-\infty$ (b) $-\infty$
   (c) $-\infty$ (d) 1
   (e) 2   (f) 2

11. (a) 0   (b) 0
    (c) 0   (d) 0
    (e) does not exist   (f) does not exist

13. all values except -4

► Exercise Set 2.5 (Page 106)

1. 7   3. π 5. 36

7. $\sqrt{109}$ 9. 14   11. 0

13. 8   15. 4   17. $-\frac{4}{5}$

19. $\frac{3}{2}$ 21. 0   23. 0

25. $-\sqrt{5}$ 27. $\dfrac{1}{\sqrt{6}}$ 29. $\sqrt{3}$

31. $+\infty$ 33. does not exist   35. $-\infty$

37. $+\infty$ 39. does not exist   41. $+\infty$

43. $-\infty$ 45. $-\frac{1}{7}$ 47. -1

49. 6   51. (a) 2   (b) 2   (c) 2   53. 4

57. $\dfrac{1}{2\sqrt{3}}$ 59. $\dfrac{a}{2}$ 61. 0

63. does not exist   65. does not exist

67. 1   69. if $r(a)$ is defined

▶ Exercise Set 2.6 (Page 118)

Note: There are other possible answers for Exercises 1–21, 27.

1. 0.05 **3.** 1/700 **5.** 0.05

7. 1/9000 **9.** 1 **11.** $\delta = \dfrac{\epsilon}{3}$

13. $\delta = \epsilon$ **15.** $\delta = \min\left(\dfrac{\epsilon}{6}, 1\right)$

17. $\delta = \min\left(\dfrac{\epsilon}{36}, \dfrac{1}{4}\right)$ **19.** $\delta = \min(2\epsilon, 4)$

21. $\delta = \epsilon$ **27.** $\delta = \min\left(\dfrac{\epsilon}{8}, 2\right)$

▶ Exercise Set 2.7 (Page 127)

1. continuous on $(1, 2)$, $[2, 3]$, $(2, 3)$; discontinuous on $[1, 3]$, $(1, 3)$, and $[1, 2]$ at $x = 2$

3. continuous on $(1, 3)$, $(1, 2)$, $(2, 3)$; discontinuous on $[1, 3]$ at $x = 1$ and $x = 3$; on $[1, 2]$ at $x = 1$, and on $[2, 3]$ at $x = 3$

5. none **7.** none

9. $x = \pm 4$ **11.** $x = \pm 3$

13. none **15.** none

17. (a) 5 (b) $\frac{4}{3}$

21. $0, \pm 1, \pm 2, \ldots$ **31.** $-1.65, 1.35$

33. (a) 2.25 (b) 2.235

▶ Exercise Set 2.8 (Page 135)

1. none **3.** $x = n\pi$, $n = 0, \pm 1, \pm 2, \ldots$

5. $x = n\pi$, $n = 0, \pm 1, \pm 2, \ldots$ **7.** none

9. $x = \dfrac{\pi}{6} + 2n\pi$ or $\dfrac{5\pi}{6} + 2n\pi$, $n = 0, \pm 1, \pm 2, \ldots$

13. $-\frac{1}{2}\sqrt{3}$ **15.** 3

17. $\frac{7}{3}$ **19.** 1

21. 2 **23.** 0

25. $-\frac{25}{49}$ **27.** $\frac{1}{2}$

29. (a) 1 (b) 0 (c) 1

31. $-\pi$ **33.** 1

▶ Chapter 2 Supplementary Exercises (Page 136)

1. $|x| \le 2$; $\sqrt{2}, 2, 1$

3. $x \ne -2, 1$; $\frac{1}{2}$, undefined, $\frac{1}{4}$

5. all x; $-1, 3, \sqrt{3}$

7. (a) $(-6 + 6x - 2x^2)/x^2$
(b) $-9/[x(x + 3)]$
(c) $(3x^2 - 9x + 3)/(x - 3)$
(d) $(4x - 3)/(3 - x)$

9. the horizontal line $y = -\pi$; domain: all x; range: $\{-\pi\}$

11. domain: $x \ne -2$; range $y \ne -2$

13. domain: $x \ge -\frac{1}{3}$; range: $y \le 0$

15. domain: $x \neq \pm 2$; range: $y \neq 0, \frac{1}{2}$

17. Some possible answers are:
(a) $h(x) = x^3$, $g(x) = x^2 + 3$;
    $h(x) = x^6$, $g(x) = x + 3$
(b) $h(x) = x^2 + 1$, $g(x) = \sqrt{x}$;
    $h(x) = x^2$, $g(x) = \sqrt{x + 1}$
(c) $h(x) = 3x + 2$, $g(x) = \sin x$;
    $h(x) = 3x$, $g(x) = \sin(x + 2)$.

19. (a) -1 (b) does not exist   (c) 1
(d) 0   (e) $-\infty$ (does not exist)   (f) 0
(g) 0   (h) $-\infty$ (does not exist)

21. 2, 1, 0, does not exist, $+\infty$ (does not exist), does not exist

23. 5, 10, 0, 10, 0, $-\infty$ (does not exist), $+\infty$ (does not exist)

25. $\dfrac{a}{b}$ **27.** does not exist

29. 0   **31.** $3 - k$

▶ Exercise Set 3.1 (Page 147)

1. (a) $\frac{7}{2}$ (b) $3, y = 3x - \frac{9}{2}$
(c)

3. (a) $-\frac{1}{6}$ (b) $-\frac{1}{4}, y = -\frac{1}{4}x + 1$
(c)

5. (b) $y = 75x - 250$ (c) $y = 3x_0^2 x - 2x_0^3$
7. (a) $2x_0 + 1$ (b) $y = 5x - 4$
(c) $y = (2x_0 + 1)x - x_0^2$
9. (a) t_0 (b) 0   (c) speeding up
(d) slowing down
11. It is a straight line with slope equal to the velocity.
13. (a) 320,000 ft   (b) 8000 ft/sec
(c) 45 ft/sec   (d) 24,000 ft/sec
15. (a) 720 ft/min   (b) 192 ft/min
17. (a) 10   (b) 4
19. (a) 3π (b) 4π

▶ Exercise Set 3.2 (Page 157)

1. $6x$ **3.** $3x^2$

5. $\dfrac{1}{2\sqrt{x + 1}}$ **7.** $-\dfrac{1}{x^2}$

9. $2ax$ **11.** $-\dfrac{1}{2x^{3/2}}$

13. $18; y = 18x - 27$ **15.** $0; y = 0$
17. $\frac{1}{6}; y = \frac{1}{6}x + \frac{5}{3}$ **19.** (a) $8x$ (b) 8
21. $8t + 1$ **23.** $6\lambda - 1$
25. (a) D   (b) F   (c) B
(d) C   (e) A   (f) E

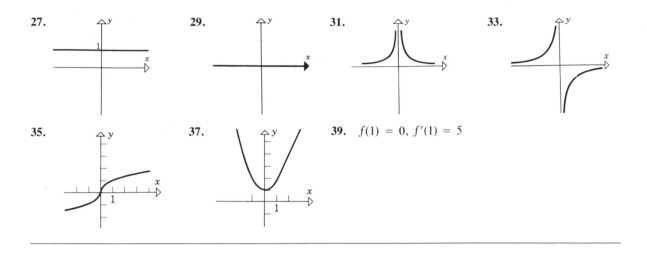

27.   **29.**   **31.**   **33.**

35.   **37.**   **39.** $f(1) = 0, f'(1) = 5$

▶ Exercise Set 3.3 (Page 170)

1. $28x^6$

3. $24x^7 + 2$

5. 0

7. $-\frac{1}{3}(7x^6 + 2)$

9. $3ax^2 + 2bx + c$

11. $24x^{-9} + 1/\sqrt{x}$

13. $-3x^{-4} - 7x^{-8}$

15. $18x^2 - \frac{3}{2}x + 12$

17. $-15x^{-2} - 14x^{-3} + 48x^{-4} + 32x^{-5}$

19. $12x(3x^2 + 1)$

21. $-5/(5x - 3)^2$

23. $3/(2x + 1)^2$

25. $7/(x + 3)^2$

27. $\left(\dfrac{3x + 2}{x}\right)(-5x^{-6}) + (x^{-5} + 1)\left(-\dfrac{2}{x^2}\right)$

29. (a) $-\frac{37}{4}$ (b) $-\frac{23}{16}$

31. $32t$

33. $3\pi r^2$

35. $\dfrac{7 - 2t^3}{(t^3 + 7)^2}$

37. $-\dfrac{2GmM}{r^3}$

39. (a) $42x - 10$ (b) 24

(c) $\dfrac{2}{x^3}$ (d) $700x^3 - 96x$

41. (a) $-210x^{-8} + 60x^2$ (b) $-6x^{-4}$ (c) $6a$

43. (a) 0 (b) 112 (c) 360

47. $(1, \frac{5}{6}), (2, \frac{2}{3})$

49. $y = 5x + 17$

51. $a = 3, b = 2$

53. $\frac{1}{2}$

55. $2 \pm \sqrt{3}$

57. $-2x_0$

63. (a) $2\left(1 + \dfrac{1}{x}\right)(x^{-3} + 7)$

$+ (2x + 1)\left(-\dfrac{1}{x^2}\right)(x^{-3} + 7)$

$+ (2x + 1)\left(1 + \dfrac{1}{x}\right)(-3x^{-4})$

(b) $(-5x^{-6})(x^2 + 2x)(4 - 3x)(2x^9 + 1)$
$+ x^{-5}(2x + 2)(4 - 3x)(2x^9 + 1)$
$+ x^{-5}(x^2 + 2x)(-3)(2x^9 + 1)$
$+ x^{-5}(x^2 + 2x)(4 - 3x)(18x^8)$

(c) $3(7x^6 + 2)(x^7 + 2x - 3)^2$

(d) $100x(x^2 + 1)^{49}$

65. $2(2x^3 - 5x^2 + 7x - 2)(6x^2 - 10x + 7)$

67. not differentiable at $x = 1$

69. (a) $x = \frac{2}{3}$ (b) $x = \pm 2$

71. (a) $n(n - 1)(n - 2) \cdots 1$ (b) 0

(c) $a_n n(n - 1)(n - 2) \cdots 1$

77. (b) f and all its derivatives up to $f^{(n-1)}(x)$ are continuous on (a, b).

▶ Exercise Set 3.4 (Page 176)

1. $-2\sin x - 3\cos x$

3. $\dfrac{x\cos x - \sin x}{x^2}$

5. $x^3 \cos x + (3x^2 + 5)\sin x$

7. $\sec x \tan x - \sqrt{2} \sec^2 x$

9. $\sec^3 x + \sec x \tan^2 x$

11. $1 + 4 \csc x \cot x - 2 \csc^2 x$

13. $-\dfrac{\csc x}{1 + \csc x}$ **15.** 0

17. $\dfrac{1}{(1 + x \tan x)^2}$ **19.** $-x \cos x - 2 \sin x$

21. $-x \sin x + 5 \cos x$ **23.** $-4 \sin x \cos x$

25. (a) $x = n\pi, \ n = 0, \pm 1, \pm 2, \ldots$

   (b) none

   (c) $x = \dfrac{\pi}{2} + n\pi, \ n = 0, \pm 1, \pm 2, \ldots$

27. (a) $y = x$ (b) $y = 2x - \pi/2 + 1$

   (c) $y = 2x + \pi/2 - 1$

29. (a) all x (b) all x

   (c) $x \neq \dfrac{\pi}{2} + n\pi, \ n = 0, \pm 1, \pm 2, \ldots$

   (d) $x \neq n\pi, \ n = 0, \pm 1, \pm 2, \ldots$

   (e) $x \neq \dfrac{\pi}{2} + n\pi, \ n = 0, \pm 1, \pm 2, \ldots$

   (f) $x \neq n\pi, \ n = 0, \pm 1, \pm 2, \ldots$

   (g) $x \neq \pi + 2n\pi, \ n = 0, \pm 1, \pm 2, \ldots$

   (h) $x \neq n\pi/2, \ n = 0, \pm 1, \pm 2, \ldots$

   (i) all x

31. $3, 7, 11, \ldots$ **33.** $\sec^2 y$

▶ Exercise Set 3.5 (Page 185)

1. $37(x^3 + 2x)^{36}(3x^2 + 2)$

3. $-2\left(x^3 - \dfrac{7}{x}\right)^{-3}\left(3x^2 + \dfrac{7}{x^2}\right)$

5. $\dfrac{24(1 - 3x)}{(3x^2 - 2x + 1)^4}$ **7.** $\dfrac{3}{4\sqrt{x}\sqrt{4 + 3\sqrt{x}}}$

9. $3x^2 \cos(x^3)$ **11.** $8x \sec^2(4x^2)$

13. $-20 \cos^4 x \sin x$ **15.** $-\dfrac{2}{x^3} \cos\left(\dfrac{1}{x^2}\right)$

17. $28x^6 \sec^2(x^7) \tan(x^7)$

19. $-\dfrac{5 \sin(5x)}{2\sqrt{\cos(5x)}}$

21. $-3[x + \csc(x^3 + 3)]^{-4}$
$[1 - 3x^2 \csc(x^3 + 3) \cot(x^3 + 3)]$

23. $\dfrac{x(10 - 3x^2)}{\sqrt{5 - x^2}}$

25. $10x^3 \sin 5x \cos 5x + 3x^2 \sin^2 5x$

27. $-x^3 \sec\left(\dfrac{1}{x}\right) \tan\left(\dfrac{1}{x}\right) + 5x^4 \sec\left(\dfrac{1}{x}\right)$

29. $\sin x \sin(\cos x)$

31. $-6 \cos^2(\sin 2x) \sin(\sin 2x) \cos 2x$

33. $12(5x + 8)^{13}(x^3 + 7x)^{11}(3x^2 + 7)$
$+ 65(x^3 + 7x)^{12}(5x + 8)^{12}$

35. $\dfrac{33(x - 5)^2}{(2x + 1)^4}$

37. $\dfrac{-64x(2x + 1)^{-3}(4x^2 - 1)^{-9} + 6(4x^2 - 1)^{-8}(2x + 1)^{-4}}{(2x + 1)^{-6}}$

39. $5[x \sin 2x + \tan^4(x^7)]^4[2x \cos 2x + \sin 2x$
$+ 28x^6 \tan^3(x^7) \sec^2(x^7)]$

41. $-25x \cos(5x) - 10 \sin(5x) - 2 \cos(2x)$

43. $y = -x$ **45.** $y = -1$

47. $\frac{2}{25}(x - 2)$

49. $-\dfrac{9 \sin^2(1/x) \cos(1/x)}{x^2}$

51. $3 \cot^2 \theta \csc^2 \theta$ **53.** $\pi(b - a) \sin 2\pi\omega$

55. (b) $\begin{cases} \cos x, & 0 < x < \pi \\ -\cos x, & -\pi < x < 0 \end{cases}$ for both

57. (a) $-\dfrac{1}{x} \cos\dfrac{1}{x} + \sin\dfrac{1}{x}$

59. 6 **61.** $\dfrac{1}{2x}$

63. $\frac{2}{3}x$ **65.** $f'\big(g(h(x))\big)\, g'(h(x))h'(x)$

▶ Exercise Set 3.6 (Page 193)

1. $\frac{2}{3}(2x - 5)^{-2/3}$ **3.** $\dfrac{9}{2(x + 2)^2}\left(\dfrac{x - 1}{x + 2}\right)^{1/2}$ **5.** $\frac{1}{3}x^2(5x^2 + 1)^{-5/3}(25x^2 + 9)$

7. $-\dfrac{15[\sin (3/x)]^{3/2} \cos (3/x)}{2x^2}$

9. $-\frac{2}{3}(2x - 1)^{-4/3} \sec^2 [(2x - 1)^{-1/3}]$

11. $-\dfrac{x}{y}$

13. $\dfrac{1 - 2xy - 3y^3}{x^2 + 9xy^2}$

15. $-\dfrac{y^2}{x^2}$

17. $-\dfrac{\sqrt{y}}{\sqrt{x}}$

19. $\dfrac{1 - 70x(x^2 + 3y^2)^{34}}{210y(x^2 + 3y^2)^{34}}$

21. $\dfrac{\frac{3}{2}x^2(x^3 + y^2)^{1/2} - y}{x - y(x^3 + y^2)^{1/2}}$

23. $\dfrac{1 - 2xy^2 \cos (x^2y^2)}{2x^2y \cos (x^2y^2)}$

25. $\dfrac{1 - 3y^2 \tan^2 (xy^2 + y) \sec^2 (xy^2 + y)}{3(2xy + 1) \tan^2 (xy^2 + y) \sec^2 (xy^2 + y)}$

27. $\dfrac{3y^2 \sin^2 (xy^2) \cos (xy^2)}{2\sqrt{1 + \sin^3 (xy^2)} - 6xy \sin^2 (xy^2) \cos (xy^2)}$

29. $-\dfrac{14}{13}$

31. 2

33. -2

35. 1

37. $\dfrac{6}{5}$

39. $-2x/y^5$

41. $-3/(y - x)^3$

43. $-\dfrac{\sin 2y + y(\sin^2 y + 1)}{(1 + x \sin y)^3}$

45. $\dfrac{2t^3 + 3a^2}{2a^3 - 6at}$

47. $-\dfrac{b^2\lambda}{a^2\omega}$

49. $-\dfrac{2y^3 + 3t^2y}{(6ty^2 + t^3) \cos t}$

51. $y = \dfrac{\sqrt{3}}{3}x, \ y = -\dfrac{\sqrt{3}}{3}x$

53. (a) $(0, 0), \left(-\dfrac{5}{\sqrt{2}}, 0\right), \left(\dfrac{5}{\sqrt{2}}, 0\right)$

 (b) $9x + 13y = 40$

▶ Exercise Set 3.7 (Page 203)

1. (a) 5 (b) 4

 (c)

3. (a) $-\frac{1}{3}$ (b) -0.5

 (c)

5. $dy = 3x^2 \, dx$

 $\Delta y = 3x^2 \, \Delta x + 3x(\Delta x)^2 + (\Delta x)^3$

7. $dy = (2x - 2) \, dx$

 $\Delta y = 2x \, \Delta x + (\Delta x)^2 - 2 \, \Delta x$

9. $dy = (12x^2 - 14x + 2) \, dx$

11. $dy = (\cos x - x \sin x) \, dx$

13. $2x$

15. -1

17. 83.16

19. 8.0625

21. 8.9944

23. 2.005

25. 0.8573

27. 0.6947

29. 0.0225

31. 0.0048

33. (a) $\pm 2 \text{ ft}^2$ (b) side: $\pm 1\%$, area: $\pm 2\%$

35. (a) opposite: $\pm 0.151''$, adjacent: $\pm 0.087''$

 (b) opposite: $\pm 3.0\%$, adjacent: $\pm 1.0\%$

37. $\pm 10\%$

39. $\pm 6\%$

41. $\pm 0.5\%$

43. 0.236 cm^3

▶ Chapter 3 Supplementary Exercises (Page 205)

1. k

3. $-2/\sqrt{9 - 4x}$ for $x < 9/4$

5. 0

7. $y + 1 = 5(x - 3)$

9. (a) 12 (b) -7 (c) 9

 (d) $-9/4$ (e) 5 (f) 21

 (g) -35 (h) -7 (i) -126

 (j) -12 (k) $3/2$ (l) $-3/2$

11. $2(x - 3)^3(x^2 + 6x + 3)/(x^2 + 2x)^2$; 3, $-3 \pm \sqrt{6}$

13. $-3(3x + 1)^2(3x + 2)/x^7$; $-1/3$, $-2/3$

15. $(7x^2 + 5x + 3)x^{-1/2}(x^2 + x + 1)^{-2/3}/6$; none

17. $-2\sqrt{2}/x^3 + 2x^{-2}/5$

19. $\sqrt{3} \ (z = \sin^2 2r)$

21. $2(x - 1)/x^3 \left(u = \left(1 - \dfrac{1}{x}\right)^2\right)$

23. 0

25. $2\ (F(x) = 2x)$ **27.** ± 1 **29.** ± 2

31. odd integer multiples of $\pi/4$

33. $\Delta x = \pi/4,\ \Delta y = 1,\ dy = \pi/2$

35. (a) $-97/48$ (b) $1 - (\pi/90)$

37. $dh = -\pi/30$ ft

39. (a) 2000 gal/min   (b) 2500 gal/min

41. $\dfrac{dy}{dx} = \dfrac{y^2 - \cos(x + 2y)}{2\cos(x + 2y) - 2xy};\ y = -\tfrac{1}{2}x$

▶ Exercise Set 4.1 (Page 213)

1. (a) $\dfrac{dA}{dt} = 2x\dfrac{dx}{dt}$ (b) 12 ft²/min

3. (a) $\dfrac{dV}{dt} = \pi\left(r^2\dfrac{dh}{dt} + 2rh\dfrac{dr}{dt}\right)$

   (b) -20π in³/sec; decreasing

5. (a) $\dfrac{d\theta}{dt} = \dfrac{\cos^2\theta}{x^2}\left(x\dfrac{dy}{dt} - y\dfrac{dx}{dt}\right)$

   (b) $-\dfrac{5}{16}$ radian/sec; decreasing

7. $4\pi/15$ in²/min   **9.** $1/\sqrt{\pi}$ mph

11. 4860π cm³/min   **13.** $\tfrac{5}{6}$ ft/sec

15. $\tfrac{8}{5}$ in²/min   **17.** $\tfrac{1}{12}$ radian/sec

19. $\dfrac{9}{20\pi}$ ft/min   **21.** 125π ft³/min

23. 250 mph   **25.** $\tfrac{36}{25}\sqrt{69}$ ft/min

27. $8\pi/5$ km/sec   **29.** $600\sqrt{7}$ mph

31. (a) $-60/7$ units/sec   (b) falling

33. -4 units/sec   **35.** 4.5 cm/sec; away

39. $\dfrac{2}{9\pi}$ cm/sec

▶ Exercise Set 4.2 (Page 220)

1. (a) (d, f) (b) $(a, d),\ (f, g)$
   (c) $(a, b),\ (c, e)$ (d) $(b, c),\ (e, g)$

3. (a) $(\tfrac{5}{2}, +\infty)$ (b) $(-\infty, \tfrac{5}{2})$
   (c) $(-\infty, +\infty)$ (d) none   (e) none

5. (a) $(-\infty, -2),\ (-2, +\infty)$ (b) none
   (c) $(-2, +\infty)$ (d) $(-\infty, -2)$ (e) -2

7. (a) $(-\infty, -\tfrac{2}{3}),\ (\tfrac{2}{3}, +\infty)$ (b) $(-\tfrac{2}{3}, \tfrac{2}{3})$
   (c) $(0, +\infty)$ (d) $(-\infty, 0)$ (e) 0

9. (a) $(1, +\infty)$ (b) $(-\infty, 0),\ (0, 1)$
   (c) $(-\infty, 0),\ (\tfrac{2}{3}, +\infty)$ (d) $(0, \tfrac{2}{3})$
   (e) $0, \tfrac{2}{3}$

11. (a) $(\pi, 2\pi)$ (b) $(0, \pi)$ (c) $(\pi/2, 3\pi/2)$
   (d) $(0, \pi/2),\ (3\pi/2, 2\pi)$ (e) $\pi/2, 3\pi/2$

13. (a) $(-\pi/2, \pi/2)$ (b) none
   (c) $(0, \pi/2)$ (d) $(-\pi/2, 0)$ (e) 0

15. (a) $(-\infty, -2),\ (-2, +\infty)$ (b) none
   (c) $(-\infty, -2)$ (d) $(-2, +\infty)$ (e) -2

17. (a) $(-1, 0),\ (0, +\infty)$ (b) $(-\infty, -1)$
   (c) $(-\infty, 0),\ (2, +\infty)$ (d) $(0, 2)$ (e) $0, 2$

19. (a)   (b)
   (c)

21. (a) 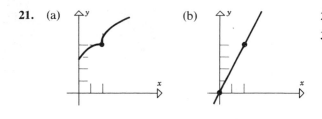 (b)

23. none

31. $f(x) = x, g(x) = 2x$ on $(-\infty, +\infty)$

► Exercise Set 4.3 (Page 227)

1. $x = \frac{5}{2}$ (stationary)

3. $x = -3, 1$ (stationary)

5. $x = 0, \pm\sqrt{3}$ (stationary)

7. $x = \pm\sqrt{2}$ (stationary)

9. $x = 0$ (not differentiable)

11. $x = n\frac{\pi}{3}$, $n = 0, \pm1, \pm2, \ldots$ (stationary)

13. $x = n\frac{\pi}{4}$, $n = 1, 2, 3, 4, 5, 6, 7$ (stationary)

15. $x = -1$ (stationary); $x = 0$ (not differentiable)

17. relative max of 5 at $x = -2$

19. relative min of 0 at $x = \pi$;
relative max of 1 at $x = \pi/2, 3\pi/2$

21. none

23. relative min of 0 at $x = 1$;
relative max of $\frac{4}{27}$ at $x = \frac{1}{3}$

25. relative min of 0 at $x = 0$;
relative max of 1 at $x = -1, 1$

27. relative min of 0 at $x = 0$

29. relative min of 0 at $x = 0$

31. relative min of 0 at $x = -2, 2$;
relative max of 4 at $x = 0$

33. relative min of 0 at $x = \pm\pi/2, \pm3\pi/2,$
$\pm5\pi/2, \ldots$;
relative max of 1 at $x = 0, \pm\pi, \pm2\pi, \ldots$

35. relative min of tan (1) at $x = 0$

37. relative min of 0 at $x = \pi/2, \pi, 3\pi/2$;
relative max of 1 at $x = \pi/4, 3\pi/4, 5\pi/4, 7\pi/4$

39. 54

41. $f(x) = -x^4$ has a relative maximum at $x = 0$,
$f(x) = x^4$ has a relative minimum at $x = 0$,
$f(x) = x^3$ has neither at $x = 0$;
$f'(0) = 0$ for all three functions.

43. (a)   (b)

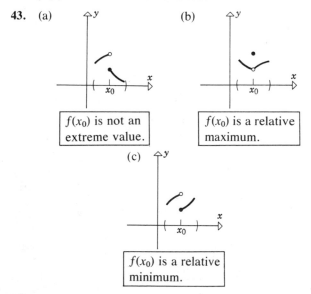

$f(x_0)$ is not an extreme value.

$f(x_0)$ is a relative maximum.

(c)

$f(x_0)$ is a relative minimum.

► Exercise Set 4.4 (Page 235)

1.

$(1, -4)$

3.

$(-1, 3)$ $(0, 1)$ $(1, -1)$

5.

$(-2, 9)$ 10 $(-1, 7)$ 5 $(0, 5)$

7.

10 $(\frac{1}{2}, 14\frac{1}{2})$ 1

9.

1 $(1, 0)$ 1

11.

$(0, -1)$ $(-1, -2)$ $(-\frac{3}{2}, -\frac{43}{16})$

13.

10 $(\frac{1}{2}, \frac{7}{16})$ $(0, 1)$ $(1, 2)$

15.

$(-1/\sqrt{2}, 7\sqrt{2}/8)$ $(-1, 2)$ $(0, 0)$ 1 $(1, -2)$ $(1/\sqrt{2}, -7\sqrt{2}/8)$

17.
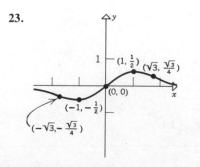
0.1 1 $(1, 0)$ $(1/2, -1/16)$ $(1/4, -27/256)$

19.
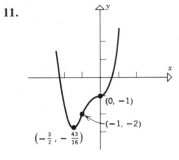
$y = 2$ $x = 3$

21.

$x = -1$ $x = 1$ $y = 1$ $(0, 0)$

23.
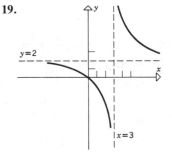
1 $(1, \frac{1}{2})$ $(\sqrt{3}, \frac{\sqrt{3}}{4})$ $(0, 0)$ $(-1, -\frac{1}{2})$ $(-\sqrt{3}, -\frac{\sqrt{3}}{4})$

25.

$y = 1$ $x = 2$

27.

$\sim (-0.8, 1.9)$

$(1, 0)$

29.

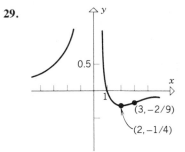

0.5

1

$(3, -2/9)$

$(2, -1/4)$

31.

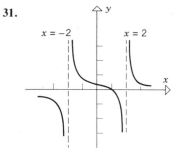

$x = -2$ $x = 2$

33.

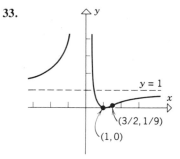

$y = 1$

$(3/2, 1/9)$

$(1, 0)$

35.

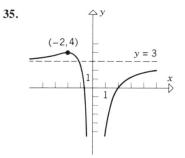

$(-2, 4)$

$y = 3$

1

1

37.

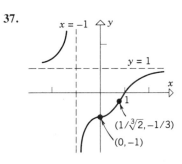

$x = -1$

$y = 1$

1

$(1/\sqrt[3]{2}, -1/3)$

$(0, -1)$

39.

$y = x$

41.

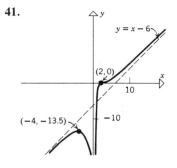

$y = x - 6$

$(2, 0)$

10

$(-4, -13.5)$

-10

43.

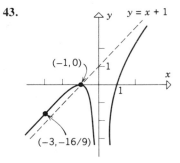

$y = x + 1$

$(-1, 0)$

1

1

$(-3, -16/9)$

▶ Exercise Set 4.5 (Page 240)

1.

1

1

3.

1

1

5.

1

-1 1

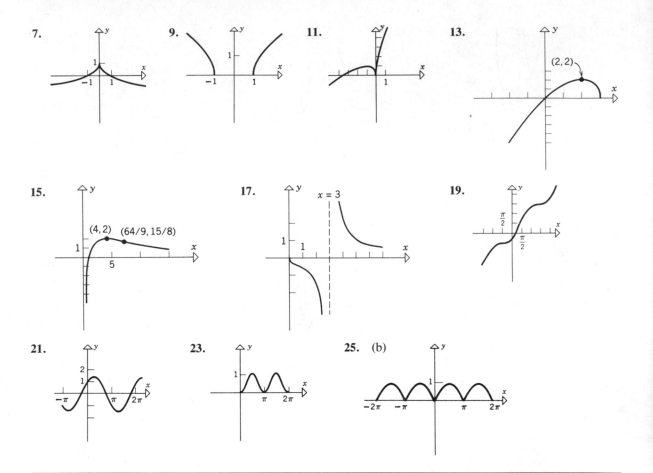

7.

9.

11.

13. (2, 2)

15. (4, 2) (64/9, 15/8)

17. $x = 3$

19. $\frac{\pi}{2}$ $\frac{\pi}{2}$

21.

23.

25. (b)

▶ Exercise Set 4.6 (Page 251)

1. maximum value 1 when $x = 0$, 1;
minimum value 0 when $x = \frac{1}{2}$

3. maximum value 27 when $x = 4$;
minimum value -1 when $x = 0$

5. maximum value $3/\sqrt{5}$ when $x = 1$;
minimum value $-3/\sqrt{5}$ when $x = -1$

7. maximum value 48 when $x = 8$;
minimum value 0 when $x = 0$, 20

9. maximum value $1 - (\pi/4)$ when $x = -\pi/4$;
minimum value $(\pi/4) - 1$ when $x = \pi/4$

11. maximum value 2 when $x = 0$;
minimum value $\sqrt{3}$ when $x = \pi/6$

13. maximum value 17 when $x = -5$;
minimum value 1 when $x = -3$

15. minimum $-\frac{13}{4}$, no maximum

17. maximum 1, no minimum

19. minimum 0, no maximum

21. no maximum or minimum

23. no maximum or minimum

25. maximum -4, no minimum

27. maximum value $3\sqrt{3}/2$ when $x = (\pi/6) + n\pi$
$(n = 0, \pm1, \pm2, \ldots)$;
minimum value $-3\sqrt{3}/2$ when $x = (5\pi/6) + n\pi$
$(n = 0, \pm1, \pm2, \ldots)$

29. maximum value 2; minimum value $-\frac{1}{4}$

31. (a) relative min of 0 at $x = a$ **33.** (b) 125
  (b) none

35. $a_0 = 9,\ a_1 = -8,\ a_2 = 2$ **43.** $\dfrac{2}{3\sqrt{3}}$

▶ Exercise Set 4.7 (Page 262)

1. $5 + 5$ **3.** (a) 1   (b) $\frac{1}{2}$

5. 500 ft by 750 ft   **7.** $10\sqrt{2}$ by $10\sqrt{2}$

9. 5 in by $\frac{12}{5}$ in   **11.** 2 in. square   **13.** $\frac{200}{27}$ ft^3

15. (a) Use all of the wire for the circle.
  (b) $\dfrac{12\pi}{\pi + 4}$ in for the circle.

17. Each side has length 4.

19. $L/12$ by $L/12$ by $L/12$

21. (a) 7000   (b) yes

23. height $L/\sqrt{3}$, radius $\sqrt{2/3}\ L$

25. height $2\sqrt{(5 - \sqrt{5})/10}\ R$,
  radius $\sqrt{(5 + \sqrt{5})/10}\ R$

27. $\dfrac{\pi}{3}$ **29.** $2\pi R^3/(9\sqrt{3})$

31. (a) 24   (b) \$24   (c) \$24.10
  (d) $R'(x) = 10,\ P'(x) = 6 - 0.2x$

35. $p/(4 + \pi)$ **37.** (b) no cube

39. (a) $\frac{3}{4}$ (b) $\frac{3}{16}$

43. (c) $\frac{1}{4}$ mi downstream from the house

▶ Exercise Set 4.8 (Page 272)

1. (a) 10, 10   (b) no minimum

3. 80 ft (\$1 fencing), 40 ft (\$2 fencing)

5. base: 10 cm square, height: 20 cm

7. ends: $\sqrt[3]{3V/4}$ units square;
  length: $\frac{4}{3}\sqrt[3]{3V/4}$ units

9. height = radius = $\sqrt[3]{500/\pi}$

11. radius: $\sqrt[6]{\dfrac{450}{\pi^2}}$ cm
  height: $\dfrac{30}{\pi}\sqrt[3]{\dfrac{\pi^2}{450}}$ cm

13. $(-\sqrt{2}, 1),\ (\sqrt{2}, 1)$

15. (a) no maximum   (b) -3

17. $1/\sqrt{5}$ **19.** $(\sqrt{2}, 1/2)$ **21.** $(-1/\sqrt{3}, 3/4)$

23. height: $4R$ **25.** $4(1 + 2^{2/3})^{3/2}$ ft
  radius: $\sqrt{2}R$

27. 30 cm from the weaker source

▶ Exercise Set 4.9 (Page 280)

1. 1.4142136   **3.** 1.8171206   **13.** 1.8954943   **15.** 4.4934095

5. -1.6716999 **7.** 1.2244395   **17.** (b) 3.1622777   **19.** -1.1653730

9. 0.5811388   **11.** -1.4526269

► Exercise Set 4.10 (Page 286)

1. 3   **3.** π **13.** (b) $\tan x$ is not continuous on $[0, \pi]$.

5. 1   **7.** 1   **29.** 8   **31.** $f(x) = x^3 - 4x + 5$

9. 5/4   **11.** $-\sqrt{5}$

► Exercise Set 4.11 (Page 293)

1.

t	s	v	$|v|$	a	direction; motion
1	-5	-9	9	-6	left; speeding up
2	-16	-12	12	0	left; neither
3	-27	-9	9	6	left; slowing down
4	-32	0	0	12	stopped
5	-25	15	15	18	right; speeding up

3. (a) 12   (b) $t = 2.2$, $s = -24.2$

5.

7.

9.

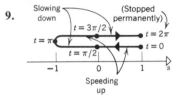

11. (a) $s = 5/16$, $v = 3/2$
(b) $s = 1$, $a = -3$

13. (b) $v = \dfrac{3}{2\sqrt{3t + 7}}$; $a = -9/500$

15. (b) 2/3 unit
(c) $0 \le t < 1$ and $t > 2$

17. (a) negative   (b) negative
(c) speeding up   (d) slowing down

► Chapter 4 Supplementary Exercises (Page 294)

1. decreasing 39π m³/sec

3. 60 ft/sec (toward pole)

5. $m = -2 = f(-1)$; $M = 27/256 = f(3/4)$

7. no m; $M = 1/\sqrt{3} = f(\sqrt{3})$

9. $m = -3 = f(3)$; $M = 0 = f(2)$

11. -2.1149075, 0.2541016, 1.8608059

13. $\left(-\frac{1}{\sqrt{3}}, \frac{3}{4}\right)$, $\left(\frac{1}{\sqrt{3}}, \frac{3}{4}\right)$ **15.**

17.   **19.**

21. at $(1, 2)$, $y = 2x$; at $(2, 3)$, $y = 3$

23. rel. max at $x = \pi/2$ and $3\pi/2$;
rel. min at $x = 7\pi/6$ and $11\pi/6$

25. rel. min at $x = 9$

27. rel. max at $x = 2\pi/3$ and $4\pi/3$;
rel. min at $x = \pi$

29. $2\sqrt{8}$ by $3\sqrt{2}$

31. $r = 2P/(3\pi + 8)$ ft   **33.** $r/h = \pi/8$

35. satisfied; $c = 0$

37. satisfied; $c = \sqrt{\pi/2}$

39. satisfied; $c = 1$

41. satisfied; $c = \frac{1}{2}, \sqrt{2}$

▶ Exercise Set 5.2 (Page 307)

1. $\frac{1}{9}x^9 + C$

3. $\frac{7}{12}x^{12/7} + C$

5. $8\sqrt{t} + C$

7. $\frac{2}{9}x^{9/2} + C$

9. $-\frac{1}{2}x^{-2} + \frac{2}{3}x^{3/2} - \frac{12}{5}x^{5/4} + \frac{1}{3}x^3 + C$

11. $28y^{1/4} - \frac{3}{4}y^{4/3} + \frac{8}{3}y^{3/2} + C$

13. $\frac{1}{2}x^2 + \frac{1}{5}x^5 + C$

15. $3x^{4/3} - \frac{12}{7}x^{7/3} + \frac{3}{10}x^{10/3} + C$

17. $\frac{1}{2}x^2 - \dfrac{2}{x} + \dfrac{1}{3x^3} + C$

19. $-4\cos x + 2\sin x + C$

21. $\tan x + \sec x + C$ **23.** $\sec x + x + C$

25. $\sec x + C$ **27.** $\theta - \cos \theta + C$

29. $\sin \theta - 5 \tan \theta + C$

31. $F(x) = \frac{3}{4}x^{4/3} + \frac{5}{4}$

33. $f(x) = \frac{4}{15}x^{5/2} + C_1 x + C_2$

35. $f(x) = 15x^2 - 3$ **37.** $\tan x - x + C$

▶ Exercise Set 5.3 (Page 313)

1. (a) $\frac{1}{24}(x^2 + 1)^{24} + C$ (b) $-\frac{1}{4}\cos^4 x + C$

(c) $-2\cos\sqrt{x} + C$ (d) $\frac{3}{4}\sqrt{4x^2 + 5} + C$

3. (a) $-\frac{1}{2}\cot^2 x + C$ (b) $\frac{1}{10}(1 + \sin t)^{10} + C$

(c) $\frac{2}{7}(1 + x)^{7/2} - \frac{4}{5}(1 + x)^{5/2}$
$+ \frac{2}{3}(1 + x)^{3/2} + C$

(d) $-\cot(\sin x) + C$

5. $-\frac{1}{8}(2 - x^2)^4 + C$ **7.** $\frac{1}{8}\sin 8x + C$

9. $\frac{1}{4}\sec 4x + C$ **11.** $\frac{1}{21}(7t^2 + 12)^{3/2} + C$

13. $\frac{2}{3}\sqrt{x^3 + 1} + C$ **15.** $-\frac{1}{16}(4x^2 + 1)^{-2} + C$

17. $\frac{1}{5}\cos(5/x) + C$ **19.** $\frac{1}{3}\tan(x^3) + C$

21. $\frac{1}{18}\sin^6 3t + C$ **23.** $-\frac{1}{6}(2 - \sin 4\theta)^{3/2} + C$

25. $\frac{1}{6}\sec^3 2x + C$ **27.** $-\frac{1}{3}\tan(\cos 3\theta) + C$

29. $\dfrac{1}{b(n + 1)}\sin^{n+1}(a + bx) + C$

31. $\frac{2}{5}(x - 3)^{5/2} + 2(x - 3)^{3/2} + C$

33. $\frac{2}{3}(y + 1)^{3/2} - 2(y + 1)^{1/2} + C$

35. $\frac{1}{3}\tan 3\theta - \theta + C$

39. (a) $\frac{25}{3}x^3 - 5x^2 + x + C$; $\frac{1}{15}(5x - 1)^3 + C$

(b) They differ by a constant.

41. $f(x) = 6x + \frac{5}{2}\cos 2x + \frac{1}{2}$

43. $\frac{1}{3}f(3x + 2) + C$ **45.** $-\frac{1}{2}f(2/x) + C$

▶ Exercise Set 5.4 (Page 322)

1. (a) 36   (b) 55   (c) 40   (d) 6

3. $\displaystyle\sum_{k=1}^{10} k$

5. $\displaystyle\sum_{k=1}^{49} k(k + 1)$

7. $\displaystyle\sum_{k=1}^{10} 2k$

9. $\displaystyle\sum_{k=1}^{6} (-1)^{k+1}(2k - 1)$

11. $\displaystyle\sum_{k=1}^{5} (-1)^k \frac{1}{k}$

13. $\displaystyle\sum_{k=1}^{4} \sin \frac{(2k - 1)\pi}{8}$

15. $\displaystyle\sum_{k=1}^{5} \frac{k}{k + 1}$

17. (a) $\displaystyle\sum_{k=1}^{5} (-1)^{k+1}a_k$ (b) $\displaystyle\sum_{k=0}^{5} (-1)^{k+1}b_k$

(c) $\displaystyle\sum_{k=0}^{n} a_k x^k$ (d) $\displaystyle\sum_{k=0}^{5} a^{5-k}b^k$

19. 5047   **21.** 2870

23. 1728   **25.** 214,365

27. $3^{17} - 3^4$ **29.** $-\frac{399}{400}$ **31.** $a_n - a_0$

33. (a) n^2 (b) -3 (c) $\frac{1}{2}n(n + 1)x$

(d) $(n - m + 1)c$

35. (a) $\displaystyle\sum_{k=0}^{14} (k + 4)(k + 1)$

(b) $\displaystyle\sum_{k=5}^{19} (k - 1)(k - 4)$

37. $\displaystyle\sum_{k=1}^{18} k \sin \frac{\pi}{k}$ **39.** Both are valid.

41. (a) $\frac{3}{2}(3^{20} - 1)$ (b) $2^{31} - 2^5$

(c) $-\frac{2}{3}\left(1 + \dfrac{1}{2^{101}}\right)$ **43.** 110

▶ Exercise Set 5.5 (Page 331)

1. (a) 46   (b) 58

3. (a) $\dfrac{\sqrt{2}\pi}{4} \approx 1.111$

(b) $\dfrac{(\sqrt{2} + 2)\pi}{4} \approx 2.682$

5. $\frac{15}{4}$ **7.** $\frac{1}{3}$ **9.** 320

11. $\frac{15}{4}$ **13.** $\frac{1}{3}$

15. (b) $\frac{1}{4}(b^4 - a^4)$ **17.** $\frac{1}{2}(b^2 - a^2)$

▶ Exercise Set 5.6 (Page 340)

1. (a) $\frac{71}{6}$ (b) 2

3. (a) $-\frac{117}{16}$ (b) 3

7. $\displaystyle\int_{-3}^{3} 4x(1 - 3x)\, dx$

9. $\displaystyle\lim_{\max \Delta x_k \to 0} \sum_{k=1}^{n} 2x_k^* \,\Delta x_k;\ a = 1,\ b = 2$

11. $\displaystyle\lim_{\max \Delta x_k \to 0} \sum_{k=1}^{n} \frac{x_k^*}{x_k^* + 1}\, \Delta x_k;\ a = 0,\ b = 1$

13. (a) 0.8   (b) -2.6 (c) -1.8 (d) -0.3

15. (a) -1 (b) $\frac{5}{2}$ (c) $\frac{21}{2}$ (d) $\frac{1}{2}k^2 - k$

17. 13/2   **19.** $\pi/2$ **21.** $25\pi/2$

▶ Exercise Set 5.7 (Page 347)

1. $\frac{65}{4}$ **3.** $\frac{81}{10}$

5. $\frac{22}{3}$ **7.** $\frac{2}{3}$

9. $-\frac{1}{3}$ **11.** $\frac{52}{3}$

13. $\frac{844}{5}$ **15.** $-\frac{55}{3}$

17. 0   **19.** $\sqrt{2}$

21. $\dfrac{\pi^2}{9} + 2\sqrt{3}$ **23.** 12

25. $\frac{9}{2}$ **27.** $\frac{203}{2}$ **29.** (b), (c)

▶ Exercise Set 5.8 (Page 352)

1. (a) $\displaystyle\int_{1}^{3} u^7\, du$ (b) $-\dfrac{1}{2}\displaystyle\int_{7}^{4} u^{1/2}\, du$

(c) $\dfrac{1}{\pi}\displaystyle\int_{-\pi}^{\pi} \sin u\, du$ (d) $\displaystyle\int_{0}^{1} u^2\, du$

(e) $\dfrac{1}{2}\displaystyle\int_{3}^{4} (u - 3)u^{1/2}\, du$

(f) $\displaystyle\int_{2}^{0} (u + 5)u^{20}\, du$

3. $\frac{121}{5}$ **5.** 10

7. $\frac{1192}{15}$ **9.** $8 - 4\sqrt{2}$

11. $-\frac{1}{48}$ **13.** $\frac{2}{3}$

15. $\frac{2}{3}(\sqrt{10} - 2\sqrt{2})$ **17.** $2(\sqrt{7} - \sqrt{3})$

19. 0   **21.** 0

23. -4 **25.** $-\frac{1}{9}$

27. $\frac{1}{3}(\sqrt{3} - 1)$ **29.** $\frac{106}{405}$

31. $3\sqrt{2}/4$ **33.** 2π **39.** (b) $\frac{3}{2}$ (c) $\dfrac{\pi}{4}$

35. $\pi/8$ **37.** 3

▶ Exercise Set 5.9 (Page 362)

1. $\frac{5}{2}$ **3.** $2 - \frac{1}{2}\sqrt{2}$ **5.** $-\frac{11}{6}$ **21.** (a) $x^3 + 1$ **23.** $\sin\sqrt{x}$

9. (a) $\frac{4}{3}$ (b) $\dfrac{2}{\sqrt{3}}$ (c)

11. $f_{\text{ave}} = 2;\ x^* = 4$

13. $f_{\text{ave}} = \frac{1}{2}\alpha(x_0 + x_1) + \beta;\ x^* = \frac{1}{2}(x_0 + x_1)$ if $\alpha \neq 0$; x^* is any point in $[x_0, x_1]$ if $\alpha = 0$.

17. 1404π lb   **19.** (b) no

25. $|x|$ **27.** $\displaystyle\int_2^x \frac{1}{t-1}\,dt$ **29.** $\displaystyle\int_0^x \frac{1}{t-1}\,dt$

31. (a) $(0, +\infty)$ (b) $x = 1$

33. (a) 0   (b) $\frac{1}{3}$ (c) 0

35. $F(x) = \begin{cases} \frac{1}{2}(1 - x^2), & x < 0 \\ \frac{1}{2}(1 + x^2), & x \geq 0 \end{cases}$

37. $F(x) = \begin{cases} \frac{1}{3}(x^3 + 1), & x \leq 0 \\ x^2 + \frac{1}{3}, & x > 0 \end{cases}$ **39.** $3/x$

43. (a) $3x^2 \sin^2(x^3) - 2x \sin^2(x^2)$ (b) $\dfrac{2}{1 - x^2}$

▶ Chapter 5 Supplementary Exercises (Page 365)

1. $-\frac{1}{2}x^{-2} + 2\sqrt{x} + 5\cos x + C$

3. $\frac{2}{9}(\sqrt{x} + 2)^9 + C$

5. $-\frac{1}{2}\cos\sqrt{2x^2 - 5} + C$

7. $2x^{3/2} + \frac{6}{17}x^{17/6} + C$

9. $\frac{1}{5}\tan(\sin 5t) + C$

11. (a) $\frac{1}{6}y^6 + y^4 + 2y^2 + C$

    (b) $\frac{1}{6}(y^2 + 2)^3 + C$

13. $\displaystyle\int_0^1 u^4\,du = 1/5$ **15.** $\displaystyle\int_1^4 \frac{u - 1}{\sqrt{u}}\,du = \frac{8}{3}$

17. $\displaystyle\int_{\pi/2}^\pi \frac{4}{\pi}\cos u\,du = -\frac{4}{\pi}$

19. $17/2$ **21.** $\frac{1}{3}$

23. (a) 20   (b) 8   (c) $4n$

    (d) $8/15$ (e) $61/24$ (f) 9

    (g) $1 + \sqrt{2}$ (h) $3(\sqrt{2} + 1)/4$

25. (a) $\displaystyle\sum_{k=1}^{9} (-1)^{k-1}\left(\frac{k}{k + 1}\right)^2 = \sum_{k=2}^{10} (-1)^k\left(\frac{k - 1}{k}\right)^2$

    (b) $\displaystyle\sum_{k=1}^{11} \frac{(-\pi)^{k+1}}{k} = \sum_{k=2}^{12} (-1)^k\frac{\pi^k}{k - 1}$

27. (a) $64\left[1 - \dfrac{(n + 1)(2n + 1)}{6n^2}\right]$

    (b) $64\left[1 - \dfrac{(n - 1)(2n - 1)}{6n^2}\right]$

    (c) $128/3$

29. (a) 12   (b) 12   (c) 12

31. (a) -4 (b) $9\pi/4$

    (c) -2 (d) 0

    (e) 0   (f) 30

    (g) 2

33. (a) 10   (b) 1

    (c) -4 (let $u = -x$)

    (d) 4

35. $f_{\text{ave}} = 7;\ x^* = -\sqrt{7/3}$

37. $f_{\text{ave}} = 13/4;\ x^* = -5/4$

▶ Exercise Set 6.1 (Page 373)

1. $\frac{32}{3}$

3. $\frac{9}{4}$

17. 24

19. $\frac{1}{2}$

5. $\frac{49}{192}$

7. $\frac{1}{2}$

21. $4\sqrt{2}$

23. $\frac{11}{2}$

9. $\frac{32}{3}$

11. $\pi - 2$

25. $\frac{2}{3}$ **27.** (a) $2(\sqrt{b} - 1)$ (b) $+\infty$

13. $\frac{9}{2}$

15. $\frac{355}{6}$

29. $9/\sqrt[3]{4}$

▶ Exercise Set 6.2 (Page 381)

1. $\dfrac{32\pi}{5}$

3. $\dfrac{373\pi}{14}$

25. $\dfrac{256\pi}{3}$

27. $\dfrac{648\pi}{5}$

5. $\dfrac{1296\pi}{5}$

7. $\dfrac{2048\pi}{15}$

29. $\dfrac{\pi}{2}$

31. $\frac{4}{3}\pi ab^2$

9. $\dfrac{\pi}{2}$

11. $\dfrac{\pi}{6}$

33. π

35. $\frac{1}{3}\pi r^2 h$

37. $V = \frac{1}{6}\pi L^3$

13. $\dfrac{3\pi}{5}$

15. 8π

39. $V = \begin{cases} 3\pi h^2, & 0 \le h < 2 \\ \frac{1}{3}\pi(12h^2 - h^3 - 4), & 2 \le h \le 4 \end{cases}$

17. 2π

19. $\dfrac{28\pi}{3}$

41. $40{,}000\pi$ ft^3 **43.** $36\sqrt{3}$

45. $\frac{1}{4}(\pi + 2)$ **47.** $\frac{2}{3}r^3 \tan\theta$ **49.** $\frac{16}{3}r^3$

21. $\dfrac{58\pi}{5}$

23. $\dfrac{72\pi}{5}$

▶ Exercise Set 6.3 (Page 389)

1. $2\pi/5$

3. $3\pi\sqrt[3]{4}(1 + 3\sqrt[3]{3})$

13. (a) $7\pi/30$

15. $9\pi/14$

5. $20\pi/3$

7. 4π

17. $\frac{1}{3}\pi r^2 h$ **19.** $\dfrac{4\pi}{3}[r^3 - (r^2 - a^2)^{3/2}]$ **21.** $b = 1$

9. $\pi/2$

11. $\pi/5$

▶ Exercise Set 6.4 (Page 394)

1. $\sqrt{5}$

3. $\frac{1}{243}(85\sqrt{85} - 8)$

9. (a)

5. $\frac{1}{27}(80\sqrt{10} - 13\sqrt{13})$ **7.** $\frac{17}{6}$

(b) dy/dx does not exist at $x = 0$

(c) $\frac{1}{27}(13\sqrt{13} + 80\sqrt{10} - 16)$

▶ Exercise Set 6.5 (Page 398)

1. $35\sqrt{2}\,\pi$ **3.** 8π **9.** 24π **11.** $\dfrac{16{,}911\pi}{1024}$

5. $\dfrac{16\pi}{9}$ **7.** $40\pi\sqrt{82}$ **15.** $S = 2\pi rh$ **19.** (b) constant functions

---

▶ Exercise Set 6.6 (Page 404)

1. $s(t) = t^2 - 3t + 7$

3. $s(t) = \frac{1}{4}t^4 - \frac{2}{3}t^3 + t + 1$

5. $s(t) = 2t^2 + t$

7. $s(t) = -\cos 2t - t - 2$

9. (a) $s = \dfrac{2}{\pi},\ v = 1,\ |v| = 1,\ a = 0$

(b) $s = \frac{1}{2},\ v = -\frac{3}{2},\ |v| = \frac{3}{2},\ a = -3$

11. $-\frac{968}{45}$ ft/sec^2

13. (a) $v(3) = 16$ ft/sec, $v(5) = -48$ ft/sec

(b) 196 ft   (c) 112 ft/sec

15. (a) 1 sec   (b) $\frac{1}{2}$ sec

17. (a) $\frac{5}{8}(1 + \sqrt{33})$ sec   (b) $20\sqrt{33}$ ft/sec

19. (a) 5 sec   (b) 272.5 m

(c) 10 sec   (d) -49 m/sec

(e) 12.46 sec   (f) 73.1 m/sec

21. $80\sqrt{10}$ ft/sec   **23.** 256 ft

25. $\frac{2}{3}$; 3   **27.** 0; 2

29. $\frac{9}{4}$; $\frac{11}{4}$ **31.** $-\frac{10}{3}$; $\frac{17}{3}$ **33.** $\frac{204}{25}$; $\frac{204}{25}$

▶ Exercise Set 6.7 (Page 410)

1. (a) 210 in-lb   (b) $\frac{5}{8}$ in-lb   **11.** 75,000 ft-lb

3. $\frac{9}{5}$ ft-ton   **5.** 20 lb/ft   **7.** $900\pi\rho$ ft-lb   **13.** (a) 96×10^9 (b) $\,{,}800{,}000$ mile-lb

9. (a) 926,640 ft-lb   (b) 0.468 horsepower

▶ Exercise Set 6.8 (Page 415)

1. (a) 2808 lb   (b) 3600 lb

3. 998.4 lb   **5.** 6988.8 lb   **7.** 5200 lb   **9.** $\dfrac{\sqrt{2}}{2}\rho a^3$

11. $14{,}976\sqrt{17}$ lb   **13.** (b) $80\,$ lb/min

▶ Chapter 6 Supplementary Exercises (Page 416)

1. (a) $\displaystyle\int_0^2 (x + 2 - x^2)\, dx$

(b) $\displaystyle\int_0^2 \sqrt{y}\, dy + \int_2^4 [\sqrt{y} - (y - 2)]\, dy$

3. (a) $\displaystyle\int_0^9 2\sqrt{x}\, dx$ (b) $\displaystyle\int_{-3}^3 (9 - y^2)\, dy$

5. (a) $\displaystyle\int_0^2 2\pi x(x + 2 - x^2)\, dx$

(b) $\displaystyle\int_0^2 \pi y\, dy + \int_2^4 \pi[y - (y - 2)^2]\, dy$

7. (a) $\displaystyle\int_0^4 2\pi x[\frac{1}{2}x - (2 - \sqrt{4 - x})]\, dx$

(b) $\displaystyle\int_0^2 \pi[(4y - y^2)^2 - 4y^2]\, dy$

9. (a) $\displaystyle\int_0^9 4\pi x^{3/2}\, dx$

(b) $\displaystyle\int_{-3}^3 \pi(81 - y^4)\, dy$

11. (a) $16/3$ (b) 8π

13. $11/4$ **15.** $\pi a^2 b/24$

17. (a) $256\pi/15$ (b) $40\pi/3$

19. $61/27$ **21.** $779/240$

23. $\frac{1}{27}\pi(145^{3/2} - 10^{3/2})$ sq. units

25. $1017\pi/5$ sq. units   **27.** $28\pi\sqrt{3}/5$ sq. units

29. 3 in-lb   **31.** $10{,}600$ ft-lb

33. 6656π ft-lb   **35.** $28\rho/3$ lb

▶ Exercise Set 7.1 (Page 428)

1. (a) yes   (b) no   (c) yes   (d) no

3. yes   **5.** no

7. no   **9.** yes

11. yes   **13.** yes

15. $x^{1/5}$ **17.** $\frac{1}{7}(x + 6)$

19. $\sqrt[3]{\dfrac{x + 5}{3}}$ **21.** $\frac{1}{2}(x^3 + 1)$

23. $-\sqrt{\dfrac{3}{x}}$ **25.** $\begin{cases} 5/2 - x, & x > 1/2 \\ 1/x, & x \le 1/2 \end{cases}$

27. $\dfrac{1}{15y^2 + 1}$ **29.** $\dfrac{1}{2\sec^2 2y}$

31. $\dfrac{1}{10y^4 + 3y^2}$

33. (b)

(c) No; $f(g(x)) = x$ for $x > 1$, but the domain of g is $x > 0$.

35. $x^{1/4} - 2$ for $x \ge 16$ **37.** $\frac{1}{2}(3 - x^2)$ for $x \le 0$

39. $\frac{1}{10}(1 + \sqrt{1 - 20x})$ for $x \le -4$

41. (b) symmetric about the line $y = x$

43. (b) $1 - \dfrac{\sqrt{3}}{3}$ **45.** (b) $\dfrac{1}{\sqrt[3]{2}}$

47.

49. $\frac{88}{7}$

▶ Exercise Set 7.2 (Page 435)

1. (a) 4 (b) -5

(c) 1 (d) $\frac{1}{2}$

(e) -3 (f) 4

(g) 3 (h) $\frac{1}{2}$

3. 0.01 **5.** e^2

7. 4 **9.** 10^5

11. $\sqrt{\frac{3}{2}}$ **17.** (b) $10^{8.2}I_0$ (c) $10{,}000$

▶ Exercise Set 7.3 (Page 445)

1. (a) $r + s$ (b) $s - r$ (c) $-r$

(d) $2s$ (e) $\frac{1}{5}s$ (f) $-2(r + s)$

5.

7.

9. $\dfrac{1}{x}$

11. $\dfrac{2\ln x}{x}$

13. $\dfrac{\sec^2 x}{\tan x}$

15. $\dfrac{1-x^2}{x(1+x^2)}$

17. $\dfrac{3x^2-14x}{x^3-7x^2-3}$

19. $\dfrac{1}{2x\sqrt{\ln x}}$

21. $-\dfrac{5\cos(5/\ln x)}{x(\ln x)^2}$

23. $-\dfrac{2x^3}{3-2x}+3x^2\ln(3-2x)$

25. $4x\ln(x^2+1)+2x[\ln(x^2+1)]^2$

27. $\dfrac{x(1+2\ln x)}{(1+\ln x)^2}$

29. $-\dfrac{y}{x(y+1)}$

31. $\frac{1}{2}\ln|x|+C$

33. $\frac{1}{3}\ln|x^3-4|+C$

35. $\ln|\tan x|+C$ **37.** $-\frac{1}{3}\ln(1+\cos 3\theta)+C$

39. $\frac{1}{2}x^2-\frac{1}{2}\ln(x^2+1)+C$ **41.** $\frac{1}{4}(\ln y)^4+C$

43. $\frac{1}{3}\ln\frac{5}{2}$ **45.** $\frac{1}{2}\ln\frac{5}{6}$

47. (a) all $x\neq 0$ (b) $x>0$

49. $\dfrac{1}{2x}+\dfrac{1}{3(x+3)}+\dfrac{3}{5(3x-2)}$

51. $\frac{3}{2}(\ln x)^2+C$ **53.** $-\frac{1}{2}(\ln x)^2+C$

55. $x\sqrt[3]{1+x^2}\left[\dfrac{1}{x}+\dfrac{2x}{3(1+x^2)}\right]$

57. $\dfrac{(x^2-8)^{1/3}\sqrt{x^3+1}}{x^6-7x+5}$
$\left[\dfrac{2x}{3(x^2-8)}+\dfrac{3x^2}{2(x^3+1)}-\dfrac{6x^5-7}{x^6-7x+5}\right]$

67. $\frac{1}{2}(1+\ln 2)$ **69.** $\pi\ln 4$ **71.** $\ln 2$

► Exercise Set 7.4 (Page 455)

1. (a) $\dfrac{1}{x},\ x>0$

(b) $x^2,\ x\neq 0$
(c) $-x^2,\ -\infty<x<+\infty$
(d) $-x,\ -\infty<x<+\infty$
(e) $x^3,\ x>0$
(f) $x+\ln x,\ x>0$
(g) $x-\sqrt[3]{x},\ -\infty<x<+\infty$
(h) $e^x/x,\ x>0$

3. (a) $\sqrt{e}$ (b) $\dfrac{\ln 2}{4\pi}$ (c) 0 (d) $\ln 2$

5. (a) 3.3 (b) 15 (c) 30

7. $-10xe^{-5x^2}$ **9.** $x^2e^x(x+3)$

11. $\dfrac{4}{(e^x+e^{-x})^2}$ **13.** $(x\sec^2 x+\tan x)e^{x\tan x}$

15. $(1-3e^{3x})e^{(x-e^{3x})}$ **17.** $\dfrac{x-1}{e^x-x}$

19. $e^{ax}(a\cos bx-b\sin bx)$

21. $3x^2$ **23.** $-3^{-x}\ln 3$

25. $\pi^{x\tan x}(\ln\pi)(x\sec^2 x+\tan x)$

27. (a) Not of the form a^x, where a is a constant.
(b) $x^x(1+\ln x)$

29. $(x^3-2x)^{\ln x}\left[\dfrac{3x^2-2}{x^3-2x}\ln x+\dfrac{1}{x}\ln(x^3-2x)\right]$

31. $(\ln x)^{\tan x}\left[\dfrac{\tan x}{x\ln x}+(\sec^2 x)\ln(\ln x)\right]$

33. $x^{(e^x)}\left[\dfrac{e^x}{x}+e^x\ln x\right]$

37. (a) k^ne^{kx} (b) $(-1)^nk^ne^{-kx}$

39. $-\dfrac{1}{\sqrt{2\pi}\sigma^3}(x-\mu)\exp\left[-\dfrac{1}{2}\left(\dfrac{x-\mu}{\sigma}\right)^2\right]$

41. $-\frac{1}{5}e^{-5x}+C$ **43.** $e^{\sin x}+C$

45. $-\frac{1}{6}e^{-2x^3}+C$ **47.** $\ln(1+e^x)+C$

49. $\frac{1}{3}(1+e^{2t})^{3/2}+C$ **51.** $\exp(\sin x)+C$

53. $\tan(2-e^{-x})+C$ **55.** $\dfrac{\pi^{\sin x}}{\ln\pi}+C$

57. $\frac{1}{2}x^2\ln 3-4\pi e^2\sin x+C$ **59.** C

61. $2e^{\sqrt{y}}+C$ **63.** -36

65. $3+e-e^2$ **67.** $\ln(\frac{21}{13})$

71. $\dfrac{\ln 3}{\ln(2/3)}$ **73.** ex^{e-1}

75. $\dfrac{-qk_0}{2T^2}\exp\left[-(q/2)\left(\dfrac{T-T_0}{T_0T}\right)\right]$

77. $e^x-3\ln(e^x+3)+C$

79. $\dfrac{1}{\ln x}$ **81.** $\frac{1}{3}e^{1-x}$ **87.** $e^{-1/e}$

▶ Exercise Set 7.5 (Page 461)

1. (a) $+\infty$ (b) 0

3. (a) $+\infty$ (b) $+\infty$

5. (a) 1   (b) 1

7.

9.

11.

13. (a) yes   (b) no

(c)

15.

17. -1 **19.** 0   **21.** 1   **23.** 1   **25.** 1

27. (a) $\lim\limits_{x \to +\infty} xe^x = +\infty$, $\lim\limits_{x \to -\infty} xe^x = 0$

(b)

29. (a) $\lim\limits_{x \to +\infty} \dfrac{x^2}{e^{2x}} = 0$, $\lim\limits_{x \to -\infty} \dfrac{x^2}{e^{2x}} = +\infty$

(b)

31. (a)   (b)

(c)

33. (a)   (b)

(c)

35. $\dfrac{3-e}{2e}$ **37.** $\tfrac{1}{2}(9 \ln 3 - 4)$

39.

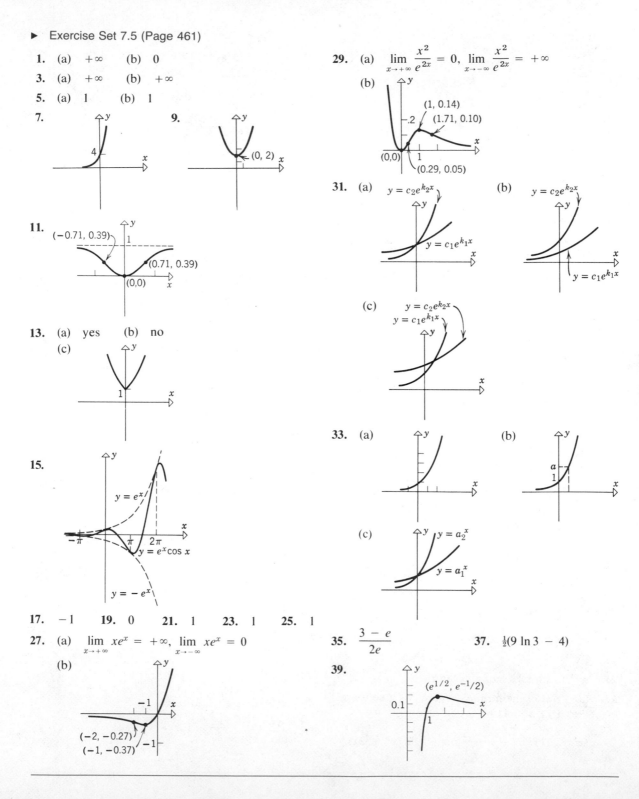

▶ Exercise Set 7.6 (Page 468)

1.

	$\sinh x_0$	$\cosh x_0$	$\tanh x_0$	$\coth x_0$	$\operatorname{sech} x_0$	$\operatorname{csch} x_0$
(a)	-2	$\sqrt{5}$	$-2/\sqrt{5}$	$-\sqrt{5}/2$	$1/\sqrt{5}$	$-1/2$
(b)	$-3/4$	$5/4$	$-3/5$	$-5/3$	$4/5$	$-4/3$
(c)	$-4/3$	$5/3$	$-4/5$	$-5/4$	$3/5$	$-3/4$
(d)	$1/\sqrt{3}$	$2/\sqrt{3}$	$1/2$	2	$\sqrt{3}/2$	$\sqrt{3}$
(e)	$8/15$	$17/15$	$8/17$	$17/8$	$15/17$	$15/8$
(f)	-1	$\sqrt{2}$	$-1/\sqrt{2}$	$-\sqrt{2}$	$1/\sqrt{2}$	-1

17. $4 \cosh(4x - 8)$ **19.** $-\dfrac{\operatorname{csch}^2(\ln x)}{x}$

21. $\dfrac{\operatorname{csch}\left(\dfrac{1}{x}\right) \coth\left(\dfrac{1}{x}\right)}{x^2}$

23. $\dfrac{2 + 5 \cosh(5x) \sinh(5x)}{\sqrt{4x + \cosh^2(5x)}}$

25. $x^{5/2} \tanh(\sqrt{x}) \operatorname{sech}^2(\sqrt{x}) + 3x^2 \tanh^2(\sqrt{x})$

27. $\frac{1}{7} \sinh^7 x + C$ **29.** $\frac{2}{3}(\tanh x)^{3/2} + C$

31. $\ln(\cosh x) + C$ **33.** $-\frac{1}{3} \operatorname{sech}^3 x + C$

35. $2 \sinh(\sqrt{x}) + C$

37. For $x > 0$:

(a) $\dfrac{x^2 + 1}{2x}$ (b) $\dfrac{x^2 - 1}{2x}$

(c) $\dfrac{x^4 - 1}{x^4 + 1}$ (d) $\dfrac{x^2 + 1}{2x}$

45. $\frac{3}{4}$ **47.** 5π

▶ Exercise Set 7.7 (Page 482)

1. $y = Cx$ **3.** $y = Ce^{-\sqrt{1+x^2}} - 1$

5. $y = \ln(\sec x + C)$ **7.** $y = e^{-2x} + Ce^{-3x}$

9. $y = e^{-x} \sin(e^x) + Ce^{-x}$

11. $y = -\frac{2}{7}x^4 + Cx^{-3}$ **13.** $y = -1 + 4e^{x^2/2}$

15. $y = 2 - e^{-t}$

17. $y = \sqrt[3]{3t - 3 \ln t + 24}$

19. $y = x + \dfrac{1}{x}$ **21.** $y^2 = -x^2 + 6x - 5$

23. (a) $y = 200 - 175e^{-t/25}$ **25.** 25 lb
   (b) 136 lb

27. (a) $y = 10e^{-0.005t}$ **29.** 196 days
   (b) 7 mg

31. 6.8 years   **33.** (a) 14,400
   (b) 38 years

▶ Chapter 7 Supplementary Exercises (Page 484)

1. (a) no   (b) yes   (c) no
   (d) yes   (e) yes

3. does not exist   **5.** $\frac{1}{2} \ln(x - 1)$

7. If $ad - bc \neq 0$, then
   $f^{-1}(x) = (-dx + b)/(cx - a)$

9. (a) $(-\infty, 5/2)$ (b) $(-2, +\infty)$
   (c) $(-\pi/3, 2\pi/3)$

11. $-\dfrac{3}{x^2}$ **13.** $\dfrac{2}{x}$

15. (a) $-(2r + s)$ (b) $2s - \frac{3}{2}r$
   (c) $\frac{1}{4}(3r - s)$

17. (a) $\ln 3/(2 \ln 5 + \ln 3)$
   (b) $\frac{1}{2}(\ln 5 - \ln 3)$

19. (a) $\sqrt{34}/5$ (b) $-3/\sqrt{34}$
  (c) $-6\sqrt{34}/25$

21. $-1/(2\sqrt{e^x})$ **23.** $(\ln x - 1)/(\ln x)^2$

25. 0   **27.** $\ln 10 - \cot x$

29. $x^4 e^{\tan x}\left(\sec^2 x + \dfrac{4}{x}\right)$ **31.** $(x^2 + a^2)^{-1/2}$

33. $6x \exp(3x^2)$ **35.** $1/(4x\sqrt{\ln \sqrt{x}})$

37. $x^{\pi-1}\pi^x(\pi + x \ln \pi)$

39. $5 \cosh(\tanh(5x)) \operatorname{sech}^2(5x)$

41. $3e^{3x} + 4e^{2x} + e^x$

45. (a) $dy = -e^{-x}\,dx$ (b) $dy = dx/(1 + x)$
  (c) $dy = 2x (\ln 2)\, 2^{x^2}\,dx$

47. (a) $\dfrac{1}{x\sqrt{4 + x}}$ (b) $5e^{5x}\sqrt{5x + e^{5x}}$

49. $Y = kt + b,\ b = \ln C$ **51.** $\ln(1 + e^x) + C$

53. $\dfrac{x^{e+1}}{e + 1} + C$ **55.** $4 \ln|x| + 3/x + C$

57. $\frac{1}{2} \ln|2 \sec x - 1| + C$ **59.** $\frac{1}{2}e^{\sin 2x} + C$

61. $\frac{1}{2}\tanh^2 x + C_1 = -\frac{1}{2}\operatorname{sech}^2 x + C_2$

63. $\ln 2$ **65.** $1/\ln 2$

69. inflection points at $0, 3 - \sqrt{3}$, and $3 + \sqrt{3}$; relative extremum at 3

73. $A = \frac{1}{2}(1 - e^{-2b}) = \frac{1}{4}$ when $b = \ln \sqrt{2}; A \to \frac{1}{2}$ as $b \to +\infty$

75. $\frac{1}{2}\pi(2 + \sinh 2)$

77. (a) about 352 million   (b) the year 2058

▶ Exercise Set 8.1 (Page 493)

1. (a) $-\dfrac{\pi}{2}$ (b) π (c) $-\dfrac{\pi}{4}$

  (d) $\dfrac{\pi}{4}$ (e) 0   (f) $\dfrac{\pi}{2}$

3. $\dfrac{1}{2}, -\sqrt{3}, -\dfrac{1}{\sqrt{3}}, 2, -\dfrac{2}{\sqrt{3}}$ **5.** $\frac{4}{5}, \frac{3}{5}, \frac{3}{4}, \frac{5}{3}, \frac{5}{4}$

7. (a) $\dfrac{\pi}{7}$ (b) 0   (c) $\dfrac{2\pi}{7}$ (d) $201\pi - 630$

9. (a) $0 \le x \le \pi$ (b) $-1 \le x \le 1$
  (c) $-\dfrac{\pi}{2} < x < \dfrac{\pi}{2}$ (d) $-\infty < x < +\infty$
  (e) $0 < x \le \dfrac{\pi}{2}, \quad -\pi < x \le -\dfrac{\pi}{2}$ (f) $|x| \ge 1$

11. $\frac{24}{25}$ **13.** $\dfrac{\pi}{2}$ **15.** $-4\sqrt{5}$

19. (a) $\dfrac{1}{\sqrt{1 + x^2}}$ (b) $\dfrac{1}{x}$
  (c) $\dfrac{\sqrt{x^2 - 1}}{x}$ (d) $\sqrt{x^2 - 1}$

21. (a)   (b)

▶ Exercise Set 8.2 (Page 499)

1. (a) $\dfrac{1}{\sqrt{9 - x^2}}$ (b) $-\dfrac{2}{\sqrt{1 - (2x + 1)^2}}$

3. (a) $\dfrac{7}{x\sqrt{x^{14} - 1}}$ (b) $-\dfrac{1}{\sqrt{e^{2x} - 1}}$

5. (a) $-\dfrac{1}{|x|\sqrt{x^2 - 1}}$ (b) $\begin{cases} 1, & \sin x > 0 \\ -1, & \sin x < 0 \end{cases}$

7. (a) $\dfrac{e^x}{x\sqrt{x^2 - 1}} + e^x \sec^{-1} x$

  (b) $\dfrac{3x^2(\sin^{-1} x)^2}{\sqrt{1 - x^2}} + 2x(\sin^{-1} x)^3$

9. (a) $-\dfrac{1}{x^2 + 1}$

  (b) $10(1 + x \csc^{-1} x)^9 \left(-\dfrac{1}{\sqrt{x^2 - 1}} + \csc^{-1} x\right)$

11. (a) $-\dfrac{1}{2\sqrt{1-x^2}}$

(b) $\dfrac{x+2x\ln x}{\sqrt{1-x^4\ln^2 x}}$

13. $\dfrac{y\sqrt{1-(x-y)^2}+\sqrt{1-x^2y^2}}{\sqrt{1-x^2y^2}-x\sqrt{1-(x-y)^2}}$

15. $\dfrac{\pi}{2}$ **17.** $-\dfrac{\pi}{12}$ **19.** $\tfrac14\tan^{-1}4x+C$

21. $\tan^{-1}(e^x)+C$ **23.** $\dfrac{\pi}{6}$

25. $\sin^{-1}(\tan x)+C$ **27.** $\sin^{-1}(\ln x)+C$

29. (a) $\sin^{-1}\left(\dfrac{x}{3}\right)+C$

(b) $\dfrac{1}{\sqrt5}\tan^{-1}\left(\dfrac{x}{\sqrt5}\right)+C$

(c) $\dfrac{1}{\sqrt\pi}\sec^{-1}\left(\dfrac{x}{\sqrt\pi}\right)+C$

31. $\dfrac{\pi}{18}$ **33.** $\dfrac{\pi^2}{4}$

35. $\dfrac{\pi}{2}-1$ **37.** $1+2\sqrt2$

39. $\dfrac{52\pi}{3}$ mi/min **41.** $2\sqrt6$ ft

▶ Exercise Set 8.3 (Page 505)

5. (a) $\ln(3+\sqrt8)$ (b) $\ln(\sqrt5-2)$

7. (a) $\dfrac{1}{\sqrt{9+x^2}}$ **9.** (a) $-\dfrac{7}{x\sqrt{1-x^{14}}}$

(b) $\dfrac{2}{\sqrt{(2x+1)^2-1}}$ (b) $-\dfrac{1}{\sqrt{1+e^{2x}}}$

11. (a) $-\dfrac{1}{|x|\sqrt{x^2+1}}$

(b) $\begin{cases}1, & x>0\\-1, & x<0\end{cases}$

13. (a) $-\dfrac{e^x}{x\sqrt{1-x^2}}+e^x\operatorname{sech}^{-1}x$

(b) $\dfrac{3x^2(\sinh^{-1}x)^2}{\sqrt{1+x^2}}+2x(\sinh^{-1}x)^3$

15. (a) $-\dfrac{1}{2x}$

(b) $10(1+x\operatorname{csch}^{-1}x)^9\left(-\dfrac{x}{|x|\sqrt{1+x^2}}+\operatorname{csch}^{-1}x\right)$

17. $\cosh^{-1}\left(\dfrac{x}{\sqrt2}\right)+C,\,x>\sqrt2$

19. $-\operatorname{sech}^{-1}(e^x)+C$ **21.** $-\tfrac13\operatorname{csch}^{-1}|x^3|+C$

23. -0.2028

25. (a)   (b)

▶ Chapter 8 Supplementary Exercises (Page 507)

1. (a) $2\pi/3$ (b) $3/4$
(c) $3/5$ (d) $3/5$

3. (a) $\pi/4$ (b) $-\pi/4$
(c) $2\sqrt6$ (d) $\pi/3$

5. (a) $33/65$ **7.** $5/4,\ -3/4,\ 34/16$
(b) $56/65$
(c) 7

9. (a)

(b) $y=f(x)=\pi/2$ (constant), $|x|\le1$

11. $-2/[x(\sec^{-1} x^2)^2\sqrt{x^4 - 1}]$

13. $(\sec x \tan x)/|\tan x|$

15. $\dfrac{1}{1 - (\ln x)^2} + \tanh^{-1}(\ln x)$

17. $3/(2\sqrt{(1 - 9x^2)} \sin^{-1} 3x)$

19. $\exp(\sec^{-1} x)/(x\sqrt{x^2 - 1})$

21. $(\pi^{\sin^{-1} x}) \ln \pi/\sqrt{1 - x^2}$

23. 1 (since $y = x$)   **25.** Show $y' = \cos^2 y$.

27. $\frac{1}{2} \ln|(1 + e^x)/(1 - e^x)| + C$

29. $\cosh^{-1}(\ln x) + C,\ x > e$

31. $\pi/18$ **33.** $2 \tan^{-1}\sqrt{x} + C$

35. $\pi/6 (u = \sqrt{x})$ **37.** $1/\sqrt{2}$

▶ Exercise Set 9.2 (Page 519)

1. $-xe^{-x} - e^{-x} + C$

3. $x \ln(2x + 3) - x + \frac{3}{2} \ln(2x + 3) + C$

5. $\frac{1}{4}x^2 \ln x - \frac{1}{8}x^2 + C$

7. $x \cos^{-1}(2x) - \frac{1}{2}\sqrt{1 - 4x^2} + C$

9. $x^2 e^x - 2xe^x + 2e^x + C$

11. $\dfrac{e^x}{2}(\sin x - \cos x) + C$

13. $\dfrac{1}{4\pi} e^{-2\pi x}(\sin 2\pi x - \cos 2\pi x) + C$

15. $\dfrac{e^{ax}}{a^2 + b^2}(a \sin bx - b \cos bx) + C$

17. $x^2 \sin x + 2x \cos x - 2 \sin x + C$

19. $-\frac{1}{3}x \cos(3x + 1) + \frac{1}{9} \sin(3x + 1) + C$

21. $x \tan x + \ln|\cos x| + C$

23. $\frac{1}{2}x[\cos(\ln x) + \sin(\ln x)] + C$

25. $\frac{1}{2}x[\sin(\ln x) - \cos(\ln x)] + C$

27. $x \tan x - \frac{1}{2}x^2 + \ln|\cos x| + C$

29. $\frac{1}{25}(1 - 6e^{-5})$ **31.** $\frac{1}{9}(2e^3 + 1)$

33. $5 \ln 5 - 4$ **35.** $\dfrac{5\pi}{6} - \sqrt{3} + 1$

37. $-\dfrac{\pi}{8}$ **39.** $\frac{1}{3}(2\sqrt{3}\pi - \frac{1}{2}\pi - 2 + \ln 2)$

41. $\frac{1}{3}(2 - \sqrt{2})$ **43.** (a) 1 (b) $\pi(e - 2)$

45. $2\pi^2$

47. (a) $-\frac{1}{3} \sin^2 x \cos x - \frac{2}{3} \cos x + C$
(b) $\frac{1}{32}(3\pi - 8)$

49. (a) $\frac{1}{15} \cos^2 5x \sin 5x + \frac{2}{15} \sin 5x + C$
(b) $\frac{1}{8} \cos^3(x^2) \sin(x^2) + \frac{3}{16} \cos(x^2) \sin(x^2)$
$+ \frac{3}{16}x^2 + C$

53. (b) $x^3 e^x - 3x^2 e^x + 6xe^x - 6e^x + C$

▶ Exercise Set 9.3 (Page 525)

1. $-\frac{1}{6} \cos^6 x + C$ **3.** $\dfrac{1}{2a} \sin^2 ax + C$

5. $\frac{1}{2}\theta - \frac{1}{20} \sin 10\theta + C$

7. $\dfrac{3}{8}x + \sin\left(\dfrac{x}{2}\right) + \dfrac{1}{8} \sin x + C$

9. $\sin\theta - \frac{2}{3} \sin^3\theta + \frac{1}{5} \sin^5\theta + C$

11. $\frac{1}{6} \sin^3 2t - \frac{1}{10} \sin^5 2t + C$

13. $-\frac{1}{5} \cos^5 x + \frac{1}{7} \cos^7 x + C$

15. $-\frac{1}{5} \cos^5\theta + \frac{2}{7} \cos^7\theta - \frac{1}{9} \cos^9\theta + C$

17. $\frac{1}{8}x - \frac{1}{32} \sin 4x + C$

19. $-\frac{1}{6} \cos 3x + \frac{1}{2} \cos x + C$

21. $-\dfrac{1}{3} \cos\left(\dfrac{3x}{2}\right) - \cos\left(\dfrac{x}{2}\right) + C$

23. $\dfrac{1}{7 \cos^7 x} + C$ **25.** $\dfrac{5\sqrt{2}}{12}$

27. 0 **29.** $\frac{1}{24}$ **33.** $\dfrac{\pi}{2}$

35. (a) $\frac{2}{3}$ (b) $\dfrac{3\pi}{16}$ (c) $\frac{8}{15}$ (d) $\dfrac{5\pi}{32}$

► Exercise Set 9.4 (Page 531)

1. $\frac{1}{3}\tan(3x+1)+C$

3. $\frac{1}{2}\ln|\cos(e^{-2x})|+C$

5. $\frac{1}{2}\ln|\sec 2x+\tan 2x|+C$

7. $\frac{1}{3}\tan^3 x+C$

9. $\frac{1}{16}\tan^4(4x)+\frac{1}{24}\tan^6(4x)+C$

11. $\frac{1}{7}\sec^7 x-\frac{1}{5}\sec^5 x+C$

13. $\frac{1}{4}\sec^3 x\tan x-\frac{5}{8}\sec x\tan x$
$+\frac{3}{8}\ln|\sec x+\tan x|+C$

15. $\frac{1}{6}\sec^3(2t)+C$

17. $\tan x+\frac{1}{3}\tan^3 x+C$

19. $\frac{1}{5\pi}\sec^4(\pi x)\tan(\pi x)+\frac{4}{15\pi}\sec^2(\pi x)\tan(\pi x)$
$+\frac{8}{15\pi}\tan(\pi x)+C$

21. $\frac{1}{3}\tan^3 x-\tan x+x+C$

23. $\frac{1}{6}\tan^3(x^2)+C$

25. $-\frac{1}{5}\csc^5 x+\frac{1}{3}\csc^3 x+C$

27. $-\frac{1}{2}\csc^2 x-\ln|\sin x|+C$

29. $\frac{2}{3}\tan^{3/2}x+\frac{2}{7}\tan^{7/2}x+C$ 31. $\frac{\sqrt3}{2}-\frac{\pi}{6}$

33. $-\frac{1}{2}+\ln 2$ 35. $\ln(\sqrt2+1)$

39. $-\frac{1}{\sqrt{a^2+b^2}}\ln|\csc(x+\theta)+\cot(x+\theta)|+C$,
where θ satisfies
$\cos\theta=\frac{a}{\sqrt{a^2+b^2}}$ and $\sin\theta=\frac{b}{\sqrt{a^2+b^2}}$

► Exercise Set 9.5 (Page 537)

1. $2\sin^{-1}\left(\frac{x}{2}\right)+\frac{1}{2}x\sqrt{4-x^2}+C$

3. $\frac{9}{2}\sin^{-1}\left(\frac{x}{3}\right)-\frac{1}{2}x\sqrt{9-x^2}+C$

5. $\frac{1}{16}\tan^{-1}\frac{x}{2}+\frac{x}{8(4+x^2)}+C$

7. $\sqrt{x^2-9}-3\sec^{-1}\frac{x}{3}+C$

9. $-2\sqrt{2-x^2}+\frac{1}{3}(2-x^2)^{3/2}+C$

11. $\frac{x}{3\sqrt{3+x^2}}+C$ 13. $\frac{\sqrt{4x^2-9}}{9x}+C$

15. $\frac{x}{\sqrt{1-x^2}}+C$ 17. $x-\tan^{-1}x+C$

19. $-\frac{\sqrt{9-4x^2}}{9x}+C$ 21. $-\frac{x}{\sqrt{9x^2-1}}+C$

23. $\frac{1}{2}\sin^{-1}(e^x)+\frac{1}{2}e^x\sqrt{1-e^{2x}}+C$

25. $\frac{2048}{15}$ 27. $\frac{1}{2}(\sqrt3-\sqrt2)$

29. $\frac{1}{243}(10\sqrt3+18)$ 31. $\frac{1}{2}\ln(x^2+4)+C$

33. $\sqrt5-\sqrt2+\ln\left(\frac{2+2\sqrt2}{1+\sqrt5}\right)$

35. $\frac{\pi}{32}[18\sqrt5-\ln(2+\sqrt5)]$

37. (a) $\sinh^{-1}\left(\frac{x}{3}\right)+C$

(b) $\ln\left(\frac{\sqrt{x^2+9}}{3}+\frac{x}{3}\right)+C$

► Exercise Set 9.6 (Page 540)

1. $\frac{1}{3}\tan^{-1}\left(\frac{x-2}{3}\right)+C$ 3. $\sin^{-1}\left(\frac{x-1}{3}\right)+C$ 7. $2\sin^{-1}\left(\frac{x+1}{2}\right)+\frac{1}{2}(x+1)\sqrt{3-2x-x^2}+C$

5. $\ln(\sqrt{x^2-6x+10}+x-3)+C$
or $\sinh^{-1}(x-3)+C$ 9. $\frac{1}{\sqrt{10}}\tan^{-1}\frac{\sqrt2(x+1)}{\sqrt5}+C$

11. $\ln(x^2 + 2x + 5) + \dfrac{3}{2}\tan^{-1}\left(\dfrac{x+1}{2}\right) + C$

15. $\dfrac{2\pi}{3} - \dfrac{\sqrt{3}}{2}$

13. $\sqrt{x^2 + 2x + 2} + 2\ln(\sqrt{x^2 + 2x + 2} + x + 1) + C$
or $\sqrt{x^2 + 2x + 2} + 2\sinh^{-1}(x + 1) + C$

▶ Exercise Set 9.7 (Page 550)

1. $\dfrac{1}{5}\ln\left|\dfrac{x-1}{x+4}\right| + C$

3. $-2\ln|x - 2| + 3\ln|x - 3| + C$

5. $\dfrac{5}{2}\ln|2x - 1| + 3\ln|x + 4| + C$

7. $-\dfrac{1}{6}\ln|x - 1| + \dfrac{1}{15}\ln|x + 2| + \dfrac{1}{10}\ln|x - 3| + C$

9. $\ln\left|\dfrac{x(x+3)^2}{x-3}\right| + C$

11. $\dfrac{1}{2}x^2 - 2x + 6\ln|x + 2| + C$

13. $3x + 12\ln|x - 2| - \dfrac{2}{x-2} + C$

15. $\dfrac{1}{2}x^2 + 3x - \ln|x - 1| + 8\ln|x - 2| + C$

17. $\dfrac{1}{3}x^3 + x + \ln\left|\dfrac{(x+1)(x-1)^2}{x}\right| + C$

19. $3\ln|x| - \ln|x - 1| - \dfrac{5}{x-1} + C$

21. $\ln\dfrac{(x-3)^2}{|x+1|} + \dfrac{1}{x-3} + C$

23. $\ln|x + 2| + \dfrac{4}{x+2} - \dfrac{2}{(x+2)^2} + C$

25. $-\dfrac{7}{34}\ln|4x - 1| + \dfrac{6}{17}\ln(x^2 + 1) + \dfrac{3}{17}\tan^{-1}x + C$

27. $\dfrac{1}{32}\ln\left|\dfrac{x-2}{x+2}\right| - \dfrac{1}{16}\tan^{-1}\dfrac{x}{2} + C$

29. $3\tan^{-1}x + \dfrac{1}{2}\ln(x^2 + 3) + C$

31. $\dfrac{1}{2}x^2 - 3x + \dfrac{1}{2}\ln(x^2 + 1) + C$

33. $\dfrac{1}{\sqrt{2}}\tan^{-1}\left(\dfrac{x+1}{\sqrt{2}}\right) + \dfrac{1}{x^2 + 2x + 3} + C$

35. $\dfrac{1}{6}\ln\left|\dfrac{\sin\theta - 1}{\sin\theta + 5}\right| + C$ **37.** $\ln\dfrac{e^x}{1 + e^x} + C$

39. (a) $\sqrt{2}, -\sqrt{2}$ **41.** $\pi(\dfrac{19}{5} - \dfrac{9}{4}\ln 5)$

43. $y = \dfrac{3 - 2Ce^x}{1 - Ce^x}$ **45.** $y = \dfrac{t-1}{Ct - t + 1}$

47. (a) $(x - 1)(x - 2)(x - 3)$
(b) $(x - 4)(x^2 + x + 5)$
(c) $(x - 2)(x - 3)(x^2 + 1)$

49. $\dfrac{1}{8}\ln|x - 1| - \dfrac{1}{5}\ln|x - 2|$
$+ \dfrac{1}{12}\ln|x - 3| - \dfrac{1}{120}\ln|x + 3| + C$

▶ Exercise Set 9.8 (Page 555)

1. $\dfrac{2}{5}(x - 2)^{5/2} + \dfrac{4}{3}(x - 2)^{3/2} + C$

3. $4 - \pi$ **5.** $4 - 6\ln\dfrac{5}{3}$

7. $\dfrac{2}{15}(x^3 + 1)^{5/2} - \dfrac{2}{9}(x^3 + 1)^{3/2} + C$

9. $2x^{1/2} - 3x^{1/3} + 6x^{1/6} - 6\ln(x^{1/6} + 1) + C$

11. $4\ln\dfrac{v^{1/4}}{|1 - v^{1/4}|} + C$

13. $2t^{1/2} + 3t^{1/3} + 6t^{1/6} + 6\ln|t^{1/6} - 1| + C$

15. $\dfrac{1}{3}(1 + x^2)^{3/2} - (1 + x^2)^{1/2} + C$

17. $-2\sqrt{x}\cos\sqrt{x} + 2\sin\sqrt{x} + C$

19. $\ln\dfrac{\sqrt{e^x + 1} - 1}{\sqrt{e^x + 1} + 1} + C$

21. $\ln\left|\tan\left(\dfrac{x}{2}\right) + 1\right| + C$ **23.** 1

25. $\dfrac{4}{\sqrt{3}}\tan^{-1}\left(\sqrt{3}\tan\dfrac{x}{2}\right) - x + C$

29. $\dfrac{2}{\sqrt{3}}\tan^{-1}\left(\dfrac{2\tanh(x/2) + 1}{\sqrt{3}}\right) + C$

31. $-\dfrac{\sqrt{3 - x^2}}{3x} + C$ **33.** $\dfrac{(x^2 - 5)^{3/2}}{15x^3} + C$

▶ Exercise Set 9.9 (Page 563)

1. (a) 0.7366   (b) 0.6532
   (c) 0.6949   (d) 0.6932

3. (a) 1.8961   (b) 1.8961
   (c) 1.8961   (d) 2.0046

5. (a) 1.1261   (b) 0.6352
   (c) 0.8806   (d) 0.8818

7. 3.1416   9. 0.6928

▶ Chapter 9 Supplementary Exercises (Page 564)

1. $\frac{1}{2}x \sin 2x + \frac{1}{4}\cos 2x + C$

3. $\frac{1}{3}\sec^3 x - \sec x + C$

5. $\frac{1}{9}\tan^3 3t + C$ 7. $x - \sin x + C$

9. $\frac{1}{6}x^3 + \frac{1}{4}(x^2 - \frac{1}{2})\sin 2x + \frac{1}{4}x \cos 2x + C$

11. $\frac{1}{4}\sec^4 x + C$

13. $-(\sin^3 2x)/48 - (\sin 4x)/64 + x/16 + C$

15. $\frac{1}{4}$ 17. $\frac{1}{2}$ 19. $\frac{1}{2}$

21. $x/\sqrt{1 + x^2} + C$

23. $\ln|\sec(e^x) + \tan(e^x)| + C$

25. $\sqrt{e^{2x} + 1} + C$

27. $\frac{1}{13}e^{3x}(3 \sin 2x - 2 \cos 2x) + C$

29. $\frac{5}{6}\pi - \sqrt{3}$ 31. $-(\ln x + 1)/x + C$

33. $\sqrt{x^2 - 9} + C$ 35. $3 \ln(3 + \sqrt{8}) - \sqrt{8}$

37. $\frac{1}{5}\sqrt{2x + 3}\,(x^2 - 2x + 6) + C$

39. $\frac{1}{2}\sin^{-1}\left(\frac{2t + 1}{2}\right) + C$

41. $-\sqrt{a^2 - x^2}/(a^2 x) + C$

43. $\frac{1}{2}[x\sqrt{a^2 - x^2} + a^2 \sin^{-1}(x/a)] + C$

45. $-\sqrt{4x - x^2} + C$

47. $\ln|(2x + 1)/(x + 1)| + C$

49. $-\dfrac{\ln|x|}{6} - \dfrac{2}{15}\ln|x + 3| + \dfrac{3}{10}\ln|x - 2| + C$

51. $\frac{1}{3}\ln|x^3 - 3x| + C$

53. $\dfrac{7}{4}\ln\left|\dfrac{x - 1}{x + 1}\right| - \dfrac{3}{2}\tan^{-1}x + C$

55. $\dfrac{1}{26}\left[\ln\dfrac{(x - 3)^2}{x^2 + 4} - 3 \tan^{-1}\left(\dfrac{x}{2}\right)\right] + C$

57. $\frac{7}{8}\tan^{-1}\left(\dfrac{x}{2}\right) - \dfrac{24 + 5x}{4(4 + x^2)} + C$

59. $\frac{1}{2}\ln(x^2 + 2x + 5) - \frac{1}{2}\tan^{-1}\frac{1}{2}(x + 1) + C$

61. $\dfrac{1}{\sqrt{2}}\sin^{-1}\sqrt{\frac{2}{3}}x + C$

63. $2(\sqrt{t} - \tan^{-1}\sqrt{t}) + C$

65. $-\frac{7}{4} + 3 \ln 2$

67. $\frac{1}{2}(x - \ln|\sin x - \cos x|) + C$

69. $\frac{1}{4}\cot^2(\frac{1}{2}x) + \frac{1}{2}\ln|\tan(\frac{1}{2}x)| + C$

73. (a) $\frac{1}{2}\tan^{-1}(1) = \pi/8$
    (b) $\pi(\pi + 2)/64$ (c) $\pi \ln 2$

75. (a) $e^{2x}(4x^3 - 6x^2 + 6x - 3)/8 + C$
    (b) $(\pi - 2)/125$
    (c) $\frac{1}{6}(\sin x \cos x)^3 + \frac{1}{16}x - \frac{1}{64}\sin(4x) + C$

77. (a) $\frac{1}{4}(\sec^4 \theta - 2 \sec^2 \theta) + C_1$
    (b) $\frac{1}{4}\tan^4 \theta + C_2,\ C_2 = C_1 - \frac{1}{4}$

79. (a) 58.9276   (b) 54.7328

81. (a) 1.8277   (b) 1.8279   83. 0.36972

85. $-\dfrac{1}{a}\ln(1 + e^{-ax}) + C$

87. $\sqrt{1 - x^2} + \ln(1 - \sqrt{1 - x^2}) + C$

89. $-\dfrac{1}{3}\left(\dfrac{x + 1}{x - 1}\right)^{3/2} + C$

91. $-\frac{1}{10}\ln|3 + 2x^{-5}| + C$

93. $\dfrac{\sqrt{2}}{3}[(x + 2)^{3/2} - (x - 2)^{3/2}] + C$

▶ Exercise Set 10.1 (Page 573)

1. 1

3. divergent

5. $\ln \frac{5}{3}$

7. $\frac{1}{2}$

9. $\dfrac{1}{2(a^2 + 1)}$

11. $-\frac{1}{4}$

13. $\frac{1}{3}$

15. divergent

17. 0

19. divergent

21. divergent

23. $\dfrac{\pi}{2}$

25. 1

27. divergent

29. $\frac{9}{2}$

31. $\dfrac{\pi}{2}$

33. (a) $\frac{1}{5}$ (b) $\dfrac{\pi}{2}$

35. (b) $\frac{1}{2}$

39. $+\infty$

41. (a) $\dfrac{1}{24}$ with $\dfrac{x}{x^5 + 1} \le \dfrac{1}{x^4}$

(b) e^{-1} with $e^{-x^2} \le e^{-x}$

43. (a) $V = \pi$ **45.** $\dfrac{2\pi NI}{kr}\left(1 - \dfrac{a}{\sqrt{r^2 + a^2}}\right)$

47. (b) 24,000,000

▶ Exercise Set 10.2 (Page 582)

1. 1

3. 1

5. 1

7. $\dfrac{1}{\pi}$

9. -1

11. 0

13. $\frac{1}{2}$

15. $\frac{3}{2}$

17. $-\dfrac{1}{2\pi}$

19. $\frac{1}{4}$

21. $-\frac{1}{12}$

23. $\frac{1}{6}$

25. $a - b$

27. 2

29. $+\infty$

31. $\frac{1}{9}$

33. (b) 2

35. $k = -1, l = \pm 2\sqrt{2}$

37. 2

39. does not exist

41. 3

▶ Exercise Set 10.3 (Page 590)

1. 0

3. $-\infty$

5. $+\infty$

7. 0

9. 0

11. π

13. 2

15. e^{-3}

17. e^2

19. 1

21. $+\infty$

23. $e^{2/\pi}$

25. 1

27. e^3

29. 1

31. e^2

33. $-\frac{1}{2}$

35. 0

37. 0

39. $+\infty$

41. 2

45. (a) 0   (b) $+\infty$
(c) 0   (d) $-\infty$
(e) $+\infty$ (f) $+\infty$
(g) $-\infty$ (h) $-\infty$

47.

49. -1

51. $\frac{1}{9}$

53. (b) $\frac{8}{5}\sqrt{2}$

55. does not exist

▶ Chapter 10 Supplementary Exercises (Page 591)

1. $\pi/2$

3. 1

9. diverges

11. 1

5. 6

7. 0

13. $\pi/4$

15. $n > -1; -1/(n + 1)^2$

17. $3/2$ **19.** $+\infty$ **29.** 0 **31.** 1

21. 0 **23.** $\frac{1}{6}$ **33.** (a) $+\infty$ (b) $3\pi/2$ **35.** 2

25. $-\dfrac{1}{4\pi^2}$ **27.** 0

► Exercise Set 11.1 (Page 601)

1. $\frac{1}{3}, \frac{2}{4}, \frac{3}{5}, \frac{4}{6}, \frac{5}{7}$; converges to 1

3. $2, 2, 2, 2, 2$; converges to 2

5. $\dfrac{\ln 1}{1}, \dfrac{\ln 2}{2}, \dfrac{\ln 3}{3}, \dfrac{\ln 4}{4}, \dfrac{\ln 5}{5}$; converges to 0

7. $0, 2, 0, 2, 0$; diverges

9. $-1, \frac{16}{9}, -\frac{54}{28}, \frac{128}{65}, -\frac{250}{126}$; diverges

11. $\frac{6}{2}, \frac{12}{8}, \frac{20}{18}, \frac{30}{32}, \frac{42}{50}$; converges to $\frac{1}{2}$

13. $\cos 3, \cos \frac{3}{2}, \cos 1, \cos \frac{3}{4}, \cos \frac{3}{5}$; converges to 1

15. $e^{-1}, 4e^{-2}, 9e^{-3}, 16e^{-4}, 25e^{-5}$; converges to 0

17. $2, (\frac{5}{3})^2, (\frac{6}{4})^3, (\frac{7}{5})^4, (\frac{8}{6})^5$; converges to e^2

19. $\left\{\dfrac{2n-1}{2n}\right\}_{n=1}^{+\infty}$; converges to 1

21. $\left\{\dfrac{1}{3^n}\right\}_{n=1}^{+\infty}$; converges to 0

23. $\left\{\dfrac{1}{n} - \dfrac{1}{n+1}\right\}_{n=1}^{+\infty}$; converges to 0

25. $\{\sqrt{n+1} - \sqrt{n+2}\}_{n=1}^{+\infty}$; converges to 0

27. (a) 3, 6, 12, 24, 48, 96, 192, 384
    (b) 1, 1, 2, 3, 5, 8, 13, 21

29. (a) $1, \frac{3}{4}, \frac{2}{3}, \frac{5}{8}$ (b) $\frac{1}{2}$

31. (a) 3   (b) 11   (c) 1001   **37.** 3

► Exercise Set 11.2 (Page 610)

1. decreasing   **3.** increasing   **25.** converges   **27.** diverges

5. decreasing   **7.** increasing   **29.** converges

9. decreasing   **11.** nonincreasing   **31.** (a) 0   (b) $+\infty$ (does not exist)

13. not monotone   **15.** decreasing   **37.** (b) converges (decreasing and bounded below

17. increasing   **19.** increasing   by 0)

21. decreasing   **23.** decreasing

► Exercise Set 11.3 (Page 618)

1. (a) converges to $\frac{5}{2}$ (b) converges to $\frac{1}{2}$ **19.** $\frac{532}{99}$ **21.** $\frac{869}{1111}$
    (c) diverges

3. $\frac{4}{7}$ **5.** 6   **23.** diverges   **31.** $\dfrac{1}{x^2-1}, \; |x| > 2$

7. diverges   **9.** $\frac{1}{3}$

11. $\frac{1}{6}$ **13.** $\frac{448}{3}$ **33.** $\dfrac{2\sin x}{2+\sin x}, \; -\infty < x < +\infty$ **35.** 1

15. $-\frac{1}{3}$ **17.** $\frac{4}{9}$

▶ Exercise Set 11.4 (Page 628)

1. $\frac{4}{3}$

3. $-\frac{1}{36}$

5. (a) converges (b) diverges
   (c) diverges   (d) diverges
   (e) converges (f) diverges
   (g) converges (h) converges

9. diverges   11. diverges

13. converges   15. diverges

17. diverges   19. diverges

21. converges   23. diverges

25. converges   27. converges

29. diverges

35. (a) diverges (b) diverges
    (c) diverges   (d) converges

37. (b) 0.005   (c) 14

▶ Exercise Set 11.5 (Page 635)

1. converges   3. inconclusive

5. diverges   7. diverges

9. converges   11. diverges

13. converges   15. converges

17. converges   19. converges

21. diverges   23. converges

25. converges   27. diverges

29. converges   31. converges

33. converges

▶ Exercise Set 11.6 (Page 644)

13. converges   15. converges

17. diverges   19. converges

21. diverges   23. converges

25. diverges   27. converges

29. diverges   31. converges

33. converges   37. converges

39. $p > 1$ 41. converges

▶ Exercise Set 11.7 (Page 655)

1. converges   3. diverges

5. converges   7. converges absolutely

9. diverges   11. converges absolutely

13. conditionally   15. divergent

17. conditionally   19. absolutely

21. conditionally   23. divergent

25. conditionally   27. absolutely

29. conditionally   31. 0.125

33. 0.1   35. 10,000   37. 40,000

▶ Exercise Set 11.8 (Page 662)

1. $1, [-1, 1)$

3. $+\infty, (-\infty, +\infty)$

5. $\frac{1}{5}, [-\frac{1}{5}, \frac{1}{5}]$

7. $1, [-1, 1]$

9. $1, (-1, 1]$

11. $+\infty, (-\infty, +\infty)$

13. $+\infty, (-\infty, +\infty)$

15. $1, [-1, 1]$

17. $1, (-2, 0]$

19. $\frac{4}{3}, (-\frac{19}{3}, -\frac{11}{3})$

21. $1, [-2, 0]$

23. $+\infty, (-\infty, +\infty)$

25. $(-\infty, +\infty)$ 27. $(a - b, a + b)$ 29. $+\infty$

► Exercise Set 11.9 (Page 672)

1. $1 - 2x + 2x^2 - \frac{4}{3}x^3 + \frac{2}{3}x^4$

3. $2x - \frac{4}{3}x^3$

5. $x + \frac{1}{3}x^3$

7. $x + x^2 + \frac{x^3}{2!} + \frac{x^4}{3!}$

9. $1 + \frac{1}{2}x^2 + \frac{5}{24}x^4$

11. $\ln 3 + \frac{2}{3}x - \frac{2}{9}x^2 + \frac{8}{81}x^3 - \frac{4}{81}x^4$

13. $e + e(x - 1) + \frac{e}{2!}(x - 1)^2 + \frac{e}{3!}(x - 1)^3$

15. $2 + \frac{1}{4}(x - 4) - \frac{1}{64}(x - 4)^2 + \frac{1}{512}(x - 4)^3$

17. $\frac{\sqrt{2}}{2} - \frac{\sqrt{2}}{2}\left(x - \frac{\pi}{4}\right) - \frac{\sqrt{2}}{4}\left(x - \frac{\pi}{4}\right)^2 + \frac{\sqrt{2}}{12}\left(x - \frac{\pi}{4}\right)^3$

19. $-\frac{\sqrt{3}}{2} + \frac{\pi}{2}\left(x + \frac{1}{3}\right) + \frac{\sqrt{3}\pi^2}{4}\left(x + \frac{1}{3}\right)^2 - \frac{\pi^3}{12}\left(x + \frac{1}{3}\right)^3$

21. $\frac{\pi}{4} + \frac{1}{2}(x - 1) - \frac{1}{4}(x - 1)^2 + \frac{1}{12}(x - 1)^3$

23. $\sum_{k=0}^{\infty} (-1)^k \frac{x^k}{k!}$

25. $\sum_{k=0}^{\infty} (-1)^k x^k$

27. $\sum_{k=1}^{\infty} (-1)^{k+1} \frac{x^k}{k}$

29. $\sum_{k=0}^{\infty} (-1)^k \frac{x^{2k}}{4^k(2k)!}$

31. $\sum_{k=0}^{\infty} \frac{x^{2k}}{(2k)!}$

33. $\sum_{k=0}^{\infty} (-1)(x + 1)^k$

35. $\sum_{k=1}^{\infty} (-1)^{k+1} \frac{(x - 1)^k}{k}$

37. $\sum_{k=0}^{\infty} (-1)^k \frac{\pi^{2k}}{(2k)!}\left(x - \frac{1}{2}\right)^{2k}$

39. $\sum_{k=0}^{\infty} \left(\frac{16 + (-1)^{k+1}}{8}\right) \frac{(x - \ln 4)^k}{k!}$

► Exercise Set 11.10 (Page 683)

1. $\frac{2^6 e^{2c}}{6!} x^6$

3. $-\frac{x^5}{(c + 1)^6}$

5. $\frac{(4 + c)e^c}{4!} x^4$

7. $-\frac{(1 - 3c^2)}{3(1 + c^2)^3} x^3$

9. $-\frac{5}{128c^{7/2}}(x - 4)^4$

11. $\frac{\cos c}{5!}\left(x - \frac{\pi}{6}\right)^5$

13. $\frac{7}{(1 + c)^8}(x + 2)^6$

15. $\frac{x^{n+1}}{(1 - c)^{n+2}}$

17. $\frac{2^{n+1} e^{2c}}{(n + 1)!} x^{n+1}$

27. $\sum_{k=0}^{\infty} (-1)^k \frac{2^k x^k}{k!}$, $(-\infty, +\infty)$

29. $\sum_{k=0}^{\infty} (-1)^k x^k$ $(-1, 1)$

31. $\sum_{k=0}^{\infty} (-1)^k 3^k x^{k+2}$, $(-\frac{1}{3}, \frac{1}{3})$

33. $1 + \sum_{k=1}^{\infty} (-1)^k \frac{2^{2k-1}}{(2k)!} x^{2k}$, $(-\infty, +\infty)$

35. $\sum_{k=0}^{\infty} (-1)^k \frac{2^{2k+1}}{(2k + 1)!} x^{2k+1}$, $(-\infty, +\infty)$

37. $\sum_{k=0}^{\infty} \frac{(-1)^k}{(2k)!} x^{4k}$, $(-\infty, +\infty)$

39. $\sum_{k=0}^{\infty} (-1)^k (x - 1)^k$, $(0, 2)$

41. $\sin \pi = 0$

43. $e^{-\ln 3} = \frac{1}{3}$

45. (a) $x^2 - 2x^4 + \frac{2}{3}x^6 - \frac{4}{45}x^8 + \cdots$
  (b) 0

► Exercise Set 11.11 (Page 692)

1. (a) 9   (b) 13

3. 1.6487

5. 0.9877

7. 0.5299

9. 0.223

11. 0.100

13. 1.0050

15. $|x| < 0.569$ **17.** $-0.144 < x < 0.137$

19. (a) 2,000,000   (b) 8   **21.** 3.140

▶ Exercise Set 11.12 (Page 700)

5. $\displaystyle\sum_{k=1}^{\infty} (-1)^{k+1} k x^{k-1}$

9. $\ln\frac{4}{3}$ **11.** $\dfrac{2}{(1-x)^3}$

13. (b) $\displaystyle\sum_{k=0}^{\infty} (-1)^k x^k$; $(-1, 1)$

15. $1 - \frac{1}{3}x - \frac{1}{9}x^2 - \frac{5}{81}x^3 + \cdots$; 1

17. $x + \frac{2}{3}x^2 + \frac{8}{9}x^3 + \frac{112}{81}x^4 + \cdots$; $\frac{1}{2}$

19. $1 - 2x^2 - 2x^4 - 4x^6 - \cdots$; $\frac{1}{2}$

21. 0.764 **23.** 0.494

25. 0.100 **27.** 0.491

29. $1 - \frac{3}{2}x^2 + \frac{25}{24}x^4 - \frac{331}{720}x^6 + \cdots$

31. $x - x^2 + \frac{1}{3}x^3 - \frac{1}{30}x^5 + \cdots$

33. $-x^3 - \frac{1}{2}x^5 - \frac{1}{3}x^7 - \frac{1}{4}x^9 - \cdots$

35. $x^2 + \frac{9}{2}x^3 + \frac{79}{8}x^4 + \frac{683}{48}x^5 + \cdots$

37. (a) 0 (b) $-\frac{3}{2}$

39. (a) $x - \frac{1}{6}x^3 + \frac{3}{40}x^5 - \frac{5}{112}x^7$

  (b) $x + \displaystyle\sum_{k=1}^{\infty} (-1)^k \frac{1\cdot 3\cdot 5\cdots(2k-1)}{2^k k!(2k+1)} x^{2k+1}$

  (c) 1

43. $\displaystyle\sum_{k=0}^{\infty} \binom{m}{k} x^k$

45. $\displaystyle\sum_{k=0}^{\infty} (-1)^k \binom{1/3}{k} x^k$; $R = 1$

47. $\displaystyle\sum_{k=0}^{\infty} \binom{2/3}{k} x^{2k}$; $R = 1$

49. $\displaystyle\sum_{k=0}^{\infty} \binom{-1/2}{k} x^{k+1}$; $R = 1$

▶ Chapter 11 Supplementary Exercises (Page 702)

1. $L = 0$ **3.** $L = 0$

5. does not exist

7. (a) $2, 3, 6$ (b) 4 (c) $1, 5$

9. (a) $|q| > \sqrt{\pi}$ (b) $q > 1/3$
  (c) none $(p = 1)$ (d) $q > e$ or $0 < q < 1/e$

11. (a) $1 + 36\displaystyle\sum_{k=1}^{\infty} (.01)^k$ (b) $15/11$

13. (a) $1/4$ (b) diverges   (c) 1

15. converges   **17.** converges   **19.** converges

21. diverges (general term does not approach zero)

23. converges absolutely

25. diverges   **27.** $R = 1; 0 \le x \le 2$

29. $R = 2; -1 < x < 3$ **31.** $R = 0; x = 1$

33. (a) $(x - 2) - (x - 2)^2/2 + (x - 2)^3/3$

  (b) $\dfrac{-1}{4(c-1)^4}(x-2)^4$, c between 2 and x

  (c) $\dfrac{(\frac{1}{2})^4}{4(\frac{1}{2})^4} = \frac{1}{4}$

35. (a) $1 + (x-1)/2 - (x-1)^2/8$

  (b) $\dfrac{(x-1)^3}{16c^{5/2}}$, c between 1 and x

  (c) $\dfrac{(5/9)^3}{16(2/3)^5} < 0.0814$

37. $\ln a + \displaystyle\sum_{k=0}^{\infty} (-1)^k \frac{(x/a)^{k+1}}{k+1}$; $R = a$

39. $\dfrac{1}{3}\left\{ 1 + \displaystyle\sum_{k=1}^{\infty} (-1)^k \frac{1\cdot 3\cdots(2k-1)}{2\cdot 4\cdots(2k)} (x/9)^k \right\}$;
  $R = 9$

41. $1 + x^2/2 + 5x^4/24 + \cdots$

43. $1 - x^2/4 - x^4/96 + \cdots$

45. $x - \frac{1}{2}x^2 + \frac{1}{6}x^3 + \cdots$

47. $-2 - 2x - \frac{8}{3}x^3 - \cdots$; -2

49. $|x| < 1.17$ **51.** 0.444

► Exercise Set 12.2 (Page 713)

1.

3.

5.

7.

9.

11.

13.

15.

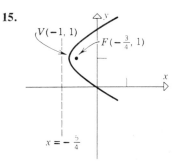

17. $y^2 = 12x$ **19.** $y^2 = -28x$

21. $y^2 = 2x$ **23.** $x^2 = -12y$

25. $y^2 = -5(x - \frac{19}{5})$ **27.** $y^2 = 6(x - \frac{3}{2})$

29. $(x - 1)^2 = 12(y - 1)$

31. $(x - 5)^2 = 2(y + 3)$

33. (a) $y = \frac{7}{6}x^2 - \frac{23}{6}x + 3$
   (b) $x = -\frac{7}{6}y^2 + \frac{17}{6}y + 2$

35. vertex: $\left(-\dfrac{B}{2A}, \dfrac{4AC - B^2}{4A}\right)$

focus: $\left(-\dfrac{B}{2A}, \dfrac{4AC - B^2 + 1}{4A}\right)$

directrix: $y = \dfrac{4AC - B^2 - 1}{4A}$

37. 16 ft   **41.** $\frac{1}{16}$ ft

▶ Exercise Set 12.3 (Page 720)

1.

3.

5.

7.

9.

11.

13.

29.
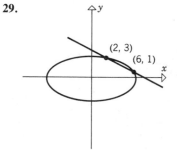

15. $\dfrac{x^2}{9} + \dfrac{y^2}{4} = 1$

17. $\dfrac{x^2}{169} + \dfrac{y^2}{144} = 1$

19. $\dfrac{x^2}{3} + \dfrac{y^2}{2} = 1$

21. $\dfrac{x^2}{16} + \dfrac{y^2}{4} = 1, \dfrac{x^2}{4} + \dfrac{y^2}{16} = 1$

23. $\dfrac{x^2}{36} + \dfrac{y^2}{81/8} = 1$

25. $(x - 1)^2 + \dfrac{(y - 3)^2}{2} = 1$

27. $\dfrac{(x - 4)^2}{32} + \dfrac{(y - 3)^2}{36} = 1$

31.

33. $k = -4$ at $(-2, -1)$
$k = 4$ at $(2, 1)$

35. πab

37. $\dfrac{4a^2 b^2}{a^2 + b^2}$

39. center: $(0, 0)$, major axis: 10, minor axis: 6

41. major axis: $r_1 + r_2$, center at the midpoint of the line segment joining the centers of C_1 and C_2

43. (b) $e \to 0$: ellipse $\to$ circle of radius a; $e \to 1$: ellipse $\to$ major axis

---

▶ Exercise Set 12.4 (Page 730)

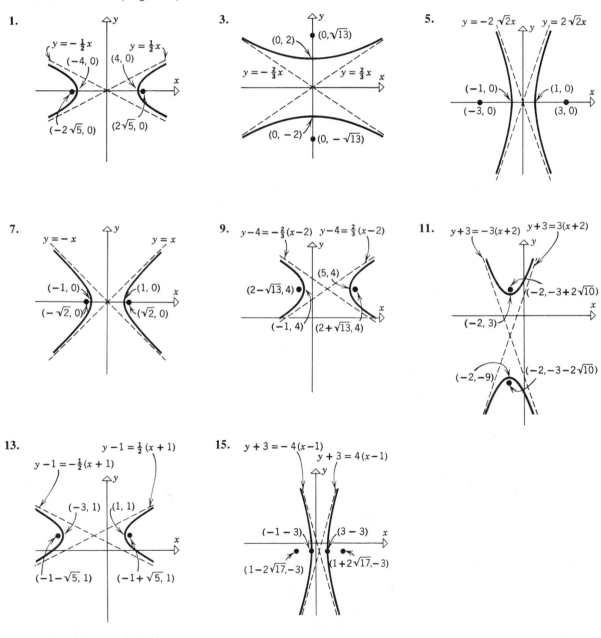

1.

3.

5.

7.

9.

11.

13.

15.

17. $\dfrac{x^2}{4} - \dfrac{y^2}{5} = 1$ **19.** $x^2 - \dfrac{y^2}{4} = 1$

21. $\dfrac{x^2}{64/9} - \dfrac{y^2}{16} = 1,\ \dfrac{y^2}{36} - \dfrac{x^2}{16} = 1$

23. $\dfrac{y^2}{56} - \dfrac{x^2}{56} = 1$ **25.** $\dfrac{x^2}{4} - \dfrac{y^2}{4/3} = 1$

27. $\dfrac{(x-2)^2}{4} - \dfrac{(y+3)^2}{5} = 1$

29. $\dfrac{(x-6)^2}{16} - \dfrac{(y-4)^2}{9} = 1$

31. $8xy - 4x - 4y + 1 = 0$

33.

35.

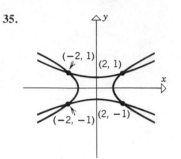

39. $(\tfrac{3}{2}\sqrt{13}, -9),\ (-\tfrac{3}{2}\sqrt{13}, -9)$

41. $(\pm 3/\sqrt{5}, 4/\sqrt{5}),\ (\pm 3/\sqrt{5}, -4/\sqrt{5})$

43. (b) $e \to 1$: hyperbola $\to$ focal axis (excluding the segment between vertices);
$e \to +\infty$: hyperbola $\to$ 2 lines perpendicular to focal axis at vertices

---

▶ **Exercise Set 12.5 (Page 740)**

1. (a) $(-1 + 3\sqrt{3}, \sqrt{3} + 3)$
  (b) $\dfrac{x'^2}{4} - \dfrac{y'^2}{12} = 1$
  (c)

3. $\theta = 45°;\ \dfrac{y'^2}{18} - \dfrac{x'^2}{18} = 1,$
  hyperbola

5. $\theta = \tan^{-1}\tfrac{1}{2};$
  $\dfrac{x'^2}{3} - \dfrac{y'^2}{2} = 1$, hyperbola

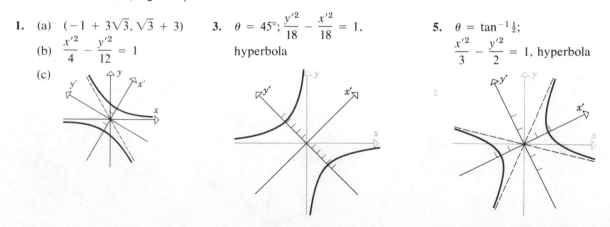

7. $\theta = 60°$; $y' = x'^2$, parabola

9. $\theta = \tan^{-1}\frac{3}{4}$;
$y'^2 = 4(x' - 1)$, parabola

11. $\theta = \tan^{-1}\frac{3}{4}$;
$\dfrac{(x' + 1)^2}{4} + y'^2 = 1$, ellipse

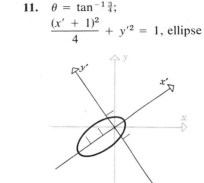

15. $x^2 + xy + y^2 = 3$

23. ellipse, point, or no graph

25. parabola, line, pair of parallel lines, or no graph

27. ellipse, point, or no graph

▶ Chapter 12 Supplementary Exercises (Page 741)

1. parabola: $V(-2, 3)$, $F(-5, 3)$, directrix $x = 1$

3. ellipse: $C(-2, 1)$, $F(-2, 1 \pm \sqrt{5})$, axis lengths 6 and 4

5. hyperbola: $C(2, 1)$, $F(2 \pm \sqrt{10}, 1)$, $V(2 \pm 3, 1)$, asymptotes $y - 1 = \pm(x - 2)/3$

7. parabola: $V(3, 1)$, $F(21/8, 1)$, directrix $x = 27/8$

9. $(y - 3)^2 = 16(x - 1)$ **11.** $y^2/9 - x^2/16 = 1$

13. $x^2/25 + y^2/16 = 1$ **15.** $(x - 3)^2 = 4(y - 3)$

21. $\pi/4$; ellipse; $(x'/\sqrt{2})^2 + (y')^2 = 1$

23. $\pi/6$; hyperbola; $4x'^2 - y'^2 = 1$

25. $\tan^{-1}(4/3)$; parabola; $y'^2 = 4(x' - 1)$

17.

19.

27.

29.

▶ Exercise Set 13.1 (Page 748)

1. $(5, \frac{2\pi}{3})$, $(3, \frac{\pi}{4})$, $(1, \frac{\pi}{2})$, $(0, \pi)$, $(4, \frac{7\pi}{6})$, $(2, \frac{4\pi}{3})$

3. (a) $(3\sqrt{3}, 3)$
(b) $\left(-\dfrac{7}{2}, \dfrac{7\sqrt{3}}{2}\right)$
(c) $(4\sqrt{2}, 4\sqrt{2})$
(d) $(5, 0)$
(e) $\left(-\dfrac{7\sqrt{3}}{2}, \dfrac{7}{2}\right)$
(f) $(0, 0)$

5. (a) $(5, \pi)$ (b) $\left(4, \dfrac{11\pi}{6}\right)$

(c) $\left(2, \dfrac{3\pi}{2}\right)$ (d) $\left(8\sqrt{2}, \dfrac{5\pi}{4}\right)$

(e) $\left(6, \dfrac{2\pi}{3}\right)$ (f) $\left(\sqrt{2}, \dfrac{\pi}{4}\right)$

7. (a) $(-5, 0)$ (b) $\left(-4, \dfrac{5\pi}{6}\right)$

(c) $\left(-2, \dfrac{\pi}{2}\right)$ (d) $\left(-8\sqrt{2}, \dfrac{\pi}{4}\right)$

(e) $\left(-6, \dfrac{5\pi}{3}\right)$ (f) $\left(-\sqrt{2}, \dfrac{5\pi}{4}\right)$

9. $y = 4$; line

11. $3x^2 + 4y^2 - 12x = 36$; ellipse

13. $x^2 + y^2 + 4x = 0$; circle

15. $3x + 2y = 6$; line

17. $r \cos \theta = 7$ **19.** $r^2 = 16 \cos 2\theta$

21. $r = 9 \sec \theta \tan \theta$ **23.** $r = 6 \sin \theta$

25.

27.

▶ Exercise Set 13.2 (Page 758)

1.
Line

3.
Circle

5.
Circle

7.
Circle

9.
Cardioid

11.
Cardioid

13.
Cardioid

15.
Limaçon

17.
Limaçon

19.
Limaçon

21.
Limaçon

23.
Limaçon

25.
Lemniscate

27.
Lemniscate

29.
Spiral

31.
Four-petal rose

33.

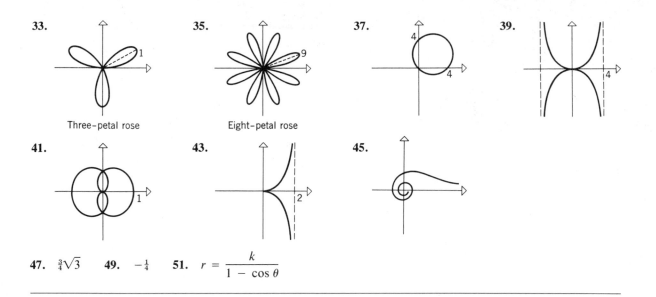

Three-petal rose

35.

Eight-petal rose

37.

39.

41.

43.

45.

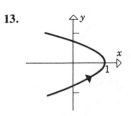

47. $\frac{3}{4}\sqrt{3}$ **49.** $-\frac{1}{4}$ **51.** $r = \dfrac{k}{1 - \cos\theta}$

▶ Exercise Set 13.3 (Page 765)

1. $\dfrac{7\pi^3}{1296}$ **3.** 6π **13.** $\dfrac{5\pi}{4}$ **15.** $100\cos^{-1}\left(\frac{3}{5}\right) - 48$

5. $\dfrac{8\pi}{3} + \sqrt{3}$ **7.** $\dfrac{9\sqrt{3}}{2} - \pi$ **17.** $\frac{4}{3}a^2$ **19.** $\frac{9}{8}(1 - e^{-4\pi})$

9. 1   **11.** 4π **21.** $\ln 6$ **23.** $2\sqrt{3} - \frac{2}{3}\pi$

▶ Exercise Set 13.4 (Page 777)

1.   **3.**   **9.** (3, 6) (5, 2)   **11.**

5.   **7.**   **13.**   **15.**

17.

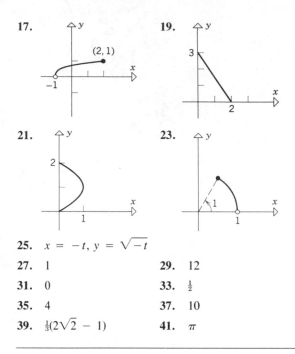

19.

21.

23.

25. $x = -t, y = \sqrt{-t}$

27. 1   **29.** 12

31. 0   **33.** $\frac{1}{2}$

35. 4   **37.** 10

39. $\frac{1}{3}(2\sqrt{2} - 1)$ **41.** π

43. $\frac{1}{27}(80\sqrt{10} - 13\sqrt{13})$ **45.** $8a$

47. $x = (2 + 3\sin\theta)\cos\theta$
$y = (2 + 3\sin\theta)\sin\theta$

49. $y = -e^{-4}x + 2e^{-2}$

51. (a) $-\frac{1}{2}$ (b) 1, 4

53. $y = 5x - 14$, $y = 6x - 17$

55. If $|a| = |b|$, a circle with center at (h, k) and radius $|a|$; if $|a| \neq |b|$, an ellipse with center at (h, k) and vertices $(h \pm |a|, k)$ if $|a| > |b|$ and vertices $(h, k \pm |b|)$ if $|a| < |b|$

57. (a)

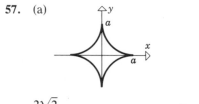

59. $\dfrac{2\sqrt{2}}{5}\pi(2e^\pi + 1)$ **61.** $\dfrac{\pi}{24}(17\sqrt{17} - 1)$

▶ Exercise Set 13.5 (Page 785)

1. $\dfrac{1}{\sqrt{3}}$

3. $\dfrac{\tan 2 - 2}{2\tan 2 + 1}$

5. -2

7. 1

9. 0

11. $-\dfrac{1}{2\sqrt{3}}$

13. $\dfrac{\sqrt{10}}{3}(e^6 - 1)$

15. $2\pi a$

17. $\dfrac{a}{3}[(\pi^2 + 4)^{3/2} - 8]$ **19.** $8a$

21. $\theta = \dfrac{\pi}{2}, \dfrac{3\pi}{2}, \sin^{-1}\frac{1}{4}, \pi - \sin^{-1}\frac{1}{4}$

23. $\tan^{-1} 3\sqrt{3}$

▶ Chapter 13 Supplementary Exercises (Page 786)

1. (a) $(1, \sqrt{3})$ (b) $(0, -2)$
(c) $(0, 0)$ (d) $(-1, 1)$
(e) $(-3, 0)$ (f) $(3/5, -4/5)$

3. (a)   (b)

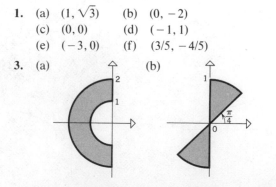

5. $2xy = 1$ (hyperbola) **7.** $y = -4$ (line)

9. $y = \sqrt{3}x$ (line)   **11.** $x = y = 0$ (point)

13. $r\cos\theta = -3$ **15.** $\tan\theta = 3$

17.   **19.**

21.   **23.**

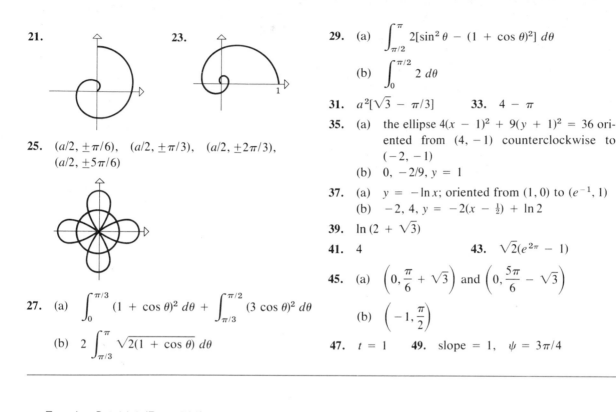

25. $(a/2, \pm\pi/6)$, $(a/2, \pm\pi/3)$, $(a/2, \pm2\pi/3)$,
     $(a/2, \pm5\pi/6)$

27.   (a) $\displaystyle\int_0^{\pi/3} (1 + \cos\theta)^2 \, d\theta + \int_{\pi/3}^{\pi/2} (3\cos\theta)^2 \, d\theta$

     (b) $\displaystyle 2\int_{\pi/3}^{\pi} \sqrt{2(1 + \cos\theta)} \, d\theta$

29.   (a) $\displaystyle\int_{\pi/2}^{\pi} 2[\sin^2\theta - (1 + \cos\theta)^2] \, d\theta$

     (b) $\displaystyle\int_0^{\pi/2} 2 \, d\theta$

31. $a^2[\sqrt{3} - \pi/3]$ **33.** $4 - \pi$

35.   (a)   the ellipse $4(x - 1)^2 + 9(y + 1)^2 = 36$ oriented from $(4, -1)$ counterclockwise to $(-2, -1)$

     (b) $0, -2/9, y = 1$

37.   (a) $y = -\ln x$; oriented from $(1, 0)$ to $(e^{-1}, 1)$

     (b) $-2, 4, y = -2(x - \frac{1}{2}) + \ln 2$

39. $\ln(2 + \sqrt{3})$

41. 4 **43.** $\sqrt{2}(e^{2\pi} - 1)$

45.   (a) $\left(0, \dfrac{\pi}{6} + \sqrt{3}\right)$ and $\left(0, \dfrac{5\pi}{6} - \sqrt{3}\right)$

     (b) $\left(-1, \dfrac{\pi}{2}\right)$

47. $t = 1$ **49.**   slope $= 1$, $\psi = 3\pi/4$

► Exercise Set 14.1 (Page 796)

1.   (a) $\sqrt{14}$; $(1, \frac{1}{2}, \frac{3}{2})$ (b) $\sqrt{11}$; $(\frac{9}{2}, \frac{3}{2}, \frac{9}{2})$
     (c) $\sqrt{30}$; $(\frac{1}{2}, -\frac{1}{2}, 4)$ (d) $\sqrt{42}$; $(\frac{3}{2}, 1, -\frac{5}{2})$

3. $(-6, 2, 1), (-6, 2, -2), (-6, 1, -2), (4, 2, 1),$
     $(4, 1, 1), (4, 1, -2), (4, 2, -2), (-6, 1, 1)$

5.   (b) $(2, 1, 6)$ (c) 49

7.   distance to x-axis is $\sqrt{y_0^2 + z_0^2}$
     distance to y-axis is $\sqrt{x_0^2 + z_0^2}$

9. $(x + 2)^2 + (y - 4)^2 + (z + 1)^2 = 36$

11. $x^2 + (y - 1)^2 + z^2 = 9$

13.   (a) $(x - 2)^2 + (y + 1)^2 + (z + 3)^2 = 9$
     (b) $(x - 2)^2 + (y + 1)^2 + (z + 3)^2 = 1$
     (c) $(x - 2)^2 + (y + 1)^2 + (z + 3)^2 = 4$

15. $(x + \frac{1}{2})^2 + (y - 2)^2 + (z - 2)^2 = \frac{5}{4}$

17. $(x - 3)^2 + (y + 2)^2 + (z - 4)^2 = 41$

19. $x^2 + y^2 + z^2 = 30 \pm 2\sqrt{29}$

21.   sphere, center $(-5, -2, -1)$, radius 7

23.   sphere, center $(\frac{1}{2}, \frac{3}{4}, -\frac{5}{4})$, radius $\frac{3}{4}\sqrt{6}$

25.   no graph

27.   largest $3 + \sqrt{6}$, smallest $3 - \sqrt{6}$

29.   all points outside the circular cylinder $(y + 3)^2 + (z - 2)^2 = 16$

31. $(2 - \sqrt{3})R$

33.   (a)   (b)   (c)

35. (a)

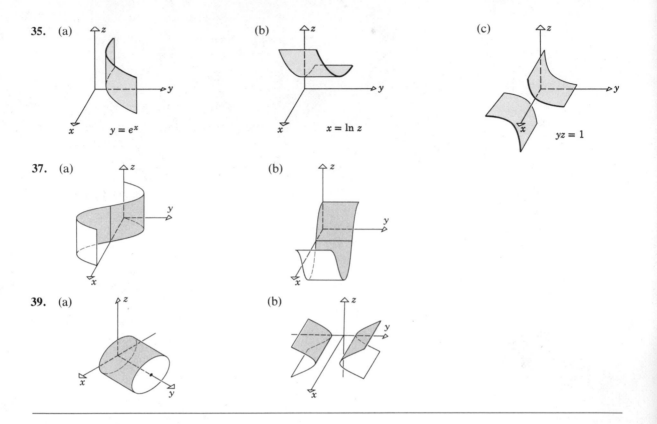

$y = e^x$

(b)

$x = \ln z$

(c)

$yz = 1$

37. (a)

(b)

39. (a)

(b)

▶ Exercise Set 14.2 (Page 808)

1.

$\langle 2, 5 \rangle$

$\langle 2, 0 \rangle$

$\langle -5, -4 \rangle$

3.

$\langle 1, -2, 2 \rangle$

$\langle 2, 2, -1 \rangle$

5.

$-5i + 3j$

$3i - 2j$

$-6j$

7.

$-i + 2j + 3k$

$2i + 3j - k$

9. (a) $\langle -1,3 \rangle$ (b) $\langle -7,2 \rangle$
(c) $\langle 2,1 \rangle$ (d) $\langle -8,7 \rangle$

11. (a) $\langle -3,6,1 \rangle$ (b) $\langle 1,-3,-5 \rangle$

13. $(4,-4)$ **15.** $(4,-4)$

17. $(8,-1,-3)$

19. (a) $-5\mathbf{i}-2\mathbf{j}$ (b) $8\mathbf{i}+10\mathbf{j}$
(c) $-2\mathbf{i}+4\mathbf{j}$ (d) $40\mathbf{i}+36\mathbf{j}$
(e) $-20\mathbf{i}+22\mathbf{j}$ (f) $-8\mathbf{i}-8\mathbf{j}$

21. (a) $-\mathbf{i}+4\mathbf{j}-2\mathbf{k}$ (b) $18\mathbf{i}+12\mathbf{j}-6\mathbf{k}$
(c) $-\mathbf{i}-5\mathbf{j}-2\mathbf{k}$ (d) $40\mathbf{i}-4\mathbf{j}-4\mathbf{k}$
(e) $-2\mathbf{i}-16\mathbf{j}-18\mathbf{k}$ (f) $-\mathbf{i}+13\mathbf{j}-2\mathbf{k}$

23. (a) $\sqrt{2}$ (b) 2 (c) 3

25. (a) $\sqrt{14}$ (b) 3

27. (a) $\sqrt{41}$ (b) 7
(c) $3\sqrt{29}+8$ (d) $3\sqrt{10}$
(e) $\langle \frac{3}{5},\frac{4}{5} \rangle$ (f) 1

29. (a) $2\sqrt{3}$ (b) $\sqrt{14}+\sqrt{2}$
(c) $2\sqrt{14}+2\sqrt{2}$ (d) $2\sqrt{37}$
(e) $\dfrac{1}{\sqrt{6}}\mathbf{i}+\dfrac{1}{\sqrt{6}}\mathbf{j}-\dfrac{2}{\sqrt{6}}\mathbf{k}$ (f) 1

31. $\langle -\frac{2}{3},1 \rangle$ **33.** $\mathbf{u}=\frac{5}{7}\mathbf{i}+\frac{2}{7}\mathbf{j}+\frac{1}{7}\mathbf{k},$
$\mathbf{v}=\frac{5}{7}\mathbf{i}-\frac{1}{7}\mathbf{j}-\frac{4}{7}\mathbf{k}$

43. $-\frac{3}{5}\mathbf{i}+\frac{4}{5}\mathbf{j}$ **45.** $-\dfrac{1}{\sqrt{14}}(3\mathbf{i}-2\mathbf{j}+\mathbf{k})$

47. $\dfrac{1}{3\sqrt{2}}(4\mathbf{i}+\mathbf{j}-\mathbf{k})$

49. $\langle -\frac{3}{2},2 \rangle$ **51.** $6\mathbf{i}-8\mathbf{j}-2\mathbf{k}$

53. circle; radius 1, center at (x_0,y_0)

55. (a) sphere; radius 2, center at $(0,0,0)$
(b) sphere; radius 3, center at (x_0,y_0,z_0)
(c) all points on and inside the sphere of radius 1, center at (x_0,y_0,z_0)

57. (a) $\left\langle \dfrac{1}{\sqrt{2}},-\dfrac{1}{\sqrt{2}} \right\rangle$, $\left\langle -\dfrac{1}{\sqrt{2}},\dfrac{1}{\sqrt{2}} \right\rangle$
(b) $\left\langle \dfrac{1}{\sqrt{2}},\dfrac{1}{\sqrt{2}} \right\rangle$, $\left\langle -\dfrac{1}{\sqrt{2}},-\dfrac{1}{\sqrt{2}} \right\rangle$

59. $(\frac{13}{4},\frac{21}{4})$

61. (a) $\left\langle \dfrac{1}{2},\dfrac{\sqrt{3}}{2} \right\rangle$ (b) $\langle -2\sqrt{2},2\sqrt{2} \rangle$

63. $(1,1),(-1,-1)$ **77.** (a) $(\frac{3}{2},\frac{1}{2})$

▶ Exercise Set 14.3 (Page 820)

1. (a) -10 (b) -3
(c) 0 (d) -20

3. (a) obtuse   (b) acute
(c) obtuse   (d) orthogonal

5. (a) $\frac{14}{13}\mathbf{i}+\frac{21}{13}\mathbf{j}$ (b) $\langle 2,6 \rangle$
(c) $-\frac{11}{13}\mathbf{i}+\mathbf{j}+\frac{55}{13}\mathbf{k}$ (d) $\langle -\frac{32}{89},-\frac{12}{89},\frac{73}{89} \rangle$

9. (a) 6 (b) 36
(c) $24\sqrt{5}$ (d) $24\sqrt{5}$

11. $\dfrac{1}{5\sqrt{2}},\dfrac{4}{\sqrt{65}},\dfrac{9}{\sqrt{130}}$

13. The right angle is at vertex B.

15. (a) the line through the origin and perpendicular to $\mathbf{r}_0$
(b) the line through (x_0,y_0) and perpendicular to $\mathbf{r}_0$
(c) circle; center at $(\frac{1}{2}x_0,\frac{1}{2}y_0)$, radius $\frac{1}{2}\|\mathbf{r}_0\|$

17. (a) $-\dfrac{3}{4}$ (b) $\dfrac{1}{7}$
(c) $\dfrac{48\pm25\sqrt{3}}{11}$ (d) $\dfrac{4}{3}$

21. (a) $\dfrac{2}{5}$ (b) $\dfrac{2}{\sqrt{5}}$ (c) $\dfrac{2}{\sqrt{5}}$

23. (a) $\frac{4}{3}$ (b) $\frac{1}{3}\sqrt{137}$

25. -12 foot-pounds   **31.** $71°$

▶ Exercise Set 14.4 (Page 830)

1. $\langle 7, 10, 9 \rangle$ **3.** $\langle -4, -6, -3 \rangle$

5. (a) $\langle -20, -67, -9 \rangle$ (b) $\langle -78, 52, -26 \rangle$
   (c) $\langle 24, 0, -16 \rangle$ (d) $\langle -12, -22, -8 \rangle$
   (e) $\langle 0, -56, -392 \rangle$ (f) $\langle 0, 56, 392 \rangle$

9. $\pm \dfrac{1}{\sqrt{5}}(2\mathbf{j} + \mathbf{k})$ **11.** $2\mathbf{v} \times \mathbf{u}$

13. (a) $\dfrac{\sqrt{374}}{2}$ (b) $9\sqrt{13}$

15. (a) $\tfrac{1}{2}\sqrt{26}$ (b) $\tfrac{1}{3}\sqrt{26}$

17. (a) $2\sqrt{\dfrac{141}{29}}$ (b) $\tfrac{1}{3}\sqrt{137}$

19. ambiguous, needs parentheses

21. 80   **23.** 1

25. (a) 16   (b) 45

27. (a) 9   (b) $\sqrt{122}$
   (c) $\sin^{-1}(\tfrac{9}{14})$

37. (a) $\tfrac{2}{3}$ (b) $\tfrac{1}{2}$

▶ Exercise Set 14.5 (Page 838)

1. $x = 3 + 2t, \; y = -2 + 3t$

3. $x = -4t, \; y = 1 + 2t$

5. $x = 5 - 3t$ **7.** $x = -t$
   $y = -2 + 6t$ $y = 6t$
   $z = 1 + t$ $z = t$

9. same as Exercise 1 with $0 \le t \le 1$

11. same as Exercise 3 with $0 \le t \le 1$

13. same as Exercise 5 with $0 \le t \le 1$

15. same as Exercise 7 with $0 \le t \le 1$

17. $x = -5 + 2t, \; y = 2 - 3t$

19. $x = 3 + 4t, \; y = -4 + 3t$

21. $x = -1 + 3t$ **23.** $x = -2 + 2t$
   $y = 2 - 4t$ $y = -t$
   $z = 4 + t$ $z = 5 + 2t$

25. $x = 3 + t$ **27.** $(-1, 1), (3, 9)$
   $y = 7$
   $z = 0$

29. (a) $(-2, 10, 0)$ (b) $(-2, 0, -5)$
   (c) does not intersect yz-plane

31. $(0, 4, -2), (4, 0, 6)$ **33.** $x = x_1 + at$
                                          $y = y_1 + bt$
                                          $z = z_1 + ct$

35. $(1, -1, 2)$ **39.** (a) no   (b) yes

41. $(1, \tfrac{14}{3}, -\tfrac{5}{3})$

45. (a) $\langle x, y \rangle = \langle 2, -1 \rangle + t\langle -7, 4 \rangle$
   (b) $\langle x, y \rangle = \langle 0, 3 \rangle + t\langle 4, 0 \rangle$

47. the line segment joining the points $(1, 0)$ and $(-3, 6)$

49. $2\sqrt{5}$ **51.** $x = t, \; y = 2 + t, \; z = 1 - t$

▶ Exercise Set 14.6 (Page 846)

1. $x + 4y + 2z = 28$ **3.** $z = 0$

5. (a) $2y - z = 1$
   (b) $x + 9y - 5z = 16$

7. (a) yes   (b) no

9. (a) yes   (b) no

11. (a) $35°$ (b) $79°$

13. (a) $z = 0$ (b) $y = 0$ (c) $x = 0$

15. $4x - 2y + 7z = 0$

17. $4x - 13y + 21z = -14$

19. $x = 5 - 2t$
   $y = 5t$
   $z = -2 + 11t$

21. $x + y - 3z = 6$ **23.** $7x + y + 9z = 25$ **31.** $\frac{5}{3}$ **33.** $\frac{137}{21}$

25. $2x + 4y + 8z = 29$

29. (a) $x = -\frac{11}{7} - 23t$ (b) $x = -5t$ **35.** $\dfrac{5}{3\sqrt{6}}$ **37.** $\dfrac{25}{\sqrt{126}}$ **39.** $\dfrac{95}{\sqrt{1817}}$

$\quad\quad\quad y = -\frac{12}{7} + t$ $\quad\quad y = -3t$ **43.** $(x - 2)^2 + (y - 1)^2 + (z + 3)^2 = \frac{121}{14}$

$\quad\quad\quad z = -7t$ $\quad\quad\quad z = 0$

▶ Exercise Set 14.7 (Page 859)

1.

Ellipsoid

3.

Hyperboloid
of one sheet

5.

Elliptic cone

7.

Elliptic
paraboloid

9.

Hyperboloid
of one sheet

11.

Hyperbolic
paraboloid

13. (a)

Hyperboloid
of one sheet

(b)

Hyperboloid
of two sheets

(c)

Paraboloid

(d)

Elliptic cone

(e)

Hyperbolic
paraboloid

(f)

Paraboloid

15.

17.

(0, 1, 0)
(1, 0, 0)

19.

(−2, 3, −9)
Circular
paraboloid

21.

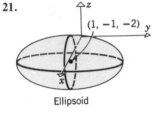

(1, −1, −2)

Ellipsoid

23. (0, −1, 5)

Circular
paraboloid

29. (a) focus: $\left(k, 0, \dfrac{k^2}{9} + 1\right)$, vertex: $\left(k, 0, \dfrac{k^2}{9}\right)$

(b) foci: $(\sqrt{5k}, 0, k)$, $(-\sqrt{5k}, 0, k)$
endpoints of major axis: $(3\sqrt{k}, 0, k)$,
$(-3\sqrt{k}, 0, k)$
endpoints of minor axis: $(0, 2\sqrt{k}, k)$,
$(0, -2\sqrt{k}, k)$

31. $z = \frac{1}{4}(x^2 + y^2)$; paraboloid

▶ Exercise Set 14.8 (Page 865)

1. (a) $(8, \pi/6, -4)$ (b) $(5\sqrt{2}, 3\pi/4, 6)$
(c) $(2, \pi/2, 0)$ (d) $(8, 5\pi/3, 6)$
(e) $(2, 7\pi/4, 1)$ (f) $(0, 0, 1)$

3. (a) $(2\sqrt{2}, \pi/3, 3\pi/4)$ (b) $(2, 7\pi/4, \pi/4)$
(c) $(6, \pi/2, \pi/3)$ (d) $(10, 5\pi/6, \pi/2)$
(e) $(8\sqrt{2}, \pi/4, \pi/6)$
(f) $(2\sqrt{2}, 5\pi/3, 3\pi/4)$

5. (a) $(2\sqrt{3}, \pi/6, \pi/6)$ (b) $(\sqrt{2}, \pi/4, 3\pi/4)$
(c) $(2, 3\pi/4, \pi/2)$ (d) $(4\sqrt{3}, 1, 2\pi/3)$
(e) $(4\sqrt{2}, 5\pi/6, \pi/4)$ (f) $(2\sqrt{2}, 0, 3\pi/4)$

11.

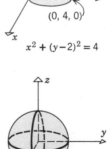

(0, 4, 0)

$x^2 + (y-2)^2 = 4$

13.

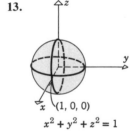

(1, 0, 0)

$x^2 + y^2 + z^2 = 1$

7.

(3, 0, 0)

$x^2 + y^2 = 9$

9.

$z = x^2 + y^2$

15.

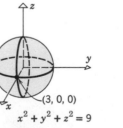

(3, 0, 0)

$x^2 + y^2 + z^2 = 9$

17.

$z = \sqrt{x^2 + y^2}$

19.

(0, 0, 2)

$x^2 + y^2 + (z - 2)^2 = 4$

21.

(1, 0, 0)

$(x - 1)^2 + y^2 = 1$

23. (a) $r^2 + z^2 = 9$ (b) $\rho = 3$

25. (a) $r(2\cos\theta + 3\sin\theta) + 4z = 1$
(b) $\rho(2\sin\phi\cos\theta + 3\sin\phi\sin\theta + 4\cos\phi) = 1$

27. (a) $r^2\cos^2\theta = 16 - z^2$
(b) $\rho^2(1 - \sin^2\phi\sin^2\theta) = 16$

29. spherical: $(4000, \pi/6, \pi/6)$
rectangular: $(1000\sqrt{3}, 1000, 2000\sqrt{3})$

31.

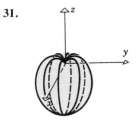

▶ Chapter 14 Supplementary Exercises (Page 866)

1. (a) $3\mathbf{i} - 4\mathbf{j}$, 5   (b) $-\mathbf{i} + 4\mathbf{j}$, $\sqrt{17}$

3. $-3\mathbf{i} + 4\mathbf{j}$ **5.** $\frac{3}{5}(4\mathbf{i} + 3\mathbf{j})$

7. $-6\mathbf{i} + 6\sqrt{3}\mathbf{j}$ **9.** $\mathbf{i} + \frac{1}{2}\mathbf{j}$

11. $\mathbf{F}_3 = -(\mathbf{F}_1 + \mathbf{F}_2) = \mathbf{i} + 5\mathbf{j}$

13. (a) $\sqrt{6}$ (b) -3
(c) $\langle 5, -5, -5\rangle$ (d) $\langle -5, 5, 5\rangle$
(e) $\frac{5}{2}\sqrt{3}$ (f) $\langle -1, 8, -9\rangle$

15. (a) $2/3$ (b) $2/5$
(c) $\cos^{-1}(-2/15)$ (d) $3/5, -4/5, 0$

17. Each side reduces to $2\mathbf{i} - 2\mathbf{j} + \mathbf{k}$.

19. $\langle -3/\sqrt{2}, 0, 3/\sqrt{2}\rangle$

21. (a) $-\frac{1}{2}\mathbf{v}$,   (b) $\frac{1}{2}(3\mathbf{i} + 5\mathbf{j} - 4\mathbf{k})$

23. $62°$ **25.** (a) $\|\mathbf{u}\| = \|\mathbf{v}\|$

27. $\pm(5\mathbf{i} + 7\mathbf{j} - \mathbf{k})/(5\sqrt{3})$

29. $x - y - z + 4 = 0$

31. $x - y - z = -1$

33. (a) $k = -2, l = 12$ (b) yes, at $(7, 0, 0)$
(c) $(136/13, -15/13, -60/13)$

35. (a) $x = 2t + 1, y = 3t - 1, z = -3t + 2$
(b) $x = 1, y = 5t - 3, z = -7t + 4$

37. (a) $x = t, y = -t + 2, z = t - 1$
(b) $60°$

39. (a) the region inside the elliptic paraboloid
$z = 4x^2 + 9y^2$
(b) the point $(0, 0, 0)$

41. elliptic cone   **43.** ellipsoid

45. hyperboloid of two sheets   **47.** 13 ft-lb

49. $\rho = \sqrt{r^2 + z^2}, \theta = \theta, \phi = \cos^{-1}\dfrac{z}{\sqrt{r^2 + z^2}}$

51. (a) $(1, 1, 1)$ (b) $(\sqrt{3}, \pi/4, \tan^{-1}\sqrt{2})$

53. (a) $z = x^2 - y^2$ (b) $xz = 1$

▶ Exercise Set 15.1 (Page 876)

1. $(-\infty, +\infty)$; $-\mathbf{i} - 3\pi\mathbf{j}$

3. $[2, +\infty)$; $-\mathbf{i} - \ln 3\mathbf{j} + \mathbf{k}$

5. $\mathbf{r} = 3\cos t\mathbf{i} + (t + \sin t)\mathbf{j}$

7. $\mathbf{r} = 2t\mathbf{i} + 2\sin 3t\mathbf{j} + 5\cos 3t\mathbf{k}$

9. $x = 3t^2, y = -2, z = 0$

11. $x = 2t - 1, y = -3\sqrt{t}, z = \sin 3t$

13. line in the xy-plane through $(2, 0)$, parallel to $-3\mathbf{i} - 4\mathbf{j}$

15. line through $(0, -3, 1)$, parallel to $2\mathbf{i} + 3\mathbf{k}$

17. $-\frac{3}{2}$

19. $(\frac{5}{2}, 0, \frac{3}{2})$

21.

23.

25.

27.

29. (0, 2, π/2) (2, 0, 0)

31.

33. $\dfrac{3}{2\pi}$

35. conical helix

► Exercise Set 15.2 (Page 886)

1. $9\mathbf{i} + 6\mathbf{j}$

3. $\mathbf{j}$

5. $2\mathbf{i} - 3\mathbf{j} + 4\mathbf{k}$

7. $\frac{1}{2}\pi\mathbf{i} + \mathbf{k}$

11. $5\mathbf{i} + (1 - 2t)\mathbf{j}$

13. $-\dfrac{1}{t^2}\mathbf{i} + \sec^2 t\,\mathbf{j} + 2e^{2t}\mathbf{k}$

15. $\langle 1, 4 \rangle$

17. $\langle -\frac{1}{2}, 8 \rangle$

19. $-2\mathbf{k}$

21. $\mathbf{j} - 2\mathbf{k}$

23. $x = 1 + 2t,\ y = 2 - t,\ z = 0$

25. $x = 1 - \sqrt{3}\pi t,\ y = \sqrt{3} + \pi t,\ z = 1 + 3t$

27. $\mathbf{r} = -\mathbf{i} + 2\mathbf{j} + t(2\mathbf{i} + \frac{3}{4}\mathbf{j})$

29. $\mathbf{r} = 4\mathbf{i} + \mathbf{j} + t(-4\mathbf{i} + \mathbf{j} + 4\mathbf{k})$

31. $2y + z = \pi/2$

33. (a) $(-2, 4, 6),\ (1, 1, -3)$

    (b) $14°,\ 19°$

37. (a) $7t^6$

    (b) $12t \sec t \tan t + 12 \sec t - \dfrac{\sin t}{t}$

          $- (\cos t)\ln t$

39. $4\mathbf{i} + 8(4w + 1)\mathbf{j}$

41. $2we^{w^2}\mathbf{i} - 8we^{-w^2}\mathbf{j}$

► Exercise Set 15.3 (Page 896)

1. $3t\mathbf{i} + 2t^2\mathbf{j} + \mathbf{C}$ 3. $\langle 0, -\frac{2}{3}\rangle$

5. $\frac{52}{3}\mathbf{i} + 4\mathbf{j}$

7. $\langle (t-1)e^t, t(\ln t - 1)\rangle + \mathbf{C}$

9. $\frac{1}{3}t^3\mathbf{i} - t^2\mathbf{j} + \ln|t|\mathbf{k} + \mathbf{C}$

11. $\frac{1}{2}(e^2 - 1)\mathbf{i} + (1 - e^{-1})\mathbf{j} + \frac{1}{2}\mathbf{k}$

13. (a) $\mathbf{F} = 3t\mathbf{i} - 2\mathbf{j} - 3t^2\mathbf{k}$
    (b) 30

15. $(1 + \sin t)\mathbf{i} - (\cos t)\mathbf{j}$

17. $(t^4 + 2)\mathbf{i} - (t^2 + 4)\mathbf{j}$

19. $2(t-1)\mathbf{i} + \frac{1}{2}\ln\frac{1}{2}(t^2 + 1)\mathbf{j} + \frac{1}{2}(t^2 - 1)\mathbf{k}$

21. $\sqrt{14}$ 23. 28

25. $e - e^{-1}$ 27. $t_0\sqrt{a^2 + c^2}$

29. $x = \frac{3}{5}s - 2, y = \frac{4}{5}s + 3$

31. $x = 3 + \cos s, y = 2 + \sin s; 0 \le s \le 2\pi$

33. $x = \frac{1}{3}[(3s + 1)^{2/3} - 1]^{3/2}$
    $y = \frac{1}{2}[(3s + 1)^{2/3} - 1]$; $s \ge 0$

35. $x = \left(\dfrac{s}{\sqrt{2}} + 1\right)\cos\left[\ln\left(\dfrac{s}{\sqrt{2}} + 1\right)\right]$

   $y = \left(\dfrac{s}{\sqrt{2}} + 1\right)\sin\left[\ln\left(\dfrac{s}{\sqrt{2}} + 1\right)\right]$

   where $0 \le s \le \sqrt{2}(e^{\pi/2} - 1)$

37. $x = 1 - 3s/\sqrt{35}, y = 5s/\sqrt{35}, z = 2 + s/\sqrt{35}$

39. $x = (\sqrt{2s + 1} - 1)\cos(\sqrt{2s + 1} - 1)$,
    $y = (\sqrt{2s + 1} - 1)\sin(\sqrt{2s + 1} - 1)$,
    $z = \frac{2}{3}\sqrt{2}[\sqrt{2s + 1} - 1]^{3/2}; s \ge 0$

41. (b) $x = a\left(\cos\sqrt{\dfrac{2s}{a}} + \sqrt{\dfrac{2s}{a}}\sin\sqrt{\dfrac{2s}{a}}\right)$, $s \ge 0$

   $y = a\left(\sin\sqrt{\dfrac{2s}{a}} - \sqrt{\dfrac{2s}{a}}\cos\sqrt{\dfrac{2s}{a}}\right)$

43. (a) $\frac{9}{2}$ (b) $9 - 2\sqrt{6}$

45. (a) $\sqrt{3}(1 - e^{-2})$ (b) $4\sqrt{5}$

47. (a) $2t + \dfrac{1}{t}$ (b) $2t + \dfrac{1}{t}$
    (c) $8 + \ln 3$

► Exercise Set 15.4 (Page 903)

1.
$T = -\frac{\sqrt{3}}{2}\mathbf{i} + \frac{1}{2}\mathbf{j}$
$\left(\frac{5}{2}, \frac{5\sqrt{3}}{2}\right)$
$N = -\frac{1}{2}\mathbf{i} - \frac{\sqrt{3}}{2}\mathbf{j}$

3.
$T = \frac{2}{\sqrt{5}}\mathbf{i} + \frac{1}{\sqrt{5}}\mathbf{j}$
$N = \frac{1}{\sqrt{5}}\mathbf{i} - \frac{2}{\sqrt{5}}\mathbf{j}$

5.
$N = -\frac{1}{\sqrt{2}}\mathbf{i} + \frac{1}{\sqrt{2}}\mathbf{j}$
$T = \frac{1}{\sqrt{2}}\mathbf{i} + \frac{1}{\sqrt{2}}\mathbf{j}$
$\left(\frac{1}{2}, \frac{1}{3}\right)$

7.
$T = -\frac{4}{\sqrt{97}}\mathbf{i} + \frac{9}{\sqrt{97}}\mathbf{j}$
$(2\sqrt{2}, 9\sqrt{2}/2)$
$N = -\frac{9}{\sqrt{97}}\mathbf{i} - \frac{4}{\sqrt{97}}\mathbf{j}$

9. $T = -\dfrac{4}{\sqrt{17}}\mathbf{i} + \dfrac{1}{\sqrt{17}}\mathbf{k}$ 11. $T = \mathbf{i}$
   $N = -\mathbf{j}$ $N = \mathbf{j}$

13. $T = -\dfrac{3}{\sqrt{10}}\mathbf{i} + \dfrac{1}{\sqrt{10}}\mathbf{k}$
    $N = -\mathbf{j}$

15. $T = \dfrac{1}{\sqrt{5}}\mathbf{j} + \dfrac{2}{\sqrt{5}}\mathbf{k}$
    $N = -\dfrac{2}{\sqrt{5}}\mathbf{j} + \dfrac{1}{\sqrt{5}}\mathbf{k}$

17. (a) $\mathbf{T} = \frac{3}{5}\mathbf{i} + \frac{4}{5}\mathbf{j}$
(b)

21. See Exercise 1. **23.** See Exercise 3.

25. See Exercise 9. **27.** See Exercise 11.

▶ Exercise Set 15.5 (Page 912)

1. $\dfrac{96}{125}$ **3.** $\dfrac{6}{5\sqrt{10}}$

5. $\dfrac{3}{2\sqrt{2}}$ **7.** $\dfrac{4}{17}$

9. 1   **11.** $\frac{2}{5}$

13. $\dfrac{2}{5\sqrt{5}}$ **17.** 1

19. $\dfrac{1}{\sqrt{2}}$ **21.** $\dfrac{4}{5\sqrt{5}}$

23. $\cos x$; maximum for $x = 0$

25. See Exercise 1.

27. See Exercise 3.

29. See Exercise 5.

31. $\kappa(0) = a/b^2$, $\kappa(\pi/2) = b/a^2$

33. 1   **35.** $\dfrac{3}{2\sqrt{2}\,a}$

37. $\kappa(t) = \dfrac{1}{4a}\csc(t/2)$

39. $\rho = 1$

41. $\rho = 2\sqrt{2}$

43. $\rho = 2\sqrt{2}$ (1, 0)

45. $\rho = 4$

47. $\rho(0) = \frac{1}{2}$ $\rho\left(\frac{\pi}{2}\right) = 4$

49. $2|p|$ **51.** $(3, 0), (-3, 0)$

53. maximum 2, minimum $1/\sqrt{2}$

55. $\kappa = \dfrac{1}{\sqrt{1 + a^2}\,r}$

57. $\sim 2.86°/\text{cm}$ **61.** $\dfrac{1}{2r}$

▶ Exercise Set 15.6 (Page 926)

1.

$\mathbf{v} = -\dfrac{3\sqrt{3}}{2}\mathbf{i} + \dfrac{3}{2}\mathbf{j}$

$\left(\dfrac{3}{2}, \dfrac{3\sqrt{3}}{2}\right)$

$\mathbf{a} = -\dfrac{3}{2}\mathbf{i} - \dfrac{3\sqrt{3}}{2}\mathbf{j}$

$\mathbf{v}(t) = -3\sin t\,\mathbf{i} + 3\cos t\,\mathbf{j}$
$\mathbf{a}(t) = -3\cos t\,\mathbf{i} - 3\sin t\,\mathbf{j}$
$\|\mathbf{v}(t)\| = 3$

3.

$\mathbf{a} = \mathbf{i} + \mathbf{j}$
$(1, 1)$
$\mathbf{v} = \mathbf{i} - \mathbf{j}$

$\mathbf{v}(t) = e^t\mathbf{i} - e^{-t}\mathbf{j}$
$\mathbf{a}(t) = e^t\mathbf{i} + e^{-t}\mathbf{j}$
$\|\mathbf{v}(t)\| = \sqrt{e^{2t} + e^{-2t}}$

5.

$\mathbf{v} = \dfrac{3}{4}\mathbf{i} + \dfrac{5}{4}\mathbf{j}$
$\mathbf{a} = \dfrac{5}{4}\mathbf{i} + \dfrac{3}{4}\mathbf{j}$
$\left(\dfrac{5}{4}, \dfrac{3}{4}\right)$

$\mathbf{v}(t) = \sinh t\,\mathbf{i} + \cosh t\,\mathbf{j}$
$\mathbf{a}(t) = \cosh t\,\mathbf{i} + \sinh t\,\mathbf{j}$
$\|\mathbf{v}(t)\| = \sqrt{\sinh^2 t + \cosh^2 t}$

7. $\mathbf{v} = \mathbf{i} + \mathbf{j} + \mathbf{k}$
$\|\mathbf{v}\| = \sqrt{3}$
$\mathbf{a} = \mathbf{j} + 2\mathbf{k}$

9. $\mathbf{v} = -\sqrt{2}\mathbf{i} + \sqrt{2}\mathbf{j} + \mathbf{k}$
$\|\mathbf{v}\| = \sqrt{5}$
$\mathbf{a} = -\sqrt{2}\mathbf{i} - \sqrt{2}\mathbf{j}$

11. $\mathbf{v} = e^{\pi/2}\mathbf{i} - e^{\pi/2}\mathbf{j} + \mathbf{k}$
$\|\mathbf{v}\| = \sqrt{2e^{\pi} + 1}$
$\mathbf{a} = -2e^{\pi/2}\mathbf{j}$

13. $\mathbf{r}(t) = -16t^2\mathbf{j}$
$\mathbf{v}(t) = -32t\mathbf{j}$

15. $\mathbf{r}(t) = (t + \cos t - 1)\mathbf{i} + (\sin t - t + 1)\mathbf{j}$
$\mathbf{v}(t) = (1 - \sin t)\mathbf{i} + (\cos t - 1)\mathbf{j}$

17. $\mathbf{r}(t) = \frac{1}{2}t^2\mathbf{i} + \mathbf{j} + \frac{1}{6}t^3\mathbf{k}$
$\mathbf{v}(t) = t\mathbf{i} + \frac{1}{2}t^2\mathbf{k}$

19. $\mathbf{r}(t) = (t - \sin t - 1)\mathbf{i} + (1 - \cos t)\mathbf{j} + e^t\mathbf{k}$
$\mathbf{v}(t) = (1 - \cos t)\mathbf{i} + \sin t\,\mathbf{j} + e^t\mathbf{k}$

21. $a_T = 0,\ a_N = 2$

23. $a_T = 0,\ a_N = \sqrt{2}$

25. $a_T = 2\sqrt{5},\ a_N = 2\sqrt{5}$

27. $a_T = \dfrac{22}{\sqrt{14}},\ a_N = \sqrt{\dfrac{38}{7}}$

29. $a_T = 0,\ a_N = 3$

31. $a_T = -3,\ a_N = 2;\ \mathbf{T} = -\mathbf{j},\ \mathbf{N} = \mathbf{i}$

33. $a_T = \frac{4}{3},\ a_N = \frac{1}{3}\sqrt{29};$
$\mathbf{T} = \frac{2}{3}\mathbf{i} + \frac{2}{3}\mathbf{j} + \frac{1}{3}\mathbf{k},$
$\mathbf{N} = \dfrac{1}{3\sqrt{29}}(\mathbf{i} - 8\mathbf{j} + 14\mathbf{k})$

37. 9×10^{10} kilometers/sec^2

41. (b) no

43. maximum 6, minimum 3

45. $15°$

47. $\mathbf{r} = -\frac{19}{16}\mathbf{i} + \frac{3}{2}\mathbf{j} + \frac{3}{16}\mathbf{k}$

49. $8\mathbf{i} + \frac{26}{3}\mathbf{j},\ \frac{1}{3}(13\sqrt{13} - 5\sqrt{5})$

51. $\mathbf{0},\ 12\pi$

53. $2\mathbf{i} - \frac{2}{3}\mathbf{j} + \sqrt{2}\ln 3\mathbf{k},\ \frac{8}{3}$

55. (a) $x = 160t$
$y = 160\sqrt{3}t - 16t^2,\ t \geq 0$
(b) 1200 ft
(c) $1600\sqrt{3}$ ft
(d) 320 ft/sec

57. $40\sqrt{3}$ ft   **59.** 800 ft/sec

61. $15°,\ 75°$

63. (a) $x = (v_0 \cos \alpha)t,$
$y = (v_0 \sin \alpha)t - \frac{1}{2}gt^2$
(c) $45°$

65. (a) $\mathbf{r}(t) = -16t^2\mathbf{k}$
$\mathbf{v}(t) = -32t\mathbf{k}$
$\|\mathbf{v}(t)\| = 32t$
(b) $\mathbf{r}(t) = \mathbf{i} + 2\mathbf{j} + (1 - 16t^2)\mathbf{k}$
$\mathbf{v}(t) = -32t\mathbf{k}$
$\|\mathbf{v}(t)\| = 32t$
(c) $\mathbf{r}(t) = (3 + t)\mathbf{i} - \mathbf{j} + (4 + t - 16t^2)\mathbf{k}$
$\mathbf{v}(t) = \mathbf{i} + (1 - 32t)\mathbf{k}$
$\|\mathbf{v}(t)\| = \sqrt{1 + (1 - 32t)^2}$

▶ Chapter 15 Supplementary Exercises (Page 928)

1. (a) $\frac{1}{2}(t+4)^{-1/2}\mathbf{i} + 2\mathbf{j}$, $-\frac{1}{4}(t+4)^{-3/2}\mathbf{i}$

(b) The graph is the part of the parabola $y = 2(x^2 - 4)$ starting at $(0, -8)$ and directed up and toward the right. At $(1, -6)[t = -3]$, $\mathbf{r}' = \frac{1}{2}\mathbf{i} + 2\mathbf{j}$, $\mathbf{r}'' = -\frac{1}{4}\mathbf{i}$; at $(2, 0)[t = 0]$, $\mathbf{r}' = \frac{1}{4}\mathbf{i} + 2\mathbf{j}$, $\mathbf{r}'' = -1/32\mathbf{i}$

3. (a) $\langle 6t^2, 3t^2 \rangle$, $\langle 12t, 6t \rangle$

(b) The graph is the straight line $x - 2y + 3 = 0$, directed to the right and up; at $(-1, 1)[t = 0]$, $\mathbf{r}' = \mathbf{r}'' = \langle 0, 0 \rangle$; at $(-5/4, 7/8)[t = -1/2]$, $\mathbf{r}' = \langle 3/2, 3/4 \rangle$, $\mathbf{r}'' = \langle -6, -3 \rangle$.

5. (a) $t(k\mathbf{i} + m\mathbf{j}) + \mathbf{C}$ (b) $\langle 4, 4 \rangle$

(c) 2   (d) $\sqrt{t^2 + 3}\,\mathbf{i} + \ln(\sin t)\mathbf{j} + \mathbf{C}$

7. (a) $1 + 1/t^2$

(b) $x = \sqrt{s^2 + 4}$, $y = 2\ln[\frac{1}{2}(s + \sqrt{s^2 + 4})]$

9. the parabola $y = x^2 + 1$ in the plane $z = 1$, traced so x increases with t

11. $\mathbf{v} = a\mathbf{i}$, $\|\mathbf{v}\| = a$, $\mathbf{a} = -a(\mathbf{j} + \mathbf{k})$, $\mathbf{T} = \mathbf{i}$, $\mathbf{N} = -(\mathbf{j} + \mathbf{k})/\sqrt{2}$, $\kappa = \sqrt{2}/a$

13. $4\pi\sqrt{37}$ **15.** $x = 1 - t$, $y = 1 + 2t$, $z = 1$

17. (a) $\langle 9, 9, -9 \rangle$ (b) $\langle 3 + 4t^3, -4t, 6t^2 \rangle$

19. (a) $(2\mathbf{i} - \mathbf{j})/\sqrt{5}$ (b) $(\mathbf{i} + 2\mathbf{j})/\sqrt{5}$

(c) $6/5^{3/2}$

21. (a) $\mathbf{j}$ (b) $\mathbf{i}$ (c) 2

23. 2/25   **25.** 1

27. $(x - 1)^2 + (y - \frac{1}{2})^2 = \frac{1}{4}$; at $(1, 0)$, $dy/dx = 0$, $d^2y/dx^2 = 2$

29. $(1/w)\langle e^t, 4e^{2t} \rangle = \langle 1, 4w \rangle$

31. $a_T = 1$, $a_N = t$

33. (a) $\langle -1, 1 \rangle$, $\langle 1, 1 \rangle$, $\sqrt{2}$

(b) $(-\mathbf{i} + \mathbf{j})/\sqrt{2}$, $(\mathbf{i} + \mathbf{j})/\sqrt{2}$, $1/\sqrt{2}$

(c) 0, $\sqrt{2}$

(d) trajectory: the branch of the hyperbola $xy = 1$ in the first quadrant, traced so y increases with t

(e) $C(2, 2)$

35. $(2t - \sin t)\mathbf{i} + (e^{2t} - 1)\mathbf{j}$

37. (a) $a_T = 0$, $a_N = 32$

(b) $a_T = -64/\sqrt{5}$, $a_N = 32/\sqrt{5}$

39. $a_T = 0$, $a_N = 8\pi^2$ (m/sec^2)

41. (a) $\sqrt{5}e^t$ (b) $2\sqrt{5}$

▶ Exercise Set 16.1 (Page 944)

1. (a) 5   (b) 3

(c) 1   (d) -2

(e) $9a^3 + 1$ (f) $a^3b^2 - a^2b^3 + 1$

3. (a) $x^2 - y^2 + 3$ (b) $3x^3y^4 + 3$

5. $x^3e^{x^3(3y+1)}$

7. (a) $t^2 + 3t^{10}$ (b) 0   (c) 3076

9. (a) 19   (b) -9 (c) 3

(d) $a^6 + 3$ (e) $-t^8 + 3$

(f) $(a + b)(a - b)^2b^3 + 3$

11. $(y + 1)e^{x^2(y+1)z^2}$

13. (a) t^{14} (b) 0   (c) 16,384

15.

17.

19.

21. all points above or on the line $y = -2$

23. all points above the line $y = 2x$

25. all points not on the plane $x + y + z = 0$

27. all points within the cylinder $x^2 + y^2 = 1$

29.

31.

33.

35.

37.

39.

41.

43.

45.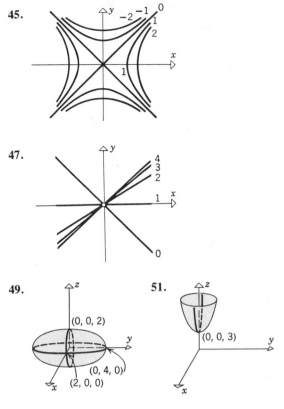

47.

49.

51.

53. concentric spheres, center at $(2, 0, 0)$

55. concentric cylinders, common axis the y-axis

57. (a) $x^2 - 2x^3 + 3xy = 0$
      (b) $x^2 - 2x^3 + 3xy = 0$
      (c) $x^2 - 2x^3 + 3xy = -18$

59. (a) $x^2 + y^2 - z = 5$
      (b) $x^2 + y^2 - z = -2$
      (c) $x^2 + y^2 - z = 0$

61.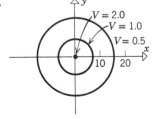

63. (a) A (b) B

65. (a) decrease   (b) increase
      (c) increase   (d) decrease

67. (a) open (b) neither
(c) closed (d) closed

69. (a) bounded (b) unbounded
(c) unbounded (d) unbounded

▶ Exercise Set 16.2 (Page 955)

1.

3.

5.

7.

9. all of 3-space

11. all points not on the cylinder $x^2 + z^2 = 1$

13. 35   **15.** -8

17. 0   **19.** does not exist

21. 1   **23.** 0

25. does not exist   **27.** 0

29. 0   **31.** $\frac{8}{3}$

33. 0   **35.** does not exist

41. $-\dfrac{\pi}{2}$ **43.** no

45. choose $\delta = \sqrt{\epsilon}$ **47.** choose $\delta = \sqrt{\epsilon}$

▶ Exercise Set 16.3 (Page 963)

1. $\dfrac{\partial z}{\partial x} = 9x^2y^2, \dfrac{\partial z}{\partial y} = 6x^3y$

3. $\dfrac{\partial z}{\partial x} = 8xy^3 e^{x^2y^3}, \dfrac{\partial z}{\partial y} = 12x^2y^2 e^{x^2y^3}$

5. $\dfrac{\partial z}{\partial x} = \dfrac{x^3}{y^{3/5} + x} + 3x^2 \ln(1 + xy^{-3/5}),$

$\dfrac{\partial z}{\partial y} = -\dfrac{3x^4}{5(y^{8/5} + xy)}$

7. $f_x(x, y) = \frac{3}{2}x^2y(5x^2 - 7)(3x^5y - 7x^3y)^{-1/2},$
$f_y(x, y) = \frac{1}{2}x^3(3x^2 - 7)(3x^5y - 7x^3y)^{-1/2}$

9. $f_x(x, y) = \dfrac{y^{-1/2}}{y^2 + x^2},$

$f_y(x, y) = -\dfrac{xy^{-3/2}}{y^2 + x^2} - \dfrac{3}{2}y^{-5/2}\tan^{-1}\left(\dfrac{x}{y}\right)$

11. $f_x(x, y) = -\frac{4}{3}y^2 \sec^2 x(y^2 \tan x)^{-7/3},$
$f_y(x, y) = -\frac{8}{3}y \tan x(y^2 \tan x)^{-7/3}$

13. (a) -6 (b) -21

15. (a) $\dfrac{1}{\sqrt{17}}$ (b) $\dfrac{8}{\sqrt{17}}$

17. $\dfrac{\partial z}{\partial x} = -\dfrac{x}{z}, \dfrac{\partial z}{\partial y} = -\dfrac{y}{z}$

19. $\dfrac{\partial z}{\partial x} = -\dfrac{2x + yz^2 \cos(xyz)}{xyz \cos(xyz) + \sin(xyz)}$

$\dfrac{\partial z}{\partial y} = -\dfrac{xz^2 \cos(xyz)}{xyz \cos(xyz) + \sin(xyz)}$

21. $f_{xx} = 8,$
$f_{yy} = -96xy^2 + 140y^3,$
$f_{xy} = f_{yx} = -32y^3$

23. $f_{xx} = e^x \cos y,$
$f_{yy} = -e^x \cos y,$
$f_{xy} = f_{yx} = -e^x \sin y$

25. $f_{xx} = -\dfrac{16}{(4x - 5y)^2},$

$f_{yy} = -\dfrac{25}{(4x - 5y)^2},$

$f_{xy} = f_{yx} = \dfrac{20}{(4x - 5y)^2}$

27. (a) $30xy^4 - 4$ (b) $60x^2y^3$
(c) $60x^3y^2$

29. (a) -30 (b) -125 (c) 150

31. (a) $\dfrac{\partial^3 f}{\partial x^3}$ (b) $\dfrac{\partial^3 f}{\partial y^2 \partial x}$

   (c) $\dfrac{\partial^4 f}{\partial x^2 \partial y^2}$ (d) $\dfrac{\partial^4 f}{\partial y^3 \partial x}$

33. $\dfrac{\partial w}{\partial x} = 2xy^4 z^3 + y,\ \dfrac{\partial w}{\partial y} = 4x^2 y^3 z^3 + x,$

   $\dfrac{\partial w}{\partial z} = 3x^2 y^4 z^2 + 2z$

35. $\dfrac{\partial w}{\partial x} = \dfrac{2x}{y^2 + z^2},\ \dfrac{\partial w}{\partial y} = -\dfrac{2y(x^2 + z^2)}{(y^2 + z^2)^2},$

   $\dfrac{\partial w}{\partial z} = \dfrac{2z(y^2 - x^2)}{(y^2 + z^2)^2}$

37. $\dfrac{\partial w}{\partial x} = \dfrac{x}{\sqrt{x^2 + y^2 + z^2}},\ \dfrac{\partial w}{\partial y} = \dfrac{y}{\sqrt{x^2 + y^2 + z^2}},$

   $\dfrac{\partial w}{\partial z} = \dfrac{z}{\sqrt{x^2 + y^2 + z^2}}$

39. $f_x = -\dfrac{y^2 z^3}{x^2 y^4 z^6 + 1},\ f_y = -\dfrac{2xyz^3}{x^2 y^4 z^6 + 1},$

   $f_z = -\dfrac{3xy^2 z^2}{x^2 y^4 z^6 + 1}$

41. $f_x = 4xyz \cosh \sqrt{z} \sinh (x^2 yz) \cosh (x^2 yz),$
   $f_y = 2x^2 z \cosh \sqrt{z} \sinh (x^2 yz) \cosh (x^2 yz),$
   $f_z = 2x^2 y \cosh \sqrt{z} \sinh (x^2 yz) \cosh (x^2 yz)$
   $\quad + \dfrac{\sinh \sqrt{z} \sinh^2 (x^2 yz)}{2\sqrt{z}}$

43. (a) -80 (b) 40 (c) -60

45. (a) $\dfrac{2}{\sqrt{7}}$ (b) $\dfrac{4}{\sqrt{7}}$ (c) $\dfrac{1}{\sqrt{7}}$

47. $\dfrac{\partial w}{\partial x} = -\dfrac{x}{w},\ \dfrac{\partial w}{\partial y} = -\dfrac{y}{w},\ \dfrac{\partial w}{\partial z} = -\dfrac{z}{w}$

49. $\dfrac{\partial w}{\partial x} = -\dfrac{yzw \cos (xyz)}{2w + \sin (xyz)}$

   $\dfrac{\partial w}{\partial y} = -\dfrac{xzw \cos (xyz)}{2w + \sin (xyz)}$

   $\dfrac{\partial w}{\partial z} = -\dfrac{xyw \cos (xyz)}{2w + \sin (xyz)}$

51. (a) $15x^2 y^4 z^7 + 2y$ (b) $35x^3 y^4 z^6 + 3y^2$
   (c) $21x^2 y^5 z^6$ (d) $42x^3 y^5 z^5$
   (e) $140x^3 y^3 z^6 + 6y$ (f) $30xy^4 z^7$
   (g) $105x^2 y^4 z^6$ (h) $210xy^4 z^6$

55. 6

57. (a) 8   (b) -2

59. (a) $\dfrac{\partial V}{\partial r} = 2\pi rh$ (b) $\dfrac{\partial V}{\partial h} = \pi r^2$
   (c) 48π (d) 64π

61. (a) $\frac{1}{5}$ (b) $-\frac{25}{8}$

63. $\dfrac{\partial z}{\partial x} = -1 - \dfrac{\cos (x - y)}{\cos (x + z)},$

   $\dfrac{\partial z}{\partial y} = \dfrac{\cos (x - y)}{\cos (x + z)},$

   $\dfrac{\partial^2 z}{\partial x \partial y} = -\dfrac{\cos^2 (x + z) \sin (x - y) + \cos^2 (x - y) \sin (x + z) + \cos (x - y) \sin (x + z) \cos (x + z)}{\cos^3 (x + z)}$

65. (a) 4 degrees per centimeter
   (b) 8 degrees per centimeter

69. (a) Both are positive; $\partial T / \partial x$.
   (b) All are negative.

▶ Exercise Set 16.4 (Page 979)

1. 0.872

3. $\frac{7}{2}$

5. $\epsilon_1 = 0,\ \epsilon_2 = \Delta x$

7. $\epsilon_1 = y\Delta x + \Delta x \Delta y,\ \epsilon_2 = 2x\Delta x$

13. $f_{xy} = f_{yx} = 12x^2 + 6x$

15. $f_{xy} = f_{yx} = -6xy^2 \sin (x^2 + y^3)$

17. (a) 4   (b) 5

19. $42t^{13}$

21. $\dfrac{3 \sin (1/t)}{t^2}$

23. $-\frac{10}{3}t^{7/3}e^{1-t^{10/3}}$

25. $\dfrac{\partial z}{\partial u} = 24u^2v^2 - 16uv^3 - 2v + 3,$

$\dfrac{\partial z}{\partial v} = 16u^3v - 24u^2v^2 - 2u - 3$

27. $\dfrac{\partial z}{\partial u} = -\dfrac{2\sin u}{3\sin v}, \dfrac{\partial z}{\partial v} = -\dfrac{2\cos u \cos v}{3\sin^2 v}$

29. $\dfrac{\partial z}{\partial u} = e^u, \dfrac{\partial z}{\partial v} = 0$

31. $\dfrac{\partial z}{\partial u} = \dfrac{2e^{2u}}{1 + e^{4u}}, \dfrac{\partial z}{\partial v} = 0$

33. $\dfrac{\partial T}{\partial r} = 3r^2 \sin\theta\cos^2\theta - 4r^3\sin^3\theta\cos\theta,$

$\dfrac{\partial T}{\partial\theta} = -2r^3\sin^2\theta\cos\theta + r^4\sin^4\theta$

$\qquad + r^3\cos^3\theta - 3r^4\cos^2\theta\sin^2\theta$

35. $\dfrac{\partial t}{\partial x} = \dfrac{x^2 + y^2}{4x^2y^3}, \dfrac{\partial t}{\partial y} = \dfrac{y^2 - 3x^2}{4xy^4}$ **37.** $-\pi$

39. $\dfrac{\partial z}{\partial r}\Big|_{r=2,\,\theta=\pi/6} = \sqrt{3}e^{\sqrt{3}},$

$\dfrac{\partial z}{\partial\theta}\Big|_{r=2,\,\theta=\pi/6} = (2 - 4\sqrt{3})e^{\sqrt{3}}$

41. $\dfrac{x^2 - y^2}{2xy - y^2}$ **43.** $\dfrac{2\sqrt{xy} - y}{x - 6\sqrt{xy}}$

45. -39 mi/hr   **47.** $-\frac{7}{36}\sqrt{3}$ radian/sec

49. $\frac{4}{3} - 16\ln 3$ °C/sec

55. (b) $z = C_1\ln r + C_2$

65. (a) $n = 2$ (b) $n = 1$
(c) $n = 3$ (d) $n = -4$

67. (a) $\dfrac{\partial w}{\partial x} = \dfrac{\partial f}{\partial x} + \dfrac{\partial f}{\partial y}\dfrac{\partial y}{\partial x}$

(b) $\dfrac{\partial w}{\partial z} = \dfrac{\partial f}{\partial y}\dfrac{\partial y}{\partial z}$

▶ Exercise Set 16.5 (Page 989)

1. $48x - 14y - z = 64,$ $x = 1 + 48t$
$y = -2 - 14t$
$z = 12 - t$

3. $x - y - z = 0,$ $x = 1 + t$
$y = -t$
$z = 1 - t$

5. $3y - z = -1,$ $x = \pi/6$
$y = 3t$
$z = 1 - t$

7. $3x - 4z = -25,$ $x = -3 + \frac{3}{4}t$
$y = 0$
$z = 4 - t$

9. $7\,dx - 2\,dy$

11. $\dfrac{y}{1 + x^2y^2}dx + \dfrac{x}{1 + x^2y^2}dy$

13. 0.10   **15.** 0.03
17. (a) all points on the x-axis and y-axis
(b) $(0, -2, -4)$
19. $(\frac{1}{2}, -2, -\frac{3}{4})$ **23.** 0.088 cm
25. 8%   **27.** $r\%$
29. 2%   **31.** 0.004 radian
33. 0.3%
35. (a) $(r + s)\%$ (b) $(r + s)\%$
(c) $(2r + 3s)\%$ (d) $(3r + \frac{1}{2}s)\%$
37. (a) $(-2, 1, 5), (0, 3, 9)$
(b) $\dfrac{4}{3\sqrt{14}}$ at $(-2, 1, 5), \dfrac{4}{\sqrt{222}}$ at $(0, 3, 9)$

▶ Exercise Set 16.6 (Page 999)

1. $4\mathbf{i} - 8\mathbf{j}$ **3.** $\dfrac{x}{x^2 + y^2}\mathbf{i} + \dfrac{y}{x^2 + y^2}\mathbf{j}$ **9.** $6\sqrt{2}$ **11.** $-\dfrac{3}{\sqrt{10}}$

5. $-36\mathbf{i} - 12\mathbf{j}$ **7.** $4\mathbf{i} + 4\mathbf{j}$ **13.** 0   **15.** $-8\sqrt{2}$

17. $\dfrac{1}{2\sqrt{2}}$

19. $\dfrac{1}{2} + \dfrac{\sqrt{3}}{8}$

21. $2\sqrt{2}$

23.

25.

27. $\dfrac{3}{\sqrt{13}}\mathbf{i} - \dfrac{2}{\sqrt{13}}\mathbf{j},\ 4\sqrt{13}$

29. $\frac{4}{5}\mathbf{i} - \frac{3}{5}\mathbf{j},\ 1$

31. $-\dfrac{1}{\sqrt{10}}\mathbf{i} - \dfrac{3}{\sqrt{10}}\mathbf{j},\ -2\sqrt{10}$

33. $\dfrac{3}{\sqrt{10}}\mathbf{i} - \dfrac{1}{\sqrt{10}}\mathbf{j},\ -\sqrt{5}$

35. $\dfrac{1}{\sqrt{5}}$ **37.** $-\frac{3}{2}e$

39. $-\dfrac{4}{\sqrt{17}}\mathbf{i} + \dfrac{1}{\sqrt{17}}\mathbf{j}$

41. (a) 5   (b) 10   (c) $-5\sqrt{5}$

43. $\dfrac{8}{\sqrt{29}}$

45. all points on the ellipse $9x^2 + y^2 = 9$

47. $36/\sqrt{17}$ **49.** (a) $2e^{-\pi/2}\mathbf{i}$

▶ Exercise Set 16.7 (Page 1008)

1. $165t^{32}$

3. $-2t \cos(t^2)$

5. 3264

7. -320

9. $-\frac{314}{741}$

11. $\dfrac{72}{\sqrt{14}}$

13. $-\frac{8}{63}$

15. $\dfrac{1}{\sqrt{2}}\mathbf{i} - \dfrac{1}{\sqrt{2}}\mathbf{j},\ 3\sqrt{2}$

17. $-\dfrac{1}{\sqrt{2}}\mathbf{i} + \dfrac{1}{\sqrt{2}}\mathbf{j},\ \dfrac{\sqrt{2}}{2}$

19. $\dfrac{1}{\sqrt{266}}\mathbf{i} - \dfrac{11}{\sqrt{266}}\mathbf{j} + \dfrac{12}{\sqrt{266}}\mathbf{k},\ -\sqrt{266}$

21. $\dfrac{3}{\sqrt{11}}$

23. $-\frac{10}{3}\mathbf{i} + \frac{5}{3}\mathbf{j} + \frac{10}{3}\mathbf{k}$

25. $3x - 2y + 6z = -49,\quad x = -3 + 3t$
$y = 2 - 2t$
$z = -6 + 6t$

27. $x - y - 15z = -17,\quad x = 3 + t$
$y = 5 - t$
$z = 1 - 15t$

31. $(1, \frac{2}{3}, \frac{2}{3}),\ (-1, -\frac{2}{3}, -\frac{2}{3})$

33. $8\,dx - 3\,dy + 4\,dz$

35. $\dfrac{yz}{1 + (xyz)^2}\,dx + \dfrac{xz}{1 + (xyz)^2}\,dy + \dfrac{xy}{1 + (xyz)^2}\,dz$

37. 0.96   **39.** 2.35 cm^3

41. 39 ft^2 **43.** 14%

▶ Exercise Set 16.8 (Page 1015)

1. $\dfrac{\partial f}{\partial v} = 8vw^3x^4y^5,\ \dfrac{\partial f}{\partial w} = 12v^2w^2x^4y^5,$

$\dfrac{\partial f}{\partial x} = 16v^2w^3x^3y^5,\ \dfrac{\partial f}{\partial y} = 20v^2w^3x^4y^4$

3. $\dfrac{\partial f}{\partial v_1} = \dfrac{2v_1}{v_3{}^2 + v_4{}^2},\ \dfrac{\partial f}{\partial v_2} = -\dfrac{2v_2}{v_3{}^2 + v_4{}^2},$

$\dfrac{\partial f}{\partial v_3} = -\dfrac{2v_3(v_1{}^2 - v_2{}^2)}{(v_3{}^2 + v_4{}^2)^2},\ \dfrac{\partial f}{\partial v_4} = -\dfrac{2v_4(v_1{}^2 - v_2{}^2)}{(v_3{}^2 + v_4{}^2)^2}$

5. $f_v(1, -2, 4, 8) = 128,\quad f_w(1, -2, 4, 8) = -512,$
$f_x(1, -2, 4, 8) = 32,\quad f_y(1, -2, 4, 8) = \frac{64}{3}$

7. $210t^{29}$

9. $\dfrac{\partial z}{\partial r} = \dfrac{2r \cos^2 \theta}{r^2 \cos^2 \theta + 1},\ \dfrac{\partial z}{\partial \theta} = -\dfrac{2r^2 \sin \theta \cos \theta}{r^2 \cos^2 \theta + 1}$

11. $\dfrac{\partial w}{\partial \rho} = 2\rho(4\sin^2\phi + \cos^2\phi),$

$\dfrac{\partial w}{\partial \phi} = 6\rho^2 \sin\phi\cos\phi,$

$\dfrac{\partial w}{\partial \theta} = 0$

13. $\dfrac{-4\cos 2y \sin 2y + 2y + 1}{2\sqrt{\cos^2 2y + y^2 + y}}$

15. $\sqrt{3} + \dfrac{\pi}{6}$ cm/sec, increasing

17. $\dfrac{\partial z}{\partial x} = -\dfrac{2x + 4z}{4x + 2z - 3y}, \dfrac{\partial z}{\partial y} = \dfrac{3z}{4x + 2z - 3y}$

27. $\dfrac{\partial z}{\partial x} = \dfrac{ye^x}{15\cos 3z + 3}, \dfrac{\partial z}{\partial y} = \dfrac{e^x}{15\cos 3z + 3}$

33. (a) $\dfrac{dw}{dt} = \dfrac{\partial w}{\partial x_1}\dfrac{dx_1}{dt} + \dfrac{\partial w}{\partial x_2}\dfrac{dx_2}{dt} + \dfrac{\partial w}{\partial x_3}\dfrac{dx_3}{dt} + \dfrac{\partial w}{\partial x_4}\dfrac{dx_4}{dt}$

(b) $\dfrac{\partial w}{\partial v_1} = \dfrac{\partial w}{\partial x_1}\dfrac{\partial x_1}{\partial v_1} + \dfrac{\partial w}{\partial x_2}\dfrac{\partial x_2}{\partial v_1} + \dfrac{\partial w}{\partial x_3}\dfrac{\partial x_3}{\partial v_1} + \dfrac{\partial w}{\partial x_4}\dfrac{\partial x_4}{\partial v_1},$

$\dfrac{\partial w}{\partial v_2} = \dfrac{\partial w}{\partial x_1}\dfrac{\partial x_1}{\partial v_2} + \dfrac{\partial w}{\partial x_2}\dfrac{\partial x_2}{\partial v_2} + \dfrac{\partial w}{\partial x_3}\dfrac{\partial x_3}{\partial v_2} + \dfrac{\partial w}{\partial x_4}\dfrac{\partial x_4}{\partial v_2},$

$\dfrac{\partial w}{\partial v_3} = \dfrac{\partial w}{\partial x_1}\dfrac{\partial x_1}{\partial v_3} + \dfrac{\partial w}{\partial x_2}\dfrac{\partial x_2}{\partial v_3} + \dfrac{\partial w}{\partial x_3}\dfrac{\partial x_3}{\partial v_3} + \dfrac{\partial w}{\partial x_4}\dfrac{\partial x_4}{\partial v_3}$

▶ Exercise Set 16.9 (Page 1029)

1. $(0, 0)$ relative min   **3.** $(1, -2)$ saddle

5. $(2, -1)$ relative min

7. $(2, 1), (-2, 1)$ saddle; $(0, 0)$ relative min

9. $(-1, -1), (1, 1)$ relative min

11. $(0, 0)$ saddle   **13.** none

15. $(0, 0), (4, 0), (0, -2)$ saddle; $(\frac{4}{3}, -\frac{2}{3})$ relative min

17. $(-1, 0)$ relative max

19. $\left(\dfrac{\pi}{2}, \dfrac{\pi}{2}\right)$ relative max

21. (b) $(0, 0)$ relative min

23. critical point $(1, 0)$

25. absolute maximum 0 at $(0, 0)$; absolute minimum -12 at $(0, 4)$

27. absolute maximum 3 at $(0, 1), (2, 1)$; absolute minimum -1 at $(1, 0), (1, 2)$

29. absolute maximum $\frac{33}{4}$ at $(-\frac{1}{2}, \pm\frac{1}{2}\sqrt{15})$; absolute minimum $-\frac{1}{4}$ at $(\frac{1}{2}, 0)$

33. $9, 9, 9$ **35.** $(\sqrt{5}, 0, 0), (-\sqrt{5}, 0, 0)$

37. $\frac{1}{27}$ **39.** length and width 2 ft, height 4 ft

41. $\frac{3}{2}\sqrt{6}$

43. length and width $\sqrt[3]{2V}$, height $\frac{1}{2}\sqrt[3]{2V}$

47. $y = 0.5x + 0.8$

▶ Exercise Set 16.10 (Page 1039)

1. max $\sqrt{2}$ at $(\sqrt{2}, 1)$ and $(-\sqrt{2}, -1)$, min $-\sqrt{2}$ at $(-\sqrt{2}, 1)$ and $(\sqrt{2}, -1)$

3. max $\sqrt{2}$ at $\left(\dfrac{1}{\sqrt{2}}, 0\right)$, min $-\sqrt{2}$ at $\left(-\dfrac{1}{\sqrt{2}}, 0\right)$

5. max 6 at $(\frac{4}{3}, \frac{2}{3}, -\frac{4}{3})$, min -6 at $(-\frac{4}{3}, -\frac{2}{3}, \frac{4}{3})$

7. max $\dfrac{1}{3\sqrt{3}}$ at $\left(\dfrac{1}{\sqrt{3}}, \dfrac{1}{\sqrt{3}}, \dfrac{1}{\sqrt{3}}\right),$

$\left(\dfrac{1}{\sqrt{3}}, -\dfrac{1}{\sqrt{3}}, -\dfrac{1}{\sqrt{3}}\right), \left(-\dfrac{1}{\sqrt{3}}, \dfrac{1}{\sqrt{3}}, -\dfrac{1}{\sqrt{3}}\right),$

$\left(-\dfrac{1}{\sqrt{3}}, -\dfrac{1}{\sqrt{3}}, \dfrac{1}{\sqrt{3}}\right);$

min $-\dfrac{1}{3\sqrt{3}}$ at $\left(\dfrac{1}{\sqrt{3}}, \dfrac{1}{\sqrt{3}}, -\dfrac{1}{\sqrt{3}}\right),$

$\left(\dfrac{1}{\sqrt{3}}, -\dfrac{1}{\sqrt{3}}, \dfrac{1}{\sqrt{3}}\right), \left(-\dfrac{1}{\sqrt{3}}, \dfrac{1}{\sqrt{3}}, \dfrac{1}{\sqrt{3}}\right),$

$\left(-\dfrac{1}{\sqrt{3}}, -\dfrac{1}{\sqrt{3}}, -\dfrac{1}{\sqrt{3}}\right)$

9. $(\frac{2}{5}, \frac{19}{5})$

11. $(1, -1, 1)$

13. $\frac{1}{2}$

15. $(3, 6)$ closest,
$(-3, -6)$ farthest

17. 9, 9, 9

19. $(\sqrt{5}, 0, 0), (-\sqrt{5}, 0, 0)$

21. length and width 2 ft, height 4 ft

▶ Chapter 16 Supplementary Exercises (Page 1039)

1. (a) 1   (b) yx (c) $e^{r+s} \ln(rs)$

3. (a) the upper nappe of the elliptical cone
$z^2 = x^2 + 4y^2$
(b) the plane with x-, y-, z-intercepts $a, b, 1$

5. $\dfrac{1}{x \sin yz} - \dfrac{3y \ln(xy) \cos yz}{\sin^2 yz}$

7. $\pi/2; 0; 1, -\pi^2/4$ **9.** 2

11. $(2x - 2y + 4r)/(x^2 + y^2 + 2z) = 2/(r + s)$

17. (a) increasing 12 newtons/m² min
(b) increasing 240 newtons/m² min

19. (a) $0 = f(0, 0)$ (b) yes

21. (a) $-(6x - 5y + y\sec^2 xy)/(-5x + x\sec^2 xy)$
(b) $-\dfrac{\ln y + \cos(x - y)}{\dfrac{x}{y} - \cos(x - y)}$

23. $\dfrac{dV}{dt} = R\left[\dfrac{dE/dt}{r + R} - \dfrac{dr}{dt}\dfrac{E}{(r + R)^2}\right]$

25. (a) $\langle 6, 45\rangle$ (b) $-174/\sqrt{17}$

27. (a) $\langle \frac{1}{3}, \frac{1}{2}, \frac{1}{6}\rangle$ (b) $\sqrt{3}/9$

29. (a) $\langle 1, 3, 0\rangle$ (b) 43/15

31. (a) $\pm(\mathbf{i} + \mathbf{j})/\sqrt{2}$ (b) $\pm(2\mathbf{i} + \mathbf{j})/\sqrt{5}$

33. $-7/\sqrt{5}$

35. (a) $\mathbf{N} = \langle 8, -6, -5\rangle$
(b) $8(x - 4) - 6(y + 3) - 5(z - 10) = 0$

37. $(1, 1, 1), (-1, -1, 1), (0, 0, 2)$

39. $(-\frac{1}{3}, -\frac{1}{2}, 2)$

41. $dV = -\dfrac{1}{3}(.2)\text{m}^3, \Delta V = -\dfrac{1}{3}(.218)\text{m}^3$

43. relative min at $(15, -8)$

45. relative min at $(3, 9)$; saddle point at $(0, 0)$

47. $\dfrac{2a}{\sqrt{3}} \times \dfrac{2b}{\sqrt{3}} \times \dfrac{2c}{\sqrt{3}}$

49. $I_1:I_2:I_3 = R_1^{-1}:R_2^{-1}:R_3^{-1}$

▶ Exercise Set 17.1 (Page 1052)

1. 7

5. 2

9. 3

13. $\frac{1}{2}(1 - \ln 2)$

17. $\frac{1}{3}$

3. 2

7. $\frac{2}{15}(31 - 9\sqrt{3})$

11. $1 - \ln 2$

15. 0

19. 1

21.

23.

25.

27. 172

29. 8

▶ Exercise Set 17.2 (Page 1062)

1. $\frac{1}{40}$

3. 9

5. $\frac{\pi}{2}$

7. 1

9. $\frac{2}{3}a^3$

11. $\frac{1}{12}$

13. 32

15. $-\dfrac{2}{\pi}$

17. 576

19. $\frac{1}{2}(\sqrt{17}-1)$

21. 0

23. $\dfrac{\pi}{4}-\dfrac{1}{2}\ln 2$

25. $\frac{50}{3}$

27. $-\frac{1}{2}$

29. $\frac{25}{2}$

31. $\sqrt{2}-1$

33. 32

35. $\frac{125}{12}$

37. $\frac{56}{15}$

39. 170

41. $\frac{27}{2}\pi$

43. $\frac{20}{3}$

45. $\dfrac{\pi}{2}$

47. $\frac{2000}{3}$

49. $\displaystyle\int_0^{\sqrt{2}}\int_{y^2}^2 f(x,y)\,dx\,dy$

51. $\displaystyle\int_1^{e^2}\int_{\ln x}^2 f(x,y)\,dy\,dx$

53. $\displaystyle\int_{-1}^1\int_{-2\sqrt{1-y^2}}^{2\sqrt{1-y^2}} f(x,y)\,dx\,dy$

55. $\displaystyle\int_0^{\pi/2}\int_0^{\sin x} f(x,y)\,dy\,dx$

57. $\frac{1}{8}(1-e^{-16})$

59. $\frac{1}{3}(e^8-1)$

61. $\frac{1}{8}\pi$

63. $\frac{1}{3}(1-\cos 8)$

65. (a) 0   (b) $-\frac{603}{40}$

67. π

69. 0

▶ Exercise Set 17.3 (Page 1072)

1. $\frac{1}{6}$

3. 0

5. 0

7. $\dfrac{3\pi}{2}$

9. $\dfrac{\pi}{16}$

11. $\dfrac{4\pi}{3}+2\sqrt{3}$

13. $\frac{4}{3}(27-16\sqrt{2})\pi$

15. $\frac{32}{9}$

17. $\dfrac{5\pi}{32}$

19. $(1-e^{-1})\pi$

21. $\dfrac{\pi}{8}\ln 5$

23. $\dfrac{\pi}{8}$

25. $\frac{16}{9}$

27. $\dfrac{\pi}{2}\left(1-\dfrac{1}{\sqrt{1+a^2}}\right)$

29. $\frac{1}{4}\pi(\sqrt{5}-1)$

31. $\frac{1}{9}(3\pi-4)a^2c$

33. $\dfrac{4\pi}{3}+2\sqrt{3}-2$

35. (b) $\dfrac{\pi}{4}$ (c) $\dfrac{\sqrt{\pi}}{2}$

▶ Exercise Set 17.4 (Page 1079)

1. 6π

3. $\dfrac{\sqrt{5}}{6}$

5. $\dfrac{\pi}{6}(5\sqrt{5}-1)$

7. $\dfrac{\pi}{18}(10\sqrt{10}-1)$

9. 8π

11. $2(\pi-2)a^2$

13. 128

▶ Exercise Set 17.5 (Page 1088)

1. 8

3. 7

5. $\frac{81}{5}$

7. $\frac{128}{15}$

9. $\frac{\pi}{2}(\pi - 3)$

11. $\frac{1}{6}$

13. 4

15. $\frac{256}{15}$

17. 9π

19. 2π

21. $\frac{\pi}{6}(8\sqrt{2} - 7)a^3$

23. (a)   (b)

25. (a) $\displaystyle\int_0^a \int_0^{b(1-x/a)} \int_0^{c(1-x/a-y/b)} dz\,dy\,dx,$

$\displaystyle\int_0^b \int_0^{a(1-y/b)} \int_0^{c(1-x/a-y/b)} dz\,dx\,dy,$

$\displaystyle\int_0^c \int_0^{a(1-z/c)} \int_0^{b(1-x/a-z/c)} dy\,dx\,dz,$

$\displaystyle\int_0^a \int_0^{c(1-x/a)} \int_0^{b(1-x/a-z/c)} dy\,dz\,dx,$

$\displaystyle\int_0^c \int_0^{b(1-z/c)} \int_0^{a(1-y/b-z/c)} dx\,dy\,dz,$

$\displaystyle\int_0^b \int_0^{c(1-y/b)} \int_0^{a(1-y/b-z/c)} dx\,dz\,dy$

27. $\frac{1}{6}a^3$

29. (a) 0   (b) $\frac{1}{2}(e^2 - 1)$

31. (a) 10   (b) 0

33. 0

▶ Exercise Set 17.6 (Page 1103)

1. 10 feet to the right of m_1

3. $\frac{13}{20}, (\frac{190}{273}, \frac{6}{13})$

21. $(\frac{1}{2}, 0, \frac{3}{5})$

23. $\left(\frac{3a}{8}, \frac{3a}{8}, \frac{3a}{8}\right)$

5. $\frac{a^4}{8}, \left(\frac{8a}{15}, \frac{8a}{15}\right)$

7. $(\frac{2}{3}, \frac{1}{3})$

25. $\frac{2}{3}\pi ka^3$

29. $\left(\frac{128}{105\pi}, \frac{128}{105\pi}\right)$

9. $(-\frac{1}{2}, \frac{2}{5})$

11. $\left(0, \frac{4(b^3 - a^3)}{3\pi(b^2 - a^2)}\right)$

33. $2\pi^2 kab$

35. $\left(\frac{a}{3}, \frac{b}{3}\right)$

13. $(0, \frac{8}{3})$

15. $\frac{a^4}{2}, \left(\frac{a}{3}, \frac{a}{2}, \frac{a}{2}\right)$

37. $\frac{1}{12}M(a^2 + b^2)$

39. $\frac{1}{4}MR^2$

41. $\frac{3}{2}MR^2$

43. $\frac{2}{3}Ma^2$

17. $\frac{1}{6}, (0, \frac{16}{35}, \frac{1}{2})$

19. $(\frac{1}{4}, \frac{1}{4}, \frac{1}{4})$

▶ Exercise Set 17.7 (Page 1118)

1. $\frac{\pi}{4}$

3. $\frac{\pi}{16}$

21. πka^4

23. $\left(0, 0, \frac{7}{16\sqrt{2} - 14}\right)$

5. $\frac{81\pi}{2}$

7. $\frac{8}{3}(10\sqrt{5} - 19)\pi$

25. $\left(\frac{3a}{8}, \frac{3a}{8}, \frac{3a}{8}\right)$

27. $\left(0, \frac{195}{152}, 0\right)$

9. $\frac{32}{9}(3\pi - 4)$

11. $\frac{2}{3}(\sqrt{3} - 1)\pi$

29. $\frac{1}{3}(1 - e^{-1})\pi$

31. 81π

13. $9\sqrt{2}\pi$

15. $\frac{4}{3}\pi a^3$

33. $\left(0, 0, \frac{2a}{5}\right)$

35. $(0, 0, \frac{11}{30})$

17. $\frac{11}{6}\pi$

19. $\frac{1}{4}(2 - \sqrt{2})\pi$

37. $\frac{63}{8}\pi$ **41.** $M(\frac{1}{4}R^2 + \frac{1}{3}h^2)$ **47.** $2\pi k\delta h\left(1 - \dfrac{h}{\sqrt{R^2 + h^2}}\right)$

43. $\frac{2}{5}MR^2$

45. (a) $2\pi k\,\delta(\sqrt{R^2 + a^2} - \sqrt{R^2 + (a + h)^2} + h)$

    (b) Other components are zero because of symmetry.

▶ Chapter 17 Supplementary Exercises (Page 1120)

1. $-\frac{1}{2}\sqrt{2}/\pi$ **3.** $-9/14$

5. $\displaystyle\int_0^1 \int_{2y}^2 e^x e^y \, dx \, dy$ **7.** $3/2$

9. (a)   (b)

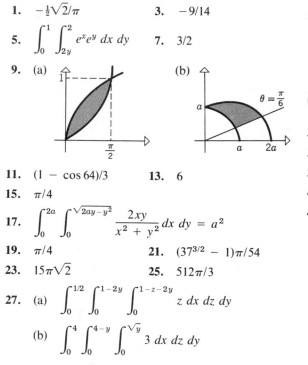

11. $(1 - \cos 64)/3$ **13.** 6

15. $\pi/4$

17. $\displaystyle\int_0^{2a} \int_0^{\sqrt{2ay - y^2}} \frac{2xy}{x^2 + y^2} \, dx \, dy = a^2$

19. $\pi/4$ **21.** $(37^{3/2} - 1)\pi/54$

23. $15\pi\sqrt{2}$ **25.** $512\pi/3$

27. (a) $\displaystyle\int_0^{1/2} \int_0^{1-2y} \int_0^{1-z-2y} z \, dx \, dz \, dy$

    (b) $\displaystyle\int_0^4 \int_0^{4-y} \int_0^{\sqrt{y}} 3 \, dx \, dz \, dy$

29. (a) $\displaystyle\int_0^{2\pi} \int_0^{\pi/3} \int_0^a (\rho^4 \sin^3 \phi) \, d\rho \, d\phi \, d\theta$

    (b) $\displaystyle\int_0^{2\pi} \int_0^{\sqrt{3}a/2} \int_{r/\sqrt{3}}^{\sqrt{a^2 - r^2}} r^3 \, dz \, dr \, d\theta$

    (c) $4\displaystyle\int_0^{\sqrt{3}a/2} \int_0^{\sqrt{(3a^2/4) - x^2}} \int_{\sqrt{(x^2 + y^2)/3}}^{\sqrt{a^2 - x^2 - y^2}} (x^2 + y^2) \, dz \, dy \, dx$

31. $8\pi a^3/3$ **33.** $\pi a^3/9$

35. $\bar{x} = 0, \bar{y} = 5a/6$ **37.** $\bar{x} = 0, \bar{y} = b/4$

39. $kabc^2/24$

41. $\bar{x} = 0, \bar{y} = 12/7, \bar{z} = 8/7$

43. $\bar{x} = \bar{y} = 0, \bar{z} = h/4$

▶ Exercise Set 18.1 (Page 1132)

1. (a) $\frac{4}{3}$ (b) 0   (c) $\frac{4}{3}$

3. $-\frac{8}{3}$ **5.** $1 - \pi$

7. 3   **9.** $-1 - \dfrac{\pi}{4}$

11. 0   **13.** $1 - e^{-1}$

15. $1 - e^3$ **17.** $\frac{23}{6}$

19. $\frac{3}{5}$

21. (a) $\dfrac{5}{4} - \dfrac{\pi}{8} + \dfrac{1}{2}\tan^{-1} 2$

    (b) $\dfrac{1}{3}\tan^{-1} 2 - \dfrac{2}{3}\tan^{-1}\dfrac{1}{2}$

    (c) $\dfrac{3}{4}$

23. $-\frac{37}{2}$ **25.** -12

27. 0

▶ Exercise Set 18.2 (Page 1143)

1. $\frac{1}{2}x^2 + \frac{1}{2}y^2 + K$ **3.** not conservative

5. $x \cos y + y \sin x + K$

7. 13   **9.** -6

11. $9e^2$ **13.** 32

15. $-\frac{1}{2}$ **17.** $\dfrac{\pi}{4}$

19. (a) $-2a$ (b) $-2a$
(c) $-2a$ (d) 0

▶ Exercise Set 18.3 (Page 1149)

1. 0   **3.** 0

5. 0   **7.** 8π

9. -4 **11.** -1

13. 0   **15.** πab

17. $\frac{1}{2}ab$ **19.** $\frac{250}{3}$

21. (b) $\left(0, \dfrac{4a}{3\pi}\right)$

23. (c) $A = \frac{1}{2}[(x_1 y_2 - x_2 y_1) + (x_2 y_3 - x_3 y_2)$
$+ \cdots + (x_n y_1 - x_1 y_n)]$

(d) 8

▶ Exercise Set 18.4 (Page 1161)

1. $\dfrac{\pi\delta_0}{6}(5\sqrt{5} - 1)$ **3.** $\frac{4}{3}\pi\delta_0$

5. $\dfrac{15\pi}{\sqrt{2}}$ **7.** $\dfrac{\pi}{4}$

9. $\dfrac{3}{\sqrt{2}}$ **11.** 9

13. $\dfrac{4\pi}{3}$ **15.** $\frac{8}{3}\pi a^4$

17. $\frac{1}{4}[37\sqrt{37} - 1]$

19. (a) $\dfrac{\sqrt{29}}{16} \int_0^6 \int_0^{(12-2x)/3} xy(12 - 2x - 3y)\, dy\, dx$

(b) $\dfrac{\sqrt{29}}{4} \int_0^3 \int_0^{(12-4z)/3} yz(12 - 3y - 4z)\, dy\, dz$

(c) $\dfrac{2\sqrt{29}}{9} \int_0^3 \int_0^{6-2z} xz(6 - x - 2z)\, dx\, dz$

21. (a) $\int_0^4 \int_1^2 y^3 z\sqrt{1 + 4y^2}\, dy\, dz$

(b) $\dfrac{1}{2} \int_0^4 \int_1^4 xz\sqrt{1 + 4x}\, dx\, dz$

▶ Exercise Set 18.5 (Page 1170)

1. $\dfrac{2}{\sqrt{29}}\mathbf{i} + \dfrac{3}{\sqrt{29}}\mathbf{j} + \dfrac{4}{\sqrt{29}}\mathbf{k}$

3. (a) $-\dfrac{2}{\sqrt{21}}\mathbf{i} - \dfrac{4}{\sqrt{21}}\mathbf{j} + \dfrac{1}{\sqrt{21}}\mathbf{k}$

(b) $\dfrac{3}{5\sqrt{2}}\mathbf{i} - \dfrac{4}{5\sqrt{2}}\mathbf{j} + \dfrac{1}{\sqrt{2}}\mathbf{k}$

(c) $\frac{3}{5}\mathbf{i} - \frac{4}{5}\mathbf{k}$

5. 2π **7.** $\dfrac{\pi}{2}$ **9.** 54π

11. $\frac{14}{3}\pi$ **13.** 0   **15.** $4\pi a^3$

19. (a) $\displaystyle\iint\limits_{R} \mathbf{F} \cdot \left(\mathbf{i} - \frac{\partial x}{\partial y}\mathbf{j} - \frac{\partial x}{\partial z}\mathbf{k} \right) dA$ $\begin{bmatrix} \text{Forward} \\ R = \text{projection on } yz\text{-plane} \end{bmatrix}$

$\displaystyle\iint\limits_{R} \mathbf{F} \cdot \left(-\mathbf{i} + \frac{\partial x}{\partial y}\mathbf{j} + \frac{\partial x}{\partial z}\mathbf{k} \right) dA$ $\begin{bmatrix} \text{Backward} \\ R = \text{projection on } yz\text{-plane} \end{bmatrix}$

(b) $\displaystyle\iint\limits_{R} \mathbf{F} \cdot \left(-\frac{\partial y}{\partial x}\mathbf{i} + \mathbf{j} - \frac{\partial y}{\partial z}\mathbf{k} \right) dA$ $\begin{bmatrix} \text{Right} \\ R = \text{projection on } xz\text{-plane} \end{bmatrix}$

$\displaystyle\iint\limits_{R} \mathbf{F} \cdot \left(\frac{\partial y}{\partial x}\mathbf{i} - \mathbf{j} + \frac{\partial y}{\partial z}\mathbf{k} \right) dA$ $\begin{bmatrix} \text{Left} \\ R = \text{projection on } xz\text{-plane} \end{bmatrix}$

21. 2π

▶ Exercise Set 18.6 (Page 1179)

1. $\operatorname{div} \mathbf{F} = z^3 + 8y^3 x^2 + 10zy$ **11.** 180π **13.** $\frac{1}{24}$

3. $\operatorname{div} \mathbf{F} = ye^{xy} + \sin y + 2 \sin z \cos z$ **15.** $\dfrac{\pi}{2}$ **17.** $\frac{4608}{35}$

5. $\operatorname{div} \mathbf{F} = \dfrac{1}{x} + xz\, e^{xyz} + \dfrac{x}{x^2 + z^2}$ **19.** $4\pi a^3$ **23.** $\pi a^2 h$

7. 8 **9.** $3\pi a^2$

▶ Exercise Set 18.7 (Page 1185)

1. $\operatorname{curl} \mathbf{F} = \mathbf{0}$ **7.** 16π **9.** 0

3. $\operatorname{curl} \mathbf{F} = -xe^{xy}\mathbf{k}$ **11.** πa^2 **13.** $\frac{3}{2}$

5. $\operatorname{curl} \mathbf{F} = -xye^{xyz}\mathbf{i} + \dfrac{z}{x^2 + z^2}\mathbf{j} + yz\, e^{xyz}\mathbf{k}$ **15.** 0 **17.** 2π

▶ Exercise Set 18.8 (Page 1192)

1. 2π **3.** 54π **11.** sources at all points except the origin; no sinks

5. $4\pi a^3$ **7.** 180π

9. no sources or sinks ($\operatorname{div} \mathbf{F} = 0$) **13.** (a) $\frac{3}{2}$ (b) -1 (c) $\mathbf{n} = -\dfrac{1}{\sqrt{2}}\mathbf{j} - \dfrac{1}{\sqrt{2}}\mathbf{k}$

▶ Chapter 18 Supplementary Exercises (Page 1193)

1. 4 **3.** 6 **17.** 32 **19.** $-45\pi/4$

5. 2 **7.** not conservative **21.** 12/5 **23.** 2

9. $\dfrac{y^2}{x} + 2y^2 + x^3 + C$ **25.** $(0, 0, \frac{149}{65})$ **27.** $\dfrac{\pi}{2}$

11. $-7\pi^2/16$ **13.** 3/2 **15.** 0

▶ Exercise Set 19.1 (Page 1201)

3. $y = c_1 e^x + c_2 e^{-4x}$

5. $y = c_1 e^x + c_2 x e^x$

7. $y = c_1 \cos \sqrt{5}x + c_2 \sin \sqrt{5}x$

9. $y = c_1 + c_2 e^x$

11. $y = c_1 e^{-2t} + c_2 t e^{-2t}$

13. $y = e^{2x}(c_1 \cos 3x + c_2 \sin 3x)$

15. $y = c_1 e^{-x/4} + c_2 e^{x/2}$

17. $y = 2e^x - e^{-3x}$

19. $y = (2 - 5x)e^{3x}$

21. $y = -e^{-2x}(3 \cos x + 6 \sin x)$

23. (a) $y'' - 3y' - 10y = 0$
     (b) $y'' - 8y' + 16y = 0$
     (c) $y'' + 2y' + 17y = 0$

25. (a) $k < 0$ or $k > 4$
     (b) 0, 4   (c) $0 < k < 4$

27. (a) $y = \dfrac{1}{|x|}[c_1 \cos(\ln|x|) + c_2 \sin(\ln|x|)]$
     (b) $y = c_1|x|^{1+\sqrt{3}} + c_2|x|^{1-\sqrt{3}}$

▶ Exercise Set 19.2 (Page 1209)

1. $y = c_1 e^{-x} + c_2 e^{-5x} + \frac{1}{16}e^{3x}$

3. $y = c_1 e^{4x} + c_2 e^{5x} - 3x e^{5x}$

5. $y = (c_1 + c_2 x)e^{-x} + \frac{1}{2}x^2 e^{-x}$

7. $y = c_1 e^{3x} + c_2 e^{-4x} - \frac{13}{216} - \frac{1}{18}x - \frac{1}{3}x^2$

9. $y = c_1 + c_2 e^{6x} + \frac{5}{36}x - \frac{1}{12}x^2$

11. $y = c_1 + c_2 x - \frac{1}{2}x^2 + \frac{1}{20}x^5$

13. $y = c_1 e^{-x} + c_2 e^{2x} - 3 \cos x - \sin x$

15. $y = c_1 e^{-2x} + c_2 e^{2x} - \frac{3}{8}\cos 2x - \frac{1}{4}\sin 2x$

17. $y = c_1 \cos x + c_2 \sin x - \frac{1}{2}x \cos x$

19. $y = c_1 e^x + c_2 e^{2x} + \frac{3}{4} + \frac{1}{2}x$

21. $y = e^{-2x}(c_1 \cos \sqrt{5}x + c_2 \sin \sqrt{5}x) - \frac{94}{729}$
     $+ \frac{19}{81}x + \frac{1}{9}x^2$

23. $y = c_1 \cos 2x + c_2 \sin 2x - \frac{1}{8}x \cos 2x$

25. (b) $-\frac{3}{16} - \frac{1}{4}x + \frac{1}{5}x e^x$

27. $y = c_1 e^{-x} + c_2 e^x - 1 + \frac{1}{2}x e^x$

29. $y = c_1 \cos 2x + c_2 \sin 2x + \frac{1}{4} + \frac{1}{4}x + \frac{1}{3}\sin x$

31. $y = (c_1 + c_2 x)e^x + \frac{1}{4}x^2 e^x + \frac{1}{8}e^{-x}$

33. $y = c_1 \cos x + c_2 \sin x + 6 - 2 \cos 2x$

35. (a) $y = c_1 \cos \mu x + c_2 \sin \mu x + \dfrac{a}{\mu^2 - b^2}\sin bx$
     (b) $y = c_1 \cos \mu x + c_2 \sin \mu x$
     $+ \displaystyle\sum_{k=1}^{n} \dfrac{a_k}{\mu^2 - k^2\pi^2}\sin k\pi x$

37. $x_0 = 1$; $y = 2x^2 - 4e^{x-1} + c$, c arbitrary

▶ Exercise Set 19.3 (Page 1213)

1. $y = c_1 \cos x + c_2 \sin x + x^2 - 2$

3. $y = c_1 e^x + c_2 e^{-2x} + \frac{2}{3}x e^x$

5. $y = c_1 \cos 2x + c_2 \sin 2x - \frac{1}{4}x \cos 2x$

7. $y = c_1 \cos x + c_2 \sin x$
    $- (\cos x)\ln|\sec x + \tan x|$

9. $y = c_1 e^x + c_2 x e^x + x e^x \ln|x|$

11. $y = c_1 \cos x + c_2 \sin x + \frac{3}{2} + \frac{1}{2}\cos 2x$

13. $y = c_1 \cos x + c_2 \sin x - x \cos x$
     $+ (\sin x)\ln|\sin x|$

15. $y = c_1 \cos x + c_2 \sin x + x \cos x$
     $- (\sin x)\ln|\cos x|$

17. $y = c_1 e^{-x} + c_2 x e^{-x} - e^{-x}\ln|x|$

19. $y = c_1 e^{-2x} + c_2 x e^{-2x} + (x - 2)e^{-x}$

21. $y = c_1 \cos x + c_2 \sin x - 1$
     $+ (\sin x)\ln|\sec x + \tan x|$

23. $y = c_1 e^x + c_2 x e^x - e^x \ln|x|$

25. $y = c_1 e^x + c_2 e^{-x} + \frac{1}{5}e^x(2 \sin x - \cos x)$

27. $y = c_1 e^{-x} + c_2 x e^{-x} + \frac{1}{2}x^2 e^{-x}\ln|x| - \frac{3}{4}x^2 e^{-x}$

▶   Exercise Set 19.4 (Page 1223)

1. (a) $y'' + 4y = 0$, $y(0) = 1$, $y'(0) = 0$
   (b) $y = \cos 2t$

3. (a) $y'' + 196y = 0$, $y(0) = -10$, $y'(0) = 0$
   (b) $y = -10 \cos 14t$

5. (a) $y = 2 \cos 8t$ (b) 2
   (c) $\pi/4$ (d) $4/\pi$

7. (a) $y = -\frac{1}{4} \cos 8\sqrt{6}t$ (b) $\frac{1}{4}$
   (c) $\dfrac{\pi}{4\sqrt{6}}$ (d) $\dfrac{4\sqrt{6}}{\pi}$

9. (a) $y'' + 4y' + 8y = 0$, $y(0) = -3$, $y'(0) = 0$
   (b) $y = -3e^{-2t}(\cos 2t + \sin 2t)$

11. (a) $y = \frac{5}{2}\sqrt{6}e^{-t/5} \cos\left(\dfrac{\sqrt{2}}{5}t - \tan^{-1}\dfrac{1}{\sqrt{2}}\right)$
   (b) $\dfrac{10\pi}{\sqrt{2}}$ (c) $\dfrac{\sqrt{2}}{10\pi}$

13. (a) $y = -\frac{1}{4} \sin 8t$ (b) $\frac{1}{4}$
   (c) $\pi/4$ (d) $4/\pi$

15. $y = \frac{1}{3}e^{-t}(\cos 7t - 2 \sin 7t)$

17. (a) $\frac{1}{32}\pi^2$ (b) $\frac{9}{4}$

19. $T = 2\pi \sqrt{\dfrac{\delta h}{\rho g}}$

▶   Exercises, Trigonometric Functions and Identities (Page A16)

1. (a) $\dfrac{3\pi}{2}$ (b) $\dfrac{13\pi}{6}$ (c) $\dfrac{\pi}{9}$
   (d) $\dfrac{23\pi}{30}$ (e) $\dfrac{\pi^2}{1800}$

3. (a) $12°$ (b) $270°$ (c) $\left(\dfrac{810}{\pi}\right)°$
   (d) $288°$ (e) $900°$

5. (a)

(b)

(c)

(d)

(e)

7.

θ	$\sin\theta$	$\cos\theta$	$\tan\theta$	$\csc\theta$	$\sec\theta$	$\cot\theta$
(a) $\dfrac{\pi}{6}$	$\frac{1}{2}$	$\dfrac{\sqrt{3}}{2}$	$\dfrac{1}{\sqrt{3}}$	2	$\dfrac{2}{\sqrt{3}}$	$\sqrt{3}$
(b) $\dfrac{-7\pi}{3}$	$\dfrac{-\sqrt{3}}{2}$	$\frac{1}{2}$	$-\sqrt{3}$	$\dfrac{-2}{\sqrt{3}}$	2	$\dfrac{-1}{\sqrt{3}}$
(c) $\dfrac{-5\pi}{4}$	$\dfrac{1}{\sqrt{2}}$	$\dfrac{-1}{\sqrt{2}}$	-1	$\sqrt{2}$	$-\sqrt{2}$	-1
(d) -3π	0	-1	0	$-$	-1	$-$
(e) π	0	-1	0	$-$	-1	$-$

9. (a) $\theta = \pm n\pi$, $n = 0, 1, 2, \ldots$
   (b) $\theta = \dfrac{\pi}{2} \pm n\pi$, $n = 0, 1, 2, \ldots$
   (c) $\theta = \pm n\pi$, $n = 0, 1, 2, \ldots$
   (d) $\theta = \pm n\pi$, $n = 0, 1, 2, \ldots$
   (e) $\theta = \dfrac{\pi}{2} \pm n\pi$, $n = 0, 1, 2, \ldots$
   (f) $\theta = \pm n\pi$, $n = 0, 1, 2, \ldots$

11. $\theta = \dfrac{5\pi}{4} \pm 2n\pi$ and
   $\theta = \dfrac{7\pi}{4} \pm 2n\pi$, $n = 0, 1, 2, \ldots$

13. $\theta = \dfrac{\pi}{3} \pm 2n\pi$ and
   $\theta = \dfrac{5\pi}{3} \pm 2n\pi$, $n = 0, 1, 2, \ldots$

15. $\theta = \dfrac{\pi}{3} \pm n\pi$, $n = 0, 1, 2, \ldots$

17. $\theta = \dfrac{4\pi}{3} \pm 2n\pi$ and

$\theta = \dfrac{5\pi}{3} \pm 2n\pi$, $n = 0, 1, 2, \ldots$

19. $\theta = \pi \pm 2n\pi$, $n = 0, 1, 2, \ldots$

21. $\theta = \dfrac{\pi}{6} \pm n\pi$, $n = 0, 1, 2, \ldots$

23. $\theta = \dfrac{7\pi}{6} \pm 2n\pi$ and

$\theta = \dfrac{11\pi}{6} \pm 2n\pi$, $n = 0, 1, 2, \ldots$

25. $\theta = \dfrac{\pi}{6} \pm 2n\pi$ and

$\theta = \dfrac{11\pi}{6} \pm 2n\pi$, $n = 0, 1, 2, \ldots$

27. (a) $\frac{1}{4}(\sqrt{2} + \sqrt{6})$ or $\frac{1}{2}\sqrt{2 + \sqrt{3}}$

(b) $\frac{1}{2}\sqrt{2 - \sqrt{2}}$

(c) $\frac{1}{2}\sqrt{2 - \sqrt{2 - \sqrt{3}}}$

29.

	$\sin\theta$	$\cos\theta$	$\tan\theta$	$\csc\theta$	$\sec\theta$	$\cot\theta$
(a)	$\dfrac{a}{3}$	$\dfrac{\sqrt{9 - a^2}}{3}$	$\dfrac{a}{\sqrt{9 - a^2}}$	$\dfrac{3}{a}$	$\dfrac{3}{\sqrt{9 - a^2}}$	$\dfrac{\sqrt{9 - a^2}}{a}$
(b)	$\dfrac{a}{\sqrt{a^2 + 25}}$	$\dfrac{5}{\sqrt{a^2 + 25}}$	$\dfrac{a}{5}$	$\dfrac{\sqrt{a^2 + 25}}{a}$	$\dfrac{\sqrt{a^2 + 25}}{5}$	$\dfrac{5}{a}$
(c)	$\dfrac{\sqrt{a^2 - 1}}{a}$	$\dfrac{1}{a}$	$\sqrt{a^2 - 1}$	$\dfrac{a}{\sqrt{a^2 - 1}}$	a	$\dfrac{1}{\sqrt{a^2 - 1}}$

31. $\frac{9}{2}\sqrt{2}$, $\frac{9}{2}(1 + \sqrt{3})$, $105°$

41. $\sin\left(\dfrac{3\pi}{2} - \theta\right) = -\cos\theta$,

$\cos\left(\dfrac{3\pi}{2} + \theta\right) = \sin\theta$

55. $\sin 3\theta = 3\sin\theta\cos^2\theta - \sin^3\theta$,
$\cos 3\theta = \cos^3\theta - 3\sin^2\theta\cos\theta$

▶ Exercises, Graphs of Trigonometric Functions (Page A25)

1. (a) $\dfrac{2\pi}{5}$ (b) 6π (c) 8π (d) $\dfrac{\pi}{7}$

(e) 2   (f) 10   (g) 1   (h) $2\pi k$

3. (a) 5   (b) $\frac{1}{3}$ (c) $\frac{1}{2}$ (d) 1

5.

7.

9.

11.

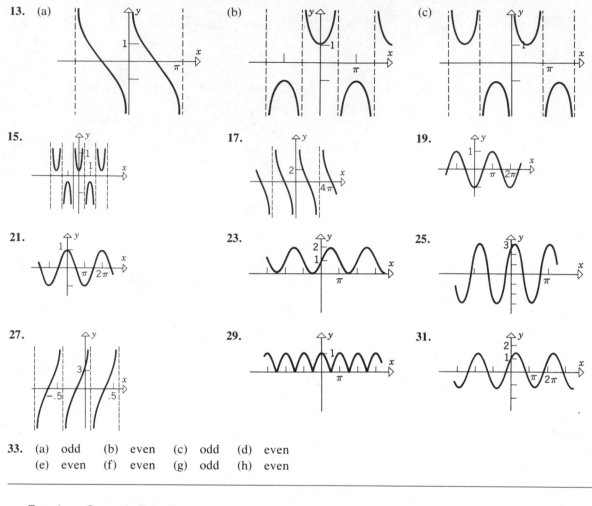

13. (a)   (b)   (c)

15.   **17.**   **19.**

21.   **23.**   **25.**

27.   **29.**   **31.**

33. (a) odd   (b) even   (c) odd   (d) even
     (e) even   (f) even   (g) odd   (h) even

► Exercises, Cramer's Rule (Page A57)

1. $x = 1, y = 2$

3. $x_1 = -\frac{7}{32}, x_2 = \frac{5}{16}$

5. $x = 2, y = -1, z = 3$

7. $x_1 = \frac{26}{21}, x_2 = \frac{25}{21}, x_3 = \frac{5}{7}$

9. $x' = x \cos \theta + y \sin \theta$
    $y' = -x \sin \theta + y \cos \theta$

---

index

44. $\int \dfrac{u^2\,du}{\sqrt{a^2-u^2}} = -\dfrac{u}{2}\sqrt{a^2-u^2} + \dfrac{a^2}{2}\sin^{-1}\dfrac{u}{a} + C$

45. $\int \dfrac{du}{u\sqrt{a^2-u^2}} = -\dfrac{1}{a}\ln\left|\dfrac{a+\sqrt{a^2-u^2}}{u}\right| + C$

$\qquad = -\dfrac{1}{a}\cosh^{-1}\dfrac{a}{u} + C$

46. $\int \dfrac{du}{u^2\sqrt{a^2-u^2}} = -\dfrac{\sqrt{a^2-u^2}}{a^2 u} + C$

47. $\int (a^2-u^2)^{3/2}\,du = -\dfrac{u}{8}(2u^2-5a^2)\sqrt{a^2-u^2} + \dfrac{3a^4}{8}\sin^{-1}\dfrac{u}{a} + C$

48. $\int \dfrac{du}{(a^2-u^2)^{3/2}} = \dfrac{u}{a^2\sqrt{a^2-u^2}} + C$

Integrals Containing $2au - u^2$

49. $\int \sqrt{2au-u^2}\,du = \dfrac{u-a}{2}\sqrt{2au-u^2} + \dfrac{a^2}{2}\cos^{-1}\left(1-\dfrac{u}{a}\right) + C$

50. $\int u\sqrt{2au-u^2}\,du = \dfrac{2u^2-au-3a^2}{6}\sqrt{2au-u^2}$

$\qquad\qquad + \dfrac{a^3}{2}\cos^{-1}\left(1-\dfrac{u}{a}\right) + C$

51. $\int \dfrac{\sqrt{2au-u^2}\,du}{u} = \sqrt{2au-u^2} + a\cos^{-1}\left(1-\dfrac{u}{a}\right) + C$

52. $\int \dfrac{\sqrt{2au-u^2}\,du}{u^2} = -\dfrac{2\sqrt{2au-u^2}}{u} - \cos^{-1}\left(1-\dfrac{u}{a}\right) + C$

53. $\int \dfrac{du}{\sqrt{2au-u^2}} = \cos^{-1}\left(1-\dfrac{u}{a}\right) + C$

54. $\int \dfrac{u\,du}{\sqrt{2au-u^2}} = -\sqrt{2au-u^2} + a\cos^{-1}\left(1-\dfrac{u}{a}\right) + C$

55. $\int \dfrac{u^2\,du}{\sqrt{2au-u^2}} = -\dfrac{(u+3a)}{2}\sqrt{2au-u^2} + \dfrac{3a^2}{2}\cos^{-1}\left(1-\dfrac{u}{a}\right) + C$

56. $\int \dfrac{du}{u\sqrt{2au-u^2}} = -\dfrac{\sqrt{2au-u^2}}{au} + C$

57. $\int \dfrac{du}{(2au-u^2)^{3/2}} = \dfrac{u-a}{a^2\sqrt{2au-u^2}} + C$

58. $\int \dfrac{u\,du}{(2au-u^2)^{3/2}} = \dfrac{u}{a\sqrt{2au-u^2}} + C$

Integrals Containing Trigonometric Functions

59. $\int \sin u\,du = -\cos u + C$

60. $\int \cos u\,du = \sin u + C$

61. $\int \tan u\,du = \ln|\sec u| + C$

62. $\int \cot u\,du = \ln|\sin u| + C$

63. $\int \sec u\,du = \ln|\sec u + \tan u| + C$

$\qquad = \ln|\tan(\tfrac{1}{4}\pi + \tfrac{1}{2}u)| + C$

64. $\int \csc u\,du = \ln|\csc u - \cot u| + C$

$\qquad = \ln|\tan\tfrac{1}{2}u| + C$

65. $\int \sec^2 u\,du = \tan u + C$

66. $\int \csc^2 u\,du = -\cot u + C$

67. $\int \sec u \tan u\,du = \sec u + C$

68. $\int \csc u \cot u\,du = -\csc u + C$

69. $\int \sin^2 u\,du = \tfrac{1}{2}u - \tfrac{1}{4}\sin 2u + C$

70. $\int \cos^2 u\,du = \tfrac{1}{2}u + \tfrac{1}{4}\sin 2u + C$

71. $\int \tan^2 u\,du = \tan u - u + C$

72. $\int \cot^2 u\,du = -\cot u - u + C$

73. $\int \sin^n u\,du = -\dfrac{1}{n}\sin^{n-1}u\cos u + \dfrac{n-1}{n}\int \sin^{n-2}u\,du$

74. $\int \cos^n u\,du = \dfrac{1}{n}\cos^{n-1}u\sin u + \dfrac{n-1}{n}\int \cos^{n-2}u\,du$

75. $\int \tan^n u\,du = \dfrac{1}{n-1}\tan^{n-1}u - \int \tan^{n-2}u\,du$

76. $\int \cot^n u\,du = -\dfrac{1}{n-1}\cot^{n-1}u - \int \cot^{n-2}u\,du$

77. $\int \sec^n u\,du = \dfrac{1}{n-1}\sec^{n-2}u\tan u + \dfrac{n-2}{n-1}\int \sec^{n-2}u\,du$

78. $\int \csc^n u\,du = -\dfrac{1}{n-1}\csc^{n-2}u\cot u + \dfrac{n-2}{n-1}\int \csc^{n-2}u\,du$

79. $\int \sin mu \sin nu\,du = -\dfrac{\sin(m+n)u}{2(m+n)} + \dfrac{\sin(m-n)u}{2(m-n)} + C$

80. $\int \cos mu \cos nu\,du = \dfrac{\sin(m+n)u}{2(m+n)} + \dfrac{\sin(m-n)u}{2(m-n)} + C$

81. $\int \sin mu \cos nu\,du = -\dfrac{\cos(m+n)u}{2(m+n)} - \dfrac{\cos(m-n)u}{2(m-n)} + C$

82. $\int u \sin u\,du = \sin u - u\cos u + C$

83. $\int u \cos u\,du = \cos u + u\sin u + C$

84. $\int u^2 \sin u\,du = 2u\sin u + (2-u^2)\cos u + C$